IMPACT AND EXPLOSION CRATERING

PLANETARY AND TERRESTRIAL IMPLICATIONS

PROCEEDINGS OF THE SYMPOSIUM ON PLANETARY CRATERING MECHANICS
Flagstaff, Arizona, September 13–17, 1976

COVER ILLUSTRATIONS

An impact crater and an experimental explosion crater which have notably affected modern research in cratering mechanics are illustrated on the covers of this book.

Front Cover: An aerial view of Meteor Crater, Arizona, one of the youngest terrestrial impact craters and one which has played major roles both in basic cratering research and as a training site for each of the Apollo astronaut crews. Field studies by E. M. Shoemaker indicate that Meteor Crater formed some 25,000 years ago, shortly before man began to inhabit the southern regions of the Colorado Plateau permanently. Shoemaker's analysis of the event indicates that an iron mass(es), traveling in excess of 11 km/second, impacted flat-lying sedimentary rocks, initiating a complex set of cratering processes and releasing some 5–10 megatons (10^{23} ergs) of kinetic energy. The result was the formation of a large bowl-shaped crater, approximately 1.1 km across and over 200 m deep, surrounded by an extensive ejecta blanket.

The nature and extent of structural deformation remained unknown until D. M. Barringer initiated a systematic exploration in the early 1900's to mine the iron meteorite. His deep drilling and excavations never located the impacting body which, unknown to him, had totally fragmented. His subsurface data, however, played a major role in Shoemaker's identification of the classic elements formed in such an impact crater, including a thick breccia lens, a fallout layer, uplifted and deformed rim strata, an ejecta blanket, and a partial sedimentary fill of talus, playa and lake beds. Vertical shafts excavated as deep as 67 m below the crater floor penetrated 30 m of fill, 10 m of fallout, and terminated in highly shocked and fused Coconino Sandstone. The white mounds seen in the floor consist of rock debris from the shafts and contain coesite and traces of stishovite.

Structural deformation in the rim of the crater consists principally of highly faulted and folded strata that have been uplifted so that they now dip away from the crater. Structural uplift, locally as great as 50 m along the upper crater walls, produced steep cliffs which expose complete sections of the Moenkopi Sandstone, 10 m thick, and the underlying Kaibab Dolomite, approximately 90 m thick. Talus on lower walls largely conceals the underlying deformed section of Coconino Sandstone.

Field studies, combined with data from over 160 holes drilled in the rim by the U.S. Geological Survey, show that approximately 175×10^6 metric tons of rock were ejected during the cratering event to form a coherent ejecta blanket with a well-ordered inversion of strata. The light-colored, hummocky terrain surrounding the crater is underlain largely by Coconino and Kaibab ejecta, locally as thick as 25 m near the rim crest. The outer edge of the ejecta blanket once extended over 2 km from the center of the crater, but erosion has reduced the range to about 1.5 km with an estimated 15% to 25% of the ejecta eroded since the time of impact. White areas on the southwest rim (left side) are abandoned silica pits in Coconino Sandstone which form the upper part of the overturned ejecta near the rim. Detailed studies by D. M. Barringer, H. H. Nininger, and E. M. Shoemaker have shown that iron fragments from the impacting body are widely but sparsely distributed on the ejecta blanket and surrounding plain, and in the crater walls, inner rim, breccia lens, and fallout layer.

The San Francisco Peaks and their associated volcanic field form the skyline 70 km to the northwest. Sunset Crater, one of the larger cinder cones in the field, has erupted as recently as about 1250 AD and locally scattered fine ash from air fallout over the Meteor Crater area. Canyon Diablo, from which the numerous meteorite fragments at Meteor Crater acquired their name, can be seen in its winding course a few kilometers northwest of the crater. Buildings on north rim are Barringer Museum and quarters. Meteor Crater as seen today averages 1186 m in diameter and 167 m in depth measured at rim crest, which rises about 50 m above surrounding plain. The crater now exhibits a squarish outline, but retains the classic bowl-shape so common to impact craters observed on other terrestrial planets and the moon. Photo by D. Roddy and K. Zeller, U.S. Geological Survey.

Back Cover: An aerial view of the Dial Pack Crater, a large experimental explosion trial formed by the detonation of a 500-ton sphere of TNT lying on alluvium. The crater exhibited a number of morphological and structural features analogous to those observed at certain large impact craters, including a very shallow bowl-shape, a floor with a broad central uplift encircled by radial and concentric ridges, faulted and folded strata in the rim, and a surrounding ejecta blanket of overturned strata. Dial Pack, 86 m in diameter and 5 m deep measured at the rim crest, is shown here about one hour after its formation with ground water flowing from fractures on the crater floor and rim and depositing light-colored patches of sediments.

The experiment, conducted in July 1970 at the Defense Research Establishment Suffield, Alberta, Canada, was one of a large number of studies related to shock wave phenomena. During this trial, the Canadian and United States Governments provided a cooperative environment for NASA and the U.S. Geological Survey to conduct a training exercise for the Apollo 15, 16 and 17 crews. Astronauts seen on the far rim of the crater include: J. P. Allen, V. D. Brand, G. P. Carr, C. M. Duke, Jr., A. W. England, R. F. Gordon, Jr., F. W. Haise, J. B. Irwin, T. K. Mattingly, W. R. Pogue, H. H. Schmidt, D. R. Scott, A. M. Worden, and T. W. Young. Employment of this crater as a simulation of the dynamics of a recent impact event was a useful addition to the field training of the Apollo 15 crew shortly before their successful mission to the cratered Hadley region on the moon. Photograph by U.S.A.F. Aerospace Audiovisual Service and courtesy U.S. Department of Defense.

IMPACT AND EXPLOSION CRATERING

PLANETARY AND TERRESTRIAL IMPLICATIONS

PROCEEDINGS OF THE SYMPOSIUM ON PLANETARY CRATERING MECHANICS
Flagstaff, Arizona, September 13–17, 1976

Edited by
D. J. Roddy, *U.S. Geological Survey*
R. O. Pepin, *The Lunar Science Institute*
R. B. Merrill, *The Lunar Science Institute*

Compiled by
The Lunar Science Institute,
Houston, Texas

This Lunar Science Institute Topical Conference was hosted by the U.S. Geological Survey, Geologic Division Branch of Astrogeologic Studies

PERGAMON PRESS
NEW YORK · OXFORD · TORONTO · SYDNEY · FRANKFURT

U.K.	Pergamon Press Ltd., Headington Hill Hall, Oxford OX3 0BW, England
U.S.A.	Pergamon Press Inc., Maxwell House, Fairview Park, Elmsford, New York 10523, U.S.A.
CANADA	Pergamon of Canada Ltd., 75 The East Mall, Toronto, Ontario, Canada
AUSTRALIA	Pergamon Press (Aust.) Pty. Ltd., 19a Boundary Street, Rushcutters Bay, N.S.W. 2011, Australia
FRANCE	Pergamon Press SARL, 24 rue des Ecoles, 75240 Paris, Cedex 05, France
FEDERAL REPUBLIC OF GERMANY	Pergamon Press GmbH, 6242 Kronberg-Taunus, Pferdstrasse 1, Federal Republic of Germany

First edition 1977

Library of Congress Cataloging in Publication Data

Impact and Explosion Cratering

Planetary and Terrestrial Implications

Proceedings of a Lunar Science Institute Topical Conference
"Hosted by the U.S. Geological Survey, Geologic
Division, Branch of Astrogeologic Studies."
Includes index.
1. Cratering—Congresses. 2. Planets—Geology—
Congresses. 3. Lunar geology—Congresses. I. Lunar
Science Institute. II. United States. Geological
Survey. Branch of Astrogeologic Studies. III. Title.
QB603.C7S95 1976 551.4'4 77–24753

PRINTED IN THE UNITED STATES OF AMERICA

Contents

Material Properties and Shock Effects

Theoretical Cratering Mechanics

EJECTA

SCALING

Preface

THIS VOLUME CONTAINS the Proceedings of the Symposium on Planetary Cratering Mechanics, held at Flagstaff, Arizona on September 13–17, 1976. The Symposium was sponsored by the Lunar Science Institute and hosted by the Astrogeology Branch of the U.S. Geological Survey. The purpose of the Symposium was to provide an organized forum for communication among scientists engaged in planetary impact and explosion cratering research. Of the 78 papers presented at the meeting, nearly all are contained among the 64 contributions to this Proceedings volume.

During the year preceding the Symposium, the Organizing Committee established the spirit of effective and cooperative interaction among individual members of historically rather separate scientific communities which later characterized the Symposium itself. Committee members were R. O. Pepin (*Lunar Science Institute*) and D. J. Roddy (*U.S. Geological Survey*), co-chairmen, and T. J. Ahrens (*California Institute of Technology*), H. Cooper (*R and D Associates*), M. R. Dence (*Department of Energy, Mines and Resources, Canada*), D. E. Gault (*NASA-Ames Research Center*), J. W. Head (*Brown University*), F. Hörz (*NASA-Johnson Space Center*), H. Masursky (*U.S. Geological Survey*), M. Settle (*Air Force Geophysics Laboratory and Brown University*), and J. Stockton (*Defense Nuclear Agency*). In particular the efforts of Drs. H. Cooper, J. Stockton, and F. Hörz were invaluable in the liaison and organizational stages of Committee activities. Drs. M. Atkins, G. Sevin, and J. Stockton of the Defense Nuclear Agency played indispensable roles in assisting in the final transmittal of a number of papers from the explosion community. P. P. Jones of the Lunar Science Institute was both administrative assistant to the Committee and director of Symposium logistics. M. S. Gibson and K. Gunter of the Lunar Science Institute prepared the volume of abstracts. M. M. Briggs of the U.S. Geological Survey, L. Mager of the Lunar Science Institute, and J. K. Roddy, D. M. Roddy, and M. R. Roddy of Flagstaff provided logistic and projection support to the Symposium.

We gratefully acknowledge these individuals, the Symposium speakers and attendees, the Editorial Staff and the Board of Associate Editors for these Proceedings, and the contributors to this volume for their efforts in creating an enthusiastic and productive interaction among workers in cratering research.

The Editors
February, 1977

Introduction

IN 1610 GALILEO GALILEI announced his discovery of craters on the moon. From these and later observations it became apparent that the dominant lunar surface features were indeed circular depressions. In the tradition of the times, Galileo chose to describe these bowl-shaped topographic features using the Greek root *κρατερ* meaning cup or bowl. However, the interesting question of the origin of such numerous features was to remain in doubt for centuries. During this long period, three questions, critical to the understanding of impact cratering, remained unanswered. One hinged on the general lack of acceptance of the existence of extraterrestrial bodies which could cause impact craters, particularly very large ones. A second more subtle problem involved the absence of technical understanding of impact cratering mechanics and its consequences. The third, and perhaps the most serious problem of all, was the psychological drawback that no one had ever observed an actual impact cratering event (or at least no one convincingly claimed that he had).

A general acceptance of the existence of extraterrestrial bodies large enough to form craters finally began to emerge during the late nineteenth century when the noted geologist G. K. Gilbert published his treatment of the impact origin of the lunar craters. During this period the concept that meteoritic fragments could at least form small craters and penetration holes became more widely accepted. Nevertheless, as late as the mid-1950's, the abundant distribution of meteoritic iron fragments around the famous Meteor (Barringer) Crater in Arizona continued to be dismissed by some as fortuitous. The modern era of impact cratering studies began in the early 1960's with the pioneering studies of R. B. Baldwin, R. S. Dietz, D. E. Gault and E. M. Shoemaker of the United States, C. S. Beals of Canada, and K. P. Stanyukovich and E. L. Krinov of the U.S.S.R. Their research, quickly amplified by a number of their students, dramatically influenced the acceptance of an impact origin for the majority of the lunar craters and their comparable terrestrial counterparts. This research, particularly the classic studies of cratering mechanics at Meteor Crater by E. M. Shoemaker of the U.S. Geological Survey and the California Institute of Technology, and the hypervelocity impact experiments of D. E. Gault at NASA-Ames, finally provided the broad conceptual framework which so strongly encouraged the rapid evolution of increasingly sophisticated studies of impact cratering. In this same decade a profound enlargement of the impact cratering data base for both lunar and terrestrial craters began with the space programs initiated by the United States and the U.S.S.R. The planetary exploration program of the National Aeronautics and Space Administration soon yielded an enormous volume of photographic and Apollo sample data showing that the inner planets Mercury, Venus, Earth,

Mars, and their satellites Moon, Phobos, and Deimos exhibit a remarkably common feature, *craters*. This vastly expanded data base, and the host of individual studies which quickly followed, have now established impact cratering as one of the dominant physical processes in the evolution of the terrestrial planets, and have led to recognition of a wide range of unanswered questions related to fundamental efforts in cratering mechanics.

A parallel development was also occurring within another active community devoted to research in nuclear and chemical explosion cratering. For almost two decades these two virtually independent groups of scientists have been engaged in research on cratering, adopting distinct but highly complementary approaches and emphases. The basic interests of the planetary cratering community have been centered on study of the origin and evolution of solid bodies in the solar system. This approach is thus largely through the geosciences, and the emphasis is primarily on the role of impact and impact cratering in accretion, physical and chemical differentiation, and morphologic evolution of planetary surfaces. On the other hand, the explosion cratering community of physical scientists and engineers is of necessity concerned with the detailed physics of shock wave propagation, crater formation, and structural deformation. As in impact cratering research, explosion studies continue to draw heavily upon empirical methods, due to the enormous complexity of the experimental and theoretical problems. The more recent interest in engineering uses of explosive cratering has prompted a rapid expansion of experimental and theoretical approaches with the result that both areas have progressed rapidly. Laboratory, field, and theoretical techniques have improved in response to increased applications demand for more explicit and reliable data bases. Here, as with impact cratering, these increased demands defined a series of new problems that centered on the dynamic *in situ* material response of geologic media and its relationship to theoretical complexities in calculation of wave propagation, fragmentation and ejection processes, crater shapes, and structural deformation effects.

The potential benefits of interaction and information exchange between the explosion and impact research communities were recognized nearly two decades ago and led to the first symposium of its kind, held in 1961 at the Geophysical Laboratory of the Carnegie Institution in Washington, D.C. The enormous wealth of experimental and theoretical data collected since that time in part provided the rationale for the 1976 Flagstaff Symposium and for the Proceedings contained in this volume. The current great breadth of mutual problem areas, however, emphasized the need for interaction among a cross section of both the explosion and impact communities, including active researchers in all types of planetary impact and cratering studies, nuclear and high explosion field trials, laboratory explosion and impact experimental studies, theoretical calculation studies, and the wide spectrum of applied groups that use impact and explosion data in allied areas. The traditional separation of these groups has been caused largely by their highly diverse and totally different goals in basic and applied research.

From the point of view of the planetary community, interest in cratering, and in impact in general, arises from investigations within three broad areas of study. The first is the very early stage of solar system history, where collisions were a fundamental process in the accretion of large planetary bodies. Here a critical question is one of the partitioning of projectile kinetic energy into heat, deformation, and kinetic energy of ejecta as a function of projectile energy and target size. In late accretionary stages, the questions of the depth of deposition of thermal and deformational energy in large-scale impact and the time-scale for temperature decay are important in assessing whether kinetic energy of the impacting body could have been a significant heat source in driving early chemical differentiation of planetary crusts.

A second area is the role of impact cratering in the morphologic evolution of planetary surfaces through time, and the study of the stratigraphy of these surfaces by observation of crater shapes and by sampling or photogeologic and multispectral mapping of ejecta. The question of the nature of the target in terms of its dynamic physical properties, and its influence on crater morphology, is one that cuts squarely across research in both impact and explosion cratering. From the planetary point of view, it is a question of what can be learned about the regolith, crust and mantle from observation of the final crater; for explosive cratering, the inverse question of what can be understood about crater shape from knowledge of the target is perhaps more central. Theoretical and computational expertise, within both the explosion and the planetary fields, is required to quantitatively address this question, and others of major interest to both communities, such as: to what extent are final crater shapes determined by *in situ* material responses and by other fundamental physical effects? From where in the vertical stratigraphy of the target do materials found at various positions in the horizontal ejecta distribution come, as functions of target composition, gravity, atmospheric effects, and impact scale sizes? What is the partitioning of initial kinetic energy into ejecta kinetic energy, thermal energy and seismic energy, particularly for cratering into unconsolidated material? These questions are fundamental in the efforts of planetary scientists to use crater ejecta as probes of the stratigraphy of planetary crusts, to understand the processes and scales of horizontal mixing, and to assess the nature and degree of modification of planetary surfaces arising from dynamic and accumulative effects of ejecta deposition and from seismically induced landslides and other mass motions of materials. The multitude of problems related to terminal positioning and dynamic cratering processes that immediately rise out of studies that address all these questions pose the really exciting areas for future research, such as reasons for transitions from one crater type to another, formation of flat-floored craters, formation of central uplifts vs. multirings, extent and distributions of uplifted sub-crater material and mascon development, overturning and formation of ejecta blankets and related ejecta distributions, origin of terraces and crater rim development, depth and range of cratering effects and mantle and crustal redistributions, as well as a host of others.

The third major area of planetary interest in cratering concerns a profoundly important event in the evolution of the inner planets of the solar system: the bombardment of the planets by a high flux of large projectiles throughout all or part of the first 500 m.y. of their history, terminating about 4 b.y. ago. All of the effects mentioned above—initial kinetic energy and coupling, materials response, crater shapes, source and distribution of ejecta, and other aspects—must be evaluated in the regime of giant cratering phenomena involving transient cavities hundreds of kilometers across and perhaps tens of kilometers deep, major redistribution of planetary crust and even upper mantle, and repositioning of crustal material by ejection to literally the opposite side of a planet. Here, in a context which so far exceeds any conceivable experimental scale, the approach must combine direct observation of the ancient giant basins with theoretical modeling.

It is clear that the problem areas of mutual interest to both communities are enormously diverse. They range from theoretical modeling and computer simulation of the cratering process itself to study of single natural or man-made cratering events; to engineering applications such as the U.S.S.R.'s program of explosive construction of reservoirs, canals and dams; and to attempts to unravel the influence of impact and cratering on the origin of planets and the evolution of their surfaces, over impact scale sizes that fracture single grains of dust to those that redistribute massive portions of a planet's crust, spanning perhaps twelve orders of magnitude in crater diameter and more than 4 b.y. of time.

The basic goal of the Symposium on Planetary Cratering Mechanics was clear from its inception: to provide a forum for exchange of data bases and state-of-the-art techniques, for active discussion of these common areas of interest, and for definition of problems within these areas—both problems that are immediately tractable by application of techniques already developed and those requiring major inter-community efforts—where impact and explosion workers can assist each other. A Symposium attendance of more than 120 scientists from Canada, France, Germany, Great Britain, South Africa, and the United States provided an assemblage of the most active researchers in cratering today. This Proceedings volume is the fruition of the Symposium papers and the active discussions and debates engendered by the speakers and audience, but even more, represents the serious interests expressed in exchanging these new data bases and ideas between the explosion and impact communities of scientists. When Galileo opened the Pandora's Box of Cratering in 1610 by announcing that one of the solar system's planets was totally saturated by craters, no one could have anticipated the dramatic and widespread importance of the physical processes associated with cratering. Hopefully, these Proceedings will express a part of the excitement and the usefulness of these ideas and data bases in further defining the evolution of our terrestrial planets and the potential man-made uses of the cratering process.

Robert O. Pepin
Houston and Minneapolis

David J. Roddy
Flagstaff

Roddy, D. J., Pepin, R. O., and Merrill, R. B., editors.
(1977) *Impact and Explosion Cratering*, Pergamon Press (New York), p. 1–10.
Printed in the United States of America

Why study impact craters?

E. M. SHOEMAKER

U.S. Geological Survey, Flagstaff, Arizona 86001 and California Institute of Technology, Pasadena, California 91125

Abstract—The thesis of this paper is that impact of solid bodies is the most fundamental process that has taken place on the terrestrial planets. The terrestrial planets were formed by this process; the last stage of accretion is still proceeding at a very slow rate. Growth of planets began when planetesimals formed as a consequence of gravitational instabilities in a particle disk in the solar nebula. These planetesimals then began to accumulate into planetary nuclei by collision. In the terrestrial planet zone this growth was accelerated by the early formation of Jupiter. Large eccentricities and inclinations acquired by the planetary nuclei in the asteroid belt, as a consequence of perturbations by Jupiter, prevented the development of a planet there. Part of the collisional debris produced in the asteroid belt was deposited on the terrestrial planets and this debris probably contributed to the late heavy bombardment of the moon. Most of the volatile constituents of Earth and Venus probably were derived from the region of the asteroid belt and more distant parts of the solar system; the volatile-rich material was deposited at a late stage of accretion. The moon evidently was too small for more than a very small fraction of this high velocity volatile-rich material to stick. It is suggested that the moon is a sample of the protoearth's mantle that was spun off when the earth had grown to about one-half its present mass. Much more research on impact craters is needed to solve nearly every aspect of the accretion process.

IN PRESENTING THIS PAPER at the Symposium on Planetary Cratering Mechanics, my goal was to identify some of the more interesting problems which the study of impact processes might elucidate. Rather than describe those problems for which solutions have been offered that are reasonably well buttressed with theory and observations, I have deliberately explored ground where much more research is needed. In an attempt to present these areas of research in a compact and provocative way, a number of conclusions that are necessarily speculative are simply stated as assertions. Full documentation of the literature bearing on these problems would result in a bibliography much longer than the present paper; in the interest of brevity, only a few relatively recent papers have been cited.

As a first speculative assertion, I submit that impact of solid bodies is the most fundamental of all processes that have taken place on the terrestrial planets. This is the central thesis of this paper. Without impact, Earth, Mars, Venus, and Mercury wouldn't exist. Collision of smaller objects is the process by which the terrestrial planets were born.

The geological record of the earliest history of impacts on the terrestrial planets has been lost. As the process is self-erasing, to a certain extent, the earliest record would have been lost even if processes of melting and internal evolution of the planets had not occurred. But much of the record of the last stages of accretion of the planets is preserved, especially on the moon, Mercury, and Mars. In fact, the last stage of accretion is still going on, albeit at a very slow rate. This is fortunate, because we can study many aspects of the processes of planetary birth by investigation of the nature of small bodies which still exist, the dynamics of their orbital evolution, and the effects which they produce when they ultimately collide with a planet. If impact and accretion were not still occurring, it would be hard to come to grips with a number of difficult problems of planetary origin and early evolution.

Now what are some of the problems of planetary accretion that we wish to know about? The first problem is how anything sticks together at all in the early stages of accretion. How is it that terrestrial planets were formed instead of ground-up debris, such as that found today in the asteroid belt, or even finer debris that might have been swept away by an early solar wind? Safronov (1969) and Goldreich and Ward (1973) have leaped the first hurdle by showing how kilometer or tens of kilometer sized bodies would have been produced from particles condensed from the solar nebula. Gas drag in the nebula caused the particles to settle rapidly to a thin disk, and gravitational instabilities in the disk caused it to collapse into small bodies prior to the dissipation of the nebula. The next question is how these initial bodies, referred to here as planetesimals, were collected into planet-sized objects.

Initially, the orbits of the planetesimals must have been nearly circular and nearly coplanar. This would be a requisite condition for the formation of planetesimals even if they were not formed by processes described by Safronov and by Goldreich and Ward. If the orbits of particles which first collected into the planetesimals had not been nearly circular and coplanar, then the particles would not have stuck together on collision. The encounter velocities would have been too high. This general requirement for low encounter velocities has been known to impact experimentalists for quite some time. In order for accretion of larger bodies to proceed, the orbits of the planetesimals must have become sufficiently eccentric after they were formed so that they could collide with one another. Orbits of growing protoplanetary nuclei formed in this way must ultimately have become still more eccentric, in order to be collected into the small number of planets found today.

Mutual scattering may have played a role in building up eccentricity at early stages of accretion. As far as most of the objects that accreted to form terrestrial planets are concerned, however, early growth of the planet Jupiter probably was critical. As soon as Jupiter grew even to a fraction of its present size, sufficient secular perturbations on the terrestrial planetesimals were produced by the gravitational attraction of Jupiter to promote rapid growth of the terrestrial planetary nuclei. I shall not inquire here how Jupiter and also Saturn were formed. We know from the abundance of helium in these two planets that, at some stage, they grew by direct gravitational collapse of parts of the solar nebula. How this collapse was initiated is a subject of controversy and current investigation. For this discussion I will take Jupiter and Saturn as given.

Jupiter did form, and as it did, considerable eccentricity was pumped into the orbits of the planetesimals in the zone of the terrestrial planets. The maximum eccentricity of the planetesimals produced by the secular perturbations was directly proportional to the eccentricity of Jupiter itself and approximately proportional to the semi-major axis of the orbit of each planetesimal. Those remaining planetesimals closest to Jupiter acquired osculating eccentricities sufficiently large for their orbits to become Jupiter-crossing. Some were then swept up by Jupiter. Those that were not swept up were deflected by a succession of close encounters with Jupiter into orbits of high eccentricity or ejected from the solar system. Some of the deflected objects probably now reside in the Oort cloud of comets (Öpik, 1973) and others ran into and smashed up other terrestrial planetesimals (Kaula and Bigeleisen, 1975). Only a small fraction of the mass of these latter objects probably now resides in the terrestrial planets (Wiedenschilling, 1975).

Farther away from Jupiter, closer to the sun, in what is now the more densely populated region of the main asteroid belt, the osculating eccentricities of the planetesimals remained too low to overlap the orbit of Jupiter. These eccentricities, nevertheless, were and still are rather large—too large to grow a planet. The present mean eccentricity of the asteroids is about 0.15. Curiously, the dispersion of the asteroids out of the proper plane of the solar system is such that the average contribution of inclination to their mutual encounter velocities is about equal to the contribution of eccentricity. The mean inclination is about 10°. It is not clear just how this distribution of inclinations came about, but evidently it is the result of mutual scattering of the asteroids, or, more particularly, of asteroid parent bodies during very close encounters.

The current r.m.s. collision velocity in the asteroid belt is about 5 km/sec. On empirical grounds, this velocity is too high for growth of a planet, as no planet is present in the belt. To put this conclusion in terms of cratering, an r.m.s. impact velocity of 5 km/sec is above the threshold of growth for the largest planetesimals or planet nuclei that might have been present in the asteroid belt. More mass is lost as crater ejecta traveling at velocities exceeding the gravitational escape velocity of the largest objects than is added to such objects by impact. The largest objects remaining in the belt, such as Ceres and Vesta, have escape velocities of several hundred m/sec. For the distribution of impact velocities among the asteroids, more mass is ejected from craters at velocities exceeding these escape velocities than is added as impacting projectiles. That, at least, is what the solar system seems to be telling us. It would be highly desirable to have experimental and theoretical demonstration of this conclusion, however. If the frequency distribution of mass as a function of ejection velocity for impact craters produced by projectiles of different velocity were accurately known, then reasonably precise bounds could be placed on the largest object that might have existed in the asteroid belt.

A first attempt to find the frequency distribution of ejection velocities from impact craters was described by Gault *et al.* (1963). At that time we were trying to determine the amount of material sprayed off the moon by impact and the steady state density of the cloud of fine impact debris in sub-orbital trajectories. The experimental results suggested that, above about 9 km/sec impact velocity, more mass is ejected from the moon at escape velocity than is added by impact. As the present r.m.s. velocity of objects hitting the moon is about 22 km/sec (Shoemaker, 1977), our interpretation of the experiment indicates that the moon is losing mass at the present time rather than growing by accretion.

On the basis of the distribution of ejection velocities found by Gault *et al.* (1963), the critical impact velocity limiting growth of an object the size of Ceres would be of the order of 1 or 2 km/sec. As the present r.m.s. impact velocity is 5 km/sec, it would appear that Ceres and the other largest asteroids must have been formed prior to complete growth of Jupiter. Are these objects simply the largest planetesimals formed by the processes described by Safronov and by Goldreich and Ward? Or did these objects grow by accretion of planetesimals while the nucleus of Jupiter was similarly growing?

Recently, O'Keefe and Ahrens (1977) have attacked the problem of ejection velocities from impact craters by computer modeling of the impact process. For objects with escape velocities between a few hundred m/sec and 1 km/sec, the critical velocities limiting growth found from computer modeling are close to those inferred from experiment. For a body with an escape velocity of 2.4 km/sec such as the moon, however, the critical impact velocity is about 21 km/sec (derived by interpolation between results obtained from computer runs for 15 km/sec and 30 km/sec impact velocity). The critical velocity found from the work of O'Keefe and Ahrens is close to present r.m.s. velocity of impacting bodies, which suggests that the moon is neither gaining nor losing much mass at the present time.

Returning to the asteroid belt, there is an unsolved theoretical difficulty concerning the mean eccentricity of the asteroids. The present mean eccentricity of these bodies is more than twice the maximum eccentricity of Jupiter. Except in low order resonances and certain secular resonances, perturbations by Jupiter alone cannot produce these high eccentricities. It is difficult to escape the conclusion that the high eccentricities as well as the high inclinations probably have been produced by mutual scattering among objects in the asteroid belt. But in order for this to happen, there must have been a great many more objects to start with about the size of Ceres and Vesta or larger. Perturbations by Jupiter would have placed these large asteroids on overlapping orbits. Then the mutual perturbations during very close encounters produced still larger eccentricities, which allowed more distant orbits to overlap which, in turn, allowed still greater perturbations. As Öpik (1963) has remarked, this process is a kind of celestial ball game. In order to build up an r.m.s. encounter velocity of about 5 km/sec, the maximum increments of velocity acquired in single encounters must be a significant fraction of the ultimate total. But the maximum increments are of the order of the escape velocity. That is why the objects must have been large. Also, there must have been many of them in order to play the ball game.

If all of these large bodies were present initially in the asteroid belt, the next question is where did they go? The answer is that most of them were destroyed by physical collision. Once the velocities reached a second critical level, collisions among even the largest asteroids would have produced catastrophic disruption rather than merger. Again, we need good experiments and good theory in order to define these critical disruption velocities as a function of the size and properties of the asteroid parent bodies and to determine the size distribution of the resulting fragments. Some of the most relevant observations to date have been contributed by Gault and Wedekind (1969).

That collisional destruction has taken place is beyond reasonable doubt. The empirical evidence is strong. As shown by Dohnanyi (1971), the size distribution of all but the largest asteroids approximately fits that which would be predicted by multiple fragmentation. More to the point, the majority of known asteroids belong to distinct orbital families, many of which were first recognized by Hirayama (1918, 1923). Except for asteroids in the Phocaea and Hungaria regions of orbit phase space, which are not true families, each of the well-defined families is almost certainly formed by collisional breakup of an individual asteroidal body. Not only are the proper elements of the orbits closely grouped, but the size distribution of the recognized members of each distinct family resembles that produced experimentally by single fragmentation events. In some cases, a major fragment of the parent object, recognizable by its size, still remains, and in other cases it has been destroyed.

There is a question, however, as to how much material has been lost from the asteroid belt. Very little mass remains in the asteroid belt now; the amount is only a small fraction of the mass of the moon. From general considerations, it appears likely that enough mass condensed from the solar nebula in this region to grow a large terrestrial planet, had the encounter velocities permitted such growth. Furthermore, we have seen that many large asteroids probably were required to produce the high mean eccentricity and inclination of the remaining objects. As the bulk of the mass left is in the few remaining large asteroids, the mass of the other large asteroids that were destroyed by collision has somehow disappeared.

Chapman and Davis (1975) have proposed that the bulk of the original asteroidal mass has been ground up very finely by repetitive impact to the point where it could be removed by radiation pressure and the Poynting–Robertson effect. If this happened, it must have occurred prior to 3.8 b.y. ago. Otherwise, we would see a much larger amount than we do of asteroidal material in the regolith of the lunar maria. This is because a large fraction of the debris would have been reduced to a broad range of sizes above the threshold for ejection by radiation pressure. This fraction would have been displaced inward by the Poynting–Robertson effect.

Another possibility is that relatively finely ground up asteroidal debris was blown away by a very strong early solar wind, which also presumably dissipated the solar nebula in the zone of the terrestrial planets. Objects up to meters across might have been removed this way, depending on the strength of the wind.

At least part of the initial mass in the asteroid belt has been deposited on the terrestrial planets. Again, the circumstantial evidence is fairly strong. Williams (1969) has discovered belts of secular resonance in proper element phase space of the asteroid orbits. For certain combinations of semi-major axis and proper inclination, secular perturbations lead to sufficiently high osculating eccentricities that any asteroids with these elements would become Mars-crossers. As pointed out by Williams (1971), the belts of secular resonance are practically devoid of asteroids. The implication is clear that the missing asteroids have been swept out.

Three things can happen to a Mars-crossing asteroid that result in its disappearance. First of all, it can impact directly on Mars. Secondly, it can be deflected into an Earth-crossing or even Venus-crossing or Mercury-crossing orbit and collide with one of the other terrestrial planets. Thirdly, it can be deflected into a Jupiter-crossing orbit, in which case it is either fairly quickly ejected from the solar system or collides with one of the Jovian planets. The relative probability of each of these general fates is very roughly similar (Wetherill, 1975). Hence, a substantial fraction of the asteroidal material lost from the belts of secular resonance now resides on Mars and a similar fraction on the other terrestrial planets. This process is still going on and must provide some of the meteoritic material now falling on the earth.

In my opinion, similar fates befell asteroidal material initially located in the Kirkwood gaps, or at least in the gap at the 2:1 resonance with Jupiter. The few asteroids remaining in the 2:1 resonance are almost Mars-crossing. Zimmerman and Wetherill (1973) have shown how material from objects on the margin of the 2:1 resonance can be placed in planet-crossing orbits by a combination of multiple collisions and secular perturbations. Single collisions, followed by secular perturbations are sufficient to place objects that are initially deep in the resonance on planet-crossing orbits. Thus, the Kirkwood gaps may also have been significant sources of material now deposited on the terrestrial planets.

Now let us take a look at what happened to the planetesimals that were formed still closer to the sun than the main asteroid belt. For planetesimals with initially zero eccentricity, the osculating eccentricities induced by the secular perturbations due to Jupiter decrease with decreasing semi-major axis, as given by the following formula (Heppenheimer, 1977):

$$e = \frac{5}{2} E\alpha K,$$

where e = maximum osculating eccentricity of planetesimal,
E = eccentricity of Jupiter,
α = ratio of semi-major axis of planetesimal to that of Jupiter,

and
$$K = 1 - \frac{1}{8}\alpha^2 - \frac{5}{128}\alpha^4.$$

As Jupiter's eccentricity varies between .027 and .062 (Brouwer and Clemence, 1961), perturbations by Jupiter alone lead to maximum eccentricities of a few hundredths or less in the vicinity of the terrestrial planets. This is enough to initiate accretion and growth of many planetary nuclei. Encounter velocities associated with these eccentricities would be down in the range of a few hundred m/sec or less. Evidently this was low enough to permit the largest planetesimals to grow by absorbing their neighbors. Again, it is crucial to know the frequency distribution of ejection velocities from impact craters in order to specify the conditions for growth.

Osculating eccentricities induced purely by secular perturbations due to Jupiter were not sufficient to collect all the planetesimals into just four planets, however. At the minimum, several dozen planetary nuclei could have formed, each sweeping out a ring on the order of a hundredth to a few hundredths of an astronomical unit wide. The fact that there are only four planets now and that the space between them has been swept clean means that either the terrestrial planetesimals or the growing planetary nuclei or both ultimately acquired eccentricities of the order of 0.1. Just as in the asteroid belt this probably occurred as a result of scattering by close encounteres with or among the larger growing objects.

As the collisional velocities increased with increasing average eccentricity (and probably with increasing inclination) a competition set in between collisional destruction and accretion. Small objects would have been fragmented and dispersed at the same time that the larger objects grew by accretion. Some objects of intermediate size could have grown for a while only to be smashed later or else to be swept up into a larger growing nucleus. Complex cycling between accretion of debris and fragmentation followed by reaccretion and refragmentation became possible. It is not surprising that most stony meteorites, the parent bodies of which probably also were formed during similar cycling between accretion and fragmentation, are compound breccias.

Near the boundary of the asteroid belt, the competition between destruction and accretion must have been close to a draw. Much of the planetesimal material near the boundary may have been pulverized and lost essentially by the same mechanisms by which most of the mass of the asteroid belt was lost. This may account for the low mass of Mars as compared with Earth and Venus. In the vicinity of the earth and still nearer the sun, the eccentricities produced by secular perturbations were smaller and the ultimate mean eccentricities and collisional velocities produced by scattering probably were smaller. The proportion of mass lost from this region probably was minor. Indeed, there may have been a net gain of mass by transfer of material from the vicinity of Mars and the asteroid belt and from more distant parts of the solar system.

Ultimately, four growing nuclei in the region of the terrestrial planets won the accretion race and consumed their competitors. As the winners grew, they did not remain at the same distance from the sun. Conservation of angular momentum required that the semi-major axis of the growing planets shifted in accordance with orbits of the material that was being added to them. The distribution of sources of material deposited on Mars, in particular, must have been asymmetric in the later stages of growth of Mars. As the larger planet Earth accreted, the space between Earth and Mars was finally swept clean. On the side of Mars toward Jupiter, however, there remained a large number of small bodies which, because of their high encounter velocities, had not coalesced into a planet. Thus, nearly all the late growth of Mars occurred by accretion of bodies with larger orbits and higher specific angular momentum. The result was that Mars was gradually driven farther from the sun as it swept up asteroids. This migration, in turn, allowed Mars to sweep up a wider region of space than it would have otherwise.

As Mars swept up asteroids, a fraction of these asteroids were perturbed by close encounters with Mars into earth-crossing orbits. Some of the earth-crossers were, in turn, perturbed by close encounters with the earth into Venus-crossers and some, by multiple encounters, became Mercury-crossers. At a late stage of accretion, a considerable amount of asteroidal material probably was handed down this way into the inner part of the solar system. This process is still going on and provides some of the earth-crossing asteroids we find today.

Between 4 and 3.3 b.y. ago the cratering rate on the moon decayed with a $1/e$ lifetime of the order of 100 m.y. (Shoemaker, 1972; Soderblom and Boyce, 1972; Neukum *et al.*, 1975). This decay time is comparable to the lifetime of possible Mars-crossing asteroids that were swept up by Mars or deflected to Earth-crossing orbits at a late stage of accretion of Mars. Therefore, it seems likely that many objects involved in the late heavy bombardment of the moon were asteroids scattered by Mars. If so, Mars was still accreting fairly vigorously at that time.

Wetherill (1975) has shown that another possible source of objects with lifetimes appropriate to account for the late heavy bombardment of the moon is the population of planetesimals and larger bodies from which Uranus and Neptune accreted. This is also a region from which many comets may have been derived. Multiple encounters with the outer planets almost certainly handed down some of this material to the region of the terrestrial planets. The impact history of the moon, in my opinion, does not require breakup at 4 b.y. ago either of a large object in the asteroid belt or of an object derived from the outer solar system, however. (See discussion by Chapman, 1976; Wetherill, 1976.)

I interpret the decipherable record of bombardment of the moon prior to 3.3 b.y. as the end stage of a period of intense bombardment that lasted throughout the moon's early history. The total mass of "asteroidal" material and "cometary" material from more distant sources that was deposited on the planets during this period constitutes the bulk of the planet Mars, perhaps as much as $\frac{1}{10}$ to $\frac{1}{4}$ of the mass of the earth and of Venus and much less of Mercury. By "asteroidal" I refer here not only to material from the present asteroid belt but also material initially condensed as close as about 1.3 a.u. from the sun. If this estimate is correct, probably most of the volatile constituents of the earth were derived from the "asteroidal" and "cometary" components. These components were not uniformly mixed in the earth to begin with. Because of the longer lifetimes against collision, they were deposited last. Thus the volatiles which produced the ocean and initial atmosphere were essentially plated on the earth in the last stages of accretion. A similar batch of volatiles was also plated on Venus.

If the above picture is correct, it may lead to a more coherent understanding of the origin of the moon. We are required to explain how the moon is simultaneously poor in iron, rich in the other refractory elements, and poor in volatile constituents. I suggest that the moon was formed by fission from the earth at a comparatively early stage of its growth. The fission occurred as a result of overspinning of the earth during accretion. At a later stage, the earth was spun down well below the stability limit during accretion of fairly large competing nuclei. Thus I view the moon as a sample of the protoearth's mantle, spun off when the earth was perhaps one-half grown. It is composed of refractory-rich material condensed probably near 0.9 a.u., minus most of the siderophile elements which separated into the protoearth's core as the earth accreted. The volatile-rich material that was plated on the earth did not accumulate on the moon because velocities of the volatile-rich objects were too high for accretion on the moon. This must be true, regardless of how the moon formed. The velocity distribution of the volatile-rich objects probably was similar to the velocity distribution of objects currently in Earth-crossing orbits. If the "cometary" component predominated, it may have been somewhat higher. As it stands now, the computer modeling of impact by O'Keefe and Ahrens just marginally supports this explanation of low volatile abundance on the moon. Clearly we need to know much more about the impact process, in addition to the celestial dynamics of the planetesimals, in order to understand the origin and composition of the terrestrial planets.

References

Bouwer, D., and Clemence, G. M.: 1961, Orbits and Masses of Planets and Satellites, *The Solar System III, Planets and Satellites* (G. P. Kuiper and B. M. Middlehurst, eds.), p. 31–94, Univ. Chicago Press, Chicago.

Chapman, C. R.: 1976, Chronology of terrestrial planet evolution: The evidence from mercury, *Icarus* **28**, 523–536.

Chapman, C. R., and Davis, D. R.: 1975, Asteroid collisional evolution: Evidence for a much larger early population, *Science* **190**, 553–556.

Dohnanyi, J. S.: 1971, *Fragmentation and Distribution of Asteroids, Phys. Studies of Minor Planets* (T. Gerhels, ed.), NASA SP-267, 263–295.

Gault, D. E., Shoemaker, E. M., and Moore, H. J.: 1963, *Spray Ejected From the Lunar Surface by Meteoroid Impact*, NASA TN-D-1767, 39 pp.

Gault, D. E., and Wedekind, J. A.: 1969, The destruction of tektites by micrometeoroid impact, *J. Geophys. Res.* **74**, 6780–6794.

Goldreich, P., and Ward, W. R.: 1973, The formation of planetesimals, *Astrophys. J.* **183**, 1051–1061

Heppenheimer, T. A.: 1977, On the Interaction between Nebular Drag, Gravitational Perturbations, and Accretion, *The Origin of the Solar System* (S. Dermott, ed.), Wiley, New York. In press.

Hirayama, K.: 1918, Groups of Asteroids Probably of Common Origin, *Astr. J.* **31**, 185–188.

Hirayama, K.: 1923, Families of Asteroids, *Jap. J. Astron. Geophys.* **1**, 55–105.

Kaula, W. M., and Bigeleisen, P. E.: 1975, Early scattering by jupiter and its collision effects in the terrestrial zone, *Icarus* **25**, 18–33.

Neukum, G., König, B., Fechtig, H., and Storzer, D.: 1975, Cratering in the earth-moon system: Consequences for age determination by crater counting, *Proc. Lunar. Sci. Conf. 6th*, p. 2597–2620.

O'Keefe, J. D., and Ahrens, T. J.: 1977, Partitioning of energy and the degree of melting and vaporization in planetary impact processes, to be submitted for publication in *Science*.

Öpik, E. J.: 1963, Survival of comet nuclei and the asteroids, *Ad. Astron. Astrophys.* **2**, 219–262.

Öpik, E. J.: 1973, Comets and the formation of planets, *Astrophys. Space Sci.* **21**, 307–398.

Safronov, V. S.: 1969, *Evolution of the Protoplanetary Cloud and Formation of the Earth and the Planets*, Nauka, Moscow, Translated by Israel Program for Scientific Translations and printed in 1972 by Keter Press, Jerusalem, 206 pp.

Shoemaker, E. M.: 1972, Cratering history and early evolution of the moon (abstract), in *Lunar Science III*, The Lunar Science Institute, Houston, 696–698.

Shoemaker, E. M.: 1977, Astronomically observable crater-forming projectiles. This volume.

Soderblom, L. A., and Boyce, J. M.: 1972, *Relative Ages of Some Near-Side and Far-Side Terra Plains Based on Apollo 16 Metric Photography*, NASA SP 315, 29-3 to 29-6.

Weidenschilling, S. J.: 1975, Mass loss from the region of Mars and the asteroid belt, *Icarus* **26**, 361–366.

Wetherill, G. W.: 1975, Late heavy bombardment of the moon and the terrestrial planets, *Proc. Lunar Sci. Conf. 6th*, p. 1539–1561.

Wetherill, G. W.: 1976, Comments on the paper by C. R. Chapman: Chronology of terrestrial planet evolution—the evidence from Mercury, *Icarus* **28**, 537–542.

Wetherill, G. W.: 1976, Where do the meteorites come from? A re-evaluation of the Earth-crossing Apollo objects as sources of chondritic meteorites, *Geochim. Cosmochim. Acta* **40**, 1297–1317.

Williams, J. G.: 1969, Secular Perturbations in the Solar System, *Ph.D. Dissertation*, University of California at Los Angeles, 1–270.

Williams, J. G.: 1971, Proper Elements, Families and Belt Boundaries, in *Physical Studies of Minor Planets* (T. Gehrels, ed.), NASA SP-267, 177–181.

Zimmerman, P. D., and Wetherill, G. W.: 1973, Asteroidal source of meteorites, *Science* **182**, 51–53.

Roddy, D. J., Pepin, R. O., and Merrill, R. B., editors.
(1977) *Impact and Explosion Cratering*, Pergamon Press (New York), p. 11–44.
Printed in the United States of America

A summary of explosion cratering phenomena relevant to meteor impact events

HENRY F. COOPER, JR.

R & D Associates, 4640 Admiralty Way, Marina del Rey, California 90291

Abstract—Theoretical and experimental studies of cratering and related phenomena from buried explosions are reviewed with emphasis on information applicable to the study of impact craters. Phenomena considered include crater volume and shape, ejecta, airblast, and ground shock. Based on a number of theoretical studies and experiments, a fairly consistent picture of the evolution of a crater (and of the mechanisms important to the cratering process) emerges.

1. INTRODUCTION

COMPARISONS OF IMPACT CRATERS and craters from buried high-explosive (HE) sources by Oberbeck (1971) coupled with studies of the crater morphology of terrestrial impact craters (such as Meteor Crater), craters from surface HE sources (e.g., Roddy, 1968, 1976) and craters from shallow-buried nuclear-explosive (NE) sources (e.g., Shoemaker, 1963; Roddy *et al.*, 1975) suggest that many late-stage impact and explosive cratering mechanisms are similar. Hence, an improved understanding of explosive cratering phenomena (where the source is known) could significantly add to the understanding of large-scale lunar and planetary impact cratering phenomena (where the source is not known). It is with this thought in mind that this summary paper was prepared to discuss explosive cratering mechanisms and related phenomena.*

Planetary and lunar impact craters are estimated to result from impact velocities between a few and several tens of km/sec which produce shock pressures up to a few 100 megabars in the earth (and impacting projectiles). Rock vaporization and melting produced by such intense pressures cannot be simulated by HE sources (which produce at most a few hundred kilobars pressure in the rock) but do result from nuclear explosions. Thus, buried NE sources most nearly simulate the close-in phenomena produced by impact events. However, most of the crater is formed after the pressure has attenuated to less than a few hundred kilobars, and buried HE and NE sources produce very similar late-time cratering and ground shock effects. Therefore, phenomena produced by both buried NE and HE sources are pertinent to understanding impact cratering events.

If a high energy density NE source is buried more than about 2 m/MT$^{1/3}$†, the prompt radiation will be deposited in the earth media, and the subsequent

*The substance of this paper was presented at the Symposium on Planetary Cratering Mechanics on 13–17 September 1976 at Flagstaff, Arizona.

†A 1-megaton (MT) explosive yield is taken to be equivalent to 4.2×10^{22} ergs.

cratering and ground motions should be essentially the same as would occur from a non-radiating hydrodynamic source (Knowles, 1973). If the NE source is sufficiently buried for the peak stress incident on the earth's surface to be less than about 100 kbar (which occurs at about 100 $m/MT^{1/3}$ depth-of-burial in hard rock), the differences in ground shock and cratering phenomena caused by NE and HE cratering bursts should be minimized. Thus, spherical HE charges detonated approximately one charge diameter below the surface should closely simulate the ground motion and cratering action of NE sources of twice the energy* buried on the order of 100 $m/MT^{1/3}$. It is fortuitous that this depth-of-burial is approximately equal to the depth-of-burst determined by Oberbeck (1971) to simulate the cratering and ejecta effects of impact cratering phenomena with HE experiments. It is also approximately half of the scaled depth-of-burial of Teapot Ess, a nuclear cratering event in dry alluvium often compared to impact cratering events (e.g., see Shoemaker, 1963).

We focus our attention on cratering and related effects from shallow-buried explosions because such events apparently involve much of the phenomena observed in impact cratering events. However, for completeness we also discuss some phenomena associated with surface and optimum depth bursts.

Figure 1 illustrates the effects of explosion cratering events discussed in subsequent sections. An explosion in the vicinity of the earth's surface releases energy into the ground and into the atmosphere.† The energy coupled into the atmosphere produces an airshock which propagates outward over the earth's surface. This attenuating and decellerating airblast wave in turn loads the ground producing what is termed as "airblast-induced ground shock". The energy

*As will be noted later, for buried explosions, HE is about twice as effective as NE in creating ground shock and cratering.

†Unless otherwise noted, our discussion of explosive cratering phenomena is constrained to explosions on the earth.

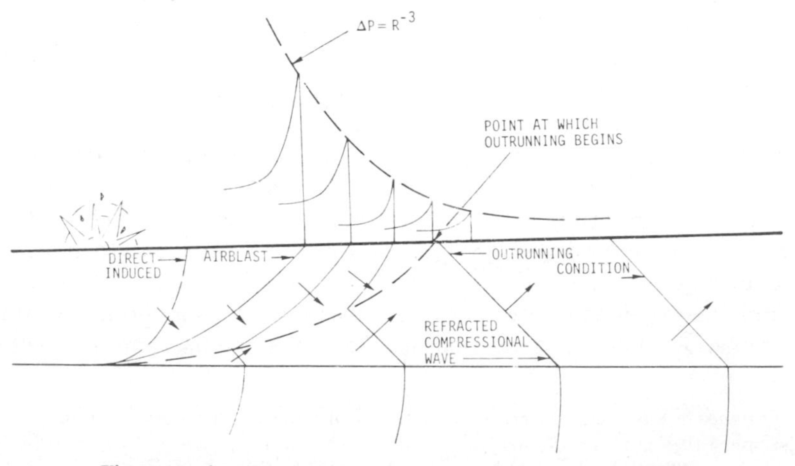

Fig. 1. Basic phenomenology from a surface burst explosion.

coupled by the nuclear source into the ground produces the crater ejecta and a hemispherically diverging "direct-induced ground shock" wave which propagates outward at the compressional wave speed of the soil or rock media. The late-time ground motions result from complex transmission, reflection and refraction of airblast-induced and direct-induced ground shock waves at the various geologic interfaces that exist at all geologic sites. The ejecta may travel on ballistic trajectories or may be significantly influenced by the blast winds and late-time airflow associated with the fireball rise phenomena—depending on the explosive yield and source geometry.

Variations in the depth-of-burst in the neighborhood of the earth's surface significantly alters the energy coupled into the ground while influencing the airblast phenomena in only a minor way. The most dramatic change in energetics occurs as the source geometry varies from slightly above the surface to slightly below the surface. For example, shallow-burial (2–3 $m/MT^{1/3}$) increases the energy coupled into the ground by at least a factor of 4 (as compared to a surface burst) while reducing the energy coupled into the air by only about 15%. As the source is buried more deeply, the energy coupled into the ground increases (and the energy coupled into the air decreases) gradually until the source is completely contained by the earth media, and no source energy is deposited in the atmosphere.

Our current understanding of these effects is derived from theoretical and experimental studies of explosive cratering phenomena. The next section briefly discusses first principal theoretical techniques and the role they play in current research to better understand explosion cratering phenomena. Then we review aspects of the experimental data base which serves to evaluate theoretical models of the cratering process. Following these discussions of theoretical and experimental studies, we describe, in qualitative terms, the various stages of the cratering process.

2. Comments on Theoretical Calculation Techniques

In recent years finite difference techniques have been developed and applied to study airblast, cratering, ground motion, and other nuclear effects from near-surface nuclear explosions. These theoretical procedures are formulated from first principles and treat radiative phenomena (transport and diffusion) coupled with material continuum response.

Calculations of weapon physics, early time fireball growth, and later-time airblast phenomena, model radiation transport phenomena and hydrodynamic motions of the bomb debris, air and ground medium. Hydrodynamic calculations have been continued to late times and have reproduced fireball rise and toroidal wind flow observed in nuclear tests (e.g., Needham, 1971). In principle, such calculations can also treat the rising dust cloud. However, credible models are lacking for treating dust pickup and the entrainment/mixing with the fireball or nuclear cloud.

The first major calculation of an explosion crater, performed by Brode and Bjork (1960), considered a nuclear surface burst and modeled the high pressure response of a porous rock as a hydrodynamic fluid. These same assumptions were used by Bjork (1961) to perform a calculation of the formation of the Arizona meteor crater. The early time phenomena in these calculations of explosion and impact cratering events had many similar features. However, important effects of material strength on low stress phenomena during the later stages of crater formation were not modeled. Since these pioneering calculations, the theoretical techniques have been extended to consider material strength and calculations of explosively produced craters have met with varying degrees of success.

In these calculations, the ground surface is usually loaded with an airblast pressure boundary condition, e.g., calculations of surface bursts usually apply the form specified by Brode (1968). Depending on the particular explosion phenomena under study, the airblast interaction with the soil may be modeled to account for gas flow into cracks, pressure equilibrium around rocks and ejecta, airblast-induced ground shock, and earth compaction (Trulio, 1970).

There are two main sources of uncertainty in applying these calculational procedures: numerical errors and the state-of-ignorance of the dynamic response of geologic media. Various first principle calculations of the close-in high pressure earth response to a given surface burst geometry have produced factor of 2 inconsistencies in calculated peak pressures and particle velocities (e.g., Cooper, 1967). The numerical treatment, particularly during re-zoning of the calculational mesh, can lead to substantial errors in the calculated stress pulse shape. Such numerical errors are usually reduced by increasing the number of zones in the calculation; however, financial constraints often limit such a brute force approach to improving accuracy. Other errors in the theoretical model could be associated with the current lack of a technique for modeling non-continuum phenomena.

Although numerical errors can be significant and must always be borne in mind, the largest errors in current calculations are thought to be associated with modeling the dynamic response of *in situ* geologic media. These errors result from known inadequacies of current theoretical constitutive relations, and from a lack of *in situ* rock property data necessary to improve the constitutive relations. This weakness is particularly apparent for post-shock material properties so important to late-stage cratering flow.

The difference between properties of small in-tact samples and the *in situ* response of geologic media is now widely acknowledged. Experiments in near-surface hard rock have demonstrated the importance of pre-existing joints and faults in determining the dynamic strength of a rock mass (e.g., Cooper and Blouin, 1971). Attempts to calculate the ground motions observed on these experiments suggest that the *in situ* strength may be an order of magnitude less than inferred from laboratory tests on small intact samples. More recent theoretical studies of the ground motions from HE surface explosions in clay shales and sandstones have had poor success in reproducing the observed ground motions—constitutive relations derived from laboratory tests have been unsuccessful in predicting even the correct arrival times of the ground shock (e.g., Port and Gajewski, 1973; Trulio and Perl, 1973; Port and Bratton, 1973; and Sandler *et al.*, 1974). Current research is emphasizing the development and application of *in situ* tests to aid in the development of credible constitutive relations, but quantitatively accurate calculations of the ground motions from surface bursts remain as a relatively rare occurrence.

After appropriately warning the reader of shortcomings of the current calculational state-of-the-art, we hasten to point out that valuable qualitative information and sometimes quantitatively accurate predictions can be obtained from the calculational procedures.

The most reliable information provided by the current ground motion calculations is probably guidance on the general phenomenology. The shapes of peak stress and particle velocity contours are probably more accurate than the precise numbers generated by a given calculation. Ejection angles and spatial variation of ejection velocities may also be more reliable than the magnitude of ejected mass—especially for surface bursts. Thus, parametric studies where the effects of changes in input parameters on such qualitative features are evaluated may lead to improved understanding of the phenomena even though the numbers calculated in any specific case may not accurately predict reality. In other words, the trends indicated by calculations may be correct even though quantitative predictions may not be credible.

In some cases, accurate quantitative predictions are possible. As illustrated by Cherry (1967), Terhune *et al.* (1970), and Glen and Thomsen (1976), calculations of cratering phenomena from deeply buried explosions have been successful in reproducing the apparent crater, ground motions and ejecta beyond the crater's edge. Furthermore, crater volumes from near-surface HE sources and from shallow-buried NE sources have been calculated to within a factor of 2—even though the crater shapes and subsequent ground motions are poorly calculated. As demonstrated by Ullrich (1976), even such late-time features as the central uplift observed on high-explosive experiments can be qualitatively if not quantitatively reproduced (if the constitutive relations are modified in a realistic although *ad hoc* way).

As discussed by Knowles and Brode (1977), the largest uncertainty exists for calculating the craters from above-surface nuclear explosions. For this case, calculated crater volumes are at least an order-of-magnitude less than predictions based on empirical generalizations of the craters produced by the large yield nuclear tests in the Pacific. This large discrepancy is thought to result from our lack of understanding of the physics of energy coupling from surface explosions and possibly the very high pressure response of rock. Trulio (1976) argues that much of the effects that contribute to the uncertainties in calculations of a surface burst are made ineffective by even shallow burial. Fortunately, our discussion emphasizes the similarities between the buried explosions and impact cratering events where many uncertainties have less impact on the quantitative predictions of cratering phenomena, and where the calculations of buried explosions have been much more successful in calculating the observed crater volumes.

No calculation to date has produced a shallow crater of the shape observed on the Pacific nuclear tests or on large impact cratering events. In fact, all calculations known to the author have produced deep bowl-shaped transient cavities, but such calculations are usually terminated before late-stage gravity-induced flow might modify the bowl shape and lead to a shallow final crater. Recently developed implicit numerical techniques may make future late-time calculations more practical.

3. Comments on the Experimental Data Base

Although first principle theoretical calculations promise to provide quantitative understanding of cratering and related phenomena, it probably will be some time before they will be completely reliable. In the interim before high confidence can be placed in the calculational procedures, our understanding of cratering phenomena will rely heavily on the analysis of experimental data. Furthermore, calculations of qualitative and quantitative features observed in various experiments constitute a necessary validation process to build confidence in the large complex theoretical codes. This section briefly highlights data from various HE and NE experiments that serve as a basis for our current understanding of explosive cratering and related effects phenomena.

HE cratering experience ranges from grams of explosive in controlled laboratory experiments up to 500 tons* of explosive in field experiments (often referred to as trials). (See Oberbeck, 1971; Piekutowski, 1975b; Vortman, 1969; Roddy, 1970, 1973; Lockard, 1974; Rooke *et al.*, 1974.) A less extensive data base is available for nuclear explosions in dry soils at the Nevada Test Site (NTS) and saturated coral at the Eniwetok and Bikini Atolls in the South Pacific (Vortman, 1968; Crawford *et al.*, 1974). Although there are significant gaps in the data base, these data allow some examination of the effects of depth-of-burst, variations in source details, geology and explosive yield.

The following subsections briefly summarize several selected effects observed on the above cratering explosions. After discussing the parameters that affect the apparent craters volume, we examine the crater shapes observed in the explosion and impact crater data base. We then summarize several features derived from studies of ejecta, ground motion and airblast data from near-surface explosions.

Apparent crater volume

With very few exceptions, inadequate data have been obtained for a given geology, explosive source and yield combination to specify crater parameters as a function of depth-of-burst. The most extensive data base exists for 256 pound

---

*One ton of TNT is equivalent to 4.2×10^{16} ergs.

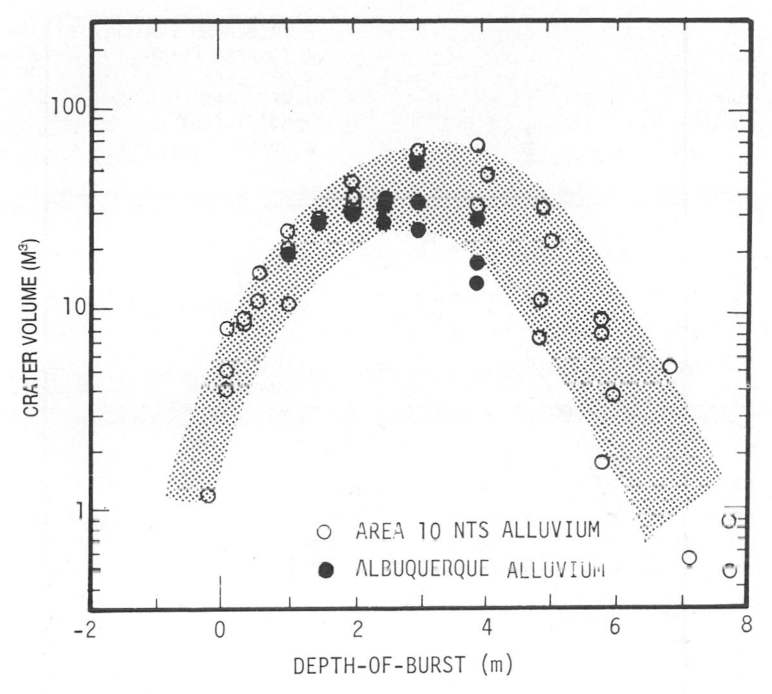

Fig. 2. Crater volume for 256 pound TNT spherical sources in dry alluvium.

TNT spheres, and this is most complete for dry alluvium (Fig. 2). The crater volume data scatter by a factor of 2 to 3 near the earth's surface and substantially more for depths greater than optimum depth-of-burst.* Any systematic difference between the cratering properties of NTS alluvium and Albuquerque alluvium is less than the data scatter for either test area. The variations in cratering efficiency (crater volume) associated with variations in explosive yield, charge placement, and measurement techniques are each estimated to be less than 5% (Vortman, 1977). Thus, the major portion of the data scatter is probably associated with geologic variations in the vicinity of the source.

It might be argued that small geologic inhomogeneities produce exaggerated effects for small charges, and that larger yields (NE or HE) would produce less data scatter. However, the scale of inhomogeneities that affect cratering phenomena also changes with explosive yield. While small inhomogeneities important for 256 pound charges might not be important for large yield sources, other large inhomogeneities unimportant for 256 pound sources may become important for larger yields. Thus, it is not obvious that the data scatter produced by geologic inhomogeneities necessarily reduces with increasing yield. Un-

*Highly controlled laboratory test conditions can reduce the crater volume variations to less than 10% (Piekutowski, 1975). However, experiment to experiment variations in the field are much more difficult to control.

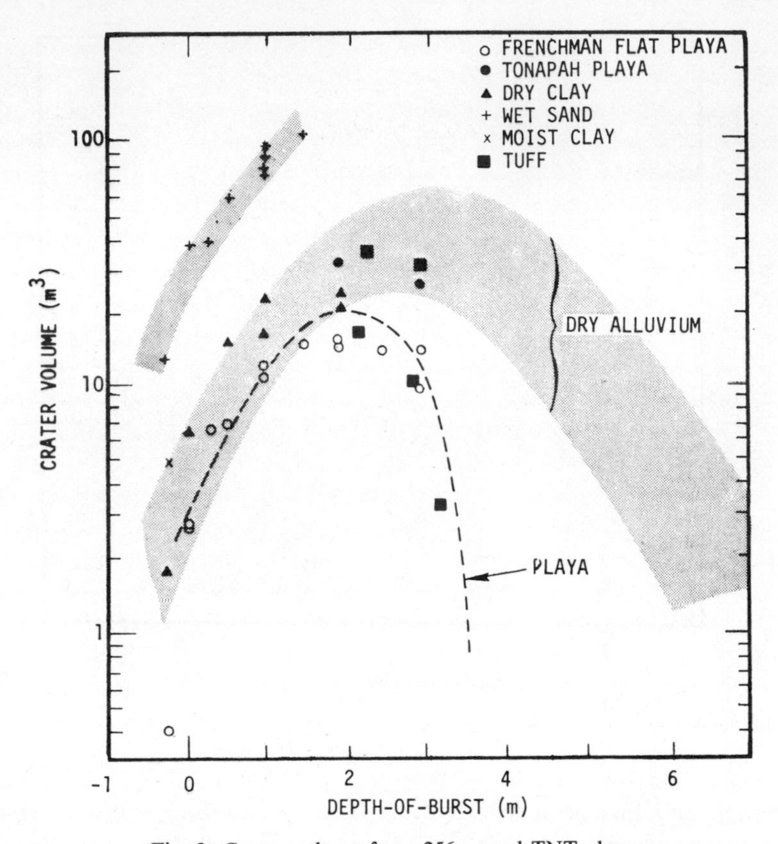

Fig. 3. Crater volume from 256 pound TNT charges.

fortunately, little or no data exist to provide significant insight into the reproducibility of large yield explosions in any given geology.

Figure 3 compares the more limited data base for 256 pound TNT spherical charges in various geologies with the alluvium data. The main systematic variation from the alluvium data results from tests in wet sand and moist clay. The reduced strength of these materials led to craters with 3 to 10 times the volume of craters in dry soils where intergranular friction opposes the late-time crater growth. This result illustrates the strong influence of water on cratering efficiency, a fact also demonstrated by other experiments and by theoretical calculations.

Data from higher yield cratering experiments are in insufficient quantity to derive height-of-burst curves for other specific yields and geologies. Thus, some method of "scaling"* is required to compare the data from various yields and

*Here the word "scaling" is italicized to denote that in a strict sense, simulitude holds for cratering phenomena only under very constrained circumstances (Crowley, 1970; White, 1971). Nevertheless, the available data are generalized in various empricial ways to provide estimates of crater parameters for conditions that interpolate and/or extrapolate the data base.

geologies. Historically, it was first suggested that apparent crater dimensions vary as $Y^{1/3}$ (where Y is the explosive yield) based on arguments from dimensional analysis (assuming the cratering process is dominated by shock phenomena which vary as $Y^{1/3}$), and early small explosive tests seemed to confirm this hypothesis (Lampson, 1946). However, additional data from larger yield buried explosions in NTS alluvium are correlated better by using a yield exponent of 1/3.4 rather than 1/3 (e.g., Violet, 1961; Chabai, 1965). Dimensional analysis might lead one to expect linear dimensions to vary as $Y^{1/4}$ if gravity effects dominate the cratering process. Thus, the observation that crater dimensions from buried explosions in alluvium vary as $Y^{1/3.4}$ suggests the importance of effects other than gravity. Other physically appealing correlation methods, which account for the resisting effects of atmospheric and lithostatic pressure and material strength, also collapse the alluvium data as well as $Y^{1/3.4}$ "scaling" (Chabai, 1965; Sun, 1970; Herr, 1971; White, 1973).

Since the cratering mechanism for deeply buried explosions is primarily one of lifting the overlying medium, it is apparent that forces that resist the lifting are important to crater formation. Thus, lithostatic forces (gravity), material strength and atmospheric pressure would be expected to resist the cratering action. On the other hand, surface and shallow-buried explosions vent to the atmosphere early in the crater formation process and the late-stage cratering process is dominated by compaction, plastic flow and scouring of media from the crater region. Experiments by Herr (1971) showed that atmospheric pressure affects the cratering from buried explosions, but has negligible influence on the cratering process for near-surface explosions. Lithostatic forces should also be less important for near-surface explosions than for optimally buried explosions. Although gravity forces are less influential in resisting early time cratering phenomena for near-surface explosions, they may still play a significant role in late-stage crater growth and afterflow which could significantly alter the maximum transient crater—especially for larger craters.

Because the cratering mechanisms from deeply buried and near-surface explosions differ, one might expect that "scaling" rules might also differ. Thus, it is not surprising that the crater volume (V) from surface explosions varies more nearly proportionally to the explosive yield (Vortman, 1968)* than implied by $Y^{1/3.4}$ "scaling"—which gives $V \sim Y^{0.882}$. The assumption that cratering efficiency (V/Y) of near-surface explosions in a given geology is independent of yield was made in developing a commonly used methodology for predicting crater parameters for near-surface explosions (e.g., Crawford *et al.*, 1974). That methodology assumes

$$V/Y = k(G)V_0(G_0, S)F(H, S), \quad (1)$$

where $k(G)$ ranks the cratering efficiency of various geologies G, $V_0(G_0, S)$ is the cratering efficiency of explosive source S in a reference geology G_0 at $H = 0$,

*In fact, HE surface bursts in several geologies produced crater volumes that increased slightly faster than Y, a feature attributed to non-scaling features such as *in situ* strength (Vortman, 1977).

and $F(H, S)$ provides for height-of-burst (H) variations for source S. Here H is the height-of-burst normalized by the characteristic length $V^{1/3}$.* Table 1 lists best estimates and ranges of values $k(G)$ for various generic geologies as derived from various HE experiments (Cooper, 1976). The reference geology G_0 is taken as saturated Pacific coral where craters have been produced by HE and two classes of NE sources. The best-estimate cratering efficiency of various generic geologies varies by a factor of 8 and the overall variation for extreme geologies (e.g., very hard rock versus very weak saturated soil) could be as large as a factor of about 25. In general, the variation in cratering efficiency appears to correlate with media strength, i.e., the weaker the medium, the larger the crater.

Table 1. HE cratering efficiency for generic geologic materials.

	$V(G, \mathrm{HE})$*-m³/ton		$k(G)$†	
Medium (G)	Range	Best estimate	Range	Best estimate
Wet soil	60–230	115	0.5–2	1
Dry soil	20–50	30	0.15–0.45	0.25
Soft rock	15–35	25	0.125–0.3	0.2
Hard rock	8.5–20	15	0.075–0.175	0.125

*$V(G, \mathrm{HE})$ is the cratering efficiency of a half-buried TNT source in geology (G).
†Based on a reference cratering efficiency for wet Pacific coral sand of 115 m³/ton.

Figure 4 illustrates the cratering effectiveness of near-surface explosions of HE and two classes of NE sources in wet coral and dry alluvium. (Note that the cratering efficiency of wet coral is about 4 times that of dry alluvium.) Equation (1) has been applied to transpose the alluvium data to the wet coral curve and vice-versa. These data show that buried HE sources have about twice the cratering efficiency of NE sources at the same scaled depth-of-burst. Surface and above-surface HE sources are many times more efficient than surface and above-surface NE sources. Furthermore, the cratering efficiency of above-surface NE sources appear to vary with energy density of the source. Although additional variations in the cratering efficiency of above-surface explosions could result from variations in source radiative characteristics (Knowles, 1973), shallow depth-of-burial ($\gtrsim 2\ \mathrm{m/MT}^{1/3}$) traps the prompt radiation in the ground, eliminating the distinction between the effects of various nuclear sources. Since cratering phenomena from buried nuclear explosions most nearly simulate impact craters, possible variations in the radiative characteristics of NE sources have little consequence in our discussion here.

*The use of $V^{1/3}$ as a characteristic length was suggested by correlation of low-frequency near-surface ground motions associated with the crater-forming process (Cooper, 1971).

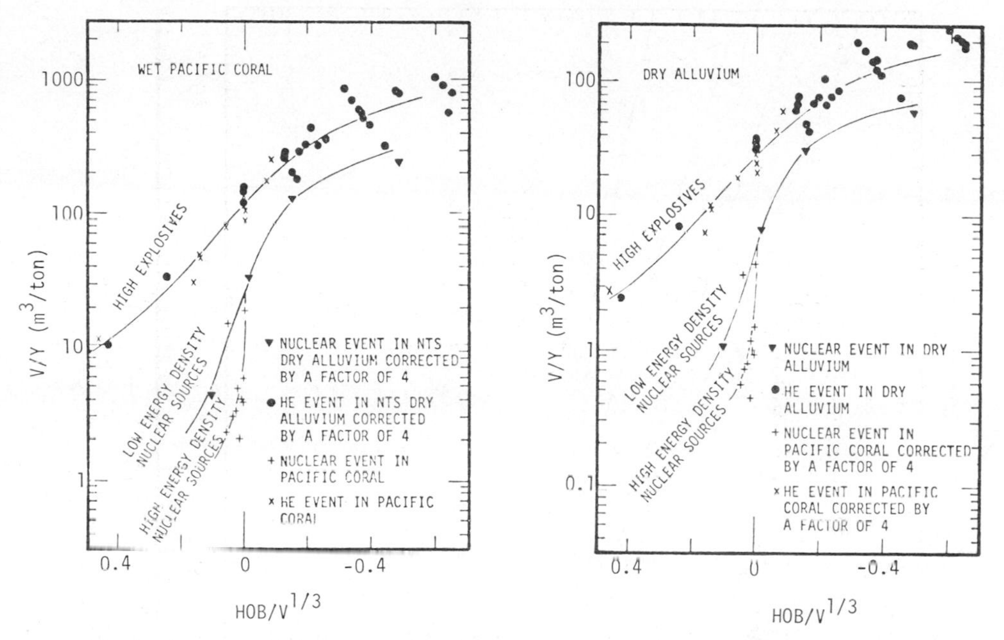

Fig. 4. Cratering efficiency of HE and NE sources in Pacific coral and dry alluvium.

Apparent crater shape

Figure 5 shows apparent crater radius (R) and depth (D) from explosions in various geologics plotted as a function of the apparent crater volume (V). Most of the cratering data are consistent with a canonical parabolic crater shape such that

$$\begin{aligned} R &\simeq 1.2V^{1/3}, \\ D &\simeq 0.5V^{1/3}, \\ V &\simeq 1.5R^2D. \end{aligned} \qquad (2)$$

Deviations from the canonical parabolic crater shape are apparent for "wet" geologies and crater volumes greater than about 10^3 m^3. In particular, the nuclear cratering experiments in saturated coral produce larger crater radii and shallower crater depths than suggested by Eq. (2). Also, several high-explosive craters in wet soils with volumes in the vicinity of about 3×10^3 m^3 have radii comparable to the canonical craters, but substantially shallower depths. In several of these cases, it is clear that late-stage inward flow significantly altered the crater shape.

Figure 6 shows crater depth versus rim diameter data for explosive* and impact cratering events on the earth, moon, and Mars (Pike, 1974; Gault *et al.*, 1975; Pike, 1977). The canonical parabolic crater shape is represented by the line

*Here, the apparent crater diameter data was converted to rim crater diameter data by using Baldwin's estimate (Baldwin, 1963) that the ratio of apparent to rim diameter is 0.83.

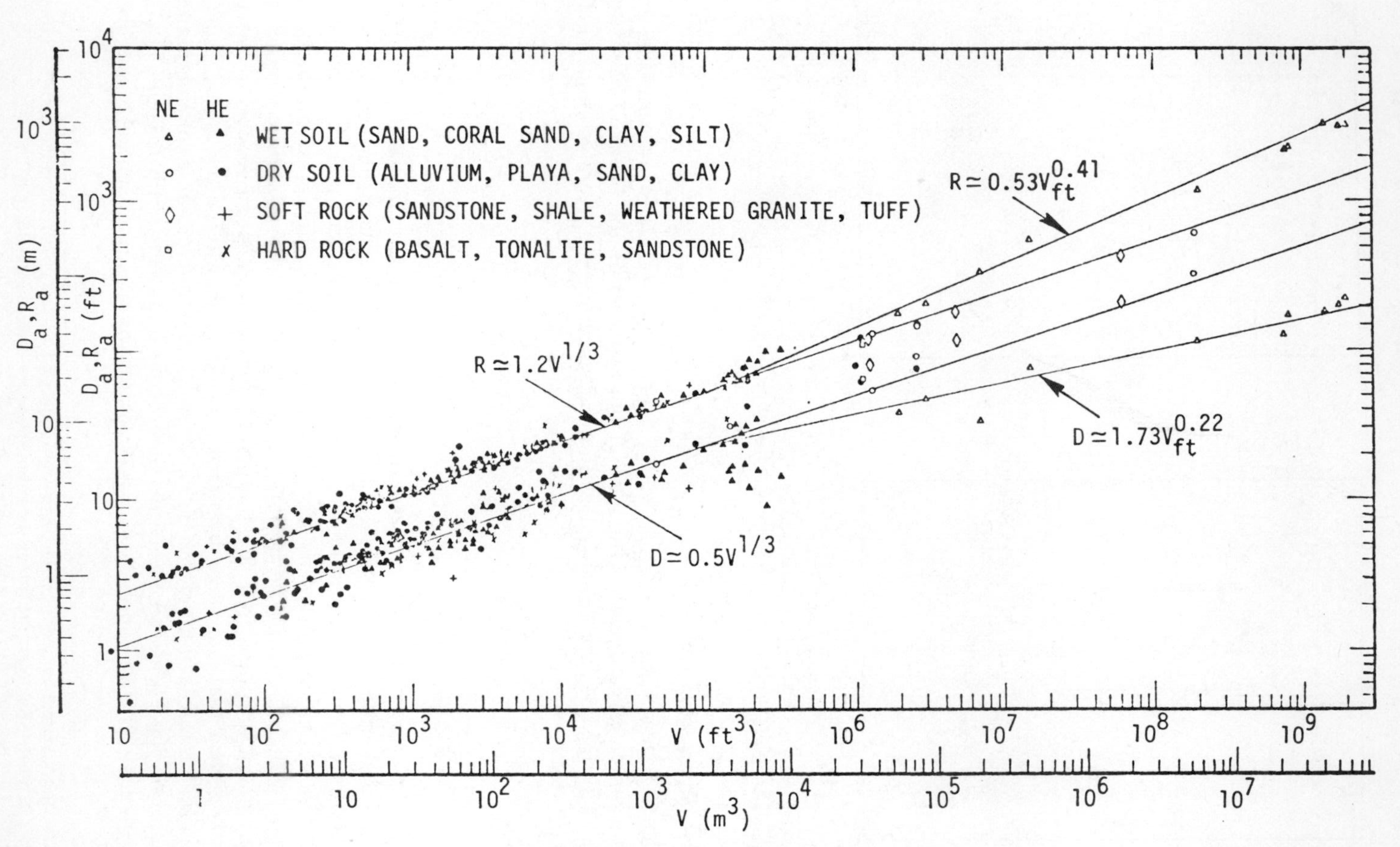

Fig. 5. Crater radii and depths as a function of crater volume.

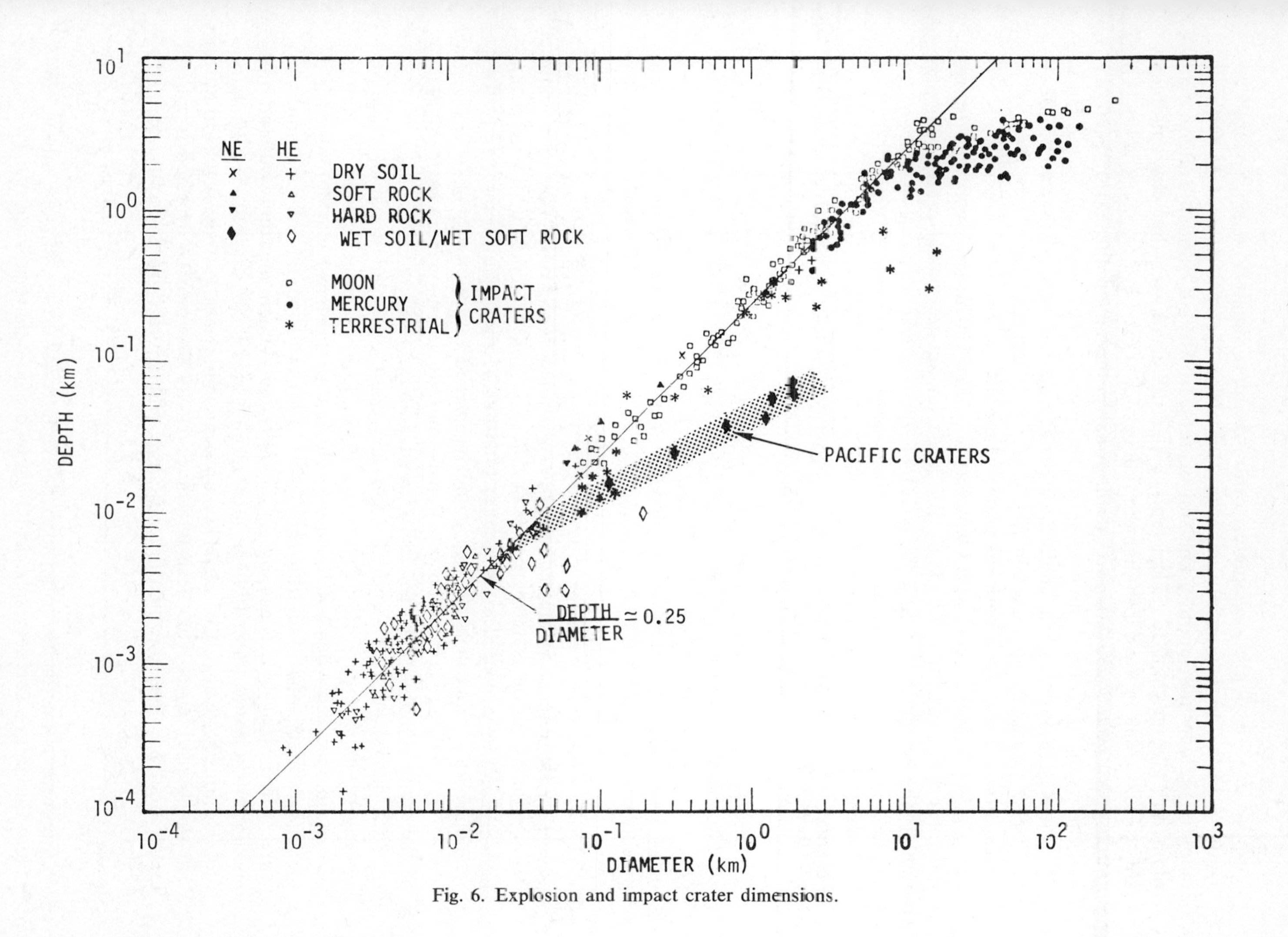

Fig. 6. Explosion and impact crater dimensions.

associated with depth-to-diameter ratio of approximately 0.25. Several deviations from the canonical shape are apparent. Terrestrial craters with diameters greater than about 2 km and craters on Mercury and the moon with diameters greater than 10–15 km tend to become shallower than the canonical shape. The break in the depth-diameter curves for these large impact craters is in rough correspondence with the gravitational potential on the earth, Mercury, and the moon; therefore, the deviation from canonical craters may be related to the gravitational forces. This hypothesis has merit, but we note that other effects must also be included to explain Fig. 6.

Deviations from the canonical crater shape also occur for terrestrial craters (fixed gravitational forces) with diameters less than 100 m; i.e., over an order-of-magnitude smaller than the break for large impact terrestrial craters. In particular, the craters from the large yield Pacific Nuclear Tests deviate from the canonical shape for diameters larger than about 30 m. Similarly, craters from high-explosive tests in soils with high water tables (Suffield Experimental Station (Roddy, 1976) and White Sands, New Mexico (Edwards, 1976)) show a marked deviation from the canonical shape for similar diameters. Furthermore, several terrestrial impact craters with ~100 m diameter appear to be significantly shallower than the canonical craters.

Thus, if deviations from the canonical depth-diameter ratio are caused by gravitational forces, some other variable, such as the *in situ* strength of the rock mass, must also be involved to account for these observations. Perhaps a meaningful parameter for determining when the deviation from the canonical crater shape will occur may be related to the ratio $\tau/\rho g V^{1/3}$ where τ is the *in situ* strength, ρ is the density, and g is the gravitational constant. Such a parameter combines the effects of strength and gravitational forces in a way that might explain the variation in terrestrial crater shapes. Soil geologies with high water tables would be expected to be weaker than harder rock. Thus, crater sizes that would be unaffected in hard rock could be altered by gravitational forces if in weaker wet soils. This hypothesis is consistent with an analysis by Melosh (1977) of the slope stability of large bowl-shaped craters.

Such late-stage crater modification has been observed at small scale in laboratory experiments. For example, Fig. 7 shows three stages of crater formation observed when a half-buried 1.7 g lead azide charge was detonated in a loose saturated sand (Piekutowski, 1976). Large horizontal outward motions were followed at late times by an inward flow which markedly altered the maximum transient crater (which, incidentally, was bowl shaped). Similar late-time motions have also been observed on large field trials. It is significant to note that the time required for these late-time gravity flows to be arrested is much longer than the time at which first principle calculations are generally stopped.

Ejecta

Figure 8 is an ejecta origin map derived from surface tangent (above and below the earth's surface) spherical 1.7 g lead azide charges in dry Ottawa sand

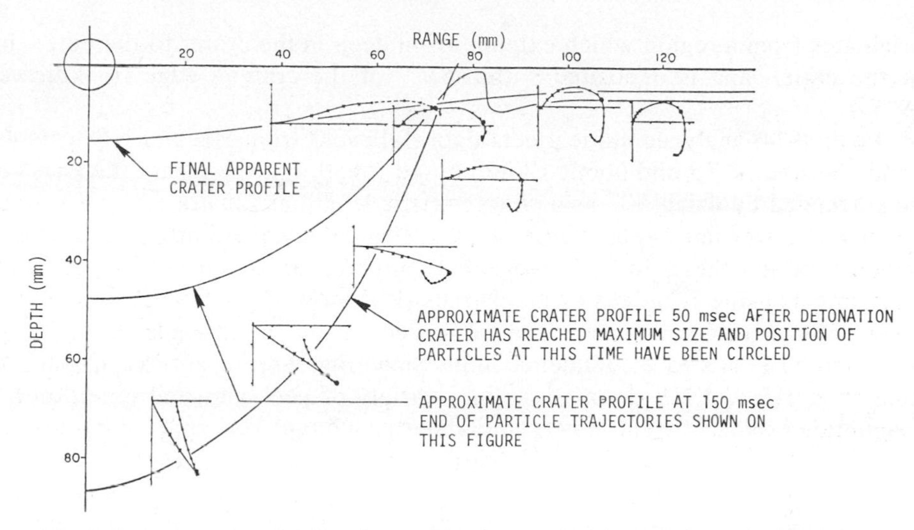

Fig. 7. Late-stage cratering flow from a 1.7 g lead azide spherical charge half buried in a loose saturated sand (Piekutowski, 1976).

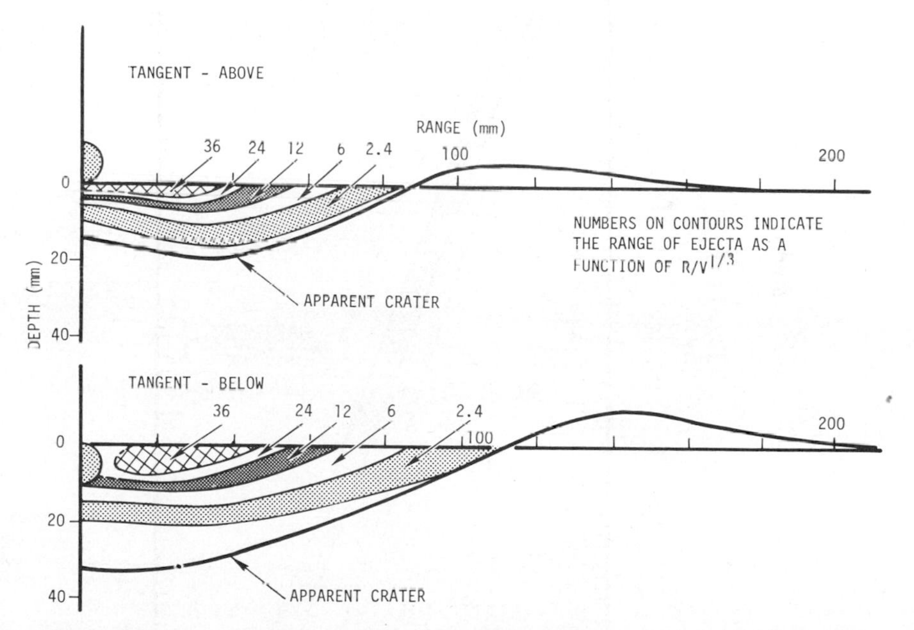

Fig. 8. Origin map of ejecta particles for surface tangent spherical 1.7 g lead azide charges in dry Ottawa sand.

(Piekutowski, 1975a). The ejecta origin contours in these laboratory experiments are qualitatively similar to those developed from high-explosive field trials with yields up to 500 tons (Rooke *et al.*, 1974). The high velocity ejecta which travels the farthest originates near the explosive source and near the earth's surface. These laboratory experiments indicate that approximately 80% of the total ejecta originates from a region which extends from deep in the crater to the outer third of the crater, and is deposited within $6V^{1/3}$ of the crater's edge (Piekutowski, 1975a).

Post (1974) analyzed static ejecta data (fallback) from HE and NE tests with yields between 1.7 g and about 1 kiloton and found that the ejecta thickness can be correlated by using $V^{1/3}$ as a characteristic length as illustrated in Fig. 9. The upper and lower lines approximate Post's 95% confidence limits and the shaded region between these limits is the data scatter observed on several laboratory experiments using 1.7 g lead azide charges (Piekutowski, 1975b). Data scatter on given field experiments has been observed to be comparable to the range indicated by Post's 95% confidence limit. Since Fig. 9 represents a correlation of data from HE and NE detonations in a variety of geologies and nine orders of magnitude explosive yield, it is remarkable that a larger variation is not observed.

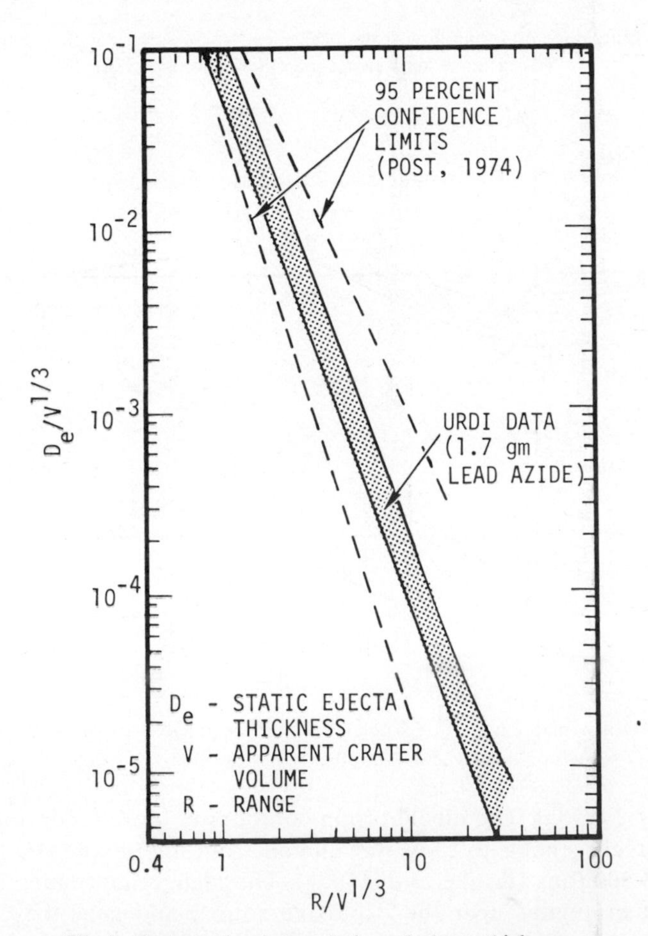

Fig. 9. Crater-volume correlation of ejecta thickness.

A best-estimate ejecta thickness (d_E) from Fig. 9 is

$$\frac{d_E}{V^{1/3}} \simeq 0.1\left(\frac{V^{1/3}}{r}\right)^3, \tag{3}$$

where V is the apparent crater volume and r is range from the burst point assumed less than about $10V^{1/3}$.

One should be cautious in using Eq. (3) to estimate ejecta thickness beyond the range of the data on which it was based ($Y \leqslant 1$ kiloton). The ejecta transport for many of these experiments is known to have been ballistic. However, blast effects from large yield surface explosions could significantly alter the ballistic trajectories, reducing the ejecta thickness close-in and increasing it further out (with respect to Eq. (3)).

Since the ejecta transport was ballistic for the experiments that make up Post's correlation, it can be used to infer a surprising characteristic of the ejection velocities as a function of yield. Post's correlation implies that debris depth scales geometrically, which requires the ejection velocities to increase as the sixth root of the explosive yield (assuming the crater volume is directly proportional to the yield and the ejection angles are independent of yield).

To illustrate this point, consider Fig. 10. Figure 10a illustrates the ballistic trajectory of a particle (ignoring drag effects) which crosses the original ground surface at range X with an initial velocity v_0 and travels a distance L down range. If the yield of this explosion is increased from Y to $8Y$, one might expect the crater dimensions to double, as illustrated in Fig. 10b. In that case, the geometrically similar particle would cross the original ground surface at a distance $2X$ with an initial velocity $\bar{v}_0$. If $\bar{v}_0 = v_0$, then the particle would again travel ballistically down range a distance L. Thus, if $\bar{v}_0 = v_0$, debris would

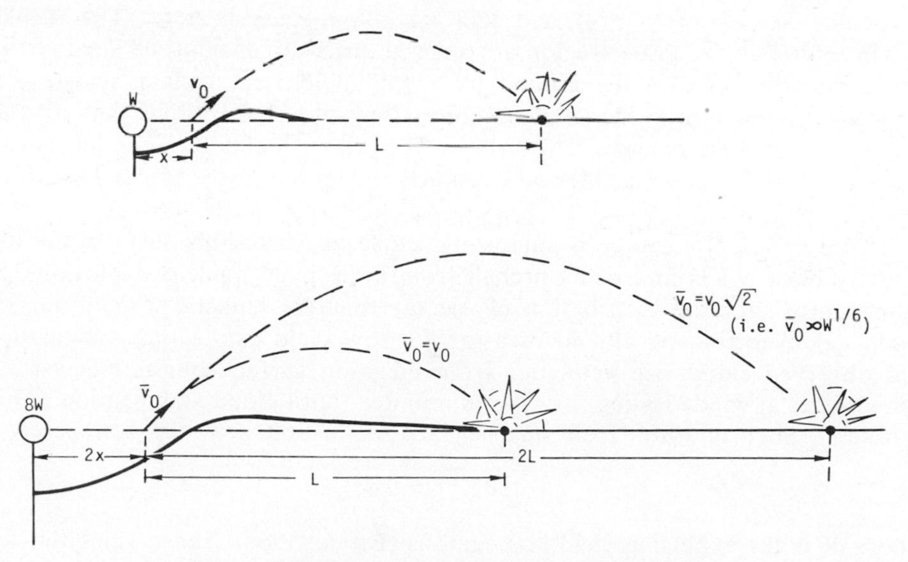

Fig. 10. Effect of ejection velocity on ejecta scaling.

accumulate at a closer-in geometric distance on the larger yield explosion. For the ballistic trajectory to be geometrically similar to the first case, its travel must increase to $2L$, implying that $\bar{v}_0 = \sqrt{2} v_0$. (The distance of travel $\Delta X = (\bar{v}_0^2/g) \sin 2\theta$ where g is the acceleration of gravity and θ is the ejection angle). In other words, the two cases would be geometrically similar if $\bar{v}_0 = Y^{1/6} v_0$ (and if the spatial variation and initial ejection angles are invariant). Thus, if Post's suggestion that the debris depth geometrically scales is literally true, the ejection velocities must vary as the sixth root of the explosive yield. Independent of these considerations, Wisotski (1977) has estimated that ejection velocities vary approximately as $Y^{1/6}$ based on the analysis of high speed photography of the large ejecta transport observed on field HE experiments with yields ranging between 0.5 and 100 tons.

That ejection velocities should vary as $Y^{1/6}$ is contary to the usual assumption that the ejection velocity at a given geometrical point is invariant under variations in explosive yield. This usual assumption is based on the observation that the conservation laws of continuum mechanics, when applied for geologic materials that are rate-independent, imply that peak shock pressure and particle velocity at geometrically scaled ranges are independent of yield. Thus, the ejection velocities are usually also assumed to be yield-independent. However, this assumption is contradicted by the above-mentioned observations.

Airblast

Above-surface and near-surface explosions produce a strong air-shock which propagates outward and loads the ground. Airblast phenomena from surface and above-surface explosions are examined in-depth by Brode (1968, 1970) and Carpenter and Brode (1974) and will not be discussed here. The airblast overpressures in the crater region are several orders of magnitude smaller than the stresses induced in the ground by a shallow-buried nuclear source—and consequently have little influence on crater formation for sources that simulate impact cratering phenomena. The airblast from more deeply buried explosions is even less intense (see Fig. 11) and even less influential in the crater excavation process.

If the buried NE source is sufficiently close to the earth's surface, the high velocity blast winds and rising fireball from large yield nuclear explosions can significantly alter the distribution of ejecta from the ballistic trajectories one might calculate for still air—or measure for low yield explosions. Calculations and observed cloud rise velocities from megaton surface bursts suggest sustained vertical winds lasting for several minutes (until cloud stabilization at high altitudes). Such updrafts from near-surface explosions scale approximately as

$$v \simeq 150 Y^{1/4} \text{ (m/sec)}, \tag{4}$$

where Y is the explosive yield in megatons (Brode, 1968). These velocities lead to a rough estimate of the weight of (spherical) ejecta chunks that can be borne

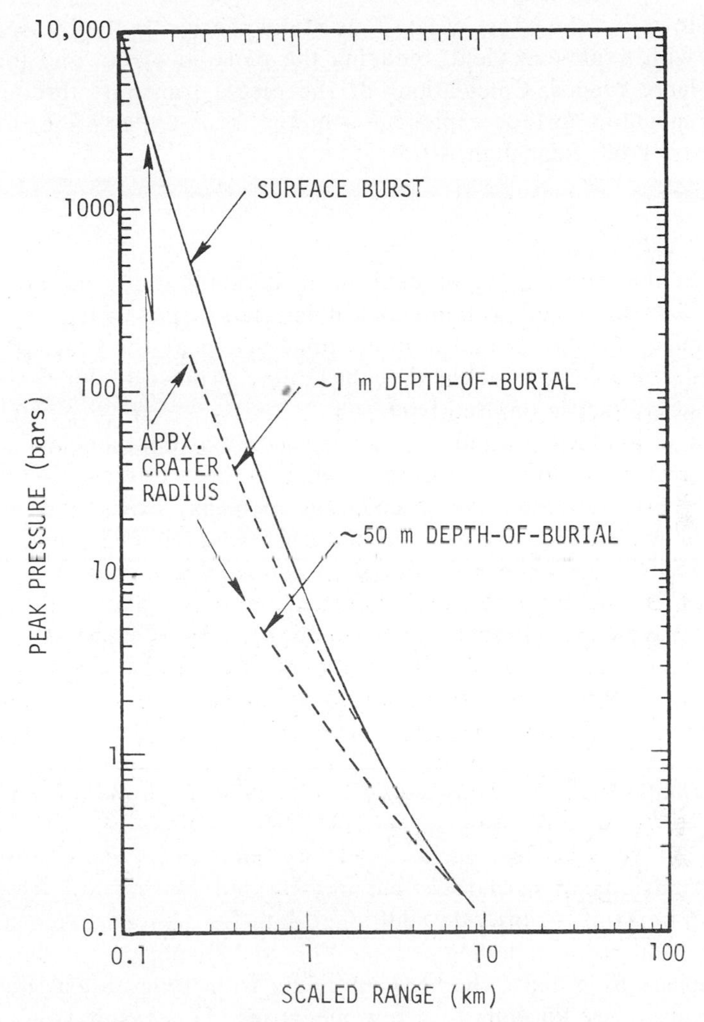

Fig. 11. Estimate of airblast peak overpressure from surface and shallow-buried megaton nuclear explosions.

aloft, viz.

$$M \simeq 150\eta^3 Y^{3/2}\,(\mathrm{kg}) \tag{5}$$

where η is the ratio of air density at altitude to that at sea level. Ejecta particles of much larger size would travel on ballistic trajectories and not be influenced by the cloud rise phenomena. Ejecta of lesser mass are likely to be lofted and distributed over a wider area than would be expected on the basis of ballistics.

The experiments from which Eq. (3) was derived involved yields less than 1 kiloton. For this yield, Eq. (5) suggests that a weight of at most .03 g could be carried aloft by the blast winds. Since most ejected particles are larger than this, Eq. (3) probably is derived from experiments where the ejecta transport was ballistic. However, the effect of late-time atmospherics on the transport of ejecta increases with explosive yield, reducing the close-in ejecta and increasing the ejecta at large ranges. Calculations of the ejecta transport through the rising fireball of megaton surface explosions support these expectations (e.g., Ganong and Roberts, 1968; Seabaugh, 1975).

Ground motions

This section comments on data used in estimating the hemispherically diverging, direct-induced ground shock illustrated in Fig. 1 and on the late-time low-frequency surface ground motions observed near the crater's edge. Since airblast-induced ground shock is thought to play at most a minor role in creating craters, that subject is omitted here.

The direct-induced ground shock propagated to depth below a surface or shallow-buried explosion has features similar to those observed on fully tamped explosions. For example, the attenuation of peak stress and peak particle velocity below a surface explosion in uniform rock media* has been observed to be approximately the same as observed on tamped explosions in the same media (Cooper, 1973). Furthermore, vertical particle velocity waveforms below surface explosions have been observed to be qualitatively the same as radial particle velocity waveforms measured on tamped explosions. For these reasons, it seems relevant to briefly discuss the ground shock observed on tamped nuclear explosions. The phenomena associated with such buried explosions are reasonably well understood, and a number of calculations compare favorably with experimental observations (e.g., Cherry, 1970; Riney *et al.*, 1972; Glen and Thomsen, 1976).

Figure 12 shows peak particle velocity and stress data from contained nuclear explosions in hard rock and unsaturated porous tuff (Cooper, 1973). Perret and Bass (1975) provide additional data for tamped explosions in these and other geologies—including acceleration and displacement data. Here $Y^{1/3}$ scaling appears to collapse the hard rock data from several test sites and yields ranging from a few kilotons to a few megatons. This result is consistent with first-principal calculations by Crowley (1970), who examined the effects of gravity on scaling the close-in ground shock and cratering from deeply buried explosions. The peak particle velocity at a given depth in hard rock is about twice the amplitude that would occur if the explosion were in unsaturated tuff. Because the impedance of hard rock is about three times that of unsaturated tuff, the peak stress at a given scaled range ($r/Y^{1/3}$ where r is range) from explosions in hard rock is about six times that which would occur from explosions in unsaturated tuff.

Figure 13 estimates the apparent craters and peak stress contours from shallow-buried (depth $\simeq 2\ m/MT^{1/3}$) nuclear explosions in dense hard and porous

---

*Geologic layering may alter this observation since, in that case, peak stresses below surface explosions may attenuate more rapidly than those from tamped explosions (Ingram *et al.*, 1975).

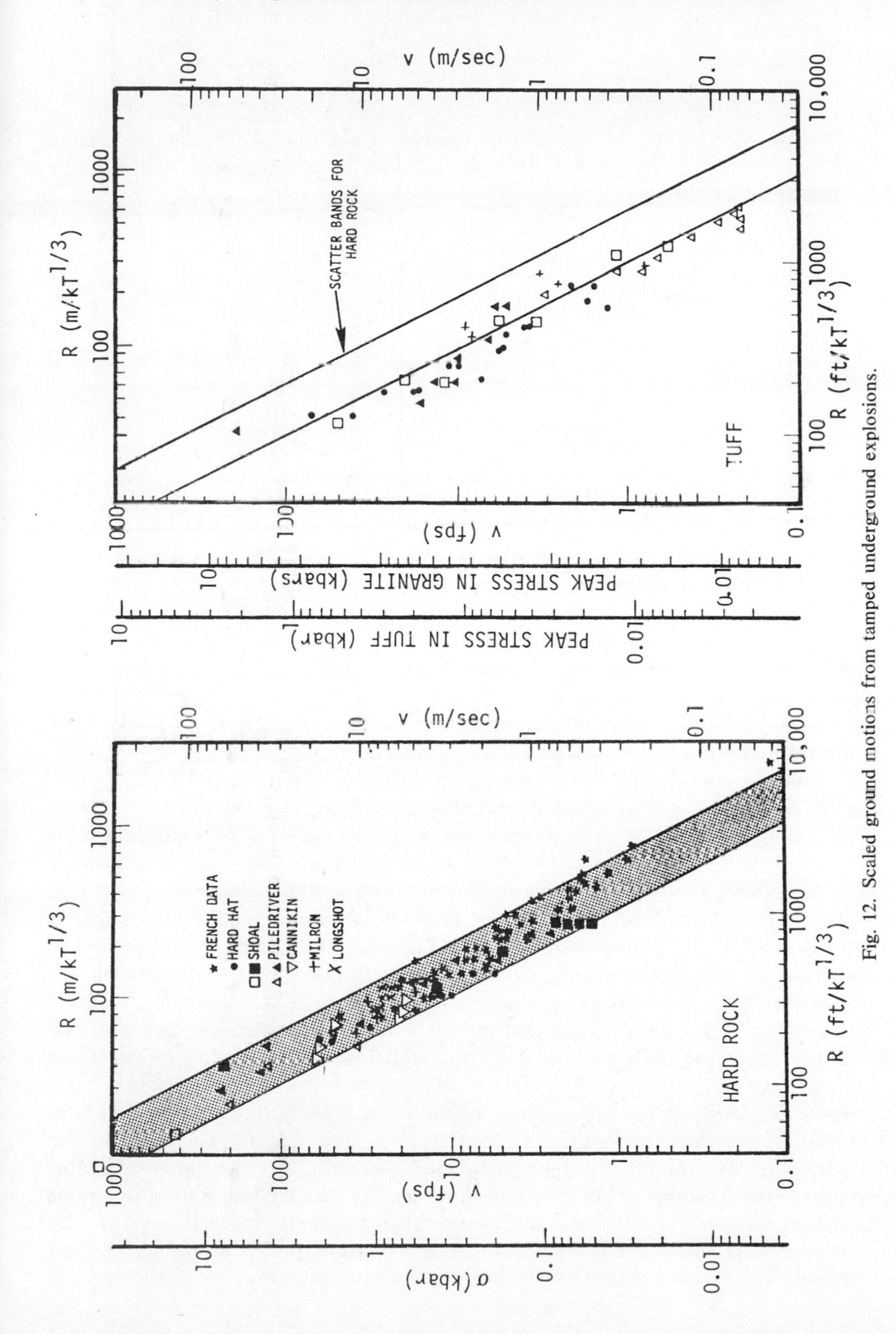

Fig. 12. Scaled ground motions from tamped underground explosions.

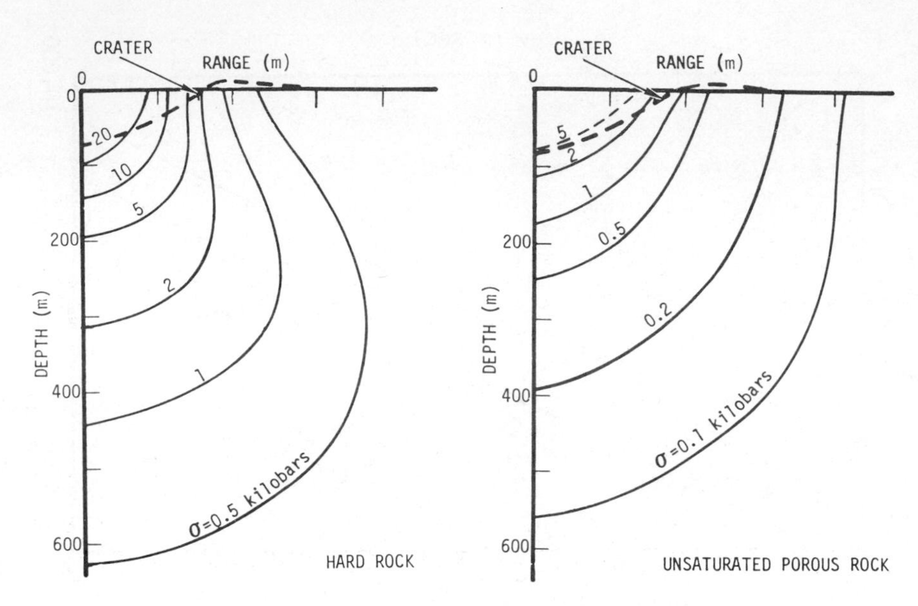

Fig. 13. Best-estimate peak stress contours for a shallow-buried (~3 m) megaton explosion in hard and unsaturated porous rock.

soft rock. These estimates were derived from analysis of NE and HE data (Cooper, 1973, 1976) coupled with the qualitative features of various theoretical calculations (Cooper *et al.*, 1972). The estimated uncertainty in these peak stress contours is about a factor of 2 at a given range and depth (i.e., from half to twice the best-estimate). Comparisons of Fig. 13 with data from tamped explosions in Fig. 12 show that such a shallow-buried nuclear explosion produces the same peak stress and particle velocity below the burst as a tamped explosion with about 16% of the yield of the shallow buried explosion. The "equivalent yield coupling factor" for creating peak stress and particle velocity is estimated to be about 4% for a contact above-surface burst and nearly 100% for a depth-of-burial of about $100\ \mathrm{m/MT^{1/3}}$. Obviously, the crater dimensions vary markedly as the depth-of-burst is varied. Figure 14 shows estimates of the apparent craters and peak stress contours from a megaton NE explosion at 100 m depth in hard rock and unsaturated porous rock. The stress contours near the earth's surface are dashed to indicate that significant uncertainties exist in that region.

The near-surface ground motions result from both the airblast loading and the shock directly transmitted from the explosive source. Close-in to the cratering region (ranges less than about $5V^{1/3}$, where V is the apparent crater volume), the low-frequency ground motions are associated with the crater forming process, and are dominated by the energy directly transmitted from the source region. Such "crater-induced" ground motions produce initially upward and outward displacements (d) whose peak amplitudes may be estimated by:

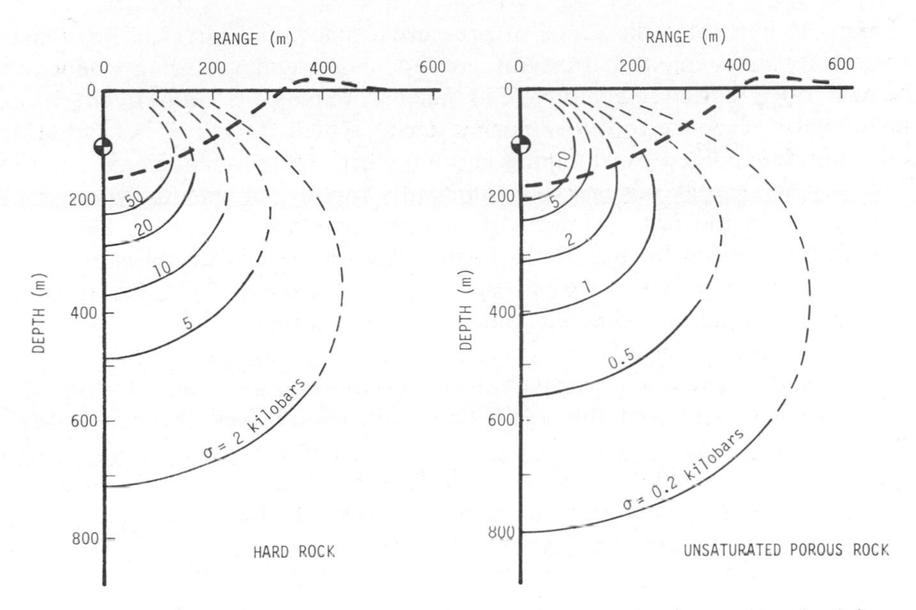

Fig. 14. Best-estimate peak stress contours for a megaton explosion at 100 m depth in hard and unsaturated porous rock.

$$\frac{d}{V^{1/3}} \simeq k\left(\frac{V^{1/3}}{r}\right)^3 \quad \text{for} \quad \frac{r}{V^{1/3}} \leqslant 5, \tag{6}$$

where r in the range from ground zero and $k \simeq 0.45$ for above-surface explosions, 0.25 for half-buried and shallow explosions and 0.1 for deeply buried explosions (Cooper and Sauer, 1977). Experiment to experiment variations, and azimuthal variations in a given experiment of a factor of 2 to 3 from Eq. (6) have been observed. Equation (6) with $0.1 \leqslant k \leqslant 0.25$, is thought to be a reasonable estimate for the peak transient displacements from impact craters. Permanent surface displacements between about half and one times the peak transient displacements estimated by Eq. (1) have been observed on various experiments.

Characteristic times associated with the crater-related ground motions have been observed to vary as $V^{1/6}$ (Cooper and Sauer, 1977). If distances vary as $V^{1/3}$, then the definition of particle velocity ($v \equiv dx/dt$) implies that crater-related peak particle velocities vary as $V^{1/6}$, consistent with the previous comments on the scaling of ejecta initial velocities. As noted earlier, such "scaling" rules are inconsistent with the usual cube-root-of-the-yield scaling procedures.

4. Discussion of Explosive Cratering Mechanisms

This section attempts to combine and supplement aspects of the previous discussion to form a qualitative model of the various stages of crater forming process. Although cratering phenomena from both shallow and deeply buried explosions are considered, the shallow-buried explosions are emphasized since they are thought to lead to phenomena most like impact craters.

Figure 15 illustrates the range of pressures, times, and physical phenomena involved in describing the transient ground shock and cratering phenomena produced by a buried megaton (MT) nuclear explosion. Subsequent to the dynamic earth response to the explosion during which the crater is formed, the crater may be modified by slumping and long-term creep phenomena.

Initially, the energy is generated sufficiently rapidly for the device to heat to temperatures on the order of tens of millions of degrees Kelvin. These high temperatures produce intense X-ray fluxes which interact with the surrounding media. At times on the order of 10^{-7} sec/MT$^{1/3}$, the energy transfer to the surrounding medium is complete, and a nearly isothermal region with temperatures on the order of 10^7°K is formed in a shell about 1.5–2 m/MT$^{1/3}$ thick around the device. In contrast, a nuclear explosion in air would produce an isothermal sphere at about this same time with a radius of about 12 m/MT$^{1/3}$.

Initial peak pressures in the ground resulting from this deposition of energy could be as high as ~1000 megabars*. The energy is subsequently redistributed by radiation diffusion and hydrodynamics. After a few μsec/MT$^{1/3}$, radiation phenomena are no longer important and the shock pressures are reduced to less

*A maximum pressure of ~100 megabars would be induced in the medium adjacent to a low energy density nuclear source, thereby significantly reducing and possibly eliminating the role of radiation diffusion phenomena in the early time physics.

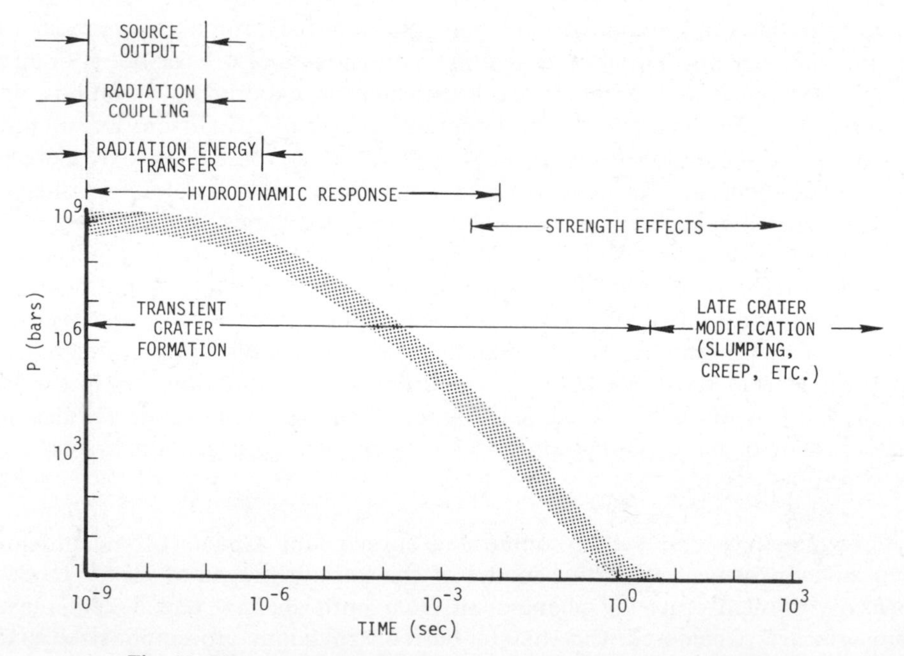

Fig. 15. Phenomenological regions for a buried megaton explosion.

than about 100 megabars. At this time, the strong shock will have propagated to a few $m/MT^{1/3}$. If the source vapors vent into the atmosphere (shallow burial), a fireball will grow via radiation diffusion processes; and, after about 10^{-3} sec, a strong airshock will break away from the fireball, propagate outward and load the ground's surface.

For times between a few $\mu sec/MT^{1/3}$ and a few $msec/MT^{1/3}$ the shock propagation phenomena are described by the equations of hydrodynamics. During this time, the ground shock is sufficiently strong to vaporize and melt earth materials as the intense shock attenuates to pressures on the order of 100 kbar. The regions of vaporization and melt extend to about 20 and 30 $m/MT^{1/3}$, respectively.

At 10^{-2} $sec/MT^{1/3}$, the peak stresses have attenuated to about 100 kbar, i.e., to the stress levels produced by HE detonations. If the depth-of-burst is ~ 100 $m/MT^{1/3}$, the shock wave is approximately incident at the ground's surface at this time. This geometry is thought to roughly simulate conditions produced in the earth by an impacting meteor. For later times (lower stresses), the mechanical properties of the earth media (porosity, strength, etc.) become important to the shock propagation and crater forming process. After the wave has propagated sufficiently for the stress to drop to several tens of kilobars, the details of the explosive source should no longer affect the ground shock signal.

The subsequent phenomena leading to crater formation is fundamentally different for deeply buried and shallow-buried explosions. Deeply buried explosions produce craters by heaving the overlying earth media upward along more or less linear trajectories centered at the explosive source. Most of the source's energy is effectively coupled to the ground, and the explosively produced gases vent to the atmosphere late in the crater forming process. On the other hand, near-surface explosions vent to the atmosphere early, and a smaller percentage of the source energy goes to create the crater. The material ejected from such craters results from stress gradients below the explosive source created by a combination of the effects of the intense shock wave moving downward below the source and free surface effects.

Based upon observations of high-explosive and nuclear events, Nordyke (1961) suggested a qualitative model of cratering mechanisms for deeply buried explosions. Subsequently, first principle theoretical calculations of cratering phenomena from buried explosions (e.g., Terhune *et al.*, 1970) verified Nordyke's model which suggests four sequential phases as summarized in Fig. 16. The role each mechanism plays in cratering varies with soil and rock properties (strength, water content, seismic velocity, etc.), the explosive source and burial depth. Near optimum depth-of-burial in soil, the gas acceleration phase is dominant. For less than optimum depths-of-burial and even for near optimum depths-of-burial in rock, other processes become dominant in crater formation.

In any case, first principle calculations of cratering phenomena from such deeply buried explosions have been very successful in reproducing apparent crater dimensions in soil and rock. For example, Fig. 17 compares calculations

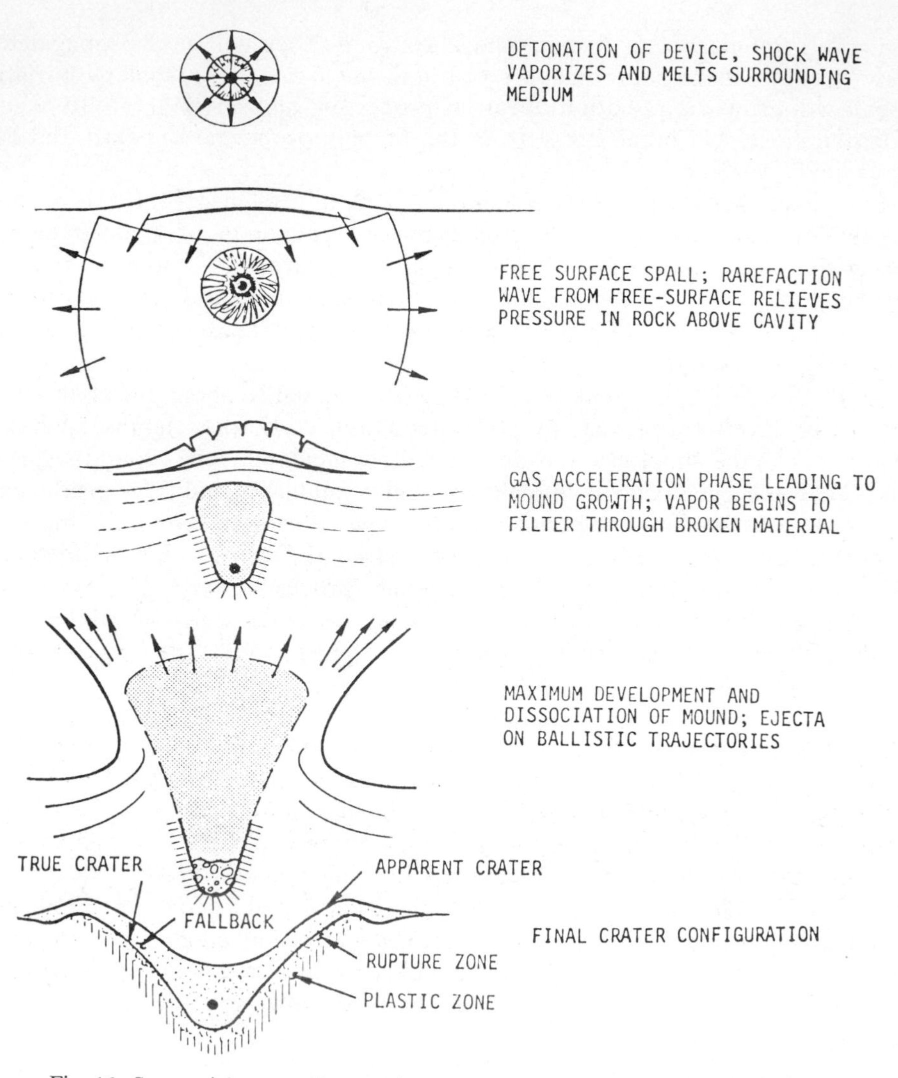

Fig. 16. Sequential stages of crater formation for deeply buried nuclear explosions.

by Terhune *et al.* (1970) and the apparent crater dimensions for Sedan (100 kilotons in alluvium) and Danny Boy (0.42 kilotons in basalt).

Shallow-buried explosions, that are more pertinent to studies of meteor impact craters, vent to the atmosphere early and the crater-forming process is dominated by compaction, plastic flow and scouring of media from the crater region. Various experimental and theoretical studies can be synthesized to provide a qualitative model of the sequential stages of the crater-forming process, as illustrated for a shallow-buried megaton explosion in Fig. 18. However, the phenomena are not so easily divided into distinct stages as for deeply-buried cratering explosions.

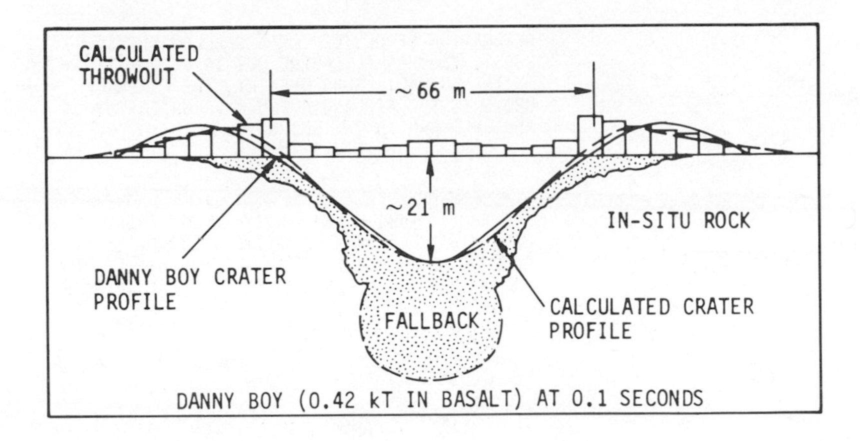

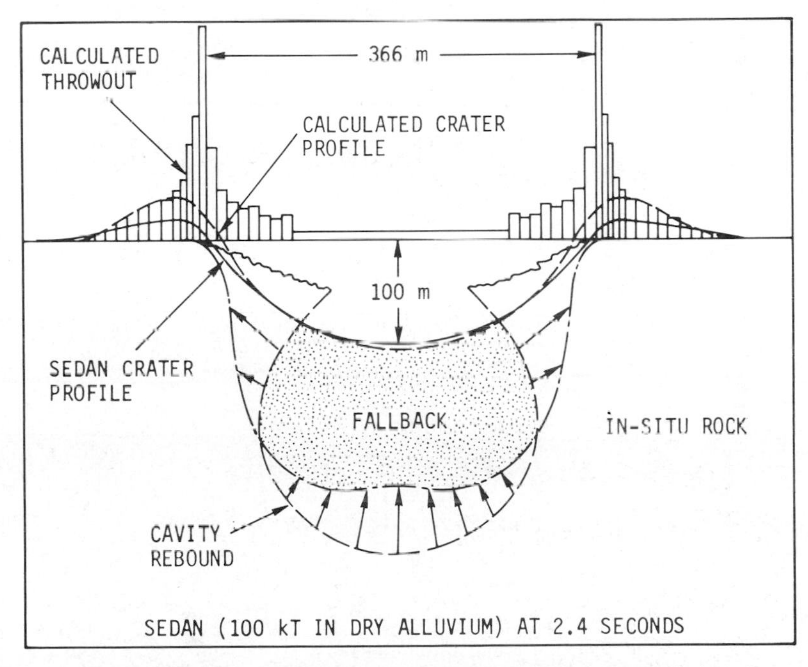

Fig. 17. Comparisons of calculated and measured craters from buried nuclear explosions. (Terhune *et al.*, 1970).

For a shallow-buried burst ($\sim 2\ \mathrm{m/MT^{1/3}}$), the vaporized material in the source region almost immediately blows upward into the atmosphere producing a strong rarefaction that moves downward from the earth's surface. For deeper explosions (say $\sim 100\ \mathrm{m/MT^{1/3}}$), the vaporized material does not intersect the earth's surface immediately, but a strong shock wave (~ 100 kbar) will interact with the free surface, sending a rarefaction back toward the source. In either case, the rarefaction wave turns the initially radial particle motion (from the burst point) toward the earth's surface producing the concave upward particle

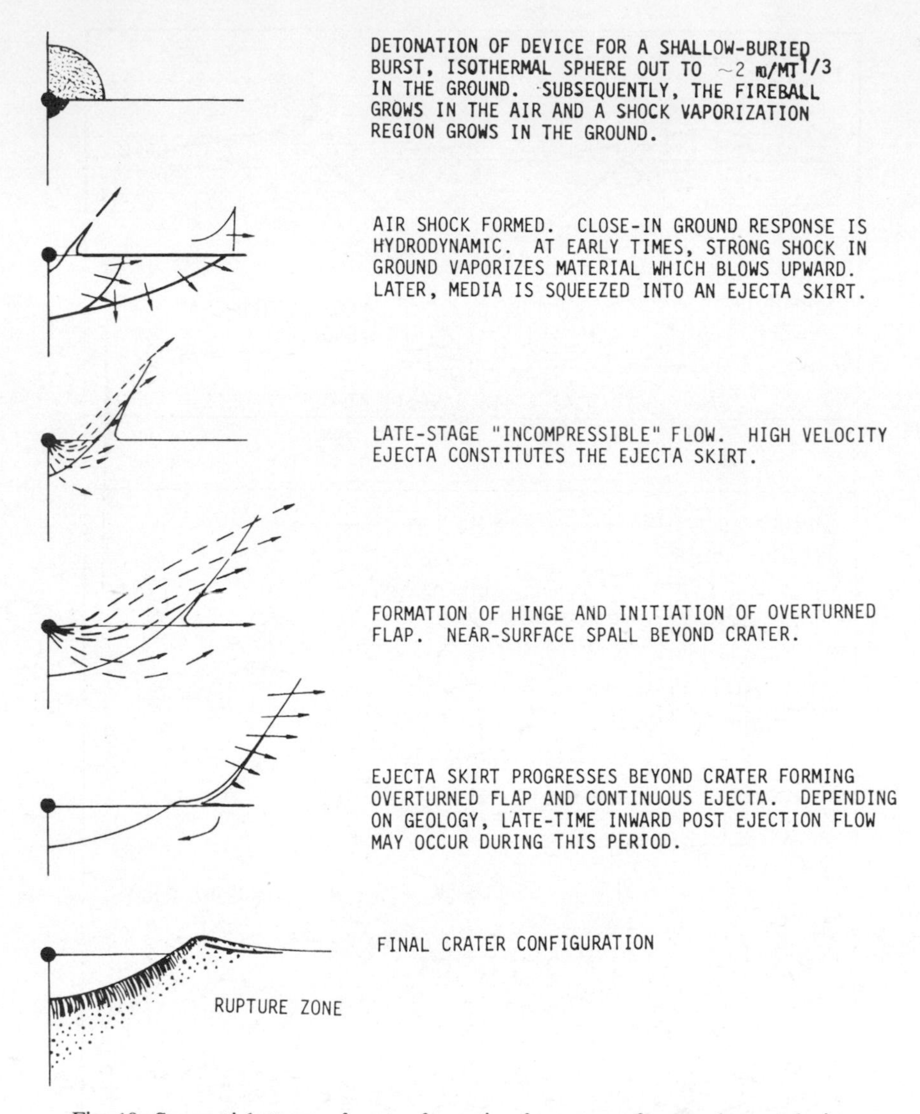

Fig. 18. Sequential stages of crater formation for near-surface nuclear explosions.

paths indicated in Fig. 18 (Gault *et al.*, 1968; Rae, 1970). An upward and outward directed conical jet sometimes called an "ejecta plume" is produced early at the edge of the transiently forming crater.

For times later than $\sim 10^{-2}/\mathrm{MT}^{1/3}$ sec (and peak stresses less than a few hundred kilobars) the mechanical properties of the earth media (strength, porosity, etc.) become important to the shock propagation, ground motion and crater-forming processes. Far behind the hemispherical direct-induced shock front in the ground, the transient crater appears to grow in a manner consistent with incompressible flow phenomena. Based on study of experiments and calculations, Maxwell and Siefert (1975) suggest concave upward particle paths ("streamlines") as indicated in Fig. 18. Since this "incompressible flow" process occurs at late times (compared to the shock compression phase), the *post-shock* material properties (rather than the virgin *in situ* material properties) control the late-stage cratering phenomena. As indicated earlier, Ulrich (1976) found such properties were also important in central peak formation from a surface HE experiment in sandstone.

Laboratory experiments (e.g., Piekutowski, 1975a, 1975b; Oberbeck, 1974), and various first principle calculations (e.g., Seabaugh, 1975), show that the angle between the ejecta skirt and the earth's surface is nearly constant throughout the crater-forming process at 40–50° for most media. Significantly steeper ejecta angles appear to result from near-surface explosions in saturated media and certain layered media (Piekutowski, 1976).

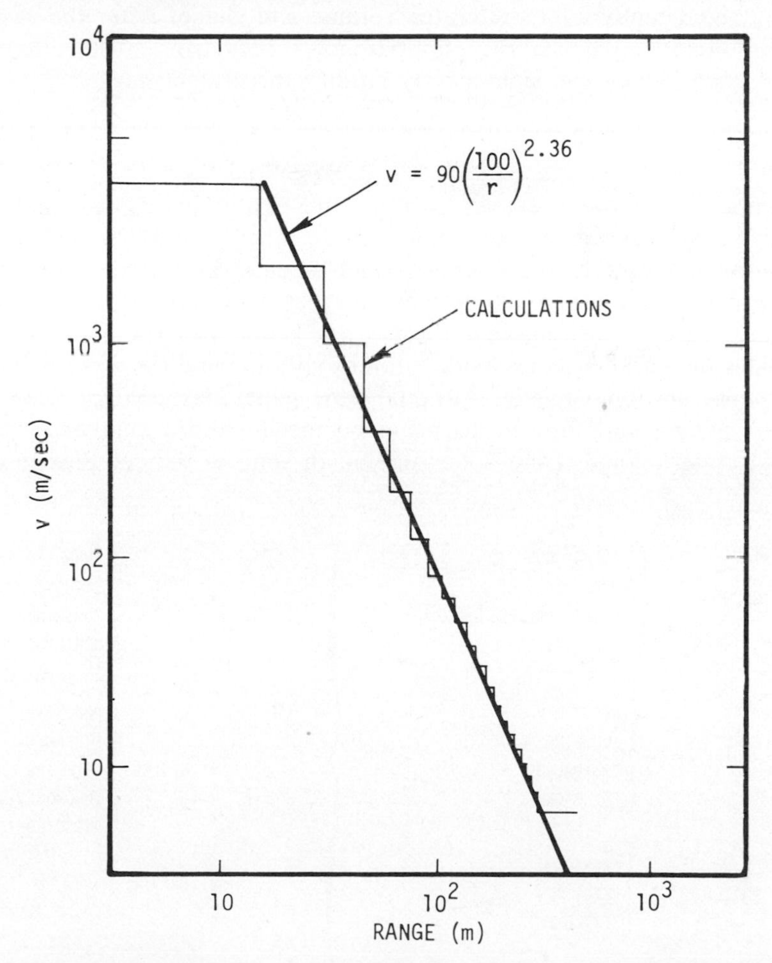

Fig. 19. Ejection velocity for a 1 megaton buried nuclear explosion in granite (Trulio, 1970).

The ejection velocities are greatest near the source and least at the crater's edge such that the ejecta forms an overturned flap and continuous ejecta field near the crater. Based on an analysis of calculations and various experiments, the ejection velocity varies approximately as $r^{-\alpha}$ where r is the range from ground zero and $2 \leq \alpha \leq 4$. For example, Fig. 19 provides results of a calculation by Trulio (1976) of a megaton explosion at 4.6 m depth-of-burial in hard rock.

The late-time ground motions continue until the post-shock material strength and gravitational forces arrest the flow. The ejected material falls back to earth along ballistic trajectories except as influenced by late-time atmospherics (blast winds, fireball rise, afterwinds, etc.).

During crater formation, reflected and refracted ground shock from underlying layers can produce a spall region that extends outward from beneath the overturned flap. During the late stages of crater formation, inward and upward flow may significantly alter the crater volume and shape. After the explosively induced motions have ceased, gravitational forces can cause slumping and late-time creep which can significantly modify the crater shape.

5. Concluding Comments

As indicated in the previous section, the evolution of nuclear explosion craters is a complex process spanning a wide range of material behavior from vaporization adjacent to the source (which occurs on a time scale of microseconds) to late-time creep and settlement (that occurs on a time scale greater than many seconds—perhaps even hours and days after the event). Figure 20 summarizes the various stages of the crater evolution and the principal variables that influence the phenomena. Although this paper has not considered all of these parameters, hopefully it has provided insight into a number of the more important cratering and related phenomena. In spite of the complexity of many

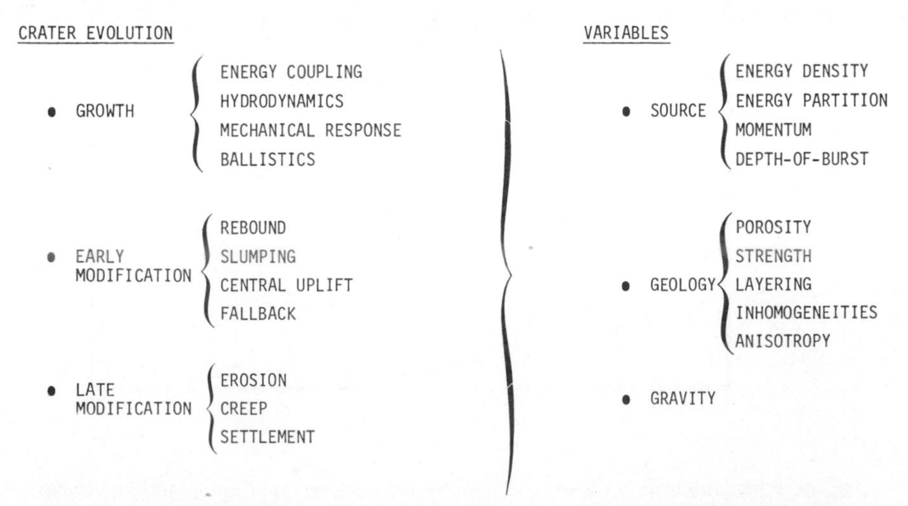

Fig. 20. Summary of mechanisms and variables important to cratering phenomena.

details which influence the cratering process, a fairly complete qualitative picture of the important mechanisms now exists.

Experiments continue to play the dominant role in our present understanding of cratering phenomena by providing both qualitative and quantitative information. Although new experiments may provide new information, it is the author's conviction that much remains to be learned from the existing data and that too little analysis has been conducted to synthesize a general picture from both experiments and theoretical studies. As an example, the hypothesis that ejecta initial velocities may vary as the sixth root of the explosive yield was inferred from examining old ejecta and ground motion data; and if true it poses important questions for theoreticians to explore in an effort to understand this unexpected result. This process of hypothesis and test is the essence of the scientific method; but has been only intermittently applied to the study of cratering mechanics.

Theoretical studies, using first principle numerical methods, can be expected to become increasingly important in future studies of cratering phenomena. Already, there is quite good confidence in the calculation of the qualitative and sometimes quantitative feature of the crater-forming process from buried explosions (and consequently from impact cratering events). The primary uncertainty that restricts confidence in these techniques is our lack of understanding of the dynamic properties of *in situ* geologic materials. Continuing research in this field can be expected to result in improved calculations of cratering, ejecta and ground shock from surface and buried explosions.

References

Baldwin, R. B.: 1963, *The Measure of The Moon.* University of Chicago Press, Chicago.

Bjork, R. L.: 1961, Analysis of the formation of meteor crater, Arizona: A preliminary report. *J. Geophys. Res.* **66**, 3379–3387.

Brode, H. L. and Bjork, R. L.: 1960, *Cratering from a Megaton Surface Burst.* The Rand Corporation, Santa Monica, California, RM-2600.

Brode, H. L.: 1968, Review of nuclear weapons effects. *Annual Review of Nuclear Science* **18**, 153–202.

Brode, H. L.: 1970, *Height-of-Burst Effects at High Overpressures.* The Rand Corporation, Santa Monica, California, DASA 2506.

Chabai, A. J.: 1965, On scaling dimensions of craters produced by buried explosions. *J. Geophys. Res.* **70**, 5075–5098.

Cherry, J. T.: 1967, Computer calculations of explosion-produced craters. *Int. J. Rock Mech. Miner. Sci.* **4**, 1–22.

Cherry, J. T.: 1970, Numerical simulation of stress wave propagation from underground nuclear explosions. In *Proc. Symposium on Engineering with Nuclear Explosives.* Las Vegas, Nevada, p. 142–220.

Carpenter, H. J. and Brode, H. L.: 1974, Height-of-burst at high overpressures. In *Proc. Fourth International Symposium on Military Applications of Blast Simulation.* Southend-on-Sea, England.

Cooper, H. F.: 1967, *Comparison Studies of Finite Difference Results for Explosions on the Surface of the Ground.* Air Force Weapons Laboratory, Kirtland Air Force Base, New Mexico, AFWL-TR-67-25.

Cooper, H. F.: 1971, *On Crater-Induced Ground Motions from Near Surface Bursts.* Air Force Weapons Laboratory, Kirtland Air Force Base, New Mexico, unpublished report.

Cooper, H. F.: 1973, *Empirical Studies of Ground Shock and Strong Motions in Rock.* R&D Associates, Marina del Rey, California, DNA 3245F.

Cooper, H. F.: 1976, *Estimates of Crater Dimensions for Near-Surface Explosions of Nuclear and High-Explosive Sources.* R&D Associates, Marina del Rey, California RDA-TR-2604-001.

Cooper, H. F. and Blouin, S. E.: 1971, Dynamic *in situ* rock properties from buried high-explosive arrays. In *Dynamic Rock Mechanics*, Twelfth Symposium on Rock Mechanics, G. G. Clark (ed.), p. 45–70. The American Institute of Mining and Petroleum Engineers, Inc., New York, N.Y.

Cooper, H. F., Brode, H. L., and Leigh, G. G.: 1972, *Some Fundamental Aspects of Nuclear Weapons*, Air Force Weapons Laboratory, Kirtland Air Force Base, New Mexico, AFWL-TR-72-19.

Cooper, H. F. and Sauer, F. M.: 1977, Crater-related ground motions and implications for crater scaling. In *Impact and Explosion Cratering*, D. J. Roddy, R. O. Pepin, and R. B. Merrill (eds.). This volume.

Crawford, R. E., Higgins, C. J., and Bultman, E. H.: 1974, *The Air Force Manual for Design and Analysis of Hardened Structures.* Air Force Weapons Laboratory, Kirtland Air Force Base, New Mexico, AFWL-TR-72-102.

Crowley, B. K. 1975, Scaling for rock dynamic experiments. *Proc. Symposium on Engineering with Nuclear Explosives*, Las Vegas, Nevada, p. 545–559.

Dillon, L. A.: 1971, *The Influence of Soil and Rock Properties on the Dimensions of Explosive-Produced Craters.* Air Force Weapons Laboratory, Kirtland Air Force Base, New Mexico, AFWL-TR-71-144.

Ganong, G. P. and Roberts, W. A.: 1968, *The Effect of the Nuclear Environment on Crater Ejecta Trajectories for Surface Bursts.* Air Force Weapons Laboratory, Kirtland Air Force Base, New Mexico, AFWL-TR-68-125.

Gault, D. E., Guest, J. E., Murrary, J. B., Dzurisin, D., and Malin, M. C.: 1975, Some comparisons of impact craters on Mercury and the moon. *J. Geophys. Res.* **80**, 2444–2460.

Gault, D. E., Quaide, W. L., and Oberbeck, V. R.: 1968, Impact cratering mechanics and structures. In *Shock Metamorphism of Natural Materials*, B. M. French and N. M. Short (eds.), p. 89–99. Mono Book Corporation, Baltimore.

Glen, H. D. and Thomsen, J. M.: 1976, Computer simulation of a high-explosive cratering experiment in a complex multilayered geology. Submitted to *Int. J. of Rock Mech. and Mining Sciences & Geomech. Abstracts.*

Herr, R. W.: 1971, *Effects of Atmospheric-Lithostatic Pressure Ratio on Explosive Craters in Dry Soil.* NASA Langley Research Center, Hampton, Virginia, NASA-TR-R-366.

Ingram, J. K., Drake, J., and Ingram, L.: 1975, *Influence of Burst Position on Airblast, Ground Shock and Cratering in Sandstone.* Miscellaneous Paper, N-75-3, U.S. Army Engineer Waterways Experiment Station, Vicksburg, Mississippi.

Knowles, C. P.: 1973, *Energy Coupling of Nuclear Weapons as It Relates to the Production of Direct-Induced Ground Shock.* R&D Associates, Marina del Rey, California, RDA-TR-3601-008.

Knowles, C. P. and Brode, H. L.: 1977, The theory of cratering and overview. In *Impact and Explosion Cratering*, D. J. Roddy, R. O. Pepin, and R. B. Merrill, (eds.). This volume.

Lampson, C. W.: 1966, Explosions in earth. In *Effects of Impact and Explosion*, Vol. **1**, Part 2, Chapter 3. Office of Scientific Research and Development, Washington, D.C.

Lockard, D. M.: 1976, *Crater Parameters and Material Properties.* Air Force Weapons Laboratory, Kirtland Air Force Base, New Mexico, AFWL-TR-74-200.

Maxwell, D. and Seifert, K.: 1975, *Modeling of Cratering, Close-In Displacements and Ejecta.* Physics International Company, San Leandro, California, DNA 3628F.

Melosh, H. J.: 1977, Crater modification by gravity: A mechanical analysis of slumping. In *Impact and Explosion Cratering*, D. J. Roddy, R. O. Pepin, and R. B. Merrill, (eds.). This volume.

Needham, C.: 1971, Airblast interaction calculations. In *Proc. of the Eric H. Wang Symposium on Protective Structures Technology*, **I**, 57–77. Air Force Weapons Laboratory, Kirtland Air Force Base, New Mexico.

Nordyke, M. D.: 1961, Nuclear craters and preliminary theory of the mechanics of explosive crater formation. *J. Geophys. Res.* **66**, 3439–3459.

Oberbeck, V. R.: 1971, Laboratory simulation of impact cratering with high explosives. *J. Geophys. Res.* **76**, 5732–5749.

Perret, W. R. and Bass, R. C.: 1975, *Free-Field Ground Motion Induced by Underground Explosions.* Sandia Laboratories, Albuquerque, New Mexico, SAND74-0253.

Piekutowski, A. J.: 1975a, *The Effect of Variations in Test Media Density on Crater Dimensions and Ejecta Distribution.* Air Force Weapons Laboratory, Kirtland Air Force Base, New Mexico, AFWL-TR-74-326.

Piekutowski, A. J.: 1975b, *A Summary of Four Laboratory-Scale High-Explosive Cratering Parametric Sensitivity Studies.* Air Force Weapons Laboratory, Kirtland Air Force Base, New Mexico, AFWL-TR-75-211.

Piekutowski, A. J.: 1976, A review of cratering and ejecta studies performed at the University of Dayton Research Institute. In *Papers Presented to the AFWL Cratering and Related Effects Review.* Kirtland Air Force Base, New Mexico.

Pike, R. J.: 1974, Depth/diameter relations of fresh lunar craters: Revision from spacecraft data. *Geophys. Res. Lett.* **1**, 291–294.

Pike, R. J.: 1977, Size-dependence in the shapes of fresh impact craters on the moon. In *Impact and Explosion Cratering*, D. J. Roddy, R. O. Pepin and R. B. Merrill, (eds.). This volume.

Port, R. J. and Gajewski, R.: 1973, Sensitivity of uniaxial stress-strain relations calculations of Middle Gust III. In *Proc. of the Mixed Company/Middle Gust Results Meeting,* **II**, 540–567. GE-TEMPO/DASIAC, Santa Barbara, California.

Port, R. J. and Bratton, J. L.: 1973, Material models, calculations and experimental results—Middle Gust IV. In *Proc. Strategic Structures Vulnerability/Hardening Long Range Planning Meeting,* **I**, 141–186. The Defense Nuclear Agency, Washington, D.C., DNA 3132P-1.

Post, R. L.: 1974, *Ejecta Distributions from Near-Surface Nuclear and HE Bursts.* Air Force Weapons Laboratory, Kirtland AFB, New Mexico, AFWL-TR-74-51.

Quaide, W. L., Gault, D. E., and Schmidt, R. A.: 1964, Gravitative effects on lunar impact structures. *Annals of the New York Academy of Sciences* **123**, 563–572.

Rae, W. J.: 1970, Analytical studies of impact-generated shock propagation. In *High-Velocity Impact Phenomena*, R. Kinslow (ed.), p. 213–291. Academic Press, New York and London.

Riney, T. D., Dienes, J. K., Frazier, G. A., Garg, S. K., Kirsch, J. W., Brownell, O. H., and Good, A. J.: 1972, *Ground Motion Models and Computer Techniques.* Systems, Science and Software, La Jolla, California, DNA 2915Z.

Roddy, D. J.: 1968, The Flynn Creek crater, Tennessee. In *Shock Metamorphism of Natural Materials,* B. French and N. Short (eds.), p. 291–322. Mono Book Corp., Baltimore, Maryland.

Roddy, D. J.: 1970, LN 303 geologic studies. *Operation Prairie Flat Symposium Report,* M. J. Dudash (ed.), p. 210–215. General Electric—TEMPO, DASIAC, Santa Barbara, California, DASA 2377-1.

Roddy, D. J.: 1973, Geologic studies of the Middle Gust and Mixed Company craters. In *Proc. of the Mixed Company/Middle Gust Results Meeting,* **II**, 79–124. General Electric Company—TEMPO, DASIAC, Santa Barbara, California.

Roddy, D. J.: 1976, High-explosive cratering analogs for bowl-shaped, central uplift and multi-ring impact craters. *Proc. Lunar Sci. Conf. 7th*, p. 3027–3056.

Roddy, D. J., Boyce, J. M., Colton, G. W., and Dial, A. L.: 1975, Meteor Crater, Arizona, rim drilling with thickness, structural uplift, diameter, depth, volume, and mass-balance calculations. *Proc. Lunar Sci. Conf. 6th*, p. 2621–2644.

Rooke, A. D., Carnes, B. L., and Davis, L. K.: 1974, *Cratering by Explosives: A Compendium and Analysis.* Waterways Experiment Station, Technical Report N-74-1.

Sandler, I. S., Wright, J. P., Baron, M. L., and Kavarna, J.: 1974, *Ground Motion Calculations for the Mixed Company Event of the Middle North Series.* U.S. Army Engineer Waterways Station, Vicksburg, Mississippi, Contract Report S-74-4.

Seebaugh, W. R.: 1975, *Studies of the Nuclear Crater Ejecta Environment.* Science Applications, Inc., McLean, Virginia, DNA 3640.

Shoemaker, E. M.: 1963, Impact mechanics at Meteor Crater, Arizona. In *The Moon, Meteorites and Comets*, B. M. Middlehurst and G. P. Kuiper (eds.), p. 301–336. University of Chicago Press.

Sun, J. M. S.: 1970, Energy counter-pressure scaling equations of linear crater dimensions. *J. Geophys.* **75**, 2003–2027.

Terhune, R. W., Stubbs, T. F., and Cherry, J. T.: 1970, Nuclear cratering on a digital computer. In *Proc. Symposium on Engineering with Nuclear Explosives.* Las Vegas, Nevada, p. 334.

Trulio, J. G.: 1970, *Calculations of Cratering, Ejecta and Dust Lofting.* Applied Theory, Inc., Los Angeles, California, DASA 2507.
Trulio, J. G. and Perl, N. K.: 1973, Limitations of present computational models of explosively-induced ground motion: Middle Gust Event 3. In *Proc. of the Mixed Company/Middle Gust Results Meeting.* GE-TEMPO/DASIAC, Santa Barbara, California, p. 568–617.
Trulio, J. G.: 1976, Ejecta formation: Calculated motions from a shallow-buried nuclear burst and its significance for high velocity impact cratering. Oral Presentation at the Symposium on Planetary Cratering Mechanics, Flagstaff.
Ulrich, G. W.: 1976, *The Mechanics of Central Peak Formation in Shock Wave Cratering Events.* Air Force Weapons Laboratory, Kirtland Air Force Base, New Mexico, AFWL-TR-75-88.
Violet, C. E.: 1961, A generalized empirical analysis of cratering. *J. Geophys. Res.* **66**, 3461–3470.
Vortman, L. J.: 1968, Craters from surface explosions and scaling laws. *J. Geophys. Res.* **73**, 3461–4636.
Vortman, L. J.: 1969, Ten years of high-explosive cratering research at Sandia Laboratory. *Nuclear Applications and Technology* **7**, 269–304.
Vortman, L. J.: 1977, Craters from surface explosions and energy dependence: A retrospective view. In *Impact and Explosion Cratering*, D. J. Roddy, R. O. Pepin, and R. B. Merrill (eds.). This volume.
White, J. W.: 1971, Examination of crater formulae and scaling methods. *J. Geophys. Res.* **76**, 8599–8603.
White, J. W.: 1973, An empirically derived cratering formula. *J. Geophys. Res.* **78**, 8623–8633.
Wisotski, J.: 1977, Dynamic ejecta parameters from high explosive detonations. In *Impact and Explosion Cratering*, D. J. Roddy, R. O. Pepin, and R. B. Merrill (eds.). This volume.

Roddy, D. J., Pepin, R. O., and Merrill, R. B., editors.
(1977) *Impact and Explosion Cratering*, Pergamon Press (New York), p. 45–65.
Printed in the United States of America

Application of high explosion cratering data to planetary problems

VERNE R. OBERBECK

NASA Ames Research Center, Moffett Field, California 94035

Abstract—Solution of many planetary cratering problems requires knowledge of processes acting during formation of impact craters. Because terrestrial meteorite craters have not been observed during formation, it was necessary in the past to observe small laboratory craters and missile impact craters. Dynamic data for larger events must be obtained from study of properly selected high explosion or nuclear events. This paper presents a review of the conditions of explosion or nuclear cratering required to simulate impact crater formation. Next, some planetary problems associated with three different aspects of crater formation are illustrated and solutions based on high-explosion data are suggested. Structures of impact craters and properly selected explosion craters formed in layered media are discussed and related to the structure of lunar basins. The mode of ejection of material from impact craters is clarified using explosion analogs and it is shown to have important implications for the origin of material in crater and basin deposits. Equally important are the populations of secondary craters on lunar and planetary surfaces. Explosion crater data aid in their recognition and prediction.

INTRODUCTION

FORMATION OF IMPACT CRATERS can be observed by impacting projectiles in the laboratory and impacting missiles at military test ranges. Such studies provide the only available dynamic data relating impact cratering effects to known conditions of impact velocities, kinetic energies, projectile properties, and target conditions. The limitation of these data is that all the events have kinetic energies less than about 10^{15} ergs. Although many terrestrial impact craters provide knowledge of the structure of large craters, few are uneroded, and the conditions of impact are unknown. Fortunately there are many recent chemical and nuclear explosion craters that are larger than missile impact craters for which the physical conditions of formation and target properties are known and dynamic data have been obtained. Studies of these explosion craters can increase our confidence in application of small scale impact experiment results. However, only those explosion events that simulate impact events can be used to solve planetary problems.

In this paper the small impact craters and explosion craters are compared in order to select the proper scaled depth of burst explosion craters as analogs of large planetary impact craters. A second objective is to illustrate explosion crater data which could have been used in the past to solve planetary geologic problems. This approach may suggest to those who study high-explosion cratering exclusively, some applications of explosion cratering to present and future planetary surface problems.

SIMULATION OF IMPACT CRATERS WITH HIGH EXPLOSIVES

Impact craters form when target material, compressed by a nearly hemispherical shock wave, is decompressed by a tensile wave and ejected. During crater formation, particle motion is first radial; then material moves nearly parallel to the crater wall just before ejection (Gault *et al.*, 1968). Explosion craters eject material by this process as well as by gas acceleration. However, the shock wave shape depends on the depth of burst: it varies from nearly hemispherical for near-surface burst events to spherical for deeply buried charges. This suggests that near surface burst explosion events may produce effects similar to impacting projectiles. Therefore, the absence of dynamic crater formation data for large impact craters and the scarcity of large, fresh impact craters led to the early application of high-explosion crater studies to many lunar crater energy scaling problems (Baldwin, 1963) and to the analysis of lunar crater ballistic problems through development of an explosion crater model for conditions of formation of meteor crater (Shoemaker, 1962). At this time, it was believed that impact events having a given kinetic energy could be simulated by explosive charges having an equivalent chemical energy provided that explosives were detonated at the proper depth of burst. Later, impact craters were compared to explosion craters formed in the laboratory. Results showed that effects of impact cratering can be simulated by shallow depth of burst explosion craters provided that variables in addition to energy are controlled (Oberbeck, 1971). Results indicated that it was necessary to control impact velocity as well as kinetic energy before it was possible to select an explosion crater analog. Figure 1 shows the relation between the mass ejected and projectile kinetic energies for impact craters formed in quartz sand targets during the simulation study. There is a separate relation between size and projectile kinetic energy for craters formed at each of the velocities indicated in Fig. 1; as a result, craters of many sizes can be produced by projectiles having the same kinetic energy. If impact velocity is specified, there is only one crater size possible for a given kinetic energy. Therefore, in order to simulate an impact crater of a given size and energy by an explosion of a given chemical energy and depth of burst, the velocity of impact also must be controlled. This required that the packing density of the explosive be selected such that the detonation pressure of the explosion-generated shock wave (determined by packing density) was the same as the pressure of the impact-generated shock wave (determined by impact velocity and equation of state of projectile and target material). Next, the charge weight was selected to produce a chemical energy equal to the projectile kinetic energy.

The conditions of formation of the laboratory impact crater and explosion craters are given in Table 1. An aluminum sphere of mass 0.43 g and kinetic energy 8.7×10^9 ergs impacted quartz sand at 2 km/sec. Charges of PETN of 150 mg having packing density equal to 1 g/cm^3 were detonated at depths between 0 and 77.5 mm. Peak shock wave pressures at the impact point and at the surface of the explosive was 83 kbar.

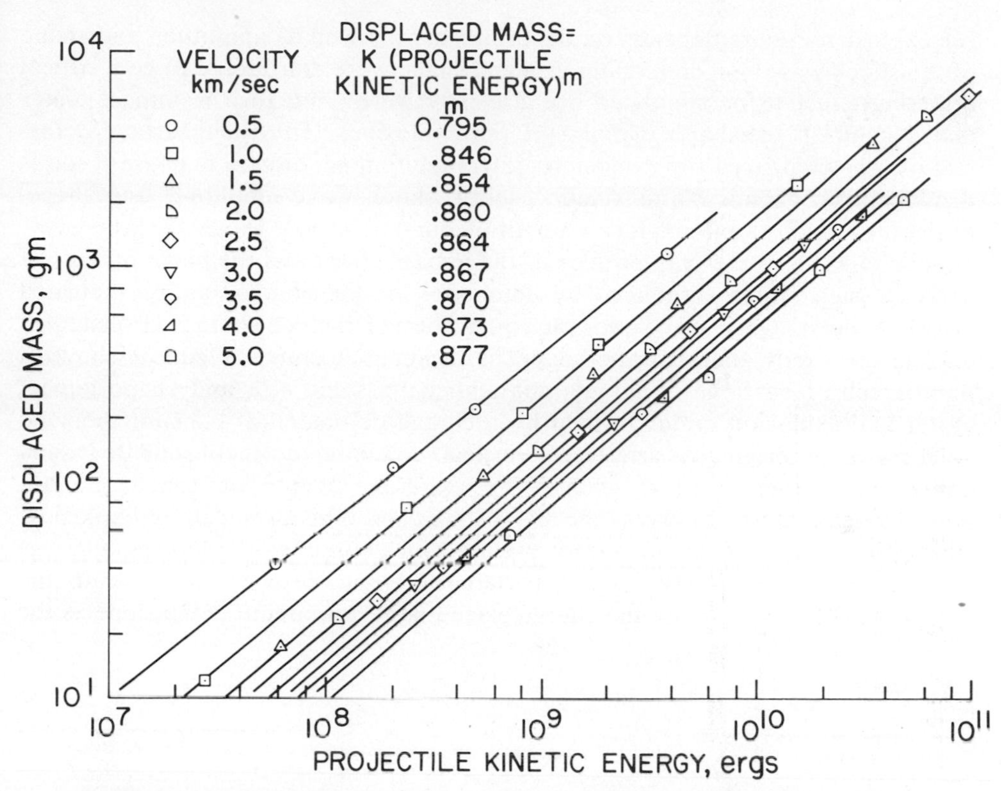

Fig. 1. Mass ejected from impact crater formed in quartz sand plotted as a function of projectile kinetic energy for each of the indicated projectile velocities.

Simulation of impact cratering effects by the detonation of explosives could be assessed by observing three effects of cratering. First, because plume growth depends on the initial particle velocity produced within the target, which depends on the amplitude and shape of the shock wave, high speed motion pictures recorded the growth of the ejecta plume of each crater. Plume growth

Table 1. Experimental conditions.

Impact craters	Explosion craters
Projectile material: aluminum	Explosive: PETN
Target material: quartz sand	Target material: quartz sand
Projectile mass: 0.43 g	Explosive mass: 150 mg
Impact velocity: 2.0 km/sec	Packing density: 1 g/cm^3
Projectile kinetic energy: 8.8×10^9 ergs	Chemical energy: 8.7×10^9 ergs
Peak impact pressure: 83 kbar	Detonation pressure: 83 kbar
	Charge depths: 0, 3.2, 6.3, 9.5, 14.3, 27.0, 39.7, 52.5, 65.2, and 77.5 mm

for explosion craters depends on depth of burst as well as amplitude and shape of the shock wave. Second, crater size and shape were considered to be a critical cratering effect to be simulated because they were fixed for the impact crater and depend on the depth of burst of the explosives. Third, subsurface deformation was compared for explosion craters and impact craters because it too is dependent on chemical and kinetic energy, shock wave amplitude and shape, projectile velocity, and explosive depth of burst.

Figure 2 shows the gray profiles of the impact crater and the black profiles of the explosion craters produced by detonation of the charges at the indicated depth of burst. Only the explosion crater formed by the charge detonated at 6.3 mm is exactly the size and shape of the impact crater. Figure 3 shows a photograph of sections of targets containing the same size and shape impact crater and explosion crater formed by a charge detonated at 6.3 mm; the sand markers were originally arrays of vertical columns of unconsolidated sand containing a thermosetting agent. The cratering events deformed the unconsolidated material in ways reflective of the conditions of impact or explosion. This deformation was preserved by heating which solidified the targets, and was revealed by cutting sections. The pattern of target deformation beneath the explosion crater formed by the charge placed at 6.3 mm depth is the same as the

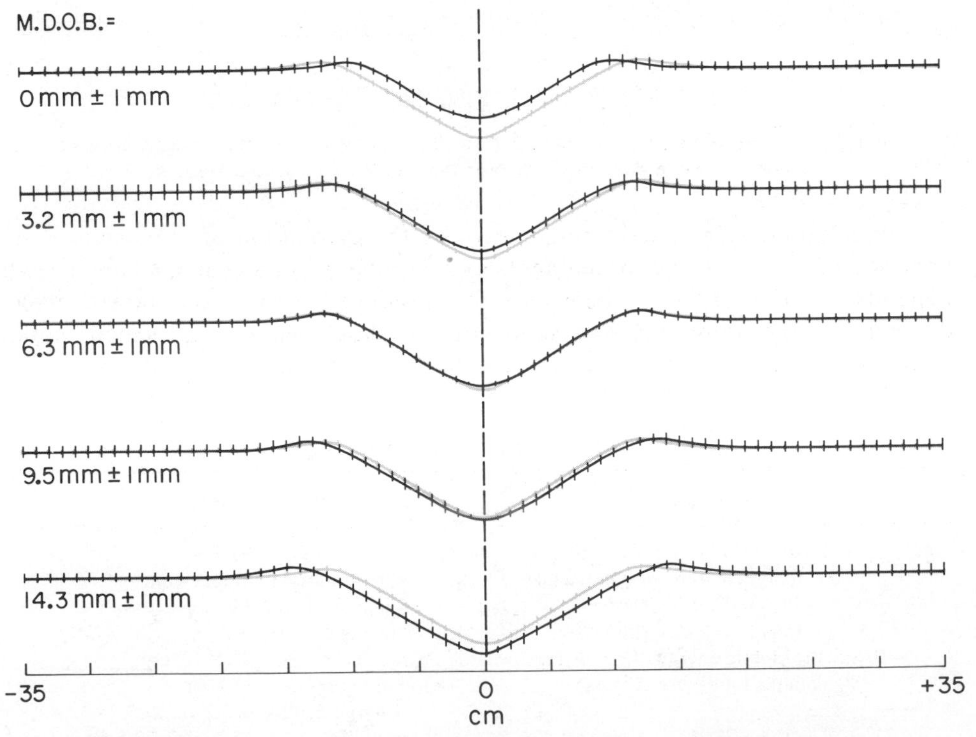

Fig. 2. Profiles of impact crater (gray) and profiles of explosion craters (black) show that the 6.3 mm depth of burst explosion is similar to the impact crater.

deformation pattern beneath the impact crater (Fig. 3). The pattern of deformation beneath explosion craters formed at other depths is much different from deformation beneath the impact crater (Fig. 4). Figure 5 shows the positions of the outer edge of the ejecta plumes of explosion craters (black) and impact craters (gray) at the indicated time in milliseconds after explosion and impact, respectively. The pattern of ejecta growth of the explosion crater formed by detonation of the charge at 6.3 mm, is most similar to the ejecta pattern of the explosion crater.

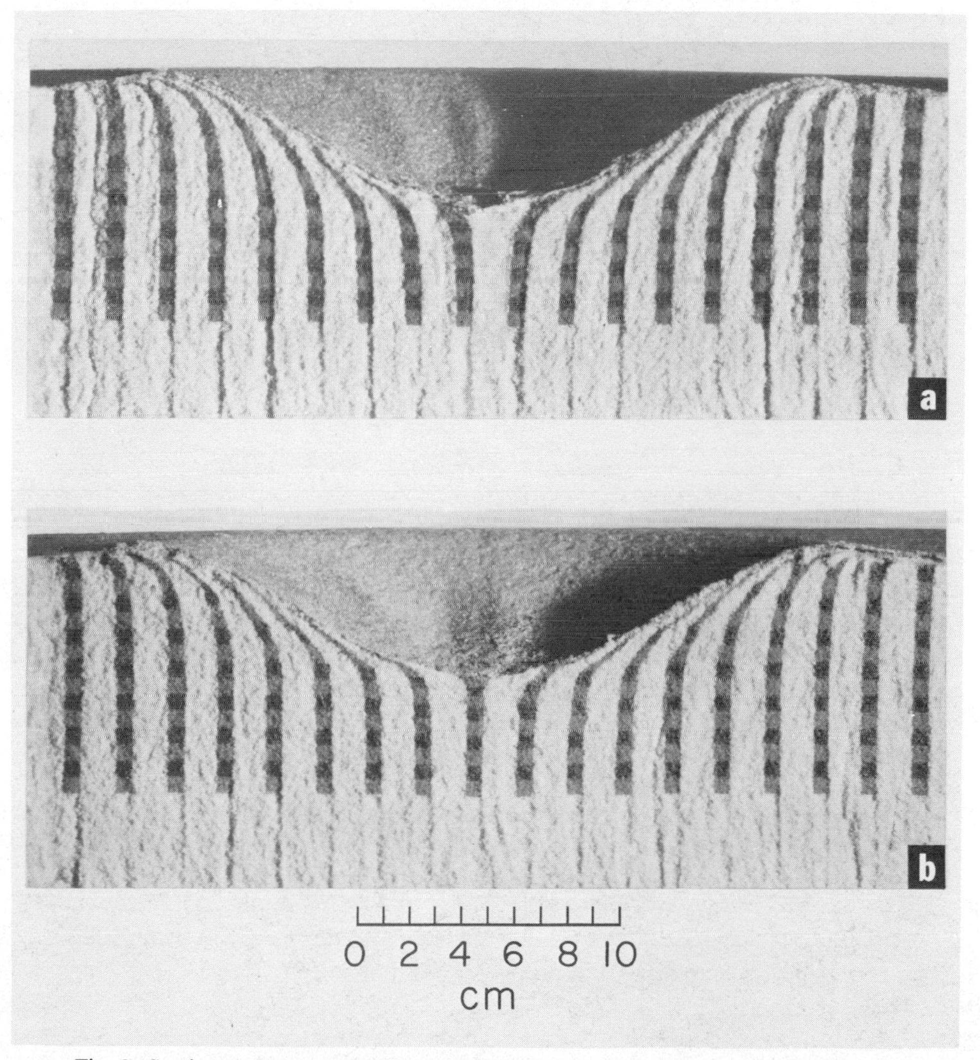

Fig. 3. Sectioned targets containing an impact crater (a) and an explosion crater (b). The vertical columns of sand were originally in an unconsolidated state but have been deformed by the cratering events. Patterns of deformation produced by impact are simulated by the 6.3 mm depth-of-burst explosion crater.

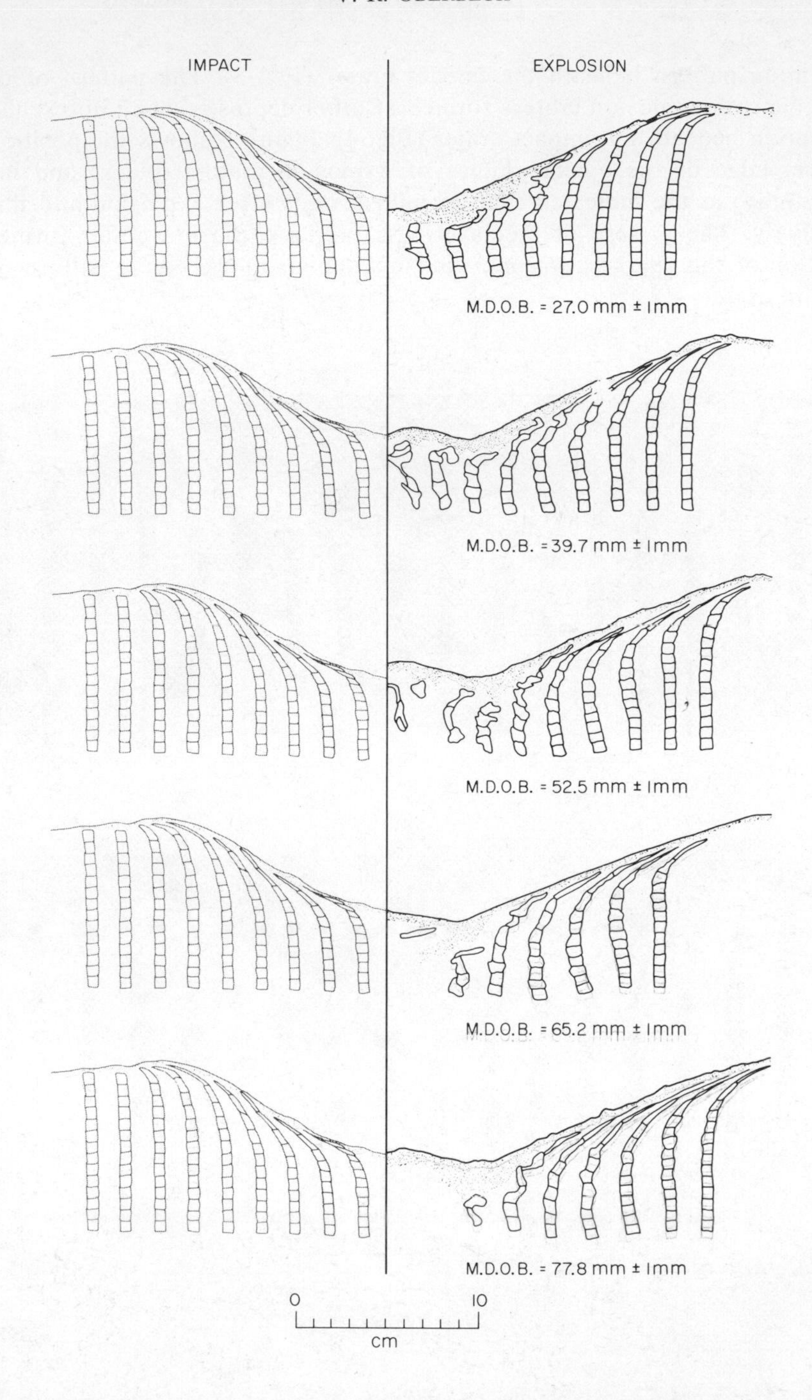

Fig. 4. Diagrams of subsurface deformation beneath impact crater and explosion craters produced by deeply buried explosives. Patterns produced by explosions are unlike those produced by impacts.

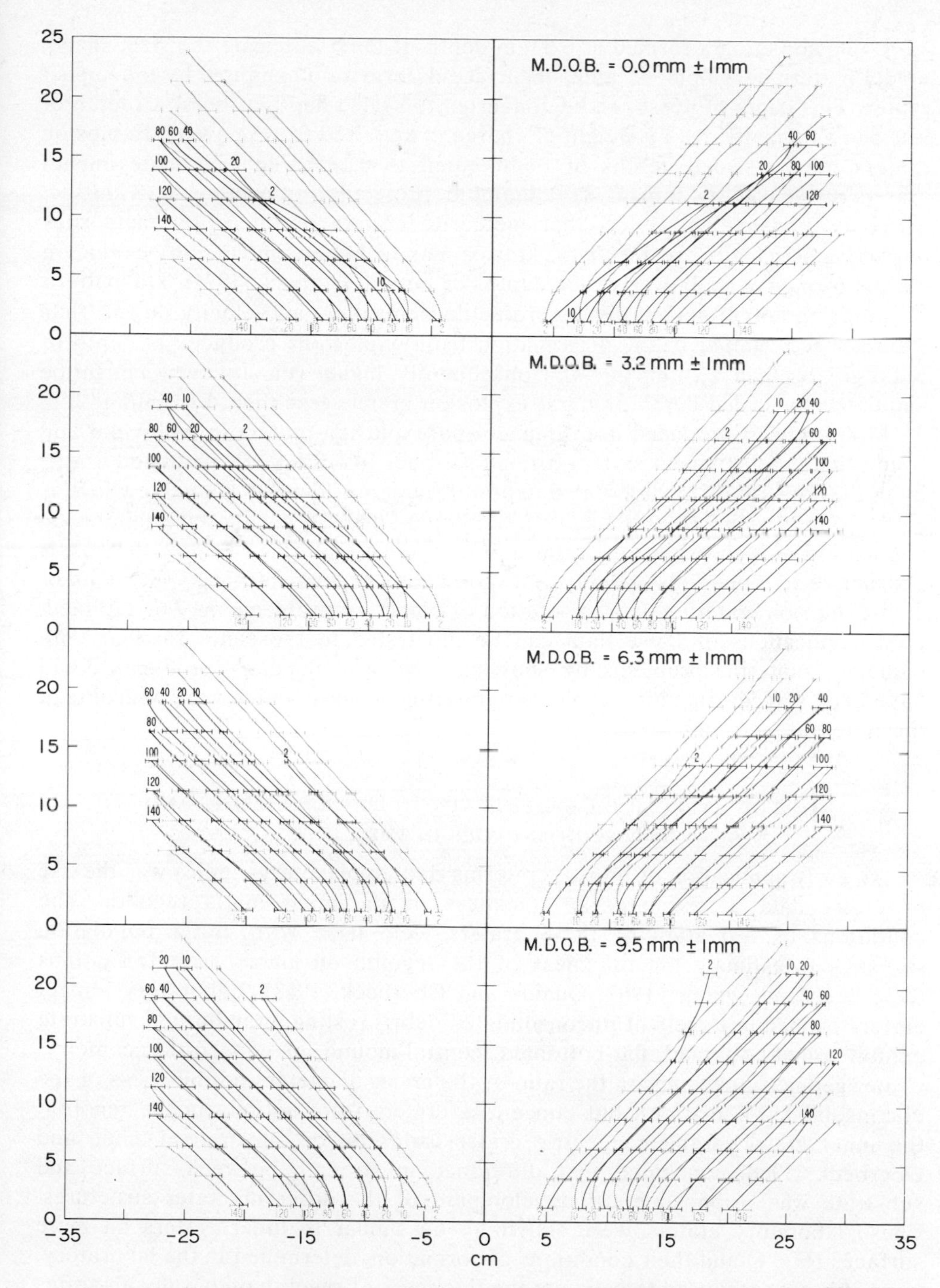

Fig. 5. Black lines show the positions of the ejecta plume of the explosion craters. Gray lines show the positions of the ejecta plumes of the impact crater. Numbers show times in milliseconds after impact or detonation.

Explosion craters formed at 6.3 mm depth of burst duplicate the size, shape, ejecta pattern development, and subsurface deformation exhibited by the impact crater. This depth of burst can be converted to scaled depth of burst λ (depth of burst in meters divided by weight of charge in kg raised to 1/3 power). Explosion craters having scaled depths of burst equal to $0.105\ m/kg^{1/3}$ simulate impact craters produced by impact of aluminum projectiles impacting at 2 km/sec. There is evidence, however, that meteorite craters produced by meteorites impacting at velocity greater than 2 km/sec may be better simulated by explosion craters formed at shallower scaled depths of burst (Oberbeck, 1971). The pattern of deformation produced by a projectile impacting at velocity lower than 2 km/sec was similar to the deformation from explosions produced at depth of burst greater than 6.3 mm (Fig. 4). Consequently, higher velocity events might be simulated by scaled depth of burst explosion craters less than $0.105\ m/kg^{1/3}$.

Moore (1976) produced both impact and explosion craters in colluvium and found that missile impact craters formed at about 10^{14} ergs were simulated in size by explosion craters having scaled depth of burst $= 0.15\ m/kg^{1/3}$, a value which is close to our value of $0.105\ m/kg^{1/3}$. Although there are still uncertainties in application of high explosion crater data to impact cratering (Moore, 1976), it is instructive to tentatively accept the experimental results so that applications of high-explosion cratering data to solution of planetary problems may be outlined. New applications of these data can be illustrated to those not familiar with planetary cratering problems by showing how past planetary problems could have been solved using high-explosion cratering analogs, which were available at the time but were not used.

USE OF EXPLOSIVE CRATER DATA TO DETECT SEISMIC AND STRENGTH DISCONTINUITIES IN THE LUNAR CRUST

An early application of impact cratering data to planetary studies was the use of crater data to estimate the thickness of the lunar maria regolith. The conditions of formation of these craters were used with crater population statistics to estimate the thickness of the regolith on mare basalt formations (Oberbeck and Quaide, 1967; Quaide and Oberbeck, 1968). Laboratory impact craters formed in targets of unconsolidated debris resting on indurated substrate exhibit round-bottomed, flat-bottomed, central-mound, or concentric geometry. Crater geometry depends on the ratio of the crater diameter and thickness of the unconsolidated debris. All but concentric craters form entirely in the regolith; the inner crater of the concentric crater forms in the substrate. Quaide and Oberbeck, 1968, concluded that difference in strength between surface and substrate was responsible for development of the different crater structures. These laboratory craters were shown to be similar to lunar craters on mare surfaces (Fig. 6) and their conditions of formation, determined in the laboratory, were subsequently used to estimate the thickness of regolith on basalt substrates (Oberbeck *et al.*, 1968). Long before these experiments were performed, high-explosive crater data were available and could have been used for both the

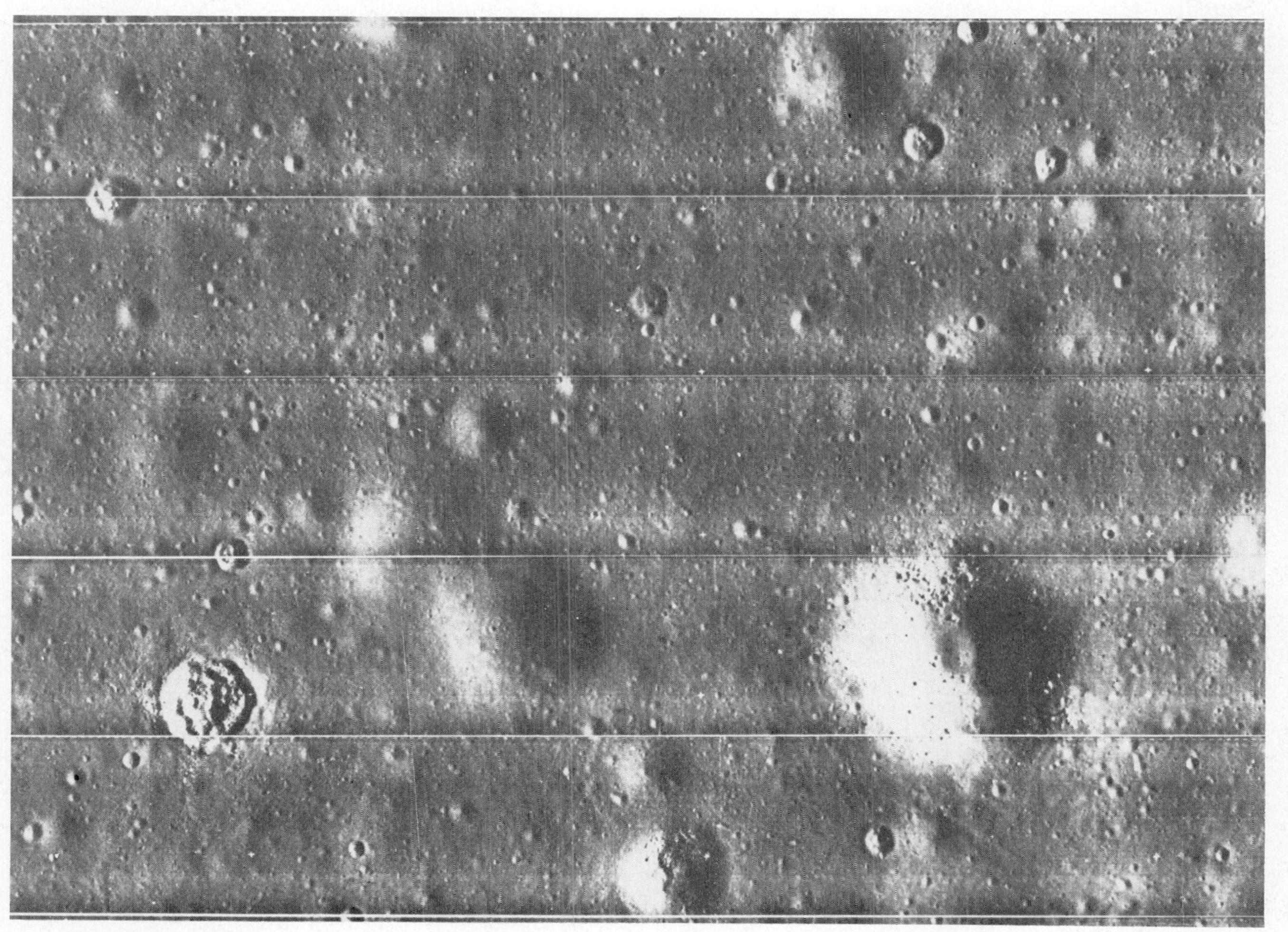

Fig. 6. Photograph of lunar mare surface showing craters having normal flat-bottomed, central mound, and concentric geometry.

detection of near-surface strength discontinuities and the interpretation of mare regolith thickness distributions. Figure 7 shows selected crater profiles adapted from experiments by Fortson and Brown (1958) to determine the effect of a soil-rock interface on crater shape. The position of the soil and concrete interface is shown on each crater profile. Twenty-seven pounds of C4 explosive

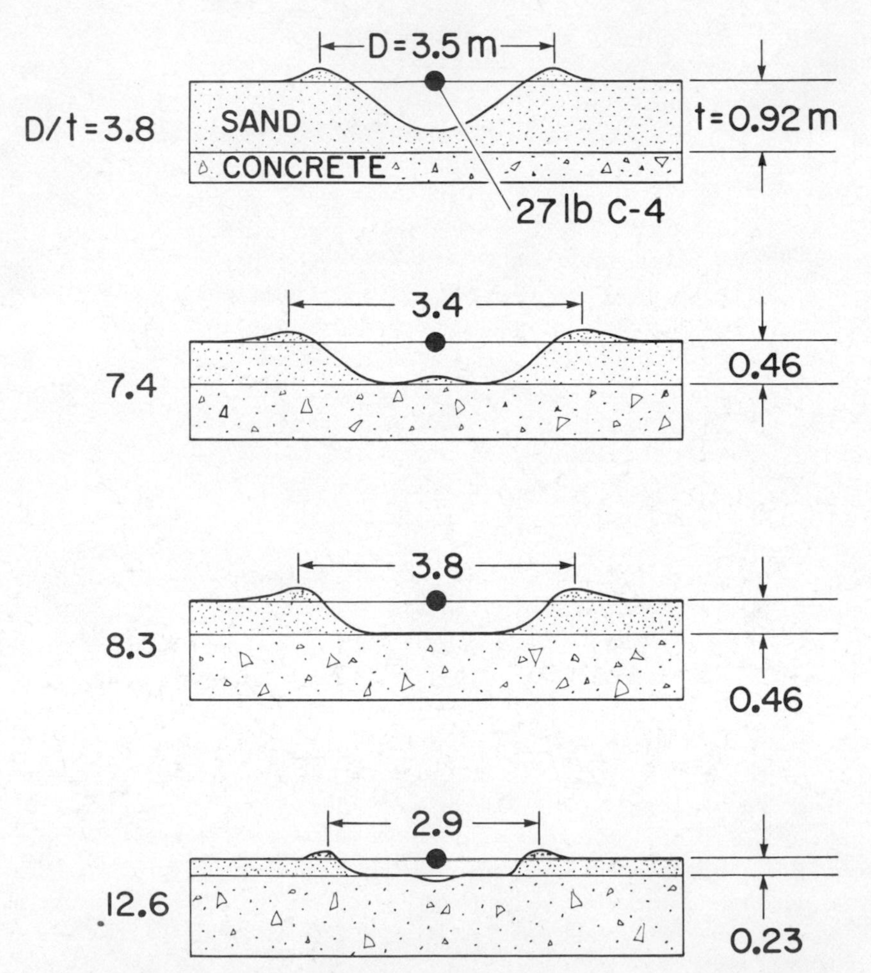

Fig. 7. Profiles of explosion craters formed in targets of soil on concrete substrates. The conditions of formation defined by the ratio of crater diameter D and soil thickness t are the same as for impact craters.

was detonated at the target surface to form each of the explosion craters. Therefore, $\lambda = 0 \text{ m/kg}^{1/3}$ for the explosion craters. This scaled depth of burst is less than that required to simulate impact craters formed by projectiles impacting at 2 km/sec but may be appropriate for lunar primary craters formed at velocities greater than 2.0 km/sec. Therefore, in hindsight, the Fortson and Brown explosion craters might have been used as appropriate analogs for some lunar impact craters produced in layered media.

Quaide and Oberbeck (1968) found that round-bottomed impact craters formed when the ratio of crater diameter and soil cover thickness was less than approximately 4. Central mound and flat-floored impact craters formed for ratios between 4 and some value between 8 and 9. Concentric impact craters formed when the ratio exceeded 9. It may be seen from Fig. 7 that the round-bottomed, flat-floored, central-mound, and concentric explosion craters formed when the ratio of crater diameter and soil thickness was 3.8, 7.4, 8.3, 12.6, respectively. The conditions of formation of the explosion craters in layered media appear to be the same as the conditions for the impact craters, thereby indicating that they are appropriate analogs for impact craters formed in layered media. If they had been used as analogs, they might have been used to detect a strength discontinuity several meters beneath the lunar mare surfaces and the regolith thickness distribution could have been determined from their conditions of formation. Piekutowski (1976) has now performed important small-scale laboratory explosion cratering experiments using layered targets and craters have the same morphologies as impact craters formed in layered media. These results support the argument that shallow depth-of-burst explosion craters formed in layered targets simulate effects of impact cratering in layered media. Moreover, they partially satisfy the requirement of Moore (1976) for further experiments to examine all target and projectile variables before it is accepted that explosion craters can simulate impact craters (Moore, 1976).

The progression of crater morphology with size for craters less than 200 m in diameter as shown in Fig. 6 is repeated for lunar impact craters with diameters between 1 and 100 km. The smallest craters are cup shaped, whereas the largest commonly are flat floored, central peaked, and terraced. In preparing their 1968 paper Quaide and Oberbeck considered—but were reluctant to propose—the hypothesis that these morphologic changes also reflected a response of cratering process to strength discontinuities as they did for the mare regolith craters. We believed, however, there would be no effective strength differences between the megaregolith and substrate for craters sufficiently large to penetrate the megaregolith. Short and Forman (1972) and Head (1976) suggested that strength discontinuities between the megaregolith and coherent subsurface formations are responsible for the observed changes in crater structure. Large terrestrial impacts are known that penetrate layered media. The Ries Crater has concentric geometry but it was first believed that the shelf separating the inner crater in the crystalline basement and outer crater in the sedimentary rocks was produced by slumping of the crater walls. However, laboratory experiments suggest that the outer crater at the Ries is larger than the substrate crater because the shock wave transmitted less energy to the substrate than would normally have been transmitted beneath the substrate surface. Rings of basins are believed to be analogous to the concentric rings of the Ries and concentric rings of laboratory impact craters (Oberbeck, 1975). The rings would, according to this hypothesis, represent craters formed in layers having different seismic velocities. Hodges and Wilhelms (1976) and Wilhelms *et al.* (1976) used this evidence from laboratory impact experiments, from high-explosion experiments, and from terrestrial impact crater data to interpret photographs of many lunar basins. They believe they find evidence for two and sometimes three main rings that could be the rims of basins of excavation in layers having different seismic properties. Their hypothesis is that rings are related to central peak rings within smaller craters and represent rims of craters produced both by excavation of material and by uplift of material from deeper layers having different seismic properties. They are the rims of craters which are not produced by slumping. For example, Wilhelms *et al.* (1976) conclude that the outer ring of the Orientale Basin, the Cordillera mountains, form the crater rim of excavation and the Outer and Inner Rook mountains together represent a major rim of a basin of excavation formed at a major seismic discontinuity.

The laboratory studies, photogeologic studies and lunar seismic profiles suggest that layering can, in fact, affect structure even on the scale of basins. However, the origin of basin rings is still uncertain. Several studies indicate other possible origins. Head (1976) and Schultz (1976) propose that large scale slumping is responsible for the outer ring of multi-ring basins and the central peak rings are modified central peaks. Roddy (1976) describes a series of high explosion craters formed by detonation of high-explosion charges and presents strong evidence for similarities of multirings produced in these craters and rings of multiring lunar basins. He concludes that surface burst charges may have been responsible for production of these features; the implication is that very high velocity impact events, which are modeled in these experiments by surface burst charges, produce ring structures. Support for this argument is our earlier discussion that produced evidence that high velocity impact craters might be simulated by smaller scale depth-of-burst explosion craters. However, some of the craters described by Roddy were produced in layered media and the water table was sometimes present at the position of the flat floor and near the depth at which the central rings arise. However, it is not certain whether water-saturated materials will behave like substrates of solid material. Further field study of these and other large high-explosion craters and code calculations of explosion cratering in layered media may prove to be valuable applications of data associated with large explosion craters to solution of important planetary problems.

MODE OF TRANSPORT OF IMPACT CRATER EJECTA

It is important for purposes of planetary stratigraphic studies to understand the mode of transport of ejecta of large impact craters and basins because the mode of transport determines the nature and origin of the deposits. Chao (1973) hypothesized that the Apollo 16 smooth plains impact breccias were Orientale ejecta transported for hundreds of kilometers in ballistic trajectories. Based in part on a study of the structure of different facies of the Orientale Basin deposits, Moore *et al.* (1974) concurred that some plains were Orientale ejecta. Oberbeck *et al.* (1974) proposed that smooth plains could not be predominantly Orientale ejecta if it was transported in ballistic trajectories as Chao maintained (Chao, 1973) and as confirmed by laboratory experiments. Velocities of ejection and impact of Orientale ejecta in ballistic trajectories would have been sufficient to cause secondary cratering and excavation of secondary crater ejecta which would mix with Orientale Basin ejecta to produce a mixed deposit. Chao (1974) concluded that formation of small laboratory impact craters that eject material in ballistic trajectories do not provide insight to formation of large lunar craters and basins. Recently Chao (1976) studied striations formed on pebbles of Ries Crater ejecta and concluded these formed when ejecta was transported from the crater rim in a non-ballistic roll glide transport mode. However, this type of transport has been observed during formation of the small laboratory craters but only after material is transported from the crater in ballistic trajectories. Figure 8 shows a side view of growth of the ejecta plume of a laboratory impact crater formed in a quartz sand target. The pattern of ejection of surficial material is the same for laboratory craters produced in layered media, which have structures like the Ries Crater. The ejecta plume has been disected by a vertical plate containing a vertical slit (Oberbeck and Morrison, 1976). We observe ejecta that moved along one radius after passing through this narrow slit. It can be seen that ejecta are in the form of a thin expanding conical sheet. Figure 9 shows motion pictures of further disection of another ejecta plume. Wires were placed across the vertical slit. Undisturbed parts of the plume pass the wires and trace out parabolic ballistic trajectories. Clearly ejecta of small impact craters is transported in ballistic trajectories. Figure 10 shows a magnification of the base of the conical ejecta plume of the laboratory impact crater shown in Fig. 9. Notice that a ground hugging debris surge builds up behind the ejecta plume after material impacts from ballistic trajectories and it follows the plume along the ground. Some believe from photogeologic studies, ballistic calculations, and terrestrial crater studies that this surge is made up of secondary crater ejecta and primary ejecta (Morrison and Oberbeck, 1975; Hörz, 1976). However, photogeologic evidence persuades others that deposits are primarily crater or basin ejecta. For example, many of the flow lobes on basin deposits (Eggleton, 1972) were believed to have been produced by a ground hugging mode of transport of primary crater ejecta. Study of ejection of material from large explosion craters may supply information required to choose between ballistic or nonballistic mode of transport for large craters. For example, Seebaugh (1976) has developed a model which should yield the conditions of ejection of different sizes of ejecta fragments from large craters. Such studies may resolve old controversies and will help to prepare solutions to new problems suggested by new imagery of planetary surfaces. For example, Carr *et al.* (1976) describe remarkable flow lobes in the continuous deposits of Martian craters and conclude that the impacting body melted permafrost which is ejected ballistically and becomes a fluidized sheet as a result of melted ice or atmospheric interaction. If it is due to melting of ice, valuable determinations of the properties of the Martian regolith may result from study of structures of Martian Crater deposits. It is not clear whether Carr *et al.* (1976) believe the ice is melted before ballistic transport or upon impact. If we consider that ice is melted before ejection, it may be possible to use high explosion craters to explain why flow patterns develop. Perhaps, craters like the one produced at Bikini can provide clues to origin of the large observed lobes. Bikini Crater was produced in water with $\lambda = .105\ \mathrm{m/kg}^{1/3}$ so that it simulates a projectile impacting in water. Figure 11 shows frames of the ejecta plume development for this event reprinted from Young (1965). As noted by Young a vertical cylindrical column of water developed and a large wave developed at the base of the column which gave rise to a base surge. Note that very high-angle ejecta are falling from the column in panel b of Fig. 11 which according to Young (1965) gives rise to a secondary base surge. The combined effects produce large amounts of water impacting near the crater rim and form a wave of water having large potential energy. Craters formed in solids have lower crater rims because the ejecta volume is conical and not as much ejecta impact near the crater rim. If impact occurs in permafrost and melts a large amount of it before ejection, a similar vertical column of water saturated soil might form above a fluidized crater rim. If so, the large amounts of potential energy of material piled at the crater rim might produce more of a flow of material away from the rim than would have been produced had no large fluidized rim or vertical ejection column formed. There may be other causes for Martian flow lobes. Schultz and Gault (1976) proposed that the atmosphere of Mars was sufficient to cause deceleration of particles in ballistic trajectories so that fragments in ballistic trajectories impact nearer the crater rim than they would on an airless body. This, also, might lead to increased flow in the vicinity of the rim after material impacts from ballistic trajectories.

Crater Populations on Planetary Surfaces

It is important to determine the relative distribution of primary and secondary craters to the total lunar crater population. Until very recently, it was widely accepted that the form of the production population of primary craters on the Moon and Mars is a power function of the form $N = KD^{-B}$ where N is the number of craters of diameter D and B is considered constant over the entire size range of craters and approximately equal to 3.5 (Chapman, 1974). The marked deficiency of small Mercurian craters has forced a reevaluation of the nature of the production population of lunar and Martian craters. In Oberbeck *et al.* (1976) morphologic evidence based on experiments was presented that lunar craters as large as 30 km have features characteristic of secondary craters. Many of the irregular clustered and chained craters mapped by Wilhelms and McCauley (1971) were identified as basin secondaries and were subtracted from the total crater population; results indicate the lunar production crater popu-

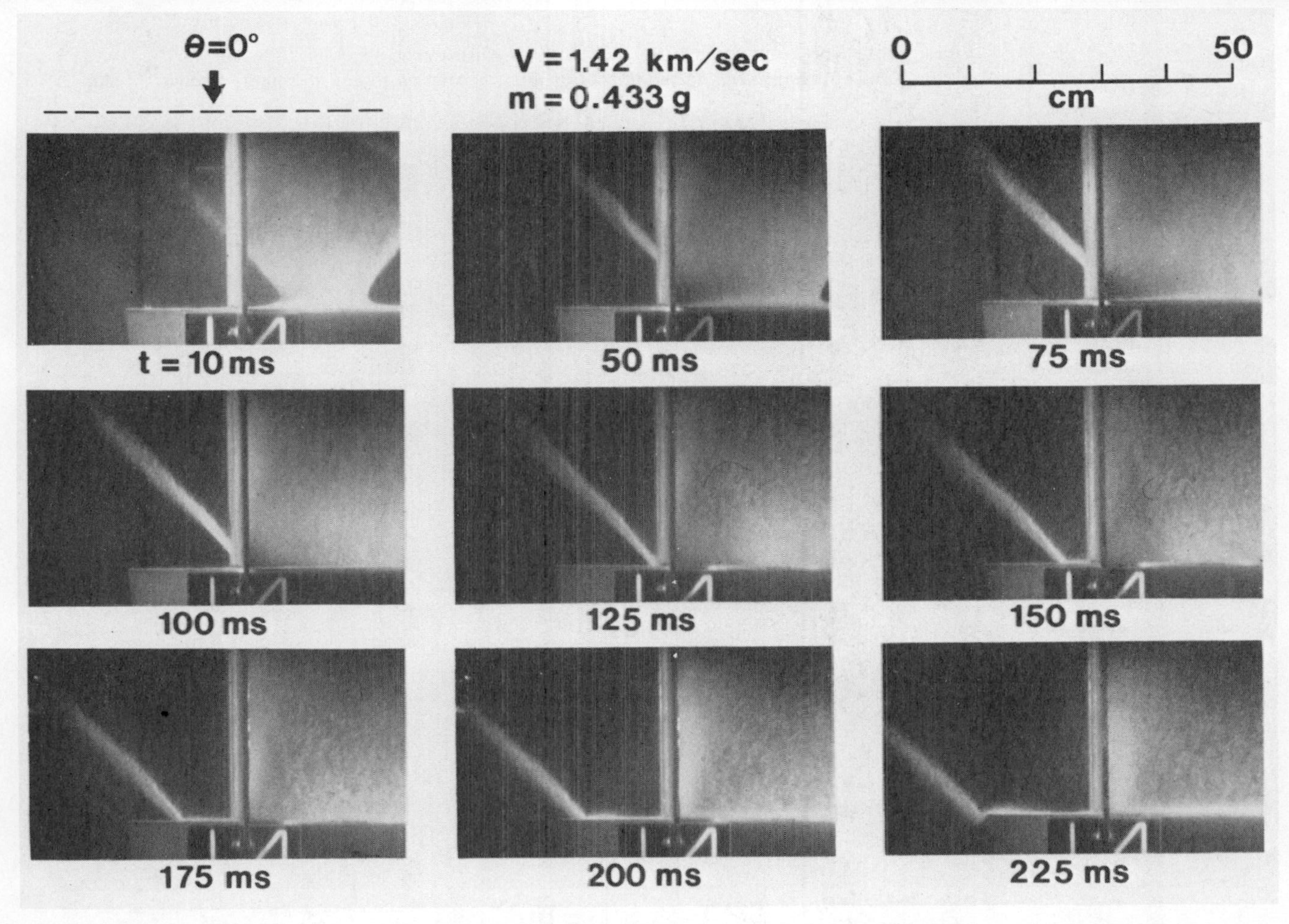

Fig. 8. Selected frames of film of an impact crater formed in quartz sand. The inclined sheet of ejecta contains material in ballistic trajectories and it moves away from the impact point as it deposits material at its base.

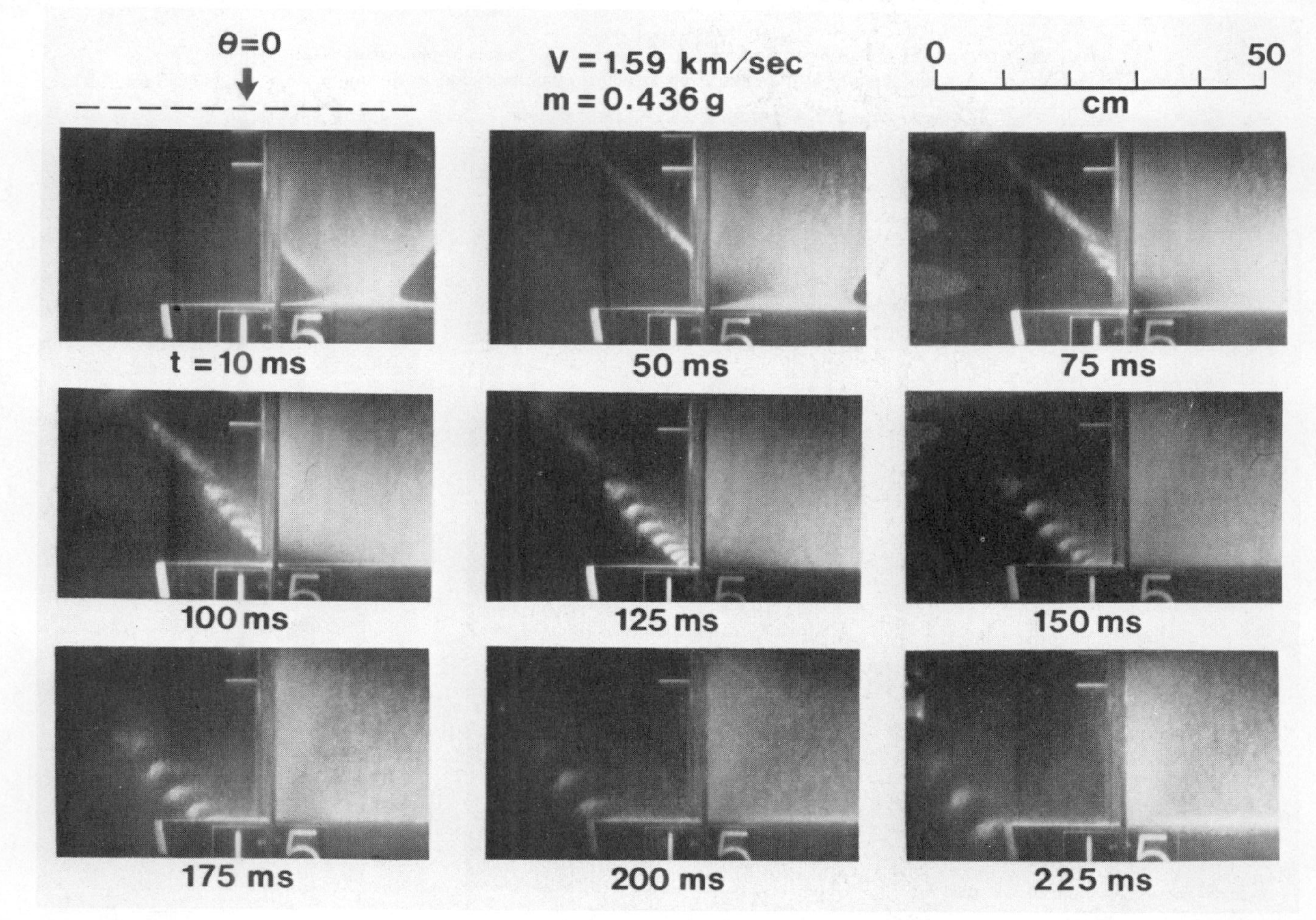

Fig. 9. Selected frames of film of an impact crater formed in quartz sand. Similar to results shown in Fig. 8 except the plume has been further disected to show material in ballistic trajectories.

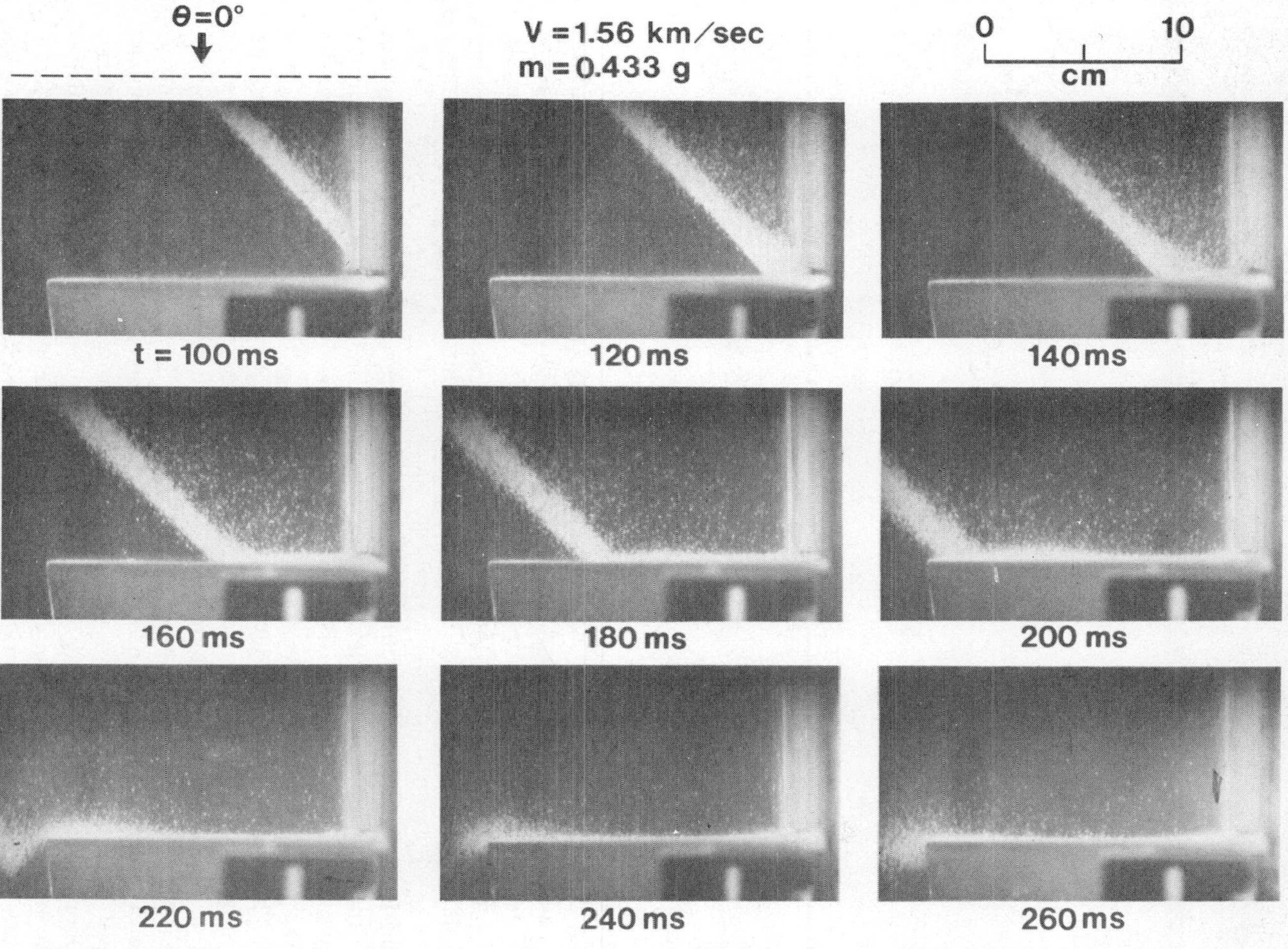

Fig. 10. Selected frames of film of ejecta plume of crater shown in Fig. 9 showing a magnified view of the base of the ejecta sheet. Notice that a ground hugging flow of debris builds up behind the ejecta sheet as it moves away from impact.

lations are not characterized by one power function. It is important to confirm that the craters mapped by Wilhelms and McCauley (1971) are basin secondaries because geologic histories of planetary surfaces have been deduced from the existence of crater populations with deficiencies in small craters. For example, Chapman (1974) assumed one power function for production of all craters and explained that the observed deficiency of small Martian craters below 30 km is due to obliteration of small craters by episodes of enhanced erosion at some time in the past. Strom (1976) concludes that a crater deficiency in certain places on the lunar uplands is due to volcanic flooding and obliteration of craters by pre-mare volcanic activity. The alternative explanation is that basin secondaries were not produced in these areas.

Therefore, it will be increasingly important to identify basin secondary craters. Secondaries have been recognized in the past by virtue of their pattern of arrangement around lunar primaries, by their subdued and irregular appearance and by their association with ray patterns. Small impact experiments were performed by firing projectiles simultaneously so that they would impact very near one another as would fragments thrown out of lunar primaries (Oberbeck and Morrison, 1974). A number of diagnostic features of secondaries became known from these experiments. For example, ridges projecting from intersection of adjacent craters are characteristic of basin secondaries. Preexisting high-explosion crater data could have provided the same information. Project Dugout consisted of simultaneous detonation of charges placed next to

Fig. 11. Photographs of the nuclear crater produced at Bikini.

one another in a row. Ridges similar to those observed near lunar secondaries were produced between the craters. The crater chain produced appears to be similar in morphology to Davy Crater chain first believed to be volcanic but now believed to be a secondary crater chain. Results suggest that explosion craters could be used to identify lunar secondary craters.

There are other applications of high explosion crater data to the secondary crater problem. For example, there now is no direct way for determining the nature of the expected size distribution of secondaries distributed globally from lunar or planetary basins. Those craters near large primaries can be assigned easily to their parent basins but secondary craters at a great distance cannot be easily associated with the parent basin. Yet it is the suspected large basin secondaries mapped by Wilhelms (1976) on the southern highlands of the moon whose origins are critical to interpretation of the geologic history of the moon, Mars, and Mercury. It would help to confirm their origin as secondaries if their size distribution was as expected for the indicated size of the parent basin. Current models for properly selected large nuclear cratering events such as those of Seebaugh (1976) may provide the expected-size distributions and velocity of ejecta from which secondary crater size distributions could be computed. It might be possible to extend such model studies to the problem of basin secondaries.

CONCLUSIONS

It is possible to simulate the ejecta plume growth, subsurface deformation, size and shape of an impact crater produced by a low-velocity impact by near surface detonation of a high-explosive charge with packing density selected to produce detonation pressure similar to the impact peak pressure and with chemical energy equal to the projectile kinetic energy. Impacts formed in layered media also may be simulated by near surface burial and detonation of high explosives in layered media.

The importance of knowing the conditions of simulation of impact craters lies in the need to select large high-explosion crater analogs for study of formation of large impact craters. This need is generated by the fact that only small laboratory impact craters may be observed during formation. Many large, high-explosion and nuclear craters have been observed, and code calculations available for these events might be applied to impact cratering problems. There has been a tendency for selection of surface burst explosion events as analogs for impact events in the papers presented at this symposium. Results presented in this paper indicate a slightly greater depth of burst explosion is required to simulate impact at low velocity, whereas others (Ivanov, 1976; Roddy, 1976) indicate that shallow depths of burst may be required to simulate impacts at velocities greater than 2 km/sec. If this is correct, a large amount of high-explosion crater data can provide valuable insight into solution of current planetary problems.

Acknowledgment—This manuscript has been improved considerably by the review of Dr. Peter Schultz.

REFERENCES

Baldwin, R. M.: 1963, *The Measure of the Moon*, University of Chicago Press, Chicago, 488 pp.

Carr, M. H., Masursky, H., Baum, W. A., Blausius, K. R., Briggs, G. A., Cutts, J. A., Duxburg, T., Greeley, R., Guest, J. E., Smith, B. A. Soderblom, L. A., Veverka, J., and Wellman, J. B.: 1976, Preliminary Results From the Viking Orbiter Imaging Experiment, *Science* **193**, 766–776.

Chao, E. C. T.: 1974, Impact Cratering Models and Their Application to Lunar Studies—A Geologist's View, *Proc. Lunar Sci. Conf. 5th*, **1**, 35–52.

Chao, E. C. T.: 1976, Mineral produced high pressure striae and clay polish: Key evidence for non-ballistic transport of ejecta from ries crater, *Science*, **194**, 615–618.

Chao, E. C. T., Soderblom, L. A., Boyce, J. M., Wilhelms, D. E., and Hodges, C. A.: 1973, Lunar light plains deposits (Cayley Formation)—A re-interpretation of origin (abstract). In *Lunar Science IV*, p. 127–128, The Lunar Science Institute, Houston.

Chapman, C. R.: 1974, Cratering on Mars, I, cratering and obliteration history, *Icarus* **22**, 272–291.

Fortson, E. P., and Brown, F. R.: 1958, Effect of Soil-rock Interface On Crater Morphology, *U.S. Army Eng. Exp. St.*, Corp. of Eng. Vicksburg, Miss. Tech. Rept. No. 20478, 28 pp.

Gault, D. E., Quaide, W. L., and Oberbeck, V. R.: 1968, Impact cratering mechanics and structures, In *Shock Metamorphism of Natural Materials*, B. M. French and N. M. Short (eds.), Mono Book Corp., Baltimore.

Head, J. W.: 1976, Significance of substrate characteristics in crater morphology and morphometry (abstract), In *Lunar Science VII*, p. 254–356. The Lunar Science Institute, Houston.

Hodges, C. A., and Wilhelms, D. E.: 1976, Formation of concentric basin rings (abstract). In *Papers*

Presented to the Symposium on Planetary Cratering Mechanics, p. 53–55, The Lunar Science Institute, Houston.
Hörz, F., Gall, H., Hüttner, R., and Oberbeck, V.: 1977, Shallow drilling in the Bunte breccia impact deposits, Ries Crater, Germany, In *Impact and Explosion Cratering*, D. J. Roddy, R. O. Pepin, and R. B. Merrill (eds.). This volume.
Ivanov, B.A.: 1976, On the mechanics of the surface explosion (abstract), In *Papers Presented to the Symposium on Planetary Cratering Mechanics*, p. 56–58, The Lunar Science Institute, Houston.
Moore, H. J.: 1976, *Missile Impact Craters (White Sands Missile Range, New Mexico) and Applications to Lunar Research*, Geol. Survey Prof. Paper 812B, 47 pp.
Oberbeck, V. R.: 1971, Laboratory simulation of impact cratering with high explosives, *J. Geophys. Res.* **76**, 5732–5749.
Oberbeck, V. R.: 1975, The role of ballistic erosion and sedimentation in lunar stratigraphy, *Rev. Geophys. Space Phys.* **13**.
Oberbeck, V. R., and Quaide, W. L.: 1967, Estimated thickness of a fragmental surface layer of Oceanus Procellarum. *J. Geophys. Res.* **72**, 4697–4704.
Oberbeck, V. R., Quaide, W. L., Mahon, M., and Paulson, J.: 1968, Monte Carlo calculations of lunar regolith thickness distributions, *Icarus* **19**, 87–107.
Oberbeck, V. R., Hörz, F., Morrison, R., Quaide, W. L., and Gault, D. E.: 1975, On the origin of the lunar smooth plains, *The Moon* **12**, 19–54.
Oberbeck, V. R., and Morrison, R. H.: 1974, Laboratory simulation of the herringbone pattern associated with lunar secondary crater chains, *The Moon* **9**, 415–455.
Oberbeck, V. R., and Morrison, R. H.: 1976, Candidate areas for *in situ* ancient lunar materials, *Proc. Lunar Sci. Conf. 7th*, p. 2983–3006.
Oberbeck, V. R., Quaide, W. L., Arvidson, R., and Aggarwal, H. R.: 1976, Comparative studies of lunar, Martian and Mercurian Craters and plains, *J. Geophys. Res.* In press.
Quaide, W. L., and Oberbeck, V. R.: 1968, Thickness determinations of the lunar surface layer from lunar impact craters, *J. Geophys. Res.* **73**, 5247–5270.
Piekutowski, A. J.: 1976, Cratering mechanisms observed in laboratory scale high explosive experiments (abstract), In *Papers Presented to the Symposium on Planetary Cratering Mechanics*, p. 102–104, The Lunar Science Institute, Houston.
Roddy, D. J.: 1976, Impact and explosion craters (abstract), In *Papers Presented to the Symposium on Planetary Cratering Mechanics*, p. 118–120, The Lunar Science Institute.
Schultz, P. H., and Gault, D. E.: 1976, Atmospheric effects on ballistic impact ejecta: Implications for Martian craters, *Trans Amer. Geophys. Union* **57**, 948.
Schultz, P. H.: 1976, Floor fractured lunar craters, *The Moon* **15**, 241–273.
Seebaugh, W. R.: 1976, A dynamic crater eject model, In *Papers Presented to the Symposium on Planetary Cratering Mechanics*, p. 133–135, The Lunar Science Institute, Houston.
Shoemaker, E. M.: 1962, Interpretation of lunar craters, In *Physics and Astronomy of the Moon*, Z. Kopal (ed.), Academic Press, New York, 538 pp.
Short, N. M., and Foreman, M. L.: 1972, Thickness of impact crater ejecta on the lunar surface, *Mod. Geol.* **3**, 69–91.
Strom, R. G.: 1976, Preliminary comparisons of the crater diameter/density distribution of lunar and Mercurian intercrater plains (abstract), In *Papers Presented to the Conference On Comparisons Of Mercury And the Moon*, p. 34, The Lunar Science Institute, Houston.
Whitaker, E. A., and Strom, R. G.: 1976, Populations of impacting bodies in the inner solar system (abstract), In *Lunar Science VII*, p. 933–934, The Lunar Science Institute, Houston.
Wilhelms, D. E.: 1976, Secondary impact craters of lunar basins (abstract), In *Lunar Science VII*, p. 935–937, The Lunar Science Institute, Houston.
Young, G. A.: 1965, The Physics of the Base Surge, NOL TR 64–103, U.S. NAV. Ord. Lab., White Oak, Md, 234 pp.

Roddy, D. J., Pepin, R. O., and Merrill, R. B., editors.
(1977) *Impact and Explosion Cratering*, Pergamon Press (New York), p. 67–102.
Printed in the United States of America

Cratering mechanisms observed in laboratory-scale high-explosive experiments

ANDREW J. PIEKUTOWSKI

University of Dayton Research Institute, Dayton, Ohio 45469

Abstract—Laboratory-scale studies of explosive crater formation and ejecta distribution were performed in cohesionless and cemented Ottawa sand. Gram-size, spherical explosive charges were detonated in carefully prepared homogeneous and two-layered test beds. Craters produced in these studies exhibited many of the morphological features which are present in larger scale explosive experiments, in laboratory-scale impact experiments, and in large impact craters. Craters produced in homogeneous test media exhibited a bowl-to-conical shape. Variations in crater shape were related to the height of burst or depth of burst of the explosive charge and the initial density of the cratering medium. Most significant variations in apparent crater morphology occurred when the cratering medium consisted of a layer of cohesionless sand above a cemented sand layer. These morphological features included: central mounds, concentric rings and terraces on the crater wall, and hummocky mounds and concentric rings in the region surrounding the crater. Formation of these features was related to the depth of the harder lower layer and the strength of the upper medium.

1. INTRODUCTION

LABORATORY-SCALE EXPERIMENTS were performed to examine mechanisms associated with the formation of explosion-produced craters. Particular attention was paid to the application of the observed mechanisms to phenomena associated with the formation of large-scale explosion craters. A variety of morphological features was observed in the craters produced by the laboratory-scale experiments. These morphological features were produced as a result of the action of a number of different cratering mechanisms.

Detailed study of the crater formation process must consider the effects of a large number of experimental variables. Because laboratory-scale operation permits careful control of experimental conditions, the number of variables to be considered could be reduced, and the effects of various parameters considered important to the cratering process systematically examined. This paper will describe and discuss the morphological features of the laboratory-scale explosion craters produced in these parametric studies and, when possible, will compare them with the morphological features of larger explosion craters. Both external and internal morphological features of the craters will be discussed. External features are those observed in the apparent crater and in the region surrounding the crater. Internal features include the structural deformation and a discussion of the dynamic processes which produce the observed deformations.

2. EXPERIMENTAL CONDITIONS

The craters described in this paper were produced by detonating small explosive charges in Ottawa sand having an average grain diameter of 0.55 mm. The size distribution of the grains was such that 95% of the grains had an average diameter which ranged from 0.42 to 0.82 mm. Certain experiments required cemented sandstones for use as a hard layer in layered media studies. Slabs of weakly cemented (4% cement) and strongly cemented (30% cement) Ottawa sandstone were cast for use in these studies as the harder layer. These sandstones were prepared by mixing Ottawa sand with the proper weight of cement and water prior to casting the slab.

Spherical, centrally-initiated charges of dextrinated lead azide were used as the explosive. The charges were 10.16 mm in diameter, weighed 1.7 g, and released 2.25×10^{10} ergs at detonation. Theoretical detonation pressure for these charges was determined to be 160 kbar. Some specialized experiments required hemispherical lead azide charges. These charges had a radius of 5.08 mm, weighed 0.85 g, and were initiated at the center of the flat surface.

3. External Morphological Features

Parametric sensitivity and various specialized cratering studies were performed in homogeneous, layered, and wet media and employed a wide range of heights-of-burst (HOB) and depths-of-burst (DOB) of the charge. Interest focused on the features and mechanisms produced by near-surface events. The various morphological features which will be described in this paper are for craters produced when the charge was surface tangent-above, i.e., the center of the charge was one charge radius above the surface (1 HOB), half-buried, or surface tangent-below (1 DOB). The features produced by the shallowly buried charges (half-buried and 1 DOB) should resemble closely those of impact craters as shown in studies performed by Oberbeck (1971).

Typical apparent crater volumes for half-buried events in the various media ranged from 90 to 1400 cm^3. Similar variations in crater volume occurred for events at the other charge configurations. As the charge configuration or media conditions were varied, the change in crater dimensions and transition from one morphological feature to another was smooth and orderly, i.e., discontinuities or abrupt changes in these features were not observed.

Two external characteristics were observed for nearly all the craters, regardless of the medium. First, well-defined raised lips or rims were formed around the craters. Second, azimuthal variations in the thickness of the ejecta blanket occurred and were manifested as rays or radially oriented mounds of debris.

Homogeneous media

A number of experiments were performed in densely packed homogeneous Ottawa sand to obtain a set of standard or benchmark crater data (Piekutowski, 1974). Additional events were detonated in other homogeneous Ottawa sandbeds which were less densely packed (Piekutowski, 1975a). *In situ* density of the sandbeds used in these studies ranged from dense (1.80 g/cm^3, 95% relative density) to loose (1.62 g/cm^3, 40% relative density). Typical craters for half-buried events in loose and dense sand are shown in Figs. 1 and 2, respectively. Large cone-shaped craters were produced in the loose sand while smaller bowl-shaped craters were produced in the dense sand. Craters produced in the intermediate density sandbeds tended to be cone-shaped for the looser sandbeds

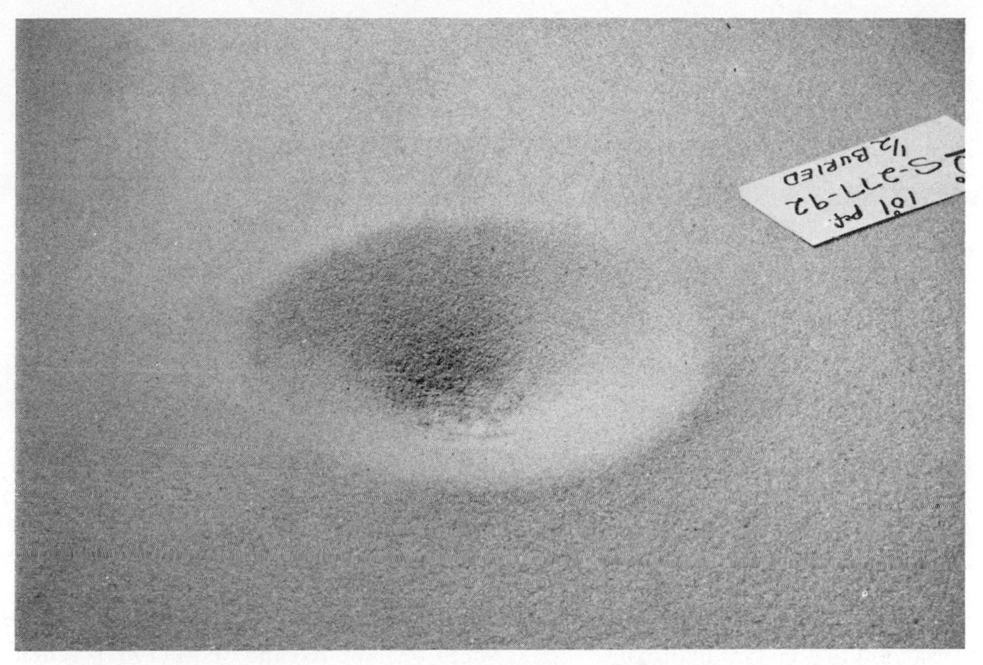

Fig. 1. Crater produced by detonation of charge half-buried in loose (1.62 g/cm^3) Ottawa sand.

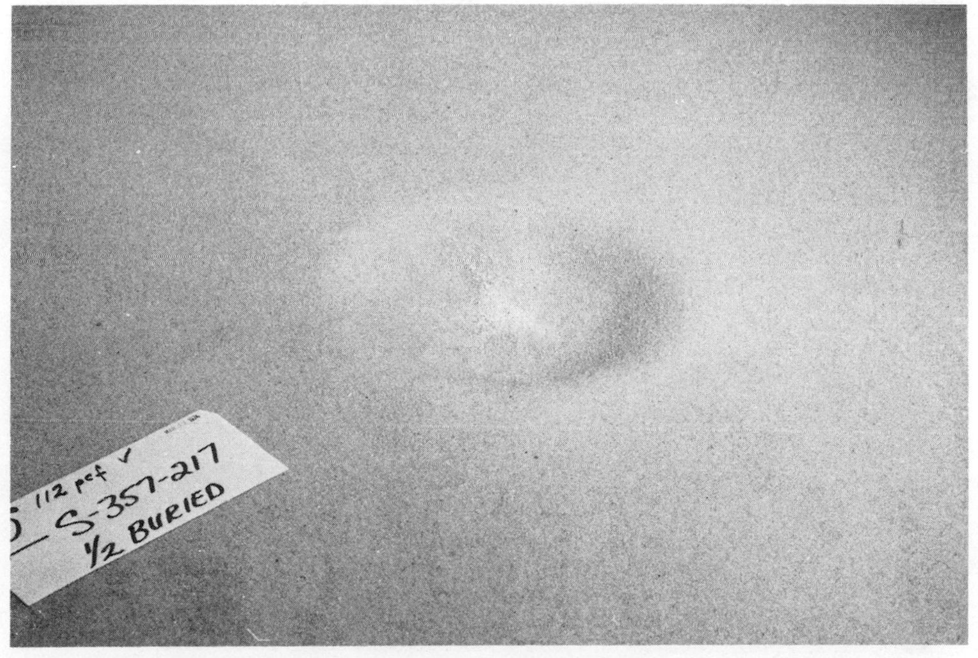

Fig. 2. Crater produced by detonation of charge half-buried in dense (1.80 g/cm^3) Ottawa sand.

and bowl-shaped for the denser sandbeds. The cone shape observed for the craters produced in the loose sandbeds probably resulted from a late-time failure of the upper portion of the crater walls since the slope of the apparent crater wall was identical to the angle of repose of the loose sand. In addition, the large number of comminuted sand grains normally distributed over the floor of the apparent craters produced in the dense sand were not visible on the floor of the craters produced in the loose sand. However, postshot examination of the cone-shaped craters revealed large numbers of comminuted sand grains scattered over a hemispherical-shaped surface just below the apparent crater floor. A significant portion of these comminuted sand grains were covered with "fresh" sand grains that apparently rolled down the crater wall during slope failure.

Several mass accountability experiments relating the weight of material actually ejected from the crater to the weight of material apparently excavated from the crater indicated that a significant fraction of the apparent crater volume produced by the events in loose sand resulted from compaction of material in the cratered region. On the other hand, some bulking of material surrounding the crater occurred for the events in dense sand. The observed behavior of the sand surrounding the crater was similar to the behavior of Ottawa sand samples subjected to constant hydrostatic-stress shear tests by Ko and Scott (1967). When loose sands were sheared in these experiments they contracted in volume until failure approached and then began to expand. Dense sands expanded from the beginning of shearing.

Fig. 3. Central mound and radially oriented ridges and valleys observed in craters produced by near-surface detonations in dense homogeneous sand.

In addition to differences in crater shape and volume, two other notable features were observed for craters produced in homogeneous media. First, small central mounds were formed in the craters produced in dense sand by near-surface detonations. Second, a system of radially oriented ridges and valleys was produced in the region surrounding the crater by near-surface detonations in all sand densities. Both features are present in the crater produced by the surface tangent-above event shown in Fig. 3. The depth from the preshot surface to the top of the mound is slightly less than one half the maximum depth of the crater. Smaller central mounds were produced by half-buried events in dense sand and were virtually nonexistent for the surface tangent-below events. Central mounds were not produced by any of the near-surface detonations in loose sand.

Layered media

Several series of experiments were performed in layered media, i.e., media containing one or more layers having different physical properties. The simplest of these layered media consisted of loose sand over dense sand. Other series of experiments (Piekutowski, 1975b) were performed in layered media consisting of dense and loose sands over cemented sandstones. Additional events were detonated in a "layered" medium which was nominally a dense homogeneous sand but contained one or more thin horizontal layers of loose sand.

Loose sand over dense sand. Photographs of several craters produced in this series of surface tangent-above events are shown in Fig. 4 for various values of the ratio D which relates the various overburden thicknesses to a "normal" crater depth. Specifically, D is the ratio of the depth of loose sand to the depth of the crater which would be produced by a surface tangent-above event in homogeneous loose sand. The floor of the crater shown in Fig. 4b has a small central mound that is surrounded by a small flat area. This flat area is the first indication of an effect of the lower layer of dense sand on the crater formation process; however, the floor of the crater did not "bottom" on the lower layer. In Fig. 4c, the flat area at the bottom of the crater had enlarged considerably, and crater profile measurements for this event indicated that the crater had "bottomed" on the dense sand lower layer. A very small central mound was formed in this crater. Crater formation proceeded into the dense sand for the crater shown in Fig. 4d and a bowl-shaped crater with a small central mound was produced. A distinct ledge was also produced in the crater at the level of the interface between the loose sand and the dense sand.

The crater produced by the surface tangent-above PRE-MINE THROW IV-6 (PMT IV-6) event and described by Jones and Henny (1976), exhibited a ledge similar to that shown in Fig. 4d. PRE-MINE THROW IV-6 was an event in which a spherical container of 100 tons of nitromethane was detonated on Yucca Flat playa. The material property profile for the Yucca Flat playa at the PMT IV-6 site exhibited a density discontinuity in the playa in the region where the ledge was formed, slightly less than 2 m below the surface. Dry density of the

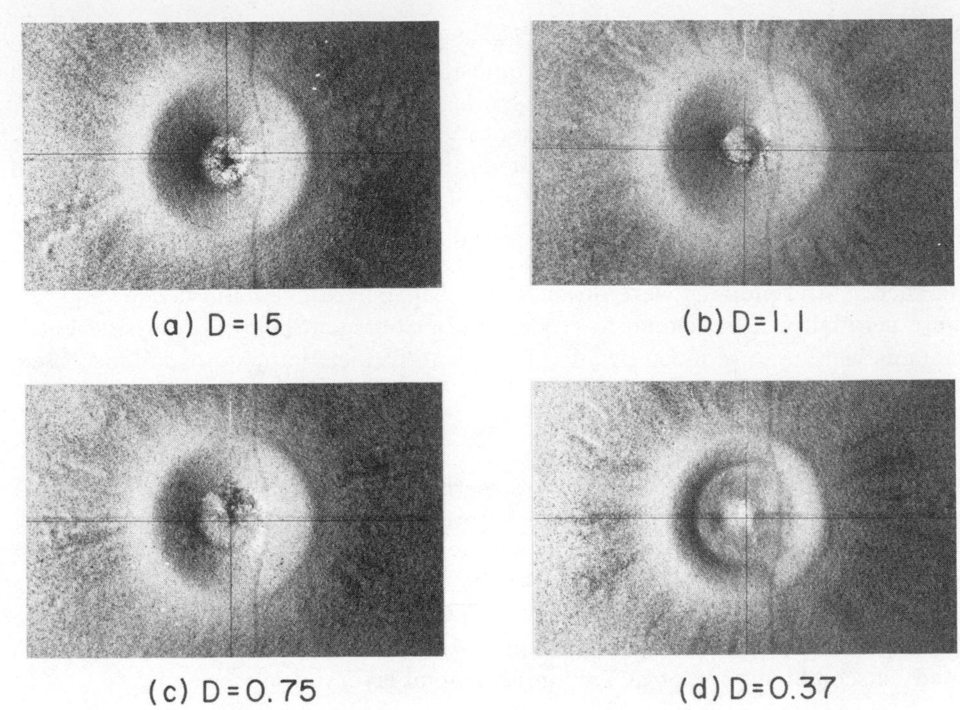

Fig. 4. Overhead views of craters produced in a layered medium (loose sand over dense sand) for various values of the ratio (D) of depth of loose sand to depth of crater in homogeneous loose sand.

playa near the surface was approximately 1.28 g/cm^3 while dry density of the playa below the ledge was approximately 1.47 g/cm^3. Moisture content of the playa in both regions ranged from 17 to 18% and increased to approximately 19% at a depth of 6.5 m below the preshot surface. Because comparable increases in media density were observed at the approximate level of the ledge in the PMT IV-6 crater, and the crater shown in Fig. 4d, the formation of the ledge almost certainly can be related to the change in material properties at the location of the ledge. It is interesting to note that a ledge was not formed in the crater produced by the next largest event in the PRE-MINE THROW IV series, PMT IV-3. This 7.1-ton nitromethane event was detonated very near the PMT IV-6 site and produced a crater which was 1.15 m deep. Since the density discontinuity was almost 2 m below the surface, it is likely that the effect it had on the formation of the PMT IV-3 crater was less than the effect a similar density discontinuity had on the formation of the crater shown in Fig. 4b.

Sand over sandstone. Use of this layered medium produced the greatest number and variety of morphological features in the apparent craters. Various thicknesses of loose and dense Ottawa sand were used as the upper medium and weakly cemented (4% cement) and strongly cemented (30% cement) Ottawa sandstone slabs were used as the lower medium. The density of the overburden, hence its resistance to shear, had the greatest effect on the shape of the apparent crater. Effects of the strength of the lower layer were not readily apparent.

Layered media with dense sand as the overburden provided the greatest variety of crater morphological features. Craters formed in a layered medium of dense sand over a weakly cemented sandstone are shown in Fig. 5 for various values of the ratio D. A brief description of the morphological features of the craters for each value of D in this figure is given in Table 1. The features which are shown in this figure and table are very similar to those observed by Oberbeck and Quaide (1967) in their laboratory-scale impact craters formed in a layered medium. In addition, these features can be compared with those described by Gault *et al.* (1975) in their summary of the classes of features found in lunar and mercurian craters.

As the lower medium began to influence the crater formation process ($D = 6$ and $D = 4$) a slight reduction in apparent crater depth occurred. Apparent crater depth increased when the value of D was 2.5 and then decreased continuously as the value of D decreased. Corresponding reductions in apparent crater radius, lip height, etc., also occurred for the craters formed in media that had values of D less than 2.5.

The transition between the various structural features appeared to be continuous and related to the thickness of the overburden. The first structures which appeared in the craters were terraces and small systems of radially oriented ridges that surrounded a central mound ($D = 6$ and $D = 4$). As the thickness of the overburden decreased ($D = 2.5$ and $D = 1.5$) the intracrater structure changed to a very well-developed central mound. For the lesser values of D, the central mound was not as evident.

High-speed photography of several of the events in a layered medium revealed that a central plume of material was formed during crater formation, and that the maximum height which the plume attained increased as the overburden thickness decreased. The axis of the plume was seldom normal to the preshot surface, and material in the plume was deposited as fallback in a raylike fashion on a portion of the region surrounding the crater. Typical distributions of comminuted sand grains and small blocks of cemented sand which were contained in the central plume are shown in Figs. 5e, f, and g. Certain of the small blocks of cemented sand formed small impact craters in the region around the crater. The raylike character of a collapsed central plume is shown more clearly in Fig. 6. High-speed films of several events for which the value of D was 2.5 show that the hemispherical mounds in these craters were formed during the collapse of central plumes which had less vertical rise but a broader cross-section. These plumes did not attain a maximum height of more than one crater depth above the preshot surface during the crater formation process.

Considerable disturbance of the region around the crater was observed for the events in which D was 0.25. These disturbances appeared to be very similar to the radially oriented ridge and valley structure which was observed around craters produced by above-surface detonations in homogeneous test beds.

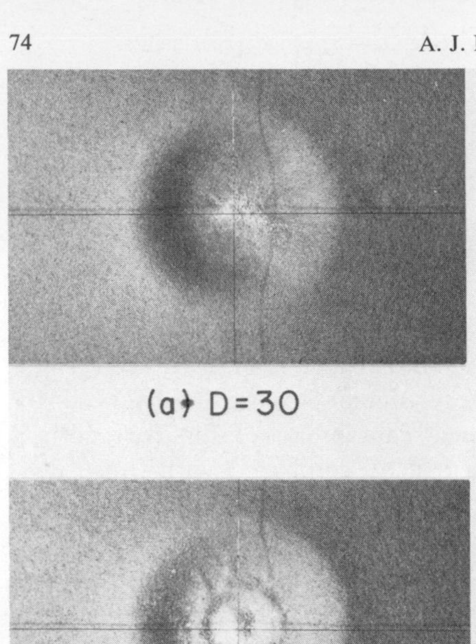

(a) D=30

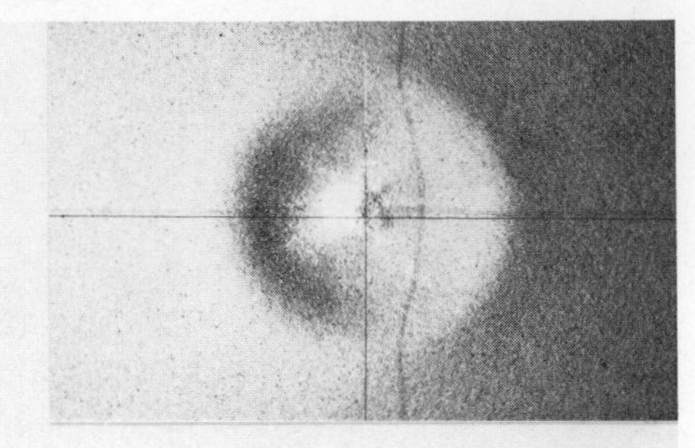

(b) D=6

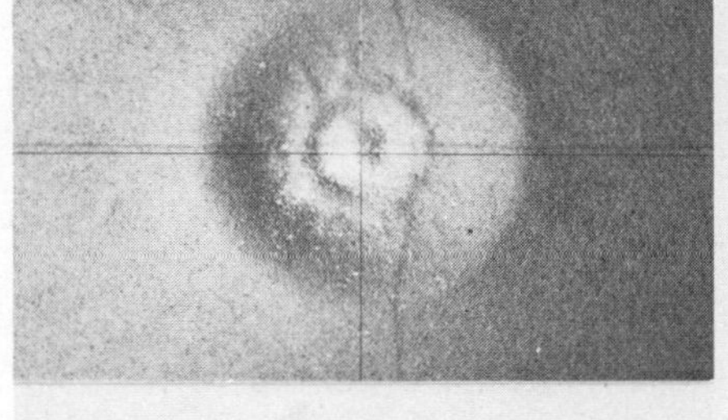

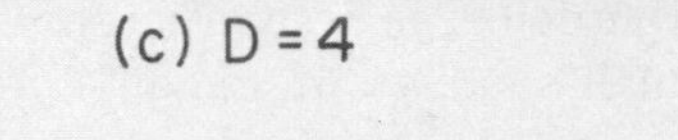

(c) D=4

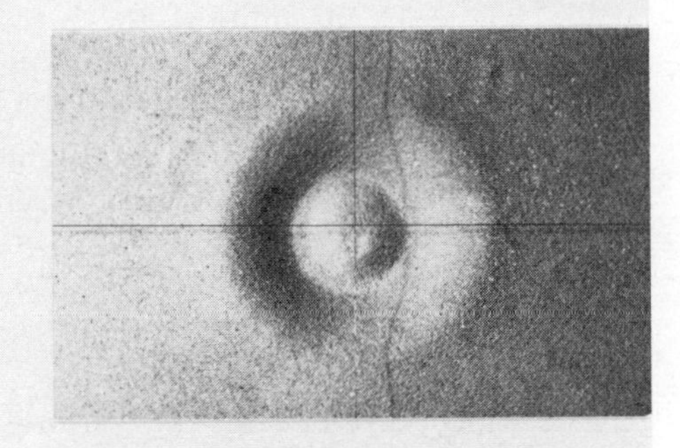

(d) D=2.5

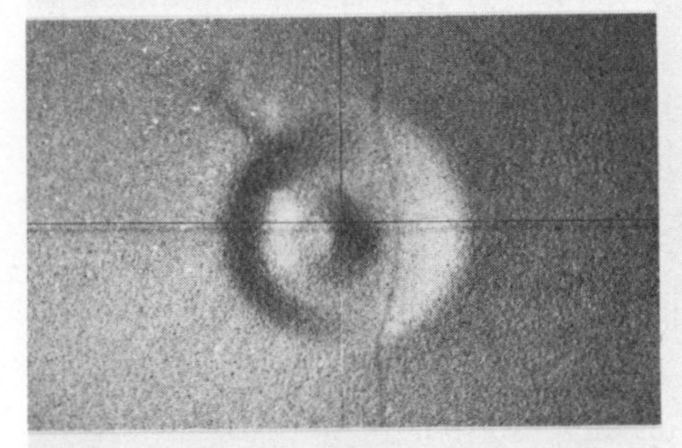

(e) D=1.5

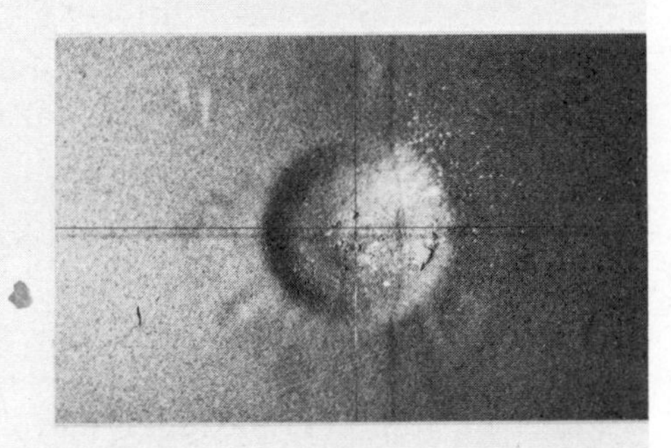

(f) D=0.75

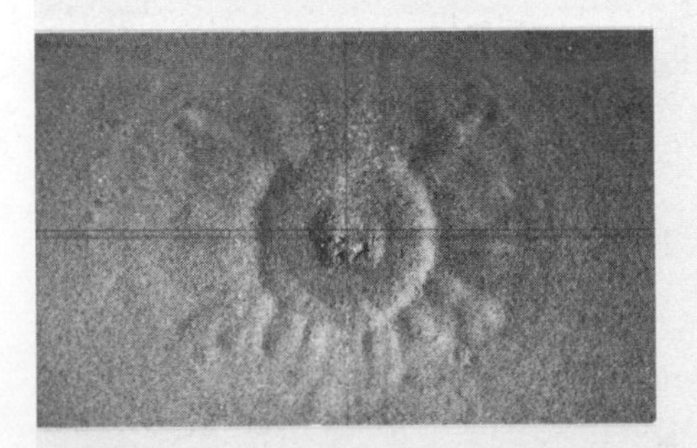

(g) D=0.25

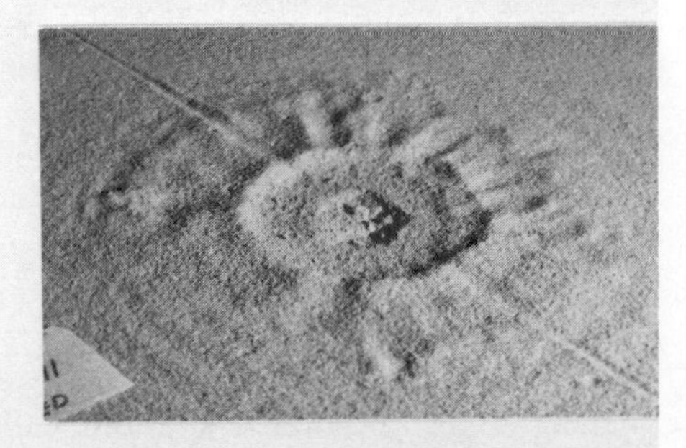

(h) Oblique View of Crater for D=0.25

However, the ridge and valley structures produced in the layered media events were much higher and more pronounced and contained hummocky mounds which appeared to define a system of concentric rings around the crater lip. The ridges and hummocky mounds were even more pronounced for the surface tangent-above event shown in Fig. 7. A number of secondary craters formed by the impact of small blocks of cemented sand are also visible in this figure. The effect of lateral inhomogeneities in the thin layer of overburden used for these events is evident as an additional system of linear structures superimposed on the ridge and valley structure and shown in Fig. 8. Preshot conditions for the event shown in Fig. 8 were identical to those of the event shown in Fig. 7 except that the overburden in Fig. 8 contained lateral zones of lower density sand which were produced when the overburden was improperly placed on the cemented sand slab.

The terrace, or inner ring structure, and central mound in the crater shown in Fig. 5c bear some similarity with the concentric ring and central mound structure which was observed in the crater produced by the 500-ton-TNT event PRAIRIE FLAT. This surface tangent-above event was detonated at the Watching Hill site of the Defence Research Establishment, Suffield, Alberta, Canada (DRES). Jones (1976) described the geology of this site as consisting of horizontally stratified silt, sands, clays, and gravels, overlying glacial till on a bedrock of shales and sandstones. He states that the depth of bedrock was variable but close to 65 m, and that water appeared in bore holes at depths ranging from 9 to 12 m. Although it is impossible to determine an appropriate value of D for this event, the depth to bedrock is such that the medium could be represented as a two-layered medium with a rather thick overburden. In addition, the probable influence of the saturated material below the water table must not be ignored when considering mechanisms that would result in the formation of the concentric rings and central mound observed in the PRAIRIE FLAT crater.

The craters shown in Figs. 5f and g have a number of features observed in certain of the craters formed in the MIDDLE GUST and MIXED COMPANY series of high-explosive experiments. In very general terms, these experiments can be considered to have been performed in a simple layered medium with a relatively thin layer of a weak material overlying a layer of more competent material. A wet and a dry site were used for the MIDDLE GUST series. Both sites consisted of approximately 2.7 m of silty to sandy clay alluvium overlying Pierre Shale. The upper 1 to 2 m of the shale was strongly weathered. A perched water table at the top of the competent shale extended to about 1.2 m below the surface of the wet site, while the water table was more than 20 m below the surface at the dry site. The MIXED COMPANY site was essentially dry and consisted of a layer of silty to clayey alluvium 1 m thick overlying Kayenta Sandstone. The upper 1 to 2 m of the Kayenta Sandstone was moderately weathered and considerably weaker than the underlying sandstone.

Fig. 5. Overhead views of craters produced in a layered medium (dense sand over weakly cemented sand) for various values of the ratio (D) of depth of cemented layer to depth of crater in homogeneous dense sand.

Table 1. Comparison of morphological features of craters produced in a layered medium. (For half-buried and tangent-below events in dense cohesionless Ottawa sand over cemented Ottawa sand.)

Ratio of depth of cemented layer to depth of crater in homogenous medium	Characteristic morphologic features	Representative crater profile
30	Normal bowl shape with small conical central mound.	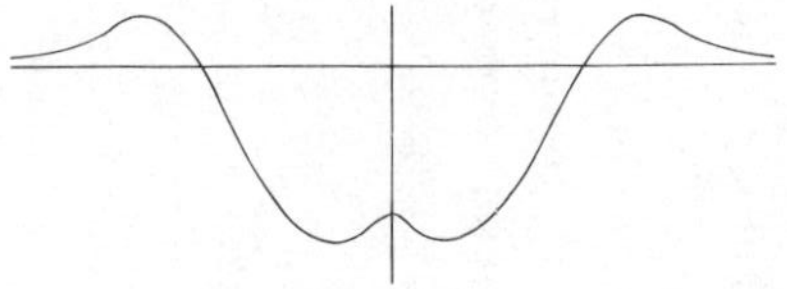
6	Slight terrace and radial ridge structure on lower wall. Mound slightly larger than normal and nonsymmetric.	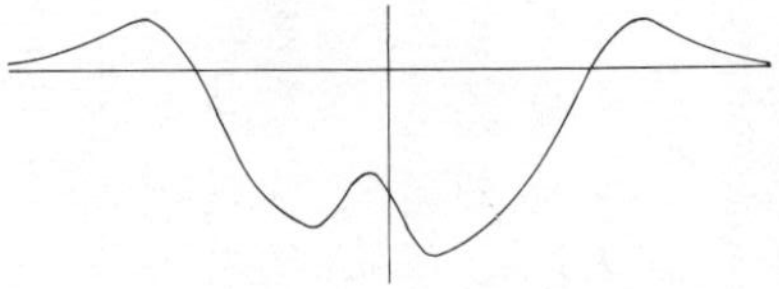
4	Well-defined terrace or concentric inner ring with concentric central mound.	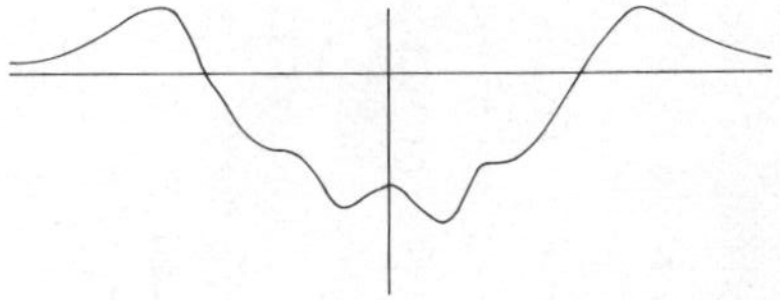

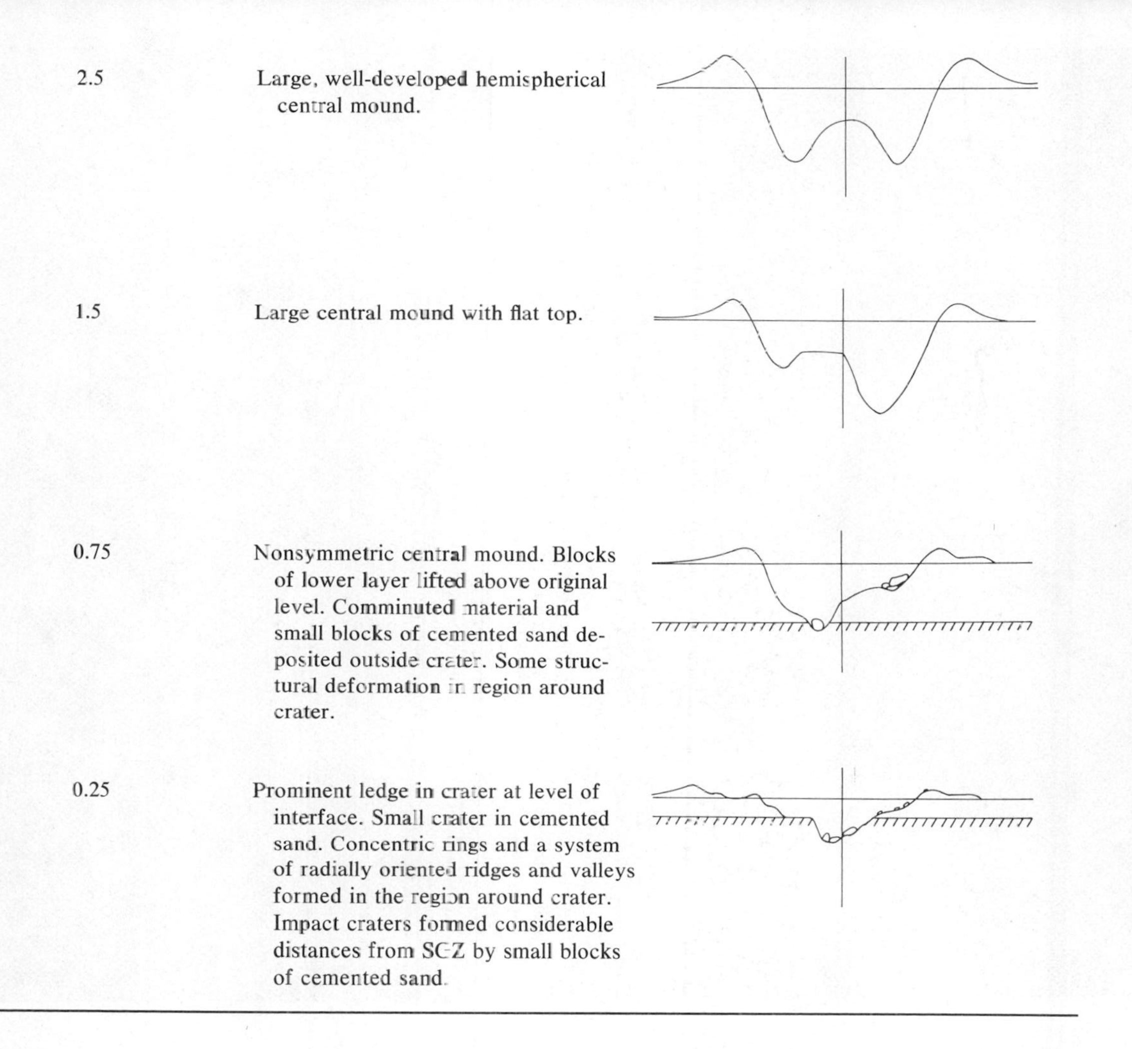

2.5	Large, well-developed hemispherical central mound.	
1.5	Large central mound with flat top.	
0.75	Nonsymmetric central mound. Blocks of lower layer lifted above original level. Comminuted material and small blocks of cemented sand deposited outside crater. Some structural deformation in region around crater.	
0.25	Prominent ledge in crater at level of interface. Small crater in cemented sand. Concentric rings and a system of radially oriented ridges and valleys formed in the region around crater. Impact craters formed considerable distances from SGZ by small blocks of cemented sand.	

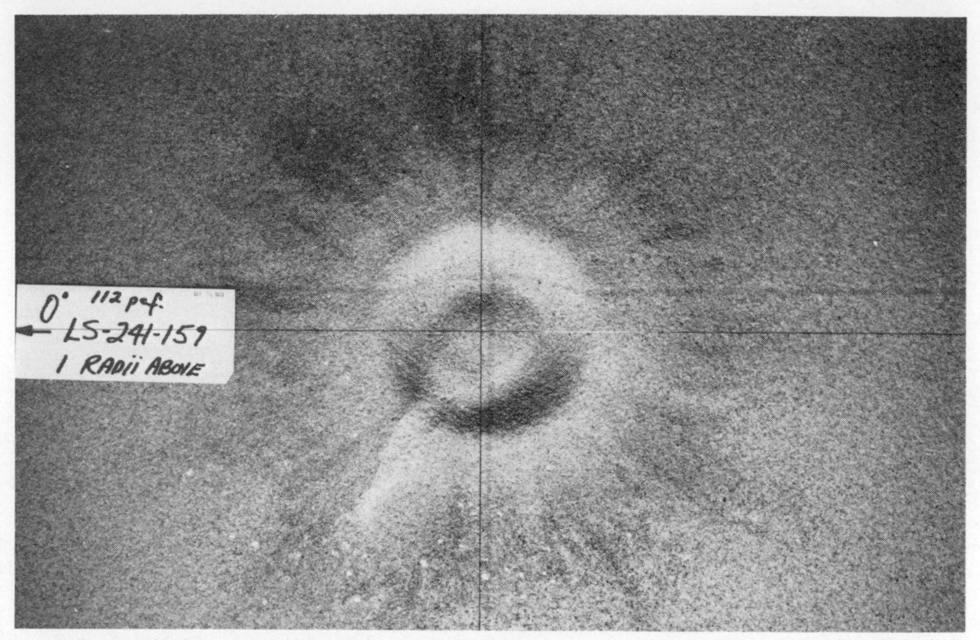

Fig. 6. Radial structure formed in ejecta blanket during deposition of material ejected from center of crater. Note the distribution of the comminuted material (white areas) around the tip of the ray-like structure. Dense homogeneous sand over cemented sand ($D = 1.5$).

Fig. 7. Crater and ridge-and-valley structure produced by near-surface detonation in a layered medium. Dense homogeneous sand over cemented sand ($D = 0.25$).

Fig. 8. Crater and ridge-and-valley structure produced by near-surface detonation in a layered medium, showing the effects of lateral inhomogeneities in the dense overburden. Dense homogeneous sand over cemented sand ($D = 0.25$).

MIDDLE GUST III and IV were 100-ton-TNT, surface tangent-above events detonated at the wet and dry sites, respectively. MIDDLE GUST III produced a crater with a ledge at the interface between the weathered and competent shale and a trough-like crater in the underlying shale (Meyers, 1973). In addition, Roddy (1973) mentions that fragmental mound material was deposited on the lower walls of this crater. The craters shown in Fig. 5f and g display both ledges and fragmental mound material distributed on the lower walls of the crater. MIDDLE GUST IV was described by Meyers to have an off-center mound similar to that shown in Fig. 5f.

Much of the description of the three MIXED COMPANY craters given by Roddy also could be used to describe the craters shown in Figs. 5f and g. MIXED COMPANY I and II were 20-ton-TNT, half-buried and surface tangent-above events, respectively. MIXED COMPANY III was a 500-ton-TNT, surface tangent-above event. All three events produced craters with central mounds or uplifts. In addition, MIXED COMPANY I and II craters exhibited benches or ledges in the crater wall at the alluvium-sandstone interface. MIXED COMPANY I had well-developed rays of crushed sandstone that were deposited beyond the overturned flaps. Similar ledges and deposition of crushed material are evident in the craters shown in Figs. 5f and g. Raylike deposition of crushed material is also shown in the crater in Fig. 6. The bulk of the crater produced by MIXED

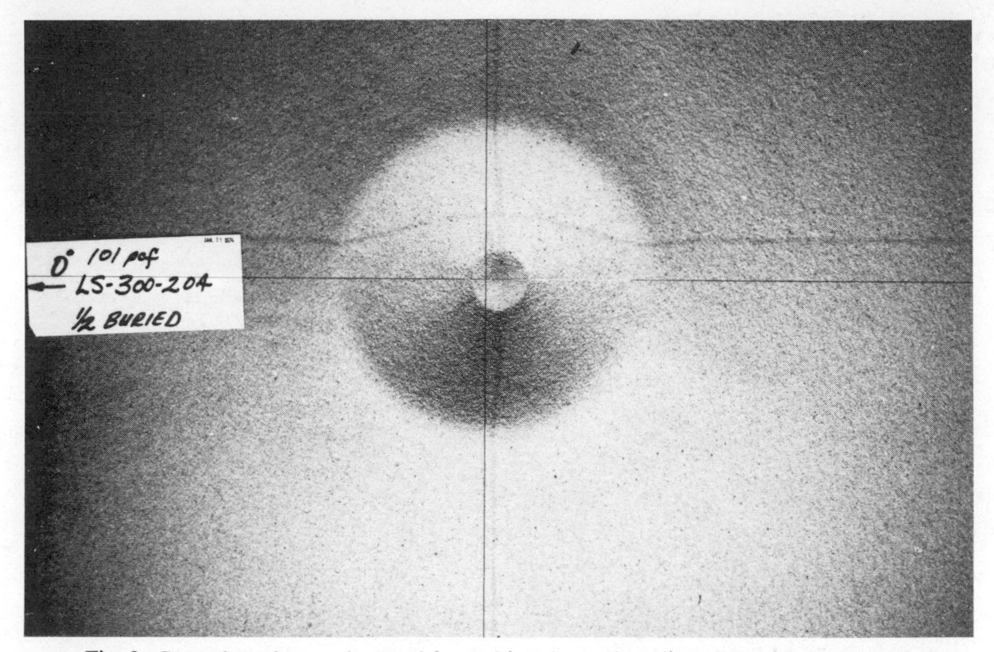

Fig. 9. Cone-shaped central mound formed in a layered medium. Loose homogeneous sand over cemented sand ($D = 1.0$).

COMPANY III was formed in the Kayenta Sandstone. Roddy describes the crater as a broad, flat-floored crater with steep walls and a sharp, well-defined central mound. He describes the central mound as a poorly formed irregular dome of massive bedded sandstone that moved upward and inward. The crater floor surrounding the central mound consisted of slabs of sandstone sloping upward to surface ground zero (SGZ). The crater shown in Fig. 5f has an off-center mound composed of brecciated cemented sand. Several blocky slabs of cemented sandstone are also in evidence in this crater. As mentioned earlier in this section, the axis of the central plume of ejected material was seldom normal to the preshot surface. Had the central plumes been ejected vertically and subsequently collapsed on themselves, the craters shown in Figs. 5f and g probably would have been flat-floored craters with central mounds.

Only limited studies were performed using loose sand as the overburden. The only unusual observed feature produced in these studies was a cone-shaped mound that formed at the bottom of the normal conical crater and is shown in Fig. 9. The crater shown in Fig. 9 closely resembles the craters with a central mound geometry described by Quaide and Oberbeck (1968).

Multilayered Media. Experiments performed in what was originally considered a homogeneous medium were, in reality, performed in a dense homogeneous sand containing one or more thin horizontal layers of loosely-packed dyed Ottawa sand. The reduction in the density of the dyed sand layer resulted

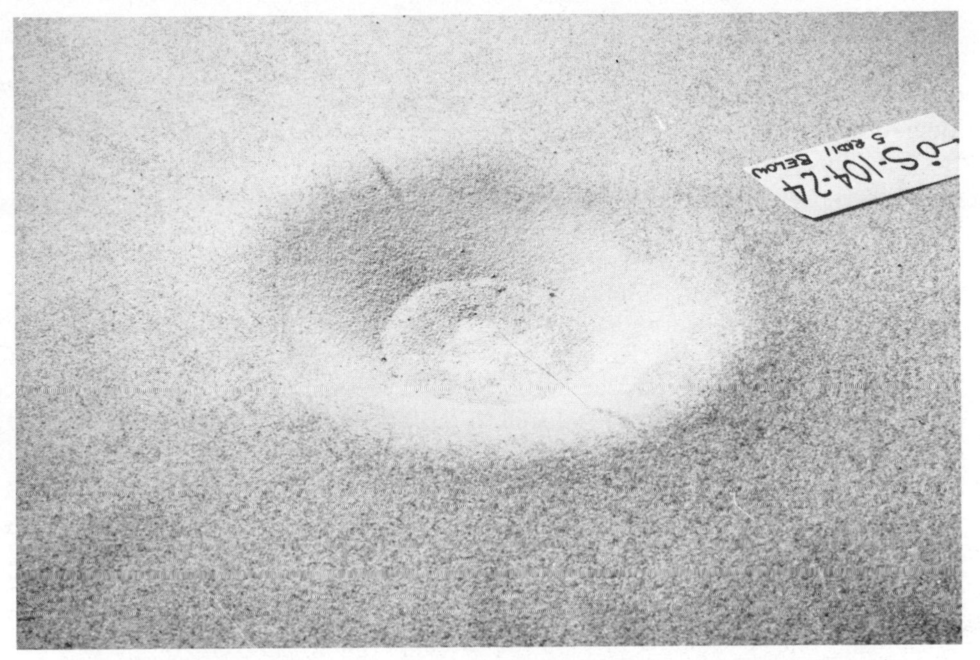

Fig. 10. Ledge formed in crater at the level of a thin layer of loose dyed sand placed in the dense "homogeneous" sand for use in exhibiting internal structural deformations.

from the technique used to place the sand in the test bed. When the layer was placed at a level below the charge but within the cratered region, a distinct ledge was formed in the crater wall at the level of the layer as shown in Fig. 10. The importance of a thin layer of foreign material in a medium should not be overlooked. Meyers described a distinct shear line along an iron-oxide-stained gypsum seam as a prominent feature of the subsurface displacement which occurred for MIDDLE GUST IV. Grout columns placed at approximately 4.6 and 6.1 m from SGZ (for this event) exhibited horizontal shear displacements of approximately 3 and 2.4 m, respectively, at a depth of 4.6 m—the level of the gypsum seam.

Wet media

A number of experiments were performed in homogeneous and layered media containing water. Two moisture contents were used for these experiments: naturally drained and saturated. The naturally drained condition was created when the media was first saturated and then drained for at least an hour before the charge was detonated. Only general observations regarding the results of these experiments will be made in this paper since the study is partially complete. The addition of water to the Ottawa sand produced two noticeable effects. First, the ejection angle (measured from horizontal) of material removed from the crater increased from that observed for similar dry test conditions. Second, material was ejected from the crater in clumpy fragments composed of numerous sand grains and water. Material originally nearest the charge was ejected at fairly high velocity as very small fragments, while material farthest from the charge was ejected at low velocity as relatively large fragments. Quite often these large fragments simply overturned and were deposited on the crater lip.

The addition of any amount of water to homogeneous dense sand resulted in a reduction of the size of the apparent crater. Craters formed in the naturally-drained homogeneous loose sand were also smaller than those produced in dry homogeneous loose sand. However, craters produced in saturated loose sand were larger than those produced in dry loose sand. Radius-to-depth ratio (aspect ratio) of the apparent craters produced in loose saturated sandbeds ranged from 6 to 20. Craters produced in dense and naturally-drained loose sandbeds had aspect ratios of 2 to 5. The depth of the crater produced in a "saturated" loose sandbed was a very sensitive function of the degree of saturation of the test bed, with the shallowest craters being produced in the most highly saturated test beds. Apparent crater radius did not vary significantly for these events. The degree of saturation of the individual test beds could be determined to within ±0.5% and ranged from 89 to 97% for the experiments performed in saturated loose sand.

The broad shallow craters formed in the loose saturated sandbeds had shapes which were similar to those formed by large-yield nuclear devices detonated at Eniwetok in the Pacific Proving Grounds. Two preshot conditions that existed in the saturated loose sandbeds—high porosity and high degree of saturation—are also found in the coral-coral sand geology of Eniwetok. High-speed films of events detonated in the saturated loose sandbeds showed that a low aspect ratio transient cavity was initially produced and that the transient cavity experienced a considerable late-time alteration of this shape. Companion experiments using saturated loose sandbeds (described later in this paper) indicate that a portion of medium below the crater apparently liquefied during crater formation and facilitated the late-time alteration of the transient cavity shape. Liquefaction of the saturated coral sand medium at Eniwetok can be used as a mechanism to explain the formation of the unusual large-aspect-ratio craters produced at that site.

A crater formed in a naturally-drained dense sandbed is shown in Fig. 11. Crater formation "bottomed" on a thin layer of loose sand which had been placed in the test bed and produced a flat-floored crater with rather steep sidewalls. A similarly shaped crater was produced by the 120-ton-ANFO event, PRE-DICE THROW II-2, which was detonated at the White Sands Missile Range. This event was detonated in a medium which consisted of layers of the following materials, starting at the surface: sandy silt; sandy silt to silt with abundant gypsum crystals; silty clay with scattered gypsum crystals; clay with very little silt; a zone of fluidlike bluish white to gray clay at almost 2.4 to 3 m; sand to sandy clay; and sand with sandstone lenses (Jones, 1975). The water

Fig. 11. Crater formed in dense homogeneous naturally drained wet sand, showing typical size and distribution of ejecta fragments produced by events in wet sand. Crater "bottomed" on thin layer of loose sand placed in sandbed.

table was located approximately 2 m below the preshot surface at the time of detonation. The crater was formed in the material which was above the 0.6-m-thick zone of fluidlike bluish white to gray clay. Crater formation apparently "bottomed" on this layer and produced a very flat-floored crater with rather steep sidewalls. Additional discussion of crater formation in a wet medium containing a thin layer is given in the next section.

4. Internal Morphological Features

Laboratory-scale operation and the small size of the explosive charge permitted the motions and processes occurring during crater formation to be observed in various ways not possible in larger experiments. In some cases processes were inferred from the displacement of tracers placed in the test medium. In other cases, the motions of the tracers were observed during crater formation with use of high-speed photography.

Permanent displacements

Small quantities of Ottawa sand used in the cratering experiments were dyed several different colors for use as tracers in particle displacement experiments.

Use of the dyed Ottawa sand provided natural tracers and generally eliminated difficulties associated with the use of artificial tracers. The dyed sand was placed in the test bed as a thin layer of a single color, as a thin layer of several colors arranged in various patterns, as vertical columns, and, on occasion, as single grains of sand. The tracers were normally placed in the test bed as the various test geologies were being constructed.

After detonation, the permanent displacements of the tracers were observed with use of several different techniques. Cross-sectional views of craters formed in dry media were obtained by "freezing" the craters. A crater was frozen by carefully pouring a thinned lacquer solution over the crater and allowing the lacquer to dry. The resulting frozen mass was removed from the test bed and sawed open to reveal the deformation of the layers, columns, etc. Cross-sectional views of craters formed in wet media were obtained by draining excess water from the media and then carefully removing a section of the crater to expose the deformed dyed sand patterns. Distortions in the plane of a layer were observed by removing the material above the layer with use of a special scraper blade.

Single and/or multiple layers of dyed sand were used in dense homogeneous media to display the general deformation of the region surrounding the crater. Typical frozen craters for events detonated in dense homogeneous media are shown in Figs. 12 and 13. The crater shown in Fig. 12 was produced by a charge detonated 5 charge radii (25.4 mm) below the surface (at the level of the fourth

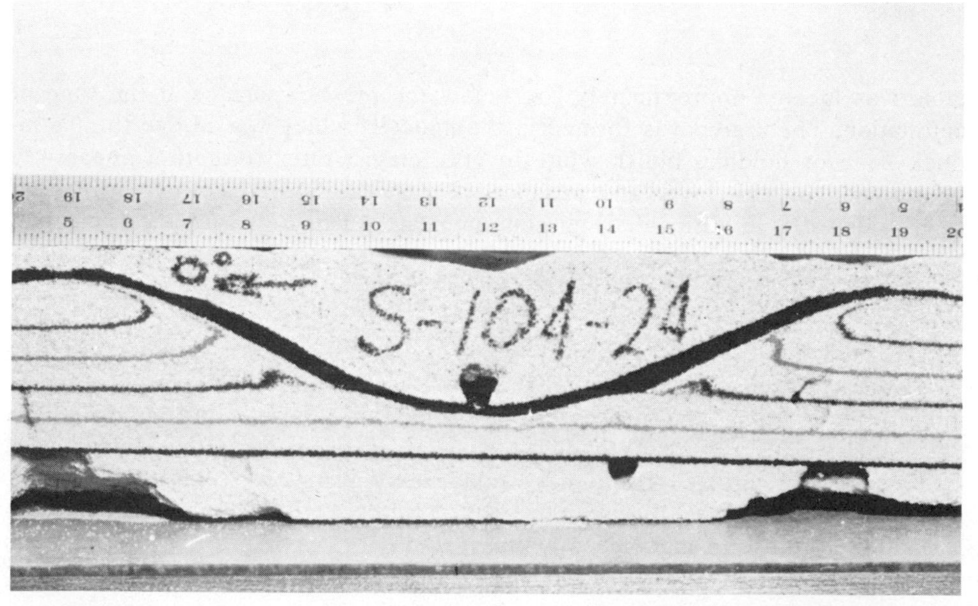

Fig. 12. Cross-sectional view of crater shown in Fig. 10. Structural deformations in region around crater are typical of events detonated in dense homogeneous sand: negligible downward displacement of layers below crater, shearing of material in region near crater walls, and overturning of material near crater lip.

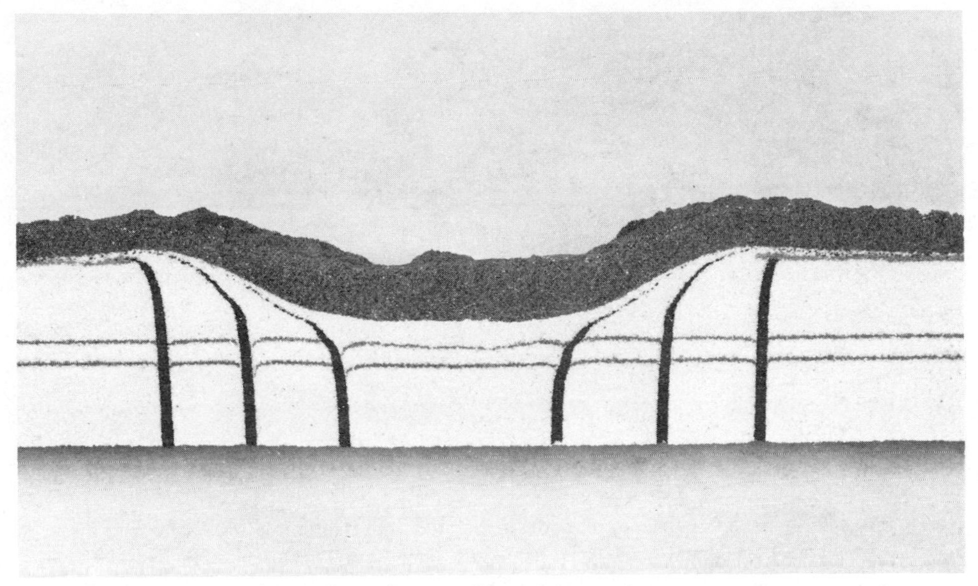

Fig. 13. Cross-sectional view of crater formed by surface tangent-above event in medium dense sand showing shear deformation of columns. Localized deformations of layers at intersection with columns occurred during installation of columns.

layer from the bottom) and the crater shown in Fig. 13 was produced by a charge which was surface tangent-above. The deformation of the dyed sand layers shown in Fig. 12 exhibit a number of the features displayed in frozen craters obtained for other events in dense media. These features are: (1) minimal downward displacement of layers below the apparent crater floor, (2) a region adjacent to the crater walls in which shearing of the medium has occurred, and (3) an overturning of material in the region of the lip. In addition, the crater wall at the level of the third layer from the bottom exhibits a ledge which was discussed previously and shown in Fig. 10.

A cross-section of a crater formed in a loose saturated sand is shown in Fig. 14. The sand used in the preparation of the test bed for this and several other events in this particular series was a mixture of equal parts (by weight) of fine-grained (0.17 mm dia) and coarse-grained (0.55 mm dia) Ottawa sand. The crater exhibits considerable downward displacement of the material below the floor. In addition, a central uplift is in evidence although a mound was not formed on the crater floor.

Several features displayed in Fig. 14 are typical of those observed in other craters formed in this wet medium when a layer of dyed sand was near the expected level of the crater floor. High-speed films of these events showed that separation of the medium along the layer occurred during crater formation, with excavation and ejection of material occurring in the medium above the layer. Considerable outward radial displacement, circumferential stretching, and thinning of material in the uplifted layer resulted and produced a rather large

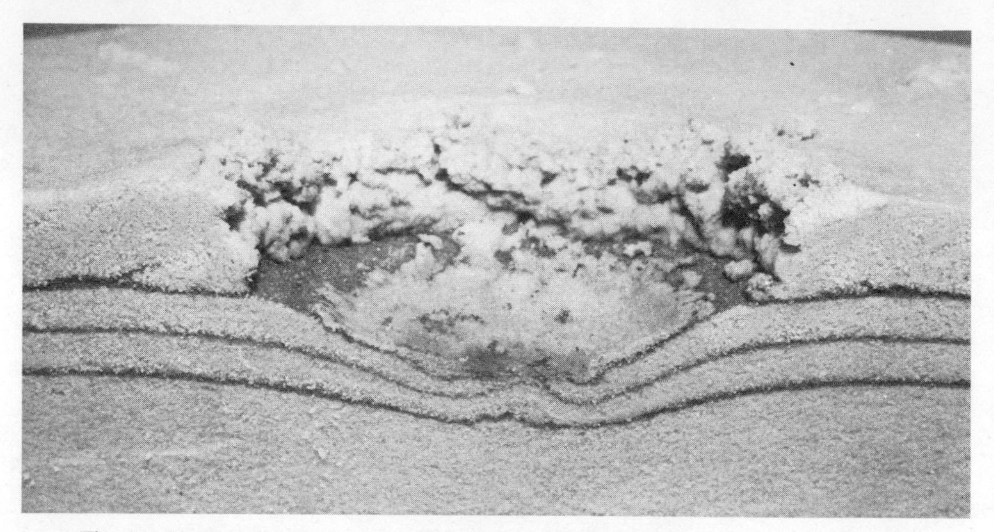

Fig. 14. Cross-sectional view of crater formed in loose saturated sand, showing the downward displacement of material below the crater and the separation of material at the level of the first layer of dyed sand.

transient lip. When ejection of material ceased, the transient lip collapsed. During collapse of the lip, considerable slumping and inward displacement of material in this region occurred. These inward and downward motions resulted in the formation of a number of circumferential and radial cracks and wrinkles in the region around the apparent lip and are very evident in the crater shown in Fig. 15. The high-speed films also show rather large "blocks" of wet sand sliding radially along the surface of the separated dyed sand layer. Quite often, these blocks were driven into the separated region. One of these blocks is clearly visible in the cross-sectional view of the crater shown in Fig. 16. The various processes observed in the high-speed films of the events detonated in these "layered" sandbeds—separation, slumping and cracking of the surface layer, sliding blocks, etc.—were not observed in films of similar events detonated in sandbeds which did not contain these layers.

Some of the features observed in the craters formed in the wet layered media were very similar to those observed in the SNOWBALL crater. Jones (1976) and Roddy (1976) have provided excellent descriptions of the surface and subsurface features of SNOWBALL and several of the other craters in the DRES series. Before comparing these features, several obvious differences between the general physical appearance of SNOWBALL crater and the laboratory-scale craters in wet sand should be pointed out. First, central mounts were not formed in the small craters. Second, SNOWBALL had a depressed rim while the small craters had raised rims. Depressed rims were occasionally formed along sections of the rims of the small craters when blocks of sand in the rim slumped into the crater. Finally, ejecta produced by the small events consisted of extremely widespread fragments rather than the more or less uniform blanket formed around SNOWBALL.

Circumferential fissures in the region around the crater lip were detected for

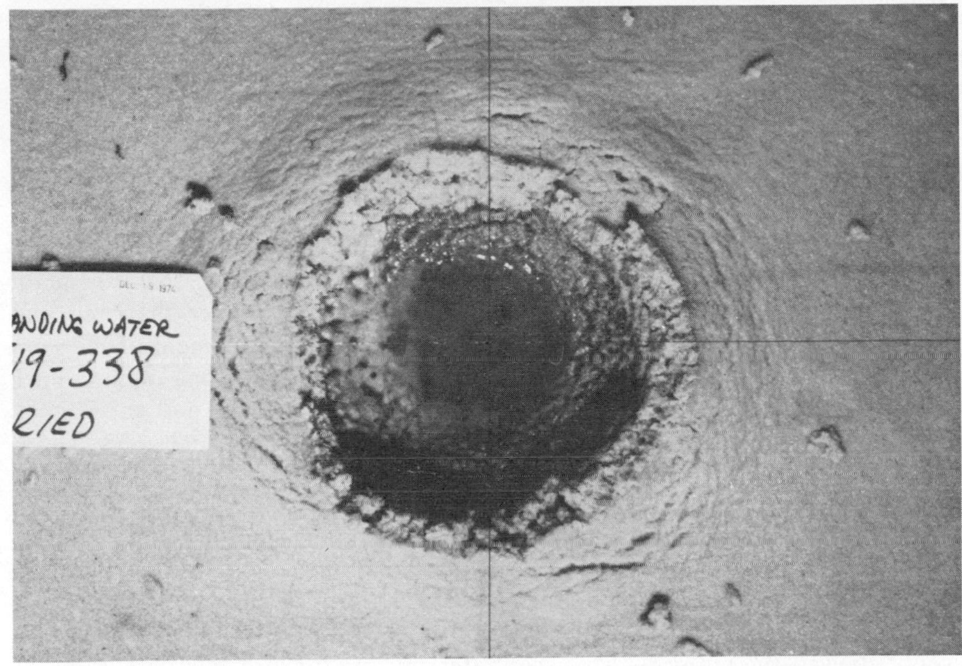

Fig. 15. Overhead view of crater formed in loose saturated sand, showing circumferential and radial cracks and wrinkles which formed during slumping of crater lip.

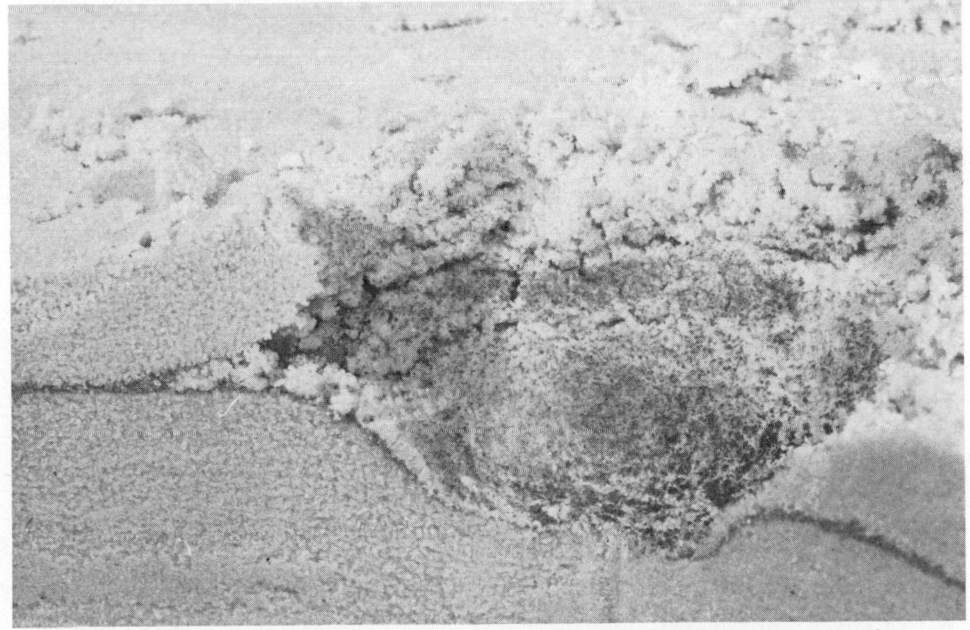

Fig. 16. Cross-sectional view of a crater formed in naturally-drained sand, showing block of sand which has been thrust into separated dyed sand layer.

the SUFFIELD 1961 and the PRAIRIE FLAT events but were most pronounced around the SNOWBALL crater. SNOWBALL also exhibited a number of radial cracks. Both types of cracks are visible in the region around the crater shown in Fig. 15.

The SNOWBALL site extended across the transition between an extensive lacustrine deposit and a gravel bar or shoal deposit. The uppermost region of the site consisted of a number of thin stratified layers of various colored silts, clays, and sands. These various layers were deposited over a bed of sand and gravel on the northern side and a layer of silty clay and a layer of coarse sand on the southern side. A continuous layer of blue clay lay under the sand and sand-and-gravel deposits. As a result of the observed displacement of markers placed in sand columns, Jones indicated that slippage and sliding of the upper strata was possible along the boundaries between the blue clay and gravel beds and the coarse delta deposit and the overlying mantle of silts and clays. These sand and sand-and-gravel layers may have had the same effect on crater formation as the dyed sand layer did in the formation of the small craters.

Although a central mound was not formed in the laboratory-scale craters, central uplift of a portion of the strata below the crater is clearly shown in Fig. 14. The central uplift observed in SNOWBALL was much more pronounced and consisted mainly of blue clay capped with clay-silt. High-speed films of the small events showed the entire floor of the crater moving upward after the transient cavity had reached maximum size. Similar late-time upward motions of the crater floor were also observed during the formation of craters in the saturated homogeneous loose sandbeds. The actual mechanism which produced the central uplift structure in the dyed sand layers and intervening strata is not clear but must be related to the readjustment of material which occurs in this region as the crater floor moves up.

Two cross-sectional views of a crater and central mound formed in a layered medium ($D = 2.5$) are shown in Fig. 17a. These views were reconstructed from a series of measurements which were taken as the overburden was being removed in 2-mm-thick lifts and clearly show the vertical deformation of layers of dyed sand which had been placed in the overburden. Figures 17b and c show the radial displacements of material within these dyed sand layers. Distortion of the striped pattern of the center layer, shown in Fig. 17b, indicated that material in this layer experienced a permanent outward radial displacement. On the other hand, distortion of the striped pattern in the lower layer, shown in Fig. 17c, indicated a considerable upward and inward flow of material had occurred in this region during crater formation.

Dynamic displacements

A special "quarter-space" test container was constructed to view the interior dynamic processes which occurred during crater formation. One wall of this quarter-space container was constructed of 5-cm-thick, clear Plexiglas® and allowed crater formation to be viewed through this wall with high-speed cameras.

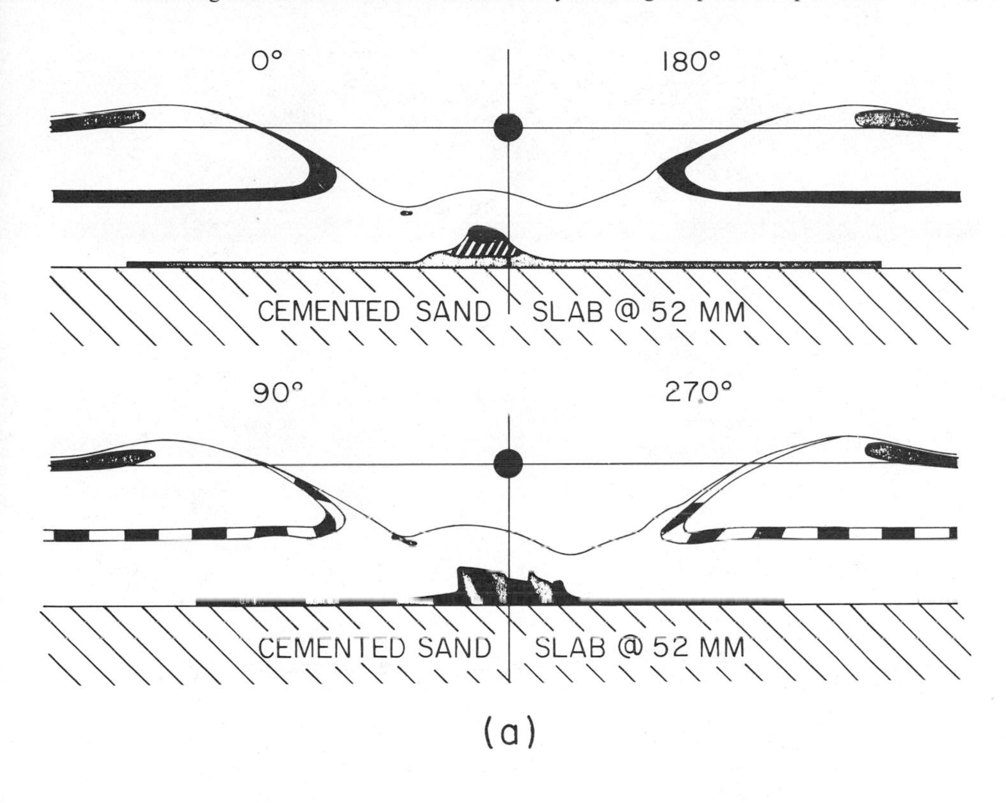

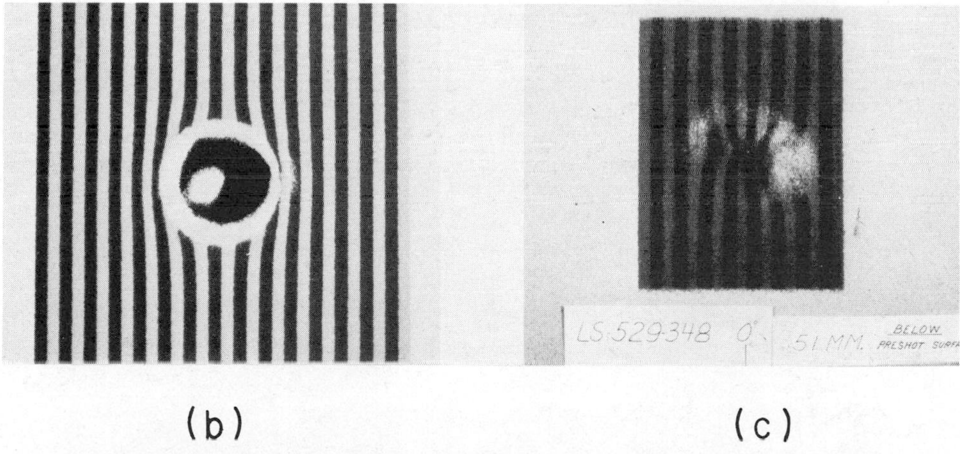

Fig. 17. Deformation of dyed sand layers placed in dense sand overburden for event in layered medium ($D = 2.5$). (a) Cross-sectional views of crater showing deformation of layers and interior of central mound. (b) Overhead view of center layer. Distortion of stripes indicates outward flow of medium in this layer. Dark sand at center of layer defines crater cross section at a depth of 26 mm below preshot surface. (c) Overhead view of lowest layer. Distortion of stripes indicates extent of inward flow of dyed sand in the layer while white areas define regions of reduced layer thickness.

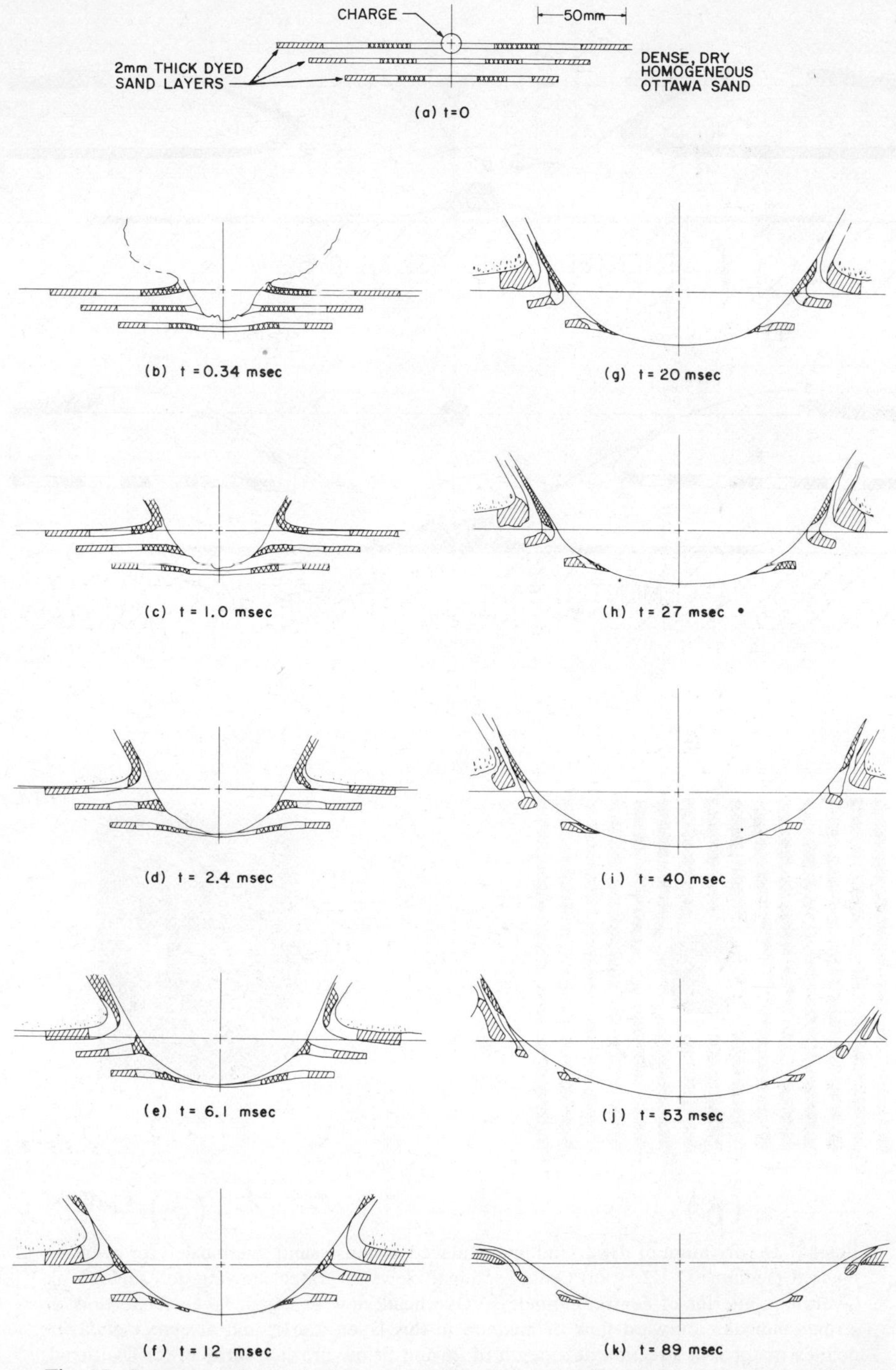

Fig. 18. Sequence of transient crater profiles and subsurface particle motions which occur during formation of a crater in dense, dry homogeneous sand.

Test beds were prepared in the quarter-space tank using the normal test bed preparation techniques, including the installation of dyed sand layers, etc. Hemispherical lead azide charges were placed in the tank with their flat surface in contact with the Plexiglas® wall.

Similar experiments were performed by Vesic *et al.* (1972) and discussed by Nelson and Taylor (1968). These experiments used the same size explosive charges to examine crater formation when the charges were deeply buried. It is important to point out that the quarter-space experiments described in this paper were performed solely to provide a means of observing the transient crater and the dynamic displacement of subsurface material. These experiments were not intended to duplicate faithfully all aspects of experiments performed in the normal half-space container.

Illustrations of the transient crater profile and attendant subsurface particle displacements at various times during the formation of a crater in dense dry homogeneous sand are shown in Fig. 18. This figure and others presented later in this section were taken from high-speed films of the various events. The test bed contained a number of dyed sand tracers similar to those used by Andrews (1975) in a study of the origin and distribution of ejecta produced in these laboratory-scale experiments. Crater formation for this event was basically an excavational process with very little compaction of material occurring in the region around the crater. Maximum cavity depth was attained approximately 12 msec after detonation (Fig. 18f) but considerable growth of the cavity radius occurred during the next 40–60 msec. Other typical features observed during the formation of craters in other media and illustrated in this figure include: (1) growth of an upthrust region proceding the main plume of ejecta, (2) spall of grains of sand from the surface of the test bed, and (3) the development of an overturned flap.

Formation of a crater in a dense saturated homogeneous sand is shown in Figs. 19b and c. The crater formed in this medium is considerably smaller than that formed in a dense dry sand. Some compaction of material below the crater was observed and, once again, the transient cavity reached maximum depth before reaching maximum radius.

Considerable compaction of the medium surrounding the crater occurred during the formation of a crater in a loose dry homogeneous sand as shown in Figs. 19d and e. A nearly hemispherical cavity existed during much of the crater formation process. A distinct change in the slope of the cavity walls was observed and the bulk of ejecta was supplied by material in the wedge-shaped region at the flanks of the cavity. Some late-time failure of the upper part of the crater walls occurred after the last time shown in Fig. 19e and resulted in the formation of a conical crater.

Individual grains of dyed sand were used to obtain particle displacements for an event detonated in loose saturated homogeneous sand. Several transient crater profiles, as well as the displacement of a number of particles, are shown in Fig. 20. The transient cavity had a nearly hemispherical shape during crater growth. After crater growth ceased, the floor of the cavity began to rise. This rise was accompanied by a settling of material in the lip region and a small inward movement

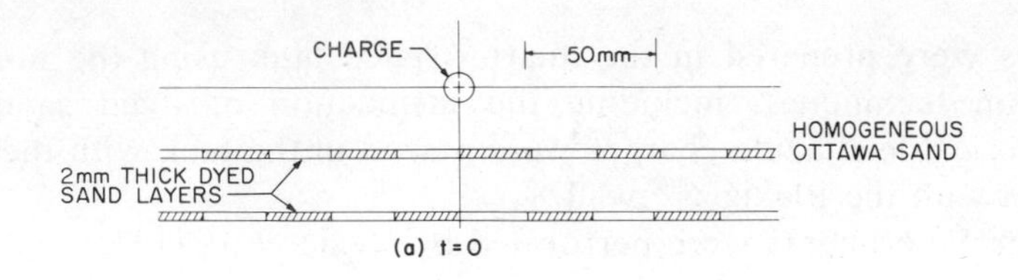

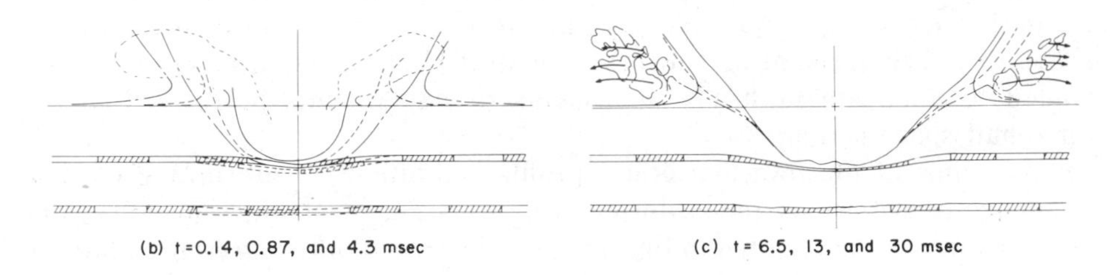

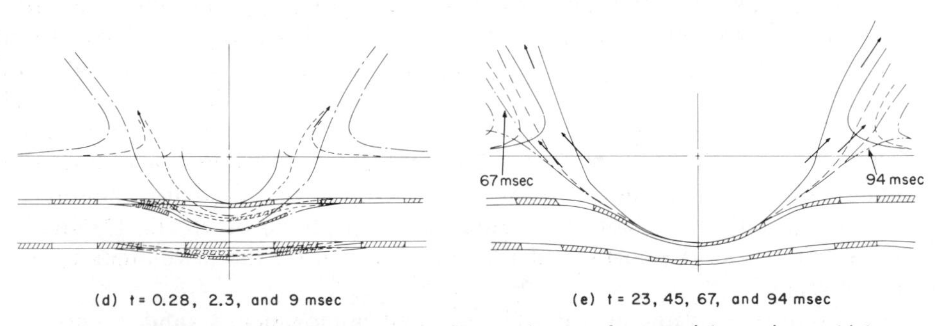

Fig. 19. Sequence of transient crater profiles and subsurface particle motions which occur during formation of a crater in (b) and (c), dense saturated homogeneous sand and (d) and (e), loose dry homogeneous sand.

of all material around the crater. These combined late-time motions resulted in the formation of a broad shallow crater.

The motions of four other sand grains not shown in Fig. 20 were analyzed using a technique described by Piekutowski *et al.* (1976). These sand grains were placed at the corners of a small square located at the intersection of the diagonal between the preshot surface and the vertical axis of the test bed and a line which connected the four deepest grains shown in Fig. 20. The results of the analysis of the motions of these four grains indicated that material in the region of the square behaved as a liquid from approximately 8 msec after detonation to 80 msec after detonation.

Thus far, the various illustrations have been for events detonated in homogeneous media. In all cases, the transient cavity was bowl-shaped. The next series of illustrations are for events detonated in layered media and exhibit several other transient cavity shapes.

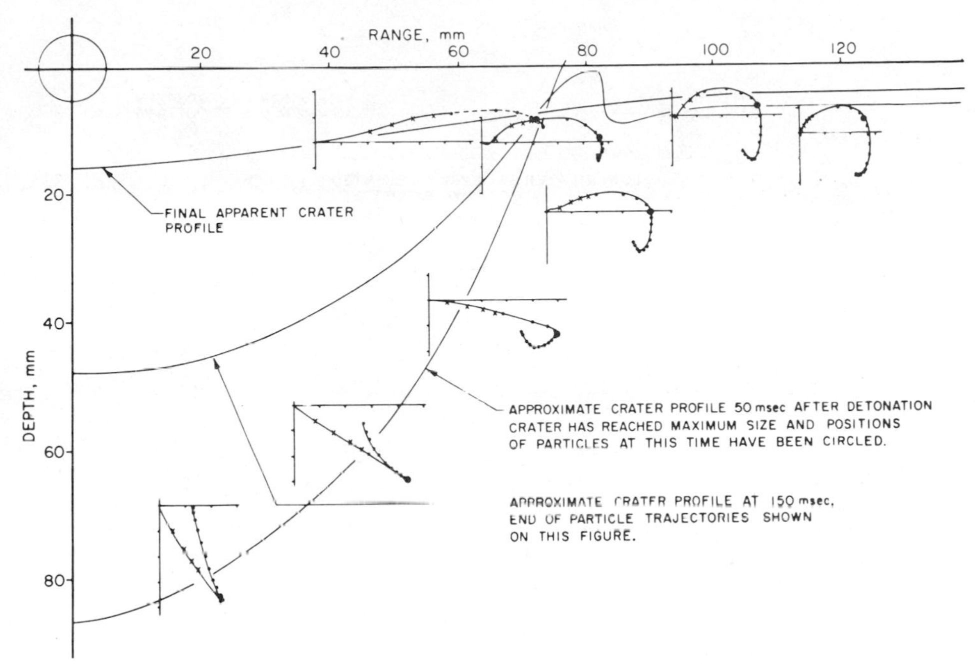

Fig. 20. Displacement time-histories of single grains of dyed sand during crater formation in a loose, saturated sand. Particle positions shown with an "X" are at 2-msec intervals. Other positions are at 10-msec intervals.

The crater formation sequence shown in Fig. 21 is for a layered medium in which the value of D was 2.5. The first deviation from a "normal" crater formation sequence was observed approximately 1.4 msec after detonation (Fig. 21c). A zone of bulked sand has formed between the "cavity" surface and the uncratered medium. The bulked zone increased in size until about 12 msec after detonation (Fig. 21e). At this time a secondary veil of ejecta is observed to form and begin leaving the cavity. In addition, the dyed sand layer in contact with the cemented sand layer begins to exhibit an inward and upward deformation. The secondary veil of ejecta continues to grow in size for the next 30 msec and then begins to fall and be deposited on the near-field ejecta blanket. The uplifted region and the bulked zone continue to increase in size until approximately 34 msec after detonation, when a distinctive Y-shaped cavity is formed. For the next 200 msec, airborne material is deposited on the transient cavity walls and a collapse of material in the cavity wall occurs. The collapse and sliding of the cavity walls continues until the slope of the walls is equal to the angle of repose of the sand. Quaide and Oberbeck (1968) described an almost identical crater formation process for certain impact craters formed in layered media.

Crater formation in a saturated layered medium, with the same thickness of dense overburden used for the event shown in Fig. 21, was identical to that shown for the event in dense, saturated homogeneous sand (Figs. 19b and c). Crater formation in a dry and a saturated layered medium with loose overburden is shown

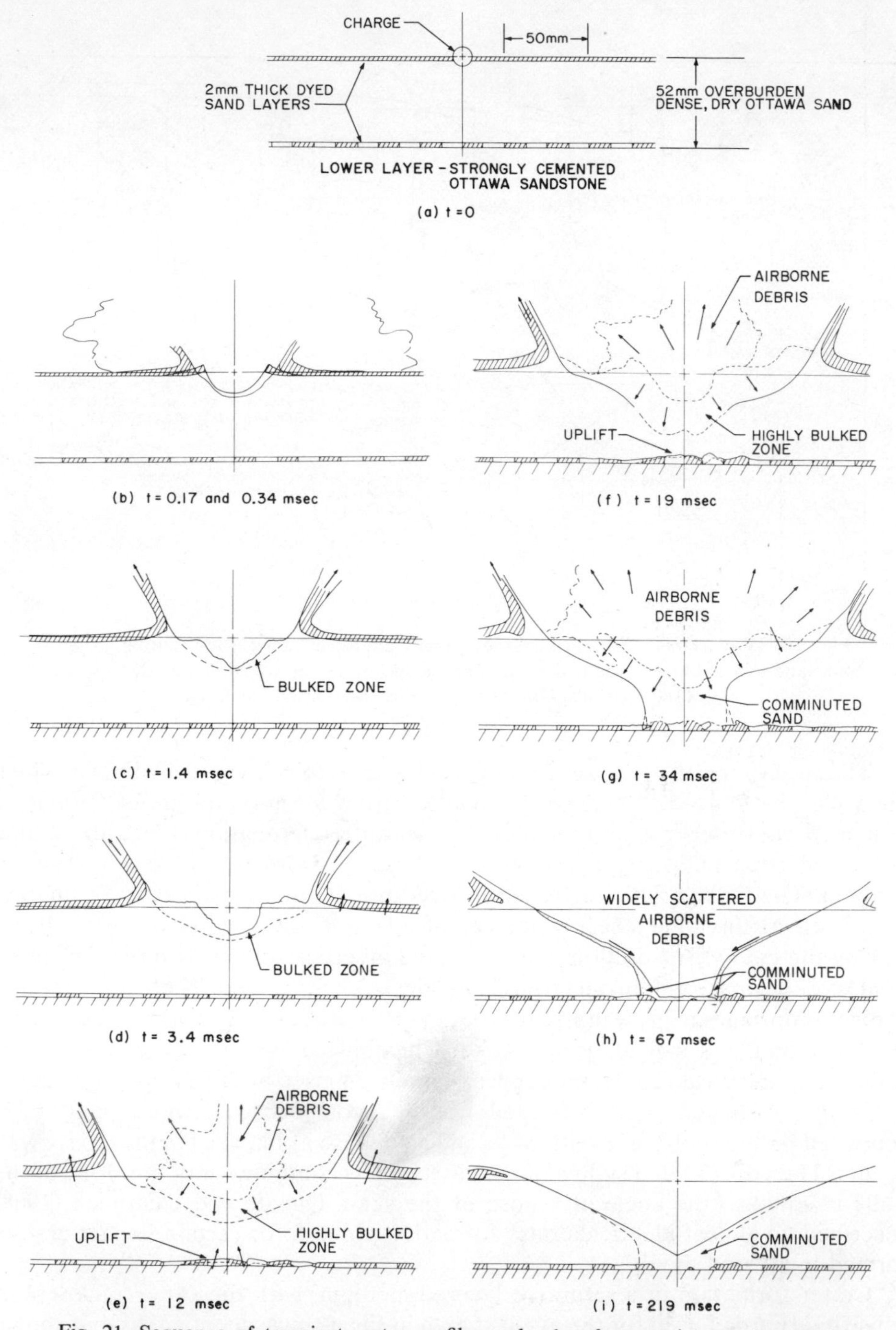

Fig. 21. Sequence of transient crater profiles and subsurface particle motions which occur during formation of a crater in a layered medium. Dense dry sand over strongly cemented sand ($D = 2.5$).

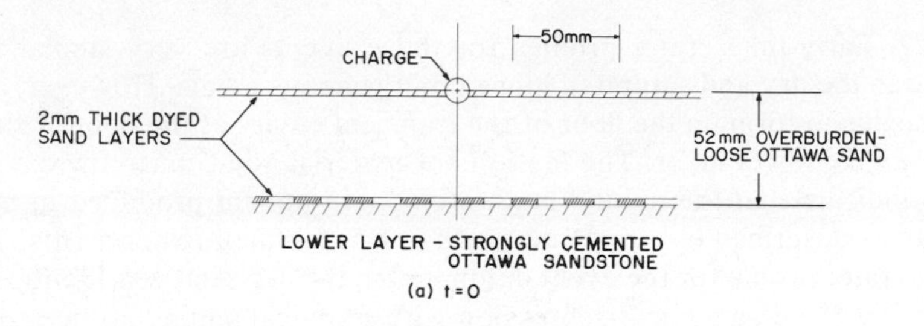

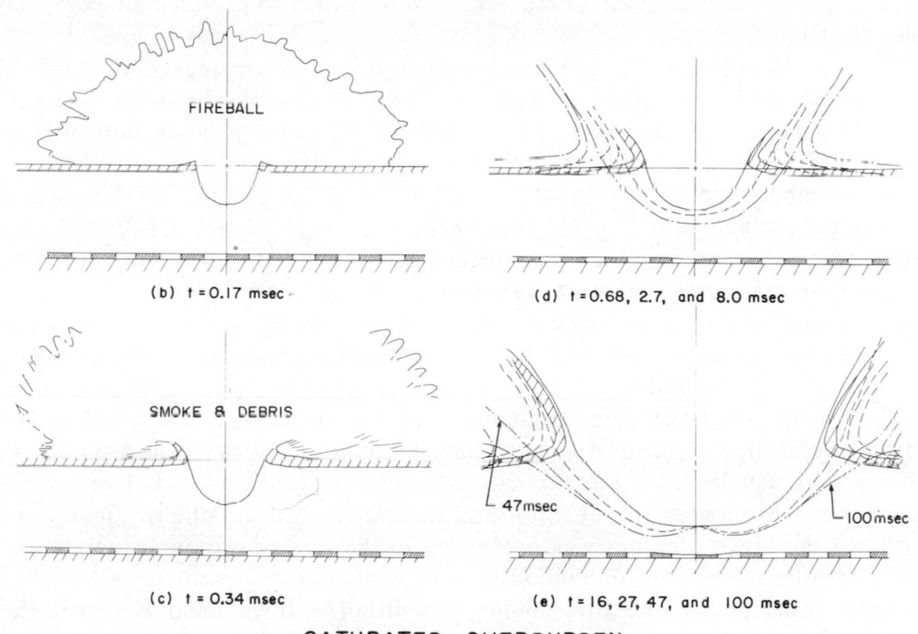

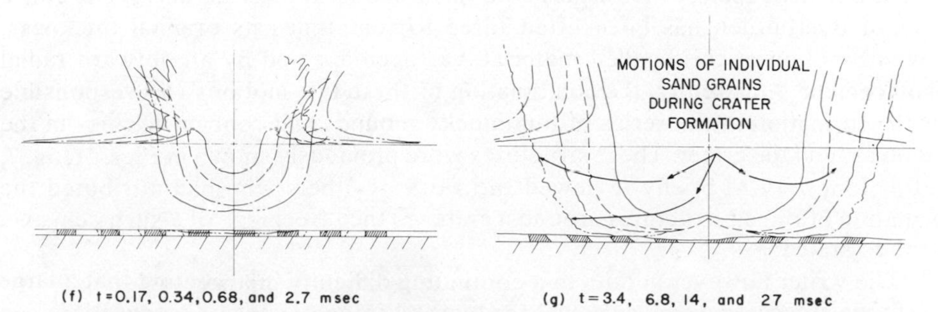

Fig. 22. Sequence of transient crater profiles and subsurface particle motions which occur during formation of a crater in a layered medium: (b) through (e), loose dry overburden ($D = 2.5$); (f) and (g), loose saturated overburden ($D = 2.5$).

in Fig. 22. Early-time crater profiles for these events are very similar to those produced in the dry and saturated loose homogeneous media. However, a central mound begins to form on the floor of the transient cavity as the floor of the cavity approaches the lower layer. The motions of material adjacent to the cavity walls strongly indicate that the mound is an erosional remnant produced in much the same way as described by Land and Clark (1965) for their experiments. The final apparent crater profile for the event detonated in the dry sand was identical to that shown in Fig. 9, i.e., a conical depression with a conical mound at the center. The trajectories of several particles in the saturated overburden (Fig. 22g) given some indication of the magnitude of the displacement which occurred during this event. These displacements coupled with a "snowplow" effect to produce an extremely high crater lip.

When the thickness of the overburden was significantly decreased, to $D = 0.25$, a flat-floored crater with a large central mound was produced. This crater was very similar in appearance to the MIXED COMPANY III crater as it was described by Roddy. Two processes which played relatively minor roles in the formation of craters in other media dominated the crater formation process in this layered medium and are shown in Fig. 23. These processes were: (1) formation of a large central plume of material which was ejected vertically from within the crater, and (2) spall of surface material in the region outside the ejecta veil.

The central plume of airborne debris developed shortly after detonation (see Fig. 23d) and was composed of brecciated cemented sandstone from the lower layer. Material in the plume was ejected and deposited in a very orderly manner. The first material to be ejected (originally closest to the charge) was lifted highest and was the last material to be deposited on the mound as fallback. The last material to be ejected, usually fairly large blocks, was merely lifted slightly before settling back down. Throughout the ejection and deposition process the original stratigraphy of the lower medium was preserved in the mound although there was a locally chaotic orientation of individual fragments due to tumbling, etc., which occurred during the time these fragments were airborne. It was clear, however, that the central mound formed in this crater was formed by fallback *and* some central uplift.

The extent of spall of the surface material is shown in Figs. 23f and g. The entire layer of overburden has been lifted three to four times its original thickness. Upward motion of this spalled material was accompanied by an outward radial displacement. Apparently, the combination of these two motions was responsible for the formation of the series of hummocky mounds and concentric rings in the region around the crater. These structures were previously shown in Figs. 5f, 5g, 7, and 8. Jones (1976) briefly reviewed the work of others who had attributed the formation of ringed structures around a crater to the "freezing" of a surface wave. He then stated:

"The writer however admits to a continuing difficulty in accepting that a large amplitude *travelling* wave, as called for by the "frozen Tsunami" suggestion, can simply stop travelling at a given point in time and space. On the other hand, it appears quite reasonable that a *standing* wave pattern—produced by interference

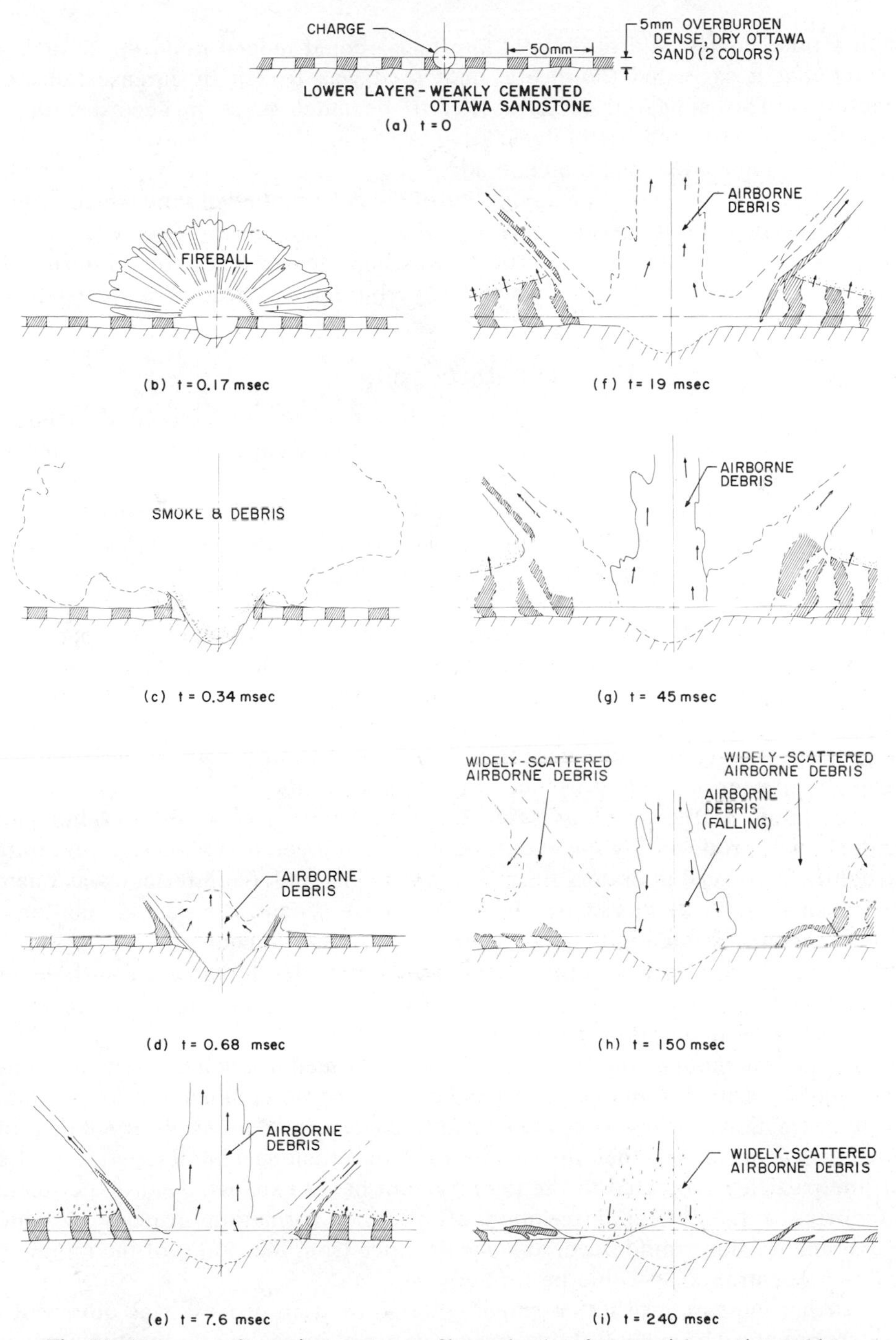

Fig. 23. Sequence of transient crater profiles and subsurface particle motions which occur during formation of a crater in a layered medium. Dense dry sand over weakly cemented sand ($D = 0.25$).

with a reflection from a horizontal interface—could indeed build up to such an extent that it exceeded the elastic limit (and was frozen by intense tensional fractures). This appeared to the writer to be much more in accord with the experimental data, and it still does."

The standing wave phenomenon advanced by Jones was perhaps responsible for the relatively coherent upward motion of the stripes of dyed sand which formed the overburden for the event shown in Fig. 23. The radially-directed, outward component of velocity almost certainly was imparted to this material during the initial phase of crater formation when the overburden was near its original density.

5. Discussion

Simple bowl-shaped craters were produced by events detonated in homogeneous media or media in which the depth to a hard layer, water table, density discontinuity, etc., was many times the depth of the crater. Jones (1976) and Roddy (1976) have provided a series of crater morphological features for events detonated in the DRES series—events performed in a medium in which the depth to the water table and the hard lower layer essentially remained constant. The nature of the features observed in these craters increased in complexity with the yield of the explosive. As Roddy stated, it was clear that changes in yield, charge shape, and HOB strongly affected the total amount of energy coupled into the alluvium and that the final crater shapes and structural deformations were sensitive to these initial conditions in terms of energy deposited in the medium. Crater formation for the smaller events was relatively free of influence from the water table but was strongly influenced by its presence in the larger events.

As shown in the previous sections, the greatest variety of morphological features was produced by the events detonated in layered media. The intracrater structures observed in craters formed in media in which the interface was one or two normal crater depths below the surface were produced by a combination of elastic rebound of material below the crater floor and fallback of airborne debris. When the interface was less than one normal crater depth below the surface and strongly influencing the crater formation process, ledges, hummocky mounds, etc. were produced in addition to intracrater mounds.

The central mounds observed in the craters formed in dry homogeneous dense sand and described in this paper are believed to be the remnants of an erosional crater formation process. They are similar to the central mounds observed by Land and Clark (1965) during their investigations of the erosional characteristics of a bed of fine particles subjected to the impingement of jet exhaust. Their experiments supported a theoretical description of a crater formation process in which maximum erosion would occur at some distance from the center of the impinging jet and a central cone would be formed.

Further indication of an erosional crater formation process was observed in craters produced by several above-surface detonations in the present studies where dyed sand patterns had been placed on the surface of the sandbed. After detonation the pattern was generally intact at the top of the central mound formed in the crater and in the region surrounding the crater. The pattern was removed, with additional material, from the cratered region. The portions of the dyed sand pattern which remained intact are consistent with the crater formation process proposed by Land and Clark in their model and demonstrated by their experiments.

Central mounds in craters formed in homogeneous media have been observed by other investigators. Carlson and Jones (1964) reported central mounds or inner cones formed in three craters produced in the AIR VENT series of experiments in Frenchman Flat playa. Shot II-1, in which a 116 kg (256 lb) sphere of TNT was detonated with its center 27 cm (0.9 ft) above the surface, produced a crater with a rather large central mound. Smaller central mounds were also observed in the craters produced by the half-buried events, Shot III-3A and Shot III-3B, which utilized 2724 kg (6000 lb) spheres of TNT. Central mounds were not observed for any of the other half-buried events in the AIR VENT series. Shot II-1 was the only above-surface detonation in the series. Carlson and Jones indicated that the inner cone structure observed in these craters may have been the result of fallback.

Fulmer (1965) also reported and discussed the formation of an inner cone in the craters produced by the detonation of 0.45 kg (1 lb) of TNT in plaster sand. Although inner cones were only observed for above-surface detonations, he noted that they may have developed when the charges were buried but were destroyed or modified by debris which fell back into the crater. The inner cones described by Fulmer exhibited the same characteristics of the central mound shown in Fig. 3, i.e., they were covered with finely divided and pressure-agglutinated sand flour.

Fulmer also described the unpublished work of Olsen performed in 1944 in which elastic rebound was suggested as the source of an inner cone formed in the crater produced in a natural silty loam by the above-surface detonation of a 45 kg (100 lb) block of TNT. The cone produced in this event was described to be relatively hard and compressed with its top at the same level as the preshot surface and was *not* made up of pieces of earth which fell back into the crater. It was theorized that the dry nature of the soil coupled with intermittent layers of lava made the ground resilient and that the mound was formed when the dense, compressed material at the bottom of the crater was broken up and piled into a cone as stresses were relieved.

Central mounds in each of the above studies were always located directly below the explosive source or the stagnation point of the impinging jet. In addition, central mounds exhibited in craters formed in the various homogeneous media were produced by detonations in which the explosive charge was detonated above the surface or half-buried.

The ridge and valley system observed around the craters produced by the above-surface bursts in this study indicate significant surface disturbance. The radial orientation of the ridges and valleys as well as blackening of individual sand grains suggests that the formation of these structures is related to an interaction between the test bed surface and jets of gas in the fireball.

A violent disturbance of the surface material surrounding the crater was reported by Carlson and Jones for Shot II-1, the only above-surface event in the AIR VENT series. Ejecta collection pads placed on the surface were destroyed by the air blast and the surface of the region around the crater was severely disturbed. Close examination of the loose disturbed material revealed that it was was almost entirely composed of broken and displaced surface material and not ejecta. Carlson and Jones attributed this disturbance of the surface to ground shock.

It is interesting to note that the ridges and valleys were not produced by the half-buried or any of the buried events in homogeneous media. The material surrounding the charge for these buried events evidently provided sufficient shielding of the test bed surface to protect it from direct contact with the fireball. Significant ridge and valley structures were produced by half-buried events in layered media with a very thin layer of overburden (see Figs. 5g, 5h, and 23). For these events, it is very probable that the production of these extracrater features was enhanced by disturbances propagated in the lower layer since the charge was resting on this layer at detonation.

6. Summary

A number of different mechanisms influenced the crater formation process and produced characteristic morphological features of the apparent craters. Crater formation was a very orderly process which began with the formation of a small transient cavity that was initially identical in shape for all geologies. The transient cavity quickly and smoothly transformed to a shape which was unique for the particular geological conditions of the cratered medium. As the rate of growth of the cavity decreased, the craters were observed to have reached maximum depth before reaching maximum radius.

Because the craters were produced using the same explosive charges, each of the characteristic mechanisms and features could be related to the preshot conditions of the cratered medium. Many features observed in the laboratory-scale craters also have been observed in other explosive and impact craters. A comparison of the geologies in which craters with similar morphological features were formed has shown that the gross features of the geologies were similar, i.e., layered, wet, etc.

Differences in the scales of the media, cratering sources, etc., for the events described in this paper suggest that a variety of morphological features could result from crater formation in a particular medium. The features produced would be a function of the source size and the general characteristics of the region of the medium exercised by the source. It is apparent that an interface, i.e., hard layer, water table, etc., in the medium exercised by the source had a pronounced effect on the morphological features of the crater produced in that medium. Detailed examination and understanding of the mechanisms, motions, etc., which are produced or occur at the interface during crater formation remains an important area of further study.

The ejection of material from the crater was also an orderly process. The initial flight conditions of ejecta fragements, i.e., size, velocity, ejection angle(s), etc., varied but were controlled by the geology of the cratered medium. Ejecta fragments generally moved outward and upward along the transient cavity wall. After leaving the cavity at the transient lip, ejecta traveled with little mixing or particle-to-particle interaction of in-flight fragments. Particles which were closely associated before crater formation were generally closely associated in the ejecta blanket but in a reverse order.

Acknwledgments—I wish to express my appreciation to Dr. G. Sevin and Capt. J. Stockton of the Defense Nuclear Agency for the opportunity and cooperation required for the preparation of this paper. I am also indebted to Dr. D. Roddy for his continued support and interest in the work described in this paper and thank him for his helpful discussions during its preparation. Special appreciation is due Mr. S. Hanchak for the meticulous care and attention to detail he exercised during the preparation of the experiments and to Mr. H. Williams for his care in assisting with the experiments and subsequent collection of data. Finally, I wish to sincerely thank Mr. D. Peterson for his suggestions, help, and assistance before and during the time this paper was being written and Drs. Fred Sauer and Peter Schultz for their thoughtful reviews.

The work described in this paper was jointly sponsored by the Defense Nuclear Agency and the Air Force Weapons Laboratory under United States Air Force Contract No. F29601-71-C-0132, No. F29601-73-C-0007, and No. F29601 74 C 0102.

References

Andrews, R. J.: 1975, Origin and Distribution of Ejecta From Near-Surface Laboratory-Scale Cratering Experiments, Air Force Weapons Laboratory Report, AFWL-TR-74-314.

Carlson, R. H. and Jones, G. D.: 1964, Ejecta Distribution Studies, Boeing Company Report, D2-90575.

Fulmer, C. V.: 1965, Cratering Characteristics of Wet and Dry Sand, Boeing Company Report, D2-90683-1.

Gault, D. E., Guest, J. E., Murray, J. B., Dzurisin, D. and Malin, M. C.: 1975, Some Comparisons of Impact Craters on Mercury and the Moon, *J. Geophys. Res.* **80**, 2444.

Jones, G. D.: 1975, Personal Communication.

Jones, G. D. and Henny, R. W.: 1976, Preliminary Results of Ejecta and Permanent Ground Displacements from PRE-MINE THROW 100-Ton Nitromethane Event (PMT IV-6), Air Force Weapons Laboratory Report, AFWL-TR-74-306.

Jones, G. H. S.: 1976, The Morphology of Central Uplift Craters, DRES Suffield Report, No. 281.

Ko, H. Y. and Scott, R. F.: 1967, Deformation of Sand in Shear, *J. of the Soil Mechanics and Foundations Division* **93**, SM5.

Land, N. S. and Clark, L. V.: 1965, Experimental Investigation of Jet Impingement on Surfaces of Fine Particles in a Vacuum Environment, NASA Technical Note D-2633.

Meyers, J.: 1973, MIDDLE GUST Crater and Ejecta Studies *Proceedings of the MIXED COMPANY/MIDDLE GUST Results Meeting*, 13–15 March 1973, Vol. II, Defense Nuclear Agency Report, DNA 3151P2.

Nelson, D. L. and Taylor, T. S.: 1968, Analysis of Vesic Crater Modeling Experiments, Nuclear Cratering Group Technical Memorandum, NCG/TM 68-14.

Oberbeck, V. R. and Quaide, W. L.: 1967, Estimated Thickness of a Fragmental Surface Layer of Oceanus Procellarum, *J. Geophys. Res.* **72**.

Oberbeck, V. R.: 1971, Laboratory Simulation of Impact Cratering with High Explosives, *J. Geophys. Res.* **76**, 5732.

Piekutowski, A. J.: 1974, Laboratory-Scale High-Explosive Cratering and Ejecta Phenomenology Studies, Air Force Weapons Laboratory Report, AFWL-TR-72-155.

Piekutowski, A. J.: 1975a, The Effect of Variations in Test Media Density on Crater Dimensions and Ejecta Distributions, Air Force Weapons Laboratory Report, AFWL-TR-74-326.

Piekutowski, A. J.: 1975b, The Effect of a Layered Medium on Apparent Crater Dimensions and Ejecta Distribution in Laboratory-Scale Cratering Experiments, Air Force Weapons Laboratory Report, AFWL-TR-75-212.

Piekutowski, A. J., Andrews, R. J., and Swift, H. F.: 1976, Studying Small-Scale Explosive Cratering Phenomena Photographically, *Proceedings of the 12th International Congress on High Speed Photography.*

Quaide, W. L. and Oberbeck, V. R.: 1968, Thickness Determinations of the Lunar Surface Layer from Lunar Impact Craters, *J. Geophys. Res.* **73**, 5247.

Roddy, D. J.: 1973, Project LN303, Geologic Studies of the MIDDLE GUST and MIXED COMPANY Craters, *Proceedings of the MIXED COMPANY/MIDDLE GUST Results Meeting*, 13–15 March 1973, vol. II, Defense Nuclear Agency Report, DNA 3151P2.

Roddy, D. J.: 1976, High-Explosive Cratering Analogs for Bowl-Shaped, Central Uplift, and Multiring Impact Craters, *Proc. Lunar Sci. Conf. 7th*, p. 3027.

Vesic, A. S., Ismael, N. M. F., and Bhushan, K.: 1972, Cratering in Layered Media, U.S. Army Engineer Waterways Experiment Station Report, E-72-31.

Roddy, D. J., Pepin, R. O., and Merrill, R. B., editors.
(1977) *Impact and Explosion Cratering*, Pergamon Press (New York), p. 103–124.
Printed in the United States of America

Nuclear cratering experiments: United States and Soviet Union

MILO D. NORDYKE

Lawrence Livermore Laboratory, Livermore, California 94550

Abstract—Both the U.S. and the Soviet Union have carried out a number of nuclear cratering experiments in a variety of geologic environments which have provided useful data for understanding the mechanics of terrestrial explosive cratering. Whereas the majority of the U.S. nuclear cratering experience has been in relatively dry unsaturated environments, all of the Soviet experience described to date is in relatively high moisture situations which were generally in a saturated condition. The range of yields studied is similar, ranging from about 100 tons to 100 *kilotons*. Inter-comparison of the data shows a good correlation when differences in the geologic conditions are taken into consideration. No significant deviation from 3.4 root scaling appears to be required to correlate the data from these two independent bodies of cratering data.

OVER THE LAST TWENTY-FIVE YEARS the U.S. and the Soviet Union have carried out a large number of nuclear cratering explosions which have served to broaden our cratering experience to include a variety of geologic conditions and explosive yield ranges. Although the procedures and conditions under which the U.S. and the Soviet Union have carried out their nuclear explosions have widely differed the results compare favorably. In this paper I intend to summarize briefly this body of cratering experience and then to offer some comments particularly with respect to the Soviet experience with which the audience may not be too familiar.

U.S. UNDERGROUND NUCLEAR CRATERING EXPERIENCE

In Table 1, all of the underground nuclear cratering experience of the U.S. and of the Soviet Union (or at least that which they have made available) has been summarized. The first three explosions under U.S. experience were carried out for the purpose of gaining data on the cratering effects of nuclear weapons. As a result they are at relatively shallow scaled depths-of-burial. The fourth explosion, Neptune, was intended to be a contained explosion but because the yield was somewhat higher than expected, it resulted in the formation of a crater (Shelton *et al.*, 1960). Interpretation of the results of this explosion was somewhat difficult because the nearest free surface to the explosion was the sloping face of Rainier Mesa with the result that most of the throwout from the explosion was thrown down the slope. However, Neptune had a very profound effect on our understanding of the cratering mechanism. Prior to the time of the Neptune explosion it was believed that the peak of the cratering curve (i.e., scaled radius versus scaled depth-of-burst) occurred at approximately the scaled depth-of-burial of the Teapot Ess event. The fact that Neptune

Table 1. Summary of nuclear cratering explosions contributing data for nuclear excavation technology.

Name	Date	Configuration	Yield (kt)	Depth-of-burst (m)	Apparent crater radius (m)	Apparent crater depth (m)	Medium	Water conditions	Ref.
U.S. explosions									
Jangle S	1951	Single	1.2	1.1	14	6.4	Alluvium	Dry	Nordyke (1961)
Jangle U	1951	Single	1.2	5.2	40	16	Alluvium	Dry	Nordyke (1961)
Teapot ESS	1955	Single	1.2	20	45	27	Alluvium	Dry	Nordyke (1961)
Neptune	1958	Single (under slope)	0.115	31	31	11	Tuff	Dry	Shelton *et al.* (1960)
Danny Boy	1962	Single	0.42	34	33	19	Basalt	Dry (<1%)	Nordyke and Wray (1964)
Johnnie Boy	1962	Single	0.5	0.53	18	9.1	Alluvium	Dry	Nordyke (1964)
Sedan	1962	Single	100	194	184	98	Alluvium	Dry (~20%)	Nordyke and Williamson (1965)
Sulky	1962	Single	0.087	27	—	—	Basalt	Dry (<1%)	Videon (1965)
Palanquin	1965	Single	4.3	85	36	24	Rhyolite	Dry (<1%)	Videon (1966)
Cabriolet	1968	Single	2.6	52	54	37	Rhyolite	Dry (<1%)	Tewes (1968)
Buggy	1968	Row of 5	1.1	41 spacing: 46 m	76[b]	21	Basalt	Dry (<1%)	Tewes (1968)
Schooner	1968	Single	35	108	130	63	Tuff	Wet (~10%) but unsaturated	Tewes (1970)
Soviet experience									
1003	—	Single	1.1	48	53.5 (62.0)[a]	31 (20)[a]	Siltstone	Saturated	Myasnikov *et al.* (1970)
1004	—	Single	~125	~178	204 (~157)[a]	100 (83)[a]	Sandstone/ shale	Saturated (12%)	Myasnikov *et al.* (1970)
T-1	—	Single	0.2	31.4	40	21	Sandstone	Saturated (12.8%)	Myasnikov *et al.* (1970)
T-2	—	Row of 3	0.2	31.4 spacing: 40 m	32.5[b]	16	Sandstone	Saturated (12.8%)	Myasnikov *et al.* (1970)
Pechora-Kama	1971	Row of 3	15	~127 spacing: ~165 m	150–170[b]	10–15	Alluvium	Saturated	Kireev *et al.* (1975)

[a]After sluffing of side slopes.
[b]Average half-width of row crater.
[c]Includes 7.6% non-water gas-forming component.

made a very sizeable crater even though it was at a scaled depth-of-burst several times greater than Teapot Ess threw serious doubt upon this belief. This led to a series of high-explosive cratering tests which confirmed that indeed the optimum depth-of-burial was about twice the Teapot Ess scaled depth-of-burial.

The Danny Boy explosion carried out in 1962 was the first nuclear cratering explosion carried out at or near what was believed to be optimum depth-of-burial (Nordyke and Wray, 1964). The medium chosen for this test was the dry basalt of Area 18 of the Nevada Test Site (NTS). Because of the very dry nature of the basalt, only a very small amount of "non-condensable"[1] gas was produced by the nuclear explosion with the result that the crater was formed almost entirely as a result of spalling of surface layers of rock in reaction to the shock wave. Figure 1 shows the Danny Boy crater and of particular note is the short range of the throwout confirming this lack of "gas acceleration".

The next explosion which was designed to explore the question of what scaling law should be used for nuclear explosions with yields in the range of a few hundreds of kilotons was the Sedan explosion (Nordyke and Williamson, 1965). The site selected for the Sedan explosion was chosen to be near the location of a series of high-explosive cratering experiments carried out over a number of years in the desert alluvium of the NTS with yields ranging from a few hundred pounds up to 500 tons of TNT. A yield of 100 kilotons was chosen for the Sedan experiment and it was judged that this would permit us to estimate what scaling should be used for explosions as large as 500 kilotons or perhaps even as large as a megaton. The results of the Sedan explosion (see Fig. 2) indicated that for all practical purposes the use of 3.4-root scaling was valid for explosions in the region of optimum depth-of-burial over the entire range from a few hundred pounds of TNT to at least 100 kilotons. Since no change in scaling was observed over at least four orders of magnitude it was believed little error would be introduced if this scaling were used up to a megaton.

The moisture content in the Sedan area was between 15 and 20% and led to a significant amount of "non-condensable" gas being produced by the nuclear explosion and assisting with the cratering action. Analysis of surface motion data on Sedan showed that gas acceleration played a major role in the formation of this crater.

I would like now to skip to the Schooner explosion which is the most recent U.S. nuclear cratering explosion and one which is of very great interest from the viewpoint of understanding cratering mechanisms (Tewes, 1970). Figure 3 shows the Schooner crater and Fig. 4 shows two cross sections through the Schooner crater. The geologic setting of the Schooner crater is of very great importance since it had a large effect on the resulting crater. The Schooner site was characterized by a layer of very strong welded tuff, about

[1]"Non-condensable" gas refers to gases that are not condensable at temperature of 1000°C and at a pressure of a few atmospheres such as water or CO_2.

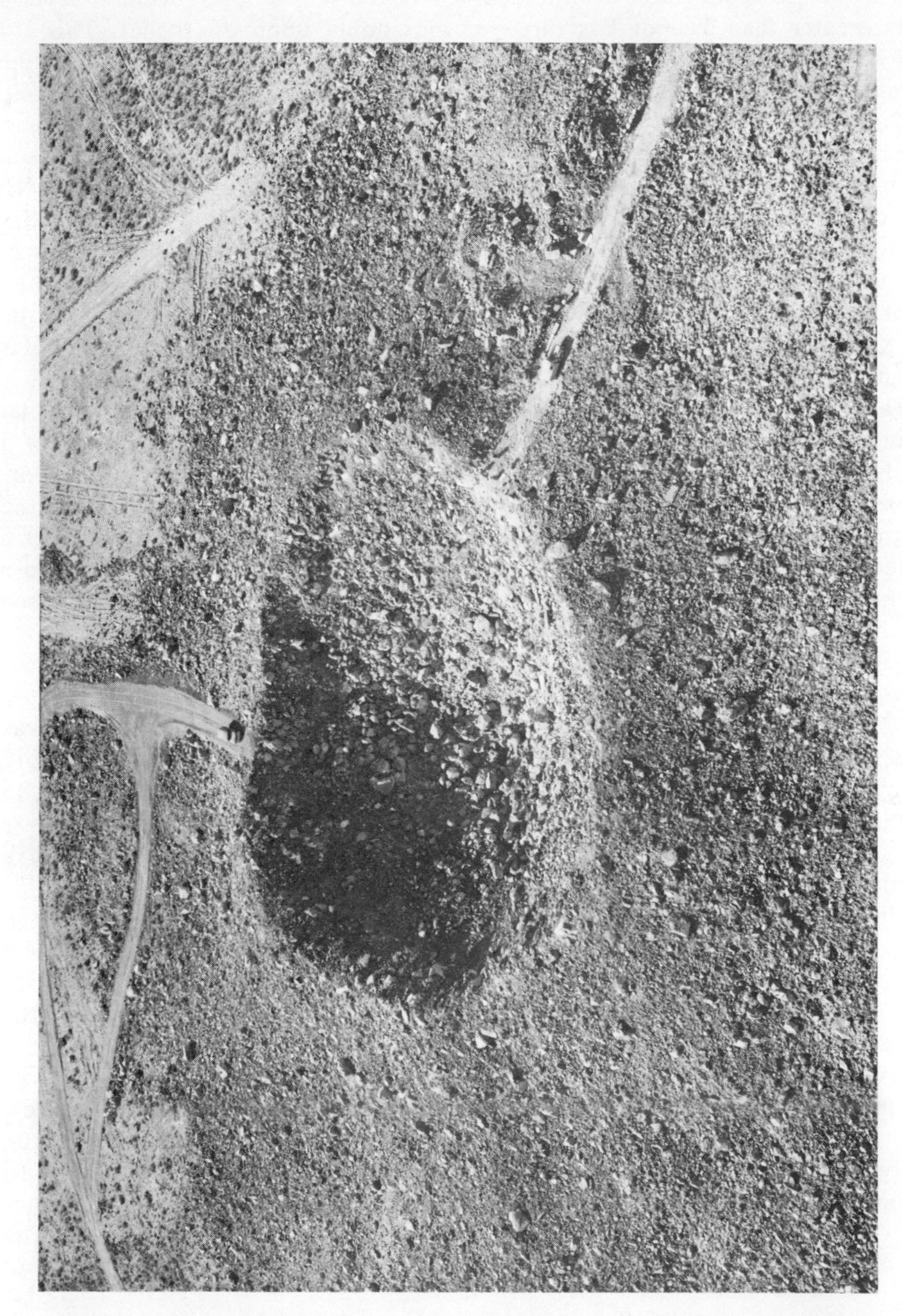

Fig. 1. Aerial view of the Danny Boy Crater in basalt.

Fig. 2. Aerial view of the Sedan Crater in NTS alluvium.

30–40 m thick, at the surface of the ground. This was underlain by a layer of weak, low-density tuff about 60–70 m thick. This was in turn underlain by a layer of strong welded tuff. The explosion was located approximately at the interface between the weak, low-density tuff and the underlying welded tuff. A perched water table existed at the bottom of the weak, low-density tuff which contributed a significant amount of water to the gases in the Schooner cavity during its formation. This resulted in a very large gas acceleration phase which together with the high-velocity characteristic of the surface layer resulted in very high velocities for the surface material. In addition, the very strong layer of tuff near the surface of the ground led to the walls of the resulting crater being very steep or even almost vertical in some locations. These features can be seen in Figs. 3 and 4.

One other U.S. nuclear explosion on which I would like to make a special comment is the Buggy event (Toman, 1968). This was a row of five 1.1 kiloton nuclear explosions in a dry basalt environment. Figure 5 shows a pre-shot geologic section along the axis of the row. Of particular significance in this Figure is the presence of the dense basalt layers which were discovered during mapping of the emplacement holes which were actually drilled to depths of 150 ft, the original planned depth-of-burial of the explosions. Consideration of the effect of these dense basalt layers on the outgoing shock wave lead to the belief that they

Fig. 3. Aerial view of the Schooner Crater in NTS alluvium.

would reflect a significant amount of the explosion energy and thereby significantly reduce the spall velocities of the surface layers. Calculations indicated that if the originally planned depth-of-burial of 150 ft were used, no crater would be formed. Therefore, on the basis of these calculations the depth-of-burial of the Buggy explosions was changed from 150 to 135 ft. Analysis of the surface motion during the explosion and the crater resulting from the explosions shown in Fig. 6 confirmed this judgment. Buggy illustrated the importance of having a good geologic understanding of the explosion environment and the need for an ability to consider the effects of geologic variations in the calculational model used to predict crater results.

U.S.S.R. UNDERGROUND NUCLEAR CRATERING EXPERIENCE

Turning now to the Soviet explosions, I will comment in more detail since I am sure they are not as familiar to the reader as the U.S. nuclear cratering explosions. The first Soviet nuclear cratering explosion is called "1003" (Myasnikov *et al.*, 1970). This explosion was fired at an unspecified date at a depth of 48 m in a siltstone environment. The yield was 1.1 kilotons. An artesian water table existed at a depth of about 14–20 m. The crater initially had an apparent

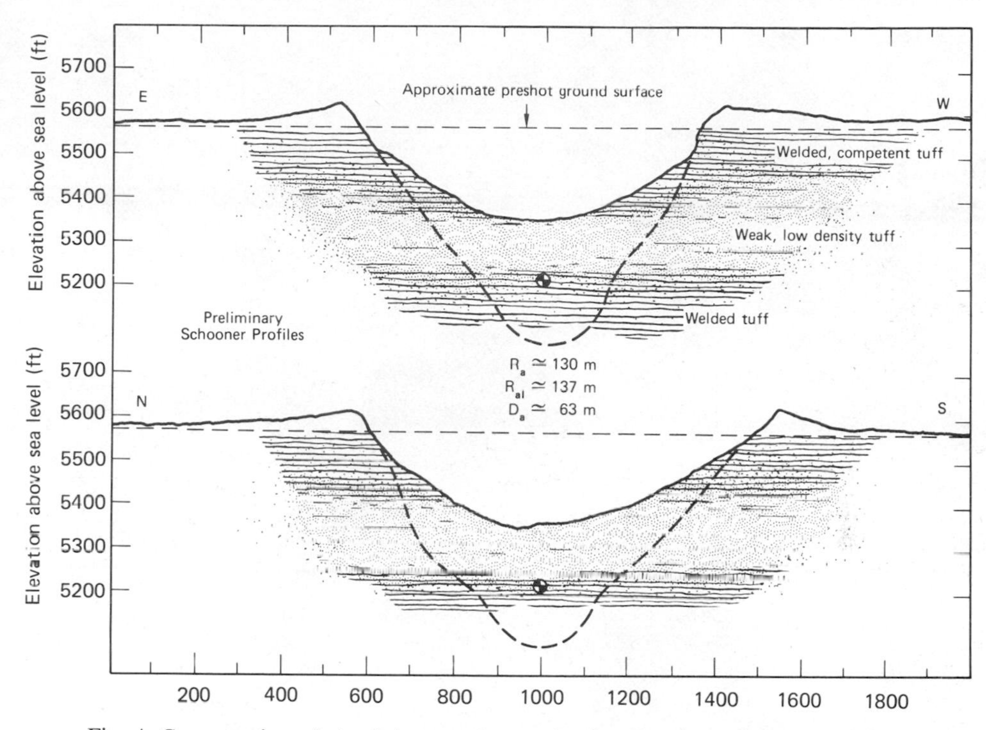

Fig. 4. Cross-section of the Schooner Crater showing the alternating beds of hard, welded tuff and weak, low-density tuff.

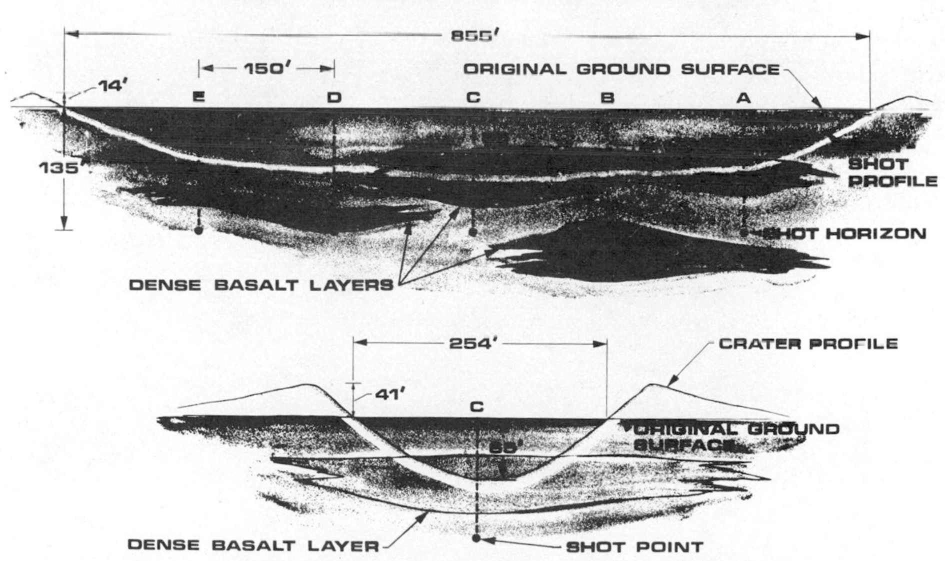

Fig. 5. Cross-section of the Buggy geologic environment showing the location of layers of hard, dense basalt.

Fig. 6. Aerial view of the Buggy Crater in basalt.

depth of 31 m and an apparent radius of 53.5 m. As a result of sluffing of the slopes of the crater under the action of artesian water flowing into the crater, over a two year period the apparent depth decreased to 20 m and the apparent radius increased to 62 m. Figure 7 shows a cross section through the "1003" crater as it existed two years after the detonation. Figure 8 shows a plan view of this same crater.

The second Soviet explosion of interest is the "1004" explosion. (Myasnikov *et al.*, 1970). Although the Soviets have never indicated a precise yield for this explosion, it has been inferred from several references to be about 125 kilotons. The depth-of-burial has similiarly not been specified but has been inferred from the geologic section shown in Fig. 9 to be about 178 m. This figure indicates the "1004" explosion was carried out in a complex geologic environment involving interbedded sandstones, siltstones, shales, and conglomerates. The Soviets have given the moisture content as 12%. Immediately after the shot the "1004" crater had an apparent crater radius of 204 m and an apparent crater depth of 100 m.

The "1004" event was placed immediately adjacent to a river bed in order that the lip of the crater would block the river and a large reservoir would be formed. Figure 10 is a composite photograph showing the "1004" crater filled with $7 \times 10^6\,m^3$ of water and a reservoir backed up behind the lip of the crater containing $10^7\,m^3$ of water. A special channel was dug to permit water from the outer reservoir to fill the crater. Filling of the crater with water resulted in substantial readjustment of some of the slopes of the crater. About $1.7 \times 10^6\,m^3$ of rock slid into the crater, mostly from the south side, reducing the apparent crater depth by 17 m to 83 m and the apparent crater volume by 25% from 6.4×10^6 to $4.7 \times 10^6\,m^3$. The crater is reported to have stabilized in this configuration and continues to stand as a useful water storage reservoir at last report.

Over the last 30 years the Soviet Union has developed a large body of experience with high-explosive row cratering explosions and has used them to construct many canals. To provide experimental data on which to base plans for future canal construction, they have conducted a number of nuclear cratering experiments. The first of these, called T-1, was a single 0.2 kiloton nuclear cratering explosion carried out in a quartzose sandstone at a depth of burst of 31.4 m (Myasnikov *et al.*, 1970). Figure 11 is a topographic map of this crater and shows the apparent crater radius to be about 40 m. The apparent crater depth is 21 m. As will be shown later this crater is considerably larger than expected based upon previous cratering experience with high explosives and "1003". The Soviets have attributed this to the presence of a large amount of gas-forming organic material. The water table was at a depth of about 2 m and the crater filled with water within several days.

The T-2 event was a row charge cratering experiment involving three 0.2 kiloton nuclear explosives spaced about 40 m apart (Myasnikov *et al.*, 1970). The geology has been described to be similar to T-1 except that the water table was at a depth of about 5.6–6.9 m. The depth-of-burst of the three explosions was 31.4 m, the same as for T-1. The spacing and depth-of-burial for

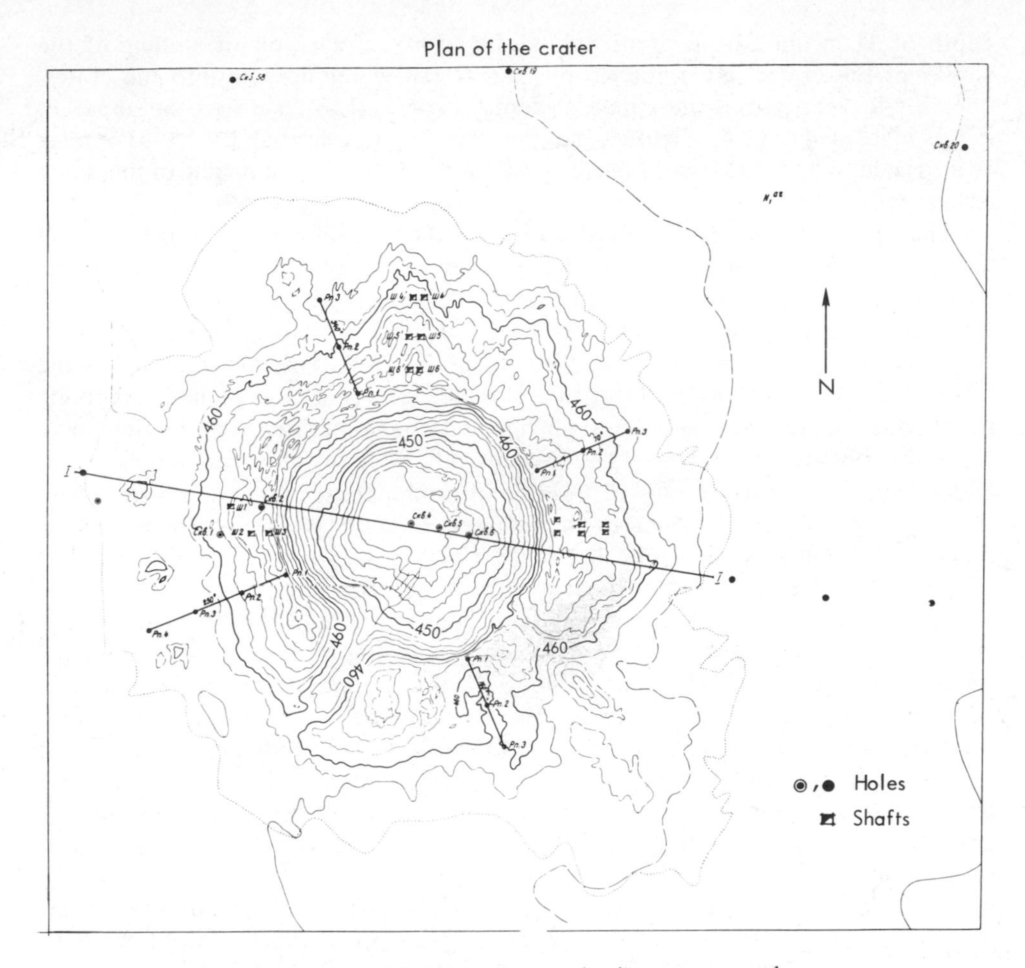

Fig. 7. Plan of Soviet 1003 crater in siltstone-type rock.

T-2 give a ratio of spacing to depth-of-burial of 1.3. Based on the results of T-1, a spacing of 40 m would correspond to a ratio of spacing to single crater radius of 1 to 1.14, depending upon whether one uses a radius for T-1 of 35 or 40 m. According to either of the Soviets' criterions, using the ratio of spacing to depth-of-burial, or U.S. criterion based upon the ratio of spacing to single crater radius, this spacing would be expected to lead to about 10% enhancement of crater radius and no enhancement of crater depth.

Figure 12 shows the results of the T-2 explosion. The width of the T-2 row crater ranged from 61 to 69 m, averaging about 65 m. The depth was about 16 m. Thus, the crater was about 20–30% narrower and about 25% shallower than expected on the basis of the T-1 crater. When judged relative to "1003" however, the T-2 crater has about the expected dimensions.

The last Soviet crater is part of a large project being studied by the Soviet Union for the diversion of water from the north-flowing Pechora River into the

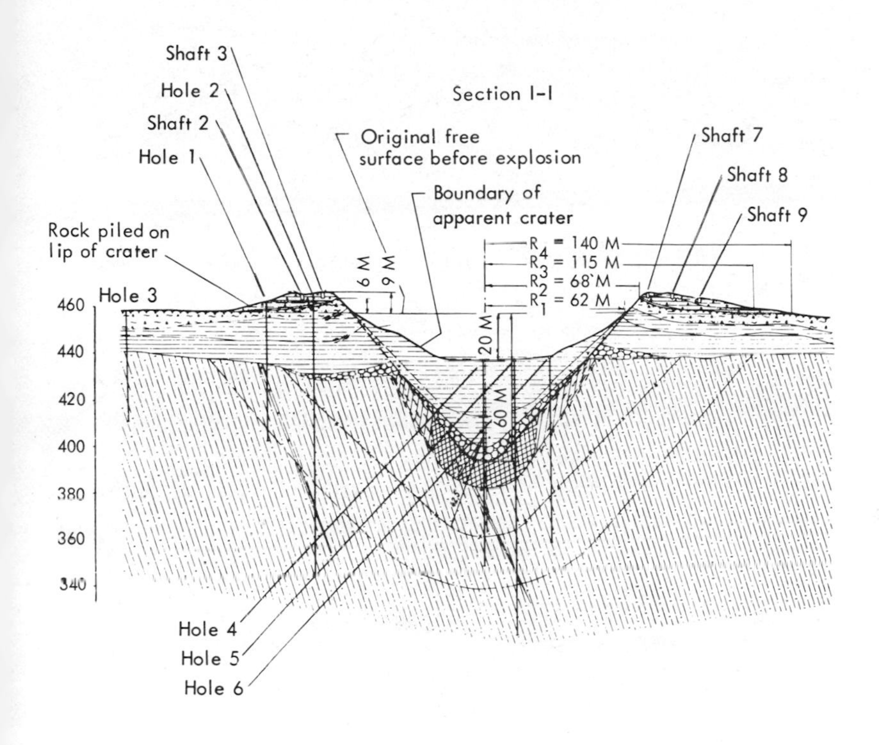

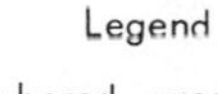

Legend

- Holes bored, preshot
- Holes bored, postshot
- Shafts
- Datum marks to study subsidence
- Boundary of fallback of ejected rock
- Clay
- Soil and plant layer with loam
- Grey-green and brown clay
- Grey-green clay
- Brown clay
- Siltstone
- Diabasic porphyrites
- Clay with chips of siltstone (more than 30%)
- Fractured siltstone
- Crushed zone
- Lower limit of zone of intense fracturing
- Lower limit of zone of block fracturing

Scale
0 20 40 60 80 100 M

Cross section of "1003" project two years after detonation.

Fig. 8. Cross-section of 1003 crater in siltstone-type rock.

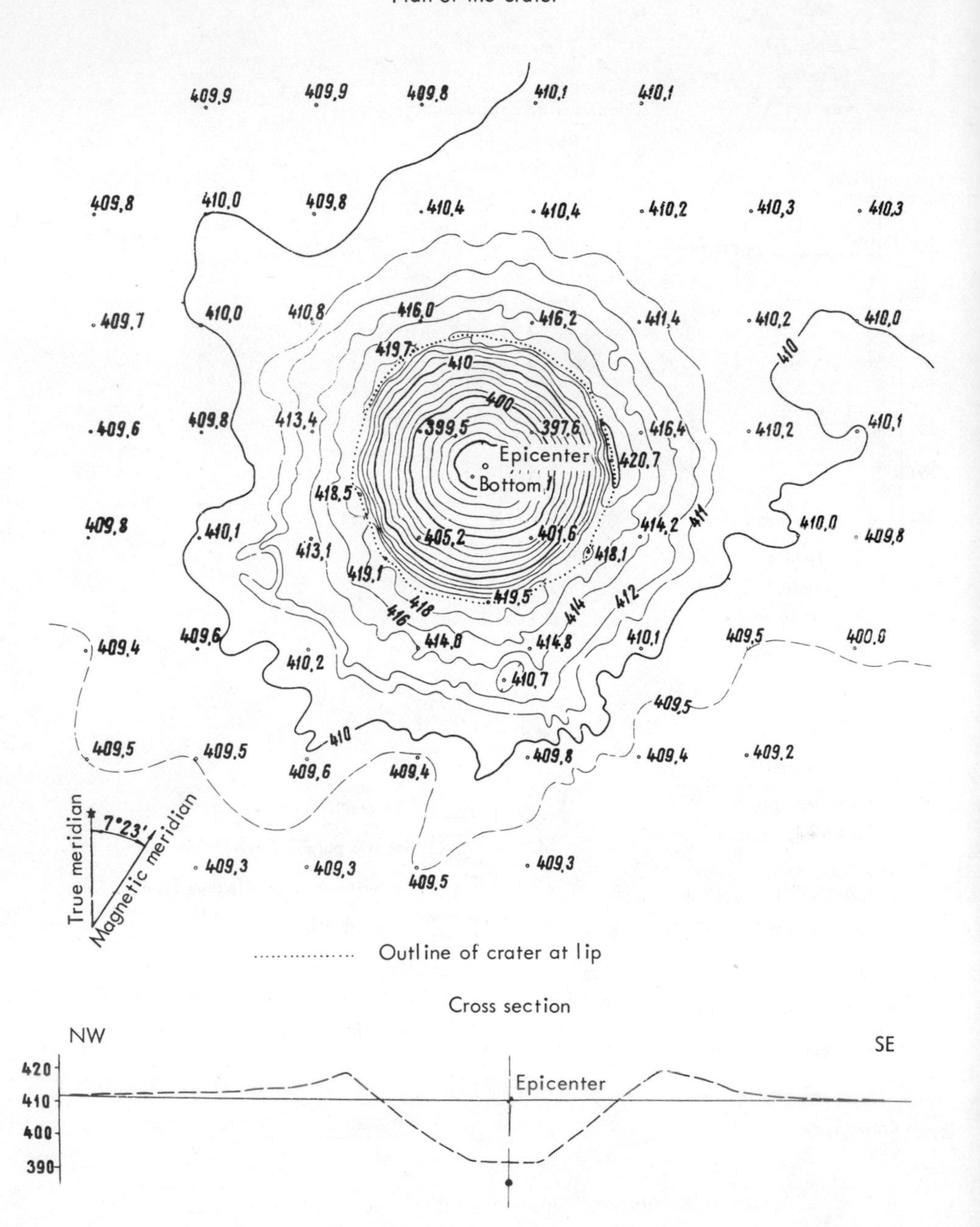

Fig. 9. Cross-section of Soviet 1004 crater in interbedded sandstones, siltstones, shales, and conglomerate.

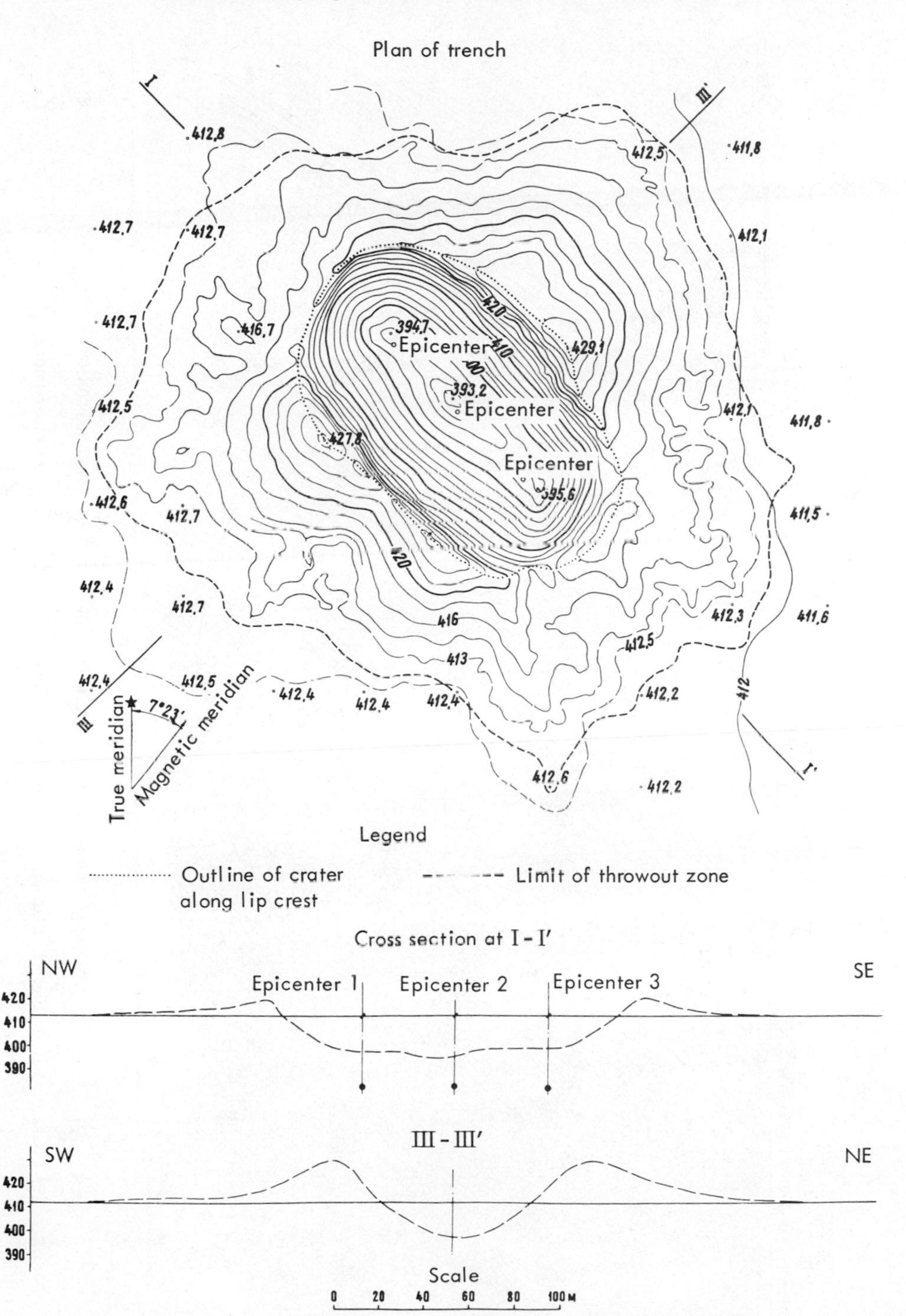

Fig. 10. Aerial view of Soviet 1004 crater showing its use as water reservoir and crater-lip dam.

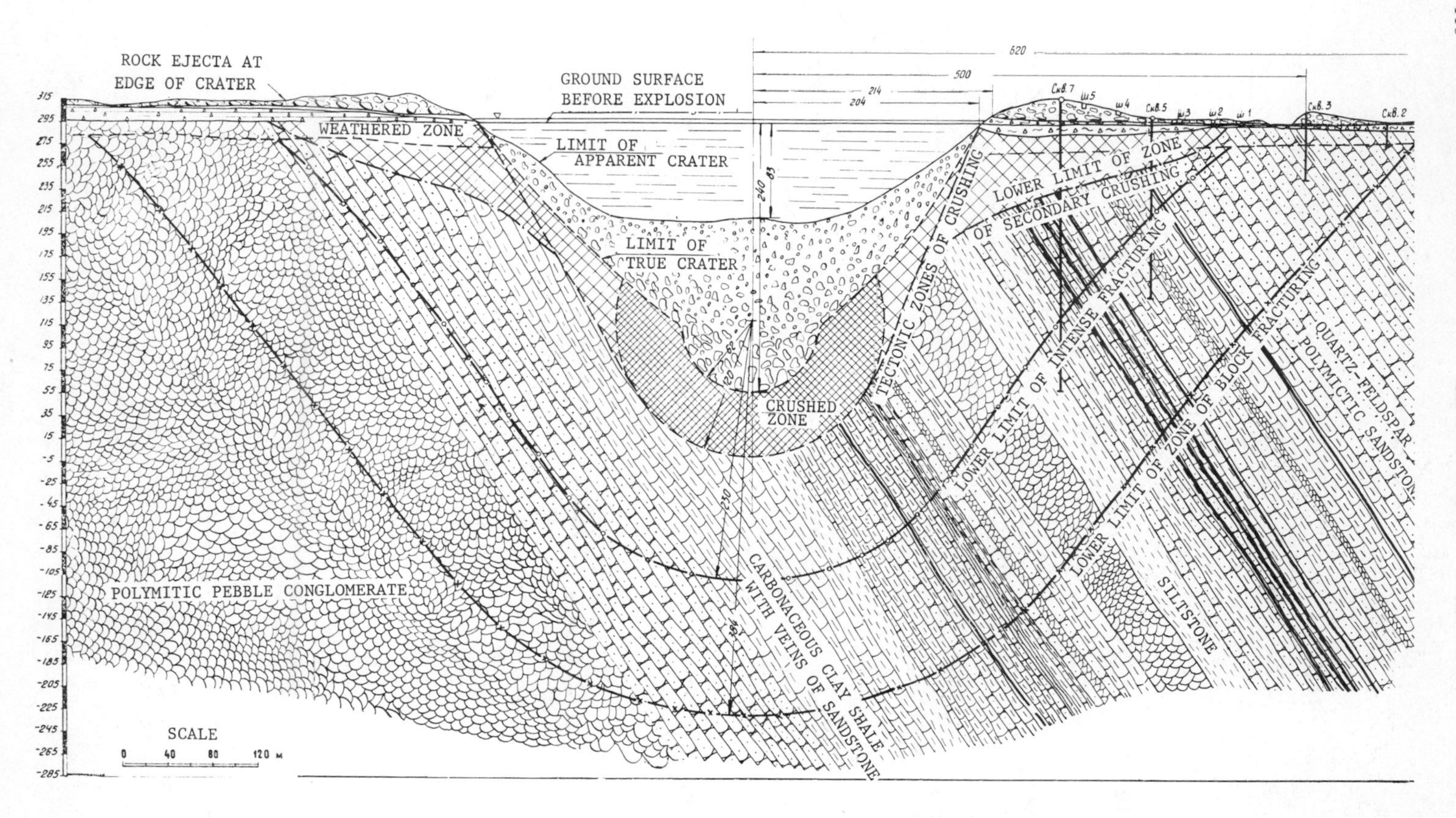

GEOLOGIC CROSS SECTION THROUGH AREA OF HOLE 1004 AFTER EXPLOSION

Fig. 11. Plan and cross-section of Soviet T-1 crater in quartzose sandstone.

Fig. 12. Plan and cross-section of Soviet 3-charge row crater T-1 in quartzose sandstone.

Kama and Volga Rivers and thence into the Caspian Sea through a 112 km long canal (see Fig. 13). The northern 65 km of this canal is being considered for construction by nuclear excavation. The northern half of the section considered for nuclear excavation is described as consisting of sandstone, siltstone, and argillite rock while the southern portion is described as consisting largely of saturated unconsolidated alluvial deposits.

Because of concern over the stability of the slopes of a nuclear canal in the southern portion of the canal a nuclear row charge cratering experiment was conducted near the southern end of the nuclear portion of the canal where the alluvial deposits were 70–135 m thick (Kireev *et al.*, 1975). Figure 14 shows a geologic cross section along the canal alignment in this region. The water table was reported to be at a depth of 5 m at the northern end of this section and 17 m at the southern end with a majority of the loose, cohensionless, alluvial sediments being saturated. No figure for water content has been provided. The underlying rock consists of interbedded sandstones, argillites, and marls.

Three 15 kiloton explosives were emplaced at depths of about 128 m which resulted in two of the explosions being within the underlying rock formation while the third was 5–10 m above the alluvium-rock interface. The scaled depths-of-burial were about 57 $m/kt^{1/3.4}$ which placed them somewhat deeper than optimum depth-of-burial for most materials. However, in view of the high water content and the weak nature of the overlying material such a depth was considered adequate by the Soviets for producing dynamic cratering action. The spacing between the charges from north to south was 163.1 and 167.5 m respectively. This spacing corresponds to a spacing to depth-of-burst ratio of about 1.3 which would be expected to enhance crater radius by about 10% but not to enhance crater depth.

Figure 15 is a schematic diagram of the crater resulting from the Pechora-Kama row charge explosion. The crater was reported to be about 700 m long, 340 m wide and 10–15 m deep. Inspection of Fig. 15 gives an average width of about 300 m. The final pan-shaped configuration of the crater resulted from extensive failure of the crater lips, with subsequent slumping of lip and fallback material into the crater, increasing its width and decreasing its depth. Such behavior in loose, unconsolidated material is quite consistent with expectations although the degree to which slumping would occur is difficult to predict. The Soviets report no further failure of the slopes since the event and they appear to have stabilized at a slope angle of 8–10°. The depth of the crater has been noted to be increasing slightly with time. The final crater is considered by the Soviets to be adequate for use as a portion of the ultimate canal if it is to be constructed by nuclear excavation and shows that it is possible to use nuclear excavation for constructing such a canal in weak, saturated, alluvial material.

To assist with comparison of the U.S. and Soviet Union data, scaled apparent crater radii and depth have been plotted against scaled depth-of-burst in Figs. 16 and 17. These plots show that the scaled apparent radius of the "1003" explosion is significantly larger than the scaled apparent radius of the Sedan explosion. This can presumably be attributed to the saturated water conditions

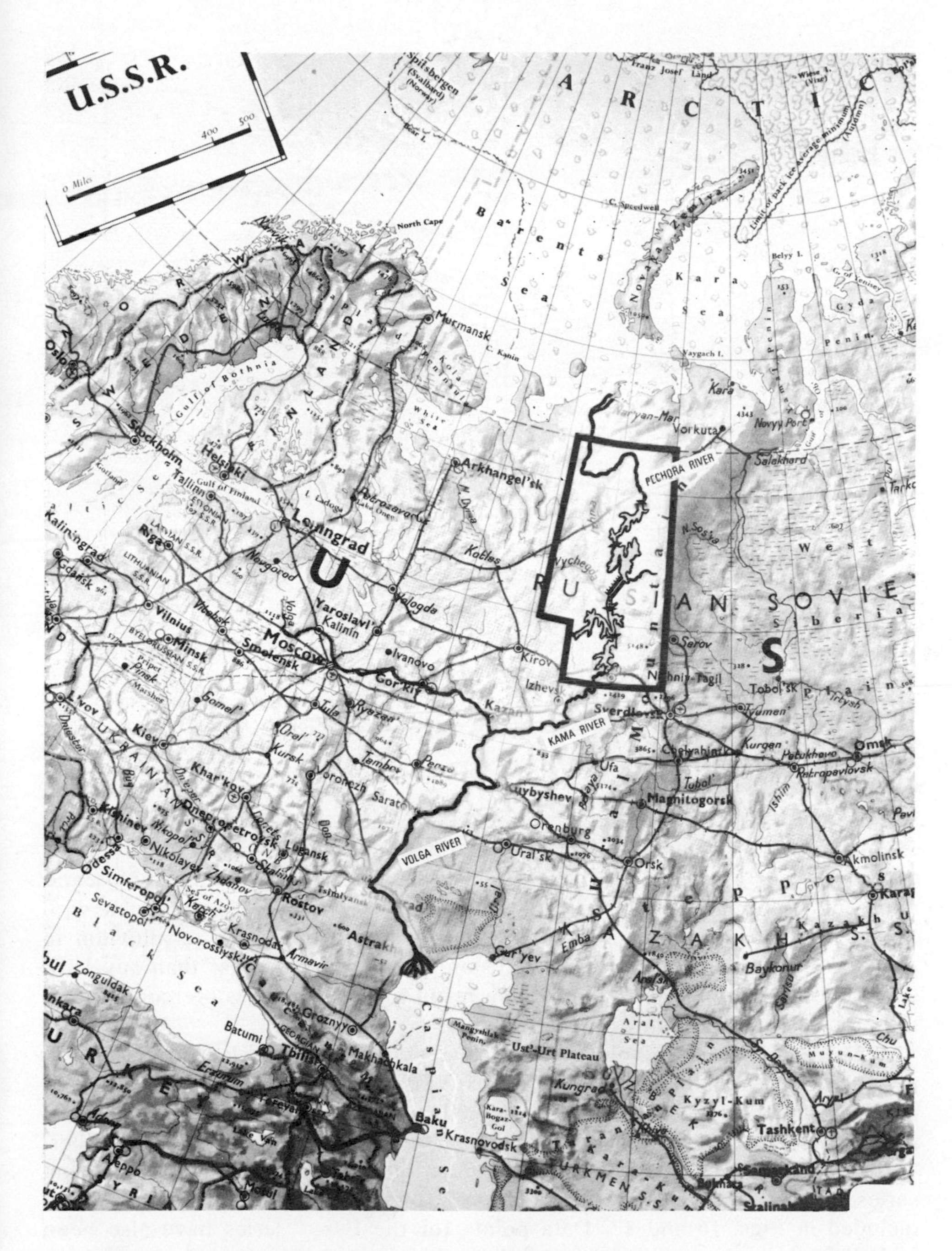

Fig. 13. Map of European-Soviet Union showing location of proposed canal to link Pechora River and Kama/Volga River.

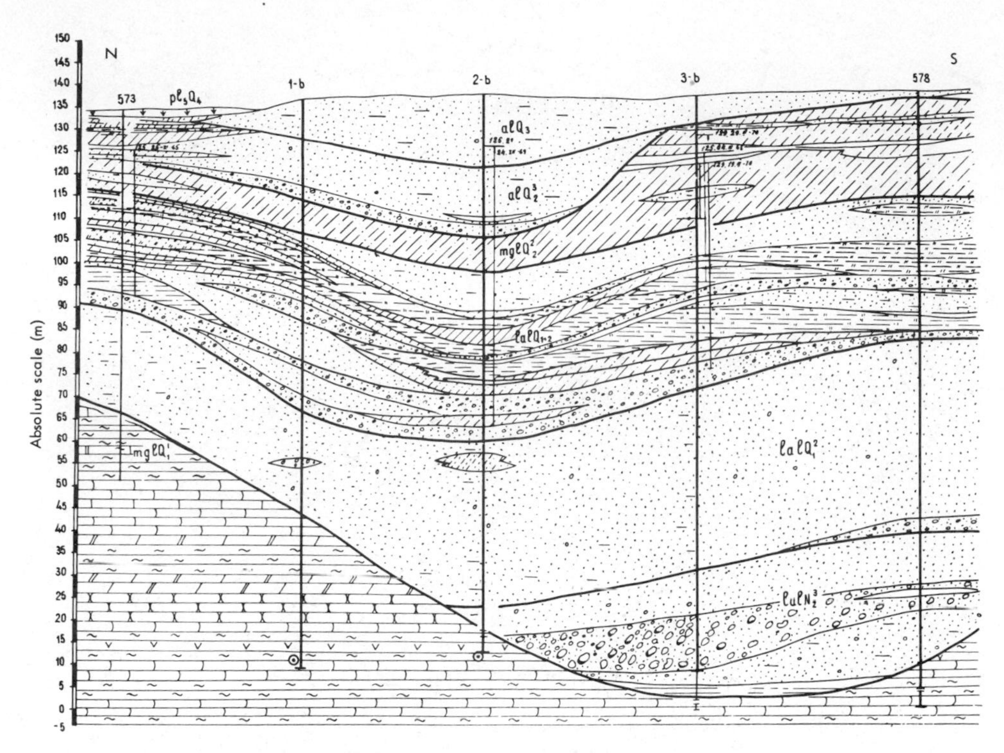

Fig. 14. Cross-section of geologic environment along alignment of proposed canal showing location of three 15 kiloton nuclear explosives used for Pechora-Kama row-charge nuclear cratering experiment (Kireev *et al.*, 1975).

and to the more competent nature of the sandstone overlying the "1003" explosion in comparison with alluvium of the Sedan event. The "1004" crater compares very favorably with the Sedan crater for both apparent radius and apparent depth reflecting the fact that the moisture content was comparable with Sedan and indicating that the medium may have been similar to alluvium in physical characteristics. The T-1 explosion is significantly larger than any U.S. experience in alluvium presumably as a result of the high moisture content and the more competent nature of the medium.

The Pechora-Kama explosion has no analog in U.S. nuclear cratering experience. The medium which would most closely approach the Pechora-Kama medium would appear to be the saturated, unconsolidated sand in which the recent Essex high-explosive cratering experiments were carried out (Miller, 1975) or the Bearpaw shale which was the site of a high-explosive cratering series several years ago. Average curves for the Bearpaw shale have been included in Figs. 16 and 17. Data points for the Essex series have also been included but no attempt was made to draw representative curves. Craters in both of these media were characterized by radii much larger than those experienced in NTS alluvium. The craters with the largest radii for the Essex series were

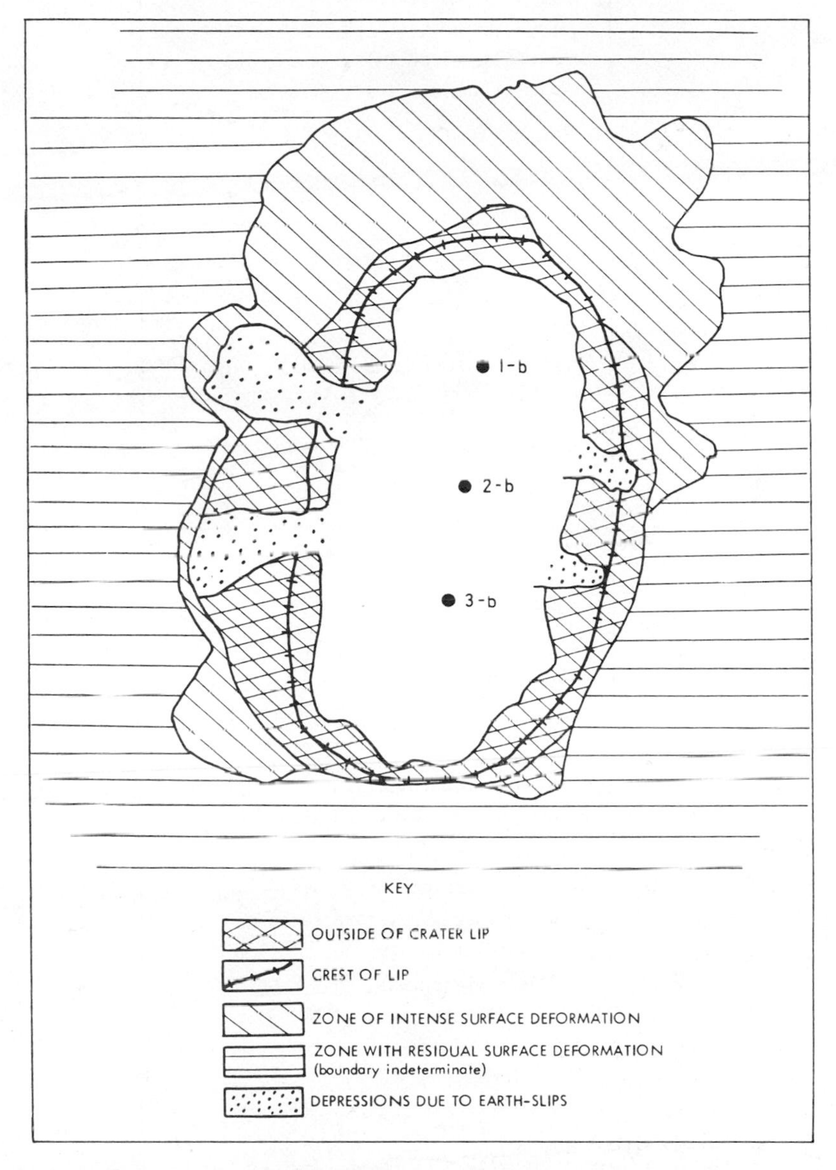

Fig. 15. Plan of crater resulting from Pechora-Kama row-charge nuclear cratering experiment (Kireev *et al.*, 1975).

generally associated with the presence of a very shallow water table which led to extensive early slope failure. Crater depths for the Bearpaw shale tended to be similar to those for high-explosive craters in NTS alluvium, whereas the depths for the Essex craters were all much smaller than NTS alluvium. The crater dimensions for the Pechora-Kama row-charge cratering experiment appear to be roughly consistent with the results of the Essex series although the paucity of

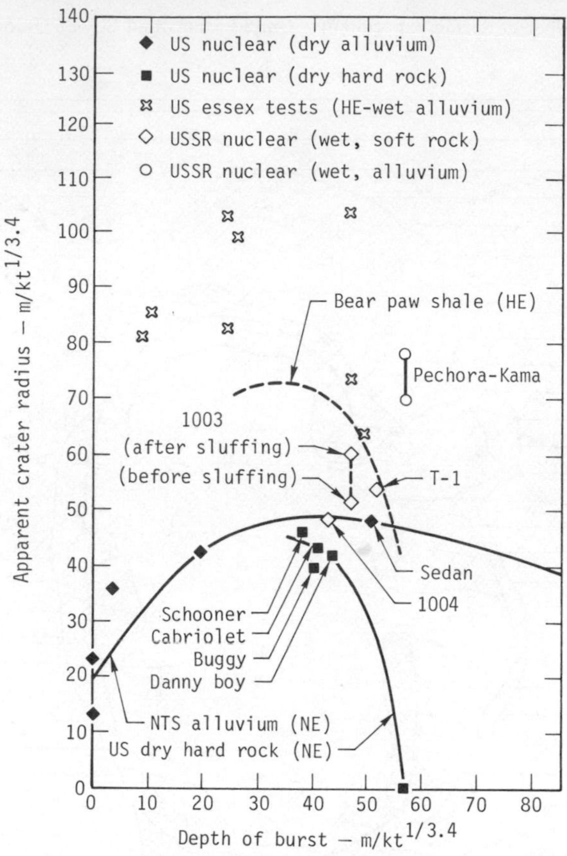

Fig. 16. Plot of U.S. and Soviet Union nuclear cratering data: Apparent scaled crater radius versus scaled depth-of-burst.

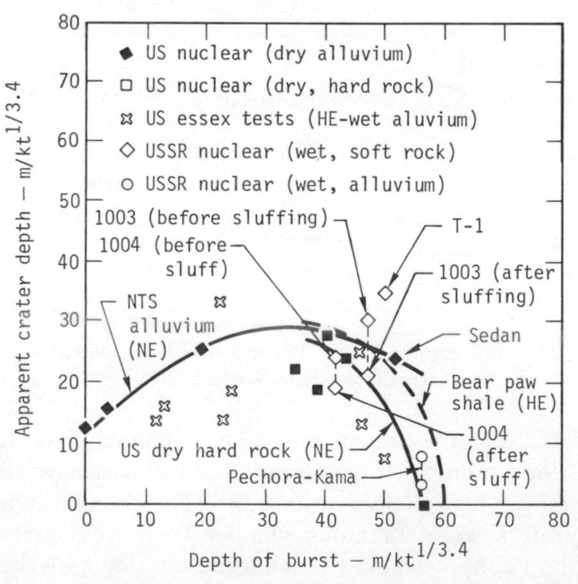

Fig. 17. Plot of U.S. and Soviet Union nuclear cratering data: Apparent scaled crater depth versus scaled depth-of-burst.

the data, the small yields of the Essex series (~10 tons), and the large scatter of the Essex data do not permit definitive conclusions to be drawn.

SUMMARY

Comparing the body of nuclear cratering experience developed by the U.S. and the Soviet Union, we see that the U.S. cratering data are characterized by explosions in the range between about 100 tons to 100 kilotons in relatively dry alluvium or hard rock media. The Soviet experience covers a similar yield range but by contrast has been entirely in saturated or partially saturated media. This has led to larger craters being formed initially, followed by varying degrees of slope failure resulting from the saturated conditions and readjustment of crater slopes. Soviet experience has shown that post-shot slope failure occurs to a limited degree and that stable slopes are produced within a relatively short length of time. Overall, the two bodies of data compare quite favorably, particularly when known geologic differences are considered. The Soviet Pechora-Kama row-charge cratering experiment in loose, unconsolidated saturated sediments is the most difficult to reconcile with U.S. experience, but its geologic environment is very different from any comparable U.S. nuclear cratering explosion.

Acknowledgments—Work performed under the auspices of the U.S. Energy Research and Development Administration under contract No. W-7405-Eng-48.

REFERENCES

Kireev, V. V., Kedrovskiy, O. L., Valentinov, Yu. A., Myasnikov, K. V., Nikiforov, G. A., Prozorov, L. B., and Potapov, V. K.: 1975, *Gruppovoi eskavatsionnyi yadernyi vzryv b allyuvial'nykh porodakh*, IAEA Technical Committee Meeting on Peaceful Uses of Nuclear Explosions. Vienna. IAEA TC 1-4/14, 418–420; see also AWRE-Trans-69, 43–69.

Myasnikov, K. V., Prozorov, L. B., and Sitnikov, I. E.: *Mechanical effects of single and multiple underground nuclear cratering explosions and the properties of the excavation dug by them, in Nuclear Explosions for Peaceful Purposes*, I. D. Morokhov, (ed.,) (Atomizdat, Moscow) Lawrence Livermore Laboratory, UCRL-Trans-10517, 70–109.

Miller, A. E.: 1975, *Preliminary combined results—Essex-I, Phases 1 and 2, US Army Explosive* Excavation Laboratory, DNA PR 0016, p. 000–639.

Nordyke, M. D.: 1961, An analysis of cratering data from desert alluvium, *J. Geophys. Res.* **67**, 1965–74.

Nordyke, M. D.: 1964, Cratering Experience with Chemical and Nuclear Explosives, *Proc. Third Plowshare Symposium*, 51–74.
Nordyke, M. D. and Wray, W. R.: 1964, A nuclear cratering detonation in Basalt, *J. Geophys. Res.* **69**, 675–690.
Nordyke, M. D. and Williamson, M. M.: 1965, *The Sedan Event*, Lawrence Livermore Laboratory, PNE-242F, p. 000–103.
Shelton, A. V., Goeckermann, R. H., and Nordyke, M. D.: 1960, *The Neptune Event: A Nuclear Explosive Cratering Experiment*, Lawrence Livermore Laboratory, UCRL-5766, p. 00–32.
Tewes, H. A.: 1968, *Results of the Cabriolet Excavation Experiment*, Lawrence Livermore Laboratory, UCRL-71196, p. 00–76.
Tewes, H. A.: 1970, Results of the Schooner excavation experiment, *Proc. Symposium Eng. with Nuclear Explosive, Conf-700101* **1** 306–333.
Toman, J.: 1968, *Project Buggy—A Nuclear Row Excavation Experiment*, Lawrence Livermore Laboratory, UCRL-71280, p. 0–6.
Videon, F. F.: 1965, *Project Sulky-Crater Measurement*, US Army Nuclear Cratering Group, PNE-601F, p. 00–66.
Videon, F. F.: 1966, *Project-Palanquin—Studies of the Apparent Crater*, US Army Nuclear Cratering Group, PNE-904, p. 00–34.

Roddy, D. J., Pepin, R. O., and Merrill, R. B., editors.
(1977) *Impact and Explosion Cratering*, Pergamon Press (New York), p. 125–162.
Printed in the United States of America

Tabular comparisons of the Flynn Creek impact crater, United States, Steinheim impact crater, Germany and Snowball explosion crater, Canada

DAVID J. RODDY

U.S. Geological Survey, 2255 North Gemini Drive, Flagstaff, Arizona 86001

Abstract—A tabular set of comparative data for 340 basic parameters and related aspects are listed for three terrestrial craters which exhibit broad, flat-floors with central uplifts. The table includes selected dimensional parameters, morphologic styles, types of structural deformation, and related features for the Flynn Creek and Steinheim impact craters and the Snowball explosion crater. The listings provide a set of parametrics derived from detailed field and laboratory studies for use in certain crater comparisons and calculations, such as dimensional and energy scaling, volume determinations, central uplifts, and ejecta masses. Dimensional parameters include diameters, depths, heights, radii, thicknesses, and volumes, each determined or estimated for rim crest, apparent, and true crater conditions. Maximum, minimum, and average values are listed for most parameters and were derived from sets of geologic cross sections constructed for each of these craters and from field and laboratory data. Dimensional parametric ratios include a variety of combinations currently used in explosion crater scaling studies, such as cratering efficiency as well as simple diameter/depth ratios. Impact kinetic energies of formation are calculated using diameter-energy, volume-energy, and equivalent length factor scaling. These calculations, using a 1/3.4 root scaling, give impact energies ranging from 1.0 to 4.7×10^{24} ergs as conservative estimates for Flynn Creek and Steinheim. The tabular listings also provide a set of data in comparative outline format that allows direct analog comparisons between the three craters with respect to morphologic and deformational styles.

The Flynn Creek Crater in the United States and the Steinheim Crater in Germany were chosen because they both exhibit the type morphology of a major class of natural hypervelocity impact craters found on each of the terrestrial planets and the moon. This large class, represented by broad flat-floored craters with central peaks, exist in a size range from a few kilometers up to hundreds of kilometers in diameter. A similar morphological class also exists with certain large-scale experimental explosion craters, as typified by the Snowball 500-ton TNT surface trial. The analog comparisons listed in tabular form demonstrate that a strong similarity exists between the two impact craters, Flynn Creek and Steinheim, and the explosion crater, Snowball.

The extent and types of morphological and structural similarities between these three craters should provide a more convincing basis for both predicting and interpreting the initial morphology and structure of certain eroded terrestrial impact sites, as well as predicting and interpreting the general structure of impact craters on the other planets and moon which are in the same morphological class and general size range. Of equal interest is the potential use of natural hypervelocity impact crater morphologies and their structural deformation to predict certain large explosion crater parameters under comparably modeled cratering conditions.

INTRODUCTION

HYPERVELOCITY IMPACT CRATERS are known now to exist in a variety of morphological and structural types on each of the terrestrial planets and their satellites. It also has been recognized that during the formation of these impact craters a broad range of dynamic physical processes were invoked which have profoundly affected the evolution of the planets and their crusts. Efforts to place adequate *quantitative bounds* on these *processes* and their *effects*, however, have been previously hindered by our limited knowledge of impact cratering mechanics, the different types of structural deformation, and the initial crater morphologies and dimensions. Fortunately, our understanding of these areas is improving with a rapidly expanding data base involving laboratory impact and explosion experiments, missile impacts, large-scale experimental explosion craters, natural impact craters, and theoretical calculational studies. A prominent requirement in any of these studies, whether they be applied to planetary or terrestrial problems, however, is the need for accurate, detailed data on the basic dimensional parameters of the initial craters and their structural deformation, cast in comparable formats. This paper addresses a limited part of this problem by presenting a tabular outline of comparative data for 340 basic dimensional, morphological, and structural parameters and related aspects for three craters of the flat-floored, central uplift type, two of which are natural terrestrial impact craters and one is a large-scale experimental explosion crater. A second intent of this paper is to list selected values that are necessary in certain numerical calculations and which might also be useful as a guide in collecting basic parametric data from other craters.

BACKGROUND

The need for dimensional data on impact craters has been increasingly manifested in recent Moon, Mars, Mercury, and Venus studies and in comparative planetology studies devoted to basic characterizations of crater classes, morphologies, structural types, ejecta distribution, and post-crater processes (Arthur, 1974; Arthur *et al.*, 1963, 1964, 1965, 1966; Baldwin, 1963; Burt *et al.*, 1976; Cintala *et al.*, 1975, 1976a, 1976b; Leighton *et al.*, 1967; Pike, 1968, 1972, 1973, 1974a, 1974b, 1976, 1977; Ronca and Salisbury, 1966; Schubbert *et al.*, 1976; Wetherill, 1975a, 1975b; and others). An example of the higher quality of planetary crater data now available is represented in Pike's (1976) newest listing of six basic dimensional parameters for over 500 lunar and terrestrial craters, a study which involves an extremely useful state-of-the-art compilation in terms of the new Apollo Lunar Topographic Orthophotomaps. The six parameters listed by Pike include values for crater diameters, depths, rim heights, flank widths, circularities, and floor diameters. These basic crater parameters, with some variations, have proven in the past to be both necessary and sufficient for most characterization schemes associated with planetary cratering and the related crustal effects. Indeed, much of our present understanding of planetary surfaces is related, in one way or another, to photogeologic data interpretations and associated calculations of cratering processes and their effects (Baldwin, 1949, 1963; Dence, 1973; French and Short, 1968; Gault *et al.*, 1975; Head 1975, 1976; Head *et al.*, 1976; Hörz, 1971; McCauley *et al.*, 1977; McGetchin *et al.*, 1973; Moore *et al.*, 1974; Oberbeck, 1977; Öpik, 1976;

Quaide *et al.*, 1965; Shoemaker, 1963; Wetherill, 1975a; Wilhelms and McCauley, 1971; and others including the many papers in the 1st through 8th Proceedings of the Lunar Science Conference). In most of these studies there have been efforts to quantify craters in some form, and as one might expect, this recognition of the wide-ranging effects of global impact cratering throughout a planet's history has come to place increasingly serious demands on the existing crater data base. Consequently, the need for information required now for studies of, (a) cratering mechanics, (b) energy sources and coupling, (c) melt and vapor formation, (d) ejecta distribution, (e) ray formation, (f) material responses, (g) modeling, (h) scaling, (i) parametric studies, and (j) computer numerical simulation, to mention a few, has expanded well beyond earlier requirements. In light of such problems, improved and upgraded tabular data sets, such as Pike's, became of critical importance, particularly if basic cratering processes are to be examined in detail. With such data sets, however, it is increasingly clear that the addition, where possible, of certain other dimensional and structural parameters would also prove to be of value. For example, it would be extremely useful in scaling to have a single quantity, or set of quantities, that has been *proven* to be functionally related to the energy of formation of the crater. Such a quantity might be what is called in the explosion literature, a *true diameter.* This is a value measured along the pre-shot ground level to the intersection with the regional post-shot ground level projected through the crater walls. This is a useful diameter because it defines at least one of the surfaces actually involved in the target rock comminution. Other critical dimensions might include the apparent and true volumes, both of which are directly related to crater energy of formation. Another approach, as described in this paper, might be the derivation of an *equivalent length factor* which can be related in some simple scaling fashion to the energy of formation.

Such dimensional information as described above, of course, cannot yet be conveniently extracted from the types of maps and other data available for the terrestrial planets. Even terrestrial erosion has degraded most of our impact craters to such an extent that initial morphologies are largely destroyed. However, a limited number of terrestrial impact craters still exist sufficiently intact to provide an improved data set for detailed tabulation, at least for certain critical dimensional and structural parameters which appear to be functionally related to the impact energy of formation. This is also true for a significant number of large-scale experimental explosion craters which have been shown to have direct morphological and structural analogs with certain impact craters (Roddy, 1968, 1976, 1977a). Consequently, a set of selected dimensions and physical parameters are listed in this paper for two, well-preserved, terrestrial impact craters and one large explosion crater of the same morphological and structural class. The dimensional data presented here, for the most part, are not available in the published literature and the structural data have not been assembled elsewhere in comparative formats.

The choice of the particular craters for this tabulation was dictated by three conditions. First, these three craters are part of a general class, in terms of their morphology and structural deformation, that is represented on each of the terrestrial planets including the Moon. The importance of this class and the widespread distribution of flat-floored craters with central uplifts has been discussed by a number of workers including, Dence (1972), Gault *et al.* (1975), Milton and Roddy (1972), Shoemaker (1962), Reiff (1977), Roddy (1968, 1976, 1977a), and others. A second reason for choosing these three craters is that a limited published and large unpublished data base already exists for each crater and the basic parametrics were readily available, particularly in the form of detailed field and laboratory data we have collected for this study. A third reason involves both their state of preservation and the simple initial structure of their target rocks. As noted earlier regarding this last point, the Earth exhibits a number of impact *structures* that fall in the class of flat-floored craters with central uplifts. The major problem is that the majority of these features have been eroded to such a degree that only the deeper structural deformation is now exposed, and the cratered regions no longer exist. Consequently, there are only a very small number of these terrestrial sites still preserved as *craters* in *flat-lying rocks* at a size small enough for complete field studies, each necessary requirements for detailed dimensional and structural studies. Both the Flynn Creek and Steinheim impact craters meet these field requirements reasonably well and exhibit essentially identical dimensions, morphologies, and structures. Fortunately, due to limited initial erosion, both of these craters also still exhibit very nearly the initial crater morphologies, except for their ejecta blankets, identified with the broader planetary class of craters which have flat-floors with central peaks. Moreover, both Flynn Creek and Steinheim have most of the major deformational styles found in other more highly-eroded, terrestrial impact sites considered to have once been in the same morphological class, even though they may be considerably larger in size.

The third crater listed in the tables, formed by a large-scale explosion experiment, also exhibits both a morphology and structure essentially identical to Flynn Creek and Steinheim. Indeed, there is a class of such craters in large-scale experimental explosion cratering that have flat floors and central uplifts (Roddy, 1968, 1976, 1977a). The explosion results are tabulated here with Flynn Creek and Steinheim to provide a comparative data base from a large-scale explosion experiment in which the energy source functions, material models, morphology, structure, ejecta distribution, and detailed physical dimensions are well known. The value of such comparisons is to establish a more convincing tie between large-scale explosion experimental cratering processes and certain very large-scale impact cratering processes. Hopefully, both research communities will find such analog approaches applicable to their individual problems where direct experimentation is not tractable. The broader value of this tabular data set, however, is to provide a well documented data base for flat-floored craters with central uplifts for explosion, impact and comparative planetary impact studies which involve basic cratering research on such aspects as, (a) cratering mechanics, (b) energy coupling, (c) melt production, (d) ejecta distribution, (e) modeling, (f) scaling, and other problems.

Previous Work

The individual geologic and cratering studies for the Flynn Creek, Steinheim, and Snowball craters have been published in a number of earlier papers and are only briefly outlined here. The field data from the Steinheim Basin was collected as part of a joint study now in progress on the comparative morphology and structural deformation with Dr. W. Reiff, Geologisches Landesmt, West Germany.

Briefly summarized, the Flynn Creek Crater was formed by a hypervelocity impact event approximately 360 m.y. ago in what is now north central Tennessee. The crater shown in Fig. 1, initially over 3.8 km across and approximately 200 m deep, appears to have been formed on a rolling coastal plain, perhaps inundated by shallow waters. The target strata, consisting of flat-lying limestones and dolomites, are over 400 m.y. old and were completely lithified at the time of impact. The impacting body, interpreted by Roddy (1968, 1976, 1977b) to be a low density mass such as a carbonaceous chondrite or a cometary mass, delivered on the order of 10^{24} ergs of kinetic energy to completely brecciate over 1 km^3 of rock to a depth of nearly 200 m and to eject over 0.8 km^3.

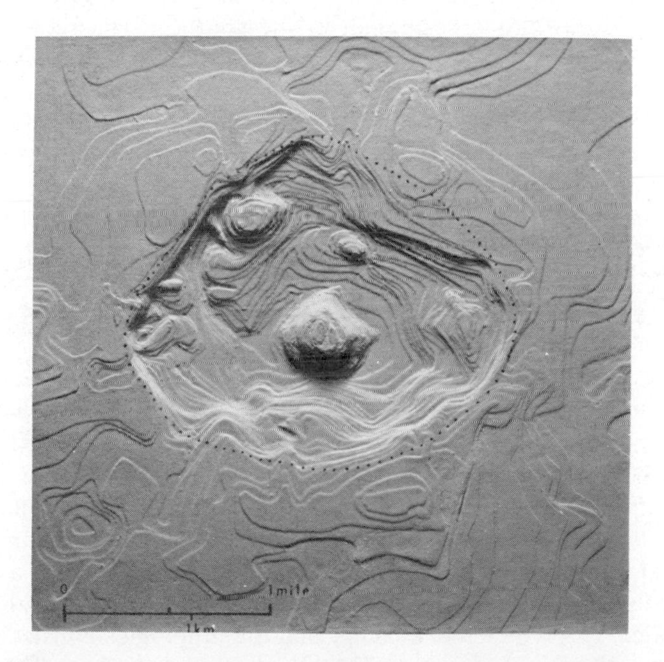

Fig. 1. Oblique view of a model of the Flynn Creek Crater constructed from the structure contour map. This represents the crater immediately before deposition of the Chattanooga Shale about 360 m.y. ago. The large hills near the crater walls are underlain by megabreccia blocks separated from the walls, and a large central uplift dominates the middle of the crater floor. The dotted line indicates the approximate tops of the crater walls as erosion has modified them. Large volumes of ejecta washed back into the crater have also modified the original outline. No vertical exaggeration and lighting is from the north.

The major morphologic and structural elements include a broad, flat-floor underlain by a breccia lens and a major uplift forming a large central hill. The breccia lens, averaging approximately 30–40 m in thickness, consists of totally fragmented limestone and dolomite. The lithologies in the upper part of the breccia lens are totally mixed and were derived mainly from the upper 150 m of strata. The lower part of the lens consists of increasingly larger blocks with a marked decrease in mixing near the base of the lens. Six drill cores on the crater floor show that limestone and dolomite beds immediately below the breccia lens are highly faulted, folded, and locally brecciated, but that deformation decreases downward with the rocks nearly flat-lying and relatively undisturbed about 100 m beneath the breccia lens.

In the middle of the crater, a sequence of steeply-dipping, folded, faulted, and locally brecciated limestone and dolomite are structurally deformed into a massive central uplift. Large subcrater floor movements towards the center have occurred with the deepest rocks raised as much as 450 m at the top of the uplift, and shatter cones in rocks uplifted over 350 m.

In the crater rim, initially flat-lying limestone and dolomite have been faulted, folded and locally brecciated, with parts of the rim uplifted and other parts locally down-dropped. A large graben in the southern rim retains part of the original ejecta blanket. Erosion began to modify the crater after its formation, washing part of the ejecta back into the crater. The surrounding hilly terrain was lowered less than 30 m maximum, with an estimated average over the entire area of 10 m, before the crater and rim were essentially immediately buried by black shale sediments. More recently, regional uplift and erosion have produced excellent exposures of the crater walls, floor, and central uplift. The geology of the Flynn Creek Crater is discussed in more detail in Roddy (1963, 1964a, 1964b, 1966, 1968, 1976), and a brief review of the morphology is given in Roddy (1977a) and of the cratering processes in Roddy (1977b). A treatment of central uplift mechanics are given in Ullrich *et al.* (1977).

The Steinheim Crater was formed by an impact event approximately 14.7 m.y. ago in what is now southwestern Germany. The crater shown in Fig. 2, initially over 4.2 km across and approximately 250 m deep, exhibits a morphology and structural deformation essentially identical to that at Flynn Creek. The target strata, over 150 m.y. old, consisted of well-lithified flat-lying limestone, dolomite, claystone, marlstone, and sandstone. The major morphological and structural elements include a broad, flat-floor underlain by a breccia lens with a major uplift forming a large central hill. The characteristics of the breccia lens are identical to Flynn Creek, except that the lithologies consist of limestone, marlstone, claystone, and sandstone. The thickness of the breccia lens averages approximately 30–40 m, as at Flynn Creek.

The central uplift in the middle of the crater has strata uplifted over 380 m and exhibits an abundance of shatter cones. Reiff (1977) has noted that drill cores in the breccia lens also have shatter cones. The writer has also identified shatter cones in fragments which were once in the ejecta blanket beyond the rim crest but have now washed back into the crater.

Fig. 2. Oblique aerial view of the Steinheim Crater in southern Germany. Note central uplift in center of crater with the town of Steinheim on the left side. The eroded crater walls surround farmed areas of rectangular fields on the crater floor. View is looking northeast. Photograph courtesy of A. Brugger, Stuttgart, December, 1968, Ltd. of Regierungspräsidium Nord-Wuerttemberg.

The strata in the crater rim, initially flat-lying, are now highly faulted, folded, and locally brecciated. A small area on the rim exhibits what may be a small remnant of the original ejecta blanket, however, we have not completed our study of the area at this writing.

Erosion began to modify the crater after its formation, washing part of the ejecta blanket back into the crater, lowering the local rim surface less than a few meters. A Tertiary-age lake formed in the crater, nearly filling it with lacustrine deposits. Erosion has subsequently breached the crater rim and removed much of the lake beds, leaving the crater as we see it today. The geology of the Steinheim Crater is discussed in more detail in Branco and Fraas (1905), Bucher (1933), Groschopf and Reiff (1966, 1967, 1969, 1970, 1971), Kranz *et al.* (1924), and Reiff (1974, 1976, 1977).

The Snowball Crater was formed by the detonation of a 500-ton TNT hemisphere on flat-lying, unconsolidated alluvium at the Defence Research Establishment Suffield, Alberta, Canada. A water table was present at about 7 m depth. The crater shown in Fig. 3, over 100 m across and approximately 9 m deep, had a broad, flat floor with a well-defined central uplift. A thin breccia lens,

Fig. 3. Oblique aerial view of the Snowball Crater formed by the detonation of a 500-ton TNT hemisphere lying on alluvium. The view is one day after formation and shows a complex central uplift extending above the lake that partly filled the crater. Note terraced walls and hummocky ejecta blanket with large concentric fracture zones. Photograph courtesy of Dr. G. H. S. Jones, Defence Research Board, Canada, and the Defence Research Establishment Suffield, Alberta, Canada.

about 1 m thick, consisting of alluvium and shock-compressed alluvium covered the floor and parts of the uplift.

The central uplift consisted mainly of a domical structure of raised clay beds which were faulted and folded into an irregular set of peaks. Initial central uplift at the level of the crater floor was over 10 m, with later subsidence reducing this to approximately 8.5 m.

Structural deformation in the rim was similar to Flynn Creek and Steinheim, with faulting, folding, and local brecciation of the clays. The rim beds however, were more consistently down-folded around the entire rim. A large, well-defined ejecta blanket with inverted stratigraphy surrounded the crater.

Within minutes after the detonation of the charge, ground water began to flow into the crater from fractures opened in the floor and rim with over a meter thickness of sands piped to the surface and deposited over the floor. This highly instrumented explosion experiment was later completely excavated by trenching for detailed subsurface studies. The Snowball Crater is discussed in more detail in Diehl and Jones (1967a, 1967b), Jones (1964, 1976, 1977), Roddy (1968, 1976, 1977a), Rooke and Chew (1965), and Rooke *et al.* (1968).

Tabular Comparisons of the Flynn Creek, Steinheim, and Snowball Craters

The physical dimensions, morphological styles, types of structural deformation and related physical features of the Flynn Creek, Steinheim, and Snowball Craters are listed in tabular form in Table 1. The cratering terminology used in this paper is illustrated in Fig. 1 in Roddy (1977a) and Fig. 4 shows a representation of selected dimensional parameters listed in Table 1. It is recognized that the large number of independent entries can prove difficult for readers to extract specific information, consequently, the tabulated data have been grouped into 12 different sections to facilitate quick review. These sections include: (A) Locations, dimensions, ratios, (B) Energies of formation, (C) Pre-impact conditions, (D) Topography, (E) Structural deformation, (F) Ejecta, (G) Shock metamorphism, (H) Field studies, (I) Geochemistry, (J) Geophysics, (K) Drilling, and (L) Other studies.

Section A, Location, Dimensions, and Ratios, contain the bulk of the parametric data and is organized with the basic dimensions, such as diameters and depths, presented first followed by heights, thicknesses, radii, volumes, and masses. The section concludes with a set of parametric ratios orientated primarily towards scaling and comparative dimensional studies. Entries are also included for the central uplifts, breccia lenses, and ejecta.

A large number of symbols were adopted of necessity to simplify the construction of Table 1 and permit internal comparisons, however, the writer does not recommend the extension of this rather arbitrary symbolization beyond the immediate use of this data set. The explanations at the beginning of Table 1 identify the parameters and associated symbols, and describe the quantitative interpretations assigned to the values listed.

The accuracy of the dimensional values are basically dependent on the quality and scale of the topographic base maps available, types of field surveys, amount of geophysical subsurface and drill data, and on the extent and quality of critical outcrops. Fortunately, both impact structures had maps especially made for each crater. A high-precision topographic map covering 256 km^2 was compiled by J. Alderman of the U.S. Geological Survey for Flynn Creek at a scale of 1:6000 with a contour interval of 3.05 m. A topographic map was also made for the Steinheim Basin by the local government at a scale of 1:8000 with a contour interval of 5 m, and an excellent outcrop map was made by Kranz (1924).

The amount and quality of rock exposures at both the Flynn Creek and Steinheim craters, in terms of critical localities for dimensional studies, are generally adequate for both craters. Several thousand specific outcrop locations were surveyed at Flynn Creek on high-resolution aerial photographs and transferred to the base map using a Kelsh topographic plotter. Field surveys at Steinheim were completed on high-resolution aerial photography and the base map directly in the field. The surface mapping, field surveys, Kelsh plotting, and the subsurface drill data were each used to construct the different geologic cross sections shown in Roddy (1977a, this volume) and Reiff (1977, this volume). The

Snowball studies involve both surface studies, photogeologic mapping, and trench excavations and wall mapping. Field survey measurements were completed by continuous survey crew operations with the geologic cross sections shown in Roddy (1977a, this volume) and Jones (1977, this volume) based upon these surveys.

The dimensional data for Flynn Creek draws heavily upon the four detailed cross sections, accompanying field data, and Kelsh topographic information plotted on the geologic map, whereas the Steinheim tabulated data is based largely on two detailed cross sections and numerous field measurements. The Flynn Creek data were corrected for regional dip of the strata, and special field surveys were completed to obtain accurate values of the crater wall slopes. The equivalent crater wall studies for Steinheim, which are still in progress, have defined the approximate boundaries of the crater; however, the slope values are presently less certain than at Flynn Creek. This uncertainty would modify the rim crest positions less than 300 m as interpreted from the graphic construction on the cross section.

Several other points might be noted regarding the use of Table 1. The listing of dimensional values in this paper are associated with three basic sets of crater profiles, i.e., values measured from the initial *rim crests*, *apparent* values, and *true* values (Fig. 4). The use of *apparent* and *true* values is derived from explosion cratering studies and has proven most useful in crater dimensional characterization and parametric tabulations. These configurations generate three different data sets for many of the dimensional parameters, which in turn, are subdivided into maximum, minimum, and averaged values. For example, the *apparent* volume of the crater includes the region as seen with the breccia lens still in place, and as used in this paper (Fig. 4), is measured from the top of the breccia lens up to the original ground level where it intersects the crater walls. The *apparent* depth is measured from the original ground level to the top of the breccia lens. The *true* diameter is measured at the intersection of the original ground level with the post-shot regional ground level *in* the crater walls or rim. The *true* depth is measured from the original ground level to the base of the breccia lens. The maximum and minimum values are taken either directly from the cross sections, where maximum or minimum values are often located due to the choice of the location of the cross section, or from the geologic map and partial cross sections. The averaged values are taken either from multiple cross sections or from the data points on the geologic map and from the field survey data.

It is customary in explosion cratering to tabulate radii rather than diameters, however, we cannot generally locate true ground zeros (GZ) for natural impact craters. This is especially true at eroded terrestrial sites as well as on the other planetary surfaces. Consequently, diameters are listed in this paper and *effective* GZ's are determined for volume computations by halfing the average diameters for Flynn Creek and Steinheim.

The rim crest positions were estimated from the best cross-section data and scaling available since these parts of both Flynn Creek and Steinheim were

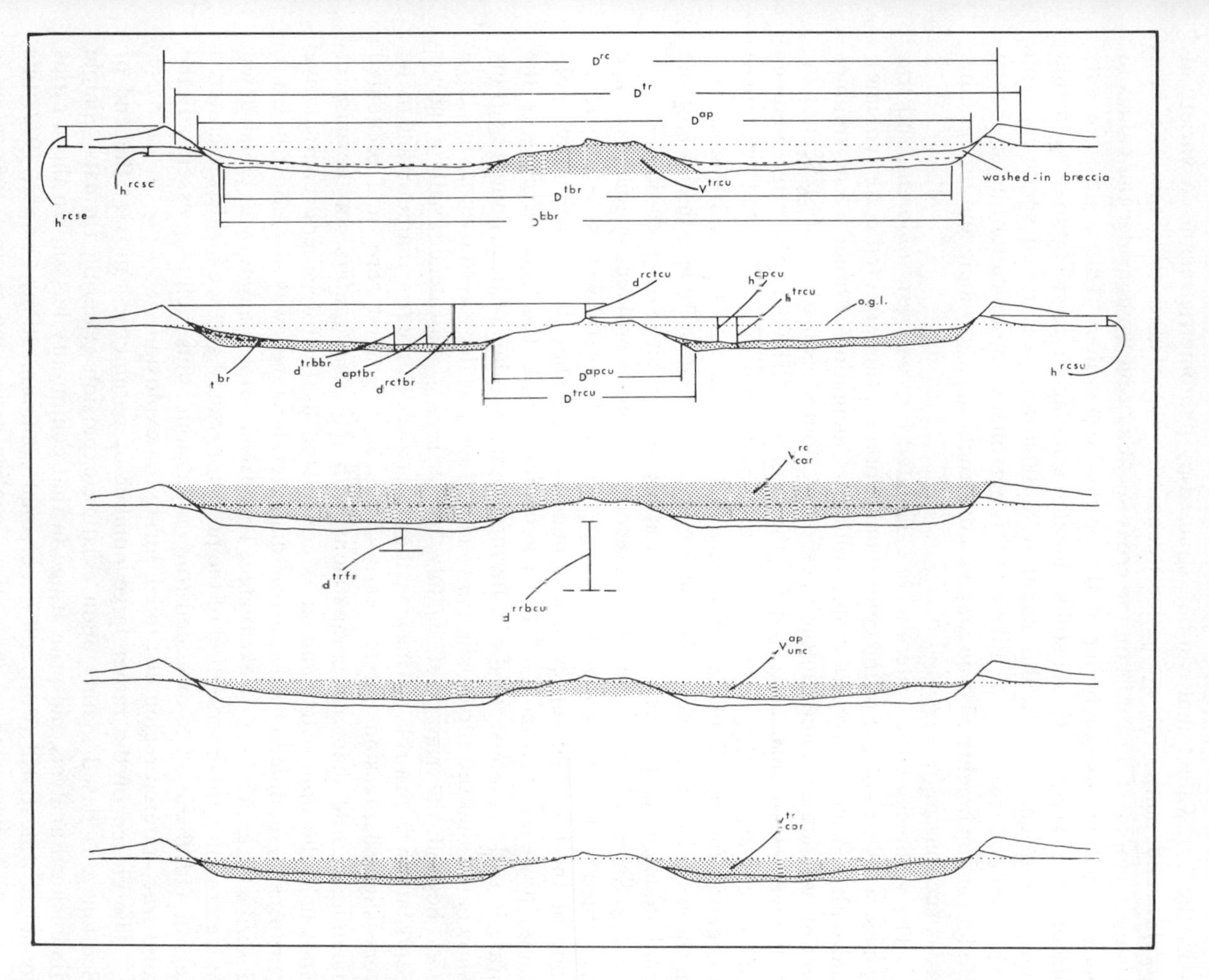

Fig. 4. Schematic representation of selected dimensional parameters listed in Table 1. The symbols are explained at the beginning of Table 1 and the crater profile is for the Flynn Creek impact crater, Section C-C′ in Roddy (1977a).

removed by erosion. The rim crest values, as well as volume and volume ratio values using rim crest estimates, are included in this paper primarily as general comparative references for lunar and planetary studies since these are essentially the *only* values that can be extracted by photometric means. A variety of graphic and numerical estimation techniques were used to obtain the best assumed rim crest positions on the cross sections, and we consider the values to be good within 90% or better of the original true values. This is possible *only* because the *crater walls* can still be located quite accurately. The total estimated rim crest height used in this paper was 100 m and was based specifically on explosion crater scaling, on Pike's (1977, personal communication) lunar scaling model for rim heights, and on graphic constructions on geologic and topographic cross sections.

The volumes of the craters were calculated for *rim crest*, *apparent*, and *true* values, with the volume of the central uplift present (uncorrected crater volume), and with it absent (corrected crater volume). Calculation of breccia lens volumes and masses are explained in their respective footnotes at the end of Section A. These values are also corrected for the central uplift volume. Crater volumes and ejecta masses for Flynn Creek and Steinheim are calculated without a correction for mass bulking and permanent true crater structural displacements in the walls and floor, however, these corrections should be less than 5% of the total values judging from such studies at Meteor Crater (Roddy *et al.*, 1975).

The Snowball explosion crater presents a special problem in that its shape continued to adjust slightly over a time period of several weeks to a year, the time of the last field survey. The immediate post-shot data is generally applicable, however, for most of the calculations since this time represents the crater's early response to the cratering event. Another critical point regarding Snowball is that the surrounding ground surface was depressed over a very large area beyond the rim crest, mainly by structural subsidence with a minor contribution from airblast. This gave the crater a very large *true* diameter due to the very slight depression out to nearly 200 m from ground zero, and prevented meaningful *true* diameter comparisons with the other craters. In the case of Snowball, the *apparent* diameter is considered useful for general comparative purposes. A modified or *pseudo-true* diameter, however, was also constructed to describe the general region excavated. This *pseudo-true* diameter was constructed by determining the intersection of the pre-shot ground surface inside the cratered region with the average extended *true* crater slope, thus including essentially the total *true* cratered region disrupted during the explosion.

The choice of the rather large number of parametric ratios at the end of Section A is based mainly upon existing explosion cratering formats. Simple dimensional analysis and prior usage in that community has shown these ratios to be of widest use (Port, 1977, personal communication; Cooper, 1977, this volume). A discussion of the various scaling ratios and comparative ratios is beyond the scope of this data tabulation; however, limited explanations are given in the footnotes at the end of Section A. The reader is also referred to the papers in this volume on scaling and to Cooper (1977, this volume) for further discussions.

The ratios are given to three figures; however, the writer recommends that these values be rounded off to the first figure for general comparison purposes. A *complete ratio set* is presented in Table 1 to allow single data point entry for scaling comparisons, i.e., where only very incomplete data is available from another crater.

One point that is of interest in the scaling ratios is the development of an *equivalent length factor*, suggested by Port (1977, personal communication) and as noted in entry (199). This involved the averaging of a series of diameter, volume, and other parametric ratios to arrive at general group parametric ratios that approximate the dimensional relationships between the Flynn Creek and Snowball craters and between the Steinheim and Snowball craters. A second approach to deriving an *equivalent length factor* involved comparisons of the craters by variable image projection of their respective cross sections. The assumption is that visual inspection of the superimposed geologic cross sections tends to scale the two craters more quantitatively than does the use of any *single* dimensional parameter. The preliminary results of this effort is an *equivalent length factor* of 31 for Flynn Creek and 32 for Steinheim, which lends strong support to the numerical averaging technique.

The kinetic energies of formation for the Flynn Creek and Steinheim impact craters are listed in Section B using three different scaling calculations, (a) volume-energy (Port, 1977, personal communication), (b) diameter-energy (Roddy *et al.*, 1975; Shoemaker, 1977, this volume), and (c) equivalent length factor (Port, 1977, personal communication, and this paper). The 1/3.4 root scaling is also used in each of the above and the footnotes following Section B further define the entry values for the energies calculated. The choices of the particular scaling laws, heights of burst equivalents, and other related critical assumptions used in these calculations require a more complete treatment than can be discussed here. Consequently, the writer suggests the kinetic energies of formation for the impact craters be interpreted only as conservative *representative* estimates, particularly because of the unknown energy partition factors between explosion and impact processes in hard rock and alluvial target medias.

The remaining Sections (C) through (L) are devoted mainly to comparative data between the three craters and are generally self-explanatory.

Table 1. Tabulated parametric values and related aspects for the Flynn Creek and Steinheim impact craters and Snowball explosion crater including: (A) Locations, dimensions, and ratios, (B) Energies of formation, (C) Pre-impact conditions, (D) Topography, (E) Structural deformation, (F) Ejecta, (G) Shock metamorphism, (H) Field studies, (I) Geochemistry, (J) Geophysics, (K) Drilling, and (L) Other studies.

(A) Locations, dimensions, ratios

Parameters: D = diameter, d = depth, h = height, r = radius, t = thickness, V = volume; M = mass; E = energy of formation using 1/3.4 root scaling.

Abbreviations: GZ = ground zero; o.g.l = original ground level at time of impact; p.g.l = post-crater regional ground level after erosion and before sedimentary filling; N/A = not applicable; ? = unknown; n.d. = not determined; n.a. = not available; est = best estimate from graphic construction or calculations where direct measurements not possible due to ejecta burial or erosion.

Numerical value prefixes: $\approx$ = approximate value considered accurate to within approximately 10% or better of actual value; $\sim$ = approximate value considered accurate to within approximately 5% or better of actual value; *no* prefix symbol to value indicates accurate to within 1% or better of actual value. Assignment of prefixes is based upon the best quality assessment available of original data sources, including topographic maps, types of field surveys, aerial photography, topographic plotter transfers, and nature of field outcrops. A value with no prefix and no abbreviation is the most accurate, whereas a value with a $\approx$ prefix and followed by an est abbreviation is a least well-known number. Values derived from direct measurement sources are quoted to nearest one meter, depending upon the exact measurement technique, and are considered as accurate as the prefixes indicate. Values based on the rim crests of the impact. craters are also quoted to the nearest one meter as measured directly on the cross sections, however, these values should be considered the least well-known as defined by the prefix $\approx$ symbol and est abbreviation, i.e., accurate to only 90% or better.

Subscripts and Superscripts listed in order of appearance in Table 1, Section A:

max = maximum
min = minimum
avg = average
rc = rim crest apparent value
ap = apparent value
tr = true value
br = breccia lens
t = top
b = base
cu = central uplift
f = filling sediments
rx = rock
rxr = rock raised in uplift
cx = crystalline or sedimentary basement
s = structural deformation
su = structural uplift
unc = uncorrected for volume of central uplift
cor = corrected for volume of central uplift
cal = calculated
erx = ejected rocked or alluvium
cs = compaction and subsidence
ds = disrupted volume
rot = rotated volume
tk = truck volume
sd = structural depression
e = ejecta
ce = continuous ejecta
de = discontinuous ejecta
efg = ejecta fragment
rfk = rim flank
dz = deformational zone

Table 1, Section (A). (*Continued*).

Parameters	Flynn Creek impact crater	Steinheim impact crater	Snowball explosion crater
Locations			
(1) Location	North central Tennessee, United States	East central Baden Württemberg, West Germany	Defense Research Establishment, Suffield, Alberta, Canada
(2) Coordinates of ~GZ	36°17′N; 85°40′W	48°02′N; 10°04′E	50°29′18″N; 110°37′42″W
(3) Present MSL for ~ original GZ	~275 m ± 15 m	~660 m ± 5 m	660.7 m
Dimensions			
(4) D^{rc}_{max} Rim crest	~4039 m (est)	≈4180 m (est)	114 m[1]
(5) D^{rc}_{min} Rim crest	3563 m (est)	≈3350 m (est)	~102 m[1]
(6) D^{rc}_{avg} Rim crest	~3830 m (est)	≈3650 m (est)	~108.5 m[1]
(7) D^{ap}_{max} Apparent	~3737 m	≈3820 m	88.0 m[1]
(8) D^{ap}_{min} Apparent	~3240 m	≈3100 m	75.0[1]
(9) D^{ap}_{avg} Apparent	~3517 m	≈3300 m (est)	~83.0 m[1]
(10) D^{tr}_{max} True	~3905 m	≈3820 m	n.a.
(11) D^{tr}_{min} True	~3423 m	≈3100 m	n.a.
(12) D^{tr}_{avg}	~3681 m	≈3300 m (est)	≈110 m[2]
(13) D^{tbr}_{max} Top br	~3499 m	≈2900 m (est)	n.d.
(14) D^{tbr}_{min} Top br	~2987 m	n.d.	n.d.
(15) D^{tbr}_{avg} Top br	~3161 m	≈2800 m (est)	≈53 ± m
(16) D^{bbr}_{max} Base br	~3414 m (est)	≈2700 m (est)	n.d.
(17) D^{bbr}_{min} Base br	~2880 m (est)	n.d.	n.d.
(18) D^{bbr}_{avg} Base br	~3029 m (est)	≈2600 m (est)	≈50 ± m
(19) D^{trcu}_{max} Base cu at bbr	~1189 m (est)	~950 m	~27.0 m
(20) D^{trcu}_{min} Base cu at bbr	~997 m (est)	≈730 m (est)	~21.5 m
(21) D^{trcu}_{avg} Base cu at bbr	~1087 m (est)	≈840 m (est)	≈24.3 m
(22) D^{apcu}_{max} Base cu at tbr	951 m	~850 m	~22.0 m
(23) D^{apcu}_{min} Base cu at tbr	~799 m	≈ 630 m (est)	~17.0 m
(24) D^{apcu}_{avg}	~899 m	≈740 m (est)	~19.5 m
(25) D^{tcu}_{max} Top cu today	402 m	~650 m	~4–15 m
(26) D^{tcu}_{min} Top cu today	~244 m	150 m (est)	sharp peaks
(27) D^{tcu}_{avg} Top cu today	~302 m	350 m (est)	~7 m to peaks
(28) d^{rctf}_{max} Rim crest to top filling sediments before compaction	N/A	N/A	7.1 m

Table 1, Section (A). (*Continued*).

Parameters	Flynn Creek impact crater	Steinheim impact crater	Snowball explosion crater
(29) d_{avg}^{rctfs} Rim crest to top filling sediments before compaction	N/A	N/A	6.6 m
(30) d_{avg}^{rctf} Rim crest to top filling sediments after compaction	N/A	N/A	~6 m
(31) d_{max}^{rctcu} Rim crest to tcu at GZ today	~90 m (est)	~185 m (est)	5.0 m (initial) 6.2 m (final)
(32) d_{max}^{rctcu} Rim crest to tcu off GZ today	~90 m (est)	n.d.	3.8 m (initial) 4.7 m (final)
(33) d_{max}^{rctbr} Rim crest to tbr	~198 m (est)	~267 m (est)	9.2 m
(34) d_{avg}^{rctbr} Rim crest to tbr	≈180 m (est)	≈250 m (est)	≈7.5 m
(35) d_{max}^{rcbbr} Rim crest to bbr	~221 m (est)	~317 m (est)	10.0 m
(36) d_{avg}^{rcbbr} Rim crest to bbr	≈220 m (est)	≈300 m (est)	≈8.5 m
(37) d_{max}^{aptf} O.g.l to top of filling sediment before compaction	≈38 ± m	≈30 ± m (est)	6.35 m
(38) d_{avg}^{aptf} O.g.l to top of filling sediment before compaction	n.d.	n.d.	≈5.8 m
(39) d_{max}^{aptf} O.g.l to top of filling sediment after compaction	~57 m	~ 30 + m	n.a.
(40) d_{max}^{aptcu} O.g.l to tcu at GZ today	~10 m (above o.g.l)	~95 m	4.3 m (initial) 5.2 m (final)
(41) d_{max}^{aptcu} O.g.l to tcu off GZ today	~10 m (above o.g.l)	n.d.	3.1 m (initial) 3.8 m (final)
(42) d_{max}^{aptbr} O.g.l to tbr	~98 m	~167 m	8.2 m
(43) d_{avg}^{aptbr} O.g.l to tbr	≈80 m	≈150	≈6.6 m
(44) d_{max}^{trbbr} O.g.l to bbr	~121 m	~217 m	9.0 m
(45) d_{avg}^{trbbr} O.g.l to bbr	≈120 m	≈200 m	≈7.6 m

Table 1, Section (A). (*Continued*).

Parameters	Flynn Creek impact crater	Steinheim impact crater	Snowball explosion crater
(46) d_{max} Regional lowering from o.g.l to p.g.l	~25 to 30 m	~20 m (est)	N/A
(47) d_{min} Regional lowering from o.g.l to p.g.l	~5 m (est)	~1 m (est)	N/A
(48) d_{avg} Regional lowering from o.g.l to p.g.l	~10–15 m (est)	~5–10 m (est)	N/A
(49) d_{max} P.g.l to tf after compaction	43 m	~20 m	N/A
(50) d_{min} P.g.l to tcu at GZ	+18 m (above p.g.l.)	65–95 m	N/A
(51) d_{max} P.g.l to tbr	88 m	167 m	N/A
(52) d_{max} P.g.l to bbr	111 m	217 m	N/A
(53) d_{min}^{aprx} O.g.l to ~ flat-lying undeformed rock below breccia lens and beyond central uplift	~221 m	n.d.	≈ 17 ± m
(54) d_{max}^{trrx} Base br to ~ flat-lying, undeformed rock below breccia lens and beyond central uplift	~100 m	n.d.	~8 ± m
(55) d_{min}^{rxrtcu} O.g.l to deepest rock raised to tcu	≈450 m	≈250–380 m	~19 ± m
(56) d_{min}^{rxrbcu} Deepest rock raised in cu to tbr	~396 m	n.d.	~11 + m
(57) d_{avg}^{cx} O.g.l to ~ top of crystalline basement	~1700 m	~1100 m	~60 m to indurated bedrock
(58) h_{max}^{apcu} Top br to tcu at GZ today	108 m	72–90 m	3.9 m (initial) 3.0 m (final)

Table 1, Section (A). (*Continued*).

Parameters	Flynn Creek impact crater	Steinheim impact crater	Snowball explosion crater
(59) h_{avg}^{apcu} Top br to tcu at GZ today using avg tbr	≈90 m	≈55 m	3.5 m (initial) 2.8 m (final)
(60) h_{max}^{apcu} Top br to tcu at GZ today off GZ	108 m	n.d.	5.1 m (initial) 4.4 m (final)
(61) h_{avg}^{apcu} Top br to tcu at GZ today off GZ using avg tbr	≈90 m	n.d.	3.5 m (initial) 2.8 m (final)
(62) h_{max}^{trcu} Base br to tcu at GZ today	~131 m	~122–140 m	4.7 m (initial) 3.8 m (final)
(63) h_{avg}^{trcu} Base br to tcu at GZ using avg tbr	≈130 m	≈105 m	3.3 m (initial) 2.4 (final)
(64) h_{max}^{trcu} Base br to tcu off GZ	~131 m	n.d.	5.9 m (initial) 5.2 m (final)
(65) h_{avg}^{trcu} Base br to tcu off GZ using average tbr	≈130 m	n.d.	4.5 m (initial) 3.8 m (final)
(66) h_{avg}^{rcsue} Rim crest including structural uplift and ejecta	100 m	100 m (est)	0.9 m
(67) h_{avg}^{rcsu} Structural uplift of rim at crater walls	10–50 m	?	N/A
(68) h_{avg}^{rcsd} Structural depression of rim at crater wall	≈12 m	≈10–15 m	≈2.5 m (4.3 m max)
(69) h_{avg}^{rce} Ejecta at rc	~50–100 m (est)	~50–100 m (est)	1.8 m (1.6 m min) (2.3 m max)
(70) t_{avg} Filling sediments over cu before compaction	~13 m (est)	≈40–50 m (est)	N/A
(71) t_{avg} Filling sediment over cu after compaction	9 m	≈35–45 m (est)	N/A

Table 1, Section (A). (*Continued*).

Parameters	Flynn Creek impact crater	Steinheim impact crater	Snowball explosion crater
(72) t_{avg} Filling sediments before compaction around cu	≈40 ± m (variable)	?	~1.0 m
(73) t_{avg} Filling sediments after compaction around cu	≈30 ± m (variable)	≈25–35 m (after erosion)	N/A
(74) t_{avg} Br plus washed-in breccia near crater walls	~70 m (variable)	≈15 + m (min est)	N/A
(75) t^{br}_{max} Br on crater floor excluding washed-in breccia	~50 + m	~70 m	~2.1 m
(76) t^{br}_{min} Br on crater floor excluding washed-in breccia	~11 m	~10 m	~0.5 m
(77) t^{br}_{avg}	~30–40 m (est)	~30–40 m (est)	~0.7–1.1 m
(78) t_{avg} Br surrounding central uplift	~20–30 m (variable)	~30–50 m (variable)	~0.6–1.7 m
(79) t_{avg} Washed-in ejecta near crater walls	≈25 + m (est) (Variable)	≈10 + m (est) (Variable)	N/A
(80) t_{avg} Washed-in ejecta on tbr in middle of crater	~3–10 m	?	N/A
(81) t_{avg} Bedded unit between washed-in ejecta and filling sediments	~1–3 m	?	N/A
(82) t^{trrd}_{avg} Deformed rock below br	~100 + m	n.d.	~8 m
(83) t^{rxtcu}_{min} Deformed rock below tcu	≈500 + m (est)	≈600 + m (est)	≈22 + m (est)
(84) t^{rce}_{avg} Ejecta near overturned flap at rc	≈50–100 m (est)	≈50–100 m (est)	1.8 m

Table 1, Section (A). (*Continued*).

Parameters	Flynn Creek impact crater	Steinheim impact crater	Snowball explosion crater
(85) t^{1De}_{avg} Ejecta in overturned flap at ~1D from GZ	≈30–50 m (est)	?	0.1–0.3 m
(86) t^{2De}_{avg} Ejecta in overturned flap at 2D from GZ	≈5 ± m (est)	?	~0.01 ± m
(87) r^{ce}_{max} Edge of continuous ejecta in overturned flap from GZ	?	?	≈300 m
(88) r^{ce}_{min} Edge of continuous ejecta in overturned flap from GZ	?	?	≈175 m
(89) r^{ce}_{avg} Edge of continuous ejecta in overturned flap from GZ	?	?	≈200–250 m
(90) r^{de}_{avg} Edge of discontinuous ejecta in overturned flap from GZ	?	?	≈350–400 m
(91) r^{efg}_{max} Single 0.1 m size ejecta fragments from GZ	?	?	1036 + m
(92) r^{rfk}_{avg} Rim flank from GZ	≈700–1700 m (est)	?	≈50–80 m
(93) r^{dz}_{avg} Folded, faulted or fractured deformational zones concentric to crater and GZ	r_1 2438 m r_2 3048 m	?	r — Δ r_1 55 m r_2 67 m — 12 m r_3 73 m — 6 m r_4 79 m — 6 m r_5 91 m — 12 m r_6 116 m — 25 m r_7 155 m — 39 m
(94) $D^{ap}_{avg}/r^{dz}_{avg}$ Ratio of (9)/(93)	1.44 — Δ 1.15 — 0.29	? — ?	1.51 — Δ 1.24 — 0.27 1.14 — 0.10 1.05 — 0.09 0.91 — 0.14 0.72 — 0.19 0.54 — 0.18

Table 1, Section (A). (*Continued*).

Parameters	Flynn Creek impact crater		Steinheim impact crater	Snowball explosion crater	
(95) $D^{tr}_{avg}/r^{dz}_{avg}$	1.51	Δ		2.00[2]	Δ
Ratio of (12)/(93)	1.21	0.30	?	1.64	0.36
				1.51	0.13
				1.39	0.12
				1.21	0.18
				0.95	0.26
				0.71	0.24
(96) $r^{dz}_{avg}/r^{dz}_{avg}$	1.25		?	1.22	
Ratios of r_2/r_1,				1.09	
r_3/r_2, r_4/r_3,				1.08	
r_5/r_4, r_6/r_5				1.15	
				1.27	
				1.34	
(97) $r^{dz}_{avg}/r^{dz}_{avg}$	N/A		?	1.33	
Ratios of r_3/r_1,				1.18	
r_4/r_2, r_5/r_3,				1.25	
r_6/r_4, r_7/r_5				1.47	
				1.70	
(98) V^{rc}_{unc} Rim crest	≈1.932 km^3		≈2.002 km^3	~27,353 m^3 [3]	
(99) V^{rc}_{cor} Rim crest	≈1.899 km^3		≈1.984 km^3	~27,133 m^3	
(100) V^{ap}_{unc} Apparent	~0.825 km^3		≈1.053 km^3	~20,193 m^3	
(101) V^{ap}_{cor} Apparent	~0.792 km^3		≈1.035 km^3	~19,848 m^3	
(102) V^{tr}_{unc} True	~1.207 km^3		≈1.365 km^3	~33,329 m^3 [4]	
(103) V^{tr}_{cor} True	~1.152 km^3		≈1.329 km^3	~32,640 m^3 [4]	
(104) V^{br}_{unc}	~0.383 km^3		≈0.312 km^3	~13,136 m^3 [4]	
(105) V^{br}_{cor}	~0.361 km^3		≈0.294 km^3	~13,432 m^3 [4]	
(106) V^{apcu}_{avg} Apparent cu above tbr	~0.033 km^3		≈0.018 km^3	~345 m^3	
(107) V^{bcu}_{avg} Central uplift below tbr and above bbr	~0.022 km^3		≈0.018 km^3	~344 m^3	
(108) V^{trcu}_{avg} True cu above bbr	~0.055 km^3		≈0.036 km^3	~689 m^3	
(109) V^{e}_{cal} Calculated ejecta	~0.825 km^3		≈1.053 km^3	≈20,193 m^3 [5]	
(110) V^{ecs}_{cal} Calculated ejecta assuming compaction and subsidence	N/A		N/A	≈14,866 m^3 [6]	
(111) V^{eds}_{cal} Calculated ejecta assuming disrupted volume	N/A		N/A	≈15,250 m^3 [7]	
(112) V^{erot}_{cal} Calculated ejecta assuming volume of revolution	N/A		N/A	≈18,085 m^3 [8]	

Table 1, Section (A). (*Continued*).

Parameters	Flynn Creek impact crater	Steinheim impact crater	Snowball explosion crater
(113) V_{cal}^{etk} Calculated ejecta from truck field values	N/A	N/A	~16,100 to [9] ~19,260 m^3
(114) V_{avg}^{e} Ejecta	~0.825 km^3	≈1.053 km^3	≈17,500 m^3 [10]
(115) M_{max}^{e} Ejecta	N/A	N/A	≈32,309 metric tons[11]
(116) M_{min}^{e} Ejecta	N/A	N/A	≈23,786 metric tons[11]
(117) M_{avg}^{e} Ejecta	~2.228×10^9 metric tons	≈2.632×10^9 metric tons	≈28,000 metric tons[11]
(118) M_{avg}^{br} Breccia lens today	~0.920×10^9 metric tons	≈0.647×10^9 metric tons	≈16,118 metric tons[12]
(119) $V\&M$ Remaining % ejecta in overturned flap	~1+%	~0.1%(?)	80%

Ratios			
(120) $D_{avg}^{rc}/d_{max}^{rc}$	19.3[13]	13.7[13]	11.8[13]
(121) $D_{avg}^{rc}/d_{avg}^{rc}$	21.3[13]	14.6[13]	14.5[13]
(122) $D_{avg}^{ap}/d_{max}^{ap}$	35.9	19.8	10.1
(123) $D_{avg}^{ap}/d_{avg}^{ap}$	44.0	22.0	12.6
(124) $D_{avg}^{tr}/d_{max}^{tr}$	30.4	15.2	12.2
(125) $D_{avg}^{tr}/d_{avg}^{tr}$	30.7	16.5	14.5
(126) $D_{avg}^{rc}/V_{unc}^{1/3rc}$	3.08	2.90	3.60
(127) $D_{avg}^{rc}/V_{cor}^{1/3rc}$	3.09	2.91	3.61
(128) $D_{avg}^{ap}/V_{unc}^{1/3ap}$	3.75	3.24	3.05
(129) $D_{avg}^{ap}/V_{cor}^{1/3ap}$	3.80	3.26	3.07
(130) $D_{avg}^{tr}/V_{unc}^{1/3tr}$	3.46	2.98	3.42
(131) $D_{avg}^{tr}/V_{cor}^{1/3tr}$	3.52	3.00	3.44
(132) $d_{max}^{rc}/V_{unc}^{1/3rc}$	0.159	0.212	0.305
(133) $d_{avg}^{rc}/V_{unc}^{1/3rc}$	0.145	0.198	0.249
(134) $d_{max}^{ap}/V_{unc}^{1/3rc}$	0.105	0.164	0.301
(135) $d_{avg}^{ap}/V_{unc}^{1/3rc}$	0.085	0.147	0.242
(136) $d_{max}^{tr}/V_{unc}^{1/3tr}$	0.114	0.196	0.280
(137) $d_{avg}^{tr}/V_{unc}^{1/3tr}$	0.113	0.181	0.236
(138) $V_{unc}^{rc}/D_{avg}^{2rc}d_{max}^{rc}$	0.665	0.563	0.253
(139) $V_{unc}^{rc}/D_{avg}^{2rc}d_{avg}^{rc}$	0.732	0.601	0.310
(140) $V_{unc}^{ap}/D_{avg}^{2ap}d_{max}^{ap}$	0.680	0.579	0.357
(141) $V_{unc}^{ap}/D_{avg}^{2ap}d_{avg}^{ap}$	0.833	0.645	0.444
(142) $V_{unc}^{tr}/D_{avg}^{2ap}d_{max}^{tr}$	0.736	0.574	0.306
(143) $V_{unc}^{tr}/D_{avg}^{2ap}d_{avg}^{tr}$	0.743	0.622	0.362
(144) $D_{max}^{rc}/h_{max}^{apcu}$ at GZ	37.4	58.1	29.2 (initial) 38.0 (final)
(145) $D_{avg}^{rc}/h_{avg}^{apcu}$ at GZ	42.6	66.36	31.0 (initial) 38.8 (final)
(146) $D_{max}^{ap}/h_{max}^{apcu}$ at GZ	34.5	53.1	22.6 (initial) 29.3 (final)
(147) $D_{avg}^{ap}/h_{avg}^{apcu}$ at GZ	39.1	60.0	23.7 (initial) 29.64 (final)

Table 1, Section (A). (*Continued*).

Parameters	Flynn Creek impact crater	Steinheim impact crater	Snowball explosion crater
(148) $D^{tr}_{max}/h^{trcu}_{max}$ at GZ	29.8	31.3	n.a.
(149) $D^{tr}_{avg}/h^{trcu}_{avg}$ at GZ	28.3	31.4	33.3 (initial)[2]
			45.8 (final)
(150) $D^{rc}_{avg}/d^{rrtcu}_{min}$	9.67	7.60–14.6	5.71 at bcu
			9.86 at bcu
(151) $D^{ap}_{avg}/d^{rrtcu}_{min}$	8.88	6.88–13.2	4.37 at bcu
			7.55 at bcu
(152) $D^{tr}_{avg}/d^{rrtcu}_{min}$	9.30	6.88–13.2	5.79 at bcu
			10.00 at bcu
(153) $d^{rctbr}_{max}/h^{apcu}_{max}$ at GZ	1.83	3.71	2.36 (initial)
			3.07 (final)
(154) $d^{rctbr}_{avg}/h^{apcu}_{avg}$ at GZ	2.00	4.55	2.14 (initial)
			2.68 (final)
(155) $d^{aptbr}_{max}/h^{apcu}_{max}$ at GZ	0.91	2.32	2.10 (initial)
			2.73 (final)
(156) $d^{aptbr}_{max}/h^{apcu}_{avg}$ at GZ	0.89	2.73	1.89 (initial)
			2.36 (final)
(157) $d^{trbbr}_{max}/h^{trcu}_{max}$ at GZ	0.92	1.78	1.91 (initial)
			2.37 (final)
(158) $d^{trbbr}_{avg}/h^{trcu}_{avg}$ at GZ	0.92	1.90	2.30 (initial)
			3.17 (final)

Equivalent Length Factor Ratios	Flynn Creek / Snowball	Flynn Creek / Steinheim	Steinheim / Snowball	Steinheim / Flynn Creek	Snowball / Flynn Creek	Snowball / Steinheim
(159) $D^{rc}_{avg}/D^{rc}_{avg}$ Ratio of (6)	35.30	1.05	33.64	0.95	0.0283	0.0297
(160) $D^{ap}_{avg}/D^{ap}_{avg}$ Ratio of (9)	42.37	1.07	39.76	0.93	0.0236	0.0252
(161) $D^{tr}_{avg}/D^{tr}_{avg}$ Ratio of (12)	33.46	1.12	30.00	0.88	0.0299	0.0333
(162) $V^{1/3rc}_{unc}/V^{1/3rc}_{unc}$ Ratio of (98)	41.34	0.99	41.83	1.01	0.0242	0.0239
(163) $V^{1/3rc}_{cor}/V^{1/3rc}_{cor}$ Ratio of (99)	41.21	0.99	41.82	1.01	0.0243	0.0239
(164) $V^{1/3ap}_{unc}/V^{1/3ap}_{unc}$ Ratio of (100)	34.44	0.92	37.34	1.08	0.0290	0.0268
(165) $V^{1/3ap}_{cor}/V^{1/3ap}_{cor}$ Ratio of (101)	34.17	0.92	37.36	1.08	0.0293	0.0268
(166) $V^{1/3tr}_{unc}/V^{1/3tr}_{unc}$ Ratio of (102)	33.09	0.96	34.47	1.04	0.0302	0.0290
(167) $V^{1/3tr}_{cor}/V^{1/3tr}_{cor}$ Ratio of (103)	32.80	0.95	34.40	1.05	0.0305	0.0291
(168) $V^{1/3br}_{unc}/V^{1/3br}_{unc}$ Ratio of (104)	30.78	1.07	28.75	0.93	0.0325	0.0348

Table 1, Section (A). (*Continued*).

Equivalent Length Factor Ratios	Flynn Creek	Flynn Creek	Steinheim	Steinheim	Snowball	Snowball
	Snowball	Steinheim	Snowball	Flynn Creek	Flynn Creek	Steinheim
(169) $V_{cor}^{1/3br}/V_{cor}^{1/3br}$ Ratio of (105)	29.96	1.07	27.97	0.93	0.0334	0.0357
(170) $V_{avg}^{1/3apcu}/V_{avg}^{1/3apcu}$ Ratio of (106)	45.82	1.22	37.44	0.82	0.0218	0.0267
(171) $V_{avg}^{1/3trcu}/V_{avg}^{1/3trcu}$ Ratio of (108)	43.15	1.15	37.47	0.82	0.0232	0.0267
(172) $V_{cal}^{1/3e}/V_{cal}^{1/3e}$ Ratio of (109)	34.44	0.92	37.37	1.08	0.0290	0.0265
(173) $V_{avg}^{1/3e}/V_{avg}^{1/3e}$ Ratio of (114)	36.13	0.92	39.20	1.08	0.0277	0.0255
(174) $M_{avg}^{1/3e}/M_{avg}^{1/3e}$ Ratio of (115)	43.01	0.95	45.47	1.05	0.0232	0.0220
(175) $M_{avg}^{1/3br}/M_{avg}^{1/3br}$ Ratio of (116)	38.50	1.12	34.24	0.88	0.0260	0.0292
(176) Equiv. Length Factor (159–175)	37.06	1.02	36.38	0.98	0.0274	0.0279
(177) $d_{max}^{rctcu}/d_{max}^{rctcu}$	18.00		37.00		0.0556	0.0270
Ratio of (31)	14.52	0.49	29.84	2.06	0.0689	0.0335
(178) $d_{max}^{rctbr}/d_{max}^{rctbr}$ Ratio of (33)	21.52	0.74	29.02	1.35	0.0465	0.0345
(179) $d_{max}^{rctbr}/d_{avg}^{rctbr}$ Ratio of (34)	24.00	0.72	33.33	1.39	0.0417	0.0300
(180) $d_{max}^{rcbbr}/d_{max}^{rcbbr}$ Ratio of (35)	22.10	0.67	31.70	1.50	0.0452	0.0315
(181) $d_{avg}^{rcbbr}/d_{avg}^{rcbbr}$ Ratio of (36)	25.88	0.73	35.29	1.36	0.0386	0.0283
(182) $d_{max}^{aptbr}/d_{max}^{aptbr}$ Ratio of (42)	11.95	0.59	20.37	1.70	0.0837	0.0491
(183) $d_{avg}^{aptbr}/d_{avg}^{aptbr}$ Ratio of (43)	12.12	0.53	22.73	1.88	0.0825	0.0440
(184) $d_{max}^{trbbr}/d_{max}^{trbbr}$ Ratio of (44)	13.44	0.56	24.11	1.79	0.0744	0.0415
(185) $d_{avg}^{trbbr}/d_{avg}^{trbbr}$ Ratio of (45)	15.79	0.60	26.32	1.67	0.0633	0.0380
(186) $d_{min}^{aprx}/d_{min}^{aprx}$ Ratio of (53)	13.00	n.d.	n.d.	n.d	n.d	0.0769
(187) $d_{max}^{trrx}/d_{max}^{trrx}$ Ratio of (54)	12.50	n.d	n.d	n.d	n.d	0.0800
(188) $d_{min}^{rxrtcu}/d_{min}^{rxrtcu}$ Ratio of (55)	23.68	1.18	20.00	0.80	0.0422	0.0500
(189) $d_{min}^{rxrbcu}/d_{min}^{rxrbcu}$ Ratio of (56)	36.00	n.d.	n.d.	n.d.	n.d.	0.0278
(190) $h_{max}^{apcu}/h_{max}^{apcu}$	27.69		18.48		0.0361	0.0542
Ratio of (58)	36.00	1.50	24.00	0.67	0.0278	0.0417

Table 1, Section (A). (*Continued*).

Equivalent Length Factor Ratios	Flynn Creek / Snowball	Flynn Creek / Steinheim	Steinheim / Snowball	Steinheim / Flynn Creek	Snowball / Flynn Creek	Snowball / Steinheim
(191) $h^{apcu}_{avg}/h^{apcu}_{avg}$	25.71		15.71		0.0389	0.0636
Ratio of (59)	32.14	1.64	19.64	0.61	0.0311	0.0509
(192) $h^{apcu}_{avg}/h^{apcu}_{avg}$	25.71					0.0389
Ratio of (61)	32.14	n.d.	n.d.	n.d.	n.d.	0.0311
(193) $h^{trcu}_{max}/h^{trcu}_{max}$	27.87		25.96		0.0359	0.0385
Ratio of (62)	34.47	1.07	32.11	0.93	0.0290	0.0311
(194) $h^{trcu}_{avg}/h^{trcu}_{avg}$	39.39		31.82		0.0254	0.0314
Ratio of (63)	54.17	1.24	43.75	0.81	0.0185	0.0229
(195) $h^{trcu}_{max}/h^{trcu}_{max}$	22.20					0.0450
Ratio of (64)	25.19	n.d.	n.d.	n.d.	n.d.	0.0397
(196) $h^{trcu}_{avg}/h^{trcu}_{avg}$	28.89					0.0346
Ratio of (65)	34.21	n.d.	n.d.	n.d.	n.d.	0.0292
(197) $t^{br}_{max}/t^{br}_{max}$ Ratio of (75)	23.81	0.71	33.33	1.40	0.0420	0.0300
(198) $t^{br}_{avg}/t^{br}_{avg}$ Ratio of (77)	38.89	1.00	38.89	1.00	0.0257	0.0257
(199) Equiv. length factor (144) through (183) excluding n.d.	29.85	0.94	31.89	1.14	0.0352	0.0354
(200) $D^{rc}_{avg}/E^{1/3.4}$ m/erg (13), (14)	2.55×10^{-4}		2.35×10^{-4}		2.25×10^{-4}	
(201) $D^{ap}_{avg}/E^{1/3.4}$ m/erg (13), (14)	2.35×10^{-4}		2.13×10^{-4}		1.72×10^{-4}	
(202) $D^{tr}_{avg}/E^{1/3.4}$ m/erg (13), (14)	2.46×10^{-4}		2.13×10^{-4}		2.28×10^{-4}	
(203) $D^{rc}_{avg}/E^{1/3}$ m/erg (13), (14)	2.82×10^{-5}		2.59×10^{-5}		3.93×10^{-5}	
(204) $D^{ap}_{avg}/E^{1/3}$ m/erg (13), (14)	2.59×10^{-5}		2.35×10^{-5}		3.01×10^{-5}	
(205) $D^{tr}_{avg}/E^{1/3}$ m/erg (13), (14)	2.72×10^{-5}		2.35×10^{-5}		3.99×10^{-5}	
(206) $d^{rctbr}_{max}/E^{1/3.4}$ m/erg (13), (14)	1.32×10^{-5}		1.72×10^{-5}		1.91×10^{-5}	
(207) $d^{rctbr}_{avg}/E^{1/3.4}$ m/erg (13), (14)	1.20×10^{-5}		1.61×10^{-5}		1.56×10^{-5}	
(208) $d^{aptbr}_{max}/E^{1/3.4}$ m/erg (13), (14)	0.65×10^{-5}		1.08×10^{-5}		1.70×10^{-5}	
(209) $d^{aptbr}_{avg}/E^{1/3.4}$ m/erg (13), (14)	0.53×10^{-5}		0.97×10^{-5}		1.37×10^{-5}	
(210) $d^{trbbr}_{max}/E^{1/3.4}$ m/erg (13), (14)	0.81×10^{-5}		1.40×10^{-5}		1.87×10^{-5}	
(211) $d^{trbbr}_{avg}/E^{1/3.4}$ m/erg (13), (14)	0.81×10^{-5}		1.29×10^{-5}		1.58×10^{-5}	
(212) $d^{rctbr}_{max}/E^{1/3}$ m/erg (13), (14)	0.15×10^{-5}		0.19×10^{-5}		0.33×10^{-5}	

Table 1, Section (A). (*Continued*).

Equivalent Length Factor Ratios	Flynn Creek / Snowball — Flynn Creek / Steinheim	Steinheim / Snowball — Steinheim / Flynn Creek	Snowball / Flynn Creek — Snowball / Steinheim
(213) $d_{avg}^{rctbr}/E^{1/3}$ m/erg (13), (14)	0.13×10^{-5}	0.18×10^{-5}	0.27×10^{-5}
(214) $d_{max}^{trbbr}/E^{1/3}$ m/erg (13), (14)	0.07×10^{-5}	0.18×10^{-5}	0.30×10^{-5}
(215) $d_{avg}^{trbbr}/E^{1/3}$ m/erg (13), (14)	0.06×10^{-5}	0.11×10^{-5}	0.24×10^{-5}
(216) $d_{max}^{trbbr}/E^{1/3}$ m/erg (13), (14)	0.08×10^{-5}	0.15×10^{-5}	0.33×10^{-5}
(217) $d_{avg}^{trbbr}/E^{1/3}$ m/erg (13), (14)	0.09×10^{-5}	0.14×10^{-5}	0.28×10^{-5}
(218) V_{unc}^{rc}/E m^3/erg	0.77×10^{-15}	0.72×10^{-15}	1.30×10^{-15}
(219) V_{cor}^{rc}/E m^3/erg	0.76×10^{-15}	0.71×10^{-15}	1.29×10^{-15}
(220) V_{unc}^{ap}/E m^3/erg	0.33×10^{-15}	0.38×10^{-15}	0.96×10^{-15}
(221) V_{cor}^{ap}/E m^3/erg	0.32×10^{-15}	0.37×10^{-15}	0.95×10^{-15}
(222) V_{unc}^{tr}/E m^3/erg	0.48×10^{-15}	0.49×10^{-15}	1.59×10^{-15}
(223) V_{cor}^{tr}/E m^3/erg	0.46×10^{-15}	0.48×10^{-15}	1.55×10^{-15}
(224) $V_{unc}^{rc}/E^{3/3.4}$ $m^3/erg^{3/3.4}$	5.73×10^{-13}	5.94×10^{-13}	2.44×10^{-13}
(225) $V_{cor}^{rc}/E^{3/3.4}$ $m^3/erg^{3/3.4}$	5.64×10^{-13}	5.33×10^{-13}	2.42×10^{-13}
(226) $V_{unc}^{ap}/E^{3/3.4}$ $m^3/erg^{3/3.4}$	2.45×10^{-13}	2.83×10^{-13}	1.80×10^{-13}
(227) $V_{cor}^{ap}/E^{3/3.4}$ $m^3/erg^{3/3.4}$	2.35×10^{-13}	2.78×10^{-13}	1.77×10^{-13}
(228) $V_{unc}^{tr}/E^{3/3.4}$ $m^3/erg^{3/3.4}$	3.58×10^{-13}	3.67×10^{-13}	3.00×10^{-13}
(229) $V_{cor}^{tr}/E^{3/3.4}$ $m^3/erg^{3/3.4}$	3.42×10^{-13}	3.57×10^{-13}	2.92×10^{-13}
(230) $V_{unc}^{br}/E^{3/3.4}$ $m^3/erg^{3/3.4}$	1.14×10^{-13}	0.84×10^{-13}	1.17×10^{-13}
(231) $V_{cor}^{br}/E^{3/3.4}$ $m^3/erg^{3/3.4}$	1.07×10^{-13}	0.79×10^{-13}	1.20×10^{-13}
(232) M_{avg}^{e}/E metric tons/erg	0.89×10^{-15}	0.94×10^{-15}	1.33×10^{-15}
(233) M_{avg}^{br}/E metric tons/erg	0.37×10^{-15}	0.23×10^{-15}	0.77×10^{-15}

(1) Averaged values from topographic map.

(2) The Snowball Crater experienced finite post-shot compression and subsidence of the alluvium giving a *true* crater diameter of ~350 m initially with a final value of ~400 m. These large diameters are due to a very small depression of the ground surface for a large distance from ground zero, and do not represent meaningful *true* diameters in the normal cratering usage of this term. A *pseudo-true* diameter was constructed from the cross-sections for these tabulations by using the intersection of the *mean slope* of the *true* crater walls with the original ground level. Irregularities in the walls gave a *pseudo-true* diameter range of ~100–120 m, with an average of ≈110 m adopted as a best approximation.

(3) The volumes and masses calculated for Snowball in (98) through (118) are averaged from topographic profiles made shortly after post-shot time and one year post-shot. These averages, therefore, represent minimum estimates since the earliest profiles have not experienced the maximum amount of subsidence.

(4) The volumes and masses calculated in (102) to (105) and (109) to (118) use a maximum *pseudo-true* diameter of 120 m. Rooke *et al.* (1968), using a *true* diameter of ~350–400 m, lists a *true* volume of 43,100 m^3 which includes the very small, but extensive, depressed ground surface well beyond the normal crater area. This region, however, does not represent the actual ejecta cratered volume.

(5)Ejecta volume calculation assumes no contribution from compaction and subsidence to apparent volume.

(6)Ejecta volume calculation assumes approximately 25% contribution from compaction and subsidence to apparent volume.

(7)Ejecta volume calculation from total disrupted volume assumes approximately 25% contribution from compaction and subsidence to true volume and uses a density for undisturbed soil of 1.6 g/cm^3 and a density for breccia of 1.1 g/cm^3. The total disrupted alluvium minus the breccia lens gives the estimated ejected alluvium.

(8)Ejecta volume calculated from volume of revolution for ejecta using cross section profiles.

(9)Ejecta volume calculated from total ejecta collected by truck continuously along two 1.83 m-wide strips from rim to ~183 m range from GZ (Diehl and Jones, 1967b). The value in entry (113) is a minimum estimate due to a real divergence and is the best field data value of the volumes listed. Measured ejecta dry densities averaged about 1.35 g/cm^3 and *in situ* densities ranged from ~1.4 to 1.6 g/cm^3.

(10)Ejecta volume averaged from all calculation techniques (5) through (9) above.

(11)The bulk *in situ* densities for Flynn Creek and Steinheim were approximated as 2.7 and 2.5 g/cm^3, respectively. Ejecta mass calculations for Snowball uses average *in situ* pre-shot density of 1.6 g/cm (G. Jackson, U.S. Army Engineer Waterways Experiment Station, 1977, personal communication). Entry (117) is average of (115) and (116).

(12)The bulk *in situ* breccia lens densities for Flynn Creek and Steinheim were approximated as 2.6 (lithified) and 2.2 g/cm^3, respectively. Breccia lens mass calculations for Snowball uses assumed average value of 1.2 g/cm^3 for breccia lens.

(13)The kinetic energies of formation for Flynn Creek and Steinheim were calculated using the *equivalent length factors* derived from averaging diameters, volumes and other parametric ratios as shown in entry (199) in Section A and entry (2) in Section B, i.e. 2.5×10^{24} ergs and 2.8×10^{24} ergs, respectively.

(14)The energy of formation of the Snowball craters was 500-tons of TNT as listed in Section B with a TNT equivalent of 2.1×10^{19} ergs.

Parameters	Flynn Creek impact crater	Steinheim impact crater	Snowball explosion crater
(B) Energies and conditions of crater formation with kinetic energies derived from 1/3.4 root scaling using Snowball Crater.[1]			
(1) Energy source	Kinetic energy of hypervelocity impact	Kinetic energy of hypervelocity impact	500-ton TNT surface tangent hemisphere
(2) Calculated energies[1] of formation	$\approx 1.0 \times 10^{24}$ ergs[2] $\approx 2.2 \times 10^{24}$ ergs[3] $\approx 4.7 \times 10^{24}$ ergs[4]	$\approx 1.2 \times 10^{24}$ ergs[2] $\approx 2.8 \times 10^{24}$ ergs[3] $\approx 3.7 \times 10^{24}$ ergs[4]	500-ton TNT
(3) HOB	Estimated ~ at surface to −100 m	Estimated ~ at surface to −100 m	At ground surface
(4) Effective HOB	Estimated surface to ~ −100 m	Estimated surface to ~ −100 m	0.0–1.92 m[5]
(5) Assumed impacting body type from published data	Comet or stoney meteorite	Comet or stoney meteorite	N/A
(6) Assumed density for impacting body and energy density for Snowball Crater	≈ 3.3 g/cm^3 [6]	≈ 3.3 g/cm^3 [6]	TNT stacked charge

Table 1, Section (B). (*Continued*).

Parameters	Flynn Creek impact crater	Steinheim impact crater	Snowball explosion crater
(7) Selected impact velocity for calculations	15 and 25 km/sec[6]	15 and 25 km/sec[6]	N/A
(8) Calculated mass of impacting body	$\approx 22.2 \times 10^{11}$ g and 8.00×10^{11}	$\approx 24.9 \times 10^{11}$ g and $\approx 8.96 \times 10^{11}$ g	N/A
(9) Calculated volume of impacting body	$\approx 6.73 \times 10^{11}$ cm^3 and $\approx 2.42 \times 10^{11}$ cm^3	$\approx 7.54 \times 10^{11}$ cm^3 $\approx 2.72 \times 10^{11}$ cm^3	N/A
(10) Calculated diameter of impacting body	$\approx$109 m and $\approx$77 m[7]	$\approx$113 m and $\approx$80 m[7]	N/A

[1]The 1/3.4 root scaling is used in this paper as the best first-order approximation for larger craters in this size range (Cooper, 1977, this volume; Roddy *et al.*, 1975).

[2]Volume scaling using average of rim crest, apparent, and true diameters of Flynn Creek and Steinheim with Snowball where $E \propto V$.

[3]Equivalent length factor, as derived in entry (199) in Section A, is ~30 for Flynn Creek and ~32 for Steinheim from technique suggested by Port (1977, personal communication). Energies calculated in (2), (3), and (4) should be considered conservative representative estimates due to unknown energy partition factors between explosion and impact processes in hard rock and alluvial target media.

[4]Diameter scaling using average of rim crest, apparent and true diameters of Flynn Creek and Steinheim with Snowball where $E \propto D^{3.4}$.

[5]The hemispherical charge was detonated in the center and at the base, however, Port (1977, personal communication) suggests that the center gravity of the charge can be considered as the effective HOB.

[6]Assumed density of 3.3 and average impact velocity of 15 and 25 km/sec from Shoemaker (1977, this volume) as a reasonable range of current estimates.

[7]Should be considered conservative representative estimates as noted in (3). Used 2.5×10^{24} ergs.

Parameters	Flynn Creek impact crater	Steinheim impact crater	Snowball explosion crater
(C) Pre-impact conditions			
(1) Age of crater formation	~360 m. y.	~14.7 m. y.	June, 1964
(2) Regional rx types	Limestone, dolomite	Limestone, dolomite, marlstone,	Unconsolidated sands, silts, clays
(3) Regional structure	Flat-lying	Flat-lying	Flat-lying
(4) Topography at impact time	Rolling hilly coastal plain and possible shallow water	Rolling hills	Flat prairie
(D) Morphology of crater and post crater modifications			
(1) Initial crater morphology	Flat-floored with central uplift	Flat-floored with central uplift	Flat-floored with central uplift
(2) Initial crater floor shape inferred from drilling or visible	Flat, hummocky	Flat, hummocky	Flat, hummocky

Table 1, Section (D). (*Continued*).

Parameters	Flynn Creek impact crater	Steinheim impact crater	Snowball explosion crater
(3) Central uplift shape	Domical	Domical	Domical, with multi peaks
(4) Initial crater wall shape	Terraced	?	Terraced
(5) Immediate erosion	Crater wall slumping and terracing	Crater wall slumping and terracing	Crater wall slumping and terracing
(6) Immediate deposition	Marine black shale	Lake deposits	Subsurface piped sands and silts
(7) Long-time erosion	Remove ejecta blanket and erode walls and peak	Remove ejecta blanket and erode walls and peak	Wind erosion minimal
(8) Long-time deposition	Fill with black shale	Partly filled with lake deposits	In fill of one meter of piped sands
(9) Morphology at time of filling	Filled	Crater with central peak	Crater with central peak
(10) Morphology today	Buried	Crater with central peak	Crater with central peak
(11) Degree of erosion and exposure today	Buried with partial exposure of valley walls and floors, moderate rock exposures	Ejecta removed crater walls and peaks slightly eroded, moderate rock exposures	No longer exists
(E) Structure of craters			
(1) Extent of breccia lens	Covers crater floor and flanks of uplift	Covers crater floor and flanks of uplift	Covers floor and most of uplift
(2) Breccia lens fragment type	Limestone and dolomite	Limestone, dolomite, marlstone, claystone	Alluvium
(3) Breccia lens fragment-size range	Sub-mm to >100 m	Sub-mm to >100 m	Sub-mm to ~1 m
(4) Breccia lens matrix	Carbonate	Carbonate	None
(5) Faulted, folded, fractured, brecciated rock below breccia lens	Yes	Yes	Yes
(6) Forced injection of breccia below breccia floor	Yes	Yes	Yes
(7) Central uplift	Yes	Yes	Yes
(8) Deformation style in central uplift	Domical, faulted, folded, brecciated	Domical, faulted, folded, brecciated	Domical, faulted, folded, brecciated
(9) Mixed stratigraphy and lithologies in central uplift	Yes	Yes	Yes

Table 1, Section (E). (*Continued*).

Parameters	Flynn Creek impact crater	Steinheim impact crater	Snowball explosion crater
(10) Galgenbergs, i.e. massive faulted or thrust blocks in breccia lens	Yes	Yes	Possible
(11) Megabreccia of massive blocks of mixed lithologies in breccia lens	Yes	Yes	Yes
(12) Authigenic breccia in crater walls and inner rim	Yes	Yes	Yes
(13) Allogenic breccia in crater walls and inner rim	Yes	Yes	Yes
(14) Injection breccia in crater wall region	Yes	Yes	Poorly developed
(15) Faulted, folded, fractured, brecciated wall and rim rx	Yes	Yes	Yes
(16) Crater average slopes	25° to 78°, 46.5° avg	?	30°
(17) Terraced crater walls	Yes	Probable	Yes
(18) Normal faulting in walls and rim	Yes	Yes	Yes
(19) Reverse faulting in wall and rim	Yes	Yes	Yes
(20) Thrust faulting walls and rim	Yes	Yes	Yes
(21) Large thrust sheets in inner wall and rim region	Yes	Yes	Moderate
(22) Max hi-angle fault displacement measured in wall and rim	>100 m	>10 m	~1 m
(23) Max thrust displacement measured in wall and rim	>50 m	>20 m	>2 m
(24) Monoclinal folding in walls and rim	Yes	Yes	Yes
(25) Anticlinal folding in walls and rim	Yes	Yes	Yes
(26) Synclinal folding in walls and rim	Yes	Yes	Yes
(27) Axial planes of folding in rim concentric to crater walls	Yes	Yes	Yes

Table 1, Section (E). (*Continued*).

Parameters	Flynn Creek impact crater	Steinheim impact crater	Snowball explosion crater
(28) Axial planes of folding in rim radial to crater walls	Yes	Yes	Yes
(29) Max shortening of rock positions radial to crater due to folding (not rock compression)	35%	>10%	>10%
(30) Concentric and radial fault, fold, or fracture system in rim	Yes	Yes	Yes
(31) Crater partly or completely filled with sediments	Filled	Nearly filled	N/A
(F) Ejecta at craters			
(1) Ejecta blanket surrounded crater	One area still remaining in graben	?	Yes
(2) Inverted stratigraphy in ejecta blanket forming overturned flap	Very poorly preserved in graben	?	Yes
(3) Fragment type	Carbonate rx	Carbonate rx	Alluvium
(4) Fragment-size range	Sub-mm to >100 m	Sub-mm to >100 m	Sub-mm to ~1 m
(5) Hummocky terrain on overturned flap of ejecta blanket	?	?	Yes
(6) Rayed extensions of outer edge of overturned flap of ejecta blanket	?	?	Yes
(G) Shock metamorphism			
(1) Fractured, shattered rock or alluvium in fragments in breccia lens, central uplift, below breccia lens, in crater walls and rim	Yes	Yes	Yes
(2) Shatter cones in central uplift	Yes	Yes	No
(3) Shatter cones in breccia lens fragments	?	Yes	No
(4) Shatter cones in ejected fragments	?	Yes	No

Table 1, Section (G). (*Continued*).

Parameters	Flynn Creek impact crater	Steinheim impact crater	Snowball explosion crater
(5) Shatter cones in limestone	Yes	Yes	N/A
(6) Shatter cones in dolomite	Yes	?	N/A
(7) Shatter cones in marlstone, sandstone	N/A	Yes	N/A
(8) Maximum size of shatter cones observed	~1 cm	~25 cm	N/A
(9) Shatter cones avg. angle	~90°	~90°	N/A
(10) Shatter cones in drill core	?	Yes	N/A
(11) Planar features in quartz	Yes	Yes	Yes
(12) Planar features on rational orientations in calcite	Yes	Yes	Probable
(13) Planar features on irrational orientations in calcite	Yes	?	Probable
(14) Enhanced microfracture on rational orientations in calcite	Yes	Yes	?
(15) Enhanced microtwin lamellae in calcite	Yes	Yes	?
(16) Enhanced normal twin lamellae in calcite	Yes	Yes	?
(17) Micro-bending and folding in calcite crystals	Yes	?	?
(18) Fused material	Probable	Possible	n.d.
(H) Field surface studies			
(1) Field mapping scales	1:6000	1:8000	1:300
(2) Contour Interval	3.05 m	5 m	0.305 m
(3) Stratigraphic sections reported	Yes	Yes	Yes
(4) Rock collections	Yes	Yes	Yes
(5) Published reports	Yes	Yes	Yes
(I) Petrographic, geochemical, and materials studies			
(1) Petrographic studies on fragments	Yes	Yes	Yes
(2) Petrographic studies on polished thin sections	Yes	Yes	No

Table 1, Section (I). (*Continued*).

Parameters	Flynn Creek impact crater	Steinheim impact crater	Snowball explosion crater
(3) Petrofabric studies	Yes	Limited	Limited
(4) Major element spectrographic studies	Yes	?	Yes
(5) Minor element spectrographic studies	Yes	?	Yes
(6) Trace element spectrographic studies	Yes	?	Yes
(7) Rock material properties studies	No	No	Yes
(J) Geophysical studies			
(1) Gravity map scale	1:6000	1:10,000	N/A
(2) Magnetic map scale	1:6000	1:10,000	N/A
(K) Core and rotary drilling			
(1) Drill core total length	~762 m	134 m	~300 m
(2) Percent drill core recovery	99.99%	~90%	~100%
(3) Combined drill core/rotary	~762 m	~2150 m	~300 m
(4) Rotary drill total length	0	~2068 m	?
(5) Cores or cuttings available	Cores	Cores & cuttings	Yes
(6) Breccia lens and underlying rocks drilled	Yes	Yes	Trench excavated
(7) Central uplift drilled		Yes	Trench excavated
(L) Other studies			
(1) Nearest volcanic or mineralized rock and type at or near surface	25 km to calcite, barite, fluorite, talena sphalerite mineralized zones zinc deposits probable originally at depth	40 km to volcanic pipes of basalt and basaltic ashes mixed with sedimentary rock and lapilli	N/A
(2) Studies on filling sediments	Yes	Yes	Yes
(3) Complete trenching of crater	No	No	Yes
(4) Subsurface terminal displacement marker studies	No	No	Yes
(5) Theoretical numerical simulation calculation	No	No	Yes
(6) Published references	>10	>15	>25

Conclusions

The tabular data in Table 1 summarizes 340 selected dimensional, morphological, structural and related features for three craters in the class which exhibit broad flat-floors with central uplifts. The basic purpose of this tabulation is to provide a detailed set of dimensional and structural comparative data on a common class of craters for use in both terrestrial and planetary studies. A second intent has been to offer an outline of selected values that have proven useful in certain numerical calculations, and which should also be useful as a guide in collecting basic parametric data from other craters.

One of the more useful results is the preliminary development of an *equivalent length factor* which provides general scaling relationships for these impact and explosion craters. Although these factors were derived specifically for this set of craters, it appears that these values could be extended as *general* scaling values to other craters of the same class, at least through a dimensional range of ~30–50. Certain of the parametric ratios calculated in Table 1 should be useful to include in future data assemblages involving scaling studies.

The complexities of the energy calculations and scaling ratios were only briefly touched upon in the footnotes to the tabulations; however, it does appear that diameter, depth, volume, and mass scaling with energy are useful in these correlations. The narrow range of 1.0 to 4.7×10^{24} ergs, derived from three different calculation techniques, supports the generalization of $\sim 10^{24}$ ergs as an acceptable minimum estimate for the formational energies of Flynn Creek and Steinheim. The problems associated with 1/3 versus 1/3.4 root scaling, height of burst considerations, energy transfer functions, and a number of other cratering process problems require more complete solutions before the simple energy scaling used in this paper can be considered other than presenting representative values.

Aside from the tabular presentation of crater data, several aspects related to crater formation might also be noted from the comparative outline. One of the more prominent points emphasized by the outline format is the close similarities between the dimensions, morphology, and structure for the two impact craters. Such close similarities argue that the initial conditions of impact and the subsequent cratering processes were not only essentially identical for these two craters, but that they represent *general* cratering processes which may be expected at other flat-floored craters with central uplifts. This should hold, at least for the general size range discussed in this paper. Furthermore, the close morphological and structural analogs between the Snowball Crater, formed by a *surface* explosion, and the Flynn Creek and Steinheim craters also implies that the center of energy of the impacting bodies had a kinetic energy transfer at or near their respective ground surfaces. This is consistent with, but does not prove, the argument that these types of impacts are the result of lower density bodies, i.e., such as carbonaceous chondrites or cometary masses (Roddy, 1968, 1976, 1977a,b).

Another consequence of the establishment of such a high degree of morphological and structural similarity is that structural deformation found at these well-preserved craters should prove helpful in predicting or interpreting terrestrial scaling and general deformational styles at other impact craters of the same morphological class. This would be useful for craters which have been greatly eroded or which exist on other planets and cannot be examined first-hand. In other words, if we have a given crater morphology, we should be able to predict with increased confidence the general styles of deformation, at least within a limited size range. This assumes essentially comparable, though not necessarily identical, media response.

One example of the use of such analog data can be inferred from the ejecta blanket at Snowball. Since the morphology and structure of Snowball are essentially identical with that at Flynn Creek and Steinheim, it follows that these and other impact craters of the same class also had overturned flaps with inverted stratigraphy as at Snowball. Certainly, Shoemaker (1960, 1963) has demonstrated this feature for Meteor Crater in Arizona. Further documentation can be found for such a generalization to larger impact craters in Gault *et al.* (1968), Oberbeck (1971, 1977), Quaide and Oberbeck (1968), Roddy *et al.* (1975), and Roddy (1976, 1977a). In this particular example, we suggest that the use of the comparative analog data substantiates the present interpretation of most planetary workers that impact craters with flat-floors and central uplifts on the other planets and the moon are surrounded by ejecta blankets of inverted stratigraphy (Head, 1976; McCauley *et al.*, 1977).

Another example of the use of the tabular parametric data is to examine the repositioning of deeper rocks in central uplifts. Each of the flat-floored craters have target materials raised through a distance equal to approximately 10% or more of their crater diameter. This was suggested in earlier studies (Milton and Roddy, 1972; Roddy, 1968, 1976), but the dimensional data in Table 1 now appears more conclusive when all three craters are compared. The implication is that one should be able to predict the range of depths from which rocks have been uplifted at other terrestrial and planetary craters of the same class. Gravitational effects would be included in this type of scaling if the proper parametric values are used.

The data presented here should be useful in testing our predicative and interpretative capabilities for *both* explosion and natural impact craters. As other workers continue to expand and refine such parametric data sets, our abilities to interpret the physical processes associated with impact cratering and how they affect planetary crustal evolution hopefully should also improve.

Acknowledgments—The author is especially indebted to Professor W. Reiff for his cooperation on the Steinheim study and for the use of certain of his data and for Fig. 2. Some of the Steinheim field measurements in this paper are part of another study by Dr. Reiff and the author.

The author wishes to express his appreciation to Dr. G. H. S. Jones, Defence Research Board, Staff, Canada, and J. Kelso and J. Lewis of the Defense Nuclear Agency, U.S. Department of Defense, who made it possible to collect the field data for the Snowball Crater. I also wish to personally thank Dr. E. Shoemaker, California Institute of Technology and U.S. Geological Survey, for his guidance during the Flynn Creek work. Special thanks are due to J. Alderman, U.S. Geological Survey, for his preparation of the special topographic map of Flynn Creek. I am also especially indebted to R. Port for his many helpful discussions on the dimensional parameters considered useful in current scaling and numerical calculations, and I am again indebted to Dr. G. Sevin and Captain J. Stockton of the Defense

Nuclear Agency for their continued cooperation in these studies. I also express my real appreciation to M., M. and M. Roddy for their continued field assistance in this work and to J. Roddy for her help in calculations and data preparation.

Reviews of this paper by Drs. G. H. S. Jones, D. Milton, R. Pike, R. Port, and W. Reiff materially assisted in improving the format and contents, and I express my thanks to each for their help. Calculations supplied by R. Port on the scaling ratios were essential in formulating that section, and discussions with R. Pike were helpful on certain aspects of the dimensional data.

This work was supported through the cooperation of National Aeronautics and Space Administration under contract W-13,130 and the Defense Nuclear Agency, Department of Defense.

REFERENCES

Arthur, D. W. G., Agnieray, A. P., Horvath, R. A., Wood, C. A. and Chapman, C. R.: 1963, The Systems of Lunar Craters, Quadrant I, *Comm. Lunar Planetary Lab.* **2**, no. 30, 72 pp.

Arthur, D. W. G., Agnieray, A. P., Horvath, R. A., Wood, C. A. and Chapman, C. R.: 1964, The System of Lunar Craters, Quadrant II, *Comm. Lunar Planetary Lab.* **3**, no. 40, 71 pp.

Arthur, D. W. G., Agnieray, A. P., Pellicori, R. H., Wood, C. A. and Weller, T.: 1965, The System of Lunar Craters, Quadrant III, *Comm. Lunar Planetary Lab.* **3**, no. 50, 158 pp.

Arthur, D. W. G., Pellicori, R. H., and Wood, C. A.: 1966, The System of Lunar Craters, Quadrant IV, *Comm. Lunar Planetary Lab.* **5**, no. 70, 207 pp.

Arthur, D. W. G.: 1974, Lunar Crater Depths from Orbiter IV Lunar Crater Depths from Orbiter IV Long-Focus Photographs, *Icarus* **23**, 116–131.

Baldwin, R. B.: 1949, *The face of the Moon*, Univ. Chicago Press, Chicago. 230 pp.

Baldwin, R. B.: 1963, *The Measure of the Moon*, Univ. Chicago Press, Chicago. 488 pp.

Branco, W., and Fraas, E.: 1905, Das kryptovulkanische Becken von Steinheim. *K. Preuss. Akad. Wiss. Abh.*, Berlin, p. 1–64.

Bucher, W. H.: 1933, Cryptovolcanic structures in the United States, *Internat. Geol. Cong., 16th, Washington, D.C., Rept., 2,* pp. 1055–1084, 1936.

Burt, J., Veverka, J. and Coole, K.: 1976, Depth-Diameter Relation for large Martian Craters determined from Mariner 9 UVS altimeter, *Icarus* **29**, 83–90.

Cintala, M. J., Head, J. W., and Mutch, T. A.: 1975, Depth/diameter relationships for martian and lunar craters (abstract), *EOS* (*Trans. Amer. Geophys. Union*) **56**, 389.

Cintala, M. J., Head, J. W. and Mutch, T. A.: 1976a, Characteristics of fresh martian craters as a function of diameter: Comparisons with the Moon and Mercury, *Geophys. Res. Lett.* **3**, 117–120.

Cintala, M. J., Head, J. W., and Mutch, T. A.: 1976b, Craters on the Moon, Mars and Mercury: A comparison of depth/diameter characteristics (abstract). In *Reports of Accomplishments of Planetology Programs, 1975–1976.* NASA TMX-3364, p. 186–187.

Cooper, H. F.: 1977, A summary of Explosion Cratering Phenomena Relevant to Meteor Impact Events. In *Impact and Explosion Cratering* (D. J. Roddy, R. O. Pepin, and R. B. Merrill, eds.). This volume.

Dence, M. R.: 1972, Meteorite Impact Craters and the structure of the Sudbury Basin, *Geol. Assoc. Canada, Special Paper 10*, p. 7–18.

Dence, M. R.: 1973, Dimensional analysis of impact structures (abstract), *Meteoritics* **8**, 343–344.

Diehl, C. H. H. and Jones, G. H. S.: 1967a, *The Snowball Crater General Background Information,* Suffield Technical Note No. 187, unclassified, 24 pp.

Diehl, C. H. H., and Jones, G. H. S.: 1967b, The Snowball Crater, profile and ejecta pattern, Defense Research Establishment Suffield, Canada, Suffield Tech. Note No. 188, unclassified.

French, B. M. and Short, N. M., eds.: 1968, *Shock metamorphism of Natural Materials.* Mono Book Corp., Baltimore, 644 pp.

Gault, D. E., Guest, J. E., Murray J. B., Dzurisin D., and Malin, M. C.: 1975, Some comparisons of impact craters on Mercury and the Moon. *J. Geophys. Res.* **80**, 2444–2460.

Gault, D. E., Quaide, W. L., and Oberbeck, V. R.: 1968, Impact Cratering Mechanics and Structures, In *Shock Metamorphism of Natural Materials* (B. M. French and N. M. Short, eds.), p. 87–99, Mono Book Corp., Baltimore.

Groschopf, P. and Reiff, W.: 1966, Ergebnisse neuer Untersuchungen im Steinheimer Becken (Württemberg). *Jh. Ver. vaterl. Naturkde. Württemberg* **121**, p. 155–168.

Groschopf, P. and Reiff, W.: 1967, Neue Untersuchungen in Steinheimer Becken, *Fortschr. Miner.* **44**, H. 1, p. 141–142.

Groschopf, P. and Reiff, W.: 1969, Das Steinheimer Becken ein Vergleich mit den Ries. *Geologica Bavarica* **61**, p. 400–412.

Groschopf, P. and Reiff, W.: 1970, Die zentrale Erhebung, Steinhirt-Klosterberg im Steinheimer Becken (Schwäbische Alb), *Jber. u. Mitt. oberrh. geol. Ver.* **52**, 169–174.

Groschopf, P. and Reiff, W.: 1971a, Vorläufige Ergebnisse der Forschungsbohrungen 1970 im Steinheimer Becken (Schwäbische Alb), *Jh. geol. Landesamt Baden-Wurttemberg* **13**, 223–226.

Groschopf, P. and Reiff, W.: 1971b, Es war ein Meteoreinschlag, Ergebnis der Bohrungen im Steinheimer Becken, *Kosmos* **12**, Stuttgart, p. 521–525.

Head, J. W.: 1975, Processes of Lunar Crater Degradation: Changes in Style with Geologic Time, *The Moon* **12**, 299–329.

Head, J. W.: 1976, Evidence for the sedimentary origin of Imbrium sculpture and lunar basin radial texture, *The Moon* **15**, 445–462.

Hörz, F.: 1971, ed. of Meteorite impact and volcanism, *J. Geophys. Res.* **76**, 5381–5798.

Jones, G. H. S.: 1964, *Preliminary Report on the Canadian Projects in the (1964)* 500 *Ton TNT Suffield Explosion*, Suffield Special Publication 45, unclassified, 140 pp.

Jones, G. H. S.: 1976, *The Morphology of Central Uplift Craters*. Defence Research Establishment, Suffield, Canada, Suffield Report No. 281, unclassified, 212 pp.

Jones, G. H. S.: 1977, Complex Craters in Alluvium. In *Impact and Explosion Cratering* (D. J. Roddy, R. O. Pepin, and R. B. Merrill, eds.). This volume.

Kranz, W.: 1924, Das Steinheimer Becken. In *Begleitworte zur geognostischen Spezialkarte von Württemberg*. Atlasblatt Heidenheim, 2. Aufl.—Württbg. Statist. Landesamt, Stuttgart, 138 pp.

Leighton, R. B., Murray, B. C., Sharp, R. P., Allen, J. D., and Sloan, R. K.: 1967, *Mariner IV pictures of Mars*, Jet Propulsion Laboratory Technical Report 32–884, Part I, Investigator's Report. 178 pp.

McCauley, J. F., Scott, D. H., Roddy, D. J., and Boyce, J. M.: 1977, *The Geology of the Orientale Basin of the Moon*. (In press).

McGetchin, T. R., Settle, M., and Head, J. W.: 1973, Radial thickness Variation in Impact Crater Ejecta: Implications for Lunar Basin Deposits, *Earth Planet. Sci. Lett.* **20**, 226–236.

Milton, D. J. and Roddy, D. J.: 1972, Displacements within impact craters, *Proc. 24th Internat. Geol. Congr., Section 15* (*Planetology*) p. 119–124.

Moore, H. J., Hodges, C. A., and Scott, D. H.: 1974, Multi-ringed basins illustrated by Orientale and associated features, *Proc. Lunar Sci. Conf. 5th*, p. 71–100.

Oberbeck, V. R.: 1971, Laboratory simulation of impact cratering with high explosives, *J. Geophys. Res.* **76**, 5732.

Oberbeck, V. R.: 1977, Application of High Explosive Cratering Data to Planetary Problems. In *Impact and Explosion Cratering* (D. J. Roddy, R. O. Pepin, and R. B. Merrill, eds.). This volume.

Öpik, E. J.: 1976, *Interplanetary Encounters*, Elsevier Scientific Publishing Co., New York, 155 pp.

Pike, R. J.: 1968, Meteoritic origin and consequent endogenic modification of large lunar craters—a study in analytical geomorphology, Ph.D. thesis, Univ. Michigan, Ann Arbor. 404 pp.

Pike, R. J.: 1972, *Apollo 16 Prelim. Sic. Report*, NASA SP-315, 29–56 to 29–61.

Pike, R. J.: 1973, *Apollo 17 Prelim. Sci. Report*, NASA SP-330, 32-1 to 32-7.

Pike, R. J.: 1974a, Craters on Earth, Moon, and Mars: Multivariate Classification and Mode of Origin, *Earth Planet. Sci. Lett.* **22**, 245–255.

Pike, R. J.: 1974b, Depth/Diameter Relations of Fresh Lunar Craters: Revision from Spacecraft Data, *Geophys. Res. Lett.* **1**, 291–294.

Pike, R. J.: 1976, Crater dimensions from Apollo data and supplemental sources, *The Moon* **15**, 463–477.

Pike, R. J.: 1977, Size-dependence in the Shape of Fresh Impact Craters on the Moon. In *Impact and Explosion Cratering* (D. J. Roddy, R. O. Pepin, and R. B. Merrill, eds.). This volume.

Quaide, W. L., Gault, D. E., and Schmidt, R. A.: 1965, Gravitative effects on lunar impact structure, *Ann. N.Y. Acad. Sci.* **123**, 563–572.

Quaide, W. L. and Oberbeck, V. R.: 1968, Thickness Determinations of the Lunar Surface Layer from Lunar Impact Craters, *J. Geophys. Res.* **76**, 5247.

Reiff, W.: 1974, Einschlag Krater Kosmischer Körpel auf der Schwäbischen and Fränkischen Alb. In *der Aufschluss* 7/8, 368 pp.

Roddy, D. J., Pepin, R. O., and Merrill, R. B., editors.
(1977) *Impact and Explosion Cratering*, Pergamon Press (New York), p. 163–183.
Printed in the United States of America

Complex craters in alluvium

G. H. S. JONES

Defence Research Board Staff, Department of National Defence, Canada.

Abstract—Experimental data are given on the morphology of several complex craters produced by the detonation of large TNT charges on the surface of alluvium at the Defence Research Establishment, Suffield, Canada. Primary attention is concentrated upon the Snowball Crater, the excavation of which is described in detail in terms of a diametral cross section along a system of displacement marker sand columns, and these displacement data are amplified by stratigraphic data. Comparable or supplementary data are given for several other explosive craters, particularly the Prairie Flat Crater.

It is shown that the craters produced in the highly compactible alluvium of the Suffield site exhibit a hierarchy which shows consistent pattern from craters with a simple, dome like central uplift to the final stage of a complex ringed crater with central uplift, marked radial, and concentric fissuring, a pseudo-volcanic extrusion of fluidized material along both radial and concentric structures, and terracing produced by late stage slumping. It is further shown that the ejecta blanket consists of a coherently overturned flap in which the original stratification is preserved, inverted, even when the stratum consist of fine, free flowing sand. The paper is essentially a summary of much more detailed individual reports which have only received limited circulation, to which reference is made in the text.

INTRODUCTION

STUDIES OF THE EFFECTS of TNT explosions with yields up to the equivalent of 1 KT nuclear were carried out over a period of many years at the Defence Research Establishment, Suffield, Canada. Many of the explosions produced complex craters exhibiting ringed structures, central uplift, and coherently overturned ejecta blankets.

Geological, seismic, and soil studies of the test sites have been given by (or cited by) Jones (1963, 1970, 1976). The present paper discusses craters produced on the Watching Hill Range, a site which consisted of horizontally stratified lacustrine silts, sands, clays, and gravels overlying glacial Till above a bedrock of softly indurated cretaceous shales and sandstones. The depth to bedrock was variable but close to 60 m, and a limited amount of free water appeared in boreholes at a depth of 10 m. However, many of the significant phenomena appear to be dependent upon the behaviour of the unsaturated but highly porous and compactible lacustrine sediments, as they were compressed to and beyond the saturation point.

The experimental data produced by the excavation of the central uplift crater called Snowball are summarized in this paper. Data from other trials, in particular the Prairie Flat ringed crater are given where these supplement the Snowball data. Full details of the excavations of both Snowball and Prairie Flat have been given by Jones (1970, 1976), both references being reports printed in limited numbers but now in the public domain.

Fig. 1. Aerial view of Snowball Crater.

The Snowball Crater was formed by the detonation of a 500 ton hemisphere of TNT on the surface, in July 1964, and Prairie Flat Crater by the detonation of a 500 ton sphere tangential to the surface on 8 August 1968. Some other trials are mentioned in the text.

Attention is drawn to Roddy (1968, 1969, 1976), which discuss many of the features of the Suffield craters using both the present writer's data and supplementary data collected by Roddy at the Suffield test site.

The Snowball Crater before excavation

Figure 1 is an aerial photograph of the Snowball Crater the day after its formation. The ground surrounding the main crater was depressed to such an extent that the ejecta did not form an elevated rim. A large structural uplift

Fig. 2. Pseudo-volcanic sand cones aligned on radial fissure.

occupied the approximate centre of the crater. From this uplift a gusher of sand-laden water rose to a height of about a metre with a stream diameter of about half a metre. Smaller gushers existed at various points around the inner slope of the crater and beyond the crater rim. By the second day the crater was partly filled with water and alluvium. The inundated areas beyond the rim, the circumferential cracks and the terracing show clearly in the photograph.

Fig. 3. Aerial view of Prairie Flat, with sand cones on concentric structures.

Many of the water sources had the appearance shown in Fig. 2, which shows a chain of small cratered sand cones situated along a fissure radial to the main Snowball crater. On the Prairie Flat crater alignments of cones were located on concentric ring structures within the main crater as shown in Fig. 3.

Circumferential and radial fissures were detected on several of the Suffield craters. In the case of Snowball the fissures were open at the ground surface, even extending through the ejecta blanket, except in the limited areas where sand and water had risen to the surface and sealed the fissures as shown in Fig. 2. At depth, however, as shown in Fig. 4, these fissures were plugged with a

Fig. 4. Section through circumferential dyke on Snowball.

close packed sand filling which penetrated the horizontal strata, including loosely packed sand strata, without spreading into "sills" along the stratification.

The sand column displacement markers

To obtain information on the origin of the ejecta blanket and to define the subsurface structural deformation the sand column technique of Perkins (1954) was modified as described by Diehl and Jones (1965) and used on the Suffield trials. In this technique numbered markers were placed at known intervals in vertical columns of mortar installed in boreholes. The pre- and post-shot locations of these columns and markers were surveyed to give information on the movement of the ground. It is stressed that the vector displacements do not indicate dynamic trajectories. Excavation of the sand columns, and the associated stratigraphical studies were completed during two consecutive summer seasons.

Section through Snowball Crater

Figures 5–9, taken together, depict a section through the Snowball Crater as determined by excavation along the main sand column line. In Fig. 5 we approach the crater from the south side, an area in which the pre-explosion stratigraphy was a series of horizontal lacustrine beds. At approximately 98 m (320 ft) from the centre of detonation (GZ) we encounter sand column S2 which exhibits significant movement in an outward and downward direction. This movement, of about 1 m at each marker location, would not have been recognized by stratigraphic study alone. Sand columns S3 and S4, show a progressive change in the nature of the residual displacement in the ground. Contrary to what might have been expected the residual displacements become less, rather than greater, as one approached the crater. The outward movement vanishes, but the vertical depression remains. This is clearly connected with the presence of the major circumferential crack at about 69 m (225 ft) from GZ.

Figure 6 is the next section sequentially in towards GZ and shows the crater wall from the region of the circumferential crack to a point well within the crater proper. Sand columns S5 to S13 all show movement inwards toward the crater, and downward. The column S5, just within the circumferential crack, shows movement which is almost a mirror image of that associated with the column S4, just outside the crack. It indicates a residual movement which is little more than the necessary inward movement to allow the crack to open. The columns progressively nearer to the crater rim, however, show increasing inward and downward movement. The stratigraphy in this area correlates with the columns and confirms the downwarping of the strata.

There were numerous sub-vertical cracks, some relatively local and others being part of circumferential systems of cracks. The dip of the upper layers increases as one approaches the rim of the crater, and the boundaries among the strata exhibit numerous examples of folding, minor slumping, and systems of

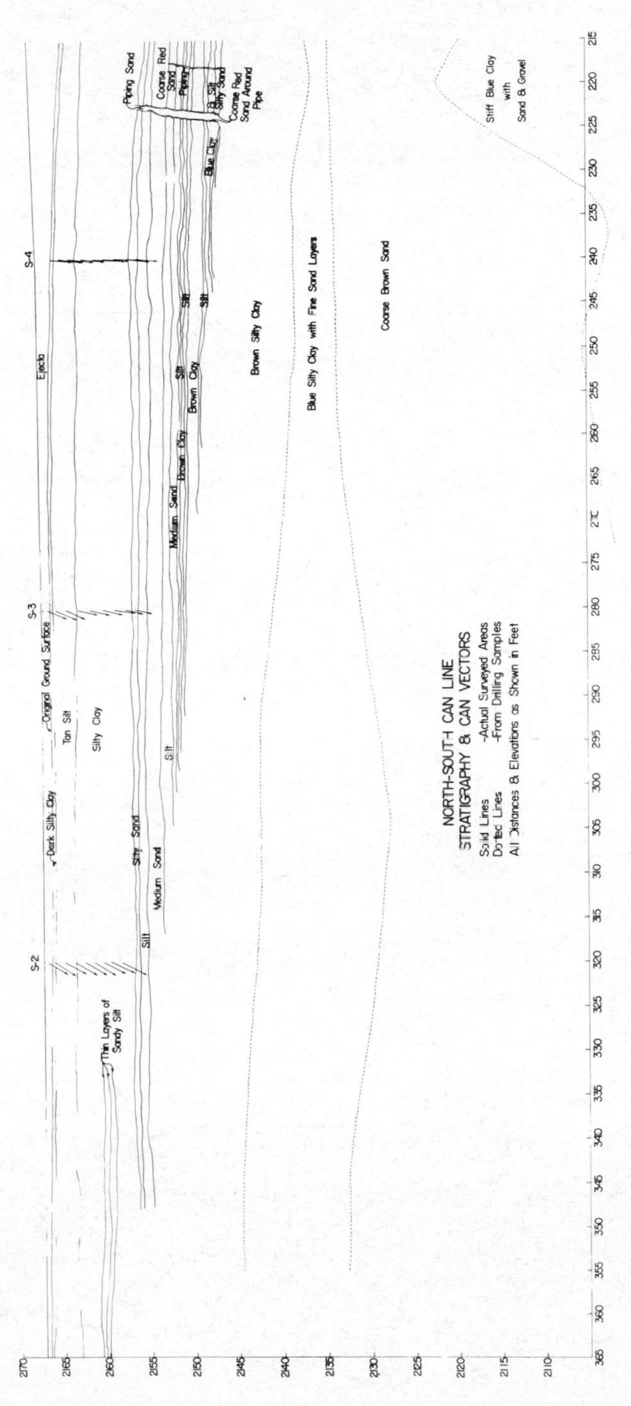

Fig. 5. Snowball section—part A.

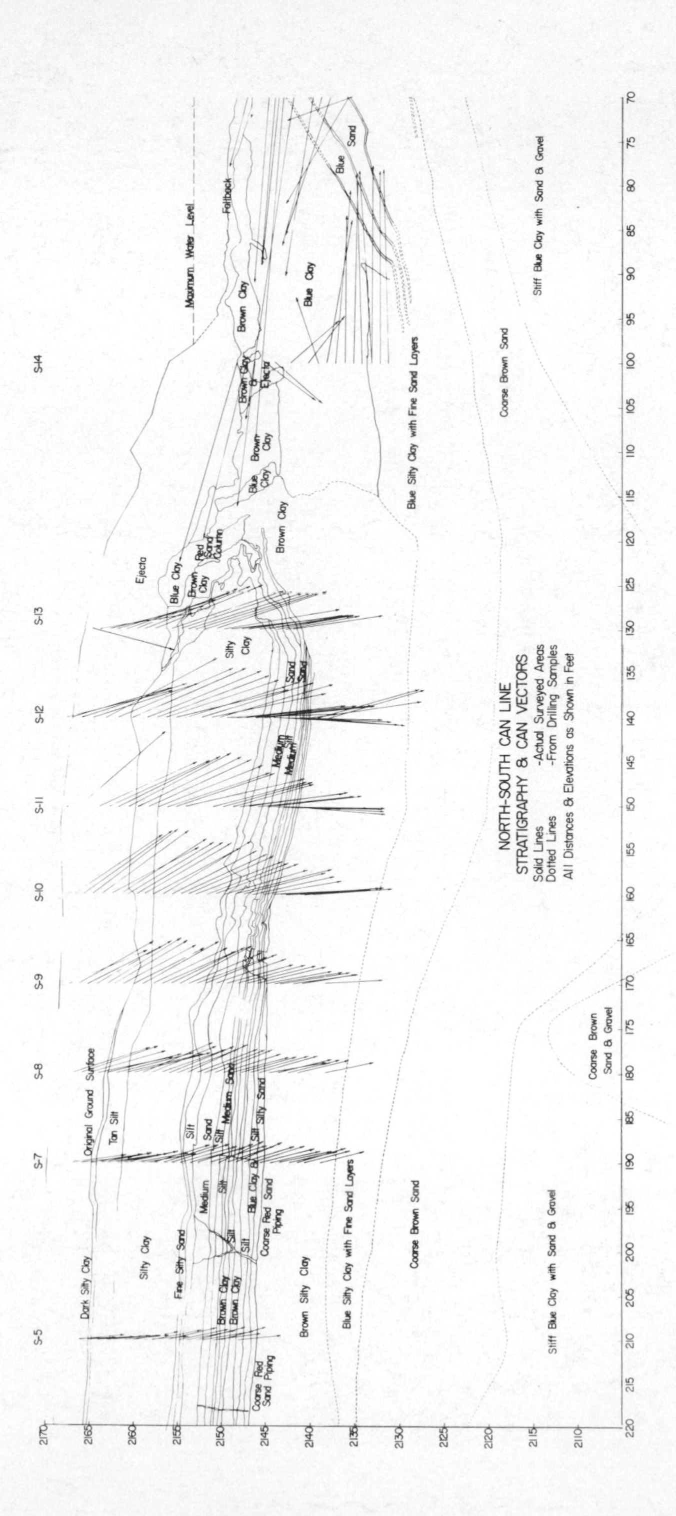

Fig. 6. Snowball section—Part B.

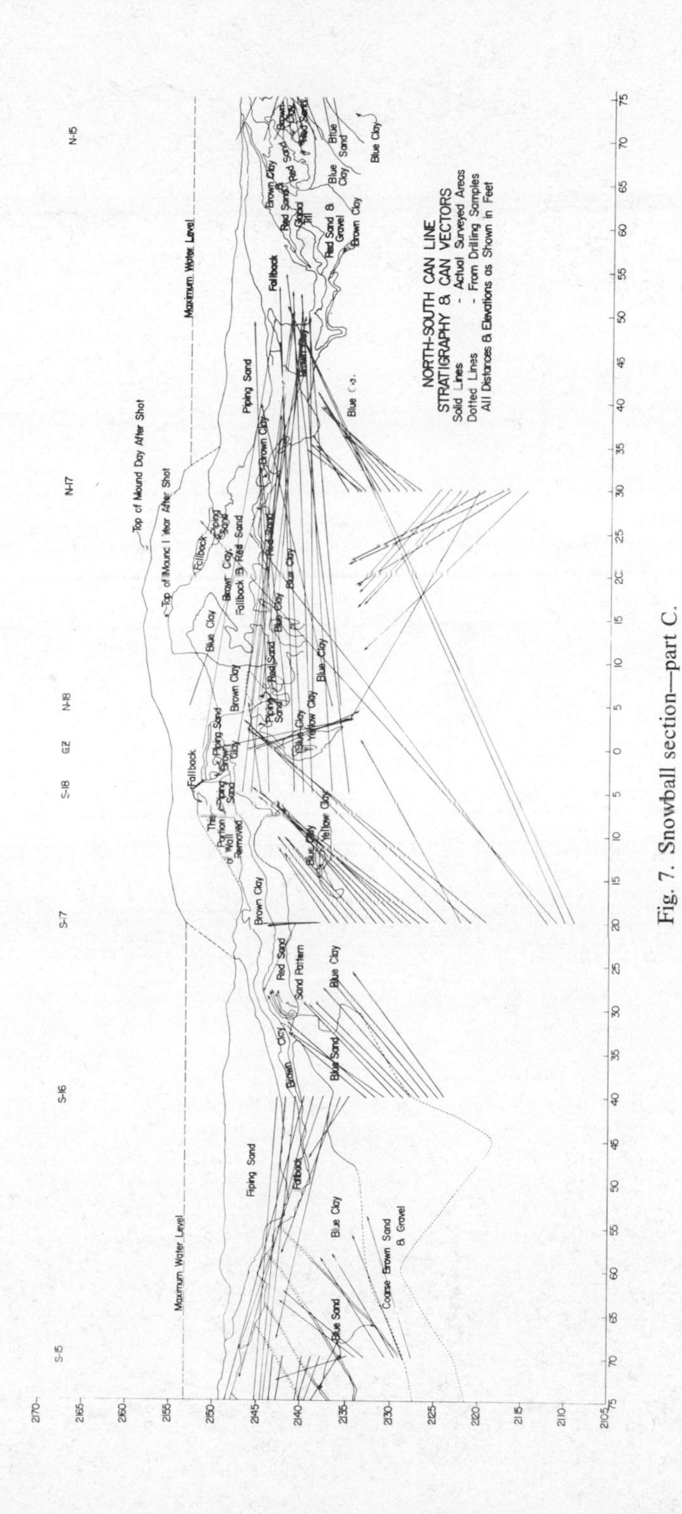

Fig. 7. Snowball section—part C.

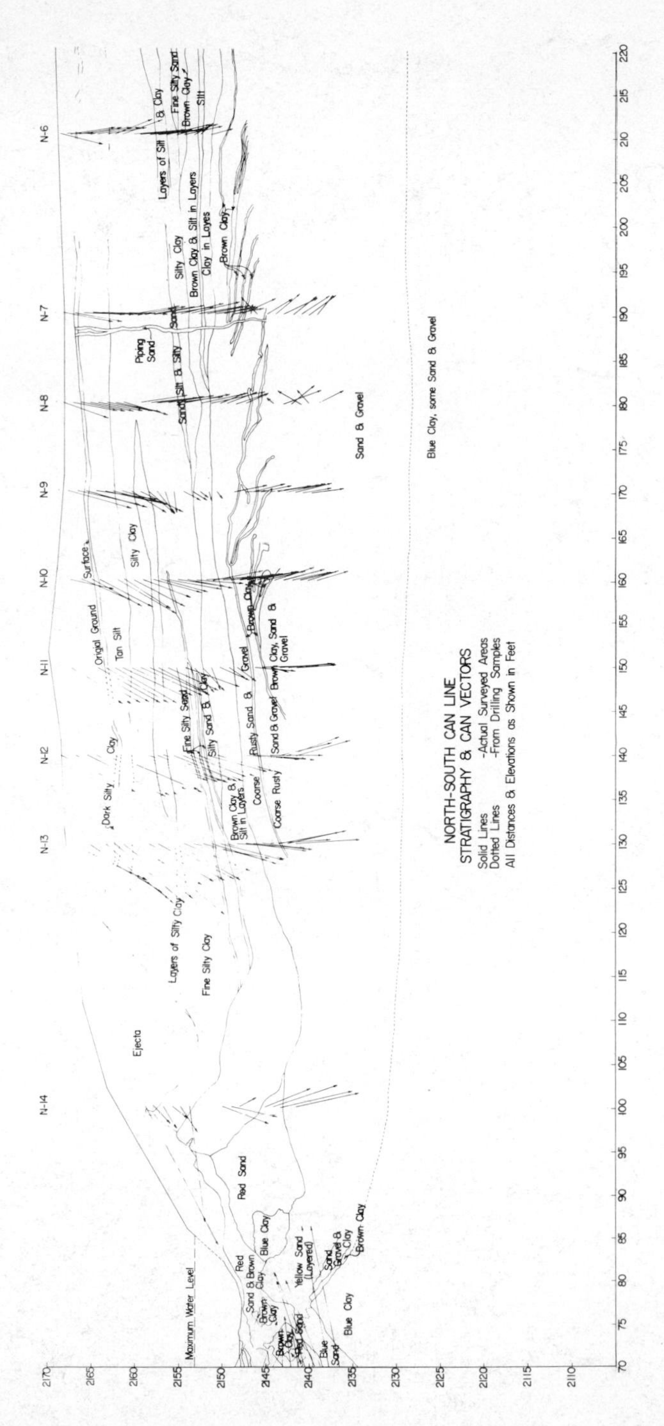

Fig. 8. Snowball section—part D.

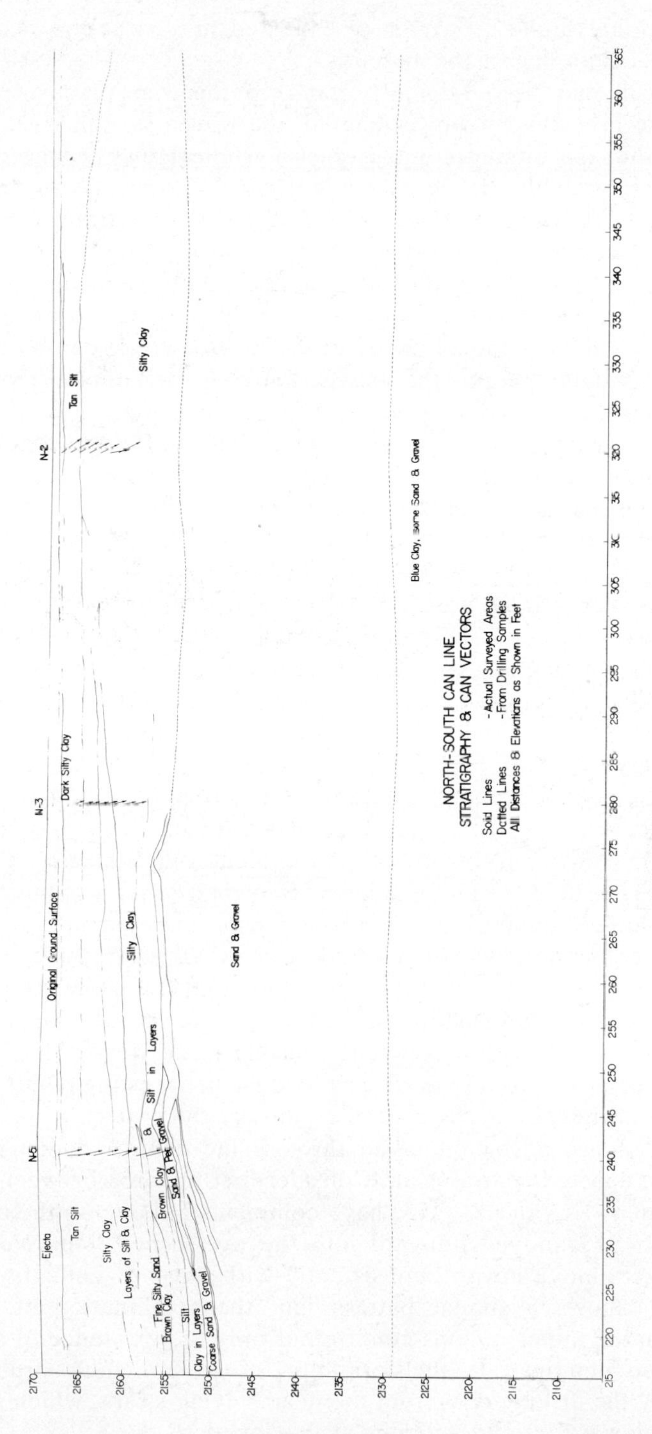

Fig. 9. Snowball section—part E.

conjugate normal faults as would be expected in a structure resulting from a stretching and slumping of the material.

If Snowball had been the only crater of this type excavated one would probably interpret this downwarping of the strata as being entirely due to slumping within the circumferential crack system. However the other two 500 ton craters at Suffield, Prairie Flat and Dialpack show no evidence at all of slumping, but do exhibit marked downwarping of the strata in this region. In both the last named craters, residual movement as indicated by the sand columns was *outward* and downward, thus denying the slump mechanism. Thus the data from Snowball represent a late phase slumping of previously downwarped strata. Figure 6 shows data obtained by core drilling to deeper depths than could be excavated, and demonstrates the downwarping of the contact between the blue clays and a coarse brown sand layer. These strata, after downwarping in this region, then rise into a major circumferential ridge within the crater proper. This downwarp beneath the rim and the upward slope within the crater is entirely consistent with the data from the other trials.

Column S13 is roughly coincident with the position of maximum downwarp. The motion of the upper markers in this column, and also of the adjacent columns, show a tendency for the top of the column to overturn outwards, and the stratigraphy shows a coherent overturning of the strata in this region despite the completely incompetent nature of the sand and silt layers. Figure 10 is a photograph of this region of Snowball, while Fig. 11 is the equivalent region for Prairie Flat. The overturned strata may be traced continuously to the limit of the ejecta blanket.

Within the crater proper, just beyond the region of coherent overturning shown in Figs. 10 and 11, there is an arch concentric with the crater, in all of the 500 ton craters. In Snowball, this is not very evident, but is clearly revealed by the drilling data. The surface manifestation on Snowball appears to have been masked by the late phase slumping and the heavy deposit of sediment.

The deeper markers of column S14 indicate very large movement inwards towards the centre, almost horizontally at this location, while the upper markers suffer large outward and slightly upward movement into the hinge region of the ejecta flap consistent with a coherent overturning of the material. Immediately below the floor of the crater we see in Fig. 6 the final resting points of markers in sand columns originally in the central regions of the crater.

Figure 7 shows a section taken through the central region of the crater, including the uplifted central peak. Consider first the area between the positions of sand columns S15 and S16. We have commented above on the upper markers from column S15, moved outward into the overturned flap. We see that the deeper markers move inward, consistently with those of columns S13 and S14. Column S16 shows a similar pattern, but the combination of S15 and S14 movement in the upper regions confirm not only the existence of an overturned flap, but also demonstrate the formation of an anticlinal structure. This is confirmed by the drilling data from the deeper lying strata, which show the blue clay forced upward into an anticline at this location.

Fig. 10. Overturned hinge region of Snowball.

The bottom series of markers from sand column S16 show the inward and upward displacement resulting in the formation of the central uplift structure. All the markers in column S17 confirm the formation of the structural uplift with the exception of the bottom four markers. The markers from the bottom four locations in the column were recovered in the anomalous position shown by the displacement vectors plotted in this figure. These vectors are quite inconsistent with the other displacement vectors, across which they pass. The markers for

Fig. 11. Overturned hinge region of Prairie Flat.

these locations were probably carried to their final resting points by the high pressure flow of water, which was most concentrated in the region of the central uplift. At some early stage in the formation of the central structure, a zone of distension, coupled with very high water pressures and hydraulic fracturing, could exist in this region and free the markers from the columns, and allow them to become entrained in the subsequent water flow.

The central uplift structure was not placed symmetrically at the ground zero of the detonation, but rather was displaced towards the north. The asymmetry is pronounced in the displacement vectors from sand columns S18 and N18. Initially these two sand columns were positioned symmetrically south and north of the GZ position respectively. Upon excavation, however, the markers from the upper levels of both of these columns were found on the north side of the crater. The recovery of the upper markers from N18 in this location is to be expected. To recover the S18 markers in this area is, however, inexplicable in terms of homogeneous material. However, the GZ of this detonation was very close to a transition zone between the horizontal lacustrine sands and silts and a system of gravel beds (a shoal or delta deposit). This factor of asymmetry in the cratered material is the only explanation that can be offered for the northward displacement of the central uplift, and the markers from sand column S18. It is, perhaps relevant that the main water source was located on the south side of the central uplift, that is to say almost exactly at GZ. The area of recovery of the markers is in a synclinal trough around the central uplift, in the area of the deepest deposit of sediment. The displacement vectors for the lower markers in column N17 are symmetrically arrayed with those from column S17, a mirror image which indicates clearly the central uplift formation of the structure.

The central structure consisted mainly of uplifted blue clay, with a capping of yellow clay-silt. Through this uplifted mass ran many fissures, which became filled with waterborne sand from considerable depth. It is probable that the material of the central mound was uplifted sufficiently to produce a tensional stress condition. In such a situation one would anticipate tension fractures as actually found in the central mound filled with waterborne sand. Excavation confirmed this structure, and also showed that the upper layers of the mound had been altered by heat or pressure to a comparatively strong material. This material shattered in the tension phase of the movement, as demonstrated in Fig. 12. This regular disruption of the material was quite different from the drying and cracking which the material would exhibit after long exposure to the air.

Figure 13 is a close-up photograph of one of the early sections cut through the central uplift. Attention is drawn to the massive up-domed blue clay (the relatively smooth bottom area of the photograph). The blocky material overlying this dome could, from this photograph, be thought to be "fall back" from the ejecta. This would definitely be a wrong interpretation as this material, as found *in situ*, was composed of large expanses of the yellow clay, essentially in conformable contact with the blue clay, but showing much greater disruption, evidence of a more frangible rather than plastic displacement. These massive "blocky" fragments of the yellow clay were not only conformably situated

Fig. 12. Tension fracturing of central uplift of Snowball.

above the blue clay but in some places the blue clay had actually flowed around the discrete blocks, completely encasing the blocks as large inclusions in the plastic blue clay. This is clearly visible in the photograph of Fig. 13, which also shows the presence of hydraulic fractures filled with sand.

Figure 14 shows the comparatively flat floor of the crater, formed by the great thickness of sedimentary infilling, overlying the thin layer of fall back material above the brecciated yellow clay which in turn overlies the plastic blue clay. The contact between the blue and yellow clays shows much relief, while the blue clay shows both included breccia of the yellow clay and some hydraulic fracturing.

Fig. 13. A sectional view of part of the Snowball uplift.

Fig. 14. Cross sectional view of part of the Snowball Crater floor.

Figure 8 is a section through the north rim of the crater. Movement of the markers in column N15 shows the outward and upward movement into the hinge of the overturned flap. The stratigraphy taken alone for this region would tend to indicate an uplifted crater rim, but the sand column markers prove that the final displacement was downward. All the data from this northern rim of the crater indicate that in this region the movement of the upper strata was controlled by the presence of gravel beds above the blue clay. The deeper strata, including the blue clay, are continuous across the crater. The stratigraphy and the upper markers of columns N13, N12, N11, N10, and N9 all show inward and downward movement, greatest near the rim and decreasing with distance outward from the rim. However, it is significant that in this wall of the crater the displacement shown by the deeper markers is consistently *outward* and down. Taken together with stratigraphic information it would appear that in this case the slumping was in the form of slippage at the boundary of the coarse delta deposit. In Prairie Flat and in the 100 ton Suffield 1961 Crater clear evidence exists of a low angle thrust in this region (Jones, 1970). Subsequent reverse slippage along a low angle thrust is not unreasonable.

In column N8, the displacement pattern is similar to that described above, but the transition between the slumping movement and the outward and downward movement is quite sharp, and located on an extension of the postulated low angle slip plane suggested above. Just beyond this column there is a deep circumferential fracture zone analogous to the fracture on the south side shown in Fig. 4, and the sand column N7 is intimately associated with this fracture. The markers show almost vertical slipping near the surface, and outward and downward movement corresponding with no inward slumping in the deeper layers. Column N6, beyond the fracture zone, shows no evidence of inward slumping but considerable outward and downward movement.

The deep lying clay layers on this northern side of the crater show no evidence of deformation. The indication is that the deep layers were to a large extent protected by the presence of the gravel. In Fig. 9 there is no visible displacement of the blue clay underlying the sand and gravel beds, but the upper layers show a gradual reduction of the shallow dip angle. Columns of markers N5, N3 and N2 show the continued presence of outward and downward residual displacement.

Conclusion

The data given in this paper (and in more detail in the cited reports) indicate that near surface explosions in the compactible sediments of the Suffield test site produce craters which are structurally consistent and are analogous to many planetary craters of the ringed or central uplift type.

The complex craters are not uniform in the overall visual surface manifestation, but do exhibit a consistent hierarchy as follows:

(a) A simple dome-like central uplift, as demonstrated by the 100 ton Distant Plain 6 tangent sphere explosion crater (Diehl *et al.*, 1968).

(b) A crater with central uplift, but with some extrusion of fluidized material from the central uplift, forming a single pseudo-volcanic cone on top of the uplift. This is illustrated by the 100 ton Suffield 1961 Crater (Jones, 1976).
(c) A crater with internal ring structures on the crater floor, with pseudo-volcanic cones superimposed on the ring structures, as illustrated by Prairie Flat (see Fig. 3 of this paper).
(d) A crater with a dominant central uplift, with a depressed rim structure, evident radial and circumferential fissuring both internal and external to the main crater, and pseudo-volcanic cones both internal and external to the main crater. The final stage of such craters includes slumping of the crater walls to form internal terracing, and to some extent masking the internal ring structure. This stage is typified by the Snowball crater, discussed in detail in this paper.

In all the observed central uplift craters the ejecta blanket consisted of a coherently overturned, stratigraphically inverted expression of the pre-existing stratigraphy. In the blanket, strata could be traced from the undisturbed position, through a hinge region under the rim, and continuously through the ejecta blanket. In the ejecta the strata, after inversion, suffer radial thinning, but they retained precisely their relative positions even when the stratum consisted of a fine, free flowing sand. This overturning is clearly *not* due to sequential fall-out of the ejected material, but is a coherent roll-back of the strata.

Even after removal of the ejecta blanket (by excavation or erosion) the existence of the inverted stratigraphy is evidenced by the "hinge region" in which strata rotate through the vertical. Associated with this hinge region there were many examples of small drape folds and block faulting.

All the Suffield craters exhibit circumferential and radial fracturing of the ground surrounding the main crater. In some cases these fractures are accentuated by the injection of fluidized material (water borne sand at Suffield). The phenomenon associated with the circumferential fissures internal and external to the main crater were tracable at depth in the form of ring dykes which penetrate the country rock without spreading into sills at the contact with even incompetent strata. In addition to the ring dykes, there was an irregular injection of fluidized material due to a form of hydraulic fracturing.

The material of the central uplift may show signs of tension fracturing, with or without the injection of fluidized material.

There may be low angle thrust faulting included beneath the coherently overturned blanket particularly in the hinge region below the rim.

The ejecta blanket may include fused material deriving from the surface materials (for details, see Jones, 1976).

The floor of the crater may be covered by a mantle of fluidized material deriving from locations deep within the structure. In extreme cases, as illustrated by Snowball, regions external to the crater may be covered locally by similar beds of ejected fluidized material. These simulate in relative position the "magmatic" material extruded in planetary counterparts, though at Suffield the deposits were sediments. However, they must not be confused with the sedimentary beds which are a feature of terrestrial astroblemes and fossil craters. Such beds would have developed at Suffield in due course if the craters had not been excavated, and would have overlain the "magmatic" deposits.

Acknowledgments—Permission to publish this paper, and to distribute the full report (Jones, 1976) to the delegates to the Symposium on Planetary Cratering Mechanics held at Flagstaff, Arizona was granted by the Department of National Defence, Canada.

Financial support and much encouragement for part of this work provided by the Defence Atomic Support Agency of the United States is gratefully acknowledged by the author.

References

Diehl, C. H. H. and Jones, G. H. S.: 1965, A tracer technique for cratering studies, *J. Geophys. Res.* **70**, 305–309.

Diehl, C. H. H., Pinnel, J., and Jones, G. H. S.: 1968, *Crater and Ejecta Studies on Distant Plain 6*, Suffield Tech. Note 208.

Jones, G. H. S.: 1963, *Strong Motion Seismic Effects of the Suffield Explosions*. Univ. of Alberta PhD thesis (and as Suffield Report 208).

Jones, G. H. S.: 1970, *Prairie Flat Crater and Ejecta Study*, DASA Report POR 2115 (WT 2115).

Jones, G. H. S.: 1976, *The Morphology of Central Uplift Craters*, Suffield Report 281.

Perkins, B.: 1954, A new technique for studying crater phenomena, Ballistic Research Labs Tech. Note 880.

Roddy, D. J.: 1968, The Flynn Creek Crater, Tennessee, *Shock Metamorphism of Natural Materials*, Mono Book Corp.

Roddy, D. J.: 1969, Geological Survey Activities, *Operation Prairie Flat Prelim. Rep.* 1, DASA 2228–1.

Roddy, D. J.: 1976, High Explosive Cratering Analogs for Bowl-shaped, Central Uplift, and Multi-Ring Craters. *Proc. Lunar Sci. Conf. 7th*, p. 3027–3056.

Roddy, D. J., Pepin, R. O., and Merrill, R. B., editors.
(1977) *Impact and Explosion Cratering*, Pergamon Press (New York), p. 185–246.
Printed in the United States of America

Large-scale impact and explosion craters: Comparisons of morphological and structural analogs

DAVID J. RODDY
U.S. Geological Survey, 2255 North Gemini Drive, Flagstaff, Arizona 86001

Abstract—Morphological and structural comparisons between natural impact and explosion craters show that a well-developed set of analogs exist. Three basic crater types, common to both impact and explosion cratering, are compared and include: (a) bowl-shaped, (b) flat-floored with central uplift, and (c) flat-floored with multiring. For each of the three different types, three separate craters are compared which include: (a) for bowl-shaped craters: the terrestrial impact Meteor (Barringer) Crater, Arizona; an explosion trial, Pre-Mine Throw IV; and a young unnamed lunar impact crater; (b) for central uplift craters: the terrestrial impact, Flynn Creek, Tennessee; an explosion trial, Snowball; and the young lunar impact Copernicus; and (c) for multiring craters: the terrestrial impact, Ries Basin, Germany; an explosion trial, Prairie Flat; and the young lunar multiring impact, Orientale Basin

The morphological analogs for the bowl-shaped craters include: (1) bowl-shape, (2) small flat floor, (3) talus-covered lower walls with blocky terrain, (4) steep upper crater walls with horizontally banded outcrops, (5) high sharp rim crest, (6) surrounding light-colored ejecta blanket, and (7) smooth to hummocky inner terrain on ejecta. The structural analogs for bowl-shaped craters include: (1) fallout layer, (2) breccia lens, (3) deep disrupted zone, (4) faulted, folded, brecciated, and fractured rim strata, (5) uplifted rim, (6) surrounding ejecta blanket, and (7) inverted strata in ejecta blanket. A number of parametric ratios were calculated for bowl-shaped craters with the depth vs apparent diameter averaging $0.2D$ and the edge of the continuous ejecta vs apparent diameter averaging $2.9D$ for uneroded craters.

The morphological analogs for the flat-floored, central uplift craters include: (1) broad flat hummocky floor, (2) large central peak, (3) locally terraced crater walls, (4) uplifted and depressed rim segments, (5) surrounding rayed ejecta blanket, and (6) smooth to hummocky inner terrain on ejecta. The structural analogs for flat-floored, central uplift craters include: (1) fallout layer, (2) breccia lens underlain by thin disrupted zone of uplifted material, (3) large central uplift underlain by a deep disrupted zone, (4) locally terraced crater walls, (5) faulted, folded, brecciated, and fractured rim strata, (6) uplifted and downfolded rim segments, (7) large rim graben, (8) surrounding ejecta blanket, (9) inverted strata in ejecta, and (10) secondary craters. Parametric ratios calculated for central uplift craters include an average maximim value of $0.05D$ for the depth vs apparent diameter and over $2.2D$ for the edge of the continuous ejecta vs apparent diameter.

The morphological analogs for flat-floored, multiring craters include: (1) broad flat-floor, (2) central basin, (3) concentric rings on floor, (4) locally terraced crater walls, (5) moderately steep walls, (6) low rim crest, (7) exterior concentric rings, (8) surrounding rayed ejecta blanket, and (9) smooth to hummocky inner terrain on ejecta. The structural analogs for flat-floored, multiring craters include: (1) fallout layer, (2) melt-breccia, (3) breccia lens underlain by disrupted zone of uplifted material, (4) uplifted material to form concentric rings on floor, (5) faulted, folded, brecciated, and fractured rim strata, (6) locally downfolded rim, (7) locally terraced crater walls, (8) surrounding ejecta blanket, (9) inverted strata in ejecta, and (10) secondary craters. Parametric ratios calculated for multiring craters include an average value of approximately $2.0D$ for the edge of the ejecta blanket vs the apparent diameter for uneroded craters.

The analog comparisons provide: (a) an outline of the major morphological and structural elements that might be expected to occur in the three crater types, and (b) form a basis for predictive and interpretative analog studies of these physical elements with respect to other craters. The arguments for a universal, well-ordered, set of general cratering processes are also supported by this broad group of physical analogs formed under such diverse conditions as at natural impact and man-made explosion events and at impact craters formed in both terrestrial and lunar environments. The common ejecta blanket formed by the overturning sequence demonstrates one of these more general types of processes which occurred at each terrestrial cratering event. Photogeologic evidence further supports the overturning processes for the lunar craters. The general applicability of analogs studies to cratering sequences, however, remains constrained by our limited understanding of initial boundary conditions and their effects on the overall cratering processes, requiring caution for specific applications.

Introduction

Natural hypervelocity impact and experimental explosion cratering research have both undergone notable expansions in the last number of years. The increased emphasis on impact cratering studies has been primarily due to the planetary program, whereas the accompanying efforts in explosion studies have been prompted by expanded interests in engineering and cratering mechanics. Difficulties inherent in each discipline, however, have inhibited data acquisition and prevented a comprehensive understanding of the full range of cratering processes. For example, natural impact events can never be simulated at larger scales due to obvious physical constraints. Consequently, scaling of experimental impacts for comparative studies is often required to extend over ranges of 10^7 or more. Equally important, the critical effects of different types of impacting bodies and different target media have proven especially difficult to assess. Explosion cratering research also continues to experience similar difficulties, especially for larger high-energy experiments. In particular, environmental concerns and restrictions on nuclear testing, among other important factors, have necessarily restricted such engineering and cratering studies.

In the past, both the explosion and impact communities have traditionally used analog techniques as one of several approaches to study problems in their own respective areas, but until recently, the exchange of cratering analog and process data between these two groups has been extremely limited and without systematic organization. Crater studies now indicate, however, that there *are* a broad set of morphological and structural analogs that occur in both impact and explosion craters. This has led to the tacit assumption that these physical analogs imply *similar*, though not necessarily *identical* cratering processes. That is, if two craters exhibit essentially identical morphologies and structures and the formational processes are well understood for one of the craters, then these same general physical processes may be inferred for the second crater. This concept has been most successful when the comparisons are restricted to surface and near-surface explosion trials which are more similar to impact conditions.

Unfortunately, the absence of published systematic documentation of impact and explosion analogs has greatly restricted their general applicability to cratering problems for either discipline. More recently, however, a limited number of analog treatments has been assembled by various workers, which documents some of the critical physical similarities and differences. This paper briefly summarizes one such study by describing impact and explosion analogs for three basic crater types, i.e., (a) bowl-shaped, (b) flat-floored with central uplift, and (c) flat-floored with multirings. The data consist of a set of published and unpublished cratering studies summarizing morphological and structural elements in an ordered format and assembled in the same comparative style. The purpose of this paper is to provide a systematic documentation of the more critical morphological and structural analogs common between large natural bowl-shaped impact and explosion craters, between central uplift impact and explosion craters, and between multiring impact and explosion craters. Hopefully, such an assemblage will increase the credibility of using such analog data for predictive aind interpretative purposes by both the impact and explosion communities.

BACKGROUND

One of the most fundamental problems confronting both impact and explosion cratering research is the intractability of *direct* observations of the complex physical processes occurring in a large-scale cratering event. In the past, the approach to this problem has been to concentrate on various independent cratering studies, such as laboratory impact and explosion experiments, field explosion trials, natural impact studies, and a limited number of theoretical studies. The integration of this wide data base, however, has not been a particularly active discipline, except for a few workers. Moore recently, though, there has been a growing recognition of the potential usefulness of exchanging the broad sets of data assembled by the two different communities. The result has been for interest to center on analog comparisions of the morphologies and structures of impact and explosion craters with the assumption that a well-defined set of physical similarities implies *comparable*, though not necessarily *identical*, physical processes of formation (Baldwin, 1963; Cooper, 1977; Curran *et al.*, 1977; Gault *et al.*, 1968; Gault and Wedekind, 1977; Johnson *et al.*, 1969; Knowles and Brode, 1977; Kreyenhagen and Schuster, 1977; Moore *et al.*, 1963; Milton and Roddy, 1972; Oberbeck, 1971, 1977; Orphal, 1977; Quaide and Oberbeck, 1968; Roddy, 1968, 1976, 1977a; Shoemaker, 1960, 1963; Trulio, 1977; Ullrich *et al.*, 1977 and others).

Despite the absence of systematic documentation, the tacit assumption of similitude has been used more recently by a number of workers to argue a wide variety of independent cratering problems in both fields, as well as to attempt to outline certain parts of the evolutionary steps in the development of the crusts on the different planets (for example, see 1st through 8th Proceedings of the Lunar Science Conferences and other papers in this volume). Indeed, the first comprehensive outline describing the sequence of events following a large natural impact cratering event was derived by Shoemaker (1960, 1963) using theoretical approaches *and* selected analog comparisons with nuclear explosion craters. The concept of analog studies has been further improved and strengthened by comparisons between morphologies, structures and formational processes of laboratory impact and natural impact craters as illustrated by the excellent studies of Gault *et al.* (1968), Gault and Wedekind (1977), Quaide and Oberbeck (1968), Oberbeck (1971, 1977), Oberbeck *et al.* (1977), and others. Use of missile impact crater data by Moore (1976) has provided another direct tie with natural impact analogs. Analog studies utilizing laboratory impact and explosion cratering experiments (Oberbeck, 1971, 1977) and large field explosion cratering trials (Roddy, 1968, 1976, 1977a) have been specifically directed towards comparisons with large natural impact craters. More recently, theoretical calculational studies by Ahrens and O'Keefe (1977) and O'Keefe and Ahrens (1975, 1976) also have begun to address the question of very large-scale impact cratering processes.

Explosion cratering research has been more successful than natural impact studies in that direct observations, extensive instrumentation, and theoretical calculations are more tractable as summarized by Cooper (1977). Theoretical studies using numerical simulation techniques, in particular, have been especially useful since each of the explosion trials are controlled events with predetermined initial conditions which permit, at least partial verification of certain processes by calculations (Knowles and Brode, 1977). Various physical *analog* approaches, however, still remain of fundamental importance in most scaling, predictive and interpretative studies of experimental explosion events (Allen, 1976; Cooper, 1977; Cooper and Sauer, 1977; Chabai, 1977; Curran *et al.*, 1977; Killian and Germain, 1977; Kreyenhagen and Schuster, 1977; Knowles and Brode, 1977; Maxwell, 1977; Orphal, 1977; Oberbeck, 1977; Seebaugh, 1977; Swift, 1977; Shock, 1977; Vortman, 1977; Wisotski, 1977). Of necessity, the direct observations and analog studies derived from such laboratory explosion experiments as those of Andrews (1977) and Piekutowski (1977), and large-scale field explosion cratering trials such as reported by Henny (1977), Jones (1977), Roddy (1968, 1969, 1970, 1976, 1977a), and others, still form a phenomological basis for many of the explosion cratering studies.

The integration of large-scale explosion data with laboratory and natural impact studies, however, has remained largely neglected, at least in terms of detailed comparative treatments of the basic elements common to both disciplines. Consequently, the purpose of this paper, is to systematically *document* the most critical morphological and structural analogs between selected large-scale experimental explosion and natural impact craters. If a broad morphological and structural set of cratering analogs can be successfully established from such diverse origins as natural impact and man-made explosions and in such diverse environments as on the earth and moon, then support would be added to the arguments for a universal, well-ordered, predictable set of cratering processes. The use of creditable analogs which can be directly compared in a single text and are described in the same comparative style should improve our understanding of the general physical processes in cratering for *both* impact and explosion sources and, hopefully, allow a more effective exchange of information in the future.

TEST SITES AND FIELD TECHNIQUES

Three terrestrial impact, three experimental explosion, and three lunar impact craters are described in the following sections. Standard geological, geophysical, and drilling techniques were employed to collect detailed cratering data at the three terrestrial impact sites, Meteor Crater, Arizona (Shoemaker, 1963; Roddy *et al.*, 1975), Flynn Creek, Tennessee (Roddy, 1968, 1977a,b), and the Ries Crater, Germany (Chao, 1977a, b; Pohl *et al.*, 1977, and others). Each of these impact craters was formed in flat-lying sedimentary rocks with the Ries being sufficiently large enough to extend downward into mafic and felsic igneous and metamorphic crystalline rocks. The three lunar craters were presumably each formed in mafic crystalline rocks overlain by regolith.

The experimental explosion craters described in this paper include the Pre-Mine Throw IV 100-ton Crater 6, Snowball 500-ton crater, and Prairie Flat 500-ton crater. The Pre-Mine Throw IV Crater 6 was formed at the Nevada Test Site in playa sediments consisting of flat-lying, unconsolidated to weakly cemented thin, interbedded clay, silt, and fine sand (Roddy, 1977c). The test site and playa were dry at shot time. The Snowball and Prairie Flat explosion trials were part of a joint research program by Australia, Canada, the United Kingdom, and the United States conducted under the TTCP and held at the Defence Research Establishment Suffield, Alberta, Canada (DRES). The U.S. Geological Survey participated in field studies of Pre-Mine Throw IV, Snowball, Prairie Flat to examine the types of morphology, structural deformation, and cratering processes involved in these trials, and to make comparisons with terrestrial and planetary impact craters (Roddy, 1968, 1976, 1977a). Both the Snowball and Prairie Flat trials were conducted at the Watching Hill test site at DRES on flat-lying alluvium with a water table at approximately 7 m depth. The target media for both craters consists basically of a thick sequence of interbedded, unconsolidated alluvial, lacustrine, and glacial till deposits. The upper part of this sequence is mainly interbedded clay, silty clay, clayey silt, silty sand, sandy silt, and gravelly clay, with beds ranging from a few millimeters to several meters thick (Jackson, 1972; Jones, 1963, 1976, 1977; Jones *et al.*, 1970). Bedrock, approximately 60 m deep, consists of thin- to medium-bedded flat-lying sandstones. Jackson (1972) describes extensive soil test data and constitutive properties for numerical calculations for cratering at this test site.

The Pre-Mine Throw IV experimental crater was formed by a 100-ton nitromethane spherical charge lying tangent on the playa surface. The Snowball Crater was formed by a 500-ton TNT hemisphere lying on the alluvial surface, whereas the Prairie Flat Crater was formed by a 500-ton TNT sphere lying tangent on the alluvium. The hemisphere was detonated at the ground surface and in the center of the charge, and the spheres were detonated at the centers of the charges. Extensive ground instrumentation including velocity and acceleration gages, airblast gages, multiple high speed cameras, colored sand columns, and numbered marker cans in sand columns, to mention a few, were utilized for these trials. No sand columns or marker cans were emplaced at Pre-Mine Throw IV. After each of the trials, extensive trench excavations were used to determine the subsurface structural deformation.

The morphologic data and structural deformation described in this paper were derived from the pre- and post-shot surveys of the sand columns, surveys of the pre- and post-shot positions of the numbered marker cans, and surveys of the open-trench excavations through the craters and along the instrument column lines. Other data sources used in this paper for summaries of the larger trials include: (a) aerial stereophotography; (b) aerial blast photography; (c) detailed ground photography; (d) geologic maps; (e) ejecta distribution maps; (f) multiple deep-trench excavations; (g) specialized topographic maps; and (h) pre- and post-shot survey data from numerous man-made installations. More detailed descriptions of these explosion cratering experiments are given in Roddy (1976, 1977c).

COMPARISONS OF MORPHOLOGIC AND STRUCTURAL ANALOGS IN IMPACT AND EXPLOSION CRATERS

Comparisons of morphologic and structural analogs are summarized in this section between three terrestrial impact craters and three experimental explosion craters. Comparisons of morphology are also briely summarized for three lunar impact craters. The craters described were chosen to represent three *different* basic morphological and structural types: (a) bowl-shaped, (b) flat-floored with central uplift, and (c) flat-floored with multirings. The choices of these three particular different types were dictated, in part, because: (a) bowl-shaped craters are the most basic and widespread of all crater forms, (b) flat-floored craters with central uplifts are the most common type in the larger size ranges and appear strongly related to initial impact and target media conditions, (c) all three impact crater types are present on each of the terrestrial planets and the Moon, (d) an understanding of the initial conditions leading to the formation of large central uplift- and multiring-type impact craters is critical to interpreting crustal structure and evolution, (e) multiring impact craters are sufficiently large enough to reposition large regions of deep crustal material nearer the surface and, therefore, have major effects on the crustal evolution and long-term thermal history, and (f) experimental explosion cratering involving all three crater types is in critical need of a data base extended to the larger sizes which are now beyond experimental tractability. The particular craters described in this section were chosen because they exhibit typical morphologies and structure and because of the availability of descriptive data. As noted earlier, the fundamental emphasis in this paper is directed toward documenting only the most common analogs between impact and explosion craters of the bowl-shaped type, analogs between impact and explosion craters of the central uplift type, and analogs between impact and explosion craters of the multiring type. To simplify these comparisons, the following descriptions are divided into morphological and structural sections for each crater, recognizing that there are certain features, such as the fallout layer or the physical character of the top of the ejecta blanket, which have dual aspects critical to both the morphology and structure. Of necessity, emphasis is placed on graphic comparisons in the figures to reduce extensive written comparisons.

Dimensions listed in this paper include, where possible, both *rim crest* values and *apparent* values, with the latter measured at the level of pre-shot or pre-impact original ground surface. Diameters and depths are given as average values unless otherwise noted. The dimensional parameters are shown graphically in Fig. 4 in Roddy (1977a, this volume) and the terminology used for the craters is shown in Fig. 1 in this paper.

Oblique aerial and space photography, geologic cross-sections, and geologic maps are included in scaled formats to allow direct visual comparisons. Comparisons of the dimensions of the various analogs over the wide range of crater sizes are facilitated by presenting both the actual dimension followed by the ratio of that dimension vs the apparent diameter. Each of the parametric ratios is written as some fraction of *D*apparent, i.e., (*xD*), to provide scaled values which can be directly compared with reference to a dominant crater parameter, the diameter. Table 1 lists the craters described in this paper with their pertinent dimensions and Table 2, at the end of the paper, provides a summary of the morphological, structural, and parametric relationships.

(a) *Bowl-shaped craters*

The craters chosen to illustrate morphological and structural analogs for bowl-shaped craters include: (1) the terrestrial impact site, Meteor (Barringer) Crater, Arizona, (2) the experimental explosion trial, Pre-Mine Throw IV Crater 6, and (3) a young unnamed lunar crater.

(1) *Meteor (Barringer) Crater, Arizona.* Meteor (Barringer) Crater, located in north central Arizona, represents the earth's most recent, large, bowl-shaped impact crater. The important morphologic elements include: (1) a bowl-shape, (2) flat-floor, (3) talus-covered lower walls with blocky terrain, (4) steep upper crater walls with horizontally banded outcrops, (5) high sharp rim crest, and (6) a surrounding light-colored ejecta blanket with a smooth to blocky, hummocky, inner terrain (Fig. 2a). The major structural elements include: (1) a fallout layer, (2) breccia lens, (3) deep disrupted zone, (4) faulted, folded, brecciated, and fractured rim strata, (5) uplifted rim, and (6) an overturned flap of inverted strata forming an ejecta blanket that surrounds the crater. A generalized cross-section and geologic map are shown in Figs. 3a and 4a, respectively.

Briefly summarized, Meteor Crater was formed 20,000 to 30,000 years ago (Shoemaker, 1975, pers. comm.) by the hypervelocity impact of an iron meteorite into nearly flat-lying sandstone and limestone. Energy scaling gives an impact energy of approximately 4.5 megatons, and calculations, assuming an impact velocity of 15 km/sec, indicate the impacting iron body would have weighed on the order of 150,000 metric tons with a diameter of approximately

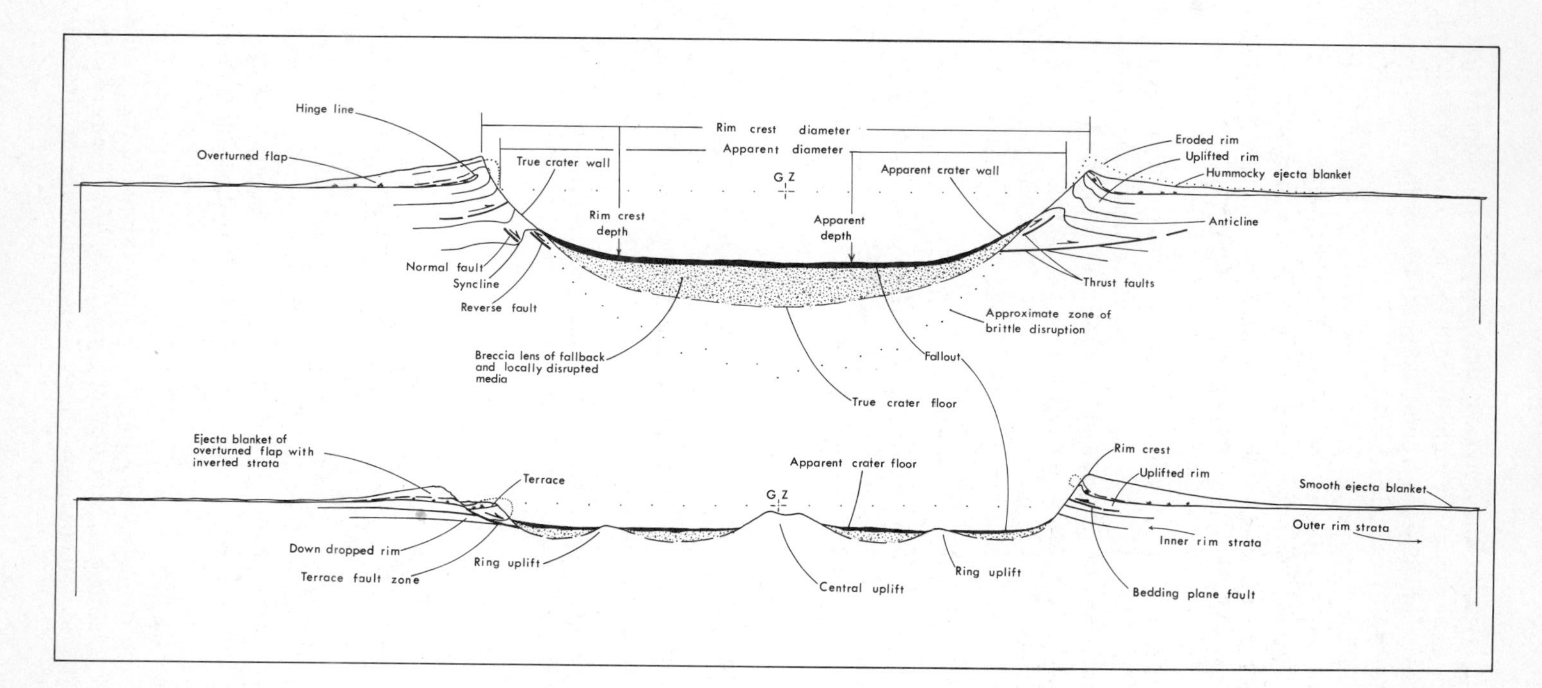

Fig. 1. Generalized geologic cross-sections of bowl-shaped and flat-floored craters showing the symbols and terminology used in this paper. Bedding planes are used to show styles of folding with close-spaced dotted lines showing projection of overturn of beds in space. The relative sense of fault movement is shown by a single half-arrow with the arrow on the side that generally had the largest absolute displacement. All of the features shown do not necessarily occur in each crater.

Table 1. Impact and experimental explosion craters described in this paper. Meteor Crater, Copernicus, and the Ries diameters are present-day mean values, which in the case of Copernicus and the Ries include the terraces. Values in parentheses are estimates of immediate post-crater rim crest and apparent diameters and depths for Meteor Crater (Roddy *et al.*, 1975), Flynn Creek (Roddy, 1977a), Copernicus and the Ries, and are derived from cross-section reconstructions. Depths for the unnamed lunar crater, Copernicus, and Orientale are to present level of floor filling, including maria, i.e., minimum values since they are not to top of fallout beneath filling material. The two diameters given for the Orientale Basin reflect two different interpretations of McCauley (1977) and Wilhelms *et al.* (1977). See text for further explanation of values in this table.

			Diameter (D)		Depth (d)		D/d		
Crater	Origin	Type	Rim crest	Apparent	Rim crest	Apparent	Rim crest	Apparent	Target media
Meteor Crater, Arizona	Impact	Bowl-shaped	1186 m (≈1100 m)	1036 m (≈1005 m)	197 m (≈215 m)	150 m 150 m	6.0 (≈5.1)	6.9 (≈6.8)	Flat-lying sandstone and dolomite
Pre-Mine Throw IV 6, Nevada	100-ton nitromethane sphere	Bowl-shaped	25.0 m	20.3 m	5.8 m	4.3 m	4.3	4.7	Flat-lying playa sediments
Unnamed crater, moon	Impact	Bowl-shaped	~2.7 km	?	~775 km	?	~3.5	?	Lunar regolith in Highlands
Flynn Creek, Tennessee	Impact	Flat-floored, central uplift	(~3830 m)	~3517 m	(~198 m)	~98 m	(~19)	~36	Flat-lying limestone and dolomite
Snowball, Canada	500-ton TNT hemisphere	Flat-floored, central uplift	108.5 m	83.0 m	7.5 m	6.6 m	15	13	Flat-lying alluvium
Copernicus, moon	Impact	Flat-floored, central uplift	~93 km (≈79 km)	~84.6 km (≈72 km)	~3.8+ km (≈4.0+ km)	~2.7+ km ≈2.7+ km	~25 (≈20)	~31 (≈27)	Lunar regolith, maria, and Highlands
Ries Crater, Germany	Impact	Flat-floored, multiring	(≈27 ± 2 km) (≈22 ± 2 km)	~25 ± 1 km (≈20 ± 1 km)	(≈1.2+ km) (≈1.1+ km)	~570 m ~570 m	(≈23) (≈20)	(≈44) (≈35)	Flat-lying sedimentary, igneous and metamorphic rocks
Prairie Flat, Canada	500-ton TNT sphere	Flat-floored, multiring	85.5 m	61.0 m	5.3 m	4.4 m	16	14	Flat-lying alluvium
Orientale Basin, moon	Impact	Flat-floored, multiring	?	≈620 km ≈860 km	?	3.5–4.5+ km	?	~177–138(?) ≈246–191(?)	Lunar Highlands and regolith

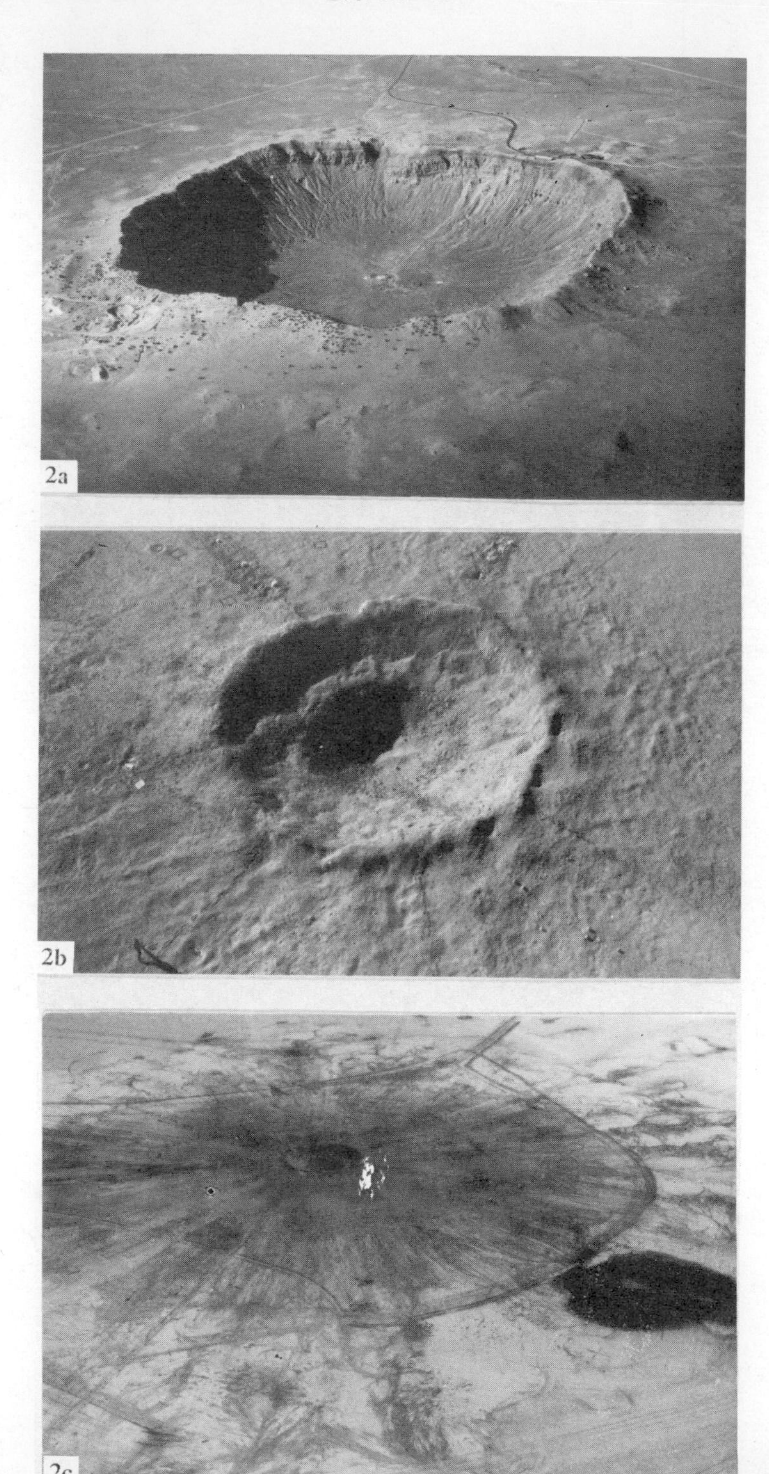
2a
2b
2c

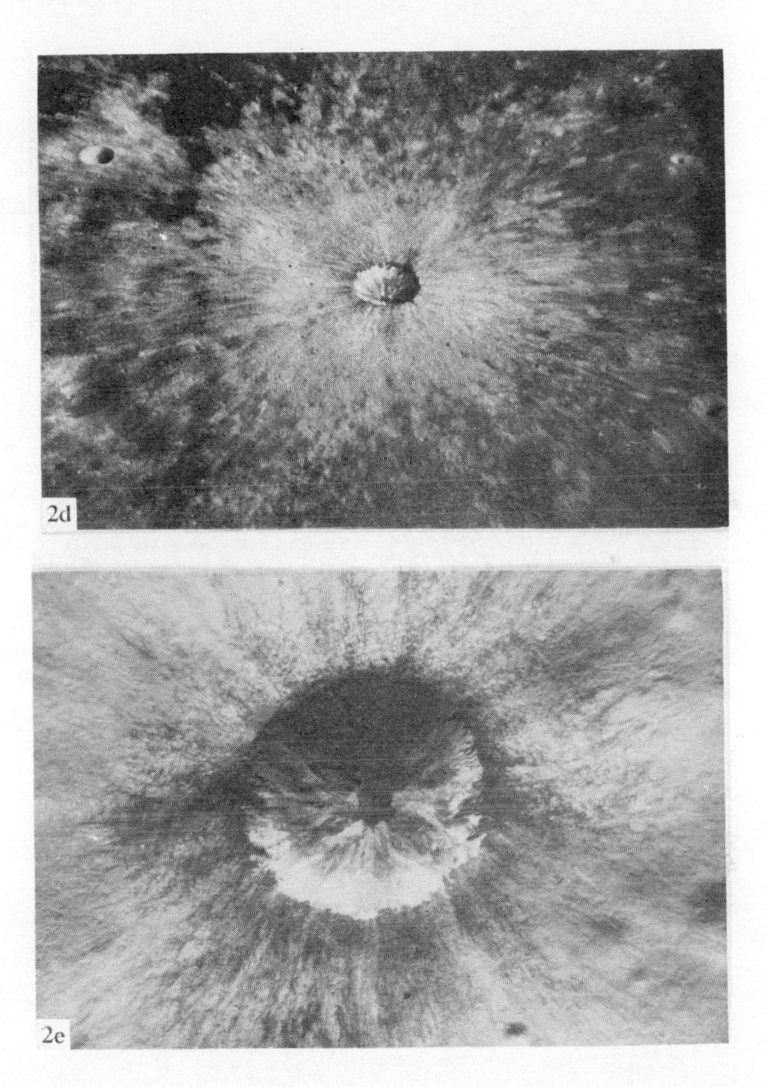

Fig. 2. Oblique aerial and space photographs of three bowl-shaped craters. The view in Fig. 2a shows the inner hummocky ejecta blanket composed mainly of Kaibab limestone and Coconino Sandstone surrounding Meteor Crater, Arizona. Uplifted, outward dipping Kaibab and Moenkopi strata are exposed in the horizontally banded rocks on the upper crater walls. For scale, note the museum buildings on the far north side of the squarish rim. U.S. Geological Survey photograph by D. Roddy and K. Zeller. The view in Fig. 2b shows an internal bench and the hummocky ejecta of the Pre-Mine Throw IV 100-ton explosion crater on Yucca Lake Playa at the Nevada Test Site, and Fig. 2c shows the rayed pattern of this crater. U.S. Geological Survev photographs by D. Roddy and R. Williamson. The view in Fig. 2d (AS17-150-23102) shows a young unnamed lunar crater with its well-developed rays, and Fig. 2e (AS15-8936P) is a near vertical view of this crater showing the small flat floor, steep lineated walls, and surrounding dark ray material locally overlying the walls and inner rim. Large blocks, up to 100 m across, are scattered near the rim crest.

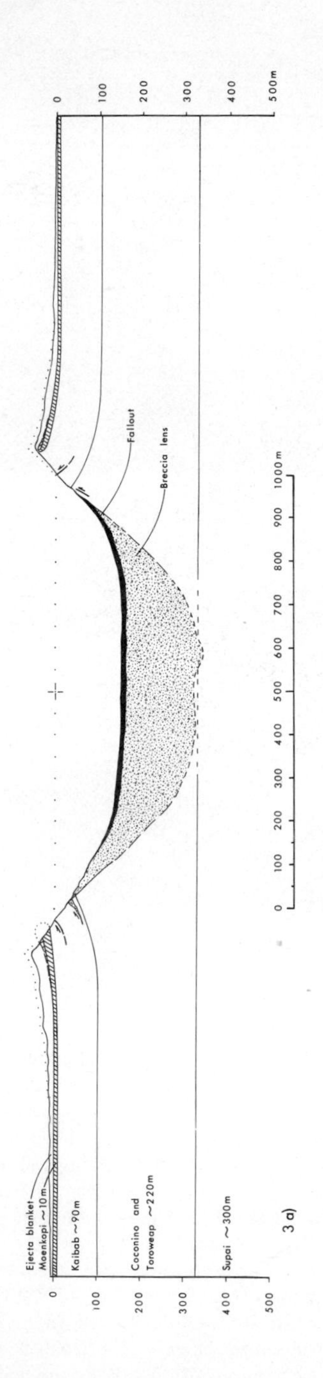
Ejecta blanket
Moenkopi ~10m
Kaibab ~90m
Coconino and Toroweap ~220m
Supai ~300m
Fallout
Breccia lens
500m
1000m

3 a)

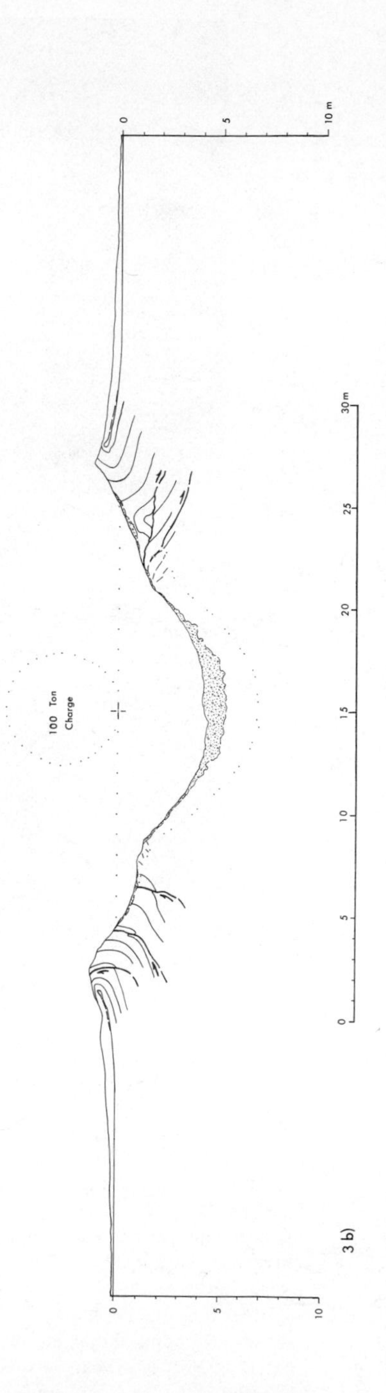
100 Ton Charge
10 m
30 m

3 b)

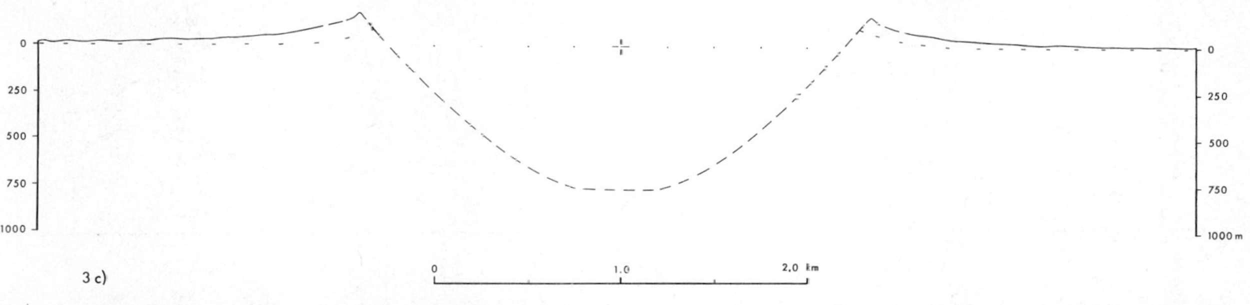

Fig. 3. Generalized geologic cross-sections of three bowl-shaped craters, with Fig. 3a of Meteor Crater, Arizona (modified from Shoemaker, 1960), Fig. 3b of the Pre-Mine Throw IV 100-ton Crater 6 (modified from Roddy, 1978), and Fig. 3c a schematic of an unnamed lunar crater. Each crater is drawn to show the apparent diameter at the same scale. The explanations for the symbols are shown in Fig. 1.

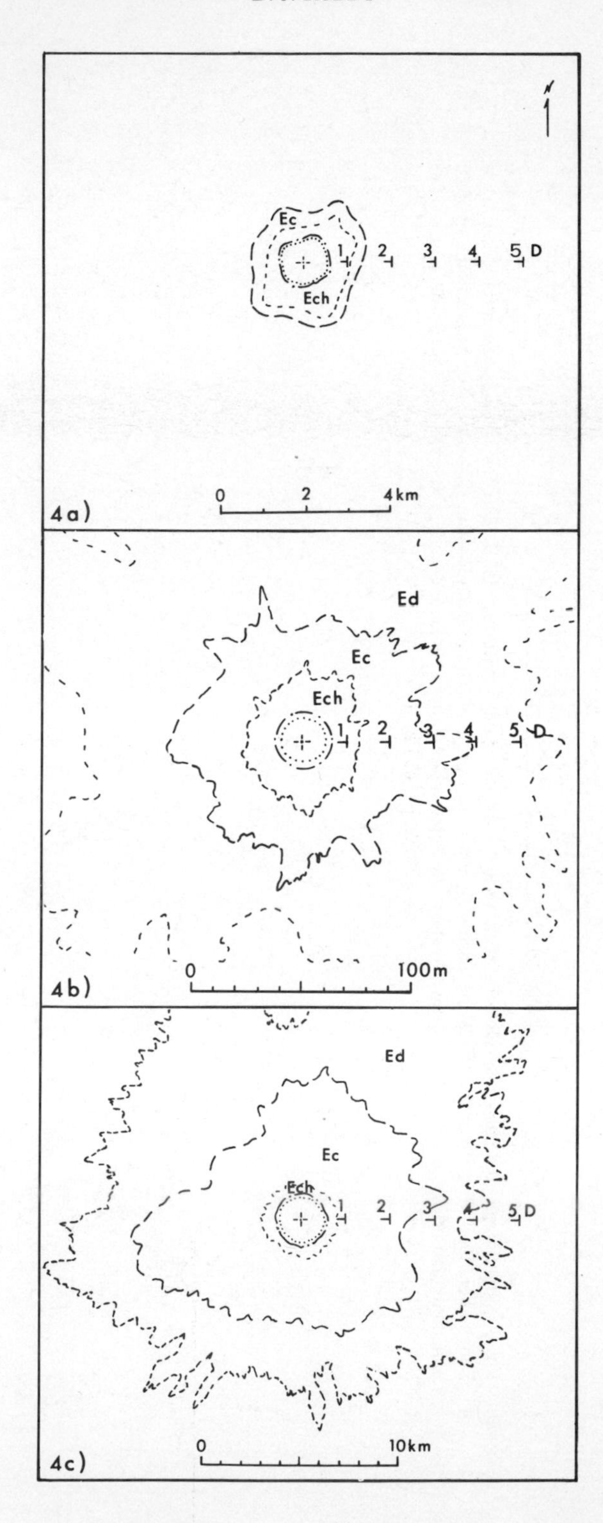
N
Ec
Ech
1
2
3
4
5
D
0
2
4km
4a)
Ed
Ec
Ech
1
2
3
4
5
D
0
100m
4b)
Ed
Ec
Ech
1
2
3
4
5
D
0
10km
4c)

The crater, as it exists today, has an average rim crest diameter of 1186 m (1036 m apparent diameter) and a depth of 167 m from the rim crest to the sedimentary deposits on the crater floor. The present depth from the rim crest to the top of the breccia lens is 197 m (150 m apparent, 0.15*D*). The rim crest now averages 47 m (0.05*D*) above the surrounding terrain. During the cratering, approximately 127×10^6 m^3 (287×10^6 metric tons) of rock were brecciated and 76×10^6 m^3 (175×10^6 metric tons) of rock ejected.

The Colorado Plateau, in the vicinity of Meteor Crater, is underlain by approximately 1200 m (1.2*D*) of nearly flat-lying Paleozoic and Mesozoic limestone, dolomite, sandstone, and shale which overlies pre-Cambrian crystalline basement (Fig. 3a). The regional structure near the crater consists of gentle monoclinal folds and wide-spaced northwest trending normal faults, commonly kilometers in length with a few meters to 30 m of displacement. Jointing consists of two mutually perpendicular sets, one of which is subparallel to the normal faults.

The geology of Meteor Crater, its analogs with two nuclear explosion craters, and the general cratering processes are described in detail by Shoemaker (1960, 1963) and Shoemaker and Kieffer (1974), and the dimensional parameters, ejecta thicknesses, crater and ejecta volumes and masses, mass balances, erosion percentages, and energies of formation are listed in Roddy *et al.* (1975). The crater, as seen today, is approximately circular in plan with slightly squarish sides oriented approximately 45° to the regional joint system. The present floor of the crater, approximately 520 m (0.5*D*) across, is underlain now by 30 m (0.03*D*) thickness of Holocene age alluvium and playa beds and Pleistocene alluvium, lake beds and talus. Beneath this sequence is a 10 m-thick (0.01*D*) unit of fallout which originally covered the crater floor, walls, and probably the entire surrounding ejecta blanket. Drilling on the crater floor shows the fallout to consist of fine to coarsely comminuted, highly shocked Coconino, Toroweap, Kaibab, and Moenkopi with minor amounts of lechatelierite and traces of meteoritic material, coesite, and stishovite (Shoemaker, 1960). Drilling on the rim located scattered remnants of a very finely comminuted, moderately shocked Coconino sandstone layer which appears to have once formed a very light-colored blanket several meters or more thick which once covered the inner ejecta blanket at least to approximately 1.4*D* from GZ (Roddy *et al.*, 1975). The finer-grained upper part of this layer was probably slightly mixed and transitional with the earliest fallout being deposited.

Inside the crater, talus deposits formed at the base of the crater walls and

Fig. 4. Generalized geologic maps of three bowl-shaped craters, with Fig. 4a of Meteor Crater, Arizona (Roddy *et al.*, 1975), Fig. 4b of Pre-Mine Throw IV 100-ton Crater 6 (modified from Roddy, 1978), and Fig. 4c of an unnamed lunar crater. Each crater is drawn to show the apparent crater diameter at the same scale. The symbols include Ec for continuous ejecta, Ech for continuous hummocky ejecta, and Ed for discontinuous ejecta. The dotted circles are the apparent diameters, the dot-dash lines are rim crests, the long-dashed lines show the approximate outer limits of the continuous ejecta, and the short-dashed lines show the very approximate limits of the continuous hummocky and the discontinuous ejecta units. The numbered symbols show multiples of the apparent diameter distances from GZ out to 5*D*.

flowed out onto the floor. The talus incorporated masses of unsorted fragmented rocks including large boulders, locally forming rough, elongated, blocky terrains along the lower parts of the crater walls. Deposition of alluvium tended to decrease the local relief and smooth the crater floor and lower walls. The crater walls average 35–45° except for the upper 20% of the walls which are locally vertical.

The structure beneath the fallout layer under the crater floor consists of a large breccia lens underlain by a thick disrupted zone (Fig. 3a). Drilling (Barringer, 1905, 1910, 1914, 1924; Shoemaker, 1960; Tilghman, 1905; Barringer, 1975, pers. comm.) and geophysical studies (Ackermann *et al.*, 1975; Regan and Hinze, 1975) indicate the breccia lens is approximately 160–200 m (0.15–0.19*D*) thick, and consists mainly of fallback and locally disrupted rock from the Kaibab and Coconino sections ranging in size from less than a millimeter to blocks meters across. The Kaibab appears most common at the top of the outer edges of the breccia lens along the crater walls, whereas the bulk of the breccia lens, according to the deeper drilling, consists mainly of Coconino sandstone breccia in a chaotic and mixed state. Shoemaker (1963) notes that lechatelierite and meteorite material were present in the breccia lens drilling and that "... cores of ordinary siltstone and sandstone of the Supai were obtained at depths of 210 m and deeper." Seismic studies by Ackerman *et al.* (1975) indicate the disrupted and fractured zone extends to at least 800 m (0.8*D*) below the crater floor, involving rocks well below the Supai. This would put the total depth of disrupted and fractured rock over 950 m (0.9*D*) below GZ. An apparent extension of the breccia lens under the south rim was encountered in a drill hole by Barringer (1924) implying that the rim was partly extended during cratering, possibly due to oblique impact or some other irregularity in the initial shock wave. The detailed structure beneath the rim, however, remains poorly defined.

The general structure in the lower crater walls consists mainly of faulted, fractured, and locally brecciated Coconino sandstone which has been slightly to moderately folded and uplifted. These strata now dip gently out from the crater at approximately 30° ± 10°. The intensity of deformation increases upward to near the tops of the crater walls where the Kaibab dolomite and Moenkopi sandstone become vertical to locally continuously overturned. Uplift accompanying the overturned rocks is as great as 47 m (0.05*D*). Complex faulting associated with the outward and upward displacement of rim strata includes thrust, normal, and reverse faulting. At least some of the high-angle fault attitudes are related to the joint orientations, whereas others occur at a variety of attitudes to the strata and jointing, with bedding plane faults being common. The four *corners* of the crater at the tops of the walls contain near vertical fault zones that extend up to the original ground surface. In the northeast corner, a single large fault has a vertical fault plane lower on the crater wall, whereas the upper end bends over into a near horizontal thrust fault extending to the original ground surface. Both authigenic and allogenic breccias are present in the wall rocks.

At the top of the crater walls, the structural deformation was dominated by the overturning process (Shoemaker, 1960, 1963; Roddy *et al.*, 1975), with strata in this region standing at near vertical angles to locally overturned. The detailed structure of the overturned flap varies locally from a smooth continuous overturn of Moenkopi sandstone in parts of the rim, to broken segmented overturned strata, to discontinuous blocks rotated progressively into an overturn. The latter is typified by the massive blocks of Kaibab dolomite up to 30 m (0.03*D*) across locally overlying parts of the rim crest. Beyond the rim crest area, the rocks rapidly disaggregate in the overturning and become fragmented unmixed breccia to form the inverted strata of the overturned flap. The author uses the term *continuous overturned flap* to represent that limited part of the structure in the crater wall region which actually has continuous unbroken overturned strata, as the Moenkopi in the northeastern rim, or certain of the overturned flaps in the explosion craters. The more general term, *overturned flap*, is used in the paper primarily to emphasize the *inverted* state of the rocks in the ejecta blanket and the nature of its formational sequences.

During the excavation phase at Meteor Crater, strata were ejected from the ground surface down to about the middle Coconino sandstone, approximately 160 m (0.15*D*) below GZ. The result was to reposition about 80–90% of the total ejecta, some 65 million cubic meters of rocks (150×10^6 metric tons), in a matter of seconds into the surrounding overturned flap within about 2–3*D* of GZ. The remaining breccia was lost to high speed ejecta, fine fragmental debris lofted into the air and carried off, and a small amount of vaporization. One of the consequences of the total ejecta process was to deposit a well-ordered *vertical* inversion of the strata in the ejecta blanket in which the shallowest rocks cratered reposition directly back upon themselves and deeper (and last) rocks cratered are positioned on top of the ejecta blanket (Fig. 3a). A second consequence of the overturning is seen in the horizontal distribution, i.e., the shallowest rocks cratered are distributed over the greatest distances as a layer lying on the original stripped ground surface. Rocks cratered from intermediate depths are redistributed as an intermediate layer in depth and in range in the ejecta blanket. Strata from still deeper levels are repositioned near the rim crest as the top ejecta blanket. The deepest and last rocks cratered, middle to lower Coconino sandstones, commonly are not ejected far enough to leave the crater and return to the breccia lens as fallback.

Erosion has continued to lower the rim crest and smooth the ejecta blanket, but drilling studies and mass balance calculations indicate that on the order of only 20–27% of the ejecta has been removed (Roddy *et al.*, 1975). The outer edge of the continuous flap that forms the ejecta blanket averages about 1543 m (1.5*D*) from GZ as seen today, but it certainly extended further immediately after impact. The inner part of the ejecta blanket exhibits a hummocky, locally blocky, surface caused predominantly by large Kaibab blocks and concentrations of Kaibab ejecta. The hummocky terrain, surrounded by smoother sections, now extends out to approximately 1.1*D* from GZ, a distance functionally related to the type of rock cratered.

(2) *Pre-Mine Throw IV Crater* 6, *NTS Nevada.* The Pre-Mine Throw

IV explosion trials were conducted at the Nevada Test Site on Yucca Lake Playa. Detonation of a 100-ton nitromethane sphere lying tangent on the dry, flat, playa surface formed Crater 6, 25.0 m across (20.3 m apparent) and 5.8 m deep (4.3 m apparent, 0.2*D*) with approximately 530 m^3 (~ 686 metric tons) of alluvium ejected. Its important morphological elements include: (1) a bowl-shape, (2) small flat-floor, (3) internal bench, (4) steep talus-covered lower walls with blocky terrain, (5) moderately steep upper walls with horizontally banded outcrops, (6) high sharp rim crest, and (7) a surrounding ejecta blanket with a smooth to blocky, hummocky inner terrain (Fig. 2b, c). The major structural elements include: (1) a thin fallout layer, (2) breccia lens, (3) deep disrupted zone, (4) internal bench, (5) faulted, folded, brecciated, and fractured rim strata, (6) uplifted rim strata, and (7) an overturned flap of inverted strata forming an ejecta blanket that surrounds the crater. A detailed cross-section and geologic map are shown in Figs. 3b and 4b, respectively and a detailed description of the trial is given in Roddy (1978).

The floor of the crater, about 5 m (0.3*D*) across, was nearly flat and surrounded by relatively steep inner crater walls averaging 35° in slope. About halfway up the walls, a prominent topographic and structural bench was formed, above which the walls become less steep (Fig. 2b). The upper 10–15% of the walls steepen abruptly again to slopes as great as 80° locally.

The floor of the crater, the walls, and surrounding ejecta blanket were irregularly covered by a finely comminuted fallout of playa material ranging in thickness from approximately 0.1 m (0.005*D*) on the floor and lower walls to less than a few millimeters at 2–3*D* from GZ. Regions of blocky terrain were prominent both inside the crater and on the inner rim, however, the greatest concentrations were below the bench. Most of the blocks, consisting of slightly shock compressed playa, as large as 0.5 m (0.03*D*) across, were derived from below the 2 m (0.1*D*) depth in the cratered region. The upper walls below the rim crest and lower walls below the bench locally exhibited slump and talus deposits that flowed out over the bench and crater floor, respectively, forming elongated blocky terrains.

The major structure beneath the crater floor, determined by deep trench excavations, consisted of a breccia lens underlain by a zone of brittle disruption. The breccia lens, consisting of fallback and locally disrupted playa, averaged about 1.0 m (0.05*D*) in thickness under the floor and thinned to a few centimeters on the outer crater walls (Fig. 3b). The breccia lens consisted of chaotically mixed playa fragments, mainly from the lower horizons cratered, i.e., the lowest 2 m (0.1*D*). Fragment sizes ranged from sub-millimeter to blocks up to 0.5 (0.03*D*) across. The zone of major brittle disruption, which extended to a depth of approximately 7 m (0.35 + *D*) below GZ, consisted of highly fractured, faulted, and locally folded playa beds separated into loose blocks of various sizes *without* stratigraphic mixing. Brittle fractures and limited faulting were present beneath the main zone but were much less intensive and well defined.

The structural deformation beneath the crater walls consisted of highly faulted, folded, fractured, and locally brecciated playa beds as shown in the cross-sections in Fig. 3b. The structure in the lowest walls was mainly that of faulted, fractured, and brecciated playa. Immediately below the bench, the playa was forced outward and moderately uplifted with beds dipping 30–60° away from the crater.

The structure forming the bench caused the only morphological deviation from a perfect bowl-shaped crater. This topographic and structural feature formed half-way down the crater walls approximately 1.3 m (0.06*D*) below the original ground surface. The bench resulted from the strong differential movements created by a strength discontinuity at about 1.8 m (0.09*D*) depth, a level at which the shear strength of the playa beds increases slightly (Peterson, 1974). Extensive low-angle thrust faulting, as well as complex normal and reverse faulting, occurred with the upper level forced back and uplifted about 1.3 (0.06*D*) (Fig. 3b). Some of the fault motions appear to have reversed in later stages, i.e., normal faults become reverse faults and early thrust faults converted to reverse faults in parts of the complex rim uplift. The beds can be repositioned at the 1.8 m (0.09*D*) depth when they are graphically unfolded back into their pre-shot positions. In this particular explosion trial, crater shape and differential strengths in layered target media exhibit an excellent cause and effect relationship in the formation of the bench.

Structural deformation above the bench and in the upper crater walls and rim consisted mainly of very complex folding and faulting associated with outward movements, rim uplift, and overturning, as shown in Fig. 3b. Development of an overturned flap caused beds nearer the surface to be uplifted and overturned to form an ejecta blanket of crudely inverted stratigraphy (Fig. 3b). Strata under the rim crest exhibit a complete range in dip from vertical to overturned. The hinge zone that was part of the overturned flap was present as a well-formed continuous overturn in every trench excavation around the crater. Structural uplift averaged 1.0 m (0.05*D*) at the hinge point and 1.5 m (0.07*D*) at the rim crest with respect to the original ground level.

The ejecta blanket forms a continuous ground cover out to an average distance of approximately 54 m (2.7*D*) from GZ with complex patterns of small dune-like deposits locally present near the outer edge (Figs. 2b, c). A hummocky surface dominates the inner part of the overturned flap extending from the rim crest out to approximately 28 m (1.4*D*) from GZ (Fig. 2b). Beyond approximately 54 m (2.7*D*), the ejecta blanket forms a discontinuous cover consisting predominantly of a very complex set of thin, linear, ribbon-like rays of fine ejecta (Fig. 2c). Scattered, small, secondary craters and secondary crater fields formed by the impact of playa ejecta blocks were present within this region of discontinuous ground cover as well as beyond its outer edge. A small number of secondary craters formed by high-angle ejecta were also present on the surface of the continuous ejecta blanket. A very limited amount of slightly fused playa fragments were deposited as final fallout on top of the overturned flap near the rim crest and downwind.

(3) *Unnamed crater, moon.* The unnamed crater shown in Figs. 2d, e is located

on the west flanks of the large crater Gagarin on the lunar farside (see Apollo frames AS15-03-0101, AS15-03-0102, AS15-03-0103, AS15-8936P, AS17-150-23102). This particular crater was chosen as a representative type because of the excellent photography, very recent age, bowl-shape and morphologic elements, even though it is somewhat deeper and more conical than the average bowl-shaped crater. It exhibits the typical morphology of a young small impact crater with: (1) deep bowl-shape, (2) small flat-floor, (3) talus-covered appearing lower walls with blocky terrain, (4) steep upper walls with local horizontally banded outcrop-like zones surrounded by a high sharp rim crest, and (5) a very light-colored, rayed, ejecta blanket with a blocky inner terrain. The crater, approximately 2.7 km across and 750–800 m deep (Pike, 1977, pers. comm.), occurs in a very rugged, intensely cratered terrain in the lunar Highlands, a region presumed to be underlain by tens of meters of regolith. An approximate profile and generalized geologic map are shown in Figs. 3c and 4c, respectively. Between approximately 1.5 and 2 km^3 of rock are estimated to have been ejected using the profile in Fig. 4c.

The floor of the crater, about 400 m (0.15*D*) across, appears flat with a covering of dark-colored material that also partly overlays the crater walls and inner rim (Fig. 2e). The lower parts of the walls exhibit irregularly lineated regions that appear to be slump and talus masses that have flowed downward and out onto the floor. Blocks at least as large as 30 m (0.01*D*) across are incorporated into the upper surfaces of these masses, with blocky terrains more prominent nearer the crater floor. The upper crater walls locally exceed approximately 60° to 70° m slope with scattered regions of very rough irregular terrain which locally has a very crude banded-like appearance of layered outcrops of either uplifted bedrock or ejecta blanket. This is especially prominent in the upper 10% of the crater walls. Two large radial breaks in the rim crest are similar to faulted or breached zones in terrestrial craters. The dark material that forms the crater floor and overlies parts of the walls, also forms a well-developed rayed pattern over the rim crest and inner rim. The distribution of this dark material indicates that it overlies the lighter-colored material on the walls, suggesting that it was encountered in the lowest units cratered, probably below 500–800 m (0.2–0.3*D*). The much larger distribution of the lighter-colored material on the crater walls and in the ejecta indicates that it was derived from the upper volumetrically larger part of the cratered region. Its position below the darker material also indicates it was ejected first, and was initially stratigraphically higher in the rocks cratered.

The sharp rim crest testifies to the freshness of this crater as does the preservation of the details of raying in the ejecta. Locally, several parts of the rim crest exhibit short, concentric ridges, 50–100 m (0.02–0.04*D*) in length, that have an appearance similar to terrestrial explosion craters in which the overturned flap of ejecta has slid outward on the original ground surface of the uplifted rim. No fallout layer was positively identified.

A light-colored, well-developed, ejecta blanket surrounds the crater and extends outward from GZ for 2.9*D*. Within this region, a continuous cover of ejecta blankets the lunar surface with blocky, fragmented rock from the crater (Fig. 2e). The very light color, which is common at certain terrestrial sites, is probably the result of highly fractured and comminuted rock which has a high reflectivity. Massive ejecta blocks over 100 m (0.04*D*) across are exposed on the ejecta blanket within 1*D* of the rim. Figure 2e shows numerous regions of extensive blocky terrain, many of which have a radial pattern to the crater. A very well-defined radial pattern of ridges and grooves, as wide as 50 m (0.02*D*) and tens of meters in length, characterized the ejecta surface, at least within 2*D* of GZ. A very poorly developed, hummocky terrain is locally present on parts of the inner ejecta blanket within 500 m (0.7*D*) of the rim crest. The absence of extensive hummocky terrain is probably due to cratering in regolith without harder layers.

The outer edge of the ejecta becomes extremely irregular in areal distribution with a pronounced complex ray pattern. Within this region, which extends to 5*D* from GZ, on the order of 50% of the lunar surface appears still covered by ejecta material. Rays can be identified, however, beyond 20*D* from GZ, although the areal distribution is less than 1% cover at those ranges.

(b) *Flat-floored craters with central uplifts*

The craters chosen to illustrate the morphological and structural analogs between flat-floored central uplift craters include: (1) the terrestrial impact site, Flynn Creek, Tennessee, (2) the experimental explosion crater, Snowball, and (3) the lunar crater, Copernicus.

(1) *Flynn Creek Crater, Tennessee.* The Flynn Creek Crater, located in north central Tennessee, was formed in a hilly coastal plain environment approximately 360 m.y. ago. The important morphologic elements included: (1) a broad flat hummocky floor, (2) large central peak, (3) locally terraced crater walls, (4) uplifted as well as depressed rim segments, and (5) surrounding ejecta blanket (Fig. 5a). The major structural features include: (1) a breccia lens underlain by a thin disrupted zone, (2) a large central uplift underlain by a deep disrupted zone, (3) locally terraced crater wall sections, (4) faulted, folded, brecciated, and fractured rim strata, (5) uplifted as well as downfolded rim segments, (6) large rim graben, and (7) the remnants of a once surrounding ejecta blanket consisting of an overturned flap of crudely inverted stratigraphy. A cross-section and structure map are shown in Figs. 6a and 7a, respectively.

The impact event occurred in flat-lying limestone and dolomite overlying a crystalline basement at about 1700 m (0.5*D*) depth. Energy balance calculations indicate approximately 100 megatons (10^{24} ergs) were involved with the impact of a lower density body (estimate 3.3 g/cm^3) approximately 100 m across and weighing some 10^6 metric tons (Roddy, 1977a). During the cratering, approximately 1.2 km^3($\sim 3.2 \times 10^9$ metric tons) of rock were brecciated with 0.8 km^3($\sim 2.2 \times 10^9$ metric tons) of rock ejected. The result was to form a crater with an estimated rim crest diameter of approximately 3830 m (3517 m apparent) with a depth of 198 m (98 m apparent, 0.03*D*). Field evidence indicates that the impact

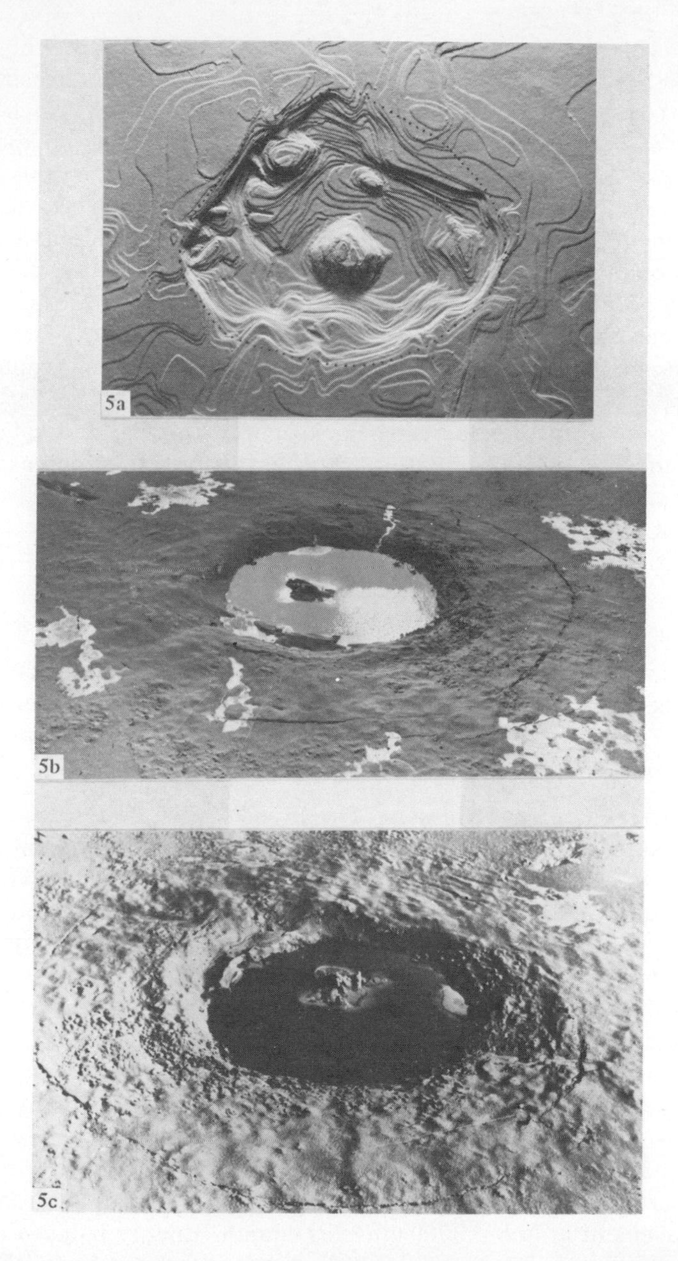

Fig. 5. Oblique aerial and model photographs of three flat-floored, central uplift craters. The view in Fig. 5a is of a model of the Flynn Creek Crater, Tennessee, constructed from the structure contour map, showing the crater after the erosion of the ejecta blanket and immediately before deposition of the Chattanooga Shale approximately 360 m. y. ago. Note the large central uplift. The large terrace-like hills near the walls are blocks slumped toward the crater. The dotted line shows the location of the approximate apparent crater diameter which averages about 3517 km across. North lighting with no vertical exaggeration.

The view in Fig. 5b,c shows the Snowball Crater formed by a 500-ton TNT hemisphere one day after the explosion. The rim crest diameter is about 108.5 m. Note the irregular central uplift in the lake, terraced walls, and surrounding hummocky ejecta blanket with concentric fractures. The light-colored patches of material in Fig. 5b are fine alluvium deposited by water flowing from the fractures. A ponded alluvial deposit can be seen in Fig. 5b in the lower left crater wall formed over a local fracture. Photographs courtesy of Dr. G. H. S. Jones, Defence Research Board, Canada, and the Defence Research Establishment Suffield, Alberta, Canada.

The view in Fig. 5e (AS17-M-2444) is across Mare Imbrium toward Copernicus in the background. The distant edges of its bright-rayed ejecta blanket include a large number of secondary craters, some over 1 km across, and crater chains in the foreground. The secondary ejecta related to Copernicus can be identified by the patterns radial to the main crater. The range from the bottom of the photograph to the center of Copernicus is about 400 km (5.5D). Figure 5d (AS 17-151-23260) shows Copernicus with its large central peaks and complex terraced walls. An irregular hummocky ejecta blanket with more recent craters surrounds the low but relatively sharp rim crest which is approximately 93 km across. Note the development of multiple terraces in Copernicus.

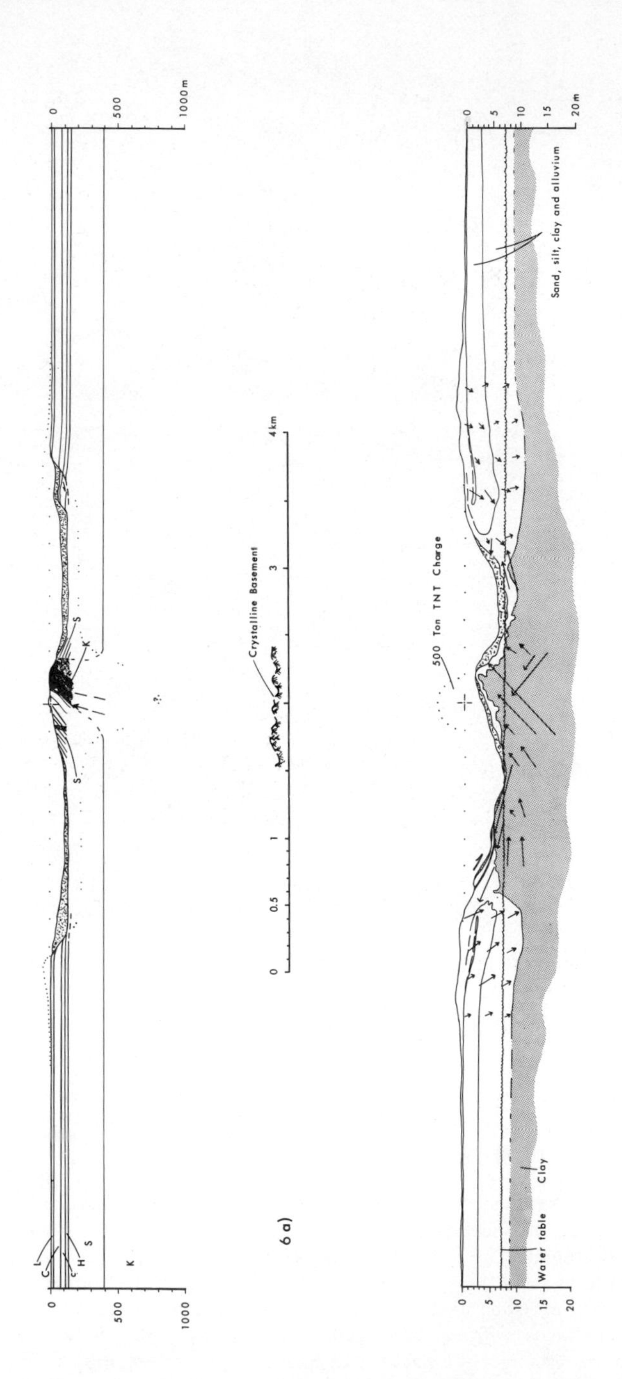

1000 m
4 km
Crystalline Basement
6 a)
500 Ton TNT Charge
Sand, silt, clay and alluvium
Clay
Water table
20 m
100 m
6 b)

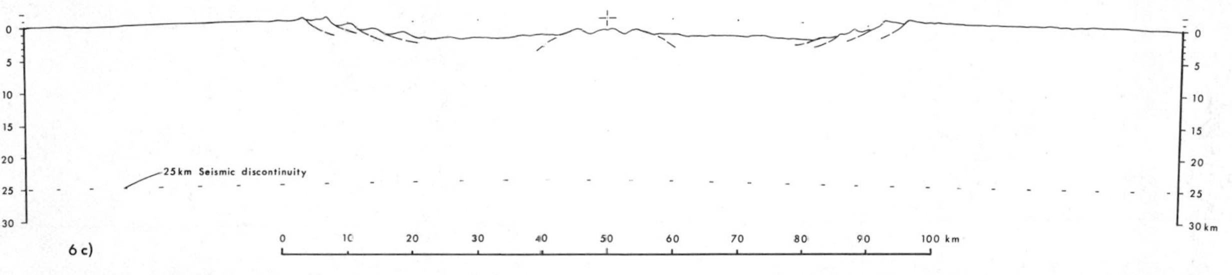

Fig. 6. Generalized geologic cross-sections of three flat-floored, central uplift craters, with Fig. 6a of Flynn Creek, Tennessee (modified from Roddy, 1968), Fig. 6b of the Snowball 500-ton TNT explosion crater (modified from Roddy, 1976), and Fig. 6c a schematic of the lunar crater, Copernicus, drawn with the actual lunar curvature. The letter symbols in Fig. 6a identify the strata described in Roddy (1968) and show the uplifted Stones River (S) and Knox (K) in the central uplift. The two dashed arrows in Fig. 6a show generalized displacements of the top of the Knox strata. The solid lines, alluvial contacts, in Fig. 6b are from open-trench excavation data and the dashed lines are from drill data. The arrows in Fig. 6b are representative terminal displacements of buried marker cans with the ends of the arrows at pre-shot positions and the arrowheads at post-shot surveyed positions as modified from Jones (1976) and Roddy (1976). Each crater is drawn to show the apparent diameter at the same scale. The explanations for the symbols are shown in Fig. 1.

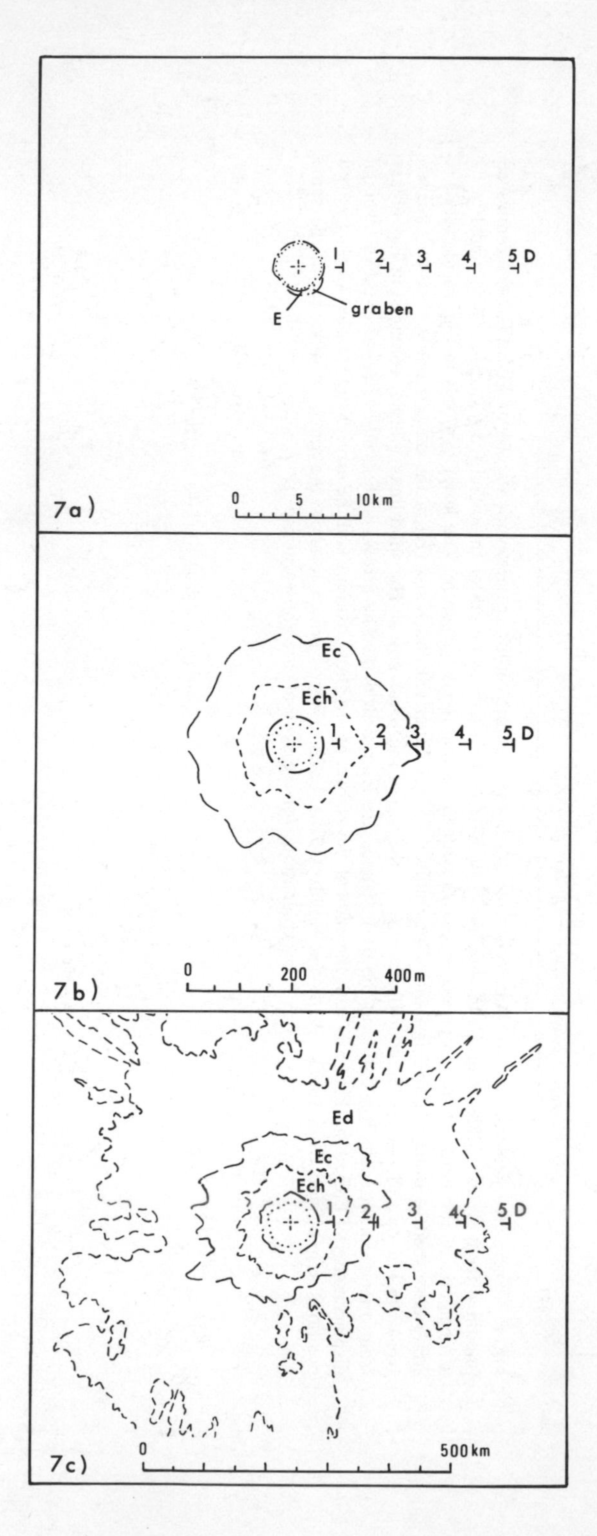
1
2
3
4
5 D
E
graben
7a)
0
5
10km
Ec
Ech
1
2
3
4
5 D
7b)
0
200
400m
Ed
Ec
Ech
1
2
3
4
5 D
7c)
0
500km

occurred on a low, rolling coastal plain or in the very shallow waters of the Chattanooga Sea. The first deposits on the crater floor, after breccia washed down from the rim, are dolomites derived from the rim crest, as well as possible reworked fallout, with early Late Devonian marine conodonts of the Chattanooga Sea. No lacustrine deposits have been identified in the crater. Areal erosion was active, however, in forming talus deposits on the rim graben. Consequently, the crater appears to have been very near local base level and experienced both limited erosion in the higher elevations as well as marine deposition at approximately the same time on the crater floor, or shortly thereafter. The important result was that the crater experienced relatively little erosion before complete burial under the fine silty muds of the Chattanooga Shale. Less than 10–15 m of regional erosion are estimated as a maximum, although this was sufficient to remove the ejecta blanket, except for that trapped in the rim graben. More recently, regional uplift has exposed the crater in the many valleys that form the Flynn Creek drainage system. The dimensional parameters, geology, and cratering sequences are described further in Roddy (1966, 1968, 1977a,b).

The morphology of the crater floor, approximately 3160 m (0.9D) across, was relatively flat on the 10 m (0.003D) scale exhibiting a hummocky surface on top of the breccia lens and a large central peak. The surface, before the deposition of breccia washed down from the rim, consisted of chaotic blocky fragments ranging in size from sub-millimeter to blocks over 100 m (0.03D) across. The larger, more blocky, terrain was most common along the base of the crater walls. A large central peak, which rose nearly 120 m (0.03D) above the floor, dominated the center of the crater. The peak, nearly 1 km (0.28D) across, had the shape of an irregular dome with flanks averaging 10–30° in slope after limited erosion, with coarse, blocky fallback covering the irregular top of the peak.

The lower crater walls were covered by fallback and reworked breccia, whereas the steep upper walls appear to have been locally clear of cover. Large blocks, over 500 m (0.14D) across, locally separated from the walls and moved toward the center of the crater forming irregular terraces. Over 50% of the inner rim strata experienced downfolding toward the crater, with other parts uplifted to form a well-defined structural rim crest. The western, crudely terraced walls were generally irregular but locally appear to follow the regional joint system, although their present forms may be the result to erosional enhancement. A large ring graben in the southern rim formed the best developed terrace (Fig. 6a). An

Fig. 7. Generalized geologic maps of three flat-floored, central uplift craters, with Fig. 7a of the eroded crater, Flynn Creek, Tennessee (Roddy, 1968), Fig. 7b a schematic map of the Snowball 500-ton TNT explosion crater, and Fig. 7c of the lunar crater Copernicus. Each crater is drawn to show the apparent crater diameter at the same scale. The symbols include Ec for continuous ejecta, Ech for continuous hummocky ejecta, and Ed for discontinuous ejecta. The dotted circles are the apparent diameters, the dot-dash lines are rim crests, the long-dashed lines show the approximate outer limits of the continuous ejecta, and the short-dashed lines show the very approximate limits of the continuous hummocky and the discontinuous ejecta units. The numbered symbols show multiples of the apparent diameter distances from GZ out to 5D.

ejecta blanket initially surrounded the crater, but was totally removed by erosion except for a large mass trapped in the rim graben.

The structure beneath the crater floor surrounding the central uplift consists of a bedded breccia, a breccia lens and a relatively thin disrupted zone. The bedded breccia sequence and depositional history are part of the post-crater events and are discussed in Roddy (1966, 1968). The breccia lens, which averages 35 m ($0.01D$) in thickness, consists of fallback and locally disrupted limestone derived mainly from the upper 100 m ($0.03D$) of rock cratered. A very limited amount of deeper rocks derived from the Stones River Group in the central uplift are also present in the breccia, added possibly by erosion and talus transport from the central peak. Fragments range in size from sub-millimeter to blocks over 100 m ($0.03D$) across. The brecciation was continuous down into the Hermitage Formation, the weakest unit in the stratigraphy (Fig. 6a), with the base of breccia in relatively sharp contact with the underlying upper part of the dolomitic limestone of the Stones River Group at approximately 140 m ($0.04D$) below GZ. The faulting, folding, brecciation, and fracturing in these underlying rocks decrease downward and are essentially absent below approximately 100 m ($0.03D$) beneath the base of the breccia lens, i.e., 235 m ($0.07D$) below the pre-impact ground surface. Injection breccias, up to 3 m thick, have been forced into weak zones along bedding planes adjacent to the central uplift at a depth of approximately 200 m ($0.06D$) below the pre-impact ground surface. This indicates that strata at these depths were temporarily opened along weaker zones (presumably in tension) during the cratering event. Other details, including a possible ring structure on the crater floor, are described in Roddy (1966, 1968).

The structure of the central uplift consists of a complex sequence of highly faulted, folded, brecciated, and fractured limestones and dolomites which have been moved in and up to form a domical-shaped central peak. The stratigraphic units exposed above the crater floor include the Stones River Group and the upper part of the Knox (Fig. 6a). The top of the Knox dolomite has been uplifted through approximately 450 m ($0.13\ D$) to the top of the central uplift and exhibits shatter cones. The termination of major disrupted strata beneath the central uplift is estimated to be on the order of 750 + m ($0.2D$) below the pre-impact ground level. Large fault zones, both high and low angle, cause numerous omissions of the Stones River strata exposed on the flanks of the central peak with most of the beds near the middle of the peak dipping 24–60° to the west and northwest. Breccia zones several meters wide cut the exposed strata at various angles and trend generally north. The available structural data indicate the totally disrupted zone beneath the central peak can be represented very generally by a broad truncated cone, although further subsurface data are definitely necessary to define the geometry.

The structure in the rim includes several types of deformational styles, with folding, faulting, brecciation, and fracturing common throughout the limestone strata. The major part of this deformation generally lies within approximately 500–750 m (0.14–$0.21\ D$) of the crater walls, exhibiting an irregular concentric distribution. The structure of the western rim is dominated by a downfold along a broad monocline dipping towards the crater, whereas the eastern rim is uplifted as much as 50 m (0.01 D) along the crests of anticlines and sharp monoclines adjacent to the crater walls. The southern rim, in contrast, consists of a complex structural sequence with a major thrust fault and a large graben. The thrust block, initially forming part of the southern crater wall, was forced out from the crater, raising strata on the order of 30 m (0.01 D) and partly overriding the large downdropped block forming the graben. The fault zone marking the southern boundary of the graben curves towards the crater with the whole complex forming a large terraced system (Figs. 6a, 7a). Absolute displacement along the fault zone is approximately 100 m (0.03 D). Large blocks along the western rim also formed irregular terrace-like structures.

A large mass of chaotic limestone breccia overlies the southern rim graben, representing the remnants of the ejecta blanket that initially surrounded the crater. The stratigraphy in this breccia is crudely inverted with respect to the normal pre-impact sequence indicating that an overturning flap process also operated here. The necessary hinge zone presumably either eroded from the crater wall region or, more likely, was totally disrupted during the cratering. The 360 m. y. old pre-impact *stripped* ground surface underlies the ejecta on top of the graben.

Anticlinal and synclinal folds with axes radial, as well as concentric to the crater locally form crenulated structural patterns adjacent to the crater walls. The concentric folds commonly have crests that plunge irregularly, and radial compression or shortening of the strata along some folds is as great as 35%. Most of the smaller normal faults and thrust faults, and the asymmetric anticlines, synclines, and monoclines trend approximately concentric to the crater walls. Dips on the fold limbs are commonly as high as 35°, with the steeper limbs dipping towards the crater. Rim faults often grade into tight anticlinal folds or into authiginic breccia or jumbled zones. Displacements on the largest faults are as great as 100 m (0.03 D). The major structural elements are shown in Fig. 2 in Roddy (1977b, this volume).

The ejecta blanket is preserved only as a remnant in the graben in the southern rim, and exhibits a crude inversion of stratigraphy indicating that the ejecta was emplaced by an overturning process. Erosion has removed all but this one set of exposures that now cover approximately 2 km^2.

(2) *Snowball Crater, DRES, Canada.* The Snowball explosion trial was conducted at the Defence Research Establishment Suffield, Alberta, Canada and consisted of a 500-ton TNT hemisphere lying on alluvium. The important morphological elements included: (1) a broad flat hummocky floor, (2) large central peak complex, locally terraced crater walls, (3) depressed rim, and (4) surrounding rayed ejecta blanket with a smooth to blocky, hummocky inner terrain (Fig. 5b,c). The major structural features included: (1) a thin fallout layer, (2) breccia lens underlain by a disrupted zone, (3) a large central uplift underlain by deep disrupted zone, (4) narrowly terraced crater wall sections, (5) faulted and folded rim with major concentric fracture system, (6) broadly

downfolded rim dipping into crater, (7) secondary craters, and (8) a surrounding ejecta blanket consisting of an overturned flap of inverted stratigraphy. Figures 6b and 7b show the cross-section and geologic map, respectively. Details of this trial are described in Diehl and Jones (1965, 1967a,b), Jones (1976, 1977), Roddy (1968, 1976, 1977a), and Rooke and Chew (1965).

The Snowball trial was conducted on unconsolidated alluvial sands, silts and clays with a water table at about 7 m (0.08*D*) depth and sandstone bedrock at approximately 60 m (0.7 *D*) depth. During the cratering, approximately 33,329 m^3 ($\sim$ 44,118 metric tons) of alluvium were totally disrupted and approximately 17,500 m^3 ($\sim$ 28,000 metric tons) of alluvium are estimated to have been ejected. The resulting crater had an average rim crest diameter of 108.5 m ($\sim$ 83 m apparent) and a rim crest depth of 7.5 m ($\sim$ 6.6 m apparent, 0.08 *D*) (Roddy, 1977a). Immediately after detonation, ground water flow from fractures in the crater floor began to form a lake depositing approximately a meter of alluvium on the floor. The crater incurred no further modifications before field studies, including trenching were completed.

The morphology of the crater floor, about 53 m (0.06*D*) across, was represented by a broad relatively flat surface surrounding a large central peak. The floor itself consisted of low hummocky terrain with relief less than about 1 m (0.01*D*) that formed the top of the breccia lens. An irregularly shaped complex of central peaks dominated the center of the crater floor with several large masses of uplifted clay standing at various angles (Fig. 5b, c). The flanks of the peaks exhibited slopes ranging from 10° to vertical with numerous color lineations and outcrops dipping at steep angles. The highest parts of the peaks rose 3.5 m (0.04*D*) above the crater floor and the average diameter at the base of the entire complex was approximately 21.5 m (0.3*D*). Blocky terrains locally covered parts of the complex with fallback covering much of the irregular tops of the peaks.

The morphology of the crater rim, locally terraced at the crater walls, was dominated by smooth to hummocky, blocky terrain of the overturned flap. Concentric faults and open fractures near the rim crest marked the outer limit of active terrace slumping. Concentric fractures in the middle and outer rim opened as a response to the shock rarefaction phase and relaxation of the high stress field. Minor tensional opening appears to have continued for several days. A thin layer of fallout tended to smooth the rougher character of the regions that were blocky. Alluvium, deposited by water flowing from a fracture on one of the crater terraces, formed ponded deposits on the terrace that eventually flowed onto the crater floor (Fig. 5b, c).

The structure beneath the crater floor consisted of a thin fallout layer, a breccia lens underlain by a deformed and disrupted zone, and a central uplift. The thin layer of fallout, which blanketed both the crater floor, walls, and surrounding rim, averaged a few centimeters thickness within the crater and thinned to millimeters at its irregular outer edges. This layer consisted of very finely comminuted alluvium initially lofted into the air by the cratering event. The fallout layer was typically transitional in grain size and texture into the top of the breccia lens and ejecta blanket, partly filling to filling the irregular depressions on the surfaces of the breccia and ejecta. The breccia lens, which averaged 0.7–1.1 m (0.01D) over the crater floor and thinned to less than a few centimeters on the upper walls, consisted of fallback and locally disrupted alluvium derived mainly from the lower 2–4 m (0.04D) of strata cratered. A limited amount of clay fragments from as deep as $\sim 10+2$ m (0.12D) were mixed in the breccia and derived from the central uplift. Fragment sizes in the breccia ranged from sub-millimeter to blocks over one meter (0.01D) across forming a chaotic mixture of unsorted to very poorly sorted breccia of alluvium of slightly shock compressed alluvium with the largest blocks commonly near or at the base of the breccia lens.

The deformed and disrupted zone beneath the breccia lens consisted of alluvium that had been strongly moved inward and upward to raise material beneath the entire crater floor as shown in Fig. 6b. The large subsurface motions, as great as 6 m (0.07D), were enhanced by the mobility of the alluvium due to the water saturation. These motions, however, were typical of all such trials with comparable initial boundary conditions. The factor that varies most is not the sense of displacement, but instead is the absolute magnitude of the displacements which are related to the specific initial conditions, i.e., dynamic late-stage strengths of the target media (Roddy, 1976).

In the center of the crater, the central uplift contained clay beds raised typically through 6–10 m (0.07–0.12D) or more with the general structure consisting of a complex dome of uplifted alluvium (Fig. 6b). The central uplift appears to be, in part, the consequence of the large overall inward motions beneath the *entire* crater floor towards the center of the crater. The large mass movements towards the center of the crater require final upward displacements around the axis of symmetry. The possible causes of such physical processes which invoke the crater floor motions and form the central uplift are discussed in Ullrich *et al.* (1977).

The major structures in the crater rim consist of a downfold sequence, locally terraced segments, and an overturned flap. The initial downfolding of the original ground surface begins approximately 55 m (0.7D) from GZ with the surface hinge at about 38 m (0.5D) from GZ. At these distances the original ground surface has been downfolded on an average of 2 m (0.02D). Locally, large terrace blocks developed and slumped towards the crater with at least 50% of the rim exhibiting some terrace development. Beds near the bases of the crater walls exhibited moderate dips out from the crater, whereas higher in the walls, the dips increased from vertical to overturned. Final downward and inward motion in the inner rim are shown in Fig. 6b and are related to the overall inward motions beneath the crater floor. Normal, reverse, and thrust faulting in the crater walls and inner rim were preserved in the more competent clay beds. Pronounced concentric fractures occurred in the rim at distances of 55, 67, 73, 79, 91, 116, and 155 m (0.7, 0.8, 0.9, 1.0, 1.1, 1.4, and 1.9D). A large radial fracture, over 50 m (0.6D) also opened on the rim. Prominent deposits of alluvium carried by water flow to the surface occurred along these concentric and radial fractures shortly after the cratering event (Fig. 5b, c).

A large ejecta blanket consisting of an overturned flap covered the rim and extended out to approximately 200+m (2.4+*D*) from GZ. A well-defined hummocky terrain formed the inner part of the blanket out to approximately 107 m (1.3*D*) from GZ. Individual as well as linear clusters of secondary craters occurred beyond the edge of the continuous ejecta blanket.

(3) *Copernicus, Moon.* The lunar crater, Copernicus, shown in Fig. 5d,e is located on the southern rim of Mare Imbrium in a complex region containing both maria and local Highland material (see, Masursky *et al.*, and Schultz, 1976 for collection of Apollo and Orbiter photography and Schmidt *et al.*, 1967 for geologic map of the Copernicus Quadrangle of the moon). It exhibits the typical morphology cf a large, young impact crater with: (1) a broad flat hummocky floor, (2) central peaks, (3) terraced and horizontally banded crater walls, (4) low rim crest, (5) a light-colored, rayed ejecta blanket, and (6) extensive secondary craters (Fig. 5d,e). Volcanic constructional forms appear locally present, including floor filling, small domes, flows, and other features.

In its present form the crater has an average rim crest diameter of 93 km (84.6 km apparent) and 3.8 km depth (2.7 km apparent, 0.03*D*) (Pike, 1976, 1977, pers. comm.) These diameters include the massive terraced rim segments that have moved downward towards the crater and consequently are larger than the immediate post-crater diameters. For comparative purposes with other craters in this paper, the diameter for the *initial* crater walls are estimated (using lunar photography and an adjustment for block slump expansion) to be approximately 15–25% less than the present observed value. Using 15% gives a conservative initial rim crest diameter of approximately 79 km (72 km apparent), and the initial rim crest depth, before lava filling and terrace adjustment, is estimated to have been slightly over 4 km. The estimated initial apparent diameter of 72 km is used for *D* ratios in this section. A generalized profile and geologic map are shown in Figs. 6c and 7c, respectively. Descriptions of the cratering processes are given in Shoemaker (1962) and Schultz (1976) provides a number of morphologic descriptions using Orbiter and Apollo photographs of Copernicus.

The broad floor of the crater, approximately 60 km (0.8*D*) across, is relatively flat on a 100 m+ relief scale except for a group of large central peaks. Locally, the floor is hummocky with what appear to be low volcanic domes and craters, irregular flows, and fractured blocky terrains which may represent both the top of the irregular breccia lens and volcanic constructional forms. The central peaks, approaching 1 km in height (0.01*D*), form irregular single mountains as well as linear ridges over 10 km (0.14*D*) in length (Fig. 5d). Slopes of several of the larger peaks average approximately 40° and locally exhibit terrains covered by massive blocks. Color lineations and rough, irregular outcrops dip steeply on the flanks of the central peaks as seen in the Orbiter high resolution photography. The central peaks give the appearance of being partially buried by the surrounding breccia lens as well as by later volcanic fill.

The crater walls consist of a complex sequence of multiple terraces that appear to be composed of large irregular to arcuate slumps of the rim strata. Material that appears to be extensive volcanic flows covering the floor tends to obscure the lower parts of the rim terraces. Other volcanic forms present on the floor, terraces, and rims are summarized in Schultz (1976). The present width of the terraced part of the crater walls is approximately 16 ± 2 km (0.2*D*), depending on the choice of the base of terraces. Figure 5d shows at least four well-developed major terraces, as well as numerous smaller irregular masses. The inward curving faults inferred for these terraces are shown in Fig. 6c and follow the pattern suggested by Dence (1968, 1971), Shoemaker (1962), and Roddy (1966, 1968) and others for these types of craters.

The rim crest, immediately beyond the present outermost terraced wall scarp, appears topographically high around the entire crater, implying (though not proving) structural uplift as well as a deposit of ejecta. Radial and concentric faults can be observed directly in the terraced section and indirectly in the rim. Large offsets in the ejecta blanket imply that some fault readjustments occurred after the emplacement of the ejecta. The surrounding ejecta blanket has a continuous ground cover with an irregular outer edge extending out to at least 1.9*D* from GZ (using initial apparent crater diameter), with hummocky terrain occupying the inner part of the blanket orbit to approximately 1.1*D* from GZ. Irregular, discontinuous to locally continuous bright-rayed ejecta continues prominently out to 4–5*D* from GZ, with thin rays from Copernicus identified at least as far as 12*D*. Secondary cratering is well developed beyond approximately 1.5*D* with elongated crater chains as well as numerous large single craters (Fig. 5c). One of the largest crater chains, over 100 km (1.4*D*) in length, contains single craters over 5 km (0.07*D*) across. The ballistic history of secondary cratering at Copernicus is discussed in detail by Shoemaker (1962).

(c) *Flat-floored craters with multirings*

The craters chosen to illustrate the morphological and structural analogs between flat-floored multiring craters include: (1) the terrestrial impact site, the Ries Crater, Germany, (2) the experimental explosion crater, Prairie Flat, and (3) the Orientale Basin on the moon.

(1) *Ries Crater, Germany*. The Ries Crater, located in southern Germany, was formed in the Swabian Alb scarp and its foothills approximately 14.8 m. y. ago. The important morphological elements *initially* included: (1) a broad flat-floor with a central basin, (2) concentric partial ring at the outer edge of the inner floor, (3) broadly terraced to hilly outer crater wall region, and (4) remnants of surrounding ejecta blanket, apparently all discontinuously to continuously covered by a layer of melt-like breccia. The major structural features include: (1) an inner crater region with a layer of melt-like breccia overlying a breccia lens of crystalline rock, (2) underlying zone of deep disrupted rock, (3) uplifted crystalline rock to form concentric structural partial ring on crater floor, (4) faulted, folded, brecciated and fractured sedimentary rim rock forming a complex, locally terraced, broad crater wall section, (5) probable buried secondary crater fields beneath a large surrounding ejecta blanket with local crude inversions, and (6) a dis-

continuous overlying melt-like breccia. The interpretations of the origin of several of these morphological and structural features remains controversial as noted in this volume by Chao and Minkin (1977) and Pohl *et al.* (1977). Regardless of these problems, however, the Ries constitutes the largest and best preserved terrestrial impact crater exhibiting a *ring* structure and still retaining a part of its initial morphology and ejecta blanket. Recent summaries by Chao (1974, 1976, 1977a,b), Chao and El Goresy (1977), Dennis (1971), El Goresy and Chao (1976, 1977), Englehardt (1974), Englehardt and Graup (1977), Englehardt *et al.* (1969), Hörz *et al.* (1975, 1977), Pohl *et al.* (1977), Stöffler (1971, 1977), Preuss and Schmidt-Kaler (1969), and others have discussed the many aspects of the Ries event and the reader is referred to those papers for more detailed analyses. The descriptions in this section draw heavily upon these and other references and a generalized cross-section and geologic map are shown in Figs. 9a and 10a, respectively.

The Ries impact event occurred in nearly flat-lying limestones, sandstones, shales, and marl and penetrated well into the crystalline basement at approximately 600 m ($0.03D$) depth. Stöffler (1977) describes the sedimentary section in this area as ranging from 480 to 780 m (0.02–$0.04D$) in thickness. Energy balance calculations indicate that between 10^{26} ergs (Port, 1977, pers. comm.) and 10^{28} ergs (Chao and Minkin, 1977) were involved in the kinetic energy of the impacting body, suggested to be a stoney meteorite by El Goresy and Chao (1976). Calculations of the body size by Port gives a diameter of about 0.4 km, and by Chao and Minkin, a diameter of about 2 km. During the cratering, Chao and Minkin (1977) estimate that on the order of 150–200 km^3 ($\sim 10^{11}$ metric tons) of rock were ejected. Stöffler (1977) in describing the Nördlingen deep drill hole, lists a depth to the melt breccia that places the apparent crater depth on the order of 570 m below the original surface depending on the choice of the surrounding rim scarp elevations. The *initial* apparent crater diameter, however, is less certain. Chao describes a cratering model in this volume which interprets the crater walls as essentially coincident with the present outermost topographic scarps, i.e., approximately 25 ± 1 km for an apparent diameter. Most of the papers on the Ries also quote only the topographic rim as the apparent diameter. Exposures, however, along the hilly southern and eastern parts of the crater suggest that the *initial* crater walls and apparent diameter may have been substantially smaller, on the order of 20 ± 1 km or less. The author's choice of the apparent diameter and crater walls rests on descriptions of proposed terracing and is discussed later in this section; however, an apparent diameter of 20 km is adopted for tentative comparative (xD) values in this paper.

The morphology of the crater shortly after its formation consisted of a broad flat, presumably hummocky, inner floor about 12 km ($0.6D$) across, a surrounding ring of crystalline hills up to a kilometer (0.05) across, and an outer zone of blocky, hilly terrain about 6 km ($0.3D$) across extending to the outermost rim scarps. This outermost zone of megablocks and local topographic terraces appears to discontinuously encircle the crater wall region. The inner part of a large ejecta blanket once apparently overlay parts or all of this hilly zone and surrounded the crater, and now exists discontinuously as far out as 40 km ($2D$)

from GZ. A layer of melt breccia underlies the crater floor, and its variants initially covered most or all of the hilly crater wall zone and inner ejecta blanket. This sequence now remains as a layer beneath the crater floor and as scattered remnants on the latter two regions out to approximately 22 km (1.1*D*) from GZ (Fig. 9a).

The interpretations of the structure beneath the crater floor are based on a limited number of surface exposures near the walls, six shallow and one deep drill core, petrologic studies and gravity, magnetic and seismic studies. Dence (1977, pers. comm.), Hörz (1977, pers. comm.), Chao and Minkin (1977), Engelhardt (1974), Stöffler (1977), Pohl *et al.* (1977), and others each interpret the Ries to have an inner basin bounded by a ring of uplifted crystalline basement and surrounded by an outer zone of complex crater floor and rim structure, although different modes of formation have been advanced for several of these features (Fig. 9a). A 1206 m deep drill core in the inner basin, approximately 3.5 km (0.18*D*) west of GZ at Nördlingen, shows this part of the floor to be underlain by 331 m of lake and alluvial sediments, which, in turn, overlie 271 m of melt breccia, called *suevite* (Stöffler, 1977; Pohl *et al.*, 1977). This unit, interpreted to include fallback and reworked fallout, is presumed on the basis of two other drill holes and geophysical studies to cover the entire floor of the inner crater. The Nördlingen drill hole shows the top of the *suevite* to lie at an average apparent depth below the original ground level of approximately 570 m (0.03*D*) with the base at an approximate apparent depth of 840 m (0.04*D*) below original ground level (Stöffler, 1977). The remaining drill core extends to a depth of over 1445 m (0.07*D*) below the original ground surface and consists of crystalline basement rocks with deformation which Pohl *et al.* (1977) interpret as indicative of a disrupted and probably locally mixed zone, the basal part of a breccia lens in this author's terms. The core below the *suevite* exhibits crystalline megablocks that have a variety of magnetic remanent orientations suggesting total block disruption and rotation. The base of this brecciated zone is not known; however, geophysical studies by Pohl *et al.* (1977) and summarized in Dennis (1971) suggest significant fracturing to a depth of over 6 km (0.3*D*). The drill core and ejecta data indicates that cratering and ejection of material extended through the sedimentary sequence and into the crystalline rocks at least to a depth of over 850 m (0.04*D*) and possibly as deep as 2500 m (0.1*D*) according to Stöffler (1977). The total nature and extent of the breccia lens, as characterized in the previous crater descriptions, remains undefined at this point.

One of the most significant structural features beneath the crater floor is the uplift of the crystalline basement and part of the overlying sedimentary section to form a large partial ring, approximately 6 km (0.3*D*) in radius, that bounds at least 80% of the inner basin. The geometry of this structure, which has a minimum uplift on the order of 350–450 m ($\sim 0.02D$), is discussed further in Pohl *et al.* (1977) in which they present arguments for the uplift of the entire crater floor. The structural nature of the ring and its mode of emplacement, however, remain unclear with respect to relative contributions from folding and faulting.

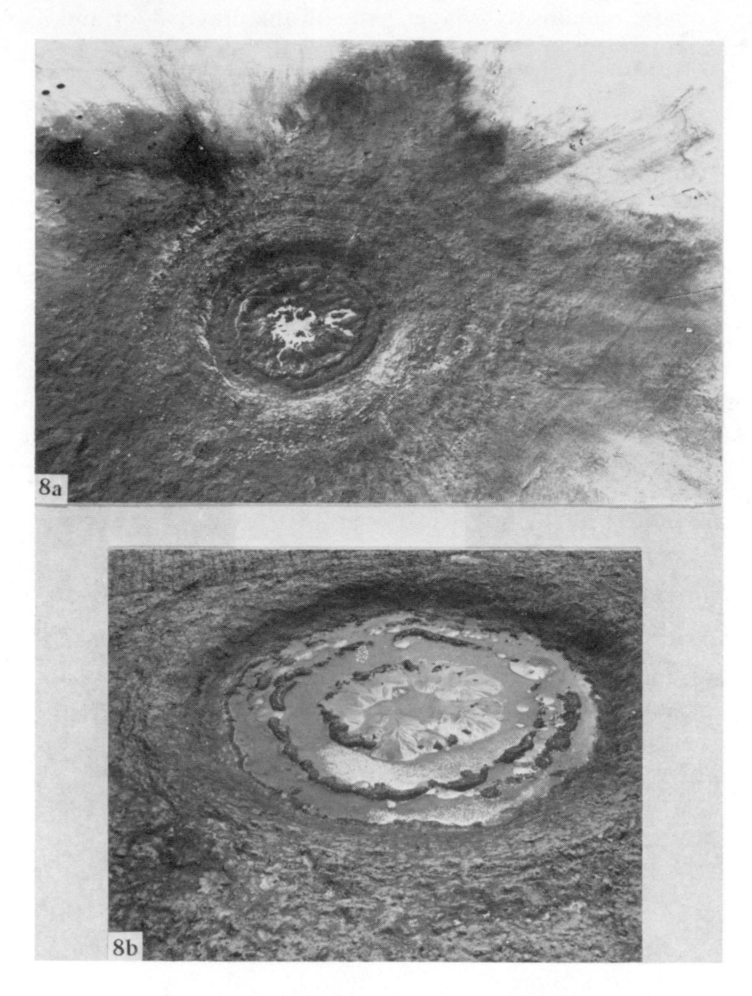

Fig. 8. Oblique aerial and space photographs of two flat-floored, multiring craters. The view in Fig. 8a shows the Prairie Flat Crater formed by a 500-ton TNT sphere about 1 hr after formation. The rim crest diameter is about 85.5 m. Note the concentric rings on the crater floor, the discontinuous ridges on the inner rim and the irregular rayed ejecta blanket. The outer edges of the hummocky terrain follow the ray-like extension as seen on the map in Fig. 10b. The view in Fig. 8b shows Prairie Flat several days later after the formation of a lake which accentuates the position of four discontinuous rings on the floor. The concentric patterns of sand cones along the rings were caused by alluvium deposited from water flow from concentric fractures along the rings. Photographs taken by U.S. Air Force Audio Visual Service, Norton AFB, California and from Roddy (1969).

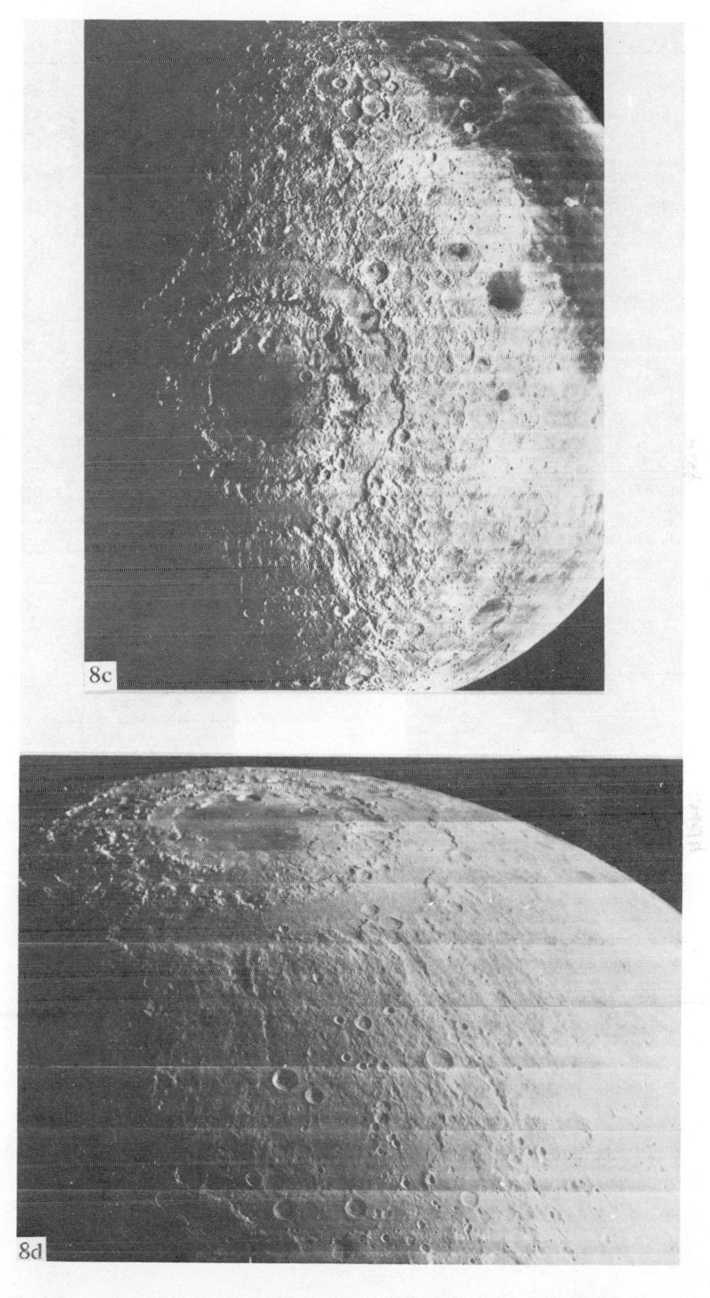

Fig. 8. (*Continued*). The views in Fig. 8c,d (LO IV-187M; LO IV-193M) show the large multiring Orientale Basin. The inner maria basin is about 300 km across with the four rings 320, 480, 620, and 920 km across. Note the rough, hummocky, radially sculptured terrain beyond the outer Cordillera Ring. Mare Imbrium is in the upper right of Fig. 8c. Note the prominent radial sculpturing beyond the outer ring with the largest trench, Bouvard, 40 km across and over 500 km long.

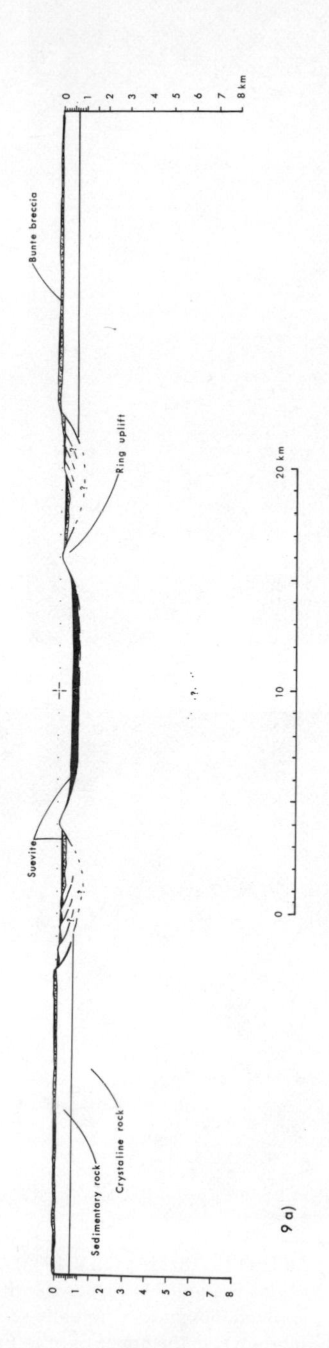
Bunte breccia
Ring uplift
Suevite
Sedimentary rock
Crystaline rock
0 10 20 km
8 km

9 a)

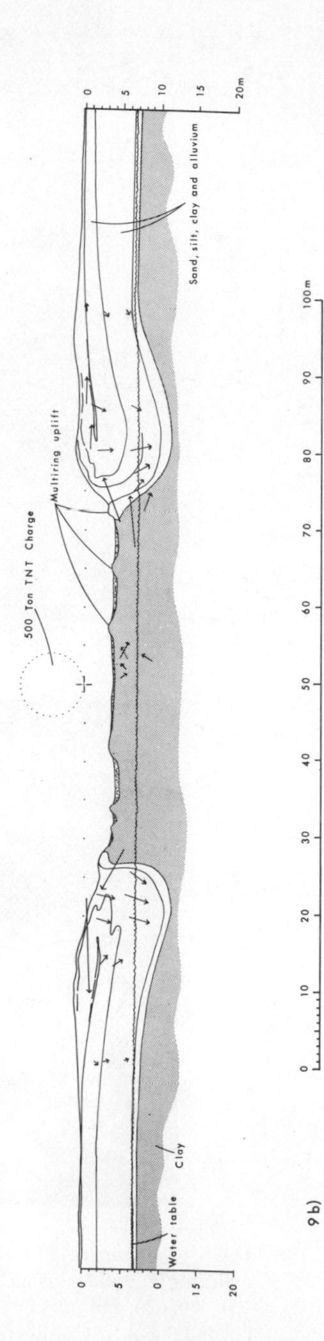
500 Ton TNT Charge
Multiring uplift
Sand, silt, clay and alluvium
Clay
Water table
20m
100m

9 b)

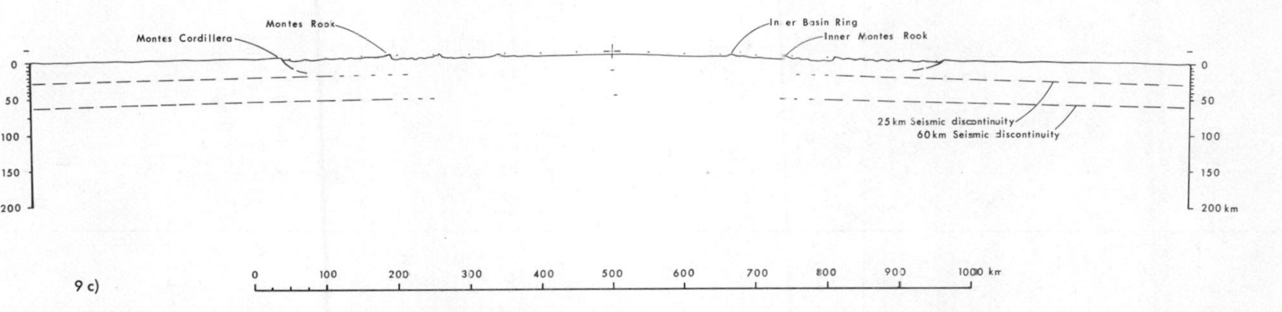

Fig. 9. Generalized geologic cross sections of three flat-floored, multiring craters, with Fig. 9a an interpretation of the Ries Crater, Germany, Fig. 9b of the Prairie Flat 500-ton TNT explosion crater (modified from Roddy, 1976), and Fig. 9c a schematic of the Orientale Basin on the moon, drawn with the actual lunar curvature. The solid lines, alluvial contacts, in Fig. 9b are from open-trench excavation data and the dashed lines are from drill data. The arrows in Fig. 9b are representative terminal displacements of buried marker cans with the ends of the arrows at pre-shot positions and the arrowheads at pre-shot surveyed positions as modified from Jones (1976) and Roddy (1976). Each crater is drawn to show the apparent diameter at the same scale. The explanations for the symbols are shown in Fig. 1.

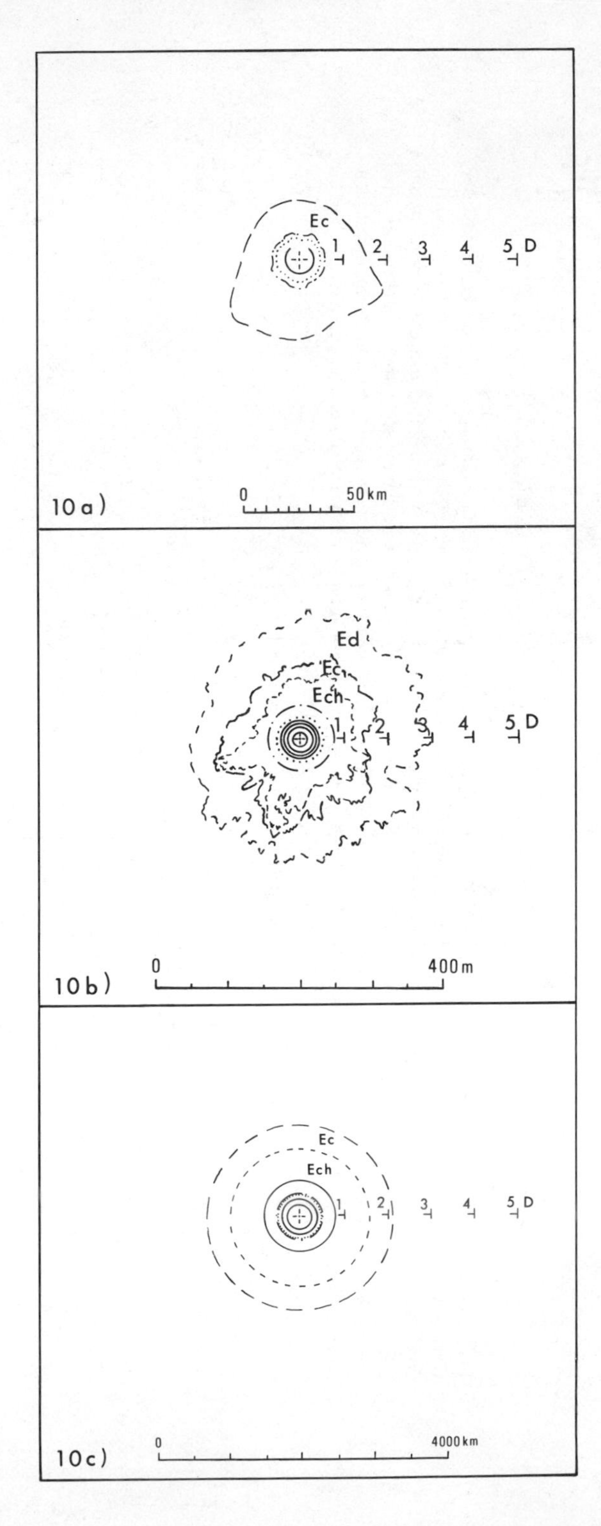
Ec
1
2
3
4
5
D
10 a)
0
50 km
Ed
Ec
Ech
1
2
3
4
5
D
10 b)
0
400 m
Ec
Ech
1
2
3
4
5
D
10 c)
0
4000 km

The structure of the outer crater region beyond the ring uplift is described by Stöffler (1977), Gall *et al.* (1977), Hüttner (1969), Wagner (1964), and others to consist basically of a megablock or klippe zone (Dennis, 1971) approximately 6 km (0.3D) wide. This zone is interpreted to consist of chaotic breccias derived mainly from the entire pre-impact sedimentary sequence and upper crystalline basement rocks with sizes ranging from sub-millimeter to megablocks over 1 km (0.05D) across (Gall *et al.*, 1977; Dennis, 1971). Hüttner (1969) assigned the name *Bunte Breccia* to collectively include all of the chaotic breccia masses too small to be mapped as separate blocks. This complex zone extends out to the outermost rim scarps at a distance of 12–13 km (0.6–0.7D) from GZ. Stöffler (1977) interprets this region to be terrace-like in structure over distances as wide as 4–6 km (0.2–0.3D) at least along parts of the eastern and southern rim. This interpretation is based mainly upon surface exposures and geophysical data. Chao and Minkin (1977), however, propose another cratering model that interprets this zone as part of the cratered region and not a terraced sequence. The structure of this critical zone and its relationship to a true crater floor remains controversial at present.

The Bunte Breccias, which occur in the outer complexity deformed wall zone and overlap the outer rim scarp areas, also form the massive ejecta blanket that now discontinuously surrounds the crater with thicknesses up to 200 m (0.01D) depending on local relief (Stöffler, 1977). Chao (1977a) describes a crude inversion with considerable local mixing near the rim in the Bunte ejecta. Stöffler (1977) further notes that the older, and therefore initially deeper, rocks are closest to the rim in the ejecta with the youngest rocks ejected further. Both of these observations suggest that an overturning process occurred in the formation of the ejecta blanket. Hörz (1977) has identified secondary cratering, at ranges of 20 km (1D) and more, which mix major percentages of the local bedrock into the ejecta blanket. Large remnants of the *suevite* melt breccia locally overlie the *Bunte Breccia* in the complex wall zone, the topographic rim scarp, and the ejecta blanket, out to ranges of 22 km (1.1D) from GZ. Thicknesses of *suevite* vary from a few meters to approximately 15 m on the ejecta (Stöffler, 1977).

As noted earlier, the basic structural elements at the Ries include a central basin surrounded by an uplifted crystalline ring and an outer complexly deformed zone and rim. The character of each of these regions and their relationships, however, require further study before a detailed analysis of the structure and

---

Fig. 10. Generalized geologic maps of three flat-floored multiring craters, with Fig. 10a of the eroded Ries Crater, Germany, Fig. 10b of the Prairie Flat 500-ton TNT explosion crater (modified from Roddy, 1969), and Fig. 10c a schematic map of the Orientale Basin on the moon. Each crater is drawn to show the apparent diameter at the same scale. The symbols include Ec for continuous ejecta, Ech for continuous hummocky ejecta, and Ed for discontinuous ejecta. The dotted circles are the apparent diameters, the dot-dash lines are rim crests, the long-dashed lines show the approximate outer limits of the continuous ejecta, and the short-dashed lines show the very approximate limits of the continuous hummocky and the discontinuous ejecta units. The numbered symbols show the apparent diameter distances from GZ out to 5D.

cratering is possible. One fundamental problem centers on what constitutes the breccia lens, i.e., does the lower suevite layer in the crater form the bulk of the lens or does the underlying brecciated region make up the real lens? Or, should an entirely new definition be constructed for breccia lenses in craters of this size and physical character, particularly with massive uplifted regions? The concept of a deep transient cavity should also be considered in terms of the central basin and its possible effects on the formation of a classical breccia lens associated with gravitational collapse of the surrounding walls.

Another critical problem is the question of uplift of the ring structure or, instead, uplift of the entire sub-crater floor region. Such movements would require physical coupling with the rim and could enhance terrace development in the complexly deformed zone between the ring and the outermost topographic scarps. The limited field evidence indicates that the highest units in the pre-impact stratigraphy are present in the tops of what appear to be massive terrace-like blocks. This is the interpretation displayed in Fig. 9a, and, if correct as shown, indicates this zone was not an ejecta zone but instead collapsed towards the crater. Consequently, the hinge zone of the overturning flap and crater walls would have been at the inner limit of this zone. A second point is that the Bunte Breccia locally appears to overlie much of this complexly deformed zone. This would also imply the hinge zone of the overturning flap to be closer to the center, that is, at the innermost terraced blocks. On the other hand, if this complex zone is the outer part of a megabreccia lens, then the outer scarps are the crater walls and another explanation is required for the emplacement of the breccia and ejecta blanket, such as suggested by Chao in this volume. A key point would be to determine if a stratigraphic inversion in the breccias are present in this complex zone, i.e., an inversion could be expected in the ejecta blanket beyond the hinge point but not in the breccia lens.

The author certainly agrees that initial terrace development is essentially contemporaneous with the final stages of such a cratering event. However, the correct choice of an apparent diameter becomes critical when scaling of any kind is applied, whether it be for a general cratering model, energy computations, melt production, or other parametric studies. In the case of the Ries, if the present field evidence of terracing is correct, then it would appear that the *initial* apparent diameter may have been on the order of 20 ± 1 km, i.e., the innermost terrace-like blocks. Scaling from Pike's (1976) lunar data would then give a rim crest diameter of approximately 22 ± 2 km and a rim crest depth on the order of 1.1 km, determined by adding the estimated rim height from Pike's data to the present known apparent depth to melt breccia. The author remains un-enthusiastic regarding his generalizations to arrive at this specific initial diameter and certainly considers the values subject to major refinement when further field data are acquired.

(2) *Prairie Flat, DRES, Canada.* The Prairie Flat explosion trial was conducted at the Defence Research Establishment Suffield, Alberta, Canada and consisted of a 500-ton TNT sphere lying tangent on alluvium. The important

morphological elements include: (1) a broad flat floor, (2) shallow central depression surrounded by four discontinuous concentric ridges on the floor, (3) moderately steep crater walls, (4) low rim crest, and (5) several subdued discontinuous concentric ridges on a surrounding, hummocky, rayed, ejecta blanket (Fig. 8a,b). The major structural features include: (1) very thin fallout layer, (2) thin breccia lens underlain by a large disrupted zone in which major inward and upward movements have uplifted alluvium beneath the entire crater floor to form concentric rings, (3) broadly downfolded rim with locally superimposed concentric faults, folds, and fractures, (4) and a surrounding rayed ejecta blanket consisting of a well-defined overturned flap of inverted stratigraphy. The cross section and geologic map are shown in Fig. 9b and 10b, respectively. Details of this trial are described in Jones *et al.* (1970), Jones (1976, 1977), Roddy (1969, 1970, 1976) and Roddy *et al.* (1977).

The Prairie Flat trial was conducted on unconsolidated alluvial sands, silts and clays with a water table at approximately 6.7 m (0.1 D) below the surface and sandstone bedrock at approximately 60 m (1.0D) depth. During the cratering, approximately 13,000 m^3 ($\sim$21,000 metric tons) of alluvium are estimated to have been ejected. The resulting crater had an average rim crest diameter of 85.5 m (61 m apparent) and a rim crest depth of approximately 5.3 m ($\sim$4.4 m apparent, 0.07 D) (Roddy, 1976). Immediately after detonation, ground water flow from concentric fractures in the crater floor began to form a lake and deposit approximately a meter of fine alluvium on the floor. The crater incurred no further modifications before field studies with trenching were completed. Concentric fractures associated with the rings allowed ground water to flow to the surface and deposit alluvial cones on the crater floor (Fig. 8b), interestingly similar to proposed volcanic extrusions on lunar crater floors.

The morphology of the crater floor, about 46 m (0.75 D) across, was represented by a relatively broad, flat, surface with a low central depression surrounded by four discontinuous concentric rings consisting of ridges and depressions. The approximate diameters of the four ridges were 17 m (0.28D), 32 m (0.52D), 37 m (0.61D), and 43 m (0.71D). The central part of the floor consisted of a shallow flat basin, about 0.5 m (0.01D) deep, surrounded by a low concentric ridge approximately 17 m (0.28D) across. This circular ridge was interrupted by numerous radially oriented topographic highs and lows, similar to the spokes of a wheel giving it a lobate or crenulated appearance (Fig. 8a). The crest of the ridge was about 1 m (0.02D) above the lowest parts of the crater floor. The lobate extensions from the inner ridge extended irregularly out to the second concentric ridge. This ridge terminated the extension of the lobate form, similar to the outer rim of a wheel, and stood about 1 m above the crater floor. In plan view, the second ridge exhibited a discontinuous scalloped outline of segmented sections around the ring. Parts of this ridge were discontinuous and folded in towards the center of the crater (Fig. 8b). The third concentric ridge was the most continuous of the rings, except for the east side of the crater, and exhibited a smooth rounded shape. The outermost ring was relatively discontinuous and occurred immediately at the base of the crater walls.

The crater walls were relatively smooth, averaging 22–25° in slope on the lower parts and increasing to 30° in the upper half. The rim was overlain by a surrounding ejecta blanket, which exhibited a series of prominent, low, discontinuous, concentric ridges over the inner rim (Fig. 10b). The continuous ejecta cover extended out to approximately 107 m (1.8*D*) from GZ with a well-defined discontinuous cover extending to approximately 170 m (2.8*D*). Lineations radial to GZ were common over much of the surface of the ejecta blanket, but were most prominent in the inner region where a blocky, hummocky surface extended out to an average distance of 83 m (1.4*D*) from GZ, with several elongated extensions (Fig. 10b). Near the rim crest, ejected clay beds formed prominent patterned terrain with a close spaced concentric character. The outer edge of the ejecta blanket was irregular with numerous ray-like to lobate extensions. Dune-like patterns and radial lineations were also common at the outer edges. The crater, its walls, and ejecta blanket were also covered by a thin layer of fallout that tended to smooth the entire area.

The structure beneath the crater floor consisted of a fallout layer and a thin breccia lens underlain by a strongly uplifted, deformed, and disrupted zone. The thin layer of fallout, which blanketed both the crater floor, walls, and surrounding rim, averaged a few centimeters thickness in the crater and thinned to millimeters at its irregular outer edges. As at other explosion trials in alluvium, this layer consisted of very finely comminuted material initially lofted into the air by the cratering event and was typically transitional in grain size and texture into the top of the ejecta blanket and breccia lens. The fallout tended to smooth the minor relief wherever it was deposited.

The breccia lens, which averages 1 m or less (0.02*D*) in thickness, consisted of fallback and locally disrupted alluvium cratered from approximately the lower 2.5 m (0.04*D*) and was generally deepest between the ridges, locally having slid from the higher parts. The tops of the ridges were commonly covered by a few centimeters (0.001*D*) or less of breccia and fallout. The breccia, which consisted of sub-millimeter fragments to shock compressed blocks 0.4 m (0.006*D*) across, tended to be thinner and more blocky towards the center of the crater. Towards the crater walls the breccia became somewhat finer and increased in thickness slightly. The crater walls exhibited an irregular covering of the breccia lens with numerous slumps and local talus development.

The disrupted alluvium beneath the breccia was moved inward and uplifted to form the ridges and depressions on the uplifted crater floor with the mobility of the clays below 6.7 m (0.1*D*) strongly enhanced by water saturation. The ridge system consisted of tight- to open-folded anticlinals and synclinals broken locally by numerous faults and fractures. The major fold axis for each ridge and depression was concentric to the crater. Minor axis radial to the center of the crater formed the radial lobate patterns in the inner crater floor. Uplift of deeper clay beds at the crest of anticlines exceeded 3.8 m (0.06*D*). During the complex uplift and anticlinal folding of the outermost ridge, open fracturing occurred allowing fragments from the breccia lens to be injected into the core of the truncated anticline. The concentric fracture system provided rapid access for ground water, under pressure due to shock increased alluvial density, to flow to the crater floor. The result was a series of alluvial cones deposited on the ridges of the floor.

The general structural deformation in the crater walls and inner rim is shown diagrammatically in Fig. 9b, and consisted of an inward dipping, faulted alluvial sequence overlain by a large overturned flap. Trench excavations showed the ground surface and underlying sand, silt, and clay units to experience gentle downfolding about 40 m (0.7*D*) from GZ. At about 30 m (0.5*D*) from GZ, the ground surface is depressed about 2 m (0.03*D*) with underlying units 20–50% thinner. High- and low-angle faults are common with up to at least 0.5 (0.01*D*) displacement. Between 25 and 30 m (0.4–0.5*D*) from GZ, approximately the upper 6.5 m (0.1*D*) of alluvium are folded upward with the upper 2.5–3 m (0.04–0.05*D*) of silty clay and sandy units fold back over the rim in a large coherent overturned flap in which the stratigraphy is well preserved in an inverted sequence. The units in it, however, are extended horizontally and thinned at least as much as 95% at the outer end. Individual horizons, such as fragile carbon lenses, are preserved remarkably intact in sand units as far as 2*D* from GZ in a rayed extension. Scattered regions of injection breccia were found in isolated lenses within the overturned flap and presumably were injected into open fractures during the overturning; however, the absence of free ballistic ejecta on the ground below the inner flap indicates that it overturned as a continuous unit.

In the innermost part of the rim, sand and silty clay units below 2.5–3 m (0.05*D*) in the upper part of the strata are folded into vertical positions and truncated by the crater wall. The more coherent strata ejected from these beds are distributed beyond the rim crest and immediately on top of the overturned flap forming discontinuous, concentric sand and clay ridges. The innermost ridges and troughs may have been formed by immediate partial subsidence of the crater walls, whereas other concentric ridges in the ejecta appear to reflect concentric folding in the underlying rim surface strata. As noted above, movement of alluvium in the region under the crater walls showed outward displacement high in the walls as part of the motion incurred during the overturning of the flap. Deeper beneath the crater walls the displacements were generally downward. Dynamic velocity and acceleration gage data show that these motions are part of the compensating volume adjustment for the inward flow under the crater floor, and occurred during the formation of the crater, i.e., in less than 2.5 sec and were not part of a late stage slump phenomena (Roddy *et al.*, 1977).

The ground surface beneath the overturned flap was locally faulted, folded, and fractured, with most faults and folds concentric to the crater wall. Both high- and low-angle faults were present, with displacements of as much as 0.4 m (0.006*D*). Over-thrusting away from the crater is common locally near the crater walls with graben and horst blocks developed a few meters further out. Excavations of surface-laid asphalt strips illustrated the complex high-angle and thrust fault movements in the inner part of the crater rim where significant outward displacements occurred (Roddy, 1969). Tightly folded anticlines exposed at the ground surface slightly beyond the crater walls were commonly fractured at their crests, exhibiting fault displacement of a few centimeters. Little to no surface deformation was observed beyond 70 m (1.2*D*) from GZ. A more complete description of this crater is given by Roddy (1969, 1976) including details of fused and shocked alluvium formed during the detonation and in Jones (1976).

(3) *Orientale Basin, Moon.* The Orientale Basin, nearly 1000 km across, is located approximately 1500 km west of Mare Humorum in the lunar highlands and represents one of the largest and freshest multiring craters on the moon. This enormous impact structure exhibits a morphology which includes: (1) a central, flat, maria-filled basin, (2) four discontinuous, concentric mountainous rings surrounded by very rough, hummocky to locally smoother lineated terrain, and (3) a surrounding radially sculptured terrain that includes the outer part of a massive ejecta blanket extending continuously to over 1300 km from the center of the basin (Fig. 8c,d).

Studies by Baldwin (1974), Hartman and Wood (1971), Head (1974, 1977), McCauley (1967), Moore *et al.* (1974), Stuart-Alexander and Howard (1970), McCauley (1977), Scott *et al.* (1978), and Wilhelms *et al.* (1977) have described the more important stratigraphic and structural features which characterize this large impact feature. In particular, the recent summaries by McCauley (1977) and the mapping by Scott *et al.* (1977) demonstrate the unique concentric patterns in the ejecta blanket and their extensive distributions. A fundamental concern noted in each of these studies is the need for the correct interpretation of the apparent crater diameter and the structural meaning of the massive rings. A brief outline of the morphology and possible structural interpretations of Orientale is summarized in this section, based largely upon the above studies, and a schematic profile and geologic map are shown in Figs. 9c and 10c, respectively.

The morphology of the inner part of the Orientale Basin consists of a relatively broad flat floor of dark colored maria, approximately 300 km across. The inner basin ring, about 320 km across, bounds the maria with a subdued, irregular, highly discontinuous zone of rounded mountains, ridges and scarps that locally exceed 1 km in height (Head, 1974). This innermost ring consists of segmented steep scarps, low elongated fracture ridges, and irregular mountain ridges. Terrain beyond the outer edges of the maria are coarsely hummocky on the 100 m and larger scale with impact craters and rough hilly, fractured regions which may consist of megabreccia and impact melt material (Head, 1974; Moore *et al.*, 1974). The surface consists of highly fractured, locally smooth, broadly rolling plains and ridges with scattered young impact craters, numerous concentric rilles, and broad areas of elongated to lobate terrain subradial to radial to the center of the basin. The material forming this unit has been called the Maunder Formation by McCauley *et al.* (1977), McCauley (1977), and Scott *et al.* (1978). This terrain includes the second concentric ring structure, the Inner Rook Mountains, and extends from the maria out to approximately the base of the third concentric ring, Montes Rook, 310 km from the center of the basin. The inner edge of the Maunder Formation is truncated by the central maria and appears to dip under it.

The second ring, the Inner Montes Rook, has a diameter of approximately 480 km, and is well defined for over 40% of its circumference. It consists of closely spaced, rounded, massifs with interior scarps reaching heights of 1.5–3.5 km (Head, 1974). McCauley (1977) notes that this scarp is the most segmented and discontinuous of the rings and is locally present only as isolated, rounded hills and massifs protruding through the Maunder unit.

The third ring, the Montes Rook, has a diameter of approximately 620 km, and represents the most rugged and continuous of the scarps with elevations reaching 4.5 km (Head, 1974). Steep interior slopes characterize much of the ring with the continuity broken only on the eastern side, as are the other inner rings.

Scott *et al.* (1978) have mapped two concentrically distributed units in the area of the Montes Rook ring consisting of an extensive Massif and Knobby Facies that McCauley *et al.* (1977) and Scott call the Montes Rook Formation. Both facies are characterized by rectilinear to equidimensional blocks and knobs, up to 100 km across in the Massif Facies, separated by smoother hummocky plains. The roughest, more hummocky, massif terrain forms the inner part of the formation adjacent to the ring of the Montes Rook, with radial lineations present but not as pronounced as they are further out from the scarp area. The material forming these facies is interpreted by McCauley and Scott as the last large ejecta to be thrown or moved from the crater, whereas, locally smoother terrain may reflect a melt blanket extending outward for hundreds of kilometers. McCauley *et al.* (1977) and Scott *et al.* (1978) have mapped the Montes Rook Formation to extend out to and locally *overlap* the outermost scarp, the Montes Cordillera Ring, which is approximately 920 km in diameter and rises as high as 3.5 km (Head, 1974). Arthur (1977, pers. comm.) indicates profile measurements show over a 5 km drop from the outer rim to the inner maria.

Beyond the Montes Cordillera Scarp is a concentric band of terrain called the Hevelius Formation which is interpreted as a continuation of the ejecta blanket (Scott *et al.*, 1978). The inner facies of the Hevelius, locally transitional with the Montes Rook Formation, consists of radial to curvilinear, elongated ridges and troughs in very rough terrain forming a unit 300–600 km wide. Massive radial sculpturing is most common to this unit with partly buried secondary craters locally prominent. McCauley interprets this widespread unit to be continuous in its cover and transitional outward into the concentric and most distant outer facies of the Hevelius Formation, placing the outer edge of continuous cover between 610 and 910 km from the center of the basin. This facies is characterized by weakly lineated terrain approximately radial to Orientale with flat-lying to rolling plains occurring in irregular depressions and on crater floors. The outer edge of this unit appears to average approximately 1300 km from the center of the basin.

Head (1974), McCauley (1977), McCauley *et al.* (1977), and Scott *et al.*

(1978) interpret the original crater to extend out to the Montes Rook ring and suggest that these scarps form essentially the initial crater wall region of Orientale. This would give the crater an apparent diameter of approximately 620 km. The maria region is viewed as an inner, deeper crater with the inner basin and inner rook scarp forming structurally raised rings, similar to the inner region of the Prairie Flat Crater (Roddy, 1976). The Maunder Formation is interpreted, in the most simplified sense of the terminology used in this paper, as a megabreccia lens overlain by impact melt breccia. The concentric massif and knobby facies of the Montes Rook Formation, irregularly bounding the proposed breccia lens as two concentric units, are interpreted by McCauley and Scott to be the massive blocky units that form the inner part of the hummocky ejecta blanket. That would place the *effective* hinge zone of the overturning ejecta in the area of the Montes Rook scarps. The overlapping relationship of the Montes Rook Formation with respect to the Cordillera ring scarp further supports the interpretation of an ejecta blanket originating further towards the center of the basin. In its simplest form, the section between the Montes Rook and Cordillera rings may be interpreted to represent a massive downdropped terrace or graben-like structure in which the region moved down and towards the center. The large radially sculptured terrain of the surrounding Hevelius Formation is defined by McCauley and Scott as the outer part of the continuous ejecta blanket which extends out to over $2D$, assuming a 620 km crater diameter. The outer facies of the Hevelius Formation, appears to the author approximately equivalent to the wide, often highly irregular, transition zone between continuous and discontinuous ejecta cover. Assuming an initial crater diameter of 620 m, the ring d/D spacings would be $0.5D$ for the inner basin ring, $0.8D$ for the inner rook rim, and $1.5D$ for the Cordillera ring.

In characterizing the distributions of the proposed impact ejecta units at Orientale, McCauley *et al.* (1977), McCauley (1977), and Scott *et al.* (1978) make the important observation regarding areal distributions that each of the formations form well-defined concentric patterns, which supports the necessary characteristic symmetry that is a consequence of the ejection processes at blanket appears to lie on top of the next unit further out, implying the general all impact events. They further indicated that the innermost unit of the ejecta overturning process which also appears common to all impact events.

It is of interest to note that the details of the inner contact of the Montes Rook Formation, however, do not appear to the author as totally consistent with a well-defined, single hinge zone, i.e., part of this unit, interpreted as innermost ejecta, lies irregularly *inside* the Montes Rook ring. The possibility exists that the Montes Rook ring may be the *outermost* scarps of a broad, terrace-like zone, thereby placing the initial crater walls much further towards the center of the basin. Perhaps such irregularities are to be expected with an event which craters over so large a region and diverse terrain, nevertheless, the definition of the distribution of the inner ejecta blanket remains a critical problem with respect to defining the exact initial location of the effective crater walls.

Wilhelms *et al.* (1977) and Hodges and Wilhelms (1976) offer an alternative cratering model for Orientale in which they interpret the inner region similar to McCauley and Scott in terms of structure and ring uplift, however, they assign the initial crater walls to the region of the outermost ring, the Montes Cordillera scarp. Terracing and local structure are interpreted in their model to give an initial diameter of approximately 860 km as discussed by Wilhelms *et al.* (1977).

The concentric patterns of the giant rings and scarps and the ejecta units, the suggestion of overlapping relationships implying an overturned flap process, the overlying relationship of the presumed ejecta on the Cordillera scarp, the massive blocky inner terrain of presumed ejecta, and the extensive radial sculpturing on the ejecta and rim, all substantiate the concept of a massive impact event. This point, however, no longer is seriously doubted. The un certainties regarding the detailed structural and cratering interpretations of Orientale, on the other hand, remain of concern. In the author's opinion the photogeologic evidence of massive, blocky, radially lineated terrain in the Montes Rook region and the overlapping relations over the Cordillera scarp presently support, though do not prove, the interpretations of Head (1974), McCauley (1977), and Scott *et al.* (1978) that the initial crater walls and hinge zone for the ejecta blanket were, at least, as close to the center as the Montes Rook region. As in the case of the Ries, further studies are necessary to resolve these questions. Since the magnitude of the Orientale event elevates it well beyond our preserved terrestrial ringed analogs in terms of direct comparisons, a new generation of experimental modeling, field studies, and theoretical calculations appear necessary for these basin-size events.

Discussion

The fundamental purpose of this paper has been to provide a systematic documentation, in the same comparative format, of the more critical *analogs* between bowl-shaped impact and explosion craters, between central uplift impact and explosion craters, and between multiring impact and explosion craters. As noted earlier, the data presented in this paper consist of a set of published and unpublished cratering studies assembled into a single ordered outline with each crater described in the same format and graphic style. The first objective has been to present physical descriptions for three different morphological and structural types of craters, both impact and explosion, with each description arranged to allow *direct* comparisons and contrasts. The second, longer range objective is to provide an improved basis for analog comparisons of dynamic, i.e., time dependent, physical processes in cratering mechanics, and, thereby hopefully, improve the level of creditability for such generalized predictions and interpretations.

The first objective was the most tractable, i.e., the morphological and structural analog comparisons. For the sake of brevity, these have been summarized in Table 2, in which the various morphological elements, structural elements, dimensional parameters and *D* ratios are presented in tabular com-

Table 2. Summary of morphological and structural crater elements, selected dimensional parameters and apparent diameter ratios written in terms of xD. Meteor Crater, Copernicus, and the Ries diameters are present-day mean values, and include terraces where present. Values in parentheses are estimates of the immediate post-crater rim crest and apparent diameters and depths for Meteor Crater (Roddy *et al.*, 1975), Flynn Creek (Roddy, 1977a), Copernicus and the Ries, as derived from cross-section reconstructions. Depths for the unnamed lunar crater, Copernicus, and Orientale are to present level of floor filling, i.e., minimum values since they are not to top of fallout beneath filling material. The two diameters given for the Orientale Basin reflect different interpretations of McCauley (1977) and Wilhelms *et al.* (1977). See text for explanation of other values in this table.

Bowl-shaped craters			
Morphological elements	Meteor Crater	Pre-Mine Throw IV, Crater 6	Unnamed lunar crater
(1) Bowl-shape	Yes	Yes	Yes
(2) Small flat-floor	Medium	Yes	Yes
(3) Talus-covered lower walls with blocky terrain	Yes	Yes	Yes
(4) Steep upper crater walls with horizontally banded outcrop	Yes	Yes	Yes
(5) High sharp rim crest	Moderate	Moderate	Yes
(6) Surrounding light-colored ejecta blanket	Yes	Yes	Yes
(7) Smooth hummocky terrain on inner ejecta blanket	Yes	Yes	Moderate
Structural elements			
(1) Fallout layer	Yes	Yes	?
(2) Breccia lens	Yes	Yes	?
(3) Deep disrupted zone	Yes	Medium	?
(4) Faulted, folded, brecciated and fractured rim strata	Yes	Yes	?
(5) Uplifted rim	Yes	Yes	Yes
(6) Surrounding ejecta blanket	Yes	Yes	Yes
(7) Inverted strata in ejecta blanket	Yes	Yes	Probable
Dimensional parameters and D ratios			
(1) Diameter, rim crest, today	1186 m	25.0 m	2.7 km
(2) Depth, rim crest, today	197 m	5.8 m	~775 km
(3) Diameter, apparent, today	1036 m	20.3	?
(4) Depth, apparent, to top fallout	150 m	4.3	?
(5) d/D	0.15 D	0.2 D	~0.29 D
(6) Rim uplift/D	0.05 D	0.06 D	?
(7) Crater floor diameter/D	0.5D	0.3D	0.15D
(8) Fallout thickness in crater/D	0.01D	0.005D	?
(9) Breccia lens thickness/D	0.15–0.19D	0.05D	?
(10) Depth of average cratering for ejecta/D	0.15 + D	0.25D	?
(11) Depth to base disrupted zone below GZ/D	0.9D	0.35 + D	?
(12) Depth to bedrock or crystalline basement below GZ/D	1.2D	20.0D	?

Table 2. (*Continued*).

Bowl-shaped craters			
Morphological elements	Meteor Crater	Pre-Mine Throw IV, Crater 6	Unnamed lunar crater
(13) Ejecta block max. diameter/D	0.03D	0.02D	~0.04D
(14) Radius of hummocky ejecta/D	1.1 + D	1.4D	0.7 + D
(15) Radius of continuous ejecta/D	1.5 + D	2.7D	2.9D
(16) Radius of average rays/D	?	5.5 + D	5D
(17) Radius maximum of rays/D	?	?	20 + D
(18) $D/V^{1/3}$	2.45	2.51	~2.30

Flat-floored central uplift craters			
Morphological elements	Flynn Creek	Snowball	Copernicus
(1) Broad flat hummocky floor	Yes	Yes	Yes
(2) Large central peak	Yes	Multiple peaks	Multiple peaks
(3) Locally terraced crater walls	Yes	Yes	Yes
(4) Uplifted and depressed rim segments	Yes	Depressed	?
(5) Surrounding rayed ejecta blanket	Eroded	Yes	Yes
(6) Smooth to hummocky terrain on inner ejecta blanket	Eroded	Yes	Yes
Structural elements			
(1) Fallout layer	?	Thin	?
(2) Breccia lens underlain by thin disrupted zone of uplifted material	Yes	Yes	?
(3) Large central uplift underlain by a deep disrupted zone	Yes	Yes	Probable
(4) Locally terraced crater walls	Yes	Yes	Yes
(5) Faulted, folded, brecciated, and fractured rim strata	Yes	Yes	?
(6) Uplifted and downfolded rim segments	Yes	Downfolded	?
(7) Large rim graben	Yes	Small	Probable
(8) Surrounding ejecta blanket	Eroded	Yes	Yes
(9) Inverted strata in ejecta	Crudely preserved	Yes	?
(10) Secondary craters	?	Yes	Yes
Dimensional parameters and D ratios			
(1) Diameter, rim crest, today	(~3830 m)	108.5 m	~93 km (≈79 km)
(2) Depth, rim crest, today	(~198 m)	7.5 m	~3.8+ km (≈4.0+ km)
(3) Diameter, apparent, today	~3517 m	83.0 m	~84.6 km (≈72 km)
(4) Depth, apparent, to top fallout	~98 m	6.6 m	~2.7+ km ≈2.7+ km

Table 2. (*Continued*).

Bowl-shaped craters			
Morphological elements	Meteor Crater	Pre-Mine Throw IV, Crater 6	Unnamed lunar crater
(5) d/D	$0.03D$	$0.08D$	$0.04D$
(6) Rim uplift/D	$0.01D$	Downfolded	?
(7) Crater floor diameter/D	$0.9D$	$0.6D$	$0.9 \pm D$
(8) Fallout thickness in crater/D	?	$0.001D$	?
(9) Breccia lens thickness/D	$0.01D$	$0.01D$	?
(10) Depth of average cratering for ejecta/D	$0.03 + D$	$0.09D$	?
(11) Depth of maximum cratering for central uplifted material/D	$0.1D$	$0.12 + D$	?
(12) Depth base disrupted zone under breccia lens from GZ/D	$0.07 + D$	$0.07 + D$	?
(13) Depth base disrupted zone under central uplift from GZ/D	$0.2 + D$	$0.2 + D$	?
(14) Height of central uplifted material/D	$0.13D$	$0.12 + D$	?
(15) Height of central uplift/D	$0.03 + D$	$0.04D$	$0.02 \pm D$
(16) Width of terraced wall zone/D	$0.1D$	$0.1 \pm D$	$0.2 \pm D$
(17) Depth to bedrock or crystalline basement from GZ/D	$0.5D$	$0.7D$	?
(18) Ejecta block max. diameter/D	?	$0.01 \pm D$	?
(19) Radius of hummocky ejecta/D	?	$1.3 + D$	$1.1 + D$
(20) Radius of continuous ejecta/D	?	$2.4 + D$	$1.9 + D$
(21) Radius of average rays/D	?	?	$4D$–$5D$
(22) Radius maximum of rays/D	?	?	$12 + D$
(23) $D/V^{1/3}$, apparent values	3.8	3.1	$\approx 2.2 + (?)$
(24) $D/V^{1/3}$, true values (Roddy, 1977a)	3.5	3.4	?

Flat-floored multiring craters			
Morphological elements	Ries	Prairie Flat	Orientale
(1) Broad flat-floor	Yes	Yes	On large scale
(2) Central basin	Yes	Yes	Yes
(3) Concentric rings on floor	One	Four	Two to three
(4) Locally terraced crater walls	Yes	?	Probable
(5) Moderately steep walls	Yes	Yes	Yes
(6) Low rim crest	Yes	Yes	Yes
(7) Exterior concentric rings	?	Yes	Probable
(8) Surrounding rayed ejecta blanket	Eroded	Yes	Yes
(9) Smooth to hummocky terrain on inner ejecta blanket	Eroded	Yes	Yes
Structural elements			
(1) Fallout layer	Yes	Yes	?
(2) Melt-breccia	Yes	Extremely limited fragments	Probable
(3) Breccia lens underlain by disrupted zone of uplifted material	Yes	Yes	?

Table 2. (*Continued*).

Bowl-shaped craters			
Morphological elements	Meteor Crater	Pre-Mine Throw IV, Crater 6	Unnamed lunar crater
(4) Uplifted material to form concentric rings on floor	Yes	Yes	?
(5) Faulted, folded, brecciated and fractured rim strata	Yes	Yes	?
(6) Downfolded rim	Flat(?)	Yes	?
(7) Locally terraced crater walls	Yes	?	Yes
(8) Surrounding ejecta blanket	Eroded	Yes	Yes
(9) Inverted strata in ejecta	Crude	Yes	?
(10) Secondary craters	Yes	Yes	Yes
Dimensional parameters and D ratios			
(1) Diameter, rim crest, today	(≈27 ± 2 km) (≈22 ± 2 km)	85.5 m	?
(2) Depth, rim crest, today	(≈1.2+ km) (≈1.1+ km)	5.3 m	?
(3) Diameter, apparent, today	~25 ± 1 km (≈20 ± 1 km)	61.0 m	≈610 km ≈910 km
(4) Depth, apparent, to top fallout	~570 m ~570 m	4.4 m	3.5 to 4.5+ km
(5) d/D	$0.03D$	$0.07D$	0.006 to $0.007 + D$ 0.004 to $0.005 + D$
(6) Rim uplift/D	Flat (?)	Downfolded	?
(7) Crater floor diameter/D	$0.9 \pm D$	$0.8D$	?
(8) Fallout thickness in crater/D	$0.1 \pm D$ (?)	$0.001D$	?
(9) Breccia lens thickness/D	$0.03 + D$ (?)	$0.02D$	?
(10) Depth of average cratering for ejecta/D	$0.04 + D$ (?)	$0.09D$	?
(11) Depth to base disrupted zone from GZ/D	$0.3 + D$	$0.2 \pm D$	?
(12) Ring spacings/D	$0.3D$	$0.3, 0.5, 0.6, 0.7D$	$0.5, 0.8, 1.5D$
(13) Ring material uplift/D	$0.02 + D$	$0.06 + D$	?
(14) Width of terraced wall zone/D	$0.3 \pm D$	—	?
(15) Depth to bedrock or crystalline basement below GZ/D	$0.03D$	$1.0D$	?
(16) Ejecta block max. diameter/D	?	$0.008 \pm D$	?
(17) Radius of hummocky ejecta/D	Eroded	$1.4D$	$1.6 \pm D$ (?)
(18) Radius of continuous ejecta/D	Eroded ($1.7 + D$)	$1.8D$	$2.1 + D$
(19) Radius of average rays/D	Eroded	$2.8 + D$	?
(20) Radius maximum of rays/D	?	$5.0 + D$	?
(21) $D/V^{1/3}$	$3.6 \pm$	2.6	?

30 m (Roddy *et al.*, 1975). Recent drilling studies further indicate that the impact occurred on a stripped, erosional surface of the Moenkopi Formation which had an average relief of about 5–10 m. The crater, estimated to have been initially about 1100 m across at the rim crest and approximately 215 m deep, has since experienced limited erosion with less than ~27% of its ejecta blanket removed (Roddy *et al.*, 1975).

parative form. Of equal importance are Figs. 3–10 which have been scaled to allow direct visual comparisons between each crater.

It was no surprise that *individual analogs* exist in that these have been noted before in individual studies (Baldwin, 1963; Oberbeck, 1977; Roddy, 1976; Shoemaker, 1963; and others). The systematic documentation of the full range of morphological and structural similarities from such a large variety of crater types, sizes, different origins and formational environments, however, has not been described previously. One clear result from such a data assemblage as shown in Table 2 and Figs. 3–10 is that each of the three different types of impact and explosion craters do indeed bear marked analog similarities within each class. A second result is that the excellent set of morphological analogs, exhibited by these craters formed by both natural impact and explosion processes and in both terrestrial and lunar environments, strongly supports the arguments for well-ordered, predictable cratering processes.

The application of these data, however, should apply to a wider spectrum of impact and explosion studies than single analog comparisons. Perhaps one of the more basic uses of the data in the text and Table 2 is to serve as an *outline* for the full range of possible major morphological and structural features at a given site. In terrestrial field studies, of course, it is necessary to accept whatever products remain after erosion. The *analog* data here suggests what levels of confidence (and absence thereof) one might expect to attach to predictions and interpretations based on a *limited* or *incomplete* set of erosional exposures and subsurface data. For example, scaling values for predictions of the base of disruption zones beneath craters are good examples of the type of data necessary to optimize a core drilling study. The need for accurate predictive capabilities are equally demanding in explosion crater studies where sound qualitative predictions still remain difficult, requiring extensive studies for each experimental trial.

Another logical use of such data sets is their use in substantiating arguments for *impact origins* of craters which exhibit analogous morphologies and surface structures. In a general sense, this has been *the* basic approach for all photogeologic studies of cratering on other planetary surfaces. In terrestrial studies, this was the basis for the initial interpretation of the origin of the Flynn Creek Crater in Tennessee and its counterpart in Germany, the Steinheim Crater (Roddy, 1968, 1977a). The development of shock metamorphic criteria, such as coesite and other shock phenomena, has tended to greatly strengthen such studies; however, the application of morphological and structural analog studies remain essential to the *proof* of an impact origin.

Another example of the application of the data in this paper is the use of the various *xD* values to predict *ranges* in certain crater dimensions, such as the expected outer limits of continuous ejecta blanket coverage. Gravity is inherently included in this type of scaling in the formation of an initial apparent diameter, and need not be considered for such simple general range scaling, whereas atmospheric effects would be critical in predicting the distances of long-range ejecta. The use of such *xD* scaling, of course, is only a first approximation and should not be expected to give other than a general range in values. A point of real concern is raised, however, by the use of apparent diameters for *xD* scaling involving craters with terraced rims. If the terraced segments cannot be accurately repositioned, *initial* apparent crater diameters cannot be determined. Consequently, careful cross-section data becomes absolutely essential for such studies. The limited terracing at Flynn Creek permitted an adequate apparent diameter to be determined, whereas the subsurface uncertainties at the Ries remain a problem with respect to the initial diameter. Another point is that the $D/V^{1/3}$ values suggest a shape factor which is related to the initial cratering conditions and the target rock, i.e., is a bowl-shaped or flat-floored crater to be expected. A better understanding of what initial parameters control the fundamental shape of a crater is of paramount importance in helping to define certain cratering processes and final morphological and structural conditions.

The second objective of this paper is directed towards an improved understanding of cratering dynamics through the use of analog processes inferred from terminal morphological and structural configurations. A full treatment of this area is beyond the scope of this descriptive paper, except to briefly note several points related to some of the important dynamic processes observed in this text. One consistent theme noted directly for most of the craters was the overturning process which formed the ejecta blankets, regardless of the size of the event. The result, where it could be examined first hand, was a repositioning of the ejecta in a general inverted pattern, with the shallowest and first material ejected forming the most widespread and lowest ejecta layer lying immediatcly on the original ground surface. The deepest and final material ejected tends to lie highest in the ejecta blanket and closest to the crater. Knowledge of this general distribution has already played a major role in the Apollo surface exploration at cratered sites. More recent studies, as of the Ries, demonstrate the effects of local secondary mixing and suggest the need for improved analog studies of secondary cratering processes. Another point can be noted in ejecta distances as displayed in Figs. 4, 7, and 10. The scaled maps suggest a relatively stable set of ranges for the hummocky ejecta blanket at each of the less-eroded craters. The ranges for the continuous ejecta, however, decrease approximately 30% from bowl-shaped craters to multiring types. The discontinuous ejecta ranges are strongly affected by the earth's atmosphere and cannot be directly compared although they appear to obey the same trend. This limited data base is insufficient to prove these trends, but in terms of the continuous ejecta, it would be reasonable to expect near-surface material to be ejected to greater scaled distances for deeper-cratered events. Further study is certainly necessary to quantify this relationship, although the overall ejection process appears quite well defined by the analogs for the craters in this paper suggesting the validity of general distribution predictions at other cratered sites.

The physical structure of the breccia lens appears more complex than the overturned flap with increasing crater size. Bowl-shaped craters tend to exhibit a fairly well-defined zone of brecciation, lenticular in shape, and an underlying zone of disruption, also generally lenticular in shape. As crater sizes increase to the flat-floored type with uplift, the breccia lens thins proportionally with uplift of the entire sub-crater floor region. As noted earlier, does this require a redefinition of the breccia lens to include the massively uplifted and repositioned sub-crater floor material, i.e., a megabreccia zone, of should this lower region be included in an enlarged disrupted zone? The Ries, which includes the complication of a melt breccia, is a good example of the problem. The point is not just one of semantics when the craters involve massive repositioning of large volumes of crusted and mantle material. A clearer understanding of sub-crater movements is becoming increasingly mandatory even for terrestrial examples, such as Sudbury, in terms of economic mineral recovery.

The complex dynamic cratering processes determined for the Snowball and Prairie Flat craters suggest a need for real caution in the interpretation that only simple movements occurred with shallow deformation. At both craters, including their terrestrial counterparts of Flynn Creek and the Ries, the subsurface material was notably moved inward and the part or all of the subsurface material below the crater floor locally uplifted. Presumably, similar structural deformation also occurred at Copernicus and Orientale, with a consequent massive repositioning of deep lunar mantle and crust. At the explosion craters the subsurface mobility of the alluvium was greatly enhanced by their water saturated condition, i.e., reduced strength and lower compressibilities. A similar rheological response must be associated with the giant lunar craters in order for them to exhibit similar ring morphologies. Consequently, a better understanding of the real dynamics involved in the formation of the explosion craters would certainly be of interest for their lunar counterparts.

It is also of interest to note that both Snowball and Prairie Flat were detonated with the same energy yield over identical media and water table depths. The differences were that one charge was a hemisphere and the other was a sphere and each initiated slightly different shaped shock waves into the target media. The center of energy for the Prairie Flat sphere was naturally higher above the ground surface, consequently, less energy was coupled into the target media. The result was to form a crater smaller than Snowball, but with an equally large raised sub-crater region. Both explosion trials formed an uplifted region, except that Snowball's uplift was more centralized, whereas Prairie Flat had a flattened central uplift surrounded by concentric rings. This would seem to indicate that a raised height of burst (HOB) uplifts a broader region but not through as great a height. The implications of these variations in terms of target layering, material properties and impacting body differences remain important areas for study (Roddy, 1976). The structural details of the ring analogs with respect to impact craters also certainly require further study.

The analog comparisons of bowl-shaped craters also presented a problem in height of burst effects. Shoemaker (1960) drew his original morphological and structural comparisons of Meteor Crater with a relatively deeply-buried nuclear cratering event. In this paper, however, a *surface* explosion sphere formed a crater closely analogous with Meteor Crater. The explosion target media, however, was totally dry and there were minimal layering effects compared with the other explosion trials which were also surface events. Since both the buried nuclear and surface chemical explosion trials formed craters similar to Meteor Crater, it is reasonable to question the significance of the HOB relationship. Apparently, within a certain range of HOB, isotropic non-layered media, and energy levels, bowl-shaped craters will occur. The conditions necessary for the transition to flat-floored crater shapes remain equally critical and undefined.

The author has previously suggested that craters which exhibit morphologies similar to Snowball and Prairie Flat were formed by *surface* impact events with a minimum of penetration, i.e., lower density bodies such as carbonaceous chondrites or comets that volatize or fragment before significant penetration occurs (Roddy, 1968, 1977b). This would seem to be supported by the surface HOB and raised HOB effect for central uplift and multiring experimental explosion craters. The physical analog suggested is that of a shock wave generated from well above the surface by an impacting body which volatizes or disintegrates very early into an event with very little penetration. This view is dampened, however, by explosion trials such as the Pre-Mine Throw IV crater which were also formed by surface explosions. If one followed the *surface impact analog concept* it would lead to the conclusion that Meteor Crater was also formed by a surface event. However, we know Meteor Crater was formed by an iron meteorite, presumably a body which has *some* penetration, as logically argued by Shoemaker (1963). It is permissible, of course, to suggest that this particular iron body was *fragmenting* at the time of impact due to atmospheric breakup or other related effects coming from fragmentation and that *effective* penetration was minimal. Again, HOB effects with various material properties need serious examination.

As noted at the beginning of this discussion, the fundamental purpose of this paper was to provide a systematic documentation that would establish that certain morphological and structural analogs *do* exist within several classes of impact and explosion craters. The multitude of questions related to terminal positioning and dynamic cratering processes that immediately arise out of such studies pose the really exciting areas for future study, such as: (a) reasons for transitions from one crater type to another type, (b) the formation of flat-floored craters, formation of central uplifts vs multirings, (c) extent and distributions of uplifted sub-crater material and mascon development, (d) overturning and formation of ejecta blankets and related ejecta distributions, (e) origin of terraces and crater rim development, (f) depth and range of cratering effects and mantle and crustal redistributions, as well as a host of others. Hopefully, simple analog studies such as described here will further encourage the explosion and impact communities to exchange their data on these mutually important cratering problems and actively participate in the future on joint cratering research.

Acknowledgments—The data presented in this paper was abstracted from a wide base of published and unpublished studies, including the author's work, and I remain indebted to the many workers whose data I have drawn upon for this paper. I also appreciate the long-term exchanges with Dr. E. M. Shoemaker on many aspects discussed in this work. The help of B. Barringer and F. Hatfield greatly assisted in making the field studies at Meteor Crater possible as did C. H. H. Diehl and Dr. G. H. S. Jones the field work at The Defence Research Establishment Suffield, Alberta, Canada. Discussions with Drs. H. Cooper, E. Chao, M. Dence, F. Hörz, J. McCauley, and D. Scott greatly assisted the writer in organizing and technically improving this study.

Reviews of the paper by Drs. H. Wilshire and R. Port materially assisted in improving the format and I express my thanks to each for their help.

I am again indebted to Dr. G. Sevin and Capt. J. Stockton of the Defense Nuclear Agency for their continued support of these studies. I most sincerely express my real appreciation to M., M., M. and N. Roddy for their continued field assistance in a number of phases of the field work and to J. Roddy for her invaluable help in data assemblage.

This work was supported through the cooperation of the National Aeronautics and Space Administration under Contract W-13,130 and the Defense Nuclear Agency, Department of Defense, Washington, D.C.

REFERENCES

Ackerman, H. D., Godson, R. H., and Watkins, J. S.: 1975, A seismic refraction technique used for subsurface investigations at Meteor Crater, Arizona. *J. Geophys. Res.* **805**, 765–775.

Ahrens, T. J. and O'Keefe, J. D.: 1977, Equations of State and Impact-Induced Shock-Wave Attenuation on the Moon. In *Impact and Explosion Cratering* (D. J. Roddy, R. O. Pepin, and R. B. Merrill, eds.), Pergamon Press. This volume.

Allen, R. T.: 1976, Late-Stage effects in Crater and Ejecta Formation (abstract). In *Papers Presented to the Symposium on Planetary Cratering Mechanics*, p. 4–5. The Lunar Science Institute, Houston.

Andrews, R. J.: 1977, Characteristics of Debris from small-scale cratering experiments. In *Impact and Explosion Cratering* (D. J. Roddy, R. O. Pepin, and R. B. Merrill, eds.), Pergamon Press. This volume.

Baldwin, R. B.: 1963, *The Measure of the Moon.* University of Chicago Press. 488 pp.

Baldwin, R. B.: 1974, On the Origin of the Mare Basins. *Proc. Lunar Sci. Conf. 5th*, p. 1–10.

Barringer, D. M.: 1905, Coon Mountain and its crater. *Proc. Acad. Nat. Sciences, Philadelphia* **57**, 861–886.

Barringer, D. M.: 1910, Meteor Crater (formerly called Coon Mountain or Coon Butte) in northern Arizona. Published by the author. 24 pp.

Barringer, D. M.: 1914, Further notes on Meteor Crater in northern central Arizona (no. 2). *Proc. Acad. Nat. Sciences, Philadelphia* **76**, 275–278.

Chabai, A. J.: 1977, Influence of Gravitational Fields and Atmospheric Pressures on Scaling of Explosion Craters. In *Impact and Explosion Cratering* (D. J. Roddy, R. O. Pepin, and R. B. Merrill, eds.), Pergamon Press. This volume.

Chao, E. C. T.: 1974, Impact cratering models and their application to lunar studies—a geologist's view. *Proc. Lunar Sci. Conf. 5th*, p. 35–52.

Chao, E. C. T.: 1976, The Ries crater of southern Germany—a model for large basins on planetary surfaces. *Fortschr. Mineral.* In press.

Chao, E. C. T.: 1977a, Mineral-produced high pressure striae and clay polish: key evidence for nonballistic transport of ejecta from the Ries crater, southern Germany. *Science.* In press.

Chao, E. C. T.: 1977b, Preliminary interpretation of the 1973 Ries research deep drill core and a new Ries cratering model. *Geol. Bavarica.* In press.

Chao, E. C. T. and El Goresy, A.: 1977, Shock attenuation and the implantation of Fe-Cr-Ni veinlets in the compressed zone of the 1973 Ries research deep drill core. *Geol. Bavarica.* In press.

Chao, E. C. T. and Minkin, J. A.: 1977, Impact Cratering Phenomenon for the Ries Multiring Structure Based on Constraints of Geological, Geophysical and Petrological Studies and the Nature of the Impacting Body. In *Impact and Explosion Cratering* (D. J. Roddy, R. O. Pepin, and R. B. Merrill, eds.), Pergamon Press. This volume.

Cooper, H. F., Jr.: 1977, A Summary of Explosion Cratering Phenomena Relevant to Meteor Impact

Events. In *Impact and Explosion Cratering* (D. J. Roddy, R. O. Pepin, and R. B. Merrill, eds.), Pergamon Press. This volume.
Cooper, H.F. and Sauer, F. M.: 1977, Crater-related ground motions and implications for crater scaling. In *Impact and Explosion Cratering* (D. J. Roddy, R. O. Pepin, and R. B. Merrill, eds.), Pergamon Press. This volume.
Curran, D. R., Shockey, D. A., Seaman, L., and Austin, M.: 1977, Mechanisms and Models of Cratering in Earth Media. In *Impact and Explosion Cratering* (D. J. Roddy, R. O. Pepin, and R. B. Merrill, eds.), Pergamon Press. This volume.
Dence, M. R.: 1968, Shock zoning at Canadian Cratering: Petrography and structural implications. In *Shock Metamorphism of Natural Materials* (B. M. French and N. M. Short, eds.), p. 169–184. Mono Book Corp., Baltimore.
Dence, M. R.: 1971, Impact Melts. *J. Geophys. Res.* **76**, 5552–5565.
Dennis, J. G.: 1971, Ries Structure, Southern Germany, A Review. *J. Geophys. Res.* **76**, 5394–5406.
Diehl, C. H. H. and Jones, G. H. S.: 1965, A tracer technique for cratering studies. *J. Geophys. Res.* **70**, 305–309..
Diehl, C. H. H. and Jones, G. H. S.: 1967a, The Snowball Crater, general background information. Defence Research Establishment, Suffield, Canada. Suffield Tech. Note No. 187, unclassified. 23 pp.
Diehl, C. H. H. and Jones, G. H. S.: 1967b, The Snowball Crater, profile and ejecta pattern. Defence Research Establishment, Suffield, Canada, Suffield Tech. Note No. 188, unclassified. 30 pp.
El Goresy, A. and Chao, E. C. T.: 1976, Evidence of the impacting body of the Ries crater—the discovery of the Fe-Cr Ni veinlets below the crater bottom. *Earth Planet. Sci. Lett.* **31**, 330–340.
El Goresy, A. and Chao, E. C. T.: 1977, The origin and significance of the discovery of the Fe-Cr-Ni veinlets in the compressed zone of the 1973 Ries research drill core. *Geol. Bavarica.* In press.
Englehardt, W. v.: 1974, Ries meteorite crater, Germany. *Fortschr. Mineral.* **52**, 103–122.
Englehardt, W. v. and Graup, G.: 1977, Ries Crater Drilling, Germany. *Geol. Bavarica* **75**. In press.
Englehardt, W. v., Stöffler, D., and Schneider, W.: 1969, Petrologische Untersuchungen in Ries. *Geol. Bavarica* **61**, 229–295.
Gall, H., Müller, D., and Pohl, J.: 1977, Zum geologischen Bau der Randzone des Ries-Kraters. In *Jb Geol. Paläont. Mh.* **2**, 65–94.
Gault, D. E. and Wedekind, J. A.: 1977, Experimental hypervelocity impact into Quartz Sand: II, Effects of gravitational acceleration. In *Impact and Explosion Cratering* (D. J. Roddy, R. O. Pepin, and R. B. Merrill, eds.), Pergamon Press. This volume.
Gault, D. E., Quaide, W. L., and Oberbeck, V. R.: 1968, Impact cratering mechanics and structures. In *Shock Metamorphism of Natural Materials* (B. M. French and N. M. Short, eds.). Mono Book Corp., Baltimore.
Hartmann, W. K. and Wood, C. A.: 1971, Moon: Origin and evolution of multi-ring basins. *The Moon* **3**, 3–78.
Head, J. W.: 1974, Orientale multi-ringed basin interior and implications for the petrogenesis of lunar highland samples. *The Moon* **11**, 327–356.
Head, J. W.: 1977, Origin of Outer Rings in Lunar Multi-ringed Basins: Evidence from Morphology and Ring Spacing. In *Impact and Explosion Cratering* (D. J. Roddy, R. O. Pepin, and R. B. Merrill, eds.), Pergamon Press. This volume.
Henny, R. W.: 1977, The effects of the geologic setting on the distribution of ejecta from a buried nuclear detonation. Unpublished Ph.D. Thesis at Michigan State University.
Hodges, C. A. and Wilhelms, D. E.: 1976, Formation of lunar basin rings (abstract). *Proc. 25th Internat. Geol. Congress*, Sydney, Australia, p. 612–613.
Hörz, F., Gall, H., Hüttner, R., and Oberbeck, V. R.: 1977, Shallow drilling in the "Bunte Breccia" Impact Deposits, Ries Crater, Germany (abstract). In *Lunar Science VIII*, p. 457–459. The Lunar Science Institute, Houston.
Hörz, F., Gall, H., Hüttner, R., Oberbeck, V. R., and Morrison, R. H.: 1975, The Ries Crater and Lunar Basin Deposits (abstract). In *Lunar Science VI*, p. 396–398. The Lunar Science Institute, Houston.
Hörz, F., Gall, H., Hüttner, R., and Oberbeck, V. R.: 1977, Shallow Drilling in the "Bunte Breccia" Impact Deposits, Ries Crater, Germany. In *Impact and Explosion Cratering* (D. J. Roddy, R. O. Pepin, and R. B. Merrill, eds.), Pergamon Press. This volume.
Hüttner, R.: 1969, Bunte Trümmermassen und Suevite. *Geol. Bavarica* **61**, 142–200.

Jackson, J. G., Jr.: 1972, Physical property and dynamic compressibility analysis of the Watching Hill blast range. *U.S. Army Engineer Waterways Experiment Station Technical Report* S-72-4, unclassified. 149 pp.

Johnson, S. W., Smith, J. A., Franklin, E. G., Moraski, L. K., and Teal, D. J.: 1969, Gravity and atmospheric pressure effects on crater formation in sand. *J. Geophys. Res.* **74**, 4838–4850.

Jones, G. H. S.: 1963, Strong Motion Seismic Effects of the Suffield Explosions. University of Alberta. Ph.D. Thesis.

Jones, G. H. S.: 1976, The Morphology of Central Uplift Craters. Suffield Report No. 281, unclassified. 212 pp.

Jones, G. H. S.: 1977, Complex Craters in Alluvium. In *Impact and Explosion Cratering* (D. J. Roddy, R. O. Pepin, and R. B. Merrill, eds.), Pergamon Press. This volume.

Jones, G. H. S., Diehl, C. H. W., Pinnell, J. H., and Briosi, G. K.: 1970, Crater and ejecta study. Defence Atomic Support Agency POR-2115 (WT-2115), unclassified. 276 pp.

Killian, B. G. and Germain, L. S.: 1977, Scaling of Cratering Experiments—Analytical and Heuristic Approach to the Phenomenology. In *Impact and Explosion Cratering* (D. J. Roddy, R. O. Pepin, and R. B. Merrill, eds.), Pergamon Press. This volume.

Knowles, C. P. and Brode, H. L.: 1977, The Theory of Cratering Phenomena, an Overview. In *Impact and Explosion Cratering* (D. J. Roddy, R. O. Pepin, and R. B. Merrill, eds.), Pergamon Press. This volume.

Kreyenhagen, K. N. and Schuster, S. H.: 1977, Review and comparison of Hypervelocity Impact and Explosion Cratering Calculations. In *Impact and Explosion Cratering* (D. J. Roddy, R. O. Pepin, and R. B. Merrill, eds.), Pergamon Press. This volume.

McCauley, J. F.: 1967, The nature of the Lunar Surface as determined by systematic geologic mapping. In *Mantles of the Earth and Terrestrial Planets* (S. K. Runcorn, ed.), p. 431–460. Interscience Publishers, London, New York, Sydney.

McCauley, J. F.: 1977, Orientale and Coloris in comparisons of Mercury and The Moon. *Phys. Earth Planet. Int.* In press.

McCauley, J. F., Scott, D. H., Roddy, D. J., and Boyce, J. M.: 1977, The geology of the Orientale basin of the Moon. U.S. Geol. Survey Prof. Paper.

Masursky, H., Colton, G. W., and El-Baz, F.: 1977, Apollo over the Moon—A Photographic View from Orbit. NASA Spec. Rept. In press.

Maxwell, D. E.: 1977, Simple Z Model of Cratering, Ejection, and the Overturned Flap. In *Impact and Explosion Cratering* (D. J. Roddy, R. O. Pepin, and R. B. Merrill, eds.), Pergamon Press. This volume.

Milton, D. J. and Roddy, D. J.: 1972, Displacements within impact craters. *Proc. 24th Internat. Geol. Congress, Section 15* (*Planetology*), p. 119–124.

Moore, H. J.: 1976, Missile Impact Craters (White Sands Missile Range, New Mexico) and Applications to Lunar Research. U.S. Geological Survey Prof. Paper 812-B. 47 pp.

Moore, H. J., MacCormack, R. W., and Gault, D. E.: 1963, Fluid impact craters and hypervelocity—high-velocity impact experiments in metals and rocks. In *Proc. Sixth Symposium on Hypervelocity Impact*, Cleveland, Ohio, April 30, May 1, 2, 1962: Cleveland, Firestone Tire & Rubber Co., vol. 2, pt. 2, p. 367–399.

Moore, H. J., Hodges, C. A., and Scott, D. H.: 1974, Multi-ringed basins—Illustrated by Orientale and associated features. *Proc. Lunar Sci. Conf. 5th*, p. 71–100.

Oberbeck, V. R.: 1971, Laboratory simulation of impact cratering with high explosives. *J. Geophys. Res.* **76**, 5732.

Oberbeck, V. R.: 1977, Application of High Explosion Cratering data to Planetary Problems. In *Impact and Explosion Cratering* (D. J. Roddy, R. O. Pepin, and R. B. Merrill, eds.), Pergamon Press. This volume.

Oberbeck, V. R., Quaide, W. L., Arvidson, R., and Aggarwal, H. R.: 1977, Comparative Studies of Lunar, Martian, and Mercurian Craters and Plains. *J. Geophys. Res.* **82**, 1681–1698.

O'Keefe, J. D. and Ahrens, T. J.: 1975, Shock effects from a large impact on the Moon. *Proc. Lunar Sci. Conf. 6th*, p. 2831–2844.

O'Keefe, J. D. and Ahrens, T. J.: 1976, Impact Ejecta on the Moon. *Proc. Lunar Sci. Conf. 7th*, p. 3007–3025.

Orphal, D. L.: 1977, Calculations of Explosion Cratering, Part I: The Shallow-Buried Nuclear Detonation JOHNIE BOY. In *Impact and Explosion Cratering* (D. J. Roddy, R. O. Pepin, and R. B. Merrill, eds.), Pergamon Press. This volume.

Peterson, R. W.: 1974, Recommended Constitutive Properties for the Pre-Mine Throw IV High Explosive Test Site. *U.S. Army Engineer Waterways Experiment Station Technical Report.* 95 pp.

Piekutowski, A. J.: 1977, Cratering Mechanisms Observed in Laboratory-Scale High-Explosive Experiments. In *Impact and Explosion Cratering* (D. J. Roddy, R. O. Pepin, and R. B. Merrill, eds.), Pergamon Press. This volume.

Pike, R. J.: 1976, Crater dimensions from Apollo data and supplemental sources. *The Moon* **15**, 463–477.

Pohl, J., Stöffler, D., Gall, H., and Ernstson, K.: 1977, The Ries Impact Crater. In *Impact and Explosion Cratering* (D. J. Roddy, R. O. Pepin, and R. B. Merrill, eds.), Pergamon Press. This volume.

Preuss, E. and Schmidt-Kaler, H. (eds.): 1969, Das Ries, Geologie, Geophysik and Genese eines Kraters. *Geol. Bavarica* **61**, 478 pp.

Quaide, W. L. and Oberbeck, V. R.: 1968, Thickness Determinations of the Lunar Surface Layer from Lunar Impact Craters. *J. Geophys. Res.* **76**, 5247.

Regan, R. D. and Hinze, W. J.: 1975, Gravity and magnetic investigations of Meteor Crater, Arizona. *J. Geophys. Res.* **80**, 776–782.

Roddy, D. J.: 1966, The Paleozoic crater at Flynn Creek, Tennessee. California Institute of Technology, unpublished Ph.D. Thesis.

Roddy, D. J.: 1968, The Flynn Creek crater, Tennessee. In *Shock Metamorphism of Natural Materials* (B. M. French and N. M. Short, eds.), p. 291–322. Mono Book Corp., Baltimore.

Roddy, D. J.: 1969, Geologic survey activities. In *Operation Prairie Flat Preliminary Report* (M. J. Dudash, ed.) vol. 1, p. 317–333. Defense Atomic Support Agency, DASA 2228-1, unclassified.

Roddy, D. J.: 1970, LN 303, Geologic studies. In *Operation Prairie Flat Symposium Report* (M. J. Dudash, ed.), vol. 1, p. 210–215. Defense Atomic Support Agency, DASA 2377-1, DASIAC SR-92, unclassified.

Roddy, D. J.: 1976, High explosive cratering analogs for central uplift and multiring impact craters. *Proc. Lunar Sci. Conf. 7th*, p. 3027–3056.

Roddy, D. J.: 1977a, Tabular Comparisons between the Flynn Creek Impact Crater, United States, Steinheim Impact Crater, Germany, and Snowball Explosion Crater, Canada. In *Impact and Explosion Cratering* (D. J. Roddy, R. O. Pepin, and R. B. Merrill, eds.), Pergamon Press. This volume.

Roddy, D. J.: 1977b, Initial pre-impact conditions and cratering processes at the Flynn Creek Crater, Tennessee. In *Impact and Explosion Cratering* (D. J. Roddy, R. O. Pepin, and R. B. Merrill, eds.), Pergamon Press. This volume.

Roddy, D. J.: 1978, Geologic, cratering, and airblast studies of the Pre-Mine Throw IV Craters, Nevada Test Site. Vol. I, Defense Nuclear Agency, Washington. POR 6838, unclassified.

Roddy, D. J., Boyce, J. M., Colton, G. W., and Dial, A. L.: 1975, Meteor Crater Arizona, rim drilling with thickness, structural uplift, diameter, depth, volume, and mass-balance calculations. *Proc. Lunar. Sci. Conf. 6th*, p. 2621–2644.

Roddy, D. J., Ullrich, G. W., Sauer, F. M., and Jones, G. H. S.: 1977, Cratering motions and structural deformation in the rim of the Prairie Flat multiring explosion crater. *Proc. Lunar Sci. Conf. 8th.* In press.

Rooke, A. D. and Chew, T. D.: 1965, Operation Snowball Project 3.1, crater measurements and earth media determinations, interim rept., Miscellaneous Paper No. 1-764, *U.S. Army Engineer Waterways Experiment Station CE*, Vicksburg, Mississippi, unclassified. 61 pp.

Schmitt, H. H., Trask, N. J., and Shoemaker, E. M.: 1967, Geologic Map of the Copernicus Quadrangle of the Moon. *U.S. Geol. Survey Misc. Geol. Inv. Map* I-515.

Scott, D. H., McCauley, J. F., and West, M. N.: 1976, Geologic Map of the West limb region of the Moon. *U.S. Geol. Survey Misc. Geol. Inv. Map* I-1034.

Shock, R. N.: 1977, The Response of Rocks to Large Stresses. In *Impact and Explosion Cratering* (D. J. Roddy, R. O. Pepin, and R. B. Merrill, eds.), Pergamon Press. This volume.

Schultz, P. H.: 1976, *Moon Morphology.* University of Texas Press, Austin, Texas. 626 pp.

Seebaugh, W. R.: 1977, A Dynamic Crater Ejecta Model. In *Impact and Explosion Cratering* (D. J. Roddy, R. O. Pepin, and R. B. Merrill, eds.), Pergamon Press. This volume.

Shoemaker, E. M.: 1962, Interpretation of lunar craters. In *Physics and Astronomy of the Moon* (Z. Kopal, ed.), p. 283–380. Academic Press, New York.
Shoemaker, E. M.: 1960, Penetration mechanics of high velocity meteorites, illustrated by Meteor Crater, Arizona. In *Structure of the Earth's Crust and Deformation of Rocks, Internat. Geol. Congress, XXI Session*, Copenhagen, Rept. **18**, 418–434.
Shoemaker, E. M.: 1963, Impact mechanics at Meteor Crater, Arizona. In *The Moon, Meteorites, and Comets* (B. M. Middlehurst and G. P. Kuiper, eds.), p. 301–336. University of Chicago Press, Chicago.
Shoemaker, E. M. and Kieffer, S.: 1974, Synopsis of the geology of Meteor Crater. In *Guidebook to the Geology of Meteor Crater*, Ann. Mtg. Meteoritical Society, Aug. 1974.
Stöffler, D.: 1971, Progression Metamorphism and Classification of shocked and brecciated crystalline rods at Impact Craters. *J. Geophys. Res.* **76**, 5541–5551.
Stöffler, D.: 1977, Structure of the Ries Crater and Distribution of Target Rocks within Different Types of Impact Breccias (abstract). In *Lunar Science VIII*, p. 908–910. The Lunar Science Institute, Houston.
Stuart-Alexander, D. E. and Howard, K. A.: 1970, Lunar Maria and circular basins—a review. *Icarus* **12**, 440–456.
Swift, R. P.: 1977, Material Strength Degradation Effect on Cratering Dynamics. In *Impact and Explosion Cratering* (D. J. Roddy, R. O. Pepin , and R. B. Merrill, eds.), Pergamon Press. This volume.
Tilghman, B. C.: 1905, Coon Butte, Arizona. *Proc. Acad. Nat. Sciences, Philadelphia* **57**, 887–914.
Trulio, J. G.: 1977, Ejecta Formation: Calculated Motion from a Shallow-Buried Nuclear Burst, and its Significance for High Velocity Impact Cratering. In *Impact and Explosion Cratering* (D. J. Roddy, R. O. Pepin, and R. B. Merrill, eds.), Pergamon Press. This volume.
Ullrich, G. W., Roddy, D. J., and Simmons, G.: 1977, Numerical Simulations of a 20-ton TNT Detonation on the Earth's Surface and Implications Concerning the Mechanics of Central Uplift Formation. In *Impact and Explosion Cratering* (D. J. Roddy, R. O. Pepin, and R. B. Merrill, eds.), Pergamon Press. This volume.
Vortman, L. J.: 1977, Craters from Surface Explosions and Energy Dependence—A Retrospective View. In *Impact and Explosion Cratering* (D. J. Roddy, R. O. Pepin, and R. B. Merrill, eds.), Pergamon Press. This volume.
Wagner, G. H.: 1964, Kleintektonische Untersuchungen in Gebiat des Nördlingen Ries. *Geol. Jahrb.* **81**, 519–600.
Wilhelms, D. E., Hodges, C. A., and Pike, R. J.: 1977, Nested-Crater Model of Lunar Ringed Basins. In *Impact and Explosion Cratering* (D. J. Roddy, R. O. Pepin, and R. B. Merrill, eds.), Pergamon Press. This volume.
Wisotski, J.: 1977, Dynamic Ejecta Parameters from High Explosive Detonations. In *Impact and Explosion Cratering* (D. J. Roddy, R. O. Pepin, and R. B. Merrill, eds.), Pergamon Press. This volume.

Roddy, D. J., Pepin, R. O., and Merrill, R. B., editors.
(1977) *Impact and Explosion Cratering*, Pergamon Press (New York), p. 247–275.
Printed in the United States of America

Terrestrial impact structures: Principal characteristics and energy considerations*

M. R. DENCE, R. A. F. GRIEVE, and P. B. ROBERTSON

Earth Physics Branch, Department of Energy, Mines, and Resources, Ottawa, Ontario, Canada K1A OY3

Abstract—The world census of craters, crater groups, and other approximately circular structures which are attributed to hypervelocity impact of extra-terrestrial bodies stands at about 80. They are so classified on a number of criteria, of which the presence of meteoritic fragments or of shock metamorphism of rocks within or around the structure are considered definitive. At least 50 additional structures appear to conform to other, less definitive criteria.

All well-investigated craters up to ~3.5 km diameter have simple bowl form. Barringer Meteor Crater (1.2 km) is the largest with identified meteorite fragments and the largest simple crater formed in sedimentary rocks. Lonar L. (1.8 km) and Brent (3.8 km) are the largest simple craters in basaltic and granitic rocks, respectively. Larger structures are complex and have shallower relative dimensions. In general those up to approximately 30 km across have distinct central uplifts (Steinheim, 3.6 km; Deep Bay, 12 km; Gosses Bluff, 24 km); some formed in heterogeneous, layered rocks (Ries, 24 km) and all more than 30 km across are ring structures with alternating topographically high and low features (Clearwater Lake West, 32 km; Charlevoix, 46 km; Manicouagan, 70 km; Popigai, 100 km). The floors of the larger structures are covered by thick sheets of impact melted rock and mixed breccias.

For both simple and complex craters formed in predominantly crystalline rocks, downward excavation is limited by rock strength to material shocked above 23 ± 7 GPa (230 ± 70 kbar). Consideration of the distribution of shock metamorphism in rocks below the excavated cavity, the volumes of impact melt and the relative diameters of the excavated cavity, the final crater rim and the outer limit of bedrock disruption, yield a series of relationships between released energy and crater size, based on nuclear explosion data. Rim diameters scale with $E^{1/3}$ in small craters, but craters in crystalline rocks with rim diameters >2.4 km scale with $E^{1/3.4}$. Excavated cavity diameters scale as $E^{1/3}$ for excavated cavities <6.4 km diameter, but as $E^{1/4}$ for larger cavities. There is an indicated change in mean rate of attenuation of shock pressure (P) with radial distance (R) from $P \sim R^{-2}$ for small events ($E < \sim 10^{17}$J), to $P \sim R^{-3}$ or more for large events ($E > \sim 10^{22}$J). Cavity growth beyond the excavation stage by approximately 50% is due to displacement of large blocks of basement rocks accelerated by the shock wave. It is suggested that resultant frictional heating and crushing between blocks lubricates movement, allowing base yielding and central uplift in large craters, with injection of frictionally melted material as pseudotachylite dikes.

INTRODUCTION

THE STUDY OF terrestrial impact craters began with the recognition of some ten craters or crater groups with which are associated abundant fragments of iron meteorites. A few, including Barringer Meteor Crater in Arizona, were known in the 19th century, but most were discovered through exploration in the desert areas of the world between 1920 and 1950 (Krinov, 1963a, 1966). At eight of these sites the craters are of modest dimensions, the largest being 185 m across.

---

*Contribution from the Earth Physics Branch No. 661.

Their origin by meteorite impact has been generally accepted. The number was augmented in 1947 by the Sikhote-Alin iron meteorite shower which created 122 craters ranging in diameter from 0.5 to 26.5 m and reproduced many of the features recognized in older craters of similar size.

There was greater reluctance in the scientific community to accept an impact origin for two craters of the order of 1 km across, Wolf Creek, Western Australia, and the Barringer Crater, until the application of the theory of hypervelocity impact as developed by Moulton (1929, 1931), Opik (1958), Shoemaker (1960), and others provided the basis for present understanding of large impact structures. Unusual rock deformation, recognized originally by Merrill (1907), was interpreted as the product of high shock pressures, and, combined with the discovery of high pressure polymorphs (Chao *et al.*, 1960, 1962), rapidly led to systematic studies of shock metamorphism (Chao, 1967a, b). It is now the principal criterion for hypervelocity impact in nature. The rationale for this view has been presented by French (1968a) and Guy-Bray (1972) and, although challenged (Currie, 1972; Nicolaysen, 1972), remains sound (Dence, 1972a; Robertson and Grieve, 1975), and is accepted here.

By 1960 several dozen approximately circular structures, the majority larger and older than the craters with meteorites, had been recognized as anomalies difficult to explain by well known geological processes. Some of these structures have the familiar bowl shape of the Barringer Crater, but many have a central high surrounded by a ring depression. They occur in a variety of geologic settings and are characterized by intensely fractured and brecciated rocks suggesting violent explosive activity. Many were attributed to cryptovolcanism (Bucher, 1936) but the idea that they could be explained by meteorite impact gradually took hold (Boon and Albritton, 1936; Daly, 1947; Dietz, 1947; Beals *et al.*, 1956). Dietz (1960, 1963) employed the more general term "cryptoexplosion structures" and used the presence of unusual fracture surfaces called shatter cones as an indication of shock wave action, and hence of hypervelocity impact. Such structures he termed "astroblemes" (star wounds), the scars of large ancient impacts. Shatterconing remains one of the mesoscopic indications of shock metamorphism, though not as yet fully understood (Milton, 1977).

No new discoveries of craters with associated meteorites have been made since 1948, but in the last 15 years the number of shock metamorphosed sites has tripled (Shoemaker and Eggleton, 1961; Dence, 1972a; Robertson and Grieve, 1975; Masaitis, 1975). At present almost 70 sites are known, with a like number of possible sites awaiting closer examination. Including those with associated meteorites there is thus a data base from nearly 80 localities with craters ranging from a few metres to over 100 km across.

Types of terrestrial impact craters

The structural division of shock metamorphosed structures into those with a regular bowl shape and those possessing an uplift in the center has led to the convention of referring to the former as *simple* and the latter as *complex* craters

(Dence, 1965). Within each of these categories further subdivisions may be recognized. In simple craters there is a transition with increasing size from impact *pits* or *holes* in which the meteorite remains intact to impact *craters* in which the impacting body is fragmented (Krinov, 1963a). This transition depends on such factors as the strength of projectile and target materials and the angle and velocity of impact (Heide, 1964; Fudali and Chapman, 1975). It corresponds approximately to the threshold of the hypervelocity regime and in the representative case of the Sikhote-Alin fall occurs at a crater diameter of 9 m, corresponding to craters formed by meteorites weighing more than 1000 kg (Krinov, 1963b). The survival of meteorites appreciably larger than 1 ton generally requires a low angle of impact and weak target materials, as at Campo del Cielo (Cassidy *et al.*, 1965; Cassidy, 1971).

Krinov (1963a, 1966) marks a further transition at craters an order of magnitude larger, which he calls *explosion craters*. In these, meteorite fragments are rare, as both meteorite and target materials are fused to form glassy ejecta. This is in accord with calculations that bodies of 1000 tons or more will impact with more than 60% of their cosmic velocity (Heide, 1964) unless the angle of impact is exceedingly low (nominally $< 5°$, Fudali and Chapman, 1975). Thus all craters 90 m or more across can be expected to contain materials showing all recognizable stages of shock metamorphism including total fusion of silicates and nickel-iron.

The change from simple craters to complex structures takes place at a diameter of approximately 1.5–2 km in layered, consolidated sedimentary rocks, and at approximately 4 km in crystalline rocks (Dence, 1972a). There appear to be no exceptions to these observations, implying that the transition is essentially a function of energy and target material properties.

Within the complex group a further distinction may be made between smaller *central uplift* craters and larger, more evolved *ring structures*. The prominence of the rings in the latter depends greatly on the level of erosion and the lithologic variation of the country rocks. Ring structure is well developed at the slightly eroded 24 km Ries Crater, where 700 m of sedimentary rocks overlie crystalline basement rocks (Pohl *et al.*, 1977). In craters formed in crystalline rocks alone, or where sedimentary rocks are little more than a veneer, rings are present in structures exceeding 30 km. The multiring appearance of the 100 km Popigai structure (Masaitis *et al.*, 1975) may represent a further structural stage represented by the largest multiring basins on the Moon (Howard *et al.*, 1974) and other terrestrial planets.

Structural characteristics of simple craters

The smaller simple craters are typified by the Odessa, Haviland, and Henbury craters, which have rim diameters between 7 and 168 m (Evans, 1961; Krinov, 1963a, 1966; Milton, 1968). Excavation or drilling has shown the craters to have a lining of fragmented and pulverized rock mixed with meteorite fragments (Fig. 1). This thin lining separates little disturbed bedrock from

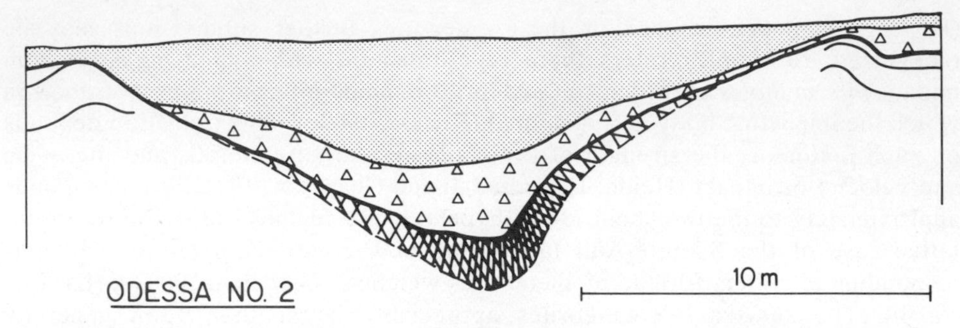

Fig. 1. Cross section of the small, simple Odessa No. 2 crater, simplified after Evans (1961), showing anticlinal structures in the rim and the zone of rock flour with meteorite fragments (cross-hatched) beneath the crater floor. Crater fill consists of coarse fragmental material in part comprising ejecta from the main Odessa Crater (triangles) and later sands (dots).

overlying crater fill. The latter generally comprises a lower layer of coarse rock fragments overlain by subsequent sedimentary fill. The rims comprise uplifted bedrock deformed into a variety of structures including low-angle thrusts, anticlines, and complex folds overlain by overturned beds and fragmental ejecta. Where there has been negligible erosion the depth to the bottom of the lining (true depth) is typically one third to one quarter the diameter of the crater. However, as crater size increases, fragment fill becomes increasingly significant so that the depth to the top of the fill (apparent depth) changes from one third to one eight of the crater diameter (Krinov, 1963b).

In the larger simple craters the structure is basically the same. Little difference is evident with rock type except in details of rim structure, which is sensitive to the influence of regional structures and the direction and angle of impact (Shoemaker, 1960; Currie and Dence, 1963). The apparent depth is generally about one eighth of the diameter; the depth to the bottom of the breccia lens is approximately twice that of the apparent depth (Fig. 2).

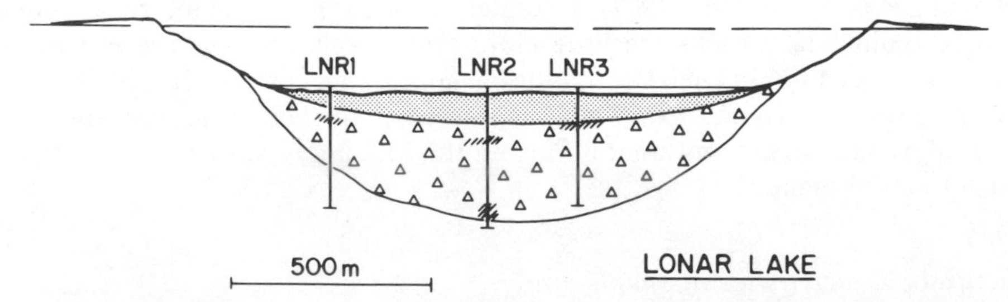

Fig. 2. Cross section of Lonar Lake Crater, India, after Fredriksson *et al.* (1973). The crater, formed in basalt, is representative of larger simple craters. Original ground surface shown by dashed line. Drill holes penetrate water, sediments derived from the eroded rim (dots), breccia (triangles), and wall rocks. Microbreccias with glass and highly shocked fragments (diagonal lines) are concentrated in the upper and basal parts of the breccia.

The early drilling at the 1.2 km Barringer Crater indicated that, as at the small craters, meteoritic fragments were present near the bottom of the breccia lens, chiefly as fine particles dispersed in glass (Shoemaker, 1960). The upper parts of the breccia comprised shattered Coconino sandstone in the center and large blocks of the overlying Kaibab limestone towards the sides, indicating that extensive slumping from the crater walls had taken place prior to the deposition of a thin layer of mixed fallout debris. Since then a number of other simple craters have been drilled: Holleford (Beals, 1960), West Hawk Lake (Halliday and Griffin, 1967; Short, 1970), Brent (Millman *et al.*, 1960; Dence, 1968; Dence and Guy-Bray, 1972) and Lonar Lake (Fredriksson *et al.*, 1973). Of these the greatest amount of structural detail is available from the core of the twelve drill holes put down in the Brent crater between 1955 and 1967.

In these craters meteorite fragments have not been identified, though glassy fragments and larger crystallized masses of melted material at Brent have Ni anomalies, which are probably indicative of admixed fused meteoritic material (Dence, 1971). Melting is attributed to intense shock. This is inferred from the compositional similarity of the melted materials to the country rocks and their intimate association with moderately to strongly shocked fragments of these rocks at both simple and complex craters (Dence, 1971; Engelhardt, 1972; Masaitis *et al.*, 1975; Grieve *et al.*, 1977). In the drill core from simple craters the thoroughly mixed, strongly shocked materials are concentrated, as at Barringer, in the upper central part and near the base of the breccia lens. In the upper part thoroughly mixed breccias are interlayered with weakly mixed, slightly shocked breccias, the latter becoming more abundant toward the crater margins. At deeper levels of the breccia lens weakly shocked materials predominate. These rocks are less brecciated and mixed and comprise a coarse breccia of internally fractured and brecciated slabs tens of metres thick, apparently derived from the wall rocks. The slabs are separated by layers 1–3 m thick of intensely crushed and smeared rock which are interpreted as the zones along which movement of the slabs took place. At these deeper levels strong brecciation and mixing are observed only in the centre of the breccia lens, where it is interpreted as the result of convergence of material from the sides (Dence, 1968; Short, 1970; Dence and Guy-Bray, 1972).

Near the base of the breccia lens the level of shock metamorphism in the breccia fragments rises and mixed breccias with strongly shocked, including shock melted, materials are again observed (Short, 1970; Fredrikkson *et al.*, 1973). At Brent, the largest of the simple craters investigated (Fig. 3), the shock melted materials at this level are not dispersed but form a lenticular mass of inclusion-bearing crystalline rock 30 m thick in the center. Heat from this concentration of molten material has thermally metamorphosed the breccias for 30 m above and below the melt layer, which, from its Ni anomaly, is taken as the deepest level of penetration of meteoritic material (Dence, 1971; Grieve *et al.*, 1977).

The breccias at Brent persist for approximately 30 m below the melt layer before giving way to fractured and locally brecciated basement rocks. Below the

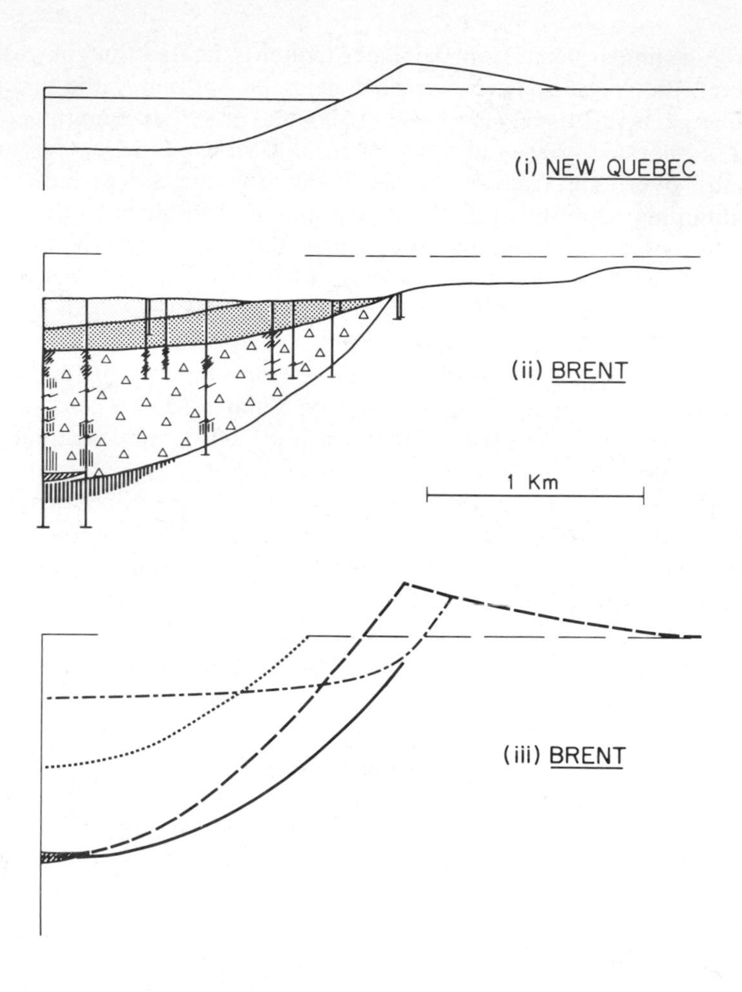

Fig. 3. Radial profiles at the same scale of: (1) New Quebec (Chubb) Crater showing lake filling crater and rim uplifted above present average ground surface (dashed line). (ii) Brent Crater, Ontario, based on surface exposures and 12 drill holes shown at their radial distance from the centre. Dashed line is estimated original surface. Ordovician sedimentary rocks filling the crater comprise restricted basin sediments (close stipple) and regional limestones and red beds (open stipple) (after Lozej and Beales, 1975). Breccias (triangles) include strongly-mixed, glass-bearing breccias of suevite type (diagonal lines) and moderately mixed and shocked crystalline breccias (vertical lines). Strongly sheared breccias are indicated by zigzag line. Lens of melt (open diagonal) is underlain by shocked basement rocks (heavy verticals). (iii) Brent Crater showing the main elements in the evolution of the crater. Dotted line outlines the volume excavated according to the model developed in the text. Short dashed line outlines the profile of the transient crater resulting from excavation and movement of ruptured basement to give an expanded cavity and uplifted rim. Slumping of the crater walls results in formation of the breccia lens above the melt zone and crater profile shown by dot-dashed line. Interface between breccia lens and fractured basement rocks indicated by solid line.

zone of thermal metamorphism the level of shock metamorphism corresponds to shock pressures of approximately 23 GPa (230 kbar), dying away systematically and rapidly with depth (Dence, 1968; Hartung *et al.*, 1971; Robertson and Grieve, 1977). Shock levels also decrease towards the rim where the uplifted and fractured basement rocks show no signs of shock pressures above 1 GPa.

These relationships have been interpreted in terms of a model in which hypervelocity impact produces a parabolic transient cavity that is immediately partly filled by slope failure and slumping of the crater walls (Dence, 1968; Grieve *et al.*, 1977). At Brent the breccias above the melt layer have a volume of 2.46 km^3. When these rocks are replaced on the crater walls, the transient cavity has a cross-section that closely approximates the formula $r^2 = 2p^2$ (Dence, 1973), where p is the depth of the transient cavity below the original plane and r is the cavity radius at original ground level. The latter is estimated to have been about 50 m above the present local height of land from an analysis of the stratigraphy and degree of compaction of the sedimentary rocks in the crater (Lozej and Beales, 1975) giving $p = 1.05$ km. Thus the depth-to-diameter ratio of the transient cavity at Brent was not greatly different from the 1 to 3 ratio observed at the smallest simple craters. However, the transient cavity volume of 3.6 km^3 does not necessarily correspond to the volume excavated, as part of the cavity is due to initial displacements in the wall rocks. This question is pursued in a later section.

Structural characteristics of complex craters

Most terrestrial craters are of complex type, as their large size favours preservation. At present, 55 shock metamorphosed structures can be placed in this category. Of these, 20 have been formed in sedimentary rocks, a like number in crystalline rocks with at most a veneer of pre-existing sedimentary rocks, and the remainder in terrain where thick sedimentary cover and crystalline basement rocks are both involved. Most of the latter exceed 30 km in diameter and are ring structures. Where well-stratified rocks are deformed they provide clear evidence of structural uplift in the centre and downdrop of rocks at the margin. There is usually little indication of rim rocks being uplifted above their initial position. Where the country rocks are crystalline, indirect evidence of uplift or downdrop is generally forthcoming from the distribution of shock metamorphism (Robertson and Grieve, 1977), and in many cases there is topographic expression of a central peak. All are much shallower relative to their diameter than simple craters.

The change from simple to complex form is relatively abrupt in craters formed in flat-lying sedimentary rocks and is illustrated by the differences between the 1.2 km Barringer crater and the 3.6 km Steinheim Basin and Flynn Creek structures (Roddy, 1968, 1977a,b; Reiff, 1977). The latter have a depth-to-diameter ratio of 1 to 24, a distinct topographic and structural peak in the centre and rim rocks largely downdropped along normal faults. Similar structural details are evident for the larger, more deeply eroded 6 km Decaturville (Offield and Pohn, 1977), 13 km Sierra Madera (Wilshire *et al.*, 1972), 14 km Wells Creek (Wilson and Stearns, 1968) and 22 km Gosses Bluff (Milton *et al.*, 1972) structures and for the deeply buried 9 km petroliferous Red Wing Creek structure (Brenan *et al.*, 1975). In all, the amount of uplift is $\frac{1}{8}$ to $\frac{1}{10}$ the overall diameter of the structure and net motion has been inwards as well as upwards in

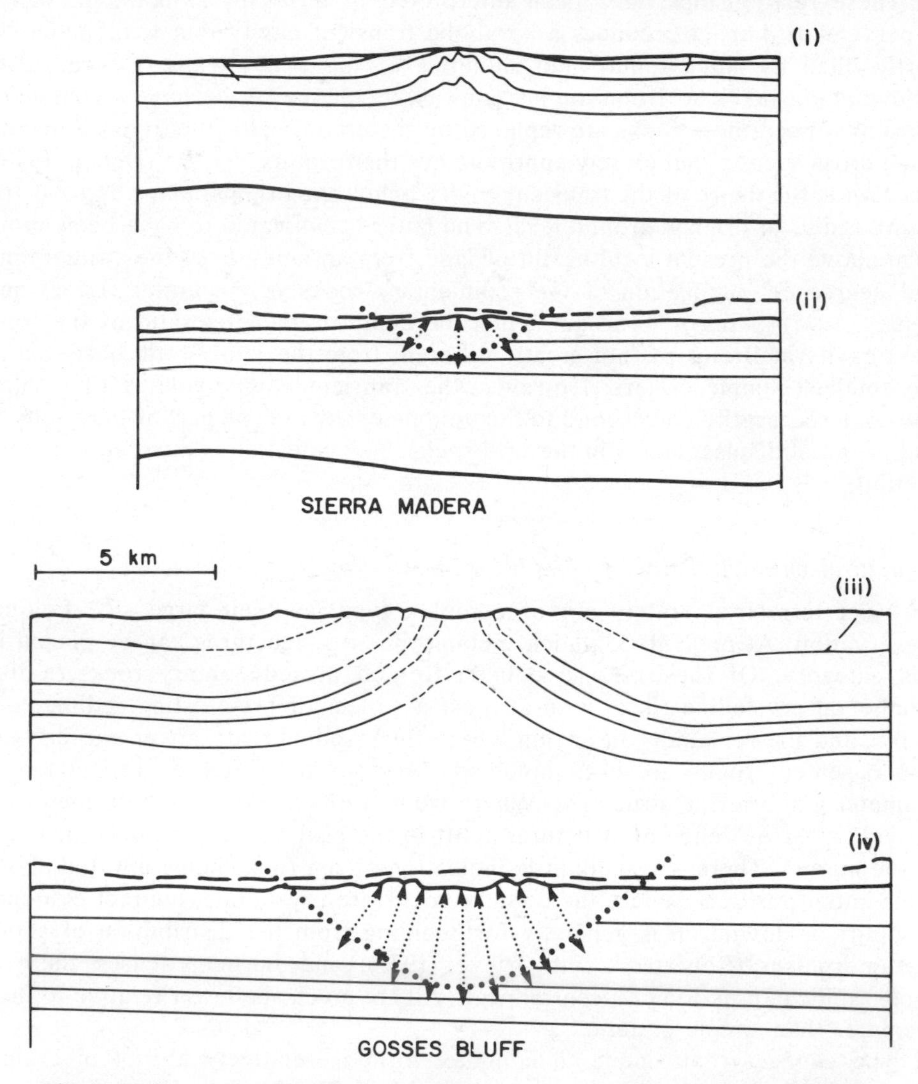

Fig. 4. Profiles of complex craters in sedimentary rocks. (i) Present profile of Sierra Madera (after Wilshire *et al.*, 1972) showing central uplift and peripheral depression. (ii) By allowing for uplift and inward motion, as indicated by circumferential shortening, marker horizons can be restored to their original positions (arrows), outlining the volume excavated (dotted line). The final, pre-erosional crater profile is indicated by a dashed line. (iii) and (iv) Similar profiles for Gosses Bluff (after Milton *et al.*, 1972). Restoration of beds in (iv) is in accord with shatter cone data.

the central uplift. In the ring depression there is also a component of motion inward, and calculations at Wells Creek (Wilson and Stearns, 1968) and Decaturville (Offield and Pohn, 1977) show that the volume of material in the central uplift is approximately balanced by the material withdrawn from the surrounding ring depression.

From the amount of circumferential shortening of key horizons (Wilshire *et al.*, 1972) or the analysis of shatter cone data (Milton *et al.*, 1972) it is possible to estimate the inward as well as upward displacements of rocks in the central uplift. Where this has been done and the rocks returned to their former position, an estimate of the volume excavated can be made. At Sierra Madera and Gosses Bluff (Fig. 4) the resulting volume corresponds to a crater with depth approximately $\frac{1}{4}$ to $\frac{1}{5}$ of its diameter, though in neither is the position of the rim of the excavated cavity well-defined. It should also be noted that, in both, the rocks in the centre of the central uplift have been shocked to 10–20 GPa (Wilshire *et al.*, 1972). More strongly shocked materials are present in mixed breccias which occur either in erosion remnants overlying the basement rocks or in irregular bodies within the central uplift. The rocks of the central uplift are generally so strongly faulted, fractured, and locally crushed that they return little seismic energy and can be considered a megabreccia. However, they retain sufficient structural coherence to allow detailed mapping where there are useful marker horizons. It is not unusual for the rocks brought up in the central uplift to be of higher density than the near-surface rocks. Thus the gravity signature of complex craters may not be the regular negative anomaly typical of simple craters (Innes, 1961) but may show modest central positive anomalies superimposed on a broad negative (Wilson and Stearns, 1968; Milton *et al.*, 1972).

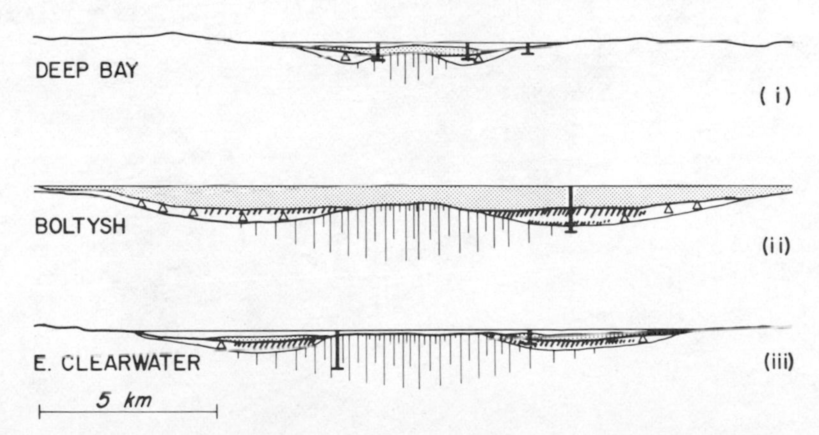

Fig. 5. Profiles of central uplift craters in crystalline rock. (i) Deep Bay (after Dence *et al.*, 1968). (ii) Boltysh (after Yurk *et al.*, 1975). (iii) East Clearwater Lake (after Dence, 1968). Only the deepest drill hole at Boltysh is shown. Vertical lines depict the estimated extent of shock metamorphism above ~5 GPa in the fractured crystalline rocks of the central uplift. Melt rocks and glass-bearing breccias in the breccia annulus (triangles) around each central uplift are indicated by diagonal lines. Stipple indicates sedimentary rocks filling the craters.

Craters formed in crystalline rocks go through a more gradual change from simple to complex form than those formed in sedimentary rocks. Examples are the well-preserved Deep Bay and Bosumtwi craters, representative of craters about 10 km across, and the East Clearwater Lake and Boltysh craters in the 20–25 km size range (Fig. 5). All but Bosumtwi are penetrated by centrally located drill holes (Dence, 1968; *et al.*, 1968; Yurk *et al.*, 1975). Sedimentary fill has largely preserved their original form which shows a change in depth to diameter ratio from 1 to 20 (Deep Bay) to about 1 to 40 (East Clearwater and Boltysh). Drill holes nearest the centre pass directly from sedimentary cover into fractured crystalline rocks without penetrating mixed breccias. Holes nearer the rim pass through thicker sedimentary strata into mixed breccias with shock melted glasses, or, in the larger craters, into substantial thicknesses of crystalline

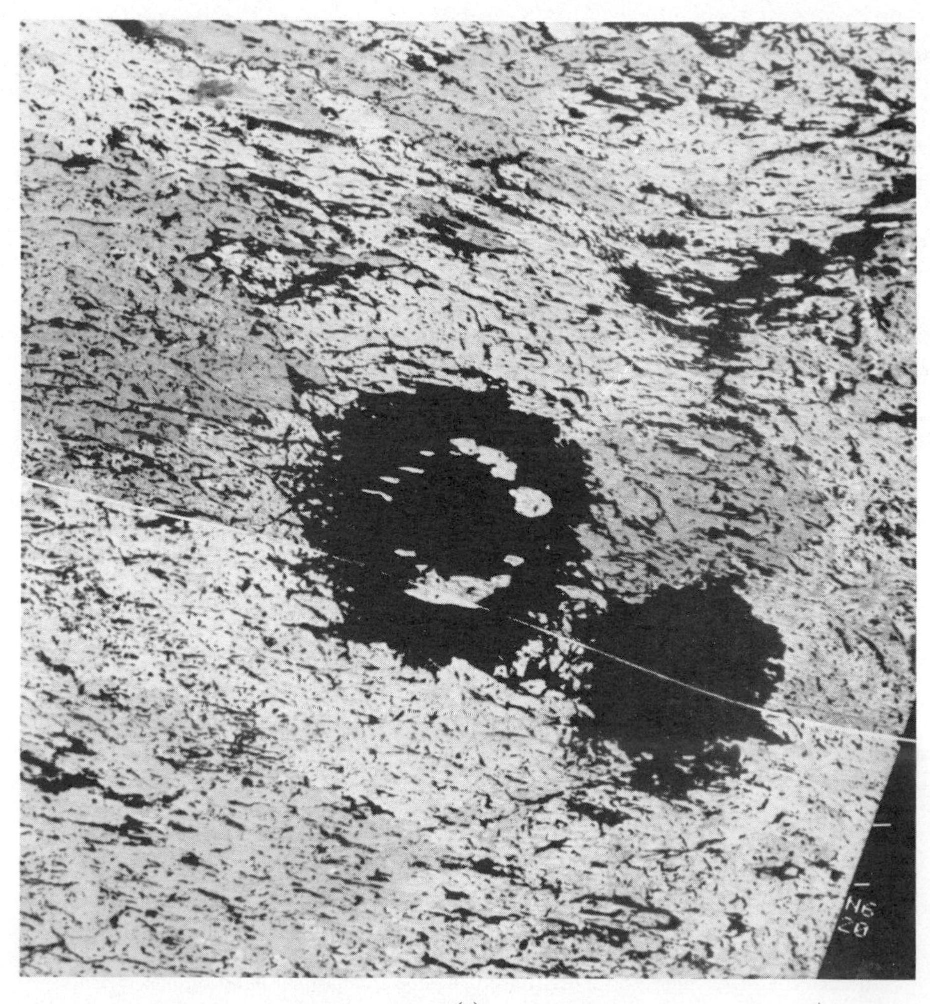

(a)

melt rocks. At Boltysh the entire succession comprises approximately 400 m of breccias and melt rocks overlying fractured basement rocks.

The annular disposition of mixed breccias and melt rocks around the central peak is characteristic of complex craters and is evident even at those 5 km across, such as Mien L. (Stanfors, 1969) and Gow L. (Robertson and Grieve, 1975; Thomas and Innes, 1977). Shock levels in the top of the central peak rocks are in the same 10–20 GPa range as in the central uplifts of craters formed in sedimentary rocks. Shock deformation dies out gradually at depth (Dence, 1968), in keeping with the interpretation that central uplift, fracturing, and tilting takes place in the crystalline rock craters in much the same way as is observed at Gosses Bluff. The rims of complex craters in crystalline rocks do show topographic evidence of uplift with downdrop along their inner margins (Dence *et al.*, 1968). Disturbance of the rim rocks extends as far as an outer depression,

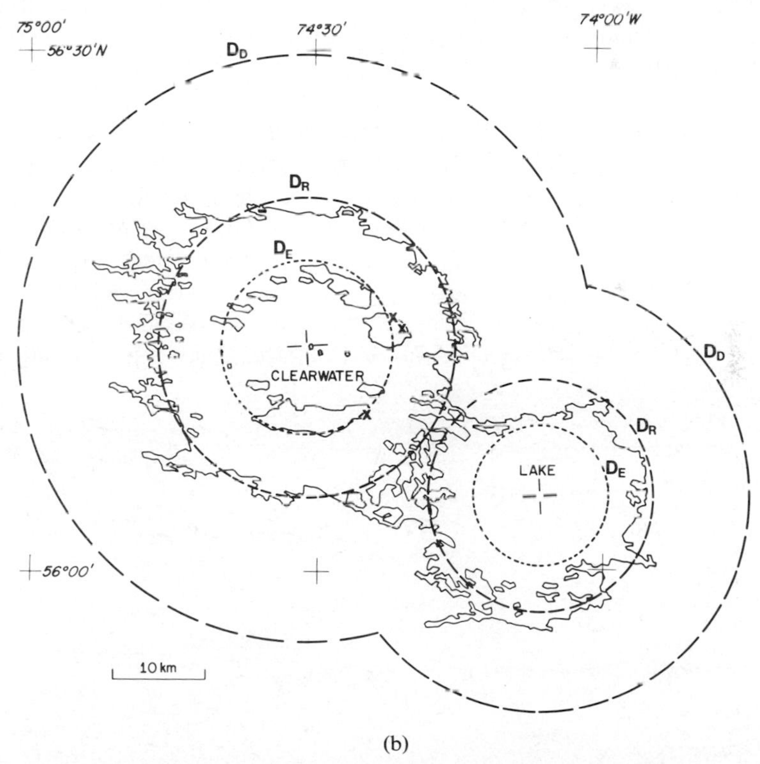

(b)

Fig. 6. (a) Mosaic from Landsat photographs of the pair of complex craters at Clearwater Lake, Quebec, formed in Archean crystalline rocks. (b) Outline map of Clearwater Lake showing the three structural rings for each crater projected down from the ground level at the time of crater formation: D_D, the outer fracture zone marking the limit of bedrock disturbance; D_R, the estimated position of the topographic rim prior to erosion; D_E, the limit of the excavated cavity. The cross marks the center of the central uplift in each crater. Limestone outcrops on the island ring of the western crater indicated by crosses.

interpreted as a fracture zone approximately twice the diameter of the crater (Innes *et al.*, 1964).

This aureole of disturbed rocks is distinct in Landsat photographs around both of the Clearwater Lake craters (Fig. 6). Formed simultaneously in crystalline rocks, they most clearly demonstrate the change in geometry from central peak to ring structure. The western crater is the larger but is shallower by at least 250 m (Dence, 1965). Its structure can be defined in terms of three rings: the outer fracture zone mentioned above, the topographic rim of the crater which corresponds approximately to the shoreline, and a conspicuous inner ring of islands. The trough between the crater rim and the island ring is a graben. The distribution of float and comparisons with other structures indicate that within the graben are preserved patches of sedimentary rocks which formed a veneer over the crystalline rocks at the time of crater formation. The innermost limestones crop out on the outer edges of the island ring marking the transition from depressed margin to uplifted centre. There is a subdued central peak marked by a few small islands on which are exposed rocks shocked to between 20 and 30 GPa. A thick sequence of breccias and melt rocks covers the islands of the ring, but not those in the center, to form an annulus of highly shocked material around the central peak as in the smaller complex craters. As melt overlies and locally engulfs limestone blocks which were near the original ground level it is concluded that melt formed a sheet lining the crater floor, riding up and over the rim of the transient cavity (Grieve *et al.*, 1977).

Similar relationships are observed at other slightly eroded, large ring structures such as the 70 km Manicouagan (Dence, 1977; Floran and Dence, 1976), 23 km Lake St. Martin (McCabe and Bannatyne, 1970), and 100 km Popigai (Masaitis *et al.*, 1975) structures (Fig. 7). At the latter there is some indication of

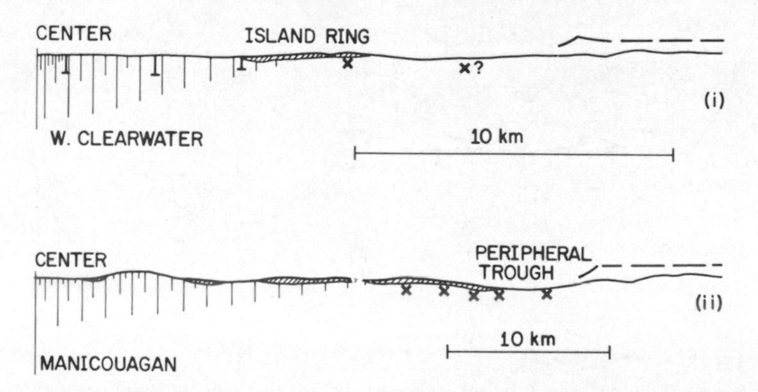

Fig. 7. Half-profiles of ring structure type of complex crater in crystalline rocks. (i) West Clearwater Lake Crater, with 3 drill holes. (ii) Manicouagan Crater. Vertical lines indicate general distribution of shock metamorphism in basement rocks, intensifying towards the center. Diagonal lines indicate remnants of crater floor lining of melt rocks and breccias. Radial distribution of pre-crater Paleozoic limestone outcrops (inferred under the peripheral trough of West Clearwater Lake Crater) indicated by crosses. Dashed line indicates position of topographic rim before erosion.

intermediate rings in the form of ridges capped with melt rocks. At Popigai and Manicouagan the centre is a topographic low as in the large lunar multiring basins such as Orientale (Howard *et al.*, 1974). The cluster of peaks at Manicouagan is displaced by 4 km or more from the centre and contains blocks of anorthosite which have been thrust up relative to the surrounding rocks.

At structures where erosion has stripped off the cover of melt rocks and breccias the topographic expression of ring structure is largely lost, though the peripheral trough generally remains conspicuous as at Charlevoix (Robertson, 1975) and Siljan (Rondot, 1975) both approximately 50 km structures (Fig. 8). At even more deeply eroded structures, the 37 km Carswell structure (Innes, 1964) and the 140 km Vredefort dome (Manton, 1965), central uplift and peripheral depression are well defined by thick stratigraphic sequences which overlie crystalline basement rocks. The sequence at Vredefort is more than 12 km thick and is amenable to the same type of structural analysis as has been applied at Gosses Bluff (Manton, 1965; Dence, 1972b). The Sudbury structure (Dietz, 1964; French 1968b, 1970; Dence, 1972b), which has been highly modified by later events, is of similar size and age to Vredefort. Sudbury is noteworthy for the exceptionally large volumes of igneous rock (Sudbury Irruptive) present in the centre, the diverse suite of breccias above and below the Irruptive and the base metal ore bodies associated with the breccias underlying the Irruptive rocks. It has yet to be determined whether all these rocks were emplaced in immediate response to the impact event, or were in part the result of subsequent igneous activity.

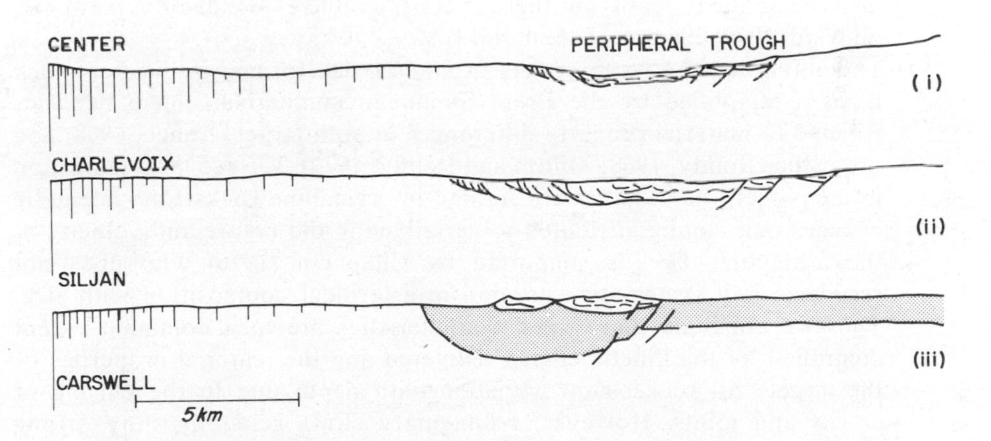

Fig. 8. Half profiles of eroded complex craters formed in crystalline rocks with one to several kilometers of overlying sedimentary rocks at time of crater formation. (i) Charlevoix Crater, with considerable topographic relief and basement rocks shocked to >20 GPa (vertical lines) in the central peak (after Robertson, 1975). (ii) Siljan, more deeply eroded, but with a thicker sedimentary rock section than Charlevoix. (iii) Carswell, where the resistant dolomites of the peripheral trough form the only positive topographic features. Wavy-lined pattern represents structures in carbonate rocks; dots represent massive sandstones (Athabasca formation at Carswell); vertical lines indicate shocked crystalline rocks as in Figs. 5 and 7.

Mechanics of crater formation

In developing a general hypothesis for the formation of both simple and complex craters certain basic observations should be emphasized.

(1) As reviewed by Grieve *et al.* (1977), the breccias within both types of crater are considered to consist largely of material disrupted, moved and mixed in the excavation process but not ejected from the cavity. In the late stages of movement this material moved laterally outwards across the expanding crater floor but in general did not go into ballistic trajectories. Except where disrupted by subsequent slumping, the interface between the breccia lining and underlying fractured basement can be mapped with reasonable precision.

(2) In both types of crater shock levels in the fractured basement rocks are similar, grading from levels between 15 and 30 GPa in the center to less than 1 GPa at the margin. However, the apparent attenuation rates differ in the two types of crater (Robertson and Grieve, 1977).

(3) In complex craters where target rocks are well stratified, it is clear from observations of the final crater form that the basement rocks underwent a net inward and upward movement in the centre and downward and inward at the margin. However, there is some indication at Red Wing Creek structure (Brenan *et al.*, 1975) that rocks at depth under the central uplift may have had a net downward displacement which dies out at greater depth. There are no stratigraphic data from simple craters on displacements at depth, but there is considerable evidence of upward and outward displacement of their rim rocks.

(4) The differences between craters of similar size formed in different rock types, exemplified by the Brent–Steinheim comparison, have been attributed to material property differences in both target (Dence, 1968) and projectile (Roddy, 1968; Milton and Roddy, 1972). We see little indication in the present data on craters formed in crystalline rocks for systematic changes that can be attributed to variations in the nature and velocity of the projectile. This is supported by Chapman (1976) who does not recognize any systematic variation in asteroidal composition with size. Thus we conclude that crater characteristics are to a dominant extent controlled by the kinetic energy delivered and the material properties of the target. All rocks show variation with depth due to the closing of cracks and joints. However, sedimentary strata generally show strong physical property variations across bedding planes, and these variations play a part in the yielding associated with central uplift formation at relatively small diameters.

Consideration of these observations suggests little change with scale or rock type between simple and complex craters at the excavation stage. The strength of rocks as diverse as anorthosite, granitic gneiss, and calcareous sandstone, places the limit of downward excavation at shock pressures in the range 23 ± 7 GPa in both types of crater. The variation in this limiting value may be a function of rock type and crater size, but also includes a considerable uncertainty due to erosion and other factors. The very considerable differences in the final crater form are thus largely a function of the late-stage movements of the basement rocks.

Basement Displacements

The contrast between the displacements in simple and complex craters are well displayed in a comparison of their apparent rates of shock attenuation (Fig. 9). In the case of the simple crater, exemplified by Brent, the apparent attenuation rate is extremely high immediately under the breccias and decreases with depth (Robertson and Grieve, 1977). In complex craters, as seen in radial traverse from the centre, the apparent rate of decrease over approximately the same range of shock pressure is extremely low near the centre and rises towards the crater margin (Fig. 9). These contrasting patterns can be reconciled if it is accepted that the Brent rocks have undergone a net downward displacement, as opposed to the net upward displacement of the rocks of the central uplift in complex craters. If it can be determined how much displacement has taken

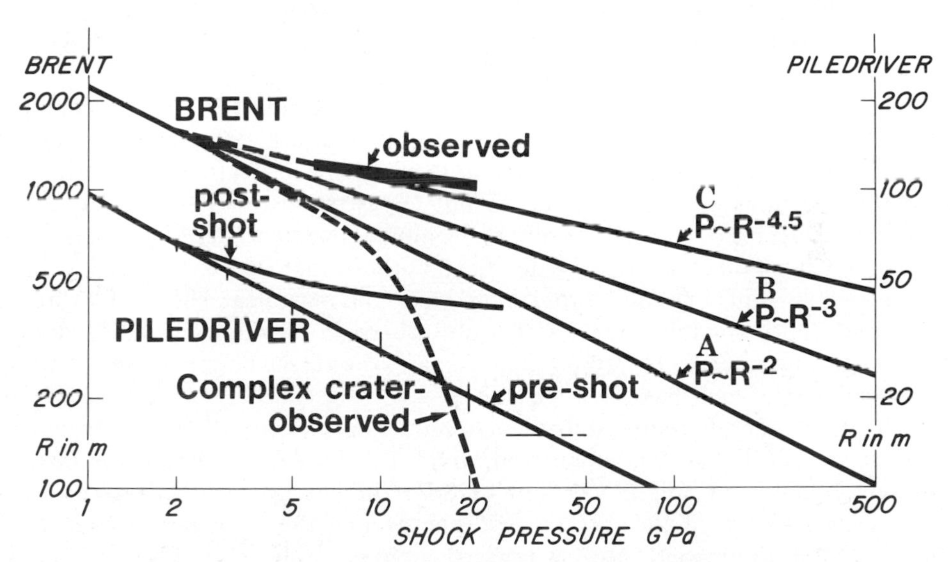

Fig. 9. Observed and inferred attenuation of shock pressures (P). Range (R) for Piledriver (after Borg, 1972) is distance from shot point. Post-shot curve for Piledriver derived from the position of stress gauges, which originally were in the positions given by the vertical lines. Rocks shocked at >27 GPa came from within the cavity (horizontal line). Restored pre-shot attenuation curve decays approximately as $P \sim R^{-2}$. Range for Brent is depth below original ground level (Fig. 3(iii)). Apparent shock pressure attenuation is observed over the range indicated. Three possible attenuation rates are indicated by cases A, B, and C, with rocks shocked at 2 GPa considered to have had negligible displacement as at Piledriver. Dashed line indicates apparent shock pressure attenuation in a complex crater such as Charlevoix, as observed on the surface. (after Robertson, 1975; Robertson and Grieve, 1977).

place, the rocks can be restored to their positions when the shock wave passed through them and the actual attenuation rate for the shock wave derived.

Such a restoration can be made in the case of the 64 kt Piledriver underground nuclear explosion in granodiorite. Borg (1972) gives the pre- and post-shot positions of rocks around the final cavity. Shock levels in these rocks measured by *in situ* stress gauges are similar to those in the rocks underlying the breccias at Brent, decreasing from 27 GPa at the glass-lined cavity wall to 1 GPa almost 100 m from the shot point. The apparent post-shot attenuation curve is similar to that observed at Brent (Fig. 9). The curve obtained when the rocks are restored to their pre-shot position shows a uniform decrease of pressure, approximately as $P \sim R^{-2}$.

It seems probable that the basement rocks at Brent underlying the breccias have undergone similar displacements to those at Piledriver. However, it is by no means certain that the shock wave followed the same attenuation law at Brent as at Piledriver, the Brent event being at least three orders of magnitude more energetic. We therefore consider three attenuation rates which may apply to Brent: Case A in which $P \sim R^{-2}$ as at Piledriver, Case B in which $P \sim R^{-3}$, and Case C in which $P \sim R^{-4.5}$ (Fig. 9). The attenuation rates are taken as constant from 400 to 2 GPa within each model, whereas, in reality the rate may vary from this mean over parts of the pressure range. In all cases the rocks shocked at 2 GPa are considered to have undergone negligible net displacement.

This leads to three possible models for the Brent Crater (Fig. 10). In each the distance from the original ground surface to the 2 GPa level is held constant at 1600 m. The shock isobars corresponding to the respective attenuation rates are drawn as portions of spheres at depth but depart from this due to more rapid attenuation near the surface, as in the computer calculations of O'Keefe and Ahrens (1975). For comparison with these authors the case is taken of an iron meteorite impacting crystalline rock at 15 km s^{-1} giving a peak shock pressure on impact of approximately 450 GPa. The difference in equation of state between the anorthositic rock used in their analysis, and the feldspathic gneisses at Brent is probably small and has been ignored.

The total energy required for each model has been calculated from the amount of rock melted or vapourized, given by the volume within the 60 GPa contour. Following Gault and Heitowit (1963), the energy absorbed in heating is approximately 20% of the total kinetic energy and the heat of fusion is 1.2×10^3 J g^{-1}. This corresponds to 700 tons of melt per kiloton of TNT equivalent energy, which is similar to the production of melt from nuclear explosions in crystalline rock (Higgins, 1970). The final position of the interface between breccias and basement rocks prior to any modification by slumping is labelled the transient crater. The position of this interface when restored to the pre-shot position in analogous fashion to Piledriver marks the limit of the excavated cavity. The difference between these two positions indicates the net displacement of the interface due to movements in the fractured basement rocks which is expressed at the surface by uplift of rim rocks.

It is clear that case A requires very large displacements of the basement

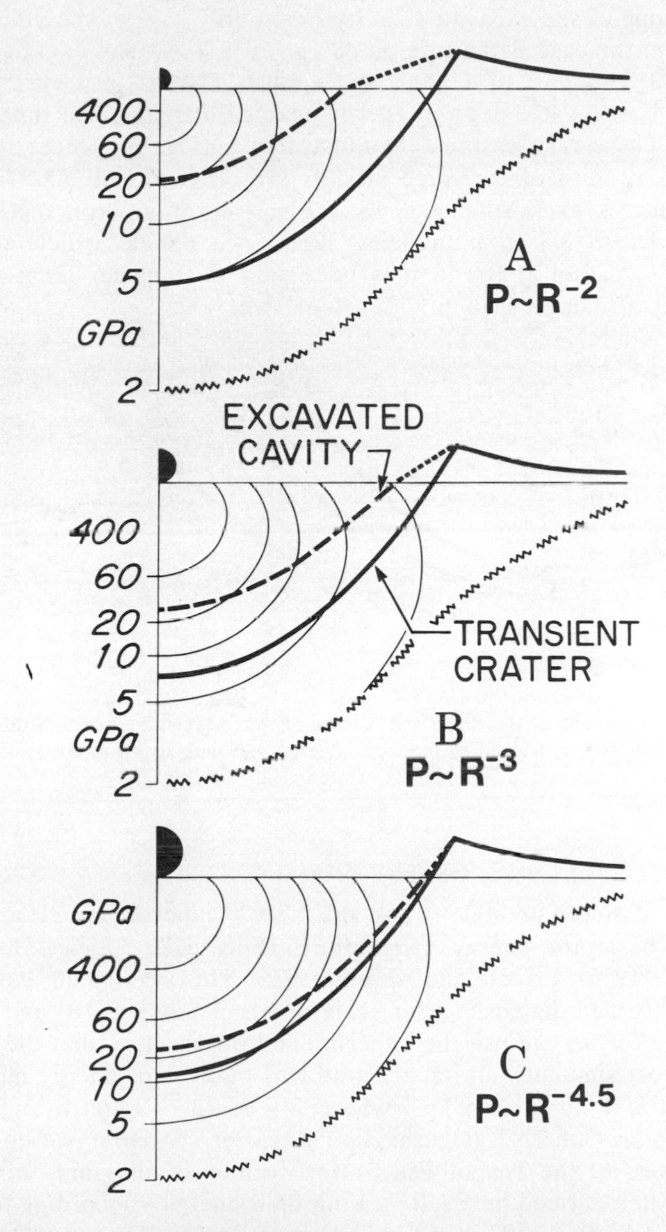

Fig. 10. Three possible models, using the Brent Crater as an example, corresponding to the three shock wave attenuation rates in Fig. 9. The black hemisphere indicates for each model the respective size of iron meteorite travelling at 15 km s^{-1} required to melt the rock volume within the 60 GPa contour. Wavy line indicates approximate limit of disrupted bedrock; solid line outlines the transient crater prior to slumping. (see text).

rocks, the volume of the excavated cavity being only 23% of the volume of the transient crater; for case B the ratio is 48%; for C 75%. The energies required are (A) 6×10^{7}J, (B) 2.5×10^{18}J and (C) 8×10^{18}J. That is, in case A twice as much material is excavated per unit energy as in case B, and four times as much as in case C. These relationships are illustrated in Fig. 11, where the slightly different crater profiles of the three models are also evident. The ratio d/D_M, where d = depth of excavation and D_M = diameter of an iron meteorite, decreases from 4.4 to 3.5 from model A to C. To decide which model best represents the situation at Brent, it is necessary to consider them within the general context of energy scaling with crater size.

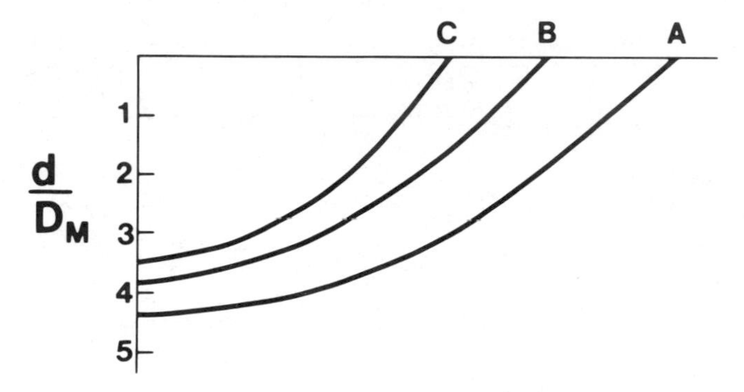

Fig. 11. A comparison of the shape and volume of the excavated cavities of the three models of Fig. 10, normalized to the same energy and with depth of excavation, d, expressed as multiples of meteorite diameter, D_M.

ENERGY SCALING

It has been standard practice to use the 1.2 kt Teapot Ess nuclear explosion crater as the basis for energy calculations (Shoemaker, 1960; Innes, 1961; Oberbeck *et al.*, 1975; O'Keefe and Ahrens, 1975). This we also do, but after first adjusting its observed dimensions for scale depth of burst (SDB) and rock type effects. For the former we use the experimental observation that the explosion crater which best simulates an impact crater of equivalent energy has a scaled depth of burst $h/W^{1/3} = 0.106$ m kg^{-1} where h = depth of burst in metres, W = energy in kilograms of TNT (Oberbeck, 1971). For the latter we compare the relative diameters of the Teapot Ess Crater, formed in alluvium, to the 0.42 kt Danny Boy Crater, formed in basalt, scaling their energies according to $W^{1/3}$. As Teapot Ess had SDB = 0.192 m kg^{-1} and Danny Boy a SDB = 0.447 m kg^{-1}, the adjustment changes the diameter of Teapot Ess from 100 to 70 m, and Danny Boy from 68 to 48 m. These are plotted on Fig. 12, where it is evident that they scale according to the relationship $E \sim D^3$, where E = energy and D = diameter. Extrapolation to the diameter of Brent passes close to the calculated energy for model A, in good accord with the argument that model A represents simple energy scaling ($E \sim D^3$) from nuclear explosions in crystalline rock.

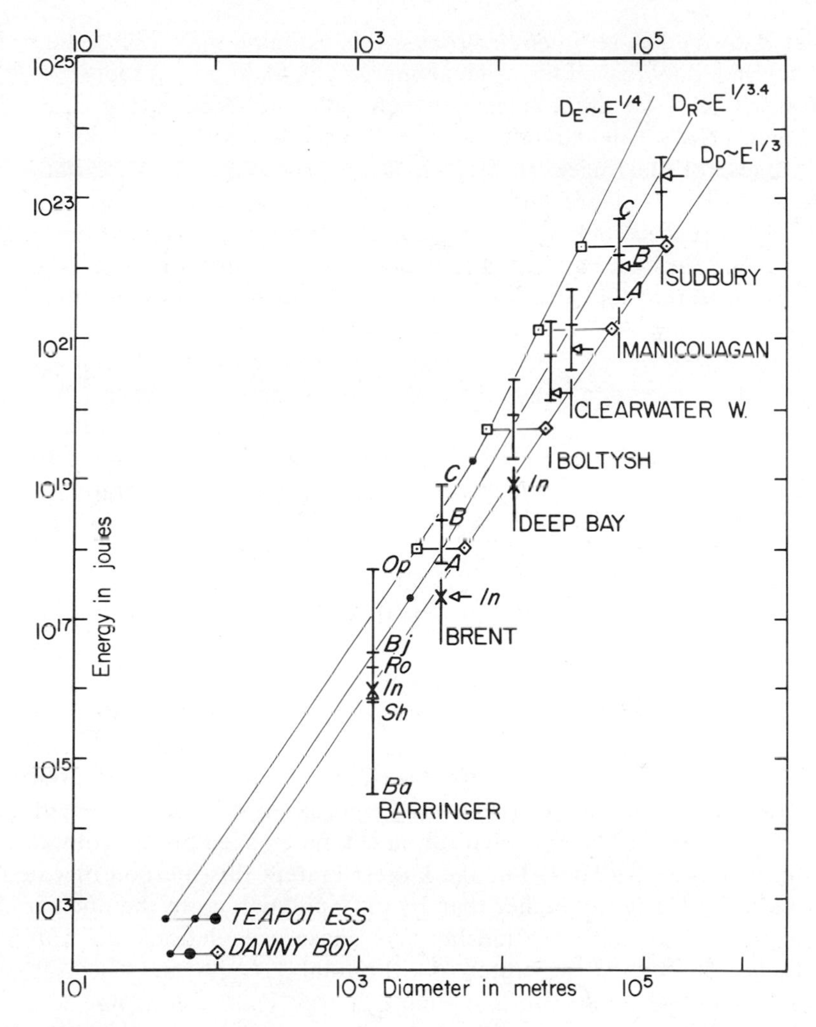

Fig. 12. Diameter versus energy plots for representative craters. The observed size of Danny Boy and Teapot Ess nuclear explosion craters is indicated by filled circles; their diameters adjusted for scaled depth of burst and media properties by dots; limits of disruption by diamonds; estimates from gravity by crosses; estimates from melt volumes by arrows; rupture limits by diamonds; excavated cavities by squares. Labelled energy estimates correspond to those in Table 1. Ba = Baldwin; Sh = Shoemaker; In = Innes; Ro = Roddy; Bj = Bjork; Op = Opik. D_D = diameter of limit of disruption; D_R = rim diameter; D_E = diameter of excavated cavity. Curves correspond to formulae (1), 2(a,b), 3(a,b).

Energy-diameter relationships for a selection of larger craters formed in crystalline rock are also shown in Fig. 12 as well as the various estimates made for the energy of formation of Barringer Meteor Crater (after Innes, 1961; Bjork, 1961; Roddy *et al.*, 1975). The estimate obtained here by extrapolation from Teapot Ess and Danny Boy with $E \sim D^3$ is slightly below that of Bjork. As

Barringer is formed in sedimentary rocks, this estimate may still be high. Bjork on the other hand increased his original energy calculation by a factor of 3.375 to allow for the material differences between tuff and sedimentary rock (Bjork, 1961). Innes (1961) calculated energies of formation using gravity data. His energy values for Barringer, Brent and Deep Bay (Fig. 12) were obtained by calculating the mass of crushed rock, and translating it into energy, using the factor 6.4 Jg^{-1} crushed rock and assuming 47% of total released energy was used in comminuting rock, as calculated for nuclear explosions (Johnson *et al.*, 1959). Innes recognized that this proportion may be a considerable overestimate in the case of impact, and indeed Gault and Heitowit (1963) calculated that only approximately 15% of the kinetic energy of impact into crystalline rock will be expended on comminution. Thus the estimates of Innes may be low by a factor of three or more.

An alternative energy calculation can be made using the volume of melted rock observed in the craters. This has been done for five of the craters in Fig. 12 using melt volumes similar to those of Grieve *et al.* (1977) and converting this thermal fraction of the energy to total energy using the factor 6×10^3 J per gram of melt (density 2.67 $g\,cm^{-3}$) as derived above. In the case of Boltysh the description of Yurk *et al.* (1975) indicates that, of a total of more than 80 km^3 of breccias in the crater, approximately 30 km^3 is melt-bearing. For this calculation we take one-third of the latter (10 km^3) to be melt. For Clearwater West and Manicouagan we estimate the observed maximum thickness of melt is approximately two-thirds the original thickness. The result of this energy calculation for Brent is not greatly different from that of Innes (1961), but for the larger craters energy estimates rise rapidly approximately in accord with gravity scaling, $E \sim D^4$, as may be expected for an estimate based on the volume of melt remaining within each crater. For the largest craters this method of calculation gives energies consistently higher than by extrapolation from the nuclear craters according to the $E \sim D^3$ relationship. As these calculations do not include materials ejected beyond the crater rim, the total energies must be still greater. There is thus a clear indication that simple energy scaling does not hold for the largest craters.

Yet another approach to energy estimation comes from an examination of the relative dimensions of the main structural elements of the craters. On general theoretical grounds (Nordyke, 1961, 1962; Gault *et al.*, 1975) the outermost limit of bedrock disruption can be expected to scale according to the energy relationship $E \sim D^3$, whereas the excavated cavity tends to conform to gravity scaling, $E \sim D^4$, at least for craters greater than 1 km across. The topographic rim of the observed crater will, in general, lie between these limits. We take the limit of disruption to be the outer limit of upheaved bedrock in the rim of simple craters and in complex craters the previously discussed outermost ring. The energy estimate is again based on the nuclear explosion craters. The diameter of outer limit of disruption, D_D, of the 68 m Danny Boy Crater is 95–100 m (Toman, 1970; Terhune *et al.*, 1970). Using the same proportion for Teapot Ess (as adjusted above for rock type and SDB) gives 98–103 m as the limit of disruption for a 5×10^{12} J event. The extrapolation from this point according to $E \sim D^3$ gives the relationship

$$D_D = 5.6 \times 10^{-6} E^{1/3} \tag{1}$$

where D_D is in kilometres and E is kinetic energy in joules. This is taken as the locus of the disruption limits for natural craters in crystalline rocks. For the particular craters under discussion the intersection of their disruption limits with this line gives a corresponding energy for each crater.

By carrying across these energy estimates, intersections are made with the rim diameters, D_R, and the estimated diameter of the excavation cavity, D_E, for each crater. When due allowance is made for erosion, the rim diameter is found to conform to the formula

$$D_R = 0.41 \times 10^{-5} E^{1/3} \quad \text{for } D_R < 2.4 \text{ km}, \tag{2a}$$

or

$$D_R = 1.96 \times 10^{-5} E^{1/3.4} \quad \text{for } D_R \geq 2.4 \text{ km}. \tag{2b}$$

Also

$$D_E = 0.24 \times 10^{-5} E^{1/3} \quad \text{for } D_E < 6.4 \text{ km}, \tag{3a}$$

or

$$D_E = 9.7 \times 10^{-5} E^{1/4} \quad \text{for } D_E \geq 6.4 \text{ km}. \tag{3b}$$

The transition from (3a) to (3b) occurs at a rim diameter of 9.2 km. These formulae do not necessarily apply to craters formed in media other than crystalline rock of normal low porosity, and will, in general, overestimate the energies where other media are involved.

Discussion

The $E \sim D_R^{3.4}$ relationship has been observed to hold for chemical and nuclear explosion craters formed in alluvium and similar materials (Vaile, 1961; Nordyke, 1962) and has been considered to express the influence of internal friction and similar forces that determine the limits of modification of the excavated cavity. Formula (2b) is almost identical to that of Oberbeck *et al.* (1975), which, when recast into the same units and applied to terrestrial craters is

$$D = 1.86 \times 10^{-5} E^{1/3.4} \tag{4}$$

However, this similarity is somewhat fortuitous. While Oberbeck *et al.* adopt Teapot Ess as their base and correct for scale depth of burst, they make no allowance for media differences and apply formula (4) to craters of all dimensions.

The energies derived for the various craters from the relative dimensions of their main structural elements are summarized in Table 1. The value adopted for Brent of 10^{18} J (240 Mt TNT) is almost five times the energy calculated by Innes

Table 1.

Crater	Rim diameter (km)	Calculated energy (J)	Basis/source
Barringer	1.2	3.0×10^{14}	Baldwin (1949)
		7.5×10^{15} (median)	Shoemaker (1960)
		9.5×10^{15}	Innes (1961)
		1.8×10^{16}	Roddy *et al.* (1975)[1]
		3.2×10^{16}	Bjork (1961)
		5.0×10^{17}	Opik (1958)
		2.0×10^{16}	This study[2]
Brent	3.8	2.06×10^{17}	Innes (1961)
		6.0×10^{17}	This study, Model A
		2.5×10^{18}	This study, Model B
		8.0×10^{18}	This study, Model C
		1.0×10^{18}	This study, calculated
Deep Bay	12	8.7×10^{18}	Innes (1961)
		2.3×10^{19}	This study
Boltysh	23	1.6×10^{20}	Estimated melt rock vol.
		4.6×10^{20}	This study
Clearwater L. West	32	7.0×10^{20}	Melt rock vol.[3]
		1.3×10^{21}	This study
Manicougan	70	1.0×10^{22}	Melt rock vol.[3]
		1.9×10^{22}	This study
Sudbury	140	2.0×10^{23}	Melt rock vol.[4]
		2.1×10^{23}	This study

[1] $E^{1/3.4}$ scaling.
[2] Assumes energy for sedimentary rocks 20% less than for crystalline rocks.
[3] Assumes $\frac{1}{3}$ of maximum thickness lost to erosion.
[4] Includes Sudbury Irruptive.

(1961) and lies midway between models A and B. Manicouagan, by this method, lies close to model B extrapolated to its dimensions according to the D^3 law. This suggests that there is a steady change in attenuation rate with increasing crater size, the indicated change being in the direction of more energy absorbed in rock deformation and less in excavation. The calculations assume no change in the proportion of energy absorbed as heat with increasing crater size. This is probably not the case, but it seems unlikely that any change in this factor would be large enough to modify significantly our conclusions.

A model for Brent intermediate between A and B gives an excavation cavity approximately 600 m deep and 1.2 km in radius, thus with depth to diameter ratio of 1:4 and volume of excavation approximately 1.3 km^3. This implies that 2.3 km^3 of bedrock under the excavated volume have been displaced as much as 500 m downward or outward to form the transient crater (Fig. 13). The relative net displacements are similar to those observed in cratering experiments using coloured sand targets (Gault *et al.*, 1968; Stöffler *et al.*, 1975), even though there are considerable differences in target material strength and porosity. The total

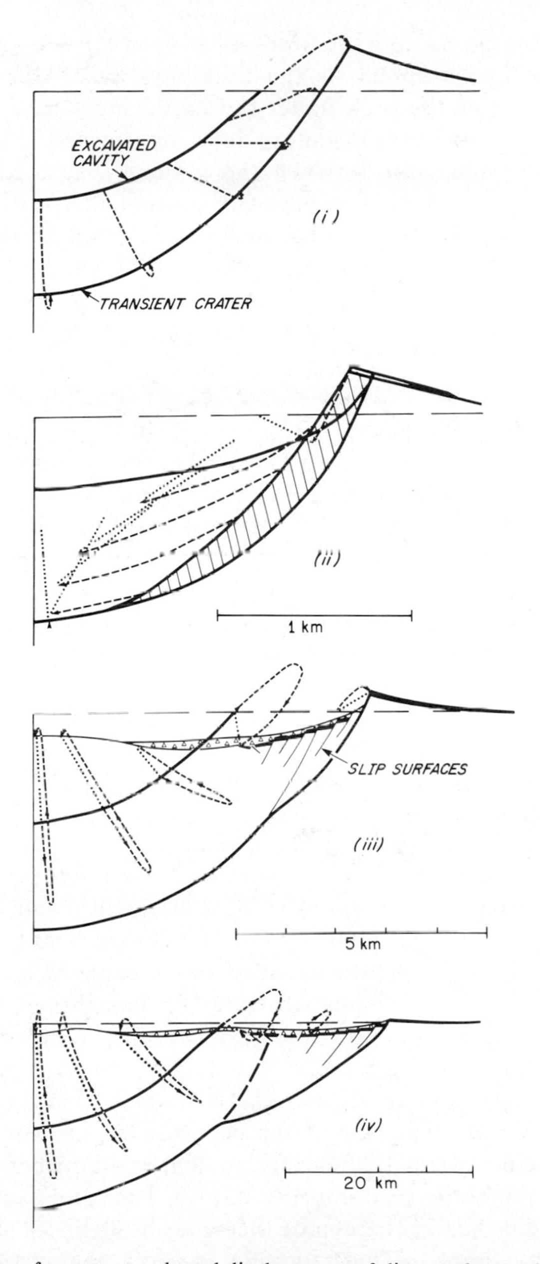

Fig. 13. Examples of apparent and total displacement of disrupted crystalline basement rocks in simple and complex craters. Dashed lines indicate approximately the total path of representative particles just below the excavated cavity. Dotted lines indicate their apparent or net displacements. (i) Transient cavity stage of simple crater formation. (ii) Modification by wall slumping of simple crater (cf. Fig. 3(iii)). Shaded volume slumps to form breccia lens. Scale is for Brent. (iii) Formation of central uplift crater. Breccia lens indicated by triangles. Scale is for Deep Bay. (iv) Formation of ring structure. Scale is for Manicouagan.

distance moved by the bedrock at Brent must have been somewhat greater than the indicated net displacement as the rock now possesses a few percent of fracture porosity. Thus the rock under the crater must have been driven down farther than its present position before relaxing upward. The formula of Pike (1967) permits a comparison between the volume displaced plus the fracture porosity induced in it and the volume of the rim of the transient crater prior to slumping. The calculation is sensitive to the rim height assigned, which can be estimated to have been 200–250 m by extrapolation from the partly eroded rim of the New Quebec Crater. The rim volume derived for Brent lies between 2 and 2.5 km^3, in reasonable agreement with the volume of displaced bedrock calculated above.

The energy calculated for Brent would be delivered by an iron meteorite with a velocity of 15 $km\ s^{-1}$ and diameter (D_M) 130 m. In the preferred model the peak pressure will begin to drop at 1.3 D_M below the contact point. This depth compares with 1.9 to 2.25 D_M calculated by O'Keefe and Ahrens (1975), and 1.8 D_M by Shoemaker (1963), who recognized that his value was probably high due to the neglect of the effect of the lateral boundaries of the projectile. Continuing excavation deepens the cavity to 4.1 D_M (Fig. 11). This additional deepening by 2.8 D_M can be compared with the enlargement by hydrodynamic flow of 3.6 D_M calculated by Shoemaker (1960, 1963). Further cavity expansion due to the displacement of bedrock below the excavated cavity gives a total transient cavity depth of 8 D_M. Immediately afterwards slumping reduces the final apparent crater depth below the original ground level to only 2.3 D_M. At Brent, compaction of breccias by later sediments has subsequently deepened the apparent crater floor to 3.4 D_M.

The crushing and frictional heat induced in the movement of bedrock blocks, which probably have dimensions of metres to hundreds of metres in larger craters such as Gosses Bluff (Milton *et al.*, 1972), will result in the formation of intensely crushed and locally melted zones which will tend to lower the strength of the rock and lubricate the late stages of movement. At a certain critical size, which for crystalline rocks is somewhat greater than that of Brent, the weakening is sufficient to allow yielding at the base resulting in uplift of the centre. The crushed and melted materials are injected into fractures that open as uplift proceeds to form pseudotachylite dikes. For craters the size of Manicouagan the initial downward displacement is of the order of 6–7 km and it seems probable that on uplift the centre rose above its present position before settling into an approximately equilibrium position (Fig. 13, iv). The total excursion of the rocks in the centre would thus have been of the order of 30 km or more in the vertical direction. The transition to ring structure in large complex craters appears to require the formation of additional surfaces of movement cutting back into the rim resulting in net downward displacement of virtually all rim rocks. However, there is apparently relative uplift of the material in the vicinity of the original rim to form an inner ring.

These concepts have some similarity to the "tsunami" model developed for large lunar basins by Van Dorn (1968) and Baldwin (1972), but differ in the appeal to frictional heat rather than direct shock-induced deformation as the main agent responsible for the weakening of target materials. A quantitative assessment is required to determine whether the model satisfies the yield criteria developed by Melosh (1977).

CONCLUSION

A general model has been developed for the mechanics of formation of terrestrial craters in crystalline rock. By scaling the limit of bedrock disruption with energy, E, the rim diameter of craters $\geqslant 2.4$ km across have been shown to obey the $E^{1/3.4}$ scaling relationship first observed empirically in explosion craters formed in alluvium. The excavated cavity accords with gravity scaling, $E^{1/4}$, for craters with final rim diameters $\geqslant 9.2$ km. The $E^{1/3.4}$ relationship appears to express the degree of crater modification by slumping of the margins, and, in larger craters, central uplift. These movements are largely controlled by rock strength and internal friction which, we suggest, is greatly decreased in the floors of large craters by frictional heating during the large initial displacements induced by shock acceleration. Heat deposited by the shock wave and from the initial depth of burial are in themselves insufficient to cause this effect, but may make a significant contribution in the case of very large craters.

The absolute energy values assigned are strongly dependent on extrapolation from the nuclear crater data. However, the relative relationships are largely independent of this base and show a reasonable consistency between energy calculations using gravity data, quantity of melt rocks and crater dimensions. According to the preferred energy estimate, approximately 75% of the volume of rock melted was ejected from the Brent Crater, whereas the proportion for Manicouagan was about 50%. In the case of Sudbury, the model indicates that all of the volume melted remained in the crater. The Sudbury melt volume estimate is crude owing to the tectonic complexity of the area, but it seems likely that there has been some addition of magma from below, as is generally accepted in current models for the Sudbury event (French, 1970; Dence, 1972b).

The subdivision of crater structure into simple and complex types in terrestrial craters parallels classifications adopted for other planetary bodies. Application of the results of the present analysis to craters on the other planets requires allowing for gravity differences and possible differences in physical properties, particularly those affected by water content. The origin of multiple rings in the largest structures remains a major question (Howard *et al.*, 1974; Head, 1977; Hodges and Wilhelm, 1976) to which the continued study of terrestrial ring structures can undoubtedly contribute. In particular, further study of the multiring morphology of the Popigai structure and its similiarities to, or differences from, Manicouagan, West Clearwater Lake and other ring structures is highly relevant.

REFERENCES

Baldwin, R. B.: 1949, *The Face of the Moon*, Univ. of Chicago Press, Chicago.

Baldwin, R. B.: 1972, The Tsunami model of the origin of ring structures concentric with large lunar craters. *Phys. Earth Planet. Interiors* **6**, 327–339.

Beals, C. S.: 1960, A probable meteorite crater of Precambrian age at Holleford, Ontario. *Pub. Dom. Obs.* **24**, 117–142.

Beals, C. S., Ferguson, G. M., and Landau, A.: 1956, A search for analogies between lunar and terrestrial topography on photographs of the Canadian Shield. *J. R. Astr. Soc. Canada* **50**, 203–211, 250–261.

Bjork, R. L.: 1961, Analysis of the formation of Meteor Crater, Arizona. *J. Geophys. Res.* **66**, 3379–3387.

Boon, J. D. and Albritton, C. C., Jr.: 1936, Meteorite craters and their possible relationship to "cryptovolcanic structures". *Field and Lab.* **5**, (1), 1–9.

Borg, I. Y.: 1972, Some shock effects in granodiorite to 270 kilobars at the Piledriver site. *Amer. Geophys. Union, Geophys. Mono* **16**, 293–311.

Brenan, R. L., Peterson, B. L., and Smith, H. J.: 1975, The origin of Red Wing Creek structure: McKenzie County, North Dakota. *Wyoming Geol. Assoc. Earth Sci. Bull.* **8**, 1–41.

Bucher, W. H.: 1936, Cryptovolcanic structures in the United States. *16th Internat. Geol. Congress Washington, D.C., 1933 Rep. 2*, p. 1055–1084.

Cassidy, W.: 1971, A small meteorite crater: Structural details. *J. Geophys. Res.* **76**, 3896–3912.

Cassidy, W. A., Villar, L. M., Bunch, T. E., Kohman, T. P., and Milton, D. J.: 1965, Meteorites and craters of Campo del Cielo, Argentina. *Science*, **149**, 1044–1064.

Chao, E. C. T.: 1967a, Impact metamorphism. In *Researches in Geochemistry* (P. H. Abetson, ed.), **2**, 204–233. Wiley, New York.

Chao, E. C. T.: 1967b, Shock effects in certain rock-forming minerals. *Science* **156**, 192–202.

Chao, E. C. T., Fahey, J. J., Littler, J., and Milton, D. J.: 1962, Stishovite, SiO_2, a very high pressure new mineral from Meteor Crater, Arizona. *J. Geophys. Res.* **67**, 419–421.

Chao, E. C. T., Shoemaker, E. M., and Madsen, B. M.: 1960, First natural occurrence of coesite. *Science* **132**, 220–222.

Chapman, C. R.: 1976, Asteroids as meteorite parent-bodies: The astronomical perspective. *Geochim. Cosmochim. Acta* **40**, 701–719.

Currie, K. W.: 1972, Geology and petrology of the Manicouagan resurgent caldera, Québec. *Geol. Surv. Can. Bull.* **198**, 153 pp.

Currie, K. W. and Dence, M. R.: 1963, Rock deformation in the rim of the New Québec crater, Canada. *Nature* **198**, 80.

Daly, R. A.: 1947, The Vredefort ring-structure of South Africa. *J. Geol.* **55**, 125–145.

Dence, M. R.: 1965, The extraterrestrial origin of Canadian craters. *Ann. N.Y. Acad. Sci.* **123**, 941–969.

Dence, M. R.: 1968, Shock zoning at Canadian craters: Petrography and structural implications. In *Shock Metamorphism of Natural Materials* (B. M. French and N. M. Short eds.), 169–184, Mono Book Corp., Baltimore.

Dence, M. R.: 1971, Impact melts. *J. Geophys. Res.* **76**, 5552–5565.

Dence, M. R.: 1972a, The nature and significance of terrestrial impact structures. *Proc. 24th Internat. Geol. Congr.* Sect. 15, p. 77–89.

Dence, M. R.: 1972b, Meteorite impact craters and the structure of the Sudbury Basin. *Geol. Assoc. Canada Spec. Paper* **10**, 7–18.

Dence, M. R.: 1973, Dimensional analysis of impact structures (abstract). *Meteoritics* **8**, 343–344.

Dence, M. R.: 1977, The Manicouagan impact structure observed from Skylab. Skylab Explores the Earth. NASA SP-380, p. 175–189.

Dence, M. R. and Guy-Bray, J. V.: 1972, Some astroblemes, craters and cryptovolcanic structures in Ontario and Québec. *24th International Geological Congress Guidebook, Field Excursion A65,* 61 pp.

Dence, M. R., Innes, M. J. S., and Robertson, P. B.: 1968, Recent geological studies of Canadian craters. In *Shock Metamorphism of Natural Materials* (B. M. French and N. M. Short, eds.), p. 339–362, Mono Book Corp., Baltimore.

Dietz, R. S.: 1947, Meteorite impact suggested by orientation of shatter cones at the Kentland, Indiana, disturbance. *Science* **105**, 42–43.

Dietz, R. S.: 1960, Meteorite impact suggested by shatter cones in rock. *Science* **131**, 1781–1784.

Dietz. R. S.: 1963, Cryptoexplosion structures: A discussion. *Amer. J. Sci.* **261**, 650–664.

Dietz, R. S.: 1964, Sudbury structure as an astrobleme. *J. Geol.* **72**, 412–434.

Engelhardt, W. von: 1972, Shock produced rock glasses from the Ries crater. *Contrib. Mineral. Petrol.* **36**, 265–292.

Evans, G. L.: 1961, Investigations at the Odessa meteor craters. In *Proceedings of the Geophysical Laboratory/Lawrence Radiation Laboratory Cratering Symposium* (M. D. Nordyke, ed.), UCRL-6438, D1-11.

Floran, R. and Dence, M. R.: 1976, Morphology of the Manicouagan ring-structure, Quebec, and some comparisons with lunar basins and craters. *Proc. Lunar Sci. Conf. 7th*, p. 2845–2865.

Fredriksson, K., Dube, A., Milton, D. J., and Balasundaram, M. S.: 1973, Lonar Lake, India: An impact crater in basalt. *Science* **180**, 862–864.

French, B. M.: 1968a, Shock metamorphism as a geological process. In *Shock Metamorphism of Natural Materials* (B. M. French and N. M. Short, eds.), p. 1–17, Mono Book Corp., Baltimore.

French, B. M.: 1968b, Sudbury structure, Ontario: Some petrographic evidence for an origin by meteorite impact. In *Shock Metamorphism of Natural Materials* (B. M. French and N. M. Short, eds.), p. 383–412. Mono Book Corp., Baltimore.

French, B. M.: 1970, Possible relations between meteorite impact and igneous petrogenesis, as indicated by the Sudbury structure, Ontario, Canada. *Bull. Volcanol.* **34**, 455–517.

Fudali, R. F. and Chapman, C. R.: 1975, Impact survival conditions for very large meteorites, with special reference to the legendary Chinguetti Meteorite. *Smithsonian Contrib. Earth Sci.* **14**, 55–62.

Gault, D. E., Guest, J. E., Murray, J. B., Dzurisin, D., and Malin, M. C.: 1975, Some comparisons of impact craters on Mercury and the Moon. *J. Geophys. Res.* **80**, 2444–2460.

Gault, D. E. and Heitowit, E. D.: 1963, The partitioning of energy for impact craters formed in rock. *Proceedings of the Sixth Hypervelocity Impact Symposium* **2**, 419–456.

Gault, D. E., Quaide, W. L., and Oberbeck, V.: 1968, Impact cratering mechanics. In *Shock Metamorphism of Natural Materials* (B. M. French and N. M. Short, eds.), p. 87–99, Mono Book Corp., Baltimore.

Grieve, R. A. F., Dence, M. R., and Robertson, P. B.: 1977, Cratering processes: as interpreted from the occurrence of impact melts. In *Impact and Explosion Cratering* (D. J. Roddy, R. O. Pepin, and R. B. Merrill, eds.), Pergamon Press, New York. This volume.

Guy-Bray, J. V.: 1972, New developments in Sudbury geology—Introduction. *Geol. Assoc. Canada, Spec. Paper* **10**, p. 1–5.

Halliday, I. and Griffin, A.: 1967, Summary of drilling at the West Hawk Lake crater. *J. R. Astr. Soc. Canada* **61**, 1–18.

Hartung, J. B., Dence, M. R., and Adams, J. A. S.: 1971, Potassium-argon dating of shock metamorphosed rocks from the Brent impact crater, Ontario, Canada. *J. Geophys. Res.* **76**, 5437–5448.

Head, J.: 1977, Origin of rings in lunar multi-ringed basins: Evidence from morphology and ring spacing. In *Impact and Explosion Cratering* (D. J. Roddy, R. O. Pepin, and R. B. Merrill, eds.), Pergamon Press, New York. This volume.

Heide, F.: 1964, *Meteorites*. University of Chicago Press. 144 pp.

Higgins, G. H.: 1970, Nuclear explosion data for underground engineering applications. *Peaceful Nuclear Explosions*. Internat. Atomic Energy Agency, p. 111–122.

Hodges, C. A. and Wilhelms, D. E.: 1976, Formation of concentric basin rings (abstract). In *Papers Presented to the Symposium on Planetary Cratering Mechanics*, p. 53–55. The Lunar Science Institute, Houston.

Howard, K. A., Wilhelms, D. E., and Scott, D. H.: 1974, Lunar basin formation and highland stratigraphy. *Rev. Geophys. Space Phys.* **12**, 308–327.

Innes, M. J. S.: 1961, The use of gravity methods to study the underground structure and impact energy of meteorite craters. *J. Geophys. Res.* **66**, 2225–2239.

Innes, M. J. S.: 1964, Recent advances in meteorite crater research at the Dominion Observatory, Ottawa. *Meteoritics* **2**, 219–241.

Innes, M. J. S., Pearson, W. J., and Geuer, J. W.: 1964, The Deep Bay crater. *Pub. Dom. Obs.* **31**, 19–52.
Johnson, G. W., Higgins, G. H., and Violet, C. E.: 1959, Underground nuclear detonations. *J. Geophys. Res.* **61**, 1457–1469.
Krinov, E. L.: 1963a, Meteorite craters on the earth's surface. In *The Moon, Meteorites and Comets* (B. M. Middlehurst and G. P. Kuiper, eds.), p. 183–207. Univ. Chicago Press.
Krinov, E. L.: 1963b, The Tunguska and Sikhote-Alin meteorites. In *The Moon, Meteorites and Comets* (B. M. Middlehurst and G. P. Kuiper, eds.), p. 208–234. Univ. Chicago Press.
Krinov, E. L.: 1966, Giant Meteorites. Pergamon Press, New York. 397 pp.
Lozej G. P. and Beales, F. W.: 1975, The unmetamorphosed sedimentary fill of the Brent meteorite crater, southeastern Ontario. *Can. J. Earth Sci.* **12**, 606–628.
Manton, W. I.: 1965, The orientation and origin of shatter cones in the Vredefort Ring. *Ann. N. Y. Acad. Sci.* **123**, 1017–1049.
Masaitis, V. L.: 1975, Astroblemes in the Soviet Union. *Sovietskaya Geol.* **11**, 52–64.
Masaitis, V. L., Mikhailov, M. V., and Selivanovskaya, T. V.: 1975, *The Popigai Meteorite Crater.* Nauka Press, Moscow (NASA Technical Translation F-16, 900).
McCabe, H. R. and Bannatyne, B. B.: 1970, Lake St. Martin crypto-explosion crater and geology of the surrounding area. *Geol. Survey Manitoba Paper 3/70.* 79 pp.
Melosh, H. J.: 1977, Crater modification by gravity: A mechanical analysis of slumping. In *Impact and Explosion Cratering* (D. J. Roddy, R. O. Pepin, and R. B. Merrill, eds.), Pergamon Press, New York. This volume.
Merrill, G. P.: 1907, On a peculiar form of metamorphism in siliceous sandstone. *U.S. Natl. Mus. Proc.* **32**, 547–550.
Millman, P. M., Liberty, B. A., Clark, J. G., Willmore, P. L., and Innes, M. J. S.: 1960, The Brent Crater. *Pub. Dom. Obs.* **24**, 1–43.
Milton, D. J.: 1968, Structural geology of the Henbury meteorite craters, Northern Territory, Australia. *U.S. Geol. Survey Prof. Paper 559-C,* C 1–17.
Milton, D. J.: 1977, Shatter cones—An outstanding problem in shock mechanics. In *Impact and Explosion Cratering* (D. J. Roddy, R. O. Pepin, and R. B. Merrill, eds.), Pergamon Press, New York. This volume.
Milton, D. J., Barlow, B. C., Brett, R., Brown, A. R., Glikson, A. Y., Manwarning, E. A., Moss, J. J., Sedmick, E. C. E., Van son J., and Young, G. A.: 1972, Gosses Bluff impact structure, Australia. *Science,* **175**, 1199–1207.
Milton, D. J. and Roddy, D. J.: 1972, Displacements within craters. *Proc. 24th Internat. Geol. Congr.*, Sect. 15, p. 119–124.
Moulton, F. R.: 1929, Report to D. M. Barringer, Lowell Observatory, Flagstaff, Arizona.
Moulton, F. R.: 1931, *Astronomy*, McMillan, New York.
Nicolayson, L. O.: 1972, North American cryptoexplosion structures. Interpreted as diapirs which obtain release from strong lateral confinement. *Geol. Soc. Amer. Mem.* **132**, 605–620.
Nordyke, M. D.: 1961, Nuclear craters and preliminary theory of the mechanics of explosive crater formation. *J. Geophys. Res.*, **66**, 3439–3459.
Nordyke, M. D.: 1962, An analysis of cratering data from desert alluvium. *J. Geophys. Res.* **67**, 1965–1974.
Oberbeck, V. R.: 1971, Laboratory simulation of impact cratering with high explosives. *J. Geophys. Res.* **76**, 5732–5749.
Oberbeck, V. R., Hörz, F., Morrison, R. H., Quaide, W. L., and Gault, D. E.: 1975, On the origin of the lunar smooth-plains. *The Moon* **12**, 19–54.
Offield, T. W. and Pohn, H. A.: 1977, Deformation at the Decaturville impact structure, Missouri. In *Impact and Explosion Cratering* (D. J. Roddy, R. O. Pepin, and R. B. Merrill, eds.), Pergamon Press, New York. This volume.
O'Keefe, J. D. and Ahrens, T. J.: 1975, Shock effects from a large impact on the moon. *Proc. Lunar Sci. Conf. 6th*, p. 2831–2844.
Opik, E. J.: 1958, Meteor impact on solid surface. *Irish Astron. J.* **5**, 14–33.
Pike, R. J.: 1967, Schroeter's rule and the modification of lunar crater impact morphology. *J. Geophys. Res.* **72**, 2099–2106.

Pohl, J., Stoffler, D., Gall, H., and Ernstson, K.: 1977, The Ries impact crater. In *Impact and Explosion Cratering* (D. J. Roddy, R. O. Pepin, and R. B. Merrill, eds.), Pergamon Press, New York. This volume.

Reiff, W.: 1977, The Steinheim Basin—An impact structure. In *Impact and Explosion Cratering* (D. J. Roddy, R. O. Pepin, and R. B. Merrill, eds.), Pergamon Press, New York. This volume.

Robertson, P. B.: 1975, Zones of shock metamorphism at the Charlevoix impact structure. Québec. *Bull. Geol. Soc. Amer.* **86**, 1630–1638.

Robertson, P. B. and Grieve, R. A. F.: 1975, Impact structures in Canada: Their recognition and characteristics. *J. R. Astr. Soc. Canada* **69**, 1–21.

Robertson, P. B. and Grieve, R. A. F.: 1977, Shock attenuation at terrestrial impact structures. In *Impact and Explosion Cratering* (D. J. Roddy, R. O. Pepin and R. B. Merrill, eds.), Pergamon Press, New York. This volume.

Roddy, D. J.: 1968, The Flynn Creek crater, Tennessee. In *Shock Metamorphism of Natural Materials* (B. M. French and N. M. Short, eds.), p. 291–322, Mono Book Corp., Baltimore.

Roddy, D. J.: 1977a, Initial pre-impact conditions and cratering processes at the Flynn Creek Crater, Tennessee. In *Impact and Explosion Cratering* (D. J. Roddy, R. O. Pepin, and R. B. Merrill, eds.), Pergamon Press, New York. This volume.

Roddy, D. J.: 1977b, Tabular comparisons of the Flynn Creek Impact Crater, United States, Steinheim Impact Crater, Germany, and Snowball Explosion Crater, Canada. In *Impact and Explosion Cratering* (D. J. Roddy, R. O. Pepin, and R. B. Merrill, eds.), Pergamon Press, New York. This volume.

Roddy, D. J., Boyce, J. M., Colton, G. W., and Dial, A. L., Jr.: 1975, Meteor Crater, Arizona, rim drilling with thickness, structural uplift, diameter, depth, volume and mass-balance calculations. *Proc. Lunar Sci. Conf. 6th*, p. 2621–2644.

Rondot, J.: 1975, Comparaison entre les astroblemes de Siljan, Suède, et de Charlevoix, Québec. *Bull. Geol. Inst. Univ. Uppsala* **6**, 85–92.

Shoemaker, E. M.: 1960, Penetration mechanics of high velocity meteorites, illustrated by Meteor Crater, Arizona. *21st. Internat. Geol. Congr. Norden*, Pt. 18, p. 418–434.

Shoemaker, E. M.: 1963, Impact mechanics at Meteor Crater, Arizona. *The Moon, Meteorites and Comets* (B. M. Middlehurst and G. P. Kuiper, eds.), p. 301–336. Univ. Chicago Press.

Shoemaker, E. M. and Eggleton, R. E.: 1961, Terrestrial features of impact origin. In *Proceedings of the Geophysical Laboratory/Lawrence Radiation Laboratory Cratering Symposium* (M. D. Nordyke, ed.), p. A-1–27.

Short, N. M.: 1970, Anatomy of a meteorite impact crater, West Hawk Lake, Manitoba, Canada. *Bull. Geol. Soc. Amer.* **81**, 609–648.

Stanfors, R.: 1969, Lake Mien—an astrobleme or a volcanic-tectonic structure? *Geol. Foren. Stockholm Forh.* **91**, 73–86.

Stöffler, D., Gault, D. E., Wedekind, J., and Polkowski, G.: 1975, Experimental hypervelocity impact into quartz sand: Distribution and shock metamorphism of ejecta. *J. Geophys. Res.* **80**, 4062–4077.

Terhune, R. W., Stubbs, T. F., and Cherry, J. T.: 1970, Nuclear cratering from a digital computer. *Peaceful Nuclear Explosions*. Internat. Atomic Energy Agency. p. 415–440.

Thomas, M. D. and Innes, M. J. S.: 1977, The Gow Lake impact structure, Northern Saskatchewan. *Can. J. Earth Sci.* **14**, 1788–1795.

Toman, J.: 1970, Results of cratering experiments. *Peaceful Nuclear Explosions*. Internat. Atomic Energy Agency, p. 345–375.

Vaile, R. B., Jr.: 1961, Pacific craters and scaling laws. *J. Geophys. Res.* **66**, 3413–3438.

Wilshire, H. G., Offield, T. W., Howard, K. A., and Cummings, D.: 1972, Geology of the Sierra Madera cryptoexplosion structure, Pecos County, Texas. *U.S. Geol. Survey Prof. Paper, 599-H*, p. H1–42.

Van Dorn, W. G.: 1968, Tsunamis on the moon? *Nature* **220**, 1102.

Wilson, C. W., Jr. and Stearns, R. G.: 1968, Geology of the Wells Creek structure, Tennessee. *Tennessee Div. Geol. Bull.* **68**, 236 pp.

Yurk, Y.-Y., Yeremenko, G. K., and Polkanov, Y. A.: 1975, The Boltysh depression—a fossil meteorite crater. *Internat. Geol. Rev.* **18**, 196–202.

Roddy, D. J., Pepin, R. O., and Merrill, R. B., editors.
(1977) *Impact and Explosion Cratering*, Pergamon Press (New York), p. 277–308.
Printed in the United States of America

Pre-impact conditions and cratering processes at the Flynn Creek Crater, Tennessee

DAVID J. RODDY

U.S. Geological Survey, 2255 North Gemini Drive, Flagstaff, Arizona 86001

Abstract—Approximately 360 m.y. ago a large, flat-floored crater, approximately 3.8 km in diameter and 200 m deep, was formed by a hypervelocity impact in what is now north central Tennessee. The initial pre-impact terrain and geologic conditions consisted of a low rolling to hilly coastal plain underlain by consolidated, nearly flat-lying, limestones and dolomites. The early Late Devonian Chattanooga Sea had innundated the general area and the impact may have actually occurred in its shallow coastal waters where depths are inferred to have been on the order of 10 m. This simple environment provided an uncomplicated set of initial conditions for the impact event.

An abbreviated sequence of events and processes are described from the entry of the impacting body in the upper atmosphere to the final late stages of cratering motions. A kinetic energy of formation of $\sim 4 \times 10^{24}$ ergs is estimated from explosion scaling, and using an assumed velocity range of 15–25 km/sec, the impacting body is estimated to have been on the order of 90–190 m in diameter, depending on the choice of density. Analog comparisons with explosion craters generated by surface and near-surface explosions indicate that Flynn Creek was formed by a comparable shock-wave set of conditions in terms of transferring all of the kinetic energy of the impacting body essentially at or near the ground surface. The interpretation is that of an impacting body which affects minimal penetration with accompanying near-total to total vaporization of itself. The exact depth of penetration and the partitioning of projectile vaporization, melting, and fragmentation depends on the choice of a stoney meteorite, such as a carbonaceous chondrite or a cometary nucleus. The impacting body appears to have been largely destroyed during its rarefaction phase since no traces are found in geochemical studies.

The cratering sequence of events and processes for Flynn Creek are inferred to represent a normal hypervelocity impact with the crater formed in 20–60 sec, depending on the scaling applied. Beyond the central uplift the brecciation and ejection processes associated with the rarefaction phase cratered to a depth of 130 m, stabilized, and continued outward to an average radius of 1760 m, terminating in less than about 60 sec after impact. During this time, over 1.2 km^3 of rock were brecciated and over 0.8 km^3 were ejected into a crudely inverted sequence forming an ejecta blanket surrounding the crater.

The results of the compression and rarefaction cratering phases were to form a broad flat-floored crater with a steeply conical, central zone of major inward and upward displacements with shatter-coned rock uplifted over 450 m. Pressures calculated for the shatter cone formation at these depths is on the order of 15–45 kbar. A possible very narrow, steeply conical, transient cavity, accompanied by a volumetric rebound phenomenon and large inward mass displacements culminating at the crater axis, appears to have contributed to the formation of the central uplift. A large bowl-shaped transient cavity is not consistent with the field and drill core data.

A breccia lens of fallback and locally disrupted target rock remained in the crater after the termination of the rarefaction phase, partly due to decreased momentum transfer at the end of the ejection process. The complex rim deformations, including the thrust faulting, terracing, and graben formation, were direct consequences of the late-stage stress conditions imposed by the decaying rarefaction field.

Erosion immediately began to degrade the crater and remove the ejecta blanket except a 2 km^2 section overlying the rim graben. Part of the rim breccia was washed back into the crater and a thin sequence of marine bedded breccias and dolomites were the first deposits on the crater floor. The rim was apparently above water for a period of time long enough to develop talus deposits, but was breached shortly thereafter and the black silty muds of the Chattanooga Sea filled the crater over the next few million years. More recently uplift and dissection of the area have exposed the once-buried crater.

INTRODUCTION

HYPERVELOCITY IMPACT CRATERING has proven to be one of the dominant physical processes affecting the surfaces and the evolution of the terrestrial planets. The extreme complexities occurring in impact events, however, continue to limit our understanding of the major physical and chemical processes associated with cratering, especially in the size range of tens of meters across and larger. More recently, critical treatments, such as Baldwin's (1963) studies on explosions and impacts, Gault *et al.* (1968) studies on laboratory impacts, and Shoemaker's (1960, 1962, 1963) studies of impact mechanics at Meteor Crater and on the moon, have greatly improved the data base for certain impact processes and provided the foundation for many of our contemporary interpretations of cratering. This has been especially true for impact craters of the simple bowl-shaped type. Parallel studies of flat-floored, central uplift, multiring, and basin impact craters, however, are considerably less advanced in terms of describing the cratering processes and deformation, primarily because of their large size and deep complex structures. This has proven particularly unfortunate in that these larger craters appear to have played a major role in the evolution of the crusts and upper mantles, and an understanding of their cratering processes is essential to any comprehensive study of the terrestrial planets (Dence *et al.*, 1977; Gault *et al.*, 1975; Gault and Wedekind, 1977; Head, 1977; Oberbeck, 1977; Wilhelms *et al.*, 1977; and others in this volume).

This paper treats one part of these broader problems in cratering mechanics by describing a preliminary outline for the cratering processes at Flynn Creek, Tennessee, a large, flat-floored, terrestrial impact crater with a central uplift. The purpose of this summary is to define; (a) the initial pre-impact conditions in terms of terrain, target media, and structure, including the possibility of impact into a shallow sea, and (b) to briefly outline the generalized sequence of events from the time of entry of the impacting body into the atmosphere through the termination of late-stage cratering motions. This abbreviated sequence of cratering events is based upon a composite of recently assembled physical constraints, including new field data, experimental laboratory impact and field explosion trials, and theoretical numerical simulation studies. A second purpose is to note several of the more critical problems yet to be solved in this type of impact cratering event with the expectation that the sequence of events and associated physical processes described here should also be generally applicable to other central uplift craters.

GEOLOGY OF THE FLYNN CREEK CRATER AND PRE-IMPACT CONDITIONS

The Flynn Creek Crater lies within central Tennessee in the extreme northeastern part of the Nashville Basin. The regional structure and stratigraphy consist of a nearly flat-lying section of limestone, dolomite, shale, and chert dipping very gently to the east away from the edge of the Nashville Dome. The sedimentary sequence is approximately 1700 m thick in this area and overlies a crystalline basement. The following abbreviated summary draws mainly upon more detailed geologic studies in Roddy (1966, 1968) and on recent morphologic, structural and parametric studies in Roddy (1977a,b).

The Flynn Creek Crater was formed in early Late Devonian time and is estimated from new cross section data to have been approximately 3830 m across and 200 m deep at the initial rim crest. Immediately after formation, the major morphologic elements of the crater included: (1) a broad flat, locally hummocky, floor surrounding a large central peak, (2) locally terraced rim segments, and (3) a surrounding ejecta blanket (Fig. 1). The major structural elements created by the impact consist of the following: (1) a breccia lens underlying the crater floor, (2) a massive central uplift, (3) an underlying zone of disruption, and (4) a complexly deformed crater rim including a major thrust block, local terraces and a large rim graben (Figs. 2, 3). The rocks involved in the deformation included the upper part of the Knox Group, Wells Creek Dolomite, Stones River Group, Hermitage Formation, Cannon Limestone, Catheys Limestone, Leipers Limestone and Sequatchie Formation, an approximately 500 m thick section of limestone and dolomite of Ordovician age (Fig. 3). Almost immediately, erosion began to remove the ejecta blanket as well as cause deposition within the crater. The terraced blocks in the rim appear to have continued to slump slightly towards the crater as evidenced by one graben fault that developed into an open fracture. The entire crater and central uplift were quickly protected from any significant erosion by the rapid disposition and complete filling by marine sediments of early Late Devonian age.

More recently, uplift and erosion during Quaternary, and probably Tertiary, time have produced the highly dissected region of steep-sided hills and valleys of the Eastern Highland Rim. Local relief, averaging 150 m along the many valley walls and floors, now contributes to the excellent exposures of the structure and stratigraphy of the Flynn Creek Crater.

The nature of the more important initial pre-impact conditions, in terms of the structure, terrain, and target media, are necessarily inferred from the post-impact stratigraphy and structural deformation. The simple regional structure, fortunately, is well exposed in this part of Tennessee and confirms that the pre-impact rocks in the Flynn Creek area were nearly flat-lying with essentially no local faulting or folding at the time of impact. The exact nature of the terrain and the possibility of the impact occurring in the shallow early Late Devonian sea, however, require more careful consideration since initial terrain extremes or a deep body of water would significantly affect the early cratering processes.

A useful line of field evidence with respect to the initial pre-crater terrain lies in the exposures along the buried rim contacts around the crater. In the southern and southeastern rim, a large graben concentric to the crater has been down-dropped over 100 m. On top of this graben, several square kilometers of a crudely inverted stratigraphic sequence of ejecta still overlie the *original* pre-

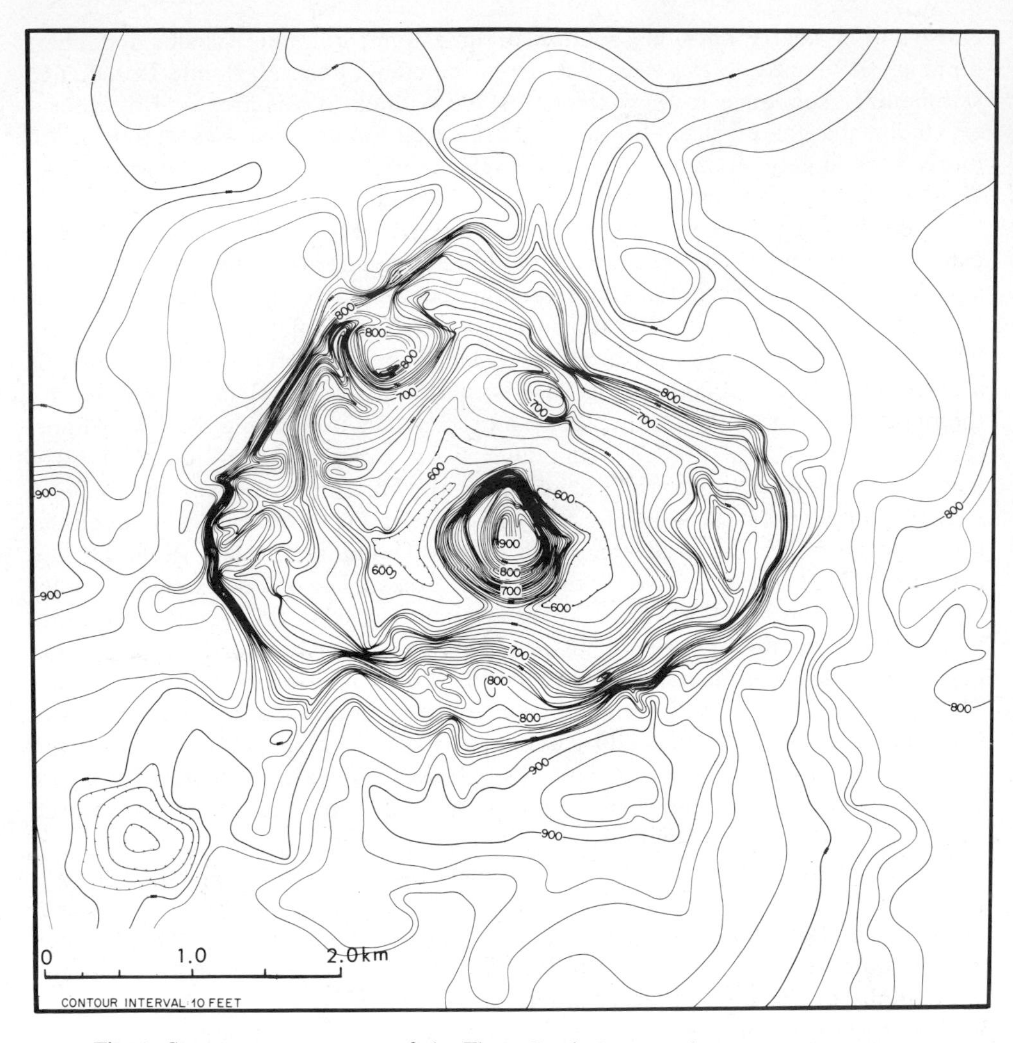

Fig. 1. Structure contour map of the Flynn Creek Crater as it appeared immediately before deposition of the Chattanooga Shale. The irregular outline of the walls is due to thick deposits of ejecta washed back into the crater and structural deformation of the rim. Modified from Roddy (1966).

impact ground surface. Two important aspects regarding the ejection of material and the initial terrain can be inferred from this graben and the surrounding field exposures of the rim. First, the presence of the large ejecta mass overlying the graben indicates that the crater was *initially* surrounded by an ejecta blanket. Second, the limited thickness and distribution of the youngest rocks involved in the cratering, and which are still exposed in the crater rim, imply a subdued type of pre-impact terrain. More specifically, the rocks underlying the original pre-impact ground surface on the graben consist of approximately a meter of basal Sequatchie dolomite which in turn is underlain by approximately 20 m of a

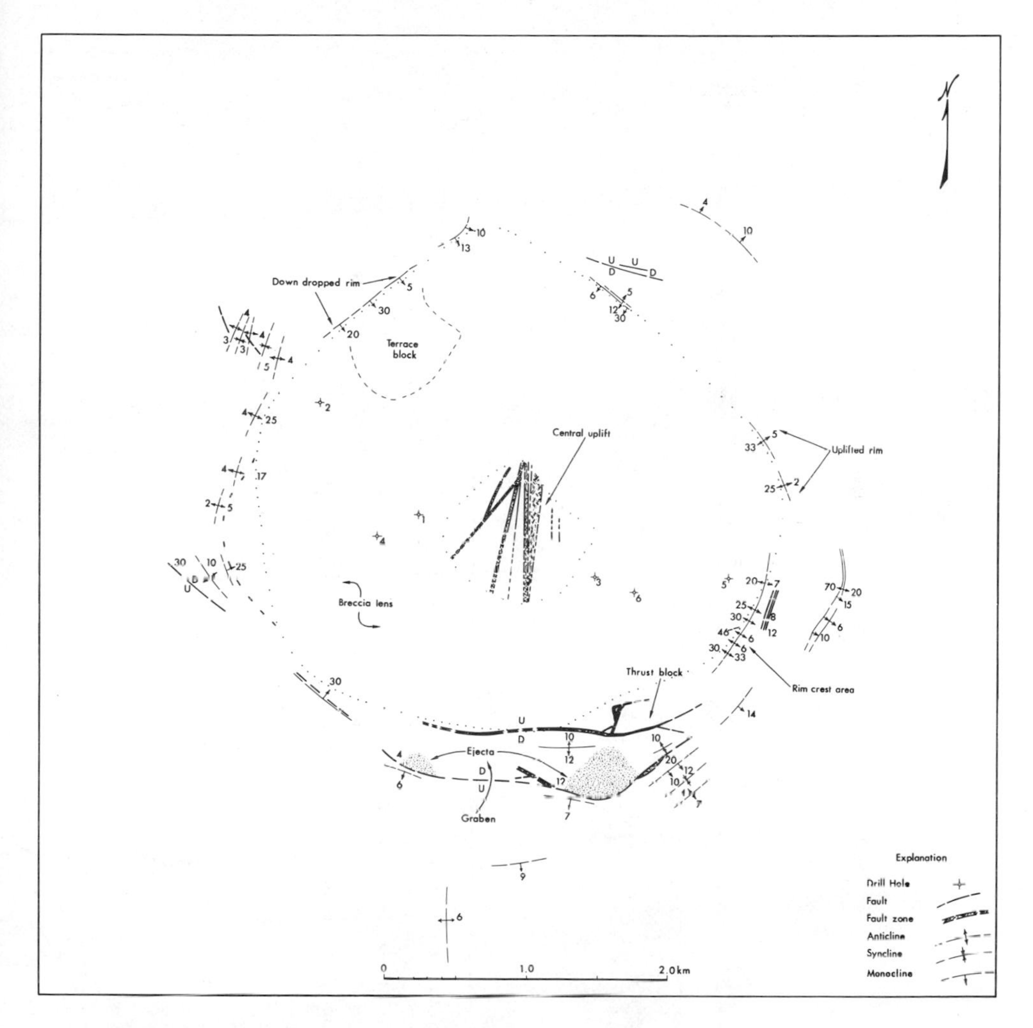

Fig. 2. Schematic map of the major structural elements formed by the impact at the Flynn Creek Crater. The location of the apparent crater walls is shown by the outer dotted line. The base of the central uplift on the crater floor is shown by the inner dotted line. Close spaced dots indicate exposed sections. Modified from Roddy (1968).

distinctive uppermost fossiliferous, dolomitic unit of the Leipers Formation, the youngest two units involved in the cratering event. The original ground surface appears to have had local relief of only a few meters since it exhibits the Sequatchie rocks along only part of its extent. The nearest continuous exposures of Sequatchie rocks are approximately 23 km north and only a few scattered areas on the outer rim retain a part of the uppermost Leipers rocks. Structural studies in these parts of the rim and the graben suggest that the original pre-impact relief did not exceed approximately 20 m and that regional lowering of the surrounding ground surface was probably less than 10 m. This limited

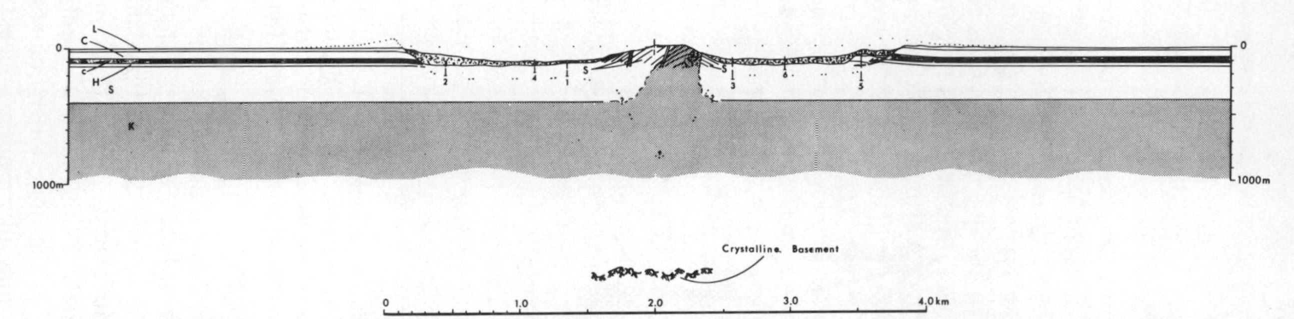

Fig. 3. Generalized geologic cross-section of the Flynn Creek Crater showing the structure immediately before deposition of sediments from the Chattanooga Sea. The numbered, vertical, dashed lines are drill hole locations projected into the plane of the cross-section. The double dots outline the approximate location of more intensely disrupted strata as inferred from drill core data and central uplift exposures. Stratigraphic symbols include: K = Knox Group; S = Stones River Group; H = Hermitage Formation; c = Cannon Limestone; C = Catheys Limestone; L = Leipers Limestone.

erosion and lowering of the rim indicate that the crater, as mapped in structure contour and outcrop today, is very close to its original gross morphologic form except for the erosion of the ejecta blanket.

The possibility of the pre-impact local relief being greater than that exposed in the graben must also be considered. However, if this were the case, one would expect a larger amount of these rocks in the breccia lens and ejecta as well as their rim erosional products to be deposited in the crater. Both of these conditions occur, but in highly limited form. Only a very small amount of fragments have been identified in the breccia lens and base of the ejecta and only a few meters of bedded dolomitic breccia derived, in part, from the upper Leipers were deposited on the crater floor. These occurrences imply that only a very limited source of these rocks was ever present, i.e., the rocks *were* as thin as the graben measurements indicate. Presumably the rim graben area, prior to impact, was a low hill or rise capped by a thin remanent of Sequatchie and upper Leipers. The other scattered locations of Leipers outcrops around the rim further suggest the pre-impact terrain was only a low rolling or very slightly hilly region with local relief on the order of 20 m or less at the time of impact.

The physical nature of the target media remains one of the final, but critical, questions in terms of the initial conditions, i.e., what was the structure and physical state of the rocks and was a standing body of water also present at the time of the impact? If the latter was the case, and the waters were moderately deep, then the impact would involve a two-layered target with the attendant terminal, but transient, result of one layer being fluid. If, instead a very shallow body of water was present, on the order of a few meters deep, the effect on an impact would be negligible.

The first part of the target media question can be easily answered from the existing field data. The type of consolidated, lithified strata present at the time of the impact is represented clearly by the limestone and dolomite section now exposed in the outcrops and core drilling at Flynn Creek. Fortunately, in terms of simplicity of pre-impact conditions and post-crater structural studies, all of these rocks were essentially flat-lying and constant in thickness throughout the area at the time of impact.

The question of an impact into a standing body of water is less definitive, but it is clearly linked with the history of the *first* sediments deposited on the crater floor and on the ejecta overlying the rim graben. These deposits show that erosion was active immediately after the formation of the crater and began to degrade the flanks of the central peak, crater walls, and ejecta blanket. As noted earlier, one of the major effects of the post-crater erosion was to remove the ejecta blanket, except for that protected in the downdropped rim graben. The thick mass of very crudely lineated breccia locally overlapping the crater walls and terrace blocks strongly suggests the inner part of the ejecta blanket was redeposited into the crater very irregularly as a chaotic mass on top of the breccia lens (Fig. 3). Presumably, the remainder of the ejecta blanket was eroded from the rim and carried away from the crater. The possible effects of an expanding shallow tidal wave created by the impact and returning after cratering are considered minimal for such shallow water depths, although this remains to be proven.

Another result of the erosional processes leads to the deposition of a variety of types of sediment in the crater and on the rim grabens. The important yet puzzling aspect of these rocks, however, is that those on the crater floor are definitely of marine origin whereas those on the higher rim graben do not appear related to marine processes. No lake or playa beds are present in either exposed sections or in drill cores anywhere on the crater floor. Instead, the first crater floor deposits are related to marine waters clearly indicating that a sea was in the area. Isolated subareal-like talus deposits on the rim graben, however, imply that the sea was quite *shallow* and *below* the uplifted rim area.

Turning first to the crater deposits, a complex series of early deposits overlie the chaotic breccia lens on the floor and exhibit a succession of stratigraphic changes. The physical and chemical details of this sequence are too lengthy to describe here; however, the general nature of these deposits is characterized by a thick non-bedded to very crudely bedded breccia transitional into overlying thinner, well bedded, dolomitic breccia which, in turn, is transitional into an overlying very thin cross-bedded dolomite. The lowest of this sequence, the first material to be deposited on the breccia lens in the crater, consists of non-bedded to very crudely lineated to bedded breccias that are similar in fragment lithologies and sizes to the ejecta on the rim graben and in the breccia lens. The unit appears prominent in non-bedded form adjacent to and locally overlying the crater walls and terraces. The distribution and character of this unit indicates that it was partly or all washed back into the crater from the inner part of the ejecta blanket on the rim. A distinctive point about this unit is that its matrix is mainly a dolomite identical to that in the overlying bedded breccias and bedded dolomite. Consequently, this first and lowest unit to be deposited appears to be the basal part of the entire post-crater depositional sequence. The only source of the dolomitic matrix is the middle and upper Leipers units exposed around the tops of the crater walls. It is significant that the ejecta blanket would have initially protected the upper Leipers dolomite unit from erosion; however, after the ejecta was stripped from the area, the upper and middle Leipers dolomitic units would have been exposed to erosion and could have provided the dolomite matrix for the ejecta washed back into the crater. This would be the proper timing necessary to supply dolomitic-rich waters for the upper part of the depositional sequence which has the dolomitic matrix in the crater. These units, including the lower non-bedded breccia, appear as thick as 50 m near the walls and thin to a few meters on the crater floor near the flanks of the central uplift.

The bedded dolomitic breccia and bedded dolomite are thickest on the lowest parts of the crater floor and thin out entirely part way up the crater walls. The bedded dolomite, up to 3 m thick locally, is the last unit to be deposited in the crater that includes very fine fragments of the underlying breccia and fragments from the upper Leipers rocks. The important point regarding these last two units is that they both contain *marine* fossil fragments of early Late Devonian age (Huddle, 1963) and consequently were deposited with *access* to the marine sea water in the area. A second critical point is that the specific marine fossil fragments in the bedded dolomite breccia and bedded dolomite are identical to those in the basal Chattanooga Shale Formation which has an extremely wide-spread distribution over several states and lies in conformable contact immediately on top of the bedded dolomite. A third critical point is the distinct change in lithology from the dolomite to the black Chattanooga Shale sediments, a transition that takes place vertically and very abruptly over a centimeter or two. Obviously, the extremely wide-spread black muds of the Chattanooga Sea were not introduced immediately onto the floor of the crater since other deposits have been identified, yet the same marine conodonts in the basal Chattanooga *were* included, at least, in the earliest bedded dolomitic breccia on the crater floor. This suggests that waters of the Chattanooga Sea were in the immediate area at the time of impact but were not deep enough to flow directly over the crater rim and ejecta blanket. Instead, it appears that the marine waters carrying the microscopic conodont fragments flowed or were initially filtered through the ejecta blanket and rim into the crater at a reduced rate such that the coarser black silty muds were initially deposited outside the crater. It is also likely that the initial environment on the crater floor was sufficiently oxidizing to eliminate the earliest Chattanooga organic material that might have been present in the waters that deposited the first dolomitic matrixes. In any event, the earliest deposition of dolomitic matrix-type units proceeded such that it received the eroded dolomite, partly in solution and partly as clastics from the upper crater walls and rim and incorporated it into a bedded dolomitic breccia and thin cross-bedded dolomite, with marine fossil fragments. Immediately thereafter, the black silty muds of the Chattanooga Shale appears to have spilled over the crater rim to eventually fill the crater over the next few million years. The conclusion one draws is that of a shallow sea with abundant black silty muds similar to that described in detail by Conant and Swanson (1961) that did not immediately flood the crater, perhaps because of the barrier of the uplifted rim and the 100 m or so thickness of ejecta blanket. After a limited period of probable wave and other types of erosion, the ejecta was removed and the black silty muds were rapidly deposited over the crater floor, walls, and rim. The details of the crater and the Chattanooga Sea interaction certainly require further study; however, it is abundantly clear that the sea *was* definitely in the area and *did* affect the earliest deposits on the crater floor.

Another line of evidence regarding the *depth* of the Chattanooga Sea at the time of impact lies in an explanation of *talus-like* deposits at the base of a cliff formed by the rim graben. This ancient talus has the character and composition of subareal deposits with no apparent marine influence of its matrix chemistry and no black, silty, mud additions. Since the presence of the Chattanooga Sea in the immediate area has been established, it would appear that the evidence of no direct communication of the talus with the sea indicates that it was formed *above* the local water level. The local relief between the talus deposits and several regions on the rim that appear to have been breached by early stream cutting (Fig. 1) was on the order of 15 m or less. This would imply that the depth of the surrounding sea was less than 15 m, a value consistent with the conditions inferred by Conant and Swanson (1961) for the coastal edges of the Chattanooga Sea. This shallow sea depth would still allow local wave action to remove the ejecta, flow over the stripped rim, and deposit marine sediments on the crater floor. In any case, the overall impression remains that of a very shallow sea, a few meters or so in depth, in this area *at* the time of impact.

The conclusions drawn from these various lines of evidence present a reasonably consistent picture of pre-impact conditions in the Flynn Creek area consisting of a low hilly to rolling coastal plain innundated by the very shallow marine waters of the Chattanooga Sea of early Late Devonian Age. The actual impact event may have occurred *in* these very shallow waters, but the depths were apparently only on the order of approximately 10 m, the minimal effects of which are discussed in the following section.

IMPACT CRATERING PROCESSES AT FLYNN CREEK

This section briefly describes a generalized set of impact conditions and cratering sequences and processes for thc formation of the Flynn Creek Crater. These interpretations are based upon the critical initial conditions derived from the field data listed in the previous section and upon selected experimental impact explosion and theoretical cratering studies. The cratering sequences and the associated physical processes are, of necessity, summarized here in a qualitative, phenomenological format since theoretical numerical simulations have yet to be completed. The interest in selecting quantitative values in this section is directed towards identifying future problem areas that will utilize calculational approaches. The choice of Flynn Creek as a representative of flat-floored craters with central uplifts was based upon the following: (1) availability of extensive field and laboratory data (Roddy, 1966, 1968, 1977a, b), (2) simple initial pre-impact conditions, (3) moderately well preserved crater with limited erosional degradation, (4) well-defined structural deformation in flat-lying target media, (5) drill data on deep structure, (6) large-scale experimental explosion data on central uplift craters (Roddy, 1976, 1977a, b), and (7) theoretical computational data on explosion craters with central uplifts (Ullrich *et al.*, 1977). This section includes: (a) Physical parameters inferred for impacting body, (b) Atmospheric passage of the impacting body, and (c) Impact cratering sequences and processes for the Flynn Creek Crater.

(*a*) *Physical parameters inferred for impacting body*

The physical parameters of the impacting body which are of concern include: (1) the energy, (2) velocity, (3) mass, (4) density and strength, and (5) dimensions. The choice of specific values for each of these parameters include a number of assumptions described below in that field and laboratory studies have as yet detected no traces of the impacting body.

The kinetic energy of an impacting body cannot be estimated *directly* with any real level of confidence from existing impact data or theoretical calculations, at least for such large-scale events as Flynn Creek. The fundamental reason for this is that there simply have been no techniques developed to experimentally *verify* calculations at such a scale. The traditional approach is to *estimate* the *energy of formation* by scaling from explosion cratering data. The argument in its simplest form is geometric, i.e., given a known explosion crater diameter (D_1) and its energy of formation (E_1), one can calculate the energy (E_2) of another crater with a known (D_2) using,

$$D_1 = D_2(E_1/E_2)^{1/3},$$

which expresses cube-root scaling. In solving this equation it is convenient to put D in metric units and E in ergs where 1 kiloton of TNT $= 4.2 \times 10^{19}$ ergs (Glasstone, 1964). As long as gravitational and other effects are negligible, cube-root scaling is reasonably accurate; however, as crater sizes increase into the tens-of-meters range new exponents have been found necessary. Cooper (1977), Cooper and Sauer (1977), Chabai (1965, 1977), Roddy *et al.* (1975), Saxe and Del Manzo (1970), Vaile (1961), Vortman (1970, 1977) and others, have discussed the choice of 1/3.4 root scaling to fit the larger craters in the kilometer size range. The argument to use 1/3.4 root scaling is based upon the best empirical fit to craters larger than a few tens of meters and is adopted for use in this paper. Two other types of scaling explained in Roddy (1977b) are also employed, volume and equivalent length factor scaling. These three types of scaling give an average energy of formation of approximately 4×10^{24} erg, i.e., on the order of 100 megatons of TNT energy, scaled from the Snowball explosion crater which has morphology and structural deformation very similar to Flynn Creek (Roddy, 1977a, b). The details of this type of scaling in terms of the critical choice of correct *true* diameters and associated calculations are given in Roddy (1977b). A simple comminution estimate of fragment crushing energies also gave 10^{24} ergs (Roddy, 1968), further suggesting that the value of 10^{24} ergs is reasonable using scaling of dynamic explosion energies.

The energy of formation determined by the explosion scaling is assumed to be approximately equal to the kinetic energy of the impacting body. There are no theoretical or experimental treatments that show this assumption to be rigorously correct at the kilometer diameter sizes, consequently, the author interprets the energy of formation listed above as an approximation correct to the nearest order of magnitude in terms of the existing uncertainties in our present energy scaling techniques. Chabai (1977), Cooper (1977), Cooper and Sauer (1977), Gault and Wedekind (1977), Killian and Germain (1977), Kreyenhagen and Schuster (1977), Trulio (1977), Vortman (1977), and others discuss this more fully in this volume. An example of the problems that still occur in scaling between explosion craters is typified by the difference in cratering efficiency between nuclear and high explosive (HE) chemical cratering, in which HE charges are twice as efficient as nuclear charges in excavating a crater. This is due, in part, to the nuclear release of other types of energy, such as radiation, that do not effectively contribute to cratering and are not present in HE processes (Cooper, 1977). In any event, the value of 10^{24} ergs is adopted in this paper as a reasonable approximation of the minimum kinetic energy of the Flynn Creek impacting body.

There are no direct techniques, at present, to determine the actual velocities of bodies which formed impact craters. The total velocity range permissible for the earth, however, is known to lie between approximately 11.2 and 72 km/sec with a maximum of ~45 km/sec for prograde impacts and ~72 km/sec for retrograde impacts. Bodies that might be traveling at less than 11.2 km/sec would be accelerated to at least that value if gravitationally attracted by the earth alone, thereby giving a lower impact limit. Bodies moving faster than approximately 45 km/sec are unlikely impact candidates in that they are moving with sufficient velocity to escape the gravitational field of the inner solar system. Shoemaker (1977a, b), Shoemaker and Helin (1977), Wetherill (1977), Buchwald (1975) and others, list velocities in the approximate range of 15–40 km/sec for bodies kilometers in diameter and crossing the earth's orbit. Dohnanyi (1972) lists 15 km/sec as a mean for some 5000 velocity observations for smaller objects entering our atmosphere, and Shoemaker (1977b) calculated 24.6 km/sec as the mean velocity for the kilometer-size Apollo asteroid bodies now crossing the earth's orbit. The impact velocities adopted as *representative* values for the calculations in this paper are 15 and 25 km/sec.

The next step, calculating the mass, is accomplished simply by substituting the kinetic energy estimated from explosion scaling and the representative velocity(s) in,

$$KE = \tfrac{1}{2}mv^2,$$

in cgs units. The masses calculated for 15 and 25 km/sec are shown in Table 1.

As noted earlier, field and geochemical studies have detected no traces of the impacting body (Roddy, 1966, 1968). This is not surprising if one looks at the normal sparse distribution of iron fragments at Meteor (Barringer) Crater, Arizona, which is relatively young at less than 30,000 yr old with limited erosion. Furthermore, the sizes of bodies creating impact craters are actually quite small with respect to the total region cratered and over which the body fragments are scattered after impact, i.e., volumetric dispersion is extreme for the fragmented, melted, and partly vaporized impacting body. Chemical degradation of iron fragments during 360 m.y. of time at Flynn Creek would

Table 1. Masses, volumes, and diameters of impacting body assuming spherical shape and impact velocities of 15 and 25 km/sec and densities of 3.3 and 1.0 g/cm³.

Impact energy (ergs)	Impact velocity (km/sec)	Mass (metric tons)	Volume for 3.3 g/cm³ body (m³)	Diameter for 3.3 g/cm³ body (m)	Volume for 1.0 g/cm³ body (m³)	Diameter for 1.0 g/cm³ body (m)
4×10^{24}	15	3.6×10^6	1.1×10^6	127	3.6×10^6	190
4×10^{24}	25	1.3×10^6	0.4×10^6	90.5	1.3×10^6	135

also be expected to severely if not totally eliminate the small number of fragments remaining after impact, *if* the impacting body were iron.

Arguments discussed by Roddy (1966, 1968, 1976, 1977a, b), however, suggest that the body was not an iron but more likely a stoney meteorite or a cometary mass. Briefly, this is inferred from morphological and structural comparisons with certain high explosive craters generated by surface and near-surface explosions, with the interpretation that the lower density types of impacting bodies would experience relatively little penetration while delivering a large shock wave at the surface and would more effectively volatilize leaving little trace of their material. This, in effect, is suggesting such an impact can be equated in a general sense with the height of burst (HOB) effects of certain surface and near-surface explosion craters (Roddy, 1976, 1977a,b). The permissibility of carbonaceous chondritic and cometary bodies is further supported by the studies of Wetherill (1975a,b, 1977) and Shoemaker (1977a,b) for earth-orbit crossing bodies. Densities in these cases range from 3.3 g/cm^3 for the carbonaceous chondritic stoney meteorites to less than 1.0 g/cm^3 for cometary nuclei. Table 1 gives the volumes and diameters, assuming a spherical shape, calculated for the Flynn Creek event for impacting bodies with densities of 3.3 and 1.0 g/cm^3. The values in Table 1 should be considered as *representative* values only, especially in view of the range in the possible velocities.

The assumption of a specific type of impacting body for calculations, as noted above, generally *implies* choosing an accompanying density and strength based upon *other* bodies of that type for which physical properties data are available. For example, if an iron meteorite is assumed, or known, to have been the impacting body, it generally follows that there is a tacit assumption that this meteorite also had the strength of iron. The problem created in such assumptions is that widely divergent *effective* bulk densities and strengths are really to be expected for large impacting bodies with different histories and angles of atmospheric passage.

At the present time, there are no *a priori* methods to determine such values, although *effective* densities and strengths remain of vital interest in both phenomenological and calculational approaches. This becomes especially important when equating concepts of line penetration bursts of an impacting body to single explosion equivalent heights of burst (HOB) or depths of burst (DOB). For example, if one holds the energy constant, by varying the mass, for two iron bodies traveling at two different velocities between 11 and 40 km/sec, the very low velocity, high strength iron will penetrate deeper (DOB) than the very high velocity, high strength iron. If instead the mass is held constant, the very high velocity iron will penetrate deeper (DOB) than the very low velocity iron. If the same conditions are imposed on two coherent stoney meteorites, the penetrations are distinctly less than with the iron bodies and also exhibit the same relative order of penetration depths as the irons in terms of high and low velocities, but they have significantly less variation in penetration depths between themselves under the various conditions. Trulio (1977) discusses these points in greater detail in this volume. These changes to shallower DOB by lower density bodies can lead to impacts that effectively transfer the bulk of their kinetic energy to the target at or near its surface. Very large, high velocity bodies that volatize quickly may actually have their effective center of kinetic energy above the surface, i.e., a HOB (Roddy, 1976, 1977a).

(b) Atmospheric passage of impacting body

Large bodies on the order of 100 m across, as calculated for Flynn Creek, are exposed to several important dynamic processes during their passage through the earth's atmosphere that can affect their final physical state prior to impact. The basic sequence of events consists of: (1) the body entering the upper atmosphere, (2) sweeping out a column of air, (3) creating a large bow shock wave, (4) decelerating, (5) losing mass by ablation, and (6) impacting the earth's surface with possible strength degradations due to the above processes including differential shock pressure buildup, heating, rotation, and fragmentation.

The entry of the Flynn Creek impacting body into the earth's atmosphere was not seriously affected until it descended below about 100 km altitude, assuming the same atmospheric density structure as we have today. Bodies of this size will sweep out a narrow column of air and form a large conical atmospheric bow shock wave initiating from the frontal edge (Martin, 1966). The physical shape of the bow wave will depend on the entry angle, velocity, and shape of the body with continuous changes occurring as the descent progresses into denser air. A compressed, heated zone of air will form the leading edge of the bow shock wave with a region of reduced pressure behind and above the wave and impacting body. If the Tunguska cometary event (Krinov, 1966) can be used as an example, excluding the contribution by the high altitude shock detonation, the bow wave would be expected to travel with the Flynn Creek body to the earth's surface. The leading edge of the bow wave adjacent to the impacting body would have nearly the same velocity as the body itself with a continuous decrease in bow wave velocity moving away from the atmospheric penetration axis. At Tunguska, the bow wave and high altitude shock detonation effects leveled a large forested area 60–80 km across. At Flynn Creek, such a bow wave would be expected to assist the air blast shock wave created on impact in stripping the target surface, or setting it in motion if it were a shallow body of water. One of the physical effects of such a process would have been to nearly instantly pre-stress the upper target material beyond the immediate impact area. This would be similar in gross effect to the air blast target loading in surface explosions (Needham, 1971). This pre-stressing normally does not seriously affect the cratering motions in large-scale explosion trials because the stress levels are several orders of magnitude lower than the direct induced cratering stresses (Cooper, 1977). The stress levels in an atmospheric bow shock wave accompanied by the possible impact air blast, however, have not been calculated and their surface pre-stress effects remain to be determined. A disturbing point was raised in this area by Maxwell and Moises (1971) in a numerical simulation which utilized a very low magnitude pre-stressed (although not from airblast) target surface below a large exploding charge. The effect appears to enhance certain cratering flow regions, although the results are still uncertain. The times of arrival of the outer parts of the bow wave on the earth's surface with respect to the earliest ejection of cratered material is also an undetermined factor. An early dismissal of possible pre-stress and ejecta transport effects over the total cratered and ejecta transport area may be premature at this time.

The interactions of the collapsing column of air above the impact also find analogs in surface explosion trials which have an atmospheric fireball expansion phase and an atmospheric collapse phase. Their subsurface cratering effects are generally negligible in these phases except for the entrapment of ejecta fines, the result of which can be a complex inward flow of lofted fine debris back towards the crater area late in the cratering ejection phase. The expansion of the impact shock-induced air blast into the low pressure region created by the swept-out air column above the impacting body also is a subject that has not been studied in terms of ejecta transport. Again, this is an area requiring numerical simulations to determine the possible interactions.

Pre-stressing the impacting body prior to impact becomes an important consequence of the high pressure atmospheric buildup at the leading edge. Transient stresses in the leading edge of the body can be raised in a matter of seconds to the 2–4 kbar level (Trulio, pers. comm., 1977) with the trailing edge experiencing essentially little pressure change. This extreme stress gradient, its rapid formation, and continuously changing stress levels across a 100 m diameter body, must certainly affect the material properties of the body. In the simplest condition, a non-rotating body could reach the earth's surface with a rapidly building high pressure gradient that could seriously degrade its *effective* strength. A rotating body, due to a continuously changing position of the leading edge, would experience a much more severe changing stress history, in seconds, with the possible effects of early breakup of even a larger body. The effect of such pre-stressing on modifying the free surface disintegrations by possible enhanced rarefactions has not been studied. The initial megabar stress level in the body at impact time, admittedly greatly exceeds the 2–4 kbar level in descent; however, if the pre-stressing causes the body to create a slightly different-shaped shock wave in the target or affects the impact energy transfer rates, then this inflight process is important. For example, a low velocity, irregularly shaped body with zones of weakness and forced into rotation in the descent, could be expected to deliver less than a well-formed initial target shock front. On the other hand, atmospheric drag and nonlaminar flow would tend to abruptly slow rotation with an attendant increase in internal strain. The leading edge would initially experience a monotonically increasing rate of fractional heating with an accompanying ablative mass loss. Flash-frothing and shedding of the accumulated thin froth would account for a very limited mass loss for a large body. The low heat conduction of these body types would create only a thin, very hot, outer skin on the leading edge, again a pre-stress condition, although probably of limited importance.

The physical conditions affecting the deceleration and ablative mass loss are related principally to the initial velocity and angle of entry into the atmosphere, and body shape. A body traveling at 15 km/sec will experience an atmospheric transient time, where deceleration and ablation became important, ranging from approximately 7 sec for a vertical entry to over twice that time for very low angle entries. Shoemaker (1962) describes probability arguments for mean entry angles averaging about 45°, with vertical and very low angles both being lower in frequency. A 15 km/sec body, therefore, entering at 45° would remain in the denser atmosphere on the order of 10 sec with the major deceleration and ablation occurring in this time. During this period the kinetic energy of the body is transferred into: (1) vaporization of body and air, (2) surface melting, (3) heating of the body, (4) heating and translation of air in front of the body, (5) formation of shock waves, (6) probable rotation of the body, and (7) possible fragmentation (Buchwald, 1975).

The critical problems of deceleration and ablative mass loss have been described in detail by Baldwin and Sheaffer (1971), Buchwald (1975), Heide (1964) and others for smaller bodies that are strongly affected by atmospheric passage. For bodies on the order of 1000 metric tons and over several meters across, however, the deceleration and mass loss for iron meteoritic are less than a few percent, except for very low angle entries. For bodies the size of Flynn Creek, extrapolations of the existing data from smaller iron bodies would indicate less than 1% losses in velocity and mass. If the Flynn Creek body was a stoney meteorite, such as a carbonaceous chondrite or cometary nucleus, mass loss with volatization would have been 10–40% greater than for iron (Heyda, pers. comm., 1973); however, this is an area that again requires a calculational study for rigorous answers. Certainly, cometary masses do enter our atmosphere as evidenced by the Tunguska event (Krinov, 1966) even though it fragmented at some 5 km above the earth's surface. If Flynn Creek *was* formed by such a body, it may be indicating the approximate lower threshold for cometary body sizes to accomplish a complete atmospheric transient and cause a cratering event. The empirical point is that Flynn Creek appears to be at the approximate lower size level at which central uplift craters are formed on earth. Perhaps lower density bodies less than about 200 m do not survive the atmospheric passage, as with Tunguska. The processes described earlier that contributed to decreasing the strength of the impacting body would play a more serious role of degradation in a mass presumably as weak as a comet.

The present uncertainties regarding the actual atmospheric passage of the impacting body detract in no way from the fact that the impact *did* occur regardless of the descent history. For example, if the projectile was extremely irregular in shape and severely fragmenting, the result was still to form the Flynn Creek Crater. In this sense the questions raised are somewhat academic. On the other hand, these points became critical when one tries to construct the detailed initial conditions necessary for numerical computer calculations of the shock wave history in both the body and target limestones. For example, if an impacting body fragments just prior to impact with particles separated by only a few meters, it would then transfer its total kinetic energy over a wider impact area. The result would be to develop a shock wave front with a significantly different shape and energy distribution than a line penetration source from a single, unfragmented body. This condition might more nearly approximate the wider areas affected by physically large high explosive charges, whereas the single point or line source would be more similar to nuclear charges, in terms of initial physical extent. The initial energy distribution function of the projectile across the impact surface and its time history, *effective* equation-of-state and other conditions will eventually have to be decided with the best pre-impact condition data available before proceeding to a full cratering calculation.

(c) *Impact cratering sequences and processes for the Flynn Creek Crater*

A review of the initial condition indicates that the Flynn Creek impact event was formed on the order of 360 m.y. ago in flat-lying carbonate rocks. The impact environment appears to have been a low rolling coastal plain, possibly in shallow waters on the order of 10 m or less deep. The impacting body is inferred to have been a stoney meteorite or cometary nucleus traveling on the order of 15–25 km/sec, for the sake of this discussion.

The following outline of impact cratering sequences and processes is determined or inferred to have operated during the formation of the Flynn Creek Crater. In simplest form, a hypervelocity impact of the type we are concerned with consists of a nearly instantaneous transfer of kinetic energy of motion into an intense compression and heating of both target and projectile by extremely high pressure shock waves. These conditions, continuously transitional in location and time, evoke further rapid changes in the shocked material by violent relaxation through complex and rapidly changing rarefaction fields. The result is the nearly instantaneous formation of a crater commonly surrounded by ejecta. Excellent treatments of laboratory impact cratering processes have been summarized by Gault *et al.* (1968) and Riney (1965), and the classic treatment of impact mechanics at Meteor (Barringer) Crater, Arizona is outlined in Shoemaker (1960, 1963). The terms projectile and impacting body are used interchangably in this section and the term target is used to indicate the limestone and dolomite bedrock involved in the cratering. No pretense is implied that the following brief outline is a comprehensive treatment of the total impact processes that occurred at Flynn Creek. The antiquity of the event, the lack of total field exposures, and the absence of numerical simulations preclude such a rigorous compilation at present. The interest in this section is instead directed toward an interpretation of the general sequence of events for a central uplift crater with emphasis on the known initial conditions and resulting structural deformation. The knowledge of a relatively simple pre-impact environment in terms of the terrain and geology greatly improves the credibility for outlining a normal impact sequence.

Turning to the Flynn Creek event, at the instant of impact a complex set of dynamic physical processes was set instantly into motion. A schematic representation of selected parts of this cratering event is shown in Fig. 4 to aid

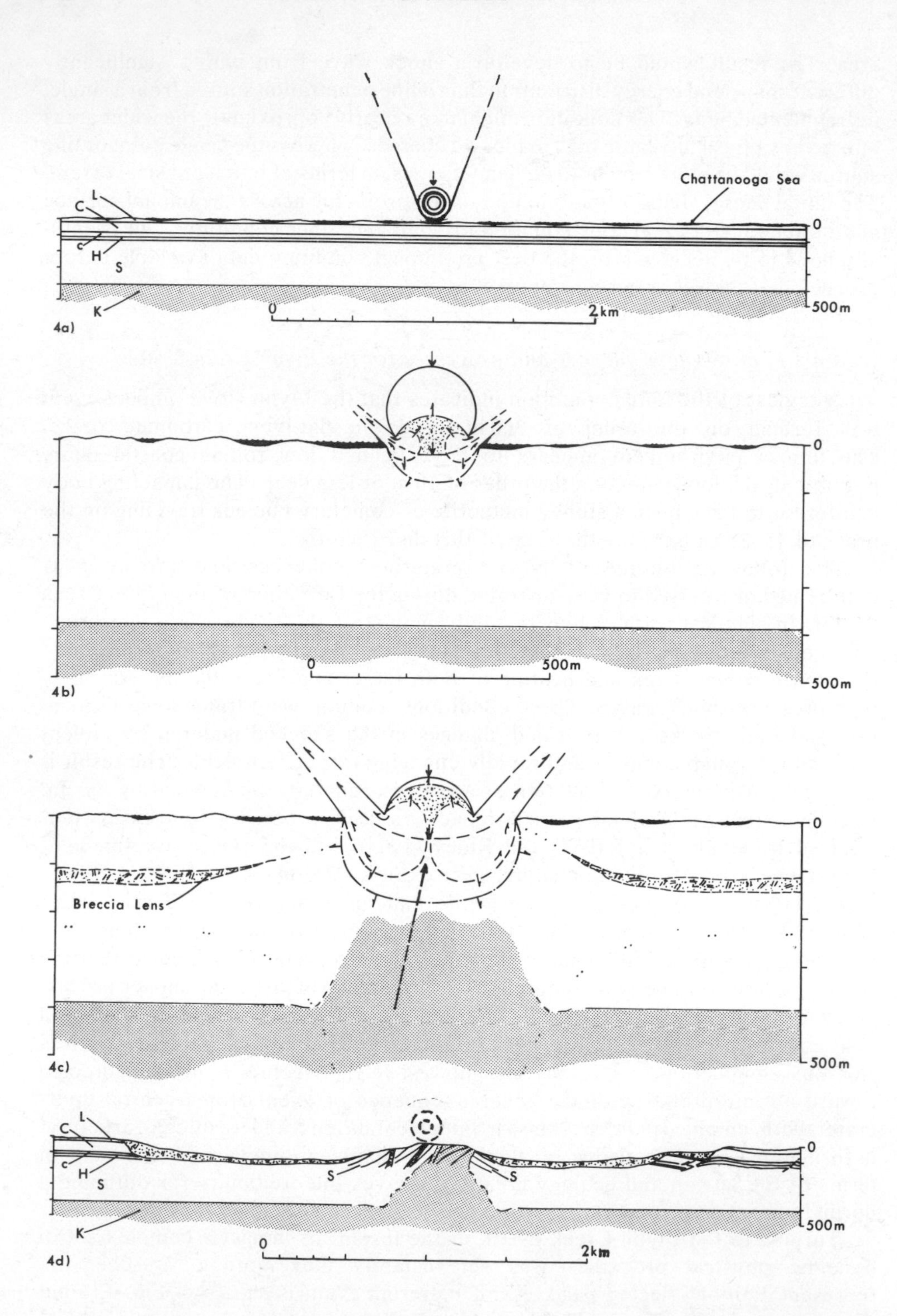
Chattanooga Sea
L
C
c
H
S
K
0
500m
0
2 km
4a)
0
500m
4b)
0
500m
0
500m
Breccia Lens
0
500m
4c)
L
C
c
H
S
S
S
K
0
500m
0
2 km
4d)

the phenomenological outline. The earliest cratering phase consisted basically of a very high pressure, high temperature period which lasted on the order of fractions of a second. Essentially, instantly after contact of the impacting body with the limestone target a well-developed set of *shock waves* was generated, one of which traveled up into the (pre-stressed?) projectile and the other down into the limestone.

Two physical aspects of these shock waves are critical in the early impact processes, their extremely high velocities and their generation of very high pressures. The shock wave velocities, inferred from Shoemaker (1960) for a 15 km/sec iron impact at Meteor (Barringer) Crater, generated back into the impacting body would have been on the order of 12 km/sec and into limestone

---

Fig. 4. Schematic sequence of selected times in the Flynn Creek cratering event. Figure 4a shows the initial geologic and inferred terrain conditions immediately before impact. The concentric spheres represent the different sizes of impacting bodies as a function of velocity and density, i.e., the innermost sphere is for 25 km/sec at 3.3 g/cm^3, next larger sphere is for 15 km/sec at 3.3 g/cm^3, next larger is for 25 km/sec at 1.0 g/cm^3, and the largest sphere is for 15 km/sec at 1.0 g/cm^3. The bow shock wave is shown extending from the leading edge of the impacting body. The stratigraphic symbols are described in Fig. 3. Figure 4b shows the impacting body, using the 15 km/sec and 1.0 g/cm^3 conditions for a 190 m diameter cometary nucleus as an example, microseconds after impact. The dash-dot lines are the expanding shock waves and the dashed lines with half-arrows represent *one* of an infinite number of pressure relaxation profiles in rarefaction field. Shocked part of projectile, shown in pattern, is being destroyed by the expanding rarefaction. The jetting stage has passed and the impacting body and limestone are now beginning to establish a steady state flow of vaporized, melted and hydrodynamically effected material. Arrows show the directions that solid surfaces are moving and half-arrows show the direction of movement of the shock waves, rarefaction fields, and flowing materials. The text describes the velocities used to draw these figures. Note that the shock wave expanding into the limestone is inferred to be slightly flat due to differential travel velocities in the vertical and horizontal directions. Figure 4c shows the continuation of 4b with the bulk of the remaining impacting body in a shocked state. The compressed leading edge has penetrated less than one-half its radius and the body is nearly totally disrupted by rarefaction. In this representation, on the order of only 0.1 sec has passed since impact and the projectile is nearly destroyed. Vaporization, melting, and hydrodynamic effects still dominate the flow. The transition to non-hydrodynamic flow, fragmentation and ejection will be completed within a second or so. The breccia lens and incipient central uplift will *not* form for another 5–20 sec, but they are drawn here to illustrate their relative terminal positions with respect to the early time sequence. Note that this type of impacting body does not penetrate to the depth of the base of the breccia lens before its destruction, although a narrow, conical transient cavity is inferred to develop within a few seconds in this central zone. The double-dot line shows the approximate zone of more intensely disrupted strata that will develop over the total period of cratering time. Long dashed arrow shows shatter-coned Knox dolomite that will be uplifted along a generalized path several seconds after impact. Figure 4d shows the final crater formed after 20–60 sec with flat-floor and massive central uplift.
The two spheres represent the maximum and minimum impacting body sizes in 4a.

would have been on the order of 17 km/sec. These velocities are obviously well in excess of the acoustic velocities of the materials allowing the term *hypervelocity impact* to be used to characterize the event (Swift, 1972). Despite the upward directed shock wave into the projectile, its particle motions are still generally downward due to the overall high impact velocity. This net downward motion, in part, explains why fragments of the impacting body finally reside in the breccia lens and walls of craters. Chou and Hopkins (1972), Gault *et al.* (1968), Kinslow (1970), Maxwell (1977), Shoemaker (1960), Stupochenko *et al.* (1967), Rinehart (1975) and others describe a wide spectrum of calculations on relative particle velocities for various natural materials.

Returning to the rapidly expanding high pressure zones in the projectile and target, physical conditions immediately in front of the shock waves remain totally undisturbed, except for a probable narrow shock or elastic precursor zone. Immediately behind the very rapidly expanding shock waves, however, the extreme shock compressions caused by the impact generate very transient, very high pressures in the low megabar range and temperatures in the $10^{3\circ}$C range, with pressures, temperatures, densities and entropies of the material engulfed by the shock wave each instantly raised to new and higher values. These extreme changes in the physical state of the impacting body and limestone are imposed in a matter of microseconds out to ranges, in these limestones, on the order of several hundred meters as scaled from explosion data. The meteorite is totally engulfed in this high pressure region at this early time. Shoemaker (1960) calculates pressures for an iron impact at Meteor Crater in rock types comparable to those at Flynn Creek to be in the 4–5 megabar range. A consequence of such extreme conditions in both the rock and impacting body is that both materials experience a very limited amount of shock-induced vaporization and melting, at impact velocities of 15 km/sec, adjacent to the growing impact cavity. At higher impact velocities the width of such a vapor and melt zone expands rapidly. Gault *et al.* (1974) and Ahrens and O'Keefe (1975) and O'Keefe and Ahrens (1976, 1977) describe calculations that infer the percent of impacting body and target material which are vaporized and melted, but in general above impact velocities of approximately 30 km/sec an iron body totally vaporizes and above approximately 25 km/sec a stoney meteorite is largely vaporized. The physical extent of such vaporization and melting, however, is actually confined to a relatively small volume involving only part of the immediately surrounding, limestone out to less than about 0.03–0.05 D, where D is the apparent diameter of the crater. In this case, the carbonates are presumably reduced to CaO and CO^2 vapor while the question of the physical character of a carbonate melt remains unanswered.

Another effect of the very high shock pressures is to create a physical state in which the impacting body, or part of it depending on the impact velocity and possible fragmentation history, and the immediately adjacent limestones respond hydrodynamically, temporarily exhibiting a fluid behavior. The strength of the projectile and immediately adjacent target rocks at this point would have been exceeded by factors of 10^3 and greater, literally causing them to flow. The width of such a stressed zone in the Flynn Creek rocks would have been on the order of about 0.06 *D*, i.e., the very high pressure, hydrodynamic stressed zone transiently occupies only a small part of the total volume cratered.

It is necessary to return to the early impact time to clarify a critical process that now becomes ultimately important in affecting the later cratering stages, i.e., the development of the complex fields of rarefactions. This involves the inevitable unloading or relaxation of the high pressure zones created by the outrunning shock waves. The basic mechanism for pressure release lies in the interactions of the shock waves with all free surfaces. Stated simply, material semi-infinitely deep in a shocked zone moves only in the direction induced by the shock wave. Near a free surface, however, material experiences a different unloading path due to the fact that an unconfined free surface cannot support stress across that surface, i.e., continuity conditions require an instant equilibration of the stress field. Consequently, the high pressure zones created in the target and projectile, together with very low surface pressures, define a decreasing stress gradient along which material can be accelerated and ejected. The practical result is that the free surface moves. The point is that impact craters, at least in hard rock systems, are *not* formed during the very high pressure compression stage. Instead, they form as a response to the later dynamic rarefaction fields developed along all free surfaces. The destruction of the projectile is also intimately involved in pressure unloading by rarefaction as shown schematically in Fig. 4. In part, these conditions explain why only a small fraction of even a low velocity iron projectile remains in a crater. It might also be noted that the instantaneous development and rapid growth of the rarefaction region behind the shock waves rapidly reduces the high pressure field. Consequently, materials actually experience an extremely rapid shock loading and a long-term time history of pressure decay. The innumerable free surfaces developed early in the cratering history only add severe complexity to the geometry of the shock and rarefaction fields.

One important consequence of the early development of the rarefaction at the interface and edges of the initial impact surface is the development of a very high velocity, high energy jet of projectile and target rock (Gault *et al.*, 1968). The material involved incorporates the vaporized, melted, and hydrodynamically affected region unloading behind the initial very high pressure shock waves. The flow is mobilized in the first milliseconds with an initial low ejection angle, on the order of a few degrees, which appear to increase with time. The velocities can be over four times the impact velocity as shown by computations (Chou and Hopkins, 1972) and experiment, although the masses of the projectile and target involved are extremely small. The jetting lasts only a fraction of a second but the extremely high velocity of the jetted material, assuming it is not all converted to vapor, would expose it to early total ablation in the earth's atmosphere. On the airless moon such jetting may account for certain early ray material.

Returning to Flynn Creek, several points should be noted regarding the initial condition and the time history up to this point. The total time elapsed since impact to the approximate end of the hydrodynamic and jetting phase is on the order of 0.01–0.10 sec as inferred from explosion cratering scaling (Cooper, 1977; Cooper and Sauer, 1977; Gault and Wedekind, 1977). In this extremely short period, the shock wave in the projectile would have traveled to the trailing edge of a 100 m body and would be under total rarefaction conditions, while the target shock wave has expanded to nearly 200 m from the impact point. If a shallow sea, on the order of 10 m deep, was present at impact time it probably would not have seriously affected this impact event. A layer of water 10 m thick would have been equivalent to a layer of rock with no effective tensile strength. Since 10 m is only one tenth of a 100 m projectile diameter, it would not have been a significant energy sink in either the penetration or cratering phases. The expected effect of such a thin water layer would have been the production of extensive steam and water vapor, but such a thin layer would have had such large volumetric dispersion that it probably did not seriously augment the cooling or deceleration of high speed ejecta.

The presence of relatively flat impact surfaces, such as a rolling coastal plain, and flat-lying target rocks greatly simplified the geometry of the initial expanding shock waves and subsequent rarefaction field. Returning momentarily to the instant of impact at the Flynn Creek ground surface, leads to an interesting conclusion regarding explosion crater analogs with similar initial crater geometries (Roddy, 1976). If the contact between the impacting body and limestone is essentially flat, the initial shape of the shock waves would have been co-planar for fractions of a microsecond, with the limestone shock wave rapidly expanding downward into the general form of a hemisphere. The exact shape, however, is critical to the geometry of the rarefaction field developed behind the shock wave and consequently the stress and flow fields that are associated with the actual ejection of the target rocks. The explosion experimental data would suggest that a near-surface to shallow penetrating projectile, as envisioned for Flynn Creek, generated a slightly flattened hemisphere that expands with this form through the upper several hundred meters of limestone. The results of the explosion experiments were to form broad flat-floored craters, with massive repositioning of sub-crater material moved inward and upward (Roddy, 1976, 1977a), structural deformation identical to that at Flynn Creek. Shock velocities and the release history inferred for the Flynn Creek limestones in the horizontal versus vertical directions suggest a similar flattening of the expanding shock wave may have occurred. Numerical calculations are crucial to this question and their similitude with the surface explosion trials.

By approximately 0.1–1.0 sec the impact has begun to stabilize, in terms of the rarefaction phase, and a continuous ejection of limestone was in progress. The shock front, at 0.1 sec, is near the position that will mark the final crater walls and pressures have now decayed to the level of a few 100 kbars near the impact point. The inner hydrodynamic zone, which extended out to about 0.05 D, has been ejected at high velocity. Carbonate vapors, melts, and possible fine fragments forming this initial, high velocity ejecta would have been severely affected by ablation and deceleration and possible interaction with the expanding sea steam front, if such existed. The mass involved in the hydrodynamic and immediately adjacent transition zone is estimated to be less than about 3% of the total ejected rock for the 15 km/sec impact velocity, and consequently is a very small part of the total rock cratered in the later stages under much lower pressures. The impacting body as noted earlier is also disintegrating under the rarefaction conditions by this time.

Time scaling indicates that the crater will continue to grow, however, for another 20–60 sec (Trulio, pers. comm. 1977). If explosion data and analog calculations can be applied, a near constant angle of ejection is achieved within several seconds (Wistoski, 1977) and remains throughout the main ejection period. Experimental explosion field data and theoretical calculations suggest that several factors combined to produce fragmentation and ejection which, in part, include: (1) rarefaction interactions behind the main shock wave with extreme unloading at the surface, (2) approximately spherical divergence of the flow field from the center of energy generated in the limestone by the impact penetration, and (3) bending of the near-surface limestones due to a decrease with range in the velocity with which those layers moved up and out (Trulio, pers. comm., 1977).

At Flynn Creek, the fragmentation and ejection processes produced a large central zone of deeply disrupted and uplifted rock surrounded by a broad, flat-floored cratered region. The geometry of the totally disrupted crater zone is similar to that of a flat saucer with a deep central cone, a configuration which is one of the more fundamental areas of concern in physical and theoretical studies for both the impact and explosion communities. Structural studies, summarized in Roddy (1968, 1977a,b), show that major deformation in the center of the crater extends to depths on the order of 500 m, and drill core data indicate that the subsurface shape of the deformed zone appears steeply conical, with a surface diameter of about 1.2 km. In the upper strata, Stones River dolomite and limestone beds have been moved inward tens of meters to form a crude domical central uplift with deeper massive units of Knox dolomite uplifted over 450 m to the level of the original impact ground surface. Besides intense faulting, folding, brecciation and fracturing, the Knox dolomites also exhibit shatter cones which were formed at depths of approximately 450 m below the impact ground surface. Roddy and Davis (1977) estimate the formational pressures for the shatter cones at Flynn Creek to have been on the order of between 15 and 45 kbars based on explosion data. Apparently, the central region beneath the impact point experienced a violent uplift to bring massive blocks over 100 m across in coherent form through these large vertical distances. Again, an identical structural configuration is found in the Snowball Crater formed by a 500-ton TNT surface explosion with ratios of uplift versus apparent diameter depth on the order of 7 to 10 (Roddy, 1977a,b). The boundaries of the central uplifted rocks appear to be rather well-defined according to drill data (Roddy, 1968), although extensive low angle faulting and brecciation may account for somewhat larger displacements than the drill cores indicate. Drill data in one core adjacent to the central uplift showed a clear example of extensive tensional opening along a weakly bedded zone of bentonite. Approximately 200 m from the edge of the uplift and 200 m below the impact surface a massive injection of fine breccia consisting of Stones River and Knox rocks were driven into the bentonite bed, forcing it from a 1.0 m thickness to over 3 m thick. Presumably, this type of tensional expansion was common elsewhere at depth around the uplifted rocks. The concept of an open, bowl-shaped, transient cavity, however, is not compatible with the structural deformation. Instead, a transient, very steep-sided conical cavity, similar to an armor piercing or jet penetration cavity may be more consistent with the observed field data. Such a region appears to have existed in the central uplift of the Snowball explosion crater and allowed material to be ejected from deeper levels near the penetration axis of symmetry as shown in Roddy (1977a). Certainly, the surface exposures at Flynn Creek show that massive Knox strata from original depths of 450 m were uplifted to the ground level. Drill core data on the flanks of the uplift further confine the uplift to a very narrow zone on the order of a kilometer or less across at the crater floor level (Roddy, 1968, 1977b). The violent decompression of the walls of such a possible narrow, transiently-opened, conical cavity would undoubtedly collapse the region in seconds. The initial movements in the central uplift are estimated to begin 5–10 sec after impact, judging from the numerical studies of Ullrich *et al.* (1977).

In the strata below the crater floor, inward movement of sub-crater floor rocks from the upper Stones River Group appear to extend out as far as the western rim. Here, the rim beds fold down instead of exhibiting the classical impact rim uplift of smaller craters. The interpretation is that deeper inward movements under the rim allowed a volume readjustment for downfolding of the upper rim rocks. Dence *et al.* (1977) and Melosh (1977) also consider similar rim failure processes. This was a common mode of structural adjustment at the large explosion craters described by Roddy (1973, 1976, 1977a). Presumably, the inward sub-crater floor movements at Flynn Creek are related to the massive inward movements to form the central uplift, as they were at the explosion craters. The reasons for these motions remain less clear. Ullrich (1976) and Ullrich *et al.* (1977) have provided the only numerical treatments of central uplift formation. These studies deal with explosion crater uplifts and indicate that several conditions were critical to the formations of uplift, which include: a surface-generated shock wave, low compactability of target, layering, and plastic volumetric increases during Mohr-Coulomb yield. The results of those numerical simulations indicate two possible mechanisms may have contributed to the explosion crater central uplifts, a gross volumetric rebound of the target rocks following the maximum peak shock pressure, followed by a possible later-stage gravitational collapse of deeper material. Unfortunately, the complexities of this model do not allow a simple direct application to Flynn Creek, at least not without a new set of numerical calculations. The general interpretation, however, is that high energy, shallow penetration impacts on layered, hard rock, systems will induce uplift. The volumetric rebound phase calculated for the explosion craters would also seem to fit the hard rock, low compactibility requirements at Flynn

Creek, as well as the massive upward expansion of the sub-crater rocks. Presumably this would be consistent with the violent volumetric release phase into the proposed transient, narrow, conical cavity inferred above. Parametric numerical calculations, such as those completed in detail by Ullrich (1976), remain one of the most critical areas of study for Flynn Creek and all other flat-floored, central uplift craters.

Beyond the flanks of the central uplift, the fragmentation and ejection processes stabilized at an effective depth of approximately 130 m and continued radially out to an average radius of approximately 1760 m. These cratering depths and radii are comparable again to scaled values for certain explosion craters formed by surface explosions (Roddy, 1976, 1977a,b). The cratering terminated at the base of the slightly weaker Hermitage limestone and penetrated a few meters irregularly into the top of the slightly denser dolomites of the Stones River Group. Presumably, the depth of cratering was partly a material property effect related to the strength of the Hermitage limestone and Stones River dolomites. Major structural deformation, including faulting, folding, local brecciation, and fracturing extended down to approximately 230 m below the impact ground surface, a relatively shallow depth considering the large diameter, and the decaying rarefaction field continued to eject material for an estimated 20–60 sec. A limited amount of breccia remained in the crater as fallback and locally disrupted limestone that did not have sufficient momentum to be ejected. As would be expected, the breccia lens material at Flynn Creek consists mainly of the lower horizons fragmented within the 20–60 sec period of ejection cratering. The result of the ejection process was to create an apparent crater volume of approximately 0.8 km^3 and a true volume of totally brecciated material of about 1.2 km^3. The total mass of brecciated material was approximately 2×10^9 metric tons and the mass of the ejecta was 0.9×10^9 metric tons (Roddy, 1977b). Approximately 80–90% of the rock ejected remained in the ejecta blanket surrounding the crater out to about 2.5 *D*, as inferred from explosion cratering data on similar crater types (Roddy, 1977a). A crude overturning process appears to have also operated in the ejection of material as evidenced by the coarsely inverted ejecta overlaying the graben. Shoemaker (1960) and Roddy (1968, 1973, 1976, 1977a) note that such overturning is a normal consequence of all ejection processes. The preservation of the inverted character in the ejecta blanket during deposition, however, depends upon the nature of the fragmental material. The ejection process terminated, with a rather irregular outline along a radius of approximately 1760 m. At this distance the shock stress had decayed to the few-kilobar level and the strength of the rocks plus confining pressure could not be further overcome, i.e., crater walls formed.

The deformation in the rim, including the rim uplift, rim downfolding, thrust faulting, terrace block and graben faulting and other minor faulting, folding, brecciation and fracturing were basically formed as part of the final, late-stage rarefaction processes. Such large deformations *do not* occur in the early cratering phases with the initial passage of the compressional shock wave, i.e., within an approximate 0.1 sec. Explosion experimental data shows that major rim deformation occurs in the same general time as that for the crater wall formation (Roddy *et al.*, 1977), consequently, the large rim thrust block, rim uplift and depression, faulting and folding are late-time effects. The thrust block was apparently activated in time before the graben formed since it partly overrides the down-dropped block. The large masses involved in these deformations must first overcome strength, inertial and gravitational effects, probably occurring on the order of at least 15–30 sec after impact. Final motions involved the late-stage slumping of the terrace and graben blocks toward the crater.

SUMMARY

Field and laboratory studies indicate that the initial field conditions at Flynn Creek consisted of a low rolling to hilly coastal plain underlain by consolidated, nearly flat-lying, limestone and dolomite beds. The impact may actually have occurred in the shallow coastal waters of the Chattanooga Sea with water depths inferred to have been no greater than about 10 m. These conditions imply that the impact event was uncomplicated by the pre-impact terrain and geology, and suggest that a normal impact sequence of events should have occurred.

The outline of the impact sequences and processes was derived from a number of interpretations, inferences, and calculations based on field and laboratory data from Flynn Creek and on laboratory impact, theoretical, and explosion crater data. On the order of 4×10^{24} ergs are estimated from explosion scaling for the kinetic energy of the event, and assuming a velocity range of 15–25 km/sec, the impacting body is estimated to have been on the order of 90–190 m in diameter, depending on the choice of density. The morphological and structural similarities of Flynn Creek with the 500-ton TNT Snowball trial and other surface and near-surface explosions has prompted the argument for a surface or near-surface effective transfer of kinetic energy by the impacting body. The implication is that of minimal penetration, on the order of the radius of the body, with accompanying near-total to total vaporization of the projectile, depending on the choice of a stoney meteorite or cometary nucleus. The absence of any traces of the body in geochemical studies (Roddy, 1966, 1968), the surface-generated explosion crater analogs (Roddy, 1976, 1977a) and the existence of high-velocity low-density, easily vaporized impacting bodies in earth-crossing orbits (Wetherill, 1977; Shoemaker, 1977a,b) suggest at least the permissibility of inferring carbonaceous chondritic or cometary impacting bodies for Flynn Creek as possible candidates for the impacting body.

The numerous complexities that may affect an impacting body during its atmospheric descent, such as large-body fragmentation and internal stress gradients, appear to be insoluble without computational studies. All that can be said with certainty is that the impact *did* occur and *appears* to have expended its kinetic energy at or near the surface, i.e., Flynn Creek is not a bowl-shaped penetration-type crater created by a normal lower-velocity iron impact.

The cratering sequence of events and processes inferred for Flynn Creek represents a normal hypervelocity impact in which the crater was formed in 20–60 sec, depending on the time scaling applied. In this period of time over 1.2 km^3 of rock were brecciated with over 0.8 km^3 ejected into a surrounding ejecta blanket. Within ~1.0 sec after impact, the initial shock waves will have traveled over 10 km from the crater and rarefaction and related processes will have destroyed the projectile and initiated the ejection process. A large inward movement of sub-crater floor rock toward the center of the crater was accompanied by a massive, violent central uplift of rock from over 450 m depth. A narrow, steeply conical transient cavity accompanied by the volumetric rebound phenomena described by Ullrich *et al.* (1977) and large inward mass movements culminating at the crater axis appear partly responsible for the uplift. The driving mechanisms for these motions at Flynn Creek, however, remain uncertain.

The brecciation and ejection process, beyond the edge of the central uplifts moved downward to approximately 130 m and continued out to about 1760 m, ending some 20–60 sec after impact, and formed a large ejecta blanket with crudely inverted strata. Explosion analog estimates indicate a radius for the continuous ejecta blanket to have been approximately 2.5 *D*. The overturning process during ejection terminated at the point where the strength of the rocks plus the confining pressure exceeded the stress conditions in the rarefaction field, and at the ground surface, this point constituted the crater walls. A large breccia lens of fallback and locally disrupted rock remained in the crater. Complex deformation in the rim, such as thrust blocks, terrace faulting and graben development was a direct response to the final stress conditions imposed by the decaying rarefaction field.

A period of erosion followed formation of the crater with the removal of the ejecta blanket, possibly by shallow wave action, and degradation of the inner rim. On the higher parts of the rim, at least in the area of the graben, talus deposits formed along a fault scarp. Within the crater, marine waters of the Chattanooga Sea of early Late Devonian age deposited a sequence of coarse breccias washed back from the rim and are overlain by progressively finer bedded dolomitic breccias and cross-bedded dolomites deposited in the same shallow marine environment. As the crater rim was breached, deposition changed abruptly to the black silty muds of the open, shallow Chattanooga Sea and filled the crater over the next several million years with its slow deposition. The Flynn Creek area remained buried under sediments through at least Early Mississippian time, when the Fort Payne sediments were deposited.

Uplift and erosion during Quaternary, and probably Tertiary, time has produced a highly dissected region of steep-sided hills and valleys with an average relief of 150 m. The structure of the Flynn Creek Crater is now well-exposed along the valley floors and walls. An outline of the generalized sequence of events at Flynn Creek is shown in Fig. 5.

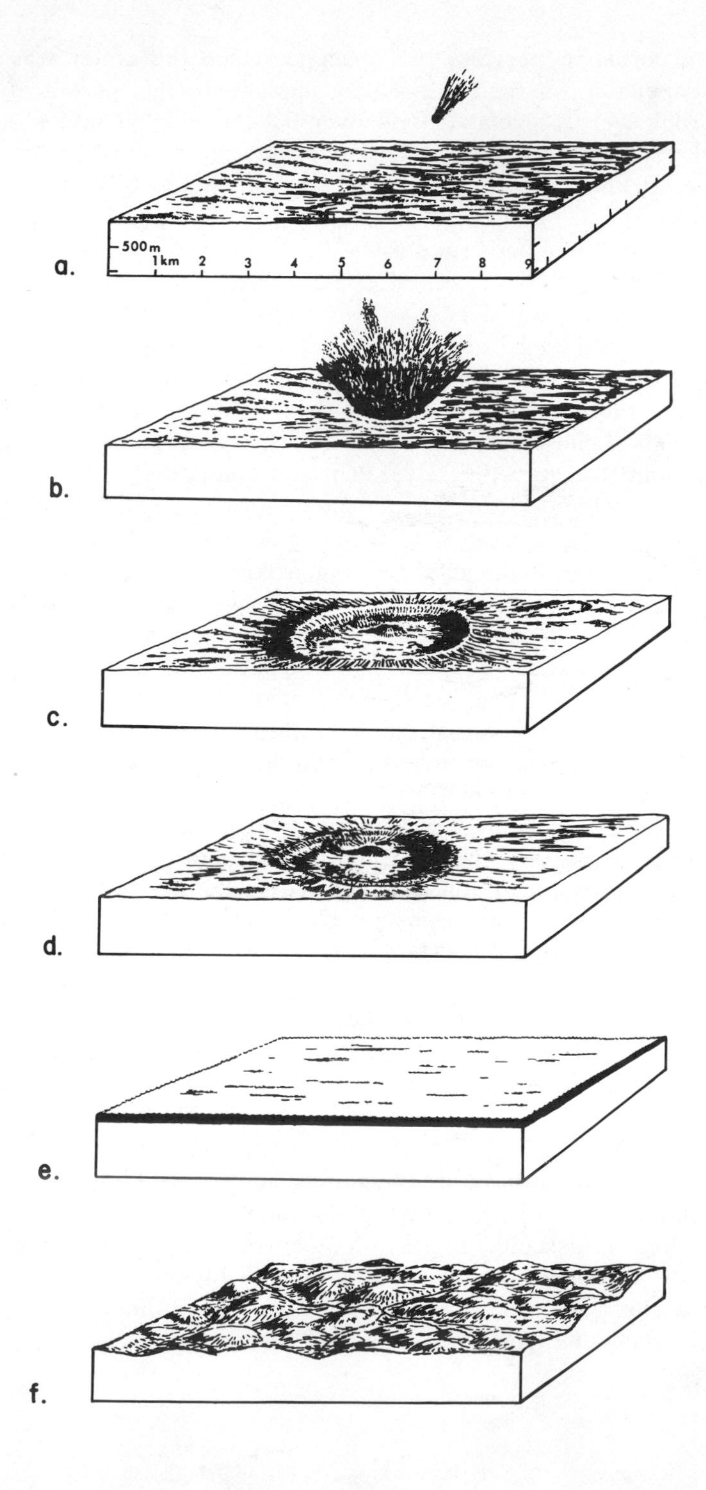

500 m
1 km
2
3
4
5
6
7
8
9
a.
b.
c.
d.
e.
f.

Acknowledgments—I wish to express my appreciation to Dr. E. M. Shoemaker, California Institute of Technology and U.S. Geological Survey for early discussions on Flynn Creek and his contagious enthusiasm for the study of craters. Numerous discussions with Dr. J. Trulio and his continued assistance and interest in certain of the calculations were invaluable in preparing the section on cratering processes.

Reviews of this paper by Drs. D. Scott, L. Soderblom, F. Hörz, G. H. S. Jones, J. Trulio, and discussions with G. W. Ullrich greatly assisted the writer in addressing a wide area of interest from atmospheric entry to the termination of the impact event, and I express my thanks to each for his help.

I am again indebted to Dr. G. Sevin and Capt. J. Stockton of the Defense Nuclear Agency for their continued support of these studies. I sincerely express my real appreciation to M., M., and M. Roddy for their field assistance at Flynn Creek and to J. Roddy for her help in data assemblage.

This work was supported through the cooperation of the National Aeronautics Space Administration under contract W13,130 and the Defense Nuclear Agency, Department of Defense, Washington, D.C.

References

Ahrens, T. J. and O'Keefe, J. D.: 1975, Shock effects from a large impact on the moon. *Proc. Lunar Sci. Conf. 6th*, p. 2831 2844.

Ahrens, T. J. and O'Keefe, J. D.: 1977, Equations of State and Impact-Induced Shock-Wave Attenuation on the Moon. In *Impact and Explosion Cratering* (D. J. Roddy, R. O. Pepin, and R. B. Merrill, eds.), Pergamon Press, New York.

Baldwin, R. B.: 1963, *The Measure of the Moon.* University of Chicago Press. 488 pp.

Baldwin, B. and Sheaffer, Y.: 1971, Ablation and Breakup of Large Meteoroids during Atmospheric Entry. *J. Geophys. Res.* **76**, 4653–4668.

Buchwald, V. F.: 1975, *Handbook of iron meteorites.* V. 1, Iron Meteorites in General. University of California Press, Berkeley.

Chabai, A. J.: 1965, On scaling dimensions of craters produced by buried explosives. *J. Geophys. Res.* **70**, 5075–5098.

Chabai, A. J.: 1977, Influence of Gravitational Fields and Atmospheric Pressures on Scaling of Explosion Craters. In *Impact and Explosion Cratering* (D. J. Roddy, R. O. Pepin, and R. B. Merrill, eds.), Pergamon Press, New York.

Chou, P. C. and Hopkins, A. K.: 1972, *Dynamic response of materials to intense impulsive loading.* Air Force Materials Laboratory, Wright Patterson Air Force Base, Ohio. 55 pp.

Fig. 5. Diagrammatic summary of the major events inferred in the history of the Flynn Creek Crater outlined as follows: (a) The Flynn Creek area, approximately 360 m.y. ago and immediately before impact, was a rolling coastal plain inferred to have been innundated by the shallow waters of the early Late Devonian Chattanooga Sea. Topographic highs were capped by rocks of Upper Ordovidian age; (b) crater formed by hypervelocity impact of stoney meteorite or cometary nucleus; (c) result of the impact was to form the Flynn Creek Crater, 3.8 km across and 200 m deep, with a large central peak, locally terraced rims, large rim thrust blocks and graben, and surrounding ejecta blanket. The very shallow sea waters, inferred to have been on the order of only 10 m deep and probably present throughout the area, are not shown in (c) and (d) for clarity of the illustration. Such waters would not have covered the 100 m high rim; (d) erosion removed the ejecta, washed a large amount back into crater, and formed small valleys in the rim. Subaqueous erosion and marine deposition of bedded breccias and bedded dolomite in the crater are basal units of Chattanooga Shale deposited in its initial very shallow waters; (e) regional coverage by the Chattanooga Sea continued filling the crater with black silty muds throughout early Late Devonian time; (f) area remained buried until uplift and erosion during Quaternary, and probably Tertiary, and time produced the highly dissected region that now exposes the crater.

Conant, L. C. and Swanson, V. E.: 1961, Chattanooga shale and related rocks of central Tennessee and nearby areas. *U.S. Geol. Survey Prof. Paper 357.* 91 pp.

Cooper, H. F., Jr.: 1977, A Summary of Explosion Cratering Phenomena Relevant to Meteor Impact Events. In *Impact and Explosion Cratering* (D. J. Roddy, R. O. Pepin, and R. B. Merrill, eds.), Pergamon Press, New York. This volume.

Cooper, H. F. and Sauer F. M.: 1977, Crater-related ground motions and implications for crater scaling. In *Impact and Explosion Cratering* (D. J. Roddy, R. O. Pepin, and R. B. Merrill, eds.), Pergamon Press, New York. This volume.

Dence, M. R. Grieve, R. A. F., and Robertson, P. B.: 1977, Terrestrial Impact Structures: Principal Characteristics and Energy Considerations. In *Impact and Explosion Cratering* (D. J. Roddy, R. O. Pepin, and R. B. Merrill, eds.), Pergamon Press, New York.

Dohnanyi, J. S.: 1972, Interplanetary Objects in Review: Statistics of their masses and dynamics. *Icarus* **17**, 1–48.

Gault, D. E., Quaide, W. I., and Oberbeck, V. R.: 1968, Impact cratering mechanics and structures. (B. M. French and N. M. Short, eds.) Mono Book Corp., Baltimore.

Gault, D. E., Hörz, F., Brownlee, D. E., and Hartung, J. G.: 1974, Mixing of the lunar regolith. *Proc. Lunar Sci. Conf. 5th*, p. 2365–2386.

Gault, D. E., Guest, J. E., Murray J. B., Dzurisin, D., and Malin, M. C.: 1975, Some comparisons of impact craters on Mercury and the Moon. *J. Geophys. Res.* **80**, 2444–2460.

Gault, D. E. and Wedekind, J. A.: 1977, Experimental hypervelocity impact into quartz sand: II, Effects of gravitational acceleration. In *Impact and Explosion Cratering* (D. J. Roddy, R. O. Pepin, and R. B. Merrill, eds.), Pergamon Press, New York.

Glasstone, S.: 1964, *The Effects of Nuclear Weapons.* U.S. Atomic Energy Commission, Washington, D.C. 730 pp.

Head, J. W.: 1977, Origin of Outer Rings in Lunar Multi-ringed Basins: Evidence from Morphology and Ring Spacing. In *Impact and Explosion Cratering* (D. J. Roddy, R. O. Pepin, and R. B. Merrill, eds.), Pergamon Press, New York.

Heide, F.: 1964, *Meteorites.* University of Chicago Press. 144 pp.

Huddle, J. W.: 1963, Conodonts from the Flynn Creek crypto-explosion structure, Tennessee. *U.S. Geol. Survey Prof. Paper 475 C*, C55–C57.

Killain, B. G. and Germain, L. S.: 1977, Scaling of cratering experiments—Analytical and heuristic approach to the phenomenology. In *Impact and Explosion Cratering* (D. J. Roddy, R. O. Pepin, and R. B. Merrill, eds.), Pergamon Press, New York. This volume.

Kinslow, R.: 1970, *High-Velocity Impact Phenomena.* Dept. of Engineering Science, Tennessee Technological University, Cookeville, Tennessee. Academic Press, New York.

Krinov, E. L. : 1966, *Giant Meteorites.* Pergamon Press, New York. 397 pp.

Kreyenhagen, K. N. and Schuster, S. H.: 1977, Review and comparison of hypervelocity impact and explosion cratering calculations. In *Impact and Explosion Cratering* (D. J. Roddy, R. O. Pepin, and R. B. Merrill, eds.), Pergamon Press, New York. This volume.

Martin, J. J.: 1966, *Atmospheric Reentry, An Introduction to its Science Engineering.* Prentice Hall, New Jersey.

Maxwell, D. E.: 1977, Simple Z model of cratering, ejection, and the overturned flap.

Maxwell, D. E. and Moises, H.: 1971, Hypervelocity impact cratering calculations. Physics International Company, PIFR-190. 201 pp.

Melosh, H. J.: 1977, Crater modification by gravity: A mechanical analysis of slumping. In *Impact and Explosion Cratering* (D. J. Roddy, R. O. Pepin, and R. B. Merrill, eds.), Pergamon Press, New York. This volume.

Needham, C.: 1971, *Airblast Interaction Calculations.* Proceedings of the Eric H. Wang Symposium on Protective Structures Technology, Vol. 1 (21–23 July 1970), p. 57–77. Air Force Weapons Laboratory, Kirtland Air Force Base, New Mexico.

Oberbeck, V. R.: 1977, Application of high explosion cratering data to planetary problems. In *Impact and Explosion Cratering* (D. J. Roddy, R. O. Pepin, and R. B. Merrill, eds.), Pergamon Press, New York. This volume.

O'Keefe, J. D. and Ahrens, T. J.: 1976, Impact Ejecta on the Moon. *Proc. Lunar Sci. Conf. 7th*, p. 3007–3025.

O'Keefe, J. D. and Ahrens, T. J.: 1977, Impact-induced energy partitioning, melting, and vaporization on terrestrial planets. *Proc. Lunar Sci. Conf. 8th.* Vol. 3.

Rinehart, J. S.: 1975, *Stress Transients in Solids.* Dept. of Mechanical Engineering, Univ. of Colorado and Technical Director Hyper Dynamics, P.O. Box 392, Santa Fe, New Mexico.

Riney, T. D.: 1965, *Developments in Mechanics* (S. Ostrach and R. H. Scranlan, eds.) p. 419. Pergamon Press, Oxford.

Roddy, D. J.: 1966, The Paleozoic crater at Flynn Creek, Tennessee. Unpub. Ph.D. thesis, California Institute of Technology, Pasadena. 232 pp.

Roddy, D. J.: 1968, The Flynn Creek Crater, Tennessee. In *Shock Metamorphism of Natural Materials* (B. M. French and N. M. Short, eds.), p. 291–322. Mono Book Corp., Baltimore.

Roddy, D. J.: 1973, Project LN303, geologic studies of the Middle Gust and Mixed Company craters. In *Proceedings of the Mixed Company/Middle Gust Results Meeting.* 13–15 March, 1973. Vol. 2, p. 79–128. Defense Nuclear Agency, Department of Defense, DNA 3151P2.

Roddy, D. J.:1976, High-explosive cratering analogs for bowl-shaped central uplift, and multiring impact craters. *Proc. Lunar Sci. Conf. 7th*, p. 3027–3056.

Roddy, D. J.: 1977a, Large Scale Impact and Explosion Craters: Comparisons of Morphological and Structural Analogs. In *Impact and Explosion Cratering* (D. J. Roddy, R. O. Pepin, and R. B. Merrill, eds.), Pergamon Press, New York. This volume.

Roddy, D. J.: 1977b, Tabular Comparisons of the Flynn Creek Impact Crater, United States, Steinheim Impact Crater, Germany, and Snowball Explosion Crater, Canada. In *Impact and Explosion Cratering* (D. J. Roddy, R. O. Pepin, and R. B. Merrill, eds.), Pergamon Press, New York. This volume.

Roddy, D. J. Boyce, J. M., Colton, G. W., and Dial, A. L.: 1975, Meteor Crater, Arizona, rim drilling with thickness, structural uplift, diameter, depth, volume, and mass-balance calculations. *Proc. Lunar Sci. Conf. 6th*, p. 2621–2644.

Roddy, D. J. and Davis, L. K.: 1977, Shatter Cones Formed in Large Scale Experimental Explosion Craters. In *Impact and Explosion Cratering* (D. J. Roddy, R. O. Pepin, and R. B. Merrill, eds.), Pergamon Press, New York. This volume.

Roddy, D. J., Ullrich, G. W., Sauer, F., and Jones, G. H. S.: 1977, Cratering motions and structural deformation in the rim of the Prairie Flat multiring explosion crater. *Proc. Lunar Sci. Conf. 8th.* Vol. 3.

Saxe, H. C. and DelManzo, D. D., Jr.: 1970, A study of underground explosion cratering phenomena. In *Proceedings, of the Symposium on Engineering with Nuclear Explosives.* Jan. 14–16, 1970, Las Vegas, Nevada, p. 1701–1725. American Nuclear Society and U.S. Atomic Energy Commission, CONF-700101.

Shoemaker, E. M.: 1960, Penetration mechanics of high velocity meteorites, illustrated by Meteor Crater, Arizona. In *Structure of the Earth's Crust and Deformation of Rocks*, p. 418–434. Internat. Geol. Congress, XXI Session, pt. 18, Copenhagen.

Shoemaker, E. M.: 1962, Interpretation of lunar craters. In *Physics and astronomy of the moon*, (Z. Kopal, ed.), 283–380. Academic Press, New York.

Shoemaker, E. M.: 1963, Impact mechanics at Meteor Crater, Arizona. In *The solar system, the moon, meteorites, and comets*, (B. M. Middlehurst and G. P. Kuiper, eds.) Vol. 4, p. 301–336. University of Chicago Press, Chicago.

Shoemaker, E. M.: 1977a, Why Study Impact Craters? In *Impact and Explosion Cratering* (D. J. Roddy, R. O. Pepin, and R. B. Merrill, eds.), Pergamon Press, New York. This volume.

Shoemaker, E. M.: 1977b, Astronomically Observable Crater-Forming Projectiles. In *Impact and Explosion Cratering* (D. J. Roddy, R. O. Pepin, and R. B. Merrill, eds.), Pergamon Press, New York. This volume.

Shoemaker, E. M. and Helin, E. F.: 1977, Populations of Planet-crossing Asteroids and the Relation of Apollo Objects to Main-belt Asteroids and Comets. In *Relationships Between Comets, Minor Planets, and Meteorites* (A. H. Delsemme, ed.) Proc. I.A.U. Colloquium 39. In press.

Stupochenko, Ye. V., Losev, S. A., and Osipov, A. I.: 1967, Relaxation in Shock Waves. In *Applied Physics and Engineering.* Springer-Verlag, New York.

Swift, H. F.: 1972, Hypervelocity Impact. In *Dynamic Response of Materials to Intensive Impulsive Loading* (P. C. Chou and A. K. Hopkins, eds.) p. 517–538. Air Force Materials Laboratory, Wright Patterson Air Force Base, Ohio.

Trulio, J. G.: 1977, Ejecta Formation: Calculated Motion from a Shallow-Buried Nuclear Burst, and Its Significance for High Velocity Impact Cratering. In *Impact and Explosion Cratering* (D. J. Roddy, R. O. Pepin, and R. B. Merrill, eds.), Pergamon Press, New York. This volume.

Ullrich, G. W.: 1976, *The Mechanics of Central Peak Formation in Shock Wave Cratering Events.* AFWL-TR-75-88, Air Force Weapons Laboratory, Kirtland AFB, New Mexico.

Ullrich, G. W., Roddy, D. J., and Simmons, G.: 1977, Numerical Simulations of a 20-Ton TNT Detonation on the Earth's Surface and Implications Concerning the Mechanics of Central Uplift Formation. In *Impact and Explosion Cratering* (D. J. Roddy, R. O. Pepin, and R. B. Merrill, eds.), Pergamon Press, New York. This volume.

Vaile, R. B., Jr.: 1961, Pacific craters and scaling laws. *J. Geophys. Res.* **66**, 3413–3438.

Vortman, L. J.: 1970, Nuclear excavation. In *Education for Peaceful Uses of Nuclear Explosives* (L. E. Weaver, ed.), p. 65–79. University of Arizona Press, Tucson.

Vortman, L. J.: 1977, Craters from Surface Explosions and Energy Dependence—A Retrospective View. In *Impact and Explosion Cratering* (D. J. Roddy, R. O. Pepin, and R. B. Merrill, eds.), Pergamon Press, New York. This volume.

Wetherill, G. W.: 1975a, Pre-mare cratering and early solar system history. In *Proc. Soviet-Amer. Conf. on Cosmochemistry of the Moon and Planets (Moscow).* NASA SP-390. In press.

Wetherill, G. W.: 1975b, Late heavy bombardment of the Moon and terrestrial planets. *Proc. Lunar Sci. Conf. 6th*, p. 1539–1561.

Wetherill, G. W.: 1977, The Nature of the Present Interplanetary Crater-Forming Projectiles. In *Impact and Explosion Cratering* (D. J. Roddy, R. O. Pepin, and R. B. Merrill, eds.), Pergamon Press, New York. This volume.

Wilhelms, D. E., Hodges, C. A., and Pike, R. J.: 1977, Nested-Crater Model of Lunar Ringed Basins. In *Impact and Explosion Cratering* (D. J. Roddy, R. O. Pepin, and R. B. Merrill, eds.), Pergamon Press, New York. This volume.

Wistoski, J.: 1977, Dynamic Ejecta Parameters from High Explosive Detonations. In *Impact and Explosion Cratering* (D. J. Roddy, R. O. Pepin, and R. B. Merrill, eds.), Pergamon Press, New York. This volume.

Roddy, D. J., Pepin, R. O., and Merrill, R. B., editors.
(1977) *Impact and Explosion Cratering*, Pergamon Press (New York), p. 309–320.
Printed in the United States of America

The Steinheim Basin—an impact structure

WINFRIED REIFF

Geologisches Landesamt Baden-Württemberg, Urbanstrasse 53,
7000 Stuttgart 1, West Germany

Abstract—The Steinheim Crater was formed by impact in flat-lying carbonate and marlstone rocks on the Swabian Alb in Southern Germany approximately 14.7 million years ago. It was a shallow crater, averaging 3.4 km across and about 220 m deep, with a large central uplift about 150 m in height. Shatter cones, first described from Steinheim, are prominent in the central uplift rocks. Major structural deformation consists of: (a) a large breccia lens in the crater, (b) central uplift with shatter cones, (c) faulted, folded and locally brecciated rim rock, (d) faulted, folded and locally brecciated rock below the breccia lens, and (e) shock metamorphic features in the breccia, such as planar features and shatter cones. The Steinheim Basin is interpreted as an impact crater exhibiting structure and morphology very similar to that of the Flynn Creek impact crater in Tennessee, United States, which was also formed in flat-lying carbonate rocks.

THE STEINHEIM BASIN is situated in Southern Germany on the Swabian Alb (latitude: 48°02′N, longitude 10°04′E; Fig. 1). In the flat upland area with low relief surface there is a flat-floored, roundish basin. The diameter is 3.4 km (max. 3.8 km, min. 3.1 km), the depth is 90 m. In the center stands a hill with a diameter of about 900 m and a height of about 50 m (Fig. 2). The strata surrounding the basin are limestones and marls of the Upper Malm. The strata in the central hill are limestones and marls of the Lower Malm and clays, limestones and sandstones of the Dogger. On the flanks of the basin and in the basin floor are Tertiary lake deposits and Quaternary (Fig. 3).

As early as the nineteenth century the Steinheim Basin was recognized as a geological anomaly. At the end of the nineteenth century some geologists thought that it was a maar lake. But they could not find volcanic material. Therefore Branco and Fraas (1905) suggested the idea of cryptovolcanism. Kranz (1924) varied this idea and interpreted the Steinheim Basin as a cryptoexplosion structure. Similar conditions and lake deposits of the same age were found 40 km ENE in the Nördlinger Ries. The conclusion was drawn, that both structures originated from the same event. As early as in 1934 and 1936 Rohleder and Stutzer compared the Steinheim Basin and the Nördlinger Ries with the Meteor Crater in Arizona—implying that the former two were impact structures. However, Rohleder and Stutzer could not prove this theory for Steinheim.

More than twenty years later, Dietz (1959) proposed that shatter cones (Fig. 4), first described in the Steinheim Basin (Branco and Fraas, 1905), and later found in other so-called cryptovolcanic structures, indicated an origin by impact. The investigations made by Shoemaker and Chao (1960) in the Ries Basin initiated a new era of research.

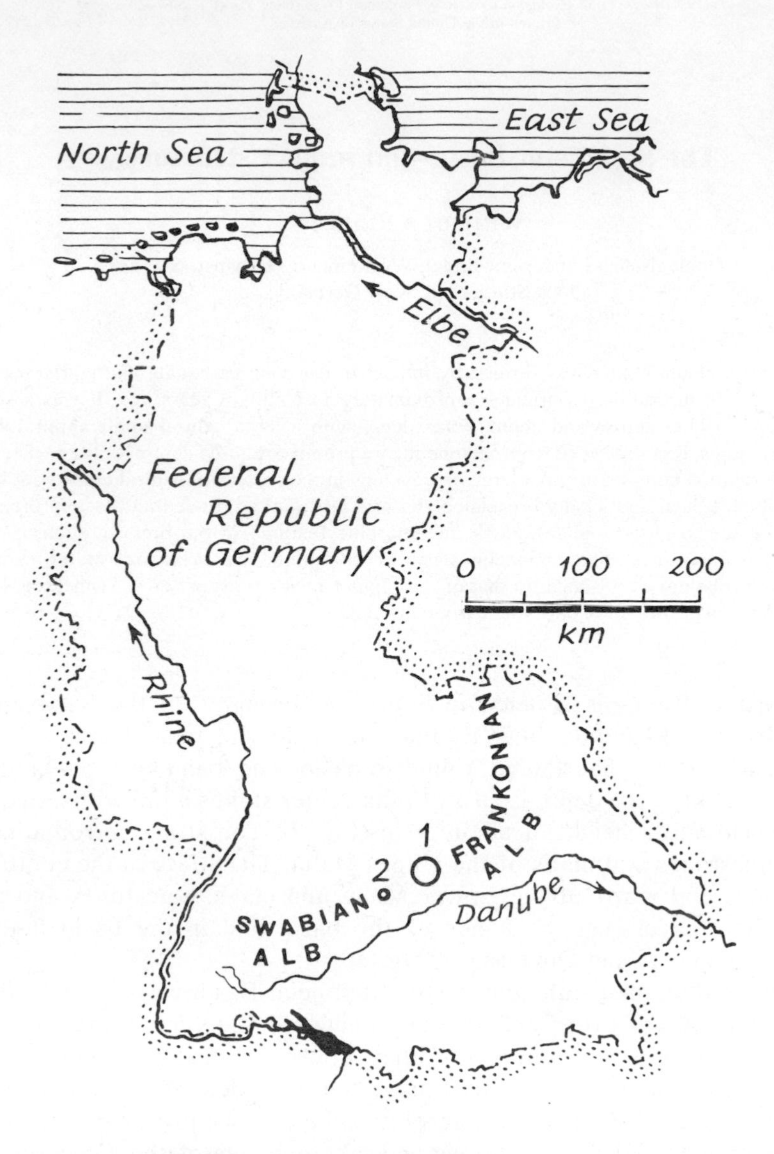

Fig. 1. Sketch of the Federal Republic of Germany with the position of Nördlinger Ries and Steinheim Basin.

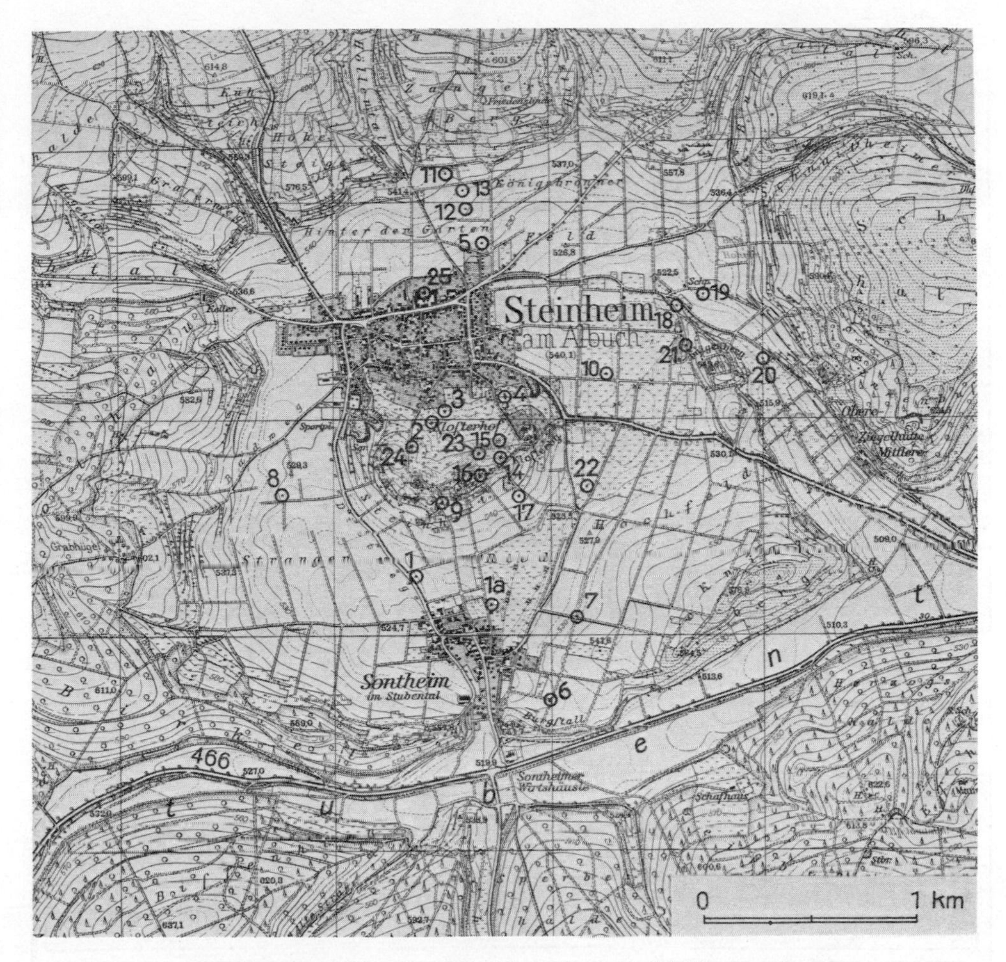

Fig. 2. Topographic map of Steinheim Basin and drilling locations.

These investigations in the sixties gave an important hint (Groschopf and Reiff, 1966; Engelhardt *et al.*, 1967) and in 1971 (Groschopf and Reiff) supplied final proof that the Steinheim Basin is an impact structure. The diameter of the crater is on the average 3.4 km, the original depth about 220 m. The diameter of the central uplift is about 900 m and the height 150 m. The Steinheim Crater has had a depth/diameter ratio of 1 : 16.

Between 1964 and 1972, in the Steinheim Basin holes were drilled (1 core drilling)—10 in the central hill, 16 in the basin floor. The drillings in the Steinheim Basin showed Quaternary deposits and Tertiary lake beds overlaying a 20–70 m thick layer of ejecta material (fallback: "Primary Basin Breccia", Groschopf and Reiff, 1966). This breccia consists of Lower Malm and Dogger, probably with some portions of Lias. The breccia is a mixture of small pieces of limestone, marls, clays, and sandstone (Fig. 5), but there are also big blocks. In this breccia we find well-developed, fine-striated shatter cones and quartz grains

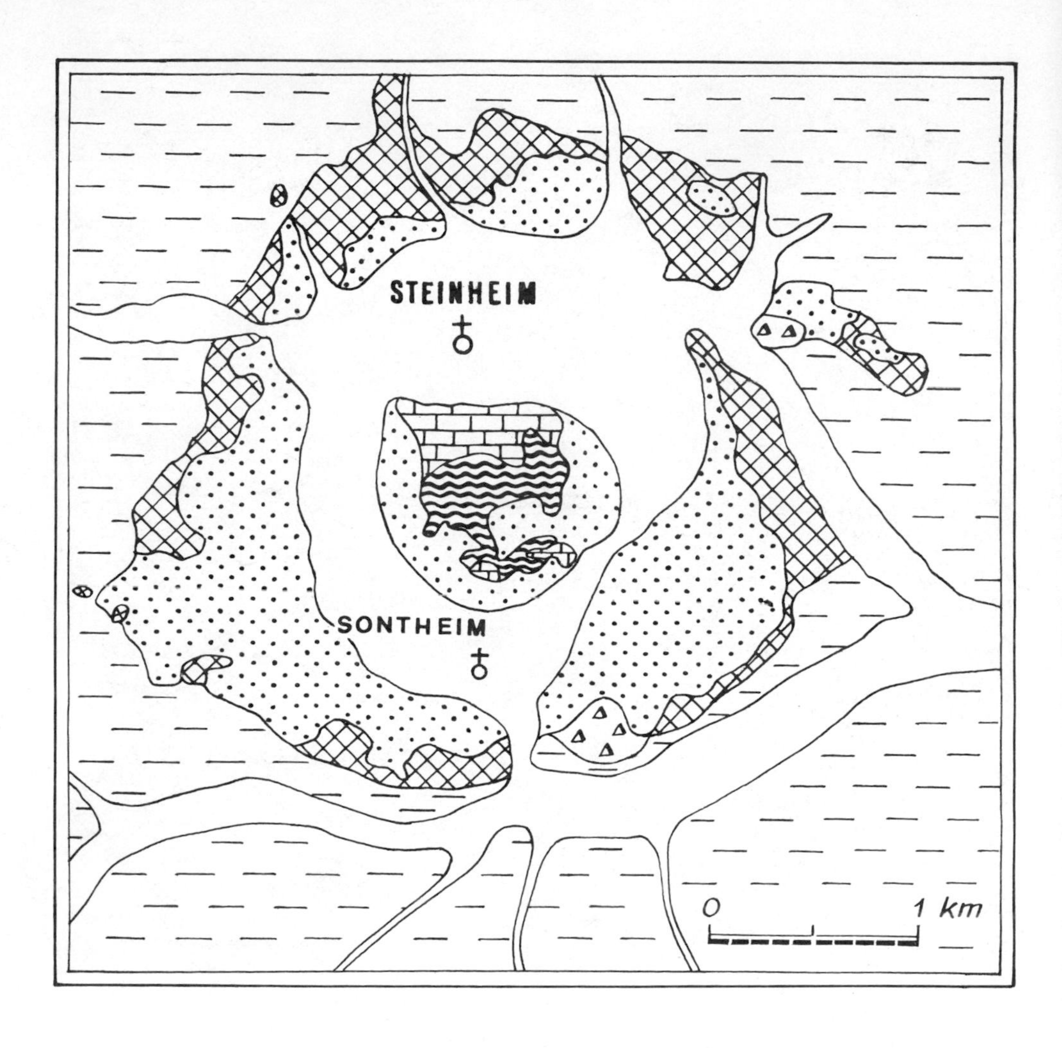

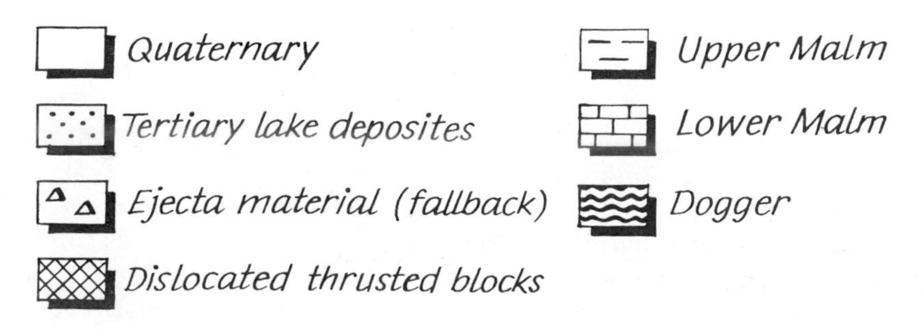

Fig. 3. Geologic sketchmap of Steinheim Basin.

Fig. 4. Shatter cones from the Steinheim Basin; north flank of the central uplift.

Fig. 5. Primary basin breccia (fallback), Steinheim Basin, core of drilling 22, SE of the central uplift, depth 77–78 m.

with planar features (60%‖{0001} and about 30%‖{10$\bar{1}$1}; Engelhardt *et al.*, 1967). The breccia covers the crater floor, part of thrusted blocks in the crater and near the rim and the flanks of the central uplift (Fig. 8).

The fact that the flanks of the central hill are covered with fallback ejecta shows that this uplift was in existence when the ejecta came down (Fig. 9). Exposures and drillings—one was 603 m deep—proved (Groschopf and Reiff, 1971; Reiff, 1974) that the central uplift has a structure of a megabreccia very similar to the megabreccia of the Wells Creek central uplift. In the flanks of the central uplift there are strata of Lower Malm with an inclination of 30–60%. In the center, stratified clays of Dogger and Lias have been offset along their layers and show a high degree of fracturing. The layers are almost vertical or have a high dip angle, they are faulted, folded and brecciated. They have been lifted some 250–380 m (Fig. 6), which is nearly the same as the Knox Dolomite uplift in the equal-sized Flynn Creek Crater, but probably in the case of Steinheim where there are clays a greater degree of plastic deformation was possible.

In the Lower Malm limestone of the central uplift flanks there are many shatter cones. In the Dogger of the central uplift, shatter cones are rare because banks of hard limestones are rare. The Dogger sandstone, which before the impact lay approximately 340 m under the surface, contains a great many quartz grains with planar features. The quartz grains with planar features in the fallback ejecta originate from this Dogger sandstone. On the contrary, quartz grains in the Triassic sandstone, which lay about 560 m under the surface, contain no planar features, but they are highly fractured. Consequently, the stress phenomena of the rocks caused by shock metamorphism decreases with depth.

Below the 80 m thick block of the Triassic sandstone layers, there is another block of Dogger and Lias. A facies of fine-grained sandstone in the Lowest Lias is very much fractured. However, the individual quartz grains are not fractured and do not show any planar features. The Lias sandstone also shows a shatter cone (Fig. 7). The shatter cones which consist of limestone of the Lower Malm and Dogger have a fine-striated surface structure. When compared with those, this particular shatter cone of the Lias sandstone has only a few broad striae. As the sandstone is very fine grained, the difference is probably not due to the rock structure. It is possible that at lower pressure the striated structure of the shatter cones is coarser than in the case of shatter cones which were formed by high pressure.

The block of Dogger and Lias strata below the Triassic sandstone has been pushed, by radial forces, towards the center and from there in a downward direction when the central uplift came into existence (Fig. 8).

The crater floor has been disturbed. It is highly fractured and partially of a brecciated structure. Also blocks have been uplifted and are now some 100 m above their original position. Possibly, the thrusted block of Galgenberg is a partially developed ring structure similar to the Inner Wall of the Ries Basin.

The Upper Malm in the center of the impact area at the site of the present central uplift could be evaporated or carried away as fine dust (David, 1969), but we have no evidence. The rest of the Upper Malm layers, which had surrounded

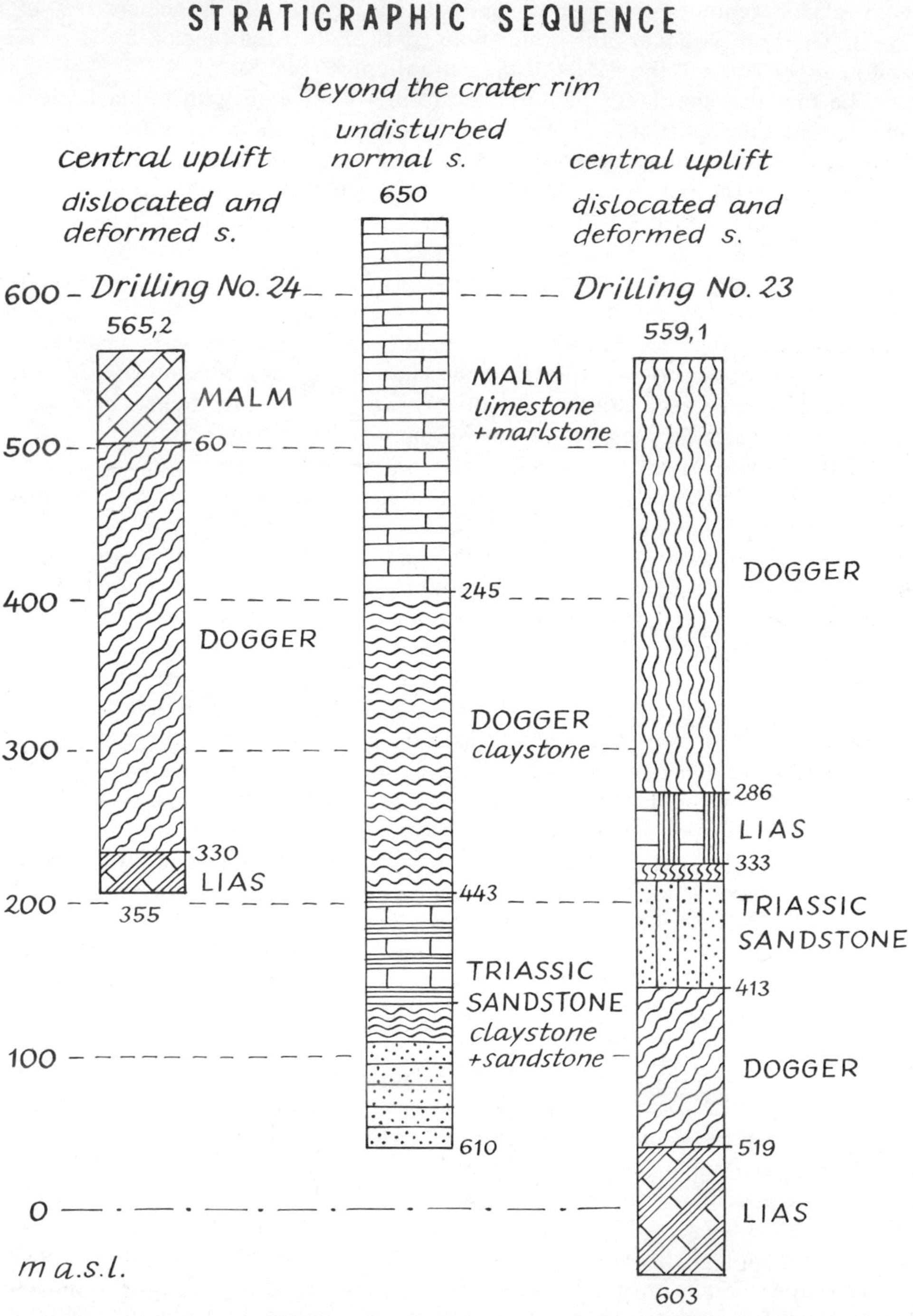

STRATIGRAPHIC SEQUENCE

beyond the crater rim

central uplift
dislocated and deformed s.
Drilling No. 24
565,2
MALM
60
DOGGER
330
LIAS
355

undisturbed normal s.
650
MALM
limestone
+marlstone
245
DOGGER
claystone
443
TRIASSIC
SANDSTONE
claystone
+sandstone
610

central uplift
dislocated and deformed s.
Drilling No. 23
559,1
DOGGER
286
LIAS
333
TRIASSIC
SANDSTONE
413
DOGGER
519
LIAS
603

600
500
400
300
200
100
0
m a.s.l.

Fig. 6. Profile of the drillings 23 and 24 in the central uplift of the Steinheim Basin.

Fig. 7. Shatter cone, Lower Lias sandstone, core of drilling 23, depth 589 m, central uplift of Steinheim Basin.

the center, were pushed in blocks, radially away from the center. While being pushed away, the blocks became highly fractured. In a former quarry at the "Burgstall"—a hill in the wall zone at the south rim—there are some parts which preserved their stratified structure, but they are in a vertical and inclined position. Some parts have been folded or show intensive breccia formation. In the massive breccia rock, chert nodules show that the rock originally was stratified, because nonbedded limestones of the Upper Malm have no chert nodules. This breccia is very similar to breccias in the same position in the Flynn Creek Crater. Thrusted blocks like the Burgstall form a circle of a width of 200–400 m making up the crater wall.

The rim zone shows rock that has been weakly fractured; but this belt with a width of approximately 300 m is of a shallow flexure structure studded with small faults. At some small exposures we see that the layers of the rim dip slightly toward the crater. There is no evidence that these rocks have been uplifted or overturned. It is possible that in small areas a slight uptilting exists, but there are no distinct exposures.

Material of the impacting body has never been found. But in size, form and structure, and in many additional details, the Steinheim crater is very similar to the Flynn Creek Crater. Moreover, structural and theoretical comparisons of the Flynn Creek Crater with known impact and explosion craters indicate that such craters were formed by low-density impacting bodies, i.e., comets (Roddy, 1968). Recently, El Goresy and Chao (1976) could find a hint that the Ries Basin was created by stone meteorite. So possibly the Steinheim Basin was also created by

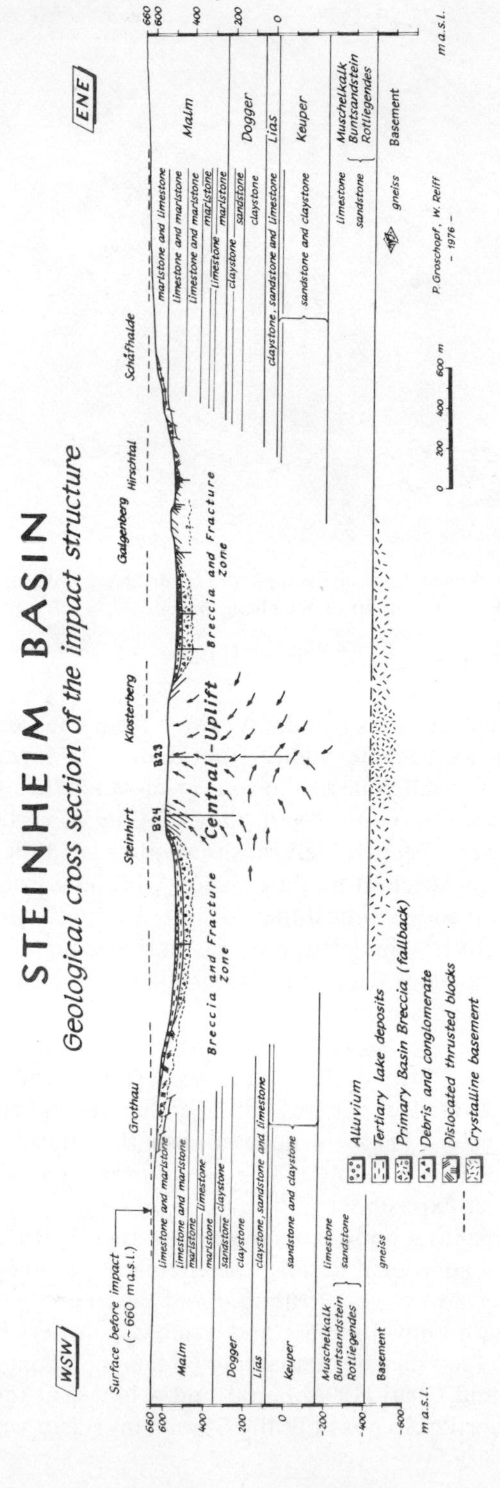

Fig. 8. Cross section through the Steinheim Basin.

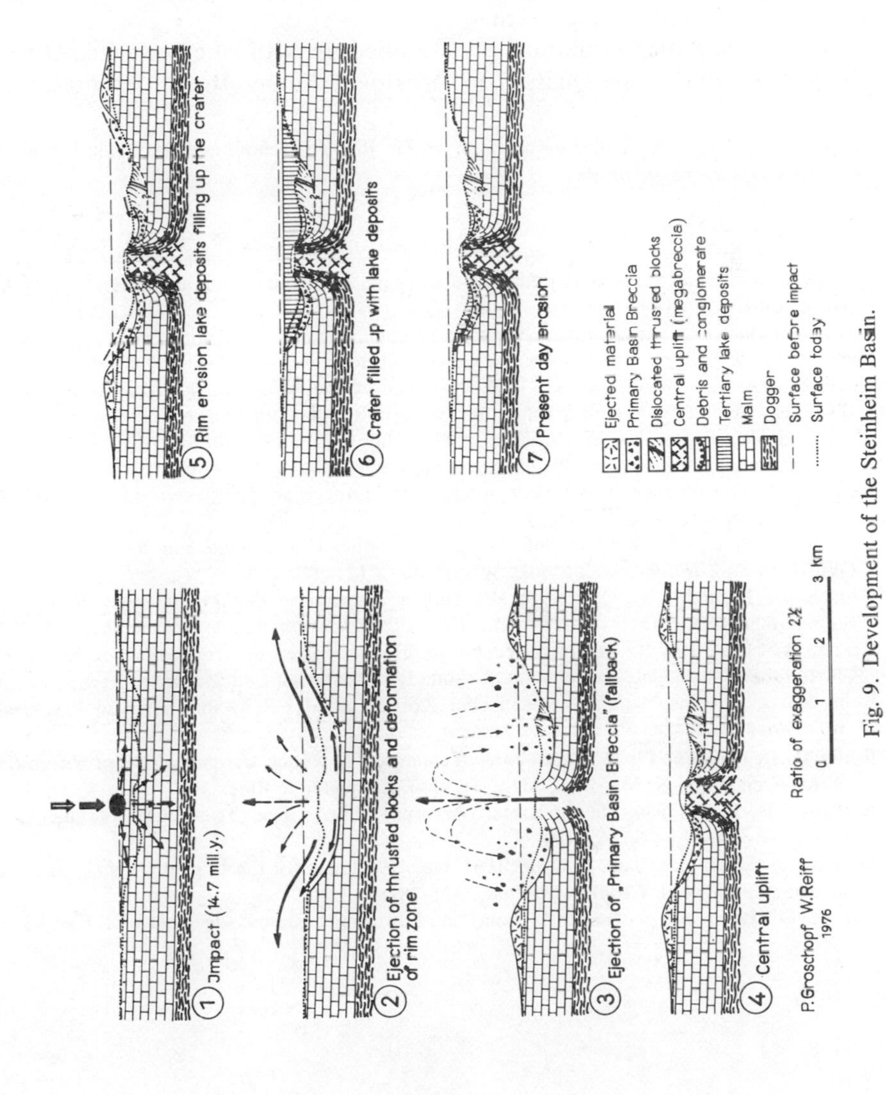

Fig. 9. Development of the Steinheim Basin.

an impact of a stony meteorite, because both craters have the same age, but there is no evidence.

By radiometry (K/Ar) and fission track (Gentner and Wagner, 1969) on suevite rock, the age of the Ries Crater was determined to be 14.7 million years. In both the Ries and the Steinheim Basin there are Tertiary lake deposits of the late Tortonian/early Sarmatian age. As a consequence, it is probable that both craters were created by the same event.

In the Sarmatian both craters were filled up with lake deposits. At the end of the Tertiary and in the Quaternary, erosion exhumed the buried craters (Fig. 9).

Acknowledgments—The author wishes to thank Dr. Paul Groschopf and Dr. David J. Roddy for their assistance and exchange of ideas.

REFERENCES

Branco, W. and Fraas, E.: 1905, Das Kryptovulkanische Becken von Steinheim. *Abh. Kgl. Preuss. Akad. Wiss.*, 1–64.

David, E.: 1969, Das Ries-Ereignis als physikalischer Vorgang. *Geologica Bavarica* **61**, 350–378.

Dietz, R. S.: 1959, Shatter cones in cryptoexplosion structures (Meteorite impact?). *J. Geol.* **67**, 496–505.

El Goresy, A. and Chao, E. C. T.: 1976, Evidence of the impacting body of the Ries Crater—the discovery of Fe-Cr-Ni veinlets below the crater bottom. *Earth Planet. Sci. Lett.* **31**, 330–340.

Engelhardt, W. v., Bertsch, W., Stöffler, D. Groschopf, P., and Reiff, W.: 1967, Anzeichen für den meteoritischen Ursprung des Beckens von Steinheim. *Die Naturwissenschaften* **54**, 198–199

Gentner, W. and Wagner, G. A.: 1969, Altersbestimmungen an Riesgläsern und Moldaviten. *Geologica Bavarica* **61**, 296–303.

Groschopf, P. and Reiff, W.: 1966, Ergebnisse neuer Untersuchungen im Steinheimer Becken (Württemberg). *Jh. vaterl. Naturkde. Württemberg* **121**, 155–168.

Groschopf, P. and Reiff, W.: 1971, Vorläufige Ergebnisse der Forschungsbohrungen 1970 im Steinheimer Becken (Schwäbische Alb). *Jh. geol. Landesamt Baden-Württemberg* **13**, 223–226.

Kranz, W.: 1924, Das Steinheimer Becken.-In: Begleitworte zur geognostischen Spezialkarte von Württemberg. Atlasblatt Heidenheim, 2. Aufl. *Württ. Statist. Landesamt*, p. 1–38.

Reiff, W.: 1974, Einschlagkrater kosmischer Körper auf der Schwäbischen und Fränkischen Alb. *Aufschluss* **25**, 12–24.

Roddy, D.J.: 1968, The Flynn Creek Crater, Tennessee. In *Shock Metamorphism of Natural Materials* (B. M. French and N. M. Short, eds.), p. 291–322. Academic Press.

Rohleder, H. P. T.: 1934, Meteor-Krater (Arizona)—Salzpfanne (Transvaal)—Steinheimer Becken. *Z. deutsch. geol. Ges.* **85**, 463–468.

Shoemaker, E. M. and Chao, E. C. T.: 1961, New Evidence for the Impact Origin of the Ries Basin, Bavaria, Germany. *J. Geophys. Res.* **66**, 3371–3378.

Stutzer, O.: 1936, Meteor-Krater (Arizona) und Nördlinger Ries. *Z. deutsch. geol. Ges.* **88**, 510–523.

Roddy, D. J., Pepin, R. O., and Merrill, R. B., editors.
(1977) *Impact and Explosion Cratering*, Pergamon Press (New York), p. 321–341.
Printed in the United States of America

Deformation at the Decaturville impact structure, Missouri

TERRY W. OFFIELD and HOWARD A. POHN

U.S. Geological Survey, Denver, Colorado 80225

Abstract—The Decaturville structure of central Missouri consists of a central uplift surrounded by a structural depression, in turn bounded by a normal fault. In the central uplift, Cambrian rocks normally 300 m or more deep are exposed, and isolated blocks of Precambrian basement pegmatite and schist occur 550 m above their normal position. Convoluted strata in the uplift indicate that inward movement and crowding of beds accompanied the upward movement. The inward movement involved folding and thrusting, and was succeeded by adjustments on steep faults with both upward and downward displacement as much as 150 m. The depressed zone around the uplift is characterized by thrusts inward and outward relative to the center, and by steep faults which formed both before and after thrusting took place.

Shock features include monolithologic and mixed breccias, shatter cones, planar elements in quartz, and intense intragranular deformation. Shatterconed rocks form a capping, relatively flat-lying layer over a megabraccia column containing blocks up to several tens of meters in size. These jumbled blocks are from formations which make up the bottom 240 m of the 550 m disturbed sequence. A mixed-breccia matrix around the blocks contains quartz grains with planar features probably indicating shock pressures as great as 70–100 kb. According to study of drill cores, shock effects decrease in intensity downward and the basement rocks do not contain shock features.

Post-structure erosion in the area is estimated at less than 50 m, which could not have destroyed a cavity estimated to be 3100 m in diameter and 550 m deep. Instead the cavity was destroyed by the immediate inward movement of beds to form the central uplift. This movement is believed to have been a response initiated during the phase of rarefaction and pressure release following passage of the shock wave, and produced the ring fault and structural depression as necessary concomitants to the development of the central uplift. As the cavity closed, material spalling from the walls and probably fallback breccia were trapped, mixed, and partly crushed to form the megabreccia mass at the center. This explanation may apply to other astroblemes where erosion is not thought to be great but where a central uplift is present without an obvious topographic crater.

INTRODUCTION

FOR MORE THAN 100 years geologists have been intrigued by the Decaturville structure, a small area of intense disturbance marked by surface exposure of Precambrian rock and minor lead-zinc mineralization. Explanations of the structure have generally invoked subterranean volcanic explosive processes (e.g., Bucher, 1963; Snyder and Gerdeman, 1965), but detailed mapping of the entire structure and study of drill core reveal a complex structural pattern not consistent with an endogenetic origin. This pattern and the presence of shock-deformation features are strong evidence that the Decaturville structure was formed by impact of an extraterrestrial body, as first suggested by Boon and Albritton (1936). The 7300 m of drill core available and excellent exposures in key areas provide insight into late-stage deformation during crater formation in layered rocks and may explain puzzling features in other astroblemes.

GEOLOGIC SETTING

The Decaturville structure is located on the Ozark Plateau of central Missouri about 13 km south of Camdenton (Fig. 1). A ring fault bounds the slightly elliptical structure (5.6–6.4 km in diameter) and separates it from the surrounding plateau of flat-lying, little-disturbed sedimentary rocks (Fig. 2). The structure consists of a central uplift and surrounding structurally depressed zone. Cambrian strata and Precambrian rocks exposed in the uplift, and Ordovician and Silurian limestones preserved in the ring depression do not crop out elsewhere in central Missouri. Figure 3 shows the stratigraphic units involved in the structure. Either the Jefferson City Dolomite or the Roubidoux Dolomite caps the adjacent plateau, and the Gasconade Dolomite is exposed in valleys.

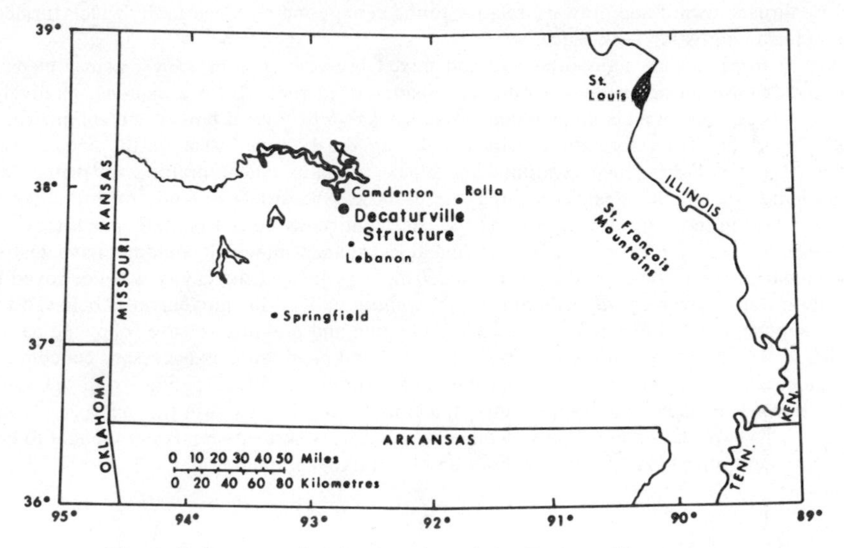

Fig. 1. Index map showing location of the Decaturville structure.

The structure lies on the crestal zone of a gentle regional upwarp called the Proctor anticline, the limbs of which dip less than 2 m per km. Within 5–6 km of the ring fault a few beds dip as steeply as 32°; they do not define folds and lie outside an undisturbed zone all around the structure and are believed to indicate either solution collapse in the carbonate terrane or the presence of small faults related to very minor regional deformation. As much as 2 km outside the ring fault a few low-amplitude undulations and one fairly sharp monoclinal flexure are oriented crudely circumferential with respect to the Decaturville structure. A few small faults probably related to the formation of the structure also are found within 2 km of the ring fault.

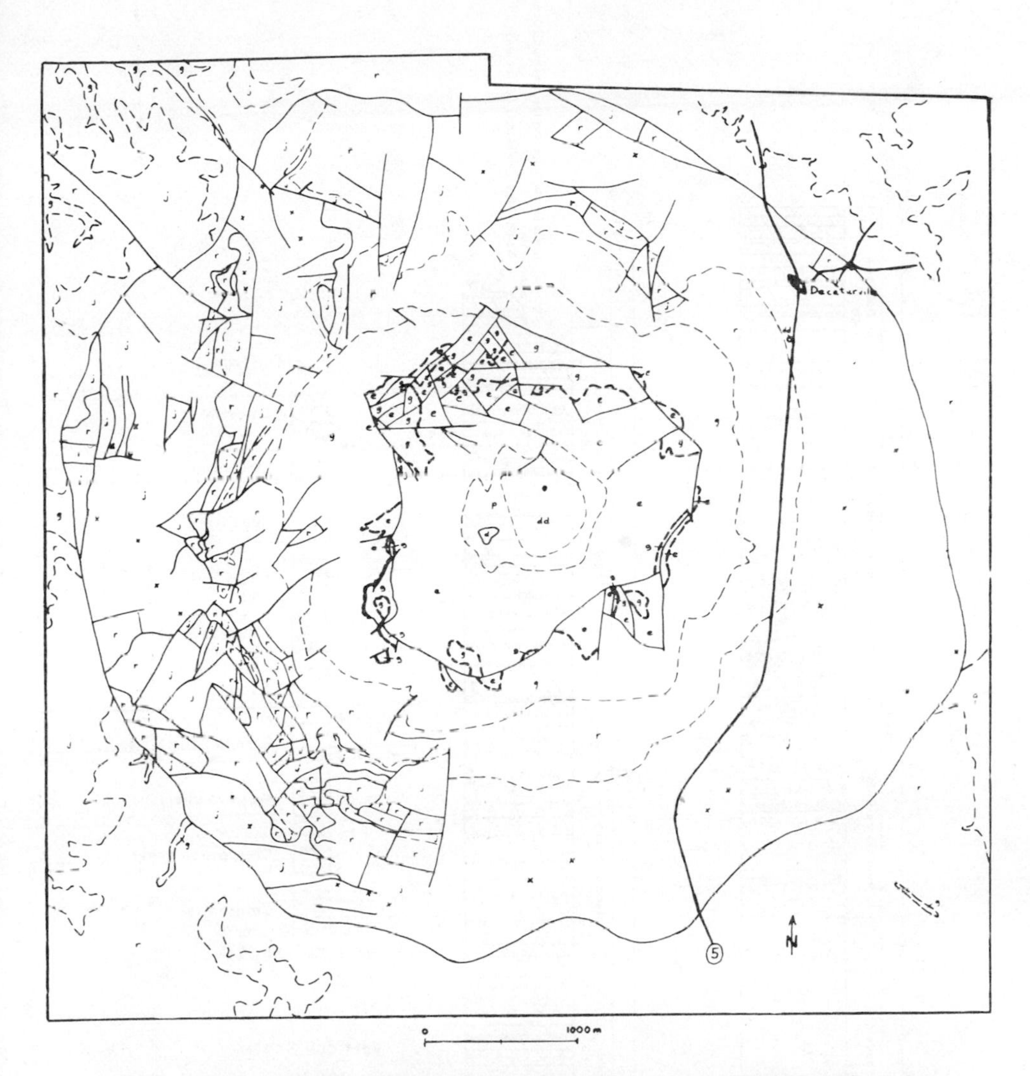

Fig. 2. Generalized geologic map of the Decaturville structure. Refer to Fig. 3 for stratigraphic section. Solid black, Precambrian; d, Davis Formation; dd, Derby and Doe Run Dolomites; p, Potosi Dolomite; e, Eminence Dolomite, all of Ordovician age; heavy dashed line, Gunter Sandstone Member of Gasconade; g, Gasconade Dolomite; r, Roubidoux Dolomite; j, Jefferson City Dolomite; x, isolated outcrops of Ordovician and Silurian limestones resting on Jefferson City. Dashed lines contacts, solid lines faults.

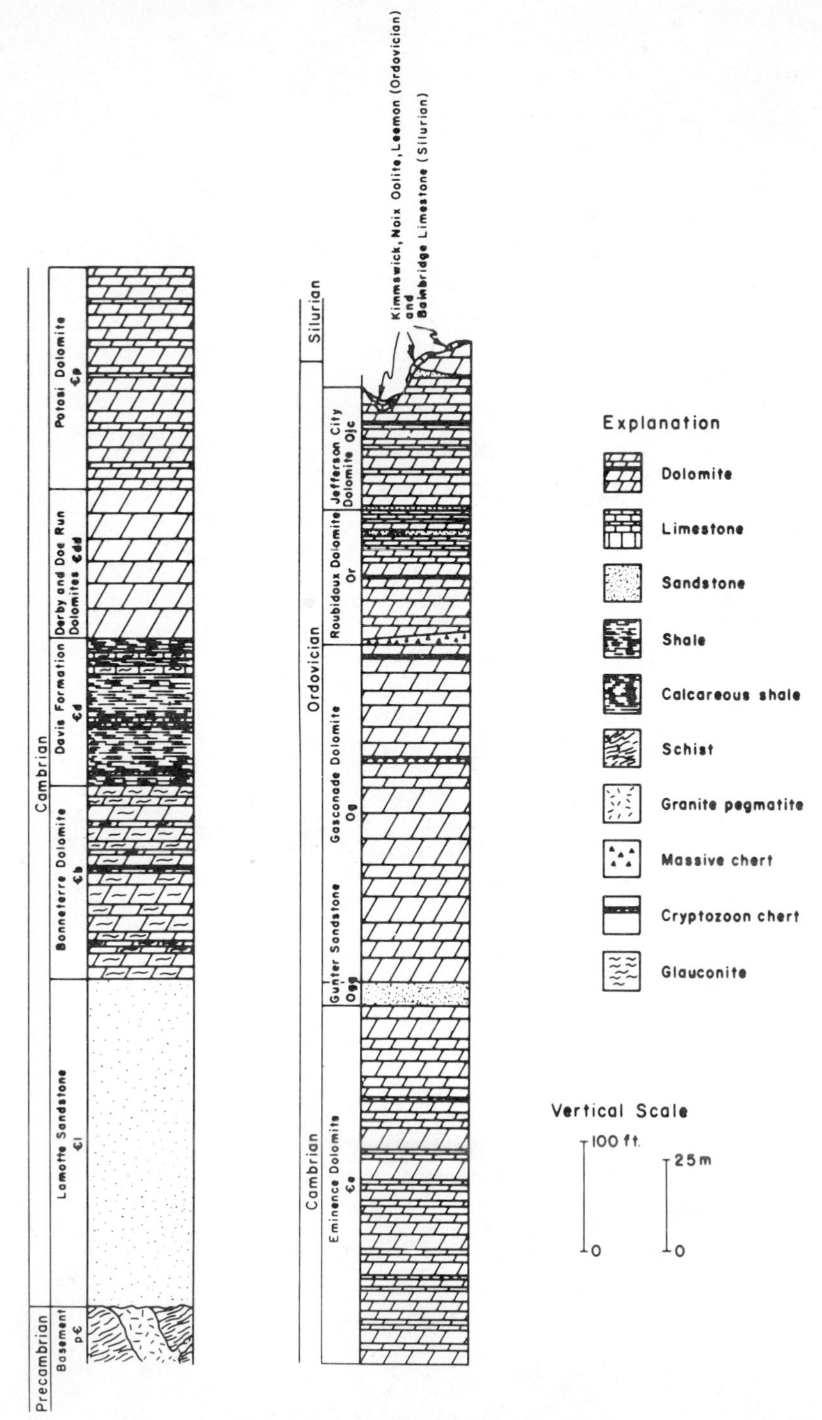

Fig. 3. Local stratigraphic sequence.

STRUCTURE

Ring fault and depressed zone

The boundary of the structure is a circular normal fault with displacements of 20–130 m down toward the center of the structure. Outcrops just outside the fault are rare, but drag on footwall beds commonly is very small and extends no more than 30–60 m outside the fault. The ring fault is broken by a few cross faults, with low dips localized along them as far as 180 m beyond the ring. The cross faults have little vertical displacement outside the ring fault and so are considered to be tear faults indicative of horizontal movements.

The ring fault is poorly exposed but its even trace indicates that it is generally steep at the present ground surface. However, a drill hole 150 m inside the fault indicated an average dip of 45°, suggesting that the steep dips seen at the surface become shallower at modest depths. We believe that in profile the fault curves inward below ground to form a continuous bowl-like boundary around the entire Decaturville structure.

A structurally depressed zone 1300–1600 m wide lies inside the ring fault. Figure 4 shows lines of equal vertical displacement throughout the structure and indicates that the depressed zone is variously monoclinal, synformal, and antiformal in cross section. The zone gives way inward to the central uplift at a hinge line generally within the Roubidoux outcrop belt. Roubidoux in normal stratigraphic sequence (Fig. 3) is exposed by erosion in relative structural highs within the depressed zone, but more commonly is exposed in windows through thrust plates of Jefferson City or in thrust plates overlying Jefferson City rocks (Fig. 2). Complex interrelated folds, steep faults, and particularly thrust faults can be mapped or reasonably inferred throughout the zone in the western half of the structure. The remainder of the zone has sparse outcrops and may be of equally complex structure, but because only the top few meters of the rock section generally are recognized in outcrop, it is believed that fairly gentle, broad undulations and small faults and folds characterize this part. In the southern and eastern areas of the ring depression, small folds typically are asymmetric, with the steeper limb toward the center of the structure, and the few inferred faults appear to be steep.

For the western sector, outcrop data are good enough to permit construction of block diagrams, which in turn provide a means to add inferred structural details that appear necessary to explain observed surface complexities. Selected block diagrams are shown in Fig. 5 to illustrate different styles of deformation around the structure.

Block 1 displays a schuppen-like arrangement of 10 thrust plates. Based on intersection and superposition relationships, it appears that the two earliest thrusts moved northward clockwise relative to the center of the structure, and the later thrusts moved northeast directly toward the center. From measured positions of marker beds, lateral displacements probably range from 30 to 90 m, with vertical throw on the order of 6–30 m. In block 2, folds oriented circumferential to the center of the structure formed first and were broken by steep

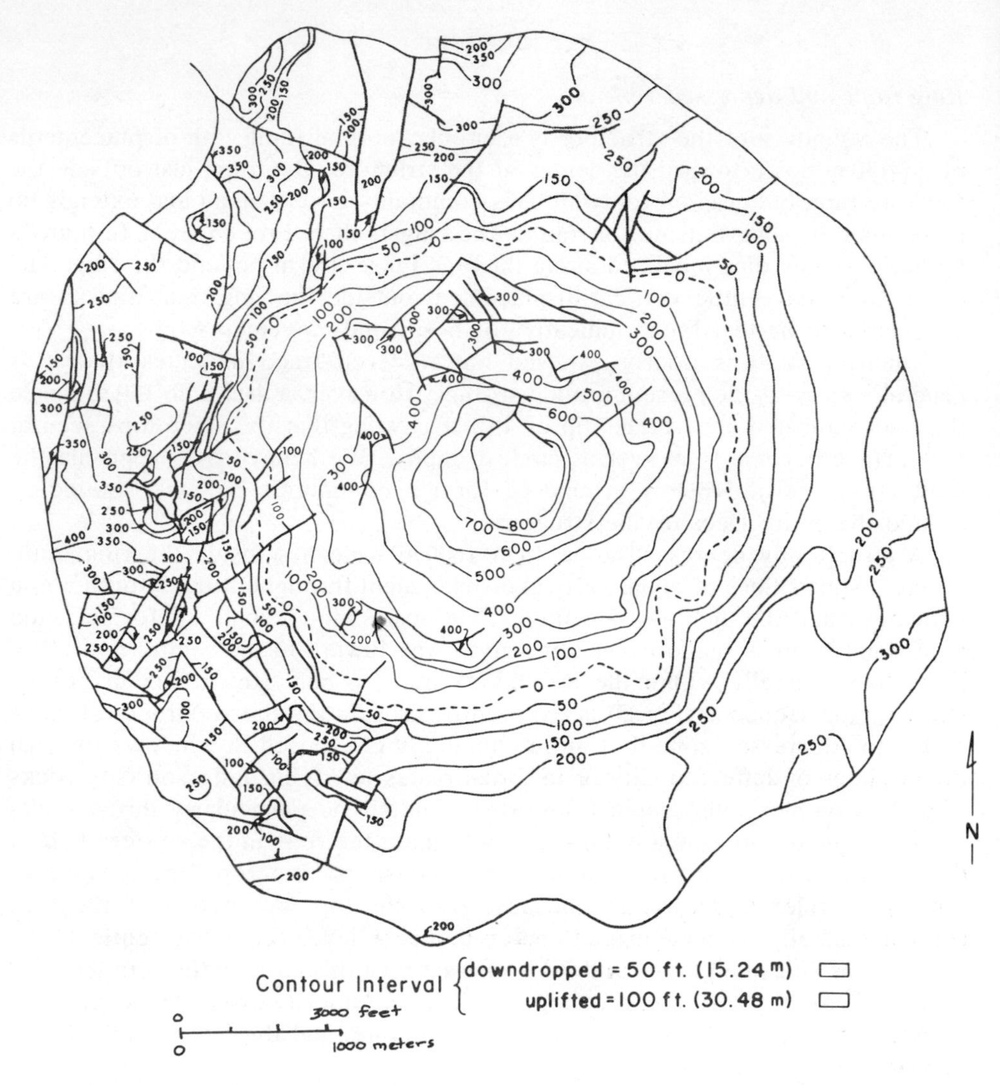

Fig. 4. Schematic map contoured to show lines of equal vertical displacement in feet. Central uplift area unshaded, ring-depression area shaded.

faults in a horst-and-graben pattern. A small thrust appears to have formed after the folds and before the steep faults, but the main thrust movement clearly was after the normal faults and involved displacement of the entire mass toward the center of the structure. Block 3 depicts a series of imbricate thrusts involving outward movement of plates. In the outermost thrust a marker sandstone in the Roubidoux evidently underwent thickening by intergranular flow during its outward thrusting of sliding; at one place it is 12 m thick instead of its usual 1.2 m. A small outlier of the sandstone shows the former minimum extent of the thrust plate. The section is virtually unfolded.

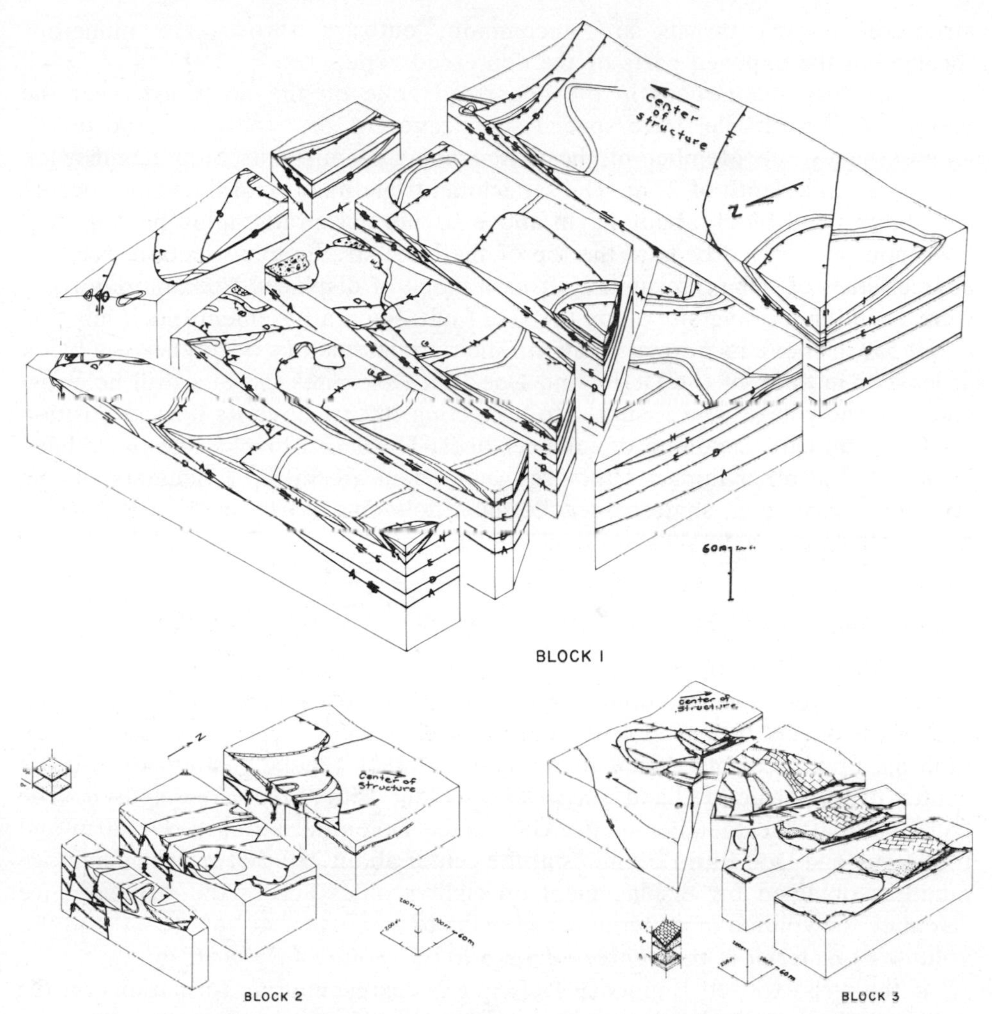

Fig. 5. Block diagrams displaying local structural complexities in the ring depression. In block 1 thrusts lettered in time sequence (A formed first, J last).

These and other structures seen in the depressed zone permit generalizations concerning the sequence of deformation: circumferential folds are common and radial folds relatively rare; folds appear to be the earliest structures in the sequence. Associated steep faults post-date these folds and generally are earlier than thrust faults. On the well-exposed west side of the structure, however, two phases of faulting on steeply dipping surfaces occurred—one associated with early folds and before thrusting, and a second one of late, post-thrust vertical adjustments along relatively long faults. In two out of three instances where outward and inward thrusts occur together, inward movement appears to have taken place earlier than outward thrusting. Except in the southwest part of the structure, inward thrusts are uncommon; outward thrusts are numerous throughout the exposed parts of the depressed zone.

Three rock occurences in the depressed zone on the northeast, near the village of Decaturville, have special significance. A 55×180 m outcrop of the Gunter Sandstone Member of the Gasconade Dolomite sits atop Kimmswick Limestone at a depth of 27 m. The sandstone, triple its normal thickness, occurs as a large tilted block about 175 m above its normal stratigraphic position and lies upon undistorted beds at the top of the local stratigraphic section. Nearby, coarse flakes of mica, presumably from a block of displaced Precambrian rock, were found in a grave site along the ring fault. Such a basement block must lie about 550 m above its normal position. The third anomalous occurence is a block at least 15 m thick of the Derby and Doe Run Dolomites cut in a drill hole just south of the village, and assumed to be resting 300 m above its normal position on Jefferson City Dolomite or younger beds. Displaced blocks like these have been noted at other impact structures and in fact are rather a hallmark of this structural class (e.g., shatterconed Permian dolomite blocks in the rim zone at Sierra Madera, Texas; Wilshire *et al.*, 1972b).

Central uplift

The line marking the change from depressed zone to central uplift is irregular but generally circular; the uplift diameter is about 10% greater in the northeast-southwest direction than in other directions (Fig. 4). Except for areas of block faulting, upward displacement increases in a fairly regular fashion toward the center of the structure, and successively older formations are exposed. The Gunter Sandstone Member of the Gasconade Dolomite is raised 90–120 m and the Derby and Doe Run Dolomites at the center about 280–300 m. Using average numbers obtained for displacement on eight profiles across the structure, we calculate the volume of material in the uplift to be about $7.4–7.6 \times 10^8$ m^3 and the volume of material in the depressed zone to be about $7.4–7.9 \times 10^8$ m^3.

If the well-exposed Eminence Dolomite is representative, formations on the uplift are intricately deformed on a small scale. However, it appears that the formations as whole units have relatively gentle dips off the dome. The undivided Derby and Doe Run Dolomites are intensely deformed and shatterconed, but as a composite unit must be nearly flat lying in order to cover the central area 490–610 m across; the unit is only 43 m thick and its base is nowhere exposed in the area.

In the entire uplifted sequence, the Gunter Sandstone Member is the only thin marker bed useful for detecting structural offsets. Its outcrop line is conspicuously lobate and defines numerous gentle folds around axes radial to the center (in contrast to circumferential folds in the depressed zone). At about the zone of Gunter outcrops, a system of apparently steep faults nearly rings the center of the structure (Fig. 2). On the east side of the dome the central area is displaced upward along these faults, and on the west side it is displaced downward. Outside this fault line, block faulting has raised Eminence and

Gunter as much as 90–120 m into the Gasconade outcrop belt. In a complex of steep faults on the north side, Roubidoux and Jefferson City blocks are down-dropped into the Gasconade, the vertical displacement of the Jefferson City being as much as 135 m. The block faulting seems to have been a late phase of major vertical adjustment.

Structures outlined by the Gunter are very complex. In the northwest area folds vary from broad, open warps to tight folds overturned outward, all cut by steep faults with scissors, hinge-like, and simple vertical displacements. In the southwest area deformation appears to have involved three thrust sheets, partly broken into schuppen during thrusting, and later steeply dipping faults along which small block movements occurred. The thrusts brought Gasconade inward over Eminence, and Gunter in two slices over Gasconade. The inward thrusts and the lobate Gunter line are believed to indicate an important phase of inward movement as the uplift formed. The total length of Gunter segments (Fig. 2) is about 30% longer than the perimeter of the somewhat triangular area which they outline. This implies a shortening of the perimeter of the Gunter at its original prestructure stratigraphic level, as a result of inward and upward movement during formation of the central uplift. Even if some lengthening of Gunter segments is due to thinning as it moved, still 20% shortening of the perimeter during imward crowding is a reasonable minimum estimate. Similar configurations of marker beds in central uplifts have been reported from Sierra Madera, Texas, Wells Creek Basin, Tennessee, Gosses Bluff, Australia, and Vredefort, South Africa (Wilshire and Howard, 1968).

Mapping of structure inside the Gunter Member is difficult. Bed truncations and breccia lenses indicate numerous faults, but displacements and types of faults are difficult to determine. In the Eminence belt most deformation involves small open basins and canoe-like folds which commonly appear to be slightly rotated and laterally dislocated short distances. The Derby and Doe Run are highly shattered but lie essentially flat across the central area. Drilling shows the Derby and Doe Run to form a nearly continuous cap layer over a column of breccia involving large (dimensions up to 60 m) blocks of rock units from Potosi down to the basement. These blocks are jumbled together, commonly displaced several tens of meters out of place stratigraphically, and surrounded by a matrix of fine-grained mixed breccia.

Near the center of the structure, blocks from other formations are mixed with Derby and Doe Run at the surface. Displaced blocks and slivers include Precambrian pegmatite and schist, Lamotte Sandstone, shale of the Davis Formation, and Potosi Dolomite. The Precambrian block is about 15 × 20 × 15 m and is some 200 m above normal stratigraphic position and 550 m above normal elevation. Only a few other small slivers of Precambrian were seen in the numerous drill cores from the central area. About 500 m southwest of the pegmatite a trench exposes shattered Potosi in normal position as host to blocks of Eminence, Derby and Doe Run, and Davis surrounded by mixed breccia containing fragments of sulfide minerals. One block of upfaulted Davis covers an area 90 × 90 m and is 40 m thick. Sandstone from the Lamotte about 150 m below this position in the Potosi has been injected to form a rind around one block of shattered Potosi.

The main column of megabreccia at the center below the Derby and Doe Run is defined by drilling as on the order of 600 m long in the northeast-southwest direction and at least 275 m wide. The subsurface complexity shown by cores would be difficult to overstate. From just below the capping layer to as deep as drill information exists, it is rare to be able to connect formations cut in adjacent holes only 15 m apart. The mixed breccia around the blocks exhibits evidence of flow and shock deformation. In four holes around the pegmatite, the mixed breccia occurs not just as matrix but makes up most of the 160–240 m penetrated in each hole.

Depth of the megabreccia column is greater than 320 m, judging from the one hole that went that deep. The basement and overlying Lamotte were disrupted during uplift, but the small amount of these rocks found in the cores suggests less mixing with other units than happened with stratigraphically higher formations. If the stratigraphic pile were uplifted without disruption, then basement would lie about 200 m deep at the center of the uplift. In fact, except for isolated blocks torn loose and lifted, both the Lamotte and the basement lie below 320 m. The lack of Lamotte mixed in breccia 320 m deep suggests that it may occur at least a few tens of meters deeper, and as the Lamotte is 90 m thick, the basement below Lamotte is likely to be 440–470 m or more deep. Normally, basement in the surrounding plateau is 490–520 m below the elevation at the center of the structure, so little general uplift of the basement surface under the structure seems indicated.

Intensity of formation mixing within the megabreccia mass is greatest at its apparent center: only there are blocks of basement found, and only there is fine-scale mixed breccia the dominant lithology rather than a filling around large blocks. The general pattern seen in the megabreccia suggests that it is of limited lateral extent and confined under beds which have slid inward and upward to cap the uplift center, and that it dominantly involved formations above the Lamotte but no higher than the Potosi.

Shock-deformation features

The Decaturville structure contains several kinds of features indicating extremely intense and/or unusual deformation: monolithologic and mixed breccias, rock granulation and small-scale rhombohedral rock cleavage, shatter cones, and intragranular features in quartz, dolomite, calcite, and mica.

Monolithologic breccias consisting of clasts and mylonitic matrix derived from single beds occur throughout the structure in most of the fine- to medium-grained carbonate rocks (Fig. 6a). They are like impact breccias described by Wilshire and others (1972a), and result from shattering and dilation of individual beds. Even in sequences of beds less than 3 cm thick, each brecciated bed maintains perfect stratigraphic coherence with no mixing of adjacent beds. For example, in the Jefferson City an 8 cm bed of clean gray dolomite typically will

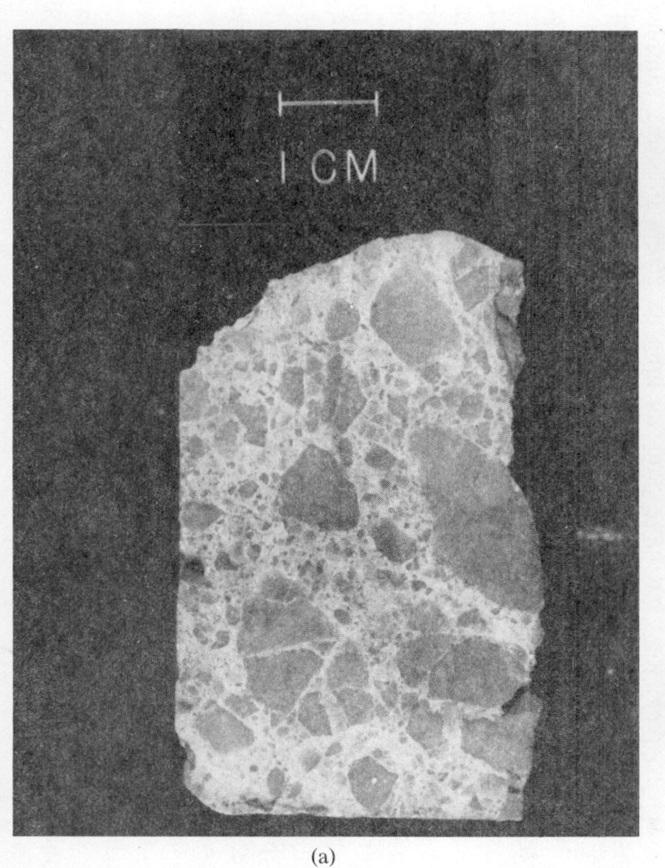

(a)

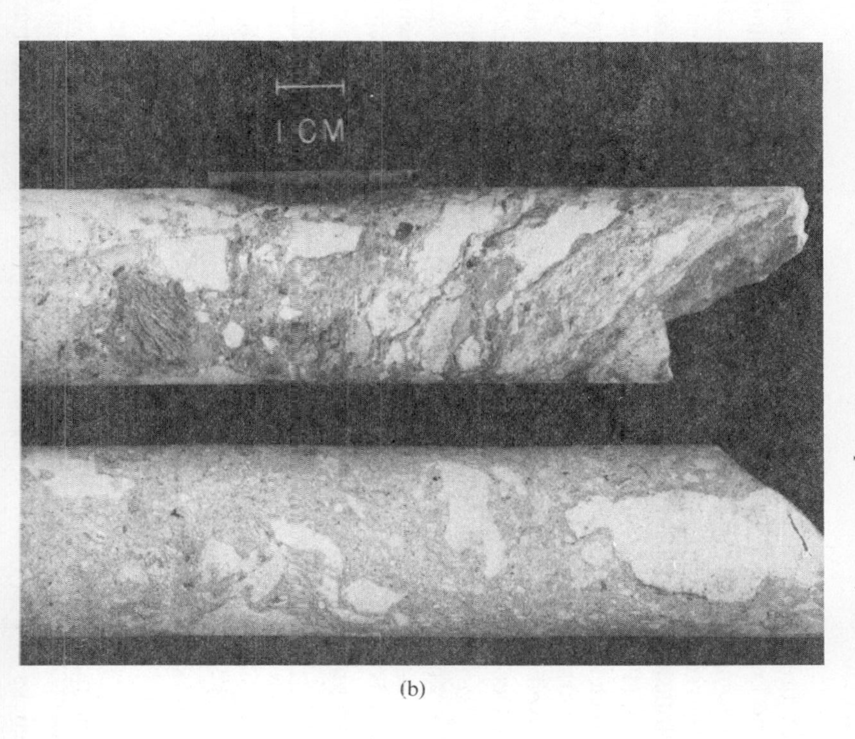

(b)

Fig. 6. (a) Sawed surface of monolithologic breccia. (b) Cores of mixed breccia showing fragment mixture and flow features.

be thoroughly brecciated and 3 cm beds of tan marly dolomite above and below it will show no brecciation whatever. Many fragments appear to have undergone rotation and some alignment of grains near the borders of fragments suggests flow of the matrix. The shattering and dilation which fragment the dolomite beds probably is followed by both flowage and crushing recompaction to produce the matrix. Roddy (1968) has described identical deformation in carbonate rocks at the Flynn Creek impact crater in Tennessee and at the Steinheim impact crater in Germany (pers. comm., 1976).

In the mixed breccia around blocks of the megabreccia column, dolomite fragments of Derby and Doe Run are markedly predominant, though glauconitic pieces of Davis also are common. Quartz may constitute as much as 10% of some samples, and some grains display planar elements. Shatter-cone segments are present in a few fragments of Potosi and Derby and Doe Run within the breccia. The mixed breccia may show fine banding in the matrix, with folds, swirls, and wrapping around sculpted fragments (Fig. 6b), and in places dolomite fragments themselves appear to be drawn out plastically to form ragged-ended schlieren. At places the breccia contains up to 10% sulfide fragments. Many fragments are botryoidal or concentrically banded, with sharp broken edges clearly showing that they predate brecciation. Overgrowths on a few sulfide fragments suggest minor deposition or redistribution of iron sulfide after the breccia formed.

Well-formed shatter cones are common in outcrops of Derby and Doe Run in a central area measuring about 460 × 520 m. Cones are not found in Derby and Doe Run outside that area nor in the surrounding outcrop belt of Potosi but are present in blocks of Potosi faulted into Derby and Doe Run at the center. Cones typically are 2.5–5 cm long but have axes as long as 25 cm in some samples. Striations plot stereographically to give apical angles ranging from 79° to 92°, with a mean of 86°. Unfortunately bedding cannot be identified in the massive shatterconed dolomites, so shatter-cone orientations cannot be used to determine a focus of energy release at Decaturville.

Planar elements in quartz include cleavage, planar features, and probable deformation lamellae such as described by Carter (1968). Open rhombohedral cleavage planes are common in grains from the Gunter Sandstone Member, and are found in a few grains from the Lamotte Sandstone but not in quartz from any other formations. A few grains from the Lamotte have well-developed basal cleavage, a type not seen in the Gunter. Both types of cleavage in Lamotte grains may show small displacement or faults (Carter, 1968). Grains with any kind of planar fractures probably make up no more than 2–3% of the quartz in Lamotte. Moreover, only Lamotte samples from the center show planar elements; samples from a trench 500 m away show intergranular flowage but no planar elements. Quartz from the Precambrian rocks shows unusual curving or chevron-like strain extinction bands, but complexly sutured boundaries are undisturbed and cleavage is absent.

One to two percent of the quartz grains of mixed-breccia matrix display rhombohedral and basal cleavages, about equally abundant and well developed.

In a very few grains planar features are oriented parallel to {10$\bar{1}$3}. Even rarer are probable deformation lamellae, planar elements parallel to {0001} and containing abundant tiny cavities or inclusions. Planar elements at Decaturville are generally similar to those from Meteor Crater and Middlesboro, Kentucky, and deficient in the {10$\bar{1}$3} relative to Sierra Madera, Texas, Clearwater Lake, Quebec, or rocks experimentally shocked at 250–300 km pressures. The stratigraphic origin of grains with various shock features in the mixed breccia is not certain; they may have come from the Lamotte, Gunter, or Davis. As no fragments from units higher than Potosi have been found in the breccia, it does not seem likely that sand grains from the Gunter would be present. Grains in recognizable fragments of the Lamotte do not show planar features. However, planar features are found in clusters of angular grains such as occur in silty portions of the Davis.

Cleavage in quartz (other than the simple open fracturing seen in the Gunter) begins to form at pressures of about 50 kb (Short, 1966; Hörz, 1968). Microfaults associated with cleavages indicate pressures in excess of 40 kb (Carter, 1968). These features are seen in at least part of the Lamotte, probably the top part judging from the absence of shock indicators in the available samples of pegmatite and schist from the underlying basement. The pegmatite does have mica with kink bands; these cannot be said definitely to have formed by shock, and if they did they could indicate pressures as low as 9 kb (Hörz, 1970). Planar features and probable deformation lamellae in quartz believed to come from the Davis are not common and are rather poorly developed, but they likely indicate deformation under pressures of about 100 kb or somewhat less, depending on the temperature and other factors (Carter, 1968; Hörz, 1968).

Permanent damage to crystal lattices is shown by asterism and line broadening. X-ray examination of single dolomite crystals from shatter cones in the Derby and Doe Run near the center of the structure has been reported by Simons and Dachille (1965) and by Dachille *et al.* (1968). One Decaturville sample showed a complete powder arc and extreme line broadening, comparable to shocked calcite from a nuclear-explosion site (Simons and Dachille, 1965). Similar X-ray patterns have been obtained from calcite crystals taken fron the crater wall and breccia at the Flynn Creek impact center in Tennessee (Roddy, pers. comm., 1976). Dachille and others (1968, Fig. 7) show that Decaturville samples have X-ray patterns similar to powders which were shocked at about 40 and 90 kb. They correspond in damage intensity to samples from Sierra Madera, Texas, and the Steinheim Basin, Germany. Shatter cones in dolomite may form at pressures as low as 15–20 kb, and they are also found at Sierra Madera in dolomite and limestone shocked to levels around 100 kb (Wilshire *et al.*, 1972b). Thus, the range of 40–90 kb for shatterconed rocks at Decaturville, suggested by the asterism study, extends across the upper two-thirds of the probable 10–100 kb range in which shatter cones form in dolomite.

Control points for reconstructing the envelope of shock pressure are: (1) probable 20 kb at the outer limit of shatter cone formation, (2) no more than

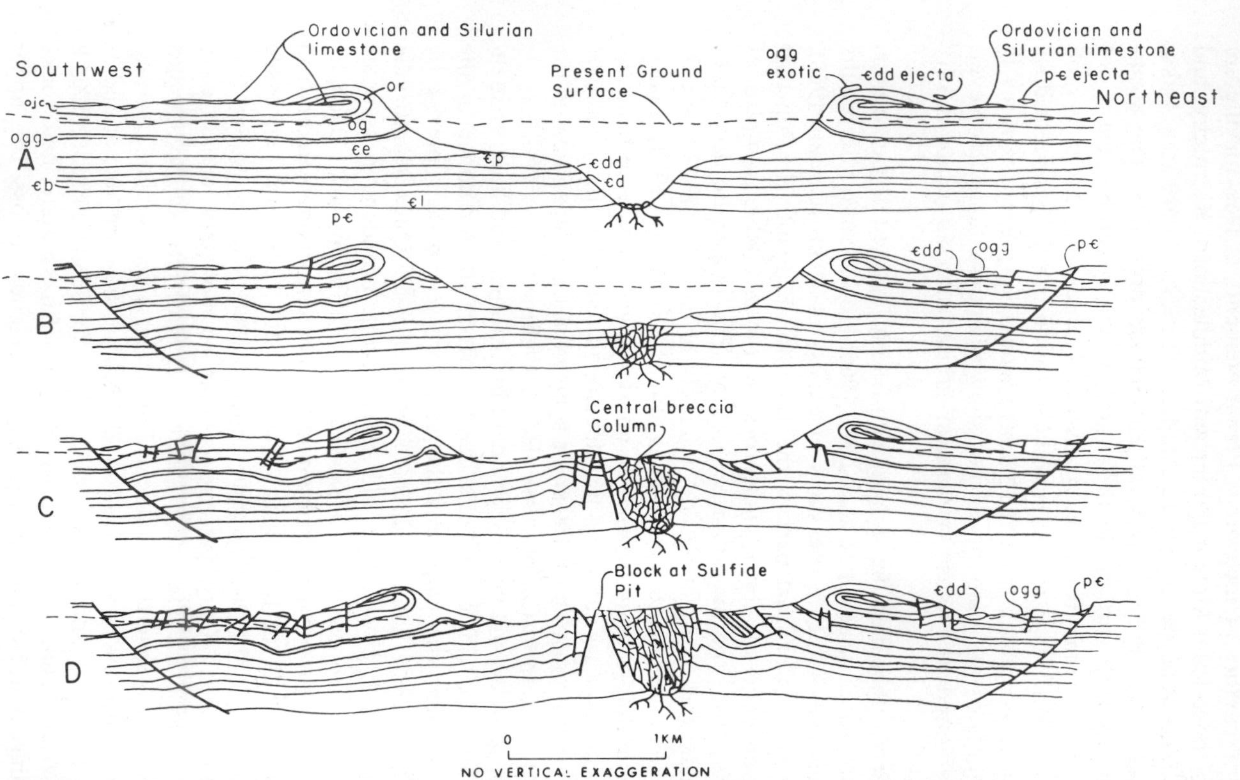

Fig. 7. Schematic cross section showing inferred time sequence of movements during the cratering event. (The entire sequence is interpreted to have occurred within minutes.) A. Transient cavity—ejecta blocks present in the structure are still airborne. B. Beginning of structural adjustments—ring fault forms; formations collide and fracture at center of structure. C. Numerous faults form in response to different centripetal movement; folding becomes more intense toward the central breccia column, giving rise to steep faults as seen at the sulfide pit where a large block has been raised. D. Beds glide over the breccia column as seen in present-day configuration. Formations represented in the cross section are as follows: Ordovician: Ojc, Jefferson City Dolomite; Or, Roubidoux Dolomite; Og, Gasconade Dolomite; Ogg, Gunter Sandstone Member of Gasconade Dolomite. Cambrian: Ꞓe, Eminence Dolomite; Ꞓp, Potosi Dolomite; Ꞓdd, Derby and Doe Run Dolomites; Ꞓd, Davis Formation; Ꞓb, Bonneterre Dolomite; Ꞓl, Lamotte Sandstone. Precambrian; pꞒ, schists and pegmatites.

10–15 kb in the basement rocks, (3) about 50 kb at the top of the Lamotte, (4) 70–100 kb for quartz grains believed to be derived from the Davis, and (5) probable 90 kb for the centermost part of the shatterconed Derby and Doe Run. In replacing these points to their pre-deformation positions, the inward and upward movement involved in forming the uplift must be considered. It is difficult to establish the deepest stratigraphic level at which shock energy was directly imparted to the rocks. Maximum pressure levels recorded in any of the rocks preserved at the center are in the Davis and Derby and Doe Run Formations, but no evidence precludes the possibility of greater shock pressures in higher strata. If the point of origin of the shock wave was in the Derby and Doe Run or Davis, it would have been 275–330 m below the inferred original surface. It was not deeper, because pressure levels clearly diminished to 40–50 kb in the Lamotte a few tens of meters downward, and to 10–15 kb or less in the basement below the Lamotte. It might have been higher, with no rocks from high enough in the section preserved in the central breccia column to indicate the true point of shock-wave origin.

Age of Structure and Depth of Erosion

Bainbridge Limestone of Middle Silurian age is involved in the deformation at Decaturville. Distribution of remnants of Mississippian and Pennsylvanian strata in the surrounding area suggests that the area was relatively high in late Paleozoic time. If the Decaturville structure formed before Pennsylvanian time, some Pennsylvanian rocks would have been deposited in topographic lows and should be found in remnants along with the patches of Ordovician and Silurian limestone. Similarly, if the structure formed when Pennsylvanian strata covered the area, remnants of those rocks should be present in the depressed zone. Thus, it seems likely that the structure formed after Pennsylvanian strata had been stripped from the area.

Other evidence supporting this is that the structure postdates sulfide mineralization in the area. The age of mineralization in central Missouri is not certain, but by comparison with similar microbanded sulfide in western Missouri (A. V. Heyl, pers. comm., 1974) it is at least post-Pennsylvanian and possibly as young as Cretaceous.

Therefore, the structure appears to have formed probably after Pennsylvanian and possibly after Cretaceous time, when the area was covered by Jefferson City strata with patches of younger limestone preserved in local topographic lows. These strata are preserved in full sequence in the depressed zone of the structure. The described exotic blocks of basement rock, Derby and Doe Run, and Gunter, lie upon these strata and are believed to be ejecta (or in the case of the Gunter block, part of an overturned flap) lying on the original ground surface. If this is the case, then it follows that post-structure erosion in the area has only been enough to remove most of the Jefferson City from the hilltops outside the ring fault, a lowering of the surface by less than 50 m. Thus, the Decaturville structure is not deeply eroded, and instead is believed to be only slightly modified from its original appearance.

INTERPRETED CRATER PROFILE AND POST-CRATER DISRUPTION

Two models involving different penetration by the impacting body, different transient cavity size and shape, and different methods of forming the central uplift have been considered to fit most of the field observations as well as what has been seen in cratering experiments. One model calls for very shallow or no penetration by a low-density impacting body. From cratering experiments, this may be expected to result in excavation of a shallow crater with central uplift, possibly containing a disrupted and brecciated core. In this model the observed breccia core would have a cap of Derby and Doe Run as those formations would not have been excavated, but merely uplifted. Lateral movements might have occurred during formation of the uplift but are likely to have been small in comparison to the vertical displacements involved. The crater would have bottomed no deeper than the top of the Derby and Doe Run and thus would have been less than 300 m deep. Brecciation below the Derby and Doe Run would represent the effect of bulking as the uplift formed during relaxation after passage of the shock wave. This explanation accounts well for the observation that except for a small amount of Potosi, all fragments in the breccia core come from Derby and Doe Run and lower formations. It also could explain the orientation of large elongate blocks pointing inward and upward. It does not seem to account for basement ejecta, unless jetting of deep material occurred along transient fractures that extended some 180 m below the crater floor. (Subcrater material has been found in ejecta in some cratering experiments.) The model also does not appear to be in accord with the inferred scale of lateral movement of strata.

Our preferred model of original cavity shape is highly inferential, but takes account of all pertinent observations. There is no direct evidence on depth of penetration, but from the configuration of probable shock envelopes, penetration seems likely to have been at least as deep as the Potosi Dolomite, and possibly as deep as the Davis Formation. Thus the cratering mechanics would have been different than those of the shallow-burst cratering experiments. Moreover, it seems clear that the basement must have been adjacent to a free surface at least transiently. With probable impacting-body penetration as deep as Potosi, or Davis, excavation to the basement at the very center of the crater is not unlikely, and we infer that the instantaneous excavation (or transient cavity) bottomed at or near the basement surface, as shown in the cross section of hypothetical post-crater deformation (Fig. 7).

This means that the Derby and Doe Run would have been excavated over the central area and the cap that now covers the center must have been emplaced from the side. Inward movement of about 300 m from all sides is indicated by the inward crowding of beds at the Gunter Member level, and by general comparison with inferences from shatter-cone orientations at Gosses Bluff

(Milton *et al.*, 1972), inward movement of deeper strata may have been even greater. Roddy (1968, 1976) has shown that inward movement is the rule in large-scale experimental cratering events of similar configuration. We thus infer that the Derby and Doe Run moved inward a minimum of 300 m and closed over the bottom one-third of the excavated cavity, now filled with blocks that spalled from the crater walls during the shock unloading phase, or were fractured and torn loose as the strata moved inward (Fig. 7c). The movement of the Derby and Doe Run and subjacent beds involved bedding-plane slip, block displacements, and considerable flowage (as shown by thinning of units and flow deformation in the Lamotte Sandstone as well as the mixed breccia). The Derby and Doe Run Dolomites and formations below them make up most of the breccia core. The Derby and Doe Run probably slid more readily and farther than lower beds because they were moving atop the relatively water-rich, limey shales of the Davis.

The blocks overridden by the Derby and Doe Run now form the breccia core, changed in shape from a filling of the lower part of the crater to a column by crushing inward movement of the crater walls. This explanation accounts for the observation in drill core that long blocks in the megabreccia core commonly stand nearly on end, pointing inward and upward in an acicular pattern. The inferred 300 m of inward movement permits us to establish the position of the crater wall at the level of the Derby and Doe Run prior to the inward move. Above that level the crater probably widened sharply as shown in Fig. 7a. Our conjectured initial crater is a shallow bowl with a deeper excavation at its center.

The inward move of Derby and Doe Run across the deep inner crater accounts for the absence of blocks from strata above the Derby and Doe Run (except for a minor amount of Potosi which probably fell in during the move). Higher strata would have been excavated over the wider area of the main crater bowl and would be present in the central column only as minor pieces of fallback material. Even fallback pieces of upper strata would not be expected to be present if the Derby and Doe Run beds had closed the lower part of the crater before ejected material could fall back. That this is feasible is indicated by the fact that at Gosses Bluff, blocks had risen more than 1830 m to stand in open space and then toppled or were thrown outward apparently before airborne ejecta landed in the area (Milton *et al.*, 1972). The time required for such dramatic movement in the uplift obviously must be measured in seconds. Most of the highly shocked material would have been ejected, but some would have been mixed with rubble on the crater floor. This material is now found only in the mixed breccia, highly diluted by the addition of grains and fragments produced by the grinding together of the large blocks spalled from the crater walls.

The profile of the upper portion of the transient crater is based primarily on three points of information. A downdropped block of Eminence Dolomite in the northeast sector of the inferred bowl indicates that Eminence was not entirely excavated there and at least in part must have been below the crater floor. A similar argument can be used to explain the presence of a Jefferson City

Dolomite block in the northwest sector, thus putting a probable limit on the diameter of the crater at the original ground surface. An additional surmise is that the tectonic style of the Gunter Member suggests involvement in the crater rim and overturned flap (such as described at Meteor Crater by Shoemaker, 1960). The only reasonable explanation for the presence of the large Gunter block at the village of Decaturville, atop rocks which would have been at ground surface, appears to be that it slid into place as a projection of the overturned flap. The Gunter commonly is overturned in the northeast sector where this block remnant is found.

The maximum depth of the inferred transient cavity shown in Fig. 7a was about 550 m and rim diameter was about 3100 m. This excavation was modified in a matter of seconds by the inward and upward move of strata in the surrounding ground. The physical impetus for such movement is not completely understood, but the movement may be considered roughly as a rebound following intense compression by the shock wave. D. J. Roddy (pers. comm., 1974) has indicated that recent large-scale cratering experiments and computer simulations of shock-wave cratering show inward crater-floor movements to be common in surface-burst events.

Figure 7 gives a highly schematic sequence of inferred stages in the development of the structure from transient cavity through post-excavation disruption. Instantly after, or even during crater excavation, the inward mass displacement involved in the initiation of the central uplift produced the ring fault (Fig. 7b). The position of such a ring fault in impact structures may depend on the depth to which strata are disturbed. Movement of strata downward and inward along the boundary fault produces a ring depression and provides further impetus for the rise of a central peak, and this push-pull mechanism appears to have acted effectively at Decaturville. As an aside, the inferred linkage of uplift and ring fault may explain the difference between Decaturville and Crooked Creek. Uplift must be supported by flow and slippage of strata inward. This movement may not be able to develop fully if the ring fault is incomplete. It seems possible that a configuration like that at Crooked Creek (a central horst with a dome collapsed around it) would occur if initial uplift were unsupported and partial collapse then ensued because a ring fault did not form to provide additional impetus for inward movement of material.

The centripetal movement of beds in the central-peak area is clear and appears to have been an early phase in a very complex structural sequence. Beds in the ring depression on the southwest side also moved inward, largely before steep faults and outward thrusts formed. Elsewhere on the west side of the structure in the ring depression, outward thrusting seems to have been more conspicuous than the other phases of movement. We believe that the outward movement occurred primarily by gravity sliding of beds from the central peak. Some sliding may have occurred as the peak was being lifted, but most probably was after the peak was essentially formed. In other sectors of the perimeter of the peak, late outward sliding of layers may have been precluded by fault-block topography. Block faulting was the dominant structural phase on the north and south sides and outward thrusts have not been identified in those areas.

The push-pull mechanism of ring fault-central uplift has been discussed for lunar craters, where the slump of large masses on crater walls is believed to have contributed energy and mass to the formation of central peaks (Dence, 1968), once inward movement has been initiated by rebound after shock-wave passage. The final configuration at Decaturville may have been much like lunar craters, with blocks unevenly downfaulted inside the ring boundary and a central peak which as it formed destroyed the transient cavity of Fig. 7a.

Implications for Other Astroblemes

No other cratering events in layered rocks are reported to have resulted in beds moving inward *over a preserved core of breccia* to form a central uplift (although subsurface data are seldom available to reveal such a core). Other facets of our proposed reconstruction, however, may have more general application and help to explain some puzzling aspects of other astroblemes. Decaturville had always been considered to be deeply eroded, not based on field evidence but on the presumption that much of the crater topography had been removed. Yet there is no reason to suppose deep erosion. Paleogeographic evidence neither requires nor provides support for having more strata atop the present section at the time of impact. Moreover, emplacement of the large block of Gunter atop the highest preserved beds in the section seems possible only as part of the flap of strata overturned outward around the transient cavity. An exotic block of Derby and Doe Run and fragments from the basement lie atop the section like the Gunter block, but are from sufficient depth that they must be ejecta. A basement fragment appears to be caught in the ring fault.

We have estimated a maximum of about 45 m of post-structure erosion. Thus the structure immediately after its formation would have shown a ring zone dropped as much as 175 m below the surrounding plateau surface. If our reconstructed initial crater profile is approximately correct, the central peak probably was at most 30–60 m higher than today. The raised rim of the crater stood on the flanks of the central peak as an easily eroded pile of broken rock perhaps as much as 180 m above the plateau level.

Erosion at Sierra Madera (Wilshire *et al.*, 1972b) is thought to have been 600 m and the present surface is considered to show the style of deformation at a deep level in the impact structure. Shatterconed Permian rocks from the central-peak area are found along ring faults in Cretaceous strata at the present topographic rim, 3.2 km outside the shatter-cone envelope. Wilshire *et al.* (1972b) conjectured that such blocks were driven outward and upward along curved faults. We suggest instead that, as with the exotic blocks near the village of Decaturville, these blocks are ejecta trapped in faults at, or not far below, original ground surface as jostling occurred in the ring-boundary zone. This explanation seems to fit well with the presence of highly shocked, even melted, mixed breccia material filling cracks and blanketing hillsides in the central uplift. If the present surface is little different from the original surface after the structure formed, then the mixed breccia may well represent material deposited in the transient cavity as the floor of the cavity was being uplifted to become part of the surface of the central peak. The highly granulated and partly fluidized breccia would naturally have drained or flowed by gravity into cracks between blocks in the rising central peak. Scaling from Decaturville, the transient cavity at Sierra Madera would have been about 7300 m across; the cavity wall would have been just outside the outermost mapped occurrences of mixed breccia.

At Gosses Bluff, the presence of impact melt at the present surface suggests that little post-crater erosion can have taken place. Milton *et al.* (1972) speculate that Mt. Pyroclast, 5 km from the center of the structure, may represent a mass of suevite-like material which was plastered against the crater wall. Other evidence suggests that the central peak formed as ejected material was still falling or moving laterally along the surface, as we now surmise for Decaturville and Sierra Madera, and which was observed at the Snowball experimental cratering event (D. J. Roddy, pers. comm., 1976). It is interesting to note that, if Mt. Pyroclast marks the position of the transient cavity wall, then the cavity diameter was approximately half the 20–22 km diameter of the final disturbed area, essentially as inferred for the ratio of transient cavity to final ring diameter at Decaturville.

The implications of terrestrial impact structures for the structure of lunar craters have been discussed by Wilshire *et al.* (1972b) and Milton *et al.* (1972). They make the reasonable analogy that, as in terrestrial impact structures, peaks in lunar craters consist of rocks brought up from deep below the crater floor. We would only add that the ratio of upward displacement to final ring (or possibly apparent crater) diameter probably varies considerably depending on layering and especially on the ability to flow or slip of the disturbed rocks in order to develop the push-pull mechanism linking central uplift and ring fault. In lunar terrain of massive basalt layers, that ratio might be much smaller than the 1:10 observed at Decaturville and several other comparable structures. In terrain where fragmental material is extremely deep and could move by granular flow, the ratio might be significantly larger than 1:10.

Acknowledgement—This work was supported by the National Aeronautics and Space Administration, under contract no. 160-75-01-43-10.

REFERENCES

Boon, J. D. and Albritton, C. C., Jr.: 1936, Meteorite craters and their possible relationship to "cryptovolcanic structures". *Field and Lab.*, **5**, 1–9.

Bucher, W. H.: 1963, Cryptoexplosion structures caused from without or from within the Earth? ("astroblemes" or "geoblemes"?). *Amer. J. Sci.* **261**, 597–649.

Carter, N. L.: 1968, Dynamic deformation of quartz. In *Shock Metamorphism of Natural Materials* (B. M. French and N. M. Short, eds.), p. 453–474. Mono Book Corp., Baltimore, Maryland.

Dachille, F., Gigl, P., and Simons, P.Y.: 1968, Experimental and analytical studies of crystalline damage useful for the recognition of impact structures. In *Shock Metamorphism of Natural Materials* (B. M. French and N. M. Short, eds.), p. 555–569. Mono Book Corp., Baltimore, Maryland.

Dence, M. R.: 1968, Shock zoning at Canadian craters: petrography and structural implications. In *Shock Metamorphism of Natural Materials* (B. M. French and N. M. Short, eds.), p. 169–184. Mono Book Corp., Baltimore, Maryland.
Hörz, F.: 1968, Statistical measurements of deformation structures and refractive indices in experimentally shock loaded quartz. In *Shock Metamorphism of Natural Materials* (B. M. French and N. M. Short, eds.), p. 243–253. Mono Book Corp., Baltimore, Maryland.
Hörz, F.: 1970, Static and dynamic origin of kink bands in micas. *J. Geophys. Res.* **75**, 965–977
Milton, D., Barlow, B. C., Brett, R., Brown, A. R., Glikson, A. Y., Manawaring, E. A., Moss, F. J., Sedmik, E. C. E., Van Son, J., and Young, G. A.: 1972, Gosses Bluff impact structure, Australia. *Science* **175**, 1199–1207.
Roddy, D. J.: 1968, The Flynn Creek crater, Tennessee. In *Shock Metamorphism of Natural Materials* (B. M. French and N. M. Short, eds.), p. 291–322. Mono Book Corp., Baltimore, Maryland.
Roddy, D. J.: 1976, Impact, large-scale high explosive, and nuclear cratering mechanics: Experimental analogs for central uplifts and multiring impact craters (abstract). In *Lunar Science VII*, p. 744–746. The Lunar Science Institute, Houston.
Shoemaker, E. M.: 1960, Penetration mechanics of high velocity meteorites, illustrated by Meteor Crater, Arizona. *Internat. Geol. Cong., 21st*, Copenhagen, Rept., pt. 18, p. 418–434.
Short, N. M.: 1966, Shock processes in geology. *J. Geol. Education* **14**, 149–166.
Simons, P. Y., and Dachille, F.: 1965, Shock damage of minerals in shatter cones (abstract). Geol. Soc. Amer. Ann. Mtg. program, p. 156–157.
Snyder, F. G. and Gerdeman, P. E.: 1965, Explosive igneous activity along an Illinois-Missouri-Kansas axis. *Amer. J. Sci.* **263**, 465–493.
Thompson, T. L., and Satterfield, I. R.: 1975, Stratigraphic and conodont biostratigraphy of strata contiguous to the Ordovician-Silurian boundary in eastern Missouri. *Missouri Geol. Survey Rept.* Inv. No. 57, pt. 2, p. 61–108.
Wilshire, H. G., and Howard, K. A.: 1968, Structural pattern in central uplifts of cryptoexplosion structure as typified by Sierra Madera. *Science*, **162**, 258–261.
Wilshire, H. G., Howard, K. A., and Offield, T. W.: 1972a, Impact breccias in carbonate rocks, Sierra Madera, Texas. *Bull. Geol. Soc. Amer.* **82**, 1009–1018.
Wilshire, H. G., Offield, T. W., Howard, K. A., and Cummings, D.: 1972b, Geology of the Sierra Madera cryptoexplosion structure, Pecos County, Texas. *U.S. Geol. Survey Prof. Paper* 599-H, 42 pp.

Roddy, D. J., Pepin, R. O., and Merrill, R. B., editors.
(1977) *Impact and Explosion Cratering*, Pergamon Press (New York), p. 343–404.
Printed in the United States of America

The Ries impact crater

JEAN POHL

Institut für Allgemeine und Angewandte Geophysik, Universität München, Theresienstrasse 41, 8 München 2, Germany

DIETER STÖFFLER

Institut für Mineralogie, Universität Münster, Gievenbecker Weg 61, 44 Münster, Germany

HORST GALL

Bayerische Staatssammlung für Paläontologie und hist. Geologie, Richard Wagner-Strasse 10, 8 München 2, Germany

KORD ERNSTSON

Institut für Geologie, Universität Würzburg, Pleicherwall 1, 87 Würzburg, Germany

Abstract—Presently available data on the Ries impact structure as far as they relate to the interpretation of the crater-forming process are summarized and discussed. The Ries projectile impacted a stratified target of horizontal layers of various sedimentary rocks (limestones, shales, sandstones), about 600–700 m thick, underlain by a basement of crystalline rocks. The main structural characteristics of the crater are: flat-floored central basin (crater depth ~0.6–0.7 km) due to uplifted crystalline basement bordered by an inner ring of uplifted basement and sedimentary megablocks (radius 6 km), zone of megablocks from all stratigraphic units between 6 and 12–13 km radius. The perimeter of the crater is defined by an outer circular system of downfaulted target rocks at 12–13 km radius (tectonic rim). Continuous crater deposits occur in the megablock zone and in the "Vorries" which extends from the tectonic rim up to a radial distance of at least 40 km. A melt-bearing top unit of suevite breccia forms a continuous fallback layer in the central cavity and isolated fallout patches outside the inner ring up to a radius of 22 km. Distant ejecta may reach as far as 350 km. Ballistic ejection and secondary mass transport by the impact of landing ejecta is considered to be the main mechanism for the emplacement of the continuous deposits beyond the megablock zone. Inside this zone most of the displaced masses were transported by gliding and overthrusting mechanisms. The subsurface structure of the crater is modelled by density and seismic velocity models indicating brecciation and fracturing as deep as 6 km. The crater reveals a negative gravity anomaly of −18 mgal. The central cavity is characterized by strong, but irregular negative magnetic anomalies due to the suevite layer of up to 400 m thickness. Estimates of the total volume of excavated rocks ranges from 124 to 200 km^3. The total melt volume is on the order of 0.1–0.5 km^3. The depth of the transient crater cavity is estimated to be about 2 km. Modification of this cavity is achieved by central uplifting of the basement and subsidence of the marginal zone of the crater (megablock zone).

1. INTRODUCTION

THE RIES (Ries of Nördlingen or Ries Kessel) is a flat circular basin, 22–23 km in diameter, which is located between the Jurassic limestone plateaus of the Swabian and Franconian Alb, about 110 km NW of München, Bavaria (Fig. 1).

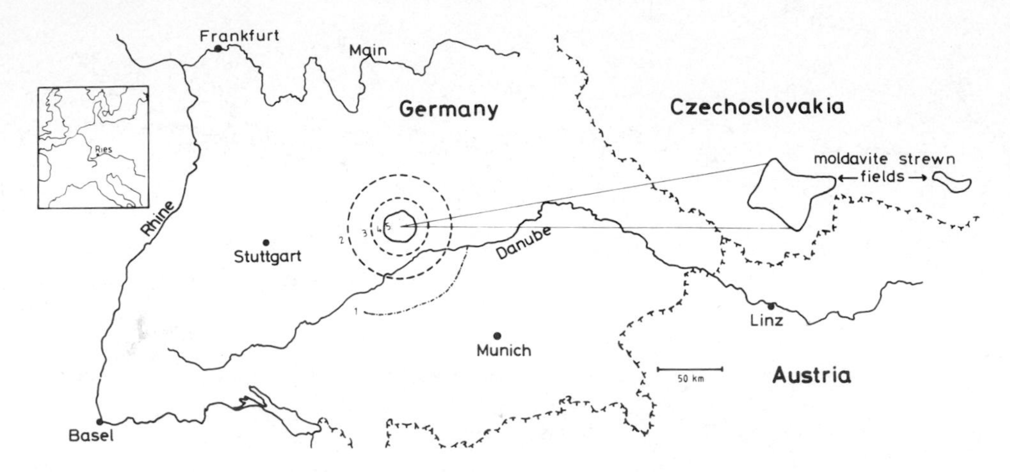

Fig. 1. Map of central Europe showing the location of the Ries and its ejecta; 1 = outer limit of distribution of Malmian Reuter blocks, 2 and 3 = maximum extent of Bunte breccia and suevite, respectively, 4 = tectonic rim of the crater, 5 = inner ring.

Geological investigations of the Ries area began as early as 1834 (v. Cotta) resulting in a variety of theories about its origin. The theory that the Ries was formed by meteorite impact was first proven by the discovery of coesite (Shoemaker and Chao, 1961) although it was already suggested by Werner in 1904. Most of the older literature is reviewed by Oberrhein. Geol. Verein (1926), Dorn (1948), Preuss (1964) and Dehm (1969). Research along an impact hypothesis was summarized and reviewed since 1961 by Preuss (1964), Engelhardt (1967a), Bayer. Geol. Landesamt (1969), Dennis (1971), Chao (1973), Engelhardt and Stöffler (1974), and Gall *et al.* (1975). The impact concept is unanimously accepted by now.

The Ries crater represents one of the best investigated, large, complex terrestrial impact structures with a well preserved "ejecta blanket." Therefore, the Ries may serve as an important reference point for the discussion of impact cratering mechanics, although the complex layered structure of the Ries target may differ from the average situation on planetary surfaces.

It is the objective of this paper to present a synopsis of field and laboratory data relevant to the mechanics of the crater-forming process. It is hoped to provide useful constraints for theoretical and experimental studies simulating impact events of the magnitude and conditions of the Ries.

2. Structure and Composition of the Target

2.1. *Pre-impact geology*

The Ries impact occurred in the upper Tortonian (upper Miocene; Table 1) 14.8 m.y. ago (Bolten and Müller, 1969; Gentner *et al.*, 1963) in a target that consisted of horizontally layered sediments of early Permian, Mesozoic, and

Tertiary age above a variety of crystalline basement rocks of pre-Permian age. The basement is part of a mountain belt formed during the Variscan orogeny and subsequently peneplained during the Permian and early Triassic.

The target consists of eight major stratigraphic units shown in a reconstructed pre-impact profile (Fig. 2). Overall thickness and lithologic facies of the sedimentary rocks are not quite constant for the entire target area. Thickness and distribution of the uppermost and the lower part of the sedimentary strata are highly irregular. The uppermost unit (0–50 m thick) consists of middle Oligocene (Müller, 1972) and upper Miocene sediments (sands, shales, fresh-water limestone, coal) formed as part of the Molasse sedimentation of the Alpine geosyncline (Table 1). These sediments covered a large part of the crater area. South of the crater, the upper Miocene occurs in a continuous layer underneath the Ries crater deposits (Fig. 3). Details of the remaining stratigraphic units can be seen from Fig. 2. The occurrence of Buntsandstein is still questionable; if at all, it could be present W to NW of the target. The Permian (sandstones and rhyolite) fills elongated troughs in the crystalline basement striking generally WSW.

The most distinct unconformity in the Ries stratigraphy is formed by the surface of the crystalline basement which had a relief of at least 100 m in the target area at the time of the impact (Section 4). The stratigraphy and structure of the basement are very complicated and not well known. A reconstruction was attempted by Graup (1975, Fig. 4). The most important feature is that most of the

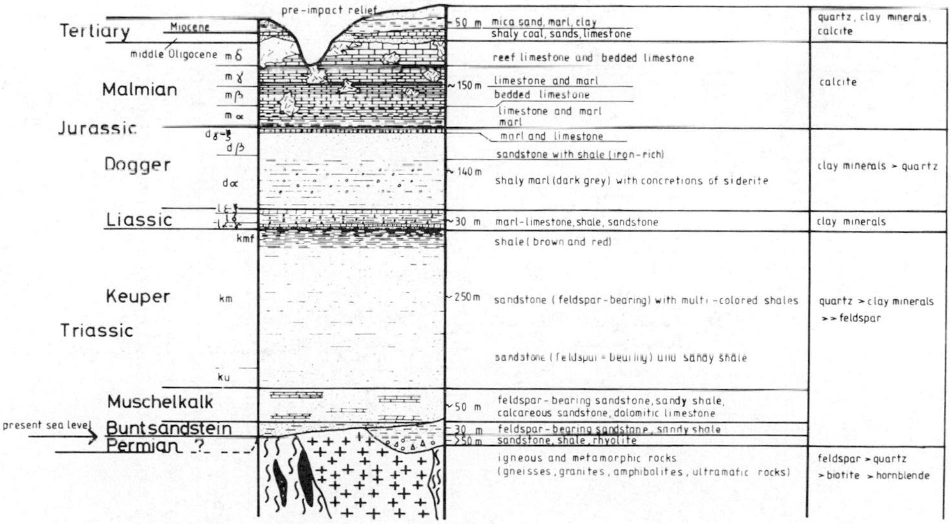

Fig. 2. Geological and mineralogical profile of the central area of the Ries crater before impact. From the area W of the crater to the area E of the crater Triassic, Liassic and Dogger decrease from 350 to 250 m, 46 to 16 m, and 160 to 125 m. Malmian decreases to about 110–120 m.

Table 1. Tertiary sedimentation in the Ries crater area*

			Crater	Vorries†	Sedimentation		Erosion state
	Pliocene		↕	↕	Post-impact Obere Süsswassermolasse	Fluviatile, limnic, max 150 m	Relicts
		Sarmatian	↕		Post-impact lake sedimentation	Lacustrine, limnic-brackish, max 500 m	Partly eroded
					Impact		
Upper Tertiary		Tortonian	↕	↕ S	Pre-impact Obere Süsswassermolasse OSM	Fluviatile-lacustrine, limnic max 80 m	Strongly eroded
	Miocene	Helvetian		↕ S	Obere Meeresmolasse OMM	Marine, max 60 m	Partly eroded
		Burdigalian					
		Aquitanian		↕ SW	Untere Süsswasser-molasse USM	Lacustrine, limnic, max 20 m	Strongly eroded
Lower Tertiary	Oligocene		↕	↕ SE	Marginal facies of Untere Meeresmolasse UMM	Fluviatile-lacustrine, limnic-brackish, max 75 m	Strongly eroded
	Eocene						
	Paleocene						

*See also Fig. 2.2.

†S, SE, SW indicate the sedimentation area relative to the crater.

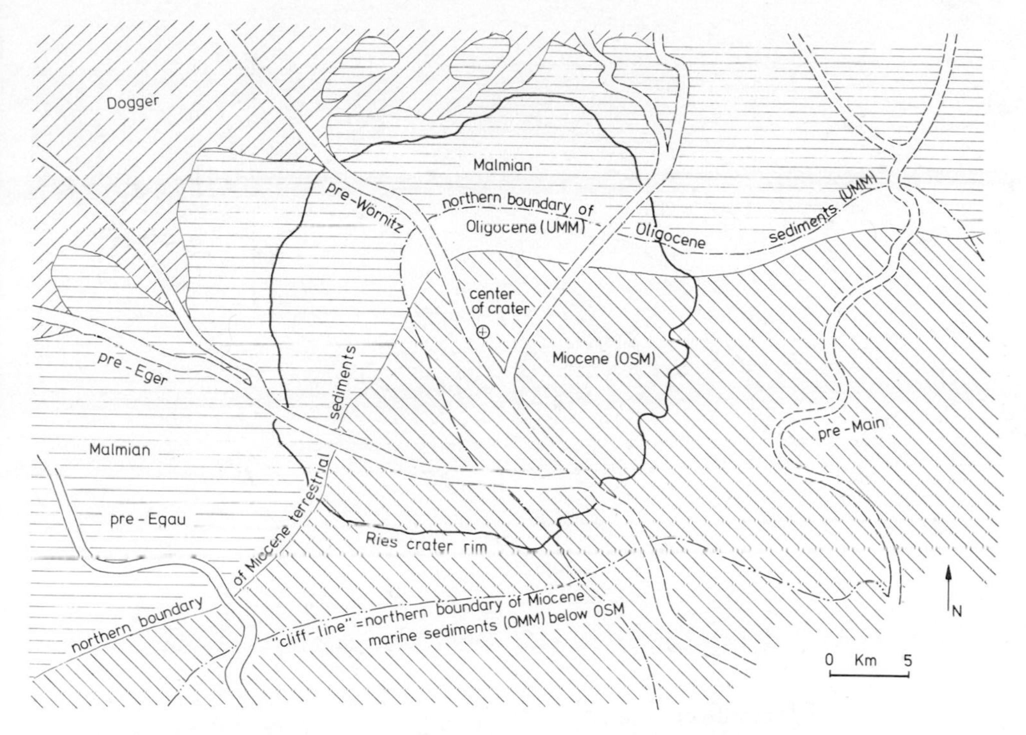

Fig. 3. Probable geological map of the Ries area before impact. Maximum extent of Tertiary sediments is indicated. UMM = Untere Meeresmolasse, OMM = Obere Meeresmolasse, and OSM = Obere Süsswassermolasse.

surface is formed by a granitic intrusion which cuts through older gneisses, amphibolites and ultrabasic rocks with steeply (?) inclined foliation planes (see also Bayer. Geol. Landesamt, 1974, 1977).

Details of the mineralogical and chemical composition of the target rocks will be discussed along with the properties of the impact formations (see section 3.2).

2.2. *Pre-impact morphology*

The Upper Malmian limestone strata were exposed as a land surface for most of the Cretaceous and Tertiary. This land surface was then covered by an incomplete sequence of Oligocene and Miocene sediments as indicated in Figs. 2, 3 and Table 1. In the Tortonian, shortly before the impact occurred, the Ries area became a region of relatively strong erosion which was drained into the Molasse basin by a number of predominantly NW-SE running rivers (Gall *et al.*, 1975, 1977; Fig. 3). The western rivers penetrated as deep as Dogger (100–150 m of relief) and the "pre-Main" river reached the lowermost Malmian (150–200 m of relief) (Bader and Schmidt-Kaler, 1977). This erosion cut the Malmian limestone plateau into a system of isolated mesas, most distinctly developed in

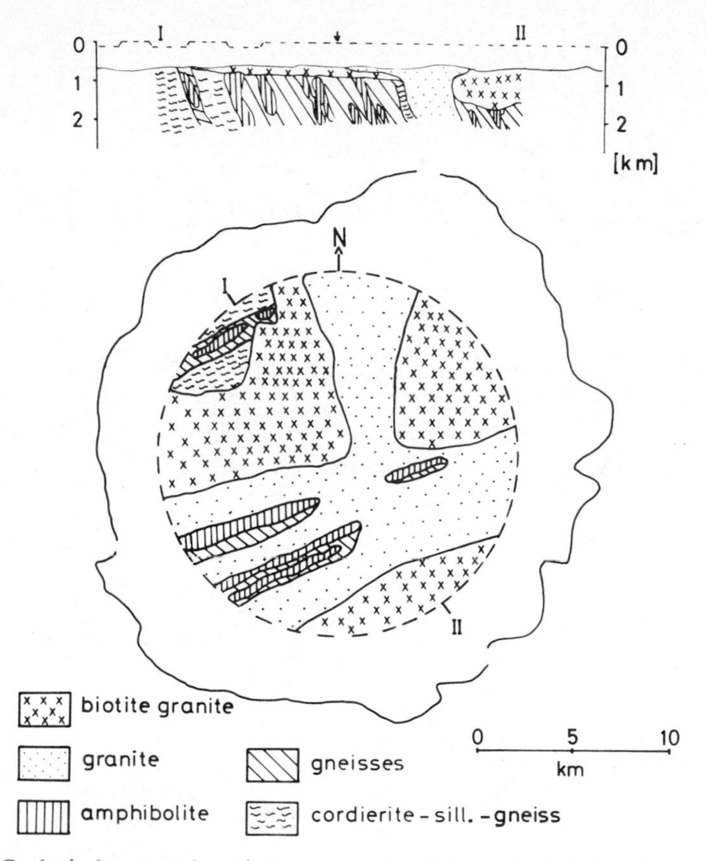

Fig. 4. Geological map and vertical cross section of the crystalline basement of the Ries crater area reconstructed by Graup (1975); modified after Graup (1977); irregular circular line = tectonic rim of the crater; I–II indicates the position of the cross section.

the northern part of the Ries (Fig. 3; Gall *et al.*, 1975, 1977). The marked escarpment of the Malmian limestone presently extending across the Ries crater from WSW to ENE at its S-rim probably was not as pronounced prior to impact as it is now, because the sedimentary strata were not yet tilted to the SE. In contrast to previous assumptions Gall *et al.* (1977) believe that most of the crater area was covered by the Malmian limestone plateau which was more thoroughly dissected into mesas in the northern part (Fig. 3). Thus the target surface displayed a maximum relief of about 150 m near to the point of impact. It is not known whether the impact occurred onto the Upper Malmian limestone (probably covered by 20–50 m of Tertiary sands, marls or shales) or onto the slope or bottom of a valley cut into Dogger sandstones.

3. Surface Formations of the Crater

3.1. *Present morphology and surface geology*

The present Ries is an almost circular, flat basin, 22–23 km in diameter, typically 410–430 above sea level (Fig. 5). It can be subdivided into a flat *central basin*, about 410–420 m in altitude and 11–12 km in diameter formed by post-impact Tertiary lake sediments. This plain is bordered in the W, SW, and E by a ring-shaped chain of isolated hills standing about 50 m above the basin. The zone between this *inner ring* and the *morphological rim* of the basin (22–23 km diameter) is formed in part by additional plains of post-impact lake sediments and in part by a hummocky relief extending into the rim area. The hummocks marginal to the plains reach elevations of 450–500 m in the northern part and 550–600 m in the southern part of the Ries.

The autochthonous rocks of the northern rim zone consist of Keuper, Liassic, and Dogger whereas the southern rim zone is formed by Malmian limestones. The hummocks of the inner ring and those extending to a radius of 12–13 km are mainly displaced megablocks up to 1 km in size of the sedimentary rock strata (Keuper, Liassic, Dogger, Malmian) and of the crystalline basement, sometimes covered by post-impact lacustrine limestone. Blocks of crystalline rocks uplifted from the basement by 450–550 m prevail in the inner ring. For reasons discussed below it is believed that at a radius of 12–13 km some concentric fault system separates displaced and down-faulted megablocks from the autochthonous, undisturbed strata marking what is called the *tectonic rim* of the crater (Figs. 6, 7; Gall *et al.*, 1977). The concentric zone between the inner

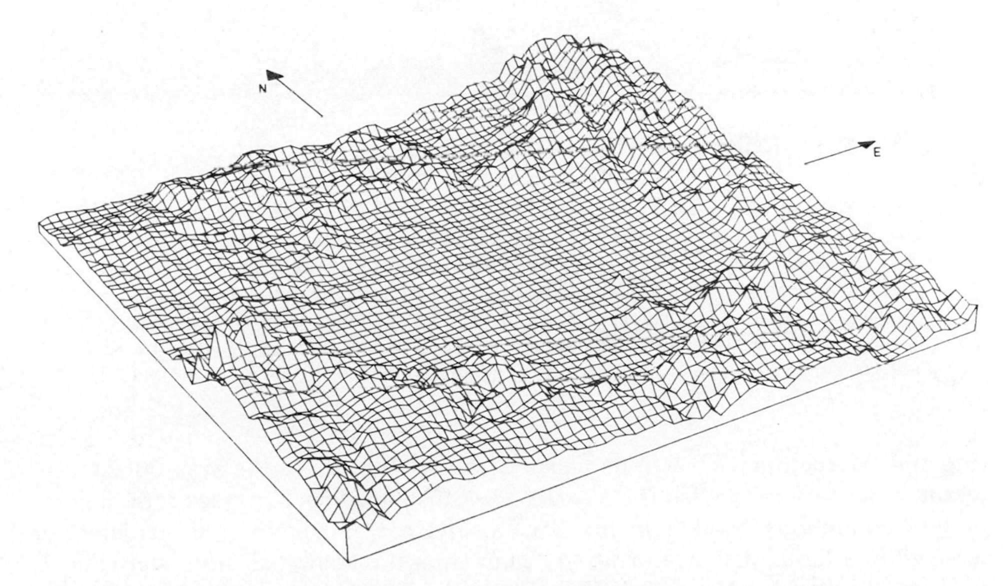

Fig. 5. Morphology of the present Ries Basin; grid distance 0.5 km, vertical exaggeration 12.5-fold, horizontal coordinate lines 400 m above sea level.

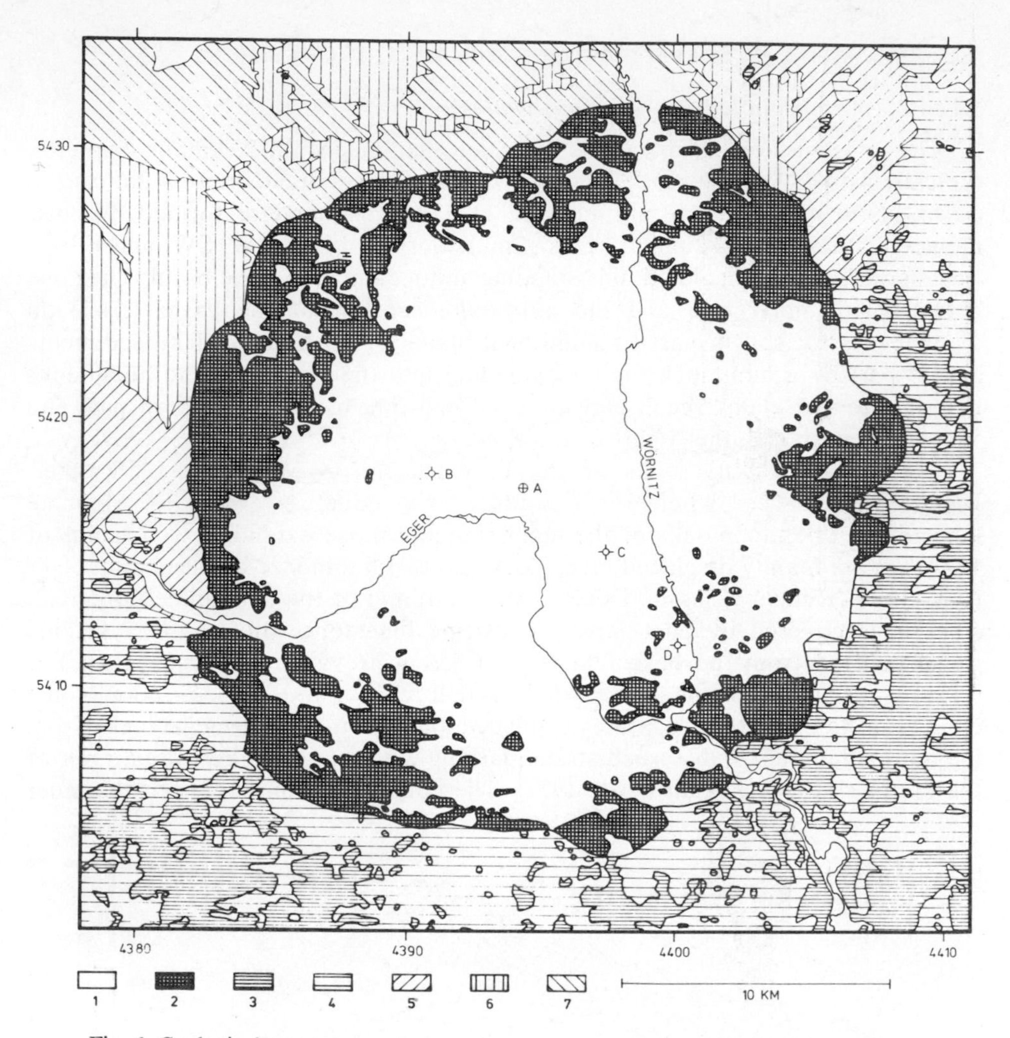

Fig. 6. Geological map of the Ries crater; A = center of crater, B = drillhole Nördlingen 1973, C = drillhole Deiningen, D = drillhole Wörnitzostheim; 1 = post-impact Tertiary and Quaternary sediments, 2 = megablock zone (out-cropping), 3 = "Vorries" continuous deposits, 4 = Malmian, 5 = Dogger, 6 = Liassic, 7 = Keuper; 2 and 3 are displaced rocks, 4–7 are autochthonous rocks; this simplified map is based on the Geological Map 1 : 100,000 (Schmidt-Kaler *et al.*, 1970), on detailed geological maps 1 : 25,000 and on a preliminary draft for the geological map 1 : 50,000 (Gall and Müller, 1977).

ring and the tectonic rim will be called *megablock zone.* In the area outside the tectonic rim which is called the *Vorries zone*, displaced rock masses form a more or less continuous blanket in the SW, S, and SE and have been geologically mapped to a radial distance of up to 42 km from the center of the crater (Fig. 7; Gall *et al.*, 1975). This formation will be referred to as continuous Vorries deposits. Beyond the edge of the continuous Vorries deposits we define a *zone*

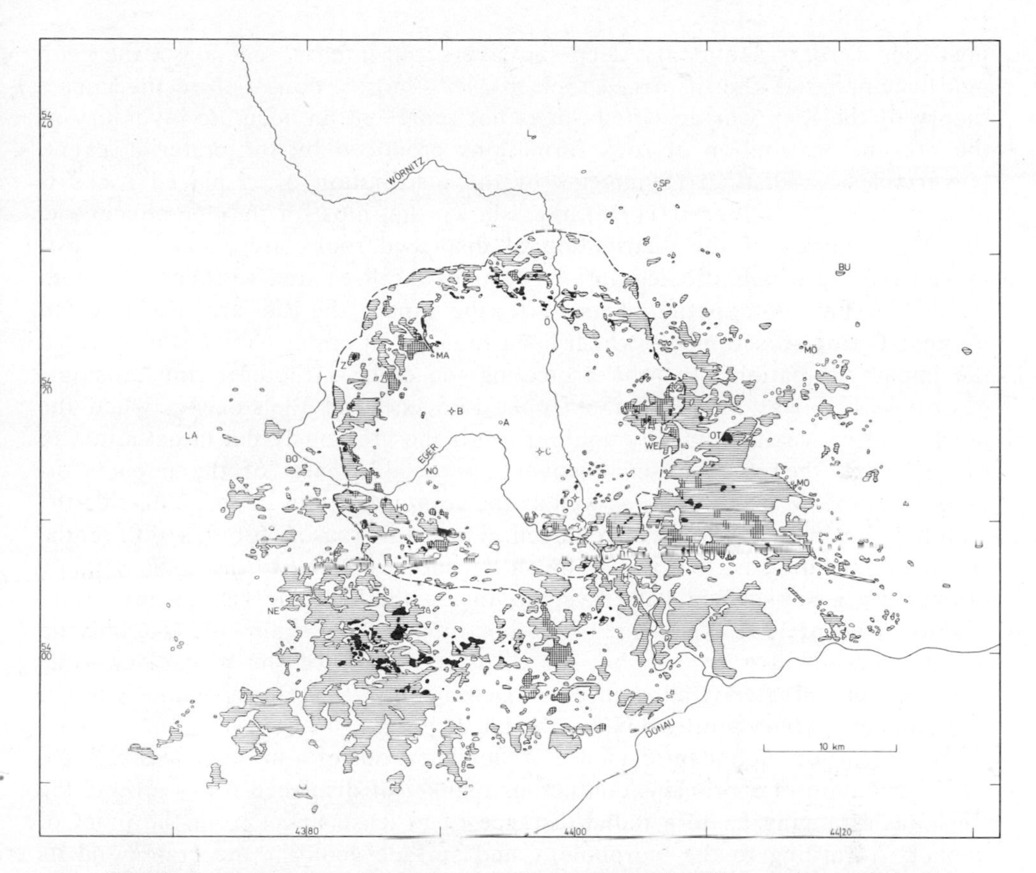

Fig. 7. Geological map of displaced rock masses at the Ries crater; horizontal hatching = Bunte breccia and megablocks; black = suevite; cross hatching = crystalline basement rocks; A = crater center; B = drilling Nördlingen 1973; C = drilling Deiningen; D = Wörnitzostheim; dashed line = tectonic rim of the crater; Bu = Bubenheim; Di = Dischingen; G = Gundelsheim; H = Harburg; Ho = Holheim; K = Kirchheim; La = Lauchheim; L = Lentersheim; Ma = Maihingen; Mö = Möhren; Mo = Monheim; Ne = Neresheim; Nö = Nördlingen; Ö = Öttingen; Ot = Otting; P = Polsingen; R = Ronheim; Sp = Spielberg; Su = Sulzdorf; U = Ursheim; We = Wemding; Z = Zipplingen; modified from Schmidt-Kaler *et al.* (1970).

of distant ejecta in which three types of rocks occur which are presumably related to the Ries event; the Reuter blocks and the Molasse bentonites south of the Danube, and the moldavite tektites of Czechoslovakia (see section 3.2.5).

The present morphology of the Vorries deposits is very smooth and relatively continuous in the E and SE where it is covered in some places by post-impact upper Miocene and Pliocene fluviatile sediments (Table 1). It is more hummocky and discontinuous in the S and SW. Only isolated remnants of ejecta were found in the W and NE (Fig. 7; Gall *et al.*, 1975). No displaced rock masses occur beyond the northwestern tectonic rim of the crater (Fig. 7), where the radial distribution of the megablocks ends abruptly. Due to the partial cover by post-Ries Tertiary sediments, deep soil layers and intense vegetation, the geological mapping of the displaced rock masses—mostly done before the impact theory of the Ries was accepted—does not represent an accurate inventory of the present distribution of rock formations produced by the cratering event. Nevertheless, a distinct asymmetry of the distribution of displaced rocks is obvious (Fig. 7). Gall *et al.* (1975) have shown that most of the non-concentric, irregular features of the distribution of displaced rocks are caused by post-impact erosion which affected various parts of the Ries area with very different intensities: For a long period of time after the impact, the Ries area was covered by post-Tortonian sediments which filled the central crater cavity, and covered the impact formations, thereby protecting the entire structure from erosional destruction. It was not until the Upper Pliocene and Pleistocene, when the northern area was tectonically uplifted to tilt the previously flat target strata to the SE, that the post-Tortonian cover, appreciable parts of the impact formations in the W, N and NE, as well as the complete crater deposits outside the tectonic rim in the NW were removed. The main reason for this differential surface denudation is, that the Malmian limestone plateau to the S was mainly drained by a subsurface karst system. Inside the tectonic rim of the crater (radius of about 12–13 km), part of the post-impact lake sediments (possibly up to 100 m) were also eroded, but the displaced rocks of the megablock zone remained less affected (Gall *et al.*, 1976; Dehm *et al.*, 1977) (for a summary of the post-impact Tertiary stratigraphy see Table 1).

In conclusion, abundant evidence argues in favour of a more or less concentric distribution of a primary, continuous blanket of displaced rocks around the Ries Basin ranging up to a radial distance of at least 40 km from the point of impact. According to the morphology and surface geology, the crater and its impact formations can be subdivided into major concentric zones and structural elements: central basin, inner ring, zone of megablocks, tectonic rim of the crater, Vorries continuous deposits, and zone of distant ejecta (Figs. 1, 5–7).

3.2. *Classification and composition of impact formations*

The main criteria used in classifying the impact formations of the Ries are the geological setting, the stratigraphic provenance, the petrographic composition, and the stage of shock metamorphism. In addition, for polymict breccia formations, textural properties have to be considered.

The impact formations fall into two major groups which may be called *allochthonous* or *displaced rocks* and *autochthonous* or *in situ rocks*. According to the geological setting the displaced rocks may be subdivided into *outer impact formations* or *outer crater deposits* deposited in the inner ring structure and outside the inner ring (radius 5–6 km) and *inner impact formations* or *inner crater deposits* occurring within the central crater cavity inside the inner ring. According to criteria of petrographic composition and grain size we classify the displaced Ries rocks into 6 major units (Table 2; compare also Hüttner, 1969; Engelhardt *et al.*, 1969; Engelhardt and Stöffler, 1974; Gall *et al.*, 1975): (1)

Table 2. Impact formations of the Ries crater (stages of shock metamorphism according to Stöffler, 1971a).

Impact formation	Particle size (m)	Stratigraphic provenance	Shock metamorphism	Geological setting	Texture
Impact melt	As inclusions < 0.2–0.5 m	Crystalline rocks	Stage IV 550–1000 kbar	As inclusions in suevite or as larger coherent bodies	Polymict (mixed with rock and mineral clasts)
Suevite	< 0.2–0.5 m	Crystalline rocks ≫ sedimentary rocks	Stages 0–IV < ~ 1000 kbar	Central crater cavity, megablock zone, and Vorries zone	Polymict
Dike breccias	< 0.2–0.5 m	Crystalline rocks ≫ sedimentary rocks	Stages 0–II < ~ 350 kbar	Crater basement megablocks, surface megablocks	Polymict
Crystalline breccia	< 0.5–1 m	Crystalline rocks	Stages 0–II < ~ 350 kbar	As irregular bodies within or on top of Bunte breccia, central crater cavity	Polymict
Bunte breccia	< 25 m	Sedimentary rocks ≫ crystalline rocks	Stages 0–II < ~ 350 kbar	Megablock zone and Vorries zone	Polymict
Megablocks	~ 25–1000 m	All stratigraphic units	Stages 0–I < ~ 50–100 kbar	Crater basement, inner ring, megablock zone and Vorries	Monomict
Brecciated and fractured autochthonous rocks	—	All stratigraphic units	Stage 0 < ~ 50 kbar	At the tectonic rim, undisplaced crater basement	Monomict

megablocks, (2) Bunte breccia, (3) crystalline breccia, (4) dike breccia, (5) suevite, and (6) impact melt rock.

For the outer impact formations it may be convenient in some cases to consider (1)–(4) throughout the megablock zone and the Vorries as one unit which represents the *bulk continuous deposits* (in German: Bunte Trümmermassen, Hüttner, 1969) as opposed to the suevite and impact melt which overlay the bulk deposits with a distinct discontinuity (Fig. 7).

3.2.1. *Megablocks and Bunte breccia.* Displaced fragments of all stratigraphic units of the Ries target rocks which are larger than 25 m in size and can be mapped geologically will be called *megablocks* (German: Schollen; they were called "klippes" by Dennis, 1971). Megablocks are composed of a coherent, though brecciated or fractured mass of one or more rock types which still show their primary stratigraphic relation (Figs. 8, and 9). They reach dimensions of more than 1 km and are predominant in the megablock zone (Fig. 6). Smaller blocks, however, were observed in the Vorries up to radial distances of more than 35 km. Analysis of the drill core Nördlingen 1973 (Bayer. Geol. Landesamt, 1974, 1977) indicates that megablocks of crystalline rocks also occur at the base of the suevite breccia layer in the central crater cavity (Engelhardt and Graup, 1977; Stöffler, 1977). The lateral dimensions of these blocks are not known (see section 4.6).

With decreasing block size a continuous transition into finer-grained breccia called *Bunte breccia* is observed in the outer impact formations (Hüttner, 1969). An intimate mixture of megablocks and Bunte breccia is most typical for the Vorries whereas in the megablock zone Bunte breccia (block size <25 m) is less abundant compared to megablocks.

Bunte breccia is a polymict breccia of clastic material derived from all stratigraphic units of the target (Figs. 10, 11). The clast size ranges from a few microns upward (Hüttner, 1969). The following *compositional characteristics* of the finer-grained variety of Bunte breccia were found (Schneider, 1971 and own analyses) from localities within a radius of about 12–18 km. Among the rock clasts Malmian limestone predominates over Dogger, Liassic, Keuper, and basement rocks. Crystalline basement rocks usually range from 3–10% of the total volume of rock clasts (Fig. 12). The main mineral clasts are quartz, alkali feldspar, plagioclase, and mica (decreasing abundance) (Fig. 13). The fine-grained matrix consists mainly of calcite and subordinate amounts of kaolinite, mixed layer illite-montmorillonite, illite, and montmorillonite. In Bunte breccia from the outer zone of the Vorries deposits, the matrix is dominated by local Tertiary sand and clay (Hörz, 1976; Schneider, 1971) and rock clasts from lower stratigraphic levels of the target are less abundant.

The *macroscopic texture* of Bunte breccia is very complicated and varies from place to place (Hüttner, 1969). There is no indication of sorting. In certain parts the breccia appears more fine-grained (<20 cm), in others many large blocks (1–25 m) are mixed with smaller amounts of the fine-grained variety. A most important observation is that many rock clasts in the Bunte breccia are

Fig. 8. Megablock of Malmian limestone near Ebermergen, in the south-eastern continuous deposits, 18.5 km from the point of impact (dimensions of block ≈ 300 m).

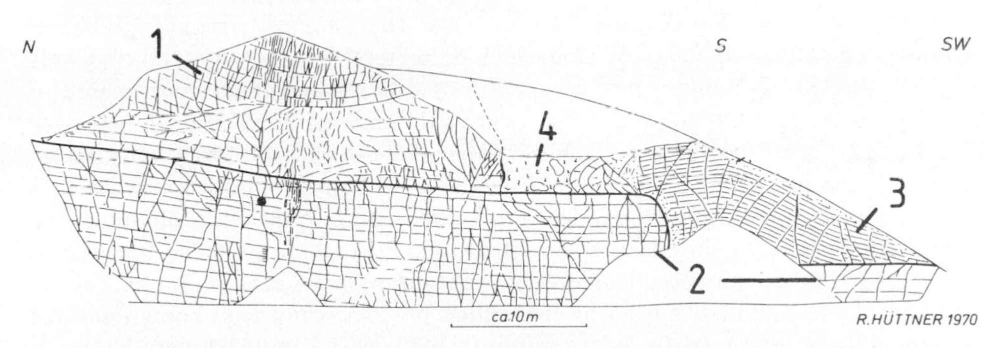

Fig. 9. Displaced Malmian megablocks (1 = Malmian δ, 3 = Malmian β) and Bunte breccia (4) overriding the original target surface (heavy solid line; 2 = Malmian δ); quarry Schneider near Ursheim; from Hüttner in Schmidt-Kaler *et al.* (1970).

monomict or sometimes polymict breccias, indicating several phases of deformation during the process of excavation, transport and deposition. Soft rock clasts (clay, shale, marl) sometimes display a strong deformation (folding, bending, elongation in one direction) and flow-like parallel alignments, especially near the target surface (Wagner, 1964; Hüttner, 1969).

Fig. 10. Bunte breccia on top of limestone (Malmian δ) which forms the polished and striated target surface; photograph courtesy of J. Kavasch; lateral dimension about 100 m; quarry Bschor near Ronheim.

3.2.2. *Crystalline breccia and dike breccias.* Occasionally irregular bodies of polymict *crystalline breccias* of some tens of meters in size, which consist only of a mixture of crystalline rock fragments of different lithology and degree of shock metamorphism (stages 0, I, II) occur in the inner ring, the megablock zone and in the Vorries (Hüttner, 1969; Stöffler, 1969; Abadian, 1972). The stratigraphic relations of these crystalline breccias to the surrounding Bunte breccia and megablocks are not always clear. In some outcrops they are obviously on top of Bunte breccia. In cases where the lateral extension of these breccias is less than 25 m they may be considered per definitionem as part of Bunte breccia. The abundance and distribution of crystalline breccia is not well known and the present data is restricted to a few outcrops such as Leopold-Meyers-Keller at Nördlingen. Some quantitative modal data are plotted in Fig. 18 (Abadian, 1972).

Dike breccias show some similarities to crystalline breccia in terms of their petrographic composition. They occur as dikes of irregular shape and various thickness in the decimeter to meter range penetrating displaced megablocks of various lithology, most frequently crystalline rocks. The dikes contain fragmental material of variable lithological and stratigraphic provenance. Results of petrographic analyses (Stöffler, 1969; Abadian, 1972; Abadian *et al.*, 1973) suggest that two types of dikes can be distinguished: I. Dikes filled with a mixed breccia of fragmental material which is predominantly derived from rocks

Fig. 11. Bunte breccia fragments of Malmian limestone, Liassic shale, Keuper shale and sandstone, crystalline rocks and Tertiary shale; quarry Bschor, Ronheim; scale = cm.

different in composition and stratigraphic provenance from the adjacent country rock. II. Dikes filled with brecciated material which is partly or predominantly derived from the adjacent country rock by some frictional process.

Type I dikes were observed only in megablocks near to the surface (Fig. 14). In most cases they consist of a mixture of moderately shocked crystalline rock fragments (see section 3.4.1). Abadian *et al.* (1973) interpreted the type I-dikes as open fractures which were filled by ejected material after the deposition of the megablocks.

Type II dikes are represented by two kinds of dikes which occur in megablocks of the surface deposits (Abadian *et al.*, 1973) and in displaced blocks of the crater basement as revealed by the drillhole Nördlingen 1973 (Stöffler *et al.*, 1977). Type II dikes of the surface megablocks consist almost exclusively of rock fragments derived from the adjacent rock by a frictional process (Abadian *et al.*, 1973). The dikes discovered in the crystalline crater basement by drilling

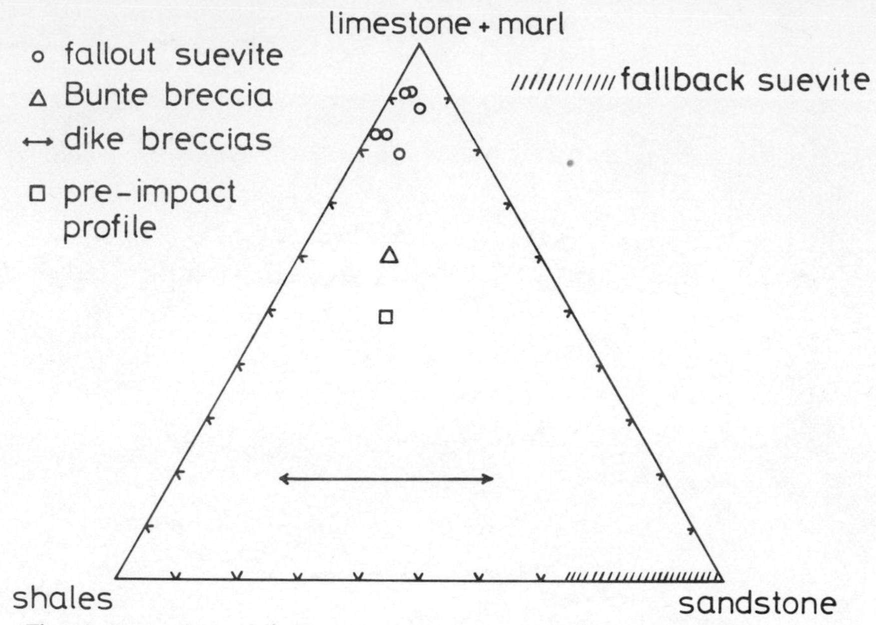

Fig. 12. Proportions of limestone + marl, shale and sandstone in the coarse (~1–10 cm) fragment population of Ries impact breccias and in the pre-impact profile of the sedimentary rock strata. Bunte breccia from Otting, fallback suevite from the drill core Nördlingen 1973 (from Stöffler *et al.*, 1977).

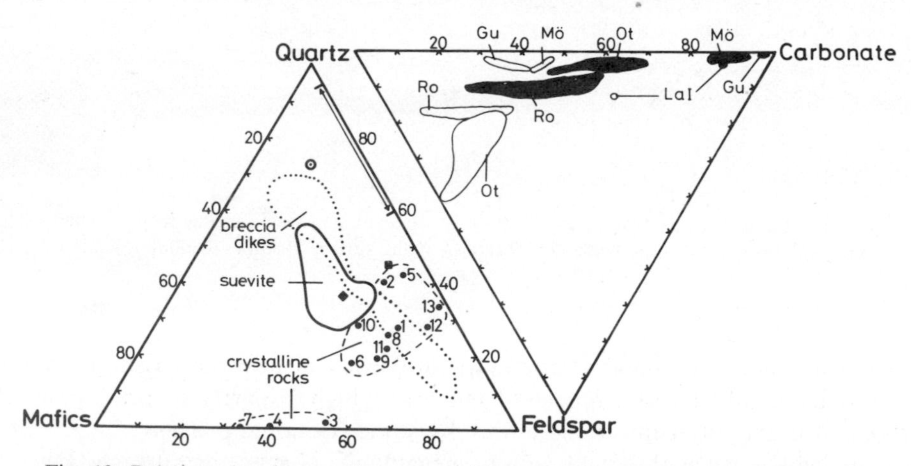

Fig. 13. Relative abundances of quartz, feldspar (alkali feldspar + plagioclase), carbonate and mafic mineral fragments (biotite + amphibole + chlorite) in suevites, dike breccias and Bunte Breccia and in various types of crystalline basement rocks of the Ries. 1, 2, 4, 5, 6, = various gneisses; 3 = amphibolite; 7 = hornblende diorite; 8–11 = granodioritic rocks; 12, 13 = granites (data for 1–13 from Graup, 1975) ⊙ Bunte Breccia of Otting; ◆ suevite of Otting; ■ suevite of Zipplingen; arrow = quartz: feldspar ratio of 10 occurrences of Bunte Breccia (Schneider, 1971); solid line and dotted line = fields of suevite and dike breccia samples, respectively, drill core Nördlingen 1973 (data of Stöffler *et al.*, 1977). Bunte Breccia samples in right triangle given for the grain size ranges. 0.063–0.125 mm (white areas) and 1–2 mm (gray areas) at the localities of Gundelsheim (Gu), Lauchheim (La I), Möhren (Mö), Otting (Ot), and Ronheim (Ro) (data from Schneider, 1971).

Fig. 14. Dike breccia in gneiss at Klostermühle, Maihingen after G. Wagner, published by Hüttner (1969); parallel and subparallel lines in gneiss indicate the primary schistosity and foliation.

contain not only fragmental country rocks but also a variable proportion of rock fragments of the sedimentary strata and other, higher shocked regions of the crystalline basement (Figs. 12, 13, 18). About 70 dikes several centimeters to about one meter in thickness were found in the drill core Nördlingen 1973 between about 600 and 1200 m depth (Figs. 21, 29). Shocked quartz fragments and a relatively high abundance of sedimentary rock fragments are restricted to the upper section of these dikes (above about 1170 m, Stöffler *et al.*, 1977). These dikes are interpreted as being formed by an injection process which took place during the early stage of crater excavation (Stöffler, 1977).

3.2.3. *Suevite.* Suevite is a polymict breccia of clastic material derived predominantly from the crystalline basement. The particle sizes of the rock and mineral clasts range from a few microns up to about 20 cm, in rare cases up to half a meter (Figs. 16, 17). The particle size distribution is typical for impactoclastic material because of a strong deficiency of fine-grained material (see Stöffler *et al.*, 1976). Most characteristic is that the rock and mineral clasts belong to all stages of shock metamorphism, including partially and completely molten rock material. The crystalline rock fragments display a great variety of different rock types such as granites, gneisses, and amphibolites. Only about 0.2–5% of the rock clast population are sedimentary rocks. The fine-grained detrital matrix mainly composed of quartz, feldspar biotite, and amphibole fragments shows no effects of welding or severe recrystallization. Montmorillonite, which is an abundant matrix constituent, probably results from a secondary alteration of fine-grained clastic or glassy (?) material.

Based on texture, composition and structural position, two types of suevite formations can be distinguished:

(a) *Fallout suevite*, deposited outside the inner ring, forms isolated patches (Fig. 7) on top of megablocks or Bunte breccia (Fig. 16).

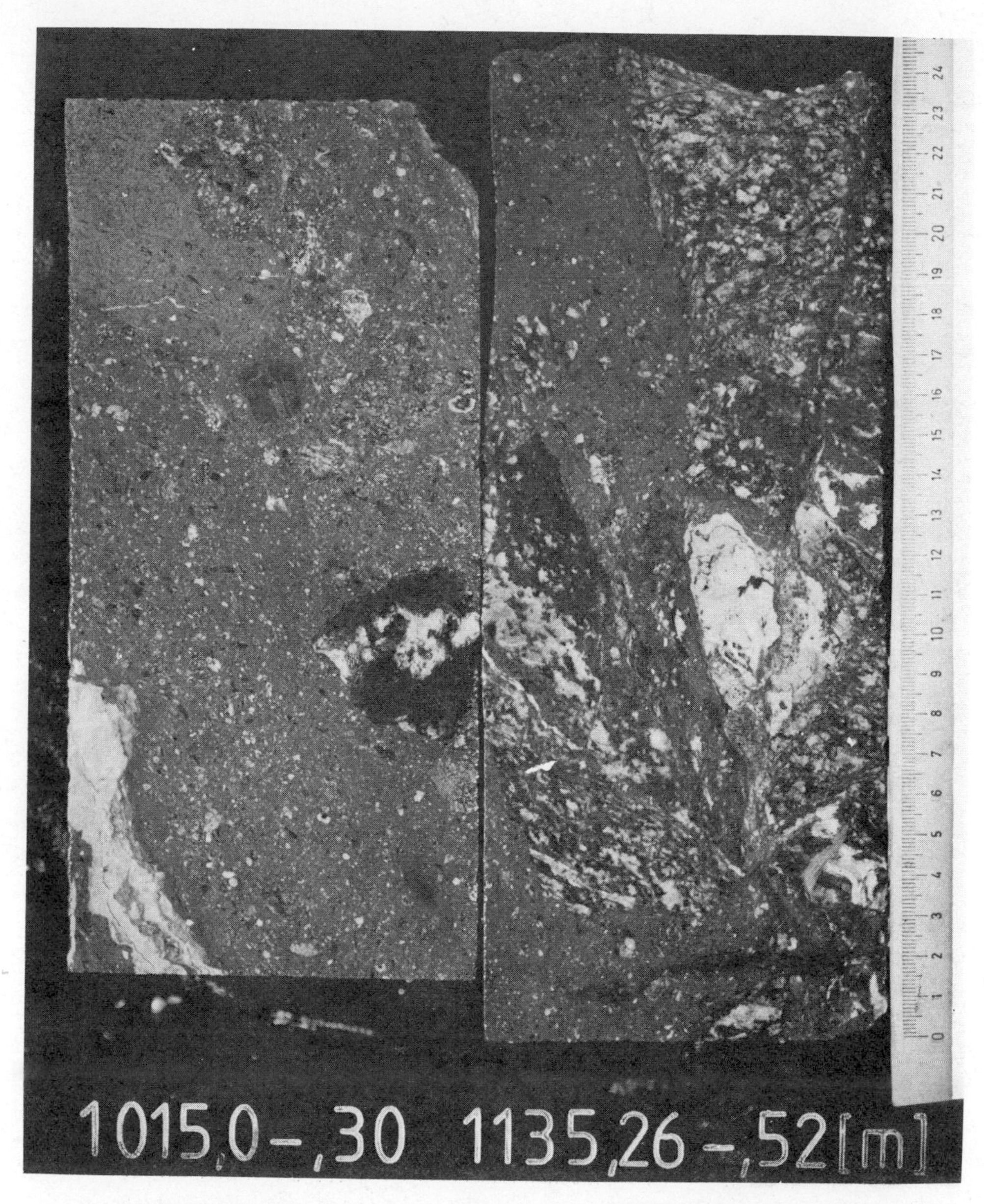

Fig. 15. Sections of dike breccias from the drill core of Nördlingen 1973. The vein-like intrusion into the biotite gneiss country rock is best visible in the right photograph. Lower left corner: brecciated fluorite dikelet. Scale = cm.

Fallout suevite has been studied thoroughly in the past 15 years (Shoemaker and Chao, 1961; Hörz, 1965; Pohl, 1965; Stöffler, 1966, 1971b; Förstner, 1967; Chao, 1967, 1968, 1973; Engelhardt *et al.*, 1969; Pohl and Angenheister, 1969). The most typical in the *modal composition* of the fallout suevite is the abundance of dark-brown to black glass bodies, mostly some mm to a few dm in size, which often have a characteristic bomb- or pancake-like shape (Hörz, 1965). The average glass content is about 15 vol.%. The chemical and petrographic composition of these glasses will be discussed in section 3.2.4. The fallout suevite differs from its fallback counterpart in the abundance and type of sedimentary rock inclusions (Ackermann, 1958; Engelhardt *et al.*, 1969). 80 to 90% of these inclusions (Fig. 12) which comprise only 0.05–1.2% of the total rock, are Malmian limestones. The modal composition, the macroscopic texture and the color vary between the widely distributed occurrences of suevite to some degree whereas the chemical composition of the glass bombs is rather constant compared to the large compositional variation of the lithic clasts (see section 3.2.4).

(b) *Fallback suevite* forms the top layer of the impact formations within the central crater cavity as revealed by the drillings of Deiningen (Förstner, 1967) and Nördlingen 1973 (Bayer. Geol. Landesamt, 1974, 1977). In the drill core Nördlingen 1973 a complete section of the suevite layer was obtained (Figs. 21, 29). This suevite profile can be subdivided into 3 major units: suevitic sandstones and conglomerates from 314 to 331.5 m, melt-rich suevite from 331.5 to about 525 m, melt-poor suevite from 525 to 602 m as intercalations within the brecciated crystalline basement.

Small, brownish, elongated to isometric bodies of extremely vesiculated and recrystallized melt particles, less than a millimeter to some centimeters in size, are the most contrasting feature of the *modal composition* and *texture* as compared to the fallout suevite (Fig. 17). In addition, the fallback suevite contains many less sedimentary rock inclusions among which only sandstone and shale, but no limestone was found. The melt particles are also less abundant compared to the fallout suevite (about 3–5 vol.% on the average for the entire suevite layer).

Some relevant properties of the *mineral and lithic clast population* in both types of suevite are presented in Figs. 12, 13, and 18 (see also Stöffler *et al.*, 1977). The most important conclusions which can be drawn from these data for the origin of fallout and fallback suevite are: (1) the lithic clast population of both suevites (Fig. 18) is dominated by gneisses and differs distinctly from the petrographic composition of the crystalline megablocks, 79% of which are granites; (2) the abundance of melt and sedimentary rock inclusions is markedly higher in the fallout suevite than in the fallback suevite; and (3) the mineral clast population of both types of suevite deviates from a mere mixture of comminuted basement rocks because of excess quartz, possibly derived from comminuted sandstone (Fig. 13).

3.2.4. *Impact melt.* Shock produced melt of the Ries target rocks is found in two types of impact formations: (1) as inclusions in suevite (Fig. 17, see section 3.2.3), and (2) as isolated bodies of impact melt rocks.

Fig. 16. Suevite on top of Bunte breccia, quarry of Aumühle, northern part of megablock zone.

Only two occurrences of *impact melt rock*, with lateral extents of 10–50 m were found near Polsingen and Amerbach in the megablock zone, about 1.5–3 km inside the tectonic rim of the crater. They overlay megablocks or Bunte breccia. The reddish melt rock consists of a vesicular, fine-grained recrystallized glass matrix with fluidal texture in which many angular fragments of crystalline rocks in different stages of shock metamorphism are embedded. The matrix glass is completely crystallized to a fine-grained aggregate of feldspar, pyroxene, hematite, and cristobalite.

The *glass bombs of the fallout suevite* most intensely investigated by Hörz (1965), Engelhardt (1976b, 1972), El Goresy (1968) and Stähle (1972) are mixtures of fluidal, vesiculated and schlieren-rich glass with fragmental inclusions of comminuted basement rocks in all stages of shock metamorphism. Quartz fragments (6–23%), 5–30% which display shock effects, are most abundant; less frequent are crystalline rock fragments (1–7%), plagioclase (0.1–0.3%), SiO_2-glass (1–3%) and accessories (Zircon, apatite, biotite comprising 0–0.7%) according to Engelhardt (1972). The decomposition products of some accessories indicate very high temperatures of the melt of at least 1950°C (El Goresy, 1968; Stähle, 1972). Ni-iron spherules are extremely rare in these glasses but Fe-sulfide spherules are more abundant (El Goresy, 1968; Chao, 1963; Stähle, 1972).

In contrast to the glass bombs of the fallout suevite, the *melt inclusions of the fallback suevite* (see section 3.2.3) have a pumice-like vesiculated texture and are completely recrystallized to montmorillonite and analcite (Förstner, 1967;

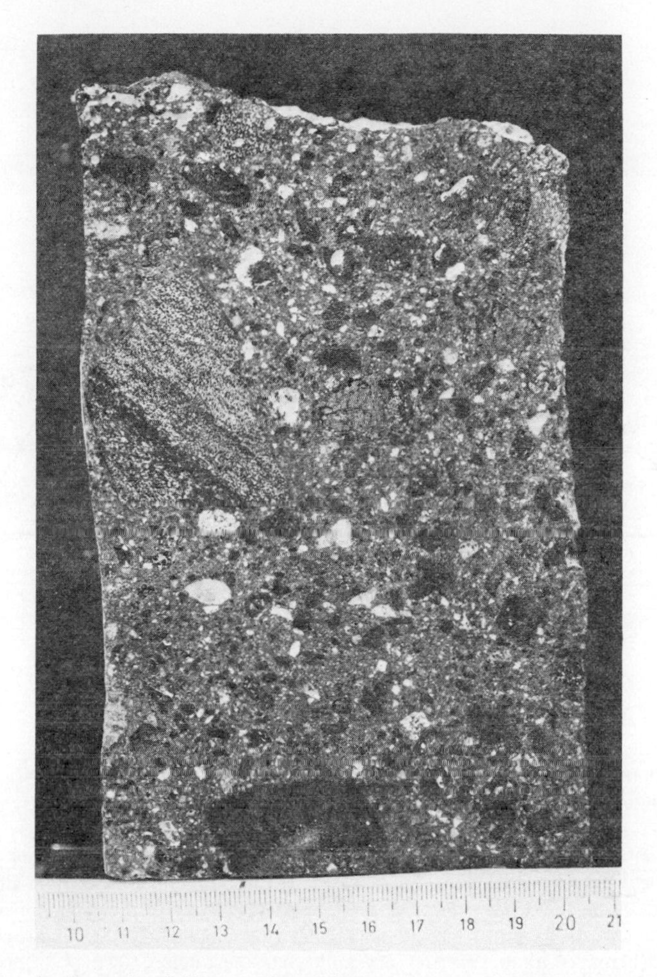

Fig. 17. Suevite from the drill core of Nördlingen (depth 417.5 m), dark inclusions are melt products, other fragments are crystalline rocks of lower shock metamorphism; scale = cm.

Stähle and Ottemann, 1977; Stöffler *et al.*, 1977). They contain many less fragmental inclusions than the glass bombs or are even free of them.

The chemical composition of the various occurrences of glassy impact melt is remarkably constant as far as the fallout suevite is concerned (Table 3, Fig. 19). The melt of the glass bombs probably originates from shock-fused crystalline basement rocks (Fig. 19), most probably from gneiss (Engelhardt, 1977) although the scatter of 88 analyses of different localities (Table 3; Stähle, 1972) is still large enough to allow for other crystalline source rocks or mixtures of source rocks. However, most of the melt volume in the Ries is contained in the fallback suevite or in a possible, yet unknown central impact melt layer. The composition of the melt in the fallback suevite is not known because its chemistry is severely

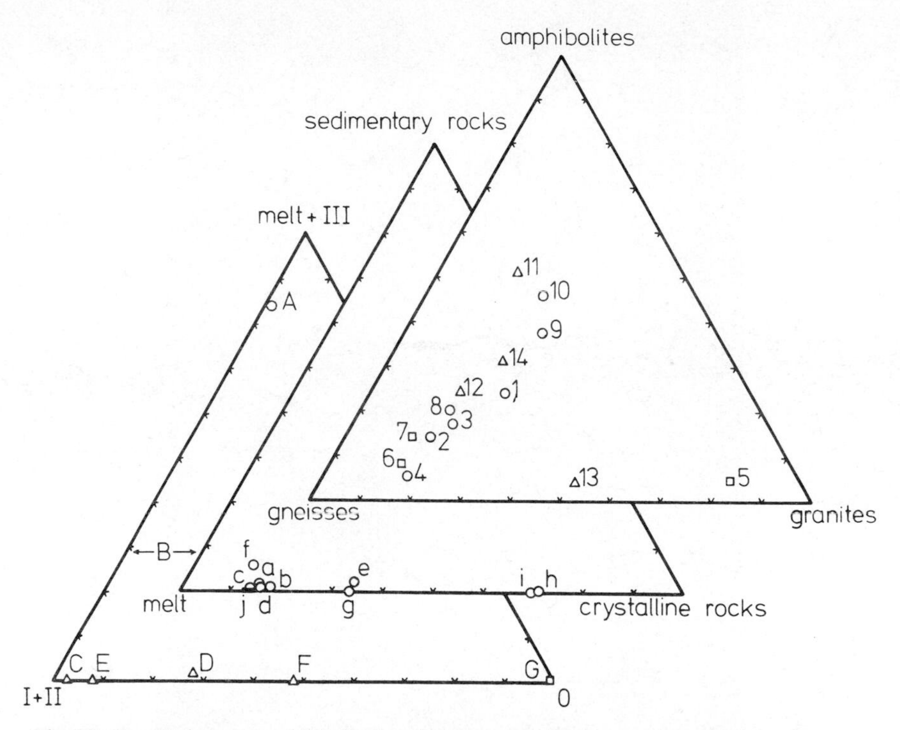

Fig. 18. Population plots of lithologies of coarse particles in polymict Ries breccias and of autochthonous and allochthonous crystalline basement rocks (circles = suevite, triangles = crystalline breccias, squares = crystalline basement megablocks).

Right plot: 1 = five fallout suevite occurrences (fragments of shock stages O and I); 2 = same as 1 but only fragments of shock stage II (Stöffler, unpublished data and data of Engelhardt *et al.*, 1969), 4 = suevite of Zipplingen (fragments of the 4–32 mm size class, Stöffler, unpublished data), 5 = large blocks and megablocks of crystalline rocks of the Ries ejecta blanket (Graup, 1975); 6 = crystalline basement below 602 m of the drill core of Nördlingen 1973, calculated from the core description of Bauberger *et al.*, 1974; 7 = same as 6 but including hornblende gneisses within amphibolites; 8 = suevite of the drill core of Nördlingen 1973 (fragments of the 8–28 mm size class, from Stöffler *et al.*, 1977); 9 = suevite of the drill core Nördlingen 1973 (fragments 60 mm, Bauberger *et al.*, 1974); 10 = same as 9 but including hornblende gneisses within amphibolites; 11, 12, 13, 14 = polymict crystalline breccias of Nördlingen, Maihingen, Appetshofen and Lierheim (data from Abadian, 1972).

Middle plot: a–g = fallout suevite occurrences (a–f = data of Ackermann, 1958); a, b = Aufhausen, c = Mauren, d = Aumühle; e = Bollstadt; f = Otting; g = Zipplingen (Stöffler *et al.*, 1977); h = fallback suevite, drill core Nördlingen 1973 (Stöffler *et al.*, 1977); i, j = fallback and fallout suevite of the drill cores of Deiningen and Wörnitzostheim, respectively (approximate values of Förstner, 1967).

Left plot: A = average of 5 fallout suevite occurrences (Engelhardt *et al.*, 1969); B = average of the drill core fallback suevite of Nördlingen 1973 (ratio stage 0 : I : II not exactly known, Stöffler *et al.*, 1977); C, D, E, F = polymict crystalline breccias (see 11, 12, 13, 14); G = large blocks and megablocks of crystalline rocks from the ejecta blanket (Graup, 1975).

Table 3. Chemical analyses of crystalline basement rocks, suevite, and glass bombs.

	1	2	3	4	5	6	7	8	9	10	11	12
SiO_2	53.39	57.50	65.62	59.09	72.13	58.31	57.12	52.63	64.01	62.66	64.03	1.33
TiO_2	0.70	0.88	0.43	0.98	0.13	0.98	0.86	0.33	0.78	0.78	0.79	0.096
Al_2O_3	17.47	17.90	15.83	17.85	15.37	14.97	13.53	10.06	16.85	15.48	15.25	0.39
Fe_2O_3	4.46	1.70	1.07	1.81	0.34	1.46	1.81	4.29	3.40	1.53	5.22	0.43
FeO	5.29	4.70	2.88	5.01	0.81	4.18	5.24	3.28	1.44	2.55		
MnO	0.24	0.10	0.08	0.16	0.03	—	—	—	—	—	0.077	0.017
MgO	4.86	3.32	1.66	3.20	0.38	4.17	5.64	2.70	1.23	1.82	3.04	0.31
CaO	7.70	4.28	3.40	4.87	0.80	3.51	6.52	4.87	2.97	2.82	3.96	0.52
Na_2O	3.55	3.91	2.90	3.35	3.45	4.52	3.26	3.62	4.01	4.70	3.02	0.46
K_2O	1.21	2.43	4.09	2.35	5.66	3.23	1.85	3.24	3.63	5.74	4.01	0.82
P_2O_5	0.20	0.18	0.57	0.29	0.24	0.31	0.26	0.33	0.40	0.33	0.21	0.059
H_2O	1.39	2.93	1.41	1.91	0.60	3.35	2.19	3.82	1.05	1.35	—	—
CO_2	0.39	0.82	1.21	0.12	0.29	0.60	1.34	—	0.54	0.93	—	—
S	—	—	—	—	—	0.71	0.48	0.28	—	—	—	—
Σ	100.85	100.65	100.85	100.99	100.23	100.38	100.23	99.55	100.31	100.69	99.61	—

1–5 = crystalline basement rocks from the drill core Nördlingen 1973 (from Graup, 1977), amphibolite (depth 625.3 m, 1), hornblende-biotite-plagioclase gneiss (711.9 m, 2), migmatic gneiss (926.5 and 1201.5 m, 3 and 4), granite (540.5 m, 5).

6–8 = drill core Nördlingen 1973 (from Stähle and Ottemann, 1977), suevite (depth 400.0 m, 6), melt-poor suevite (597.0 m, 7), dike breccia (737.1 m, 8).

9 = suevite (depth 61.1 m) from the drill core Wörnitzostheim (from Förstner, 1967).

10 = suevite (depth 340 m) of the drill core Deiningen (from Förstner, 1967).

11 = 88 microprobe analyses of 51 matrix glasses from glass bombs of fallout suevite occurrences (Amerdingen, Aufhausen, Aumühle, Bollstadt, Fünfstetten, Grossorheim, Mauren, Otting, and Zipplingen (from Stähle, 1972).

12 = standard deviation σ of 88 analyses of No. 11.

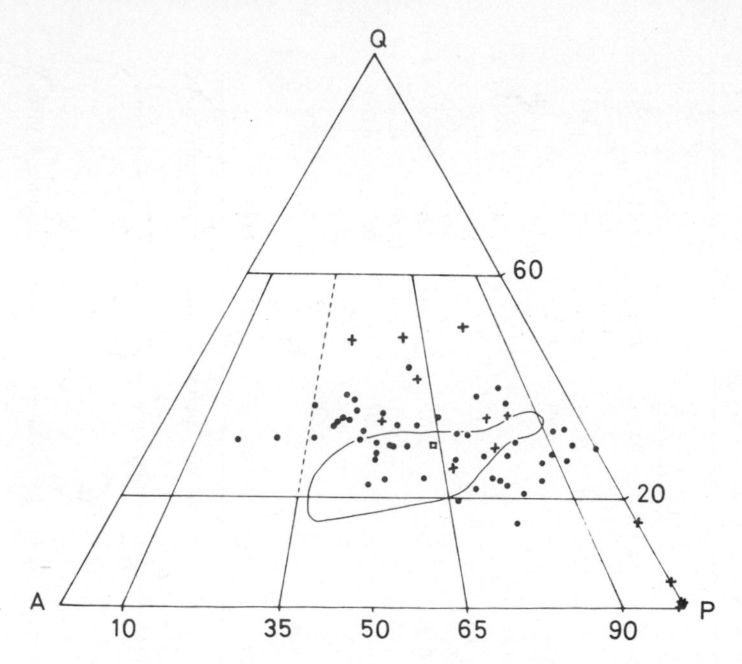

Fig. 19. Modal composition of crystalline basement rocks after Graup (1975) as compared to the normative composition of 52 samples of matrix glass (solid line; 88 microprobe analyses) from the fallout suevite of Amerdingen, Aufhausen, Aumühle, Bollstadt, Fünfstetten, Großsorheim, Mauren, Otting, and Zipplingen (data from Stähle, 1972); calculated according to the plutonic norm of Rittmann (1973); A = alkali feldspar, P = plagioclase, Q = quartz.

altered now. It is possible that the extremely vesiculated melt particles are derived in part from fused sedimentary rocks or from a more or less homogenized mixture of sedimentary and crystalline rocks and that they originate from a part of the melt zone which is different from that of the fallout glass bombs (for details see Stöffler, 1977).

Neither the Ries melt nor other Ries impact formations is markedly enriched in siderophile elements (Ir, Os, Re, Au, Pd) so that any contamination by the impacting body is uncertain (Morgan, 1976), although the Ni-content of the glass-bombs is slightly higher than that of the crystalline basement rocks (Hörz, 1965; Stähle, 1972).

3.2.5. *Distant ejecta.* There are three possible types of distant ejecta which might be related to the Ries impact event: (1) single rock fragments of Malmian limestone (Malmian δ and younger), generally less than 20 cm in size, were found south of the Danube at radial distances up to 70 km within fine-grained Molasse sediments of Ries age and at the boundary of Tertiary and Quaternary (Fig. 1; Reuter, 1926; Scheuenpflug, 1973; Engelhardt, 1975; Gall *et al.*, 1975). Part of these so called *Reuter blocks* are Ries ejecta (Gall and Müller, 1975; Gall *et al.*, 1975). (2) Thin layers of *bentonite* occur in the same area where the Reuter blocks are found. The bentonites are tuff-like layers consisting mainly of fine glass particles recrystallized to montmorillonite. The Ries origin of these sediments is still controversial (Harr, 1976). (3) The *moldavite tektites* of the Czechoslovakian strewn field (Fig. 1) are commonly considered as Ries ejecta (Gentner *et al.*, 1963).

The strongest argument in favour of the Ries origin of the bentonites and moldavites is the common age of 14.7 m.y. which is the age of the Ries event (Gentner and Wagner, 1969). If the bentonites and moldavites originate from the Ries, they might represent fused silica-rich Tertiary sediments from the top of the target which were ejected with extremely high velocities at the very beginning of the impacting process (see Gault *et al.*, 1968; O'Keefe and Weiskirchner, 1970).

3.3. *Structure, stratigraphy and thickness of the outer impact formations*

Regional variations of the composition, thickness and structure of the Ries outer impact formations have been reviewed recently by Gall *et al.* (1975). The regional variation in the distribution of the fallback material inside the central crater cavity could be evaluated only by indirect geophysical observations and by extrapolation from drill core data. It will be discussed in section 4. Essential characteristics of the outer impact formations deposited outside the inner ring are schematically represented in Fig. 20. They will be discussed here following vertical and horizontal sections.

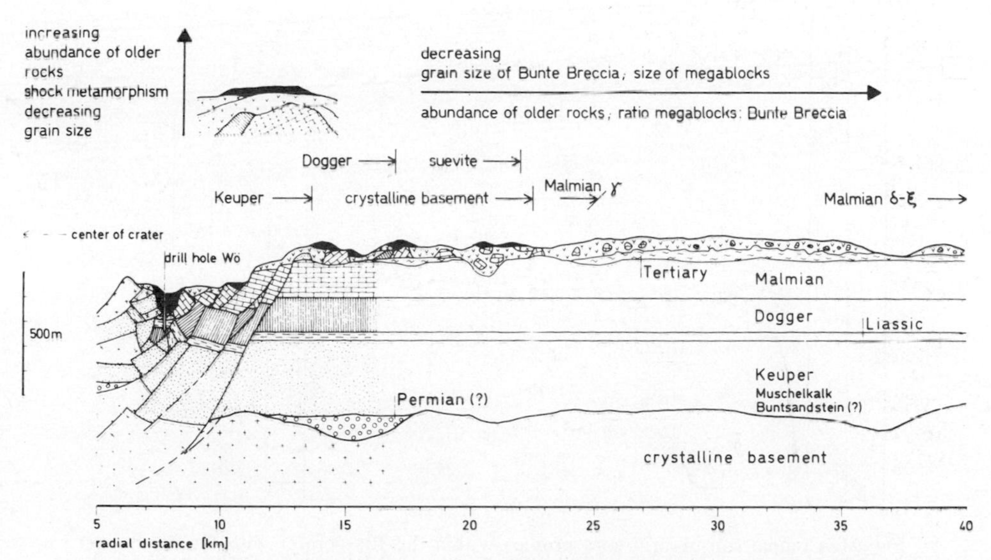

Fig. 20. Schematic profile of the megablock zone and southern "Vorries" continuous deposits; Wö = Wörnitzostheim; small arrows indicate the maximum radial extent of megablocks from different stratigraphic levels and of the suevite.

Within the *megablock zone* complete vertical sections of the displaced rock masses are not exposed and the information obtained from the surface geology has to be supplemented by geophysical and drilling data (e.g., drill holes of Wallerstein 1948, Nördlingen 1950, 1955, and Wörnitzostheim 1965; Fig. 21). Near the inner ring, the section of displaced rocks consists predominantly of megablocks, which have been uplifted mainly from the crystalline basement and the lowest section of the sedimentary strata (Triassic), occasionally covered by a layer of suevite of limited lateral extension. The drill core of Wörnitzostheim gives one of the few good examples of an inverted stratigraphy in the megablock zone, near the inner ring, with suevite, granite, and an inverted block of Keuper, Liassic and Dogger sediments going from top to bottom (Fig. 21, Gall *et al.*, 1976; Dressler and Graup, 1974). Quite commonly the primary stratigraphic relations of megablocks (e.g., megablock of Keuper, Liassic, and Dogger strata near Harburg described by Müller, 1969) is disturbed by shear faulting. Approaching the tectonic rim of the crater (Fig. 20) the number of megablocks of higher stratigraphic levels (Dogger and lower Malmian) increases and their structural position is less disturbed. Displacement by slumping dominates over tilting and overturning (Hüttner, 1969; Gall *et al.*, 1977). It is typical that fine-grained Bunte breccia is subordinate in the megablock zone, i.e., block sizes below about 10 m are rare. On the average, the particle size seems to decrease towards the top of the displaced masses. Other typical structural features are the polished and striated surfaces (Schliff-Fläche) on top of displaced Malmian

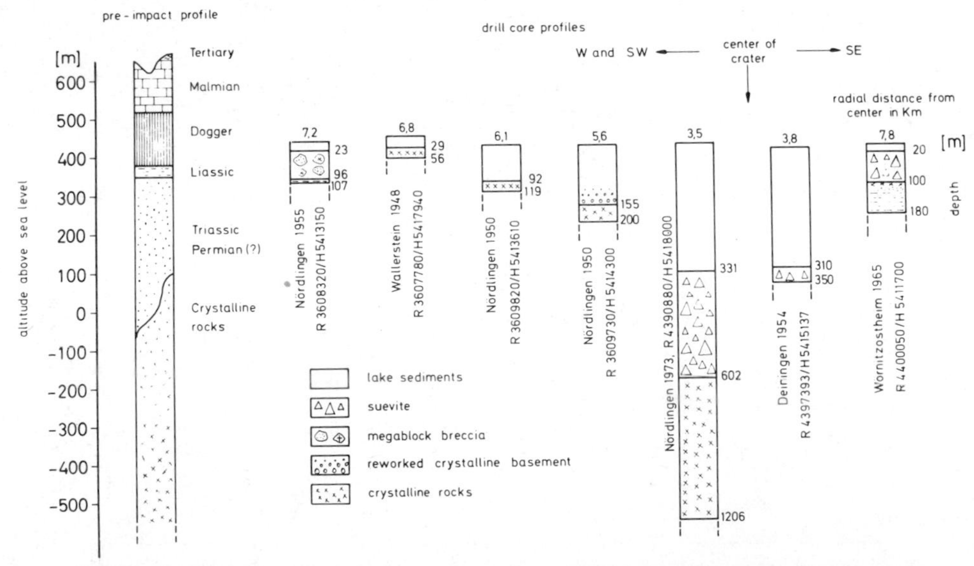

Fig. 21. Comparison of drill core profiles within the Ries crater with the pre-impact stratigraphy (left). Figures on top of the drill cores indicate the radial distance from the center of crater; R and H values refer to the grid system of the topographic maps 1 : 25,000 of the states of Baden-Württemberg and Bayern (Bavaria).

megablocks obviously produced by the violent gliding of masses of Bunte breccia after or during the displacement of the blocks (e.g., Holheim, Siegling quarry; Ronheim, Bschor quarry, Fig. 10). Polished and striated surfaces on top of autochthonous Malmian extend far beyond the tectonic boundary of the crater, e.g., exposures at Gundelsheim or Möhren, up to 24 km from the point of impact.

A remarkable observation in vertical sections of the continuous deposits of the megablocks and the *Vorries zone* is the lack of any regular, layered stratigraphy (Fig. 20). Although the gross principle of inverted stratigraphy appears to hold as far as it can be checked in field outcrops and drill holes, the sequence of inverted rocks is generally incomplete and irregular. The contacts between various stratigraphic units are often steeply inclined. Near the tectonic rim of the crater all main stratigraphic units may be found at the top of the continuous deposits if a cross section circumferential to the rim is considered. However, suevite, if present, is always found on top of other displaced rocks, mostly on Bunte breccia but also on large blocks of all stratigraphic levels (see sketch inset in Fig. 20). The contact to the underlaying polymict breccias (Bunte breccia) which usually have a very strong relief, is extremely sharp. Laminated structures were observed at this contact. Schneider (1971) found that in finer grained Bunte breccia (mostly analyzed in short profiles of the top section) the abundance of rock fragments of lower stratigraphic levels increases from the bottom to the top. This holds also for the abundance of shocked minerals (quartz), whereas the grain size decreases in the same direction.

A genetically important observation was made by field geologists as early as 1905 (v. Ammon) and recently discussed in detail by Hüttner (1958, 1969), Gall (1969), and Hörz *et al.* (1975): at some radial distance from the crater, Bunte breccia contains an appreciable amount of local material excavated from the ground surface and incorporated into the breccia probably by secondary cratering of landing ejecta (Fig. 22). This effect is most conspicuous in places where the pre-impact surface is formed by soft Tertiary sediments (clay, sand). The local material which was horizontally transported in some cases over a distance 1.5–3 km (Hüttner, 1969) comprises 20–80% of the breccia (Hüttner, 1969; Hörz, 1976). Schneider (1971) confirmed this observation by a detailed petrographic analysis of Bunte breccia from several localities.

As discussed by Hüttner (1969), Schneider (1971), and Gall *et al.* (1975) the following radial variations of various properties of the continuous deposits of the Ries appear relevant for the interpretation of the crater-forming process:

(a) The size or volume of megablocks decreases as a function of range, most conspicuously observed with Malmian limestone blocks (Fig. 23).

(b) The abundance of megablocks (>25 m) from the uppermost strata of the target increases as a function of range (Figs. 20, 24); the largest radial range of ejected blocks decreases with their primary depth in the target (Fig. 20). The radial extent of blocks from the crystalline basement and also the suevite does not follow this rule; they extend to a larger radial distance than Keuper or even Dogger megablocks. Their pattern of

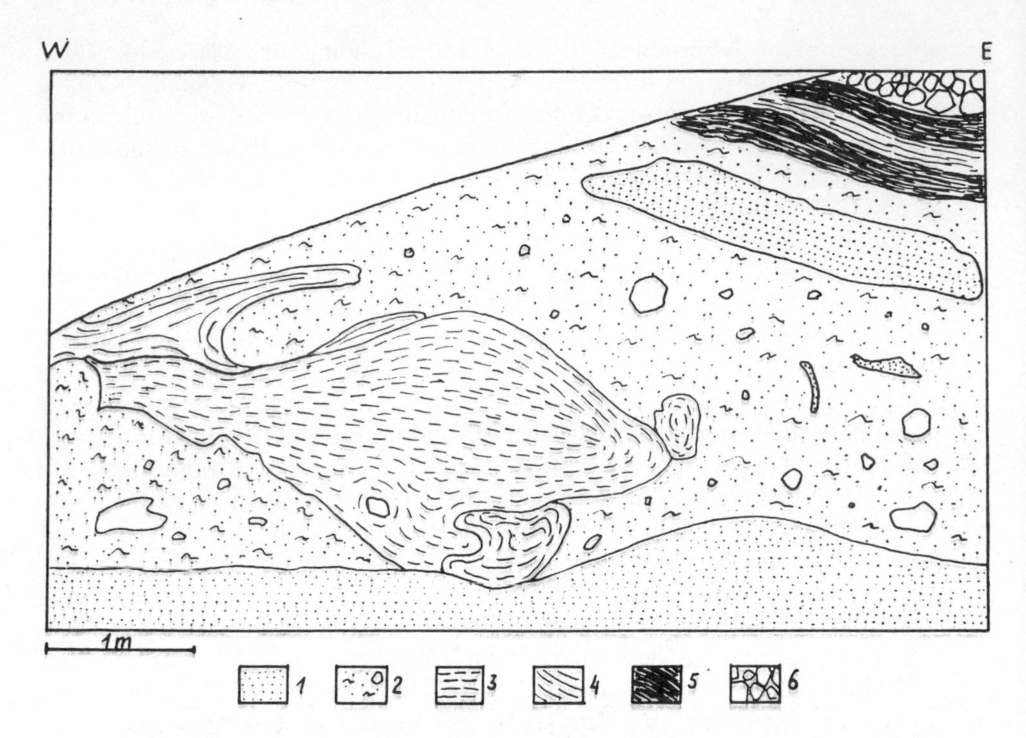

Fig. 22. Bunte breccia on top of Molasse sediments (Tertiary sandstone) with inclusions of large blocks of local Molasse sediments; quarry of Guldesmühle near Dischingen; 1 = Miocene (OMM = Obere Meeresmolasse) sand; 2 = matrix mainly derived from 1; 3 = Miocene (OSM = Obere Süsswassermolasse) marl; 4 = weathering products; 5 = Tertiary clay (?); 6 = blocks of Malmian limestone; from Hüttner (1969).

distribution is less concentric to the crater (Fig. 7, note possible ray-like extension of crystalline blocks).

(c) The ratio of fine-grained Bunte breccia to megablocks increases as a function of range.

(d) In areas of the megablock and Vorries zone where the effect of post-impact erosion is known or negligible, it is found that the thickness of the continuous deposits (Bunte breccia and megablocks taken as one unit) does not show a regular decrease as a function of range. This is probably due to the relief of the pre-impact surface and to mass transport and redistribution by secondary cratering (Oberbeck, 1975; Hörz, 1976). Local depressions of the target contain very thick deposits. In the pre-existing, former Main River valley thicknesses of up to 200 m were found at a distance of about 20 km from the crater center (Birzer, 1969; Bader and Schmidt-Kaler, 1977). On the average the thickness varies between 100 and a few meters.

(e) Suevite reaches a thickness of 84 m in the drill hole of Wörnitzostheim (Figs. 20, 21; Förstner, 1967; Pohl and Angenheister, 1969) in the

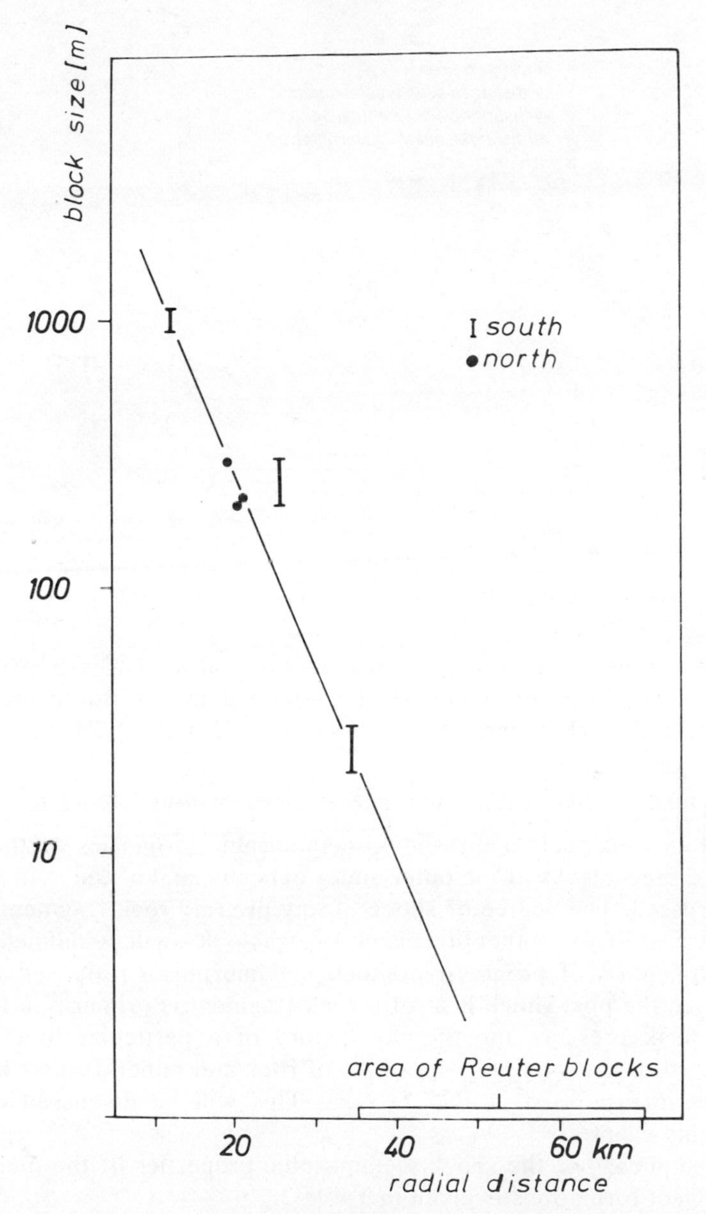

Fig. 23. Size of Malmian megablocks in the Ries outer impact formations as a function of range (after Gall *et al.*, 1975).

megablock zone. In the Vorries the thickness of suevite ranges between 25 and 5 m. Since the upper quenched zone of suevite (see section 3.4) is still preserved in many suevite outcrops (Engelhardt *et al.*, 1969) the intensity of erosion was very weak in some occurrences. Therefore 25 m is considered a maximum primary thickness in the Vorries.

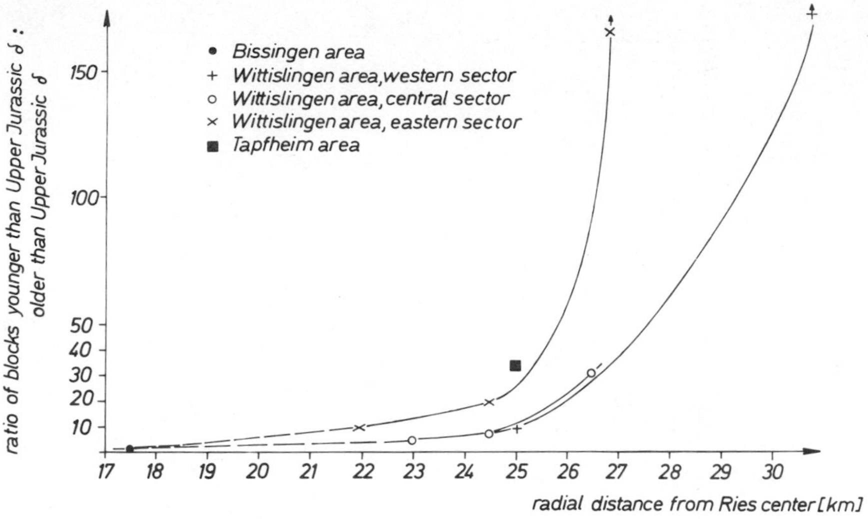

Fig. 24. Stratigraphic provenance of Malmian megablocks as a function of range at various radial sections of the Vorries continuous deposits (after Gall *et al.*, 1975).

(f) The volume fraction of local ground material increases as a function of range, reaching some 80% of the total volume of Bunte breccia at its outer margin (Hüttner, 1969; Schneider, 1971; Gall, 1974; Hörz, 1976).

3.4. *Shock metamorphism and thermal history of various impact formations*

In the following section the shock metamorphic properties of the polymict breccias, the megablocks of the outer crater deposits, and of the crater basement will be discussed. The degree of shock of any discrete rock fragment—whether it is a small clast in a polymict breccia or a megablock—will be defined according to the classification of progressive shock metamorphism proposed by Stöffler (1971a). Since the post-shock heat of a rock fragment is primarily a function of the shock peak pressure the thermal history of a particular breccia will be closely related to the relative proportions of rock and mineral clasts of different shock stage incorporated in that breccia. This will be discussed in the last section of this chapter.

A brief summary of the shock metamorphic properties of the main types of the Ries impact formations is given in Table 2.

3.4.1. *Shock metamorphism of suevite and bulk continuous deposits*

Impact melt representing the highest stage of shock metamorphism is found in most terrestrial craters (e.g., Dence, 1971; Grieve *et al.*, 1977) as coherent sheets of melt rocks in the central area of the crater and as discrete smaller particles in suevite breccias. In the Ries only two smaller bodies of coherent melt in the eastern megablock zone are known. Geophysical data do not exclude the possibility that a coherent layer of impact melt rock could be discovered in the central crater basin by drilling. *Suevite* contains the remaining fraction of melt along with rock and mineral clasts of all stages of shock metamorphism. Petrographic and chemical properties of the melt inclusions have been discussed in section 3.2.3. Quantitative data with respect to the distribution of shock stages in the total clast population are given in Fig. 18 (see also Engelhardt *et al.*, 1969; Stöffler *et al.*, 1977). The fraction of melt and rock clasts of shock stage III is distinctly higher in the fallout suevite than in the fallback suevite of the drill hole Nördlingen 1973. The abundance of clasts of stage 0 (planar deformation structure lacking) is remarkably low in the fallout suevite. Quenched high pressure polymorphs like coesite and stishovite are present in this type of suevite (Shoemaker and Chao, 1961; Stöffler, 1971b). In the fallback suevite of the drill hole Nördlingen 1973 (Fig. 21) a vertical variation in the abundance of shock effects was found (Stöffler *et al.*, 1977): the volume fraction of melt inclusions and of shocked quartz fragments decreases continuously with depth in the lower part of the suevite section, below about 380 and 450 m respectively (Fig. 25) indicating a complex mixing and deposition history of the fallback suevite. Unequivocal shock effects as defined by stages I–IV in the sedimentary rock clasts have not been observed in the fallout suevite although they cannot be excluded with certainty in the fallback suevite. Microscopically observable shock effects in the mineral and rock fragments of the *Bunte breccia* are restricted to quartz and crystalline rocks. Planar elements in quartz fragments and shock stages 0, I, II in crystalline rock clasts (e.g., Schneider, 1971) indicate that this material is derived from zones of the crater basement which were affected by peak shock pressures below about 400 kbar. Occasionally these rock clasts display shatter cones. Shock effects in the sedimentary rock fragments (e.g., sandstones, shales) are consistently lacking. Since sedimentary rocks comprise more than 90–95% of the volume of Bunte breccia it must be concluded that the bulk of the rock material in the Bunte breccia was affected by shock pressures much less than 50 kbar. The clast population of the *crystalline* breccias and *dike breccias* belongs to shock stages 0, I and II (Fig. 18); stage III is extremely rare (Abadian, 1972). Coesite and stishovite are common constituents of these rocks (Stöffler, 1971b). The rocks penetrated by dikes are usually very weakly shocked (stage 0). This holds also for all *megablocks* in the megablock and Vorries zone. The mode of brecciation and fracturing of these blocks, however, is quite characteristic. It is basically different in competent and incompetent rocks. Limestone, expecially massive reef limestone, and crystalline rocks failed by brittle fracturing; marl, shale, and clay display severe plastic deformation but unusual brittle fracturing is also frequently observed (Hüttner, 1969). Different types of deformation can be distinguished in limestone and crystalline rocks (Wagner, 1964; Hüttner, 1969): (a) extremely intense *fracturing*: the fractures are irregularly oriented and have an extreme but variable spacing in the millimeter to centimeter range, in contrast to the preexisting tectonic fracture systems, which are spaced in the order of decimeter to meter. (b) *brecciation*: highly angular rock fragments produced by brittle fracturing

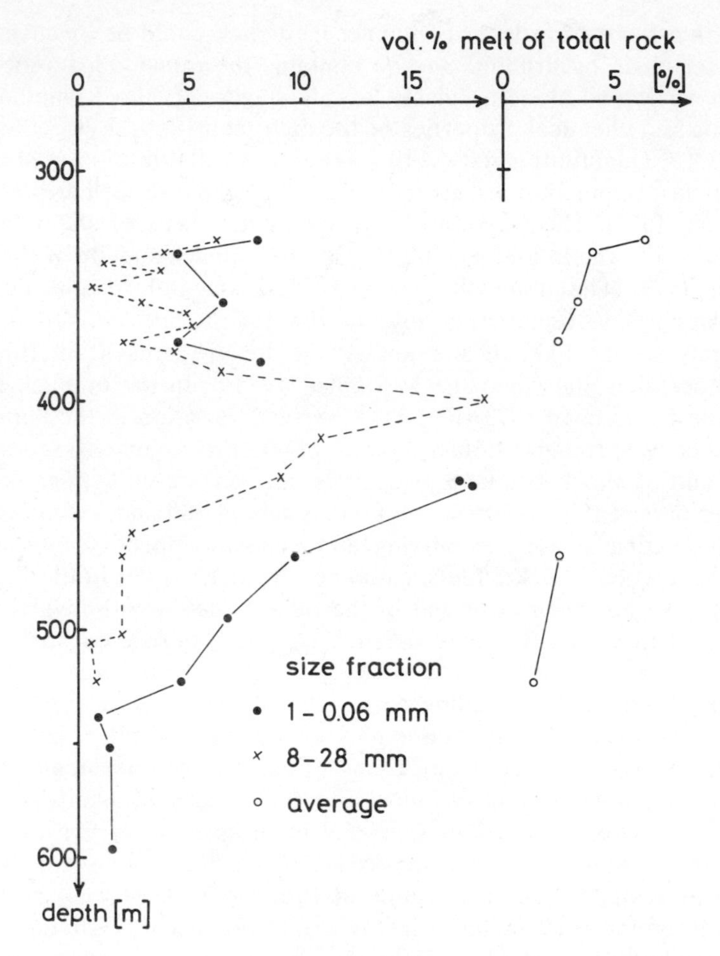

Fig. 25. Abundance of melt products in the suevite of the drill core Nördlingen 1973 as a function of depth (modified after Stöffler, 1977), average calculated on the basis of the grain size distribution.

moved and rotated such that larger fragments (up to some centimeters) are embedded in a more fine-grained matrix (Figs. 26, 27). This may be called a monomict breccia ("Gries").

Within one megablock, regions of both types of deformation may alternate within the range of a decimeter to several meters (Fig. 28), producing complicated megatextures of fractured rocks within monomict breccias or vice versa. Only weak shock effects have been observed in the minerals of the crystalline megablocks (stage 0). Kink bands in biotite are most typical; planar deformation structures in quartz indicating peak pressures of more than 90 kbar, are mostly absent (Stöffler, 1969; Graup, 1975). In brecciated limestone a change of the thermo-luminescence was observed (Engelhardt, 1975).

3.4.2 *Shock metamorphism of the crater basement.* The crystalline rocks at the base of the suevite layer in the drill core Nördlingen 1973 (see Figs. 21, 29), which can be interpreted as displaced megablocks, reveal shock effects in quartz, feldspar, biotite, hornblende, and other constituents (Engelhardt and Graup, 1977). These authors derived an attenuation of the peak pressure from about 160 kbar at a depth of 505 m to about 90 kbar at 670 m on the basis of shock effects in quartz. In this depth range a high intensity of shatter coning was found (Bayer. Geol. Landesamt, 1977) which decreases below about 600 m. The attenuation of the peak shock pressures in the whole profile is discontinuous (Fig. 29). At 1206 m the peak pressure may have been less than 10 kbar (kinking in biotite).

3.4.3 *Thermal history.* Thermal annealing of breccias or of their substrates after deposition is restricted to suevite. All other impact formations lack any mineralogical or other effects which could be assigned unequivocally to post-shock heat. This means that the bulk continuous deposits of the megablock and Vorries zone never experienced any shock-induced increase of temperature. In contrast, suevite was deposited as a relatively hot impact formation on top of all other types of displaced rock masses (Fig. 16). The top and bottom zones of the *fallout suevite* were thermally quenched upon deposition of the suevite, leaving a central hot zone in which shock-produced glasses recrystallized to a fine-grained aggregate of plagioclase and clinopyroxene (Engelhardt *et al.*, 1969). Vertical vents indicate outgassing of the hot suevite and hydrothermal formation of secondary minerals after deposition, mainly montmorillonite, quartz and zeolites. The highly vesiculated glasses of the fallback suevite are completely recrystallized to montmorillonite and analcite (Förstner, 1967; Stähle and Ottemann, 1977; Stöffler *et al.*, 1977). The vesicles are filled with euhedral zeolite minerals (erionite, clinoptilolite, harmotome, phillipsite). The zeolite assemblage varies in composition as a function of depth (Stöffler *et al.*, 1977). These minerals indicate a maximum range of post-depositional temperatures of 250–750°C. More accurate temperature estimates are based on the remanent magnetization (Pohl, 1977) and on fission track measurements (Wagner, 1977) from which temperatures of at least 600°C are inferred. Depending on the thickness of the hot suevite layers cooling will take a rather long time. Calculations made for the 200 m thick suevite layer found in the drill core Nördlingen 1973, assuming an infinite horizontal extension, show that cooling from 600 to 100°C takes about 2000 years (Pohl, 1977). Simultaneously the layers below the hot suevite blanket are heated to a certain extent. The high post-depositional heat of suevite resulted in a strong *remanent magnetization* (Angenheister and Pohl, 1964; Pohl and Angenheister, 1969; Pohl, 1974, 1977). The intensity of magnetization is often several orders of magnitude higher than the mean magnetization of unshocked basement rocks from which the suevite was derived. Data from various suevite localities are summarized in Table 4 and Fig. 29. All measured samples have a reversed natural remanent magnetization with a constant direction. The intensity of remanent magnetization varies between a few gammas and several hundred gammas.

The remanence is carried mainly by magnetite which is present in all con-

Fig. 26. Brecciated Malmian limestone ("Gries"), Dirgenheim, southern continuous deposits; scale = cm.

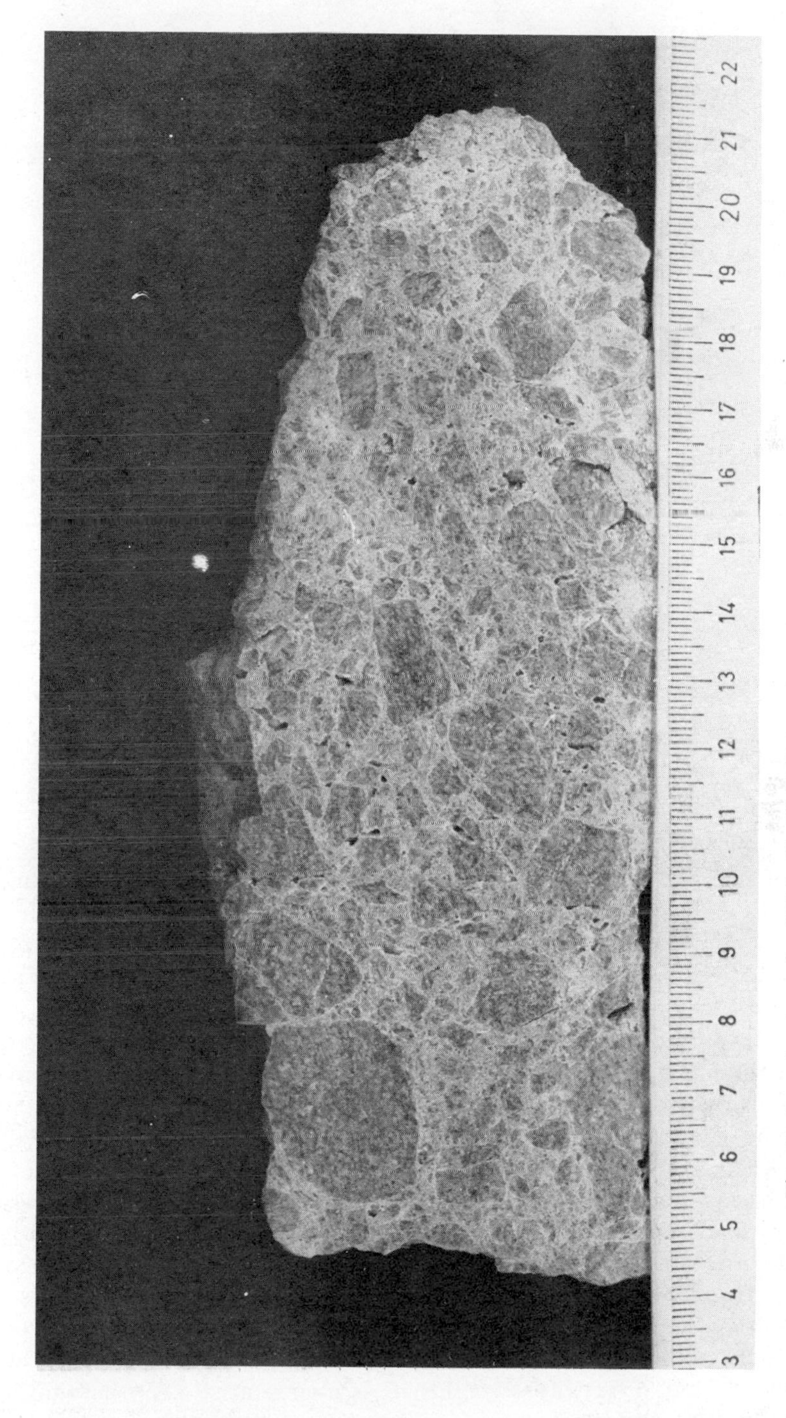

Fig. 27. Brecciated granite ("Gries"); W of Schmähingen, southern megablock zone; scale = cm.

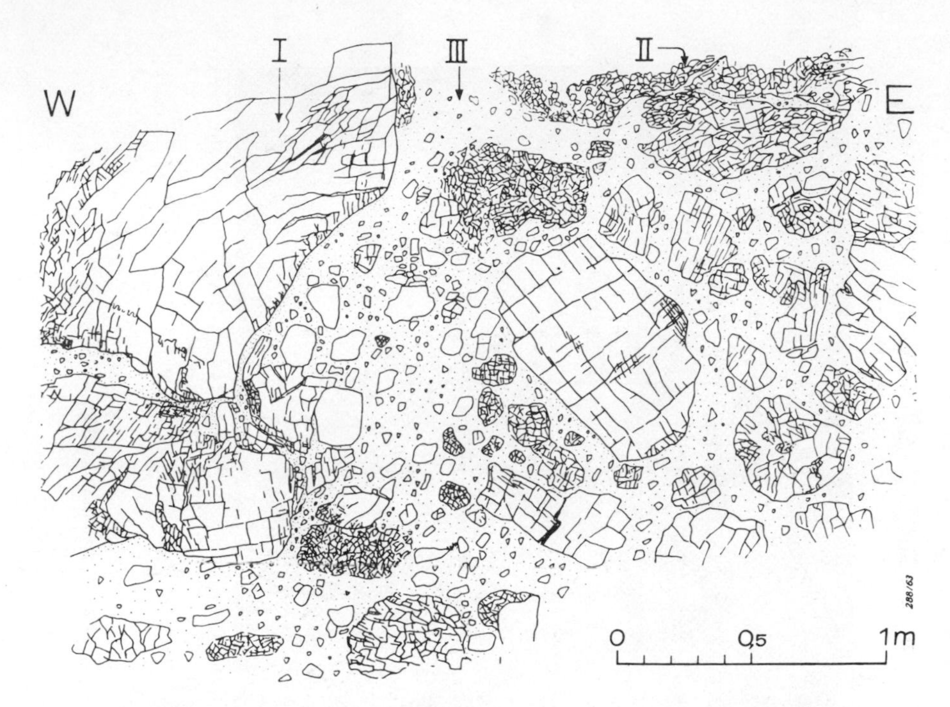

Fig. 28. Malmian megablock near Eglingen, southern continuous deposits, with three different types of deformation (from Wagner, 1964), 21 km from the point of impact; I = fracturing but primary joints and bedding planes preserved, II = intense fracturing, III = brecciation.

stituents of suevite: Rock fragments, glass and matrix. There is no predominance for particular carriers. Maghemite, as a second magnetic phase which carries an important fraction of the NRM, has been identified in the lower part (c. 450–525 m) of the high-temperature suevite and in the low-temperature suevite (525–650 m) from the drill core Nördlingen 1973.

The main part of the natural remanent magnetization which is carried by magnetite with high temperatures is best explained as a thermoremanent magnetization. Therefore the mean temperatures of suevite at the time of deposition were on the order of 600°C, in agreement with temperature estimates from mineralogical investigations (Hörz, 1965; El Goresy, 1968; Stöffler *et al.*, 1977) and from fission track dating (Wagner, 1977). The increase of magnetization of the suevite in comparison with the magnetization of the constituent rocks is mainly due to the generation of magnetite by shock metamorphism and/or subsequent thermal metamorphism, especially in the high-temperature suevite. The production of large quantities of magnetite by these processes from Fe-bearing mafic minerals and melts has been described by many authors (references in Pohl, 1971; Stähle, 1972). It is characteristic for suevitic impact breccias on earth.

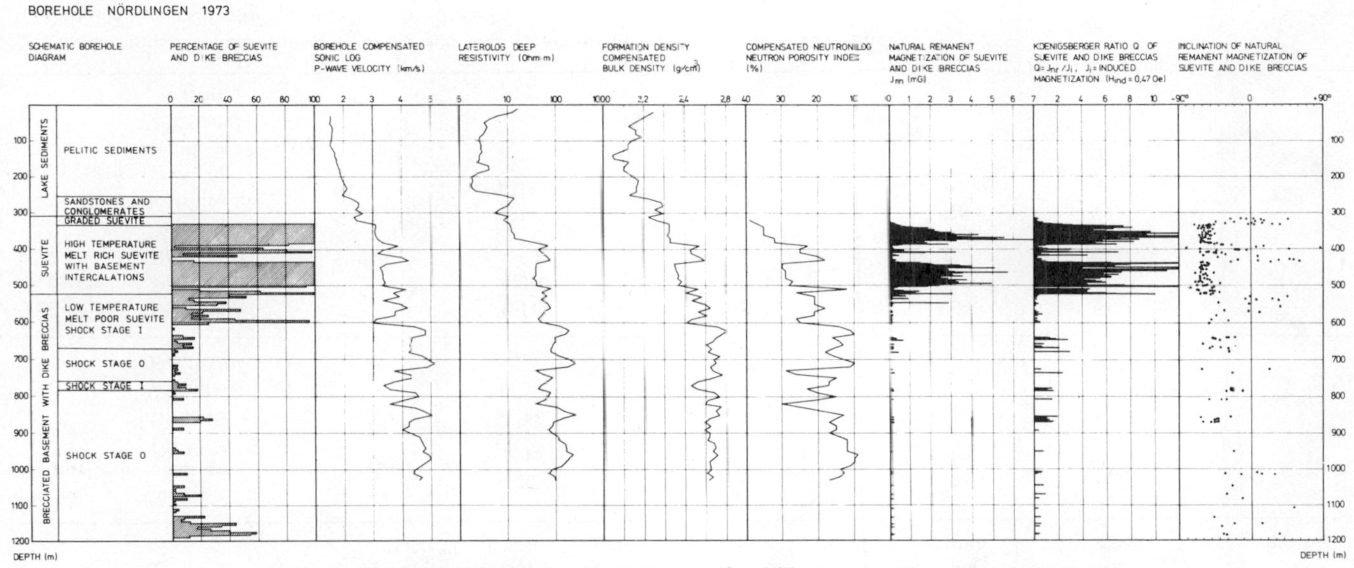

Fig. 29. Drillhole Nördlingen 1973. Schematic drill core profile and geophysical measurements. In the schematic drill core profile the indicated shock stage refers to the basement rocks, not to the suevite and dike breccias. Magnetic properties measured on basement rocks are not shown except for basement intercalations in suevite between 331.5 and 525 m. Positive inclinations between 331.5 and 525 m are due to these intercalations. Positive inclinations in suevitic breccias below 525 m are due to unstable components of the natural remanent magnetization.

Table 4. Natural remanent magnetization of suevite (high-temperature suevite).

		Natural remanent magnetization J_{nr}(mG)		Susceptibility (10^{-6} cgs)		Koenigsberger ratio $Q = J_{nr}/J_i$ (J_i induced in 0.47 Oe)		Declination (°)	Inclination (°)
		mean	max min	mean	max min	mean	max min	mean	mean
Fall-out suevite	Surface outcrops 12 sites*	0.59	1.37 0.04	260	510 30	6.4	30 1.3	194	−57
	Drillhole Wörnitzostheim Depth 20–100 m	3.00	14.30 0.10	560	2000 64	11.3	30 1.5	—	−59
Fall-back suevite	Drillhole Deiningen I† Depth 332–350 m	0.04	0.11 0.006	60	156 30	1.5	11 0.6	—	−61
	Drillhole Nördlingen 1973 Depth 331.5–390 and 435–525 m	2.30	6.95 0.05	800	2600 80	5.9	20 0.9	—	−56

*max and min are mean values of sites, Dec. and Inc. of surface outcrops after af-cleaning.
†Only the upper 20 m of the suevite were drilled.

4. Subsurface Structure of the Crater

The subsurface structure of the crater has been investigated by geophysical methods complemented by several drillholes. The following sections will review the measurements. An interpretive summary of the present knowledge of the crater structure is given at the end of the chapter.

4.1 *Drillholes*

Of the major drill sites which provided information on the subsurface structure of the crater, two are located inside the central crater (Nördlingen 1973, Deiningen 1953), four are in the area of the inner ring (Wallerstein, 1948, Nördlingen 1950, 1955) and one in the megablock zone (Wörnitzostheim 1965). The location of the main drillholes is shown in Figs. 6 and 32 and a summary of the drilled profiles is given in Fig. 21. The most important is the deep drillhole Nördlingen 1973 (Bayer. Geol. Landesamt, 1974, 1977). Figure 29 shows a schematic core diagram and measurements of physical parameters. Some results of the encountered impact formations, which are considered to be typical for the central crater cavity, have already been discussed in section 3. Further results will be mentioned in the following sections.

4.2 *Geoelectric measurements*

Geoelectric methods (DC-depth sounding with Schlumberger and dipole electrode configurations) have been used in the Ries crater to investigate the low resistivity upper layers down to a depth of 600–800 m (Ernstson, 1974; Engelhard and Hansel, 1976; Blohm, Friedrich, and Homilius, 1977). An electrical resistivity profile in the central crater obtained from the drill hole Nördlingen 1973 is shown in Fig. 29.

From the geoelectric measurements the thickness of the post-impact Tertiary lake sediments in the Ries crater, the location of the boundary of the central crater and the inner ring and some details of the complex structure of the inner ring were obtained. Figure 30 shows the thickness of the lake sediments near the area of the inner ring (Ernstson, 1974). From the circular form of this central basin bordered almost completely by the structure of the inner ring the center of the crater can be fixed with great accuracy. Figure 30 also gives an approximate idea of the morphology of the central part of the Ries crater shortly after the deposition of the fall-back ejecta. Not shown is the lake sediment's maximum thickness in the central crater, which may reach 400 m according to geoelectric and seismic data. The relief of the surface of the fall-back suevite material all over the central crater probably was not much greater than about 100–200 m. Geoelectric methods also yielded an indication about the highly varying thickness of the suevite layer and thus information concerning the relief of the brecciated crystalline rock masses below the suevite. Specifically it could be shown near the center that the basement rocks form a ring-like uplift with a diameter of 4–5 km, consistent with seismic, magnetic and gravity data. The maximum relief of this uplift is about 300 m (Fig. 40).

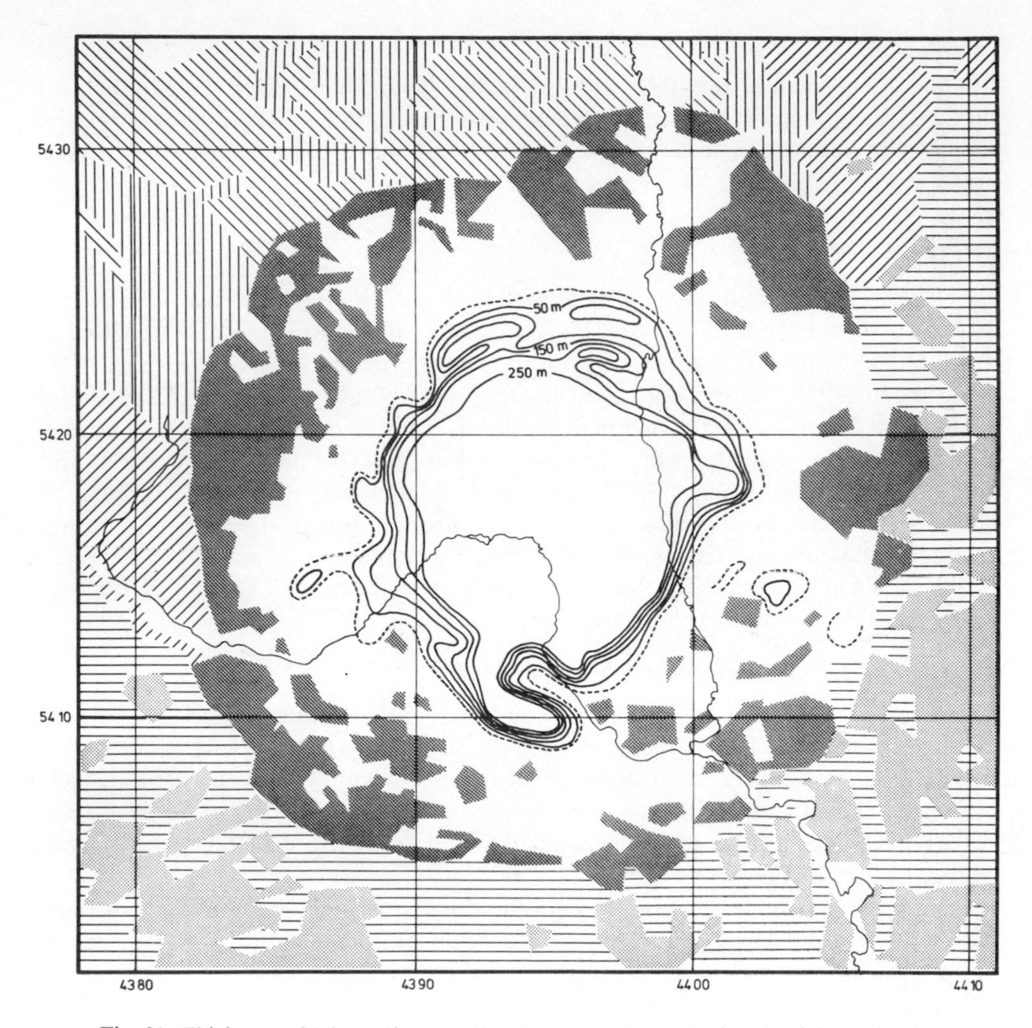

Fig. 30. Thickness of lake sediments (after Ernstson 1974). Maximum thickness is not shown. Schematic geology according to Fig. 6.

4.3 *Seismic measurements*

Early refraction seismic work by Reich and Horrix (1955) with refraction lines 3–4 km long established many of the main features of the crater: central crater cavity, inner ring, and megablock zone. The most detailed information on the near surface structure is provided by an EW reflection profile from the center of the crater to about 6 km W of the tectonic crater rim (Angenheister and Pohl, 1969, Fig. 32). A redrawing of the seismic record section is shown in Fig. 31. Good reflections were obtained from the bottom of the lake sediments in the central crater and from a mesozoic layer and the surface of the basement outside the crater. The crater boundary in the basement is indicated by the disap-

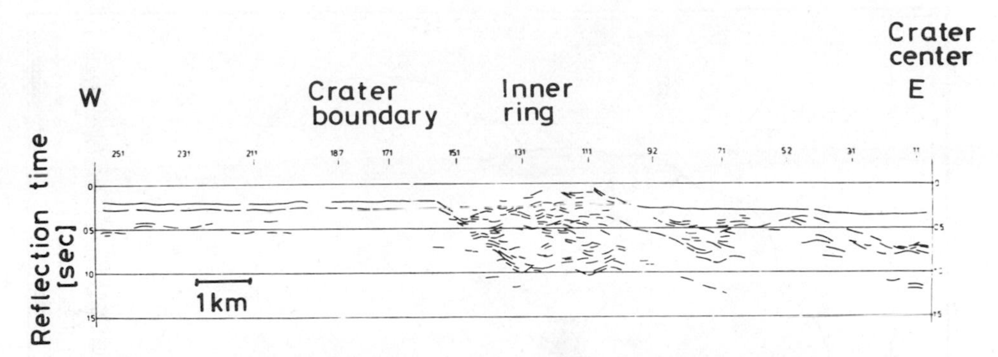

Fig. 31. Seismic reflection profile (RR in Fig. 32). Simplified diagram showing important events on the record section. Datum plane is 400 m above sea level (Angenheister and Pohl, 1969).

pearance of reflections from the undisturbed basement. It is located about 2.5 km inside the tectonic rim of the crater (Fig. 32). Towards the center the basement is fractured and has subsided, in part together with sedimentary layers, c. 100–200 m (Fig. 32), possibly as a series of down-faulted blocks which no longer form a continuous reflecting horizon. Still closer to the inner ring megablock crater deposits, with thicknesses up to several hundred meters, may also contribute to the scattering of the seismic signals and the "chaotic" aspect of the reflected signals. The inner ring is mainly characterized by a steep contact to the lake sediments (Fig. 40).

Velocity-depth soundings with symmetrical arrangement of shot points and receivers with respect to the sounding point were made at the center and outside of the crater in order to determine the depth of the brecciation zone (Pohl and Will, 1974). It could be shown that at depths of about 3 km the P-wave velocities are still considerably lower than in the undisturbed basement outside the crater. The velocity-depth distribution shown in Fig. 34 down to 2.5 km is taken from these measurements. The velocity distribution obtained from the surface measurements was confirmed for the upper 1000 m by the velocity measurements in the drill hole Nördlingen 1973 (Fig. 29).

Using the data obtained from the above described measurements, a new interpretation of two 40 km long refraction profiles (6, Harburg, and 7, Holheim, Fig. 32) was recently suggested (Ernstson and Pohl, 1977) utilising a ray tracing program for two-dimensional velocity distributions. The underground structure is modelled by velocity isolines (Fig. 33). Results show a 20–25 km large bowl-shaped zone with reduced velocities. At the center lowered velocities can still be identified at depths of about 6 km, which is in agreement with the interpretation of gravity measurements described in section 4.4. The velocity distributions inside and outside the crater obtained from these models and their differences are shown in Fig. 34. Also shown is the density difference for density model II of Fig. 37. The agreement between the density and the velocity models can be explained by the correlation of the velocities and the densities measured

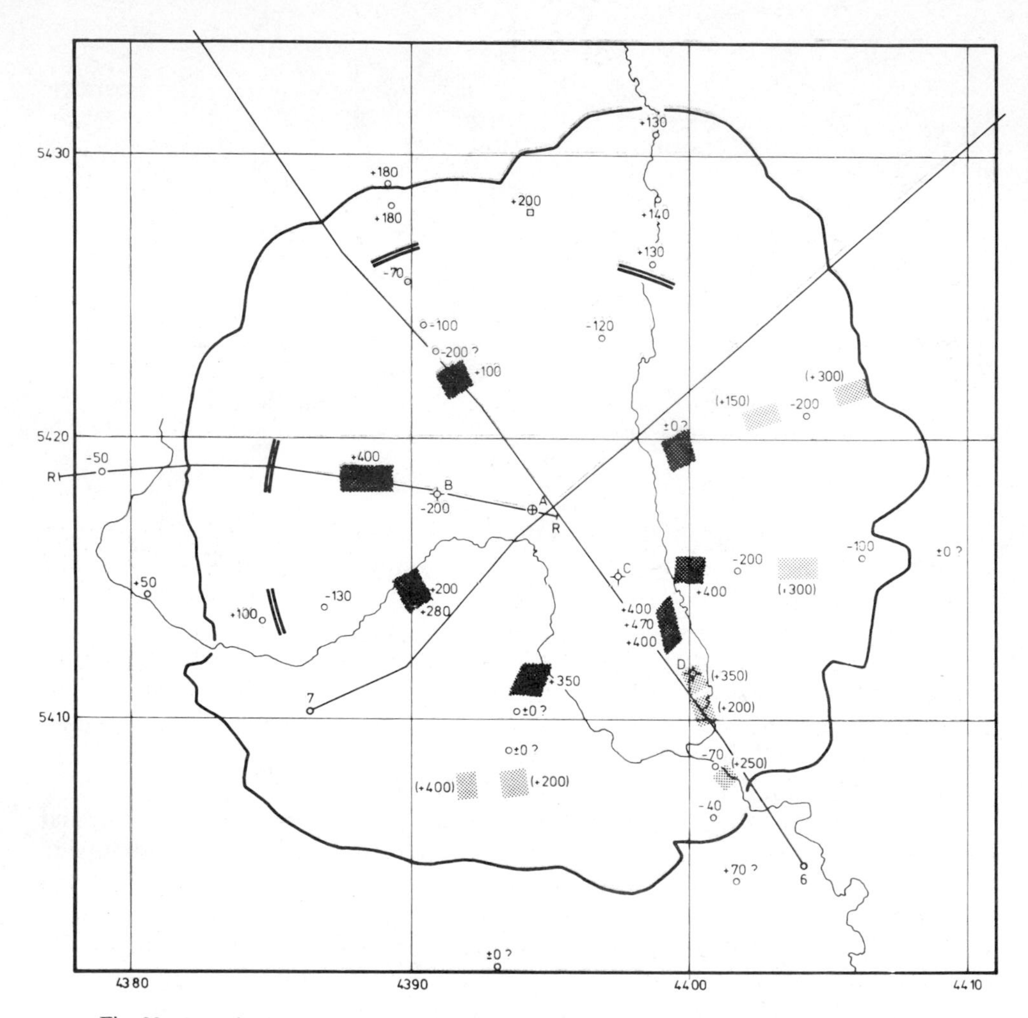

Fig. 32. Compilation of available elevation above sea level (in m) data of the crystalline basement and of subsurface crystalline basement and Malmian limestone (?) megablocks in the area of the inner ring and the megablock zone. Dark cross-hatching indicates location and depth of crystalline basement blocks in the area of the inner ring. Light cross-hatching indicates blocks with high seismic velocity in the marginal zone of the crater between the inner ring and the tectonic rim, probably Malmian limestone, possibly basement blocks. Heavy double line indicates approximate location of the crater boundary within the crystalline basement (○ seismic data, □ geoelectric data). Also shown is the location of the reflection profile (RR, see Fig. 31) and of two refraction profiles (6 and 7, see Fig. 33).

in the brecciated basement rocks in the drill hole Nördlingen 1973 (Ernstson and Pohl, 1974).

4.4 *Gravity measurements*

Figure 35 shows the Bouguer anomaly map published by Jung and Schaaf

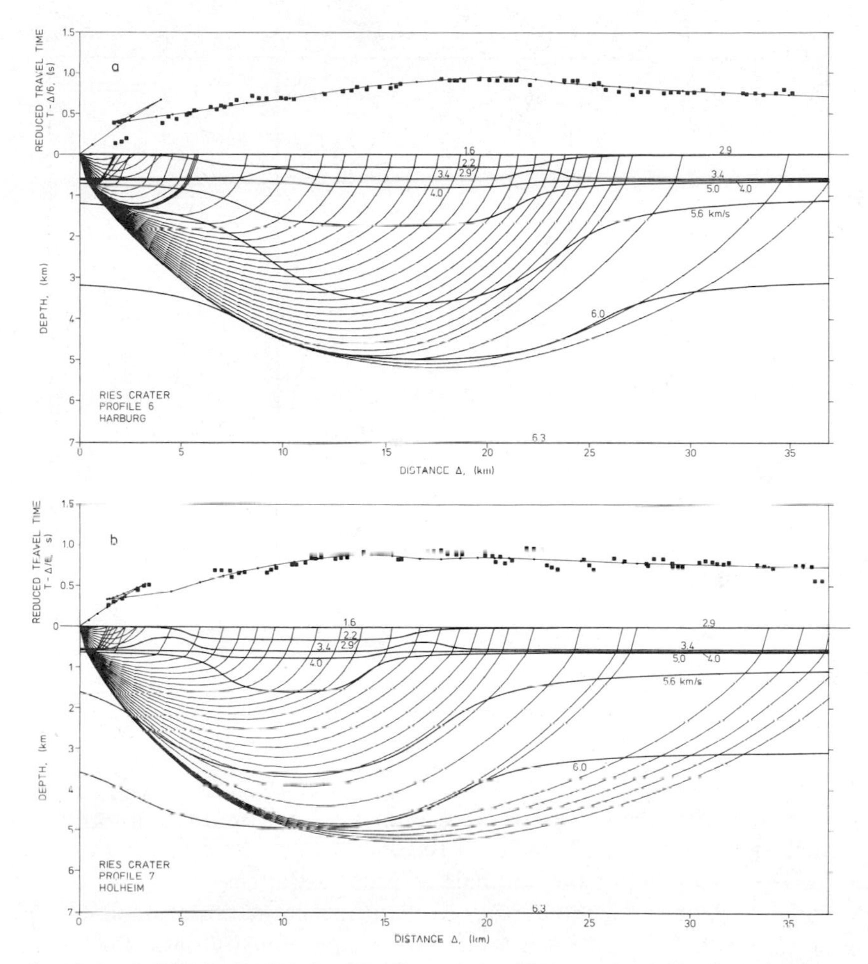

Fig. 33. Velocity distribution in the Ries crater calculated for refraction profiles 6 (Harburg) and 7 (Holheim) (see Fig. 32). Top: Full squares indicate measured travel times. Calculated travel times for models shown below are indicated by a continuous line. Bottom: Velocity distribution shown by velocity isolines in km/s (thick lines). Thinner lines are seismic ray paths. Vertical exaggeration 2×.

(1967). The datum plane is 400 m above sea level. For the determination of the anomaly due to the crater structure, various regional fields have been constructed by Kahle (1969). A residual anomaly which is thought to be due solely to the impact structure is shown in Fig. 36 (Kahle, 1969). The anomaly has a maximum negative value of about −18 mgal and its location is concentric to other structures within crater. On radial profiles the inner ring is marked by a relative maximum at a radial distance of c. 5–6 km from the center. A second kink at a radial of about 10 km can be associated with the crater boundary in the crystalline basement (see section 4.3). The residual anomaly has been used to calculate the total mass deficiency below the datum plane by Jung and Schaaf

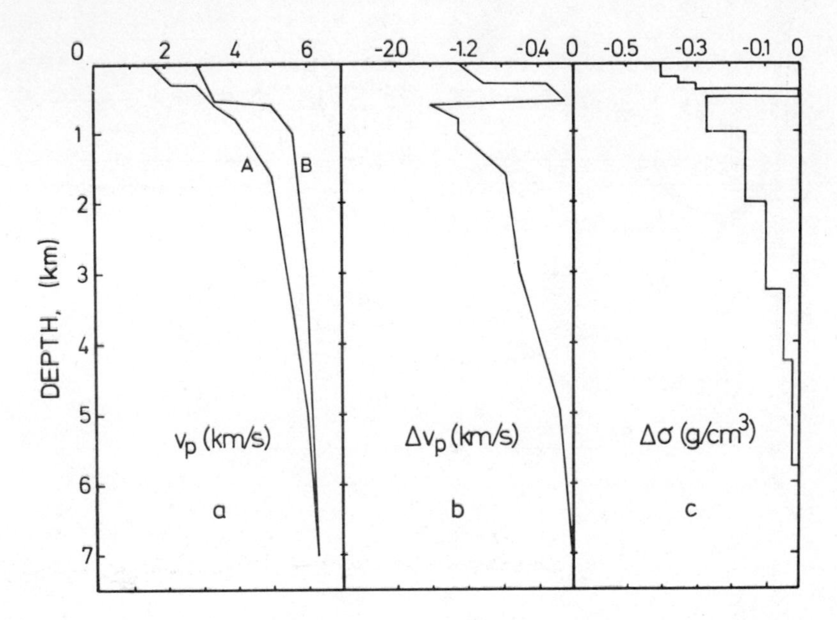

Fig. 34. (a) Velocity-depth-distribution (p-wave velocity) at the center of the crater (A) and outside (B) the crater, according to models shown in Fig. 33, (b): velocity differences $\Delta v_p = v_{pA} - v_{pB}$ at the center of the crater, (c) density differences $\Delta\sigma$ at the center of the crater between crater rocks and off-crater rocks according to the density model II (Fig. 37).

(1967), Kahle (1969), and Ernstson and Pohl (1977). The integration gives mass deficiencies between about 70,000 and 100,000 Mt, depending on the assumptions made for the zero level of the anomaly at great distances.

Model calculations for the density distribution in the crater made before 1974 (Jung *et al.*, 1969) suffered from a lack of reliable density data. Since that time the drill hole Nördlingen 1973 and several other drillholes outside the crater provided better data on the density distribution for the upper 1000 m (Table 5 and Fig. 29). This enabled new and refined model calculations (Ernstson and Pohl, 1977). Some results for density bodies with circular symmetry are shown in Fig. 37. Details of the near-surface structure have been omitted. The density difference between the undisturbed rocks outside the crater and the brecciated crater filling as a function of depth is given in Fig. 34. The models indicate the existence of a bowl-shaped zone of reduced density with a depth of at least 5–6 km at the center of the crater, in agreement with results of seismic modelling discussed in section 4.3.

4.5 *Magnetic field measurements*

Magnetic surveys have been made by Reich and Horrix (1955), Angenheister and Pohl (1969) and by Bundesanstalt für Geowissenschaften und Rohstoffe (1971). Figure 38 shows an airborne total intensity map (Bundesanstalt für

Geowissenschaften und Rohstoffe, 1971) and Fig. 39 a ground magnetic total intensity map (Pohl and Angenheister, 1969). The central part of the crater is characterized by negative anomalies ($-300\,\gamma$ on the ground magnetic map) which are due to the thick layer of reversely magnetized high-temperature fall-back suevite below the lake sediments (see also Figs. 29, 40 and Table 4). The presence of larger coherent impact melt bodies is not excluded by this interpretation.

Some negative anomalies of smaller extent in the megablock zone are due to local fallout suevite patches below shallow lake sediments as was shown, e.g., by the drillhole Wörnitzostheim (Fig. 21). Similar anomalies are measured above outcropping fallout suevite in the Vorries.

The large positive anomalies in the SW of the crater and outside the crater are not related to the impact structure. They are caused by basic or ultrabasic paleozoic rocks within the basement. Amphibolites with high magnetic susceptibility (Pohl, 1974) which are found in the ejecta and in the drillhole Nördlingen 1973 are probably part of these rocks.

4.6. *Subsurface structure*

From the geophysical and drillhole data, the following picture of the structure of the crater can be drawn. In agreement with the surface morphology and geology (section 3.1) the crater is divided (Fig. 40) into a central crater cavity surrounded by the rampart-like structure of the inner ring (diameter c. 12 km) and the marginal megablock zone between the inner ring and the tectonic rim (diameter 25–26 km). The *central crater* is filled with post-impact lake sediments with a thickness up to about 400 m. The lake sediments can be divided into an upper series of pelitic sediments (thickness 200–350 m) deposited under quiet water conditions and a lower series of sandstones and conglomerates deposited under turbulent conditions shortly after the impact (thickness 50–100 m?).

Below the lake sediments a graded suevite layer about 15 m thick was found in the drillhole Nördlingen which may represent a preserved part of the latest fallback material in the crater. This layer in turn is underlain by a high-temperature suevite fallback breccia with a highly variable thickness ranging from 0 to more than 400 m. The large basement blocks found within the suevite of the drillhole Nördlingen (Fig. 29) and the variation of petrographic characteristics (see sections 3.2.3, 3.2.4, and 5.3.2) indicate a multiphase process for the deposition of the fallback suevite. The petrographic character of the suevite may also show important lateral variations within the central crater.

The basis of the suevite layer consists of brecciated basement rocks probably all over the central crater. In the drillhole Nördlingen the brecciated basement consists of several thick rocks units (megablocks) showing, alternatively, shock metamorphism stage I and stage 0 (Fig. 29, Engelhardt and Graup, 1977). The upper unit (525–670 m) (mainly amphibolite and some granite) is characterized by the presence of numerous glass-poor, low-temperature suevitic dikes, by shatter cones and by traces of meteoritic material (El Goresy and Chao, 1976). The

Fig. 35. Bouguer gravity anomaly in the Ries crater area (after Jung and Schaaf, 1967) in mgal. Schematic geology according to Fig. 6.

lower units contain also numerous breccia dikes as discussed in section 3.2.2. Part of these dikes contain sedimentary material.

The morphology of the surface of the basement in the central crater shows an important relief (Fig. 40). Towards the center the basement rocks form a series of peaks in a ring-like arrangement with a diameter of 4 to 5 km. The points of some of the peaks are probably at higher elevation than the precrater basement surface.

For depths greater than 1200 m both the reduced velocity and density indicate brecciation and fracturing down to about 6 km depth at the center of the crater. Fracturing and brecciation may extend still deeper, because microcracks are partly closed at this depth ($p \approx 2$ kbar) and cementation and healing have

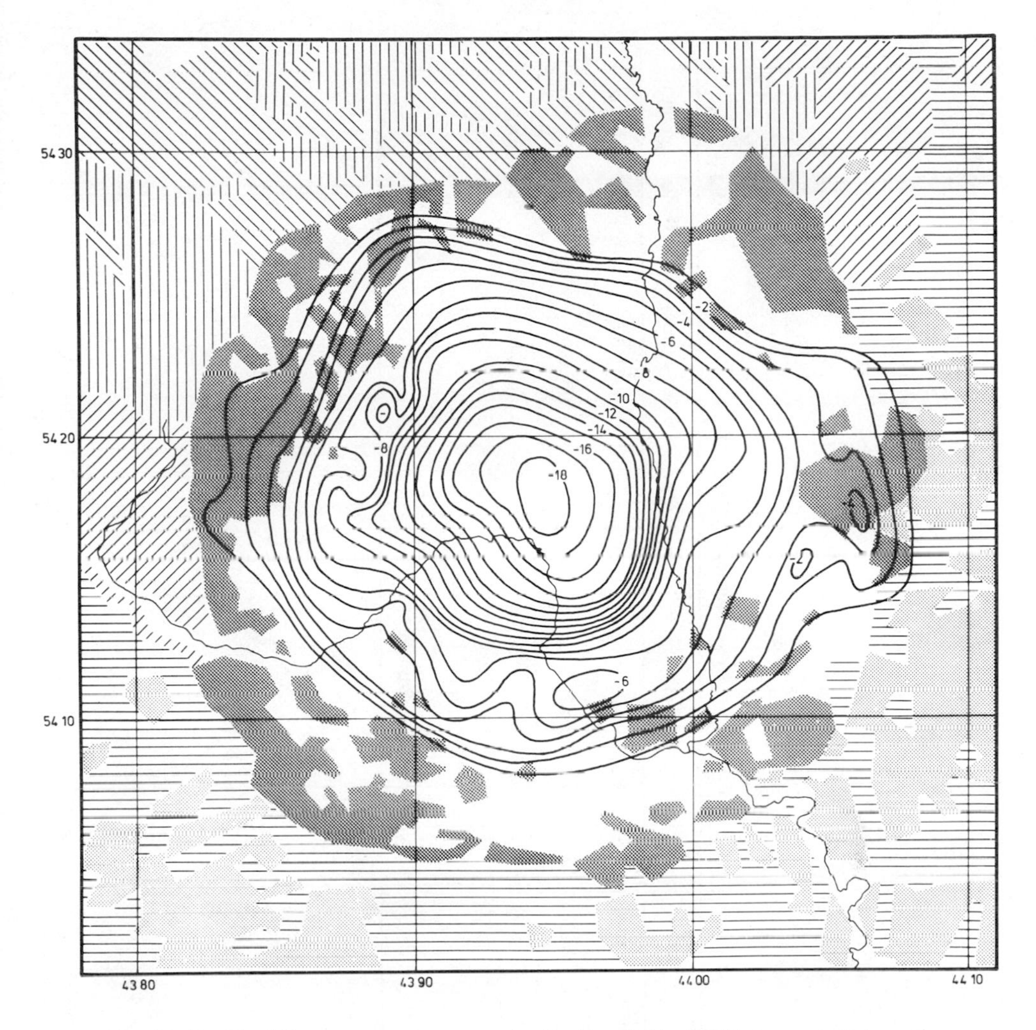

Fig. 36. Residual Bouguer gravity anomaly II after Kahle (1969) in mgal. Schematic geology according to Fig. 6.

probably taken place (Simmons *et al.*, 1975). In the basement section of the drill core Nördlingen 1973 the decrease in density and velocity is due to the presence of dike breccias and to brecciation and fracturing of basement rocks. For larger depths it is not yet possible to decide whether this is solely due to *in situ* brecciation by the shock wave or possibly also due to subsequent mass-movement during readjustment and relaxation of the crater.

The *inner ring* is a rampart like structure which encircles most of the central crater. Towards the center of the crater it forms a steep, well-defined contact to the lake sediments and the suevite (Figs. 30, 40). In contrast the exterior side of the inner ring is poorly defined. In some places the inner ring is interrupted and relatively thick lake sediments are found outside such gaps (Fig. 30). In the

Table 5. Seismic velocities and densities of pre-crater and crater rocks.

Pre-crater rocks	Velocity (m/s)	Density (g/cm^3)	Crater rocks	Velocity (m/s)	Density (g/cm^3)
Malmian limestones	4600–5000	2.52	Pelitic lake sediments	1700–2200	1.85–1.9
Dogger-Malm	2500–3000	2.25–2.35	Psammitic-psephitic lake sediments	2700–3000	2.15
Keuper-Dogger	2500–2800	2.25–2.35	Suevite*	3100–3400	2.25–2.30
Basement*	5000–6000	2.65–2.95	Brecciated basement* (including dyke breccias)	3000–6000	2.45–2.90
			Inner ring	3300–4000	—
			Megabloc zone	2500–4000	—

*See Figs. 29 and 34.

RIES CRATER

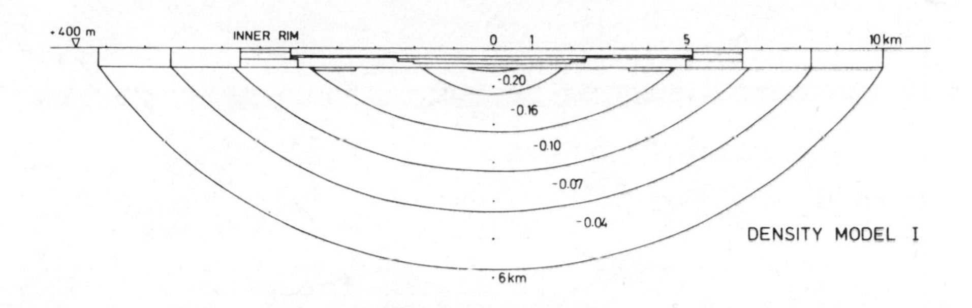

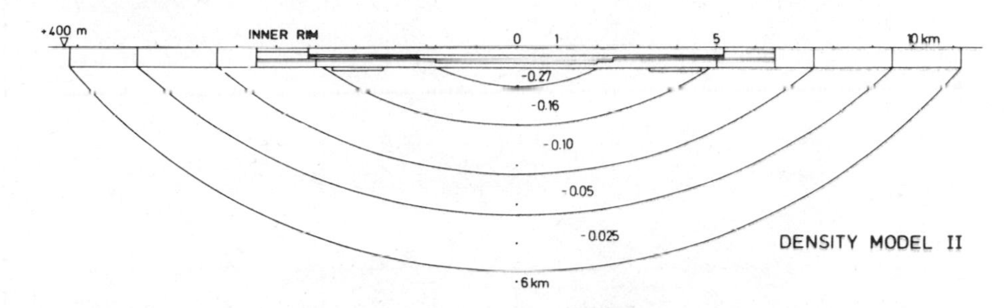

Fig. 37. Density distribution models for the Ries crater. Density differences between off-structure rocks and rocks within the crater in g/cm^3. Density differences within the upper 500 m not shown. No vertical exaggeration (Ernstson and Pohl, 1977).

north, the inner ring structure is apparently divided into several approximately concentric highs and troughs in a steplike arrangement. Outcrops, drillholes and seismic data show that the inner ring structure is formed in part of weakly shocked crystalline basement rocks (Fig. 32), and in part of sedimentary rocks (Fig. 20). Shallow drillholes at Nördlingen show in particular that sediments are underlying the basement rocks, indicating an inverted stratigraphy (Fig. 21).

Just outside the inner ring the *megablock zone* consists largely of ejected rock masses down to a depth of several hundred meters as indicated by the drillhole Wörnitzostheim and geophysical data. These ejecta and the lake sediments which fill several troughs outside the inner ring have replaced the pre-crater sedimentary strata in these areas. However, it is not yet proven if the pre-crater sediments have been removed by ejection or if they have subsided together with the ejecta blanket. The ring depression in the basement up to a radial distance of about 10 km (section 4.4) is considered strong evidence for subsidence, since ejection of basement rocks from outside the central crater is most unlikely. Subsidence would also explain the presence of high-velocity megablocks (probably Malmian limestones) in this zone at depths of 200–300 m

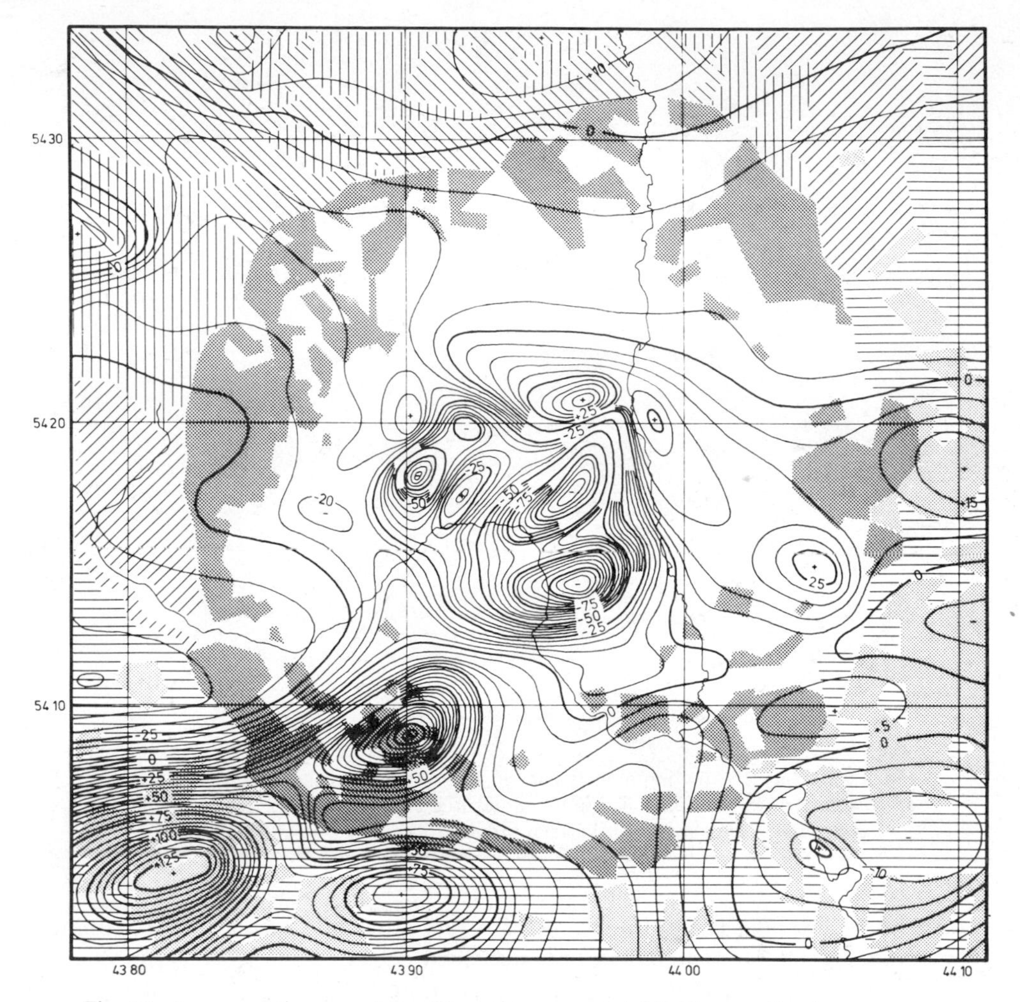

Fig. 38. Aeromagnetic map of the Ries crater area: total field anomalies in gamma. Flight height 1000 m above sea level (Bundesanstalt für Geowissenschaften und Rohstoffe, 1971). Schematic geology according to Fig. 6.

below their normal stratigraphic position. At radial distances >10 km subsidence movements are confined mainly to the sedimentary cover (Gall *et al.*, 1977). The tectonic rim is defined as the outer limit of these blocks.

5. CRATERING MODEL

5.1. *Volume estimates of excavated rocks and impact energy*

Calculations of masses and volumes of rocks which have been displaced by the impact process are subject to large uncertainties because not only are the

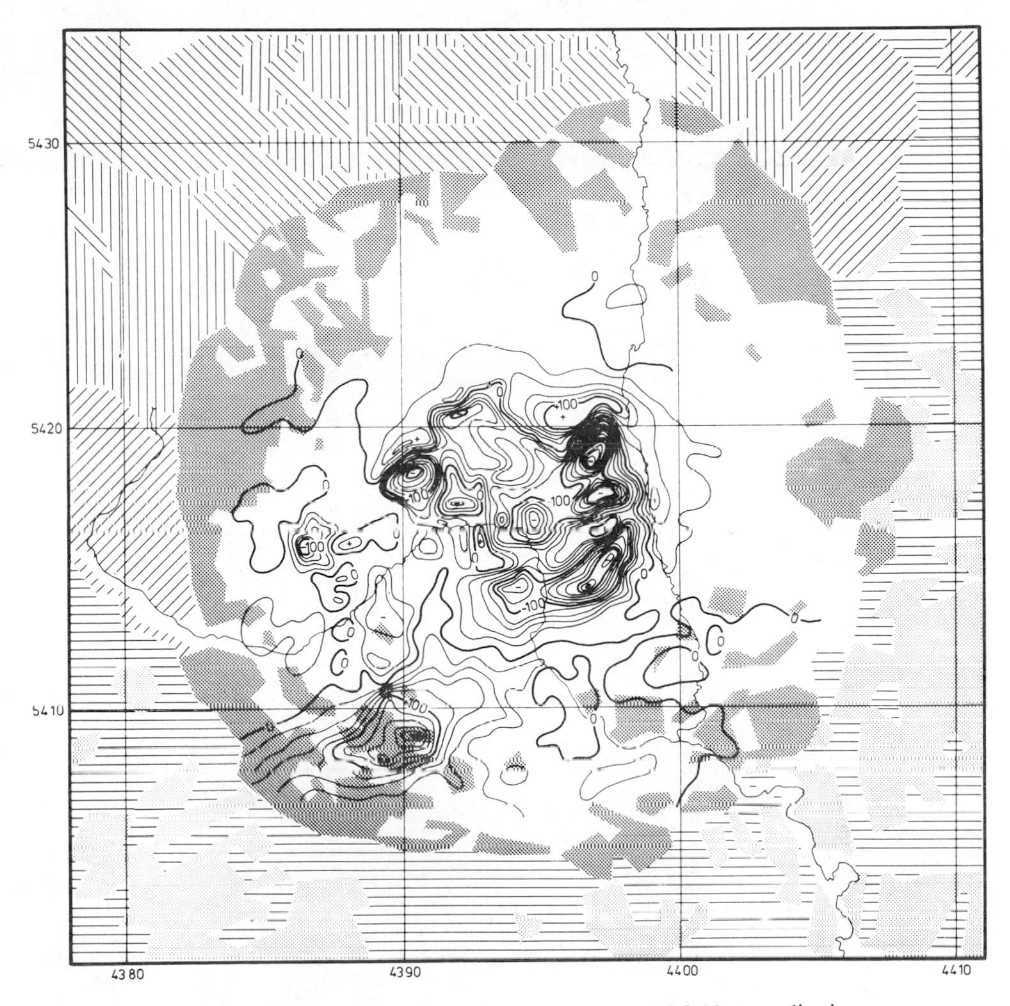

Fig. 39. Ground magnetic map of the Ries crater area; total field anomalies in gamma (Pohl and Angenheister, 1969). Schematic geology according to Fig. 6.

structure and composition of the marginal crater zone, between the inner ring and the tectonic rim, not well enough known, but also the structure and composition of the inner ring, the mass-transport by readjustment and finally the erosional processes since crater formation.

We nevertheless performed some volumetric calculations in order to constrain estimates of possible maximum and minimum values. For such calculations the following general assumptions were made: Horizontal layering of sedimentary strata, sediment thickness before the impact 600 m, pre-impact basement surface at sea-level.

5.1.1. *Estimate of the volume of excavated sediments.* To estimate a maximum value we assume that all sediments which occupied the present topographic

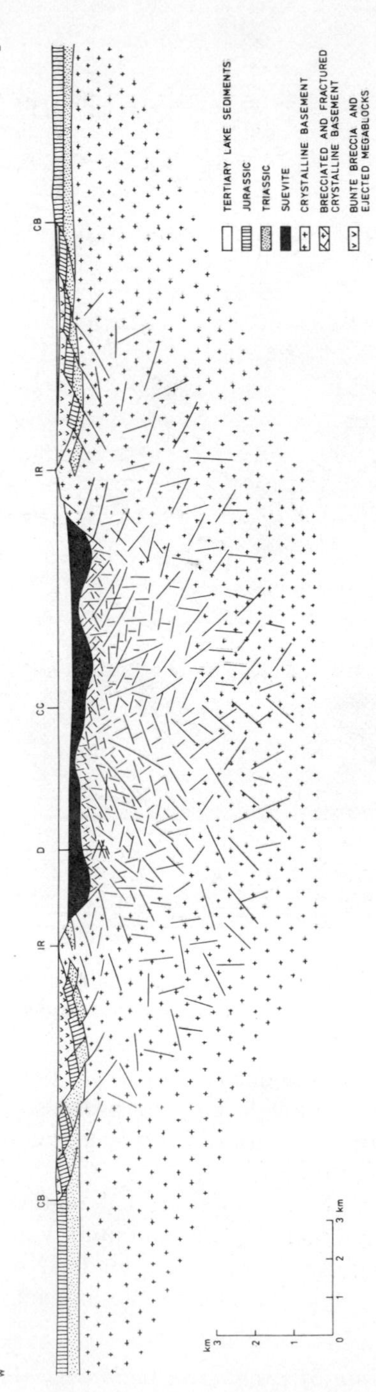

Fig. 40. Schematic cross section of the Ries crater. No vertical exaggeration; CC = crater center, IR = inner ring, CB = crater boundary (tectonic rim).

crater basin (Fig. 5) below the 600 m level were ejected (84 km^3). To this we add the volume of lake sediments (28 km^3), the volume of fallback suevite above sea level (7 km^3, out of a total fallback suevite volume of 13 km^3), and the volume of sediments replaced by basement rocks in the inner ring (5–10 km^3 ?) and in the megablock zone (5–10 km^3 ?). This gives a maximum total volume of about 129–139 km^3 for sedimentary rocks which could have been ejected beyond the tectonic rim (diameter ~ 25 km). It is possible to assume that half of the lake sediments (14 km^3) are derived from sedimentary crater deposits within the tectonic rim. We can further take into account the subsidence in the megablock zone discussed in section 4.3. A subsidence of 0.1 km only between radial distances of 6 and 10 km corresponds to a volume of c. 20 km^3. Thus the maximum ejected sediment volume would be decreased by about 34 km^3 (provided that ejected basement rocks corresponding to the subsidence volume have not all been deposited in the megablock zone). In addition, the possibility that erosion has removed a large amount of crater deposits in the megablock zone completely out of the crater, has to be considered. In conclusion we feel that the volume of sediments ejected *beyond the tectonic rim* (diameter 26 km) can hardly exceed 100 km^3, which is somewhat less than previous estimates (Gall *et al.*, 1975). The total volume of *excavated* sediments could be somewhat larger, because we have to take into account the volume of sedimentary deposits in the megablock zone, which may range up to 30 km^3 (?).

Some minimum value for the volume of excavated sediments can be estimated by projecting the central crater cavity (diameter 12 km) to 600 m above sea level. This yields about 71 km^3, taking the same values as above for the volumes of lake sediments, suevite and inner ring structure. However, the 71 km^3 would have been deposited in part in the megablock zone and in part outside the crater, so that the volume ejected beyond the tectonic boundary would be much less. This model assumes that the present topographic depression in the area of the megablock zone between the inner ring and the tectonic rim was formed exclusively by subsidence and erosion.

5.1.2. *Estimates of the volume of excavated basement rocks.* The mass deficiency (section 4.1) indicates that the presently missing volume of crystalline rocks in the basement (below −100 m) ranges between 22 and 30 km^3 (calculated for an original density of 2.8 g/cm^3). To this we add the volume of missing basement rocks between sea level and −100 m for a radial distance of 10 km

5.1.3. *Estimate of the volume of impact melt.* If we assume that all impact (excavation in the central cavity and subsidence in the megablock zone, 31 km^3). Thus the total volume of excavated basement rocks could range between 53 and 61 km^3. Part of these rocks is found in the fallback suevite (9 km^3, corrected for porosity) and part in the inner ring (5–10 km^3). This gives estimated volumes between 34 and 47 km^3 for basement rocks ejected beyond the inner ring. Volumes of 53 and 61 km^3 correspond to flat spherical cavities of a depth of 2.0 and 2.1 km and a diameter of 8.0 and 8.4 km. A subsidence of about 0.15 km between the inner ring and a radial distance of 10 km would fill up these cavities.

melt produced in the Ries event is contained in the suevite, the melt volume can be calculated from the volume of suevite and the volume fraction of melt in the suevite. According to Stöffler (1977) the volume of fallback and fallout suevite could be on the order of 8–10 km^3 and 0.2–0.4 km^3, respectively. If 75% of the original fallout suevite has been removed by erosion (rough estimate), the total primary volume of suevite ranges then from about 9 to 11 km^3. The melt content in the fallout suevite is 15 vol.%, in the fallback suevite 3.5–5 vol.% as measured in the drill core of Nördlingen 1973 (Stöffler *et al.*, 1977). From these figures we obtain a total volume of melt of 0.3–0.55 km^3. Because of the porosity of the melt particles which is on the order of 70% in the fallback suevite and 40% in the fallout suevite, the volume of the melt itself will be in the range of 0.1–0.2 km^3. The highest estimates of the total fallback suevite volume (Bayer. Geol. Landesamt, 1977) which are on the order of 15 km^3, would increase this figure to 0.2–0.6 km^3. Depending on the total volume of excavated rocks of the Ries estimated above (124 km^3 minimum, 200 km^3 maximum, 150 km^3 probable value) the melt volume in the Ries ranges between 0.27 and 0.44% of the total volume of excavated rocks for the maximum estimate of the melt volume. This is much less than melt volume estimates for Canadian craters (Grieve *et al.*, 1977) which range from 1 to 5% of the transient cavity volume which might correspond to our minimum estimate of the crater volume (124 km^3). The Ries impact melt volume would be appreciably higher if a coherent layer of impact melt was present in the central area of the crater.

5.1.4. *Impact energy.* A rough estimate of the impact energy can be made from the volume of displaced masses. The total excavation mass will include the mass deficit below the elevation of 400 m obtained by gravity measurements (75,000–100,000 Mt) plus the mass of the lake sediments (56,000 Mt), the mass of fallback suevite (32,000 Mt) and the mass of ejected sediments above 400 m (145,000 Mt for a radius of 10 km and a thickness of 0.2 km). This adds up to masses between 308,000 and 333,000 Mt. These are not maximum masses since fallback has only partly been taken into account. To lift a mass of 300,000 Mt up to an elevation of 5 or 10 km (max. horizontal transport 10–20 km) an energy of $1.5 \cdot 10^{26}$ erg and $3 \cdot 10^{26}$ erg respectively is needed. As only part of the impact energy is converted into kinetic energy of ejection, these figures may be multiplied by a factor of 2 to 3 in order to obtain estimates for the kinetic energy of the impacting body.

Using the scaling laws of Roddy *et al.* (1975) for Meteor Crater ($D_1 = D_2(E_1/E_2)^{1/n}$, D = diameter of simple bowl-shaped crater, E = impact energy), the following table for the crater diameter was calculated for impact energies of 10^{26} and 10^{27} erg and different scaling exponents n(D in km).

n	3	3.4	3.5	3.6	4
10^{26} erg	11.8	6.7	5.9	5.3	3.5
10^{27} erg	25.3	13.1	11.4	10.1	6.2

For the more probable scaling exponents between 3.4 and 3.6 there is a rough agreement with the diameter of the central crater cavity in the Ries for an energy between 10^{26} and 10^{27} erg.

There is also a rough agreement with the diameter of a primary crater (before readjustment) calculated with the scaling equation of Gault (1974) for craters larger than 1 km which gives $D = 7.1$ km and $D = 12.6$ km for impact energies of 10^{26} and 10^{27} erg respectively.

5.2. *Interpretation of the subsurface structure*

We think that the subsurface structure of the crater described in the previous sections can be explained by the following cratering model. A primary cavity about 2–3 km deep is formed by the impacting body. The diameter of this transient crater probably was a few kilometers less than the diameter of the present inner ring. The rampart-like structure of the inner ring can be interpreted as the remnants of the uplifted rim of this primary crater (Angenheister and Pohl, 1974; Engelhardt, 1975) or possibly rim deposits of inverted stratigraphy. This is suggested by the inverted sequences of rocks in drillholes in the inner ring area (basement rocks overlying sedimentary rocks) and the overturned 80 m thick sediment block in the drillhole Wörnitzostheim about 2 km outside of the inner ring.

The primary cavity was filled by fallback ejecta, mainly suevite, and by readjustment movements bringing the brecciated basement up to the present elevation in the central crater. The readjustment consisted of downward and inward movements in the megablock zone producing the ring depression in the basement and inward and upward movements in the central crater giving rise to the innermost peak ring. Part of the rim of the transient crater may have collapsed into the central crater. This could explain the interruptions of the inner ring structure. Large relative horizontal movements in the central crater explain the alternating crystalline basement units of various shock stages found in the drill core Nördlingen 1973 (Fig. 21; Engelhardt and Graup, 1977; David, 1977). The upper unit (525–670 m, shock stage I, shatter cones, traces of meteoritic material, El Goresy and Chao, 1976) probably originates from a boundary region of the primary cavity, but it may have been displaced vertically and horizontally over a distance of 1–3 km. According to Stöffler (1977) the melt-poor suevitic dikes have been intruded during the formation of the primary cavity.

As a consequence of these readjustment movements the strata outside the inner ring subsided together with the outer crater deposits in the megablock zone. The amount of subsidence decreases with increasing distance from the center and terminates at the tectonic rim at a radial distance of about 12–13 km.

5.3. *Interpretation of the impact formations*

5.3.1. *Continuous deposits of the megablock and Vorries zone.* The outer impact formations of the Ries crater have been subdivided into a lower unit represented by megablocks and Bunte breccia and an upper unit, the fallout suevite (section 3.2). Three major types of transport and deposition are considered to be involved in the formation of this lower unit, the bulk continuous deposits:

(1) over- and under-thrusting of megablocks by gliding and shearing
(2) ballistic transport of ejecta without major secondary mass transport upon landing of the ejecta
(3) ballistic transport of ejecta with major secondary mass transport.

Types (1) and (2) are characteristic of the megablock zone, types (2) and (3) are recognized in the "Vorries" continuous deposits.

In addition to the material transported and deposited according to (1) and (2), megablocks near the tectonic rim of the crater and probably at greater depth of the megablock zone were displaced by down-faulting and slumping of autochthonous target rocks (Figs. 20, 40). This was probably induced by subsidence of the marginal crater zone during readjustment of a transient crater cavity (see section 5.2).

We believe that the complex structure of the megablock zone is primarily the result of the radially decreasing shock wave energy. When the growing crater reached the radius of the later formed inner ring, the mode of excavation gradually transformed from a ballistic to a gliding mode of excavation (thrust faults) which is more and more restricted to the uppermost strata (Malmian) with increasing range. Consequently most of the allochthonous megablocks originate from the target area which is close to the region of the present inner ring. Many of the characteristics of the impact formations in the megablock zone continue into the Vorries continuous deposits beyond the tectonic rim of the crater. The decrease in the size of megablocks and of the abundance of blocks from the lower part of the pre-impact stratigraphy as a function of range is most likely the result of the particle velocity decreasing radially away from the point of impact in horizontal and vertical direction (e.g., Gault *et al.*, 1968; papers in this volume).

The texture of the Bunte breccia in which the megablocks are embedded, reveals a number of genetically important properties (see sections 3.2.1 and 3.3) which indicate the nature of several subsequent phases in the process of its formation: (a) *shock compression* resulting in shock metamorphism of target rocks ranging from stage II (about 350–400 kbar) to low pressure monomict brecciation (less than about 50 kbar), (b) *comminution, deformation and mixing* of fragmental rock material from all stratigraphic levels and from a broad zone of variable shock compression. This is achieved during the excavation process by which the downward and outward moving material is deflected into an upward moving flow of particulate material which shears along the slope of the growing crater cavity. It is plausible to assume that part of the deformation of the fragments and most of the Bunte breccia takes place during this process. (c) *impacting of the ejecta* on the ground and *horizontal movement* on the target surface. Certain textural and deformational effects of the Bunte breccia may be assigned to this last phase such as the formation of the polished and striated surfaces on limestone, flow textures around large blocks and near to the base of the breccia deposits, and plastic deformation of incompetent rocks such as shale and clay (Wagner, 1964; Hüttner, 1969; Chao, 1976, 1977). The increasing incorporation of local material into Bunte breccia with increasing range is explained by the increasing impact velocity of the landing ejecta as described by the model of Oberbeck (1975). The horizontal and radial flow of the primary and secondary ejecta also explains the preferential filling of local depressions of the target surface which results in an extremely variable thickness of the continuous deposits in the Vorries (see section 3.3).

Chao (1976, 1977) has developed a different model for the emplacement of the Bunte breccia and megablocks. He suggests that they were emplaced predominantly by a roll and glide mode of transport rather than ballistically. To our opinion all observed textural and deformational features can be assigned to one of the phases (a), (b), or (c) discussed above. There is no unequivocal field evidence that excludes ballistic transport. In general, horizontal gliding requires preceding ballistic transport. Only where the particle velocities are insufficient for ballistic ejection, that is in the marginal zone of the crater, gliding may be an important primary mode of displacement.

5.3.2. *Suevite and impact melt rocks.* Some genetically important properties of suevite as well as the characteristic differences between fallout and fallback suevite have been discussed in section 3.2.3. According to these data and theoretical considerations about cratering mechanics (e.g., Gault *et al.*, 1968; and papers in this volume) the following model for the origin of suevite is proposed (see also Stöffler, 1977). Upon impact a shell of melt which surrounds the central region of vaporized rock is formed in the target by the propagating shock wave. The region of the melt zone which is nearest the free surface of the transient cavity, is first moving downward and sideward and then accelerated upward into ballistic trajectories. This gives a plausible mechanism for the mixing of melt and rock fragments of decreasing intensity of shock metamorphism. The uppermost part of the shell of melt which involved mainly sedimentary rocks, was probably ejected with very high velocity beyond the present outer margin of the continuous deposits. This had already happened before the stagnation point of the penetrating projectile was reached. The lower part of the melt zone, preferably the outer shell of it, which was mainly confined to the crystalline basement rocks had a smaller particle velocity and was ejected late in the excavation process, after the Bunte breccia was formed. This is the most probable source of melt from which particles already mixed with shocked rock debris during the downward movement were disrupted by the upward acceleration and subsequently shaped into bomb-like forms, which are characteristic of the fallout suevite. Fragment-laden melt particles and rock fragments were probably mixed with expanding vaporized target rock and deposited as a high density current. A few very large lumps of melt were ejected to the eastern crater rim forming isolated bodies of impact melt rock (see section 3.2.4). It is an open question how the discontinuous distribution of fallout suevite and the admixture of sedimentary rock fragments from the upper section of the target can be explained in the scope of this schematic model.

The source region of the melt inclusions in the fallback suevite must be a deeper part of the melt zone in which the velocity and vector of particle movement did not result in an ejection beyond the rim of the primary crater cavity. Also the innermost, hottest, and less viscous shell of the melt zone should have been incorporated into the suevite since it cannot be ejected beyond the crater due to the vertical downward movement. Thereby fused sedimentary strata (depending on the penetration depth of the projectile and extent of the zone of vaporized rock) could be mixed into the melt. This might explain the smaller size of the melt inclusions and their extreme vesiculation which could result from the high content of water in the sedimentary rocks.

The question is whether the material of the fallback suevite was actually ejected vertically to relatively large heights or whether it moved turbulently more or less like a surge in the crater. The latter seems to be required to explain the intrusive character of the lowermost suevitic breccia (below 515 m) and the "suevitic" dike at 642 m in the drill core Nördlingen 1973. On the other hand the "sorted" suevite above 331 m of this drill core speaks in favor of a late genuine fallback phase which interacted with the atmosphere. The main layer of the melt-rich suevite within the central crater might be explained by a vertical ejection and fallback mechanism rather than by ground surging (see continuous decrease of melt content with depth; Fig. 25). An earlier ground surge derived from an area at some lateral distance from the projectile could have formed the melt-poor, cold suevite intrusively. On top of it the true fallback suevite originating from the vicinity of the projectile was deposited later in time.

Acknowledgment—The authors are indebted to various colleagues who provided useful information and unpublished data, especially to Prof. W. v. Engelhardt, and Dr. G. Graup, Tübingen, and Dr. F. Hörz, Houston. They should like to thank Mrs. G. Grant, Miss F. Möllers, Mrs. U. Ewald, H.-D. Knöll, U. März, R. Ostertag, and W.-U. Reimold for their valuable technical assistance in the preparation of the manuscript. We thank the German Research Association (Deutsche Forschungsgemeinschaft) for its generous financial support over the past decade. We acknowledge the very helpful comments of Dr. F. Hörz, Houston, and Dr. H. C. Halls, Toronto, who reviewed the manuscript with great care.

REFERENCES

Abadian, M.: 1972, Petrographie, Stoßwellenmetamorphone und Entstehung polymikter kristalliner Breccien im Nördlinger Ries. *Contr. Mineral. Petrol.* **35**, 245–262.

Abadian, M., Engelhardt, W. v., and Schneider, W.: 1973, Spaltenfüllungen in allochthonen Schollen des Nördlinger Ries. *Geologica Bavarica* **67**, 229–237.

Ackermann, W.: 1958, Geologisch-petrographische Untersuchungen im Ries. *Geol. Jb.* **75**, 135–182.

Ammon, L. v.: 1905, Die Scheuerfläche von Weilheim in Schwaben. Ein Beitrag zur Riesgeologie. *Geogn. Jh.* **18**, 153–176.

Angenheister, G. und Pohl, J.: 1964, The remanent magnetization of the suevite from the Ries area (Southern Germany) *Z. Geophysik*, **30**, 258–259.

Angenheister, G. and Pohl, J.: 1969, Die seismischen Messungen im Ries von 1948–1969. *Geologica Bavarica* **61**, 304–326.

Angenheister, G. and Pohl, J.: 1974, Beiträge der angewandten Geophysik zur Auswahl des Bohrpunktes der Forschungsbohrung Nördlingen 1973. *Geologica Bavarica* **72**, 59–63.

Bader, K. and Schmidt-Kaler, H.: 1977, Der Verlauf einer präriesischen Erosionsrinne im östlichen Riesvorland zwischen Treuchtlingen und Donauwörth. *Geologica Bavarica* **75**, 401–410.
Bauberger, W., Mielke, H., Schmeer, D., and Stettner, G.: 1974, Petrographische Profildarstellung der Forschungsbohrung Nördlingen 1973 (von Meter 263 an bis zur Endteufe im Maßstab 1 : 200). *Geologica Bavarica* **72**, 33–34.
Bayer. Geol. Landesamt (ed.): 1969, Das Ries. *Geologica Bavarica* **61**, 478 pp.
Bayer. Geol. Landesamt (ed.): 1974, Die Forschungsbohrung Nördlingen 1973. *Geologica Bavarica* **72**, 98 pp.
Bayer. Geol. Landesamt (ed.): 1977, Ergebnisse der Ries-Forschungsbohrung 1973: Struktur des Kraters und Entwicklung des Kratersees. *Geologica Bavarica* **75**, 470 pp.
Birzer, F.: 1969, Molasse und Ries-Schutt im westlichen Teil der Südlichen Frankenalb. *Geol. Bl. NO-Bayern* **19**, 1–28.
Blohm, E.-K., Friedrich, H., and Homilius, J.: 1977, Ein Ries-Profil nach geoelektrischen Tiefensondierungen. *Geologica Bavarica* **75**, 381–393.
Bolten, R. and Müller, D.: 1969, Das Tertiär im Nördlinger Ries und in seiner Umgebung. *Geologica Bavarica* **61**, 87–130.
Bundesanstalt für Geowissenschaften und Rohstoffe, Hannover: 1971, Aeromagnetic Map of the German Federal Republic 1 : 100 000.
Chao, E. C. T.: 1963, The petrographic and chemical characteristics of tektites. In *Tektites* (J. A. O'Keefe, ed.), p. 51–94. Univ. of Chicago Press.
Chao, E. C. T.: 1967, Impact metamorphism. In *Researchers in Geochemistry* (P. H. Abelson, ed.), **2**, p. 204–233. John Wiely and Sons, Inc., New York.
Chao, E. C. T.: 1968, Pressure and temperature histories of impact metamorphosed rocks—based on petrographic observations. In *Shock Metamorphism of Natural Materials* (B. M. French and N. M. Short, eds.), p. 135–158. Mono Book Corp. Baltimore.
Chao, E. C. T.: 1973, Geologic implications of the Apollo 14 Fra Mauro breccias and comparison with ejecta from the Ries Crater, Germany. *J. Research U.S. Geol. Survey*. **1**, 1–18.
Chao, E. C. T.: 1976, The Ries Crater, a model for the interpretation of the source areas of lunar breccia samples (abstract). In *Lunar Science VII*, p. 126–128. The Lunar Science Institute, Houston.
Chao, E. C. T.: 1977, Preliminary interpretation of the 1973 Ries drill core. *Geologica Bavarica* **75**, 421–441.
Cotta, B. v.: 1834, Geognostische Beobachtungen im Riesgau und dessen Umgebungen. *N. Jb. Miner. usw*, p. 307–318.
David, E.: 1977, Abschätzung von impaktmechanischen Daten aufgrund von Ergebnissen der Forschungsbohrung Nördlingen 1973. *Geologica Bavarica* **75**, 459–470.
Dehm, R.: 1969, Geschichte der Riesforschung. *Geologica Bavarica* **61**, 25–35.
Dehm, R., Gall, H., Höfling, R., Jung, W., and Malz, H.: 1977, Die Tier- und Pflanzenreste aus den obermiozänen Riessee-Ablagerungen in der Forschungsbohrung Nördlingen 1973. *Geologica Bavarica* **75**, 91–109.
Dence, M. R.: 1971, Impact melts. *J. Geophys. Res.* **76**, 5552–5565.
Dennis, J. G.: 1971, Ries structure, Southern Germany, a review. *J. Geophys. Res.* **76**, 5394–5406.
Dorn, P.: 1948, Ein Jahrhundert Riesgeologie, *Z. dt. geol. Ges.*, **100**, 348–365.
Dressler, B. and Graup, G.: 1974, Gesteinskundliche Untersuchungen am Suevit der Bohrung Wörnitzostheim I im Nördlinger Ries. *Der Aufschluß* **25**, 404–411.
El Goresy, A.: 1968, The opaque minerals in impactite glasses. *Shock Metamorphism of Natural Materials* (B. M. French and N. M. Short, eds.), p. 531–554. Mono Book Corp., Baltimore.
El Goresy, A. and Chao, E. C. T.: 1976, Evidence of the impacting body of the Ries crater—the discovery of Fe-Cr-Ni veinlets below the crater bottom. *Earth Planet. Sci. Lett.* **31**, 330–340.
Engelhard, L. and Hansel, J.: 1976, Ein Beitrag zur Erkundung der Struktur des Nördlinger Rieses auf Grund geoelektrischer Schlumberger-Sondierungen. *Abh. Braunschw. Wiss. Ges.* **26**, 1–19.
Engelhardt, W. v.: 1976a, Neue Beobachtungen im Nördlinger Ries. *Geol. Rundschau* **57**, 165–188.
Engelhardt, W. v.: 1976b, Chemical composition of Ries glass bombs. *Geochim. Cosmochim. Acta* **31**, 1677–1689.

Engelhardt, W. v.: 1972, Shock produced rock glasses from the Ries Crater. *Contr. Mineral. Petrol.* **36**, 265–292.

Engelhardt, W. v.: 1975, Some new results and suggestions on the origin of the Ries basin. *Fortschr. Miner.* **52**, 375–384.

Engelhardt, W. v.: 1977, personal communication.

Engelhardt, W. v., Stöffler, D., and Schneider, W.: 1969, Petrologische Untersuchungen im Ries. *Geologica Bavarica* **61**, 229–296.

Engelhardt, W. v. and Stöffler, D.: 1974, Ries meteorite crater, Germany. *Fortschr. Miner.* **52**, Beih. 1, 103–122.

Engelhardt, W. v, and Graup, G.: 1977, Stoßwellenmetamorphose im Kristallin der Forschungsbohrung Nördlingen 1973. *Geologica Bavarica* **75**, 255–271.

Ernstson, K.: 1974, The structure of the Ries crater from geoelectric depth soundings. *J. Geophys.* **40**, 639–659.

Ernstson, K. and Pohl, J.: 1974, Einige Kommentare zu den bohrlochgeophysikalischen Messungen in der Forschungsbohrung Nördlingen 1973. *Geologica Bavarica* **72**, 81–90.

Ernstson, K. and Pohl, J.: 1977, Neue Modelle zur Verteilung der Dichte und Geschwindigkeit im Ries-Krater. *Geologica Bavarica* **75**, 355–371.

Förstner, U.: 1967, Petrographische Untersuchungen des Suevit aus den Bohrungen Deiningen und Wörnitzostheim im Ries von Nördlingen. *Contr. Mineral. Petrol.* **15**, 281–307.

Gall, H.: 1969, Geologische Untersuchungen im südwestlichen Vorries. Das Gebiet des Blattes Wittislingen. Diss. Univ. München, 166 pp.

Gall, H.: 1974, Geologische Bau- und Landschaftsgeschichte des südöstlichen Vorrieses zwischen Höchstädt a.d. Donau und Donauwörth. *N. Jb. Geol. Paläont. Abh.* **145**, 1, 58–95.

Gall, H. und Müller, D.: 1975, Reutersche Blöcke—außeralpine Fremdgesteine unterschiedlicher Herkunft in jungtertiären und quartären Sedimenten Südbayerns. *Mitt. Bayer. Staatsamml. Paläont. Hist. Geol.* **15**, 207–228.

Gall, H., Müller, D., and Stöffler, D.: 1975, Verteilung, Eigenschaften und Entstehung der Auswurfs massen des Impaktkraters Nördlinger Ries. *Geologische Rundschau*, **64**, 915–947.

Gall, H., Hollaus, E., and Trischler, J.: 1976, Obermiozäne Seesedimente und Bunte Trümmermassen der Forschungsbohrung Wörnitzostheim I im Nördlinger Ries, *Geolog. Blätter NO-Bayern.* **26**, 188–206.

Gall, H. and Müller, D.: 1977, 4. Stratigraphie. *Erläuterungen zur Geologischen Karte des Rieses 1 : 50 000.* Bayerisches Geologisches Landesamt, München, in press.

Gall, H., Müller, D., and Pohl, J.: 1977, Zum geologischen Bau der Randzone des Impaktkraters Nördlinger Ries. *N. Jb. Geol. Paläont.* **1977**, 65–94.

Gault, D. E.: 1974, Impact cratering, *A Primer in Lunar Geology*, (R. Greeley and P. Schultz, eds.). NASA TM-X-62, 359, 137–176.

Gault, D. E., Quaide, W. L. and Oberbeck, V. R.: 1968, Impact cratering mechanics and structures. In *Shock Metamorphism of Natural Materials* (French, B. M. and Short, N. M., eds.), p. 87–99, Mono Book Corp., Baltimore.

Gentner, W., Lippolt, H. J., and Schaeffer, O. A.: 1963, Argonbestimmungen an Kaliummineralien—XI. Die Kalium-Argon-Alter der Gläser des Nördlinger Rieses und der böhmischmährischen Tektite. *Geochim. Cosmochim Acta* **27**, 191–200.

Gentner, W. and Wagner, G. A.: 1969, Altersbestimmungen an Riesgläsern und Moldaviten. *Geologica Bavarica* **61**, 296–303.

Graup, G.: 1975, *Das Kristallin im Nördlinger Ries.* Diss. Univ. Tübingen, Germany.

Graup, G.: 1977, Die Petrographie der kristallen Gesteine der Forschungsbohrung Nördlingen 1973. *Geologica Bavarica* **75**, 219–229.

Grieve, R. A. F., Dence, M. R., and Robertson, P. B.: 1977, Cratering processes: as interpreted from the occurrence of impact melts. In *Impact and Explosion Cratering* (D. J. Roddy, R. O. Pepin, and R. B. Merrill, eds.), Pergamon Press, New York. This volume.

Harr, K.: 1976, Petrographische Untersuchungen an Glastuffen und Bentoniten der südwestdeutschen Molasse, Diss. Tübingen.

Hörz, F.: 1965, Untersuchungen an Riesgläsern *Beitr. Miner. Petrol.* **11**, 621–661.
Hörz, F.: 1976, Personal Communication.
Hörz, F., Gall, H., Hüttner, R., and Oberbeck, V. R.: 1975, The Ries crater and lunar basin deposits. *Proc. Lunar Sci. Conf. 7th*, p. 396–398.
Hüttner, R.: 1958, *Geologische Untersuchungen im SW-Vorries auf Blatt Neresheim und Wittislingen.* Diss. Univ. of Tübingen, Germany. 347 pp.
Hüttner, R.: 1969, Bunte Trümmermassen und Suevit. *Geologica Bavarica* **61**, 142–200.
Jung, K. and Schaaf, H.: 1967, Gravimetermessungen im Nördlinger Ries und seiner Umgebung. Abschätzung der gesamten Störungsmasse. *Z.f. Geophysik* **33**, 319–345.
Jung, K., Schaaf, H., and Kahle, H. G.: 1969, Ergebnisse gravimetrischer Messungen im Ries. *Geologica Bavarica* **61**, 337–342.
Kahle, H. G.: 1969, Abschätzung der Störungsmasse im Nördlinger Ries. *Z. Geophysik* **35**, 317–345.
Morgan, J. A.: 1976, Personal Communication.
Müller, D.: 1969, Ein neues Profil vom Mittelkeuper bis zum Unterdogger bei Harburg nahe dem Nördlinger Ries. *Mitt. Bayer. Staatssamml. Paläont. Hist. Geol.* **9**, 73–92.
Müller, D.: 1972, *Die Oligozän-Ablagerungen im Gebiet des Nördlinger Rieses.* Diss. Univ. München, Germany. 249 pp.
Oberbeck, V. R.: 1975, The role of ballistic erosion and sedimentation in lunar stratigraphy. *Rev. Geophys. Space Phys.* **13**, 337–362.
Oberheinischer Geologischer Verein (ed.): 1926, Das Problem des Rieses. *Verlag der Stadt Nördlingen*, Nördlingen, Germany. 291 pp.
O'Keefe, J. A. and Weiskirchner, W.: 1970, Die Tektite als natürliche Gläser. *Glastechnische Berichte* **43**, 199.
Pohl, J.: 1965, Die Magnetisierung der Suevite des Rieses. *N. Jb. Miner. Abh.* **H.9–11**, 268–276.
Pohl, J.: 1971, On the origin of the magnetization of impact breccias on Earth. *Z.f. Geophysik*, **37**, 549–555.
Pohl, J.: 1974, Magnetisierung der Bohrkerne in der Forschungsbohrung Nördlingen 1973. *Geologica Bavarica*, **72**, 65–74.
Pohl, J.: 1977, Paläomagnetische und gesteinsmagnetische Untersuchungen an den Kernen der Forschungsbohrung Nördlingen 1973. *Geologica Bavarica* **75**, 329–348.
Pohl, J. and Angenheister, G.: 1969, Anomalien des Erdmagnetfeldes und Magnetisierung der Gesteine im Nördlinger Ries. *Geologica Bavarica*, **61**, 327–336.
Pohl, J. and Will, M.: 1974, Vergleich der Geschwindigkeitsmessungen im Bohrloch der Forschungsbohrung Nördlingen 1973 mit seismischen Tiefensondierungen innerhalb und außerhalb des Ries. *Geologica Bavarica* **72**, 75–80.
Preuss, E.: 1964, Das Ries und die Meteoritentheorie. *Fortschr. Mineral.* **41**, 271–312.
Reich, H. and Horrix, W.: 1955, Geophysikalische Untersuchungen im Ries und Vorries und deren geologische Deutung. *Beih. Geol. Jb.* **19**, 119 pp.
Reuter, L.: 1926, Die Verbreitung jurassischer Kalkblöcke aus dem Ries im südbayerischen Diluvial-Gebiet. *Jber. u. Mitt. Oberrh. Geol. Ver. N.F.* **14**, 191–218.
Rittmann, A.: 1973, *The stable mineral assemblage of igneous rocks: a method of calculation.* Springer, Berlin–Heidelberg–New York.
Roddy, D. J., Boyce, J. M., Colton, G. W., and Dial A. L., Jr.,: 1975, Meteor crater, Arizona, rim drilling with thickness, structural uplift, diameter, depth, volume, and mass-balance calculations. *Proc. Lunar Sci. Conf. 6th*, p. 2621–2644.
Scheuenpflug, L.: 1973, Zur Problematik der Weißjuragesteine in der östlichen Iller-Lech-Platte. *Eiszeitalter und Gegenwart* **23/24**, 154–158.
Schmidt-Kaler, H.: 1969, Versuch einer Profildarstellung für das Rieszentrum vor der Kraterbildung. *Geologica Bavarica* **61**, 38–40.
Schmidt-Kaler, H., Treibs, W., and Hüttner, R.: 1970, Geologische Übersichtskarte des Rieses und seiner Umgebung 1:100 000. Exkursionsführer zur Geologischen Übersichtskarte des Rieses 1:100 000. *Bayerisches Geologisches Landesamt*, München, 68 pp.
Schneider, W.: 1971, Petrologische Untersuchungen der Bunten Breccie im Nördlinger Ries. *B. Jb. Miner. Abh.* **114**, 136–180.

Shoemaker, E. M. and Chao, E. C. T.: 1961, New evidence for the impact origin of the Ries Basin, Bavaria, Germany, *J. Geophys. Res.*, **66**, 3371–3378.

Simmons, G., Siegfried, R., and Richter, D.: 1975, Characteristics of microcracks in lunar samples. *Proc. Lunar Sci. Conf. 6th*, p. 3227–3254.

Stähle, V.: 1972, Impact glasses from the suevite of the Nördlinger Ries. *Earth Planet. Sci. Lett.* **17**, 275–293.

Stähle, V. and Ottemann, J.: 1977, Petrographische Studien am Suevit und an den Gangbreccien der Forschungsbohrung Nördlingen 1973, *Geologica Bavarica* **75**, 191–217.

Stöffler, D.: 1966, Zones of impact metamorphism in the crystalline rocks of the Nördlinger Ries crater *Contr. Miner. Petrol.* **12**, 15–24.

Stöffler, D.: 1969, Kristalline Trümmermassen. in Petrologische Untersuchungen im Ries, Engelhardt, W. v. Stöffler, D. and Schneider, W. *Geologica Bavarica*, **61**, 285–288.

Stöffler, D.: 1971a, Progressive metamorphism and Classification of Shocked and Brecciated Crystalline Rocks at Impact Craters. *J. Geophys. Res.* **76**, 5541–5551.

Stöffler, D.: 1971b, Coesite and Stishovite in Shocked Crystalline Rocks. *J. Geophys. Res.* **76**, 5474–5488.

Stöffler, D.: 1977, Research drilling, Nördlingen 1973: polymict breccias, crater basement, and cratering model of the Ries impact structure. *Geologica Bavarica*, **75**, 443–458.

Stöffler, D., Knöll, H.-D., Reimold, W.-U. and Schulien, S.: 1976, Grain size statistic, composition and provenance of fragmental particles in some Apollo 14 breccias. *Proc. Lunar Sci. Conf. 7th*, p. 1965–1985.

Stöffler, D., Ewald, U., Ostertag, R., and Reimold, W.-U.: 1977, Ries deep drilling: I. Composition and texture of polymict impact breccias. *Geologica Bavarica*. **75**, 163–189.

Wagner, G. A.: 1977, Spaltspurendatierungen an Mineralien aus kristallinen Riesgesteinen. *Geologica Bavarica*, **75**, 349–354.

Wagner, G. H.: 1964, Kleintektonische Untersuchungen im Gebiet des Nördlinger Rieses. *Geol. Jb.* **81**, 519–600.

Werner, E.: 1904, Das Ries in der Schwäbisch-fränkischen Alb. *Bl. Schwäb. Albvereins*, **16**, 153–167.

Roddy, D. J., Pepin, R. O., and Merrill, R. B., editors.
(1977) *Impact and Explosion Cratering*, Pergamon Press (New York), p. 405–424.
Printed in the United States of America

Impact cratering phenomenon for the Ries multiring structure based on constraints of geological, geophysical, and petrological studies and the nature of the impacting body

E. C. T. CHAO with energy considerations by JEAN A. MINKIN

U.S. Geological Survey, National Center 929, Reston, Virginia 22092

Abstract—In this paper a model is proposed which outlines the cratering phenomenon for the Ries multiring structure. This schematic reconstruction conforms to constraints imposed by geological, geophysical, and petrological studies and the nature of the postulated impacting body. Among these constraints are: (1) the shallow geometry of the crater (depth-to-diameter ratio 1:33); (2) the predominance of the nonballistic mode of transport of most of the sedimentary ejecta apparently under locally very high confining pressures; and (3) evidence that the most strongly shocked rocks are the crystalline rock ejecta and suevite. The model is also based on the impact of a stony meteorite with a suggested diameter of about 3 km, and an assumed impact velocity of 15 km/sec. These constraints were derived in part from earlier data and in part from new observations made in the summers of 1975 and 1976 by Chao.

In contrast with experimentally produced simple bowl-shaped craters, the Prairie Flat 500-ton high explosive experiment produced a shallow multiring structure with predominant horizontal or lateral displacement of target material. In these respects, it is similar to the Ries, although the cratering conditions are by no means the same.

The effects of layering of geologic formation, and the role of water are briefly discussed. The mechanism of base surge is concluded to be of no significance. Energy-mass relationships implied by the proposed model are also briefly treated in an appendix.

INTRODUCTION

FROM PRESENTATIONS and discussions at the Symposium on Planetary Cratering Mechanics held in Flagstaff, Arizona, September 13–17, 1976, it seems very probable that two distinct types of cratering mechanisms and phenomena exist: one for simple bowl-shaped craters with or without central uplifts, and one for flat-floored multiring structures. This is in contrast with the proposal of Dence *et al.* (this issue) that all craters (simple bowl-shaped, craters with central uplift, and multiring structures) are produced by one and the same mechanism, dominated by ballistic ejecta transport. Dence *et al.* called on post-cratering adjustments, such as subsidence, faulting, deep sliding, and uplift due to rebound, to account for the morphology of complex craters. I do not think the proposal of Dence *et al.* satisfactorily explains the origin of multiring structures.

The Ries multiring basin of southern Germany is the best preserved of the very large terrestrial impact structures, and is now probably the most thoroughly studied. On the basis of field and laboratory data obtained to date an attempt is made here to delineate the phenomenon of formation of the Ries Basin. Calculations by J. A. Minkin show that the estimated kinetic energy of impact is compatible with the estimated total ejecta volume based on the shallow crater geometry. Energy partitioning valid for laboratory experimental craters (Gault and Heitowit, 1963) appears not to be applicable to large craters. The reconstruction presented in this paper meets the detailed constraints of geologic, geophysical, and petrologic observations itemized and summarized below. It is hoped that this will serve to aid others in the development of a theoretical basis for understanding the mechanics of formation of shallow multiring structures, and serve as a starting point for interpreting multiring basins on the moon, Mercury, and on Mars.

CONSTRAINTS FOR THE SCHEMATIC RECONSTRUCTION OF THE DEVELOPMENT OF THE RIES BASIN

The Ries Crater of southern Germany has been studied for more than 100 yr, more intensively during the last 15 yr. Many new data have been obtained during the last few years, particularly after the 1973 Ries research deep drill core became available. Important earlier geological, geophysical, and petrological observations by numerous investigators (citcd in Chao, 1977a), together with new data on the crater rim and diameter, crater depth, and the evidence of nonballistic ejecta transport are summarized and discussed in a comprehensive paper by Chao (1977a). Readers interested in obtaining detailed background information regarding the Ries structure are urged to consult that paper which is the principal source for the constraints enumerated below. Additional new data concerning mineral-produced high pressure striae and clay polish as evidence for nonballistic ejecta transport (Chao, 1976), position of the present crater floor (Chao, 1977b) and the nature of the impacting body (El Goresy and Chao, 1976, 1977) were obtained within the last year. All of these data put stringent constraints on the nature of the cratering phenomenon of the Ries multiring structure.

The principal parameters and constraints to be considered are summarized and itemized below:

1. *Origin*

The Ries is an impact crater: Evidence based on the occurrence of coesite, stishovite, other shock features, and armalcolite (see references cited in Chao, 1977a).

2. *Impacting body*

(a) Probably a stony meteorite with density about 3.5 g/cm^3 (El Goresy and Chao, 1976, 1977).
(b) Probably a large body, up to about 3 km in diameter (estimate in part based on the occurrence of Fe–Cr–Ni veinlets in the compressed zone in the 1973 Ries drill core 3.5 km, N75°W from the geometric center of the crater, Chao and El Goresy, 1977; also see appendix).
(c) Impact velocity—assume to be 15 km/sec.

3. *Morphology and dimensions* (Chao, 1977a)
   (a) The morphological rim is the crater rim, 25 km in diameter.
   (b) Crater depth: 750 m.
   (c) Depth–diameter ratio: 1:33.
   (d) Crater volume: 184 km^3.

4. *Structural constraints*
   (a) Geophysical studies indicate the presence within the crater of a 12 km diameter ring (partly exposed) and the possible presence of a completely buried 4 km ring. Based on thickness variation of suevite deposits within the crater, the crater floor within the 12 km ring is probably 100–200 m deeper than outside the 12 km ring (Reich and Horrix, 1955; Pohl and Angenheister, 1969; and other references on geophysical studies in Chao, 1977a).
   (b) Crater floor lies between the base of the fallback ejecta and a clearly defined compressed zone at shallow depths. There is no evidence of a melt layer (Chao, 1977b).
   (c) Refraction profiles show that reduced seismic velocities extend to depths of about 5 km below the crater (Pohl *et al.*, this issue). Fracturing therefore extends to depths of at least 5 km.
   (d) Crater wall rocks adjacent to the morphological rim are highly to intensely fractured, but down-faulting is negligible. There is no evidence of extensive post-event adjustment by concentric faulting (Bannert, 1969; Chao, 1977a).

5. *Geologic constraints*

Ejecta volume, mode of ejecta transport, and distribution and overlap relationship (Chao, 1977a).
   (a) Volume—(estimated on the basis of proportions of the crater volume related to depths of excavation into the sedimentary section and into the basement crystalline rocks respectively, see Chao, 1977a)—multicolored sedimentary ejecta (Bunte Trümmermassen and bunte breccia): 166–176 km^3; suevite and crystalline ejecta: 18–8 km^3.
   (b) Nonballistically transported sedimentary ejecta are greatly dominant over ballistically transported sedimentary ejecta (Chao, 1977a). Transport of the multicolored sedimentary ejecta mainly in the nonballistic roll-glide mode resulted in intimate mixing locally of stratigraphically widely separated lithologies. The ejecta were transported at high velocity and apparently under confining pressure, as suggested by plastic deformation of limestone concretions enclosed in shale in the sedimentary ejecta (Fig. 1) and mineral-produced striae and clay polish on sharply angular limestone fragments in the ejecta (Fig. 2). The striae and polish, more fully

Fig. 1. Photograph showing a plastically deformed Lias delta limestone concretion collected from within dark gray shale in the sedimentary ejecta from the Siegling quarry near Holheim. Note the ruptured slab shown by arrows A and B. The displacement of this slab at B is about 4 mm whereas there is no visible displacement at A, so that the slab was foreshortened between A and B by plastic flow under very high confining pressure. Scale in millimeters.

described in Chao, 1976, were not produced during the initial phase of acceleration by shock near the impact point, since the plastically deformed or striated limestone fragments show no sign of high strain-rate shock deformation.

(c) Evidence of downward pressure transmitted by the ejecta mass of the order of 3 kbar (i.e. greater than the compressive strength of limestone under a confining pressure of 0.5 kbar (Robertson, 1955, 1972)) was found in ruptured limestones on top of the crater wall (Fig. 3).

(d) Nonballistically transported sedimentary ejecta extend over both disturbed and undisturbed autochthonous substrate to distances as great as

Fig. 2. Photograph of a sharply angular fragment of Malm limestone from the sedimentary ejecta from Bschor quarry near Ronheim. Note the closely spaced deeply incised mineral-produced striae on both surfaces crossing fracture ridges. The striae are shown at higher magnification in the inset. Areas between the striae are slightly polished. The limestone interior is completely free of evidence of shock deformation. Both bar scales 1 cm.

one crater diameter beyond the crater rim. Ballistically transported sedimentary breccia overlies nonballistically transported sedimentary ejecta near and within a crater radius beyond the crater rim. Stratigraphically higher blocks are found farthest from the rim.

(e) Crystalline rock ejecta and suevite, formed mainly at high pressures and temperatures late in the cratering process and then ejected ballistically, overlie the sedimentary ejecta outside the crater. Within the crater, high temperature suevite overlies displaced sedimentary blocks and fallback ejecta beyond the 12 km ring and close to the rim, and inside the 12 km ring high temperature suevite overlies low temperature suevite.

6. *Petrologic constraints*

(a) Most strongly shocked ejecta is suevite, containing melt products and consisting mostly of basement crystalline rocks.

(b) Sedimentary ejecta are largely unshocked; moderate to strongly shocked sedimentary rocks are present in suevite but they are rare.

Fig. 3a. Photograph of the highly fractured crater wall at Siegling quarry near Holheim. Note the striations (schliffläche) on the top of the bedrock surface trending SW towards the viewer. The surface dips about 10° away from the crater. Note the areas of ruptured limestone near the hammers at the lower right and upper left of the photograph. Lower hammer is 32.2 cm in length.

CRATERING MODELS

Various models have been proposed for interpreting the phenomenon of the development of the Ries Crater. Gall *et al.* (1975) were convinced that the simple bowl-shaped model based on laboratory impact experiments in sand can be applied to the Ries Crater. During a workshop discussion meeting of the participants of the Flagstaff symposium, it became clear that if the simple bowl-shaped model is to be applied to the Ries, the 12 km diameter ring with a transient crater cavity reaching a depth of about 2 km should be considered as the starting point of such a model. Areas immediately adjoining the outside of this 12 km diameter ring should show evidence of the overturned flap of such a model. Hence no material could have been ejected from areas between the 12 km ring structure and the morphological crater rim. Furthermore, the morphological rim with a diameter of 25 km must then be accounted for by either uplifting of the rim or down-dropping of the inner ring, since the morphological southern rim now stands about 200 m higher in elevation than the 12 km ring. It is necessary for such a model to rely extensively on central uplift and rim subsidence to account for present shallow crater geometry with depth–diameter ratio 1 : 33.

Fig. 3b. Closeup photograph of the ruptured limestone shown in the lower right corner of Fig. 3a. The fractures are hinged with zero displacement at both ends and maximum of about 8 mm in the middle. Such ruptures are interpreted as produced when the stress upon this limestone exceeded the confining pressure exerted by the ejecta mass moving over this area. Since the compressive strength of limestone increases with increasing confining pressure according to Robertson (1955, 1972), and assuming the Malm massive limestone is similar to the Solenhofen limestone tested, the estimated compressive pressure was about 3 kbar here.

Another possible choice would be a model based on the experimental impact in a layered sand model (cohesionless sand over cemented sand) as reported by Piekutowski (this issue). In this experiment, a bowl-shaped crater was produced in the lower layer of bonded sand, surrounded by a flat shelf in the upper layer, extending into a crater rim. Ballistically expelled ejecta surround the crater. The layered stratigraphic section of the Ries vaguely resembles the target of this experiment, assuming the basement crystalline rocks as analogous to the cemented sand and the Mesozoic section as comparable to the loose sand. This layering effect has also been suggested for the Ries by Oberbeck *et al.* (1975a).

Unfortunately neither of the models described above can account for the mixing and overlap relationship of the Ries ejecta, and most importantly, for the dominance of nonballistically over ballistically transported ejecta.

The third experimental model available for comparison is the Prairie Flat ring structure produced by the detonation of a 500-ton sphere of TNT lying tangent on the ground surface (Roddy, 1969, 1970; Jones *et al.*, 1970). Figure 4 is a photograph of this structure. Cross sections of the Prairie Flat Crater (Roddy,

Fig. 3c. Closeup photograph of the rupture shown at the upper left corner of Fig. 3a. Here the fracture displacement ranges from zero to a maximum of about 2 cm.

1969, 1973, 1976; Jones, 1976) show the predominance of horizontal or lateral displacement vectors of the target materials, with a coherent overturned flap. In his oral presentation at the Flagstaff cratering symposium, Dr. Jones also described evidence of nonballistic movement. The phenomenon and the theoretical basis for the formation of such a multiring structure are not yet clear. I find this example to be a possible analogue to the Ries structure, in agreement with Roddy (1976), if the Prairie Flat ejecta can be demonstrated to have been truly transported nonballistically.

A schematic reconstruction of the Ries cratering phenomenon, to be realistic and accurate, must meet all the constraints itemized above. The model introduced below has been derived after thorough consideration of the evidence, and is believed to satisfy all of the observational data.

DESCRIPTION OF STAGES OF THE CRATERING PHENOMENON WHICH COULD HAVE PRODUCED THE RIES MULTIRING BASIN

T_0. Pre-impact geologic setting of the Ries based on reconstruction of the stratigraphic section (see Chao, 1977a). This is a north (N)–south (S) section. Gently dipping Malm limestone (specific gravity (s.g.) about 2.4) (M), thickness about 100 m near the center to about 200 m near the southern rim. Below this shales (s.g. about 2.2) with some sandstones of Dogger and Lias (D), and shales

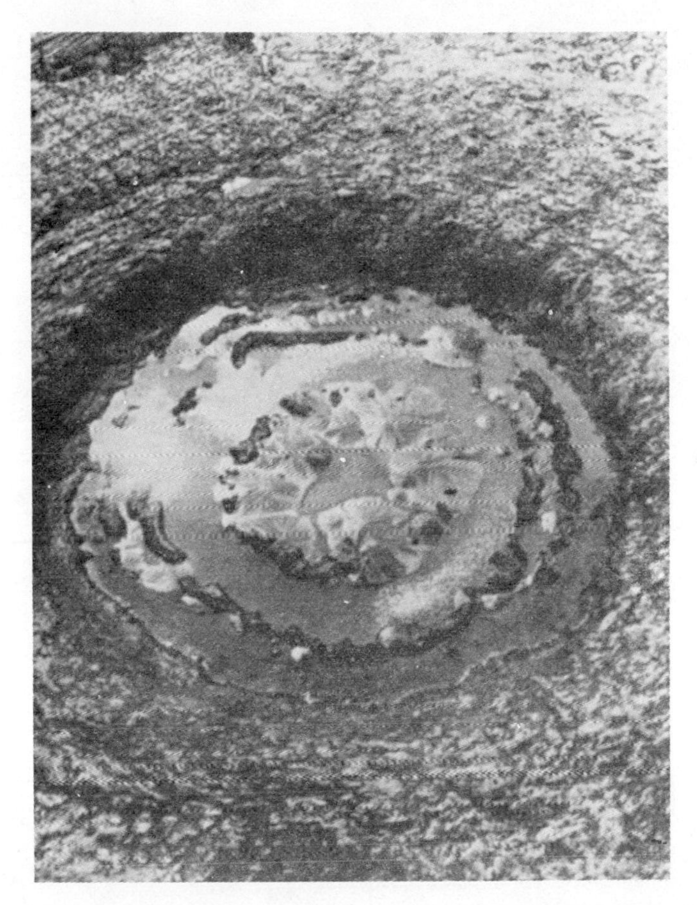

Fig. 4. Photograph of the Prairie Flat multiring crater structure produced by the detonation of a 500-ton TNT sphere lying tangent on the ground surface. The diameter of the outer rim is about 61 m (Roddy, 1976).

(with some sandstones) of the Triassic Keuper (T) with a total thickness of about 450 m. The sedimentary sequence about 600 m thick was unconformably (U) underlain by basement crystalline rocks (s.g. about 2.6) with foliation trending NE–SW and dipping SE. The water table was probably located at the depth of the Albtrauf (AT). The impacting body was probably a stony meteorite (density about 3.5 g/cm^3) estimated to be about 3 km in diameter, impacting at an assumed velocity of 15 km/sec. T_1. Stage of initial penetration. With an impact velocity of 15 km/sec, the stony meteorite (m) penetrated down into the crystalline basement rocks, a distance of about 650 m, in about 100 msec. Due to the lag time for material response, the amount of material vaporized and fused (short dashes) in this short time span is probably negligible. This should account for the negligible amounts of fused limestone and other Mesozoic sediments in the Ries. As the result of penetration the lower sedimentary beds were compressed downward and the upper sedimentary beds were ruptured and upturned

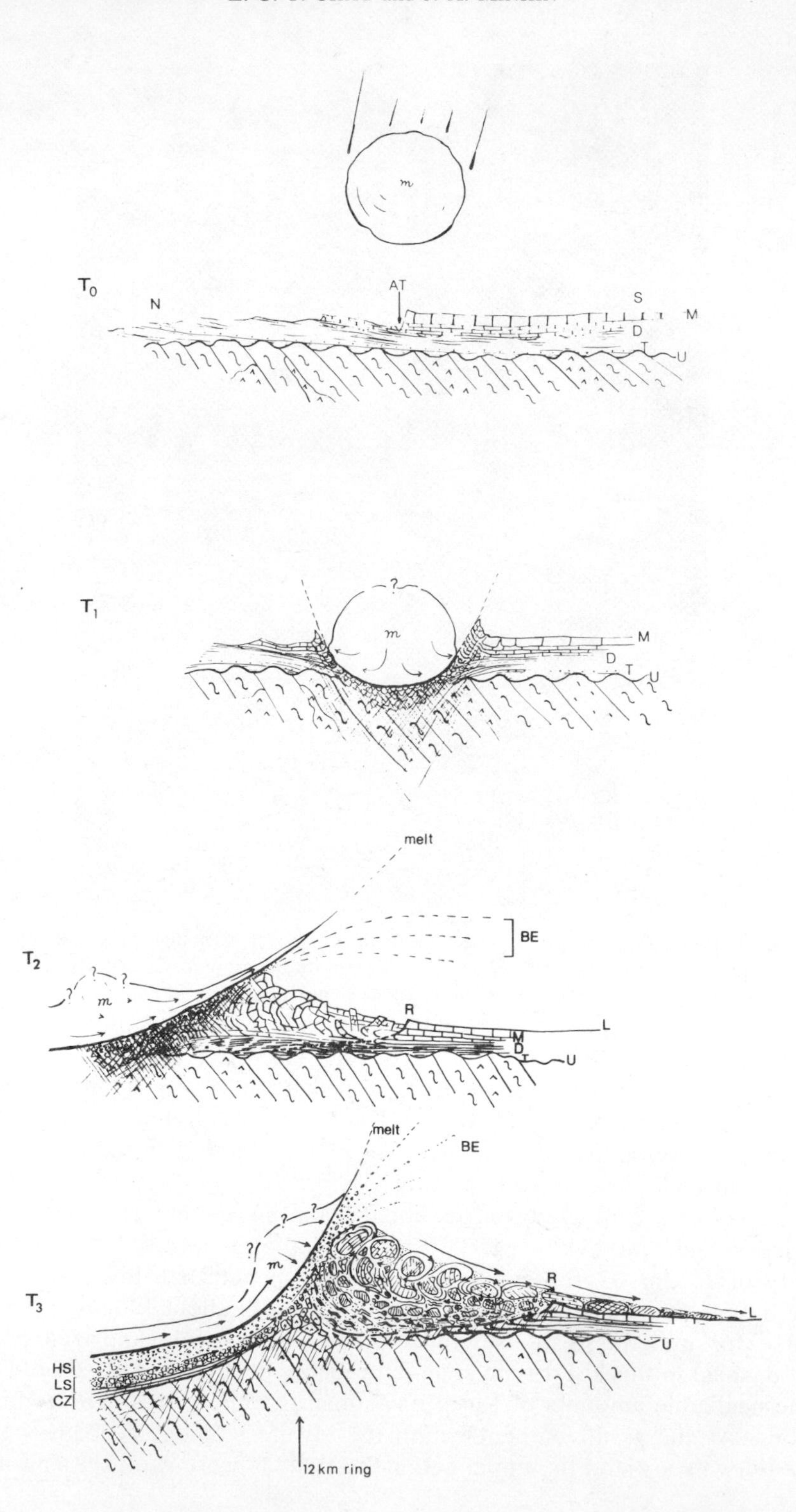

m
T0
AT
N
S
M
D
T
U
T1
?
m
M
D
T
U
melt
BE
T2
?
m
R
L
M
D
T
U
melt
BE
?
m
T3
R
L
U
HS
LS
CZ
12 km ring

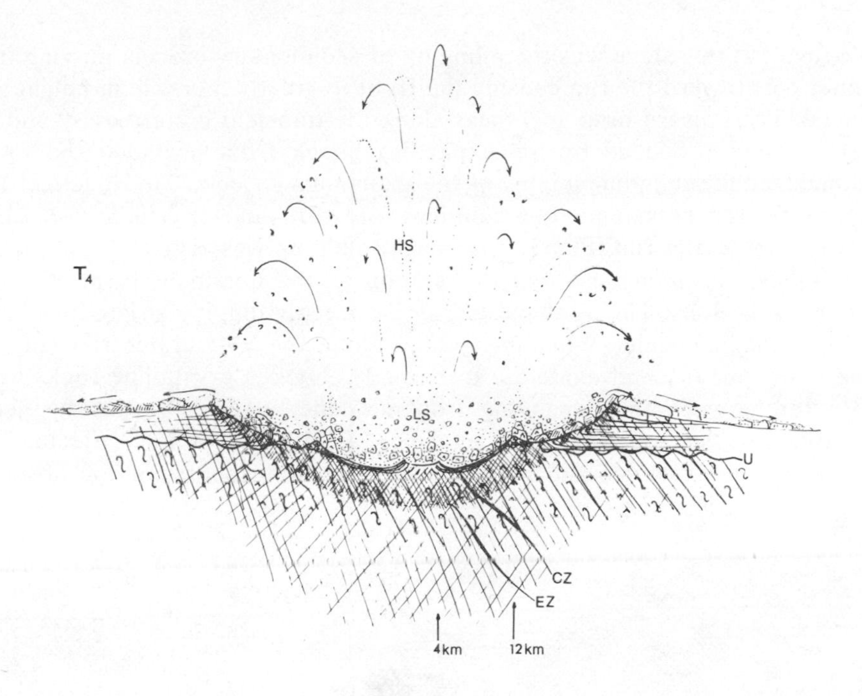

and displaced upward and outward. The depressed lower sedimentary beds near the center were crushed and brecciated. A very small fraction of these, not much more than a percent, was incorporated in suevite. Another very small fraction was incorporated in fine breccias and was injected into fractures below the floor of the crater to form injection dikes and dikelets during the cratering process. Most was later expelled with the crushed crystalline rocks as shown in T_2 and T_3. Peak pressure and temperature were reached in the basement rocks, producing suevite as the most intensely shocked and partly fused ballistic ejecta. (The back surface shown for the impacting body for this stage and for T_2 and T_3 is highly speculative and is assumed to be related to the differential non-uniform flow and deformation of a large stony meteorite upon impact.) T_2. Details of the southern half of the crater cross section developed during this stage. Lateral enlargement with onset of lateral flow of the impacting body. Lateral displacement of the sedimentary beds and their rupture and upturn increased as the result of drag and shear forces produced by the flow of the impacting body. A small amount of suevite, crystalline rocks and bunte breccia was ejected ballistically (BE). Mixing between the shale and limestone beds barely began as the sedimentary beds were displaced from the central area outward piling up on previously ruptured and displaced beds. L—land surface; R—transient crater rim; m, M, D, T, and U same as above. Long dashed line delineates the crater floor. T_3. Continued development of the southern half of the crater cross section. Crater development after several seconds had elapsed. The crater was slightly deeper, causing steepening of the laterally advancing front lined with partial melt and fragments. The major development at this stage was the piling up of sedimentary breccia moving from the inner part toward the rim causing the front to greatly increase in height as it advanced. The upward drag and shear along the transient crater cavity and the lateral pressure produced by the expanding shock front provided the forces which induced the roll-glide mixing of the sedimentary ejecta. This mechanism is indispensible for explaining the confining pressure and the indicated 3 kbar pressure at the crater rim (Fig. 3). The height build-up was critical for producing the *roll-glide, cascading, nonballistic* motion as the dominant form of ejecta transport in the Ries. This is suggested by the steeply dipping striated wall rock surface at the Schneider–Wemding quarry along the east crater rim (Fig. 5). Furthermore, suevite and moderate to strongly shocked crystalline rocks acted as a buffer between the expanding front and the mixed sedimentary ejecta, accounting for the lack of shock evidence among sedimentary ejecta. This upturned mixing sequence and nonballistic advance accounts for the observed

Fig. 5. Photograph of a steeply dipping slab of Malm delta thick-bedded limestone showing well developed schliffflâche (arrow). The strike of this surface is NS, dipping about 60° away from the crater or eastward. The relative movement as based upon the striae is downward so that ejecta moved outward and downward on this surface. On the left are more gently dipping Malm beta and gamma limestones overlying the striated Malm delta limestone and separated by a few inches of sedimentary breccia (bunte breccia) with Triassic Keuper purple shale. There is no evidence of faulting. Hence the steeply inclined surface may result from the steep advancing front shown in the illustration for T_3. Schneider–Wemding quarry, 1 km south of Wemding.

ejecta distribution, where the uppermost limestone beds in general travelled the farthest. It also allows the small amounts of crystalline and sedimentary breccia and the suevite ejected ballistically to overlie the essentially nonballistically transported sedimentary ejecta (Bunte Trümmermassen) close to the crater rim. Hence the phenomenon shown schematically in T_3 satisfies the constraints imposed by geologic and petrologic observations for the Ries. Arrows illustrate the shear and drag effects and direction of movement. The large vertical arrow points to the location of the 12 km diameter ring. It was produced by the combined shear and drag forces of the laterally advancing meteorite and the upward drag induced in the basement as the overlying materials were turned upward and outward. HS—high temperature suevite; LS—low temperature suevite; CZ—compressed zone. Other notations same as above. T_4. Final stage of the cratering process after most of the sedimentary ejecta had been expelled nonballistically outside of the crater, and the impacting body was completely dissipated by vaporization, melting, and ejection. At this time much of the suevite was produced, with high temperature suevite (HS) ejected to great heights and low temperature suevite (LS) mixed with large blocks of weakly shocked crystalline rocks slightly lifted (only as much as several tens of meters) before falling back into the crater. There is no separate melt layer. A compressed zone (CZ) full of shatter cones underlay the fallback low temperature suevite. Shock pressures, probably less than 20 kbar at the top of the compressed zone (see Chao and El Goresy, 1977), attenuated gradually downward. Dilation-injection dikes produced during stages T_2–T_4 are common under the crater floor. A zone of expansion (EZ) at depths above 1186 m (Chao, 1977a) perhaps lifted the entire block of crystalline rock below the crater floor some 15 m. The inner 4 km ring, uplifted some 150 m, is possibly the result of rebound. The transient crater floor may have extended a few hundred meters below the present crater floor. The extent of fracture shown below the crater floor is to conform generally to the reduced velocity shown by seismic velocity studies (Pohl and Will, 1974) and by the two refraction profiles across the crater (Ernstson and Pohl, 1977). Some fractures may extend beyond 10 km.

DISCUSSION

The early stages T_0–T_2 of the schematic reconstruction of the Ries cratering phenomenon are basically in harmony with the simple bowl-shaped model, except that in T_2 the ballistically transported portion of the ejecta is limited to that part immediately next to the thin melt layer. In both stages T_2 and T_3 emphasis is placed on lateral displacement and the piling up of the disrupted sedimentary rocks of the upper part of the section. The steepening of the expanding front is the result of the slight deepening of penetration and the increasing amount of material to be displaced. Beginning at stage T_2, the process departs drastically from that which produces a bowl-shaped crater with a simple overturned flap by ballistic ejection.

Although I stated above that I favor the Prairie Flat multiring structure as a possible analogue to the Ries, it must be pointed out that the analogy cannot be treated too literally, since the two craters differ in many important details. These include:

1. The Ries is about 400 times as large in diameter as the Prairie Flat Crater. 2. The nature and geologic setting of the target materials and the origin for producing the shock are different. An impacting body is not involved in the Prairie Flat event, hence the manner and the depths of penetration are different. There is no effect of the spreading and flow of the impacting body over the target materials. 3. The Prairie Flat structure has a coherent overturned flap where the inverted stratigraphic units, although thin, are essentially continuous and intact. It has a distinct hinge area located approximately under the position of the crater rim. The Ries structure has no coherent overturned flap and hence no corresponding hinge area. Instead, the nonballistically transported ejecta blanket which directly overlies the area near and beyond the crater rim, consists of intensely fractured, broken and mixed lithologic units that only at best crudely display an inverted stratigraphic relationship (Chao *et al.*, 1977). As shown in T_2 and T_3, such overturning and mixing probably began to take place nearly half a distance of the radius within the morphological crater rim rather than at the crater rim. 4. There is a much more pronounced uplift or rebound of the crater floor in the Prairie Flat structure than in the Ries. 5. At the Ries, evidence of locally high confining pressure in the ejecta blanket and high pressures exerted on the crater wall are significant. Such items of evidence have not been reported for the Prairie Flat event.

In spite of the differences outlined above, which to a large extent may be a matter of scaling, the important similarity is that lateral and radially outward forces appear to have been major factors in producing both shallow multiring structures. In the case of the Ries I postulated that these lateral and radially spreading forces occurred behind the interface between the fragmented and flowed impacting body or shock front (?) and the ejecta-target materials at stage T_3. I have called on these radially spreading lateral forces to produce the roll-glide mixing action of the brecciated, crushed and broken Malm limestone and the Dogger and Keuper shales that were caught between the transitional crater rim, the shallow crater floor and the advancing interface front. As a result of this process shearing and drag effects outward along the advancing interface and inward and upward along the crater floor, as seen in T_3, produced the ring structure with a 12 km diameter. The inner 4 km ring seen in T_4 was attributed to effects of rebound. I suggest that the coherent overturned flap for the small Prairie Flat event is possibly representative of a transitional phase to the overturned ejecta blanket of the Ries produced by the nonballistic roll-glide cascading mode with much more mixing. Perhaps in principle, a similar mechanism is operating in the two cases.

I have also assumed that a large, low density impacting body, such as a stony meteorite, rather than a small, high density iron body, with the same impacting velocity, will favor the production of a large shallow crater. A large, shallow structure such as the Imbrium Basin on the moon with a diameter of 1300 km could result from the impact of a stony body with a size close to about 100 km in diameter. On the basis of the reconstruction above, hydrodynamic flow of both the large impacting body and the target rock should be considered jointly. It is possible that the interaction of fluid flow between the impacting body and the target rocks could have produced expanding ring waves. Each wave could have exerted unequal pressures over the excavated target area to produce the multiring structure. In this concept, the energy coupling between the impacting body and the target rocks will be much more effective and efficient than the detonation of high energy explosives as in the Prairie Flat event. A very shallow basin coupled with dominant lateral propagation or flow seems most effective in producing nonballistically transported ejecta.

Both experimental and theoretical cratering investigations have pointed out the importance of material characteristics, e.g., whether the target material is saturated with water or is layered (Quaide and Oberbeck, 1968; Oberbeck *et al.*, 1975a; Roddy, 1976; Piekutowski, this issue). The Ries geologic setting indicates that a predominantly shale layer saturated or partially saturated with water occurs between brittle limestone above and denser, tougher crystalline rocks below. The impacting body definitely penetrated into the basement crystalline rocks. The 12 km ring may mark the boundary of the unconformity, perhaps reflecting the sharp change in material properties between the wet shale formation and the basement crystalline rocks. The shale formation is very important in providing a medium for producing the confining pressure within the mixed ejecta mass. Furthermore it is an effective lubricant for nonballistic transport of ejecta to distances of a crater diameter or more from the crater rim. In the cases of impact into the saturated shale formation of the Ries, areas of permafrost on Mars, or any water-saturated formation, ejecta should more readily flow nonballistically and fluid-dynamically under pressure. During the initial penetration phase, there is no significant layering effect between the brittle Malm limestone and the underlying wet shale formation. I believe that the above proposed schematic reconstruction would have produced a multiring structure even without a layered target, that the shallow penetration is most importantly the effect of the low density of the impacting body pointed out above.

In the schematic reconstruction, the effect of air shock was omitted because it has very little to do with the transport of the ejecta. The insignificant role of base surge (principally due to gaseous expansion where only fine particulate materials are transported) in the transport of ejecta mass was pointed out some time ago (Chao, 1973; Oberbeck, 1975). The huge mass of sedimentary ejecta from the Ries is unshocked, therefore there was no production of steam or water vapor by shock-induced heating. Hence "fluidization", as occurs in base-surge transport, is also not a realistic mode of ejecta transport involved in cratering mechanics.

In summary, two key observational facts have dominated the above reconstruction of the Ries cratering phenomenon. These are (1) the shallow configuration of the geometry of the crater with a depth–diameter ratio of 1:33, and (2) the predominance of ejecta transported nonballistically under locally high confining pressures. The schematic reconstruction shows that critical to the production of a shallow crater is shallow impact penetration (shallow depth of burst). This and the nonballistic ejection of excavated materials appear to be genetically related, i.e., if extensive nonballistic transport of ejecta is recognized then the associated crater must be a shallow structure and vice versa. This also means that the shallow configuration of a crater may not have anything to do with post cratering readjustment.

IMPLICATIONS

The schematic reconstruction presented in this paper is intended to illustrate the cratering phenomenon responsible for producing the Ries multiring structure. The dominance of nonballistic ejecta transport in the Ries goes hand in hand with the phenomenon of producing a shallow basin type crater. By analogy, large multiring basins observed on planetary surfaces, e.g., the Mare Imbrium Basin and the Orientale Basin on the moon, the Caloris and other basins on Mercury, and the numerous Mars craters that show ejecta blankets with radial flow structures are all basically shallow basins which may be interpreted as produced by impacts of large stony bodies of relatively low density. It is most interesting to note that possibly because of the abundance of water or ice on Mars as shown by the Viking results, nonballistically transported ejecta masses on Mars appear prevalent (Carr, oral communication).

Future investigations may further demonstrate the importance of a cratering phenomenon drastically different from that derived from the simple bowl-shaped model.

Appendix

Energy and mass considerations of the proposed reconstruction of the Ries cratering phenomenon (by Jean A. Minkin).

The validity of the model proposed above must ultimately be tested by theoretical and experimental cratering investigators to determine whether the phenomenon as described is consistent with the principles of cratering mechanics. Here we present a preliminary quantitative test of the reasonableness of this model by examining some of the energy-mass relationships implied by the parameters observed and deduced.

1. *Characteristics of the impacting body*

Several equations are available for calculating the kinetic energy (K.E.) of the impacting body on the basis of the diameter of the crater it produced. For a terrestrial crater the size of the Ries, Baldwin (1963) derived the relationship:

$$D_{km} = 0.3284E_e - 7.9240 \quad (1)$$

where D_{km} = log of the diameter in km, and E_e = log of the K.E. in ergs.

For the Ries with a diameter of 25 km, K.E. is equal to approximately 2.5×10^{28} ergs. Oberbeck *et al.* (1975b) used the equation

$$D_r = kE^{1/3.4}, \quad (2)$$

where D_r = the rim diameter in km, E = K.E. in ergs, and $k = 1.617 \times 10^{-7}$ for the earth.

For the Ries we get from this calculation a value for the K.E. of 7.0×10^{27} ergs. Still another estimate, based on the determined crater volume of approximately 200 km^3 and assuming a cratering efficiency at the Ries similar to the rock craters at the Nevada Test Site of 1.5×10^{-24} km^3/erg (R. Port, written communication), yields a value for the K.E. of 1.3×10^{26} ergs.

Averaging these three determinations gives a value for the K.E. of the impacting body of the order of 10^{27} ergs.

An impacting body with this K.E., a density of 3.5 g/cm^3 and an impacting velocity of 15 km/sec has a diameter of about 0.8 km if it is a sphere. If it has the shape of an oblate spheroid with major axis four times as great as minor axis, its major axis is about 1 km and its minor axis 0.25 km.

The question may be raised as to whether the parameters used above to determine the K.E. are valid when applied to a multiring structure such as the Ries. We may get some answer to this question by examining what changes in the dimensions of the impacting body would be dictated by variations in the K.E. For each order of magnitude by which the K.E. changes, the dimensions of the impacting body will change by a factor of approximately 2. Thus, if the K.E. of the impacting body is of the order of 10^{28} ergs the diameter of the impacting sphere is about 1.7 km and for the spheroid the major axis is 2.0 km and the minor axis 0.5 km. A K.E. of 10^{26} ergs corresponds to a sphere less than 0.4 km in diameter or a spheroid with major axis about 0.4 km and minor axis 0.10 km, dimensions which appear to be rather small to be compatible with the 4 km diameter inner ring and the presence 3.5 km away from the center of the crater of the Fe–Cr–Ni veinlets believed to be relicts of the vapor-condensed impacting body. (Such a judgement is of course subject to further knowledge and understanding of the nature and rate of flow of the target and the projectile during the impact process.) In light of the present indications it seems reasonable to assume that the K.E. of the impacting body of the Ries was greater than 10^{27} ergs, and probably of the order of 10^{28} ergs. Thus Eqs. (1) and (2) cited above seem at least approximately valid for the Ries Crater and more suitable than the approximation based on the value of the cratering efficiency measured at the Nevada Test Site.

2. *Energy–mass relationships for the material ejected from the crater*

Gault and Heitowit (1963) determined on the basis of small cratering experiments that the energy required for comminution of the target is $5.5–6.5 \times 10^7$ ergs per gram of ejecta. They further estimated that the energy expended in this fracturing and crushing of the target is equal to 10–24% of the initial K.E. of the impacting body. For the Ries, if we accept 10^{28} ergs as the K.E. of impact, the energy expended in comminution would have been of the order of 10^{27} ergs, corresponding to a predicted mass of ejecta $\geqslant 1.5 \times 10^{19}$ g. If we use 200 km^3 as an approximate value for the volume of the ejecta, and 2.3 g/cm^3 as the average density of the ejecta (Ernstson and Pohl, 1974), we see that the mass of the ejecta from the Ries is more probably $\approx 5 \times 10^{17}$ g. Conversely, if the energy expended in comminution at the Ries was 10^{27} ergs and the mass of ejecta is 5×10^{17} g, then the calculated energy required for comminution of the target would be of the order of 10^9 ergs/g of ejecta. Thus there is a discrepancy of 1 to 2 orders of magnitude between the energy–mass relationship predicted by Gault and Heitowit and the values determined for the Ries.

This corroborates the earlier conclusion by Chao (1974) that estimates based on results of small-scale cratering experiments cannot be extrapolated successfully to structures of the magnitude and complexity of the Ries. Relatively more energy must be involved in target comminution in large, shallow craters than in the simpler, smaller bowl-shaped types. This also supports Innes' finding (1961) that a smaller proportion of the fragmented material produced by the impact event is ejected as the craters become larger.

Attempts at calculation of the energy partitioning and expenditure in the transport of the ejecta must await further analysis and understanding of the complex thrust forces and turbulent motion Chao envisions in his schematic mechanism of nonballistic ejecta transport.

Acknowledgments—We wish to thank O. B. James, D. J. Roddy, R. Port, and G. H. S. Jones for their helpful reviews and suggestions. Gratitude is expressed to G. H. S. Jones and the Defence Research Establishment, Suffield, Alberta, Canada, for permission to use the photograph of the Prairie Flat Crater in Fig. 4. We are also grateful for support of this work by NASA Contract W-13, 130.

REFERENCES

Baldwin, R. B.: 1963, The Measure of the Moon. Univ. of Chicago Press, Chicago.

Bannert, D.: 1969, Luftbildkartierung des Lineationsnetzes von. Ries und seiner Umgebung. *Geol. Bavarica* **61**, 379–384.

Chao, E. C. T.: 1973, Geologic implications of the Apollo 14 Fra Mauro breccias and comparison with ejecta from the Ries crater, Germany. *J. Res.* **1**, 1–18.

Chao, E. C. T.: 1974, Impact cratering models and their application to lunar studies—a geologist's view. *Proc. Lunar Sci. Conf. 5th*, 35–52.

Chao, E. C. T.: 1976, Mineral-produced high pressure striae and clay polish: Key evidence for nonballistic transport of ejecta from the Ries crater, southern Germany. *Science* **194**, 615–618.

Chao, E. C. T.: 1977a, The Ries crater of southern Germany—a model for large basins on planetary surfaces. *Geol. Jahrb.* In press.

Chao, E. C. T.: 1977b, Preliminary interpretation of the 1973 Ries research deep drill core and a new Ries cratering model. *Geol. Bavarica* **75**. In press.

Chao, E. C. T. and El Goresy, A.: 1977, Shock attenuation and the implantation of Fe-Cr-Ni veinlets in the compressed zone of the 1973 Ries research deep drill core. *Geol. Bavarica* **75**. In press.

Chao, E. C. T., Hüttner, R., and Schmidt-Kaler, H.: 1977, Vertical section of Ries sedimentary ejecta blanket as revealed by 1976 drill cores from Otting and Itzing. In *Lunar Science VIII*, The Lunar Science Institute, Houston, 163–165.

Dence, M. R., Grieve, R. A. F., and Robertson, P. B.: 1977, Terrestrial impact structures: Principal characteristics and energy considerations. In *Impact and Explosion Cratering* (D. J. Roddy, R. O. Pepin, and R. B. Merrill, eds.). This volume.

El Goresy, A. and Chao, E. C. T.: 1976, Evidence of the impacting body of the Ries crater—the discovery of Fe-Cr-Ni veinlets below the crater bottom. *Earth Planet. Sci. Lett.* **31**, 330–340.

El Goresy, A. and Chao, E. C. T.: 1977, The origin and significance of the discovery of the Fe-Cr-Ni veinlets in the compressed zone of the 1973 research drill core. *Geol. Bavarica* **75**. In press.

Ernstson, K. and Pohl, J.: 1974, Einige Kommentare zu den Bohrlochgeophysikalischen Messungen in der Forschungsbohrung Nördlingen 1973. *Geol. Bavarica* **72**, 81–90.

Ernstson, K. and Pohl, J.: 1977, Neue Modelle zur Verteilung, der Dichte und Geschwindigkeit im Ries-Krater. *Geol. Bavarica* **75**. In press.

Gall, H., Müller, D., and Stöffler, D.: 1975, Verteilung, Eigenschaften und Entstehung der Auswurfsmassen des Impaktkraters Nördlinger Ries. *Geol. Rundschau*, **64**, 915–947.

Gault, D.E. and Heitowit, E. D.: 1963, The partition of energy for hypervelocity impact craters formed in rock. *Proc. Sixth Symposium on Hypervelocity Impact* **2**, 419. Natl. Tech. Info. Service, Springfield, Va, U.S.A., Document no. AD 423064.

Innes, M. J. S.: 1961, The use of gravity methods to study the underground structure and impact energy of meteorite craters. *J. Geophys. Res.* **66**, 2225–2239.

Jones, G. H. S., Diehl, C. H. W., Pinnell, J. H., and Briosi, G. K.: 1970, Crater and ejecta study. Defence Atomic Support Agency POR-2115 (WT-2115), unclassified.

Jones, G. H. S.: 1976, The morphology of central uplift craters. Defence Research Establishment Suffield. Canada. Suffield Report 281.

Oberbeck, V. R.: 1975, The role of ballistic erosion and sedimentation in lunar stratigraphy. *Rev. Geophys. Space Phys.* **13**, 337–362.

Oberbeck, V. R., Quaide, W. L., and Arvidson, R. E.: 1975a, Secondary cratering on Mercury, the Moon and Mars. Oral presentation at the International Colloquium of Planetary Geology, Rome, Italy, September, 1975.

Oberbeck, V. R., Hörz, F., Morrison, R. H., Quaide, W. L., and Gault, D. E.: 1975b, On the origin of the lunar smooth-plains. *The Moon* **12**, 19–54.

Piekutowski, A. J.: 1977, Cratering mechanisms observed in laboratory-scale high explosive experiments. In *Impact and Explosion Cratering* (D. J. Roddy, R. O. Pepin, and R. B. Merrill, eds.). This volume.

Pohl, J. and Angenheister, G.: 1969, Anomalien des Erdmagnetfeldes und Magnetisierung der Gesteine im Nördlinger Ries. *Geol. Bavarica* **61**, 327–336.

Pohl, J. and Will, M.: 1974, Vergleich der Geschwindigkeitsmessungen im Bohrloch der Forschungbohrung Nördlingen 1973 mit seismischen Tiefensondierungen innerhalb und ausserhalb des Ries. *Geol. Bavarica* **72**, 75–80.

Pohl, J., Stöffler, D., Gall, H., and Ernstson, K.: 1977, The Ries impact crater. In *Impact and Explosion Cratering* (D. J. Roddy, R. O. Pepin, and R. B. Merrill, eds.). This volume.

Quaide, W. L. and Oberbeck, V. R.: 1968, Thickness determinations of the Lunar surface layer from Lunar impact craters. *J. Geophys. Res.* **73**, 5247–5270.

Reich, H. and Horrix, W.: 1955, Geophysikalische Untersuchungen im Ries und Vorries und deren geologische Deutung. *Beih. Geol. Jb.* **19**.

Robertson, E. C.: 1955, Experimental study of the strength of rocks. *Geol. Soc. Amer. Bull.* **66**, 1275–1314.

Robertson, E. C.: 1972, Strength of metamorphosed graywacke and other rocks. In *The Nature of the Solid Earth*, McGraw-Hill, New York.

Roddy, D. J.: 1969, Geologic survey activities. In *Operation Prairie Flat preliminary report* (M. J. Dudash, ed.), vol. 1, Defence Atomic Support Agency DASA 2228-1, unclassified, 317–333.

Roddy, D. J.: 1970, LN 303, Geologic studies. In *Operation Prairie Flat Symposium Report* (M. J.

Dudash, ed.), vol. 1, pt. 1, Defence Atomic Support Agency, DASA 2377-1, DASIAC SR-92, unclassified, 210–215.

Roddy, D. J.: 1973, Geologic studies of the middle Gust and Mixed Company Craters. *Proceedings of the Mixed Company/Middle Gust Results Meeting, 13–15 March* 1973. vol. 11 Defense Nuclear Agency, DNA 3151P2, unclassified, 79–124.

Roddy, D. J.: 1976, High-explosive cratering analogs for bowl-shaped, central uplift, and multiring impact craters. *Proc. Lunar Sci. Conf. 7th*, p. 3027–3056.

Roddy, D. J., Pepin, R. O., and Merrill, R. B., editors.
(1977) *Impact and Explosion Cratering*, Pergamon Press (New York), p. 425–448.
Printed in the United States of America

Shallow drilling in the "Bunte Breccia" impact deposits, Ries Crater, Germany

FRIEDRICH HÖRZ

Geology Branch, NASA Johnson Space Center, Houston, Texas 77058

HORST GALL

Bayerische Staatssammlung, Richard Wagnerstr. 10, D8 München 2, Germany

RUDOLF HÜTTNER

Geologisches Landesamt Baden-Württemberg, Albrechtstr. 5, D75 Freiburg, Germany

V. R. OBERBECK

NASA Ames Research Center, Moffett Field, California 94035

Abstract—This is a field report concerning a shallow core drilling program in the "Bunte* Breccia" deposits, which constitute ≈90% of all impact breccias beyond the outer rim of the Ries (Germany), an ≈26 km diameter impact crater. A total of 11 locations having radial ranges between 16.5 and 35 km from the crater center were drilled and ≈480 m of core were recovered.

The cores consist of breccias whose components are derived from both the terrain outside the crater ("local") and the crater cavity itself ("crater"). The local components completely dominate the breccias at the larger ranges and possibly constitute >90% of the breccia volume at the greatest distances examined. Clast sizes, frequency, and lithologies as well as matrix character are extremely variable. Breccia matrices vary between, as well as within, specific localities indicating that thorough mixing is omnipresent but variable in intensity. There are various stages of mixing at different locations and/or times prior to final emplacement because "breccias within breccias" are common. The overall texture of the matrix is "massive"; the clast orientation is highly irregular and many twisted and swirly deformation structures are observed, indicating that the mixing and emplacement process was predominantly turbulent. The Bunte Breccia is surprisingly deep (e.g., 84 m at 27 km range). This great depth of Bunte Breccia, together with the preponderance of local components necessitate an emplacement mechanism that ploughed up and mixed the crater surroundings to depths greater than 50 m.

These new studies as well as previous observations are strong evidence for a ballistic emplacement mechanism of the Bunte Breccia deposits at distances greater than 5 km from the Ries rim.

INTRODUCTION

"EJECTA" BLANKETS associated with impact craters of all sizes attest to the importance of the cratering process in redistributing planetary materials. Depending on size and depth of any given crater, such impact deposits may

*Bunt = multicolored.

originate from the uppermost surface or from crustal depths measured in kilometers, if not tens of kilometers. Detailed understanding of these deposits which dominate cratered, planetary surfaces may only come forward, after quantitative assessment of the nature and emplacement mechanism(s) of the ejecta. Of particular interest are relatively large craters because their deposits may contain deep seated, possibly crustal materials.

Many current and fundamental problems in lunar surface evolution depend on a better understanding of large scale ejecta distribution: the source area(s) of the Apollo 14 and 16 rocks (Oberbeck *et al.*, 1973, 1974, 1975; Head, 1974; Chao *et al.*, 1973; Moore *et al.*, 1974; Morrison and Oberbeck, 1975), the assessment of the enrichment of siderophile elements indicative of meteoritic contamination (Ganapathy *et al.*, 1973), and the absolute formation ages of major basins (Schaeffer and Husain, 1974) are some examples requiring an understanding of the mechanics of ejecta emplacement.

Knowledge concerning the ejecta of relatively small laboratory experiments is fairly detailed (e.g., Gault *et al.*, 1968; Oberbeck, 1971, 1975; Stöffler *et al.*, 1975) and is also available for Meteor Crater, Arizona (Shoemaker, 1963; Roddy *et al.*, 1975) and explosive or nuclear craters (e.g., Andrews, 1976; Wisotski, 1976; Carlson and Roberts, 1963). Our understanding of large scale crater ejecta, however, is based on an extrapolation of these relatively small scale events and detailed photogeologic studies.

Present hypotheses based on such extrapolations and especially photogeologic studies which focus on the interpretation of a variety of geomorphic land forms are, however, controversial. A ballistic ejection mechanism leading to significant incorporation of local materials and final emplacement of the deposits via a debris surge of the combined crater—and locally derived materials has been proposed by Oberbeck *et al.* (1973, 1974, 1975), Oberbeck (1975), Morrison and Oberbeck (1975), and others. In contrast, Chao *et al.* (1974), Chao (1976a,b), and Moore *et al.* (1974) propose one massive, groundhugging, flow regime emanating from the crater; they consider ballistic transport and associated secondary cratering of minor importance for continuous deposits. Implications of these hypotheses are best illustrated in interpreting the potential source area(s) of the Apollo 14 and 16 samples. While Oberbeck *et al.* (1975), Morrison and Oberbeck (1975), Head (1974) and Head and Hawke (1975) postulate a largely local derivation of these samples, Moore *et al.* (1974) and Chao *et al.* (1974) suggest derivation from the Mare Imbrium if not Mare Orientale basins. Chao (1974) specifically challenged the prediction of the ballistic hypothesis that significant amounts of local materials are incorporated into continuous deposits.

The Ries Crater (Germany) is the largest terrestrial crater with significant parts of the breccia deposits beyond the outer crater rim still preserved (see Pohl *et al.*, 1977). Thus the Ries offers the opportunity to examine directly the continuous deposits and thereby to evaluate the emplacement mechanism(s) of large scale crater deposits. A shallow drilling program was therefore pursued in August–October, 1976. The following is a brief description of some field observations.

THE RIES CRATER

Pohl *et al.* (1977) present a detailed summary of the entire structure including target stratigraphy, lithologies, target morphology prior to impact and the various, principal breccia deposits. The following report addresses exclusively the so called "Bunte Breccia", i.e., those deposits derived predominantly from the upper stratigraphic units of the Ries target, mainly comprising sedimentary rocks (see Pohl *et al.*, 1977). Because these deposits constitute approximately 90% of all breccias beyond the outermost crater rim they are considered a valuable analogue for continuous deposits observed around large scale, planetary impact craters (Fig. 1).

The Ries not only offers the opportunity to study such deposits in general, but it also is ideally suited—owing to a fortunate geological circumstance—to discriminate between ballistic and nonballistic ejecta deposition modes: A cliff line striking approximately east-west (Gall, 1974a) occurs to the south of the crater rim as illustrated in Fig. 1. This cliff line marks the northernmost extent of widespread shallow marine sediments (middle Miocene; "Obere Meeres-

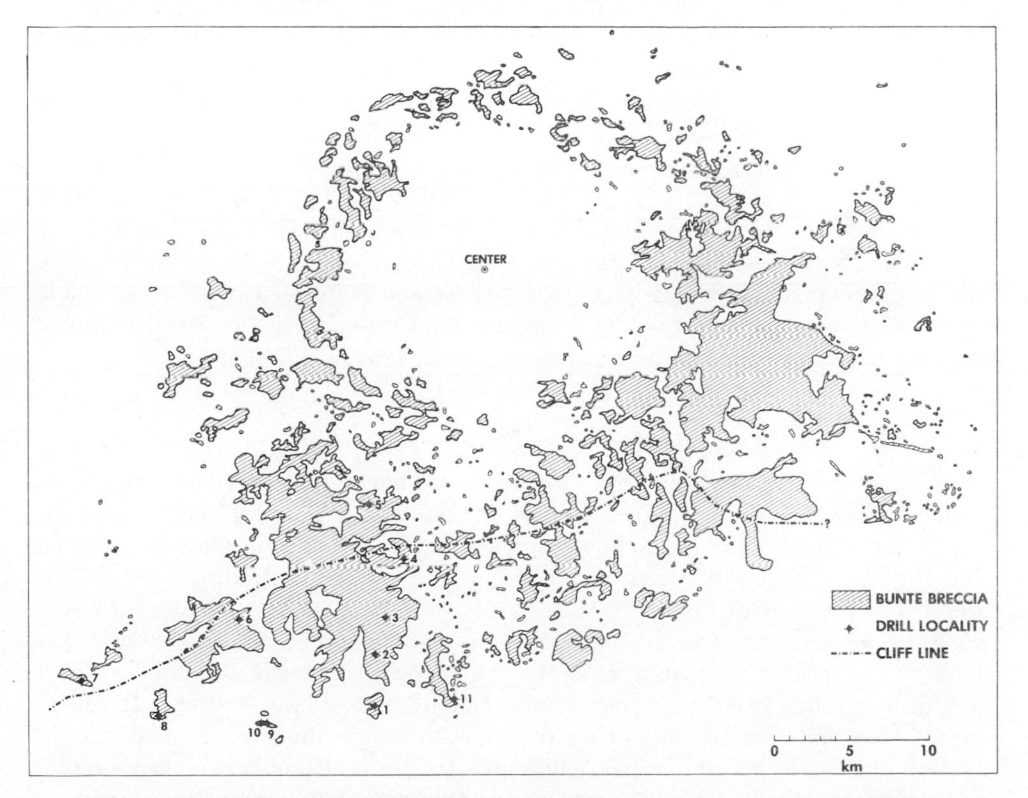

Fig. 1. Distribution of "Bunte Breccia" around the Ries Crater, Germany. Drilling locations are indicated as well as position of Miocene cliff line marking the northernmost invasion of marine OMM sediments.

molasse" (=OMM); generally medium to fine grained sands). In addition, fresh-water deposits on top of OMM occur also *predominantly* south of this cliff line, though specific facies did extend farther to the north, i.e., into the area of the crater cavity as thin (<50 m), probably discontinuous, eroded deposits (upper Miocene; "Obere Süßwassermolasse" (OSM), fine grained sands, marls, fresh-water limes, and especially clays). These conditions, i.e., cliff line and distribution of OMM and OSM deposits existed prior to the Ries event (Bolten and Müller, 1969; see also Fig. 2.2 of Pohl *et al.*, 1977). Accordingly, *most* of the OSM and *all* of the OMM sediments long known to make up a substantial part of the Ries' continuous deposits, can be taken as diagnostic indicators of "local" materials from the terrain outside the crater cavity and yet incorporated into the breccias (Ammon, 1905; Hüttner, 1958, 1969; Gall, 1969; Schneider, 1971; Gall *et al.*, 1975, and many others). It is possible therefore to distinguish unambiguously breccia components of "local" origin versus those that were genuinely derived from the crater cavity (Jurrassic limestones, marls, clays, and sandstones; triassic clays, shales and sandstones; and a few crystalline components). The stratigraphic-lithologic control of crater derived materials on one hand and a different, equally well documented set of lithologies of the surrounding target surface on the other hand (see also Pohl *et al.*, 1977), enable therefore detailed assessment of the source area of the breccia components.

Little quantitative data exist on the absolute amount of local materials in the "Bunte Breccia" deposits. Their vertical distribution within a given profile is unknown; hardly any data exist concerning their abundance as a function of radial range (Schneider, 1971; Hüttner, 1969; Gall, 1969; Gall *et al.*, 1975). Indeed, accurate thickness determinations for the entire deposit are scarce and confined to a few drill holes (e.g., Hüttner, 1969; Birzer, 1969) and geophysical measurements (Reich and Horrix, 1955; Bader and Schmidt-Kaler, 1977); thickness estimates were traditionally based on field criteria, i.e., reconstruction of detailed target morphology (e.g., Hüttner, 1958, 1969; Gall, 1969).

SELECTION OF DRILL LOCALITIES

Though breccia deposits with "local" components are described from many localities, the drilling program focussed on the "south-Vorries", because of the above mentioned cliff line and its consequences concerning OMM sediments and because detailed, modern geologic maps are available (Hüttner, 1958; Gall, 1969). The breccias of the south-Vorries and their field relations are best known.

A radial traverse was accomplished by drill localities 1–5 (see Fig. 1 and Table 1) to obtain information about the amount of local components as a function of range from the crater center. Locations 1, 2, and 3 were selected on top of local hills for two major reasons: (a) to obtain the least eroded vertical section and, (b) to determine to what degree the Bunte Breccia is merely draping a preexisting relief as a thin veneer or whether the deposits themselves are controlling the present-day relief. Clarification of this question is paramount in volumetric estimates of the Bunte Breccia deposits and also bears on the emplacement mechanism. Drill holes number 4 and 5 are placed in areas where suevites are known to overly the "Bunte Breccia" (see e.g., Pohl *et al.*, 1977) and thus yielded demonstrably uneroded profiles of Bunte Breccia. With the exception of core number 5, all cores were obtained from south of the cliff line, for most reliable identification of local components; number 5 was just north of the cliff line and serves as a tie point for potential changes of breccia character south of the cliff line.

Locations 7–11 were chosen to assess areal variations of Bunte Breccia between ≈2 and 3 crater radii range. They explore Bunte Breccia essentially at the greatest radial range for any given azimuth. They are all located on top of modest local topographic highs, the typical mode of occurrence for Bunte Breccia at such distances.

Locality number 6, also on a local hill, served two purposes: first it is a classic locality for the discussion of ejecta thickness and its relation to preexisting relief (e.g., Hüttner, 1958) and the outcrop "Guldesmühle" in its vicinity has received recent attention for the interpretation of the emplacement mechanism of Bunte Breccia (e.g., Hüttner, 1969; Schneider, 1971; Hörz *et al.*, 1974; Chao, 1974). Second, core number 6 may be considered part of a (poorly documented) radial subtraverse in conjunction with localities 5, 7, and 8.

Drill cores were obtained using a truck-mounted drilling rig; core diameter was 101 mm; core recovery in Bunte Breccia was ≈95%; some drilling through large megaclasts (>5 m) was destructive because of rotary drilling for economic reasons. A total of 560 m was drilled and 480 m of core was recovered.

Table 1 sumarizes some first order geographic, topographic, and geologic data. The observations we report here were all obtained in the field during initial core description, aided by a hand lens. The present report contains only major findings, which we feel confident will remain valid also after laboratory analyses because they address large scale features only.

The Depth of Bunte Breccia

Thickness of the Bunte Breccia deposits is highly variable (Table 1) and the deepest deposit was penetrated at locality 11 (Lutzingen, Goldbergalm). Because the drill hole was not placed on the very top of a hill, a total depth of >100 m of Bunte Breccia may be postulated for that locality at a distance of ≈27 km from the crater center (Fig. 1). Gall (1974b) describes Bunte Breccia from Tapfheim (≈29 km) overlain by fluviatile deposits of the Danube at an elevation ≈410 m a.s.l. Together with other reported Bunte Breccia deposits at many intermediate locations, this would indicate that Bunte Breccia may have covered substantial parts of the south and southeast Vorries to depth in excess of 50 m. This is one of the most important results of the drilling program because traditional interpretations had postulated that most of the deposits were draping a preexisting relief as a relatively thin veneer, thus mimicking the preimpact relief. However, the depths obtained now are comparable to those obtained from the east/southeast Ries by Birzer (1969) and Bader and Schmidt-Kaler (1977), who

Table 1.

Drill location	Range[1] (km)	Azimuth	A.s.l. (m) Top[2]	Bottom[3]	Total thickness[4] of Bunte Breccia	Bottom of Bunte[5] Breccia a.s.l. (m)	*In situ*[6] material
1	28.5	194	495	480	>15	<480	??[7]
2	25.5	196	534	471	52	482	OMM
3	23	196	553	469	76	477	OSM
4	19	195	502	460	34	468	Upper Jurassic Lime
5	16.5	207	568	488	>80	<488	??[8]
6	27	115	558	486	47	511	OSM
7	36.5	225	492	446	17	475	Upper Jurassic Lime
8	35	216	560	528	28	532	OSM
9	32	206	506	499	>7	<499	??[9]
10	32	206	505	482	21	484	OMM
11	27	184	503	401	84	419	OSM

[1]Range: From crater center (see Pohl *et al.*, 1977).
[2]Elevation of drill point above sea level (= top of profile).
[3]Elevation of final drill depth above sea level (= bottom of profile).
[4]Total thickness of Bunte Breccia encountered.
[5]Elevation above sea level of contact between Bunte Breccia and undisturbed country rock.
[6]Undisturbed country rock in contact with Bunte Breccia; because "local" components may form "megaclasts" of considerable size (> 10 m) it was necessary to penetrate substantially into the country rock for reliable identification of the Bunte Breccia substrate contact.
[7]Exploratory drill hole; terminated in OSM megaclast.
[8]Contact misinterpreted initially; terminated in upper Jurassic limestone megaclast.
[9]Exploratory drilling revealed Bunte Breccia >7 m; drill hole no. 10 is at same locality (≈10 m distance away) and fully cored.

report "unusual" depths of up to 200 m for deposits filling a localized depression, i.e., the preexisting Main valley (at a radial range of ≈28–30 km). Since our results were obtained during drilling of hills, our thicknesses encountered cannot be attributed to trapping of ejecta in depressions.

The deposits of Bunte Breccia encountered at our localities are mostly deeper than expected and do not mimick a preexisting relief. Indeed, where possible, i.e., where reliable geologic maps exist, the breccia depth may be qualitatively predicted by simply projecting the OSM (OMM) substrate as a horizontal datum plane (e.g., localities 2, 3, 4, 7, 8, 9, 10). Thus, the new drilling results suggest that the preexisting target morphology in the south-Vorries is not necessarily related to the present-day morphology; small scale present-day relief appears to have little in common with the morphology just prior to the Ries event. No detailed field data exist to determine whether the present relief and distribution of Bunte Breccia is a primary feature associated with the emplacement mechanism (e.g., analogous to dunes, ridges, and other morphologic landforms observed around large scale lunar craters) or whether the relief is exclusively determined by erosion since crater formation some 15×10^6 yr ago (Gentner and Wagner, 1969).

Breccia Components

The deposits to the south of the cliff line are composed of a large number of crater derived and local components (Fig. 2). Clast sizes may vary from tens of meters to <1 mm; clast size appears to be independent of source area because clasts >10 m from both local and crater sources are observed. These clasts are embedded into a relatively fine grained, clastic matrix. The vol.% of matrix is variable, generally, however, >50% of the entire volume. Because of the abundance of clays and unconsolidated sands, the entire breccia deposit is relatively unconsolidated and parts of the recovered cores disintegrate rather readily.

The following field terms apply: "matrix" is everything <1 cm in grain size. "Clasts" occur on all scales and will be qualified by size; "megaclasts" are monolithic inclusions >1 m by definition. In the following, we will describe the nature of the breccias from large to small scales, i.e., from megaclasts to matrices.

1. *Clasts*

The extreme variability in abundance of megaclasts is striking. Figure 3 attempts to illustrate this point by plotting volumes of local and crater derived megaclasts as well as the remainder of the core (= *all* materials <100 cm). Not only is the total volume of megaclasts highly variable, but also the ratio of local versus crater derived megaclasts.

We also recorded the frequency of occurrence of all clasts between 5 and 100 cm in the field, starting with drill hole number 3 (Fig. 4). The observations

Fig. 2. General view of cores from drill hole No. 8. Note the differing clast sizes and different color shades, indicating different lithologies. Also note the relatively unstructured, massive character of the breccia matrix and the irregular orientation of clasts (length of core container: 1 m; core diameter; 101 mm).

are at present confined to only that half of the core surface which was readily accessible for inspection after the cores were placed in their respective containers. Thus the data are by necessity of a survey type nature, though valid in comparing relative frequencies of local versus crater derived clasts. No volume percentages are given because this will be possible only after a more thorough planimetric study of the cores. However, the qualitative data contained in Fig. 4 serve to illustrate two points which will certainly hold after more detailed analyses: first, the ratio of local versus crater derived clasts of 5–100 cm diameter varies in an irregular fashion with respect to radial range and second, the frequency of these clasts per running core meter (= unit surface ≈ unit volume) is also variable.

Hüttner (1969) and Gall *et al.* (1975) report decreasing mean diameters for megablocks >25 m of crater derived materials (in particular of upper Jurassic limestones) as a function of range. The data contained in Figs. 3 and 4 indicate that significant variations may exist within these trends. Referring to Fig. 3, the breccia encountered at locality 3 is much more "fine grained" than that of locality 2, despite the fact that core 3 was taken somewhat closer to the crater. In keeping with the large scale trends, cores 7, 8 and 10, however, are of a more fine grained nature, if not devoid of megaclasts (see also Fig. 2). Frequency of crater derived megaclasts appears indeed to decrease with increasing range (cores 10, 8, 7) though localities 4, 3, and 2 also contain either little or no crater material >1 m. Data of Fig. 4 show that relatively small crater clasts may occur in variable abundance in all localities; no systematic trends are obvious within the core materials, in contrast to Schneider (1971). Though not illustrated, clasts of all sizes and both of local and crater derived origin may occur along the entire profile; their distribution appears to be irregular. There are no apparent trends and concentrations of either source toward the top or bottom of the profile, again in contrast to Schneider (1971).

In conclusion the field observations of clast populations present strong

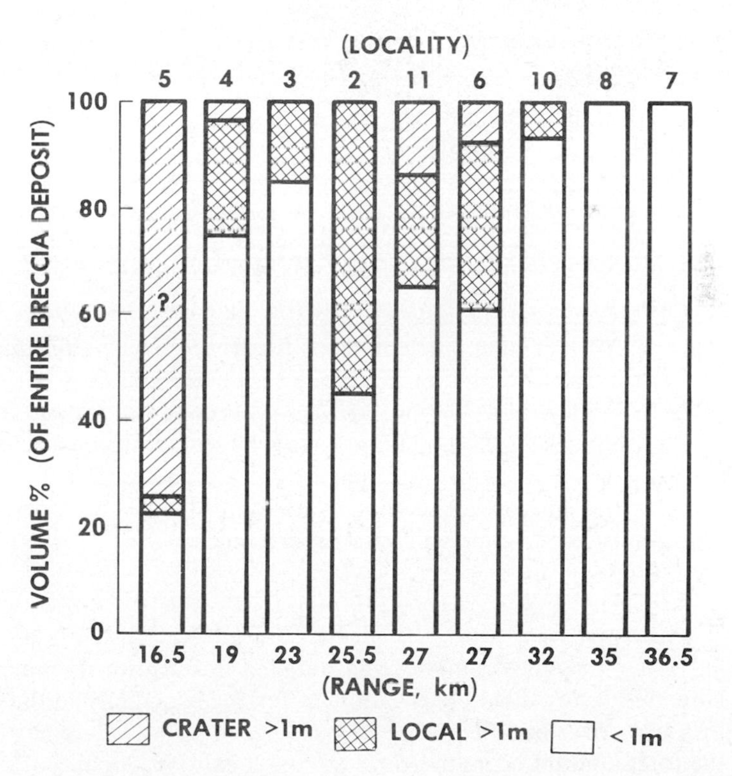

Fig. 3. Vol.% of local and crater derived "megaclasts" (>1 mm) and cumulative volume of all other components including genuine matrix (<1 m) as a function of radial range from the crater center. The source area of all megaclasts encountered in No. 5 is difficult to evaluate, because of its position north of the cliff line. The radius of the Ries Crater is ≈13 km.

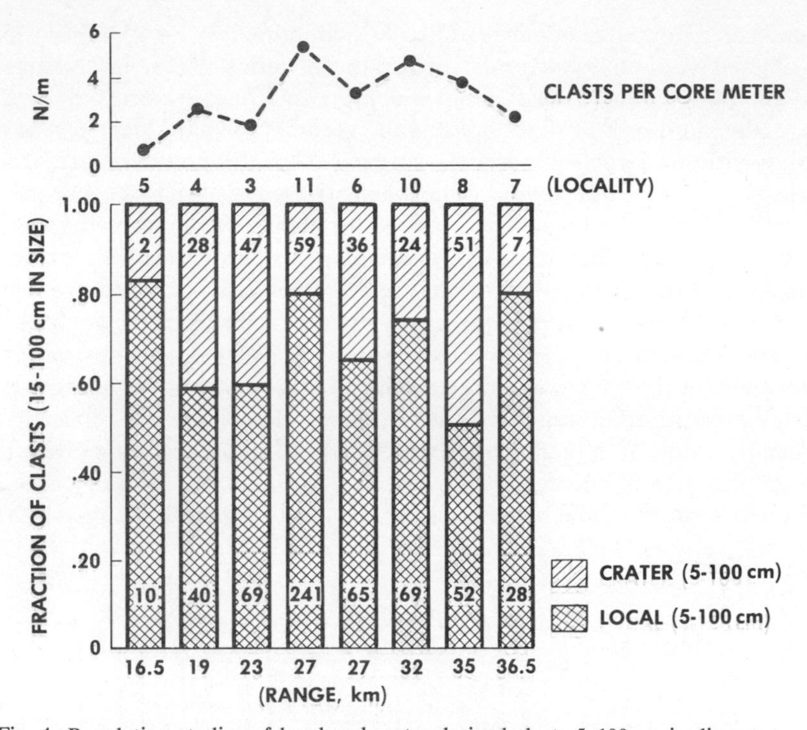

Fig. 4. Population studies of local and crater derived clasts 5–100 cm in diameter as a function of range. "Fraction" simply refers to numbers of clasts with observed absolute numbers inserted in each column. N/m refers to total number of clasts 5–100 cm in size per running core meter of matrix.

evidence that the emplacement mechanism of the Bunte Breccia produced highly variable clast sizes, incorporated large, though variable amounts of local clasts and mixed both crater and locally derived components rather thoroughly. Significant variations in clast populations may be encountered over distances measured in a few kilometers, i.e., the spacings of e.g., holes 5, 4, 3, and 2, along a radial traverse, though natural or man made outcrops in the Vorries suggest that such variations occur over still smaller distances.

2. *The Matrix*

The matrix is extremely complex and detailed description is not possible in the field. However, a few field observations referring especially to the *variability* of the matrix are described below:

First, the total amount of matrix differs from locality to locality. It may range from approximately 20% (locality 4) to >80% (localities 8 and 7). The amount present at each locality is roughly approximated by the <1 m category in Fig. 3, because the total volume occupied by clasts from 5–100 cm diameter is generally <20% of the entire core.

Second, the matrix encountered in different localities is not uniform as evidenced by different color shades, variable grain size, different proportions of clay and sand components, and various populations of discrete clasts between .1 and 1 cm in diameter. To the naked eye—aided by a hand lens—the matrix is largely made up of locally derived materials (see also Schneider, 1971), i.e., soft clays and unconsolidated sands of OMM or OSM, which are intimately mixed on centimeter and smaller scales, though frequently either one of them may dominate if not—on occasion—constitute the entire matrix.

Third, the variations in matrix character described above do even occur within one profile, though generally on a more subtle scale (Fig. 5). Within a given profile the various types of matrices may have dimensions on the order of a few meters, though variations on smaller and larger scales are also observed. The transition from one matrix type to the other may either be gradual, typically measured in decimeters or it may be rather abrupt (<1 cm) (Fig. 5). Most frequently, different types of matrices are sandwiched in between megaclasts, with the character of the matrix remaining fairly homogeneous *between* clasts, but different on either side of the clasts.

Fourth, the .1 and <1 cm clasts, considered part of the matrix, are also variable both with respect to frequency and lithological character; especially frequency and component lithologies of crater derived sources appear to be independent of the <.1 cm ground mass comprising the bulk of the matrix.

The above field observations concerning the matrix and its variability lead to the following conclusions: the matrix is to a large degree locally derived. It is extremely well mixed on centimeter scales, though pronounced heterogeneities measured on decimeter and meter scales exist within any locality. The frequent change in matrix character on either side of megaclasts and the occasionally observed, knife edge sharp contacts between two neighboring matrix types (Fig. 5) suggest that a large proportion of the matrix may be deposited as discrete, extremely polymict "megaclasts". The variable lithologies and thoroughly mixed nature of these matrix clasts seems to indicate that various matrix types were formed initially at distinctly differing locations and subsequently transported to a common, final resting place. It is unknown at present how far such breccia clasts were transported and how far apart their respective source areas were. However, all are derived from south of the cliff line (excepting cores 5 and 7), which limits both distances to less than ≈2 km for at least locality 4; similar distances may be postulated for locality 8 also. Hüttner (1958) and Gall (1974a) reported on limestone clasts in the Bunte Breccia which are characterized by pholade bore-holes and which, therefore, were dislodged from the Miocene cliff line. Such limestones were found as far as 10 km to the south of their closest source area, indicating significant lateral transport of local components.

Textural Observations

All core materials were inspected for preferred orientation of clasts, lineations in the matrix, style of deformation of soft clays, and noncohesive sands,

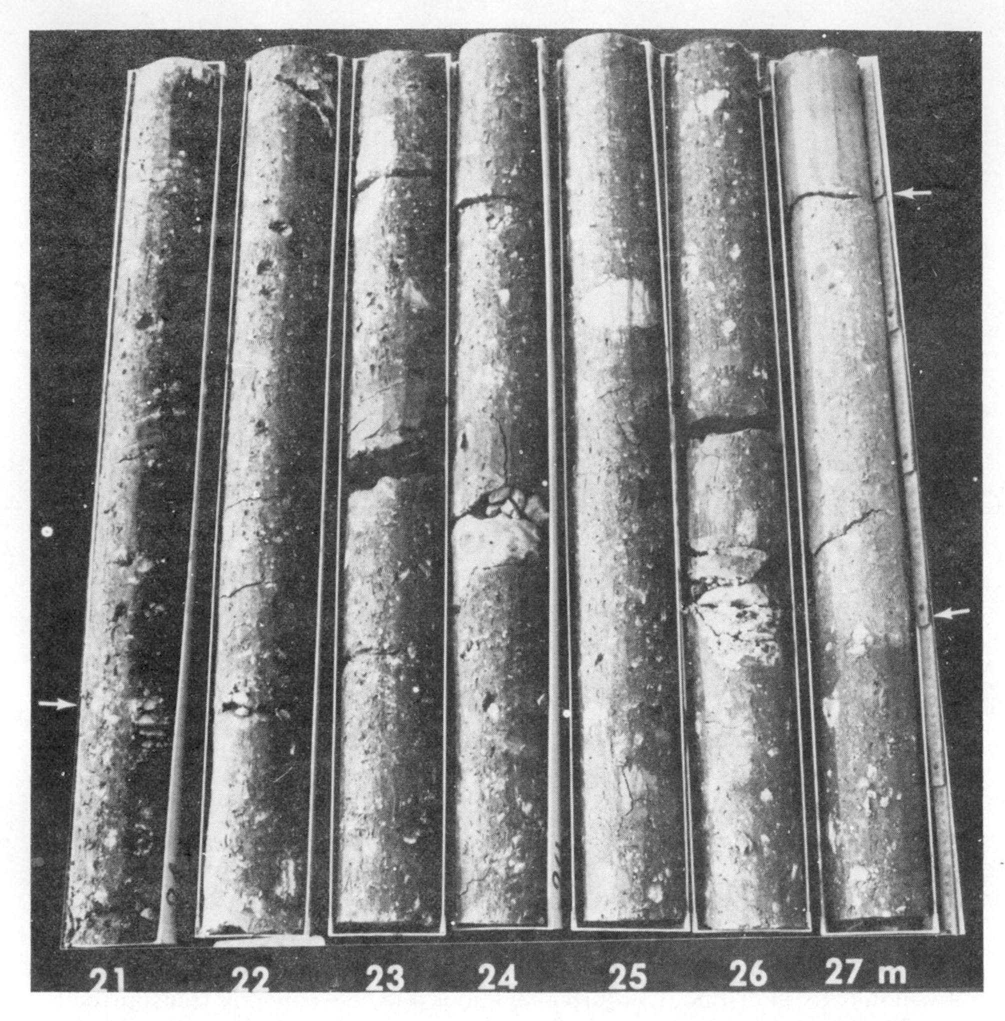

Fig. 5. Cores from locality No. 8 illustrating various matrix characters: Note distinct change of matrix character at ≈21.30 and at 27.35 m. Though more difficult to observe, the character of the matrix between 21.30 and 27.30 m also changes *gradually* from a more clay-rich (OSM), darker matrix to a lighter, more sandy (OSM/OMM) matrix. Note also irregular distribution and orientation of clasts and the equant shapes of many clay clasts in contrast to some deformed inclusions. Furthermore light colored, monomict, cataclastic limestones are readily identified. At 27.80 m, a *sharp contact to the in situ OSM substrate* (= fine grained sands) occurs.

etc., in an attempt to determine whether the depositional environment was largely turbulent or laminar. For the first we expect a rather random orientation of linear textural elements, while the latter should result in more or less lineated textures. The observations were performed on the cylindrical core surfaces occasionally aided by orthogonally cut samples.

1. *Orientation of monolithic clasts* <50 *cm*

These observations generally refer to clasts up to 50 cm in size, with the bulk of the observations, however, based on 1–10 cm clasts. The majority of all clasts appears to be oriented irregularly as illustrated in Fig. 5. The field evidence argues strongly for a random orientation of clasts. These observations apply not only to the orientation of a clast's longest axis, but also to primary sedimentary textures such as rare bedding planes in some sand clasts, foliation of shales and primary layering in clay clasts. On occasion, however, sections of matrix were encountered, which did contain some uniformly oriented clasts, distributed over 10–50 cm of core length. Such occurrences are rare and they are confined to individual matrix clasts, i.e., do not transgress into an adjoining matrix type. Common dip angles of such oriented clast groups are variable from group to group, i.e., matrix type to matrix type, thus strengthening our observations that discrete breccia matrices are deposited as megaclasts.

2. *Lineations in the matrix*

Two types of "lineations" were searched for: one among the abundant .1–1 cm clasts embedded in the matrix and a second among the fine grained sands and clays in the groundmass proper. In general, no preferred orientation of .1–1 cm clasts is observed; their orientations are highly irregular, if not random both with respect to longest axis and primary sedimentary textures. In general, the fine grained sandy and clayey groundmass also displays no lineations. Thus the overall appearance of the matrix is that of a thoroughly mixed, rather "massive" deposit, devoid of linear structural and textural elements (Fig. 6).

Though the above "massive" character is by far the most abundant, there are also matrix sections which display irregular swirls and other highly contorted configurations of alternating and interfingering sandy and clayey matrix components as well as individual sand and clay clasts. Such textures are indicative of turbulent forces during ejecta emplacement.

In contrast to these "massive" and "swirly" matrix types there are however, rare instances where good and even extremely well developed matrix lineations occur. In these sections the clay and sand clasts are extremely deformed, elongated and smeared out, quite frequently >5 cm in length and <.5 cm in width. Such matrix lineations are preferably horizontally oriented though other orientations occur; furthermore they are particularly abundant in locality 2 and extremely rare to nonexistent in the other localities.

3. *Deformation structures of clasts* >1 cm

Two types of deformation may be observed: "external" deformation referring to overall shape of the easily deformed clay and sand clasts and "internal" deformation as evidenced by primary, sedimentary features.

Many examples exist for external deformation, especially well displayed by

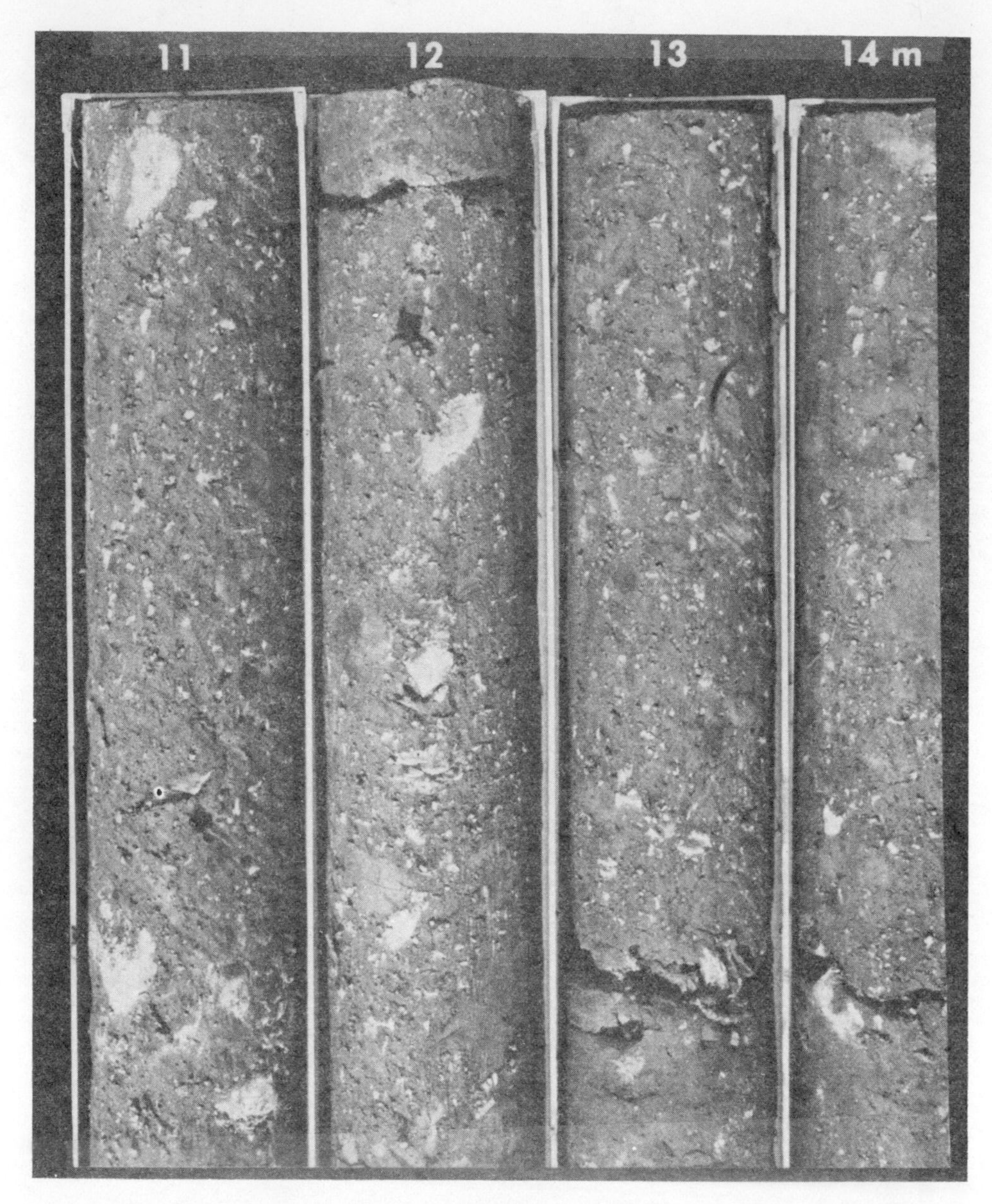

Fig. 6. Closeup of Bunte Breccia (locality 8). Note irregular orientation of clasts and "massive" character of matrix.

highly contorted, twisted, and elongated clay clasts (Fig. 7). Very commonly, however, one may also observe clay, and in particular, sand clasts that are relatively equant and undeformed, as also illustrated in Fig. 7. Notice the close spatial association of deformed and undeformed clasts in Fig. 7. Clasts >5 cm

Fig. 7. Unusual concentrations of relatively large clay clasts (locality 11). Note various degrees of deformation of relatively soft clay clasts and irregular direction of deformation force vector. Note also "mixed clast" at 54.09–54.31, representing "breccia within breccia", and the relatively sharp contacts with surrounding breccia matrix.

appear to be more severely and more frequently deformed than those <5 cm, where relatively equant shapes seem to dominate (Fig. 7).

Primary sedimentary structures of clay, shale, and sand clasts are commonly deformed, ranging from modestly wavy, possibly folded textures to extremely complex, contorted, and twisted configurations. Again, however, undeformed clasts are commonly observed, though not as frequently as the deformed ones and furthermore predominantly in the size range <5 cm (Fig. 7).

The "local" megaclasts (>1 m) are generally only modestly deformed, if at all. Complexly contorted, twisted, and swirly textures are extremely rare in the megaclasts. In many cases original sedimentary textures are so well preserved and so consistent along the entire megaclast that dip angles of the entire inclusion may be readily obtained; these dip angles are highly variable.

We thus conclude from these textural observations that the depositional environment of the Bunte Breccia was dominated by turbulent forces. Furthermore, utterly deformed, contorted and twisted clasts may be in close spatial association with completely undeformed inclusions of the same lithology, indicating highly variable energy distributions during breccia formation. Laminar flow is evidenced by lineated matrix types but appears to be of subordinate importance.

Evidence for Extensive Mixing

General

As already partly described and illustrated in Figs. 5–7, the core materials are extensively mixed on all scales. Clasts ranging from meter to probably submicroscopic sizes may occur anywhere along the core profiles. Though variable, the preponderance of mixing on centimeter and millimeter scales in essentially all matrix sections is truly extraordinary. Local and crater derived materials participate equally well in this mixing process and may be encountered on all scales of naked eye and hand lens inspection. Component populations with respect to size frequency and lithology of clasts are variable, however, any particular component may never be excluded *a priori*, though any specific component may very well be absent in some cases. The intensity of the mixing process must have been such that the above described matrix differences only represent variations of an extremely efficient, omnipresent and thorough mixing process operating south of the cliff line.

This mixing process was not only thorough but also extremely energetic. The energetic nature of the mixing process is illustrated by small limestone clasts (<1 mm in size) which occur in the center of decimeter sized sand or clay inclusions (Fig. 8). Even more startling are similar, small inclusions of clays within clays or sands; almost incomprehensible are discrete sand inclusions lodged inside pure clay clasts. These inclusions have no obvious connection to neighboring sand clasts. The relatively unconsolidated sands must have penetrated the very viscous clays while maintaining their integrity and without significantly deforming the host material either. These phenomena are truly startling, but attest to very high relative velocities between clast and host materials during the mixing process.

Qualitatively similar arguments can be made for some core sections which have matrices of predominantly one lithology, e.g., almost pure OSM clays or OMM sands. The clastic components in these sections are evently distributed and display a large variety of lithologies and/or clast sizes. However, the primary sedimentary textures in such "matrices" do not indicate large scale deformation, much less wholesale auto brecciation and mixing of the host material. Again it appears that penetration was accomplished rather nondestructively by *all* clast components, irrespective of local or crater derived origin. Particular emphasis is placed on such observations which involve exclusively locally derived host and clast materials because they demonstrate collisions at relatively high velocities of local components only.

Fig. 8. Closeup of Bunte Breccia (locality 5, ≈70.90 m deep). One of the best examples of deformed, i.e., folded clay clast. Close inspection will reveal some clastic, brecciated material within this clast, thus illustrating three "breccia within breccia" generations.

In contrast to this relatively energetic and violent mixing, there is, however, evidence for less energetic conditions in close spatial association: many clasts are undeformed and must have been embedded rather gently into the polymict matrix. Furthermore, the contacts between megaclasts and matrix or contacts between various matrix types may be exceedingly sharp without evidence of mixing. Clasts <10 cm of noncohesive sands and soft clays more often display knife sharp contacts than evidence of mixing with the surrounding, polymict breccia (Figs. 5–8). These observations imply that some, if not most, of the clastic material was embedded into the polymict matrix relatively gently.

In conclusion, the mixing process operating south of the cliff line was extremely efficient, thorough, and omnipresent. The intensity of mixing may range from extremely energetic, requiring high relative velocities, to rather gentle forces. Evidence for both violent and relatively gentle forces is so abundant and in such close spatial association that one must postulate a mixing process of dramatically different energy levels in close spatial and temporal proximity.

Breccias within breccias. The above described clastic materials within monolithic host materials that are embedded in some matrix can be viewed as "breccia within breccia". The same applies to relatively small sections (<1 m) of one matrix type breccia bounded by some other matrix type(s) (e.g., Fig. 5). Frequently one observes also distorted, yet distinct "clasts" composed of two (or more) intimately mixed and smeared out lithologies (Fig. 7); again they can be viewed as breccias within breccias. Thus breccias within breccias are common on a large variety of scales and up to four "generations" of breccias within breccias were encountered, e.g., a specific matrix type (breccia 4) containing a mixed clast composed of two lithologies (breccia 3), one of them containing clastic material (breccia 2) if not even brecciated (breccia 1) components (Fig. 8). While the term "generation" may be valid for one specific set of observations, such relative "timing" is not applicable to others. We envision the formation of such breccias within breccias as individual stages of a continuous mixing and recycling process; neither the number of "generations" nor their order of formation may be of general significance, as they represent minor and unpredictable excursions of a more general mixing process. They do indicate, however, that individual constituents of the Bunte Breccia and thereby the entire deposit have a complex history of mixing with discrete mixing events separated in space and time and occurring on a variety of scales (see also Hüttner, 1969).

THE CONTACT BUNTE BRECCIA/COUNTRY ROCK

Previous studies of the Guldesmühle outcrop (Hüttner, 1969), suggested a generalized "transition zone" enriched in local components at the contact of Bunte Breccia and its substrate. At Guldesmühle, Bunte Breccia rests on relatively unconsolidated OMM sands which are also observed in decreasing abundance over a vertical distance of 3 to 4 meters above the relatively sharp contact. Thus Guldesmühle served as a classical site to study the incorporation of local materials (e.g., Hüttner, 1968, 1969; Oberbeck *et al.*, 1975; Chao, 1974). Such "transition" zones are not observed in any of the new core materials. With the exception of locality 8, all cores displayed extremely sharp contacts measured over <5 cm; most are indeed knife edge sharp (Fig. 9). The contact at locality 8 may have a transition zone of 50 cm width (see Fig. 5). Such sharp contacts are surprising because in many cases the substrate is formed by unconsolidated OSM or OMM sands (see Table 1), i.e., by exactly those materials that in turn make up a large fraction of the matrix and clasts. Thus the outcrop Guldesmühle needs re-evaluation and cannot be generalized to the entire Bunte Breccia deposit.

The substrate itself lacks any deformation features (except at locality 4). At distances measured in less than decimeters the primary sedimentary textures, e.g., bedding planes, etc., are not only horizontal, but also completely undisturbed. At locality 4, the substrate is formed by upper Jurassic limestones which are thoroughly fractured and shattered over a distance of ≈ 90 cm; there is a distinct gradient of decreasing deformation with increasing depth along this 90 cm zone.

Furthermore, none of the contacts observed represents the old land surface prior to impact, because they lack any pre-Ries weathering horizon. Such weathering horizons are found as discrete clasts in the Bunte Breccia deposit, however. The lack of any weathering products at the present contacts implies complete stripping of the old land surface to depths of $\gg 5$ m; the present contact

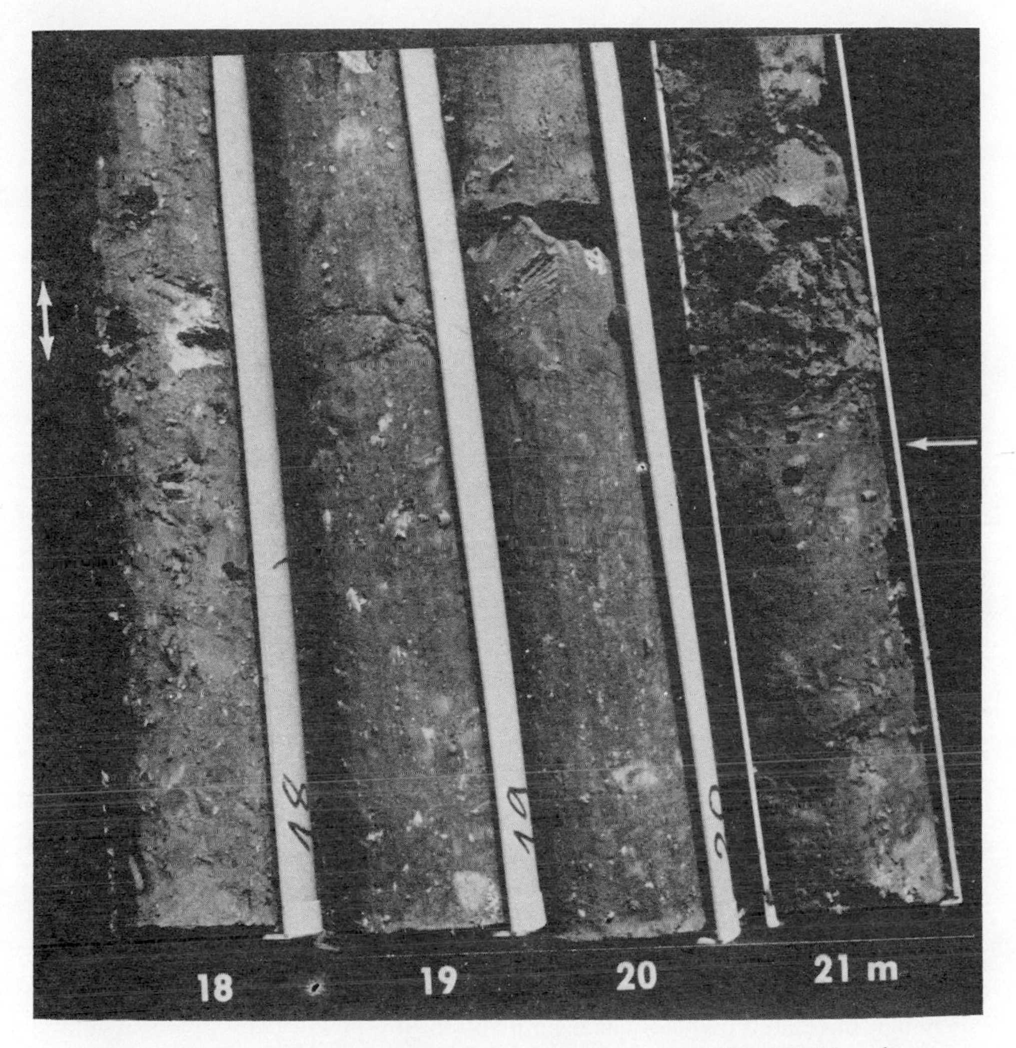

Fig. 9. Sharp contact of Bunte Breccia and underlying unconsolidated OMM sands at locality 10 and 21.35 m depth. Again note a "matrix" clast between 21.02 and 21.35 m and a gradual change in matrix character at ≈18.45 m

surface was created by the ejecta emplacement mechanism prior to final deposition of the Bunte Breccia.

We therefore conclude from the observations concerning the contact of allochthonous/authochtonous masses that most contacts are rather sharp and the old target surroundings were stripped to depths below the old weathering horizon before final deposition of the breccia layer.

Percentage of Local Materials

Oberbeck *et al.* (1974) postulated that the percentage of locally derived components in large scale crater deposits increases with radial range. Therefore an attempt was made in the field to estimate the ratio of locally derived materials versus primary crater components for the Bunte Breccia. Because of the fine grained matrices that constitute as much as 80% of the core materials, accurate ratios may be obtained only after careful examination of the matrix and detailed volumetric assessment of all clasts both of which must be completed in the laboratory.

Presently available field estimates, however, indicate that local components dominate the character of Bunte Breccia south of the cliff line. Local materials may vary between 50 and 80% of the entire deposit for most cores; cores number 7 and 10 may contain >90% local components. Schneider (1971) arrived at similar observations based on the analysis of heavy mineral populations and he actually identified "breccias that do not contain components from the crater cavity". This latter finding, however, may apply only to rare matrix types and does not hold for an entire breccia profile; crater components—though subordinate—may be found essentially anywhere in our cores.

Although no precise volume fractions for the local components can be given at present, the important and fundamental prediction of the ballistic ejecta hypothesis is strengthened by present field estimates: continuous deposits around large impact craters contain local materials which may dominate the entire deposits at sufficient ranges; at the Ries, local components may constitute as much as ≈90% of the total deposits at ranges of 2–3 crater radii.

Discussion

The reported field observations contain important new results for the Ries itself and by analogy, for planetary, large scale, continuous crater deposits beyond the crater rim.

Though of limited planetological implication, the newly determined thickness of Bunte Breccia is significantly deeper than previous estimates and thus of great importance to the Ries. Bunte Breccia is not a thin veneer mimicking a preexisting landscape. Thus, a variety of previous estimates of Bunte Breccia volumes need serious reconsideration as well as the reconstruction of preimpact target morphology and relief. It is unclear at present, to what degree thickness variations and present-day relief are the result of erosion or to what extent they may reflect primary features of the emplacement mechanism.

A variety of new observations concerning the incorporation of local materials and the mixing of the entire breccia unit were described. These lead to important observational insight and constraints about the formation of Bunte Breccia and—by analogy—other planetary, large scale crater deposits:

(1) Incorporation of local materials is significant and may lead to widespread

deposits that contain only subordinate amounts of crater derived materials.

(2) Judging from the total thickness of Bunte Breccia and from the percentage of local components present (>90%), the emplacement mechanism must have excavated and disturbed the surrounding crater terrain to depths in excess of 50 m.

(3) As evidenced by the sharp contacts of Bunte Breccia with the underlying substrate and by the undisturbed nature of the substrate itself it is mandatory to invoke a process that excavated and stripped the crater surroundings to significant depths prior to terminal emplacement of the breccias.

(4) Random orientation of clasts and the massive matrix character indicate a predominantly turbulent environment, rather than laminar flow.

(5) Multiple generations of breccias within breccias indicate discrete mixing and brecciation events at discrete locations and/or times, thus attesting to a complex, multi-stage mixing process.

(6) Various matrix types are generated during the intense mixing process and may be incorporated into the overall deposit as discrete clasts.

(7) Evidence for energetic mixing is present, requiring significant relative velocities of both crater and locally derived materials to penetrate each other.

(8) Evidence for "violent" and "gentle" forces is so abundant and in such close spatial association that an environment of dramatic energy variations over small distances must be postulated.

These observations lead to the following major constraints: any ejecta emplacement mechanism must be capable of excavating local terrain, incorporating and thoroughly mixing these materials with crater derived components and transporting such components laterally after having stripped or excavated the local crater surroundings.

Two rather different emplacement mechanisms are presently suggested for the Ries deposits: Oberbeck *et al.* (1974), Oberbeck, (1975), and Morrison and Oberbeck (1975) suggest a ballistic regime whereby primary crater ejecta is transported in ballistic trajectories and impacts the surrounding terrain with sufficient velocities to cause secondary cratering. As a consequence local materials are excavated, mixed with the primary ejecta and both primary and secondary ejecta combine into a terminal, ground-hugging flow. In contrast, Chao (1976a,b) argues against a ballistic transport mode and suggests a ground-hugging, roll-glide type ejecta transport, in which the ejecta leave the crater cavity essentially as one massive ground-hugging flow that must spill over tens of kilometers.

Many of the features observed are predicted by the ballistic hypothesis, most importantly the incorporation of local materials. Judging from the sizes of crater derived "megablocks" in the Vorries (see Pohl *et al.*, 1977; Hüttner, 1969; Gall, 1969) which may reach tens, if not hundreds, of meters in dimension at distances of 1–2 crater radii, excavation of the local substrate to tens of meters is perfectly plausible and expected. Indeed, Hüttner (1958) elaborates already on the possibility that such large blocks may have shattered and excavated local terrain north of the cliff line after impacting from ballistic trajectories, setting a debris surge in motion. New observations, e.g., breccias within breccias, the turbulent mixing environment, the existence of various breccia matrices formed at distinctly different locations, and, finally, the evidence of proximal "violent" and "gentle" forces are all compatible with secondary cratering followed by formation of a ground hugging debris surge for final breccia emplacement. Though velocity vectors of secondary ejecta are predominantly downrange, i.e., radially away from the primary crater, they *can* be distributed essentially over the entire 360° range, thus leading to a large array of collision—and mixing conditions over a relatively large span of time. We therefore consider all observations consistent with a ballistic ejection mechanism and resulting debris surge.

We note that the observations of Chao (1976a,b) may also be entirely consistent with ballistic transport followed by a debris surge, inasmuch as the observed striation of breccia components indicate nothing else but relative movement of particles; similarly the large striae (Schliffflächen) frequently observed in areas where the Bunte Breccia substrate is Malm limestone (Wagner, 1964) prove only relative movement. Such relative motions, however, are nonspecific for the presently competing emplacement mechanisms, because they are produced easily during secondary cratering and the subsequent ground-hugging debris surge. The observations of Chao (1976a,b) are not suitable to distinguish between a "primary" or a "secondary" debris surge.

We thus conclude that our present and previous observations are consistent with a ballistic emplacement concept. The energy expanded in excavating local materials followed by intense mixing and final emplacement via a debris surge is derived from the kinetic energy contained in ballistic ejecta. Unless other mechanisms are suggested by which up to 90% local materials are incorporated into a crater's continuous deposits, the ballistic hypothesis remains the most viable one.

Acknowledgment—It is a pleasure to thank Messrs. K. Camann, H. Arndt, L. Schuler, H. Jahn, and G. Karg of Prakla Seismos GmbH for the successful completion of the drilling program. The generous assistance of President H. Vidal and Drs. H. Schmidt-Kaler and H. Gudden are gratefully acknowledged as well as helpful discussions with Profs. D. Stöffler and W. v. Engelhardt and Drs. J. Pohl and G. Graup.

REFERENCES

Ammon, L.: 1905, Die Bahnaufschlüsse bei Fünfstetten am Ries und an anderen Punkten der Donauwörth-Treuchtlingen Linie, *Geognostische Jahreshefte* **16**, 145–185.

Andrews, R. J.: 1976, Characteristics of debris from small scale cratering experiments, In *Impact and Explosion Cratering*, D. J. Roddy, R. O. Pepin, and R. B. Merrill, (eds.). This volume.

Bader, K. and Schmidt-Kaler, H.: 1977, Der Verlauf einer präriesischen Erosionsrinne im östlichen Riesvorland zwischen Treuchtlingen und Donauwörth, *Geologica Bavarica*. In Press.

Birzer, F.: 1969, Molasse und Ries-Schutt im westlichen Teil der südlichen Frankenalb, *Geol. Bl., No. Bayern* **19**, 1–28.

Bolten, R. and Müller, D.: 1969, Das Tertiär im Nördlinger Ries und seiner Umgebung, *Geologica Bavarica* **61**, 87–130.

Carlson, R. H. and Roberts, W. A.: 1963, *Mass distribution and throwout studies, Project Sedan*, PNE-217, F, Boeing Co., Seattle, Washington. 144 pp.

Chao, E. C. T.: 1974, Impact cratering models and their application to lunar studies—a geologist's view, *Proc. Lunar Sci. Conf. 5th*, p. 35–52.

Chao, E. C. T.: 1976a, The Ries Crater, a model for the interpretation of the source areas of lunar breccia samples (abstract). In *Lunar Science VII*, p. 126–128. The Lunar Science Institute, Houston.

Chao, E. C. T.: 1976b, Mineral produced high pressure striae and clay polish: Key evidence for nonballistic transport of ejecta from Ries Crater, *Science* **194**, 615–618.

Chao, E. C. T., Soderblom, L. A., Boyce, J. M., Wilhelms, D. E., and Hodges, C. A.: 1973, Lunar light plains deposits (Caley Formation)—a reinterpretation of origin (abstract). In *Lunar Science IV*, p. 127–128. The Lunar Science Institute, Houston, Texas.

Gall, H.: 1969, Geologische Untersuchungen im SW-Vorries; das Gebiet des Blattes Wittislingen, Ph.D. Thesis, München, Germany.

Gall, H.: 1974a, Neue Daten zum Verlauf der Klifflinie der Oberen Meeresmolasse (Helvet) im Südlichen Vorries, Mitt. Bayer. Staatssamml. Paläont, *Hist. Geol.* **14**, 81–101.

Gall, H.: 1974b, Geologischer Bau und Landschaftsgeschichte des SE Vorrieses zwischen Höchstädt a. d. Donau und Donauwörth, *N. Jb. Geol. Paläont. Abbh.* **145**, 58–95.

Gall, H., Müller, D., and Stöffler, D.: 1975, Verteilung, Eigenschaften und Entstehung der Auswurfmassen des Impakt Kraters Nördlinger Ries, *Geologische Rundschau* **64**, 915–947.

Ganapathy, R., Morgan, J. W., Krahenbuhl, U., and Anders, E.: 1973, Ancient meteoritic components in lunar highland rocks: Clues from trace elements in Apollo 15 and 16 samples, *Proc. Lunar Sci. Conf. 4th*, p. 1238–1261.

Gault, D. E., Quaide, W. L., and Oberbeck, V. R.: 1968, Impact cratering mechanics and structures. In *Shock Metamorphism of Natural Materials*, B. M. French and N. M. Short, (eds.), Mono Book, Baltimore.

Gentner, W. and Wagner, G. A.: 1969, Altersbestimmungen an Riesgläsern und Moldaviten, *Geologica Bavarica* **61**, 296–303.

Head, J. W.: 1974, Stratigraphy of the Descartes Region at Apollo 16; implications for the origin of samples, *The Moon*, **11**, 77.

Head, J. W. and Hawke, B. R.: 1975, Geology of the Apollo 14 region (Fra Mauro): Stratigraphic history and sample provenance, *Proc. Lunar Sci. Conf. 6th*, p. 2483–2501.

Hörz, F., Oberbeck, V. R., and Morrison, R. H.: 1974, Remote sensing of the Cayley plains and Imbrium Basin Deposits (abstract). In *Lunar Science V*, p. 357–359. The Lunar Science Institute, Houston.

Hüttner, R.: 1958, Geologische Untersuchungen in SW-Vorries auf Blatt Neresheim und Wittislingen, Ph.D. Thesis, Tübingen, Germany.

Hüttner, R.: 1969, Bunte Trümmermassen und Suevit, *Geologica Bavarica*, **61**, 142–200.

Morrison, R. H. and Oberbeck, V. R.: 1975, Geomorphology of crater—and basin deposits: Emplacement of the Fra Mauro formation, *Proc. Lunar Sci. Conf. 6th*, p. 2503–2530

Moore, H. J., Hodges, C. A., and Scott, D.: 1974, Multiring basins—illustrated by Orientale and associated features, *Proc. Lunar Sci. Conf. 5th*, p. 71–100.

Oberbeck, V. R.: 1971, Laboratory simulation of impact cratering with high explosives, *J. Geophys. Res.* **76**, 5732–5749.

Oberbeck, V. R.: 1975, The role of ballistic erosion and sedimentation in lunar stratigraphy, *Rev. Geophys. Space Phys.* **13**, 337–362.

Oberbeck, V. R., Hörz, F., Morrison, R. H., and Quaide, W. L.: 1973, *Emplacement of the Cayley Formation*, NASA TMX, p. 62–302.

Oberbeck, V. R., Hörz, F., Morrison, R. H., Quaide, W. L., and Gault, D. E.: 1975, On the Origin of Lunar Smooth Plains, *The Moon* **12**, 19–54.

Oberbeck, V. R., Morrison, R. H., Hörz, F., Quaide, W. L., and Gault, D. E.: 1974, Smooth plains and continuous deposits of craters and basins, *Proc. Lunar Sci. Conf. 5th*, p. 111–136.

Pohl, J., Stöffler, D., Gall, H., and Ernstson, K.: 1977, The Ries Impact Crater, In *Impact and Explosion Cratering*, D. J. Roddy, R. O. Pepin, and R. B. Merrill, (eds.). This volume.

Reich, H. and Horrix, W.: 1955, Geophysikalische Untersuchungen im Ries und Vorries und deren geologische Deutung, *Beit. geol. Jb.* **19**, 119 pp.

Roddy, D. J., Boyce, J. M., Colton, G. W., and Dial, A. L.: 1975, Meteor Crater, Arizona, rim drilling with thickness, structural uplift, diameter, volume and mass balance calculations, *Proc. Lunar Sci. Conf. 6th*, p. 2621–2644.

Schaeffer, O. A. and Hussain, L.: 1974, Chronology of lunar basin formation, *Proc. Lunar Sci. Conf. 5th*, p. 1541–1555.

Schneider, E.: 1971, Petrologische Untersuchungen der Bunten Breccie im Nördlinger Ries, *B. Jb. Miner. Abh.* **114**, 136 180.

Shoemaker, E. M.: 1963, *Impact Mechanics at Meteor Crater, Arizona, The Moon, Meteorites and Comets*, B. M. Middlehurst and G. P. Kuiper, (eds.), University of Chicago Press, Chicago.

Stöffler, D., Gault, D. E., Wedekind, J., and Polkowski, G.: 1975, Experimental hypervelocity impact into quartz sand: distribution and shock metamorphism of ejecta, *J. Geophys. Res.* **80**, 4062–4077.

Wagner, G.: 1964, Kleintektonische Untersuchungen im Gebiet des Nördlinger Rieses, *Geol. Jb.* **81**, 519–600.

Wisotski, J.: 1976, Dynamic ejecta parameters from high explosive detonations. In *Impact and Explosion Cratering*, D. J. Roddy, R. O. Pepin, and R. B. Merrill, (eds.). This volume.

Roddy, D. J., Pepin, R. O., and Merrill, R. B., editors.
(1977) *Impact and Explosion Cratering*, Pergamon Press (New York), p. 449–460.
Printed in the United States of America

Rochechouart impact crater: Statistical geochemical investigations and meteoritic contamination

P. LAMBERT

Service Géologique National, B.R.G.M., BP. 6009, 45018 ORLEANS, Cedex, FRANCE

Abstract—Analyses by atomic absorption spectrometry of Rochechouart samples confirm the Ni anomaly of allochthonous breccias and the meteoritic origin of this Ni enrichment. The average Ni content of impact melt rocks is 10 times that of the target, suevite breccias 5 times, polymict breccias 2.5 times. Metallic particles are not detected in the glass fraction and are probably now all destroyed. The particular setting of Ni in secondary minerals reflects the Ni mobility. Ni moved from the melt and was trapped in the clast fraction, even probably outside of the melt area. Consequently Ni content of impact melts is too low and Ni content of polymict breccias too high. The Ni mobility is restricted to the fall back breccias unit. The disymmetric distribution of the Ni anomaly may be due to an inclined trajectory of the Rochechouart projectile. Geochemical investigations for Fe/Mg indicate that the recovered melt of Rochechouart was probably generated in the deep part of the target, not far from the actual ground level corresponding to the crater floor at the time of impact.

INTRODUCTION

THE IMPACT ORIGIN of the Rochechouart structure (France = 45°50′ N, 000°56′ E) was first suggested by Kraut (1967) and confirmed later by Kraut and French (1971), Raguin (1972), and Lambert (1974a). Country rocks are Hercynian crystalline rocks. Middle Jurassic is the most probable date of the impact (Hurley, 1969; Lambert, 1974b). The Rochechouart crater is deeply eroded and lacks any visible circular morphology. The present exposure level is near the bottom of the original crater, and remnants of fall back breccias and melt rocks are preserved on top of basement rocks. The breccia unit (comprising fall back breccias and autochthonous breccias) is no more than 100 m thick, drawing a continuous cartographic blanket in the inner zone (12 km diameter: see Fig. 1), surrounded by a crown of scattered patches (22 km diameter as maximum extent (Fig. 1)). Impact melts occur only in the inner zone. The probable impact point was deduced from the zoneographic study of shock effects intensity of both into the ejected mass and the target (Lambert, 1977). Crater size was also deduced: 20 km diameter may be considered as the minimum value (Lambert, 1977) (Fig. 1).

Meteoritic contamination

Preliminary studies of 50 Rochechouart samples by atomic absorption spectometry (A. A. S.) for Fe, Na, K, Mg, Co, and Ni indicated a strong Ni enrichment of impact melts with respect to metamorphic and granitic rocks of the basement (Lambert, 1975).

190 new analyses were performed, using the same method (Table 1). The Ni enrichment is confirmed: impact melts (E breccias) have the highest Ni content,

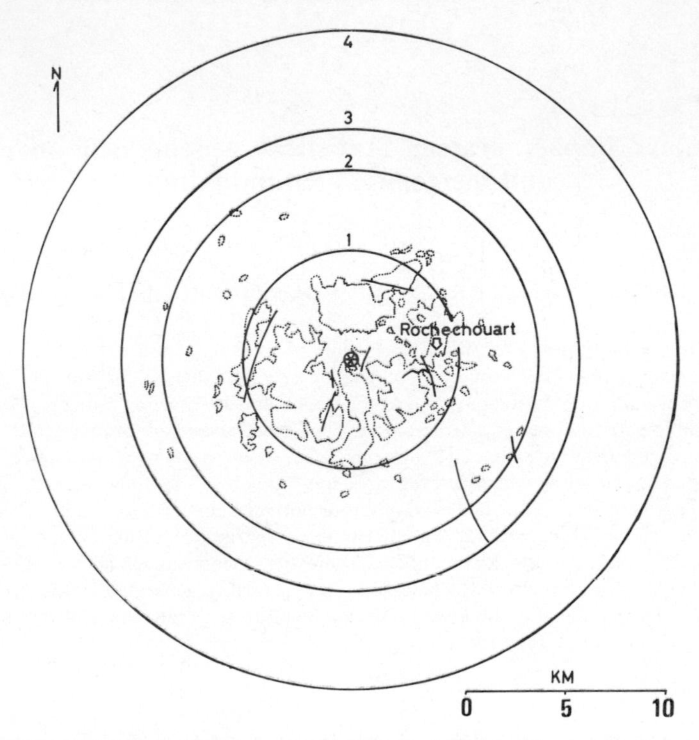

Fig. 1. Extent of Rochechouart breccias: circle 1: impact melts occurrence; circle 2: minimum crater diameter; circle 3: most probable size; circle 4: maximum crater diameter.

Table 1. A. A. S. analyses: Fe, Mg, Ni average contents of rock types from Rochechouart basement and from fall back breccias (breccias C for polymict rocks without glass, D for suevites). s: standard deviation. Rock source is determined from the actual extent of each rock type of the basement, in the estimated area of the crater (circle 2 or 3, Fig. 1): gneiss: 50%, granite ≃ 25%, orthogneiss ≃ 25%. As shown by Fig. 4 the rock source for impact melts was probably near this theoretical composition. For D and C breccias, Fig. 4 suggests a more gneissic contribution (see also the discussion).

	N	$\overline{Fe}$%	s	$\overline{Mg}$%	s	$\overline{Ni}$ ppm	s
gneiss	22	3.51	1.00	1.01	0.36	<14	8
orthogneiss	8	1.12	0.44	0.20	0.12	∽2	
granite	15	0.77	0.65	0.17	0.17	∽2	
rock source		2.50		0.60		∽9	
breccias C	43	2.70	1.12	0.70	0.32	26	16
breccias D	49	2.85	1.17	0.73	0.63	48	34
impact melts E	50	2.54	1.25	0.41	0.34	105	135
impact melts E (Ni > 30ppm)	27	2.91	0.46	0.60	0.38	202	169

some values reaching 600 or 700 ppm. The average Ni content of E is ten times that of the target. It is five times larger for suevite* (D breccias). Polymict breccias without glass (C breccias) (as "Bunte" breccias from Ries) show also a weak Ni enrichment (factor 2.5). In the Rochechouart structure, monomict breccias (B breccias) are very common below the crater floor. Their Ni content is normal, relative to the basement rock type from which they are formed. Shattered rocks of the target (A) are not enriched. Gneiss are usually the richest Ni rocks (between <10 ppm and 25 ppm). Few occurrences of serpentinites may be noted (Ni content about 2,500 ppm). The Ni enrichment of impact formation by such crystalline rocks cannot be accepted, for the following reasons. Their occurrences actually in the basement are exceptional. They have never been observed as clasts in the breccias. Moreover, because of their high Mg content (>20%), the Ni enrichment would be associated with a Mg enhancement. Table 1 and Figs. 2 and 3 clearly indicate that there is no change for Mg contents in impact breccias. For Fe, Mg, the composition of breccia units (C. D. E.) reflects that of the basement quite closely (Figs. 2 and 3). Ni enrichment cannot be due to the contamination from some Ni rich crystalline rocks or from any endogenic contamination. There was no more than 10 or 20 ppm as indigenous Ni in the basement rock source. Therefore, as yet suggested by preliminary results (Lambert, 1975), the hypothesis of meteoritic origin for Ni enrichment is the most plausible.

Nature of Rochechouart projectile

Eight selected samples of highest Ni content from C. D. E. breccias, as well as 2 samples of the basement were studied by Anders' team for siderophile elements by radiochemical neutron activation. The Ni enrichment of impact rocks is confirmed and the nature of the projectile is deduced in this last work (Janssens *et al.*, 1976, 1977). The Rochechouart meteorite was apparently a II A iron (Janssens *et al.*, 1976, 1977). Its composition would then be Fe = 94.5%; Ni = 5.5%, judging from the average of 47 II A iron meteorites.

Distribution of Ni in Rochechouart breccias

Figure 4 gives the distribution of Ni in the whole fall back unit. Ni enrichment is the highest in the center of the structure but the distribution is very heterogeneous. The highest values occur in a band oriented NE-SW. The Ni distribution may be correlated with the distribution of impact melts, and the distribution of the highest shock effects in the target (see Lambert, 1977).

Variations of Ni content in analysed rocks

Selected impact melts (E) and suevites (D) were sampled with respect to the glass fraction and clasts of basement rocks. One hundred and forty analyses

---

*Suevite is a polymict breccia with glass debris and clastic matrix.

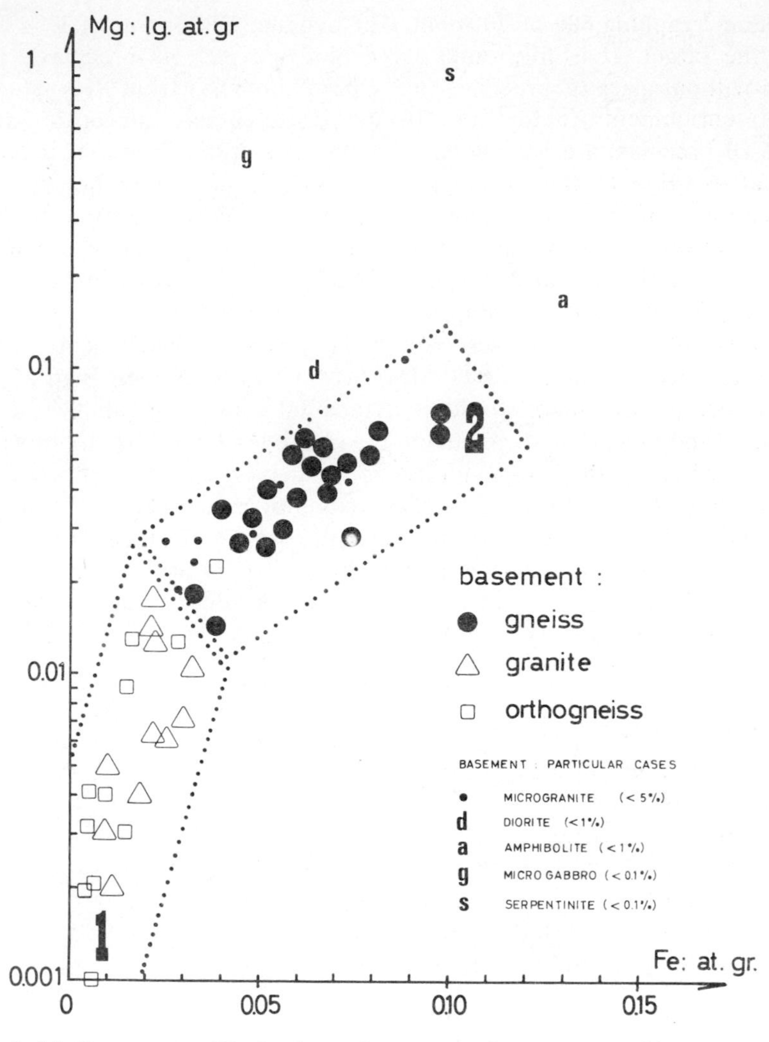

Fig. 2. Mg-Fe contents of Rochechouart basement rocks: zone 1: granitic area (granite + orthogneiss); zone 2: gneissic area.

were performed by A. A. S. but only for Ni (Table 2). Results clearly indicate that Ni distribution is also very heterogeneous on the scale of one hand specimen. Ni is usually more concentrated in the glass fraction, but clasts always show a higher Ni content in comparison with the autochthonous basement. Sometimes (in impact melts only), Ni is more abundant in the clast fraction than in glasses (Table 2). The same rock type from the basement (Ni = few ppm) may reach several hundred ppm when sampled in impact melts. These results imply that originally meteoritic material was mostly dispersed in the glass fraction. Later Ni moved and was trapped also in the clast fraction.

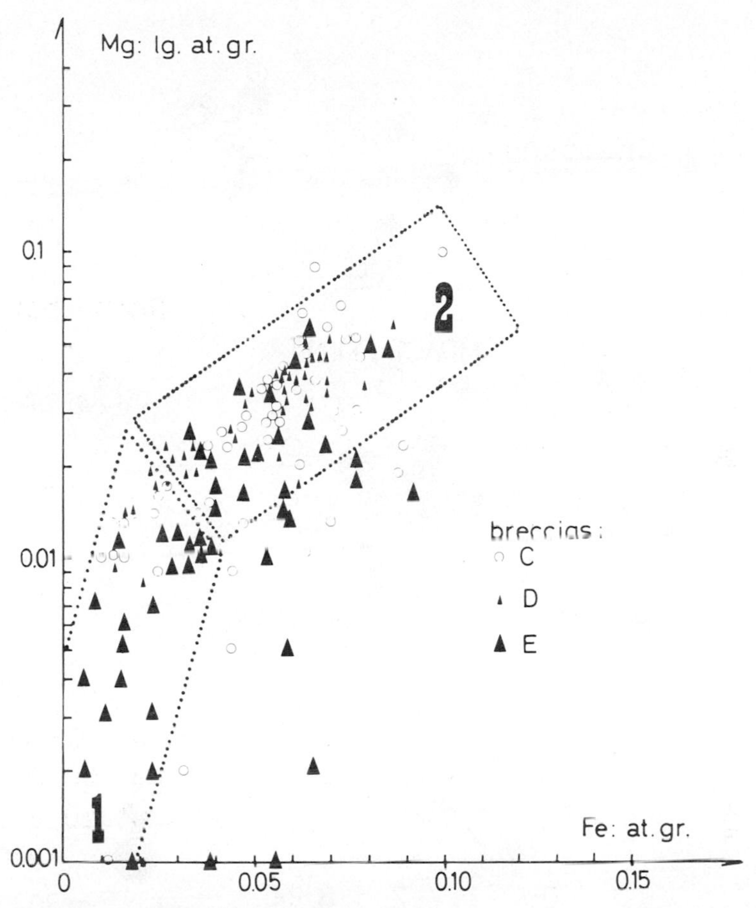

Fig. 3. Mg-Fe contents of Rochechouart fall back breccias: C = polymict breccias without glass; D = suevites; E = impact melts. E values are equally scattered in the granitic area (1) and gneissic one (2). It is not the case for C and D concentrated in the gneissic area (2) or only at the top of the granitic one (1).

Mineralogical setting of Ni into Ni-rich Rochechouart samples

At this time Ni-Fe spherules or other metallic particles have never been observed in polished sections from Ni rich breccias. Microprobe analyses of glasses were performed in order to determine the mineralogical setting of Ni and to detect the manner of Ni dispersion and concentration. All Rochechouart glasses from impact melts are recrystallised, essentially in a mixture of K feldspar and quartz (Table 3). Ni was not detected in these minerals. In most cases Ni is relatively concentrated (about 0.5%) in secondary chlorites filling cavities and bubbles. Higher Ni contents (1–3%) are sometimes detected in secondary limonites (Table 3). It is interesting to underline that all hydrated limonites analysed have an Fe/Ni ratio similar to that for projectiles. Excep-

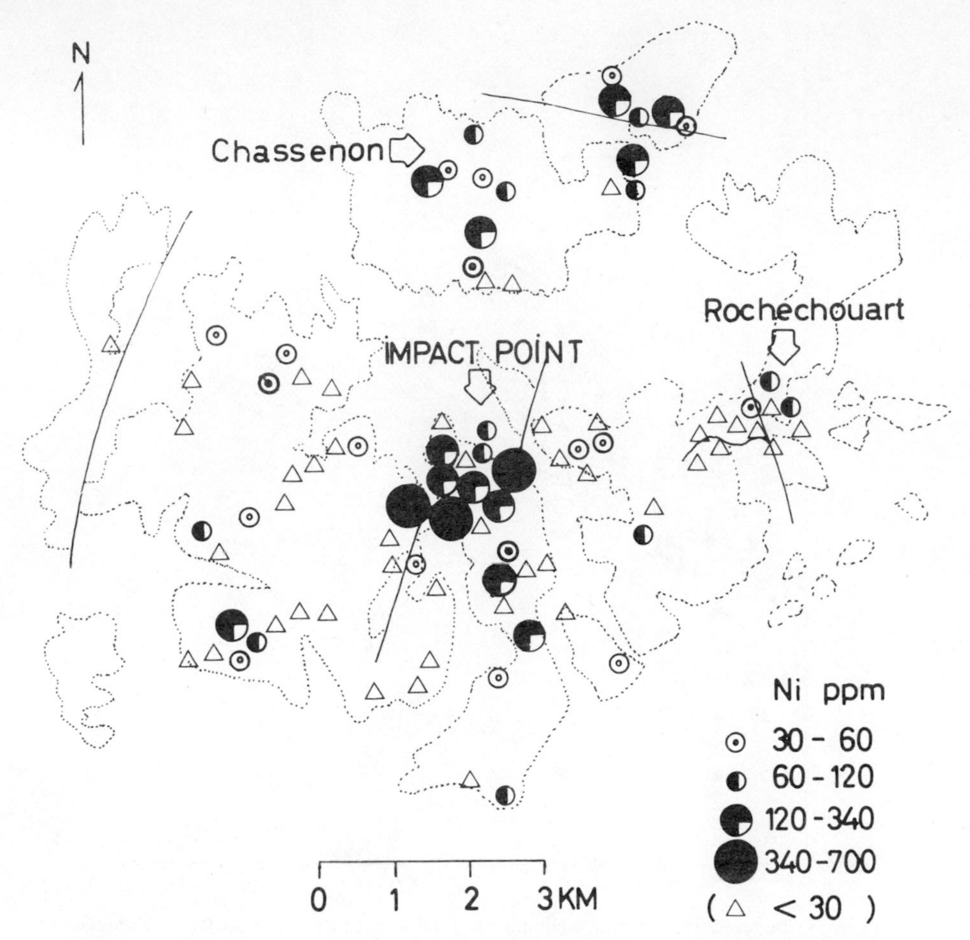

Fig. 4. Ni distribution in fall back breccias of the Rochechouart structure (dashed line: breccias extent, full line: faults).

tionally an automorphous bravoite was found (Fig. 5) (Ni 13%, Co = 0.7%) (Table 3) surrounded by limonite. The Ni/Co ratio of this bravoite is near 19.5 which is close to the cosmic ratio (20:1).

Discussion

The disymmetric distribution of the Ni enrichment in allochthonous breccias is mainly due to the distribution of melt rocks which could be related to nothing more than erosion. But this hypothesis is not entirely convenient (see Lambert, 1977). The area covered by the allochthonous breccia unit is more than 100 km^2 and is nearly circular. The maximum thickness of this unit is 50 m, and in most cases it is only a few meters. Therefore, the role of any erosion pattern

Table 2. A. A. S. analyses. Variation of Ni (in ppm) in the glass fraction and basement clasts of suevites (D) and impact melts (E).

reference	breccias	glass %	glass only			glass except		
			N	$\overline{Ni}$ ppm	s	N	$\overline{Ni}$ ppm	s
4A	D	30	9	160	38	6	76	58
101	D	30	2	250	140	3	77	64
377	D	10	2	150	90	2	82	4
919	D	10	2	45	14	2	25	7
948	D	5	2	183	25	3	75	30
950	D	15	3	58	25	2	37	4
9	E	60	6	98	50	6	43	50
203 A	E	65	1	260		1	210	
203 E	E	85	4	567	50	1	20	
229 B	E	50	10	254	143	3	492	114
447 A	E	90	3	480	200	2	180	5
483 B	E	50	2	98	3	3	205	131
931 B	E	40	3	83	6	2	75	14
932 B	E	80	6	276	40	3	218	148
939 A	E	90	4	121	52	1	35	

inside the fall-back unit is very limited. On the other hand, the distribution of Ni in allochthonous breccias, except all breccias containing more than 5% of glass, shows the same trend. A possible explanation for Ni, melts and high shock level distributions could be an inclined trajectory of the Rochechouart projectile coming from NE or SW.

Usually the meteoritic contamination in terrestrial crater is deduced from the recovery of Ni–Fe spherules. They are too rare to be detected or do not exist at Rochechouart. It is supposed that 165 Myr of weathering and oxidation have led to their entire destruction. Even in fresh glasses from a young crater (Aouelloul) where spherules are found, the Ni content of spherules vary from <2 to 9% (Morgan *et al.*, 1975). Metallic spheroids within impactite from the Barringer meteorite crater have Ni content from 13–22% representing enrichment factors of 2–3 in spheroids compared to analyses of the bulk meteorite (Kelly *et al.*, 1974).

The particular setting of Ni (in secondary minerals) and its concentration in basement rock clasts reflect the Ni mobility. Ni in chlorites represents probably the latest stage of Ni–Fe distribution. Limonites which are "near" the composition of meteorical material could be due to the oxidation of Ni–Fe spherules.

Table 3. Microprobe analyses of several phases in the recrystallised impact melt n° 447A—($\bar{N}i$ = 480 ppm, $\bar{K}$ = 8.60%)—(H_2O calculated).

	secondary K-feldspar				secondary chlorite	
	%	c.f.	%	c.f.	%	c.f.
SiO2	63.74	2.990	65.00	3.029	42.32	8.1298
TiO2	–	–	–	–	0.06	0.0087
Al2O3	18.26	1.0090	17.49	0.9604	12.78	2.8938
Fe2O3 FeO	0.21	0.0087	0.06	0.0022	10.55	1.6950
Cr2O3	0.01	0.0006	–	–	–	–
MnO	–	–	–	–	0.23	0.0374
MgO	0.04	0.0026	–	–	18.68	5.3493
CaO	0.11	0.0056	0.02	0.0012	1.08	0.2223
Na2O	0.23	0.0210	0.11	0.0096	–	–
K2O	15.98	0.9564	16.60	0.9867	0.64	0.1569
H2O	–	–	–	–	12.47	8.0000
NiO	0.02	0.0008	–	–	0.51	0.060
TOTAL	98.59	4.9945	99.27	4.9891	99.32	26.55

	secondary limonites (crystallochemical formula of goethite)					
	%	c.f.	%	c.f.	%	c.f.
Fe2O3 FeO	85.95	1.9424	87.19	1.9659	89.30	1.9813
NiO	3.44	0.0864	2.04	0.0511	1.14	0.0281
H2O	9.97	1.0000	10.00	1.0000	10.16	1.0000
TOTAL	99.36	3.0288	99.23	3.0170	100.60	3.0094

	secondary bravoite %
S	51.05
Fe	29.90
Co	0.69
Cu	0.20
Ni	13.27
Total	95.11

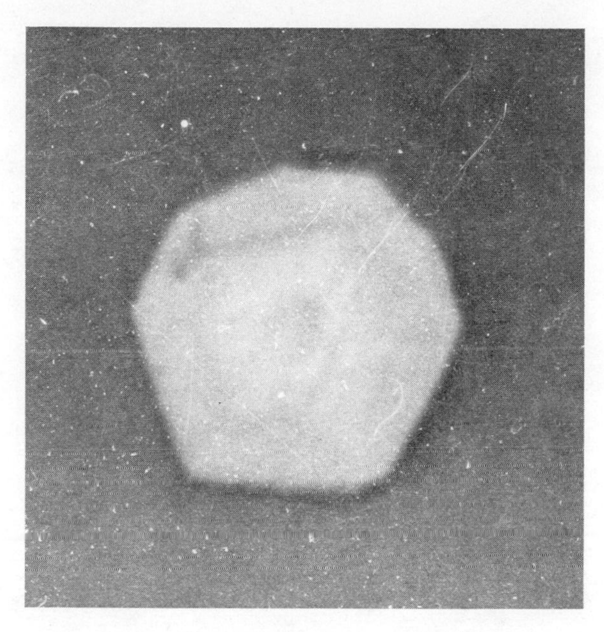

Fig. 5. Automorphous (30 μ size) bravoite in the n° 477 impact melt (see chemical analyses Table 3).

In the Ni–Co poor milieu, as in the granite-gneiss melt, Ni, Co could have been trapped by sulfides as bravoite with preservation of the meteoritic Ni/Co ratio. Changes in equilibrium conditions involved new formation of limonite.

Considering the composition of the Rochechouart meteorite, the Ni content of each breccia type, and the one of each related rock source, the proportion of meteorite in each breccia type may be deduced. The whole impact melt formation contains, about 0.2% of projectile material, whereas Ni rich impact melts reach 0.4%. In suevites the proportion falls to 0.085% and 0.045% for C breccias.

As fall back breccias from Rochechouart are mainly constituted by C breccias, the most part of the recovered material of the Rochechouart projectile is actually located in these low shocked ejecta.

In the case of a large impact crater meteoritic material is often suspected to be concentrated in the melt fraction. Rochechouart does not seem to confirm this assumption.

However this result may not reflect the original proportion of meteoritic material located in C breccias, as Ni movements from glasses to clast fractions (demonstrated in section entitled *Variations of Ni content in analysed rocks* at a small scale) are probably not restricted to the melt area (see Fig. 6).

Nevertheless it is not possible to determine if these movements are responsible for most (or all) of the Ni in C breccias, or if migration is a minor effect. In any case such movements are restricted to the fall back breccias, as rocks lying under the crater floor show a regular Ni content (even autochthonous breccias).

Figures 2 and 3 suggest some interesting aspects concerning the nature of

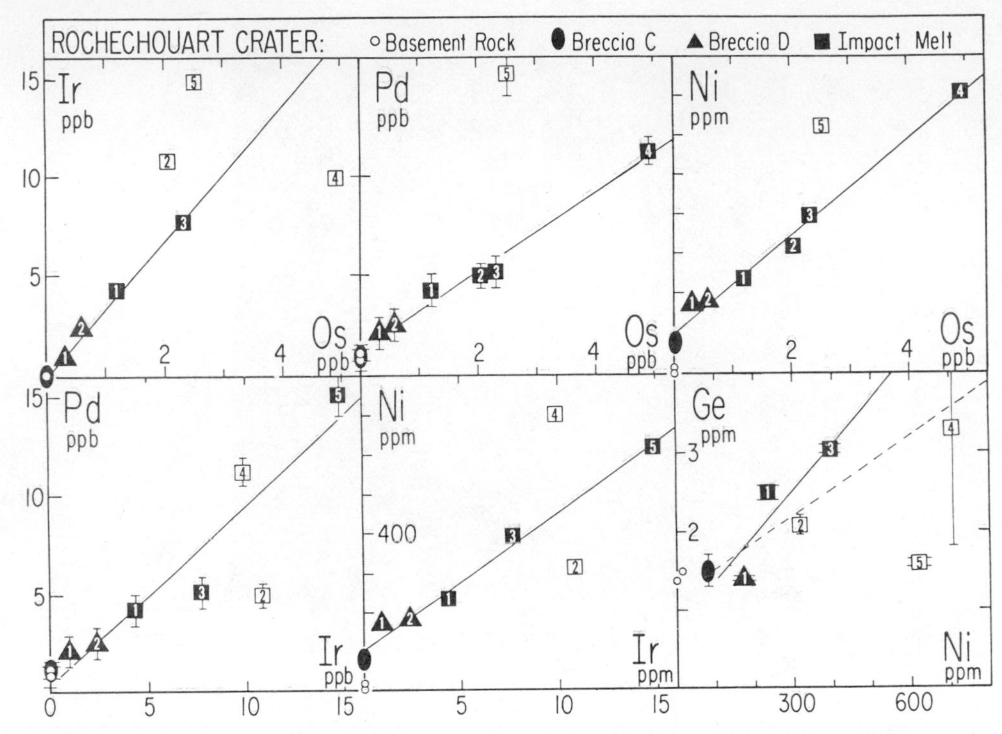

Fig. 6. Interelement correlations for diagnostic siderophiles (Janssens *et al.*, 1976, 1977)—Slope gives the ratio in the Rochechouart meteorite, while intercepts give indigenous abundances. The high intercept for Ni (100 ppm), consistently above the basement rocks (< 10 ppm) suggest Ni movements even outside of the melt area. Impact melts are depleted in Ni whereas D and C breccias are enriched, consequently Os/Ni and Ir/Ni slopes are too low explaining high Ni intercept.

rock source for E. D. and C. breccias. Figure 3 clearly indicates that impact melts (lying just above the crater floor) reflect the composition of the basement just below the crater (see definition of rock source in Table 2). C and D values are concentrated only in the gneiss area, or in the top of the granitic one, suggesting a much more gneissic rock source for C and D breccias (gneiss: about 75% in D, near 85% in C). These observations seem to indicate that the actual extent of the three main rock types of the basement, in the area of the crater, was different at the pre-impact level.

Probably intrusive granites were deeper. But even if the limit of granite intrusions was just above the actual ground level, there is always too much gneiss in C breccias. As orthogneiss occur on all the eastern part of the actual basement, it is suggested that gneiss is lying above orthogneiss and that the whole metamorphic features are inclined to the west. These two deductions are in agreement with the structural scheme of this part of the French "Massif Central" (A. Autran, pers. comm.). In any case we have to consider that the Rochechouart target was essentially constituted by gneiss, except for the deeper part where granitic rocks were more abundant, reaching 50% at the level of the crater floor. The recovered melt of Rochechouart does not reflect the average composition of the target; it is more granitic.

There are two possibilities to explain this result:

(1) The shock wave melted granites preferentially,
(2) The recovered melt came from the deep part of the target.

The first hypothesis has to be ruled out, as suggesting a reaction occurring only under strict equilibrium conditions. This leaves alternative (2) as the most probable.*

Consequences

During the impact process, in the center of the structure, the melt ends in the target nearly at the same level that excavation growing ends: in other words, in the target, the limit of melt is close to the limit of excavation (below the impact point).

For the melt produced in the deepest part of the target, axial and radial velocities are very low as such a melt is recovered in the crater.

Melt occurring in the upper level of the target is not mixed with the one produced deeper. This suggests a quick deflection of axial particle motions or a quicker decrease of axial velocity than radial one.

The meteoritic contamination of Rochechouart melt rocks implies that some projectile material reached the deep part of the target, suggesting that the axial particle motion in the projectile ended nearly at the same time and level than axial particle motion in the target.

References

Hurley, P. M.: 1969, Rb, Sr isotopic investigation of crater rocks of possible impact origin: Variation in isotopic abundances of Stronium, Calcium and Argon and related topics. DOC-MIT 1381, 17. Annual progress report for 1969 U.S.A., E.C. contract +(30-1-1381), p. 27.

Janssens, M. J., Hertogen, J., Takahashi, H., Anders, E., Lambert, P.: 1976, Meteoritic material in the Rochechouart crater and prevalence of irons among crater forming meteorites (abstract). In *Papers Presented to the Symposium on Planetary Cratering Mechanics*, p. 62–63. The Lunar Science Institute, Houston.

Janssens, M. J., Hertogen, J., Takahashi, H., Anders, E., Lambert, P.: 1977, Rochechouart meteorite crater: Identification of projectile. *J. Geophys. Res.* in press.

Kelly, W. R., Holdsworth, E., Moore, C. B.: 1974, The chemical composition of metallic spheroids and metallic particles within impactite from Barringer Meteorite Crater Arizona. *Geochim. Cosmochim. Acta* **38**, 533.

Kraut, F.: 1967, Sur l'origine des clivages du quartz dans les brèches "volcaniques" de la région de Rochechouart. *C. R. Acad. Sci. Fr.* **264**, 2609.

---

*This assumption deduced from observation of natural phenomenon will have to be compared to results obtained from theoretical calculations. New information, where phase changes have been taken into account, is beginning to be available (O'Keefe and Ahrens, 1976) and seems in agreement.

Kraut, F., French, B. M.: 1971, The Rochechouart meteorite impact structure, France; Preliminary geological results. *J. Geophys. Res.* **76**, 5407.

Lambert, P.: 1974a, Etude geologique de la structure impactitique de Rochechouart (Limousin-France) et son contexte. *Bull. B. R. G. M. Fr.*, section 1, n° 3, p. 153.

Lambert, P.: 1974b, La structure d'impact de météorite géante de Rochechouart. Doc. spec., Univ. Paris-Sud, p. 148.

Lambert, P.: 1975, Nickel enrichment of impact melt rocks from Rochechouart. Preliminary results and possibility of meteoritic contamination. *Meteoritics*, **10**, 433.

Lambert, P.: 1977, The Rochechouart crater = Shock zoning study. *Earth Planet. Sci. Lett.* **35**, 258–268.

Morgan, J. W., Higuchi, H., Ganapathy, R., Anders, E.: 1975, Meteoritic material in four terrestrial meteorite craters. *Proc. Lunar Sci. Conf. 6th*, p. 1609–1623.

O'Keefe, J. D., Ahrens, T. J.: 1976, Impact ejecta on the Moon—Contribution number 2763, Division of Geological and Planetary Sciences. California Institute of Technology, Pasadena, California.

Raguin, E.: 1972, Les impactites de Rochechouart (Haute-Vienne). *Bull. B. R. G. M. Fr.*, section 1, n° 3, p. 1.

Stöffler, D.: 1975, Ries deep drilling results: Implications for the substructure of the crater basement and the distribution of excavated masses. *Meteoritics* **10**, 495.

Roddy, D. J., Pepin, R. O., and Merrill, R. B., editors.
(1977) *Impact and Explosion Cratering*, Pergamon Press (New York), p. 461–480.
Printed in the United States of America

Buried impact craters in the Williston Basin and adjacent area

H. B. SAWATZKY

10th Floor, 630–6th Ave. S. W., Calgary, Alberta, Canada

Abstract—Five subsurface structures, either within or adjacent to the Williston Basin, reveal some of the essential ear-marks of terrestrial impact craters. One (Viewfield, Saskatchewan) is a simple bowl-shaped crater and the remainder are complex models, which, in addition to the structurally elevated outer rim, also display a positive highly disturbed central core. Two of these features appear to be Late Cretaceous, or younger, in age and three are dated as Jurassic-Triassic.

The Red Wing Creek, North Dakota and Viewfield, Saskatchewan geophysical anomalies (interpreted as buried impacts) have yielded commercial oil production; the former from the central area and the latter from the outer rim. Some interesting hydrocarbon occurrences are indicated at the other sites. These too may be potential commercial oil producers but they have been inadequately explored to date.

INTRODUCTION

THE MAIN PURPOSE of this paper is to briefly outline, with the aid of pertinent maps, cross-sections based on drilling logs and seismic records (Sawatzky 1972, 1975, 1976), the basic parameters which allow us to identify five subsurface features in the Williston Basin and adjacent area as buried impact craters (Viewfield, Saskatchewan, T. 7, R. 8, W2M; Hartney, Manitoba, T. 5, R. 24, W1M; Red Wing Creek, North Dakota, T. 148N, R. 101W; Dumas, Saskatchewan, T. 11, R. 1 and 2, W2M; Eagle Butte, Alberta, T. 8 and 9, R. 4 and 5, W4M). Additionally this paper is intended to show that impact craters are potential commercial sites for the accumulation of hydrocarbons.

The locations of these structures are listed in Table 1 and shown in Fig. 1 which illustrates the distribution of craters and shatter cone sites in North America. Because of the rapid rate at which new sites have been examined and confirmed during the past few years, some of the more recent discoveries probably are missing from this diagram.

VIEWFIELD (SASKATCHEWAN) CRATER

A bowl-shaped depression approximately $1\frac{1}{2}$ miles (2.4 km) in diameter, with a present rim elevation of approximately 200 ft (60 m) and a minimum depth of 600 ft (185 m) has been identified in Saskatchewan, Canada, on the basis of drilling and geophysical information. Fig. 2 shows structure contours on the top of the Mississippian. However, the wells near the center of the feature (see Fig. 3, 13–29–7–8W2 and 15–29–7–8W2) did not reach the Mississippian. Detailed seismic data, as well as theoretical crater shape calculations suggest that wells may still be several hundred feet short of this target.

Figure 3 is an electric log section based on electric drill logs (spontaneous

Table 1. Meteorite impact confirmed sites.

Map ref.	Map ref.
1. New Quebec Crater, Quebec	18. Gow Lake, Saskatchewan
2. Brent, Ontario	19. Lac La Moinerie, Quebec
3. Manicouagan, Quebec	20. Haughton Dome, Devon Island
4. Clearwater Lakes, Quebec	21. Slate Islands, Lake Superior
5. Holleford, Ontario	22. Ile Rouleau, Quebec
6. Deep Bay, Saskatchewan	23. Barringer, Arizona
7. Carswell, Saskatchewan	24. Odessa, Texas
8. Lac Couture, Quebec	25. Crooked Creek, Missouri
9. West Hawk Lake, Manitoba	26. Decaturville, Missouri
10. Pilot Lake, Mackenzie Dist.	27. Flynn Creek, Tennessee
11. Nicholson Lake, Keewatin Dist.	28. Kentland, Indiana
12. Steen River, Alberta	29. Manson, Iowa
13. Sudbury, Ontario	30. Middlesboro, Kentucky
14. Charlevoix, Quebec	31. Serpent Mound, Ohio
15. Lake Mistastin, Labrador	32. Sierra Madera, Texas
16. Lake St. Martin, Manitoba	33. Wells Creek, Tennessee
17. Lake Wanapitei, Ontario	
Probable sites	
A. Elbow, Saskatchewan	D. Red Wing Creek, North Dakota
B. Viewfield, Saskatchewan	E. Eagle Butte, Alberta
C. Hartney, Manitoba	F. Dumas, Saskatchewan

potential and induced resistivity) pertaining to the wells shown in the previous figure. On the periphery of the rim disturbed Mississippian limestone and dolomite are sandwiched between Lower Watrous redbeds which are Lower Jurassic–Upper Triassic. Hence, the impact is considered to be of Jurassic–Triassic age (Fig. 4).

A seismic record section on an east-west line crossing the crater is shown in Fig. 5. The datum for the section is the Second White Speckled Shale (Upper Cretaceous) at 0.650 sec. However, this horizon displays a maximum depression over the crater of approximately 35 ft (11 m). The top of the Blairmore is also essentially flat, but the basal part and the Jurassic rocks show pronounced slumping into the crater. This is considered to be the result of post-impact Middle Devonian salt solution and collapse, as well as a certain amount of differential compaction of the clastic section over the crater. The seismic event labelled "Mississippian" depicts the top of this horizon well enough (except through the crater depression, where it more adequately represents the top of the Lower Watrous). The precise attitude of the Mississippian is difficult to follow through the crater but is suggested by strong dips displayed between this horizon and the Devonian Birdbear horizon. In addition to the general crater profile, the Birdbear shows several distinct dislocations. These are evident on all of the seismic lines associated with the impact area. The Prairie Evaporite (P.E.) can be mapped continuously across the crater and probably owes its configuration to a certain amount of plastic flow at the time of impact plus post-impact

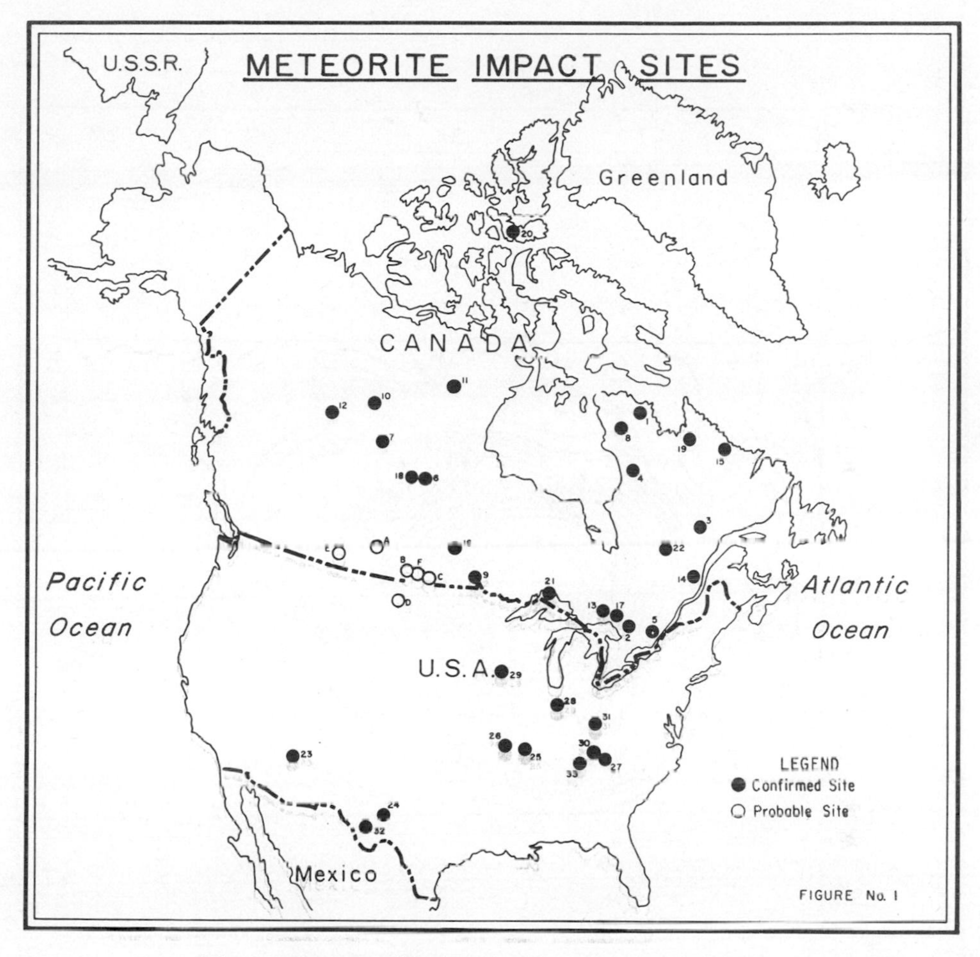

Fig. 1. Index map of impact sites in Canada and the U.S.A.

solution. Some of the sag is no doubt the result of the replacement of high-seismic velocity Mississippian rocks in the crater by relatively low-seismic velocity fill. The seismic reflection from Cambrian-Ordivician strata can be correlated quite successfully and shows its regional character. Minor "wrinkles" are attributed to differential seismic velocity effects in the overlying section.

There are presently some 30 wells producing oil from Mississippian rocks along the crater rim at initial rates as high as 300–400 bbls. per day, from gross producing layers of approximately 200 ft in thickness (60 m) at the best rim positions. Detailed, multifold seismic data were the key to development of this pool and greatly aided in the interpretation of this structure. For a more detailed description of the Viewfield anomaly, the reader is referred to an earlier paper (Sawatzky, 1972).

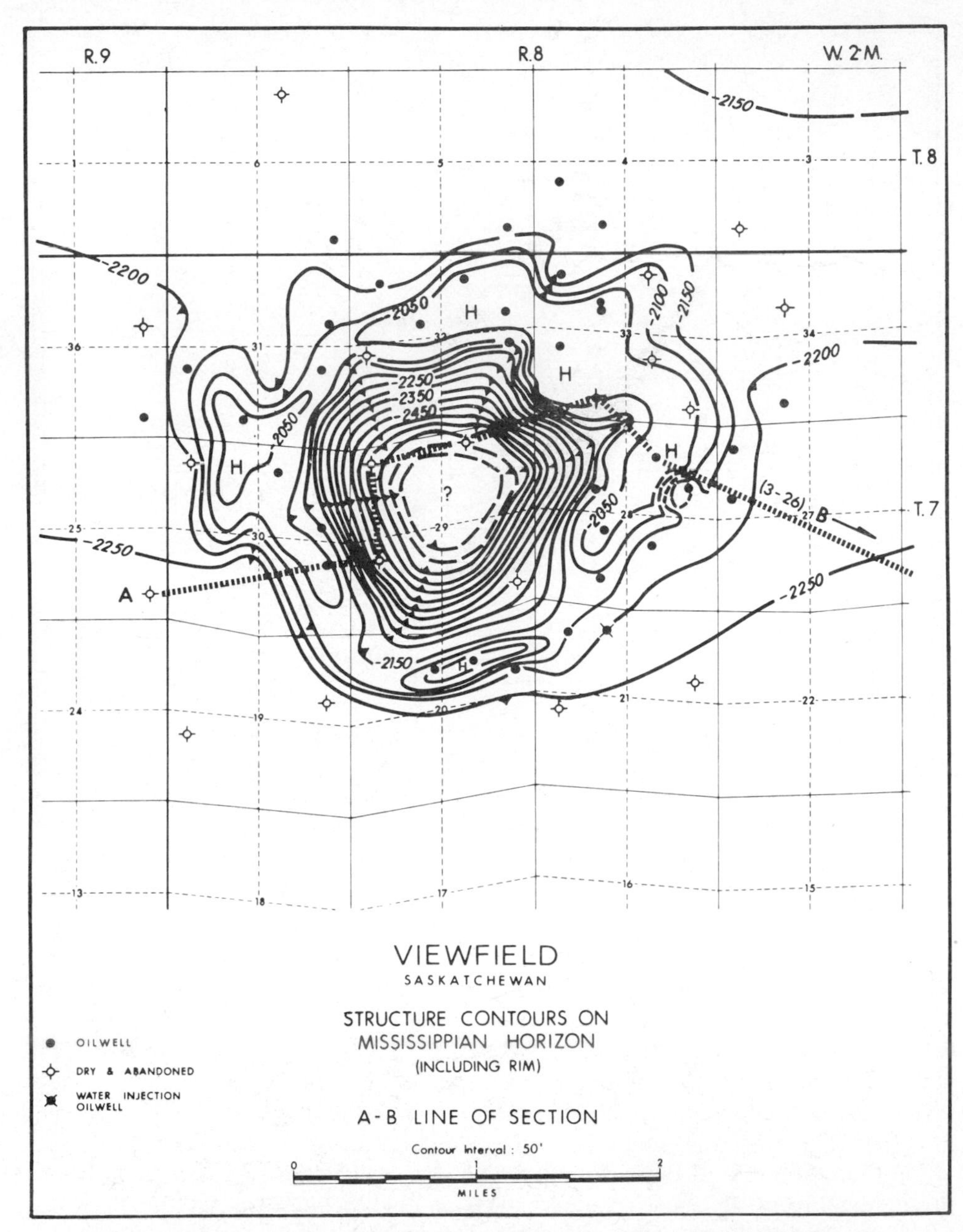

Fig. 2. Structure contours on Mississippian horizon of the Viewfield Crater, Saskatchewan. See numbers of holes for cross-section A-B in Fig. 3. North is at the top.

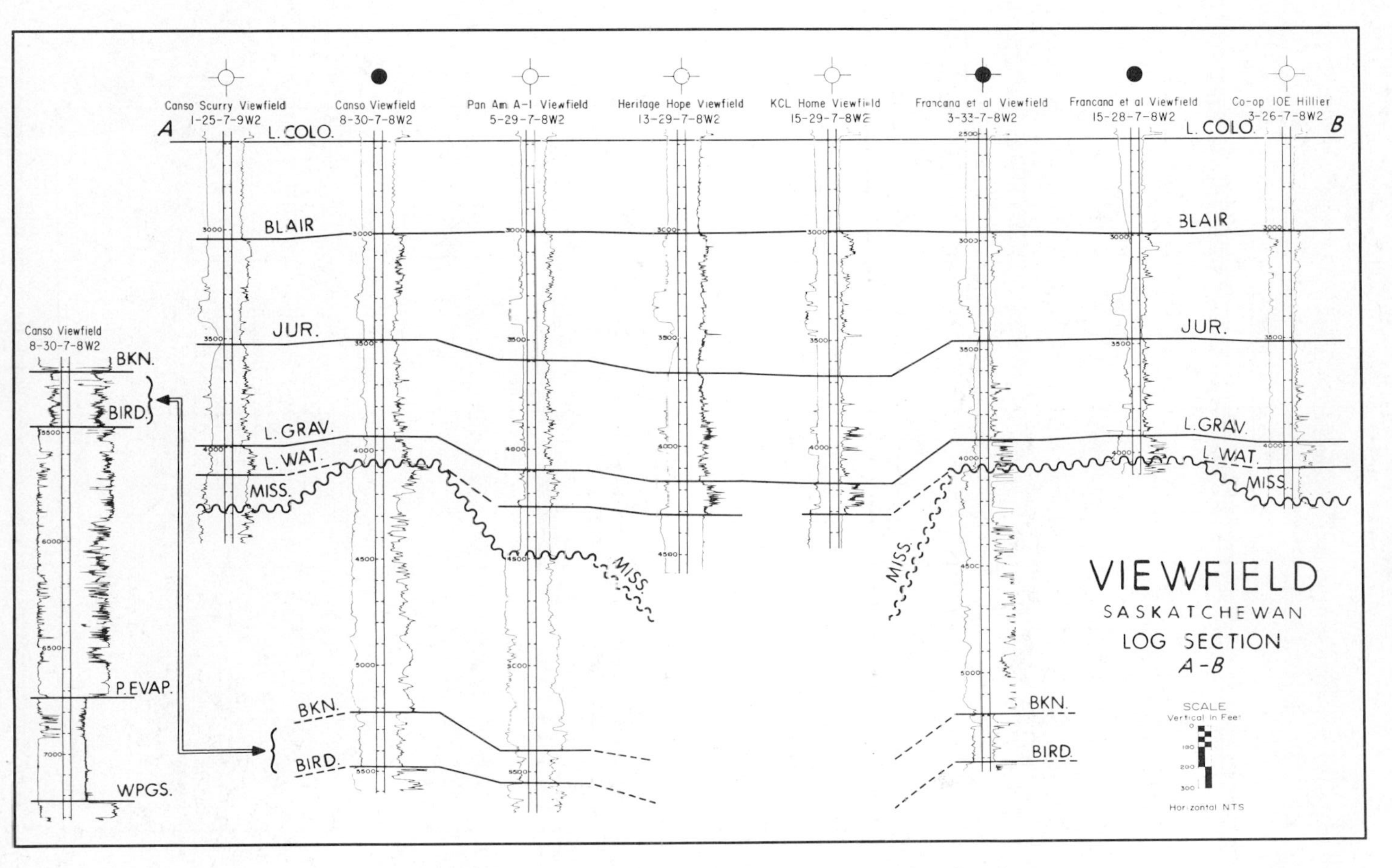

Fig. 3. Electric-log section of Viewfield structure, Saskatchewan, Canada.

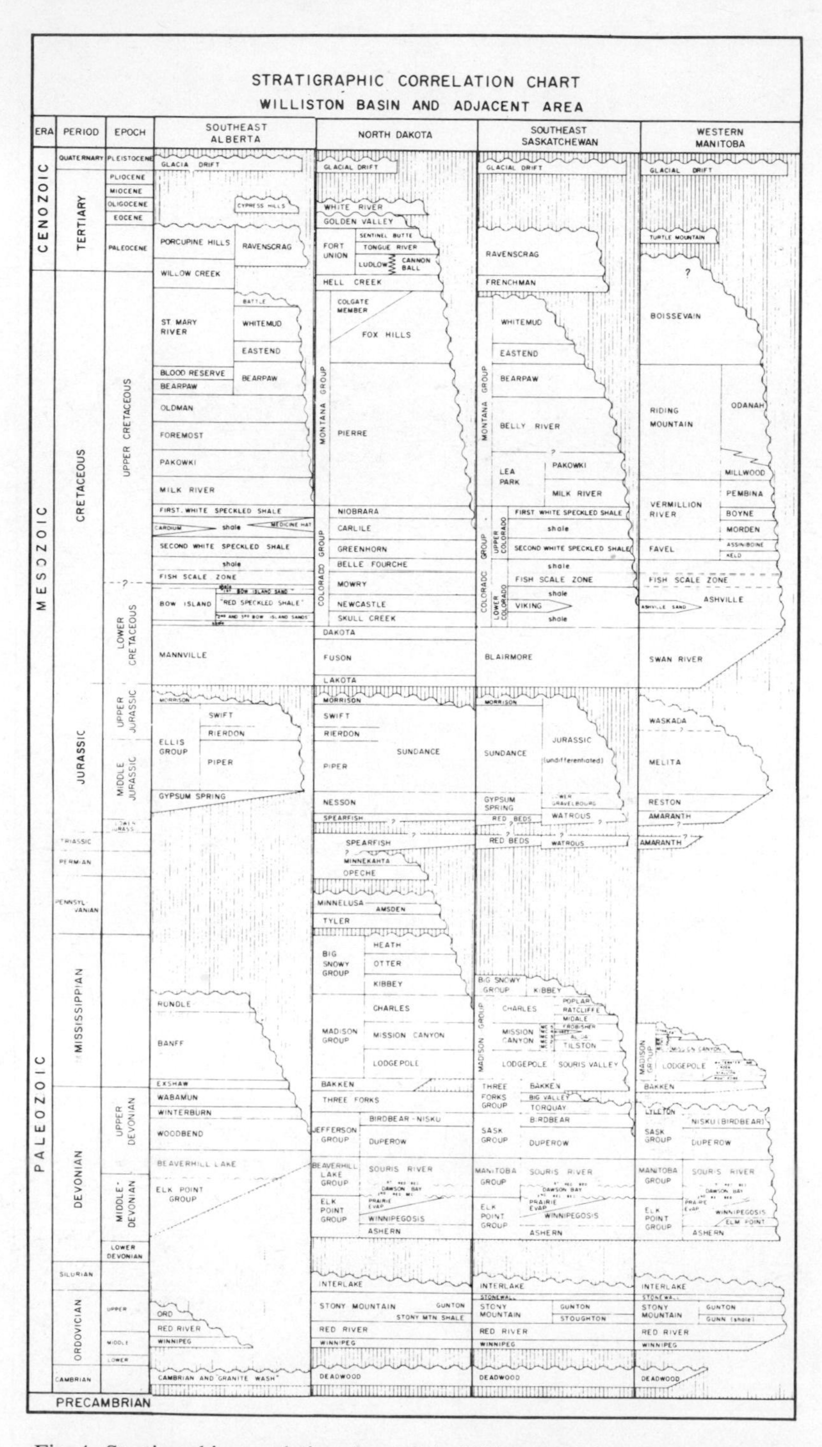

Fig. 4. Stratigraphic correlation chart of the Williston Basin and adjacent area.

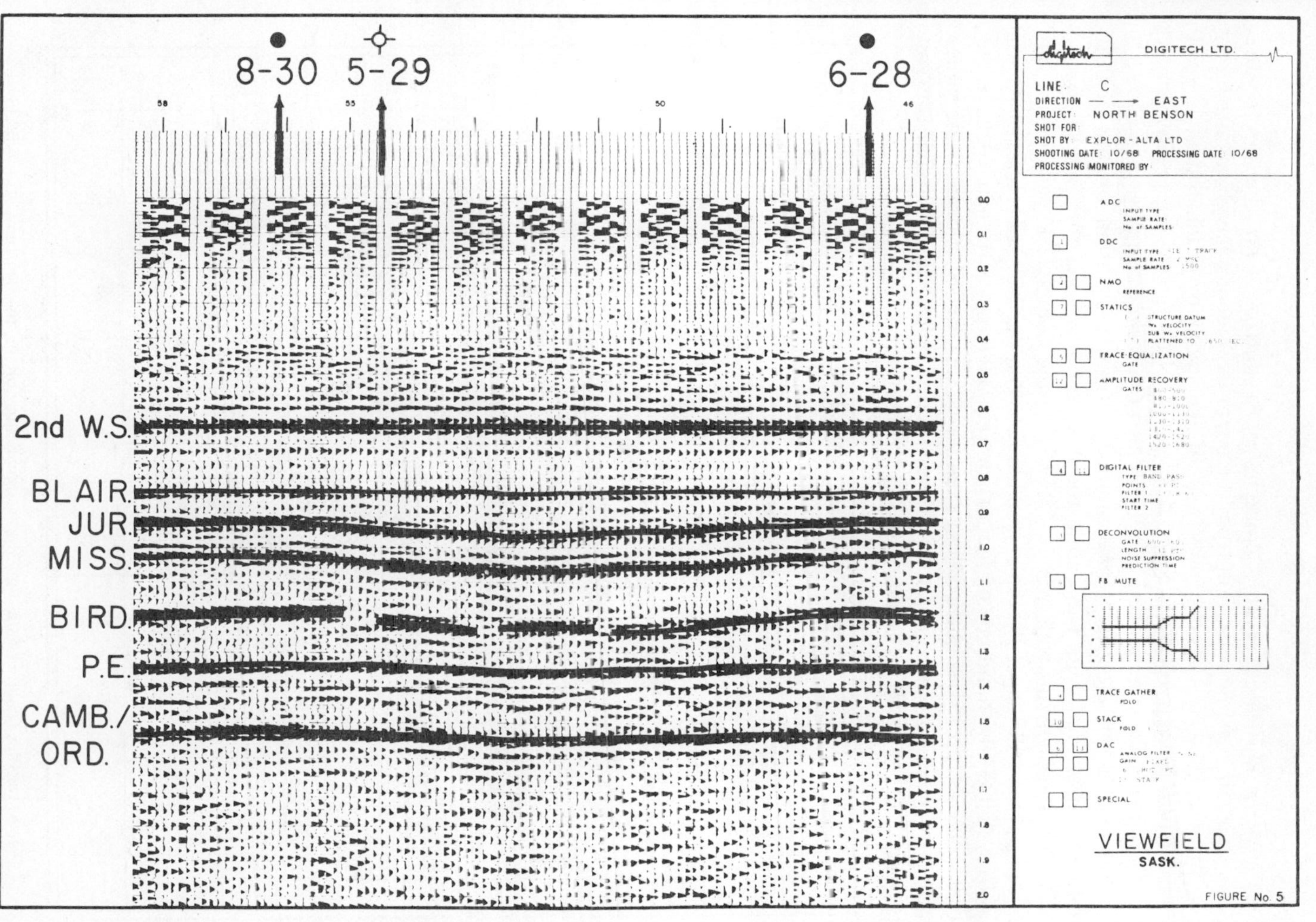

Fig. 5. Seismic section of the Viewfield structure, Saskatchewan, Canada.

HARTNEY (MANITOBA) CRATER

Figure 6 illustrates seismic contours of a horizon near the top of the Mississippian. Seismic reflection from the upper Cretaceous Favel was used as a datum at 0.350 sec. Although only half of the anomaly is illustrated, a good appreciation of its size and shape are possible. Note the strong positive outer rim, the inner ring syncline and the highly disturbed central uplift.

Figure 7 is an electric log section A-A′ from a position on the central uplift to an off-rim well. The datum for this section also is the Favel horizon. Note: (1) the high rim well at 7-27-5-24Wl; (2) the complete excavation of the Mississippian and Upper Devonian section between the 7-27-5-24Wl and 16-33-6-24Wl wells; (3) the hundreds of feet of stacked Devonian carbonate rocks at the 1-29 well (interpreted as an excellent example of repeated section); (4) the rather straightforward correlation of the geologic section below the Devonian Dawson Bay in all of the wells penetrating this horizon or deeper.

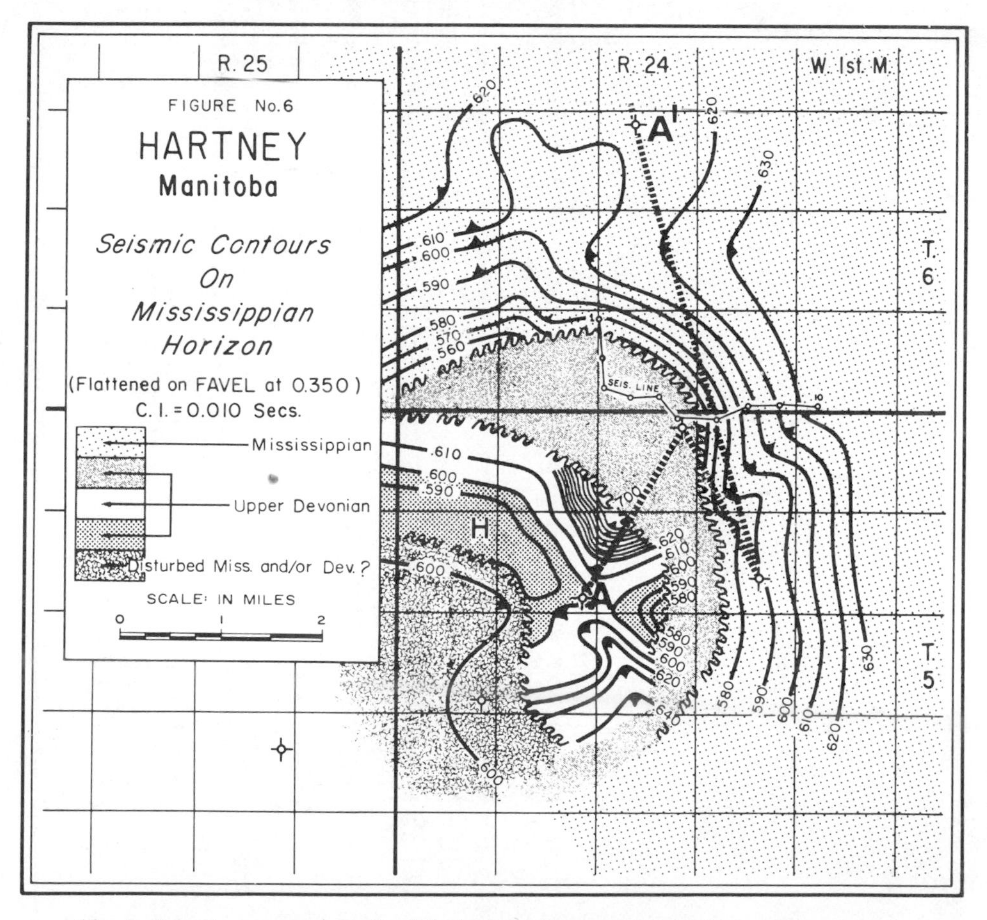

Fig. 6. Seismic contours on the Mississippian horizon of the Hartney structure in Manitoba, Canada. North is at the top.

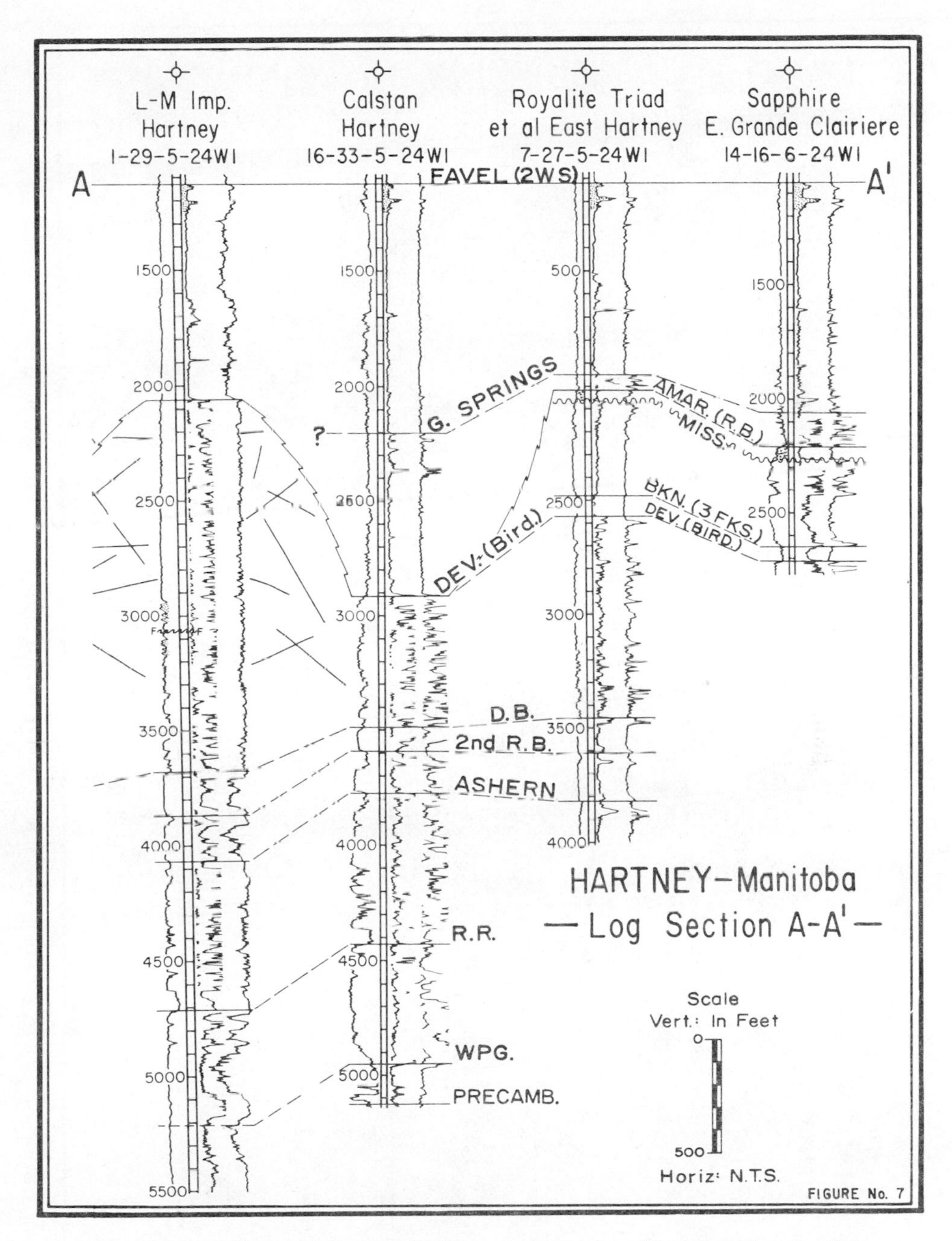

Fig. 7. Electric-log section through Hartney structure in Manitoba, Canada. Hole locations are along line A-A' shown in Fig. 6.

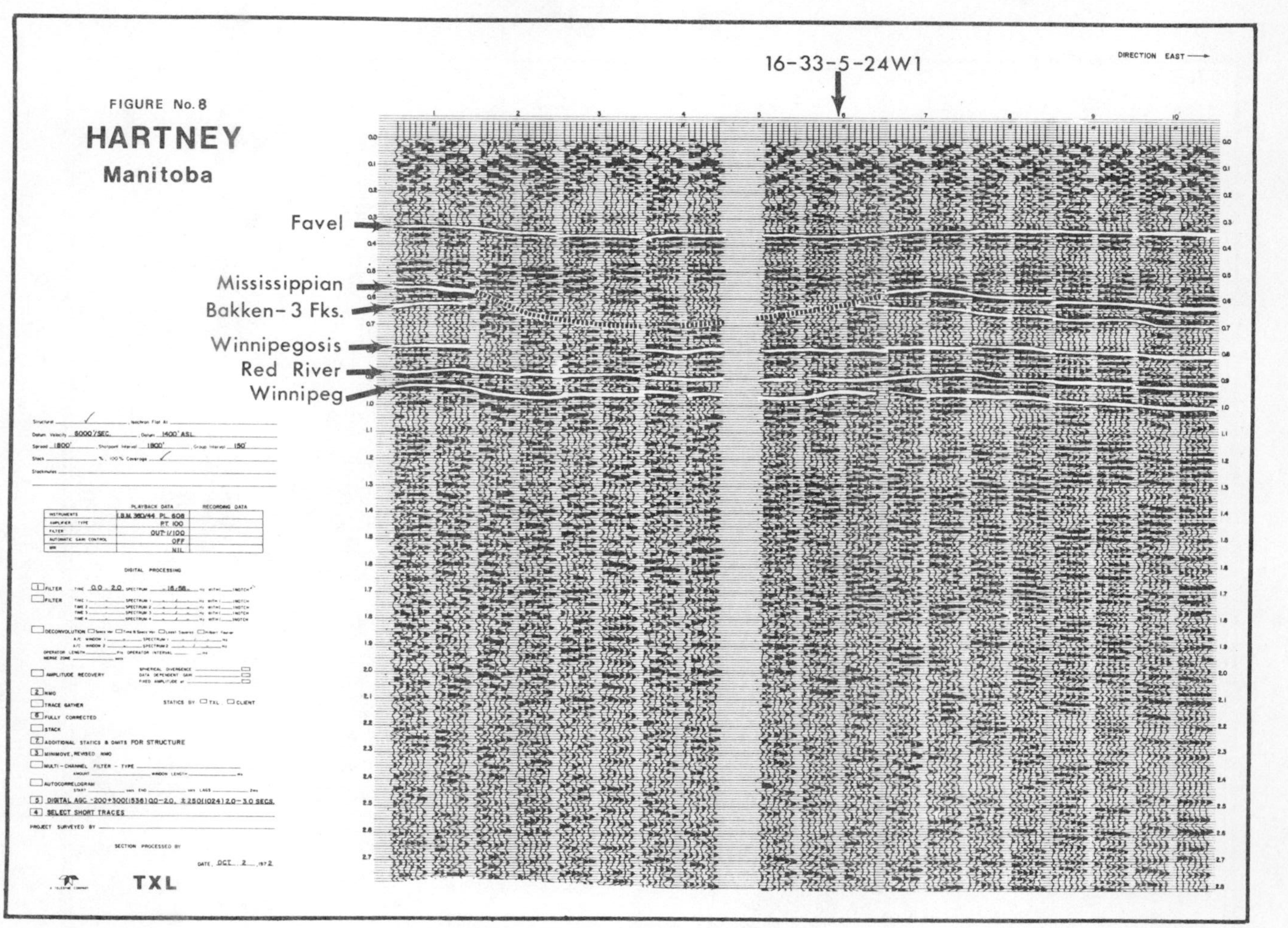

Fig. 8. Seismic section which was taken across the northeast corner of the Hartney structure.

Figure 8 is a seismic section composited from complete coverage crossing the northeast corner of the crater. This is a structural section with a 1400 ft (approximately 430 m) ASL datum. Note: (1) the slight depression on the Favel horizon; (2) the complete absence of the Mississippian-Three Forks reflections in the Synclinal area. The crater also appears to be Jurassic–Triassic in age.

Red Wing Creek (North Dakota) Crater

The location of wells pertinent to this feature are shown in Fig. 9 which also depicts the line of section B-B′ illustrated in Fig. 10. Commercial oil production is obtained from wells located on the complex central uplift as indicated on the electric log section. The discovery well (True 22–27 Burlington Northern, SE, NW Sec. 27, T. 148N, R. 101W) contains approximately 3000 ft (860 m) of producing rocks in what has been described as steeply dipping, intensely thrust-faulted Mission Canyon (?) carbonate rock. The inferred outer rim and crater profile on the left side of Fig. 10 are based on seismic information.

Since this structure appears to be pre-Piper—post-Minnekahta in age, its origin falls into approximately the same time interval as Viewfield and Hartney. The Piper and younger sediments indicate drape over this feature; however, some of the differential relief appears to be related to thickness variations in Charles salts that were affected at the time of impact, as well as during the post-impact period. Although some deformation extends into pre-Mississippian rocks, correlations are readily made at and below the Bakken-Three Forks interval. It is interesting to compare Fig. 10 with Fig. 7 and note the similarity in structural style. Shock metamorphic evidence has been recognized in cores obtained from wells located on the central uplift (State of North Dakota Industrial Commission Hearing, September, 1973).

Eagle Butte (Alberta) Canada

Figure 11 is based on bore hole data and shows the configuration of this anomaly at the base of Fish Scales (Upper Cretaceous). Note the pronounced central uplift and rim syncline. Inadequate well control prevents us from properly mapping the outer rim of this feature. However, existing seismic data (not available for publication) should be helpful in delineating the outer rim and better define the complex central core.

The electric log section illustrated in Fig. 12 pertaining to the wells shown in Fig. 11 should again be compared with similar sections related to the complex craters at Hartney and Red Wing Creek (Fig. 7 and Fig. 10, respectively). The similarities are again striking. The Milk River Formation is present immediately below the well casing at the surface in the 7-5 well, which appears to be drilled near the crest of the central uplift. Detailed examination of the interval between the Milk River and basal Colorado in the 7-5 and 7-32 wells indicates consider-

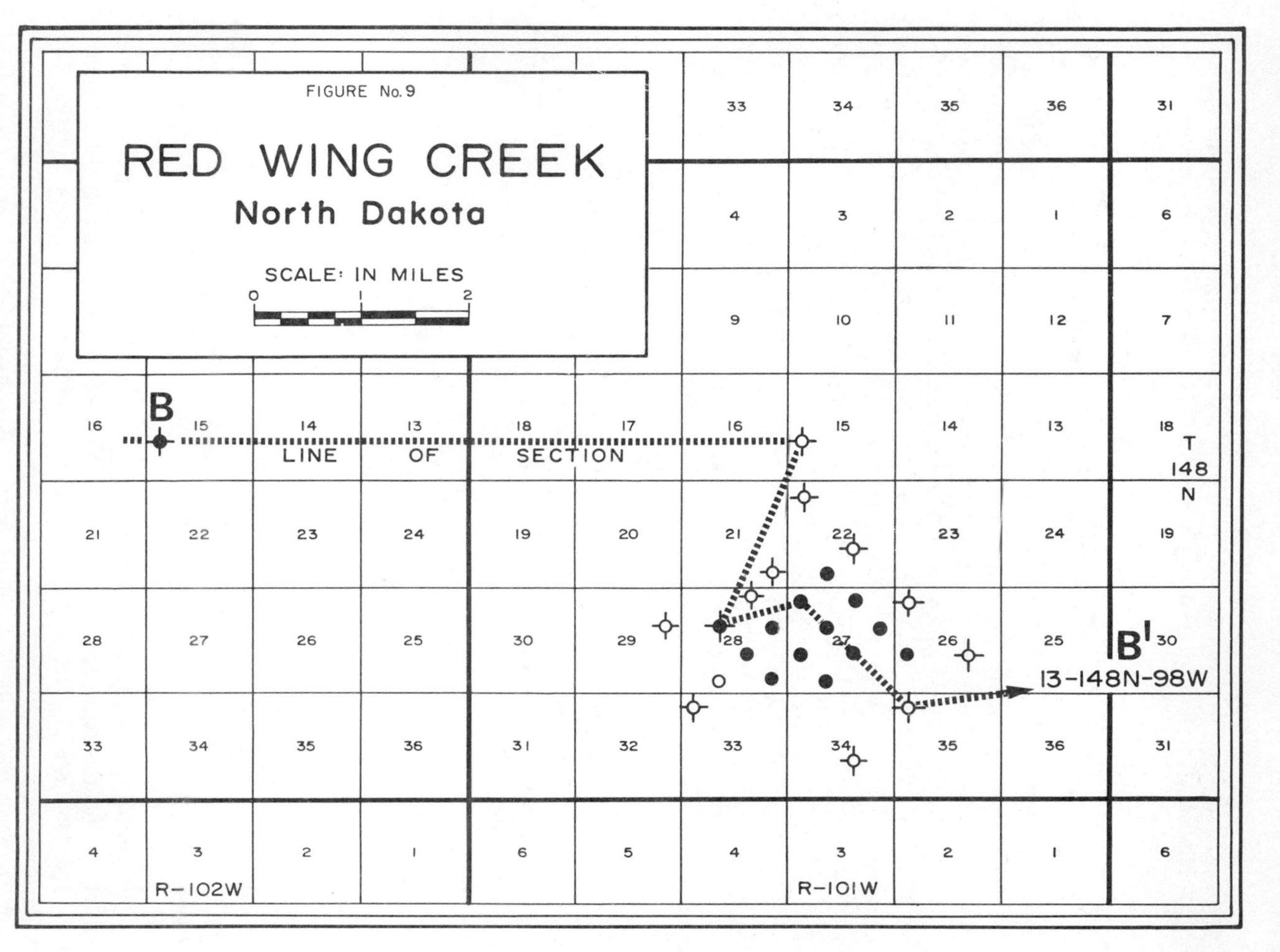

Fig. 9. Location of wells at the Red Wing Creek structure in North Dakota.

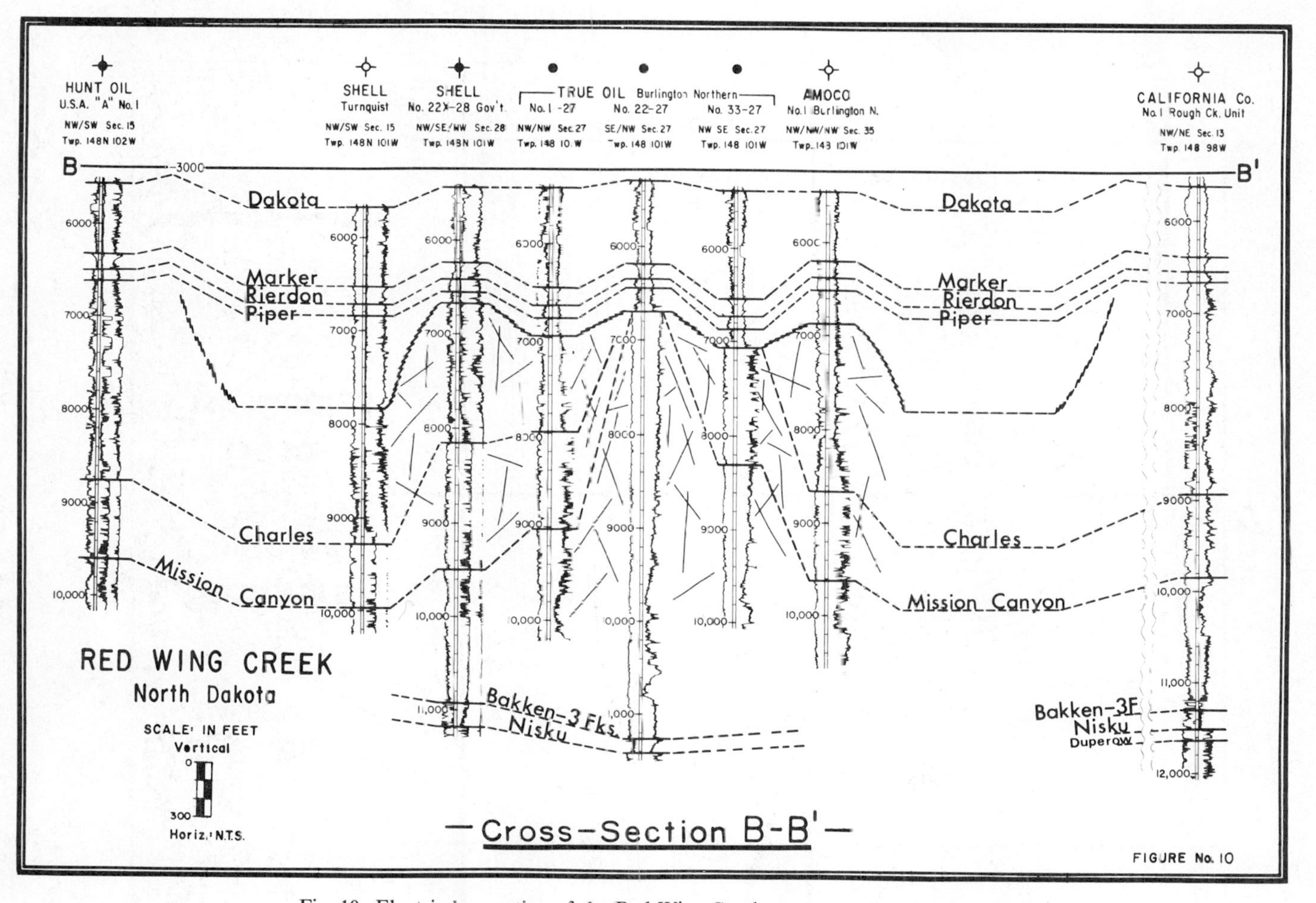

Fig. 10. Electric-log section of the Red Wing Creek structure, North Dakota.

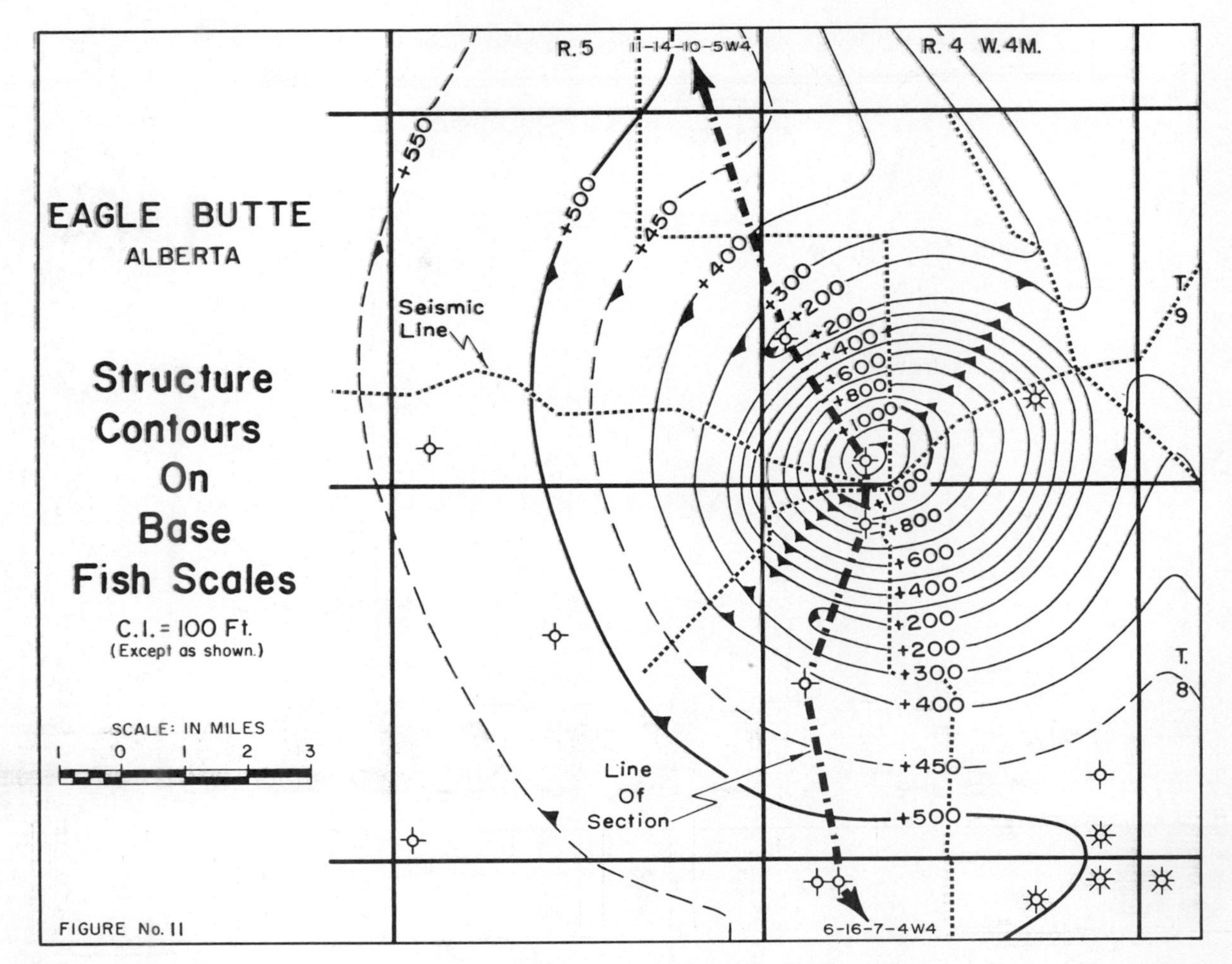

Fig. 11. Structure contours on the base Fish Scales horizon of the Eagle Butte structure, Alberta, Canada. North is at the top.

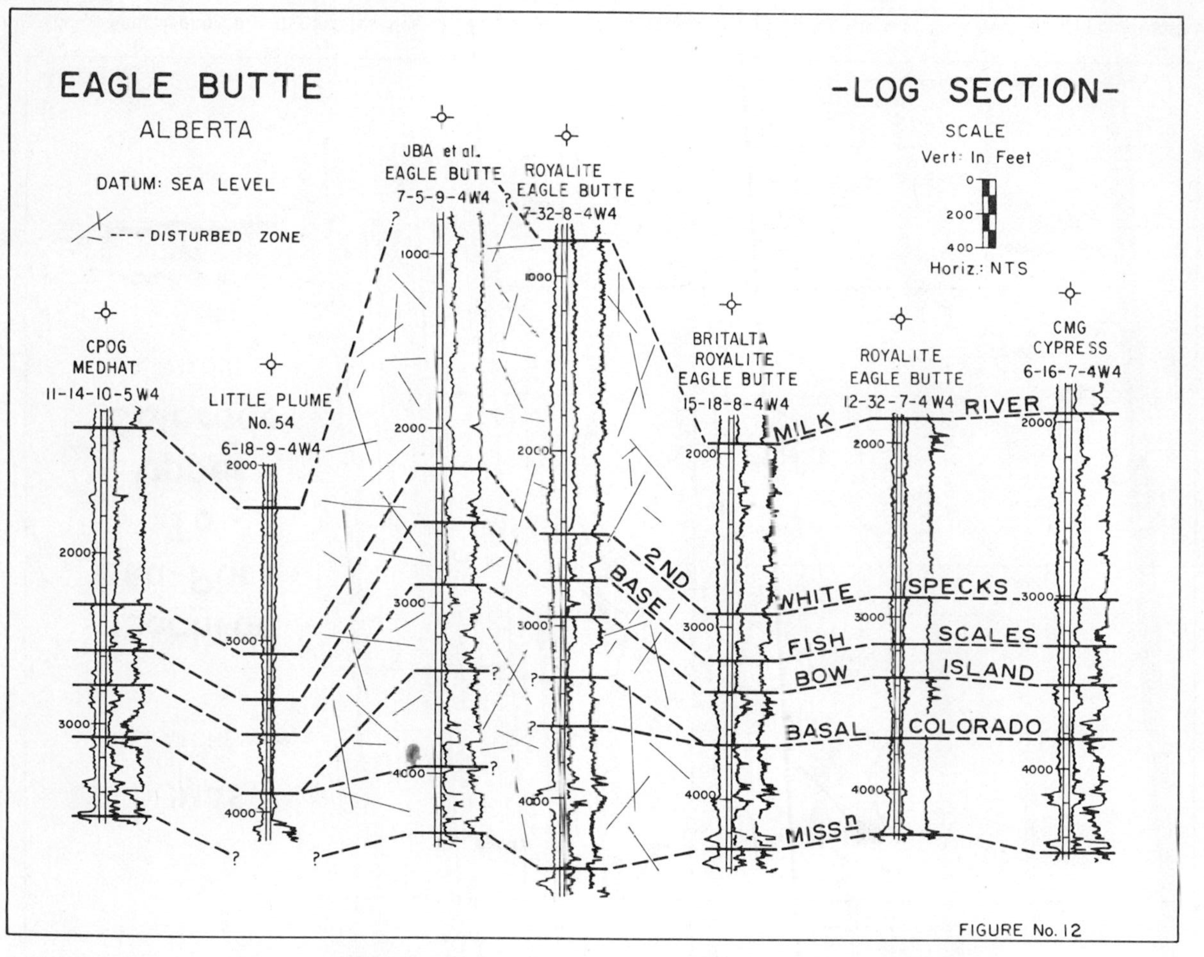

Fig. 12. Electric-log section of the Eagle Butte structure, Alberta, Canada.

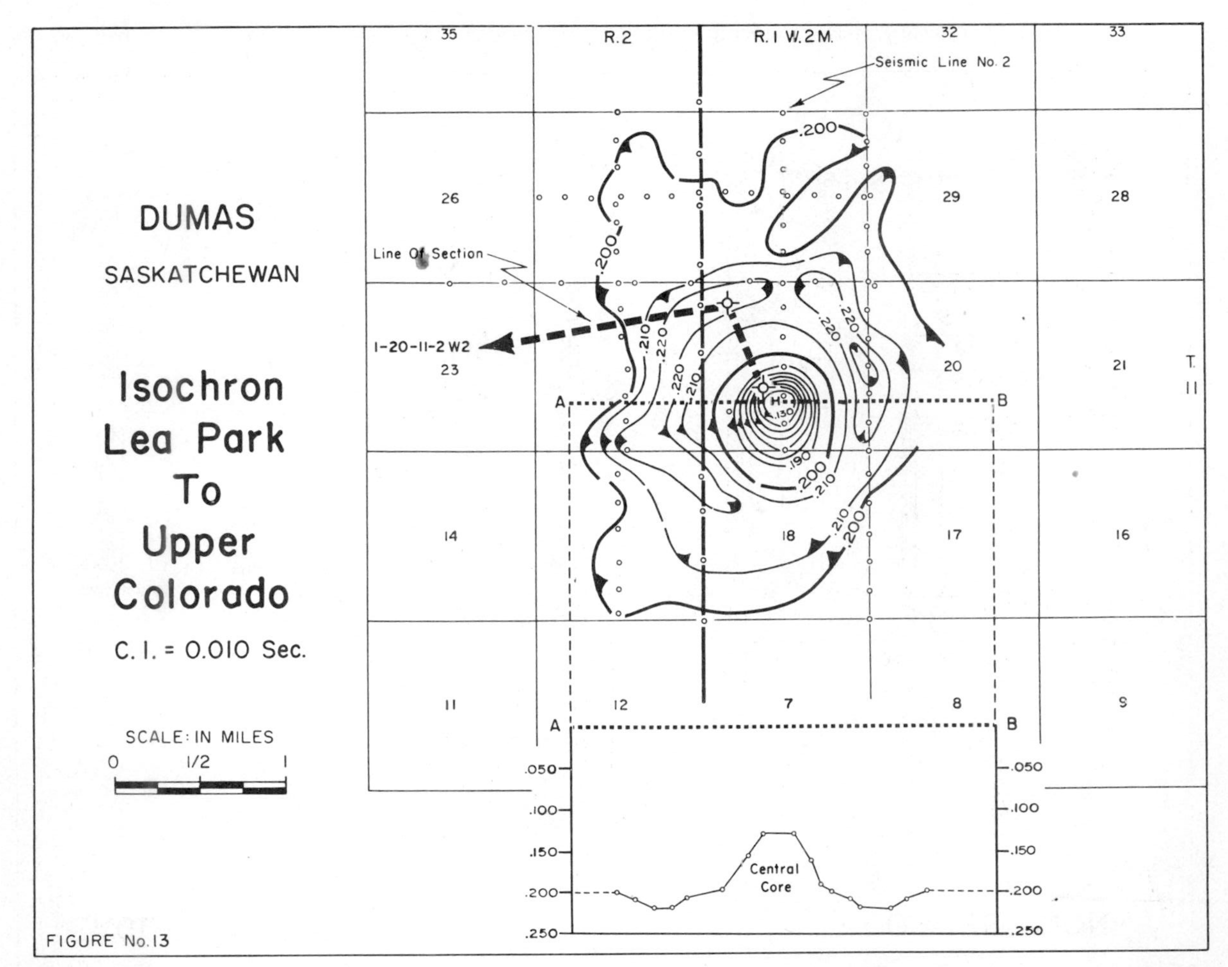

Fig. 13. Seismic Isochron map of difference of time of travel of seismic waves between the Lea Park and the Upper Colorado. North is at the top.

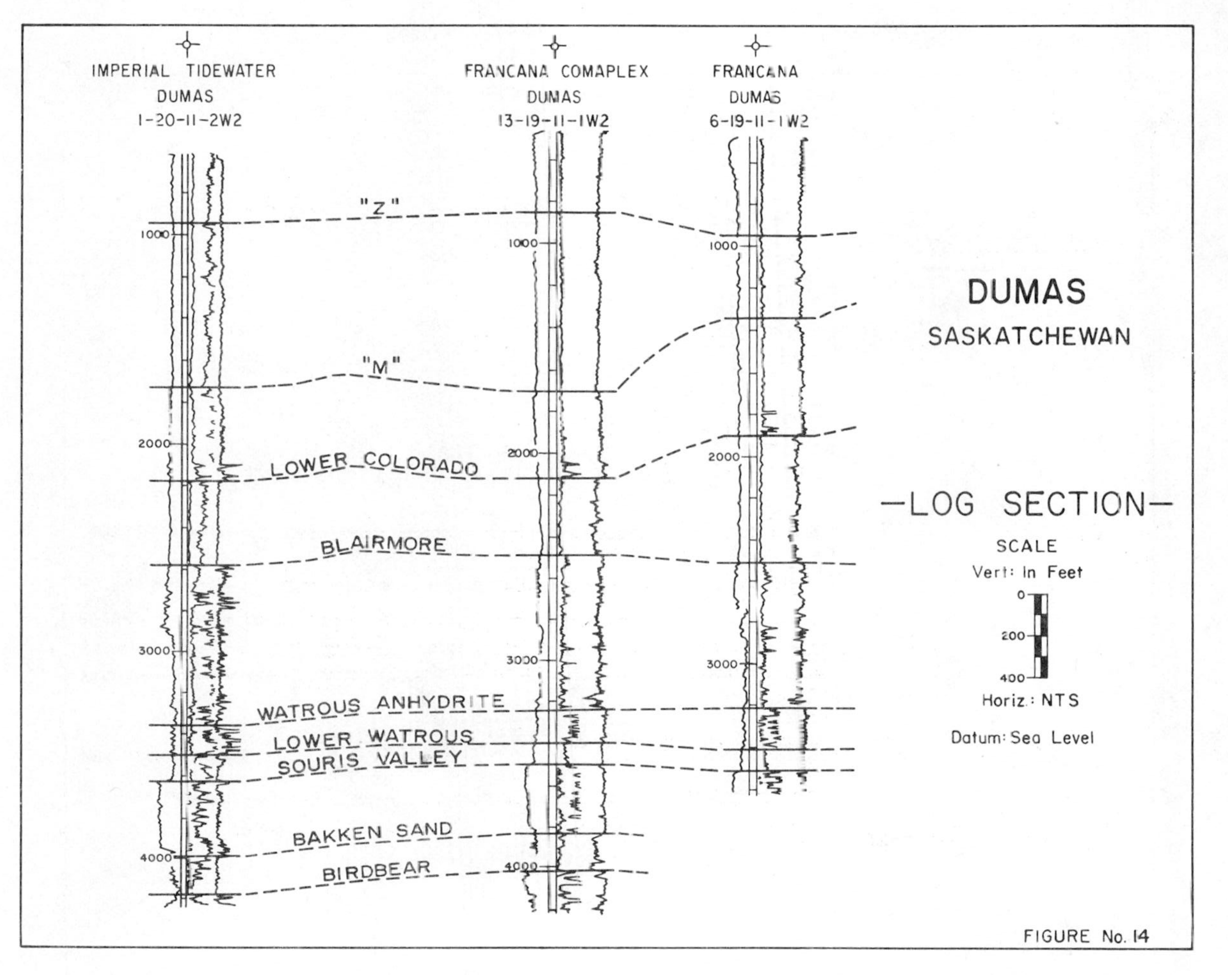

Fig. 14. Electric-log section of the Dumas structure, Saskatchewan, Canada.

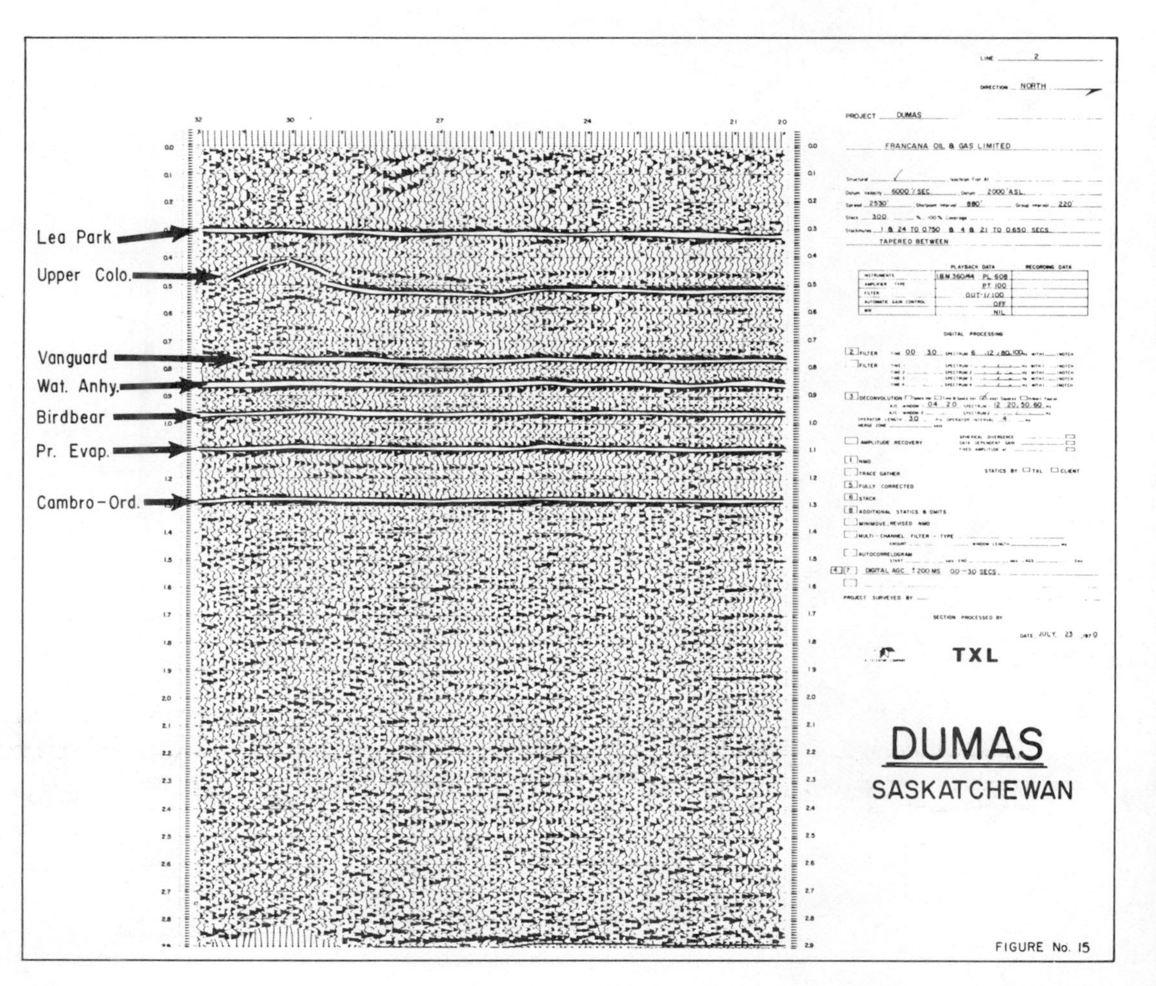

Fig. 15. Seismic section across the Dumas structure, Saskatchewan, Canada.

able faulting and repetition of section. Most of the structural relief has disappeared at the base of the Colorado Group. The age of the Eagle Butte structure is either Late Cretaceous or perhaps even Tertiary.

DUMAS (SASKATCHEWAN) CRATER

The Dumas Crater in Saskatchewan, Canada, is identified by the variations in seismic time interval between the Lea Park and Upper Colorado events (Fig. 13). The structure embraces an area of approximately 9 square miles (23 km^2) centered in Section19-11-1 W2M. Again a pronounced positive central uplift, a rim syncline and the subtle indication of a positive outer rim can be identified.

An electric log section showing half of the crater profile is illustrated in Fig. 14. Note in particular the additional thickness of strata between the Blairmore and "M" marker in the 6-19 well. This is interpreted to be part of the central uplift.

Figure 15 is a seismic structure section pertaining to Line #2 (Fig. 13) and it indicates the following:

(1) A relatively flat and undisturbed pre-Jurassic section. Note in particular that there is no disruption at the Prairie Evaporite level.
(2) The faulted and unusually thick section between the Jurassic Vanguard and Upper Colorado horizons in the vicinity of SP #30.
(3) The synclinal area on the Colorado events between SP #26 and SP #29.
(4) The slightly positive outer rim on the Upper Colorado at SP #25.
(5) The relatively flat-lying Lea Park horizon.

Seismic and well data indicate that the Dumas structure was formed in Late Cretaceous time (during final deposition of Upper Colorado or early Lea Park sediments).

For additional details and support information on the "buried impact crater" origin for the subsurface structures described in this paper (as opposed to alternate explanations such as tectonic deformation and/or salt solution) the reader is referred to earlier articles (Bishop, 1954; Dence, 1965, 1972; Haites and Van Hees, 1962; Parker, 1967; Sawatzky, 1972, 1975, 1976; Smith and Pullen, 1967, Swensen, 1967; Wilson *et al.*, 1963) listed under selected references.

CONCLUSION

Two of the five features discussed (Viewfield and Red Wing Creek) have resulted in commercial oil production and some interesting oil and gas shows were encountered at several other sites (Eagle Butte and Dumas). Hence, this type of structure has a very respectable record of commercial hydrocarbon production. We, therefore, urge both the pure science research community and the mineral explorationists to direct their efforts toward a greater degree of cooperation in this particular endeavour. The results would be mutually beneficial.

Acknowledgments—The writer thanks the Management of Francana Oil and Gas Ltd. for permission to publish this paper and extends his appreciation, in particular, to Mr. W. E. Nicholson for drafting the illustrations.

REFERENCES

Baldwin, R. B.: 1949, *The Face of the moon.* University of Chicago Press, Chicago. 239 pp.

Bishop, R. A.: 1954, Saskatchewan exploratory progress and problems. In *Western Canada Sedimentary basin, AAPG,* p. 474–485.

DeMille, G., Shouldice, J. R., and Nelson, H. W.: 1964, Collapse structures related to evaporites of the Prairie Formation, Saskatchewan. *Geol. Soc. Amer. Bull.*, **75**, 307–316.

Dence, M. R.: 1965, The Extraterrestrial Origin of Canadian Craters. *Ann. New York Acad. Sci.*, **123**, 941–969.

Dence, M. R.: 1972, The nature and significance of terrestrial impact structures. *Proc. 24th Intern. Geol. Cong., Canada,* sec. **15**, p. 77–89.

Dietz, R. S.: 1961, Astroblemes. *Sci. Amer.* **205**, 50–58.

Haites, T. B. and Van Hees, H.: 1962, The Origin of Some Anomalies in the Plains of Western Canada. *J. Alberta Soc. Petrol. Geologists* **10**, 511–533.

Milner, R. L.: 1956, Effects of salt solution in Saskatchewan (abstract). In *1st Williston Basin Symposium, North Dakota Geol. Soc.*, p. 111.

Milton, D. J., Barlow, B. C., Brett, R., Brown, A. R., Glickson, A. Y., Manwaring, E. A., Moss, F. J., Sedmick, E. C. E., Van Son, J., and Young, G. A.: 1972, Gosses Bluff Impact Structure. Australia. *Science* **175**, 1199–1207.

Parker, J. M.: 1967, Salt solution and subsidence structures, Wyoming, North Dakota, and Montana. *AAPG Bull.* **51**, 1929–1947.

Roberston, P. B., and Grieve, R. A. F.: 1975, Impact Structures in Canada. Their Recognition and Characteristics. *The J. R.A.S.C.* **69**, 1–21.

Sawatzky, H. B.: 1972, Viewfield—a producing fossil crater? *Canadian Soc. Exploration Geophysicists J.* **8**, 22–40.

Sawatzky, H. B.: 1975, Astroblemes in Williston Basin. *AAPG Bull.* **59**, 694–710.

Sawatzky, H. B.: 1976, Two Probable Late Cretaceous Astroblemes in Western Canada, Eagle Butte, Alberta and Dumas, Saskatchewan, *Geophysics.* In press.

Smith, D. G., and Pullen, J. R.: 1967, Hummingbird structure of southeast Saskatchewan. *Bull. Canadian Petroleum Geology* **15**, 468–482.

Swenson, R. E.: 1967, Trap mechanics in Nisku Formations of northeast Montana. *AAPG Bull.* **51**, 1948–1958.

Wilson, W., Surjik, D. L., and Sawatzky, H. B.: 1963, Hydrocarbon Potential of the South Regina area, Saskatchewan. *Saskatchewan Dept. Mineral Resources Rept.* **76**, 17 pp.

Roddy, D. J., Pepin, R. O., and Merrill, R. B., editors.
(1977) *Impact and Explosion Cratering*, Pergamon Press (New York), p. 481–487.
Printed in the United States of America

Crater morphometry from bistatic radar*

RICHARD A. SIMPSON, G. LEONARD TYLER, and H. TAYLOR HOWARD

Center for Radar Astronomy, Stanford, California 94305

Abstract—Bistatic radar data can be used to identify and measure the dimensions of anomalously scattering regions on planetary surfaces. A simple case in which spectral features in Apollo 14 echoes were correlated with parts of the lunar crater Lansberg has been investigated quantitatively. Crater diameter was found to be approximately 35 km, compared with 40 km obtained from maps. The central peak was detected and appears to be surrounded by a relatively flat floor. Crater walls are quite rough and, on the west side at least, are on the order of 12 km from base to rim (in the horizontal dimension). Rough areas match quite well those which one would expect from lunar charts, with all the roughest surface (presumably the interior wall) being contained within the mapped rim. The ejecta blanket extends 30 km beyond the rim on the east side of the crater. In neither the ejecta nor the crater itself is there any evidence of enhanced unpolarized power, implying that the detected radar anomalies are caused by greater roughness of the surface and not by an increase in the concentration of blocks and irregular scatterers on or near the surface. Generalization of this technique will permit detection and measurement of surface regions which scatter anomalously (as a result of centimeter to meter scale roughness) on Mars, where high altitude photography does not have sufficient resolution, on Venus, where orbital photography is impossible, and on other targets.

1. INTRODUCTION

CRATER DIMENSIONS, including detailed limits of ejecta blankets, can be estimated for fresh craters using radar techniques. Monostatic observations of such features as Tycho and Copernicus have been reported before (Zisk *et al.*, 1974; Thompson, 1974). Here we discuss briefly bistatic identification techniques, which have been introduced previously (Tyler and Howard, 1973) but not covered quantitatively.

Using monostatic techniques, crater dimensions can be estimated by studying the excess brightness in radar imagery. Since craters such as Tycho and Copernicus are viewed from earth at oblique (20–50°) angles, the excess backscattered energy recorded in the image indicates the presence of boulders, blocky material, and surface roughness at a higher concentration than that of the environs. The upturned crater rim and inner wall facing the radar scatter energy preferentially back toward the source, so these features will also appear brighter. In principle the slope of these tilted surfaces could be estimated based on the increase in signal strength; in practice, the complex geometry makes this exceedingly difficult.

In bistatic radar the transmitter is located aboard a moving spacecraft. The continuous wave emitted interacts with the planetary surface and the scattered wave is received on earth. Because the spacecraft is moving, the signal is

*This research was supported by NASA grant NSG 7029. The Center for Radar Astronomy is operated with partial support from NASA grant NGL 05-020-014.

Doppler shifted by each scattering element. Since the velocity with respect to each element is different, Doppler shifts are different and the received signal is a broadened version of the original. No ranging information is obtained in bistatic experiments, so imaging (in the sense used for the monostatic case) is impossible and all analysis must be done in the frequency domain. The interested reader is referred to Tyler and Ingalls (1971) for more detail on the mechanics of bistatic radar.

In this paper we re-examine bistatic radar data obtained during an Apollo 14 command module orbit of the moon. Although many anomalously scattering regions have been identified in the data, we have selected an example which can readily be associated with the crater Lansberg for illustration of the morphometric potential of bistatic radar. Crater dimensions, including width of interior walls and extent of the ejecta blanket have been estimated in this simple case where the target lies directly along the radar ground track. The technique can be generalized for scattering areas at arbitrary positions with respect to the ground track and could easily be inverted for characterization of a "blind" target, such as Venus.

2. EXPERIMENTAL CONDITIONS

For purposes of the experiment described here (Apollo 14 command module transmissions at 2287.5 MHz), the earth and moon may be considered fixed; all motion can be incorporated into the spacecraft velocity vector. The locus of points defined by

$$\frac{\bar{v}_{sc} \cdot \bar{r}_i}{\lambda_0 |\bar{r}_i|} = \Delta_i \tag{1}$$

is the source of all contributors to Doppler shift Δ_i where $\bar{v}_{sc}$ is the spacecraft velocity vector, $\bar{r}_i$ is the spacecraft-to-scatterer vector, and λ_0 is the radio wavelength (13 cm). The condition which applied for the following analysis allow (1) to be simplified such that

$$\Delta_i = \frac{|\bar{v}_{sc}| \sin \phi}{\lambda_0} \tag{2}$$

where ϕ is the angle of incidence on the mean surface.

As the spacecraft moves over the surface, most of the received signal comes from a small region around the specular point—that point on the mean surface where angle of incidence equals angle of reflection (and where $\phi = \phi_0$). Anomalously scattering regions will generally appear in the spectrum displaced from the mean response because $|\bar{v}_{sc}| \sin \phi$ is different from $|\bar{v}_{sc}| \sin \phi_0$ (Fig. 1). In most cases these features appear in the spectrum at relatively high frequencies as the specular point approaches the anomaly. They drift through the mean frequency at closest approach and then continue on to negative frequencies. This drift is illustrated in Fig. 2a for two features which have been identified as the east and west interior walls of the crater Lansberg.

A sequence of spectra, plotted in more detail than Fig. 2 (see Fig. 4 of Tyler and Howard, 1973), has been studied to identify the parts of Lansberg. As a subject for analysis, this crater is ideal because of the favorable geometries which obtained. The spacecraft orbit was circular at 110 km altitude and the specular point track bisected the crater, so that simple calculations were sufficient to locate anomalously scattering parts of the surface. In terms of the variables shown in Fig. 1, displacement of a scatterer from the specular point (measured along the ground track) is simply

$$d = h\left[\tan \phi_0 - \tan \left\{\arcsin\left(\sin \phi_0 - \frac{\lambda_0 \Delta}{v_{sc}}\right)\right\}\right] \tag{3}$$

where Δ is Doppler offset of the anomaly from that of the specular point.

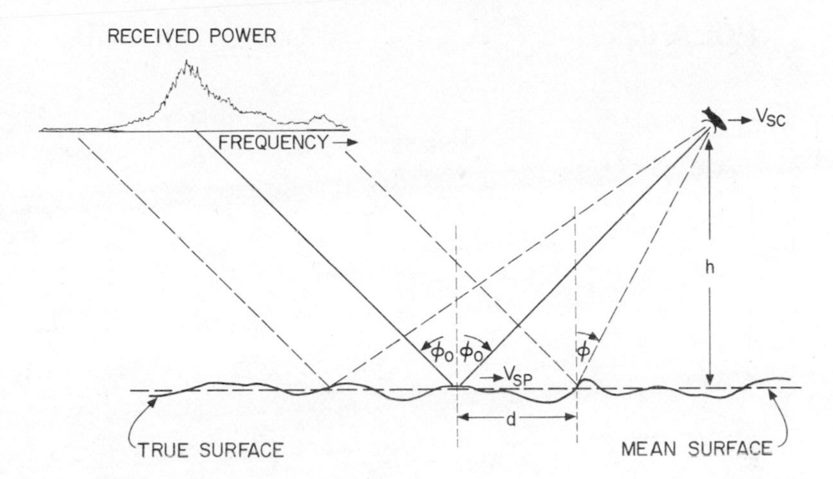

Fig. 1. Geometry for bistatic radar probing of a planetary surface. For this paper all definitions are in the orbital plane; in general, this simplification is not possible. h and v_{sc} are spacecraft altitude and speed, respectively; v_{sp} is speed of the specular point as it moves along the mean surface (dashed line); ϕ is angle of incidence on the mean surface, ϕ_0 is angle of incidence at the specular point. An anomalous region (e.g., a crater rim) displaced d from the specular point scatters energy at a different Doppler frequency and is seen in the resultant spectrum as a spur offset from the main echo.

3. Crater Analysis

Four discrete features were identified in the spectra (Fig. 3) and correlated with lunar maps. An inner and outer boundary on the far (west) wall were defined, corresponding to the lower and upper edges (respectively) facing the receiver. At a slightly lower frequency an increase in received energy, corresponding to the earth-facing side of the central peak, can be identified. The floor surrounding this peak appears relatively smooth (perhaps similar to that of the ejecta blanket); a quantitative estimate of roughness would be very difficult to make, however. Finally, a break in the slope of the spectrum is inferred to be the boundary between the ejecta blanket and the more rugged wall within the near (east) rim. The success with which these four features were located is shown in Fig. 4. Six estimates of position are indicated, corresponding to every second specular point shown (the lines are so closely spaced in a few cases that it may be impossible to resolve them in the figure). Comparison of inner and outer limits on the far wall gives an average base to rim width of 11.4 km. By subtracting the near rim estimate from the outer limit of the far wall, an estimate of crater diameter averaging 35 km is obtained; this compares with 40 km derived from LAC 76 (1964). Since the radar value is based on detection of the rough inner walls and not the rim proper, it is not surprising that the diameter is slightly underestimated.

In addition to identifying the discrete features, one can set an approximate bound on the eastern edge of the ejecta blanket, which is assumed responsible for distortion of the expected (approximately bell) shape of the spectrum (Fig.

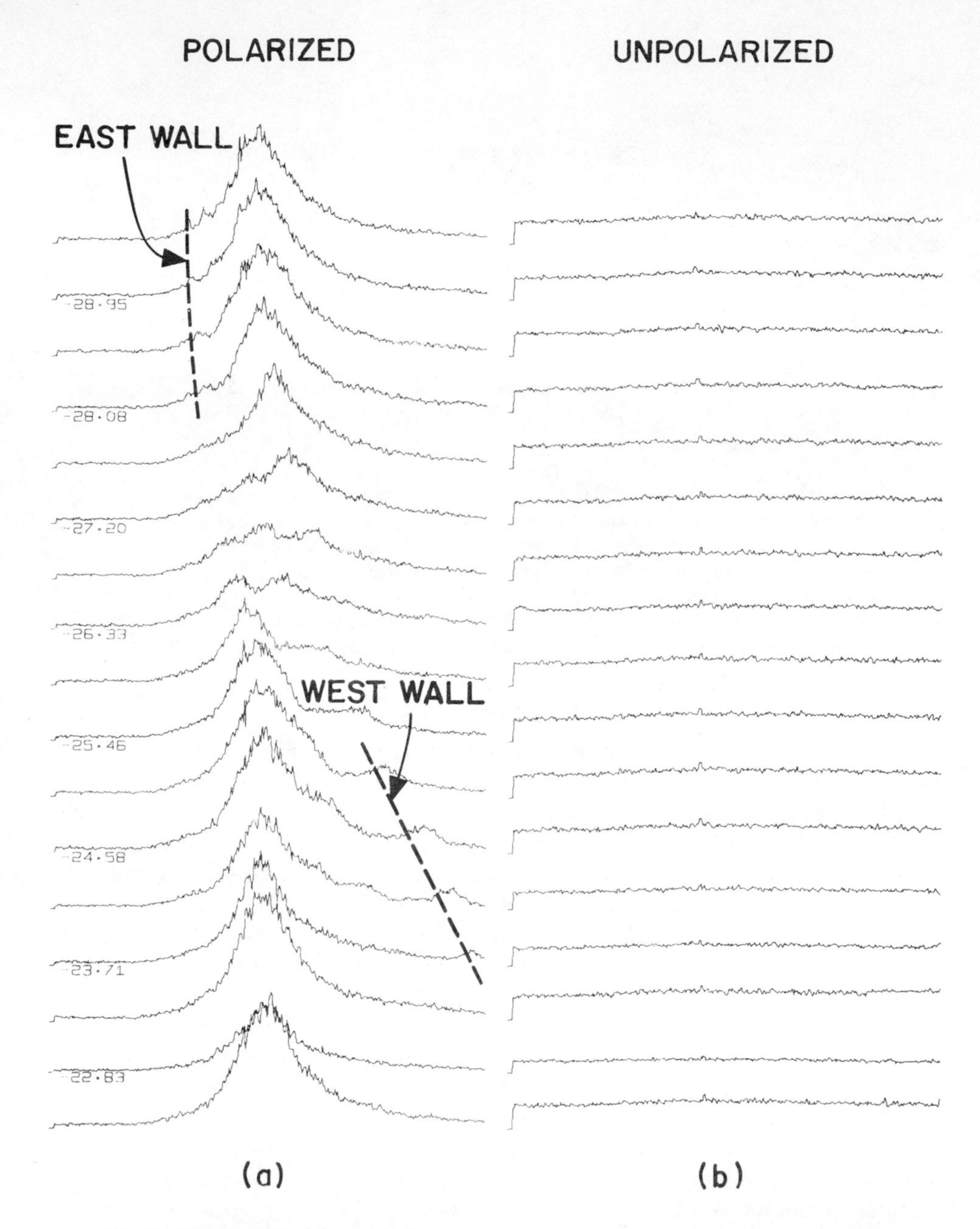

Fig. 2. Polarized (a) and unpolarized (b) echo power spectra plotted versus longitude (°E) of the specular point. Lower dashed line indicates drift of Lansberg west interior wall as specular point approaches crater; upper dashed line shows drift of east interior wall as specular point moves away. Absence of drifting features in (b) implies concentration of blocks and small scale structure is uniform over Lansberg region. Slight leftward skew of spectrum peak beginning at 25°W in (a) is symptomatic of detection of ejecta blanket.

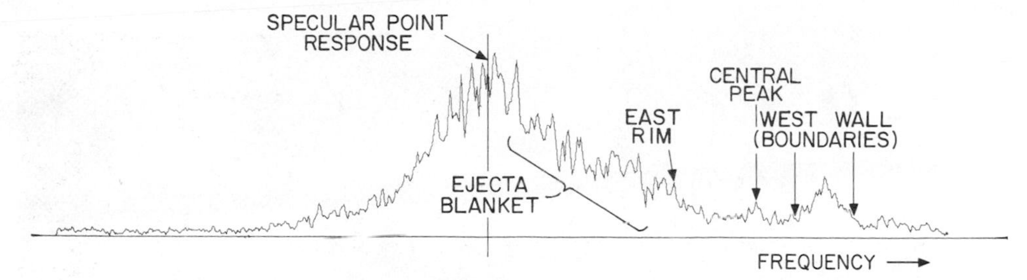

Fig. 3. Spectral features labeled according to source. Ejecta blanket is not so sharply defined as discrete features. Spectrum extends to ±10 kHz about the center frequency. Specular point position is (0.75°S, 24.53°W).

3). Using (3), one can locate the boundary approximately where the specular point track crosses 25°W longitude (Fig. 4). In addition, a skew of the spectrum toward negative frequencies is noted after the specular point moves beyond about 25°W (Fig. 2a), indicating a change in scattering properties of the surface in that vicinity. This radar boundary is 30 km beyond the crater rim, somewhat outside the ejecta limits as mapped by Eggleton (1965) but not inconsistent with his results.

In the above analysis the specular point was always east of and moving

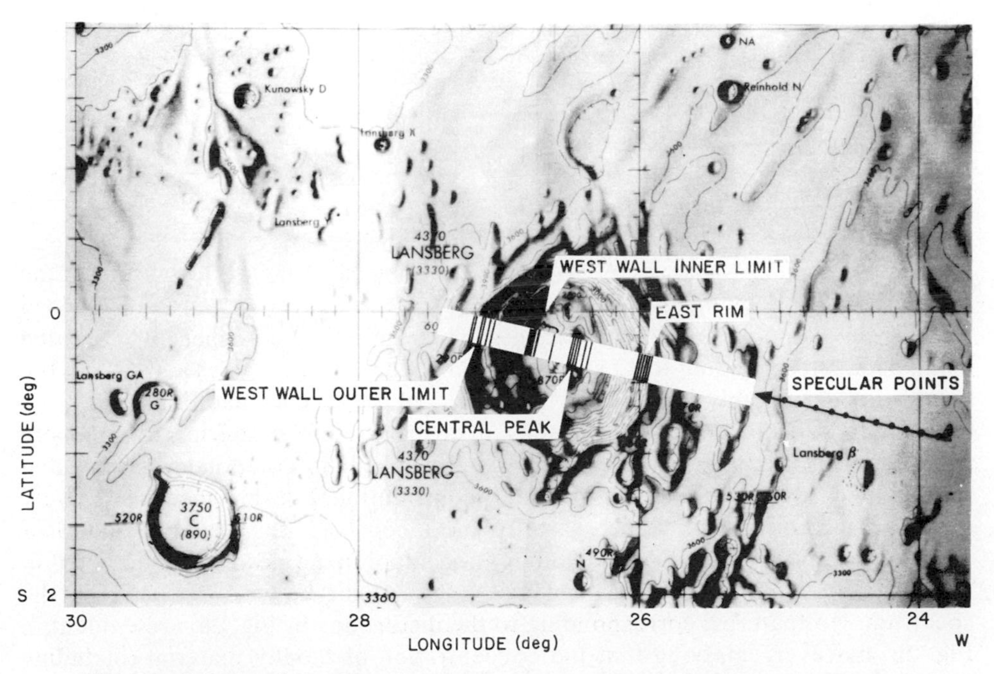

Fig. 4. Positioning of Lansberg rims and central peak from data during specular point approach. Six estimates for each of four located features are shown as straight lines across the ground track. Specular point track corresponding to these estimates is indicated to right of crater. Ejecta blanket extends to where track crosses 25°W.

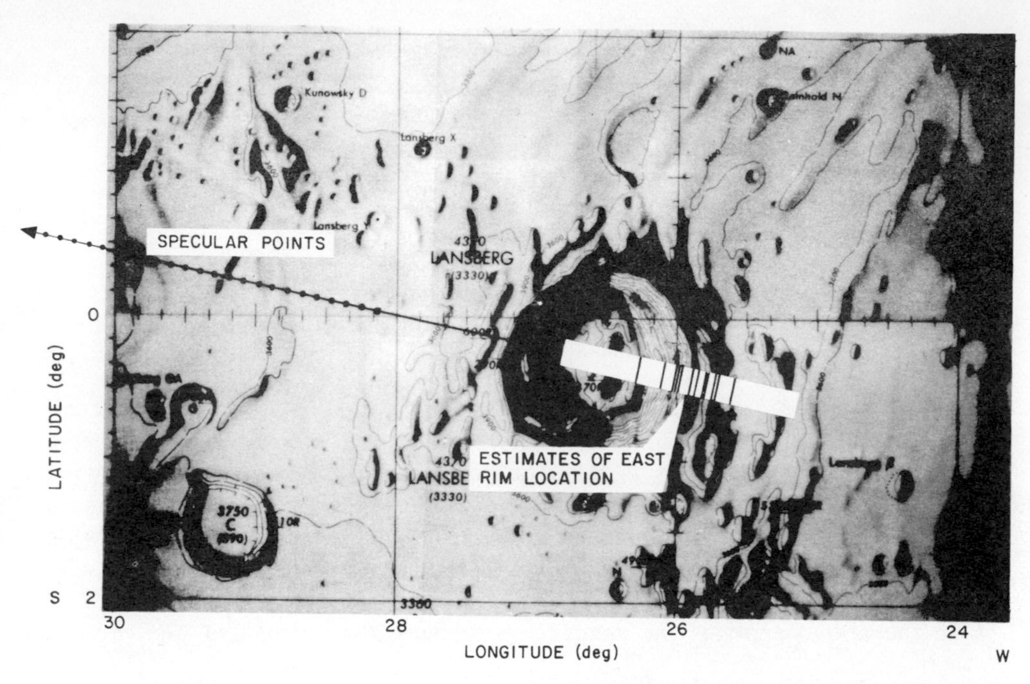

Fig. 5. Positioning of Lansberg east rim from data taken as specular point moved west of the crater. Wider spread in estimates results from lower signal strength and greater sensitivity of d to Δ for large incidence angles ϕ.

toward the anomaly. The same techniques may be applied as the specular point moves away from the crater; in this case, however, results were not so satisfactory. The distance between transmitter and anomaly increases and scattered power diminishes as $|\bar{r}_i|^{-2}$ (Tyler and Ingalls, 1971). Furthermore, the incidence angle ϕ changes more slowly (Fig. 2a) so that errors in reading Δ lead to relatively larger errors in estimating d. The east rim of Lansberg was located using data corresponding to 12 specular points west of the crater (Fig. 5), but the scatter is considerably larger than that obtained under the conditions of Fig. 4.

Unpolarized power, which is believed to result from scattering by boulders and irregular structures, is plotted in Fig. 2b. This echo component is expected to be much weaker than the specular contribution but has been positively identified in returns from other parts of the moon (see, for example, Fig. 5 of Tyler and Howard, 1973). A small amount of unpolarized power was detected in the Lansberg region, as evidenced by a very slight upward bow across the spectrum. No features, corresponding to the distortions in Fig. 2a, are evident in Fig. 2b, however, implying that the concentration of blocky material (including fragments and edges that are sharp on the scale of $\lambda_0 = 13$ cm) in and around Lansberg is not significantly different from that in the surrounding area. The enhanced power which appears in the polarized signal is thus believed to arise from an increase in the density of high angle slopes at Lansberg. A more detailed discussion of the unpolarized component has been given by Tyler and Howard (1973).

4. Concluding Remarks

Data such as these have been available for some time, but this is believed to be the first instance in which actual measurements of a surface feature have been made using the radar spectra. The ideal geometry of the circular spacecraft orbit and bisection of Lansberg by the track of the specular point have made this analysis quite straightforward. The technique, of course, can be generalized; some of the ground work has already been laid out by Tyler and Ingalls (1971). Detection of a large off-axis anomaly some 1200 km either north or south of Mars' equator at approximately 0°W longitude (without identification of its source) has been made using similar techniques (Simpson *et al.*, 1977).

In many respects the imagery which can be obtained from monostatic observations is superior for determining dimensions or radar anomalies. The processing required to obtain these results can be quite involved in comparison to that used here, however. Further, the requirement that both transmitter and receiver be co-located on earth restricts aspect angles to essentially one for the moon. A bistatic configuration, on the other hand, is limited only by the spacecraft orbit. For a target such as Venus, where radar may be the only means to obtain surface imagery on a global scale, the ability to obtain obliquely forward scattered echoes will be important to interpretation of the surface structure.

Overall we believe both monostatic and bistatic techniques can be used in conjunction with other remote sensing tools to obtain a better understanding of planetary cratering. The sensitivity of radar to surface structure on centimeter to hundred meter scales makes it especially valuable as an adjunct to high altitude photography. Detection of the Lansberg ejecta blanket 30 km from the rim and failure to detect an excess of blocky material anywhere in the crater's vicinity are important examples of this.

References

Eggleton, R. E.: 1965, Geologic Map of the Riphaeus Mountains Region of the Moon, U.S. Geological Survey.

Montes Riphaeus: 1964, LAC 76, U.S. Air Force—Aeronautical Chart and Information Center, St. Louis.

Parker, M. N. and Tyler, G. L.: 1973, Bistatic-Radar Estimation of Surface-Slope Probability Distributions with Applications to the moon, *Radioscience* **8**, **3**, 177–184.

Simpson, R. A., Tyler, G. L., and Campbell, D. B.: 1977, Arecibo Radar Obervations of Martian surface characteristics near the equator, *Icarus*, in press.

Thompson, T. W.: 1974, Atlas of Lunar Radar Maps at 70-cm Wavelength, *The Moon* **10**, 51–85.

Tyler, G. L. and Ingalls, D. H. H.: 1971, Functional Dependences of Bistatic-Radar Frequency Spectra and Cross Section on Surface Scattering Laws, *J. Geophys. Res.* **76**, 4775–4785.

Tyler, G. L. and Howard, H. T.: 1973, Dual-Frequency Bistatic-Radar Investigations of the moon with Apollos 14 and 15, *J. Geophys. Res.* **78**, 4852–4874.

Zisk, S. H., Pettengill, G. H., and Catuna, G. W.: 1974, High-Resolution Radar Maps of the Lunar Surface at 3.8-cm Wavelength, *The Moon* **10**, 17–50.

Roddy, D. J., Pepin, R. O., and Merrill, R. B., editors.
(1977) *Impact and Explosion Cratering*, Pergamon Press (New York), p. 489–509.
Printed in the United States of America

Size-dependence in the shape of fresh impact craters on the moon

RICHARD J. PIKE

U.S. Geological Survey, Menlo Park, California 94025

Abstract—Eleven changes in the shape of fresh lunar craters occur within a diameter range of 10–30 km (average, 17.5 km), and mark the transition from small, simple craters to large, complex or modified craters. Seven ratio-level variations—those of rim-crest diameter with depth, rim height, flank width, rimwall slope, floor diameter, circularity, and rim-crest evenness—all are defined for fresh-appearing craters from the new Apollo data and expressed mathematically where practicable. These relations constitute a shape model for interpreting fresh craters on the moon. The size-dependent changes reflect the occurrence of central peaks, rimwall terraces, and a flat floor within craters over 10–20 km across. Interpretation of these features remains speculative. The threshold diameter probably is the maximum size that can be attained by a stable crater-landform having only a simple morphology. The changes in crater shape may depend upon the manner in which the moon's gravity and rock strength interact with the energy of impact to control flowage, collapse, or elastic recoil of the target materials at the time a crater forms.

INTRODUCTION

FRESH IMPACT CRATERS on the moon comprise a morphologic sequence over at least eleven orders of magnitude, within which they display greater complexity with increasing size (Hörz and Ronca, 1971; Howard, 1974). The sequence is not perfectly continuous, however, in that craters of adjacent size-ranges differ significantly with respect to specific characteristics (Gilbert, 1893; Quaide *et al.*, 1965; Stuart-Alexander and Howard, 1970; Pohn and Offield, 1970; Howard, 1974). Lunar craters exhibit changes for no fewer than eleven attributes of shape at a diameter of roughly 15 km as smaller, simple craters undergo a transition to larger, more complex craters. The changes usually are described as being either morphologic (qualitative) or geometric or morphometric (quantitative), but the distinction is artificial: some attributes simply are measured more exactly than others (Griffiths, 1960).

Seven of the size-dependent changes in crater shape are defined here at the ratio level of measurement (Griffiths, 1960): graphs of rim diameter against crater depth (Pike, 1974a), rim height, flank width, floor diameter, rimwall slope, circularity of the rim crest, and profile evenness of the rim crest all inflect at values of rim diameter ranging over about 10–30 km. The other four changes are documented elsewhere (Pike, 1975) at the nominal and interval levels of measurement (Griffiths, 1960): the restriction of swirl-textured floors mostly to craters bridging the simple-to-complex transition, and presence-or-absence or elaborateness of central peaks, flat interior floors, and terraces on the rimwall. Analogous size-morphology thresholds have been identified in terrestrial impact structures (Dence, 1964, 1968; Losej and Beales, 1975) and in craters on Mercury (Gault *et al.*, 1975a) and Mars (Pike, 1971a; Hartmann, 1972).

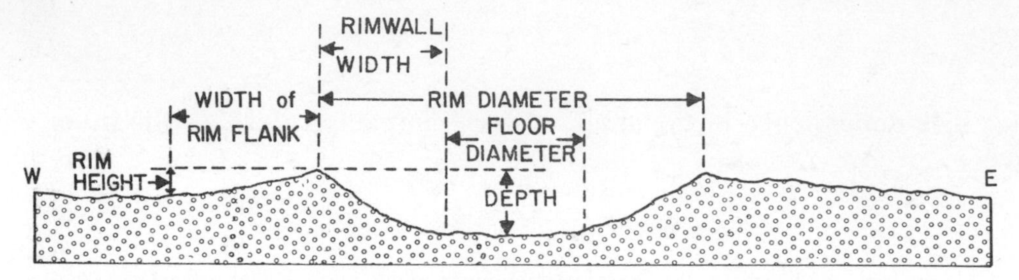

Fig. 1. Six crater dimensions. Explicit descriptions of crater parameters and their measurement given in Pike (1976). Topographic profile across the crater Proclus from photogrammetry of Apollo 17 Mapping Camera photographs. Corrected for planetary curvature. No vertical exaggeration. Compass directions are approximate only. Data have been compiled for nearly 500 lunar craters (Pike, 1976).

Ratio-Level Variations

The relation of depth and other geometric attributes of lunar craters (Fig. 1) to their overall size, expressed here by the rim-crest diameter, can best be ascertained on graphs in which one or both quantities are plotted as logarithms (Baldwin, 1949). The high quality of the new Apollo photogrammetric data, which are listed in full in Pike (1976), enables these plots to be determined for craters between 400 m and 300 km across with more accuracy than they could from the old shadow-length data. Most measurements—the linear dimensions are averages—come from the 1:250,000 Lunar Topographic Orthophotomaps (LTOs). Explicit definitions and detailed techniques of mensuration are given in Pike (1976). Fresh craters, Copernican and Eratosthenian in age as mapped by the U.S. Geological Survey, predominate in all graphs to avoid contaminating resulting trends with the effects of crater degradation.

Seven of the ten ratio-level relations of crater-rim diameter to other dimensions are linear regressions of the form $\log y = \log b + a \log x$; Four of them—depth/diameter, height/diameter, width/diameter, and rim diameter/floor diameter—require two such expressions, and three—ejecta-width/diameter, central peak relief/diameter, and length of rays/diameter—require only one. This form does not apply to the semi-logarithmic plots of rim slope, circularity, and evenness against diameter—although the constituents of rimwall slope (crater depth and rimwall width) do plot exponentially as two linear fits. Calculated least-squares fits to data in the graphs are presented where applicable. All equations written for the new data are summarized in Table 1, which gives values of a (slope) and b (intercept on the y axis at one km rim diameter) for eleven exponential expressions (variables in km), along with sample size and the standard error of estimate of y on x (x always being the rim diameter).

The new equations (Table 1) constitute fairly complete models of the geometry of both simple and complex fresh craters on the moon that may be helpful in estimating the initial dimensions of multi-ring basins and large craters as well as the volume and distribution of their ejecta. They also serve as

Table 1. Data for shape model of fresh lunar craters.

Relation	Size-range	N	Slope (a)	Intercept at 1 km diam. (b)	Standard error	Source
Depth/diameter	<15 km	171	1.010	0.196	+0.038 −0.027	(Pike, 1974a)
Depth/diameter	>15 km	33	0.301	1.044	+0.067 −0.063	(Pike, 1974a)
Rim height/diam.	<15 km	124	1.014	0.036	+0.0075 −0.0062	
Rim height/diam.	>15 km	38	0.399	0.236	+0.036 −0.031	
Rim width/diam.	<15 km	117	1.011	0.257	+0.055 −0.046	
Rim width/diam.	>15 km	46	0.836	0.467	+0.081 −0.069	
Floor diameter/ rim diameter	<20 km	38	1.765	0.031	+0.011 −0.009	
Floor diameter/ rim diameter	>20 km	53	1.249	0.187	+0.012 −0.011	
Width continuous ejecta/diameter	all	84	1.001	0.674	+0.162 −0.157	(Moore *et al.*, 1974)
Relief of central peak/diam.	>27 km	22	0.900	0.032	+0.011 −0.008	
Length (radius) of rays/diam.	all	50	1.25	4.41	+2.94 −1.76	(Moore *et al.*, 1974)

All variables in km; equations of the form $\log y = \log b + a \log x$, where a is slope, b is intercept, x is crater diameter, and y is the dependent variable (depth, etc.).

standards of comparison for the shapes of craters on Earth, Mars, Mercury, and other planets and their satellites. The equations update or supplant older models of crater shape (Baldwin, 1963; Pike, 1972) and are not expected to change significantly with additional data.

Depth

The relation of depth to diameter for craters on the moon has a long history that was reviewed in detail elsewhere (Baldwin, 1949; Pike, 1968). The depth/diameter plot recently was revised from Apollo photogrammetric data for 204 craters between 60 m and 275 km across (Pike, 1974a). The new graph (not reproduced here) shows a sharp diminution in the rate of increase of depth as crater diameter increases beyond 20 km. The two resulting subgroups of craters overlap somewhat between 12 km and about 25 km diameter. The distribution of

171 craters smaller than about 15 km across is described by the expression

$$R_i = 0.196 D_r^{1.010}, \tag{1}$$

where R_i is crater depth (km) and D_r is rim-crest diameter (km). Crater terminology and symbols are discussed in Pike (1968). Equation (1) applies equally to fresh mare and to fresh upland craters. The depth/diameter distribution of the 33 craters over about 15 km in diameter follows the expression

$$R_i = 1.044 D_r^{0.301}. \tag{2}$$

Large mare craters do not appear to differ systematically from large upland craters. The two linear fits intersect at a crater diameter of about 10.6 km. Equation (1), which resembles depth/diameter expressions for experimental explosion and meteoritic impact craters on earth (Pike, 1974a), reveals that small lunar craters are about 50% deeper than older telescopic data showed them to be. The difference between old and new depth data is much less for the larger craters. The new depth/diameter equations have been compared with earlier expressions and discussed in greater detail in Pike (1974a).

Rim height

Before the Apollo missions, the relation of rim height to crater size on the moon (Baldwin, 1949; Pike, 1967) was less certain than the depth/diameter curve because rim height could not be measured as accurately as crater depth. The height/diameter relation has been revised from Apollo photogrammetric data, including most of the craters plotted in Pike (1974a). The resulting graph (Fig. 2) establishes a well-defined inflection of rim height. The 162 craters divide into two groups at a diameter of about 17 km, the two linear fits intersecting at a diameter of about 21.3 km. The 124 smaller craters follow the expression

$$R_e = 0.036 D_r^{1.014}, \tag{3}$$

where R_e is crater rim height and all units are in km. Fresh upland and mare craters do not appear to differ. The height/diameter distribution of the 38 larger craters follows the expression

$$R_e = 0.236 D_r^{0.399}. \tag{4}$$

Neither of these expressions differs radically from the visually-fitted regression lines given previously for fresh craters on the moon (Pike, 1967, 1972), although the slope of Eq. (3) now is essentially 1.0, the value expected from dimensional theory (Pike, 1967, 1968).

The rim-height/diameter relation described for fresh lunar craters in Fig. 2 by Eq. (3) is quite consistent with the height/diameter characteristics of fresh experimental and meteoritic impact craters on earth. Equation (3) is very much like an expression derived previously for 28 of the most freshly formed ter-

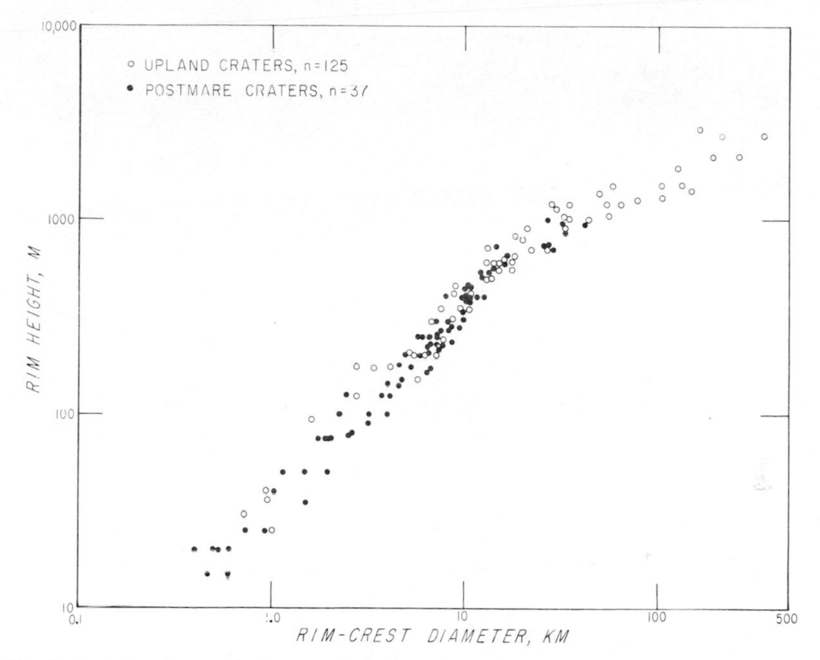

Fig. 2. Relation between rim height, R_e, and diameter, D_r, for 162 fresh lunar craters. Data from photogrammetry of Apollo 15–17 pictures. Distribution inflects at a diameter of about 15 km, in the same manner as the depth/diameter plot (Pike, 1974a).

restrial meteorite craters (Pike, 1972),

$$R_e = 0.033D_r^{0.94}, \tag{5}$$

and also resembles an expression derived for a sample of 100 experimental explosion craters (Pike, 1972),

$$R_e = 0.042D_r^{0.98}; \tag{6}$$

variables for both equations are in km. Equation (3) differs only somewhat more from the height/diameter expression describing laboratory impact craters (Gault *et al.*, 1966),

$$R_e = 0.022D_r, \tag{7}$$

variables converted to km.

Width of rim flank

According to prior studies over a wide size-range of craters on the moon, the relation of rim-flank width to diameter was linear throughout (Baldwin, 1963; Pike, 1968; Guest and Murray, 1969; Pike, 1972). The width/diameter relation has been reexamined, using the same 162 craters as in Fig. 2. The resulting graph

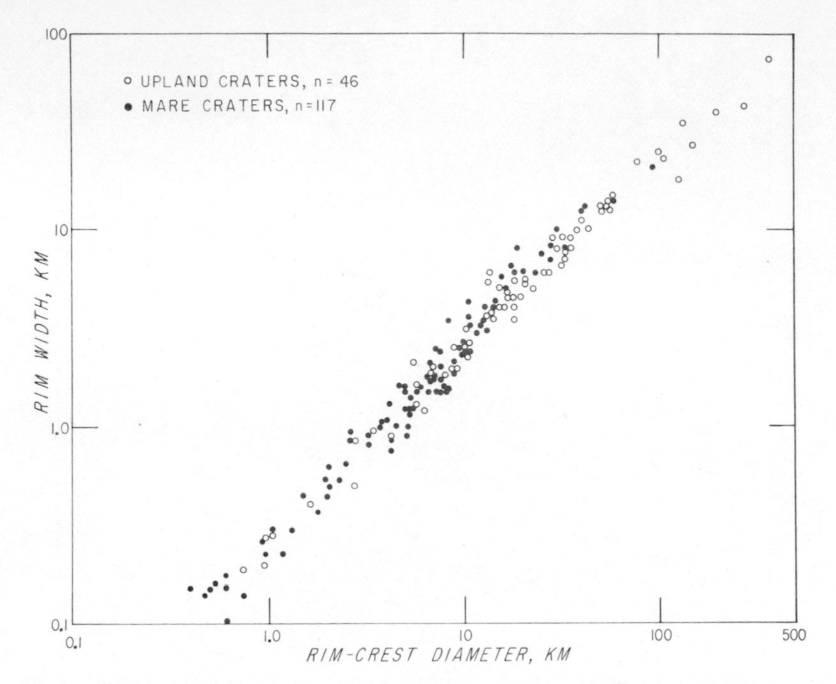

Fig. 3. Relation between width of rim flank, W_e, and rim-crest diameter, D_r, for the same 162 fresh lunar craters in Fig. 2. Distribution inflects at a diameter of about 15 km.

(Fig. 3) shows a perceptible diminution of rim-flank width as crater diameter increases beyond 20 km, although the inflection is not as strong as those evident for depth/diameter and rim height/diameter. Evidently prior studies did not show this subtle inflection because flank width is susceptible to large errors in visual measurements on photographs.

The 162 craters in Fig. 3 divide by inspection into two linear segments at a crater diameter of about 17 km. The 117 smaller craters follow the trend

$$W_e = 0.257 D_r^{1.011}, \tag{8}$$

and the 46 larger craters are described by the expression

$$W_e = 0.467 D_r^{0.836} \tag{9}$$

where W_e is rim-flank width and both variates are in km. The two linear fits intersect at a crater diameter of about 30.4 km, although visually the inflection seems closer to 20 km. This discrepancy probably reflects some nonlinearity in the data, particularly those describing smaller craters. As in the case of depth/diameter and rim height/diameter, fresh upland craters do not appear different from mare craters.

Equation (8) shows that rim flanks of fresh lunar craters under about 20 km in diameter are between 25% and 30% wider than had been ascertained from pre-Apollo observations by Baldwin (1963, Fig. 22) for about 100 craters

$$W_e = 0.206D_r, \tag{10}$$

and by Pike (1972) for 400 craters

$$W_e = 0.170D_r. \tag{11}$$

The latter measurements all were visual estimates from photographs; this tends to result in narrower rim-flanks because terrain of the outer rim is not sufficiently steep or rugged to cast perceptible shadows on photographs. The new values of W_e for smaller craters on the moon also are systematically higher by 30% or so than rim-flank widths of 15 terrestrial meteoritic impact craters

$$W_e = 0.200D_r^{0.97}, \tag{12}$$

and 20 experimental-explosion craters

$$W_e = 0.190D_r^{0.98}, \tag{13}$$

(both from Pike, 1972). The latter discrepancies may be explained by the lower lunar gravity, which would tend to produce broader blankets of ejecta on the topographic rim and on subjacent annular zones of uplifted bedrock.

Rimwall slope

Philip Fauth (1894) discovered that walls (between rim crest and the floor) of lunar craters under about 30 km across did not vary significantly in slope with changing diameter, whereas the slope angles decrease appreciably with increasing diameter for craters above this size. The latest graph of rimwall slope against diameter, for 106 fresh craters from the new Apollo data (Fig. 4), shows the tangent of slope increasing gradually from about 0.34 (19°) in craters 0.5 km across to a maximum of about 0.55 (29°) at a diameter of about 10–20 km, then dropping sharply for larger craters to a value of about 0.25 (14°) at 50–60 km diameter, and then declining somewhat less rapidly to a value of 0.12 (7°) for the largest crater. The revised relation confirms the thrust of Fauth's original findings, but places the change in slope angle at a crater diameter of about 15 km rather than 30 km.

Floor diameter

The size of the flat floor within the larger craters on the moon has been studied in little detail (Warner, 1961; Pike, 1968). Measurements from LTOs and Apollo photographs show that the floor diameter/rim diameter plot for fresh craters (Fig. 5) inflects perceptibly at a rim diameter of about 20 km and a floor diameter of about 8 km. Craters smaller than this threshold value have increasingly narrow floors, down to about a kilometer across in craters 7 km or so in diameter. Flat floors comprise increasingly large fractions (up to 75%) of the rim diameter in fresh craters over 20 km across.

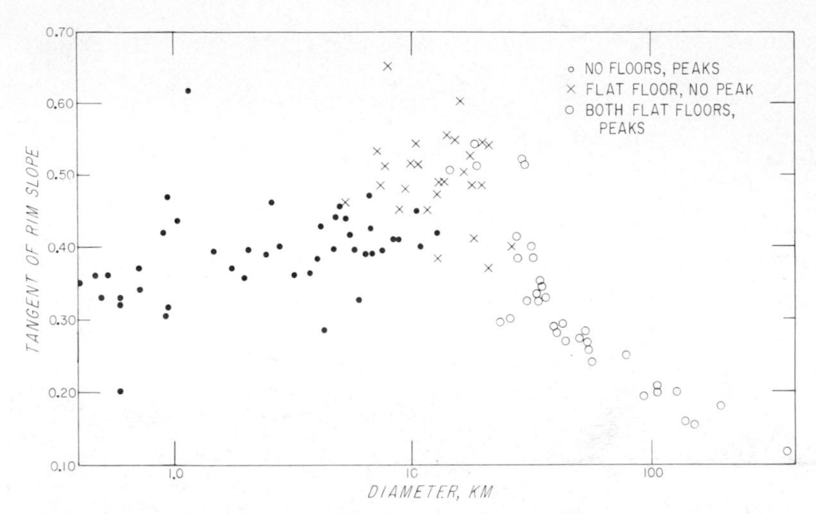

Fig. 4. Relation between tangent of inner slope of the rimwall, tan W_i, and rim-crest diameter, D_r, for 106 fresh lunar craters. Data from photogrammetry of Apollo 15–17 pictures. Slope increases with crater diameter to a maximum of about 29° (0.55) for craters about 15 km across, and then decrease sharply (see text). Dots—craters with no flat floors; crosses—craters with flat floors but no central peaks; circles—craters with both peaks and flat floors (cf. Fig. 8).

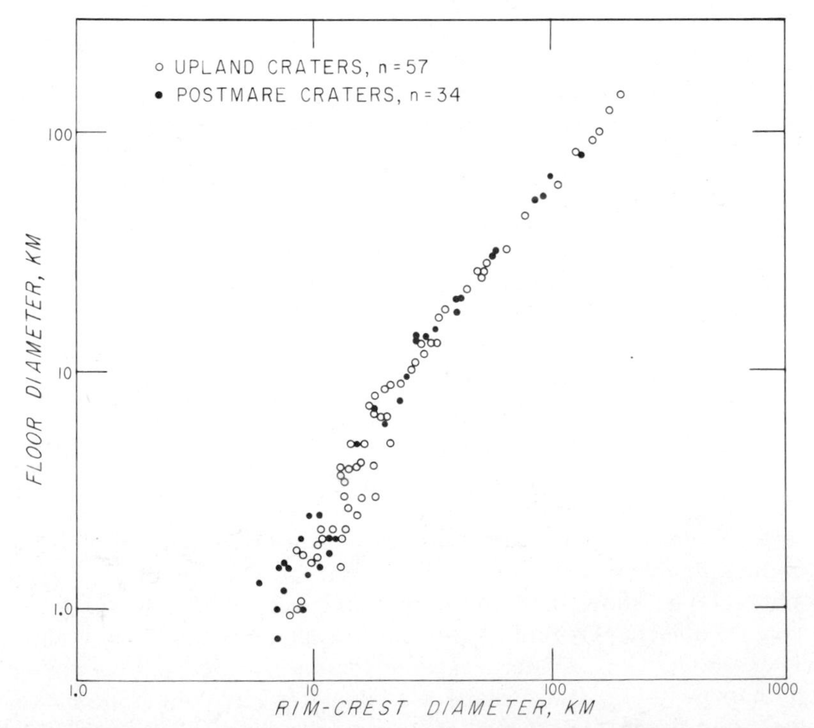

Fig. 5. Relation between diameter of the inner flat floor, D_f, and rim-crest diameter, D_r, for 91 fresh lunar craters. All but a half-dozen points from photogrammetry of Apollo 15–17 pictures. Distribution inflects at a diameter of about 20 km.

The 91 craters in Fig. 5 were divided by inspection into two linear segments at a diameter of roughly 20 km. The 38 smaller craters are described by the trend

$$D_f = 0.031 D_r^{1.765}, \quad (14)$$

and the 53 larger craters follow the expression

$$D_f = 0.187 D_r^{1.249}, \quad (15)$$

where D_f is diameter of the flat floor and the two variables are in km. The equations intersect at a crater diameter of 33 km, again a higher figure than visual inspection would lead one to expect. The difference is ascribed to curvature in the field of smaller craters (Fig. 5). Equation (14) does not extend to craters less than 5 km across (approximately where D_f goes to zero); Eq. (15) also seems to fit data for the flooded mare basins up to a diameter of 900 km. Upland craters have been distinguished from mare craters in Fig. 5, but again, no difference in the two subsamples is evident in the plot.

Rim-crest circularity

The planimetric symmetry of lunar craters—expressed by circularity or polygonality—has been examined qualitatively and quantitatively by several investigators, but the findings disagree. Ronca and Salisbury (1966), Adler and Salisbury (1969), and Murray and Guest (1970) found no functional relation between crater size and values of a circularity index. Similar negative results are evident for Mars from Mariner 6 and 7 pictures (Oberbeck *et al.*, 1972). However, Fielder (1961) and Quaide *et al.* (1965) both observed that craters more than 20 km across are more polygonal than are smaller craters, and Schultz (1976) judged the rims of most craters over 15 km across to be either scalloped or polygonal. Pike (1968) found increases, statistically, in the frequency of rim-crest polygonality and in strength of polygonality with increasing crater size. Pohn and Offield (1970), on the other hand, judged fresh craters between 16 km and 48 km to have markedly polygonal rim crests, but saw the rim crests of craters both larger and smaller than this size range as more nearly circular. Although circularity is not the same as polygonality, these conflicting results can be tested to some extent quantitatively, using an index of crater circularity defined as the ratio of the area of an inscribed circle (fitted to the planimetric outline of the rim crest) to the area of a circumscribed circle.

Rim-crest circularity was calculated using data from Apollo photogrammetric maps and orthophotos. The outcome for 200 fresh lunar craters between 400 m and 370 km across is a graph (Fig. 6) showing that circularity is not constant with crater size but attains a maximum at a rim diameter of 10 km or so (cf. Fig. 4). Circularity declines in both larger and smaller craters, but the decrease is more perceptible among the larger craters. Fresh craters on the upland do not appear to differ significantly from fresh mare craters in circularity, save where rims of upland craters have been distorted by uneven subjacent terrain.

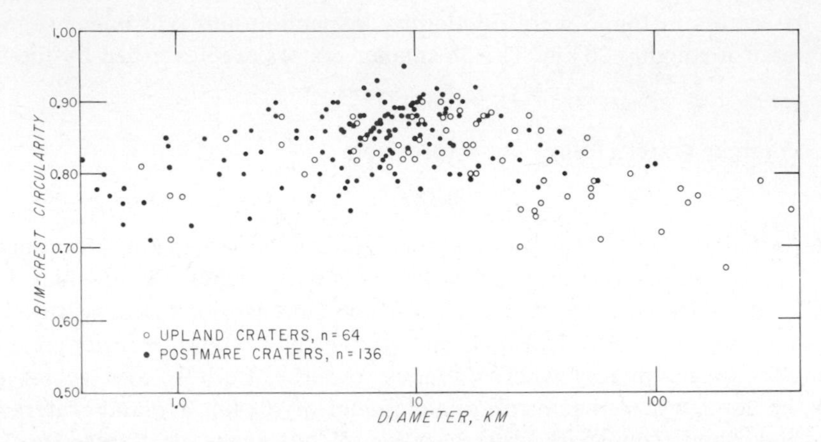

Fig. 6. Relation between rim-crest circularity (see text for definition) and diameter, D_r, for 200 fresh lunar craters. Data from photogrammetry of Apollo 15–17 pictures. Circularity increases with diameter to a maximum of about 0.90 in craters 10 km across, and then decreases (see text).

Evenness of rim crest

A common telescopic observation has been that rim crests of lunar craters below a certain diameter appeared to be relatively even and smooth, whereas rim crests of larger craters were significantly rougher (Baldwin, 1949). The critical diameter has been placed at 16 km by Firsoff (1959) and at 50–60 km by Shoemaker (1965), but excepting the work of MacDonald (1929) evaluations of rim-crest evenness for craters have been qualitative. The standard deviation of elevation along the rim crest was calculated for 35 representative, fresh-appearing lunar craters to test the supposed dependency of rim-crest roughness upon crater size. Thirty-six elevations were determined along the rim crest of each crater at intervals of 10 degrees from due north. Because the 35 sampled craters lie at different lunar elevations, each standard deviation was divided by its mean elevation to measure relative rather than absolute dispersion. The resulting statistic, the coefficient of variation (V_c), is the index of rim-crest evenness.

A plot of V_c against the logarithm of crater diameter (Fig. 7) indicates a significant increase in rim-crest roughness for craters over 10–20 km across. For 13 of the 21 mare craters (15 km in diameter and less), V_c remains relatively constant, at a low value of about 0.006, whereas the statistic increases systematically to a maximum of about 0.05 for the eight mare craters over 15 km across. The 14 upland craters have less even rim crests than the mare craters. V_c values for the upland craters also increase with the crater size, but the relation is not very systematic and any size-dependent inflection is much less evident than that shown by the mare craters. The sample of small upland craters may not be large enough to show an inflection, if one exists. Most of the scatter among the larger upland craters in Fig. 7 is attributed to the inherently less even surface upon which they formed.

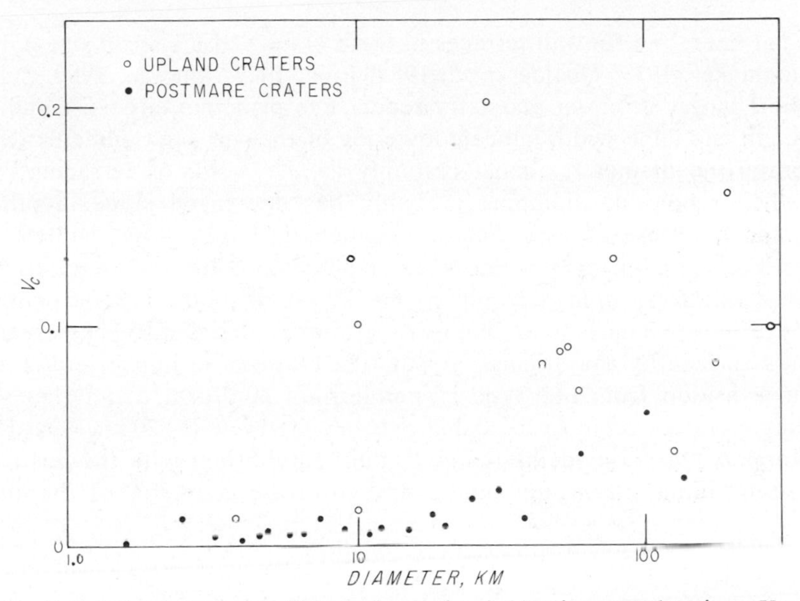

Fig. 7. Relation between coefficient of variation for 36 rim-crest elevations, V_c, and diameter, D_r, for 35 fresh lunar craters. Data from photogrammetry of Apollo 15–17 pictures. Upland craters have much more irregular rims than mare craters. Values of V_c for mare craters remain constant at small diameters, but begin to increase with crater size at a diameter of about 10–20 km.

Non-Ratio Variations

At least four discontinuities, or transitions, in more qualitatively-expressed aspects of shape also interrupt the morphologic continuum of lunar craters at the same size-range indicated by the seven changes in crater geometry. The graphs and a full discussion have appeared elsewhere (Pike, 1975). First, using semiquantitative index numbers that reflect the degree of complexity of crater floors, rimwall terraces, and central peaks (Smith and Sanchez, 1973), Pike (1975) documented a size-threshold that divides the sample of fresh craters into two contrasting shape categories at a diameter of about 10–20 km. In the same paper, the replotted frequency data of Smith and Sanchez (1973) revealed that the transition from 100% absence to 100% presence (0.5 frequency of occurrence) of peaks, terraces, and flat floors in lunar craters is centered within a restricted size-range, about 9–22 km in diameter, and averages 15 km for the three features. Finally, Pike (1975) found that a crater diameter of 15 km exactly bisects the frequency distribution derived by Smith and Sanchez (1973) for the occurrence of "swirl texture" on the floors of certain fresh lunar craters.

Discussion

It is likely that all eleven shape changes—which occur at an average crater diameter of 17.5 km—have a common origin related to formation of central peaks, a flat floor, and rimwall terraces in fresh craters of a critical size (Gilbert, 1893; Shoemaker, 1959; Quaide *et al.*, 1965; Pike, 1967; Mackin, 1969; Schultz, 1976). The changes in crater geometry reflect two principal effects. Diminution of rim height and flank width indicate lowering of the rim crest and enlargement of the crater-rim diameter, almost certainly as the result of terracing of the rimwall. This wholesale slumping probably has decreased slope, depth, circularity, and evenness as well. Second, changes in slope, floor diameter, and again depth reflect emplacement and widening of the flat floor in larger craters—in part probably from uplift accompanying formation of the central peak. The net result of these changes is to diminish the topographic disturbance created at the moon's surface by the original impact. The changes in impact crater shape mark the transition from one type of equilibrium landform to another (Pike, 1967): simple craters seem to be stable only up to about 15–20 km across. The larger, complex craters evidently cannot attain equilibrium with the postimpact environment without departing significantly from the geometry of the smaller craters.

Depth vs. rimwall width

Further insight into the simple-to-complex transition in lunar craters may be obtained by examining the inflected relation between depth and rimwall width (Fig. 8) in some detail. The presence of flat floors and central peaks, which generally are accompanied by rim terraces, is indicated for each crater. The depth/rimwall-width data divide into three relatively discrete fields, group I (no special features), group II (excepting four craters, small flat floors only), and group III (excepting three craters, prominent flat floors and central peaks). Group I meets group III at a rimwall width of about 5 km and a depth of about 2.5 km, but group II (craters attaining depths up to 3.9 km) spans this inflection without any change in slope on the graph. Although flat floors (group II) are the first complication of the simple shape as craters increase in size, depth does not diminish with increasing width of the wall until central peaks begin to appear (group III). The change from group II to group III craters marks the change from simple to complex, or modified craters on the moon.

Group I contains 44 craters and exhibits a positive slope of about 1.0 (Fig. 8). These craters range from 400 m to 12.8 km in diameter and have no terraces, peaks, or flat floors. Wall slopes typically are between 18° and 24° averaged from rim crest to crater center, but reach 30° and more on upper slopes of the wall. Cauchy, at 2675 m, is the deepest crater in the group; the slope of its wall (6.4 km long) averages about 23°. The depth of Cauchy approaches the 3000 m upper limit proposed by Quaide *et al.* (1965) for the depth of a stable crater landform on the moon. The symmetrical, simple shapes of group I craters and their similarity to much smaller experimental impact and explosion craters suggests that they differ little in form from the transient craters that developed at the climax of the original impact event, except for the addition of some fallback ejecta and minor rock slides on the walls.

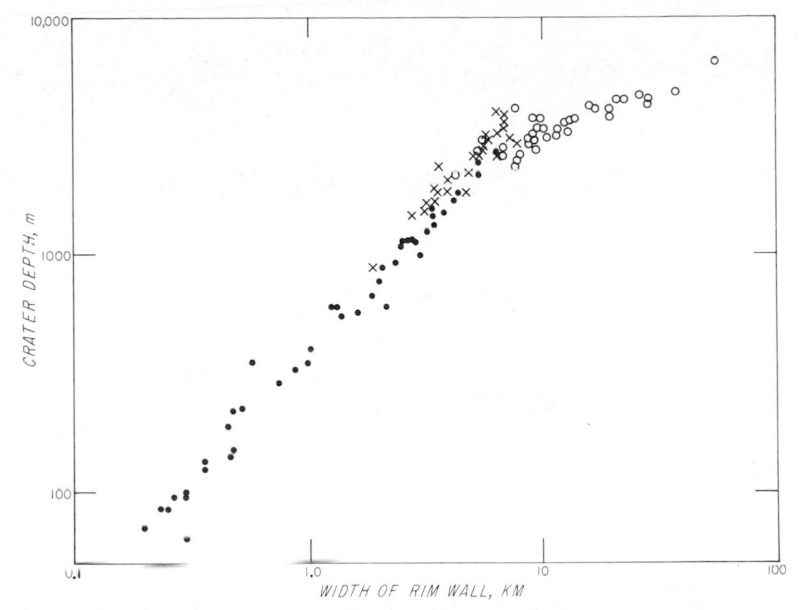

Fig. 8. Relation between crater depth, R_i, and width of the rimwall, W_i, for 106 fresh lunar craters. Data points and crater types same as in Fig. 4. Craters divide into three groups (see text). Dots—no floors or peaks (all group I craters); crosses—flat floors but no peaks (all but four group II craters); open circles—both peaks and flat floors (all but three group III craters). Terraces usually present in craters with peaks. Transition from group II to group III craters marks change from simple to complex or modified craters.

Group II in Fig. 8 contains 27 craters and is above and parallel to group I. The craters range over about 5–29 km in diameter and have no prominent central peaks or rim terraces. Group II craters do have very small flat floors (the same craters less than 20 km across in Fig. 5) and are comparatively deep for their size; their rimwalls are narrow and steep. Evidently moderate infilling, not accompanied by extensive terracing, has formed a small floor that reduces the rimwall width substantially without much reducing crater depth. In the four anomalous craters (shown by open circles in Fig. 8) that contain central peaks—Proclus, Peirce, Diophantus, and an unnamed farside crater 14.6 km across—the peaks are abnormally small for the size of the crater (Fig. 9). The rimwalls of these four craters also are among the steepest in the sample. Other craters within the group II size range—such as Messier B (Pike, 1971b) and Taruntius E—have small flat floors, but only low hummocks instead of well-developed central peaks. These observations and floor frequency data in Pike (1975) suggest that emplacement of small flat floors is the first manifestation (with respect to increasing crater size) of the change in shape at the 10–20 km diameter threshold. The small peaks within larger craters in the group may be large blocks of fallback ejecta or rocks that fell from the upper crater wall onto the crater bottom but were not buried by the flat floor. Alternatively, these

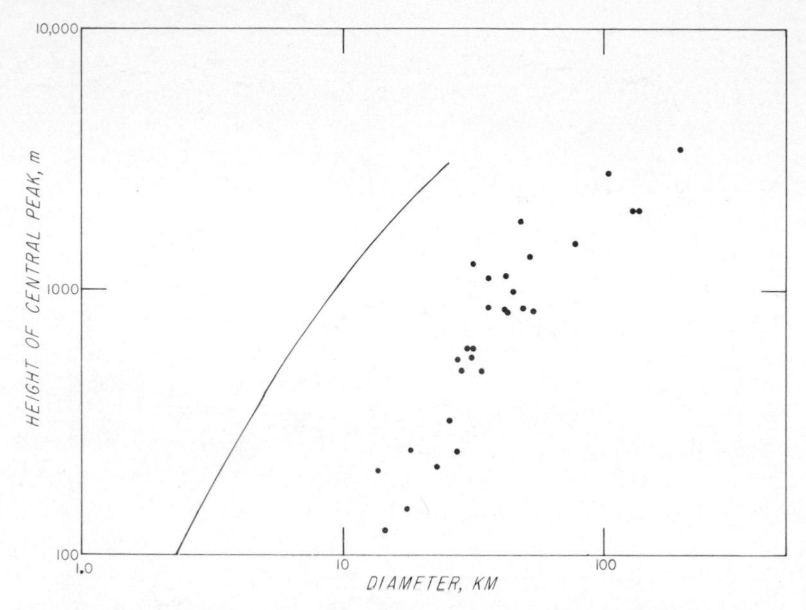

Fig. 9. Relation between relative height of central peak, R_{cp}, and rim diameter, D_r, for 29 fresh lunar craters. Data all from photogrammetry of Apollo 15–17 pictures. Peaks in seven craters under 28 km across are systematically lower than peaks in the 22 larger craters, for which an exponential equation has been fit (Table 1), supplanting the arithmetic fit to shadow data (Wood, 1973). Solid curve is Wood's (1973) fit to amount of central uplift in terrestrial impact structures.

protuberances may be blocks of material that have been thrust up from depth by recoil from the impact. The flat floor in these smaller craters is interpreted as a more or less levelled-out deposit of impact fallback and loose rock debris that moved rapidly in sheetlike slides down the crater walls, probably right after impact, until the slope stabilized (Pike, 1971b). The deposit probably is not impact melt in such small craters (Howard and Wilshire, 1975). Typically, the floor forms a gently-sloping conical surface that is not quite flat and level, but inclined 4° or 5°. The maximum depth for group II craters is about 4 km, corresponding to a crater diameter of about 20 km (Eq. 1), and the limiting slope angle of the wall is about 27–31°. The largest group II craters appear to contain the longest steep slopes on the moon, 6 km to nearly 8 km. Thus the limiting conditions for a stable geometry of slopes on the moon may be about 7 km and 30° (cf. Pike, 1971b).

Group III in Fig. 8 contains 35 craters and slopes positively away from the other two fields at about 0.40. Most craters in group III, which range in diameter from 18 to 370 km, are full-fledged complex or modified craters, with well-developed terraces, flat floors, and central peaks. The larger floors in group III craters are more complicated topographically than the smaller floors in group II craters, and may consist of uplifted material veneered with impact melt. Compared with group II craters, the floors are wider, the peaks (if present at all in group II) are larger (Fig. 9), and the depths are significantly less per unit diameter. Walls slope at only 17–22° for small group III craters and are much less in large craters (Fig. 4). Small craters in group III are on the order of only 2700 m deep; the large group III craters are all over 4000 m deep (the maximum for group II), once rimwall slopes get wider than 15,000 m, with a maximum depth of 6000 m in the largest crater. There are three anomalous craters in the group III field without well-developed central peaks—Conon, Vitruvius A, and an unnamed crater on the farside—that appear to mark the transition from group II to group III. These three (shown by crosses in Fig. 8) have low hummocks or proto-peaks on their small flat floors, tend to be deeper than other group III craters of similar size, and are the three smallest craters in group III; their walls, which are only 6500 m to 8000 m long slope at 20–22°.

If the simple group I craters do not differ too much in geometry from the form of the transient cavity reached at the height of the impact, then Fig. 8 might provide one way to estimate the shape and size of transient cavities for larger, more complex craters. Group II craters differ only slightly from those in group I. Hence, the group II field probably could be shifted up about 500 m and to the right on Fig. 8 until it is aligned with the upper part of group I, in order to obtain a fair approximation of the shape of the transient cavities for group II craters. An analogous restoration of group III craters would involve rotating the field counter-clockwise about an origin marked by a rim width of 6 km and a depth of 2500 m. Such a restoration for the larger group III craters would involve very great differences between observed and hypothesized depths (Pike, 1968). Whether or not this practice is justified remains one of the more intriguing questions about cratering on the moon because of possible extrapolation to the initial cavities of excavation of large basins and the radial distribution of their ejecta (Pike, 1967, 1974b; McGetchin *et al.*, 1973).

Two alternative explanations

Various hypotheses to account for the discontinuity in crater shape—including the cometary-impact model developed later by Roddy and others (1969), isostasy (Baldwin, 1963), and variable depth of impact focus (Baldwin, 1963)—were reviewed by Quaide *et al.* (1965) and more recently by Allen (1975). Currently the two main alternatives are centripetal collapse and elastic recoil, although conceivably the two mechanisms act together (Gilbert, 1893), and none of the other processes has been excluded.

Quaide *et al.* (1965) concurred with Shoemaker's (1959, 1962) view that deep-seated slumping of the crater rim to form terraces converged at the crater center to thrust up a peak (see also Gilbert, 1893). Subsequently this model was adopted by Dence (1968), Mackin (1969), and by Gault *et al.* (1968, 1975a). Centripetal collapse of the rim and uplift of the central peak, which were likened to formation of a central jet in a liquid splash (e.g., Harlow and Shannon, 1967), are presumed to result from gravitational instability of lunar rock materials comprising long slopes (Fig. 8) that exceed some critical value, shown here to be about 30°, in transient craters larger than 10–20 km across and deeper than 4 km (Fig. 8). According to the collapse hypothesis, peak-forming energy is potential energy that is stored momentarily in the mass of the upthrust rim of the transient crater and then released in a late stage of the cratering event. This mechanism would produce most of the size-dependent changes in crater shape, with a veneer of impact melt perhaps accounting for the level floors in larger craters such as Tycho and Aristarchus (Howard and Wilshire, 1975). Explanations for the increase of rim-crest circularity and rimwall slope in craters up to 10–20 km across are less evident and may not even be related to the main shape discontinuity. Quantitative evaluation of the gravity-collapse hypothesis currently is underway (Melosh, 1977).

Geologic evidence from impact structures on earth, however, suggests that central peaks did not result from deep-seated centripetal collapse of the crater rim, but are rebound phenomena as Baldwin (1963) suggested (see also Gilbert, 1893). Elastic recoil depends upon crater-forming energy—as manifested by compression of the ground surface by the shock wave—rather than upon potential energy stored in the rim of the transient cavity. Data of Wood (1973) imply that the amount of uplift estimated to have occurred in the centers of terrestrial impact craters is directly proportional to the size of the crater and hence to impact energy (Fig. 9). Theoretical calculations by Dent (1973) are consistent with this view.

Study of the Gosses Bluff and Sierra Madera structures on earth shows that rocks of the central uplifts came from stratigraphic horizons well below those that could have been affected by rim slumping (Milton and Brett, 1968; Milton *et al.*, 1972; Wilshire *et al.*, 1972). Additionally, the central uplift appears to develop in early, not late, stages of a cratering event (Milton and Roddy, 1972). The centripetal movement that seems to accompany formation of a central uplift may have caused the rims of these structures to collapse into slump terraces with attendant loss of crater depth, not vice-versa. If the terrestrial analogy can be extended to large craters on the moon, then rim terraces and flat floors with "swirl-textured" material (Smith and Sanchez, 1973) probably develop consequent to central peaks rather than the other way around. Much of the shallowness of large craters may result from accompanying (and instantaneous) uplift of the floor area between the peaks and the rim (Offield and Pohn, 1977), rather than exclusively from terracing of the rim. This structural shallowing also would enable a comparatively small volume of impact melt and fallback material to form a broad, flat floor in large craters.

Although the exact mechanisms responsible for central uplifts in lunar craters according to the recoil, or rebound, hypothesis remain conjectural, the geology of terrestrial impact craters and analyses of experimental explosion craters imply that peaks result from reflection of the impact-rarefaction waves by target material at the excavated ground surface (Milton and Roddy, 1972). In very large impacts, there also may be interaction of these waves with material-strength discontinuities at depth, in the case of the moon perhaps with that indicated at a 25 km depth by seismic data (Toksöz *et al.*, 1972). The latter mechanism bears some similarity to that proposed by Oberbeck and Quaide (1967) to explain morphologic differences in small mare craters excavated in various thicknesses of lunar regolith (see Wegener, 1975, for the genesis of this idea). The presence of discontinuities may play a role in formation of concentric rings in lunar basins (Hodges and Wilhelms, 1976), and also may (Sabaneyev, 1962; Pike, 1971a; Head, 1976) or may not (Allen, 1975) be a likely mechanism for forming peaks in intermediate-size craters.

Critical to explaining the existence of central peaks or uplifts according to any process is their obvious dependence upon crater size. Why peaks form only in craters above a particular diameter is not clear, but two influences, an energy threshold and gravitational acceleration, may be involved. The role of increased energy in changing crater form was addressed implicitly by Gilbert (1893), and still is speculative. A threshold in the relation between impact energy and strength of the target materials could provide physical circumstances that favor development of central peaks and other structural displacements. It is suggested

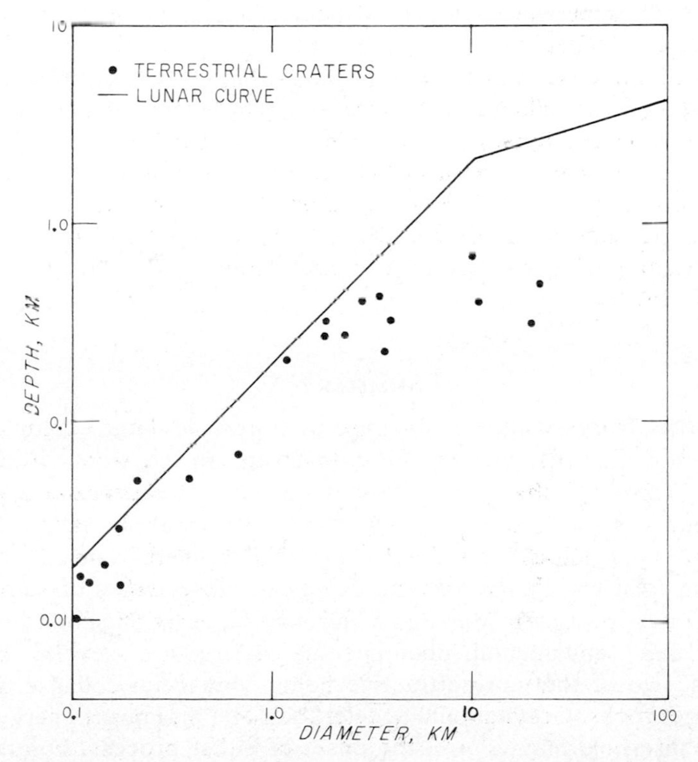

Fig. 10. Depth/diameter plot for 21 terrestrial impact craters (some unpublished data, but most are from Pike, 1972 and 1976). Most craters are relatively fresh; otherwise, depths were determined from results of drilling or gravity work. Solid curves are Eqs. (1) and (2) for fresh lunar craters, simple and complex, respectively. Depth data for terrestrial craters inflect at about one-sixth of the diameter at which depth inflects for lunar craters.

that the mechanics of impact cratering change qualitatively once energy exceeds some critical value. Perhaps beyond this level, a significant fraction of the target material—some of which becomes impact melt—behaves as a fluid rather than as fractured and comminuted rock. The energy level at which the postulated change in shock-hydrodynamic regime occurs on the moon is unknown, but it may not be far from the 10^{24}–10^{25} ergs (Innes, 1961; Baldwin, 1963) thought to be required for development of a central uplift in terrestrial meteorite craters. Chao's (1976) suggestion that the basic mechanism producing shallow (complex) impact craters such as the Ries may differ substantially from that responsible for bowl-shaped (simple) craters such as the Arizona Meteor Crater is consistent with the existence of a critical energy threshold.

Gravity is thought to affect the way energy couples with the target materials to form peaks, because central uplifts (Dence, 1968) and diminished depth/diameter ratios (Fig. 10; see also Losej and Beales, 1975) first appear in terrestrial impact craters 2–4 km across, or about one-sixth the crater size at which similar changes take place on the moon. Some influence of gravity also is indicated for the appearance of central peaks in craters on Mars (Hartmann, 1972) and Mercury (Gault *et al.*, 1975a,b) and possibly for the relative height of peaks and uplifts in terrestrial and lunar craters (Fig. 9; Wood, 1973; Gault *et al.*, 1975a,b). Although the influence of gravity has been interpreted in the context of the centripetal-collapse hypothesis (Quaide *et al.*, 1965; Gault *et al.*, 1975a,b), it has not yet been evaluated in terms of elastic recoil. Currently the role of elastic recoil in forming peaks is being reexamined by computer-modelling large experimental craters from more elaborate hydrodynamic and material-strength assumptions than have been used in past simulations of this type (Ullrich, 1976; Ullrich *et al.*, 1977).

SUMMARY

A transition from small, simple impact craters to large, complex craters occurs at a threshold diameter of about 10–20 km on the moon. Eleven of the constituent changes in shape are summarized here: the appearance of central peaks, flat floors, rimwall terraces, and swirl texture; and the variation with rim diameter of crater depth, rim height, rim-flank width, rimwall slope, floor diameter, rim-crest circularity, and rim evenness. Observation of some of these changes in craters on Earth, Mars, and Mercury suggests that the transition is a widespread and fundamental phenomenon. There are several contending explanations, two of the more attractive being centripetal collapse and elastic recoil of target rocks. Gravitational acceleration and an impact energy/strength-of-materials threshold may control the onset of either process. Future research should concentrate on developing criteria for testing all alternative hypotheses. A solution to the problem may emerge from some combination of theoretical modelling of the impact process, interplanet comparison of crater shapes, and the geologic investigation of terrestrial impact structures and experimental explosion craters.

Acknowledgments—Work was supported by NASA Contract W13,130. Various drafts of this paper benefited from comments by P. H. Schultz, H. J. Moore, D. E. Wilhelms, C. A. Hodges, and Mark Settle.

References

Adler, J. E. and Salisbury, J. W.: 1969, Circularity of lunar craters. *Icarus* **10**, 37–52.

Allen, R. C.: 1975, Central peaks in lunar craters. *The Moon* **12**, 463–474.

Baldwin, R. B.: 1949, The *Face of the Moon*. Univ. Chicago Press, 239 pp.

Baldwin, R. B.: 1963, *The Measure of the Moon*. Univ. Chicago Press, 488 pp.

Chao, E. C. T.: 1976, Mineral-produced high-pressure striae and clay polish: key evidence for nonballistic transport of ejecta from Ries crater. *Science* **194**, 615–618.

Dence, M. R.: 1964, A comparative structural and petrographic study of probable Canadian meteorite craters. *Meteoritics* **2**, 249–270.

Dence, M. R.: 1968, Shock zoning at Canadian craters: petrography and structural implications. In *Shock Metamorphism of Natural Materials* (B. M. French and N. M. Short, eds.), p. 168–184. Mono Book Corp., Baltimore.

Dent, B.: 1973, Gravitationally induced stresses around a large impact crater (abstract). *Amer. Geophys. Union Trans.* **54**, 1207.

Fauth, P.: 1894, Neue Beiträge zur Begründung einer modernen Selenologie. *Astron. Nachrich.* **137**, 17–26.

Fielder, G.: 1961, *Structure of the Moon's Surface*. Pergamon Press, New York. 266 pp.

Firsoff, V. A.: 1959, *Strange World of the Moon*. Basic Books, 226 pp.

Gault, D. E., Quaide, W. L., Oberbeck, V. R., and Moore, H. J.: 1966, Luna 9 photographs: evidence for a fragmental surface layer. *Science* **153**, 985–988.

Gault, D. E., Quaide, W. L., Oberbeck, V. R.: 1968, Impact cratering mechanics and structures. In *Shock Metamorphism of Natural Materials* (B. M. French and N. M. Short, eds.), p. 87–99. Mono Book Corp., Baltimore.

Gault, D. E., Guest, J. E., Murray, J. B., Dzurisin, D., and Malin, M. C.: 1975a, Some comparisons of impact craters on Mercury and the Moon. *J. Geophys. Res.* **80**, 2444–2460.

Gault, D. E., Wedekind, J. A., Nakata, G., and Jordan, R.: 1975b, Effects of gravitational acceleration on hypervelocity impact craters formed in quartz sand (abstract). *Amer. Geophys. Union Trans.* **56**, 1015.

Gilbert, G. K.: 1893, The Moon's face—a study of the origin of its features. *Phil. Soc. Washington Bull.* **12**, 241–292.

Griffiths, J. C.: 1960, Some aspects of measurement in geosciences. *Mineral Industries* **29** (no. 4), 1, 4, 5, 8.

Guest, J. E. and Murray, J. B.: 1969, Nature and origin of Tsiolkovsky Crater, lunar farside. *Planet. Space Sci.* **17**, 121–141.

Harlow, F. H. and Shannon, J. P.: 1967, Distortion of a splashing liquid drop. *Science* **157**, 547–550.

Hartmann, W. K.: 1972, Interplanet variations in scale of crater morphology—Earth, Mars, Moon. *Icarus* **17**, 707–713.

Head, J. W.: 1976, The significance of substrate characteristics in determining morphology and morphometry of lunar craters. *Proc. Lunar Sci. Conf. 7th*, p. 2913–2929.

Hodges, C. A. and Wilhelms, D. E.: 1976, Formation of lunar basin rings. Submitted to *Icarus*.

Hörz, F. and Ronca, L. B.: 1971, A classification of impact craters. *Modern Geol.* **2**, 65–69.

Howard, K. A.: 1974, Fresh lunar impact craters: review of variations with size. *Proc. Lunar Sci. Conf. 5th*, p. 61–69.

Howard, K. A. and Wilshire, H. G.: 1975, Flows of impact melt at lunar craters. *U.S. Geol. Survey J. Research* **3**, 237–251.

Innes, M. J. S.: 1961, The use of gravity methods to study underground structures of meteorite craters. *J. Geophys. Res.* **66**, 2225–2239.

Lozej, G. P. and Beales, F. W.: 1975, The unmetamorphosed sedimentary fill of the Brent meteorite crater, southeastern Ontario. *Canadian J. Earth Sci.* **12**, 606–628.

MacDonald, T. L.: 1929, The altitudes of lunar craters. *British Astron. Assoc. J.* **39**, 314–324.

Mackin, J. H.: 1969, Origin of lunar maria. *Geol. Soc. Amer. Bull.* **80**, 735–748.
McGetchin, T. R., Settle, M., and Head, J. W.: 1973, Radial thickness variation in impact crater ejecta: implications for lunar basin deposits. *Earth Planet. Sci. Lett.* **20**, 226–236.
Melosh, H. J.: 1977, Crater modification by gravity: a mechanical analysis of slumping. In *Impact and Explosion Cratering* (D. J. Roddy, R. O. Pepin, and R. B. Merrill, eds.), Pergamon Press. This volume.
Milton, D. J. and Brett, R.: 1968, Gosses Bluff astrobleme, Australia—the central uplift region (abstract). *Geological Society of America, 64th Annual Meeting*, Tucson, p. 82.
Milton, D. J. and Roddy, D. J.: 1972, Displacements within impact craters. *Proc. 24th Internat. Geol. Cong., Montreal*, Sec. 15, 119–124.
Milton, D. J., Barlow, B. C., Brett, R., Brown, A. R., Glikson, A. Y., Manwaring, F. A., Moss, F. J., Sedmik, E. C. E., VanSon, J., and Young, G. A.: 1972, Gosses Bluff impact structure, Australia. *Science* **175**, 1199–1207.
Moore, H. J., Hodges, C. A., and Scott, D. H.: 1974, Multiringed basins—illustrated by Orientale and associated features. *Proc. Lunar Sci. Conf. 5th*, p. 71–100.
Murray, J. B. and Guest, J. E.: 1970, Circularities of craters and related structures on Earth and Moon. *Modern Geol.* **1**, 149–159.
Oberbeck, V. R. and Quaide, W. L.: 1967, Estimated thickness of a fragmental surface layer of Oceanus Procellarum. *J. Geophys. Res.* **72**, 4697–4704.
Oberbeck, V. R., Aoyagi, M., and Murray, J. B.: 1972, Circularity of martian craters. *Modern Geol.* **3**, 195–199.
Offield, T. W. and Pohn, H. A.: 1977, Deformation at the Decaturville impact structure, Missouri. In *Impact and Explosion Cratering* (D. J. Roddy, R. O. Pepin, and R. B. Merrill, eds.), Pergamon Press. This volume.
Pike, R. J.: 1967, Schroeter's Rule and the modification of lunar crater impact morphology. *J. Geophys. Res.* **72**, 2099–2106.
Pike, R. J.: 1968, "Meteoritic origin and consequent endogenic modification of large lunar craters." Unpublished Ph.D. dissertation, 404 pp.
Pike, R. J.: 1971a, Genetic implications of the shapes of martian and lunar craters. *Icarus* **15**, 384–395.
Pike, R. J.: 1971b, Some preliminary interpretations of lunar mass-wasting processes from Apollo 10 photography. *Analysis of Apollo 10 Photography and Visual Observations*, NASA SP-232, 14–20.
Pike, R. J.: 1972, Geometric similitude of lunar and terrestrial craters. *Proc. 24th Internat. Geol. Cong., Montreal*, Sec. 15, 41–47.
Pike, R. J.: 1974a, Depth/diameter relations of fresh lunar craters: revision from spacecraft data. *Geophys. Res. Lett.* **1**, 291–294.
Pike, R. J.: 1974b, Ejecta from large craters on the Moon: comments on the geometric model of McGetchin *et al. Earth Planet. Sci. Lett.* **23**, 265–271.
Pike, R. J.: 1975, Size-morphology relations of lunar craters: discussion. *Modern Geol.* **5**, 169–173.
Pike, R. J.: 1976, Crater dimensions from Apollo data and supplemental sources. *The Moon* **15**, 463–477.
Pohn, H. A. and Offield, T. W.: 1970, Lunar crater morphology and relative age determination of lunar geologic units—Part 1, Classification. *U.S. Geol. Survey Prof. Paper 700-C*, p. C153–C162.
Quaide, W. L., Gault, D. E., and Schmidt, R. A.: 1965, Gravitative effects on lunar impact structures. *New York Acad. Sci. Annals* **123**, 563–572.
Roddy, D. J., Jones, G. H. S., and Diehl, C. H. H.: 1969, Similarities of 100- and 500-ton TNT explosion craters and proposed comet impact craters (abstract). *Trans. Amer. Geophys. Union* **50**, 220.
Ronca, L. B. and Salisbury, J. W.: 1966, Lunar history as suggested by the circularity index of lunar craters. *Icarus* **5**, 130–138.
Sabaneyev, P. F.: 1962, Some results deduced from simulation of lunar craters. In *The Moon—Symposium No. 14 of I. A. U.* (Z. Kopal and Z. K. Mikhailov, eds.), p. 419–431. Academic Press.
Schultz, P. H.: 1976, *Moon Morphology*. University of Texas Press, 626 pp.
Shoemaker, E. M.: 1959, Address to earth sciences session. *Proc. Lunar and Planetary Exploration Colloquium* 2, no. 1, 20–28.

Shoemaker, E. M.: 1962, Interpretation of lunar craters. In *Physics and Astronomy of the Moon* (Z. Kopal, ed.), p. 283–359. Academic Press.
Shoemaker, E. M.: 1965, Preliminary analysis of the fine structure of the lunar surface. *NASA Tech. Rep.* 32-700, p. 75–134.
Smith, E. I. and Sanchez, A. G.: 1973, Fresh lunar craters: morphology as a function of diameter, a possible criterion for crater origin. *Modern Geol.* **4**, 51–59.
Stuart-Alexander, D. E. and Howard, K. A.: 1970, Lunar maria and circular basins—a review. *Icarus* **12**, 440–456.
Toksöz, M. N., Press, F., Anderson, K., Dainty, A., Latham, G., Ewing, M., Dorman, J., Lammlein, D., Nakamura, Y., Sutton, G., and Duennebier, F.: 1972, Velocity structure and properties of the lunar crust. *The Moon* **4**, 490–504.
Ullrich, G. W.: 1976, The mechanics of central peak formation in shock wave cratering events. *U.S. Air Force Rep.* AFWL-TR-75-88. 138 pp.
Ullrich, G. W., Roddy, D. J., and Simmons, G.: 1977, Numerical simulations of a 20-ton TNT detonation on the earth's surface and implications concerning the mechanics of central uplift formation. In *Impact and Explosion Cratering* (D. J. Roddy, R. O. Pepin, and R. B. Merrill, eds.), Pergamon Press. This volume.
Warner, B.: 1961, Accretion and erosion on the surface of the Moon. *Planet. Space Sci.* **5**, 321–325.
Wegner, A.: 1975, The origin of lunar craters. Translation of 1921 paper by A. M. Celâl Şengör. *The Moon* **14**, 211–236.
Wilshire, H. G., Offield, T. W., Howard, K. A., and Cummings, D.: 1972, Geology of the Sierra Madera cryptoexplosion structure, Pecos County, Texas. *U.S. Geol. Survey Prof. Paper* 599-H. 42 pp.
Wood, C. A.: 1973, Moon: central peak heights and crater origins. *Icarus* **20**, 503–506.

Roddy, D. J., Pepin, R. O., and Merrill, R. B., editors.
(1977) *Impact and Explosion Cratering*, Pergamon Press (New York), p. 511–526.
Printed in the United States of America

Fourier analysis of planimetric lunar crater shape—Possible guide to impact history and lunar geology

DUANE T. EPPLER, DAG NUMMEDAL, and ROBERT EHRLICH

Department of Geology, University of South Carolina, Columbia, South Carolina 29208

Abstract—Fourier analysis in closed form of the rim crest shape of 247 nearside lunar craters greater than 18 km in diameter suggests that crater shape is affected by age and regional physiography. Young craters are more circular than old craters, possibly as a result of crustal deformation early in the moon's history. Mare craters are more circular than highland craters, probably as a reflection of the relative structural and lithologic simplicity of thick flood basalt filling mare basins as compared with more complex geologic relationships that exist in highland regions. Crater size does have a marked effect on planimetric shape over the size range investigated. Statistical analysis indicates that although many shape components (harmonics) are independent of crater age, physiography and size, many of these harmonics, such as the eleventh, display polymodal amplitude frequency distributions suggesting that other factors contribute to crater shape. Craters comprising each shape family of the eleventh harmonic typically are located in the same general geographic region of the moon. Data obtained from Fourier analysis of crater shape indicates that the method shows promise as a tool for probing lithologic and structural characteristics of lunar bedrock.

INTRODUCTION

PREVIOUS CRATER STUDIES seeking to relate cratering mechanics to crater morphometry have described craters in terms of physical parameters such as depth, rim crest diameter, ejecta blanket diameter, ejecta blanket thickness, terrace width, rim height, central peak height, and relative circularity (Baldwin, 1963; Ronca and Salisbury, 1966; McGetchin *et al.*, 1973; Murray and Guest, 1970; Pike, 1972, 1973, 1974, 1976; Short and Forman, 1972). Observations from such studies have supported theories favoring impact nature of crater origin (Baldwin, 1949, 1963; Pike, 1974), established a framework within which hypervelocity impact models can be built (Gault, 1973; Wood, 1973; Croft, 1976), provided a basis for estimates of regolith and ejecta thickness (Oberbeck and Quaide, 1967; McGetchin *et al.*, 1973), and revealed relationships between crater age, crater size, and present and past meteorite fluxes (Marcus, 1966; Ronca, 1972; Soderblom and Lebovsky, 1972; McGill, 1974).

Such studies have concentrated on the cratering history of the moon and the physical response of lunar crust to cratering. Because major craters completely penetrate lunar regolith, crater shape may carry information concerning the lithologic or structural aspects of bedrock. If bedrock is regionally variable, then this may be expressed by regional variation in crater shape. Because all craters excepting those produced by extremely oblique trajectories are highly circular in plan view, any information that shape might carry concerning lunar geology must involve small deviations from this dominant circular shape. In order to adequately evaluate relationships between variation in small scale lunar crater shape and variation in lunar crustal properties, a method that precisely measures these deviations must be used.

This paper introduces a new technique for analysis of crater shape which can detect small-scale deviations from circularity. The method employs Fourier analysis in closed form to describe planimetric crater shape in mathematical terms. Fourier analysis was originally adapted to shape analysis by Ehrlich and Weinberg (1970) and has since been applied successfully to a wide variety of geologic problems (Ehrlich *et al.*, 1972, 1974; Nummedal and Boothroyd, 1976; Pryzgocki *et al.*, 1976; Yarus *et al.*, 1976a, 1976b). Since Ehrlich and Weinberg (1970) give a detailed technical description of the method, only a brief outline of the technique is given here. Instead, we emphasize means by which data generated by the Fourier analysis technique is interpreted and applied to planetary problems.

Fourier Shape Analysis

Any closed shape can be thought of as a composite of many less complex shapes. For example, the four-sided form with rounded corners shown in Fig. 1 is the sum of a circle and a square. Aspects of both component parts are clearly expressed in the total form. The degree to which the shape's corners are rounded depends upon the circle's relative contribution to the total form. As the circle increases in size over the square, the composite shape takes on an increasingly rounded appearance.

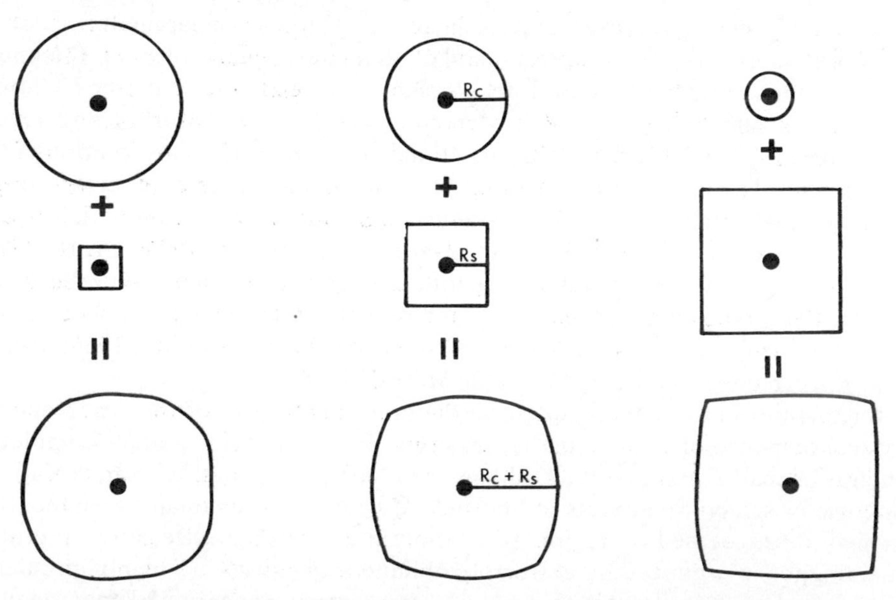

Fig. 1 Composite shapes formed by adding successive vector radii R_c and R_s for a circle and a square. Increased dominance of circular shape over quadrate shape results in successively more rounded forms.

The bell-shaped forms in Fig. 2 result from combining a square and a triangle. In this case, the composite figure's shape is determined by two factors. First, it is a function of the relative *dominance* of the triangle with respect to the square. Second, it is a function of the triangle's *orientation* with respect to the square. Rotating the triangle in relation to the square significantly alters the spacing of points on the periphery thereby changing the composite shape's appearance.

Basic shapes used in Fourier analysis differ from common shapes discussed above (Fig. 3) and are used because of the ease with which they can be mathematically described and compared with composite shapes. The number of bumps on the periphery of each harmonic shape is the same as the harmonic number of that shape. For this reason, the dominance of the fourth harmonic shape in a composite shape is a measure of the tendency for that shape to have four bumps. It is an indication of how square a shape is. Dominance of the third harmonic is a measure of triangularity, and so on. Although low order harmonics, for example two through six, contain information about gross shape, higher order harmonics carry information about small-scale details on the surface of the composite shape.

The contribution that each basic shape or harmonic makes to total shape in

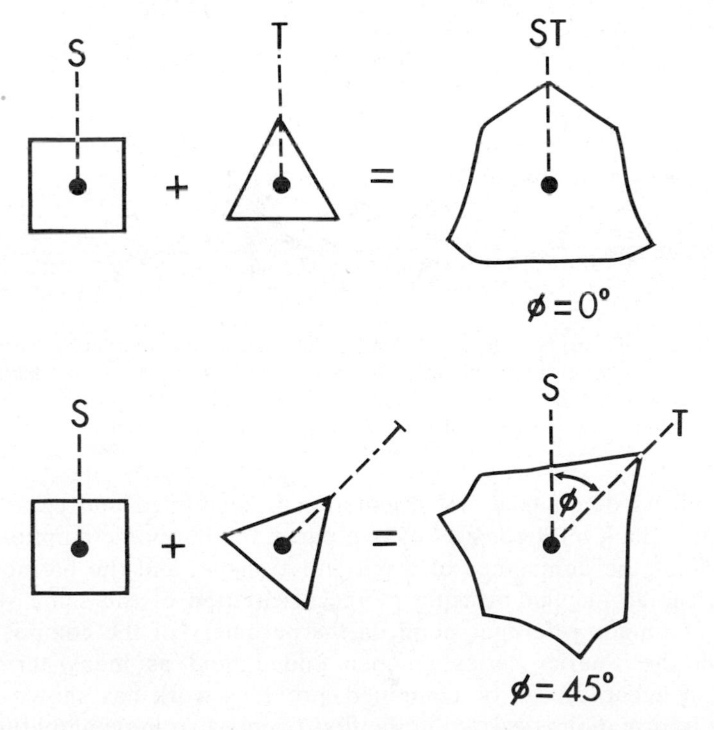

Fig. 2 Composite shapes formed by addition of a square and a triangle. Rotation of the triangle with respect to the square results in distinctly different composite shapes. ∅ is the angular distance in degrees through which the triangle has been rotated as measured between two arbitrarily placed reference lines *S* and *T*.

Fig. 3. Shapes of some basic forms used in Fourier analysis of closed forms. Note that the number of lobes characteristic of each shape is the same as the shape's harmonic number. Low order harmonics contribute to gross shape. High order harmonics contribute to small scale irregularities on the shape periphery.

terms of both its dominance and orientation is quantified and expressed in a Fourier series. Each term consists of two parts: the harmonic amplitude, which is a measure of the dominance of a particular shape, and the harmonic phase angle, which is an angular measure of the orientation of the same shape with respect to a common reference point on the periphery of the composite shape.

Although the Fourier series is open ended, and as many terms as are necessary for accuracy can be computed, previous work has shown that sufficient shape information is carried in the first twenty harmonic amplitude values. Consequently, interpretation of shape variation commonly is made on the basis of the first twenty harmonic amplitude values. Harmonic phase angles provide useful interpretive data when one or more harmonic shapes show a tendency to be oriented either with respect to each other, with respect to an external parameter such as the azimuth of an impacting projectile, or with respect to the structural grain of the impacted medium.

The sample used for this study consists of the rim crest outlines of 247 nearside lunar craters that were traced directly from USGS 1:1,000,000 lunar geologic maps (Fig. 4a and b). The coordinates of 48 points on the periphery of each crater outline were digitized, recorded on magnetic tape, and read into an IBM system 370/168 computer. Then craters were normalized with respect to size, and harmonic amplitudes describing the contribution that each of the first twenty harmonic shapes makes to total crater shape were calculated for each crater.

The smallest image size from which harmonic data can be accurately calculated is approximately three quarters of an inch. At a scale of 1:1,000,000, this places the lower limit of accuracy at approximately 18 km. Consequently, only craters with a diameter greater than 18 km are included in the sample. In addition, craters with rim crests severely degraded by overlapping impact events or partially obscured by mare flood basalt have been eliminated from the sample. Virtually all mapped craters not eliminated on the basis of the above criteria are included to give a sample that represents the total nearside crater population in terms of age, morphology, and both physiographic and geographic location.

Harmonic Amplitude Spectra

Harmonic amplitudes generated for a given crater can be displayed best by plotting harmonic amplitude as a function of harmonic number to produce the amplitude spectrum of a given shape. In this format, easy comparison of Fourier shape data for different craters is accomplished. Shape spectra for lunar craters Aristillus and Cavalerius are shown in Fig. 4c. Careful analysis of the curves reveals real similarities and differences in shape between the two craters. For example, the value of the second harmonic amplitude is virtually the same for both craters indicating that they each have an equal tendency to have two bumps. In other words, they are equally elongate. Examination of the traced outline for each crater bears this out (Fig. 4b). At both the third and fourth harmonics, the spectra diverge significantly. Cavalerius displays greater harmonic amplitudes over this interval indicating that it has a greater tendency to have both three- and four-fold symmetry. Looking again at the crater outline, this fact is borne out by the quadrate and, to a lesser degree, triangular nature of Cavalerius as compared to Aristillus, which is essentially circular. Even more dramatic divergence of the curves occurs at the eleventh, twelfth, fourteenth, and fifteenth harmonics. This indicates that one crater has a greater or lesser tendency to have eleven, twelve, fourteen, or fifteen bumps on its periphery. Although these reflect smaller scale irregularities than lower order harmonics, harmonics eleven, twelve, fourteen, and fifteen still measure bumps of appreciable size. For example, the wavelength of the tenth harmonic oscillations on a 30 km crater is approximately 10 km. Such variation is real and can be

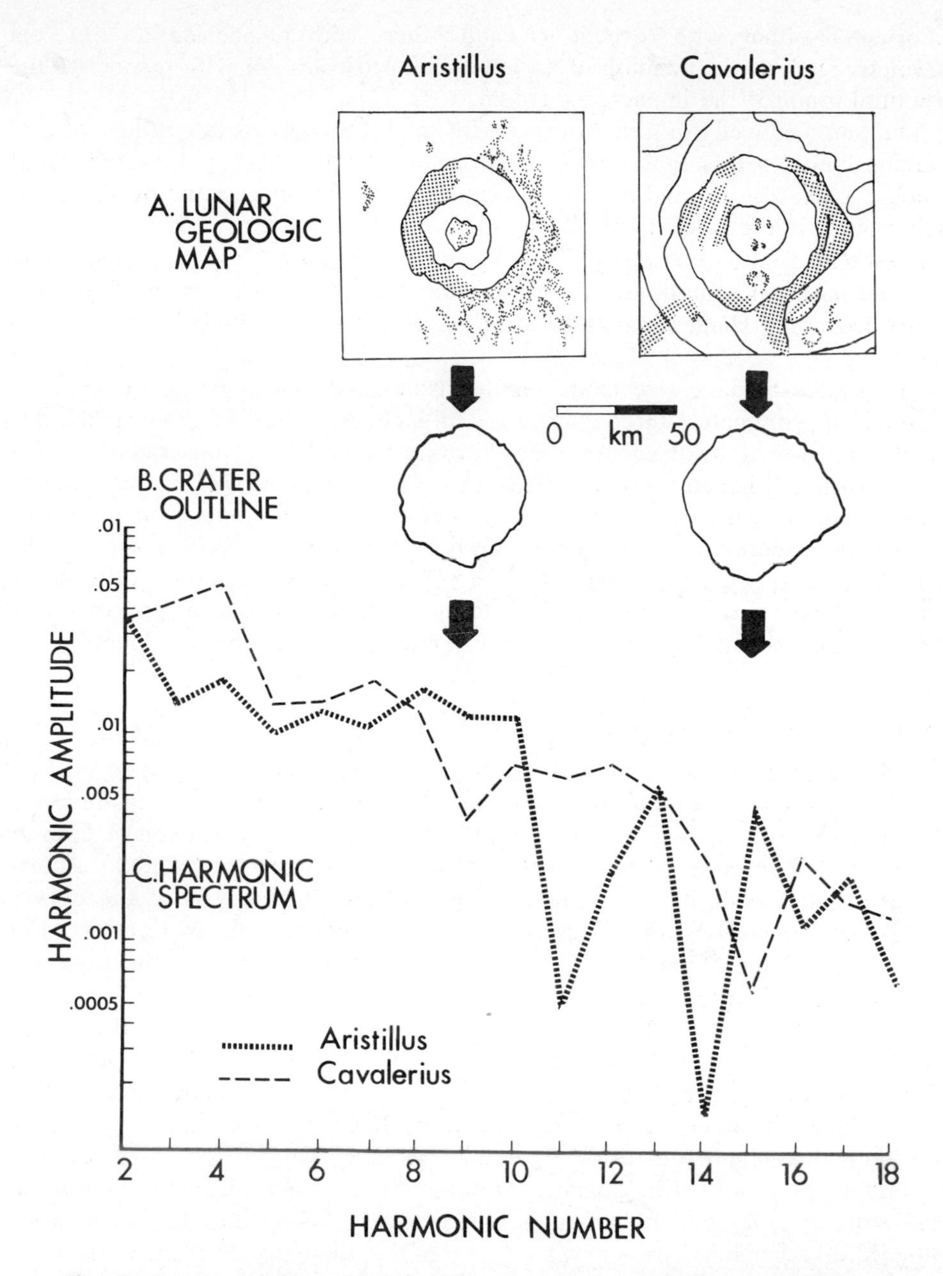

Fig. 4a. Craters Aristillus and Cavalerius as shown on USGS 1:1,000,000 lunar geologic maps. b. Crater outlines traced from geologic maps. c. Harmonic spectra for lunar craters Aristillus and Cavalerius. Harmonic amplitude is plotted as a function of harmonic number.

repeated when crater outlines are retraced, coordinates redigitized, and harmonic amplitudes recomputed.

The relative position of one harmonic spectrum with respect to another on a harmonic amplitude diagram is a reflection of the gross circularity displayed by a crater. If a crater is perfectly circular, the only harmonic shape contributing to its composite shape is the zeroth harmonic (Fig. 3). Harmonic amplitudes of all other harmonic numbers will equal zero. It follows that as a crater approaches perfect circularity, harmonic amplitude values for all but the zeroth harmonic decrease. Consequently, if the harmonic amplitude spectrum of one crater falls consistently below that of another crater, then it can be said that the first crater is more circular than the second crater. Whereas the harmonic amplitude spectrum for Aristillus does not fall invariably below that of Cavalerius, it does so with a fair degree of consistency indicating that it is slightly more circular than Cavalerius.

Spectra of Mean Harmonic Amplitude

In order to determine how planimetric crater shape is affected by age, size, physiographic location, and geomorphology, lunar craters are divided into groups on the basis of these characteristics. For age investigations, all pre-Imbrian craters form one group, all Imbrian craters another, the Eratosthenian craters a third, and the Copernican craters a fourth. The mean harmonic amplitude spectrum is computed for each group by averaging harmonic amplitude values at respective harmonic numbers for all craters in a given group. At the same time, the frequency distributions for craters in related groups are subjected to a chi-square test to determine where statistically significant shape variation occurs. Finally, mean harmonic spectra for related crater groups are plotted as a function of harmonic number and harmonics that show differences significant at the 0.05 level are indicated.

Age

Shape variation resulting from crater age is shown in Fig. 5a in which mean harmonic amplitude is plotted as a function of harmonic number. Inspection of mean harmonic spectra for both Copernican and Eratosthenian craters through the first six harmonics reveals that mean spectra for both groups fall *consistently* lower on the diagram than do mean spectra of Imbrian and pre-Imbrian crater groups. This suggests that younger craters consistently approach circularity more than older craters, an effect first reported by Ronca and Salisbury (1966). Conversely, Imbrian and pre-Imbrian craters display shapes that are more irregular than Copernican and Eratosthenian craters. That this relationship is not due to chance is indicated by significant differences between frequency distributions of the four groups. In the second, fifth and sixth harmonics, chi-square tests show significant differences to be present at the 0.05 level. It remains possible that, because the mean size of Imbrian and pre-Imbrian craters is

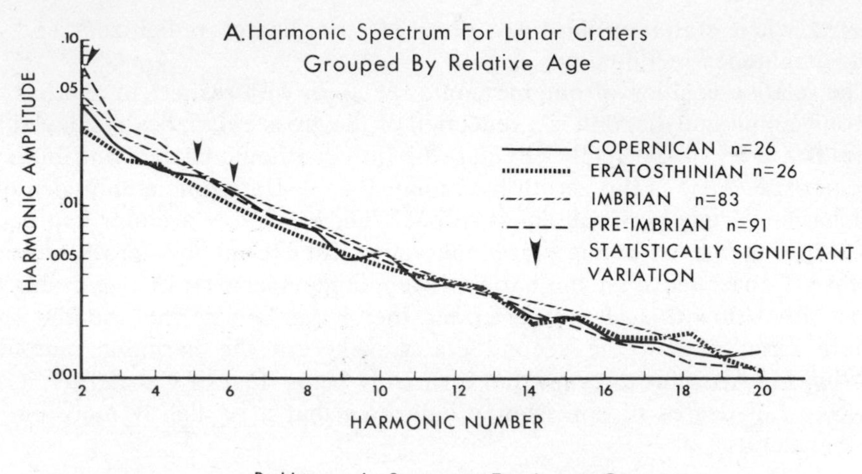

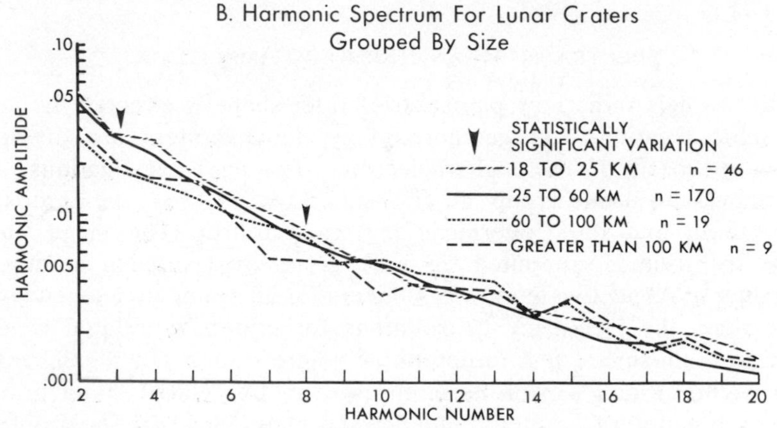

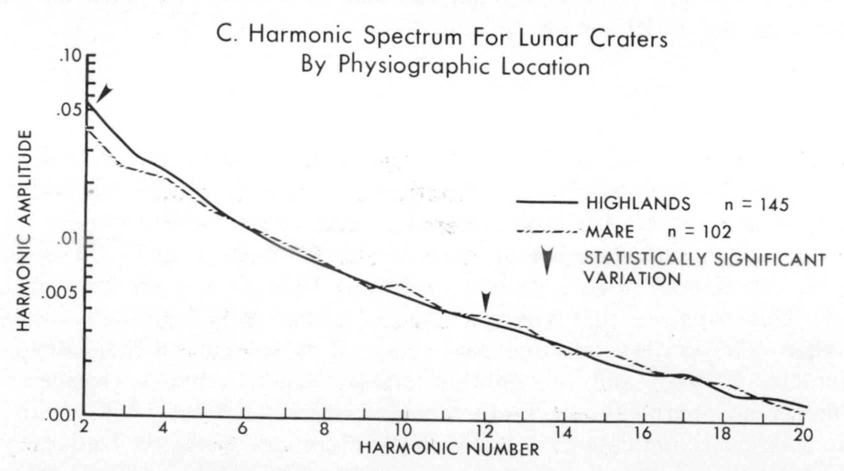

Fig. 5. Mean harmonic spectra for craters grouped by age (a), size (b), and physiographic location (c). Mean harmonic amplitude is plotted as a function of harmonic number. Harmonics significant at the 0.05 level are indicated.

greater than that of Copernican and Eratosthenian craters, changes in shape with age may be due to differences in size. However, inspection of mean amplitude spectra for craters grouped by size (Fig. 5b) indicates that large craters approach circularity more than small craters. In addition, chi-square tests of frequency distributions for craters grouped by size indicate significant differences in the third and eighth harmonics, a sharp contrast to the second, fifth, and sixth harmonics which are significant for craters grouped by age (Table 1).

The observed increase in circularity with decreasing crater age may be evidence of crustal deformation during the moon's history. As Ronca and Salisbury (1966) suggest, both compression and extension of lunar crust would distort craters that initially were circular. The resulting elongation will increase amplitude values of the second harmonic, producing the effect observed in Fig. 5a.

Size

For analysis of size-shape relationships, lunar craters were grouped according to the following diameter ranges: 18–25 km, 25–60 km, 60–100 km, and greater than 100 km. Previous work has shown that depth-diameter ratios of lunar craters change abruptly at approximately 15 km (Quaide *et al.*, 1965; Pike, 1967, 1974). The lower limit of crater sizes encompassed by this study falls slightly above the size range over which depth-diameter ratios abruptly change in Pike's (1974) curves. Through the size range investigated, our findings appear to be consistent with those of most previous workers (Ronca and Salisbury, 1966; Murray and Guest, 1970), in that there are no dramatic shape differences between the large and small craters, except for the fact that larger craters tend more toward circularity (Fig. 5b).

Mare vs. Highland

Figure 5c shows mean harmonic amplitude spectra plotted for craters grouped by physiographic location. The spectra for mare and highland craters coincide closely at almost all harmonic numbers. Most marked divergence occurs at the second harmonic where mare craters have a lower mean amplitude value than do highland craters, indicating that the rim crest shape of mare craters is typically less elongate than that of highland craters. Statistical significance in chi-square test of the second harmonic supports this conclusion (Table 1). Since upland surfaces are older than mare surfaces, the former will contain a larger number of old craters. As discussed above, old craters tend to be elongate probably as a result of large-scale structural modification. Consequently, upland craters should have larger second harmonic amplitudes than mare craters, as observed. However, this difference may also be a reflection of relative bedrock homogeneity in mare basins as compared with complex structural and lithologic relationships in highland regions. It is conceivable that impact events in highly faulted terrain will excavate craters that display elon-

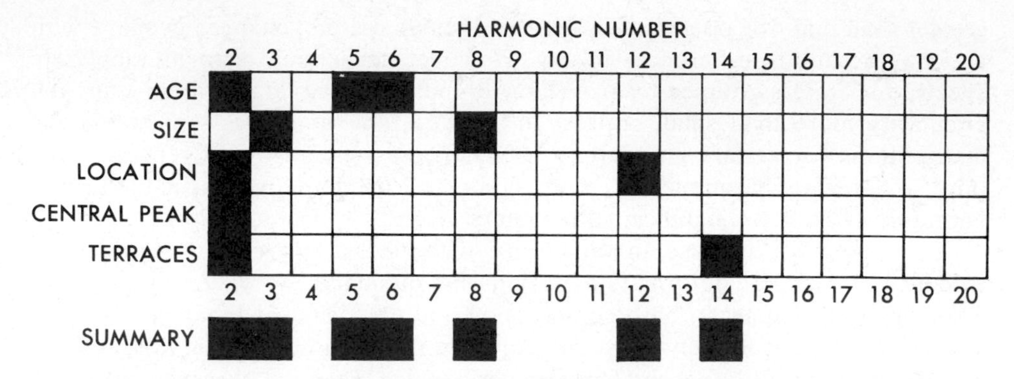

HARMONIC NUMBER	2	3	4	5	6	7	8	9	10	11	12	13	14	15	16	17	18	19	20
AGE	■			■	■														
SIZE		■					■												
LOCATION	■										■								
CENTRAL PEAK	■																		
TERRACES	■												■						
SUMMARY	■	■		■	■		■				■		■						

Table 1. Harmonic numbers carrying significant shape information for lunar craters at the 0.05 level. Darkened blocks indicate harmonic numbers at which significant differences occur in harmonic amplitude frequency distributions when craters are grouped on the basis of criteria shown on the vertical axis. Shape variation that occurs in harmonic numbers not darkened in the summary row is independent of these variables and is a function of other processes attendant to impact events.

gation in a direction parallel to regional faulting. Such elongation is expressed in a Fourier series by an increase in the second harmonic amplitude value.

The twelfth harmonic, in addition to the second harmonic, shows statistical significance in a chi-square test for craters grouped by physiographic location (Fig. 5c and Table 1). Although marked separation of mean spectra occurs at the second harmonic, relatively minor divergence is observed at the twelfth harmonic. Such small separation does not suggest that the chi-square test is in error, but rather indicates that significant shape variation does not manifest itself in the sample mean. For this reason, it is important to realize that mean harmonic amplitude diagrams must be used circumspectly as a diagnostic tool for determining the nature of shape variation.

HARMONIC AMPLITUDE FREQUENCY DISTRIBUTIONS

Chi-square contingency tables

For many crater groups, harmonic amplitude frequency distributions, also known as shape frequency distributions, are polymodal at some harmonics. In such cases, shape differences commonly are expressed not in variation of sample means, but rather in the presence, absence, or shifting of modes in shape frequency distributions. When this occurs, means of compared frequency distributions may not be affected significantly. Accordingly, shape differences between crater groups are evaluated at each harmonic by comparing entire shape frequency distributions in chi-square contingency tables. Such analyses indicate harmonics in which non-random differences between frequency distributions occur and permit subsequent inspection of shape frequency histograms to determine the nature of statistically significant variation.

Results of chi-square tests performed on all crater groups are summarized in

Table 1. Various groups into which craters have been divided are shown vertically and harmonic number is shown horizontally. Blocks that are shaded indicate harmonics in which significant shape variation occurs between one or more respective crater groups. For example, shading of the second and twelfth harmonics opposite physiographic location signifies shape variation discussed with regard to Fig. 5c.

Shape frequency distributions of crater groups analysed do not display significant differences for most harmonics evaluated (Table 1). Significant variation occurs predominantly in low order harmonics. This could be either a real effect or a result of lack of small-scale resolution of the crater rim with concomitant loss of high order detail typically expressed in harmonics fifteen through twenty. Future work will resolve such ambiguity by comparing crater shape traced from 1:1,000,000 scale maps with shape of the same crater traced directly from rectified high resolution photographs. All crater groups excepting size express significant shape differences at the second harmonic. Except for the second harmonic, all harmonic numbers in which significant differences occur are mutually exclusive for each category. In other words, every crater group studied shows significant variation for at least one harmonic that does not display variation related to any other crater group. For craters grouped by age, it is the fifth and sixth harmonics, for craters grouped by the particular size classification shown in Fig. 5c, it is the eighth, and for craters grouped by physiographic location, it is the twelfth. Thus, through detailed examination of shape frequency distributions of harmonic amplitudes for any of these harmonics, it is possible to observe the effect that either age, size, location, or geomorphic factors have on crater shape.

Shape frequency histograms

Figure 6 shows shape frequency histograms at the twelfth harmonic for craters grouped as mare or highland craters. Frequency is plotted as a function of harmonic amplitude. Variation in harmonic amplitude along the x-axis can be thought of in terms of a continuous change in shape between two end-points. At the left side of the histogram are craters with extremely low amplitude values for the twelfth harmonic. These have virtually no tendency to have twelve bumps on their periphery. In terms of the twelfth harmonic alone, they are virtually featureless. Craters at the other end of the histogram have such a strong tendency to have twelve bumps that they appear as twelve-sided polygons.

Since the twelfth harmonic is affected significantly *only* by crater location (Table 1), frequency distribution differences between the histograms of mare and highland craters are a physiographic effect that can be attributed *only* to crater location. Highland craters show a unimodal distribution skewed slightly toward lower amplitude values. In contrast, amplitude values of mare craters are bimodally distributed. A marked deficiency of craters occurs between amplitude values of .0026 and .0032, an interval coincident with the highland crater mode. Addition of mare craters with amplitude values within this narrow range will

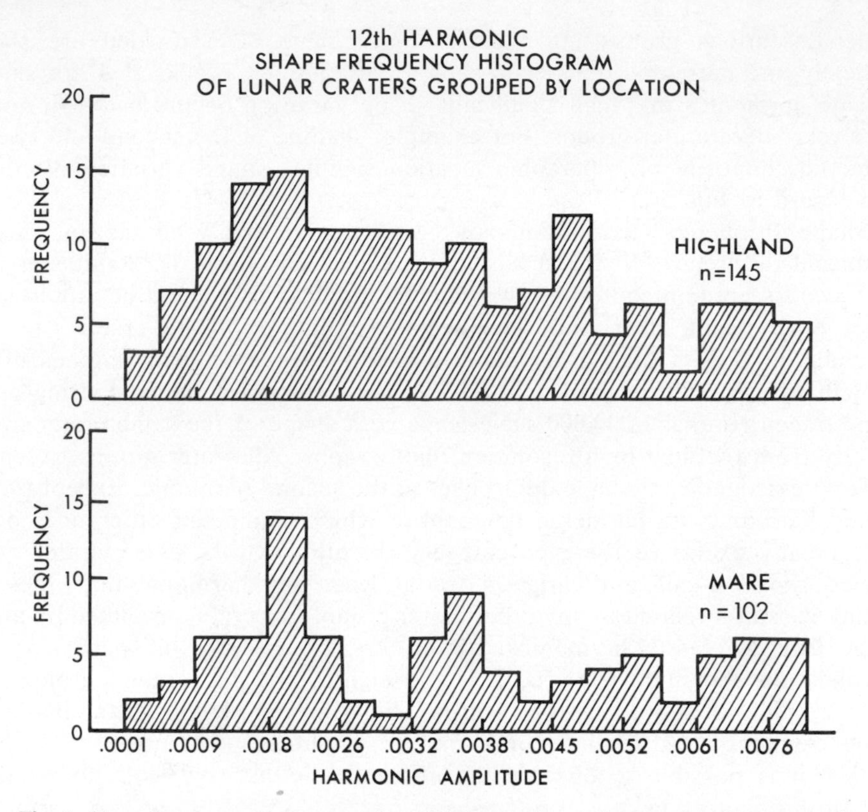

Fig. 6. Shape frequency histograms of twelfth harmonic amplitude values for mare and highland craters.

produce a unimodal distribution roughly mirroring that of the highland craters, indicating that mare and highland crater populations differ only in the presence or absence of craters with an intermediate but sharply defined tendency to have twelve bumps.

Why the twelfth harmonic should carry shape information related to crater location is unexplained, as are the reasons for the nature of harmonic amplitude differences that result. Our analysis of shape has not yet reached the level of these intricacies. However, it can be concluded that observed differences in the twelfth harmonic are tied to physiographic location and may result either from variation in lunar crustal properties or the sharp differences in cratering density between undersaturated mare and saturated highland regions.

Results of chi-square analyses (Table 1) point to those harmonics where significant differences ascribed to age, size, physiographic location and geomorphic factors exist between shape frequency distributions. Most harmonics do not reflect these differences. Thus, any variation that exists between craters at harmonics unaffected by any of the above crater groups must be either random variation or the result of previously unexplored factors contributing to crater shape. The effects of variables such as crustal composition and geologic structure have not been evaluated previously because conventional crater shape sampling techniques have been unable to measure regional variations in lunar crust buried beneath thick sequences of regolith. Impact events forming virtually all craters included in this study penetrate lunar regolith to bedrock providing us with a potential sampling of widespread crustal characteristics across the moon's surface. If bedrock characteristics affect crater shape, each lithology might produce a characteristic shape frequency distribution. Similarly, regional patterns formed by intersecting fault trends, joint sets, or fractures formed prior to impact events may be expressed in planimetric crater shape.

In order to determine whether or not additional harmonics contain shape information unaffected by previously considered variables, shape frequency distributions for the entire lunar sample of craters greater than 18 km in diameter were examined at those harmonics where significant shape differences were reflected in chi-square results. Table 1 shows that harmonic numbers four, seven, thirteen, nine through eleven, and fifteen through twenty are unaffected by either age, size, physiographic location, or geomorphic crater characteristics. Shape frequency histograms for the ninth, eleventh, thirteenth, eighteenth, and twentieth harmonics show varying degrees of polymodality, an indication that previously unaccounted for factors may be influencing frequency distributions. Polymodality displayed by the eleventh harmonic is most pronounced with amplitudes clustering into at least four modes (Fig. 7a). Once polymodal harmonics have been identified, shape frequency histograms so affected can be subdivided into shape families that are defined by these individual modes (Fig. 7b). The planimetric shape of craters in any given shape family might be related to a previously unstudied characteristic common to all craters in that family such as lithology of the impacted medium or patterns of regional joint sets in the crater vicinity. Preliminary study of craters in each of the four shape families defined in Fig. 7b indicates that craters common to a given shape family are not randomly distributed over the moon's surface, but rather from geographic clusters, suggesting that crustal characteristics of a regional nature do indeed affect planimetric crater shape. Work in progress is directed toward both accurately delineating these regions and subsequently determining the nature of physical characteristics significantly affecting crater shape for each shape family.

Conclusions

The relationship between lunar crater shape and crater size, age, physiographic location, and geomorphology has been studied previously by conventional morphometric methods of measuring and comparing physical parameters such as depth, rim crest diameter, rim height, central peak height, terrace width, and ejecta blanket diameter. Although these techniques have given us valuable information regarding both cratering history and the response of lunar crust to impact events, they have not been shown to be sensitive to regional changes in

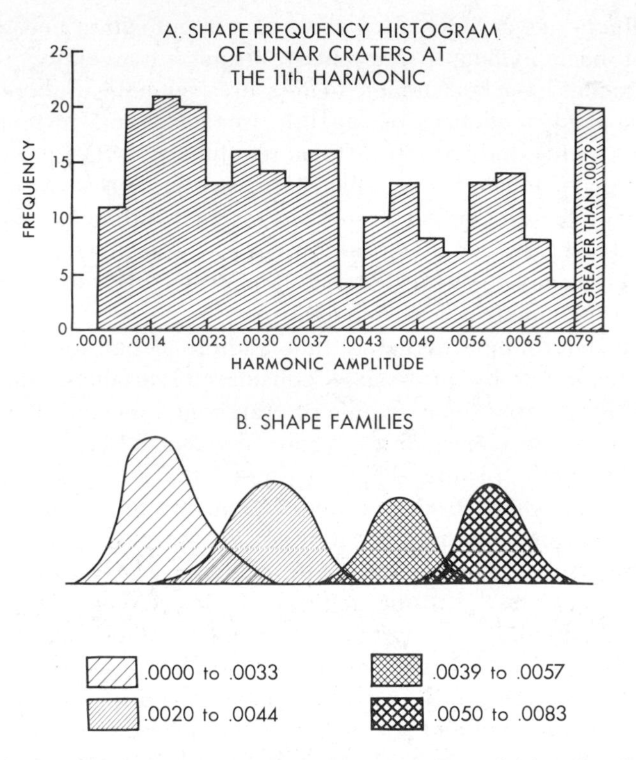

Fig. 7a. Shape frequency histogram of eleventh harmonic amplitude values for the total crater sample. b. Hypothetical component shape families of the eleventh harmonic shape frequency histogram.

bedrock composition. If the lithology of lunar crust influences impact crater morphology, a method of analysis that is sensitive to small-scale changes in crater shape must be used. Fourier analysis in closed form is a new technique that can provide detailed information regarding planimetric crater shape. Preliminary analysis of the rim crest outline of 247 nearside lunar craters larger than 18 km in diameter reveals the following information:

(1) Imbrian and pre-Imbrian craters are more elongate, and therefore less circular, than younger craters, possibly as a result of widespread crustal deformation early in the moon's history.
(2) Crater size does not affect the planimetric shape of craters greater than 18 km in diameter.
(3) Highland craters are less circular than mare craters, probably in response to greater structural and lithologic complexity of highland crust as compared to the relative simplicity of homogeneous mare flood basalt.
(4) Craters comprising each shape family of the eleventh harmonic typically are located in the same general geographic region of the moon. Chi-

square tests show that crater shape variation as expressed by the eleventh harmonic is unaffected by crater age, size, or physiographic location, indicating that the eleventh harmonic amplitude values may be controlled by bedrock characteristics.

Acknowledgments—Mr. Joseph Berger of the Computer Services Division, University of South Carolina, provided invaluable assistance in the computational aspects of this study. This work is funded by NASA grant #NSG7076.

References

Baldwin, R. B.: 1949, *The Face of The Moon.* University of Chicago Press, Chicago, 239 pp.

Baldwin, R. B.: 1963, *The Measure of The Moon.* University of Chicago Press, Chicago, 488 pp.

Croft, S. K.: 1976, Ejection energy-diameter scaling laws for giant impacts (abstract). In *Papers Presented to the Symposium on Planetary Cratering Mechanics*, p. 22–24. The Lunar Science Institute, Houston.

Ehrlich, R., Orzeck, J., and Weinberg, B.: 1974, Detrital quartz as a natural tracer—Fourier grain shape analysis. *J. Sed. Petrology* **44**, 145–150.

Ehrlich, R., Vogel, T. A., Weinberg, B., Kamille, D. C., Byerly, G., and Richter, H.: 1972, Textural variation in petrogenic analyses. *Bull. Geol. Soc. Amer.* **83**, 665–676.

Ehrlich, R. and Weinberg, G.: 1970, An exact method for characterization of grain shape. *Jour. Sed. Petrology* **40**, 205–212.

Gault, D. E.: 1973, Displaced mass, depth, diameter, and effects of oblique trajectories for impact craters formed in dense crystalline rocks. *The Moon* **6**, 32–44.

Marcus, A. H.: 1966, A stochastic model for the formation and survival of lunar craters, II—approximate distribution of diameter of all observable craters. *Icarus* **5**, 165–177.

McGetchin, T. R., Settle,, M., and Head, J. W.: 1973, Radial thickness variation in impact crater ejecta: implications for lunar basin deposits. *Earth Planet. Sci. Lett.* **20**, 226.

McGill, G. E.: 1974, Morphology of lunar craters: A test of lunar erosion models. *Icarus* **21**, 437–447.

Murray, B. C. and Guest, J. E.: 1970, Circularity of craters and related structures on earth and moon. *Modern Geol.* **1**, 149–159.

Nummedal, D. and Boothroyd, J. C.: 1976, Fourier analysis of landforms: A comparison of features in the Washington scablands and selected erosional forms on Mars (abstract). NASA TM X-3364, p. 148.

Oberbeck, V. R. and Quaide, W. L.: 1967, Estimated thickness of a fragmental layer of Oceanus Procellarum. *J. Geophys. Res.* **72**, 4697–4704.

Pike, R. J.: 1967, Schroeter's rule and the modification of lunar crater impact morphology. *J. Geophys. Res.* **72**, 2099–2106.

Pike, R. J.: 1971, Height-depth ratios of lunar and terrestrial craters. *Nature* **234**, 56–57.

Pike, R. J.: 1972, Geometric similitude of lunar and terrestrial craters. *Proc. 24th Internatl. Geol. Congr.*, Sec. 15, p. 41–47.

Pike, R. J.: 1973, Lunar crater morphometry. In *Apollo 17 Prelim. Sci. Rep.*, NASA SP-330, p. 32-1 to 32-7.

Pike, R. J.: 1974, Craters on Earth, Moon, and Mars: Multivariate classification and mode of origin. *Earth Planet. Sci. Lett.* **22**, 245–255.

Pike, R. J.: 1976, Simple to complex impact craters: The transition on the moon (abstract). *Lunar Science VII*, The Lunar Science Institute, Houston, p. 700–702.

Pryzgocki, R. S., Yarus, J. M., Ehrlich, R., and Onofryton, J. K.: 1976, The nature of shape frequency distributions of primary and detrital quartz and the relationships between size and shape (abstract). *Geol. Soc. Amer. Northeast-Southeast Section Abst. w/Prog.* **8**(2), p. 250.

Quaide, W. L., Gault, D. E., and Schmidt, R. A.: 1965, Gravitative effects on lunar impact structures. *N.Y. Acad. Sci. Ann.* **123**, 563–572.

Ronca, L. B.: 1972, The geomorphic evolution of the lunar surface. In *The Moon*, D. K. Runcorn and H. C. Urey (eds.), p. 43–54. D. Reidel, Dordrecht.

Ronca, L. B. and Salisbury, J. W.: 1966, Lunar history as suggested by the circularity index of lunar craters. *Icarus* **5**, 130–138.

Short, N. M. and Forman, M. L.: 1972, Thickness of impact crater ejecta on the lunar surface. *Modern Geol.* **3**, 69–91.

Soderblom, L. A. and Lebovsky, L. A.: 1972, Technique for rapid determination of relative ages of lunar areas from orbital photography. *J. Geophys. Res.* **77**, 279–296.

Wood, C. A.: 1973, Moon: Central peak heights and crater origins. *Icarus* **20**, 503–506.

Yarus, J. M., Pryzgocki, R. S., and Ehrlich, R.: 1976a, Fourier grain shape analysis identifies bedrock from saprolite and stream sediment—Carolina piedmont and Blue Ridge (abstract). *Geol. Soc. Amer. Northeast-Southeast Sect. Abst. w/Prog.* **8**(2), p. 305–306.

Yarus, J. M., Van Nieuwenhuise, D. W., Pryzgocki, R. S., and Ehrlich, R.: 1976b, Sources of shoaling in Charleston harbor—Fourier grain shape analysis as a natural tracer (abstract). *Geol. Soc. Amer. Northeast-Southeast Section Abst. w/Prog.* **8**(6).

Roddy, D. J., Pepin, R. O., and Merrill, R. B., editors.
(1977) *Impact and Explosion Cratering*, Pergamon Press (New York), p. 527–538.
Printed in the United States of America

A stratigraphic model for Bessel Crater and southern Mare Serenitatis

RICHARD A. YOUNG

Department of Geological Sciences, S.U.N.Y., College at Geneseo, Geneseo, New York 14454

Abstract—Stratigraphy visible in the walls of Bessel Crater in Mare Serenitatis permits construction of models for impact mechanics and regional geology. Evidence for layering and shallow subsurface compositional differences is supported by published results of several remote sensing experiments as well as crater morphologies at all scales. The younger brownish-gray flows of the central basin appear to be underlain at depths ≤350 m by the dark mantle(s) that occurs around the basin margin. Crater depth/diameter ratios appear to be significantly affected by the subsurface stratigraphy.

There may be no lunar mare surfaces where large Copernican-age craters have impacted homogeneous (nonlayered) materials. This implies that there may be no unique model for impact phenomena which is related in a simple or direct way to various commonly used crater parameters. Conversely, comparative studies of craters with similar diameters in different mare regions may provide insight into variable regional stratigraphies.

INTRODUCTION

BESSEL CRATER in Mare Serenitatis is located in a region which is better covered by a diversity of earth-based and Apollo orbital data sets than any comparable area on the moon. It is also one of the few areas where visual, photographic, and remote sensing observations have all recorded obvious evidence of local and regional stratigraphic boundaries within the Serenitatis Basin. The quantity and quality of information available for this region allow construction of a reasonably detailed model for the subsurface stratigraphy which can be seen in the wall of Bessel Crater on Apollo 15 and Apollo 17 panoramic camera frames. The model can be applied to the analysis of cratering mechanics for a lunar crater diameter (15.3 km) comparable to the terrestrial Ries Crater (25 km), for which there is relatively good stratigraphic information. However, the very existence of a unique stratigraphic setting in southern Mare Serenitatis may preclude the general application of these measurements to other lunar areas. Bessel lies within the transition interval defined by Pike (1976) as a region in which crater parameters cannot be accurately predicted by simple mathematical relationships.

GEOLOGIC SETTING AND PREVIOUS WORK

The concentric filling of the Mare Serenitatis Basin has been discussed in numerous publications. A recent summary is contained in Howard *et al.* (1973). Immediately prior to the Apollo 17 Mission serious consideration was given to the possibility that the high albedo central fill in Serenitatis was younger than the darker material around the edge of the basin (Howard *et al.*, 1973; Lucchitta,

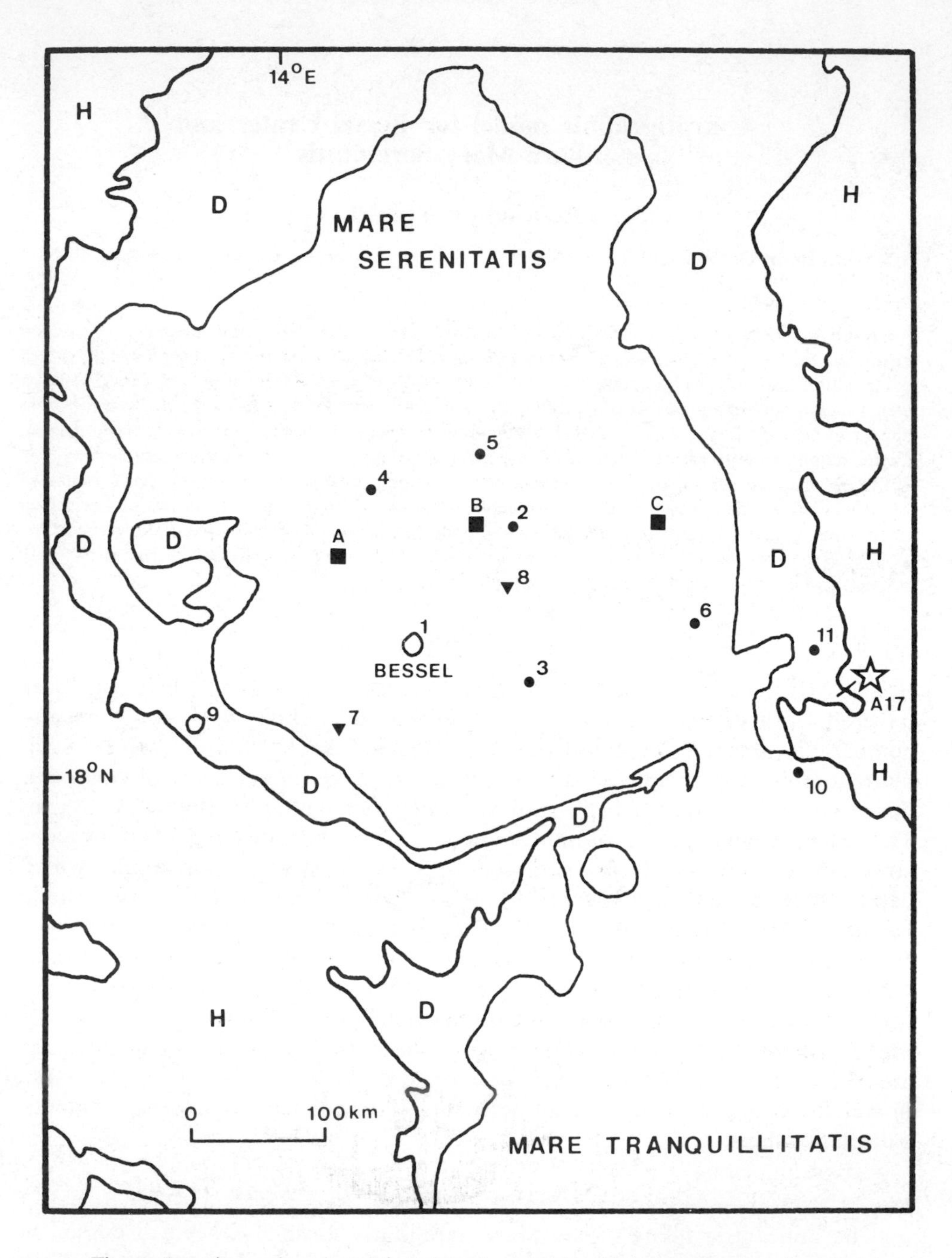

Fig. 1. Locations of craters (circles) and crater frequency count areas (squares) referred to in tables and text. Triangles are craters flooded by lavas. Highlands (H) and low albedo mare units (D) are generalized from U.S.G.S. Geologic Atlas of the Moon Quadrangles I-489, 463, 705, 510, 722, 799.

1973). Basic geologic information resulting from analyses following the Apollo 17 Mission is contained in Muehlberger (1974), Adams *et al.* (1974), Head (1974), and Young *et al.* (1974).

Schmitt (Muehlberger, 1974) observed layering in Bessel (Fig. 1) from lunar orbit and described its ejecta as bluish gray in contrast to the brownish-gray color of the surface flows in the central basin. Because the floor of Bessel is only 1100 m below the level of the surrounding mare surface, Schmitt's description is the first clear evidence of a distinctive shallow discontinuity in central Serenitatis. However, Muehlberger (1974) interpreted Schmitt's observations as evidence for a 2–4-km-thick surficial fill of brownish-gray basalt (prior to the completion of accurate topographic maps).

Young and Brennan (1976), Young *et al.* (1974), and Young (1975) have presented evidence that the unique stratigraphy of central Serenitatis has produced measurable differences in the size distributions, shapes, and ejecta characteristics of craters less than 1 km in diameter. The significance of the size distributions has been discussed by Young (1975), and the data in Table 1 are

Table 1. Crater distributions for three 350 km^2 areas in Mare Serenitatis (locations on Fig. 1).

Center of 17 m interval	Ave. number of craters[b]				% Decrease
m	A	B	C	Ave.	
117	479	420	351	417	—
134	350	270	232	284	32
151	221	195	163	193	32
168	138	138	108	128	33
185	100	94	80	91	29
202	56	60	59	58	36
219	49	43	43	45	25
236[a]	38	39	46	41	8
253	33	34	35	34	17
270	18	23	21	21	38
287	14	20	21	18	14
304	12	13	17	14	22
322	10	11	12	11	21
338	7.1	10	10	9.0	18
355	6.2	8.4	4.3	6.3	33
373	5.4	6.5	5.0	5.6	11
390	2.7	4.7	7.3	4.9	13
407	3.6	6.3	4.3	4.7	4
424	1.8	2.8	3.3	2.6	23
441	2.7	3.7	1.3	2.5	4
458	1.8	3.7	1.0	2.2	12

Sun elevation 24°–33°.
Apollo panoramic frames AS15-9310, 9329, 9341.
Anomalous distributions underlined.
[a]Beginning of anomalous diameter range.
[b]A, B, and C are averages of 3 individuals.

included to emphasize the distinct relative increase of craters with diameters near 240 m. Although crater distributions with similar characteristics are present in some other mare regions, Serenitatis showed the most uniform populations and the most distinct anomalies of 34 test areas widely distributed throughout the lunar maria (Young and Brennan, 1976). The size-frequency anomaly near 240 m diameters might be caused by a subsurface horizon which produces a relative increase in crater diameters for a particular size range of impacting objects as compared with deeper or shallower events. This effect can most readily be seen in Table 1, area C. The forty-six craters centered near 236 m may include several which would have been distributed in larger or smaller intervals if there were a random distribution of objects impacting a homogeneous medium. In this example a buried regolith near a depth of 50 m might be producing the anomalous distribution by a complex effect on depth/diameter ratios. It is difficult to speculate on the precise mechanism because several other independent lines of evidence indicate the presence of several discrete layers at relatively shallow depths.

Examination of all Apollo panoramic frames in south-central Serenitatis has also revealed an unusually high proportion of relatively fresh craters with diameters smaller than 700 m that have flat or hummocky floors and concentric wall terraces (Young and Brennan, 1976). Craters at the upper end of this size range are too deep for their floors to be merely reflecting the transition from regolith to coherent bedrock. This phenomenon implies a number of layers at different depths, which influence the impact process for a range of crater diameters analogous to the experiments by Oberbeck and Quaide (1968) for 2-layer models. If the layering in Serenitatis were not unique, other mare regions should show similar crater shape distributions (allowing for age differences).

The craters Finsch and Bobillier (4.5 and 6.5 km), in the vicinity of Bessel, have obviously been flooded by lavas and provide some limits on the lava filling which followed their formation but preceded the Bessel impact. Reasonable assumptions for rim height/diameter ratios indicate surficial flooding of 150–200 m at distances of 90–100 km northeast and southwest of Bessel. The Apollo 17 Lunar Sounder Experiment (Phillips *et al.*, 1974) has demonstrated the probable existence of a subsurface reflector near 100 m in depth 80 km southwest of Bessel. More recent information from the Apollo Lunar Sounder Experiment (May *et al.*, 1976) and the 15 MHz radar data (Brown *et al.*, 1976) indicates a strong regional subsurface boundary at 900–1200 m depths overlain by 12–20 thinner layers. This would imply flows, ejecta blankets, and/or regolith layers with maximum thicknesses of 45–100 m.

Additional confirmation of the general picture that is emerging from these several independent lines of evidence comes from the infrared spectral reflectance data of Thompson *et al.* (1975). The experimental spectral map they present for the south-central Serenitatis basin can be interpreted as evidence of a shallow subsurface compositional difference exposed in the ejecta of craters smaller than 2 km in diameter scattered over much of the central basin. Thompson *et al.* (1975) interpret their data as indicating low TiO_2 soils in the central basin with a rock-to-glass ratio higher than the dark mantle south of Bessel. A possible interpretation of the SRR unit on their map, which they interpreted as "dark mantle" material "mixed with" the soils of the central basin, could involve impact excavation of a glass-rich deposit from a shallow depth, similar to the dark mantle along the basin margin. Their most intense SRR color values are in the general vicinity of Deseilligny Crater (Fig. 1), where the surface elevations are 150–200 m lower than at Bessel, consistent with thinner surface flows. These color spectral maps by Thompson *et al.* also show that the larger craters (>5 km) penetrate material with properties more like the lunar uplands.

Several investigators have presumed that the radar and Lunar Sounder data which define the deeper discontinuity at 900–1200 m are detecting the dark mantle dipping beneath the brownish flows of the central basin. Recent gravity modeling (Sjogren, 1976) suggests that relatively thin mare fills can account for observed lunar mascons, a view that could be consistent with Lunar Sounder and radar data for Mare Serenitatis. However, there is no compelling reason to assume either that the dark mantle deposit(s) which rings the southern Serenitatis basin coincides with the most prominent subsurface discontinuity, or that it coincides with the base of the mare fill. Existing evidence permits models with 2 or more major subsurface discontinuities which might be regional ejecta blankets, buried regoliths, dark mantles, flow surfaces, and/or compositional differences.

Examination of strata exposed in the walls of Bessel Crater to depths of 1100 m may be one means of choosing among presently viable alternatives as to whether various "remotely sensed" horizons might be the floor of the Serenitatis basin, the "dark mantle", or other surfaces.

Stratigraphy Visible at Bessel Crater

Figure 2 shows the different layers which are present in the western wall of Bessel on two Apollo orbital frames. Part A shows the entire wall with a prominent dark layer in the upper half about 500 m down from the rim.

The random impact erosion, slumping, and seismic shaking that must occur on such slopes (34°) make it unlikely that this relatively horizontal dark layer is only a surficial deposit which slid uniformly down to its present position from the rim crest. Material from the dark layer itself is mass wasting from the outcrop into irregular dark stripes on the slope below. A slight darkening near the rim crest (not accurately represented in this dodged photograph) may be a thin dark mantle on the surface formed when the buried layer was stratigraphically inverted in the rim ejecta.

The low sun elevation view (Fig. 2B) shows only that part of the wall immediately above the dark horizon. Portions of the dark layer can be seen just above the shadow cast by the east rim. Several prominent ledges can be traced for several kilometers around the crater wall. The detail in this degraded print is even more striking when viewed on second or third generation transparencies with 20× magnification. The most continuous (uppermost) layer can be traced

Fig. 2. Panoramic frames of Bessel at moderate and low sun elevations. (A) AS15-9330 (26° sun): South-looking oblique view of western rim. Dodged mosaic to bring out details. (B) AS17-2345 (3° sun): Layers in west rim above dark mantle layer. Crater wall is 1100–1200 m high on west side. Crater superimposed on rim (a) has 0.5 km diameter. Letters b, c, indicate corresponding points in both views. Dark mantle layer (d). See Fig. 3 for interpretation.

Table 2. Dimensions for Bessel Crater measurements and estimates from LTO Map 42D2 and Orbital Photography (Fig. 2).

Elevation of original surface	4700 m
Diameter	15.3 km
Rim height above original surface (ave. 5350 m elev.)	650 m
Rim height west wall (layers visible)	450 m
Interior relief using ave. rim height	1750 m
Floor elevation	3600 m
Original thickness of strata above dark mantle, west wall (estimated)	350 m
Thickness of overturned flap at rim (west wall)	180 m
Uplift of west rim (estimated)	380 m
Dip of upturned rim strata (assumed)	10°

for at least 10 km around the western wall. Other layers, although less continuous, are parallel to the upper layer and uniformly spaced. The evenness of these relatively thin layers suggests they probably lie below the overturned rim ejecta flap. It is unlikely they could have survived being overturned and still have remained so continuous and relatively intact.

Measurements on these photographs and topographic orthophotomap LT042D2 were used to estimate or calculate the values in Table 2 and to construct Fig. 3. It was assumed that if the continuous ledges are flows beneath the overturned ejecta flap, then the dark layer near the middle of the wall is in a normal stratigraphic sequence below the flows. If the dark layer had been overturned in the rim ejecta flap, it should be visible (repeated) somewhere near or below the rim crest. However, if the ejecta flap formed mainly by overturning

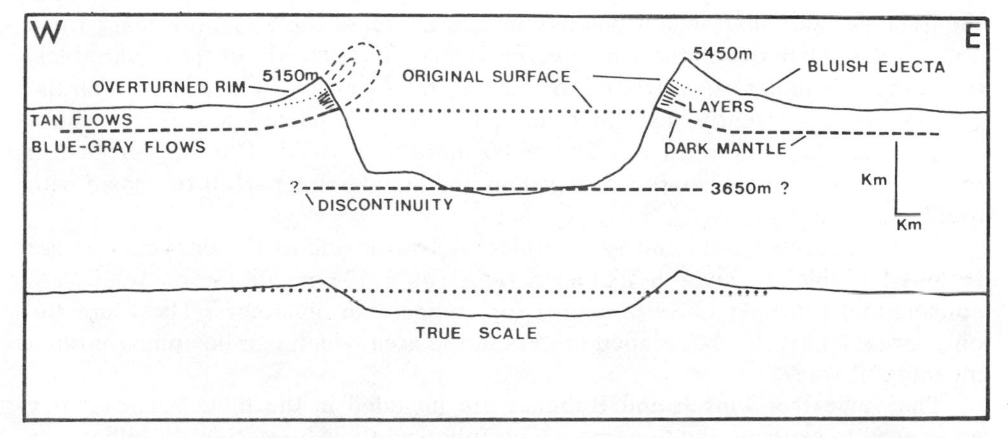

Fig. 3. Profiles and interpretative cross-sections of Bessel Crater from LTO Map 42D2. Table 2 lists dimensions to accompany this figure. Vertical exaggeration is 3× in top view.

of the strata above the dark layer, only a small amount of the dark layer material might be present along the rim crest. Some ejecta from depths below the dark layer (bluish-gray lavas?) would occur outside the rim crest, as observed by Schmitt (Muehlberger, 1974).

The assumption that the strata on the rim in the overturned flap come from above the dark layer is consistent with the proportional thicknesses of ejecta compared with interior relief for craters like Meteor Crater and the Ries Crater. If the uppermost continuous layer represents the original (pre-impact) surface flows, allowing for a thin regolith, the dark layer would have been at a depth of approximately 350 m. This is reasonable if craters like Finsch and Bobillier, now flooded by 150–200 m of flows, are assumed to have impacted on the buried dark mantle deposits ringing the Serenitatis basin and seen in the walls of Bessel Crater.

REGIONAL STRATIGRAPHIC EVIDENCE

Bessel D, 175 km northeast of Bessel, also shows evidence of competent ledges near its rim, and a faintly discernible irregular dark layer is present starting about 300 m below its rim (Apollo Frame AS15-9320). The crater is only 5 km in diameter but is 1100 m deep. Its floor is 950 m below the adjacent mare surface, nearly equivalent to the depth and mean floor elevation of Bessel. Deseilligny and several other craters show ledges near their rims and faint dark stripes on their inner walls, but none of the photos of these craters available to the author has sufficient contrast to be worth reproducing for this paper.

Table 3 is a comparison of some significant dimensions of all relatively fresh, large craters in the vicinity of Bessel and on the dark mantle to the south. The craters between 5 and 7.5 km near Bessel in the central basin have floor elevations generally within 100 m of Bessel's. The two extremes, Borel and Banting, impacted into surfaces which are correspondingly higher or lower than the average. Sarabhai, which belongs in this group, is a crater on a mare ridge about 150 m higher than the surrounding surface. Even with all these variables, including a diameter range from 5 to 15.3 km, the inner floor elevations of all the craters are consistently near 3600 m. Only Borel impacted a surface whose original elevation differed from 4700 m by more than 100 m. Borel is furthest to the east across a major mare ridge system and may be in a part of the basin with a different stratigraphy.

For comparison and contrast, 3 craters on or outside of the dark mantle are included (Table 3). Over a diameter range of 6–12 km, all these craters are conspicuously deeper (30% or more) for comparable diameters. These are the only large, relatively undegraded craters in the area which can be compared in a meaningful way.

The values for Finsch and Bobillier are included in the table because they were used to estimate the flooding which followed their formation. Bobillier was only flooded around its exterior, whereas Finsch has an almost completely filled interior as well. It is interesting that the estimated 200 m of flooding around

Table 3. Values of selected crater parameters of the Serenitatis basin region.

Crater (dia. in km)		Floor elevation (m)	Original surface elevation (m)	Interior relief (m)	Rim height (m)	Location
1. Bessel	(15.3)	3600	4700	1700	600	Central basin
2. Sarabhai[a]	(7.5)	3400	4750	1700	300	Central basin
3. Deseilligny	(6)	3600	4600	1275	250	Central basin
4. Banting	(5)	3825	4700	1150	250	Central basin
5. Bessel D	(5)	3700	4650	1100	150	Central basin
6. Borel[b]	(5)	3510	4400	1100	200	Central basin
7. Bobillier[c]	(6.5)	3400	4700	1260	50	Central basin
8. Finsch[c]	(4.5)	4560	4650	100	0–25	Central basin
9. Sulpicius Gallus	(12)	3100	4900	2250	300	Dark mantle
10. Fabbroni	(11)	3570	5250	2150	450	Outside dark mantle
11. Clarke	(6)	3850	5000	1600	400	Dark mantle

[a]Crater impacted on local rise, probable mare ridge.
[b]Borel is on opposite side (east) of major mare ridge system where stratigraphy may be significantly different.
[c]Crater flooded by younger flows; impact occurred on buried surface.

Bobillier corresponds rather closely with the 200-ft lower elevation of its floor, and it has nearly the same depth as Deseilligny.

The flooding around Bobillier and Finsch was estimated by comparison with other crater rim heights for unflooded craters in Table 3. If the stratigraphy can have an effect on the rim heights as well as the diameters of the younger, unflooded craters, perhaps Clarke would be a better choice for estimating the original rim height of Bobillier. This alternative would give an apparent flooding of 300–400 m, in better agreement with the estimated thickness of pre-impact flows above the dark mantle in Bessel. Craters outside the central basin do appear to have higher rims than comparable craters in central Serenitatis, but the numbers involved are insufficient for meaningful comparisons.

Although Pike (1976) has shown there is a lack of strict proportionality among crater parameters (i.e., depth/diameter ratios) over much of the diameter range discussed here, it may be more than coincidental that all prominent craters with diameters from 5 to 15 km in the central brownish-gray flows have similar floor elevations, whereas both large and small craters (6–12 km) on the dark annulus have significantly greater depths at comparable diameters. The subsurface discontinuity defined by remote sensing at depths from 900 to 1200 m may be exerting a significant degree of control over the depth/diameter ratios of craters in south central Serenitatis.

Although the larger craters, 5–15 km, may have been influenced by the major discontinuity at 900–1200 m in the central basin, the small craters (<2 km) were influenced by the multi-layered nature of the surficial flows above the dark layer visible at Bessel. If the dark layer in Bessel is continuous with the dark mantle around the basin margin, then the discontinuity at 900–1200 m may be the mare basin floor beneath the entire lava fill of the basin. Alternatively, there might be two or more dark layers coincident with regional unconformities. Lucchitta and Schmitt (1974) have shown that the dark mantle and pyroclastic(?) materials of the basin edge and the Sulpicius Gallus region are of two different ages and may be up to 50 m thick.

It seems likely that the floor of Bessel, which is relatively flat with no central peak, is coincident with one of these many possible subsurface horizons. More precise spectral reflectance data in this region may eventually allow a more detailed analysis of the subsurface stratigraphy. Existing infrared analyses (Thompson *et al.*, 1975) show that craters larger than 5 km in Serenitatis are excavated into material with a spectrum similar to the lunar uplands, allowing for mixing with surface lavas.

CONCLUSIONS AND IMPLICATIONS

The values in Table 2 and Fig. 3 allow comparisons of Bessel Crater parameters with terrestrial craters and models of impact mechanics. If the simplest stratigraphic interpretation is correct, it suggests that the dark mantle material of the annulus in the southern part of the basin dips northward very gradually under the brownish-gray lavas of the interior basin with a slope less than 0.15 degrees to the place where it is intersected by Bessel.

A relatively thin brownish-gray surface fill for Serenitatis has implications for the nature of mare ridges in this basin, which are demonstrably younger than the surface flows. Many postmare craters in Mare Serenitatis exhibit evidence of filling by flows along mare ridge margins. This demonstrates that ridges continued to form long after the surface lavas cooled and had been cratered to a significant degree. Some mare ridges in the basin on the central lavas have a local relief of between 400 and 500 m, greater than the apparent thickness of the individual lava flows in which they formed. This relationship is more suggestive of a late-stage intrusive origin than of autointrusion of crusts on lava lakes or structural deformation, as has been suggested (Hodges, 1973; Bryan, 1973). The dark mantle(?) horizon at a depth of approximately 350 m would provide an obvious horizon for sill-like lateral intrusions (broad arches) on the flanks of mare ridges which appear so prominent in Serenitatis.

The thinness of the surficial lavas in the central basin may also explain why no appreciable postmare isostatic subsidence has occurred in this basin since the youngest flows formed. Such subsidence would seem to be a requirement of the model proposed by Muehlberger (1974), who estimated a 2–4-km-thick fill of brownish-gray lavas.

The eastern part of Mare Serenitatis is 400 m lower than the region near Bessel Crater. An east-west profile across the basin at right angles to the north-trending mare ridges shows the lavas to be at distinctly different levels between adjacent ridges. It seems possible that early ridge volcanism controlled the filling of the basin in parallel segments, whereas late stage volcanism, or continued activity, developed the detailed relief on the youngest ridges. A thin uniform surface fill is also more consistent with the form and distribution of the ridge systems. If the brownish lavas were 2–4 km thick in the central basin, with a subsurface contact sloping upward toward the margins, the ridges might be expected to show a related trend in their morphology or distribution resulting from the superposed stress field.

If the proposed stratigraphic boundaries in Mare Serenitatis have influenced crater dimensions as proposed in this discussion, it is very likely that some of the scatter of data which shows up in Pike's (1976) crater relationships near the transition interval from small to larger diameters (10–30 km) may be a reflection of a number of unique stratigraphic sequences in several regions on the moon. The evidence near Bessel suggests that crater parameters, especially depth/diameter ratios, are significantly affected by near-surface stratigraphy. The precise relations are probably complex and cannot be isolated or quantitatively treated without arbitrary assumptions. Conversely, published crater parameter interrelationships based on lunar measurements may include significant stratigraphic influences not readily apparent in randomly selected craters from different maria. Comparisons of crater dimensions between coherent portions of individual maria may provide more insight into subsurface structure or regional inhomogeneities.

References

Adams, J. B., Pieters, C., and McCord, T. B.: 1974, Orange glass: Evidence for regional deposits of pyroclastic origin on the moon, *Proc. Lunar Sci. Conf. 5th*, p. 171–186.

Brown, W. E., Jr., Saunders, R. S., and Kobrick, M.: 1976 Lunar subsurface structure in the Sulpicius Gallus region (abstract), In *Lunar Science VIII*, p. 97–98, The Lunar Science Institute, Houston.

Bryan, W. B.: 1973, Wrinkle ridges as deformed surface crust on ponded mare lava, *Proc. Lunar Sci. Conf. 4th*, p. 93–106.

Head, J. W.: 1974, Lunar dark mantle deposits: Possible clues to the distribution of early mare deposits, *Proc. Lunar Sci. Conf. 5th*, p. 207–222.

Hodges, C. A.: 1973, Mare ridges and lava lakes, *Apollo 17 Prelim. Sci. Rep.*, NASA SP-330, Section 31, Part B, p. 31-12 to 31-21.

Howard, K. A., Carr, M. H., and Muehlberger, W. R.: 1973, Basalt stratigraphy of southern Mare Serenitatis, *Apollo 17 Prelim. Sci. Rep.*, NASA SP-330, Section 29, Part A, p. 29-1 to 29-12.

Lucchitta, B. K.: 1973, Geologic setting of the dark mantling material in the Taurus-Littrow region of the Moon, *Apollo 17 Prelim. Sci. Rep.*, Section 29, Part B, p. 29-13 to 29-25.

Lucchitta, B. K. and Schmitt, H. H.: 1974, Orange material in the Sulpicius Gallus Formation at the southwestern edge of Mare Serenitatis, *Proc. Lunar Sci. Conf. 5th*, p. 223–234.

May, T. W., Peeples, W. J., Maxwell, T., Sill, W. R., Ward, S. H., Phillips, R. J., Jordan, R., and Abbott, E.: 1976, Subsurface layering in Maria Serenitatis and Crisium: Apollo lunar sounder results (abstract), In *Lunar Science VII*, p. 540–541. The Lunar Science Institute, Houston.

Muehlberger, W. R.: 1974, Structural history of southeastern Mare Serenitatis and adjacent highlands, *Proc. Lunar Sci. Conf. 5th*, p. 101–110.

Oberbeck, V. R. and Quaide, W. L.: 1968, Genetic implications of lunar regolith thickness variations. *Icarus* **9**, 446–465.

Phillips, R. J., Adams, G. F., Brown, W. E., Eggleton, R. E., Jackson, P. L., Jordan, R., Linlor, W. I., Peeples, W. J., Porcello, L. J., Ryu, J., Schaber, G. G., Sill, W. R., Thompson, T. M., Ward, S. H., and Zelenka, J. S.: 1974, The Apollo 17 Lunar Sounder Experiment, *Apollo 17 Prelim. Sci. Rep.*, NASA SP-330, Section 22, p. 1–26.

Pike, R. J.: 1976, Simple to complex impact craters: The transition on the Moon (abstract), In *Lunar Science VII*, p. 700–702, The Lunar Science Institute, Houston.

Sjogren, W. L.: 1976, Quantitative mass distribution model for Mare Orientale (abstract), In *Lunar Science VII*, p. 818–820, The Lunar Science Institute, Houston.

Thompson, T.V., Matson, D. L., Phillips, R. J., and Saunders, R. S.: 1975, Vidicon spectral imaging: Color enhancement and digital maps, *Proc. Lunar Sci. Conf. 6th*, p. 2677–2688.

Young, R. A.: 1975, Mare crater size-frequency distributions: Implications for relative surface ages and regolith development, *Proc. Lunar Sci. Conf. 6th*, p. 2645–2662.

Young, R. A., Brennan, W. J., and Nichols, D. J.: 1974, Problems in the interpretation of lunar mare stratigraphy and relative ages indicated by ejecta from small impact craters, *Proc. Lunar Sci. Conf. 5th*, p. 159–170.

Young, R. A. and Brennan, W. J.: 1976, Selected aspects of lunar mare geology from Apollo orbital photography, NASA CR-147424, 139 pp.

Roddy, D. J., Pepin, R. O., and Merrill, R. B., editors.
(1977) *Impact and Explosion Cratering*, Pergamon Press (New York), p. 539–562.
Printed in the United States of America

Nested-crater model of lunar ringed basins

DON E. WILHELMS, CARROLL ANN HODGES, and RICHARD J. PIKE

U.S. Geological Survey, Menlo Park, California 94025

Abstract—We propose a model for the origin of impact-basin rings whereby the main topographic rim of a basin approximates the limit of excavation and inner rings approximate the rims of craters formed inside the transient crater by some perturbation in the cratering process. The cause of this complexity in transient cavities may be the presence of discontinuities in the target material. The second inward ring may have formed at the seismic discontinuity about 20 km deep in the lunar crust, and the third, innermost ring of a few large basins at the crust-mantle interface about 60 km deep. Slumping increased the original diameters of many rings and split some initially coherent rings into subsidiary or partial rings. Deformation outside the transient crater produced external arcs. This model differs from prevalent hypotheses of ring formation whereby an inner ring approximates the transient crater rim and major faulting of the flank produced the outer ring structures.

INTRODUCTION

THE NATURE AND ORIGIN of multiple concentric rings in large lunar and planetary impact excavations are controversial. Such cavities are called ringed basins, multi-ringed basins, or simply, basins, but they are actually large, complex craters. Especially disputed is identification of the basin ring that corresponds to the rim crest of a smaller, simpler crater. The rim crest of a lunar crater is generally acknowledged as the boundary of ejecta excavation, except as modified by slumping. Estimates of volume and source depth of basin ejecta now diverge widely and would be greatly improved by correct identification of the excavation boundary.

Interpretations in this paper are based mostly on observations of lunar craters and basins, discussed in order of increasing complexity. We conclude that the outer rim of relatively small, double-ring basins corresponds to the rim crest of simpler craters and thus approximates the boundary of excavation, whereas the inner ring is the rim of an inner—or nested—crater. Evidence from terrestrial and theoretical craters presented at the Symposium on Planetary Cratering Mechanics (Flagstaff, September, 1976) appeared to support this previously developed hypothesis (Hodges and Wilhelms, 1976a,b,c) and, in particular, our preferred explanation for the formation of nested craters: layered target materials. We extrapolate our interpretations of double-ring basins to the more complex multi-ringed basins and conclude that the main topographic rim also defines the cavity of excavation of these large basins. The next-inward major ring is a subsidiary crater rim formed at a lunar seismic discontinuity, and the few existing third, innermost rims developed at a second, lower seismic discontinuity. Slumping caused much of the complexity of ring structures in these basins, and bedrock deformation produced arcuate ridges outside the

Fig. 1. Schrödinger Basin (center) and Antoniadi (lower right). Rough ejecta and linear secondary-impact valleys surround Schrödinger. Rugged discontinuous blocks of the inner ring of Schrödinger resemble those of single central peaks of craters. This ring is unlike the terraces on the crater wall from which it is widely separated by smooth plains material, probably impact melt in part. Antoniadi is unusual in that it has both a central peak typical of craters and a ring typical of basins. North at top in this and all subsequent photographs. Lunar Orbiter IV 8 M.

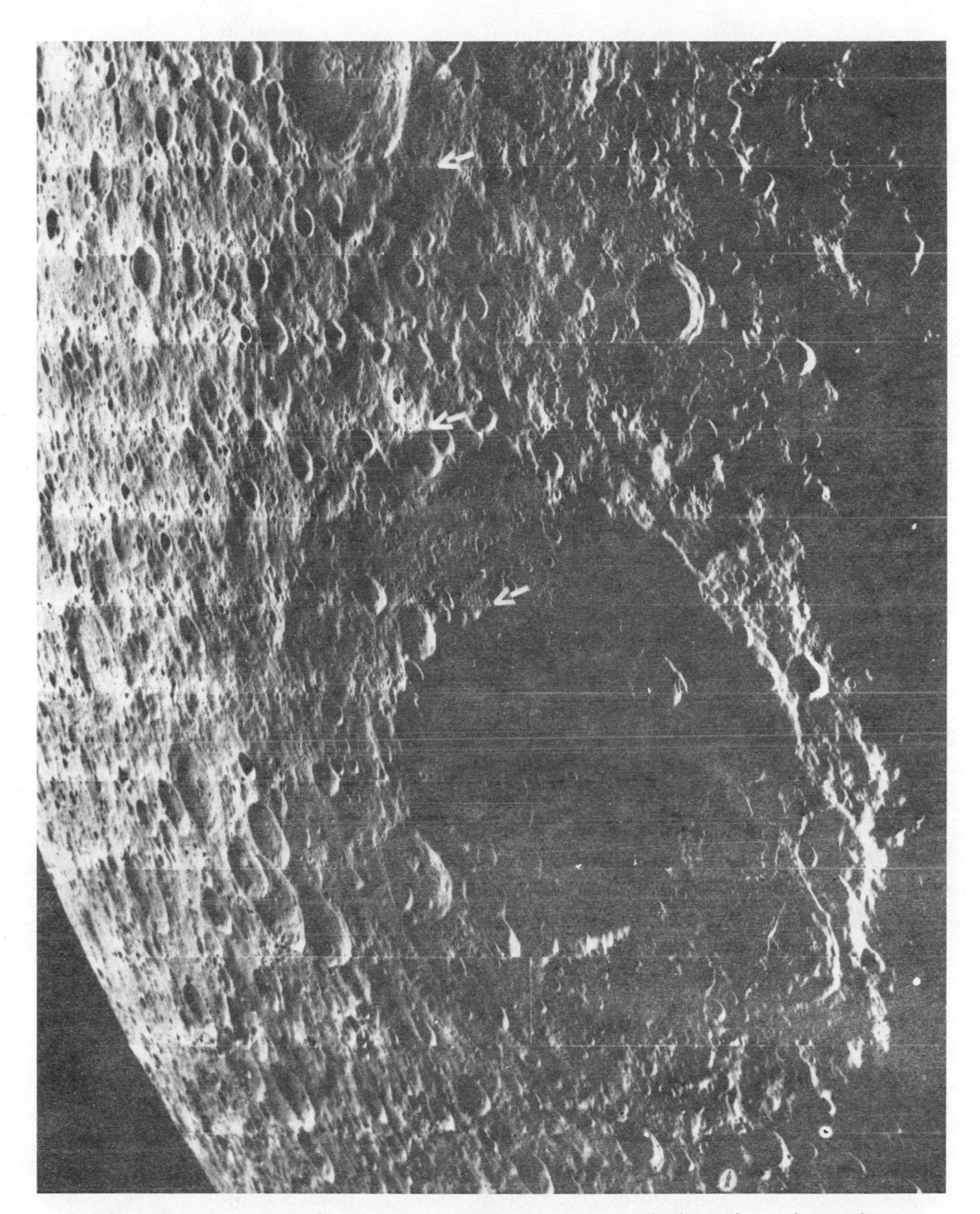

Fig. 2. Moscoviense Basin on the lunar far side. Lower arrow indicates inner ring 200 km in diameter (partly flooded by mare in northeast). A broad bench extends outward to the next ring (middle arrow). In the nested crater model, both rings are crater rim crests. Upper arrow indicates outermost arcuate structure believed associated with the basin; this arc may have been produced by local deformation outside the transient crater. Lunar Orbiter V 124 M.

excavated cavity. The topographic rim of large basins is identified as the limit of excavation mostly on the basis of ejecta textures, ring height, ring continuity, and interpretation of terrestrial analogs. Further evidence includes new measurements of ejecta and secondary crater distribution.

Aspects of our model have been proposed by other workers. Baldwin (1974) and Chao *et al.* (1975) also regarded the topographic basin rim as the limit of ejecta excavation. Oberbeck (1975) proposed a nested-crater model for the Ries Crater (Germany) and suggested that layering might also cause lunar ring formation. Hartmann and Wood (1971), McCauley (1976), Head (1976a), and others also suggested that lunar subcrustal layering or structure could influence ring formation. McCauley (1976) proposed that one ring might form at a seismic discontinuity and represent a topographic bench within which excavation occurred. McCauley (1976) and Head (1976b) also suggested that shock deformation exterior to transient cavities produced concentric structures, as in terrestrial craters formed by impact (Roddy, 1968) or chemical explosions (Roddy, 1976; Jones, 1976). Faulting has been invoked by many workers to create rings (McCauley, 1968, 1976; Mackin, 1969; Hartmann and Wood, 1971; Short and Foreman, 1972; McGetchin *et al.*, 1973; Head, 1974, 1976b; Gault, 1974; Schultz and Gault, 1975; Head *et al.*, 1975; Floran and Dence, 1976). These and other basin topics have been reviewed by Hartmann and Kuiper (1962), Baldwin (1963), McCauley (1967, 1968), Stuart-Alexander and Howard (1970), Wilhelms (1970), Wilhelms and McCauley (1971), Hartmann and Wood (1971), Hartmann (1972a,b), Mutch (1973), Moore *et al.* (1974), Howard *et al.* (1974), and Wood and Head (1976).

SMALL DOUBLE-RING BASINS

The outer rings that bound fresh double-ring basins (Fig. 1) closely resemble the rim crests of central-peak craters like Copernicus (Hartmann and Wood, 1971). Outside the crest of a typical crater or small basin, rugged, generally concentric, massive or hummocky-textured ejecta grades outward into a radially textured deposit that extends about a basin or crater diameter from the crest. Inside the crest are arcuate, step-like terraces or blocky masses that obviously have slumped from the rim crest (Gilbert, 1893). Thus the original rim-crest diameter was smaller than that presently observed. By restoring terraces upward and outward along inferred slump planes, we estimated that the original diameter of the double-ring basin Mendeleev (partly covered by Defense Mapping Agency topographic maps) was about 12–14% smaller than the present 320 km. Other basins of the Schrödinger or Mendeleev type include Korolev (420 km, 5°S, 158°W; Hodges and Wilhelms, 1976a) and Birkhoff (370 km, 59°N, 147°W). The morphologic similarities of the outer rims of these basins to the rims of craters indicate that, before slumping, they marked the outer limit of ejecta excavation.

In contrast, Dence (1968) suggested that the outer rims are fault scraps along which the rim crest collapsed to produce the inner ring. We think terracing of this magnitude unlikely because the inner rings are separated from the nearest

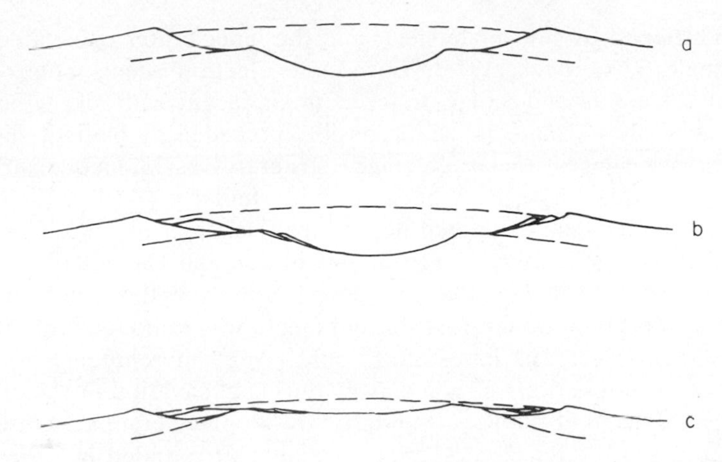

Fig. 3. Hypothetical development of typical double-ring basin 360 km in diameter with an inner ring 180 km in diameter. Upper dashed line represents original ground surface; lower dashed line represents 20 km seismic discontinuity. (a) Immediately after impact. (b) After slumping: small crater-like terraces (Schrödinger type) on right, large terraces on left. Inner crater also shown as slumped on left. (c) After post-impact recoil and isostatic adjustment. Vertical scale not exaggerated; true lunar curvature.

true terraces of the outer rim by a topographic bench that is much wider than any of the terrace benches (Figs. 1, 2, 3). Jagged rings like those of Schrödinger (Fig. 1) are morphologically unlike terraces, and there are no transitional landforms between them. We propose instead that the inner rings are the uplifted rims of subsidiary cavities that formed within the transient crater (Fig. 3; Hodges and Wilhelms, 1976a,b,c).

Terrestrial Analogs

A nested-crater origin for double-ring basins is supported by the concentric structures that develop in terrestrial meteorite-impact craters, experimental impact and explosion craters, and computer-modeled explosion craters. Most such craters formed in layered materials.

Concentric craters were produced by laboratory impacts into a weakly cohesive layer overlying a more resistant substrate. The weak layer was preferentially excavated by low-angle shear, and the resulting crater contained a small inner crater in the substrate (Oberbeck and Quaide, 1967; Quaide and Oberbeck, 1968; Oberbeck, 1975). R. T. Allen (oral communication, 1976) simulated benched craters in computer calculations involving the differential behavior of layered materials. Piekutowski (1976) produced internal and external rings in layered target materials by small surface explosions.

The Ries is probably a double-ring (or possibly three-ring) structure (Dennis, 1971; Engelhardt, 1974; Stöffler, 1975; Chao, 1976a) consisting of an inner crater (12 km across) in crystalline basement surrounded by a shallower, broader crater (24 km) excavated in overlying sedimentary rock. The sedimentary strata could have been stripped by low-angle shear and the inner crater excavated at higher ejection angles (Oberbeck, 1975). Low-angle ejection seems compatible with textures of the ejecta and radial striations in subjacent bedrock, which indicate ground-hugging flow (Chao, 1976a,b) possibly preceded by ballistic flight (Oberbeck, 1975). We suggest that such ringed structures as Manicouagan and West Clearwater, Canada, may have been formed similarly so that their transient cavities were larger than estimated heretofore (Dence *et al.*, 1965; Dence, 1968; Floran and Dence, 1976). Dence (1972) and Floran and Dence (1976) concluded that the limestones between the bounding scarps and the inner ring of both craters were in place and therefore that no ejecta was removed from this region; alternatively, however, the limestones could represent remnants of strata that were mostly but not entirely stripped away during excavation of the outer crater, as probably occurred at the Ries. Similarly, the Popigay complex crater, Siberia, comprises an inner crater in crystalline basement surrounded by an outer crater in sedimentary rocks. The sedimentary rocks were likely susceptible to excavation across a broader diameter (Masaytis *et al.*, 1972, 1975).

Craters produced by experimental chemical explosions at Suffield, Alberta, also contain concentric uplifted structures within the excavation cavity (Roddy, 1976; Jones, 1976). In the Prairie Flat test, fractured anticlines of substrate material formed three concentric rings on the crater floor. The Pre-Mine Throw IV crater contains a bench at a weak stratigraphic discontinuity.

LARGE BASINS

Ring origin is most controversial for the large Orientale, Imbrium, Nectaris, Crisium, and Humorum basins, whose multiple-ring structural style was first systematically described by Hartmann and Kuiper (1962). There have been many attempts to compare basins by ring spacing and morphology (for example, Hartmann and Kuiper, 1962; Stuart-Alexander and Howard, 1970; Hartmann and Wood, 1971; Wilhelms and McCauley, 1971; Wilhelms, 1973; Howard *et al.*, 1974). As many as five rings have been mapped at Orientale, Imbrium, and Crisium. We propose that the large basins are dominated by two or three rings that are the rim crests of nested craters, analogous to the two rings of double-ring basins like Schrödinger or the Ries. Other rings are subsidiary features derived from these crater rims by slumping.

The principal competing explanations for rings are the tsunami-wave mechanism (Van Dorn, 1968, 1969; Baldwin, 1972, 1974) and various megaterrace models whereby slumping or related types of inward, downward, or rotational movement took place on a very large scale and divided the originally simple transient crater rim into two or more concentric rings (McCauley, 1968; Mackin, 1969; Hartmann and Wood, 1971; Short and Foreman, 1972; McGetchin *et al.*, 1973; Head, 1974; Gault, 1974; Schultz and Gault, 1975; Head *et al.*, 1975; Floran and Dence, 1976; Head, 1976b; McCauley, 1976). Tsunami-wave (or water-drop) mechanisms may produce spaced ripples (Van Dorn, 1968, 1969; Baldwin, 1972), but it has not been established that these can "freeze" in place or otherwise produce the observed concentric rings. Although we agree that slumping is a major factor in the origin of partial rings, we believe that it cannot account for the major rings.

Orientale

Montes Cordillera, the outermost conspicuous ring that bounds the east side of the Orientale Basin (Fig. 4; Table 1) is generally acknowledged as the topographic limit of the basin (for example, Head, 1974). In the western part of the basin poor photography makes it unclear whether the continuation of the Cordillera ring is complete. Two rings, about 480 and 620 km across, collectively called Montes Rook, lie inward from the Cordillera ring (940 km). An additional ring inside the Rook rings is indicated by the steep mountainous scarp that bounds the northwest shore of Mare Orientale; continuation of this scarp to lower scarps in other sectors makes a circle about 330 km in diameter (Fig. 4). Outside the Cordillera rim, about 650 km from the basin center (Hartmann and Kuiper, 1962; McCauley, 1968), are discontinuous arcuate ridges consisting mainly of pre-Orientale crater rims (Fig. 4).

We consider the Cordillera ring to mark the approximate boundary of excavation of the Orientale Basin because of its resemblance to crater rims. It is topographically prominent and marks the beginning of the obviously thick ejecta blanket and the secondary craters of the Orientale Basin (Fig. 4). Raised massifs along part of the ring indicate that, like crater rims, the Cordillera ring probably includes uplifted "bedrock" as well as ejecta in an "overturned flap".

The significance of the Cordillera as a primary structure in our model contrasts with the passive or secondary origin of the scarp most commonly proposed. Head (1974) and McCauley (1976) believe that the Cordillera ring marks the limit of a zone of deformation outside the transient crater and that the scarp formed when the zone collapsed by rotational, inward, and downward movement, uplifting Montes Rook, the original basin rim. Variants of this megaterrace hypothesis place the movement either before (McCauley, 1976) or after (Head, 1974) completion of crater excavation. We suggest instead that the arcuate ridges outside the Cordillera represent the deformation outside the transient crater.

Many workers have considered one or the other of the Rook rings to approximate the outer limit of ejecta excavation (Hartmann and Wood, 1971; McGetchin *et al.*, 1973; Moore *et al.*, 1974; Howard *et al.*, 1974; Head, 1974, 1976; Schultz and Gault, 1975; Head *et al.*, 1975; McCauley, 1976). The inner Rook ring also has been considered an expanded central peak (Head, 1974; McCauley, 1976). Short and Forman (1972), Dence (1973), and Gault (1974) proposed that the innermost 330-km ring most closely approximates the limit of basin excavation. This is highly unlikely, however, for the ring is the same size as the Schrödinger outer rim—yet is part of a vastly greater basin.

We agree with many of the arguments in favor of a crater-rim origin for Montes Rook. They are very prominent rimlike structures with steep inward

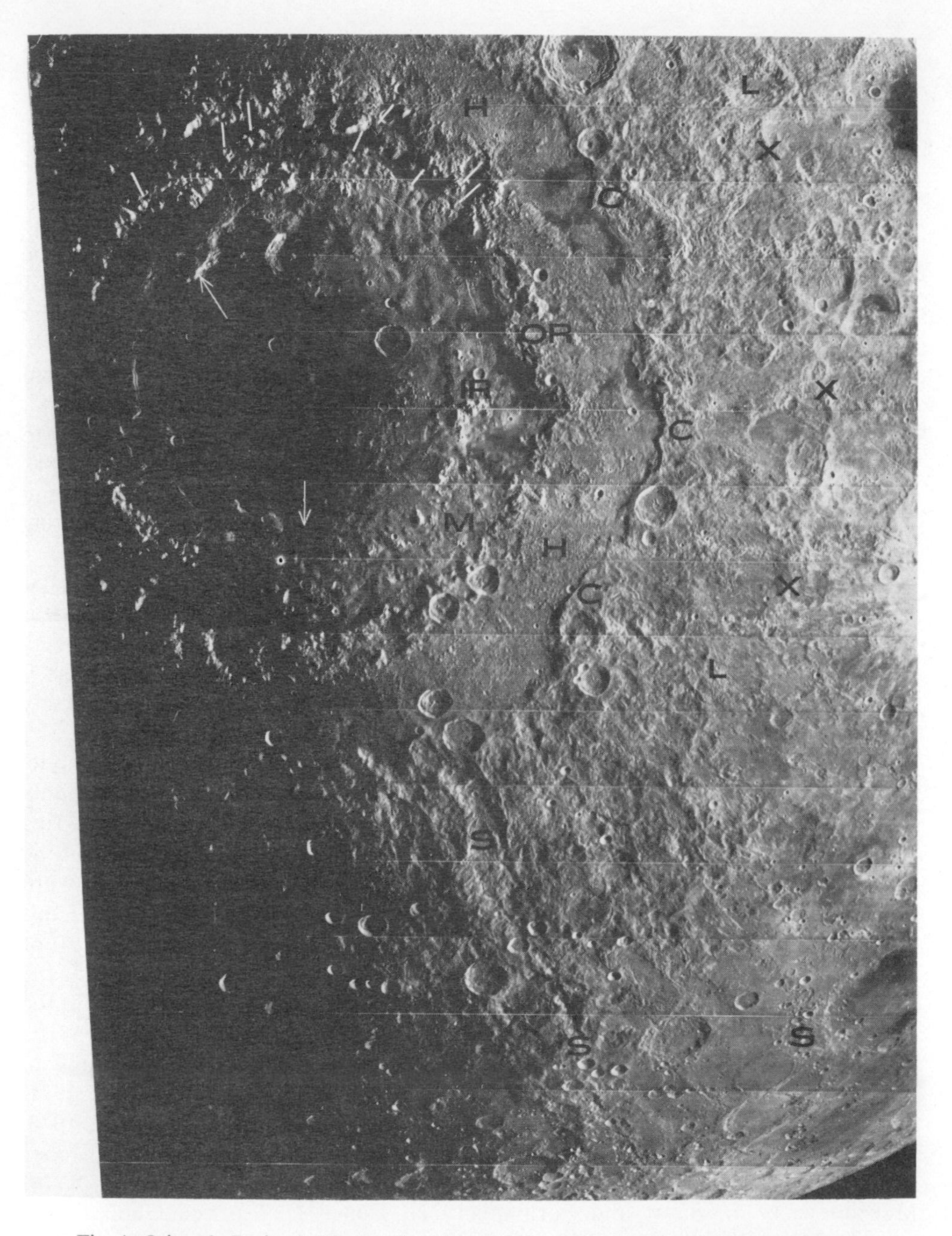

Fig. 4. Orientale Basin. C, Montes Cordillera (470 km radius); OR, outer Montes Rook (310 km radius); IR, inner Montes Rook (240 km radius); lines in north indicate terrain believed separated by slumping; arrows, innermost ring (165 km radius); M, probable impact melt; H, hummocky ejecta; L, lineated ejecta; X, external deformation ridges; S, secondary impact craters. Lunar Orbiter IV 187 M.

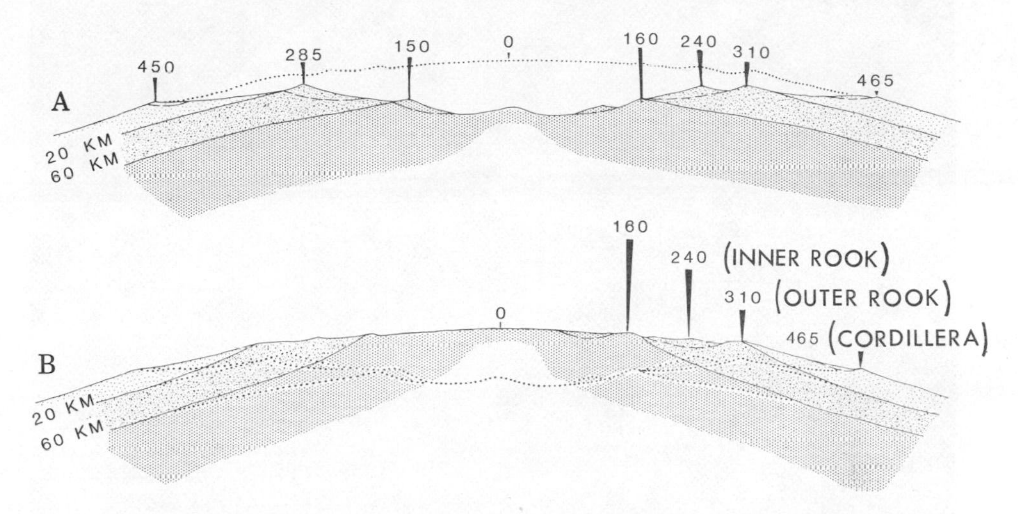

Fig. 5. Inferred profiles of Orientale Basin immediately after impact (A) and at present, after rebound (B). Dashed lines indicate slump faults. Dotted lines in each drawing correspond with solid or dashed lines in the other. Figures (kilometers) refer to radii of present rings and to depths to major seismic discontinuities observed in Oceanus Procellarum (see Fig. 12). Patterned areas include bedrock, disrupted bedrock, and ejecta; impact melt, fallback, and post-basin mare and crater materials are not shown. Pre-slumping configuration shown in left side of (A); post-slumping configuration shown in right side of (A) and all of (B). Slumping of ring originally 285 km from basin center has produced two Rook rings at 240 and 310 km. True lunar curvature; no vertical exaggeration.

scarps and a gentler outer flank. They also seem to be the inner limit of an ejecta deposit (McCauley, 1976). In contrast to the other hypotheses, however, we do not believe Montes Rook is the *outer* boundary of ejecta excavation; it is the boundary of a crater excavated inside the transient crater. Its ejecta is hummocky and blocky, quite unlike the lineated, thick material that originates at Montes Cordillera. The hummocky Rook ejecta extends from Montes Rook to the Cordillera scarp in all sectors and overlaps that scarp and the outer ejecta in several places (McCauley, 1976). This superposition precludes formation of the Cordillera scarp by terracing after the cessation of cratering.

The two Rook rings seem to have been derived from a single rim by downfaulting or slumping of the inner ring towards the basin center (Fig. 5). In the north, the outer edge of the inner ring lies at the foot of the outer ring scarp, and in many places protrusions in one ring match cavities in the other (Fig. 4). The attitude of the presumed faults that separate the two rings is conjectural. Hartmann and Wood (1971) believed that they are steep and served as conduits for the mare materials that fill the trough between the rings. We, however, believe that they are more likely to be shallow by analogy with craters, whose wall terraces are very likely low-angle gravitational slides (Gilbert, 1893). A similar split of an original Rook ring may have occurred in the southeast, but the depression between ring crests is obscured by a wide belt of undulatory, cracked

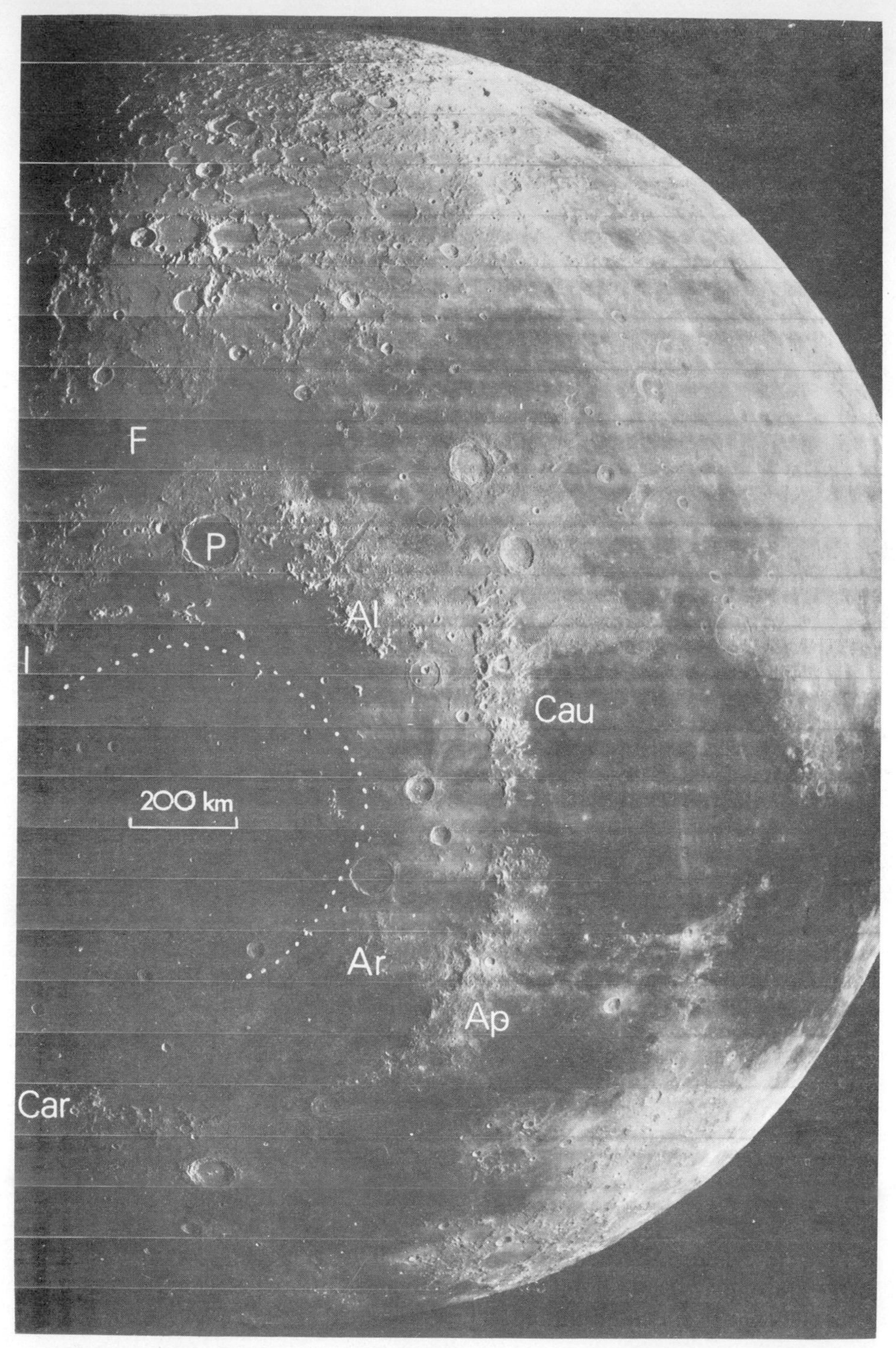

Fig. 6. Imbrium Basin. Car, Montes Carpatus; Ap, Montes Apenninus; Cau, Montes Caucasus; Al, Montes Alpes; Ar, Montes Archimedes; F, Mare Frigoris; P, Plato; I, Iridum crater flooded by mare basalts of Sinus Iridum. Dotted line, inferred original inner Imbrium crater rim. Proposed outer crater rim consists of Car, Ap, southern Cau, Al. Lunar Orbiter IV 115 M.

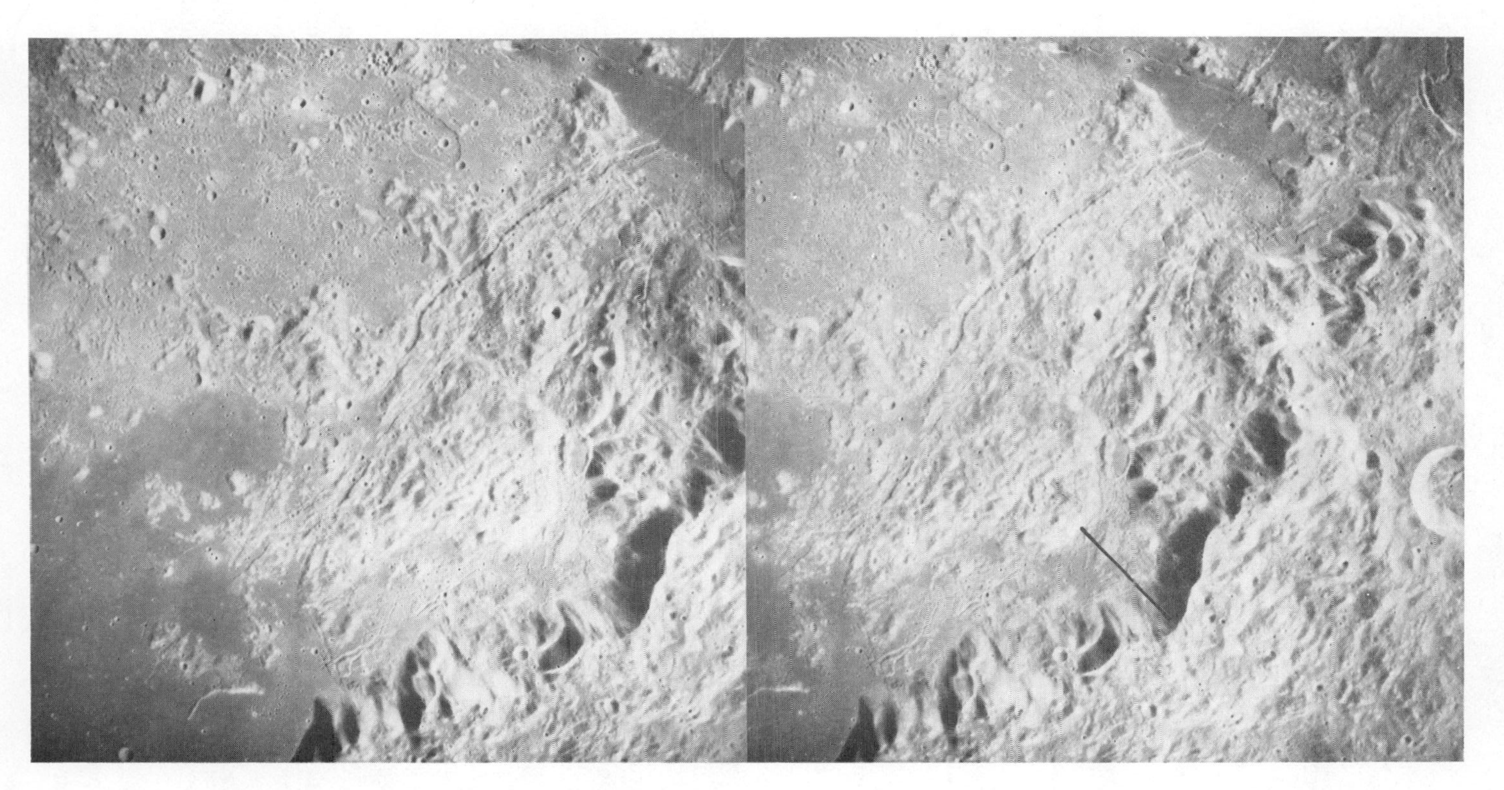

Fig. 7. Montes Apenninus front near crater Conon (right); head of Rima Hadley, upper right. Line indicates matching features of slump mass and re-entrant of mountain front from which it was derived; movement about 30 km northwest. Part of Montes Archimedes in upper left, embayed by planar Apennine Bench Formation. Cracked material between slump mass and mountain front is probably impact melt emplaced after slumping; conspicuous NE-SW grabens parallel to mountain front indicate that basinward movement continued after emplacement of impact melt. Stereo pair; Apollo 17 mapping frames 2709 (right) and 2710.

material that is likely impact melt (Moore *et al.*, 1974; Howard *et al.*, 1974; Head, 1974). This relation shows that slumping preceded solidification of the impact melt. To the west, the division into two rings seems less marked or nonexistent. The crest of the reconstructed original Rook ring had a diameter of about 570 km, between the observed rings.

The innermost, 330 km ring of Orientale may also be a subsidiary crater rim nested inside the Rook and Cordillera rings. This hypothesis is shared by otherwise divergent models (McCauley, 1976).

Imbrium

The main topographic rim of Mare Imbrium comprises Montes Carpatus in the south, Montes Apenninus in the southeast, part of Montes Caucasus in the east, Montes Alpes in the northeast, and a broad arc whose surface is mostly covered by deposits of the younger craters Plato and Iridum in the north and northwest (Fig. 6). Montes Apenninus, the Moon's largest mountain range, consists of many major massifs, overlain by hummocky, concentrically oriented ejecta deposits that grade outward into radial deposits (Figs. 6, 7). No rimlike structures exist outside Montes Apenninus. This same pattern occurs on crater rims, and in our model, Montes Apenninus approximates the initial, outer Imbrium rim crest.

Attempts have been made to identify interior rings by fitting circles to the large islands in Mare Imbrium, to segments of the topographic basin rim, to Montes Archimedes (Fig. 6), or to mare ridges (Hartmann and Kuiper, 1962; Wilhelms and McCauley, 1971). The islands and ridges do seem to form two subcircular structures. We suggest that these structures, like the two Rook rings of Orientale, may have originally comprised a single inner crater rim crest (Figs. 6, 8; Table 1). Montes Archimedes may be a remnant of the flank of that rim. If Imbrium has a third, innermost ring like the 330-km ring of Orientale, it is completely obscured by mare.

Several large arcuate features lie outside of and concentric with the Carpatus-Apenninus-Alpes mountains (Figs. 6, 8). The most conspicuous are southeast of the basin and bound Sinus Medii (Wilhelms, 1964). The north shore of Mare Frigoris is also concentric with Imbrium and lies on an arc with the northern part of Montes Caucasus (Fig. 6). These features have been considered part of the Imbrium crater rim (Wilhelms and McCauley, 1971), but the northern part of Montes Caucasus contains grooves that cross the mountain front at an oblique angle. The grooves presumably were formed by secondary impact or ground flow of ejecta, so should be approximately perpendicular to a crater rim. Their obliquity suggests that the northern Caucasus and the Mare Frigoris trough are external to the Imbrium transient crater. Thus we believe that only the southern part of Montes Caucasus, which lies on a circle with Montes Apenninus and Alpes, is part of the crater rim. This apparently different origin for grossly similar features illustrates the problem of identifying the position of basin rims.

Table 1. Radii (km) of major features of seven lunar basins.

Basin	Center coordinates	Third ring	Intermediate ring		Topographic rim of basin		Exterior arcs	Limits of continuous ejecta†		Limit of secondary crater concentrations†	
			Present	Pre-slumping	Present	Pre-slumping					
Imbrium	34°N, 17°W	—	330,410	350	590	530	950,1150	1250	1400 1200	2300	2700 1400
Orientale	20°S, 95°W	165	240,310	285	470	450	650	1000	1050 900	1450	1800 1100
Nectaris	16°S, 34°E	140	200,300	265	435	415	550	750	950 650	1250	1500 700
Crisium	18°N, 59°E	—	175*	—	270*, 340*	315*	500	550	650 500	1100	1300 650
Humorum	24°S, 39°W	—	150	—	205, 280	245	370	620	800 550	950	1000 750
Moscoviense	26°N, 146°E	—	100	—	205	205	320	400	450 300	—	
Schrödinger	75°S, 134°E	—	80	—	160	140	—	300	400 250	500	950 250

*Averages of east-west and north-south dimensions.
†Left—average maximum radius; right—radii of major outlines and re-entrants.

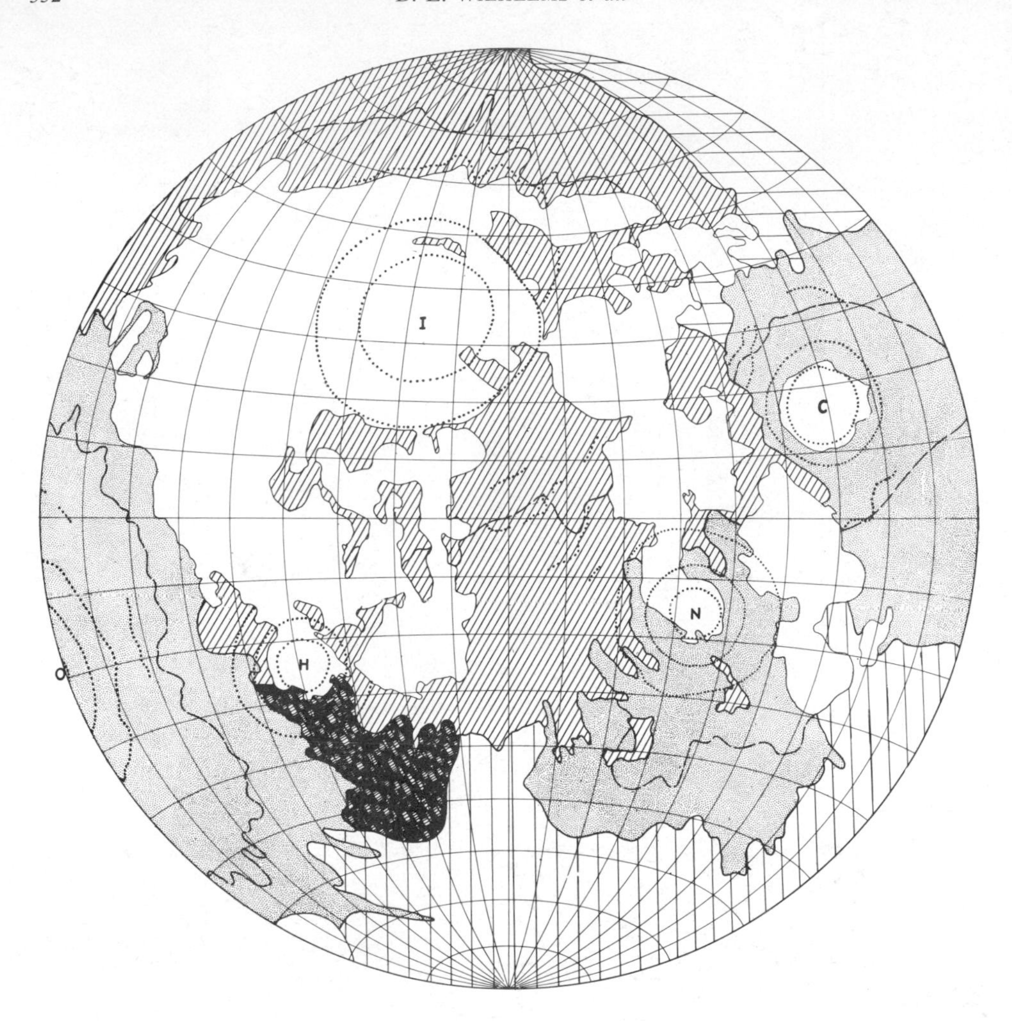

Fig. 8. Map of near side of Moon showing the five large basins discussed in text: Orientale (O), Imbrium (I), Nectaris (N), Crisium (C), Humorum (H). Humorum—dark pattern; Imbrium—diagonal lines; others—light shading. Dots show major rings and external arcs in their inferred pre-slumping position (Table 1). Dashed line separates continuous basin deposits (closer to basin center) from concentrations of secondary impact craters; additional secondaries occur closer to and farther from basin. Horizontal lines in northeast (upper right) represent materials of Nectarian basins Serenitatis and Humboldtianum. Vertical lines at bottom (south), miscellaneous materials older than the basins. Blank areas are mare and crater deposits that obscure unit boundaries or rings. Equal area projection.

Nectaris, Humorum, and Crisium

The Nectaris Basin was considered as the prototype multi-ringed basin in the earliest systematic basin study (Hartmann and Kuiper, 1962) and is readily compared with Orientale. Rupes Altai, a Cordillera-like ring 435 km west of the basin center, clearly is the topographic rim of the Nectaris Basin and marks the inner limit of the lineated basin ejecta blanket. Two partial rings inside Rupes Altai and its continuations are observed in the southern and eastern sectors of the basin, but merge in a broad, high uplift in the northwest and north. We propose that this ring, coherent in places and split in others, is the analog of Montes Rook and marks a second crater rim crest inside the main Altai rim. Arcuate shelves in Mare Nectaris may represent a third, innermost crater rim about 280 km across, analogous to the innermost Orientale ring.

Crisium (Fig. 9) and Humorum (Fig. 10) are less easily compared with Orientale or other basins. The main mare-bounding topographic rims of both basins are complex, but we believe that each is a major crater rim that has been divided by Rook-like slumping. Stereoscopic photographs available for Crisium show that the highest part of the western basin rim lies outside the more coherent ring that borders the mare; thus inward slumping of the inner ring is

Fig. 9. Crisium Basin. A massive ring 540 km in diameter borders the mare in the north, west, and south. An inner ring is suggested by the conspicuous shelf in the mare. Arrows indicate incomplete ring 680 km in diameter; central, double arrow in crater Cleomedes is 140 km long. Rectified telescopic photograph courtesy of E. A. Whitaker and W. K. Hartmann.

Fig. 10. Humorum Basin. Lower dots to right of highest part of southern scarp; center dots, least modified part of topographic basin rim; inner dots, mare structures that may mark inner ring (well seen only on telescopic photographs taken at low sun angle). S, probable or possible (S?) secondary impact craters. Sets of dots each 35 km long. Lunar Orbiter IV 136 M.

indicated. Salients and re-entrants in massifs and other features in the rings comprising the basin rims commonly match (Figs. 9, 10) and suggest segmentation by faulting. Crisium (Wilhelms, 1973; DeHon and Waskom, 1976) and Humorum each contain mare shelves that suggest innermost rings. The greatest uncertainties about both basins revolve about partial rings outside the topographic basin rims. Those of Crisium (Wilhelms and McCauley, 1971; Olson and Wilhelms, 1974; DeHon, 1975; Wilhelms and El-Baz, 1976) consist of irregular troughs or scarps that are probably external deformation features and not crater rims; but 370 km south of Humorum's center is a scarp about 300 km long intermediate in morphology between Montes Cordillera and the Crisium external arcs (Fig. 10). This scarp apparently is nearly as high as the compound ring that encloses Mare Humorum. Therefore it cannot be confidently classified as a partial crater rim or exterior arc without additional data.

Extent of Ejecta and Secondary Craters

To test our identification of the topographic basin rims as the boundaries of basin excavation and to help resolve ambiguities in basins like Humorum, we mapped the continuous (ground-hugging) ejecta and the secondary impact craters around seven basins and plotted the radii of these features against radii of the basin rings (Figs. 8, 11; Table 1). Comparable data for large central-peak craters have been fitted with least-squares regressions, and we compare the candidate basin rings with these curves. Despite uncertainties in these extrapolations and in the photogeologic interpretation of the older basins, some conclusions can be drawn from the graphs. First, the topographic basin rims of Schrödinger and Moscoviense clearly fit the curves for craters better than do the inner rings (as might be expected for these small basins). Second, the innermost basin rings and the outermost arcs for all basins but Humorum can be excluded as likely limits of basin excavation. Third, for Orientale and Imbrium, the Cordillera and Apennine rings, respectively, best fit the ejecta curve (see also Moore *et al.*, 1974), and the Apennine ring is the best fit for Imbrium to the secondary-crater curve. Although more definitive conclusions may be unwarranted at this time, these data clearly are compatible with large excavations delimited by the main topographic basin rims.

The ring radii used on the graphs (Fig. 11) are observed, post-slumping measurements. Like all crater rims over 10–20 km across, basin rims probably have been enlarged somewhat by slumping. The amount of enlargement at Imbrium is measurable along the Apennine Front covered by Apollo mapping stereo photographs (Fig. 7). Slivers, blocks, and chaotic masses of material lying northwest of the Apennine Front fit well with topographic re-entrants in the present front. A former rim crest about 30–60 km closer to the basin center than the present Apennine crest is reconstructed by restoring these slumps. The Cordillera and Altai rings are nearly continuous for at least half the circumference of their respective basins and probably have enlarged less. Some short arcuate ridges seem to have moved inward as much as 10–30 km. The diameters measured along the crests of the Cordillera (940 km) and the Altai (870 km) rings thus are probably only about 20–60 km larger than the original transient crater rim crests. The restored ring radii (Table 1) do not substantially alter the quantitative results (Fig. 11) or their interpretation.

The ratio of terrace width to crater radius decreases with increasing crater

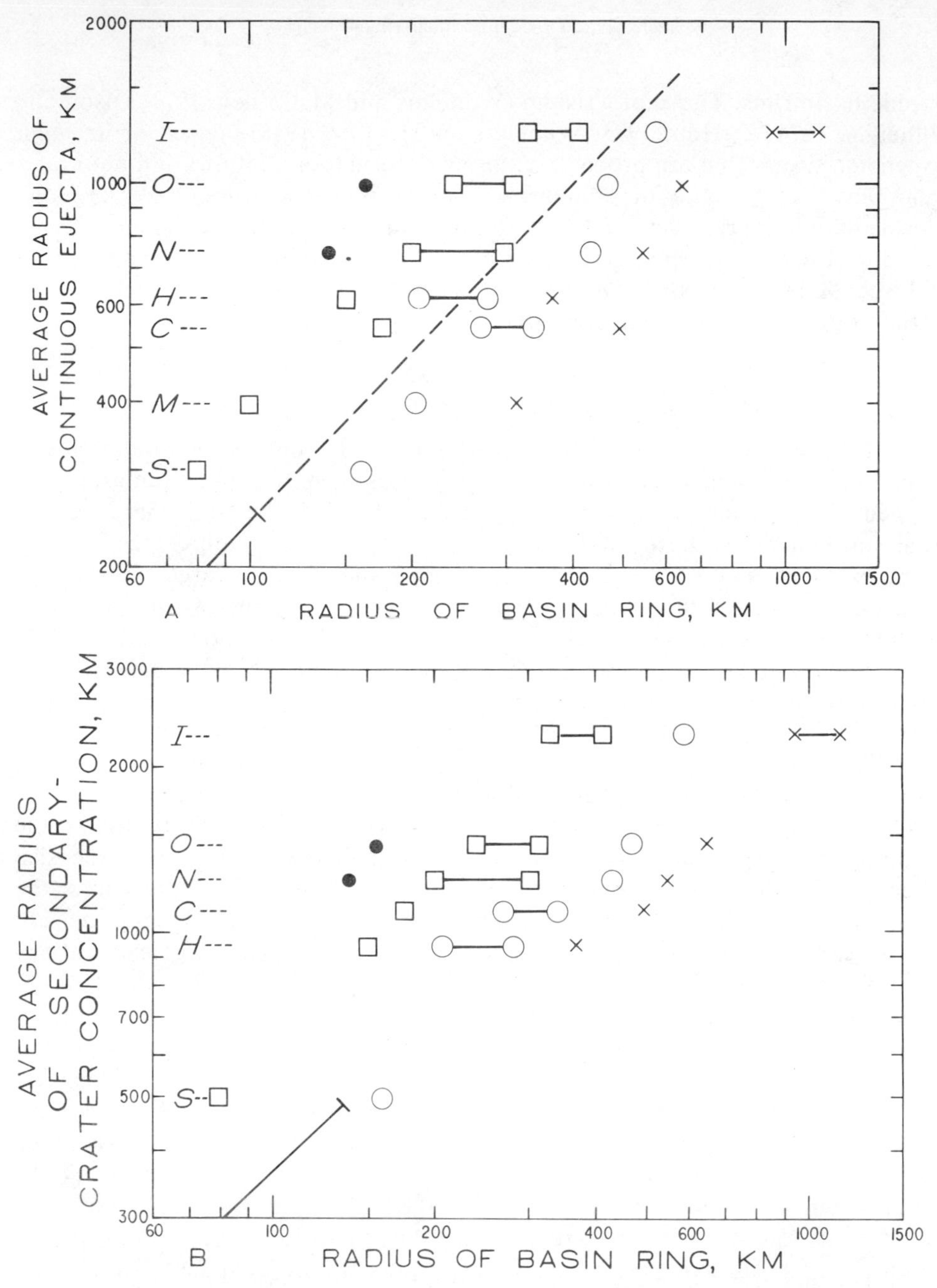

Figure 11. Average radius of continuous ejecta blanket (A) and secondary-crater concentrations (B) as a function of basin-ring radius, for four classes of rings: Exterior arcs (crosses); topographic rims (open circles); inner rings or intermediate rings (squares); innermost rings (dots). I, Imbrium; O, Orientale; N, Nectaris; C, Crisium, H, Humorum; M, Moscoviense; S, Schrödinger. Horizontal bars connect presumably split segments of same ring. Solid lines are least-squares fits to normal lunar craters (A, $n = 84$, from Moore *et al.*, 1974; B, $n = 17$, this paper, where $R_{sc} = 5.2\ R_r^{0.9}$, and R_r (km) is average observed rim radius and R_{sc} (km) is average radial limit of concentration of secondary craters for 17 impact primaries 18–260 km across).

size, so that the terrace widths observed in craters are compatible with the amount of slump observed on these basin rings but are not compatible with slumping on the order required to create megaterraces (Hodges and Wilhelms, 1976a).

Discussion

Although correlations of rings among basins and the diameters of transient craters of some basins are still not certain, we believe that a nested-crater origin best explains the major rings. The topographic basin rims, the most conspicuous and continuous rings in all basins, probably are the outer crater rims. Smaller buried rings in Humorum and Crisium and exposed inner rings in the smallest basins represent inner rims. The three largest basins (Imbrium by inference) probably have three crater rims; their intermediate rims are composite, slump-divided rings. Arcs outside the topographic basin rims probably are external deformation features.

Although the formational mechanism for nested craters is beyond the scope of our study, we suggest that it relates in some way to layering in the target media. This suggestion stems from the apparent cause of concentric ring formation in some terrestrial craters (cited above and elaborated on by Hodges and Wilhelms, 1976a) and on an apparent consistency of ring spacing with lunar crustal layering. Seismic discontinuities have been detected near the Apollo 12 and 14 sites at about 20 km and 60 km depth (Fig. 12) (Toksöz, 1974; Anderson, 1974). Interpretive profiles of small double-ring basins (Fig. 3) seem to us consistent with formation of the inner rings at the 20 km discontinuity, below which more coherent crustal rock with higher seismic velocity was uplifted to form the inner crater rims. The apparently brecciated, fractured rock above (Simmons *et al.*, 1973) was excavated across a wider diameter. Original depth-diameter ratios necessary to create this geometry are about 1/12. Interpretive profiles of Orientale having depth-diameter ratios of 1/10 to 1/15 are compatible with formation of the Rook ring from the more coherent rock below the 20 km discontinuity and the 330 km ring at the crust-mantle discontinuity (Fig. 5). We suggest therefore that the second rings of both double-ring and larger three-ring basins form at the 20 km discontinuity and the few observed third rings form at the 60 km discontinuity.

The gradual morphologic transition from central-peak craters to ringed basins (Stuart-Alexander and Howard, 1970; Hartmann and Wood, 1971; Hartmann, 1972a; Howard, 1974) could mean that peaks and rings form by basically similar mechanisms (Hodges and Wilhelms, 1976a). We believe that evidence from terrestrial craters (Roddy, 1968, 1976; Jones, 1976; Wilshire *et al.*, 1972; Milton *et al.*, 1972) accords with elastic rebound models of peak formation (Milton and Roddy, 1972) better than with passive models (Shoemaker, 1959; Quaide *et al.*, 1965; Dence, 1968). In elastic rebound, material at depth responds immediately to the excavation by flowing forcefully toward the center and then upward, whereas in passive models peaks are pushed up by converging slides induced by

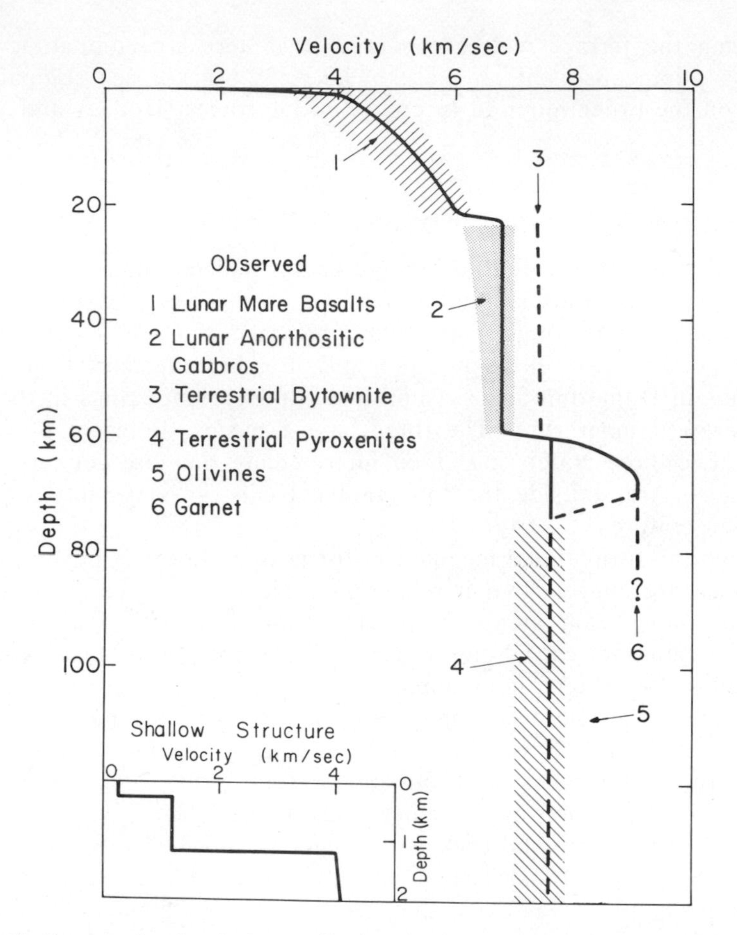

Fig. 12. Compressional velocity profile for the lunar crust and upper mantle and laboratory data on velocities of lunar and terrestrial rock. The velocity model is shown by a heavy line (or dashed heavy line where the model is uncertain). Inset shows shallow structure at Apollo 17 landing site. From Toksöz (1974).

gravity. We envision (Hodges and Wilhelms, 1976a) that as in central peaks, material of a lower resistant layer was uplifted, and, if sufficient energy was involved, was ejected at the core of the uplift. This process was suggested by field relations at Gosses Bluff, Australia, which is a central uplift whose center was excavated by ejection as well as erosion (Milton *et al.*, 1972). The size of the resulting craterlike inner ring should therefore be a function of impact energy (and thus size of transient crater), lithology, and depth to the layers. The larger the ring, the more material is ejected, and the more nearly the ring resembles a normal crater rim. Because this ring forms inside the larger transient cavity of excavation, the result is a crater inside a crater.

However, the apparent peak-to-ring transition can also be explained by passive models of peak formation. If peaks form by inward slide of material in craters too small to penetrate a subsurface discontinuity, rings might form as nested crater rims in craters large enough to penetrate it. In this case rings and peaks would form by very different processes.

A peak-to-ring transition might also occur in homogeneous target materials—large peaks becoming rings as their centers collapsed (Ullrich, 1976; Schultz, 1976) or were ejected (Head, 1976a; Hodges and Wilhelms, 1976a). However, this explanation seems inapplicable to departures from the general size-morphology transition, such as the ring-to-peak basins Antoniadi (Fig. 1) and Compton (175 km, 56°N, 105°E), which are smaller than many craters with peaks alone. Local heterogeneities in stratigraphy, on the other hand, could account for these features.

By all mechanisms of peak and ring formation, the crater floor must have rebounded substantially by elastic recoil (Gilbert, 1893) and slower plastic (isostatic) recovery (Baldwin, 1949; Wise and Yates, 1968; Chao *et al.*, 1975). The amount of uplift would be greatest in the basin center, so that any original outward slope of the interior nested rims would be increased (Figs. 3, 5).

The genetic relation between peaks and rings clearly needs further investigation. Our ring model, however, rests chiefly on interpretation of the lunar observations and does not depend entirely on any one mode of origin for central peaks.

Conclusions

The two rings of small, double-ring basins and the three rings of a few large basins like Orientale are basically analogous to crater rim crests. The outermost ring is the outer limit of excavation (except as modified by slumping). The one or two inner rings are rims of craters nested within the excavation cavity that were formed by perturbations in the cratering process. The perturbations probably resulted from intersection of horizontal discontinuities in the lunar target materials. Additional partial rings were produced by slumping of segments of these nested crater rims. Passive slump mechanisms are not adequate, however, to produce major rings such as the Cordillera and the combined Rook rings of Orientale.

Possible explanations of the way in which planetary stratification produces rings and central peaks include reflection of the shock wave, refraction of the shock wave, interaction of the rarefaction front and the free surface, or a combination of these or other processes. Cratering specialists should explore these possibilities, together with the role of material strength and gravitational acceleration in peak and ring formation.

Acknowledgments—We thank J. M. Boyce, W. K. Hartmann, J. F. McCauley, D. J. Milton, H. J. Moore, and V. R. Oberbeck for helpful reviews and discussions. The work was supported by the Lunar Programs Office, NASA, under contract No. W13,130.

REFERENCES

Anderson, D. L.: 1974, The interior of the Moon. *Physics Today*, March 1974, 45–49.

Baldwin, R. B.: 1949, *The Face of the Moon.* Univ. Chicago Press, Chicago.

Baldwin, R. B.: 1963, *The Measure of the Moon.* Univ. Chicago Press, Chicago.

Baldwin, R. B.: 1972, The tsunami model of the origin of ring structures concentric with large lunar craters. *Phys. Earth Planet. Interiors* **6**, 327–339.

Baldwin, R. B.: 1974, On the origin of the mare basins. *Proc. Lunar Sci. Conf. 5th*, p. 1–10.

Casella, C. J. and Binder, A. B.: 1972, Geologic map of the Cleomedes quadrangle of the Moon. U.S. Geol. Survey Map I-707.

Chao, E. C. T.: 1976a, The Ries crater of Southern Germany—A model for large basins on planetary surfaces. *Geologisches Jahrbuch* (In press).

Chao, E. C. T.: 1976b, Mineral-produced high-pressure striae and clay polish: Key evidence for nonballistic transport of ejecta from Ries crater. *Science* **194**, 615–618.

Chao, E. C. T., Hodges, C. A., Boyce, J. M., and Soderblom, L. A.: 1975, Origin of lunar light plains. *U.S. Geol. Survey Jour. Res.* **3**, 379–392.

DeHon, R. A.: 1975, Mare Spumans and Mare Undarum: Mare thickness and basin floor. *Proc. Lunar Sci. Conf. 6th*, p. 2553–2561.

DeHon, R. A. and Waskom, J. D.: 1976, Geologic structure of the eastern mare basins. *Proc. Lunar Sci. Conf. 7th*. p. 2729–2746.

Dence, M. R.: 1968, Shock zoning at Canadian craters, petrography and structural implications. In *Shock Metamorphism of Natural Materials* (B. M. French and N. M. Short, Eds.), p. 169–184. Mono Book Corp., Baltimore.

Dence, M. R.: 1972, Meteorite impact craters and the structure of the Sudbury basin. *Geol. Assoc. Canada Spec. Pap. 10*, 7–18.

Dence, M.R.: 1973, Dimensional analysis of impact structures (abstract). *Meteoritics* **8**, 343–344.

Dence, M. R., Innes, M. J. S., and Beals, C. S.: 1965, On the probable meteorite origin of the Clearwater Lakes, Quebec. *Royal Astron. Soc. Canada, Jour.* **59**, 13–22.

Dennis, J. G.: 1971, Ries structure, southern Germany, a review. *J. Geophys. Res.* **76**, 5394–5406.

Engelhardt, W. v.: 1974, The Ries structure and its impact formation. *Fortschr. der Miner.* **52**, Special Issue 1, 103–109.

Floran, R. J. and Dence, M. R.: 1976, Morphology of the Manicouagan ring-structure, Quebec, and some comparisons with lunar basins and craters. *Proc. Lunar Sci. Conf. 7th*, p. 2845–2865.

Gault, D. E.: 1974, Impact Cratering. In *A Primer in Lunar Geology* (R. Greeley and P. H. Schultz, eds.), p. 137–175. Ames Res. Center.

Gilbert, G. K.: 1893, The moon's face, A study of the origin of its features. *Phil. Soc. Wash. Bull.* **12**, 241–292.

Hartmann, W. K.: 1972a, Interplanet variations in scale of crater morphology—Earth, Mars, Moon. *Icarus* **17**, 707–713.

Hartmann, W. K.: 1972b, *Moons and Planets: An introduction to planetary science.* Bogden and Quigley, Tarrytown-on-Hudson, New York.

Hartmann, W. K. and Kuiper, G. P.: 1962, Concentric structures surrounding lunar basins. *Ariz. Univ. Lunar Planet. Lab. Commun.* **1**, 51–56.

Hartmann, W. K. and Wood, C. A.: 1971, Moon: Origin and evolution of multi-ring basins. *The Moon* **3**, 3–78.

Head, J. W.: 1974, Orientale multi-ringed basin interior and implications for the petrogenesis of lunar highland samples. *The Moon* **11**, 327–356.

Head, J. W.: 1976a, Significance of substrate characteristics in determining crater morphology and morphometry (abstract), In *Lunar Science VII*, p. 354–356, Lunar Science Institute, Houston.

Head, J. W.: 1976b, Origin of rings in lunar multi-ringed basins. In *Papers Presented to the Symposium on Planetary Cratering Mechanics*. p. 47–49. Lunar Science Institute, Houston.

Head, J. W., Settle, M., and Stein, R. S.: 1975, Volume of material ejected from major lunar basins and implications for the depth of excavation of lunar samples. *Proc. Lunar Sci. Conf. 6th*, p. 2805–2829.

Hodges, C. A. and Wilhelms, D. E.: 1976a, Formation of lunar basin rings. Submitted to *Icarus*.
Hodges, C. A. and Wilhelms, D. E.: 1976b, Formation of lunar basin rings (abstract). In *25th International Geological Congress*, Sydney, Australia. Abstracts **2**, Sec. 15, 612–613.
Hodges, C. A. and Wilhelms, D. E.: 1976c, Formation of concentric basin rings (abstract). p. 53–55. In *Papers Presented to the Symposium on Planetary Cratering Mechanics*. Lunar Science Institute, Houston.
Howard, K. A.: 1974, Fresh lunar impact craters: Review of variations with size. *Proc. Lunar Sci. Conf. 5th*, p. 61–69.
Howard, K. A., Wilhelms, D. E., and Scott, D. H.: 1974, Lunar basin formation and highland stratigraphy. *Reviews Geophys. and Space Phys.* **12**, 309–327.
Jones, G. H. S.: 1976, The morphology of central uplift craters. *Suffield Rpt. No. 281*. Defence Research Establishment Suffield, Ralston, Alberta, Canada.
Mackin, J. H.: 1969, Origin of lunar maria. *Geol. Soc. America Bull.* **80**, 735–748.
Masaytis, V. E., Mikhaylov, M. V., and Selivanovskaya, T. V.: 1972, The Popigay meteorite crater. *Internat. Geologic Rev.* **14**, 327–331.
Masaytis, V. L., Mikhaylov, M. V., and Selivanovskaya, T. V.: 1975, *The Popigay meteorite crater*. Nauka Press, Moscow (NASA Tech. Translation F-16900).
McCauley, J. F.: 1967, The nature of the lunar surface as determined by systematic geologic mapping. In *Mantles of the Earth and Terrestrial Planets* (S. K. Runcorn, ed.), p. 431–460, Interscience, New York.
McCauley, J. F.: 1968, Geologic results from the lunar precursor probes. *AIAA Jour.* **6**, 1991–1996.
McCauley, J. F.: 1976, Orientale and Caloris (abstract). In *Papers Presented to the Conference on Comparisons of Mercury and the Moon*, p. 24. Lunar Science Institute, Houston.
McGetchin, T. R., Settle, M., and Head, J. W.: 1973, Radial thickness variation in impact crater ejecta: Implications for lunar basin deposits. *Earth Planet. Sci. Lett.*, **20**, 226–236.
Milton, D. J. and Roddy, D. J.: 1972, Displacements within impact craters. *Intl. Geol. Congr. Rep. Ses. 24th*, 119–124.
Milton, D. J., Barlow, B. C., Brett, R., Brown, A. R., Glikson, A. Y., Manwaring, E. A., Moss, F. J., Moss, E. C. E., Van Son, J., and Young, G. A.: 1972, Gosses Bluff impact structure. *Science* **175**, 1199–1207.
Moore, H. J., Hodges, C. A., and Scott, D. H.: 1974, Multiringed basins—illustrated by Orientale and associated features. *Proc. Lunar Sci. Conference 5th*, p. 71–100.
Mutch, T. A.: 1973, *Geology of the Moon, A Stratigraphic View*, Princeton Univ. Press, Princeton, N.J. (2nd edition).
Oberbeck, V. R.: 1975, The role of ballistic erosion and sedimentation in lunar stratigraphy. *Reviews Geophys. and Space Phys.* **13**, 337–362.
Oberbeck, V. R. and Quaide, W. L.: 1967, Estimated thickness of a fragmental surface layer of Oceanus Procellarum. *J. Geophys. Res.* **72**, 4697–4704.
Oberbeck, V. R., Morrison, R. H., Hörz, F., Quaide, W. L., and Gault, D. E.: 1974, Smooth plains and continuous deposits of craters and basins. *Proc. Lunar Sci. Conf. 5th*, p. 111–136.
Olson, A. B. and Wilhelms, D. E.: 1974, Geologic map of the Mare Undarum quadrangle of the Moon. U.S. Geol. Survey Map I-837.
Piekutowski, A. J.: 1976, Cratering mechanisms observed in laboratory-scale high-explosive experiments (abstract). In *Papers Presented to the Symposium on Planetary Cratering Mechanics*, p. 102–104. Lunar Science Institute, Houston.
Pike, R. J.: 1967, Schroeter's rule and the modification of lunar crater impact morphology. *J. Geophys. Res.*, **72**, 2099–2106.
Pike, R. J.: 1976, Crater dimensions from Apollo data and supplemental sources. *The Moon* **15**, 463–477.
Quaide, W. L., Gault, D. E., and Schmidt, R. A.: 1965, Gravitative effects on lunar impact structures. *Annals N.Y. Acad. Sc.* **123**, 563–572.
Quaide, W. L. and Oberbeck, V. R.: 1968, Thickness determinations of the lunar surface layer from lunar inpact craters. *J. Geophys. Res.* **73**, 5247–5270.
Roddy, D. J.: 1968, The Flynn Creek structure, Tennessee. In *Shock Metamorphism of Natural Materials*. (B. M. French and N. M. Short, eds.), p. 291–322. Mono Book Corp., Baltimore.

Roddy, D. J.: 1976, High-explosive cratering analogs for bowl-shaped, central uplift, and multiring impact craters. *Proc. Lunar Sci. Conf. 7th*, p. 3027–3056.

Schultz, P. H.: 1976, *Moon Morphology*. Univ. Texas Press, Austin.

Schultz, P. H. and Gault, D. E.: 1975, Seismic effects from major basin formations on the Moon and Mercury. *The Moon* **12**, 159–177.

Shoemaker, E. M.: 1959, Address to earth sciences session. *Proc. Lunar Planetary Exploration Colloq.* **2** (1), 20–28.

Short, N. M. and Forman, M. L.: 1972, Thickness of impact crater ejecta on the lunar surface. *Modern Geol.* **3**, 69–91.

Simmons, G., Todd, T., and Wang, H.: 1973, The 25-km discontinuity: Implications for lunar history. *Science* **182**, 158–161.

Stöffler, Dieter: 1975, Ries deep drilling results: Implications for the structure of the crater basement and the distribution of excavated masses (abstract). *Meteoritics* **10**, 495–497.

Stuart-Alexander, D. E. and Howard, K. A.: 1970, Lunar maria and circular basins—A review. *Icarus* **12**, 440–456.

Toksöz, M. N.: 1974, Geophysical data and the interior of the Moon. In *Annual Review of Earth and Planetary Sciences* **2**, 151–177.

Ullrich, G. W.: 1976, The mechanics of central peak formation in shock wave cratering events. Air Force Weapons Laboratory, AFWL-TR-75-88.

Van Dorn, W. G.: 1968, Tsunamis on the Moon. *Nature* **220**, 1102–1107.

Van Dorn, W. G.: 1969, Lunar maria: Structure and evolution. *Science* **165**, 693–695.

Wilhelms, D. E.: 1964, Major structural features of the Mare Vaporum quadrangle. *Astrogeologic studies ann. prog. rept.* July 1963–July 1964, pt. A., 1–16 (U.S. Geol. Survey open-file report).

Wilhelms, D. E.: 1970, *Summary of lunar stratigraphy—Telescopic observations.* U.S. Geol. Survey Prof. Paper 599F.

Wilhelms, D. E.: 1973, Geologic map of the northern Crisium region. *Apollo 17 Prelim. Sci. Rept.*, p. 29-29 to 29-34. NASA SP-330.

Wilhelms, D. E. and McCauley, J. F.: 1971, *Geologic map of the near side of the Moon.* U.S. Geol. Survey Map I-703.

Wilhelms, D. E. and El-Baz, Farouk: 1976, *Geologic map of the east side of the Moon.* U.S. Geol. Survey Map I-948.

Wilshire, H. G., Offield, T. W., Howard, K. A., and Cummings, D.: 1972, *Geology of the Sierra Madera cryptoexplosion structure, Pecos County, Texas.* U.S. Geol. Survey Prof. Paper 599-H.

Wise, D. U. and Yates, M. T.: 1968, Mascons as structural relief on a lunar 'moho'. *J. Geophys. Res.* **75**, 261–268.

Wood, C. A. and Head, J. W.: 1976, Comparison of impact basins on Mercury, Mars, and the Moon: *Proc. Lunar Sci. Conf. 7th*, p. 3629–3651.

Roddy, D. J., Pepin, R. O., and Merrill, R. B., editors.
(1977) *Impact and Explosion Cratering*, Pergamon Press (New York), p. 563–573.
Printed in the United States of America

Origin of outer rings in lunar multi-ringed basins: Evidence from morphology and ring spacing

JAMES W. HEAD

Department of Geological Sciences, Brown University, Providence, Rhode Island 02912

Abstract—The concentric rings associated with lunar basins have been studied with particular emphasis on the location of basin features equivalent to the crater rim crest in smaller craters, and on the mode of formation of additional basin rings. Examination of the morphology and morphometry of features spanning the transition from craters to three-ring basins (diameters about 140–435 km) shows that the two rings in peak-ring basins represent a ring of expanded central peaks and a crater rim crest modified (enlarged) by formation of imbricate terraces. Central peak ring diameter (Y) shows a linear relation to crater rim crest diameter (X) such that $Y = 0.56X - 17.55$. Examination of the morphology and morphometry (primarily ring spacing) of features spanning the transition from peak-ring basins to multi-ring basins suggests that in three-ring basins the central peak ring is maintained and the intermediate ring closely corresponds to the crater rim crest. The outer ring apparently forms as a fault near the edge of intense structural uplift of the crater rim crest, and a large portion of the rim flank collapses inward forming a wide megaterrace. Formation of the basin outer ring scarp appears to result from a specific change in the style of the modification stage of the cratering event, distinct from that of smaller peak-ring basins and craters.

INTRODUCTION

LUNAR BASINS are large circular depressions with distinctive concentric rings and have been of major importance in the evolution of the surface features and crust of the Moon (Hartmann and Wood, 1971; Head, 1974; Howard *et al.*, 1974; Moore *et al.*, 1974; Wood and Head, 1976). Although virtually all workers agree that basins are of impact origin, there has been no consensus on the origin of the specific multiple rings and their correspondence to features in smaller craters. The purpose of this study was to examine both the morphology *and* morphometry (primarily ring spacing) of several of the freshest lunar basins, including Orientale, Imbrium, Nectaris, Crisium, and Humorum, and to compare the characteristics of their three most prominent rings to features in smaller craters. Specific questions include: (1) What is the location of the basin feature that is equivalent to the crater rim crest in smaller craters?, and (2) What is the mode of formation of additional basin rings?

GENERAL MORPHOLOGY AND FACIES DISTRIBUTION

The morphologic and morphometric sequence of craters has been established by workers such as Baldwin (1963), Pike (1972), and Howard (1974). Craters less than about 15 km diameter are characterized by circular outlines, generally smooth walls, and bowl-shapes often with small flat floors (Fig. 1). Larger craters show polygonal outlines, terraced walls, wide flat floors, and central peaks or

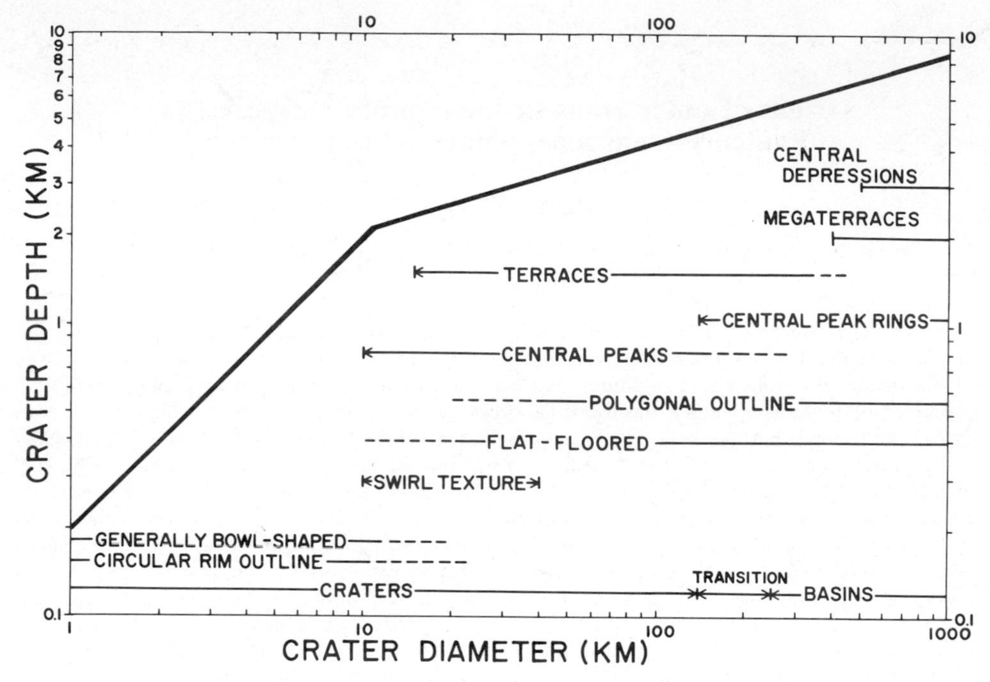

Fig. 1. Major morphologic features in craters and basins and depth–diameter relationship for fresh craters (solid line). (Solid line from Pike, 1974.)

central peak clusters. In craters in this size range, inward slumping of the wall terraces causes enlargement of the crater rim crest in the terminal stages of the cratering event.

The first concentric ring to form with increasing crater size in addition to the crater rim crest is the central peak ring (or CPR for abbreviation) (Fig. 1) (Hartmann and Kuiper, 1962; Hartmann and Wood, 1971). This is so named because of its morphologic similarity to central peaks and is first seen at diameters of about 150 km. At transition diameters central peak rings often surround a central peak. Such structures (Compton, for example) are known as central peak basins. The central peak disappears in larger structures, such as Schrödinger, and these features are known as peak-ring basins (Hartmann and Wood, 1971; Wood and Head, 1976). For the purposes of this discussion, central peak basins are included with peak-ring basins. Above diameters of about 250 km, an additional concentric ring forms (Fig. 1) to produce the characteristic multi-ringed basin (Hartmann and Wood, 1971; Wood and Head, 1976). The characteristics of the rings associated with peak-ring basins, and multi-ringed basins are illustrated in Fig. 2. A significant question is which of the major rings in multi-ringed basins corresponds to features observed in smaller basins and craters (Fig. 2).

The distribution of facies shows distinctive relationships to diameter and ring development. Craters between 10 and 150 km diameter show a hummocky and

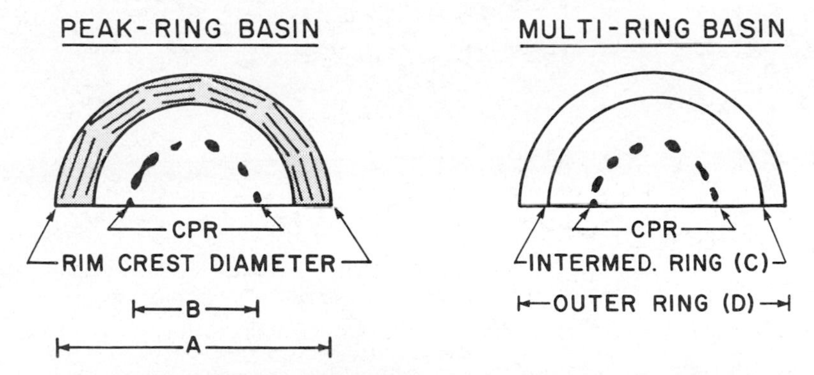

Fig. 2. Diagrammatic representation of major basin features.

radial zone extending from the rim crest outward (Howard, 1974), which represents the textured ejecta deposit. A series of step-like terraces forms the wall and extends from the rim crest to the edge of the flat crater floor. The distinctive terraced wall slopes inward at 15°, and may span as much as 50% of the crater radius. The regionally flat crater floor is often distinctly hummocky and is draped and mantled with a rough-textured crenulated deposit which has been interpreted to be partly impact melt (Howard and Wilshire, 1975; Howard, 1974; Hawke and Head, 1977).

Peak-ring basins show a similar facies development, with a textured ejecta deposit exterior to the rim crest, a distinctively terraced crater wall, a rough-textured hummocky or knobby floor, and the addition of light plains deposits (Howard *et al.*, 1974, Fig. 4 in particular).

Since multi-ringed basin interiors are often obscured by mare material, the relatively young sparsely flooded Orientale Basin provides an excellent example of facies development in this size range (McCauley, 1968; Head, 1974; Howard *et al.*, 1974; Moore *et al.*, 1974). The distribution of facies mapped by Head (1974) is shown in Fig. 3. Textured ejecta deposits surround the outer ring and a distinctive domical facies is found almost exclusively between the intermediate and outer ring. Portions of the radially textured exterior deposits have been traced locally across the domical facies to the vicinity of the intermediate ring (Head, 1974, pp. 337–338 and Fig. 4). A crackled and corrugated facies is restricted to the interior part of the basin inside the second (Outer Rook) ring, is often draped over a very hummocky floor, and appears similar in origin to the impact-melt facies in smaller craters (Head, 1974; Howard, 1974). The non-mare plains facies in the basin interior also appear to be related to impact-melt deposits.

The Inner Rook ring is morphologically similar to the central peak rings in peak-ring basins (Head, 1974; Howard *et al.*, 1974; Wood and Head, 1976), whereas the intermediate ring (Outer Rook) and the outer ring (Cordillera) appear to represent inward dipping fault scarps each bearing some resemblance to crater rim crests. Extensive step-like wall terraces analogous to those in

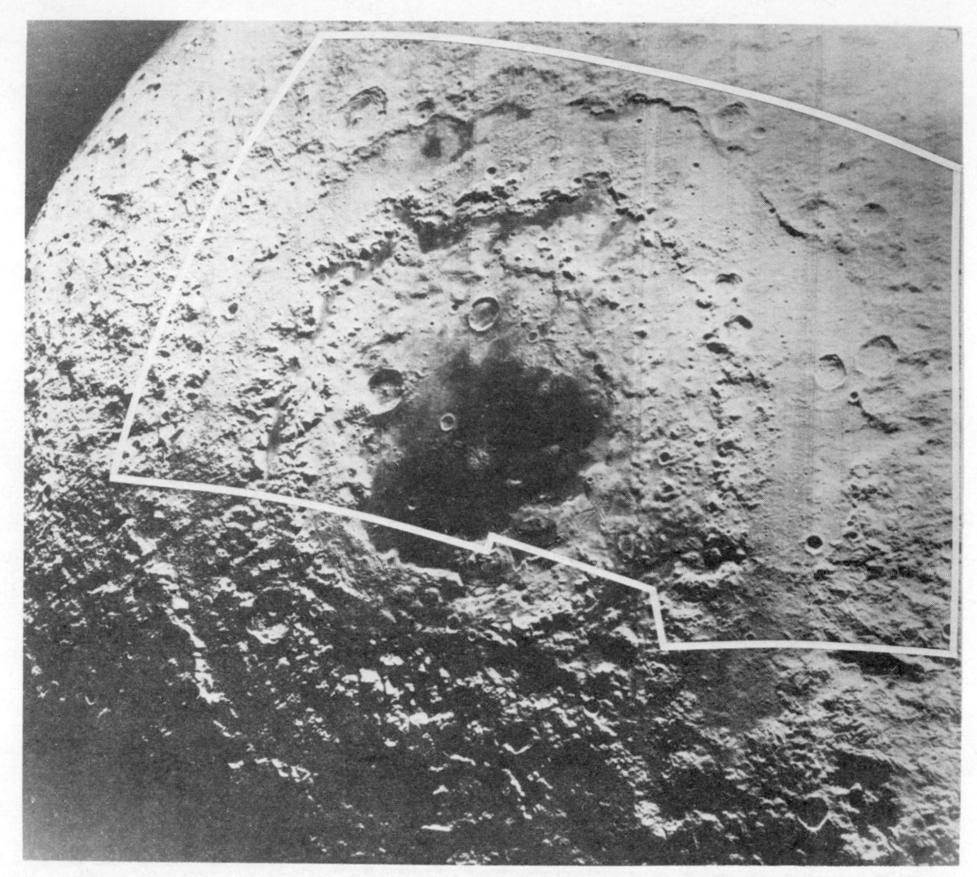

Fig. 3. Major facies of the Orientale Basin interior (after Head, 1974). Solid black spots mark peaks of central peak ring.

peak-ring basins and smaller craters are not observed. The intermediate ring appears to have a more sinuous outline than the outer ring (Head, 1974, Fig. 2).

On the basis of deposit morphology and distribution, and the progression from smaller craters, through peak-ring basins, to multi-ring basins, the intermediate ring (Outer Rook) is interpreted as the closest approximation to the crater rim. Impact melt-related facies and the central peak ring lie within this ring, as they do within the crater rim crest of smaller structures, and radial rim facies can occasionally be traced inward to this ring. According to this model (Head, 1974), the outer ring (Cordillera) is a fault scarp that formed outside the crater rim crest in the terminal stages of the cratering event as a large portion of the crater exterior collapsed inward toward the recently excavated crater, forming a megaterrace. This collapse pushed the wall of the initial Orientale Crater inward, distorting it and slightly decreasing its radius. The domical facies is interpreted as radially textured ejecta which was disrupted and modified to a jumbled domical texture by movement associated with the formation of the megaterrace.

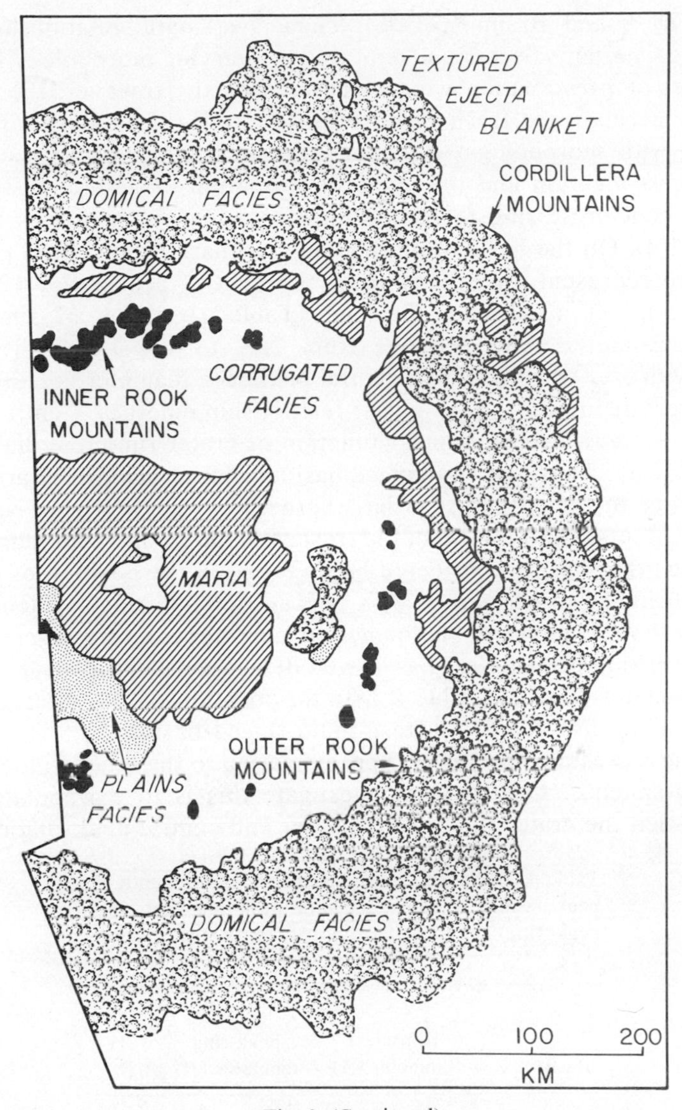

Fig. 3. (Continued)

Basin Ring Spacing

Hartmann and Kuiper (1962) observed that the ratio of successive ring diameters is usually about $\sqrt{2}$ or in some cases 2. Hartmann and Wood (1971) used a larger sample of basins, and compared ring diameters with respect to the diameter of the most prominent ring. They concluded that the special role of $\sqrt{2}$ ratios was confirmed (see their Fig. 27). However, Howard *et al.* (1974) re-measured the basins and found only a crude correlation of ring ratios with $\sqrt{2}$

(see their Fig. 6 and Brennan, 1976). Thus, systematic relationships do not clearly emerge, perhaps because several rings of varying morphology in basins in various states of preservation were being compared. However, the spacing of major rings in relatively fresh peak-ring and multi-ringed basins (Fig. 2), in combination with morphologic evidence, may provide some indication of the crater rim crest location and the mode of formation of additional basin rings.

The first concentric ring to form with increasing crater size is the central peak ring (CPR). On the basis of morphologic similarity the central peak ring is interpreted to represent an expansion of the central peaks. For 12 peak-ring basins between 140 to 435 km diameter (Table 1), ratios of rim crest diameter/CPR diameter show a range from 2.19 to 1.85, generally inversely correlative with size (Fig. 4). When central peak-ring diameter is plotted against crater rim crest diameter (Fig. 5) a clear relationship emerges. Central peak-ring diameter (Y) is shown to be a linear function of crater rim crest diameter (X): $Y = 0.56X - 17.55$. Thus, for two-ringed basins, central peak rings are generally relatively closer to the rim crest in large structures. Central peak rings can be identified in Orientale and Imbrium, the freshest large basins with more than two rings. For the other basins considered here (Table 2), the position of the central peak ring is believed to be marked by a concentric mare ridge system. Imbrium exhibits both the ridge system and numerous central peak-ring segments. Orientale, about 900 km in diameter, has a well-developed central peak ring about 480 km in diameter (Fig. 3). Table 2 lists the location of the intermediate and outer rings for the five relatively fresh multi-ringed basins.

Which ring represents the closest approximation to the crater rim crest on the basis of ring spacing? One way to investigate this is to extrapolate the relationship between the crater rim crest diameter and central peak-ring diameter in

Table 1. Peak-ring basin dimensions (km). Central peak basins (containing a central peak and a central peak ring) are included and are indicated by an asterisk.

Basin	Rim crest diameter (A)	Central peak-ring diameter (B)	A/B
*Antoniadi	140	65	2.15
*Compton	175	80	2.19
Milne	240	120	2.00
Bailly	300	145	2.07
Schrödinger	320	155	2.06
Lorentz	330	160	2.06
Birkhoff	320	170	1.88
Near Schiller	350	180	1.94
Poincare	335	180	1.86
Korolev	405	200	2.03
Grimaldi	410	220	1.86
Apollo	435	235	1.85

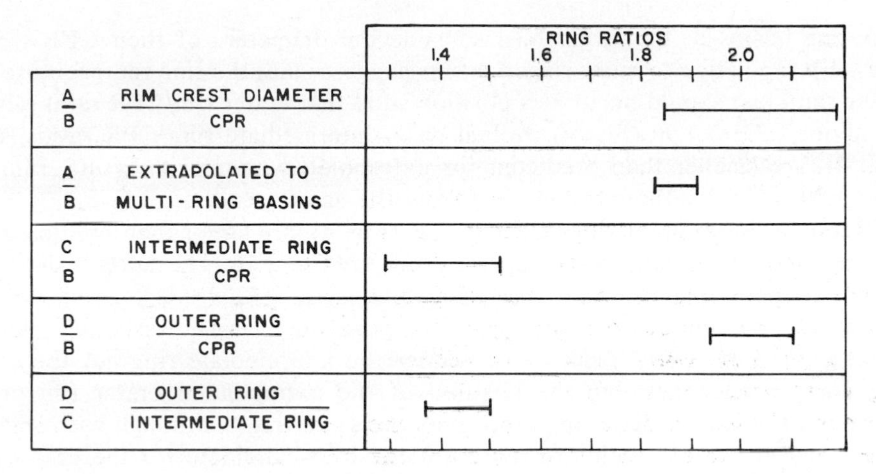

Fig. 4. Ratios of various basin and crater parameters.

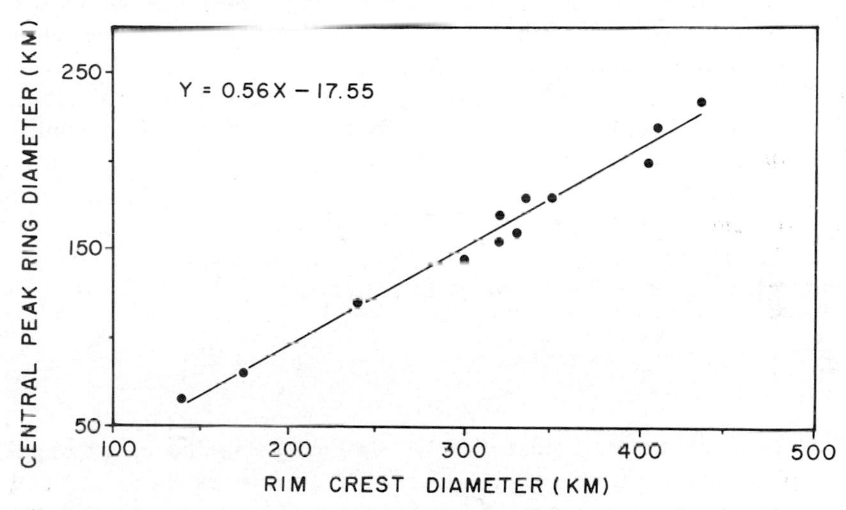

Fig. 5. Relation between central peak-ring diameter (Y) and crater rim crest diameter (X) for the 12 peak-ring basins listed in Table 1. The straight line is a least squares fit to the data.

Table 2. Multi-ring basin dimensions (km).

Basin	Outer ring diameter (D)	Intermediate ring diameter (C)	Central peak-ring (or equivalent) diameter (B)
Imbrium	1340	970	670
Orientale	930	620	480
Nectaris	840	600	400
Crisium	670	450	330
Humorum	560	410	270

peak-ring basins to the larger basins, using the diameters of their CPR's as a basis. Extrapolation to multi-ringed basins suggests that the ring representing the crater rim crest should occur at a position such that crater rim/CPR ratio ranges are about 1.83 to 1.90 (Fig. 4). Actual basin intermediate ring/CPR ratios, C/B (Fig. 4), are smaller than predicted for extrapolated crater rim crests, ranging from 1.29 to 1.52, although they do follow the same trend.

Ratios of basin outer rings/CPR's, D/B (Fig. 4), are larger than predicted for extrapolated crater rim crests, ranging from 1.94 to 2.10. The latter values are typical of peak-ring basins in the diameter range of 200–350 km and do not follow the size-related trend observed for peak-ring basins. Assuming proportional growth of central peak rings, neither the intermediate ring nor the outer ring correspond exactly to the position of the extrapolated crater rim crest. However, the intermediate ring represents the same trend, although with smaller ratios, and at present seems to represent the best candidate for the crater rim crest on the basis of ring spacing (Fig. 4). Therefore, on the basis of morphologic and morphometric evidence, the intermediate ring represents the *closest approximation* to the location of the crater rim crest, and the outer ring represents a ring which formed concentric to, but outside of, the crater rim crest. The circular nature of central peak rings and their similarity to peak-ring basin features and central peaks in craters suggests that they formed by a similar, but poorly understood process (Head, 1976), and that they have not been moved substantially toward the basin interior. The morphometric relationships noted (Figs. 4, 5) suggest that some process may cause the crater rim crest (1) to form closer to the central peak ring or (2) to be displaced closer to the central peak ring, than predicted from peak-ring basin extrapolation.

DISCUSSION

If the intermediate ring most closely corresponds to the initial crater rim crest, then to what, if any, crater-associated feature does the outer ring correspond? The range of outer ring/intermediate ring ratios for the basins considered here, D/C, is 1.37 to 1.50 (Fig. 4). Using the same outer ring dimensions, but plotting them against the intermediate ring position extrapolated from peak-ring basins (Fig. 4), yields a range of 1.05 to 1.13. Pike (1976) has documented the pronounced rim exterior topography around craters over a wide range of diameters and has shown that crater rim width/crater diameter ratios range from 1.4 to 1.6. Rim flank topography is made up of ejecta and structural uplift components. In terrestrial experimental craters, structural uplift is most important toward the crater rim crest, usually within about 1.50 crater radii from the crater center (Carlson and Jones, 1965). Pike (1968) and Settle and Head (1977) have suggested that structural uplift in lunar craters lies between the rim crest and the break in slope at the base of the rim flank. On the basis of these comparisons, it is concluded that the outer basin ring forms within the region where significant structural uplift of the basin rim is to be expected. Therefore, the formation of the outer ring scarp may be closely associated with structural uplift of the inner portion of the crater rim flank.

If the outer ring scarp results from failure along the edge of significant structural uplift, what effect might this have on the intermediate ring and its relation to the central peak ring? Several characteristics of the intermediate ring are relevant: (1) although the ring is morphologically similar to crater rims, the abundant step-like terraces seen in smaller craters are not prominent, (2) the ring is composed locally of straight segments but is regionally crenulated, as if the circumference had undergone shrinkage (Head, 1974). Two factors appear to be important in explaining the deviation between extrapolated and actual crater rim/CPR ratios noted previously (Fig. 4): (1) the terracing that causes up to 20% rim crest diameter enlargement in craters does not seem to be as significant in three-ringed basins, and (2) the formation of the megaterrace and its inward movement compresses and crenulates the crater rim crest, decreasing its radius. Both of these factors tend to decrease the intermediate ring diameter from the extrapolated values and readily account for the magnitude of difference observed between actual and extrapolated crater rim crest/CPR ratios.

These characteristics and the ring ratios suggest the following model for origin of the outer two rings: The cratering event formed two inner rings, a central peak ring and an uplifted crater rim crest, with deposition of ejecta

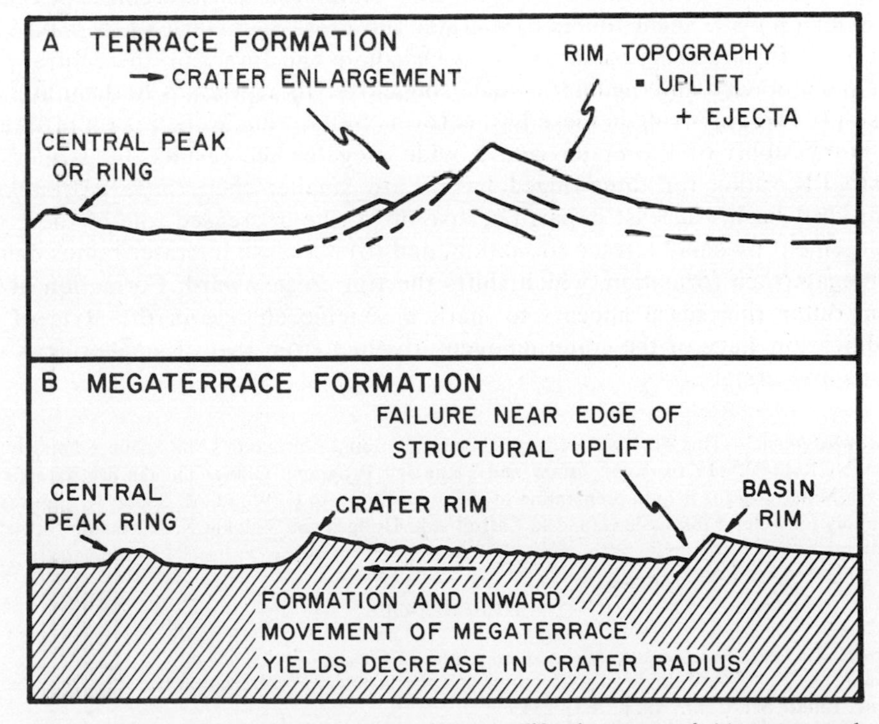

Fig. 6. Models for contrasting styles of the modification stage of (a) craters and peak-ring basins and (b) multi-ringed basins. Not to scale.

during the process. Spacing of these two rings shows a systematic relationship in peak-ring basins between about 140 and 435 km diameter (Fig. 5). In basins of this size range, the rim crest is enlarged by imbricate slumping during the modification stage and a series of slump terraces are produced on the crater wall. However, in the terminal stages of the event in structures over about 400 km diameter, a large segment of the rim (megaterrace) collapsed, approximately at the edge of intensive structural uplift, and deformed the recently deposited ejecta (Fig. 6). Megaterrace formation is in contrast to the style of the modification stage of smaller peak ring basins and craters where a smaller portion of the rim is destroyed by imbricate terrace formation, and the rim crest diameter is enlarged. Potential factors contributing to this change in style of the modification stage are discussed elsewhere (Head, 1977) and include: (a) changes in the characteristics of the excavation stage of the cratering event (particularly depth/diameter relationships of the cavity), (b) rim loading by ejecta, and (c) variations in the characteristics of the substrate.

CONCLUSIONS

(1) Peak-ring basins are characterized by two rings, a central peak ring and a crater rim crest enlarged by slumping. The central peak ring corresponds to enlarged central peaks and is not the true crater rim crest. Central peak ring diameter (Y) is a linear function of crater rim crest diameter (X) ($Y = 0.56X - 17.55$). (2) In three-ring basins, such as Orientale and Imbrium, the central peak ring is maintained and the intermediate ring closely corresponds to the crater rim crest. (3) The outer ring in these basins forms by faulting in the region of intense structural uplift of the crater rim: a wide megaterrace results. (4) Crater rim crest/CPR ratios for three-ringed basins are smaller than those extrapolated from CPR basins at least in part because of (a) the decreased role of the crater enlargement by small terrace formation, and (b) decrease in crater radius caused by megaterrace formation, which shifts the rim crest inward. Formation of the basin outer ring scarp appears to mark a specific change in the style of the modification stage of the cratering event, distinct from that of smaller peak-ring basins and craters.

Acknowledgments—This work was performed under National Aeronautics and Space Administration Grant NGR-40-002-116 from the Lunar and Planetary Programs Office. Thanks are extended to Stefana Matarazza for help in preparation of the manuscript, to C. Wood, M. Settle, R. Hawke, and P. Spudis for helpful discussions, and to Carroll Ann Hodges and William K. Hartmann for helpful reviews.

REFERENCES

Baldwin, R.: 1963, *The Measure of the Moon*. Univ. of Chicago Press, 488 pp.

Brennan, W. J.: 1976, Multiple ring structures and the problem of correlation between lunar basins. *Proc. Lunar Sci. Conf. 7th*, p. 2833–2843.

Carlson, R. H. and Jones, G. D.: 1965, Distribution of ejecta from cratering explosions in soils. *J. Geophys. Res.* 70, 1897.

Hartmann, W. K., and Kuiper, G. P.: 1962, Concentric structures surrounding lunar basins. *Comm. Lunar and Planet. Lab.* **1**, 51–66.

Hartmann, W. K. and Wood, C. A.: 1971, Moon: Origin and evolution of multiring basins. *The Moon* **3**, 2–78.

Hawke, B. R., and Head, J. W.: 1977, Impact melt in lunar crater interiors (abstract). In *Lunar Science VIII*, p. 415–417. The Lunar Science Institute, Houston.

Head, J. W.: 1974, Orientale multi-ringed basin interior and implications for the petrogenesis of lunar highland samples. *The Moon* **11**, 327–356.

Head J. W.: 1976, The significance of substrate characteristics in determining morphology and morphometry of lunar craters (abstract). In *Lunar Science VII*, p. 354–356. The Lunar Science Institute, Houston.

Head, J. W.: 1977, Characteristics and Mode of Formation of Impact Basins. Unpublished manuscript.

Howard, K. A.: 1974, Fresh lunar impact craters: Review of variations with size. *Proc. Lunar Sci. Conf. 5th*. p. 61–69.

Howard, K. A., Wilhelms, D. E., and Scott, D. H.: 1974, Lunar basin formation and highland stratigraphy. *Rev. Geophys. Space Phys.* **12**, 309–327.

Howard, K. A., and Wilshire, H.: 1975, Flows of impact melt at lunar craters. *Jour. Research U.S.G.S.* **3**, 237–251.

McCauley, J. F.: 1968, Preliminary photogeologic map of the Orientale basin region, Advanced Systems Traverse Research Project Report, Astrogeology 7, U.S.G.S. Interagency Report, p. 32–33.

Moore, H., Hodges, C. A., and Scott, D. H.: 1974, Multiringed basins—illustrated by Orientale and associated features. *Proc. Lunar Sci. Conf. 5th*, p. 71–100.

Pike, R. J.: 1968, Meteoritic origin and consequent endogenic modification of large lunar craters: Ph. D. Thesis, Univ. Michigan, 404 pp.

Pike, R. J.: 1972, Geometric similitude of lunar and terrestrial craters. Proc. 24th. Int. Geol. Congress, Montreal, Canada, August 1972, Sect. 15, Planetology, p. 41–47.

Pike, R. J.: 1976, Simple to complex impact craters: The transition on the Moon (abstract). In *Lunar Science VII*, p. 700–702. The Lunar Science Institute, Houston.

Settle, M., and Head, J. W.: 1977, Radial variation of lunar crater rim topography, *Icarus*, In press.

Wood, C. A., and Head, J. W.: 1976, Multi-ring basins of Mars, Mercury, and the moon (abstract). In *Lunar Science VII*, p. 950–52. The Lunar Science Institute, Houston.

Roddy, D. J., Pepin, R. O., and Merrill, R. B., editors.
(1977) *Impact and Explosion Cratering*, Pergamon Press (New York), p. 575–591.
Printed in the United States of America

Martian fresh crater morphology and morphometry—a pre-Viking review

MARK J. CINTALA

Department of Geological Sciences, Brown University, Providence, Rhode Island 02912

Abstract—Martian craters exhibit a wide variety of shapes, morphologies, and degrees of degradation. Fresh martian craters, observed at Mariner 9 A-frame resolution (1–3 km), display a regular progression of increasing morphologic complexity with size. Wall terraces first appear in craters larger than 50 km; central peaks, found in some craters as small as 5 km, also increase in frequency of occurrence with crater size. Small bowl-shaped craters make up the vast majority of craters classified as fresh, but flat floors rapidly increase in frequency of occurrence with increasing crater diameters. Both primary ejecta emplacement and crater erosion models have been proposed to account for the distinctive pedestals surrounding some martian craters. Depth/diameter ratios of fresh-appearing martian craters, determined through Mariner 9 ultraviolet spectrometer topographic measurements, are less than those found for fresh lunar craters. The sparsity of depth values for fresh martian craters prevents close comparisons between martian and mercurian fresh crater morphometries. Viking orbiter photography will be valuable in obtaining: (1) Statistically significant morphologic data for small (<15 km) martian craters, (2) accurate high resolution morphometry through shadow measurements and/or photogrammetry, and (3) higher quality albedo data for small features.

INTRODUCTION

THE PREDICTIONS OF Öpik (1950) and Tombaugh (1950) was substantiated in 1965 when the American spacecraft Mariner 4 provided photographic evidence of significant numbers of craters on the surface of Mars (Leighton *et al.*, 1965). Since then, seven American and six Soviet spacecraft have had encounters with Mars (summarized in Mutch *et al.*, 1976); not only have they confirmed the existence of craters as prominent landforms on both Mars and its moons, but they also provided a substantial data base with which to perform first-order statistical analyses of martian crater morphology and morphometry.

The purpose of this paper is to present a brief review of martian crater data and to provide a short summary of martian fresh crater morphology and morphometry based on pre-Viking information. Craters studied herein are considered to be of impact origin on the basis of arguments outlined in McCauley *et al.* (1972) and Hartmann (1973).

PRE-VIKING MARS MISSIONS

1. *Early spacecraft*

Mariner 4 flew by Mars on 15 July 1965, returning 22 photographs of the planet from a minimum distance of about 10,000 km (see Table 1). Although the best spatial resolution was on the order of 3 km, craters of varying degrees of

Table 1. Mariner—Mars photographic summary.

Spacecraft	Encounter date	Closest approach	Cameras	Best resolution	Number of photographs[a]	Areal coverage
Mariner 4	14 July 1965	9540 km	1	3 km	22	1%
Mariner 6	31 July 1969	3430 km			25	
			A	3 km		
			B	0.3 km		10%
Mariner 7	5 August 1969	3500 km			35	
Mariner 9	14 November 1971[b]	1390 km	A	1 km	7329	100%
			B	0.1 km		

[a]Near-encounter photography only.
[b]Date of orbital insertion.

degradation and morphology were recognized (Leighton *et al.*, 1967). Favoring an impact origin for the observed craters, these investigators concluded that, due to the relatively small percentage of "sharp, fresh-appearing craters of any size", surface modifying processes were more effective on Mars than on the Moon. Nevertheless, martian craters were observed to have morphologic characteristics similar to those observed in lunar craters: terraces, central peaks, polygonal outlines, and recognizable ejecta deposits around some of the larger craters.

The flyby of Mariner 6 on 31 July 1969, followed by that of Mariner 7 on 5 August 1969, provided greater areal coverage and much higher spatial resolution than that obtained by Mariner 4 (Table 1). These spacecraft provided further evidence for similarities between Mars and the Moon, with the generally more degraded appearance of the martian craters being the major difference between martian and lunar craters (Leighton *et al.*, 1969). Murray *et al.* (1971) pointed out that two distinct crater populations were visible in these photographs: small, relatively fresh, bowl-shaped craters and large, flat-floored craters, more modified, for the most part, than lunar upland craters of similar size. When coupled with other considerations, these two distinct morphology/size populations suggested to Murray *et al.* that the large, more modified craters dated back to the earliest periods of martian history, while the small, fresh craters were formed after the most intense erosion of the large craters had taken place. No large, fresh craters "comparable to Kepler and Copernicus on the Moon" were observed by these investigators on the Mariner 6 and 7 photographs of Mars.

2. *Mariner 9*

Mariner 9 was inserted into martian orbit on 14 November 1971. During its 349 days of active operations in near-Mars space, the spacecraft transmitted to Earth 7329 photographs of the planet and its moons. The result was global coverage of Mars at resolutions of 1–3 km (A-frames) and photography of select targets at resolutions approaching 100 m (B-frames; Table 1). This extensive set of images provides a substantial basis for formulating a large body of data on crater morphometry, morphology, and location. Unless otherwise indicated, Mariner 9 information serves as the source of all data presented here (Arvidson *et al.*, 1974).

Fresh Martian Craters

1. *Morphology*

Fresh lunar craters larger than several kilometers in diameter generally have a crisp appearance, and are characterized by rays, satellitic craters, radial rim facies, sharp rim crests, and smooth or terraced walls, depending on size (Pohn and Offield, 1970; Arthur *et al.*, 1963; Wood, 1972; Smith and Sanchez, 1973; Head, 1975; Howard, 1974; Pike, 1974). Fresh craters are of Copernican age (with rays) or Eratosthenian age (fresh morphologies without rays) (Wilhelms and McCauley, 1971); large Imbrian craters exhibit only slightly higher levels of degradation, but pre-Imbrian craters are usually highly modified (Head, 1975). Class I craters, the freshest class of Arthur *et al.* (1963), are characterized by sharp rim crests, while Class II craters have blurred or broken rims. Class I craters are of Copernican, Eratosthenian, and even Imbrian age (C. A. Wood, personal communication).

In the following discussions, a martian crater is classified as unmodified or nearly unmodified if it falls into one of three categories: (1) deep, bowl-shaped, with smooth walls; (2) deep, bowl-shaped, with terraced walls; or (3) deep, flat-floored, with terraced walls. The "deep" category is a visual impression of depth, as compared to others such as shallow and extremely shallow. The validity of this classification scheme is supported by the facts that (1) less than 1% of the fresh craters so defined have breached rims or superposed craters, compared to more than 25% of the remaining (degraded) population; and (2) more than 98% of this fresh crater population have raised rims, as opposed to less than 25% of the degraded craters. Further discussion regarding this classification scheme can be found in Arvidson (1974) and Cintala *et al.* (1976b).

In general, fresh martian crater morphology follows a diameter-dependent progression (Hartmann, 1973). The simplest morphology is found in the smallest craters (Fig. 1): a bowl shape, with sharp, raised rims, usually lacking flat floors at Mariner 9 A-frame resolution. This morphology remains essentially constant through diameters of approximately 10 km. The first indication of a deviation from the simple bowl shape occurs in significant numbers of craters between diameters of about 10 and 15 km, where the precursors of wall terraces begin to appear and flat floors become more prominent (Fig. 2).

Morphologies rapidly become more complex (Dence *et al.*, 1968) in the diameter range of 10–40 km. Central peaks are found more often, well-developed terraces become more frequent, and flat floors grow in proportion to the rim diameter (Fig. 3). Above diameters of about 40 km, craters generally maintain a constant overall morphology: well-developed terraces, prominent, often mul-

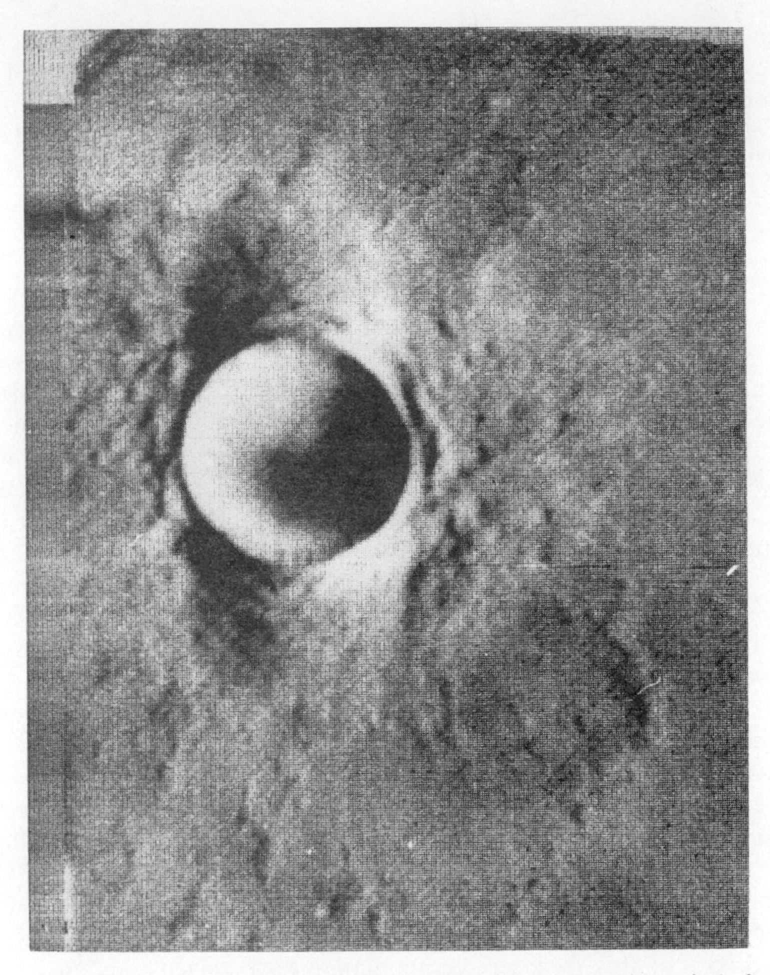

Fig. 1. Small (9 km) crater, displaying a bowl shape and the sharp rim typical of most small, fresh martian craters (Mariner 9 DAS 08838834). Apparent vertical concentration of ejecta is a computer processing artifact.

tiple, central peaks, and large, flat floors, often with low albedo splotches on them (Fig. 4). High resolution images have shown at least some of these dark patches to be fields of sand dunes (Cutts and Smith, 1973).

These morphologic changes are reflected in the crater statistics. About 34,000 martian craters classified as fresh have been analyzed for variation in morphologic characteristics as a function of diameter (Cintala *et al.*, 1976a,b). Based on a somewhat smaller data set, a similar analysis of central peaks and wall terraces has been carried out by Smith (1976).

The size-frequency distribution (Fig. 5) for fresh craters illustrates the preponderance of small craters in the fresh crater population. Comparison with a similar plot for bowl-shaped craters (Fig. 6a) shows the large percentage of fresh martian craters with apparent bowl shapes. The floor-type histograms (Figs.

Fig. 2. Fresh martian crater (11 km diameter) exhibiting the precursors of wall terraces and a small, flat floor. Note rim strata (Mariner 9 DAS 08297874). Vertical pattern is an enhancement effect.

6b,c) display this bowl-shaped crater distribution—as is apparent in Fig. 6a, no craters without flat floors are found above 50 km; indeed, there is a marked decrease in the percentage of bowl-shaped craters between 20 and 30 km diameter. Some craters maintain a bowl-shaped appearance at diameters in excess of 40 km, persisting in much larger craters than on the Moon (Howard, 1974; Head, 1976). Whether this is a real or photographic/visual effect is not known; perhaps better resolution, more suitable viewing and lighting geometries, and/or photogrammetry based on Viking orbiter photography will help explain this apparent anomaly.

The flat floor distribution is the complement of that for the bowl-shaped craters, with 100% occurrence taking place at 50 km. It is interesting to note that wall terraces, which make their first appearance between 10 and 20 km, tend to

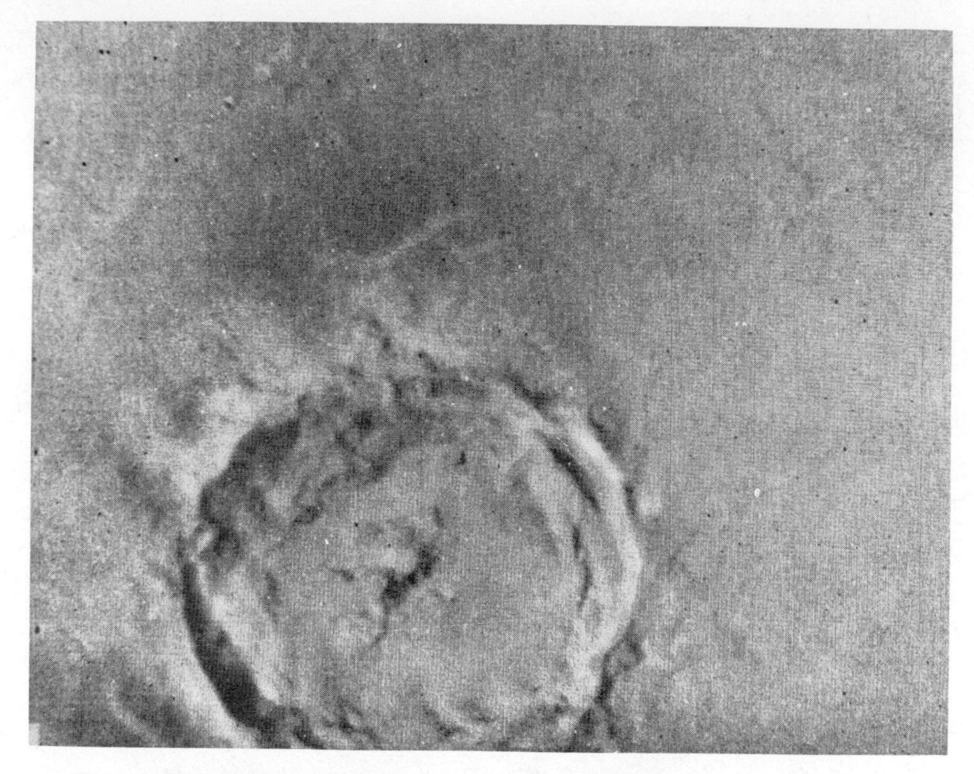

Fig. 3. A 27 km crater with central peaks, wall terraces, and a flat floor, illustrating increasingly complex morphology as larger diameters are approached (Mariner 9 DAS 11480749).

follow the flat-floor distribution (Fig. 6e). Several possible mechanisms can be advanced in an attempt to explain the apparent relationship of flat floors to wall terraces. First, a layered target could account for flat floors due to a change in the cratering process as a competent substrate is encountered by the expanding shock wave; at the same time, the reflection of this wave from the stronger layer may combine with gravity in mobilizing large blocks of material from the crater walls (Head, 1976; Cintala and Head, 1977). Second, wall slumping during the terminal stages of the event would move material from the crater walls toward the floor; the "pooling" of this debris at the bottom of the crater would result in a flat floor (Quaide *et al.*, 1965; Cintala and Head, 1977.) A combination of these two mechanisms could account for the observed trends; ejecta fallback within the crater may also play a role in determining final crater floor morphology.

Central peaks are observed at diameters as small as 5 km, but the actual percentage is too small to show at the scale of Fig. 6d. The frequency plot of central peaks exhibits a uniform increase with size until a diameter of about 70 km is reached. The fluctuations in the distribution above 70 km may be due to: (1) the small statistical sample at these diameters (Fig. 5), (2) variations in the target

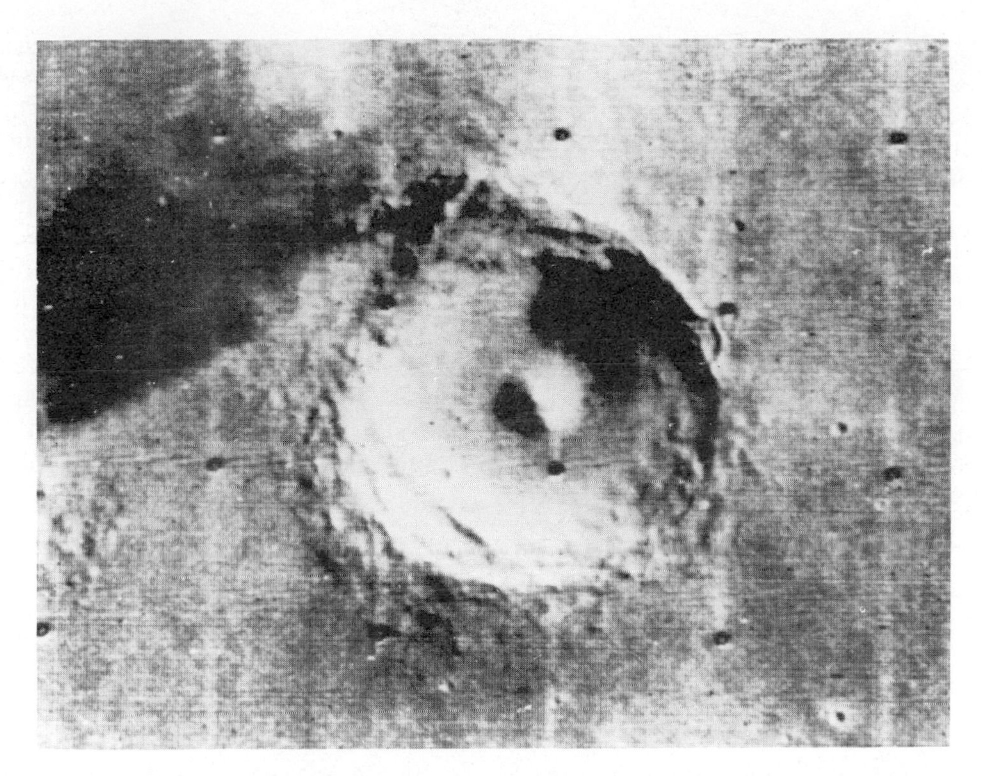

Fig. 4. The morphology of craters larger than about 50 km is well illustrated in this 60 km crater. The well-developed wall terraces, central peak (often multiple), and flat floor are typical of large, fresh martian craters. The dark splotch on the floor, probably a sand dune field (see Cutts and Smith, 1973), is a feature not uncommon in fresh craters in this size range (Mariner 9 DAS 07651038).

material's response to the impact process due to local or regional substrate variations (Head, 1976), (3) small amounts of eolian material trapped inside the crater, helping to obscure the central peaks (Cintala *et al.*, 1976b; Smith, 1976), (4) latitude-dependent, nonimpact mechanisms of central peak formation, such as volcanism or pingo formation (Cordell *et al.*, 1974), or (5) some combination of these effects.

The floor hummocks histogram (Fig. 6f) shows very low frequencies. Again, there may be a number of causes for this effect. Floor hummocks in lunar craters are very often below 1 km in diameter; for example, the largest hummock on the floor of the fresh lunar crater Copernicus (95 km diameter), as mapped by Howard (1975), is 4 km in diameter. These linear dimensions approach or are within the limits of resolution of Mariner 9 A-frames. Head (1974) has demonstrated that relatively small volumes of material would be required to cover the hummocks on the floor of Copernicus. In the martian case, otherwise insignificant amounts of eolian material may serve to obscure hummocks on crater floors.

Exceptions to these statistical trends may be due to a number of effects, such as statistical spread in impact velocities, variations in projectile density, composition, and strength, poorly understood atmospheric effects, subsurface water content and target/substrate characteristics. More detailed work must be carried out in order to determine the relative importance of these effects.

Finally, Fig. 7 illustrates the empirically derived distribution relations for central peaks, wall terraces, flat floors, and bowl-shaped craters. These curves show, in effect, the probability of finding a particular feature at a given diameter. For instance, about 25% of all bowl-shaped craters considered here are between 3 and 4 km in diameter; integrating over the entire range will give 100% of all bowl-shaped craters. This type of display aids in emphasizing the relative abundances of various features: from Figs. 6b and c, it is observed that flat floors are about as frequent as bowl-shaped craters between 30 and 40 km; the distribution relations for these two features show that it takes less than 1% of all

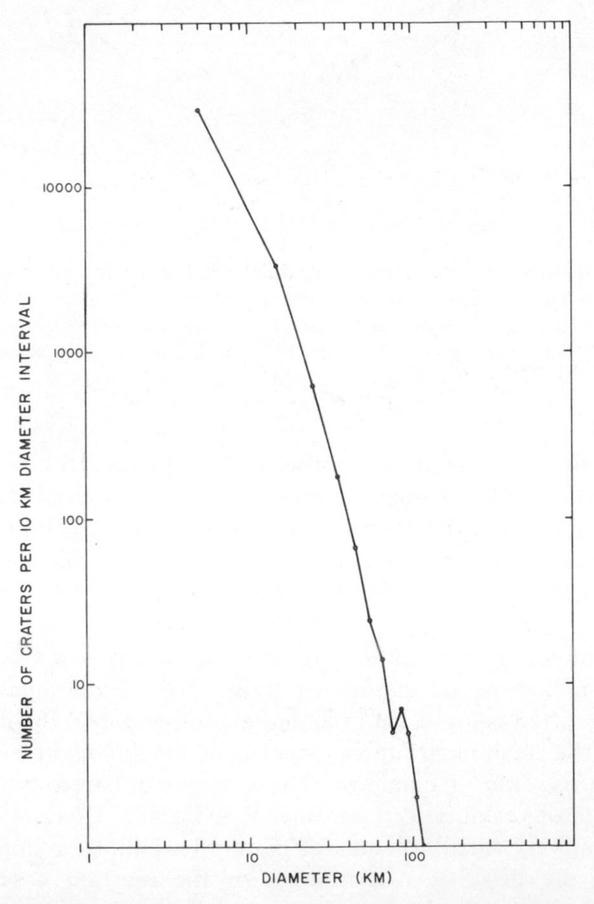

Fig. 5. Incremental plot for fresh martian craters. The small irregularity at 80–90 km is due to poor statistics for the larger craters. (From Cintala *et al.*, 1976b.)

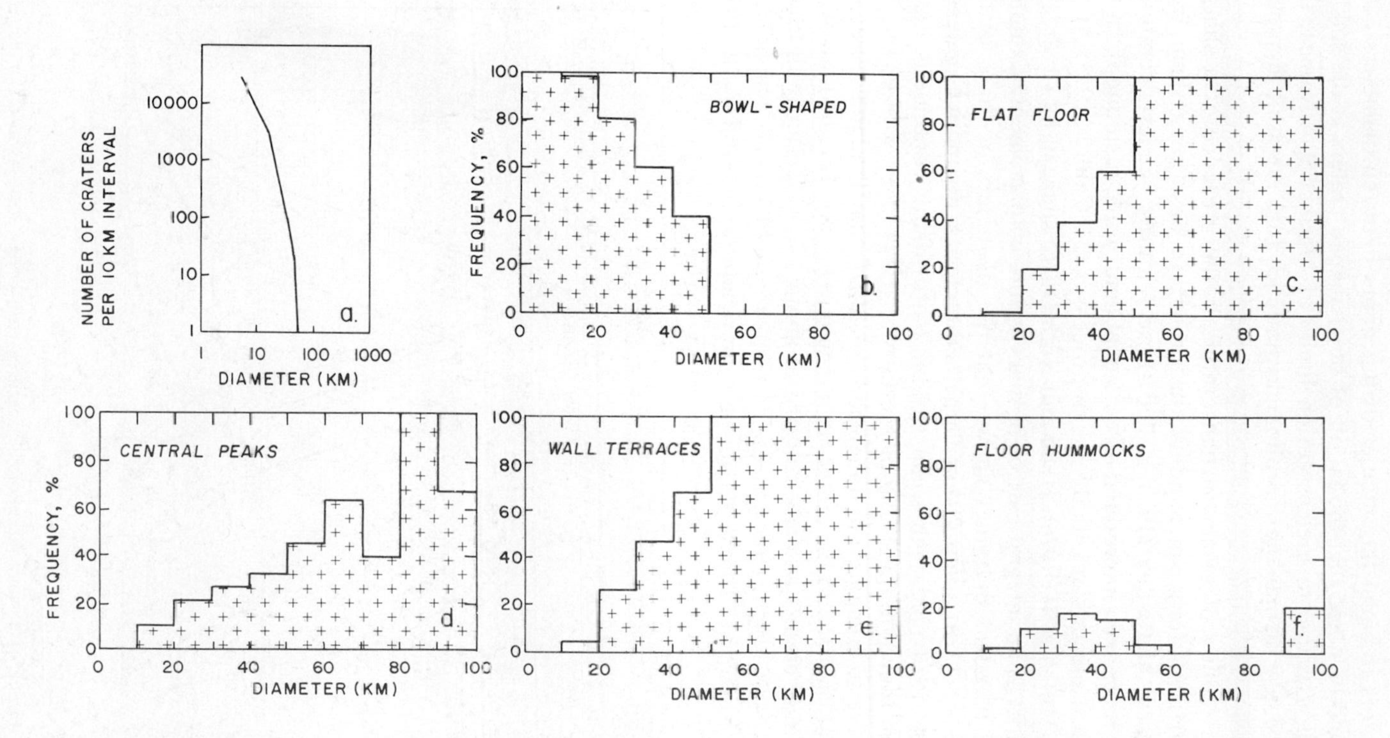

Fig. 6. (a) Incremental plot of fresh martian bowl-shaped craters. Compare with Fig. 6; (b–f) percentage of craters with designated feature as a function of size (e.g., about 67% of all fresh craters between diameters of 40 and 50 km have terraced walls); (b) bowl-shaped craters; (c) flat floors; (d) central peaks; (e) wall terraces; (f) floor hummocks. (From Cintala *et al.*, 1976b.)

bowl-shaped craters to equal the number of flat floors between diameters of 30 to 40 km. It should be noted that the falloff of bowl-shaped craters at the smaller diameters in Fig. 7 is due to the resolution limits of Mariner 9 A-frames; this factor may also cause incomplete morphologic statistics for craters at diameters less than 10 km.

A distinctive feature of some martian craters is the presence of a sharply defined concentric platform surrounding the crater; in describing them from Mariner 9 imagery, McCauley (1973) called these features "pedestal craters." These pedestals have been interpreted as erosional landforms, resulting from the gradual removal of ejecta blankets by eolian processes (McCauley, 1973; Arvidson *et al.*, 1976; Mutch *et al.*, 1976). Head and Roth (1976) suggest, however, that the pedestals form as an integral part of primary ejecta emplacement. Examination of Mariner 9 A and B frames shows that rim diameters of these craters range from less than 1 to about 20 km; larger craters lack distinctive pedestals (Head and Roth, 1976), although some exhibit more lobate continuous ejecta deposits than are found on the Moon or Mercury.

It is instructive to compare the dimensions of lunar, martian, and mercurian ejecta deposits normalized to the crater rim diameter (Fig. 8). The distinctive rim topography of lunar craters extends outward to about 1.3–1.7 crater radii (Settle and Head, 1976), while the textured ejecta deposits reach to 1.8–2.8. The fields of dense satellitic craters are found out to 4.5 crater radii (Head and Roth, 1976).

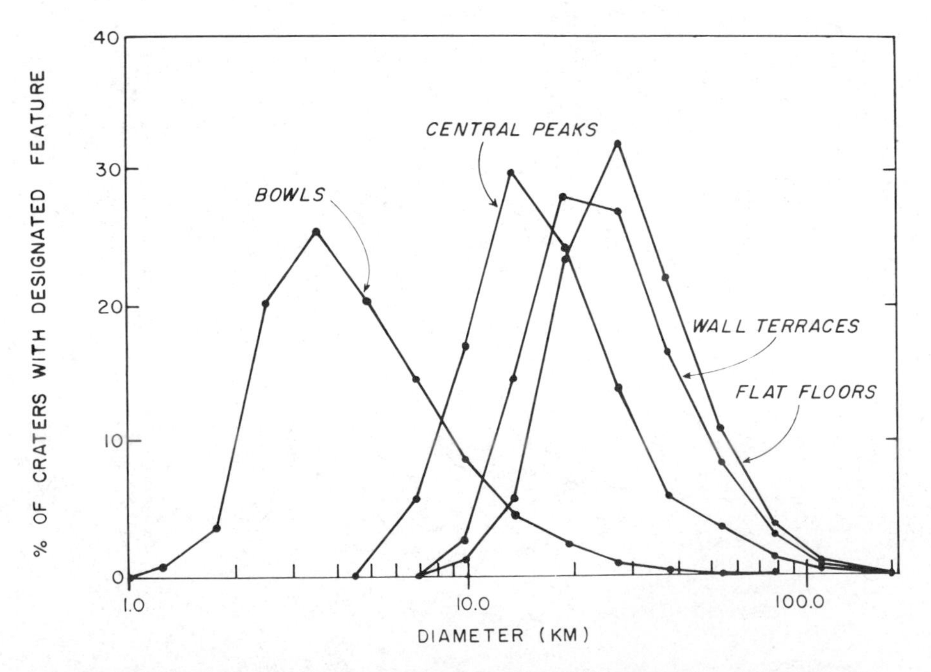

Fig. 7. Distribution of various morphologic features with crater size. As an example, about 30% of all central peak craters are found at diameters near 15 km. (From Cintala *et al.*, 1976b.)

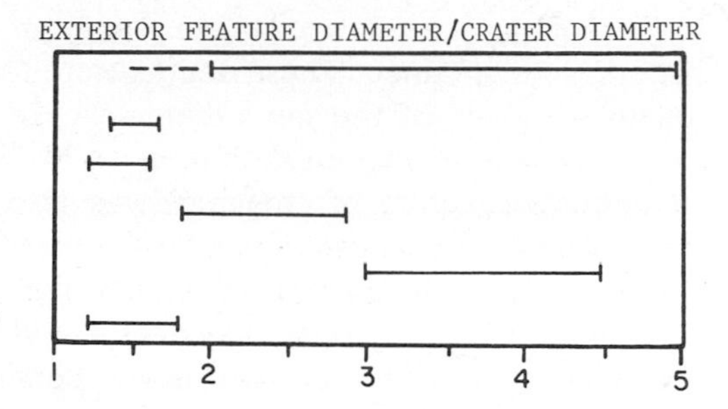

Fig. 8. Normalized measurements of external morphologic features of martian, lunar, and mercurian craters. (Adapted from Head and Roth, 1976.)

On Mercury the edge of the continuous ejecta deposits extends to 1.2–1.8 crater radii (derived from Gault *et al.*, 1975). Martian pedestals, however, cover a range of about 1.5 to 5.0 crater radii, values inconsistent with gravitational control of the extent of the continuous ejecta (Mars' surface gravity is nearly identical to that of Mercury) (Head and Roth, 1976).

Among the arguments presented by Head and Roth against the erosional model of pedestal formation are: (1) the undegraded appearance of the pedestals coupled with the lack of superposed craters imply relative youth, (2) their symmetry would require very even variations in wind direction, (3) secondary craters identified near some pedestals are evidence against local stripping, (4) some pedestals exhibit ridges or ramparts on their margins, features difficult to reconcile with an erosional origin of the margins, and (5) the apparent interaction of some pedestal margins with each other also argues against an erosional origin.

Pedestal craters seem to be evenly distributed in latitude, but the normalized pedestal diameters may be greater in the high latitudes (Arvidson *et al.*, 1976). Although the evidence appears to favor a primary origin for most martian pedestals, Viking imagery will be necessary to make the final determination of their origin, small-scale morphology, volumes, and locational dependence.

Features that are rare or unrecognized in Mariner photography of martian craters are secondary craters associated with a definite primary crater (McCauley *et al.*, 1972), high or low albedo rays commonly associated with fresh lunar and mercurian impact craters (McCauley *et al.*, 1972), and impact melt deposits on crater rims, walls, and floors. Again, the higher resolution and better photometric response of the Viking orbiter cameras (Carr *et al.*, 1972) may provide more numerous examples of the features mentioned above.

2. *Morphometry*

The first attempt at gathering and interpreting depth/diameter information for martian craters was presented by Leighton (1966). Through photoclinometric processing (the measurement of reflectivity as a function of the slope of the scattering surface relative to the source of illumination), depths for 41 martian craters were obtained. These depths were found to be less than those determined for lunar craters of the same diameters. Leighton attributed this difference to a more active erosional environment on Mars as compared to that on the Moon; an unknown source of 'fogging' was also cited as a possible reason for the shallow depths. Because the method utilized in reducing the photographic data to crater depths is dependent on relative light intensities, any scattering or abnormal light levels in the optical system would result in incorrect crater depths. The existence of this defect was never confirmed, although subsequent martian photography tends to strengthen this hypothesis (Leighton *et al.*, 1969).

Pike (1968, 1971) later plotted Leighton's depth/diameter values on log-log axes; this method of data presentation was found earlier to be an effective method of detecting changes in crater morphometry (Baldwin, 1949; Pike, 1967). A moving average technique enhanced a possible "inflection" in the data for martian craters, occurring near the diameter at which a similar break in slope exists for fresh lunar craters (Fig. 9; Pike, 1971, 1974); this inflection was not apparent in Leighton's arithmetic depth/diameter plot. Pike (1971) attributed the difference in depth/diameter ratios between the martian and lunar craters to a more active martian surface erosional and sedimentary environment, along with more intense endogenic processes operative on Mars.

Topographic profiles generated through reduction of Mariner 9 ultraviolet spectrometer (UVS) data (Barth *et al.*, 1974) have been utilized as a source of martian crater depth measurements by numerous workers (Cintala *et al.*, 1975, 1976c,d; Burt *et al.*, 1976a,b). By determining the amount of atmospheric scattering at ultraviolet wavelengths, the UVS measured, in essence, the surface atmospheric pressure; because the atmospheric pressure is controlled directly by local topography, a profile of the terrain directly below the UVS footprint can be obtained. Detailed treatment of the theory behind this technique can be found in Hord *et al.* (1972, 1974).

Depth/diameter values for 139 martian craters between the diameters of 15 and 200 km are plotted in Fig. 10. The large scatter in this diagram can be attributed to a number of effects. Since the technique depends upon scattering by atmospheric constituents (mainly CO_2), scattering caused by suspended or entrained atmospheric dust or undetected atmospheric components would contribute to anomalous measurements, as would low and/or high ultraviolet albedo areas on the surface. The scatter in the plot results, to a large degree, from the incorporation of craters of all degrees of degradation into the population. Pike (1968, 1971) has shown, for example, that the amount of scatter in depth/diameter relationships for lunar craters increases with apparent crater age. This same phenomenon is, without a doubt, applicable to martian craters, in light of impact weathering as in the lunar case (Chapman *et al.*, 1969; Murray *et al.*, 1971; Cintala *et al.*, 1975) and the active surface erosional regime (e.g., Sagan, 1973).

Varying interpretations have been applied to these data. Burt *et al.* (1976a,b) argue that the deepest martian craters measured are as deep as those found on

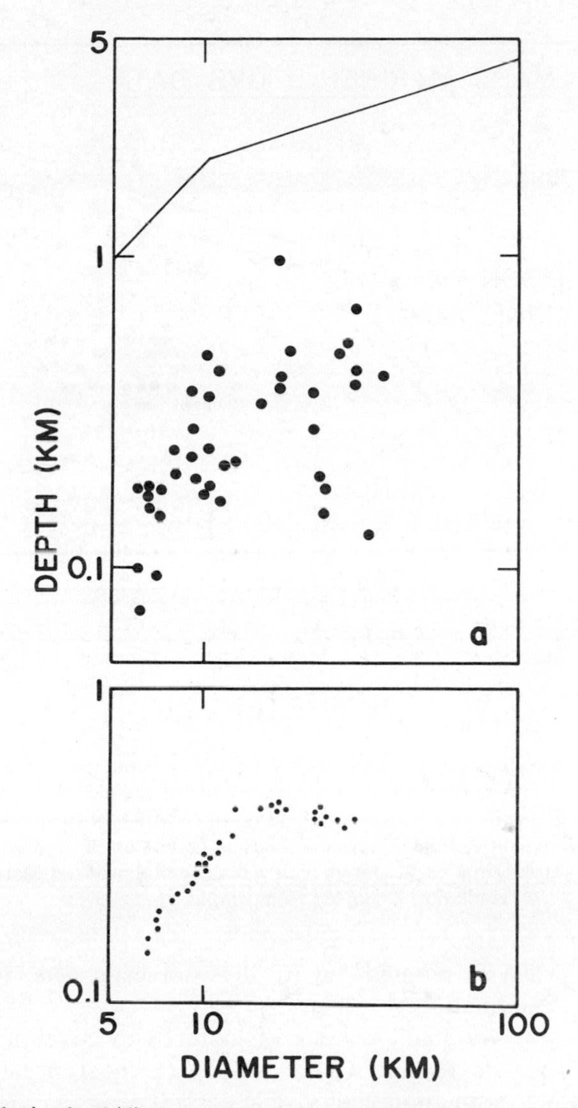

Fig. 9. Logarithmic depth/diameter plot for 41 martian craters of all degrees of degradation; data obtained through photoclinometric processing of Mariner 4 photographs (Leighton, 1966). (a) Original data, with Pike's (1974) lunar depth/diameter fit included for comparison. (b) Martian data after application of a 9-point moving average. (Adapted from Pike, 1971.)

Mercury (Gault *et al.*, 1975), a planet with a surface gravity essentially equal to that of Mars; this would be expected if gravity were the major controlling factor in determining the final crater morphology (Hartmann, 1972; Gault *et al.*, 1975) and morphometry (Gault *et al.*, 1975). A possible correlation between crater depth/diameter ratio and elevation of the surrounding terrain was also identified

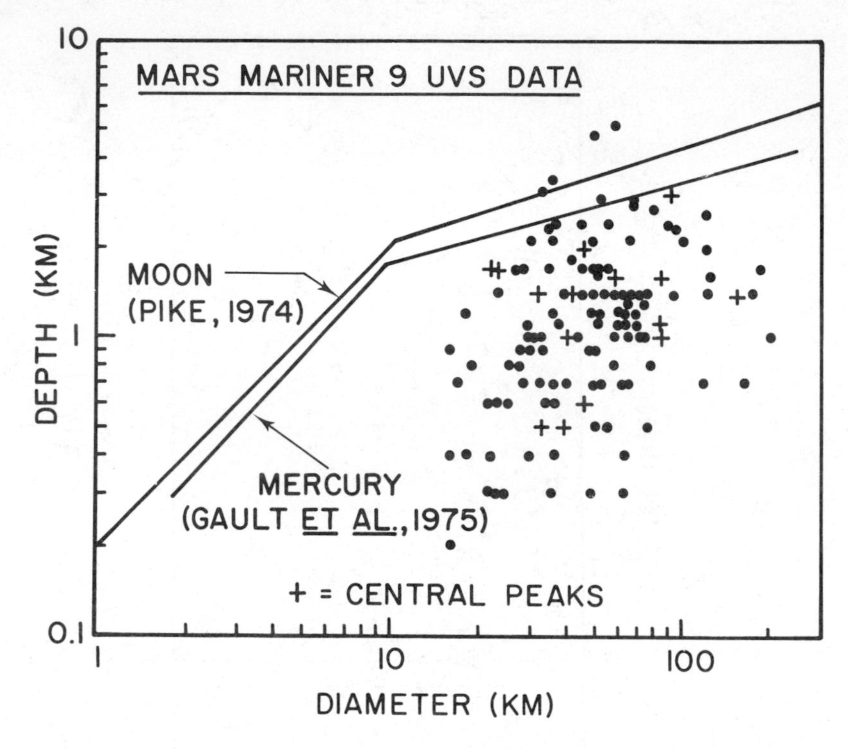

Fig. 10. Logarithmic depth/diameter plot for 139 martian craters between the diameters of 15 and 200 km; data obtained from measurements of Mariner 9 UVS profiles (Barth *et al.*, 1974). Shown for comparison are the least-squares fits for fresh lunar (Pike, 1974) and mercurian (Gault *et al.*, 1975) craters. Martian craters of all degress of degradation are included; crosses represent craters with visible central peaks at Mariner 9 A-frame resolution. (Adapted from Cintala *et al.*, 1976d.)

by Burt *et al.* This, as pointed out by these investigators, requires more and better data for confirmation.

It takes a relatively small volume of material to cover lunar central peaks (Head, 1975); for this reason, Cintala *et al.* (1976c,d) treated those martian craters in the UVS population with visible central peaks as the least modified. It can be seen in Fig. 10 that these craters generally fall below both the lunar and mercurian depth/diameter relationships. On this basis, it was suggested by Cintala *et al.* that factors other than gravity might also play a role in determining the final crater depth/diameter ratios; among such agents proposed were impact velocity (projectile energy), projectile/target properties, and atmospheric effects.

The number of martian craters with available depth/diameter information is too small, unfortunately, to permit more detailed statistical depth/diameter analysis on the basis of discrete morphologic criteria or degree of degradation. Viking orbiter photography should provide much more accurate, precise, and abundant topographic data through both shadow measurements and photogrammetric analysis.

Phobos and Deimos

Both known martian moons exhibit craters with a wide spectrum of planimetric shapes and apparent degress of degradation (Pollack *et al.*, 1972, 1973). Well-defined morphologic features such as central peaks and wall terraces have not been positively identified in Mariner 9 photographs of craters on Phobos and Deimos. One or more of the following possibilities may account for the apparent deficiency of such morphologic features: (1) such structures are below the resolution of available Mariner 9 photography, (2) most of these craters are below the onset diameter for such features found on Mars, the Moon, and Mercury (Smith and Sanchez, 1973; Gault *et al.*, 1975; Cintala *et al.*, 1976a,b; Smith (1976); Wood *et al.*, 1976), and/or (3) the gravitational fields of the two satellites are too small to give rise to these morphologies (e.g., Hartmann, 1972). Higher resolution and better viewing geometries should give the Viking orbiters a better chance of detecting possible morphologic variations associated with the craters of Phobos and Deimos.

Acknowledgments—This summary was carried out under funding provided by the Planetology Program Office, Office of Space Science, NASA Headquarters, Grant NGR-40-002-088, which is gratefully acknowledged. Helpful comments were supplied by J. W. Head and P. D. Spudis; thanks go to Alex Woronow for his critical review of the manuscript. S. H. Bosworth, S. M. Matarazza, A. W. Gifford, and R. Roth provided valuable assistance in the preparation of the manuscript. Support of the author by the William F. Marlar Memorial Foundation in the form of a fellowship is greatly appreciated.

References

Arvidson, R. E.: 1974, Morphologic Classification of Martian Craters and Some Implications, *Icarus* **22**, 264.

Arvidson, R. E., Jones, K. L., and Mutch, T. A.: 1974, Craters and Associated Aeolian Features on Mariner 9 Photographs: An Automated Data Gathering and Handling System and Some Preliminary Results, *The Moon* **9**, 105.

Arvidson, R. E., Coradini, M., Carusi, A., Coradini, A., Fulchignoni, M., Federico, C., Funiciello, R., and Salomme, M.: 1976, Latitudinal Variation of Wind Erosion of Crater Ejecta Deposits on Mars, *Icarus* **27**, 503.

Arthur, D. W. G., Agnierary, A. P., Horvath, R. A., Wood, C. A., and Chapman, C. R.: 1963, *Comm. Lunar Planet. Lab.*, Univ. Arizona (Tucson), 71.

Baldwin, R. B.: 1949, *The Face of the Moon*, Univ. Chicago Press, Chicago.

Barth, C. A., Hord, C. H., Stewart, A. I., Lane, A. L., Dick, M. L., Shaffner, S. H., and Simmons, K. E.: 1974, *An Atlas of Mars: Local Topography*, Laboratory for Atmospheric and Space Physics publication, Univ. Colorado, Boulder.

Burt, J., Veverka, J., and Cook, K.: 1976a, Depth/Diameter Relation for Large Martian Craters Determined from Mariner 9 UVS Altimetry (abstract), *Reports of Accomplishments of Planetology Programs*, 1975–1976, NASA TM X-3364.

Burt, J., Veverka, J., and Cook, K.: 1976b, Depth/Diameter Relation for Large Martian Craters Determined from Mariner 9 UVS Altimetry, *Icarus* **29**, 83.

Carr, M. H., Baum, W. A., Briggs, G. A., Masursky, H., Wise, D. W., and Montgomery, D. R.: 1972, Imaging Experiment: The Viking Mars Orbiter, *Icarus* **16**, 17.

Chapman, C. R., Pollack, J. B., and Sagan, C.: 1969, An Analysis of the Mariner 4 Cratering Statistics, *Astron. J.* **74**, 1039.

Cintala, M. J., Head, J. W., and Mutch, T. A.: 1975, Depth/Diameter Relationships for Martian and Lunar Craters (abstract), *EOS* **56**, 389.

Cintala, M. J., Head, J. W., and Mutch, T. A.: 1976a, Characteristics of Fresh Martian Craters as a Function of Diameter: Comparison with the Moon and Mercury (abstract), *Reports of Accomplishments of Planetology Programs*, 1975–1976, pp. 188–189 NASA TM X-3364.

Cintala, M. J., Head, J. W., and Mutch, T. A.: 1976b, Characteristics of Fresh Martian Craters as a Function of Diameter: Comparison with the Moon and Mercury, *Geophys. Res. Lett.* **3**, 117.

Cintala, M. J., Head, J. W., and Mutch, T. A.: 1976c, Craters on the Moon, Mars, and Mercury: a Comparison of Depth/Diameter Characteristics (abstract), *Reports of Accomplishments of Planetology Programs*, 1975–1976, NASA TM X-3364.

Cintala, M. J., Head, J. W. and Mutch, T. A.: 1976d, Martian Crater Depth/Diameter Relationships: Comparison with the Moon and Mercury, *Proc. Lunar Sci. Conf. 7th* 3575.

Cintala, M. J. and Head, J. W.: 1977, Relationship of Morphology and Morphometry in Fresh Lunar Craters, unpublished manuscript.

Cordell, B. M., Lingenfelter, R. E., and Schubert, G.: 1974, Martian Cratering and Central Peak Statistics: Mariner 9 Results, *Icarus* **21**, 443.

Cutts, J. A. and Smith, R. S. U.: 1973, Eolian Deposits and Dunes on Mars, *J. Geophys. Res.* **78**, 4139.

Dence, M., Innes, M., and Robertson, P.: 1968, Recent Geological and Geophysical Studies of Canadian Craters, in *Shock Metamorphism of Natural Materials* (B. French and N. Short, eds.), Mono Book Corp., Baltimore.

Gault, D. E., Guest, J. E., Murray, J. B., Dzurizin, D., and Malin, M. C.: 1975, Some Comparisons of Impact Craters on Mercury and the Moon, *J. Geophys. Res.* **80**, 2444.

Hartmann, W. K.: 1972, Interplanet Variations in Crater Morphology, *Icarus* **18**, 707.

Hartmann, W. K.: 1973, Martian Cratering, 4, Mariner 9 Initial Analysis of Cratering Chronology, *J. Geophys. Res.* **78**, 4096.

Head, J. W.: 1975, Processes of Lunar Crater Degradation: Changes in Style with Geologic Time, *The Moon* **12**, 299.

Head, J. W.: 1976, The Significance of Substrate Characteristics in Determining Morphology and Morphometry of Lunar Craters, *Proc. Lunar Sci. Conf. 7th*, 2913.

Head, J. W. and Roth, R.: 1976, Mars Pedestal Crater Escarpments: Evidence for Ejecta-Related Emplacement, in *Papers Presented to the Symposium on Planetary Cratering Mechanics*, pp. 50–52, The Lunar Science Institute, Houston.

Hord, C. W., Barth, C. A., Stewart, A. I., and Lane, A. L.: 1972, Mariner 9 Ultraviolet Spectrometer Experiment: Photometry and Topography of Mars, *Icarus* **17**, 443.

Hord, C. W., Simmons, K. E., and McLaughlin, L. K.: 1974, Mariner 9 Ultraviolet Spectrometer Experiment: Pressure-Altitude Measurements on Mars, *Icarus* **21**, 492.

Howard, K. A.: 1974, Fresh Lunar Impact Craters: Review of Variations with Size, *Proc. Lunar Sci. Conf. 5th*, 61.

Howard, K. A.: 1975, Geologic Map of the Crater Copernicus, *U.S. Geological Survey Misc. Investigations Series*, Map I-840.

Leighton, R. B.: 1966, The Photographs from Mariner IV, *Scientific American* **214**, 54.

Leighton, R. B., Murray, B. C., Sharp, R. P., Allen, J. D., and Sloan, R. K.: 1965, Mariner 4 Photography of Mars: Initial Results, *Science* **149**, 627.

Leighton, R. B., Murray, B. C., Sharp, R. P., Allen, J.D., and Sloan, R. K.: 1967, Mariner 4 Pictures of mars, *Jet Propul. Lab. Tech. Rep.*, 32–884.

Leighton, R. B., Horowitz, N. H., Murray, B. C., Sharp, R. P., Herriman, A. H., Young, A. T., Smith, B. A., Davies, M. E., and Leovy, C. B.: 1969, Mariner 6 and 7 Television Pictures: Preliminary Analysis, *Science* **166**, 49.

McCauley, J. F.: 1973, Mariner 9 Evidence for Wind Erosion in the Equatorial and Mid-Latitude Regions of Mars, *J. Geophys. Res.* **78**, 4123.

McCauley, J. F., Carr, M. H., Cutts, J. A., Hartmann, W. K., Masursky, H., Milton, D. J., Sharp, R. P., and Wilhelms, D. E.: 1972, Preliminary Mariner 9 Report on the Geology of Mars, *Icarus* **17**, 289.

Mutch, T. A., Arvidson, R. A., Jones, K. L., Head, J. W. and Saunders, R. S.: 1976, *Geology of Mars*, Princeton Press, Princeton.

Murray, B. C., Soderblom, L. A., Sharp, R. P. and Cutts, J. A.: 1971, Surface of Mars, 1. Cratered Terrains, *J. Geophys. Res.* **76**, 313.

Öpik, E. J.: 1950, Mars and the Asteroids, *Irish Astron. J.* **1**, 22.

Pike, R. J.: 1967, Schroeter's Rule and the Modification of Lunar Crater Impact Morphology, *J. Geophys. Res.* **72**, 2099.

Pike, R. J.: 1968, Meteoritic Origin and Consequent Endogenic Modification of Large Lunar Craters—A Study in Analytical Geomorphology, Ph.D. thesis, Univ. of Michigan, Ann Arbor.

Pike, R. J.: 1971, Genetic Implications of the Shapes of Martian and Lunar Craters, *Icarus* **15**, 384.

Pike, R. J.: 1974, Depth/Diameter Relationships of Fresh Lunar Craters: Revision from Spacecraft Data, *Geophys. Res. Lett.* **1**, 291.

Pohn, H. A. and Offield, T. W.: 1970, Lunar Crater Morphology and Relative-Age Determination of Lunar Geologic Units—Part 1. Classification, *U.S. Geol. Survey Res. Prof. Paper 700-C*, C153.

Pollack, J. B., Veverka, J., Noland, M., Sagan, C., Hartmann, W. K., Duxbury, T. C., Born, G. H., Milton, D. J., and Smith, B. A.: 1972, Mariner 9 Television Observations of Phobos and Deimos, *Icarus* **17**, 394.

Pollack, J. B., Veverka, J., Noland, M., Sagan, C., Duxbury, T. C., Acton, C. H., Jr., Born, G. H., Hartmann, W. K., and Smith, B. A.: 1973, Mariner 9 Television Observations of Phobos and Deimos, 2, *J. Geophys. Res.* **78**, 4313.

Quaide, W. L., Gault, D. E., and Schmidt, R. A.: 1965, Gravitative Effects on Lunar Impact Structures, *Annals N.Y. Acad. Sci.* **123**, 563.

Sagan, C.: 1973, Sandstorms and Eolian Erosion on Mars, *J. Geophys. Res.* **78**, 4155.

Settle, M. and Head, J. W.: 1976, Radial Variation of Lunar Crater Rim Topography, *Icarus*, in press.

Smith, E. I.: 1976, Comparison of the Morphology-Size Relationship for Mars, Moon, and Mercury, *Icarus* **28**, 543.

Smith, E. I. and Sanchez, A. G.: 1973, Fresh Lunar Craters: Morphology as a Function of Diameter, A Possible Criterion for Origin, *Mod. Geol.* **4**, 51.

Tombaugh, C. W.: 1950, cited in *Sky and Tel.* **9**, 272.

Wood, C. A.: 1972, The System of Lunar Craters, Revised, *The Moon* **3**, 408.

Wood, C. A., Cintala, M. J. and Head, J. W.: 1976, Morphological Characteristics of Fresh Craters: Mercury, Moon, and Mars (abstract), in *Papers Presented to the Conference on Comparisons of Mercury and the Moon*, The Lunar Science Institute, Houston.

Wilhelms, D. E. and McCauley, J. F.: 1971, Geologic Map of the Near Side of the Moon, *U.S. Geological Survey, Map I-703.*

Roddy, D. J., Pepin, R. O., and Merrill, R. B., editors.
(1977) *Impact and Explosion Cratering*, Pergamon Press (New York), p. 593–602.
Printed in the United States of America

Distribution and emplacement of ejecta around martian impact craters

MICHAEL H. CARR

U.S. Geological Survey, Menlo Park, California 94025

Abstract—The pattern of ejecta around most martian craters is distinctively different from that for Mercury and the moon. Fresh-appearing martian craters are typically surrounded by layers of ejecta, each with a low ridge or escarpment at its outer edge. Small craters (<10 km diameter) may have only one ejecta layer; larger craters may have many layers with complexly lobed outer margins. Deflection of the ejecta by pre-existing obstacles and the large distance to which the continuous ejecta extends from the source crater suggest final emplacement as a surface debris flow, probably following ballistic ejection. The unique character of the martian craters is tentatively attributed to gases entrained in the ejecta either as a result of permafrost in the target medium or from the atmosphere.

DISTRIBUTION AND EMPLACEMENT OF EJECTA AROUND MARTIAN IMPACT CRATERS

THE PATTERN of ejecta around martian impact craters is, in most cases, distinctively different from the pattern around lunar and mercurian craters. This unique pattern, although suspected from Mariner 9 data, could not be documented in detail until Viking Orbiter photographs were acquired early in the Viking mission. At that time the atmosphere was extremely clear. This combined with the higher resolution Viking cameras resulted in views of the martian surface that were far more detailed than anything available previously. These views revealed complex lobate patterns of ejecta around most fresh-appearing martian craters. Relatively few craters have the ejecta pattern that is characteristic of lunar and mercurian craters—coarse disordered texture close to the rim which becomes finer further out, grading imperceptibly into dense fields of secondary craters and finally discrete secondaries and rays. Most fresh-appearing martian craters have ejecta that commonly consists of several layers each with a distinct outer edge marked by a low ridge or escarpment. This type of crater was recognized from the Mariner 9 images and was termed a rampart crater (McCauley, 1973). The peculiar morphology was attributed to modification of the ejecta by wind action (McCauley, 1973; Arvidson, 1976). While this may be true in some cases, it now appears (Carr *et al.*, 1977) that many of the unique characteristics of the ejecta of martian craters are primary and not due to secondary modifications. They may result from movement of ejecta along the ground as a debris flow after ballistic deposition (Carr *et al.*, 1977). In this paper we shall look briefly at the patterns of ejecta around martian craters and discuss possible implications concerning the mechanism of ejecta emplacement.

Rampart craters occur in a wide range of morphologies from those that seemingly have only one layer of ejecta with a simple, almost circular, outline to those with several ejecta layers, each complexly lobed. The simple rampart crater of Fig. 1 shows most of their general characteristics. This crater has many features typical of lunar craters such as terraced walls and a central peak partly surrounded by a flat floor. The ejecta pattern, however, is distinctly non-lunar. A layer of ejecta appears to extend from the crater rim out to approximately one crater radius. The outer edge of the layer is marked either by

Fig. 1. A 9 km rampart crater in Chryse Planitia. An inner ejecta layer is clearly demarcated by a low ridge or escarpment. Beyond the edge of the continuous ejecta are clusters of secondary craters, low hills and faint radial striae. The dark ring to the northwest is a camera artifact. (Viking Orbiter frame 10A56)

a low ridge or an escarpment that gives the rampart crater its name. Over most of the ejecta surface is a concentric pattern of low ridges and grooves but part has a fine radial pattern. Beyond the edge of the ejecta layer are low hills, lines of shallow craters, and faint radial striae, all reminiscent of the distal edges of lunar and mercurian craters. These features appear to be transected by the edge of the main ejecta as though overridden.

The description given above applies in general to craters in 5–20 km size range. Larger craters commonly have more ejecta layers. Typical relations are shown in Fig. 2 where an upper layer clearly overlies a lower, more lobate layer, indicating successive deposition. In the bottom right of the picture the upper layer appears to have buckled and flowed around a protrusion. The buckled surface and the fact that the ejecta is nearly as thick as the protrusion is high, indicates very little volume reduction of the ejecta after deposition and suggests the ejecta was not deposited from a low density, solid-gas mixture such as a base surge but rather was emplaced as a debris flow. In general, the edges of the ejecta layers are more lobate for larger craters. The crater Yuty (Fig. 3) for example, has several layers of ejecta with complex lobate outer margins. A faint radial pattern is visible in places. A small crater close to the rim of Yuty is partly covered with ejecta, yet is clearly visible, indicating that the ejecta is thin. In another more complex example the ejecta cannot so easily be separated into discrete layers as at Yuty, although the emplacement mechanics (near-surface outward flow) of the ejecta appears similar (Fig. 4). Northwest of the crater, a low mesa forms an obstacle to radial flow. Very little ejecta is visible on top of the mesa while it is present on either side, indicating that at this distance ballistic deposition is negligible and radial flow is the dominant emplacement mechanism.

The crater Arandas (Fig. 5) is one of the least modified large craters yet observed. Several tiers of ejecta are clearly visible. The most distant lobes appear to have ridden over the polygonal fractures of the surrounding plains (Fig. 6), while the inner lobes appear to have ridden over just-deposited ejecta. To the south and southwest the flow of ejecta has been diverted by craters. The ejecta extends almost to the rim crest of the crater (Fig. 7) to the south but did not flow over and into it again suggesting little volume change after deposition. On the Arandas side of the crater to the southwest (Fig. 6) is a series of ridges, that could be pressure ridges caused by the crater acting as a barrier to flow. Each of the layers has a strong radial pattern. In places, small ridges parallel the edge of individual ejecta blankets. Lines of secondary craters are visible to the west and southwest.

Studies of lunar and terrestrial impact craters suggest that the final mode of emplacement of impact ejecta may be either simply ballistic or by outward radial flow following ballistic ejection from the craters. Many aspects of lunar craters suggest that relatively small amounts of radial flow can follow ballistic emplacement (Howard, 1972). The avalanches around larger craters such as Tsiolkovsky (Guest and Murray, 1969), Tycho (Shoemaker *et al.*, 1968) and Aristarchus (Guest, 1973) have been interpreted as part of the impact process, although this is not universally accepted. Products of ejecta flows are also seen in the ejecta

Fig. 2. Detail of ejecta around a 15 km diameter crater in Chryse Planitia. An upper ejecta layer transects a lower, more lobate layer. In the bottom right of the picture the ejecta has flowed around a low obstacle. (Viking Orbiter frame 10A66)

blankets of fresh basins on the moon such as Orientale (Moore *et al.*, 1974). Thus it appears that, although the final distribution of ejecta around lunar craters predominantly reflects ballistic emplacement, in rare cases modifications do result from subsequent radial transport.

On Mars, flow after ballistic emplacement appears to be the rule rather than the exception. The pattern of ejecta around obstacles to radial flow, such as older craters and hills, strongly supports the flow mechanism. The ejecta is

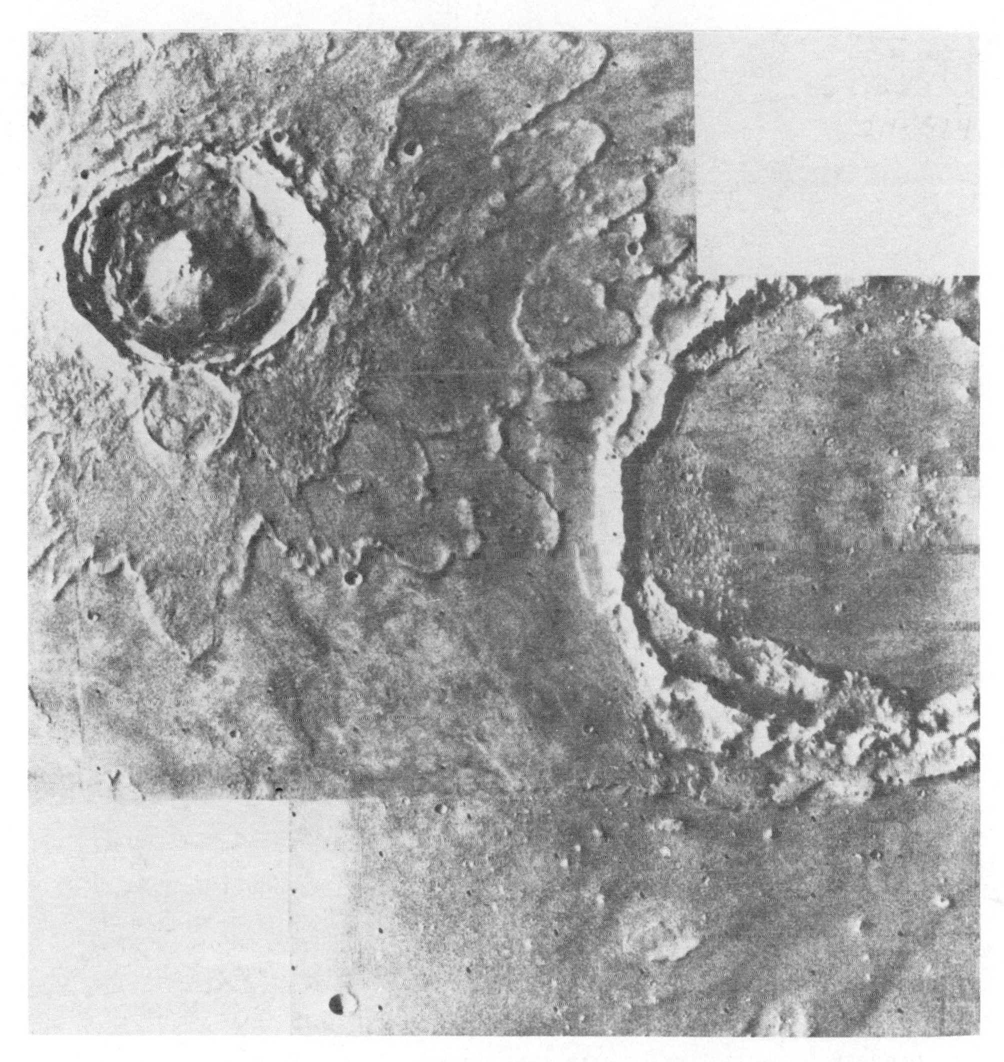

Fig. 3. The 19 km diameter crater Yuty (22°N, 34°W) with several ejecta layers each complexly lobed. A pre-Yuty, crater close to the rim of Yuty, is buried by ejecta, yet is still visible, indicating a thin ejecta blanket. (Viking Orbiter frame 3A07)

deformed into ridges which wrap around the obstacles as though the ejecta flowed around them. "Shadow zones" appear on the lee sides. The lack of ejecta on the obstacles themselves and in the lee zones rules out subaerial deposition. Additional support for the flow mechanism is the strong radial pattern and transection of older features. The ridges at the distal ends of each flow are probably pressure ridges caused by buildup of ejecta at the edge of the flow as the flow front slows and finally stops. Some ridges have been previously attributed to selective erosion (McCauley, 1973), but the preservation of very

Fig. 4. An 11 km diameter, multi-layered ejecta crater at 21°N, 36°W. Little ejecta is visible on the mesa to the northwest of the crater. Ejecta is however present adjacent to the mesa on the north side, at a comparable distance from the crater, suggesting that final emplacement of ejecta is by surface flow not by ballistics. (Viking Orbiter frame 34A77)

fine primary textures, such as shallow impact craters and rays immediately adjacent to the ridge (Fig. 1) indicates that the ridge is a primary texture.

Additional evidence supporting flow is the larger radial extent of continuous ejecta. On the moon, continuous ejecta occurs up to 0.75 crater diameter from the rim and on Mercury only up to 0.4 crater diameter (Gault *et al.*, 1975). In contrast, martian craters have continuous ejecta up to 2 crater diameters from the rim (Carr *et al.*, 1977) despite the fact that the gravity fields on Mercury and Mars are similar. The radial extent of the ejecta, together with the morphologic characteristics described above, suggested to Carr *et al.* (1977) that during the formation of most martian craters ejecta is emplaced by a combination of ballistics and surface flow. They postulated that the early phases of crater formation on Mars are similar to those on the moon and Mercury. The later stages differ, however, in that on Mars the ejecta continues its radial outward

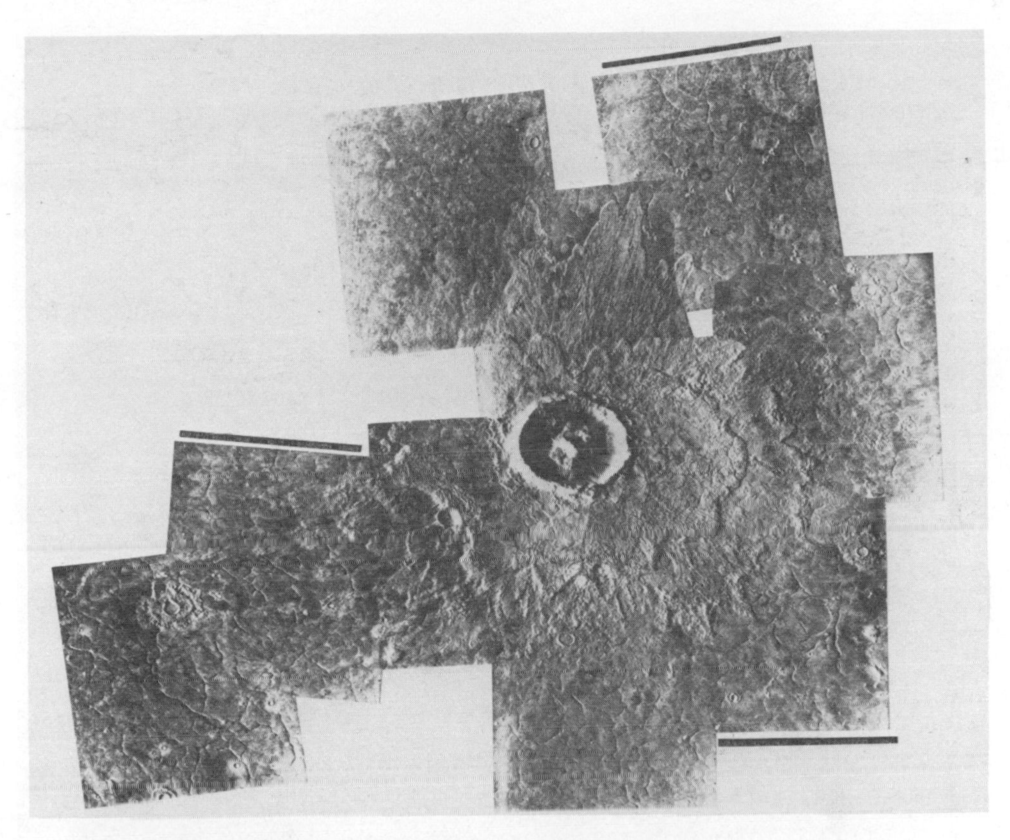

Fig. 5. Mosaic of Arandas, a 28 km diameter crater at 43°N, 14°W. Several layers of ejecta are clearly visible. To the southwest of the crater, low ridges on the surface of the ejecta wrap around a pre-existing crater, suggesting resistance to flow.

motion as a debris flow following ballistic deposition. This may happen to a limited degree on the moon and Mercury but is the general case on Mars. They attributed the predominance of flow features in the Mars case to either permafrost in the target medium or to the presence of an atmosphere.

This paper has described the unique morphology of impact ejecta on Mars and suggested possible causes. The suggestions are, however, very tentative and much systematic work must be done before we can be more definitive about the precise mechanisms. Of particular importance is comparison between different regions of Mars. Lunar-like craters do occur. If their properties relative to the rampart craters are regionally dependent this would strongly suggest the target materials are strongly influencing the ejecta morphology. Terrestrial impact craters should also be examined for evidence of the debris flows postulated for Mars. Finally, experimental work badly needs to be done to simulate the Martian feature and determine what factors control the final distribution of ejecta.

Fig. 6. Detail from Fig. 5 showing the sharp contrast between the outermost ejecta lobes and the surrounding fractured plains. Note also the strong radial texture and details of the peripheral "rampart". (Viking Orbiter frame 9A42)

Fig. 7. Detail of inner ejecta layers from Fig. 5. At the bottom of the picture ejecta has flowed around a pre-existing crater causing reduced deposition on the downstream side. (Viking Orbiter frame 32A30)

References

Arvidson, R. H., Carusi, A., Coradini, A., Coradini, M., Fulchignoni, M., Frederico, C., Funicello, R., and Salomone, M.: 1976, Latitudinal variation of wind erosion of crater ejecta deposits on Mars. *Icarus* **27**, 503–516.

Carr, M. H., Crumpler, L. S., Cutts, J. A., Greeley, R., Guest, J. E., and Masursky, H.: 1977, Martian impact craters and emplacement of ejecta by surface flow. *J. Geophys. Res.* **81**. In press.

Gault, D.E., Guest, J.E., Murray, J. B., Dzurisin, D., and Malin, M. C.: 1975, Some comparisons of impact craters on Mercury and Moon. *J. Geophys. Res.* **80**, 2444–2460.

Guest, J. E.: 1973, Stratigraphy of ejecta from the lunar crater Aristarchus. *Geol. Soc. Amer. Bull.* **84**, 2873–2893.

Guest, J. E. and Murray, J. B.: 1969, Nature and origin of Tsiolkovsky Crater, lunar farside. *Planet. Space Sci.* **17**, 121–141.

Howard, K. A.: 1972, Ejecta blankets of large craters exemplified by King Crater. *Apollo 16 Prelim. Sci. Rep.* NASA SP-315, 29-70 to 29-79.

McCauley, J. F.: 1973, Mariner 9 evidence for wind erosion in the equatorial and mid-latitude regions of Mars. *J. Geophys. Res.* **78**, 4123–4137.

Moore, H. J., Hodges, D. A., and Scott, D. H.: 1974, Multiringed basins—illustrated by Orientale and associated features. *Proc. Lunar Sci. Conf. 5th*, p. 71–100.

Shoemaker, E. M., Batson, R. M., Holt, H. E., Morris, E. C., Remilson, J. J., and Whitaker, E. A.: 1968, Television observations from surveyor VII. *Surveyor VII: A Preliminary Report*, NASA SP-173, p. 13–81.

Roddy, D. J., Pepin, R. O., and Merrill, R. B., editors.
(1977) *Impact and Explosion Cratering*, Pergamon Press (New York), p. 603–612.
Printed in the United States of America

Probable distribution of large impact basins on Venus: Comparison with Mercury and the Moon

GERALD G. SCHABER and JOSEPH M. BOYCE

U.S. Geological Survey, Branch of Astrogeologic Studies, Flagstaff, Arizona 86001

Abstract—Low resolution (80 km) 12.5 cm wavelength radar maps have been used to identify probable large impact basins on Venus. Here we define basins as any simple crater or multi-ringed structure. It is assumed that, like the Moon, large basins are characterized by relatively flat floor materials with relatively low radar reflectivity and mountainous rim deposits of high radar reflectivity. A total of twelve basins (>600 km diameter) have been tentatively recognized on 8% of the planet's surface. Four circular, weakly reflecting areas, ranging in diameter from 1000 to 1700 km, are considered good prospects for impact-type basins while eight other features ranging in size from 600 to 1130 km diameter are identified with lower confidence. A basin size-frequency distribution for Venus is compared with similar data for the moon and Mercury.

Normalizing the total number of lunar basins larger than 600 km diameter per unit area to 1.0, the ratio of Venus basins is 1.2 and the ratio of basins on Mercury is 0.4.

Wetherill (1975) and Hartmann (1976) have found that the total integrated flux of all families of known objects (asteroids and comets) appears to be nearly the same on all terrestrial planets to within a factor of three of the lunar rate. Our preliminary data for the Venus large basin distribution are in general agreement with their findings.

Preservation of impact basin morphology suggests that plate tectonics has not operated to any appreciable extent to recycle the Venus crust and destroy the basins, and that the atmosphere of Venus has not significantly eroded these features. The radar data indicate that the cratering record preserved on Venus may be very similar to that of the moon, Mars, and Mercury. Verification of this fact would have significant implication regarding a similar evolution of crustal viscosity on all four planets.

INTRODUCTION

EARTH-BASED RADAR IMAGES covering 22% of the surface area of the planet Venus are currently available at resolutions of 80 km or better. Radar resolution is approximately one-half that of the "line pair" resolution used in planetary television systems. A total of 8% of Venus has been imaged at about 20 km resolution while 19% has been imaged at 80 km resolution. These data were obtained at 12.5 cm wavelength at the Goldstone Tracking Station (California) and at the Arecibo Astronomy and Ionosphere Center (Arecibo, Puerto Rico) (Rumsey *et al.*, 1974; Goldstein *et al.*, 1976; Campbell *et al.*, 1976) using interferometry techniques described by Rogers and Ingalls (1969). Earlier, lower resolution Venus radar images were obtained at 3.8 and 70 cm wavelengths (Rogers and Ingalls, 1970; Rogers *et al.*, 1974; Campbell *et al.*, 1970). In the present study we address the 12.5 cm wavelength data of Rumsey *et al.* (1974) (Fig. 1). This low resolution (80 km) radar image covers an area equivalent to 19% of the surface of Venus. Our objective was to map potential large (>600 km) impact structures and relate their size frequency distribution to those of

Mercury and the moon. In this paper we regard a "basin" as any large simple or multi-ringed impact crater. These preliminary data suggest that the cloud-covered planet may be cratered at least as densely as the moon by very large impact structures.

VENUS RADAR REFLECTIVITY

Regions of high (radar-bright) or low (radar-dark) radar reflectivity on the Venus radar images have been described as possibly resulting from variations in small scale surface roughness, mean slope, intrinsic radar absorptivity, or dielectric constant (Carpenter, 1966; Rogers and Ingalls, 1970; Goldstein and Rumsey, 1970; Jurgens and Dyce, 1970).

Although increases in local roughness are apparently the major cause of anomalously radar-bright features such as Alpha, and Beta (Fig. 1), higher intrinsic reflectivity must be contributing to the returned signal strength (Rogers and Ingalls, 1970). Jurgens and Dyce (1970) reported that the existence of large enhancements in the depolarized portion of the returned power for some of the radar-bright regions demonstrates that these features are caused primarily by a greater abundance of small scale surface debris and larger than average slopes.

The overall specular nature of the scattering properties for the surface of Venus suggested to Jurgens and Dyce (1970) that the average surface of the planet does not contain an appreciable number of either large vertical slopes or boulder-sized debris. The recent Soviet Venera 9 and 10 surface panoramas (Avduyevskiy *et al.*, 1976), on the other hand, did reveal large numbers of sharp, angular blocks and eroded, rocky outcrops. However, it is not known how these two landing sites relate to the average surface of Venus.

The presence of nearly circular areas of low radar reflectivity on Venus were first reported by Rogers and Ingalls (1969, 1970), Goldstein and Rumsey (1970), and Campbell *et al.* (1970) at 3.8, 12.5, and 70 cm wavelengths. The large 2000 km diameter radar dark region centered near 30°S–25°W (International Astronomical Union coordinates) was the first such feature identified on all three radar maps. Rogers and Ingalls (1969) and Campbell *et al.* (1970) suggested that the large, circular, radar-dark features may be possible analogues to the lunar maria. Goldstein and Rumsey (1970) suggested that the lack of echoes from these regions could be attributed either to intrinsic radar absorptivity of the material or to unusual smoothness of the surface. The studies of Pollack and Whitehill (1972) and Schaber *et al.* (1975) support the importance of high intrinsic radar absorptivity of mare surface materials in producing weak returns by reflected radar signals from the moon. The question as to whether the weak returns from Venus are due to such intrinsic absorptivity (mare-like materials) and/or unusual smoothness or some other factor may be answered in the future by detailed observations of the depolarized radar signal (Campbell *et al.*, 1970).

In addition to the low radar reflectivity of the areas which we interpret to be basin floors on Venus, there appears to be closely associated a highly reflective annulus that may represent rim mountains and ejecta deposits (Campbell *et al.*,

1976). The association of poorly reflecting basin floor and highly reflecting basin rim deposits was also noted during preliminary studies of lunar radar maps (Thompson and Dyce, 1966). We have adopted this interpretation for the present analysis.

Venus Craters and Basins

The recognition of distinct crater morphology on the high resolution 12.5 cm wavelength radar maps of Rumsey *et al.* (1974) and Goldstein *et al.* (1976) has spurred interest in Venus as another terrestrial planet on which evidence of the early solar system flux history may be preserved (Saunders and Malin, 1977; Malin and Saunders, 1977; Science News, 1976; Campbell *et al.*, 1976) (Fig. 2).

Earth-based radar elevation data (Rumsey *et al.*, 1974; Goldstein *et al.*, 1976) have shown that craters on Venus are extremely shallow with respect to their diameters, suggesting either an erosion and infilling mechanism or efficient isostatic compensation. Saunders and Malin (1977) have noted a deficiency of craters with diameters smaller than about 80 km on the high resolution radar maps covering portions of the Venus equatorial region.

The thirty or so craters (40–350 km diameter) recognized thus far, on the high resolution Venus radar maps have been characterized by Saunders and Malin (1977) as having a size-frequency distribution similar to that of the most heavily cratered terrains on Mars and the moon. The present Earth-based radar resolution is not yet sufficient to clearly distinguish between craters of volcanic and impact origin (Malin and Saunders, 1977). However, the preliminary assessment of a lunar and martian crater size-frequency distribution on Venus strongly supports an impact origin for the great majority of the observed venusian features. It follows then, by analogy with the lunar, martian, and mercurian crater data, that the surface of Venus should also contain the scars of numerous large basins of the size of Imbrium (Moon), Caloris (Mercury) and Hellas (Mars).

The best evidence for major impact basins on Venus was reported by Campbell *et al.* (1976). They discovered an elongate basin-like feature (65°N, 65°W IAU coordinates) 1000×1500 km in size with a well defined, radar-bright rim and a radar-dark floor. The rim material reflects from two to three times as much radar power as the surrounding terrain and up to ten times as much as the "basin" floor. Campbell *et al.* (1976) noted that the proposed annulus of basin deposits extends from 400 to 1000 km away from the rim crest. Such an ejecta width is consistent with similar deposits mapped around major lunar basins such as Mare Imbrium and Mare Orientale.

Results of Present Study

In the present study the 80 km resolution Venus radar map (Fig. 1) was analyzed using a color television film density slicer system for enhancement of subtle changes and gross patterns of contrasting radar reflectivity. Analysis by this technique permitted the recognition of 12 possible basins (>600 km diameter) on about 8% (3.5×10^7 km^2) of the total surface area of the planet (Fig.

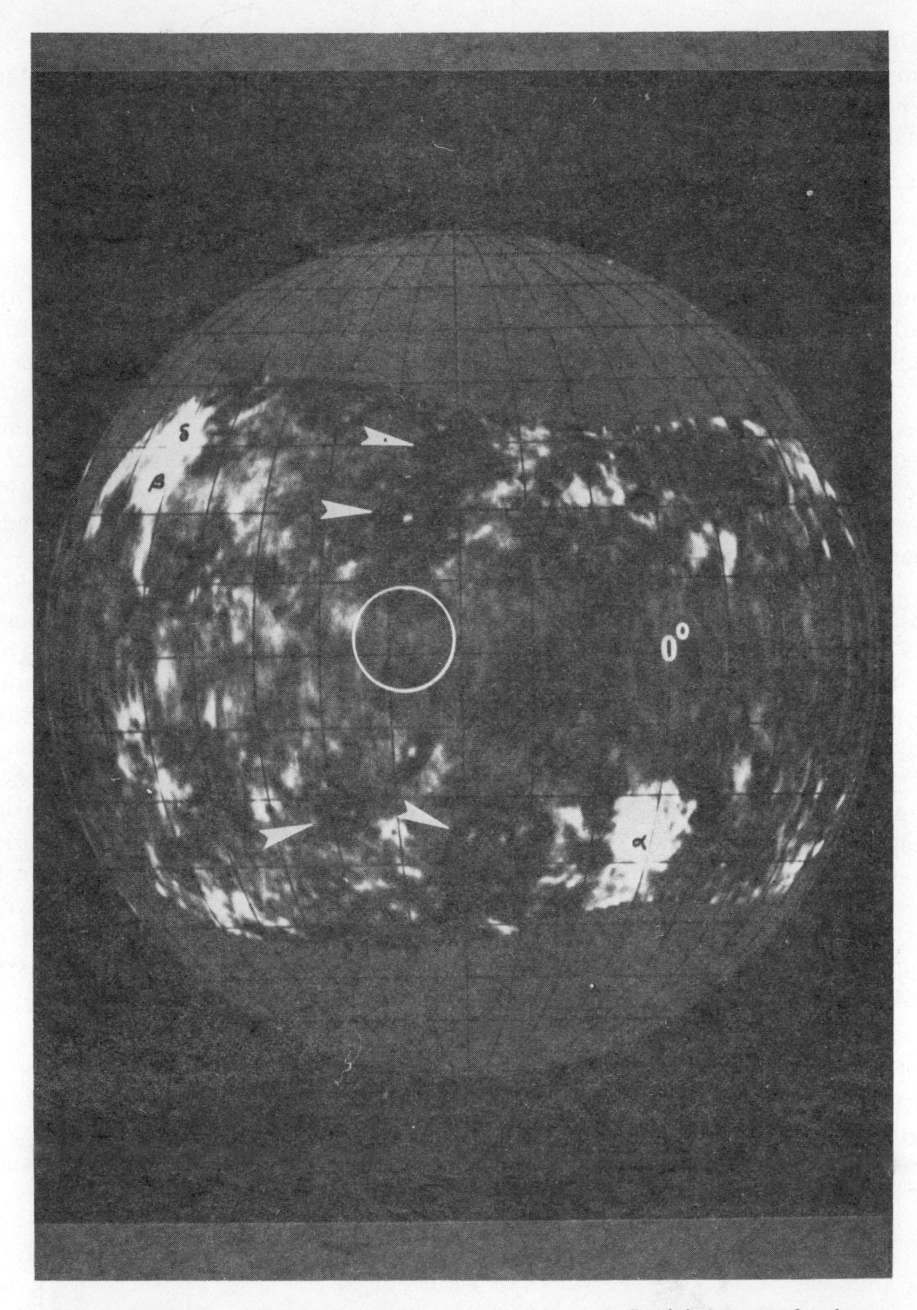

Fig. 1. Overview 12.5 cm wavelength radar brightness (reflectivity) map of a large portion of Venus. Resolution is about 80 km. The coordinate grid is spaced 10 degrees using the convention adopted by the International Astronomical Union. The circled area is shown in high resolution in Fig. 2. Arrows indicate radar dark floors of four most probable basin features. Map is a compilation of time-Doppler data taken during the 1969, 1970 and 1972 conjunctions; after Rumsey *et al.* (1974).

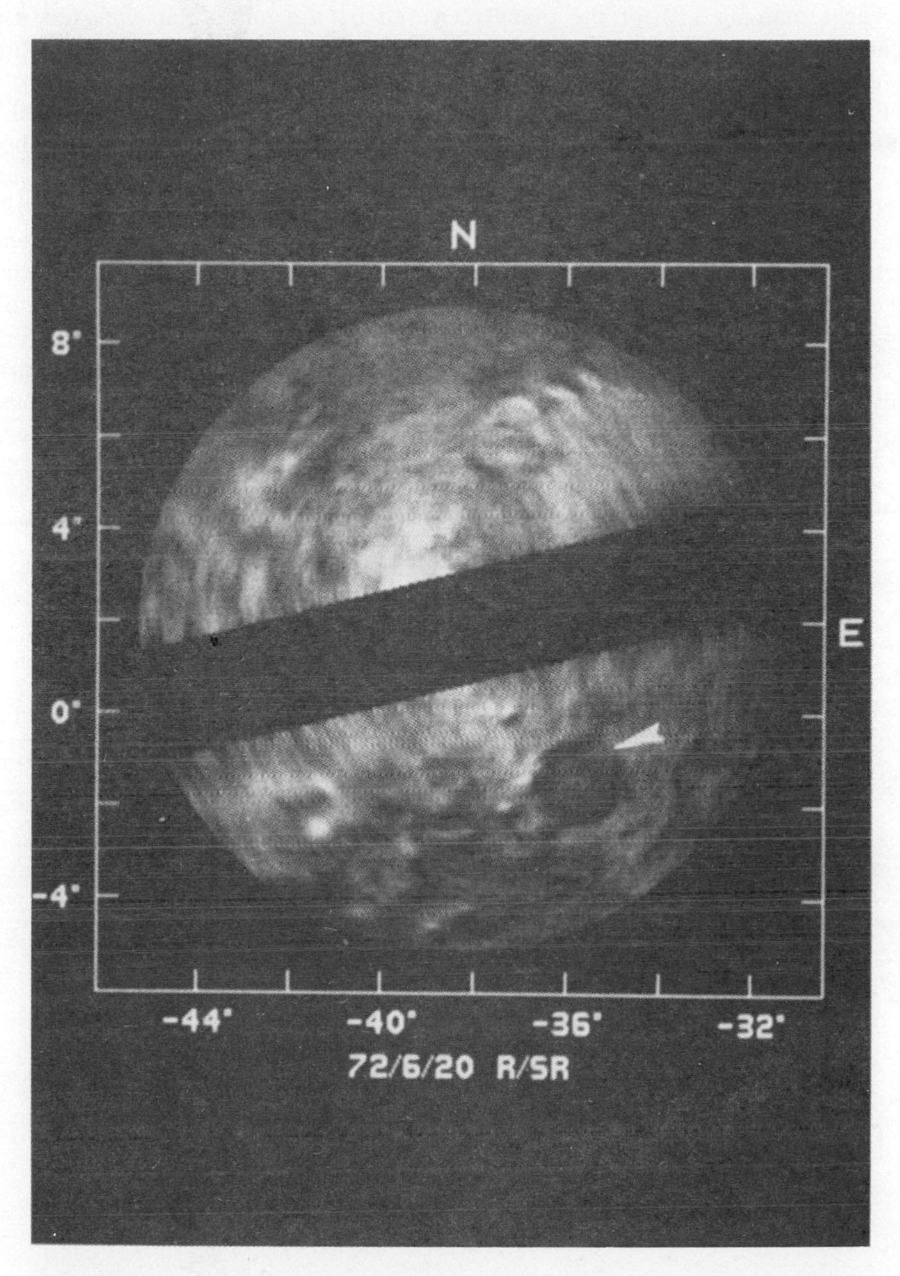

Fig. 2. High resolution (~20 km) radar brightness map of a portion of Venus showing a heavily cratered surface. The largest crater seen (arrow) is about 160 km diameter; the smallest craters that can be seen are about 40 km. The map was produced by an interferometric technique; after Rumsey *et al.* (1974).

2). The remaining 11% of the planet covered by the radar map was not used because of distortion by the north-south ambiguity smear at the equator (Rumsey *et al.*, 1974) and the effects of limb curvature.

Four circular, weakly reflecting features ranging in diameter from 1000 to 1700 km are considered very good prospects for major impact features. These structures have center coordinates at 29°S–26°W; 20°S–50°W; 19°N–37°W; and 28°N–31°W (Fig. 3). Eight other crudely circular features of low reflectivity, ranging in size from 600 to 1130 km, are identified, but with a lower confidence level. All such mapped features are characterized by weakly reflecting interiors and some by rings of above average radar reflectivity extending at least one basin radius from the rim.

A preliminary size-frequency distribution of the Venus basins (>600 km) is shown in Fig. 4a and is compared to the statistically more reliable basin frequency data for the moon and Mercury (Fig. 4b, 4c). The data show an

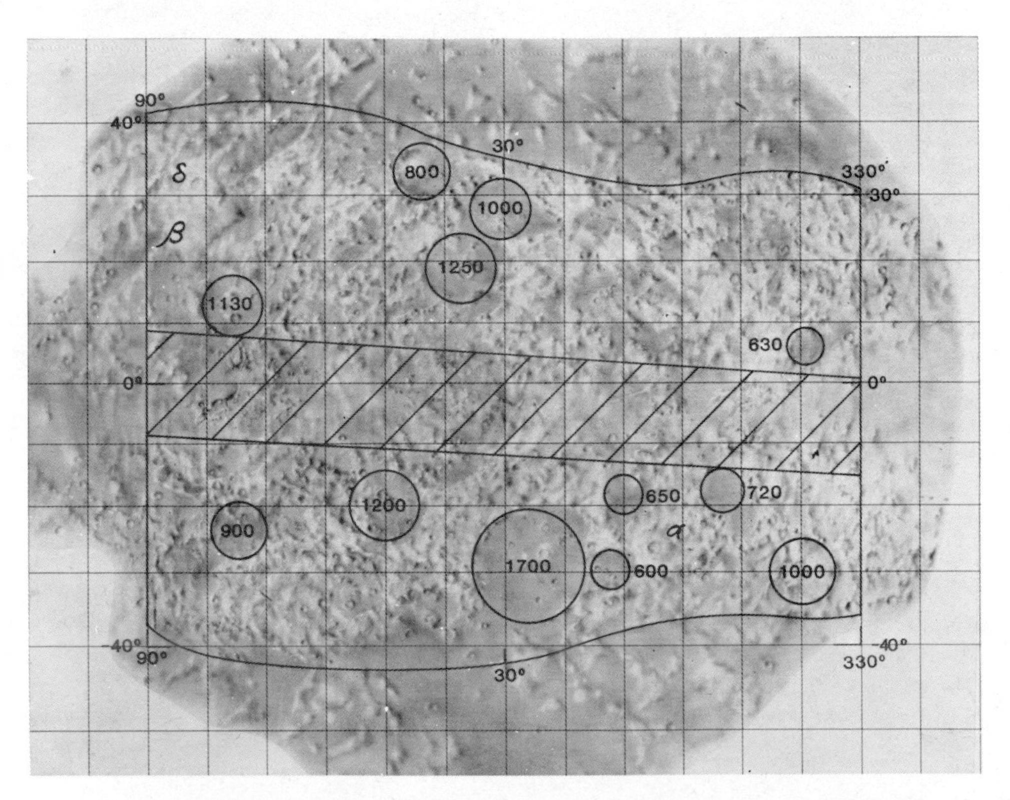

Fig. 3. A portion of Venus showing the location of twelve possible impact basin structures over 600 km in diameter. Locations for basins derived during analysis of radar map shown in Fig. 1. Basin mapping restricted to area outlined by solid line. Cross-hatched area not analyzed due to image streaking, an artifact of image aquisition. Base map from a preliminary mercator map of Venus interpreted from the Fig. 1 radar map by airbrush cartographers of the U.S. Geological Survey (Flagstaff, Arizona).

increasing number/unit area of large basins for Mercury, Moon, and Venus, in that order. Normalizing the total number of lunar basins/unit area (>600 km diameter) to 1.0, the ratio of Venus basins to lunar basins is 1.2 and Mercury basins is 0.4.

Relative Impact Probabilities on Venus

Wetherill (1975), Hartmann (1976), and Shoemaker and Helin (1976) have calculated the relative impact probabilities and the relative crater production rates on the terrestrial planets. Wetherill (1975) has shown that the relative impact probability/unit area (normalized to the Moon = 1.0) for Venus can vary from 0.94 to 3.02 times the lunar rate depending on the orbits of any single family of modern comets or asteroids.

Hartmann (1976), extending Wetherill's analysis to account for velocity and gravity differences, compared crater production rates for terrestrial planets, including Venus. He found that for early planetesimals in nearly circular orbits, Venus can have relative production rates (normalized to the Moon) between 1.9 and 2.7. Hartmann's calculation of relative crater production rates using Wetherill's list of modern comet and asteroid orbits showed Venus to range between 0.6 and 1.8 times the lunar rate. Hartmann (1976) suggests that crater production rates are susceptible to many variables associated with the family of impacting bodies and an individual's interpretation of the published data.

Shoemaker and Helin (1977) have recently calculated the rates of loss of Apollo asteroids by collision with the terrestrial planets. Their model of the space distribution of the Apollos is derived from the orbital elements of 19 known Apollos for which orbits of usable precision are available. Shoemaker and Helin (1977) have calculated the loss of Apollo asteroids to absolute magnitude 18 per 10^9 yr to be as follows: Mercury (~ 100), Venus (~ 1500), Earth (~ 3100), Moon (~ 200), and Mars (~ 300). These data, normalized to the moon, indicate that Mercury sustains about one-half the Apollo impacts per unit areas as does the moon with Venus and the earth collecting the great majority of impacts at a rate 7.5 and 15.5 times the lunar rate, respectively.

Our preliminary data on the Venus basin population relative to the moon, are within the limits (factor of three) suggested by Wetherill (1975) and Hartmann (1976). Additionally, the orbits of bodies similar to those required to explain the orbital distribution of the Prairie Network fireballs and observed chondritic meteorites appear to best fit our relative basin frequencies for Mercury, Venus, and the moon (Wetherill, 1975, Table 1) (Figs. 4a, b, and c). On the other hand, if the data of Shoemaker and Helin (1977) could be extrapolated back to the early solar system flux ratios for the terrestrial planets, then our preliminary counts of large Venus basins may be underestimated by a factor of five or more. This would, of course, be making the additional assumption that the average crustal conditions in the planets were similar.

The persistence of numerous ancient impact craters on the moon, Mars, and Mercury has been used to show that these bodies have had a more rigid and harder (higher viscosity) crust than that of Earth for at least the past 4×10^9 yr

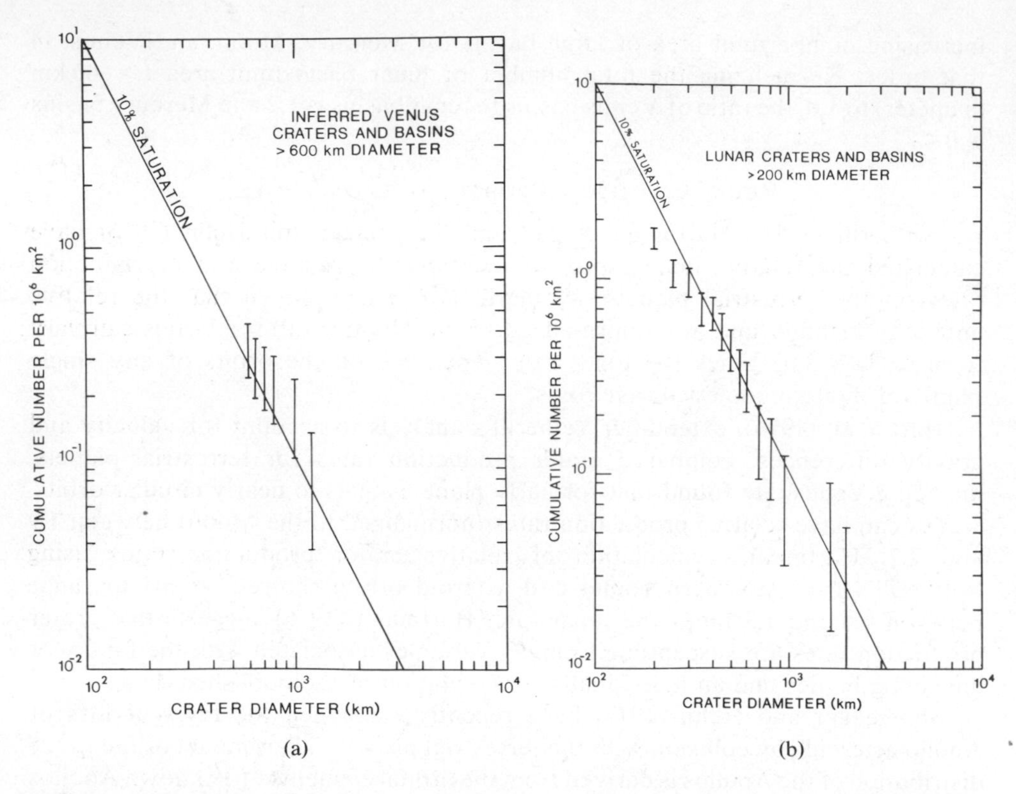

(a) (b)

(Baldwin, 1963, 1971; Schaber *et al.*, 1977). Verification of a crater population on Venus like that of the moon, Mars, and Mercury would have significant implications regarding the similar evolution of crustal viscosity on all four planets. Although the density and size of Venus is nearly identical to that of Earth, any model of thermal evolution of this planet must take into consideration that its crust may have been distinctly non Earth-like for at least the past 4×10^9 yr. The suggestion of Weertman (1970) that the viscosity of the outer surface layer of Venus is remarkably low compared to that of Earth may be incorrect, despite the extremely high surface temperatures.

Conclusions

In the present study the 80 km resolution 12.5 cm wavelength radar map of Rumsey *et al.* (1974) was analyzed for evidence of Venus basin structures, characterized by regions of nearly circular weak radar reflectivity, surrounded by rings of anomalously high radar reflectivity. A total of 12 probable basins were identified ranging in size from 600 to 1700 km on an area equivalent to approximately 8% of the planet. The preliminary basin size frequency distribution determined for Venus from these low resolution data suggests that the cloud-covered planet could be more cratered/unit area by basins >600 km diameter than either the moon (1.2 times) or Mercury (2.7 times).

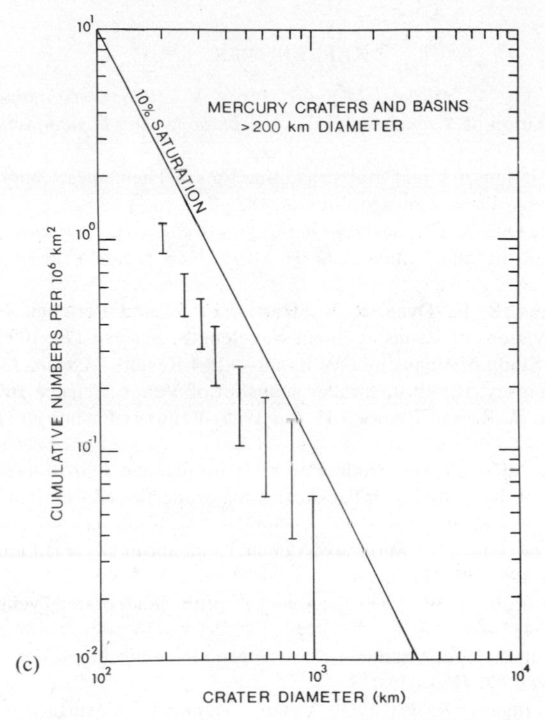

Fig. 4. Log–log large crater size-frequency diagrams for (a) Venus, (b) Moon, and (c) Mercury. Moon and Mercury basin data from work of Schaber *et al.* (1977). Venus data represent 12 craters >600 km diameter counted on 8% (3.5×10^7 km^2) of planet. Lunar data from 62 craters (>200 km) mapped on 100% (3.8×10^7 km^2) of surface. Mercury data from 35 craters (>200 km) mapped on 45% (3.38×10^7 km^2) of surface.

Our estimate of the large basin distribution on Venus is in good agreement with the distribution of 40–300 km size craters as determined by Saunders and Malin (1977) from analysis of the high resolution Venus radar maps. Our result is also in general agreement with probabilities of impact and relative crater production rates reported for the terrestrial planets using total integrated flux of all known asteroids and comets (Wetherill, 1975; Hartmann, 1976).

The fact that the radar data indicate a cratering history on Venus that is very similar to that of the moon, Mars, and Mercury is of significant importance. The existence of numerous craters on Venus also implies that the surface of the planet, like that of the moon, Mars, and Mercury, has not been significantly disturbed by plate-tectonics as it operates on Earth. Additionally, preservation of abundant large craters on the Venus surface suggests that atmosphere-related erosion has been orders of magnitude slower than on Earth, or that the dense atmosphere is a comparatively recent feature of the planet.

Acknowledgments—The paper presents the results of one phase of research carried out by the U.S. Geological Survey, under NASA Contract W, 13-576, sponsored by the Planetology Office, Office of Space Sciences, National Aeronautics and Space Administration. The authors express their thanks to M. C. Malin and C. A. Wood for critical review of the manuscript.

REFERENCES

Avduyevskiy, V., Ishevskiy, V., Markov, M., and Moroz, V.: 1976, Outstanding Success of Soviet Cosmonautics; Translation of *Vydayushchiysya uspekh sovetskoy kosmonavtiki*, Moscow, Pravda No. 52 (21021), p. 3–4.

Baldwin, R. B.: 1963, Variations in Lunar Craters as Functions of Their Ages, Chapt. 9 in *The Measure of the Moon*, Univ. Chicago Press, Chicago, Illinois, 488 pp.

Baldwin, R. B.: 1971, The Question of Isostasy On the Moon, *Phys. Earth Planet. Interiors* **4**, 167–179.

Campbell, D. B., Dyce, R. B., and Pettengill, G. H.: 1976, New radar image of Venus, *Science* **193**, 1123–1124.

Campbell, D. B., Jurgens, R. F., Dyce, R. B., Harris, F. S., and Pettengill, G. H.: 1970, Radar interferometric observations of Venus at 70-cm wavelength, *Science* **170**, 1090–1092.

Carpenter, R. L.: 1966, Study of Venus by CW Radar—1964 Results, *Astron. J.* **71**, 142–152.

Goldstein, R. M. and Rumsey, H.: 1970, A radar snapshot of Venus, *Science* **169**, 974–977.

Goldstein, R. M., Green, R. R., and Rumsey, H. G.: 1976, Venus radar images. *J. Geophys. Res.* **81**, 4807–4817.

Hartmann, W. K.: 1976, Relative crater production rates on planets. *Icarus*, in press.

Jurgens, R. R. and Dyce, R. B.: 1970, Radar backscattering properties of Venus at 70 cm, *Astron. J.* **75**, 297–314.

Malin, M. C. and Saunders, R. S.: 1977, Surface of Venus: Evidence of Diverse Land-Forms from Radar Observations. *Science* **196**, 987–990.

Pettengill, G. H. and Thompson, T. W.: 1968, A radar study of the lunar crater Tycho at 3.8 cm and 70 cm wavelength, *Icarus* **8**, 457–471.

Pollack, J.B. and Whitehill, L.: 1972, A multiple-scattering model of the diffuse component of lunar radar echoes, *J. Geophys. Res.* **77**, 4289–4303.

Rogers, A. E. E. and Ingalls, R. P.: 1969, Venus: Mapping the surface reflectivity by radar interferometry, *Science* **165**, 797–799.

Rogers, A. E. E. and Ingalls, R. P.: 1970, Radar mapping of Venus with interferometric resolution of the range—Doppler ambiguity, *Radio Science* **5**, 425–433.

Rogers, A. E. E., Ingalls, R. P., and Pettengill, G. H.: 1974, Radar map of Venus at 3.8 cm wavelength, *Icarus* **21**, 237–241.

Rumsey, H. C., Morris, G. A., Green, R. B., and Goldstein, R. M.: 1974, A radar brightness and altitude image of a portion of Venus, *Icarus* **23**, 1–7.

Saunders, R. S. and Malin, M. C.: 1977, Venus: Geological Analysis of Radar Images, *Geologicia Romano*, in press.

Schaber, G. G. and Boyce, J. M.: 1976, Moon-Mercury: Basins, secondary craters and early flux history (abstract), in *Papers Presented to the Conference on Comparisons of Mercury and the Moon*, The Lunar Science Institute, Houston.

Schaber, G. G., Boyce, M. J., and Trask, N. J.: 1977, Moon-Mercury: Large Impact Structures, Isostasy and Average Crustal Viscosity, *Phys. Earth Planet. Interiors* **15**, in press.

Schaber, G. G., Thompson, T. W., and Zisk, S. H.: 1975, Lava flows in Mare Imbrium: An evaluation of anomalously low earth-based radar reflectivity, *The Moon* **13**, 395–423.

Shoemaker, E. M. and Helin, E. F.: 1977, Populations of planet-crossing asteroids and the relationship of Apollo objects to main-belt asteroids and comets, *Int. Ast. Union Trans., Colloquim* **39**, in press.

Thompson, T. W. and Dyce, R. B.: 1966, Mapping of lunar radar reflectivity at 70 centimetres, *J. Geophys. Res.* **71**, 4843–4853.

Venus: Hints of a dynamic planet, (editorial), *Science News* **109**, 228–229.

Weertman, J.: 1970, The Creep Strength of the Earth's Mantle, *Rev. Geophys. Space Phys.* **8**, 145–168.

Wetherill, G. W.: 1975, Late heavy bombardment of the Moon and terrestrial planets, *Proc. Lunar Sci. Conf. 6th*, p. 1539–1561.

Roddy, D. J., Pepin, R. O., and Merrill, R. B., editors.
(1977) *Impact and Explosion Cratering*, Pergamon Press (New York), p. 613–615.
Printed in the United States of America

The nature of the present interplanetary crater-forming projectiles

G. W. WETHERILL

Department of Terrestrial Magnetism, Carnegie Institution of Washington, Washington D.C. 20015

THE INTERPLANETARY OBJECTS available for laboratory analysis are meteorites ranging in mass up to $\sim 10^{11}$ g, but only rarely $>10^6$ g. The composition of these objects varies, and includes soft friable carbonaceous objects ($\rho \sim 2.2$ g/cm^3), both relatively undifferentiated stones (ordinary chondrites) and stony bodies more similar to differentiated igneous rocks (achondrites) with $\rho \sim 3.5$ g/cm^3, as well as metallic iron–nickel objects ($\rho \sim 8$ g/cm^3). Only the largest of these bodies, i.e., those $\gtrsim 10^8$ g will produce terrestrial craters, as smaller bodies lose their interplanetary kinetic energy while traversing the atmosphere. The relative abundance of these meteorite types is quite uncertain because variation in the probability of surviving atmospheric passage is a function not only of their size, but also of their strength and entry velocity.

Somewhat indirect evidence of this relative abundance is given by the photographic Prairie Network (Ceplecha and McCrosky, 1976) which provides data on meteoroids in the range 100–10^6 g even if they do not survive ablation in the atmosphere. These data show that approximately one-third of these meteoroids have similar ablation characteristics to those of two photographically observed falls of ordinary chondrites, Pribram and Lost City. The remainder are weaker and less dense objects, many of which may be similar to carbonaceous chondrites. No iron objects were observed by this network indicating that they are relatively rare, at least in this size range. This is also suggested by the fact that only ~5% of the recovered meteorite falls are irons, in spite of the fact that their greater mechanical strength and more unusual appearance should facilitate their recovery.

Terrestrial craters are produced by larger bodies, up to $\sim 10^{18}$ g in mass. Large bodies in earth-crossing orbits similar to those of the meteoroids will be subject to collisional fragmentation while traversing the part of their orbit which lies in the asteroid belt. These collision fragments can impact the earth as meteoroids, thereby providing a link between telescopically observed bodies and at least a portion of the population of photographed and analyzed meteoroids and meteorites. This link is strengthened by spectrophotometric data on earth-crossing bodies ("Apollo objects" in the ~2 km diameter (10^{16} g) range (Chapman, 1976)) which indicates similarities between the composition of these bodies and chondritic meteorites. Estimates of the present number of these Apollo objects are similar to the number required to explain the surface density of terrestrial craters and post-3.3 b.y. lunar craters (Shoemaker *et al.*, 1975;

Wetherill, 1976). Putting all these data together leads to the conclusion that many, perhaps almost all, of the observed terrestrial impact craters are produced by silicate objects similar in composition to chondritic meteorites of various types. The Apollo objects may themselves be derived from the asteroid belt or may be residual, volatile-poor cores of short-period comets (Öpik, 1963; Wetherill, 1976).

In addition to Apollo objects, impacts between the earth and ordinary active comets will also occur. The relative importance of these impacts is very difficult to estimate because of lack of knowledge regarding the mass of comets. The nuclei of relatively inactive short-period comets have been observed (Roemer, 1966, 1972) and subject to uncertainties imposed by assumption of the albedo, their mass is similar to those of the Apollo objects. Taking into account their relative numbers, these data can be used to plausibly argue that the number of terrestrial craters produced by these short-period comets is small compared to those produced by Apollo objects. The same cannot be said of the very active long-period and near-parabolic comets. The nucleus of these comets is obscured by the coma, and it is unlikely that the magnitude of the nucleus has been reliably determined for any of these bodies. Furthermore, the high velocities (up to 72 km/sec) associated with the commonly retrograde orbits of these bodies will produce larger craters for the same impacting mass of lower velocity (15–40 km/sec) directly revolving Apollo objects and short-period comets. Considering these uncertainties, it is possible that as many as one-half the terrestrial impact craters could be produced by retrograde active comets. It would be quite valuable if theoretical or experimental criteria could be developed for distinguishing the effects of these very high velocity impacts from those produced by low velocity impacts of the same kinetic energy.

In any case almost all the impacting material is likely to be of low density, ranging from that of silicate rocks ($\sim 3\ g/cm^3$) to cometary material (probably $\sim 1\ g/cm^3$), rather than strong high density iron objects. This could raise a problem insofar as only iron meteorites are definitely known to be associated with terrestrial impact craters. Possible explanations of this apparent discrepancy are:

(1) Iron objects predominate in the size range ($\sim$50 m) responsible for the production of $\sim$1 km craters such as Meteor Crater, Arizona, associated with iron meteorites. This size range falls in between the $\leqslant$1 m objects observed as the Prairie Network meteoroids and those ($\geqslant$500 m) observed as Apollo objects.

(2) Craters are preferentially formed by the strong, dense iron bodies, resulting in the number of craters being unrepresentative of the actual flux of bodies of this composition.

(3) Silicate projectiles are completely destroyed by their impact and consequently are not found in the vicinity of terrestrial craters.

All three of these explanations can qualitatively explain the discrepancy, but quantitative resolution of this problem will require a better understanding of the nature of the incoming bodies as well as of cratering mechanics.

REFERENCES

Ceplecha, Z. and McCrosky, R. E.: 1976, Fireball end heights: A diagnostic for the structure of meteoric matter. Center for Astrophysics, *Preprint Series*, No. 442.

Chapman, C. R.: 1976, Asteroids as meteorite parent-bodies: The astronomical perspective. *Geochim. Cosmochim. Acta* **40**, 701–719.

Öpik, E. J.: 1963, Survival of comet nuclei and the asteroids. *Advan. Astron. Astrophys.* **2**, 219–262.

Roemer, E.: 1966, The diameter of cometary nuclei. In *Nature et Origine des Cometes, Mem. Soc. Roy. Sci. Liege, Ser. 15, 12,* 23–28.

Roemer, E.: 1972, Periodic comet Encke (1970–1). Centr. Bur. Astron. Telegrams; I. A. U., (B. Marsden, ed.) Circular 2435.

Shoemaker, E. M., Helin, E. F., and Gillett, S. L.: 1975, Populations of the planet-crossing asteroids. Abstract of paper presented at the International Colloquim of Planetary Geology in Rome, Italy, September 22–30, 1975.

Wetherill, G. W.: 1976, Where do the meteorites come from? A reevaluation of the earth-crossing Apollo objects as sources of stone meteorites. *Geochim. et Cosmochim. Acta* **40**, 1297–1317.

Roddy, D. J., Pepin, R. O., and Merrill, R. B., editors.
(1977) *Impact and Explosion Cratering*, Pergamon Press (New York), p. 617—628.
Printed in the United States of America

Astronomically observable crater-forming projectiles

E. M. SHOEMAKER

U.S. Geological Survey, Flagstaff, Arizona 86001 and California Institute of Technology, Pasadena, California 91125

THE FLUX OF OBJECTS large enough to produce impact craters on earth is so low that millennia would be required to obtain a useful estimate of the flux by direct observation of actual encounters. If we wish to estimate this flux, we have the choice of astronomical search for potential crater-forming projectiles or of examining the geologic record of impact cratering. It turns out that these alternate methods are about equally difficult, and the results from each method, at the present stage of investigation, have roughly comparable uncertainties.

Astronomically observable objects which are in orbits that permit eventual collision with the earth, the moon, or the other terrestrial planets include the planet-crossing asteroids and the comets. Probably the smallest such object found by telescopic observation, and certainly the faintest, is the asteroid PL-6344, discovered with the 48-inch Schmidt Camera of the Hale Observatories on Palomar Mountain in the course of the Palomar-Leiden Survey of faint minor planets (Van Houten *et al.*, 1970). With an absolute blue photographic magnitude estimated at 22.92, the diameter of PL-6344 is close to 100 m, if its Bond albedo is about 0.1, as found for several other earth-crossing asteroids. The encounter velocity with earth of PL-6344 would be 15.8 km/sec, neglecting atmospheric drag.

Assuming this object has a density like that of common chondritic meteorites, it would produce a crater about $1\frac{1}{2}$ times the diameter of Meteor Crater, Arizona. Its kinetic energy on entry into the earth's atmosphere, about 2×10^{24} ergs, would be about four times greater than that of a small nucleus or fragment of the nucleus of a comet which produced the great Tunguska meteoric fireball in 1908 (Krinov, 1966). My estimate of the energy of the Tunguska object is derived from the period of the fundamental mode of the atmospheric gravity wave recorded at stations in England. There is almost an overlap in impact energy between the smallest object observed at the telescope and the largest object for which there is a direct observational record of its encounter with the earth.

Unfortunately, as the precision of the orbits obtained from the Palomar-Leiden survey is insufficient to recover a fast-moving object beyond a few years after the original plates were exposed, PL-6344 is lost. The smallest earth-crossing asteroid for which an orbit adequate for recovery has been obtained is 1976 UA, discovered in the month following that of the Symposium on Planetary Cratering Mechanics (Helin and Shoemaker, 1977). This object is about 200 m in diameter, again assuming a Bond albedo of 0.1 or a visual geometric albedo of

0.2; it missed the earth by a little more than a million km on its discovery apparition. A few well-observed periodic comets may have solid nuclei of about the same size as 1976 UA.

Objects which are classified as asteroids by the usual telescopic criteria dominate the flux of crater-forming projectiles on earth. It should be noted, however, that the distinction between an asteroid and a comet is based on the absence or presence of an observable coma. An object is called an asteroid if it has been stellar in appearance at all times that it has been observed. As some comets exhibit comas only sporadically, the distinction clearly is arbitrary; it depends upon completeness of observation. Generally, asteroids and comets have distinctly different types of orbits, but some planet-crossing asteroids have orbits nearly like the orbits of a few short-period comets. The present weight of evidence indicates that many and probably most earth-crossing asteroids are the degassed nuclei of former short-period comets (Öpik, 1963; Wetherill, 1976; Shoemaker and Helin, 1977).

Our knowledge of planet-crossing asteroids, particularly of the earth-crossing objects, is extremely fragmentary. At the time of writing, only 20 earth-crossers had been observed sufficiently well for determination of orbits useful for statistical studies. About half of these objects were found accidentally, in the course of astronomical observations directed toward quite different objectives. The difficulty of finding them lies in the fact that nearly all earth-crossers are so small that they are observable by telescopes of modest aperture only when they are relatively close to the earth. Large aperture, wide-field instruments, such as the 48-inch Schmidt camera, could be used to find most of the Apollo objects to absolute visual magnitude 18 (about 0.7 km diameter) or brighter, but the few existing instruments of this class are dedicated primarily to other research and have not been available for the massive effort that would be required.

From the discovery of seven earth-crossing asteroids found in the course of three different systematic surveys of the sky, it is possible to make a rough estimate of the total population. Besides information on the amount of sky photographed and the statistical distribution of the direction of this coverage with respect to the position of the sun, such an estimate requires a model of the characteristic photometric function or distribution of such functions for the earth-crossing asteroids, a model of their average distribution in space, and a model of their magnitude-frequency distribution. Data needed for these models can be obtained from photometric observations of a limited number of the known earth-crossers, from the orbital elements of the 20 known objects, and from the magnitude-frequency distribution of main-belt asteroids and the crater size distributions on relatively young surfaces on the moon, Mercury, and Mars. In addition, it is necessary to establish the threshold for detection of fast-moving objects in each survey. When all this information is combined, the best estimate of the number of earth-crossing asteroids to absolute visual magnitude 18 is $(0.8 \pm 0.3)10^3$. The uncertainty indicated here is the standard deviation of the mean for a very small sample ($n = 7$). When all the model dependent uncertainties are included, the precision of the estimate is about $\pm 50\%$, i.e., the population is estimated to be

800 ± 400. In addition, there may be a systematic error in estimating the magnitude threshold at which discovery may be considered complete.

Wetherill (1976) has obtained a somewhat similar estimate of the population of earth-crossing asteroids by means of a very different but not totally independent statistical argument. From the fact that 16 earth-crossers brighter than absolute blue magnitude 18 have been discovered and that none have been rediscovered accidentally, Wetherill finds that the most probable number of earth-crossers to absolute blue magnitude 18 is about 700. This estimate employs an adopted upper bound of 1200 for the population of earth-crossers to absolute blue magnitude 18, together with a lower bound of about 200 derived from the fact that no earth-crossers have been accidentally rediscovered. Given an average difference between blue and visual magnitudes for asteroids of about 0.8 and an exponential form for the magnitude-frequency distribution,

$$N_v = Ke^{bv}, \tag{1}$$

where N_v is the cumulative number of objects to absolute magnitude v, v is the absolute visual magnitude, and where b is close to 1, a lower bound of about 400 is found for earth-crossers to absolute visual magnitude 18. If an upper bound of 1200 is retained for the population of earth-crossers to absolute visual magnitude 18, the most probable number is still close to 700.

At most, a few percent of the earth-crossing objects to visual magnitude 18 have been discovered so far. On the basis of Eq. (1) and b close to 1, there are about 10^5 asteroidal objects the size of PL-6344 (about 100 m) and larger in earth-crossing orbits, of which just a few have been recognized.

The flux of earth-crossing objects in the vicinity of the earth can be found from the estimated population by utilizing information about the orbits of the known objects. Assuming that the orbits of the known objects are representative of the entire population, the present rate of impact of objects to a specified absolute magnitude, I_v, is given by

$$I_v = \frac{N_v}{4R^2\pi} \left(\frac{\sum_{j=1}^{j=n} P_j}{n} \right) \tag{2}$$

where R is the radius of planet, P_j is the probability of impact for known individual planet-crossing object, and n is the number of known planet-crossing objects. With the equations of Öpik (1951 and 1956), the probability of impact for each known object, P_j, may be computed from the semi-major axis, eccentricity, and inclination of its orbit (Table 1). It is explicitly assumed in the derivation of Öpik's equations that, over a sufficiently long period of time, the argument of perihelion of the planet-crossing object is randomly distributed, as a result of secular perturbations.

The impact rate on earth calculated from Eq. (2), using $N_{18} = 800 \pm 400$ and the mean probability of impact for 20 earth-crossing asteroids (Table 1) is $(0.7 \pm 0.35)10^{-14}\ \text{km}^{-2}\ \text{yr}^{-1}$. Impact velocities of these objects, which may be calculated with the aid of Öpik's equations, range from 15.0 to 40.3 km/sec (Table 1). The rms impact velocity on earth is 24.6 km/sec.

If reasonable assumptions are made about the albedo and density of the earth-crossing asteroids, the rate of production of impact craters on earth can be calculated from the estimated population of earth-crossers and the data provided in Table 1. The albedos of four earth-crossers have been investigated by polarimetry or thermal infrared radiometry (Dunlap, 1974; Dunlap *et al.*, 1973; Gehrels *et al.*, 1970; Morrison *et al.*, 1976). Visual geometric albedos of all four objects are close to 0.2. Spectrophotometric studies of these four objects show that they are roughly similar in spectral reflectance to common chondritic meteorites (Chapman *et al.*, 1973; Dunlap, 1974; Gehrels *et al.*, 1970; Gradie, 1976). Indeed, a fairly strong case can be made that the earth-crossing asteroids are sources of at least some of the chondrites (Wetherill, 1976). Hence, it is plausible to assign densities to the earth-crossers that are comparable to chondrite densities. Adopting 0.2 as the mean visual geometric albedo for all earth-crossing asteroids, a mean density of $3.3\,gm/cm^3$, and a previously used scaling relationship between crater diameter and kinetic energy of the projectile (Table 2), the current rate of production of impact craters on earth is found to be 3 ± 1.5 craters 10 km in diameter or larger per million years.

Reduced to unit area, the production rate on earth for craters 10 km in diameter and larger is $(0.7 \pm 0.35)10^{-14} km^{-2} yr^{-1}$. This estimate probably is conservative, as a fairly high mean albedo has been adopted, which leads, in turn, to a small estimate of mean area and volume for an asteroid of a given magnitude. The asteroid (1580) Betulia, which is earth-crossing during most of its cycle of secular perturbation (Wetherill and Williams, 1968), has recently been found to be somewhat darker than the other earth-crossers studied to date (L. A. Lebofsky, G. J. Veeder, B. H. Zellner, and M. J. Lebofsky, pers. comm., 1976). Its visual geometric albedo, inferred from polarimetry, is about 0.1. If half of the earth-crossers should turn out to be darker objects like Betulia, the estimated production rate for 10 km and larger craters would rise to about $(0.8 \pm 0.4)10^{-14} km^{-2} yr^{-1}$.

For the land areas of earth the rate of production of craters 10 km across and larger, as estimated from the Apollo data, is about one per million years. It is of interest, then, that at least one 10 km impact crater approximately one million years old, the Bosumtwi crater of Ghana, has been recognized (Gentner *et al.*, 1964; Fleischer *et al.*, 1965). A statistically more meaningful comparison with the geological record of impact can be made by counting ancient impact craters and eroded impact structures on the shield or stable areas of the continents. In the Mississippi lowland of east central United States, a region underlain principally by nearly flat-lying Paleozoic sedimentary rocks, about a dozen so-called cryptoexplosion structures have been found. On the basis of personal examination of these structures and detailed studies by a number of investigators, I now interpret four of the cryptoexplosion structures as deformed rocks that once underlaid impact craters 10 km in diameter or larger. The average age of the beds exposed in a $700{,}000\,km^2$ area in which most of the cryptoexplosion structures are found is estimated at 235 m. y. (Shoemaker *et al.*, 1963); the range of age is from late Cambrian to the present, or roughly, the last 500 m. y. The average production

Table 1. Impact probabilities and velocities of impact of Earth-crossing asteroids with the terrestrial planets and the moon[1].

Asteroid	Impact on Mercury		Impact on Venus		Impact on Earth		Impact on Moon		Impact on Mars	
	Impact Velocity (km/sec)	Probability of impact (10^{-9}/yr)	Impact Velocity (km/sec)	Probability of impact (10^{-9}/yr)	Impact Velocity (km/sec)	Probability of impact (10^{-9}/yr)	Impact Velocity (km/sec)	Probability of impact (10^{-9}/yr)	Impact Velocity (km/sec)	Probability of impact (10^{-9}/yr)
1566 Icarus	46.8	1.37	37.7	2.72	31.8	1.75	29.9	.11	21.1	0.24
1974 MA	42.5	.26	36.3	1.01	32.7	0.57	30.9	.04	25.0	0.06
1936 CA Adonis	22.2	.57	27.8	7.08	27.3	10.99	25.1	.68	21.6	0.76
1976 UA	13.7[2]	1.33[2]	19.5	15.78	23.2	14.84	12.9	.64	—	0
1864 Daedalus	—	0	25.3	2.34	24.8	1.26	22.2	.08	18.3	0.14
1965 Cerberus	—	0	20.6	5.28	20.0	3.12	16.7	.16	11.0	0.39
1937 UB Hermes	—	0	19.6	6.05	21.9	3.47	19.0	.19	17.2	0.34
1973 EA	—	0	32.4	1.53	30.1	0.64	28.1	.04	23.3	0.07
1862 Apollo	—	0	17.5	8.33	20.3	4.29	17.1	.23	15.1	0.41
1685 Toro	—	0	—	0	17.2	4.18	13.3	.18	12.1	0.36
1976 AA	—	0	—	0	15.7	7.98	11.3	.30	—	0
PL-6743	—	0	—	0	17.0	4.20	13.1	.18	13.8	0.31
1620 Geographos	—	0	—	0	16.1	4.46	11.8	.18	9.8	0.41
1959 LM	—	0	—	0	15.1	14.90	10.4	.52	10.2	0.99
1950 DA	—	0	—	0	17.7	2.50	13.9	.11	14.6	0.19
1866 Sisyphus	—	0	—	0	27.8	0.95	25.5	.06	22.0	0.07
1973 NA	—	0	—	0	40.3	0.60	38.8	.04	31.4	0.04
1863 Antinous	—	0	—	0	19.5	1.20	16.2	.06	17.7	0.08
1975 YA	—	0	—	0	35.5	2.46	33.7	.16	25.3	0.25
PL-6344	—	0	—	0	15.8	4.81	11.4	.18	16.3	0.22
rms impact velocity	34.2		27.3		24.6		21.7		19.0	
mean probability of impact		0.18		2.51		4.46		0.21		0.27

[1]Impact probabilities and impact velocities listed here are calculated with the use of the equations of Öpik (1951, 1976). For a listing of the orbital elements required for these calculations, exclusive of the elements of 1976 UA, see Wetherill (1976).

[2]Impact velocity and probability of impact of 1976 UA with Mercury are based on a statistical model of the secular variation of eccentricity and inclination of Mercury derived from the 10 m.y. histories of planetary elements presented by Cohen *et al.* (1973).

rate of impact craters 10 km and larger indicated by these data is $(2.2 \pm 1.1)10^{-14}\,km^2\,yr$.

An independent estimate of the cratering rate on earth has been made by Dence (1972) on the basis of ancient craters and deeply eroded structures interpreted to be of impact origin that are found on the Canadian shield. Assuming that discovery of ancient impact structures is complete at diameters of 20 km and above, Dence finds that the average rate of production of craters 20 km in diameter and larger was $10^{-15}\,km^{-2}\,yr^{-1}$, from eight structures distributed over the entire shield, or about $0.5 \times 10^{-14}\,km^{-2}\,yr^{-1}$, from four structures in the Quebec-Labrador area alone. These estimates are equivalent to production rates for craters 10 km in diameter and larger of $0.3 \times 10^{-14}\,km^{-2}\,yr$ and $1.6 \times 10^{-14}\,km^{-2}\,yr^{-1}$. The rate based on the distribution for the entire shield probably is too low, owing to loss of impact structures by erosion (Dence, 1972). The impact rate based on four structures in the Quebec-Labrador area is comparable to that found for the Mississippi lowland and has a similar statistical uncertainty.

The geological record of impact on North America for the last half billion years indicates a rate of crater production about two to three times higher than that calculated from observations of the earth-crossing asteroids. While this difference is not statistically significant, particularly in view of some of the uncertainties involved in deriving crater production rates, it is nevertheless of interest to see whether the difference might arise from an error in the crater scaling relation adopted for calculation of crater diameters (Eq. (3) in Table 2). Equation (3) is an empirical relationship that has been found from experimental explosion craters formed by ejection of material and by outward and upward displacement of the material in the crater walls. The diameter of probably all craters 10 km across and larger on the earth, however, has been substantially increased by collapse of the crater walls. (See, for example, Melosh, 1977.) Examination of collapsed lunar craters suggests that the final crater diameter may be as much as 35–40% larger than the initial crater formed by excavation (Shoemaker, 1962). A 40% increase in crater diameters would lead to an increase in the estimated production of 10 km diameter craters by a factor of 1.8. Hence, the geological record of cratering may be in even closer agreement with the cratering rate implied by the astronomical data than is suggested by Table 2. A more realistic estimate for the present production on the earth of craters 10 km in diameter and larger is $(1.2 \pm 0.6)10^{-14}\,km^{-2}\,yr^{-1}$.

Comparison with the cratering record on the lunar surface affords an independent check on the cratering rates derived from the estimated earth-crossing asteroid population. This comparison largely avoids the problem of diameter enlargement by collapse of the crater walls, as almost no lunar craters at 10 km diameter show evidence of collapse (Smith and Sanchez, 1973; Howard, 1974). Problems do arise, however, in identifying the ages of extensive mare surfaces which must be examined to obtain good statistics for craters that are 10 km in diameter and larger. Relatively young mare surfaces have crater densities for $D \geqslant 10$ km which range from 15 to 50 per $10^6\,km^2$ (Trask, 1966); older mare surfaces have densities near 100 per $10^6\,km^2$. The rate of cratering clearly declined during the early period of extrusion of mare lava flows (Shoe-

Table 2. Impact rates and rates of crater production by impact of Earth-crossing asteroids with the terrestrial planets and the moon.

	Mercury	Venus	Earth	Moon	Mars
	(10^{-14} km^{-2} yr^{-1})	(10^{-14} km^{-2} yr^{-1})	(10^{-14} km^{-2} yr^{-1})	(10^{-14} km^{-2} yr^{-1})	(10^{-14} km^{-2} yr^{-1})
Rate of impact of Earth-crossing asteroids to absolute visual magnitude 18[(1)]	0.2 ± 0.1	0.4 ± 0.2	0.7 ± 0.35	0.4 ± 0.2	0.15 ± 0.07
Rate of production of craters 10 km in diameter and larger by impact of Earth-crossing asteroids[(2)]	0.4 ± 0.2	0.5 ± 0.25	0.7 ± 0.35	0.6 ± 0.3	0.15 ± 0.07

[(1)]Based on an estimated population of 800 ± 400 for Earth-crossing asteroids to absolute visual magnitude 18 and on the mean probabilities of impact given in Table 1.

[(2)]Based on adopted mean visual geometric albedo for Earth-crossing asteroids of 0.2 and mean density of 3.3 gm/cm^3; includes corrections for effect of difference of surface gravity on diameters of craters produced on each planet. Diameters of craters have been calculated from the following formula determined from terrestrial cratering experiments

$$D_e = (74 \text{ m/kt TNT equivalent}^{1/3.4})\ W^{1/3.4} \quad (3)$$

where D_e is diameter of crater formed on Earth, W is kinetic energy of projectile in kilotons TNT equivalent (Shoemaker *et al.*, 1963); one kt TNT equivalent is 4.185×10^{19} ergs. This equation is valid only for bowl-shaped craters. The diameters given by equation (3) are then scaled for gravitational acceleration at the surface of other planets or the moon by

$$D_p/D_e = (g_e/g_p)^{1/6} \quad (4)$$

where D_p is diameter of crater on other planet or moon, g_e is surface gravity on Earth, and g_p is surface gravity on other planet or moon (Gault and Wedekind, 1977). The cumulative frequency of the craters produced is assumed to be proportional to $D^{-1.7}$, consistent with the observed distribution of post-mare lunar craters larger than 3 km (Shoemaker *et al.*, 1963; Baldwin, 1971).

maker, 1971 and 1972; Soderblom and Boyce, 1972; Boyce *et al.*, 1974; Neukum *et al.*, 1975). A moon-wide study of large post-mare craters reveals an average density at $D \geq 10$ km of about 40 per 10^6 km^2 (Shoemaker *et al.*, 1963). I estimate that the density of craters at $D \geq 10$ km on 3.3 b.y. old surfaces is 20 ± 10 per 10^6 km^2. This gives an average rate of crater production over the last 3.3 b.y. of (0.6 ± 0.3) 10^{-14} km^{-2} yr^{-1}, which is essentially identical with the present cratering rate derived from the earth-crossing asteroids (Table 2).

It should be noted that Neukum *et al.* (1975b) have estimated the density of craters at $D \geq 10$ km on 3.3 b. y. old mare surfaces as only 6 per 10^6 km^2. Their estimate is not obtained by counting craters 10 km in diameter and larger, however. Instead, they count small craters and infer the number of large craters from a "calibration" curve of crater size-frequency distribution. The calibration curve has been obtained by fitting the shapes of crater size distribution curves found for several surfaces of different ages (Neukum *et al.*, 1975a). This procedure predicts substantially fewer craters at $D \geq 10$ km than are actually found on the lunar maria. Part of the discrepancy may be accounted for by the fact that some large craters are older than the lava flows which immediately surround them (Neukum *et al.*, 1975b).

Following procedures identical with those used for estimating the crater production rate on earth and on the moon, the current rate of impact of earth-crossing asteroids and the rate of crater production from these impacts may be estimated for each of the other terrestrial planets (Table 2). The estimated cratering rates on Mercury and Venus are roughly two thirds the rate on Earth (neglecting corrections for collapse of crater walls). It should be borne in mind that these estimates utilize the orbits of a relatively small number of Earth-crossers, especially the estimate for Mercury. Because objects on different orbits have large differences in probability of impact with a given planet, the estimated cratering rates are strongly dependent on the observational sample of orbits. If the next Earth-crossing asteroid to be discovered should have an orbit like that of Icarus, the estimated cratering rate on Mercury would be increased by about 20%. The uncertainties in the cratering rates attributable to the small samples of orbits are not included in the estimated errors shown in Table 2.

In addition to the Earth-crossing asteroids, there must be another class of asteroids that produce craters on Mercury and Venus. Near encounters of Earth-crossers with Mercury and Venus will deflect a certain fraction of these objects into orbits with aphelion distances less than the perihelion distance of the earth. These deflected asteroids are then no longer earth-crossing; they are virtually precluded from astronomical discovery because the phase angle is never less than 90° and they are always extremely faint. From Monte Carlo simulations of the deflection process, Wetherill (pers. comm., 1977) finds that about 5–15% of the earth-crossing asteroids are ultimately deflected into non-earth-crossing small orbits, if their orbits are highly eccentric to start with. This would be the case for objects which were once the nuclei of periodic comets. For Earth-crossing asteroids with initial orbits of low eccentricity, such as the objects derived by deflection of Mars-crossers, probably about 25% are ultimately deflected into orbits too small to cross the orbit of the earth. The asteroids which are

Table 3. Impact rates and rates of crater production by impact of asteroids on Mars.

Asteroid class	Perihelion distance (AU)	Number of asteroids to absolute visual magnitude 18	Rate of impact on Mars to absolute visual magnitude 18 (10^{-14} km^{-2} yr^{-1})	Rate of production of craters 10 km in diameter or larger (10^{-14} km^{-2} yr^{-1})
Earth- and Mars-crossing (Apollo asteroids)	$q < 1.02$	750 ± 300	0.15 ± 0.07	0.15 ± 0.07
Deep Mars-crossing (Amor asteroids)	$1.02 \leq q \leq 1.30$	600 ± 400[1]	0.08 ± 0.05[2]	0.06 ± 0.04[3]
Moderate Mars-crossing	$1.30 < q \leq 1.68$[4]	$15{,}000 \pm 9000$[5]	1.4 ± 0.8[6]	0.8 ± 0.5[7]
Shallow Mars-crossing	$1.68 < q \leq 1.73$[4]	$10{,}000 \pm 7000$[8]	0.2 ± 0.15[9]	0.1 ± 0.1[9]
Total		$26{,}000 \pm 17{,}000$	1.8 ± 1.1	1.1 ± 0.7

(1)Based on discovery of 3 objects in systematic surveys.
(2)Based on estimated population and orbital elements for 17 known objects.
(3)Based on data indicated in footnotes (1) and (2) and Eqs. (3) and (4) in Table 2.
(4)As noted by Wetherill (1974), secular variations of eccentricity will preclude collision with Mars for many objects with perihelion distances in this range.
(5)Based on discovery of 3 objects in systematic surveys.
(6)Based on estimated population and orbital elements for 31 known objects.
(7)Based on data indicated in footnotes (4) and (5) and Eqs. (3) and (4) in Table 2.
(8)Based on discovery of 2 objects in systematic surveys.
(9)Rough estimate only.

exclusively Mercury- and Venus-crossing must have relatively high probabilities of collision with Mercury and Venus and must contribute significantly to the production of impact craters on these two planets. Very roughly, the crater production rates shown for Mercury and Venus in Table 2 should be increased about 10–20%. Furthermore, the crater production rate on Venus at $D \geq 10$ km should be increased still more to account for collapse of crater walls. This correction is comparable to that for Earth. Less than 10% of the craters on Mercury have collapse terraces at $D \leq 10$ km (Gault *et al.*, 1975), however, so that the scaling laws given by Eqs. (3) and (4) (Table 2) are assumed to be approximately correct for Mercury.

In the case of Mars, the impact of Earth-crossing objects probably accounts for about $\frac{1}{10}$–$\frac{1}{5}$ of the current production of craters 10 km across and larger. Most of the craters on Mars are produced by impact of Mars-crossing asteroids that do not overlap the orbit of Earth. The ratio of Mars-crossing asteroids to Earth-crossing asteroids is estimated to lie between 10 and 60 (Shoemaker and Helin, 1977). The average probability of impact on Mars for these asteroids is less than half the average probability of impact for Earth-crossers, however, and their rms impact velocity is less than half that of the Earth-crossers. Adopting $25{,}000 \pm 16{,}000$ as a best estimate of the number of Mars-crossing asteroids to visual magnitude 18, the rate of production of craters 10 km in diameter and larger by impact of Mars-crossers is estimated to be $(0.95 \pm 0.65)10^{-14}$ km^{-2} yr^{-1} (Table 3). When added to the crater production by impact of asteroids that are both Earth- and Mars-crossing, this gives a total cratering rate on Mars of $(1.1 \pm 0.7)10^{-14}$ km^{-2} yr. Almost no craters on Mars show collapse terraces at $D = 10$ km (Smith, 1976); thus no correction of the crater scaling relations is indicated. The total cratering rate on Mars is comparable to the cratering rate on Earth after correction of the rate on Earth for crater wall collapse.

I wish to thank S. L. Gillett for programming most of the calculations used in preparing Tables 1, 2, and 3. G. W. Wetherill contributed many helpful suggestions and generously shared unpublished results from his Monte Carlo studies of orbital evolution of Earth-crossing asteroids.

REFERENCES

Baldwin, R. B.: 1971, On the History of Lunar Impact Cratering: The Absolute Time Scale and the Origin of Planetesimals. *Icarus* **14**, 36–52.

Boyce, J. M., Dial, A. L., and Soderblom, L. A.: 1974, Ages of the Lunar Nearside Light Plains and Maria. *Proc. Lunar Sci. Conf. 5th*, p. 11–23.

Cohen, C. J., Hubbard, E. C., and Oesterwinter, C.: 1973, Planetary Elements for 10,000,000 Years. *Celest. Mech.* **7**, 438–448.

Chapman, C. R., McCord, T. B., and Pieters, C.: 1973, Minor Planets and Related Objects. X. Spectrophotometric Study of the Composition of (1685) Toro. *Astr. J.* **78**, 502–505.

Dence, M. R.: 1972, The Nature and Significance of Terrestrial Impact Structures. *24th Internat. Geol. Cong. Sect.* **15**, 77–89.

Dunlap, J. L.: 1974, Minor Planets and Related Objects. XV. Asteroid (1620) Geographos. *Astr. J.* **79**, 324–332.

Dunlap, J. L., Gehrels, T., and Howes, M. L.: 1973, Minor Planets and Related Objects. IX. Photometry and Polarimetry of (1685) Toro. *Astr. J.* **78**, 481–501.

Fleischer, R. L., Price, P. B., and Walker, R. M.: 1965, On the Simultaneous Origin of Tektites and Other Natural Glasses. *Geochim. Cosmochim. Acta* **29**, 161–166.

Gault, D. E., Guest, J. E., Murray, J. B., Dzurisin, D., and Malin, M. C.: 1975, Some Comparisons of Impact Craters on Mercury and the Moon. *J. Geophys. Res.* **80**, 2444–2460.

Gault, D. E. and Wedekind, J. A.: 1977, Experimental Hypervelocity Impact into Quartz Sand: II, Effects of Gravitational Acceleration. In *Impact and Explosion Cratering* (D. J. Roddy, R. O. Pepin, and R. B. Merrill, eds.), Pergamon Press. This volume.

Gehrels, T., Roemer, E., Taylor, R. C., and Zellner, B. H.: 1970, Minor Planets and Related Objects. IV. Asteroid (1566) Icarus. *Astr. J.* **75**, 186–195.

Gentner, W., Lippolt, H. J., and Müller, O.: 1964, Das Kalium-Argon-Alter des Bosumtwi-Kraters in Ghana und die Chemische Beschaffenheit Seiner Gläser. *Zeit. Naturf.* **19a**, 150–153.

Gradie, J. C.: 1976, Physical Observations of Object 1976 AA (abstract). *Amer. Astron. Soc. Bull.* **8**, 458–459.

Helin, E. F., and Shoemaker, E. M.: 1977, 1976 UA: Second Asteroid with Orbit Smaller than Earth's (abstract). To be published in *Amer. Astron. Soc. Bull.*

Howard, K. A.: 1974, Fresh Lunar Impact Craters: Review of Variations with Size. *Proc. Lunar Sci. Conf. 5th*, p. 61–69.

Krinov, E. L.: 1966, *Giant Meteorites.* Pergamon Press, New York. 397 pp.

Melosh, H. J.: 1977, Crater Modification by Gravity: a Mechanical Analysis of Slumping. In *Impact and Explosion Cratering* (D. J. Roddy, R. O. Pepin, and R. B. Merrill, eds.), Pergamon Press. This volume.

Morrison, D., Gradie, J. C., and Reike, G. H.: 1976, Radiometric Diameter and Albedo of the Remarkable Asteroid 1976 AA. *Nature* **260**, 691.

Neukum, G., König, B., and Arkani-Hamed, J.: 1975a, A study of Lunar Impact Crater Size-Distributions. *The Moon* **12**, 201–229.

Neukum, G., König, B., Fechtig, H., and Storzer, D.: 1975b, Cratering in the Earth-Moon System: Consequences for Age Determination by Crater Counting. *Proc. Lunar Sci. Conf. 6th*, p. 2597–2620.

Öpik, E. J.: 1951, Collision Probabilities with the Planets and the Distribution of Interplanetary Matter. *Proc. Roy. Irish Acad.* **54**, Sect. A, 165–199.

Öpik, E. J.: 1963, The Stray Bodies in the Solar System. Part 1. Survival of Cometary Nuclei and the Asteroids. *Advan. Astron. Astrophys.* **2**, 219–262.

Öpik, E. J.: 1976, *Interplanetary Encounters.* Elsevier, New York. 155 pp.

Shoemaker, E. M.: 1962, Interpretation of Lunar Craters. In *Physics and Astronomy of the Moon*, (Z. Kopal, ed.), p. 283–359. Academic Press, New York.

Shoemaker, E. M.: 1971, Origin of Fragmental Debris on the Lunar Surface and the History of Bombardment of the Moon. *Inst. Invest. Geol. Diput. Prov. Barcelona* **25**, 27–56.

Shoemaker, E. M.: 1972, *Cratering History and Early Evolution of the Moon* (abstract). In *Lunar Science III*, p. 696–698. The Lunar Science Institute, Houston, Texas.

Shoemaker, E. M., Hackman, R. J., and Eggleton, R. E.: 1963, Interplanetary Correlation of Geologic Time. *Adv. Astronaut. Sci.* **8**, 70–89.

Shoemaker, E. M., and Helin, E. F.: 1977, Populations of Planet-crossing Asteroids and the Relation of Apollo Objects to Main-belt Asteroids and Comets. To be published in *Internat. Astron. Union Trans. Colloq.* **39**.

Smith, E. I.: 1976, Comparison of the Crater Morphology-Size Relationship for Mars, Moon, and Mercury. *Icarus* **28**, 543–550.

Smith, E. I., and Sanchez, A. G.: 1973, Fresh Lunar Craters: Morphology as a Function of Diameter, A Possible Criterion for Crater Origin. *Mod. Geol.* **4**, 51–59.

Soderblom, L. A., and Boyce, J. M.: 1972, Relative Ages of Some Near-Side and Far-Side Terra Plains Based on Apollo 16 Metric Photography. NASA SP 315, p. 29-3 to 29-6.

Trask, N. J.: 1966, Size and Spatial Distribution of Craters Estimated From the Ranger Photographs. *Jet Propulsion Lab Tech. Rept.* 32–800. p. 252–263.
Van Houten, C. J., Van Houten-Groeneveld, I., Herget, P., and Gehrels, T.: 1970, The Palomar-Leiden Survey of Faint Minor Planets. *Astron. Astrophys. Suppl.* **2**, 339–448.
Wetherill, G. W.: 1974, Impact Rates on Mars and the Moon. *The Moon* **9**, 227–231.
Wetherill, G. W.: 1976, Where do Meteorites Come From? A re-evaluation of the Earth-crossing Apollo Objects as Sources of Chondritic Meteorites. *Geochim. Cosmochim. Acta* **40**, 1297–1317.
Wetherill, G. W., and Williams, J. G.: 1968, Evaluation of the Apollo Asteroids as Sources of Stone Meteorites. *J. Geophys. Res.* **73**, 635–648.

Roddy, D. J., Pepin, R. O., and Merrill, R. B., editors.
(1977) *Impact and Explosion Cratering*, Pergamon Press (New York), p. 629–633.
Printed in the United States of America

Historical variations in the density and distribution of impacting debris in the inner solar system: Evidence from planetary imaging

LAURENCE A. SODERBLOM

United States Geological Survey, 2255 North Gemini Drive, Flagstaff, Arizona 86001

INTERPLANETARY CORRELATIONS

THE UBIQUITOUS IMPACT CRATERS found scattered over the surfaces of Mercury, Mars, and the moon record the historical variations in the distribution of impacting debris in the inner solar system for the last several billion years. The traditional method for establishing relative ages of various surfaces on an individual planet has been to compare the areal density of impact craters superposed on these surfaces; older surfaces, having been exposed longer, are presumed to have more craters. Not until the Apollo missions returned samples from the moon for radioisotope age dating did we have a reasonably accurate knowledge of the historical variations in the flux of debris impacting the moon. We now know that about 4 b.y. ago the flux decreased rapidly from a torrential rate, in which surfaces were saturated with large (>10 km diameter) craters in a few hundred million years, to a much lower level which has remained roughly constant for the last 3–4 b.y. (Hartmann, 1970; Soderblom and Lebofsky, 1972; Boyce *et al.*, 1977). Younger surfaces only rarely display craters in this size range.

Now that we have a reasonable understanding of the flux history for the moon, the next step is to extend this knowledge to Mercury and Mars in order to develop flux histories and time scales for the surface evolutions of those planets. The first of two ways to approach this problem is to examine the current distribution of debris in the solar system using telescopic data for comets and asteroids and meteorite falls to develop dynamic models consistent with that current distribution (cf. Wetherill, 1974; Shoemaker, 1977). As these dynamic models contain expected lifetimes for various families of objects, it is possible to infer relative flux histories for different terrestrial planets. Major drawbacks with this approach include the following: (1) the observational data pertaining to current distribution of asteroids and meteorites in the solar system are far from complete, (2) the inferred relative fluxes at Mercury, the moon, and Mars are extremely sensitive to the complexities and assumptions in the dynamic modeling, (3) families of objects that were important in early planetary history had short half-lives and are now nearly gone.

The second method, the subject of this paper, is to examine the global cratering records using planetary images (Lunar Orbiter and Apollo mission for the Moon; Mariner 10 for Mercury; Mariner 4, 6, 7, 9, and Viking Orbiters I and

II for Mars) to establish for each planet the global succession of stratigraphic units and the variations in crater densities among these global provinces. The next step is to examine the resulting maps of global crater density for large differences in crater density between adjacent units in the stratigraphic succession—differences that may be correlated from planet to planet. Such differences suggest large changes in the cratering flux over relatively short periods of geologic time. This method is very similar to that of tree-ring dating. Using the lunar flux history it is then possible to assign absolute ages to similar changes in cratering flux on Mercury and Mars and thereby to establish gross impact flux histories for those planets.

As previously mentioned, the lunar flux abruptly declined about 4 b.y. ago. Older surfaces are so intensely cratered with large craters that they appear ancient. Younger surfaces (the light plains and maria) have far lower crater densities and appear extremely young by comparison. Radiometric ages of Apollo samples tell us that this tremendous difference in apparent age is not real, it is simply the result of the extremely rapid decline in the cratering flux a few hundred million years after accretion. It is this large change in the cratering flux that we attempt to identify in martian and mercurian cratering records permitting a correlation with the lunar time scale.

Mariner and Viking images of Mars reveal a dichotomy in the crater density of various geologic terrains similar to that seen for the moon. The southern hemisphere is characterized by a dense population of large (20–100 km diameter) craters. These cratered terrains have been overlain by a variety of younger plains that have partly buried the large craters so that many areas are now only 30–40% saturated with these large craters (Soderblom *et al.*, 1974). The martian cratered terrains are similar in appearance and crater density to the lunar highlands. By contrast, the plains on Mars as on the moon are sparsely cratered; large craters (20–100 km) superposed on these surfaces are rare. The tremendous difference in crater density between the oldest martian plains and martian highlands is inferred to have resulted from the same rapid fall in flux known to have occurred on the moon. This is the "tree-ring" we correlate between Mars and the moon.

The same correlation exists for Mercury. The intercrater plains and highlands are densely cratered, and the smooth plains sparsely cratered (Trask and Guest, 1974; Soderblom, 1977). The key to the model proposed here is that (1) the rapid decrease in flux that occurred 3–4 b.y. ago can be correlated from planet to planet and (2) the oldest plains on these planets record the cratering flux averaged over the last 3–4 b.y.

SUMMARY OF EXISTING DATA AND CONCLUSIONS

In order to establish the range of variation of crater density of post-highland plains for each of the planets (Mercury, Moon, and Mars) it is first necessary to *map* crater density on a global scale on each of the planets. For this purpose a narrow diameter-range was chosen (5–10 km). Crater abundance in this size range has been continuously mapped for available coverage for each of the three planets. Craters in this size range superposed on the plains usually display pristine morphology suggesting that most were formed after the plains; evidently, most pre-existing craters in this size range were erased by the plains-forming processes. Secondly these craters are large enough to have withstood impact and eolian erosion processes which have affected smaller craters since plains formation[1]. It is the uniform fresh appearance of the 5–10 km diameter craters that leads us to believe that they best record the integrated cratering flux since plains formation. Global crater density maps for the 5–10 km diameter range have been collected for Mars[2], the moon[3], and Mercury (Soderblom *et al.*, 1974; Soderblom, 1977). Figure 1 compares the frequency distribution of plains by crater density (5–10 km diameter) for Mercury, the moon, and Mars. Assuming that the oldest plains record the accumulation of impact craters in this size range over the last ~4 b.y., rough estimates of the average crater production rates (crater density of oldest plains divided by 4×10^9 yr) are given in Table 1.

A possible flaw in the model proposed here is the assumption of a roughly continuous evolution of new plains surfaces during and following the rapid decline in flux. In particular, the model assumes that there was not a long hiatus (half a billion years or longer) in the evolution of plains following the early decrease in flux. For the moon we know from Apollo data that this was not the case. For Mars we can make strong geologic arguments that such a void did not occur as we find abundant volcanic structures which both predate and postdate the decline in flux. For Mercury the argument is weaker, the "oldest plains" could be much younger than 4 b.y. If this were the case, however, the cratering flux at Mercury relative to the moon would be even higher than in Table 1, an observation that would be very difficult to explain. Rather it suggested that the higher values and variation in crater density on the mercurian plains indicate that the decline in flux and tapering off of plains-formation occurred nearly at the same time.

It should be realized that the values given in Table 1 are production rates of craters of the same *diameter range* on the planets, *not* impact flux rates of objects in the same *mass range.* Relative to the moon typical encounter velocities are higher by roughly 50% at Mercury and lower by about the same factor at Mars (cf. Shoemaker, 1977; Neukum and Wise, 1977). As the

---

[1]On Mars the density of smaller craters ($\leqslant 1$ km) appears to be controlled by latitudinal belts of wind erosion and deposition; such small craters exhibit a wide range of morphologies and their density reflects the intensity of eolian activity, not the age of the plains-forming events. On Mercury and the moon such small craters (< 1 km) are being continuously eroded by impact processes, their density represents a balance between crater formation and erosion and may not reflect the age of the plains.

[2]Data actually were acquired for 4–10 km diameter craters and have been converted to 5–10 km diameter range.

[3]Neukum and Wise (1977) have pointed out an error in our earlier work for the moon. Originally we did not count 5–10 km craters but erroneously estimated their densities from a morphological erosion parameter (Soderblom *et al.*, 1974). The new data presented here was acquired by C. D. Condit by actually mapping craters in the 5–10 km diameter range.

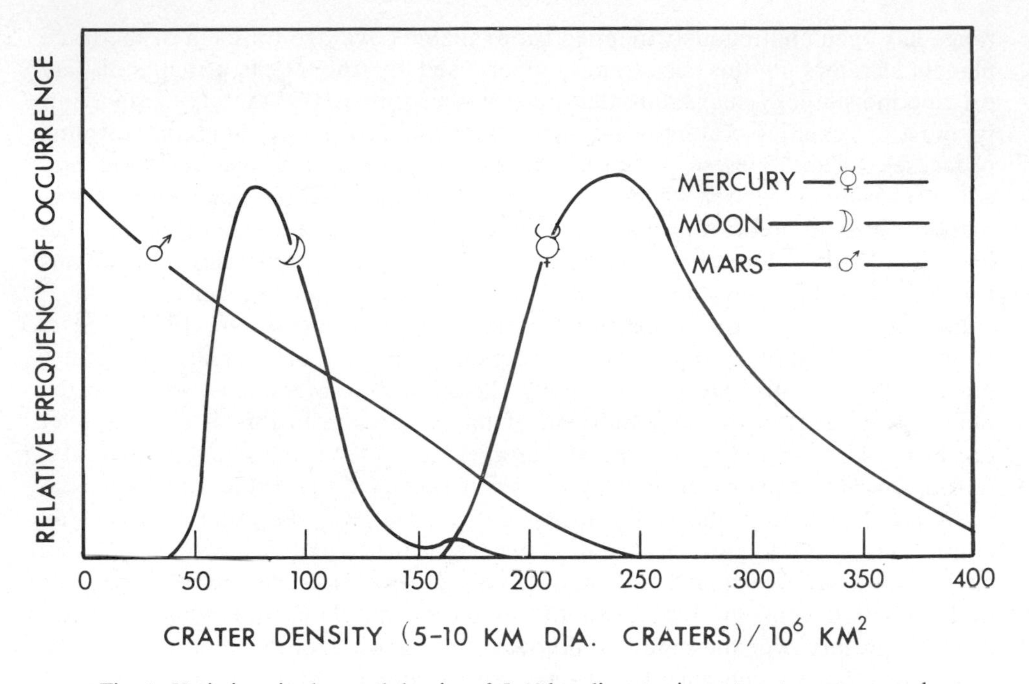

Fig. 1. Variations in the areal density of 5–10 km diameter impact craters superposed on mercurian, lunar, and martian plains. These relative frequencies of occurrence by crater density of plains (maria and light plains on the moon; volcanic plains on Mercury, and volcanic, eolian and other sedimentary plains on Mars) were acquired by continuously mapping the density of 5–10 km craters on a global scale. See text for data sources.

Table 1.

Planet	Crater production rate (5–10 km diameter)	Equivalent production rate (>10 km diameter)*
Mercury	$\sim 10 \times 10^{-14}\ km^{-2}\ yr^{-1}$	$\sim 3 \times 10^{-14}\ km^{-2}\ yr^{-1}$
Moon	$\sim 4 \times 10^{-14}\ km^{-2}\ yr^{-1}$	$\sim 1.3 \times 10^{-14}\ km^{-2}\ yr^{-1}$
Mars	$\sim 6 \times 10^{-14}\ km^{-2}\ yr^{-1}$	$\sim 2 \times 10^{-14}\ km^{-2}\ yr^{-1}$

*These values are provided for comparison with data of Shoemaker (this publication). They are calculated assuming a cumulative frequency distribution $\propto D^{-2}$ (D is diameter) for craters in this size range. As these values are *averages* for the last ~4 b.y., they are expected to be somewhat higher (~ a factor of 2) than modern production rates. This arises because the impact fluxes were still gradually decaying between 3 and 3.5 b.y. ago (cf. Soderblom and Lebofsky, 1972).

diameter of a crater scales roughly as $v^{2/3}$ (v is velocity), objects of the same mass will produce craters about 30% larger on Mercury and 30% smaller on Mars. Hence, to compare the flux of equivalent masses we should compare, for instance, the density of 6.5–13 km diameter craters on Mercury, with 4–10 km diameter on the moon, with 3.5–7 km diameter craters on Mars. Using published crater-size-frequency curves for Mercury and Mars (cf. Murray *et al.*, 1974; Neukum and Wise, 1977) the density of 6.5–13 km craters on Mercury is lower by about a factor of 2 than the density of 4–10 km craters. On Mars the density of 3.5–7 km craters is higher than that of 4–10 km craters by about a factor of 2. Therefore, based on the data of Table 1, the relative fluxes of objects of the same mass range (10^{13-14} g) has been roughly equal for Mercury and the moon and a factor of 2 to 3 higher for Mars.

Acknowledgment—This work was carried out under the sponsorship of S. E. Dwornik, Chief of Planetary Geology, NASA contract W13-576.

References

Boyce, J. M., Schaber, G. G., and Dial, A. L., Jr.: 1977, Age of Luna 24 Mare Basalts Based on Crater Studies, *Nature*, **265**, 38–39.

Hartmann, W. K.: 1970, Lunar Cratering Chronology, *Icarus* **13**, 299–301.

Murray, B. C., Belton, M. U. S., Danielson, G. E., Davies, M. E., Hapke, B., O'Leary, B. T., Strom, R. G., Suomi, V. E., and Trask, N. J.: 1974, Mercury's Surface: Preliminary Description and Interpretation from Mariner 10 Pictures, *Science* **185**, 169–179.

Neukum, G. and Wise, D. U.: 1977, Mars: A Standard Crater Curve and Possible New Time Scale, *Science* **194**, 1381–1388.

Shoemaker, E. M.: 1977, Astronomically Observable Crater-Forming Projectiles. This volume.

Soderblom, L. A.: 1977, Relative Ages of Mercurian Plains, *Abs. 8th Planetary Program Investigators Meeting*, St. Louis, May 1977, NASA Memorandum.

Soderblom, L. A., Condit, C. D., West, R. A., Herman, B. M., and Kreidler, T. J.: 1974, Martian Planetwide Crater Distribution: Implications for Geologic History and Surface Processes, *Icarus* **22**, 239–263.

Soderblom, L. A. and Lebofsky, L. A.: 1972, Technique for Rapid Determination of Relative Ages from Orbital Photography, *J. Geophys. Res.* **77**, 279–296.

Trask, N. J. and Guest, J. E.: 1975, Preliminary Terrain Map of Mercury, *J. Geophys. Res.* **80**, 2461–2477.

Wetherill, G. W.: 1974, Problems Associated with Estimation of the Relative Impact Rate on Mars and Moon, *The Moon* **9**, 227.

Roddy, D. J., Pepin, R. O., and Merrill, R. B., editors.
(1977) *Impact and Explosion Cratering*, Pergamon Press (New York), p. 635–637.
Printed in the United States of America

Cratering mechanics and future Martian exploration

HAROLD MASURSKY

U.S. Geological Survey, 2255 North Gemini Drive, Flagstaff, Arizona 86001

A FOLLOW-ON MARS MISSION probably will be launched in 1984 followed by a Mars Surface Sample Return mission in 1988 or 1990. Planning for these missions is in progress. An understanding of crater mechanics, and the geologic and topographic implications of various crater morphologies, will be even more useful for planning future orbiter, lander, rover, and penetrator missions than they were for the Viking landings in 1976. In order to plan most effectively for future missions, additional research in the following areas would be most helpful:

(1) Conducting additional studies of various types of craters and crater ejecta (lobate, blocky, rectilinear, etc.) based on theoretical and experimental work and comparative studies with terrestrial, lunar and Mercurian craters; from these studies more accurate landing hazard and rover trafficability analyses can be made at the scale of the spacecraft.
(2) Conducting similar studies of the mechanics of secondary cratering.
(3) Conducting studies of cratering in periglacial, permafrost, eolian, mass-wasted, and lava flow areas on Earth, to help define and explain some of the observed crater morphologies on Mars.
(4) Conducting studies of large-block populations, to aid in planning penetrator targets.
(5) Using Viking photos to derive a new Martian flux curve from crater counts of a large number of geologically diverse areas. The present Martian flux curve is derived from the lunar flux curve; it is based on Mariner 9 photography. Relative age determinations derived from a recent Martian flux curve (Neukum and Wise, 1977) indicate that geologic events on Mars have all occurred within the first billion years of solar system history, which seems unlikely. Crater counts made of a larger variety of geologic units would probably provide data from which a flux curve could be constructed that was similar (Soderblom *et al.*, 1974) but not exactly like the lunar curve. More reasonable ages for geologic units and predictions of the physical nature of the surfaces studied could then be obtained from the new curve.

The research and detailed studies outlined above will provide much useful new information for: (1) planning candidate sites for future lander and rover missions, (2) planning future targets for a possible penetrator mission, and (3) defining areas where high resolution imaging by future orbiter missions would add to our knowledge of the planet.

Hazard analyses, for a Lander mission, would include determinations of the size and distribution of ejecta blocks and secondary craters in the candidate landing area. Viking lander photographs provide estimates at the appropriate scale of the landing of block size and distribution for the two limited areas where Viking spacecraft in 1976 landed. Estimates for the rest of the planet must be based on studies of experimental craters, and theoretical model studies and inferences based on geologic maps of the planet. The two orders of magnitude difference between the present orbital imaging scale and the scale at which blocks are a hazard to the lander must be bridged by these data. The Viking extended mission is acquiring data that are a factor of five higher resolution; a follow-on mission would require data at a factor of three higher resolution than this. If periapsis of the Viking Orbiters is lowered from 1500 to 300 km, then higher resolution picture (factor of 4 to 5) will help close this gap.

Crater counts from Viking photographs will also be used to determine the homogeneity or heterogeneity of the surface; mantling of a surface by wind laid deposits or stripping of a surface, so that hazardous blocks are covered or exposed, can be deduced from breaks in slope of the crater size-frequency curve and from photo interpretation. Imaging resolution can be determined by the point on the curve where rollover occurs; imaging discrimination is often indicated by the general shape of the curve. Surface characteristics, and the reliability of visual imaging of the surface, as defined by crater size-frequency curves, are important factors in choosing landing sites.

Crater counts also will be useful in planning the possible locations for penetrators. The probable thickness of mantles that would affect the depth of penetration can be determined from the crater curve. The type of ejecta blanket around craters located in a possible penetrator site will also give a clue to characteristics of the substrate, and therefore the depth of penetration. Peculiar lobate flows around some Martian craters may be an indication of permafrost. Penetrators targeted for such areas would add valuable information about the presence and depth of permafrost on Mars. Study of terrestrial experimental craters may be the best clue to the interpretation of the nature of the target material. Blocks large enough to deflect a penetrator can cause considerable loss of information and even failure of the penetrator mission. Large-block population studies are therefore essential before candidate sites for a penetrator mission are selected.

Estimates of block size and distribution, and the size and areal extent of secondary craters, will be useful in making trafficability studies for planning rover traverses. Rates of travel (kilometers per day) for the rover would be constrained, in part, by large populations of ejecta blocks or secondary craters. Because the rover will have a limited traverse capability, the allowable distance between a proposed landing site and the nearby area to be examined in detail by the rover will be determined, in part, by these data.

Sample collection sites along rover traverses will be defined, in part, by studies of various types of ejecta blankets, and by crater counts used to define geologic units and to determine the relative ages of these units. Samples collected on ejecta blankets will provide information about the subsurface materials brought to the surface as ejecta blocks or as central peak material. Studies of ejecta patterns and central peaks of experimental explosion craters and terrestrial impact craters will allow the design of effective rover traverses to analyze and return samples.

We have learned from past planetary missions that an understanding of the processes involved in both crater formation and degradation provides clues to the age and geologic history of an area. Additional studies of the mechanics of crater formation and degradation derived from Earth-analogue studies and higher resolution photographs will help to define the geologic age relationships of the various geologic units on Mars; from these studies a more detailed history of the planet can be developed. Such studies, combined with higher resolution photographs obtained by an orbiter will be useful in planning candidate landing sites for future lander missions.

References

Neukum, G. and Wise, D. U.: 1976, Mars: A standard crater curve and possible new time scale. *Science* **194**, 1381–1387.

Soderblom, L. A., West, R. A., Herman, B. M., Kreidler, T. J., and Condit, C. D.: 1974, Martian planetwide distribution: Implications for geologic history and surface processes. *Icarus* **22**, 239–263.

Roddy, D. J., Pepin, R. O., and Merrill, R. B., editors.
(1977) *Impact and Explosion Cratering*, Pergamon Press (New York), p. 639–656.
Printed in the United States of America

Equations of state and impact-induced shock-wave attenuation on the moon*

THOMAS J. AHRENS

Seismological Laboratory, California Institute of Technology, Pasadena, California 91125

JOHN D. O'KEEFE†

Department of Earth and Space Sciences, University of California, Los Angeles, California 90024

Abstract—Current equation-of-state formulations, used for finite-difference cratering flow calculations, are cast into a framework permitting comparison of peak pressures attained upon impact of a sphere, with a half-space, along the impact symmetry axis, to one-dimensional impedance match solutions. On the basis of this formulation and application of thermochemical data, the regimes of melting and vaporization are examined. For the purpose of identifying material which will, upon isentropic release from the impact-induced shock state, result in a solid just brought to its melting point, i.e., incipiently melted (IM), completely melted (CM), just brought to its boiling point, i.e., incipiently vaporized (IV), and completely vaporized (CV) state, the pressures at which the critical isentropes intersect the Hugoniots of iron and gabbroic anorthosite (GA) are examined in detail. The latter rock type is assumed to be representative of the lunar highlands. The Hugoniot pressures, for which IM, CM, IV, and CV will occur upon isentropic expansion, are calculated to range from 2.2 to 16.8 Mbar, respectively for iron. For the high-pressure phase (hpp) assemblage of GA, modelled as a mixture of plagioclase in the hollandite structure and pyroxene in the perovskite structure, IM, CM, IV, and CV are calculated to occur upon isentropic expansion from Hugoniot states ranging from 0.43 to 5.9 Mbar, respectively. The spatial attenuation of shock pressure along the impact axis is found to be clearly represented by two regimes, if the peak pressure, P, and radius normalized to that of the projectile, r, are fitted to expressions of the form $P \propto r^a$. At distances from 2.2 to 5.6 projectile radii into a GA target, the constant, a, is on the order of -0.2. This low-attenuation rate, near-field regime, extends further into the target at the slower impact velocities and arises because of the slightly divergent flow associated with the penetration of a spherical projectile. For the near-field impact regime, an impact at 5 km/sec of an iron object with a GA surface will induce CM for GA but the iron will remain solid. At 15 km/sec, partial vaporization (PV) occurs for both GA and iron, whereas at 45 km/sec, CV occurs in both materials. Similar calculations are summarized for a GA meteoroid striking a GA surface at velocities ranging from 5 to 45 km/sec. At greater radii, in the far-field regime, the exponent, a, varies systematically from -1.45 to -2.15 for impacts of GA onto GA as the impact velocity is increased from 5 to 45 km/sec. For an iron projectile impacting at speeds of 5–45 km/sec, the exponent, a, varies from -1.67 to -2.95. By comparison, the equivalent value of a, reported for both contained and surface explosions in various rocks is ~ -2. It is suggested that, given field data on shock attenuation (based on identification of various shock metamorphic features versus distance), overall crater size, and some chemical data as to the type of meteoroid which produced a crater, quantitative bounds on the impact velocity of the meteorite may be obtained.

---

*Contribution Number 2844, Division of Geological and Planetary Sciences, California Institute of Technology, Pasadena, California 91125.

†Present address: Seismological Laboratory, California Institute of Technology, Pasadena, California 91125.

INTRODUCTION

THE FORMULATION of an equation-of-state for a rock type which we believe is typical of the lunar highland province, gabbroic amorthosite (GA), provides an opportunity to investigate the following two important and related problems regarding impact phenomena:

(1) The impact velocities, and hence shock pressures, required to bring the meteorite and target material to the melting point, i.e., incipient melting (IM), induce complete melting (CM), and to produce a liquid at the boiling point, i.e., incipient vaporization (IV), and completely vaporize (CV) upon isentropic release is examined. We consider a highland terrane cratered by meteorites having equations-of-state ranging from iron to GA.
(2) The spatial attenuation of peak shock-pressures in a GA composition, again, as the result of impact of objects having equations-of-state similar to iron and GA.

It has been only recently determined that, even in very large and ancient astroblemes, geochemical evidence (via analyses of the minor siderophile element contents of impact melts and ejecta (Morgan *et al.*, 1975; Lambert, 1976)) allows specification of the causative meteorite. Moreover, recent quantitative studies of the effect of shock on minerals and their application to the spatial distribution of shock metamorphic features within *in situ* shocked rock, can be used to infer distinct bounds on the spatial shock attenuation rate (Robertson and Grieve, 1977). If in addition the spatial attenuation rate can be related to the meteorite size and velocity, then improved estimates can be obtained of the chemistry, mass, and velocity, of the meteoroids which have bombarded the earth's zone as a function of time (Dence, 1972). Much of the required data is not now available for lunar craters. However, if the moon formed and evolved near the earth, the late cratering history of the earth should, with minor modification, be applicable to the moon.

In the present paper we obtain, within the framework of a two-phase Tillotson equation-of-state model the impedance-match solutions specifying the peak pressure upon impact of a GA lunar surface by meteoroids of GA and iron in the 5–45 km/sec range (O'Keefe and Ahrens, 1975, 1976). We then apply the available thermochemical data for the low-pressure phase (lpp) assemblage and a theoretical model for the high-pressure phase (hpp) assemblage to construct critical release adiabats which define regimes in the pressure (P)–volume (V) plane which specify IM, CM, IV, and CV. Finally we fit the results of a series of hypervelocity impact, finite-difference, flow calculations specifying the spatial on-axis peak pressure attenuation due to the shock wave and compare these to previous calculations and measurements of pressure attenuation from explosions.

CONDITIONS FOR ONE-DIMENSIONAL IMPACT

It is useful for the purposes of (a) providing an analytic comparison with finite-difference impact calculation and (b) obtaining an estimate of the impact velocity required to induce melting and vaporization in the target and impactor, to obtain one-dimensional impedance match solutions for an assumed equation-of-state. As in our previous treatments of impact flows produced upon the impact of a meteorite with a planetary surface ((O'Keefe and Ahrens, 1975, 1976), we employ a $P-V-E$ (specific internal energy), dependent equation-of-state of the form first proposed by Tillotson (1962). In the compressed region this is given by

$$P = \left[a + \frac{b}{\left(\frac{E}{E_0\eta^2} + 1\right)} \right] \frac{E}{V} + A\mu + B\mu^2, \tag{1}$$

where $a = 0.5$ is the polytropic constant minus 1, at high temperature. The constant, b, is defined such that $(a+b)$ is the STP Grüneisen parameter, $\gamma = V(\partial P/\partial E)_V$, and A is the bulk modulus. Here $\eta = V_0/V$ and $\mu = \eta - 1$, where V_0 is the STP specific volume. The parameters b, B, and E_0 are obtained by fitting to Thomas-Fermi calculations of the equation-of-state in the 10^2 Mbar range, and Hugoniot data. The parameters for GA (Table 1) are based on previous shock-wave measurements on lunar sample 15418 (Ahrens *et al.*, 1973). As in O'Keefe and Ahrens (1976), we consider the equilibrium equation-of-state for GA to have discreetly different Tillotson parameters which specify the lpp and hpp assemblage. For the latter, hpp, assemblage, stable above ~150 kbar, we assume plagioclase has the density corresponding to the hollandite structure and pyroxene has a density corresponding to the perovskite structure. The effect of applying this relatively complex, but realistic equation-of-state to the description of the impacts onto GA is examined in some detail by O'Keefe and Ahrens (1976).

For convenience we define, $E = 0$, for the lpp at STP. To invert Eq. (1) to the $P - u_s$ (particle velocity) plane, the energy term for a Hugoniot state in the hpp regime is given by

$$E = P(V_{00} - V)/2 - E_{TR}, \tag{2}$$

where V_{00} is the lpp specific volume and E_{TR} is the increase in internal energy in going from the lpp to the hpp assemblage at STP. The latter term is zero when referring to the lpp.

It should be noted that if the E_{TR} term is not included in Eq. (2), the resultant Hugoniot curve generated would be with respect to the hpp at STP and hence be that of the metastable Hugoniot (McQueen *et al.*, 1967). Moreover we note that in applying the definition of η and μ for describing the hpp, the initial specific volume used, V_0, is that of the hpp. In calculating the Hugoniot curve for GA (lpp) and for iron, $V_{00} = V_0$ and $E_{TR} = 0$. Upon eliminating E between Eqs. (1) and (2), the Hugoniot pressure, $P = P_H$, is given analytically upon solution of the following quadratic equation where V (and hence η and μ) is the independent variable

$$a'P_H^2 + b'P_H + c' = 0. \tag{3}$$

Here

$$a' = \frac{s}{k}\left[\frac{as-1}{V}\right], \tag{4a}$$

$$b' = (a+b)s/V - 1 + E + R(1-2as/V)/k + ls/k, \tag{4b}$$

$$c' = E_{TR}^2 a/(kV) - E_{TR}(a+b)/V + l(1-E_{TR}/k), \tag{4c}$$

where

$$s = (V_{00} - V)/2, \tag{4d}$$

$$k = E_0\eta^2, \tag{4e}$$

$$l = \mu A + \mu^2 B. \tag{4f}$$

The negative sign is used in the usual quadratic formula. Once $P_H(V)$ along the Hugoniot, centered at V_{00}, is calculated, the corresponding value of u_p is simply calculated from the Rankine-Hugoniot equation

$$u_p = [P_H(V_{00} - V)]^{1/2}. \tag{5}$$

By solving Eq. (3) at a series of specific volumes, the $P_H - u_p$ relations given in Fig. 1 are obtained from the parameters in Table 1. Straightforward application of the method of impedance matching (Duvall and Fowles, 1963) (matching pressure and particle velocity), and application of the Rankine-Hugoniot equations yield the peak shock states defined in Tables 2 and 3 for impact of GA and iron projectiles on a GA half-space.

To calculate the amount of melting and vaporization that occurs in an impact event, the release isentropes for IM, CM, IV, and CV were calculated using the Tillotson equation-of-state (Eq. (1)). The calculation of the isentropes requires that at least one state along the desired isentrope be known initially. These initial states are typically the internal energy or temperature at one atmosphere pressure. To obtain the internal energy contents, at one atmosphere pressure of iron required for IM (1809°K), CM, IV, (3145°K), and CV, we have utilized enthalpy and entropy data tabulated in the JANAF Tables (1965). In the case of GA, the above melting and vaporization processes are incongruent, and an approximate treatment was used for simplicity. Following the rock modelling scheme of Ahrens and O'Keefe (1972), GA was assumed to comprise a mixture of 0.714 (mass fraction) anorthite, and 0.286, enstatite. Using the JANAF thermochemical data, through the $MgSiO_3$ melting point (1798°K) and into the liquid, and extrapolating using the entropy systematics described in Ahrens and O'Keefe (1972), the energy and entropy for incipient and complete vaporization is inferred. Assumptions regarding volatilization are made on the basis of oxide volatility and are discussed by Ahrens and O'Keefe (1972). In the case of $CaAl_2Si_2O_8$, a similar but even less certain extrapolation was employed using the thermochemical data tabulated in Robie and Waldbaum (1968) and, again, the systematics of Ahrens and O'Keefe. The densities, at one atmosphere, given in Table 1 for IM, CM, and IV for the lpp and hpp assemblages of GA and iron are obtained from Eq. (1).

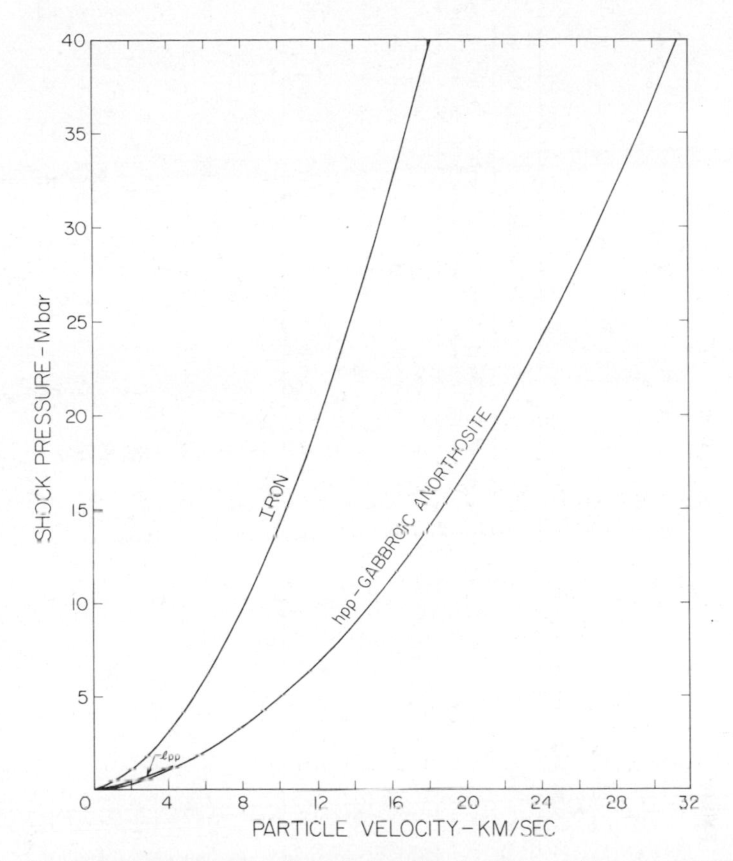

Fig. 1. Pressure-particle velocity Hugoniot relation for iron, and low-pressure phase (lpp) and high-pressure phase (hpp) assemblage of gabbroic anorthosite. Gabbroic anorthosite Hugoniots are both centered at specific volume of lpp and calculated using the Tillotson parameters of Table 1.

Only in the case of iron can meaningful comparisons be made with other theoretical treatments and the actual densities at CM and IV. Table 1 gives densities of 7.36 and 6.10 g/cm^3 for CM and IV, respectively. Interpolating, the results of the Mie-Grüneisen reduction of the iron data by McQueen *et al.* (1970) to 1809° yields a considerably lower density, 6.696 g/cm^3, for CM. The CRC Handbook (1970) lists density values ranging from 6.99 to 7.24 g/cm^3 for iron at the liquidus, slightly lower, but in good agreement with those calculated from Eq. (1). For IV, at 3145°K, the CRC Handbook gives density values ranging from 5.84 to 5.93 g/cm^3 which again is close to, but still slightly lower than, the 6.10 g/cm^3 value indicated in Table 1.

The onset of IM, CM, IV, and CV upon expansion (assumed isentropic) from Hugoniot states even for the one-dimensional case is of interest. We have calculated the $P - V$ isentropes which pass through the states IM, CM, and IV specified in Table 1 at atmospheric pressure and high temperatures using Eq. (1)

Table 1. Complete equation-of-state parameters.

Material	Normal density (g/cm³)	b	A (Mbar)	B (Mbar)	E_0 (10^{12} erg/g)	Incipient melting: Energy[β] (10^{12} erg/g)	Incipient melting: Density[α] (g/cm³)	Incipient melting: Entropy[c] (10^7 erg/g °K)	Complete melting: Energy[β] (10^{12} erg/g)	Complete melting: Density[α] (g/cm³)	Complete melting: Entropy[c] (10^7 erg/cm °K)	Incipient vaporization: Energy[β] (10^{12} erg/g)	Incipient vaporization: Density[α] (g/cm³)	Incipient vaporization: Entropy[c] (10^7 erg/cm °K)	Complete vaporization: Energy[β] (10^{12} erg/g)	Complete vaporization: Entropy[c] (10^7 erg/g)
Iron	7.80	1.5	1.279	1.05	0.095	0.0105	7.43	d	0.0132	7.36	d	0.024	6.10	d	0.0867	d
Gabbroic anorthosite (lpp)	2.936	0.145	0.705	0.751	4.89	0.0176	2.80	2.663	0.02065	2.77	2.837	0.0472	2.57	3.517	0.1816	7.10
Gabbroic anorthosite (hpp)	3.965	0.128	2.357	1.258	18.0	0.00575_b	3.89	2.663	0.00964	3.88	2.837	0.0319	3.79	3.517	0.16836	7.10

[α] At 1 atmosphere and high temperature.
[β] With respect to gabbroic anorthosite, lpp at STP.
[c] With respect to gabbroic anorthosite, lpp at STP value of $S_{P=1\,\mathrm{atm}}^{298°\mathrm{K}} = 0.534 \times 10^7$ erg/g °K.
[d] Not utilized in calculations.

Table 2. Peak shock (impedance match) states for gabbroic anorthosite impacting gabbroic anorthosite.

Impact velocity (km/sec)	Shock pressure (Mbar)	Particle velocity (km/sec)	Shock velocity (km/sec)	Density (gm/cm^3)	Internal[a] energy density (10^{10} erg/g)
5	0.62[b]	2.50	8.45	4.17	3.13
7.5	0.99[b]	3.75	8.99	5.04	7.03
15.0	3.04[c]	7.50	13.81	6.43	28.13
30.0	10.11[d]	15.0	22.96	8.47	112.50
45.0	21.29[d]	22.50	32.23	9.73	253.13

[a]With respect to low-pressure phase of gabbroic anorthosite at STP.
[b]Partial melting upon isentropic release.
[c]Partial vaporization upon isentropic release.
[d]Complete vaporization upon isentropic release.

and found their intersection with the Hugoniot curves (Table 4). In the case of iron and the lpp of GA, this is a straightforward procedure which merely involves numerically calculating the locus of V, P, and E_s (energy) states reached by an isentropic process which satisfies the energy integral along the isentrope

$$E_s = -\int_{V(\mathrm{IM,CM,\ or\ IV})|_{P=0}}^{V} V(P, E)\, dP. \tag{6}$$

In principle, the actual value of the entropy along one of the isentropes, so calculated, is not explicitly specified but, in fact, requires an additional assumption equivalent to a thermodynamic model for C_P or C_v, the specific heat at constant pressure or volume. Instead, for the lpp we have used the available thermodynamic data for enstatite and anorthite, to separately estimate the entropy associated with the critical isentropes for IM, CM, and IV using the method outlined in Ahrens and O'Keefe (1972). (These data are utilized in calculating the critical isentropes for the hpp assemblage below.)

In the case of CV, the pressure is obtained by using the distended formulation of the Tillotson (1962) equation-of-state:

$$P = \frac{aE}{V} + \left\{\frac{bE/V}{(E/E_0\eta^2)+1} + A\mu e^{-\beta[(V/V_0-1)-1]}\right\} e^{-\alpha[V/V_0-1]^2} \tag{7}$$

in the regime where

$$1 > V/V_0 \qquad \text{for } E > E_{\mathrm{CV}},$$

where $\alpha = \beta = 5$, and E_{CV} is the CV energy at standard pressure. In the partial vaporization regime

$$V_0 > V \quad \text{and} \quad E_{\mathrm{IV}} < E < E_{\mathrm{CV}},$$

where E_{IV} is the energy associated with IV. We utilize the interpolation relation

Table 3. Peak shock (impedance match) states for iron impacting gabbroic anorthosite.

Impact velocity (km/sec)	Shock pressure (Mbar)	Particle velocity[a] iron (km/sec)	Shock velocity[a] iron (km/sec)	Density iron (g/cm^3)	Internal energy iron (10^{10} erg/g)	Particle velocity anorthosite (km/sec)	Shock velocity anorthosite (km/sec)	Density anorthosite (g/cm^3)	Internal energy density[b] anorthosite (10^{10} erg/g)
5.0	0.825[c]	1.59	6.60	10.35	1.26	3.41	8.24	5.01	5.81
7.5	1.531[d]	2.51	7.76	11.62	3.15	4.99	10.45	5.62	12.45
15.0	4.807[e]	5.21	11.74	14.13	13.57	9.79	16.72	7.08	47.92
30.0	15.852[f]	10.70	18.85	18.18	57.25	19.30	27.98	9.47	186.25
45.0	33.556[g]	16.35	26.11	21.02	133.66	28.65	39.89	10.42	410.41

[a]With respect to iron, at rest.
[b]With respect to low-pressure phase of gabbroic anorthosite at STP.
[c]Iron solid upon isentropic release; gabbroic anorthosite completely melted upon isentropic release.
[d]Iron solid upon isentropic release; gabbroic anorthosite partially vaporized upon isentropic release.
[e]Both iron and gabbroic anorthosite partially vaporized upon isentropic release.
[f]Gabbroic anorthosite vaporized upon isentropic release; iron partially vaporized.
[g]Both iron and gabbroic anorthosite completely vaporized upon isentropic release.

Table 4. Hugoniot pressures (Mbar) inducing a change of state upon isentropic release to atmospheric pressure.

	Iron	Gabbroic anorthosite (lpp)	(hpp)
Incipient melting	2.2	0.97	0.43
Complete melting	2.6	1.1	0.52
Incipient vaporization	4.2	1.9	1.02
Complete vaporization	16.8	7.9	5.9

(Allen, 1967; Hageman and Walsh, 1970)

$$P = \frac{(E - E_{IV})P_E + (E_{CV} - E)P_C}{(E_{CV} - E_{IV})}, \tag{8}$$

where P_E refers to the pressure calculated from Eq. (7), the distended region, and P_C refers to the pressure calculated from Eq. (1), the compressed region.

The straightforward calculation of the isentrope (Eq. (6)) using the distended formulation Tillotson equation-of-state and starting at a one atmosphere pressure state at complete vaporization leads to unreasonable pressures ($\simeq 10^2$ Mbar) for the intersection of the complete vaporization isentrope with the Hugoniot. The reason for this is that the second term in Eq. (7) does not have the correct form in the distended regime, where μ is negative. The correct cohesive energy, $\simeq E_{CV}$, is not obtained by taking $|\int_{V=V_{IV}}^{V=\infty} P\,dV|$. This problem was partially circumvented by choosing a reference state for complete vaporization, which did not involve the distended region of the equation-of-state. To this end, we adopted the approach of Zeldovich and Raizer (1967), who demonstrate, for single phase materials, that the isentrope that passes through the vapor–liquid critical point has a specific internal energy equal to approximately twice the binding energy at standard volume. Because of this assumption the curves shown in Figs. 2 through 4 for CV should be considered approximate.

The construction of release isentropes for the hpp assemblage (Fig. 4) which is used as a model for GA, requires two independent assumptions in addition to the parameters of the P, V, E equation-of-state:

(1) The entropy difference between the lpp and hpp assemblage; no data presently exists giving the slope of pertinent phase boundaries.
(2) A model for the specific heat for the assemblage.

Although more elaborate methods have recently been suggested for estimating entropies of phases based on detailed knowledge of crystal structure (e.g., Saxena, 1976), we have used the method of Fyfe (1958) to obtain the entropy of the hpp assemblage. This method sums the entropies of the equivalent oxide mixture (data obtained from Robie and Waldbaum, 1968). By using the STP entropy of stishovite, this procedure probably provides a good approximation, as the major effect of the assumed phase changes in both pyroxene and plagioclase is the increase of Si^{+4} coordination from four to six O^{-2} ions. Upon mass weighting entropies for the plagioclase and pyroxene hpp mixture, the transition entropy at STP is given by:

$$\text{lpp} \rightarrow \text{hpp}; \quad \Delta S_{TR, P=0}^{298°\text{K}} = -0.188 \times 10^7 \text{ erg/g °K}. \tag{9}$$

Recognizing the shortcomings of the Debye theory for describing C_V, when applied to silicates (Kieffer, 1977), we have assumed an equivalent STP, Debye temperature, Θ_D, for the hpp assemblage. Using the density-Debye temperature relation proposed for oxides by Anderson (1965), a value of $\Theta_D = 1029°$ K is obtained. The entropy, at high temperature at atmospheric pressure for the hpp assemblage, is then calculated from

$$S_{\text{hpp}}^{T} = S_{\text{hpp}}^{298°\text{K}} + \int_{298°\text{K}}^{T} (C_p/T)\, dT, \tag{10}$$

where, taking into account the entropy of the known lpp assemblage, and Eq. (9), $S_{\text{hpp}}^{298°\text{K}} = 0.534 \times 10^7$ erg/g °K.

From the thermodynamic relations:

$$C_P = C_V\left(\frac{\Theta_D}{T}\right) \Big/ \left[1 - C_V\left(\frac{\Theta_D}{T}\right)\left(\frac{\partial V}{\partial E}\right)_P \frac{\gamma}{V} T\right], \tag{11}$$

and from Eq. (1), it follows that

$$\left(\frac{\partial V}{\partial E}\right)_P = \left\{\frac{a}{V} + \frac{b}{V\chi} - \frac{bE}{V\chi^2}\left[\frac{1}{E_0}\left(\frac{V}{V_0}\right)^2\right]\right\}\left\{\frac{aE}{V^2} + \frac{bE}{\chi V^2} + \frac{2bE^2}{\chi^2 V_0^2 E_0} + \frac{AV_0}{V^2} + 2B\left(\frac{V_0}{V} - 1\right)\frac{V_0}{V^2}\right\}^{-1}, \tag{12}$$

and

$$\gamma = a + \frac{b}{\chi} - \frac{bE}{\chi^2}\left[\frac{1}{E_0}\left(\frac{V}{V_0}\right)^2 + 1\right], \tag{13}$$

where $\chi = E/E_0$.

The variation of Θ_D with volume follows from

$$(\partial \Theta_D / \partial T)_P = -\gamma\alpha, \tag{14}$$

where

$$\alpha = \frac{C_V}{V}\left(\frac{\partial V}{\partial E}\right)_P \left[1 + \frac{1}{V}\left(\frac{\partial V}{\partial E}\right)_P \gamma T\right]. \tag{15}$$

In order to obtain the internal energies, relative to the lpp at STP, for the hpp at one atmosphere, such that the entropy value associated with the critical isentropes for IM, CM, IV, and CV are the same as for the lpp, we have simultaneously solved Eqs. (10)–(15), using T, the absolute temperature as the independent variable. Thus upon isentropic release from a shock state either in the lpp or hpp regime to one atmosphere and high temperature, and achieving states corresponding to IM, CM, IV, or CV, the entropy of the material is the same, however, the isentropic paths from high-pressure differ for the lpp and hpp cases, inferring that reverse transformation from the hpp occurs.

We have not yet compared the results of the Debye theory, when used with

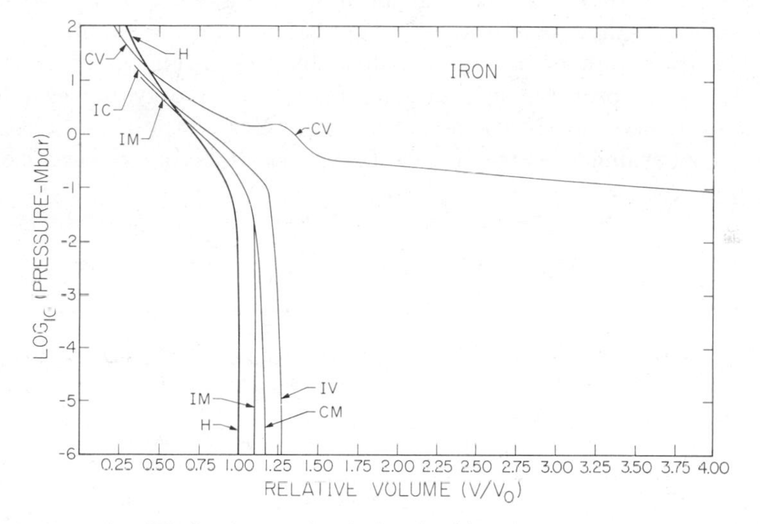

Fig. 2. Hugoniot (H) for iron and calculated critical isentropes, corresponding to incipient melting (IM), complete melting (CM), incipient vaporization (IV), and complete vaporization (CV). The H and IM curves agree closely with similar results obtained by Tillotson (1962).

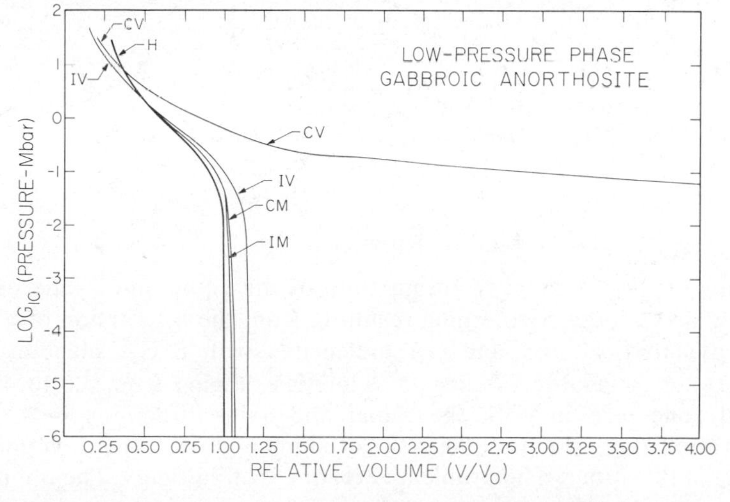

Fig. 3. Hugoniot and critical isentropes for lpp, gabbroic anorthosite. Curve labeling similar to that in Fig. 2.

the Tillotson formulation as described in Eqs. (12)–(15), when the high temperature entropy has been explicitly measured.

To describe the low cohesion of iron or GA, we have assumed $P_C = 0$, when $V_0/V < 0.99$ and $V_0/V < 0.995$ respectively. This corresponds to a dynamic tensile strength of 0.64 and 0.71 kbar, respectively. The latter physical quantities are both poorly constrained experimentally, although recent measurements of the tensile strength of novaculite and quartzites (Shockley *et al.*, 1973) report some data in this range. By assuming $P_C = 0$ upon tensile failure, a physical model of the formation of a fragment dust and coexisting vapor is implicitly assumed. The total pressure upon failure results from only the gaseous (P_E) component of a mixture of dust and gas. Using Eqs. (1), (6), and (7), and the assumption concerning dynamic brittle failure to dust, the release isentropes shown in Figs. 2, 3, and 4 are calculated.

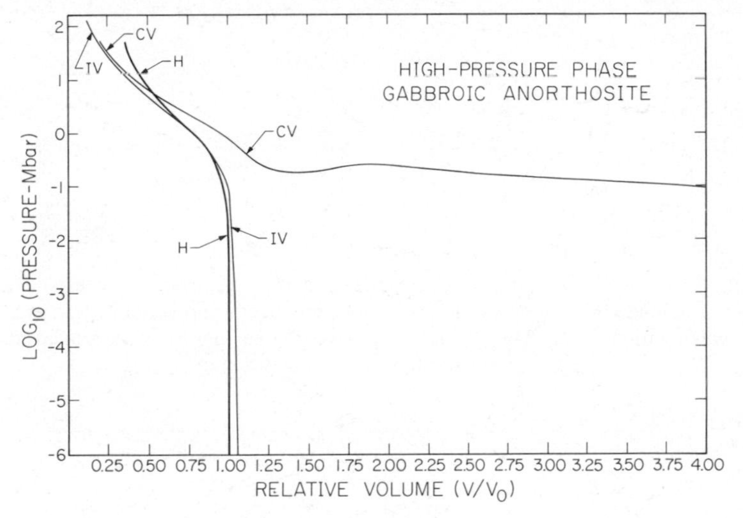

Fig. 4. Hugoniot and critical isentropes for hpp, gabbroic anorthosite. Curve labeling similar to that in Fig. 2. Compression, V/V_0, corresponds to zero-pressure specific volume of hpp.

RESULTS

Applying the above detailed formulation of the equation-of-state of GA, the impact flow and energy partitioning resulting from the interaction of a series of spherical hypothetical iron and GA meteorites with a GA planetary surface (half-space) are calculated for impact velocities ranging from 5.0 to 45 km/sec. The initial zone size in both the radial and axial directions is 0.5 cm. The problem is run from ~300 to ~500 μsec of simulated time at which point all the stresses are ~10^{-3} Mbar. The problem is rezoned 4 to 6 times. The boundaries of the meteoroid and half-space are constrained to have zero normal stresses. The finite difference algorithms used are given in detail in Hageman and Walsh

(1970). Previous results for energy partitioning and ejecta distributions upon impact of a 10 cm diameter, 15 km/sec iron object are given in O'Keefe and Ahrens (1976). Ejecta (0.5 cm, initial zone size) distributions and energy partitioning versus impact velocity will be presented elsewhere (O'Keefe and Ahrens, in preparation). Because the shape of the isentropes for the hpp assemblage of GA strongly controls peak pressure decay, cf., O'Keefe and Ahrens (1976), and field data relating to spatial shock attenuation are beginning to become available, cf., this volume (Robertson and Grieve, 1977), initial calculations of shock attenuation along the impact, symmetry axis are presented in Figs. 5 and 6 and Table 5. The steep release adiabats which are associated with release from shock states above 0.15 Mbar, give rise to the rapid attenuation rates indicated in the far-field regimes in Figs. 6 and 7. The effect of

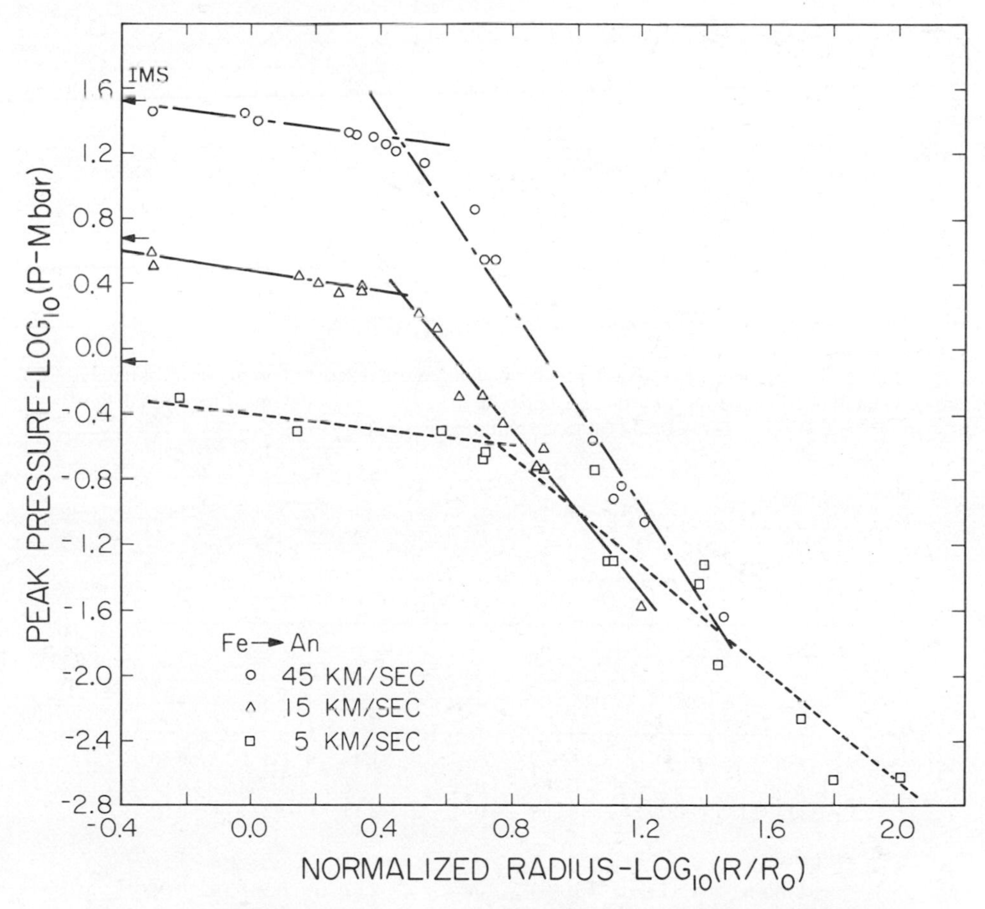

Fig. 5. Log_{10} peak shock pressure versus $\log_{10}$ normalized radius at various impact velocities. R_0 is radius of impactor for iron object. Parameters of lines fit through calculation are given in Table 5. Arrows (IMS) indicate impedance-match solutions (Table 3) valid for one-dimensional flow.

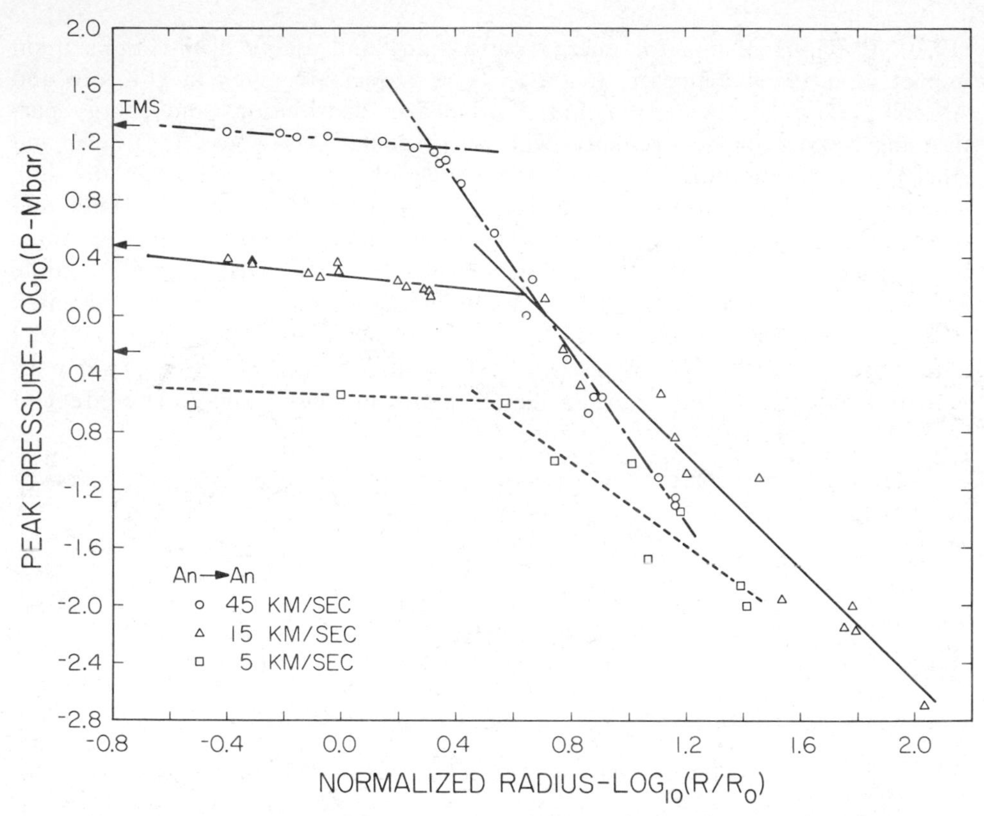

Fig. 6. Log_{10} peak shock pressure versus $\log_{10}$ normalized radius at various impact velocities for gabbroic anorthosite impactors. Arrows indicate one-dimensional flow pressures (Table 2). Notation is similar to Fig. 5.

Table 5. Peak centerline pressure attenuation; $\log_{10}(P - \text{Mbar}) = a \log_{10}(R/R_0) + b$.

	Near-field			Far-field		
An → An	a	b	$r^{2\,(a)}$	a	b	$r^{2\,(a)}$
5 km/sec	—	—	—	−1.45	0.151	0.85
15 km/sec	−0.222	0.285	0.83	−1.97	1.399	0.95
45 km/sec	−0.293	1.208	0.92	−2.15	2.373	0.92
Fe → An						
5 km/sec	−0.232	−0.398	0.69	−1.67	0.681	0.91
15 km/sec	−0.295	0.470	0.88	−2.49	1.475	0.97
45 km/sec	−0.149	1.220	0.89	−2.95	2.082	0.99

[a]Linear regression correlation coefficient.

including the phase change is explicitly discussed by O'Keefe and Ahrens (1976). The scatter is largely the result of the rezoning procedure in the calculation. As previous workers have noted (e.g., Gault and Heitowit, 1976; Bjork and Rosenblatt, 1965; Heyda and Riney, 1965), two regimes of attenuation are to be expected. These are indicated as near- and far-field in Table 5. Within the precision of the present calculations, centerline peak pressure decay in the near- and far-field is described in terms of:

$$\log_{10}(P - \text{Mbar}) = a \log_{10}(R/R_0) + b, \qquad (15)$$

where P is the centerline pressure, R is the distance from the point of impact and R_0 the initial meteoroid radius. In the near-field regime, the peak pressure may be obtained from the one-dimensional impedance match solution (Tables 3 and 4). The agreement with impedance match solutions indicated in Figs. 5 and 6 appears to be closer, the higher the impact speed, as $R \rightarrow 0$. This may be due to shorter time steps in the calculation at the higher impact speeds. The slow spatial decay rate observed in the near-field is insensitive to both impact velocity and impactor type. In the present case of a spherical projectile, the near-field shock attenuation undoubtedly results from the slightly divergent flow induced in the target as the projectile imbeds itself into the target. In the case of the normal impact of a flat-nosed cylinder or rod, or in the assumed constant-hemisphere energy model of Gault and Heitowit, no attenuation, i.e., $a = 0$, is expected in the near-field regime. The onset of the characteristic far-field attenuation rate begins at depths corresponding to ~2.6 to ~5.6 projectile radii for the case of iron impacting GA as the impact velocity decreases from 45 to 5 km/sec. In the case of GA impacting GA, the far-field rates are established at depths of from ~2.2 to ~3.5 projectile radii, as the impact velocity decreases from 45 to 5 km/sec. Dienes and Walsh (1970) have shown as part of their late-stage equivalence studies of hypervelocity impact that late-stage far-field flows are insensitive to the exact shape of the impactor, provided the axial to radial dimensions of the impactor are roughly comparable.

Comparison of the present results with other attenuation calculations and measurements are of interest. Dienes and Walsh found good agreement between their computer solution with experimental data, over a range of ~10–1000 kbar, for the impact at 7.3 km/sec of an aluminum sphere with an aluminum half-space. The centerline particle velocity in this case, which is to first-order proportional to shock pressure, decays as, $a \cong -2$, which is comparable to the value of "a", listed in Table 5, for the far-field, in this impact velocity range. A larger set of experimental impact data, up to ~8 km/sec, for various aluminum projectiles is shown to satisfy a relation comparable to present results, where $a \cong -1.6$ (Billingsley, 1969). Gault and Heitowit (1963), on the basis of their constant energy, hemisphere model, calculate, for the impact at 6.35 km/sec of an aluminum sphere onto basalt, (similar to the present An $\rightarrow$ An calculation) an initial pressure decay rate with $a \cong -3$ to -4, which is considerably greater than predicted here. However, at the point where the shock has decayed to pressures on the order of $\sim 10^2$ kbar, their indicated attenuation rate is considerably slower than is predicted here.

Since the cratering effects of large explosions are often used for scaling of impact flows to greater dimensions than those available from results of experiments, it is of interest to compare pressure decay rates from near-surface and contained explosions. A large number of such data are summarized by Cooper (1973), who finds a decay coefficient of the centerline pressure of $a \cong -2$ is compatible with field data for both contained and surface explosions. This is close to the value found, for example, for An$\rightarrow$An at 15 km/sec. Also, Butkovich and Borg (1974) demonstrated that the decay of the experimentally observed shock pressure in the range 650 to ~ 1 kbar surrounding the 5 kton, Hardhat nuclear explosion in granodiorite also gave an $a \cong -2$ dependence. Moreover, they point out that in general, calculational results such as presented here, always give, to a first approximation, the decay rates in terms of pre-flow, i.e., Eulerian coordinates, whereas attenuation inferred from post-flow shock metamorphic features, must for meaningful comparisons, be corrected such that material motion, after the shock wave has passed, is taken into account.

DISCUSSION

The major new result described here is that different impact velocities and meteorite shock impedances give rise to significantly different spatial peak shock attenuation rates in the far-field.

We anticipate that present calculational results will prove useful in placing bounds on the shock pressures, experienced by both *in situ* rocks and ejecta, in the vicinity of impact craters, which petrologic and geochemical analyses indicate are partially melted, melted, or, incongruently vaporized. We are well aware of the difficulties in determining the pre-impact position of different rock units in other than structures formed in simple sedimentary terranes, e.g., Gosses Bluff. Moreover, it should be pointed out that hypervelocity impact on a laterally inhomogeneous or anisotropic terrane will necessarily influence the shock attenuation rate. Aside from those difficulties, (a) given the size of a crater, and hence a measure of the total projectile energy, (b) some knowledge of the chemistry, and hence shock impedance, of the impactor, and (c) a measure of the spatial rate of peak shock-pressure attenuation (probably inferred from shock metamorphic features) we believe it will be possible, on the basis of the present results, to estimate the impact velocity associated with a given crater. Although the required knowledge of crater size, meteoroid impedance, and spatial shock decay rate is now limited to only a few terrestrial craters, we hope the methodology presented will be useful in inferring impact velocities and cratering histories for other terrestrial planets.

Acknowledgments—This research supported under NASA Grant, NSG 7129. We appreciate the computational assistance of M. Lainhart, José Helu, and J. Huber and the opportunity to present this material at the Symposium to a critical audience. We have profited from critical comments on this manuscript offered by Raymond Jeanloz, G. Wayne Ullrich, and Robert N. Schock.

REFERENCES

Ahrens, T. J., and O'Keefe, J. D.: 1972, Shock melting and vaporization of lunar rocks and minerals. *The Moon*, **4**, 214–249.

Ahrens, T. J., O'Keefe, J. D., and Gibbons, R. V.: 1973, Shock compression of a recrystallized anorthositic rock from Apollo 15, (*Suppl. 4*) *Geochim. Cosmochim. Acta*, **3**, 2575–2590.

Allen, R. T.: 1967, Equation of state of rocks and minerals, General Dynamics, General Atomic Division, Special Nuclear Effects Laboratory Report, GAMD-7834A, p. 25.

Anderson, O. L.: 1965, Determination and some uses of isotropic elastic constants of polycrystalline aggregates using single-crystal data, In *Physical Accoustics 3 B*, W. P. Mason (ed.), Academic Press, 43–95

Billingsley, J. P.: 1969, Comparison of experimental and predicted axial pressure variations for semi-infinite metallic targets. *Proc. AIAA Hypervelocity Impact Conference, Paper #69–361, Vol. III, AIAA*, p. 10.

Bjork, R. L., and Rosenblatt, M.: 1965, Hypervelocity impact of end-oriented rods. *Proc. Seventh Hypervelocity Impact Symposium, IV*, Martin Marietta Corp., Tampa, Florida, Vol. 4, p. 195–211.

Butkovich, T. R., and Borg, I. Y.: 1974, Peak radial pressures associated with the hard hat underground nuclear explosion especially as they pertain to shock effects in recovered grandiorite, Lawrence Livermore Laboratory Report UCLR-75594, p. 31.

Cooper, H. F., Jr.: 1973, Empirical studies of ground shock and strong motions in rock, R&D Associates, Santa Monica, Ca, Rpt. DNA 3245F, p. 85.

Dence, M. R.: 1972, The nature and significance of terrestrial impact structures. *Proc. 24 Int. Geol. Cong., Sect. 15*, 77–89.

Dienes, J. K., and Walsh, J. M.: 1970, Theory of impact: Some general principles and the methods of Eulerian codes, In *High Velocity Impact Phenomena*, R. Kinslow (ed.), Academic Press, 46–104.

Duvall, G. E., and Fowles, G. R.: 1963, Shock waves, In *High Pressure Physics and Chemistry 2*, R. S. Bradley (ed.), Academic Press, 209–292.

Fyfe, W. S., and Verhoogen, J.: 1958, General thermodynamic considerations, In *Metamorphic Reactions and Metamorphic Facies*, W. S. Fyfe, F. J. Turner, and J. Verhoogen (eds.), *Mem. 73, Geol. Soc. Am.*, 21–51.

Gault, D. E., and Heitowit, E. D.: 1963, The partition of energy for hypervelocity impact craters formed in rock. *Proc. Sixth Hypervelocity Impact Symposium*, Cleveland, Ohio, Vol. 2, p. 419–456, (unpublished).

Hageman, L. J., and Walsh, J. M.: 1970, Help, a multimaterial eulerian program for compressible fluid and elastic-plastic flows in two space dimensions and time, Systems, Science, and Software Report 3SR-350, Vol. 1, p. 139.

Heyda, T. F., and Riney, T. D.: 1965, Peak axial pressures in semi-infinite media under hypervelocity impact. *Proc. Seventh Hypervelocity Impact Symposium, IV*, Martin Marietta Corp., Tampa, Florida, Vol. 3, p. 75–122.

JANAF Thermochemical Tables: 1976, Thermal Research Laboratory, Dow Chemical Company, Midland, Michigan , Clearinghouse for Federal Scientific and Technical Information, PB 168–370.

Kieffer, S. W.: 1977, Thermodynamics and lattice vibrations of minerals: I. Mineral heat capacities and their relationships to simple lattice vibrational models. Submitted to *J. Geophys. Res.*

Lambert, P.: 1977, The meteoritic contamination in the Rochechouart crater: statistical investigations. In *Impact and Explosion Cratering*, D. J. Roddy, R. O. Pepin, and R. B. Merrill (eds.). This volume.

Long, G.: 1970, Density of liquid elements, In *Chemical Rubber Handbook of Chemistry and Physics, 50th Ed.*, B 225–226.

McQueen, R. G., Marsh, S. P., and Fritz, J. N.: 1967, Hugoniot equation of state of twelve rocks. *J. Geophys. Res.*, **72**, 4999–5036.

McQueen, R. G., Marsh, S. P., Taylor, J. W., Fritz, J. N., and Carter, W. J.: 1970, The equation of state of solids from shock wave studies, In *High-Velocity Impact Phenomena*, R. Kinslow (ed.), Academic Press, p. 294–416.

Morgan, J. W., Higuchi, H., Ganapathy, R., and Anders, E.: 1975, Meteoritic material in four terrestrial meteorite craters. *Proc. Lunar Sci. Conf. 6th*, 1609–1623.

O'Keefe, J. D., and Ahrens, T. J.: 1975, Shock effects from a large impact on the moon. *Proc. Lunar Sci. Conf. 6th*, 2831–2844.

O'Keefe, J. D., and Ahrens, T. J.: 1976, Impact ejecta on the moon. *Proc. Lunar Sci. Conf. 7th*, 3007–3026.

Robertson, P. B., and Grieve, R. A. F.: 1977, Shock wave attenuation: Apparent variation with crater dimensions. In *Impact and Explosion Cratering*, D. J. Roddy, R. O. Pepin, and R. B. Merrill (eds.). This volume.

Robie, R. A., and Waldbaum, D. R.: 1968, Thermodynamic properties of minerals and related substances at 298.15°K (25.0°C) and one atmosphere (1.013 bars) pressure and at high temperature. *Geol. Survey Bull.*, **1259**, 265.

Saxena, S. K.: 1976, Entropy estimates for some silicates at 298°K from molar volumes. *Science*, **143**, 1241–1242.

Shockley, D. A., Petersen, C. F., Curran, D. F., and Rosenberg, J. T.: 1972, Dynamic tensile failure of rocks, Stanford Research Institute Report, PYU-1087, pp. 70.

Tillotson, J. H.: 1962, Metallic equations of state for hypervelocity impact. *General Atomic Report GA* 3216, p. 137.

Zeldovich, Y. B., and Raizer, Y. P.: 1966, *Physics of Shock Waves and High Temperature Phenomena*, Academic Press, New York, Vol. II.

Roddy, D. J., Pepin, R. O., and Merrill, R. B., editors.
(1977) *Impact and Explosion Cratering*, Pergamon Press (New York), p. 657–668.
Printed in the United States of America

The response of rocks to large stresses

R. N. SCHOCK

University of California, Lawrence Livermore Laboratory, Livermore, California 94550

Abstract—To predict the dimensions and characteristics of impact- and explosion-induced craters, one must know the equation of state of the rocks in which the crater is formed. Recent experimental data shed light upon inelastic processes that influence the stress/strain behavior of rocks. We examine these data with a view to developing models that could be used in predicting cratering phenomena. New data are presented on the volume behavior of two dissimilar rocks subjected to tensile stresses.

INTRODUCTION

WHEN A BODY IMPACTS or explodes in rock or soil, some of the energy is converted into heat and some into mechanical work on the surrounding material. The relative magnitude of heat and work as well as the form of the work itself (i.e., elastic, inelastic, fracture, compaction) are directly dependent on the mechanical response of the surrounding material to the stress conditions. Thus, knowledge of the nature of the relation between stress and mechanical response (strain) is needed to predict the effect of an impact or explosion. The crater form can be predicted if source parameters are known; the source energy and boundary conditions can be predicted if the crater form is known. Terhune and Stubbs (1970) have given an excellent description of the effect of material parameters (such as strength and compressibility) on crater dimensions. They have also compared calculations and observations of explosion-created craters. The purpose of this paper is to review recent experimental work on the response of rock to stress. We seek constitutive relations that can be incorporated into computer codes whose function is to predict the phenomenology of explosions or impacts. The underlying purpose of these experiments has been to increase understanding of the physical processes responsible for observed behavior, so that models developed can be applicable to a broad range of stress and strain conditions (i.e., so that they can be truly predictive, rather than simply fitted to experimental data).

To develop inelastic constitutive models, we must determine the stress and the strain tensor and the tensor that couples them (Schock, 1970) over the range of conditions encountered. Unlike those for elastic materials, the moduli (or stress/strain coupling coefficients) are not independent of stress state. The range of conditions in rock is commonly described by mean pressure (or the invariants of stress), shear stress (or the invariants of stress deviation), and strain rate. Constitutive relations may be considered in terms of stress and strain or, if strain-rate effects are considered, in terms of stress and strain rates. Experimental arrangements incorporating this range of conditions have become common.

See, for example, the papers by Brace *et al.* (1966), Scholz (1968), Swanson and Brown (1971), Schock and Duba (1972), Schock *et al.* (1973), and Scholz and Kranz (1974). The answers to such questions as, "How complex must the models be to accurately describe the desired behavior?" and "What kinds of approximations must be used to make them usable in a time-limited computer code?" are of interest.

In the discussion that follows, brittle, ductile, and porous rocks will be discussed in order. Then, important effects such as fluid saturation and sample size will be considered. Within each classification, behavior in compression, tension, and at high strain rate will be considered. In almost all of the experimental work considered, there was a stress geometry such that the intermediate principal stress was equal to either the maximum or minimum principal stress. Other experimental conditions are difficult to achieve and will be given only brief mention. However, it should be noted that for point source explosions or impacts, shock waves and resultant stresses closely approximate this geometry over significant times and distances.

Brittle Rocks

For our purposes, the failure of brittle rocks may be characterized by a through-going fracture that propagates at sonic or nearly sonic speeds. Inelastic behavior may be observed in the axial stress/strain relation, but total axial strain before failure is usually less than 1% (Griggs and Handin, 1960). Granite, limestone and dolomite at low confining pressure, and quartz-cemented sandstones and metamorphic rocks exhibit this behavior. Failure stress in these materials is strongly dependent on and increases with confining pressure (Jaeger and Cook, 1969). It is not uncommon to observe changes of an order of magnitude with 0.1 GPa confining pressure. This is due primarily to the strong effect of pressure, which increases friction and thereby inhibits sliding on inter-granular crack surfaces.

One of the most striking characteristics of low-porosity, brittle rocks is that before they fail in compression, there is a pronounced nonlinear behavior in the axial-stress/radial-strain relation. This results from inelastic volume dilatancy (Brace *et al.*, 1966), which characteristically precedes failure in these rocks (Scholz, 1968; Schock *et al.*, 1973). This behavior has been ascribed to the opening and the propagation of cracks whose major axes are oriented parallel to the active principal stress. Such cracks open with a tensile stress in the region of the crack tip, even though all of the macroscopic stresses are compressive. Cracks with other orientations are compressed shut at much lower pressures.

Swanson and Brown (1971) determined that at constant strain rate, the curve that describes compressive failure in granite as a function of confining pressure is independent of loading path. A similar observation was made by Schock *et al.* (1973) for the onset of dilatant behavior as a function of confining pressure. If one considers the success of critical-strain-energy criteria for failure (Griffith, 1921; Sih and MacDonald, 1974), this uniqueness in behavior suggests a uniqueness in strain at a given mean pressure and shear stress. This hypothesis has been tested on several brittle rocks (Schock, 1976; Costantino and Schock, 1976) and has been found to be true for the stress conditions prescribed. This allows the construction of a constitutive relation that expresses dilatant strain in the form,

$$\epsilon_d = \exp\left[\frac{dP}{x(\tau)} - A(\tau)\right],$$

where dP is an increment of mean pressure, τ is shear stress, and x and A are material constants. The production of dilatant volume thus appears to follow an exponential law. This form of constitutive relation is not only simple, but it expresses the rock behavior in terms of experimentally measurable and thermodynamically definable parameters. The relation has the additional advantage of being able to predict failure shear stress accurately (Schock, 1976). The physical meaning of the exponential form is not yet clear.

After loading in compression due to an impact or an explosion, one or more of the principal stresses in the rock medium may become tensile during unloading. Brittle rocks typically fail in tension at stress levels an order of magnitude or more below those in compression, again presumably because of the lack of friction on grain boundary cracks when the active stress is tensile. Thus, significantly larger volumes of rock may be affected by inelastic phenomena resulting from tensile stresses than compressive stresses. Investigation of this stress regime has been carried out by Brace (1964), who monitored the axial strain and, more recently, in our laboratory (Schock and Louis, 1974), where both axial and radial strain were monitored. These latter results on Westerly granite were obtained on "dog-bone-shaped" samples in the apparatus described by Schock and Duba (1972).

The experimental stress paths are shown in Fig. 1. The circles represent failure points and collectively describe failure in tension as a function of confining pressure. The two points on the ordinate are extension data (all principal stresses are compressive), for which the minimum principal stress was atmospheric pressure. The coincident strain data are shown in Fig. 2 in terms of mean pressure and volume strain. Significantly, the amount of dilatant behavior is a function of the ratio of tensile to compressive stress in the particular test. Apparently, oriented cracks that will not open until they propagate through the specimen when no stress is tensile, are pulled open by the tensile stress when the compressive stress is lowest. The rock also becomes substantially weaker in these instances, perhaps because of the presence of open cracks. The dilatant behavior as a function of the tensile-stress/compressive-stress ratio (as shown in Fig. 2) would seem to lend itself to a simple relation useful in a constitutive equation. More work is required to define the exact form of this relation.

Failure in brittle rocks is a strong function of strain rate (Green and Perkins, 1968; Logan and Handin, 1970; Green *et al.*, 1972). Increases of failure stress of about 5% per order-of-magnitude increase in strain rate generally are observed. This is a significant amount, which, when considered over 8 to 10 orders of

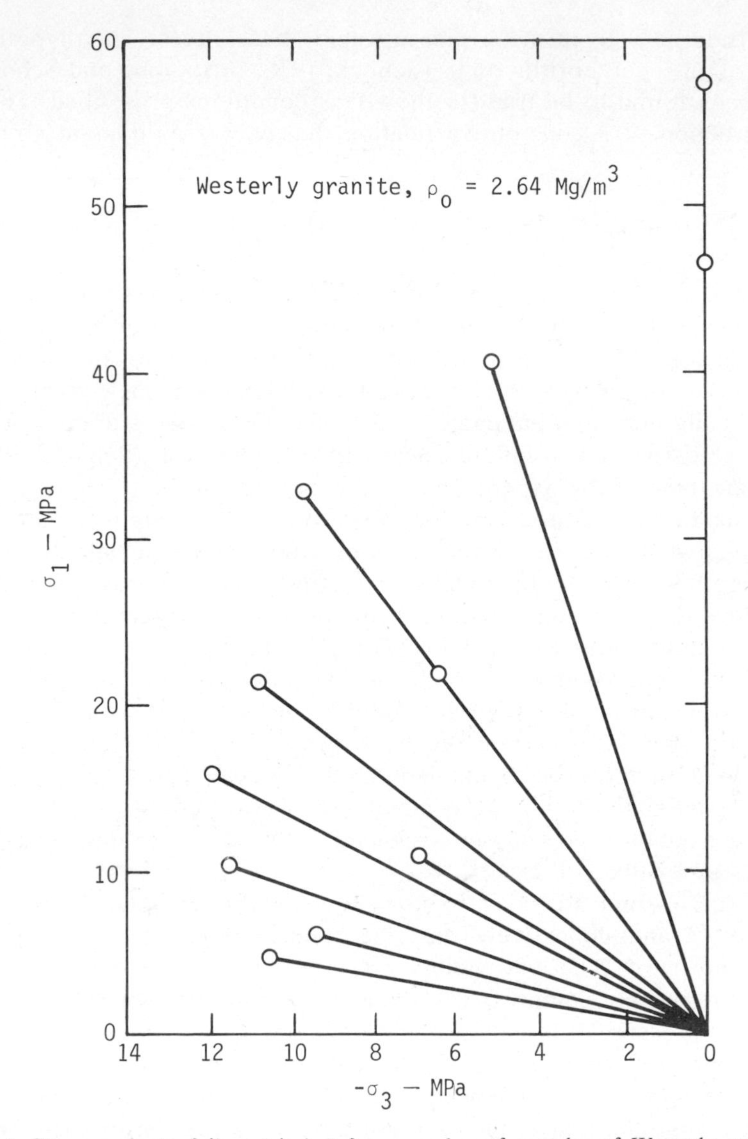

Fig. 1. Stress paths to failure (circles) for a number of samples of Westerly granite (initial density 2.64 Mg/m^3) in terms of the tensile stress (σ_3) and the maximum compressive stress (σ_1). The intermediate principal stress (σ_2) was in all cases equal to σ_1.

magnitude of strain rate, must be accounted for in calculations of the effect of dynamic impulses on rocks. In addition, there is evidence in the combined results of static and shock-wave experiments that the onset of dilatant behavior is suppressed (occurs at higher stress) and that the dilatant strain is reduced as the strain rate increases (Schock and Heard, 1974). This evidence, together with the observation of Scholz (1968) that microfracturing becomes localized only

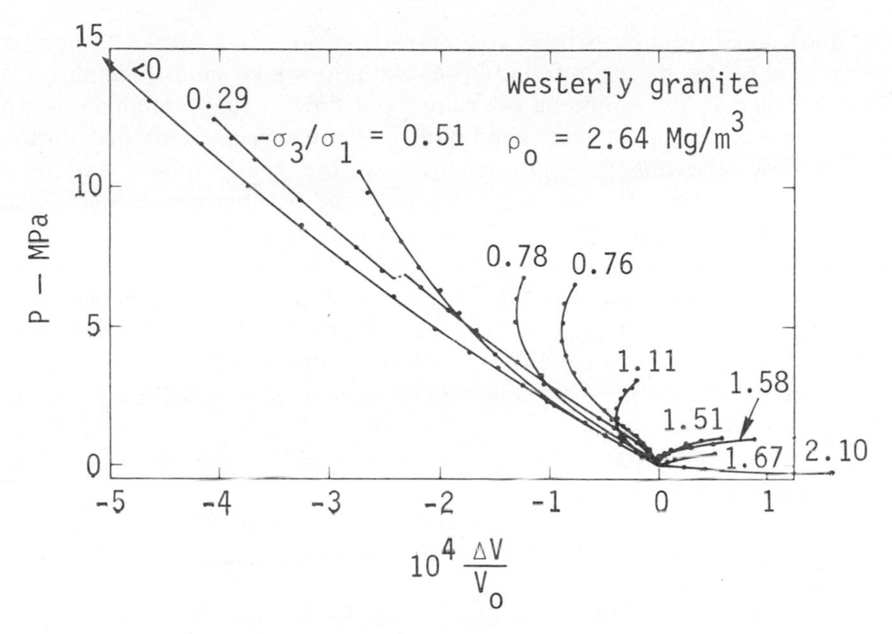

Fig. 2. Behavior of Westerly granite during the loadings shown in Fig. 1 in terms of mean pressure $(2\sigma_1 + \sigma_3)/3$ and sample volume strains.

near the failure stress, suggests that brittle failure at very high strain rates ($>10^4$/s) may be a much more disruptive process involving more of the rock volume than commonly observed visually in the laboratory at low strain rates. At low strain rates, microfractures have time to terminate and relieve local stress concentrations. For example, some brittle rocks are observed to strain for periods of greater than two weeks at constant stress below their fracture strengths (Kranz and Scholz, 1976). Laboratory specimens failed at strain rates of $\sim 10^{-4}$/s commonly show one or two through-going fractures. On the other hand, there is evidence of "pulverized" rock at the edges of nuclear-explosion-induced cavities (Borg, 1972), where extremely high strain rates (perhaps $>10^5$/s) were achieved.

Ductile Rocks

With increased confining pressure, many of the mineral constituents in rock undergo a transition from brittle to ductile behavior (Handin *et al.*, 1967). In addition, some rocks contain minerals that are ductile at normal pressures. The resulting behavior is distinguished from brittle failure in that the rock does not achieve a maximum shear stress at a fixed strain. Many ductile rocks exhibit work-hardening; stress and strain continue to increase in a highly nonlinear manner, with the result that a unique failure surface does not exist.

Since dilatancy is related to the opening of microcracks, it is expected to be an inherent property of brittle rocks and to be absent in ductile materials, where flow and creep reduce stress concentrations at crack tips. Experimental confirmation of this has been found in several graywacke sandstones that exhibit brittle fracture at low confining pressures and flow at high confining pressures (Schock *et al.*, 1973). At high confining pressure, argillaceous and carbonate cements flow, allowing for rearrangement of the brittle quartz and feldspar grains and suppressing the dilatancy characterized by microfracturing.

A diminishing of dilatant behavior also is seen in these rocks when they are subjected to tensile stresses. Graywacke sandstone loaded in a similar manner to the granite in Figs. 1 and 2, exhibits little or no tendency to dilate (Fig. 3). Since at these low confining pressures, the rock fails by brittle fracture, the explanation for this behavior must lie in the nature of the cracks themselves. In order for tensile stress of the order of megapascals (tens of bars) to open cracks in granite, aspect ratios must be very small ($<10^{-5}$) (Walsh, 1965). Thus, the average aspect ratio of the cracks present in this sandstone is large enough so that they do not open before the material fails.

The effect of an increasing strain rate is: (1) to raise the stress level for a given amount of strain (higher deformation modulus), and (2) to raise the pressure at which rocks go from brittle to ductile behavior (Handin *et al.*, 1967; Schock *et al.*, 1973). This decrease in ductility with increasing strain rate amplifies the importance of brittle deformation phenomena in explosive and impact events in rock. Even though there are rocks that behave in a ductile manner at the highest strain rates, most common rock types, if ductile at

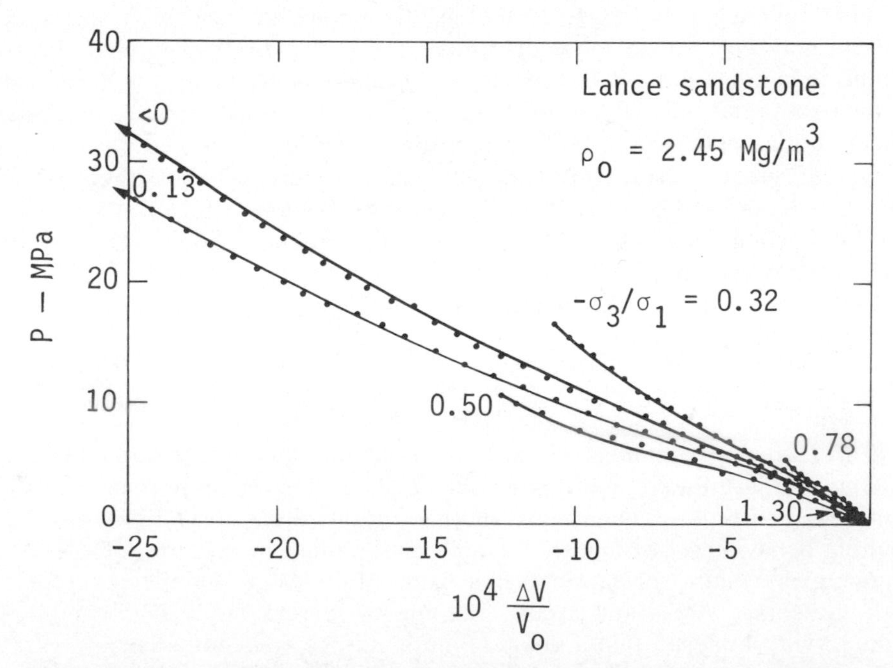

Fig. 3. Behavior of Lance sandstone during loading with one principal stress tensile, in terms of mean pressure and volume strain.

atmospheric confining pressure, show a ductile/brittle transition with increasing strain rate (Handin *et al.*, 1967).

Porosity

Early observation (Schock *et al.*, 1973; Schock and Heard, 1974) indicated that granites and graywacke sandstones did not fail when compressed quasi-statically in uniaxial strain (constant radial strain) to simulate plane-wave shock-loading conditions. On the other hand, loading to failure did take place in very porous, brittle rocks, such as tuff, subjected to the same conditions (Heard *et al.*, 1971). Furthermore, there appeared to be little or no dilatancy prior to failure when these tuffs were loaded at constant confining pressure. This suggests that at least in some rocks, catastrophic pore collapse rather than a through-going fracture was the dominant failure mode. This idea is supported in part by a curvature of the failure envelope concave to the shear stress axis. The subsequent work of Duba *et al.* (1974a) on a sandstone with 26% gas-filled porosity verified these conclusions by demonstrating that the failure envelope was effectively depressed by pore collapse from that for the matrix material without pores.

Another significant observation is that the compressibility of a material in the pore-collapse region is a function of the shear stress (Schock *et al.*, 1971; Schock *et al.*, 1973; Shipman *et al.*, 1974; Schock *et al.*, 1976). The volume strain in the pore collapse region, unlike that in the dilatant region previously discussed, is stress path dependent. This shear-enhanced compaction is not incorporated in most constitutive relations derived to treat inelastic pore collapse (Herrmann, 1969; Carroll and Holt, 1972a; Carroll and Holt, 1972b; Bhatt *et al.*, 1975). Instead, only properties under the hydrostatic or presumed hydrostatic conditions of most experiments are treated.

One of the more successful forms of constitutive relations is

$$P = 2/3\tau(\ln 1/\eta),$$

where τ is yield stress $(\sigma_1 - \sigma_3)$ η porosity, and P pressure in the yield region. This form results from a consideration of the ideally plastic deformation of a hollow sphere. τ can be made to vary with porosity. For rocks, Bhatt *et al.* (1975) considered τ in terms of a Mohr–Coulomb material.

To date, two methods have been used to treat shear-enhanced pore collapse specifically. Shipman *et al.* (1974) fitted data on porous uranium metal. Data on other porous metals (Johnson *et al.*, 1974; Kuhn and Downey, 1971) and on some porous rocks and soils (Nelson *et al.*, 1971) can be fitted with models using movable failure surfaces and computing strain through "associated" flow rules. These models are complex and require a large number of tests to define an equally large number of parameters.

Fluid Saturation

When water is allowed into the pore space in a dry rock, it can introduce large departures from the response to stress. The collapse of pore space in water-saturated rocks is controlled, not only by the strength of the pore wall, but by the compressibility of the water. The pressure on the pore fluid controls failure by dictating the effective stress, i.e., the difference between applied stress and pore pressure (Terzaghi, 1943). When rocks remain completely saturated to failure, the failure stress is observed to decrease with increasing fluid pore pressure (Heard, 1960). During dilatant behavior, brittle rock will behave almost as if it were dry, if the total volume of water is fixed so that the rock becomes unsaturated. As the microcracks open, the resulting volume increase is such that the pore pressure drops until the rock becomes undersaturated (Duba *et al.*, 1974b). If that pore space is connected so as to allow the fluid pressure to increase, strength will decrease. This is the mechanism of the suggested dilatancy model of earthquake generation (Scholz *et al.*, 1973). The brittle/ductile transition is also controlled by the effective stress (Heard, 1960).

Wherever movement of fluid is possible, strong strain-rate effects on behavior are expected (Martin, 1972). The movement of fluid through pore space is a strong function of its viscosity. Since viscosity is the relation between stress and strain rate, it follows that the pressure in the fluid is a function of time, at a given strain.

Water may also introduce complications through its behavior as a thermodynamic fluid during shock loading (Stephens, 1969). Consider a saturated rock that has been adiabatically shocked such that the temperature is above 100°C. On isentropic unloading, the pressure will drop faster than the temperature, and water may convert to steam with a volume increase and a release of energy associated with the latent heat of vaporization. This energy will be added to that from the impact or explosion to enhance crater formation.

SUMMARY

These observations lead to the development of simplified constitutive models that can be used to predict the response of rock to impact loading by identifying the important parameters that determine that response. The importance of shear stress and mean pressure in defining regions of behavior has been demonstrated. It has been shown experimentally that dilatancy is related to failure in low-porosity, brittle rocks. The onset of dilatant behavior can be defined in terms of mean stress and shear stress, and, once begun, it can be described by a simple constitutive relation involving these two system variables. For rocks that exhibit flow instead of fracture, little or no dilatancy is observed.

Compaction of pore space can be an important process in rock behavior in terms of enhanced compression and decreased failure shear strength, both of which absorb energy that might otherwise be used in the cratering process. In addition, compaction is influenced by shear stress. These observations may be summarized in a schematic diagram (Fig. 4). Here, the axes are the two system variables, shear stress and confining pressure (function of mean stress). The failure envelope, which defines the limit of shear stress in terms of mean stress for low-porosity material, the dilatant and compaction region boundaries, and pore

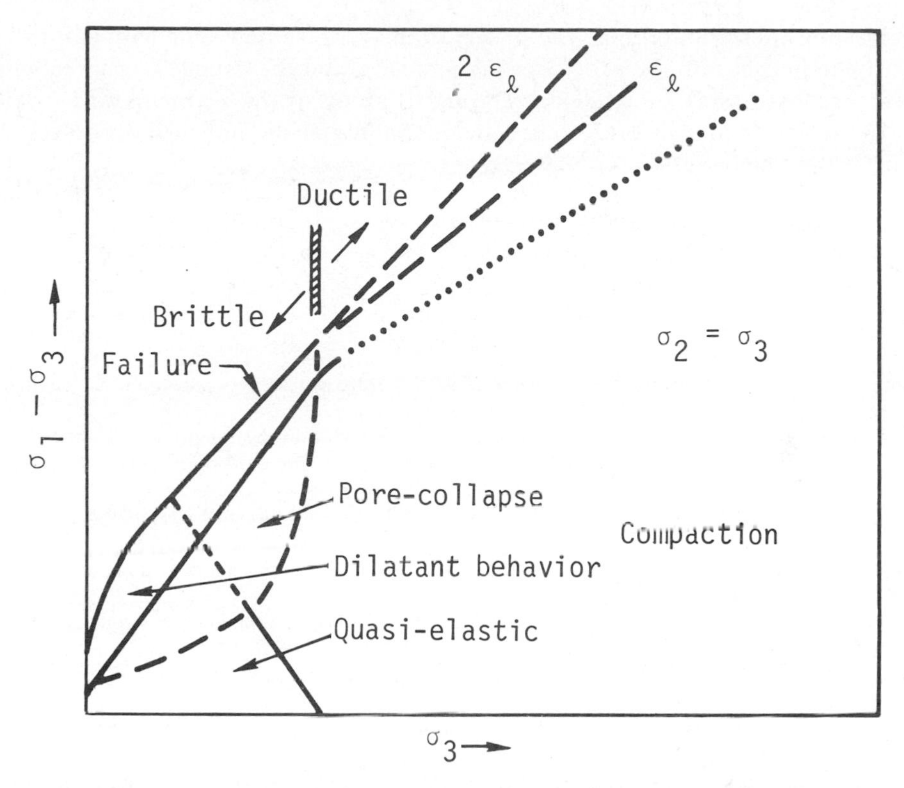

Fig. 4. Schematic representation of boundaries in shear-stress/confining-pressure space for noncyclically loaded rock. Axial strains shown for ductile failure are values of permanent strain. After Schock *et al.* (1973).

collapse and ductile failure envelopes, are all defined in terms of the system variables. In Fig. 4, all of these boundaries are shown in a general, not a rigorous sense. That is, they are movable in both coordinates and may not even exist for a given rock type (e.g., compaction in low-porosity rock). Within a given region, a third system variable may be used to define a constitutive relation, such as has been shown for dilatant and compacting material.

Some qualifications not shown in Fig. 4 must be considered. The influence of water has been mentioned, but it has not been quantified to allow treatment in this sense. More studies are needed. The effect of strain rate likewise is not shown. Cyclical loading affects strength (Peng *et al.*, 1974), dilatant behavior (Scholz and Kranz, 1974), as well as compaction (Schock *et al.*, 1976). The effect of the intermediate principal stress may also be important (Handin *et al.*, 1967; Mogi, 1972), the symmetry of the cratering process at early times notwithstanding. More studies to define and describe this effect are called for. Finally, sample size can be a serious problem, in terms of critical phenomena excluded by the limited size of laboratory samples. Pratt *et al.* (1976) show that failure shear stress can be a strong function of sample size.

Despite these limitations, the observations represented schematically in Fig. 4 provide insight into the processes taking place, and they suggest ways in which these processes may be modeled. Through a study of the mathematical form of the model, we may achieve a better understanding of the physical processes that control the behavior.

Acknowledgment—This work was performed under the auspices of the U.S. Energy Research and Development Administration under Contract No. W-7405-Eng-48.

REFERENCES

Bhatt, J. J., Carroll, M. M., and Schatz, J. F.: 1975, A spherical model calculation for volumetric response of porous rock. *J. Appl. Mech.* **42**, 363.

Borg, I. Y.: 1972, Some shock effects in granodiorite to 270 kbar at the Piledriver site. In *Flow and Fracture of Rocks* (H. C. Heard *et al.*, eds.). Am. Geophys. Union Monograph 16, Washington, D.C.

Brace, W. F.: 1964, Brittle fracture of rocks. In *State of Stress in The Earth's Crust* (W. R. Judd, ed.), Elsevier, New York.

Brace, W. F., Paulding Jr., B. W., and Scholz, C.: 1966, Dilatancy in the fracture of crystalline rocks. *J. Geophys. Res.* **71**, 3939.

Carroll, M. M. and Holt, A.: 1972a, Suggested modification of the Pα-model for porous materials. *J. Appl. Phys.* **43**, 759.

Carroll, M. M. and Holt, A.: 1972b, Static and dynamic pore-collapse relations for ductile porous materials. *J. Appl. Phys.* **43**, 1626.

Costantino, M. S. and Schock, R. N.: 1976, A constitutive relation for compressive loading in Nugget sandstone. Lawrence Livermore Laboratory Report UCRL-52036, Livermore, California.

Duba, A. G., Abey, A. E., Bonner, B. P., Heard, H. C., and Schock, R. N.: 1974a, High-pressure mechanical properties of Kayenta sandstone. Lawrence Livermore Laboratory Report UCRL-51526, Livermore, California.

Duba, A. G., Heard, H. C., and Santor, M. L.: 1974b, Effect of fluid content on the mechanical properties of Westerly granite. Lawrence Livermore Laboratory Report UCRL-51626, Livermore, California.

Green, S. J. and Perkins, R. D.: 1968, Uniaxial compression tests at varying strain rates on three geologic materials. In *Proc. of 10th Symposium on Rock Mechanics* (K. E. Gray, ed.), pp. 35–54, American Institute of Mining, Metallurgical, and Petroleum Engineer, Austin, Texas.

Green, S. J., Lesia, J. D., Perkins, R. D., and Jones, A. H.: 1972, Triaxial stress behavior of Solenhofen limestone and Westerly granite at high strain rates. *J. Geophys. Res.* **77**, 3711.

Griffith, A. A.: 1921, The phenomena of rupture and flow in solids. *Phil. Trans. Royal Society, London* **A221**, 163.

Griggs, D. T. and Handin, J.: 1960, Observations on fracture and a hypothesis of earthquakes. In *Rock Deformation* (D. T. Griggs and J. Handin, eds.), pp. 347–364, Geological Society of America Memoir 79, New York.

Handin, J., Heard, H. D., and Magourik, J. M.: 1967, Effects of the intermediate principal stress on the failure of limestone, dolomite, and glass at different temperatures and strain rate. *J. Geophys. Res.* **72**, 611.

Heard, H. C.: 1960, Transition from brittle to ductile flow in Solenhofen limestone as a function of temperature, confining pressure and interstitial fluid pressure. In *Rock Deformation* (D. T. Griggs and J. Handin, eds.), pp. 193–226, Geological Society of America Memoir 79, New York.

Heard, H. C., Schock, R. N., and Stephens, D. R.: 1971, High-pressure mechanical properties of tuff from the Diamond Mine site. Lawrence Livermore Laboratory Report UCRL-51099, Livermore, California.

Herrmann, W.: 1969, Constitutive equation for the dynamic compaction of ductile porous materials. *J. Appl. Phys.* **40**, 2490.

Jaeger, J. C. and Cook, N. G. W.: 1969, *Fundamentals of Rock Mechanics*. Methuen and Co. Ltd., London.

Johnson, J. N., Shipman, F. H., Green, S. J., and Jones, A. H.: 1974, The influence of deviatoric stress in the compaction of porous metals at high pressure. In *Proc. 4th Intl. Conf. on High Pressure*, pp. 130–137, The Physio-Chemical Society of Japan, Kyoto.

Kranz, R. L. and Scholz, C. H.: 1976, Critical dilatant volume at the onset of tertiary creep (abstract). *Trans. Amer. Geophys. Union* **57**, 330.

Kuhn, H. A. and Downey, C. L.: 1971, Deformation characteristics and plasticity theory of sintered powder materials. *Intl. J. Powder Metall.* **7**, 15.

Logan, J. M. and Handin, J.: 1970, Triaxial compression testing at intermediate strain rates. In *Proc. of 12th Symposium on Rock Mechanics* (G. B. Clark, ed.), pp. 164–194, American Institute of Mining, Metallurgical, and Petroleum Engineers, Rolla, Missouri.

Martin III, R.: 1972, Time-dependent crack growth in quartz and its application to the creep of rocks. *J. Geophys. Res.* **77**, 1406.

Mogi, K.: 1972, Effect of triaxial stress system on fracture and flow of rock. *Phys. Earth Planet. Interiors* **5**, 318.

Nelson, I., Baron, M. L., and Sandler, I.: 1971, Mathematical models for geologic materials for wave propagation studies. In *Shock Wave and the Mechanical Properties of Solids* (J. J. Burke and V. Weiss, eds.), pp. 289–351, Syracuse University Press, Syracuse, New York.

Peng, S. S., Podnieks, E. R., and Cain, P. J.: 1974, Study of rock behavior in cyclical loading. *J. Soc. Petrol. Engr.* **14**, 19.

Pratt, H. R., Johnson, J. R., Swolfs, H. S., Black, A. D., and Brechtel, C.: 1976, Experimental and analytical study of the response of earth materials to static and dynamic loads. Air Force Weapons Laboratory Report AFWL-TR-75-278, Kirtland Air Force Base, New Mexico.

Schock, R. N.: 1970, Dynamic elastic moduli of rocks under pressure. In *Proc. of a Symposium on Engineering with Nuclear Explosives, January 14–16, 1970*, pp. 110, American Nuclear Society, Hinsdale, Illinois.

Schock, R. N., Heard, H. D., and Stephens, D. R.: 1971, Mechanical Properties of graywacke sandstones and granodiorite (abstract). *Trans. Amer. Geophys. Union* **52**, 345.

Schock, R. N. and Duba, A. G.: 1972, Quasi-static deformation of solids with pressure. *J. Appl. Phys.* **43**, 2204.

Schock, R. N., Heard, H. C., and Stephens, D. R.: 1973, Stress-strain behavior of a granodiorite and two graywacke sandstones on compression to 20 kilobars. *J. Geophys. Res.* **78**, 5922.

Schock, R. N. and Heard, H. C.: 1974, Static mechanical properties and shock loading response of granite. *J. Geophys. Res.* **79**, 1662.

Schock, R. N. and Louis, H.: 1974, Unpublished results.

Schock, R. N.: 1976, A constitutive relation describing dilatant behavior in Climax Stock granodiorite. *Intl. J. Rock Mech. Mining Sci. and Geomech. Abstr.* **13**, 221.

Schock, R. N., Abey, A. E., and Duba, A. G.: 1976, Quasistatic deformation of porous beryllium and aluminum. *J. Appl. Phys.* **47**, 53.

Scholz, C.: 1968, Microfracturing and the inelastic deformation of rock in compression. *J. Geophys. Res.* **73**, 1417.

Scholz, C. H., Sykes, L. R., and Aggarwal, Y. P.: 1973, The physical basis for earthquake prediction. *Science* **181**, 803.

Scholz, C. and Kranz, R.: 1974, Notes on dilatancy recovery. *J. Geophys. Res.* **79**, 2132.

Shipman, F. H., Johnson, J. N., Green, S. J., and Jones, A. H.: 1974, The mechanical response of porous uranium alloy. In *Proc. AEC/AMMRC Conf. on the Physical Metallurgy of Uranium Alloys, February 12–14, 1974,* Vail, Colorado.

Sih, G. C. and MacDonald, B.: 1974, Fracture mechanics applied to engineering problems—strain energy density fracture criterion. *Eng. Fract. Mech.* **6**, 361.

Stephens, D. R.: 1969, Personal communication.

Swanson, J. R. and Brown, W. J.: 1971, An observation of loading path independence of fracture in rock. *Intl. J. Rock Mech. Mining. Sci.* **8**, 277.

Terhune, R. W. and Stubbs, T. F.: 1970, Nuclear cratering on a digital computer. In *Proc. of a Symposium on Engineering with Nuclear Explosives, January 14–16, 1970*, pp. 334–359. American Nuclear Society, Hinsdale, Illinois.

Terzaghi, K.: 1943, *Theoretical Soil Mechanics*, Wiley, New Work.

Walsh, J.: 1965, The effect of cracks on the compressibility of rocks. *J. Geophys. Res.* **70**, 381.

Roddy, D. J., Pepin, R. O., and Merrill, R. B., editors.
(1977) *Impact and Explosion Cratering*, Pergamon Press (New York), p. 669–685.
Printed in the United States of America

On fracture mechanism of rocks by explosion

V. M. Tsvetkov, I. A. Sisov, and N. M. Syrnikov

O. J. Schmidt Institute of Physics of the Earth, USSR Acad. Sci., Moscow, USSR

Abstract—The problem of fracture mechanism of rocks by the explosion under spherical symmetry conditions is concerned. It is the rosin that is chosen as a model medium, transparency of which allows to use the optical methods for recording compressional wave and crushing fronts. The displacement velocities are measured by the electromagnetic sensors in the medium.

The studies performed have confirmed the law of geometric similarity for explosion wave and the crushing front by the charge explosion of various weight. At a distance of $r > 4a_0$ (a_0—charge radius) the crushing front is shown to lag behind the wave front; the crushing onset being determined by the process behind the wave front.

The results calculated with using the relations of elastic theory allowed to draw a conclusion of two fracture mechanisms of medium (shear and tensile fractures) which are realized at various distances from the center.

Introduction

The investigation of rock fracture process by the explosion is considered to be of particular interest. The range of practical questions associated with the rock fracture by the explosion is rather wide, however, the fracture process itself has now been studied insufficiently. This situation is accounted for by the lack of reliable experimental data which provide recording the processes of fracture formation and development of natural rocks by the explosion loading effect. Therefore in fact nothing is known about the moment of the fracture onset and what parameters of mechanical state define its onset and following development. In the particular paper the results of model laboratory investigations and calculations are presented which give an idea of crushing mechanism of rocks by the explosion.

Experimental Studies

Rosin is chosen as a model experimental medium (density $\rho = 1.08\,\mathrm{g/cm^3}$, the velocity of propagation of longitudinal and transverse waves $C_l = 2.37 \cdot 10^3$ m/s and $C_t = 1.04 \cdot 10^3$ m/s. The rosin is considered to be rather uniform medium. Like the majority of rocks the rosin has no grain structure or any characteristic parameter with the length dimension. In the initial stage of investigation this particular feature of rosin has an advantage as it decreases the number of parameters effecting the fracture process. It is the rosin transparency that provides the use of the optical methods to record the fronts of compressional wave and crushing medium. In the rosin blocks, enormous enough to eliminate the boundary effect, the explosions of spherical charges of PETN by weight $0.16 \div 2.4$ g were performed. The compressional wave front has been recorded as a narrow dark line on the photographs due to the short-term change of the medium refraction coefficient; the crushing front having been determined by the irreversible loss of the rosin transparency as a result of forming a great number of cracks. The photochronogram obtained in the process of one of those tests is shown in Fig. 1. A detailed description of the optical technique is given in Lukishov *et al.* (1975). The displacement velocity of the compressional wave was measured by the electromagnetic sensors spaced at various distances from the center on the same plane of charge. The sensors were made from the copper wire of diameter $0.05 \div 0.1$ mm. The rosin block was enclosed into the field of constant electromagnet so that the sensor plane appeared to be normal to the field direction. The typical oscillograms of displacement velocity are shown in Fig. 2.

The numerous experiments displayed the high stability of the compressional wave front and, of particular importance, the fracture front. The latter shows one to note the propagation not separate cracks but the front of a great number of cracks, i.e., the crushing. If the fracture were followed by the separate cracks it should not be possible for the stable fracture front to take place. In the photograph of block in section (after the experiment; Fig. 3) is seen that the central zone is crushed into a lot of small fragments, the size of which is less the distance to the charge. In future we shall be of particular interest in the crushing process and it is this term that will be used. Figure 4 illustrates the time curve locuses of the compressional wave and crushing fronts for all the charges used (a_0—charge radius). The variation of the results for the charges of different weight falls within the experimental errors and does not exceed 5%. This confirms the law of geometric similarity by the explosion.

Let's consider the medium at different distances from the center. The photochronogram of Fig. 1 shows that in the vicinity of the explosion cavity the rosin remains transparent behind the wave front. This zone extends to distances of $r = 2.7a_0$. As seen from Figs. 4 and 5 the stress of the wave front at this distance can be estimated by the time curve locus of the front propagation and the velocity of the medium displacement; it makes up $\sigma_r = -11 \cdot 10^3$ kg/cm^2. It is likely to consider that at stresses of wave front more than $11 \cdot 10^3$ kg/cm^2 the rosin is deformed not brittly (i.e., with crack formation resulting in the transparency loss) but plastically. One has already observed the identical picture with the experiments of plane shock waves (Lukishov *et al.*, 1975). Note that later on with the stresses decreasing behind the wave front there also occur the cracks in this zone—this phenomenon is followed by an abrupt darkening in the photochronogram of Fig. 1.

At a distance from $2.7a_0$ to $4a_0$ the medium loses its transparency at the wave front irreversibly; this pointing to the brittle fracture of rosin. At a point of $r = 4a_0$ the stress makes up $\sigma_r = -4 \cdot 10^3$ kg/cm^2 at the wave front. At distances exceeding $r = 4a_0$ (as seen from the photochronogram) the fracture at the wave front does not take place. The crushing front starts lagging behind the wave front and changes the velocity abruptly. Then the velocity of crushing front changes loosely. Over the distance range of $4a_0 < r < 10a_0$ it equals 1.2 km/s and over the range of $12a_0 < r < 22a_0$—0.96 km/s. The latter value practically coincides with the velocity of Rauleigh waves for the rosin which equals $C_R = 1$ km/s. At distances of $r = 22a_0$ the velocity of crushing front starts falling abruptly and at $r = 25a_0$ the crushing front stops. The zone having the radius of $r = 25a_0$ can be identified with the zone size of rosin crushing by the contained explosion. To check the latter assertion special experiments were performed with retaining the block fractured in metallic bomb of cubic shape. The study of the block after the explosion (Fig. 3) shows that the radius of fracture zone coincides with the distance at which the crushing front stop occurs.

Figure 5 illustrates the maximum displacement velocity of the rosin as a function of distance to the center. The attenuation of the maximum displacement velocity obtained in the experiment can be given by

$$v_m = 1330(a_0/r)^{1.65} \qquad \text{for } 2 \leqslant \frac{r}{a_0} \leqslant 8.4$$

$$= 330(a_0/r) \qquad \text{for } \frac{r}{a_0} > 8.4, \tag{1}$$

where v_m—maximum displacement velocity, m/s; r—distance from the center, a_0—radius of the charge.

The experimental data of the maximum displacements of medium at various distances can be represented by the following relations:

$$u_m = 10.5\left(\frac{a_0}{r}\right)^{1.37} \qquad 3 \leqslant \frac{r}{a_0} \leqslant 30 \tag{2}$$

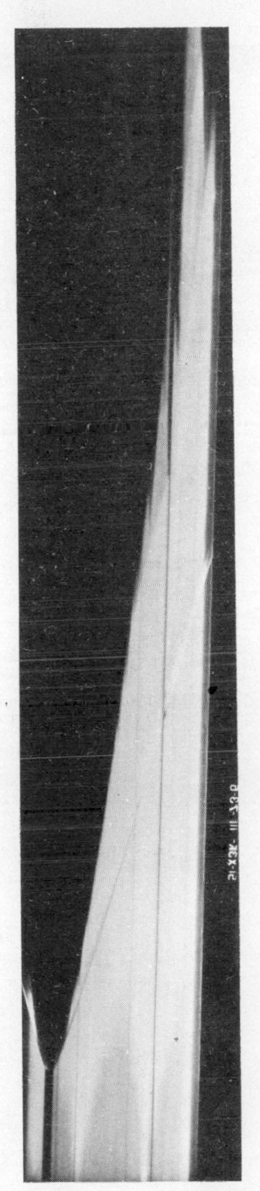

Fig. 1. Photochronogram of the explosion development in the rosin block.

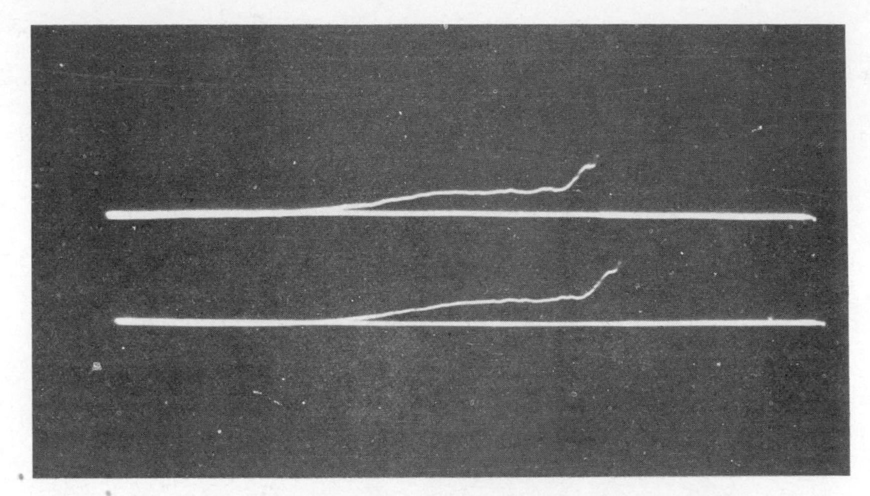

Fig. 2. Typical oscillograms of the medium velocity ($r = 15a_0$).

Fig. 3. Photography of rosin block in section after the explosion.

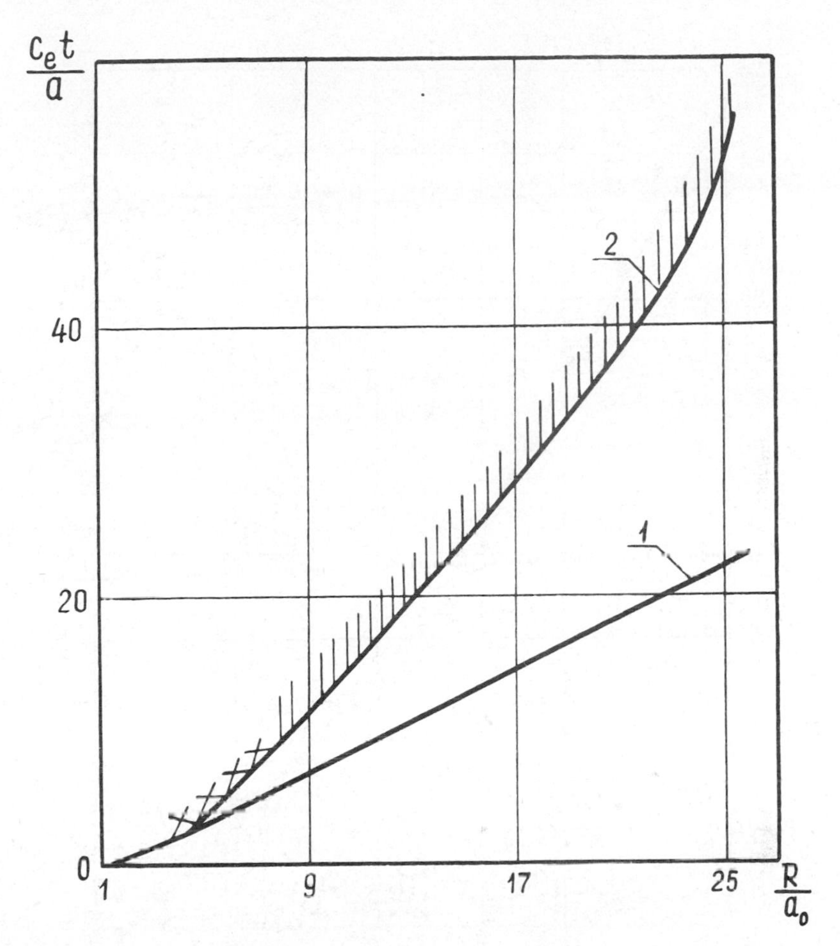

Fig. 4. Time curve locuses of the compressional wave front (1) and crushing front (2).

where u_m—maximum displacement, mm. Within the crushing the maximum displacement coincides with the residual ones; the backward motion occurring to the center at distances exceeding the crushing zone size. This fact allows to estimate the finite density of the medium after the explosion. In Table 1 the parameters $\eta = (\rho_0 - \rho)/\rho$ are given which are calculated by the Eq. (2).

The direct measurement of the mean density of samples withdrawn at various distances from the center and consisting of the majority of fragments of rosin fractured has confirmed the calculated values obtained. At the same time the measurements showed that the density of a separate rosin fragment fractured remains equal to the initial one. This fact permits us to confirm that the above decreasing of the mean density is due to the voids between the fragments of the medium fractured.

The integral volume of voids to be estimated within the crushing zone is of particular interest.

Table 1.

r/a_0	3	5	7	10	15	20	25
η%	4.5	2.9	1.5	0.68	0.27	0.13	0.08

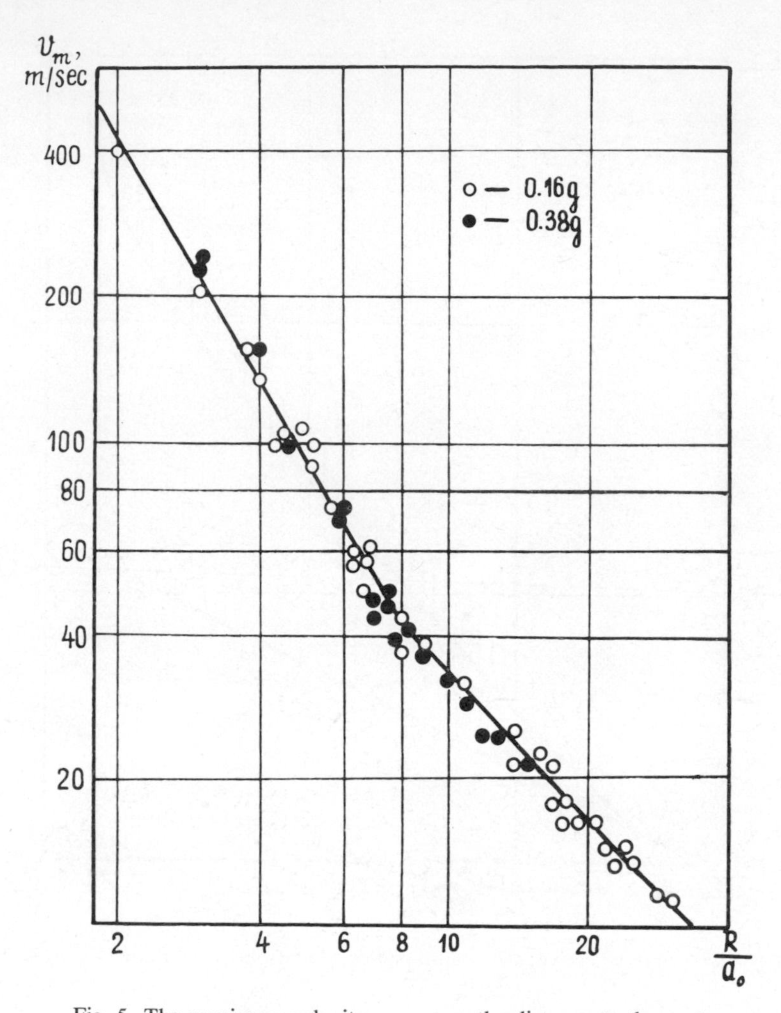

Fig. 5. The maximum velocity v_m versus the distance to the center.

Figure 6 shows the total volume values for voids inside the crushing zone in radius area related to the volume of the explosion cavity the size of which is determined experimentally and equals $a = 2.8a_0$. The analysis of such a relationship shows that the volume of void within the crushing zone exceeds three times the cavity volume. The latter is considered to be very important from the viewpoint of studying the fracture medium state. A considerable volume of void also results in exceeding the volume displaced at the elastic boundary three times the cavity volume. The volume displaced into the elastic zone characterizes the radiated elastic signal (Rodionov *et al.*, 1971). The identical relationship of voids and the cavity volumes was observed while performing the large-scale underground explosions (Rodionov and Tsvetkov, 1971).

CALCULATIONS

The experimental relations (1) show that beginning with the distance $\approx 8.5a_0$ the attenuation of the maximum velocity occurs in accordance with the law

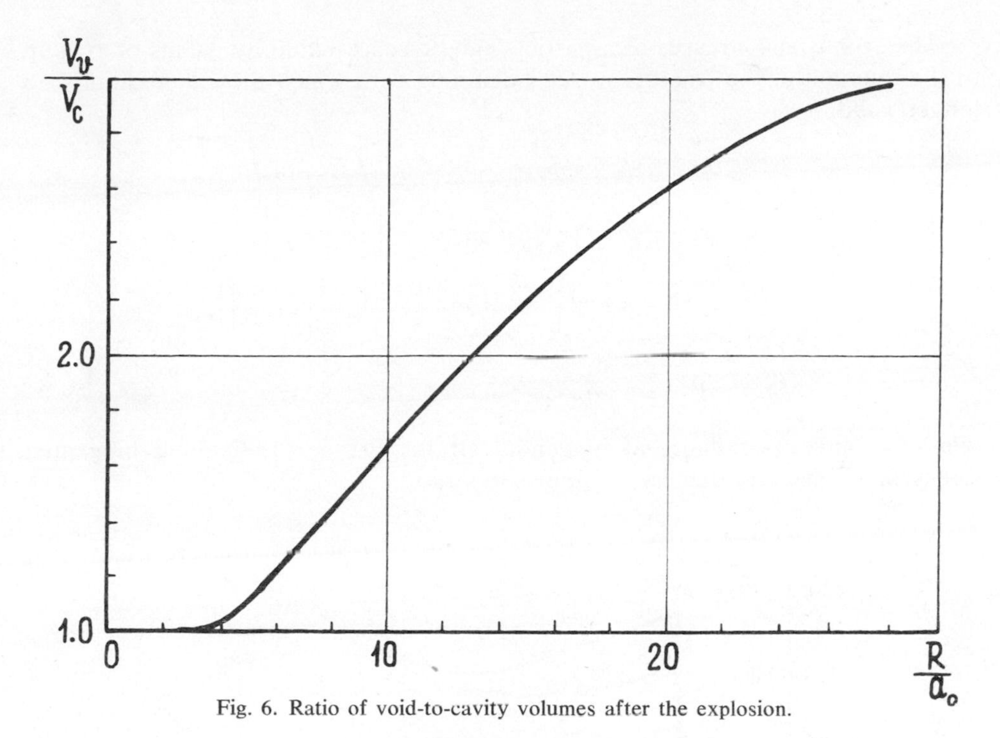

Fig. 6. Ratio of void-to-cavity volumes after the explosion.

$v_m \sim 1/r$. Besides in this zone the compressional wave front propagates with the velocity of sound. The facts mentioned assure that between the wave and crushing fronts the medium behavior is considered to be elastic; allowing it to calculate stresses and strains in the compressional wave up to the fracture moment.

The equation of elastic medium motion in spherical coordinates is represented by

$$\frac{\partial^2 u}{\partial r^2}+\frac{2}{r}\frac{\partial u}{\partial r}-\frac{2u}{r}=\frac{1}{c_L^2}\frac{\partial^2 u}{\partial t^2},$$

where u—medium displacement. The curves of displacement velocity $v(t)$ recorded in the process of experiments permit us to plot the elastic displacement potential for the symmetric wave $f(\zeta)$ expanding spherically, where $\zeta = t - r/c$.

If the velocity of the medium displacement with time $v(t)$ and the displacement $u(t) = \int_0^t v\, dt$ are known at any distance from the center R_0 one can write the following relation based on the elastical potential definition

$$u(t) = \frac{f'}{c_L R_0} + \frac{f}{R_0^2}. \tag{3}$$

The equation calculated takes the form

$$f(\zeta) = R_0 c_L\, e^{-C_L \zeta / R_0} \int_{\zeta_0}^{\zeta} e^{C_L t / R_0} u(t)\, dt. \tag{4}$$

The strains and stresses of a particle can be represented by means of $f(\zeta)$ up to the moment of the fracture onset, beginning with which the calculation is no longer valid.

$$\epsilon_r = \frac{\partial u}{\partial r} = -\frac{f''}{c_L^2 r} - \frac{2f'}{c_L r^2} - \frac{2f}{r^3}, \quad \epsilon_\varphi = \frac{u}{r} = \frac{f'}{cr^2} + \frac{f}{r^3},$$

$$\sigma_r = -\rho_0 c_L^2 \left\{ \frac{f''}{c_L^2 r} + \frac{2(1-2\nu)}{1-\nu} \left[\frac{f'}{c_L r^2} + \frac{f}{r^3} \right] \right\},$$

$$\sigma_\varphi = -\rho_0 c_L^2 \left\{ \frac{\nu}{1-\nu} \frac{f''}{c_L^2 r} - \frac{1-2\nu}{1-\nu} \left[\frac{f'}{c_L r^2} + \frac{f}{r^3} \right] \right\},$$

$$\rho = \frac{\rho_0}{1 - f''/c_L^2 r},$$

where ϵ_r and ϵ_φ—radial and tangential strains, σ_r, σ_φ—radial and tangential stresses, ρ—medium density, ν—Poisson's ratio.

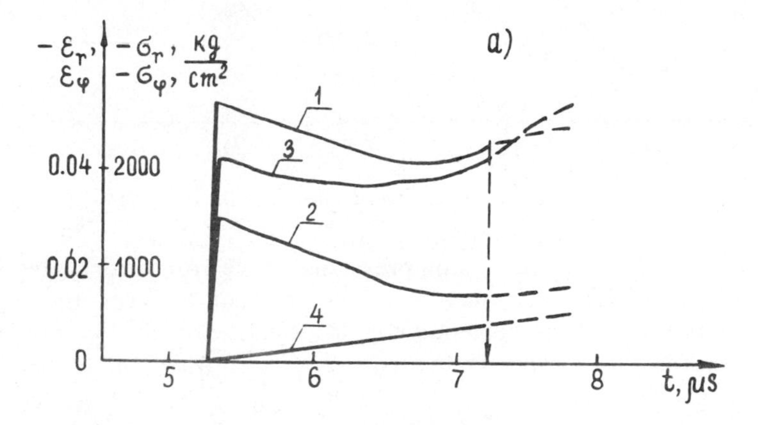

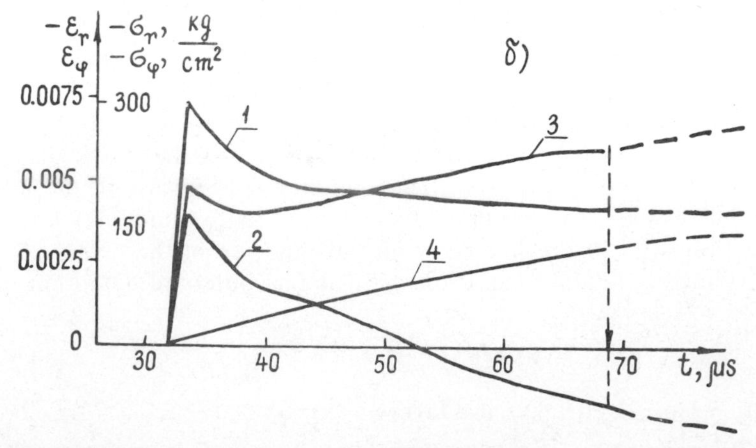

Fig. 7. The change of radial (1) and azimuthal (2) stresses; radial and azimuthal (4) strain in time in the particle with the initial coordinate $a/R_0 = 5.25a_0$, $b/R_0 = 22a_0$.

The calculations were performed with the Eqs. (4)–(5) by using the experimental oscillograms of the velocity with time at various distances. For control the oscillograms obtained at one distance from the center were scaled at another one and compared with the experimental oscillograms. A remarkable quantitative coincidence obtained confirms the elastic nature of the medium behavior up to the moment of the fracture onset.

As seen from Fig. 7 the calculation results are illustrated by the stresses and strains versus time at two fixed points. The moment of the fracture onset shown in these curves up to which the elastic relationships are valid is marked by a vertical line with an arrow. As the plots show the radial and azimuthal component behavior is various. The azimuthal stresses and strains behind the wave front are considered to change monotonely: (the strains increase while the stresses decrease); in the far zone to the moment of fracture onset the azimuthal stresses which were compressional become extensive ones.

The above method of elastic calculation, using generalized time curve locuses of wave and crushing fronts, allows us to define the parameters of stress-strain rosin state at the moment proceeding the fracture onset. The calculation results are shown in Figs. 8, 9 which illustrate the character of the stresses changing σ_r, σ_φ and the strain ϵ_r and ϵ_φ at the crushing front versus its position. These plots also show that the radial stresses and strains at the crushing front are always compressional but the azimuthal strain—always extensive. The behavior of the azimuthal stresses is quite different. At distances less than $8a_0$ they are con-

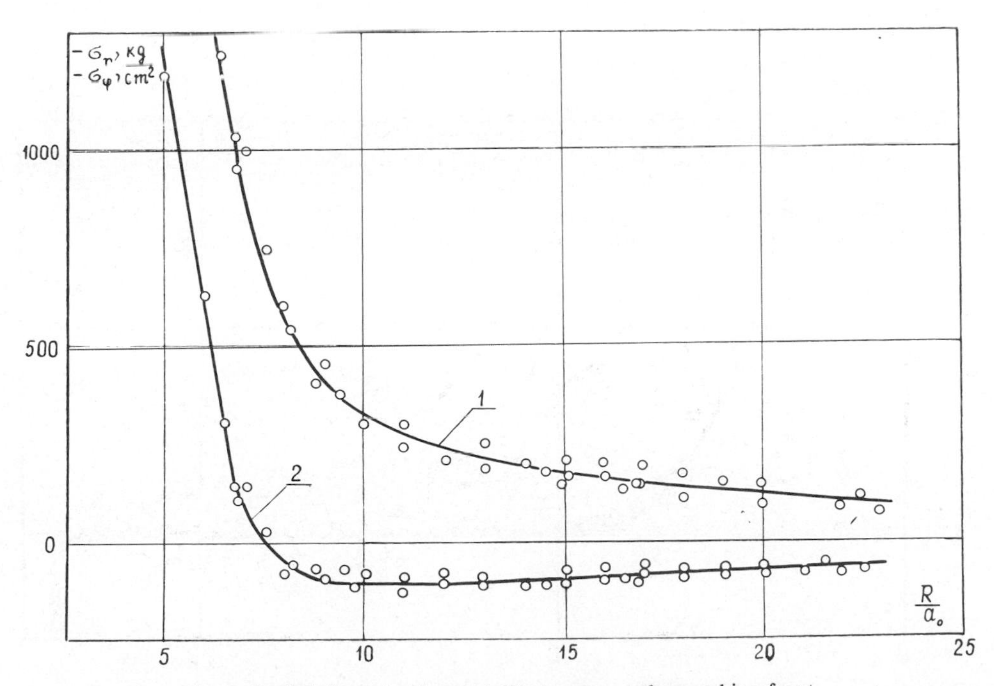

Fig. 8. Radial (1) and azimuthal (2) stresses at the crushing front.

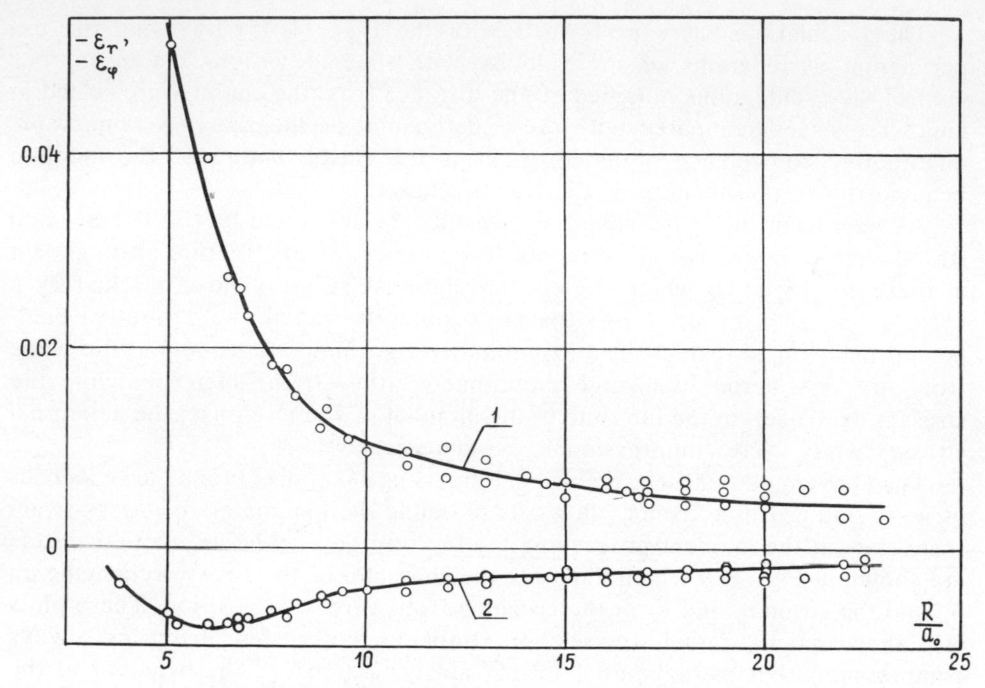

Fig. 9. Radial (1) and azimuthal (2) strains at the crushing front.

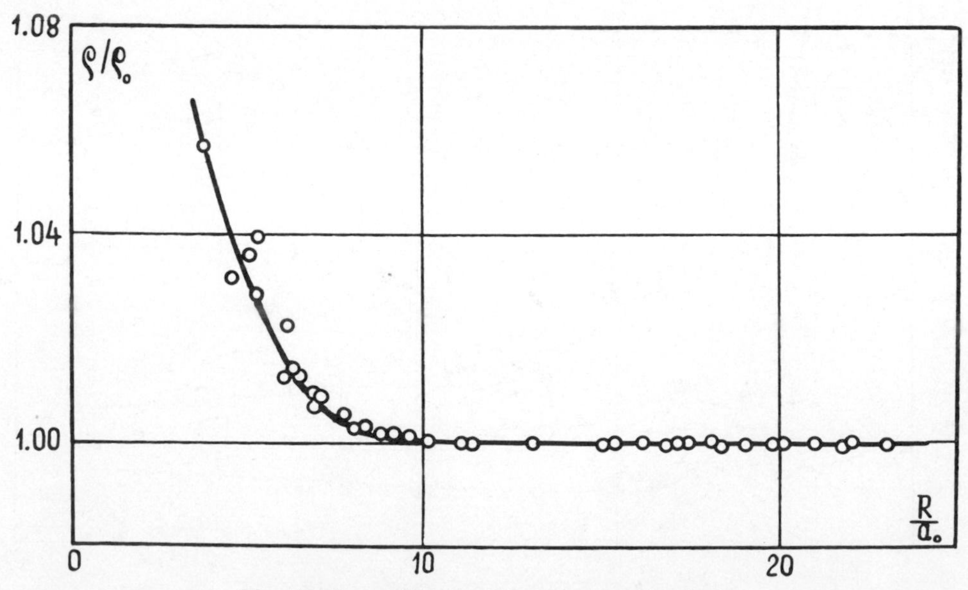

Fig. 10. The medium density at the crushing front.

sidered to be compressional and at distances more than $8a_0$—extensive. Furthermore beginning with distance equal to $r = 10a_0$ all parameters of the medium are slightly changed at the crushing front and the rosin density appears to be equal to the origin one (Fig. 10). In this zone the fragments of the fracture medium are stretched in the radial direction.

Discussion of the Results

The above facts suggest that in the vicinity of the boundary $r = (8 \div 10)a_0$ at the crushing front there takes place the fracture mechanism replacement; closer to the center the fracture is likely to occur as a result of shear; at more remote distances—due to forming the radial grid cracks. The replacement of fracture mechanism is also indicated by the following fact: if at distances of $r < 10a_0$ the velocity of propagation of crushing front exceeds the velocity of Rauleigh wave C_R, in the range of $10a_0 < r < 20a_0$ the velocity of crushing front equal to $0.4C_L$ practically coincides with the velocity of Rauleigh waves which is equal to $C_R = 0.41C_L$ in our case.

The limiting velocity of the extensive crack propagation as some transverse disturbance is known to approach the velocity of Rauleigh waves (Petch, 1973). The above results allow us to consider that in the remote zone the crushing front represents the front of extensive radial cracks which propagate with limiting velocity as a result of extensive azimuthal stress effects. Simultaneously with crack propagation their branching out is not solely to occur. Such a possibility was noted by Joffe (1951) whose calculations show that with the velocity of crack propagation exceeding $0.6C_L$, the extensive stresses near the crack vertex over two symmetric inclined planes are assumed to be more than on the crack plane itself. These calculations were confirmed by the experiments on the glass. Kuznetsov (1968) showed that the number of cracks might decrease at the extensive crack front propagated. This gives rise to the fact that the cracks of different length possess various velocities of propagation therefore more quick cracks will be ahead of slow ones and unload the stress field in front of them; the slow cracks coming to a halt. These two mechanisms (branching out and crack absorption) are likely to occur. It is extremely difficult to appreciate their relative effect directly in the experiment due to the absence of the method of the fragment size estimation (or the number of cracks) at the crushing front of propagation. The final picture which one can observe in the explosion block in section demonstrates not only the fracture at the crushing front but the additional fracture which suffers the medium behind the crushing front. The additional fracture is associated with the fact that behind the crushing front the medium fractured continues moving from the center and deformed for some time interval. This additional deformation increases while approaching the center from the boundary of crushing zone. Figure 11 plots the maximum shear strains and the strains at the crushing front which the medium suffers at various distances. The maximum shear strain values $\gamma_m = \epsilon_r - \epsilon_\varphi$ were estimated on the basis of the experimental dependence for the maximum displacements in the rosin (2).

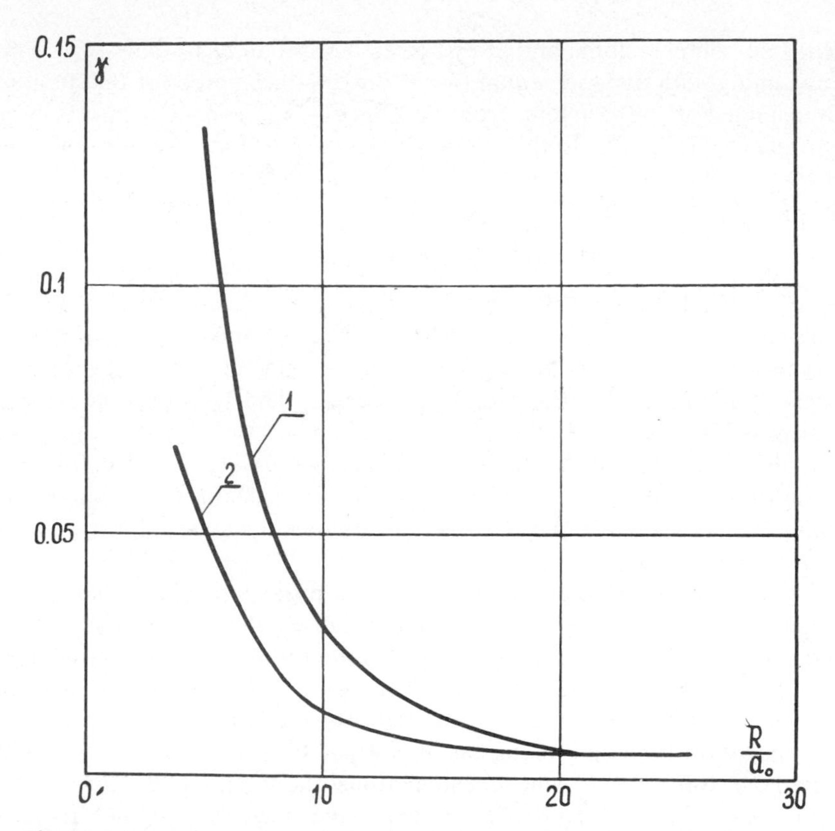

Fig. 11. The maximum shear strain (1) and shear strain at the crushing front (2).

The considerable shear strains of the medium behind the crushing front are responsible for subsequent additional fracture which one may call "volume" as it occurs simultaneously all over the volume of the fracture medium. This volume fracture appears to be responsible for causing the "refragment" zone not far from the explosion cavity and often stands out by more light color. In Fig. 3 the "refragment" zone stands out sharply. The identical refragment zone is observed by the explosions in other rock media (Rodionov *et al.*, 1968).

Thus the investigations performed show that with the contained explosion in rocks the stable fracture front is realized where the medium is crushed, i.e., is divided by the majority of fragments, the size of which is small as compared with the distance to the center. There one observes two zones where the fracture takes place at the crushing front according to different mechanisms: shear and tensile fractures. The second zone assumes to be a zone where the medium is divided by the radial cracks and to exceed two and more times the first zone in size and more than ten times—in volume. In its turn in the first zone—shear fracture zone—one may distinguish two regions: the region where the fracture occurs at the wave front ($2.7a_0 < r < 4a$) and the region where the fracture occurs behind the wave front ($4a_0 < r < 8a_0$). Here it is the second zone (the zone of fracture behind the wave front) that in volume exceeds ten times the first one. The fracture at the crushing front is supplemented by the subsequent volume fracture due to the prolonged deformation of the medium.

The above results of the stressed state of the medium at the crushing front enables one to put a criterion for describing the limiting state of the medium before the fracture onset. This state being actually limited is evidenced by the plots of Figs. 7 and 12 which display the changes of stresses and strains in a particle in time. As is seen from the plots the shear stresses and strains in the nearest zone ($r < 8a_0$) as well as the extensive azimuthal stresses and strains in the remote zone ($r > 8a_0$) do attain the maximum values at the moment of fracture.

There exist some approaches to describing the limiting state of rocks. Rather

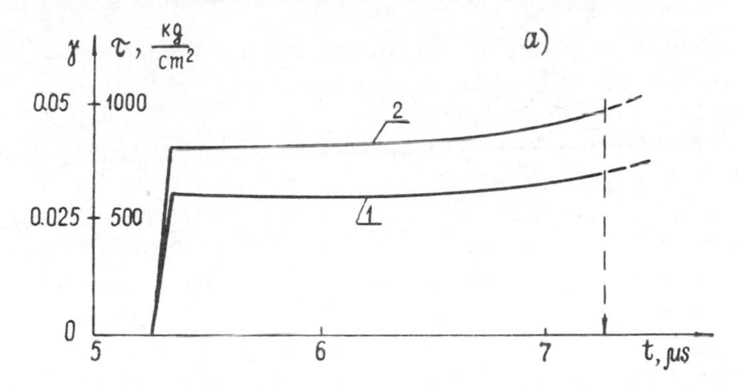

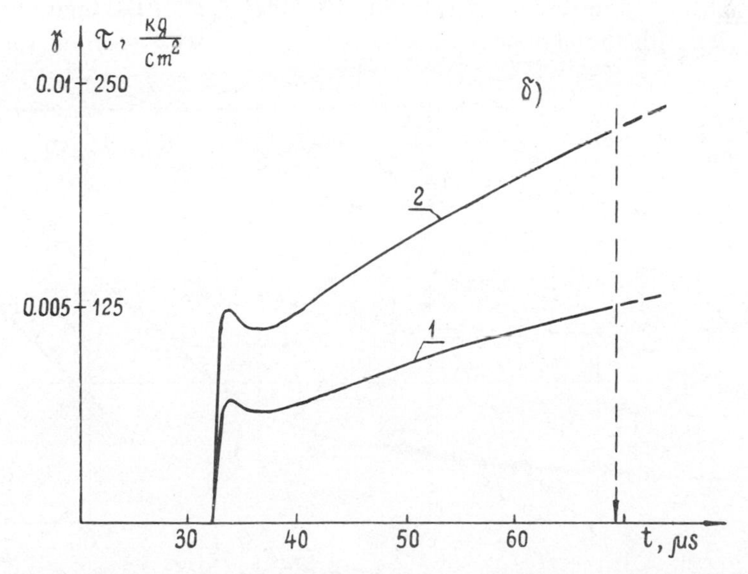

Fig. 12. The change in time of shear stresses (1) and shear strains (2) in the particle with the initial coordinate $a/R_0 = 5.25a_0$ and $b/R_0 = 22a_0$.

often to describe the fracture the maximum shear stress τ (in our case $-\tau = |\sigma_r - \sigma_\varphi|/2$) is plotted versus the parameter that is characterized by the stresses state of the medium before causing the fracture. Stavrogin (1969) suggests to use the minimum-to-maximum ratio of main stresses, i.e., σ_φ/σ_r ratio, as such a parameter. The data processes are shown in Fig. 13. As is seen from the plot the curve possesses two distinct regions the boundary between which lies in the vicinity of $\sigma_\varphi/\sigma_r = -0.6$. In the region to the left of the boundary the tensile fracture takes place at the crushing front and in the region to the right the shear fracture is observed. In the region of the shear fracture the strength results from the parameter σ_φ/σ_r by the exponential law

$$\tau \sim \tau_0 \cdot \exp\left(A \cdot \frac{\sigma_\varphi}{\sigma_r}\right),$$

as well as in the static experiments (Stavrogin, 1969) and changes ten times. In the zone of tensile fracture the strength changes slightly but there is observed an enormous spread of experimental points.

The latter affirms the well-known experimental fact: the tensile strength of brittle fracture solids (and especially rocks) is rather unstable and depends on the presence of defects severely.

Figure 13 also illustrates the relationships typical for the rosin strength under static conditions. This relationship was obtained with the proportional loading and strain rate $\dot{\gamma} = 10^{-3}\ 1/S$. The similarity of the above relationships of static and dynamic experiments in the first part of the plot points to the general fracture mechanism by shear. In the static experiments there is no possibility to perform the tensile fracture due to the procedure difficulties associated with the state type tension-compression. The numerical difference of the limiting value of shear stresses with the similar parameters σ_φ/σ_r suggests the strain rate effect.

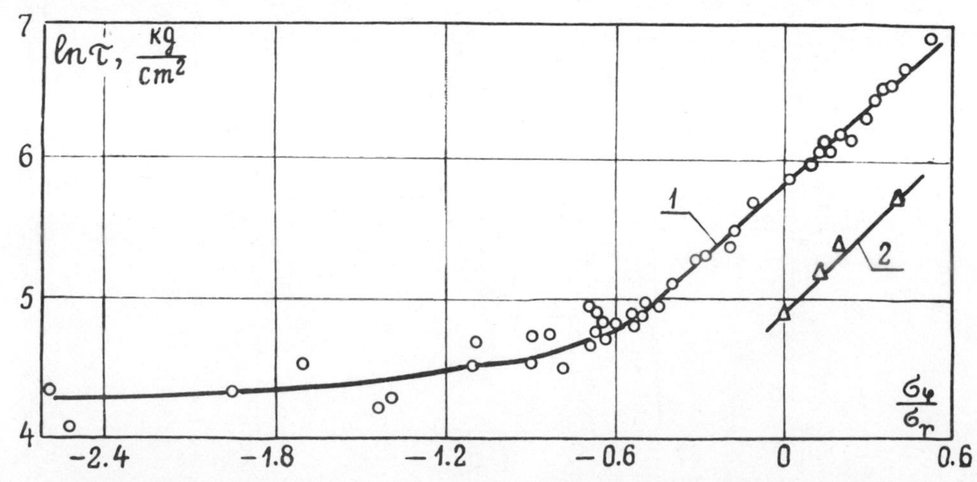

Fig. 13. The limiting shear stress versus stressed state under dynamic (1) and static (2) conditions.

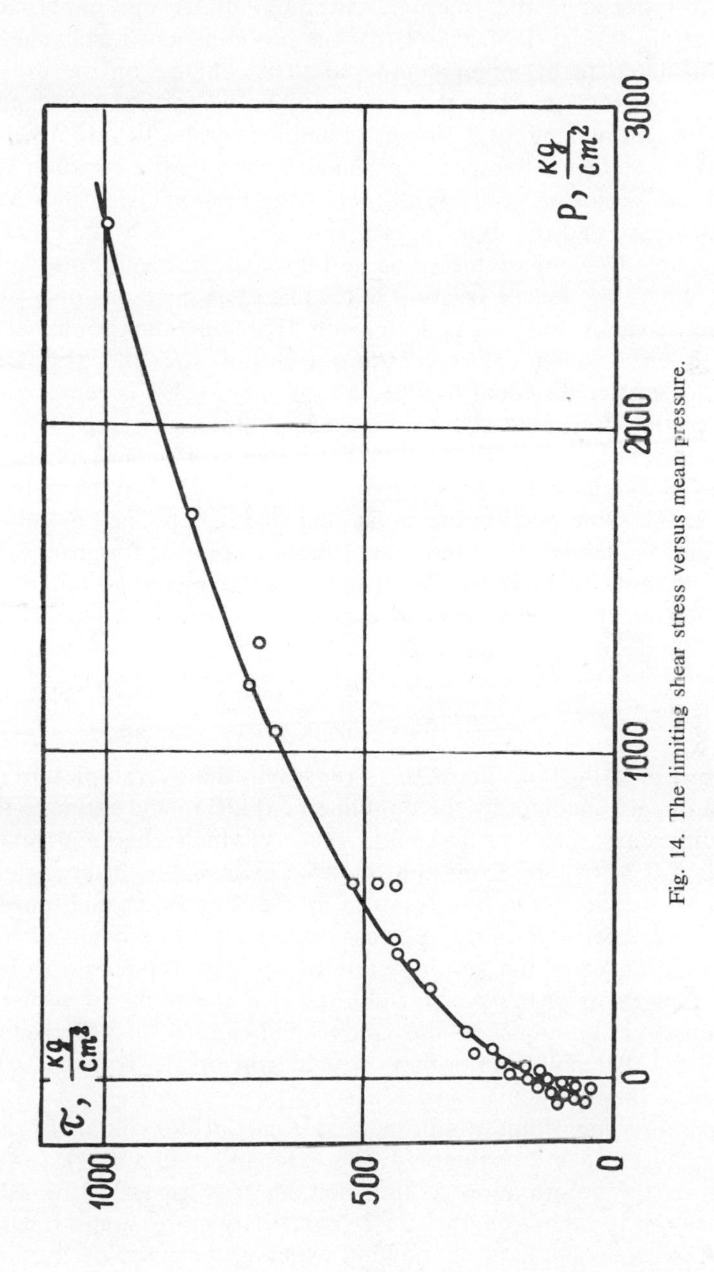

Fig. 14. The limiting shear stress versus mean pressure.

The velocities of shear strain in the explosion experiments estimated in accordance with the strain as a function of time (Fig. 7) lie within $(10^2 \div 10^3)1/S$ at the moment proceeding the fracture. Although under our experimental conditions the strain rate at the fracture front is not constant, its change of tenfold may be considered to be minor compared to its change on the order of $5 \div 6$ while static experiments. Thus the difference between the static and dynamic curves of strength making $3 \div 3.3$ times is defined by the strain rate changing on the order of $5 \div 6$. This difference is assumed not to be the constant value but a function of the explosion scale. With increasing the explosion power the strain rates will decrease and the dynamic strength will approach static ones.

One is unable to compare the static and dynamic curves of strength in detail in the zone where the tensile fracture takes place; as in statics only one point is known—the strength with uniaxial tension (Kuzmin and Pooh, 1959) equals $30\,kg/cm^2$. Such a small value of static fracture strength compared to the dynamic tensile strength equal to $80 \div 130\,kg/cm^2$ (Fig. 8) is represented by the real result of the strain rate effect and stressed state.

The fracture criterion is often plotted (Cherry *et al.*, 1968) by the maximum shear stress τ versus mean pressure $p = -\frac{1}{3}(\sigma_r + 2\sigma_\varphi)$. Our experimental data processed in the same manner are presented in Fig. 14. The smooth-increasing curve obtained is identical with the same functions plotted for granite, limestone, and salt in statics (Cherry *et al.*, 1968). However, it should be noted that such a representation conceals the difference in the fracture nature which is displayed on the plot of Fig. 13 very distinctly.

CONCLUSION

The above investigations allow us to represent the overall picture of fracture mechanism of rock medium by the contained explosion and estimate the relative effect of different aspects. All the parameters which characterize the stress-strain state at the crushing front appear to be changeable. There is no constant critical value to characterize the fracture moment onset. In the tensile fracture region one can observe the tensile stress and strain to be changed the least.

The plotted curve of the limiting state of the rosin appeared to be identical with the functions given by A. N. Stavrogin for the rocks of some tens. This result permits us to consider the above indications of the rosin fracture mechanism by the explosion rather typical for brittle fracture solid media including rocks than particular.

It seems rather important that in the major part of the volume of the crushing zone the fracture process begins with the extensive radial crack formation and proceeds with the deformation of the medium fractured. Taking into account that in rocks the tensile strength is less than the compression strength and real rock massifs possess the natural cracking which reduces the fracture strength to the lithostatic pressure value the above results are indicative of cracking and lithostatic pressure effect on the explosion crushing by the large-scale natural explosions.

References

Cherry, J. F., Larson, D. B. and Rapp, E. G.: 1968, A Unique Description of a Failure of Brittle Materials, *Int. J. Rock Mech. Min. Sci.* **5**, 455.

Joffe, E.: 1951, *Philis. Magazine* **42**, 739.

Kuzmin, E. A. and Pooh, V. P.: 1959, Velocity of Brittle Crack Growth in Glass and Rosin. In col. *Some Problems of Solid Strength*, M-L. Publ. of Academy of Science, USSR.

Kuznetsov, V. M.: 1968, On Unsteady Propagation of Crack Systems in Brittle Body, *PMTF* No. 2.

Lukishov, B. G., Rodionov, V. N. and Tsvetkov, V. M.: 1975, Model Research on the Crushing Effect of Explosions. In *Peaceful Nuclear ExplosionIV*, Int. Atom. Energy Agency, Vienna.

Petch, N. J.: 1968, Metallographic Aspects of Fracture. In col. *Fracture* **I**. Academic Press, New York and London.

Rodionov, V. N., Adushkin, V. V., Kostjuchenko, V. N., Nikolaevsky, V. N., Romashov, A. N. and Tsvetkov, V. M.: 1971, *Mechanical Effect of Underground Explosion*. M. Nedra Press, Moscow.

Rodionov, V. N., Sizov, I. A. and Tsvetkov, V. M.:1968, Study of Cavity Development by Contained Explosion. In col. *Explosion Engineering* (Vzryvnoe Delo), N. 64/21, M. Nedra Press, Moscow.

Rodionov, V. N. and Tsvetkov, B. M.: 1971, Some Results of Observations on Underground Nuclear Explosions, *Atomic Energy* **30**, No. 1.

Stavrogin, A. N.: 1969, Study of Limiting States and Strains of Rocks, *Izv. Academy USSR. Physics of the Earth*, No. 12.

Roddy, D. J., Pepin, R. O., and Merrill, R. B., editors.
(1977) *Impact and Explosion Cratering*, Pergamon Press (New York), p. 687–702.
Printed in the United States of America

Shock attenuation at terrestrial impact structures*

P. B. ROBERTSON and R. A. F. GRIEVE

Earth Physics Branch, Department Energy, Mines and Resources, Ottawa, Canada K1A OY3

Abstract—Based on the development of planar deformation features in quartz, four levels (A to D) of shock metamorphism have been quantitatively established for up to fifty grains per sample in autochthonous rocks from the central uplifts of two complex craters—the Slate Islands and Charlevoix impact structures—and in drill core samples from the base of one simple crater—Brent, Canada. Applying earlier experimental results for the various shock levels, pressures of 8.8, 12, 15, and 23 GPa are assigned to quartz grains of types A through D deformation, respectively, and an average shock pressure calculated for each sample from the mean of the individual grain pressures. Average shock pressures in samples examined at the three structures range from 5 to 23 GPa. Equal-pressure shock contours defined from surface samples are concentric about the crater center at the complex craters, and contours defined from subsurface samples in the simple crater are roughly hemispherical beneath the postulated point of impact.

An attenuation rate of $P \sim R^{-20}$ calculated from quartz deformation at Brent is excessively high and may be due to the section in which shock pressures were determined having been shortened by approximately 50%, the upper portions being displaced downward in the cratering process approximately 550 m, with proportionally less displacement of the lower sections. A rate of R^{-2}–R^{-3} results if this section is expanded and returned to its original, pre-crater position.

In a comparison of shock attenuation at the surface in the two complex craters, shock pressure was plotted against radial proportion (the ratio of sample distance from the crater center to the transient crater radius). Statistically the slopes can be considered equivalent so that shock attenuation is believed to be similar at impact structures which do not differ in size by more than a factor of two. An attenuation rate for complex craters was established by comparing the inferred and hypothetical excavated cavities for modelled attenuation rates of R^{-2}, R^{-3}, and $R^{-4.5}$. The two cavities coincide more closely as attenuation rate increases so that a value of $\sim R^{-5}$–$R^{-5.5}$ has been estimated as representative for complex craters. Shock attenuation rates appear, therefore, to be more rapid in the large, complex craters in comparison with the smaller, simple craters.

As a consequence of this more rapid attenuation, for craters formed in target materials with similar strength properties, the original excavated cavity of a complex crater must be proportionally shallower than that of a simple crater. In addition, the zone of autochthonous material beneath the crater floor, in which tectosilicates display planar features, must also be proportionally narrower at large, complex craters.

INTRODUCTION

SHOCK PRESSURES generated by hypervelocity meteorite impact produce a variety of characteristic, high-strain-rate, high stress effects in the target rocks and meteorite. Such shock effects range from total vaporization and melting near the point of impact, where shock pressure is several hundred GPa (1 GPa = 10 Kbar), through partial melting or conversion to dense high-pressure phases, to the development of solid state glasses and various planar deformation features especially in tectosilicates. The latter occur where shock pressures have attenuated

*Contribution from the Earth Physics Branch No. 688

to values just above Hugonoit elastic limits (approximately 5–10 GPa) which, in a large event, can be several kilometres from the point of impact.

Because the various shock effects are produced over different regimes of the shock-pressure spectrum, their distribution in the target rocks at an impact site can provide estimates of the pressure levels outward from the point of impact, and thus allow shock-wave attenuation rates to be calculated. However, target materials shocked to pressures above approximately 30–40 GPa are excavated and removed from the crater, or redistributed and mixed to form the allochthonous breccias within the crater. Although some system to the distribution of shock grades has been noted in the allochthonous breccias (Dence, 1968) they are for the most part heterogeneous in terms of shock level. Materials representative of the major portion of the shock spectrum, therefore, cannot be used to estimate the outward diminution of shock pressures.

Rocks of the walls and floor of the crater, although fractured and brecciated, to a large extent retain their relative positions in terms of radial distance from the impact point, and the succession of shock metamorphic effects in these rocks can provide estimates of shock levels, and rate of change of shock levels with distance, for a portion of the shock-wave decay curve.

In the large complex craters the rocks of the central uplift have received a net upward displacement from their pre-crater levels (Dence, 1968; Milton *et al.*, 1972) although their motion to reach this position is still a matter of discussion. The uplifted material preserves only relative radial distances with respect to the shock point, and calculation of shock-wave attenuation cannot be done directly. In simple craters there is conflicting evidence regarding displacements in the autochthonous rocks which form their transient crater floors. The central portion of the floor of Barringer Crater is formed in the Supai Formation at a depth which would indicate no measurable displacement of these rocks (Roddy *et al.*, 1975). In contrast, in the Piledriver nuclear crater, twenty-five times less energetic, radial displacements of up to 20 m have been measured for the cavity walls (Borg, 1972; Dence *et al.*, 1977). Calculation of shock-attenuation rates in material beneath the transient crater floor of a simple crater could provide a means to assess these apparently different behaviours.

The present study examines shock effects in samples from beneath a simple crater and from the central uplifts of two complex impact structures, and shock pressures are established for these samples. Shock-wave attenuation rates are determined by comparison of the distribution of shock pressures with the distribution predicted in simple and complex craters modelled according to different attenuation rates.

THE CRATERS

Brent, Ontario, is a simple crater in acidic to mafic gneisses with an erosional surface diameter of 2.9 km. The true crater contains 610 m of breccias overlain by 260 m of younger sediments (Dence and Guy-Bray, 1972). Samples for which shock pressures have been determined are basement gneisses from drill holes 1-59, 1-67, and 2-67 (Fig. 1). Few measurements were made in the latter two holes and the

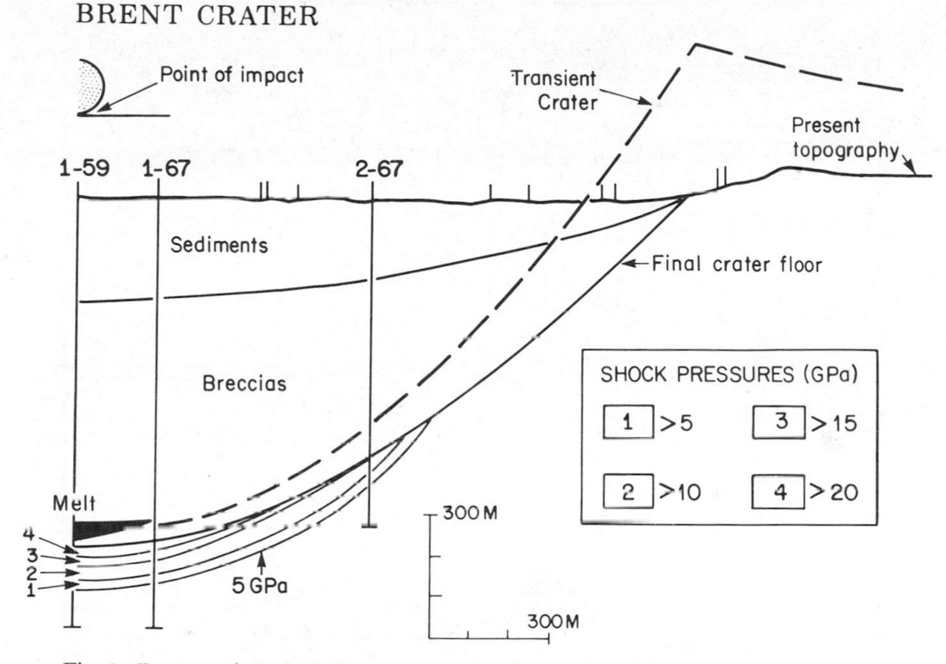

Fig. 1. Cross-section of the Brent crater showing zoning of shock pressures based on quartz deformation in 3 drill holes which penetrate the crater floor (positions of additional holes are indicated at the surface). The "final crater floor" configuration results from slumping of the transient crater.

sequence of shock pressures used in modelling attenuation rates is from hole 1-59. In the 30-m section underlying the crater floor in this hole, reliable indications of shock metamorphism have been largely obliterated through secondary thermal metamorphism from an overlying thin layer of impact melt, and sampling is from depths below this thermal zone.

The Slate Islands, Lake Superior, represent a portion of the central uplift of a largely submerged complex impact structure (Halls and Grieve, 1976). Bedrock geology of the islands is complex with the principal units being Archean age, felsic to mafic metavolcanics, plus Keweenawan diabase and minor Keweenawan basalts and sedimentary rocks (Sage, 1974).

Charlevoix, Quebec, is a complex impact crater in Grenville age gneisses and anorthosite and minor Ordovician limestones. Descriptions of the geology, structure and shock metamorphism have been detailed by Robertson (1968) and Rondot (1972), among others.

Shock deformation of a given mineral is influenced not only by magnitude of the shock pressure, but also by the mineral's textural and mineralogic environment. On the basis of planar feature development Grieve and Robertson (1976) estimated pressures of 14.3 GPa for large quartz grains compared with only 12.5 GPa for smaller quartz grains within the same sample. Shock pressures in quartz of a Slate Islands diabase are approximately 4 GPa lower than in quartz of nearby acid metavolcanics, and planar feature development in quartz within poikilitic garnets

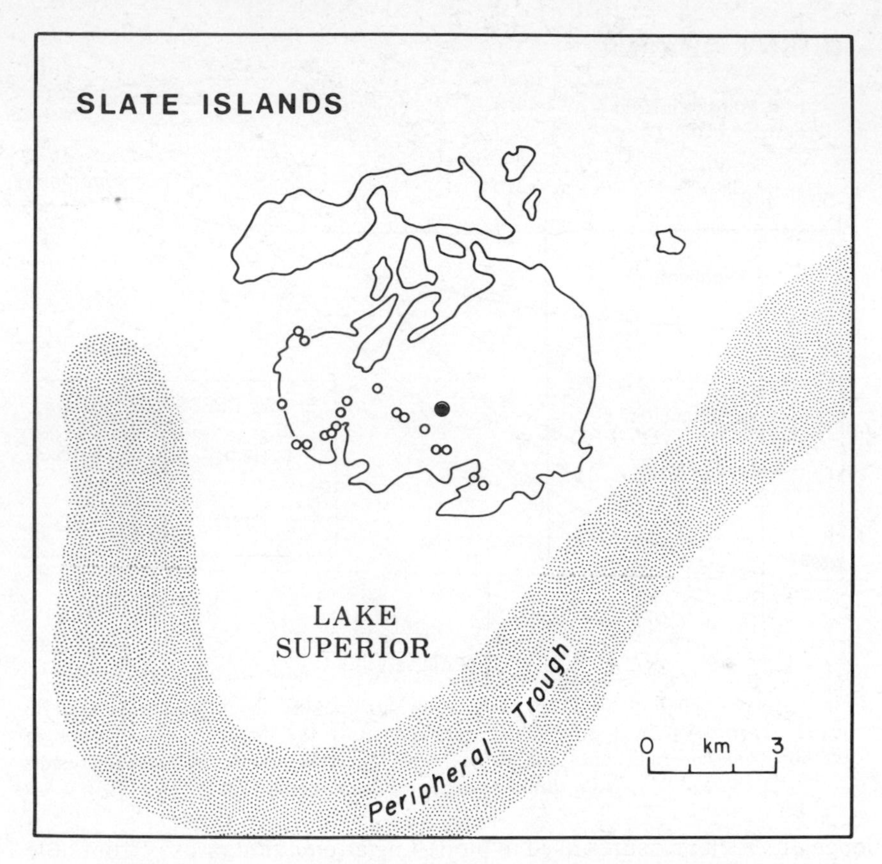

Fig. 2. Sample locations at the Slate Islands for which shock pressures have been determined from quartz deformation (Table 4, Fig. 5). Center of impact structure (circle with dot) determined by Halls and Grieve (1976) from shatter cone orientations.

has been observed to be weaker than in matrix quartz of the same sample (unpublished observations). To minimize the effects of lithologic variation at the Slate Islands, average pressures are from acid metavolcanics (Fig. 2), and observations on diabase which give anomalously low pressures have been excluded. Although lithologic variation at Charlevoix is not extreme, samples for which average shock pressures have been determined (Fig. 3) were taken largely from the charnockitic gneiss unit. Also, because earlier studies (Robertson, 1968, 1975) indicated anomalies in shock distribution on the St. Lawrence River side of the structure, probably connected with post-crater tectonic disturbances, samples were selected predominantly from the north and west of the central peak.

SHOCK-PRESSURE DETERMINATIONS

The shock levels encountered in the basement rocks of simple craters and in rocks forming the central uplifts in complex impact structures are characterized

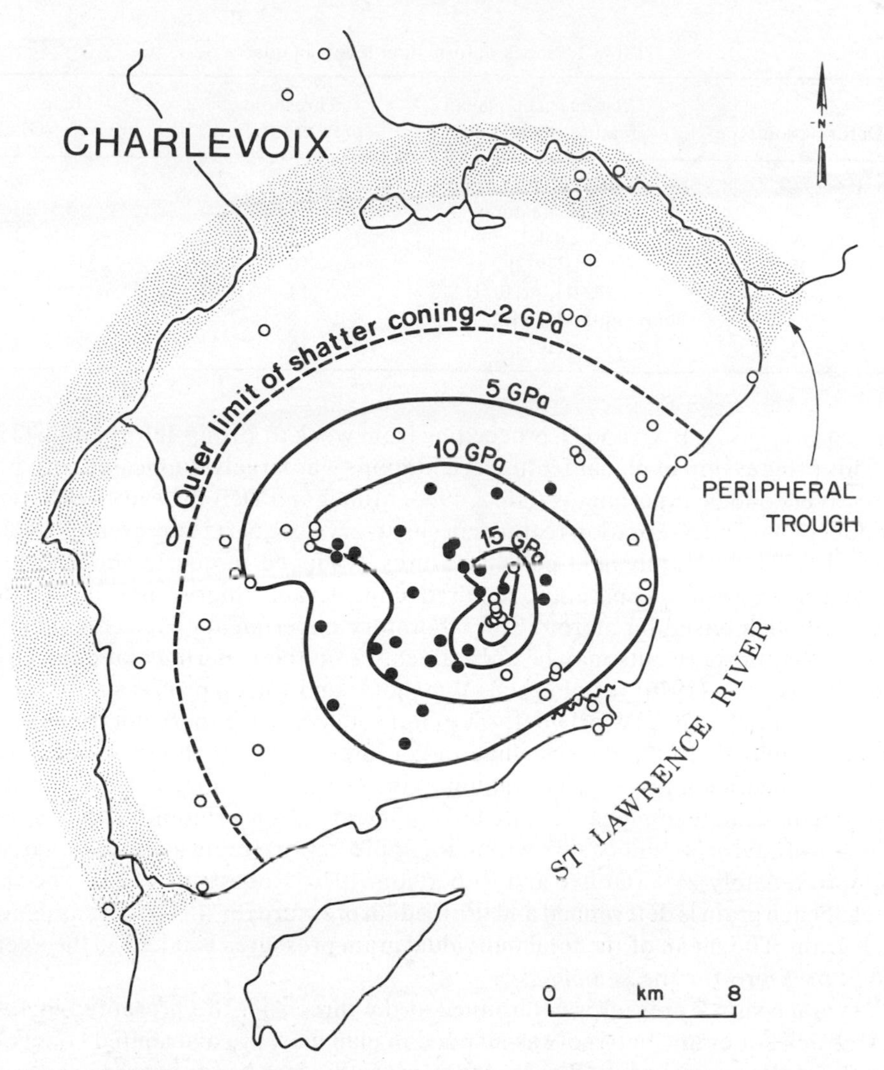

Fig. 3. Shock pressure contours at the Charlevoix complex impact structure. Contours were established from shock pressure data for all locations plotted, and are based on both quartz and feldspar deformation (Robertson, 1975). Quartz deformation data used in the present attenuation study (Table 3, Fig. 5) are located by solid circles.

by microscopic planar deformation features in quartz and feldspars. These features which have been termed "planar features", "planar elements", "shock lamellae", have been described extensively from many terrestrial impact structures (Bunch, 1968; Carter, 1968; Engelhardt and Bertsch, 1969; Robertson *et al.*, 1968). It has been observed that planar deformation features in quartz, and to a lesser extent in feldspars, form parallel to a relatively small number of rational, low-index crystallographic planes. Robertson *et al.* (1968) further noted that specific orientations of planar features in quartz are characteristic of successive grades of shock metamorphism. On this basis they recognized four shock levels, which they

Table 1. Shock deformation levels in quartz.

Deformation type	Characteristic planar feature orientations	Threshold pressure (GPa)	Mean pressure (GPa)
—	Shatter cones but no planar features	3.5	5.5
A	c{0001}	7.5	8.8
B	$\omega\{10\bar{1}3\}$	10	12
C	$\{22\bar{4}1\}$, $r\{10\bar{1}1)$ $z\{01\bar{1}1\}$, $\xi\{11\bar{2}2\}$	14	15
D	$\pi\{10\bar{1}2\}$	16	23

defined as types A, B, C, and D, proceeding from weak to strong deformation (Table 1). This progression of planar feature orientations was largely duplicated in two sets of recovery shock experiments (Hörz, 1968; Müller and Défourneaux, 1968) which produced planar deformation features in single-crystal quartz at pressures over the range 10–17 GPa. Equivalent planar features produced in quartz shocked in the Piledriver explosion apparently required considerably higher pressures (Borg, 1972) although based on more recent laboratory experiments, Hanss *et al.* (1977) believe that these results may be anomalous. From the experimental data, Grieve and Robertson (1976) established threshold and median pressures for the development of types A to D deformation produced by hypervelocity meteorite impact (Table 1). In a polycrystalline target interactions of the shock wave with phase boundaries and irregular grain margins produce localized stress variations resulting in a range of planar feature development within a sample. Typically, A to C or B to D deformation occurs within a sample, representing a pressure variation of approximately 25% (Grieve and Robertson, 1976; Robertson, 1975). The shock level of each grain is determined and the median pressure for this level is assigned to each grain. The mean of the total individual grain pressures is taken as the average shock pressure for the sample.

Average shock pressures determined at the three sites are presented in Tables 2–4. Shock zones at Charlevoix as mapped in plan view are distributed concentrically about the central peak (Fig. 3). At Brent, equivalent zones in cross-section are practically hemispherical beneath the crater (Fig. 1).

SHOCK-WAVE ATTENUATION

Simple craters

Attenuation of shock pressures in the basement rocks of the Brent crater (Table 2) is illustrated in a logarithmic plot of pressure versus depth (Fig. 4). Depth is calculated from the original plane, estimated to have been about 200 m above the collar of drill hole 1-59 (Dence, 1968). Over this 80 m interval the calculated average rate of shock-pressure attenuation is $P \sim R^{-20}$. Such a rate is orders of magnitude higher than typical rates of approximately R^{-2} to R^{-3} (Cherry and Petersen, 1970; Dence *et al.*, 1977; O'Keefe and Ahrens, 1976) or extreme rates of $\sim R^{-7}$ (Porzel,

Table 2. Shock pressures at Brent from quartz deformation in drill hole 1-59.

Depth[1] (m)	% quartz with[2] planar features	% type A	% type B	% type C	% type D	Average pressure (GPa)
1103	100	—	—	—	100	23.0
1106	100	—	5	5	90	22.1
1107	100	—	—	15	85	21.8
1108	100	—	5	10	85	21.7
1109	100	—	10	10	80	21.1
1110	100	—	20	35	45	18.0
1112	100	—	35	15	50	18.0
1113	88	—	12	38	38	16.5
1114	95	—	85	5	5	12.4
1116	100	—	95	5	—	12.2
1117	95	—	75	10	—	13.1
1122	95	—	90	5	—	11.8
1123	95	—	70	25	—	12.4
1130	85	25	60	—	—	10.2
1137	50	35	10	5	—	7.8
1140	50	45	5	—	—	7.3
1142	58	58	—	—	—	7.4
1147	55	10	45	—	—	8.8
1159	40	35	5	—	—	7.0
1161	31	23	8	—	—	6.8
1169	40	35	5	—	—	7.0
1176	30	30	—	—	—	6.5
1182	15	15	—	—	—	6.0
1188	5	5	—	—	—	5.7

[1]Depths are from impact point, 220 m above present ground surface.
[2]Twenty grains measured per sample.

1958) calculated for nuclear and explosion craters, or predicted by modelling impact processes. Extrapolation to the surface at Brent of an R^{-20} attenuation rate would require initial shock pressures of 10^{17} GPa at the point of impact, pressures unattainable from the impact of meteoritic materials even at maximum geocentric velocities.

This observed rate of shock-pressure decrease can be reconciled with the lower rates for nuclear craters if the length of the section over which shock effects have been measured has been reduced from its pre-impact extent, either by differential vertical compaction or through removal of portions of the section. Dence *et al.* (1977), using the observed Brent attenuation data presented here, calculated the amount of shortening which took place for craters modelled for attenuation rates of R^{-2}, R^{-3}, and $R^{-4.5}$. In each case they assumed that material shocked at 2 GPa experienced no net vertical movement, and this pressure was fixed at 1600 m depth, whereas the upper part of the section was allowed to expand differentially to conform to the modelled attenuation rates. On this basis the top of the section, where shock pressures are calculated as 23 GPa, lay originally at depths of 480, 690 and 920 m for the three attenuation rates, which in turn requires post-impact reductions

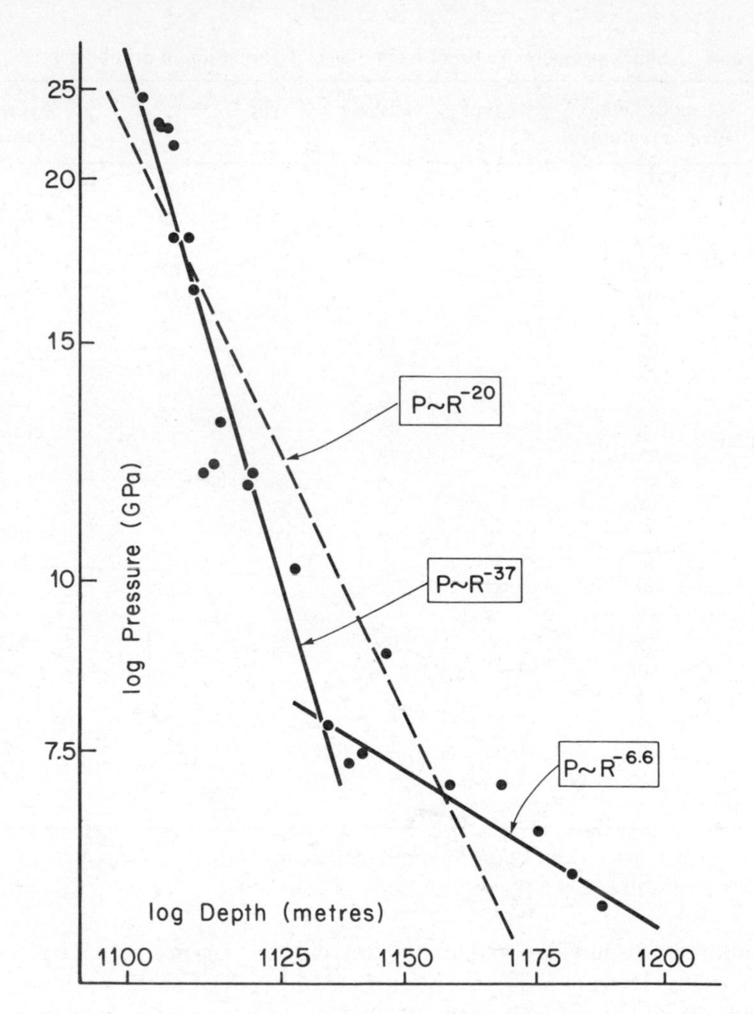

Fig. 4. Log pressure versus log depth below impact point (Table 1) in Brent drill hole 1-59 (Fig. 1). An attenuation rate of R^{-20} is calculated from all data, whereas rates of R^{-37} and $R^{-6.6}$ are found for deformation above 1145 m and below 1135 m, respectively. This apparent discontinuity may result from gaps in the preserved shock record.

of the length of this section of 56%, 45%, and 27%, respectively. Dence *et al.* (1977) concluded, on the basis of elaborate energy and scaling considerations, that the Brent crater closely approximates a model with an attenuation rate between R^{-2} and R^{-3}. Accepting this model, material shocked to 23 GPa lay initially at approximately 550 m, and the section of basement rocks under discussion was reduced by roughly 50%.

Gravity studies at simple craters, and particularly Brent (Innes, 1961), show the rocks of the basement to be slightly lower than average density, so that there is no permanent densification beneath the crater floor. Rather, reduction in the length of the shocked section must result from lateral displacements, either as a general integrated spreading of the section or by removal of discrete segments of the shock record. Although petrographic observations of the basement rocks at Brent do not reveal any distinct gaps in the spectrum of shock effects, a possible discontinuity is apparent in Fig. 4 where attenuation rates of R^{-37} and $R^{-6.6}$ more closely fit the data respectively above and below 1140 m, than does the average attenuation rate of R^{20}.

Complex craters

Distribution of pressures derived from shock effects is given in Fig. 3 for rocks of the central uplift at the Charlevoix structure. Concentric zones of shock pressure diminish outward from 23 GPa at the central peak, to 5 GPa at approximately

Table 3. Shock pressures at Charlevoix from quartz deformation.

Sample	Radial[1] proportion	% quartz with[2] planar features	% type A	% type B	% type C	% type D	Average pressure (GPa)
150	.59	20	20	—	—	—	6.2
39	.50	35	35	—	—	—	6.7
61	.55	36	36	—	—	—	6.7
84	.55	48	48	—	—	—	7.1
52	.39	62	58	4	—	—	7.7
139	.48	68	60	8	—	—	8.0
81	.34	80	44	36	—	—	9.3
86	.57	76	28	44	4	—	9.7
87	.50	100	52	42	—	—	10.3
137	.42	80	20	60	—	—	10.1
5	.28	82	24	58	—	—	10.1
71	.36	92	36	56	—	—	10.3
73	.28	100	44	56	—	—	10.6
82	.40	88	16	64	8	—	10.9
77	.26	96	36	52	4	—	11.1
76	.19	100	32	64	4	—	11.1
106	.42	92	0	80	12	—	11.8
136	.48	96	20	72	4	—	11.2
6	.30	100	12	88	—	—	11.6
104	.20	100	4	88	8	—	12.1
109	.31	100	12	68	20	—	12.2
141	.27	100	—	88	12	—	12.4
126	.09	100	—	84	16	—	12.5
142	.26	100	—	84	16	—	12.5
25	.18	100	—	77	5	18	14.1
118	.20	100	—	56	28	16	14.6
102	.16	100	—	56	4	40	16.5
53	.18	100	2	58	6	34	15.9
74	.13	100	—	44	8	48	17.5

[1]Ratio of distance from crater center to transient crater radius.
[2]Minimum of twenty-five grains measured per sample.

Table 4. Shock pressures at Slate Islands from quartz deformation.

Sample	Radial proportion[1]	% quartz with planar features[2]	% type A	% type B	% type C	% type D	Average pressures (GPa)
5	.52	10	10	—	—	—	5.8
18	.48	35	21	14	—	—	7.1
19	.45	65	29	36	—	—	8.8
49	.33	70	15	45	10	—	9.9
13	.51	100	23	77	—	—	11.3
112	.15	100	40	40	20	—	11.3
127	.18	90	—	70	20	—	12.0
28	.33	100	5	80	15	—	12.3
15	.47	100	—	80	20	—	12.6
27	.33	100	—	65	35	—	13.1
36	.30	100	—	64	24	12	14.0
52	.30	100	—	25	55	20	15.7
24	.36	100	—	15	70	15	15.8
116	.11	100	—	30	40	30	16.5
120	.10	100	6	29	18	47	17.5
34	.33	100	—	35	10	55	18.4
124	.13	100	—	15	25	60	19.4
128	.19	100	—	10	25	65	19.9

[1]Ratio of distance from crater center to transient crater radius.
[2]Twenty grains measured per sample.

8–9 km distance, or roughly halfway to the peripheral trough. At Slate Islands, shock pressures determined for the central uplift rocks are approximately 5–6 GPa at the shoreline and increase inwards in a regular manner to 20 GPa approaching the inferred shock center (Fig. 2 and Table 4). As fewer data are available from this impact site a shock-zoning map equivalent to that for Charlevoix has not been prepared.

To compare the radial distribution of shock pressures at the two impact sites of different size, distances from the inferred shock center must be normalized in terms of a structural parameter common to both craters. The peripheral trough would appear to be an obvious choice for this parameter. However, in the model for complex crater formation proposed by Dence (1968 and personal communication), the peripheral trough forms near the transient crater rim at small craters, but moves outward with increasing magnitude of the impact event. Therefore, the structural parameter to which radial distances are normalized is the radius of the transient crater, rather than the peripheral trough. At Charlevoix, distribution of the Ordovician rocks indicates that the margin of the transient crater lies inside the peripheral trough, with a radius at the original ground surface of 13.5 km (Robertson, 1975). Direct evidence for position of the transient crater at Slate Islands is not visible, as the critical region is underwater. Slate Islands is, however, approximately equivalent in form to the Nicholson Lake structure where indications are that the transient crater margin and peripheral trough coincide (Dence *et al.*, 1968; Dence, personal communication). Thus the radius of the Slate Islands transient crater is taken as 7 km, the distance to the peripheral trough.

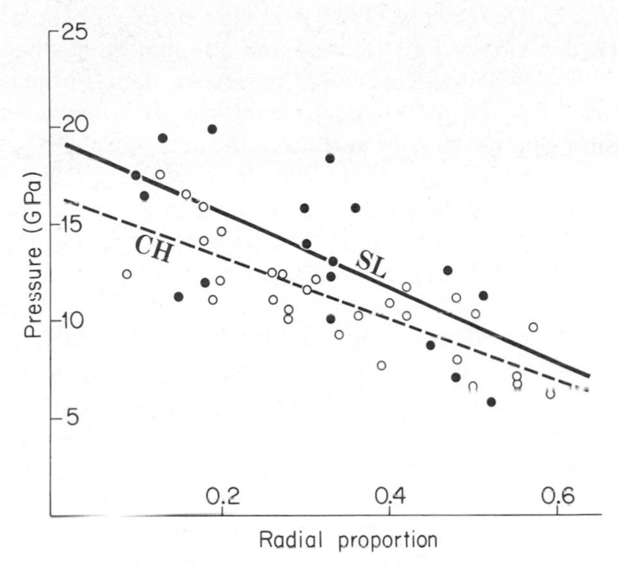

Fig. 5. Shock attenuation in the Slate Islands (SL) and Charlevoix (CH) central uplifts. Charlevoix data are shown by open circles: dots indicate Slate Islands data. Radial proportion is the ratio of sample distance from the crater center to the transient crater radius. Although data from Charlevoix are available inward to a radial proportion = 0, only data from equivalent radial proportion as sampled at Slate Islands are plotted for comparison.

For each sample from the two complex craters, the ratio of its radial distance from the shock center to its respective transient crater radius is calculated as its radial proportion (Tables 3 and 4). Average shock pressure has been plotted against radial proportion for each crater (Fig. 5), and a linear regression curve fitted to each data set. The slope of the Slate Islands attenuation curve (– 193.6) is somewhat steeper than that for Charlevoix (– 160.2) but statistical analysis of the two data sets using Student's t test indicates that the slopes are not significantly different. In other words, the normalized attenuation of shock effects in the central uplift is equivalent for complex craters which do not differ in dimensions by more than a factor of approximately two. This conclusion supercedes our earlier findings (Grieve and Robertson, 1976) which were based on fewer data for the Slate Islands. Actual values for the shock attenuation have not been calculated as they would not be meaningful in terms of the original pressure attenuation because of subsequent structural adjustment to the central uplift configuration.

The intercept of the Slate Islands curve on the pressure axis (Fig. 5), and in fact the entire portion of the curve for which field data are available, lies at higher pressures than for corresponding proportional distances at Charlevoix. The different positions are taken to indicate a lower degree of erosion at the Slate Islands structure (Grieve and Robertson, 1976).

An attempt to derive the original attenuation rate at complex craters, for comparison with the rate at Brent, has been made through modelling studies similar to those employed by Dence *et al.* (1977) using the surface observations described above. As shock deformation distribution and attenuation has been shown to be similar at the surface of Slate Islands and Charlevoix, data for the latter have been used as the model for complex impact structures. It is assumed, allowing for post-crater erosion, that rocks at the surface of the central uplift originally lay at the floor of the excavated cavity. At impact they were initially driven radially downward and outward by the advancing shock front, and subsequently inwards and upwards to form the central uplift. The trajectories along which the upward movement occurred in the modification stage have been defined from the analysis by Milton *et al.* (1972) of shatter cone orientations and seismic data at the Gosses Bluff structure, and are taken as applying to Charlevoix and complex craters in general.

Models of the excavated cavity configuration and distribution of equal pressure contours in the advancing shock front are presented in Fig. 6, for attenuation rates of R^{-2}, R^{-3}, and $R^{-4.5}$ (based on Fig. 10, Dence *et al.*, 1977). Dimensions of the excavated cavity diameter are defined from field observations and the depth is defined by a pressure of 25 GPa, which represents shock pressures typical of the highest grade shock effects found in central uplifts of complex craters or immediately beneath the floor of simple craters. The position of shock-pressure contours calculated from quartz deformation in the surface rocks at Charlevoix (cf, Fig. 3) has been plotted on a horizontal plane 1 km below the impact point to take into account the estimated combined effects of erosion and position of the central uplift below the original ground surface. These observed surface contours of shock pressure were then rotated downward and outward along trajectories equivalent to those at Gosses Bluff to intersect the corresponding pressure contours in each of the three models. The locus of these intersections should define the floor of the excavated cavity. In all cases the excavated cavity thus defined (the inferred cavity, Fig. 6) lies within the cavity, based on the model (hypothetical cavity), but as attenuation rate increases the two cavities become coincident over a larger proportion. It is predicted that the inferred and hypothetical excavated cavities will be superimposed at an attenuation rate between approximately R^{-5} and $R^{-5.5}$, and therefore the surface distribution of shock pressures determined from Charlevoix and Slate Islands is in agreement with a modelled complex crater with such an attenuation rate. From arguments based on energy and crater dimensions Dence *et al.* (1977) concluded that attenuation rates for complex craters lie between R^{-3} and $R^{-4.5}$. Although this rate is somewhat slower than that established above, in both the above analysis and that of Dence *et al.*, the attenuation rate for complex craters (Charlevoix) is higher than that calculated for simple craters (Brent).

DISCUSSION

The arguments and models used to reconcile the observed R^{-20} attenuation rate at Brent with an original attenuation rate of R^{-2} to R^{-3} necessitate the section beneath the excavated cavity to have been permanently reduced or shortened by approximately 50%. Evidence for compression of a similar magnitude is found in

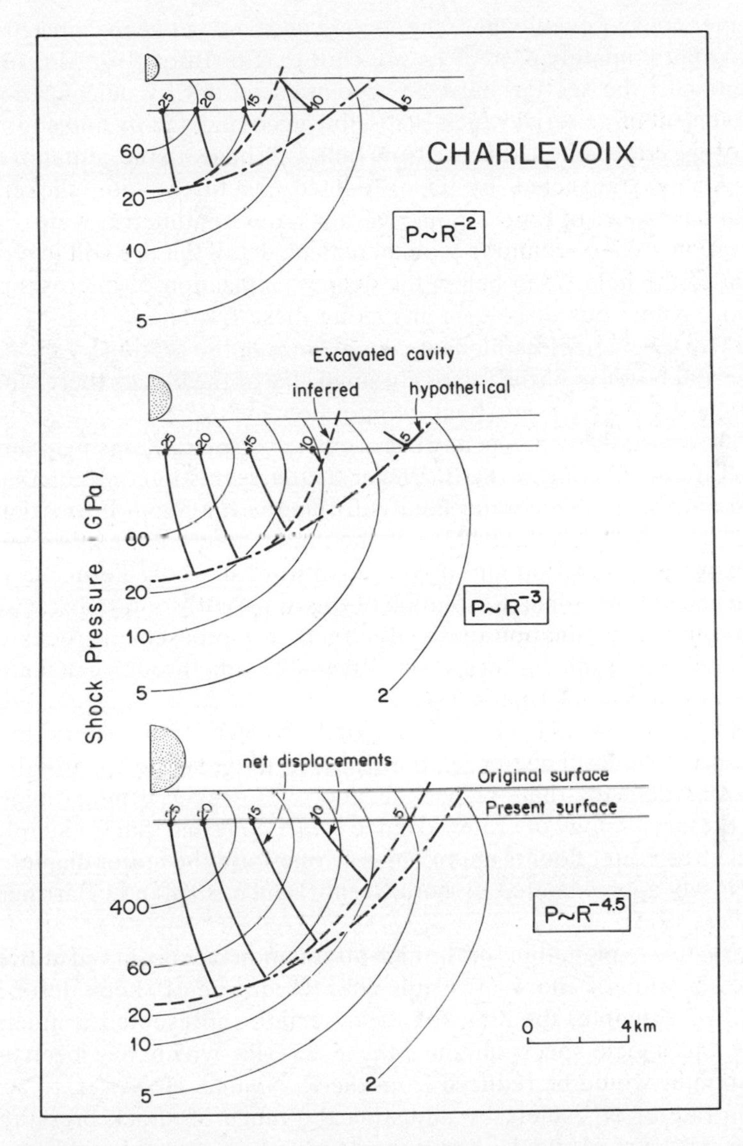

Fig. 6. Models of the excavated cavity at Charlevoix based on attenuation rates of R^{-2}, R^{-3}, and $R^{-4.5}$ (cf. Dence *et al.*, 1977, Fig. 10). The hypothetical excavated cavity shape is controlled by pressure of 25 GPa at the base, and a radius determined from field observations. Points representing shock-pressure contours at the present surface (Fig. 3) are plotted on a horizon approximately 1 km below the point of impact to account for erosion and failure of the central peak to be uplifted to the original ground surface. These points originally lay at the floor of the excavated cavity, and therefore have been rotated downward in the diagram (along trajectories determined from Gosses Bluff) to intersect the equivalent shock pressure contours. The locus of points defined by these intersections is the inferred excavated cavity. The hypothetical and inferred cavities come into closer correspondence as the attenuation rate increases.

the Piledriver nuclear event where the section shocked between 2 and 20 GPa was reduced to approximately 45% of its pre-shot length (Borg, 1972; Fig. 3).

Shortening of the section must be accomplished not by densification but by lateral movement of material which is possibly accomodated by and contributes to upraising of the crater rim. Drill core from hole 1-59 beneath the center of the Brent crater is highly transected by chlorite-filled fractures with slickenside-like surfaces, and by zones of brecciation generally a few centimetres wide. Fracturing and brecciation are less common with increasing depth but are still in evidence to the bottom of the hole, 75 m below the deepest indication of microscopic shock deformation. Numerous displacements along these fractures of blocks less than one metre to a few metres in thickness could shorten the section by the calculated percentage, but because of the relatively small size of the blocks there would be no obvious gaps in the observed shock progression.

There are few data available, however for such displacements in natural events in stratified rocks. Drilling at the Barringer Crater seems to contradict large scale downward movement. The crater floor is formed in the Supai Formation near its upper boundary (Roddy *et al.*, 1975). If the crater floor had been excavated at a higher stratigraphic position and driven downward it would lie in the Coconino Formation and not the Supai. Resolution of this apparently contradictory evidence could come from examination of shock effects in the basement rocks of simple craters where stratigraphic information is available, and through calculation of the shock attenuation rate at such a site.

It might be argued that either Brent or Barringer is anomalous in terms of morphology and depth-diameter relationships, resulting in the apparently contrasting situations beneath their respective floors. However, summarizing several studies of the morphology of craters, Dence *et al.* (1977) find that for simple craters, depth to the true crater floor is approximately one-third the crater diameter, a ratio which is closely approximated by both Brent (Dence, 1968) and Barringer Crater (Roddy *et al.*, 1975).

An alternative explanation for the high attenuation rate observed at Brent could lie in the calibration of shock pressures with the degree of shock deformation in quartz. If, for example, the shocked 85 m section represented a much smaller portion of the shock spectrum than the 5–23 GPa which has been assumed, attenuation rates would be reduced from the R^{-20} value. However, to achieve an attenuation rate of R^{-3} solely by adjusting the range of shock pressures would require the spectrum of shock effects observed to be produced within a spread of less than 1 GPa, a requirement unsupported by laboratory data.

The higher pressures at a given proportional distance at Slate Islands compared with Charlevoix are likely a consequence of a lesser degree of erosion, although there is no independent evidence to assess this assumption. Both structures have formed in crystalline rocks near the margin of the Canadian Shield, and are presumed to have had similar erosional histories, so that their relative degrees of erosion should provide an estimate of their relative ages. Determinations of the age of the Slate Islands impact event from paleo-magnetic studies (Halls, 1975) and radiometric methods (R. Bottomley, personal communication) are regarded as inconclusive. Our belief is that Slate Islands could be younger than Charlevoix, which has been dated radiometrically at 350 ± 25m.y. (Rondot, 1971).

The conclusion that attenuation rates at small, simple craters are less rapid than at large, complex craters has implications for the relative dimensions of the excavated craters, and for the volume and distribution of the various characteristic products of shock metamorphism. If the position of the crater floor is controlled largely by the minimum pressure which exceeds the yield strength of the target rocks, then for rocks of similar strength properties and where the initial impact pressure is equivalent this depth is a direct function of attenuation rate. Therefore in complex craters with the more rapid attenuation rate, the floor of the excavated cavity is proportionally shallower than for simple craters.

Comparisons of the volume and distribution of melt, and the efficiency of cratering between simple and complex craters due to this difference in attenuation rates are given by Grieve *et al.* (1977) and Dence *et al.* (1977). An increased attenuation rate at complex craters also controls distribution in the autochthonous basement rocks of the type of shock deformation discussed here; planar deformation features in tectosilicates. Although the absolute distance over which planar features are produced is greater at larger craters, because of the attenuation rate this distance represents a smaller proportion in terms of the overall crater dimensions compared with smaller, simple craters.

References

Borg, I. Y.: 1972, Some shock effects in granodiorite to 270 kilobars at the Piledriver site. *Flow and Fracture of Rocks.* (H. C. Heard *et al.*, eds.), p. 293–312, Am. Geophys. Union, Geophys. Monograph 16.

Bunch, T. E.: 1968, Some characteristics of selected minerals from craters. *Shock Metamorphism of Natural Materials*, (B. M. French and N. M. Short, eds.), p. 413–432, Mono Book Corp., Baltimore.

Carter, N. L.: 1968, Meteorite impact and deformation of quartz. *Science* **160**, 526.

Cherry, J. T. and Petersen, F. L.: 1970, Numerical simulation of stress wave propagation from underground nuclear explosions. *Peaceful Nuclear Explosions*, p. 241–325, Internat. Atomic Energy Agency, Vienna.

Dence, M. R.: 1968, Shock zoning at Canadian craters: petrography and structural implications. *Shock Metamorphism of Natural Materials*, (B. M. French and N. M. Short, eds.), p. 169–183, Mono Book Corp., Baltimore.

Dence, M. R., Grieve, R. A. F., and Robertson, P. B.: 1977, Terrestrial impact structures: principal characteristics and energy considerations. This volume.

Dence, M. R. and Guy-Bray, J. V.: 1972, Some astroblemes, craters and cryptovolcanic structures in Ontario and Quebec. *24th Internat. Geol. Cong. Guidebook, Field excursion A65*, 61 pp.

Dence, M. R., Innes, M. J. S., and Robertson, P. B.: 1968, Recent geological and geophysical studies of Canadian craters. *Shock Metamorphism of Natural Materials*, (B. M. French and N. M. Short, eds.), p. 339–362, Mono Book Corp., Baltimore.

Engelhardt, W. von and Bertsch, W.: 1969, Shock induced planar deformation structures in quartz from the Ries crater, Germany. *Contrib. Mineral. Petrol.* **20**, 203.

Grieve, R. A. F., Dence, M. R., and Robertson, P. B.: 1977, Cratering processes as interpreted from the occurrence of impact melts. This volume.

Grieve, R. A. F. and Robertson, P. B.: 1976, Variations in shock deformation at the Slate Islands impact structure, Lake Superior. *Contrib. Mineral. Petrol.* **58**, 37.

Halls, H. C.: 1975, Shock induced remanent magnetization in late Precambrian rocks from Lake Superior. *Nature*, **225**, 692.

Halls, H. C. and Grieve, R. A. F.: 1976, The Slate Islands: a probable complex meteorite impact structure in Lake Superior. *Can. Jour. Earth Sci.* **13**, 1301.

Hanss, R. E., Montague, B. R., Galindo, C., and Hörz, F.: 1977, X-ray diffraction studies of shocked materials. (abstract), In *Lunar Science VIII*, The Lunar Science Institute, Houston.

Hörz, F.: 1968, Statistical measurements of deformation structures and refractive indices in experimentally shock loaded quartz. *Shock Metamorphism of Natural Materials*, (B. M. French and N. M. Short, eds.), p. 243–254, Mono Book Corp., Baltimore.

Innes, M. J. S.: 1961, The use of gravity methods to study the underground structure and impact energy of meteorite craters. *Jour. Geophys. Res.*, **66**, 2225.

Milton, D. J., Barlow, B. C., Brett, R., Brown, A. R., Glikson, A. Y., Manwaring, E. A., Moss, F. J., Sedmick, E. C. E., Van son, J., and Young, G. A.: 1972, Gosses Bluff impact structure, Australia. *Science* **175**, 1199.

Müller, W. F. and Défourneaux, M.: 1968, Deformationsstrukturen in Quarz als Indikator für Stosswellen: Eine experimentelle Untersuchung an Quarz-Einskristallen. *Zeitschr. Geophysik.* **34**, 483.

O'Keefe, J. D. and Ahrens, T. J.: 1976, Partitioning of energy from impact cratering on planetary surfaces (abstract). In *Papers Presented to the Symposium on Planetary Cratering Mechanics*, p. 93–95. The Lunar Science Institute, Houston.

Porzel, F. B.: 1958, A new approach to heat and power generation from contained nuclear explosions. In *Second United Nations Conf. on Peaceful Uses of Atomic Energy*, p. 293–299, A/CONF./P/2178.

Robertson, P. B.: 1968, La Malbaie structure, Quebec—a Palaeozoic meteorite impact site. *Meteoritics* **4**, 89.

Robertson, P. B.: 1975, Zones of shock metamorphism at the Charlevoix impact structure, Quebec. *Bull. Geol. Soc. America* **86**, 1630.

Robertson, P. B., Dence, M. R., and Vos, M. A.: 1968, Deformation in rock-forming minerals from Canadian craters. *Shock Metamorphism of Natural Materials*, (B. M. French and N. M. Short, eds.), p. 433–452, Mono Book Corp., Baltimore.

Roddy, D. J., Boyce, J. M., Colton, G. W., and Dial, A. L. Jr.: 1975, Meteor Crater, Arizona, rim drilling with thickness, structural uplift, diameter, depth, volume and mass-balance calculations. *Proc. Lunar Sci. Conf. 6th*, 2621.

Rondot, J.: 1971, Impactite of the Charlevoix structure, Quebec. *Can. Jour. Geophys. Res.*, **76**, 5414.

Rondot, J.: 1972, Géologie de la structure de Charlevoix. *Proc. 24th Internat. Geol. Congress, Section 15 (Planetology)*, 140.

Sage, R. P.: 1974, Geology of the Slate Islands, District of Thunder Bay, *Summary of field work*, (V. G. Milne, D. F. Hewitt, and K. D. Card, eds.), p. 80–86. Geological Branch, Ont. Div. Mines.

Roddy, D. J., Pepin, R. O., and Merrill, R. B., editors.
(1977) *Impact and Explosion Cratering*, Pergamon Press (New York), p. 703–714.
Printed in the United States of America

Shatter cones—An outstanding problem in shock mechanics

DANIEL J. MILTON

U.S. Geological Survey, Reston, Virginia 22092

Abstract—Shatter cones are the characteristic form of rock fracture in impact structures. They apparently form as a shock front interacts with inhomogeneities or discontinuities in the rock. Shatter cones are produced within a limited range of shock pressures, not precisely known but extending from about 20 to perhaps 250 kbar—possibly the range in which shock waves decompose into elastic and deformational fronts. Apical angles range from less than 70° to over 120°; controlling factors are not known. Shatter coning as a physical process has received little attention; theoretical explanations that have been advanced are at best incomplete, if not erroneous.

INTRODUCTION

SHATTER CONES are a prominent and diagnostic feature of impact structures. Geologists, despite the confident use they make of shatter cones in their reconstruction of the development of impact structures, have at best a vague idea of their immediate origin and mechanical significance. Several physicists attending the Planetary Cratering Mechanics Conference expressed an interest in pursuing theoretical or experimental studies of shatter coning. At their request, I have prepared this introduction to the problem, extracting aspects that seem particularly relevant from an extensive literature on shatter cones as a geologic phenomenon (see particularly Dietz, 1968, 1972).

MORPHOLOGY

Shatter cones are conical fracture surfaces striated radially from the apex. Although rare oddities in other geologic environments may resemble them (Dietz, 1968), shatter cones occur in nature only in crypto-explosion structures, and few such structures, if sufficiently exposed and explored, lack them. One must be unusually broadminded to entertain hypotheses for shatter coning other than that it results from shock generated by impact.

Shatter cones range in size from less than 1 cm to over 10 m. The surfaces are cleanly cut, passing through grains in coarse indurated rocks, but there is no relative displacement along the cone surface, at least on a megascopic scale. However, powdery material suggestive of fault gouge is not uncommon on freshly exposed cone surfaces and recently scanning electron microscopy has revealed micrometer-size spherules which the discoverer interprets as fused rock (Gay, 1976). The striations are sharp narrow grooves between broader convexities. Characteristically, subsidiary or parasitic cone segments lie with their apices on the flanks of the master cone, so that the striations have an overall "horse-tail" pattern. Complete cones are less common than partial cone segments. Surfaces of different cones rarely intersect; usually one terminates against the other. Some inhomogeneity, such as a shale chip in sandstone or a small fossil may be found at the apex of a cone. Commonly the apices of a multitude of cones will lie along the contact of two lithologies or even along particular bedding planes in otherwise homogeneous rock. It may be assumed that cones initiate at inhomogeneities or discontinuities within the rock, although the feature responsible may not always be seen. Variations in the abundance and distribution of such inhomogeneities are probably principally responsible for the great variability of habit shown by shatter cones—large or small, complete or partial segments, widely spaced or overlapping.

A wide variety of rock types is shatter coned at one or another of the world's impact structures. Locally, shatter cones may occur in one rock type and not in another. For example, in the Government Reef Formation at Vredefort, shatter coning is generally restricted to shaly beds and is absent in the enclosing quartzite, although other quartzite units at Vredefort are extensively coned (Manton, 1965). In general, dense rocks are more susceptible to shatter coning than associated porous or weakly consolidated rocks. Cones in fine-grained rocks, especially carbonates, have the sharpest and most closely spaced striations.

The often cited value of 90° for the apical angle of shatter cones is at best a generalization, as angles between 66° and 122° have been measured (Table 1). Apical angles clearly differ from one structure, but no clear relation has emerged that would indicate whether these differences are a function of lithology, distance from the point of impact, or depth beneath the surface. Even individual cones may not have constant apical angles—cones with noncircular cross sections and cones flaring outward or toeing inward occur. Nevertheless, at nearly half the shatter cone stations at Gosses Bluff (most of which included several cone segments in one outcrop rather than a single cone) the standard deviation of striation directions from the cone of best fit was less than 4° (Fig. 1).

Conditions of Formation

A hypervelocity impact event consists of a compression stage during which a shock front expands into the ground from the area of impact, followed by an excavation stage during which gross displacements of material produce the crater bowl itself, the tilted rim, the central uplift in larger craters, breccias in and beneath the crater, and the ejecta deposits. Shatter cones form during the compression stage, as is indicated by the occurrence of broken cones in breccia and also by the orientation of cones in place in the crater floor and central uplift. As measured, orientations show little pattern, but at those craters that formed in horizontal strata, displacements during the excavation stage can be determined and if shatter-coned outcrops are restored to their pre-impact position, cone axes point inward and upward toward a point near the original ground surface at the center of the structure. This is striking evidence for, beyond the basic hypothesis

Figs. 1–6. A gallery of shatter cones in sandstone, Gosses Bluff, Australia, showing the variety that can be found at a single structure with a limited range of rock types.

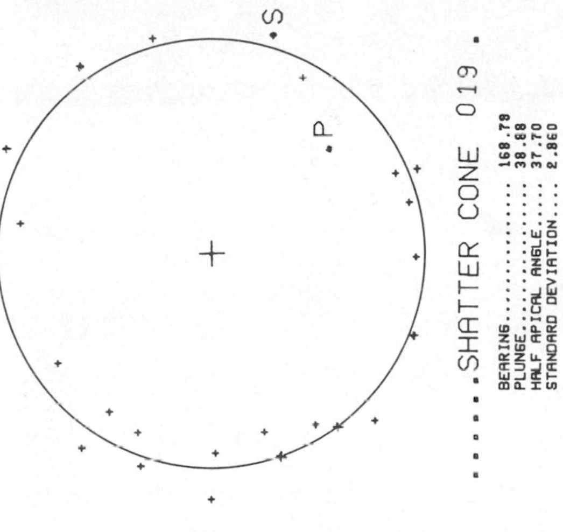

Fig. 1. Measurement of striations and computer printout of results for large cones in massive Mereenie sandstone. The cone segments near the instrument describe two-thirds of the cone, shadowed segments at right center and upper right complete it. The pole of the stereogram is the cone axis, P is lower pole of bedding (face at left in photo approximates the base of a bed), S is south vector, V is vertical vector.

Fig. 2. Pervasively shatter-coned Harajica sandstone. Rock is somewhat heterogeneous with scattered shale chips, but bedding is weakly developed.

that cryptoexplosion structures are caused by impact, the formation of shatter cones during the compression stage with their axes normal to an advancing hemispherical shock front. Most cones point toward the focus, but at most localities a few percent point in the opposite direction (Fig. 6). Measurements at Gosses Bluff show that the reversed cones and nearby normal cones have precisely opposite orientations and identical apical angles. This is evidence that they represent the negative branch of the mathematical cone and are not produced by a distinct reflected shock wave, as has been suggested.

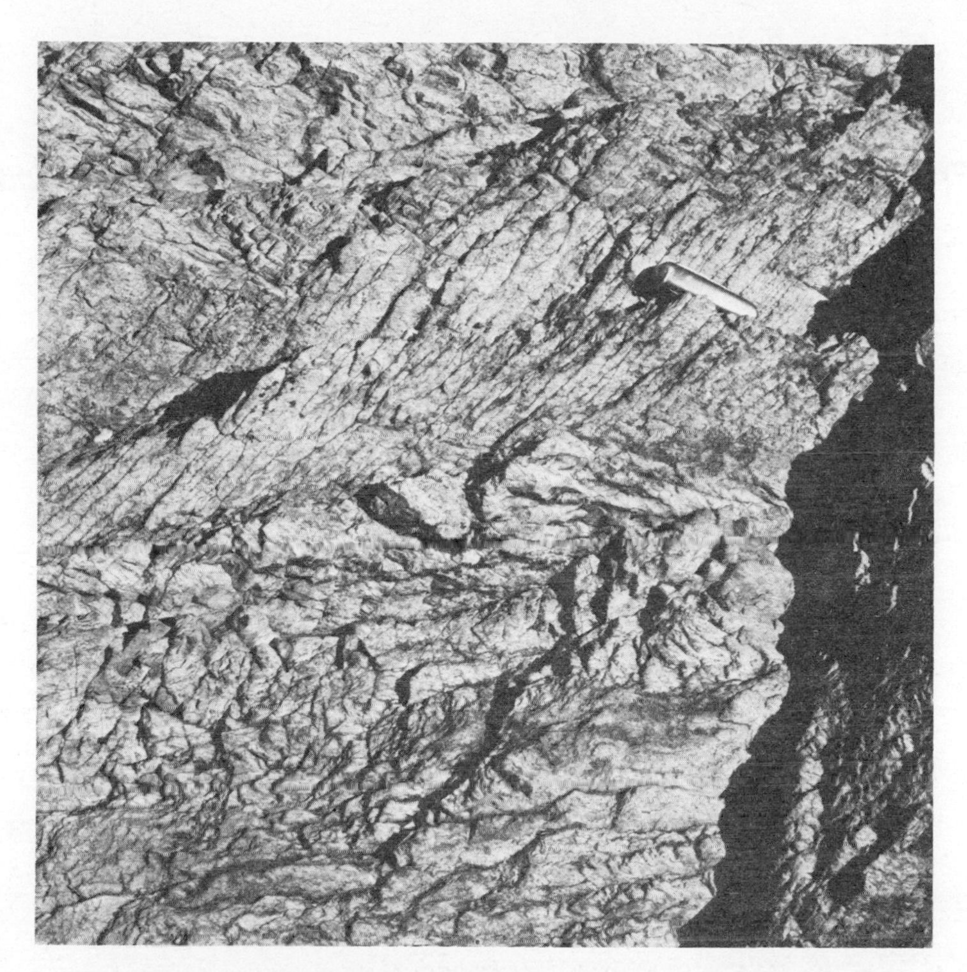

Fig. 3. "Shatter cleavage" in strongly silicified upper Carmichael sandstone. Fine short striations, almost requiring a hand lens to be seen, show that the cleavage planes are composed of flat cone segments preferentially developed in certain orientations.

Shock Pressure of Formation

Minute features that appear to be shatter cones have been produced by laboratory hypervelocity impact into dolomite (Shoemaker *et al.*, 1961) and limestone (Schneider and Wagner, 1976), but determination of the local shock pressure in these small scale experiments is difficult. Excellent shatter cones have been produced in quartz diorite in 100 and 0.5 ton TNT cratering experiments at shock pressures of 25–45 kbar (Roddy and Davis, this vol.). Shatter cones are conspicuously absent in ordinary quarry blasting; fractures radiating from shot holes may define rude cones but they lack the characteristic striations.

The microscopic deformation features in mineral grains can be used to establish shock zones in impact structures and, as these features can be

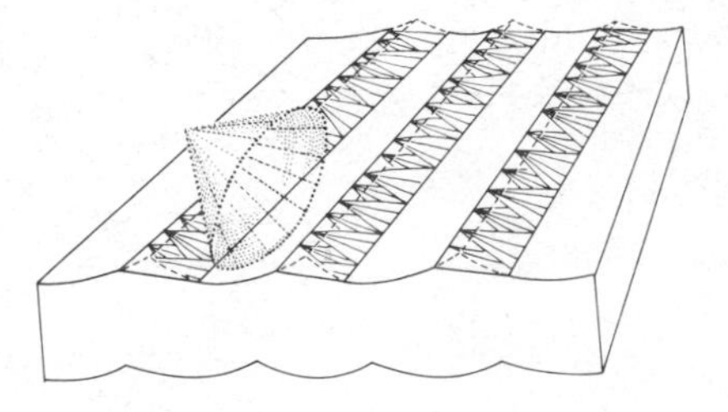

Fig. 4. Ripple-marked Harajica sandstone. Only a small part of the cone developed, forming surfaces that start on the near side of each ripple crest and terminate on the far side. The photograph views from the *bottom* the nearly planar parting surface composed of broad ripple troughs and shatter cone surfaces that cut off the ripple crests.

reproduced in accurately calibrated experiments, shock pressure levels can be assigned to successive zones. This has been done at several impact structures, notably at Charlevoix, Quebec (Robertson, 1975). Shatter cones are found in rock where the shock intensity was insufficient to cause any microscopic effects, so that the minimum requisite shock can only be roughly estimated. At Charlevoix, the weakest planar features in quartz grains, corresponding to about 50 kbar, are found 8 or 9 km from the crater center, whereas shatter cones are found as far as 14 km where, by extrapolation, shock pressures would have attenuated to about 20 kbar. This limit is consistent with the presence of shatter cones beneath the floors but absence in the walls of craters with diameters as diverse as Kaalijarvi, Estonia (110 m), Lonar, India (1800 m), and Ries, Germany

Fig. 5. Coning initiated along Scolithus tubes, casts of vertical burrows made by sand-dwelling organisms, in Mereenie sandstone "pipe-rock". Tubes about 20 cm long.

(24 km). At Charlevoix, virtually all outcrops within a radius of 12 km of the center (shock about 225 kbar) are shatter coned, but Robertson (1968) notes a maximum in the quality of cones (more complete cone segments, striations sharply defined, closely spaced and convergent to the apex) in both gneisses and limestone at about 7 km (corresponding to 50 or 100 kbar) with progressive decrease in quality inward and outward.

The upper limit of shatter coning must be determined at impact structures where zones of more intense shock have survived erosion. Shatter cones are absent in rocks that show bulk maskelynitization (shock vitrification of feldspar) but in some shatter cones from Manicouagan and perhaps the Ries, certain twin lamellae in plagioclase are maskelynitized (M. Dence, pers. comm.). This suggests a limit of about 250 kbar.

Theory of Formation

A theoretical explanation of shatter coning developed by Johnson and Talbot (1964) unfortunately has never been formally published and, although widely

Table 1. Apical angles of shatter cones.

Structure / Unit	Range	Mean and std. dev.	Predominant lithology	Restored distance from focus (km)	Reference
Gosses Bluff (78)[1]	66–96°	80.2 ± 6.1°	Sandstone	3.5–6.5	Milton *et al.* (1972 and unpublished)
Hermannsburg Ss. (7)	71–83°	78.1 ± 6.1°	Sandstone	5	
Harajica Ss. (34)	66–93°	78.4 ± 5.9°	Sandstone	3.9	
Mereenie Ss. (27)	71–93°	82.4 ± 5.4	Well-cemented sandstone	3.8	
Carmichael Ss. (5)	75–96°	80.5 ± 9.3°	Sandstone	3.8	
Stairway Ss. (5)	81–89°	85.2 ± 2.9°	Calcareous sandstone	3.8	
Charlevoix (63)	70–112°			12–15?	D. W. Roy (unpublished)
Ordovician (17)	70–98°	87.2°	Limestone		
Ordovician (7)	82–108°	90.6°	Arkosic sandstone		
Ordovician (7)	82–102°	91.8°	Quartzite		
Precambrian (22)	82–112°	94.2°	Charnockite		
Precambrian (10)	84–122°	99.6°	Metasediments and granitic gneisses		
Vredefort (23)	90–122°	—	Quartzite	46–57	Manton (1965)
Kimberley-Elsburg Series (8)	90–97°	—	Gritty quartzite		
Jeppestown Series (5)	94–100°	—	Gritty quartzite		
Jeppestown Series (1)	90°	—	Epidiorite		
(Intrusive) (1)	98°	—	Alkali granite		
Government Reef Series (1)	100°	—	Pure quartzite		

Government Reef Series (1)	91°	—	Argillaceous quartzite		
Hospital Hill Qzt.	106–122°	—	Fine-grained quartzite		
Orange Grove Qzt. (3)	93–103°	—	Fine-grained quartzite		
Sierra Madera (27)	76–108°	88.5°	Dolomite	1–5	Howard and Offield (1968)
Tessey Ls. (1)	82°	—	Dolomite		
Gilliam Ls. (11)	77–108°	—	Dolomite		
Word Fm. (6)	83–101°	—	Siltstone and sandstone		
Upper Hess Fm. (5)	76–82°	—	Dolomite		
Lower Hess Fm. (4)	82–93°	—	Dolomite		
Ile Rouleau (5)	86–98°	94°	Dolomite	~2?	Caty *et al.* (1976)
Sudbury (1)	87°	—	Quartzite	~50	French (1972)
Manicouagan (1)	85°	—	Gneiss	~20?	Currie (1972)
Lonar (1)	72°	—	Basalt	0.5	Fredriksson *et al.* (1973)
Wells Creek[2] (886)	(35–85°)	(56°)	Dolomite	~2?	Wilson and Stearns (1968)

Units at Gosses Bluff, Vredefort and Sierra Madera in sequence from top to bottom.
[1]Number of cones (or assemblages of cone segments) measured.
[2]Photographs and actual specimens from Wells Creek suggest apical angles of about 30°. The reported measurements may be the spread of striations in partial cone segments rather than full apical angles.

Fig. 6. Segment of the positive branch of a cone (right) and exterior cast of a segment of the negative branch (left) apex to apex. Striations have been marked for measurement. Other positive segments at far left top and bottom. Mereenie sandstone.

accepted, has received little critical scrutiny. Briefly, they proposed that shatter coning occurs when the elastic precursor of a multiple shock front encounters a small region of anomalous density or compressibility, producing a scattered wave. The direct and the scattered waves interact constructively to give stresses which exceed the elastic limit of the rock within a conical region, while stresses remain below the elastic limit outside this volume. As the stress is removed, the material outside which returns elastically to its original state separates along the conical boundary from the permanently deformed material within. Some explanation is required, however, for the removal of stress before the arrival of the deformational wave puts the entire rock well into the plastic regime. Material can be driven past its equilibrium elastic limit and will relax immediately behind the elastic wave (Ahrens and Duvall, 1966) but it is not clear that such relaxation produces conditions suitable for operation of the Johnson-Talbot mechanism. A two-wave structure will occur during strong unloading from the plastic state also; perhaps a mechanism analogous to Johnson and Talbot's may operate during relaxation from the peak compression, rather than before.

Johnson and Talbot's hypothesis places the minimum and maximum stress for shatter coning at definite (if not always known) levels: the Hugoniot elastic limit and the critical stress for decomposition into a two-wave shock for the rock involved. Correlation of the range in which these stresses fall for rocks in general and the estimates for the occurrence of shatter cones appears plausible, but has not been examined more specifically. It might be possible, for example, to relate the presence or absence of shatter cones to the Hugoniots of specific rock types in interbedded lithologies with and without cones.

An alternate hypothesis proposed by Gash (1971), has shatter cones produced by the interaction of the compressive wave and the tensile wave reflected from the free surface. It is not clear that Gash, in dealing with stress fields macroscopically, has accounted for the formation of actual cones, rather than simply fractures whose overall orientation fit a conical pattern. In any case, the prediction from his theory that cones form beneath the area of impact with axes vertical or fanned through a small angle is contradicted by the presence of shatter cones in the outer zones of some impact structures and the very low angles between cone axes and original horizontal bedding in these cones.

At most sites, the presence of shatter cones has simply been noted. Clearly, much could be learned from more detailed, and perhaps different types of observations. The development of a tentative physical theory incorporating the facts now known could provide the stimulus for further field studies.

Acknowledgments—I thank D. W. Roy for furnishing data on Charlevoix shatter cones in advance of publication. C. A. Hodges, P. B. Robertson, and an anonymous individual gave very helpful reviews. Work supported by NASA contract W13130.

References

Ahrens, T. J. and Duvall, G. E.: 1966, Stress relaxation behind elastic shock waves in rocks, *J. Geophys. Res.* **71**, 4349–4360.

Caty, J-L., Chown, E. H., and Roy, D. W.: 1976, A new astrobleme: Ile Rouleau structure, Lake Mistassini, Quebec, *Can. J. Earth Sci.* **13**, 824–831.

Currie, K. L.: 1972, Geology and petrology of the Manicouagan resurgent Caldera, Quebec. *Geol. Survey Can. Bull.* **198**, 153 pp.

Dietz, R. S.: 1968, Shatter cones in cryptoexplosion structures. In *Shock Metamorphism of Natural Materials* B. M. French and N. M. Short, (eds.), p. 267–285. Mono Book Corp., Baltimore.

Dietz, R. S.: 1972, Shatter cones (shock fractures) in astroblemes. *Internatl. Geological Congress*, 24th session, sect. 15, 112–118.

Fredriksson, K., Dube, A., Milton, D. J., and Balasundaram, M. S.: 1973, Lonar Lake, India: An impact crater in basalt, *Science*, **180**, 862–864.

French, B. M.: 1972, Shock-metamorphic features in the Sudbury Structure, Ontario: A review, *Geol. Soc. Can.*, Spec. Pap. **10**, 19–28.

Gash, P. J. S.: 1971, Dynamic mechanism for the formation of shatter cones, *Nature Physical Science*, **230**, 32–35.

Gay, N. C.: 1976, Spherules on shatter cone surfaces from the Vredefort structure, South Africa, *Science* **194**, 724–725.

Howard, K. A. and Offield, T. W.: 1968, Shatter cones at Sierra Madera, Texas, *Science* **162**, 261–265.

Johnson, G. P. and Talbot, R. J.: 1964, A theoretical study of the shock wave origin of shatter cones, Air Force Inst. Tech., M. S. thesis, Wright-Patterson AFB, Ohio, GSF/Mech 64–35, 92 pp.

Manton, W. I.: 1965, The orientation and origin of shatter cones in the Vredefort Ring, *New York Acad. Sci. Annals* **123**, 1017–1049.

Milton, D. J., Barlow, B. C., Brett, R., Brown, A. R., Glikson, A. Y., Manwaring, E. A., Moss, F. J., Sedmik, E. C. E., Van Son, J., and Young, G. A.: 1972, Gosses Bluff impact structure, Australia, *Science* **175**, 1199–1207.

Robertson, P. B.: 1968, La Malbaie structure, Quebec—A Palaeozoic meteorite impact site, *Meteoritics* **4**, 89–112.

Robertson, P. B.: 1975, Zones of shock metamorphism at the Charlevoix impact structure, Quebec. *Bull. Geol. Soc. Amer.* **86**, 1630–1638.

Schneider, E. and Wagner, G. A.: 1976, Shatter cones produced experimentally by impacts in limestone targets, *Earth and Planet. Sci. Lett.* **32**, 40–44.

Shoemaker, E. M., Gault, D. E., and Lugn, R. V.: 1961, Shatter cones formed by high speed impact in dolomite. U.S. Geol. Survey Prof. Paper 424-D, 365–368.

Wilson, C. W., Jr. and Stearns, R. G.: 1968, Geology of the Wells Creek structure, Tennessee, *Tenn. Div. Geol. Bull.* **68**, 236 pp.

Roddy, D. J., Pepin, R. O., and Merrill, R. B., editors.
(1977) *Impact and Explosion Cratering*, Pergamon Press (New York), p. 715–750.
Printed in the United States of America

Shatter cones formed in large-scale experimental explosion craters

DAVID J. RODDY

U.S. Geological Survey, Branch of Astrogeologic Studies, 2255 North Gemini Drive, Flagstaff, Arizona 86001

L. KIM DAVIS

U.S. Army Engineer Waterways Experiment Station, Vicksburg, Mississippi 39180

Abstract—Shatter cones were formed in tonalite in several large-scale explosion cratering experiments near Cedar City, Utah. In these experiments, a series of 0.5-ton and 100-ton TNT spherical charges were detonated at various heights of burst, forming craters from a few meters to over 21 m across at their rim crests. All of the shatter cones had conical fracture surfaces with poorly to well-developed striae fanning out from the apex. Three of the 0.5-ton trials exhibited complete single shatter cones, as well as small groups *in situ* on the crater floor, commonly within a cone of approximately 110° under ground zero. Cone segments were also present in their ejecta. Stereographic measurements of the *in situ* cones on the crater floors showed they pointed toward ground zero, and that their axes were approximately normal to the shock front and passed through ground zero ±20°. Apical angles of these cones averaged 86° with a range from 78° to 94°. Maximum lengths of striations on some cones were as great as 0.6 m.

Detonation of the 100-ton TNT surface sphere formed several large ejecta blocks with nested and elongated well-developed shatter cone segments with striations up to nearly 1 m in length. A 100-ton TNT sphere raised above the ground surface formed a single shatter cone which remained *in situ* in a grout column below the crater.

Formational pressures of the shatter cones, estimated from stress attenuation curves constructed for these experiments, ranged from approximately 20 to 60 kb in the tonalite. The formational stress range of shatter cones at the Flynn Creek impact structure, a 3.6 km diameter crater in Tennessee, was estimated from our attenuation data to lie between approximately 15 and 45 kb, depending on the assumed height of burst.

The explosion cratering experiments conclusively establish that shatter cones can be formed by shock wave processes in cratering, and that the cones normally point in the direction of the energy source. These experiments allow a more confident use of shatter cones as criteria for shock wave processes and direction of propagation, generated by either explosion or hypervelocity impact, and describe a formational stress range in crystalline rocks in the 40 ± 20 kb range.

INTRODUCTION

IN 1968, A SERIES of 0.5-ton and 100-ton TNT explosion experiments were conducted in granitic rock near Cedar City, Utah as part of a basic research program on cratering and shock wave propagation. These trials, titled the Mine Shaft Series, formed craters ranging from a few meters to over 21 m across at their rim crests. Structural deformation consisted of the normal faulting, shattering, local brecciation, shearing, and minor folding typical of cratering at this scale in crystalline rocks.

Of special interest, however, was the formation of an important type of shock metamorphic feature, *shatter cones*, previously unreported from high explosion experiments. Shocked rocks with these features characteristically exhibit conical fracture surfaces with striae radiating outward from the apex in horsetail patterns. More recently, these unusual features have received increased attention in hypervelocity impact studies due to their widespread occurrence at many larger terrestrial impact sites, with the result that it has become common practice by many workers to accept their presence as sufficient criteria for proof of an impact event. This has naturally prompted an increased interest in identifying the overall physical conditions and mechanics of formation of shatter cones. This paper addresses a limited part of this problem by describing the first reported occurrence of shatter cones in high explosion trials and by briefly summarizing calculations of the approximate pressure range for their formation.

Background to Shatter Cone Studies

Shatter cones were first described by Branco and Fraas (1905) at the Steinheim Basin, a 3.6 km diameter crater, in southern Germany. Lacking any direct evidence, they inferred that the crater, and the abundant shatter cones, were formed by some type of buried volcanic explosion and proposed the term cryptovolcanic, i.e., hidden or covered volcanism. This was certainly not an unreasonable inference at that time considering the general lack of understanding of hypervelocity impact phenomena. Since their early work, however, a dramatic increase in the number of shatter cone localities has been reported from around the world with over 36 new locations identified in association with proven or suspected impact structures ranging in size from a few kilometers to over 100 km across. The mounting field and laboratory evidence from these terrestrial sites has steadily grown to now clearly indicate a shock wave origin for shatter cones with formational pressures at least on the order of 10 kb or greater *in surface* and *near-surface* rocks. Dietz, who has been responsible for many of the initial field discoveries of shatter cones (Dietz, 1947, 1959, 1960, 1961a, 1961b, 1963a, 1963b, 1964, 1966a, 1966b, 1966c, 1967), nicely defined the state of knowledge in two summaries (Dietz, 1968, 1972) which indicates: (a) "Shatter cones are conical fracture surfaces with striae which fan outward from the apex in horsetail-like packets. The striae are sharp grooves between intervening, rounded ridges—or the reverse on negative mold faces. Small, parasitic half-cones are shingled on the face of the master cone. Cones range greatly in length from less than 1 cm to more than 12 m for one shale cone at Kentland. ...They are best developed in aphanitic and dense rocks, especially carbonates, but are also known from shale, sandstone, quartzite, granite, gneiss, and other lithologies" (Dietz, 1972). (b) Shatter cones occurring in impact structures in the few kilometer diameter range tend to be concentrated in the central uplift rocks (Milton *et al.*, 1972; Howard and Offield, 1968; Rieff, 1977; Roddy, 1968). In the 100 km diameter range, as at Vredefort and Sudbury, shatter cones tend to occur further toward the edge of the deformed rocks and are arranged essentially symmetrically around the center (Dietz and Butler, 1964; Manton, 1965). (c) Shatter cones commonly appear to point in random directions, but after the host rocks are stereographically rotated back to their original position, the cones invariably point toward the shock wave source area and indicate the source came from above rather than below ground (Milton *et al.*, 1972; Howard and Offield, 1968). Cones in an individual rock unit tend to be parallel and to point in the same direction. Occasionally a few cones will point in the opposite direction of the source area. (d) Apical angles are reported to vary commonly from approximately 60° to 120°, with an overall average close to 90°. Lithologies and proximity to ground zero appear to play a serious role in controlling the angles (Milton, 1977). (e) An inhomogeneity in the rock, such as a clastic quartz grain in limestone or a nodule, may be at the apex of a cone. (f) Rocks of the same lithology and at the same distance from the shock wave source tend to have cones of the same size and orientation. Adjacent, but different, rock types may or may not have shatter cones. (g) Shatter cones which formed last, i.e., millimeters to centimeters farther from the shock wave source, truncate earlier shatter cones producing a stacked appearance on a thoroughly shatter coned rock. (h) Gay (1976) has noted that microspherules on shatter cones at the Vredefort show local melting and considerable dilation across the cone surfaces. (i) Shock metamorphic features formed by high-pressure shock waves, such as certain planar features, coesite, stishovite, shock-formed glass, and numerous other solid state deformational features, formed *closer* to the shock wave source, whereas shatter cones are commonly located further *out* from the shock source area.

The foregoing field and laboratory observations obviously place a number of important physical constraints on the formation of shatter cones. Dietz (1968, 1972) and Milton (1977) discuss the above aspects in greater detail, however certain critical points deserve brief attention here. One problem is that, until recently, well-developed shatter cones have not been identified at impact craters with meteoritic fragments. Consequently, the proof of an impact origin by *direct association* with a meteorite impact had been subject to legitimate question. This problem has been partly alleviated by the recent work of El Goresy and Chao (1976) who have reported traces of *stoney meteoritic* material at the Ries impact crater in Germany, which does have shatter cones. Both shatter cones and meteoritic material also have been reported from the impact craters of Kaalijarv (Dietz, 1968) and Lonar (Milton, 1977).

There are three other sources of evidence now available that also conclusively show shatter cones are formed by shock waves, one involves nuclear cratering, a second involves the high explosion cratering described in this paper, and the third involves laboratory experimental impact cratering. Dietz (1960) reported shatter cones in strongly shocked beds of tuff around the Rainier nuclear explosion chamber at the Nevada Test Site. Bunch and Quaide (1968) also reported shatter cones formed in subandesite ejecta at the Danny Boy nuclear crater at the Nevada Test Site. Neither experiments, however, yielded mechanisms or formational pressures for the shatter cones, although they did clearly demonstrate a shock wave origin. The high explosion cratering described in this paper is the first to combine field location, orientation, apex angles, and formational pressures for a single event (Roddy and Davis, 1969). Shoemaker *et al.* (1961), and more recently, Schneider and Wagner (1976) have formed millimeter-sized shatter cone segments and fragmental cones in laboratory impact cratering experiments. Shoemaker *et al.* (1961) estimated shock pressures on the order of 4.7 kb at the furthest edge of their minute shatter cone surfaces on the floor of an experimental impact crater in sandy dolomite. Schneider and Wagner (1976) are less definitive in their estimates of pressures but note that their shatter cones must have formed below 300–400 kb in limestone targets. The pressures for both these laboratory experiments cannot be applied quantitatively to many of the large natural impacts for a number of reasons, such as different initial conditions, Hugoniots and rock types. However, the important point is that these tests demonstrate *unequivocally* that shatter cones are formed by shock waves in laboratory experimental hypervelocity impacts. The combination of all the new field, laboratory, and experimental data, appears to clearly establish that shatter cones can also be formed by the shock waves generated from a natural hypervelocity impact event, conditions long ascribed to by Dietz (1947, 1972).

The critical questions that remain now involve determining the actual physical conditions of formation of shatter cones such that this data can be used to further define the overall impact process. For example, are shatter cones found only in the central uplifts of craters in the few kilometer diameter range and in outer concentric zones of very large impact structures or are these zones part of extended uplift? Another question is how accurate are shatter cone orientations for prediction of the center of the shock wave source area, i.e., can a real height, depth, or line of burst be determined? What do apex angles, different lithologies, and varying orientations mean in terms of rock properties and shock histories? And, of course, there is the fundamental question of what is the range of pressures required for formation and why does a range exist.

Of the aforementioned problems needing solution, that of determining formational pressures as a function of distance is of major concern in terms of solving field relationships at impact sites. That is, if shatter cones are formed in a given rock type for which equation of state data can be obtained, what can be determined about the overall impact conditions? Fortunately, a limited amount of field and laboratory data are available on this particular problem. At each of the proven impact structures other shock metamorphic features are now known to also be present which, coupled with laboratory experiments, have helped establish the prior existence of transient high shock pressures (Chao, 1968, 1976; Chao and Minkin, 1977; Dence, 1968, 1972; Dence, *et al.*, 1977; Engelhardt, 1972, 1974a, 1974b; Engelhardt and Bertsch, 1969; Engelhardt *et al.*, 1969; Hörz, 1971; Hörz and Quaide, 1973; Robertson, 1975; Stöffler, 1966, 1971, 1972; Pohl and Stöffler, 1977). These extensive field studies and laboratory data show a general trend of *decreasing* high-pressure levels existing in concentric patterns out from the center of large concentricly deformed structures. Laboratory experiments, such as those of Hörz and Ahrens (1969) and others, clearly show that the shock pressures at these sites were well in excess of that permissible for normal near-surface or surface volcanic explosion processes. That is, 200+ kb shock pressure levels *in* surface and near-surface rocks, arranged symmetrically in concentric zones of decreasing pressure and structural deformation outward from a center, is certainly inconsistent with any *known* volcanic or endogenic processes operating near the earth's surface. The close proximity of shatter cones in these shocked rocks clearly ties their formation to the overall deformational process that formed the entire structure. It may be noted for completeness that such unique associations and pressure conditions have not convinced a limited number of workers who still express skepticism of an impact origin for both the structures and shatter cones (Amstutz, 1965; Bucher, 1963; Currie, 1968; Gay, 1976; Nicolaysen, 1976; Zimmerman, 1971). None have yet, however, described any plausible endogenetic mechanisms that would account for the high shock pressures, various shock metamorphic features including shatter cones, circularity of the structures, and the types of structural deformation, i.e., *in surface* and *near-surface* rocks. Indeed, contemporary studies of rock mechanics by Baer and Norris (1969), Biot (1965), Means (1976), Handin and Hager (1957, 1958), Handin *et al.* (1963), Jaeger and Cook (1969), Johnson (1970), and Turner and Weiss (1962), all describe the extreme difficulty in generating high pressures in near-surface rocks under low confining pressures. Simple calculations on compressional rim folding by Roddy (1968) further suggests the great unlikelihood of any of the observed types of structural deformation in the rim rocks being formed by volcanic gas expansion processes. The important point, as noted in the comprehensive summaries in French and Short (1968) and Hörz (1971), is that the majority of shock metamorphic features found at these terrestrial structures, which include numerous solid state deformational features, require formational pressures normally in excess of 50 kb in *near-surface* rocks. An equally important point is that the available field evidence for larger structures indicates that it is most common for shatter cones, at least well-developed cones, to also occur in a generally concentric zone immediately surrounding the inner zone of higher pressure features. The conclusion from this field relationship, as noted in the summaries by Dietz (1968, 1972) and Milton (1977), is that well-developed shatter cones form at shock pressures generally *lower* than the majority of the other shock metamorphic features although overlapping certainly does occur (Dietz and Butler, 1964; Hargraves, 1961; Howard and Offield, 1968; Manton, 1965; Milton *et al.*, 1972; Wilson and Stearns, 1968). Robertson (1968, 1975) and Rondot (1968, 1971, 1972, 1975) describe one exception in which shatter cones at the eroded, 35+ km diameter Charlevoix impact structure in Canada, occur from the center of the structure outwards to about 12–14 km. Shock metamorphic features, formed in the inner part of this structure, experienced pressures on the order of 200+ kb. They note, however, that the shatter cones are best developed at a range of about 7 km, where pressures are estimated to be in the 50 kb range, and are increasingly poorly developed both inward and outward from the 7 km zone. Carter's (1965) comprehensive study of basal deformation lamellae in quartz also indicate shock formational pressures between 35 and 60 kb for shatter cones.

A final source of information for shatter cone formation, of course, lies in theoretical studies which, to date, are limited to only those by Johnson and

Talbot (1964) and Gash (1971). Unfortunately, neither study has yet been tested rigorously by field and laboratory observations and experiments, and therefore have not been broadly useful. Milton (1977) discusses both of these studies and the reader is referred to his paper for review.

With our present state of knowledge, we now know with certainty that shatter cones *can be formed* by shock waves generated during a hypervelocity impact and that the formational pressures appear to exceed at least 10 kb. What is not known are the mechanics of shatter cone formation, the formational pressure range, and what can be deduced about the overall impact process from the presence of shatter cones in various rock types at different ranges. These problems mark some of the more important research areas to attack in the immediate future with respect to shatter cones.

TEST PROGRAM

The experimental program, titled the Mine Shaft Series, consisted of 0.5-ton and 100-ton TNT trials detonated at various heights of burst (HOB) above and below the target rock surface. The 0.5-ton charges formed a calibration cratering series consisting of ten events in which 454 kg (1000 lb) cast TNT spheres were detonated at several HOB's ranging from above ground level to partly buried (Fig. 1). One of the 100-ton TNT trials which formed shatter cones, titled Mine Ore, was constructed by stacking individual 15 kg blocks of TNT to approximate a spherical shape. The target rock for this trial was excavated in the form of a shallow bowl to allow the TNT sphere to lie one-tenth of the charge radius below the ground surface, i.e., 0.24 m deep (Fig. 1). A second 100-ton TNT charge, titled Mine Under, was placed on top of a platform supported by wooden timbers to provide a HOB of approximately 4.8 m. Each of the 0.5-ton and 100-ton spheres were detonated at the center of the charge. A complete discussion of the overall cratering experiments is beyond the scope of this paper, however, the reader requiring further information will find extended treatments in Davis (1970a, 1970b) and Meyer and Rooke (1969).

GEOLOGY AND PHYSICAL PROPERTIES OF TEST MEDIUM

The test site was located 15 km northwest of Cedar City, Utah on the eastern slope of The Three Peaks, a small north–south trending mountain formed by an igneous intrusive body of granitic-like rock. The intrusion is partially encircled on the lower flanks of the mountain by a sedimentary cover of rocks of Jurassic, Cretaceous, and Tertiary age. Mackin (1947) originally classified the crystalline rocks in this region as quartz monzonite. In the area of the test site; however, an absence of potash feldspar, an essential mineral for quartz monzonite, indicates the rock is better termed quartz diorite. Using Shand's (1947) classification, the rock type is further defined as tonalite (Kolb *et al.*, 1970; Farrell and Curro, 1968; Saucier, 1969). The rock in outcrop consists of very light gray to light gray,

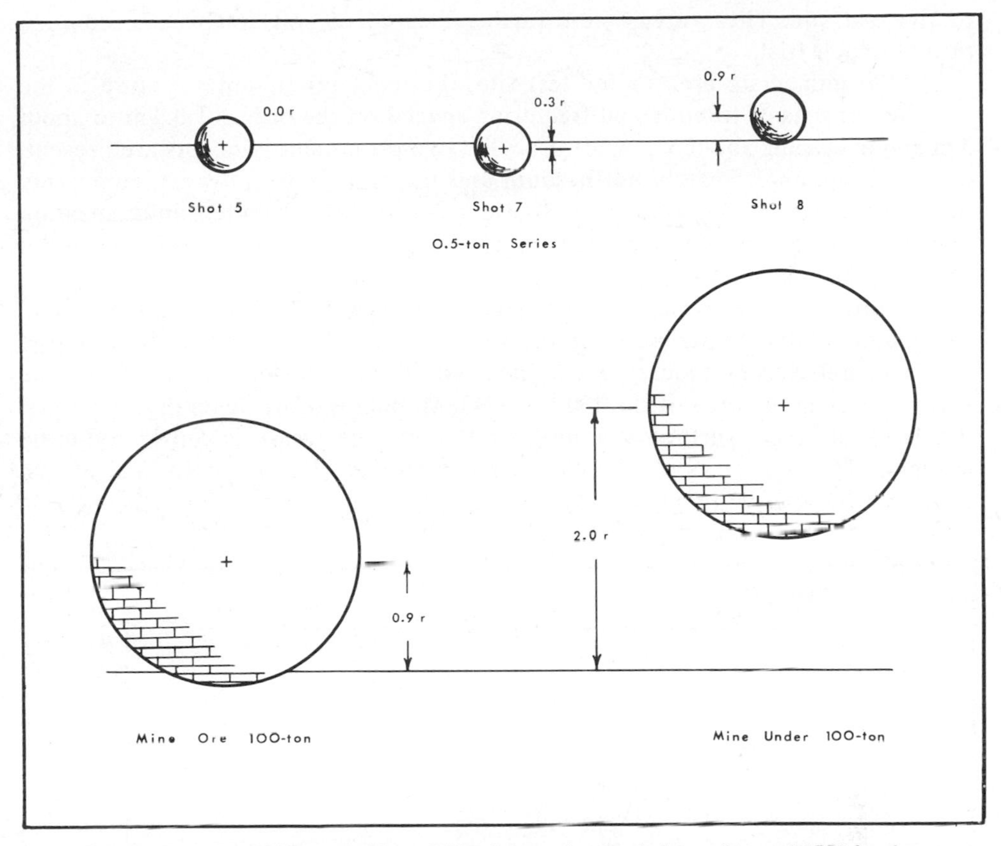

Fig. 1. The 0.5-ton TNT and 100-ton TNT charges showing heights of burst (HOB) for the five trials in which shatter cones were formed. The HOB is defined with respect to the center of the charge and the ground surface, where r is the radius (2.4 m for 100-ton).

fine- to medium-grained crystalline tonalite. The degree of weathering and friability appears related to its proximity to the ground surface and to adjacent joints and fractures. Petrographic examination of thin section shows the rock to consist of a fine-grained matrix of quartz, plagioclase feldspar, pyroxene, and biotite with lesser amounts of magnetite and other trace accessories. Clay is the principle alteration product associated with the mica. Phenocrysts of plagioclase feldspar, pyroxene, and biotite are reported as common in the samples examined (Saucier, 1969). The determination of pyroxene versus hornblende complicates the identification of the rock as tonalite; however, such distinctions in classification are not critical in terms of the rock properties and the formation of shatter cones, and the rock type is hereafter referred to as tonalite in this paper.

The upper part of The Three Peaks exhibits a continuous exposure of the tonalite, whereas the lower flanks of the mountain are partly to completely covered by alluvium and colluvium ranging from a few centimeters to over 3 m in thickness. Locally, however, large outcrops of tonalite are exposed in the area of the test site. These areas were further cleared of soil and weathered rock prior to each trial.

In the immediate area of the test site, the most prominent structure in the tonalite consists of jointing and fracturing spaced on the order of 0.3 m to about 5 m and averaging about 3 m apart. Locally, two prominent joint sets are present, one striking approximately north–south and the other set east–west, commonly terminating against the more pervasive north–south set. Limited mineralization consisting of thin films of hematite, magnetite, and calcite are locally present on some of the joint faces.

A third joint set consists of curvilinear surfaces formed by weathering and exfoliation of the upper parts of the tonalite near the ground surface. Large concave and convex blocks of tonalite complicated the local structure of the surface rock in the area of the 100-ton TNT Mine Ore trial. Generally, however, the depth of severe surface weathering in the test site area was considered to be minimal. This aspect as well as the detailed geological and geophysical results are given by Kolb *et al.* (1970) and Farrell and Curro (1968).

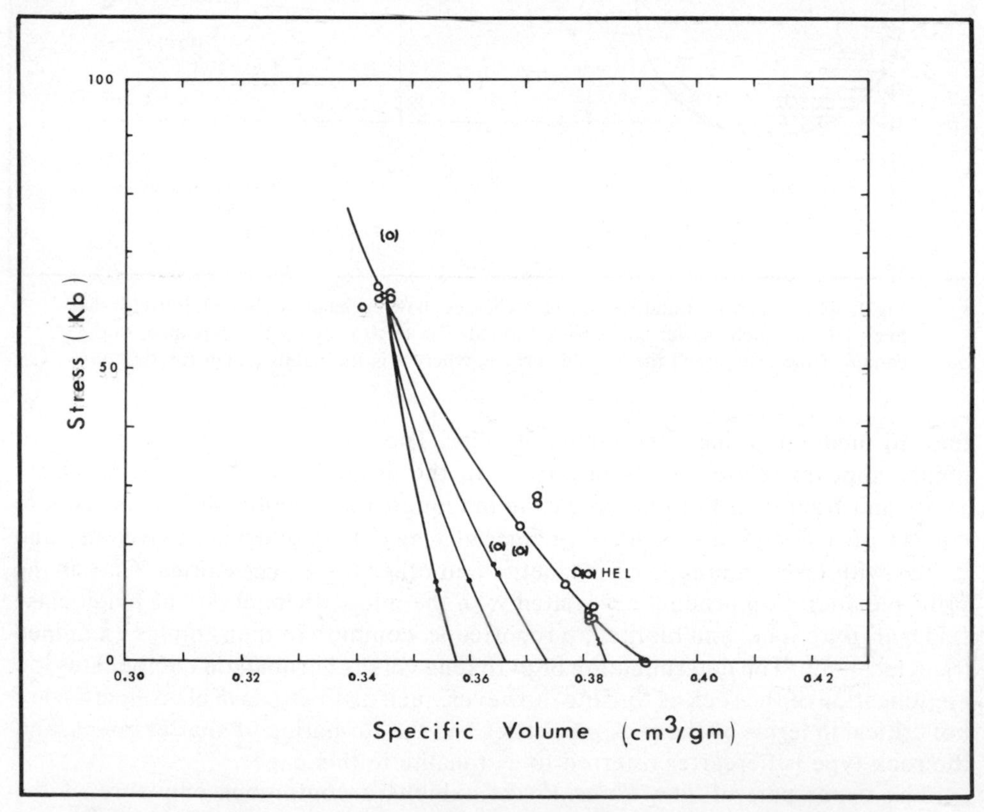

Fig. 2. Preliminary stress-specific volume data for Cedar City tonalite from McFarland (1968). Solid line ——— is Hugoniot; —●— is release adiabat; —●●— bounds to release adiabat; ( ) less reliable data.

The seismic velocity of the rocks exposed at the test site ranged from approximately 2900 to 3700 m/sec with a compressional wave velocity of approximately 3965 m/sec and a shear wave velocity of 2287 m/sec. Hugoniot data, with release adiabats, for the tonalite is shown in Fig. 2, and the particle velocity–pressure relation is shown in Fig. 3 from McFarland (1968). Laboratory tests indicate that the unconfined rock strength increases rapidly with increasing strain rates with normal viscoelastic responses. The average unconfined compressive strength was 1035 kg/cm^2 and the Young's modulus and Poisson's ratio

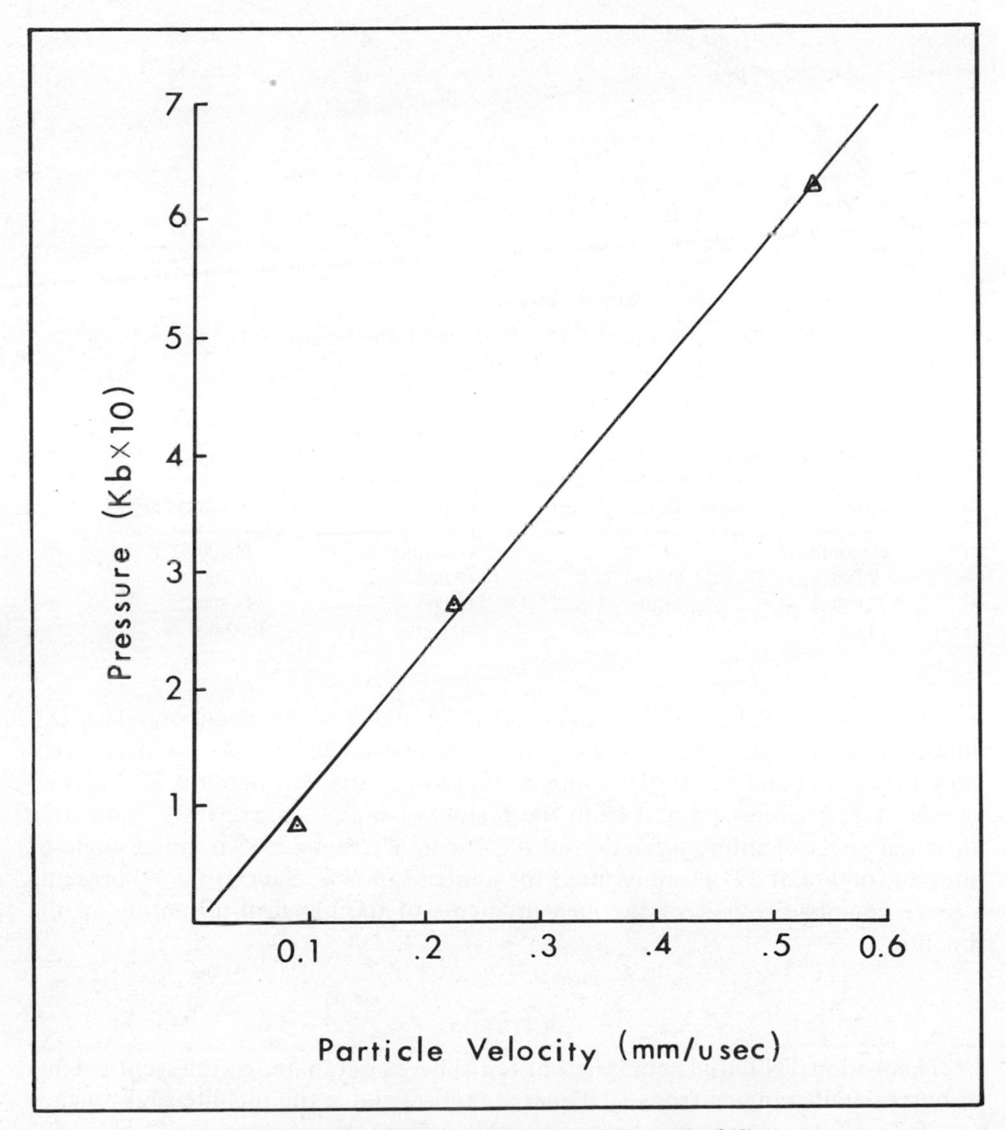

Fig. 3. Hugoniot of tonalite from McFarland (1968).

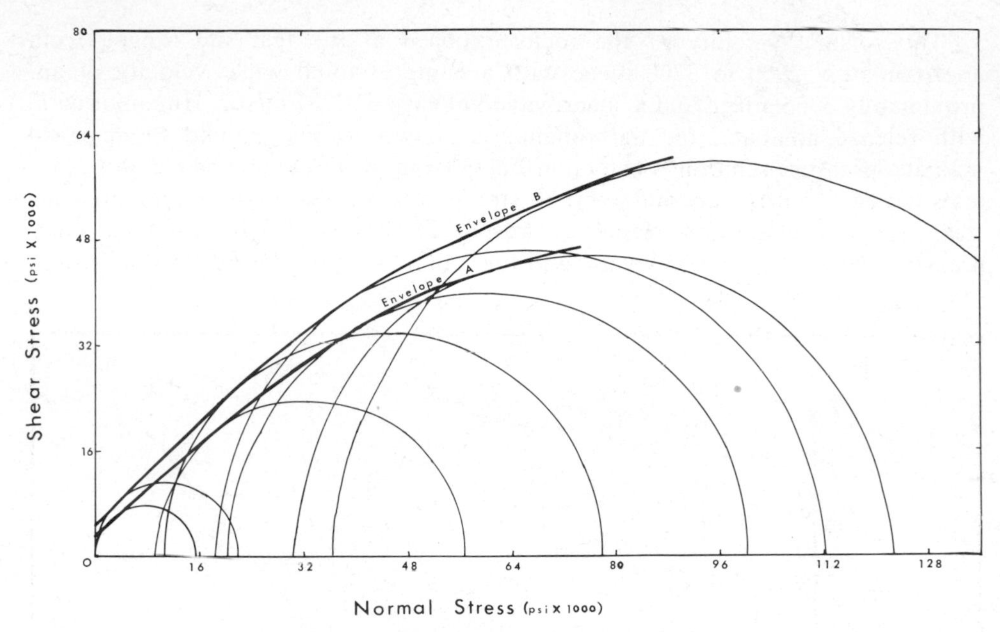

Fig. 4. Mohr's circles for Cedar City tonalite for high-pressure tests from Saucier (1969).

Envelope A for average rock		Envelope B for stronger rock	
σ_3,psi	Deviatory, psi	σ_3,psi	Deviatory, psi
Unconfined	14,700	Unconfined	22,000
9,000	49,100	10,000	67,700
20,000	81,000	18,000	93,300
30,000	92,200	36,000	120,000

(measured at a 352 kg/cm^2 loading) were 2.79 and 0.17, respectively. The calculated bulk modulus, from hydrostatic compression tests, was approximately 7.0×10^4 kg/cm^2, and the static, confined Poisson's ratio was about 0.25. Figure 4 shows Mohr's circles plotted from the results of high-pressure tests, indicating an initial angle of internal friction of 45° for intact samples. An initial angle of internal friction of 37° was measured for jointed samples. Saucier (1969) presents a more complete review of the measurements of the physical properties of the tonalite.

CRATERING

The 0.5-ton TNT trials consisted of ten spheres detonated at different heights of burst (HOB) ranging from +2.0 charge radii(r) above the tonalite rock surface to $-0.3r$ partly contained. As one would expect, the charges well above the surface produced no cratering while the deepest buried charge formed the largest crater. A blocky, fragmental ejecta blanket surrounded each crater. Topographic profiles of craters which formed shatter cones or shatter cone surfaces, Craters 2, 5, 7, and 8, are shown in Fig. 5. Average apparent crater dimensions for these four trials ranged from a radius of 1.19–2.87 m across and 0.27–0.52 m deep. The true crater dimensions ranged from 1.19 to 3.38 m across with maximum depths of 0.34–1.25 m. The apparent crater values represent the dimensions measured from ground zero (GZ) to the upper surface of the breccia lens as it initially existed in the crater. The apparent crater, as used in this paper, represents diameters and depths measured from the original ground surface with the breccia lens still in place. The true crater, as used in this paper (see Figure 4 in Roddy, 1977a) represents the crater dimensions measured from the original ground surface with the breccia lens removed. The variabilities in the crater dimensions were strongly related to the pronounced joint system and the degree of weathering along the joints and rock fractures.

The 100-ton TNT Mine Ore sphere, buried one-tenth charge radius into the tonalite rock surface, formed an irregularly shaped crater which was also strongly affected by the local joint and fracture pattern. The average apparent crater radius was 7.0 m with a 2.74 m maximum depth. The average true crater radius was 8.84 m with a 3.84 m maximum depth (Fig. 6). A very blocky, fragmental ejecta blanket with a highly irregular areal distribution surrounded the crater.

The 100-ton TNT sphere, Mine Under, detonated on the tower formed a *bulked* mound of fractured and brecciated tonalite about 1 m in height. Removal of this fragmental rock showed a true crater on the order of 0.7 m deep and less than 4 m in radius (Fig. 6). Detailed treatments of the cratering and ejecta studies are given by Davis (1970a, 1970b) and Meyer and Rooke (1969).

Shatter Cone Occurrences

Shatter cones were positively identified in three of the 0.5-ton TNT craters and in their ejecta, in ejecta from the Mine Ore 100-ton TNT crater, and in one grout column beneath the Mine Under 100-ton TNT tower shot. Striated surfaces similar to poorly developed shatter cone segments were also present in a fourth 0.5-ton crater. The major distinction in occurrence between the 0.5-ton and the Mine Ore surface 100-ton event is that all the shatter cones were ejected from the crater in the larger event. In the smaller trials, they were found in ejecta, as well as remaining *in situ* or only slightly dislodged on the crater floor, and pointed toward ground zero. In the grout column, the cone remained in place well below the crater floor. Naturally, *in situ* orientation measurements could only be made on the cones remaining in place on the crater floors in the 0.5-ton trials. The original orientations of the ejected cones and the grout column shatter cone could not be determined. Another difference between the trials is that cone lengths were commonly a factor of 4 to 10 times longer in the 100-ton Mine Ore

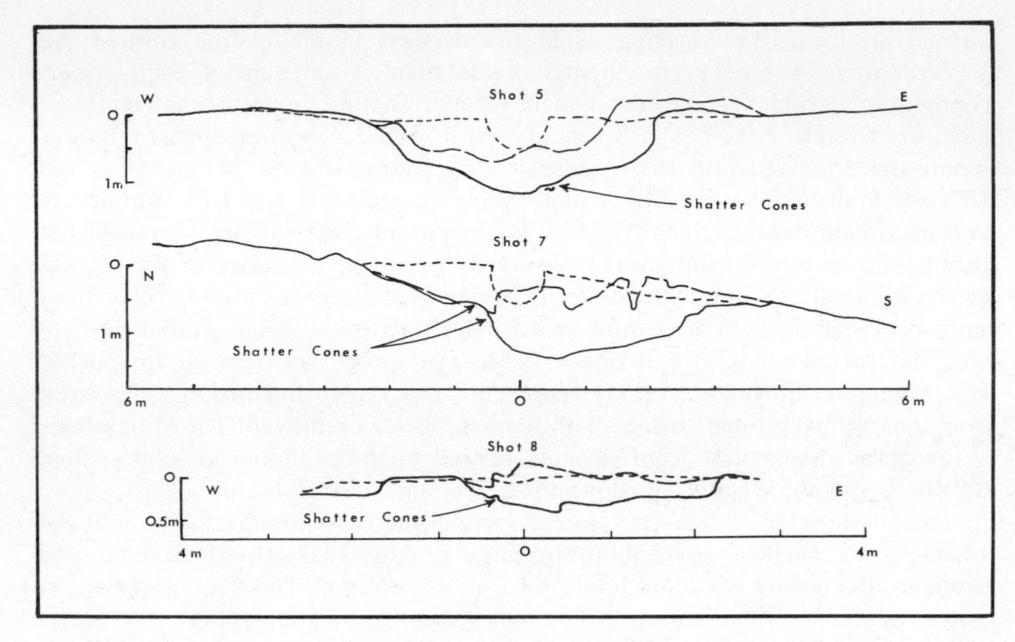
Shot 5
W
E
O
1m
Shatter Cones
Shot 7
O
N
S
1m
Shatter Cones
6m
O
6m
Shot 8
O
W
E
Shatter Cones
0,5m
4m
O
4m

Fig. 5.

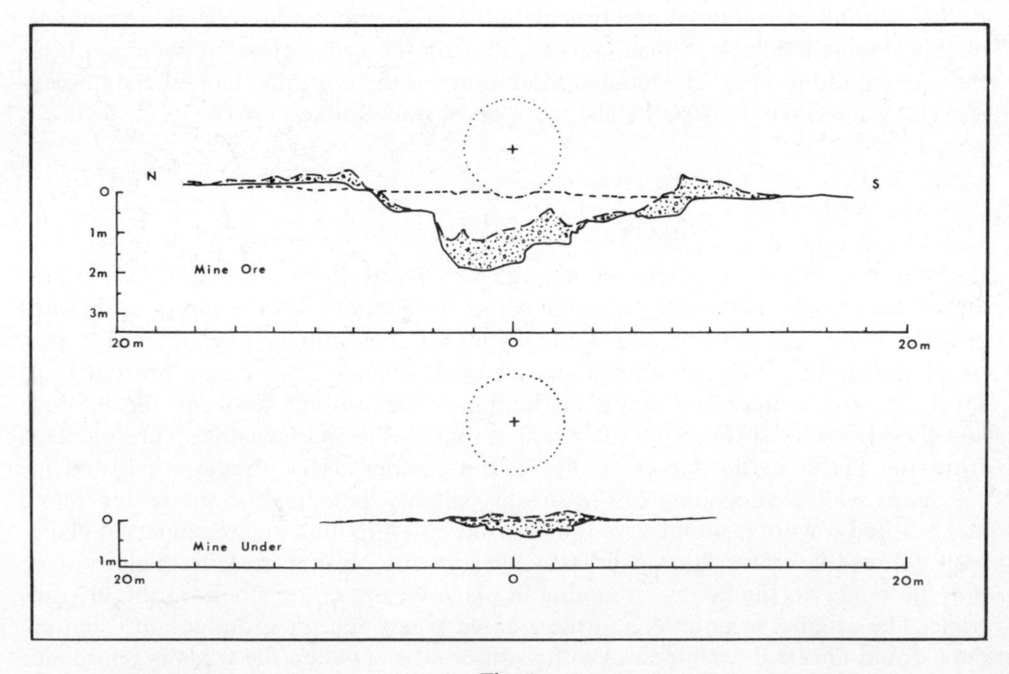
N
S
O
1m
2m
Mine Ore
3m
20m
O
20m
O
Mine Under
1m
20m
O
20m

Fig. 6.

trial as compared with the smaller trials. Apex angles could be measured only for the 0.5-ton trials and ranged between 78° and 94°.

(a) *Shatter cones in the 0.5-ton TNT trials*

Moderately to well-developed shatter cones were found *in situ* on the crater floors and in the ejecta of three of the 0.5-ton trials, i.e., Craters 5, 7, and 8. Crater 2 also showed shear surfaces with striations which resembled poorly developed shatter cone segments. Figure 5 shows the locations of the shatter cones on the floors of Craters 5, 7, and 8. In Craters 7 and 8 the shatter cone positions were offset from the plane of the topographic profiles by approximately 0.3 m. Their locations on the profiles in Fig. 5 therefore do not represent their actual position, but instead, is a superimposed position with respect to GZ, i.e., the distance from the cone apex to GZ is the actual measured field distance. The shatter cones in Crater 5 were broken loose and slightly dislodged before their *in situ* position could be measured with sufficient accuracy to reconstruct their exact position on the topographic profile. Shatter cone fragments were also found in the ejecta surrounding each of these craters, but their individual positions were not recorded due to required immediate post-crater excavation studies. Shatter cones may also have been formed in one or more of the other 0.5-ton craters and not observed due to the very limited field time available in this particular experimental series.

Crater 2 was formed by an 0.5-ton sphere detonated at an HOB of 0.9*r*. Complex jointing caused a very irregular crater to be formed with an apparent radius of 1.37 m and depth of 0.27 m. The average true crater radius measured 1.49 m with a depth of 0.46 m. A number of rock surfaces on the floor of the crater exhibited shear lineations similar to the striations found on poorly developed shatter cones in coarse-grained crystalline rock found in certain impact structures.

Crater 5 was formed by an 0.5-ton sphere detonated at an HOB of 0.0*r*, i.e., half buried. The crater was relatively symmetrical with an apparent radius of 1.65 m and depth of 0.52 m. The average true crater radius was 2.5 m with a depth of 1.13 m. Several small shatter cone segments were found in the ejecta and one block with moderately well-developed nested cone segments was found slightly dislodged on the crater floor. This block was about 0.5 m in length and

---

Fig. 5. Topographic profiles of the 0.5-ton TNT craters. Shatter cones are shown in generalized *in situ* locations although then were not in the planes of the profiles. Short-dashed line is original ground surface. Long-dashed line is the top of breccia lens and forms the surface of the apparent crater. Solid line is the true crater below the breccia lens. Modified from Davis (1970).

Fig. 6. Topographic profiles of the Mine Ore and Mine Under 100-ton TNT craters. The short-dashed line is the original ground surface. The long-dashed line in the crater marks the top of the breccia lens and forms the surface of the apparent crater. The long-dashed line outside of the crater marks the inner ejecta. The solid line is the true crater below the breccia lens and ejecta. Modified from Davis (1970).

had nested cone segments that curved over 45° from apex to tail. The approximate distance from the cone apex region to GZ was 0.87 ± 0.1 m. Shear surfaces were less evident at this crater and commonly occurred lower on the true crater walls than at the other craters.

Crater 7 was formed by an 0.5-ton sphere detonated at an HOB of $-0.3r$, i.e., partly buried. This crater, the most deeply buried of the series, exhibited very few shear surfaces and was somewhat squared due to the effects of the vertical joints. The apparent crater radius was 2.87 m with a depth of 0.27 m and the true crater radius was 3.38 m with a depth of 1.25 m. In the crater, several moderately well-developed shatter cones and segments remained in place on a steep wall on the floor. The apex of some of the cones protruded out from the crater floor surface and were badly damaged or destroyed by the fallback and turbulent rock movements during the cratering process. Their location in the crater was inside a cone of approximately 110° or less measured from the vertical at the center of the charge. Stereographic measurements show that the cones point within approximately ±20°, toward the center of the charge or slightly lower. The measured distances from the damaged apexes to GZ averaged about 0.69 ± 0.1 m. The furthest cone surface from GZ that could be measured averaged approximately 1.29 m in distance. The cone surfaces extended below the crater floor but their total lengths could not be determined. Limited excavation tended to indicate the cone surfaces terminated within a few centimeters of the floor level, but this generalization may not be valid for all the cones.

The cone surfaces were generally roughly striated with crudely radiating, broadly spaced ridges and grooves (Fig. 7). Several cone surfaces exhibited curved or horsetailed striations. Complex intersections of some surfaces gave an unusual transected character to many of the cones and segments. Most of the shatter cones and segments, when broken out of place by hand, exhibited pronounced shear surfaces with slickensides on one or more of the sides away from the cone surface. The average length of the cone segments *in situ* in the crater was between 10 and 25 cm. One exception was a large cone segment that, when partly excavated by hand, exhibited striations 0.6 m in length.

Apical angles for the cone segments in place ranged from 78° to 94° and averaged 86°. Figure 8 shows a nearly complete cone which has a flaired shape with lower apical angles nearest the apex. The rock surfaces tended to disintegrate easily with a rapid reduction in the integrity of the finer striations. Cone nesting was present but not well developed. The larger cones had irregularly curved surfaces that tended to give the cones a distorted shape. Some of the cone surfaces, however, appeared to have secondary slickensides developed on the cone faces, tending to partly obliterate the cone striation. In a number of cases the cone surfaces that had slickenside development also had shatter cone casts in the nearby fallback or ejecta. This suggests that the cones were formed first and their adjacent blocks with cast surfaces were ejected with a shearing motion. Normally, no direct evidence of motion could be determined on most of the shatter cone surfaces. Figures 9 and 10 show two other shatter cones and cone segments with coarsely developed striae.

Crater 8 was formed by an 0.5-ton sphere detonated at an HOB of $0.9r$, the same as Crater 2. The crater shape was also influenced by the local jointing but considerably less than was Crater 2. The apparent crater radius was 1.19 m with a depth of 0.18 m and the true crater radius was 1.19 with a depth of 0.34 m. The shatter cones in Crater 8 were generally identical to those in Crater 7, except that the largest, most complete cone in those trials was formed on this crater's floor (Figs. 11, 12, 13). This cone was 0.3 by 0.2 m at the base and consisted of an irregular curving surface that formed a broad, flat cone, i.e., the entire block had a crenulated surface. The length of the striation on the sides of this cone was 10–15 cm for the part exposed above the crater floor. The total length of the cone surface below the floor could not be determined. Stereographic measurements, however, gave a surprisingly consistent set of apical angles that ranged from 78° to 94°, averaging 86°. These are the same apical values determined for three other cones in this crater. The main cone on the crater floor pointed toward GZ and its axis passed within approximately ±20° of GZ.

The measured distance from the approximate apex of the cone to GZ was between 39.4 and 44.5 cm, depending on the exact choice of axis on the damaged apex. At the level of the crater floor, the furthest cone surface from GZ averages about 55 cm in distance, although this surface certainly extended further below the crater floor.

In general, all the shatter cones had the same surface characteristics as those in Crater 7 except that crude nesting was somewhat more common. Development of secondary slickensides occurred on certain cone faces with large, well-developed, fault-formed slickensides on other sides of the cone. Some of these fault slickensides have faces parallel to the shatter cone surface on the opposite side of the cone or cone segment implying the possibility of some type of parallel displacement. Obviously, a number of complex movements were induced on each of these blocks.

Fig. 7. Moderately well-developed complete shatter cone found *in situ* on the floor of the 0.5-ton TNT Crater 7 trial. Apex is largely destroyed. Bar scale is approximately 15 cm long.

Fig. 8. Shatter cone found *in situ* on the floor of the 0.5-ton TNT Crater 7 trial. Note decrease in apex angle toward top of cone. Dark lines were drawn on cone to outline individual striae.

Fig. 9. Shatter cone found *in situ* on the floor of the 0.5-ton TNT Crater 7 trial showing part of a complete cone with irregular surface, vertical fracture, and slickensides. Bar scale is 10 cm long.

Fig. 10. Shatter cone segment found *in situ* on the floor of the 0.5-ton TNT Crater 7 trial. This block was part of a nearly complete cone that fragmented when it was taken from the crater floor. Poorly developed or incipient nesting is seen on the bottom half. Bar scale is 5 cm long.

Fig. 11. Oblique view of single large shatter cone on floor of 0.5-ton TNT Crater 8. Arrow points to cone apex region and intersection of white cord is ground zero. Apex of cone is offset to the left in photograph from ground zero by about 22 cm. True crater shown here is 3.84 m across and 1.25 m deep.

Fig. 12. Close-up of single large shatter cone on floor of 0.5-ton TNT Crater 8 shown in Fig. 11. Solid arrow points to cone apex region. Dashed arrow points toward ground zero at top of photograph. Cone points toward ground zero with apex approximately 40–45 cm from GZ. Note striae and crude nesting on irregular curving surface of cone. Apex of cone was destroyed during cratering. Pen is 13.5 cm long.

Fig. 13. Single, irregularly shaped shatter cone formed on floor of 0.5-ton TNT Crater 8. This is the same cone shown in Figs. 11 and 12. Poorly developed nesting is seen on left side of block. Striae are approximately 10 cm long on sides of cone surface. Apex damaged during cratering and totally destroyed during excavation. Bar scale is 5 cm long.

(b) *Shatter cones in the 100-ton TNT Mine Ore trial*

This 100-ton TNT crater was formed by the detonation of a sphere at an HOB of 0.9*r*. Ejecta from this trial formed a very blocky, irregular blanket surrounding the crater. During the ejection process, three large blocks of tonalite were thrown out of the crater to distances ranging from 15 to 17 m from GZ along a south–southwest radial. These three blocks exhibited the only shatter cone segments and casts observed at this trial (Fig. 14). They also represent the largest and best developed shatter cones of all the trials.

Ejecta block A, a large barrel-shaped fragment shown in Figs. 14, 15, 16, 17, weighed about 400 kg and exhibited excellent, well-developed, nested cone segments extending over a single, unbroken surface for 90 cm. The cone segments were terminated at the top and bottom of the block by fracturing. The shatter coning on this surface consisted of millimeter- to centimeter-deep grooves separated by rounded cone segments forming slightly irregular nesting. Individual cone striations are as long as 80 cm or more. Crudely to non-shatter coned regions occur locally toward the top of block, i.e., in the direction the

Fig. 14. Shatter cone surfaces on large blocks ejected from the 100-ton TNT Mine Ore trial. Block A, weighing over 400 kg, has shatter cone surface about 1 m in length with nested cones. Individual striae are up to 0.8 m in length. Block B, weighing over 200 kg, exhibits more prominent horsetail striae than on block A. The surface of block C seen in the photograph is the cast from the shatter-coned surface of block B. See broom lying on block B for scale.

Fig. 15. Shatter cone surface on block A ejected from the 100-ton TNT Mine Ore trial. Block is 1 m long with individual striae up to 0.8 m in length. Cone nesting can be seen as elongated, narrow-angle cones. Relief between the grooves and ridges is up to 2 cm locally. Note left end of block is poorly to non-shatter coned. The shatter cone surface does not extend around more than 30% of the circumference of the block.

cones pointed. At this end of the block the cones become increasingly poorly developed, as though this was the beginning of the shatter cone development in the rock. The shatter cone surface on block A extended around the curved surface for at least 30% of the total circumference, and the cones segments continued to point in the same direction. No slickensides could be determined on the cone surfaces and both ends of block A showed only flat, fractured surfaces. Cone apex angles were not measured due to other trial experiments in progress.

Ejecta block B shown in Figs. 14 and 18 also exhibited well-developed, nested cone segments as on block A; however, horsetail curving of the striae was more common. Block C shown in Fig. 14 is the cast broken from the shatter cone surface of block B. Blocks B and C weighed about 200 and 100 kg respectively. Some of the cone segments on block B extended the length of the block, i.e., 80 cm. The apparent cone segment angles on the rock surface of block A tended to be slightly smaller than those of block B, with an average apparent angle in plan view of 22°, although this may be related to the variations produced by the horsetail curvatures. Shatter coning was not present on the sides of blocks B or C, and the nesting on block B was also less common than in block A.

Fig. 16. Shatter cone surface on block A ejected from the 100-ton TNT Mine Ore trial. Pen is 13.5 cm long.

Fig. 17. Shatter cone surface on block A ejected from the 100-ton TNT Mine Ore trial. Pen is 13.5 cm long.

Fig. 18. Shatter cone surface on 200+ kg block B ejected from the 100-ton TNT Mine Ore trial. Note horsetail striations and nesting. Large curvature of striae is most common on this block. Pen is 13.5 cm long.

The pre-shot location of the tonalite that fractured into blocks A, B, and C can be estimated with some certainty even though they were thrown from the crater. Ejecta tends to follow a ballistic path from its initial point to its final position along a single radial, as shown in numerous explosion trials. Using the radial passing through the blocks and GZ, the crater topographic profiles along this radial and the block distances from GZ suggest a general launch region for the blocks. This position is further refined by using independent data determined from subsurface markers ejected from the crater (Davis, 1970). This unique marker data allows a general set of subsurface crater depths to be translated into ejecta ranges as plotted in Fig. 19. For example, rock located between the 15 and 30 m contours was found to be ejected to a range of between 15 and 30 m. It is important to note; however, these are only general ranges valid for this experiment and should not be extrapolated to craters with different initial conditions. In the case of this 100-ton trial, blocks A, B, and C were ejected to distances ranging from 15 to 17 m from GZ, and using Fig. 19, we find a pre-shot minimum apex distance from GZ of 1 m and a maximum apex distance of 6.5 m. Since the blocks were about 1 m in length, the furthest end of the cone surfaces would have been 2 and 7.5 m, respectively. It seems much more reasonable, however, that the shatter cones would have formed nearer the approximate 110° cone region common in the 0.5-ton craters. Assuming this was the case, the ap-

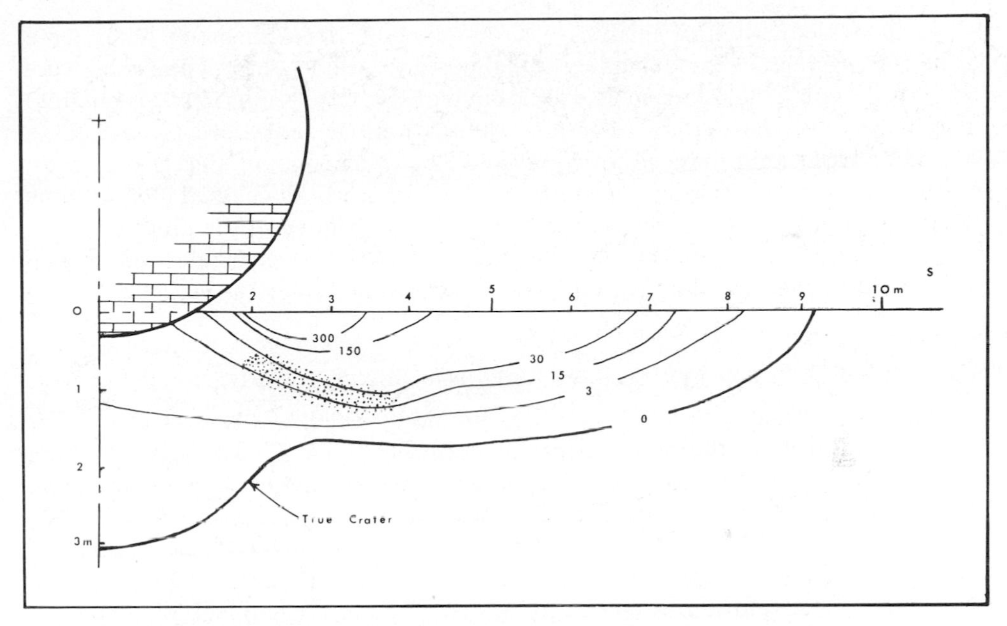

Fig. 19. Generalized south radial profile of true crater formed by the Mine Ore 100-ton TNT event. The numbered contours indicate approximate average ranges in meters at which discrete fragments of ejecta were deposited. These approximate average ranges were determined from color-coded grout fragments ejected during the cratering with the terminal position measured from the pre-shot column location. The three large blocks A, B, and C, exhibiting shatter cone surfaces had a most probable pre-shot location in the shaded area as described in the text. Profile and range contours modified from Davis (1970).

proximate distance from GZ to the apex of the shatter cones would have been on the order of 2–3 m, and the farthest cone surface would have been 3–4 m.

The crater and its ejecta field were thoroughly searched but no other shatter cones were located. A large number of shear surfaces were present in the crater walls and ejecta, some with cone-like striations, but no other definitive cones were identified. The blocky chaotic nature of the ejecta and fallback in the crater could, however, easily have masked a limited number of other shatter-coned fragments.

(c) *Shatter cones in the 100-ton TNT Mine Under tower trial*

This 100-ton TNT event was formed by the detonation of a sphere supported on a tower at an HOB of $2r$. The raised charge actually caused the tonalite to fracture and bulk upwards forming a 1 m high mound and to scatter a limited amount of ejecta. Shear surfaces were present on many of the fragmented blocks in the bulked mound and ejecta, but no shatter cone surfaces were observed at the ground surface.

Prior to detonation of this charge, a number of 20 cm diameter holes were drilled along three lines extending radially from ground zero. The holes were backfilled with colored cement grout having strength properties designed to match those of the granite. After the detonation, the holes were redrilled to examine explosion-produced fracture and shear damage in the grout cores. Figure 20 shows the original location of a small, poorly developed shatter cone formed in a grout column near ground zero. Cores from all of the grout columns shown in Fig. 20 had numerous slickenside surfaces on shear planes that angled downward in the direction away from the explosion center (Davis, 1970).

FORMATIONAL PRESSURES FOR SHATTER CONES

Various calculational techniques are commonly employed to predict the free field stress levels in the target media at cratering trials. These predictions are then used in the planning of the actual instrumentation layouts. Fortunately, two sets of predictions were made for the stress attenuation below the 100-ton TNT Mine Ore sphere, one set for the near-in and intermediate ranges (Schuster, 1977) and another set for the intermediate and far-out ranges (Donat, 1969). These data have allowed estimates to be made of the formational pressures for shatter cones in four of the cratering experiments.

It is beyond the scope of the paper to describe each calculational technique used to determine the stress predictions, consequently the reader requiring further information is referred to the original references. What is presented, however, is the composite stress attenuation curves predicted for the 100-ton Mine Ore crater as derived from Schuster (1977) and Donat (1969). We have normalized the original pressure data in such a way as to allow attenuation values for both the 0.5-ton and 100-ton events to be read from the single data plot in Fig. 21. This figure shows the predicted stresses in terms of their attenuation as a function of *scaled depth* in the rock, i.e., actual depth in meters divided by the cube root of the charge weight in kilograms. This type of normalization is valid only for comparable shot geometries, with the depths measured from the intersection of the original ground surface with the vertical axis of the charge sphere to selected calculated stress points in the rock. A large number of positions were calculated such that attenuations for both on-axis and off-axis positions were predicted (Schuster, 1977; Donat, 1969). Due to the complexity of displaying such data, however, Fig. 21 displays only the averaged attenuations of all near on- and off-axis stresses by a long-dashed line. The three separate branches of stress attenuation indicated by the dotted lines show more detailed values for off-axis regions increasing in range to the right and exhibiting the lowest stresses. Use of these three branches, however, requires the off-axis angle and specific rock location. In the region where shatter cones appear to form, the maximum stress differences for intermediate off-axis locations generally are on the order of 30% and more commonly average 10% or less for decreasing off-axis positions at comparable depths.

The stress attenuation curves for Fig. 21 have an upper pressure limit

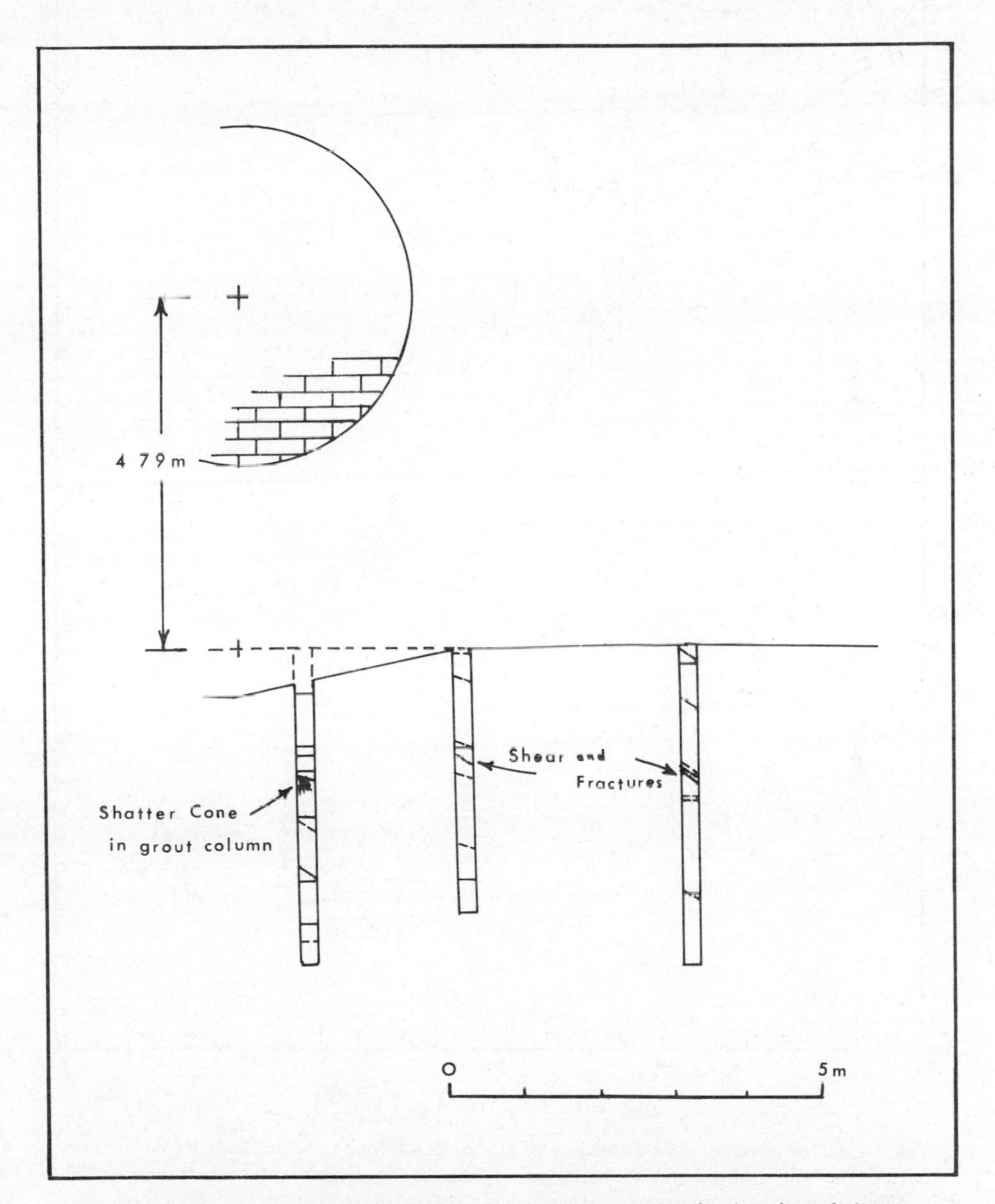

Fig. 20. Mine Under 100-ton TNT charge and true crater showing location of shatter cone, shears, and fractures in grout columns. Columns were redrilled after cratering trial to determine deformation. Cores from all grout columns had numerous slickensides on shear surfaces that cut columns at different angles. Modified from Davis (1970).

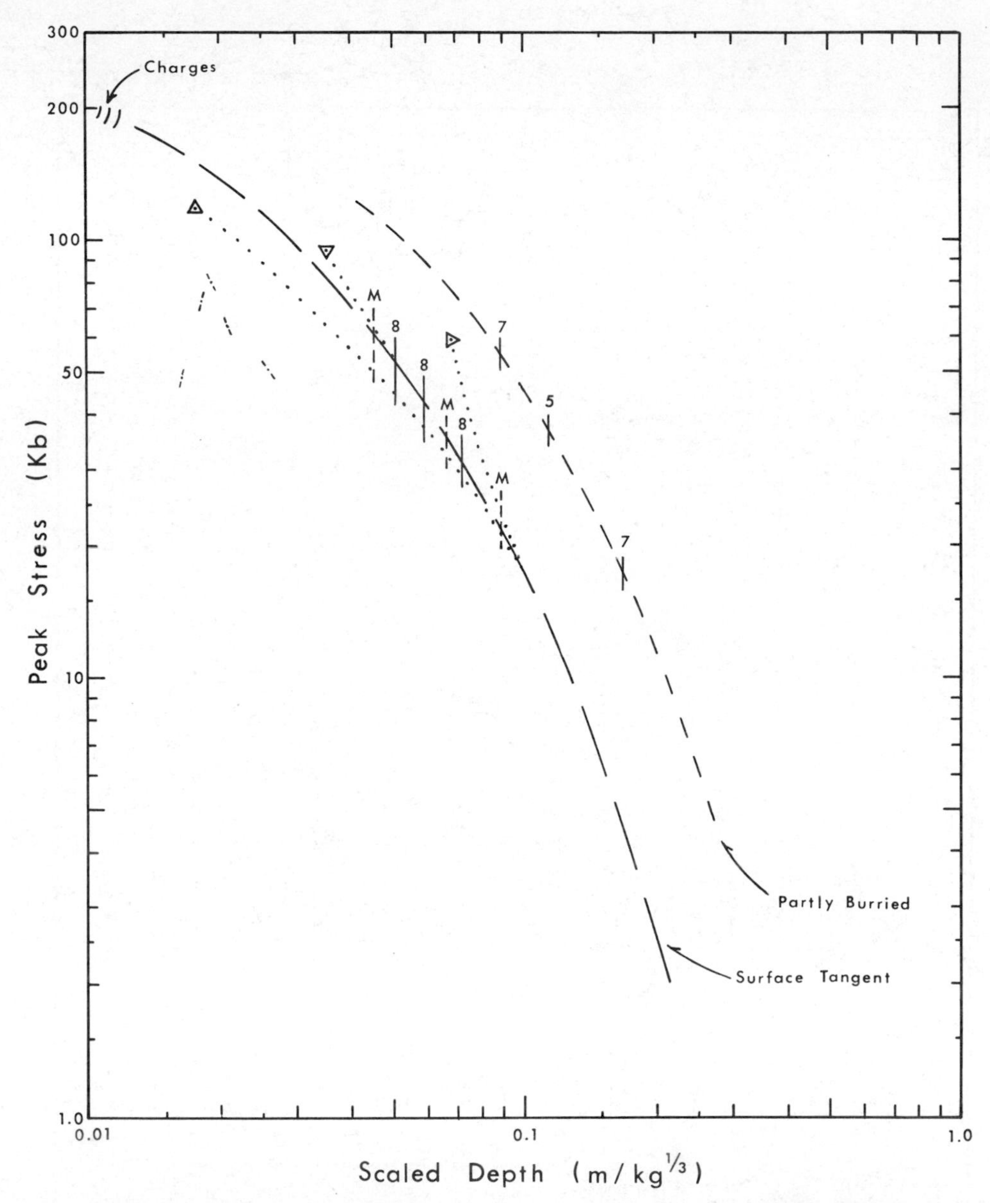

Fig. 21. Stress attenuation as a function of scaled depth for the Mine Shaft Explosion Series. The long-dashed curve shows the average stress for charges with surface tangent geometries in terms of near on-axis to far off-axis positions in the tonalite. The curve above about 10 kb is based on theoretical calculations for the 100-ton Mine Ore trial by Schuster (1977), and below 10 kb it is based on theoretical calculations from Donat (1969). The long-dashed curve is normalized for different charge energies by (m/kg$^{1/3}$) for surface and near-surface tangent geometries such that Crater 8 and the 100-ton Mine Ore Crater can be plotted on the same curve. The dotted curves show three separate branches of stress attenuation exhibiting more detailed values for off-axis regions surrounding ground zero. From left to right, the three dotted branches represent increasing off-axis positions of

determined by the Chapman–Jouget pressure for TNT, i.e., approximately 200 kb at the rock–TNT interface. A series of pressure and displacement gauges were also installed in the rock below the 100-ton Mine Ore trial (Lieberman *et al.*, 1969), and gave values shown by the —•— symbols in Fig. 21. The gauge data, although consistently lower than the predicted curve, shows the same stress attenuation slope indicating that they were responding in a comparable manner. The absolute gauge values, however, are considered to be low by a factor approximately of 30% due to reduced energy coupling.

The half-buried TNT spheres created a set of stress attenuation conditions different from the one-tenth radius buried events. To correct for this configuration an energy coupling factor was derived from Ingram (1977) which provided the basis for the construction of the short-dashed curve in Fig. 21. The scaled depth for the partly buried charges, is calculated again from the ground zero point. The absolute values of this stress attenuation curve are less certain than the long-dashed curve for surface charges due to uncertainties in the coupling factor; however, in the region of shatter cone formation we consider it to be reasonably accurate based on the available experimental data.

The estimated formational pressures for the shatter cones found *in situ* in the different trials are listed in Table 1. The scaled depths for the shatter cones use the distances measured from the intersection of the original ground surface with the vertical axis of the charge to the apexes of the shatter cones and to the farthest cone surfaces. Ranges of distances are necessary where apexes were highly damaged or where some dislodgement of shatter cone blocks had occurred. In the case of the ejected shatter cones at Mine Ore, the best estimates from all of the field data and topographic profiles were used to establish the most likely pre-shot range. A minimum and maximum range are also tabulated, although the large size of blocks A, B, and C strongly suggest, as described earlier in this paper, that they were derived from the shaded area in Fig. 19 and not from the two end extremes.

---

stresses calculated for vertical columns at ranges from ground zero of 0.61 m, 1.52 m, and 3.05 m, respectively. Use of these three stress branches, rather than the long-dashed curve, requires the exact off-axis angle and slant range from ground zero. The short-dashed curve is constructed for partly-buried charge geometries from energy coupling data from Ingram (1977). This curve is also normalized for different charge energies by ($m/kg^{1/3}$) such that craters 5 and 7 can be plotted on the same curve, but remains only valid for partly buried charges which do not exceed approximately $-0.5\ r$ HOB. The vertical dashed-lines, designated M, represent the most-likely shot positions of the shatter cones ejected from the 100-ton Mine Ore Crater. The vertical solid lines, designated 5, 7, and 8, represent the measured positions to apexes and ends of shatter cones found *in situ* in these three craters. Note that stress ranges for a single crater are due to the stress drop over the length of individual shatter cones, as listed in Table 1. The five symbols —•— indicate pressure and displacement gage data from Lieberman *et al.* (1969) and are orientated to show the trend of their plotted gage data. The variations in scaled sizes of the various charges are shown in the region of the Chapman–Jouget pressure at approximately 200 kb.

Table 1. Formational stress levels of shatter cones determined from stress attenuation curves in Fig. 21. Stress ranges are listed where shatter cones are located over broad regions. Impact energy of formation for Flynn Creek Crater is estimated between approximately 2.5 and 4.7×10^{24} ergs from Roddy (1977a) and for Gosses Bluff it is estimated at approximately 10^{28} ergs by Milton (1977, pers. comm.)

Crater	Charge	HOB	Shatter cone	Undisturbed Distance to GZ(m)		Scaled depth (m/kg$^{1/3}$)	P(kb)
Crater 5	0.5-ton TNT	0.0*r*	Slightly dislodged, apex of cone	~0.87		~0.113	~37
Crater 7	0.5-ton TNT	0.0*r*	*in situ*, apex of cone	0.69		0.089	55
			in situ, end of cone	1.29		0.167	17
Crater 8	0.5-ton TNT	0.9*r*	*in situ*, apex of cone	0.39		0.051	53
			in situ, apex of cone	0.45		0.059	43
			in situ, end of cone	0.55		0.072	31
Mine Ore	100-ton TNT	0.9*r*	ejecta	1.0 m	Min. estim.	0.022	130
			ejecta	2.0 m	Most likely range	0.045	62
			ejecta	3.0 m	Most likely range	0.067	36
			ejecta	4.0 m	Most likely range	0.089	22
			ejecta	5.0 m	Maximum estimate	0.111	14
			ejecta	6.0 m	Maximum estimate	0.134	8.5
			ejecta	7.5 m	Maximum estimate	0.168	4.5
Flynn Creek	~10^{24} ergs	near surface(?)	uplifted, displaced	450	Surface tangent	~0.104	~17
				480	Surface tangent	~0.111	~14
				450	Partly buried	~0.104	~45
				480	Partly buried	~0.111	~38
Gosses Bluff	~10^{28} ergs	near surface(?)	uplifted, displaced	3500	Surface tangent	~0.056	~46
				6500	Surface tangent	~0.105	~15
				3500	Partly buried	~0.56	~95
				6500	Partly buried	~0.105	~42

The formational pressures for the shatter cones in Crater 8 and the 100-ton Mine Ore event were determined using the long-dashed curve in Fig. 21, whereas the short-dashed curve was used for the 0.5-ton half-buried events of Craters 5 and 7.

The formational pressure for the shatter cone formed in the grout core under the 100-ton sphere tower shot is more difficult to determine. An extrapolation of data given by Carpenter and Brode (1974) indicates the peak airblast pressure on the rock surface directly under a raised charge would be about 35 kb. We estimate that peak stresses in the region of the grout shatter cone would have been on the order of 15–20 kb. However, the strength differences between the tonalite and grout, the different attenuation rates, coupling variations at the core interface, and other conditions present an uncertainty to this pressure estimate that is difficult to evaluate.

Formational pressures for shatter cones were also estimated for two natural impact structures, Flynn Creek and Gosses Bluff. Both were chosen because of the excellent field control on the pre-impact distances to their shatter-coned rocks. Flynn Creek, a 3.6 km diameter crater in north central Tennessee, had a pre-impact distance ranging from approximately 450 to 480 m from the original ground level at GZ to the pre-shot location of the shatter cones. Gosses Bluff, a 20+ km diameter crater in central Australia, had a pre-impact range of approximately 3500–6500 m from the original ground level at GZ (Milton, 1977). We recognize fully the very real number of differences between the explosion cratering trials and these large impact events. However, since estimates of the impact cratering energies were available for the formation of both impact structures, we felt it would be instructive to estimate the formational pressures for these natural shatter cones. The kinetic energy for the impact at Flynn Creek has been calculated to be 10^{24} ergs (Roddy, 1977a) and $\sim 10^{28}$ ergs for Gosses Bluff (Milton, 1977, personal communication). The choice of a height of burst was less certain, although Roddy (1968, 1976, 1977a,b, c) has offered arguments for a surface burst geometry for Flynn Creek. Consequently, we used both the surface tangent and partly buried curves in Fig. 21. As seen in Table 1, the formational pressure for shatter cones at Flynn Creek average approximately 15 kb for surface tangent and about 45 kb for partly buried. For Gosses Bluff, the pressures average 46 and 15 kb for the 3.5 and 6.5 km ranges, respectively, for surface tangent, and 95 and 42 kb for the same ranges under a half-buried configuration. The reader, naturally, is encouraged not to consider these as final values until there is adequate confirmation from experimental studies on the same material. The general pressure range, however, appears quite consistent with the available petrofabric and X-ray studies on these sedimentary rocks.

CONCLUSIONS

These high explosion trials demonstrate now, beyond any doubt, that shatter cones can be formed by shock wave processes during cratering and that average formational pressures in these crystalline rocks are in the 20–60 kb range. Although the mechanics of formation were not a part of this study, a number of physical conditions related directly to shatter cone formation were identified. The more important of these conditions, which tend to confirm Dietz's (1968, 1972) list described earlier in this paper, include the following. (a) Shatter cones in these trials have normal conical fracture surfaces with striae which fan outward from the apex with nesting common on large cone segments. (b) Single, large, complete shatter cones occur in 0.5-ton crater floors near centers of craters, commonly within a 110° cone under GZ. (c) *In situ* cones point in the direction of the shock wave source with their axes normal to direction of shock wave propagation and passing through GZ ± 20°. (d) Stereographic measurements show apical angles averaged 86 ± 8° and can exhibit this full range in a single large cone. The tonalite was relatively homogeneous in the shatter-coned rocks, although material properties varied with depth from the ground surface. Weathered rock exhibited reduced strengths. (e) Apexes were commonly damaged on the large cones and could not be examined for apex inhomogenieties. (f) Large ejecta blocks exhibited shatter cone surfaces with all cones pointing in same direction. (g) Some cones, which were nearer GZ in Mine Ore, are truncated by cones a few centimeters further away. (h) Shatter coned rocks commonly exhibit no faults and little or no fractures except for the cone surface themselves, and otherwise appear megascopically relatively undamaged. (i) Small, shatter cone segments in the 0.5-ton trials remained both *in situ* on the true crater floors, commonly within a 110° cone under GZ, and in the ejecta. (j) The 100-ton surface charge ejected large blocks with nested and elongated cone segments, but no cones were found *in situ* in the crater. (k) Some cone surfaces extended beyond the crater floor into the non-cratered rock. (l) Pre-shot jointing modified the crater shapes but did not appear to affect the cone formation. (m) Estimates of average formational pressures in the tonalite range from approximately 17 to 62 kb. (n) Formational stress attenuation along large single cone surfaces is as great as 38 kb. (o) The minimum formation stresses determined for an *in situ* cone was 17 kb with all other cone locations having higher stresses. The HEL for Cedar City tonalite is listed at about 15 kb (McFarland, 1968); however, Schuster (1977, pers. comm.) indicates it may be higher.

As noted earlier, many of the conditions listed above have direct application to shatter cones found at natural impact structures. The most fundamental point, of course, is the direct confirmation that shatter cones are clearly formed by shock wave processes during cratering. Moreover, the average orientation geometry can now more confidently be used to determine the shock source area. Also of interest, is that these trials confirmed that apical angles tend to remain fairly constant in the same rock type, exhibiting a relatively narrow spread of only 16°. These trials also confirm that shatter cones can be found both *in situ* and as ejecta, as we find at the Steinheim impact crater in Germany.

Perhaps some of the most interesting results are the estimates made of the formational pressure ranges for the shatter cones in the tonalite. The range of 20–60 kb appears quite reasonable when compared with shock metamorphic evidence from a number of natural impact structures, i.e., shatter cones are not found in association with other higher pressure features. As noted earlier, previous workers on impact structures have suggested pressure ranges of 20–80 kb on the basis of shock metamorphic studies (Dietz, 1972) and 20–40 kb on the basis of theoretical studies (Gash, 1971). Stress ranges do indeed exist as evidenced by the stress attenuation values calculated at different distances along individual shatter cone surfaces observed in these explosion trials.

We offer a strong word of caution, however, on extrapolation of these results to natural impact events before further experimental data are completed. The stress attenuation curves shown in Fig. 21, are based solely on *theoretical calculations* to *predict* a stress field under a given set of initial conditions. Furthermore, the calculational effort was originally completed to examine ground motion and was not designed especially for the close-in cratered region, i.e., the calculated zoning was not tailored exclusively for a cratering study of stress attenuation (Schuster, 1977, personal communication). The stress attenuation curves shown in Fig. 21, however, do give predictive results that are internally consistent at the intermediate stress levels for two independent sets of theoretical calculations (Donat, 1969; Schuster, 1977). The fact that these two studies overlap with common values at the intermediate pressures, using different calculational approaches, suggests that the predicted stresses are generally correct for these trials. The gauge data described earlier and shown on Fig. 21 also tend to confirm the general validity of the predicted stresses. They average within 30% of their respective predicted calculational values, which is reasonable for this particular experiment and the requirements for these early, close-in gauges. Although the theoretical calculation and the gauge data are considered reasonable for these trials, *absolute* pressure data for the formation of shatter cones in other media and under other initial conditions must still await instrumented tests.

Several other factors are also important in any study of the formational pressures of shatter cones, such as possible fundamental differences between shock waves propagated by an impact versus high explosives. Are the pressure, time, and energy transfer histories the same and is the Hugoniot of the rocks modified by shock conditioning? How do different rock types with different pre-shock stress histories, different confining pressures, and different structural attitudes respond to shock wave passage at different angles? The explosion trials described in this paper show that shatter cone surfaces can experience differential pressures as large as 38 kb. What does such a large stress drop over short distances actually imply in terms of the initiation of shatter coning and how are the final cone sizes related to such differential stress fields?

Another point of concern is that some workers have argued that the formation of shatter cones involves an elastic/plastic wave front interaction (Johnson and Talbot, 1964; Dietz, 1972). The HEL shown in Fig. 3, however, is below the average stress level of cone formation. In fact, it is below the lowest level of shatter cone formational stresses by two or more kilobars. The mechanics of shocked rock material remain quite unclear in terms of their elastic/plastic behavior and precursor interactions, consequently we are uncertain as to what the lowest shatter cone stresses actually can be with respect to such HEL data.

The *total* possible stress range in which shatter cones may have formed in

Mine Ore is shown in Table 1 to lie between 4.5 and 130 kb. As noted earlier in this paper, the Mine Ore cones most likely were formed in a much more *restrictive* range of between 22 and 62 kb. Robertson (1975) notes, however, that *poorly developed* shatter cones can occur with higher pressure shock metamorphic features. Were these cones formed at the same time as the other features or was there a later state of lower shock overpressure?

The question of energy coupling and its rates of transfer also remains prominent. As noted earlier, different stress attenuation values exist for raised, surface, and partly buried explosions. The solution to the different attenuations lies in determining valid *coupling factors*, among other considerations. Coupling values, however, are generally determined only through adequate experimentation. The coupling factor used to construct the curve in Fig. 21 is, at best, a generalization based upon a number of experiments, mostly in other rock types. Consequently, the stress attenuation for the partly buried configuration, although constructed with our best information, remains an approximation without detailed experiments in the same media. The extrapolation of impact data to our pressure curves presupposes a knowledge of the coupling factor, a quantity that is poorly known. That is, does an impact approximate a line or a point energy source and what is its *effective* HOB? These points remain of real concern in any such extrapolations of stress data.

Obviously, a number of questions have yet to be answered concerning the mechanics of shatter coning and their formational pressures, i.e., if these unique features are to play a real quantitative role in shock metamorphism and impact cratering. This remains true for both explosion and hypervelocity impact craters. Fortunately, a number of the more important parameters have been determined by other workers, and as noted in this paper, certain of the basic conditions of formation have been confirmed or were established in these explosion experiments. Indeed, we felt sufficiently encouraged with this data base to attempt two stress determinations for the large impact craters of Flynn Creek in Tennessee and Gosses Bluff in Australia, and were happily rewarded with formational stress which appear at least consistent with the limited petrofabric and X-ray studies of those rocks. Recognizing the many problems inherent with such extrapolations of the explosion data, we suggest the 40 ± 20 kb range appears to be most reasonable, particularly for the Flynn Creek event. We take more comfort, however, with respect to the general applicability of this explosion data, in the fact that the shatter cones formed in these explosion experiments exhibit essentially *all* of the characteristics of shatter cones formed in natural impact craters, and that their formational pressures are presently *consistent* with those interpreted from hypervelocity impact structures.

Acknowledgment—We wish to express our appreciation to Dr. J. K. Ingram, U.S. Army Engineer Waterways Experiment Station for his assistance during the explosion trials. We are especially indebted to S. Schuster, California Research and Technology, for the use of his unpublished calculational data on stress attenuation at the Mine Ore trial and for his continued advice regarding our calculations. D. J. R. is again indebted to Dr. G. Sevin and Captain J. Stockton of the Defense

Nuclear Agency for their continued cooperation in these studies. D. J. R. also expresses his real appreciation to M., M., and M. Roddy for their continued field assistance and to J. Roddy for her help in data assemblage.

Reviews of this paper by Drs. P. DeCarli, R. Dietz, J. Ingram, D. Milton, and A. Rooke materially assisted us in its improvement and we express our thanks to each person. Drs. H. Cooper, F. Sauer, and J. Trulio also gave advice in the computational work.

This work was supported by the National Aeronautics and Space Administration under contract W-13,130 and the Defense Nuclear Agency, Department of Defense.

References

Amstutz, G.: 1965, A morphological comparison of diagnetic cone-in-cone structures and shatter cones. *Ann. N.Y. Acad. Sci.*, 1050.

Baer, A. J., and Norris, D. K.: 1969, *Proceedings, Conference on Research in Tectonics* (*Kink Bands and Brittle Deformation*). Queen's Printer, Ottawa, Canada.

Branco, W., and Fraas, E.: 1905, *Das kryptovulkanische Becken von Steinheim.* Akad. Wiss. Berlin, Phys.-math. Kl. Abh. 1.

Biot, M. A.: 1965, *Mechanics of incremental deformations.* John Wiley & Sons, Inc., N.Y..

Bucher, W. H.: 1963, Cryptoexplosion structures caused from without or from within the earth? (Astroblemes or Geoblemes?). *Am. J. Sci.* **261**, 597.

Bunch, T. E., and Quaide, W. L.: 1968, Shatter cones in the Danny Boy nuclear crater. In *Shock Metamorphism of Natural Materials* (B. M. French and N. M. Short, eds.), Mono Book Corp., Baltimore.

Carter, N.: 1965, Basal quartz deformation lamellae—a criterion for recognition of impactites. *Am. J. Sci.* **263**, 786.

Carpenter, H. J., and Brode, H. L.: 1974, Paper No. H3, Height of burst blast at high overpressure, presented at 4th International Symposium on Military Applications of Blast Simulations, South-end-on-Sea, England. R & D Assoc., Santa Monica, California.

Chao, E. C. T.: 1968, Pressure and temperature histories of impact metamorphosed rocks—based on petrographic observations. In *Shock Metamorphism of Natural Materials* (B. M. French and N. M. Short, eds.), p. 135–158, Mono Book Corp., Baltimore.

Chao, E. C. T.: 1976, The Ries crater in southern Germany—a model for large basins on planetary surfaces. *Fortschr. Mineralogie,* in press.

Chao, E. C. T., and Minkin, J.: 1977, Impact Cratering Phenomenon for the Ries Multiring Structure Based on Constraints of Geological, Geophysical and Petrological Studies and the Nature of the Impact Body. In *Impact and Explosion Cratering* (D. J. Roddy, R. O. Pepin, and R. B. Merrill, eds.). This volume.

Currie, K.: 1968, Mistastin Lake, Labrador, a new Canadian crater. *Nature* **22**, 776.

Davis, L. K.: 1970a, Mine Shaft Series—Subtask N 123, Calibration Cratering Series, TR N-70-4, U.S. Army Engineer Waterways Experiment Station, Vicksburg, Mississippi.

Davis, L. K.: 1970b, Mine Shaft Series, Events Mine Under and Mine Ore, Subtask N 121, Crater Investigations, TR N-70-8, U.S. Army Engineer Waterways Experiment Station, Vicksburg, Mississippi.

Dence, M. R.: 1968, Shock zoning at Canadian craters: Petrography and structural duplications. In *Shock Metamorphism of Natural Materials* (B. M. French and N. M. Short, eds.), p. 169–184, Mono Book Corp., Baltimore.

Dence, M. R.: 1972, Meteorite impact craters and the structure of the Sudbury Basin. *Geol. Assoc. of Canada, Special Paper 10.*

Dence, M. R., Grieve, R. A. F., and Robertson, P. B.: 1977, Terrestrial impact structures: Principal characteristics and energy considerations. In *Impact and Explosion Cratering* (D. J. Roddy, R. O. Pepin, and R. B. Merrill, eds.). This volume.

Dietz, R. S.: 1947, Meteorite impact suggested by orientation of shatter cones at the Kentland, Indiana disturbance. *Science* **105**, 42.

Dietz, R. S.: 1959, Shatter cones in cryptoexplosion structures (meteorite impact?). *J. Geol.* **67**, 496.
Dietz, R. S.: 1960, Meteorite impact suggested by shatter cones in rock. *Science* **131**, 1781.
Dietz, R. S.: 1961a, Vredfort ring structure: meteorite impact scar? *J. Geol.* **69**, 499.
Dietz, R. S.: 1961b, Astroblemes. *Sci. Amer.* **205**, 2.
Dietz, R. S.: 1963a, Cryptoexplosion structures: a discussion. *Am. J. Sci.* **261**, p. 650–664.
Dietz, R. S.: 1963b, Astroblemes: ancient meteorite-impact structures on the earth. In *The Solar System*, v. 4, 'The moon, meteorites, and comets', (B. M. Middlehurst and G. P. Kuiper, eds.), p. 285–300, Univ. of Chicago Press, Chicago.
Dietz, R. S.: 1964, Sudbury structure as an astrobleme. *J. Geol.* **72**, 412.
Dietz, R. S.: 1966a, Shatter cones at the Middlesboro Structure, Kentucky. *Meteoritics* **3**, 27.
Dietz, R. S.: 1966b, Striated surfaces on meteorites; shock fractures, not slicken-sides. *Meteoritics* **3**, 31.
Dietz, R. S.: 1966c, Shatter cones and astroblemes. *Proc. Oregon Lunar Geological Field Conference*, Bend, Oregon.
Dietz, R. S.: 1967, Shatter cone orientation at Gosses Bluff astrobleme. *Nature* **216**, 1082.
Dietz, R. S.: 1968, Shatter cones in cryptoexplosion structures. In *Shock metamorphism of natural materials* (B. M. French and N. M. Short, eds.), p. 267–285, Mono Book Corp., Baltimore.
Dietz, R. S.: 1972, Shatter cones (shock fractures) in Astroblemes. In *sec. 15, Planetology of 24th International Geological Congress*, Montreal, Canada.
Dietz, R. S., and Butler, L.: 1964, Orientation of shatter cones at Sudbury, Canada. *Nature* **204**, 280.
Donat, K.; 1969, Director, Program 2—Earth Motions and Material Properties, Mine Shaft Series, U.S. Army Engineer Waterways Experiment Station, Vicksburg, Mississippi.
Englehardt, W. V.: 1972. Shock produced rock glasses from the Ries Crater. *Contr. Mineral. and Petrol.* **36** 265.
Englehardt, W. V.: 1974a, Ries meteorite crater, Germany. *Fortschr. Miner.* **52**, 103.
Englehardt, W. V.: 1974b, Meteoritenkrater. *Naturwissenschaften* **61**, 413.
Englehardt, W. V., and Bertsch, W.: 1969, Shock induced planar deformation structures in quartz from the Ries crater, Germany. *Contrib. Mineral. and Petrol.* **20**, 203.
Englehardt, W. V., Stöffler, D., and Schneider, W.: 1969, Petrologische untersuchungen im Ries. *Geologica Bavarica* **61**, 229.
El Goresy, A., and Chao, E. C. T.: 1976, Evidence of the impacting body of the Ries crater—the discovery of Fe-Cr-Ni veinlets below the crater bottom. *Earth Planet. Sci. Lett.* **31**, 330.
Farrell, W. J., and Curro, J. R.: 1968, Site selection investigation for the Mine Shaft Series. Misc. Paper S-68-18, U.S. Army Engineer Waterways Experiment Station, Vicksburg, Mississippi.
French, B. M., and Short, N. M., eds.: 1968, *Shock metamorphism of natural materials*. Proc. 1st Conf., NASA, Goddard Space Flight Center, April 14–16, 1966. Mono Book Corp., Baltimore.
Gash, P. J. Syme: 1971, Dynamic mechanism for the formation of shatter cones. *Natural Physical Science* **230**, 32.
Gay, N. C.: 1976, Spherules on shatter cone surfaces from the Vredefort structure, South Africa. *Science* **194**, 724.
Handin, J. W., and Hager, R. V., Jr.: 1958, Experimental deformation of sedimentary rocks under confining pressure: tests at high temperature. *Am. Assoc. Petroleum Geol. Bull.* **42**, 2892.
Handin, J. W., and Hager, R. V., Jr.: 1957, Experimental deformation of sedimentary rocks under confining pressure: tests at room temperature on dry samples. *Am. Assoc. Petroleum Geol. Bull.* **41**, 1.
Handin, J. W., *et al.*: 1963, Experimental deformation of sedimentary rocks under confining pressure: pore pressure tests. *Am. Assoc. Petroleum Geol. Bull.* **47**, 717.
Hargraves, R. B.: 1961, Shatter cones in rocks of Vredefort Ring. *Geol. Soc. S. Africa Trans.* **64**, 147.
Hörz, F. (ed.): 1971, 'Meteorite impact and volcanism'. *J. of Geophys. Res.* **76**, 5381.
Hörz, F., and Quaide, W. L.: 1973, Debye-Scherrer investigations of experimentally shocked silicates. *The Moon* **6**, 45
Hörz, F., and Ahrens, T. J.: 1969, Deformation of experimentally shocked biotite. *Am. Jour. Sci.* **267**, 1213.
Howard, K., and Offield, T.: 1968, Shatter cones at Sierra Madera, Texas. *Science* **162**, 261.

Ingram, J. K.: 1977, Summary and analysis of CENSE data, U.S. Army Engineer Waterways Experiment Station, Vicksburg, Mississippi, in press.

Jaeger, J. C., and Cook, N. G. W.: 1969, *Fundamentals of rock mechanics*, Methuen & Co., Ltd., London.

Johnson, G. P., and Talbot, R. J.: 1964, A theoretical study of the shock wave origin of shatter cones. M. S. thesis, Air force Inst. of Technol., Wright-Patterson AFB, Ohio.

Johnson, A. M.: 1970, *Physical processes in geology*. Freeman Cooper & Co., San Francisco.

Kolb, C. R., Farrell, W. S., Hunt, R. W., and Curro, S. R.: 1970, Operation mine shaft: Geological investigation of the mine shaft sites, Cedar City, Utah. U.S. Army Engineer Waterways Experiment Station, Vicksburg, Mississippi, unclassified, 322 pp.

Lieberman, P., Nagumo, G., Miller, D., Knox, R.: 1969, Close-in Pressure Displacement Measurements in Mine Ore. Defense Atomic Support Agency, DASA 2321, unclassified.

Mackin, J. H.: 1947, Some structural features of the intrusions in the Iron Springs District. U.S. Geol. Survey Guidebook to the Geology of Utah, no. 2. Utah Geological Society, Salt Lake City, Utah.

Manton, W. I.: 1965, The orientation and origin of shatter cones in the Vredefort Ring. *Ann. N.Y. Acad. Sci.* **123**, 1017.

McFarland, C. B.: 1968, Mine Shaft Series—Subtask N201, Material Properties. Interim Data Report, December, unclassified, unpublished report.

Means, W. D.: 1976, *Stress and strain, basic concepts of continuum mechanics for geologists*. Springer-Verlag, N.Y., 339 pp.

Meyer, J. W., and Rooke, A. D., Jr.: 1969, Mine Shaft Series, Events Mine Under and Mine Ore, Ejecta Studies. U.S. Army Engineer Waterways Experiment Station, Vicksburg, Mississippi, 92 pp.

Milton, D. J.: 1977, Shatter cones—an outstanding problem in shock mechanics. In *Impact and Explosion Cratering* (D. J. Roddy, R. O. Pepin, and R. B. Merrill, eds.). This volume.

Milton, D. J., Barlow, B. C., Brett, R., Brown, A. R., Glikson, A. Y., Manwarning, E. A., Moss, F. J., Sedmick, E. C. E., Van son J., and Young, G. A.: 1972, Gosses Bluff impact structure, Australia. *Science* **175**, 1199–1207.

Nicolaysen, L. O.: 1976, The Vredefort structure: A review of recent studies on its constitution and origin. In *Papers Presented to the Symposium on Planetary Cratering Mechanics*, p. 78–80. The Lunar Science Institute, Houston.

Pohl, J., Stöffler, D., Gall, H. and Ernstson, K.: 1977, The Ries impact crater. In *Impact and Explosion Cratering* (D. J. Roddy, R. O. Pepin, and R. B. Merrill, eds.). This volume.

Rieff, W.: 1977, The Steinheim basin—an impact structure. In *Impact and Explosion Cratering* (D. J. Roddy, R. O. Pepin, and R. B. Merrill, eds.). This volume.

Robertson, P. B.: 1968, La Malbaie structure, Quebec: a Paleozoic meteorite impact site. *Meteoritics* **4**, 1.

Robertson, P. B.: 1975, Zones of shock metamorphism at the Charlevoix impact structure, Quebec. *Bull. Soc. America* **86**, 1630.

Roddy, D. J.: 1968, The Flynn Creek Crater, Tennessee. In *Shock Metamorphism of Natural Materials* (B. M. French and N. M. Short, eds.), p. 291–322, Mono Book Corp., Baltimore.

Roddy, D. J.: 1976, High-explosive cratering analogs for bowl-shaped, central uplift and multiring impact craters. *Proc. Lunar Sci. Conf. 7th*, 3027.

Roddy, D. J.: 1977a, Tabular comparisons of the Flynn Creek impact crater, United States, Steinheim impact crater, Germany and the Snowball explosion crater, Canada. In *Impact and Explosion Cratering* (D. J. Roddy, R. O. Pepin, and R. B. Merrill, eds.). This volume.

Roddy, D. J.: 1977b, Large-Scale Impact and Explosion Craters: Comparisons of Morphological and Structural Analogs. In *Impact and Explosion Cratering* (D. J. Roddy, R. O. Pepin, and R. B. Merrill, eds.). This volume.

Roddy, D. J.: 1977c, Pre-impact conditions and cratering at the Flynn Creek Crater, Tennessee. In *Impact and Explosion Cratering* (D. J. Roddy, R. O. Pepin, and R. B. Merrill, eds.). This volume.

Roddy, D. J., and Davis, L. K.: 1969, Shatter cones at TNT explosion craters. *Trans. AGU* **50**, 220.

Rondot, J.: 1968, Nouvel impact météoritique fossile?: La structure semicirculaire de Charlevoix. *Can. J. Earth Sci.* **5**, 1305.
Rondot, J.: 1971, Impactite of the Charlevoix structure, Quebec, Canada. *J. Geophys. Res.* **76**, 5414.
Rondot, J.: 1972, La transgression Ordovicienne dans le Comté de Charlevoix, Quebec. *Canadian J. Earth Sci.* **9**, 1187.
Rondot, J.: 1975, Comparaison entre les astroblèmes de Siljan, Suède et de Charlevoix, Quebec. *Bull. Geol. Inst. Univ., Uppsala, N.S.* **6**, 85.
Saucier, K. L.: 1969, Properties of Cedar City tonalite. MPC-69-9, U.S. Army Engineer Waterways Experiment Station, Vicksburg, Mississippi, unclassified.
Schuster, S.: 1977, Mine ore stress attenuation calculations. Written communication. Calif. Research and Technology, Woodland Hills, Calif.
Schneider, E., and Wagner, G. A.: 1976, Shatter cones produced experimentally by impact in limestone targets. *Earth Planet. Sci. Lett.* **32**, 40.
Shand, S. J.: 1947, *Eruptive rocks.* Revised 3rd ed., John Wiley & Sons Inc., N.Y.
Shoemaker, E. M., Gault, D. E., and Lugn, R. V.: 1961, Shatter cones formed by high speed impact in dolomite. U.S. Geol. Survey Prof. Paper 424-D, p. 365–368.
Stöffler, D.: 1966, Zones of impact metamorphism in the crystalline rocks of the nordlinger Ries crater. *Contrib. Mineral. Petrol.* **12**, 15.
Stöffler, D.: 1971, Progressive metamorphism and classification of shocked and brecciated crystalline rocks at impact structures. *J. Geophys. Res.* **73**, 5541.
Stöffler, D.: 1972, Deformation and transformation of rock-forming minerals by natural and experimental shock processes. *Fortschr. Miner.* **49**, 50.
Turner, F. J., and Weiss, L. E.: 1962, *Structural analysis of metamorphic tectonites.* McGraw Hill, N.Y.
Wilson, C. W., and Stearns, R.: 1968, Geology of the Wells Creek structure, Tennessee. *Bull. 68, Tenn. Div. Mines,* Nashville.
Zimmermann, R. A.: 1971, Formation of shatter cones in the Steinheim Basin. *N. Jb. Miner. Mh.* **1**, 19.

Roddy, D. J., Pepin, R. O., and Merrill, R. B., editors.
(1977) *Impact and Explosion Cratering*, Pergamon Press (New York), p. 751–769.
Printed in the United States of America

Impact conditions required for formation of melt by jetting in silicates

SUSAN WERNER KIEFFER

Department of Geology, University of California. Los Angeles, California 90024

Abstract—The velocities required to produce the abundant impact glass found on the surface of the moon and in meteorites are a subject of controversy. In this paper it is demonstrated that the process of jetting which occurs when particles collide at oblique angles may produce melt at much lower velocities than are required for melt production in head-on collisions. The minimum velocities of impact required for jetting in aluminum, bronzitite, dunite and quartz are calculated by the method of shock polars. Conditions of impact velocity and angle which give rise to the regular (jetless) regime are calculated and the critical angle-velocity relations at which the regular regime breaks down and jetting arises are given. Shock-velocity particle-velocity equations-of-state with three shock regimes (low-pressure, mixed-phase, and high pressure) are used. The results depend on the validity of two assumptions: (1) that pressures attained are sufficiently high that the material behaves hydrodynamically; (2) that, in spite of possible complications by a double shock-wave structure under mixed-phase, regular-regime conditions, jetting occurs upon breakdown of regular regime conditions when pressures are within the mixed phase regime. Jetting should arise in bronzitite, dunite and quartz at relative velocities of impact as low as 1–2 km sec^{-1}. At such velocities material which passes near the stagnation point in the jet-forming region is subjected to sufficiently high pressures that it is probably melted. Thus melt may be formed by the collision of meteoritic particles at relative velocities of 1–2 km sec^{-1}, less than one-half the values obtained from one-dimensional theory head-on collisions.

1. INTRODUCTION

ABUNDANT GLASS is found on the lunar surface and in ancient meteorites and it is generally accepted that much of this glass was formed during the impact of particles in space or into planetary regoliths. Because calculations by one-dimensional analyses (e.g., Ahrens and O'Keefe, 1973) show that impact velocities on the order of 4 to 6 km sec^{-1} are required for the production of glass from nonporous silicates by impact, it has been assumed that velocities of this magnitude were required for production of the relatively abundant glass (including devitrified glass) which is observed on the moon and in the meteorites.

However, dynamical theories for the origin of solid matter from the primitive nebular gas cloud predict that particle velocities would have been relatively low (on the order of 1 km sec^{-1} or less) until the accreting bodies attained nearly asteroidal size (Goldreich and Ward, 1973; Cameron, 1973). For this reason, it has generally been concluded that velocities of ~4 km sec^{-1}—at which glass production commences in a one-dimensional model—were not attained until bodies of asteroidal size or larger, with significant gravitational fields, existed. It then follows that impact glass must have been produced on bodies which were a minimum of asteroidal size (Urey, 1956; Fredriksson, 1963; Kurat, 1967;

Wlotzka, 1969; Dodd, 1971). Thus, the apparent requirement of velocities in excess of 4 km sec^{-1} for impact melting places severe constraints on the origin of any impact glasses found in the solar system.

Primitive chondritic meteorites—specifically, the unequilibrated ordinary chondrites—contain a unique mixture of glass spherules (or devitrified remnants thereof) which are called *droplet chondrules* and relatively unshocked fragmental *matrix.* (Note: there are other rounded fragments in chondritic meteorites which are also called chondrules, but this study pertains only to those chondrules which appear to have formed as the result of cooling of liquid droplets and, hence, are called droplet chondrules.) The volatile trace element geochemistry (Larimer and Anders, 1967) of the droplet chondrules and their generally droplet-like shapes suggest a high-temperature origin (about 1000°C). The matrix is fine-grained, fragmental and of generally the same composition as the droplet chondrules, except for a higher trace-element volatile content, which suggests that at least some fraction of the matrix equilibrated at lower temperatures (Larimer and Anders, 1967).

On the basis of textural arguments, I have suggested that droplet chondrules and some of the matrix in chondritic meteorites may have been produced by collisions of relatively *small* particles in space and *not* on bodies of asteroidal size (Kieffer, 1975). The motivation for the study of jetting presented here was to inquire if the minimum velocities at which molten silicate could be produced by impact are consistent with the relatively low velocities predicted by dynamical theories for the early stages of accretion.

The argument for the origin of chondrules from small body impacts may be summarized as follows: High velocity impacts of meteoritic particles onto large bodies produce craters and ejecta with a wide variety of impact products. I have called such impacts in which one particle is much larger than the other "particle-to-parent" impacts (Kieffer, 1975). Ejecta from such events vary widely in size, texture and composition and contain a variety of glasses and weakly, moderately and strongly shocked crystalline fragments. For example, a wide spectrum of glasses (agglutinates, shards, spherules, and diaplectic glasses which vary in size from microns to centimeters) is seen in ejecta from terrestrial craters, in lunar surface samples and in the achondrites, which are believed to have been part of large parent bodies. Our understanding of impact processes and products is heavily weighted by the many studies of craters and their ejecta which have been done in the past 25 years. Neither the chondrules alone nor the chondrules-plus-matrix assemblage contains the spectrum of shock products seen as the result of high velocity particle-to-parent impacts.

By considering the nature of grain-to-grain impacts in naturally shocked Coconino Sandstone I have presented a qualitative model for the impact of two small particles of roughly comparable size (Kieffer, 1975). I have called these impacts "particle-to-particle impacts" to contrast them with "particle-to-parent" impacts in which one particle is much larger than the other. Upon impact of two approximately equidimensional spherical particles, a roughly hemispherical compressive shock wave is propagated into both particles. In a uniform half space (for example, the larger particle of a particle-to-parent impact) the hemispherical shock would decay uniformly away from the center of impact. However, during collision of small spherical particles the shock wave is disturbed by reflection from boundary surfaces before it has attenuated to seismic strengths and a complex system of rarefaction waves is set up. The tensile stresses arising from the rarefaction waves cause spallation and fracturing at the free surfaces of the spheres and a great deal of pulverized material is produced. Material in the immediate vicinity of the point of collision is subjected, however, to much higher pressures than material in the central and far edge regions because high stress concentrations result from the geometry of the particles. If the impact velocities are relatively low, this material may merely be fractured; if impact velocities are high, the material may be melted and ejected from the point of impact through a process known as *jetting.*

The process of jetting is well known in the field of military ballistics (e.g., Birkhoff *et al.*, 1948); however, its occurrence in silicates has only been inferred from microscopic textural relations in naturally shocked Coconino Sandstone (Kieffer, 1975; Kieffer *et al.*, 1976). Material which was jetted into collapsing pores in shocked Coconino Sandstone from Meteor Crater, Arizona, forms "cores" of cryptocrystalline coesite (see Fig. 1). These "cores" are located in collapsed pores of the original unshocked Sandstone (Kieffer, 1971). The grain boundaries of coesite in these cores show an equilibrium texture characteristic of high-temperature recrystallization; from this texture it was inferred that these cores were formed from jets of molten silica (Kieffer *et al.*, 1976). The cores probably reached temperatures in excess of 3000°K during part of their shock history, although they lie within 50–100 μm of crushed quartz whose temperature never exceeded a few hundred degrees Kelvin. The jetting mechanism is a process which allows extremely nonequilibrium textures to develop within shocked porous materials.

I have proposed (see Kieffer, 1975, for detailed model) that droplet chondrules were formed by a similar jetting process during collision of meteoritic particles with diameters ranging in order of magnitude from 0.5 mm to 20 cm. (The range of diameters is estimated from the size range of chondrules.) Wasson (1972) has argued that geochemical evidence suggests an origin from small particles and Cameron (1973) has suggested dynamical conditions under which chondrules might originate from small bodies. The chondrules were formed as jets extruded from the collision points of high stress concentrations and the fine-grained matrix observed in the chondrites may be the fraction of impacting grains which did not enter the jet; it experienced a relatively low pressure and low temperature history.

Since the most likely time for the collision of *small* particles to be the *dominant* process during accretion is before any large bodies have accreted, the formation of chondrules may have occured very early in the accretion of the planets from the nebular material. Dynamical models (Goldreich and Ward, 1973) would restrict this to the time during the collapse of the nebular cloud to a flat disk and prior to the formation of gravitationally stable bodies on the order

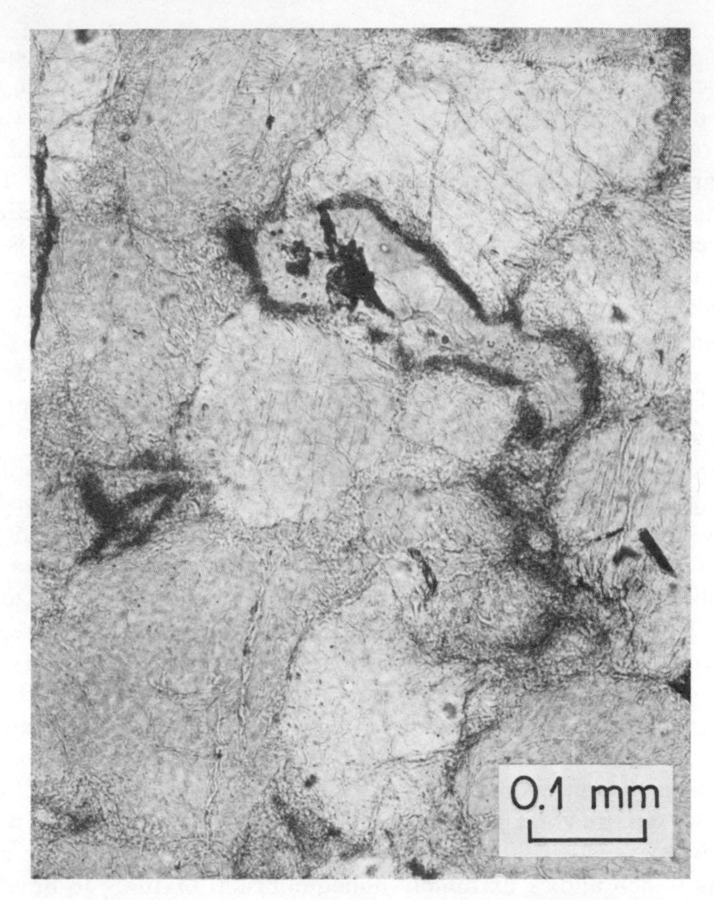

Fig. 1. A coesite core within Class 3 Coconino Sandstone from Meteor Crater, Arizona, formed as a molten jet injected into a pore cooled from high temperature during rarefaction. The core is surrounded by an opaque rim which contains coesite, stishovite and cryptovesicular amorphous froth (see Kieffer *et al.*, 1976 for description). Photo by M. W. Wegner. Plane polarized light.

of 5 km diameter. If it could be demonstrated that chondrules formed during the earliest stages of accretion, and if minimum velocities of formation can be stated, a unique tie can be made between dynamical models for the origin of the planets and geochemical theories and data for their evolution.

The object of this study then was to inquire at what velocities shock melting may occur when the effect of natural irregularities in the shape of particles is accounted for in producing stress concentrations. Although it would be desirable to analyze the collision of two spheres, for example, shock wave theory for impacting spheres is complex and can probably be developed quantitatively only with the aid of large computer programs. An instructive approximation can be made by considering the oblique impact of planar surfaces (Walsh *et al.*, 1953; referred to as WSW in this paper) and much can be learned about the velocities at which high pressures are generated from consideration of such a geometry. Even for this simplified case, a complete thermodynamic analysis is not currently feasible. In this paper calculations of the velocity-angle relationships at which jets are produced are given for *quartz*, the mineral for which the most complete shock wave equation-of-state data are available and for *dunite* and *bronzitite*, two materials characteristic of meteorites. Calculations are also given for *aluminum*, the material used as part of many shock experiments in order to allow comparison with the previous calculations of oblique shock conditions done by WSW. It will be shown that molten jets are probably produced at very low velocities of impact.

2. Oblique Impacts

In oblique impacts the behavior of the material depends on the angle of impact, 2θ (Fig. 2). For a given impact velocity, v_p, specified normal to the free surfaces of the impacting plates, the material exhibits one of two possible behaviors. For half-angles less than a certain critical angle, denoted θ_{cr}, the shock configuration is such that the plates simply close along their interface (Fig. 2a). The shock configuration which exists under such conditions is called the

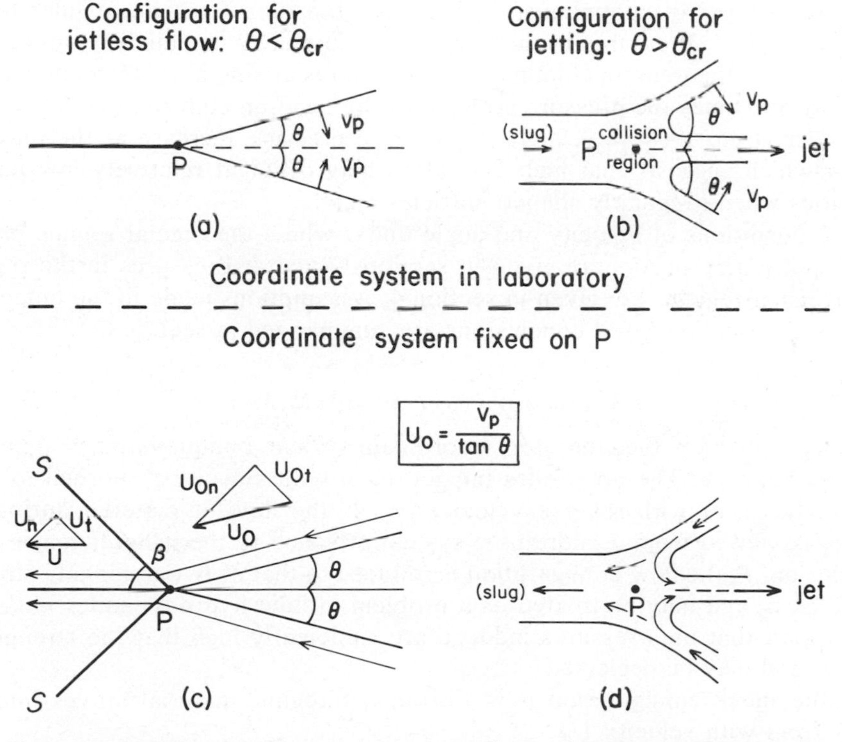

Fig. 2. Configuration for jetless flow (regular regime) and jet-forming flow (irregular regime).

regular regime. For half-angles greater than the critical angle, the shock configuration is such that a stream of material is ejected forward from the point of collision (Fig. 2b). The ejected material is called a *jet*, and the shock configuration which produces the jet is called the *irregular regime.* Excellent computer generated schematics of the steady-state fluid flow configuration in the vicinity of the jet are given by Harlow and Pracht (1966). The phenomenon of jetting was first analyzed with a hydrodynamic model developed by Birkhoff *et al.* (1948). Quantitative descriptions of the shock wave geometry, prediction of the critical angle, θ_{cr}, and verification with experiments on several metals were given by WSW and by Allen *et al.* (1959). Similar work has been done by the Russian physicists Al'tshuler *et al.* (1962). The analysis and notation of WSW are followed in this paper.

Relative to head-on collisions, very high pressures are generated under conditions of oblique impact, particularly under jet-forming conditions. (Comparison of a head-on with an oblique collision is made at the same particle velocity behind the shock. In the head-on collision this is taken to be equal to the impact velocity of the plates, v_p, specified normal to the free surfaces.) Pressures throughout the regular regime are calculated from the mechanical shock jump conditions of conservation of mass and momentum. In aluminum, the pressure at θ_{cr} is twice that attained in the head-on collision (WSW, p. 351). The shock pressure changes discontinuously across the transition from the regular to the irregular regime. Pressures in the irregular regime are calculated by application of Bernoulli's theorem: in aluminum, the pressures arising in jet formation are as much as five times the pressure attained in the head-on collision (WSW, p. 352; Al'tshuler *et al.*, 1962, p. 989). It is this large pressure increase at the onset of jetting which suggests that melt formation may occur at relatively low impact velocities when irregularly shaped particles collide.

The conditions of velocity and angle under which the regular regime breaks down and jetting occurs are given in sections 3 and 4. Pressures in the regular and irregular regimes are given in section 4. Assumptions made in the model are discussed in section 5 and conclusions are summarized in section 6.

3. Formulation of the Model

When viewed in the laboratory coordinate system, oblique impacts appear as in Figs. 2a and b. The projectiles impact, each with velocity v_p normal to their free surfaces, or with relative velocity $2v_p$. If the flow of material during the impact is viewed from a coordinate system attached to the instantaneous point of collision, P, the flow configuration is reduced to that of two impinging streams (Figs. 2c, d) and may be treated as a problem in fluid hydrodynamics *under the assumption* that the pressures induced are sufficiently high that the strength of the material may be neglected.

If the shock configuration is in the regular regime material moves into the shock front with velocity U_0

$$U_0 = v_p/\tan\theta \qquad (1)$$

It is conventional to convert the shock velocity U_0 into a pseudo-"Mach number" according to

$$M = U_0/c_0 \tag{2}$$

where

$$c_0 = \left(\frac{\partial P}{\partial \rho}\right)_S^{1/2}\Bigg|_{P=0,\ \rho=\rho_0} \tag{3}$$

If $\theta < \theta_{cr}$ material enters the shock with the supersonic velocity U_0 relative to the coordinate system centered on P. The flow is turned discontinuously through the angle θ, called the *wedge angle*. The oblique shock front, S, connects two zones *assumed to be in steady state thermodynamic equilibrium*. The shock front stands at angle β, called the *wave angle*, to the plane of symmetry. The Rankine–Hugoniot conditions may be derived for flow through the shock front by considering components of the flow vectors parallel and perpendicular to the shock front (Figs 2c, d). Conservation of mass and momentum require that

$$\rho_0 U_{0n} = \rho U_n \tag{4}$$

and

$$P - P_0 = \rho_0 U_{0n}(U_{0n} - U_n) \tag{5}$$

Henceforth it shall be assumed that $P_0 \approx 0$, because pressures of interest are much larger than ambient pressure. Continuity across the shock front requires that $U_{0t} = U_t$.

From the geometry in Fig. 2 and the Rankine–Hugoniot equations, the flow vector U is given by

$$U = \left[\frac{\rho_0 U_0^2 - (1 + \rho_0/\rho)P}{\rho_0}\right]^{1/2} \tag{6}$$

The wave angle, β, is obtained from

$$\tan\beta = \left[\frac{P}{\rho_0 U_0^2 (1 - \rho_0/\rho) - P}\right]^{1/2} \tag{7}$$

and the wedge angle θ is obtained from

$$\tan\theta = \frac{\{P[\rho_0 U_0^2 (1 - \rho/\rho_0) - P]\}^{1/2}}{(\rho_0 U_0^2 - P)} \tag{8}$$

The flow velocities U_t and U_n are, respectively,

$$U_t = \frac{\rho_0 U_0^2 - P}{\rho_0 U_0} \tag{9}$$

$$U_n = \frac{\{P[\rho_0 U_0^2 (1 - \rho/\rho_0) - P]\}^{1/2}}{\rho_0 U_0} \tag{10}$$

These velocities can be plotted in a *shock polar hodograph*, which is the two-dimensional equivalent of the one -dimensional $U_s - u_p$ Hugoniot curve. In order to construct the hodograph, however, a shock equation of state, the Hugoniot, must be specified.

For metals, which have a simple behavior uncomplicated by phase changes, the equation of state can be specified as a rather simple function of the density:

$$P = A\left(\frac{\rho}{\rho_0} - 1\right) + B\left(\frac{\rho}{\rho_0} - 1\right)^2 + C\left(\frac{\rho}{\rho_0} - 1\right)^3 + \cdots \quad (11)$$

which may be extrapolated to pressures in the megabar range. The equations of state of silicates are generally complex, however, due to phase transitions, and such a simple equation of state, based on low-pressure parameters gives answers seriously in error at high pressure.

Typically, Hugoniots of silicates are characterized by three regimes: (1) A *low-pressure regime* in which the material retains the same structure as the starting material. In a one-dimensional geometry there are usually two waves associated with shock compression in this regime: a compressional wave which travels with the longitudinal wave velocity and takes the material up to the yield pressure. At this pressure the material relaxes toward its hydrodynamic equilibrium state at higher pressure (McQueen *et al.*, 1967, p. 5015). It is a common approximation for both metals and silicates to use the bulk sound speed to represent the whole shock process in this regime; (2) a *mixed-phase regime* in which the low-pressure phase is partially transformed to a high-pressure phase or phases; and (3) a *high-pressure phase regime* in which the low-pressure phase is completely transformed to a dense phase.

This multiphase behavior of silicates can most easily be described by specifying the shock velocity-particle velocity ($U_s - u_p$) relation for each phase. The $U_s - u_p$ curves are nearly linear over the region of pressure and particle velocity characterizing each phase, n (Ruoff, 1968):

$$U_{s,n} = U_{0,n} = c_n + s_n u_{p,n} \quad (12)$$

for $n = 1, 2, 3$ for the low-pressure, mixed, and high-pressure phase regimes, respectively. A representative equation-of-state of this form is shown in Fig. 3a. In most representations, the longitudinal elastic wave is ignored and the linear curves shown on such a diagram represent an "effective" hydrostatic compression (Wackerle, 1962). With such an approximation, the U_s-axis intercept is equal to the bulk sound speed for the low pressure phase. From Eq. (12) and the Rankine–Hugoniot equations, $P - V$ curves of the form

$$P_H = \frac{c_n^2(V_0 - V)}{(V_0 - s_n V_0 + s_n V)^2} \quad (13)$$

are obtained (Fig. 3b). In this equation V_0 is the initial specific volume of the zero pressure phase of the material. The coefficients c_n and s_n for the "equilibrium curves" are given in Table 1 (see Wackerle, 1962, and section 5 of this paper for a discussion of the meaning of the equilibrium curve).

This paper represents, to my knowledge, the first attempt to account quantitatively for jetting in materials which exhibit high-pressure polymorphism;

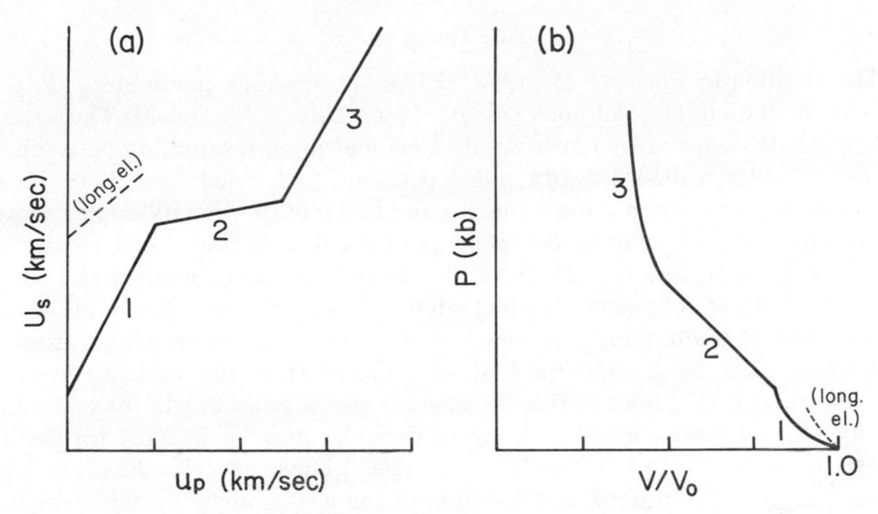

Fig. 3. (a) Schematic shock velocity-particle velocity Hugoniot and (b) schematic pressure-compression Hugoniot, showing three regimes: (1) low-pressure phase, (2) mixed-phase, and (3) high-pressure phase regimes.

although calculations have been performed for iron, the α-ϵ transition has been ignored. As will be discussed in a later section, several simplifying assumptions must be made about the multiple wave structure arising from elastic waves and/or high-pressure polymorphism. These complications are generally present at the lower pressures (regimes 1 and 2 of the equation-of-state) where the assumption of effective hydrostatic compression is most strongly violated.

Table 1. Material parameters.

Material	Regime	Equation of state (U_s, u_p in km sec^{-1})	Velocity limits (km sec^{-1})	Pressure limits, P (kbar)	Compression limits, $x = (\rho_0/\rho)$	ρ_0 (g cm^{-3})
Aluminum[1] (2024 alloy)		$U_s = 5.355 + 1.345\,u_p$	—	>200	—	2.785
Quartz[2] (single crystal)	1	$U_s = 3.68 + 2.12\,u_p$	$u_p < 0.95$	<150	>0.83	2.65
	2	$U_s = 5.56 + 0.14\,u_p$	$0.95 < u_p < 2.45$	$150 < P < 380$	$0.83 > x > 0.59$	
	3	$U_s = 1.74 + 1.70\,u_p$	$u_p > 2.45$	>380	<0.59	
Bronzitite[3] (Stillwater, Montana)	1	$U_s = 6.0 + 1.1\,u_p$	$u_p < 1.4$	<398	>0.80	3.28
	2	$U_s = 7.3 + 0.2\,u_p$	$1.4 < u_p < 2.1$	$398 < P < 600$	$0.80 > x > 0.70$	
	3	$U_s = 5.2 + 1.2\,u_p$	$u_p > 2.1$	>600	<0.70	
Dunite[3] (Twin Sisters Peaks, Wash.)	1	$U_s = 6.6 + 0.9\,u_p$	$u_p < 1.65$	<440	>0.80	3.32
	2	$U_s = 7.8 + 0.2\,u_p$	$1.65 < u_p < 2.44$	$440 < P < 730$	$0.80 > x > 0.68$	
	3	$U_s = 4.4 + 1.5\,u_p$	$u_p > 2.44$	>730	< 0.68	

1. Rice *et al.*, 1958
2. Wackerle, 1962
3. McQueen *et al.*, 1967

4. Results

The solution to Eqs. (4), (5), (9)–(13) is a set of shock parameters (P, ρ, U, U_n, U_t) for the initial conditions (P_0, ρ_0, U_0 or $U_0/c_0 = M$, and θ). The solution is most easily obtained by constructing a geometric representation between two of the variables, called a *shock polar* (Courant and Friedrichs, 1948, p. 306; WSW, 1953). A commonly used shock polar is a plot of velocity components; u and v, which are the velocity components of the flow in the x- and y-directions respectively, or U_n and U_t, which are the components perpendicular and parallel to the shock front. The curve so generated is the geometric locus of all velocity components characterizing the shock state which can be attained from the unshocked state by a stationary shock, and thus is the equivalent of the one-dimensional Hugoniot curve. A separate shock polar exists for each initial density, ρ_0, and initial velocity U_0. Other variables may be selected for the axes of the shock polar, e.g., Al'tshuler *et al.* (1962) chose the $(\theta - P)$-plane which shows the pressure attained as a function of the wedge angle through which the flow is turned.

In Fig. 4 a generalized shock polar in the (u-v) plane is shown (Courant and Friedrichs, 1948, p. 311). The coordinate system is chosen so that the incoming flow is parallel to the x-axis. The loop, called a Descartes loop, is obtained by

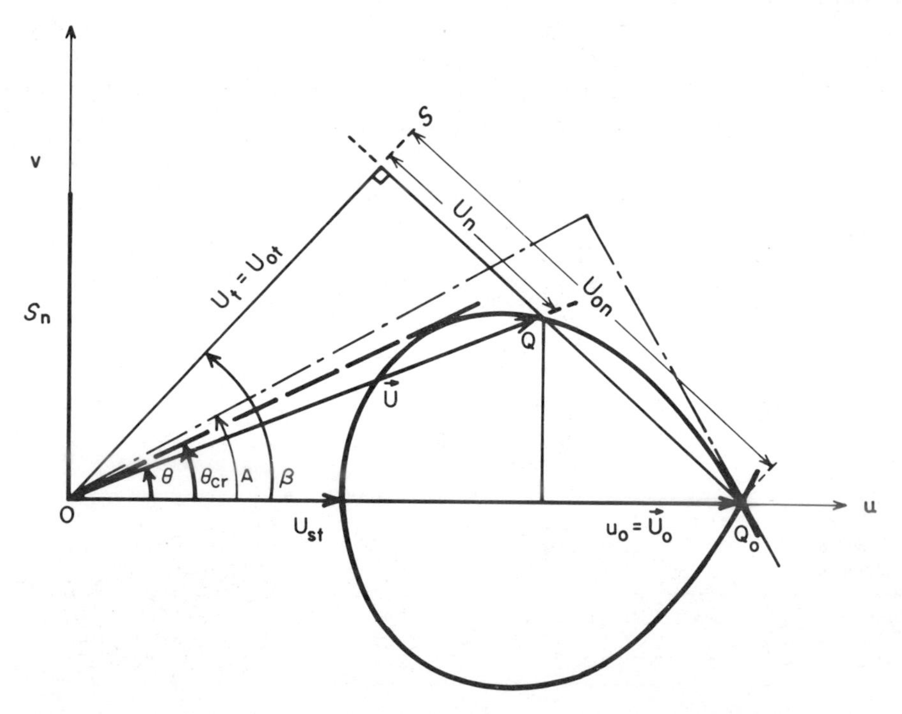

Fig. 4. Schematic hodograph with horizontal (u) and vertical (v) velocity components as axes. Notation is explained in text.

considering the variation of the density ρ. The density increases from ρ_0 at the double point, Q_0, to a maximum at the left intercept of the curve with the u-axis. The shock polar has a double or isolated point, Q_0, at $U_n = U_0$, $U_t = 0$, which is the end point of the vector U_0.

For any point Q on the shock polar, the flow is turned through the wedge angle θ between OQ and the u-axis. The velocity $\vec{U}$ behind the shock front is given by the vector OQ. The components U_{0n}, $U_{0t} = U_t$, and U_n of the velocity vectors $\vec{U}_0$ and $\vec{U}$ can be obtained by considering the components of OQ_0 and OQ perpendicular and parallel to the shock line. The following general conclusions can be obtained from such a diagram (Courant and Friedrichs, 1948, p. 312):

1. The wave angle, β, is always greater than the wedge angle, θ.
2. A point Q on the polar near Q_0 represents a weak shock with small velocity change and weak pressure. As $Q \to Q_0$ the shock becomes sonic.
3. The shock configurations which exist when $\theta < \theta_{cr}$ are those of the *regular regime* described above. For $\theta < \theta_{cr}$ there are two possible shocks; a weak and a strong shock. The wave angle β (obtained by construction as above) becomes the Mach angle, A_0, for the weak sonic shock, and becomes 90° for a normal shock. The weak and strong shocks coincide at $\theta = \theta_{cr}$. The existence of an extreme wedge angle, θ_{cr}, implies the existence of an extreme wave angle, β_{cr}. There is a critical angle $\theta = \theta_{cr}$ above which no regular shock transition exists.

Jetting arises when θ exceeds θ_{cr} and the "irregular regime" is entered (WSW; see also Chou *et al.*, 1976). The wave configurations which exist when $\theta > \theta_{cr}$ may be analyzed by methods developed by Courant and Friedrichs (1948) for air shocks or Al'tshuler *et al.* (1962) for solids, but these procedures are complex. However, since jetting is initiated at $\theta = \theta_{cr}$, it is possible to solve for the *onset* of jetting conditions by constructing shock polars of the *regular* regime to obtain the critical angle, θ_{cr}, as a function of impact velocity.

The shock polars of aluminum, quartz, bronzitite and dunite are shown in Fig. 5 for various Mach numbers, M. Shock states in the low-pressure phase regime are shown as dotted lines, in the mixed-phase regime as dashed lines, and in the high-pressure phase regime as solid lines. The polars shown are the mathematical solutions to the equations above, without, at this point, regard to restrictions imposed by the use of Eq. (12), an "equilibrium equation-of-state." These restrictions are discussed in the next section.

The critical angle, $2\theta_{cr}$, is shown as a function of relative impact velocity $2v_p$, in Fig. 6. The critical angle at which jetting onsets along a polar for a given Mach number is obtained from the tangent to the polar, $\tan\theta = u/v$. The impact velocity v_p is:

$$v_p = U_0 \tan\theta_{cr} = Mc_0 \tan\theta_{cr} \tag{14}$$

Shock polars for aluminum are given so that a comparison of these results can be made with the original WSW calculations and because aluminum is commonly used as a driver or target material in laboratory shock experiments. The shock polars of aluminum are smooth and lobate, a form characteristic of

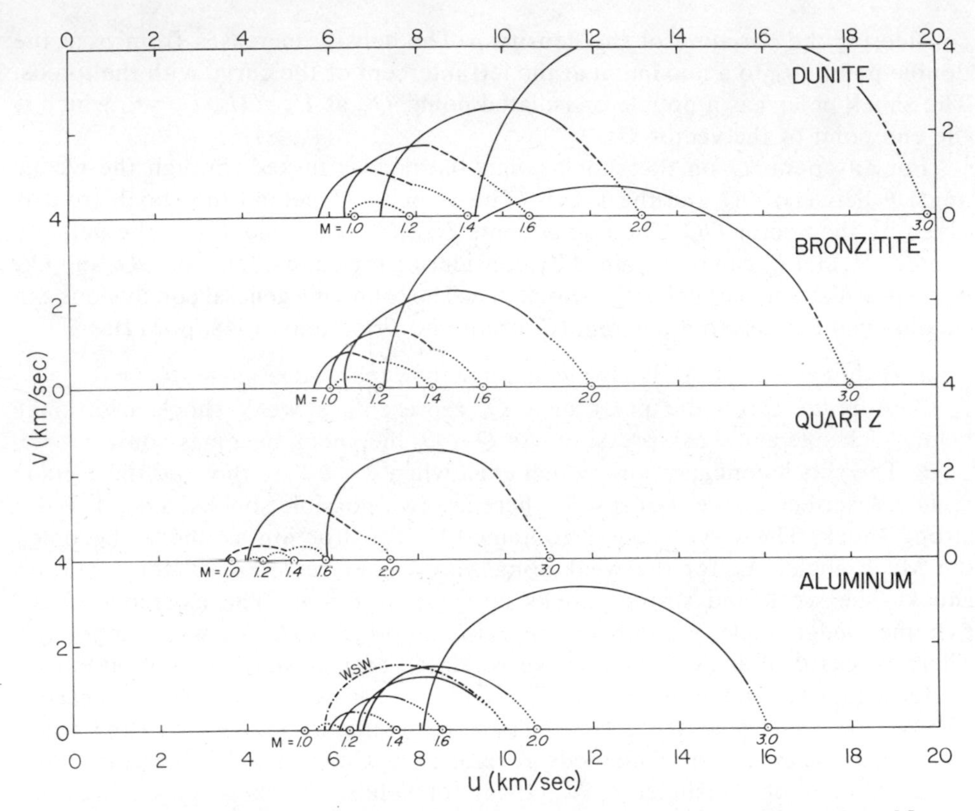

Fig. 5. Shock polars for aluminum, quartz, bronzitite and dunite. Mach numbers M for each polar are given on the abscissa. On each polar the dotted section represents shock conditions in the low-pressure phase regime, the dashed polar represents conditions in the mixed phase regime, and the solid line represents conditions in the high-pressure phase regime. The dash-dot curve labelled WSW for aluminum is explained in the text.

simple materials which do not undergo a phase change in the shock front. These shock polars were calculated from the parameters given in Table 1. WSW used a polynomial equation-of-state of the form $P(\mu)$, which can be closely approximated by choosing a value of $s = 1.1$ in Eq. (11), instead of $s = 1.345$, which is the more recent value of Rice *et al.* (1958). The sensitivity of the polars to changes of slope, s, of the $U_s - u_p$ curves can be seen by comparing the two curves centered at $M = 1.87$, corresponding to $U_0 = 10$ km/sec. The dash-dot curve corresponds closely to the original WSW result approximated here by $s = 1.1$; the solid curve results from the use of the newer value of $s = 1.345$. The predicted values of θ_{cr} required for jetting (Fig. 6) are in nearly as good agreement with the WSW experimental as is the original WSW curve.

In contrast to the smooth lobate polars of aluminum, the polars of silicates are characterized by a discontinuous change of slope in the curves at the onset

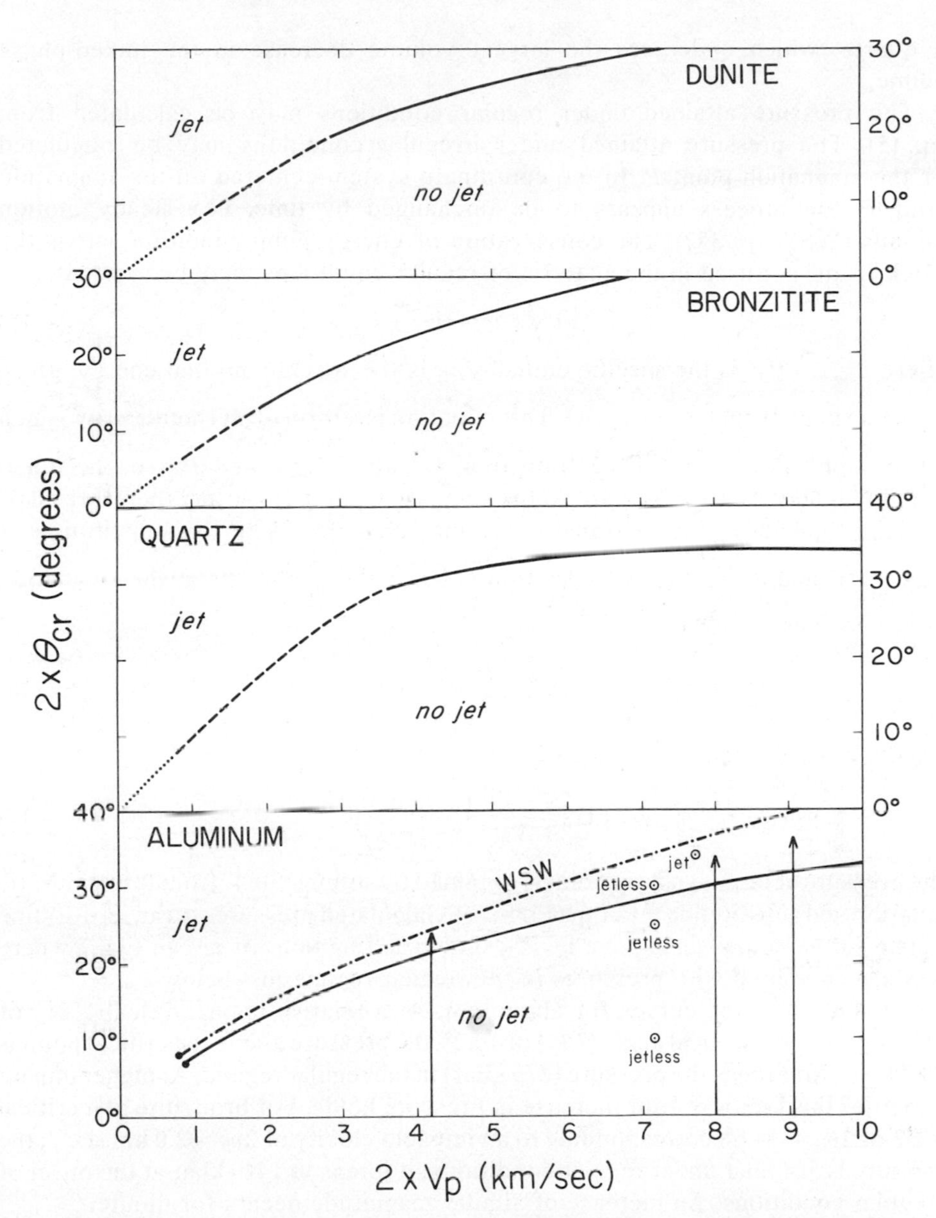

Fig. 6. Theoretical curves for transition from regular jetless conditions to irregular jet-forming collisions.

of the mixed-phase region. Geometrically, this "shoulder" is a region of shock states where the vertical velocity, v, does not change as rapidly with respect to horizontal velocity changes as in the low- or high-pressure regimes. It reflects directly the small velocity changes in the mixed phase regions of the $U_s - u_p$ Hugoniot (Fig. 3a). The shoulder is most pronounced at low Mach numbers and in quartz, which undergoes the largest volume decrease in the mixed-phase regime.

The pressure attained under regular conditions may be calculated from Eq. (5). The pressure attained under irregular conditions may be calculated for the stagnation point P. In the coordinate system centered on the stagnation point P the process appears to be unchanged by time, i.e., steady motion prevails (WSW, p. 352). The conservation of energy jump condition across the shock front, not used in the analysis of regular conditions, may be applied:

$$\tfrac{1}{2}U_0^2 + i_0 = \tfrac{1}{2}U^2 + i \tag{15}$$

where $i = e + PV$ is the specific enthalpy, e is the specific internal energy, given across a shock front by $\frac{P}{2}(V_0 - V)$. This equation is a form of Bernoulli's law which can be applied across a shock transition. Because $e_0 \approx 0$ and $P_0 \approx 0$, the initial enthalpy i_0 may be taken as zero. At the stagnation point $U = 0$ and therefore all of the initial kinetic energy is transformed into enthalpy. With the substitution of $i = e + PV$ and $e = \frac{P}{2}(V_0 - V)$, equation (15) for the conditions at the stagnation point becomes

$$\tfrac{1}{2}U_0^2 = \frac{P}{2}(V_0 + V) \tag{16}$$

For the equation of state (12) this gives

$$\tfrac{1}{2}U_0^2 = \frac{\tfrac{1}{2}c_n^2(V_0^2 - V)^2}{V_0(1 - s_n) + sV^2} \tag{17}$$

The pressure at the stagnation point is obtained by solving for V from this quadratic equation and substituting into Eq. (13) for P. Calculated pressures in the jet-forming regime for velocities just above v_{cr} are shown as functions of $2v_{cr}$ in Fig. 7 where they are compared with pressures in the regular regime just below v_{cr}.

Consider first the curves for aluminum. At a relative impact velocity $2v_p$ of 1.43 km sec^{-1} and critical angle ($2\theta_{cr}$) of 10.2°, the pressure above the critical point is 1087 kbar, four times the pressure (273 kbar) in the regular regime. At higher impact velocities this factor of four increase in pressure holds. For bronzitite at a critical angle of $2\theta_{cr} = 14.6°$, corresponding to an impact velocity of $2v_p = 2.0$ km sec^{-1}, the pressure is 524 kbar under regular conditions; it jumps to 1440 kbar at the onset of irregular conditions. An increase of similar magnitude occurs for dunite.

The behavior of quartz under oblique impact is somewhat different than the behavior of bronzitite and dunite because of the relatively lower impedance and the great difference in density and compressibility between the low- and high-pressure phases. At a given critical angle or impact velocity, the pressures induced under either regular or irregular conditions are much lower than the pressures induced in bronzitite and dunite. At an impact velocity of $2v_p$ of 3.6 km sec^{-1}, the critical angle 2θ is 29.2°. The pressure attained under regular conditions is 380 kbar; it jumps to 838 kbar under irregular conditions. In quartz

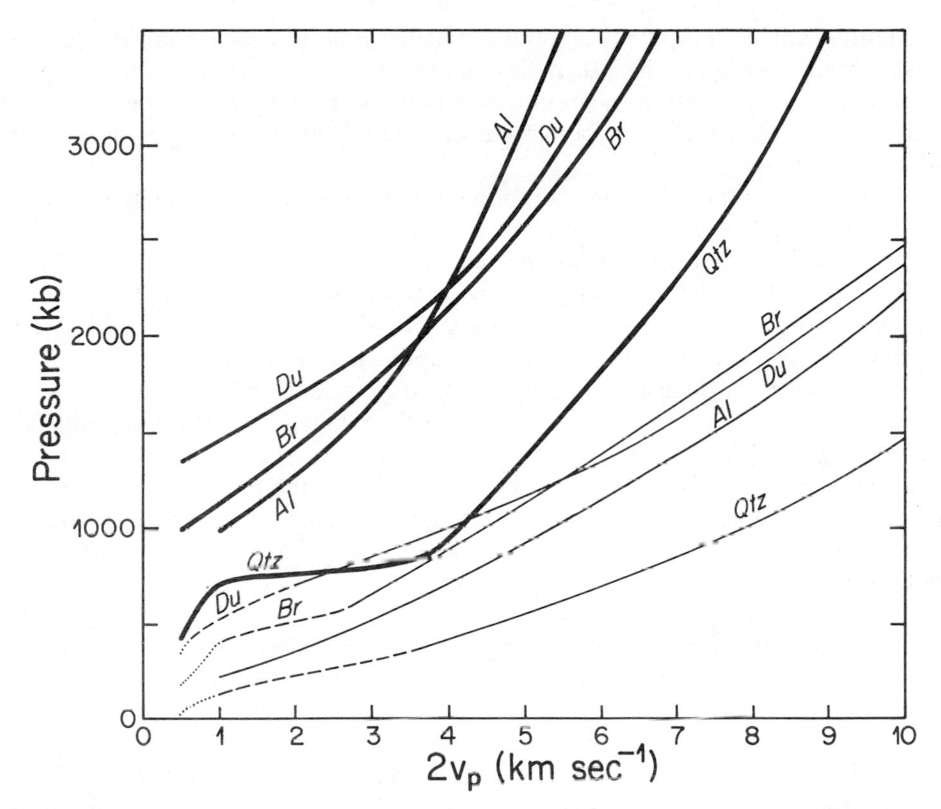

Fig. 7. Pressures attained under critical conditions in the regular regime (jetless conditions; light curves) and at the stagnation point in the irregular regime (jet-forming conditions; heavy curves).

relatively high angles of collision are required for production of jets of SiO_2 for a given impact velocity.

The high pressures attained upon the break down of regular mixed-phase regime conditions are adequate to cause melting *if* they can be directly compared to the pressures required to cause melting under one-dimensional conditions. Melting of aluminum occurs upon release from pressures in excess of 1 mbar (Urlin, 1966; Mineev and Savinov, 1967) or may actually occur in the shocked state at 1400 kbar (Mitchell and Shaner, 1976). Measured release adiabats of minerals (e.g., quartz, Ahrens and Rosenberg, 1968, p. 73) and thermodynamic calculations of energy changes with pressure (Ahrens and O'Keefe, 1972) suggest that melting of minerals occurs upon release from pressures in the high-pressure phase regime. The pressures at the onset of this regime are: 380 kbar for quartz, 600 kbar for bronzitite and 730 kbar for dunite. Pressures attained in jets upon the breakdown of regular mixed-phase regime conditions are well in excess of these values (Fig. 7). Thus, consideration of the pressure attained at the stagnation point suggests that silicate which passes through a stagnation point into a jet at impact velocities of 1.0 km sec^{-1} for quartz, 0.9 km sec^{-1} for bronzitite, and 0.7 km sec^{-1} for dunite will be molten. Since material does not pass directly through the stagnation point but rather to the sides where it experiences somewhat lower pressures, a more conservative statement would be that molten silicate jets are probably produced at impact velocities in the range of 1–2 km sec^{-1}.

There is, however, a consideration of the fluid flow characteristics which cautions against use of pressure as a simple measure of thermal state. The material which ultimately forms the jet passes through several stages of compression (J. M. Walsh, private communication; Harlow and Pracht, 1966): (1) shock compression as the material passes across a detached shock which stands across the entry flow; (2) adiabatic compression in continuous flow as the material passes through (or near) the stagnation point; and (3) rarefaction as it expands into the jet. Under the assumption that stages (2) and (3) are adiabatic, the only irreversible heating which occurs is due to the initial deceleration of material across the detached shock. Any pressure increase which is attained between the detached shock and the stagnation point does not contribute to irreversible heating and, hence, to melting of the material. It is, unfortunately, not simple to calculate the pressure increase corresponding to the shock jump alone, because the position of the detached shock depends on the collision angle and the shock itself is not a simple planar front, but rather is curved because of the influence of rarefactions emanating from the free surfaces of the colliding plates.

Numerical calculations of jet-forming flow configurations for substances with an ideal gas and a stiffened gas equation-of-state suggest that a large fraction of the internal energy change is accomplished as the material passes through the detached shock. Figures 2 and 4 of Harlow and Pracht (1966, p. 1952 and 1953) show isotherms in the region of the stagnation point. Their results represent only two specific cases, but allow comparison of a typical "metallic" substance (represented by the stiffened gas equation-of-state) and a more compressible substance (the perfect gas). Minerals, which are effectively more compressible than metals because of their phase changes, might be expected to show behavior between these two extremes. The positions of the isotherms for both materials demonstrate that more than 80% of the total internal energy increase is accomplished across the detached shock, and less than 20% in the continuous flow between the detached shock and the stagnation point. The energy increase across a shock wave is simply the area under the Rayleigh line in a P-V plot and the energy increase in an adiabatic process is the area under the adiabat. A simple consideration of P-V curves for silicates (Fig. 3b) shows that the above constraint on internal energy jumps requires *either* that the shock pressure is approximately equal to the stagnation pressure *or* that both the shock pressure and the stagnation pressure lie in regime 3 where the Hugoniot is so steep that relatively large pressure changes cause only small perturbations to the energy. If the first case is true, then the pressure gives a reasonable criterion for melting. If the second case is true, the argument is weaker, but since the pressure is in regime 3 in this case, melting will probably be attained.

In summary, then, a consideration of shock pressures attained in jets suggests that molten jets may be produced by collision of silicates at relative velocities of collision of 1–2 km sec^{-1}.

5. Discussion of Assumptions

The applicability of the above analysis to minerals depends on the validity of two assumptions for the mechanical stability of the shock configuration in the regular (jetless) regime:

(1) The pressure attained must be sufficiently high that the hydrodynamic conditions implied in the use of the equilibrium equation-of-state are attained.

(2) The pressure obtained must be sufficiently high that a single-wave shock structure exists.

If these conditions are fulfilled, the above analysis should give the velocity and critical angle at which the jetless regime breaks down and, hence, the conditions under which jetting must arise.

Assumption 1: The rheological properties of materials under shock deformation are not well known. For aluminum, WSW specify 200 kbar as the minimum pressure required for hydrodynamic conditions to be obtained within the low-pressure regime. For quartz, the onset of hydrostatic conditions appears to be above the Hugoniot elastic limit (Wackerle, 1962). In general, it may be assumed that hydrodynamic conditions are attained in the mixed phase regime.

Assumption 2: A single-wave shock structure exists when pressures are sufficiently high that the shock wave which raises the material to the final pressure can overrun elastic or plastic precursor waves. WSW show that a single-wave structure exists above 200 kbar in aluminum. As a general rule, single wave conditions in minerals are obtained at pressures on the order of a few hundred kilobars to a half megabar, approximately at the onset of the high-pressure regime (Wackerle, 1962; Ahrens and Gregson, 1964; Ahrens and Rosenberg, 1968).

These two criteria, if strictly interpreted, restrict the validity of the theory to conditions where pressures attained under regular conditions are within the *high-pressure regime.* The minimum relative velocities of impact, $2v_p$, which produce these pressures at the critical angle are: 0.9 km sec^{-1} for aluminum (200 kbar pressure); 3.6 km sec^{-1} for quartz, 1.9 km sec^{-1} for bronzitite and 2.8 km sec^{-1} for dunite.

In view of the uncertainties in assumptions 1 and 2 and the qualitative nature of the considerations of melting, laboratory verification of the conclusions should be attempted. The existence of jetting in silicates has been inferred from petrographic properties of shocked porous Coconino Sandstone from Meteor Crater, and a semi-quantitative argument has been advanced above that they formed under relatively low-velocity impact conditions, but laboratory experiments on silicates would greatly advance our understanding of this process in complex substances.

6. Conclusions

The jetting model described above gives results which are consistent with inferred conditions in the Coconino Sandstone, namely, that melted material is found intermixed with shocked quartz which cannot have seen average pressures greater than 200 or 250 kbar. The shock polars calculated for conditions in the regular regime and the critical angles calculated for this regime are probably not correct in detail because of possible complications of elastic or plastic precursor waves, but the calculations provide estimates of the velocities at which jets might arise and can be used to provide estimates of the pressures induced in the jetting material. The calculations lead to the conclusion that molten jets may be produced by collision of silicates at relative impact velocities of 1–2 km sec^{-1}.

Acknowledgments—J. M. Walsh called the author's attention to complications in the flow which prevent a simple interpretation of melting by consideration of pressures. I appreciate reviews and helpful comments by G. R. Fowles, E. Gaffney, P. Juda and J. M. Walsh. This work was supported by NASA NSG 7052.

References

Ahrens, T. J. and Duvall, G. E.: 1966, Stress relaxation behind elastic shock waves in rocks. *J. Geophys. Res.* **71**, 4349.

Ahrens, T. J. and Gregson, V. G.: 1964, Shock compression of crustal rocks: Data for quartz, calcite, and plagioclase rocks. *J. Geophys. Res.* **69**, 4839.

Ahrens, T. J. and O'Keefe, J. D.: 1972, Shock melting and vaporization of lunar rocks and minerals. *The Moon* **4**, 214.

Ahrens, T. J. and Rosenberg, J. T.: 1968, Shock metamorphism: Experiments on quartz and plagioclase. In *Shock Metamorphism of Natural Materials* (B. M. French and N. M. Short, eds.), pp. 59–82, Mono Press, Baltimore.

Allen, W. A., Morrison, H. L., Ray, D. B. and Rogers, J. W.: 1959, Fluid mechanics of copper. *Phys. Fluids* **2**, 329.

Al'tshuler, L. V., Kormer, S. B., Bakanova, A. A., Petunin, A. P., Funtikov, A. I. and Gubtin, A. A.: 1962, Irregular conditions of oblique collision of shock waves in solid bodies. *Sov. Phys. JETP* **14**, 986.

Birkhoff, G., MacDougall, D. P., Pugh, E. M. and Taylor, G.: 1948, Explosives with lined cavities. *J. Appl. Phys.* **19**, 563.

Cameron, A. G. W.: 1972, Models of the primitive solar nebula. In *Symposium on the Origin of the Solar System* (H. Reeves, ed.), pp. 56–70, Cen. Nat. Rech. Sci., Nice.

Cameron, A. G. W.: 1973, Accumulation processes in the primitive solar nebula. *Icarus* **18**, 407.

Chou, P. C., Carleone, J. and Karpp, R.: 1976, Criteria for jet formation from impinging shells and plates. *J. Appl. Phys.* **47**, 2975.

Courant, R. and Friedrichs, K. O.: 1948, *Supersonic Flow and Shock Waves*, Interscience, New York.

Dodd, R. T.: 1971, The petrology of chondrules in the Sharps meteorite. *Contr. Min. Pet.* **31**, 201.

Fredriksson, K.: 1963, Chondrules and the meteorite parent bodies. *Trans. N. Y. Acad. Sci.* **25**, 756.

Goldreich, P. and Ward, W. R.: 1973, The formation of planetesimals. *Astrophys. J.* **183**, 1051.
Harlow, F. H. and Pracht, W. E.: 1966, Formation and penetration of high-speed collapse jets. *Phys. Fluids* **9**, 1951.
Kerridge, J. F. and Kieffer, S. W.: 1977, A constraint on impact theories of chondrule formation. *Earth Planet. Sci. Lett.*, in press.
Kieffer, S. W.: 1971, Shock metamorphism of the Coconino Sandstone at Meteor Crater, Arizona. *J. Geophys. Res.* **76**, 5449.
Kieffer, S. W.: 1975, Droplet chondrules. *Science* **189**, 333.
Kieffer, S. W., Phakey, P. P. and Christie, J. M.: 1976, Shock processes in porous quarzite: Transmission-electron microscope observations and theory. *Contr. Min. Pet.* **59**, 41.
Kurat, G.: 1967, Formation of chondrules. *Geochim. Cosmochim. Acta* **31**, 491.
Larimer, J. W. and Anders, E.: 1967, Chemical fractionations in meteorites. II. Abundance patterns and their interpretation. *Geochim. Cosmochim. Acta* **31**, 1239.
Mineev, V. N. and Savinov, E. V.: 1967, Viscosity and melting point of aluminum, lead and sodium chloride subjected to shock compression. *Sov. Phys. JETP* **25**, 411.
McQueen, R. G., Marsh, S. P. and Fritz, J. N.: 1967, Hugoniot equation of state of twelve rocks. *J. Geophys. Res.* **72**, 4999.
Mitchell, A. C. and Shaner, J. W.: 1976, Melting of aluminum at 1.4 Mbar (abstract). *Amer. Phys. Soc. Winter Mtg.*, Stanford, 1286.
Rice, M. H., McQueen, R. G. and Walsh, J. M.: 1958, Compression of solids by strong shock waves. In *Solid State Physics* **6** (F. Seitz and D. Turnball, eds.), pp. 1–63, Academic Press, N. Y.
Ruoff, A. L.: 1967, Linear shock-velocity-particle-velocity relationship. *J. Appl. Phys.* **38**, 4976.
Shapiro, A. H.: 1953, *The Dynamics and Thermodynamics of Compressible Fluid Flow*. Ronald Press Co., N. Y.
Urey, H.: 1956, Diamonds, meteorites and the origin of the solar system. *Astrophys. J.* **124**, 623.
Urlin, V. D.: 1966, Melting at ultrahigh pressures in a shock wave. *Sov. Phys. JETP* **22**, 341.
Wackerle, J.: 1962, Shock wave compression of quartz. *J. Appl. Phys.* **33**, 922.
Walsh, J. M., Shreffler, R. G. and Willig, F. J.: 1953, Limiting conditions for jet formation in high velocity collisions. *J. Appl. Phys.* **24**, 349.
Wasson, J.: 1972, *Meteorites*. Springer-Verlag, Berlin.
Wlotzka, F.: 1969, On the formation of chondrules and metal particles by "shock melting". In *Meteorite Research*, (P. M. Millman, ed.), pp. 174–184, Reidel, Dordrecht.

Roddy, D. J., Pepin, R. O., and Merrill, R. B., editors.
(1977) *Impact and Explosion Cratering*, Pergamon Press (New York), p. 771–790.
Printed in the United States of America

Dynamical implications of the petrology and distribution of impact melt rocks

WILLIAM C. PHINNEY

NASA Johnson Space Center Houston, Texas 77058

CHARLES H. SIMONDS

The Lunar Science Institute, Houston, Texas 77058

Abstract—Studies of natural impact craters provide constraints that can be applied to theoretical models and experimental production of craters. In particular, the petrographic and structural relations of rocks resulting from melting of target materials during impact place constraints on five aspects of crater formation: (1) Motion of shock melted material must commence in the central zone of the target, be underway before ejection of some of the lowermost parts of the target, allow incorporation of fragmental material from kilometers away from the zone of shock melting, allow for outward flow over the edge of the transient cavity in large craters, provide for settling into depressions and cracks after formation of significant thicknesses of fragmental breccia on the base of the cavity, and be completed within tens of seconds to a couple of minutes after impact. (2) Melts must be mixed by a process that is efficient enough to homogenize chemical compositions, lithologies, and shock features to a scale of cubic millimeters from scales that originally varied by kilometers. Furthermore, this mixing mechanism must proceed to completion within tens of seconds to a few minutes after impact. (3) Energies required for heating and melting of the necessary volumes of target rocks imply impact energies ranging from 10^{25} ergs for kilometer sized craters to 10^{30} ergs for 50 km sized craters. (4) Depths of excavation for simple craters of a few kilometers diameter show diameter to depth ratios of 3 to 6 but for larger, more complex craters this ratio increases to 8 to 12. (5) Formation of structural modifications such as central uplifts, uplifted rings, and large scale slumping of crater walls occurs within tens of seconds to a couple of minutes after impact.

INTRODUCTION

TO INCREASE our understanding of the formation of impact craters and the properties of their ejecta there must be an integration of data from theoretical models, cratering experiments, and observations of natural impact craters. Constraints provided by observations of natural craters can be applied to theoretical models as well as to the variables and scaling assumptions in experimentally produced craters. Studies of natural craters generally have concentrated on shape, size, and form on the moon, Mars, and Mercury or have concentrated on structures and petrographic shock effects on Earth. This paper summarizes the petrologic and chemical data for samples of impact-melt rocks and associated breccias from various structural locations in and around eight terrestrial craters. These data provide constraints on the following five aspects of crater formation: (1) trajectories of melt, (2) extent of mixing in melt, (3) thermal energy from impact, (4) depth of excavation, and (5) timing of formation of structural modifications.

1. *Trajectories of melt.* Because melt forms only within the small fraction of a crater that experiences very high shock pressures, the initial location of melted material can be identified as a restricted volume near the axis of the projectile through the target volume (O'Keefe and Ahrens, 1975, 1977). Furthermore, the identification of the present location of melt deposits can be accomplished easily by petrographic and field studies (Grieve, 1975; Floran *et al.*, 1976; Dence, 1971). Therefore, the initial and final positions of various types of melted material place constraints on the trajectories that may be assigned to this material.

2. *Extent of mixing in melt.* Knowledge of the chemical and lithologic compositions of the target rocks as well as their distribution allows comparison with the chemical composition of resulting melt rocks and the lithologies of included fragments. Extent of homogenization of initial chemical and lithologic heterogeneities, departures from predicted average compositions, and distribution of included fragments with respect to size, lithologies, and shock features provide constraints on the extent of mixing and the process of mixing.

3. *Thermal energy from impact.* Several studies of the energy partitioning resulting from impact events suggest that about 1/4 to 1/3 of the kinetic energy of the projectile is converted to thermal energy (Braslau, 1970; Gault and Heitowit, 1963; O'Keefe and Ahrens, 1975). Because of the low specific heats and large latent heats of fusion for the usual silicate target rocks the bulk of this thermal energy is utilized for heating and fusing that part of the target that is melted. If the volume of impact-melted rock can be estimated, the energy expended to heat and fuse the target rocks can be calculated to define a lower limit for the thermal energy in an impact event. These values combined with estimates of the percentage of melt in the total excavation and partitioning of energy can provide constraints on the total energy and partitioning of energy in natural impact events.

4. *Depth of excavation.* Knowledge of the vertical distribution of rock units in the target area prior to impact can be compared with the rock types which occur as fragments in the ejecta to determine the depth of excavation. For some of the large craters that may have excavated rocks from several kilometers depth, determination of the vertical distribution of rock units may require interpretation of geophysical data (gravity, seismic, magnetic, and heat flow studies). From this information come constraints on depth of excavation and extent of uplifting of central peaks.

5. *Timing of formation of structural modifications.* Because impact melts have chilled rapidly to form rocks with glassy or very fine-grained igneous textures, it is possible to place limits of seconds to minutes on the length of time required for these melts to solidify (Simonds, 1975; Simonds *et al.*, 1976). The structural relations of these melt-rocks with structural modifications of the crater, such as central peaks, uplifted rings, and large slump blocks, can place constraints on the timing of the formation of the structural modifications. If the chilled units are faulted over such structures, the structures formed later than the solidification of the melt units. If the chilled units are draped over such

structures, however, the structures must have formed before the solidification of the melt units.

EVALUATION OF CONSTRAINTS

Petrologic studies and structural relations of impact-melt rocks associated with terrestrial craters should provide constraints for the five listed aspects of craters. To evaluate these constraints we first, review briefly some recent models of thermal history for impact-produced melt-clast mixtures and, second, discuss for several terrestrial craters the features that are applicable to the five aspects listed above.

Petrologic studies of lunar highlands samples and intensive investigation of the melt sheet at the Manicouagan impact structure have provided a thermal model for breccia lithification. This model describes the ejecta as a two component mixture of superheated melt and relatively cool fragments of minerals and rocks. Studies of interaction of melt with SiO_2 and ZrO_2 clasts (Carstens, 1975; Stähle, 1972; El Goresy, 1965), degree of oxidation in glass (Hörz, 1965), as well as digestion of clasts by melt (Simonds, 1976) suggest melt temperatures of 1600 to 1800°C or more. In contrast the lack of shock features in most fragmental material suggests an upper limit of about 200°C for shock heating of most clasts (Ahrens and O'Keefe, 1972). This apparent bimodal temperature distribution in the excavated material is attributed to the rapid radial attenuation of shock-heating in the target. Materials with intermediate levels of shock, hence intermediate post-shock temperatures, are observed in minor amounts in nature; they are neglected in the simplified model. Mixing of clasts and melt occur in all proportions. The resulting petrographic textures are largely a function of the relative proportion of the two components and the resulting temperature to which the mixtures equilibrate. In descriptions of field occurrences the lithologic terminology varies somewhat from one worker to another. Thus, it is necessary to develop a reasonably uniform terminology throughout this paper. Although there is a complete gradation in textural features among impact-produced rocks it is possible to cluster the resulting rocks into two very general groups based on their matrix characteristics: (1) fragmental, or clastic, breccias and (2) melt rocks. Both types may display a large variety of clasts. Although the fragmental breccias may contain fragments of glass that were initially melted, the melt rocks consist of masses of igneous-textured, crystalline rocks that extend continuously for at least several meters as sheets, dikes, sills, etc. Within each of these two groups there are two more rather general subgroups: (1a) polymict fragmental breccias, (1b) monomict fragmental breccias, (2a) clast-laden melt rocks, and (2b) clast-free melt rocks. The polymict fragmental breccias containing no glass are referred to as allogenic breccias. The polymict fragmental breccias containing glass as clasts or as part of the matrix and clasts are referred to as suevites. The monomict fragmental breccias of this paper are referred to as authigenic breccias or brecciated basement. Terminology for the melt rocks should be self-evident.

Before discussing the relevant data for each of several craters it should be pointed out that variations in the structure of target-rock units, differences in composition and strength, variations in porosity and water content of target rocks, and variations in projectile energies may drastically affect the extent of melting, types of impact deposits produced, and resulting structural relations. Also the extent of erosion influences the degree to which extrapolation and drilling data are required. Such differences must be considered in any detailed comparison of natural crater characteristics. Eight craters are discussed: the first five are primarily in crystalline rocks with very little or no overlying sedimentary cover, the next two are in crystalline rocks with a significant overlying sedimentary cover, and the last one is entirely in sedimentary rocks.

Manicouagan Crater, Quebec, Canada. 65 km diameter; complex crater with central peak and two or three rings; target consisted of wide variety of Precambrian crystalline, amphibolite facies, metamorphic rocks with thin (<50 m) covering of limestone; crater has been highly eroded to expose only the lowest 300 m of fill and the sub-floor target rocks; age is 214 million years (Triassic) (Currie 1972; Wolfe, 1971).

1. A sheet of impact-melt rock a few hundred meters thick rests directly on a brecciated basal gneiss over the entire crater (except for the central peak) to a diameter of about 60 km. The sheet extends over both an inner crater of 40 km diameter and an outer crater which presumably formed by the slumping of large blocks into the excavation (Floran *et al.*, 1976; Floran and Dence, 1976). The total thickness of the sheet is unknown because the highest exposed outcrops, 230 m above the base, are still part of the melt sheet. The melt-rock fills in around large blocks of dislodged gneiss at the base of the crater cavity where the blocks produce an irregular surface with relief of several tens of meters. Most inclusions in the melt-rock are small (<3 cm) unshocked fragments but there are also a few large inclusions tens of meters across. There are very few included fragments in the melt-rock that occur in the first few meters above the base and adjacent to large boulders. There are abundant fragments throughout the next higher few tens of meters and then a continual decrease in fragments higher in the sheet. The textures of the crystallized melt rock indicate that the lowermost 20 to 40 meters of the sheet cooled to subliquidus temperatures in an interval that could range from tens of seconds to a few minutes (Simonds, 1975; Simonds *et al.*, 1976.

Thus the melt which formed near the center of the crater must have moved in such a way that it mixed extensively with unshocked material that was originally a few kilometers away in the target. Final movement of this mixture developed an extensive sheet of fragment-laden melt at the bottom of the crater, where it permeated the space between large disoriented blocks.

2. The melt rocks display a very limited range of chemical composition and although the total amount of included fragments may vary with height in the sheet, the fragments are evenly distributed in each sample throughout a few hundred km^3 of melt. To illustrate the chemical homogeneity, all analyzed samples of melt rock contain $57.3 \pm 2.1\%$ (2σ) SiO_2 whereas the target rocks

contain from 42 to 72% SiO_2 (Floran *et al.*, in press). Grieve *et al.* (in press) and Floran *et al.* (1976) have calculated the proportions of various target rocks required to produce the observed contents of major and trace elements in the melt. A mixing model consistent with these compositions requires components to be added from all of the major target rocks. Thus the melted material from widely separated sources must be thoroughly mixed and homogenized. In addition the mineral and lithic debris from all of these units is so evenly distributed throughout the melt that no 1 mm^2 of a thin section is free of clasts. Most of the preserved clasts lack features indicative of shock pressures over 100 to 200 kb, many show no indication of shock pressures greater than 50 kb, that is, they lack planar elements in quartz and feldspar. Thus, most of the presently visible clasts come from a zone where shock pressure is much lower than in the fused zone, probably a few to several kilometers away from the zone of total fusion. To smooth out the centimeter to kilometer heterogeneitics in the chemistry, clast types, and shock effects that occurred in the target, material separated by several kilometers must have been mixed into every gram of melt. This evidence for the intimate mixing and homogenization of both melt and clasts from widely separated points within the target suggests that flow regimes are intensely turbulent on a wide range of scales. Thermal equilibration between cold clasts and hot melt takes only a matter of seconds once the materials are firmly in contact (Onorato *et al.*, 1976). If enough clasts are present, the mixture equilibrates to temperatures so low that crystallization ceases within seconds leaving glass in the matrix; a higher proportion of melt typical of most of the melt rocks is sufficient to initiate crystallization, raising the melt's viscosity greatly and preventing further flow. Thus the process that mixes clasts and melt must not only be thorough but also extremely rapid. Further evidence for rapid mixing exists in the glassy to very fine-grained clast-free zones immediately adjacent to large blocks of gneiss near the base of the melt sheet. The drag along these surfaces removed inclusions from the zone of intense shear next to the blocks to the zone of minimum shear away from the blocks but the melt was cooled quickly enough to produce glass and skeletal crystals.

3. Because the grain size of the melt rock still appears to be increasing at its highest exposures, the thickness of the sheet is believed to be significantly greater than shown by the actual outcrop pattern. Estimates of volume of melt range up to 600 km^3 (Grieve *et al.*, 1976; Floran *et al.*, 1976; Simonds, 1975). This estimate does not include the extensive volumes that may have been present as dikes, sills, tongues, and caps in the higher levels of the breccias and suevites as at Popigay Crater (Masaytis *et al.*, 1975). Nor does it include the volume in now-eroded suevites. Inclusions of these units would probably double the volume of melt. Six hundred km^3 represents between 3 to 4% of the excavated volume of the transient cavity (Grieve *et al.*, 1976). This volume of melt would require about 3 to 4×10^{28} ergs of energy to heat and fuse the target rocks, and, therefore, close to 10^{30} ergs for the kinetic energy of the impacting projectile.

4. Although the pre-impact stratigraphy is not unique enough to provide any limits to the depth of excavation, the lack of granulite or other inferred lower

crustal rocks as clasts suggests that depths of 10 to 15 km would be a maximum. Furthermore, gravity and seismic data (Sweeney, in press) suggest that the depth of excavation is from 3 to 8 km. If the transient cavity diameter is placed between 35 and 45 km the diameter to depth ratio is between 5 and 15.

5. Large disoriented blocks of gneiss hundreds of meters across occur on the floor of the crater. It is not certain how many of these blocks result from slumping but very fine-grained, rapidly chilled melt fills the space between them and intrudes cracks in them suggesting that the blocks were in their present location within tens of seconds to a few minutes after impact. If the diameter of the transient cavity is placed between 35 and 45 km as argued by Floran and Dence (1976) then the melt sheet, which extends continuously to about 60 km diameter, must cover a slumped wall that dropped at least several hundred meters (perhaps over a kilometer) as indicated by the presence of Ordovician limestone outcrops between diameters of 45 and 60 km (Currie, 1972). The limestones which formed a veneer over the Precambrian gneisses at the time of impact could have existed only at much higher elevations outside the crater and would have been ejected from inside the transient cavity. Thus the slumping of the walls to bring the limestones below the rapidly chilled, clast-laden melt sheet in the outer part of the crater occurred within the first few tens of seconds to a few minutes after formation of the melt.

West Clearwater Lake Crater, Quebec, Canada. 38 km diameter; complex crater with central uplift and one or two rings; target consisted of wide variety of Precambrian, crystalline, high grade metamorphic rocks with thin (<50 m) covering of limestone; crater has been highly eroded to expose only the lowest 100 m of fill and the sub-floor target rocks; age is 300 million years (Bostock, 1969; Dence *et al.*, 1964).

1. A melt sheet at least 130 m thick rests on either a basal authigenic breccia of gneiss or on a thin layer of suevitic breccias that also rests on the basal authigenic breccia (Fig. 1). The total thickness of the sheet is unknown because the highest exposed outcrops consist of melt rock whose grain size is still coarsening. The aerial extent of the melt sheet is undefined but it is best exposed over an uplifted ring of basement that forms a circular set of islands in West Clearwater Lake. Although this melt sheet has not received the same degree of detailed study as that at Manicouagan, the similarities suggest a nearly identical history for the motion of melt.

2. Similar chemical relationships are noted here as at Manicouagan. The melt rocks display a very limited range of composition while the target rocks display a wide range (Bostock, 1969). For example the melt rocks contain $59.3 \pm 1.9 (2\sigma)$ SiO_2 whereas the target rocks range from 49 to 73% SiO_2. Simple mixing models require the melt to be a mixture of the various target rocks. As at Manicouagan abundant fragments of target rocks are evenly dispersed in the fine-grained melt rocks. Thus the mixing requirements discussed for the melt sheet at Manicouagan appear to apply equally well to the melt sheet at West Clearwater Lake.

3. Estimates of the volume of the existing melt sheet at West Clearwater

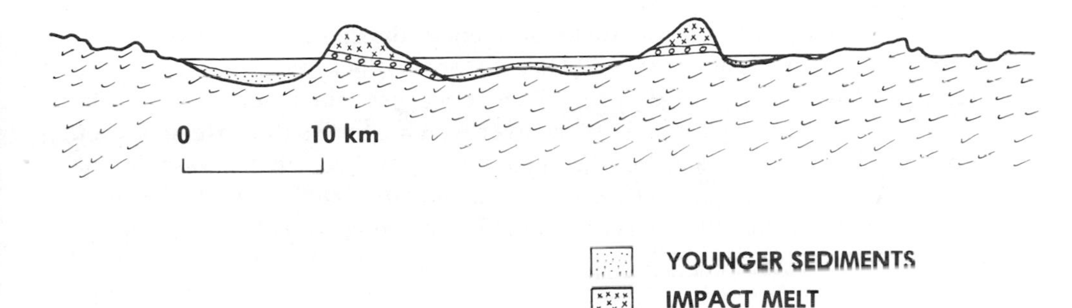

Fig 1. Cross-section of the crater at West Clearwater Lake, Canada (after Dence *et al.*, 1964).

Lake range from 34 to 50 km^3 (Grieve *et al.*, 1976). These estimates must represent a minimum volume because the grain size of the melt rocks continues to coarsen through the uppermost part of the exposed section. Again, as at Manicouagan this estimate ignores the melt that might have been present as dikes, sills, tongues, and caps in the now-eroded, overlying breccias. It also ignores the melt in now-eroded suevites. The minimum estimates of volume of melt represent about 5% of the volume of the assumed transient cavity and would require 2 to 3×10^{27} ergs of energy to heat and fuse the necessary target rocks (Grieve *et al.*, 1976; Dence *et al.*, 1964). Thus, between 10^{28} to 10^{29} ergs of kinetic energy would be required for the impacting projectile.

4. Pre-impact stratigraphy is not unique enough to allow studies of included fragments to provide limits on the depth of excavation.

5. A massive melt sheet occurs on the uplifted ring at about 20 km diameter. It is not clearly established that the rapidly cooled melt is draped over the ring and is continuous with melt outside or inside the ring. The occurrence of limestone below the melt sheet inside the crater, however, is well-established. This relationship provides the same arguments as it did at Manigouagan: the veneer of limestone on the basal gneiss is far below its normal stratigraphic level (Bostock, 1969) and must occur on a slumped block. Otherwise the limestone would have been excavated from the surface of the transient cavity. Further-

more, the presence of the rapidly chilled, clast-laden melt sheet above the limestone requires that the slumping took place within a few tens of seconds to a few minutes after formation of the melt.

Mistastin Lake Crater, Labrador, Canada. 20 km diameter; complex crater with central uplift; target consisted of wide variety of Precambrian, crystalline, high grade metamorphic rocks; crater has been highly eroded to preserve only some isolated patches of fill and the sub-floor target rocks; age is 40 million years (Grieve, 1975; Currie, 1971).

1. Patchy occurrences of melt rocks throughout the central part of the crater allow reconstruction of a melt sheet at least 80 meters thick and extending over a diameter of between 12 and 13 km. The sheet lies either directly on the brecciated basement or on a thin breccia layer which overlies the basement rock (Grieve, 1975). The grain-size of the melt rock continues to increase in the highest exposures suggesting that the sheet was initially significantly thicker than it is at present. A crude subhorizontal foliation of elongate inclusions suggests horizontal flow in the melt. The inclusions display features that range from unshocked to highly shocked. Also the clast content decreases upward in the melt sheet. The similarities to characteristics displayed by the Manicouagan melt sheet suggest nearly identical movements of melt.

2. The melt rocks display a restricted range of composition compared to the country rock. Analyses of samples from the major unit of the melt sheet indicate $58.4 \pm .87\%$ SiO_2 (1σ) while analyses of the target rocks indicate a range from 51 to 68% SiO_2. The lower, fine-grained units contain somewhat lower SiO_2 ($\sim 54\%$) but again display a restricted overall range (Grieve, 1975). Mixing models for the main mass of the melt require inputs from all target rocks. Abundant fragments in the melt range from unshocked to highly shocked, thereby indicating derivation of fragments from well outside the central, intensely shocked, zone of fusion. Again the similarities with Manicouagan suggest that the same mixing requirements apply. Grieve (1975) suggests that turbulent homogenization of the melt occurred to distances of at least 10 km from the source of the melt.

3. An estimate of the minimum volume of melt at Mistastin Lake is 12 km^3 (Grieve *et al.*, 1976). In that the initial melt sheet may have been significantly thicker than the present remnant, and additional melt may have been present in suevite or as dikes, sills, tongues, and caps in the overlying breccias (Masaytis *et al.*, 1975), the volume of melt may have been significantly larger. The 12 km^3 estimate represents about 5% of the assumed transient cavity and requires about 1×10^{27} ergs for heating and fusing the target rocks. Thus the kinetic energy of the impacting projectile must have been on the order of 10^{27} ergs as a minimum.

4. Data on pre-impact stratigraphy does not appear to be unique enough to provide limits on the depth of excavation from studies of included fragments.

5. There do not appear to be any clear-cut arguments to delineate the timing of structural events.

Boltysh Crater, Ukraine, Soviet Union. 25 km diameter; complex crater with central uplift; target consisted of Precambrian granitic rocks; crater has been moderately eroded but covered with 500 m of Cretaceous and Tertiary sediments (all data are from drilling information); age is 70 million years (Yurk *et al.*, 1975; Masaytis, 1975).

1. Overlying an authigenic breccia of granitic basement are allogenic and suevitic breccias that contain two layers of melt-rock: a lower layer, 20 m thick, and an upper layer, 200 m thick (Fig. 2). The lower layer has a glassy matrix and contains up to 30% shocked and unshocked clastic fragments (Yurk *et al.*, 1975). The upper layer has a more crystalline igneous texture and contains clastic fragments similar to those in the lower unit. In contrast to the three previously described craters where the melt sheets rest directly on granitic basement the bases of the melt sheets at Boltysh occur at heights of 80 and 180 m above the granitic basement and rest on allogenic or suevitic breccias. Barring misinterpretation of observations on the one deep drill core that penetrates all of these units, the melt motions must have been different from those at Manicouagan. In the Boltysh Crater there must be significant deposition of fragmental material as allogenic and suevitic breccias before the more massive melt sheets settled into their final positions.

2. The breccias as well as the impact melts are all reported to be "extremely similar in chemical composition" which is very much like that of the granitic basement (Yurk *et al.*, 1975). The included fragments of crystalline basement are

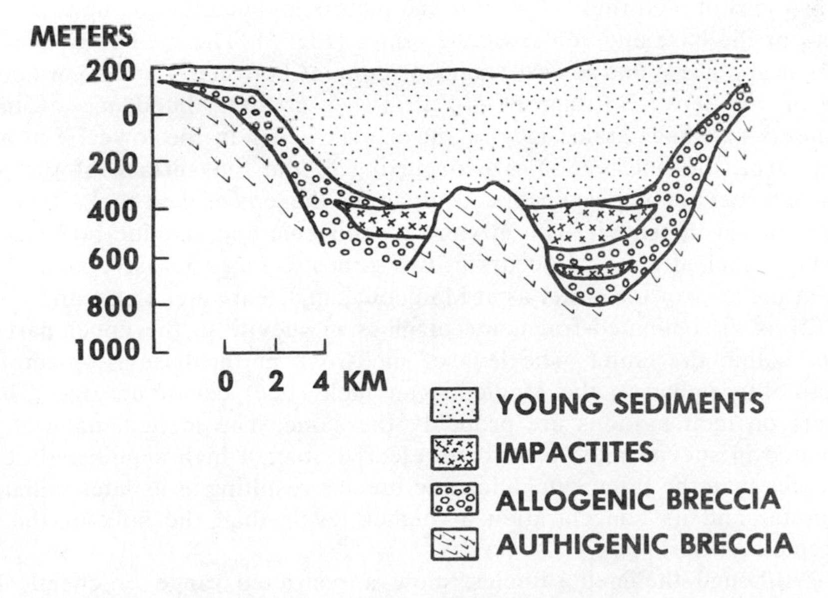

Fig. 2. Cross-section of Boltysh Crater, Ukraine, USSR (after Yurk *et al.*, 1975).

very abundant and show varying degrees of shock effects. These observations are not inconsistent with the mixing requirements discussed for the Manicouagan Crater.

3. From the available cross section of the melt units (Yurk *et al.*, 1975) a minimum estimate of about 20 km^3 can be calculated for the volume of melt. In that the size of the transient cavity is not known it is unclear what percentage of the excavated material consisted of melt. This would require about 1×10^{27} ergs of thermal energy to heat and fuse the target material. Thus, the kinetic energy of the impacting projectile must have been between 10^{27} and 10^{28} ergs.

4. In that all of the clastic fragments appear to be granitic, as would be expected from the granitic basement, no depth of excavation can be estimated from the fragment population. Drilling data, however, indicate a present depth of about 1 km. Allowing for erosion, central uplift, and bulking, the excavation may have been as deep as 2 km.

5. The lack of any melt units above the central peak suggests that the rapidly cooled melt sheets abut the uplifted central peak. This relationship would require the uplift to have been present when the melt solidified and, therefore, to have formed in tens of seconds to a few minutes after impact.

Brent Crater, Ontario, Canada. 3 km diameter; simple crater (no rings or central peak); target consisted of wide variety of Precambrian, crystalline, high grade metamorphic rocks; crater has undergone several hundred meters of erosion leaving about 560 m of impact-produced fill overlain by 270 m of Ordovician sediments within the crater and deposited soon after crater formation providing a protective cap (drilling data provide the major source of information); age is 414 million years (Hartung *et al.*, 1971).

1. A lens of melt rock a few hundred meters in diameter and up to 35 m thick occurs at the base and center of the crater (Fig. 3). The melt rock displays its coarsest grain size and lowest fragment content between 5 and 15 m above the base of the lens (Hartung *et al.*, 1971). A greater abundance of included fragments and finer grain sizes of the matrix occur in the lowest 5 m and the upper 20 m, the finest grain size occurring at the top surface of the lens. A glass-rich suevitic breccia occurs 500 m above the lens of melt rock. Between the melt lens and the suevite is a mixture of allogenic and suevitic breccias. Thus, the shock melted material occurs in two zones: (1) as a massive melt sheet, or lens, at the base of the crater as at Manicouagan, Clearwater Lake, and Mistastin and (2) as disseminated fragments of glass in suevite in the upper part of the ejecta within the crater. The lens of melt-rock at the base is texturally and structurally similar to the Manicouagan melt sheet except in size. The constraints on melt motions are probably the same. The melted material that is deposited in suevite must have been ejected at very high angles and at higher velocities than the fragmental allogenic breccia resulting in its later fallback into the crater and its concentration at higher levels than the bulk of the clastic allogenic breccias.

2. Although the melt samples show a restricted range of chemical composition (Hartung *et al.*, 1971), an impact event of this size requires only a small

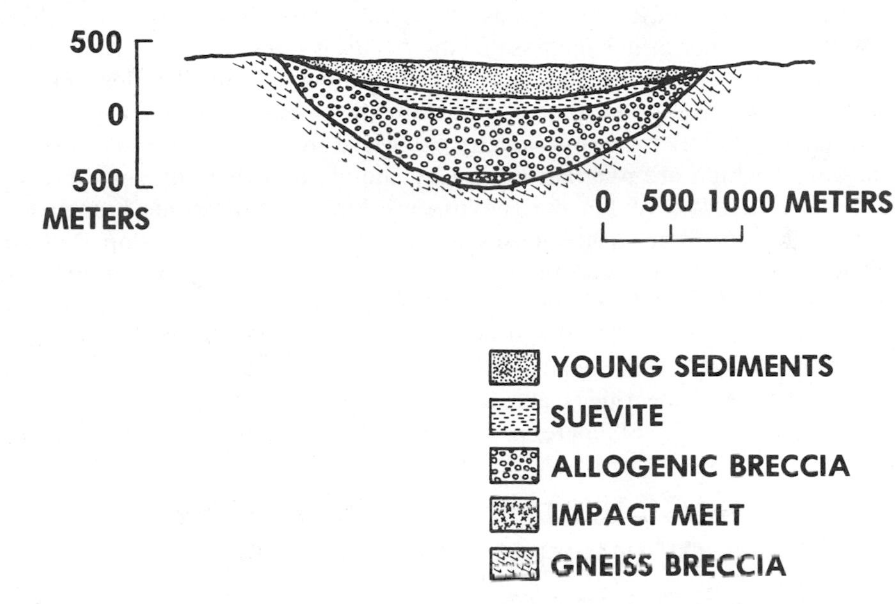

Fig. 3. Cross-section of Brent Crater, Canada (after Hartung *et al.*, 1971).

volume of the target to be melted. In that the initial rock compositions within such a small volume may not vary much, a homogeneous melt in this case does not demand extensive mixing. The abundance of evenly distributed granitic clasts throughout the melt, however, suggests that extensive mixing did occur.

3. The minimum volume of the melt lens has been calculated as 5×10^{-2} km^3 which is about 1% of the volume of the transient cavity (Grieve *et al.*, 1976). Additional melt in the suevite or small intrusive units might increase these values slightly. The thermal energy required for heating and fusion is about 2×10^{24} ergs or about 10^{25} ergs for the kinetic energy of the impacting projectile.

4. From drilling data and from estimates of erosion, the depth of excavation in the transient cavity has been estimated as 1.05 km and the diameter is about 3 km giving a diameter to depth ratio of about 3 (Dence *et al.*, 1976). Data on pre-impact stratigraphy are not unique enough to allow clast populations to provide limits on the depth of excavation.

5. There do not appear to be any data for clear-cut arguments on the timing of structural events.

Popigay Crater, Northern Siberia, Soviet Union. 100 km diameter with 70 km inner crater; complex crater with central uplift and 3 rings; target consisted of wide variety of Archean crystalline, high-grade, metamorphic rocks overlain by less than 1 km of Proterozoic sandstones and dolomites which in turn, were overlain by less than 1 km of younger sediments; crater has been slightly eroded by glaciers and rivers whose channels have been filled with a few tens of meters of sediment; sub-floor rocks exposed only on uplifted rings and outside the crater; probably 2 to 3 km of fill remain in central part of crater, some remnants of ejecta remain outside crater; age is 30 million years (Masaytis *et al.*, 1975).

1. Within the inner crater the base of the cavity is exposed on uplifted ridges which appear to form three concentric rings around a central uplift (Fig. 4). Where exposures are good the ridges show a basal authigenic breccia of Precambrian crystalline rocks overlain by extensive sheets of melt rock, referred to as tagamite, which are overlain in turn by suevite and then allogenic breccia (Masaytis *et al.*, 1975). Within the suevites are lenses of allogenic breccia and within the allogenic breccia are lenses of suevite. On the inner slopes of the ridges the suevite rests directly on authigenic breccia while on the outer slopes tagamite rests on the authigenic breccia. Dikes of tagamite 5 to 7 m thick extend into the basal authigenic breccia and pinch out downwards. Similar dikes occur in large blocks of gneiss that occur as huge clasts in the allogenic breccia. In the depressions between the ridges are suevitic units at least 700 to 800 m thick (possibly up to a few kilometers) and allogenic breccias of undetermined

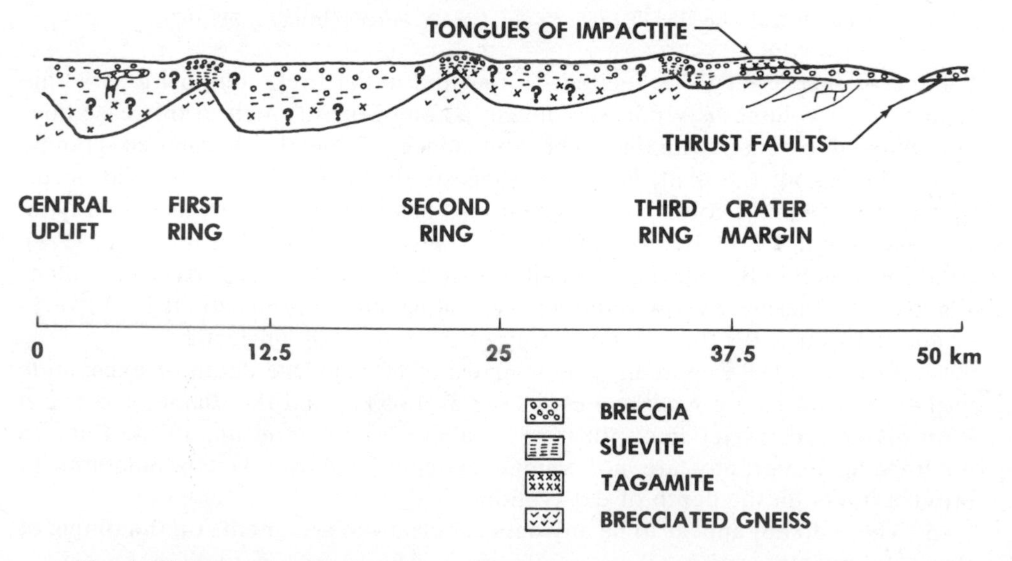

Fig. 4. Diagrammatic cross-section of the Popigay Crater, USSR. The section is interpreted from descriptions given in the text by Masaytis *et al.* (1975). Vertical relations are not to scale.

thickness. Within these suevites and breccias are numerous tagamite intrusions including dikes up to 50 m across, sills greater than 70 m thick, and cylindrical necks up to 150 m in diameter, the latter feed caps of tagamite that apparently formed as flows as much as tens of meters thick over the upper surfaces of the suevites and breccias. It is assumed that thick sheets of tagamite occur at the base of the crater to form the source of the dikes, sills, and caps.

At several locations around the outer edge of the inner crater, tongues of tagamite appear to spill out into the outer crater where they extend for distances of several kilometers. The tagamite usually occurs as an extensive network of thick lenses or as sub-radial dikes, 15 to 20 m wide, that form conduits to the lenses. These lenses are overlain by suevite and underlain by allogenic breccia which in turn overlies authigenic breccia of the basement. Rarely suevite rests directly on the authigenic breccia in these tongues. Occasionally caps of tagamite up to 60 m thick occur on the upper surface of the tongues. In the remainder of the outer crater and also outside the outer crater, there are no other tagamite units, but suevite consistently overlies allogenic breccia.

Throughout the crater the tagamites contain abundant clastic fragments among which Archean gneisses predominate, although most other rocks of the target are also present. Thus, the bulk of the shock melted material originates in the gneisses which occurred in the deeper part of the target in contrast with the experimental work of Stöffler *et al.* (1975), who found that the melted material was derived from the uppermost layer. The included fragments indicate various degrees of shock metamorphism. Xenoliths of allogenic breccia are especially prevalent in the tagamites of the dikes and caps. Such xenoliths are to be expected in rocks formed from melts that have traversed several hundred meters through relatively unconsolidated allogenic breccias. Porous tagamites occur in thin dikes and at the upper and lower margins of the caps. The pores in many cases are elongated paralled to the contacts.

Melt-rock that exists as sheets on the uplifted ridges and as dikes, sills, and caps in or on the breccias contain fragments from a wide range of target rocks and indicate a wide range of shock intensity. Thus the trajectories of melted material that end up in the melt sheets must have a similar history as discussed for Manicouagan. The tagamite dikes in the gneisses of both the crater floor and the large blocks in allogenic breccia indicate that the melt was initially mobilized and injected into the gneissic basement before the large blocks were ejected along with material that eventually come to rest well up in the fallback breccias. The melt must then have moved to mix with fragments derived from kilometers away from the zone of melting and then settled back into the lowest parts of the crater where it forms thick sheets. The inferred pool of melt at the base was squeezed upward to form dikes, sills, and surface flows, probably from the force developed by the weight of the thick pile of ejecta deposited over the pool of melt. The melt was also forced outward over the rim of the transient cavity as dikes, lenses, and caps. Thus, massive melt was restricted to the base of the crater although later mobilization may have carried it elsewhere by intrusion and extrusion.

Melted material occurs also as glassy material in the suevites which result from fallback of ejecta both inside and outside the crater. The consistent occurrence of suevite above the allogenic breccia outside the inner crater suggests that the melted fragments were ejected at higher angles and/or higher velocities than the less shocked material. Within the inner crater the suevites and allogenic breccias are interlayered and the overall stratigraphic sequence is unclear.

2. Once again the similarity of chemical compositions for all analyzed melt rocks (tagamites and suevites) is striking. For example, analyses indicate $63.1 \pm 1.2 (1\sigma)$ % SiO_2 for 22 samples of tagamites and 63.1 ± 4.0% SiO_2 for 14 samples of suevites derived from a target containing carbonates, quartzites, and a wide variety of crystalline gneisses of both mafic and salic compositions (Masaytis *et al.*, 1975). Included fragments range from quartzites and limestones to gneisses, the last of which predominate. Thus, the same mixing arguments apply to these occurrences as at Manicouagan.

3. Lack of data on the volume of a basal melt sheet makes thermal energy requirements impossible to estimate. In any case the extent of melt appears to surpass that at Manicouagan indicating an impact energy greater than 10^{30} ergs.

4. From both the clast population in the breccias and the nature of the basal material in the uplifted ridges it is clear that the approximately 1.5 km thick sequence of sedimentary rocks was completely excavated within the crater. In addition a significant amount of the crystalline basement was excavated. The lack of granulites or other lower crustal rocks among the clastic fragments appears to preclude very deep excavation below 10 to 15 km. Furthermore, fragments in the allogenic breccias consist of more than 50% sedimentary rocks while gneissic rocks are less abundant (Masaytis *et al.*, 1975). In the melt rocks the fragment population is reversed with gneissic rocks being predominant. In that melt generally makes up only about 5 to 10% of the total of excavated material, the clasts suggest that most of the excavated material consisted of sedimentary rocks. Therefore, excavation depths on the order of 3 to 5 km seem reasonable. If the 72 km diameter of the inner ring represents the diameter of the transient cavity, the diameter to depth ratio is between 14 and 24.

5. The uplift rings display sheets of rapidly cooled melt-rock resting directly on uplifted basement. It is not clear whether the melt sheets extend continuously over the basement within the depressions between the rings. If so the uplifted rings must have formed within tens of seconds to a few minutes after impact. If the 72 km diameter inner crater represents the margin of the transient cavity as is the case at Manicouagan (Floran and Dence, 1976), Ries (Pohl *et al.*, 1977), and West Clearwater (Dence, 1964), then the tongues of rapidly cooled, clast-laden melt-rocks that extend over parts of the outer crater would suggest that the down-dropping of the crater walls to form the outer crater must have occurred within tens of seconds to a few minutes after formation of the melt.

Ries Crater, Bavaria, Germany. 24 km diameter; complex crater with central uplift and a ring; target consisted of crystalline granitic and gneissic rocks overlain by 500 to 600 m of limestone and sandstone; crater has been only slightly eroded leaving most of the original ejecta both inside and outside the crater; age is 15 million years (Stöffler *et al.*, 1976; Pohl *et al.*, 1977).

1. No massive melt units are yet recognized although it is possible that they may be intersected in deep drill holes. The deepest drill hole that has been completed inside the crater penetrates first suevite, then large masses of crystalline rock but it is not clear whether these masses represent displaced blocks on the floor of the crater or basal units beneath the floor of the crater. This drill hole is 3.8 km from the center of the crater and did not intersect any massive melt sheets. However, it is possible that the drilling site may be on the marginal slope of an uplifted, circular central rise (Pohl, 1976; Ernstson, 1974) rather than in the deepest part of the crater. There are two relatively small occurrences of melt rocks a few tens of meters across on the eastern rim of the crater. It is not known whether these are connected by dikes with larger masses of melt as appears to be the case at the Popigay Crater. Melted material also occurs as fragments within suevite which occurs both as thick units within the crater and as a thin unit over allogenic breccia outside the crater. Although the allogenic breccia extends to more than 25 km outside the crater, the overlying suevite appears to be restricted to within 12 km of the crater. The greatest concentration of suevite, however, occurs inside the crater. Crater ejecta in the allogenic, or Bunte, breccia consist of 90% or more sedimentary rocks derived from the upper strata of the target. In contrast, the rock fragments within the suevites consist of 98 to 99% crystalline gneisses derived from the lower strata of the target. The remaining 1 to 2% of rock fragments in the suevites display different populations outside the crater than inside (Pohl *et al.*, 1977). The suevites outside the crater contain many limestone fragments from the uppermost layers of the target whereas the suevites inside the crater contain no limestone fragments from the upper 200 m of the target. The aerodynamically shaped forms of much of the melted material in suevitic breccias indicate that suevites result from fallback of ballistically ejected material. The composition of this melted material is that of the crystalline rocks found in the lower units of the target. The preceding relationships suggest that: (a) the melted material is derived from deep in the target, (b) the melted material was concentrated in high angle trajectories, and (c) the material in suevites underwent a different set of angles and velocities of ejection than the material in allogenic breccias.

2. Homogeneity of the glass composition in suevite has been emphasized by von Engelhardt (1967). The composition of the glass does not match any given unit of the crystalline basement but does seem to match a mixture of crystalline basement rocks with essentially no admixed sediment.

3. Assuming that there are no massive impact melts and that all of the melt occurs as glass in suevites there is only about 0.1 km^3 in the ejecta (Stöffler *et al.*, 1976). This represents between 0.05 to 0.1% of the excavated transient crater and about 10^{25} ergs of thermal energy. Because the estimated proportion of melt is nearly two orders of magnitude less than for other large craters it is clear that this represents an anomaly and an estimate of impact energy cannot be made easily from the thermal energy.

4. Comparison of pre-impact stratigraphy with that of the deep drill hole

suggests that at least 825 m of excavation occurred at this location 3.8 km from the center of the structure (Stöffler, 1976). Below 825 m are large blocks of gneiss with thin units of suevite separating them. If these blocks are similar to those that were dislodged and displaced at the base of the transient cavity at Manicouagan then the excavated depth may be significantly greater. Also the eccentric position of the drill hole 3.8 km from the center of the crater, which should be the deepest part of the excavation, indicates that the maximum depth should exceed 825 m. Bulking should add a few more tens of meters. Thus the excavated transient cavity would be expected to be at least 1 km and probably 2 to 3 km (Pohl *et al.*, 1977). If the 12 km diameter inner ring represents the diameter of the transient cavity, the diameter to depth ratio is between 4 and 12.

5. There are no definitive data for delineating the timing of structural features within the crater.

Meteor Crater, Arizona, USA. 1.2 km diameter; target consisted of 15 m of red sandstone over 80 m of limestone over white quartz sandstone, lower part of target was water-saturated; crater has been only slightly eroded; some remnants of ejecta still remain outside crater, raised rim up to 60 m high still present as an overturned flap, impact debris inside the crater is covered by up to 30 m of lake and alluvial sediments; age is 20,000 to 30,000 years (Roddy *et al.*, 1975).

1. Although no massive melt rock or suevite occur some melted fragments of quartz sandstone occur at the base of the crater (Kieffer, 1971). In addition, many small "bombs" of glass derived from the limestone occur in the ejecta blanket outside the crater (Nininger, 1954).

2. No melt units occur to provide data for mixing models, and no systematic data are published on the composition of glass fragments.

3. Inasmuch as no melt units or suevites occur the melted volume cannot be estimated. It is probably very small.

4. From drilling data and corrections for bulking a depth of excavation of 173 m has been estimated for the transient cavity (Roddy *et al.*, 1975). A 1 km diameter transient cavity gives a diameter to depth ratio of about 6.

. There are no major structural modifications such as a central peak or uplifted ridges.

Summary

From the descriptions and discussions of the eight selected terrestrial craters it is possible to summarize some of the constraints for the five aspects of crater formation that were outlined in the introduction.

1. *Trajectories of melt.* Extensive shock melting occurs in the central zone of most craters. The melt is intensively mixed and then mobilized and injected as dikes into the surrounding rocks before some of them are ejected as large blocks. Much of the melt ends up as massive melt sheets that remain close to the base of the crater. Most of the remaining melt is ejected and is incorporated as

glass fragments in suevitic breccias. At Ries and Popigay where the initial target stratigraphy is well known, the suevitic materials are derived from deep levels of the target.

The melt which forms sheets at the base of the excavation must move in such a way as to allow extensive mixing of melts and incorporation of clastic fragments that were separated initially by kilometers and derived from kilometers away from the zone of shock melting. The fragments are especially prevalent in the lowest and highest parts of the melt sheet. The melt also must move in such a way to concentrate on the outer slopes of uplifted ridges as at Popigay, fill in around large blocks on the base of the transient crater as at Manicouagan, and settle into large depressions, while occasionally allowing for deposition of a significant thickness of allogenic breccia below the melt sheet as at Boltysh. Motion of the melt must also allow it to spill out over the rim of the transient crater to emplace significant volumes of clast-laden melt in the outer part of the crater as at Manicouagan and West Clearwater. The occurrences of melt rocks as dikes, sills, caps, and tongues as at Popigay probably result from later intrusions of still fluid melt that was forced from the basal melt sheets. Furthermore, all of this motion must be completed within tens of seconds to a few minutes after formation of the melt.

The most possible mechanism for the formation of the massive melt sheets is the coalescence of liquids that line the expanding transient cavity (O'Keefe and Ahrens, 1975, 1977; Grieve *et al.*, 1976). The major problems to be evaluated experimentally and theoretically are: (1) The mechanism by which the melt becomes thoroughly homogenized, (2) the mode of incorporation of a large volume of initially widely separated, clastic, fragmental material so that it becomes evenly distributed throughout the entire melt sheet, (3) the method by which units of fragmental material are deposited below the melt sheet as at Boltysh and (4) the apparent lack of melt sheets in some large craters as the Ries. Further constraints on the motion of the melt are imposed by summary item 2 below.

The melt ejected ballistically is deposited mostly as fragments in the suevites that form extensive units within the crater where they may be interlayered with fragmental allogenic breccia. Thus most of these shock melted fragments were ejected at high angles. The remainder of the ejected, shock-melted fragments are deposited in suevites outside the crater where they consistently overly the fragmental allogenic breccias. Their occurrence above the allogenic breccias constrains the combinations of angles and velocities of transport of the most highly shocked material compared to the less shocked or unshocked materials of the allogenic breccias. Further work is necessary to determine the combinations of angles and velocities of the trajectories required to explain the suevite occurrences. Such work seems particularly appropriate in view of the deep origin of suevites indicated by the natural occurrences in contrast to the shallow origin suggested by Stöffler's experiments (Stöffler, 1975). An additional variable that may relate to the extent to which melt may form and coalesce is the amount of water present in the target rocks. Large impact generated steam explosions may produce more suevitic and allogenic breccias and less of a coherent melt sheet (Simonds *et al.*, 1977).

2. *Extent of mixing in melt.* Homogenization to a scale of cubic millimeters of widely diverse chemical compositions, lithologies, and shock features that vary on a scale of kilometers in the target requires an extremely efficient mixing mechanism. Furthermore this mechanism must proceed to completion within tens of seconds to a couple of minutes after impact, depending on the size of the crater. In some of the fragment-rich melts the cooling to a non-viscous state may occur within a few seconds after incorporation of the fragments. Among the possible mixing mechanisms to be evaluated experimentally and theoretically are: (1) high velocity melt overtaking lower velocity fragments along the base of crater (Grieve, 1975). The melt may form a thick film that lines the expanding cavity and moves outward to mix with ejecta, (2) turbulence that follows the impacting projectile in its very low pressure wake, (3) rising currents behind the rising fireball causing centripetal winds that pull fragments into central zone of crater (Jones and Sandford, 1976), (4) rebound of brecciated material from the walls and base of the crater into the melt.

3. *Thermal energy from impact.* Impacting projectiles must have kinetic energies ranging from 10^{25} or 10^{26} ergs for relatively small simple craters such as Brent to more than 10^{30} ergs for large craters such as Popigay and Manicouagan. Large volumes of melt are associated with the larger craters. The percentage of the transient cavity converted to melt appears to increase with the size of the crater (Grieve, 1976; Dence, 1971).

4. *Depth of excavation.* The depth of excavation does not appear to increase linearly with the diameter of craters. The smaller craters such as Meteor and Brent, whose transient cavities are in the range of 1–3 km, are excavated to depths of a few hundred meters to a kilometer and display diameter to depth ratios of 3–6. Larger craters such as Manicouagan and Popigay, whose transient cavities are at least an order of magnitude larger in diameter than Meteor and Brent, are not excavated to more than an order of magnitude greater depth but rather only four or five times greater and display diameter to depth ratios of 8–20. A similar conclusion holds for the lunar highlands where the fragments in breccias and clast-laden melt rocks show a general lack of plutonic rocks and an abundance of previously existing breccias and melt rocks. This distribution of fragments indicates that most craters did not penetrate more than a few km below a surface that is nearly saturated with 50–100 km diameter craters (Hörz *et al.*, 1976).

5. *Timing of formation of structural modifications.* The structural relations of rapidly-cooled, clast-laden melt rocks suggest the uplifting of rings, large scale slumping of walls, and uplifting of central hills within tens of seconds to a couple of minutes after the impact event. Such rapid timing may occur during a period of complex, high magnitude ground motions that occur during crater formation and allows failures of low angle slopes (Melosh, 1976).

Hopefully these constraints will provide input for further experimental and theoretical research on craters.

Acknowledgment—This paper constitutes Contribution No. 273 of the Lunar Science Institute, which is operated by the Universities Space Research Association under Contract No. NSR-09-051-001 with the National Aeronautics and Space Administration.

References

Ahrens, T. J. and O'Keefe, J. D.: 1972, Shock melting and vaporization of lunar rocks and minerals. *The Moon* **4**, 214–219.

Bostock, H. H.: 1969, The Clearwater Complex, New Quebec. *Geol. Surv. Can. Bull.* **178**, 63 pp.

Braslau, D.: 1970, Partitioning of energy in hypervelocity impact against loose sand targets. *J. Geophys. Res.* **75**, 3987–3999.

Carstens, H.: 1975, Thermal history of impact melt rocks in the Fennoscandian Shield. *Contrib. Mineral. Petrol.* **50**, 145–155.

Chao, E. C. T.: 1976, Mineral-produced high-pressure striae and clay polish: key evidence for nonballistic transport of ejecta from the Ries Crater. *Science* **194**, 615–618.

Currie, K. L.: 1971, Geology of the resurgent cryptoexplosion crater at Mistastin Lake, Labrador. *Geol. Surv. Can. Bull.* **207**.

Currie, K. L.: 1972, Geology and petrology of the Manicouagan Resurgent Caldera, Quebec. *Geol. Surv. Can. Bull.* **198**.

Dence, M. R., Innes, M. J. S., and Beals, C. S.: 1964, On the probable meteorite origin of the Clearwater Lakes, Quebec. *Royal Astron. Soc. Can. J.* **59**, 13–22.

Dence, M. R.: 1971, Impact melts. *J. Geophys. Res.* **76**, 5552–5565.

Dence, M. R., Grieve, R. A. F., and Roberston, P. B.: 1976, Terrestrial impact craters: Principal characteristics and energy considerations (abstract). In *Papers Presented to the Symposium on Planetary Cratering Mechanics*, p. 28–29. The Lunar Science Institute, Houston.

El Goresy, A.: 1965, Baddeleyite and its significance in impact glasses. *J. Geophys. Res.* **70**, 3453–3456.

Ernston, K.: 1974, The structure of the Ries Crater from geoelectric depth soundings. *Zeit f. Geophysik.* **40**, 639–659.

Floran, R. J., Simonds, C. H., Grieve, R. A. F., Phinney, W. C., Warner, J. L., Rhodes, M. J., Jahn, B. M., and Dence, M. R.: 1976, Petrology, structure and origin of the Manicouagan melt sheet, Quebec, Canada: A preliminary report. *Geophys. Res. Lett.* **3**, 49–52.

Floran, R. J. and Dence, M. R.: 1976, Morphology of the Manicouagan ring-structure, Quebec, and some comparisons with lunar basins and craters. *Proc. Lunar Sci. Conf. 7th*, p. 2845–2865.

Floran, R. J., Grieve, R. A. F., Phinney, W. C., Warner, J. L., Simonds, C. H., Blanchard, D. P., and Dence, M. R.: in press, Manicouagan impact melt, Quebec. Part I: Stratigraphy, petrology, and chemistry. *J. Geophys. Res.*

Gault, D. E. and Heitowit, E. D.: 1963, The partition of energy for hypervelocity impact craters formed in rock. *Proc. Sixth Symp. on Hypervelocity Impact*, p. 420–456.

Grieve, R. A. F.: 1975, Petrology and chemistry of the impact melt at Mistastin Lake Crater, Labrador. *Bull. Geol. Soc. Amer.* **86**, 1617–1629.

Grieve, R. A. F., Dence, M. R., and Robertson, P. B.: 1976, The generation and distribution of impact melts: Implications for cratering processes (abstract). In *Papers Presented to the Symposium on Planetary Cratering Mechanics*, p. 40–42. The Lunar Science Institute, Houston.

Grieve, R. A. F. and Floran, R. J.: in press, Manicouagan impact melt, Quebec, part II: Chemical interrelations with basement and formational processes. *J. Geophys. Res.*

Hartung, J. B., Dence, M. R., and Adams, J. A. S.: 1971, Potassium-Argon dating of shock metamorphosed rocks from the Brent Impact Crater, Ontario. *J. Geophys. Res.* **76**, 5437–5448.

Hörz, F.: 1965, Untersuchungen an Riesgläsern. *Beitrage Mineral. Petrog.* **11**, 621–661.

Hörz, F., Gibbons, R. V., Hill, R. E., and Gault, D. E.: 1976, Large scale cratering of the lunar highlands: Some Monte Carlo model considerations, *Proc. Lunar Sci. Conf. 7th*, p. 2931–2945.

Jones, E. M. and Sandford, M. T.: 1976, Numerical simulation of a very large explosion at the earth's surface with possible applications to tektites (abstract). In *Papers Presented to the Symposium on Planetary Cratering Mechanics*, p. 64. The Lunar Science Institute, Houston.

Kieffer, S. W.: 1971, Shock metamorphism of the Coconino Sandstone at Meteor Crater, Ariz. *J. Geophys. Res.* **76**, 5449–5473.

Masaytis, V. L.: 1975, Astroblemes in the Soviet Union. *Soviet Geology*, no. 11, p. 52–64.

Masaytis, V. L.:, Mikhaylov, M. V., and Selivanovskaya, T. V.: 1975, *The Popigay Meteorite Crater.* Nauka Press, Moscow. 124 pp.

Melosh, H. J.: 1976, Crater modification by gravity: a mechanical analysis of slumping (abstract). In *Papers Presented to the Symposium on Planetary Cratering Mechanics*, p. 76–78. The Lunar Science Institute, Houston.

Nininger, H. H.: 1954, Impactite slag at Barringer Crater. *Amer. J. Sci.* **252**, 277–290.

Oberbeck, V. R., Morrison, R. H., and Hörz, F.: 1975, Transport and emplacement of crater and basin deposits. *The Moon* **13**, 9–26.

O'Keefe, J. D. and Ahrens, T. J.: 1975, Shock effects from a large impact on the moon. *Proc. Lunar Sci. Conf. 6th*, p. 2831–2844.

O'Keefe, J. D. and Ahrens, T. J.: 1977, Impact cratering on the moon. *Phys. Earth Planet. Int.* In press.

Onorato, P. I. K., Uhlmann, D. R., and Simonds, C. H.: 1976, Heat flow in impact melts: Apollo 17 Station 6 Boulder and some applications to other breccias and xenolith laden melts. *Proc. Lunar Sci. Conf. 7th*, p. 2449–2468.

Pohl, J.: 1976, Geophysical measurements in the research drill hole Nordlingen 1973 and the deep structure of the Ries Crater. Oral Presentation at the Symposium on Planetary Cratering Mechanics, Flagstaff.

Pohl, J., Stöffler, D., Gall, H., and Ernstson, K.: 1977, The Ries Impact Crater. In *Impact and Explosion Cratering* (D. J. Roddy, R. O. Pepin, and R. B. Merrill, eds.), Pergamon Press. This volume.

Roddy, D. J., Boyce, J. M., Colton, G. W., and Dial, A. L.: 1975, Meteor Crater, Arizona, rim drilling with thickness, structural uplift, depth, volume and mass—balance calculations. *Proc. Lunar Sci. Conf. 6th*, p. 2621–2644.

Simonds, C. H.: 1975, Thermal regimes in impact melts and the petrology of the Apollo 17, Station 6 boulder. *Proc. Lunar Sci. Conf. 6th*, p. 641–672.

Simonds, C. H., Warner, J. L., and Phinney, W. C.: 1976, Thermal regimes in cratered terrain with emphasis on the role of impact melt. *Amer. Mineral.* **61**, 569–577.

Simonds, C. H., Warner, J. L., Phinney, W. C., and McGee, P. E.: 1976, Thermal model for impact breccia lithification: Manicouagan and the moon. *Proc. Lunar Sci. Conf. 7th*, p. 2509–2528.

Simonds, C. H., Phinney, W. C., and Warner, J. L.: 1977, Effect of water on cratering: A review of craters and impactites on the earth, moon and Mars (abstract). In *Lunar Science VIII*, p. 874–876. The Lunar Science Institute, Houston.

Stähle, V.: 1972, Impact glasses from the suevite of the Nördlinger Ries. *Earth Planet. Sci. Lett.* **17**, 275–293.

Stöffler, D., Gault, D. E., Wedekind, J., and Polkowski, G.: 1975, Experimental hypervelocity impact into quartz sand: Distribution and shock metamorphism of ejecta. *J. Geophys. Res.* **80**, 4062–4077.

Stöffler, D.: 1976, Ries deep drilling: Fallback breccia profile and structure of the crater basement (abstract). In *Papers Presented to the Symposium on Planetary Cratering Mechanics*, p. 136–138. The Lunar Science Institute, Houston.

Stöffler, D., Pohl, J., and Gall, H.: 1976, The Ries impact crater. Oral Presentation at the Symposium on Planetary Cratering Mechanics, Flagstaff.

Sweeney, J. F.: in press, Gravity study of great impacts. *J. Geophys. Res.*

von Englehardt, W.: 1967, Chemical composition of Ries glass bombs. *Geochim. Cosmochim. Acta* **31**, 1667–1689.

Wolfe, S. H.: 1971, Potassium-Argon ages of the Manicouagan-Mushalagan Lakes structure. *J. Geophys. Res.* **76**, 5424–5436.

Yurk, Yu. Yu., Yeremenko, G. K., and Polkanov, Yu. A.: 1975, The Boltysh depression: a fossil meteorite crater. *Internat. Geol. Rev.* **18**, 196–202.

Roddy, D. J., Pepin, R. O., and Merrill, R. B., editors.
(1977) *Impact and Explosion Cratering*, Pergamon Press (New York), p. 791–814.
Printed in the United States of America

Cratering processes: As interpreted from the occurrence of impact melts*

R. A. F. Grieve, M. R. Dence, and P. B. Robertson

Earth Physics Branch, Department of Energy, Mines, and Resources, Ottawa, Canada K1A OY3

Abstract—Impact melts are a common product at hypervelocity impact structures and as such provide information on the cratering process. Their principal occurrences are as small bodies in mixed breccia deposits, either as ejecta or within the final cavity, and as coherent sheets lining the crater floor. Both types of melt are compositionally homogeneous and can be modelled as a mixture of the target lithologies close to the point of impact. A model based on field and petro-chemical observations on impact melts but considered consistent with theory and experiment is presented for the cratering process. Compositional homogeneity is achieved during the excavation stage and is due to the dynamic conditions accompanying melt generation and movement, with the turbulent flow of super-heated, low viscosity melt containing internal velocity gradients produced by pressure attenuation and differential target response. At the termination of excavation the bulk of the melt has been ejected to be widely dispersed or form suevite breccias. The remainder forms a lining to the transient crater and has an internal structure resulting from variations in the number and type of lithic inclusions incorporated during movement. These relationships are disrupted by slope failure, during modification in simple craters, to give rise to a basal melt pool and overlying mixed breccia deposits interspersed with allochthonous basement blocks and clastic breccias derived from the collapse of the transient cavity wall. Base failure in complex structures allows the preservation of the original melt stratigraphy and the major modifications occur due to minor slope failure associated with the development of a central uplift and/or basement rings. This produces a thin mixed breccia deposit overlying the melt sheet within the trace of the transient cavity. The genesis of these breccias is considered equivalent to that of the slumped mixed breccias within simple craters and not the product of fallback as interpreted by other workers. Comparisons between the melt volumes and their relative distributions at simple and complex structures indicate apparent anomalies. These may be explained in terms of the presented model and result from greater particle velocity attenuation in large structures, with consequently decreasing excavation efficiency with increasing size, and to changes in the style of post-growth modification with cavity size.

Introduction

Impact melted lithologies constitute approximately 30% of the returned sample from the heavily cratered lunar highlands (Simonds *et al.*, 1976a) and have been recognized at over 60% of the known terrestrial impact structures. They are associated with both simple bowl-like craters and complex central uplift or ring structures and, with the possible exception of carbonate lithologies, are produced in targets of all compositions (Dence, 1971). Their apparent absence at some terrestrial structures can be generally attributed to extensive erosion or the current reconnaissance status of investigations at a particular site. Consequently, as a common lithology, the observed character, distribution, and

*Contribution from the Earth Physics Branch No. 660.

relative volumes of impact melt provide important information on the processes operating during a hypervelocity impact event.

In this work we examine some of the general properties of impact melts and in combination with experimental, theoretical and other field observations develop a model for their formation and movement during the compressional, excavation, and modification stages of crater formation (Gault *et al.*, 1968).

OCCURRENCE AND DISTRIBUTION

The volumetrically important modes of occurrence of impact melted material are as discrete glassy to very fine-grained crystalline masses in mixed breccia deposits and as fine- to medium-grained, igneous-textured, sub-horizontal sheets. The overall field relations and petrographic character of these types of melt are described in Dence (1971) and only the major points pertinent to constraining models on cratering mechanics are reemphasized or amplified here.

Coherent melt sheets or their remnants are exposed at moderate to deeply eroded complex structures, such as West Clearwater, Mistastin, and Manicouagan, and are known from drilling at well preserved simple craters such as Brent (Table 1). The field relations at the different structural types differ principally in that at simple structures the melt forms a pool at the base of the cavity, whereas at complex structures it has the form of an annulus around the central uplift (Fig. 1a,b).

In both structural types, the general upward succession is fractured and/or shocked country rock followed by a layer of mixed breccia, varying in thickness from a few centimeters to meters within a single structure, which terminates in a sharp contact with overlying very fine-grained basal melt (Fig. 2a) containing country rock inclusions, some of which show shock features (Dence, 1964; Grieve, 1975; Floran *et al.*, 1976, 1977). At structures with an extensive exposed melt sheet,

Table 1. Principal facts on impact structures mentioned in text.

	Lat.	Long.	Present diameter (km)	Age (m.y.)	Structural type
Monturaqui, Chile	23°56′S	68°17′W	0.5	<1	Simple
Lonar Lake, India	19°59′N	76°51′E	1.8	0.05	Simple
Tenoumer, Mauritania	22°55′N	10°24′W	1.8	2.5 ± 5	Simple
West Hawk, L., Manitoba	49°46′N	95°11′W	3	100 ± 50	Simple
Brent, Ontario	46°05′N	78°29′W	4	450 ± 40	Simple
Deep Bay, Saskatchewan	56°24′N	102°59′W	12	100 ± 5	Complex—central uplift
E. Clearwater L., Quebec	56°05′N	74°07′W	22	290 ± 20	Complex—central uplift
Ries, W. Germany	48°53′N	10°37′E	24	14.8 ± 0.7	Complex—ring
Mistastin L., Labrador	55°53′N	63°18′W	28	38 ± 4	Complex—central uplift
W. Clearwater L., Quebec	56°13′N	74°30′W	32	290 ± 20	Complex—ring
L. St. Martin, Manitoba	51°47′N	98°33′W	38	225 ± 40	Complex—ring
Manicouagan, Quebec	51°30′N	68°30′W	75	214 ± 5	Complex—ring
Popigai, U.S.S.R.	71°30′N	111°30′E	100	30 ± 7	Complex—ring

the basal unit is observed to grade upwards into fine–medium grained melt with well-developed cooling fractures (Fig. 2b) which passes into coarser grained, weakly jointed melt. Inclusions are less conspicuous in these upper units but are nevertheless recognizable microscopically as relicts of refractory unshocked minerals as high as 105 m above the base and as occasional meter-sized blocks of weakly shocked basement tens of meters above the base of the melt (Fig. 2c) (Grieve, 1975; Masaitis *et al.*, 1975; Simonds *et al.*, 1976a).

The melt sections at West Clearwater, Mistastin, and Manicouagan are erosional remnants with preserved thicknesses of 130, 80, and 230 m, respectively (Dence, 1971; Floran *et al.*, 1976; Grieve, 1975). Drilling at East Clearwater and Lake St. Martin (Dence *et al.*, 1965; McCabe and Bannatyne, 1970) indicates that the coarse-grained, inclusion-poor melt unit is capped by an uppermost melt unit (approximately 30 m thick at E. Clearwater) which shows a reduction in grain size and complimentary increase in the number of megascopic inclusions; the reversal of the sequence observed at the base of the melt sheet. The original lateral extent of the melt sheets at terrestrial complex structures is generally unknown because of erosion. However, it is apparent from the present outcrop distribution that the melt forms not only a lining to the cavity floor, but also extends beyond the trace of the transient cavity rim (Fig. 1a), as estimated from morphological and structural data (Dence, 1971; Floran and Dence, 1976).

Significant volumes of suevite, a melt bearing mixed breccia, are exposed at slightly eroded structures. The type locality is the Ries (Hörz, 1965; von Engelhardt *et al.*, 1969), although suevite deposits are also well developed at the Popigai structure (Masaitis *et al.*, 1975). Another variety of melt-bearing mixed breccia, which is confined to within the transient cavity and represents moved but not ejected material, occurs as dike-like bodies in the cavity floor (Currie, 1970, 1972), as a layer between the basement rocks and the melt sheet (Grieve, 1975; Yurk *et al.*, 1975), and as complex intercalations with weakly to unshocked allochthonous clastic breccias and basement blocks. The latter is well documented from drilling at Brent (Dence, 1965), West Hawk Lake (Short, 1970), Deep Bay (Dence *et al.*, 1968), and the Ries (Stettner, 1974), and with the associated clastic debris constitutes the bulk of the material filling the crater. The melt-bearing mixed breccias are volumetrically less important than the clastic breccias and basement blocks and although occurring at all depths, are concentrated in the upper portion of the crater fill, which in the case of Brent overlies an identifiable melt layer (Fig. 1b).

Mixed breccias of this type, which occur within the cavity and both overlie and underlie the melt sheet, have also been called suevite (Currie, 1970, 1972; Bostock, 1969; Masaitis *et al.*, 1975; Phinney and Simonds, 1976; von Engelhardt and Graup, 1977). They have many similarities to suevite ejecta, the most obvious of which is the occurrence of twisted glass fragments with inclusions of country rock, some of which show shock damage, in a more clastic matrix (Fig. 2d). However, there are important differences. They lack melt bombs of the "fladen" type with true aerodynamic shapes (Hörz, 1965; Masaitis *et al.*, 1975) and are relatively melt-poor. In the case of the Ries, they contain only 8% by

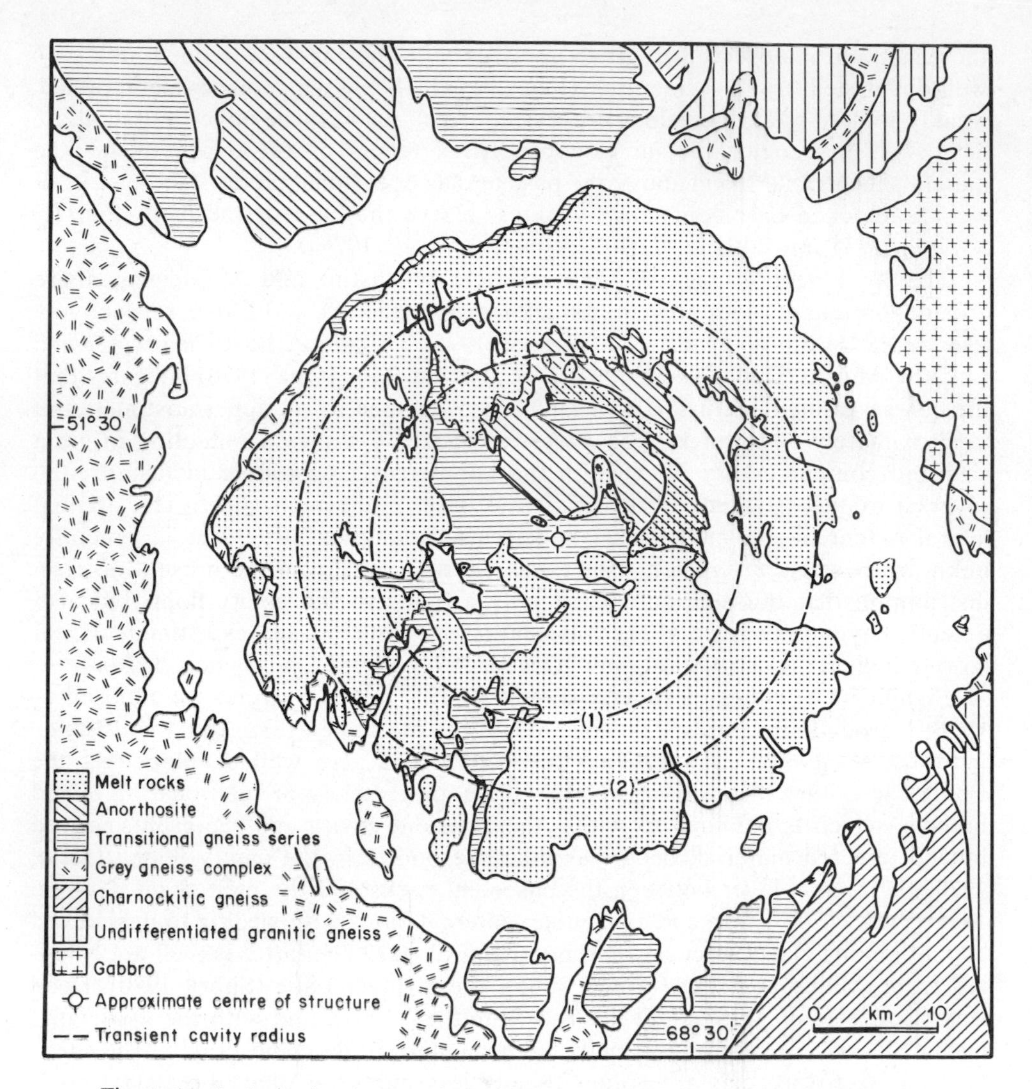

Fig. 1. (a) Geologic sketch map of Manicouagan complex impact structure, Quebec, modified from Currie (1972), Murtaugh (1975). Coherent melt sheet overlies basement and extends beyond rim of transient cavity, estimated radius (1) 15 km, (2) 22 km (Floran and Dence, 1976).

volume of impact melt, approximately 50% of that observed in suevite ejecta (Stöffler, 1977; Stöffler *et al.*, 1977).

In this work and the classic occurrences at the Ries suevite throw out ejecta and superficially similar suevite material forming mixed breccias within the cavity have been discriminated (Chao, 1976; Stöffler, 1977). However, the unqualified use of the term suevite at other impact structures has resulted in what are considered to be unwarranted arguments and conclusions regarding the physical conditions accompanying the formation and distribution of melt-bearing

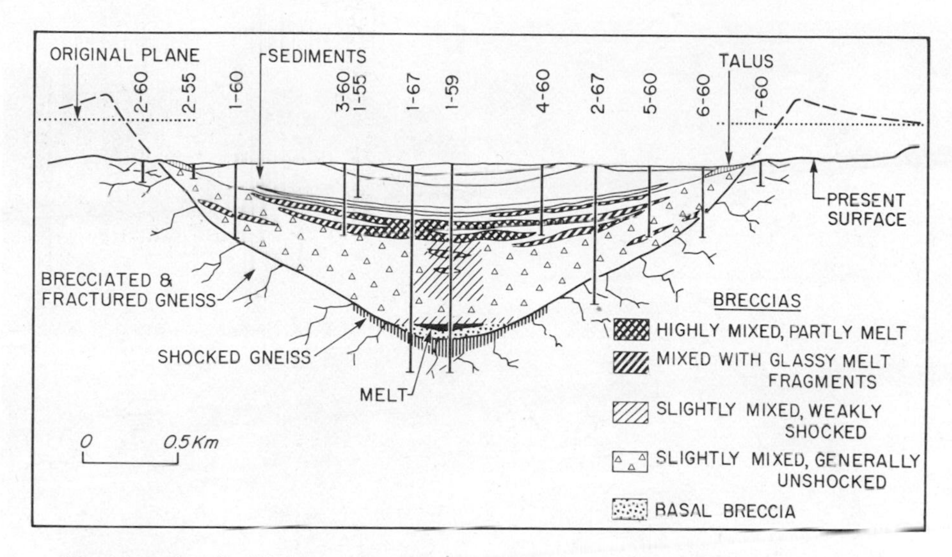

Fig. 1 (*continued*). (b) Schematic geologic cross-section of Brent simple crater, Ontario. Section based on drill hole information from locations shown. Note small basal melt pool and melt-bearing mixed breccias, which associated with essentially clastic material constitute the crater fill products.

breccias within the cavity (Currie, 1970, 1972; Bostock, 1969). Typical is the work of Masaitis *et al.* (1975) on Popigai, where both melt-rich, fladen-bearing and melt-poor, lithic or mineral clast-rich, fladen-absent breccias are recognized. These breccias occur as true ejecta outside the trace of the transient cavity, underlie and overlie the melt sheet (known as tagamites) within the cavity and are intimately associated with allochthonous clastic breccias and basement blocks. However, the unqualified designation of suevite to all of these various mixed breccias (Masaitis *et al.*, 1975) with their different modes of occurrence results in problems in the interpretation of the detailed stratigraphy of the impact products in terms of a cratering model (Phinney and Simonds, 1976). The petrographic and textural differences between suevite breccias: as ejecta and deposits within the cavity (Stöffler, 1977), reflect differences in their physical movement during the excavation and modification stages of crater development and, as discussed later, the recognition of these differences is fundamental to interpreting these melt-bearing lithologies in terms of a consistent model of cratering processes.

Compositional Characteristics

Analyses of fresh samples from extensive coherent melt sheets indicate a high degree of compositional homogeneity (Table 2). Compositional divergence from the average can generally be traced to local variations in the relative abundance of various types of basement inclusions and the degree of assimilation they have undergone (Floran *et al.*, 1976, 1977; Grieve, 1975). As expected, these

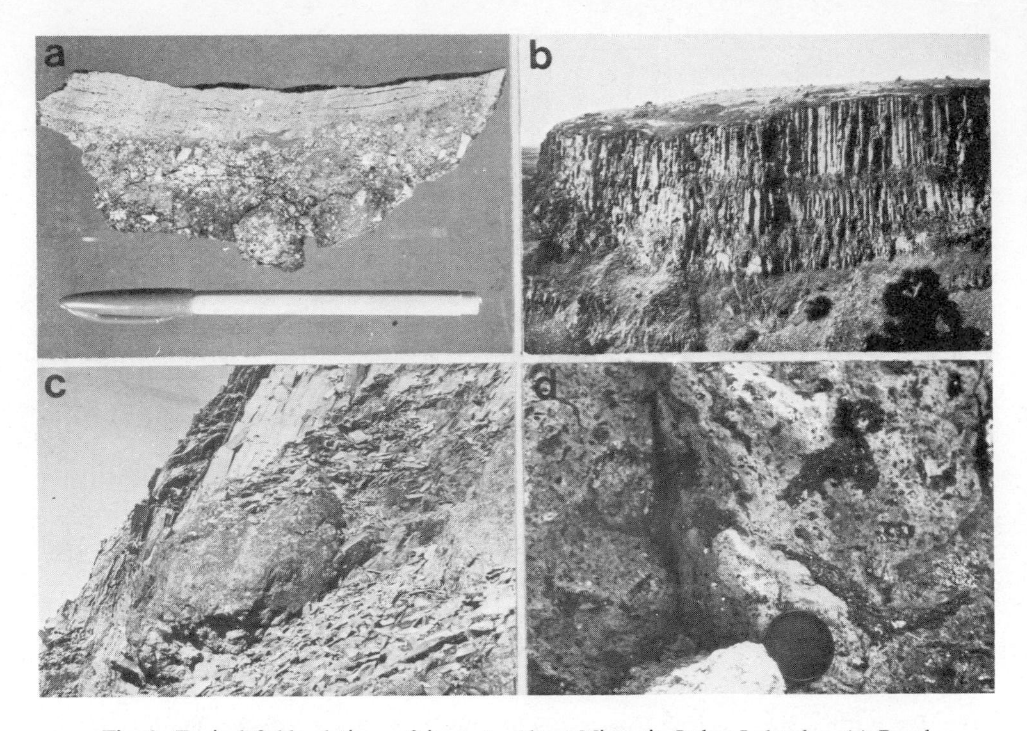

Fig. 2. Typical field relations of impact melt at Mistastin Lake, Labrador. (a) Basal contact. Slightly vesicular, very fine-grained melt (light gray) with sub-horizontal cooling fractures overlying mixed breccia. Melt is chilled (dark gray selvage) against breccia, which contains both clastic debris and inclusion-rich melt clasts (light gray). Pen length is 15 cm. (b) Fine–medium grained coherent melt sheet with vertical cooling fractures. Outcrop is 80 m high. (c) Weathered-out inclusions of basement (center, far right–center) several meters in max. dimension, 40 m above exposed basement of cavity floor. (d) Mixed breccia with contorted melt glass fragments (dark gray) in a clastic (light gray) matrix. Breccia never achieved free-flight and is in form of 1–2 m wide, sub-vertical dike cutting basement and lying topographically below nearest exposed melt. Breccias such as this are termed "suevite" by some workers.

variations are most common in the inclusion-rich basal melt units in which chemical equilibrium was arrested due to rapid cooling. Compositional variations in melt from simple structures with only a small amount of melt, such as Lonar Lake, Monturaqui, and Tenoumer, have also been noted and ascribed to incomplete melting and mixing due to reaction kinetics and variations in the local concentration of particular relict mineral and lithic clasts (Bunch and Cassidy, 1972; Fudali, 1974; Schaal *et al.*, 1975). As indicated in Table 2, these variations are minor and melt sheets are in general remarkably homogeneous in composition, given the volumes of melt involved and the postulated diversity of rock types in the target (Grieve, 1975; Simonds *et al.*, 1976a).

Compositional variance due to alteration is encountered in some melt rocks. It generally takes the form of hydration, oxidation, or alteration to minerals such as montmorillonite clays, with the resultant changes in composition being highly

Table 2. Variation in composition of melt at some impact structures.

Structure	Manicouagan		W. Clearwater		Mistastin		Popigai				Ries	
Form	Sheet		Sheet		Sheet		Sheet		Suevite		Suevite	
No. samples	50		14		75		22		14		46	
	$\bar{X}$	s.d.	$\bar{X}$	s.d.	$\bar{X}$	s.d.	$\bar{X}$	s.d.	$\bar{X}$	s.d.	$\bar{X}$	s.d.
SiO_2	57.75	1.56	59.35	1.88	56.15	2.01	63.13	1.22	63.11	4.03	63.99	1.31
TiO_2	0.77	0.08	0.82	0.13	0.94	0.19	0.76	0.15	0.84	0.23	0.79	0.09
Al_2O_3	16.51	0.96	16.06	0.58	20.11	1.94	14.68	0.53	13.39	1.12	15.32	0.34
FeO*	5.87	0.45	3.83	0.41	5.53	1.36	6.75	0.87	7.02	2.04	5.14	0.40
MnO	0.11	0.03	0.06	0.02	0.09	0.03	0.08	0.03	0.07	0.04	0.07	0.01
MgO	3.50	0.60	2.62	0.57	1.29	0.56	3.82	0.52	3.17	0.66	3.01	0.27
CaO	5.92	0.91	4.51	0.51	7.54	0.97	3.43	0.99	3.70	1.14	3.85	0.38
Na_2O	3.82	0.24	3.89	0.40	4.39	0.37	1.96	0.29	1.73	0.31	3.01	0.39
K_2O	3.03	0.37	3.47	0.45	1.82	0.47	2.72	0.23	2.54	0.38	4.20	0.60

*Total iron recalculated as FeO.

Data sources: Manicouagan—Currie, 1972; Floran *et al.*, 1976, 1977, data standardized to remove inter-laboratory bias (Grieve and Floran, 1977): W. Clearwater—Bostock, 1969: Mistastin—Currie, 1970; Grieve, 1975; Marchand, 1976, relatively high variation result of sample weighted towards inclusion-rich basal melt and inter-laboratory bias: Popigai—Masaitis *et al.*, 1975: Ries—von Engelhardt, 1972; Stähle, 1972.

variable (Dence *et al.*, 1974; von Engelhardt, 1967). A more severe form of alteration is evidenced at Brent where melt plagioclase is partially replaced by potash feldspar. This alteration is also recognized in the overlying mixed breccias and is attributed to the interaction of solutions heated by the high residual temperatures present following melt emplacement and crater formation (Hartung *et al.*, 1971). The overall effect of these "late" alteration phenomena is to complicate the chemical relationships of the melt. They introduce a cautionary note to conclusions about the compositional characteristics of impact melts in general and have been used as an argument against the impact origin for the melt at some shock-metamorphosed structures (Currie, 1971).

As with coherent melt sheets, the melt in the Ries suevite ejecta is characterized by little compositional variance (Stähle, 1972; von Engelhardt, 1972). The "suevite" at Popigai is compositionally more diverse (Table 2). This may be the result of differences in analytical methods, with the Ries data coming from microprobe analyses of inclusion-free melt glass (Stähle, 1972) and the Popigai data from wet chemical analyses on glass with 15–20% crystalline inclusions (Masaitis *et al.*, 1975). In general, the composition of the melt in suevite and other mixed breccias is similar to that of the associated coherent melt sheet (Bostock, 1969; Grieve, 1975; Masaitis *et al.*, 1975). However, occasional glass fragments in the mixed breccias underlying the melt sheet and ejected glass bombs have compositions more representative of individual analyses of basement lithologies (Table 3) than the average composition of the melt sheet (Grieve, 1975; Grieve *et al.*, 1976).

The relationship of the average major element composition of the melt with the surrounding and underlying basement rocks is shown in Fig. 3 for the melt

Table 3. Compositional comparison of rare melt fragments in mixed breccias with individual basement rocks at Mistastin Lake impact structure.

	Melt (1)	Mangerite	Melt (2)	Anorthosite
SiO_2	61.8	61.8	54.1	56.4
TiO_2	1.2	1.2	0.7	0.4
Al_2O_3	15.1	14.3	24.8	22.9
FeO*	8.0	8.6	2.8	2.6
MnO	0.1	0.1	0.04	0.08
MgO	0.7	1.0	0.8	1.4
CaO	4.5	4.5	8.2	10.0
Na_2O	4.2	3.7	5.6	4.9
K_2O	3.6	3.2	1.2	1.1

*Total iron recalculated as FeO.

Melt(1)—Float sample of glassy melt with aerodynamic shape. Melt(2)—Glassy melt clast in mixed breccia underlying melt sheet (Grieve, 1975, unpublished data). Mangerite—average. Anorthosite—average from central uplift (Marchand, 1976).

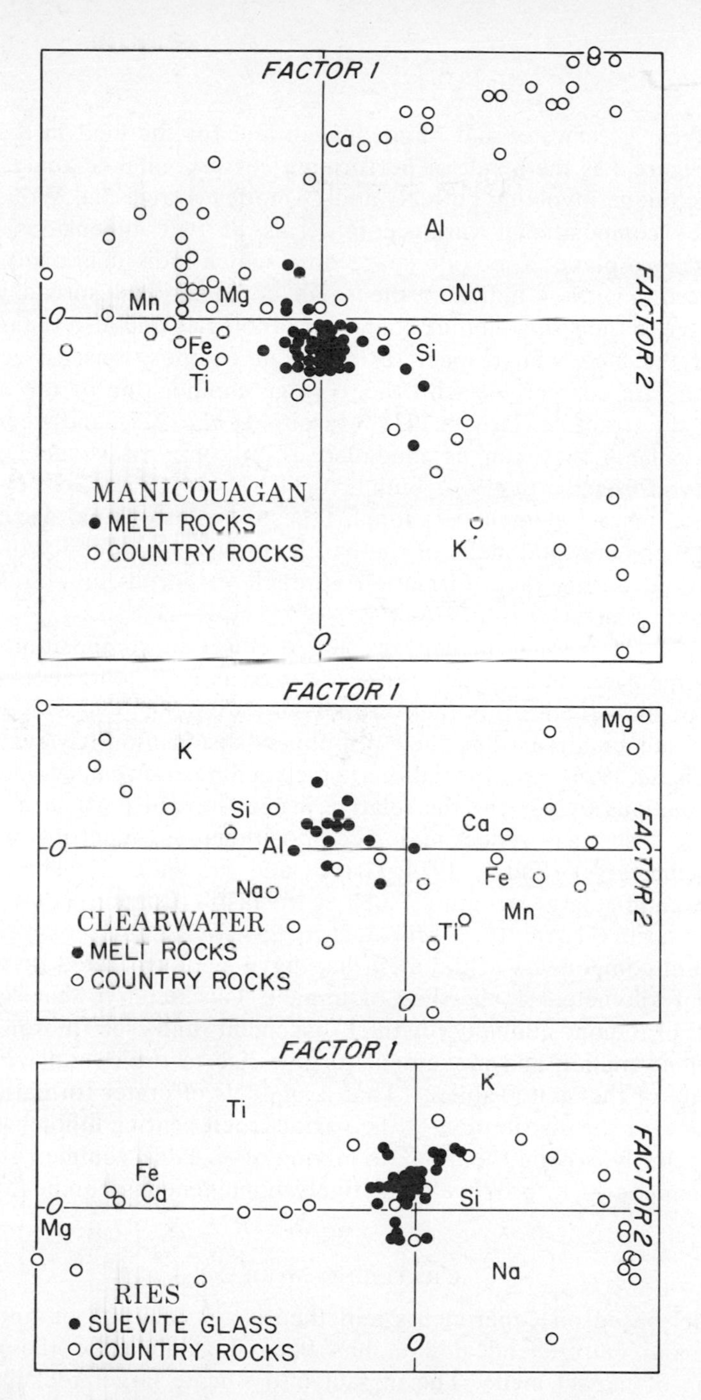

Fig. 3. Results of correspondence analysis on melt and associated basement at Manicouagan, W. Clearwater and Ries. Data from Currie (1970), Floran *et al.* (1977), Bostock (1969), von Engelhardt (1972), and Stähle (1972). Note close compositional clustering of melt and their intermediate position with respect to basement compositions.

sheets at West Clearwater and Manicouagan and for the melt in the suevite at the Ries. Figure 3 is the result of performing correspondence analysis, a factor analysis technique involving both *R*- and *Q*-modes (David and Woussen, 1973). It illustrates compositional variance in terms of two dimensions, which are represented as a plane in n-oxide space onto which individual analytical points are projected. Figure 3 indicates the small compositional spread within melt rocks relative to their surrounding basement lithologies and also the intermediate position of the melt with respect to basement compositions. In general, melt compositions are comparable with the average composition of the basement in the area of the structure (Dence, 1971; Masaitis *et al.*, 1975), and where sufficient data are available they can be modelled as mixtures of particular basement lithologies(von Engelhardt, 1972; Fudali, 1974; Grieve, 1975; Grieve and Floran, 1977). Rb/Sr and trace element data support these conclusions and also indicate that the impact melts are total melts of the basement with no evidence of substantial partial melting during their formation (French *et al.*, 1970; Marchand, 1976; Rondot, 1971; Winzer *et al.*, 1976).

The results of mixing calculations at structures in compositionally diverse basement lithologies indicate that those required as melt components outcrop at the center of the structure and their proportions are in keeping with geologic and structural arguments regarding their pre-impact distribution (Grieve and Floran 1977; Marchand, 1976). Local variations in melt composition can be accommodated in the calculations by varying the relative proportions of particular components and in some cases may reflect major compositional asymmetries in the melted volume of the target (Fudali, 1974; Grieve and Floran 1977). The present outcrop distribution at large structures such as Mistastin (Currie, 1970; Grieve, 1975) and Manicouagan (Currie, 1972; Murtaugh, 1975; Fig. 1a, this work) suggests that the basement components of the melt may have been separated by distances of the order of kilometers at the time of impact. This requires that several cubic kilometers of compositionally distinct basement units be thoroughly mixed during melt formation and movement to give rise to the overall compositional homogeneity of the melt (Table 2). That is, models of crater formation must not only account for the distribution of the various melt bearing lithologies, but must also satisfactorily explain the intimate mixing of sizeable volumes of chemically distinct components to produce a relatively homogeneous composition.

CRATERING MODEL

A model based on experiments and theoretical calculations and consistent with field and petro-chemical data has been developed for the genesis and distribution of impact melts. The impact into silicate target rocks of either an iron or stony projectile travelling at typical velocities of 15–20 km/sec produces an attenuating shock wave with initial pressures of several hundred GPa, 1 GPa = 10 kb, (Shoemaker, 1960). Thermodynamic calculations and Hugoniot data indicate that the increase in internal energy accompanying compression and subsequent pressure release by rarefaction is sufficient to melt and/or vaporize a small volume of the target close to the point of impact (Ahrens and O'Keefe, 1972). The lower limit of melting, based on numerous shock recovery experiments, corresponds to peak pressures of 50–80 GPa and residual temperatures of the order of 1600°K (DeCarli and Jamieson, 1959; Gibbons *et al.*, 1975; Stöffler and Hornemann, 1972; Wackerle, 1962). From nuclear explosion data, the upper limit of melting and onset of vaporization is considered to occur at pressures above approximately 200 GPa in sialic crystalline rocks (Butkovich, 1965; Cherry and Petersen, 1970) with corresponding temperatures in the range 3000–4000°K (Stöffler, 1971; Ahrens and O'Keefe, 1972).

In addition to melting and vaporizing a portion of the target, the shock wave imparts a particle velocity to the shocked material. During compression, velocities are radial in direction but are subsequently deflected outwards and upwards by the interaction of rarefaction waves. It is these deflected particle motions that are responsible for the growth of the transient cavity during the excavation stage of crater development (Gault *et al.*, 1968). Velocities are a function of shock pressure and rock type (Fig. 4) and for an impact into a granitic target initial velocities of the order of 6.5 and 3.2 km/sec are expected at pressures of 200 and 60 GPa, respectively (Altschuler and Sharipdzhanov, 1971). In a heterogeneous target containing, for example, ultramafic rocks in addition to granite, velocities in the ultramafic rocks are approximately 1 km/sec lower than in

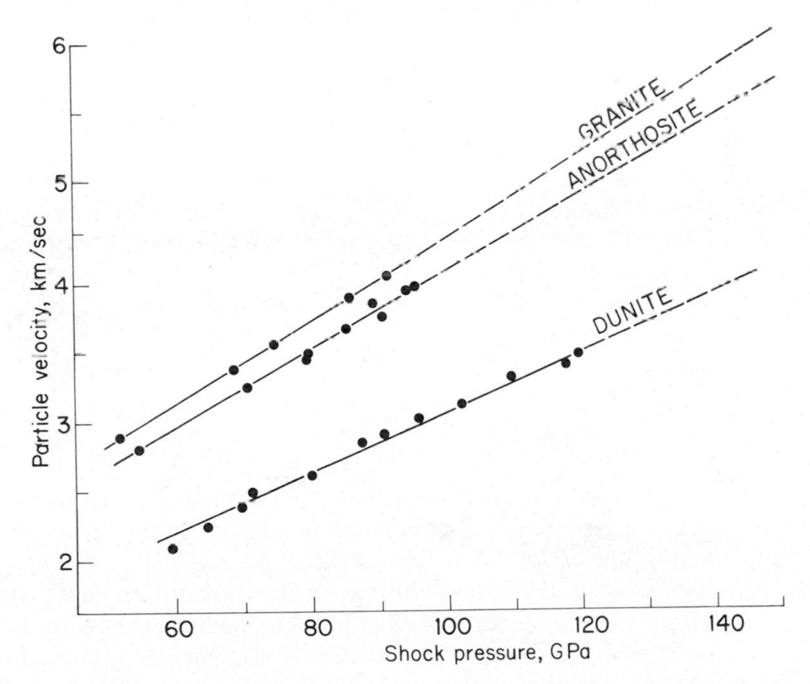

Fig. 4. Variation in particle velocity with shock pressures above those considered necessary for melting of granite, anorthosite, and dunite. Data from McQueen *et al.* (1967). Note variation in velocity with pressure and rock type.

the granite at pressures of 60 GPa (Fig. 4) (McQueen *et al.*, 1967). Therefore pressure attenuation and lithologic heterogeneities cause differential initial particle velocities of several km/sec in the highly shocked part of the target. After rarefaction, both the direction and speed of the particles are changed. However the interaction of the rarefaction wave will not remove the relative differences in velocity within the now melted/vaporized volume.

Complex numerical models suggest that the melt/vapor volume may have the approximate geometry of a truncated sphere with its center below the original ground surface (O'Keefe and Ahrens, 1975). However, a simpler model which considers a hemispherical volume (Dence, 1971; Gault and Heitowit, 1963) is presented for purposes of illustration (Fig. 5). For an iron projectile travelling at 15 km/sec peak shock pressures are approximately 450 GPa. Although the pressure attenuation rate is not constant throughout the total shocked volume, decreasing with increasing distance from the shock point, a minimum rate of the order of the third root of the radial distance has been assumed in the presented model for the melt/vapor volume (Dence *et al.*, 1977; Gault and Heitowit, 1963; Shoemaker, 1960; Stöffler, 1977). Differential velocities of several kilometers per second occur on compression and following rarefaction temperature variations of several thousand degrees Kelvin are established in the now molten and partially vaporized volume (Fig. 5a). The average temperature of the melt is conservatively estimated at 2500°K. This is on the basis of a post-shock temperature attenuation rate of r^{-1} derived from shock recovery experiments (lower temperature-pressure limit) and nuclear explosion data (upper temperature-pressure limit) and the contribution of 25% of the overlying high-temperature, >4000°K, silicate vapor which is moving with high relative velocity and is being incorporated into the melt during movement into the growing transient cavity (Fig. 5a). Temperature decay rates as high as r^{-2} can be calculated from the data of Wackerle (1962) and if the lower limit of melting is considered as the best documented point on the decay curve, then the average temperature of the melt is well in excess of 2500°K. At these decay rates, the volume of vapor relative to the melt also increases. Irrespective of the mean temperature, it is apparent that on formation the melt is highly super-heated. This is consistent with the occurrence of lechatelierite, baddeleyite, and other oxides, which have formation temperatures of approximately 2000°K (Levin *et al.*, 1956), as unmixed relict phases in quenched glassy melts (Chao, 1968; El Goresy, 1968).

Density and viscosity gradients will also be established and extremely low average viscosities of less than 10 poise would be expected in a silicate melt with an average temperature of 2500°K (Shaw, 1972). Low viscosities in combination with high velocities and the large volume of melt result in large Reynolds numbers and thus the movement of the melt is highly turbulent. It is this dynamic condition of super-heated melt, with low viscosity and internal velocity gradients, moving in a turbulent flow field that produces rapid chemical homogenization from a heterogeneous target. Rapid homogenization is evidenced by the small compositional variation within melt ejected during cavity growth (suevite glass) and its chemical equivalence to melt that remained within the structure throughout all stages of crater development (coherent melt sheets) (Table 2). The rare fragments with compositions similar to individual target rocks that have been observed (Table 3) probably represent material that separated from the bulk of the melt early in the cratering process.

A portion of the mixed and melted/vaporized material is ejected from the growing cavity along the projections of the paths indicated in Fig. 5a and is widely dispersed or forms suevite ejecta deposits. The remainder, probably originating from a truncated, roughly conical volume vertically beneath the point of impact, is in turbulent flow and follows overall resultant paths which are originally below the horizontal and generally downward (Fig. 5a). This material is also deflected upwards and outwards parallel to the base of the cavity (Fig. 5a) but it travels over a long path length and particle velocities attenuate to values insufficient for ejection. At this point the excavation stage of crater development has effectively terminated. The melt forms a lining to the transient cavity and the basic stratigraphy of melt overlying mixed breccia with both clastic and melt components and/or basement is established.

In the movement of the melt during cavity growth, its leading edge at any particular time overruns less strongly shocked and therefore slower moving crystalline target material and incorporates it as inclusions showing various degrees of shock deformation. The ejected melt contains 20–30% inclusions, some of which may have been incorporated on landing. Five to ten percent of the inclusions have characteristic shock deformation features up to and including the highest sub-solidus grades (von Engelhardt, 1972). Of the melt that remains within the cavity, the basal unit contains an average of approximately 15% inclusions, which are considered to represent 50% of the original clast content (Simonds *et al.*, 1976b). The addition of 30% inclusions, the bulk of which have a modest temperature derived from the pre-impact geothermal gradient and a weak super-imposed post-shock temperature, to melt at 2500°K or higher effectively lowers the melt temperatures to close to the liquidus. Thermal calculations indicate that, when firmly in contact, equilibration between cold clasts and super-heated melt is an extremely rapid process, occurring in a matter of seconds (Onorato *et al.*, 1976); a time span less than that required to complete the excavation stage at an impact event capable of producing a large complex structure (Gault *et al.*, 1968).

Thus as cavity growth terminates, portions of the basal unit of melt have temperatures more in keeping with those of endogenic silicate liquids and relatively high viscosities due to lowered temperature and the presence of undigested crystalline clasts. The further the melt travels the greater the number of weakly shocked and unshocked inclusions it incorporates and as a result this inclusion-rich basal melt unit is best developed and thickest towards the outer edge of cavity. Melt which is vertically under the point of impact and is originally driven downwards also picks up inclusions. However, these inclusions come from the volume between the melt and the base of the cavity, which was subjected to pressures of 20–30 GPa (Dence *et al.*, 1977), and have post-shock

Fig. 5. Cratering model. (a) Compressional (right) and excavation (left) stages. Impact propagates a shock wave attenuating as r^{-3} and imparts a particle velocity to target. Shock contours are shown and relative differences in particle velocity, U_p, indicated by length of arrows. Deflection of target material by rarefaction waves results in excavation of vapor, melt, and crystalline material. Some of melt forms lining to cavity and overlies mixed breccias. Internal stratigraphy of melt established due to inclusions incorporated during movement. See text for details. Stratigraphic thicknesses exaggerated for illustration. (b) Modification stage. Simple structure (left) undergoes slope failure. Wall of transient cavity slumps inwards to disrupt melt-mixed breccia lining, overrides small basal melt pool, and forms intimate mixture of melt-bearing mixed breccia, clastic breccia and allochthonous basement blocks. Complex structure (right) collapses by base failure with downdrop of rim and uplift of central peak. Melt has a passive role and retains gross stratigraphy. Note order of magnitude difference in scale with simple structure. (c) Final configuration. Simple bowl-like depression (left) with small basal melt layer and filled by melt-bearing mixed breccias and clastic debris from transient cavity wall. Complex structure (right) with shallow depth/diameter. Melt forms lining to floor with basal inclusion-rich melt overlying breccia at outer edge replaced by inclusion-poor basal melt overlying basement towards center. True suevite ejecta is confined to outside the transient cavity.

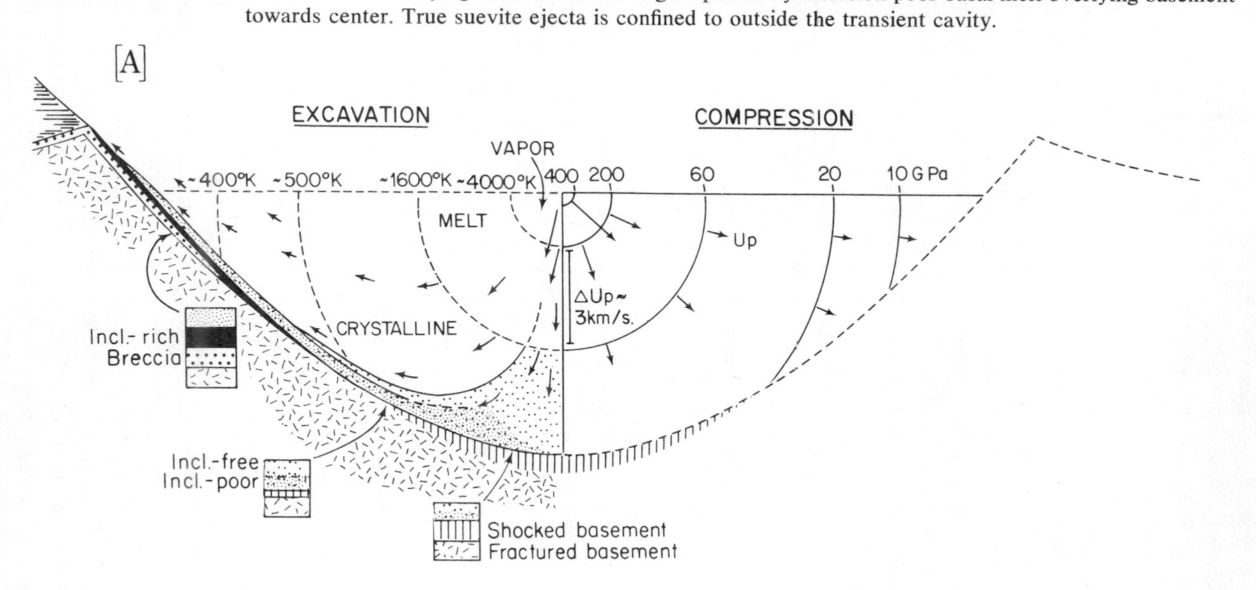

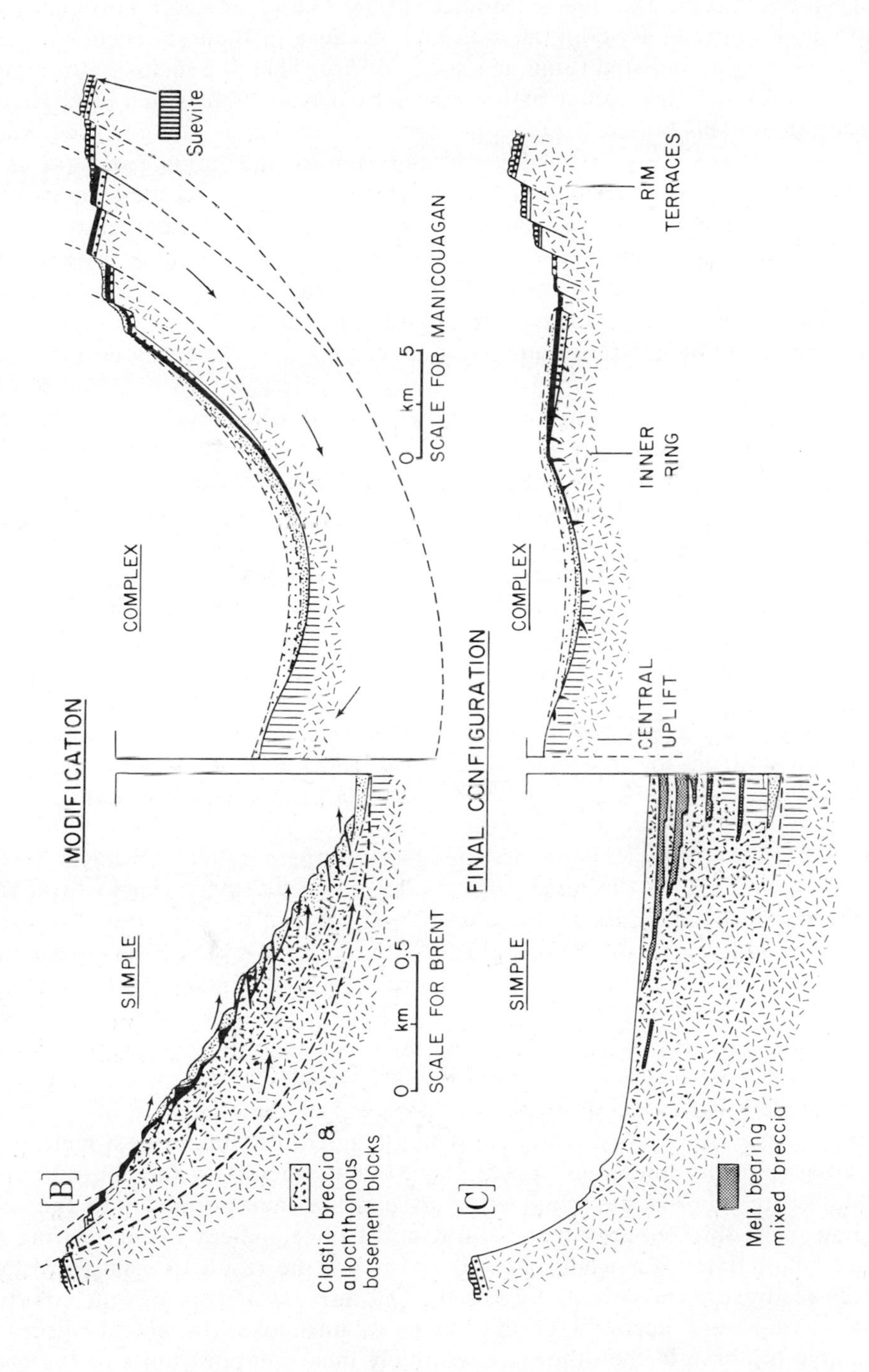
[B]
MODIFICATION
SIMPLE
COMPLEX
Suevite
Clastic breccia &
allochthonous
basement blocks
0 km 0.5
SCALE FOR BRENT
0 km 5
SCALE FOR MANICOUAGAN
[C]
FINAL CONFIGURATION
SIMPLE
COMPLEX
CENTRAL
UPLIFT
INNER
RING
RIM
TERRACES
Melt bearing
mixed breccia

temperatures in excess of 500°K (Stöffler, 1971). These relatively hot inclusions do not significantly under-cool the melt and, because of their shocked state, are relatively easily assimilated (Simonds *et al.*, 1976b). Thus the inclusion-rich melt is replaced towards the center of the cavity by a basal unit which is relatively inclusion-poor. This hotter inclusion-poor melt overrides the inclusion-rich more viscous basal unit developed towards the rim of the cavity and forms a stratigraphically higher layer. Melt from the center and behind the leading edge incorporates relatively few inclusions, retains its super-heated temperatures and is thus able to digest the bulk of the inclusions. It forms a thick layer at the center of the cavity, thinning towards the outer edge and overlying the other inclusion-bearing melt units. Thus a gross lateral and vertical stratigraphy, which is a function of the relative number of inclusions and their degree of shock damage, forms in the melt lining to the transient cavity.

As these internal relationships are being established crater development passes from the excavation to modification stage and the transient cavity undergoes rapid readjustment to produce the final crater form (Fig. 5). For small terrestrial structures (<4 km diameter) in crystalline rocks this takes the form of slope failure. The walls of the transient cavity slump inwards and the melt lining and underlying mixed breccias are swept into the center of the cavity along with weakly to unshocked basement rocks of the transient cavity wall (Fig. 5b). This material forms the bulk of the crater-fill at structures such as Brent (Dence, 1965) and the melt lining the cavity wall no longer has the form of a coherent sheet. It is now intimately mixed with clastic debris to form melt-bearing mixed breccias which are associated with clastic breccias and allochthonous basement blocks derived from the transient cavity wall. Melt lining the base of the cavity is over-ridden by this slumped material and remains as a small basal pool (Fig. 5c).

At larger structures, collapse is principally by base failure (Melosh, 1976), with the down drop of the outer edge of the transient cavity rim to form rim terraces and/or distinct annular grabens and the uplift of the center of the cavity floor as a more or less rigid plug (Fig. 5b). With increasing size of the cavity, this type of collapse leads to the formation of complex central uplift and ring structures (Dence, 1968; Pike, 1971). During collapse, the melt and mixed breccia lining of the transient cavity retains the gross stratigraphy established during cavity growth. The dominant modifications are the piercement of the melt by a horst of underlying basement in the development of the central uplift, and the run back of the low viscosity, relatively high temperature, uppermost melt units into topographically low areas. Local slope failure occurs on the central uplift and blocks of shocked basement spall off and are incorporated into the surrounding melt; this requires that central uplift development occurs during or within a short time after cavity growth and is not the result of relatively slow isostatic readjustment. Slope failure of the rim terraces and, if present, uplifted basement rings may account for the blocks of unshocked basement observed well above the base in the otherwise relatively inclusion-poor units of the melt sheet (Grieve, 1975; Simonds *et al.*, 1976b). Additional mass wasting produces an inclusion-rich uppermost melt unit, such as that observed at E. Clearwater and Lake St. Martin (Dence, 1965; McCabe and Bannatyne, 1970), and overlying mixed clastic-melt breccias. Melt-bearing mixed breccias overlying the melt sheet within the cavity are present at Popigai (Masaitis *et al.*, 1975) and may be represented on the moon by the so-called corrugated facies, which is interpreted as a layer of mixed melt and clastic debris partially capping impact melt, in the interior of complex structures such as Copernicus, Orientale, and Tycho (Head, 1974; Moore *et al.*, 1974). Thus the general sequence: basement, clastic breccia, mixed breccia, melt, mixed breccia, which is preserved at only the least eroded terrestrial complex structures, such as Popigai, is developed. The origin of the overlying mixed breccias at Popigai is considered equivalent to similar breccias found in simple structures, that is they are slump breccias and not fallout ejecta. A similar situation may occur at the Ries with suevite breccias overlying slumped allochthonous basement blocks and clastic breccias inside the trace of transient cavity (von Engelhardt and Graup, 1977; Stöffler, 1977). Above these mixed breccias is a 25 m layer of "sorted suevite", which shows grain size ordering (Stettner, 1974) and may represent true fallback material. Between the "sorted suevite" and the post-impact lake sediments there are a further 50 m of mass wasted and reworked suevite breccia, which slumped into the crater due to minor slope failure and degradation after the deposition of a surficial fallback layer (Stöffler, 1977).

As mentioned earlier, it is believed that the basic stratigraphy of the melt sheet is only slightly modified during the collapse of complex structures. The basic stratigraphy is well documented at Manicouagan. At the outer edge of melt sheet, now lying beyond the original transient cavity rim (Floran and Dence, 1976; Fig. 1a), the unshocked and fractured basement is overlain by up to 30 m of breccia grading upwards from essentially clastic to melt-bearing mixed breccia which passes, with a sharp contact, into an extremely fine-grained, inclusion-rich, basal melt unit (Currie, 1972; Murtaugh, 1975). Towards the center of the structure, both the mixed breccias and inclusion-rich melt are absent and inclusion-poor (<15%) fine-grained or essentially inclusion-absent (<2%) medium-grained melt rests directly on shocked basement. The upper unit is absent at the extreme outer edge of the sheet and is thickest in the center of the structure. The melt which intrudes the central uplift as dikes and sills is of the inclusion-poor, fine-grained variety and the only major reversal in the basic lateral and vertical variation in clast content and grain-size occurs in melt adjacent to the central peaks of anorthosite. This melt contains numerous shocked anorthosite inclusions and is relatively finer grained. However, it contains few inclusions of other basement lithologies and represents high-temperature, inclusion-absent melt modified by the incorporation of anorthosite blocks, produced by spalling during and immediately after central uplift formation.

VOLUMES OF MELT: IMPLICATIONS FOR EXCAVATION EFFICIENCY

Comparison between small simple craters (<4 km) and larger complex structures suggests that the relative volume and distribution of melt varies with crater size. As developed above, simple and complex structures have undergone basically different modification stages. Therefore, the volume data have been normalized to the excavation stage of crater development, on the assumption that all structures initially had an excavated cavity with a parabolic cross-section (Melosh, 1976) with the equation of the type $r^2 = 4ap$, where r is radius and p depth. The volume of melt at Brent (Table 4) is based on drilling results (Dence, 1971) and, as the structure was protected soon after formation by a sedimentary cover (Lozej and Beales, 1975), is considered to represent a *realistic estimate* of the volume of melt remaining within the transient cavity. Estimates of the volumes at the complex structures: W. Clearwater, Manicouagan, and Mistastin (Table 4), are derived from the present outcrop distribution with some allowance for erosion of the melt sheet. They are considered *minimum estimates*, as no allowance has been made for the melt contained in overlying mixed breccia deposits, such as observed at Popigai (Masaitis *et al.*, 1975).

The radius of the transient cavity at the complex structures has been estimated from structural data, such as the position of remnants of pre-impact sedimentary cover, and morphology. However, considerable uncertainty is attached. For example, compare the minimum and maximum estimates given in Table 4 for Manicouagan (Floran and Dence, 1976). The depth of the excavated cavity, and thus the volume, has been calculated using pressure attenuation rates of r^{-3} and $r^{-4.5}$ (Table 4) and a peak shock pressure of 25 GPa for the base of the cavity, based on the observed level of shock metamorphism (Robertson and Grieve, 1977). A full discussion of the derivation of the attenuation rates and their implications is given in Dence *et al.* (1977).

The data in Table 4 indicate that, for a given attenuation rate, complex

Table 4. Relative volumes of melt at various structures.

	Brent	Mistastin	W. Clearwater	Manicouagan	
Radius EC, km	1.5	6	8.5	15(min)	22(max)
Vol. melt in EC, km^3	5×10^{-2}	8	24	80	320
Model 1; $P \alpha r^{-3}$					
Vol. EC, km^3	2.7	170	486	2670	8427
Vol. melt/Vol. EC	0.02	0.05	0.05	0.03	0.04
Model 2; $P \alpha r^{-4.5}$					
Vol. EC, km^3	3.6	231	658	3614	11400
Vol. melt/Vol. EC	0.01	0.03	0.04	0.02	0.03

EC—Excavated cavity with form $r^2 = 4ap$, see text for details on models 1 and 2. EC radii and melt data taken or calculated from Bostock (1969), Currie (1970, 1972), Dence (1971, 1973), Floran and Dence (1976), and Grieve (1975).

structures as a group contain relatively more melt within the transient cavity. This suggests that cavity excavation is less efficient at large hypervelocity impact events. It does not necessarily imply that more of the kinetic energy of impact is partitioned into increasing the internal energy of the target. However, this is suggested by the conclusion that the attenuation of shock deformation zones is more rapid at large structures (Robertson and Grieve, 1977) and by changes in the energy-size scaling laws for large impact events (Dence *et al.*, 1977). Additional support is found in the observation that the relative volumes of melt tend to converge when a low attenuation rate (r^{-3}) is used for simple craters and a high rate ($r^{-4.5}$) for complex structures (Table 4).

The apparent discrepancy between the volumes of melt at different structural types has been used to criticize the interpretation that the relatively large volumes of igneous-textured rocks at complex structures are in fact impact melts (Chao, 1974). Although sufficient petro-chemical data exist to dismiss this criticism (Dence, 1971; Grieve, 1975; Floran *et al.*, 1976 and references therein), this anomaly warrants further examination. The discrepancy could result from an underestimate of the initial cavity volumes of complex structures. However, this is considered unlikely. For as in the case of Manicouagan, the depth of the cavity is well constrained by geophysical (Sweeney, 1977) and shock zoning data (Murtaugh, 1972), which indicate an excavated cavity with, if anything, a shallower depth to diameter ratio than is given by the relation $P \alpha r^{-4.5}$. Therefore the cavity volumes given in Table 4 for complex structures may be considered maximum values.

The most probable explanation is that, the relative amount of energy required to achieve comminution of the less shocked, still crystalline, portion of the target and ejection of this material and the melt is a function of cavity size and increases with cavity size (Gault *et al.*, 1975). In terms of the physical movement of the melt, this can be envisioned as follows. Irrespective of the size of the event, the melt has the same particle velocity after interaction with the rarefaction wave. However, in the following movement, into and finally out of the cavity, the melt in a large event, which must result in a correspondingly larger cavity (although not as large as would be expected from using the size-energy relationships determined from small impact events), moves a smaller portion of the final transient cavity radius in a given time. It must travel further in terms of actual distance before it is ejected and is therefore subjected to deceleration forces for a longer period of time. Thus, in a large relative to a small cavity, a greater proportion of the shock melted material has insufficient velocity for ejection and remains within the cavity. Dence *et al.* (1977) calculate that for a pressure attenuation rate of r^{-3} a minimum of 14% of the transient cavity volume at Brent was raised above its melting point and of this approximately 10% was not ejected. This can be compared with Manicouagan where even an attenuation rate of $r^{-4.5}$, which results in approximately 35% of the cavity volume being melted, requires that at least 15% of the melt produced has insufficient velocity for ejection to account for the present distribution of the melt sheet.

At Brent, only 1% of the melt is presently in the form of a sheet, the bulk is dispersed as glassy fragments in the overlying mixed breccias (Dence, 1971). If this distribution relationship held for all structures, then the data in Table 4 suggest that the total volume of melt at complex structures (as a sheet and in mixed breccia deposits) exceeds by a factor of 2–3 the total volume of the initial excavated cavity. Although relatively more melt remains within the cavity during the excavation stage and is available to form a larger melt sheet at complex structures, the final distribution of the melt in the cavity is a function of the modification stage of the cratering process. The approximate inverse relationship in the partitioning of the melt into coherent melt sheet and mixed breccias between simple and complex structures is a result of the basic difference in the forms of transient cavity collapse in the crater model outlined earlier. The volumetrically dominant occurrence of melt as a coherent sheet at complex structures emphasizes the essentially passive role the cavity lining plays in base failure collapse, as opposed to its intimate involvement in slope failure at simple bowl-like structures.

Summary and Concluding Remarks

A cratering model has been presented which accounts for the principal characteristics of impact melts and is consistent with experimental and theoretical studies of hypervelocity impact. One of the distinctive features of impact melts, whether as ejecta or large coherent sheets on the crater floor, is their overall chemical homogeneity and their compositional correspondence to a mixture of target lithologies. It is proposed that these properties are a function of the dynamic conditions existing in the melt during the early stages of the excavation process. Compositional mixing and resultant homogeneity are achieved by the turbulent flow of super-heated, low viscosity silicate liquid, containing large velocity gradients established by pressure attenuation and the differential response of the various target lithologies to the passage of the shock wave. Although, it is believed that at this stage, the model is consistent with theory, it is hoped a more rigorous numerical treatment of the attendant physical conditions can be undertaken to test whether they are consistent with the achievement of compositional homogeneity in the short time span, a few seconds to minutes, available during cavity growth.

By the termination of the excavation stage the bulk of the melt has left the cavity and some of it forms suevite ejecta. The remainder, 10–15% of the original volume melted, forms a lining to the cavity and overlies shocked and fractured basement and a layer of melt-bearing mixed breccias and clastic breccias, which thickens towards the rim. An internal structure is established in the melt lining as the result of the incorporation of inclusions of various shock levels and their effect upon melt temperature. At this point, the modification stage becomes the dominant factor in cavity evolution and the further development of simple and complex structures diverges. In simple craters, slope failure disrupts the cavity lining and the melt is intimately mixed with material slumped off the crater wall to form crater fill products: melt-bearing mixed breccia, clastic breccia and allochthonous basement blocks. It is only at the base of the cavity that a small discrete pool of melt remains and the original stratigraphy of the cavity lining partially maintained.

By contrast, complex structures are modified by deep centripetal sliding; a process that has little effect on the basic stratigraphy of the cavity lining or the internal relationships in the melt. The major disruptions occur due to relatively minor slope failure of the uplifted central peak, annular rings or the collapsing cavity rim. This results in the formation of a thin inclusion-rich melt and mixed breccia capping to the melt sheet. These uppermost mixed breccias, which occur within the cavity, have a mode of origin equivalent to the slump mixed breccias within simple craters and it is considered erroneous to ascribe their origin to fall back ejecta. Although not denying the existence of a relatively thin layer of true melt-bearing fallback material, which may overlie these mixed breccias, it is believed that the melt component of the bulk of the mixed breccias occurring above the melt in both simple and complex structures never achieved free-flight. The apparent discrepancy in the relative volumes of melt at small and large structures is a consequence of greater velocity attenuation in the melt, due to increased path length, within the cavity at large structures and indicates that as cavity size increases the efficiency of the excavation stage of crater formation decreases.

This work has stressed the common characteristics of impact melts. It has as a thesis that, although the relationships of the melt-bearing rocks at any particular structure may appear unique, they are in fact consistent and may be logically accommodated in a single cratering model. The apparent variation between the melt at different structures arises principally through the modification stage of crater formation, with alteration effects, variations in level of erosion and differences in interpretation between workers recognized as additional complications. We would emphasize that studies of the melt at any one structure contribute only a portion of the evidence on which the model is based and further detailed field and laboratory work are required at as many melt occurrences as possible to test and/or further refine the model. There is also a need for theoretical calculations to better define the physical conditions existing in the melt/vapor volume during the excavation stage of a hypervelocity impact into a heterogeneous crystalline target.

Acknowledgments—We thank C. H. Simonds and D. Stöffler for constructive reviews which hopefully have resulted in improvements in the paper.

References

Altschuler, L. V. and Sharipdzhanov, I. I.: 1971, Additive equations of state of silicates at high pressures. *Izvestiya, Physics of the Solid Earth*, No. 3, 11–28.

Ahrens, J. T. and O'Keefe, J. D.: 1972, Shock melting and vaporization of lunar rocks and minerals. *The Moon* 4, 214–249.

Bostock, H. H.: 1969, The Clearwater Complex, New Quebec, *Geol. Surv. Canada Bull.* **178**, 63 pp.

Bunch, T. E. and Cassidy, W. A. 1972, Petrographic and electron microprobe study of the Monturaqui impactite. *Contrib. Mineral. and Petrol.* **36**, 95–112.

Butkovich, T. R.: 1965, Calculation of the shock wave from an underground nuclear explosion in granite. *J. Geophys. Res.* **70**, 885–892.

Chao, E. C. T.: 1968, Pressure and temperature histories of impact metamorphosed rocks—based on petrographic observations. *Shock Metamorphism of Natural Materials* (B. M. French and N. M. Short, eds.), pp. 135–168, Mono Book Corp., Baltimore.

Chao, E. C. T.: 1974, Impact cratering models and their applications to lunar studies—a geologist's view. *Proc. Lunar Sci. Conf. 5th.*, pp. 35–52.

Chao, E. C. T.: 1976, Physical parameters, ejecta overlap relationship and transport characteristics, and nature of impacting body as constraints for the cratering model of the Ries. In *Papers Presented to the Symposium on Planetary Cratering Mechanics*, pp. 19–21. The Lunar Science Institute, Houston.

Cherry, J. T. and Petersen, F. L.: 1970, Numerical simulation of stress wave propagation from underground nuclear explosions. In *Peaceful Nuclear Explosions*, 241–326, International Atomic Energy Agency, Vienna.

Currie, K. L.: 1970, Geology of the resurgent cryptoexplosion crater at Mistastin Lake, Labrador, Canada, *Geol. Surv. Canada Bull.* **207**, 62 pp.

Currie, K. L.: 1971, Origin of igneous rocks associated with shock metamorphism as suggested by geochemical investigations at Canadian craters. *J. Geophys. Res.* **76**, 5575–5585.

Currie, K. L.: 1972, Geology and petrology of the Manicouagan resurgent caldera, Quebec. *Geol. Surv. Canada Bull.* **198**, 153 pp.

David, M. and Woussen, G.: 1973, Correspondence analysis, a new tool for geologists. *Proc. Mining Pribram, Czechoslovakia, No. 1*, 41–65.

DeCarli, P. S. and Jamieson, J. C.: 1959, Formation of an amorphous form of quartz under shock conditions. *J. Chem. Phys.* **31**, 1675–1676.

Dence, M. R.: 1964, A comparative structural and petrographic study of probable Canadian meteorite craters. *Meteoritics* **2**, 249–270.

Dence, M. R.: 1965, The extraterrestrial origin of Canadian craters. *Annals of New York Acad. Sci.* **123**, 941–969.

Dence, M. R.: 1968, Shock zoning at Canadian craters: Petrography and structural implications. *Shock Metamorphism of Natural Materials* (B. M. French and N. M. Short, eds.), pp. 169–183, Mono Book Corp., Baltimore.

Dence, M. R.: 1971, Impact melts. *J. Geophys. Res.* **76**, 5552–5565.

Dence, M. R.: 1973, Dimensional analysis of impact structures (abstract). *Meteoritics* **8**, 343–344.

Dence, M. R., Innes, M. J. S., and Beals, C. S.: 1965, On the probable meteorite origin of the Clearwater Lakes, Quebec. *J. Roy. Astron. Soc. Canada* **59**, 13–22.

Dence, M. R., Innes, M. J. S., and Robertson, P. B.: 1968, Recent geological and geophysical studies at Canadian craters. *Shock Metamorphism of Natural Materials* (B. M. French and N. M. Short, eds.), pp. 339–362, Mono Book Corp., Baltimore.

Dence, M. R., Engelhardt, W. von, Plant, A. G., and Walter, L. S.: 1974, Indications of fluid immiscibility in glass from West Clearwater Lake impact crater, Quebec, Canada. *Contrib. Mineral. and Petrol.* **46**, 81–97.

Dence, M. R., Grieve, R. A. F., and Robertson, P. B.: 1977, Terrestrial impact structures: Principal characteristics and energy considerations. This volume.

El Goresy, A.: 1968, Opaque minerals in impactite glasses. *Shock Metamorphism of Natural Materials* (B. M. French and N. M. Short, eds.), pp. 531–554, Mono Book Corp., Baltimore.

Engelhardt, W. von: 1967, Chemical composition of Ries glass bombs. *Geochim. Cosmochima. Acta.*, **31**, 1677–1689.

Engelhardt, W. von: 1972, Shock produced rock glasses from the Ries crater. *Contrib. Mineral. and Petrol.* **36**, 265–292.

Engelhardt, W. von and Graup, G.: 1977, Stosswellenmetamorphose im Kristallin der Forschungsbohrung Nördlingen 1973. *Geologica Bavarica* **75**, in press.

Engelhardt, W. von, Stöffler, D., and Scheider, W.: 1969, Petrologische Untersuchungen im Ries. *Geol. Bavaria* **61**, 229–295.

Floran, R. J. and Dence, M. R.: 1976, Morphology of the Manicouagan ring-structure, Quebec, and some comparisons with lunar basins and craters. *Proc. Lunar Sci. Conf. 7th.*, pp. 2845–2865.

Floran, R. J., Simonds, C. H., Grieve, R. A. F., Phinney, W. C., Warner, J. L., Rhodes, M. J., and Dence, M. R.: 1976, Petrology, structure and origin of the Manicouagan melt sheet, Quebec, Canada: A preliminary report. *Geophys. Res. Lett.* **3**, 49–52.

Floran, R. J., Grieve, R. A. F., Phinney, W. C., Warner, J. L., Simonds, C. H., Blanchard, D. P. and Dence, M. R.: 1977, Manicouagan impact melt, Quebec. Part I: Stratigraphy, petrology and chemistry. (Submitted to *J. Geophys. Res.*)

French, B. M., Hartung, J. B., Short, N. M., and Dietz, R. S.: 1970, Tenoumer Crater, Mauritania: Age and petrologic evidence for origin by meteorite impact. *J. Geophys. Res.* **75**, 4396–4406.

Fudali, R. F.: 1974, Genesis of the melt rocks at Tenoumer crater, Mauritania. *J. Geophys. Res.* **79**, 2115–2121.

Gault, D. E. and Heitowit, E. D.: 1963, The partition of energy for hypervelocity impact craters formed in rock. *Proc. 6th Symposium on Hypervelocity Impact, Cleveland, Ohio*, **11**, part 2, 419–456.

Gault, D. E., Quaide, W. L., and Oberbeck, V. R.: 1968, Impact cratering mechanics and structures. *Shock Metamorphism of Natural Materials* (B. M. French and N. M. Short, eds.), pp. 87–99. Mono Book Corp., Baltimore.

Gault, D. E., Guest, J. E., Murray, J. B., Dzurisin, D., and Malin, M. C.: 1975, Some comparisons of impact craters on Mercury and the Moon. *J. Geophys. Res.* **80**, 2444–2460.

Gibbons, R. V., Keiffer, S. W., Schaal, R. B., Hörz, F., and Thompson, T. D.: 1975, Experimental calibration of shock metamorphism of basalt (abstract). In *Papers Presented to the Conference on the Origins of Mare Basalts and Their Implications for Lunar Evolution*, pp. 44–47. The Lunar Science Institute, Houston.

Grieve, R. A. F.: 1975, Petrology and chemistry of the impact melt at Mistastin Lake Crater, Labrador. *Bull. Geol. Soc. Amer.* **86**, 1617–1629.

Grieve, R. A. F., Dence, M. R., and Robertson, P. B.: 1976, The generation and distribution of impact melts: Implications for cratering processes (abstract). In *Papers Presented to the Symposium on Planetary Cratering Mechanics*, pp. 40–42, The Lunar Science Institute, Houston.

Grieve, R. A. F. and Floran, R. J.: 1977, Manicougan impact melt. Quebec. Part II: Formational processes and chemical interrelations with basement lithologies. (Submitted to *J. Geophys. Res.*)

Hartung, J. B., Dence, M. R., and Adams, J. A. S.: 1971, Potassium-Argon dating of shock metamorphosed rocks from the Brent impact crater, Ontario, Canada. *J. Geophys. Res.* **76**, 5437–5448.

Head, J. W.: 1974, Orientale multi-ringed basin interior and implications for the petrogenesis of lunar highland samples. *The Moon* **11**, 327–356.

Hörz, F.: 1965, Untersuchungen an Riesgläsern. *Beitr. Mineral. Petrogr.* **11**, 621–661.

Levin, E. M., McMurdie, H. F., and Hall, F. P.: 1956, *Phase diagrams for ceramists*. American Ceramic Society, Columbus, Ohio, 286 pp.

Lozej, C. P. and Beales, F. W.: 1975, The unmetamorphosed sedimentary fill of the Brent meteorite crater, south-eastern Ontario. *Can. J. Earth Sci.*, **12**, 629–635.

Marchand, M.: 1976, A geochemical and geochronologic investigation of meteorite impact melts of Mistastin Lake, Labrador and Sudbury, Ontario. Ph.D. thesis, McMaster Univ., Hamilton, Ontario, 142 pp.

Masaitis, V. L., Mikhaylov, M. V., and Selivanovskaya, T. V.: 1975, *Popigayskiy Meteoritnyy Krater*. Nauka Press, Moscow, 124 pp.

McCabe, H. R. and Bannatyne, B. B.: 1970, Lake St. Martin crypto-explosion crater and geology of the surrounding area. *Geol. Surv. Manitoba*, Geol. Paper 3/70.

McQueen, R. G., Marsh, S. P., and Fritz, J. M.: 1967, Hugoniot equation of state of twelve rocks. *J. Geophys. Res.* **72**, 4999–5036.

Melosh, H. J.: 1976, Crater modification by gravity: A mechanical analysis of slumping. In *Papers Presented to the Symposium on Planetary Cratering Mechanics*, pp. 76–78. The Lunar Science Institute, Houston.

Moore, H. J., Hodges, C. A., and Scott, D. H.: 1974, Multiringed basins—illustrated by Orientale and associated features. *Proc. Lunar Sci. Conf. 5th*, pp. 71–100.

Murtaugh, J. G.: 1972, Shock metamorphism in the Manicouagan crypto-explosion structure, Quebec. *24th Internat. Geol. Congr.*, Sect. 15, 133–139.

Murtaugh, J. G.: 1975, Geology of the Manicouagan crypto-explosion structure. Ph.D. thesis, Ohio State Univ., 299 pp.

O'Keefe, J. D. and Ahrens, T. J.: 1975, Shock effects from a large impact on the Moon. *Proc. Lunar Sci. Conf. 6th*, pp. 2831–2844.

Onorato, P. I. K., Uhlmann, D. R., and Simonds, C. H.: 1976, Heat flow in impact melts: Apollo 17, Station 6 boulder. *Proc. Lunar Sci. Conf. 7th*, pp. 2449–2467.

Phinney, W. C. and Simonds, C. H.: 1976, Dynamical implications of the petrology and distribution of impact produced rocks. In *Papers Presented to the Symposium on Planetary Cratering Mechanics*, pp. 99–101. The Lunar Science Institute, Houston.

Pike, R. J.: 1971, Genetic implications of the shapes of martian and lunar craters. *Icarus* **15**, 384–395.

Robertson, P. B. and Grieve, R. A. F.: 1977, Shock attenuation at terrestrial impact structures. This volume.

Rondot, J.: 1971, Impactite of the Charlevoix structure, Quebec. *J. Geophys. Res.* **76**, 5414–5423.

Schaal, R. B., Hörz, F., Gibbons, R. V., and Keiffer, S. W.: 1975, Impact melts of well-characterized lunar and terrestrial basalts (abstract). In *Papers Presented to the Conference on the Origins of Mare Basalts and Their Implications for Lunar Evolution*, pp. 144–148. The Lunar Science Institute, Houston.

Shaw, H. R.: 1972, Viscosities of magmatic silicate liquids, an empirical method of prediction. *Amer. Jour. Sci.* **272**, 870–893.

Shoemaker, E. M.: 1960, Penetration mechanics of high velocity meteorites, illustrated by Meteor crater, Arizona. *21st Internat. Geol. Congr.*, Sect. 18, 418–434.

Short, N. M.: 1970, Anatomy of a meteorite impact crater: West Hawk Lake, Manitoba, Canada. *Bull. Geol. Soc. Amer.* **81**, 609–648.

Simonds, C. H., Warner, J. L., and Phinney, W. C.: 1976a, Thermal regimes in cratered terrain with emphasis on the role of impact melt. *Amer. Mineral.* **61**, 569–577.

Simonds, C. H., Warner, J. L., Phinney, W. C., and McGee, P. M. 1976b, Thermal model for breccia lithification: Manicouagan and the Moon. *Proc. Lunar Sci. Conf. 7th*, pp. 2509–2528.

Stähle, V.: 1972, Impact glasses from suevite of the Nördlinger Ries. *Earth Planet. Sci. Lett.* **17**, 275–293.

Stettner, G.: 1974, Das Grundgebirge in der Forschungsbohrung Nördlingen 1973 im regionalen Rahmen und seine Veränderungen durch den Impackt. *Geol. Bavarica* **72**, 35–51.

Stöffler, D.: 1971, Progressive metamorphism and classification of shocked and brecciated crystalline rocks at impact craters. *J. Geophys. Res.* **76**, 5541–5551.

Stöffler, D.: 1977, Research drilling Nördlingen 1973: Polymict breccias, crater basement, and cratering model of the Ries impact structure. *Geologica Bavarica* **75**. In press.

Stöffler, D., Ewald, V., Ostertag, R., and Reimold, W.-U.: 1977, Research drilling Nördlingen 1973, Ries: Composition and texture of polymict impact breccias. *Geologica Bavarica* **75**. In press.

Stöffler, D. and Hornemann, V.: 1972, Quartz and feldspar glasses produced by natural and experimental shock. *Meteoritics* **7**, 371–394.

Sweeney, J.F.: 1977, Gravity study of great impact. (Submitted to *J. Geophys. Res.*)

Wackerle, J.: 1972, Shock wave compression of quartz. *J. Appl. Phys.* **33**, 922–937.

Winzer, S. R., Lum, R. K. L., and Schumann, S.: 1976, Rb, Sr and Strontium isotopic composition, K/Ar age and large ion lithophile trace element abundances in rocks and glasses from the Wanapitei Lake impact structure. *Geochim. Cosmochim. Acta.* **40**, 51–58.

Yurk, Yu. Yu., Yeremenko, G. K., and Polkanov, Yu. A.: 1975, The Bolytsh depression—a fossil meteorite crater. *Internat. Geology Rev.* **18**, 196–202.

Roddy, D. J., Pepin, R. O., and Merrill, R. B., editors.
(1977) *Impact and Explosion Cratering*, Pergamon Press (New York), p. 815–841.
Printed in the United States of America

Impact melt on lunar crater rims

B. RAY HAWKE and JAMES W. HEAD

Department of Geological Sciences, Brown University, Providence, Rhode Island 02912

Abstract—Deposits of lava-like material around relatively fresh lunar craters are interpreted as impact melt on the basis of deposit distribution, lack of volcanic sources, morphology of the material, and time of emplacement. Exterior melt deposits have three modes of occurrence: thin veneer, flows, and ponds. Exterior melt ponds and flows were detected around 55 craters ranging from 4 to 300 km in diameter. Hard rock veneers and very small ponds are the dominant mode of occurrence around the smallest craters. At craters larger than about 10 km, small ponds near the rim crest, and flow lobes and channels are prominent. Flows are more conspicuous than ponds up to crater diameters of about 50 km. Changes in the mode of occurrence of rim melt deposits are correlated with major morphologic changes in crater interior structure. Melt deposits first become abundant on crater rims in the same size range that wall terraces and central peaks become common in the crater interior. There is a tendency for melt deposits to occur at greater relative maximum distances from the rim crest with increasing crater size, at least up to diameters of 50 km. Proportionally larger amounts of the total shock-melted material appear to have been emplaced on the rims of the larger craters.

Two factors are important in controlling the distribution of exterior melt deposits: (1) the pre-impact topography of the target site and (2) the angle of incidence and approach direction in an oblique impact as inferred by ray pattern, secondary crater field, or ejecta blanket asymmetry. Melt deposits are often concentrated on crater rims adjacent to topographic lows in the rim crest and commonly occur on the rim opposite zones of maximum wall slumping. The development of both rim crest lows and maximum wall slump zones appears to be strongly influenced by pre-event topography.

Evidence is presented indicating that the melt which is concentrated on the rim was emplaced during and slightly after the modification stage of the impact cratering event. On the basis of the above observations a model is suggested for the emplacement of large amounts of impact melt on crater rims. At the end of the excavation stage, the transient cavity is lined with shock-melted material and the greatest melt concentration is in the lower portions of the cavity. The modification stage begins with the initial collapse and rapid shallowing of the transient cavity due to rebound and wall slumping. As the crater shallows molten material is moved upward and is often given a lateral component of movement by terrace formation. With continued rebound and wall slumping, some of the molten material is removed from the crater at low velocities and emplaced on the crater rim. The molten material that does not escape the crater generally settles onto the crater floor but some remains as ponds, flows, and veneer on the interior crater walls. After the modification is complete, impact melt flows in response to the local topography to form the deposits observed today.

I. INTRODUCTION

LAVA-LIKE DEPOSITS that were emplaced in a fluid state occur in and around many fresh lunar craters. Early workers commonly ascribed such material to comparatively recent lunar volcanic activity (Strom and Fielder, 1970, 1971; Strom and Whitaker, 1971; El-Baz, 1971, 1972). In recent years, in large part due to the photogeologic and petrographic data generated by the Apollo program and intensive study of terrestrial impact structures, these deposits have been interpreted to be of impact melt origin (Dence, 1971; Howard and Wilshire, 1973, 1975; Guest, 1973; Hawke, 1976). In particular, Howard and Wilshire (1975) have presented strong evidence that these lava-like materials formed by impact melting rather than volcanism. They noted that the characteristics of the materials, the lack of apparent sources, and the distribution patterns are most easily explained if the lava-like deposits are impact melts with variable proportions of unmelted inclusions. A major objection to a volcanic origin for these materials has been their association with Copernican-age craters generally less than about 1.5 b.y. old. There is little evidence for internally derived surface volcanism in this time period (see Head, 1976a). However, Hulme (1974) has suggested that these deposits may be volcanic, having been melted by post-impact viscous creep.

Howard and Wilshire (1975) presented data for twenty fresh craters with melt deposits and concluded that the distribution of melt deposits conformed to asymmetries of other ejecta from the same craters. They suggested oblique impact as a control in melt distribution and found that melt was concentrated downrange to distances as great as one crater radius from the rim. Several possibilities for the mode of emplacement were suggested: (1) the melt may have been erupted from dikes, (2) the melt may have fallen as a rain of hot clots that coalesced and ran together on the surface, (3) molten material may have splashed out as a mass at the end of the cratering event, and (4) melt may have been mixed with the rest of the ejecta and later sweated out. Still, the processes controlling the formation and distribution of lava-like deposits are not well understood.

The purpose of this study was to determine the origin, distribution, and modes of occurrence of lava-like deposits and the factors responsible for their emplacement. Lava-like deposits around a large number of lunar craters were investigated in order to obtain information on the variation in morphology and the distribution as a function of crater size. The results should have bearing on lunar and terrestrial impact cratering processes and the provenance of the abundant impact melt rocks in the lunar sample collection (Grieve *et al.*, 1974; Simonds, 1975).

II. METHOD

The criteria used in this study to identify lava-like or melt deposits are similar to those described and illustrated by Howard and Wilshire (1975). These include the various indications of fluid flow (flow lineations, leveed channels, and the ponding of material to a level surface), cooling cracks in ponds, tension cracks in veneer, and gradational relationships among the various melt morphologies. The albedo of rim material is often useful in locating lava-like deposits, since rim ponds, veneer, and the thinner flows tend to have a lower albedo than the associated ejecta deposits (Howard and Wilshire, 1975). Blocky outcrop ledges as well as blocky ejecta from superposed craterlets demonstrate that the veneer solidified to hard rock. The morphology of superposed craterlets can be useful for verifying that certain ponds and flows are not deposits of fine-grained debris (Quaide and Oberbeck, 1968).

A variety of Lunar Orbiter, Apollo, and Earth-based photography was employed to locate and characterize lunar melt deposits. Lava-like material exhibits a number of distinctive morphologies including flow lobes and channels, hard rock veneer over irregular surfaces, complexly fractured ponds on crater floors, and smaller ponds on crater walls and rims. Craters which exhibited these characteristic melt morphologies in their exterior deposits were studied to determine the manner in which these morphologies varied as a function of crater size. Special emphasis was placed on exterior ponds and flows since they could also be recognized in low resolution photography.

To date, a total of 55 craters with *ponds* or *flows* of lava-like material on their rims have been identified. Nineteen of the craters are from the population presented by Howard and Wilshire (1975) and include such well-known craters as Copernicus, Tycho, Aristarchus, King, and Theophilus. Many of the remaining 36 craters have exterior melt deposits which have not been previously recognized or described.

A variety of data was collected concerning the size and location of these craters as well as the morphology and morphometry of the crater and melt deposits. Where photographic coverage and resolution permitted, we determined the locations of the most extensive ejecta and melt deposits, the position of the lowest portion of the rim crest, the maximum and minimum wall width, the location of the most abundant wall pools, the melt distribution on the crater floor, and the location of any pre-existing topographic lows exterior to the crater. Extensive use was made of the new NASA 1:250,000 Lunar Topographic Orthophotomaps and large scale lunar maps, as well as the Lunar Astronautical Charts. The data relevant to this study are presented in Table 1.

III. Results and Discussion

A. Origin of lava-like deposits

The results of this study support an origin of these deposits by impact melting (Howard and Wilshire, 1975) on the basis of the following evidence:

Deposit distribution. Melt distribution patterns on crater exteriors are asymmetrical and are controlled by pre-impact topography and oblique impact (see Section III.D). The distribution of lava-like material appears linked to the impact cratering process. It is unlikely that volcanic processes would consistently produce extensive deposits of lava-like material in the downrange direction of an oblique impact or in the direction in which the parent crater has a lower rim crest segment. Lava-like materials are often seen draping the crater walls and rim crest and extending for distances up to approximately 1.5 radii from the crater lip. There are clear indications that the crater was the source of this material and that it was emplaced by the impact cratering event. The uphill-facing flow front and the thin flow lobes which have apparently topped a hill north of King crater (see Fig. 5 of Howard and Wilshire, 1975) strongly suggest that these materials were propelled radially away from the crater. Lava-like materials are commonly found near crater rim crests and some flows appear to have originated in this vicinity (Figs. 2 and 3). Concentration on the crater rim seems an unlikely style of volcanic activity since craters which have been partly flooded with mare basalt do not exhibit evidence of extensive extrusive activity high on their rims.

Lack of apparent sources. The lack of sources has been a common objection to a volcanic origin for lava-like material (El-Baz, 1972; Howard, 1972). The observations made in this study lend support to this objection. The few sources that have been suggested can generally be explained by lunar processes other

Table 1. Data for craters with exterior melt deposits. Asymmetry in ejecta distribution was used to determine the downrange direction. Maximum wall slumping was usually taken as the widest portion of the terraced wall. When two directions are given, the one listed first was estimated to be dominant and was used in the correlation. Further discussion is given in the text.

Crater	Approximate diameter (km)	Maximum distance of melt from rim (crater radii)	Most extensive melt deposits	Topographically low rim crest segment	Maximum wall slumping	Most extensive ejecta, rays and secondaries[1]
Schrödinger	300	0.70	E	S	E	NE
Humboldt	201	0.96	SE, ENE	S, ENE	NNW	SSE
Tsiolkovsky	190	0.49	SE	W	SE	SSE
Petavius	177	0.68	S	—	S	SSE
Hausen	170	0.41	WSW	SSW	—	—
Langrenus	132	0.68	SSE	SSW, NW	—	SE
Pythagorus	128	0.72	NE	NE	S	S
Theophilus	100	0.94	NE	NE	SW	NE
Copernicus	95	0.66	NW	N, S	S	NW
Aristoteles	87	0.50	S	—	E	N
Tycho	85	0.88	E	E	SW	E
Fabricius	80	0.31	S, WSW	S	N	SW
O'Day	76	0.89	ESE	SE	W	SE
Sharonov	75	—	—	—	—	—
Von Neuman	72	0.17	NE	NE	SW	—
King	71	0.83	NE	S, NE	SSE	NNW
Philolaus	70	1.00	NNW	NNW	W	NE
Cavalerius	60	0.11	NE	N, S	SSW	ESE
Zucchius	60	0.25	E, N	—	N	—
Maunder	54	0.16	W	SW	—	SSE
Anaxagoras	51	0.65	ESE	E	NNW	ESE
Rutherfurd	50	1.33	NNW	N	SSW	NNE
Crater on SW rim of Mandel'shtam	49	0.90	NNE	N	SW	—
Crater near Steno (35°N, 162°E)	45	1.20	SSE	SE	N	—
Aristarchus	45	0.65	N, SE	NNE, E	NNE	SE
Olbers A	37	1.58	NNW	NNW	NNW	NE

Mairan	35	0.47	SW	—	E	—
Crater S of Green	35	0.08	NNW	—	N	—
Necho	33	1.03	NE	NE	SW	N
Crater N of Zhukovsky	32	—	W	—	—	—
Crater near Weiner (41°N, 149°E)	30	0.85	N	N	S	S, N
Crater NE of Saha (0°30′S, 108°E)	27	0.46	NNW	NNW	SE	—
Crater on rim of Bečvář	26	0.31	ENE	ENE	SW	NE
Proclus	26	—	SW	—	SW	NE
Crater near Kovalsky	25	0.15	NNE	NNE	SSW	NE
Lalande	24	0.40	NE	N	SW	N
Crater W of Firsov (5°30′N, 108°E)	22	0.28	W	SSW	—	—
Dawes	18	0.21	SE	SE, N	N	SW
Crater on W rim of West Bond	18	0.48	W	W, E	—	SSW
Crater W of Fermi (19°30′S, 116°30′E)	18	0.89	NNE	NNE	NW	NNE
Crater on SE rim of Skoldowska	15	0.53	NW	NW	—	—
Crater near Mandel'shtam (5°N, 167°E)	15	0.47	W	—	—	NW
Crater NE of Cyrano	14	—	NNE	NNE	—	NNE
Crater NE of Meshchersky (13°N, 127°30′E)	14	0.46	SE	—	—	NW
Crater on rim of O'Day	13	—	N	N	—	—
Crater N of Aitken (14°S, 173·30′E)	13	—	N	N	—	—
Crater on W rim of Papaleski	10	—	W	E	—	—
Crater near Curie (23°30′S, 88°E)	10	0.32	N	N, S	—	—
Crater on NW rim of Sklodowska	9.5	1.20	SSW	SSE	—	SW

Table 1. (*Continued*).

Crater	Approximate diameter (km)	Maximum distance of melt from rim (crater radii)	Most extensive melt deposits	Topographically low rim crest segment	Maximum wall slumping	Most extensive ejecta, rays and secondaries[1]
Crater near Chauvenet (13°S, 137°E)	8	0.32	N, S	—	—	W
Crater NW of Valier	7	0.48	SW	SW	—	—
Römer Y	6.6	0.82	SW	SSW	—	SSW
Crater near McKellar	5.4	0.02	S	—	—	SE
Crater at 21°S, 100°E (near Kovalsky)	4.7	0.09	W	N, ESE	—	ESE
Crater on rim of Gibbs	4	0.30	NNE	SW	—	SSW
Small crater on W wall of Lobachevsky	—	—	E	E	W	E

[1](i.e., inferred downrange direction.)

than volcanism. It has been commonly suggested that the fractures formed by the cratering event served as channelways for lava extrusion on the crater exterior (e.g., El-Baz, 1972). However, some melt deposits are found at large relative distances from the rim crest (Table 1) where evidence for fractures is lacking. Fracturing is probably not an important process at large distances from the rim crest (see Head, 1976b).

Morphology of the deposits. Evidence has been cited at Tycho (Shoemaker *et al.*, 1968), King, and Copernicus (Howard and Wilshire, 1975), that the various types of fluid material are gradational and interrelated. Similar relationships are exhibited by the Necho crater exterior melt deposits (Figs. 4 and 10). There are many examples where rim ponds appear to have been formed by runoff from the surrounding areas (e.g., Hawke, 1976). The draping of a hard rock veneer over subjacent terrain is particularly difficult to account for by volcanic processes but could easily be explained by the deposition of material rich in impact melt.

Age and time of emplacement of lava-like deposits. French (1972a, 1972b) suggested that large scale magmatic activity within an impact crater would be likely only if the crater penetrated a region where there is either active magmatic activity or a high geothermal gradient at depth. Most of the craters in our sample are small and relatively fresh. Since there is little evidence for internally derived surface volcanism in the last 1.5 b.y. (Head, 1976a), and in light of our current understanding of the thermal history of the lunar crust (see Taylor, 1975), it appears unlikely that such a region would have been penetrated by the fresh craters in our population. Hulme (1974) has recently suggested that minor amounts of melt can be generated by melting induced by viscous creep within large young craters. While it may be theoretically possible to produce melt in this process, mechanisms for concentrating this melt in the places it is commonly observed on crater exteriors (from high on the rim out to distances of approximately 1.5 crater radii) appear to be lacking. According to Hulme (1974) the melt is emplaced at some interval of time after crater formation (10^4–10^7 yr), but considerable evidence exists that the melt is emplaced during the terminal stage of the cratering event (see Section III.E and Howard and Wilshire, 1975).

B. Size range of craters with exterior melt deposits

Howard and Wilshire (1975) stated that lava-like flows are generally absent from the rims of craters less than about 30 km in diameter, but this survey revealed 18 craters with diameters less than 20 km that have rim ponds and or flows. Ponds and flows of probable impact melt have been identified around 55 craters ranging from 4 to 300 km in diameter but most of the craters (88%) are less than 100 km (Fig. 1). The smallest crater in the sample is a 4 km crater superposed on the rim of Gibbs. An even smaller crater with possible melt flows on its rim is located on the west wall of Lobachevsky crater (Mattingly *et al.*, 1972). Thus, the processes acting to generate and emplace these deposits operated at crater diameters as small as 4 km. The 300 km Schrödinger basin is

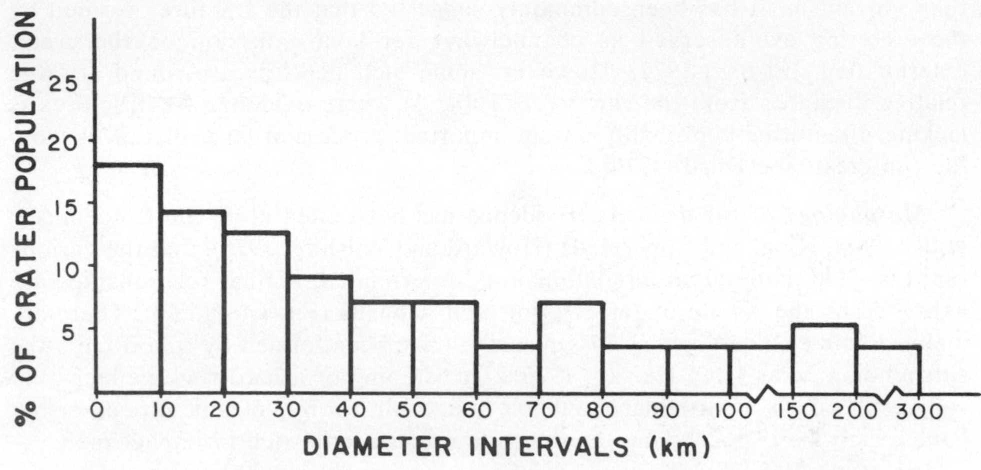

Fig. 1. Size distribution of the 55 craters with exterior melt deposits investigated in this study.

the largest impact structure in our sample with probable melt deposits on the rim. Exterior melt deposits associated with such large craters are of particular interest because of their possible implications for the distribution of melt around larger multiringed basins. Deposits of probable impact melt origin have been noted in (Head, 1974) and around (Moore *et al.*, 1974) the 900 km Orientale basin.

C. Morphology of deposits as a function of crater size

Melt around small craters (4–10 km) typically occurs as a cracked, hard rock veneer. The melt has in places coalesced in local lows to produce very small ponds on, or very near, the rim crest. Short, thin flows may also be present. Hard rock veneers are much more common than either rim ponds or flows around fresh craters of this size (Fig. 5a). The presence of melt deposits may help in distinguishing primary from secondary impact craters in this size range since very little melt would be expected to form at the impact velocities commonly associated with secondaries (<1 km/sec for secondaries from a Copernicus-sized primary crater; Shoemaker, 1962; Oberbeck, 1975).

Craters from 10 to 20 km in diameter have exterior ponds and flows that are larger and more extensive relative to the associated crater than those around smaller craters. Significantly greater amounts of melt have been emplaced on the rims of craters in the 10–20 km size range. The mean maximum distance of recognizable ponds and flows from the rim crest is $0.51R$ (R = crater radius) for craters between 10 and 20 km, as opposed to $0.39R$ for the melt deposits associated with craters less than 10 km in diameter (Fig. 5a). Typical of the crater in the 10–20 km size range is the 15 km crater near Mandel'shtam (Fig. 2) on the lunar farside (Moore, 1972; Howard and Wilshire, 1975). Small ponds occur near the rim crest and flow lobes and channels extend as far as 3.5 km ($0.47R$) from the

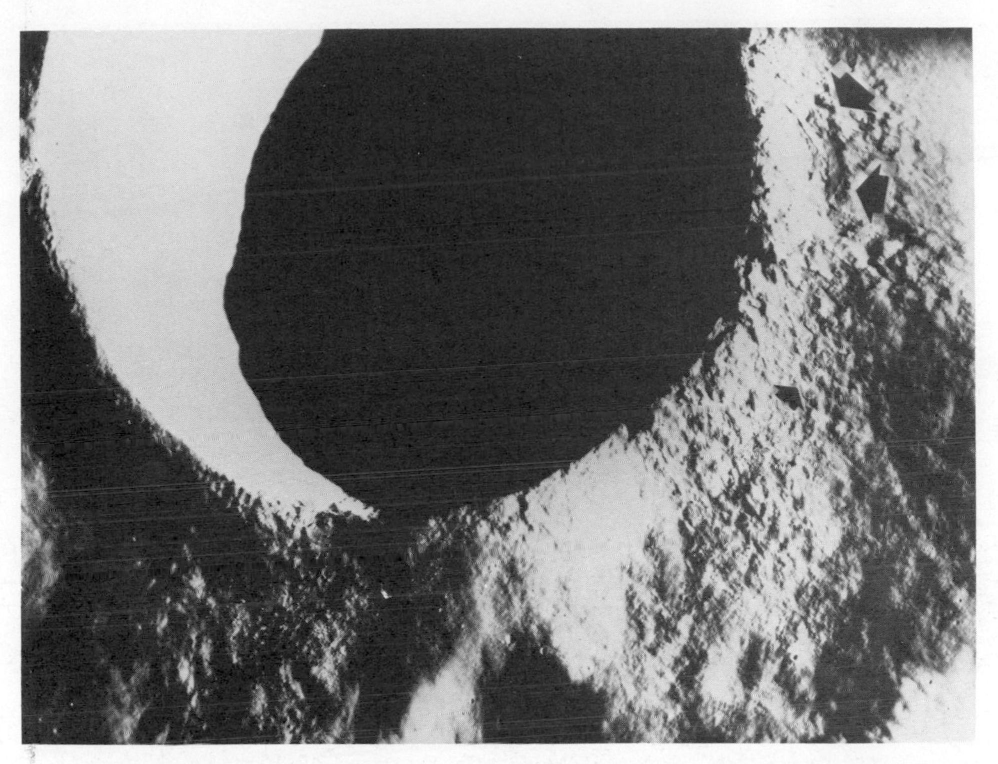

Fig. 2. Exterior melt deposits on the west rim of a 15 km crater near Mandel'shtam (5°N, 167°E). Flow lobes and channels are well developed and a few small ponds can be seen near the rim crest. North is toward the bottom of the photograph (AS-16-4136).

crater. Similar features can be seen less distinctly on the rim of the 14 km crater northeast of Meshchersky (Fig. 3).

Around craters in the 20–50 km diameter range, the melt deposits appear to be relatively more extensive than those around smaller craters. The mean maximum distance of ponds and flows from the rim crest is $0.64R$. Flows are the dominant mode of rim occurrence but ponds are locally numerous and are in places intimately related to the flows (Fig. 5b). The melt deposits around Necho crater (33 km) are generally representative of those associated with craters in this size range (Fig. 4). Thin flows of impact melt extend approximately one crater radius from the rim crest. Several ponds can be seen in the area of the flows. Aristarchus (45 km) is another example of a crater in this size range where flows are the dominant mode of rim occurrence. The more degraded craters in this range commonly exhibit only ponds, probably because of the more rapid destruction of the fine-textured features indicative of melt flows.

At crater diameters larger than about 50 km, flows are less significant and ponds appear to be the dominant mode of rim occurrence. Howard and Wilshire (1975) noted that the rim ponds of these larger craters represented the greatest concentrations of lava-like material and are the most recognizable melt features

Fig. 3. Probable melt flows on the rim of a 14 km crater northeast of Meshchersky (13°N, 127°30′E). North is toward the top of the photograph (AS-16-4974).

on low-resolution photographs or at old, degraded craters. The mean maximum distance of this ponded material from the crater rim crest is 0.62*R* and individual values are generally between 0.5 and 1.0*R* as noted by Howard and Wilshire (1975). Ponds continue to be the dominant melt morphology around such large impact structures as Tsiolkovsky, Humboldt, and Schrödinger.

In summary, several general trends can be discerned (Figs. 5a,b). Around the smallest craters with recognizable melt deposits, hard rock veneers and very small ponds are predominant. At craters larger than about 10 km, small ponds near the rim crest and flow channels and lobes are prominent. Flows are generally more conspicuous than ponds around fresh craters up to 50 km in diameter. Large ponds are dominant at craters greater than 50 km in width. There is a tendency for the relative maximum distances that melt deposits occur away from the parent crater rim crest to increase with crater size, at least up to diameters of 50 km. The change in melt morphology and distance of maximum melt occurrence, as well as observations at individual craters suggest that relatively greater amounts of shock-melted material have been emplaced on the rims of the larger craters. There are, however, individual exceptions such as Crookes crater (50 km) which has interior melt deposits but no recognizable rim flows or ponds. Additional work is in progress to establish more firmly the volumes of melt present in and around lunar craters, and changes in melt distribution with crater size (Hawke and Head, 1977). An observation with

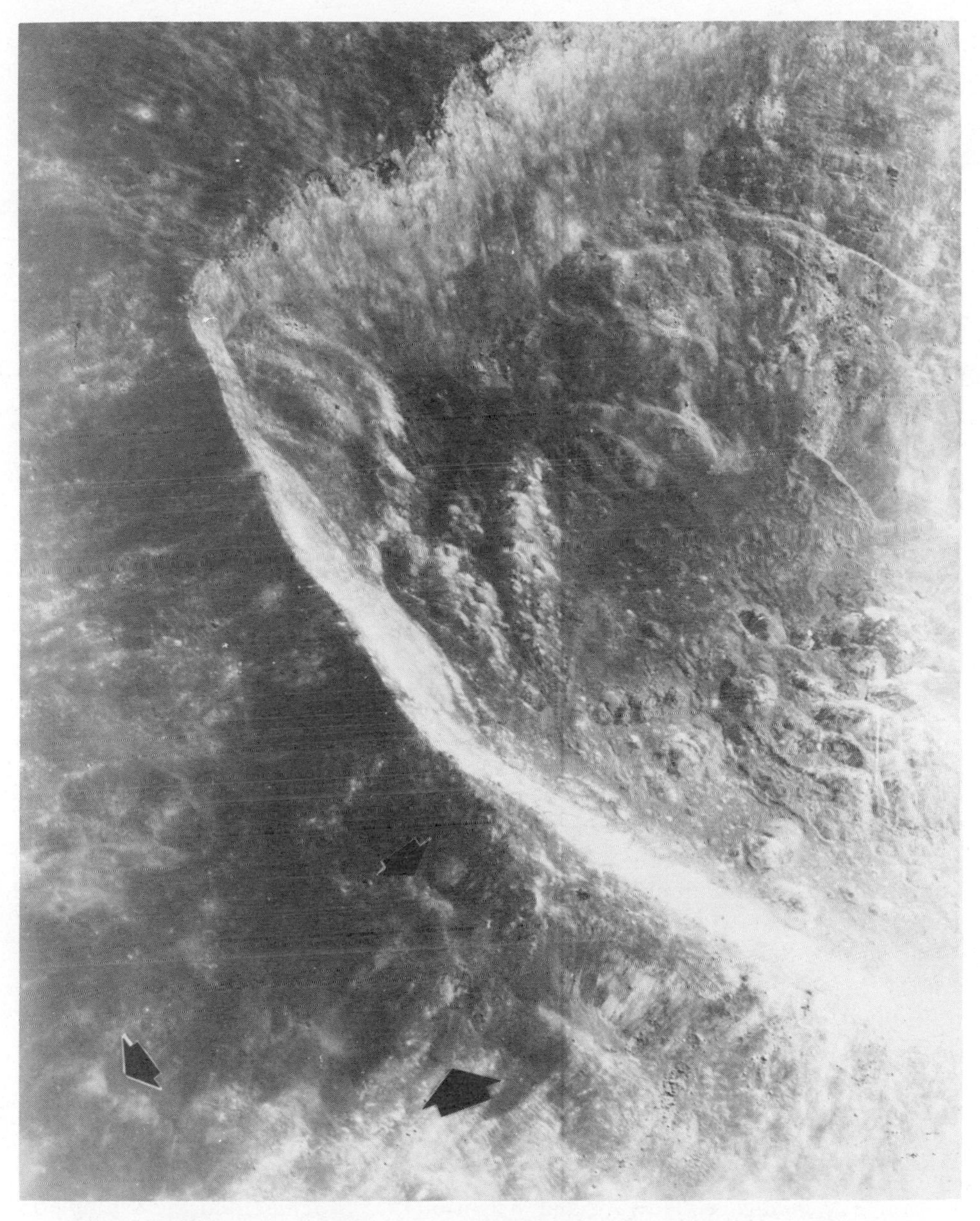

Fig. 4. Oblique pan photograph of the melt deposits on the northeast rim of the fresh crater Necho (33 km in diameter). The area in the foreground is adjacent to the low portion of the rim crest and exhibits a variety of melt morphologies including flows, small ponds and a thin veneer. The arrows indicate a leveed channel and two small rim ponds (AS-17-2049).

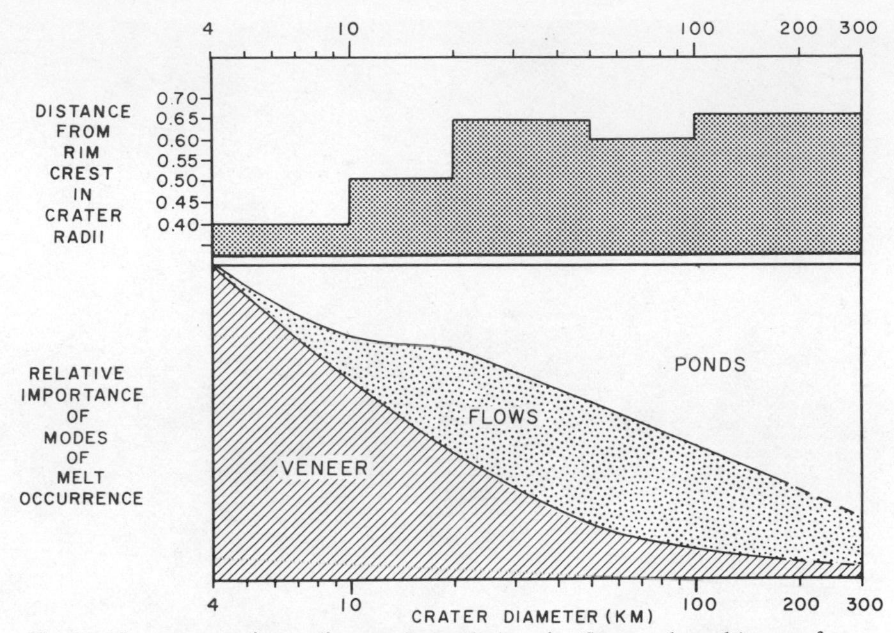

Fig. 5a. The mean maximum distance that melt deposits (flows and ponds) occur from the crater rim crest has been plotted for a variety of diameter intervals. The intervals are 4–10 km, 10–20 km, 20–50 km, 50–100 km, and 100–300 km. The trend suggests a small increase with crater size up to diameters of about 50 km. The bottom portion of the figure illustrates the change in the relative importance of modes of exterior melt occurrence with crater size. This part of the figure is schematic and should not be taken as a quantitative indication of either the volumes of melt present in each morphology or the relative areas covered.

potentially important implications for the processes responsible for melt emplacement is that while melt deposits are observed around craters less than 10 km in diameter, *large* amounts of impact melt are first seen on the rims of craters between 10 and 20 km in diameter. This is the size range in which profound changes occur in the morphology of crater interiors (Fig. 5b). At larger diameters, the modification processes responsible for these changes become more intense and melt deposits are more extensive. A relationship between crater modification and melt distribution is suggested.

D. Factors responsible for melt deposit asymmetry

Angle of incidence. Howard and Wilshire (1975) documented the asymmetry of melt deposits around numerous lunar impact craters. They noted that when oblique impact is suggested by the asymmetry of rays, secondary crater field, or ejecta blanket, the most extensive melt deposits are often in the inferred downrange direction. The data set assembled for this study was used to test this hypothesis. It was possible to determine both the direction of most extensive melt deposits and the inferred downrange direction based on crater ejecta asymmetry for 39 of the craters in the sample. Of these, the direction vectors coincide in 23% of the cases and are within 22.5° in another 23% (Fig. 6).

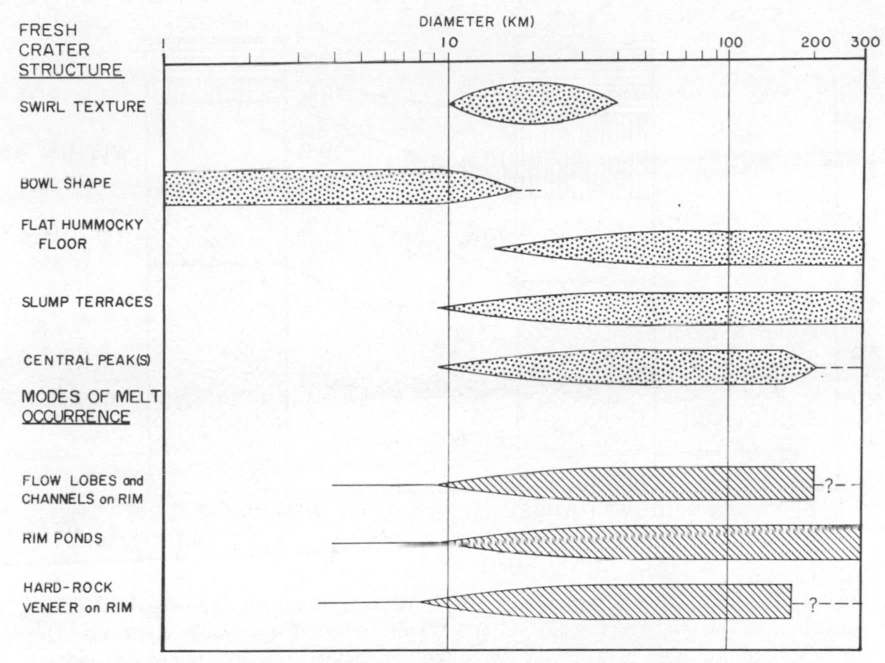

Fig. 5b. Change in crater structure and morphology of melt deposits as a function of crater diameter. Note that the exterior melt deposits become common in about the same size range as do central peaks and terraces, structures thought to be formed during the modification stage of the cratering event. Data are from Howard (1974), Smith and Sanchez (1973), and our own observations.

However, a significant number of craters show no obvious correspondence between projectile approach direction and melt distribution. Angle of incidence appears to be important in some cases (Fig. 6) as suggested by Howard and Wilshire (1975), especially at very large crater diameters ($D > 130$ km) as will be discussed later. However, additional factors appear to be necessary to explain the observed range of melt asymmetries.

Role of pre-event topography in controlling melt distribution. Hawke (1976) has recently presented the results of a study of the melt deposits associated with King crater which emphasized the influence of pre-impact topography in producing the observed melt distribution. King (71 km) is a fresh farside crater that has extensive melt deposits on its northern rim. Figure 7a shows King crater and the surrounding area and Fig. 7b is a sketch map of the region. Two large pre-existing craters (117 and 40 km in diameter) can be seen north of the King target site. King is superposed on the rims of these two pre-existing topographic lows, producing a segment of the King rim crest which is abnormally low. A plot of rim crest elevation as a function of position around the crater (Fig. 8) shows that much of King's northern rim stands 1–2 km above the floor as opposed to 4–5 km for the remainder. All of King's rim flows and ponds are located in the area adjacent to this rim crest low. Shock-melted material

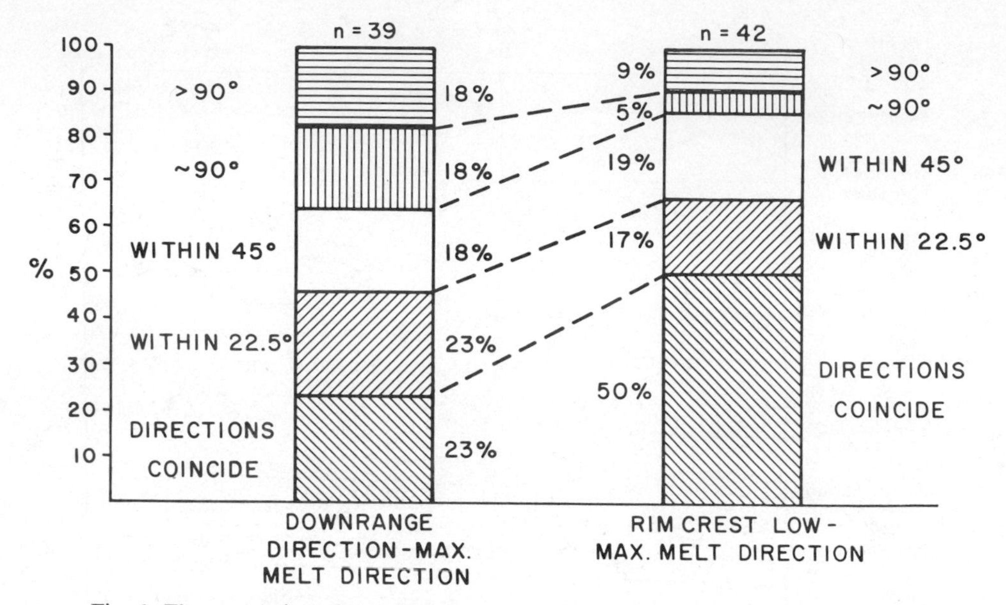

Fig. 6. The correlation of the direction of most extensive melt deposits with (1) the direction of low rim crest segments and (2) the inferred downrange direction. The angles listed are those between the two direction vectors. N is the number of craters used in each case.

appears to have escaped preferentially over the breached northern wall late in the cratering event. An additional effect of the smaller pre-existing crater was to provide the basin in which molten material collected.

Since the existence of a rim low appears to have been important in producing the observed King melt distribution and since other workers (Howard, 1972; Schultz, 1976) have noted a melt deposit–rim low correspondence at individual craters, the craters in this study were examined for any correlation of melt deposit location and rim crest elevation. The direction of both the most extensive melt deposits and the lowest rim crest segment could be determined with confidence for 42 of the craters under consideration. The results are shown in Fig. 6. In 50% of the cases the directions coincide exactly. An additional 17% have directions which are within 22.5°. The direction vectors are separated by 90° or more in only 14% of the population. In almost every instance, the lowest segment of the crater rim crest occurs where the younger crater intersected a pre-existing topographic low, usually a crater.

The pre-event topography is important not only in controlling the direction of the most extensive melt deposits but also may influence the distance that the molten material can travel. It seems significant that in those instances where melt deposits are found at large distances from the rim crest ($>0.80R$) there is usually a major pre-existing topographic low in the same direction and this low was intersected by the parent crater (87% of the cases).

Study of the farside crater Necho (33 km) provided additional insight into the relationship between crater morphometry and melt distribution. Abundant ponds and flows of impact melt can be seen on the northeast rim as well as within the crater itself (Figs. 4 and 10). A low rim crest segment was produced where Necho intersected the rim of a pre-existing 200 km crater (Fig. 9) and the most extensive melt deposits are found adjacent to this rim low. Ejecta distribution and ray patterns suggest that the projectile approached from the south. Figure 11 illustrates the relationship of the melt deposits to the rim crest elevation and the inferred downrange direction of an oblique impact. The melt deposits correspond more closely to the low rim crest segment than to the inferred downrange direction.

One of the most striking features of Necho crater is its asymmetrically terraced wall. There is considerable variability in the width of the terraced wall as can be seen in Fig. 11, where width is plotted as a function of position around the crater. The terraced wall is widest on the topographically high southwest side of the crater. Wall slumping appears to have been most extensive in areas where the pre-impact topography was high (Fig. 11). The exterior melt deposits are adjacent to the lowest segment of the rim crest and to that portion of the wall where minimum slumping occurred. The melt deposits are also directly opposite the region of maximum slumping as indicated by the width of the terraced wall.

This configuration was commonly observed in the craters analyzed and the relationship between maximum slumping (generally indicated by the width of terraced walls) and the distribution of impact melt was evaluated. The direction of the most extensive melt deposits and that of the maximum width of the terraced wall could be determined for 32 of the craters in the population. In 47% of the cases, the directions are essentially opposite. In only 22% of the cases is the maximum slumping on the same side of the crater as the most extensive deposits of impact melt (Fig. 12). It could be argued that melt deposits were initially abundant in the area where maximum wall slumping occurred, but that they were subsequently carried into the crater cavity with the slump terraces, thus creating an apparent melt deficiency on this area of the rim. However, melt deposits generally extend far beyond the area affected by wall slumping and examination of individual craters shows that wall ponds and flows are commonly not as abundant in the area where slumping has been more extensive (Fig. 10).

On the basis of the above correlations, both the rim crest elevation and the location of the maximum slump zone are related to the direction of most extensive melt deposits on crater rims. More extensive slumping was commonly observed on the side of the crater which had been topographically higher prior to the impact event. Higher topography probably acted to promote slumping and the formation of a wider terraced wall.

Topographic effects on ejecta asymmetry. The use of asymmetries of rays, secondary crater chains, and ejecta blankets to infer the direction of projectile approach for an oblique impact may not be valid in all cases. While obliquity of the impact is a common cause of asymmetrical ejecta distribution (Moore, 1971, 1972), El-Baz and Worden (1971) suggested the importance of additional factors. These include topographic shadowing by positive features, differences in the

Fig. 7. (a) Low sun-angle metric photograph of the King crater region (Apollo 16 metric photograph 3002). (b) Sketch map of the region shown in (a). A portion of the King rim crest is low because the crater intersected two pre-existing craters. The 20 km wide exterior melt pond is adjacent to the low rim crest segment.

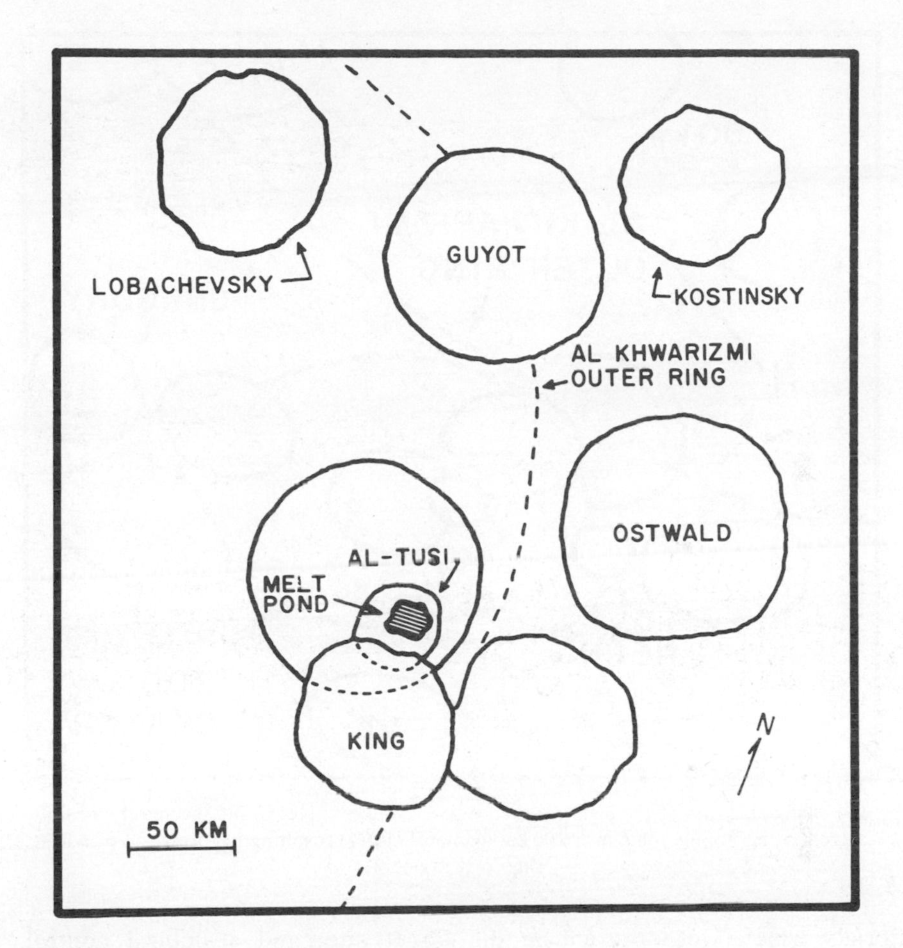

7(b)

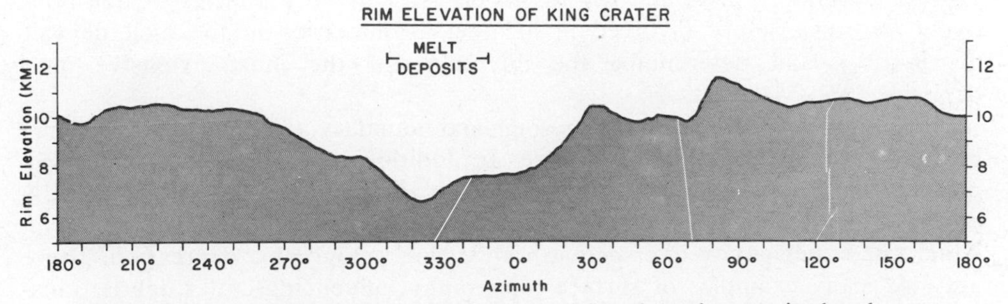

Fig. 8. Rim crest profile of King crater. The elevation of the rim crest is plotted as a function of position around the crater. Much of the rim crest stands 4–5 km above the crater floor as opposed to 1–2 km for the rim segment adjacent to the most extensive exterior melt deposits.

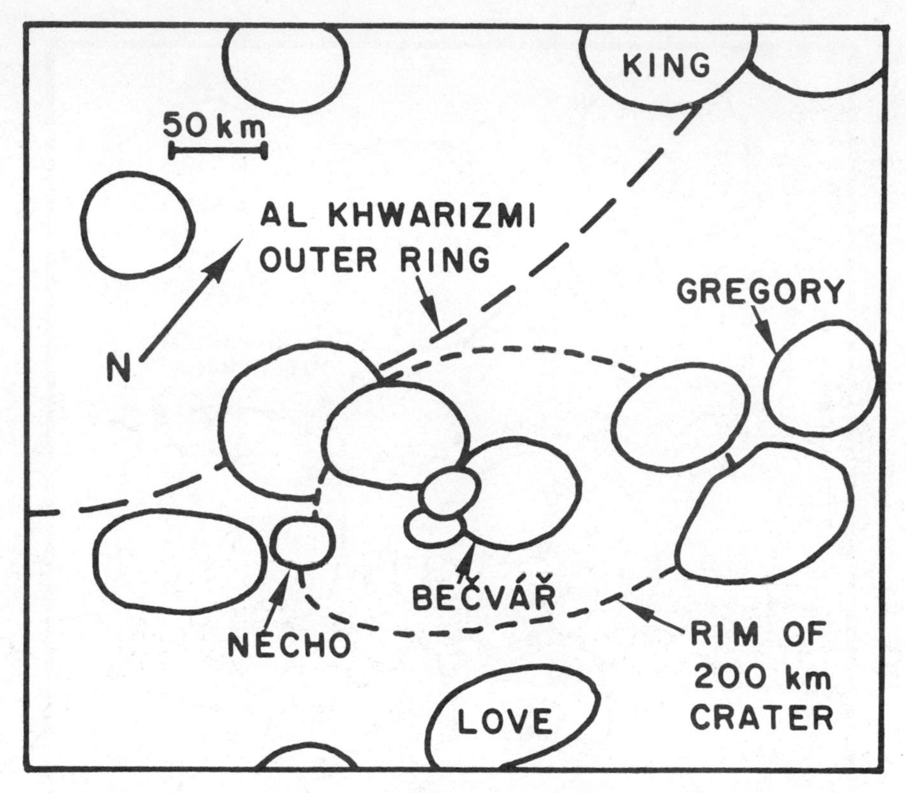

Fig. 9. Sketch map of the Necho crater region. Note that Necho intersected the rim of a pre-existing topographic low (200 km diameter crater) resulting in formation of a low rim crest segment.

materials which were present in the target site, and structural control. In addition, Soderblom (1970) has discussed the downslope displacement of the center of mass of the ejected material that results from impact on a sloping surface. There are also uncertainties involved with the determination of ejecta blanket, secondary field, and ray asymmetries. This is in contrast to the relatively low uncertainty involved in locating asymmetries in the melt deposit distributions and determining the directions of the most extensive melt deposits.

If a crater is situated near a mare-highland boundary, the ejecta and rays may be more visible on the mare surface. In addition to topographic shadowing, pre-existing topography may influence ejecta distribution in other ways. It is possible that ejecta may be more extensively distributed in the direction in which the crater cavity intersects a pre-existing topographic depression. There are also many examples of surface topography influencing crater debris transport. Howard (1972) suggested that the lower north rim of King would allow easier egress of ejecta, and he also gave several illustrations of continuous ejecta having been greatly influenced by surface topography. Many of the craters in our population show correspondence between the apparent direction of the most extensive ejecta deposits and the low rim crest segment. Additional studies are in progress to further investigate topographic control of ejecta distribution. It seems that while oblique impact is the dominant factor in producing asymmetrical ejecta, ray, and crater chain distributions, a variety of topographic effects can produce extensive ejecta deposits adjacent to rim lows. Thus, both melt deposit and ejecta asymmetry appear to be controlled in part by pre-existing topography. This may account for at least some of the instances in which the direction of the most extensive melt deposits corresponds to the inferred downrange direction for an oblique impact and would suggest an even smaller role for oblique impact in controlling the distribution of impact melt.

The influence of crater size. While the data indicate the pre-impact topography is the most important factor in producing the observed melt asymmetries (Fig. 6), oblique impact may nevertheless play some role. Some craters exhibit exterior melt deposits which appear to correspond only to the inferred downrange direction, and there is evidence that topographic effects may not be important at large crater diameters. The six largest craters under study (130–300 km) seem to show little correlation between the direction of the most extensive melt deposits and rim height or pre-impact topography but do correlate well with the inferred downrange direction for an oblique impact. Above a certain size (100–130 km), an impact crater may fail to be influenced by topographic variations on the scale generally seen in the lunar highlands (approximately 2–4 km).

The majority of craters smaller than 130 km show melt distribution–crater morphometry relationships very similar to those described at Necho where the melt deposits are most extensive adjacent to a rim low and are opposite the area of maximum wall slumping. Familiar lunar craters which exhibit similar relations are Lalande, King, Rutherfurd, O'Day, Theophilus, Fabricius, and Anaxagoras.

E. Time of melt emplacement

A key question in understanding the origin of impact melt deposits is their time of emplacement. Knowledge of the time of melt deposition is important, not only in the interpretation of lava-like material as impact melt but also for the clues it offers as to the processes responsible for melt emplacement. We refer to the *excavation stage* of the cratering event, in the sense of Gault *et al.* (1968), as the time period when the greatest bulk of the material is removed from the expanding cavity under conditions of lower stresses and ejection velocities than typical of the initial *compression stage.* Gault *et al.* (1968) used the term *modification stage* to refer to both short term and long term modification and degradation processes. However, we use the term in a more restricted sense, as the *modification stage of the cratering event* (see Settle and Head, 1977) to refer to those processes, primarily wall slumping and rebound, operating to modify the transient crater cavity in the terminal stages of the event.

(a)

Fig. 10. (a) Mosaic of Apollo 17 pan frames showing Necho crater (33 km in diameter). (b) Sketch map of Necho crater.

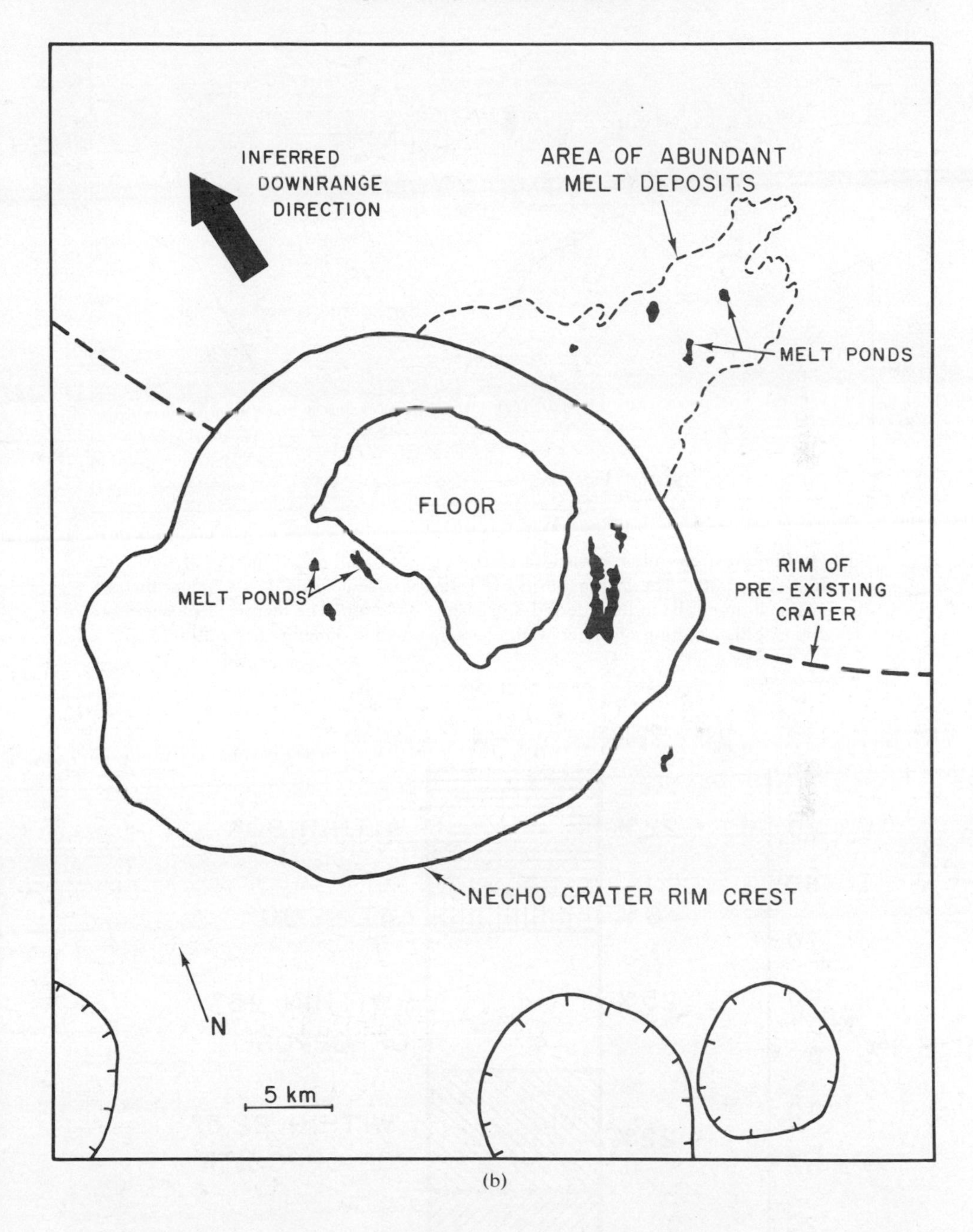

(b)

It has been noted by many workers that melt deposits commonly overlie the continuous ejecta deposits of the parent crater (e.g., El-Baz, 1972; Howard, 1972; Howard and Wilshire, 1975). The results of this study support these observations. The stratigraphic relationships demonstrate that the melt was emplaced after the bulk ejecta, either because it was removed from the crater

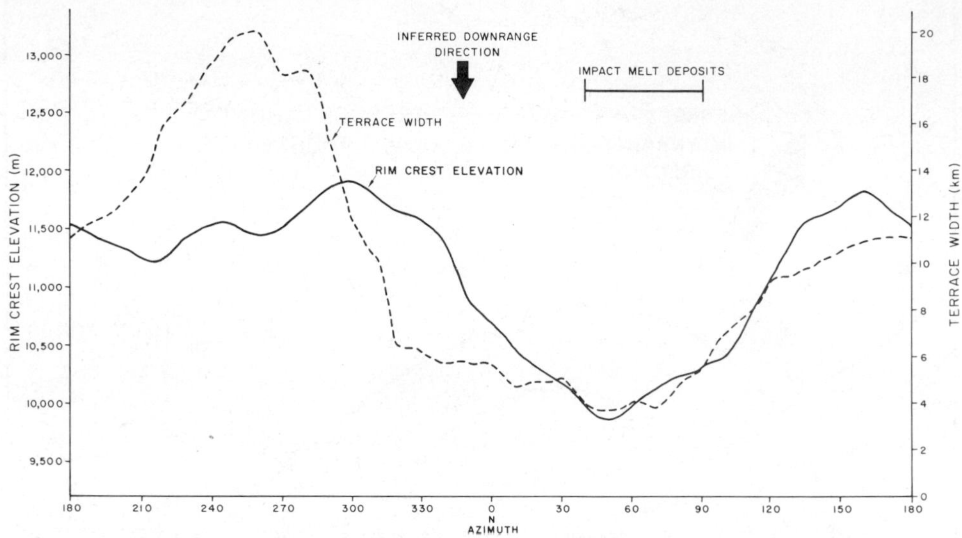

Fig. 11. Rim crest elevation and width of the terraced wall as a function of position around Necho crater. The melt deposits are found adjacent to the low segment of the rim crest and do not lie in the inferred downrange direction. The melt is also generally opposite that section of crater wall where maximum slumping has occurred.

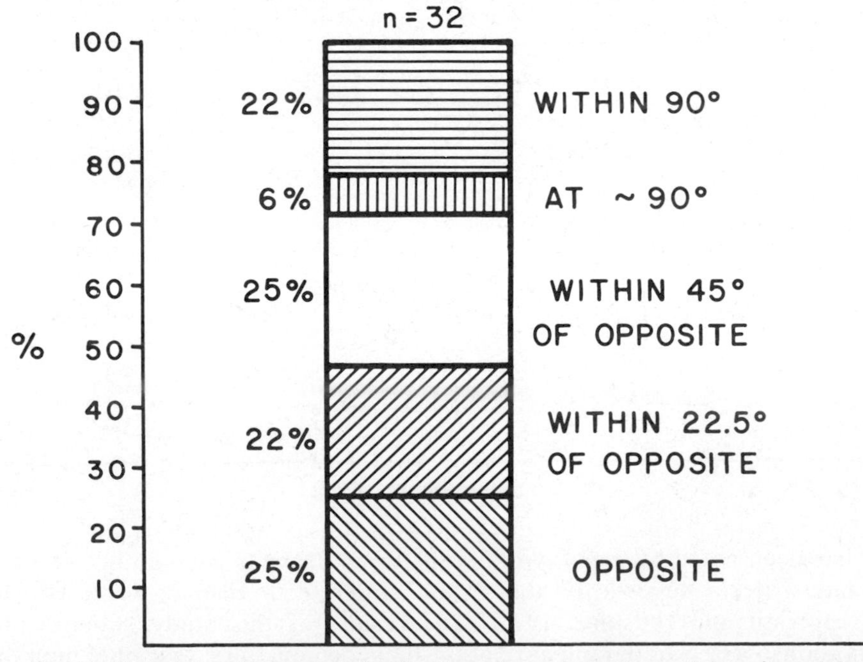

Fig. 12. The relationship between the direction of most extensive melt deposits and the direction of maximum wall slumping, generally indicated by the maximum width of the terraced wall, for 32 lunar craters.

after the bulk ejecta or because it was ejected at higher angles. Flow features have been located which appear to have been cut by the concentric faults which bound the crater lip and terrace segments (for example, note the truncated flow on the rim of the crater near Weiner shown in Fig. 10 of Howard and Wilshire, 1975). If these truncated flow features have been cut by the faults which formed during the later part of the crater modification stage of the impact cratering event and are not simply the result of more recent mass wasting, they would indicate that the melt had been emplaced and had flowed before the termination of the modification stage. Such features are rare and most crater modification must have ceased a short time after the impact event because most melt deposits on the walls and rim have flowed and ponded in response to subjacent wall and rim topography. The above relationships strongly suggest that the melt deposits were emplaced after the bulk ejecta and flowed to form the observed occurrences after the majority of the modification stage of the cratering event. The occurrence of some flow features which appear to have been truncated by faults which formed late in the modification stage of the cratering event suggests that some flows must have solidified quickly and implies a close association in time between the impact event and the emplacement of melt deposits. In addition, if some of the flows around Tycho have been impacted by late-arriving secondaries as suggested by Shoemaker *et al.* (1968) and Howard and Wilshire (1975), these flows must have formed and solidified (probably due to a high content of cold, unmelted inclusions; see Simonds, 1975) within a matter of minutes after the impact event. The above observations are difficult to explain by the viscous creep melting model of Hulme (1974) which requires a considerable interval between crater formation and melt emplacement.

Dence (1971) has suggested that the sheet of impact melt which lines the transient cavity may have extended over the cavity rim. This model implies that the melt was ejected during the terminal stages of the excavation stage. While some melt may have been ejected at this time, the following evidence suggests that most of the observed melt was emplaced on the walls and rim *during the modification stage* and then flowed to assume the observed morphologies just after this stage: (1) Melt deposits are often seen superposed on the swept zones of impact craters. The ejecta deposited on the crater rim is the last material excavated and the first to be deposited (Oberbeck, 1975) and the swept zone is thought to have been formed by the radial flow of the upper portion of this rim material after deposition (Guest, 1973; Howard, 1974). The most likely explanation for the presence of melt deposits on swept zones is that the molten material was emplaced at some later time, probably during the modification stage. Any early melt that was deposited immediately on the rim material should have been removed with the upper portion of the rim material during the radial flow interval. (2) Melt deposits are often draped over extensive portions of crater walls (Hawke and Head, 1977) and often cover topography formed during the modification stage. While such wall deposits may be explained as the result of disruption and flowage of the molten material present on the rim of the transient cavity, it seems more likely that much of this melt was emplaced during the crater modification stage. At numerous craters, the amounts of melt that appear to have moved down the walls seem much greater than the amounts which could have reasonably drained from the surfaces of visible slump terraces. (3) Howard (1974) has described dark rays crossing crater walls and rims which he interpreted as trains of glassy ejecta emplaced very late in the cratering event.

IV. Model for the Formation of Impact Melt Deposits on Crater Rims

A major problem in the study of lunar impact craters has been understanding the mechanism by which significant quantities of molten material were emplaced on crater rims (see Oberbeck, 1975). Experimental cratering studies (Stöffler *et al.*, 1975) and models based on the study of terrestrial impact craters (Dence, 1968) suggested that the melt material that was ejected would have high velocities and would be found at great distances from the crater. Some molten material is almost certainly being ejected from the transient cavity throughout its growth as suggested by the cratering experiments of Stöffler *et al.* (1975). This molten material is either widely dispersed at great distances from the parent crater or thoroughly mixed with the bulk ejecta, so little photogeologic evidence for its existence is observed. Large concentrations of impact melt are observed on crater rims and any model which attempts to explain these occurrences must account for the following:

(1) The common occurrence of exterior melt deposits adjacent to topographically low segments of the rim and opposite regions of the crater wall which have experienced maximum slumping.
(2) The change in melt deposit morphology with crater size (Fig. 5a).
(3) The relatively greater amounts of melt present on the rims of the larger craters.
(4) The increase in relative maximum distance that melt occurs from the rim crest with increasing crater size (Fig. 5a).
(5) The observation that exterior melt deposits become more abundant around craters in the same size range that profound changes occur in crater interior morphology (Fig. 5b). Increases in melt amounts and distances of maximum occurrence parallel increases in the degree of crater modification.
(6) The occurrence of the melt deposits overlying features formed during the modification stage of the cratering event.

Various models for melt emplacement on crater rims have been proposed. Oberbeck (1975) showed that in layered targets with coherent substrates the last material removed during the excavation stage exits the cavity at very high angles and is therefore deposited in and around the crater. Grieve (1975) and Grieve *et al.* (1976) employ the results of particle motion studies to explain the distribution of melts at terrestrial impact craters. They suggest that accelerated target materials are deflected outwards and upwards by the interaction of rarefaction waves, and some of the melt leaves the expanding cavity to form suevite-type deposits. The remainder has insufficient velocity for ejection due to particle velocity attenuation and remains within the impact structure. While there is evidence to support each model and some melt may be emplaced by the processes described, neither adequately accounts for the melt distribution–crater morphology correlations and relationships described in this paper.

On the basis of our observations the following model is proposed for the concentration of melt on lunar crater rims: At the end of the excavation stage of the cratering event, the transient cavity is lined with shock-melted material. Greater concentrations of melt are present in the bottom of the cavity. The modification stage begins with the initial collapse and rapid shallowing of the transient cavity. This rapid decrease in depth is due largely to wall slumping and rebound. As the crater shallows due to this activity, molten material is moved upward. With continued wall slumping and rebound, some of the molten material is removed from the cavity in a direction commonly opposite the area of maximum wall slumping and over a topographically low portion of the crater rim. The material is often given a lateral component of movement by terrace formation. The melt is ejected at low velocities and is emplaced on the crater rim. The molten material that does not escape the crater generally settles onto the crater floor but some remains as ponds, flows, and veneer on the interior crater walls. After the modification stage is complete, impact melt material flows in response to the local topography to form the deposits we see today.

Acknowledgments—This work was carried out under NASA Grant NGR-40-002-116 from the Office of Space Science Lunar and Planetary Programs. Thanks are also extended to the National Space Science Data Center for many of the photographs used in this study and to NASA and the Defense Mapping Agency for the topographic maps (Lunar Topographic Orthophotomaps) used in this study. Special appreciation is due to Mark Cintala for his useful suggestions and the identification of some of the melt deposits used in this study, and to C. Wood and P. Spudis for helpful discussions. A. Gifford, S. Matarazza, and D. Roth aided in the preparation of manuscript and figures. B. K. Lucchitta and H. J. Moore provided helpful reviews.

References

Dence, M. R.: 1968, Shock zoning at Canadian craters: Petrography and structural implications (B. M. French and N. M. Short, eds), *Shock Metamorphism of Natural Materials*, p. 169–184, Mono Book Corp., Baltimore.

Dence, M. R.: 1971, Impact melts, *J. Geophys. Res.* **76**, 5552–5565.

El-Baz, F.: 1970, Lunar igneous intrusions, *Science* **167**, 49–50.

El-Baz, F.: 1972, King Crater and its environs, NASA SP-315, p. 29-62–29-70.

El-Baz, F. and Worden, A. M.: 1971, Orbital science investigations, NASA SP-289, p. 25-1–25-27.

French, B. M.: 1972a, Production of deep melting by large meteorite impacts: The Sudbury Structure, Canada, *Proc. 24th Internat. Geol. Congr.*, Sect. 15, 125–132.

French, B. M.: 1972b, Shock-metamorphic features in the Sudbury Structure, Ontario: A review, *Geol. Association Canada, Special Paper No. 10*, 19–28.

Gault, D., Quaide, W., and Oberbeck, V.: 1968, Impact cratering mechanics and structures, (B. M. French and N. M. Short eds.), *Shock Metamorphism of Natural Materials*, p. 87–99, Mono Book Corp., Baltimore.

Grieve, R. A. F.: 1975, Petrology and chemistry of the impact melt at Mistastin Lake Crater, Labrador, *Geol. Soc. Amer. Bull.* **86**, 1617–1629.
Grieve, R. A. F., Plant, A. G., and Dence, M. F.: 1974, Lunar impact melts and terrestrial analogs: The characteristics, formation and implications for lunar crustal evolution, *Proc. Lunar Sci. Conf. 5th*, 261–273.
Grieve, R. A. F., Dence, M. F., and Robertson, P. B.: 1976, The generation and distribution of impact melts: Implications for cratering processes, (abstract). In *Papers Presented to the Symposium on Planetary Cratering Mechanics*, p. 40–42. The Lunar Science Institute, Houston.
Guest, J. E.: 1973, Stratigraphy of ejecta from the lunar crater Aristarchus, *Geol. Soc. America Bull.* **84**, 2873–2894.
Hawke, B. R.: 1976, Ponded material on the north rim of King Crater: Influence of pre-event topography on the distribution of impact melt, *EOS*: *Trans. Amer. Geophys. Union* **57**, 275.
Hawke, B. R. and Head, J. W.: 1977, Impact melt in lunar crater interiors, (abstract). In *Lunar Science VIII*, p. 415–418. The Lunar Science Institute, Houston.
Head, J. W.: 1974, Orientale multi-ringed basin interior and implications for the petrogenesis of lunar highland samples, *The Moon* **11**, 327–356.
Head, J. W.: 1976a, Lunar volcanism in space and time, *Rev. Geophys. Space Phys.* **14**, 265–300.
Head, J. W.: 1976b, Evidence for the sedimentary origin of Imbrium sculpture and lunar basin radial texture, *The Moon* **15**, 445–462.
Howard, K. A.: 1972, Ejecta blankets of large craters exemplified by King Crater, NASA SP-315, 29-70–29-77.
Howard, K. A.: 1974, Fresh lunar impact craters: Review of variations with size, *Proc. Lunar Sci. Conf. 5th*, p. 61–69.
Howard, K. A. and Wilshire, H. G.: 1973, Flows of impact melt at lunar craters, (abstract). In *Lunar Science IV*, p. 389–390. The Lunar Science Institute, Houston.
Howard, K. A. and Wilshire, H. G.: 1975, Flows of impact melt at lunar craters, *J. Res. U.S. Geol. Survey*, **3**, no. 2, 237–251.
Hulme, G.: 1974, Generation of magma at lunar impact crater sites, *Nature*, **252**, 556–558.
Mattingly, T. K., El-Baz, F., and Laidley, R. A.: 1972, Observations and impressions from lunar orbit, NASA SP-315, 28-1–28-16.
Moore, H. J.: 1971, Craters produced by missile impacts, *J. Geophys. Res.* **76**, 5750–5755.
Moore, H. J.: 1972, Ranger and other impact craters photographed by Apollo 16, NASA SP-315, 29-45–29-51.
Moore, H. J., Hodges, C. A., and Scott, D. H.: 1974, Multi-ringed basins—illustrated by Orientale and associated features, *Proc. Lunar Sci. Conf. 5th*, p. 71–100.
Oberbeck, V. R.: 1975, The role of ballistic erosion and sedimentation in lunar stratigraphy, *Rev. Geophys. Space Phys.* **13**, 337–362.
Quaide, W. L. and Oberbeck, V. R.: 1968, Thickness determinations of the lunar surface layer from lunar impact craters, *J. Geophys. Res.* **73**, 5247–5270.
Schultz, P. H.: 1976, *Moon Morphology*, University of Texas Press, 626 pp.
Settle, M. and Head, J. W.: 1977, The role of rim slumping in the modification stage of impact crater formation. To be submitted to *J. Geophys. Res.*
Shoemaker, E. M.: 1962, Interpretation of lunar craters, (Z. Kopal, ed.), *Physics and Astronomy of the Moon*, Academic Press, N.Y.
Shoemaker, E. M., Batson, R. M., Holt, H. E., Morris, E. C., Rennilson, J. J., and Whitaker, E. A.: 1968, Television observations from Surveyor VII, NASA TR 32-1264, 9–76.
Simonds, C. H.: 1975, Thermal regimes in impact melts and the petrology of the Apollo 17 Station 6 boulder, *Proc. Lunar Sci. Conf. 6th*, p. 641–672.
Smith, E. I. and Sanchez, A. G.: 1973, Fresh lunar craters: Morphology as a function of diameter, a possible criterion for crater origin, *Mod. Geol.* **4**, 51–59.
Soderblom, L. A.: 1970, The distribution and ages of regional lithologies in the lunar maria, Ph.D. thesis, California Institute of Technology.
Stöffler, D., Gault, D. E., Wedekind, J., and Polkowski, G.: 1975, Experimental hypervelocity impact into quartz sand: distribution and shock metamorphism of ejecta, *J. Geophys. Res.* **80**, 4062–4077.

Strom, R. G. and Fielder, G.: 1970, Multiphase eruptions associated with the craters Tycho and Aristarchus, *Arizona Univ. Lunar Planetary Lab. Commun.*, v. 8, pt. 4, no. 150, 235–288.

Strom, R. G. and Fielder, G.: 1971, Multiphase eruptions associated with the craters Tycho and Aristarchus, In *Geology and Physics of the Moon* (G. Fielder, ed.), p. 55–92. Elsevier.

Strom, R. G. and Whitaker, E. A.: 1971, An unusual far-side crater, NASA SP-323, 20–24.

Taylor, S. R.: 1975, *Lunar Science: A Post-Apollo View*, Pergamon, 372 pp.

Roddy, D. J., Pepin, R. O., and Merrill, R. B., editors.
(1977) *Impact and Explosion Cratering*, Pergamon Press (New York), p. 843–859.
Printed in the United States of America

Some peculiarities of selective evaporation in target rocks after meteoritic impact

O. V. PARFENOVA

Moscow State University, Department of Petrography

O. I. YAKOVLEV

V. I. Vernadsky Institute of Geochemistry and Analytical Chemistry,
USSR Academy of Science, Moscow

Abstract—Selective evaporation causes chemical alteration of impact products during impact metamorphism. A comparison between chemical alteration of crater rock and of evaporated experimental products indicates the existence of distinct similarities. The following peculiarities of selective evaporation of crater rocks were found: (1) As a result of impact melting and evaporation, the ratio K_2O/Na_2O increases in glass impact rocks. This phenomenon may be explained by more intensive removal of Na_2O in vapor phase; (2) the ratio K_2O/Na_2O in glass impact rocks depends on the acid-base properties of target rocks. The highest values of K_2O/Na_2O correspond, as a rule, to targets with the highest acidity. This relationship also exists in experimental results; (3) the content of SiO_2 decreases in some glass impact rocks relative to the target rocks as well as in balance calculations. This may be explained by rather high fugacity of SiO_2. The SiO_2 fugacity in ultra-acid melts may even be higher than the fugacity K_2O. These relationships of compositional variations from target rock to glass impact rock are important petrochemical criteria of impact explosive craters.

STUDY OF IMPACT EXPLOSION and its products is something new and unusual for petrologists. There is no other known geological phenomenon in which, for a very short time, such high pressures (up to several megabars) and temperatures (up to hundreds of thousand degrees) can develop. The target substance, depending on the epicentrum of the explosion, is converted to plasma or vaporized, melted, deformed, fragmented, and scattered over large distances. Moreover, the short duration of the impact explosion apparently gives rise to highly imbalanced alteration processes, which makes things yet more complicated.

Despite these difficulties, the study of impact craters has been expanded and intensified in recent years. A primary objective in present research is the identification of reliable structural-geological and mineral-petrographic features unambiguously indicative of an impact explosion. This is indeed necessary, for there is no consensus of opinion as to the way the explosive craters are formed. The meteorite impact hypothesis generally is poised against the possibility that the structures were formed by explosions accompanying volcanic eruptions. A great number of studies have compared the characteristics of explosion craters of a doubtful origin with structures of authenticated volcanic origin. A major difficulty in deciding whether a ring structure is an impact or a volcanic feature is the great mineral-petrographic similarity of the rocks. For example, both authenticated volcanic and impact structures contain similar glassy rocks occurring as covers or flows. In different craters, these rocks are either wholly glassy or contain inclusions of rocks and mineral fragments. Various authors often classify these rocks as lavas, tuffs, or ignimbrites. In spite of all the apparent similarities between volcanic and impact products, the two processes of formation of the ring structures are utterly different in explosion mechanism and physical parameters. This is bound to be reflected in the mineralogical, petrographic, and chemical features of the crater rocks.

A specific feature of an explosion produced by a high-speed object is that it forms a high-temperature melt capable of complete, or, at lower temperatures, selective vaporization. This study is concerned with the chemical properties of glass-containing rocks resulting from such impacts and subjected to selective vaporization. It is based on the similarities between rocks of impact craters and experimental high-temperature evaporation in vacuum melts.

I. Experimental and Theoretical Data of the Impact Explosion and High-Temperature Vaporization

Before we proceed with an interpretation of the composition of impactites, let us briefly review some physical properties of a shock-induced explosion and the physico-chemical laws of the vaporization process established experimentally.

Short characteristics of high-speed impact

The physical theory of high-speed shock has been developed by Stanyukovich (1971), Sedov (1957), and Zel'dovich and Reizer (1966). According to this theory, when a meteorite hits the surface of a planet at a speed of as much as several tens of kilometres a second, an explosion-like event takes place as a result of the abrupt deceleration of the meteorite against the ground and conversion to heat of the kinetic energy. In the process, the meteorite and some of the target substance undergo alterations, from simple mechanical fragmentation to transition into another aggregate state (melt and vapour). Theoretic calculation carried out by Zel'dovich and Reizer (1966) shows that in the case of an impact of a meteorite moving at a speed of 15 km/sec, the proportions of the mass of a granite target that will be vaporized, melted, and fragmented are respectively, 1:16:100 (in meteorite mass units).

From the impact explosion theory it also follows that from the energy release site, a shock wave will spread with a front pressure of over one million atmospheres.

As the dense high-temperature gas expands (the initial temperature may be as high as 10,000°K), or, to be more exact, as the reflected wave spreads, the meteorite and some target material scatter in space in the form of vapour, melt, and solid fragments thus forming the typical impact feature, the crater. One feature of the process is the extremely abrupt drop in gas density from the edge of the semi-sphere to the explosion centre. Practically the whole gas mass gathers into a thin layer near the front surface; in the middle of the semi-sphere the gas density declines by a factor of tens and hundreds of thousands. The pressure, however, decreases only by two-thirds owing to the high temperature. Immediately after the explosion, the density of the gas in the sphere confined by the high-pressure shell is infinitesimal and tantamount to vacuum conditions (Sedov, 1957).

Thus, as a result of a high-speed meteorite impact, the substance of bedrock near the blast undergoes a strong rarefaction and a high temperature. These are conditions fit for chemical and physical transformations—vaporization of matter followed by condensation of some vapour into fluid, as well as selective volatilization of the shock-produced melt and impoverishment of the residual product in volatiles.

General physico-chemical characteristics of vaporization

Vaporization is a process depending on a great many factors, including temperature, pressure (general and partial pressure of individual gases), oxidizing-reducing conditions, the acid-basic features of the melt being evaporated, etc. The contribution of each factor to the alteration of the original rock composition has not been completely investigated. Yet experiments in the USSR and abroad in recent years have enabled some general laws of silicate melt vaporization to be established and be used for analysis of natural shock-metamorphic products. Some experimental data on vaporization of melts are given below. Special attention is given to vaporization of alkalies (K_2O, Na_2O) and silica (SiO_2) whose variation in natural shock processes is readily observable.

The experiments show that the volatility of any i-component in the melt depends on the concentration of that component in the melt, its individual features, the general chemical composition of the melt (or influences of all other components in the melt), the ambient pressure, and temperature. The general formula for the volatility of the i-th component is given by generalized Raul–Henry equation:

$$f_i = f_i^0 x_i \gamma_i, \tag{1}$$

where f_i—equilibrium volatility of the i-th component over the melt;
f_i^0—volatility of the saturated vapour of pure substance;
x_i—concentration in the melt; and
γ_i—activity coefficient.

The temperature is implicitly incorporated in (1), being functionally connected with f_i^0 and γ_i. The activity coefficient is a gross indicator of the interaction of the i-th melt component; it is determined by gross or generalized chemical characteristics of the medium. We shall now discuss the main physico-chemical factors influencing the vaporization process.

(a) *Temperature.* With increasing temperature, the pressure of the vapours of the components of the melt grows and the mass lost by vaporization is increased. With rising melt temperature, the individual components get vaporized more and more with a resultant reduction of their final-contents (Yakovlev and Kosolapov, 1976). Figure 1 shows the variation of Na_2O content in oligoclase melt and of K_2O content in potash feldspar melt as a function of temperature. The alkali content in melts of the minerals is seen to decline with growing temperature.

Silicate melts vaporize selectively, i.e., the melt components, due to their different physico-chemical properties, are vaporized in different fashions according to their volatilities, which mainly depend on the temperature. An experimental investigation of the basalt composition dependence upon the vaporization temperature at a constant pressure has shown that up to $T = 1300°C$, there is an intensive vaporization of K_2O and Na_2O; above that temperature, the melt noticeably loses FeO, and after 1400°C, SiO_2. The melt became sharply impoverished in silica at the temperature of 1600°C and higher. Meanwhile the contents of CaO, Al_2O_3, and TiO_2, owing to their low volatilities, were relatively growing in the temperature interval of 1300–1600°C. These components also begin to get vaporized above about 1600°C (Yakovlev *et al.*, 1973).

(b) *Gas pressure.* Experimental data show that vaporization depends on the general pressure of the gas medium (Yakovlev and Kosalopov, 1976). The more rarefied the gas the greater the alteration of the chemical composition of the melt being vaporized. As external pressure drops, the basalt melts during vaporization are relatively enriched in low-volatile TiO_2, CaO, Al_2O_3, and MgO while losing the high-volatile K_2O, Na_2O, and medium-volatile Fe_2O_3, SiO_2 (Fig. 2).

(c) *Initial concentration.* The initial concentration of a component in the melt is essential for evaluating its volatility during vaporization. A low concentration reduces the volatility even of a

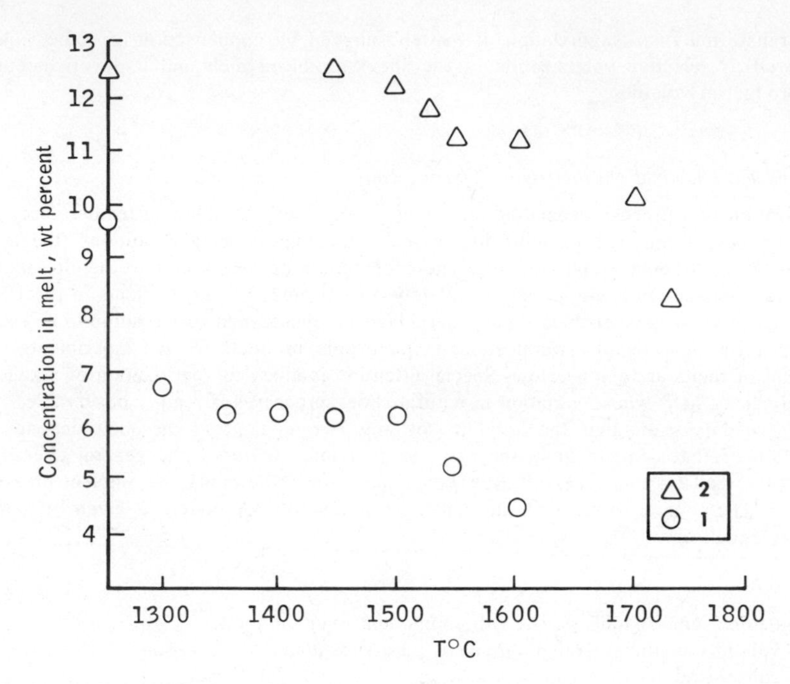

Fig. 1. Variation of the contents Na_2O in oligoclase and K_2O in K-feldspar versus temperature after vaporization of their melts in pressures 10^{-6} torr. 1—oligoclase (for Na_2O), 2—K-feldspar (for K_2O). The initial concentrations shown on the concentration axis.

high-volatile component, hampering its release from the melt (Yakovlev, 1974). Figure 3 shows that the end concentration of K_2O and Na_2O relative to the initial concentration remains practically unchanged through a wide range of temperatures of basalt melt being vaporized. The end concentration of K_2O and Na_2O in the residual melt averaged 0.19 and 0.30 wt.%, respectively. The initial concentration also is essential for the relative volatility of the substance. For example, in basalt with initial content of $K_2O = 1.93$ wt.%, and $Na_2O = 1.30$ wt.% and with $K_2O/Na_2O = 1.48$, vaporization in vacuum at 1300°C produced $K_2O/Na_2O = 0.51$. So, in this case the volatility of K_2O may be said to be higher than that of Na_2O. For basalt with initial content of $K_2O = 0.68$ wt.% and that of $Na_2O = 4.59$ wt.% and $K_2O/Na_2O = 0.15$, the residual melt after vacuum vaporization at $T = 1300$°C had $K_2O/Na_2O = 0.66$. The volatility of Na_2O in this case was, therefore, greater than that of K_2O (Yakovlev, 1974).

It should be borne in mind that a formal comparison of a component's concentration before and after vaporization may produce the wrong idea about its relative volatility. This is particularly true of medium-volatile substances, e.g., SiO_2. During basalt vaporization, silica content dropped against the relatively less volatile components depending on general pressure and temperature (Fig. 4). But a chemical analysis of the residual vaporization products, expressed in weight per cent of oxides, sometimes showed an increased SiO_2 content. For instance, basalt vaporized at $T = 1300$°C and $P = 5 \times 10^{-4}$ torr contained 2% more SiO_2 than the starting material (Yakovlev, 1974). Increased SiO_2 content, despite the component's high volatility, is due to a comparatively greater vaporization of other substances leading to ultimate increased weight share of silica in the residual melt.

(d) *Activity coefficient.* The role of this factor becomes clear if one looks at the vaporization products of melts of different compositions. The activity coefficient determines the activeness of the substance's behaviour in the melt, and it depends on the general chemical composition of the

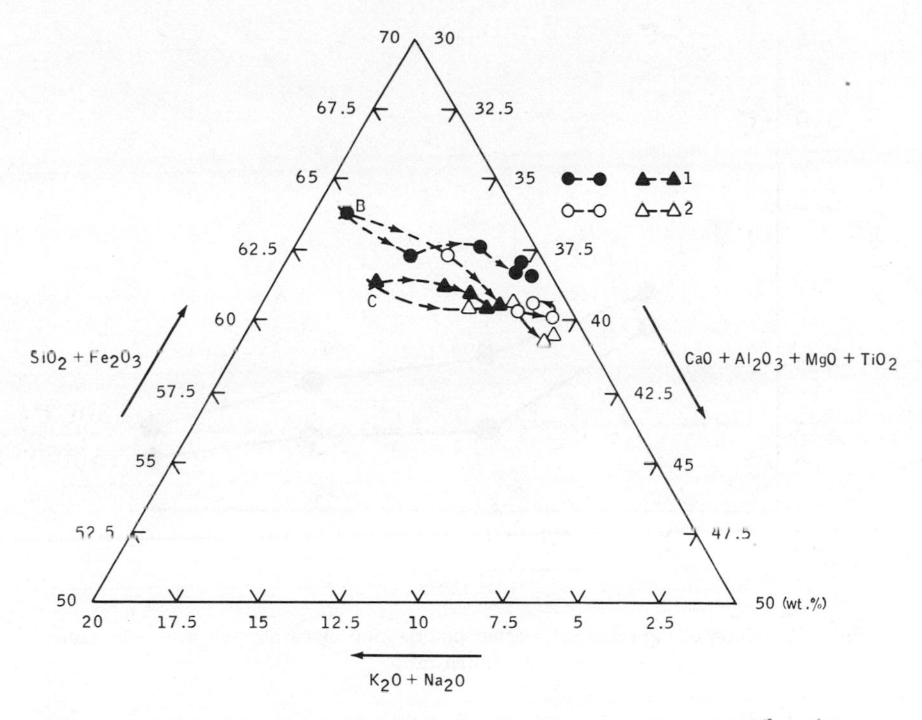

Fig. 2. Variation of the composition of the basalt melts versus pressure (10^{-3}–10^{-6} torr). Arrows show the direction of the alteration of composition from the initial value (basalts Band C) to compositions corresponding to pressures 10^{-3}, 10^{-4}, 10^{-5} and 10^{-6} torr respectively. 1—for $T = 1300°C$, 2—for $T = 1500°C$.

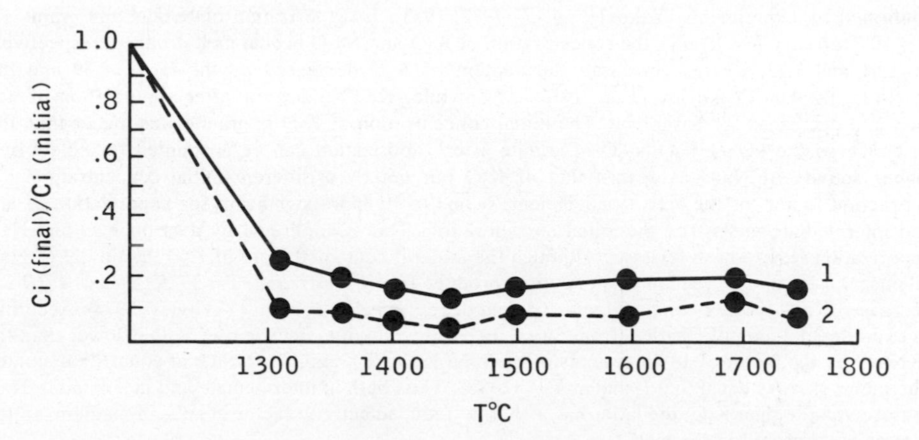

Fig. 3. Variation of the ratio C_i final/C_i initial in basalt melt versus temperature. $P = 10^{-6}$ torr. 1—for Na_2O, 2—for K_2O.

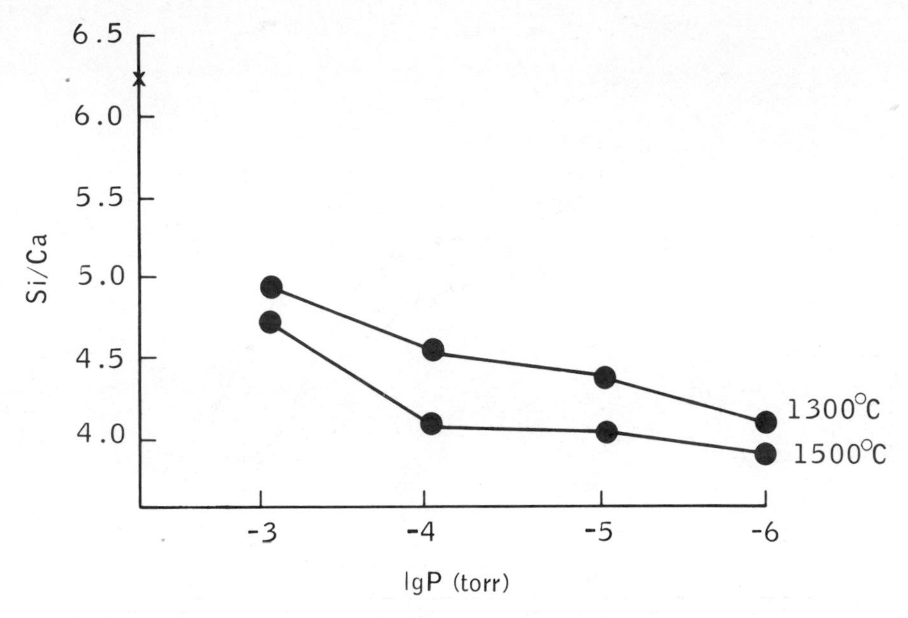

Fig. 4. Variation of the ratio Si/Ca after vaporization of basalt melt versus pressure. *x*—initial ratio.

medium being vaporized. For indicator of the medium's chemistry, one can take the activity of oxygen ions [O^{-2}], which will serve as indicator of the melt's acidity-basicity (Korzhinsky, 1963a). In view of the established dependencies of a component's activity on that of oxygen ions, the vaporization pattern of, for example, alkalies (potassium and sodium) and silica must be different for acid and basic melts. In a basic melt, the lost mass of potassium and sodium must be greater than in an acid melt (specifically, greater for potassium than for sodium), while the loss of silica must be greater in an acid melt than in a basic one. The conclusions, following from the general theory of Korzhinsky (1963a, b; 1966), about the acid-basic interaction of components in melt have been confirmed by experiments (Yakovlev *et al.*, 1972, 1973). In vaporization of basalts and granites at $P = 10^{-6}$ torr and $T = 1300°C$, the concentration of K_2O and Na_2O in acid melt dropped, respectively, by 1.01 and 1.13, whereas in basalt the content of K_2O decreased by the factor 8.39 and that of Na_2O, by 4.06 (Yakovlev *et al.*, 1973). Meanwhile, K_2O/Na_2O grew after vaporization of acid melt and decreased for basic melt. The initial concentration of K_2O in granite was higher than that of Na_2O, so the higher K_2O/Na_2O in granite after vaporization can be accounted for solely by a higher activity of Na_2O as against that of K_2O but not by a different initial concentration. The interaction factor, or the activity coefficient, seems to be more significant for vaporization of acid and intermediate melts than the initial concentration. This is confirmed by data on K_2O and Na_2O vaporization from potash feldspar. Although the mineral contained more of K_2O than it did of Na_2O (almost 4.4-fold), the vacuum vaporization product (at 10^{-5} torr and $T = 1725°C$) had a 2.2-fold decreased Na_2O content compared with 1.5-fold decreased K_2O content (Yakovlev, 1974). According to experimental data of Charles (from Story, 1973), the activity coefficient of K_2O is lower than that of Na_2O by the factor of 10^4 in systems of the type K_2O–SiO_2 and Na_2O–SiO_2 at equal alkali content (the molar share is equal to 0.1) and at $T = 1200°C$. Thus both in intermediate and in acid melt, Na_2O is more volatile than K_2O; the latter has a sharply reduced activity factor in an acid medium, so that it remains "locked" in the melt.

Experimental data proving high volatility of silica in acid melts are given by Walter and Giutronich (1967). As a result of high-temperature vaporization ($T \approx 2800°C$) of ultra-acid granite glass at $P = 1$ atm, SiO_2 concentration dropped from 82.20 to 45.00 wt.%. Compared with the volatilities of other components in ultra-acid melt, silica was inferior only to Na_2O.

Relative series of component volatilities in basic and acid melts

The joint action of the preceding vaporization factors determines for each component its place in the order of relative volatilities of petrogenic components of the melt. Vaporization study on basalt melts established this order of volatility of the main components: K > Na > Fe > Si > Mg > Ca > Al > Ti. Analysis of data given by Walter and Giutronich (1967),* on the high-temperature vaporization of acid melts (which we have confined to SiO_2 contents ranging from 82 to 70 wt.%) produces a different order of relative volatility: Na > Si > Al > [K, Fe, Mg, Ca, Al]. The positions of elements in brackets in the volatility order remain obscure.

The established high volatility of Na_2O, SiO_2, and low volatility of K_2O in acid melts are important experimental bases for analysis of the change suffered by high-temperature melts of an impact origin.

II. Brief Petrochemical Characteristics of Meteorite Crater Rocks

We have mentioned earlier that melted rocks of crater structures are represented by vitric or glass-containing units in which rock and mineral fragments are present alongside glass. A great amount of published analyses of this kind of rocks are available. As for analyses of the country rock enclosing the crater (target rocks), they are much fewer, due to the great variety of target rock compositions. Calculating the composition of an "averaged" target requires a great number of analyses of each variety to be carried out as well as knowledge of the volume ratios of these rocks. Table 1 gives analyses of target rocks and of glassy rocks of ten craters from published data (Bunch, 1972; Currie and Shafigullah, 1967, 1968; Engelhardt, 1972; Taylor and Kolbe, 1964). The craters Brent, Carswell, East and West Clearwater, and Manicouagan are in Canada, Lonar is in India, Monturaqui in Chile, the Henbury group of craters in Australia, crater Nordlinger Ries in West Germany, and Popigai in the USSR.

Melted rocks in most craters are composed of glass and mineral fragments and the analyses are for the whole rock. In the cases of Ries, Lonar, Monturaqui, and Henbury we had microprobe analyses of glasses where FeO and Fe_2O_3 contents are given as FeO. In Table 1 analyses for all craters are given in such form for convenience of their comparison. Rock analyses of Henbury and Monturaqui craters have been adjusted for the substance of iron meteorites found there by its subtraction.

Henbury, Lonar, and Monturaqui craters are situated, respectively, in gray wacke sandstones, traps, and granites, while in all other craters the target is composite.

As seen from Table 1, despite the heterogeneous composition of the melt

*Walter and Giutronich (1967) give the relative volatilities order as K > Si > Na > Fe > Mg ≥ Al ≥ Ca. But this series was built without any regard to the change in acid-basic properties of melt. Walter and Giutronich failed to take account of a sharp change in SiO_2 content dropping from 82 to 45 wt.% of melt being vaporized, i.e., they disregarded that the medium from acid had turned basic.

Table 1. Composition of target rocks (1) and vitreous rocks (2) of several craters.

Crater / Weight per cent of oxide	Brent		E. Clearwater		W. Clearwater		Carswell		Nordlinger Ries		Popigai		Henbury		Monturaqui		Lonar		Manicouagan	
	1	2	1	2	1	2	1	2	1	2	1	2	1	2	1	2	1	2	1	2
SiO_2	66.1	61.2	63.2	62.8	60.8	60.4	66.7	60.8	66.07	64.03	63.14	62.85	77.3	75.6	72.9	71.83	50.56	51.6	54.7	54.7
TiO_2	0.57	0.77	0.51	0.50	0.73	0.71	0.32	0.13	0.43	0.79	0.59	0.76	0.82	0.89	0.20	0.39	2.78	2.9	0.63	0.73
Al_2O_3	15.7	15.7	15.6	15.8	16.7	15.9	17.0	18.9	15.25	15.25	15.2	14.85	10.85	12.3	15.2	14.91	12.79	13.7	21.6	19.3
FeO_{sum}	4.86	6.84	4.5	4.14	4.73	4.8	5.22	3.35	2.96	5.22	6.26	6.74	3.71	3.96	2.2	2.16	15.7	13.8	5.67	4.9
MgO	1.08	2.08	3.1	2.9	2.38	2.67	1.1	3.72	2.04	3.04	3.06	3.57	2.01	2.33	0.78	1.42	5.40	5.40	2.92	3.92
CaO	1.55	1.38	3.3	3.8	4.82	4.48	1.2	2.26	2.61	3.96	3.77	3.23	0.83	0.59	1.1	3.98	10.29	9.7	7.0	5.70
MnO	0.09	0.15	0.08	0.10	0.09	0.05	0.11	0.02	0.07	0.08	0.08	0.08	n.d.	n.d.	0.05	0.08	n.d.	n.d.	0.09	0.11
K_2O	4.08	7.60	3.10	3.67	2.38	3.60	5.17	8.88	3.03	4.01	2.39	2.75	2.95	3.27	3.47	2.86	0.59	0.6	1.80	2.95
Na_2O	3.60	1.86	3.80	3.05	4.31	3.78	2.17	1.25	3.70	3.02	3.02	2.22	1.00	0.87	4.18	2.88	2.55	2.2	4.52	4.02
K_2O/Na_2O	1.1	4.1	1.1	1.2	0.6	1.0	2.4	7.1	0.8	1.3	0.8	1.2	2.95	3.8	0.83	1	0.23	0.22	0.4	0.7
ΔZ^0_{1200}	−0.316	0.130	0.056	0.019	0.205	0.286	−0.411	0.402	−0.307	0.064	−0.081	−0.122	−1.267	−1.125	−0.637	−0.612	1.306	1.043	0.795	0.802

rocks and, in many cases, the layered nature of the target rocks the value of K_2O/Na_2O invariably is larger for glassy rocks than for the unaffected target rocks. Only for the Lonar Crater, the ratio is practically equal.

Comparison with volcanogene units

Glasses and glass-containing rocks of crater structures are chemically equivalent to basalts, or high-potassium andesite-dacites, or liparites, if the terminology of effusive rocks can be used. When recalculated by usual petrographic techniques, the compositions coincide with those of the high-potassium effusive series. By this feature they are distinguished from volcanites of type I (lavas, tuffs, subvolcanic and neck facies [Marakushev and Yakovleva, 1975]), but resemble volcanites of type II, which include a large group of rocks described by various authors as ignimbrites, welded tuffs, automagmatic breccias, etc., and referred by Yakovlev and Yakovleva (1973), to the fluidoporphyries; they are marked by a heightened K_2O/Na_2O.

It is to this class that most of the opponents of the meteorite hypothesis refer the melt rocks of the craters. Without going into the petrographic distinctions between ignimbrites and glassy rocks caused by impact (referred to as impactites), we shall only compare their chemical features. With few exceptions, each rock group has a prevalence of K_2O over Na_2O. But the variation range of K_2O/Na_2O for the two groups is different. Numerous analyses of type II volcanites (Marakushev and Yakovleva, 1975), show the ratio to vary from 0.8 to 1.4, with the most common values between 0.9 and 1.2. As seen from Table 1, for glassy rocks and glasses of crater structures the value varies in a very wide range, from 0.22 to 7.1.

Different compositions of melt rocks of craters and target rocks

A comparison of the compositions of melt rocks and target rocks for most of the craters under consideration (Table 1) was previously carried out by Dence (1971), who computed rated compositions for rock pairs (melt rocks and target rocks). Dence showed that there are craters with a similar composition of the pair and others where it is different. The latter units differ by one of the following features: (1) melt rock contains less quartz; (2) it is richer in potash feldspar; and (3) melt rocks are enriched in potash feldspar and impoverished in quartz. Dence points out that melt rocks of the craters are more similar in composition to their parent rock than to one another. Dence attributes the differences to the addition of the meteoritic contamination, or to the selective fusion of components with relatively low melting points; secondary effects introduced during cooling and crystallization by circulating vapours and solutions may lead to significant changes in composition.

The most interesting of the explanations of the changes of composition given by Dence seems to us impact selective fusion, but this explanation does not seem plausible, since selective melting would primarily release the more fusible melts, i.e., melts enriched in SiO_2 and alkalies. But the melt rocks of craters are, on the contrary, usually impoverished in SiO_2 and Na_2O (Table 1). Besides, selective melting implies identity between the composition of the starting rock and the gross composition of restite and the melts. Glassy rocks often contain incompletely melted fragments of target rocks and minerals and yet differ in composition from target rocks. Brecciated rocks containing a small amount of glass usually present in craters differ from target rocks in the same manner.

Currie and Shafigullah (1967, 1968) also noted the potassium enrichment of melt rocks and their sodium impoverishment as compared with the country units. Citing a vaporization experiment by Walter and Carron (1964), where the amount of alkalies changed very slightly, Currie and Shafigullah believed that meteorite impact vaporization cannot lead to the considerable loss of alkalies which is recorded in melt rocks as compared with target rocks. Currie's conclusion is suspect since in all the rest of Walter's experiments (Walter and Carron, 1964), which were carried out in vacuum, the amount of alkalies changed quite noticeably.

Currie and Shafigullah are certainly right in saying that the regularities observed in the change of compositions of melt rocks and target rocks invalidate both the volcanogene hypothesis and the mixed volcanogene-shock hypothesis according to which a meteorite impact caused subsequent volcanic eruption. Currie believes that under these hypotheses it would be hard to explain the regularly observed determinate petrochemical relations between the country rocks and the rocks generated at a great enough depth.

III. COMPARISON WITH EXPERIMENTAL DATA

(a) *Ratio* K_2O/Na_2O. As mentioned, there is a steady tendency for K_2O/Na_2O to be greater in melt crater rocks as compared with target rocks. This constant trend in the variation of the value suggests that there must be the same cause behind it. We believe it to be the selective vaporization of target substance, which is only possible at the high temperatures of meteorite impact. That no such vaporization takes place during volcanic eruption is confidently confirmed by similar composition of effusives and intrusives.

It seems that thus far, the only criterion to confirm vaporization in meteorite craters must be the comparison of the preceding experimental data on substance vaporization at various T and P.

Figure 5 shows the variation pattern of K_2O/Na_2O in meteorite crater rocks (Table 1) and in experimental samples (Table 2). The ratio is evidently higher in melt rocks. The only experimental point below the line of equal ratios is that of Yakovlev's experiment for basaltic rocks. The deviation is explained by what has been said earlier about the different course of vaporization from acid and basic melts.

For study of K_2O/Na_2O variation during vaporization, an experiment was conducted on unaltered rocks of the Popigai Basin, where K_2O/Na_2O is equal to 0.8. The ratio in the mixture was 0.6. The vaporization experiment was carried out by A. Ulyanov (pers. comm.); it produced an increase of K_2O/Na_2O up to

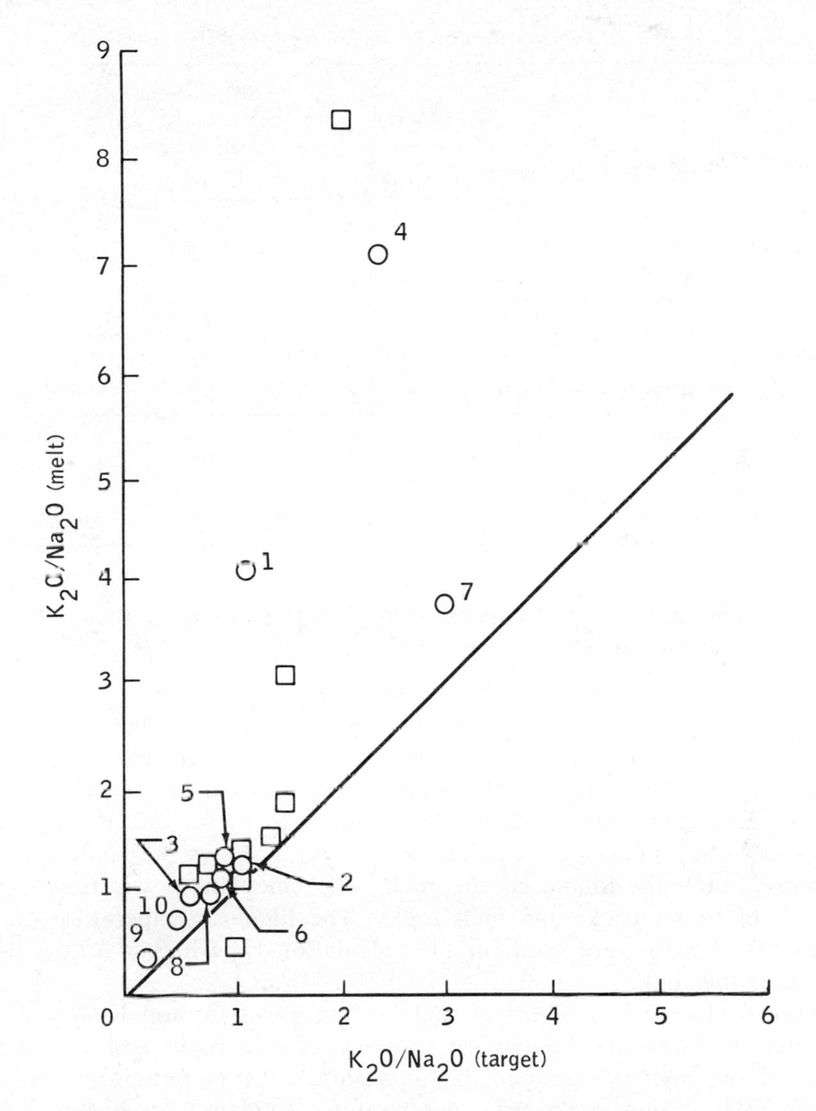

Fig. 5. Variation of the ratio K_2O/Na_2O in crater rocks (circle) and in experimental produced glasses (square). 1. Brent, 2. East Clearwater, 3. West Clearwater, 4. Carswell, 5. Nordlinger Ries, 6. Popigai, 7. Henbury, 8. Monturaqui, 9. Lonar Lake, 10. Manicouagan.

1.3. In the melt rocks of the Popigai Basin (called tagamites) the ratio is 1.2 (Table 1). The experiment demonstrated that by selective vaporization the same change in potassium-sodium ratio can be produced as is observed in meteorite crater rocks.

It should be noted that so far there is no accepted notion as to the mechanism of the vaporization induced by meteorite shock. We believe that vaporization

Table 2. K_2O/Na_2O changed by vaporization (experimental data).

T°C	P torr	K_2O/Na_2O Initial	K_2O/Na_2O Melt	SiO_2 content in initial rock (weight per cent)	Source
2060–2090	$\sim 10^{-3}$	1.9	9.0	71	Walter and Carron (1964)
2010–2080	1.5×10^{-3}	1.9	15.1	71	
1780	9×10^{-4}	1.9	3.1	71	
2050	760	1.4	1.9	71	
2600–2800	760	1.0	1.2	71	Walter and Giutronich (1967)
		1.0	1.1	70	
1400	10^{-6}	1.3	1.5	75	Yakovlev and Kosolapov (1973)
	10^{-6}	1.5	0.5	43	
1600	10^{-6}	0.6	1.3	63	Ulyanov (1974)

was not from melt "lakes". Vaporization from the shock-produced cloud seems much more probable.

(b) *The dependence of* K_2O/Na_2O *value on target acidity and basicity.* It has been already stated and confirmed by experimental data that vaporization of alkalies, and hence the K_2O/Na_2O value, is different in acid and basic melt. With the purpose of investigating this feature on rocks of meteorite craters, we applied the method proposed by Marakushev (1976) for calculating the ΔZ^0_{1200} values. ΔZ is an index of the acidity-alkalinity; it shows the contribution of all normative minerals composed the rock. The calculation was based on composition of target rocks and melt rocks. The highest temperature of the ΔZ values ($T = 1200°$) were used for the calculation. The higher ΔZ_{1200}, the more alkaline is the rock.

Figure 6 plots ΔZ_{1200} of target rock on the abscissa, and K_2O/Na_2O on the ordinate. Dots indicate the average values for target rocks and melt units. The length of the interval between the dots shows the amplitude of variation of K_2O/Na_2O in target rocks and glassy units. Evidently, a higher K_2O/Na_2O generally corresponds to relatively more acid targets. This is in full agreement with what follows from experimental data, and hence this relationship is due to a low volatility of K_2O in acid melts, which is the lower compared with Na_2O the greater the melt acidity. Deviations are possible and are indeed observed (e.g., Monturaqui; see Fig. 6), since K_2O/Na_2O, as has been stated in the experimental section of the study, depends not only on the target rock composition but also on temperature and pressure which may vary considerably from case to case.

(c) *Variation of rock acidity-alkalinity.* An analysis of the variation of ΔZ_{1200} shows that the value increases (heightening alkalinity) from target rocks to melt rocks in the majority of craters. The increase is most obvious in Carswell (by

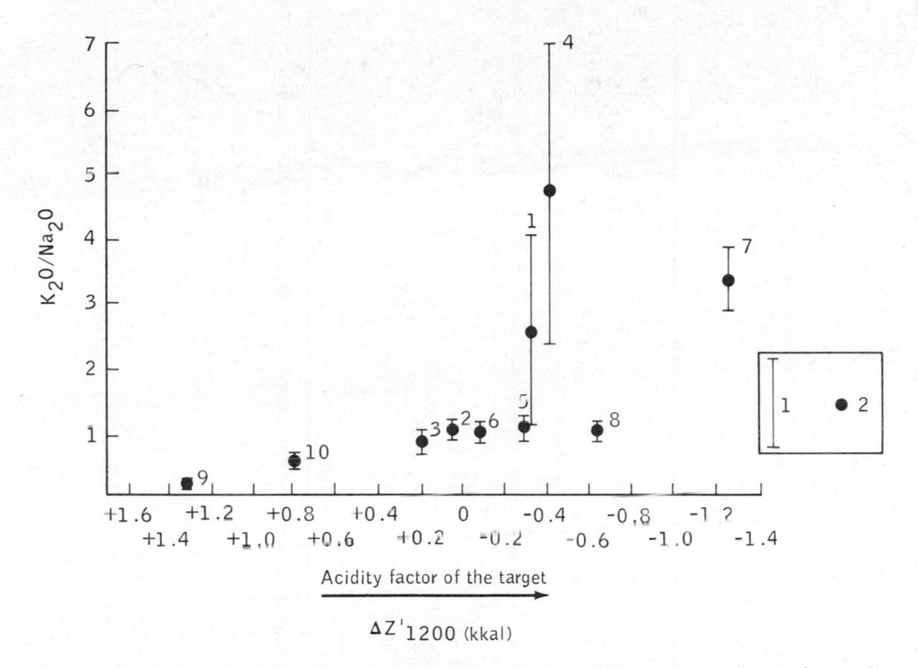

Fig. 6. Relating of K_2O/Na_2O versus the acidity factor of the target rocks. 1—interval of the meanings K_2O/Na_2O in target and impact melt. 2—Middle meaning of this ratio. 1. Brent, 2. East Clearwater, 3. West Clearwater, 4. Carswell, 5. Nordlinger Ries, 6. Popigai, 7. Henbury, 8. Monturaqui, 9. Lonar Lake, 10. Manicouagan.

0.813), and also in Brent (by 0.446) and Ries (0.371); for Western Clearwater, Monturaqui, and Henbury, the alkalinity rise is slight; for Manicouagan ΔZ remains practically invariable; and for Popigai and Eastern Clearwater, there is a minute drop in alkalinity. Only in Lonar Crater is a significant increase in acidity observed. The greatest alkalinity (ΔZ) apparently falls on rocks of acid craters (Table 1). This is due to vaporization features of acid components (mainly SiO_2), whose volatility, according both to experimental data and to the theory of acid-basic interaction, grows with increasing melt acidity.

(d) *Relative volatility.* Comparison of the contents of petrogenic elements (Table 1) between target rocks and melt rocks produced by meteorite shock shows that the content of certain elements in melt rocks is decreased as compared with the original, while other elements remain unchanged or even increase in quantity. The former group includes as a rule Na_2O, SiO_2, and the latter, K_2O, MgO, TiO_2, CaO.

It has been shown earlier that for an analysis of relative volatility of elements the concentration in weight per cent is not always a reliable indicator, since the apparent increase of an element may be in fact due to the relatively greater evaporation of other components. This explains the apparent enrichment of melt rocks in several components, including K_2O. For an objective quantification of the altered content of components due to vaporization, we made use of a method

Table 3.

Weight per cent of oxides \ Crater	Brent		Western Clearwater		Nordlinger Ries		Henbury		Manicouagan		Carswell		Monturaqui		Popigai	
	1 wt.% ≡ gram	2 gram	1 wt.% ≡ gram	2 gram	1 wt.% ≡ gram	2 gram	1 wt.% ≡ gram	2 gram	1 wt.% ≡ gram	2 gram	1 wt.% ≡ gram	2 gram	1 wt.% ≡ gram	2 gram	1 wt.% ≡ gram	2 gram
SiO_2	66.1	31.7	60.8	53.9	66.07	43.0	77.3	65.2	54.7	40.8	66.7	17.99	72.9	39.47	63.14	53.72
TiO_2	0.57	0.41	0.73	0.65	0.43	0.53	0.82	0.77	0.63	0.54	0.32	0.04	0.20	0.21	0.59	0.65
Al_2O_3	15.7	8.13	16.7	15.1	15.25	10.23	10.85	10.6	21.6	14.4	17.0	5.59	15.2	8.19	15.2	12.69
FeO_{sum}	4.86	3.54	4.73	4.3	2.96	3.5	3.71	3.4	5.7	3.66	5.22	1.0	2.2	1.19	6.26	5.76
MgO	1.08	1.08	2.38	2.38	2.04	2.04	2.01	2.01	2.92	2.92	1.1	1.1	0.78	0.78	3.06	3.06
CaO	1.55	0.73	4.82	4.00	2.61	2.66	0.83	0.51	7.0	4.25	1.2	0.67	1.1	2.19	3.77	2.76
K_2O	4.08	3.94	2.38	3.2	3.03	2.7	2.95	2.84	1.8	2.2	5.17	2.63	3.47	1.57	2.39	2.35
Na_2O	3.60	0.96	4.31	3.4	3.70	2.0	1.00	0.75	4.52	3.00	2.17	0.37	4.18	1.58	3.02	1.90

1—composition of target rocks; 2—composition of vitreous rocks calculated assuming MgO content constant during vaporization (weight per cent per 100 g of substance).

similar to scaling by an inert component used in analysis of metasomatic change. Namely, in the experimentally obtained series of low-volatile elements Ti, Ca, Mg, we assume that the content of Mg, which is apparently the least volatile element in the meteorite craters under study, remains constant during vaporization. Then we carried out the appropriate recalculations for other elements. An example of recalculation for some craters is given in Table 3. In this way the effect of volatilization of a number of compounds like SiO_2, Na_2O is brought in evidence for most craters. The effect of the vaporization of Al_2O_3 becomes more distinct; the amount of K_2O in these recalculations remains virtually constant, so that, in most meteorite craters, potassium behaves like a low-volatile component, due to its very low activity in acid and intermediate melts.

A comparison of the ratios of starting and ultimate quantities of a component thus enabled the following volatility order to be established for most craters: $Na > Si > [Al, K, Ca, Ti, Mg]$. The order is similar to that obtained by us from Walter's experimental data for acid melts.

By these recalculations, craters can be divided into two groups. One group is characterized by intensive vaporization, with Na_2O, SiO_2, and Al_2O_3 being volatilized. The starting amount of silica may be reduced by a half, and that of Na_2O by more than two-thirds. The high volatility of silica in natural craters located amid intermediate and acid rocks agrees well with Walter's experiment on vaporization of ultra-acid melts, where SiO_2 dropped from 82.2 to 45%. This group includes craters Brent, Carswell, Ries, Monturaqui.

The second group includes Eastern and Western Clearwater, Henbury, and Lonar. Popigai occupies an intermediate position. For Eastern and Western Clearwater craters, only alkalies have a statistically significant difference, but, as can be seen from Table 1, the differences are very small and comparative to sodium loss in the first group.

The existence of two groups of craters seems to indicate a difference in impact parameters, especially temperature. At times these differences were sufficient for a major alteration of target rock composition, in other cases these differences produced by lower temperatures, caused just a slight (less than 10%) volatilization of several elements.

Conclusion

The similarity of the alteration of the chemical composition of rocks as observed in experiments and in analyses of crater rocks is evidence of the importance of selective vaporization, which we believe to be a major factor behind chemical transformations produced by shock explosion. The comparison of experimental and natural data warrants at least three conclusions:

1. Increased K_2O/Na_2O value in vitreous rocks of meteorite craters is a result of selective vaporization of sodium from acid and intermediate high-temperature melts. K_2O has a low volatility in such melts, so the residual product is potassium-enriched.

2. K_2O/Na_2O value of the vitreous rocks of meteorite craters depends on the acidity-basicity of the country rocks (target rocks). The habitually higher K_2O/Na_2O value is observed in rocks of more acid craters.

These regular features of composition alteration, especially the regular increase in K_2O/Na_2O in vitreous rocks as against target rocks, are important indicators of the craters' shock-explosive origin.

3. Reduced silica content in vitreous rocks of some craters is a result of SiO_2 vaporization at high temperatures. Of the petrogenic components of an ultra-acid melt, SiO_2 has a high relative volatility, though inferior to Na_2O.

Further experiments on vaporization and comparison with regular features observed in nature may elucidate the chemical transformations produced by shock explosions and approach an evaluation of their physical parameters.

Acknowledgements—The authors are particularly thankful to V. A. Zharikov, Associate Member of the USSR Academy of Sciences, to Professor A. A. Marakushev for critically reading the manuscript and giving helpful suggestions, and also to Dr. C. P. Florensky for consultations.

References

Bunch, T. E.: 1972, Petrographic and electron microprobe study of the Monturaqui impactite, *Contrib. Mineral. Petrol.* **36**, 95–112.

Currie, K.L. and Shafigullah, M.: 1967, Carbonatite and alkaline igneous rocks in the Brent Crater, Ontario, *Nature* **215**, 725.

Currie, K. L. and Shafigullah, M.: 1968, Geochemistry of some large Canadian craters, *Nature* **218**, 457.

Dence, M. R.: 1971, Impact melts, *J. Geophys. Res.* **76**, 5552–5565.

Engelhardt, W.: 1972, Shock produced rock glasses from the Ries Crater, *Contrib. Mineral. Petrol.* **36**, 265–292.

Korzhinsky, D. S.: 1963a, Relationship between oxygen activity, acidity, and reducing potential during endogene mineral formation. *Izv. AN SSSR, Ser. Geol.*, No. 3.

Korzhinsky, D. S.: 1963b, Thermodynamic potentials of open systems, whose acidity and reducing potential are determined by external conditions. *Dokl. AN SSSR*, **152**, No. 2.

Korzhinsky, D. S.: 1966, Acid-base interaction in melts. *Issledovanie Prirody I Tekhniki Mineraloobrazovaniya*, Nedra Press, Moscow (in Russian).

Marakushev, A. A.: 1976, Method of thermodynamic calculation of basicities of bedrocks and minerals, *Bulleten' MOIP*, No. 1.

Marakushev, A. A. and Yakovleva, E. B.: 1975, Genesis of acid lavas. *Vestn. MGU, Ser. Geol.*

Sedov, A. I.: 1957, *Methods of Similarity and Dimensionality in Mechanics*, Gostekhizdat Publishers, Edition 4.

Stanyukovich, K. P.: 1971, *Unstabilized Motion of a Solid Medium*, Nauka Press, Moscow (in Russian).

Story, W. C.: 1973, Volatilization studies on a terrestrial basalt and their applicability to volatilization from the lunar surface, *Nature Phys. Sci.* **241**, 154

Taylor, S. R. and Kolbe, P.: 1964, Herbury impact glass: parent material and behaviour of volatile elements during melting, *Nature* **203**, 309–391.

Walter, L. S. and Carron, M. K.: 1964, Vapor pressure and vapor fractionation of silicate melts of tektite composition. *Geochim. Cosmochim. Acta* **28**, 937–952.

Walter, L. S. and Giutronich, J. E.: 1967, Vapor fractionation of silicate melts at high temperatures and atmospheric pressures, *Solar Energy* **11**, 163–168.

Yakovlev, O. I.: 1974, In dissertation submitted for the scientific degree of candidate in Geological-Mineralogical Sciences. MGU, Moscow.

Yakovlev, O. I. and Kosolapov, A. I.: 1976, Vaporization of melts in vacuum, *XXV Session of International Geological Congress*, Nauka Press, Moscow (in Russian).

Yakovlev, O. I., Kosolapov, A. I., *et al.* 1972, Findings of basalt melt vapor fractionation in vacuum, *Dokl. AN SSSR*, **204**, No. 4.

Yakovlev, O. I., Kosolapov, A. I., *et al.* 1973, Vaporization of K and Na from melts in vacuum, *Vestn. MGU, Ser. Geol.*, **28**, No. 5, 85–88.

Yakovlev, G. F. and Yakovleva, E. B.: 1973, Ore-bearing fluidoporphiry complexes of south-western Altai, *Vest. MGU, Ser. Geol.* **28**, No. 2.

Zel'dovich, B. Ya. and Reizer, Yu. P.: 1966, *Physics of Shock Waves and High-Temperature Hydrodynamic Events*, Nauka Press, Moscow (in Russian).

Roddy, D. J., Pepin, R. O., and Merrill, R. B., editors.
(1977) *Impact and Explosion Cratering*, Pergamon Press (New York), p. 861–867.
Printed in the United States of America

Shock wave, a possible source of magnetic fields?

B. A. IVANOV[1], B. A. OKULESSKY[2], and A. T. BASILEVSKY[3]

USSR Academy of Science, Moscow

Abstract—One possible effect due to impact cratering is the transformation of part of the kinetic energy of the projectile to energy in the electromagnetic field. Shock-induced polarization of piezo- and dielectric material could be the mechanism by which this is accomplished. Although the estimates show that the fraction of kinetic energy transformed by this mechanism is rather small, the effect could be sufficient to form some magnetic features of planets that have even a slight planetary magnetic field of their own. The possible limits of this effect are still uncertain. The mechanism suggested is attractive because it gives a possible explanation for the high values of remanent magnetization displayed by some lunar samples, without calling for the hypothesis of a strong global magnetism for the moon in the past epochs.

CRATERING IS the most prominent but probably not the most interesting effect of impacts or explosions. Other physical and chemical processes are also induced by shock waves. Results of these processes should be found during investigations of planets that have impact-cratered surfaces.

Several effects due to shock wave are well known, such as shock metamorphism, shock melting, chemical differentiation in impactites, etc. (see for example Dence, 1971; Masaytis *et al.*, 1975; Parfenova and Yakovlev, 1976). One more possible effect due to impact cratering is the transformation of part of the kinetic energy of the projectile to electromagnetic energy. The possibility of the latter transformation was first recognized during the analysis of the magnetic profiling data taken by Lunokhod-2 at Le Monier area (Ivanov *et al.*, 1976).

Lunokhod-2 profiling revealed quasisinusoidal magnetic anomalies associated with impact craters with impact diameters 50–400 m (see Fig. 1). These anomalies represent the quasisinusoidal variations of magnetic field intensity around the certain level which is the local background. In searching for the mechanism of formation of these anomalies the suggestion was made that crater excavation and anomaly induction are both caused by the same phenomenon—shock wave propagation, as the result of a magnetic field impulse produced by shock-induced electric polarization in the target rocks.

Shock-induced polarization can be considered as an electric current flowing through the shock wave front. This current is capable of causing some magnetic effects on adjacent materials. When a shock wave is passing through materials containing shock-polarized and ferromagnetic components, the induced elec-

[1]Schmidt Institute of Physics of the Earth.
[2]Shirshov Institute of Oceanology.
[3]Vernadsky Institute of Geochemistry and Analytical Chemistry.

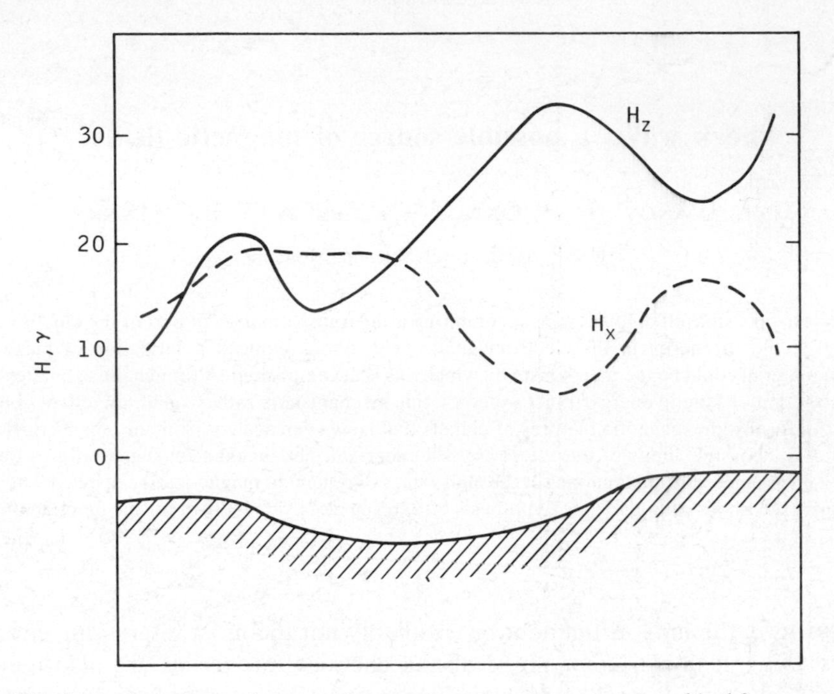

Fig. 1. Typical shape of magnetic anomaly across the crater. Lunokhod-2 measurements, bottom of Le Monier, March 16, 1973. H_x—horizontal component of magnetic field, H_z—vertical one.

tromagnetic field could be fixed by the ferromagnetic components. It may also be preceded or followed by direct shock effects on ferromagnetic components.

Lunar rocks consist of plagioclase, pyroxene, and olivine and other minerals. Between these components silica minerals occur. It is known that plagioclase, pyroxene, and olivine are dielectrics and silica minerals are piezoelectrics. Under the shock wave propagation shock-polarization occurs in piezoelectric (Graham *et al.*, 1965) as well as in some dielectric materials (Hawer, 1965; Mineev *et al.*, 1968). A review of theoretical and experimental works in this field can be found in Mineev *et al.* (1976). The ferromagnetic component of lunar rocks is represented mainly by Fe-metal at abundance up to 0.1% (Fuller, 1974). This content is sufficient to produce the observed local magnetic anomalies of the Moon (Guskova *et al.*, 1974).

This mechanism allows the evaluation of the amplitude and geometry of the magnetic field induced during the propagation of a semispherical shock wave into the semispace occupied by shock polarized dielectric. Experimental data show that shock compression of polycrystalline rocks induces polarization in the direction normal to the shock wave front (Mineev *et al.*, 1968). To simplify the evaluation let us suppose, as a first approximation, that the shock wave generates in a unit of volume a dipole momentum P and that the time of relaxation of this polarization is infinitely large (i.e., the current of depolarization is negligible). A more accurate estimation needs additional information on target mineralogy and character of the polarization mechanism. The assumption used is rather correctly applicable if a width of polarized zone is comparable to shock wave radius.

Let us consider a volume

$$dV = R^2 \sin\theta \, d\theta \, d\rho \, dR \tag{1}$$

in spheric coordinates. The shock wave front passes through this volume during the time

$$dt = \frac{dR}{D}, \tag{2}$$

if the shock front has a hemispheric shape, and D is a velocity of this front. After the shock wave has passed, rocks in the volume (1) have been polarized and the value of its dipole momentum is

$$d\mathcal{P} = P(R)\, dV \tag{3}$$

directed normal to the front surface ($P(R)$ = specific density of charge, coulomb/cm^2 in cgs).

Then, according to (1) and (2), the rate of changing in time is

$$\frac{d\mathcal{P}}{dt} = PDR^2 \sin\theta \, d\theta \, d\rho. \tag{4}$$

The derivative $d\mathcal{P}/dt$ have the meaning of an electric current through the shock front per unit of the shock front surface.

Let us consider a value and direction of magnetic field, generated by polarization current (4) between points A_0 at (R, θ, ρ) (see Fig. 2) and A_1 at $(r, \theta, 0)$. According to Landau and Lipshitz (1967) this field is determined as

$$dH = DP\left[\frac{\bar{R}}{|R|} \times \operatorname{grad} \frac{1}{\delta}\right] R^2 \sin\theta \, d\theta \, d\rho \tag{5}$$

where δ is a vector which connects the points A_0 and A_1.

Supposing that D and P are constant at any point of the shock front surface, then at any given moment of the time we can integrate (5) throughout the front surface:

$$\begin{aligned} H_y &= DP\,\mathrm{I}(R, A), \\ H_x &= H_z = 0, \end{aligned} \tag{6}$$

where H_x, H_y, and H_z—are components of magnetic field at A_1 and

$$\mathrm{I}(R_1 A_1) = \int_0^{2\pi} d\rho \int_0^{\pi/2} d\theta R^2 \sin\theta \frac{z \cos\rho \sin\theta - x\cos\theta}{[x^2 + z^2 + R^2 - 2R(x \sin\theta \cos\rho + z\cos\theta)]^{3/2}}, \tag{7}$$

(x and z are coordinates of A_1 point, $y = 0$), i.e., $\mathrm{I}(R, A)$ is a function of mutual position of shock wave and point of measurements.

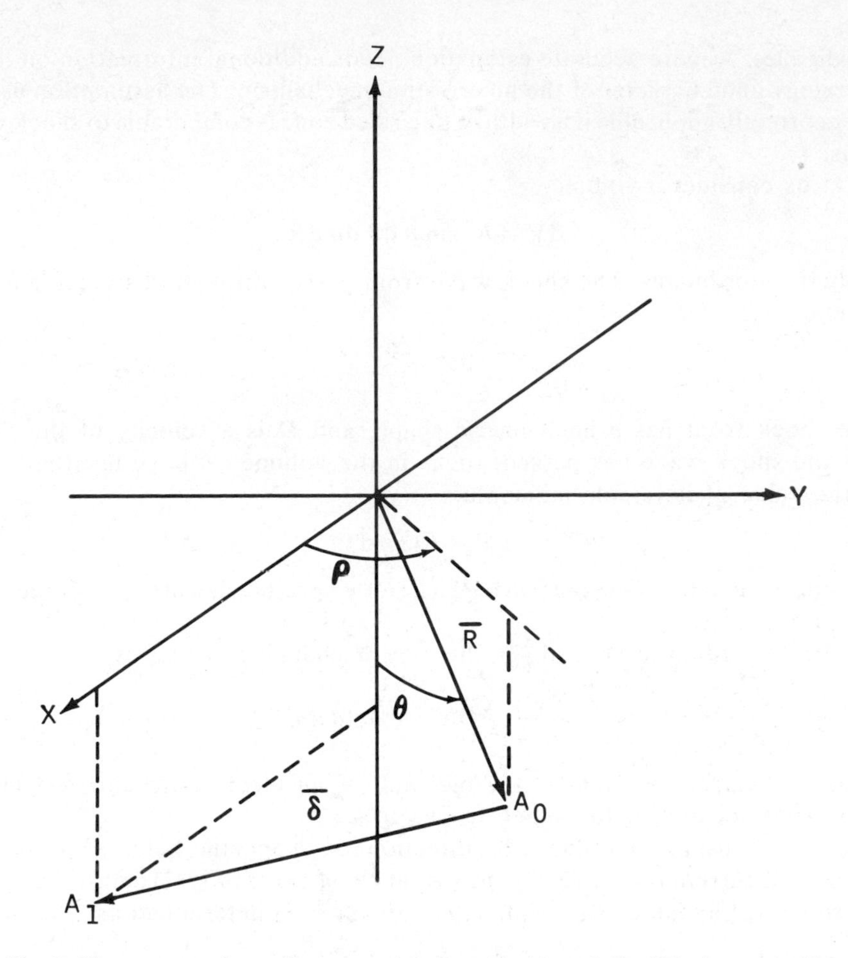

Fig. 2. The systems of coordinates used in Eqs. (4), (6), and (7). XY plane coincides with ground surface. A_0 is a point on the shock front which generates magnetic field in A_1.

Let us designate

$$H_0 = DP. \tag{8}$$

H_0 has the dimension of the intensity of magnetic field. If the units of D are km/s and the units of P are coulomb/cm^2 then

$$H_0 = 0.4\pi \times 10^5 PD \text{ (oersted)}. \tag{9}$$

For $D = 8$ km/s and $P = 10^{-9}$ coulomb/cm^2

$$H_0 = 10^{-3} \text{ Oe} = 100\gamma, \tag{10}$$

(we used the P value for enstatite from Mineev *et al.*, 1968). The P value is very high for quartz (Gracham *et al.*, 1965) but its polarization depends on the orientation of the crystal axes. The total H then depends on the percentage of quartz in the rocks.

The geometric factor $I(R_1A_1)$ in (6) can be estimated for a crater of conical shape with ratio diameter/depth = 8 (see Fig. 3) and for a moment of time when the shock wave radius equals 0.6 depth. $I(R_1A_1)$ for this case is shown on Fig. 3 for A_1 points placed on the level of future crater profile. Then the maximum magnetic intensity consists of about $0.5H_0$ and is achieved at a distance as large as the shock wave radius from the center. If one assumes that ferromagnetic components of rocks beneath the crater bottom were magnetized by this field and that they kept some part of the remanent magnetization (in proportion to the amplitude of the influenting field) the geometry of the horizontal component of the magnetic field across the crater is shown on Fig. 3. Comparing Fig. 1 and Fig. 3 shows the qualitative agreement between the calculated impact-induced anomalies and the magnetic anomalies revealed by Lunokhod-2, some craters in Le Monier area. Comparing these figures one can take into account the presence of the background at natural phenomena and the absence of it at calculated model.

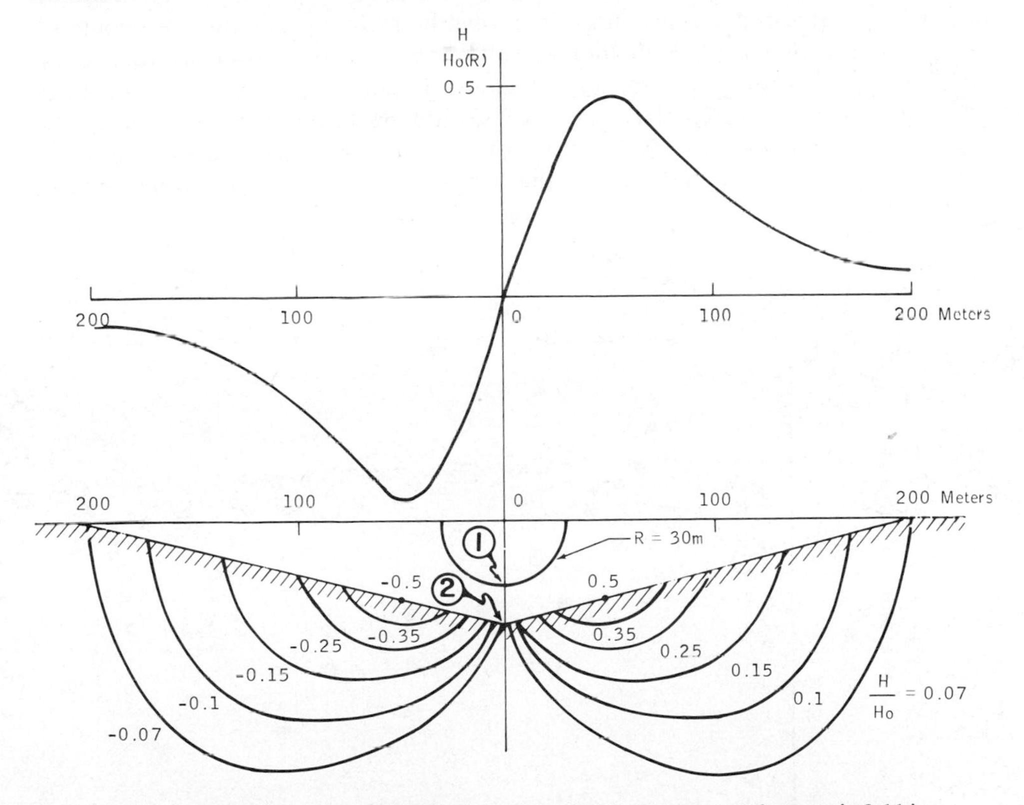

Fig. 3. Calculated shape of magnetic field: (a) calculated character of magnetic field in the bottom of future crater at the moment of time when radius of shock wave front is equal to 0.6 crater depth; (b) geometry of lines of equal intensity of magnetic field at the same moment of time: 1—position of shock-wave front, 2—bottom of future crater; vector of magnetic intensity is directed normal to the plane of the figure. To make this figure more obvious crater radius is adopted as large as 200 m.

This agreement probably indicates that the near-crater magnetic anomalies observed by Lunokhod-2 were produced by the above described mechanism of shock-induced polarization. The possible limits of this effect are now uncertain.

The possibility of formation by this mechanism of the magnetic anomalies associated with large craters is actually determined by the time of depolarization of rocks after the shock wave. If this time is rather small the effect of shock wave polarization would be localized on inhomogeneities in the rock massif and would be negligibly small at the distance as large as one crater diameter. Because of this circumstance the application of this effect to large impact craters and basins is unclear. The relations between the remanent magnetization of returned lunar samples and proposed mechanism are still not clearly understood. To resolve this problem the experimental simulation is now in progress.

If this mechanism of rock magnetization was indeed realized on the moon and other planets, its role should be much more large in the epoch of intense bombardment. We suppose that at any given impact event the percentage of intensely magnetized target rocks is probably rather small but at epoch of intense bombardment the high frequency of impacts could result in rather large cumulative percentage of intensely magnetized samples. If this mechanism is real then intensely magnetized samples should be found between young impacities of the moon also. It is interesting to note that the highest remanent magnetization is characteristic for the very lunar rocks which were formed

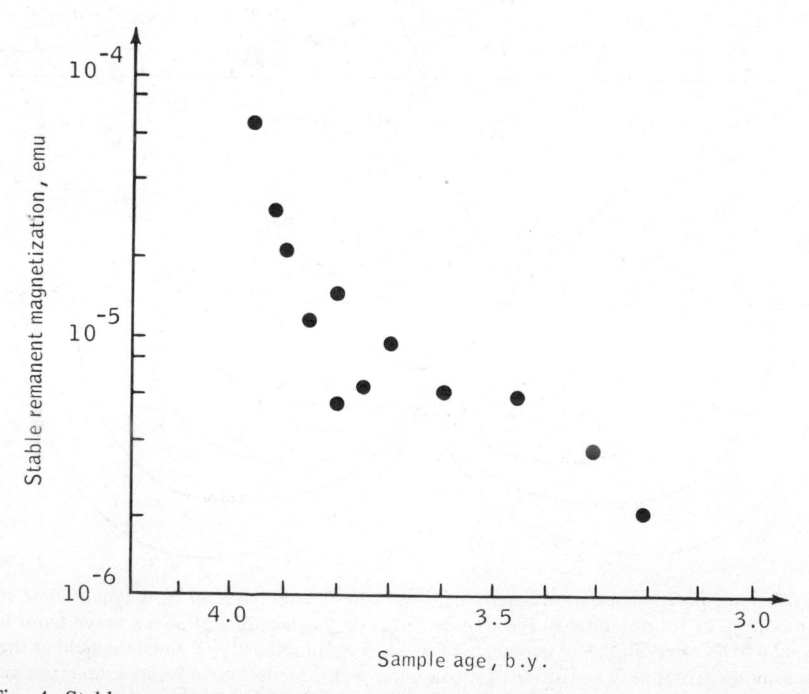

Fig. 4. Stable remanent magnetization of lunar samples vs. their absolute age. According to Okulessky (1976).

during this intense-bombardment epoch (Fuller, 1974; Okulessky, 1976); see Fig. 4. Simple estimations show that the fraction of kinetic energy transformed by this mechanism into electromagnetic energy is rather small and therefore the effect could be sufficient to form some magnetic features only on planets that have slight proper planetary magnetic field.

Acknowledgments—The authors are indebted to L. G. Bolkhovitinov for prompting them on the idea on magnetic effect of shock polarization. Authors also thank L. L. Vanjan and E. A. Eroshenko for fruitful discussions. Sincere thanks to L. B. Ronca from Wayne State University, Detroit, U.S.A., for editing the text.

REFERENCES

Dence, M. R.: 1971, Impact melts. *J. Geophys. Res.*, **76**, N 23, 5552.

Fuller, M.: 1974, Lunar magnetism. *Rev. Geophys. Space Phys.*, **12**, 123.

Gracham, R. A., Neilson, F. W., and Benedict, W. B.: 1965, Piezoelectric current from shock-loaded quartz—a submicrosecond stress gauge *J. Appl. Phys.*, **36**, 1775.

Hawer, G. E.: 1965, Shock-induced polarization in plastics. II. Experimental study of plexiglas and polystyrene. *J. Appl. Phys.*, **36**, 2113.

Ivanov, B. A., Okulessky, B. A., and Basilevsky, A. T.: 1976, Impulse magnetic field due to shock-induced polarization in rocks as a possible cause of magnetic field anomalies on the Moon, related to craters. *Pisma v Astronomicheskij Journal*, **2**, N 5, 257–260 (in Russian).

Landau, L. D. and Lipshitz, E. M.: 1967, *Theory of Fields*. Nauka Press, Moscow (in Russian).

Masaitis, V. L., Mikhaylov, M. V., and Selivanovskaya, T. V.: 1975, *Popigay meteoritic craters*. Nauka Press, Moscow (in Russian).

Mineev, V. N., Tjunuaev, U. P., Ivanov, A. G., Lisitzin, U. V., and Novicski, E. Z.: 1968, Shock polarization of water and enstatite. *Fizika Zemli*, N 4, 33 (in Russian).

Mineev, V. N. and Ivanov, A. G.: 1976, EMF due to shock compression of materials. *Uspekhy Fisicheskih Nauk*, **119**, N 1, 75–110 (in Russian).

Okulessky, B. A.: 1976, On the formation of magnetism of the Moon, *Tectonics and Structural Geology. Planetology*. Nauka Press, Moscow, 235–242 (in Russian).

Parfenova, O. V. and Yakovlev, O. I.: 1976, Selective evaporation in target rocks after meteoric impact. In *Papers Presented to the Symposium on Planetary Cratering Mechanics*, p. 96–98. The Lunar Science Institute, Houston.

Roddy, D. J., Pepin, R. O., and Merrill, R. B., editors.
(1977) *Impact and Explosion Cratering*, Pergamon Press (New York), p. 869–895.
Printed in the United States of America

The theory of cratering phenomena, an overview

C. P. Knowles and H. L. Brode

R&D Associates, 4640 Admiralty Way, Marina Del Rey, California 90291

Abstract—The essential features of detailed continuum mechanics solutions of cratering action associated with explosive or hypervelocity impact sources are explained. The initial energy coupling factors include radiation transport for nuclear explosive sources. The physical factors which must be accounted for in such calculations include material properties from ionization, dissociation, vaporization, melting, and crystal phase changes. The hydrology and geologic details of a crater site are equally important. The mathematical models and the numerical techniques are reviewed. The most successful cratering calculations either use a combined Lagrangian and Eulerian grid system or a hybrid moving mesh. Comparison with observed craters reveals that calculated craters agree fairly well for impact and chemical explosive craters, but calculations still seriously underestimate the volume of high yield nuclear surface burst craters.

Introduction

This paper has two objectives. One is to provide some insight into the physical processes of cratering. That is, what mechanisms are important and which ones seem to dominate crater formation. The second objective is to review the current theoretical capabilities for crater prediction; to examine what can be done from first principles using mathematical models, and what factors seem still to be beyond the present state-of-the-art.

The review is in eight sections: (1) Initial conditions and source phenomena; (2) Energy coupling; (3) Physical models; (4) Mathematical models; (5) Numerical programs; (6) Equations of state and material models; (7) Comparisons of theoretical results with observed craters; and (8) Conclusions.

Initial Conditions and Controlling Phenomena

Figure 1 represents the span of both time and pressure involved in cratering. This particular curve is for a nuclear explosive crater, and so pressures start at about 1000 Mbar. At these pressures, all important physical processes are occurring in times on the order of nanoseconds (10^{-9} sec). Calculations must then start with time steps much less than a nanosecond, and with pressures of the order of 10^9 bars. To complete the cratering action for large planetary craters, the calculations must continue all the way out to times of the order of tens of seconds, when the pressure has returned to ambient and the main motion has subsided. It is necessary to span roughly ten orders-of-magnitude in time and nine orders-of-magnitude or more, in pressure. Adequate treatment of all the pertinent physics and relevant material properties throughout this range presents a rather formidable problem.

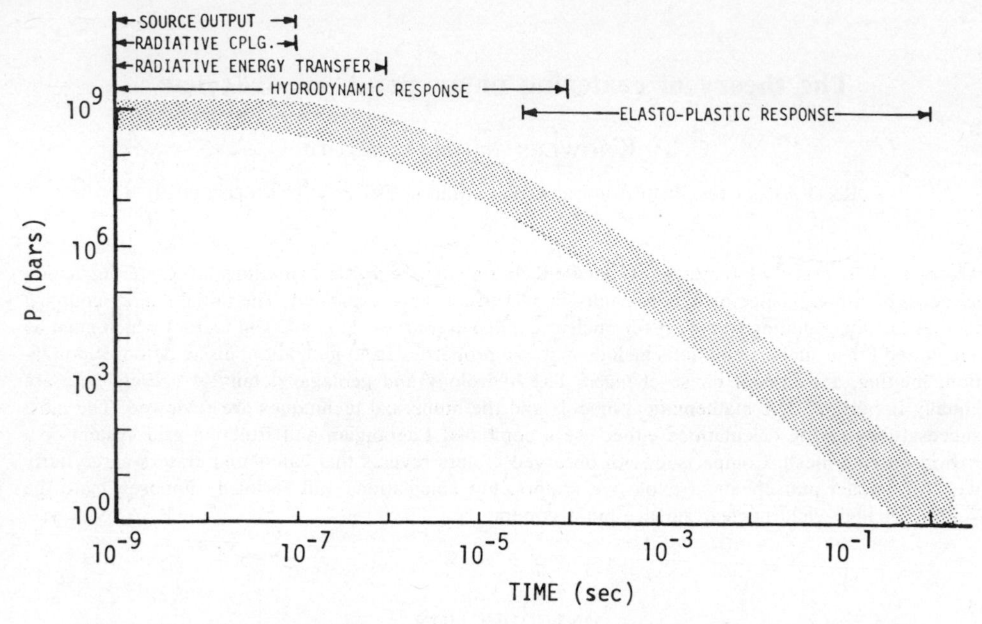

Fig. 1. Typical pressure–time sequence for energy coupling, cratering and ground shock.

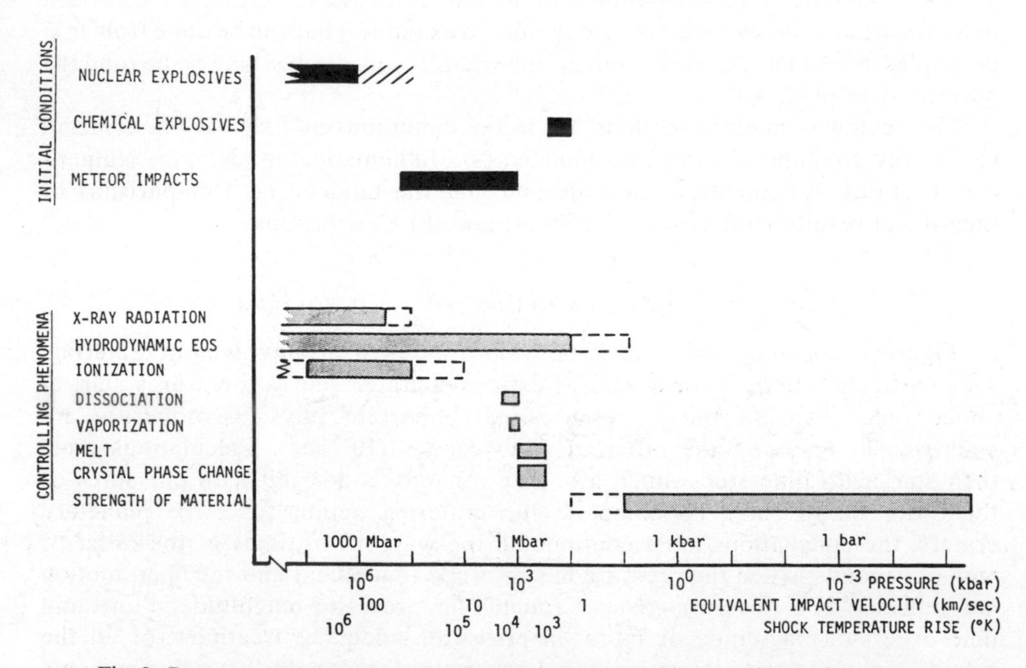

Fig. 2. Pressure, temperature, and equivalent impact velocities associated with sources and relevant physical phenomena for cratering.

Figure 2 is a plot with a multiple abscissa showing the pressure levels of interest and the equivalent impact velocity (in this example, for a nickel impact). For an ice meteorite impact, the pressure is roughly a factor of three or four lower. Also indicated on the abscissa is a temperature rise in the shocked material. At the high end, pressures of a 1000 Mbar are associated with shock heating temperatures of something of the order of a million degrees Kelvin. The initial pressure, velocity, and temperature ranges for nuclear or chemical explosive cratering and for meteor impact cratering are suggested at the top of this figure.

In the lower portion of Fig. 2 are listed some of the important physical phenomena to be followed in pursuing a theoretical calculation of a cratering event. At the very highest energy density level, radiation (photon) flow represents the most significant energy transport mechanism. It is important for temperatures above a few hundred thousand degrees Kelvin and pressures of a few hundred megabars, surpassing hydrodynamics or physical mass transport, conduction, turbulent diffusion, etc.

As the temperature falls, hydrodynamic motions soon transport energy more effectively than does radiation diffusion. At these high temperatures, the material becomes a highly ionized plasma, where electrons are completely stripped from the atoms. Ionization becomes an important process at fairly high energy density levels, as shown here, but quickly becomes unimportant once the pressures drop below 100 Mbar or so. Partial ionization may still contribute to equation-of-state changes below this 100 Mbar level, but it is no longer an important energy mechanism at material densities of interest to cratering.

Further down in energy density, one must account for molecular dissociation and vaporization, each occurring roughly around the range of a few megabars, depending on the material and its density state. Melting occurs somewhat below a megabar for shock heated natural materials. In addition, there often are significant crystalline phase changes that go on in this same pressure region. Finally, the strength of the material, its resistance to motion or distortion brings the whole cratering process to a stop and determines the crater shape. Material strength begins to be important in the range somewhere below 100 kbar and, in most materials, becomes a dominant mechanism below 10 kbar. Strength and gravity are the two influences which finally halt the crater formation.

Energy Coupling into a Surface from Nuclear, Chemical, and Impact Sources

Appropriate initial conditions for modeling a cratering event differ considerably with the type of explosive or the impact under consideration. For nuclear explosives, the energy densities are generally highest. Historically, U.S. nuclear tests have led to craters from two classes of nuclear sources: high energy density sources, and a group of lower energy-density sources. The high energy-density nuclear explosives can generate pressures in excess of 1000 Mbar in ground material and lead to shock temperatures greater than 10 million degrees Kelvin. At such temperatures, blow-off velocities can exceed 100 km/sec.

The low energy-density nuclear sources generate pressures and temperatures an order of magnitude lower; typically, 100 Mbar or so in initial pressure and around a million degrees Kelvin. In this case, radiation plays much less of a role. In fact, for most of these low energy-density sources, radiation was not a dominant energy transport mechanism in the coupling of energy to the ground. The impact velocities for meteor impacts start at an upper limit of roughly 70 km/sec and span velocities down to a few kilometers per second. Meteor impacts tend to stay in the energy-density range somewhat below nuclear explosives and above chemical explosives. The latter typically generate pressures of only 100–200 kbar.

Some of the large yield Pacific nuclear tests characterize the high energy-density, surface burst nuclear explosion, the radiation-dominated case. There are two phases in the process of energy flow into the ground from such nuclear sources. The first is the radiation output of the source itself shining onto the ground and being absorbed by it. The nuclear reactions which generate the energy take placé on the scale of a few tens of nanoseconds, very rapidly leading to extremely high (>100 Mbar) pressures in the nuclear device itself, and temperatures somewhere between 5 million and 50 million degrees Kelvin. As a consequence, exceedingly high velocities of the source materials are induced (>100 km/sec).

In a fraction of a microsecond, radiation energy flows from the source into the ground. For this short time, only radiation transport is important; hydrodynamic motions, even at 10^3 km/sec, do not move far in 10^{-8} sec. The ground heats up to approximately 10 million degrees Kelvin and immediately re-radiates most of the energy back into the air (or void) above the surface.

In all nuclear surface burst cases, only a very small amount of the total explosion energy is ultimately involved in the cratering action. That is true for a surface burst chemical charge also, but less dramatically. In addition to the energy directly deposited in the ground, the atmosphere around the source also heats up by radiation diffusion. This heated atmosphere re-radiates energy into the ground, creating a fairly thin layer of very hot plasma (Fig. 3a). This depth grows to the order of 25 cm directly under a "typical" megaton contact burst. At this stage, before any appreciable mass motions have occurred, there exists at the ground surface a rather thin layer of very hot, high energy/density material.

At temperatures below 5 million degrees Kelvin, the air becomes quite transparent, and passes any further radiation quite freely, allowing only Compton scattering from the free electrons and ions. The source itself continues to radiate freely above a few tens of millions of degrees Kelvin, while the ground absorbs and "burns out" or becomes a fully-ionized plasma around ten million degrees Kelvin. A sphere of fully-ionized air will grow to several tens of meters, but the heated ground layer will still be quite thin—a hot plasma layer of normal density ground material, tapering off to less than a centimeter in thickness at a range of 10 m or so. The normal density and very high temperature creates very

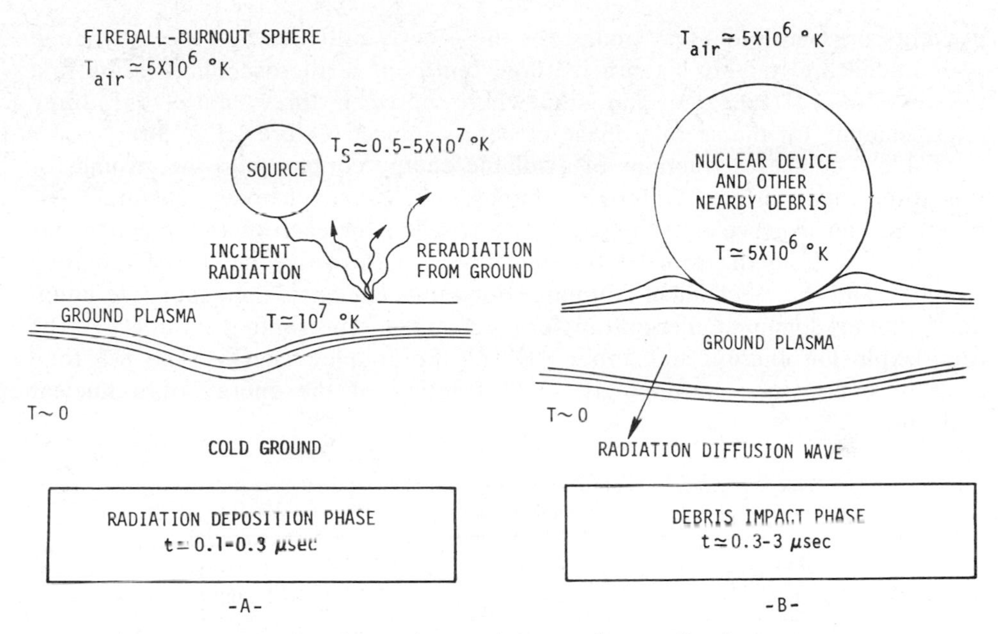

Fig. 3. High yield surface nuclear cratering coupling features.

high pressures in this layer of vaporized earth. These conditions are quite in contrast to those from either a hypervelocity impact (kinetic energy) or a chemical explosive. Both of the latter sources, as well as nuclear sources, strike the surface and generate high compressions and high pressures by means of a shock wave, but only the nuclear sources are hot enough to be dominated by radiation transport prior to shock dynamics.

In the high yield nuclear explosion, just after this initial radiation phase, the vaporized debris from the nuclear device and its associated material impact the ground surface and add more energy to the earth (Fig. 3b). This debris slap may take up to 3 or 4 μsec, depending on the burst height, and may inhibit some re-radiation out of the ground by occluding and recompressing it.

The net energy coupled to the ground from such a surface burst is about 8% of the nuclear yield. The energy from debris impact represents only about 2% of the nuclear device yield. The chief reason why the debris slap transfers so little energy (~ 2%), when it represents around 25% of the yield, is that the velocities of impact are so high that the resulting stagnation temperatures are in a region where radiative transfer again allows much of the energy to radiate to the atmosphere. That is, the stagnation energies on impact are such that the material again becomes completely ionized, and the material can no longer absorb radiation; Compton scattering (photons deflected by electrons or ions) is all that impedes the escape of energetic photons. Consequently, the nuclear device debris impact is a very inelastic impact, yet one which never generates the high stagnation pressures that may be expected in lower velocity impacts. The rapid escape of stagnation energy by radiation prevents pressure build-up. Timing for these early radiative and hydrodynamic phenomena are roughly a tenth to three tenths of a microsecond for the first radiative phase (Fig. 3a), and somewhere between three tenths and three microseconds for the second phase or debris impact features (Fig. 3b).

Table 1 compares fractions of available energy coupled into the ground for the three near-surface cratering sources, nuclear explosives, chemical explosives, and hypervelocity impacts. The fraction for each of two mechanisms for coupling from the three different sources is shown; namely, the radiative coupling and the mechanical coupling. For a nuclear explosion, radiative coupling is the predominant energy transfer mechanism, representing about 6% of the total explosion energy, and about 80% of the coupled energy. The 8% total coupled energy is a surprisingly small fraction of the energy of a nuclear explosion.

Table 1. Typical theoretical energy coupling for surface burst configurations.

Source	Radiative coupling	Mechanical coupling	Total energy coupling
Nuclear explosive	~0.06	~0.02	~0.08
Chemical explosive	—	0.05–0.1	0.05–0.1
Hypervelocity impact	—	0.5–1.0	1.00

Chemical explosives create a low enough energy density that radiation coupling is not important. In that case, the energy transferred into the ground is solely by mechanical impedance coupling. The fraction is somewhere between 5 and 10% of the total energy released in the chemical detonation of a sphere of explosive tangent to the ground surface (e.g., Ialongo, 1973). Again, only a small amount of the detonation energy is coupled from such a surface burst. In this case, the impedance match is poor between expanding explosive vapors and the earth. In a hypervelocity impact, however, most of the available energy starts off in the ground material. Much of it may come back out again as splash of vaporized material expanding above the original impact surface or into the surrounding atmosphere (if there is one), but, initially at least half the energy is coupled into the ground. Since in most nearly vertical impacts the hypervelocity object sinks below the surface before its motion is dissipated, it could be viewed as fully coupled except for low angles of impact. A somewhat more detailed description of impact coupling is provided in the paper by Trulio (1977).

PHYSICAL MODELS FOR THE THEORY OF CRATERING

What follows is a brief discussion of the principal theoretical techniques used to model cratering events. The physical model suggests what physical phenomena are pertinent, but then a mathematical model of those physical features is needed before a numerical model can be constructed for solution by high-speed electronic computers. The computer codes that are employed have been developed over the past twenty years and have found application in a wide variety of dynamics problems. In virtually all cases, the codes used for cratering are continuum mechanics finite difference codes. That is, the material is assumed to be a continuum rather than discrete particles. Usually, no attempt is made to model the small scale jointing, inclusions, voids, inhomogeneities, or stratifications. It is much more common to assume average material properties in the large, either as homogeneous solids, fluids, or gases, and characterize very complicated geologic materials in terms of uniform bulk properties. While most convenient, and almost a necessity for simple computer calculation, such gross treatment of natural materials may well be the most glaring deficiency in current theoretical calculations.

The material is usually assumed to be isotropic within large layers. Most codes can treat simple horizontal layering and can allow different properties in each layer. Unfortunately, the frictional forces transmitted across layer boundaries is the subject of considerable controversy; and only a limited amount of work has been done to characterize anisotropic behavior. The ejecta, which is often the subject of considerable interest, is seldom treated realistically, but is assumed to be continuous in cylindrical sheets or rings in these finite difference codes (Trulio, 1977).

In general, the finite difference codes treat only the early part of the problem with acceptable verisimilitude. Various ballistic extrapolations have been attempted, some of which include the effects of atmospheric drag forces on ejecta particles (with assumed simple shapes) in attempts to model the final impact points and trajectories of ejected fragments after they leave the cratering region (Seebaugh, 1976). These ballistic extrapolations rely heavily on sparse experimental data and on computer calculation results which provide either maximum or average velocities of material emerging from a cratering event.

Mathematical Models and Calculational Techniques

As mentioned, virtually all computer codes treat the medium as a continuum. This convenient assumption allows a mathematical model of the dynamics to be expressed in differential equations representing the conservation of mass, momentum, and energy (Fig. 4). In order to integrate these questions, numerical techniques are invented to deal with finite approximations to the infinitesimals characterizing the partial differential equations. In either Lagrangian (mass) coordinates or Eulerian (space) coordinates, the numerical methods assume finite-sized cells, numerical regions over which the materials are assumed to be uniform.

Typically, the codes are symmetric in a cylindrical coordinate system which assumes that there are no gradients in the circumferential direction, and that for any given sized cell, the material can be represented as a uniform material within that ring-shaped cell dimension (Fig. 4). As a result, gradients or changes in properties in the flow field cannot be any sharper or finer than the cell size. For

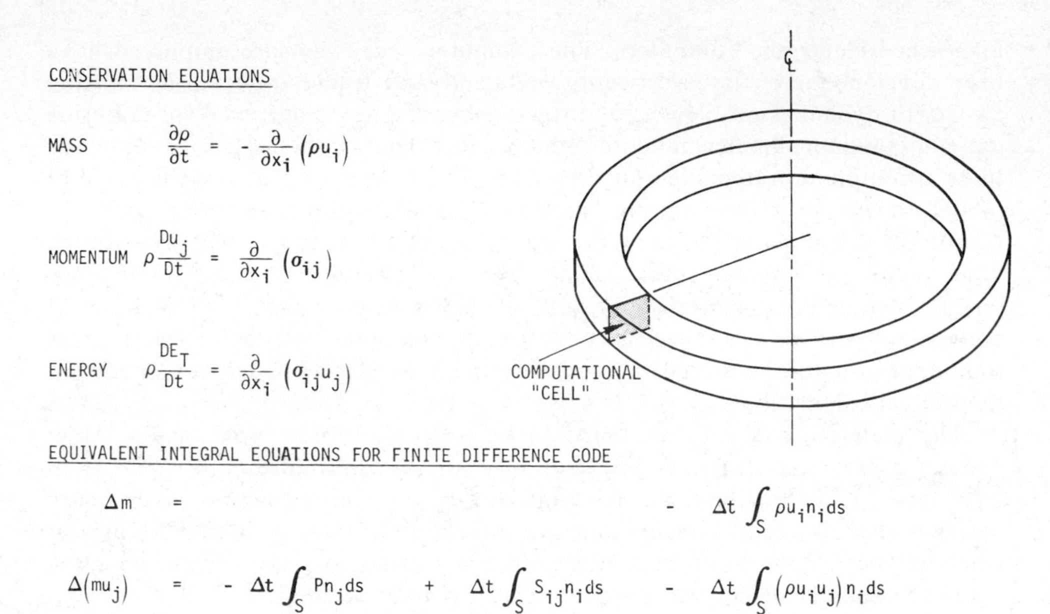

Fig. 4. Typical differential and difference equation characteristics.

practical numerical reasons, one cannot characterize gradients or flow features on a scale any smaller than three to six cell dimensions.

Shown in the lower half of Fig. 4 is an integral form of these differential equations. These integral expressions more appropriately represent the kinds of equations that are numerically integrated in most time-marching finite difference calculations. A small mass of material (ΔM) is defined, and forces are allowed to act on that incremental mass to distort and move the cell (in Lagrangian form) or to allow material to move through a fine grid (in Eulerian form).

We have made reference to two basic kinds of calculational techniques. The schemes are called Lagrangian and Eulerian. There are some hybrid schemes which combine properties of both, and which work particularly well for problems where the shear motions can be reasonably anticipated.

A Lagrangian code is one in which all motions are expressed in terms of small mass segments, i.e., the coordinates are fixed on the mass elements in the material, and follow these masses as they move and distort under the forces present. The mass of each cell remains constant always, and the cell is allowed to deform under stress gradients over the surface of the cell. Such a coordinate system has the following advantage. There can be no artificial mixing of materials, i.e., the mass of each cell is clearly defined initially, and is not altered by the motions generated in the calculation. Of course, zones of different material may be combined to yield mixtures, but only in units of mass cells. In addition, Lagrangian codes allow somewhat more accurate finite differencing than is possible in a space-coordinate (Eulerian) code, and it is easier to zone and follow special regions of interest such as small regions near a reflecting boundary or around a critical point.

But there are also some serious disadvantages to Lagrangian coordinate codes. Probably the worst of these is the inability of such codes to treat problems with severe shear distortion. Unfortunately, cratering action represents one such case where severe distortion is essential to the motions to be modeled (see Fig. 5). Consequently, before the end of a cratering calculation, a very tangled set of mass cells or Lagrangian grid would result. Without careful stability control, zones can turn inside out. Thus, masses in the flow field can become negative, and result in nonsense because of the failure of the numerics. Even if careful stability controls limit time steps so that zone boundaries do not cross each other, a serious problem remains when zones distort. As any zone distorts, one or more of its dimensions can shrink and become much smaller than originally specified. But as zone dimensions collapse, the time steps must become proportionately smaller* and the number of time cycles or the running time of such a calculation can soon become excessive.

- MASS OF ZONE CONSTANT
- ZONES DISTORT UNDER STRESS GRADIENT

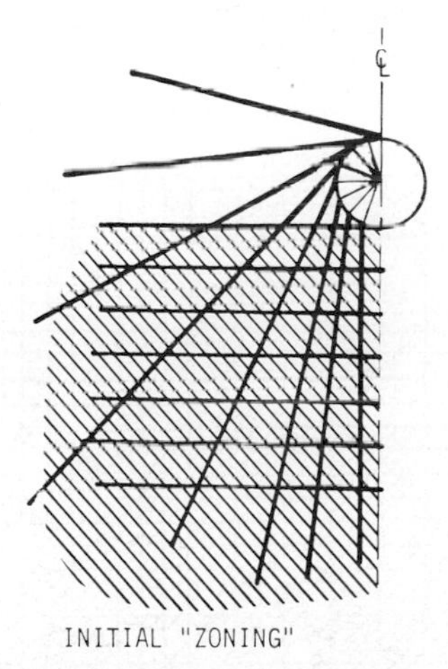

INITIAL "ZONING"

ADVANTAGES

- NO ARTIFICIAL MIXING OF MATERIALS
- MORE ACCURATE FINITE DIFFERENCING
- EASY TO ZONE SPECIAL REGIONS OF INTEREST

-A-

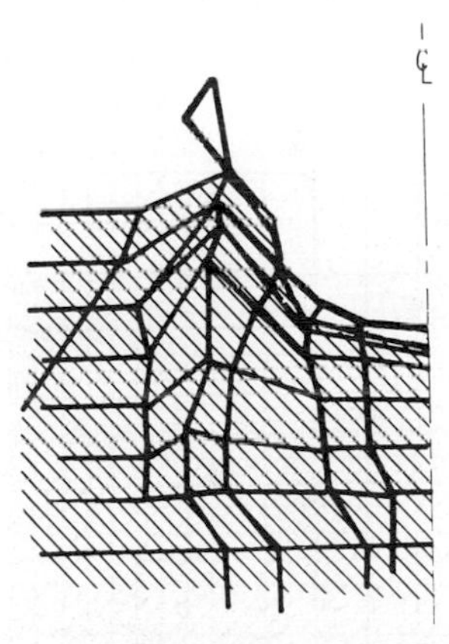

DISTORTED LAGRANGIAN GRID

DISADVANTAGES

- IN PROBLEMS WITH SEVERE DISTORTION (I.E., CRATERING) ZONES CAN TANGLE (TURN INSIDE OUT)
- ZONES CAN COMPRESS OR DISTORT TO SMALL DIMENSIONS CAUSING VERY SMALL ALLOWABLE TIME STEP

-B-

Fig. 5. Lagrangian calculation features.

---

*The Courant stability condition for numerical stability requires that each time step be so small that no signal can propagate beyond the next zone, i.e., $\Delta t \leq \Delta R_{min}/C$, where the largest allowable time increment (Δt) is less than the smallest zone dimension ΔR_{min} divided by the local sound speed.

Eulerian coordinate schemes avoid this zone distortion and time step shrinkage problem. In a Eulerian system, the zones are spacial coordinates whose size is fixed and does not follow the flow. That is, the material is allowed to flow through the grid of cells. Such a coordinate system has the obvious advantage that it can handle motions with large shears. Since the grid is not allowed to distort, the time steps determined by the Courant stability condition do not change due to zone dimension changes, and the running time of the code can be better controlled.

On the other hand, a Eulerian system suffers from a problem not present in Lagrangian coordinate systems; namely, as material flows through the Eulerian grid, interfaces between materials are lost, and the materials rapidly and artificially mix. For example, as in Fig. 6, starting with a small number of different materials in layers or well-defined regions, in a short time, that material will be spread out throughout the grid. In fact, without special treatment, every cell in the grid will soon have a mixture of every material in it. While some computer codes rely on artificial inhibitors to reduce the rate of mixing, there may still be a good deal of unrealistic mixing involved. That mixing as well as the numerical

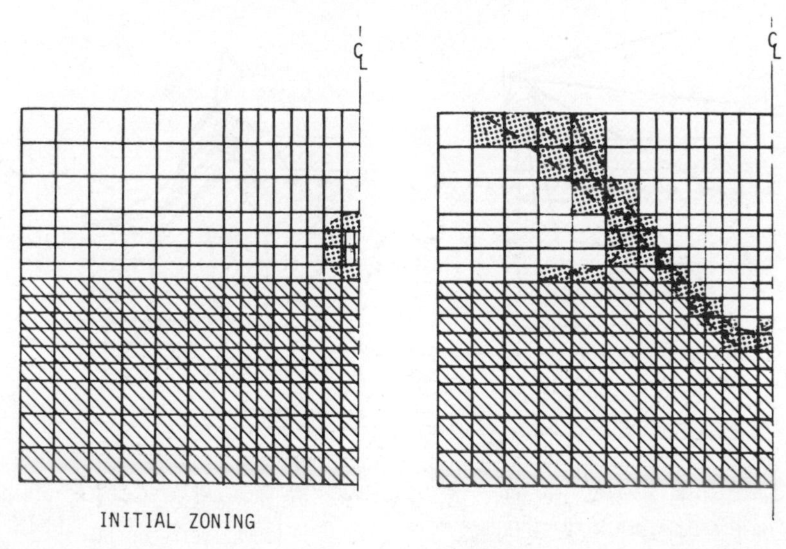

Fig. 6. Eulerian calculation features.

effects of streaming through cells results in more artificial diffusion than might be expected in a Lagrangian formulation.

Frequently, a combination of Eulerian and Lagrangian coordinate systems is invoked in order to minimize the deficiencies noted above. What is usually employed is a coupled Eulerian/Lagrangian grid in which, in regions of high distortion (where the Lagrangian technique will tend to fail), a Eulerian grid is used. Such a region might be the zones within one or two diameters of an impact body (as in Fig. 7a). At the boundaries of this Eulerian region, some coupling to a Lagrangian grid is needed, then all the mass at larger distances is carried in mass coordinates. By such a combined system a calculation can benefit, where the distortions are small, from having the greater accuracy of the Lagrangian treatment, but, where the distortions are large, the Eulerian scheme keeps the calculation from being halted by numerical stability limits.

Another factor in explosion-like dynamic problems is the fact that the source at the beginning of the problem may be quite small, yet, as the disturbance spreads, the active mesh grows to very substantial dimensions. For example, in a nuclear cratering calculation, the bomb itself has dimensions on the order of a meter, and the initial grid system might span only 10–15 m. However, at the end of the cratering process, a grid of hundreds of meters is needed in order to handle the crater and some of the ground shock. It is necessary, then, either to rezone the problem from time to time, or to use an expanding grid, in which the

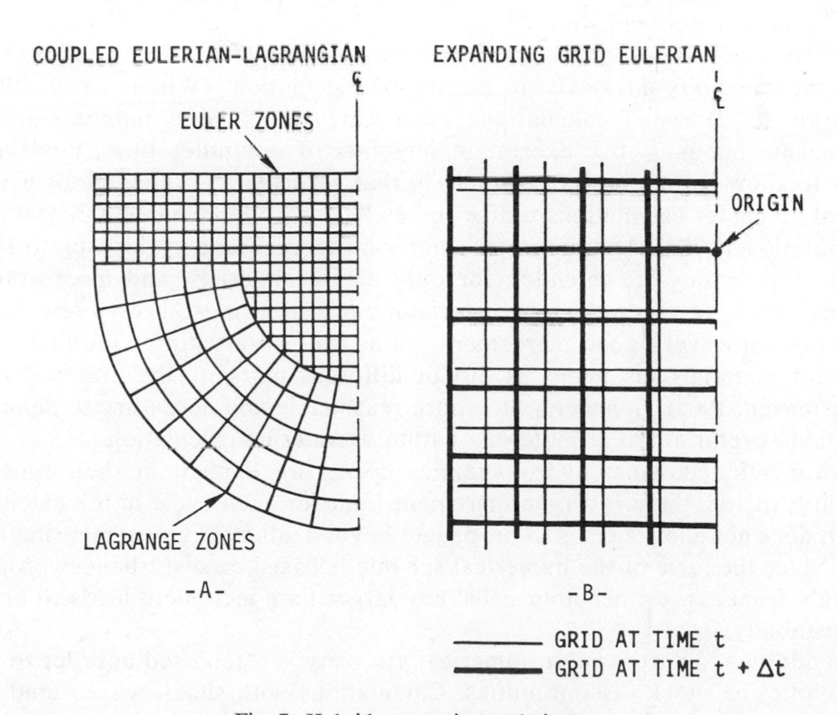

Fig. 7. Hybrid computing techniques.

zones are forced to expand each time step. Such an accordian grid, as in Fig. 7b, can be made to move out with time in a way that just keeps pace with expanding shocks or with moving masses.

CALCULATIONAL PROGRAMS—NUMERICAL MODELS

Typical problems in two-dimensional (2D) modeling of hypervelocity impact (metal on metal) are characterized by the work of Walsh and Tillotson (1963), Bjork and Olshaker (1965), Bjork, Kreyenhagen and Wagner (1967), or Rosenblatt (1971). A more recent example involving natural materials may be found in the work of Maxwell and Moises (1971). An early example is the calculation of Arizona Meteor Crater by Bjork (1961).

Some examples of 2D calculations of chemical explosive cratering are: Christensen (1970), Schuster (1973), Buckingham *et al.* (1973), Ialongo (1973), and Maxwell and Moises (1971).

Since the first nuclear surface burst calculation (Brode and Bjork, 1960), many such calculations have been reported. A representative few are Ialongo *et al.* (1974) and Orphal *et al.* (1975).

Shallow-buried nuclear calculations are somewhat simpler than surface burst calculations, but more demanding than chemical explosive cratering calculations. Examples of such calculations are those reported by Maxwell *et al.* (1973) and McDonald and Reid (1974).

While all of these cratering calculations are limited to cylindrical symmetry (2D), the capability does exist to do full 3D calculations (Wilkins *et al.*, 1974 or Matuska, 1972). Such calculations must carry very large numbers of cells (~100,000) and must use exorbitant amounts of computer time, however, in order to show resolutions comparable to that for usual 2D calculations. Current typical computer calculations utilize somewhere between 5000 and 20,000 active computational cells. Most programs run for a few thousand cycles—up to 10,000.

Most programs are intended for only a few materials, and most cratering calculations have used only one material model, or, at most, a very few. On one occasion, however, good agreement with test observations required many different materials (as many as 50) in different parts of the cratered region (Christensen, 1968). In general, it is both reasonable and necessary to define and maintain several different materials within a cratering calculation.

As mentioned, most hydrodynamics codes are explicit in their time step stability; that is, there is a time increment limit for each cycle in the calculation which does not allow signals to propagate beyond adjacent zones in a single time step. Since the logic of the numerical scheme is based on disturbances propagating only from nearest neighbor cells, any larger time increment leads to numerical instability.

In addition, an artificial or numerical viscosity is often used in order to avoid instabilities at shock discontinuities. Calculations with shock waves tend to be unstable in the discontinuity of the shock front itself, and the result often is violent numerical instabilities or significant artificial oscillation of hydrodynamic variables just behind the shock front, and wherever steep velocity or pressure gradients exist. The artificial viscosity technique forces a limit on such gradients. While all the conservation relations are still observed, and shock conditions are preserved, the smeared shocks become quite a handicap, if very rapidly changing phenomena are of interest, since those are the very regions in which an artificial viscosity distorts the representation. However, on the scale of craters, and after the initial shock expansion, these velocity and pressure gradients are usually rather small, and so an artificial viscosity seldom causes significant alterations.

Cratering calculations typically are limited to two spatial dimensions. For an explosion crater, initial or early-time overall grid size might be of the order of 20 m in radius and 10 m in depth. Each cell in this grid is a ring in cylindrical symmetry in most 2D codes. Since stability dictates that shocks be spread over 3–6 zones, no gradients sharper than 30–60 cm can be resolved in a grid with cells of 10 cm size. In fact, it is generally meaningless to think about any features whose dimensions are as small as 10 cm or even 20 or 30 cm in such a calculation. As it becomes necessary to expand the grid, larger cells are generated to keep the number of cells from growing unmanageable, and, as a consequence, the resolution limit becomes even larger. Cells can grow to be very substantial (tens to hundreds of meters) in cases where it is important to follow the ground shock well beyond the crater itself.

As for temporal resolution, at the early times in an explosion calculation, a time step of something of the order of a few nanoseconds per cycle is dictated by the source size and energy or pressure levels. It is generally true that these time steps are smaller than the time scale over which interesting phenomena in the calculation are occurring. At later times, when the problem has gotten much larger, the time step may increase and go to something in the range of a millisecond per cycle. Typically, the calculation will run several thousand cycles up to ten or fifteen thousand cycles in order to follow a crater into a regime where simple ballistic extrapolations can be used to estimate the ejecta motions.

Equation of State and Material Properties

At sufficiently high temperatures, all materials can be characterized by a thermodynamic state and treated as a fluid or gas. For most dynamics problems in normal density materials, thermodynamic equilibrium can be assumed and a full description of the energy and compressional properties of the matter can be provided, using any two of the independent thermodynamic variables such as internal energy and density. Although reducible to two parameters, the described matter often is not simple. A lot of complicated phenomena occur at high pressures. It is necessary in some regimes to treat ionization, dissociation, and phase changes. The compressibility and heat capacity of high temperature fluids are closely related to the degree of dissociation or ionization. At a level when electrons are boiling out of a K, L, or M shell, the atoms behave in a more compressible and heat-absorbent fashion than a fully-ionized plasma, and the effective ratio of specific heats goes from close to $\gamma = 5/3$ in the fully-ionized plasma down to something less than $\gamma = 1.2$ in the transition regions. The ratio between pressure and energy density is represented by the factor $(\gamma - 1) \simeq P/(\rho E)$, and this ratio changes appreciably with the degree of ionization or molecular dissociation.

Illustrated in Fig. 8, compression histories can differ due to crystalline phase change. When a material is shocked to a few hundred megabars and is rapidly released from the shocked state to about 700 kbars, the material undergoes a phase transition, a crystalline reorganization, which causes the material to move from one principal adiabat on the left (Fig. 8) onto another at larger specific volume. Such a time-dependent transition becomes a very complicated phenomenon when necessary to treat in detail under the assumption of thermodynamic equilibrium. For very small samples and particularly for the kinds of

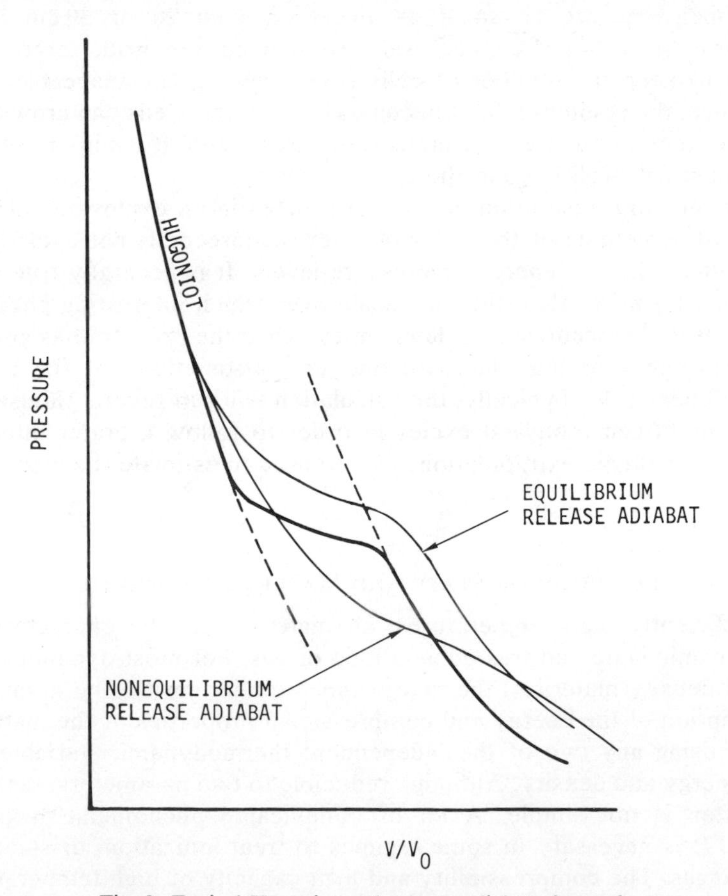

Fig. 8. Typical Hugoniot and release adiabats for rock.

experiments carried out in the laboratory, the loading and unloading times are small enough that it is quite likely that the material will be out of equilibrium for some of the time. That is, in trying to change from one crystalline phase structure to another, the expansion goes on faster than the material can follow, and so comes back down a release adiabat with different crystalline states than equilibrium would dictate.

This kind of crystal structure nonequilibrium appears to be important only in very small scale, i.e., on scales of the order of a few millimeters up to a centimeter. On larger scales, it is quite likely that the material will behave as in local thermodynamic equilibrium. However, at lower stress levels, some soils or rocks exhibit stress or strain rate effects that could be significant for cratering (Knott, 1973). For a clay shale studied by Knott, the dynamic failure strength increased by as much as 3.5 times that of the static failure strength, and the initial modulus increased nearly 10 times.

The lower energy regime is also very complicated. In Fig. 9 is shown the mean stress as a function of relative compaction of a typical natural material. The range of stresses or pressures appropriate here are from a few bars (at Point A) up to 100 kbar or so at Point B. In compressing from A to B, quite a variety of phenomena take place. In general, loading a material up to some mean stress and then allowing it to re-expand results in permanent compaction of the material, i.e., the pores, voids, cracks, and joints in the material will collapse or close and not all will reopen. If one compresses the material beyond a point where all the pores are closed, and then relieves the stress, under some conditions, the pores open up again, and then the load and unload curves will begin to coincide, representing a simple, single, compression curve. Of course, some materials, instead of showing permanent compaction under stress, result in a bulking or a net increase in volume after being shocked. Furthermore, if one unloads and goes into tension, the material reaches a point where it fails in tension. Tensile failure is seldom well-treated in any detail in calculations. Typically, a tension cutoff is invoked when a tensile pressure exceeds some arbitrary value at which the material is assumed to be unable to resist any further tensile forces.

The definition of failure criteria is one of the most significant aspects of constitutive properties. Shown in Fig. 10 are just three of a variety of failure criteria presently in use. In Fig. 10, the relation between the mean stress and the unbalanced or deviatoric stress is plotted. The J_1, representing the mean compressive stress in the material, is positive for compressions and negative for tensions. The J_2' or $\sqrt{J_2'}$ is a measure of the shear stress in the material.

The Von Mises ideally plastic material is illustrated in Fig. 10a. As the material is subjected to shear, it remains elastic up to some shear stress level at which failure occurs. This failure level is assumed independent of the confining stress. Unfortunately, this Von Mises model does not reproduce the behavior of real materials very well, and so more detailed models have developed. In Fig. 10b, a combination of the fixed failure level with a region where the failure level is proportional to the confining stress is illustrated. This model is referred to as a

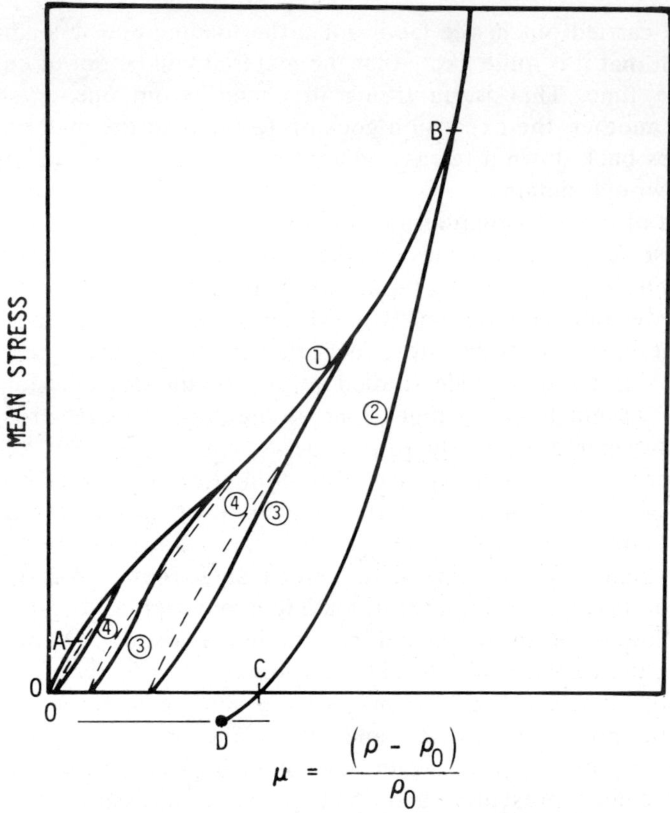

① VIRGIN LOADING CURVE ② UNLOADING CURVE AT HIGH PRESSURE ③ UNLOADING CURVES AT INTERMEDIATE PRESSURES

④ RELOADING CURVES

AB - NONLINEAR PORTION OF LOADING CURVE

B - POINT AT WHICH ALL MICROCRACKS IN MATERIAL ARE CLOSED

OC - PERMANENT COMPACTION AT HIGH PRESSURE

D - TENSION CUT-OFF

Fig. 9. Typical hydrostat for rock materials.

Von Mises plus Mohr–Coulomb model, and allows for some increase in the failure criteria with confining stress up to a limit, taking into account the fact that as the material is confined, it becomes harder for it to fail.

Each of these criteria is aimed at defining the behavior of a material as it yields or becomes plastic—as it ceases to be elastic. A most complicated model is the Cap model; developed after a considerable amount of work, and still under

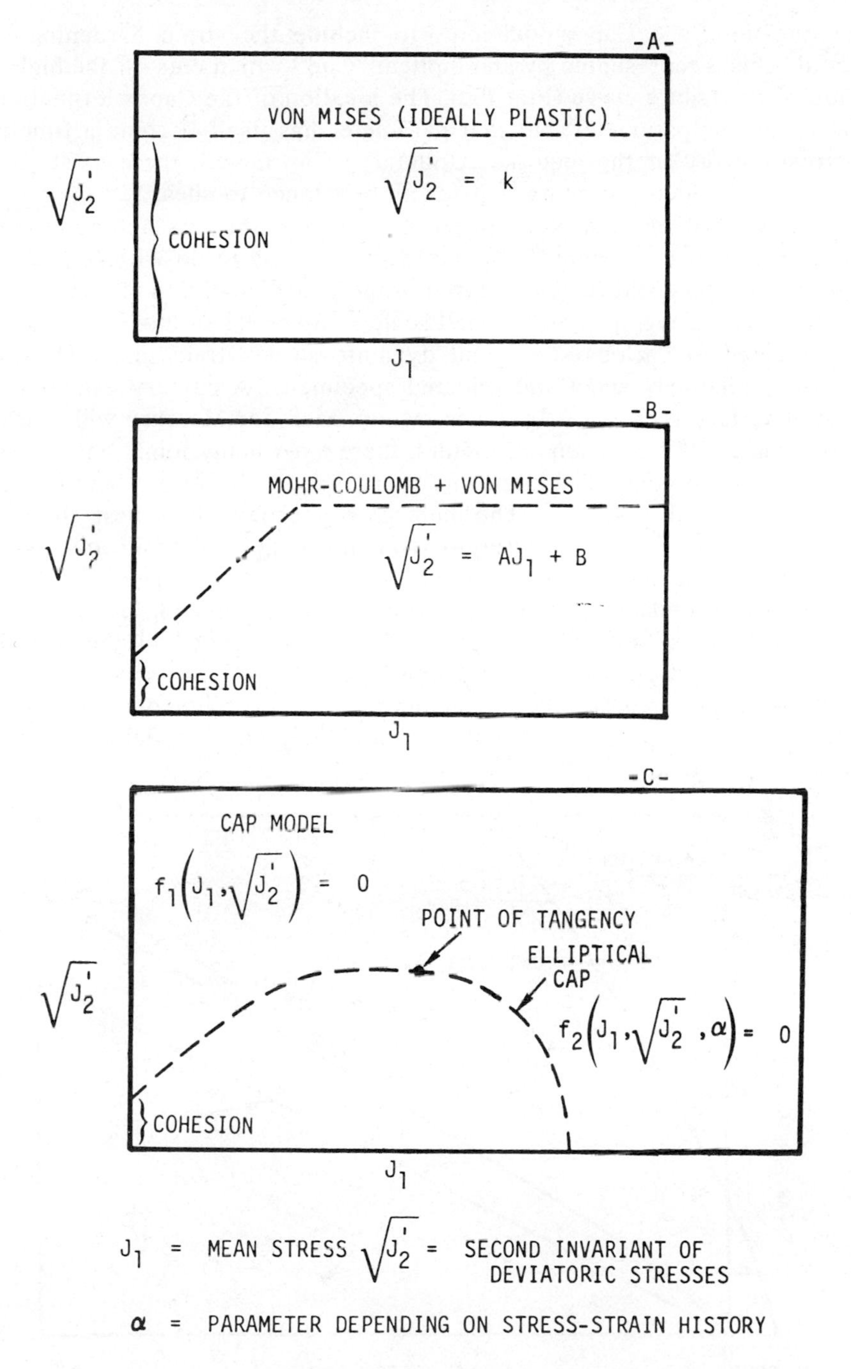

Fig. 10. Yield surfaces for plastic failure of soil or rock materials.

some question. The Cap model aims to include the strain hardening of the material. This is represented by an elliptical "Cap" which cuts off the high stress portion of the failure curve (Fig. 10c). The location of the Cap is a function of a strain hardening parameter, but that parameter has itself become a function of the stress history of the material. Under the Cap model, there exists a mean stress beyond which the material loses all resistance to shear.

For terrestrial or planetary cratering, however, there is another aspect of material properties that should be considered, having to do with the difference between laboratory and *in situ* material properties. Usually, in the past, material properties were developed from small-scale samples taken into the laboratories and subjected to various static and dynamic stress–strain tests. These were necessarily relatively small and selected specimens. A cursory examination of any solid surface section of the Earth, Moon, Mars, or Mercury will reveal that on the scale of meters or tens of meters, there exist many joints, many fissures, stratifications and voids. The kinds of craters that are of interest here cover very large areas of such a surface. The material properties which best characterize such large regions are very different from those derived from any particular small sample, just because of the joints, inclusions, inhomogenieties, etc. One might reasonably develop a fairly strong and stiff model, based on intact laboratory samples, but an acre or more of the area from which the lab samples came might be better characterized as much weaker—having no tensile strength, much less shear strength, and much more compressibility due to the *in situ* voids, open cracks, and inhomogeneities. Such an altered model is illustrated in Fig. 11.

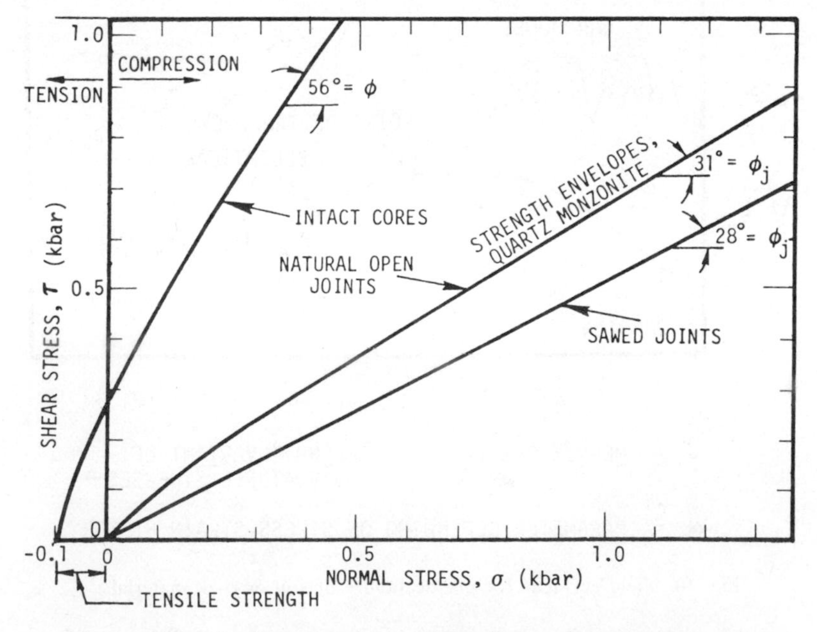

Fig. 11. Strength of intact and jointed specimens of quartz monozonite.

One of the most dramatic influences on soil or rock behavior under shear loads is that of saturation. As shown in Fig. 12 (Duba *et al.*, 1974), a dry rock may exhibit an order of magnitude greater shear resistance than a saturated rock. Much of the differences noted between dry and wet soil or rock craters may be attributable to such drastic reductions in failure strength with water content. The phenomenon involves lubrication, the reduction of shear resistance by filling cracks and voids with fluids, by minimizing crushing while promoting slippage.

In addition to hydrology and joint spacing problems, one must also give consideration to shock conditioning of the material. Once a strong shock has passed through a material, it may be shattered or so altered as to be weakened from its initial *in situ* condition. Such a shocked material is represented in Fig.

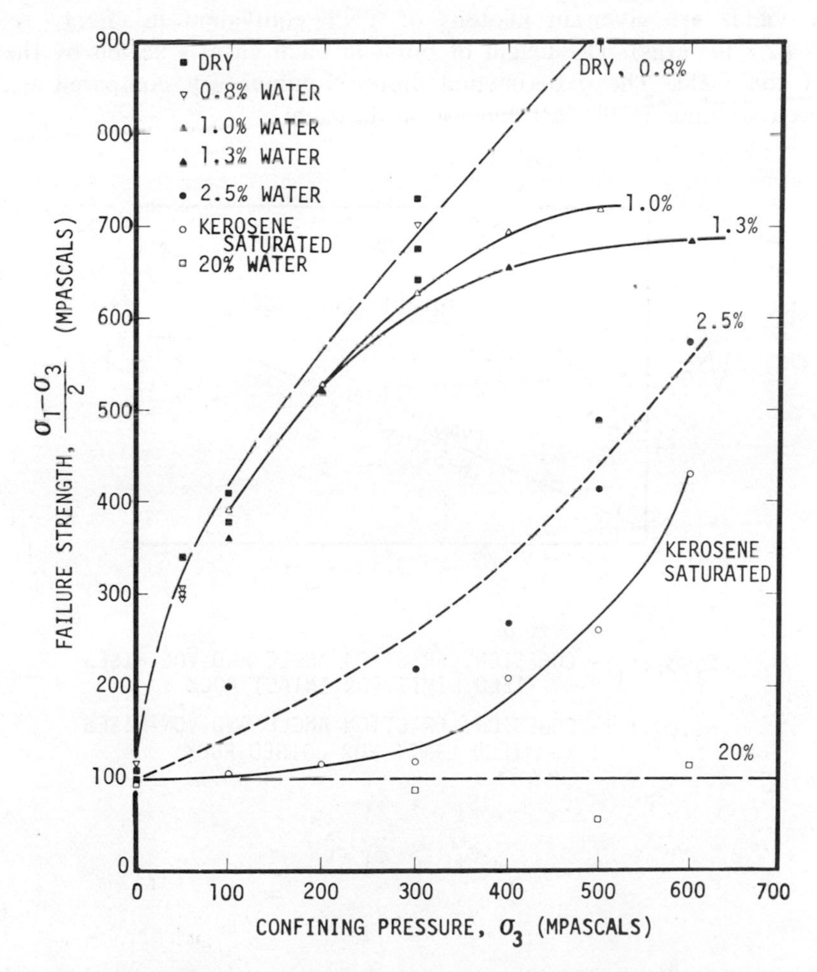

Fig. 12. Failure strength for westerly granite as a function of confining pressure, fluid type, and fluid content (Duba *et al.*, 1974).

13. That weakening can lead to significantly larger motions in late stages of crater formation, and so to larger craters than predicted on the strength of undisturbed rock. A calculation by Orphal *et al.* (1975) for a large yield in the Pacific atoll geology used such a weakening feature, and got an enlarged crater. A convenient illustration of this shock altering is discussed for rock salt by Short (1960).

COMPARISON OF THEORY WITH OBSERVED CRATERS

The crater parameters from two groups of calculations are listed in Table 2—the first group models specific high-explosive and buried nuclear craters; the second group is theoretical predictions for surface burst high-yield nuclear explosion craters.

The yields are given in kilotons of TNT equivalent in energy released ($1\ \text{kT} = 4.2 \times 10^{19}$ ergs). The height of burst in each case is scaled by the cube root of the yield. The experimental crater volumes are compared with the calculated volumes in the last column on the right.

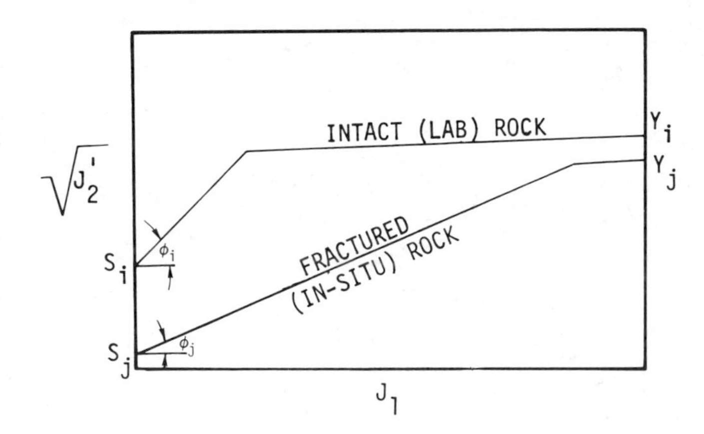

S_i, ϕ_i, Y_i - COHESION, FRICTION ANGLE AND VON-MISES YIELD LIMIT FOR INTACT ROCK

S_j, ϕ_j, Y_j - COHESION, FRICTION ANGLE AND VON-MISES YIELD LIMIT FOR JOINED ROCK

$$S_i > S_j$$

$$\phi_i > \phi_j$$

$$Y_i > Y_j$$

Fig. 13. Strength behavior of laboratory (intact) rock specimen compared with fractured *in situ* rock.

Table 2. Comparison of calculated and experimental crater dimensions.

Event	Yield (kT)	Shob ($m/kT^{1/3}$)	Experimental R(m)	D(m)	V(m^3)	Agency	Calculated R(m)	D(m)	V(m^3)	$\frac{V_{Exp}}{V_{Calc}}$
High explosive and buried nuclear craters										
Johnie Boy	0.5	0.73	22	12	6.0×10^3	PI	21	14	9.6×10^3	0.62
Middle Gust I	0.02	0.0	15	5.0	2.0×10^3	ATI	12	5.0	1.0×10^3	1.86
Middle Gust II	0.1	10.5	13	4.6	1.4×10^3	ATI	7.0	3.0	4.0×10^2	3.43
Middle Gust III	0.1	5.3	16	7.3	2.8×10^3	ATI	17	4.9	1.6×10^3	1.71
Middle Gust IV	0.1	5.3	13	4.6	1.2×10^3	ATI	11	7.9	9.9×10^2	1.26
Mixed Company	0.5	5.1	18	4.0	2.0×10^3	SHI	21	4.0	1.6×10^3	1.21
						PWA	24	6.1	2.8×10^3	0.69
Surface burst high yield nuclear craters										
Theoretical	1000	0.6	170	70	2.8×10^6 *	PI	38	33	7.9×10^4	35
High yield			340	20	2.8×10^6	PI	67	40	3.0×10^5	9.3
Nuclear surface burst (scaled to 1 MT)	1000	0.6	120	52	1.1×10^6	SHI	35	17	2.3×10^4	50
KOA	1380	0.1	700	52	2.3×10^7	AFWL	220	91	9.5×10^6	2.2

The experimental values for 1 MT surface bursts are extrapolations or interpolations from Pacific proving ground tests to other soils, scaled to 1 MT.

When all the approximations in the theoretical calculations are considered along with all the experimental and geological (rheological) uncertainties, the comparison of the crater dimensions for the HE and buried nuclear cases is quite gratifying. Crater volumes seem at variance by less than a factor of two in all but one case.

Unfortunately, for the surface burst high-yield nuclear craters, the agreement is much worse. The initial calculations miss by a factor of 35 to 50. The second listed high-yield case is a reduced strength calculation with shock conditioning and other effects invoked to weaken the material to shearing forces and so to allow late-time crater motions to go further and to result in a larger crater. Even here, the expected crater volume still exceeds by almost an order of magnitude the calculated volume. The concept that the initial shock fractures the rock or breaks up the cementation in soils or weathered rock may be realistic in assessing its resistance. However, no good experimental confirmation of this factor exists.

Another large-yield calculation, by Ullrich (1977), does much better on volume for the KOA crater, but the extrapolated shape is still very bowl-shaped unlike the actual shallow saucer-shaped crater (Table 2). As discussed by Port (1977), KOA was an event complicated by the highly stratified and saturated geology of the Eniwetok Atoll and may have had significant late-time redistribution of material that was not included in the calculation. However, the calculation did use a low Von Mises shear failure limit (14 bar) and no tensile strength for the material.

The high yield explosive source cratering problem introduces a richer set of physical phenomena of importance to be modeled. Failure to get good agreement in this case with limited test observations should not detract from the relatively good agreement for less energetic sources. Current theoretical tools are most applicable for hypervelocity impact crater studies, as well as for chemical explosives and buried nuclear shots.

Emphasis has been on comparing crater volumes, but how well do calculations do on such details as crater shape? Cooper (1976) shows the variations in observed crater dimensions and shapes. Schuster (1973) has calculated the crater for Middle Gust I. Figure 14 compares his result with an experimental crater profile. The calculated volume was almost half the observed volume from this high explosive shot. However, the cusp in the calculated profile can be imagined to account for the central rise in the real crater if it has been allowed to slide back down to the crater floor. The comparison is reminiscent of snapshots of splash waves in a water crater, one earlier than the other. One more slosh might have made the theory more like the final actual crater.

While in gross terms the kind of comparison illustrated in Fig. 14 is not bad, there seems to be more to be learned about the final stages of cratering. Furthermore, shape comparison between calculated and observed craters in the high-yield nuclear case seem much worse. Cooper (1976) also discusses the increasing aspect ratio for meteor craters and for the Pacific large-yield craters. These craters do not fit the constant ratio rule for radius versus diameter that is appropriate for small explosive craters. The ratio of diameter to depth is about

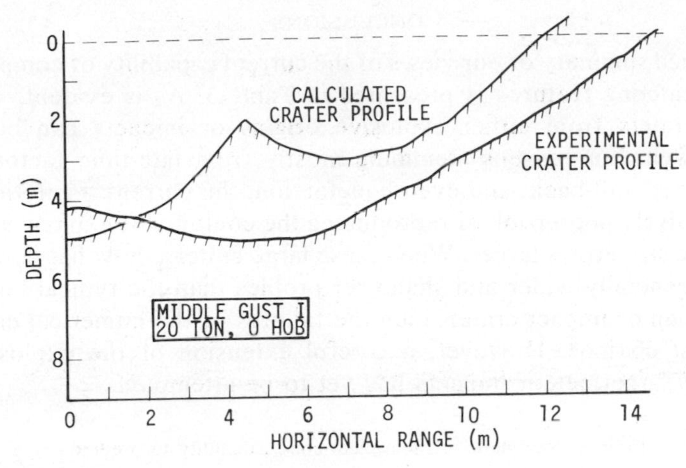

Fig. 14. Comparison of theoretical and experimental crater profiles from a chemical explosion.

2.5 for these smaller craters. For the large explosion craters in wet and stratified media or the large lunar, Martian, or Mercurian craters show an increasing ratio with increasing size. Crater shape, then, becomes for these large craters, a feature to be simulated in cratering calculations. Figure 15 illustrates this shape factor for scaled craters typical of the wet coral sand large-yield nuclear bursts of the Pacific, and a similar scaled crater from the low-yield Nevada Proving Ground dry alluvium shots. The comparison with theoretical calculations shows how poorly the fit is to both shape and volume. Even the degraded strength material calculation does not reproduce either shape or volume.

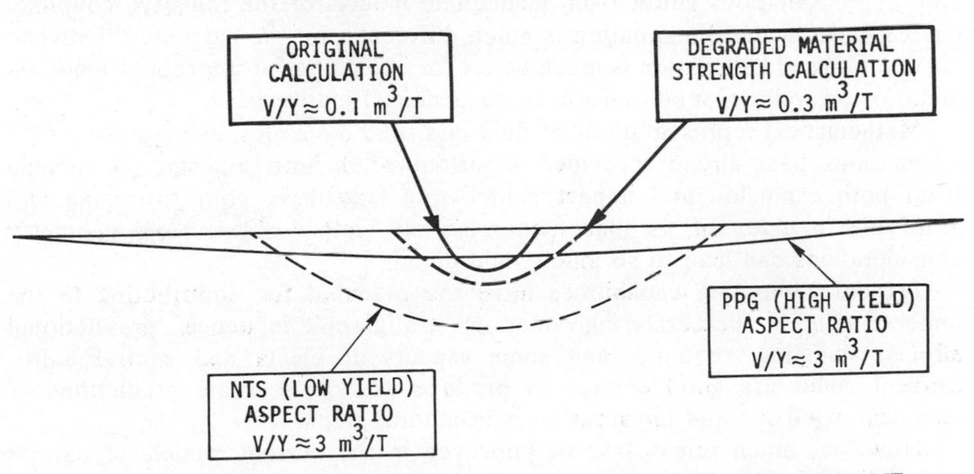

Fig. 15. Comparison of calculated and empirically predicted crater profiles for 1 MT theoretical nuclear surface burst.

CONCLUSIONS

A simplified summary of our views of the current capability of computer codes to predict cratering features is presented in Table 3. As is evident, we believe that small craters from either explosive origins or impacts can be modeled reasonably well. For reasons stemming mostly from late-time factors such as slumping, creep, fall-back, and even liquefaction, the current theoretical models do a progressively poorer job of reproducing the contours of explosion or impact craters as the size grows larger. When these large craters show flat floors, central uplifts, and generally wider and shallower profiles than the typical bowl-shaped small explosion or impact crater, then the failures of the numerical calculations become most obvious. However, a careful extension of the explosive crater work to very large meteor impacts has yet to be attempted.

Table 3. Summary of theoretical code capability to predict cratering.

Source	Volume	Shape
Chemical explosive and non-radiating low yield nuclear	OK	OK
High yield non-radiating nuclear	OK	NO
Small hypervelocity impact	OK	OK
Large hypervelocity impact (modern techniques)	OK?	NO?
High yield surface nuclear	NO	NO

Most of these calculational deficiencies stem from gravitational and stratification influences or reflect complex material properties not adequately modeled for craters in natural geologic materials. It may also be argued that the high-yield cratering calculations suffer from inadequate models of the radiative coupling. On a smaller scale, the situation is much different, and the agreement between observation and calculation is much better for explosions or impacts of metal on metal or on well-behaved uniform (homogeneous) materials.

Mathematical representations of fluid and solid dynamics, as modeled by 2D calculations, have already provided important insight into cratering phenomena from both explosion and impact sources. In fact, it is both surprising and gratifying to many of us that such relatively simple physics and geometry considerations can lead to so much valid detail.

Present computing capabilities have the potential for contributing to the understanding of such cratering factors as stratigraphic influences, gravitational effects, slumping, rebound, and some aspects of ejecta and central uplift. Current codes are good enough to produce usefully-accurate predictions of chemical explosive and impact craters in natural media.

However, much remains to be improved in the current models of natural material. While Von Mises and Mohr–Coulomb models of continuum mechanics are a significant step better than hydrodynamics alone, the modeling of such complex structures as occur in natural geologies demands much more sophistication. Without a model of jointing, faulting, and large rock mass failure and dynamic response, cratering action calculations are unreliable in their predictions of crater dimensions, and cannot generate ejecta size and velocity descriptions.

The use of laboratory sample rock properties in large-scale crater calculation models has been shown to be grossly incorrect. The pertinent dynamic properties of large volumes of geologic material are dominated by such features as void ratios, the amount of joint cementation, shear resistance of jointed rock, and spacing of such joints and faults, their orientation, anisotropy, and variability all are of greater pertinence than the bulk modulus of an intact rock sample when dealing with large-scale craters. The properties of the imperfections in the rock are more important than the "ideal" properties.

Hydrology, or its Martian equivalent, is obviously of great significance. The crater and the ejecta field is much influenced by fluid content even at levels short of saturation. A saturated medium is dramatically different in dynamic response than a dry one. Such factors as closing or collapsing of fluid-filled voids, joint lubrication, and saturation induced by compression can be most important.

Stratification and non-uniform material properties can influence craters markedly, but models of such features have not always been very sophisticated. The dynamics at the interfaces, and the variability of properties across interfaces are seldom correctly controlled or modeled. Without fairly detailed models, the observed benching, terracing, and flat-bottoming of craters and possibly the growth of central uplifts may not be adequately predicted.

Such late-time effects as slope stability under gravity clearly play an important part in crater formation, as do the ballistics for ejecta and does the sliding of rubble up the crater side or the folding over of the lip. Slumping, or late-time changes in crater dimensions are controlled by material properties as well as gravity. The shear resistance under gravity for freshly cratered materials may be important, and yet difficult to determine.

The proper description of a material will depend on the load levels it must respond to as well as its geologic and cratering history. Some mixtures of materials can behave like one or the other constituent at some loads, but other mixtures will respond quite unlike any of its constituents under shock loads, resulting in shock speeds, and dynamic properties which are not an average in any sense of the properties of the elements of the mixture. For example, gaseous inclusions (bubbles) in a material can lead to shock or sound speeds in the mixture lower than that in either the solid or the gas, and can lead to shock properties surprisingly unlike either (Parkin *et al.*, 1961). Models for natural materials under the wide range of compressive and shear loads experienced in cratering actions are by no means complete at present, and several improvements may be necessary before predictions to better than a factor of two to five in volume for large-scale planetary craters or large-yield explosion craters are possible.

The shock-conditioning phenomena, or the concept of crushing a medium under the initial compressive shock loads and then allowing it to be swept away by subsequent shear motions as a weakened material is an attractive possibility among the likely mechanisms responsible for the larger-than-computed craters. The details of such mechanisms are not generally available from experimental investigations, as yet.

The computer codes in current use are also less than ideal. Some obvious improvements are already appreciated and under development or consideration. One aspect which needs improvement is the way in which the need for global coverage and shock front detail is met. The whole flow field is necessary to the description, since it is all in communication and any part can influence any other part. At the same time, the importance of following in accurate detail the rapid compression and expansion near the shock fronts cannot be slighted, and requires fine zoning as well as extensive zone coverage. The need for more sophisticated means of covering the essentials in the flow field without greatly increasing the number of cells or zones is obvious. The potential for more efficient means of handling shock fronts and interfaces is also high. The need for faster running codes remains.

Beyond the development of faster, better codes, there are new machines which promise to handle larger problems better. The parallel-processing machines (like ILLIAC-IV) or the pipe-line machines (like STAR or CRAY) offer real advantage for such calculations as are involved in cratering predictions. It is likely that other special-purpose computers will be developed which can accommodate more detail and more ambitious calculations at less cost.

It requires no special insight to predict improved computer capabilities, even when the specific improvements are as yet unanticipated. What should be appreciated is the competitive balance between the time and cost involved in theoretical modeling (computer calculations) and experimental modeling, and the degree of reliability or extrapolatability of each. While both tests and computations become more expensive and time-consuming as their sophistication grows, the need for both in a mutually-supporting program is obvious. The theoretical results become more detailed and exact, but will always require calibration and verification because of their complex and obscure derivations. The impracticality of creating large-scale test craters leads to the fielding of smaller experimental programs, but without the confidence of theoretical extension, scaling to the larger planetary craters would remain most uncertain and impractical.

References

Bjork, R. L.: 1961, Analysis of the Formation of Meteor Crater, Arizona: a preliminary report, *J. Geophys. Res.* **66**, 3379–3387.

Bjork, R. L., Kreyenhagen, K. N., and Wagner, M. H.: 1967, *Analytical Study of Impact Effects as Applied to the Meteoroid Hazard*, U.S. Government Printing Office, Washington, D.C., NASA CR-757.

Bjork, R. L. and Olshaker, A. E.: 1965, *The Role of Melting and Vaporization in Hypervelocity Impact*, The Rand Corporation, Santa Monica, California, RM-3490-PR.

Brode, H. L. and Bjork, R. L.: 1960, *Craters from a Megaton Surface Burst*, The Rand Corporation, Santa Monica, California, RM-2600.

Buckingham, A. C., Hancock, S. L., McKay, M. W., and Orphal, D. L.: 1973, *Calculations of Cactus Ground Motion and Design of the Mine Throw II Charge*, Physics International for Defense Nuclear Agency, Washington, D.C., DNA 3131F-3.

Christensen, D. M.: 1968, Theoretical Ground Motion and Crater Calculations Event 6, *Operation DISTANT PLAIN Symposium II Proceedings*, Defense Atomic Support Agency, Washington, D.C., DASA 2207.

Christensen, D. M.: 1970, *ELK-40 Prediction Calculation of Ground Motion for DISTANT PLAIN Event 6*, Defense Nuclear Agency, Washington, D.C., DASA 2471.

Cooper, H. F., Jr.: 1976, *Estimate of Crater Dimensions for Near-Surface Explosion of Nuclear and High Explosive Sources*, R & D Associates, Marina Del Rey, California, RDA-TR-2604-001.

Duba, A. G., Heard, H. C., and Santor, M. L.: 1974, *Effect of Fluid Content on the Mechanical Properties of Westerly Granite*, Lawrence Livermore Laboratory, University of California, Livermore, California, UCRL-51626.

Ialongo, G.: 1973, *Prediction Calculations for the MIXED COMPANY III Event*, Shock Hydrodynamics Division, Whittaker Corporation for Defense Nuclear Agency, Washington, D.C., DNA 3206T.

Ialongo, G., McDonald, S. W., and Reid, J. B.: 1974, *A Two-Dimensional Calculation of Large Burst Phenomenology*, Defense Nuclear Agency, Washington, D.C., DNA 3397F.

Knott, R. A.: 1973, *Effect of Loading Rate on the Stress-Strain Characteristics of a Clay Shale in Unconsolidated-Undrained Triaxial Compression*, U.S. Army Engineer Waterways Experiment Station, Soils and Pavement Laboratory, Vicksburg, Mississippi, S-73-68.

Matuska, D.: 1972, Personal Communication, (Air Force Weapons Laboratory, Kirtland Air Force Base, New Mexico, and DNA High Altitude Symposium).

Maxwell, D. E. and Moises, H.: 1971, *Hypervelocity Impact Cratering Calculations*, Physics International for NASA Manned Spacecraft Center, Houston, Texas, PIFR-190.

Maxwell, D. E., Reaugh, J., and Gerger, B.: 1973, *JOHNIE BOY Crater Calculations*, Physics International for Defense Nuclear Agency, Washington, D.C., DNA 3048F.

McDonald, J. W. and Reid, B.: 1974, *Crater, Ejecta and Ground Motion Calculating from JOHNIE BOY*, Defense Nuclear Agency, Washington, D.C., DNA 3168T.

Orphal, D. L., Maxwell, D. E., Reaugh, J. E., and Borden, W. F.: 1975, *A Computation of Cratering and Ground Motion from a 5 MT Nuclear Surface Burst over a Layered Geology*, Defense Nuclear Agency, Washington, D.C., DNA 3711F.

Parkin, R. B., Gilmore, F. R., and Brode, H. L.: 1961, *Shock Waves in Bubbly Water*, The Rand Corporation, Santa Monica, California, RM 2795-PR (Abridged).

Port, R. J.: 1976, Cratering from Large Explosive Events, (abstract). In *Papers Presented to the Symposium on Planetary Cratering Mechanics*, p. 108. The Lunar Science Institute, Houston.

Rosenblatt, M.: 1971, *Numerical Calculations of Hypervelocity Impact Crater Formation in Hard and Soft Aluminum Alloys*, Air Force Materials Laboratory, Wright-Patterson Air Force Base, Ohio, AFWL-TR-70-254.

Schuster, S.: 1973, *Results of Pre-Test Prediction Calculation OF MIDDLE GUST I, II, and III*, Air Force Weapons Laboratory, Kirtland Air Force Base, New Mexico.

Seebaugh, W. R.: 1976, A Dynamic Crater Ejecta Model (abstract). In *Papers Presented to the Symposium on Planetary Cratering Mechanics*, p. 133–135. The Lunar Science Institute, Houston.

Short, N. M.: 1960, *Fracturing of Rock Salt by a Contained High Explosive*, Lawrence Radiation Laboratory, University of California, Livermore, California, UCRL-6054.

Trulio, J. G.: 1977, *Ejecta Formation: Calculated Motion from a Shallow-Buried Nuclear Burst, and its Significance for High Velocity Impact Cratering*, this volume.

Ullrich, G. W.: 1977, Personal Communication, (Air Force Weapons Laboratory, Kirtland Air Force Base, New Mexico).

Walsh, J. M. and Tillotson, J. H.: 1963, *Proceedings of the Sixth Symposium on Hypervelocity Impact*, Vol. 2, p. 59–73. Cleveland, Ohio.

Wilkins, M. L., Blum, R. E., Cronshagen, E., and Granthan, P.: *A Method for Computer Simulation of Problems in Solid Mechanics and Gas Dynamics in Three Dimensions and Time*, Lawrence Livermore Laboratory, University of California, Livermore, California, UCRL-51574.

Roddy, D. J., Pepin, R. O., and Merrill, R. B., editors.
(1977) *Impact and Explosion Cratering*, Pergamon Press (New York), p. 897–906.
Printed in the United States of America

Calculations of explosion cratering—I The shallow-buried nuclear detonation JOHNIE BOY

D. L. ORPHAL

Physics International Company, 2700 Merced Street, San Leandro, California 94577

Abstract—JOHNIE BOY was a nominal 0.5 kiloton nuclear event conducted in Nevada Test Site alluvium with the detonation point approximately 58 cm below the ground surface. The cratering dynamics for JOHNIE BOY were computed to a real time of 0.8 sec, when motion in the crater region had nearly terminated. The crater volume was computed to be $9.6 \times 10^3\ m^3$ as compared to the measured value of $6 \times 10^3\ m^3$. Computed values for the crater depth and radius were 14 m and 21 m, respectively, compared to measured values of 12.1 m and 22 m.

INTRODUCTION

THERE HAVE BEEN very few theoretical calculations of impact planetary cratering (Bjork, 1961; Maxwell and Moises, 1971; O'Keefe, 1976; O'Keefe and Ahrens, 1975, 1976). With the exception of the recent work by O'Keefe and Ahrens, these calculations were performed a number of years ago. Since the time when these early calculations were performed, a number of significant advances have been made in both the computer hardware and numerical techniques used for theoretical cratering calculations. In particular, the larger and faster computers now available and advances in numerical rezoning techniques allow computation of cratering dynamics to very late times as compared to most earlier computations, while retaining reasonable spatial and temporal resolution. Computation of crater formation to much longer times allows detailed study of cratering dynamics throughout most or all of the crater formation process.

With the exception of the work by O'Keefe and Ahrens, and recent research on microcratering of selected nongeologic materials by hypervelocity impact of small projectiles, most theoretical cratering calculations performed since the mid or late 1960s have been performed for explosions. One purpose of this paper is to illustrate current capability for computing cratering dynamics for surface or shallow-buried high explosive and low yield (i.e., essentially nonradiative) nuclear explosions. This will be done in Part I by presenting the essential results of a recent calculation of cratering for the JOHNIE BOY nuclear explosion.

Analysis of the results of the JOHNIE BOY and other recent calculations have revealed a number of basic characteristics of the explosion cratering process. Part II of this paper briefly reviews these results.

BACKGROUND

JOHNIE BOY, a nominal 0.5 kt (kiloton) nuclear explosion, was detonated in Area 18 alluvium at the Nevada Test Site with the initial point of energy release located approximately 58.5 cm below the ground surface. The cratering dynamics for the JOHNIE BOY detonation were computed to a time of 0.8 sec when motion in the crater region had nearly terminated. Details of the various numerical techniques and material models used in the calculation are omitted here, but may be found in the report by Maxwell *et al.* (1973). The JOHNIE BOY calculation was performed using the nonlinear explicit finite difference computer program PISCES 2D ELK (Hancock, 1976).

Figure 1 illustrates the initial geometry used for the calculation. The computation assumed axial symmetry through the center of the detonation. Initial computational zones were 3 cm square throughout a region extending from a height of 54 cm above the ground surface to 180 cm below the ground surface and to a radius of 150 cm.

The first 300 μsec of the calculation was performed in Eulerian coordinates. At a time of 300 μsec, a Lagrangian computational grid extending in space beyond a range and depth of 3.6 m was coupled to the existing Eulerian grid. At a time of 7.84 ms (milliseconds) coupling of energy from the Eulerian grid to the

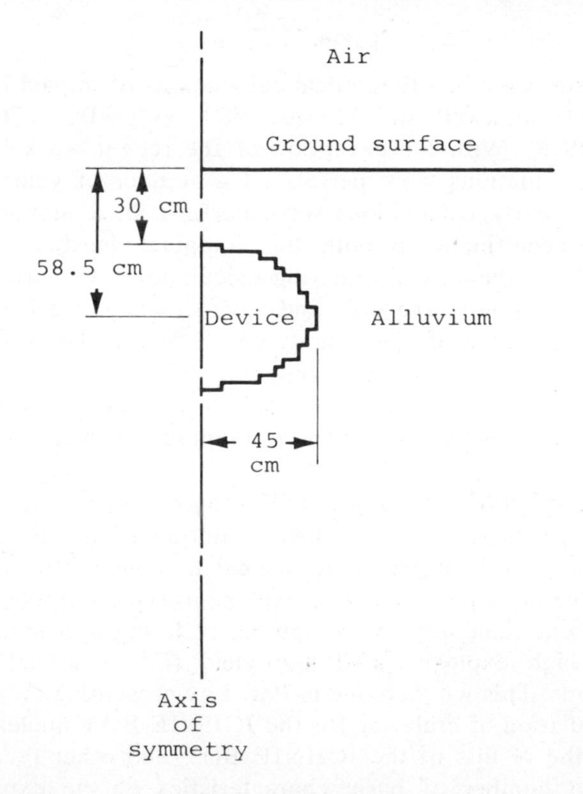

Fig. 1. Initial geometry.

Lagrangian grid was complete and the Eulerian grid was removed from the calculation. Thus for times greater than 7.84 ms the calculation was performed solely in Lagrangian coordinates. The nominal zone in the crater region at termination of the calculation was 320×275 cm.

The air initial density and pressure were 1.225×10^{-3} g/cc and zero pressure respectively. The alluvium had a nominal initial density of 1.8 g/cc. Gravitational forces were included in the calculation resulting in an increase of the alluvium initial density and hydrostatic pressure with depth. The device region was modeled using the equation-of-state for alluvium and an initial energy density of 90.6167×10^{12} ergs/cc, corresponding to 0.5 kt total energy. The initial pressure in the device region was 45.3084 Mb (Megabars).

For times greater than 300 μsec, the airblast was not computed explicitly. Rather a time-and space-dependent pressure boundary condition was applied to the ground surface. This boundary condition was modeled after Brode (1968) but with overpressure reduced to account for degredation of the air shock due to source burial. More specifically, the airblast overpressure boundary condition was taken as

$$P(R,t) = P_{\text{Brode}}(R,t)\, F(t), \quad t \geq 50\mu\text{sec}, \tag{1}$$

where $F(t) = 0.725(t_{\mu\text{sec}}/t_{\mu\text{sec}} + 200)$.

The function $F(t)$ was constructed from earlier calculations to correspond to the fraction of the original explosion yield above the ground plane as a function of time for times greater than 50 μsec. Any error in using Eq. (1) to model the JOHNIE BOY airblast is thought to have negligible effect on the computed crater formation.

Computational Results

The maximum energy coupled to the ground by the explosion was computed to be about 0.15 kt or 30% of the total yield. This energy is coupled in a small volume immediately surrounding the device. Figure 2 shows the total (internal plus kinetic) energy versus time propagated beyond the hemispherical surfaces designated by the original radii, R_0, and measured from the intersection of the original ground surface with the axis of symmetry. Also shown in Fig. 2 is the percentage of the total explosion energy (0.5 kt or 2.09×10^{19} ergs) propagating beyond the various hemispherical surfaces and the approximate peak pressure corresponding to these hemispherical radii. As can be seen from Fig. 2, the maximum energy coupled beyond the 3.6 m radius is achieved by a time of about 1.5 ms and slightly more than 6% of the total explosion energy is propagated beyond this range. Somewhat less than 3% of the explosion energy is propagated beyond the 5.2 m hemisphere and full coupling at this range is not achieved until a time of about 5 ms.

Figure 3 displays the time-dependent partitioning of the energy propagated in the ground beyond the 3.6 m hemisphere. In this plot, *IE* is the internal energy and is equal to the sum of the energy due to pdV work on the ground material

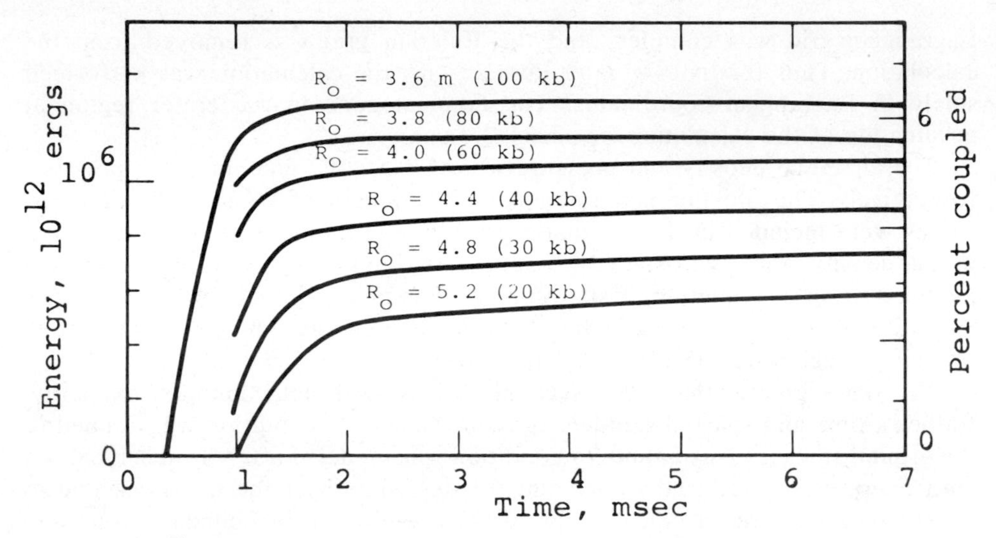

Fig. 2. Computed energy coupling versus time and for specific ranges.

and the distortional energy of the ground material resulting from work done against the shear strength of the alluvium.

Figures 4 and 5 display contours of peak pressure below the ground surface. The device region at an initial pressure of 45 Mbar is shown in Fig. 4. The effect of the degraded airblast is evident in the near surface portion of the 0.3 kb and 0.2 kb contours shown in Fig. 5.

Figure 6 displays the peak particle velocity versus depth along the axis of symmetry.

The computed growth of the JOHNIE BOY crater as a function of time is shown in Fig. 7 by a series of crater profiles. The crater achieves a maximum

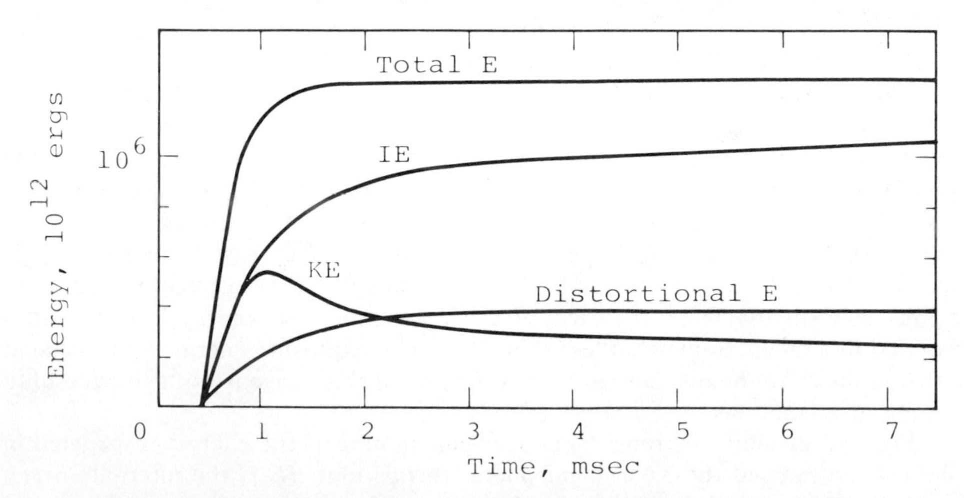

Fig. 3. Partition of energy versus time in the ground beyond the range of 3.6 m.

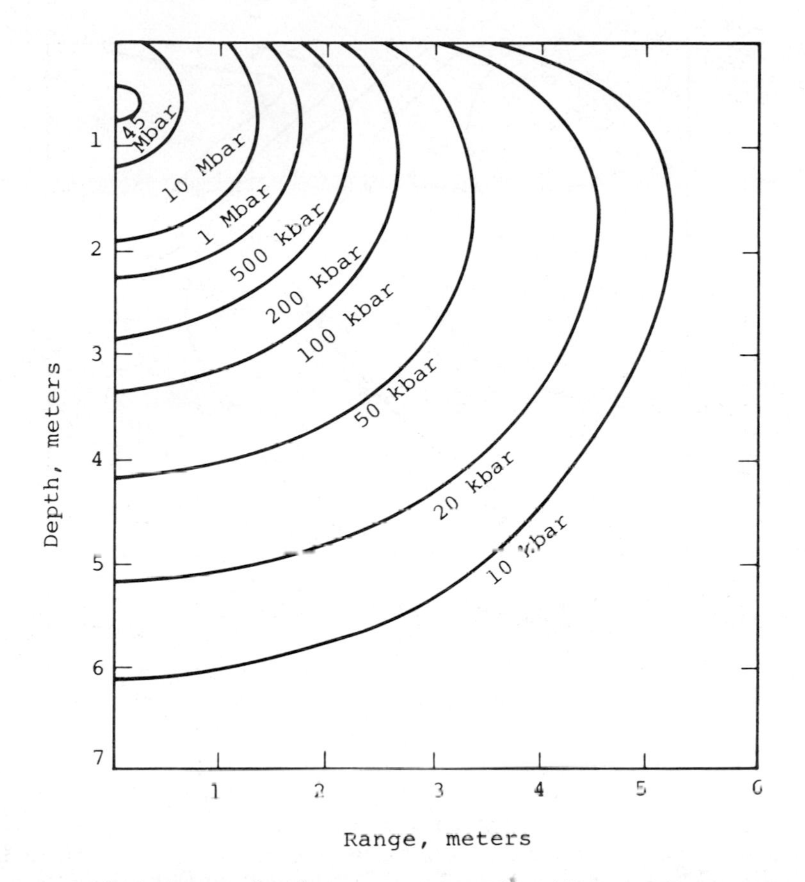

Fig. 4. Computed peak pressure contours from 45 Mb to 10 kb.

depth of about 14 m at a time of about 0.2 sec. The computation shows a small, very localized, rebound of the crater floor near the axis of symmetry. This rebound is not thought to be physical but rather the result of inaccuracies in the numerical finite difference equations on the axis of symmetry.

The final crater radius of 21 m is achieved at a time of 0.8 sec. The computation shows no significant changes in crater depth or radius for times greater than 0.8 sec.

An interesting feature of the JOHNIE BOY crater growth is evident from the crater profiles shown in Fig. 7. It is noted that until the crater has achieved maximum depth (i.e., a time of about 0.2 sec) crater growth is nearly hemispherical. Beyond a time of 0.2 sec, crater depth remains nearly constant while the crater radius continues to grow, resulting in an increase in the radius to depth ratio from 1 to about 1.5.

The "true" and apparent crater profiles measured for JOHNIE BOY are compared to the computed profile in Fig. 8. The computed profile is expected to compare most closely with the "true" crater (i.e., the crater profile that would

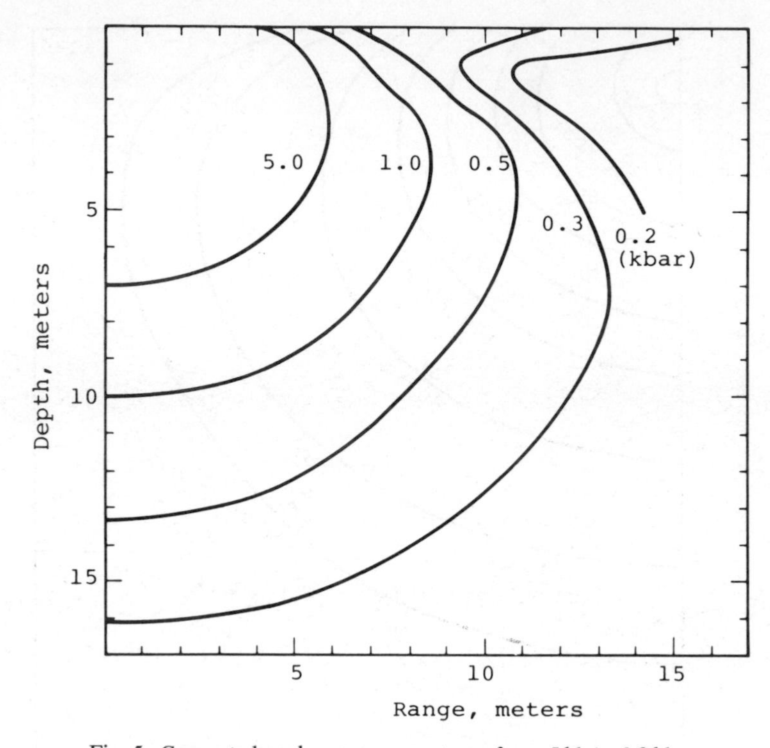

Fig. 5. Computed peak pressure contours from 5 kb to 0.3 kb.

have existed in the absence of ejecta fallback, wall slumping, etc.). The measured true crater depth of 12.1 m compares reasonably well with the calculated depth. The computed crater radius of 21 m also compares well with the measured value of 22 m. However, the computed and measured crater profiles differ significantly in shape resulting in a computed crater volume of 9600 m^3 as compared to 6000 m^3 measured.

Essentially all the ejected material was computed to leave the crater region from the transient crater lip. Figure 9 shows selected trajectories for reference mass particles located in the crater region. It can be seen that much of the mass ejected was computed to be lofted on trajectories that passed close to the final crater lip (the final computed crater profile is shown as the dashed line). In this figure the "kink" in the upper particle trajectory shown is not physical, but is due to an error introduced during rezoning.

Examination of Fig. 9 also reveals the interesting feature that the trajectories of the ejected mass intersect the original ground surface with a nearly constant angle of 38–40° (measured counterclockwise from the horizontal). This feature will be discussed in somewhat more detail in Part II of this paper.

Because the principal objectives of the JOHNIE BOY calculation were to compute cratering dynamics, the final crater size and geometry, and the explosion induced ground motions, detailed studies of the ejecta field were not performed. Therefore, no further data concerning the JOHNIE BOY ejecta will be presented.

Summary and Discussion

A theoretical calculation of cratering for the JOHNIE BOY shallow-buried nuclear explosion was performed to a time of 0.8 sec when motion in the crater region had nearly terminated. The computation showed no significant changes in crater depth or radius for times greater than 0.8 sec. The computed crater volume was 9.6×10^3 m^3 compared to the experimental value of 6×10^3 m^3. The computed crater depth and radius were 14 m and 21 m respectively, compared to measured values of 12.1 m and 22 m.

The results of the JOHNIE BOY calculation are considered representative of the state-of-the-art for theoretical calculations of cratering for surface and

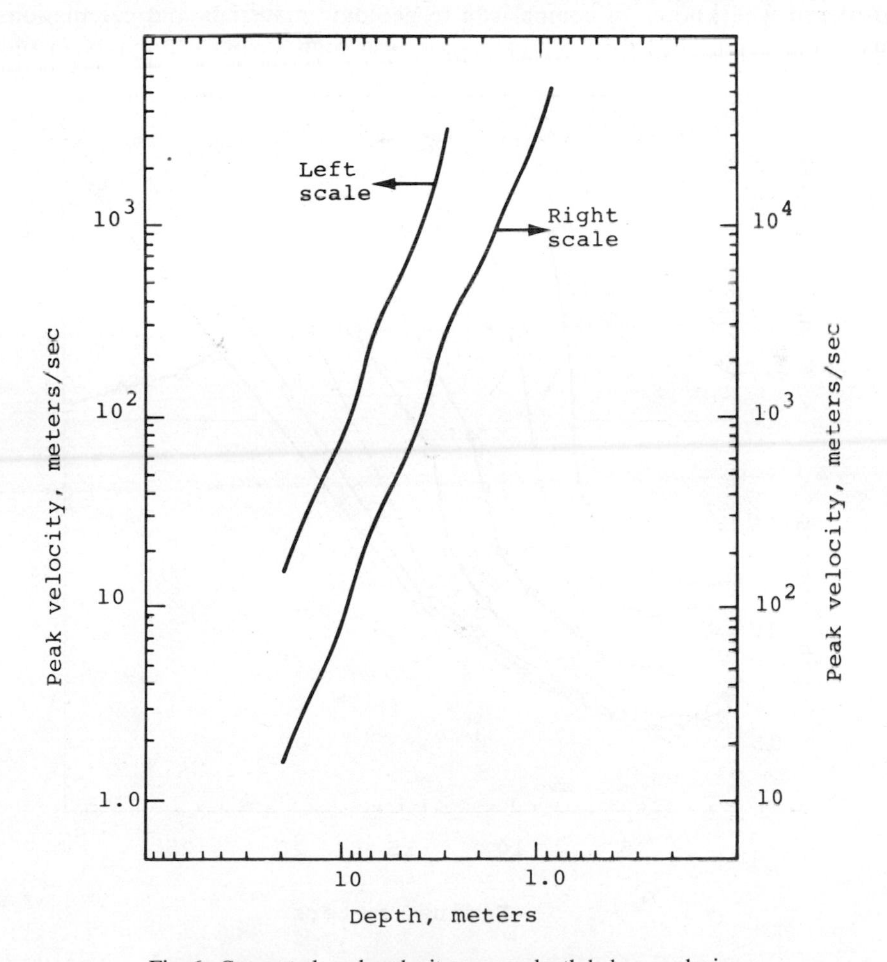

Fig. 6. Computed peak velocity versus depth below explosion.

shallow-buried high explosive and low-yield (i.e., non radiative) nuclear detonations. That is for surface and shallow-buried high explosive and low-yield nuclear detonations, computations generally result in crater volumes that are within approximately a factor of two of observed values (Knowles and Brode, 1977). However, in many instances, even when crater volume is accurately computed, the calculated crater shape differs significantly from observations.

For the case of cratering from a high yield (i.e., radiative) nuclear surface burst, recent calculations have resulted in crater volumes that are a factor of 30 to 50 smaller than is estimated on the basis of empirical scaling of data from lower yield nuclear and high explosive detonations (Knowles and Brode, 1977).

The reasons for the discrepancies between calculated and observed explosion crater volumes and shapes are not well understood. It seems likely at this time that one major source of error lies in the modeling of the stress-strain response of geologic materials. For instance, the stress-strain characteristics of metals are considered well known in comparison to geologic materials and calculations of crater volumes and shapes for explosion and high velocity impacts in metals

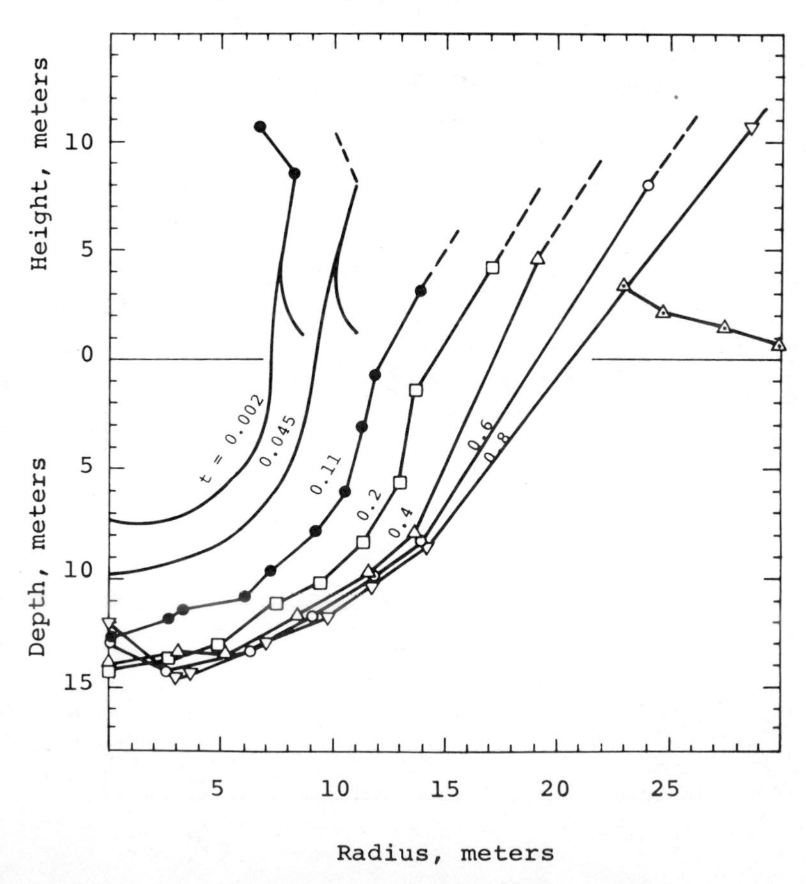

Fig. 7. Computed crater profiles at specific times (sec).

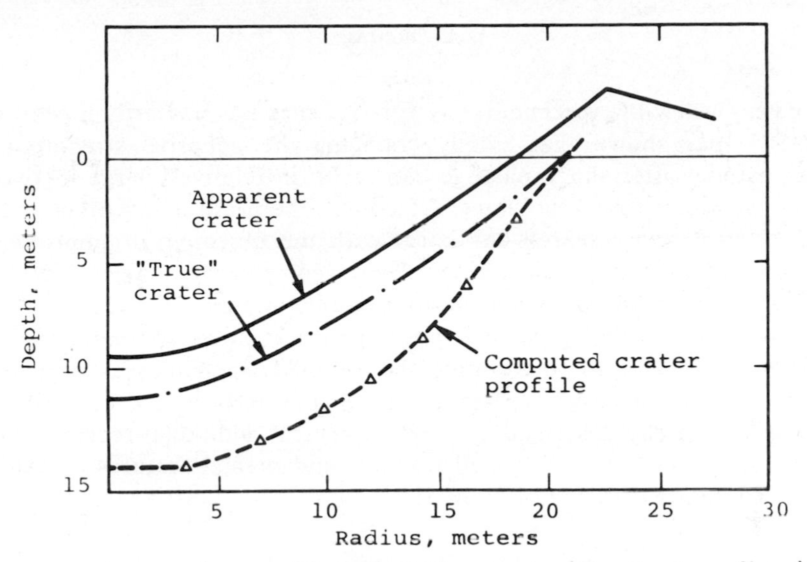

Fig. 8. Comparison of measured JOHNIE BOY apparent and "true" crater profiles with computed profile.

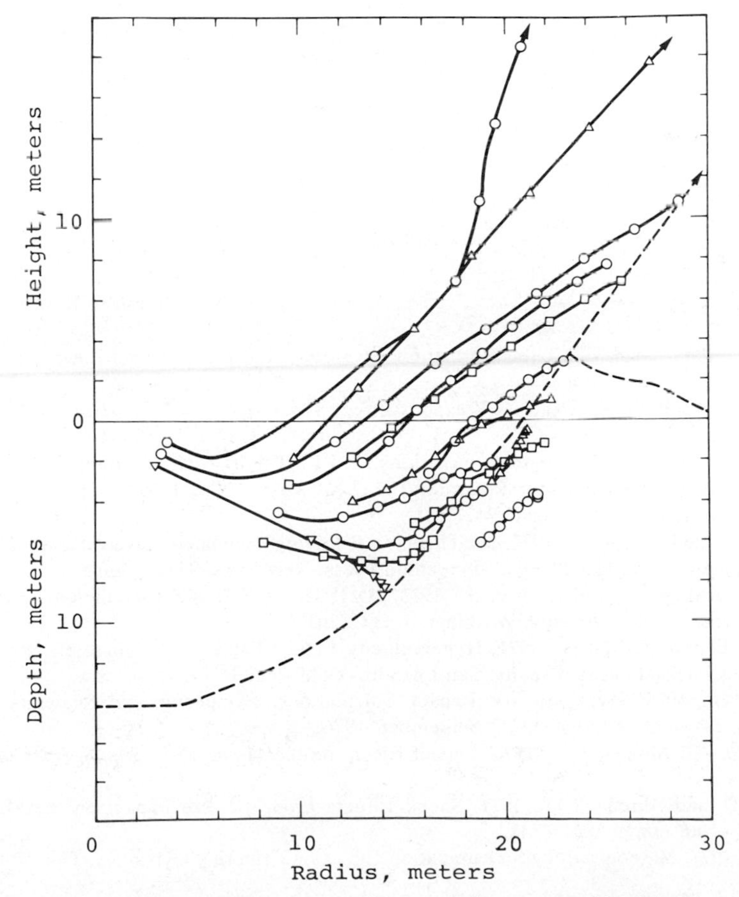

Fig. 9. Selected particle trajectories (symbols are spaced by 0.1 sec).

compare very well with experiments (see for a recent example Bertholf *et al.*, 1975). Swift (1977) has shown that simply changing the cohesive strength of the geologic material after shock passage can result in relatively large increases in computed explosion crater volumes. Of course there are many other potential sources of error and research continues with the objective of improving our capability to accurately predict the size and geometry of explosion craters as well as related phenomena such as ground motion and ejecta fields.

In addition to calculating the size of an explosion crater, theoretical cratering calculations may be used to examine the detailed mechanics associated with crater formation. Analyses of the results of the JOHNIE BOY calculation, which was just briefly described, as well as several additional recent explosion cratering calculations, have revealed several fundamental features of explosion cratering. These are discussed in Part II.

Acknowledgments—A major theoretical cratering calculation such as JOHNIE BOY results from the efforts of a number of people. The JOHNIE BOY calculation was performed by members of the Applied Mechanics Department of Physics International Company. A similar calculation was performed by Kreyenhagen and his coworkers at Shock Hydrodynamics, Inc. At Physics International, principal credit for the JOHNIE BOY calculation is due to Don Maxwell, John Reaugh and Bence Gerber.

The JOHNIE BOY calculation was performed under a contract from the Defense Nuclear Agency, Washington, D.C.

REFERENCES

Bertholf, L. D., Buxton, L. D., Thorne, B. J., Byers, R. K., Stevens, A. L., and Thompson, S. L.: 1975, Damage in Steel Plates from Hypervelocity Impact. II. Numerical Results and Spall Measurement, *J. Appl. Phys.* **46**, 3776–3783.

Bjork, R. L.: 1961, Analysis of the Formation of Meteor Crater, Arizona. *J. Geophys. Res.* **66**, 3379–3387.

Brode, H. L.: 1968, Review of Nuclear Weapons Effects, *Annual Review of Nuclear Science, Volume 18* (E. Segre, ed.).

Hancock, S. L.: 1976, Finite Difference Equations for PISCES 2D ELK, A coupled Euler-Lagrange Continuum Mechanics Computer Program, Technical Memo TCAM 76-2, Physics International Company, San Leandro, California 94577.

Knowles, C. P. and Brode, H. L.: 1977, The Theory of Cratering Phenomena, an Overview. (D. J. Roddy, R. O. Pepin, and R. B. Merrill, eds.), Pergamon Press, New York. This volume.

Maxwell, D., Reaugh, J., and Gerber, B.: 1973, JOHNIE BOY Crater Calculations, Report DNA 3048F, Defense Nuclear Agency, Washington, D.C. 20305.

Maxwell, D. E. and Moises, H.: 1971, Hypervelocity Impact Cratering Calculations, Report PIFR-190, Physics International Company, San Leandro, California 94577.

O'Keefe, J. D.: 1976, Hypervelocity Impact Calculations, Symposium on Planetary Cratering Mechanics, Flagstaff, Arizona, 13–17 September, 1976.

O'Keefe, J. D. and Ahrens, T. J.: 1976, Impact Ejecta on the Moon, *Proc. Lunar Sci. Conf. 7th.*, p. 3005–3025.

O'Keefe, J. D. and Ahrens, T. J.: 1975, Shock Effects From a Large Impact on the Moon, *Proc. Lunar Sci. Conf. 6th*, p. 2831–2844.

Swift, R. P.: 1977, Material Strength Degradation Effect on Cratering Dynamics. This Volume.

Roddy, D. J., Pepin, R. O., and Merrill, R. B., editors.
(1977) *Impact and Explosion Cratering*, Pergamon Press (New York), p. 907–917.
Printed in the United States of America

Calculations of explosion cratering—II
Cratering mechanics and phenomenology

D. L. ORPHAL

Physics International Company, 2700 Merced Street, San Leandro, California 94577

Abstract—Detailed examination of recent cratering calculations for surface and shallow buried high explosive and nuclear detonations has revealed several important features of explosion cratering dynamics and phenomenology. Principal among these are:

(1) Prior to the time the crater achieves its maximum dynamic depth, crater growth is nearly hemispherical.
(2) Crater growth subsequent to time of maximum dynamic depth is achieved principally by "shearing" material from the crater wall.
(3) The cratering flow field, defined loosely as the material flow field in the cratering region but behind the outgoing shock wave, closely approximates steady-state incompressible flow.
(4) The radial velocity characterizing the cratering flow field, $\dot{R}$, is approximately independent of angle.
(5) $\dot{R}$ may be approximated by the equation $\dot{R} = \alpha R^{-Z}$ where to a good first approximation α and Z are constants.
(6) Ejecta is lofted from near the transient crater lip resulting in a thin sheet of ejecta in the shape of an inverted cone. Except at very early and very late times the angle at which ejecta leaves the ground surface is nearly constant and is usually 40–60° from the horizontal.
(7) Overturned flaps are observed and are computed to occur at late times.

INTRODUCTION

THEORETICAL CALCULATIONS provide one of the means of studying the basic mechanics and phenomenology associated with explosion and impact crater formation. Detailed analyses of a number of recent calculations of explosion cratering, including the JOHNIE BOY calculation discussed in Part I, have revealed several interesting features of the explosion cratering process. The purpose of this paper is to review these results. For the sake of brevity only selected computational results will be presented. The following discussion is somewhat artifically sub-divided into examinations of the characteristics of: (a) overall crater growth as a function of time, (b) the flow field of the material in the crater region, and (c) the ejection process.

CRATER GROWTH

Define t_D to be the time at which an explosion crater achieves its maximum depth. Figure 1 shows computed material velocity fields in the crater region for a 5 Mt (megaton) nuclear surface explosion over a layered shale-sandstone geology. The top portion of Fig. 1 shows the cratering flow field at a time of about $t_D/3$; i.e., well before the crater has achieved maximum depth. At this time

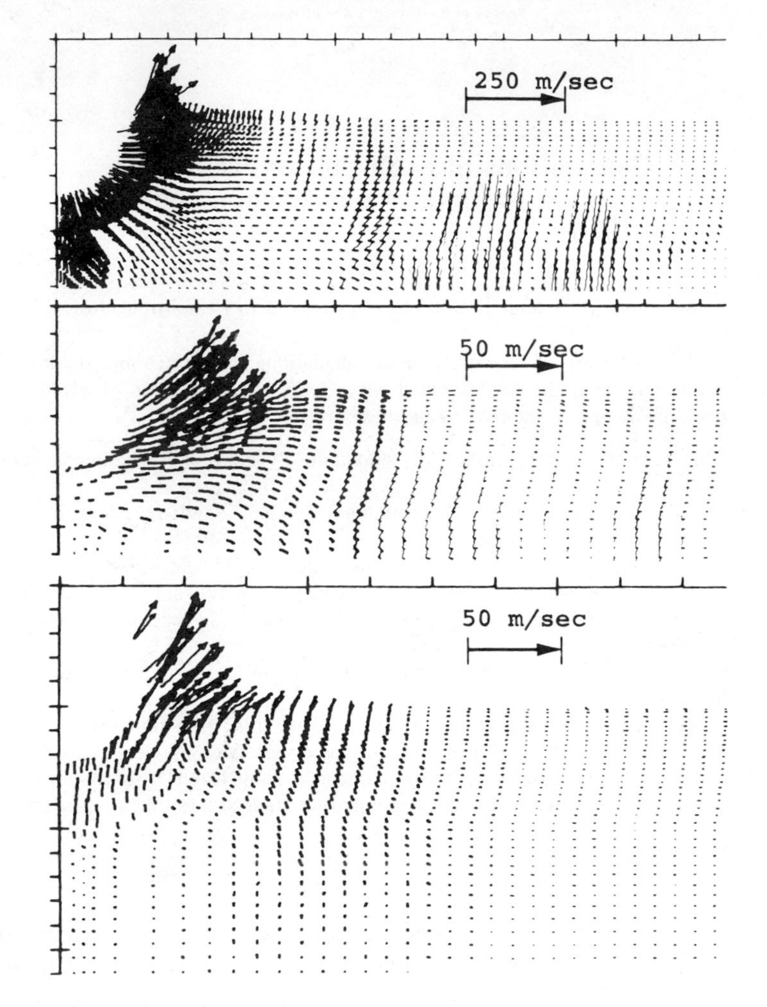

Fig. 1. Calculated vector velocity fields in crater region for a 5 Mt nuclear surface explosion at times $t_D/3$, t_D and $2t_D$ where t_D is time crater achieves maximum depth.

the crater is growing essentially hemispherically and the transient crater depth and radius are nearly identical.

The middle portion of Fig. 1 shows the cratering velocity field at a time nearly equal to t_D. At this time the transient crater radius and depth are still nearly equal. However, the character of the cratering flow has changed significantly. Instead of an outward hemispherical motion, the cratering flow field has altered to what will be called here a "shearing flow" along the transient crater wall. This shearing flow continues until very late times and essentially all crater growth subsequent to time t_D is due to the shearing of material from the transient crater wall, forming a relatively thin sheet, or cone, of ejecta.

The bottom portion of Fig. 1 shows the cratering flow field at a time $2t_D$. The motion is still dominantly a shearing flow. Note, however, the rebounding of the

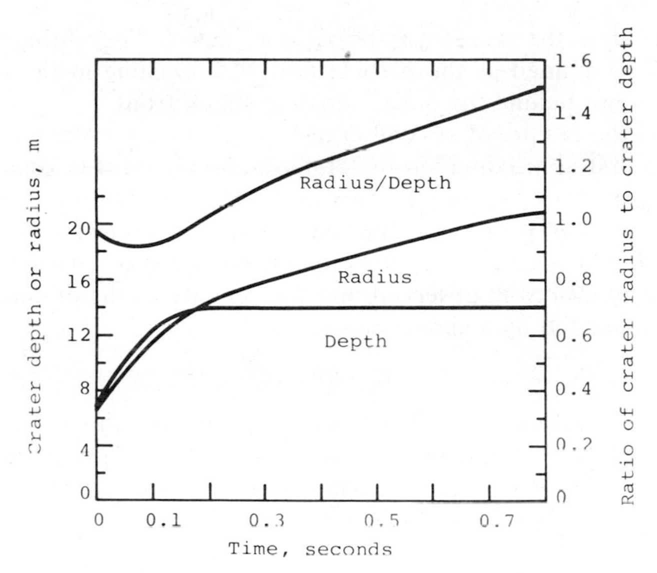

Fig. 2. Computed JOHNIE BOY crater radius, depth, and radius to depth aspect ratio versus time.

crater floor. Such rebounding is seen in most calculations of explosion cratering. In nearly all cases the computed upward displacement of the floor is small. However, under particular circumstances of layering, material properties, location of the water table, etc., central uplift or central peak formation is computed. This phenomenon is discussed in more detail by Ullrich (1976) and Ullrich *et al.* (1977).

In summary, although it is certainly an oversimplification, crater growth may be conveniently visualized as a two-stage process. Before the time that the crater reaches its maximum depth crater growth is essentially hemispherical with a transient crater radius to depth aspect ratio nearly equal to one. For times after maximum crater depth has been achieved, the crater radius continues to grow, principally by the "shearing" of material from the transient crater wall. During this second stage of crater growth the radius-to-depth ratio increases to some value greater than one. Also during this stage of cratering a central peak may be formed. This two-stage process is illustrated again in Fig. 2 which shows the computed crater radius, depth, and radius-to-depth aspect ratio calculated for the JOHNIE BOY shallow-buried nuclear explosion (see Part I).

Cratering Flow Field

Some very general remarks regarding the cratering flow field were made above. However, several more fundamental features of the cratering flow field have been revealed by the extensive work of Maxwell and his co-workers (Maxwell and Seifert, 1975; Maxwell, 1977). Several of Maxwell's observations and results will be briefly reviewed here. For this purpose it is necessary to at least loosely define the words "cratering flow field". Therefore, the cratering flow field will be defined as the particle flow field existing in the region of the transient crater but behind the initial outgoing shock front.

In studying the results of several cratering calculations Maxwell discovered that to a good first approximation the cratering flow field was steady-state, i.e., time independent during a major portion of the crater forming process. Furthermore, again to a good first approximation, the cratering flow field was independent of angle (the coordinate system used in this discussion is illustrated in Fig. 3). Finally, Maxwell observed that the velocity of the cratering flow field could be approximated by a simple power law

$$\dot{R} = \alpha R^{-Z}, \tag{1}$$

where $\dot{R}$ is the radial velocity, R is a radial distance as shown in Fig. 3 and α and Z are parameters characterizing the strength and shape of the flow field, respectively. To a good first approximation α and Z are constants. For surface and shallow-buried explosions $Z = 3$ seems to be a good approximation. (Note that constant α and Z is only a first approximation. Maxwell's work shows that Z varies with θ and α must be a function of time in order that total energy is conserved.)

Figure 4 shows a plot of spatial averages of $\dot{R}$ versus R from the JOHNIE BOY calculation at the three times indicated. As can be seen, at a time of 2 ms,

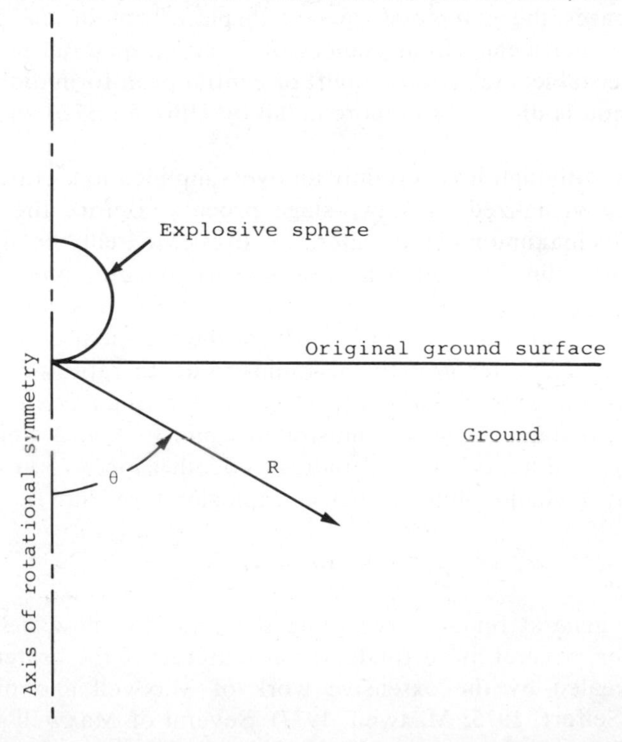

Fig. 3. Coordinate system used for describing cratering flow field.

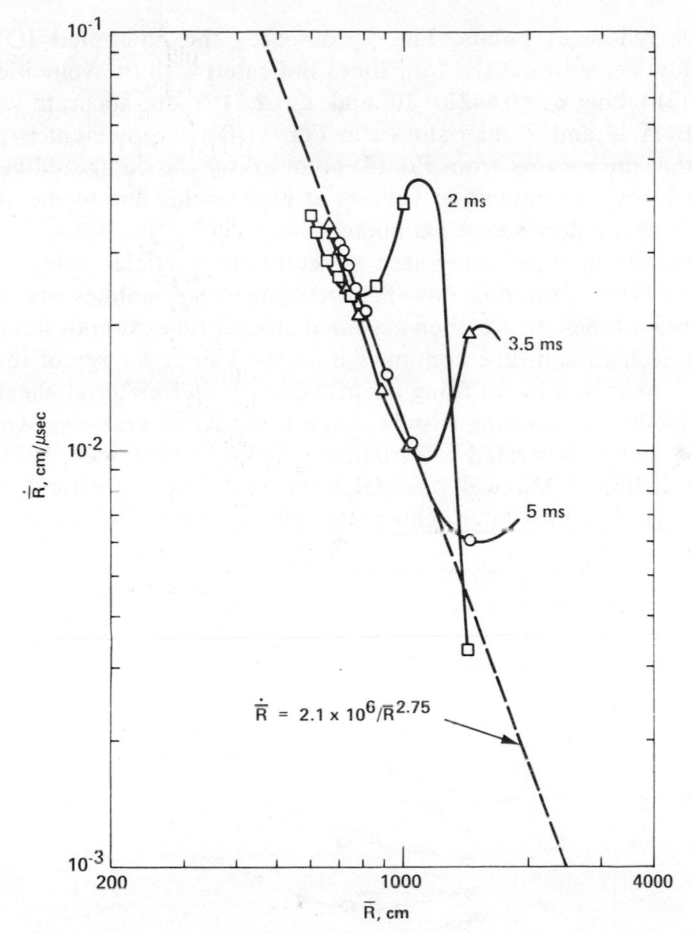

Fig. 4. Spatial average of radial material velocity, $\dot{R}$, versus range at times of 2, 3.5, and 5 ms.

the shock front is located just beyond the 1000 cm range. The approximation of the material velocities behind the shock front to the form of Eq. (1) is evident.

In his studies Maxwell made a further observation of basic importance. Namely that the calculated cratering flow fields approximated incompressible flows and thus

$$\nabla \cdot \mathbf{u} = 0. \tag{2}$$

where $\mathbf{u}$ is the flow velocity.

Equations (1) and (2) may be used to define the full vector velocity field characterizing the cratering flow:

$$\mathbf{u}(R, \theta) = \left[\alpha R^{-Z}\right]\hat{R} + \left[\alpha R^{-Z}\frac{(Z-2)\sin\theta}{1+\cos\theta}\right]\hat{\theta}, \tag{3}$$

where $\hat{R}$ and $\hat{\theta}$ are unit vectors in the R and θ directions.

For selected mass points, Fig. 5 compares the computed JOHNIE BOY cratering flow velocities at the four times indicated with the velocities calculated using Eq. (3) taking $\alpha = 0.6422 \times 10^6$ and $Z = 2.71$ (more accurate values for the JOHNIE BOY α and Z than shown in Fig. 4). This agreement is good. (Note, however, that the vectors from Eq. (3) point above the computed vectors at low angles and below the computed vectors at high angles due to the slight angular dependence of the flow shape parameter Z).

Equation (3) may be integrated to compute particle trajectories or flow stream lines in the cratering flow field. Adjacent streamlines are visualized as forming stream tubes. If it is then assumed stream tubes do not interact and that the flow in each stream tube continues until the kinetic energy of the material in the tube is consumed by working against gravity and material shear strength, a simplified model of cratering results. Such a model of cratering was developed by Maxwell and is illustrated schematically in Fig. 6 (Maxwell, 1973).

One prediction of Maxwell's model is the nearly hemispherical growth of the crater until the time the crater achieves maximum depth. Subsequent to this time

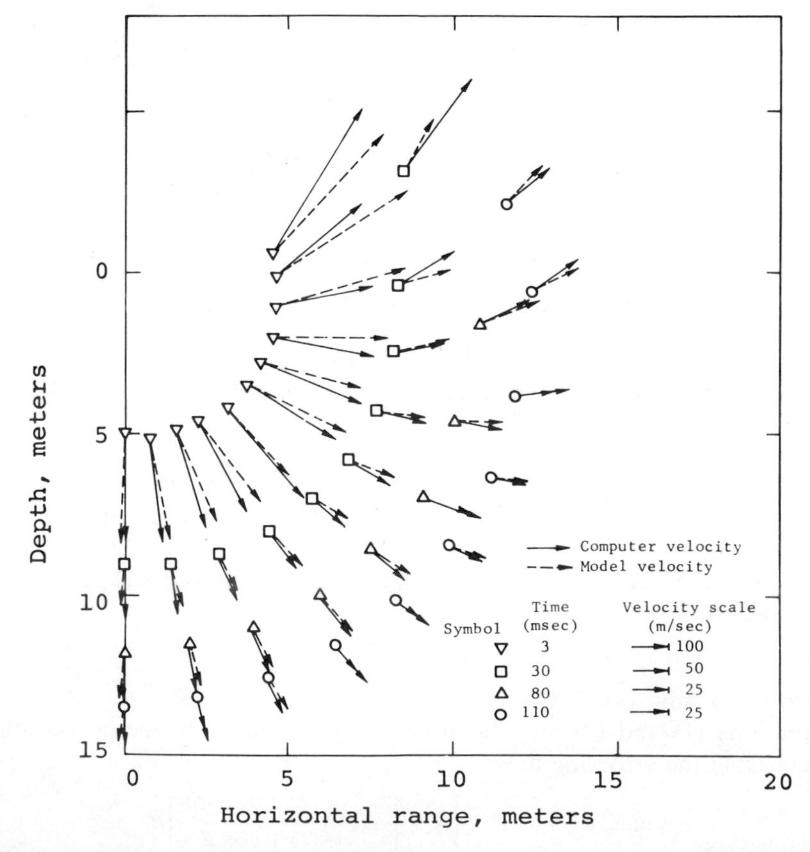

Fig. 5. Computed cratering flow velocities for selected mass points compared to Eq. (3) model.

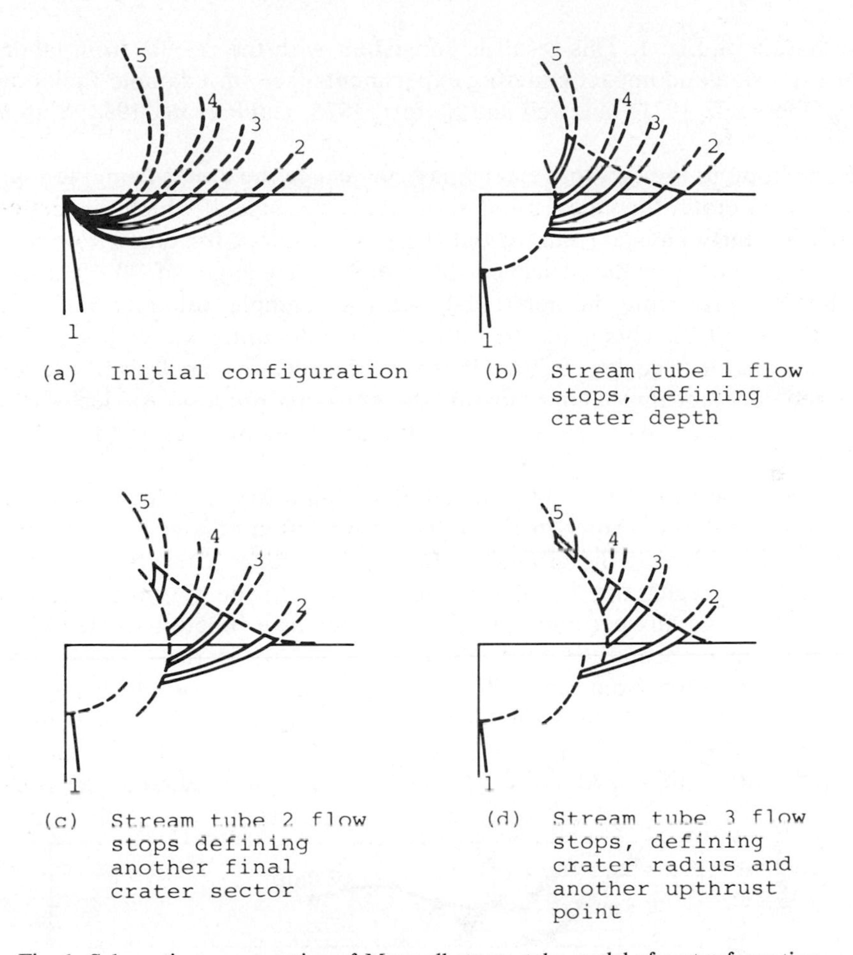

Fig. 6. Schematic representation of Maxwell stream tube model of crater formation.

the crater continues to grow with points forming the final crater wall stopping sequentially in time from the bottom of the crater to the point defining the crater radius which is the last point in the transient crater wall to come to rest (see Fig. 6). These features of crater growth history are in qualitative agreement with the results of detailed calculations (see Fig. 1 for example).

Maxwell's observations and his consequent model have proved very useful in illuminating some of the basic mechanics involved in explosion cratering, including some interesting features of the ejection process which will be addressed next.

EJECTA

Calculations of explosion cratering show that the ejecta material leaves the crater region from the lip of the transient crater. This can be seen from an examination of Fig. 1. This result is consistent with the results from laboratory scale explosion and impact cratering experiments (see for example Piekutowsky, 1975; Oberbeck, 1971; Maxwell and Seifert, 1975; Gault *et al.*, 1968; Seifert and Maxwell, 1974).

In addition to the fact that ejecta material leaves the crater from the region of the transient crater lip, calculations show that the angle at which this material is ejected is nearly constant throughout the time required for crater formation. In most cases, ejecta material leaves the crater at an angle of 40–60° measured counterclockwise from the horizontal (see for example, Ialongo, 1973; Ullrich, 1976; Trulio, 1977). This is illustrated by the results obtained from the JOHNIE BOY calculation (see Fig. 9, Part I) where the ejection angle is about 40° and remained constant throughout most of the cratering process. Angles of ejection tend to be higher at very early times in the cratering process and lower at very late times.

Figure 7 shows some of the detailed ejecta data for the 5 Mt nuclear surface explosion calculation shown in Fig. 1. The top portion of Fig. 7 shows the angle at which ejecta leaves the ground surface versus range from the center of the crater for the specific times (milliseconds) indicated. The bottom portion of Fig. 7 shows the mass flux per unit area being ejected as a function of range at these times. As can be seen, at 0.3 ms which is very early in the cratering process, ejecta angles range from about 30–80°. However, most of the mass is being ejected at a range of about 7.5 m which corresponds to the location of the crater lip at a time of 0.3 ms. The mass being ejected at the 7.5 m range is being ejected at an angle of about 55°. At a time of 3 ms, the range of ejection angles is smaller

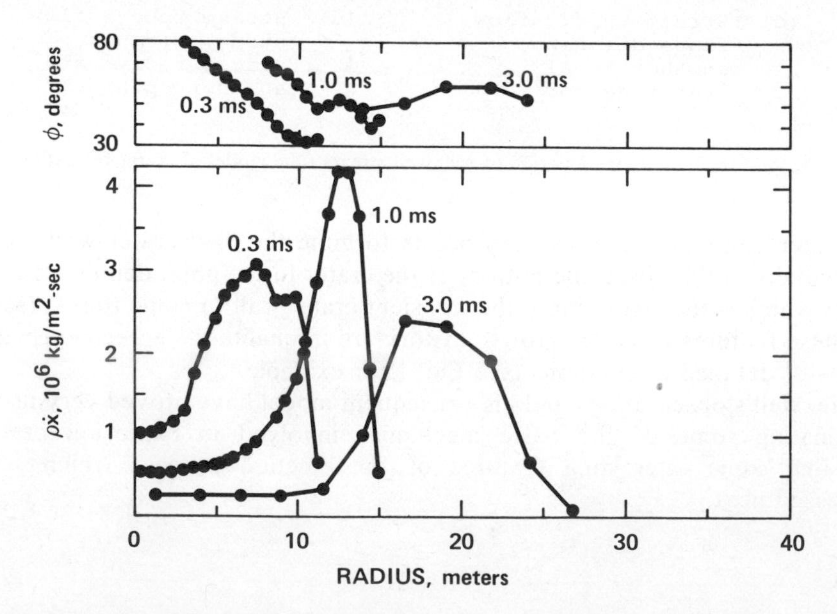

Fig. 7. Computed ejecta mass flux per unit area, $\rho\dot{x}$, and ejection angle, Ø, versus range at indicated times (msec) for 5 Mt nuclear surface explosion.

and, again, the major portion of the ejecta mass is being ejected at an angle of about 55°. This constancy of ejecta angle maintains until very late times. At very late times when ejection velocities are low, ejection angle decreases and, in this calculation, an overturned flap (Roddy *et al.*, 1975) is formed (Fig. 8). Overturned flaps are observed features of most explosion craters and many terrestrial impact craters (Roddy, 1976, 1977). However, most calculations of explosion cratering have not been performed to sufficiently late times to compute a coherent overturned flap.

The ejection process is simply a result of the fundamental material flow field existing in the cratering region during crater formation. As shown by Maxwell, a consequence of the description of the cratering flow field given by Eq. (3) is that ejecta material passes through the original ground surface at an angle given by

$$\emptyset = \tan^{-1}(Z - 2). \tag{4}$$

Most flow fields analyzed for surface or near-surface explosions show $Z \approx 3\text{–}4$ near the free surface and thus $\emptyset \approx 45\text{–}65°$.

Also as a consequence of Eq. (3), the vertical and horizontal components of

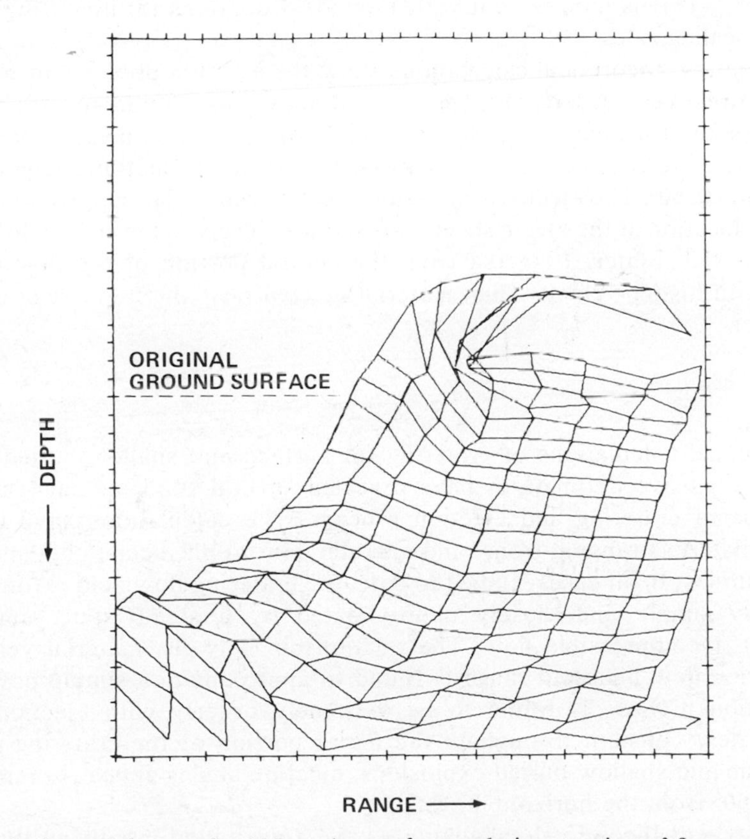

Fig. 8. Computed crater for 5 Mt nuclear surface explosion at a time of five seconds showing formation of overturned flap.

the velocity with which ejecta passes through the ground surface can be shown to be

$$\dot{x} = \frac{(Z-2)\,\alpha}{R^Z} \qquad \text{(vertical)},$$

$$\dot{y} = \frac{\alpha}{R^Z} \qquad \text{(horizontal)}. \tag{5}$$

Thus material is ejected at higher velocity and therefore to higher altitudes and larger ranges (ignoring air drag, etc.) from the central portion of the final crater than from the outer regions of the final crater. This is consistent with laboratory experimental results (see, for example, Piekutowsky, 1975) as well as studies of terrestrial impact and explosion craters and ejecta fields and theoretical calculational results.

It must be noted that Eqs. (5) are only a first approximation to the relationship between ejection velocity and range. Recent experiments (Cooper and Sauer, 1977) suggest that times for crater formation scale as the sixth root of crater volume, V, or explosion yield, W. If particle displacements scale as $V^{1/3}$ as believed based on experimental data, this suggests particle velocities scale as $V^{1/6}$ or $W^{1/6}$. This is inconsistent with Eqs. (5). Equations (5) imply the familiar cuberoot scaling rules.

In summary, theoretical calculations show the ejection process for explosion cratering to be very orderly. Ejecta is lofted above the ground surface from the region near the transient crater lip. The angle of ejection is nearly constant for most of the cratering and ejection process. The angle of ejection seems to range from about 40–60°. The ejecta forms a thin sheet or veil in the shape of an inverted cone. The location of the ejecta sheet corresponds closely to the current location of the crater wall. Material ejected from the central portion of the final crater is ejected with higher velocity than material ejected near the outer regions of the final crater.

SUMMARY

Theoretical calculations of cratering for surface and shallow-buried nuclear and high explosive detonations have revealed several fundamental features of the calculated cratering and ejection process. The calculations have involved "target" materials ranging from sand and alluvium to plasticene clay and shales and sandstones. In all cases studied to date, the cratering flow field is found to be remarkably simple and closely approximated by a steady-state, angular-independent, incompressible flow. The relationship between material velocity in the cratering flow field and range is found to approximate a simple power law. The ejection process is found to be extremely orderly with ejection angles approximately constant throughout the major portion of the cratering process. For surface and shallow-buried explosions, ejection angles appear to range from about 40–60° from the horizontal.

While recent theoretical calculations have contributed greatly to the understanding of the basic mechanics associated with explosion cratering, many questions remain. For instance, how the strength and shape of the cratering flow field is related to such parameters as explosion yield and geometry, site geology (particularly material properties and stratigraphy), gravity, etc., is not yet well understood.

Existing experimental and computational results strongly suggest that the mechanics of crater formation is similar for shallow-buried explosions and high velocity impacts (see, for example, Oberbeck, 1971; Trulio, 1977), although this remains to be fully demonstrated from a theoretical standpoint. If high velocity impact cratering exhibits a first order simplicity comparable to that observed for explosion cratering, it seems reasonable that such questions as the origin of the rings in lunar multi-ringed basin structures may be resolvable on the basis of appropriate theoretical calculations.

Acknowledgments—The author has benefited greatly from numerous conversations and informal discussions with a large number of researchers in impact and explosion cratering. However, I wish to express particular appreciation to Don Maxwell. Most of my understanding of cratering mechanics was either first born or clarified as a result of our numerous discussions and joint efforts on computations.

This work was supported under a contract from the Defense Nuclear Agency, Washington, D.C.

References

Cooper, H. F. and Sauer, F. M.: 1977, Crater-Related Ground Motions and Implications for Crater Scaling. In *Impact and Explosion Cratering.* This volume.

Gault, D. E., Quaide, W. L., and Oberbeck, V. R.: 1968, Impact Cratering Mechanics and Structure, *Shock Metamorphism of Natural Materials* (B. M. French and N. M. Short eds.), Mono Book Corporation, Baltimore.

Ialongo, G.: 1973, Prediction Calculations for the Mixed Company III Event, DNA 3206T, Defense Nuclear Agency, Washington, D.C.

Maxwell, D. E.: 1973, Cratering Flow and Crater Prediction Methods, Technical Memorandum TCAM 73-17, Physics International Company, San Leandro, CA.

Maxwell, D. E.: 1977, Simple *Z* Model of Cratering, Ejection and the Overturned Flap. In *Impact and Explosion Cratering.* This volume.

Maxwell, D. and Seifert, K.: 1975, Modeling of Cratering, Close-in Displacements and Ejecta, DNA 3628F, Defense Nuclear Agency, Washington, D.C.

Oberbeck, V. R.: 1971, Laboratory Simulation of Impact Cratering with High Explosives, *J. Geophys. Res.*, **76**, 5732–5749.

Piekutowsky, A. J.: 1975, The Effect of a Layered Medium on Apparent Crater Dimensions and Ejecta Distribution in Laboratory Scale Cratering Experiments, AFWL-TR-75-212, Air Force Weapons Laboratory, Kirtland Air Force Base, NM.

Roddy, D. J.: 1976, High-Explosive Cratering Analogs for Bowl-Shaped Central Uplift, and Multiring Impact Craters, *Proc. Lunar Sci. Conf. 7th.*, p. 3027–3056.

Roddy, D. J.: 1977, Large scale Impact and Explosion Craters: Comparisons of Morphological and Structural Analogs. In *Impact and Explosion Cratering.* This volume.

Roddy, D. J., Boyce, J. M., Colton, G. W., and Dial, A. L. Jr.: 1975, Meteor Crater, Arizona, Rim Drilling with Thickness Structural Uplift, Diameter, Depth and Volume and Mass Balance Calculations, *Proc. Lunar Sci. Conf. 6th*, p. 2621–2644.

Trulio, J. G.: 1977, Ejecta Formation: Calculated Motion from a Shallow-Buried Nuclear Burst, and its Significance for High Velocity Impact Cratering. In *Impact and Explosion Cratering.* This volume.

Ullrich, G. W.: 1976, The Mechanics of Central Peak Formation in Shock Wave Cratering Events, AFWL-TR-75-88, Air Force Weapons Laboratory, Kirtland Air Force Base, NM.

Ullrich, G. W., Roddy, D. J., and Simmons, G.: 1977, Numerical Simulations of a 20-ton TNT Detonation on the Earth's Surface and Implications Concerning the Mechanics of Central Uplift Formation, In *Impact and Explosion Cratering.* This volume.

Roddy, D. J., Pepin, R. O., and Merrill, R. B., editors.
(1977) *Impact and Explosion Cratering*, Pergamon Press (New York), p. 919–957.
Printed in the United States of America

Ejecta formation: Calculated motion from a shallow-buried nuclear burst, and its significance for high velocity impact cratering

JOHN G. TRULIO

Applied Theory, Inc. ATI

Abstract—A study was made of axisymmetric motion calculated to a time of 80 milliseconds (ms) after a 1 megaton (MT) nuclear explosion at a depth of 15 feet (ft) in granite. Free-flying ejecta at 80 ms will lead to a crater 45×10^6–65×10^6 ft^3 in volume. Vaporization of granite is responsible for the presence of so much ejecta so soon after the burst. Specifically, from about 15 ms on, granite vapor fills a hemispherical bowl about 100 ft deep, and its pressure decays only slowly relative to stresses in solid granite on the crater region. As a result, vapor in the bowl creates, and maintains below it, an adjacent near-axis island of high compressive stress. Solid granite not already fragmented cracks as it is driven upward and outward from that island—which at 80 ms extends obliquely upward from a depth of about 450 ft on-axis, to 250 ft of depth and range.

It is then shown that iron and granite meteroids normally incident upon granite half-space, produce deeper gas reservoirs than the shallow-buried nuclear burst, and at least as quickly. Previous work on meteroid penetration implies in addition that the region over which impact would cause granite to vaporize is shallow relative to the depth at which an explosive charge of equal energy would produce its largest crater ("optimum" burial).

In many geologic media, ejecta fragments account for more than half the volume of the crater generated by a burst between optimum depth and 15 ft/MT$^{1/3}$. The mass of fragments produced by the burst is then dictated by just a few properties of the medium, of which low tensile strength is most important. As a result, the volumes of explosively formed ejecta craters have so far proven calculable to within about a factor of two. The volumes of ejecta craters produced by meteroid impact should be calculable with similar accuracy using present computational models of crater formation. How well shapes can be predicted remains to be seen.

SEVERAL YEARS AGO a calculation was made at ATI of ground motion during the ejecta-forming stages of a 1 MT nuclear explosion at a depth of 15 feet (ft) in granite. That calculation, and its relation to craters formed by high velocity impact events of a kind entered into by meteroids, are the subject of this paper.

1. THE CALCULATION AND ITS MAIN RESULTS

Principal features of the computational model

A burial depth of 15 ft was chosen as an example of a shallow burst whose ensuing motion could safely be considered free of important radiation-hydrodynamic effects. Accordingly, radiative flow of energy was neglected in the calculation, as were rate processes (e.g., heat conduction).

The constitutive equations used in the calculation to describe granite were fairly typical of the present state of the art. Specifically: (a) Material strength was taken into account. At confining pressures of .1 kilobar (kb) the postulated granite medium could support in equilibrium a maximum of .13 kb of shear stress, and .4 kb of shear stress at 1 kb of confining pressure. Its limiting strength in shear was 17 kb and was reached at 56.3 kb of confining pressure. (b) All components of stress were set to zero when the density of material fell to values at which the mean stress became tensile; non-zero mean stresses were allowed to develop on recompression. (c) Tension could be sustained in one direction, if in other directions stresses were sufficiently compressive to give an overall compressive mean stress; otherwise the material's tensile strength was zero. (d) Also computed were nonlinear elastic deformation, and thermomechanical effects over the entire range of states from cold dense solid to hot dilute gas.

Parameters of the constitutive equations, such as those used to characterize variable compressibility and strength, were chosen mainly to fit data from laboratory tests of NTS granite (Piledriver site; Schuster and Isenberg, 1970).

Principal features of the computed field: The axial stagnation point, its downward motion, and the high-stress region below it

Using the computational model just described, and with 1 MT of energy released over 1.5×10^{-8} s in a small sphere 15 ft below the ground surface, the spherical shock driven outward from the center of energy release reached the ground surface in about .030 milliseconds (ms). The pressure was then about 60 megabars (mb) at the shock front and 18.5 mb at the center of symmetry. Immediately thereafter, vaporized granite began to vent to the atmosphere and the shock spread outward and downward in the earth. Velocity vector plots of the field of motion after .6 and 6.9 ms are shown in Fig. 1.

One feature of the fields shown in Fig. 1 merits special attention, namely, downward motion of the point at which the flow stagnates on the symmetry axis. That point reached depths of 23, 69, and 100 ft after about .6, 6.9, and 15 ms of motion, respectively, and—eventually—110 ft.

The flow stagnation point on the axis is the end-point of a curve C of zero vertical particle velocity in the half-plane of motion. Clearly, pieces of rock that contribute to the ejecta field must move upward for a while, and each ejecta fragment must therefore be found at some time above the time-dependent curve C. Also, if the curve C becomes stationary, then ejecta can only originate thereafter above C. Hence, with respect to computed ejecta production, it is a key fact that (i) not only did the stagnation point come to rest at a depth equal to many times its initial depth of 15 ft, but that (ii) its depth became nearly constant at a time when the outermost disturbance in granite had propagated only a fraction ($<\frac{1}{3}$) of the anticipated crater radius (700–1000 ft).

More significant still is the fact that the mean stress just below the stagnation point remained high relative to mean-stress values at almost all points of equal or smaller depth. In fact, only the main outgoing compression wave presented higher mean-stress values. However, while particles gained sizable initial velocities on being struck by the main wave, its duration at any given particle was relatively short; high mean stresses persisted only in the region just below the

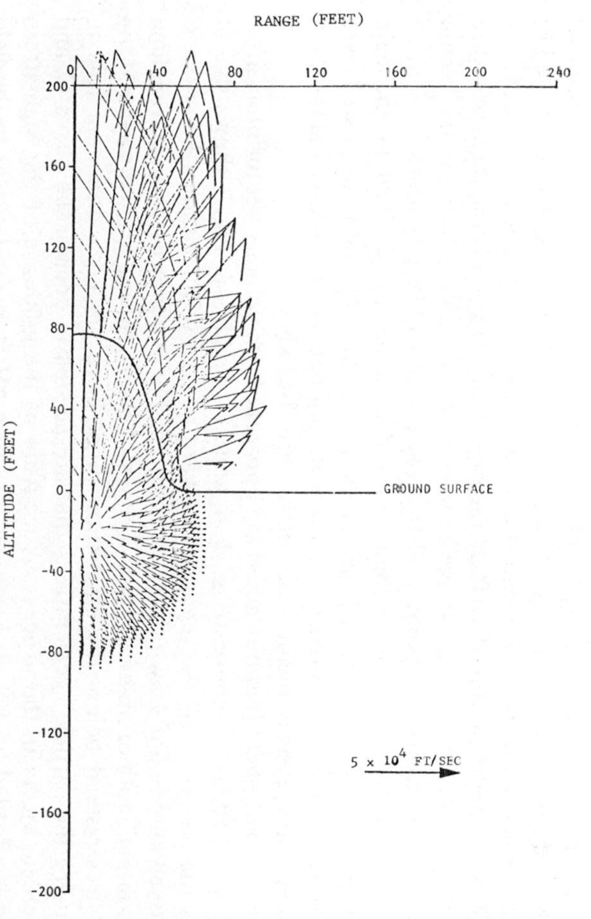

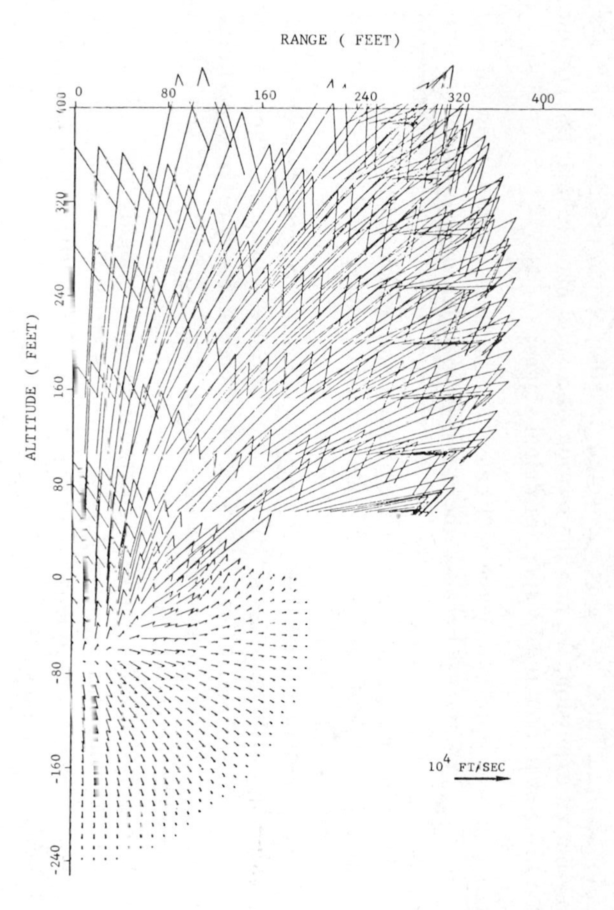

Fig. 1. Velocity vector fields for a shallow-buried burst after .6 ms of motion (left) and 6.9 ms of motion (right). Stagnation points of the flow along the symmetry axis lie at respective depths of 23 and 69 ft.

stagnation point. Thus, at ranges covered by the crater produced during a shallow-buried explosion, we find the motion of a solid particle near the ground surface to be characterized by three principal stages:

(i) A first compression wave will cause the particle to move upward and outward—outward because the wave propagates in that direction, and upward owing to stress relief at the ground surface (peak pressures in the air at the ground surface are about 4 and .65 kb, respectively, at 6.9 and 15 ms; peak mean-stress values in the main ground wave are then about 40 and 15 kb).

(ii) The rarefaction that follows the main-wave peak will reduce the horizontal (outward) component of particle velocity. However, the vertical velocity component will continue to increase for rarefying particles, because stresses in the air are much lower than in the ground, even behind the main outgoing wave.

(iii) The particle will then be accelerated along a roughly radial line extending from the region of persistently high stress (below the stagnation point), with consequent increases in both its vertical and horizontal velocity components.

The features of the mean-stress field just noted are evident in contour plots of mean stress; contours at 6.9, 15, and 80 ms are shown in Figs. 2, 3, and 4. Incipient separation of the two high-stress domains (sub-stagnation point and main-wave) is noticeable at 6.9 ms (Fig. 2), when mean stresses in excess of 40 kb are found near the symmetry axis at a depth of 160 ft, and also in a small oblique region running from 120–140 ft of depth. At 15 ms, the depth interval 120–140 ft presents a more clearly defined near-axis island of high stress ($\sim$ 8 kb) than at 6.9 ms; along the axis, mean stresses as large as 8 kb appear again at 15 ms only below a depth of about 240 ft, in the rarefaction that trails the main pressure wave (Fig. 3). By 80 ms (Fig. 4), peak stresses of .4–.5 kb are found in the region capped by the stagnation point, and that region runs obliquely upward from an on-axis depth of about 450 ft, to about 250 ft of depth and range; at 80 ms, the main wave also carries peak mean-stress values of .4–.5 kb, but at distances well over 1000 ft from the high-stress island. In the wide gulf between the island and the main outgoing wave, typical mean-stress values are .1–.2 kb.

The velocity-field features noted are apparent from velocity vector plots after 6.9, 15, and 80 ms of motion (Figs. 5, 6, and 7). In addition, Fig. 7 shows that at 80 ms the boundary of the region of upward and outward motion (the curve C discussed above; not drawn) increases appreciably in depth with range, reaching a maximum depth of about 200 ft at a range of 500 ft. Stress gradients around the high-stress island below the stagnation point have played a dominant role in producing the particle accelerations that are needed to deepen the curve C, and that create ejecta in the process. A measure of the influence of the high-stress island is afforded by Fig. 4. At ranges less than 550 ft, mean stresses in the field of Fig. 4 would not exceed 100 bars above 750 ft of depth, were it not for the presence of the island at $\frac{1}{4}$ to $\frac{1}{2}$ of that depth. Moreover, there is virtually no other

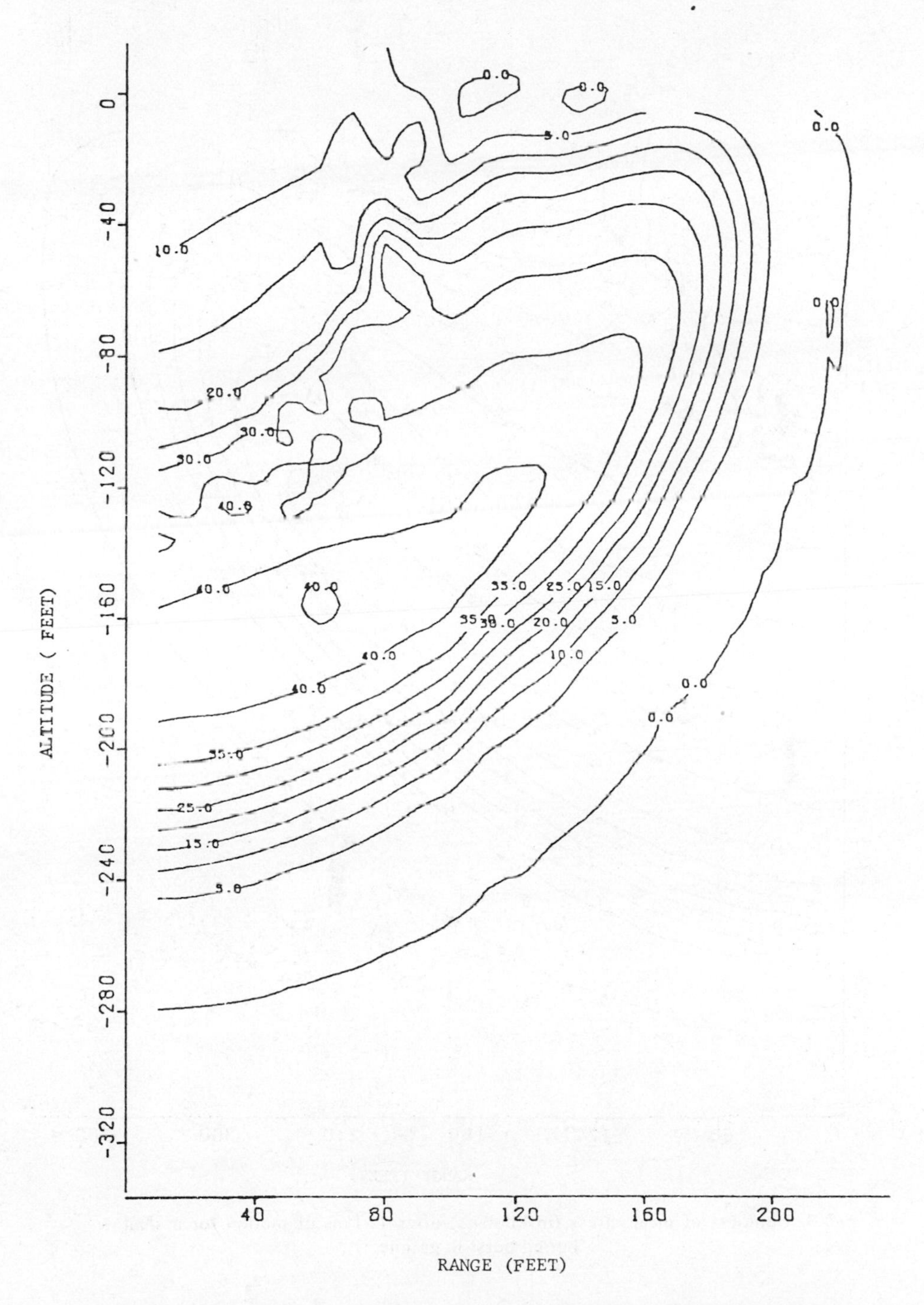

Fig. 2. Contours of mean stress (in kilobars) after 6.9 ms of motion for a shallow-buried burst in granite.

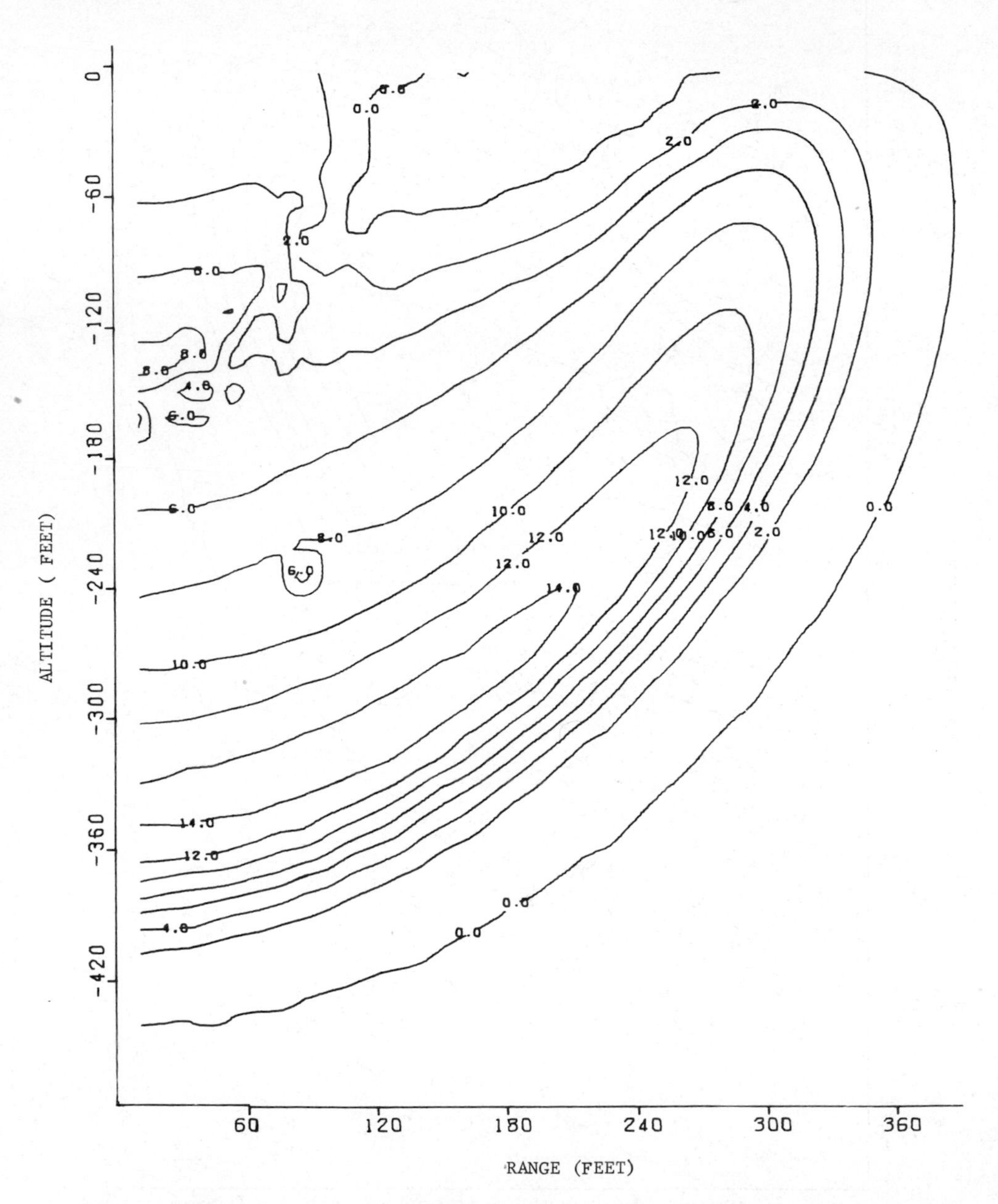

Fig. 3. Contours of mean stress (in kilobars) after 15.1 ms of motion for a shallow-buried burst in granite.

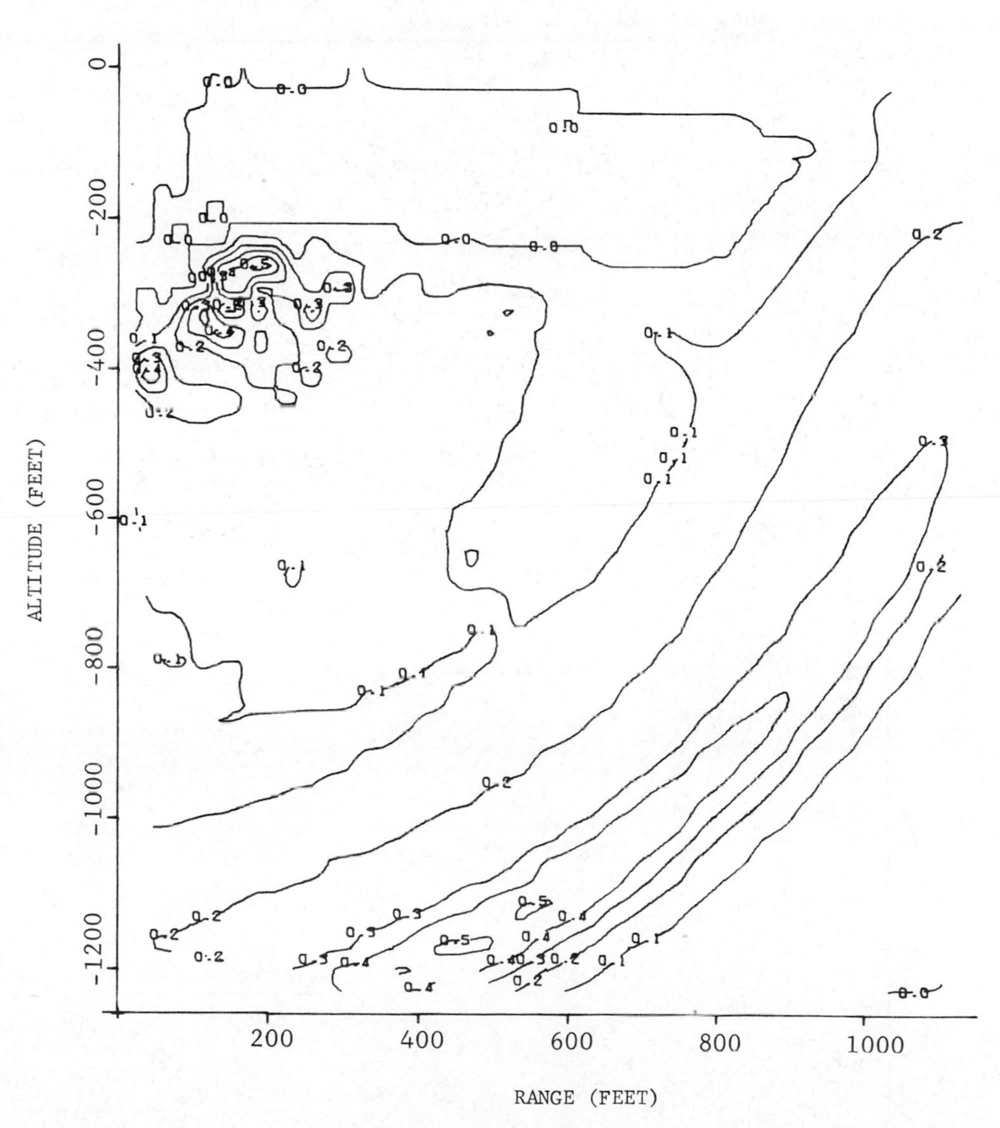

Fig. 4. Contours of mean stress (in kilobars) after 80 ms of motion for a shallow-buried burst in Granite.

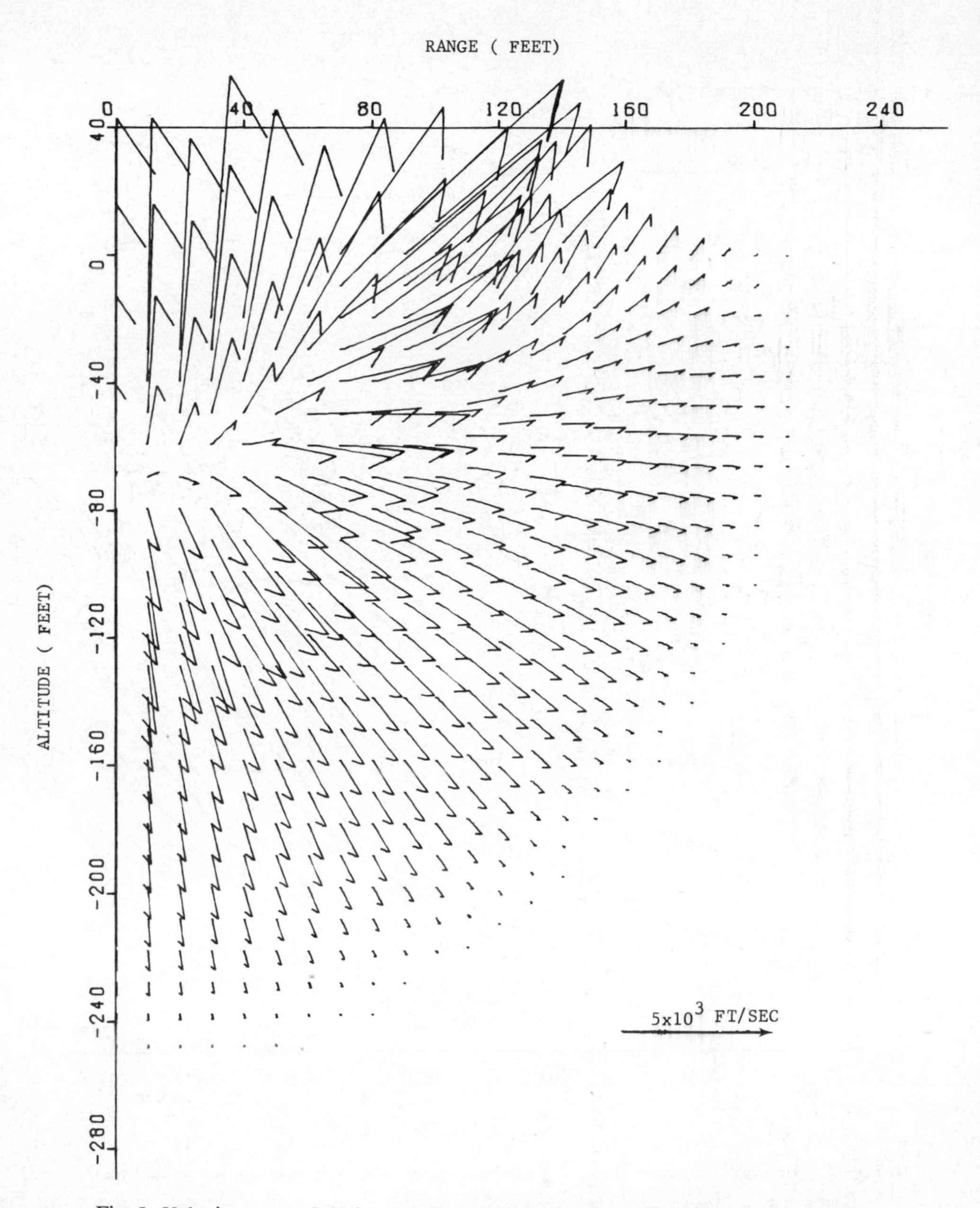

Fig. 5. Velocity vector field for a shallow-buried burst in granite after 6.9 ms of motion.

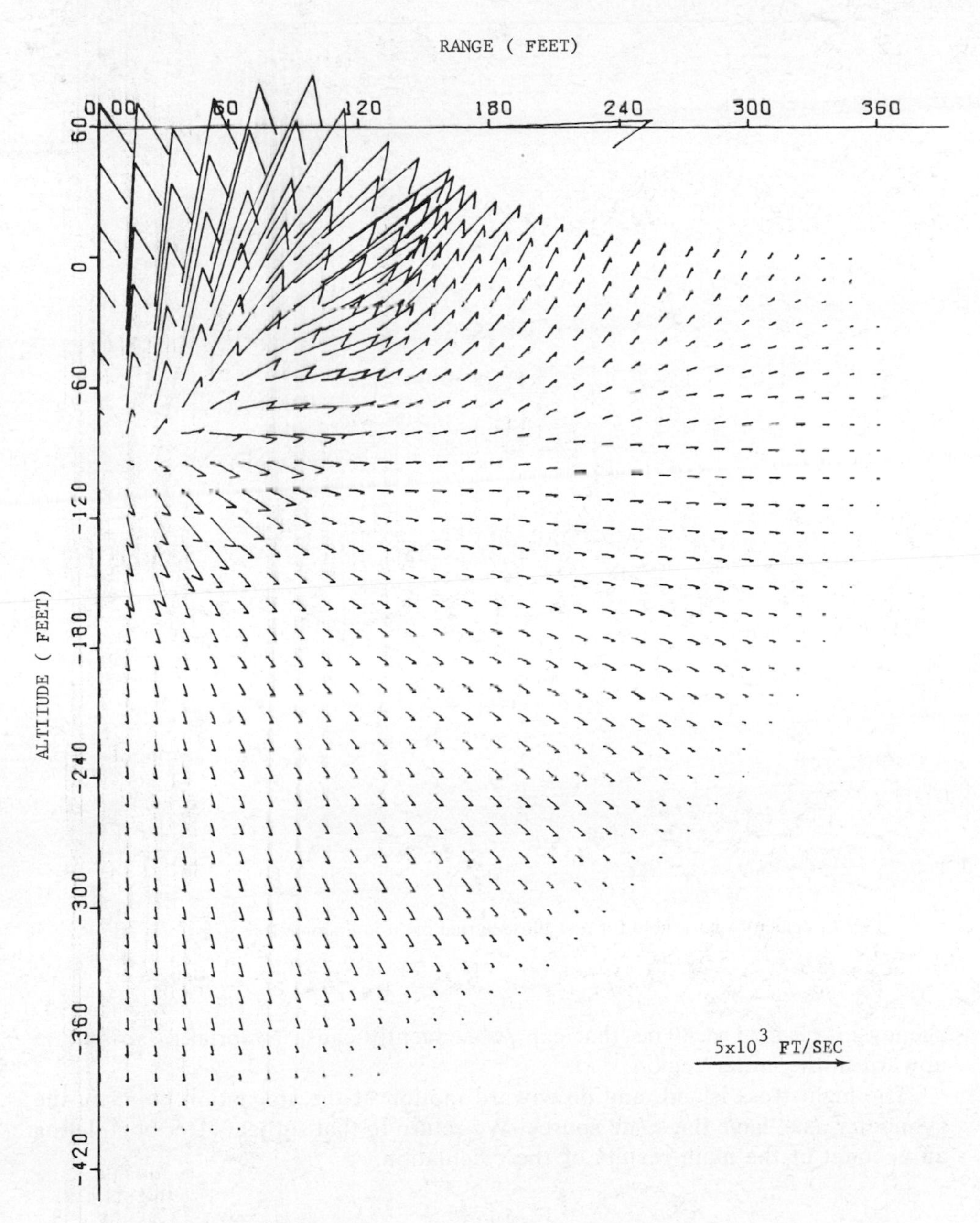

Fig. 6. Velocity vector field for a shallow-buried burst in granite after 15.1 ms of motion.

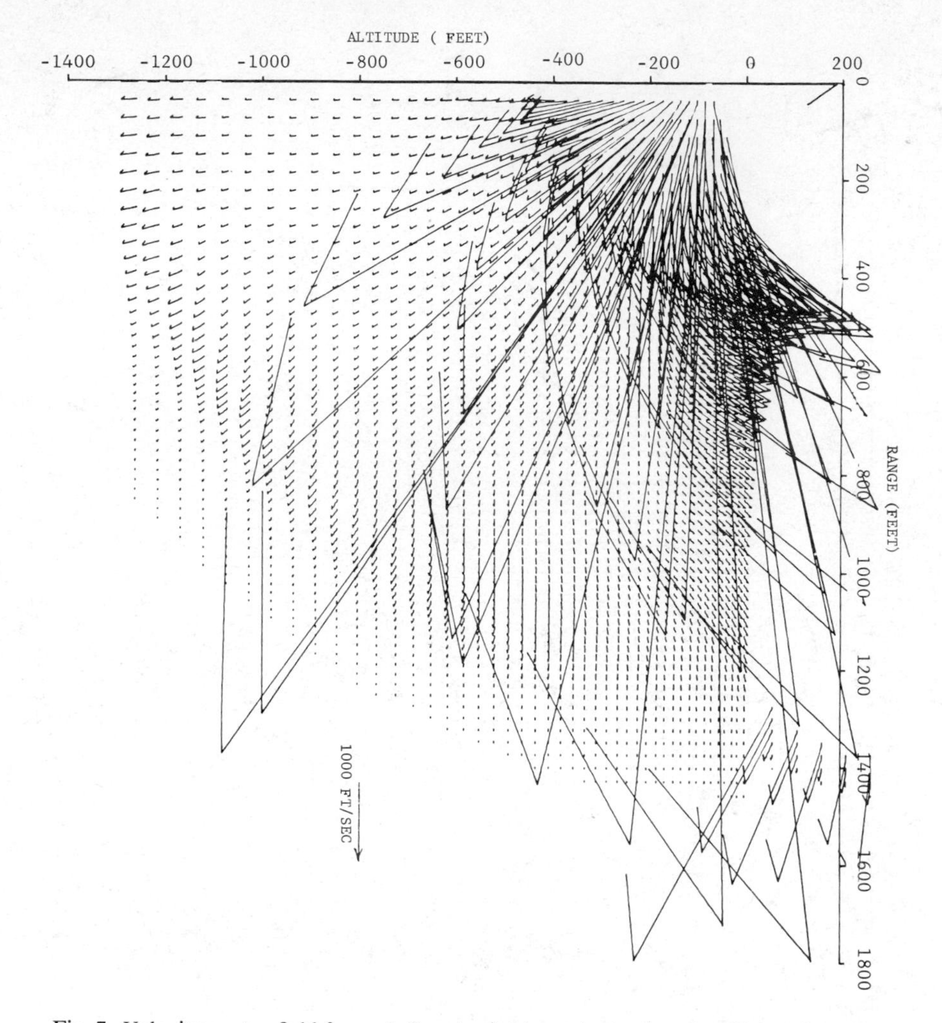

Fig. 7. Velocity vector field for a shallow-buried burst in granite after 80 ms of motion

agency in the field at 80 ms that can subsequently cause material to accelerate upward in the crater region.

The high-stress island, and downward motion of the stagnation point on the symmetry axis, have the same source. We return to that subject after completing an account of the main results of the calculation.

The ejecta field: Estimated total ejecta mass and ejecta crater volume

The speed of upward-outward motion on the crater region diminishes with range and depth. Hence, in the two-dimensional deformation that attends such motion, tensile principal stresses develop and the material cracks as it moves (the interacting rarefactions of uniaxial spall are not required to create ejecta fragments). At 80 ms, cracked rock occupies almost the entire region to a range of 900 ft and a depth of 300 ft (Fig. 8). Representative histories of the velocities with which pieces of that rock come through the original ground-surface plane are shown in Figs. 9 and 10 for the range intervals 300–350 ft and 600–650 ft. Evidently, a sizable crater will form as rock (a) moves upward and outward from the high-stress island, and then (b) cracks as it deforms during that motion.

Ballistic extrapolation of the velocity field at 80 ms yields an estimated ejecta-crater volume of 45×10^6 ft^3 to 65×10^6 ft^3, with the precise value depending mainly on the assumed shape of the final apparent crater. Allowance has been made in that estimate for bulking, both of cracked rock shallower than the crater but not ejected (5% volume increase; Roddy et al., 1975), and ejecta fallback (22% volume increase; Ramspott, 1970). The crater volumes quoted are probably underestimates in that (a) except for ejecta, they include no contribution at all from the inelastic deformation of ground material, nor (b) do they allow for any impulse delivered to the crater region after 80 ms by the high stress island. All ejecta (5.7×10^6 metric tons) fall outside the crater, whose estimated radius is 450–550 ft; within a range of 500 ft, 1.0×10^6 metric tons of ejecta have passed through the ground plane by 80 ms (over 12 ft^3 per ton of yield).

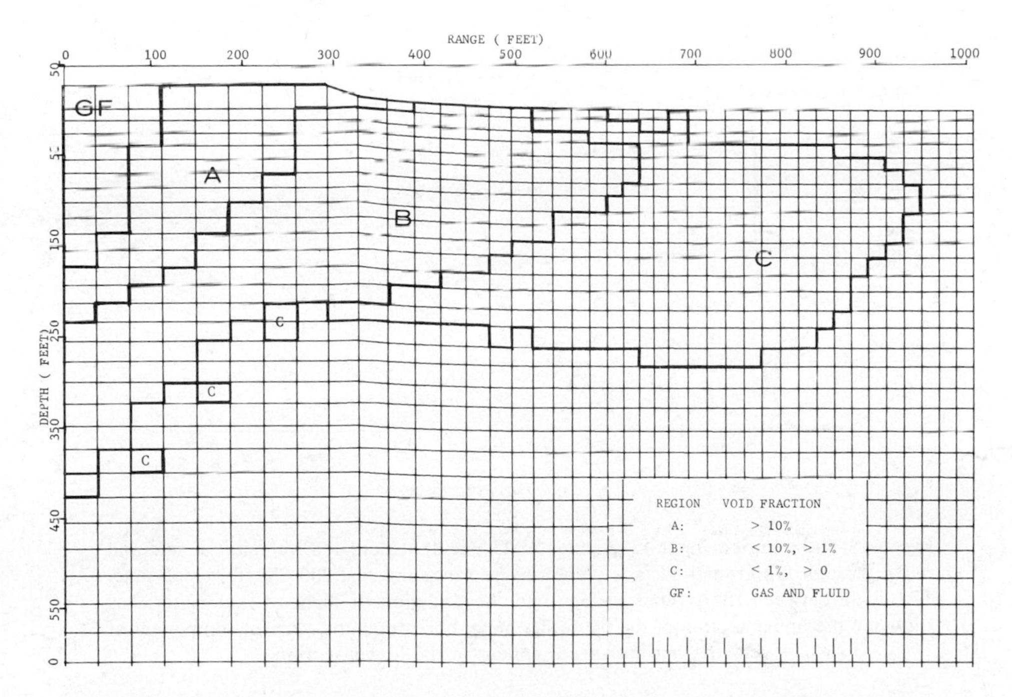

Fig. 8. Calculated region of cracked rock 80 ms after a shallow-buried burst in granite (1 MT yield).

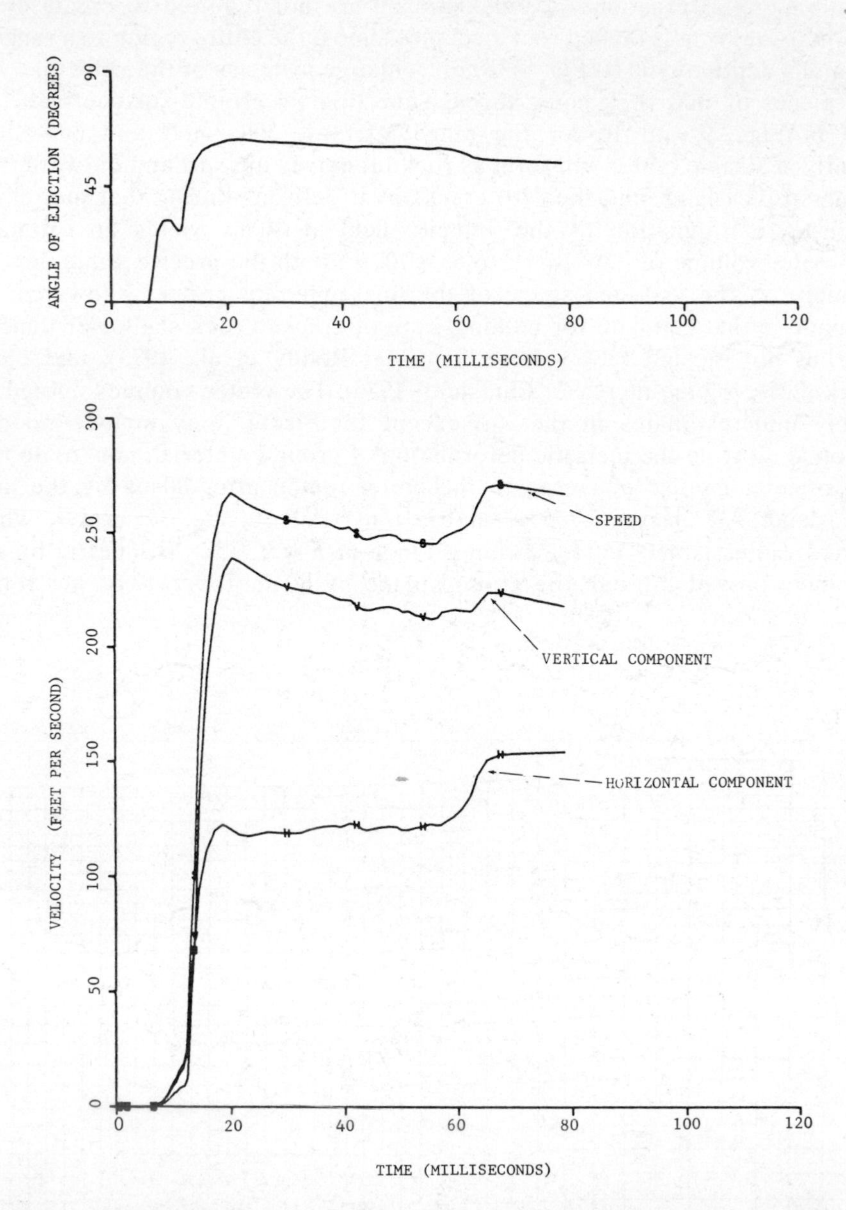

Fig. 9. Shallow-buried burst in granite (1 MT yield): calculated velocity of material rising through the ground surface plane on the range interval 300–350 ft, versus time. The angle between the ground surface and the direction of flight of that material is shown in the upper section of the figure. Its mean horizontal and vertical component of velocity, and its speed, appear in the lower section.

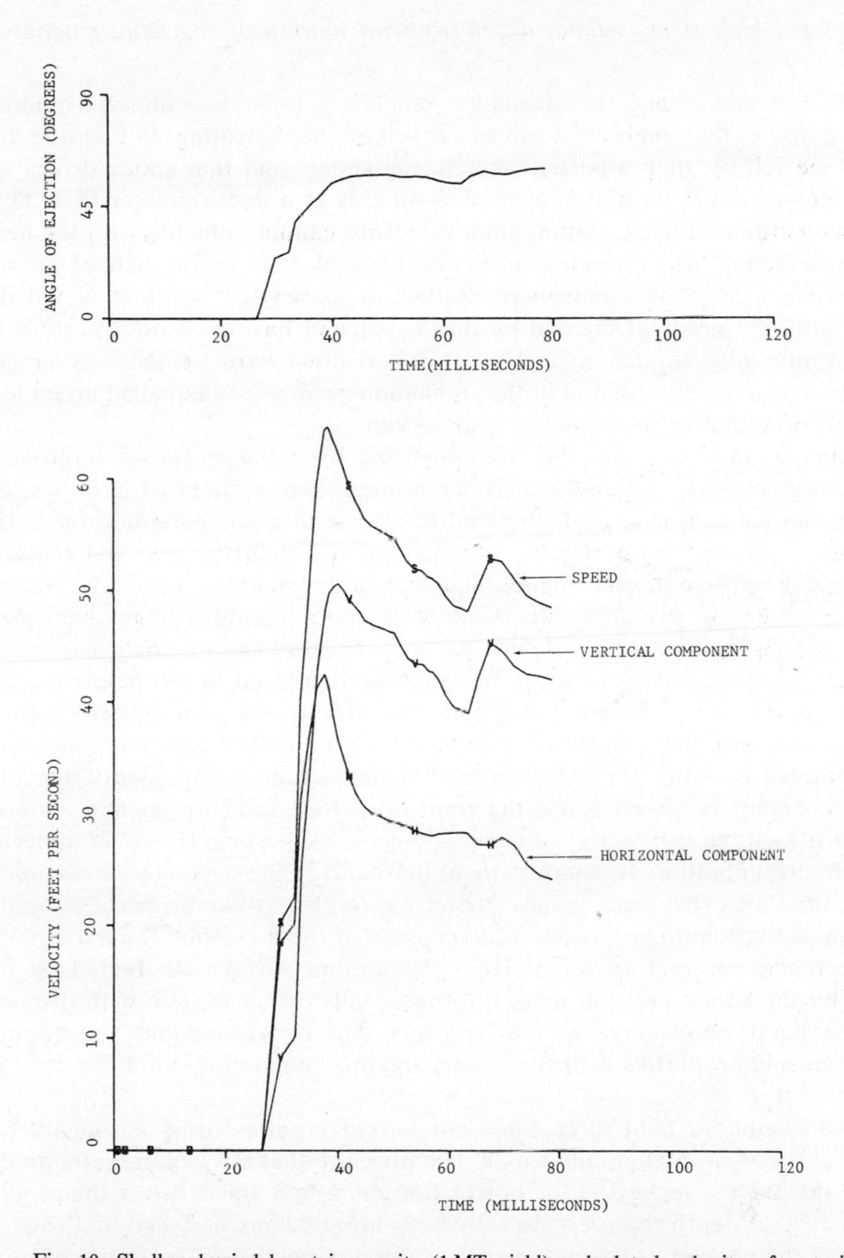

Fig. 10. Shallow-buried burst in granite (1 MT yield): calculated velocity of material rising through the ground surface plane on the range interval 600–650 ft, versus time. The angle between the ground surface and the direction of flight of that material is shown in the upper section of the figure. Its mean horizontal and vertical component of velocity, and its speed, appear in the lower section.

Origin of the high-stress island, and downward motion of the axial stagnation point

At .6, 6.9, and 15 ms, the stagnation point is a point in vaporized granite. Granite particles that vaporize do so as a result of shock heating. In fact, the first disturbance felt by such a particle is a strong shock, and that shock drives the particle downward if its initial location is on-axis at a depth greater than 15 ft. The downward-moving stagnation point therefore cannot coincide with the head of the rarefaction wave created when the original 15 ft radius ball of granite vapor expands into the atmosphere. Rather, it marks the point at which the vertical pressure gradient created by that expansion has acted on a particle of granite vapor long enough to nullify its initial downward velocity. Averaged along the axis to a depth of 100 ft, the stagnation-point speed equalled about half the speed of sound in just-shocked granite vapor.

In almost any gas, lowering the pressure by a large factor requires a many-fold increase in volume, and hence a large displacement of particles. By contrast, the volume-increase-factor that gives rise to a comparable drop in the mean stress of condensed material is much smaller than for gas, and requires much smaller particle displacements. Thus, when the stagnation point of present interest ran out of gas, and into condensed granite, only a slight additional upward expansion took place against gas above; more than a slight expansion would have caused the mean stress in condensed material to fall below the gas pressure, giving rise to downward acceleration. Hence, the point of zero particle velocity came to a rest a short distance below the boundary between condensed and vaporized granite. A precise thermodynamic criterion for identifying that boundary cannot be given, since the transition from solid to gas-like material behavior is continuous in terms of thermodynamic properties. However, in terms of depth, its definition is sharper than thermodynamic considerations might suggest; the shock that heats granite, thereby giving it gas-like properties, rapidly decays in strength both as a result of divergence of the flow from the burst point, and overtaking rarefaction waves from the ground surface. In fact, heat deposited by the shock per unit mass of granite falls off so rapidly with distance from the burst point (inverse third power) that no reasonable variation in thermodynamic properties is likely to alter the final stagnation-point depth by as much as 25 ft.

In the computed field after 1 ms, condensed granite forms a roughly hemispherical bowl around granite vapor. The pressure that the vapor exerts on the walls of the bowl is highest at the bowl's bottom, about 100 ft below the ground surface. At that depth the pressure falls to 34 kb at 6.9 ms and 9 kb at 15 ms; in both cases the pressure of granite vapor at the ground surface is about a fourth as large as at 100 ft of depth. By 80 ms, expansion of vapor has proceeded so far that the bowl is filled with a low-density mixture of gaseous and condensed granite ("smoke"), at a uniform pressure of about 40 bars.

The numbers just quoted help to establish one basic point, to wit: the drop in gas pressure is slow relative to the decay of stress in solid ground. Of course, compared to the time of formation of a crater (or even the ejecta field) from a

1 MT burst, 15 ms is a short period. However, in 15 ms, unloading the surface of a uniformly compressed half-space of solid granite would cause almost complete stress relief to a depth well over 200 ft. By contrast (as just noted), the pressure 100 ft down the symmetry axis is 9 kb at that time, when the maximum mean stress in the entire field is only 15 kb. Thus, the high-stress island below the stagnation point forms because (a) the pressure of vaporized granite decays only slowly as the vapor expands (primarily upward), while (b) stress relief is relatively rapid near the ground surface in solid granite. In short, the presence of the island is explained qualitatively by the fact that freely expanding gas delivers much more impulse to an adjoining object than does freely expanding solid, given equal initial stresses on the object. Quantitatively, however, formation of the island, and its properties, would be difficult to describe without the aid of a high-speed computer.

It should be noted that the island of high stress is not a wave, nor does it radiate compression waves. It is instead a region of compressed material, pushed steadily from above, whose slow expansion (a rarefaction process) continually lowers its own stress and drives crater material upward and outward (and deeper material downward and outward). Granite is accelerated away from the high-stress island along the roughly radial lines from its center that define the direction of the stress gradient.

2. High Velocity Impact as a Means of Producing Ejecta and Ejecta Craters

Basic considerations

We now examine the early stages of vertical impact of iron and granite spheres on a granite half-space, for incoming speeds between 1 and 10 cm/μs (10–100 km/sec). Note that a ball of granite 15 ft in radius and containing a megaton of energy has a mean energy (relative to its undisturbed state) of 3.95×10^{13} erg/g and 1.05×10^{14} erg/cc. The same energy/g appears in kinetic form in a meteor of any density moving at 89 km/s, while an object of density 7.85 g/cc has a kinetic energy of 1.05×10^{14} erg/cc at a speed of 52 km/s; the latter speed becomes 89.5 km/s at a density of 2.65 g/cc. Energy per unit mass is a rough measure of temperature–an unimportant property in the absence of energy transport by radiation, thermal conduction and viscosity. Energy per unit volume measures pressure, and is therefore a key property with or without transport. Thus, on the very crude source-strength criterion of equal energy per unit volume, meteoroid speeds of 1–10 cm/μs cover the range of interest in comparing high velocity impact sources to the explosive source assumed in our calculation of a shallow-buried burst.

Some properties of the field created by impact can be deduced by elementary methods (Courant and Friedrichs, 1948), given single-shock Hugoniot data for iron and granite (those data (Trunin *et al.*, 1972; Trunin, 1970), and the other experimental data cited below, are taken as exact for purposes of this discussion). For example, we find that in the impact of an iron sphere on granite, iron will not vaporize (AIP Handbook, 1957) at all at an incoming speed less than about 1.3 cm/μs; the corresponding speed for a granite sphere is about .6 cm/μs. Slightly less rigorous is the conclusion that at a speed of .9 cm/μs for iron, and 1.2 cm/μs for granite, the topmost point of the sphere will not change its velocity until that point has sunk below the surface of undisturbed earth (Shoemaker, 1963). The speeds .9 and 1.2 cm/μs would apply if the shock created by impact were driven through each sphere at the pressure produced at initial contact. Since divergence of the flow below the sphere will weaken the shock reflected through it, the speeds quoted are almost certainly overestimates ("almost", because the shock pressure can be somewhat higher in oblique collision along the sides of the sphere than in the head-on collision at its bottom; see below). Thus, an almost certain underestimate of the depth to which a sphere will penetrate a granite half-space is obtained by computing the depth of its topmost point when the impact shock first reaches that point, assuming no change in shock pressure relative to its value at first contact. The stagnation point of flow along the symmetry axis should reach at least the depth so computed. That depth is shown in units of the initial radius of the incoming sphere in Fig.11. It is also shown in Fig. 11 in feet, for incoming spheres with a megaton of kinetic energy.*

A better underestimate of the maximum depth reached by the stagnation point

To obtain a closer estimate (again almost certainly an underestimate) of the depth reached by the axial stagnation point, the original impact problem will now be replaced by yet another problem of vertical uniaxial motion. Specifically, let us

(1) confine the spherical missile to a vertical cylindrical tube of radius equal to that of the sphere before impact;
(2) up to the time when the missile becomes half buried, assume it undergoes uniaxial vertical compression behind a shock driven by the pressure at initial contact;
(3) after the missile becomes half buried, treat it as an incompressible right circular cylinder having the same initial radius as the undisturbed sphere;
(4) consider shocked granite a polytropic gas on subsequent expansion, taking the pressure at the shock front equal always to its value at first impact;
(5) let the pressure P_I calculated at the missile-earth interface (step (4)) act over a horizontal circular area whose radius increases with time; to compute the change in that radius, assume that the pressure in the impacting sphere falls at a constant rate from P_I on the symmetry axis to zero at the edge of the circle.

*Material strength is neglected in Section 2. Also, the linear fit used to relate shock and particle velocities covers the pressure range $2 \leq P \leq 20$ mb for granite and $7 \leq P \leq 35$ mb for iron (Trunin *et al.*, 1970, 1972). Considerably larger pressures are found at the higher missile speeds considered (55 mb for an iron sphere at 6 cm/μs).

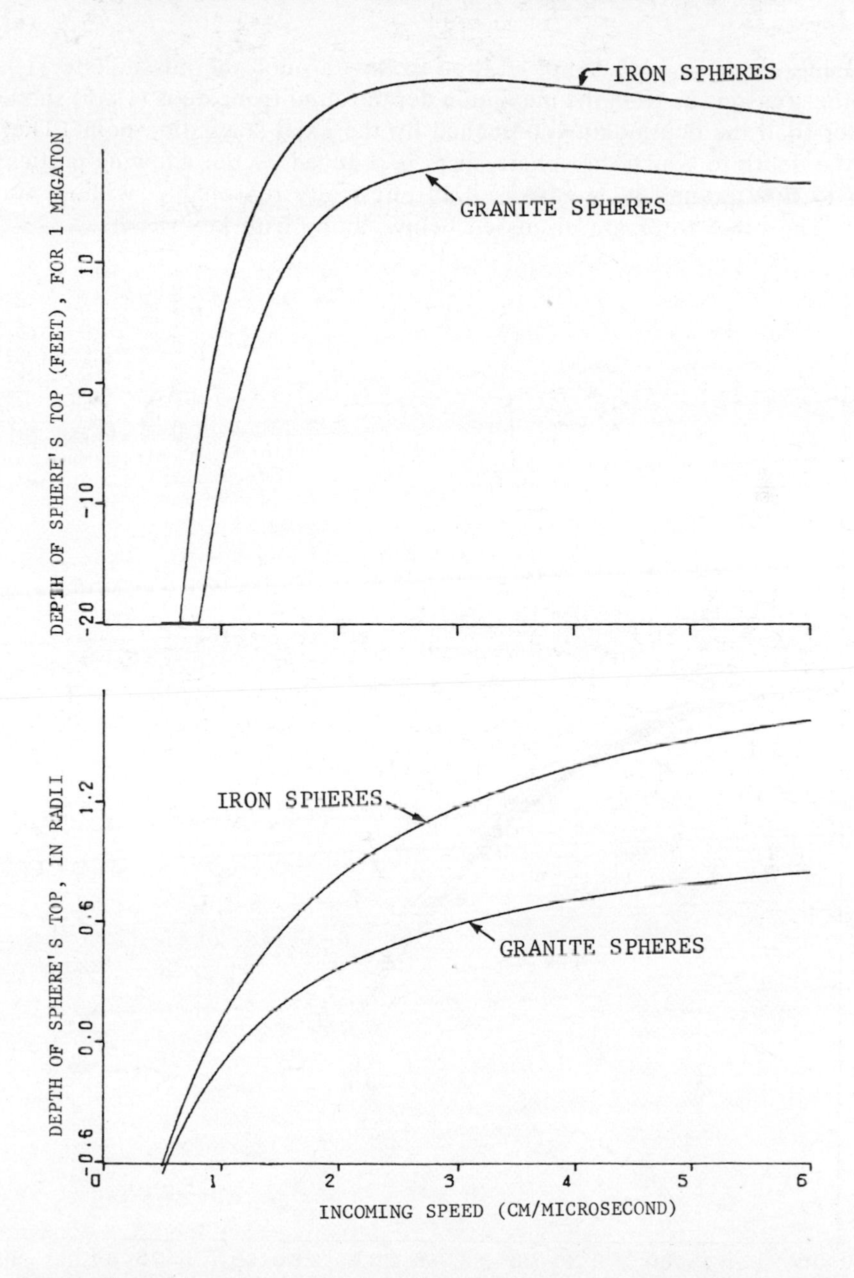

Fig. 11. Normal impact of iron and granite spheres on a flat granite half-space. For incoming speeds of .5–6 cm/ms (5–60 km/s), the figure shows the approximate depth attained by the topmost point (P_{top}) of a given sphere at a time (t_{top}) when the disturbance created by impact first reaches that point. The speed of the first disturbance that moves upward through the sphere is taken as that of a steady shock driven by the pressure created on initial impact. The depth of P_{top} at t_{top} is expressed both in undisturbed-sphere-radii (lower half of figure), and in feet for incoming spheres with 1 MT of kinetic energy (4.184×10^{22} erg; upper half of figure).

Mainly because earth is not allowed to flow around the missile (step(1)), but for other reasons as well, the maximum depth found from steps (1)–(5) should be smaller than the depth actually reached by the axial stagnation point. The idea that the depth to which the sphere sinks is reduced by not allowing particles of earth to flow around it, is advanced as physically reasonable, without further proof. The other steps are discussed below, along with key results.

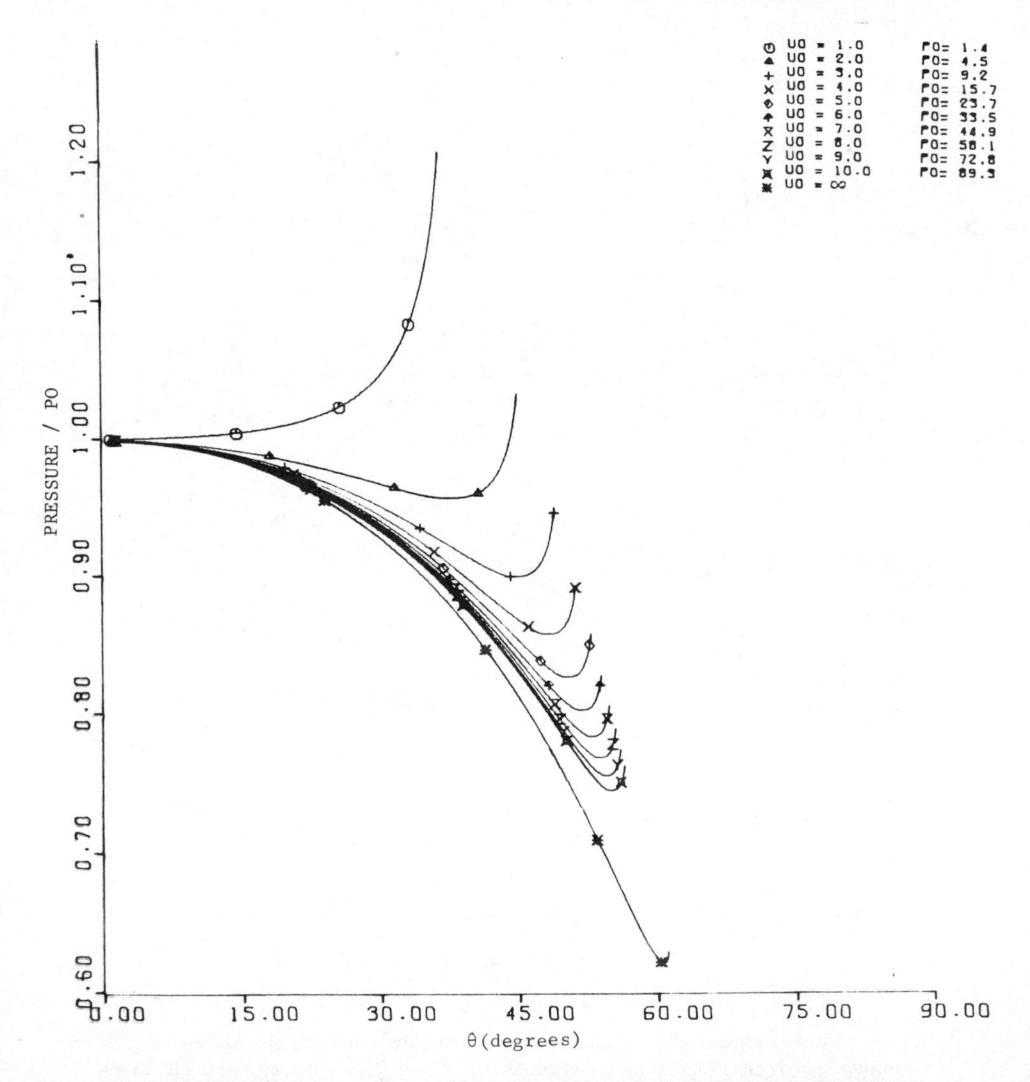

Fig. 12. Initial pressure at points on the surface of a spherical granite missile striking a granite half-space, for all possible angles of strong regular reflection. Position on the spherical surface is measured by the angle θ between the ground surface and the tangent to the undisturbed sphere at that surface. Pressure is given in units of the pressure *PO* created by normal impact at each of ten missile speeds, *UO*·(*UO*, *PO*)-values are tabulated in the figure (centimeters/microsecond, megabars).

Initial impact pressures around spherical missiles

Relative to step (2), the equations for strong regular shock formation at interfaces where granite and iron spheres collide with a flat granite half-space, have been solved exactly for normal incidence, up to the limiting impact angle for such shock formation. Results are presented in Figs. 12 and 13. Those figures show that for most conditions of present interest, the maximum pressure around the surface of the sphere–up to the limiting angle–is generated in the head-on collision that occurs at its bottom (the impact angle θ is zero in a head-on collision, while $\theta = 90°$ for a half-buried sphere). Limiting angles between the

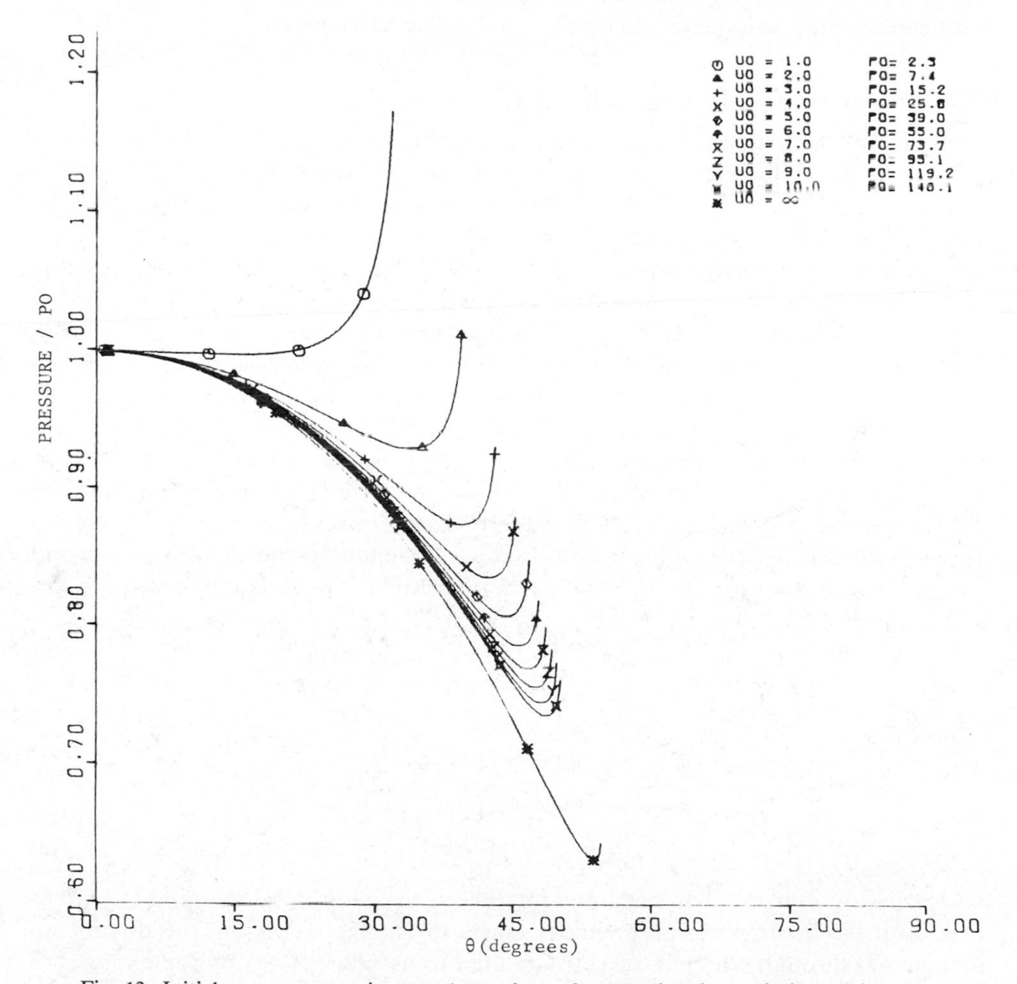

Fig. 13. Initial pressure at points on the surface of a spherical iron missile striking a granite half-space, for all possible angles of strong regular reflection. Position on the spherical surface is measured by the angle θ between the ground surface and the tangent to the undisturbed sphere at that surface. Pressure is given in units of the pressure *PO* created by normal impact at each of ten missile speeds, *UO* · (*UO*, *PO*)-values are tabulated in the figure (centimeters/microsecond, megabars).

undisturbed surfaces of sphere and ground (a) run from 32° to 51° for iron spheres and 36° to 56° for granite spheres, (b) increase with the speed of the incoming missile, and (c) are reached when the shock in the ground deflects particles passing through it by the maximum angle possible. A greater impact angle than the limit for a strong regular shock is presumably accomodated by formation of a Mach stem (perhaps curved) in the earth, and a further increase in pressure as θ reaches a value at which the shock breaks away from the sphere.

The oblique-shock equations on which Figs. 12 and 13 are based, and the method used to solve them, can be applied with little change to the collision between an arbitrarily directed sphere and a flat half-space.

Uniaxial motion without radial edge effects

In accord with step (3), we take motion within the cylindrical tube of step (1) to be uniaxial at the time $t_{1/2}$ when the sphere becomes half buried. Below the collision interface, granite is then found in a uniformly shocked state consistent with step (2). The cylindrical missile of step (3) is also considered uniformly shocked, although for incoming speeds greater than 1.2 cm/μs, more than $\frac{1}{6}$ of the cylinder would be undisturbed at time $t_{1/2}$; we neglect downward displacement that occurs as the shock travels the rest of the way to the missile's free surface. Accordingly, the cylinder's velocity at time $t_{1/2}$ is equal to that of just-shocked granite. Its subsequent motion as an incompressible object is readily described under the polytropic-gas assumption of step (4), provided that decay of the shock in the earth is ignored; in so doing, the estimated depth of penetration can be further reduced, but not increased. We then find that the density ρ_I, sound speed C_I, particle velocity U_I and pressure P_I of adiabatically expanding earth at the collision interface, are related to the density ρ_e^s, sound speed C_e^s, particle velocity U_e^s (<0) and pressure P_e^s of just-shocked earth, by the equation

$$(\rho_I/\rho_e^s)^{\gamma_e a} = (C_I/C_e^s) = (1-b) + b(U_I/U_{se}^0) \equiv \xi$$
$$= \{1 - [1 - (P_I/P_e^s)]/B\}^a \approx (P_I/P_e^s)^a, \tag{1}$$

where

$$a = (\gamma_e - 1)/(2\gamma_e), \tag{2}$$

$$b = -\tfrac{1}{2}(\gamma_e - 1)(U_e^s/C_e^s) > 0, \tag{3}$$

$$B = \rho_e^s (C_e^s)^2/(\gamma_e P_e^s), \tag{4}$$

and γ_e is a polytropic gas constant for the earth medium. Equation (1) implies that until the incompressible cylinder starts to change direction, the downward distance D through which it moves is related to its velocity U_I by the expression

$$\left(\frac{3S_e\rho_e^0}{4U_e^s\rho_m^0}\right)\frac{D}{R_0} = \frac{1}{b^2}\int_\xi^1 \frac{(x-1+b)\,dx}{[1-B(1-x^{1/a})]}$$
$$\approx \frac{[\alpha\xi - (\alpha-1)(1-b)]\xi^{-a} - [1+(\alpha-1)b]}{\alpha(\alpha-1)b^2}. \tag{5}$$

In Eq. (5), R_0 is the radius of the undisturbed sphere, $S_e(<0)$ is the velocity of the shock created in the earth on initial impact, ρ_e^0 and ρ_m^0 are the respective densities of undisturbed earth and missile, and

$$\alpha \equiv (1-a)/a = (\gamma_e + 1)/(\gamma_e - 1). \tag{6}$$

Solution of the uniaxial problem defined by steps (3)–(5) is completed by the following relation between ξ and t (note that $S_e < 0$):

$$-\left(\frac{3S_e\rho_e^0}{4R_0\rho_m^0}\right)(t - t_{1/2}) = \frac{1}{b}\int_\xi^1 \frac{dx}{[1 - B(1 - x^{1/a})]} \approx \frac{(\xi^{-\alpha} - 1)}{\alpha b}. \tag{7}$$

Curves of D/R_0 vs. U_0, where U_0 is the speed of an incoming iron or granite missile, have been developed from Eq. (5) for a granite earth. Those curves are presented in Fig. 14 for one value of γ_e (1.524). In generating the curves shown, C_e^s was computed from the formula:

$$C_e^s = c_e + [\tfrac{1}{2}\gamma_e(\gamma_e - 1)]^{1/2} U_e^s. \tag{8}$$

The constant c_e in Eq. (8) was obtained from an empirically based linear expression (also used for Fig. 14; Trunin *et al.*, 1970) relating S_e to U_e^s;

$$S_e = c_e + \lambda_e U_e^s \tag{9}$$

The constant γ_e was set equal to $2\lambda_e - 1$, in keeping with the fact that 60–92% of the maximum volume change that granite can undergo in a single shock (according to Eq. (9)) takes place under the conditions of impact of Fig. 14. In the limit of maximum single-shock compression, γ_e becomes equal exactly to $2\lambda_e - 1$ for many materials, given Eq. (9). However, γ_e probably falls somewhat on adiabatic expansion from single-shock states of interest here, since (i) at densities $< .02$ g/cc, and at appropriate temperatures, $\gamma_e \approx 1.2$ for SiO_2 (Krieger, 1965), but (ii) during the cylinder's deceleration, density decreases by only a small factor (<7.7 when $U_I = 0$, if P_I, C_I, U_I satisfy Eq. (1); $\gamma_e = 1.524$) relative to the 250–500 fold decrease needed to reach a density of .02 g/cc. As Eq. (5) implies, a small decrease in γ_e during expansion would probably reduce D/R_0, but only slightly.

To see whether the depth reached by the CoM is reduced by treating the missile as incompressible, and by ignoring shock decay in the earth, uniaxial motion was computed without those restrictions. To do so, discrete equations and a computer were used—the simplest method. The resulting depths are shown as isolated points in Fig. 14 for $U_0 = 1$, 3.5 and 6 cm/μs. To obtain each point, computation was begun at the time when the cylinder was uniformly shocked, and with the shock in the earth at a depth appropriate to collision with a cylindrical missile. Pressure was computed in granite exactly as required by Eqs. (1), (8), and (9). Also, apart from a change in the values of c_e, λ_e (Eq. (9)) those same formulas were used to relate pressure and density in iron (constants c_i and λ_i, analogous to c_e and λ_e, are known for iron; Trunin *et al.*, 1972). However, the assumption that shocked missile-material expands as a polytropic gas with $\gamma_i = 2\lambda_i - 1$ is probably less valid for iron than granite; under the impact

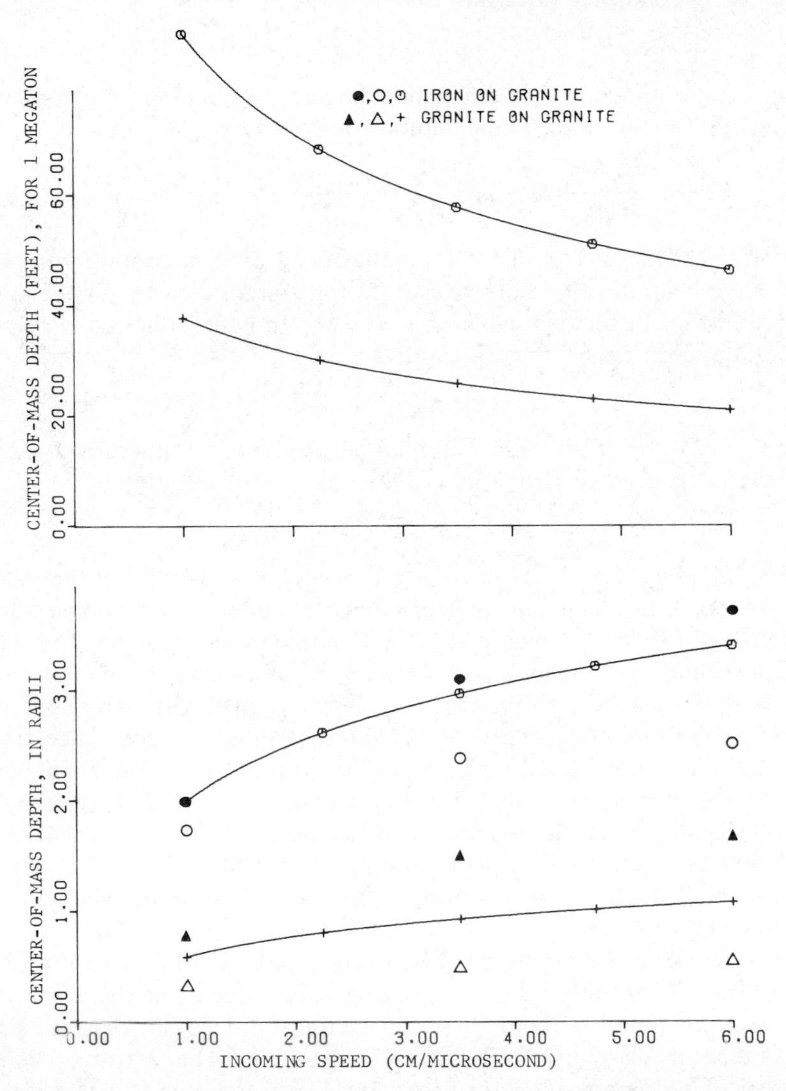

Fig. 14. Normal impact of iron and granite spheres on a flat granite half-space. A given sphere was converted to a uniformly-shocked cylinder of the same radius when the impact shock reached its center. Motion was taken as uniaxial (vertical) thereafter in computing the maximum depth reached by the cylinder's center of mass. The solid curves give that depth for an incompressible cylinder, ignoring decay of the shock in the ground. For six isolated points (△, ○) compressibility and shock decay were taken into account; also shown for those six cases are peak displacements of the cylinder's bottom face (▲, ●).

conditions of Fig. 14, iron achieves from 32% to 84% of the maximum volume decrease it can undergo in a single shock.

Figure 14 shows that Eq. (5) does understate the depth of penetration of the missile, but not that of its CoM, for the uniaxial problem of steps (1)–(4). Hence, the chief error in Eq. (5) as a source of lower-bound values of D/R_0 lies in neglect of lateral expansion of the missile.

Uniaxial motion, overcorrected for radial expansion of the missile

Initial stress distributions around the sphere-earth interface (Figs. 12 and 13) strongly suggest that significant lateral expansion starts no sooner than time $t_{1/2}$ (when the sphere becomes half buried), and at a distance R_0 from the sphere's vertical axis. Naturally, lateral expansion begins at radius R_0 and at time $t_{1/2}$ for the cylindrical missile and uniaxial field of steps (3) and (4). In accord with step (5) (above) the depth reached by the cylinder's CoM will now be calculated taking into account growth of the impact surface from time $t_{1/2}$ on.

We assume that pressure decays linearly with radius within the cylinder at any instant, falling to zero at the time-varying radius R of the cylinder's outer edge. At the cylinder's axis, the pressure is computed according to Eq. (1) from its instantaneous CoM velocity. The cylinder is again considered incompressible and uniform in thickness; its height is then inversely proportional to R^2. In contrast to the radial fall-off of pressure that causes R to increase, the pressure of Eq. (1) is applied uniformly over the bottom face of the thinning cylinder in computing its vertical motion.

Under the assumptions stated, vertical acceleration of the cylinder is governed by the equation

$$4\rho_m{}^0 R_0(R_0/R)^2(dU_I/dt) = 3P_I. \tag{10}$$

Radial acceleration of the cylinder's outer edge is computed using the density of the missile before impact (a smaller acceleration would result from use of the higher, but transient, density produced by the impact shock). The radial velocity V of the cylinder's edge then satisfies the equation

$$dV/dt = P_I/(\rho_m{}^0 R). \tag{11}$$

In addition, U_I and V are equal to $-dD/dt$ and dR/dt, respectively. Hence

$$dD/dR = -U_I/V. \tag{12}$$

Equations (10)–(12) are more conveniently written in terms of variables μ, ν, ζ, and σ, defined, respectively, as U_I/U_e^s, V/U_e^s, D/R_0 and R/R_0. We then find from Eqs. (10)–(12) that

$$A^2\nu d\nu = \{1-[1-(1-b+b\mu)^{1/a}]B\}\sigma^{-1}d\sigma, \tag{13}$$

$$d\mu = \tfrac{3}{4}\sigma^3\, d\nu, \tag{14}$$

$$d\zeta = -(\mu/\nu)\, d\sigma, \tag{15}$$

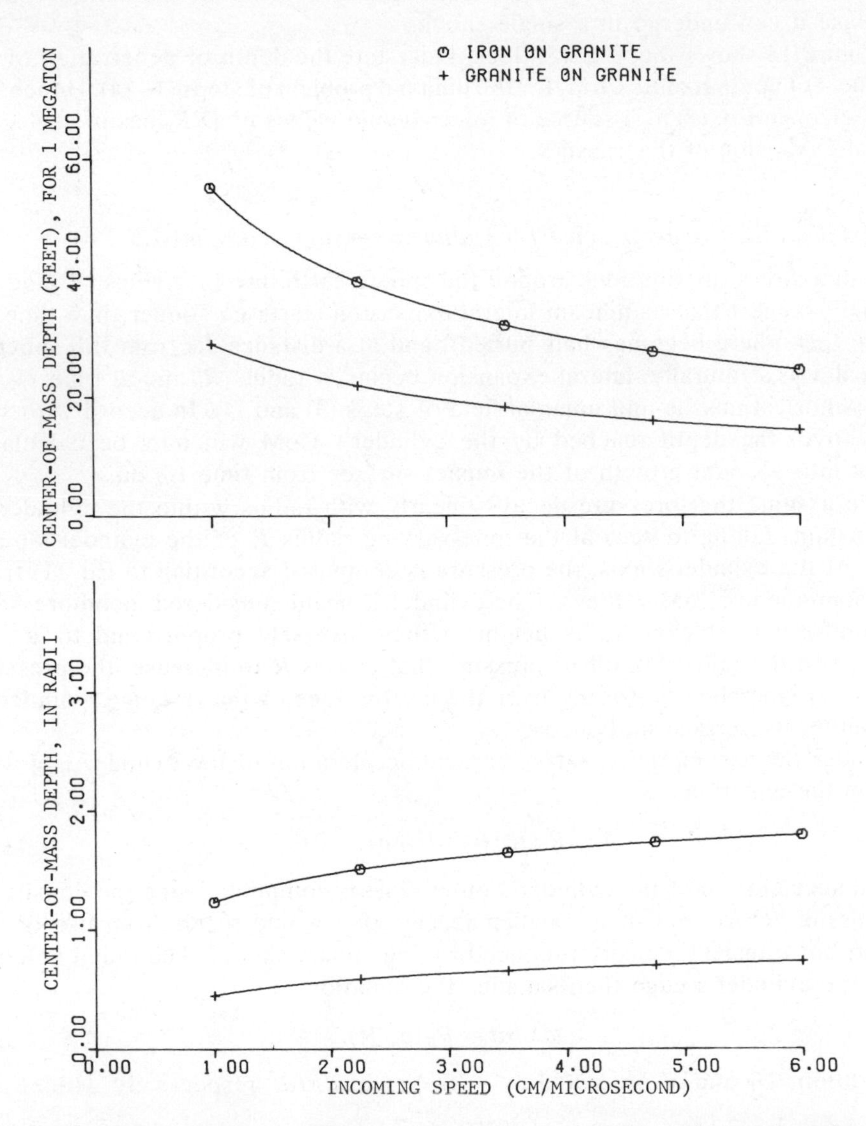

Fig. 15. Underestimates of depths reached by bottom surfaces of iron and granite spheres in normal impact on a flat granite half-space. When the impact shock reached the center of a given sphere, it was converted to a uniformly shocked imcompressible cylinder of equal radius. Vertical motion of the cylinder was opposed by a pressure P_I on its entire bottom face, with P_I related to interface velocity as in purely uniaxial (vertical) motion. The cylinder's outer edge had an outward acceleration $P_I/\rho_m{}^0R$, as if the pressure were always zero there (R and $\rho_m{}^0$ are the missile's instantaneous radius and its density before impact, respectively). The bottom of the sphere reaches greater depths than the cylinder's center of mass.

where

$$A \equiv (\rho_m{}^0 U_e^s / \rho_e{}^0 S_e)^{1/2}. \quad (16)$$

Treating the reduced radius σ as an independent variable, and with $\mu = 1$ and $\nu = 0 = \zeta$ when $\sigma = 1$, the simple task of integrating Eqs. (13)–(15) numerically has been carried out. The resulting values of reduced CoM depth, ζ, appear in Fig. 15.

The calculation that led to Fig. 15 has just a few key features. Most importantly, no radial expansion has been allowed below the cylinder-earth interface in computing the pressure P_I at that interface. Hence, for any interface velocity, that pressure has almost certainly been overestimated. Then, since radial expansion of the cylinder is driven (in the calculation) by an axial pressure P_I, with no confinement at all by surrounding earth, radial displacement of the cylinder's edge has very likely taken place at too high a rate. The area of the impact interface has therefore grown too rapidly and the cylinder has thinned too rapidly.

Applying a uniform pressure greater than any that would actually be found at the cylinder's bottom for a given bottom-surface velocity, to an area greater than the surface would have at that velocity, must lead to an underestimate of the depth reached by the cylinder's CoM. In addition: (i) under uniaxial conditions, the bottom of a shocked compressible cylinder moves farther downward than the CoM of an incompressible one (Fig. 14), but (ii) not as far down as the bottom of a compressible sphere of the same mass and radius in axisymmetric (not uniaxial) motion. Hence, the depths shown in Fig. 15 are in all likelihood smaller than the depths to which normally incident iron and granite spheres would actually penetrate a flat granite half-space. In view of Fig. 14, Fig. 15 quite probably also understates the depths to which centers-of-mass descend, at least for iron missiles.

Underestimates of meteroid penetration, and best estimates

With respect to the depths attained by energy epicenters, the penetration depths of Figs. 11 and 14 are lower-bound estimates only—and they are indeed much smaller than present best estimates (not intended as bounds) of energy-epicenter depth (Shoemaker, 1963). In particular, for a value of U_0 (1.5 cm/μs) that appears appropriate to Meteor Crater, the latter depth is 6–8 times that of Fig. 15. Before weighing that fact, however, note that the question originally addressed has been answered with some rigor: the shot-point for the explosion of Section 1 is shallower (15 ft/MT$^{1/3}$) than the centers of energy created in most collisions of granite and iron meteoroids with earth-like solids.

In comparing present lower-bound penetration depths to best estimates of the depths reached by energy epicenters, note that the former are obtained by computing CoM displacements, while the latter are not. Instead, the latter depth refers to the bottom of the hole created in the earth by a missile, at the time when its deepest-lying particles come to rest. The walls of the hole are then plated with the rest of the missile, whose CoM is therefore shallower (perhaps by far) than the bottom of the hole. The present CoM depth should be smaller still, since—with one minor exception—the steps taken to simplify the impact problem uniformly reduce that depth. (The exception—missile incompressibility —is minor because it has little effect for iron (Fig. 14), and because, in making best estimates, incompressibility is assumed during part of a missile's descent—the greater part for Meteor Crater.)

It is noteworthy that, in uniaxial flow, even greatly exaggerated radial expansion of a missile (Fig. 15) does not lower the depth reached by its CoM as much as a factor of two, vis-à-vis no expansion at all (Fig. 14; for an iron missile with $U_0 = 1.5\,\text{cm}/\mu\text{s}$, the ratio of the two depths is .6). That fact has the following implication: divergent earth-flow around a missile must quickly cause a large reduction in the pressure on its bottom surface, if that surface is to reach depths required by best estimates of penetration. In fact, without such flow, depths of penetration cannot much exceed those of Fig. 14.

More directly related to the best-estimate formulas is the use of a cylinder, in place of an incident sphere of equal radius, when early motion is approximated as uniaxial. As the first step in deriving those formulas, the sphere is considered part of an infinite plane slab whose thickness is equal to the sphere's diameter. On that basis, an energy-epicenter depth of 4–5L was ultimately found for Meteor Crater, where L is the thickness of the unshocked slab. However, the diameter of the sphere evidently sets an upper limit to L. The average thickness of the sphere when viewed in the vertical direction ($\frac{2}{3}$ of its diameter), might be more appropriate for the slab. Put somewhat differently, as the shock progresses through the upper half of the sphere-in-slab, its top surface receives downward impulse at the full pressure of the impact shock; the actual pressure above a bare sphere is more nearly zero. The added downward impulse amounts to $\frac{1}{6}$ of that delivered to the earth as the shock traverses the missile. Of greater importance is relief of interface pressure sooner near the edge of a sphere than along its axis, even in uniaxial motion; that effect is ignored in choosing the length-scale, L, of the best-estimate formulas. Within the framework of those formulas, the depth reached by the center of energy in Meteor Crater might be better estimated as 3–4 missile diameters (or somewhat less) than 4–5 diameters.*

Other approximations used to obtain best estimates have less clear effects. Specifically, in the case of Meteor Crater, more than half the ultimate depth is found from a formula for the penetration of solids by short incompressible jets. The jet-length is taken as the height of the column of disturbed earth and missile, at a time when the rarefaction from the missile's free surface interacts strongly with the missile-earth interface. That choice of jet-length is somewhat arbitrary with respect to both the time at which the formula for a short jet is applied, and

*When present shock-particle velocity relations are used (Trunin *et al.*, 1970, 1972), the changes in quantities contributing to the original best estimate of center-of-energy depth (Shoemaker, 1963, Table 1 and pp. 325, 326) are all negligible.

the extent to which shocked rock is considered part of the jet. Shocked rock actually diverges from the missile in continuous fashion; the point in its flow where divergence effects become strong enough to halt the jet-formation process, is not clearly marked. For example, it seems natural to ask whether treating the incoming missile itself as a jet of incompressible fluid would set a probable upper bound of about 1.7 diameters to the depth reached by the center of energy. Moreover, some error might be expected when the theory of incompressible jets is applied under conditions of partial or total vaporization.

It is not clear whether current best estimates of energy-epicenter depth will prove too large or too small when the uncertainties in those estimates are eliminated. The point to be recognized and acted upon is that major sources of uncertainty can be removed using computational tools presently at hand. Calculation of the motion attending normal incidence of a sphere on a layered medium, at least during the period of energy-epicenter descent, lies well within the scope of present computational methods.

3. Discussion and Conclusions

The results shown in Figs. 11 and 15 establish that the impact sources examined will produce energy epicenters at greater depths than that of our shallow-buried explosive source. Hence, unless penetration takes place to depths as great as those considered optimum for buried explosions, which would conflict with current best estimates (Shoemaker, 1963), we expect larger ejecta masses and ejecta craters to be produced by the impact systems just examined than by the shallow-buried burst discussed above. Meteor Crater provides direct support for that conclusion.

Meteor crater

Studies of Meteor Crater and its environs (Roddy *et al.*, 1975) have shown that ejecta account for about $\frac{2}{3}$ of the crater volume (as measured from the original ground level), and that the crater was produced by impact of an iron-nickel meteoroid. The missile penetrated the earth to an estimated depth of 320–400 ft (Shoemaker, 1963). Estimates of its energy at impact range from 1.3 to 18 MT, with a value of about 4.3 MT perhaps most widely accepted at present. On that basis the energy epicenter reached a scaled depth of 122–358 $\text{ft/MT}^{1/3}$, with a most likely value of about 220 $\text{ft/MT}^{1/3}$. Optimum burial depth (i.e., the depth at which a buried explosion excavates the largest possible crater) is 800–1300 $\text{ft/MT}^{1/3}$ for explosive events from 10^{-7} MT to perhaps 10^2 MT, depending on the type of explosive (chemical or nuclear) and the medium in which the burst occurs.

The ejecta crater from the shallow-buried burst, and actual craters

For the actual nuclear burst (Johnie Boy) most like that of Section 1, the crater volume was $280 \times 10^6\ \text{ft}^3/\text{MT}$. However, the scaled shot-point-depth was

22 ft/MT$^{1/3}$ for Johnie Boy rather than 15 ft/MT$^{1/3}$, and the burst occured in dry soil rather than granite. Given present empirical evidence (Cooper, 1976), a change in burial depth from 22 ft/MT$^{1/3}$ to 15 ft/MT$^{1/3}$ would lead to a reduction in crater volume of about 30%, but the effect of the medium-difference (dry soil versus granite) is harder to assess. Data are available to show that a change from dry soil to dry hard rock would have caused the crater volume to decrease by about a factor of 2.1 for Johnie Boy (Cooper, 1976). However, if actually found in nature, the hypothetical granite medium of Section 1 would probably be considered wet [in present computational models of hard rock, "wetter" or "dryer" translate mainly into lower or higher shear-strength; to account for anomalously large displacements in the Piledriver event, the granite of Section 1 was made weaker than most of the samples of NTS granite tested]. As it happens, nuclear experience is not broad enough to define the effect on crater volume of a change from "dry hard rock" to the "wet" variety. On the other hand, for nuclear explosions at depths of 15–22 ft/MT$^{1/3}$, craters in wet soft rock are about 2.2 times as large as in dry soft rock (Cooper, 1976). If we adopt that factor (2.2) for hard rock as well, then we are led to "expect" that the crater from a nuclear burst at a depth of 15 ft/MT$^{1/3}$ would have a volume of about 95×10^6 ft^3/MT in dry granite, and 205×10^6 ft^3/MT in wet granite.

For the burst of Section 1, the hole left by ejecta has a calculated volume from .46 to .68 times that of the "expected" crater in wet granite. Evidently however (Fig. 7), a quite significant mass of granite will leave the crater at times later than the final time of calculation—although much of that material will eventually fall back into the crater. At a guess, ejecta alone will ultimately account for a crater volume of 65–85 ft^3/MT (guesswork could be replaced in this case by a relatively simple calculation of the trajectories of fragments in free flight at 80 ms, but below the ground surface). The hole left by ejecta would then account for .68–.89 of the "expected" crater volume in dry granite, and .32–.41 for wet granite.

Ejecta production

The individual volume-fractions just quoted must be taken as rough estimates, because of the incompleteness of the data on which "expected" crater volumes have been based, the scatter of those data, and the early time at which the calculation ended. Rather, the significance of those fractions follows from the observation that the computed field (ejecta masses and velocities included) would be virtually identical to that of Section 1 if weak ("wet") granite were replaced by strong ("dry") granite. That assertion rests on the fact that rock breaks into pieces when its tensile strength (not its shear strength) is exceeded.

The tensile strength of *in situ* granite, both computationally and in reality, is low relative to the stresses driving the computed motion of the medium up to the time (<80 ms) of formation of the extensive region of cracked granite seen in Fig. 8. Failure of material in shear ("plastic flow") prior to cracking is a brief and unimportant stage in its history, and hence in the production of free-flying ejecta fragments. Instead, some of those fragments form "[when the shock reaches the free surface of the ground [and] is reflected as a tensional wave" (Shoemaker, 1963). At the shallow burial depth of 15 ft/MT$^{1/3}$, downward migration of the burst epicenter, and creation of an island of high stress many times deeper than the original burst point (Section 1), also comprise a key initial phase of the ejecta-production process. By either route, "momentum is trapped in the material above the cavity [or high-stress island], and it continues to move upward and outward, following ballistic trajectories" (Shoemaker, 1963). In either case, ejecta formation is a simple dynamical process, occuring too early during an explosive or impact event to be much influenced by relatively subtle and little understood properties of geologic materials at low stresses (< 100 bars).

The volume-fractions quoted above help to establish two points, to wit: for a 1 MT burst 15 ft below the ground surface (i) the computed volume of the hole vacated by ejecta is an appreciable fraction of the volume of the crater that would actually be produced in any hard rock, and (ii) that fraction varies widely with the wetness of the medium. Because those two points are important, so are the volume fractions quoted. In turn, as discussed below, (i) and (ii) are important points because of the simplicity of the process of ejecta production.

Plastic flow

Given the simplicity of the early ejecta-production process, the variability of crater size with wetness (item ii) must be taken as evidence that other mechanisms contribute to crater formation. While several such mechanisms have been identified (Trulio, 1975), plastic flow has generally been considered most important. The crater volumes that can be accounted for by plastic flow are potentially quite large. In a liquid (water, for example) only gravity limits the depth that a crater can attain (on earth, to 32 ft of depth for each bar of overpressure). As the example suggests, crater growth by plastic flow takes place over much longer times than ejecta formation. The underlying reason for that fact is that flowing material remains simply connected to the rest of the earth, continually transferring momentum and mechanical energy to surrounding, more static, material; momentum is not trapped in flowing solid. More specifically, fast-moving material that does not break away from the earth to form an ejecta fragment, quickly loses its energy to other material that either (a) blocks its path or (b) develops shear stresses that oppose its motion. As a result, the shear strengths that must be postulated if crater volumes due to plastic flow are to approach actual crater volumes, are lower than those normally associated with geologic solids; shear stresses in excess of about ten bars cannot be allowed to develop during most of the plastic-flow process (Trulio, 1975). Whether and when material around the floor of a developing crater might be that weak has yet to be established.

For computational purposes, plastic flow presents a further, and more basic, problem: The equations used to relate stress and strain in states of shear failure represent an extension of current theories of metal plasticity. Their validity for deformations in which inelastic volume changes attend shear failure is, mildly put, in serious doubt. Hence, except for geologic media that fail mainly in ductile fashion (like a metal), there is little reason to trust the predictions of present computational models for craters formed mainly by plastic flow.

Unfortunately, under the conditions created by explosive and impact events, most near-surface geologic solids undergo significant volume changes as they fail in shear. However, there would appear to be some significant exceptions. Crater dimensions might be predicted accurately for explosions in mud or wet sand, for example, or more generally in saturated fine-grained materials. In the present context, a more interesting possibility is presented by impact events on material hot enough to occupy molten, plastic, or at least very weak, states. Large lunar craters come first to mind, since a solid crust tens of kilometers thick over magma then becomes a thin and probably unimportant outer layer. Crater-formation-by-gravity-waves (van Dorn, 1969) might well be an apt description for those craters, depending on the radial distribution of the moon's temperature at the time of impact. We note also that the bottom of the vapor bowl created by the burst of Section 1 would not have become stationary if the material below were molten; in the absence of shear strength, that material would have continued to move slowly downward under the pressure of gas above.

Prospects for calculating the dimensions of craters formed by meteoroid impact

The simplicity of the ejecta-production process makes it clear that the volume of the hole vacated by ejecta (including fallback and bulking; Section 1), will increase rapidly with burial depth in the medium of Section 1. The process is also simple enough so that at greater depths of burial, ejecta masses should be as readily calculable as at 15 $ft/MT^{1/3}$; in fact, the expulsion of cracked material by expanding cavity gases (Shoemaker, 1963), should increasingly replace the more complex mechanism of burst-epicenter-migration. At the same time, the ejecta produced in the calculation of Section 1 do suffice to account for most of the volume expected for a crater in dry granite (item (i) of the section on "Ejecta Production"), to which plastic flow should contribute little. These facts justify some confidence in the principal conclusion reached here: The volumes of ejecta craters, of which Meteor Crater stands as one example, can be calculated to within a factor of two for both impact and explosive events, using present computational models.

The shape of an ejecta crater is not established until most of the ejecta have come to rest under gravitational forces. That phase of ejecta-crater formation lasts at least ten times as long as its ejecta-producing phase. As a result, calculation of the settling process is too expensive to be practical without modifying the tools at hand. Until then, there is little point in estimating the accuracy with which ejecta-crater shapes might be calculated.

Acknowledgment—Motion attending a shallow-buried nuclear explosion in granite was calculated under Contract No. F04701-68-C-0192 between the Air Force Space and Missile Systems Organization and Applied Theory, Inc. Much credit is due Mr. Sheldon Schuster, who ably exercised the AFTON 2A code to compute that motion, and to Mr. Eugene Stokes and Mr. Neil Perl for their assistance with the other calculations reported here. To Dr. David Roddy for many helpful discussions of the Meteor Crater complex, and to Prof. Thomas Ahrens for references to high-pressure Hugoniot data on iron and granite, I wish also to express sincere thanks.

References

American Institute of Physics Handbook: 1957, p. 4–42 and 4–140, McGraw-Hill, N.Y.

Cooper, H.: 1976, Estimates of Crater Dimensions for Near-Surface Explosions of Nuclear and High-Explosive Sources, RDA-TR-2604-001, p. 24, 34.

Courant, R. and Friedrichs, K.: 1948, *Supersonic Flow and Shock Waves*, p. 123–131, 178, 179. Interscience, N.Y.

Krieger, F.: 1965, *The Thermodynamics of the Silica/Silicon-Oxygen Vapor System.* RAND Corp.Memorandum RM-4804-PR.

Ramspott, L.: 1970, *An Empirical Analysis of the Probability of Formation of a Collapse Crater by an Underground Nuclear Explosion.* (U) UCRL-50883, p. 11.

Roddy, D., Boyce, J., Colton, G., and Dial, A.: 1975, Meteor Crater, Arizona, rim drilling with thickness, structural uplift, diameter, depth, volume, and mass-balance calculations *Proc. Lunar Sci Conf. 6th*, p. 18.

Schuster, S. and Isenberg, J.: 1970, *Free Field Ground Motion for Beneficial Facility Siting*, (U) SAMSO-TR-70-88, **2**, 17–30.

Shoemaker, E.: 1963, Impact Mechanics at Meteor Crater, Arizona. *The Moon, Meteorites, and Comets* (B. Middlehurst and G. Kuiper, eds.), p. 313, 316, 326. University of Chicago Press. Words in brackets[ ] signify changes or additions to the original material.

Trulio, J., Perl, N., and Latham, R., 1975, *Limitations on the Size of Ejecta Craters.* Final Report for Contract No. DNA001-75-C-0298 p. 1, 2, 27, 28, (to be published).

Trunin, J., Podurets, M., Simakov, G., Popov, L., and Moiseyev, B., An Experimental Verification of the Thomas-Fermi Model for Metals under High Pressure *Soviet Phys.*, JETP **35**, 550.

Trunin, J., Simakov, G., Podurets, M., Moiseyev, B., and Popov, L.: 1970, Dynamic Compressibility of Quartz and Quartzite at High Pressure. *Izv. Earth Phys.* No. 1, p. 8–11.

van Dorn, W.: 1969, Lunar Maria: Structure and Evolution. *Science*, **165**, 693–695.

Appendix A: Further Discussion of Steps (2)-(4)

For small impact velocities especially ($U_0 \leq 2\ \mathrm{cm}/\mu\mathrm{s}$), considerably higher initial pressures can be expected along nearly vertical portions of a sphere's surface than at its bottom, during initial impact of iron and granite spheres with a granite medium (Figs. 12, 13). However, the possibility that the first disturbance might reach the center of the sphere along a path other than the symmetry axis itself, is quite remote; up to the limiting angles of Figs. (12) and (13), the shock created by head-on collision ($\theta = 0^0$) (a) runs at least as swiftly through the sphere as the shocks generated at most other angles, (b) starts sooner than any other shock, and (c) travels the same distance to the sphere's center. Moreover, at every impact angle, the initial shock reflected into the sphere will in reality weaken somewhat due to flow divergence, a process prevented by step (1) of Section 2. Hence, when the center of the actual sphere is first disturbed, its bottom will probably lie close to the position computed in step (2), and will almost certainly lie at greater depth. On the other hand, deceleration of its CoM could be more rapid than if the impact pressure at $\theta = 0$ were found at all angles. Perhaps more probable and interesting is a pinching-downward of material below the sphere's mid-horizontal plane, and pinching-upward of material above the plane, as the sphere sinks below the ground surface—a toothpaste-tube effect that runs counter to the dominant tendency of an incoming missile to flatten on collision.

After the sphere becomes half-buried (and even earlier) expanding granite vapor will impart downward impulse to its upper surface. That impulse is difficult to compute, but is surely much smaller than the impulse delivered to the missile's bottom surface during the period when it becomes half buried. Thus, it is legitimate to compute a minimum time for the shock to reach the topmost point of the sphere by assuming the sphere to be part of an infinite slab that collides uniaxially with a given half-space. However, under the same assumption, the downward speed of the CoM at that time could be overestimated (Section 2). We have therefore neglected all downward impulse delivered to the missile on its upper surface, and have thereby further underestimated the ultimate depth of the axial stagnation point.

To progress further, motion within the cylindrical tube of step (1) has been taken as uniaxial after half-burial of the sphere. The sphere is then replaced by a cylindrical piston of equal mass and CoM velocity. As a result, motion normal to the axis of the tube is suppressed. Adding the degree of freedom represented by such motion should increase the rate of dissipation of the missile's kinetic energy into heat, and therefore reduce its depth of penetration. However, large differences in penetration in the two cases (sphere versus cylinder in tube) are intuitively unlikely. Accordingly, as for step (1) of Section 2, we make the following assertions on grounds of physical reasonableness: Let a sphere and a right circular cylinder of the same material, radius, mass, and momentum strike the earth from directly above, and let all motion be confined to a vertical cylindrical tube whose radius equals that of the two missiles. The maximum CoM and bottom-surface displacements of the two bodies will then differ by no more than a small fraction (perhaps $<.1$) of either.

Confining the cylinder in question to a tube prevents earth from flowing around it, and also prevents its lateral expansion. The former aids penetration while the latter inhibits it. To obtain a lower bound on the depth reached by the CoM of the cylinder, we have therefore tried to exaggerate both its rate of lateral expansion and the attendant increased opposition to its downward motion—while still prohibiting flow around it. A procedure for so doing is presented, discussed and applied in Section 2.

APPENDIX B: OBLIQUE-SHOCK EQUATIONS AND THEIR SOLUTION

General considerations

At the time t_0 when a spherical missile first strikes a flat half-space from above, only the point at the bottom of the sphere—its "south pole"—makes contact with the half-space. However, viewed on smaller and smaller scales of distance (as under a microscope whose magnification can be increased at will), the south polar region appears flatter and flatter, differing as little as desired from a plane parallel to the surface of the half-space. The conditions created in missile and half-space on initial impact (time t_0) are therefore assumed to be those produced by the normal collision of two flat half-spaces.

Consistent with near-flatness of the south polar region (when highly magnified) is the fact that at time t_0 the distance to the half-space from any particle on that region is tiny compared with its distance from the (south) pole. Indeed, for particles closer and closer to the pole, the ratio of those two distances (particle-to-halfspace versus particle-to-pole) approaches zero. Hence, recalling that mechanical disturbances travel at finite speed, the gap between such a particle and the ground surface at time t_0 must close thereafter at the speed of the incoming sphere—there isn't time for a shock to reach either a near-polar particle on the missile's surface, or the halfspace-particle it strikes, before the two particles collide.

It is physically reasonable that in both materials after time t_0, shock speeds vary continuously from their values for normal incidence (i.e., at time t_0), and we make that assumption; its strict justification lies after the fact in its consistency with initial-boundary conditions and the laws of motion. The conclusion reached above then has a finite range of application: The gap between the half-space and a particle on the missile's surface closes at the speed of the incoming sphere, for particles within some finite distance of the sphere's south pole. Hence, until a later time than t_0 (as yet unspecified), the conditions created by impact are locally determined; the first "signal" to arrive at a point of the half-space-surface is actually a particle on the surface of the missile. As a result, when highly magnified (Fig. 16), the field around a point where missile and half-space are just starting to make contact becomes identical to the field produced by oblique impact of two homogeneous half-spaces. Calculation of that field, and definition of its range of validity for the present impact problem, are taken up below. We note first that the remarks above apply to any body whose bottom surface turns smoothly, and for any direction of incidence.

Oblique impact of flat homogeneous half-spaces: Background, geometry, definitions, and material properties

As the sphere-halfspace collision proceeds, the initial point of contact grows quickly to a finite area ("contact surface"), bounded at any instant by a horizontal circle about the system's vertical axis of symmetry. Above that circle and outside it, respectively, lie the separated, undisturbed boundary surfaces of the missile and halfspace. Hereafter we refer to a point of that circle as an "impact point". In an azimuthal half-plane, the impact surface is represented by a curve (also termed the "impact surface"); the missile's south pole lies at one end of that curve, and an impact point ("the" impact point) at the other.

As noted above, the impact point and its immediate neighborhood are equivalent to a system of two flat homogeneous half-spaces in collision. The angle between the free surfaces of the pertinent half-spaces is termed "the impact angle"; the impact angle varies with the instantaneous position of the impact point on the missile.

We require that in the limit of zero impact angle, the field resulting from collision of two flat homogeneous half-spaces approach the uniaxial field for normal incidence (in which steady shocks run away from the impact surface in both directions). Under that condition, and up to a finite impact angle to be determined, (i) oblique impact gives rise to a steady field of motion in any frame stationary with respect to the impact point, and (ii) the shocked state in each material is reached by "strong regular reflection" (Courant and Friedrichs, 1948). To show how the properties of shocked material (particularly its pressure, P) depend in regular reflection on the speed U_0 of the incoming missile, and the impact angle θ, we adopt the following definitions:

$P \equiv$ impact point.
$B_U \equiv$ line traced in the plane of flow by the undisturbed boundary surface of the upper medium (a half plane).
$B_L \equiv$ line traced in the plane of flow by the undisturbed boundary surface of the lower medium (a half plane).
$C \equiv$ line traced in the plane of flow by the interface (a half plane) on which the two shocked materials make contact.
$S_U \equiv$ line traced in the plane of flow by the shock front in the upper material (a half plane).
$S_L \equiv$ line traced in the plane of flow by the shock front in the lower material (a half plane).
(x_U, y_U)-axes $\equiv$ Cartesian axes in the plane of flow, with P as origin, and B_U as the positive x_U-axis.
(x_L, y_L)-axes $\equiv$ Cartesian axes in the plane of flow, with P as origin, and with B_L as the positive x_L-axis.

$\theta \equiv$ angle spanned in a clockwise rotation about P from B_U to $B_L \equiv$ impact angle.
$\gamma \equiv$ angle spanned in a clockwise rotation about P from the negative x_U-axis to C; particles of upper material turn through angle γ on crossing S_U.
$\varphi \equiv$ angle spanned in a clockwise rotation about P from C to the negative x_L-axis; particles of lower material turn through angle φ on crossing S_L.
$\alpha \equiv$ angle spanned in a clockwise rotation about P from C to S_U.
$\beta \equiv$ angle spanned in a clockwise rotation about P from S_L to C.

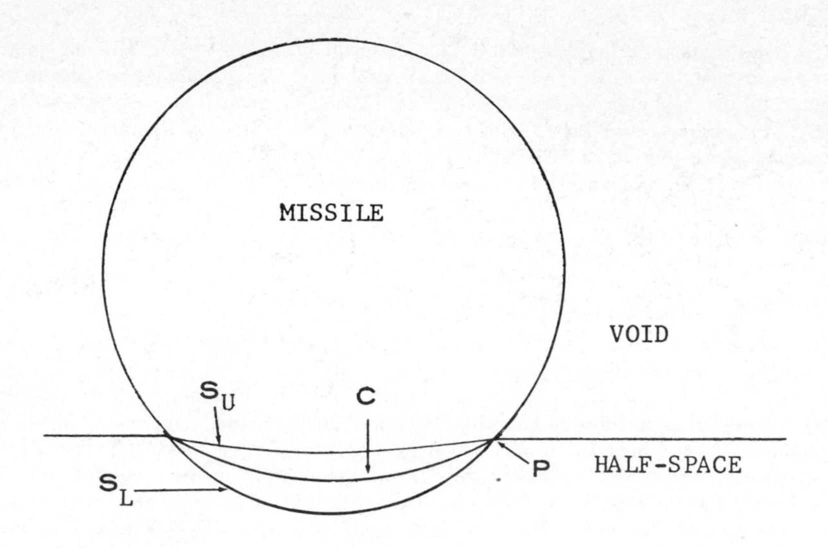

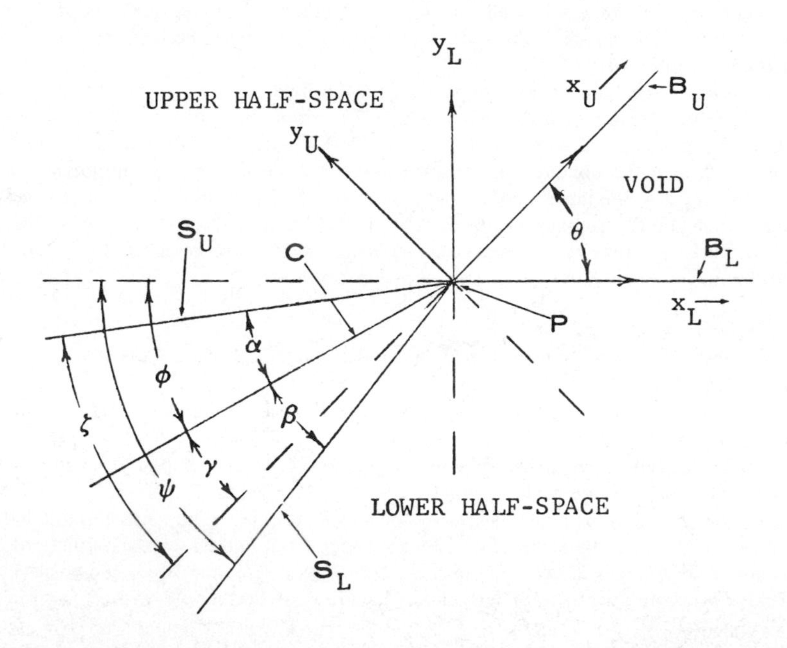

Fig. 16. Cross-section through the vertical axis of a sphere during normal impact with a flat half-space, in the period of strong regular shock reflection. Shocks in the missile (S_U) and half-space (S_L), as well as their contact surface (C), are shown on the scale of missile dimensions in the upper schematic. A point P at the edge of the contact surface is identified. Greatly magnified, P and its neighborhood present flow conditions identical to those created by oblique collision of two flat homogeneous half-spaces (lower schematic).

$\zeta \equiv$ angle spanned in a clockwise rotation about P from the negative x_U-axis to S_U (note that $\zeta = \alpha + \gamma$).

$\psi \equiv$ angle spanned in a clockwise rotation about P from S_L to the negative x_L-axis (note that $\psi = \beta + \varphi$).

Evidently the angle $(\gamma + \varphi)$ is traversed in a clockwise rotation about P from the negative x_U-axis to the negative x_L-axis, or from B_U to B_L. Hence $\gamma + \varphi$ equals θ. In calculating oblique-impact states, we make direct use of that identity in the form

$$(\tan \gamma + \tan \varphi)/(1 - \tan \gamma \tan \varphi) = \tan \theta. \tag{17}$$

Frequent reference is made below to velocity components tangential to, and normal to, the lines S_U, S_L, and C. As used here, "tangential" and "normal" denote directions in which the x_L- and y_L-axes would point if, after rigidly rotating those axes, the positive x_L-axis coincided with S_U, S_L or C. Accordingly, the tangential direction always runs outward from the impact point P along a given line. Also, the normal to S_U points from unshocked to shocked material, while the reverse is true of the normal to S_L.

Three further conventions are adopted: (i) the tangential and normal components of a vector are denoted respectively by subscripts "t" and "n"; (ii) overhead symbols $\sim$, $-$, and $\wedge$, respectively, are used to refer to S_U, S_L, and C; and (iii) a superscript "0" signifies a property of unshocked material. Thus, we define

$\tilde{\nu}_{tx}, \tilde{\nu}_{ty} \equiv$ components along the x_L- and y_L-axes, respectively, of a unit vector tangential to S_U.

$\hat{\nu}_{nx}, \hat{\nu}_{ny} \equiv$ components along the x_L- and y_L-axes, respectively, of a unit vector normal to c.

etc. Also:

$U^0_{Ux}, U^0_{Uy} \equiv$ components of velocity of unshocked upper-half space-material, along the x_L- and y_L-axes, respectively.

$U_{Lx}, U_{Ly} \equiv$ components of velocity of shocked lower-half space-material, along the x_L- and y_L-axes, respectively.

$\bar{U}^0_{Lt}, \bar{U}^0_{Ln} =$ components of velocity of unshocked lower-half space-material, tangential and normal, respectively, to S_L.

Components of the various unit normal and tangential vectors are readily expressed in terms of the angles defined above. In compact (matrix) notation, we find that

$$\begin{pmatrix} \tilde{\nu}_{tx} & \tilde{\nu}_{ty} \\ \tilde{\nu}_{nx} & \tilde{\nu}_{ny} \end{pmatrix} = \begin{pmatrix} -\cos(\varphi - \alpha) & -\sin(\varphi - \alpha) \\ \sin(\varphi - \alpha) & -\cos(\varphi - \alpha) \end{pmatrix}, \tag{18}$$

$$\begin{pmatrix} \bar{\nu}_{tx} & \bar{\nu}_{ty} \\ \bar{\nu}_{nx} & \bar{\nu}_{ny} \end{pmatrix} = \begin{pmatrix} -\cos \psi & -\sin \psi \\ \sin \psi & -\cos \psi \end{pmatrix}, \tag{19}$$

$$\begin{pmatrix} \hat{\nu}_{tx} & \hat{\nu}_{ty} \\ \hat{\nu}_{nx} & \hat{\nu}_{ny} \end{pmatrix} = \begin{pmatrix} -\cos \varphi & -\sin \varphi \\ \sin \varphi & -\cos \varphi \end{pmatrix}. \tag{20}$$

Data on the properties of the two materials most conveniently take the form of relations between shock velocity and particle velocity for uniaxial shock propagation. More specifically, relative to unshocked material, the shock velocity ΔS is a single-valued function of the velocity ΔU of shocked particles. For generality, that function could be introduced symbolically into the oblique-shock equations. More valuable than generality, however, are the concreteness and simplicity gained by using linear relations like those actually found for iron and granite (Trunin *et al.*, 1970, 1972); in fact, casting nonlinear ΔS–ΔU relations into multilinear form may be the best approach to the general oblique-shock problem. Accordingly, we write

$$|\Delta S| = c + \lambda |\Delta U|. \tag{21}$$

The constants (λ, c) in Eq. (21) are replaced by (λ_U, c_U) and (λ_L, c_L), respectively, for material of the upper and lower half-spaces. In addition, the pressure increase ΔP occasioned by crossing the shock is given by the following Rankine-Hugoniot relation:

$$\Delta P = \rho_0|\Delta U||\Delta S| \equiv P - P_0, \tag{22}$$

where ρ_0 is the density, and P_0 the pressure, of unshocked material; values of ρ_0 for the upper and lower materials are denoted ρ_{0U} and ρ_{0L}, respectively, while P_0 is negligible in both.

Basic equations

For the impact angles considered, the void between half-spaces closes at the speed of unshocked upper material relative to unshocked lower material (see above). Hence, relative to the latter, the impact point translates uniformly in the direction of the positive x_L-axis at speed $U_0 \cot \theta$; unshocked material in the upper half-space then has velocity-component $-U_0$ in the y_L-direction and zero in the x_L-direction (zero in both directions for the lower material). Accordingly, relative to (x_L, y_L)-coordinates with an origin at the impact point P, we have

$$(U^0_{Ux}, U^0_{Uy}) = (-U_0 \cot \theta, -U_0), \tag{23}$$

$$(U^0_{Lx}, U^0_{Ly}) = (-U_0 \cot \theta, 0). \tag{24}$$

The component of particle velocity tangential to S_U in unshocked upper material is equal to the scalar product of the vector (U^0_{Ux}, U^0_{Uy}) with the unit vector $(\tilde{\nu}_{tx}, \tilde{\nu}_{ty})$. Hence, since material experiences no tangential acceleration on crossing S_U, the velocity along S_U in both unshocked and shocked states is given by the formula

$$\tilde{U}^0_{Ut} = U_0[\cot \theta \cos(\varphi - \alpha) + \sin(\varphi - \alpha)] = \tilde{U}_{Ut}. \tag{25}$$

Similarly, we find that

$$\tilde{U}^0_{Un} = U_0[-\cos \theta \sin (\varphi - \alpha) + \cos (\varphi - \alpha)]. \tag{26}$$

Recalling that (i) relative to a shock front, material slows down on crossing the front, and (ii) S_U is a stationary shock in the frame of the (x_L, y_L)-axes, it follows from Eq. (21) that

$$\tilde{U}_{Un} = [(\lambda_U - 1)\tilde{U}^0_{Un} + c_U]/\lambda_U. \tag{27}$$

Eqs. (22) and (27) imply in addition that

$$\Delta P = (\rho_{0U}/\lambda_U)\tilde{U}^0_{Un}(\tilde{U}^0_{Un} - c_U). \tag{28}$$

Relative to the frame defined by the unit vectors $(\tilde{\nu}_{tx}, \tilde{\nu}_{ty})$ and $(\tilde{\nu}_{nx}, \tilde{\nu}_{ny})$, unit vectors tangential and normal to the contact surface C are given (Eqs. (18), (20)) by (cos α, sin α) and (–sin α, cos α). From Eq. (26), it then follows that

$$\hat{U}_{Ut} = \tilde{U}^0_{Ut} \cos \alpha + \tilde{U}_{Un} \sin \alpha, \tag{29}$$

$$\hat{U}_{Un} = -\tilde{U}^0_{Ut} \sin \alpha + \tilde{U}_{Un} \cos \alpha. \tag{30}$$

In order for the contact surface C to be stationary, the component of particle velocity normal to C must vanish ($\hat{U}_{Un} = 0$), whence (Eq. (30))

$$\tilde{U}^0_{Ut} \sin \alpha = \tilde{U}_{Un} \cos \alpha. \tag{31}$$

Eqs. (25)–(29) and (31) comprise a set of basic relations for the upper half-space. By exactly parallel arguments, the following analogous set of equations (differently ordered) is obtained for the lower half-space:

$$\bar{U}^0_{Lt} \sin \beta = -\bar{U}_{Ln} \cos \beta, \tag{32}$$

$$\Delta P = (\rho_{0L}/\lambda_L)(-\bar{U}^0_{Ln})(-\bar{U}^0_{Ln} - c_L), \tag{33}$$

where

$$\bar{U}^0_{Lt} = U_0 \cot \theta \cos \psi, \tag{34}$$

$$-\bar{U}^0_{Ln} = U_0 \cot \theta \sin \psi, \tag{35}$$

$$-\bar{U}_{Ln} = [-(\lambda_L - 1)\bar{U}^0_{Ln} + c_L]/\lambda_L. \tag{36}$$

Also:

$$\hat{U}_{Lt} = \bar{U}^0_{Lt} \cos \beta - \bar{U}_{Ln} \sin \beta. \tag{37}$$

Equations for oblique impact, in a form useful for computation

Since $\zeta = \gamma + \alpha$ and $\theta = \gamma + \varphi$ (above), $\varphi - \alpha = \theta - \zeta$ and (Eqs. (25), (26))

$$\tilde{U}^0_{Ut} = U_0 \csc \theta \cos \zeta, \tag{38}$$

$$\tilde{U}^0_{Un} = U_0 \csc \theta \sin \zeta. \tag{39}$$

Also, since $\alpha = \zeta - \gamma$, Eq. (31) can be written in the form

$$\tan \gamma = (\tilde{U}^0_{Ut} \sin \zeta - \tilde{U}_{Un} \cos \zeta)/(\tilde{U}^0_{Ut} \cos \zeta + \tilde{U}_{Un} \sin \zeta). \tag{40}$$

Eqs. (38)–(40) hold for any fluid.

The many equations now before us are equivalent to a single P–θ–U_0 relation. To see why, note that for the upper material $|\Delta S|$ and $|\Delta U|$ are equal, respectively, to $\tilde{U}^0_{Un}$ and $(\tilde{U}^0_{Un} - \tilde{U}_{Un})$. Hence, taken with Eq. (22), the ΔS–ΔU relation makes it possible in principle to find both $\tilde{U}^0_{Un}$ and $\tilde{U}_{Un}$ as functions of P for any given medium (independently of U_0 and θ). Further, ζ depends in a known way on U_0, θ, and $\tilde{U}^0_{Un}$ (Eq. (39)). Evidently, therefore, $\tan \gamma$ is a function of P, θ, and U_0. Similar reasoning shows that $\tan \varphi$ is also determined by P, θ, and U_0. Hence, Eq. (17) becomes—in principle—a single P–θ–U_0 relation. With ΔS a linear function of ΔU (Eq. (21)), that relation assumes a rather simple form; in fact, an explicit (but computationally inconvenient) solution can then be obtained for the entire problem.

By combining Eq. (27) with Eqs. (38), (39), and (40), we obtain the result

$$\tan \gamma = \cos \zeta (U_0 \csc \theta \sin \zeta - c_U) \div [\lambda_U U_0 \csc \theta - \sin \zeta (U_0 \csc \theta \sin \zeta - c_U)]. \tag{41}$$

Also, Eqs. (28) and (39) imply that

$$(\rho_{0U}/\lambda_U) U_0 \csc \theta \sin \zeta (U_0 \csc \theta \sin \zeta - c_U) = \Delta P, \tag{42}$$

or

$$U_0 \csc \theta \sin \zeta = \tfrac{1}{2} c_U [1 + (1 + 4\lambda_U \Delta P/\rho_{0U} c_U{}^2)^{1/2}] \equiv B_U. \tag{43}$$

Using Eq. (43) to eliminate θ from Eq. (42), we find that

$$\tan \gamma = \cos \zeta \sin \zeta/(A_u - \sin^2 \zeta), \tag{44}$$

where

$$A_U \equiv \lambda_U B_U/(B_U - c_U). \tag{45}$$

Similarly, recalling that $\psi = \alpha + \varphi$ (and $\theta = \gamma + \varphi$), we conclude from Eqs. (32)–(36) that

$$\begin{aligned} \tan \varphi &= \cos\psi (U_0 \cot \theta \sin \psi - c_L) \div [\lambda_L U_0 \cot \theta - \sin \psi (U_0 \cot \theta \sin \psi - c_L) \\ &= \cos \psi \sin \psi/(A_L - \sin^2 \psi), \end{aligned} \tag{46}$$

and

$$(\rho_{0L}/\lambda_L) U_0 \cot \theta \sin \psi (U_0 \cot \theta \sin \psi - c_L) = \Delta P, \tag{47}$$

or

$$U_0 \cot \theta \sin \psi = B_L, \tag{48}$$

where

$$B_L \equiv \tfrac{1}{2} c_L [1 + (1 + 4\lambda_L \Delta P/\rho_{0L} c_L{}^2)^{1/2}], \tag{49}$$

$$A_L \equiv \lambda_L B_L/(B_L - c_L). \tag{50}$$

Moreover, Eqs. (43) and (48) can be combined to give

$$\sin \zeta = (B_U/B_L) \sin \psi \cos \theta. \tag{51}$$

Through Eq. (48), ψ becomes an explicit function of P, θ, and U_0, and, hence (Eq. (51)), so does ζ. As a result, $\tan \gamma$ and $\tan \varphi$ are known as functions of P, θ, and U_0 (Eqs. (44) and (46)), and Eq. (17) therefore yields an explicit P–θ–U_0 relation. For a given value of U_0, iterative methods could be used to find P from that relation for each of a series of values of θ. Instead, as suggested by the most complete reduction of Eq. (17) that we achieved (below): (i) Eq. (17) was divided by $U_0 \sin \psi$, (ii) uniform steps were taken in ψ for a given U_0, and (iii) at each step the quantity $\tan \theta / U_0 \sin \psi$ ($= B_L^{-1}$; Eq. (48)) was varied to satisfy Eq. (17) as modified. Accordingly, we write

$$z = \tan \theta / (U_0 \sin \psi) = 1/B_L, \tag{52}$$

$$\Delta P = \rho_{0L}(1 - c_L z)/(\lambda_L z^2), \tag{53}$$

and, with B_U given by Eq. (43):

$$\sin \zeta / \sin \psi = (zB_U)/[1 + (zU_0 \sin \psi)^2]^{1/2}. \tag{54}$$

Also:

$$\tau_L = \tan \varphi / \sin \psi = \cos \psi / (A_L - \sin^2 \psi). \tag{55}$$

Hence (Eqs. (44), (52), and (54)):

$$\tau_U \equiv \tan \gamma / \sin \psi = (\sin \zeta / \sin \psi) \cos \zeta / (A_U - \sin^2 \zeta). \tag{56}$$

Eq. (17) then becomes

$$U_0^{-1}(\tau_U + \tau_L)/(1 + \tau_U \tau_L \sin^2 \psi) = z. \tag{57}$$

For a head-on collision ψ and θ are both zero, and, given U_0, the pressure jump ΔP^* can be computed from relatively simple formulas for uniaxial shock reflection; for $1/z$, Eq. (57) then yields the value $(c_L/2)[1 + (1 + 4\lambda_L \Delta P^* / \rho_{0L} c_L^2)^{1/2}]$. Starting from that value for a given U_0, Eq. (57) is solved by iteration for each of a closely-spaced set of ψ's, with z equated on the first iteration to its last value for the previous ψ.

By treating as basic flow variables the respective speeds $U_0 \csc \theta$ and $U_0 \cot \theta$ of unshocked upper and lower material, Eq. (57) has been reduced to the form

$$r^2(1 + \tan \varphi \cot \theta)^2 - r^4(1 + \tan^2 \varphi)$$
$$= A_U[A_U(1 + \cot^2 \theta) - 2r^2](\tan \varphi - \cot \theta)^2, \tag{58}$$

where

$$r = B_U/s_L, \tag{59}$$

$$\tan \varphi = (B_L/s_L)[1 - (B_L/s_L)^2]^{1/2}/[A_L - (B_L/s_L)^2], \tag{60}$$

and

$$s_L = U_0 \cot \theta = \text{speed of unshocked lower material}. \tag{61}$$

Eq. (58), being a quartic in $\cot \theta$, can be solved explicitly for $\cot \theta$ as a function of the other parameters it contains. In that sense, $\cot \theta$ has been explicitly determined as a function of P and s_L. Complexity aside, however, the solution is not of much use here; with s_L fixed, U_0 is only known after $\cot \theta$ has been evaluated ($U_0 = s_L \tan \theta$), and θ or ψ needs to be chosen at will rather than P.

The strong-shock limit; other comments

Equation (46) holds regardless of the cause of the shock in the lower material. The same is true for Eq. (44) and the upper material. For proof, derive Eq. (46) with an arbitrary speed s_L in place of

$U_0 \cot \theta$; only single-shock material properties (Eq. (21)), a Rankine–Hugoniot relation (Eq. (22)), and stationary directions of the unshocked flow and shock front, are needed.

The fact that the two shocks are caused by a collision of half-spaces is expressed by (a) the simple geometric relation $\theta = \gamma + \varphi$ (Eq. (17)) between the impact angle and the two angles through which flow is turned, and (b) a kinematic condition (equivalent to Eqs. (43) and (48)) specifying the relative velocity of the half-spaces prior to collision. Evidently, (a) and (b) are independent; the angle between the two half-spaces says nothing about their relative velocity, and vice versa. Further, the angles made by the relative-velocity vector with the two half-space surfaces cannot be changed by rotating or translating the reference frame. One of those angles is 90° here, while the other is not (except for a head-on collision). Hence, even if the same homogeneous material fills both half-spaces, no mix of translation and rotation can yield a reference line across which the velocities of the half-spaces, and their surfaces as well, are mirror images of one another; if it did, the relative-velocity vector would make equal angles with the two half-space surfaces.

The oblique-shock relation exemplified by Eq. (46), whose shock-source-independence gives it general character, has important properties. As background, we note that to express physically reasonable material behavior at high pressures ($\lambda \Delta P/\rho_0 c^2 \gg 1$), λ must be >1 in Eq. (11) (the values 1.262 and 1.32 fit granite and iron). Hence, for $0 \leq \psi < \pi/2$, we have $\tan \varphi > 0$, while $\tan \varphi = 0$ both at $\psi = 0$ and $\psi = \pi/2$. Thus, in a plot of φ versus ψ at constant P (a "shock polar"), $\tan \varphi$ rises to a maximum value, and then decreases, as ψ is increased. Similar statements hold for γ as a function of ζ, but φ reaches its maximum before γ in the cases considered (when granite strikes granite, $B_U/B_L = 1$ and Eq. (5) shows that $\gamma < \varphi$; γ is also $<\varphi$ when iron strikes granite). It follows from Eq. (46) that $\tan \varphi$ reaches a maximum value ($\tan \varphi_{max}$) of $\frac{1}{2}[A_L(A_L - 1)]$ at $\psi = \psi_m$, where $\cos \psi_m = [(A-1)/(2A-1)]^{1/2}$. Furthermore, φ_{max} increases monotonically with ΔP to a limiting value (41.0° for granite and 37.6° for iron).

Evidently, when $\psi \neq \psi_m$, two different shocks will turn a flow through one and the same angle at the same shock pressure. For $0 \leq \psi < \psi_m$ we speak of "strong regular reflection", and of "weak regular reflection" when $\psi > \psi_m \geq \pi/2$ (the shock can be viewed as "reflected" from a plane parallel to the direction of flow of shocked material; unshocked material is then obliquely "incident" upon that plane); if $\psi > \pi/2$, the flow is not steady. The difference between weak and strong reflections for the same φ and P is apparent when $\varphi = 0$ ($\psi = 0$ versus $\psi = \pi/2$). As $\psi \to 0$, unshocked material moves toward the impact point with a speed that increases without bound, and the pressure P appears everywhere at once on the surface S_L of the lower half-space. With $\psi = \pi/2$, the pressure P drives a shock (relative to unshocked material) parallel to S_L—as if that pressure were applied uniformly over a plane normal both to the plane of flow and the surface of the half-space. The latter case ($\psi = \pi/2$) is plainly incompatible with the shock produced by collision of two half-spaces, while the former ($\psi = 0$) simply represents a head-on collision. Thus, strong reflection is the proper choice for $\theta = 0$. Moreover, no transition can occur to weak reflection from a shock angle $\psi < \psi_m$; a discontinuous jump in ψ implies motion of the shock at infinite speed relative to unshocked material. However, when ψ becomes equal to ψ_m, a continuous transition can be made to weak reflection, and that transition is not easily ruled out as inconsistent with initial or boundary conditions. Transition to weak reflection would of course be accompanied by a decrease in φ; thus, permitting that transition, we found that $\theta(= \gamma + \varphi)$ reached a maximum value not much larger than those shown in Figs. 11 and 13. Since transition to Mach reflection is also possible, occurs widely, and seems more plausible here, the figures show results for regular reflection in the "strong" case only.

Roddy, D. J., Pepin, R. O., and Merrill, R. B., editors.
(1977) *Impact and Explosion Cratering*, Pergamon Press (New York), p. 959–982.
Printed in the United States of America

Numerical simulations of a 20-ton TNT detonation on the earth's surface and implications concerning the mechanics of central uplift formation

GILBERT WAYNE ULLRICH

Air Force Weapons Laboratory, Kirtland Air Force Base, New Mexico 87118

DAVID J. RODDY

U.S. Geological Survey, Flagstaff, Arizona, 86001

GENE SIMMONS

Department of Earth and Planetary Sciences, Massachusetts Institute of Technology, Cambridge, Massachusetts 02139

Abstract—Central uplifts are common features observed in craters on the Earth, the Moon, Mars, and Mercury. Since these uplifts do not occur in all craters, they should be useful in providing strong constraints on both planetary evolution and numerical cratering simulations. Unfortunately, because the mechanics of central uplift formation have been poorly understood, little use of those constraints has yet been made.

Therefore, a series of numerical simulations of the ground response to a high-explosive detonation was accomplished to examine the influence of model conditions on calculated central-uplift formation. Data from a previous numerical simulation of a high-explosive detonation were used as a surface boundary condition, and the ground response was simulated by a computer code that modeled two-dimensional, axisymmetric problems of continuum-mechanics with elastic-plastic material models. A calculation that modeled the 20 ton high-explosive detonation designated Mixed Company II showed that, when ballistically extrapolated, the computed motions at a simulated time of 23.9 msec were consistent with the observed crater and formation of a central uplift. The results of a series of calculations in which compaction, layering, and material-yield models were varied indicated: (1) the calculated upward-motions below the crater were eliminated when material compactibility was increased, (2) the model of test-site layering in the Mixed Company II numerical simulation only slightly influenced the upward velocities below the crater, (3) plastic volumetric-increases of material during Mohr–Coulomb yield contributed significantly to the magnitude of upward motions, (4) upward velocities for points on the axis of symmetry were first calculated where strength effects were important, and (5) the inclusion of a lower, "fluid" layer modified the calculated response in an overlying, "dry" layer in a manner that *may* have eventually resulted in upward motions.

A mechanical model of central uplift formation was developed with the results of the numerical calculations and work by others as guides. In this model, the two principal causes of central-uplift formation are the volumetric rebound of material following the maximum shock-pressure and a later gravitational-collapse of crater walls where the apparent material strength is insufficient to support the crater walls. Each mechanism may be completely dominant in specific cases; however, the two mechanisms need not be mutually exclusive. Finally, since the mechanical model is only as broad as the information on which it is based, the extrapolation of the model should be made with caution and on a case-by-case basis.

Introduction

Central uplifts are common features of craters. Such uplifts have been observed in meter-sized craters (Roddy, 1968, 1976) and in craters measured in tens of kilometers (Baldwin, 1963; Hartmann, 1973). They occur in craters produced by chemical explosives (Roddy, 1968, 1973, 1976) and in ancient impact structures on the earth (Howard *et al.*, 1972; Roddy, 1968). They have been seen in craters on the Moon (Baldwin, 1963), on Mars (Carr *et al.*, 1976; Hartmann, 1973), and on Mercury (Murray *et al.*, 1974). However, while several authors have advanced hypotheses as to the cause of central uplifts (Baldwin, 1963; Dence, 1968; Milton and Roddy, 1972; Short, 1965), a satisfactory explanation of the mechanics of central-uplift formation has not been demonstrated.

An understanding of the mechanics of central-uplift formation is important for several reasons. First, since central uplifts are associated with shock-wave-produced craters, a satisfactory theory of cratering mechanics must include an explanation of central-uplift formation. Second, since central uplifts are observed in some, but not all, hypervelocity impact craters on planetary surfaces, the occurrence of a central uplift must provide information on the conditions during the impact event. Third, the central uplifts of craters represent important sampling sites for any extraterrestrial landing or remote sensing mission because the material in the uplift is the deepest material exposed by the cratering event (Howard *et al.*, 1972; Milton *et al.*, 1972; Roddy, 1968; Stearns *et al.*, 1968). An understanding of central uplift formation might aid the determination of the preimpact stratigraphic location of the material.

We used a series of numerical simulations of a high-explosive detonation on a layered halfspace to examine causes of central-uplift formation. The AFTON-2A computer code (Niles *et al.*, 1971; Trulio, 1966; Trulio *et al.*, 1969, 1976), that is actively used for ground-shock calculations, was used with changes only in material models. These models (Ullrich, 1976) were limited to the simplest possible expressions to demonstrate which factors were most important to the numerical results. One model of the detonation of 20 tons of TNT, arranged in a surface-tangent, spherical charge, was used for all the numerical simulations. The basic assumptions for the calculations were axial symmetry, isotropic materials, and no energy transfer by conduction or radiation.

The generalization and application of the results of any numerical simulation must be done cautiously because the results strictly apply only to definite models with specific input parameters. In order to generalize our specific numerical results to mechanisms of central uplift formation, we use the following procedure. First, the available information on central uplift formation is reviewed to provide the broadest possible base of data. Then, a numerical simulation of a high-explosive experiment is described to demonstrate the applicability of the numerical results to that one experiment. The results of additional numerical-simulations in which material models were varied are then described to show which mechanisms were important to the formation of a central uplift in that experiment. These mechanisms are generalized based on the available information described in the initial review.

We define the terms "shock-wave-cratering event" and "central uplift" to clarify their uses in this paper. A "shock-wave cratering event" is an event that transfers a large amount of energy to a small volume of a halfspace by sending a shock wave into the halfspace and forms a crater at the surface. This term applies to both a hyper-velocity impact and a surface chemical or nuclear explosive detonation. Further, as Shoemaker (1961) suggests, this term is more basic than explosion cratering event because, even if the characteristics of craters are controlled by the expansion of gases near the source of the event (Baldwin, 1963), the passage of the shock wave through the material is the mechanism which establishes the conditions for such expansion. A "central uplift" is a local feature that is topographically high, physically near the center of a crater, and consists of material that was displaced upward during the cratering event. This term applies to a definite structural feature, and is not meant to include the possibility that material ejected from the crater may subsequently fall into the crater and form a hill at the center.

Background Information

Three types of information are available concerning central uplifts in shock-produced craters. First, observations of the occurrence of central uplifts and of the structure in central uplifts provide constraints on any explanation of central uplift formation. Second, numerical simulations of shock-wave-cratering events and dynamic field measurements constrained by observations provide guides to the physical processes involved in central-uplift formation for specific conditions. Third, previously-proposed hypotheses concerning the causes of central-uplift formation provide an important framework for research effort.

Observations of the structure of central uplifts suggest that similar sets of physical principles apply to both hypervelocity impact events and explosive detonations. Extensive field studies have shown that the material in central uplifts of both ancient impact structures on earth and high-explosive craters is displaced upward from its original position (Carnes, 1975; Howard *et al.*, 1972; Roddy, 1968, 1976; Stearns *et al.*, 1968). Horizontal displacements of the material that form central uplifts are probably inward in the deeper regions (Howard *et al.*, 1972; Milton *et al.*, 1972; Roddy, 1976), and outward in the shallower regions (Howard, *et al.*, 1972; Milton, *et al.*, 1972). Deformation of the material in the central uplift range from complete brecciation or rubble (Roddy, 1973), through folding and brecciation (Milton *et al.*, 1972; Roddy, 1968, 1973), to a tightly folded dome (Roddy, 1968).

The peak-shock-pressures experienced by material in central uplifts may be estimated on the basis of the occurrence of shock effects. One macroscopic shock-produced feature that occurs frequently in central uplifts (Dietz, 1968) is the shatter cone. A preliminary theoretical study of shatter-coning (as reported by Dietz, 1968) suggested that shatter cones are shock fractures formed along a traveling boundary between the plastic and elastic response of a material, with the plastic domain moving relative to an elastic domain. The analysis is consistent with the observations that shatter cones appear to be formed prior to significant material displacement, and very high-pressure effects have not been found associated with shatter cones. Thus, shatter cones, and by association central uplifts, appear to be formed in material where shock stresses were near, but above, the Hugoniot elastic limit.

Observations of the occurrence of central uplifts on the Earth, Mars, and the Moon have been interpreted to show that gravity has an influence on the occurrence of central uplifts (Hartmann, 1972, 1973). Hartmann (1972) suggested that data on the size distribution of craters with central uplifts as a function of crater diameter indicated that the smallest crater-diameter for central-uplift formation was inversely proportional to the gravitational acceleration. An extensive study of impact structures in Canada has led Dence (1968) to a similar conclusion. Hartmann's data showed, however, that the inferred minimum diameter of craters with central uplifts was significantly different from the diameter at maximum frequency, indicating that gravitational stress was not the only parameter important to central-uplift formation.

Observations of the structure and occurrence of central uplifts in shock-produced craters provide information on the final positions of material, but do not allow unique interpretations of the complete motion of material during the event. Insight into the dynamic motion can only be obtained through theoretical studies and active measurements during individual cratering events. Such information, however, bears directly on the processes that are important to central uplift and cratering mechanics.

One such theoretical study was a numerical simulation, accomplished by Maxwell and Moises (1971), of the event which may have formed the Sierra Madera formation (Howard *et al.*, 1972). For this simulation a sphere with a radius of 100 m and a velocity of 30 km/sec was assumed to impact vertically on a halfspace. Both the sphere and the halfspace were assumed to be composed of the same material, which had an assumed density of 2.7 g/cc. The parameters of the material equation-of-state were based on Hugoniot data only for basalt. The yield model was a 20 MPa von-Mises limit until a calculational zone experienced zero pressure, after which that zone was assumed to have no shear strength. The calculation resulted in upward velocities below the impact point after a simulated time of 5.5 s with a toroidal flow pattern developed by 9.5 s that continued until the calculation was terminated at 30 s. The toroidal flow was contained within a region that was dominately hydrodynamic with violent upward and inward motions near the axis of symmetry accompanied by downward motions near the surface from one to two crater radii. The downward flow at the calculated crater edge began between the simulated times of 5 and 10 s after a period of upward and outward motions. The entire flow pattern after 5.5 s could be explained by the flow of a liquid under the influence of gravity.

Dynamic measurements of motions in the crater-wall region of the Prairie Flat 500-ton TNT detonation at Suffield, Canada showed motions similar to the motions calculated during the numerical simulation of Sierra Madera. In this experiment velocity gages were buried at various ranges and depths (Hoffman *et al.*, 1971). One set of velocity gages was located at a range of 25.6 m and depths of 1.52, 3.04, and 5.18 m. Displacements of these gages after 2.5 s, obtained by integration of the velocity data, were consistent with the permanent displacement markers (Roddy, 1976) in the same region. The motion of these gages was upward and outward for 1.2 s followed by downward and inward flow to 2.5 s. The velocity changes after 0.4 s were consistent with motion in a gravitational field. Significantly, central uplifts and indications of a fluidized substrate occurred in this experiment and the other 500-ton experiments at this test site.

The mechanisms of central-uplift formation that have been postulated previously may be divided broadly into stress-wave effects and gravitational adjustments of a transient crater. The importance of this division is that the stress-wave effects should occur in at least an order-of-magnitude shorter period of time than the gravitational adjustments. The first of these broad divisions includes rebound (Baldwin, 1963; Boon and Albritton, 1938), reflection of stress waves from material discontinuities (Short, 1965), shear-wave effects (Maxwell and Moises, 1971) and combined effects of stress-wave interaction with the free boundary (Roddy, 1968, 1976). Several authors (Dence, 1968; Gault *et al.*, 1968; Pike, 1971; Shoemaker, 1963) have suggested that a cause of central-uplift formation is a deep sliding, or base failure, resulting from the gravitational stresses produced by the difference in height between the rim and the center of a crater. Melosh (1976a, 1976b) has investigated the static stability of craters with diameter to depth ratios of 6 and found that if a Mohr-Coulomb failure criterion is used, the craters are always stable; but if a von-Mises failure criterion is used deep failures will occur for sufficiently large craters. However, in order for 30 km-diameter craters on the moon to have central uplifts, this analysis results in a requirement that the shear strength of lunar material be less than 6 megapascals. Thus, Melosh concludes that either gravitational failure must accompany dynamic processes or *in situ* lunar materials have small ultimate-strength characteristics.

Numerical Simulation of Mixed Company II

High-explosive cratering experiments provide excellent opportunities to examine the mechanisms of central-uplift formation through numerical simulation because: (1) the preshot material properties of the medium were extensively tested, (2) the test conditions are known, and (3) strong constraints of the numerical results are provided both by careful documentation of the structure of the crater and central uplift and by dynamic ground-motion measurements. One high-explosive event, designated Mixed Company II served as a particularly useful experiment because a very large central uplift (Fig. 1), relative to the crater size, was produced. This large size clearly demonstrated that at least one mechanism of central-uplift formation was particularly effective and also reduced resolution problems associated with numerical calculations.

The Mixed Company II experiment (Carnes, 1975) was the detonation of 20 tons of TNT, arranged in a spherical charge placed above, and tangent to, the

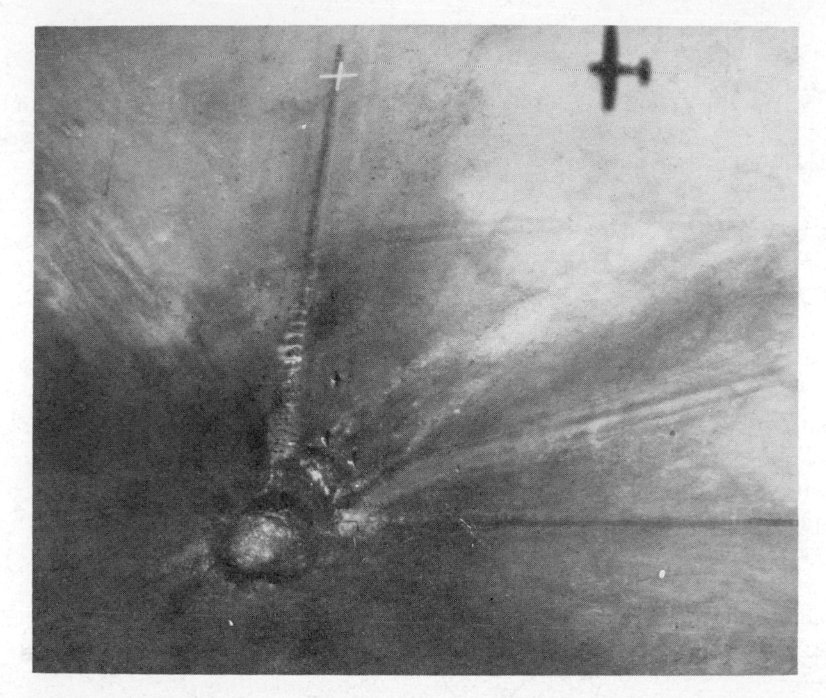

(a)

(b)

Fig. 1. (a) Oblique aerial view of Mixed Company II crater showing the prominent central uplift. Note men on the crater rim for scale. The spacing between ground crosses is approximately 61 meters. (b) Close-up view of the large central uplift showing the blocky surface. The stake on top of the uplift is approximately 0.4 meters in length. (U.S. Geological Survey aerial photograph by D. Roddy and R. Williamson).

ground surface. The Mixed Company test site consisted of a thin deposit of sandy clayey silt over a 20.7 m thick section of Kayenta formation with no significant water content. The silt, which was approximately 0.6 m thick at the Mixed Company II test site, appeared to become slightly cemented at depth (Ehrgott, 1973). The Kayenta is a fluvial deposit that consists of lenticular to irregularly bedded layers of fine-to-medium grained sandstone, siltstone, and conglomerate with occasional layers or lenses of shale. The material in the Kayenta formation had a density of 2.35–2.42 g/cc and unconfined compressive strengths of 13.8–34.5 MPa. Lenticular interfaces at the Mixed Company II test site occurred at depths of 3.5 and 7.1 m. The top 1.2 m of the Kayenta formation were weathered significantly, which resulted in a lower cohesive strength than the underlying rock.

Observed crater morphology, structural information (Carnes, 1975), field studies by one of the authors (DJR), and records from active ground-motion gages (Ingram, 1976), provided the constraints on the numerical simulation of the Mixed Company II experiment. Profiles (Fig. 2) of the crater that was formed (Carnes, 1975) showed that the apparent crater extended a maximum of 1.2 m below the original ground level at a radius from ground zero of 3.7 m. The crater was only approximately symmetric, with radii at the original ground level varying from 5.6 to 6.6 m. The central uplift, represented by true crater dimensions, extended a distance of 2.4 m from the vertical axis through ground zero and was 2.1 m high. The mound was composed of uplifted and brecciated sandstone (Roddy, 1973). A poorly-developed overturned flap and thin blanket of ejecta surrounded the crater, and no fused material was found. Deformation in the crater wall and rim consisted mainly of shattering and local brecciation. A 3.7 m deep sand column with displacement markers had been placed at ground

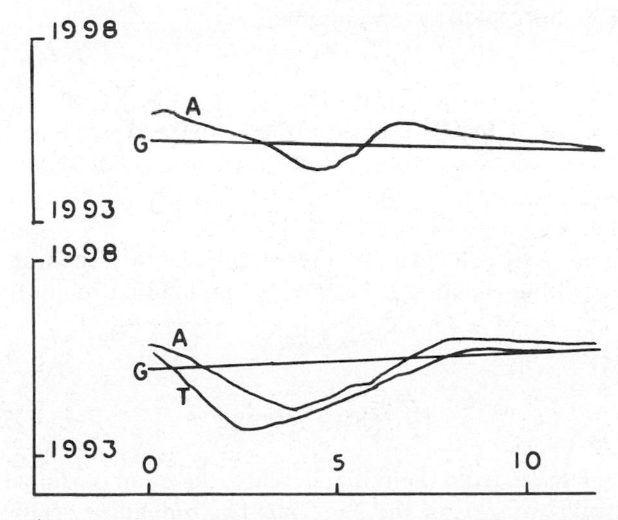

Fig. 2. Apparent (a) and true (T) crater profiles of the 20 ton TNT event, Mixed Company II, compared to the original ground profile (G). The top profile represents the north radial, and the bottom profile represents the south radial.

zero (Carnes, 1975). During excavation, the bottom displacement marker, originally placed near 3.1 m depth, was found near the top of the mound, indicating an upward displacement of at least 3.1 m. The remainder of the column below the marker was stretched out continuously for several feet in the central uplift. Ground shock instrumentation (Ingram, 1975) recorded a prominent horizontal-pulse propagating outward from ground zero at a depth of 1.5 m. The parameters of this pulse, recorded at ranges of 16.5, 21.3, and 28.3 m, also provided constraints on the numerical simulation.

The numerical simulation of this event included the use of three mathematical models of the physical processes that were assumed to be important (Ullrich, 1976). The first was a model of the surface-pressure boundary condition to simulate the high explosive detonation. The second was the AFTON-2A computer code (Niles *et al.*, 1971; Trulio *et al.*, 1976) that modeled the initial response of the ground to the surface boundary condition. This code included approximations to physical relations and the properties of the materials at the test site. By a simulated time of 23.9 msec, the material in and below the crater region was calculated to be separated and moving ballistically (Ullrich, 1976) and useful comparisons could be made with the ground-motion gages. The third model was a ballistic extension of the conditions calculated with the first two models at 23.9 msec.

The material model used in the numerical simulations was divided into relations describing the hydrostatic pressure-density characteristics and the elastic-plastic characteristics of the test-site materials. The general hydrostatic relation for the material, that expressed pressure (P) as a function of material density (ρ), was:

$$P = f(\mu, \mu^*), \tag{1}$$

where the excess compression, μ, was defined as:

$$\mu = \frac{\rho - \rho_i}{\rho_i}, \tag{2}$$

with ρ_i the initial material density and μ^* the maximum excess-compression ever calculated in that calculational zone. This functional relationship was divided into a low-pressure region, for μ less than or equal to the volume fraction of air-filled voids, and a high-pressure region.

The low-pressure hydrostat was further divided into a loading relation, for $\mu = \mu^*$, and an unloading relation for $\mu < \mu^*$. The loading relation was:

$$P = K_L\mu, \tag{3}$$

where:

$$K_L = \tfrac{1}{3}\rho_i C_p{}^2 \frac{(1+\nu)}{(1-\nu)}, \tag{4}$$

defined K_L for each layer from the initial density, the compressional wave speed, C_p, and the Poisson's ratio, ν, of the material. The unloading relation was:

$$P = \frac{K_u[(\beta - 1)\mu_x + \mu]^2 - K_v(\mu_x - \mu)^2 + K_v(\beta\mu_x)^2}{2\beta\mu_x}, \tag{5}$$

for

$$(1-\beta)\mu_x < \mu < \mu_x$$

where

$$\mu_x = \min\begin{Bmatrix}\mu^* \\ \mu_3\end{Bmatrix}; \qquad \beta = \frac{2K_L}{K_u + K_v};$$

and μ_3 was the volume of air-filled voids. The parameter K_u was defined by relation (4) where C_p was replaced by the sonic velocity at the initial release of pressure, C_u; and K_v was defined by the same relation with the sonic velocity as the pressure approaches zero, C_v.

The form of relation (5) was chosen solely to provide a linear transition in $dp/d\mu$ as μ varied between μ_x and $(1-\beta)\mu_x$ in unloading. For $\mu < (1-\beta)\mu_x$ the pressure-density relation was assumed to be:

$$P = -K_v[(1-\beta)\mu_x - \mu], \tag{6}$$

with the material in tension.

The low-pressure region thus allowed for a constant ratio of volumetric recovery to volumetric compression after a complete cycle of loading and unloading. The amount of this ratio was defined by the value of β, with a $\beta = 1$ resulting in complete volumetric recovery. The parameter defined as:

$$C = (1-\beta) \tag{7}$$

was the compactibility of a material.

The high-pressure hydrostat was assumed to be independent of μ^*, which resulted in one relation describing both loading and unloading. This relation was

$$P = K_L\mu_3 + K_m(\mu - \mu_3) - (K_m - K_u)\mu_s\left[1 - \exp\frac{(\mu_3 - \mu)}{\mu_s}\right] \tag{8}$$

where K_m and μ_s are material parameters describing an exponential hydrostat above μ_3.

The elastic-plastic relations consisted of a shear-modulus relation, a yield-surface description, and a plastic-flow rule. The model for the shear modulus, G, was a hybrid constant G-constant ν relation. In the low pressure region, the shear modulus was determined from the relation:

$$G = \frac{dp}{d\mu}\left[\frac{3(1-2\nu)}{2(1+\nu)}\right] \tag{9}$$

where $dp/d\mu$ varied depending on whether the loading (3) or unloading (5, 6) relations were being used. Relation (9), therefore, resulted in a constant Poisson's ratio, but a variable shear modulus. In the high-pressure region the shear modulus was assumed to be a constant defined by:

$$G = K_u\left[\frac{3(1-2\nu_L)}{2(1+\nu_L)}\right], \tag{10}$$

where ν_L was the Poisson's ratio of the low-pressure region.

A modified Mohr-Coulomb and von-Mises yield surface (Fig. 3) that was independent of the third invariant of the deviator stresses was assumed for all materials. The yield relation, Y, was described by the relation:

$$Y = \begin{cases} T_0 - P \tan \phi & P < P_{YLD}, \\ Y_{MAX} & P \geq P_{YLD}, \end{cases} \tag{11}$$

with T_0 the cohesion, ϕ, the angle of internal friction, Y_{MAX}, the von-Mises yield strength, and P_{YLD} defined by:

$$T_0 + P_{YLD} \tan \phi = Y_{MAX}. \tag{12}$$

Material spall was assumed to occur when the value of the yield surface for a calculational zone reached zero. At the locations of material spall, all forces except gravity were assumed to be removed. Further, in order to achieve motions consistent with the observed terminal deformations in the crater and measured ground-motions, the yield surface had to include a loss of cohesion in each calculational zone during plastic flow in that zone. This loss of cohesion was accomplished by use of a parameter, S, evaluated for each zone, which-modified the expressions containing T_0 to:

$$T_0(1 - S), \tag{13}$$

with S initially zero. The value of S was incremented by 0.04 for the first twenty-five cycles of plastic flow for each calculational zone. The shift was accomplished in increments to avoid a drastic change in the yield surface description during one calculational cycle, which might result in calculational instability, and was always completed in less than 0.8 msec of simulated time.

Finally, a relation, called a flow rule, was required to describe the inelastic strain that occurred during flow with stress conditions limited by the yield surface. The associated flow rule was used in all calculations except for a numerical simulation to evaluate the influence of the flow rule on the numerical

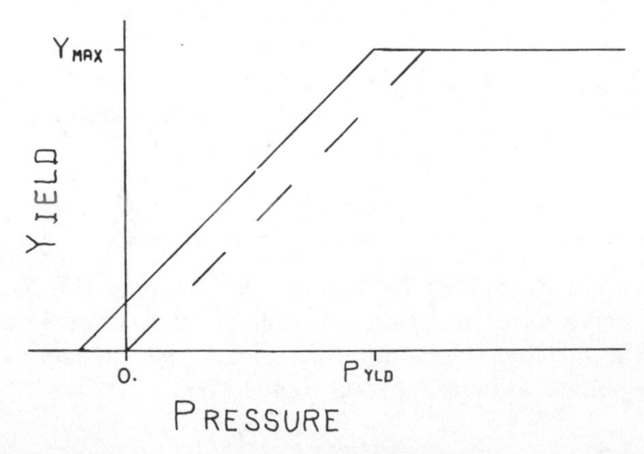

Fig. 3. Schematic yield surface for materials. Solid line indicates original yield condition as a function of pressure. Dashed line indicates the shifted yield condition. No quantitative relation is expressed.

results. This flow rule was derived with the Method of Plastic Potential (Trulio, *et al.*, 1969) and resulted in a plastic volumetric increase, called "bulking", when the yield surface was a function of pressure. When the yield surface was independent of pressure, this flow rule reduced to the Prandtl-Reuss flow rule. Also, the Prandtl-Reuss flow rule was used if: (1) the material was in tension, (2) the plastic volumetric strain had reached 0.1, or (3) the value of the yield surface was less than 0.5 T_0. The first of the conditions was caused by the description of soil response in tension; the second condition limited the amount of bulking; and the third condition was caused by a singularity in the expression for the flow rule when the third invariant of the deviator stresses was ignored and the value of the yield surface was near zero.

The material property relations just described require the specification of 12 parameters for each given material. The values for these parameters were developed through 21 attempts to accomplish a satisfactory simulation of the Mixed Company II experiment. These 21 attempts were divided into 12 attempts to obtain calculated motions consistent with the observed crater (Ullrich, 1976) and 9 more attempts to obtain calculated motions consistent with the measured ground-motions. The site model and material parameters used in the final numerical simulation, designated MC 4.10, are contained in Table 1.

The numerical simulation, MC 4.10, resulted in calculated flow-field conditions which were consistent with the formation of a central uplift by a simulated time of 16 msec (Figs. 4 and 5). The material within a range of 4 m and a depth of 8 m had achieved upward velocities with the maximum vertical velocities near the vertical axis. Also, all the material within that region had separated and was in ballistic motion. The numerical simulation was continued to 24 msec only to obtain comparisons with measured ground-motions.

After the simulated time of 24 msec, the numerical simulation of the motion in the crater region was extrapolated with ballistic relations. Each calculation

Table 1. Material parameters for Mixed Company II numerical simulation.

	Layer number				
Parameter	I	II	III	IV	V
Depth (m)	0–.55	.55–1.77	1.77–3.47	3.47–7.13	7.13–20
ρ_i(g/cc)	1.89	2.35	2.35	2.47	2.47
ν	.2	.2	.2	.2	.2
c_p(m/sec)	152	1676	1676	1981	3048
μ_3	.24	.051	.051	.023	.023
c_μ(m/sec)	381	1829	1829	2134	3200
c_v(m/sec)	152	1524	1524	1829	2896
β	.276	.992	.992	.994	.998
K_m(GPa)	68	68	68	68	68
μ_s	.40	.30	.30	.27	.27
T_0(MPa)	.069	.35	6.90	5.17	5.17
tan ϕ	.466	.7	1.0	.75	.75
Y_{max}(MPa)	51.7	759	759	207	207

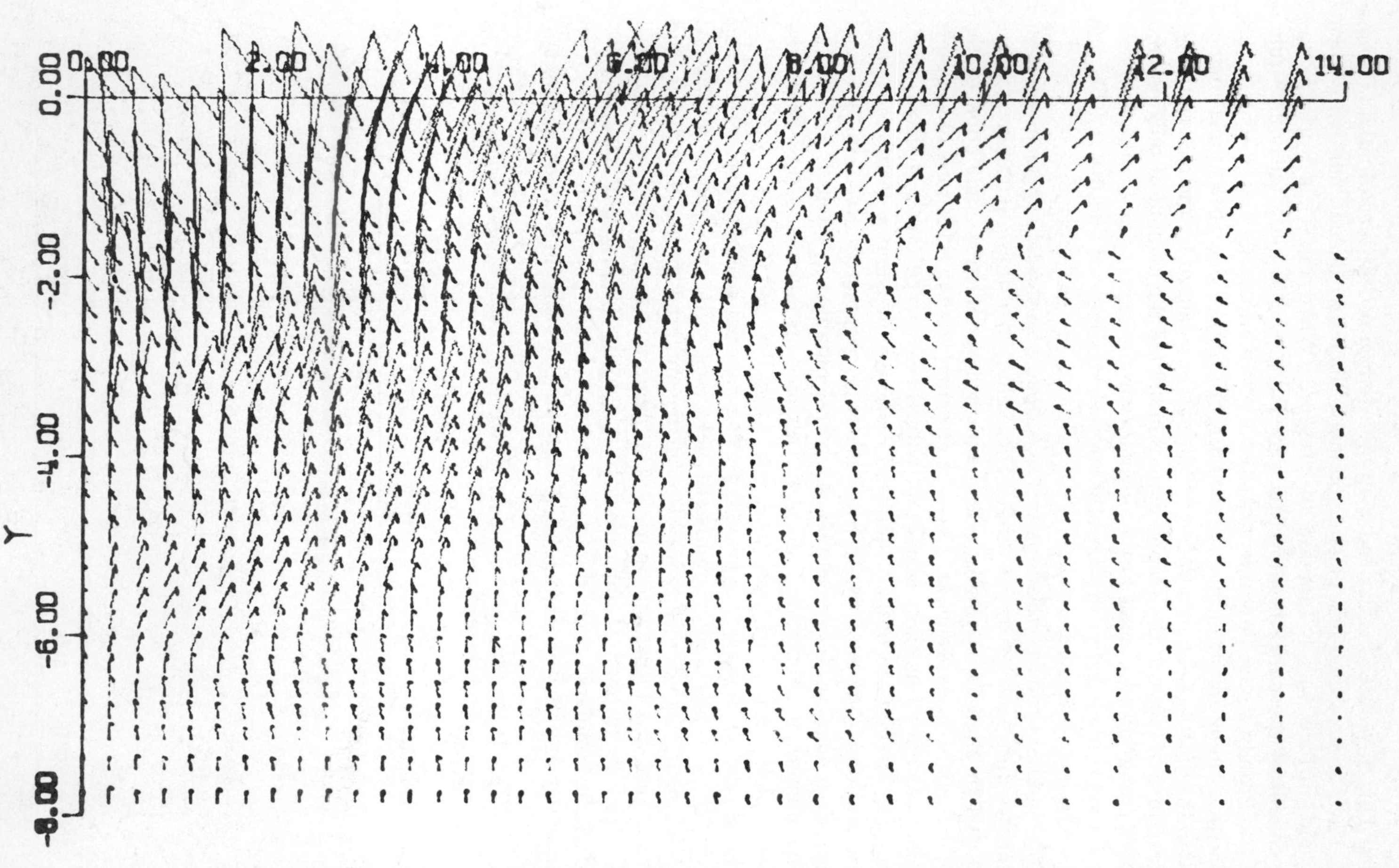

Fig. 4. Velocity vector plot for MC4.10 at 16 msec after detonation. Axes provide distance scale in meters. Velocity is proportional to the length of the vector with the distance between axis marks also representing a velocity scale of 20 m/sec.

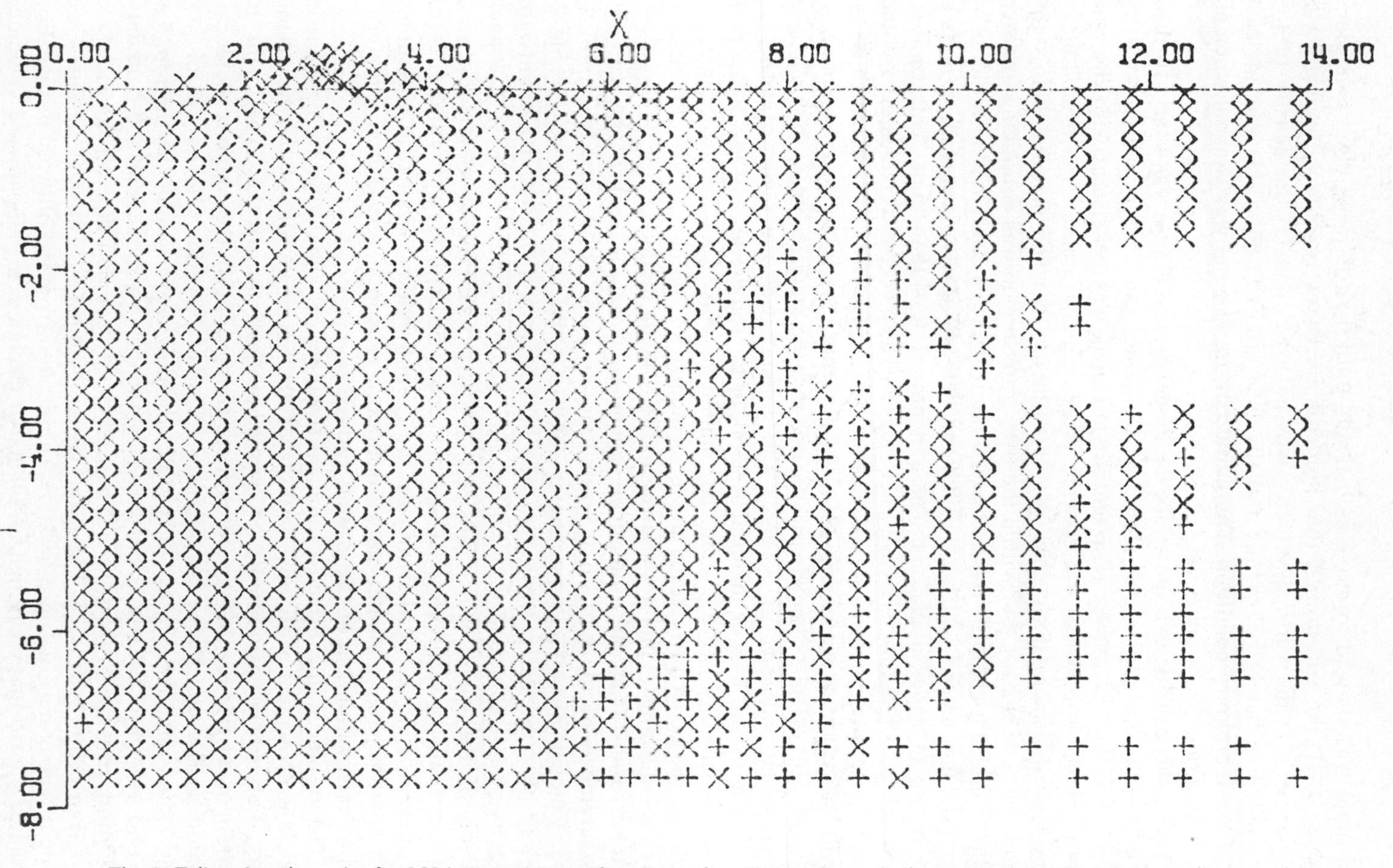

Fig. 5. Failure locations plot for MC4.10 at 16 msec after detonation. The marks, ×, indicate material spall and the marks, +, indicate plastic flow. Figure axes are in meters.

grid point was allowed to move ballistically until the following three conditions were met: (1) the grid point either immediately below or radially outward was not moving, (2) the vertical velocity was negative, and (3) the density of the material immediately below and radially outward (Ullrich, 1976) was at least 1.5 g/cc, an arbitrary condition. The motion of a grid point was stopped when all three conditions were once met.

Using this ballistic extrapolation, the model crater (Fig. 6) was nearly formed by 620 msec. The radius and slope of the model true crater wall, defined by motionless material between 3.6 and 7.6 m range, was consistent with the true crater profile. Within that crater wall, the calculated motion was beginning to bring fallback material to form the apparent crater. The material below the original ground surface and within the true crater region was continuing to move ballistically from an uplift region toward the true crater wall. Within a radius of 3 m, a definite mound had been calculated with the grid point on the axis, that had begun at a depth of 2.9 m, at a depth of less than .6 m. This upward displacement of 2.3 m is nearly 75% of the displacement indicated by the permanent-displacement marker described by Carnes. However, as indicated by the downward velocity vectors near 2.5 m depth, the uplift was beginning to collapse. This calculated collapse indicated that the material below the uplift had not obtained a density of 1.5 g/cc. The collapse was the result of either an insufficient inward flow of material at depth or the requirement of too high a bulk density to stop downward motion. The limited amount of inward flow of material could be the result of material model inadequacies, numerical inaccuracies, or gravitational adjustments not allowed in the ballistic model.

The results from the numerical simulation were also consistent with the measurements of the horizontal ground-motion pulse propagated from ground zero. Comparisons of five parameters of that pulse at three ranges (Table 2) showed that the early portion of the wave, from arrival to peak outward velocity, was characterized to within 10% of the measured values. The back portion of the wave, which included positive phase duration and outward displacement, was characterized less accurately but still always within 62% of the measured value. Further, during the ground-motion parametric variations, we found that, for calculated outward-motions to stop, a very cohesive material had to exist within 2 m of the ground surface.

The results of the Mixed Company II numerical simulation were, then, in sufficient agreement with observations of the crater produced by the 20-ton detonation and measurements of ground-motions during the experiment to warrant a description of the conditions that produced the velocity field calculated at 16 msec. These conditions were shown by the velocity vector plot (Fig. 7) at a time of 5.5 msec. In this plot the downward velocity in the top .5 m beyond a range of 10 m was produced by the airblast model surface boundary condition. However, this wave was not efficiently coupled through the interface at 0.55 m depth because of the large acoustic-impedance mismatch. The large velocities at a range of approximately 9 m resulted from the outward propagation of the compressional wave that originated at ground zero. The upward velocities

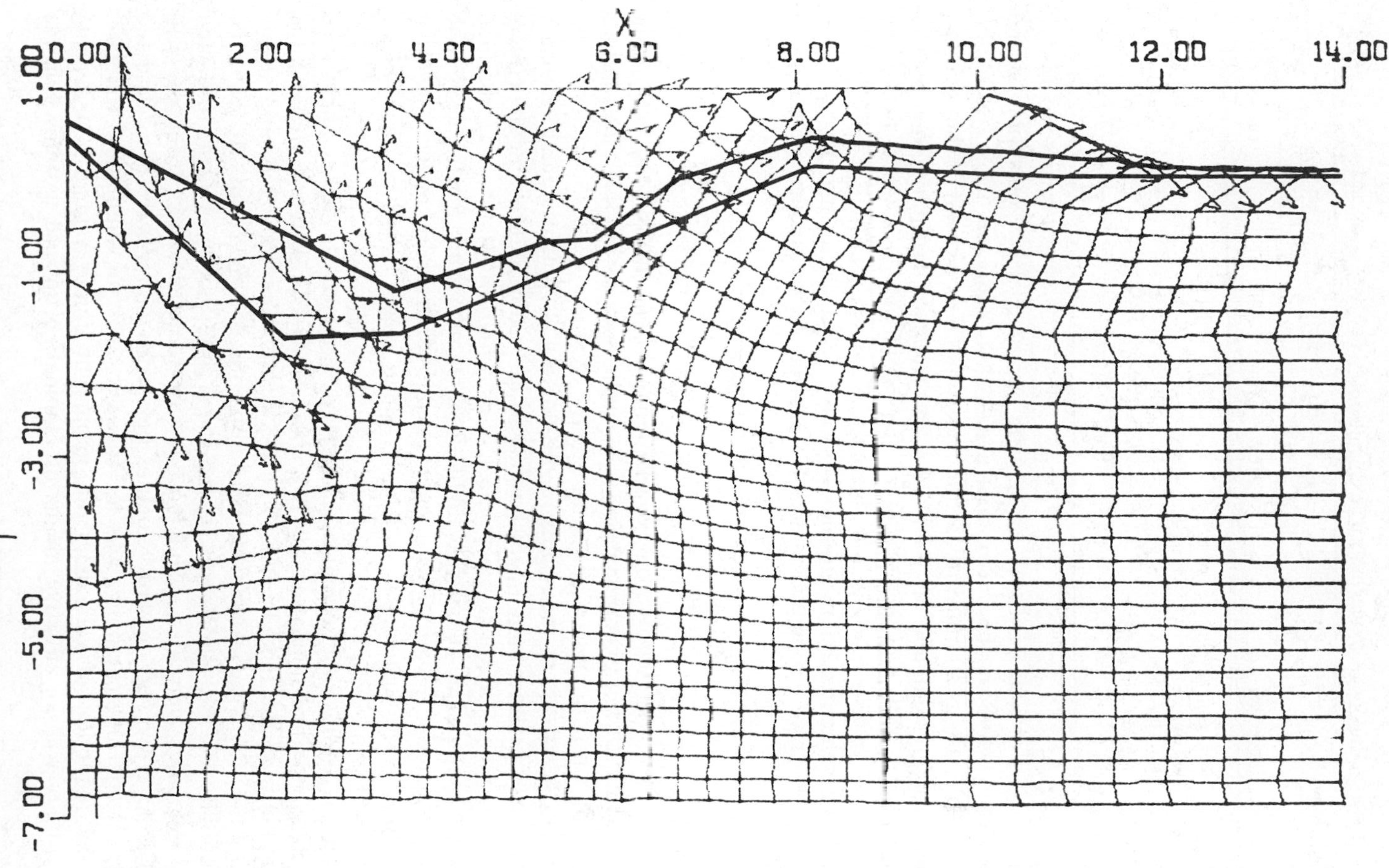

Fig. 6. Calculation grid plot for MC4.10 at 620 msec after detonation compared to the apparent and true crater profiles along the southern radial. Axes provide distance scale in meters. Velocity is proportional to the length of the vector with the distance between axis marks also representing a velocity scale of 10 m/sec.

Table 2. Horizontal ground motion at 1.5 m depth, simulated versus measured.

	16.5		21.3		28.4	
Range (m)	Sim.	Meas.	Sim.	Meas.	Sim.	Meas.
Time of arrival (m/sec)	8.5	8	11	11	14	14
Time of peak (m/sec)	10	10.5	13	12	16	16
Positive phase duration (m/sec)	11.5	8	13	8	10	11
Maximum velocity (m/sec)	3.5	3.2	2.3	2.1	.9	1.0
Displacement (cm)	1.3	1.2	1.1	0.8	.45	.58

at a range of 5 m and depth of 5.8 m were produced by a reflection of the compressional wave at the 7.1 m interface. At a radius of 6.1 m below the 1.8 m interface almost all outward motion had stopped as a result of the rebound from peak compression and the reflected stress waves. The rotational motion at a radius of 4.9 m indicated the location of the calculated shear wave. Behind all of these waves, further rebound and bulking was calculated which resulted in the large upward velocities below ground zero. By 5.5 msec, all material within a radius of 2 m from ground zero was calculated to be spalled and moving ballistically.

PARAMETRIC NUMERICAL SIMULATIONS

The results of the Mixed Company II numerical simulation, MC 4.10, were applicable only to *that* experiment and were influenced by complex interactions of rebound, strength, reflections, and shear. Evaluation of the relative importance of the effects to the results of the numerical simulation and extension of those results to other target media required information concerning the effects of material models on central uplift mechanics. This information was obtained through 13 additional numerical simulations summarized in Table 3. The influence of parametric variations in compaction, strength, and material layering were evaluated through model variations which were described in detail by Ullrich (1976).

Influences of material compaction were determined through comparisons among five numerical simulations. In two simulations, MC 4.10 and MCP-02, the compactibility of the models below the .55 m interface was zero, and strong upward velocities were calculated. In simulations MC 2.12 and MCP-09, the compactibility of the models below the .55 m interface was near 10%, and upward velocities were also calculated but with reduced values from MC 4.10 and MCP-02. However, upward velocities below the crater were not calculated when the model compactibility was increased to 30% for simulation MC 2.13. Thus, the reduction of rebound from maximum compression by increasing compactibility was an effective means of eliminating upward motions below a simulated crater.

Comparisons between MC 2.12 and MCP-09 showed that the shear-wave

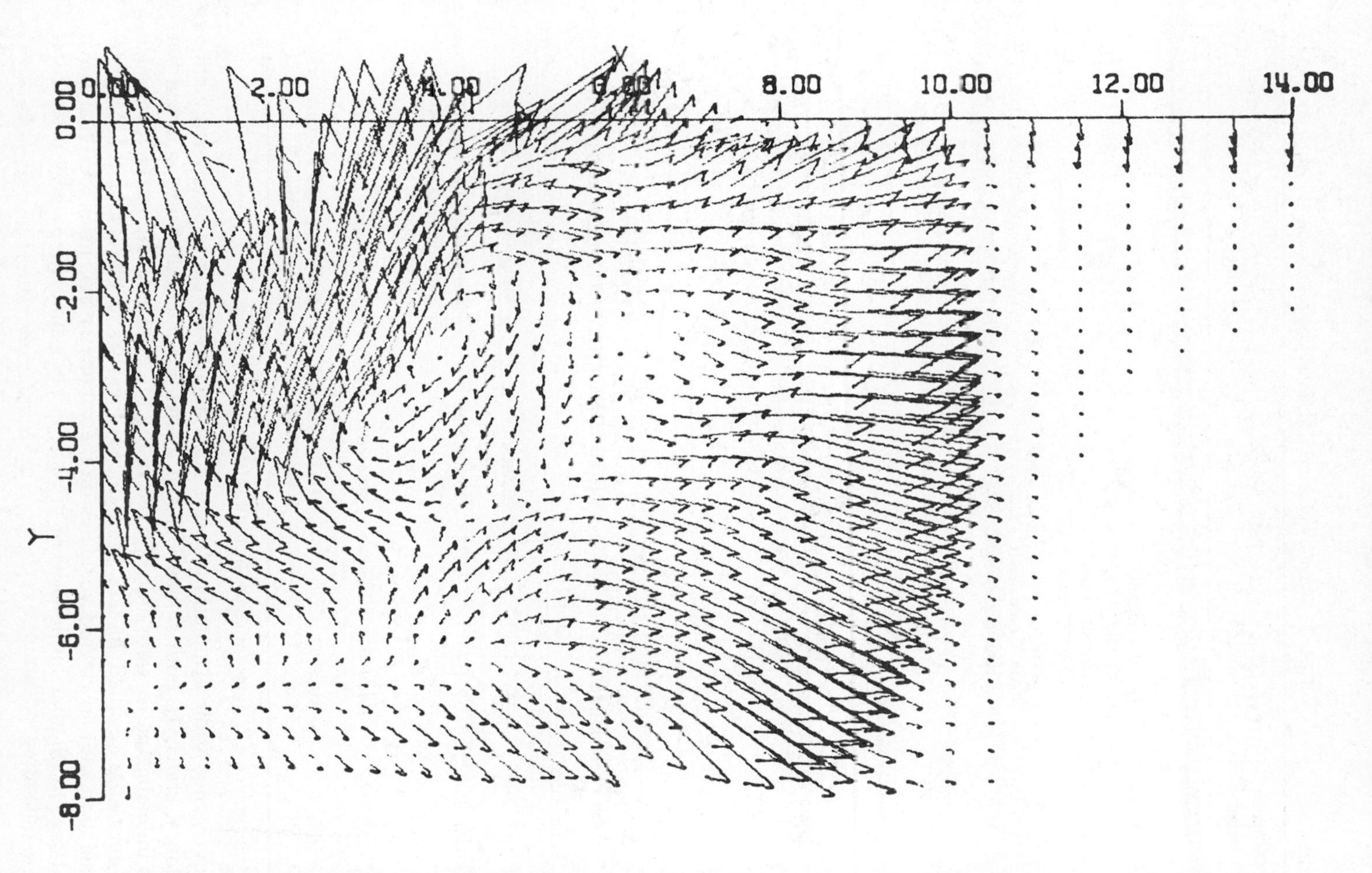

Fig. 7. Velocity vector plot for MC4.10 at 5.5 msec after detonation. Axes provide distance scale in meters. Velocity is proportional to the length of the vector with the distance between axis marks also representing a velocity scale of 20 m/sec.

Table 3. Summary of numerical simulation results.

	Model designation	Model baseline	Model change	Major result
	Baseline			
1.	MC2.12*	Mixed Company II Crater, material model with 10% compaction		Generally consistent with crater but quantitative difference with ground motion
2.	MC4.10*	Mixed Company II Crater and Ground Motions		Consistent with crater and motions
3.	MCP-02*	No compaction-two layers		Central mound
	Compaction			
4.	MC2.13*	MC2.12	30% compaction	No central mound
5.	MCP-09*	MC2.12	Different release path	Like MC2.12
		MC2.13	10% compaction	
	Layering			
6.	MCP-01	MCP-02	No soil	Little change in central mound
7.	MCP-03	MCP-02	4 layers	Little change
		MC2.12	No compaction	Increased velocities
	Strength			
8.	MC2.15	MC2.12	Cohesion retained	No central mound
9.	MCP-05*	MCP-02	No bulking	Completely different
10.	MCP-06*	MCP-02	No yield	Large transient velocity
11.	MCP-07	MCP-02	Altered weakening	No change
12.	MCP-08	MCP-02	Altered weakening	No change
13.	MCP-21*	MCP-02	21MPa yield	Deeper upward motions
14.	MC4.11*	MC2.13	"Fluid" layer	Probable splash

*Important implications for central-uplift formation.

convergence at the symmetry axis was not a dominating cause of upward motion below a crater. In both simulations the model compactibility was identical but the unloading paths were more curved in the MCP-09 calculation. This variance in curvature resulted in less pronounced shear-wave effects in the MCP-09 simulation with downward velocities on the symmetry axis after the shear wave had passed. Similar motions were present in both simulations only after complete unloading.

The influence of varying models of material layering was also evaluated by comparisons among five numerical simulations. Results from three (MCP-01, MCP-02, and MCP-O3) indicated that the inclusion of layers with property discontinuities similar to the discontinuities of MC 4.10 changed the calculated upward velocities on the symmetry axis at depths near 3 m less than a factor of two with only small changes in the entire motion field. Also, even in a numerical simulation of motions in a homogeneous, incompactible halfspace, MCP-01, upward motions were calculated for the region below the crater, indicating that stress-wave reflections were not necessary for central-uplift formation. Comparison between two numerical simulations, MC 2.12 and MC 2.13, in which the same layering and different compaction models were used showed that reflected waves would not overcome the effects of increasing compactibility. However, since no major strength increase or acoustic-impedance increase was ever modeled below .55 m depth, the possibility that an important material change at a specific depth could be influential in some instances was not eliminated.

The results from four numerical simulations indicated the importance of strength and plastic-flow models to the numerical simulation of the Mixed Company II experiment. The calculated motions in a simulation, MC 2.15, where the strength model retained cohesion, but not tensile strength, during plastic flow were inconsistent with the observed displacements in the crater. This result, combined with the requirement for a cohesive layer near 2 m depth in order for simulated and measured ground motions to be consistent, indicated that either materials below the ground zero of the Mixed Company II experiment were much weaker than where the ground motions were measured or that a loss of cohesion during plastic deformation was a necessary part of the material model. The results from a second numerical simulation, MCP-06, indicated that upward velocities greater than 40 m/sec were calculated at significant depth when a no-yield condition was applied. These high velocities were transient, however, and resulted in no permanent displacement of material. In calculation MCP-05, we examined the influence of eliminating the material "bulking" described in the associative flow-rule. The calculated motions in this simulation were much different from the Mixed Company II simulation and were inconsistent with the observed crater in that no prominent upward-velocities below the crater were calculated. Thus, the plastic volumetric-increase described in the associated-flow rule during Mohr-Coulomb plastic flow, was also an important part of the model used in the MC 4.10 simulation. The von-Mises strength was reduced to 21 MPa in a calculation, designated MCP-21, in order that the calculated peak pressures of 2000 MPa would be much greater than the von-Mises yield condition. In this simulation, upward motions were calculated first at a depth of 3.66 m on the axis of symmetry where the von-Mises limit was nearly 10% of the peak stress. Flow at shallower depths was more like a fluid with insufficient strength to overcome the initial outward-velocities.

Finally, one numerical simulation was accomplished to demonstrate the influence of a lower "fluid" layer on crater formation. This calculation, MC 4.11, simulated a 30% compactible material that was 4.88 m thick over a fluid-like halfspace. The fluid was modeled by an incompactible material with an initial density of 2.47 g/cc, a compressional wave speed of 1829 m/sec, a Poisson's ratio of .47, and a constant yield strength of .1 MPa. The calculated flow at 23.1 msec (Fig. 8) showed that a bowl-shaped depression was forming in the "fluid" that was being filled by downward moving material from the upper layer. All compression in the outward-moving fluid and overlying material had been relaxed, eliminating volumetric rebound from compression as a driving mechanism for upward motions. However, the motion in this "fluid" was similar to that calculated by Harlow and Shannon (1967) for the splash of a liquid drop into a deep pool, and to the motion calculated by Maxwell and Moises (1971) in the Sierra Madera simulation. Also, the geometry of the MC 4.11 simulation was similar to the test site for the Snowball and Prairie Flat experiments where significant central uplifts occurred. Thus, a long extension of the MC 4.11 simulation would be expected to develop large upward velocities near the axis of symmetry as the fluid recovered under gravitational flow. The amount of this recovery would be controlled by properties of the fluid, which is *speculated* to control also the development of a central uplift formed by the material in the upper layer as that material is redirected upward by the fluid.

DISCUSSION

Numerical modeling of physical events is subject to two basic causes of error. One of these causes is the numerical error caused by replacing the space and time continuums with a discrete grid on which calculations are accomplished at specific moments in time. This form affects the accuracy of the calculated numerical values and usually, but not always, can be reduced by decreasing the grid spacing and calculational time increments. We recognize that very small calculational zones can actually increase numerical error; however, the zoning of all the numerical simulations described in this paper was probably sufficiently large that such numerical problems were unimportant. Evaluation of numerical error in complex calculations can only be inexact; however, from previous investigations of numerical error in the AFTON-2A code (Cooper, 1971; Trulio *et al.*, 1967), such errors are not expected to affect seriously relations between calculated values. The second and more serious cause of error is the use of invalid mathematical models for the actual physical processes. This type of error can result not only in numerical values that are incorrect, but also in a calculated response that is completely invalid.

A necessary, but not sufficient, condition for a model to simulate the actual physical processes of a cratering event is that calculations reproduce the observed data—both dynamic measurements and observed final conditions for the physical event. The results of the Mixed Company II numerical simulation do not completely meet this condition but are sufficiently close to observed and measured data to support the contention that at least most of the phenomena important in the experiment were reproduced in the simulation. Thus, the results

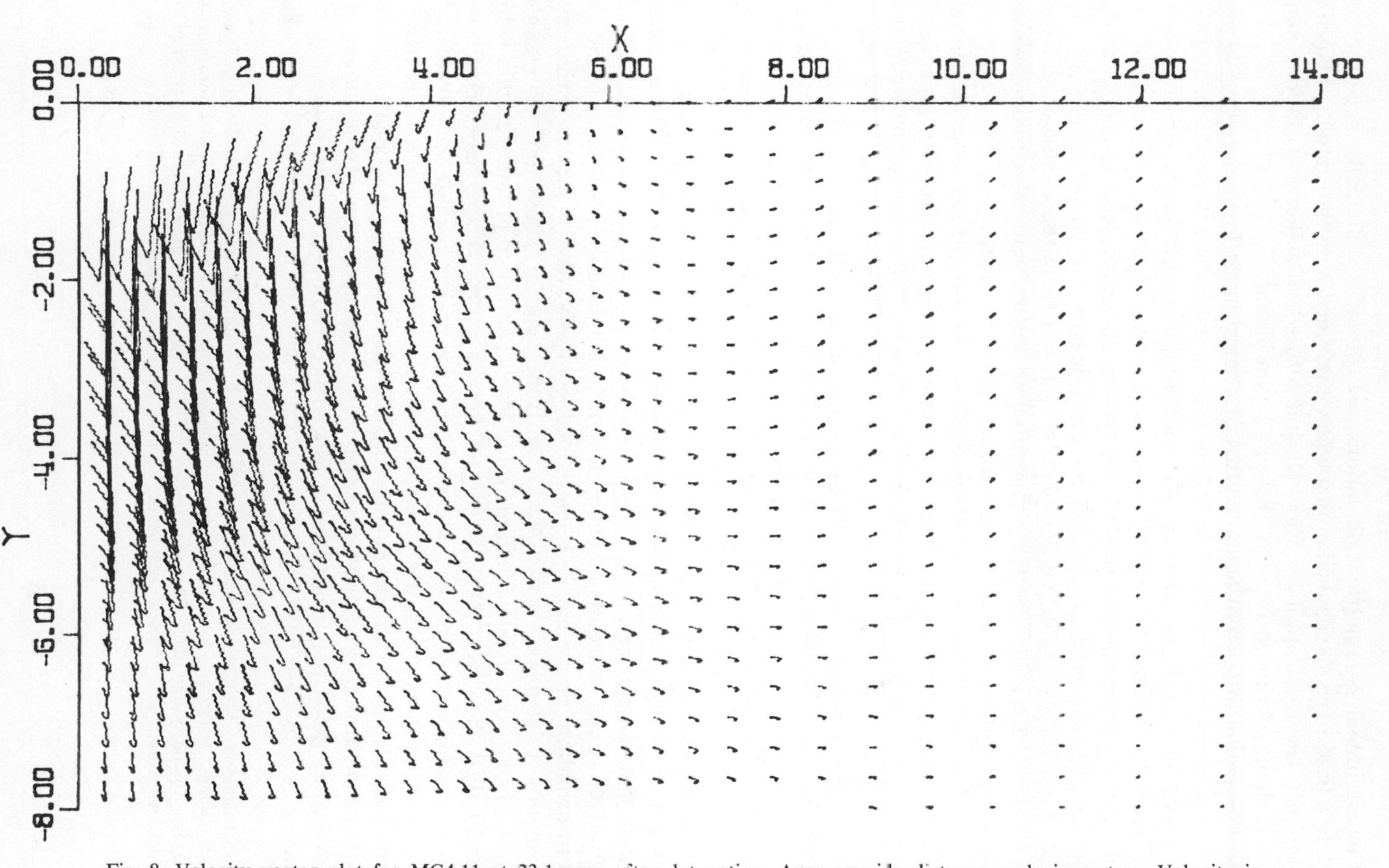

Fig. 8. Velocity vector plot for MC4.11 at 23.1 msec after detonation. Axes provide distance scale in meters. Velocity is proportional to the length of the vector with the distance between axis marks also representing a velocity scale of 30 m/sec.

of the additional parametric calculations are thought sufficiently accurate to provide a base for a mechanical model when combined with the stability analysis by Melosh, the calculation by Maxwell and Moises, and the observations and measurements of Prairie Flat and Snowball experiments.

In this model the two principal causes of central-uplift formation are: (1) the volumetric rebound of material following the maximum shock pressure, and (2) later gravitational collapse of crater walls where the apparent material strength is insufficient to support the crater walls. Each mechanism may be completely dominant in specific cases; however, the two mechanisms need not be mutually exclusive. The rebound mechanism is controlled by the compactibility of the target materials with sufficient compactibility resulting in the complete elimination of this mechanism. Further, this rebound must occur in a region where material strengths are sufficient to retard some of the outward displacements caused by the initial compressional wave. The formation of the rebound central uplift is enhanced by the bulking of material producing a rubbly, brecciated structure. The gravitational mechanism is controlled by a balance between the apparent maximum strength of the material below a dynamically-forming crater and the gravitational stresses acting in the crater-wall region. If sufficient material strength can be mobilized at precisely the time the crater walls require support, then the gravitational mechanism can be eliminated. This requirement for material strength below the crater allows the rebound and gravitational mechanisms to interact. For, at the time when the material strength is required to support the crater walls, the rebound mechanism *may* result in a reduced confinement or spall which may significantly reduce the effective material strength below the crater.

Finally, we recognize the mechanical model just described is only as broad as the information on which it is based. Specifically, the model is based on information related to craters produced from near or above-surface energy sources. These sources have demonstrated the capability to produce craters with central uplifts; however, the capability of deeper explosive sources or penetrating impact sources to produce craters with central uplifts has not been evaluated and the depth-of-event question remains open. Further, the effect of extreme material discontinuities at depth on central-uplift mechanics has also not been evaluated in this work. For these reasons, the extrapolation of the model of central-uplift mechanics described in this article should be made with caution and on a case-by-case basis.

Acknowledgments—Financial support during this study was obtained from U.S. Air Force Institute of Technology, the Defense Nuclear Agency, through Subtask SB144, and the National Aeronautics and Space Administration under Contract NGL-22-009-187. The computations were accomplished at the Air Force Cambridge Research Laboratory and the Air Force Weapons Laboratory.

References

Baldwin, R. B.: 1963, *The Measure of the Moon*, The University of Chicago Press.

Boon, J. D. and Albritton, C. C., Jr.: 1937, Meteorite Scars in Ancient Rocks, *Field and Lab*, **5**, 53–64.

Carr, M. H., Masursky, H., Baum, W. A., Blasius, K. R., Briggs, G. A., Cutts, J. A., Duxbury, T., Greeley, R., Guest, J. E., Smith, B. A., Soderblom, L. A., Veverka, J., and Wellman, J. B.: 1976, Preliminary Results from the Viking Orbiter Imaging Experiment, *Science* **73**, 766–776.

Carnes, B. L.: 1975, *Middle North Series, Mixed Company Event: Craters and Ejecta from Near-Surface Bursts on Layered Media*, POR-6612 (WT 6612), U.S. Army Engineer Waterways Experiment Station, Weapon Effects Laboratory, Vicksburg.

Cooper, H. F.: 1971, *On the Application of Finite Difference Methods to Study Wave Propagation in Geologic Materials*, Technical Report AFWL-TR-70-171, Air Force Weapons Laboratory, Kirtland AFB, New Mexico.

Dence, M. R.: 1968, Shock Zoning At Canadian Craters: Petrography and Structural Implications, *Shock Metamorphism of Natural Materials*, (B. M. French and N. M. Short, eds.). Mono Book Corp., Baltimore.

Dietz, R. S.: 1968, Shatter Cones in Cryptoexplosion Structures, *Shock Metamorphism of Natural Materials*, (B. M. French and N. M. Short, eds.), p. 267–284. Mono Book Corp., Baltimore.

Ehrgott, J. Q.: 1973, Preshot Material Property Investigation for the Mixed Company Site: Summary of Subsurface Exploration and Laboratory Test Results, *Proceedings of the Mixed Company/Middle Gust Results Meeting*. **2**, 491–539 General Electric Company—TEMPO (DASIAC), DNA 31S1P2.

Gault, D. E., Quaide, W. L., and Oberbeck, V. R.: 1968, Impact Cratering Mechanics and Structures, *Shock Metamorphism of Natural Materials*, (B. M. French and N. M. Short, eds.), p. 87–100. Mono Book Corp., Baltimore.

Harlow, F. H. and Shannon, J. P.: 1967, The Splash of a Liquid Drop, *J. Appl. Phys.*, **38**, 3855–3866.

Hartmann, W. K.: 1972, Interplanet Variations in Scale of Crater Morphology—Earth, Mars, Moon, *Icarus* **17**, 707–713.

Hartmann, W. K.: 1973, Martian Cratering, 4, Mariner 9 Initial Analysis of Cratering Chronology *J. Geophys. Res.*, **78**, 4096–4116.

Hoffman, H. V., Sauer, F. M., and Barclay, B.: 1971, *Operation Prairie Flat, Project Officers Report—Project LN-308: Strong Ground Shock Measurements*, POR-2108 (WT-2108), Stanford Research Institute, Menlo Park.

Howard, K. A., Offield, T. W., and Wilshire, H. G.: 1972, Structure of Sierra Madera Texas, as a Guide to Central Peaks of Lunar Craters, *Geol. Soc. Amer. Bull.*, **83**, 2795–2808.

Ingram, J. K.: 1975, *Middle North Series Mixed Company Event: Ground Shock from a 500-ton High-Explosive Detonation on Soil over Sandstone*, POR-6613, U.S. Army Engineer Waterways Experiment Station Weapons Effects Laboratory, Vicksburg.

Maxwell, D. and Moises, H.: 1971, *Hypervelocity Impact Cratering Calculations* PIFR-190, Physics International Company, San Leandro.

Melosh, H. J.: 1976a, The Mechanical Stability of Very Large Craters, (abstract). In *Lunar Science VII*, p. 549–551. The Lunar Science Institute, Houston.

Melosh, H. J.: 1976b, Crater Modification by Gravity: A Mechanical Analysis of Slumping (abstract). In *Papers Presented to the Symposium on Planetary Cratering Mechanics*, p. 76–78. The Lunar Science Institute, Houston.

Milton, D. J. and Roddy, D. J.: 1972, Displacements Within Impact Craters, *Proceedings, International Geological Congress, Twenty-Fourth Session*, Section 15, p. 119–124.

Milton, D. J., Barlow, B. C., Brett, R., Brown, A. R., Glikson, A. Y., Manwaring, E. A., Moss, F. J., Sedmik, E. C. E., Van Son, J., and Young, G. A.: 1972, Gosses Bluff Impact Structure, Australia, *Science*, **175**, 1199–1207.

Murray, B. C., Belton, M. J. S., Danielson, G. E., Davies, M. E., Gault, D., Hapke, B., O'Leary, B., Strom, R. G., Suomi, V., and Trask, N.: 1974, Mariner 10 Pictures of Mercury: First Results, *Science*, **184**, 459–461.

Niles, W. J., Germroth, J. J., and Schuster, S. H.: 1971, *Numerical Studies of AFTON-2A Code Development and Applications Vol. II*, AFWL-TR-70-22, Air Force Weapons Laboratory, Kirtland AFB, New Mexico.

Pike, R. J.: 1971, Genetic Implications of the Shapes of Martian and Lunar Craters, *Icarus* **15**, 384–395.

Roddy, D. J.: 1968, The Flynn Creek Crater, Tennessee, *Shock Metamorphism of Natural Materials*, (B. M. French and N. Short, eds.), Mono Book Corp., Baltimore.

Roddy, D. J.: 1973, Geologic Studies of the Middle Gust and Mixed Company Craters, *Proceedings of the Mixed Company/Middle Gust Results Meeting*, **2**, 79–128. General Electric Company, TEMPO (DASIAC).

Roddy, D. J.: 1976, Impact, Large-Scale High Explosive, and Nuclear Cratering Mechanics: Experimental Analogs for Central Uplifts and Multiring Impact Craters (abstract). In *Lunar Science VII*, p. 744–746. The Lunar Science Institute, Houston.

Shoemaker, E. M.: 1961, Interpretation of Lunar Craters, *Physics and Astronomy of the Moon*, (Z. Kopal, ed.), p. 283–359. Academic Press, New York.

Short, N. M.: 1965, A Comparison of Features Characteristic of Nuclear Explosion Craters and Astroblemes, *Annals of the New York Academy of Sciences*, **123**, 573–616.

Stearns, R. G., Wilson, C. W., Jr., Tiedemann, H. A., Wilcox, J. T., and Marsh, P. S.: 1968, The Wells Creek Structure, Tennessee *Shock Metamorphism of Natural Materials*, (B. M. French and N. Short, eds.), p. 323–338. Mono Book Corp., Baltimore.

Trulio, J. G.: 1966, *Theory and Structure of the AFTON Codes*, AFWL TR-66-19, Air Force Special Weapons Center, Kirtland AFB, New Mexico.

Trulio, J. G., Germroth, J. J., Niles, W. J., and Carr, W. E.: 1967, *Study of Numerical Solution Errors in One- and Two-Dimensional Finite Difference Calculations of Ground Motion*, AFWL-TR-67-27, Vol. I, Air Force Weapons Laboratory, Kirtland AFB, New Mexico.

Trulio, J. G., Carr, W. E., Germroth, J. J., and McKay, M. W.: 1969, Ground Motion Studies and AFTON Code Development, *Numerical Ground Motion Studies*, Vol. III, AFWL-TR-67-27, Air Force Weapons Laboratory, Kirtland AFB, New Mexico.

Trulio, J. G., Perl, N. K., and Balanis, G. N.: 1976, *The AFTON-2A Computer Code Revised User's Manual Parts 1 and 2*, AFWL-TR-75-111, Air Force Weapons Laboratory, Kirtland AFB, New Mexico.

Ullrich, G. W.: 1976, *The Mechanics of Central Peak Formation in Shock Wave Cratering Events*, AFWL-TR-75-88, Air Force Weapons Laboratory, Kirtland AFB, New Mexico.

Roddy, D. J., Pepin, R. O., and Merrill, R. B., editors.
(1977) *Impact and Explosion Cratering*, Pergamon Press (New York), p. 983–1002.
Printed in the United States of America

Review and comparison of hypervelocity impact and explosion cratering calculations

K. N. KREYENHAGEN and S. H. SCHUSTER

California Research and Technology, Inc., 6269 Variel Avenue, Suite 200,
Woodland Hills, California 91367

Abstract—Two-dimensional (axisymmetric) finite-difference numerical code calculations of several hypervelocity impacts, near-surface nuclear bursts, and near-surface high explosive cratering events are reviewed and compared. Impacts in the 20 km/sec range couple 50–80% of their energy to the target; surface explosions couple only 1–10%. Once this difference in coupling efficiency is taken into consideration, the subsequent motions or flow patterns in cratering from impacts and explosions are similar, with differences being due more to gravity and material property variations than to the nature of the sources. Calculations agree well with experimental observations of small and intermediate scale craters in various media, but have not been able to predict the wide, shallow shape of very large craters.

1. INTRODUCTION

CRATERING BY HYPERVELOCITY IMPACTS and by explosions was first modeled in two-dimensional (axisymmetric) numerical solutions in analyses of meteoroid impacts on space vehicles by Bjork (1959), of cratering from a megaton surface burst by Brode and Bjork (1960), and of the Arizona meteorite crater by Bjork (1961). In these finite difference particle-in-cell code solutions, materials were assumed to behave hydrodynamically, based on the expectation that the important stages of cratering occur under pressures which greatly exceed nominal material strengths. The same assumption was made in subsequent studies of hypervelocity impacts into metal targets by Bjork (1963), by Walsh and Tillotson (1963), by Bjork and Olshaker (1965), and by Bjork *et al.* (1967). Because of the hydrodynamic assumption, these numerical analyses were essentially confined to the early, high pressure stage of cratering.

Hydrodynamic-elastic-plastic constitutive relations formulated in a 2-D Lagrangian code by Wilkins (1969) provided the basis for extending numerical analyses to later stages of impact cratering in metals by Dienes and Walsh (1970), by Kreyenhagen *et al.* (1970), and by Rosenblatt (1970, 1971). Similar techniques have been applied to cratering by small high explosive charges in aluminum by Rosenblatt *et al.* (1970). Elastic-plastic constitutive relations have also been developed for relatively weak, brittle materials, and code analyses have been made of small-scale impacts into graphite (Kreyenhagen *et al.*, 1973), and of large high explosive surface events, including DISTANT PLAIN-6, a 100-ton sphere detonated on clay (Christianson, 1970), MIDDLE GUST-3, a 100-ton sphere detonated on layered clay-shale (Schuster, 1977), and MIXED COMPANY-3, a 500-ton sphere detonated on sandstone (Ialongo, 1973). Cratering calculations have also been made for a number of surface or near-surface nuclear bursts, as exemplified by JOHNIE BOY, a 500-ton shallow burst in alluvium (Maxwell *et al.*, 1973).

The present paper reviews and compares some results of these impact and near-surface explosion cratering calculations, particularly in the early-time stages emphasized by most of the analyses. Extension of an analysis of an impact into brittle graphite is described to illustrate some late-stage cratering processes. Comparisons of calculated craters with experimental observations are made for different types of cratering events.

2. STAGES IN CRATER FORMATION

It is convenient to divide cratering processes into two stages—a relatively short initial high pressure stage and a longer cratering flow stage. (In weak materials and in very large craters, there may be a third stage when relaxation processes slowly modify the final form of the crater. However, this stage has not been modeled in finite difference calculations.)

In the initial high pressure stage, hypervelocity impact or explosive energy deposition produces a strong shock which propagates outward from the source, creating an expanding region of extreme pressure. Where the high pressure region interacts with free surfaces (including the back surface of an impacting projectile), pressure relaxation occurs as rarefaction waves propagate into the region. Pressure in material behind the shock front thus rapidly equilibrates at a low level, and the high pressure phase ends when the shock pulse separates from the vicinity of the impact or energy deposition and propagates on into the target. The pulse propagating outward from the source (and upward, in the case of buried explosions), continues to interact with the free surface to produce upward acceleration. Beneath the source, the separated pulse does not further influence crater excavation in a direct way. It can, however, exert indirect effects by heating or otherwise altering the properties of the medium. In porous materials, the separated pulse can also directly affect crater volume and shape by crushing material, thereby allowing subsidence.

During the subsequent cratering flow stage, pressures are no longer extreme in the crater region, but material therein continues to flow because of the momentum acquired during the initial shocking and pressure relief processes. This cratering flow is responsible for excavation of the crater.

3. INITIAL HIGH PRESSURE STAGE

Pressure levels during the initial high pressure phase are determined by the nature of the impact or explosive source. Table 1 shows some representative values for initial pressures in impacts and explosions.

The energy in a hypervelocity impact is of course, initially kinetic, and it is directed into the target. As a consequence, a major fraction of this energy is transferred into the target as kinetic and internal energy at early times. Figure 1

Table 1. Peak initial pressures in impact and explosion cratering events.

Source	Peak pressure, Mb
Hypervelocity impacts:	
Iron–iron, 5.5 km/sec	1.8
Iron–iron, 20 km/sec	15
Iron–iron, 72 km/sec	143
Aluminum–aluminum, 5.5 km/sec	.7
Aluminum–aluminum, 20 km/sec	4.9
Aluminum–aluminum, 72 km/sec	47
Granite–granite, 10 km/sec	1.4
Granite–granite, 20 km/sec	4.5
Granite–granite, 30 km/sec	9.4
Nuclear surface bursts	10–1000
Chemical high explosions	0.2–0.4

shows a typical early flow field produced by an impact. Most of the flow is still downward, but pressure relaxation around the edges of the impact diverts an increasing part of the flow upward. A portion of the energy is thus removed from the target by early ejecta. At sufficiently high velocities (above ~9 km/sec for iron–iron impacts) some projectile and target material will also vaporize and blow off.

In nuclear surface explosions, energy is initially deposited in the ground to various depths by radiation, but reradiation and blow-off of vaporized material carries much of this energy out of the ground in a few microseconds. Figure 2 illustrates the early flow field from a calculation of a 1 Mt surface explosion.

In high explosive events, peak pressures are much lower than in most impacts or nuclear explosions. Most of the energy goes into expansion of the detonation products and into the air above the surface as an expanding blast wave, as seen in the early-time flow field calculated for MIXED COMPANY-3 in Fig. 3. Only part of the energy is incident on the ground beneath the explosion, and of that, only a small fraction becomes coupled to the ground for this burst geometry.

Calculations of chemical and nuclear explosions have generally assumed that energy above the surface does not significantly affect cratering. Hence, the continuing expansion of detonation products, fireball, and air blast are usually simulated by application of a boundary pressure condition which is independent of the motions of the target surface.

Table 2 compares the energy coupled to the ground from several calculations of representative cratering events at the end of the initial high pressure stage, i.e., at the approximate time when the shock pulse separates from the cratering flow. As is seen, within each type of cratering event, there can be substantial variations in the coupling efficiencies because of variations in source height, cratering medium, and impacting material. Nonetheless, impacts clearly tend to

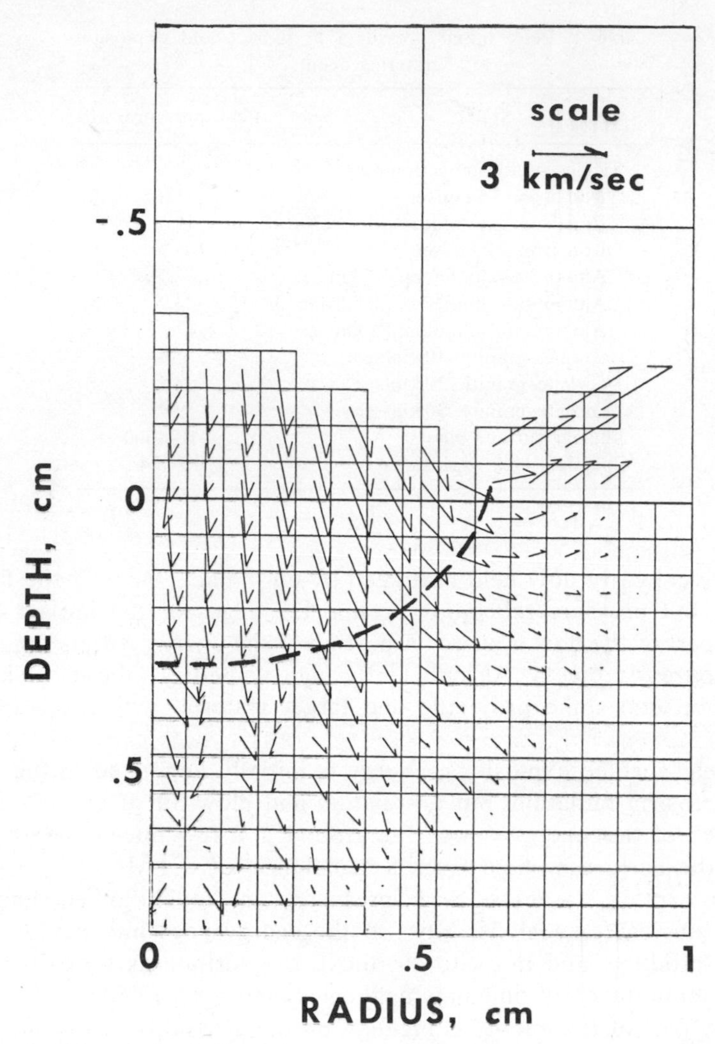

Fig. 1. Particle velocity field at 2.4 μsec after 3 km/sec impact of 1 cm diameter water drop on graphite surface. Dashed line is water-graphite interface. Left boundary is axis of symmetry (Kreyenhagen *et al.*, 1973).

be more efficient than explosions. Between different impacts, lower velocities (which give lower energy densities) and higher ratios of projectile density to target density favor more efficient coupling. Between different explosions, shallow burial, lower energy densities, and softer target media favor more efficient coupling.

Further partitioning of energy at the end of the initial high pressure stage is available from the calculations by Bjork *et al.* (1967) of several hypervelocity impact problems. For these cases, kinetic and internal energy were partitioned between the separated shock pulse, the cratering flow, and the ejecta regions, as

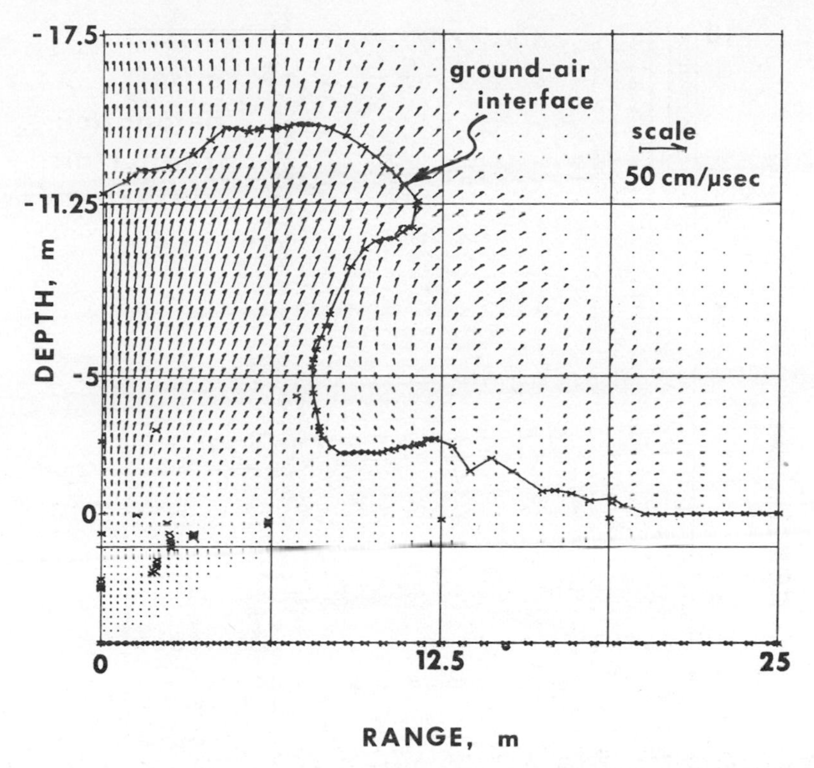

Fig. 2. Particle velocity field at 39 μsec after 1 Mt nuclear surface burst (Ialongo *et al.*, 1974).

defined in Fig. 4a. Figure 4b is a particle velocity field showing the actual partitioning in a typical case; the shock pulse is clearly separated from the crater flow by a quiescent region. Table 3 shows the fractions of the initial kinetic energy which are found as kinetic energy and internal energy in these regions after shock detachment.

The kinetic energy in the crater flow region is the primary driver of subsequent crater excavation. Higher impact velocities, and lower ratios of particle-to-target density reduce the fraction of the projectile's kinetic energy remaining in this flow after separation of the initial shock pulse.

4. Cratering Flow and Late-Stage Processes

The crater continues to form after shock pulse separation as the kinetics of the crater flow continue to distort and/or fracture additional material and to throw out additional ejecta. The processes which lead to the final crater are strongly influenced by low-stress material properties and, for large craters, by gravity. Because these processes take so long in weak materials, finite-difference cratering calculations have generally not been extended until crater formation is

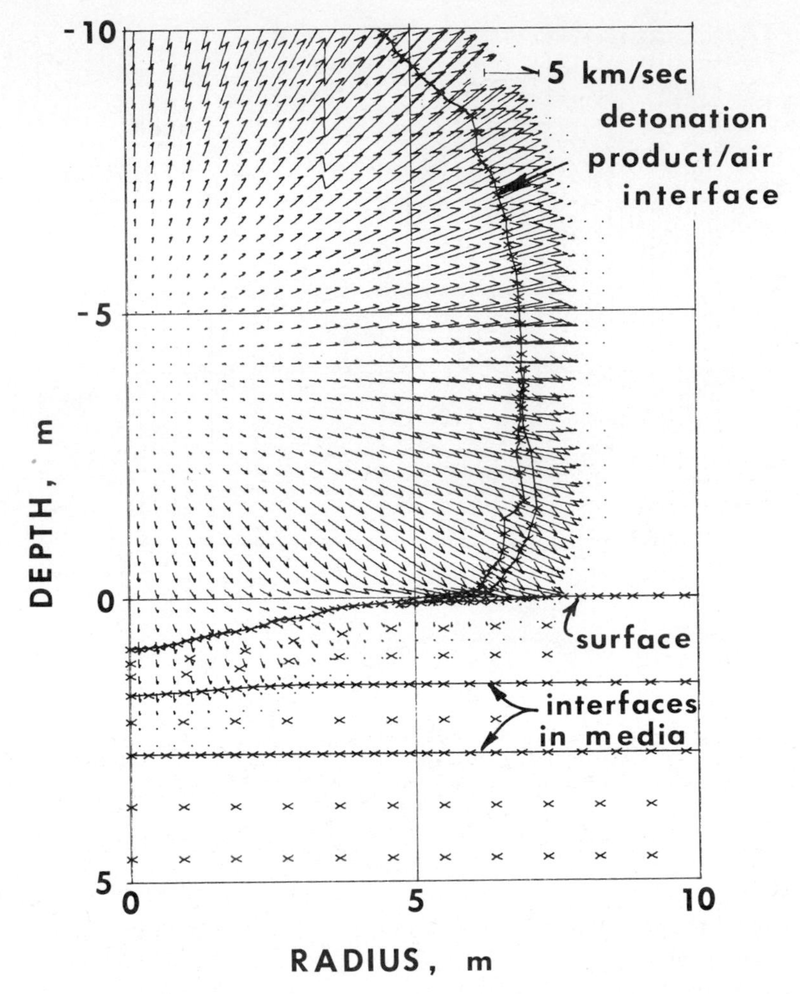

Fig. 3. Particle velocity field at 1.5 msec after detonation of MIXED COMPANY-3, a 500-ton high explosive sphere tangent above siltstone/sandstone (Ialongo, 1973).

complete. However, some calculations have gone far enough to include portions of the late-stage flow and fracture processes. One of these calculations will be used to illustrate the processes.

Figure 5 shows the particle velocity field 7.2 μsec after impact of a water droplet at 3 km/sec into graphite, which was modeled as a weak, brittle, porous material. The flow is initially radial, i.e., downward and outward, with an upward component near the free surface due to the pressure relief. At 41 μsec (Fig. 6) this flow pattern persists, and the developing crater remains essentially hemispherical.

The stresses beneath the crater in Fig. 6 are compressive. In the region around the sides of the crater, radial stresses are compressive. However, in the

Table 2. Comparison of percentages of initial source energy coupled to target after shock detachment.

Event—source	Energy below surface after shock detachment %		Reference
Nuclear			
1 Mt surface burst	1		Orphal (1975)
JOHNIE BOY—0.5 kt shallow (0.5 m) burst in alluvium	30		Maxwell *et al.* (1973)
High explosive			
MIXED COMPANY-III, 500-ton sphere (tangent above) surface burst on siltstone/sandstone	5.6		Ialongo (1973)
MINE ORE, 100-ton sphere (center 0.9 radii above surface) on strong granite	2.6		Maxwell and Moises (1971)
DISTANT PLAIN-6, 100-ton sphere (tangent above) surface burst on silty clay	7.3		Christianson *et al.* (1970)
Hypervelocity impact			
	20 km/sec impact, %	72 km/sec impact, %	
Iron into aluminum	83	65	
Aluminum into iron	61	63	
Porous aluminum ($\rho = 0.44$) into aluminum	56	48	Bjork *et al.* (1967)
Porous aluminum into iron	41	38	

region above the dashed line, axial and hoop stresses are tensile, due to upward and outward displacements in this region. These stresses result in rebound and/or material yielding or failure at late times. For this impact into graphite, compressed material beneath the crater rebounded upward, and material around the sides of the crater rebounded inward. Tensions beyond the immediate region of the crater also produced downward rebound near the free surface. Close to the crater, the tensile stresses exceeded a failure criterion. When this occurred in a computational cell, a crack was assumed to form and the stress tensor was appropriately adjusted in the continuing analysis. A subsequent flow field in Fig. 7 shows rebound occurring all around the crater. The discontinuity around the crater at about 0.5 cm depth is a horizontal crack propagating out into the target.

Figure 8 shows the final computational grid for the water impact into graphite. The calculation was stopped at this point because stresses everywhere had dropped below the material strength. The basic open crater is still hemispherical, but it is surrounded by a region of shattered material with sufficient upward and outward momentum to be ejected. One of the tensile cracks in the surrounding material connects to the surface, and an annular "spall" slab is thus separated from the target. In this small-scale impact, the annular slab will

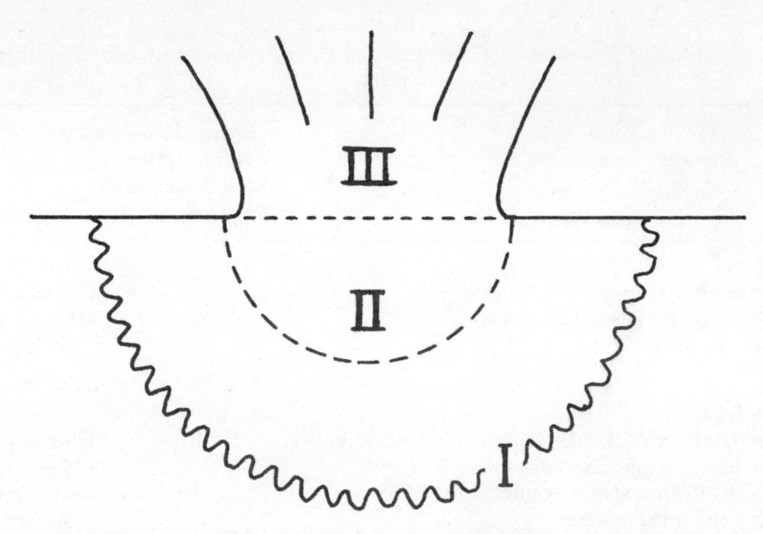

Fig. 4a. Definition of regions for energy partitioning after shock detachment. Region I: separated shock pulse, Region II: crater flow, Region III: ejecta above original surface.

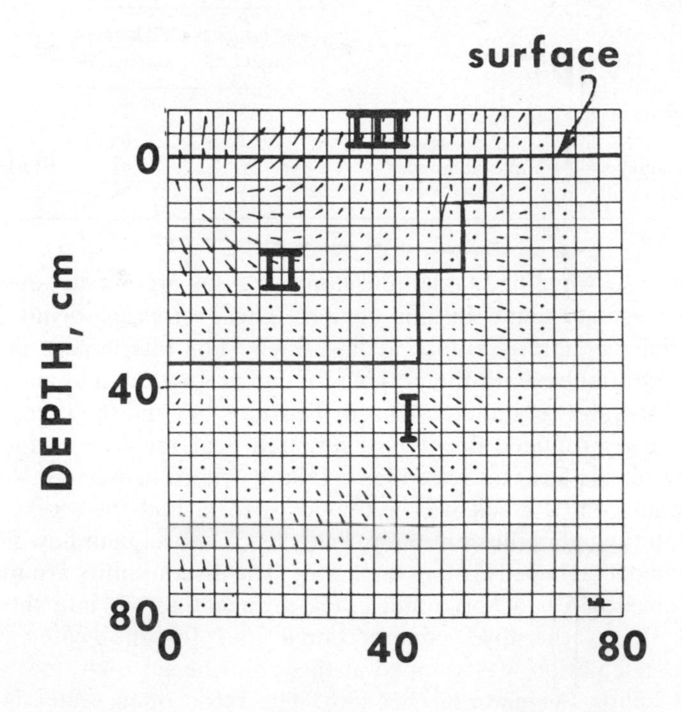

Fig. 4b. Illustration of regions in velocity field at 135 μsec after 10×10 cm aluminum cylinder impact on iron at 20 km/sec (Bjork *et al.*, 1967).

Table 3. Percentage of initial projectile kinetic energy in different regions after shock detachment.

	Region I, separated shock pulse		Region II, crater flow		Region III, early ejecta: kinetic and internal energy, %
	kinetic energy, %	internal energy, %	kinetic energy, %	internal energy, %	
Iron into aluminum					
20 km/sec	45	11	22	5	17
72 km/sec	28	12	17	8	35
Aluminum into iron					
20 km/sec	22	10	10	19	39
72 km/sec	12	7	5	39	37
Porous aluminum into aluminum					
20 km/sec	30	7	18	1	44
72 km/sec	23	6	18	1	52
Porous aluminum into iron					
20 km/sec	10	8	7	15	60
72 km/sec	5	7	5	21	62

continue to fly off. In a large impact where gravity is significant, it would probably fall back.

These same late-state flow processes, modified by gravity, media properties, geology, and surface overpressures, are seen in calculations of cratering from explosive events, including the analyses of small explosive charges in aluminum (Rosenblatt *et al.*, 1970), and the 20-ton surface explosion in MIXED COMPANY-2 (Ullrich, 1976). Figure 9 shows the velocity field from an analysis of DISTANT PLAIN-6 (Christianson, 1970), in which material beneath the crater rebounds. Surface material in this case did not rebound because it had no tensile strength to inhibit upward displacement.

5. Experimental Comparisons

The credibility of numerical calculations of cratering events is relatively good for some classes of events, and poor for others. A brief review of this situation is useful in establishing areas where understanding of cratering processes is inadequate.

Predictions of hypervelocity impacts into ductile metals give results which are in good agreement with experiments. Figure 10 shows an example of such a comparison between calculations by Rosenblatt (1971), and dynamic observations of crater growth by Prater (1970). Similarly, predictions of explosive cratering in ductile metals have been successful. Figure 11 compares the calculations by Rosenblatt *et al.* (1970) with experimental observations by Kratz and Hartenbaum (1970). These comparisons indicate that early-time processes

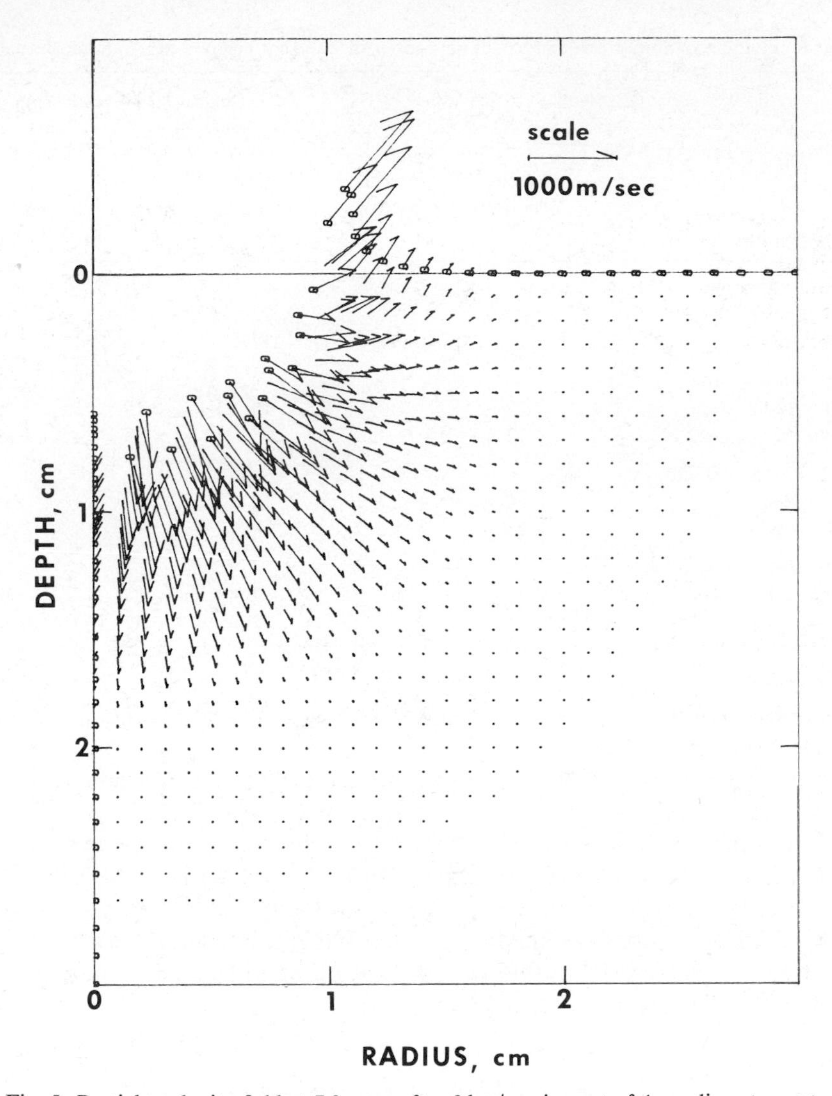

Fig. 5. Particle velocity field at 7.2 μsec after 3 km/sec impact of 1 cm diameter water drop on graphite (Kreyenhagen *et al.*, 1973).

for impacts and chemical explosions are either treated correctly, or that the final craters are insensitive to these conditions. The comparisons also indicate that the elastic-plastic models used for ductile materials like soft aluminum are adequate.

Recently, with the introduction of failure, cracking, and post-failure models, accurate predictions of impact cratering by small particle impacts into weak, brittle materials have been made. Figure 12 compares experimental and numerically calculated craters for a glass sphere impact into a composite graphitic material. The calculation used a material model which simulated the reinforcing

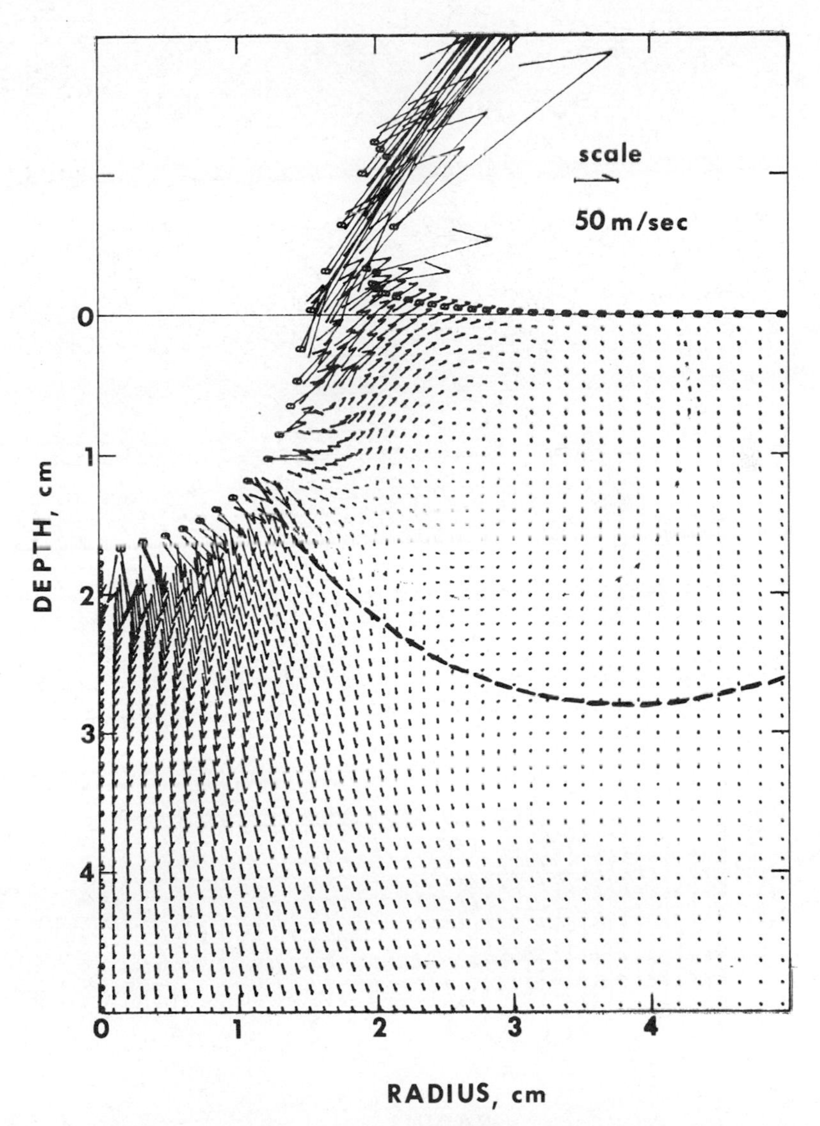

Fig. 6. Particle velocity field at 40.8 μsec after 3 km/sec impact of 1 cm diameter water drop on graphite. Material above dashed line is under axial and hoop tension.

structure of the composite by anisotropic sub-regions. The predicted crater profile conforms to the details of the observed profile.

In the foregoing comparisons, the craters were the result of small-scale impacts and explosions in media which were well characterized and controlled, and where precise measurements of crater dimensions were feasible. Predictions of 10–1000-ton events involving near-surface explosions on geologic media have been somewhat less precise than for small events, but they nonetheless indicate

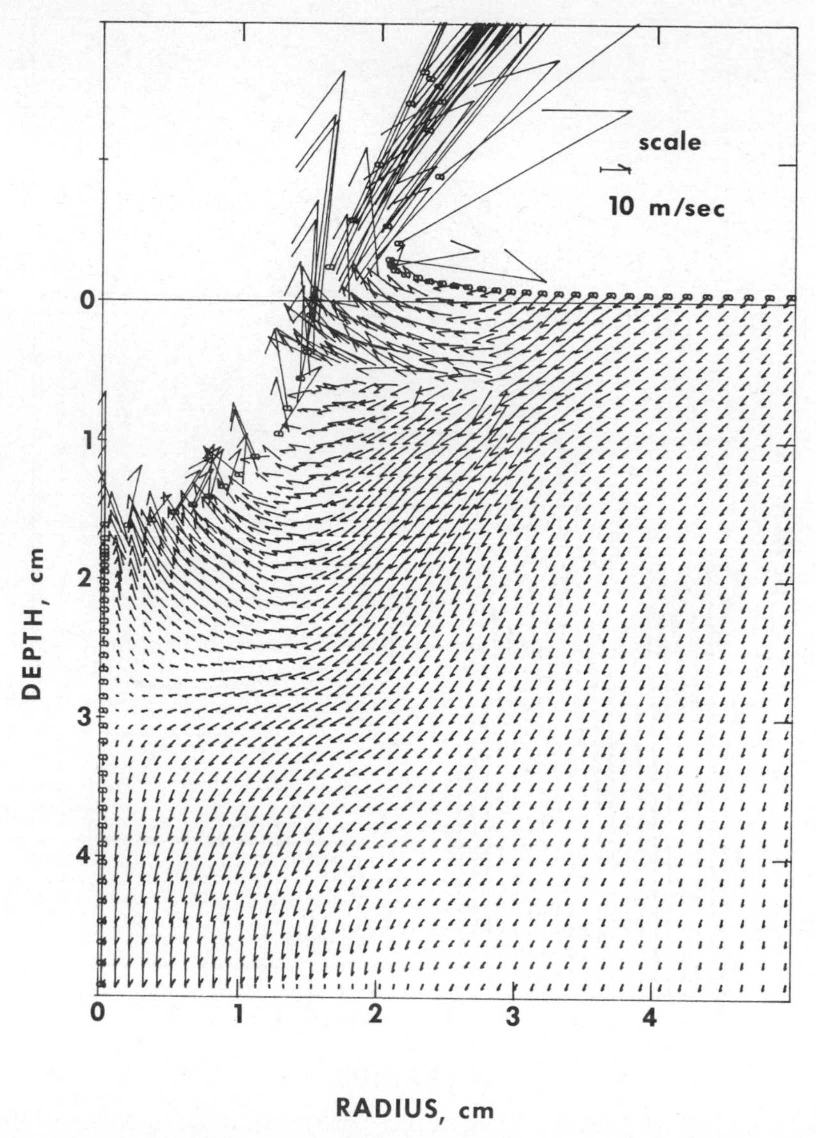

Fig. 7. Particle velocity field at 51.9 μsec after 3 km/sec impact of 1 cm diameter water drop on graphite.

adequate modeling of the basic processes and properties involved. Figure 13 shows pretest crater calculation profiles for two explosive events, the 500-ton MIXED COMPANY-3 burst (Ialongo, 1973) and the 100-ton MIDDLE GUST-3 burst (Schuster, 1977), and compares these with experimental observations of the craters. Figure 14 shows a comparison between a post-test calculation (Maxwell *et al.*, 1973) of JOHNIE BOY, a 500-ton shallow-buried nuclear burst,

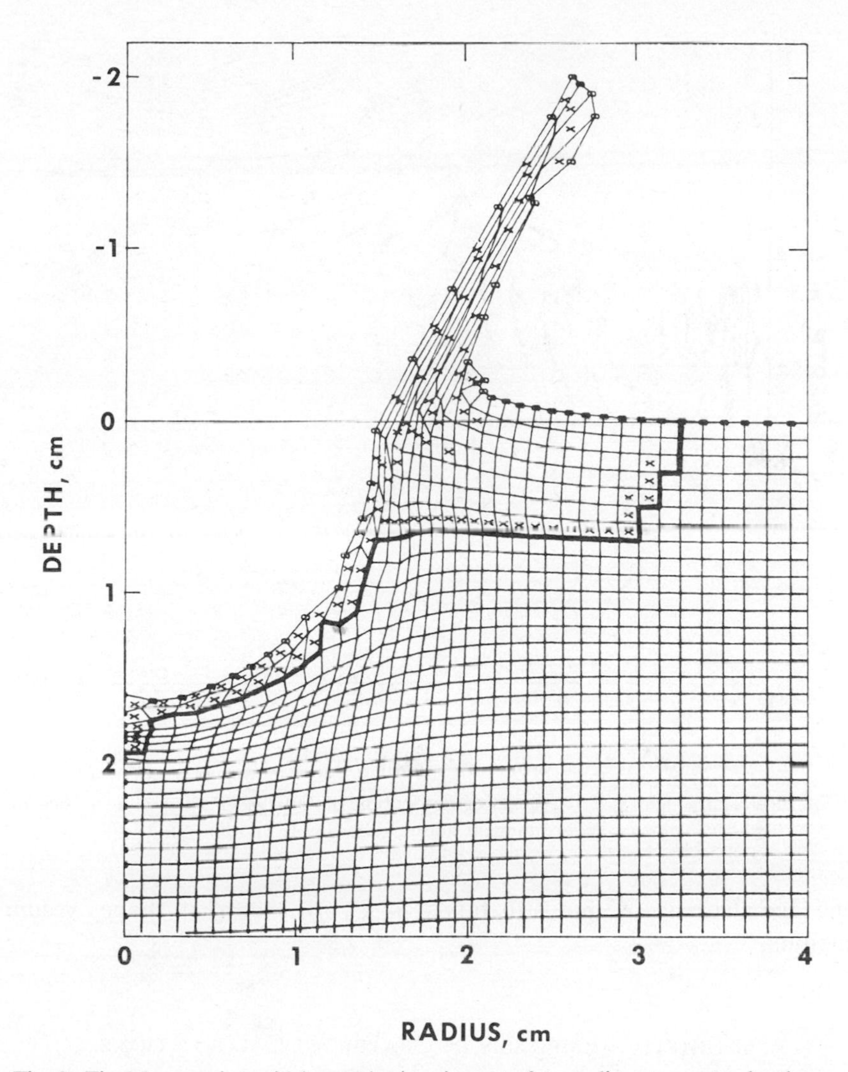

Fig. 8. Final Lagrangian grid from 3 km/sec impact of 1 cm diameter water droplet on graphite. Symbol "x" in cell denotes that failure has occurred. Heavy line delineates predicted crater profile.

with the measured crater. Although the dimensions are somewhat overpredicted, the crater profiles are in reasonable qualitative agreement.

Code analyses of large megaton-scale explosions have been both qualitatively and quantitatively poor, as seen in Fig. 15. The experimental profile is obtained by nominal cube-root scaling to 1 Mt of large nuclear explosions at the Pacific Proving Ground. The calculated profiles are scaled from Swift (1977). The larger calculated profile was obtained by substantially degrading material strength after the shock wave passage. This increased the predicted crater

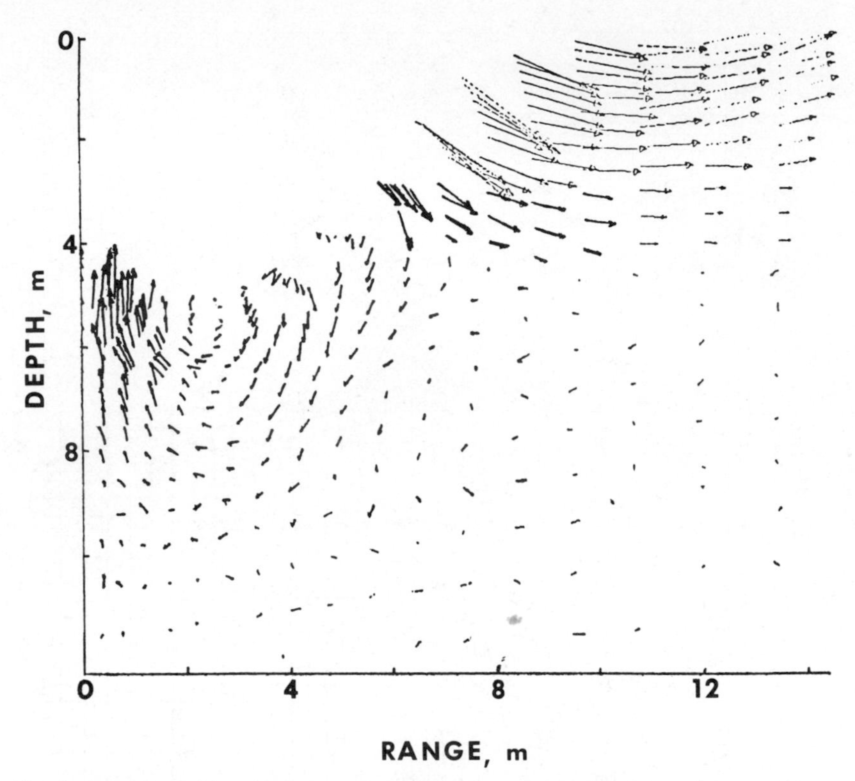

Fig. 9. Velocity field at 220 μsec after detonation of DISTANT PLAIN-6, a 100-ton high explosive sphere tangent above silty clay (Christianson, 1970).

volume and dimensions, but still failed to give the correct shape, volume, or dimensions.

6. Shifting Emphasis in Cratering Calculations

The current inability to calculate large craters is apparently due to mechanisms and/or material properties which become increasingly important during the late stages of larger-scale events. These may include:

(a) temporary weakening of a saturated medium due to dynamic pore pressure effects (liquefaction),
(b) slumping or collapse of crater walls which become unstable beyond certain heights (a function of gravity and of the properties of the medium),
(c) simultaneous slapping of the surface around the crater by the falling ejecta blanket and by fallback of annular slabs spalled off earlier in the crater event (e.g., the slab seen in Fig. 8), and
(d) converging rebound flow at late times.

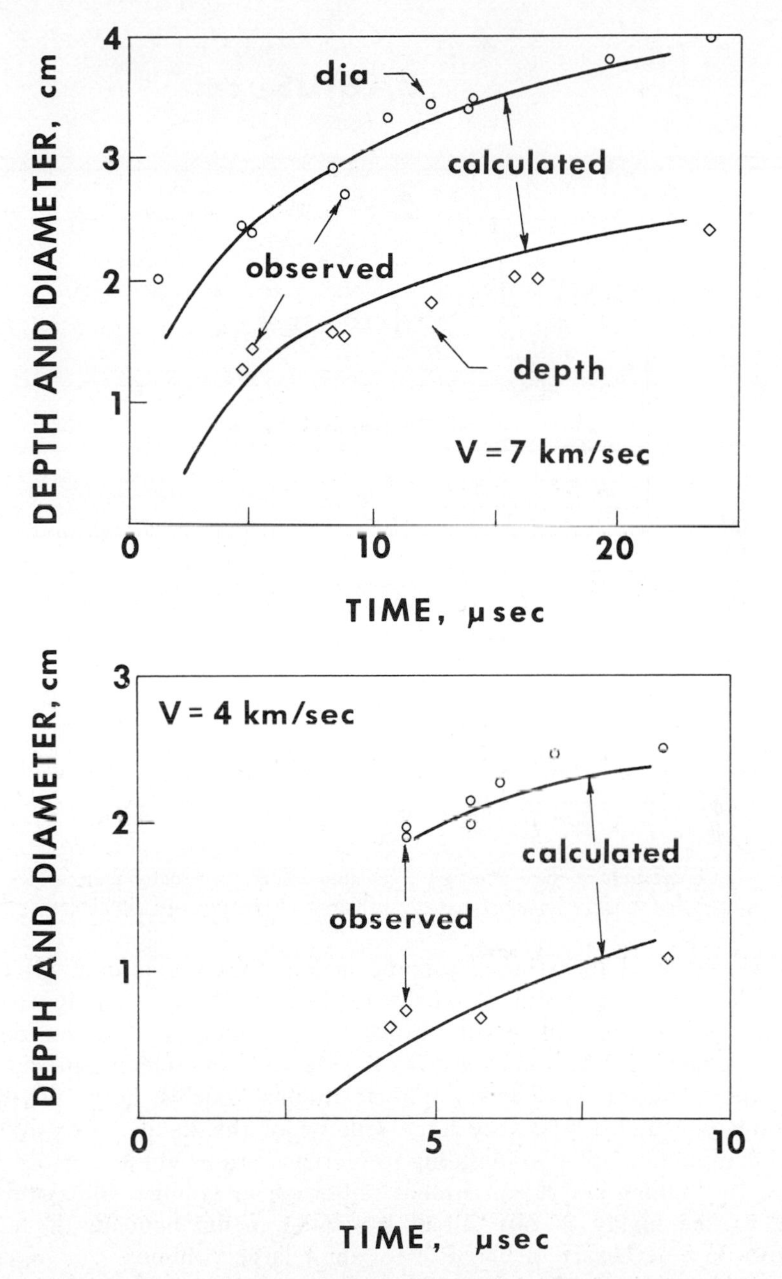

Fig. 10. Comparison of pre-test calculations of crater growth from aluminum sphere impacts into 1100 Aluminum (Rosenblatt, 1971) with flash radiograph observations (Prater, 1970). Results are normalized to 1 cm sphere impacts.

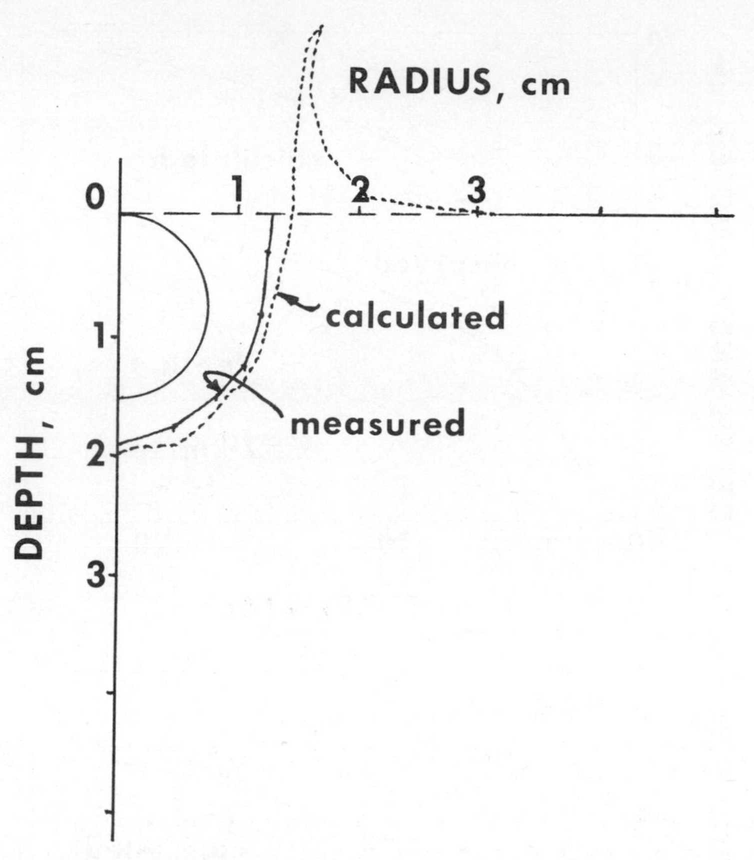

Fig. 11. Comparison of measured and calculated crater from detonation of 1.5 cm diameter sphere of explosive (tangent below surface) in soft aluminum.

A combination of these processes and interactions seems qualitatively consistent with most crater profiles (impact, nuclear, and high explosive) observed in various media, and with major trends in the changes of crater quantities (dimensions, volumes) with impact or explosive yield parameters, media properties, layering, and gravity. There are some inconsistencies, however, the most bothersome of which is the very large volume of the Pacific Proving Ground craters. Slumping of walls is not going to increase crater volume. More likely it will result in bulking and a consequent reduction in volume. Substantial compression of the highly porous, albeit saturated media beneath these craters appears to be a necessary mechanism for their large volumes.

The candidate mechanisms listed above are coupled and sometimes competing, and they are further influenced by layering. Ballistic, and incompressible processes can of course occur over a wide region during the late-stage cratering, but other regions affecting the final crater experience compressible processes, fracture, hysteretic or dilatant behavior, and dynamic changes in properties. These can best be treated by improved numerical models.

Fig. 12. Photomicrograph of crater from 4 km/sec glass sphere impact into carbon–carbon composite material. Solid line is profile as numerically calculated by author and associates.

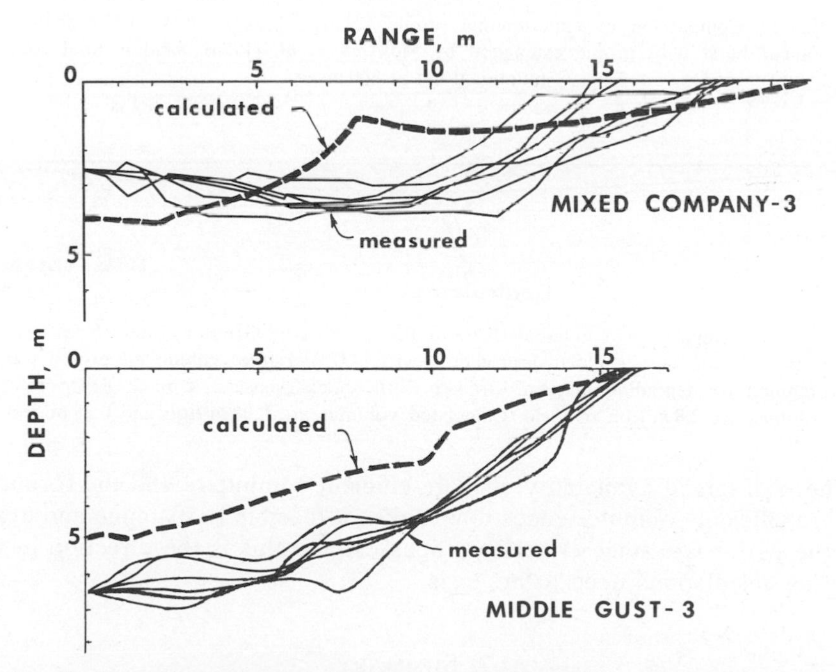

Fig. 13. Comparison of experimental and pre-test calculations of crater profiles for MIXED COMPANY-3 (Ialongo, 1973) and MIDDLE GUST-3 (Schuster, 1977). Experimental profiles were measured along several radials.

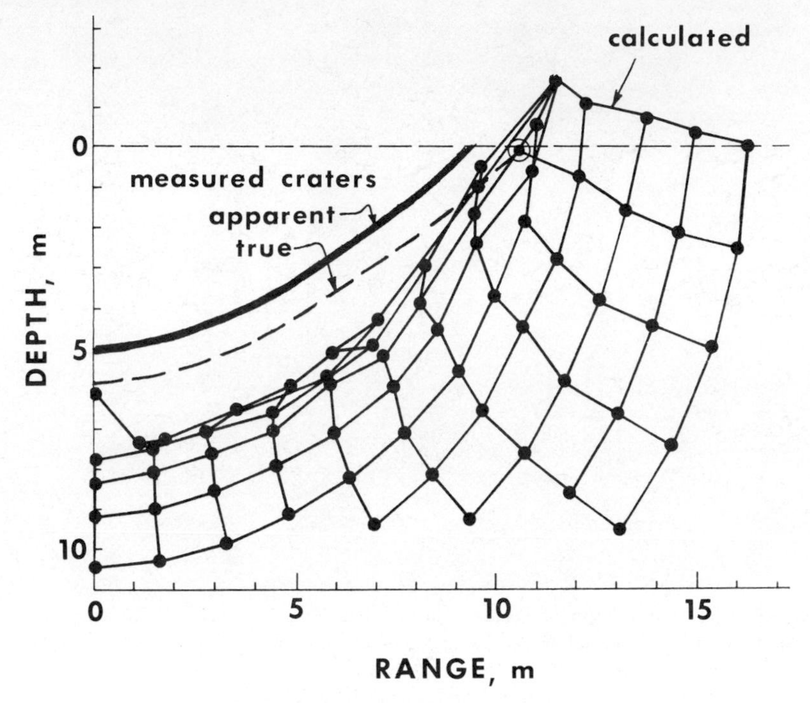

Fig. 14. Comparison of experimental profiles from JOHNIE BOY 0.5 kt shallow nuclear burst with profile calculated by Maxwell *et al.* (1973). Grid is final computational net at 800 msec.

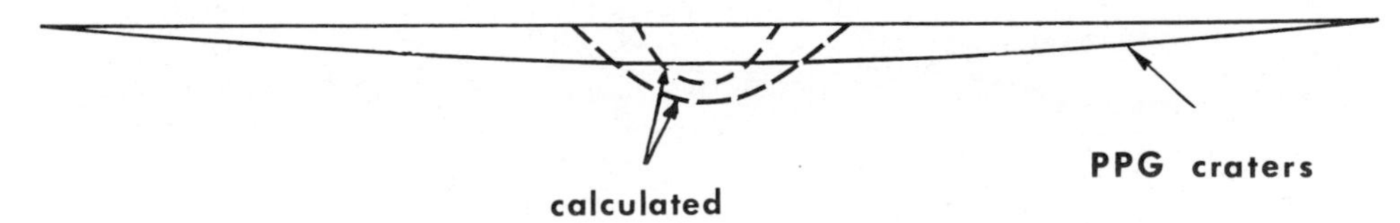

Fig. 15. Comparison of nominal shape of Pacific Proving Ground craters from large nuclear bursts with calculated profiles (Swift, 1977). Larger calculated profile was obtained by degrading material strength with shock passage. Typical PPG crater volumes are 2.8 m^3/ton of yield. Calculated volumes are 0.08 m^3/ton and 0.28 m^3/ton.

The widespread availability of more efficient computers and the formulation of more efficient computer codes now makes it feasible to examine and evaluate hypotheses for late-stage cratering processes, and this is the direction in which cratering calculations need to progress.

7. SUMMARY

Calculations show substantial differences between energy coupling and mass flow during the initial high pressure stage of cratering by hypervelocity impacts, nuclear explosions, and chemical explosions. Impacts in the 20 km/sec range couple 50–80% of their energy to the target, while surface explosions couple only 1–10%. Once the initial coupling efficiencies are taken into consideration, the subsequent flow patterns in cratering from impact and explosion craters are similar, with differences being due more to gravity and material property variations than to the nature of the sources. Final craters are largely determined by that fraction of the initial energy which is converted to kinetic energy of the crater flow, and by material properties which establish accommodation to the stresses induced by that flow.

Calculations of small impact and explosion craters in ductile and brittle materials correlate well with experiments. Calculations of cratering by intermediate-scale explosions (10–1000 tons) also agree fairly well with experiments. However, calculations of craters from very large explosive events (megatons) fail to predict the broad, shallow nature of the Pacific nuclear craters, apparently because very late relaxation processes are not modeled or considered in the calculations.

Acknowledgment—This work was partially supported by the Defense Nuclear Agency.

References

Bjork, R. L.: 1959, Effects of a Meteoroid Impact on Steel and Aluminum in Space, *Proc. Xth Intl. Astron. Congress, London*, Springer-Verlag, 505–514.

Bjork, R. L.: 1961, Analysis of the Formation of Meteor Crater, Arizona: A Preliminary Report, *J. Geophys. Res.* **66**, 3379–3388.

Bjork, R. L.: 1963, *Review of Physical Processes in Hypervelocity Impact and Penetration*, RM-3529-PR, Rand Corporation, Santa Monica, 51 pp.

Bjork, R. L., and Olshaker, A. E.: 1965, *The Role of Melting and Vaporization in Hypervelocity Impact*, RM-3490-PR, Rand Corporation, Santa Monica, 27 pp.

Bjork, R. L., Kreyenhagen, K. N., and Wagner, M. H.: 1967, *Analytical Study of Impact Effects as Applied to the Meteoroid Hazard*, NASA CR-757, 186 pp.

Brode, H. L., and Bjork, R. L.: 1960, *Cratering from a Megaton Surface Burst*, RM-2600, Rand Corporation, Santa Monica.

Christianson, D. M.: 1970, *ELK 40 Prediction Calculation of Ground Motion for DISTANT PLAIN Event 6*, DASA-2471, Defense Nuclear Agency, Washington, 109 pp.

Dienes, J. K., and Walsh, J. M.: 1970, Theory of Impact: Some General Principals and the Method of Eularian Codes, *High Velocity Impact Phenomena*, (R. Kinslow, ed.), Academic Press, New York, p. 45–104.

Ialongo, G.: 1973, *Prediction Calculations for the MIXED COMPANY-III Event*, DNA 3206T, Defense Nuclear Agency, Washington, 158 pp.

Ialongo, G., McDonald, J. W., and Reid, J. B.: 1974, Personal Communication.

Kratz, H. R., and Hartenbaum, B.: 1970, *Measurement of Surface Waves Produced by Small Explosions in Aluminum*, DASA 2594, Defense Nuclear Agency, Washington.

Kreyenhagen, K. N., Wagner, M. H., Piechocki, J. J., and Bjork, R. L.: 1970, Ballistic Limit Determination in Impacts on Multi-Material Laminated Targets, *J. AIAA* **8**, 2147–2157.

Kreyenhagen, K. N., Wagner, M. H., and Goerke, W. S.: 1973, *Direct Impact Effects in Hypersonic Erosion*, SAMSO-TR-73-339, Air Force Space & Missile Systems Organization, Los Angeles, 107 pp.

Maxwell, D., Reaugh, J., and Gerber, B.: 1973, *JOHNIE BOY Crater Calculations*, DNA 3048F, Defense Nuclear Agency, Washington, 58 pp.

Orphal, D. L.: 1975, Personal Communication.

Prater, R. F.: 1970, *Hypervelocity Impact—Material Strength Effects on Crater Formation and Shock Propagation in Three Aluminum Alloys*, AFML-TR-70-295, Air Force Materials Laboratory, Wright-Patterson AFB, 290 PP.

Rosenblatt, M.: 1970, *Analytical Study of Strain Rate Effects in Hypervelocity Impacts*, NASA CR-61323, 194 pp.

Rosenblatt, J., Piechocki, J. J., and Kreyenhagen, K. N.: 1970, *Cratering and Surface Waves Caused by Detonation of a Small Explosive Charge in Aluminum*, DASA 2495, Defense Nuclear Agency, Washington, 76 pp.

Rosenblatt, M.: 1971, *Numerical Calculations of Hypervelocity Impact Crater Formation in Hard and Soft Aluminum Alloys*, AFML-TR-70-254, Air Force Materials Laboratory, Wright-Patterson AFB, 68 pp.

Schuster, S.: 1977, *Results of Pre-Test Prediction Calculations of MIDDLE GUST I, II, and III*, AFWL-TR-76-284, Air Force Weapons Laboratory, Kirtland AFB, 109 pp.

Swift, R. P.: 1977, Material Strength Degradation Effects on Cratering Dynamics, In *Impact and Explosion Cratering*, D. J. Roddy, R. O. Pepin, and R. B. Merrill (eds.), Pergamon Press, New York. This volume.

Ullrich, G. W.: 1976, *The Mechanics of Central Peak Formation in Shock Wave Cratering Events*, AFWL-TR-75-88, Air Force Weapons Laboratory, Kirtland AFB, 138 pp.

Walsh, J. M. and Tillotson, J. H.: 1963, Hydrodynamics of Hypervelocity Impact, *Proc. Sixth Symp. Hypervelocity Impact* **22**, p. 59–73.

Wilkins, M. L.: 1969, *Calculation of Elastic-Plastic Flow*, UCRL-7322 Rev. 1, Lawrence Livermore Laboratory, Livermore, 101 pp.

Roddy, D. J., Pepin, R. O., and Merrill, R. B., editors.
(1977) *Impact and Explosion Cratering*, Pergamon Press (New York), p. 1003–1008.
Printed in the United States of America

Simple Z model of cratering, ejection, and the overturned flap

DONALD E. MAXWELL

Science Applications, Incorporated, 8201 Capwell Drive, Oakland, California 94621

Abstract—Previous studies of explosive cratering have led to models based on simple approximations to describe the dominant features of the cratering and ejecta process. These so-called Z models are based on the observations that: (1) a relatively simple and persistent cratering flow field accounts for almost all of the excavation of the non-vaporized material; (2) the conversion of this flow kinetic energy to distortional energy and gravitational potential dominates the final crater size and shape. Though a computer program is required to handle the general Z models, the simpler models can be computed by hand to illuminate the dominant mechanisms. A very simple Z model is described and applied to cases in which gravity dominates the final crater radius. This example starts with the so-called constant α, Z flow field to describe the early crater growth. A subsidiary assumption is then applied in lieu of precise energy conservation to terminate the flow. Conservation of momentum is employed as a constraint to obtain the scaling rules for final crater radius as a function of energy release and the acceleration of gravity.

THE CRATERING OF EARTH MEDIA by surface explosions is a complex phenomenon of combined effects that are difficult to untangle. However, computer simulations and controlled experiments have uncovered general features that can be approximated by the so-called Z models to provide simple and illuminating descriptions.

The initial studies were sponsored by the Defense Nuclear Agency and are reported in the references. These references provide data that support the general statements that will be given. The correlations of these data with the Z models will not be shown and many details are omitted for the sake of brevity.

These studies are not complete. The objective of this paper is to expose the methods in hopes that they will provide a useful framework for others to build upon. An example will be given at the end.

The Z models are based on the following observations:

(1) The transit of the ground shock through the incipient cratering region initiates a cratering flow field that persists long after the impulsive stresses have decayed.
(2) The associated cratering process can be approximated as incompressible flow along stationary streamlines.

The response of both porous and non-porous media could be described by these observations in the ten cases that were examined in the references. It should be emphasized that the incompressibility approximation applies only after the stresses have decayed, and does not apply to the initial porosity or the ejecta.

Figure 1 displays a typical cratering flow field at early times when the crater volume is perhaps a few percent of its final volume. The crater wall is nearly hemispherical and mass is being ejected in an expanding lip. The ejection angle

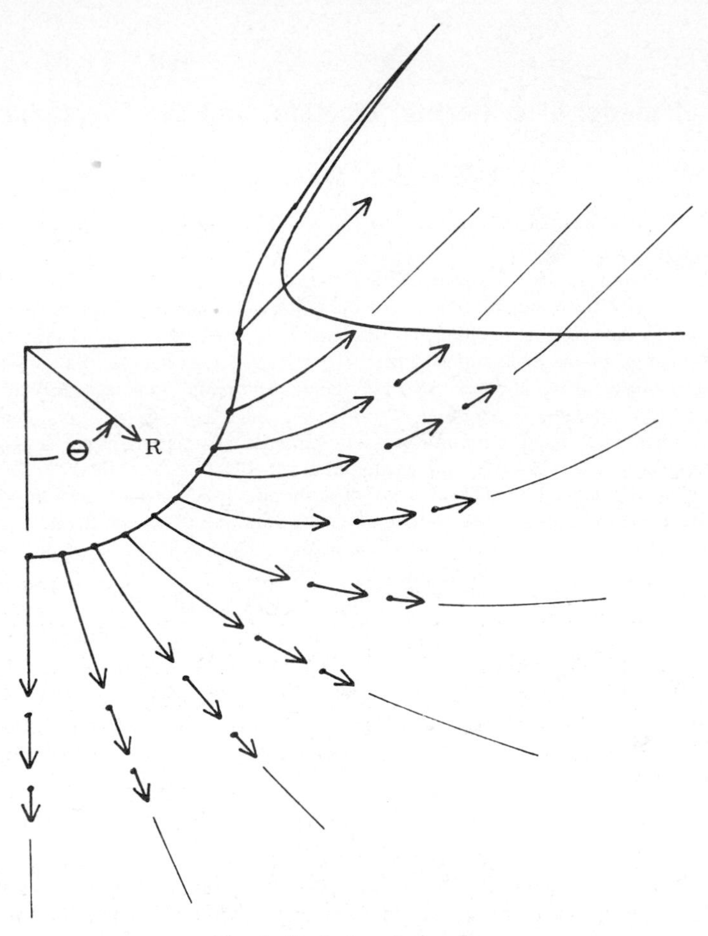

Fig. 1. Typical cratering flow.

is nearly constant during the crater growth of a given case, but varies between about 40 and 60 degrees from the horizontal for different cases.

The flow field is approximated by three basic assumptions as the starting point for all of the current Z models:

(1) Spall occurs at or near the ground plane, which subsequently allows independent ballistic trajectories.

(2) $\nabla \cdot \mathbf{u} = 0$ below the ground plane, corresponding to incompressible flow in the velocity field $\mathbf{u}$.

(3) $u_R = \alpha(t)/R^Z$, corresponding to the radial component of velocity below the ground plane (in spherical polar coordinates; see Fig. 1).

Assumptions (2) and (3) lead to stationary streamlines and a variety of other expressions that apply below the ground plane. Some of these are:

(4) $u_\theta = u_R(Z-2)\tan(\theta/2)$ for the other velocity component.

(5) $1-\cos\theta=(1-\cos\theta_0)(R/R_0)^{Z-2}$ for the streamlines.

(6) $R^{Z+1}=R_0^{Z+1}+(Z+1)\int_0^t \alpha(t)\,dt,$

(7) $R_c^{Z+1}=(Z+1)\int_0^t \alpha(t)\,dt,$

where $\alpha(t)$ is the time dependency of the velocity field at a fixed point, and is defined by Eq. (3). $R(t)$ and $\theta(t)$ are the time-dependent coordinates of a point that originated at R_0 and θ_0. $R_c(t)$ is the time-dependent radial distance to a crater wall point defined by the condition $R_0=0$.

Figure 2 displays particle displacements and crater growth features predicted by the previous expressions for the case $Z=3$ with ballistic trajectories and zero gravity applied above the ground plane. An initial test particle configuration is shown in Fig. 2A. Figures 2B and 2C display the corresponding particles and the growth of the crater and ejecta lip as time proceeds. The general features have been confirmed in the studies of the references. The rotational velocities of the ejected particles can also be evaluated if it is desired to construct models of particle size by balancing centrifugal force with surface tension or tensile strength.

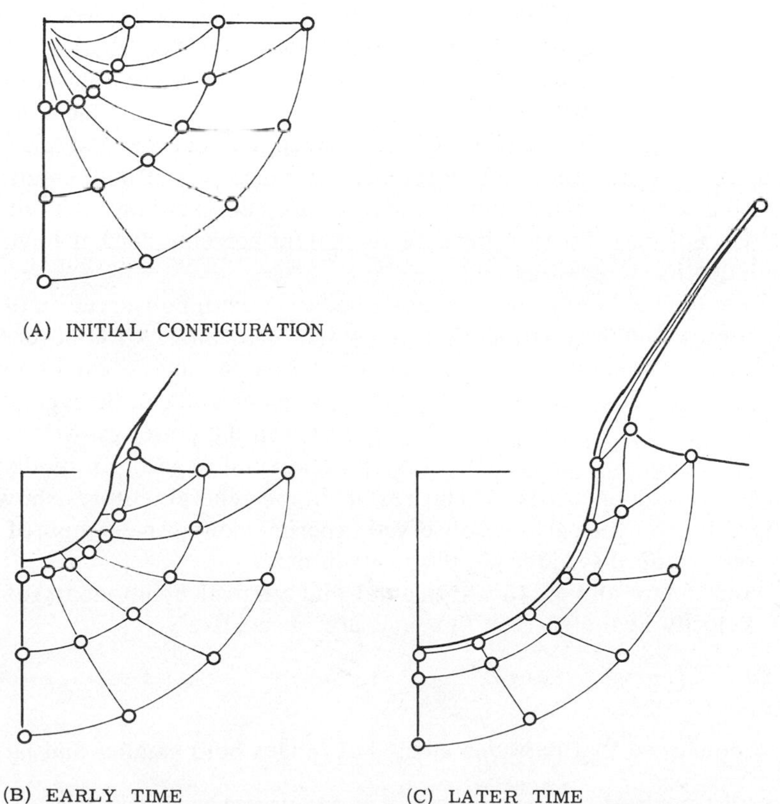

Fig. 2. Particle displacements and crater growth for $Z=3$, gravity $=0$.

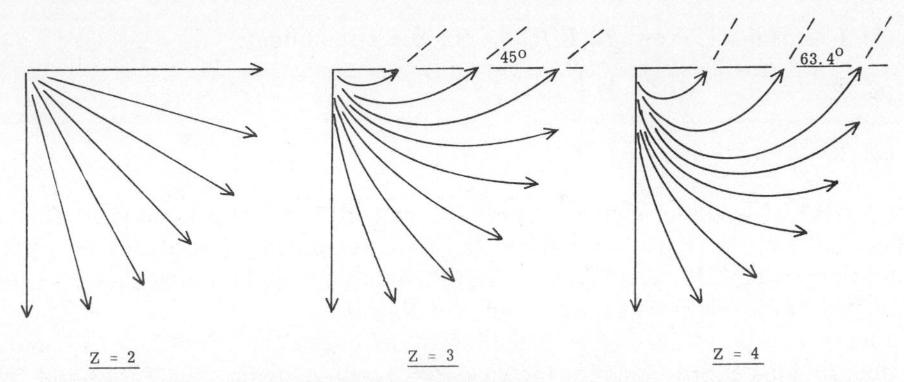

Fig. 3. Streamlines for $Z = 2$, 3, and 4.

Figure 3 displays streamlines for values of $Z = 2$, 3, and 4 applied uniformly below the ground plane. More realistic cratering flow fields have values $Z \approx 2$ near the downward axis, $Z \approx 4$ near the surface, with an average value $Z \approx 3$ applying to the average flow below the ground plane. The velocity of a particle depends on the Z value of its streamline and the time-dependent strength of the flow, defined by $\alpha(t)$, which may be different for different streamlines and must be established by other methods. One of these methods is described next.

In a computer program called the CRAZY code, the flow field is subdivided into stream tubes, defined by adjacent streamlines. Energy conservation is then applied independently in each tube, with the contained flow slowing and stopping as its kinetic energy is converted into distortional energy and gravitational potential. A computer approach is required because of the mathematical complexities that are associated with precise energy conservation.* Results of this method will not be presented because of certain corrections that have not yet been included in the existing code.

The case of constant α and Z is not compatible with conservation of energy, but provides a simple method that gives fair agreement with observed data during much of the crater growth. In this case, Eqs. (6) and (7) can be integrated directly. Then the displacement field, including the ejecta, can be evaluated at all times. An additional assumption is required to stop the crater growth to provide the final crater depth (which occurs first experimentally) and to provide the final crater width (which occurs last). This results in the general features shown in the diagrams of Fig. 4, which also are observed experimentally. An example of a simple assumption to stop the crater growth is given next.

For constant α and Z, the horizontal and vertical components of surface ejection velocity at the surface range, y, are, respectively,

(8) $u_H = \alpha / y^Z$,

(9) $u_V = (Z - 2) u_H$.

It can be confirmed that particles ejected at ranges both smaller and larger than

*The special case in which distortional energy is approximated by plastic yielding at a constant yield stress can be solved by hand.

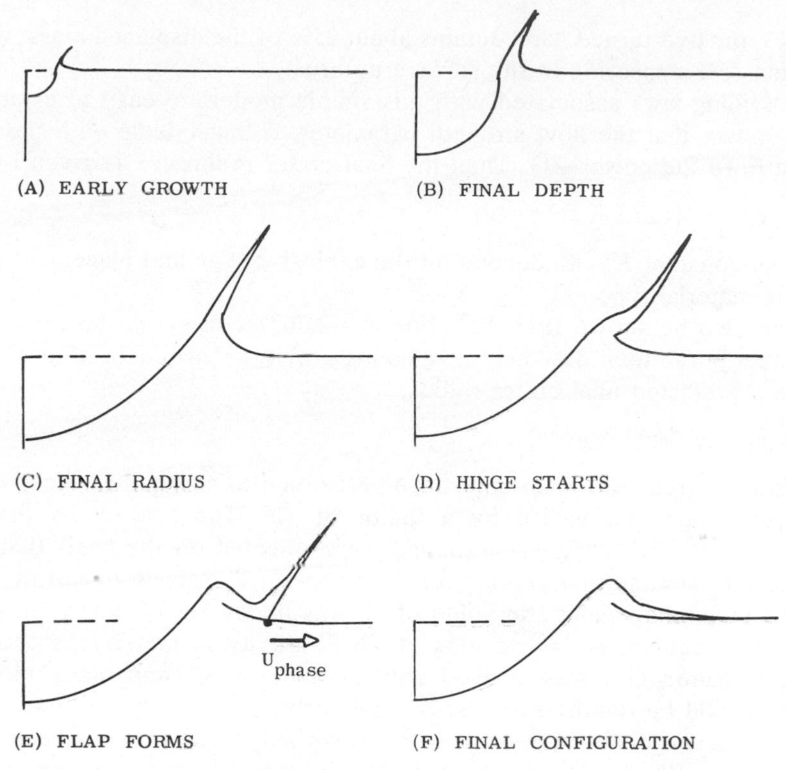

Fig. 4. Crater development.

y_1 return to the original surface plane at ranges larger than y_2, where

(10) $y_1 = [4Z(Z-2)\alpha^2/g]^{1/(2Z+1)}$,

(11) $y_2 = y_1[1 + 1/2\,Z]$,

where g is the acceleration of gravity. If the crater size is limited by gravity, then it is reasonable to approximate the final crater radius by y_1 and to freeze the flow below the ground plane when $R_c = y_1$. This leads to hemispherical final craters, in contrast to the CRAZY code method that generally produces more realistic bowl or shallow dish-shaped craters. In any case, the ejected particles continue on ballistic paths. Without air drag, the locus of the lofted particles at any time lies in a coherent sheet, forming the overturned flap of Fig. 4C, with its hinge point at y_2 and its nearly constant* surface phase velocity, u_{phase}. Figure 4D displays the final configuration when rebound and slumping effects can be ignored. The displaced crater mass, M_c, defined by the product of the *in situ* density and the volume excavated below the original ground surface, now lies above the surface because of the incompressibility assumption. It can be shown that the lofted ejecta (overturned flap) contains the mass $M_c[1-3/(Z+1)]$. Thus,

*Actually, the predicted phase velocity is $u_{\text{phase}} = \sqrt{2g(y_I - y_e)/(Z-2)}$ where y_I and y_e are the impact and ejection ranges.

for $Z \approx 3$, the overturned flap contains about 25% of the displaced mass, with the remaining 75% appearing in the surface upthrust.

The scaling laws associated with this simple model are easy to establish. It can be shown that the flow strength parameter, α, must scale with the energy release, E, to the power $Z/3$. Then the final crater radius, y_1, is given by

$$(12)\quad y_1 = K' [4Z(Z-2)E^{2Z/3}/g]^{1/(2Z+1)},$$

where the constant K' can depend on the explosive type and placement and the cratering material type.

It can also be shown that the value $Z = 2.7071$ conserves the total vertical momentum in the total flow field (neglecting gravity). This chain of assumptions leads to a predicted final crater radius,

$$(13)\quad y_1 = K \cdot E^{(1/3.55)}/g^{(1/6.4)}.$$

Cursory correlations with data were performed to scope the values of K in an energy range that varied by a factor of 10^8. The four cases that were examined were near-surface bursts and were selected on the basis that it was reasonable to assume that gravity controlled the final crater size and that crater wall slumping was small. The value of K was about 5 ± 0.5 when energy was expressed in equivalent metric tons of TNT, gravity in meters/sec^2, and y_1 in meters. Equation (13) was derived only as an example and many more correlations would be required to assess it properly.

The particular chain of approximations that led to Eq. (13) has been presented as an example of one of many ways to apply the basic Z assumptions of (1), (2), and (3). The CRAZY code method is another. Many variations and correlations with data have not been presented. Although the Z method of approach is not perfected, it is claimed that it provides a useful and convenient framework for viewing mechanisms that otherwise are obscured by complexity.

References

Maxwell, D., Seifert, K., Reaugh, J., and McKay, M.: 1973, Advanced method to predict surface-burst cratering response. *Report PIFR*-388, Physics International Company, San Leandro, California.

Maxwell, D.: 1973, Cratering flow and crater prediction methods. *Technical Report TCAM* 73-17, Physics International Company, San Leandro, California.

Maxwell, D. and Seifert, K.: 1974, Modeling of cratering, close-in displacements, and ejecta. *Report DNA* 3628*F*, Defense Nuclear Agency, Washington, D.C.

Roddy, D. J., Pepin, R. O., and Merrill, R. B., editors.
(1977) *Impact and Explosion Cratering*, Pergamon Press (New York), p. 1009–1024.
Printed in the United States of America

Numerical simulation of a very large explosion at the earth's surface with possible application to tektites

ERIC M. JONES and MAXWELL T. SANDFORD II

University of California, Los Alamos Scientific Laboratory, Los Alamos, New Mexico 87545

Abstract—A two-dimensional, radiation transport/hydrodynamics calculation has been done of a 2.1×10^{25} erg point explosion at the earth's surface. The crater and ejecta are ignored and the ground is treated as a flat, free-slip, reflective boundary. This high energy release produces a fireball comparable in size to an atmospheric scale height. The fireball rises buoyantly to 40 km altitude in 60 sec. Thereafter the bow-wave accelerates and breaks out of the atmosphere, pulling the fireball to greater altitude. Particle velocities of 2.4 km/sec at 170 km altitude occur at 120 sec.

The rising fireball creates radially converging afterwinds up to 300 m/sec occurring 3.5 km from ground zero at 40 sec.

If high energy impacts do create a fireball-like entity, hot gases in the rising fireball may contribute to the formation and/or dispersal of some kinds of impact melts and tektites. At 60 sec there is gas at 3500°K, 4×10^{-6} g/cm^3 rising at 1.2 km/sec which could, by virtue of drag, support 7 cm radius rocks against gravity.

1. INTRODUCTION

VERY LARGE NATURAL EXPLOSIVE EXPANSIONS of the atmosphere may occur at or near the earth's surface [resulting from the impact of meteorites, cometary bodies (Urey, 1957), or even compact black holes (Jackson and Ryan, 1973; see, however, Wick and Isaacs, 1974)]. The mechanics of meteorite impacts are discussed in some detail by Stanyukovich (1961, see also Brode, 1968). Shoemaker (1963) described in detail the formation of Meteor Crater, Arizona, by the process of shock propagation following impact of an iron meteorite. He used field relations and scaling laws from nuclear explosions of appropriate depth of burst to determine the energy release and an estimate of the mass and radius of the incoming meteorite. Bjork (1961) was the first worker to attempt numerical computer modeling of meteorite impact. The principal distinction between meteorite impact and explosive sources is that, during an impact, energy is delivered to the target along a (decaying) line source over some finite time, whereas an explosion provides energy almost instantaneously at a point. The excavation of a crater in both cases, however, is largely the consequence of the interaction of a strong shock with a free surface, and the resulting momentum which is trapped as a rarefaction is reflected off the surface. At significant distances from the shot or impact point, processes are so similar that nearly identical craters will be produced *provided* proper depth of burst is selected. The structural *details* of Teapot Ess Crater at the USERDA Nevada Test Site, for example, are nearly identical to those observed at Meteor Crater, Arizona, although in a smaller scale.

In addition to the cratering and seismic phenomena accompanying impacts, there may be important atmospheric effects as illustrated by the leveling of about 2200 km^2 of forest around the craterless site of the 1908 Tunguska explosion (Zotkin and Tsikulin, 1966). Other examples of atmospheric effects may be the tektite fields associated with Ries Crater and with Ashanti Crater (Lake Bosumtwi) (Cohen, 1963). In these cases one might postulate that the tektites, formed during the impact, were carried aloft by the rising fireball and deposited as the cloud drifted in the stratospheric winds. It is even possible that some kinds of tektites are formed in the fireball.

In this paper we model the atmospheric effects of a large impact under the following assumptions:

(1) The crater does not affect the airflow,

(2) there is no dynamically significant particulate loading, and

(3) at some early time, the fireball produced in the atmosphere is describable as a Sedov–Taylor blast wave (Taylor, 1950; Sedov, 1969).

In particular, we used a one-dimensional (spherical) radiation transport/hydrodynamics program, RADFLO (Zinn, 1973) to compute the evolution of a 4.2×10^{25} erg explosion (in air) to 0.25 sec when the fireball radius was 2.6 km and the central temperature had dropped to 11 eV (1.3×10^5 K). Because the fireball is so large, further one-dimensional calculations ignore important two-dimensional effects. At 0.25 sec the fireball was cut in half and upper half placed on the ground. The subsequent two-dimensional (cylindrical) evolution was calculated with the radiation transport/hydrodynamics program SN-YAQUI (Anderson and Sandford, 1975; Sandford *et al.*, 1975).

II. Initial Conditions

We used the upper hemisphere of a spherical, 4.2×10^{25} erg fireball at 0.25 sec for initial conditions.

This procedure is valid as an approximation to a 2.1×10^{25} erg point-explosion on a rigid, flat surface. The spherical calculation was terminated when the fireball had become comparable in size with an atmospheric scale height. Figure 1 shows the initial distributions of density, specific internal energy, and velocity at 0.25 sec. The variations of flow field variables with distance from the explosion site are in excellent agreement with the predictions of the blast wave solution (Brode, 1955; Sedov, 1959; see also Zel'dovitch and Raizer 1966, p. 93).

If a significant amount of energy is deposited in the air in the first few milliseconds after a massive impact, the air-flow will resemble a blast wave. Knowles (1976) has reviewed the energy partitioning in large impacts versus the nuclear case. While it is clear that the air above a hypervelocity impact will be heated to at least a few thousand degrees Kelvin, the exact heating is a function of impact velocity. The fireball we have calculated describes the nuclear case. It is expected that a large hypervelocity impact will create a similar fireball and cloud rise.

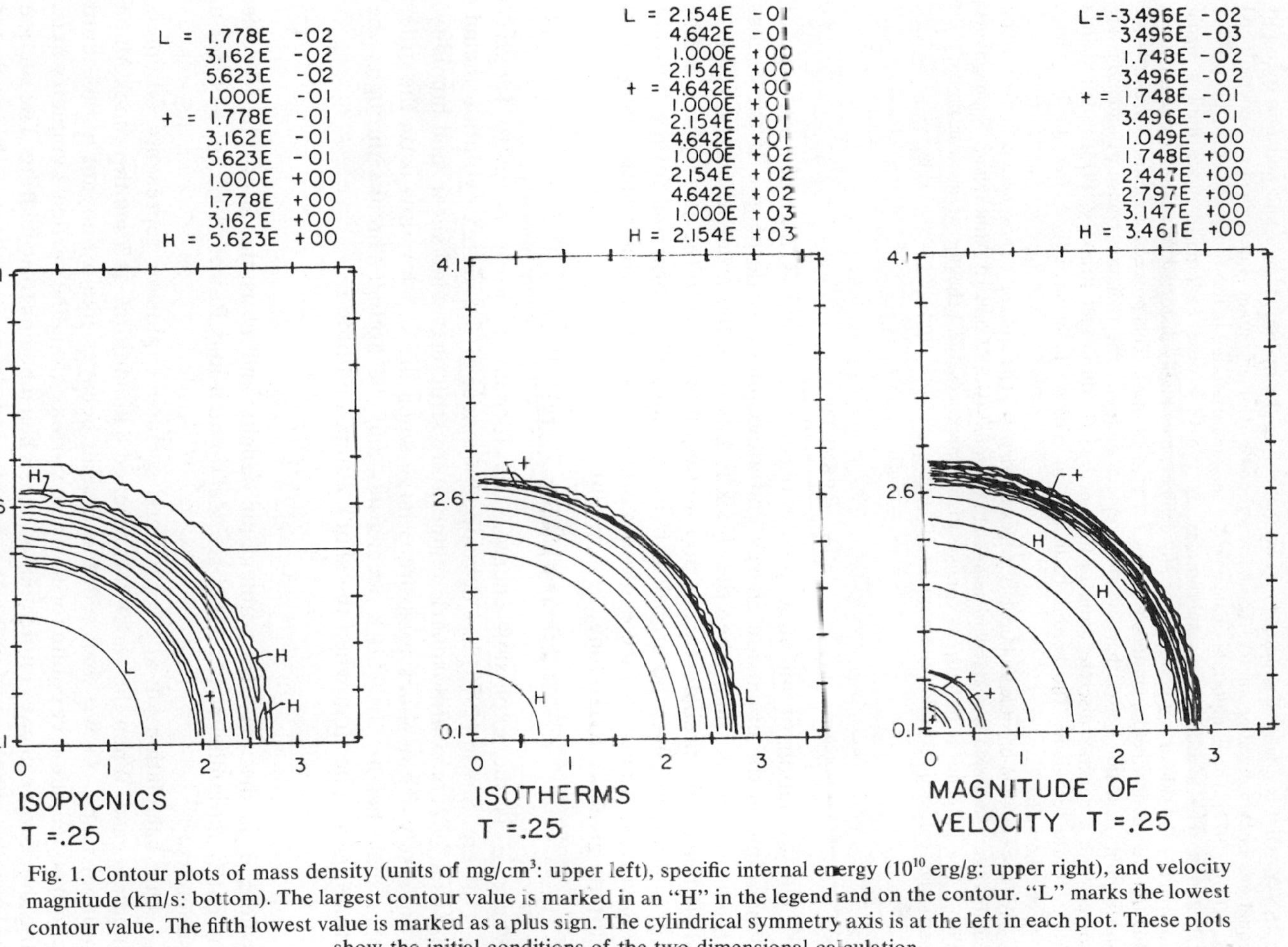

Fig. 1. Contour plots of mass density (units of mg/cm^3: upper left), specific internal energy (10^{10} erg/g: upper right), and velocity magnitude (km/s: bottom). The largest contour value is marked in an "H" in the legend and on the contour. "L" marks the lowest contour value. The fifth lowest value is marked as a plus sign. The cylindrical symmetry axis is at the left in each plot. These plots show the initial conditions of the two dimensional calculation.

III. Two-dimensional Method

SN-YAQUI is a large computer program that solves the hydrodynamics and radiation transport equations in cylindrical geometry. The basic structure of our program is described in Anderson and Sandford (1975). Three important changes have been incorporated in the version used for the present calculation. The hydrodynamics subroutines are run in a pure Lagrangian mode until a signal approaches the side or top boundary or any cell becomes highly distorted. We then rezone the mesh and transfer all flow field quantities to a new rectangular mesh. This procedure seems to preserve the shock jump condition better than did the previous method. The radiation transport equations are solved with a nongray discrete ordinate (S_n) method (Lathrop and Brinkley, 1973) rather than a Monte Carlo technique. The bottom boundary, in addition to being a rigid, free-slip boundary, reflects photons.

We use equation-of-state and opacity tables for dry air. The ambient atmosphere is in hydrostatic equilibrium and employs a standard temperature profile (U.S. Standard Atmosphere Supplements, 1966). The program in its current configuration reproduces the behavior of nuclear fireballs. In particular, calculations of lower energy ($\lesssim 10^{23}$ erg) explosions produce buoyant ring vortices.

IV. Results

The evolution of an atmospheric blast wave follows one of two courses depending on the explosion energy and atmospheric scale height. If the excess shock pressure is small when the shock radius is on the order of an atmospheric scale height, the fireball reaches pressure equilibrium with the atmosphere. Being hot and underdense, the fireball is buoyant and will deform and form a buoyant ring vortex. A near-surface explosion produces a vortex which stops rising at an altitude given by (Glasstone, 1962, p. 36)

$$h(\mathrm{km}) = 9.4 \log E(\mathrm{erg}) - 191.2, \tag{1}$$

where E is the explosion energy. The formula is fairly accurate for $E > 4 \times 10^{21}$ erg. Sowle (1975) has developed a model of the vortex evolution including such effects as atmospheric temperature structure, wind shear, and humidity. Conversely, if the shock pressure is large when the shock radius is on the order of a scale height, the shock accelerates and the fireball "breaks through" the atmosphere. The time when the shock reaches altitude is

$$\tau = 24(\rho_0 H^5/E)^{1/2}, \tag{2}$$

where ρ_0 is the sea-level ambient air density and H is the atmospheric scale height (Andriankin *et al.*, 1962; see Zel'dovitch and Raizer, 1966, p. 849, for a summary).

Our calculation of a 2.1×10^{25} erg surface explosion represents an intermediate case. When the shock has reached 8 km altitude at 3 sec the shock Mach number is 4.4. As we shall see, the fireball behaves like a buoyant bubble until about 60 sec but never quite forms a ring vortex. Along the vertical symmetry axis, the shock wave decays until about 60 sec, when it begins to accelerate. The suction behind the shock accelerates the bubble and drags it well past the predicted stabilization altitude of 47 km.

The following discussion concentrates on three aspects of the flow field:

motion of the shock wave, surface winds, and flow field structure on the vertical axis.

Figure 2 shows the distance travelled by the shock and its Mach number at various times. We show both the surface shock and the vertical (axial) shock. Initially the horizontal and vertical shock are equally strong, but the horizontal shock eventually decays into an acoustic wave (Mach number = 1). The vertical shock decays more slowly. During the first 10 sec (the hemispherical expansion phase of shock wave), the vertical component shows approximately normal decay, but the exponential atmosphere eventually dominates. Between 60 and 90 sec the vertical shock accelerates from $M = 2.7$ to $M = 5.7$.

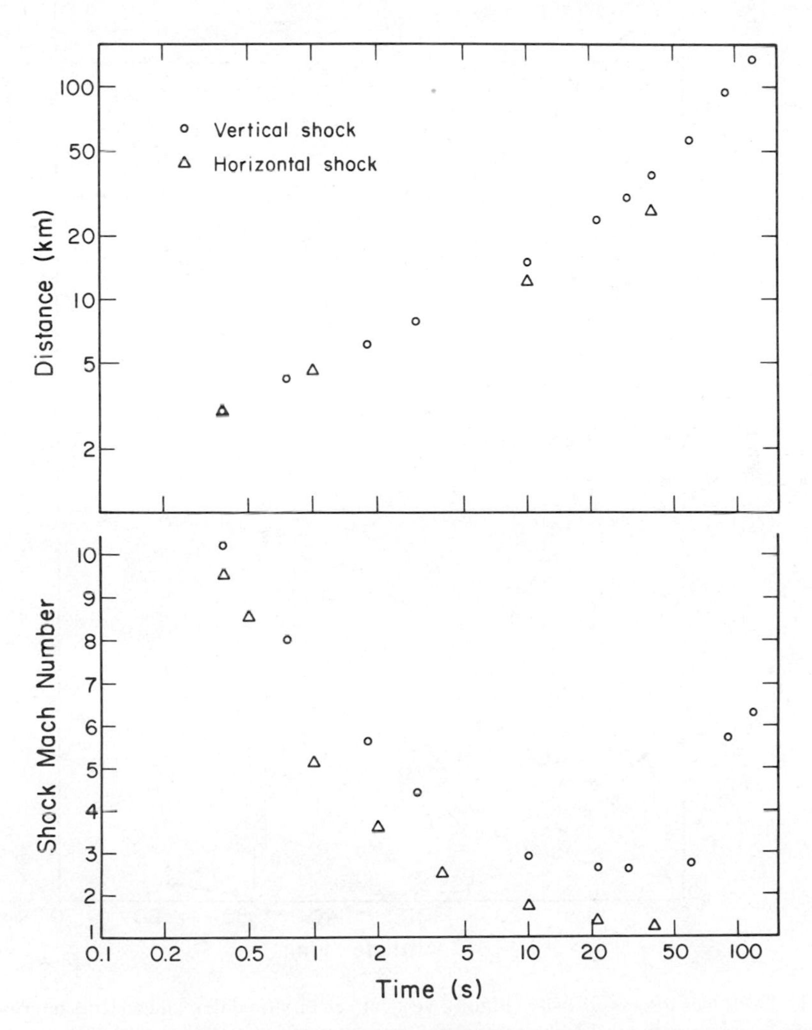

Fig. 2. Position of the vertical and horizontal shock waves (top) and the mach number = shock propagation speed/ambient sound speed.

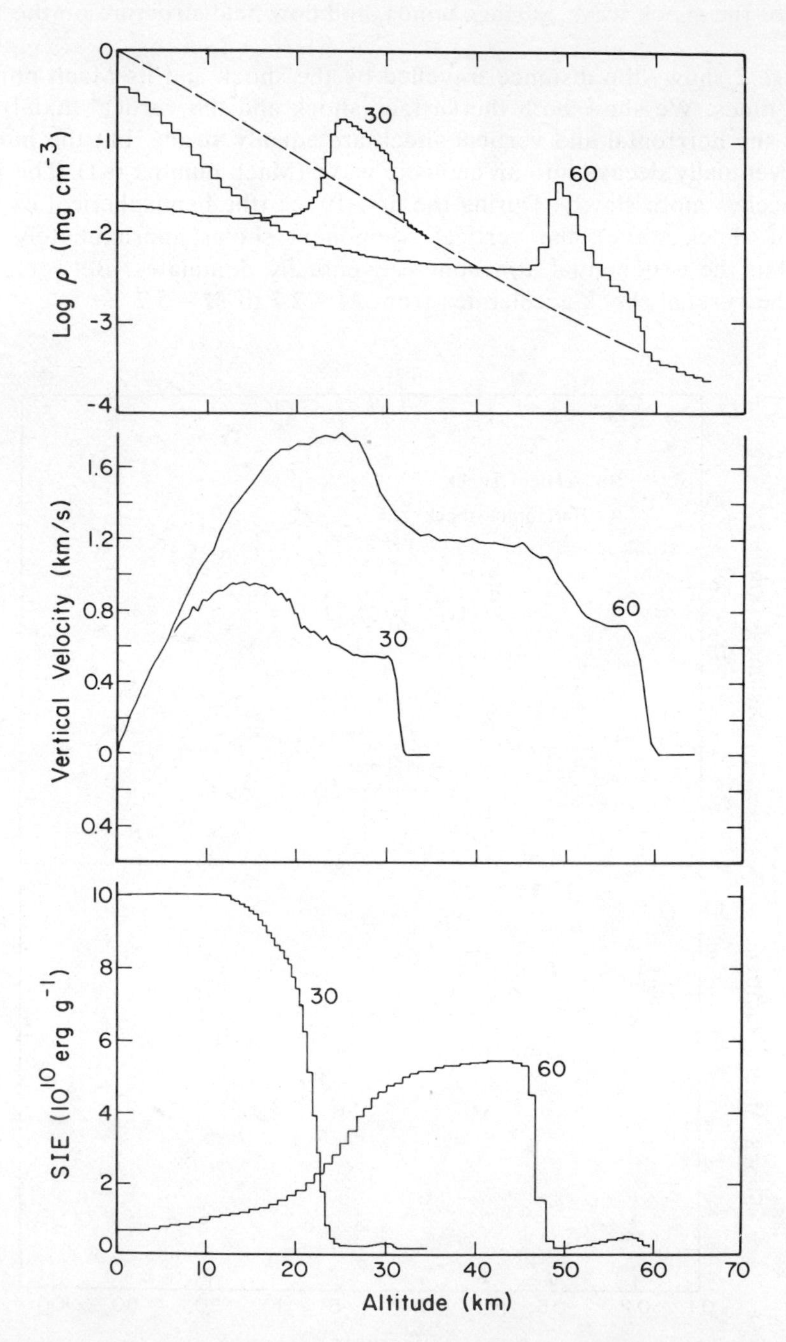

Fig. 3. Profiles of mass density (ρ: top), vertical velocity (middle), and specific internal energy (SIE: bottom) along the symmetry axis at 30 and 60 sec.

Figures 3 and 4 show profiles of mass density, ρ, velocity, v, and specific internal energy along the symmetry axis at 30, 60, 90, and 120 sec. At 30 and 60 sec (Fig. 3) the shock shows weak compressive heating, but by 120 sec (Fig. 4) the shock temperature is about 3.3 times ambient.

The vertical motion of the fireball changes over from buoyant motion to accelerated rise at about 60 sec. At this time, the hottest material on the axis is at

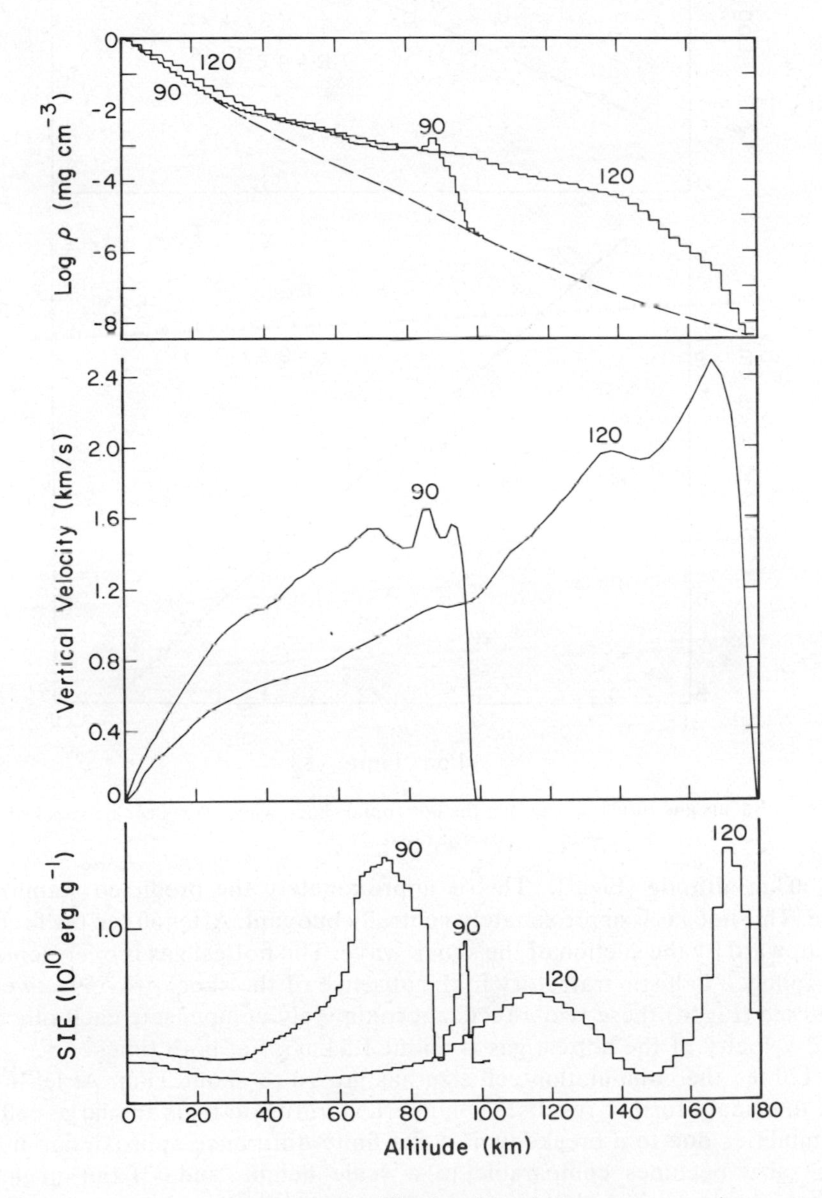

Fig. 4. Same as Fig. 3; at 90 and 120 sec.

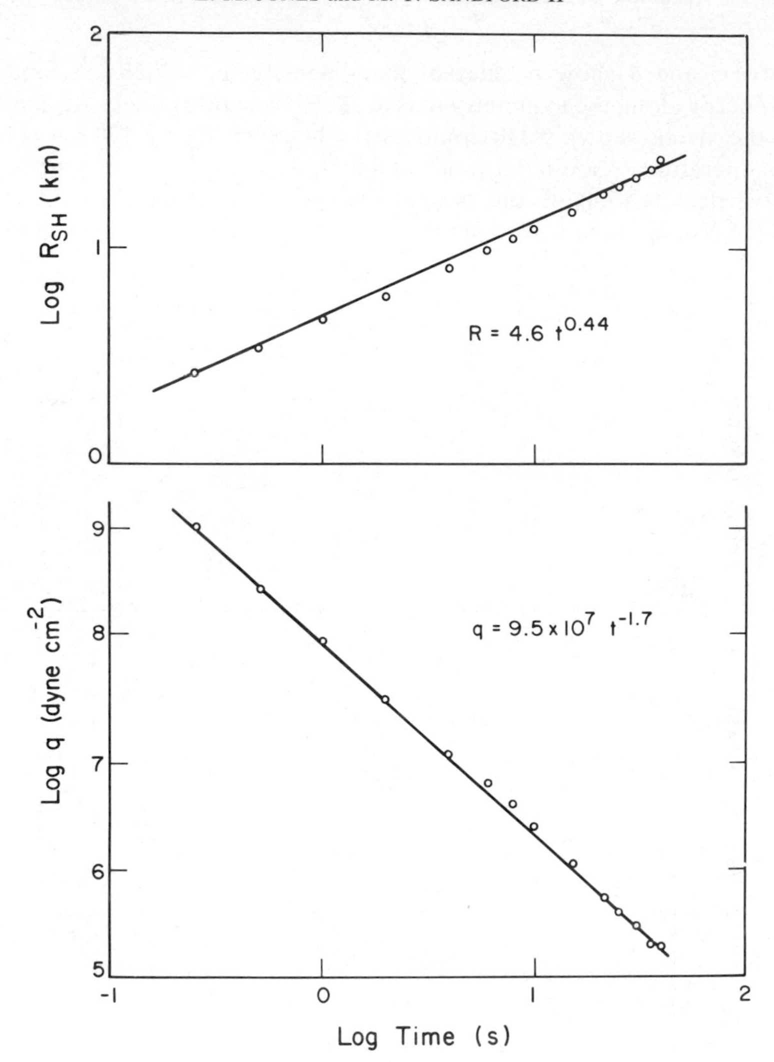

Fig. 5. Radius and thrust ($q = \rho v^2$) of the horizontal shock wave. Trees can be knocked over if log $q \geq 5.5$.

about 40 km altitude (Fig. 3). This is approximately the predicted stabilization altitude. This hot gas is approximately neutrally buoyant. After 60 sec the fireball is drawn upward by the suction of the shock wave. The hottest gas is over-dense and would follow a ballistic trajectory in the absence of the shock wave. Between 90 and 120 sec (Fig. 4) these two effects approximately compensate each other; the vertical velocity of the hottest gas is about 1.5 km s^{-1} at both times.

At 120 sec the computation cell size has grown to about 3 km. At least three effects invalidate further results: (1) numerical diffusion from the large cell size, (2) instabilities due to a breakdown of the finite difference approximation when the cell size becomes comparable to a scale height, and (3) our neglect of magnetohydrodynamic and plasma effects.

Surface winds are dominated by two effects: the shock wave and the afterwinds created by the suction of the rising fireball. The outward flow of the shock wave can knock down trees as long as the velocity thrust, $q = \rho u^2$, exceeds a threshold of approximately 3.4×10^5 dyne cm^{-2} (see Zotkin and Tsikulin, 1966). The Tunguska explosion leveled trees over a 26.5 km radius area (Zotkin and Tsikulin, 1966). Figure 5 shows shock radius and peak velocity thrust along the surface as functions of time from our calculation. Approximate relations fitting these results are:

$$R = 4.6\, t^{0.44}\ \text{km}, \tag{3}$$

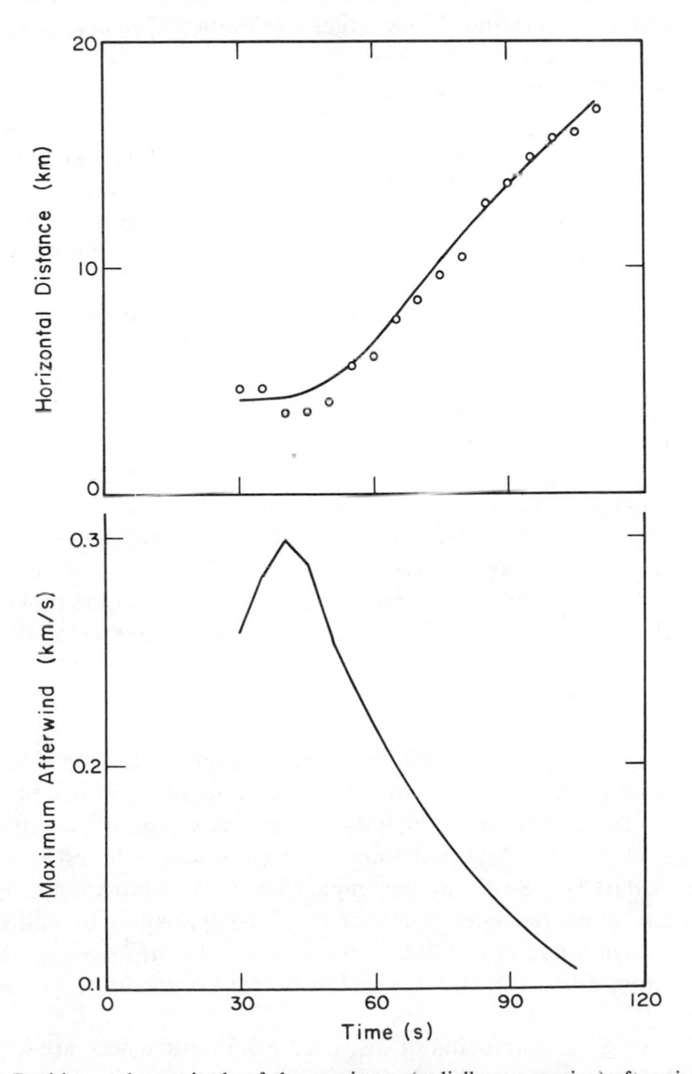

Fig. 6. Position and magnitude of the maximum (radially converging) afterwinds.

and

$$q = 9.5 \times 10^7\, t^{-1.7} \text{ dyne cm}^{-2}. \tag{4}$$

There is an unresolved paradox here. Our calculation indicates that at 2.1×10^{25} erg surface explosion produces of $\geqslant 3.4 \times 10^5$ dyne cm^{-2} at $R \leqslant 27.5$. The Tunguska explosion probably occurred at several kilometer attitude and the energy may have been 5.5×10^{23} erg (Boyarkina and Bronshten, 1976), yet trees were felled over a 26.5 km radius area. Perhaps $q = 3.4 \times 10^5$ dynes cm^{-2} is not the appropriate threshold.

As the fireball rises, large afterwinds build up when air is drawn into the space vacated by the fireball. Figure 6 shows the position and value of the maximum surface afterwinds. The largest velocities, 300 ms^{-1}, are found about 4 km from the impact point at 40 sec.

The overall structure of the fireball is illustrated by the sets of contour plots showing mass density (mg/cm^3), specific internal energy (10^{10} erg/g) and particle speed (km/s). The cylindrical symmetry axis is at the left and the ground is the horizontal axis. In the contour plots a table of contour values in included at the upper right of each plot. The lowest (L) contour value and highest (H) are labeled with the appropriate symbol. The fifth smallest value contour is labeled with a cross (+).

Figure 7 shows plots at 30 sec. The shock is clearly delineated and the fireball shows considerable vertical distortion. The velocity structure intersecting the surface at 15 km radius marks the stagnation point separating the shock from the afterwind region closer to ground zero.

Plots at 60 sec are shown in Fig. 8. The fireball shows the beginnings of buoyant distortion. There has been considerable flattening of the bottoms of the innermost density and internal energy contours. However, there is no evidence that the formation of a ring vortex occurs. Calculations of lower energy explosions do produce ring vortices.

At 90 sec (Fig. 9) shock acceleration is clearly evident in the upward elongation of the top of the fireball. The process continues at 120 sec (Fig. 10).

V. IMPLICATIONS

In the lunar and mercurian environments, impacts can be evaluated largely on the basis of rock mechanics and ballistic dynamics. Transient atmospheres may influence the dispersal of small ejecta particles. The presence of permanent atmospheres on Earth, Mars and Venus can complicate the effects of the inpact process considerably. Some of the impacting body's energy is imparted to the air and, if the imparted energy is large and is deposited in milliseconds, high temperature, high velocity gas flows are created. The atmospheric flow may alter the deformation, dispersal, and deposition of impact products smaller than about 10 cm.

Tektites are glass fragments of disputed origin but which are known in some cases, to be associated with certain large terrestrial impacts. Our analysis

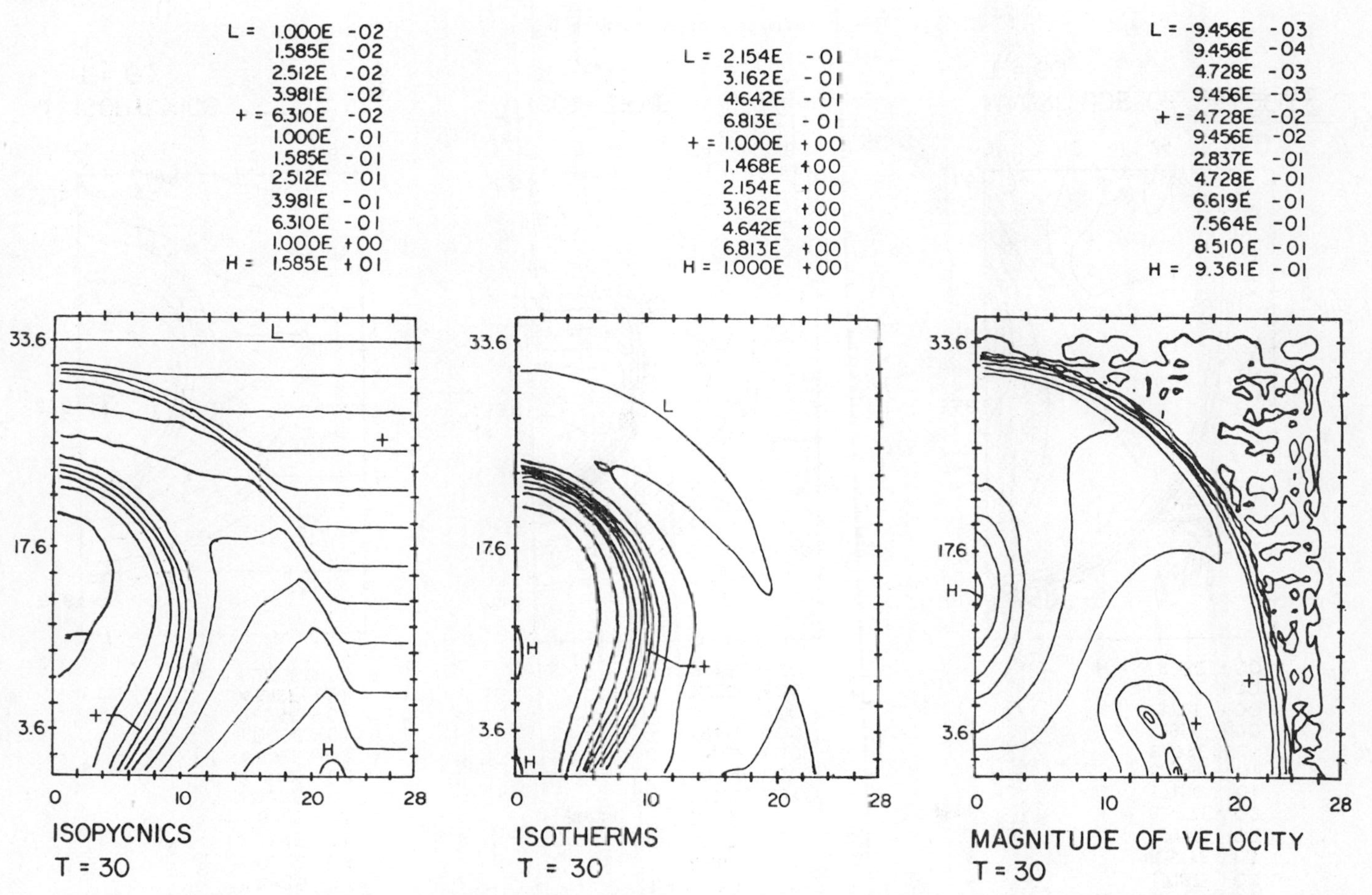

Fig. 7. Contour plots of mass density (left), specific internal energy (middle), and velocity magnitude (right) at 30 sec. See Fig. 1.

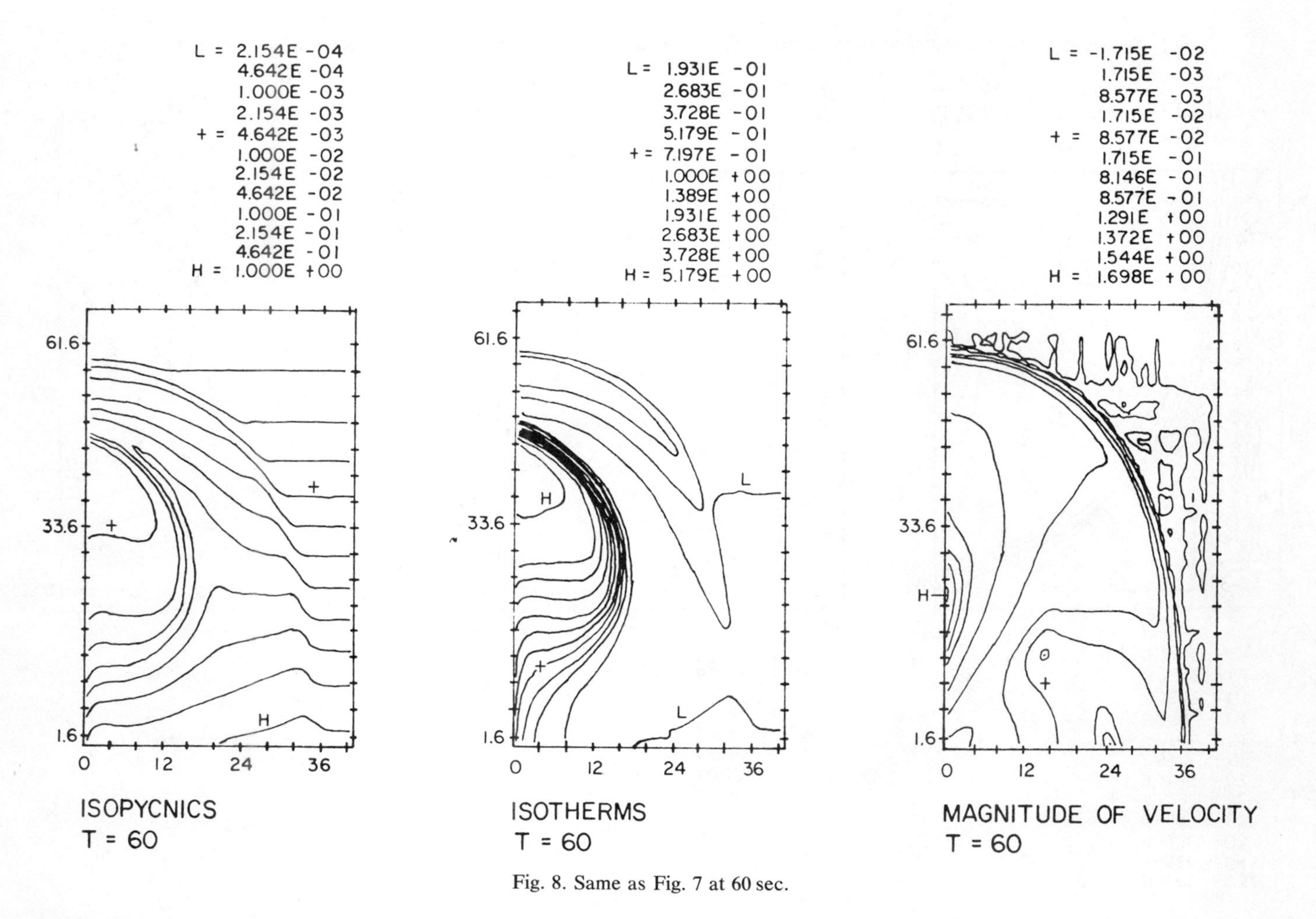

Fig. 8. Same as Fig. 7 at 60 sec.

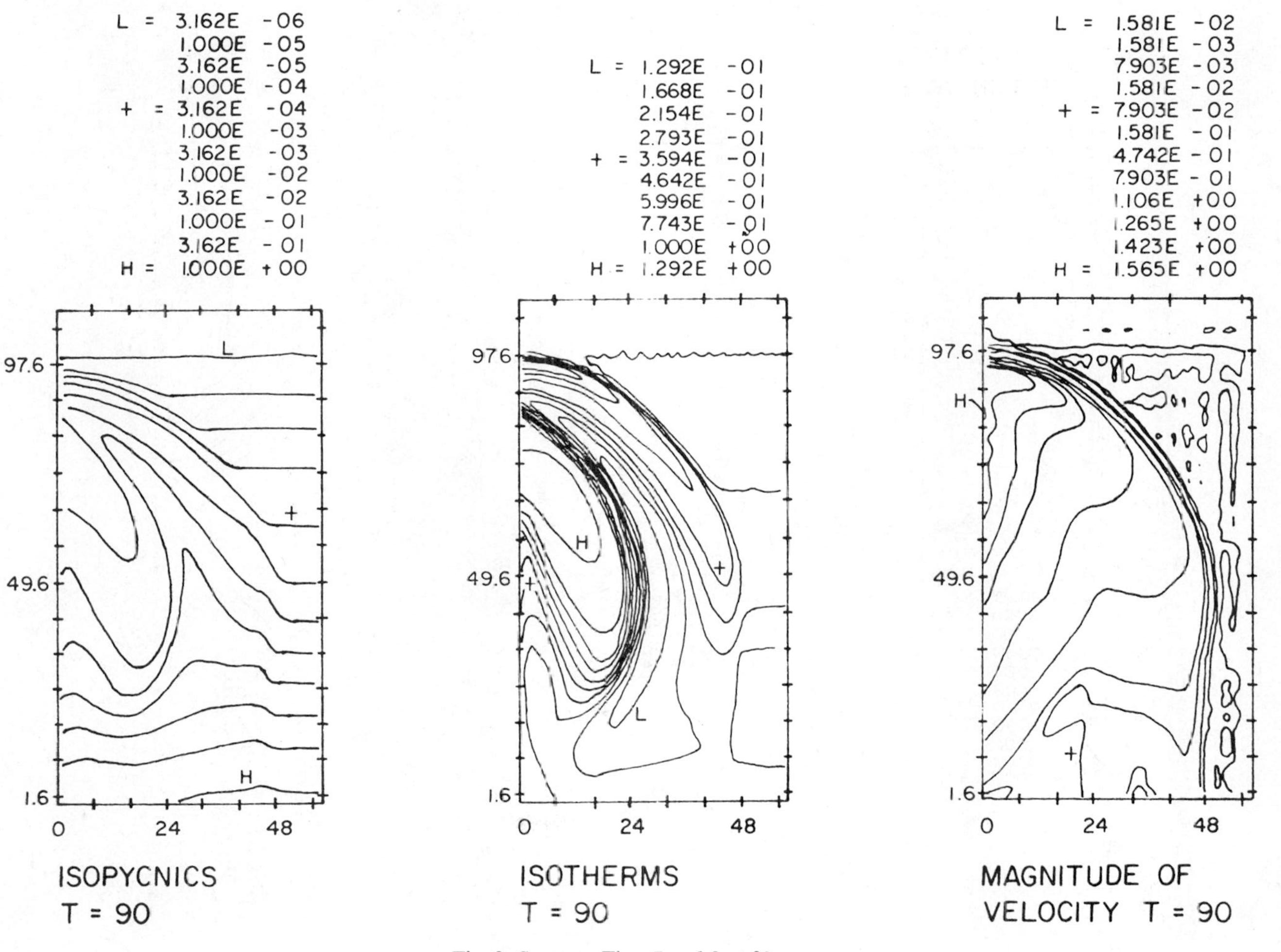

Fig. 9. Same as Figs. 7 and 8 at 90 sec.

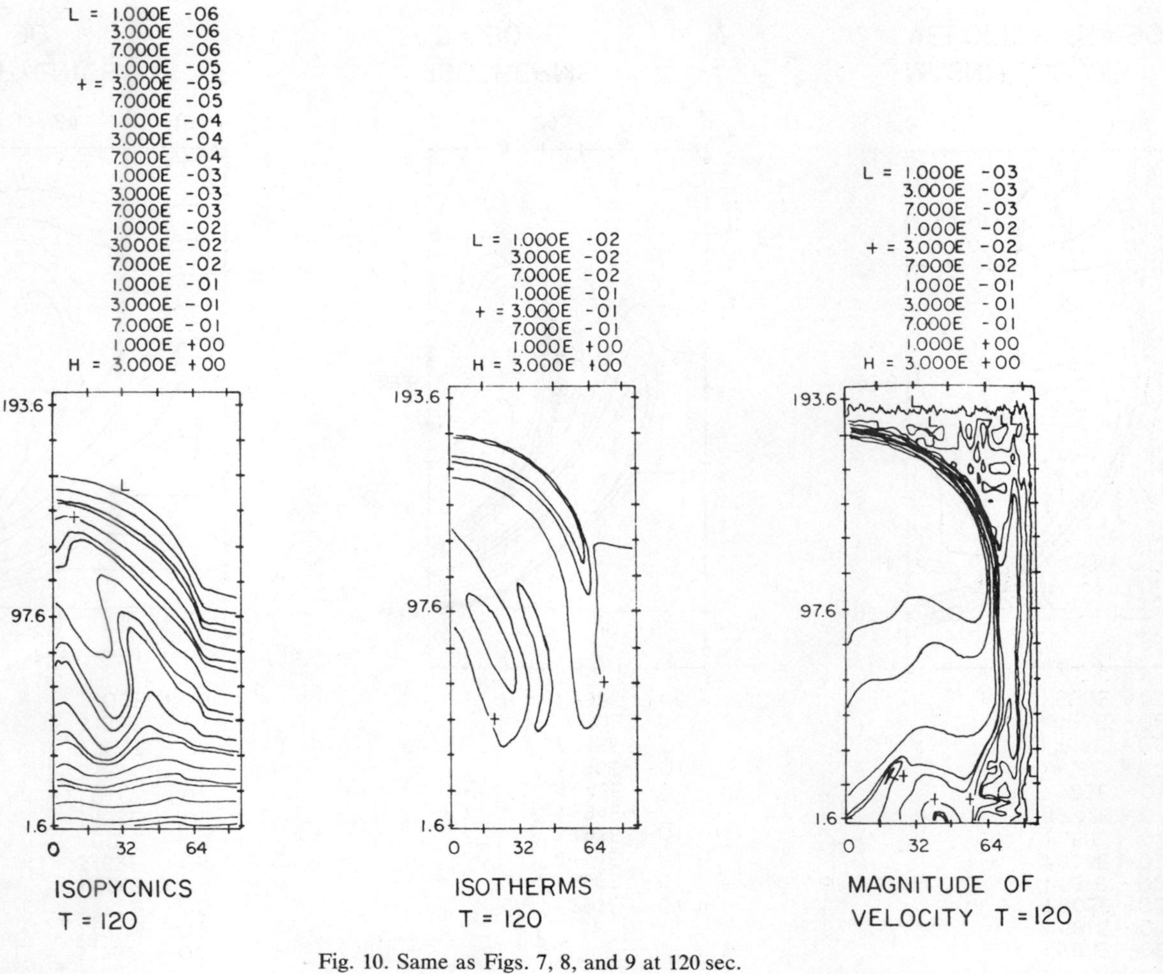

Fig. 10. Same as Figs. 7, 8, and 9 at 120 sec.

suggests that the final distribution of impact glass may be strongly influenced by the atmospheric phenomena. While we do not propose any new theories of tektite origin, we would like to point to some implications of our calculation for consideration by serious students of the tektite problem. We suspect that the class of objects called tektites is heterogeneous: some may be fully explained only in terms of entry into the earth's atmosphere at velocities near or exceeding the terrestrial escape velocity creating a variety of aerodynamic forms (Chapman *et al.*, 1962) while others may have never left the atmosphere. Let us conjecture that the spectrum of tektite types can be loosely divided into two categories: I, *atmospheric tektites* formed in terrestrial impacts, modified by motion through the fireball, and deposited by gravity after transport over some large distance by fireball motion under the influence of stratospheric winds and II, *exo-atmospheric* tektites ejected into the atmosphere from above.

The exo-atmospheric tektites may be further divided into two subgroups: those of terrestrial origin on the one hand and the rest. Chapman *et al.* (1962) and Adams (1963) have argued that the class of terrestrial exo-atmospheric tektites is empty owing to the impossibility of accelerating fused or solid tektites from the earth's surface to space without violating observational constraints provided by the "australites". We would like to point out that the present calculation produces a 60 km diameter hole in the atmosphere. Impacts depositing at least 2.1×10^{25} erg of energy to the atmosphere provide a possible means of ejecting terrestrial impact products from the atmosphere. This mechanism deserves further attention.

The atmospheric tektites may be represented by the tektites associated with Reis and Ashanti Craters. Cohen (1963) has described the fan shaped dispersal patterns (apparently oriented downwind in both cases) and indications that at least some of the Reis tektites show decreasing size with distance from the crater. We would like to point out that both of these observations are consistent with the hypothesis that each of these impacts produced a fireball which stabilized at a finite altitude, drifted in the prevailing stratospheric winds and deposited smaller and smaller tektites as time passed. "Ordinary" atmospheric sorting (in the absense of fireballs) would produce increasing tektite size with distance from the impact. In each case, the energy deposited in the air was much less than 2.1×10^{25} erg.

We suspect that careful consideration of the effects of the fireball environment seen by some tektites may help unravel the puzzle of their formation. We note that in our calculation there is gas at 60 sec with a density of 4×10^{-6} g/cm^3 and temperature 3500 K moving upward at 1.2 km/sec. Spherical rocks of approximately 7 cm radius could be supported against gravity by drag.

Acknowledgments—This work has been supported by the Los Alamos Scientific Laboratory which is operated by the University of California under contract to the U.S. Energy Research and Development Administration. We would like to thank T. R. McGetchin for helpful discussions and H. L. Brode for suggesting improvements to the manuscript. Peter Goldreich suggested this problem to us in early 1974. The first attempt revealed serious numerical difficulties. A six-month effort by J. W.

Kodis, H. G. Horak, and R. C. Anderson produced the new rezoning algorithm and permitted successful completion of the calculation. The rebuilding work was supported, in part, by the Defense Nuclear Agency.

References

Adams, E. W.: 1963, Aerodynamic Analysis of Tektites and Their Hypothetical Parent Bodies. In *Tektites.* (J. A. O'Keefe, ed.), p. 150. University of Chicago Press, Chicago.

Anderson, R. C. and Sandford, M. T., II: 1975, Yokifer: A Two-Dimensional Hydrodynamics and Radiative Transport Program. Los Alamos Scientific Laboratory Report LA-5704-MS.

Andriankin, E. I., Kogan, A. M., Kompaneets, A. S., and Krainov, V. P.: 1962, Propagation of a Strong Explosion in an Inhomogeneous Atmosphere. *Zh. Prikl. Mekhan. i Tekhn. Fiz.*, No. 6, p. 3.

Boyarkina, A. P. and Bronshten, V. A.: 1976, The Energy of the Tunguska Meteorite Explosion and Consideration of the Atmospheric Inhomogeneity. *Astron. Vestniks* **9**, 172.

Brode, H. L.: 1955, Numerical Solutions of Spherical Blast Waves. *J. Applied Phys.* **26**, 766.

Brode, H. L.: 1968, Review of Nuclear Weapons Effects. *Ann. Rev. Nuclear Sci.* **18**, 163.

Chapman, D. R., Larson, H. K., and Anderson, L. A.: 1962, NASA Tech. Report R-134.

Cohen, A. J.: 1963, Asteroid-or Comet-Impact Hypothesis of Tektite Origin. In *Tektites.* (J. A. O'Keefe, ed.), p. 189. University of Chicago Press, Chicago.

Glasstone, S. (ed.): 1962, *The Effects of Nuclear Weapons.* U.S. Govt. Printing Office, Washington.

Jackson, A. A., IV and Ryan, M. P.: 1973, Novel Explanation of Tungus 'Meteorite' Incident. *Nature* **245**, 88.

Knowles, C. P.: 1976, The Theory of Impact Cratering Phenomena, An Overview. This volume.

Lathrop, K. D. and Brinkley, F. W.: 1973, TWOTRAN II: An Interfaced Exportable Version of the TWOTRAN Code for Two-Dimensional Transport. Los Alamos Scientific Laboratory Report LA-4848-MS.

Sandford, M. T., II, Anderson, R. C., Horak, H. G., and Kodis, J. W.: 1975, Improved, Implicit Radiation Hydrodynamics. *J. Comp. Phys.* **19**, 280.

Sedov, L. I.: 1959, *Similarity and Dimensional Methods in Mechanics.* Academic Press. New York.

Shoemaker, E. M.: 1963, Impact Mechanics at Meteor Crater, Arizona. In *The Moon, Meteorites, and Comets* (B. M. Middlehurst and G. P. Kuiper, eds.), p. 301. University of Chicago Press, Chicago.

Sowle, D. H.: 1975, *Implications of Vortex Theory for Fireball Motion.* U.S. Defense Nuclear Agency Report DNA-3581F.

Stanyukovich, K. P.: 1961, Elements of the Theory of the Impact of Solid Bodies with High Velocities. In *Artificial Earth Satellites.* (L. V. Kurnosova, ed.), p. 292. Plenum Press, New York.

Urey, H. C.: 1957, Origin of Tektites. *Nature* **179**, 556.

U.S. Standard Atmosphere Supplements: 1966, U.S. Government Printing Office, Washington.

Wick, G. L. and Isaacs, J. D.: 1974, Tungus Event Revisited. *Nature* **247**, 139.

Zel'dovitch, Ya. B. and Razier, Yu. P.: 1966, *Physics of Shock Waves and High Temperature Hydrodynamic Phenomena.* Academic Press, New York.

Zinn, J.: 1973, A Finite Difference Scheme for Time-Dependent Spherical Radiation Hydrodynamics Problems. *J. Comp. Phys.* **13**, 569.

Zotkin, I. T. and Tsikulin, M. A.: 1966, Simulation of the Explosion of the Tungus Meteorite. *Soviet Physics-Doklady* **11**, 183.

Roddy, D. J., Pepin, R. O., and Merrill, R. B., editors.
(1977) *Impact and Explosion Cratering*, Pergamon Press (New York), p. 1025–1042.
Printed in the United States of America

Material strength degradation effect on cratering dynamics

R. P. SWIFT*

Physics International Company, San Leandro, California 94577

Abstract—An assessment of the effect of shock-induced degradation of material strength on cratering formation is made by comparing two numerical cratering computations for a large scale nuclear surface burst over layered rock. In one of the calculations, ELK 76, the failure strength model based on laboratory data remained unchanged throughout the calculation. In the other calculation, ELK 76-DEG, the failure strength model was modified to allow the cohesive strength and pressure dependent portion of the failure surface to be reduced as a function of peak mean stress attained during shock passage. The computation results show that crater shape and formation times are strongly dependent on material strength; with increasing radius to depth ratio, larger volume, and longer formation time associated with the reduction in strength behind the outgoing shock wave. The increase in volume and formation time is the consequence of a large gain in kinetic energy at the expense of distortional energy. The increase in radius to depth ratio is a result of the major portion of the gain in kinetic energyshowing as increased material velocity in the region of the crater lip. Finally, the results of these two calculations clearly illustrate the need for a greater understanding of the post shock-conditioned response of rock.

INTRODUCTION AND SUMMARY

FOR THE PAST DECADE OR SO, several computations of cratering, ejecta, and ground motion associated with low yield nuclear and high explosive events have been performed. These calculations have usually been performed with either Lagrangian, Eulerian, or coupled Euler-Lagrange finite difference codes that implement complex material models to describe the nonlinear response of geological media. Overall, these calculations have predicted several features that compare favorably with available experimental data. Recently, effort has been made to compute high-yield nuclear surface cratering phenomena. The computed results obtained indicate that the crater volumes are over 10 times smaller than predictions based on large yield cratering data from the Pacific Proving Ground. Furthermore, the computed crater shape, being approximately hemispherical, is not in agreement with existing crater geometries produced by megaton size explosions in the Pacific where some radius to depth ratios (R/D) are on the order of 16.

In the past several months considerable effort has been expended to identify mechanisms that could possibly account for discrepancies between the values of crater volume and shape that results from empirical extrapolation of low yield data to large yield surface explosions or from computations of large yield events with those characteristic features of the megaton size Pacific craters. Several feasible mechanisms have been identified, such as source configuration, water

*Present address: Lawrence Livermore Laboratory, Livermore, California 94550.

washing, rebound, slumping, post-shocked material behavior, and many more. None of these alone appear to be able to totally account for the discrepancies in crater volume and shape, but if taken collectively, may be able to do so. For the present paper the effect of the "post-shocked" material behavior on the computed crater characteristics is considered.

Because the major portion of crater formation occurs with the material in the crater region residing at a low mean stress state near its cohesive limit, it was indicated that the treatment of the low mean-pressure portion of the material failure surface behind the outgoing shock wave generated by an explosion may have a significant influence on computed crater growth. Analyses based on the flow field modeling technique for cratering and ejecta, Maxwell and Seifert (1975), suggested that an improved treatment of this portion of material models used in calculations may result in crater volumes about an order of magnitude larger than previously computed. This hypothesis stimulated an investigation of the effects on cratering caused by competent material strength characteristics being degraded by the passage of the shock wave.

In this paper a comparison is made of the results for two large-yield nuclear surface burst cratering computations, one with and one without shock-induced degradation of the material strength. Both computations were performed with ELK, a two-dimensional, coupled Euler-Lagrange finite difference code. The geology employed for both computations was identical and is representative of a layered shale-sandstone media. In the calculation without material strength degradation, denoted as ELK 76, the rock material was assumed to have the same failure surface characteristics after shock-wave passage as the original competent rock. A characteristic of the strength models used was a cohesive strength (i.e., yield strength at zero confining pressure) that was about 0.5–0.7 of the high pressure von Mises limit. It was postulated that the cohesive strengths in the ELK 76 calculation were too high to represent the true characteristics of the material in and near the crater region following shock passage, and as a consequence, crater growth was retarded.

In the calculation with shock-induced material strength degradation, denoted as ELK 76-DEG, the failure strength model was somewhat arbitrarily modified to allow the cohesive strength and pressure dependent portion of the failure surface to be reduced as a function of the peak mean stress that the material experiences during the passage of the outgoing shock wave. This modification mainly affected the low-pressure portion of the material failure surfaces where the material resides during most of the crater formation time. The von Mises limits remained unchanged. The cohesive strength of the material experiencing peak shock pressures of 10 MPa or more was reduced to 0.2 MPa. A lower limiting value for the degraded cohesive strength could have been used, for example, a value of zero. However, this would have necessitated a major modification to the models employed, and thus to avoid this, the value of 0.2 MPa was used. Cohesive strengths for material experiencing weaker peak shock pressures are reduced less, so that material near the elastic transition ranges essentially had no change in its failure model. Aside from this alteration to the failure surface model, the material models and all other calculational aspects for both computations were the same.

A comparison of the results from ELK 76, the calculation in which strength of material remained unchanged throughout the entire calculation, with the ELK 76-DEG results that are based on the degradation strength characteristics, show the following features:

(1) The reduction in material strength behind the shock wave causes a drop in distortional energy in the ELK 76-DEG Lagrange grid as compared to the ELK 76 Lagrange grid which is first noticeable after 100 msec. This difference in distortional energy continues to increase until about 300 msec where, thereafter, it remains constant at about 9.5% of the total distortional energy. Forty percent of this energy change occurs in the zones adjacent to the crater boundary with the rest spread throughout the active part of the grid, decreasing with increasing distance from the crater.

(2) The loss in distortional energy is offset by gains in kinetic energy and dilatational energy with a ratio of about 3:2 in favor of kinetic energy. The major increase in kinetic energy in confined early to the entire near crater region and eventually distributed to the crater side, near surface, and lip regions. The gain in dilatational energy appears uniformly distributed throughout the near and far crater field.

(3) The dynamic growth of the crater remains essentially unchanged prior to 100 msec. Thereafter, the ELK 76-DEG crater grows at a higher rate with corresponding longer formation times for the crater depth and radius than those observed for ELK 76.

(4) The strength degradation effect leads to an increase in final crater depth of 25% and to an increase in final crater radius of 77%. This results in an increase in radius-to-depth ratio of 1.42 and an increase in volume of 3.3 for ELK 76-DEG over those computed for ELK 76.

(5) The influence of the enhanced crater growth is reflected in the displacement response beyond the immediate crater region in a manner compatible with the distribution of the change in kinetic energy. This influence down along the axis of the symmetry extends only to a little beyond twice the crater depth, while near the surface the influence is noticeable out to about five crater radii.

In the following sections the degradation strength model is presented and the results above are illustrated further with selected numerical results from the ELK 76 and the ELK 76-DEG calculations being displayed and compared.

Computational Configuration and Material Models

Before discussing the strength degradation model used in ELK 76-DEG, a brief discussion of the computational configuration and the constitutive models used in the ELK 76 and ELK 76-DEG calculations is given. The geology simulated is representative of a layered shale-sandstone media. Figure 1 is a schematic of the computational configuration showing the Eulerian and Lagrangian regions and the models used in these regions. Gravity was imposed on both the Euler and Lagrange grids throughout the calculation. The Euler grid was dropped at 60 msec when energy coupling to the Lagrange grid was complete and activity in the Euler region was negligible. After the Euler grid was dropped airblast loading on the surface was produced according to the Brode airblast function (Brode, 1970).

The constitutive models included the Tillotson–Allen equation-of-state for rocks (Allen, 1967), for regions experiencing high shock pressures, an elastic-plastic soil Cap model (Nelson, *et al.*, 1971), for the intermediate stress region where strength effects are important, and an elastic model for the seismic region. In the Euler soil region, see Fig. 1, where the stress levels were 10–20 GPa and higher, only the Tillotson–Allen model was used. In the Lagrange region the Tillotson–Allen model was coupled to the Cap model when the rock material

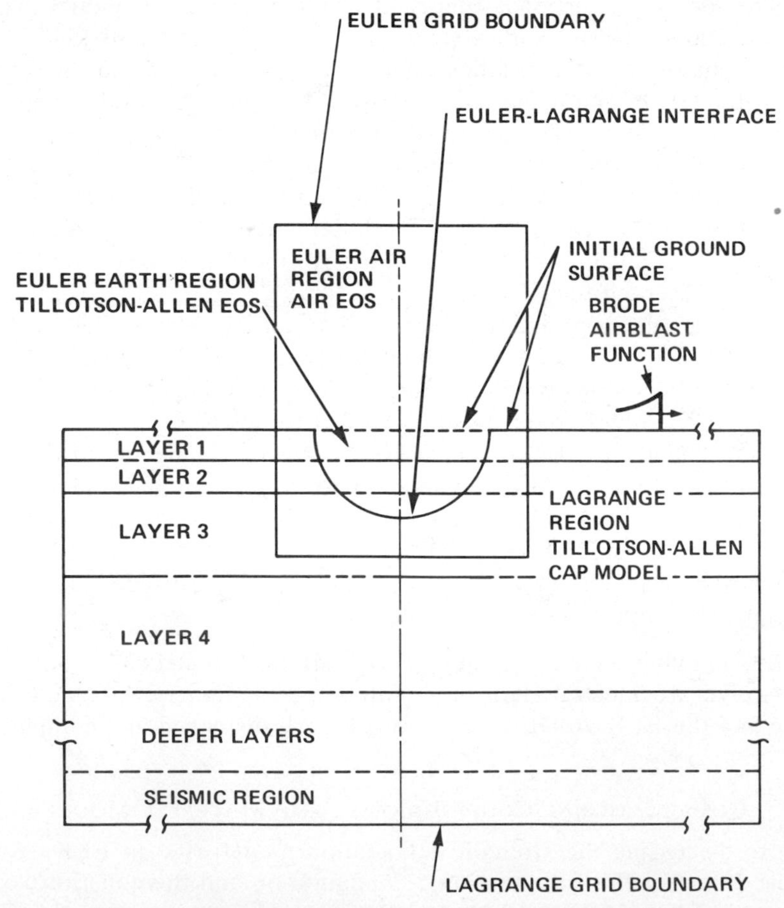

Fig. 1. Computational schematic for ELK 76 and ELK 76-DEG.

became fully compacted. The Cap model provided for compaction of unfilled voids and allowed shear strength dependence on pressure. In the fully compacted region the shear strength was independent of pressure. The effect of water in partially filled voids was accounted for in the fully compacted region by mixing the hydrostats of quartz and water to construct a hydrostat for the high pressure behavior. For the ELK 76-DEG calculation, only the strength part of the Cap model was altered with all other constitutive properties the same as those in ELK 76.

Strength Degradation Model

Rocks under shock-type loading may upon unloading from the shocked state and subsequent reloading be in a partially or completely pulverized state. The failure strength characteristics would have a tendency to change from that of a competent rock material to those similar to a soil. In effect, the cohesive strength and the Mohr–Coulomb failure limit of the competent rock would degrade in proportion to the degree of pulverization. Although the upper failure limit, the von Mises limit, is probably not very sensitive to the degree of rock competency, the onset of the von Mises limit (i.e., the lowest mean hydrostatic pressure associated with the von Mises limit) is probably sensitive on the same order as the cohesive strength and the Mohr–Coulomb limit.

One way to account for the degradation of failure strength of a competent rock is to relate the amount of degradation to the mean hydrostatic pressure associated with the shock loading. If the pressure attained is below some pressure associated with the onset of compressive pulverization, the strength characteristics of competent rock remain unchanged. For higher pressures, the cohesive strength and the Mohr-Coulomb surface degrade as some function of pressure to a lower limit, beyond which the pulverized rock material is considered to have a cohesive strength of soil.

The failure surface employed in ELK 76 is defined in the form

$$\sqrt{J_2'} = A - C \exp(BJ_1), \tag{1}$$

where $J_2' = 1/2\, S_{ij}S_{ij}$ is the second deviatoric stress invariant, $J_1 = \sigma_{ii} = -3P$ is the first stress invariant, S_{ij} and σ_{ij} are components of the deviatoric stress tensor and total stress tensor respectively, P is the mean hydrostatic pressure, and A, B, and C are constants. The cohesive strength is defined at $J_1 = 0$ as

$$\sqrt{J_2'} = A - C,$$

while the von Mises limit is approached as $J_1 \to \infty$ and is

$$\sqrt{J_2'} = A.$$

For ELK 76-DEG, the degradation of the cohesive strength with pressure is expressed in the form

$$A - C = \zeta(\bar{P})(A - C_0) \tag{2}$$

where $A - C_0$ is the cohesive strength of the competent rock material and $\bar{P}$ is the peak pressure the material has thus far experienced. The function $\zeta(\bar{P})$ is assumed to be a piecewise linear function of P and is expressed for different limits of pressure as follows:

$$\zeta(\bar{P}) = \begin{cases} 1 & \text{for } \bar{P} \leq P_L, \\ 1 - \delta\left(\dfrac{\bar{P} - P_L}{P_u - P_L}\right) & \text{for } \bar{P}_L \leq \bar{P} \leq P_u, \\ 1 - \delta & \text{for } \bar{P} \geq P_u, \end{cases} \tag{3}$$

where P_L is the mean hydrostatic pressure associated with the onset of compressive pulverization and P_u is the mean hydrostatic pressure associated with complete pulverization under compression. The value of the constant δ is

$$\delta = 1 - \frac{(A - C)_{P_u}}{A - C_0}, \tag{4}$$

where $(A - C)_{P_u}$ is the cohesive strength of the shock degraded material. Combining Eqs. (2) and (3) with Eq. (1) gives the failure surface used in ELK 76-DEG as

$$J'_2 = \begin{cases} A - C_0 \exp(BJ_1) & \text{for } \bar{P} \leq P_L, \\ A - \left[C_0 + \delta(A - C_0)\dfrac{\bar{P} - P_L}{P_u - P_L}\right] \exp(BJ_1) & \text{for } P_L \leq \bar{P} \leq P_u, \\ A - [C_0 + \delta(A - C_0) \quad \exp(BJ_1)] & \text{for } \bar{P} \geq P_u. \end{cases} \tag{5}$$

An illustration of how linear pressure dependent degradation affects the cohesive strength and the failure surface is shown in Fig. 2. It is noted that during unloading the value of $\zeta(\bar{P})$ does not vary with mean pressure, but retains the minimum value it has thus far attained. A zero cohesive strength for $\bar{P} \geq P_u$ is given by $\delta = 1$, see Fig. 2a. For cohesive strengths greater than zero, $\delta \leq 1$, and the cohesive strength and failure surface are unchanged if $\delta = 0$. The failure surfaces corresponding to the competent material, partial degradation and total degradation are shown in Fig. 2b.

The values of the degradation model parameters δ, P_L, and P_u were given in the following manner. A value of δ was given such that the degraded cohesive strength, $(A - C)_{P_u}$, for any zone is reduced to 0.2 MPa when the peak pressure has exceeded P_u. To ensure the same seismic behavior for ELK 76-DEG as that for ELK 76, the lower limit P_L was assumed to be $P_L = P_{\text{seismic}}$, the seismic pressure. P_{seismic} is taken as 2.5 times the overburden pressure in a layer P_{OB}. Mueller and Murphy (1971) have shown that good correlations with underground explosions are obtained if peak overpressure at the elastic radius is taken to be 1.5 times the overburden. Since gravity acts explicitly in these ELK calculations, the seismic pressure is 2.5 P_{OB} for comparison with total pressure. Depending on the depth of the material, P_{seismic} varied from about 0.2–20 MPa. Below P_{seismic} the material behaved elastically. The material was considered to be fully degraded when the peak pressure exceeded an upper limit $P_u = (P_L + 10)$ MPa.

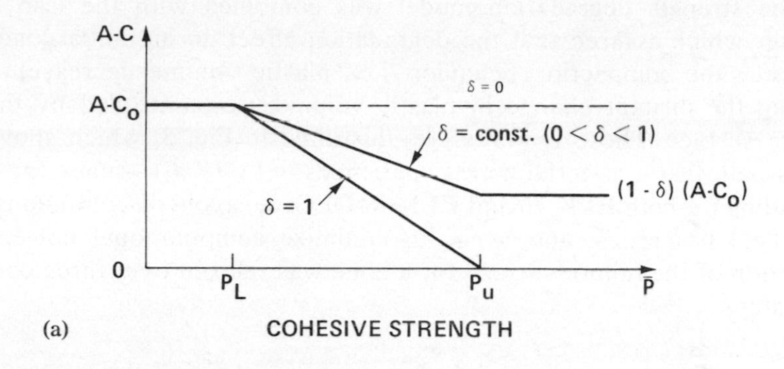

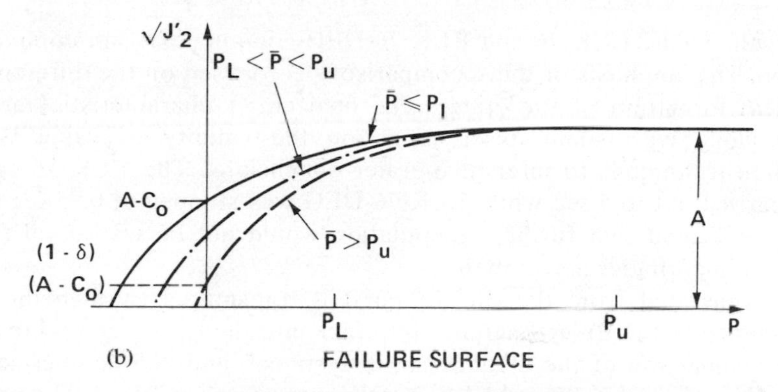

Fig. 2. Strength degradation model.

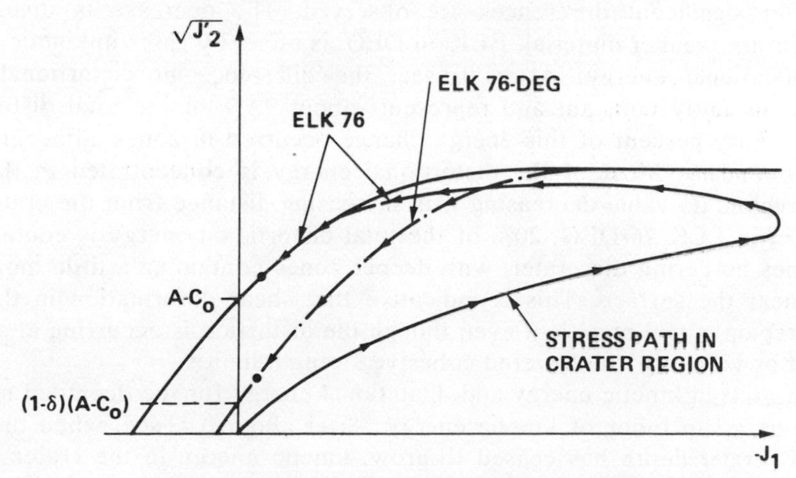

Fig. 3. Failure surfaces with typical stress paths.

The strength degradation model was combined with the Cap model in a manner which assured that the degradation effect during initial loading did not influence the compaction behavior (i.e., plastic volume decrease). During unloading the dilatant change in plastic volume was controlled by the degraded failure surface. These features are illustrated in Fig. 3, which shows a typical stress path that a material zone experiences in ($\sqrt{J_2'}$, J_1)—space for loading and unloading for both ELK 76 and ELK 76-DEG. To avoid discontinuous changes in flow field properties, and hence to minimize computational noise, each finite reduction of the failure surface for a zone was relaxed over three computational time steps.

Discussion of Computational Results

The results of the ELK 76 and ELK 76-DEG computations are compared in this section. The emphasis of these comparisons is focused on the differences in the transient formation of the crater. The final crater characteristics are also compared along with some observations on the validity of using ballistic extrapolation techniques to infer final crater dimensions. The ELK 76 calculation was carried out to 5 sec while ELK 76-DEG was stopped at 0.95 sec where the results indicated that further computation would not be warranted for the purpose of comparing crater growth.

Energy generated from the surface burst is transmitted through the Euler region, then across the Euler-Lagrange interface into the Lagrange grid (e.g., see Fig. 1). A comparison of the dilatational, distortional, and kinetic energies as a function of time for ELK 76 and ELK 76-DEG are shown in Fig. 4. The energies are plotted as a function of the total energy transmitted into the Lagrange grid. Prior to 0.1 sec, virtually no difference in the energy partitioning was noted, while thereafter significant differences are observed. The decrease in distortional energy in the weaker material, ELK 76-DEG, is offset by gains in kinetic energy and dilatational energy. After 0.3 sec, the difference in distortional energy remains fairly constant and represents about 9.5% of the total distortional energy. Forty percent of this energy change occurred in zones adjacent to the crater boundary. Most of the distortional energy is concentrated in the near crater region, its value decreasing with increasing distance from the crater. For example, in ELK 76-DEG, 20% of the total distortional energy is contained in the zones bordering the crater, with deeper zones containing a little more than those near the surface. This is indicative that shear deformation in the near crater region is still prevalent even though the distortion is occurring at a stress level at or very near the lowered cohesive strength limit.

The gains in kinetic energy and dilatational energy for the degraded material are about 3:2 in favor of kinetic energy. After about 0.43 sec, when the ELK 76-DEG crater depth has ceased to grow, kinetic energy in the crater bottom region is solely caused by a slight rebound and is very small in comparison to the kinetic energy in the expanding crater walls. About 40% of the total gain in kinetic energy is concentrated in the lateral and near-surface region of the ELK

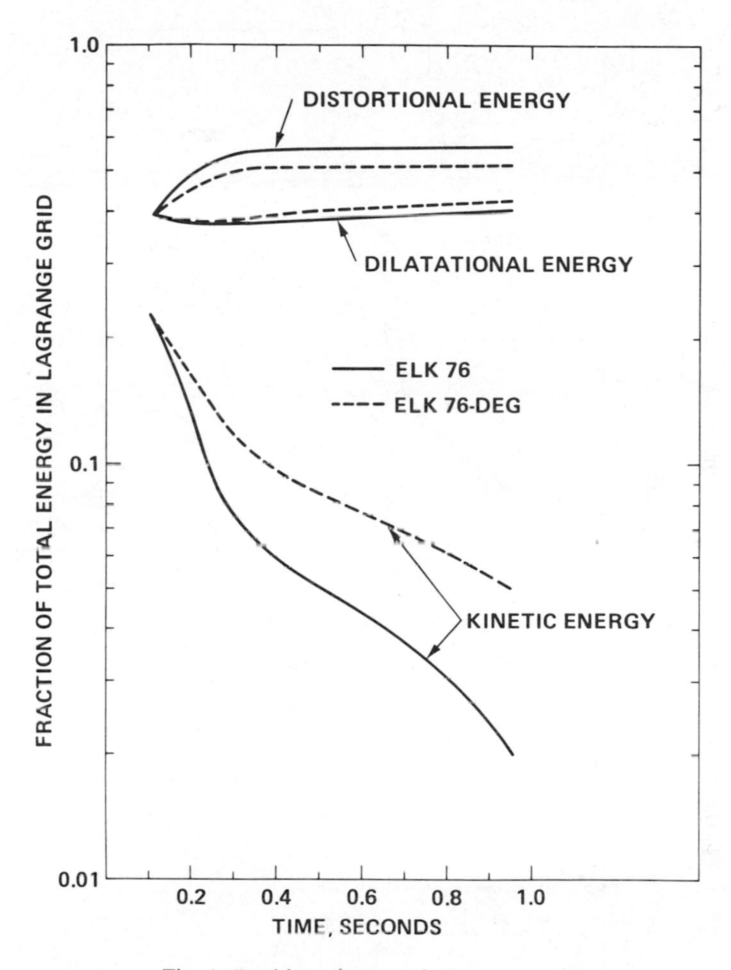

Fig. 4. Partition of energy in Lagrange grid.

76-DEG crater. This increase of kinetic energy results directly in an increase in crater radius and volume as illustrated below. At 0.95 sec, the degraded material has about 2.5 times more kinetic energy than the non-degraded material. In the ELK 76-DEG crater, 28% of the total kinetic energy in the Lagrange grid remains in the near surface region at 0.95 sec, while only 8.1% remains in the same region in ELK 76. Finally, the gain in dilatational energy in ELK 76-DEG represents only about 5% of the total dilatational energy. In contrast to the gain in kinetic energy and the loss in distortional energy, the increase in dilatational energy is distributed more uniformly throughout the near and far cratering regions.

Consistent with the temporal variations in energy partitioning, there is virtually no difference observed between the ELK 76 and ELK 76-DEG crater contours prior to 0.1 sec, while considerable differences are observed thereafter. This is illustrated in Figs. 5–7 which show a comparison of the crater contours

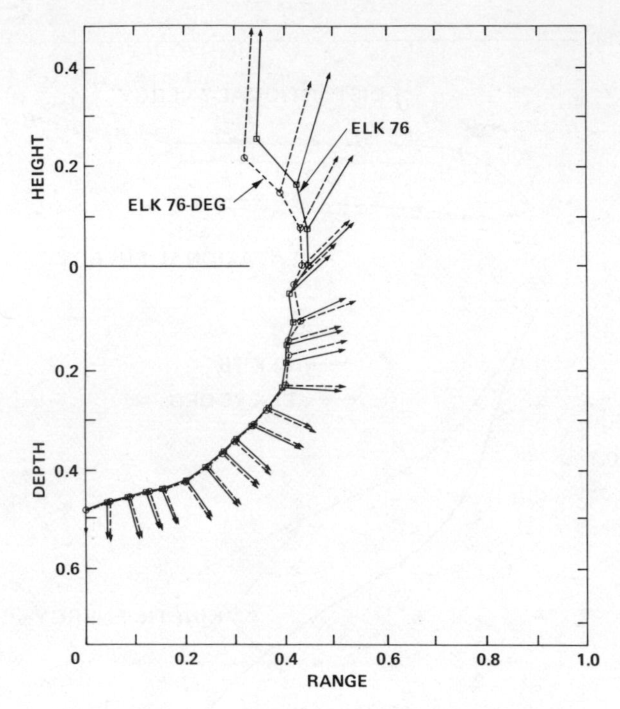

Fig. 5. Crater contour and velocity vectors at 0.1 sec.

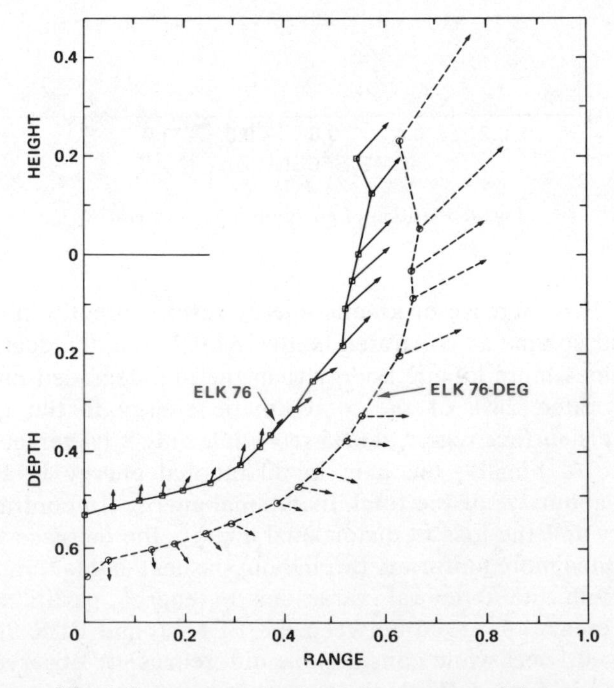

Fig. 6. Crater contour and velocity vectors at 0.4 sec.

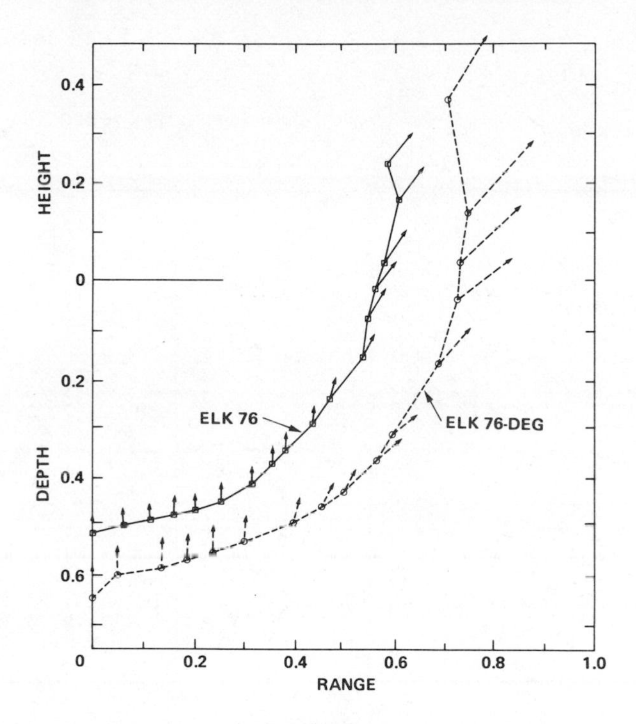

Fig. 7. Crater contour and velocity vector at 0.55 sec.

and the velocity vectors on the contours at 0.1, 0.4, and 0.55 sec. The scale of the contour is normalized with respect to the final crater radius obtained for ELK 76-DEG. At 0.4 sec and 0.55 sec, the crater volumes for ELK 76-DEG are larger than those for ELK 76 by factors of 1.76 and 2.12, respectively. By 0.4 sec, Fig. 6, the bottom of the ELK 76 crater is rebounding while the ELK 76-DEG crater depth is still increasing. By 0.55 sec, the ELK 76-DEG crater has started to rebound. A comparison of the velocity fields at 0.2 and 0.55 sec are shown in Figs. 8 and 9, respectively. At 0.2 sec, the observed difference in velocity field is fairly uniform throughout the near crater region, while at 0.55 sec the major difference is the near surface and lip regions of the crater. Differences in the far field crater region are quite small and are indistinguishable for the scale shown.

A time history comparison of the normalized crater depth is shown in Fig. 10 and illustrates the retardation and subsequent rebounding of the crater bottom. The crater depth is normalized with respect to the maximum computed depth for the ELK 76-DEG crater occurring at 0.43 sec. A maximum crater depth of 0.8 occurs at 0.27 sec for ELK 76. A comparison of the normalized crater radius as a function of time is shown in Fig. 11 where the normalized value is the final radius obtained for ELK 76-DEG by ballistic extrapolation from 0.95 sec. While the ELK 76 crater radius reaches a maximum of 0.58 at 0.8 sec and essentially remains there, the ELK 76-DEG crater radius is continuing to grow at 0.95 sec where it is 0.82 of its final value. Figure 12 illustrates how the crater radius-to-depth ratios vary with time. Initially, the crater forms in a nearly

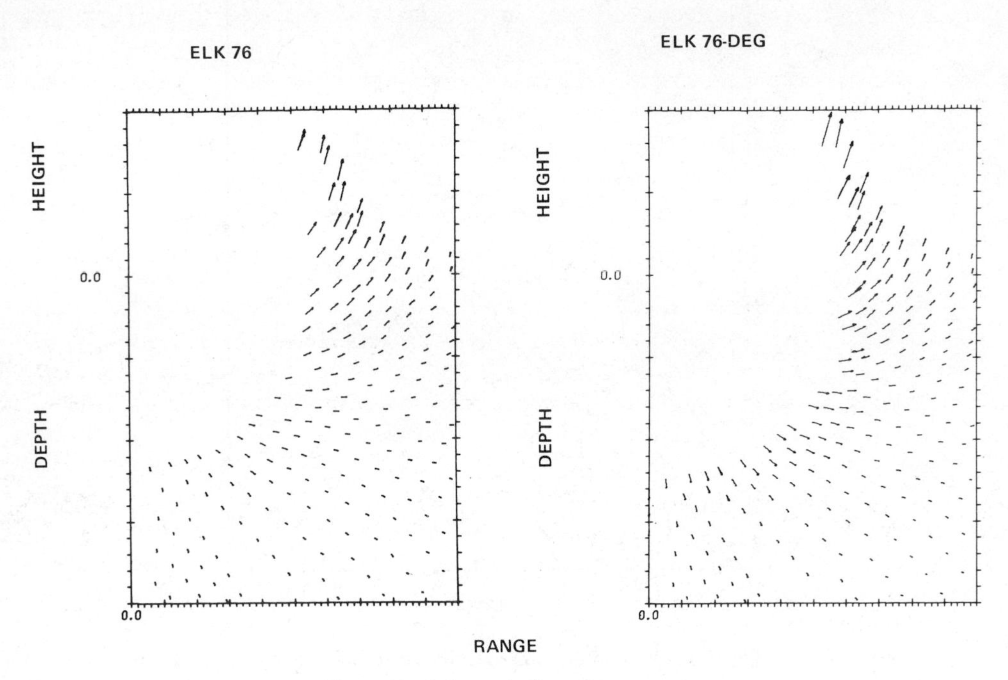

Fig. 8. Cratering velocity field at 0.2 sec.

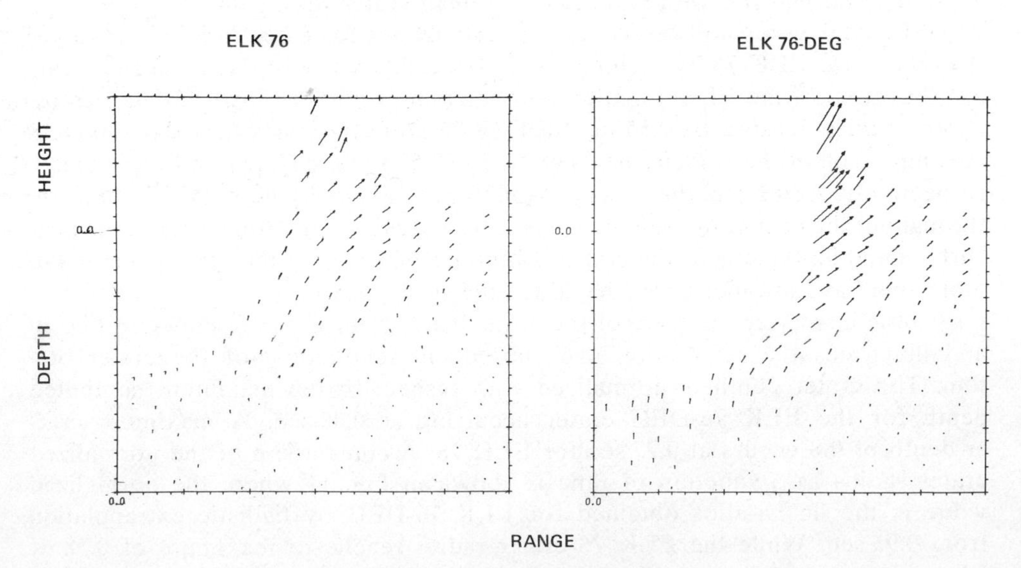

Fig. 9. Cratering velocity field at 0.55 sec.

hemispherical manner, with the crater depth growing slightly faster than the crater radius. After a period of time the combined effects of dissipation and redistribution reduce the kinetic energy in the bottom region of the crater to levels greatly below that in the side region. From there on, the radius-to-depth ratio exceeds unity and the formation of the crater is controlled by the radius growth. The longer period of hemisphericity for the ELK 76-DEG crater shown in Fig. 12 is attributed to the gain in the kinetic energy in the crater bottom region resulting from the material unloading from the shock state along the degraded failure surface as was illustrated in Fig. 3.

The time history growth of the ELK 76-DEG crater volume relative to the ELK 76 crater volume is shown in Fig. 13. The actual computed results are only valid to 0.95 sec when the ELK 76-DEG calculation was terminated. The two final crater volumes ratios of 4.9 at 4.99 sec and 3.3 at 4.7 sec are based on the final computed crater dimensions out to 5 sec for the ELK 76 crater along with ballistic extrapolation values of the ELK 76-DEG crater radii from computed conditions at 0.55 and 0.95 sec, respectively.

The final crater contours are compared in Fig. 14. The ELK 76 contour is the computed value at 5 sec, while the ELK 76-DEG contour is based on a ballistic extrapolation from the computed conditions at 0.95 sec. The final crater characteristics based on the contours in Fig. 14 are given in Table 1, and compared to values that might be expected based on empirical scaling (Cooper, 1973). Note, all the values in Table 1 are normalized with respect to the value of the parameters for ELK 76-DEG. Thus, for example, the volume of the ELK 76-DEG crater is a factor of 3.3 greater than the volume of the ELK 76 crater, but it is a factor of 4.8 less than that expected from scaling.

It is of interest to point out that although the material region existing in a fully degraded state in ELK 76-DEG is many times larger than the crater region, the effects of strength degradation are confined to a smaller region near the crater. For example, the paraboloid region of complete degradation extends on the surface out to a range of 8.7 crater radii and down to a depth of 3.6 times the crater depth. However, comparisons of differences in ground motion response were observed along the surface only out to a range of 5 crater radii and down along the axis of symmetry to about twice the crater depth.

The differences observed in ground motion response occur late, long after the direct induced shock has passed. Comparisons show that the attenuation of the direct-induced shock wave is not affected by the degradation in material strength. From these observations, it can be inferred that, over a certain bound, the value of pressure for total strength degradation, P_u, used in the degradation model has little influence on the final crater dimensions. The peak mean stress attained at twice the crater depth is 0.36 GPa, while near the surface at a range of five crater radii, it is 0.06 GPa. The values used for P_u range from about 0.01 GPa at the surface to 0.03 GPa at great depths. Thus, if $P_u = 0.06$ GPa was used, negligible changes in the results would be expected. However, the influence of the other parameters in the degradation model cannot be directly

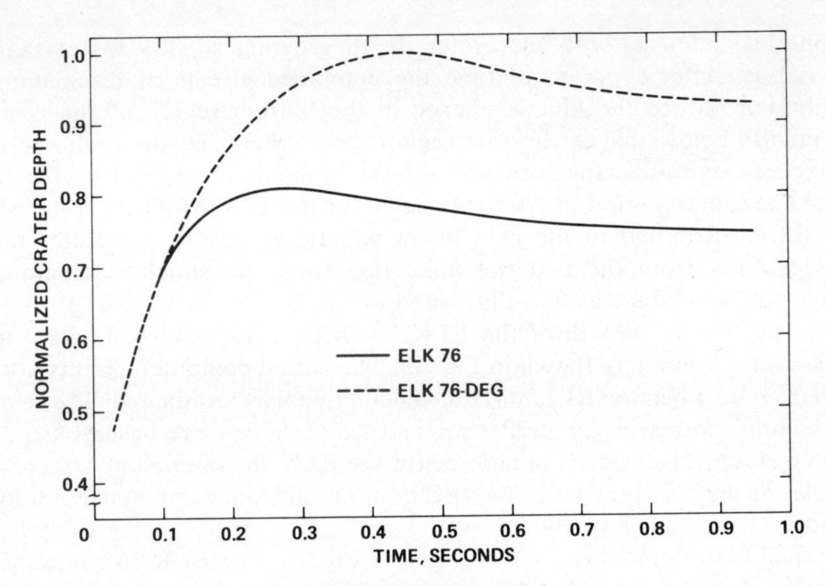

Fig. 10. Crater depth versus time.

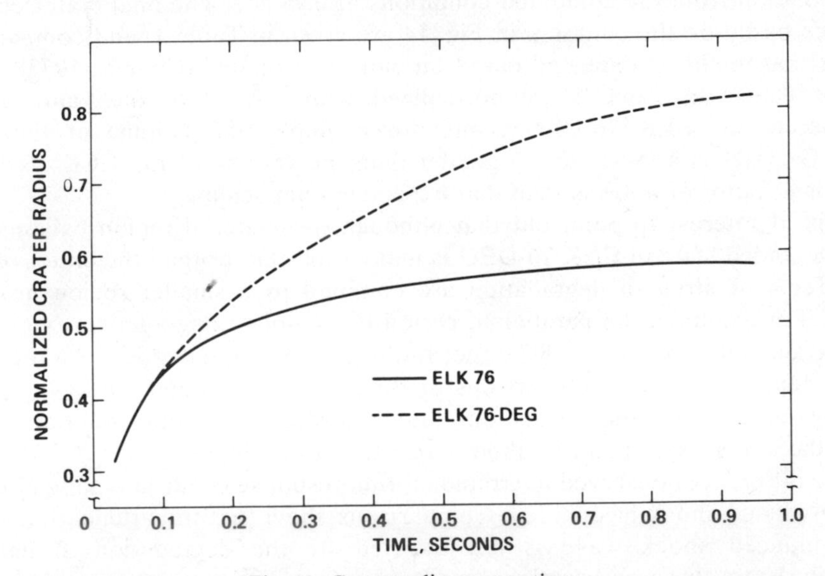

Fig. 11. Crater radius versus time.

inferred from the present results. Their effect on crater dimensions and ground motion would require further computations.

The use of ballistic extrapolation to determine final crater radius has been used mainly in the past to avoid the expense of calculating to late times. It is seen from the results shown in Fig. 13 that caution is warranted when inferring the actual crater radius from extrapolated values at too early a time, for misleading values of crater volume and shape can result. Furthermore, the

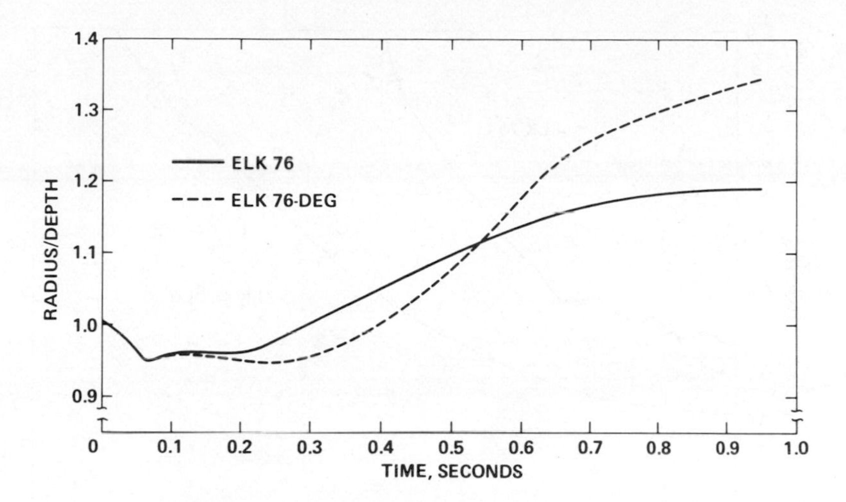

Fig. 12. Crater radius to depth ratio versus time.

anticipated cost savings by not calculating out to late times is somewhat questionable, because experience indicates that most of the expense in a large cratering calculation occurs in the early time stage where zone size and time steps are small.

Used properly, however, ballistic extrapolations are useful in that they can provide a measure of the gravitational influence as compared to the strength influence in retarding the crater formation. This is illustrated in Fig. 15 where extrapolated final values of crater radius for both ELK 76 and ELK 76-DEG are compared to the computed radius values at times used for extrapolation. The early time extrapolated values are significantly larger than the later values and

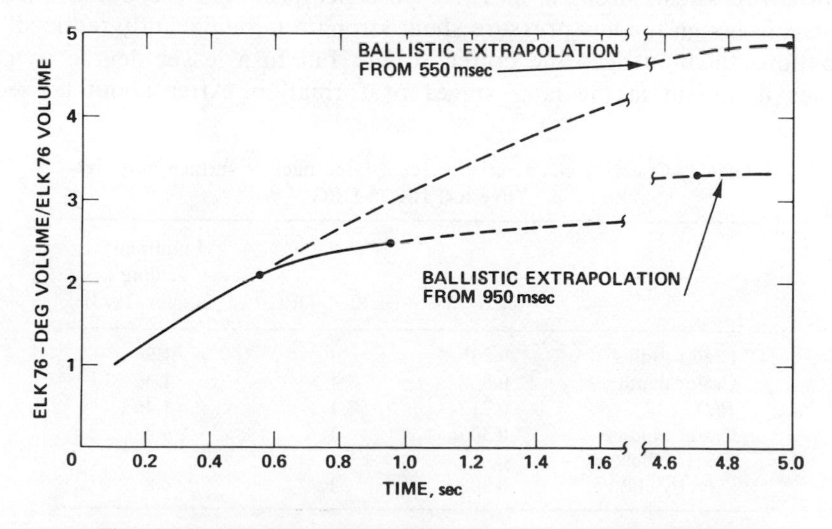

Fig. 13. ELK 76-DEG to ELK 76 crater volume ratio versus time.

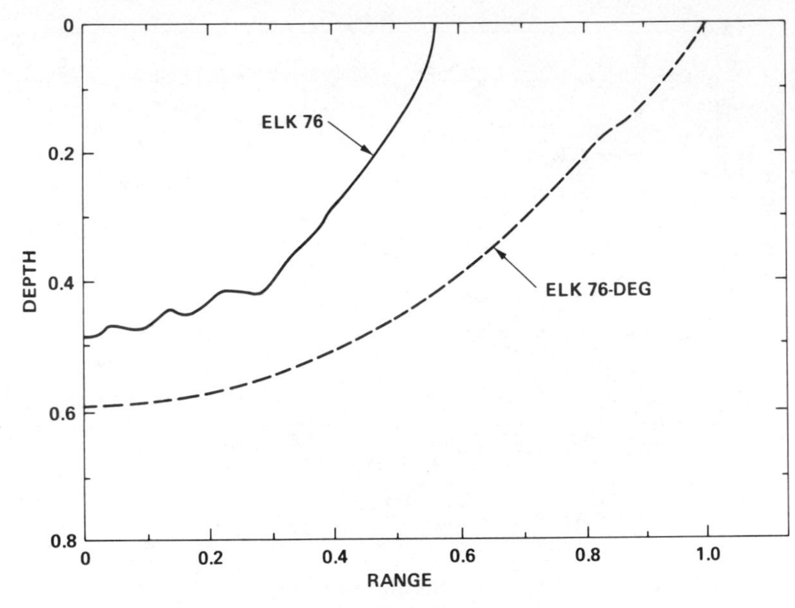

Fig. 14. Final crater contours.

indicate that much of the early loss in kinetic energy is associated with shear strength dissipation. Furthermore, the rate and amount of kinetic energy loss through strength dissipation is in proportion to the value of the shear strength, increasing or decreasing as the shear strength increases or decreases. For example, in ELK 76 the low pressure shear strength dominates the gravity influence, and the crater growth is retarded almost completely by 0.55 sec where the computed crater radius has attained about 98% of its final value at 0.8 sec. The effect of reducing the low pressure shear strength in ELK 76-DEG prolongs the crater formation time. Even though the low pressure shear strength is significantly reduced, shear dissipation still influences the crater growth but to a lesser degree as gravity becomes dominant in the later stages of formation. After about 0.7 sec, the

Table 1. Cratering characteristics for a large nuclear surface burst relative to ELK 76-DEG.

	ELK 76	ELK 76-DEG	Empirical* scaling (Cooper, 1973)
Crater radius	0.565	1	1.98
Crater depth	0.8	1	1.36
R/*D*	0.71	1	1.46
Crater volume	0.306	1	4.8
Crater efficiency (vol/yield)	0.306	1	4.8

*Assuming geology is described as dry soil.

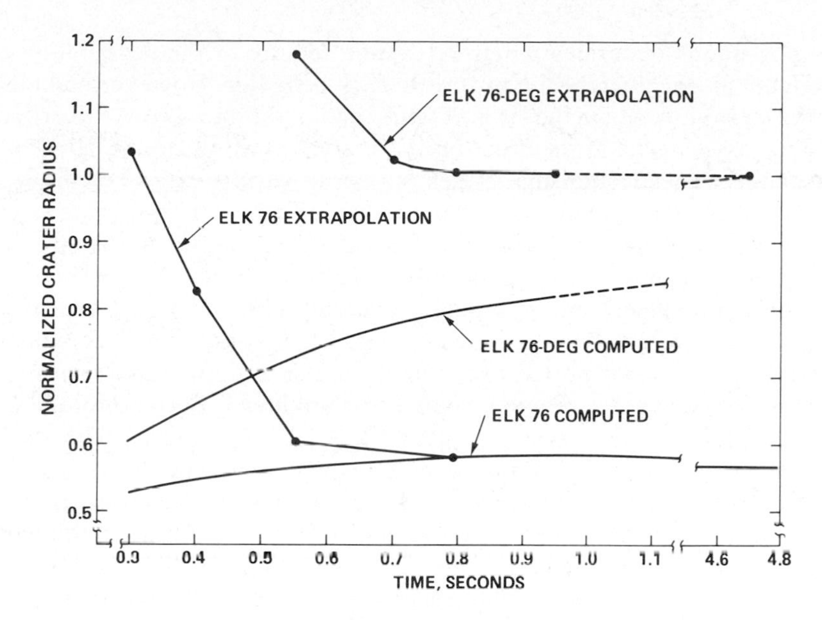

Fig. 15. Computed radius versus ballistic extrapolated radius.

extrapolated value of the crater radius remains essentially constant while the computed radius is still growing at 0.95 sec, having only 82% of its indicated final extrapolated value. The formation time of the ELK 76-DEG crater is about a factor of 5.9 longer than that for the ELK 76 crater. Note this value is almost twice the increase in volume value. This result is in fair agreement with what would be expected by empirical prediction of crater formation times (Sauer and Cooper, 1976).

Finally, it may be implied from the observed results of the two cratering calculations that when the ballistic extrapolated radius value becomes constant with time, that value is a good approximation of the final crater radius. Note, this is only an indication derived from the immediate results and has yet to be verified in general.

Conclusions

The comparison of the computed cratering formation characteristics for a large yield nuclear surface burst show that crater shape and formation time are strongly dependent on material strength, with larger R/D, larger volume, and longer formation time associated with a reduction in the low pressure failure strength (i.e., cohesive limit) after passage of the direct-induced ground shock. The increase in volume and formation time are a consequence of the gain in kinetic energy at the expense of distortional energy, and the increase in R/D is attributed to the eventual distribution of this kinetic energy to the crater side, near-surface, and lip regions.

The post-shock material strength relaxation feature of the degradation model is considered physically realistic for brittle rock behavior. However, the idealization of this behavior used in the ELK 76-DEG computation was done to provide an expeditious assessment of its effect on the crater resulting from a large nuclear surface burst. The question that arises is, would further reduction in material strength or altering parameter values in the model lead to larger gains in kinetic energy and hence, larger craters with increased R/D's? Unfortunately the answer to this question cannot be conclusively ascertained from the present results. Undoubtedly more kinetic energy would result from a further reduction in material strength, and the volume would increase. But, whether the trend for this kinetic energy to be mainly transported to the upper crater and lip regions with an R/D continuing to increase as strength is reduced, is unknown. Furthermore, it cannot be directly inferred that a zero strength hydrodynamic material with the same compressibility characteristics and layering would result in the largest crater. Layering features could greatly influence the redirecting of kinetic energy from the lower crater region to the upper crater region. In any event, the partitioning and distribution of energy in the crater flow field appears strongly dependent on the post-shocked material strength. This uncertainty associated with the influence of post-shock material behavior on cratering dynamics represents an area where more basic computational efforts are needed.

REFERENCES

Allen, R. T.: 1967, Equation of State of Rocks and Minerals, General Atomic Report GAMD-7834, San Diego, California.

Brode, H. L.: 1970, Height of Burst Effects at High Overpressures, Defense Atomic Support Agency DASA 2506, Rand Co., Santa Monica, California.

Cooper, H. F.: 1973, Empirical Studies of Ground Shock and Strong Motions in Rock, Defense Nuclear Agency, DNA 3245F, R&D Associates, Santa Monica, California.

Cooper, H. F. and Sauer F. M.: 1977, Crater-Related Ground Motions and Implications for Crater Scaling. In *Impact and Explosion Cratering* (D. J. Roddy, R. O. Pepin, and R. B. Merrill, eds.), Pergamon Press, New York. This volume.

Maxwell, D. E. and Seifert, K. S.: 1975, Modeling of Cratering, Close-in Displacements and Ejecta, DNA 3628F, Physics International Company, San Leandro, California.

Mueller, R. A. and Murphy, J. R.: 1971, Seismic Characteristics of Underground Nuclear Detonations Part I. Seismic Spectrum Scaling. *Bulletin of the Seismology Society of America* **61**, 1675–1692.

Nelson, I. M., Barron, M. and Sandler, I.: 1971, Mathematical Models for Geological Materials and Wave Propagation Studies. In *Shock Waves: the Mechanical Properties of Solids; Proceedings of the 17th Sagamore Army Materials Research Conference.* (J. J. Burke and V. Weiss, eds.) Syracuse University Press, Syracuse, N.Y.

Roddy, D. J., Pepin, R. O., and Merrill, R. B., editors.
(1977) *Impact and Explosion Cratering*, Pergamon Press (New York), p. 1043–1056.
Printed in the United States of America

A dynamic crater ejecta model

WILLIAM R. SEEBAUGH

Science Applications, Inc., McLean, Virginia 22101

Abstract—A dynamic model for the high explosive and nuclear crater ejecta environments has been developed. The model considers the entire history of the processes of crater formation and ejecta deposition, and provides both the time-dependent airborne ejecta environment and the ejecta blanket thickness as a function of range from the apparent crater radius to the maximum ejecta fragment range.

The ejecta blanket thickness was predicted for a 1-megaton nuclear surface burst on hard rock. The thickness at the crater lip is 3.2 m; this value is in good agreement with empirical results derived by previous investigators. The overall thickness distribution decays as the inverse 3.5 power of the range from the burst; this decay law reproduces the average distribution for experimental nuclear craters. The maximum ejecta fragment range for this case is 6.5 km.

1. INTRODUCTION

THE EJECTA BLANKET CONTAINS a substantial fraction of the total mass excavated during the formation of an explosive or impact crater. A number of empirical models for the thickness of the ejecta blanket have been developed; see, for example, Post (1974) and McGetchin *et al.* (1973). These models are based upon measurements obtained primarily from a large number of high explosive (HE) and nuclear craters. Limited data are also available for the Arizona Meteor Crater, and estimates of the ejecta thickness at the crater rim have been made for several lunar craters.

An alternative method of obtaining estimates of the ejecta blanket thickness considers the entire history of the crater formation and ejecta deposition. In this approach, which is the subject of this paper, the crater formation and ejecta throwout processes are modeled, the trajectories of representative ejecta fragments are calculated, and the ejecta blanket thickness is obtained by summing the contributions of the fragments impacting the surface. The current model was developed for HE and nuclear explosions near the earth's surface; these events are characterized by buoyant fireballs which rise through the atmosphere, perturbing the trajectories of the ejecta fragments. The problem is delineated schematically in Fig. 1. The left side of the sketch illustrates the positions of the representative fragments a few seconds after the burst. These fragments impact the surface within about a minute of the burst, forming the ejecta blanket as shown on the right side of the sketch. The inner region of the ejecta blanket is characterized by a continuous distribution of material that completely covers the original ground surface. At larger ranges, the ejecta mass is insufficient to cover the surface, and the distribution is discontinuous. The ejecta mass is defined as the amount of crater material that crosses the original ground surface and

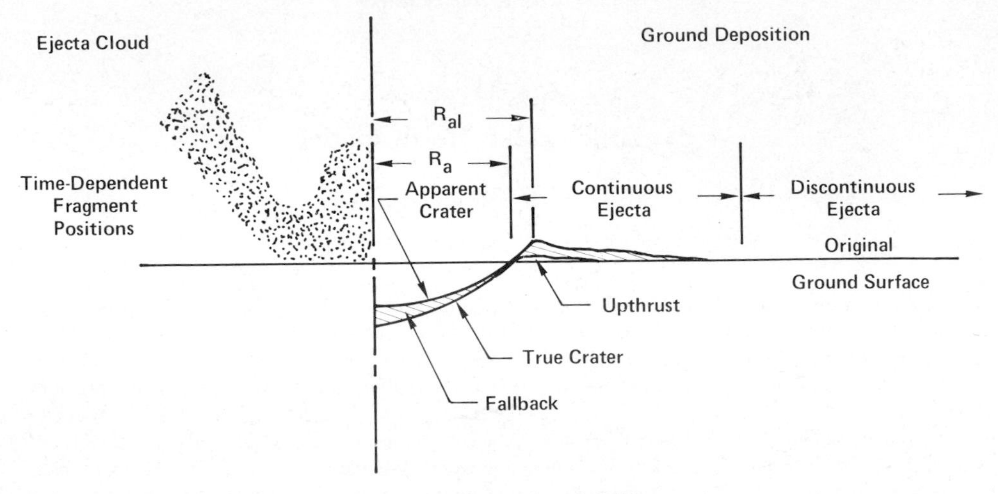

Fig. 1. Ejecta problem definition.

impacts beyond the apparent crater radius (R_a in Fig. 1); this definition excludes both fallback and upthrust. The crater lip radius is denoted by R_{al}.

2. General Description of Ejecta Model

The ejecta model is based upon recent crater calculations, selected low-yield ejecta data (primarily from HE events) and current models of the buoyant fireball and the associated wind field. The model is illustrated in flow chart form in Fig. 2. The inputs to the model are:

(1) Theoretical velocity distribution for continuous hydrodynamic material emerging from the crater.
(2) Empirical models for ejecta mass, fragment size distribution, and maximum fragment size.
(3) Experimental observations of fragment velocity for HE shots.
(4) Vortex flow model (VORDUM*) for the fireball and winds which affect ejecta fragment trajectories.
(5) Compressible flow aerodynamic drag model.

Items (1) through (3) above define the ejecta source. Ejecta fragments produced by this source are transported through the flowfield defined by the VORDUM program [Item (4) above] using the aerodynamic drag model for ejected crater material [Item (5) above]. The source model and the trajectory calculation form the complete ejecta model as shown in Fig. 2. The deposition results, which are considered herein, are obtained in the form of the cumulative mass and number of fragments impacting the ground plane per unit area, the ejecta depth, the

*VORDUM is an acronym for a vortex dust model that was developed to describe the fluid and particulate motion associated with the rise of a buoyant fireball through the earth's atmosphere.

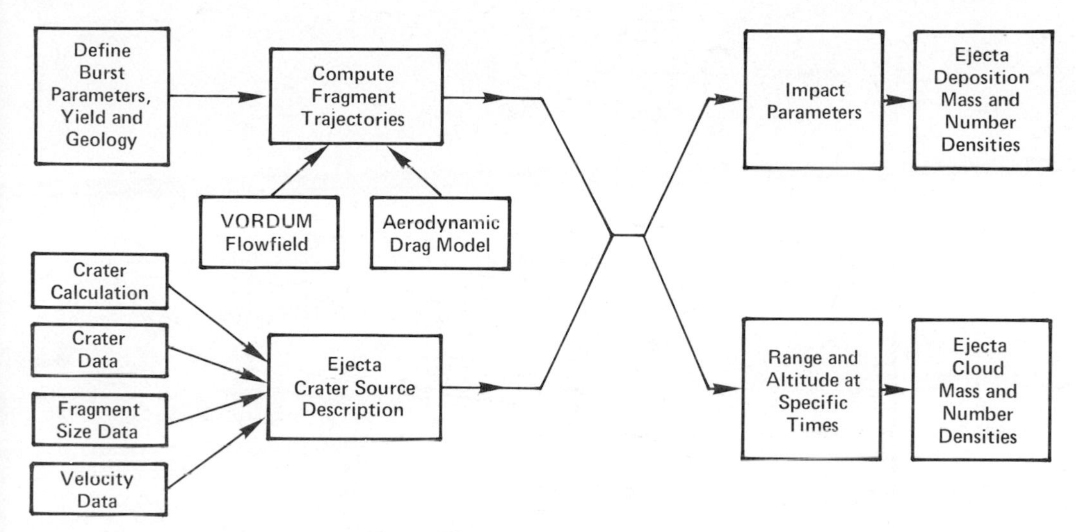

Fig. 2. Ejecta environment analysis.

fragment size class (minimum and maximum fragment diameters), and the minimum and maximum values of the impact time, velocity, angle, momentum, and kinetic energy as functions of range from the burst point. The ejecta environment model may also be used to predict the early-time airborne ejecta environment, which is termed the ejecta cloud. During the VORDUM trajectory calculations (Fig. 2), trajectory parameters in the form of range and altitude versus time are recorded for each representative fragment. This information is combined with the ejecta source description to give airborne ejecta fragment mass and number densities. These results may then be used as a means of model verification by comparing the predicted airborne fragment distribution to similar results obtained from high speed photography of test events (Wisotski, 1977). The ejecta cloud model is not considered in this paper.

3. The Ejecta Source

The most important assumptions of the ejecta model are related to the ejecta source. The source description must include the mass ejection rate, the ejection velocity (speed and angle), and the fragment size distribution as functions of radius and time. Empirical correlations have been derived for the total ejecta mass (W. M. Layson, pers comm., 1969) and for the relative origins of ejecta fragments within the crater (Davis and Carnes, 1972). The latter correlation is shown in Fig. 3 for event MINERAL ROCK, a 100-ton TNT shot on granite. As shown, the ejecta missiles originate from the upper portion of the crater away from the center of the charge. Very little ejecta comes from the region near the axis. Although cratering studies have produced correlations such as those shown in Fig. 3, attempts to construct an ejecta source model completely from experimental data have been unsuccessful. Data do not exist on the rate of ejection of material

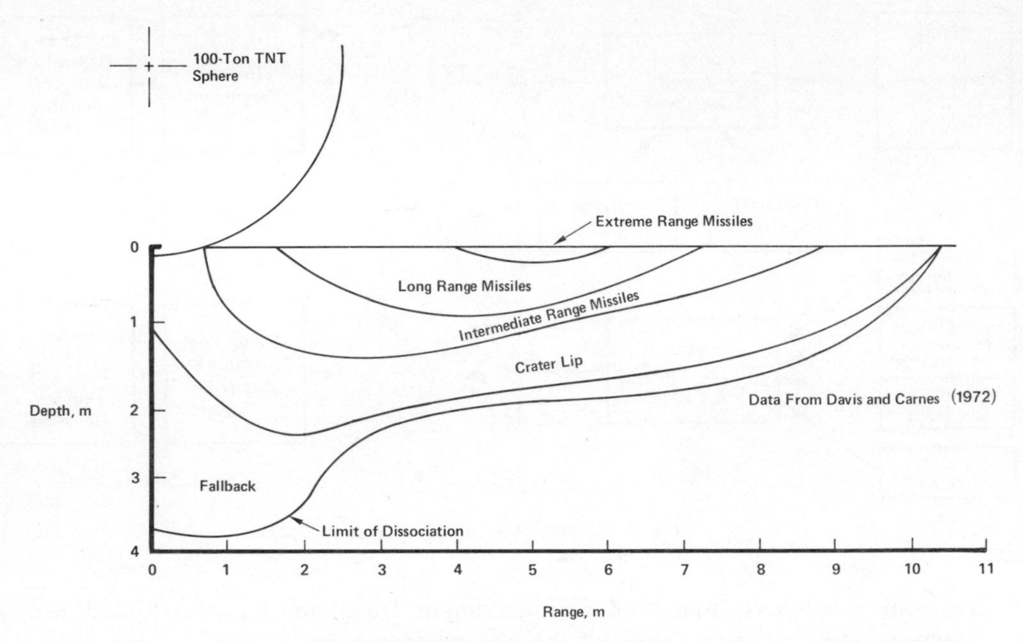

Fig. 3. Zones of origin of crater ejecta.

from the crater. The in-flight photography of ejecta fragments may be analyzed to give representative fragment ejection velocities, but such analyses do not relate mass to velocity. It is, therefore, necessary to incorporate results of theoretical calculations into the ejecta source model to complete the source description.

A number of two-dimensional hydrodynamic calculations have been performed recently using advanced elastic-plastic material models. The methods generally have not been successful in predicting the final crater dimensions. It appears that the poor agreement for the final crater volume is a result of failure to properly predict the late-time motion, which occurs at low velocities. The mass of ejecta leaving the crater with higher velocities is not as dependent on the material model as the low-velocity ejecta, and consistent results have been obtained for the high-velocity material by assuming that the ejecta motion is driven by the amount of energy coupled to the ground (defined as the total energy below the ground surface at 1 μsec after burst). The results of an analysis of the properties of the crater material crossing the original ground surface for a number of calculations are presented in Fig. 4 in the form of average vertical velocity (defined as the ratio of the momentum flux to the mass flux) versus the cumulative ejecta mass with greater average vertical velocity. Curve 1 represents the average relationship for high strength materials (wet tuff and granite), whereas Curve 2 is the average for two calculations of the nuclear event JOHNIE BOY (0.5-kiloton yield at a depth of burst of 0.6 m in alluvium). These curves suggest the source relationship given by the dashed line in Fig. 4 as

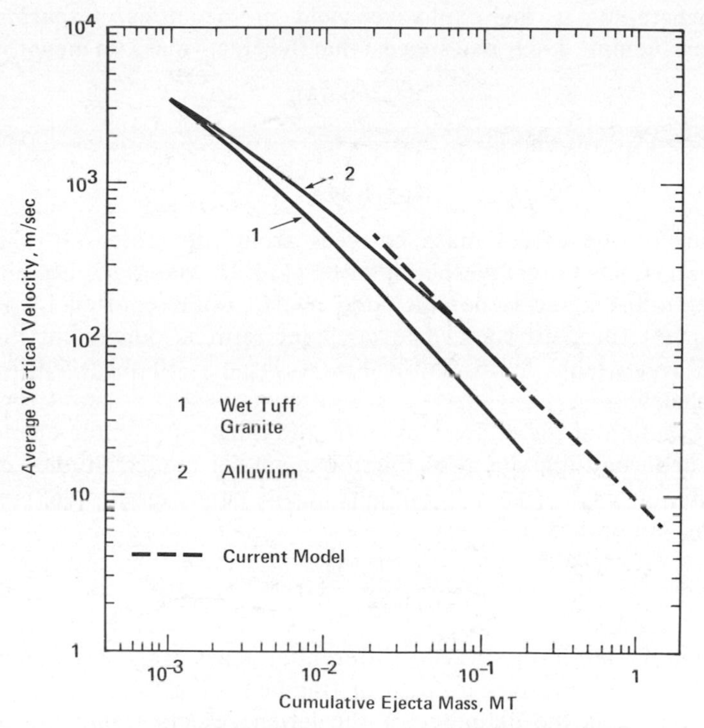

Fig. 4. Cumulative ejecta mass from crater calculations.

an upper velocity bound for a nuclear surface burst. The equation of this linc may be represented by the function

$$\frac{\partial M}{\partial V} = M_e \frac{V_{\min,z}}{\sin \theta} V^{-2}, \tag{1}$$

where $\partial M/\partial V$ is the ratio of the differential ejecta mass δM associated with differential ejection velocity δV, M_e is the ejecta mass, θ is the ejection angle relative to the horizontal, and V is the ejection velocity. This function is normalized to give the empirically determined ejecta mass (see below) at a minimum value of the vertical component of the ejection velocity $V_{\min,z}$ of 6.4 m/sec. The minimum velocity considered is then

$$V_{\min} = \frac{(V_{\min,z})}{\sin \theta} = \frac{6.4 \text{ m/sec}}{\sin \theta}. \tag{2}$$

The total ejecta mass is determined from correlations of experimental cratering data. Layson (pers. comm., 1969) has quoted the relationship

$$M_a = 2.4 W^{0.9} \tag{3}$$

for the apparent crater mass (M_a, megatons) for megaton-range surface bursts on rock, where W is the explosive yield in megatons. Experimental data (Layson, pers. comm., 1969) also suggest that the ejecta mass (in megatons) is given by

$$M_e = 0.6M_a \tag{4}$$

or, combining Eqs. (3) and (4)

$$M_e = 1.44W^{0.9}. \tag{5}$$

A large part of the ejecta mass emerges from the crater at relatively low velocities and comes to rest on the crater lip (Fig. 1). Material with initial vertical velocity below 6.4 m/sec is not included in M_e. Data reported by Post (1974) confirm Eq. (5) for hard rock. The constant term is dependent on the local geology and water table depth, with values as large as 10 observed for saturated clays and shales.

The distribution of the crater mass over the range of possible ejecta fragment sizes is an important ingredient of the ejecta source model. Studies of cratering in rock media such as basalt and granite (Layson, pers. comm., 1969) indicate that the size distribution has the form

$$\frac{\partial M}{\partial a} = \frac{0.5M_e}{\sqrt{a_m}} a^{-0.5}, \tag{6}$$

where $\partial M/\partial a$ is the ratio of the differential mass δM associated with the differential diameter δa, over a range of fragment diameters from about 10^{-4} to 1 m. The term a_m is the diameter of the largest ejecta fragment. Gault *et al.* (1963) compiled data for explosive and impact cratering events and observed that the correlation for the mass of the largest fragment

$$m_m = 6.6 \times 10^5 M_e^{0.8}, \tag{7}$$

was a reasonable representation of the data for a variation of 14 orders of magnitude of M_e. The constant of proportionality gives the maximum fragment mass m_m in kg when M_e is expressed in megatons. Assuming spherical fragments, the maximum fragment diameter $a_m(m)$ is given by

$$a_m = \left(\frac{6m_m}{\pi\rho}\right)^{1/3}, \tag{8}$$

where ρ is the density of the ejected material (the *in situ* density is used in the current model). The size distribution given by Eq. (6) is assumed for all fragment sizes with the upper limit given by a_m.

The crater mass-velocity relationships derived from the cratering calculations and the experimentally derived fragment size distributions cannot be truly coupled because the process of comminution of the crater material has not been described either theoretically or experimentally. The approach taken in the current ejecta model is to distribute the ejecta mass over the velocity-size spectrum using the aforementioned relationships, and to exclude combinations of velocity and size that appear unreasonable based upon analysis of additional experimental data.

The basis for a limiting velocity-size relationship is provided by the results of an extensive in-flight ejecta analysis using photography of HE events (Wisotski, 1977). The analysis gives the sizes and ejection conditions for a number of large ejecta fragments. The results for event MIDDLE GUST III, a 100-ton surface tangent TNT shot on a layered (clay over shale) medium, are presented in Fig. 5.* The fragment diameters were obtained by assuming spherical fragments. The data are bounded by two limits:

$$V_{\max} = 566\,W^{1/6}, \tag{9}$$

and

$$V_m = C V_{\max} \frac{a_m}{a}, \tag{10}$$

where velocities are in m/sec. The material ejected at the highest hydrodynamic velocities near the burst (Fig. 4) must either be vaporized or pulverized by the shocks measured in hundreds of kilobars. It is not reasonable to associate these high velocities with fragments of significant size. The absolute maximum velocity, Eq. (9), performs the required limiting function, with the value of the constant determined from the data of Fig. 5. The 1/6-power dependence of velocity on yield was suggested by Cooper and Sauer (1977) and observed experimentally by Wisotski (1977). Qualitative observations of results of field surveys and aerial photography for HE and nuclear events indicate that the

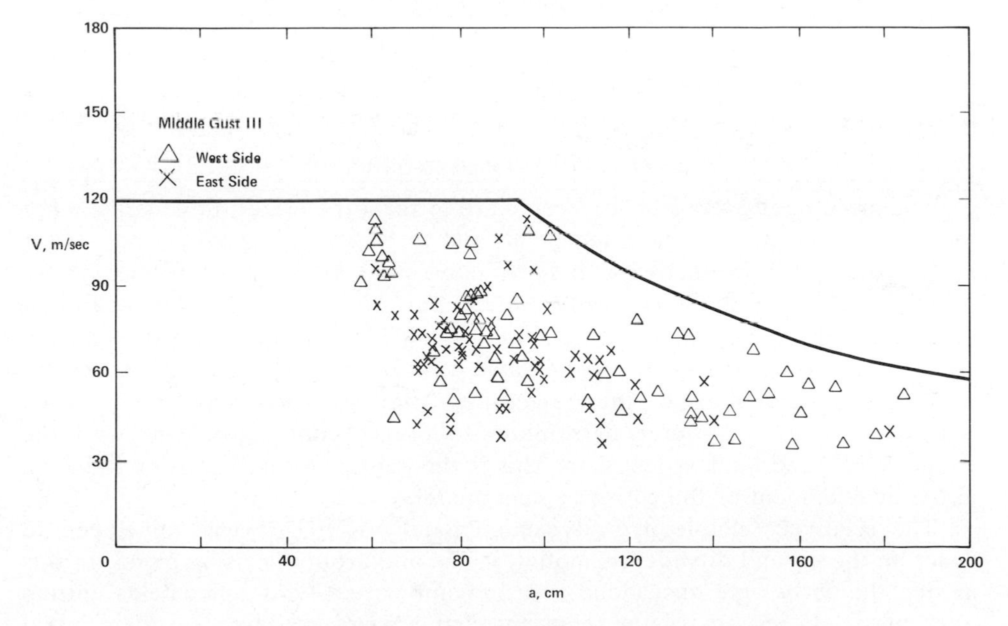

Fig. 5. Correlation of ejecta velocity and fragment diameter.

*A small number of fragments (9 out of a total of 151) which required large corrections to obtain the true azimuthal planes of the trajectories were eliminated from the sample. See Wisotski (1977) for an explanation of the correction procedure.

largest ejecta fragments do not reach the longest ranges from the burst. The use of a size-dependent maximum velocity, Eq. (10), produces the observed results. The form of this limit was also obtained from the data of Fig. 5. The parameter C is a function of the explosive type (HE or nuclear) and the depth of burst. Tentative values for this parameter are 0.5 for surface tangent HE (for example, event MIDDLE GUST III), 0.3 for half-buried HE, and 0.1 for megaton-range nuclear surface bursts. Note that the size-dependent velocity limit shown in Fig. 5 was calculated using the experimentally determined ejecta mass in Eq. (7) since the correlation of Eq. (5) does not apply to HE craters.

The relationships given in Eqs. (1)–(10) define the overall characteristics of the ejecta source. In order to complete the source model, the points of origin of the ejecta fragments within the crater must be specified. The experimental ejecta origin description illustrated in Fig. 3 is modeled effectively by a crater model developed by M. Rosenblatt and G. E. Eggum (pers. comm., 1975). This model simulates the time-phased development of the crater as described by the same hydrodynamic cratering calculations used to determine the mass-velocity relationships shown inFig. 4. The ejection and mass flux across the original ground surface are specified as functions of time (consistent with the dashed line in Fig. 4). Each increment of mass is partitioned into a range of fragment sizes according to Eq. (6), consistent with the limits specified by Eqs. (9) and (10). The result is a matrix of representative ejecta fragments, with each fragment characterized by its ejection velocity (speed and angle), diameter, ejection time, location within the developing crater, and the ejecta mass that it represents. This matrix provides the initial conditions for the ejecta trajectory calculations.

4. Ejecta Trajectories

The primary effects of the explosion are to form the crater, to generate a hot low density region near the burst point, and to create strong afterwinds which perturb the ejecta trajectories. It is of interest to know more of the motion through the very early fireball environment, but it is beyond the state of the art to describe the details of the interaction of the ejected earth material and the developing early fireball. The most reliable way to circumvent this difficulty is to make good calculations of the trajectories after pressure equilibration while adjusting the ejecta source description to obtain good agreement with the available HE and nuclear test data. This is the approach that has been followed in the development of the current ejecta model.

The relatively simple hydrodynamic model VORDUM was developed to describe the air and dust/debris motion inside and around a rising cloud. In this model, the early time dust cloud is a buoyant vortex whose flowfields entrain dust particles and distribute them through space according to their initial position, mass, and aerodynamic drag characteristics. Within the fireball radius, the aerodynamic drag is calculated for air at one-tenth ambient density. Trajectories of larger ejecta fragments are perturbed by the wind field. The boundary condition at the ground—no normal velocity component at the ground surface—is satisfied by imposing an image vortex system that moves away from the ground in the negative field. As the cloud rises, it expands adiabatically and also mixes with the ambient air. The rates at which the cloud rises and expands were obtained from test data and hydrodynamic cloud calculations. The VOR-DUM model provides the flowfield that influences the ejecta fragment motion.

Perhaps the most important assumption in the ejecta trajectory calculation relates to the drag coefficient. Previous studies have used values measured for smooth spheres, but as explained below, this assumption is not appropriate for ejecta fragments, and the present model employs a different drag model. The standard drag curves for flow over a smooth sphere are given in Hoerner (1965). The range of Reynolds numbers* of interest in the ejecta trajectory calculations is from about 10^3–10^6. For incompressible flow, the drag coefficient C_D maintains a nominally constant value of about 0.5 in the region of laminar boundary-layer flow from $Re = 10^3$ to about 3×10^5. In this regime, a substantial portion of the total drag is the pressure drag engendered by the separation of the flow away from the sphere near its equator. At a Reynolds number of about 3×10^5 for a smooth sphere the boundary layer undergoes transition to a turbulent state and the separation point moves aft over the downstream surface of the sphere, substantially reducing the pressure drag and hence the total drag. Values of C_D from 0.1–0.2 are obtained at Reynolds numbers greater than about 5×10^5. Compressibility effects on the drag coefficient are large and nonlinear throughout the Reynolds number range of interest. The influence of transition to a turbulent boundary layer is evident only for Mach numbers† below the critical value of about 0.6. At higher Mach numbers, the shock waves formed in the flow near the equator separate the boundary layer at this point for all Reynolds numbers of interest, and no drag reduction occurs. For Mach numbers greater than 1.5, the drag coefficient is approximately 1.0, independent of Reynolds number.

The aforementioned discussion applies to smooth spheres and spheres roughened to about 3% of the diameter (about the roughness of a golf ball). The ejecta fragments under consideration, in particular, the larger fragments, are not spherical and the values for smooth or slightly roughened spheres do not apply. Drag coefficients for bluff shapes such as short blunt cylinders, wedges, and so forth, with lengths nearly equal to the diameter or width, range from about 0.8–1.2 for Reynolds numbers greater than 10^3. Of even greater significance than the generally higher level for bluff shapes, as compared to smooth spheres, is the observation that no drag decrease occurs in the critical Reynolds number range for bodies with sharp edges or blunt aft faces. This is due to the "fixing" of the point of boundary-layer separation at the sharp corner of the body. These observations apply to data obtained for cubes tumbling in an uncontrolled

---

*Reynolds number is the ratio of inertial to viscous forces: $Re = \rho Va/\mu$, where ρ = atmospheric density, V = fragment velocity relative to the atmosphere, a = fragment diameter, and μ = dynamic viscosity. Drag coefficient is the ratio of the drag force to the product of the dynamic pressure and the fragment cross-sectional area, $C_D = D/(1/2\rho v^2)(\pi a^2/4)$.

†Mach number is the ratio of the fragment velocity relative to the atmosphere to the speed of sound.

manner (data summarized by Hoerner, 1965). A value of C_D of 0.75 was obtained in the incompressible range for cubes. The influence of compressibility on the drag coefficients for bluff shapes and cubes is similar to that for smooth spheres, with C_D increasing to 1.2 for a Mach number of about 1.4, and remaining nearly constant at that value for higher Mach numbers.

Since ejecta fragments have been observed to be very rough and non-spherical in shape, the smooth sphere data were not incorporated into the ejecta model. The most important effect is the elimination of the drag decrease at the critical Reynolds number. The simplest approach employs an incompressible drag law of the form

$$C_D = 0.6 + \frac{36}{Re}. \tag{11}$$

The next step is to consider the effects of the air compressibility. A cross-plot of the data given by Hoerner (1965), delineating the dependence of the smooth sphere drag coefficient on Mach number for Reynolds numbers of interest in the ejecta study, is shown in Fig. 6. The range of drag coefficients given by Eq. (11) for $10^3 < Re < 3 \times 10^5$ is from 0.60–0.64; thus, the incompressible drag coefficients are too low even for smooth spheres above a Mach number of about 0.7. Calculations are currently performed using a compressible drag law shown by the dashed lines in Fig. 6. This compressible draw law has the following characteristics:

(1) It obeys Eq. (11) for incompressible flow, defined here as $M \leq 0.4$ for all Reynolds numbers.

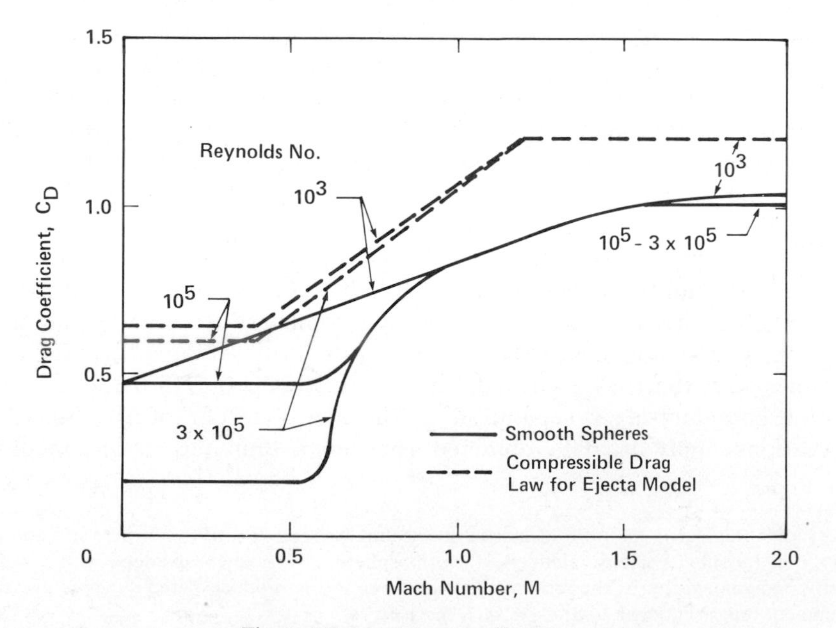

Fig. 6. Fragment drag coefficients.

(2) It obeys Eq. (11) for slow viscous flow at all Mach numbers, defined here as $Re \leqslant 60$.
(3) It fits data from Hoerner (1965) for supersonic flow over bluff objects at high Reynolds numbers, defined here as $M \geqslant 1.2$ for $Re \geqslant 60$.

For the intermediate ranges of Reynolds and Mach numbers, the drag coefficient is determined by interpolating linearly between Eq. (11) at $M = 0.4$ and $C_D = 1.2$ at $M = 1.2$. This procedure is shown in Fig. 6.

5. Results and Discussion

The ejecta blanket thickness was calculated using the current ejecta model for a 1-megaton nuclear surface burst on hard rock. The model parameters were as specified by Eqs. (1)–(11); the factor C in Eq. (10) was set equal to 0.1. The predicted ejecta blanket thickness or ejecta depth is shown in Fig. 7 (current model). The distribution begins at the apparent crater radius ($R_a/R_{al} = 0.8$, where $R_{al} = 0.17$ km). The maximum depth of 3.2 m occurs at the crater lip radius. The

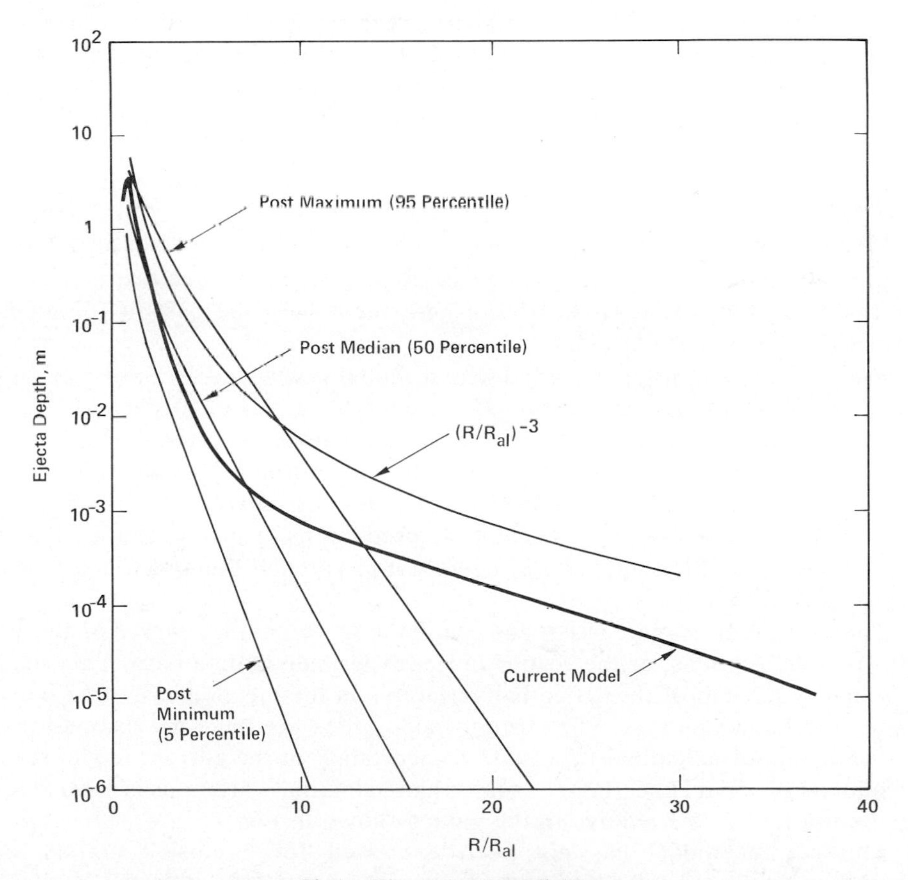

Fig. 7. Ejecta depth predictions.

depth then decreases with increasing range out to the maximum range (R/R_{al} = 37.4 or R = 6.5 km).

The thickness distribution law proposed by McGetchin *et al.* (1973) for lunar craters is also shown in Fig. 7. The ejecta thickness at the crater lip (or rim crest) is given there by the expression

$$T = 0.14R_{al}^{0.74} \tag{12}$$

where all dimensions are in meters. This relationship is based on data from the nuclear craters TEAPOT ESS and JANGLE U, data from the Arizona Meteor Crater, and estimates for the lunar craters Copernicus and Imbrium. The crater lip radius for a 1-megaton surface burst falls between the JANGLE U and Meteor Crater lip radii. The thickness predicted by Eq. (12) is 6.3 m, about twice that obtained using the current model. The higher value is expected for the shallow depths of burial represented by Eq. (12). The depth of burst question and the issue of impact crater equivalence to HE and nuclear bursts are addressed in detail by Pike (1977). The $(R/R_{al})^{-3}$ decay law of McGetchin *et al.* (1973), shown in Fig. 7, was selected as representative of distributions for lunar craters, which are formed in the absence of an atmosphere. The presence of the earth's atmosphere and the inward and upward directed air motion associated with the rising fireball suggest a preferential deposition of ejecta at the shorter ranges. This is observed in the predictions obtained with the current model between $R/R_{al} = 1$ and 6, where an $(R/R_{al})^{-4}$ law results. The predicted depth between $R/R_{al} = 6$ and 18 decreases by about $(R/R_{al})^{-2.5}$. A substantial number of the larger ejecta fragments (diameters from about 0.2 to 5 m) impact in this region. At longer ranges the size range of impacting fragments narrows to about 0.5 to 1.5 m, and the mass distribution varies as $(R/R_{al})^{-5}$. The average distribution law for the current model is $(R/R_{al})^{-3.5}$, in agreement with the best fit for nuclear craters (McGetchin *et al.*, 1973).

Results obtained using the exponential model developed by Post (1974) are given in Fig. 7 for the maximum (95 percentile), median (50 percentile), and minimum (5 percentile) ejecta thicknesses. Post's median correlation is within a factor of 2 of the results obtained using the current model to about $R/R_{al} = 8$. Beyond this range, the basic characteristics of the distributions are different, and Post's median correlation drops below the predictions of the current model. The exponential model thus appears to severely underpredict the ejecta depth at the large ranges.

The maximum (95 percentile) and minimum (5 percentile) curves of the Post (1974) model account for the scatter in the experimental data base. This scatter is primarily a result of the azimuthal variations in the ejecta distribution pattern commonly known as rays. The current model does not have provisions for the determination of azimuthal effects. The uncertainty in the current model results introduced by model uncertainties other than azimuthal variations is estimated to be about a factor of 3 relative to the values shown in Fig. 7.

The current model has not been exercised for explosive craters with diameters larger than 0.5 km; thus further comparisons with the model of

McGetchin *et al.* (1973) are not available for the range of crater sizes of interest in lunar ejecta studies. Conclusions based upon comparisons with Post's model for craters with radii from about 0.05 to 0.5 km are generally similar to those drawn from Fig. 7.

The current model may be adapted to calculate ejecta blanket thicknesses for impact craters by developing relationships analogous to Eqs. (1)–(10) for impact events. Several hydrodynamic calculations would be required to develop the relationships for ejecta mass and velocity. For planetary craters, the gravity and atmospheric density terms in the ejecta trajectory calculations would be modified accordingly. For lunar craters, a simple ballistic extrapolation of the ejecta motion as the fragments are created would be sufficient.

6. Summary and Conclusions

A dynamic model for the high explosive and nuclear crater ejecta environments has been formulated. This model, which is based on the results of the most recent hydrodynamic cratering calculations and explosive test programs, considers the entire history of the crater formation and ejecta deposition. The model provides both the ejecta blanket thickness distribution, which is considered in detail in this paper, and the airborne ejecta cloud environment, which may be used to verify the model by comparing predictions to results obtained from high-speed photography of explosive events.

The ejecta blanket thickness distribution was predicted for a 1-megaton nuclear surface burst on hard rock. The thickness at the crater lip crest is 3.2 m; this result compares favorably with empirical values derived by previous investigators. The overall ejecta thickness distribution varies as the inverse 3.5 power of the range from the burst, a result that reproduces the average distribution for experimental nuclear craters. The detailed distribution is more correctly represented by a continuous function that initially decays as range to the inverse 4.0 power, displays an intermediate −2.5 law, and finally decreases as range to the inverse 5.0 power near the maximum ejecta range of 6.5 km.

Acknowledgments—The program under which the ejecta model was developed was sponsored by the Shock Physics Directorate of the Defense Nuclear Agency; the technical representative of the sponsor was Capt. Jerry R. Stockton. The assistance of Mr. Elwood E. Zimmerman and Ms. Susan J. Rose in performing the computer analyses is acknowledged. The contribution of Dr. William M. Layson, who suggested the modeling approach and contributed to the program through frequent discussions, is especially acknowledged.

References

Cooper, H. F. and Sauer, F. M.: 1977, Crater-Related Ground Motions and Implications for Crater Scaling. In *Impact and Explosion Cratering* (D. J. Roddy, R. O. Pepin, and R. B. Merrill, eds.), Pergamon Press, New York. This volume.

Davis, L. K. and Carnes, B. F.: 1972, *Operation Mine Shaft: Cratering Effects of a 100-Ton TNT Detonation on Granite*, U.S. Army Engineer Waterways Experiment Station report MS-2151.

Gault, D. E., Shoemaker, E. M., and Moore, H. T.: 1963, *Spray Ejected from the Lunar Surface by Meteoroid Impact,* National Aeronautics and Space Administration technical note TND-1767.
Hoerner, S. F.: 1965, *Fluid-Dynamic Drag,* Published by the Author.
McGetchin, T. R., Settle, M., and Head, J. W.: 1973, Radial Thickness Variation in Impact Crater Ejecta: Implications for Lunar Basin Deposits, *Earth Planet. Sci. Lett.* **20**, 226.
Pike, R. J.: 1976, Ejecta thickness, Rim Uplift, Energy Type, and Depth of Energy Release (abstract). In *Papers Presented to the Symposium on Planetary Cratering Mechanics*, p. 105–107. The Lunar Science Institute, Houston.
Post, R. L., Jr.: 1974, *Ejecta Distributions from Near-Surface Nuclear and HE Bursts,* Air Force Weapons Laboratory report TR-74-51.
Wisotski, J.: 1977, Dynamic Ejecta Parameters from High Explosive Detonations. In *Impact and Explosion Cratering* (D. J. Roddy, R. O. Pepin, and R. B. Merrill, eds.), Pergamon Press, New York. This volume.

Roddy, D. J., Pepin, R. O., and Merrill, R. B., editors.
(1977) *Impact and Explosion Cratering*, Pergamon Press (New York), p. 1057–1087.
Printed in the United States of America

Mechanisms and models of cratering in earth media

D. R. Curran, D. A. Shockey, L. Seaman, and M. Austin
Stanford Research Institute, Menlo Park, California 94025

Abstract—The formation of craters and ejecta in solid materials by impact or explosive detonation can occur by a wide variety of mechanisms. Viscous or plastic flow and compaction are the usual mechanisms in soils, brittle fracture is a principal mechanism in hard rock, and shear banding with subsequent failure is a primary mechanism in many metals. In any given material, cratering usually occurs by the simultaneous operation of a combination of mechanisms. Progress in understanding these mechanisms is reflected by the success of our attempts to model them computationally. In this review of recent computational modeling work, we attempt to assess our present level of understanding of cratering phenomena. Emphasis is placed on cratering by fracture and comminution. Examples are shown of experiments and computations of impact and explosive cratering in a well-characterized competent rock. A model that treats the activation of inherent flaws and faults and the growth and coalescence of the resulting cracks to form fragments was used in a finite difference wave propagation code to predict crater sizes and shapes, subcrater fracture damage, and ejecta size and velocity distributions. Good agreement with measured values was obtained.

I. Introduction

The formation of craters and ejecta in geologic materials by impact or explosive detonation can occur by a wide variety of mechanisms. Viscous or plastic flow and compaction are usual mechanisms in soils, whereas brittle fracture and comminution are the principal mechanisms in hard rock. Shear banding with subsequent failure is a primary mechanism in many metals. In any given material cratering usually occurs by the simultaneous operation of a combination of mechanisms. Progress in understanding and modeling these mechanisms is reflected by the success of our attempts to compute cratering behavior. In this paper, we review recent computational modeling work for cratering by fracture and comminution.

Fracture behavior in brittle materials like hard rock is strongly dependent on the load duration. This dependence is a result of the fact that fracture is a dynamic process during which inherent flaws in the material are activated, grow, and finally coalesce to cause full separation or fragmentation. For each material there is thus a characteristic time for fracture and comminution that is associated with the fracture kinetics. If the load duration is long compared with this time, the fracture will be, by comparison, instantaneous, and static fracture mechanics concepts will apply. On the other hand, in impact or explosive cratering the load duration is of the same order as the fracture time, and the fracture kinetics must be accurately modeled if the fracture and associated cratering behavior are to be successfully predicted.

However, the fracture kinetics and the resulting size and velocity distributions of the fragments that make up the ejecta in an explosive or impact cratering event in rock have not been previously predictable from rock properties. Instead a fragment size distribution of the form

$$\rho_f(R_f) = kR_f^n, \quad (1)$$

is assumed (Anthony, 1966; Gault *et al.*, 1963; Moore *et al.*, 1963) where $\rho_f(R_f)$ is the number of fragments of radius R_f per unit volume, and where k and n are constants chosen on the basis of observed ejecta. The value of n is usually taken to be -3.5 for hard rock and -4 for soils. Since these numbers are empirical and based on limited data, no great confidence can be assigned to them when a new type of rock or soil is being considered. What has been lacking is an understanding of how rock properties determine the fragment size and initial velocity distributions in the ejecta.

The same situation has been true for predicting crater dimensions in hard rock from high explosive or impact events. Hydrocode computations have tended to predict craters that are deeper and of smaller diameters than those observed (Ferritto and Forrest, 1975). These codes usually do not treat realistically (or at all) the fracture and fragmentation processes that help determine the final crater dimensions in hard rock. Thus, to understand the dependence of the crater dimensions and the ejecta size distributions it is necessary to have a computational model that correctly simulates fracture and fragmentation of the materials of interest.

Over the past four years, Shockey *et al.* (1973a, 1974) have developed such a computational model (NAG/FRAG) that simulates the activation, growth, and coalescence of preexisting flaws in rock and other solids to form fragments. When inserted into a wave propagation code, NAG/FRAG generates cracks with size and velocity distributions that depend on the constitutive relations and the preexisting flaw distribution in the material.

Before the beginning of the present work this model had been checked against experimental data only under conditions of one-dimensional strain impact in Arkansas novaculite, a naturally occurring polycrystalline quartzite rock. For this case, the agreement was good (Shockey *et al.*, 1973a, 1974). More recently, the model was written in a form suitable for use in two-dimensional wave propagation codes and was thus available for testing against data from cratering or axially symmetric penetration experiments. This paper presents the results of such tests. As will be seen, good agreement was obtained between computed and observed cratering behavior in Arkansas novaculite and basalt. Similar agreement can be expected in other competent rock.

The objectives of the work reported here were to (1) generate by experiment simple, unambiguous crater size data and ejecta size distributions in a well-characterized hard rock and compare these data with the predictions of the NAG/FRAG computational model, and (2) simulate computationally a large scale cratering event, compare calculated and experimental results, and draw conclusions about the feasibility of using NAG/FRAG to predict cratering behavior in rock.

II. Background

The formation of craters and ejecta in earth media by impact or explosive detonation can involve fluid and plastic flow, compaction, fracture and fragmentation, shear banding, and other material response modes. However, because of the difficulties and costs of modeling these underlying cratering mechanisms, cratering predictive capabilities are usually developed using hydrodynamic codes and assuming simple fracture or fluid-flow cratering mechanisms. Such approaches cannot predict ejecta size distributions and tend to predict hemispherical craters in rock that do not agree well with the flat, shallow craters observed in specific tests such as those by Ferritto and Forrest (1975). It thus appeared that significant improvements in predictive capabilities could be obtained through more accurate modeling of the operating mechanisms.

To test this hypothesis, we performed calculations and experiments on a hard, dense, quartzite rock, Arkansas novaculite. This material was chosen because crater and ejecta were expected to be formed almost solely by fracture processes, thereby eliminating the need to model plastic flow and other possible mechanisms. Furthermore, a computational model for fracture under dynamic loads had already been developed by the present authors and applied to this rock under one-dimensional strain conditions. In this section we discuss the phenomenology of crater and ejecta formation in competent rock under dynamic surface loads and then describe the NAG/FRAG computational fracture model.

Upon dynamic load application at the surface of a rock (such as from projectile impact or explosive detonation), compressive stress waves run into the rock and give rise to subsequent tensions. At locations in the rock where tensile stresses are high enough, inherent flaws in the material become unstable and begin to grow. As the cracks continue to extend, they begin to encounter one another and link up. As crack coalescence continues, chunks of material are isolated from the main rock body and, if near a surface, may be ejected at considerable velocity.

The extent to which the fracture/fragmentation process proceeds at a given location depends on the stress history there. Thus material at some distance from the load source, where considerable attenuation of the stress has occurred, may experience various stages of crack nucleation, growth, and coalescence and remain intact as a damaged rock mass of reduced strength. Nearer to the point of load application, however, the stress amplitude and duration may be such that free fragments are generated and leave the site, thereby forming the crater ejecta and determining the size and shape of the crater.

This sequence of failure events leading to fragmentation and ejection of rock was modeled computationally by Shockey *et al.* (1973a, 1974) and applied to a case of one-dimensional strain impact. The model simulation of a plate slap experiment on Arkansas novaculite predicted a fragment size distribution in reasonable agreement with that measured. This encouraged us to insert the fracture/fragmentation model into a two-dimensional wave propagation code and to try to simulate computationally the cratering behavior in the same rock under conditions of projectile impact or in-contact surface explosion.

This computational fracture/fragmentation model, known as NAG/FRAG, treats the four stages of the dynamic fragmentation process (crack nucleation, crack growth, crack coalescence, and fragment formation) in a consecutive manner, describing each as a rate process depending on the instantaneous value of tensile stress and the material properties that govern dynamic fracture response. A wave propagation code is used to compute the stresses at each location in the material at each increment in time. Crack nucleation is allowed to occur at a given location (computational cell) when the stress in that cell (in any direction) exceeds a certain threshold value. This value is measured in dynamic fracture experiments and is considered a material property. When the threshold stress is reached, cracks begin to nucleate at a rate given by

$$\dot{N} = \dot{N}_0 \exp \frac{\sigma - \sigma_{no}}{\sigma_1}, \tag{2}$$

and once nucleated, grow according to the following growth law

$$\dot{R} = \frac{\sigma - \sigma_{go}}{4\eta} R, \tag{3}$$

where σ is the current local tensile stress, R is the half size (radius) of a crack, and $\dot{N}_0$, σ_{no}, σ_1, σ_{go}, and η are the empirically determined material properties governing dynamic fracture response.

At each time step, the code uses Eqs. (2) and (3) to compute the number and sizes of cracks in each computational cell, allowing additional cracks to nucleate, if desired, and previously nucleated cracks to grow.* After each time step, the eroding effect of the fracture damage on the stresses is computed and accounted for in computing the fracture damage during the subsequent time step.

In only mildly damaging dynamic load situations (where the stress pulse is such that significant crack coalescence and fragmentation do not occur), the final damage is a distribution of isolated microfractures of various sizes. The fracture nucleation and growth (NAG) equations (2) and (3) have been successful in predicting in quantitative detail the number, sizes, orientation, and locations of such microfractures in a wide variety of materials, including quartzite rock (novaculite), soft aluminum and copper (Barbee *et al.*, 1972), iron (Barbee *et al.*, 1972), steels (Shockey, 1973b; Seaman and Shockey, 1973), beryllium alloys (Shockey *et al.*, 1973c), and a polymer (Curran *et al.*, 1973).

More severe stress pulses actuate the last two stages of dynamic fracture response—crack coalescence and fragmentation. Crack coalescence is treated computationally by assuming that, as two cracks approach one another, interaction occurs when their strain fields overlap. Coalescence is assumed to be complete within a computational cell when the sum of the crack tip strain fields is equal to the cell volume. The volume of individual crack tip strain fields is

*For novaculite, the initial flaw size distribution was measured directly as explained by Shockey *et al.* (1973a, 1974). All preexisting flaws were activated when the tensile stress exceeded the threshold stress σ_{no}; no new flaws were subsequently activated.

assumed proportional to the cube of the crack half length (radius). The code computes at each time step the crack size distribution in each cell and also computes the total volume of crack tip strain fields for comparison with the cell volume.

The fragment size distribution in a cell is computed directly from the crack size distribution when the coalescence criterion is fulfilled. The number of cracks involved in forming a fragment was taken to be four for the case of novaculite under one-dimensional impact conditions.* Large cracks are assumed to form large fragments, smaller cracks to form smaller fragments. The procedure used to transform crack size distributions to fragment size distributions is described in more detail by Shockey *et al.* (1974).

III. Experimental

Specimen material

Arkansas novaculite, a naturally occurring polycrystalline quartzite, was the hard rock chosen for this study because much is known about its fracture and fragmentation behavior. Novaculite is pure, dense, and homogeneous, consisting of approximately equisized, equiaxed, and randomly oriented quartz grains roughly 10 μm in diameter. A population of flat, approximately circular flaws, the largest of which are about a millimeter in diameter, existed on roughly parallel planes; planar macroscopic faults traversed the rock mass every 10 to 30 cm.

In previous work by Shockey *et al.* (1973a), the size distribution of the planar flaws was determined by counting and measuring the flaw traces revealed on a polished cross section. The data were converted by a statistical transformation to obtain the inherent flaw size distribution. The fitted analytical expression is

$$N_g = \delta(R)\exp[11.1 - (3.7\times 10^2\ \text{cm}^{-1})R + (0.42\times 10^4\ \text{cm}^{-2})R^2]$$

where $\delta(R) = 1\ \text{cm}^{-3}$ for $0 < R \leq 0.05$ cm,
$\delta(R) = 0$ for $R > 0.05$ cm,
N_g = number of cracks/cm^3 with radii greater than R.

The structural and mechanical properties of this novaculite were reported by Shockey *et al.* (1973a).

Specimen configuration

For the cratering experiments cylindrical specimens approximately 12 cm in diameter by 5 cm thick were ground from large blocks of Arkansas novaculite. Each specimen was surrounded on the back and sides by about 10 cm of aluminum to (1) delay the arrival and allow attenuation of release waves originating at free surfaces and (2) confine any sections of the rock that might spall.† The sides of the rock specimens were tapered slightly to allow them to be press-fitted into the center of an aluminum annulus. The top and bottom surfaces of the aluminum-novaculite package were then lapped, and the side and bottom surfaces were coated with a thin layer of light grease to minimize air

*Individual fragments were observed to have an average of about eight sides. Each side must have been formed by one crack, but each crack contributes a side to two (adjacent) fragments. Therefore the relationship between number of cracks and number of fragments was taken as four.

†Aluminum has about the same shock impedance as novaculite.

gaps between rock and aluminum surfaces. Two additional 5-cm-thick aluminum plates were placed under the specimen. This design, shown in Fig. 1, was intended to simulate an infinite medium during the period of crater formation.

In-contact explosive experiments

The fragment catching scheme used for the explosive experiments is depicted in Fig. 2. The confining water not only provides for soft deceleration of the fragile rock fragments, but also provides for rapid cooling of the expanding hot gases. Moreover, the fragments are readily recoverable from the water. This catching scheme, however, does not provide a way to obtain fragment angle and velocity information.

As can be seen from the figure, the specimen package is placed in a 55-gal steel drum that is set in a wider but shallower plastic tank. A watertight plastic bag is placed around the specimen and covered by an inverted plastic can that is subsequently filled with water. A wash tub partially filled with water sits atop the plastic can to help counteract the upward thrust when the explosive is

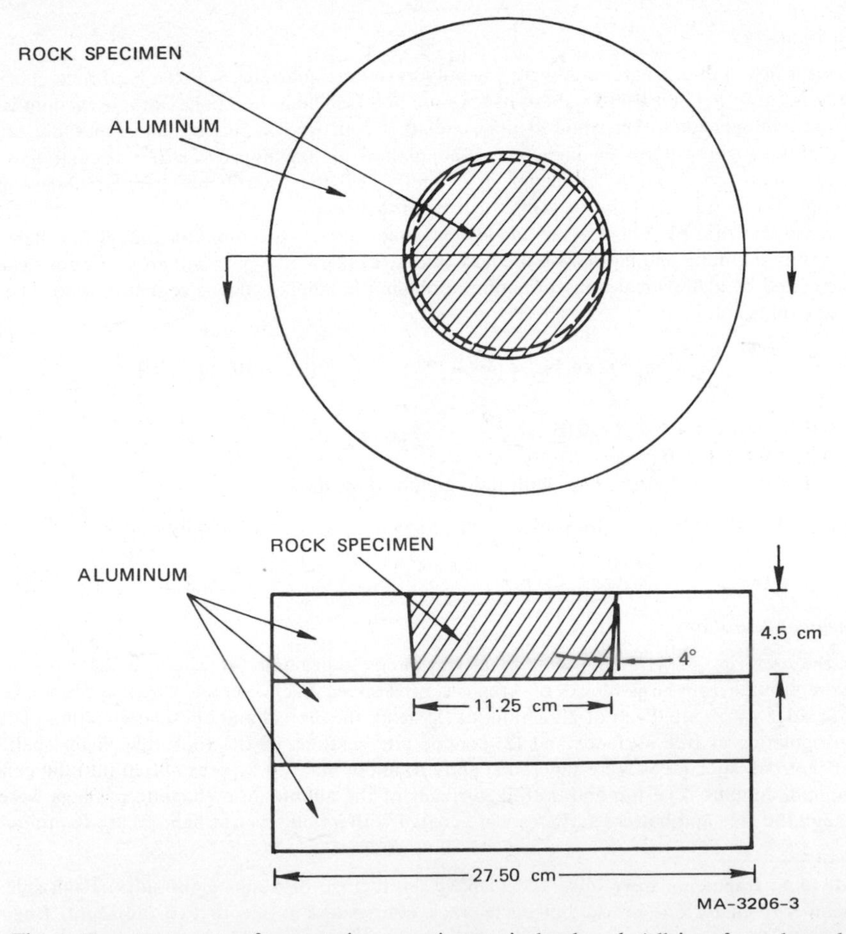

Fig. 1. Target geometry for cratering experiments in hard rock (all interfaces lapped and coated with grease).

detonated. All water and fragments were contained in the drum or the plastic tank when the experiments were performed.

Projectile impact experiments

The projectile impact experiments were performed on a specimen package (aluminum-encased novaculite specimens) having the same configuration used for the in-contact explosive experiments. The projectiles were right circular cylinders of drill rod steel, 1.15 cm in diameter by 2.54 cm long, heat treated to a hardness of R_c 58. A thin layer of copper was electroplated on all surfaces to reduce friction and seizing during transit down the gun barrel.

The experiments were conducted with a remotely fired, propellant-activated gun. The gun consisted of a 1917 Enfield action, fitted with a heavy barrel that was chambered for the 1.16-cm Winchester magnum cartridge and smooth-bored to 1.16-cm diameter. Projectile velocities were measured by three contact pins near the gun muzzle. To ensure normal impact, the specimen was positioned so close to the muzzle extension that the projectile impacted the rock before it completely exited the gun. Therefore a muzzle extension was used to prevent impinging rock and projectile fragments from damaging the gun muzzle. The muzzle extension contained slots to allow gases to escape from behind the projectile. A high speed camera was used during several impact experiments to obtain data on fragment ejection angles and velocities.

IV. Analysis and Results

Summary of cratering experiments

Fragment size distributions were determined for Experiment E1, an in-contact C4 explosive experiment, and for Experiment G1, a steel projectile impact at about 500 m/sec. Incipient stages of cratering and insight into the process of crater formation were attained in Experiment E2, which used a lower power

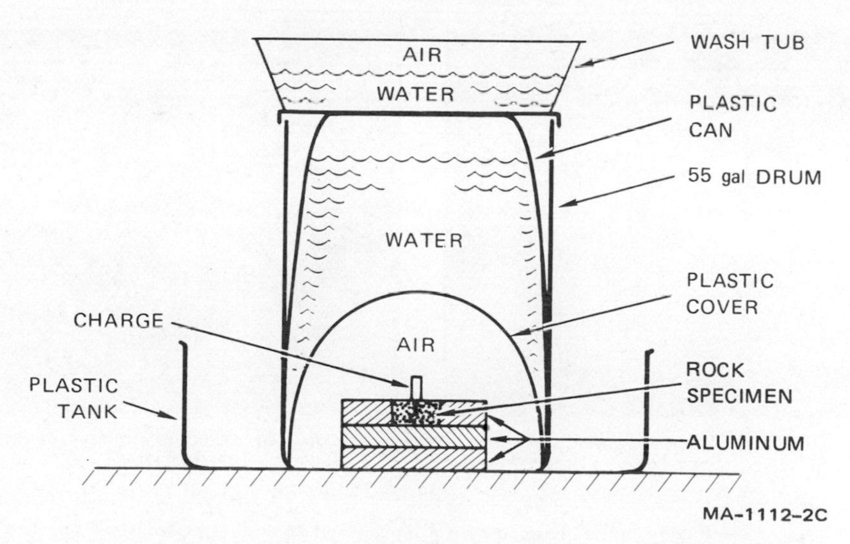

Fig. 2. Experimental arrangement for in-contact explosive ejecta recovery experiments.

explosive mixture and in G2, which used a lower impact velocity. Experiment G3 was performed at 650 m/sec to show the increase in damage at higher projectile velocities. Six additional experiments were performed in an effort to obtain information on ejection angles and velocities resulting from projectile impact at 500 and 230 m/sec.

Crater appearance

Craters produced by rod impact and in-contact explosive are shown in Fig. 3. In both cases the craters had shallow saucer-shaped configurations with aspect ratios (diameter/depth) of about 10.* The cracks are of essentially two orien-

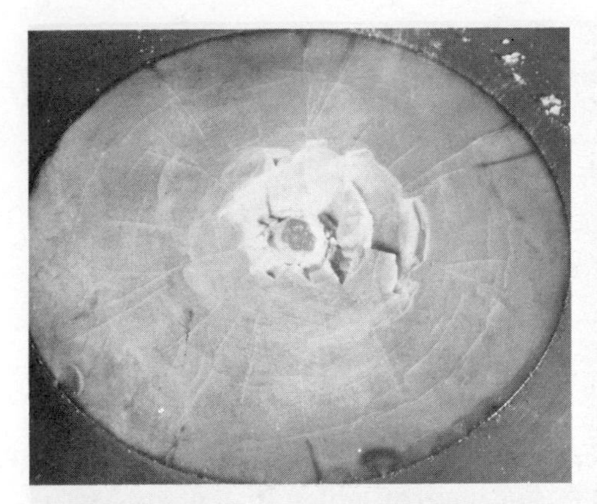

ROD IMPACT AT 140 m/sec; EXPERIMENT G2

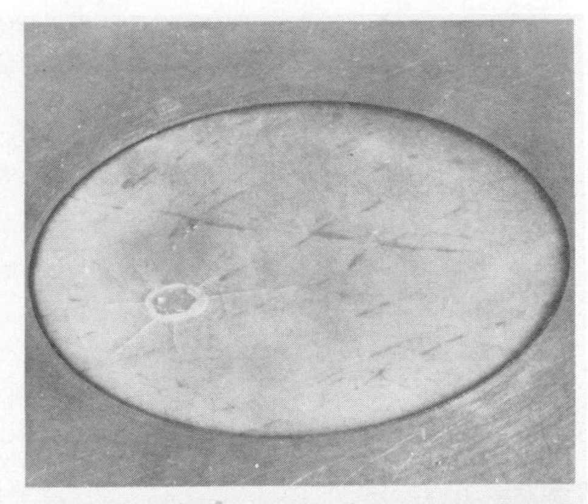

IN-CONTACT EXPLOSION OF PETN/MICROBALLOON MIXTURE; EXPERIMENT E2

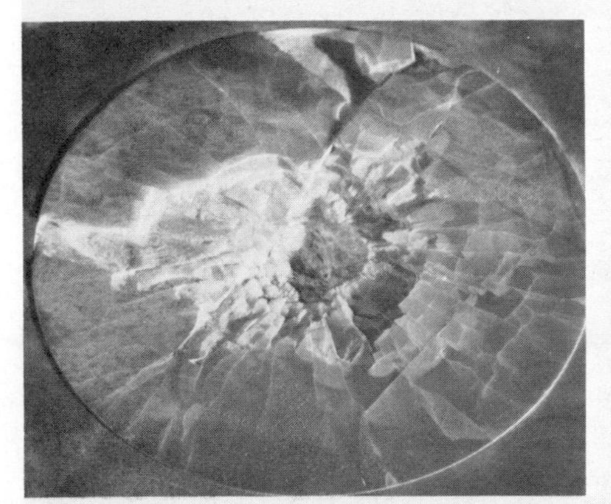

ROD IMPACT AT 500 m/sec; EXPERIMENT G1

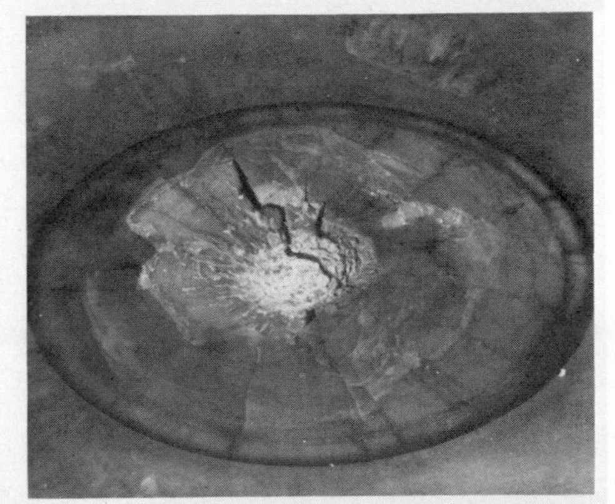

IN-CONTACT EXPLOSION OF C-4 PLASTIC EXPLOSIVE; EXPERIMENT E1

MP-3206-12

Fig. 3. Craters formed by rod impact and in-contact explosive.

*Crater profiles are best seen in Fig. 4.

tations: ring cracks concentric to the load site and radial cracks emanating from the load site.

Low velocity impacts (Experiment G2) and weak explosive mixtures (Experiment E2) left a relatively undamaged volume of rock remaining at the crater center. At higher impact velocities and with more powerful explosives, this central cone was not observed.

Subcrater fracture damage

Internal fracture damage was revealed by sectioning the specimens and lapping the sectioned surfaces. Figure 4 shows the crack patterns beneath the craters. Long cracks extend downward and outward from the load site in a fan-like array reaching entirely through the specimen thickness directly under the crater. The fracture patterns on the rear surface and on the midsection slice parallel to the rear surface are shown in Fig. 5. The subcrater fracture damage is reminiscent of the Hertzian cone fractures typically observed in quasi-static cylindrical punch experiments.

The mechanism of crater formation can be inferred from the cross section shown in Fig. 4 of the low velocity impact experiment G2, where only incipient stages of the cratering process were realized. Here portions of the rock have been ejected, but others have not. The crack pattern that leads to fragment formation and ejection is clearly shown in the left-hand side of the specimen. Large crater ejecta originating near the surface are formed by intersecting cracks of the Hertzian cone and ring types. The material adjacent to the impact site is the first to be ejected; more remote material is then ejected. The material directly under the impact site is damage-free to a depth comparable to the projectile radius, but at higher velocities acquires cracks and is ejected also.

Fragment examination

Significant changes in fragment shape for various sizes of fragments were not obvious. Figure 6 shows rock fragments from Experiment G1 retained on several sieves. Fragments were generally rather equiaxed, although fragments having one dimension about twice that of the other dimensions were commonly observed.

The extent to which uncoalesced cracks existed within fragments was investigated by examining polished cross sections of individual fragments. A number of rock fragments in several size ranges were mounted in a plastic compound and ground on abrasive papers and metallographic polishing cloths to provide a smooth section through the fragment interior suitable for microscopic examination. Many of the fragments were seen to contain cracks; some cracks appeared to be completely contained within the fragment (internal cracks), and some extended in from the fragment periphery. No evidence of shear instabilities or shock-induced phase changes was noted, although a careful,

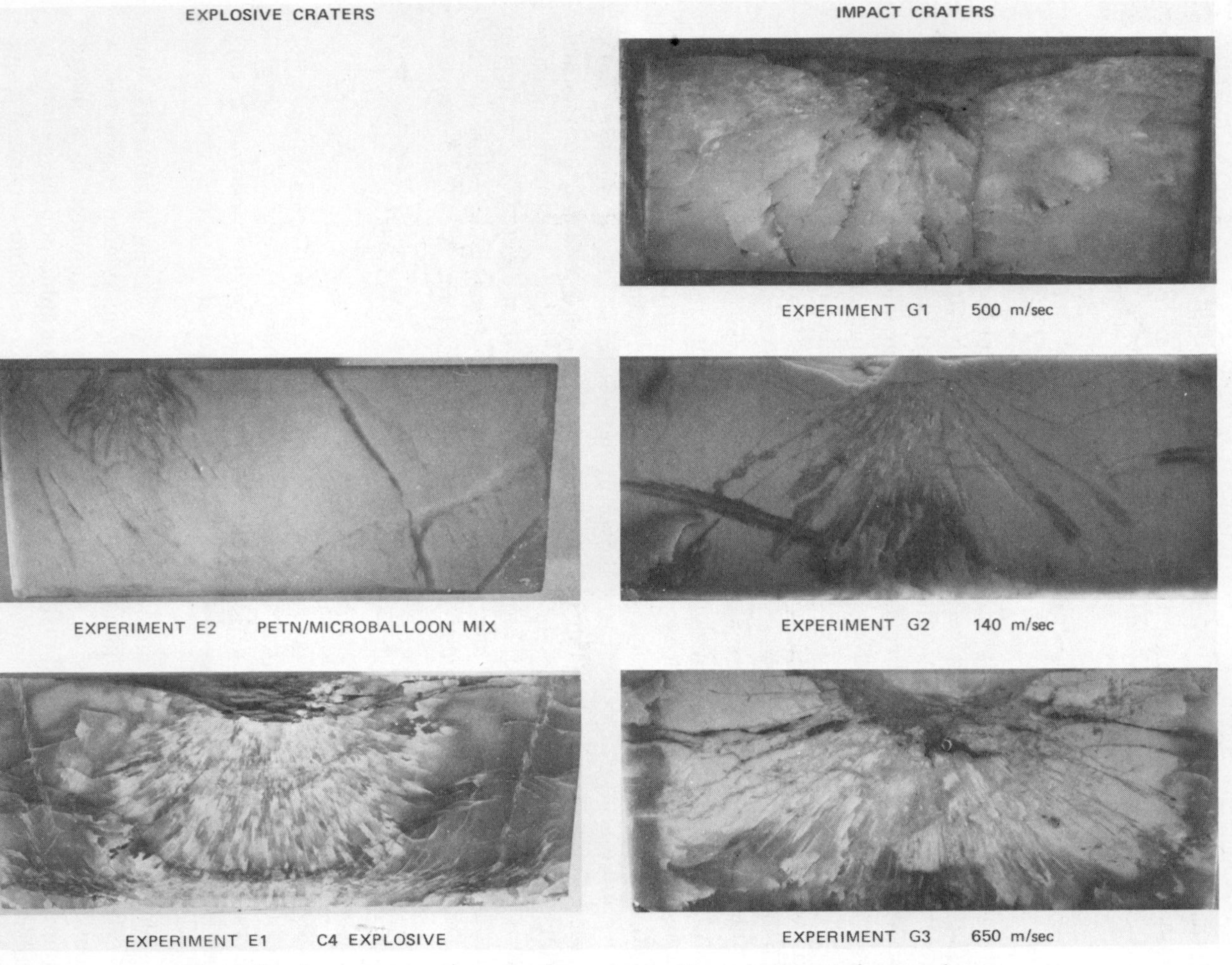

Fig. 4. Specimen cross sections showing crater profiles and subcrater fracture damage.

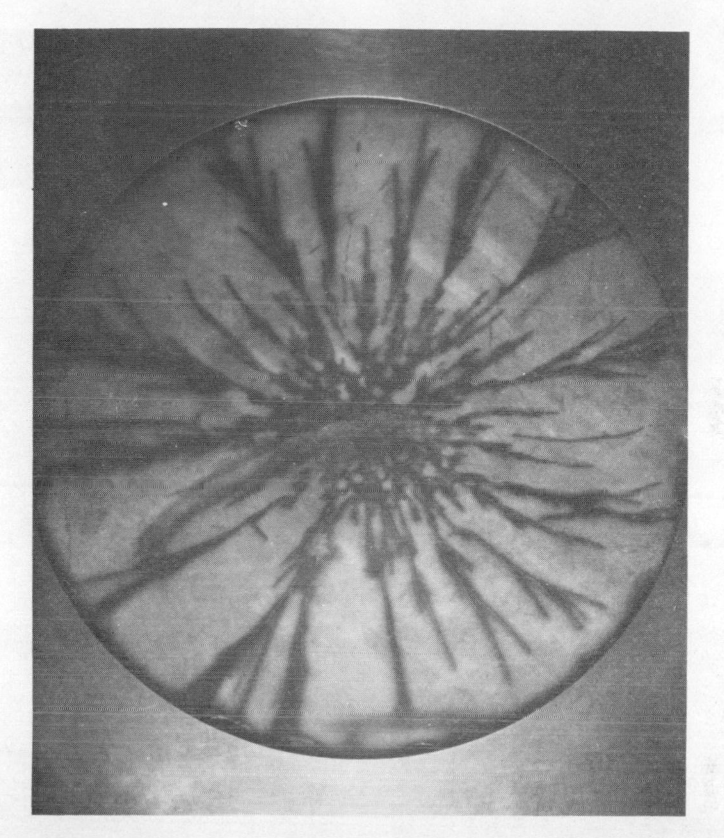

REAR SURFACE

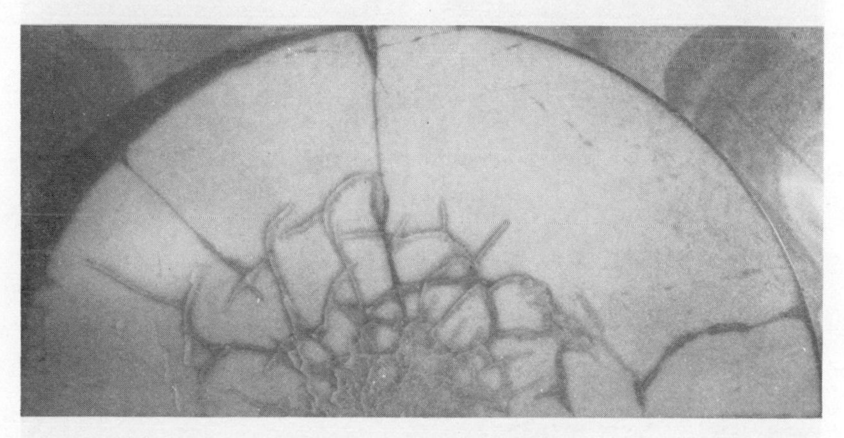

MIDTHICKNESS SECTION

MP-3206-14

Fig. 5. Fracture pattern on rear surface and at midthickness for specimen G1.

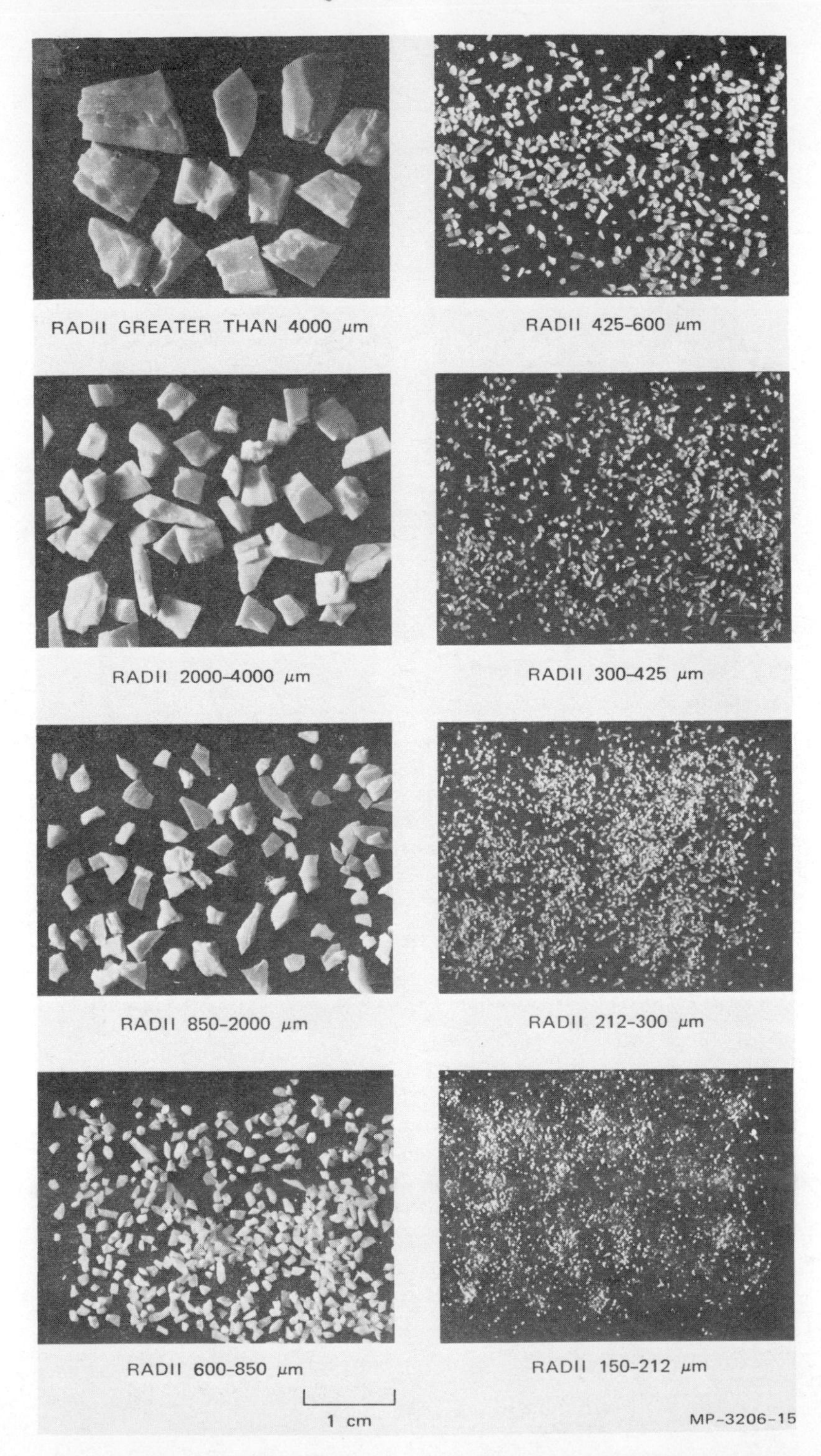

Fig. 6. Photomicrograph of various sized fragments from experiment G1.

detailed search using X-ray and transmission electron microscope techniques was not performed.

Fragment size distributions

The rock fragments produced in projectile impact Experiment G1 and in in-contact explosive cratering Experiment E1 were collected and partitioned according to size by running them through a series of sieves. The fragments retained on each sieve were photographed and then counted to obtain the data reported in Table 1. When very large numbers of small fragments were produced, direct counting of individual fragments was impractical. In these cases, a small representative sample was counted and weighed, and used to estimate the total number of fragments on the basis of the total weight. The cumulative number of rock fragments is plotted versus the fragment radius (half sieve opening) in Fig. 7 for the two cratering experiments. The similarity between the fragment size distributions produced by 500 m/sec impact and a C4 surface explosion is obvious and agrees with previous experience, since these curves are well fit by Eq. (1) with $n = -3.5$ ($k = 0.354$; ρ_f measured in cm^{-4} and R_f measured in cm). It implies that the tensile impulses produced in the rock by the two loading conditions are similar and that variations in the inherent flaw size distribution and other properties of novaculite are small. These implications are confirmed in part by previous mechanical tests and petrographic examinations performed by Shockey *et al.* (1973a, 1974).

Table 1. Measured fragment size distributions for an explosive and an impact cratering experiment.

		No. of fragments on sieve		Cumulative no. of fragments	
Sieve opening (cm)	Fragment radius (cm)	Explosive experiment E1	Impact experiment G1	Explosive experiment E1	Impact experiment G1
0.400	0.200	—	29	—	—
0.238	0.119	68	—	68	—
0.200	0.100	—	98	—	127
0.170	0.085	100*	—	168	—
0.118	0.059	266*	—	434	—
0.0850	0.0425	651*	1,474	1,085	1,601
0.0600	0.0300	1,439*	1,806	2,524	3,407
0.0425	0.0212	4,619*	5,386	7,143	8,793
0.0300	0.0150	7,210*	8,714	14,353	17,507
0.0212	0.0106	17,784*	21,988	32,137	39,495
0.0150	0.0075	53,207*	67,062*	85,344	106,557
0.0106	0.0053	141,516*	88,317*	226,860	194,874
0.0075	0.00375	266,241*	—	493,101	—
0.0053	0.00275	510,928*	—	1,004,029	—

*Determined by sampling method.

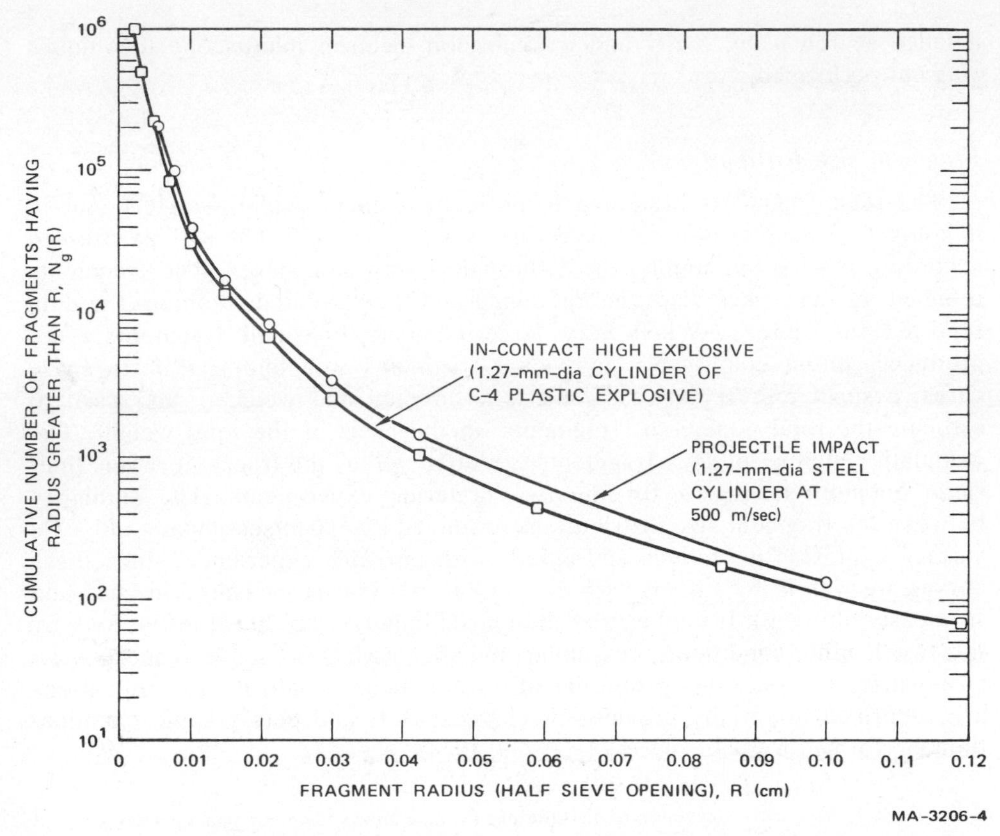

Fig. 7. Size distribution of rock fragments for an explosive and an impact cratering experiment.

Ejecta angles and velocities

The framing camera experiments indicated that fragments ejected at larger angles to the impact surface have higher velocities and are probably ejected earlier than those ejected at lower angles. In the 500 m/sec projectile impact experiments, maximum velocities of more than 400 m/sec, or about 0.85 times the impact velocity, were indicated for fragments propagating at angles to the impact surface in the range of 40–60°. For fragments ejected at about 10°, velocities of about 350 m/sec, or 0.7 times the impact velocity, were inferred.

V. COMPUTATIONAL SIMULATIONS AND RESULTS

Two in-contact explosive cratering laboratory experiments and two rod impact cratering experiments were simulated computationally using the NAG/FRAG dynamic fracture subroutine in the TOODY3 two-dimensional wave propagation code. The predicted crater shapes, ejecta distributions, and subcrater damage are presented in this chapter and compared with observations.

TOODY and SRI NAG/FRAG

Cratering calculations were performed with the two-dimensional Lagrangian wave propagation code, TOODY3, developed at Sandia Laboratories and described in detail by Bertholf and Benzley (1968). TOODY3 is well suited for these calculations because it contains a sophisticated slide line option useful in computing the large deformations occurring at the projectile-rock and explosive-rock interface.

The TOODY3 code calculates for each computational cell at each point in time the deviator stress, pressure, energy, density, and particle velocity. TRI Q artificial viscosity, as described by Shockey *et al.* (1975), was inserted into TOODY3 and used to damp numerical cell distortion. To calculate the fracture activity and hence the cratering behavior, the NAG/FRAG fracture/fragmentation subroutine was incorporated into the code in the form of a special equation of state subroutine BFRACT, developed by Seaman and Shockey (1973). This subroutine is first called for an individual cell when the tensile stress in that cell (in any direction) exceeds the threshold stress for flaw activation. Once called for that cell, it is called at every succeeding time step, and computes the fracture activity caused by the current tensile stress. Thus, in the most general form of BFRACT, the growth of previously activated flaws as well as the activation of additional flaws are computed at each increment in time, according to the crack nucleation and growth laws described in Section II. However, for novaculite, Eq. (2) is replaced by an algorithm that activates all preexisting flaws when $\sigma > \sigma_{no}$; no additional flaws are activated thereafter. A derivation of the growth parameter for novaculite is presented by Shockey *et al.* (1975). The eroding effect of the developing fracture damage on the stresses is also accounted for after each time step.

Setting up the calculations

A computational grid was constructed to conform to the geometry of the specimen package shown in Fig. 1. A sufficient number of cells were provided to obtain the necessary resolution in the novaculite near the load application. In the interests of eliminating undue computer costs, however, the cells were made as large as possible. The grid consisted of small cells near the loading site and larger cells at remote boundaries. The cell sizes varied as a geometric ratio, each successive cell having one dimension 1.05 times that of its neighbor with the other dimension constant. This cell layout kept the cells reasonably square and consistent in size with their neighbors. The projectiles and explosives were assumed to load the specimens at their centers* to provide an axisymmetric geometry.

---

*The explosive charge in Experiment E2 was inadvertently placed off center. However, because of the low strength of the explosive, edge effects were minimal, and the axisymmetric assumptions led to little error.

Computational simulations

Computational simulations of two impact and two surface explosive cratering experiments were performed. The loading conditions and the peak pressures produced at the loading site for the four simulated experiments are given in Table 2.

The wide range in initial loads was intended to produce a wide range in rock damage. The computed crater profiles, subcrater fracture damage, and ejecta size, velocity and angle distributions for these experiments are now described.

Crater profiles and subcrater fracture damage

The NAG/FRAG model calculates the extent of fracture and fragmentation in each cell, and this information is printed out after each time step. Thus the development of the crater and subcrater fracture damage can be followed. Figures 8–11 show the state of fracture and fragmentation for each of the four experiments at selected times during the cratering process. Here black cells indicate completely fragmented material, blank cells indicate material with no fracture damage, and gray cells indicate intermediate degrees of damage.

In all four simulations, cells near the circumference of the projectiles and explosive cylinders are the first to acquire fracture. The boundary between damaged and undamaged cells soon attains an approximately hemispherical shape and retains this general shape as it expands with time. The boundary between ejected and nonejected cells, however, is not hemispherical. Material in cells near the surface is more easily ejected than that in interior cells. The criteria used to compute which cells would be ejected were: cells in the first four rows that were nearly 100% fragmented, and that were unrestrained on the upper side by cells less than about 100% fragmented. These criteria were chosen before any computations of ejecta distributions were made. Furthermore the calculations were made just once; no iterations with other criteria were made to try to improve agreement with measurements.

The fourth cross section in each series shows the total computed damage and compares it with a schematic of the observed damage. The computed saucer-shaped craters and the hemispherical damaged area beneath the crater are in

Table 2. Initial pressures produced in rock specimens at load interfaces.

Experiment number	Description	Initial pressure in rock (MN/m^2)*
G1	500 m/sec impact	5,370
G2	140 m/sec impact	2,140
E1	C4 explosive cylinder	27,800
E2	PETN/microballoon explosive cylinder	733

*$1\ MN/m^2 = 10^{-2}$ kbar.

good agreement with the observed high-aspect ratio crater profiles and hemispherical damaged areas.

For the low impact velocity Experiment G2 and the low-powered explosive Experiment E2, the rock material directly beneath the load area was relatively undamaged. The computations are in accord with these observations, predicting less fracture damage in those cells.

Ejecta size distributions

The calculated ejecta size distributions in experiments G1 and E1 were obtained by computing for each cell contributing to the crater* the total number of fragments and the number of fragments having radii equal to 0.025, 0.050, 0.075, and 0.100 cm. These values were then summed and plotted as number per cubic centimeter. The smooth curve drawn through these points represented the computed crack size distribution. The results are presented in Figs. 12 and 13 along with the normalized measured distributions for comparison.

The computed curves are very good averages of the measured distributions, but do not exhibit their curvature. The computed curves are nearly straight lines because the crack size distribution in each cell is always represented as an exponential: the inherent flaw size was taken as an exponential approximating the observed flaw size distribution. During loading, growth occurs according to Eq. (3), which allows the distribution to remain an exponential. Fragmentation then occurs when the distributions reach a certain shape. Inclusion of fragment distributions from partially fragmented cells did not cause the overall distribution to vary significantly from a straight line.

It should be again emphasized that the computed distributions are the predictions of the NAG/FRAG model and do not involve assumptions of the type expressed by Eq. (1).

Ejecta velocity and angle distribution

Velocities inferred from the framing camera experiments were about three times higher than those computed.

Synopsis

The following statements summarize the results of our attempts to calculate the cratering behavior observed in the laboratory experiments on novaculite:

*It is likely that not all cells contributing to the crater contribute to the ejecta. That is, the volume of the crater is probably larger than the volume of the ejecta because some of the fragmented material beneath the load site is pushed upon by the projectile or explosive and hence does not fly out of the crater. Thus a slightly different criterion should be used for ejecta computations than for crater formation. In addition to requiring that a cell be completely fragmented and bounded on its upper side by a completely fragmented cell, it should also be required that, for the cell to contribute to the ejecta, it must have an upward component of velocity.

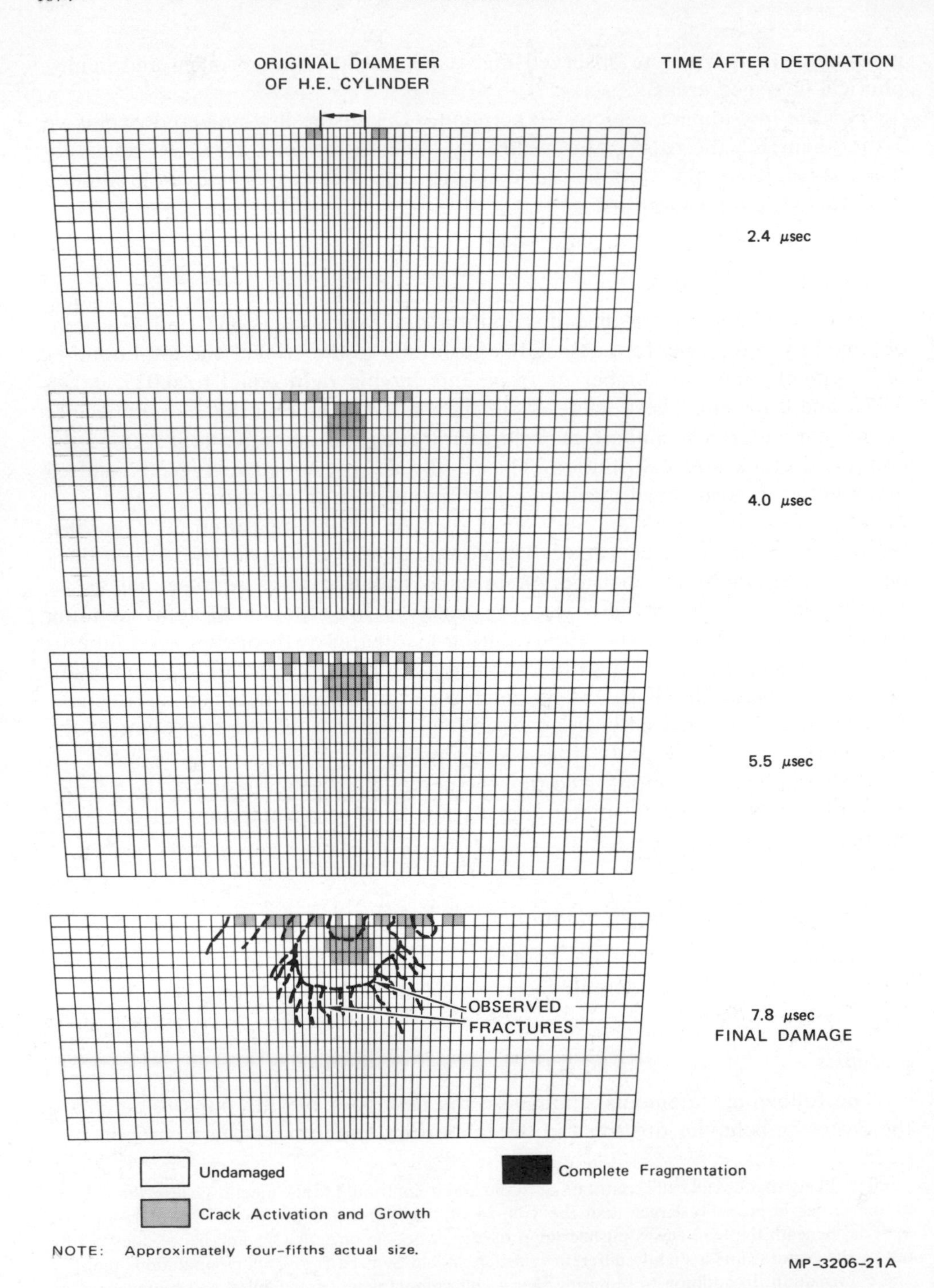

Fig. 8. Successive stages of subsurface fracture damage computed for in-contact PETN/microballoon mix explosive experiment E2.

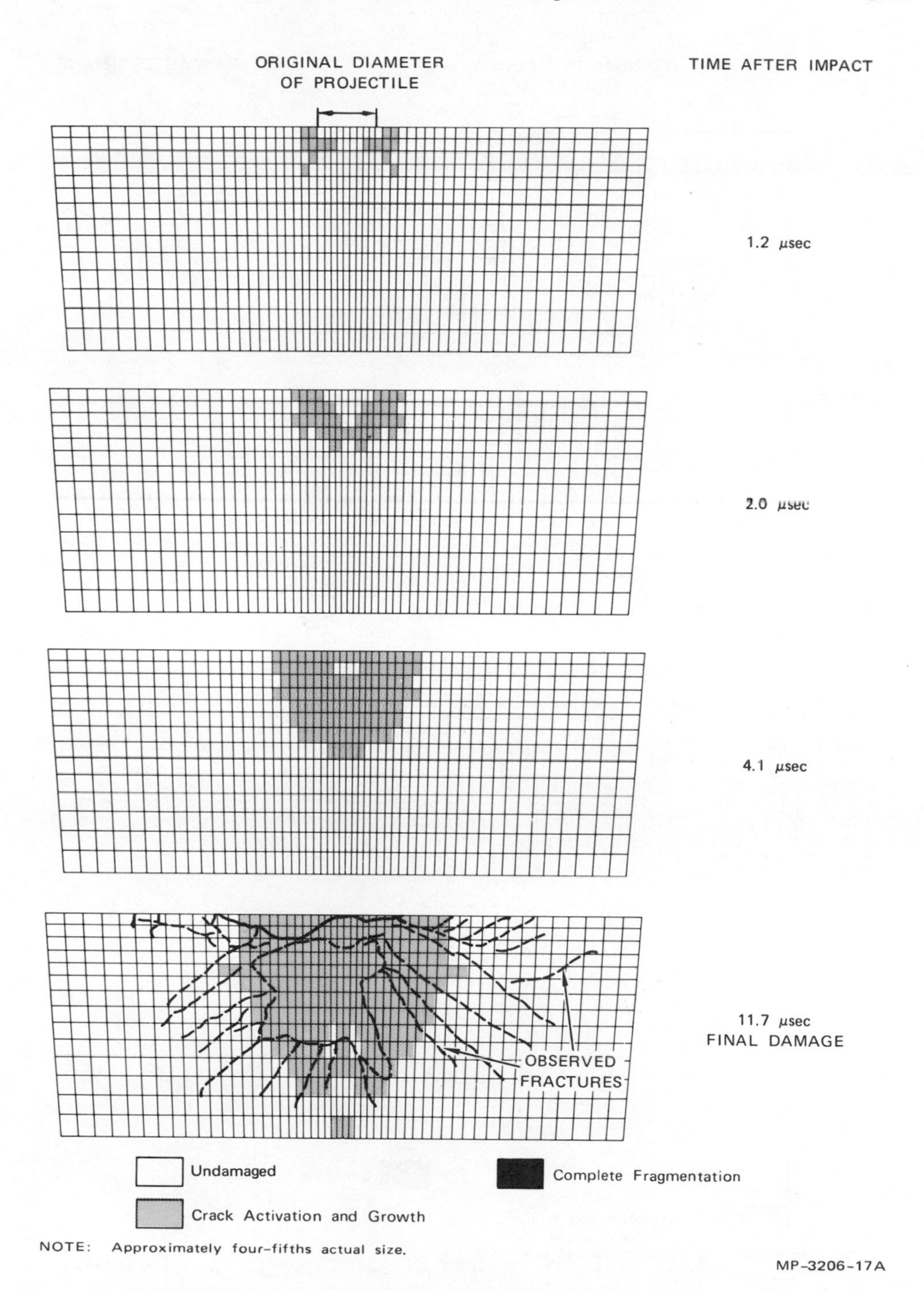

Fig. 9. Successive stages of crater formation and subcrater fracture damage computed for 140 m/sec projectile impact experiment G2.

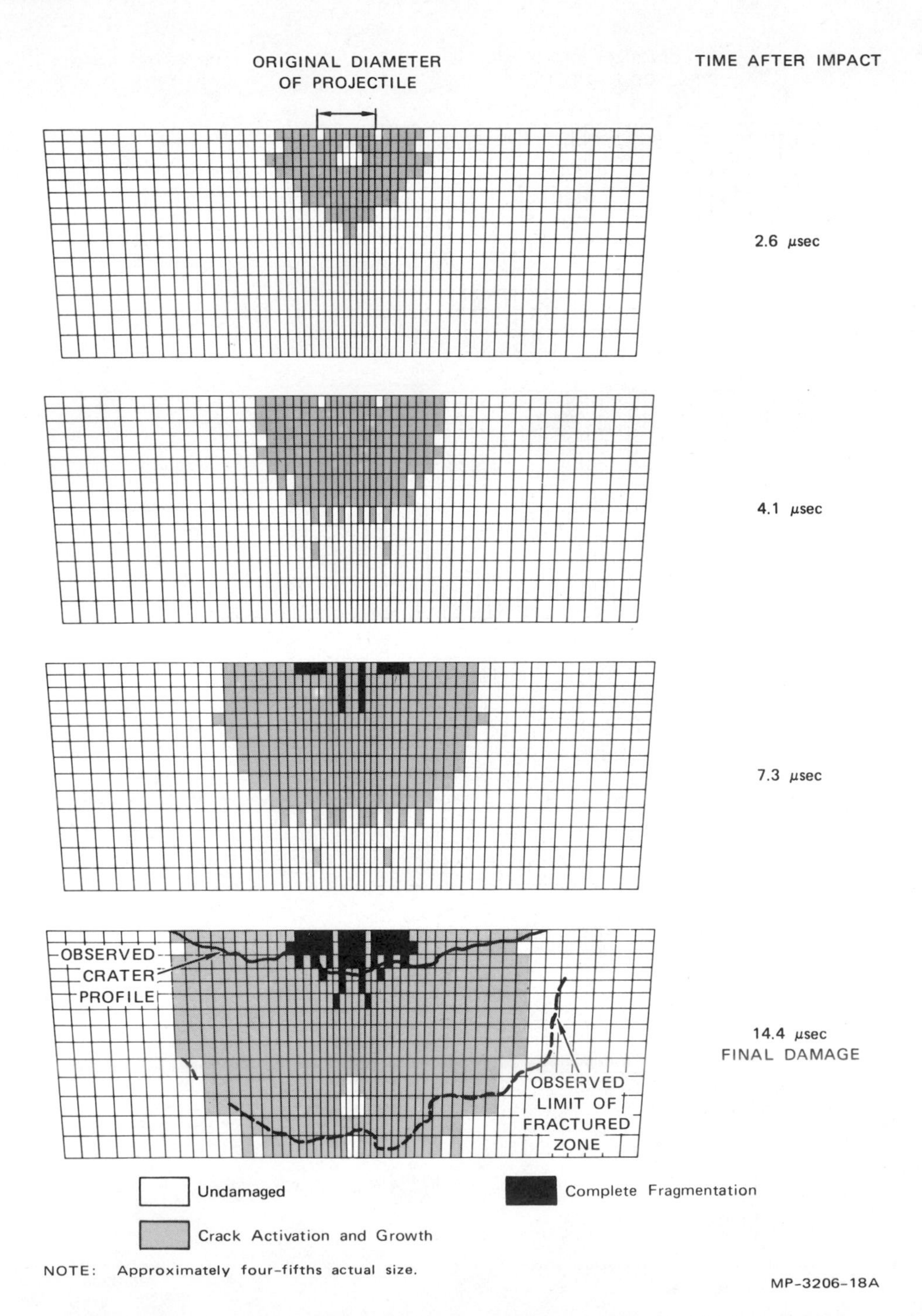

Fig. 10. Successive stages of crater formation and subcrater fracture damage computed for 500 m/sec projectile impact experiment G1.

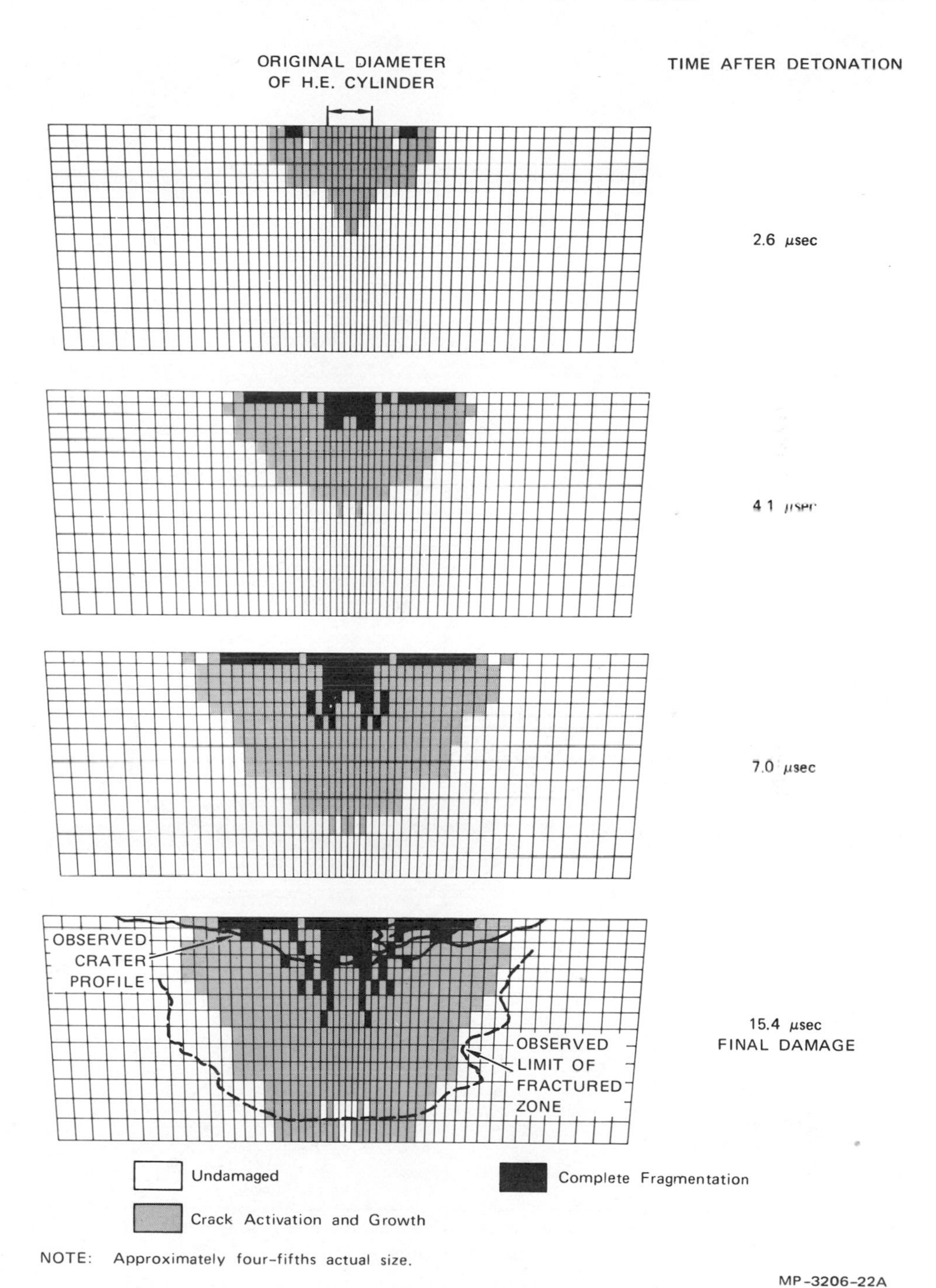

Fig. 11. Successive stages of crater formation and subcrater fracture damage computed for in-contact C4 explosive experiment E1.

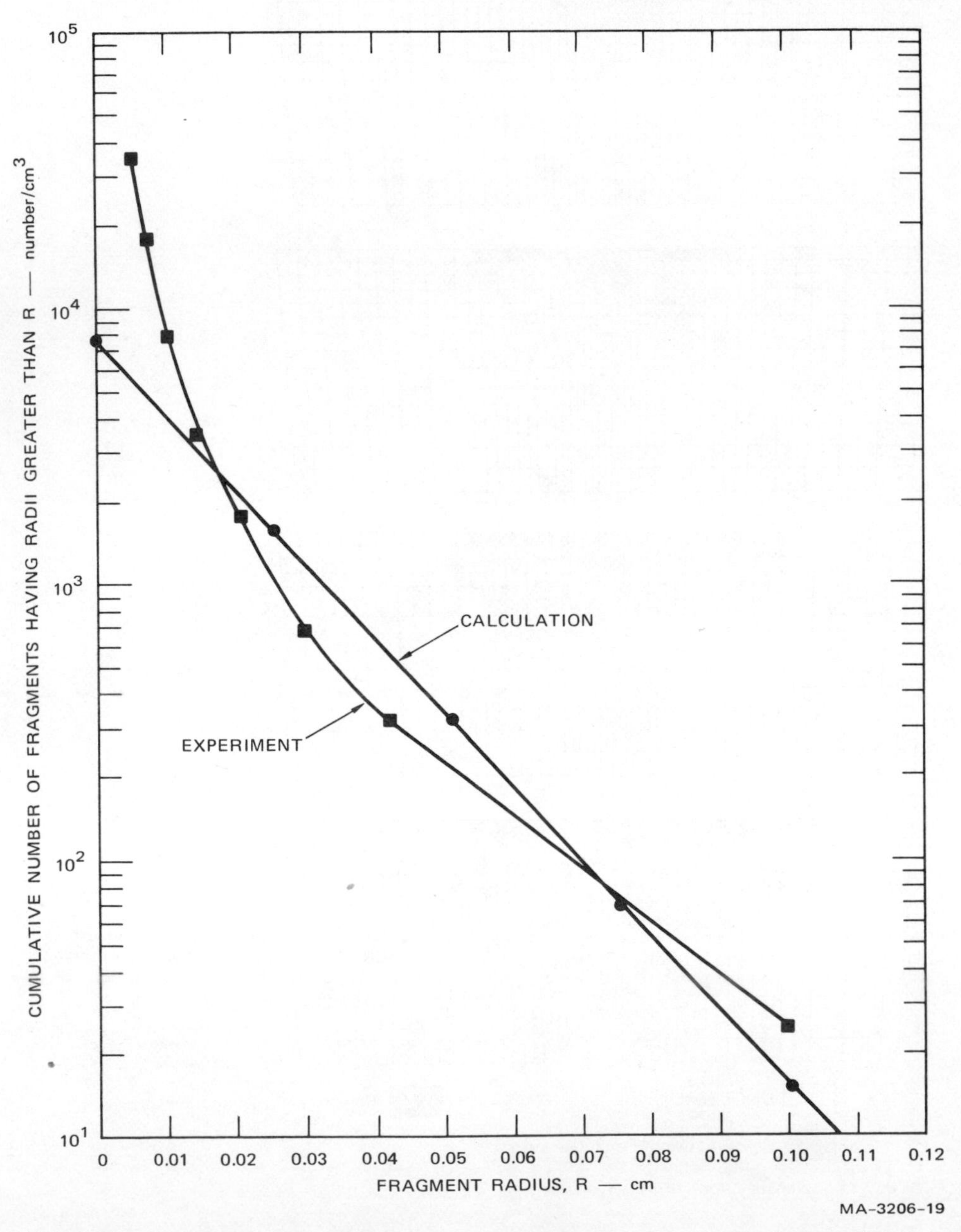

Fig. 12. Comparison of experimental and computed ejecta size distributions for 500 m/sec impact experiment G1.

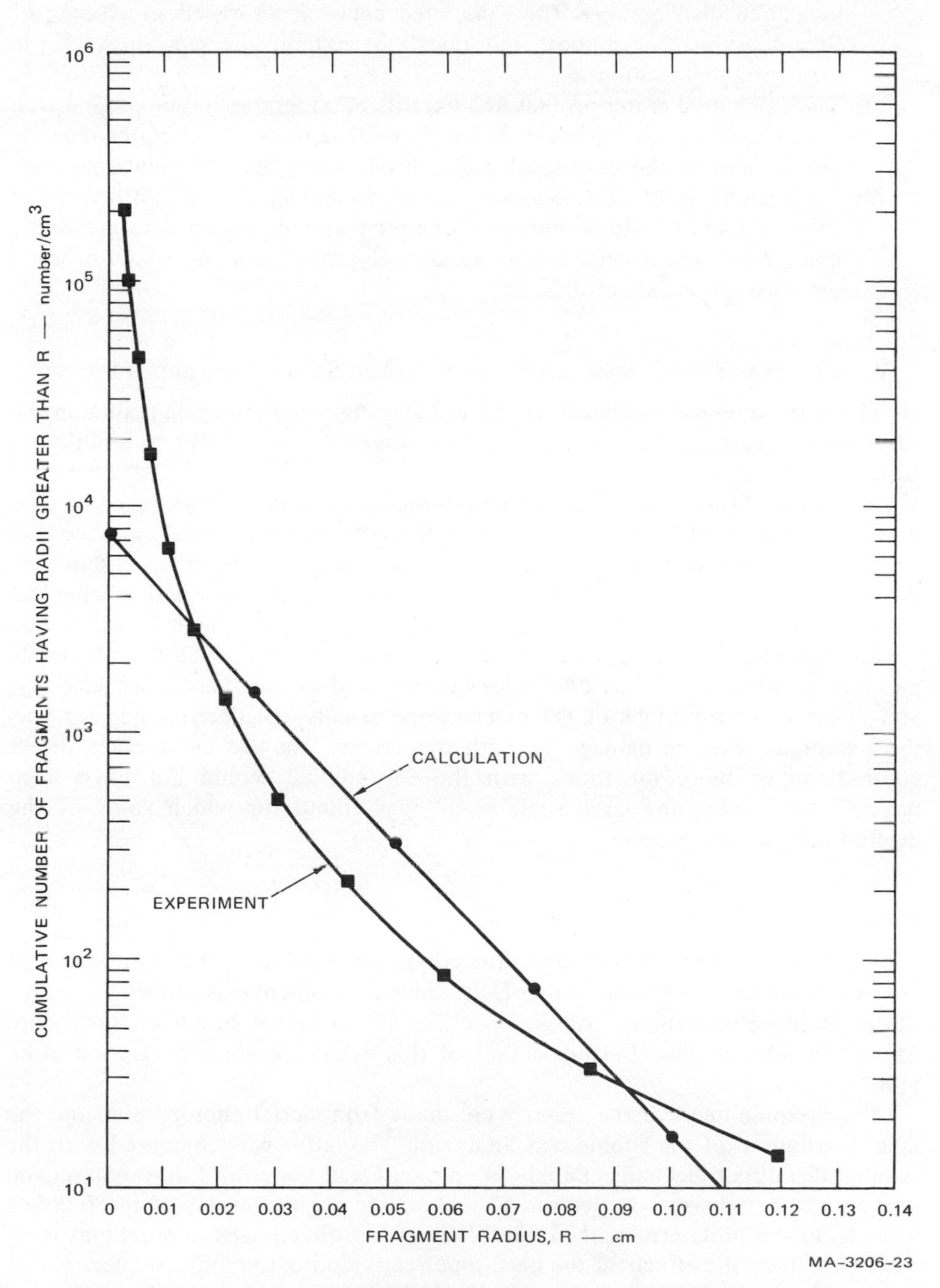

Fig. 13. Comparison of experimental and computed ejecta size distributions for in-contact C4 explosive experiment E1.

(1) The calculated ejecta size distributions are in good agreement with measured distributions. The calculated curve is an excellent average of the measured distribution, but does not exhibit the curvature of the measured distribution.
(2) The calculated crater profiles and extents of subcrater fracture agree well with observations. Although hemispherical regions of subcrater damage are computed, the calculated crater profiles are flat and saucer-shaped.
(3) Calculated ejecta velocities and angles do not agree well with framing camera data. Although more and clearer framing camera data are desirable, it appears that calculated ejecta velocities are lower than observed velocities by a factor of three.

VI. COMPUTATIONAL SIMULATION OF A LARGE SCALE CRATERING EVENT

The good agreement between computed and observed cratering results in the laboratory experiments on novaculite encouraged us to attempt to simulate a large scale event.

A review of previous large scale man-made cratering events revealed few that were performed in well-characterized, hard, dense rock, and only a few that were performed in material for which even a rough idea of the inherent flaw size distribution or grain size was available. Other microstructural and mechanical property information was likewise found to be generally lacking.

Furthermore, in only a few cases was quantitative information on crater parameters other than crater dimensions gathered after the event. Size, velocity, and direction distributions of the ejecta were usually not determined, nor was the extent of fracture damage beneath the crater. Thus in most cases direct comparison of these quantities with those predicted would not have been possible. However, one large scale event was found for which some of the desired data were available.

Pre-Schooner Delta

The large scale event best suited for computer simulation and comparison of results appeared to be Pre-Schooner Delta. This experiment was carried out with a 20-ton liquid nitromethane charge buried in dry vesicular basalt at Buckboard Mesa, Nevada, in 1964. Further details of this event are given by Lutton *et al.* (1967).

Topographic maps of the crater were made from aerial photographs, and the size distribution of the rubble was analyzed. The latter was compared with the results of a direct mechanical analysis on excavated material. Lip trenching and crater excavation were carried out to investigate the properties of the fallback and ejecta and to determine the crater dimensions. Five postshot boreholes were drilled to study the effects of the blast on effective porosity, joint frequency, and joint width; a borehole camera was used to obtain additional information on subcrater fracture damage.

The explosion, which occurred about 14 m below the surface, produced a crater that was approximately circular when viewed from above, with an approximately paraboloidal cross section profile. The true depth and radius were about 19 m and 21 m, respectively. The ejecta size distributions indicated by a photo-grid technique and by a mechanical technique are in good agreement. The average 50% grain size for the crater lip was about 0.5 m by either method.

Postshot boreholes showed an anomalous zone of fractures extending horizontally into the rock from a prominent niche in the true crater, roughly along the interface between vesicular and more dense basalt. Laboratory tests on postshot borehole samples revealed no obvious alterations of the physical properties of the basalt by the blast, but significant changes in the effective porosity, generally increasing with decreasing distance from the zero point. The greatest change in effective porosity tended to be localized in a zone flanking the crater and flaring toward the surface, but anomalous zones of high effective porosity extended along certain strata.

Rock properties

Buckboard Mesa was formed by viscous lava flow and consists of layers of basalt of various densities and mechanical properties. The porosity varied from about 20% near the surface to about 2% at 30 m. Correspondingly, saturated surface dry bulk specific gravities ranged from 2.35 to 2.75, unconfined compressive strengths from 3000–21,000 psi, and Young's moduli from 1×10^6 to 7×10^6 psi. Poisson's ratio by dynamic methods was about 0.18, and dynamic tensile splitting tests averaged near 2700 psi.

For purposes of cratering calculations, the rock was treated as a homogeneous medium with average properties. The density was taken as 2.64 gm/cm^3, the bulk modulus K as 860 kbar, the shear modulus G as 185 kbar, and the compressive yield and dynamic tensile strengths as 0.83 and 0.20 kbar, respectively. From the block size distribution of the basalt determined before the event, we inferred the inherent flaw size distribution used in the computations as discussed later.

No experimental information was available for the crack growth velocities in the Pre-Schooner Delta basalt, so we chose a simple growth law, which when inserted into NAG/FRAG, gave a reasonable fit to the observed ejecta size distribution. (Thus the ability of the NAG/FRAG approach to calculate large scale cratering behavior must be assessed from the agreement between computed and observed crater dimensions and subcrater fracture damage.) The growth law chosen and the reasoning behind its choice are discussed later.

The computational mesh consisted of rectangular cells. The two cells at the surface above the zero point were 1.4×2.7 m. Each adjacent cell in the horizontal direction was 5% wider; each successive cell in the vertical direction was 5% longer. A sufficient number of cells were incorporated to prevent reflections from computational boundaries before 9 msec.

Results

Figure 14 shows the computed fracture and fragmentation damage on a cross section taken through the center of the buried charge about 9 msec after detonation. The calculation was run for an additional 3 msec to ensure that no further growth of the crater or damaged zone occurred. The observed true crater profile and the zero fracture damage boundary determined by boring experiments are indicated for comparison. Clearly, the computed results are in good agreement with the limited observations.

The material surrounding the charge to a distance of about 5 m and the material directly above the charge have been severely fragmented, as indicated by black cells. The gray cells have undergone fracture damage and partial fragmentation. Blank cells are unfractured; thus, the interface between blank cells and gray cells may be taken as the boundary of subcrater fracture damage. It is obvious from Fig. 14 that this computed boundary agrees well with the boundary deduced by borehole observations.

The slight asymmetry of the measured crater profile and subcrater fracture damage boundary is a manifestation of the layered rock structure, the inhomogeneous properties, and the anisotropy of the cratering medium.

The material contributing to the ejecta and hence determining the crater profile must be at least partially fragmented and have some minimum upward velocity component. The extent of fracture and fragmentation is displayed in Fig. 14, and the velocity vector field is shown in Fig. 15. Comparison of the computed velocity vectors with the observed true crater profile suggests that all material with an upward velocity component exceeding about 3 m/sec was ejected.

A calculated ejecta size distribution was obtained by summing the fragments in all cells having an upward velocity component greater than 3 m/sec. The results are shown in Fig. 16. Comparison with the measured distribution shows that significantly fewer fragments of all sizes were predicted. This is attributable at least in part to inaccuracies in the assumed inherent flaw size distribution and in the assumed growth law.

The incipient flaw size distribution was inferred from the data of Fisher (1968), who determined the *in situ* block size distribution by joint analysis from pre-shot borehole camera data. The *in situ* blocks appeared to be irregularly shaped with an aspect ratio of 3 to 4. The size distribution of Fisher was based on the longest dimensions of the blocks. However, we felt that these blocks would quickly break into smaller, more regularly shaped blocks because of crack propagation through the short dimensions. Thus, the longest cracks would not really contribute to fragments with similar dimensions. Therefore, we chose the *in situ* flaw size distribution corresponding to the smaller *in situ* block dimensions (assumed equal to one-fourth the reported longest dimensions). This distribution is shown in Fig. 16.

We expect the maximum crack velocities in the Pre-Schooner Delta basalt to be about one-third the longitudinal sound speed, in accordance with theory. However, we do not expect all the cracks to travel at this maximum velocity. As

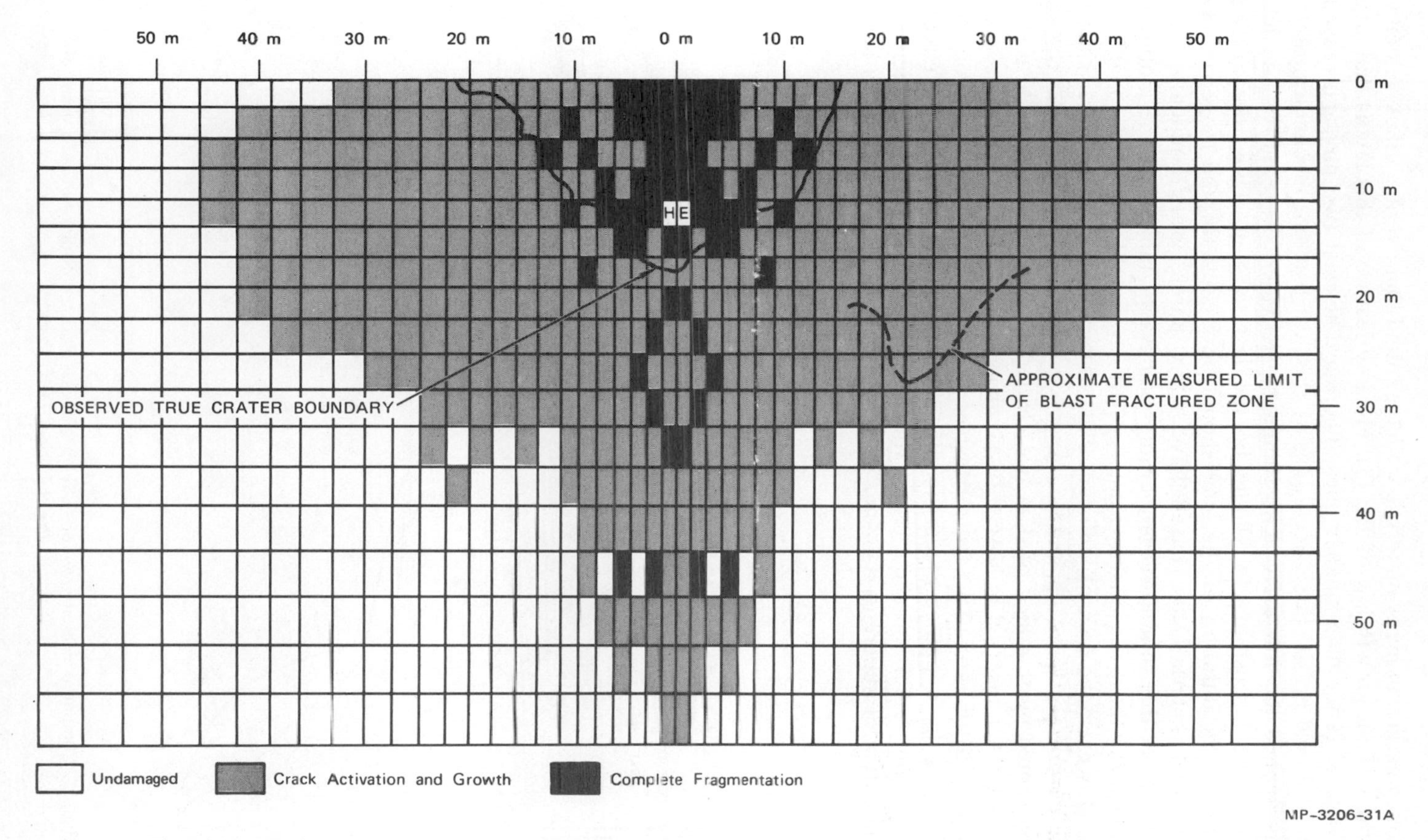

Fig. 14. Comparison of computed and observed crater profiles and subcrater fracture damage for the Pre-Schooner event.

mentioned earlier, in many materials we have found that average crack velocities obey the viscous growth law of Eq. (3). The key feature of this law is that large cracks travel faster than smaller ones. There are at least two physical reasons why this should be so. First, according to classical fracture mechanics, as the load is applied the longest cracks should be activated first, and as the load decreases the longest cracks should be the last to stop. Thus, the average velocity for the duration of the load should be greater for the longer cracks. Second, real materials like basalt have many pores, grain boundaries, and heterogeneities that inhibit the motion of small cracks more than the motion of large ones. For these reasons, a viscous growth law would appear reasonable for basalt. However, the lack of sufficient experimental data made it impossible to choose *a priori* values of σ_{go} and η.

Confronted by this lack of data, we elected to choose the simplest growth law, which when inserted in NAG/FRAG, gives a reasonable fit to the observed ejecta size distribution. The critical test of our computational simulation then becomes not the computed ejecta size distribution, since that would be circular reasoning, but rather the accuracy of the predictions of the crater dimensions and the extent of subcrater damage. That is, if choosing a crack growth law to

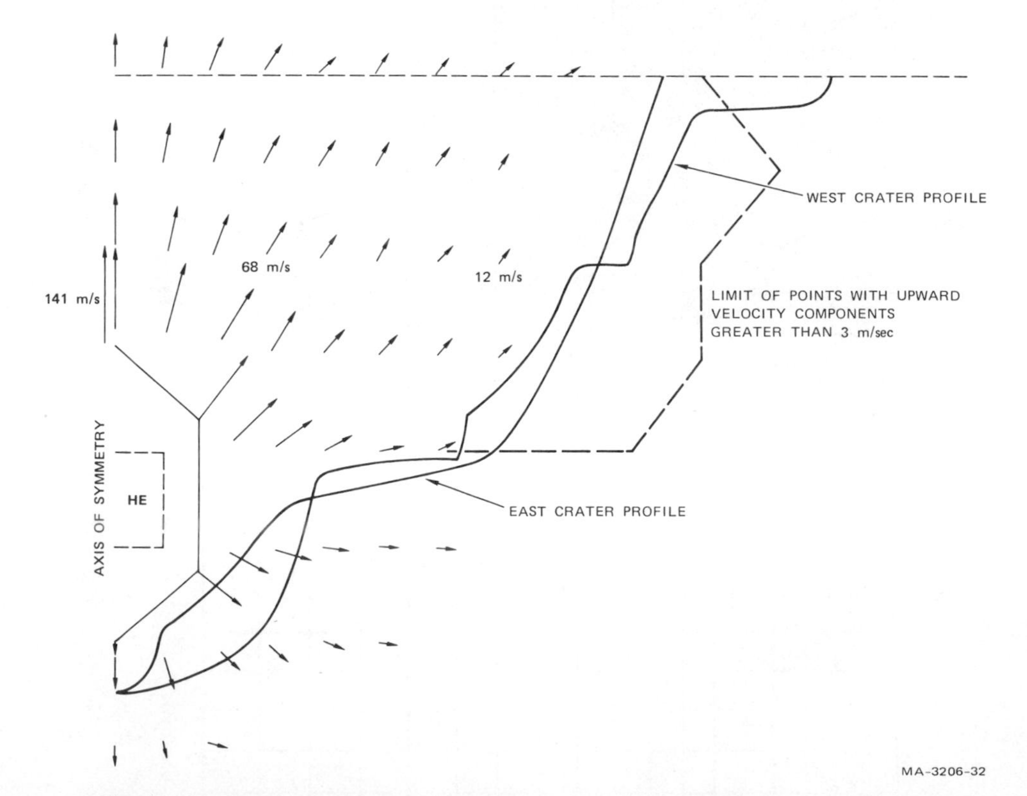

Fig. 15. Computed velocity vector field on a cross section through the zero point of Pre-Schooner Delta.

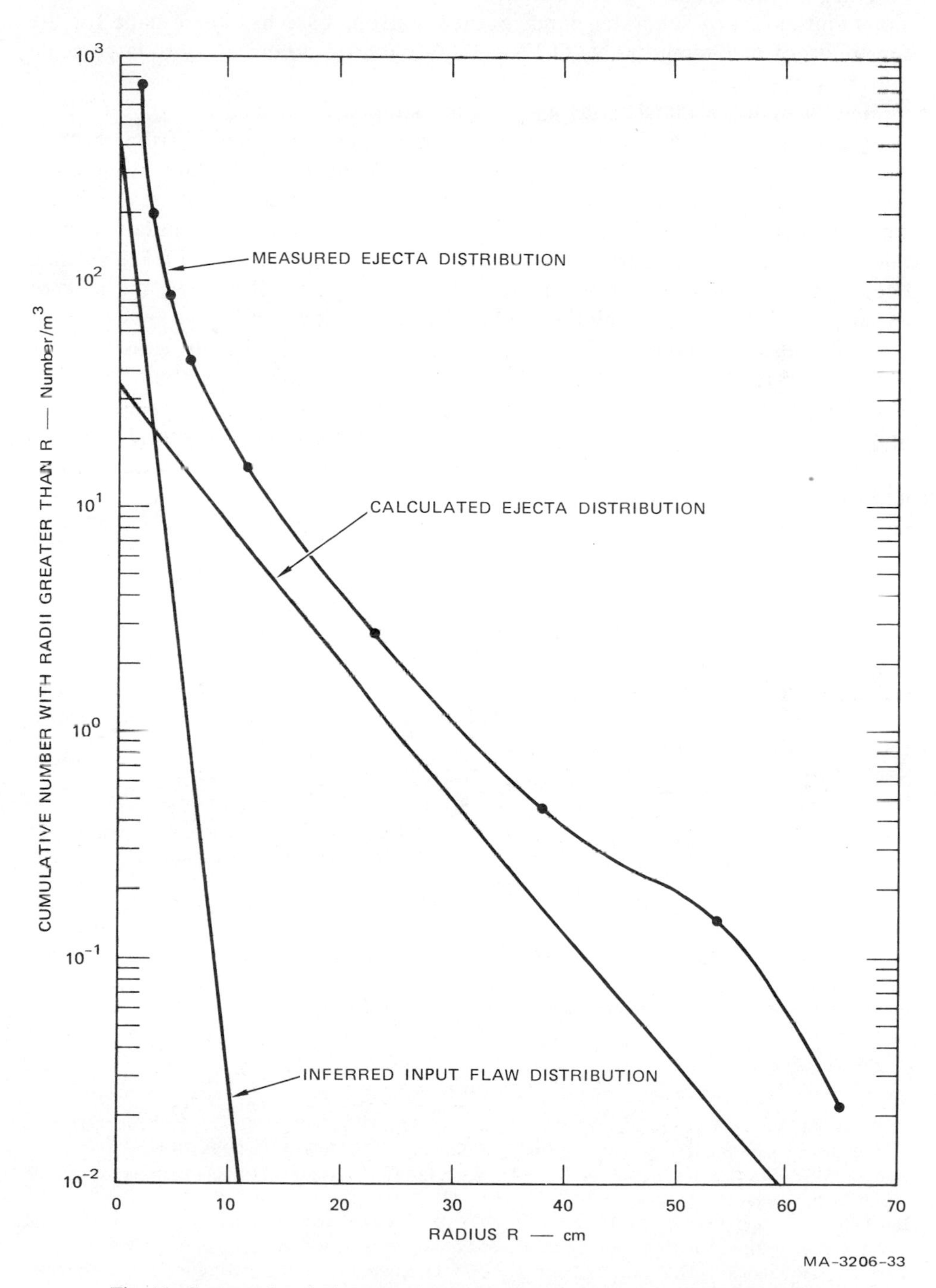

Fig. 16. Computed and observed ejecta size distribution for Pre-Schooner Delta.

give the correct ejecta size distribution also produces the correct crater dimensions and extent of subcrater damage, then a strong case has been made for the feasibility of extending the NAG/FRAG approach from laboratory to large scale cratering events.

The growth law chosen was a somewhat simpler form than that of Eq. (3); it assigns to the activated flaws constant velocities that vary linearly from zero for zero length flaws to one-third longitudinal sound speed for the largest incipient flaws. This growth law was found to give the agreement with the observed ejecta size distribution shown in Fig. 16. A better fit could have been obtained with different values of the parameters, or if the viscous growth law of Eq. (3) had been used, with a value of η on the order of 10^4 poise. However, the lack of experimental data made such a complication unjustified.

It is clear that controlled laboratory experiments to directly establish the growth law for basalt would be extremely valuable, since only then could we attempt a truly *a priori* prediction of both crater morphology and the ejecta size and velocity distributions for this large scale event. However, as was shown, the present calculation did give valid *a priori* predictions of the crater dimensions and extent of subcrater damage.

VII. CONCLUSIONS

It appears that in hard rock the initial flaw size distribution and activated flaw kinetics govern the cratering behavior. In soils or in porous rock we expect that the cratering will be strongly modified by plastic flow, pore compaction, and degree of water saturation. Nevertheless, this approach, in which the material rate dependent processes are modeled computationally, appears to hold promise for predicting cratering behavior in more detail than has been previously possible.

Acknowledgments—This work was conducted for the Defense Nuclear Agency under Contract No. DNA001-74-C-0195, DNA Project/Task/Subtask NWED Y992AXSA001 Work Order Unit 55. The Technical Monitors were Lt. Col. Judd Leech and Captain J. Stockton.

A debt of thanks is owed P. S. De Carli and J. F. Kalthoff for performing the cratering experiments, to C. F. Petersen for consultations during the course of the project, and to D. Petro and P. DuBose for technical support. The cooperation of Professors W. Goldsmith and J. L. Sachman and graduate students M. Kabo and R. Benson of the University of California at Berkeley in providing ejecta velocity information is especially appreciated.

REFERENCES

Anthony, M. V.: 1966, An Estimate of the Mass and Size Distribution of Fine Particulate from a Megaton Surface Burst on Rock. The Boeing Company, D2-125066-1 (October 1966).

Barbee, T. W., Seaman, L., Crewdson, R., and Curran, D.: 1972, Dynamic Fracture Criteria for Ductile and Brittle Metals. *J. Materials, JMLSA*, **7**, 393–401.

Bertholf, L. D. and Benzley, S. E.: 1968, TOODY II, A Computer Program for Two-Dimensional Wave Propagation. Sandia Laboratories Research Report SC-RR-68-41 (November 1968).

Curran, D. R., Shockey, D. A. and Seaman, L.: 1973, Dynamic Fracture Criteria for a Polycarbonate. *J. Appl. Phys.* **44**, 4025–4038.

Ferritto, J. M. and Forrest, J. B.: 1975, Ground Motions from Pacific Cratering Experiments 1,000-Pound Explosive Shots. Technical Report No. R808, Civil Engineering Laboratory, Naval Construction Battalion Center, Port Hueneme, California 93043 (January 1975).

Fisher, P. R.: 1968, Engineering Properties of Craters: Description of Crater Zones and Site Investigation Methods. PNE 5012-1, U.S. Army Engineer Nuclear Cratering Group, Livermore, California (February 1968).

Gault, D. E., Shoemaker, E. M. and Moore, H. J.: 1963, Spray Ejected from the Lunar Surface by Meteoroid Impact. NASA TN D-1767 (April 1963).

Lutton, R. J., Girucky, F. E. and Hunt, R. W.: 1967, Project Pre-Schooner: Geologic and Engineering Properties Investigations. PNE-505F, Nuclear Cratering Group, U.S. Army Engineer Waterways Experiment Station, Vicksburg, Mississippi (April 1967).

Moore, H. J., Gault, D. E. and Lugn, R. V.: 1963, Experimental Impact Craters in Basalt. Society of Mining Engineering Transaction (September 1963).

Seaman, L. and Shockey, D. A.: 1973, Models for Ductile and Brittle Fracture for Two-Dimensional Wave Propagation Calculations. Final Report to AMMRC DAAG46-72-C-0182 (December 1973).

Shockey, D. A., Petersen, C. F., Curran, D. R. and Rosenberg, J. T.: 1973a, Failure of Rocks Under High Rate Tensile Loads. In *New Horizons in Rock Mechanics, Proceedings of the 14th Symposium on Rock Mechanics*. (H. R. Hardy, Jr., and R. Stefanko, Eds.). American Society of Civil Engineers, New York.

Shockey, D. A., Curran, D. R., Austin, M. and Seaman, L.: 1975, Development of a Capability for Predicting Cratering and Fragmentation in Rock. SRI Final Report DNA 3730F. Prepared for Defense Nuclear Agency, Washington, D.C., 20305 (May 1975).

Shockey, D. A., Curran, D. R., Seaman, L., Rosenberg, J. T. and Petersen, C. F.: 1974, Fragmentation of Rock Under Dynamic Loads. *Int. J. Rock Mech. Sci. Geomech. Abstr.*, **11**, 303–317.

Shockey, D. A., Seaman, L. and Curran, D. R.: 1973c, Dynamic Fracture of Beryllium Under Plate Impact and Correlation with Electron Beam and Underground Test Results. Air Force Weapons Laboratory Report No. AFWL-TR-73-12.

Shockey, D. A., Seaman, L., Curran, D. R., De Carli, P. S., Austin, M. and Wilhelm, J. P.: 1973b, A Computational Model for Fragmentation of Armor Under Ballistic Impact. U.S. Army Ballistic Research Laboratories, Aberdeen, Maryland, Contract DAAD05-73-C-0025 (December 1973).

Roddy, D. J., Pepin, R. O., and Merrill, R. B., editors.
(1977) *Impact and Explosion Cratering*, Pergamon Press (New York), p. 1089–1100.
Printed in the United States of America

Characteristics of debris from small-scale cratering experiments

ROBERT J. ANDREWS

University of Dayton Research Institute, Dayton, Ohio U.S.A.

Abstract—Geology-induced changes in ejecta characteristics have been studied by conducting laboratory-scale, high-explosive cratering experiments. Changes in the general characteristics of the ejecta plume, the angle and velocity of exiting particles, the origin of material within the cratered region, depth of ejecta as a function of ground range, and the occurrence of rays in the ejecta blanket are presented and discussed. Static and dynamic data indicate that both the ejecta exit angle and ejecta exit velocity are functions of the medium density, medium moisture content, and layering of the medium. Ejecta origin within the cratered region was determined in various homogeneous and layered geologies. Origin maps have been constructed and these maps detail the geology-induced effects. Ejecta depth as a function of ground range for craters formed in dry dense medium is described by a decaying power function. The ejecta blanket of near-surface events has been studied to determine the length and number of ejecta rays produced as a function of charge configuration. Ray length increases with increasing depth of charge burial; however, the maximum number of rays occurs when the charge is half-buried. Also presented are several observations which suggest that ray patterns may be at least partially related to anomalies in the fireball and expanding detonation products.

INTRODUCTION

THE CHARACTERISTICS OF EJECTA and the parameters which effect ejecta lofting and deposition processes are of interest to investigators of explosive and impact craters. Two primary reasons for this interest are to obtain a basic understanding of the ejection and deposition processes and provide insights into the cratering mechanisms active during crater formation. Laboratory-scale, high-explosive cratering experiments have been conducted under controlled conditions to obtain crater and ejecta data at resolutions not possible outside the laboratory. The effect of *in situ* density, moisture content, and layering of media on ejecta characteristics were investigated. This paper presents a broad-based summary of results obtained from approximately 300 experiments performed over a period of three years.

Unconsolidated Ottawa sand having an average grain diameter of approximately 0.5 mm was used as the cratering medium for all events. To obtain data in layered media, various thicknesses of Ottawa sand were placed above cemented Ottawa sand layers. Centrally initiated charges of dextrinated lead azide were used as the explosive material for the experiments. The charges were formed by pressing 1.7 g of lead azide to a density of 3.1 g/cm^3 to form spheres 10.16 mm in diameter.

Although data have been collected from events detonated over a large range of burst heights, this paper presents the data from only near-surface detonations, particularly the half-buried charge configuration. Ejecta characteristics presented include: general characteristics of the ejecta plume, the angle and velocity of exiting particles, the origin of material within the cratered region, depth of ejecta as a function of ground range, and the occurrence of rays in the ejecta blanket.

EJECTA PLUME CHARACTERISTICS

High speed photographs taken during crater formation reveal that ejected material exits the crater as a thin conical veil which travels across the preshot surface at a constant angle to that surface. An upthrust region develops on the preshot surface immediately ahead of the advancing ejecta veil. The velocity of the veil, parallel to the preshot surface, is between 1.0 m/s and 1.5 m/s for a half-buried event in dry, dense (1.80 g/cm^3) sand. A fully developed plume, accentuated with squares of dyed sand, is shown in Fig. 1.

The highest portion of the veil contains material which is ejected at the highest velocity, immediately after detonation, and is deposited farthest from the crater. The trailing edge of the veil contains material which leaves the crater at low velocity and forms the overturned flap of debris near the crater lip. This low-velocity material is the last debris to be ejected and the first to be deposited on the preshot surface. Investigators of laboratory-scale impact phenomena (Oberbeck, 1976) observe essentially the same characteristics for plumes of ejecta from impact craters. Also visible in the conical veil of near-surface events

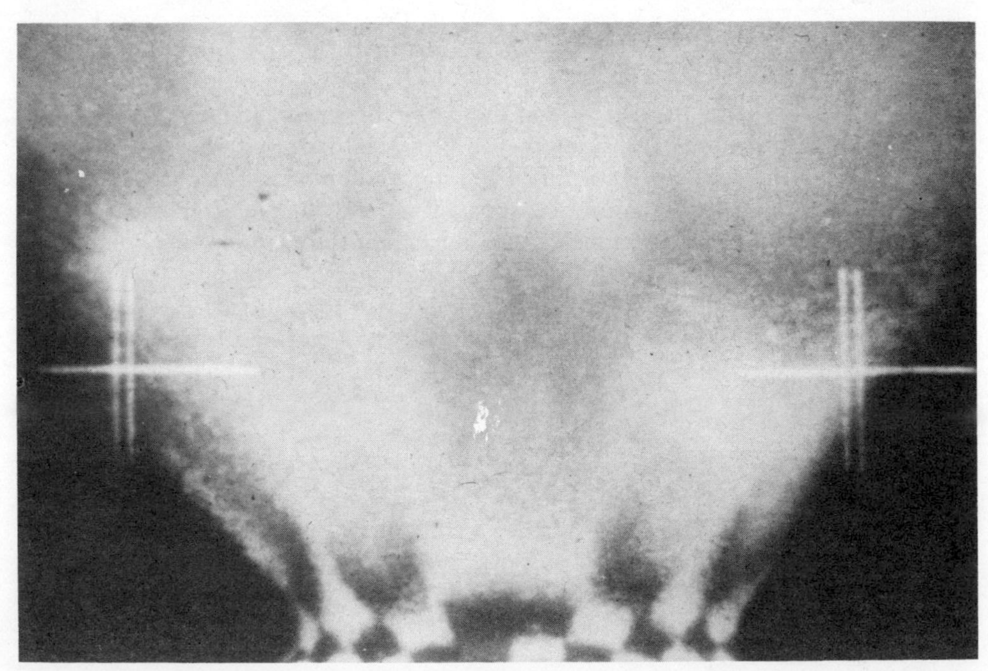

Fig. 1. Fully developed ejecta plume. The square pattern is a grid of dyed sand placed on the preshot surface.

are airborne spires of debris which form shortly after detonation and move outward at a higher velocity than the remainder of the ejecta plume. The spires are deposited on the preshot surface slightly ahead of the advancing veil as rays of ejecta.

Particle ejection angles and velocities were measured with use of high-intensity light sources and a still camera. The high-intensity lights were focused into a thin vertical plane above the test bed and normal to the preshot surface. Ejecta moving within the plane of light would normally produce streak images on the film in the still camera. A rotary shutter placed in front of the camera converts the streak images into a series of dashed trajectories. An auxiliary shutter, released a fixed time after detonation, was used to create an interruption in the dashes to relate particle position to a particular time after detonation. A "chopped" photograph showing the dashed trajectories and timing mark (interruption) is presented in Fig. 2.

The ejection angle (measured between the veil and the preshot surface) for a typical half-buried event detonated in dry dense sand is between 44° and 48°. Particles originally located near the charge are the first particles to leave the crater and are ejected at the highest velocities. Velocities as high as 8 m/s were recorded for half-buried events detonated in dry, dense sand; however, the ejection velocity for the bulk of the particles is on the order of 2.5 m/s. Decreases in the *in situ* density or increases in moisture content result in an

Fig. 2. Typical photograph used to record the motion of ejecta. This event was detonated in saturated sand and the auxiliary timing mark is visible in the particle trajectories.

Fig. 3. Chopped photograph from event in dry layered media with shallow overburden. (Note the dual-angle ejection of debris.)

increase in ejection angle. In dry, low density sand (1.57 g/cm^3), the ejection angle is between 50° and 52°. In saturated low density sand the angle increases to between 72° and 76°. Maximum particle velocities of almost 9.5 m/s have been measured for craters formed in saturated dense sand; however, maximum exit velocities are lower for events detonated in saturated low density sand (approximately 4.5 m/s).

The ejection angle is also a function of layers present in the cratering medium. In dry layered media, the ejection angle decreases as the depth of overburden decreases. However, an unusual phenomenon occurs for shallow depths of overburden in dry layered media. Half-buried events with depths of overburden less than 30 mm produce ejecta which is ejected at two distinct angles; one shallow angle (30–40°) and a near vertical angle. This dual angle can be seen clearly in Fig. 3. In saturated layered media the effect of saturation, which increases ejection angle, overshadows any layering effect. Also, no dual-angle ejection of debris occurs in saturated layered media.

ORIGIN OF EJECTA PARTICLES

Relating the original and final positions of ejecta particles is usually accomplished by placing easily identifiable tracers in the medium and recovering them after detonation. Carlson and Newell (1970) used plastic beads and

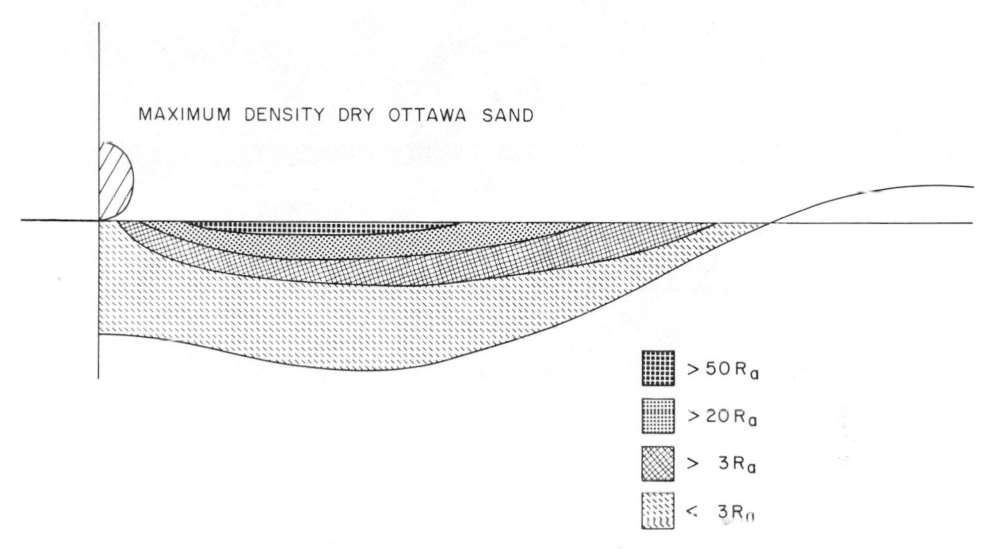

Fig. 4. Ejecta origin map for craters in dense dry (1.80 g/cm^3) sand. Patterns designate the final ejecta range as a function of apparent crater radius.

aluminum pellets for this purpose. Other investigators (Diehl and Jones, 1964; Perret *et al.*, 1963; Harvey *et al.*, 1975) have used such items as film cans, oil cans, cement balls, and radioactive pellets. The density and size of these items can differ significantly from the surrounding medium and hence, their trajectories may differ from the trajectories of the naturally occurring ejecta particles. Tracers of two different types have been used in the laboratory. Dyed Ottawa sand which has the same grain size and density as the cratering medium has been placed in the medium as multicolored layers (Andrews, 1975). A refinement of this technique involved placing small (2–3 mm diameter) glass beads at discrete locations in the region to be cratered. The postshot range and azimuth of the tracers were determined and these data were used to map the cratered region according to final debris range. The events described in this section of the paper were part of a study to determine the effect of geology on debris origin. The charge was detonated tangent above the surface for each event.

Figure 4 shows the origin map for a crater formed in dry dense sand. The maximum range missiles originate near the preshot surface in a region centered at approximately 0.4 crater radii from surface ground zero (SGZ). Mid-range ejecta particles originate from intermediate depth positions within the cratered region. The material which originates in the region forming the apparent crater wall is deposited as the overturned flap of debris in the crater lip region and the material originally beneath the charge moves down and outward and is distributed along the floor of the apparent crater.

The origin map for a crater formed in dry, low density sand is shown in Fig. 5. The maximum range missiles originate from approximately the same region as

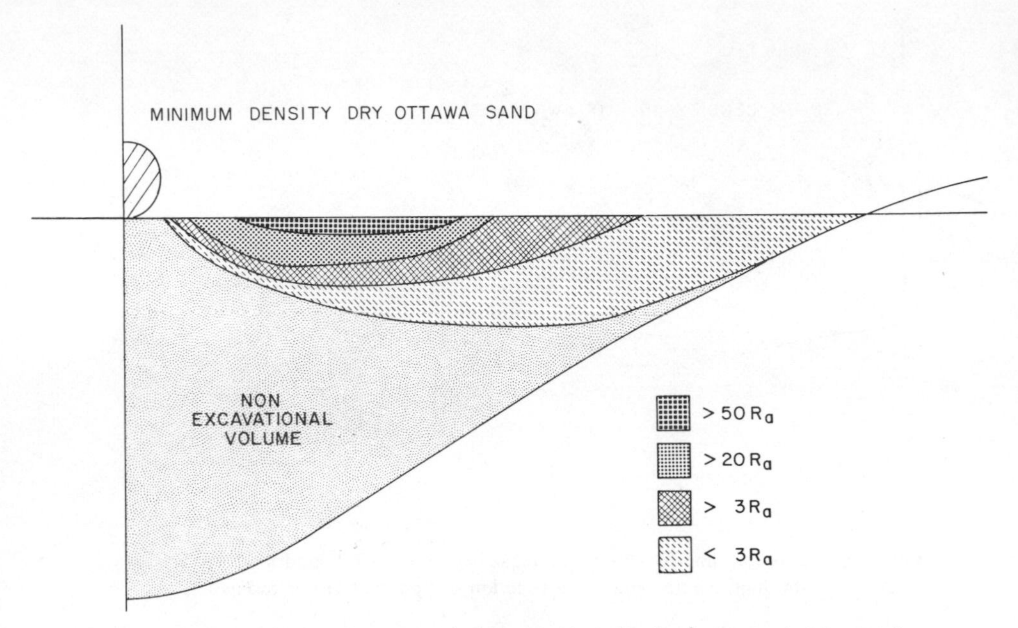

Fig. 5. Ejecta origin map for craters in low density (1.57 g/cm^3) dry sand. (Note region of compacted material.) Patterns designate the final ejecta range as a function of apparent crater radius.

for craters formed in dry, dense sand. However, nearly 40% of the crater volume is formed by compaction of material within and below the crater. Most of the material originally located below 0.3 crater depth is compacted rather than ejected. Postshot locations of tracers placed within this region indicate the material moved down and slightly outward during crater formation.

Near-surface events detonated in saturated, dense sand produce small craters; however, the entire crater volume is the result of ejection of material. Debris is ejected to greater scaled ranges in this medium than in any of the media investigated. A large percentage of the debris is ejected beyond 50 crater radii with some particles deposited beyond 130 crater radii. The origin map for a crater formed in saturated, dense sand is presented in Fig. 6.

The effect of a layered medium on the origin of ejecta particles was investigated by detonating events with 30 mm of dense, dry overburden above a cemented sand layer. Comparison of this origin map, shown in Fig. 7, with the map for the homogeneous medium (Fig. 4) reveals a larger maximum range missile region extending along the preshot surface from adjacent to the charge to approximately 0.4 crater radii. This is the only medium in which material directly beneath the charge was ejected beyond a few crater radii. A second effect of the presence of the layer is that material originally located beneath the apparent crater floor (below SGZ) is deposited between 3 and 20 crater radii from the center. High speed photographs of events detonated in layered media with shallow depths of overburden reveal a secondary ejecta veil. This secondary veil

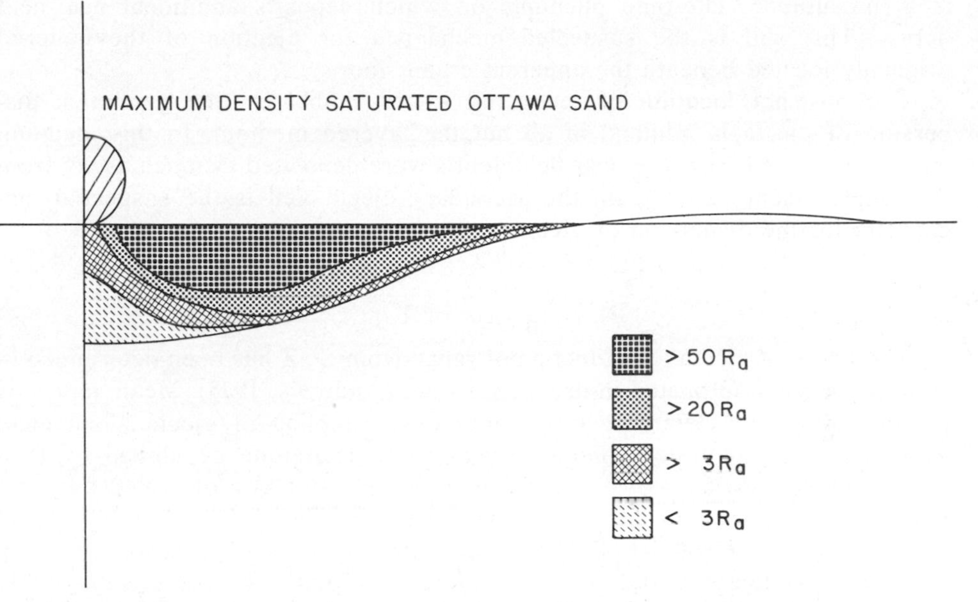

Fig. 6. Ejecta origin map for craters in dense (1.80 g/cm^3) saturated sand. Patterns designate the final ejecta range as a function of apparent crater radius.

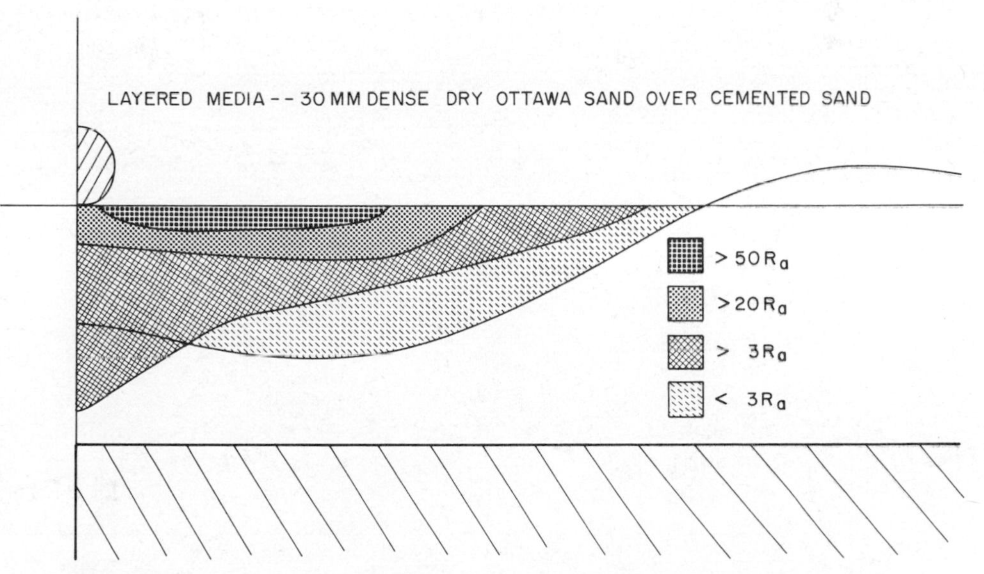

Fig. 7. Ejecta origin map for craters in layered medium with 30 mm of dense dry (1.80 g/cm^3) sand above cemented sandstone. Patterns designate the final ejecta range as a function of apparent crater radius.

is a low-altitude, late-time phenomenon which deposits additional near-field debris. This veil is the suspected mechanism for ejection of the material originally located beneath the apparent crater floor.

The postshot location of recovered tracers indicates that azimuthal dispersion of ejecta is minimal in all but the layered medium. In this medium, tracers recovered from the near-field debris were deposited as much as 90° from the emplacement axis. Again the secondary ejecta veil is the suspected mechanism for the dispersion of ejecta.

DISTRIBUTION OF EJECTA

The depth of ejecta as a function of range from SGZ has been determined in detail for events detonated in dry, dense sand (Andrews, 1975). Mean values of ejecta depth were determined by continuous sampling of ejecta along eight azimuths. Using a crater volume normalization technique developed by Post (1974), the scaled depth of ejecta was plotted as a function of scaled range. A plot of the data for half-buried events is shown in Fig. 8. The data between 2 crater radii and 40 crater radii is best described by a decaying power function. As seen in the figure, correlation between the data and fitted curve is extremely good. Figure 9 shows a comparison of this data with ejecta data from the AIR VENT series (Carlson and Jones, 1964) detonated in playa. The data shown in this figure are from events where the explosive yield spans more than six orders of magnitude. The similarity of the ejecta depth functions is apparent.

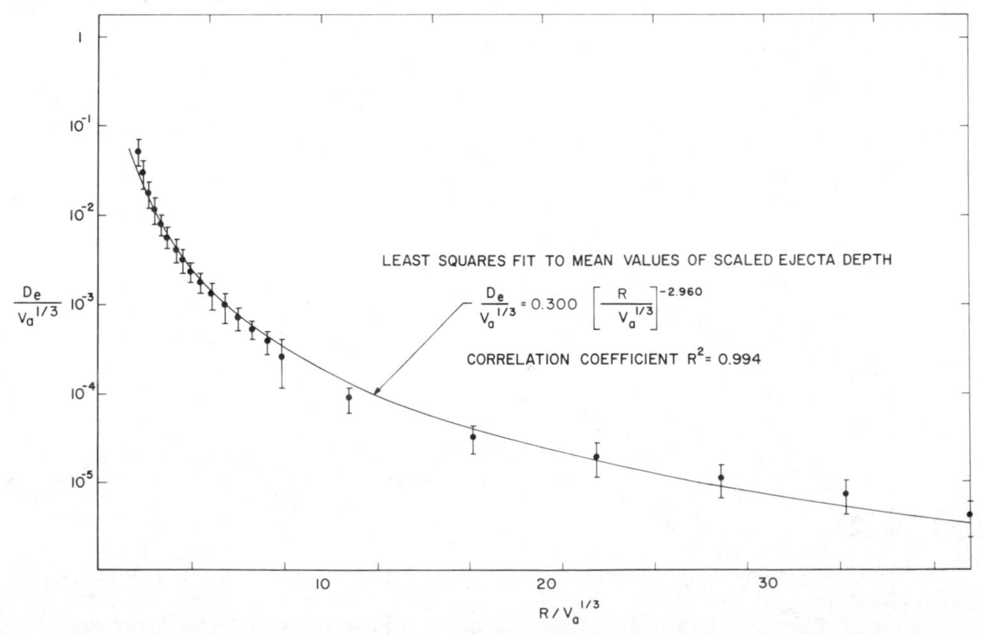

Fig. 8. Ejecta depth as a function of ground range for half-buried events in dry dense sand.

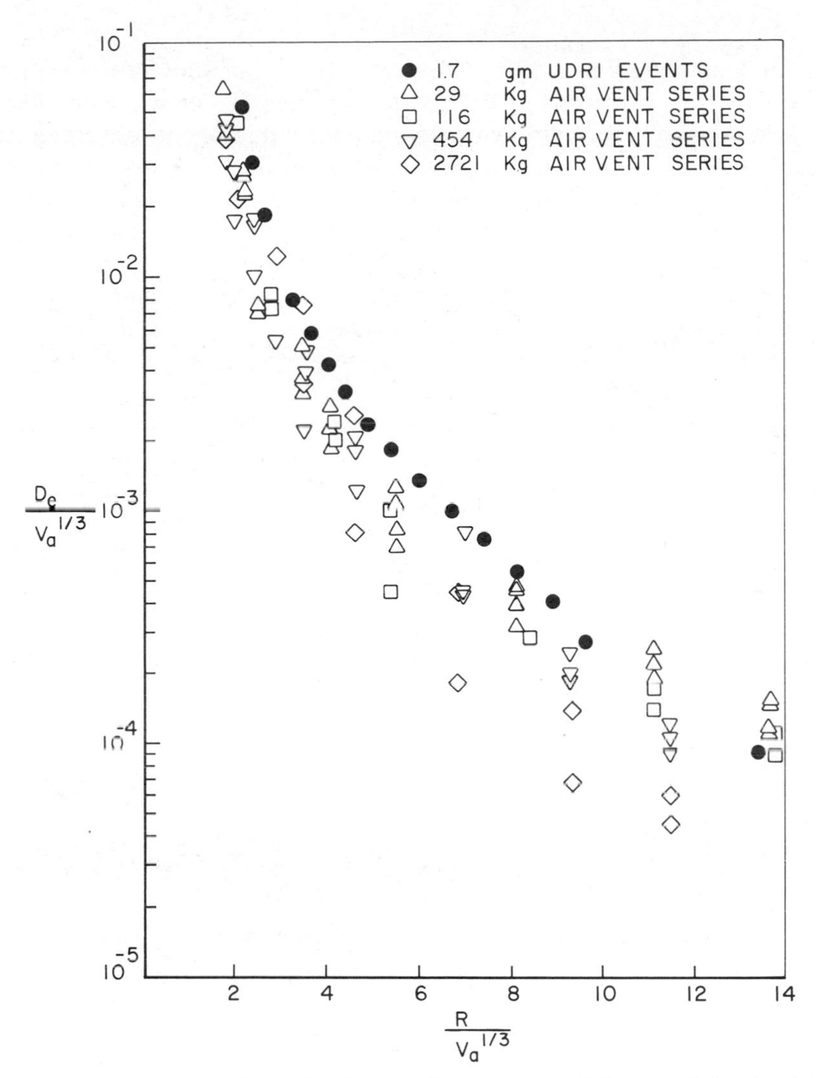

Fig. 9. Comparison of ejecta distributions from events with large variation in yield.

Changes in *in situ* density result in similar distributions (Piekutowski, 1975); however, changes in moisture content result in marked changes in the ejecta distribution. Ejecta distributions from craters in both high and low density saturated sand exhibit an order of magnitude more scatter in the data at a given range from ground zero and a tendency for the mean values to have less range dependency than dry sand distributions. Craters formed in saturated sand do not produce a continuous ejecta blanket. Rather, the ejecta is deposited as small discrete mounds of wet sand.

EJECTA RAYING

A recent study (Andrews, 1976) was performed in an attempt to characterize the rays which result from near-surface detonations in Ottawa sand. The rays resulting from laboratory-scale experiments resemble those observed in the ejecta blankets surrounding larger, field-scale explosion and impact craters and can be defined as radially oriented mounds of debris. Ray lengths extend from two to seven crater radii depending on geology and height-of-burst. The rays are not, in all cases, truly radial to SGZ, but are seldom skewed more than 10°. A few rays were observed to have multiple tails. Carlson and Jones (1964) described a primary and secondary set of rays for some of the AIR VENT craters; however, our craters exhibit only a single pattern.

The characteristics of ray length, angle of separation, and number of rays per event were determined for several near-surface charge configurations. Figure 10 shows the number of rays as a function of height-of-burst in dry, dense sand. The maximum number of rays for craters formed in dry, dense sand occurs near the half-buried charge configuration with approximately 70 rays identified. As shown in Fig. 11, average ray length is also a function of height-of-burst. As the height-of-burst decreases, average ray length increases; however, for craters

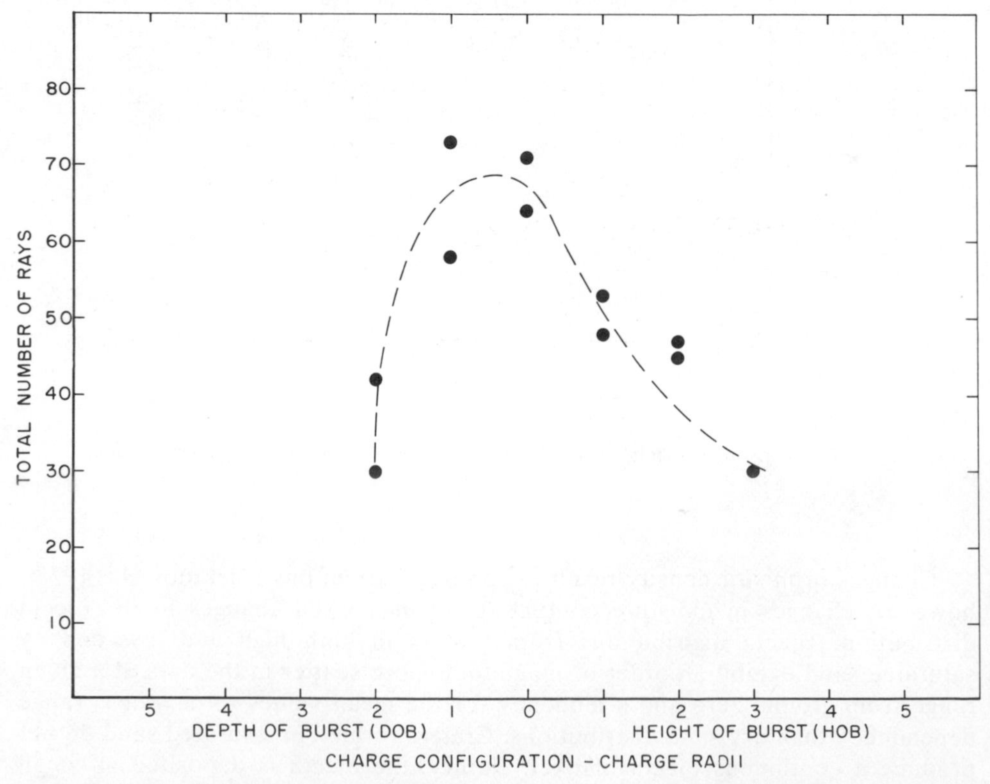

Fig. 10. Number of rays produced at near-surface charge configurations.

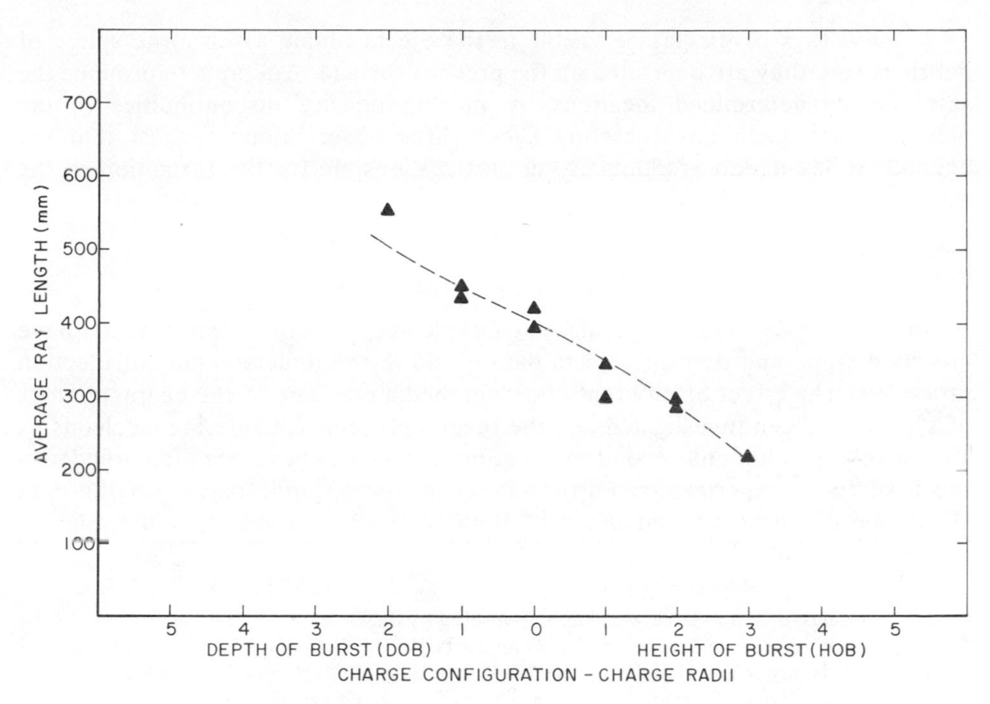

Fig. 11. Average ray length as a function of charge configuration.

formed at charge configurations deeper than 5-charge radii below the surface, the rays are indistinguishable from the surrounding ejecta.

A small amount of data on ejecta rays produced in a layered geology indicates that ray length increases as depth of overburden decreases with fewer rays produced as the depth of overburden decreases. While approximately 70 rays are produced by a half-buried event in homogeneous media, only 29 rays were identified for a half-buried event in layered media with 30 mm of overburden. No correlation between the length of a ray and its position relative to neighboring rays could be made; i.e., ray length appears to occur randomly around the crater. Also in dry, dense sand, maximum depth of ejecta on a ray is approximately four times the depth of ejecta between rays.

Three observations have been made concerning the origin of ejecta rays. High-speed photographs record fireball anomaly patterns which appear to correlate with ray patterns. However, the fireball gas jet pattern is three-dimensional with some of the jets directed upward and away from this medium. Therefore, while the correlation of rays to anomalies cannot be made with one-to-one accuracy, the gross features of the patterns are similar.

Two events were detonated with a conical airblast deflector protecting one-half of the cratering medium surface. Shielding of the preshot surface from airblast significantly reduced the number and magnitude of rays without interfering with normal cratering formation. The expanding detonation gases (and associated anomalies) apparently have an effect on the formation of rays.

Finally, rays of ejecta are visible in the ejecta plume as airborne spires of debris before they are deposited on the preshot surface. Attempts to produce the spires at predetermined locations by placing density discontinuities in the medium have been unsuccessful. These three observations suggest that the expanding detonation products are in part responsible for the formation of the rays.

CONCLUDING REMARKS

In summary, laboratory-scale, high-explosive cratering experiments have provided static and dynamic ejecta data to aid in the understanding of ejection processes. The effect of different cratering media on some of the characteristics of ejecta has been investigated and the results presented. Cratering mechanisms such as compaction and secondary ejection of material have been identified as a result of these experiments. Further experimentation and analysis of the data should be of interest to members of both the impact cratering and explosive cratering communities.

Acknowledgments—This research was jointly sponsored by the Defense Nuclear Agency and the Air Force Weapons Laboratory under Air Force Contracts F29601-73-C-0007 and F29601-74-C-0102. I thank Steve J. Hanchak and Henry L. Williams for their diligent effort in performing the laboratory experiments and Dart G. Peterson Jr. for his review of the manuscript.

REFERENCES

Andrews, R. J.: 1975, Origin and distribution of ejecta from near-surface laboratory-scale cratering experiments, Air Force Weapons Laboratory Report AFWL-TR-74-314.

Andrews, R. J.: 1976, Characteristics of ejecta rays from near-surface laboratory-scale cratering experiments, Air Force Weapons Laboratory Report, in press.

Carlson, R. H. and Jones, G. D.: 1964, Ejecta distribution studies, The Boeing Company Report D2 90575.

Carlson, R. H. and Newell, R. T.: 1970, Ejecta from single-charge cratering explosions, Sandia Laboratories Report SC-RR-69-1.

Diehl, C. H. H. and Jones, G. H. S.: 1964, A tracer technique for cratering studies, *J. Geophys. Res.* **70**, 305.

Harvey, W. T., Dishon, J. F., and Tami, T. M.: 1975, Near-surface cratering experiments, Fort Polk, Louisiana, Air Force Weapons Laboratory Report AWFL-TR-74-351.

Oberbeck, V. R. and Morrison, R. H.: 1976, Effect of basin secondaries and debris surges on survival of post-accretional lunar and planetary surfaces, (abstract) In *Lunar Science VII* p. 642–644. The Lunar Science Institute, Houston.

Perret, W. R., Chabai, A. J., Reed, J. W., and Vortman, L. J.: 1963, Project Scooter final report, Sandia Laboratories Report SC-4602 (RR).

Piekutowski, A. J.: 1975, The effect of variations in test media density on crater dimensions and ejecta distributions, Air Force Weapons Laboratory Report AFWL-TR-74-326.

Post, R. L.: 1974, Ejecta distributions from near-surface nuclear and HE bursts, Air Force Weapons Laboratory Report AFWL-TR-74-51.

Roddy, D. J., Pepin, R. O., and Merrill, R. B., editors.
(1977) *Impact and Explosion Cratering*, Pergamon Press (New York), p. 1101–1121.
Printed in the United States of America

Dynamic ejecta parameters from high-explosive detonations

JOHN WISOTSKI

Denver Research Institute, University of Denver,
Denver, Colorado 80208

Abstract—Technical information pertaining to the dynamic ejecta throwout from ton-size, high-explosive cratering events was obtained photographically from the MIDDLE GUST Series. Parametric data obtained from ejecta trajectories gave fragment velocities of over 300 ft/sec, with flight times of over 13 sec and ranges of over 2300 ft for fragments weighing over 1300 lb. Comparison of velocities and times from 1/2-ton events to those obtained from 20-ton and 100-ton events in or on similar geologies scale by the ratio of the charge weights to the 1/6 power of scaled distances of the ratio of charge weights to the 1/3 power.

I. INTRODUCTION

1. *Objectives*

THIS PAPER PRESENTS a review of dynamic ejecta data from selected high-explosive (HE) events. Dynamic ejecta fragments were photographed and analyzed to present data obtained from the analysis of some of these photographic records, especially those from the MIDDLE GUST Series. The Denver Research Institute (DRI) has photographed dynamic ejecta from over thirty-eight HE events since the MINE SHAFT Series. The sponsoring agencies were the Defense Nuclear Agency (DNA) and the Air Force Weapons Laboratory (AFWL).

2. *Background and theory*

Dynamic ejecta data presented in this paper pertains to ejecta material observed beyond the fireball debris during the detonation and cratering process. Dynamic ejecta information from HE detonations prior to the MINE SHAFT Series was very sparse. DRI participated in the MINE SHAFT Series from which limited dynamic ejecta data were obtained (Wisotski, 1969, 1970). Due to triggering and power failures, the photographic data were limited to 2.2 sec of recording. Since that time, over 38 separate events, ranging from 256 lb to 500 tons of HE, have been recorded for durations of at least 25 sec. Some of the procedures developed during the analysis of the MINE SHAFT data have been incorporated in the analysis of the data presented here (Wisotski, 1977).

II. EXPERIMENTAL PROCEDURES

1. *Experimental setup*

The detonation sources for the HE cratering events were either center-initiated ton-size spherical charges constructed from 32.6-lb (average) blocks of cast trinitrotoluene (TNT) or spherical charges (256 and 1000 lb) made as one piece of cast TNT. The charges were placed at different heights-of-burst (HOB) ranging from two times the charge radius ($2R_c$) to $-13R_c$.

The geological profiles for these detonations ranged from coral sand located on Aomon Island at Eniwetok Atoll (PACE SERIES), to a sandy alluvium over a sandstone base during the MIXED COMPANY Events at Grand Junction, Colorado. Both wet and dry geologies were employed in several of the events. (See Table 1.)

2. *Instrumentation and field operation*

Dynamic ejecta information was obtained with at least four 70 mm Hulcher cameras operated at approximately 20 frames per second. All four cameras were located at one main camera position so that reference markers placed on both sides of the charge in a plane through surface ground zero (SGZ) could be photographed. These reference markers were generally placed at 200, 500, and 1000 ft from and on both sides of SGZ for ton-size events and 100 and 200 ft for 1/2-ton events. The close-in markers were used to determine the scale for a given distance from SGZ to the camera station. The other markers were used for positioning the cameras and as reference in the calculations of trajectories.

3. *Position-time measurements*

Each position of an ejecta trajectory path was given by its X–Y coordinates referenced to SGZ or some other marker in the field-of-view of the camera; whereas, the time was represented by the time period from first-light. The exact timing of each frame was determined from 100 cycles per second timing marks.

The criterion for obtaining raw trajectory data was to track particles that were on the outer extreme of the mass of throwout near the point of impact. This criterion generally allowed the tracking of ejecta whose trajectories were within 20° or 30° of the plane-of-view of the camera. As an example, the ejecta mass thrown out during MIDDLE GUST Event I is shown in Fig. 1.

The position-time measurements of specific ejecta particles were made with a 70 mm Vanguard Motion Analyzer, Model C-11. The X–Y coordinates were measured to three decimal places. The scale in the fields-of-view as measured on the motion analyzer varied from about 100 to 140 ft per inch for ton-size detonations to 25–35 ft per inch for 1/2-ton detonations depending on the distances from SGZ to the camera station.

4. *Mathematical analysis*

The computer program that was developed to mathematically analyze the position-time data from the MIDDLE GUST Series differs from that which was used on the sparse data from the MINERAL ROCK Event (Wisotski, 1970). This was done to take advantage of the abundance of raw position-time data from each trajectory path. Each specific particle was backtracked from a few feet of impact (limited by surface dust) to a position just outside the fireball or dust cloud. The horizontal distances covered by these measurements varied from approximately 40% of the total trajectory for the close-in ejecta to 70% for the far-out ejecta. The available position-time data points using 0.5 sec intervals varied from 200 to 10 depending on the projected distances of the ejecta.

Second order least-squares fits were made to the raw position data. Three quadratic expressions were used to obtain the initial and terminal parameters for each trajectory. These quadratic expressions were in the form $Y = f(X)$, $Y = f(t)$, and $X = f(t)$.

Fig. 1. Ejecta west of surface ground zero from MIDDLE GUST I, frame 100, 4.35 sec.

Table 1. Description of events photographically covered by DRI.

No.	Event	Weight (ton)	HOB (R_c)	Geological environment
1	MIDDLE GUST I	20	0	9 ft sandy clay over shale—water table −4 ft
2	MIDDLE GUST II	100	2	9 ft sandy clay over shale—water table −4 ft
3	MIDDLE GUST III	100	1	9 ft sandy clay over shale—water table −3 ft
4	MIDDLE GUST IV	100	1	2.5 ft clay over shale—no near surface water table
5	MIDDLE GUST V	20	0	2.5 ft clay over shale—no near surface water table
6	MIDDLE GUST 2	1/2	−13[1]	9 ft sandy clay over shale—water table −3 ft
7	MIDDLE GUST 3a	1/2	0	No overburden removed in clay siol—water table −4 ft
8	MIDDLE GUST 5	1/2	1	4 ft overburden removed—in sandy clay soil—water table −1 ft
9	MIDDLE GUST 6	1/2	0	4 ft overburden removed—in sandy clay soil—water table −1 ft
10	MIDDLE GUST 7	1/2	1	9 ft overburden removed—in weathered shale—water table on surface
11	MIDDLE GUST 8	1/2	0	9 ft overburden removed—in weathered shale—water table on surface
12	MIDDLE GUST 9	1/2	0	16 ft overburden removed—in Pierre shale—saturated
13	PACE A	1/2	2	Eniwetok Atoll—coral sand—water table at surface
14	PACE B	1/2	1	Eniwetok Atoll—coral sand—water table at surface
15	PACE C	1/2	1	Eniwetok Atoll—coral sand—water table −0.5 to −1.0 ft
16	PACE D	1/2	1	Eniwetok Atoll—coral sand—water table at surface
17	PACE E[2]	1/2	0	Eniwetok Atoll—coral sand—water table −0.5 ft, 1 ft layer beach rock at 1 ft
18	PACE F	1/2	0	Eniwetok Atoll—coral snad—water table at surface
19	PACE G	1/2	0	Eniwetok Atoll—coral sand—water table at surface
20	PACE H	1/2	−1[3]	Eniwetok Atoll—coral sand—water table at surface
21	PACE I	1/2	1/2	Eniwetok Atoll—coral sand—water table at surface
22	PACE J	1/2	−1/2[3]	Eniwetok Atoll—coral sand—water table at surface
23	PACE K	1/2	0	Eniwetok Atoll—on sand spit—water table −0.2 ft
24	PACE O	1/2	0	Eniwetok Atoll—coral sand—water table at surface
25	MIXED COMPANY I	20	0	4.5 ft alluvial sandy soil over sandstone
26	MIXED COMPANY II	20	1	2.5 ft alluvial sandy soil over sandstone
27	MIXED COMPANY III	500	1	5.5 ft alluvial sandy soil over weathered sandstone −12 ft to hard sandstone layer

28	Pre-DICE THROW II-1	100	1	Queen 15 site—layered silty soil, gypsum and clay—water table −6.5 ft
29	Pre-DICE THROW II-2	120[4]	1	Queen 15 site—layered silty soil, gypsum and clay—water table −5.5 ft
30	CENSE III-2	1/10[5]	1	MIXED COMPANY site—silty soil over sandstone base—soil 5 ft thick one side and 5 ft thick other side of SGZ
31	CENSE III-4	1/10[5]	1	MIXED COMPANY site—silty soil over sandstone base—soil 3 ft thick one side and 4 ft thick other side of SGZ
32	CENSE III-6	1/10[5]	1	MIXED COMPANY site—silty soil over sandstone base—soil 1 ft thick one side and 2 ft thick other side of SGZ
33	CENSE III-7	1/10[5]	1	MIXED COMPANY site—silty soil over sandstone base—soil 0.33 ft thick
34	MX VALLEY 1	1/2	0	Sand dunes, fine blow sand, moist −1.5 ft
35	MX VALLEY 3	1/2	0	Queen 15 site, silty clay soil with sand lenses—water table −1.5 ft
36	MX VALLEY 4	1/2	0	Unconsolidated fine to medium grus sand
37	MX VALLEY 4A	1/10[6]	0	Unconsolidated fine to medium grus sand
38	MX VALLEY 5	1/2	0	Unconsolidated pebbles, cobbles, boulders in grus sand

[1]Depth of burial 17 ft.
[2]Premature detonation.
[3]Depth of burial—one charge radius (20), 1/2 charge radius (22).
[4]100-ton TNT equivalent AN/FO hemispherical-ended cylindrical charge.
[5]200 lb nitromethane cylinders ($L/D = 1$), 226 # TNT equivalent.
[6]256 # cast TNT sphere.

5. *Computer procedures*

The computer program was written to perform arithmetic and trigonometric manipulations on the raw position-time data so as to obtain the initial and terminal parameters for each tracked ejecta fragment. These parameters are escape velocity and angle; maximum height; impact velocity, angle, distance and position; total, verified, and function times; drag constant and coefficient; apparent size and weight; Reynolds number; kinetic energy and apparent fireball density. The great number of formulas associated with these parametric calculations are not presented in this paper, but are presented in Wisotski (1977). An outline of steps used in the calculations is presented below.

Since the cameras were rotated to view away from and on both sides of SGZ and also upwards toward the sky, calculations were made on the $X-Y$ coordinate data to compensate for the cameras' angular positions.

The assumed plane-of-view for any of the cameras was a plane passing through a horizontal line which passed through SGZ and perpendicular to the horizontal center line of the camera. It was assumed that there were no perturbations in the ejecta flights so that their trajectories were always in vertical planes. A plan view of the assumed plane-of-view is presented in Figs. 2 and 3.

Since most of the particles' trajectories were not exactly in the plane-of-view of the recording camera, the computer program was written to adjust the trajectories until they were in the correct plane with respect to the camera position. Examples of the apparent distortion in the horizontal positions of two different ejecta trajectories presented at even increments of distance-time, i.e., no drag, are given in Figs. 2 and 3. The fields-of-view (FOV), distances, and angles are exaggerated to emphasize the apparent distortion.

For a particle which has its trajectory on the far side of the plane-of-view of the camera (Fig. 2), the total horizontal distance covered by the particle appears to be much less than its true value and the distance-time increments are uneven when its trajectory is projected into the plane-of-view of the camera. If this same trajectory were to exist between the camera and its plane-of-view (Fig. 3), the apparent distance covered by the ejecta would be greater and the distance-time increments also uneven when the trajectory is projected into the plane-of-view of the camera.

For the paths represented in Figs. 2 and 3 the apparent distortions in their X and Y coordinates, as they would appear if they were projected into the camera's plane-of-view (no tilt), are presented in Fig. 4. The only portions of these apparent trajectories which are photographed by the camera are those outside the fireball or dust cloud. Least-squares parabolic curve fits through the X and Y coordinates which depict these paths near the fireball end and an inverse-drag routine through the fireball tend to place the particles' origins outside the crater ($X > R_a$ at $Y = 0$ where R_a is the apparent crater radius) for paths away from the plane-of-view and inside the crater ($X < R_a$ at $Y = 0$) for paths toward the camera.

An outline of the relative sequence of the major setups in the computer routine used to determine the initial and terminal parameters is as follows:

(a) Scale in SGZ.
(b) Scale in the plane-of-view.
(c) Correction of the Y coordinate for the camera's tilt.
(d) Determination of the interim least-square parabolic curve fits of $Y = f(X)$, $X = f(t)$, and $Y = f(t)$ (all or 60% of position-time data points) (Wisotski, 1977).
(e) Determination of the average drag constant (K) for a fixed increment (D), generally 2 ft on both sides of the peak (Y is maximum) from the change in velocity using $X = f(t)$ (Linnerud, 1976).
(f) Inverse-drag routine from a point on the $Y = f(X)$ curve starting outside the fireball toward SGZ until $Y = 0$. See Fig. 5.
(g) Determine the drag coefficient (C_D) from the particle's weight and presented area, the air density and the drag constant (K).
(h) Determine the Reynolds number from the particle's characteristic length or dimension, its escape velocity, and the air density and its dynamic viscosity. For a specific particle's geometry and the Reynolds number value the C_D of the particle should compare to a certain expected value (C_{DR}). See Figs. 6 and 7. Most ejecta dimensions and velocities were of such

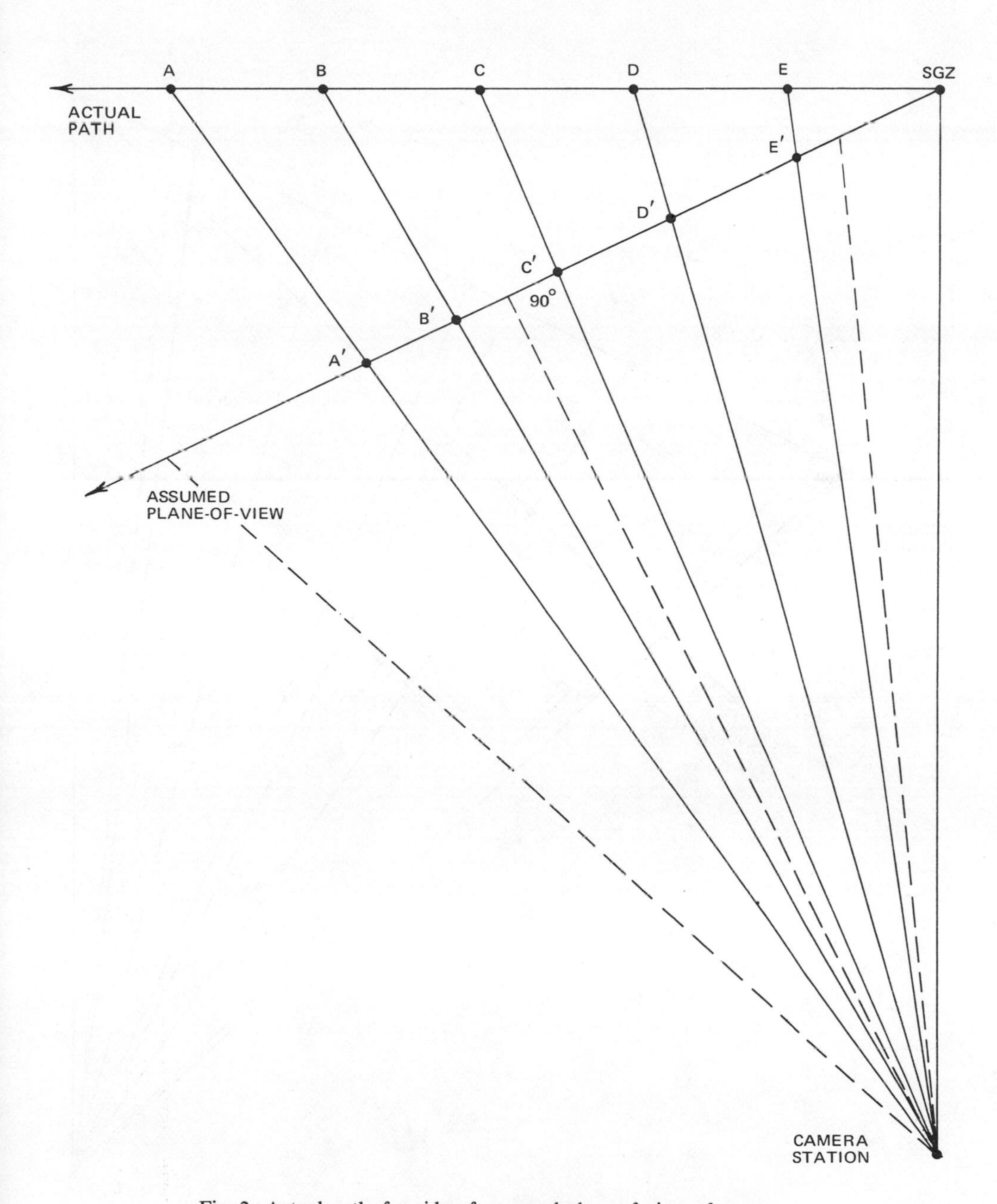

Fig. 2. Actual path, far side of assumed plane-of-view of camera.

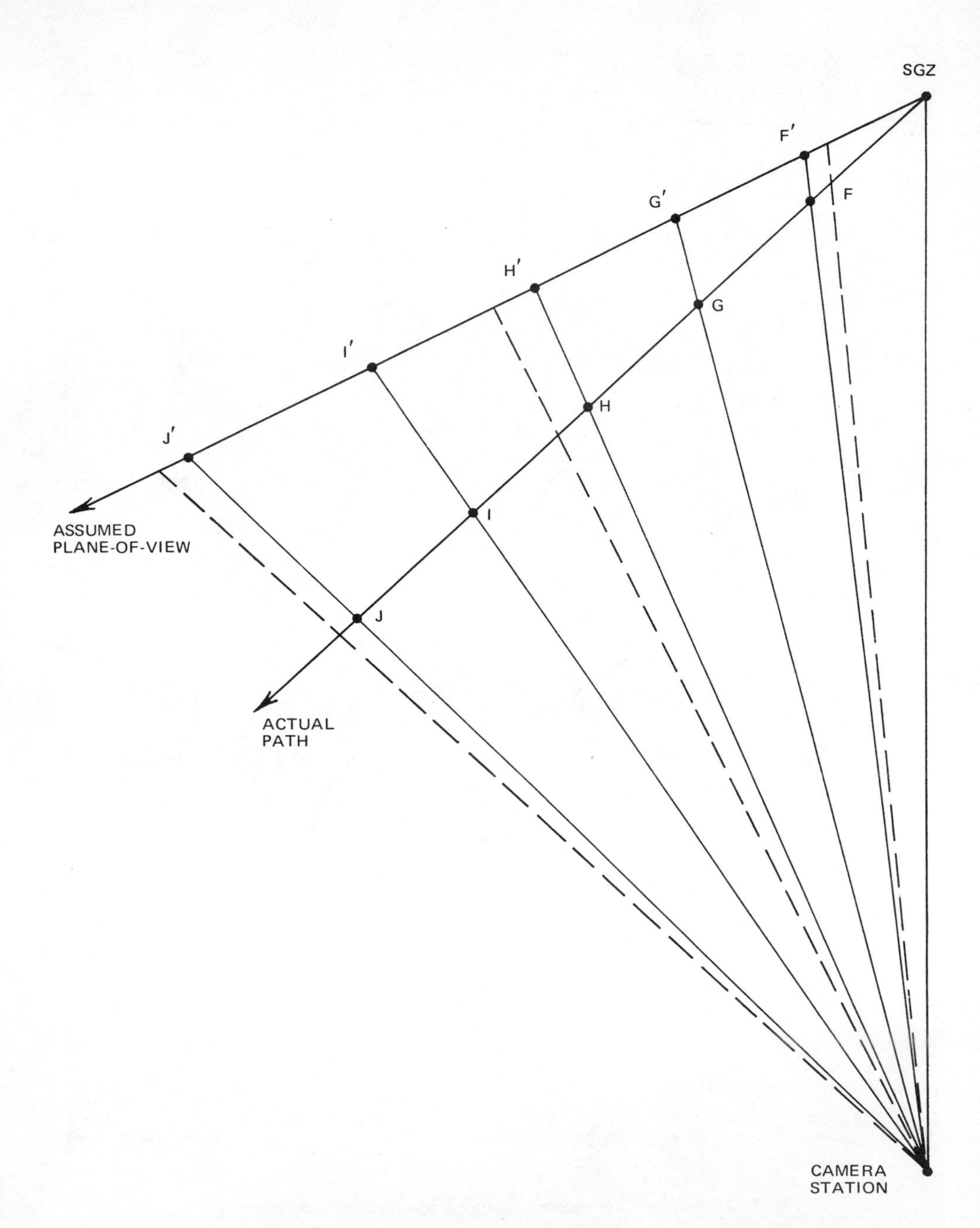

Fig. 3. Actual path near side of assumed plane-of-view of camera.

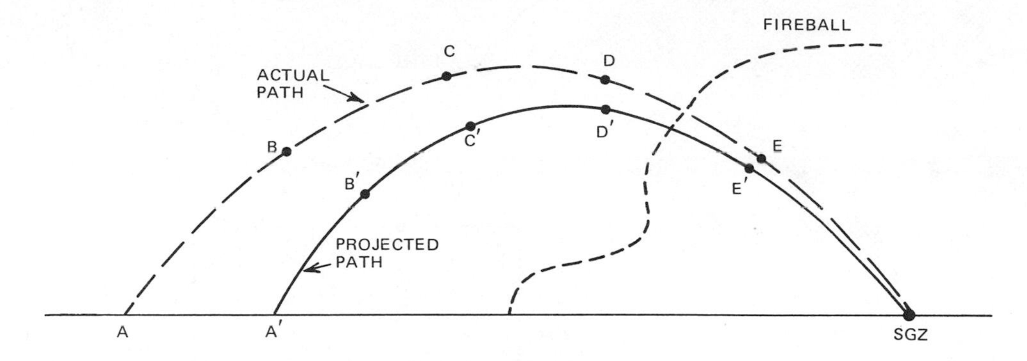

FAR SIDE OF PLANE

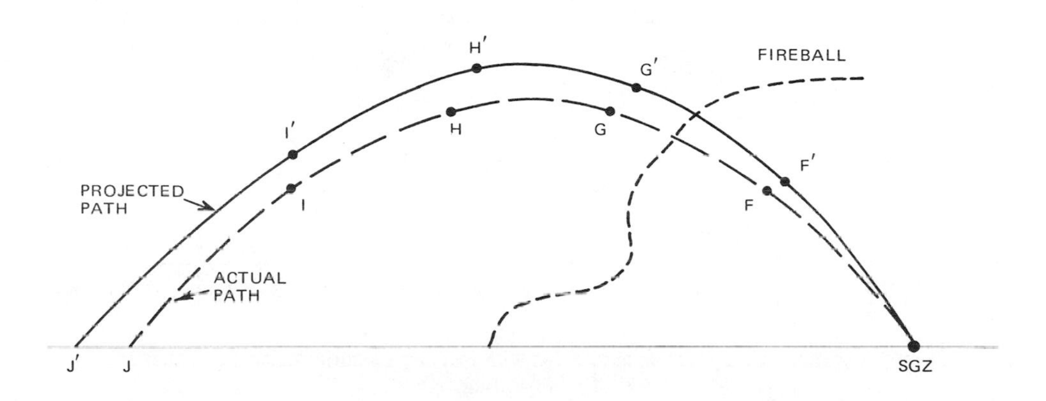

NEAR SIDE OF PLANE

Fig. 4. Actual and projected paths when their planes are and are not in the plane-of-view of the camera.

magnitude as to make $0 < C_{DR} < 0.075$ for ton-size detonation and $0 < C_{DR} < 0.175$ for 1/2-ton detonations (Binder, 1956; Scherich and Kitchin, 1967).

(i) If $C_D < C_{DR}$ rotate the ejecta's trajectory plane away (far side) from the plane-of-view in a one degree (+1°) increment and repeat steps d through h until $C_D \simeq C_{DR}$. Proceed to step j. If $C_D < C_{DR}$ rotate ejecta's trajectory away (near side) from the plane-of-view in a one degree (−1°) increment and repeat steps d through h until $C_D \simeq C_{DR}$. Proceed to step j.

(j) Determination of the final least-squares parabolic equations of $Y = f(X)$, $X = f(t)$, and $Y = f(t)$ (all or 60%).

(k) Determination of all initial (escape) and terminal (impact) parameters.

The following relationships exist between escape and impact parameter when the ejecta trajectory is in the plane-of-view of the camera: the escape velocity (V_E) is slightly greater than the impact velocity (V_I), the escape angle (θ_E) is less than the impact angle (θ_I) measured from the horizontal, the drag constant (K) is less than one, and the escape time (t_E) should be greater than zero.

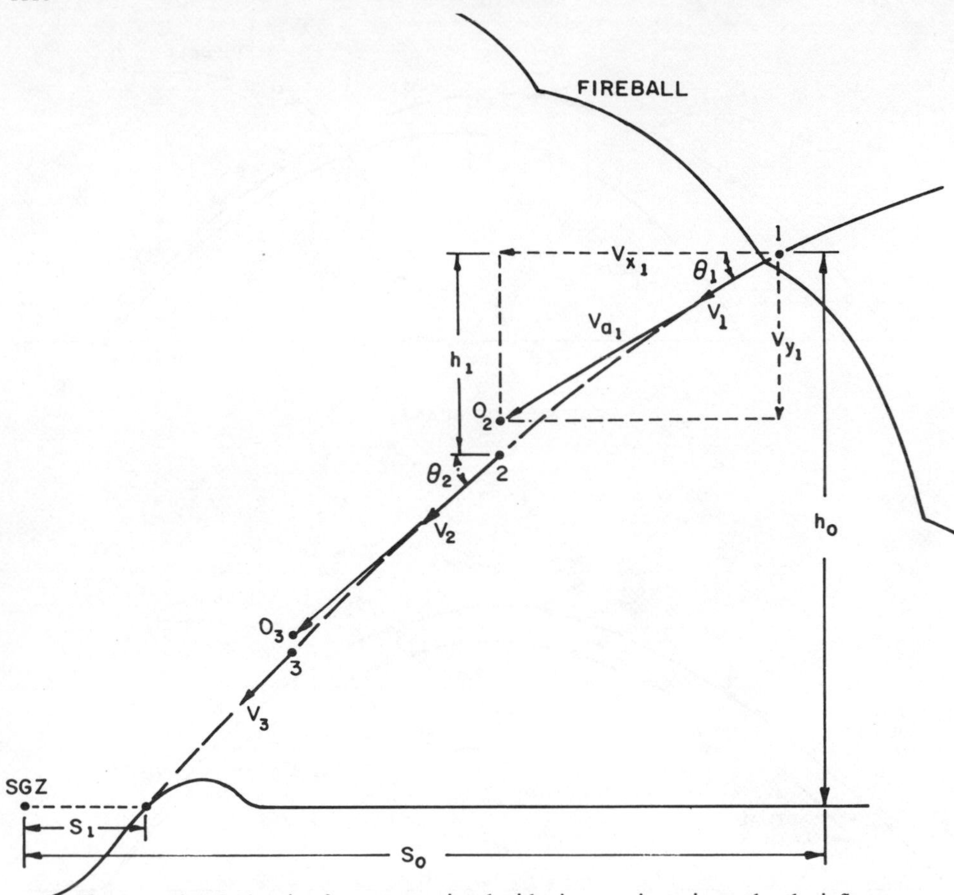

Fig. 5. A sketch showing factors associated with ejecta trajectories under the influence of inverse drag.

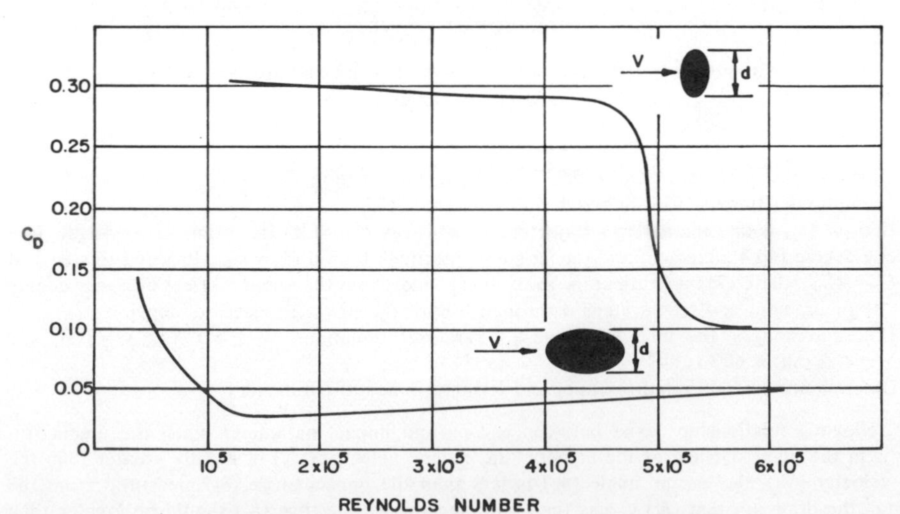

Fig. 6. Drag coefficient for ellipsoids of revolution.

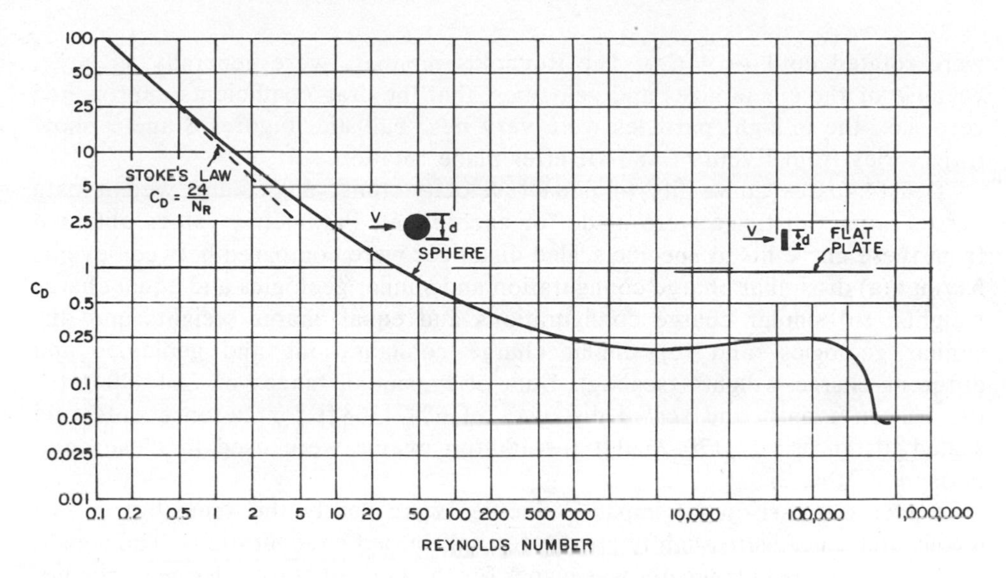

Fig. 7. Drag coefficients for sphere and flat plate.

III. Experimental Data and Analysis

Data presented here comprise a summary from AFWL Report 75-104 and stress the scalability of the impact velocities and times from 1000-lb to 20-ton and 100-ton events.

Dynamic ejecta, photographed during the detonation of 1/2-, 20-, and 100-ton events in the MIDDLE GUST Series, were analyzed to give impact and escape parameters. The TNT spheres were detonated at heights-of-burst (HOB) of 0, 1, and 2 charge radii. The HOB = 2 data are excluded from this paper.

There were two test sites, wet and dry. The wet-site geological profile was composed of a sandy clay soil approximately 9 ft thick overlaying approximately 15 ft of weathered, fractured shale; this in turn overlies a more competent shale. The water table was at a depth of approximately 4 ft (Melzer, 1971; Windham *et al.*, 1973). Five of the six 1/2-ton events (5 thru 9) discussed in this paper were detonated in a specially prepared test pit at the wet site where the overburden was removed to depths 4, 9, and, 16 ft. Water was prevented from seeping into the pit by sheet piles emplaced along the perimeter. Event 3a was detonated at the undisturbed wet site. The geologic profile at the dry site comprises an average of 28 ft of weathered, fractured shale overlying a more competent, relatively impervious shale. The craters formed at the wet site in the natural geology were two to three times the size formed at the dry site in its natural geology.

The largest ejecta fragments impacting any regions of interest were tracked back toward the fireball and cloud. Approximately one hundred of these large particles were tracked for each event. The position-time data obtained from the ejecta trajectories were adjusted for camera tilt. The ejecta trajectory planes were rotated until $C_D \simeq C_{DR}$. The Reynolds numbers were generally so large, because of the ejecta sizes and velocities, that the drag coefficients approached zero, i.e., the in-flight particles were very near ballistic. Figures 8 and 9 show trajectories from Events I and III after plane rotation.

Least-squares curve fits to impact velocity, time, angle, and weight data versus impact distance were made for each event. Parametric values obtained from these curve fits at specific scaled distances were compared between events having: (a) dissimilar charge configuration and similar geologies and equal charge weights; (b) similar charge configurations and equal charge weights and dissimilar geologies, and (c) similar charge configurations and geologies and different charge weights (scaling). Data at impact distances of 200–400 ft for the 1/2-ton events, and scaled distances of 684–1368 ft for 20-ton events, and scaled distances of 1170–2340 ft for 100-ton events were used in these comparisons (Table 2).

Under comparison (a) impact velocities were nearly the same but impact angles and times were slightly greater for half-buried configurations. The weight of the largest ejecta fragment was higher for the tangent above charge configuration when the surface overburden was soil. When the surface overburden was weathered, fractured shale, the half-buried configuration produced larger ejecta fragments.

Comparison (b) indicates that ejecta fragments from the wet site were larger and had higher impact velocities, longer times of flight, and steeper impact angles than the ejecta fragments from the dry site.

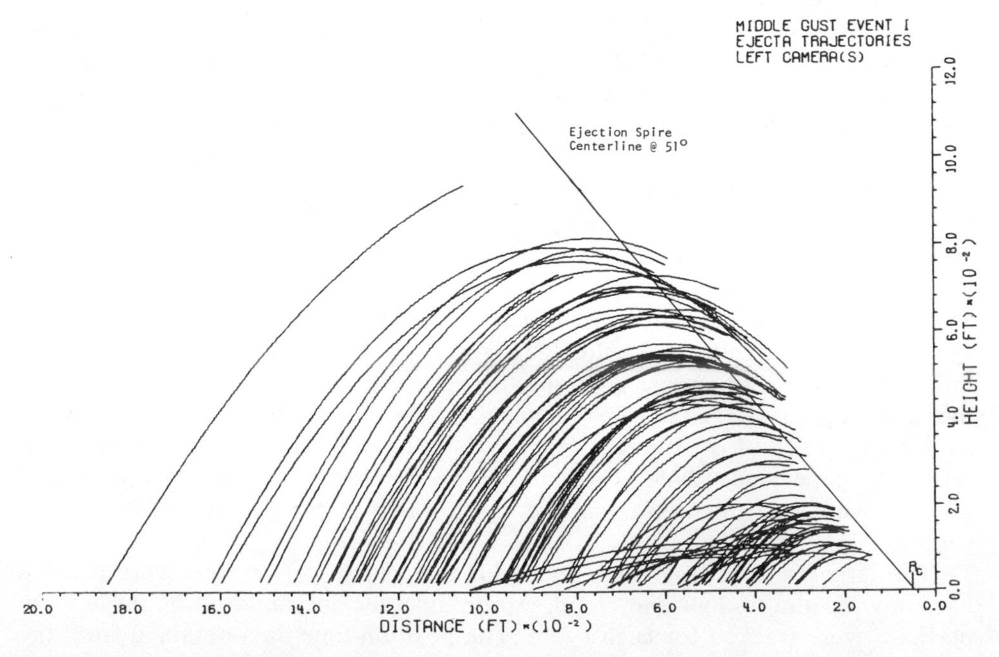

Fig. 8. Final trajectories from all data of MIDDLE GUST EVENT I, left cameras.

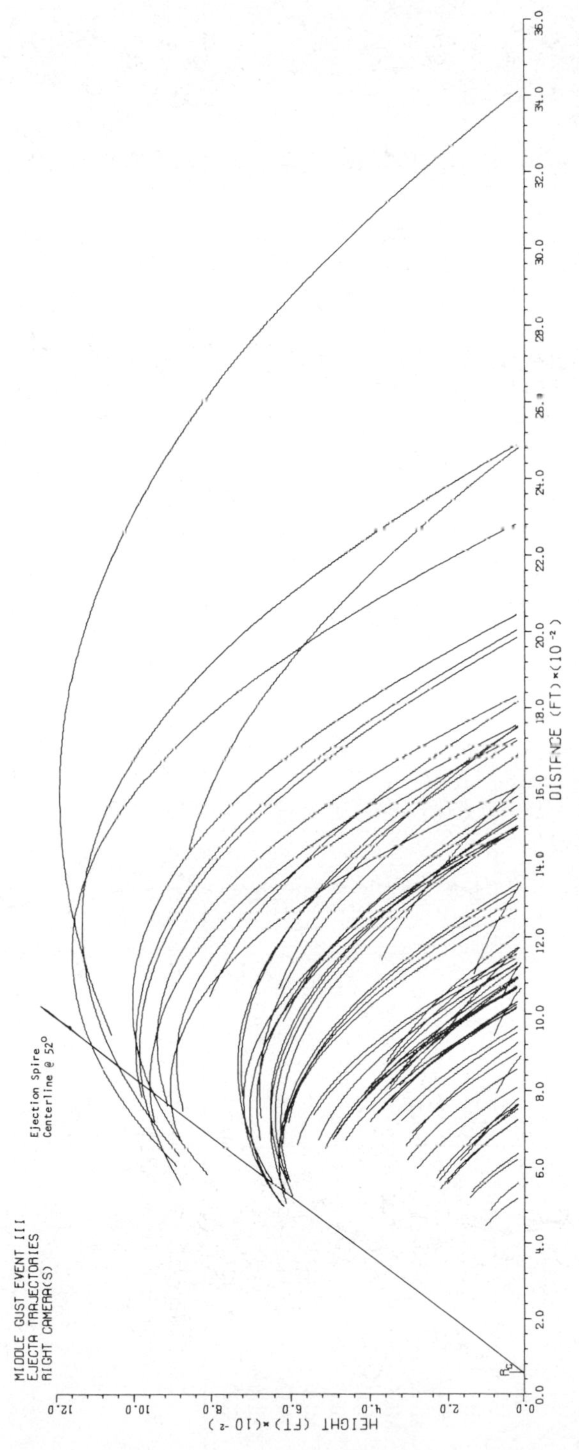

Fig. 9. Final trajectories from all data of MIDDLE GUST EVENT III, right cameras.

Table 2. Impact velocities, angles, times, and weights at three scaled distances*.

Event	Weight (tons)	Charge conf.	Site	Velocity (ft/sec)			Time (sec)			Angle (degrees)			Weight (pounds)		
				S_1	S_2	S_3	S_1	S_2	S_3	S_1	S_2	S_3	S_1	S_2	S_3
MGI	20	½-buried	Wet	159	204	240	7.8	10.4	12.4	58.5	62.0	64.5	1300	1000	810
MGV	20	½-buried	Dry	154	200	237	6.8	8.2	9.5	51.5	49.5	49.5	830	890	930
MGIII	100	Tangent	Wet	208	267	322	9.5	11.9	13.5	55.5	58.5	59.0	3600	2100	1300
MGIV	100	Tangent	Dry	206	263	311	8.1	9.4	10.0	48.5	47.0	44.0	500	490	480
MG3a†	1/2	½-buried	Wet	90	115	—	4.1	4.3	—	58.5	52.0	—	34	15	—
MG5	1/2	Tangent	−4 ft	82	106	126	3.7	4.6	5.4	49.5	52.5	54.0	71	45	29
MG6	1/2	½-buried	−4 ft	83	106	127	3.9	5.1	5.7	54.5	59.0	59.5	31	22	16
MG7	1/2	Tangent	−9 ft	83	107	133	3.5	4.4	5.3	48.5	52.5	57.0	21	23	26
MG8	1/2	½-buried	−9 ft	88	111	128	4.3	5.4	6.0	59.5	61.0	60.0	71	35	16
MG9	1/2	½-buried	−16 ft	90	116	140	4.2	5.2	6.0	58.5	61.0	64.0	82	46	24

*Scaled distances (S_n): 1/2 ton: 200, 300, and 400 ft. 20 ton: 684, 1026, and 1368 ft. 100 ton: 1170, 1755, and 2340 ft.

†Events 3a, and 5 through 9, wet site.

It was found under comparison (c) that the impact velocities and times of the dynamic ejecta approximately scale by $(w_1/w_2)^{1/6}$ (Cooper, 1976).

Since the trajectories were near ballistic outside the fireball, the maximum range any ejecta fragments would have travelled (neglecting drag forces) is:

$$S_{max} = (V_E^2 \sin 2\theta_E)/g,$$

where V_E = escape velocity, θ_E = escape angle and g = acceleration due to gravity.

It can be assumed since the trajectories are near ballistic for the greatest portion of the trajectory path, that impact velocity (V_I) and impact angle (θ_I) would be nearly equal to their escape counterparts, i.e., $V_E \simeq V_I$ and $\theta_E \simeq \theta_I$.

Assume $S_2 = (w_2/w_1)^{1/3} S_1$, where w_1 and w_2 are the TNT charge weights.

$$S_1 = (V_1^2 \sin 2\theta_1)/g \text{ and } S_2 = (V_2^2 \sin 2\theta_2)/g, \text{ and } \theta_1 = \theta_2.$$

Then $(w_2/w_1)^{1/3}$ $V_1^2 = V_2^2$, therefore $V_2/V_1 = (w_2/w_1)^{1/6}$. Since the geology for Calibration Events 5 and 6 compare nearly on a scaled basis to the geology for Events I and III, i.e., water table 1 ft below is 4 ft below the surface, comparisons of impact, velocity and impact time were made to Events III and I, respectively.

Events I and 6, $(w_I/w_6)^{1/6} = 1.85$

Ratio	S_1	S_2	S_3
V_I/V_6	1.92	1.92	1.89
t_I/t_6	2.00	2.04	2.18

Events III and 5, $(w_{III}/w_5)^{1/6} = 2.42$

Ratio	S_1	S_2	S_3
V_{III}/V_5	2.54	2.52	2.56
t_{III}/t_5	2.57	2.59	2.50

The above tabulations show good agreement of the ratios of velocity and time to the charge weight ratios to the 1/6 power. The differences are mainly due to the slightly lower (4–5°) angle values for the 1/2-ton detonations which are indicative of slightly lower velocities and slightly lower times. These lower values would increase the ratios slightly.

There are indications that ejecta weights scale directly to the charge weight at scaled distances $[(w_1/w_2)^{1/3}]$; therefore, it would follow that kinetic energies scale to the charge weights to the 4/3 power at scaled distances since their velocities scale to the 1/6 power.

Figures 10 through 14 present photographic sequence showing fireball and ejecta throwout from a number of events including those covered in the above analysis.

IV. Conclusion and Recommendations

The ten cratering events covered in this paper afforded the opportunity of comparing events where only the charge configurations were changed, where only the characteristics of the geology were changed, and where only the charge weights were changed. The dynamic ejecta impact parameters derived from these events indicated near ballistic trajectories.

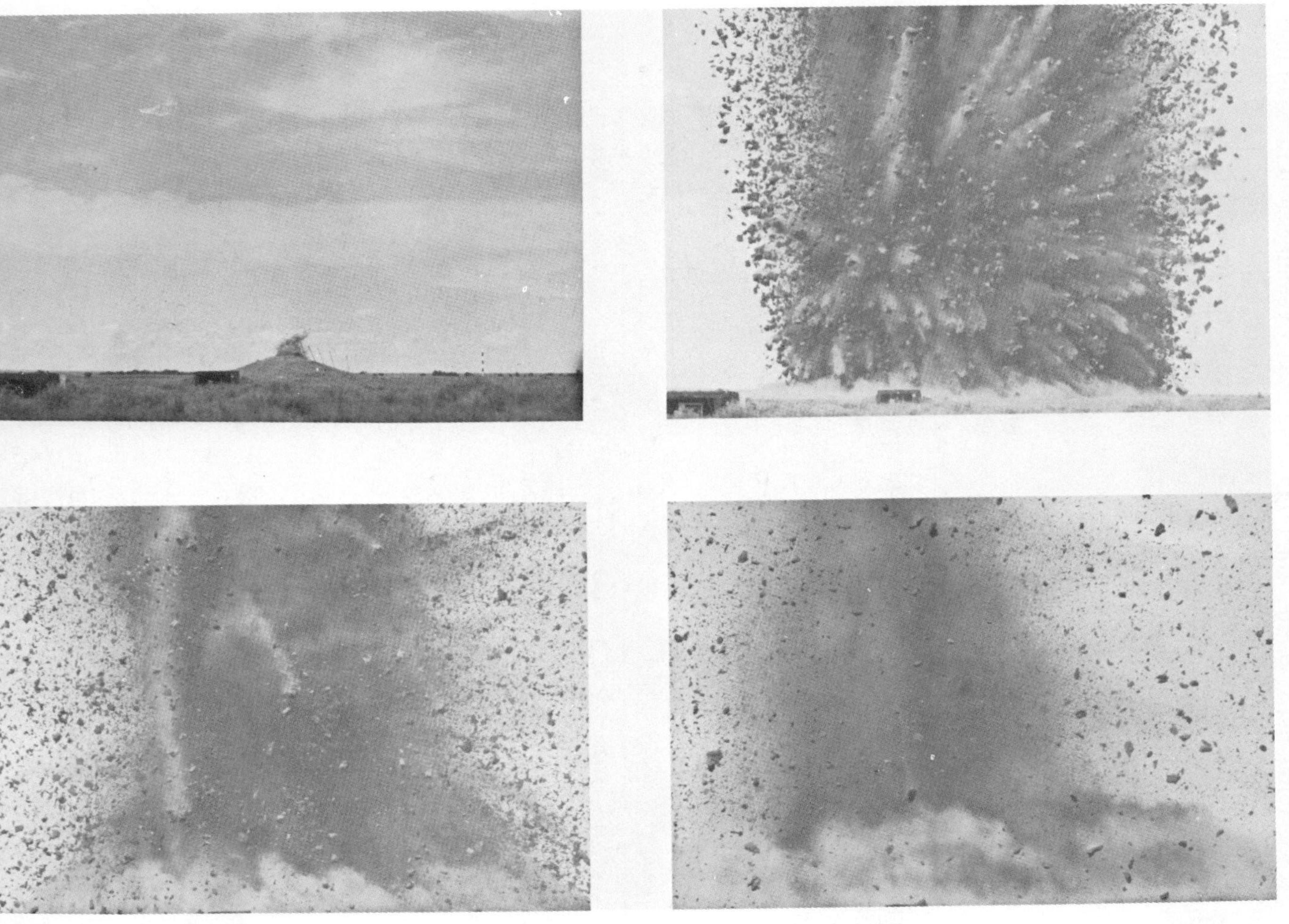

Fig. 10. MIDDLE GUST 2, 1/2 ton, DOB $= -13R_c$, frames 2, 10, 20, 40; approximate times of 0.1, 0.5, 1.0, and 2.0 sec.

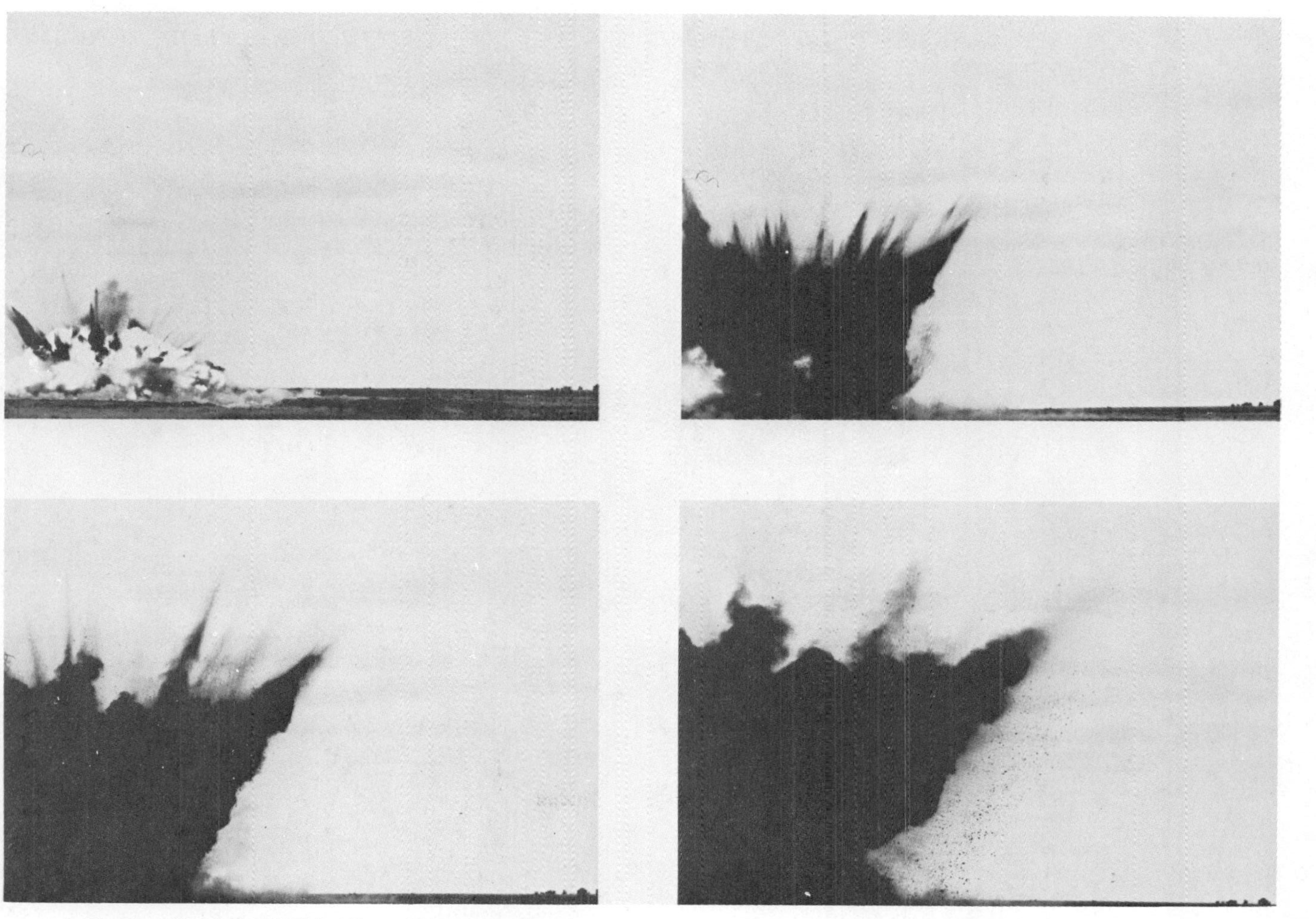

Fig. 11. MIDDLE GUST 5, 1/2 ton, HOB = $1R_c$, frames 2, 10, 20, 40; approximate times of 0.1, 0.5, 1.0, and 2.0 sec.

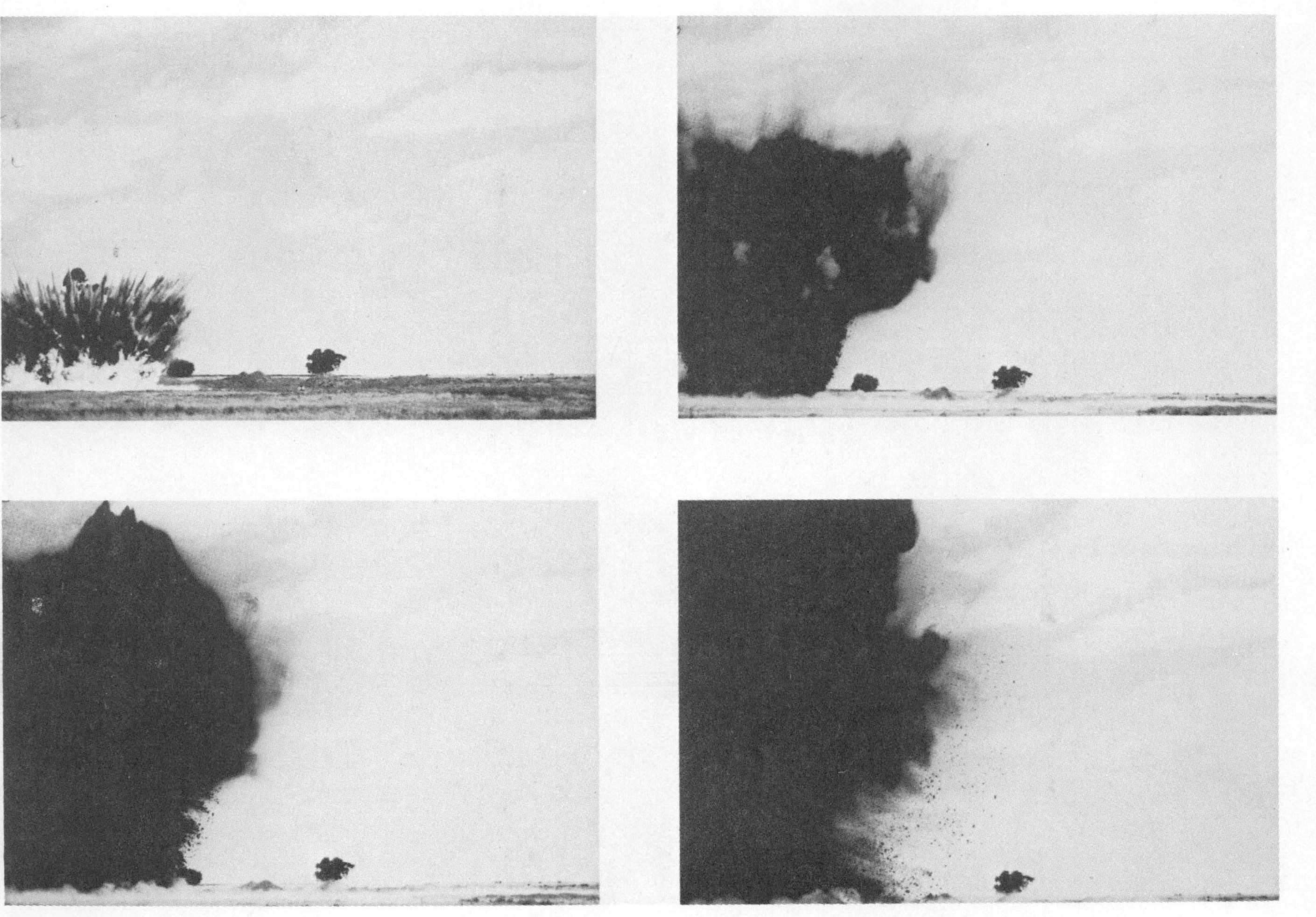

Fig. 12. MIDDLE GUST 6, 1/2 ton, HOB = $0R_c$, frames 2, 10, 20, 40; approximate times of 0.1, 0.5, 1.0, and 2.0 sec.

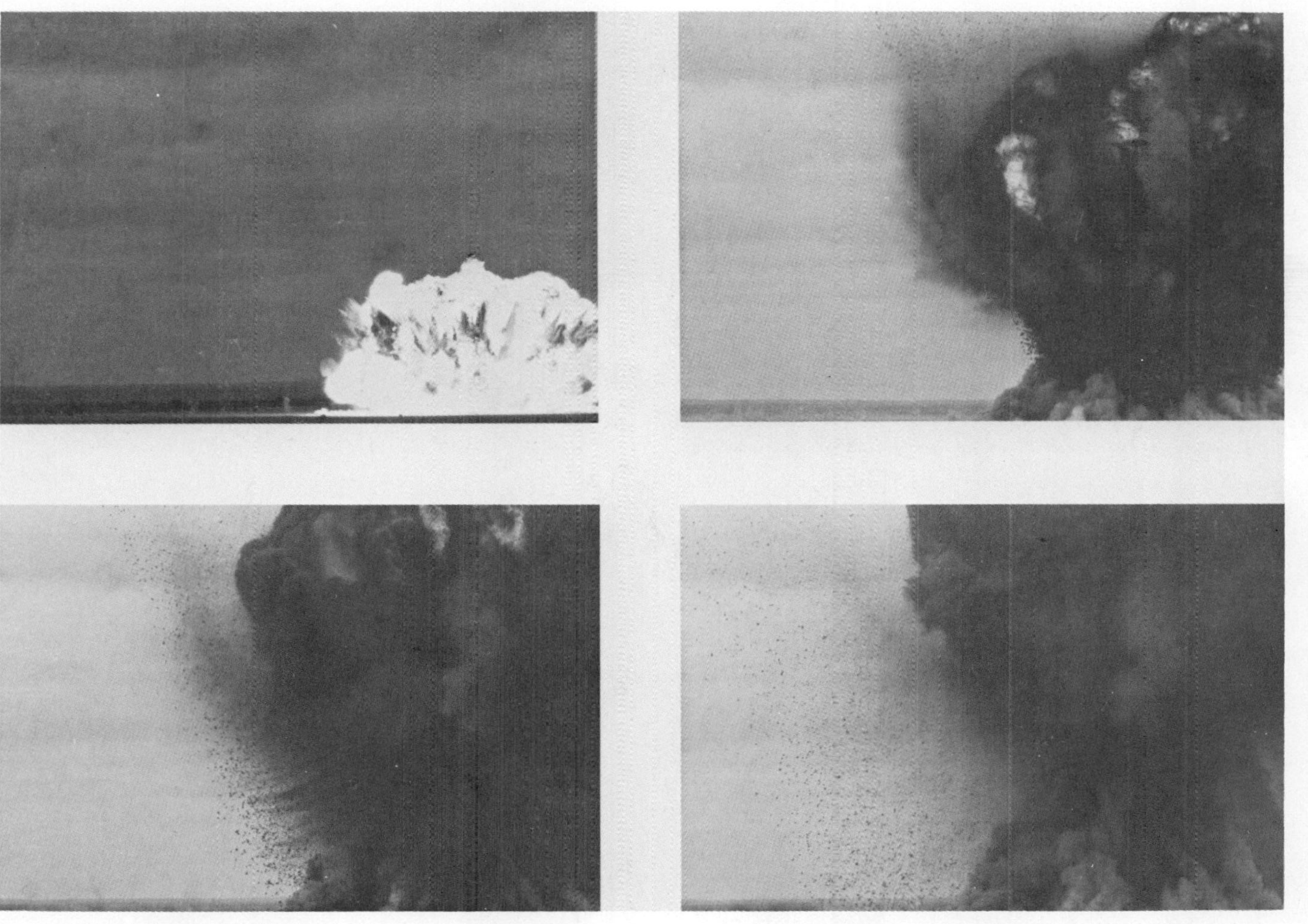

Fig. 13. MIDDLE GUST I, 20 ton, HOB = $0R_c$, frames 2, 40, 80, 120; approximate times 0.1, 2.0, 4.0, and 6.0 sec.

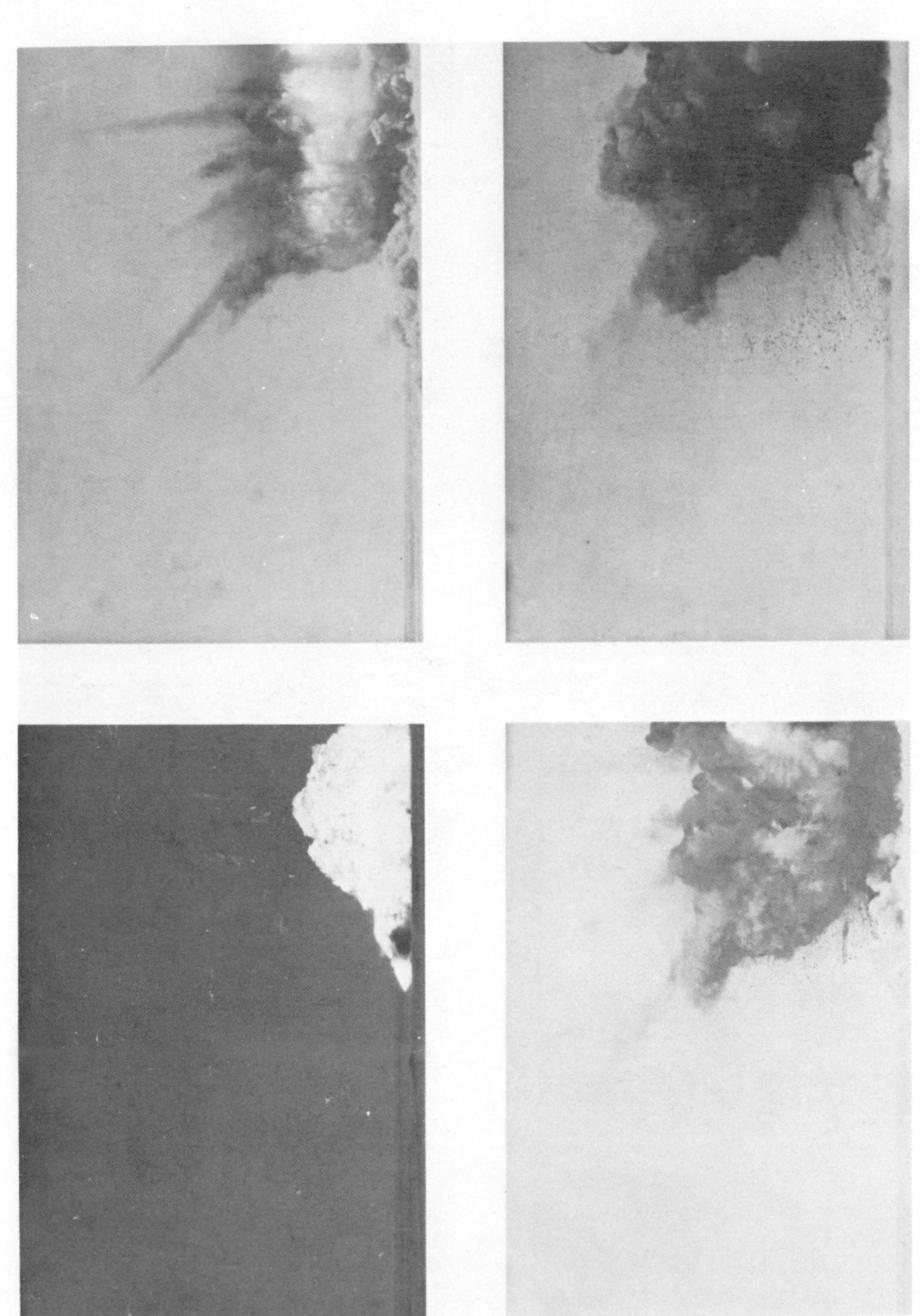

Fig. 14. MIDDLE GUST III, 100 ton, HOB = $1R_c$, frames 2, 40, 80, 120; approximate times 0.1, 2.0, 4.0, and 6.0 sec.

For half-buried charge configurations the dynamic ejecta impact velocities and angles increased with the more competent geological materials.

The wet-site geology produced higher ejecta impact velocities and higher ejecta impact angles than the dry site with similar charge weights and configurations. There are indications that dynamic ejecta velocities and times scale to the 1/6 power when the geological conditions are nearly scaled, i.e., water table position scaled.

There is a probability that dynamic ejecta weights scale directly to the charge weight at scaled distances to the charge weight to the 1/3 power. There is also a probability that kinetic energies scale to the charge weight to the 4/3 power at scaled distances. These last two conditions should be substantiated by conducting a series of scaling tests in a geology that will produce large dynamic ejecta particles and lend itself to scaling as far as layering thicknesses and water table positions are concerned. A series of 1, 8, 64, 256, and 1000-lb charges should be conducted in such a scalable geology with at least two hundred to three hundred particles tracked and analyzed.

References

Binder, R. C.: 1956, *Fluid Mechanics* (3rd ed.), Prentice Hall.

Cooper, H. F., Jr.: 1976, Personal communication.

Linnerud, H. J.: 1976, MIDDLE GUST I Ejecta Sizing and Examination of Ejecta Ballistics. DNA 3926F.

Melzer, L. S.: 1971, MIDDLE GUST Test Series Events I, II, III, Test Plan and Prediction Report. AFWL-TR-71-147.

Scherich, E. L. and Kitchin, T.: 1967, Weapon Optimization Techniques. AFATL-TR-67-128.

Windham, J. E., Knott, R. A., and Zelasko, J. S.: 1973, Geology and Material Property Comparisons for the MIDDLE GUST Test Series. Proc. MIXED COMPANY/MIDDLE GUST Results Meeting.

Wisotski, J.: 1969, Technical Photography from MINE UNDER and MINE ORE Events of the MINE SHAFT Series. DASA 2268. DRI 2495.

Wisotski, J.: 1970, Analysis of In-flight Ejecta from Photography of a 100-Ton TNT Detonation on Granite. MS 2167. DRI 2538.

Wisotski, J.: 1977, Dynamic Ejecta Measurements for Selected MIDDLE GUST and PACE Cratering Events, Denver Research Institute, Air Force Weapons Laboratory Report, AFWL-TR-75-104.

Roddy, D. J., Pepin, R. O., and Merrill, R. B., editors.
(1977) *Impact and Explosion Cratering*, Pergamon Press (New York), p. 1123–1132.
Printed in the United States of America

A model for wind-extension of the Copernicus ejecta blanket

D. E. Rehfuss, D. Michael, J. C. Anselmo, and N. K. Kincheloe

Department of Physics, San Diego State University, San Diego, California 92182

Abstract—The interaction between crater ejecta and the transient wind from impact-shock vaporization is discussed. Based partly on Shoemaker's ballistic model of the Copernicus ejecta and partly on Rehfuss' treatment of lunar winds, a simple model is developed which indicates that, if Copernicus were formed by a basaltic meteorite impacting at 20 km s^{-1}, then 3% of the ejecta mass would be sent beyond the maximum range expected from purely ballistic trajectories. That 3% mass would, however, shift the position of the outer edge of the ejecta blanket more than 400% beyond the edge of the ballistic blanket. For planetary bodies lacking an intrinsic atmosphere, our model indicates that this form of hyper-ballistic transport can be very significant for small ($\lesssim 1$ kg) ejecta fragments.

1. Introduction

Early in the crater formation process induced by meteorite infall, a cloud of vaporized material may be present, caused by high initial shock pressures between meteorite and target. If the target planet is without an atmosphere, this vapor cloud expands relatively freely into vacuum, impeded only by the presence of the non-gaseous materials being ejected from the forming crater. To a first approximation one might expect this gas to dissipate so quickly that there is negligible interaction between the wind and the ejecta particles. However, we are suggesting that this effect is not always negligible, especially at impact velocities significantly above 15 km s^{-1}, which is the approximate threshold for vaporization.

In this paper we present a calculation which implies that, for reasonable impact parameters, there may indeed be a significant interaction between vaporization wind and mechanical ejecta, yielding an extension of particle ranges beyond those expected from purely ballistic paths. We have modeled the formation of the ejecta blanket of the lunar crater Copernicus through the juxtaposition of Shoemaker's (1962) ballistic model for that crater and Rehfuss' (1972) "Lunar Winds" treatment. To allow a quantitative estimate for the ejecta mass distribution, an analytical shape is assumed for the transient crater. Equipartition into nine particle sizes is assumed for the ejected mass.

2. The Model

Based on his analysis of the secondary crater distribution, Shoemaker (1962) devises a model for the ballistic formation of the Copernicus ejecta blanket. The model specifies simultaneous values of three variables: initial ejection speed V_0, initial ejection angle $\propto$, and initial horizontal position x_0. Shoemaker's analysis includes the correlation of the secondary crater distribution around Copernicus with probable initial positions in the forming crater; he proposes that the major part of the continuous ejecta blanket was formed for ejection angles in the range 6–22° above the horizontal, whereas ejection in the range 23–90° resulted in widely dispersed material or loss from the moon's gravitational field. Oberbeck (1975) has shown that the Shoemaker model gives a concave upward ejecta curtain during the crater formation process, in agreement with laboratory cratering experiments.

The Shoemaker model does not specify how much mass is ejected at given velocities, although such information is necessary in quantitatively modeling the ejecta blanket. After first choosing the dimensions of the transient crater, we have then subdivided it according to ejection angle. We have adopted a simple cylindrical shape for our computer model of the transient crater, shown superimposed on the presently observed crater profile in Fig. 1. The transient crater radius, 30.45 km, is from Shoemaker's model. The depth estimate is based on Pike's (1967) discussion of the crater Tycho, which is only about 3% smaller in diameter than Copernicus. Pike estimated that Tycho's present depth was 29% of its original depth. A nominal depth for Copernicus could be Baldwin's (1963) 2.3 km or Shoemaker's (1962) 2.5 km. Pike's 29% ratio thus yields a transient crater depth in the range 7.9–8.6 km; we have used 8.0 km.

Having chosen the transient crater dimensions, it is then necessary to decide how much of the total crater volume would reasonably be ejected at a given initial position and velocity. We have chosen the simplest possible relation to the Shoemaker model: as shown in Fig. 2, vertical lines are placed at even-numbered ejection angles (we recall that a given angle is tied uniquely to a certain initial ejection position x_0). Then we assume that all the mass, in the annular ring between any two ejection positions, is ejected at the outermost of the two corresponding angles. For example, for a 2° spacing in the Shoemaker model, the two outermost ejection positions are at 22.8 km, corresponding to 0.25 km s^{-1} at

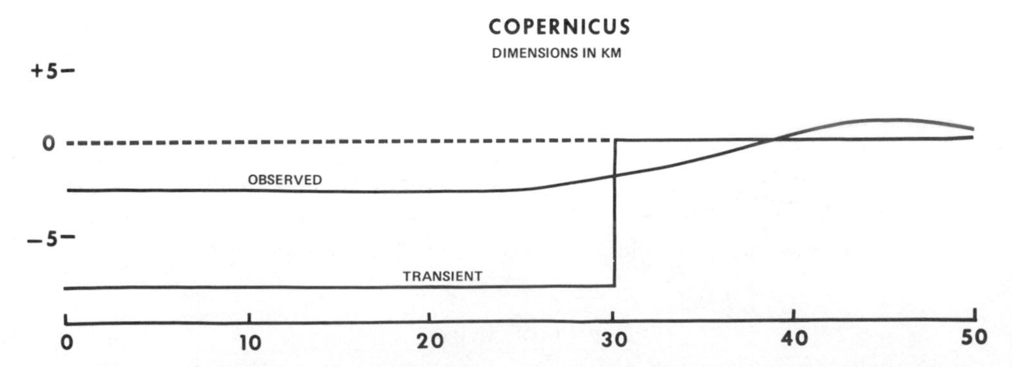

Fig. 1. Present shape of the lunar crater Copernicus, compared with the cylindrical shape utilized in the model as an estimate of the transient cavity volume involved in the formation process.

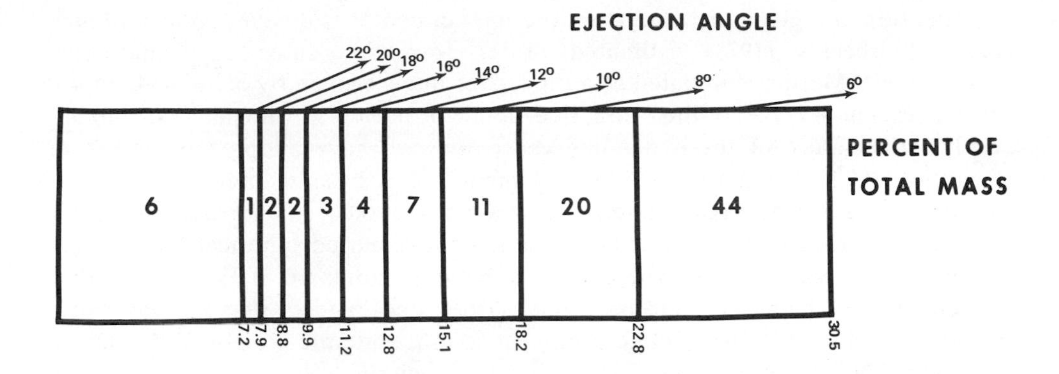

Fig. 2. Representation of the mass percentages ejected at particular angles from the transient crater. Simultaneous values of ejection position, speed, and angle are specified from the Shoemaker (1962) model. The unique speed (km/s) at a given angle (degrees) is as follows: 1 at 22, .87 at 20, .75 at 18, .63 at 16, .52 at 14, .43 at 12, .30 at 10, .25 at 8, and .17 at 6.

8°, and at 30.5 km (0.17 km s^{-1} at 6°). Between 22.8 and 30.5 km the annular mass would amount to 44% of the total transient crater mass. That entire 44% is now assigned an ejection velocity of 0.17 km s^{-1} at 6°. Similarly, the next annular ring, between 18.2 and 22.8 km, is all ejected at 0.25 km s^{-1} and 8°. The unrealistic vertical partitioning of the ejecta is not essential to the model, however. We only assume that, for example, 44% of the mass is ejected at the position and with the velocity indicated above. Instead of the Fig. 2 scheme one could have any number of other more complex geometrical arrangements, including semi-horizontal partitioning, which would be more in accord with laboratory experiments (Stöffler, 1975).

It is also necessary to consider ejecta particle sizes. We have chosen a simple particle mass distribution which nevertheless gives a wide enough sampling range to test the influence of the mass parameter on the possible interaction between vapor wind and ejecta particles. Featuring tri-decade gaps across 24 orders of magnitude, the distribution is listed in Table 1. All of the ejecta mass is assumed to be equally divided into the nine particle sizes listed in that table.

Table 1. Ejecta is divided equally into these nine sizes of cubes, in units of grams/particle.

10^{15}	10^{0}
10^{12}	10^{-3}
10^{9}	10^{-6}
10^{6}	10^{-9}
10^{3}	

Ejection at a given distance from the impact center is delayed in accordance with Oberbeck's (1975) estimated crater formation rate for Copernicus, $0.454\,\text{km s}^{-1}$. Motion is initiated at a time after impact given by $3.2/(0.454 \tan \propto)$, where 3.2 km represents the "effective depth-of-burst" in Shoemaker's model.

For the nature of the expanding gas cloud, we have adopted values from Rehfuss (1972), listed for a $20\,\text{km s}^{-1}$ impact of a basaltic meteorite upon a basaltic target surface, both materials at a density of $3.0\,\text{g cm}^{-3}$. From Vortman's (1968) empirical equations, based on data from half-buried spherical HE charges in basalt, we estimate the meteorite to have a radius of 0.73 km. For that meteorite size, from the Rehfuss model (1972) one expects that the gas cloud consists of 1.1×10^{38} monatomic atoms of average atomic weight 22.6. During post-shock adiabatic relaxation, at the instant when the bulk elastic pressure has fallen to zero, the cloud is of hemispherical radius 0.69 km, density $3.0\,\text{g cm}^{-3}$, temperature 7100 K, dynamic pressure 0.23 Mbar, and has a constant expansion velocity, $U = 3.95\,\text{km s}^{-1}$. The cloud is assumed to be homogeneous at all times and to expand adiabatically, except for its interaction with the ejecta particles.

Another feature requiring discussion is the mechanism of the interaction between wind and particle. We assume the force of the wind to be given by the product of dynamic pressure P_D, particle cross-sectional area A, and drag coefficient C_D. The dynamic pressure is a continuously varying quantity,

$$P_D = \rho U^2/2, \tag{1}$$

where U is the wind velocity. Since the density (ρ) of the cloud falls off as the inverse cube of its outer radius R, the dynamic pressure also decays rapidly, as R^{-3}, or for a fixed distance from impact center, as t^{-3}, where t is the time since impact.

The drag coefficient C_D depends on particle shape and upon the Reynolds number of the gas. For Reynolds numbers of the order $R_e \approx 10^5$, we expect the drag coefficient to be almost an order of magnitude larger for a cubical shape as compared to a spherical shape (Li and Lam, 1964; Hughes and Brighton, 1967; Whittaker, 1968). We have assumed a value $C_D = 2.0$ for cubes and $C_D = 7/(0.2 + 5 \log_{10} R_e)$ for spheres.

3. CALCULATION METHOD

Our computer program employs an iterative procedure in which both the purely ballistic (no wind) and wind-influenced trajectories are calculated at each step. The iterative procedure is necessary for the wind-influenced case because the acceleration of the particle at any moment depends on its position $d(x, y)$ in space, which is determined by the acceleration calculated in the previous step. That is, the particle is acted on by the instantaneous dynamic pressure, which is a function of the vector difference between wind velocity (U) and instantaneous particle velocity (V), and the angle of attack of the wind is related to the instantaneous position of the particle. The x- and y-accelerations are given by

$$a_x = \frac{P_{DA} C_D A x}{md}\left(\frac{x}{d} - \frac{V_x}{U}\right)^2, \tag{2}$$

$$a_y = \frac{P_{DA}C_DAy}{md}\left(\frac{y}{d} - \frac{V_y}{U}\right)^2 - g, \tag{3}$$

where the average dynamic pressure between times t_1 and t_2 is given by

$$P_{DA} = \frac{P_{Dg}R_g^{\,3}\,(t_1 + t_2)}{2U^3(t_1t_2)^2}, \tag{4}$$

and where the instantaneous particle position is $d^2 = x^2 + y^2$, its velocity components are V_x and V_y, g is the local acceleration of gravity, A/m is the area-to-mass ratio of the particle, and R_g is the initial radius of the gas cloud. P_{Dg} is the initial dynamic pressure.

4. Preliminary Parameter Study

Before implementing the wind-enhanced model, we checked the individual influences of some of the more important parameters. That is, we performed a preliminary study in which we temporarily ignored the fact that Shoemaker's model uniquely ties together ejection position, speed, and angle. The results are shown in Figs. 3, 4, and 5.

In Fig. 3 we show cases at constant ejection angle 14°, ejection speed 0.52 km s^{-1}, and initial position $x_0 = 0.69$ km from the impact center. The windless case is compared with wind-affected trajectories of 1 kg cubes and spheres. The range is almost doubled for the wind-accelerated cube, while the perfect sphere is hardly influenced at all by the wind, illustrating the effect of the drag coefficient C_D.

In Fig. 4 we present other cases for cubes ejected at the same $x_0 = 0.69$ km, but of different masses and for three different launch velocities. The ordinate is the ratio of wind-enhanced range to windless ballistic range. The wind's influence is seen to be significant for cubical masses of up to 10 kg.

In Fig. 5 we show the effect of the initial horizontal position x_0, for cubes of 1, 10, and 100 g and a fixed launch velocity, 0.52 km s^{-1} at 14°. For 1 g particles the range enhancement is seen to be significant out to almost 4 km, about a tenth of the final crater radius, or about an eighth of the transient crater radius.

5. Results

As indicated in Fig. 2 and Table 1, our model has nine different ejection velocities (in terms of both speed and angle, and associated with a particular ejection position) and nine different particle sizes. Thus there are 81 different combinations of those parameters to be calculated. Figures 6, 7, and 8 each illustrate the horizontal ranges for nine of those 81 combinations, namely those at the extremes and at the middle of the ejection angle range. Fragmentation processes are likely to result in irregular shapes which are better approximated as cubical than as spherical. Hence for the results shown in Figs. 6 through 9 we have used only the cube shape.

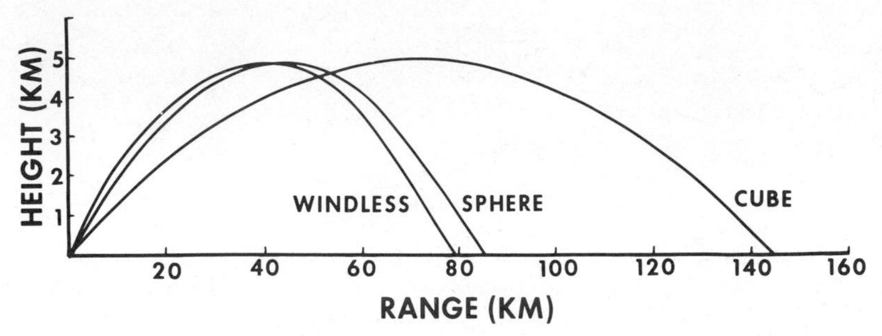

Fig. 3. Trajectories of 1 kg spheres and cubes ejected at 0.69 km from impact center at 14° at 0.52 km s^{-1}, compared to the windless case.

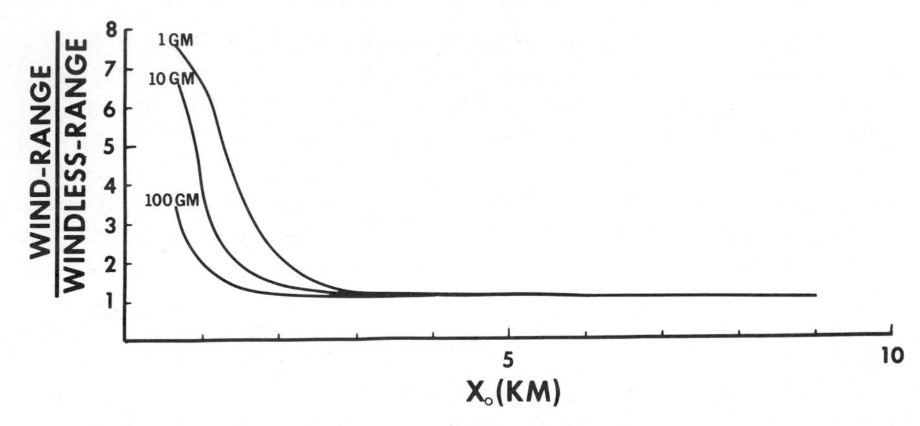

Fig. 4. Wind-accelerated ranges of cubes of various masses ejected at various velocities at 0.69 km from impact center.

In Fig. 6 we present in histogram form the final landing positions of the nine particle sizes in one "mass unit" of 9×10^{15} g launched at 1.0 km s^{-1} and 22°, from a position $x_0 = 7.9$ km from the center of the impact, with a time delay of 18 s after impact. The five most massive particle sizes, 10^3 through 10^{15} g, are not significantly influenced by the wind and land between 437 and 439 km, after following essentially ballistic trajectories. The 1 g particles are moved more than 30 km farther, out to 461 km. The smallest particles, 10^{-9} g each, are sent out to 1930 km. Figures 7 and 8 are for progressively greater distances x_0 from the impact center, and since the wind has had more time to dissipate before the initial ejection of these particles, the force of the interaction is progressively less.

When we combine together the results from all 81 combinations of velocity, ejection position, and particle mass, we obtain the results summarized in Fig. 9. Therein we have graphed the percent of ejecta mass yet undeposited versus distance from the impact center. All of the mass in the ballistic trajectory has fallen to the lunar surface by 439 km, whereas at that distance 3% of the

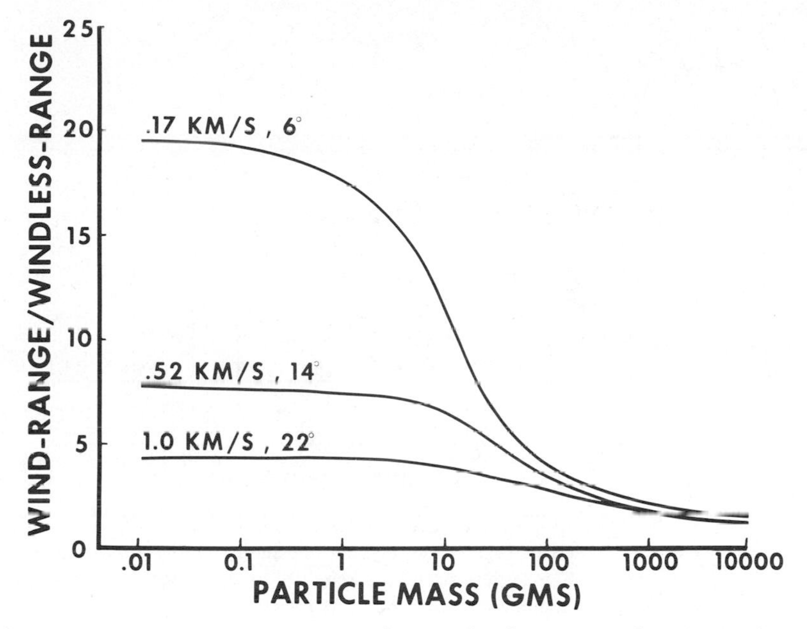

Fig. 5. Relative wind-accelerated ranges of cubes of various masses ejected at 0.52 km s^{-1} from different launch positions x_0 from impact center.

wind-accelerated material is still in flight. The last of the wind-accelerated particles do not fall until they attain a range of 1930 km from the center of the crater.

6. Discussion

We have shown that the force exerted by the transient wind resulting from impact-shock vaporization may not be negligible in all cases. According to this model of wind-accelerated ejecta, 3% of the total ejecta mass from the crater Copernicus would have been given horizontal ranges greater than those in a purely ballistic case, if the impact took place at 20 km s^{-1}.

Although 3% is a small fraction of the total mass, the redistribution of that mass is significant: the radial limit of ejecta deposition is displaced outward more than 4 times as far as in the purely ballistic case (1930 km compared to 439 km). Such an extreme outward displacement of a small amount of mass could be a contributor to the rays around Copernicus or Tycho.

Another possible effect produced by this wind-acceleration phenomenon is that it could produce a dust cloud about the moon in orbital and sub-orbital conditions (Shoemaker, pers. comm., 1976). For an event the size of Copernicus or larger, such a dust cloud suspended over the moon might persist for a time long enough to cause abnormal temperature conditions at the moon's surface, either by not allowing as much sunlight to hit the surface or by reducing the rate at which the moon radiates energy into space.

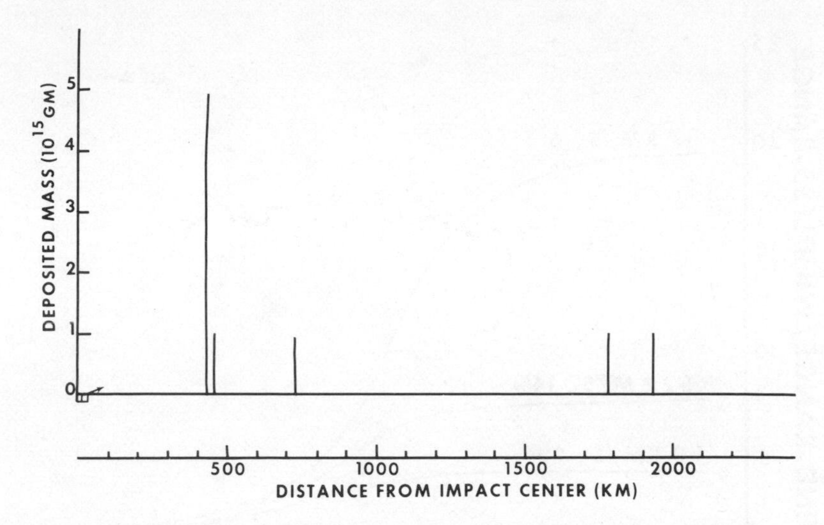

Fig. 6. Range distribution of particles ejected at 22° and 1 km s^{-1}, comprising 1% of the total crater mass, as shown in Fig. 2. The highest line indicates the almost identical landing positions of the five largest particle sizes, 10^3–10^{15} g, which have not been significantly affected by the wind.

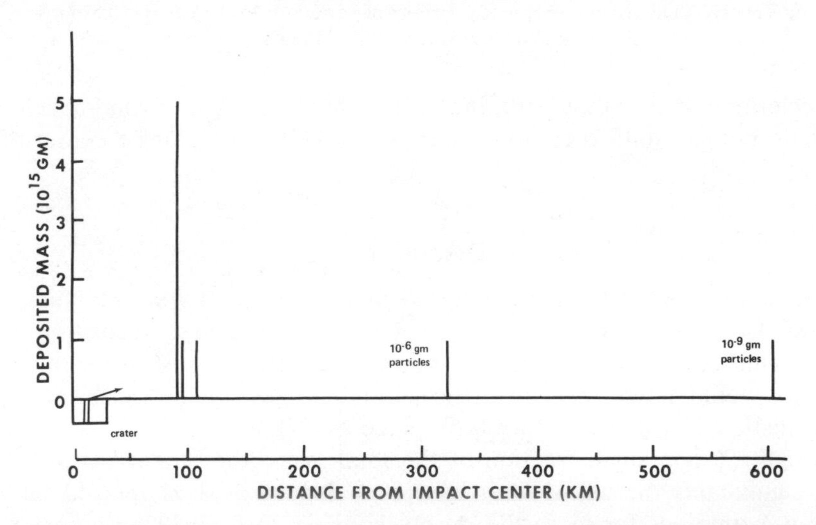

Fig. 7. Range distribution of particles ejected at 14° and 0.52 km s^{-1}, comprising 4% of the ejected mass, as shown in Fig. 2.

Although almost every detail of this model might be improved upon, it is sufficiently plausible to indicate that the effect of the vaporization wind upon ejecta material cannot always be neglected. We certainly would be surprised if a more refined model would also show that precisely 3% of the mass would be affected in the wind-accelerated case. However, for a 20 km s^{-1} impact velocity there are more reasons to consider this 3% as an underestimate than there are to

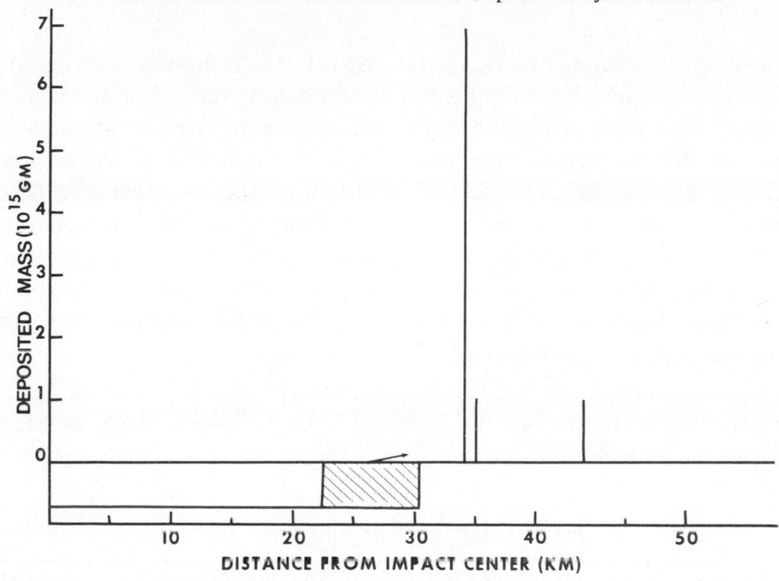

Fig. 8. Range distribution of particles ejected at 6° and 0.17 km s^{-1}, comprising 44% of the ejected mass, as shown in Fig. 2.

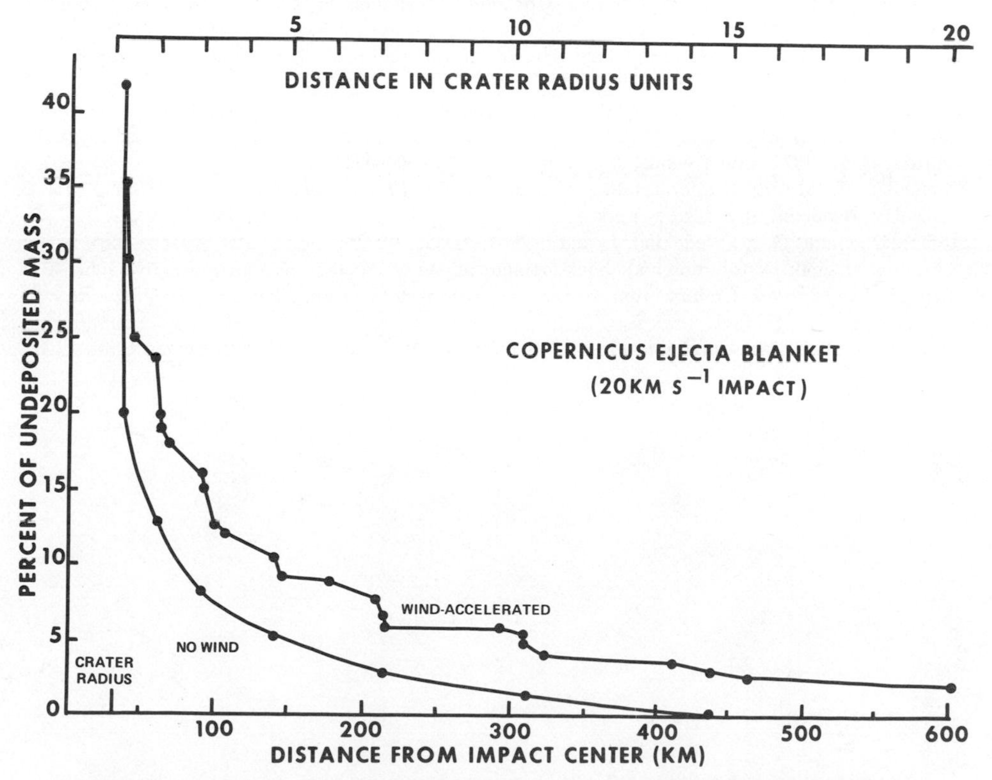

Fig. 9. Percent of crater ejecta mass not deposited versus distance from impact center (range), for the windless ballistic trajectories and for the wind-accelerated case.

consider it an overestimate. In the latter regard, C_D is poorly documented for irregular shapes at high R_e, and the gas cloud might have initial macroscopic kinetic energy. Factors making this a possible underestimate include delay of gas cloud emission due to physical occlusion by non-vaporized meteorite material. Also the crater formation rate might be faster than the one assumed here, and our meteorite size may be too small according to some scaling laws (Shoemaker, pers. comm., 1976); a larger meteorite would yield more gas and a greater interaction with ejecta material. We are continuing to investigate these possibilities, and also are considering the effects of jetting phenomena, surface curvature, lift forces, and non-homogeneous models of particle size.

Acknowledgment—This study was supported by NASA Grant NSG-07026. Much appreciated were helpful comments by E. M. Shoemaker and B. K. Germain.

REFERENCES

Baldwin, R. B.: 1963, *The Measure of the Moon*, p. 128–152, The University of Chicago Press, Chicago.

Hughes, W. F. and Brighton, J. A.: 1967, *Schaum's Outline of Theory and Problems of Fluid Dynamics*, p. 85, McGraw-Hill, San Francisco.

Li, W. H. and Lam, S. H.: 1964, *Principles of Fluid Mechanics*, p. 24, Addison-Wesley, Reading, Massachusetts.

Oberbeck, V. R.: 1975, The role of ballistic erosion and sedimentation in lunar stratigraphy, *Rev. Geophys. Space Phys.* **13**, 337–362.

Pike, R.J.: 1967, Schroeter's Rule and the modification of lunar crater impact morphology, *J. Geophys. Res.* **72**, 2099–2106.

Rehfuss, D. E.: 1972, Lunar winds, *J. Geophys. Res.* **77**, 6303–6315.

Shoemaker, E. M.: 1962, Interpretation of lunar craters, *Physics and Astronomy of the Moon*, p. 283–359, Academic Press, New York.

Stöffler, D., Gault, D. E., Wedekind, J., and Polkowski, G.: 1975, Experimental hypervelocity impact into quartz sand: Distribution and shock metamorphism of ejecta, *J. Geophys. Res.* **80**, 4062–4077.

Vortman, L. J.: 1968, Craters from surface explosions and scaling laws, *J. Geophys. Res.* **73**, 4621–4636.

Whitaker, S.: 1968, *Introduction to Fluid Mechanics*, p. 305, Prentice-Hall, Englewood Cliffs, N.J.

Roddy, D. J., Pepin, R. O., and Merrill, R. B., editors.
(1977) *Impact and Explosion Cratering*, Pergamon Press (New York), p. 1133–1163.
Printed in the United States of America

Crater-related ground motions and implications for crater scaling

HENRY F. COOPER, JR.

R & D Associates, 4640 Admiralty Way, Marina del Rey, California 90291

FRED M. SAUER

Physics International Company, 2700 Merced Street, San Leandro, California 94577

Abstract—Ground motion data from cratering explosions in a variety of geologic media suggests that

$$d/V^{1/3} \simeq k(V^{1/3}/r)^3 \text{ for } 1 \leqslant r/V^{1/3} \leqslant 5$$

where d is the transient peak surface displacement, V is the apparent crater volume, r is the range from the explosion and $k \approx 0.45$ for above-surface explosions, $k \approx 0.25$ for half-buried explosions, and $k \simeq 0.1$ for buried explosions. Thus, the use of $V^{1/3}$ as a characteristic length integrates the effects of source parameters and geologic material properties on the surface displacements produced by cratering explosions. Because of the similarity with buried explosion cratering phenomena, the above expression with $0.1 \leqslant k \leqslant 0.25$ is suggested to estimate the surface displacements from planetary and lunar impact cratering events.

Dimensional analysis procedures are applied to suggest appropriate dimensionless groups for peak particle velocity and characteristic times consistent with the assumption of $V^{1/3}$ as a characteristic length. Then, ground motion data from buried and surface explosions are examined to evaluate alternative "scaling" rules.

1. INTRODUCTION

THE PHYSICAL PARAMETERS that affect crater formation (e.g., energy and momentum coupled to the ground, and geologic properties) also affect the neighboring close-in ground motions. A variation in these parameters that leads to larger (or smaller) craters also likely leads to correspondingly larger (or smaller) crater-related ground motions.* This simple notion and analysis of near-surface horizontal displacement data from numerous near-surface cratering explosions motivated Cooper (1971) to hypothesize that the cube-root of the apparent crater volume† ($V^{1/3}$) is a characteristic length common to the crater-forming process and the subsequent crater-related ground motions (see also Crawford *et al.*, 1974). This paper reviews and extends the crater-volume correlation procedures and discusses the implications for understanding crater-forming processes and related phenomena.

It is thought that much of our explosive cratering experience should be useful

*We exclude deeply buried explosions and airbursts significantly above the ground surface. As these limits (where no crater is formed) are approached, ground motions are not easily related to the crater-forming process.

†Actually, the true crater volume would be a preferred correlating factor, but it was not measured on many experiments for which ground motion data exist.

in understanding ground motion and ejecta phenomena from impact cratering events. However, airblast-related phenomena associated with explosive cratering events have no counterpart in the impact cratering case. Thus, we first discuss the features of ground motions produced by near-surface cratering explosions so that differences from impact cratering events might be better appreciated.

We then discuss the basis for the crater volume correlation which was proposed as a means of relating the near-surface particle displacements from cratering explosions in different media. Data from several experiments are used to illustrate the extent to which the correlation procedures appear to work. We apply the methods of dimensional analysis in conjunction with the use of $V^{1/3}$ as a characteristic length to suggest dimensionless time, velocity, acceleration, and stress parameters and examine experimental data to suggest explicit forms relating the non-dimensional parameters. Finally, we summarize the key results of the paper.

Throughout our discussion we shall emphasize how parameters are observed to vary rather than how they "scale". The late time crater-related ground motions which are our main subject probably are influenced by gravity; hence, in a strict sense, "scaled" experiments (even in a given medium) cannot be conducted unless gravity is also varied (Crowley, 1970). The correlations to be discussed reflect observations from various experiments, hopefully displayed in a convenient and useful format, but they are not claimed to be scaling laws. Nevertheless, they do appear to collapse the data from many experiments and are suggestive of a relationship between crater-forming processes and the subsequent ground motions.

2. NEAR-SURFACE GROUND MOTION PHENOMENA PRODUCED BY CRATERING EXPLOSIONS

As illustrated by Fig. 1, the near-surface ground motions from near-surface cratering explosions result from two loading mechanisms. As discussed by Sauer (1964) and Brode (1968) the attenuating, decelerating airblast loads the ground producing airblast-induced ground motions. Second, the energy directly coupled to the ground by the explosive source produces direct-induced ground shock which propagates outward at the compressional wave speed of the geologic medium.

As the height-of-burst is varied, the influence of these loading mechanisms on the subsequent motions varies. At a sufficient height-of-burst, the ground is loaded by the airblast only, and no direct-induced ground motion exists. At a sufficient depth-of-burst, the explosion is totally contained maximizing the direct-induced ground shock, and no airblast and no airblast-induced ground motions will result. Even a relatively shallow depth-of-burial produces a marked enhancement in the energy that is directly coupled to the earth and reduces the close-in airblast loading on the ground.

Near the earth's surface and close-in to ground zero from a near-surface burst, the airblast arrives first and produces a downward and outward ground motion response referred to as the "airslap" in the near-field waveforms in Fig.

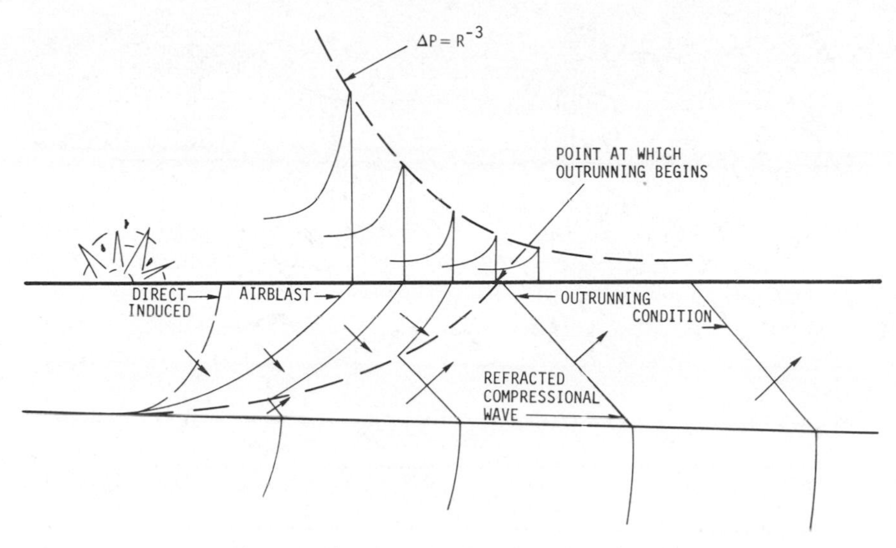

Fig. 1. Basic phenomenology from a surface burst explosion.

2. This airblast-induced signal is shortly followed by a low-frequency upward and outward particle motion referred to as "crater related" in the near-field waveforms in Fig. 2. As the explosive source is buried, the airslap portion of the signal reduces markedly due to the reduced airblast; the crater-related portion increases because of the increase in directly coupled energy to the ground; and time between the arrival of airblast and direct-induced (or crater-related) signals reduces.

Sufficiently far from the explosive source, ground motion signals precede the airblast arrival, and an "outrunning" signal condition exists. As suggested by

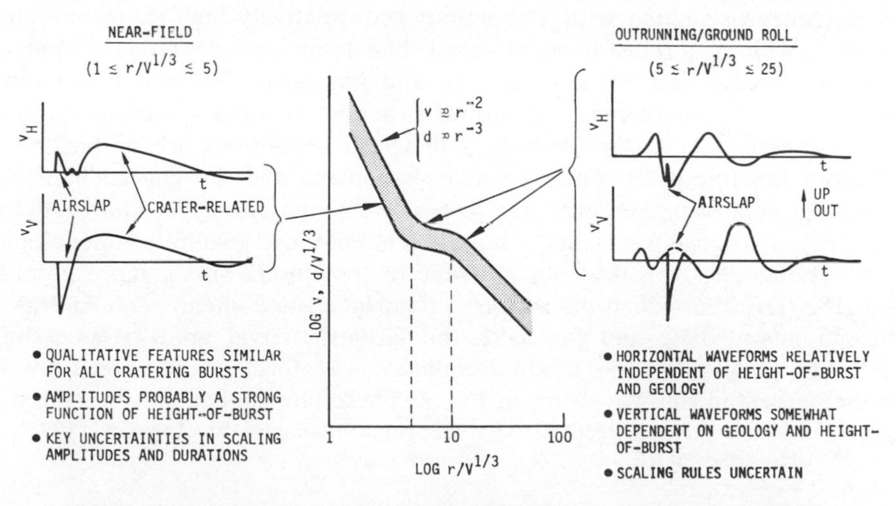

Fig. 2. Near-surface ground motion phenomenology (surface burst geometry).

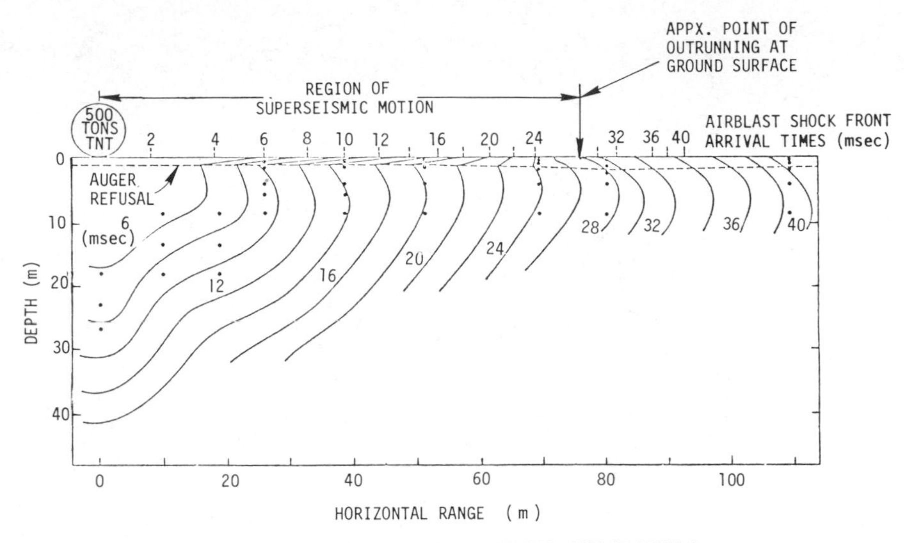

Fig. 3. Shock front profile—MIXED COMPANY 3.

Fig. 1, the outrunning signal can be associated with refracted energy from high seismic velocity layers which underlie the surface media. In an unlayered medium (such as uniform hard rock) outrunning would be initiated at the range where the airblast decelerates below the seismic velocity. For illustrative purposes, consider the arrival time contours derived from the MIXED COMPANY event (Ingram, 1975; Chisolm, 1975), which involved detonation of a 500-ton* tangent sphere of TNT on a layered soil/sandstone geology (Fig. 3).

The near-surface ground motion data from cratering explosions suggest three phenomenological regions. As indicated in Fig. 2, close-in to the crater, the particle motion waveforms are simple in character and usually consist of a single low-frequency oscillation with a superimposed, relatively high-frequency spike associated with the airblast-induced signal. The dominant low-frequency particle motion is initially upward and outward and has been observed to be either clockwise or counterclockwise on various experiments—and at various ranges in a given experiment. At distant ranges the particle motions are dominated by oscillatory low-frequency surface wave phenomena and the characteristics of the motion appear to be relatively insensitive to variations in source details (yield, depth- or height-of-burst, etc.). Between these two phenomenological regions the simple near-field wave structure transitions into a more complex wave structure from which the farther out surface wave phenomena emerge. In both the intermediate and far field, the airblast arrival appears as a high-frequency signal riding on a low-frequency waveform as suggested by the outrunning/ground roll waveform in Fig. 2. These phenomena are illustrated by the near-surface particle velocity waveforms measured by Ingram (1975) on

*One ton of TNT is equivalent to 4.2×10^{16} ergs.

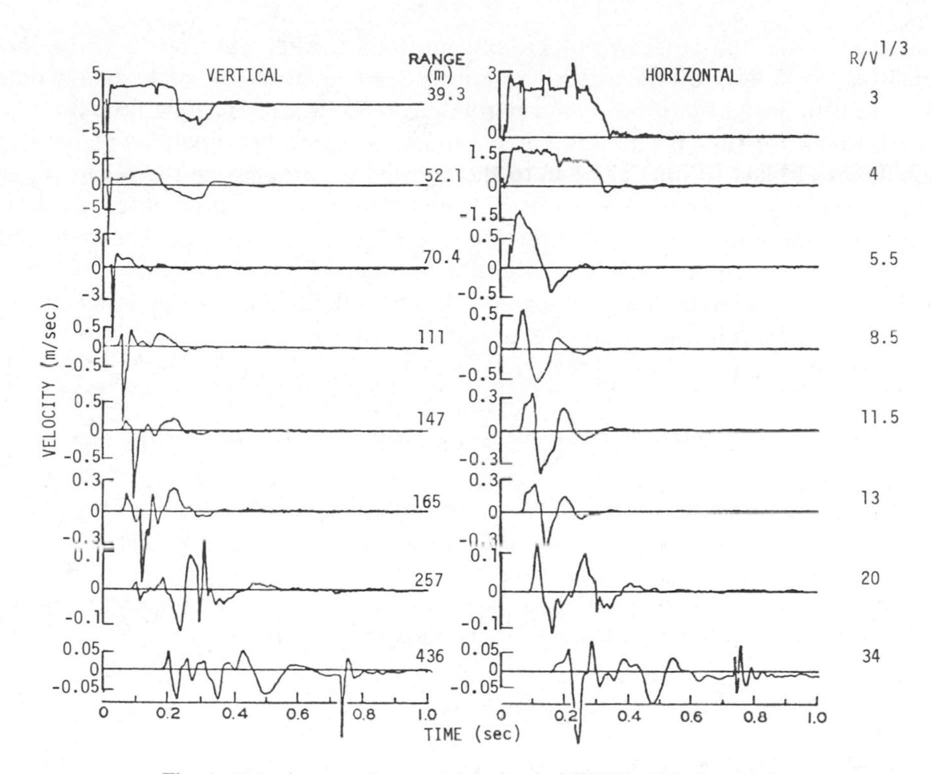

Fig. 4. Velocity waveforms, 1.5 ft depth, MIXED COMPANY 3.

MIXED COMPANY which involved the detonation of a 500-ton surface tangent spherical charge of TNT on a layered soil sandstone medium (Fig. 4).

Variations in the attenuation of peak particle velocities and displacements can be identified with the above three phenomenological regions. Although there are experiment-to-experiment deviations, close-in particle velocities and displacements attenuate approximately as r^{-2} and r^{-3}, respectively.* The attenuation rate usually reduces markedly in the transition region, which often corresponds to the initial outrunning region for surface-burst explosions (where the ground motions precede the airblast arrival). In fact, as discussed by Sauer (1964) and Bratton (1974) an increase in the ground motion amplitudes with increasing range has been observed on some experiments. Beyond the transition region, a more rapid attenuation rate is expected; however, it is not so rapid as that which occurs in the close-in region.

This paper is primarily concerned with observations of the close-in, near-surface crater-related motions. As indicated by Fig. 2, this region generally extends from the source to about 5 $V^{1/3}$ where V is the apparent crater volume.

*The exponents have been observed to vary from test to test. The exponent for close-in peak particle velocity has been observed to vary between about 1.5 to something over 2. The exponent for peak particle displacements has been observed to vary from about 2.5 to about 3.5.

The magnitude and structure of ground motions farther out appear to be less sensitive to details of the source region; so an analysis of such data would provide little insight into mechanisms related to the crater-forming process.

Although we shall discuss data from above-surface experiments which had a significant airblast loading close-in to the source, we emphasize the results from buried explosions where the close-in ground response is dominated by low-frequency motions associated with the crater-forming process. These results should be relatively independent of airblast-induced effects and should be most instructive in understanding the ground motions produced by impact cratering events (where airblast-induced effects are absent).

3. GROUND MOTION—CRATER VOLUME CORRELATION

The ground motions produced by an explosion near the surface of the earth are a function of at least the explosive yield (Y), height-of-burst (h), range from ground zero (r), and depth of the observation points (z). In addition to these independent variables, a general formulation must also account for variations in the type of explosive source and geologic medium. Thus, for example, the peak particle displacement field (**d**) might be represented by a function

$$\mathbf{d} = \mathbf{d}(G, S, Y, h, r, z), \tag{1}$$

where the parameters G and S denote some as yet undefined set of properties of the geology and source.

Equation (1) implicitly assumes the displacement field is axially symmetric. Although test data usually contradict this assumption, no procedure has yet been developed to quantitatively predict azimuthal variations, except, perhaps, in some statistical sense. The best predictions that can be made today provide bounds for the expected azimuthal variations, and possibly worst-case directions related to pre-shot *in situ* geologic conditions (joints, faults, etc.). For our purposes, Eq. (1) will be taken to represent the average displacement field, e.g., the logarithmic mean of the displacements corresponds to a 360° azimuthal variation.

If the explosive burst is near the earth's surface, a crater will be produced, the physical dimensions of which are also functions of the source type, yield, height-of-burst, and geology. For example, the apparent crater volume (V) might be expressed as

$$V = V(G, S, Y, h). \tag{2}$$

Thus, the crater volume and the displacements at a given range and depth (r, z) from a cratering burst are functions of the same independent variables describing the geology and burst configuration. Therefore, it is reasonable to expect that some explicit relationship between the crater volume and the ground motions might exist. Based on studies of ground motion data from cratering bursts in "uniform"* media, Cooper (1971) suggested that the cube-root of the apparent crater volume is a characteristic length, and that close-in near-surface peak transient crater-induced particle displacements can be represented by a function of the form

$$\mathbf{D} = \mathbf{K}(H, Z)R^{-3} \quad \text{for } R \leqslant 5^* \tag{3}$$

where $\mathbf{D} = d/V^{1/3}$, $H = h/V^{1/3}$, $Z = z/V^{1/3}$, and $R = r/V^{1/3}$. Here, $K(H, Z)$ is a function of the scaled height-of-burst and observation depth that tends to increase with decreasing Z and increasing H (with the main variation occurring for very near-surface explosions). Equation (3) is written as a vector equation because the particle displacement crater volume correlation, which was originally hypothesized for horizontal peak particle displacements, may also apply for vertical crater-induced peak particle displacements.

Obviously, Eq. (3) can only be applied to predict crater-related displacements from explosions that crater, e.g., from $h < h_0$ where h_0 is the height-of-burst beyond which no apparent crater is produced. The restriction $R \leqslant 5$ automatically constrains the spatial domain of applicability ($r \leqslant 5\,V^{1/3}$) as V tends to zero. Equation (3) is not applicable for farther out particle displacements where most of the energy is carried by surface modes and low-frequency motions reflected from deep geologic layers. Beyond $R \simeq 5V^{1/3}$, the peak particle displacements attenuate less rapidly than R^{-3}. Because of this smaller rate of attenuation, extrapolation of Eq. (3) underestimates the peak particle displacements for $R \geqslant 5$.

An important feature of Eq. (3) is that the source and geology do not appear as explicit independent variables. Therefore, such a rule allows for a considerable generalization of the existing ground motion data to estimate the ground motions in the neighborhood of explosively produced craters where ground motion measurements were not made and where the geology is unknown (provided the crater volume is known). Furthermore, it may be possible to apply the explosive crater and ground motion data base to estimate the crater-related motions produced by hypervelocity impact for those cases where the crater volume is known and the source is unknown.

The complete way of testing the hypothesized crater-volume correlation

---

*The term "uniform" is chosen in a loose qualitative sense to refer to the case where no major discontinuity significantly influences the crater formation. Possible exceptions considered by Cooper (1971) were tests at the Suffield Experimental Station where the water table may have introduced significant layering effects. However, the strength mismatch usually introduced by a soil-rock interface was in no way represented in these experiments. More recently, the MIDDLE GUST and MIXED COMPANY experiments have considered the effects of significant geologic layering on cratering and ground shock; and the resulting data suggest that some alteration of Eq. (3) may be required to model "non-uniform" or layered media (e.g., Bratton, 1973).

*The limit of applicability of Eq. (3), $R \leqslant 5$, may be medium-dependent. It was derived from observations by Cooper (1972a, 1972b, 1972c) that the close-in crater-related ground motions at the Nevada Test Site begin to transition into a more complex wavetrain of surface waves at about $5\,V^{1/3}$. Within this transition region, the peak transient displacements attenuate less rapidly than R^{-3} and may even increase with increasing range.

procedures is to conduct experiments that vary the key independent variables one at a time—yield and type of explosive, geology, burst geometry (e.g., height-of-burst), and measurement geometry (e.g., range and depth of gages). The following paragraphs review our understanding of effects of variations in these independent variables on the crater volume correlation procedures.

Source effects

Heuristic arguments suggest that the qualitative phenomena associated with late-time crater formation and the resulting ground motions can be decoupled from details of the explosive source. Such arguments are most convincing for buried cratering bursts. Recent theoretical studies by Maxwell and Seifert (1975) of first-principle theoretical calculations and small explosive experiments support the idea that late-stage cratering flows can be decoupled from details of the explosive source in a manner consistent with the hypothesized crater volume correlation. According to Maxwell, the late-stage flow is reasonably modeled as incompressible, and final crater dimensions and peak amplitude near-surface motions are determined by media strength and gravitational effects.

In principle, companion high-explosive and nuclear experiments would be desirable to verify that systematic variations in the crater volume correlation do not result from variations in the explosive source. Although such companion experiments might be conducted with buried explosions (and in fact some limited data exist as will be discussed later), surface nuclear cratering tests are prohibited by the Limited Nuclear Test Ban. Some other means of testing the effect of energy density variations (possibly via lasers as suggested by Kuhl and Wuerker (1974)) might be instructive. Also, the ground motions from hypervelocity impact experiments and buried explosions might be compared to evaluate the effects of dramatically different sources of cratering and related phenomena.

If we accept the premise that details of the source do not significantly affect the crater volume correlation procedures, then, although the crater volume is a dependent variable, it may be taken as an appropriate measure of the important source region and geology parameters affecting crater-related ground motions. An appropriate test of the crater volume correlation procedures would then require several orders of magnitude variation in V—holding other variables constant. We shall discuss such a comparison shortly.

Effect of geologic variations

Figure 5 illustrates that the crater volume correlation collapses the near-surface particle displacement data from similar above-surface explosions in very different "uniform" geologies. Sources for these data are Sauer (1970), Murrell (1970), Hoffman *et al.* (1971), Murrell (1972), Murrell (1973), Joachim (1973), and Murrell and Carleton (1973). The source and crater parameters summarized in Table 1 are given by Rooke *et al.* (1974) and Lockard (1974). The magnitude of

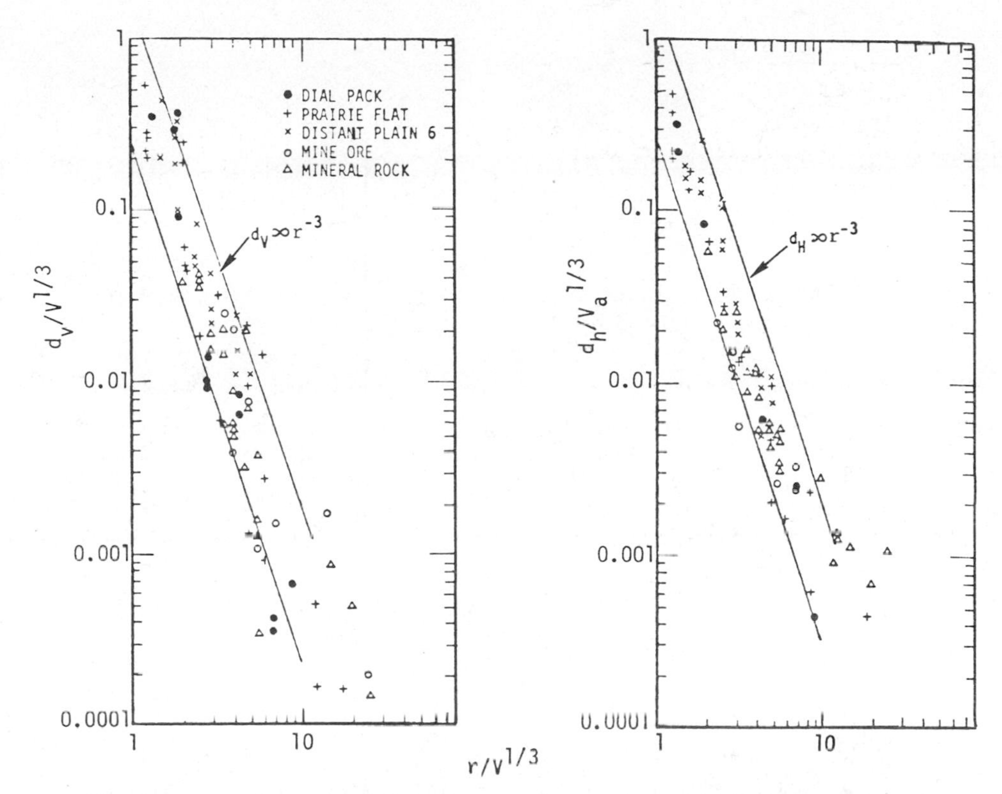

Fig. 5. Crater volume correlation of near surface peak particle displacements from large surface-tangent HE events in "uniform" media.

peak transient displacements on these several events would differ by more than an order of magnitude if they were compared directly in an unscaled sense. If Cauchy (cube-root of the yield) scaling rules were applied, the PRAIRIE FLAT, DIAL PACK, and DISTANT PLAIN 6 data would collapse, but they would still systematically differ from the Cauchy scaled MINERAL ROCK and MINE ORE data. Thus, it is seen that the crater-volume correlation procedure apparently integrates the effects of differing geologies, and therefore offers advantages over the usual yield scaling procedures.

Although the data in Fig. 5 scatter by an order of magnitude, no particular experiment-to-experiment systematic variation is apparent. Figure 6 illustrates that azimuthal variations in particle motions were observed on MINERAL ROCK (Cooper, 1973). Presumably, this particular variation was associated with a variation in *in situ* rock properties in two distinct directions with respect to the major joints, and therefore is primarily a systematic variation. Such observations of azimuthal variations have been common on most well-instrumented explosive cratering experiments, even in supposedly uniform soil geologies. For example, Fig. 7 summarizes the azimuthal variation in particle velocity waveforms at 17-ft (5.2 m) depth and 65-ft (19.8 m) range from the explosion of a half-buried 20-ton

Table 1. Crater parameters from several high explosive events (spherical charges of TNT).

Event	Yield (tons)	Geometry	Geology	Radius (m)	Depth (m)	Volume (m^3)
Mineral rock	100	0.9 charge radius height-of-burst	Cedar City tonolite	8.87	3.14	179
Mine ore	100	0.9 charge radius height-of-burst	Cedar City tonolite	6.92	2.74	97.6
Distant Plain 6	100	Above-surface tangent	Suffield silt	13.0	4.9	575
Prairie Flat	500	Above-surface tangent	Suffield silt	31.3	4.18	5160
Dial pack	500	Above-surface tangent	Suffield silt	30.5	3.93	4110

*One ton is equivalent to 4.2×10^{16} ergs.

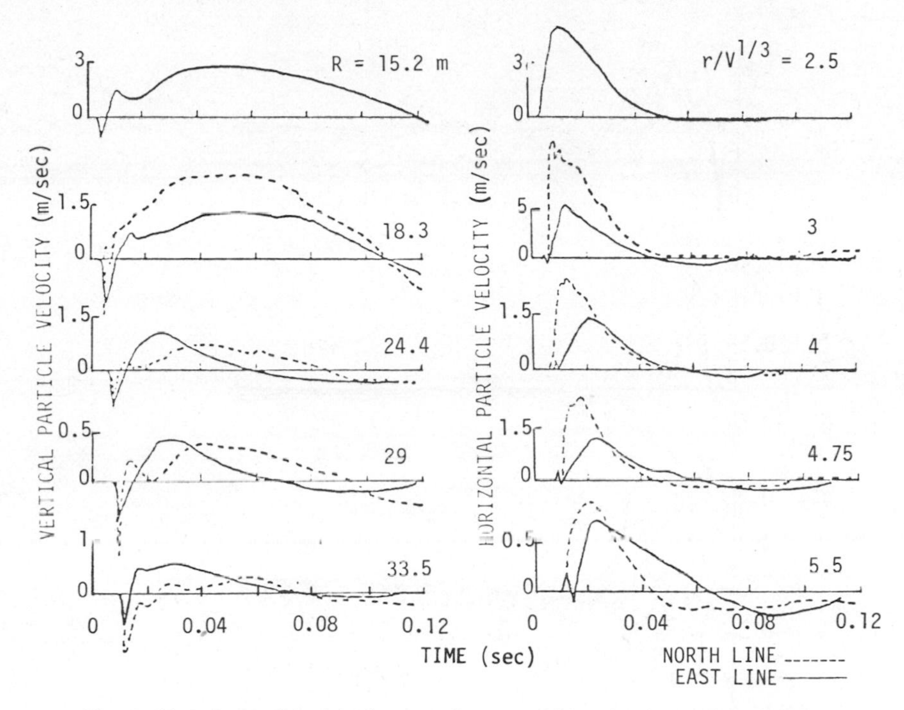

Fig. 6. Close-in particle velocity waveforms at 0.61 m depth on mineral rock.

TNT sphere in Nevada Test Site (NTS) playa (Sauer and Vincent, 1967). The site was presumed to be "homogeneous" before the experiment, but analysis of the resulting data indicates reflected waves from an unexpected inhomogeneity.

Major geologic layering introduces characteristic lengths other than $V^{1/3}$ that can influence the ground motions. With one exception, recent high-explosive tests in layered clay/shale [MIDDLE GUST Series (Bratton, 1974)] and sand/sandstone [MIXED COMPANY Series (Ingram, 1975; Chisolm, 1975)] provide data consistent with the correlation in Fig. 5. The exception was for explosions in a wet clay/shale geology where the crater volume correlation over-predicted the peak transient displacements by a factor of three.

Burst geometry

Cooper (1971) noted a height-of-burst effect in the crater-volume correlation of horizontal displacements summarized in Eq. (3). These studies suggested that for near-surface horizontal motions, $K(H, Z) \simeq 0.45$ for $H > 0$ and $K(H, Z) \simeq 0.1$ for $H \leqslant 0$ as illustrated in Fig. 8. More recent studies suggest that if the data from half-buried high-explosive sources are examined separately (Fig. 9), then $K(H, Z) \simeq 0.25$ for $H \simeq 0$. (Presumably, this height-of burst effect is related to basic differences in cratering phenomena for buried and above-surface explosions, e.g., a larger percentage of the crater volume may be ejected from buried bursts. Also, the airblast may be more important in the crater-forming

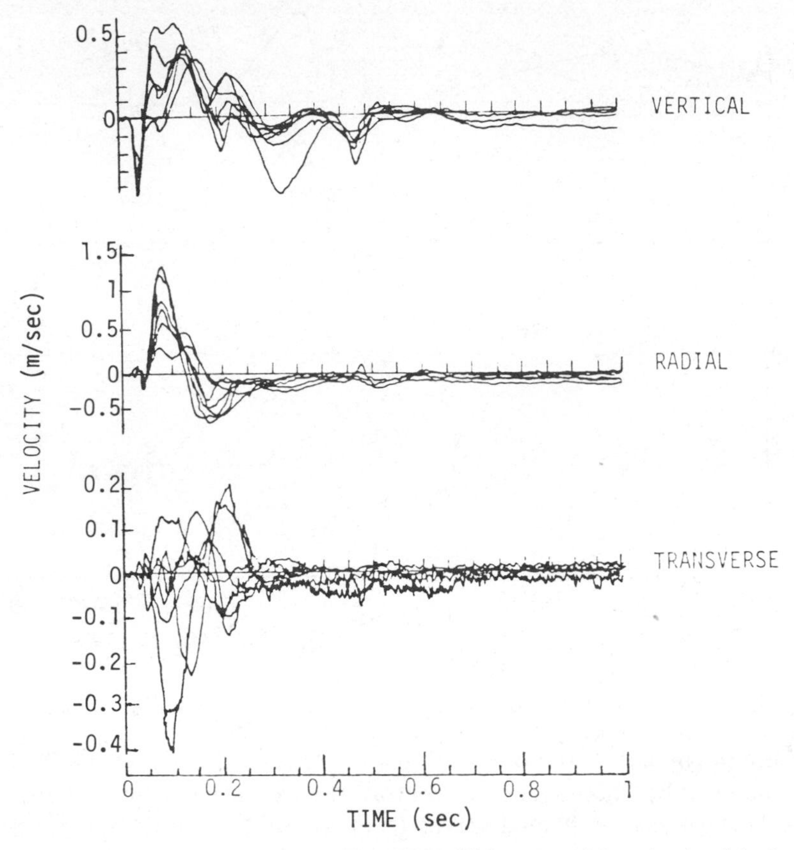

Fig. 7. Symmetry experiment on FLATTOP III (range = 19.8 m, depth = 5.2 m).

process for surface and above-surface bursts than for buried bursts.) If Oberbeck's (1971) equivalence between explosion and impact craters is assumed, then $0.1 \leqslant K \leqslant 0.25$ would be appropriate for horizontal peak transient displacements from impact cratering events.

Range and depth of gages

As indicated earlier, a change in surface motion phenomena occurs at some range beyond about 5 $V^{1/3}$ where surface waves begin to form. Hence, the crater volume correlation procedures are limited to ranges less than about 5 $V^{1/3}$. Because surface effects sometimes attenuate rapidly with depth, care should be taken to compare measurements at comparable "scaled" depths. While the "scaling" is difficult to quantify, it is clear that data from "shallow" and "deep" gages should be analyzed separately. Based on studies of crater-induced ground motions in hard rock, Cooper (1973) suggested that, in "uniform" media, the horizontal crater-induced motions have a characteristic waveform that does not

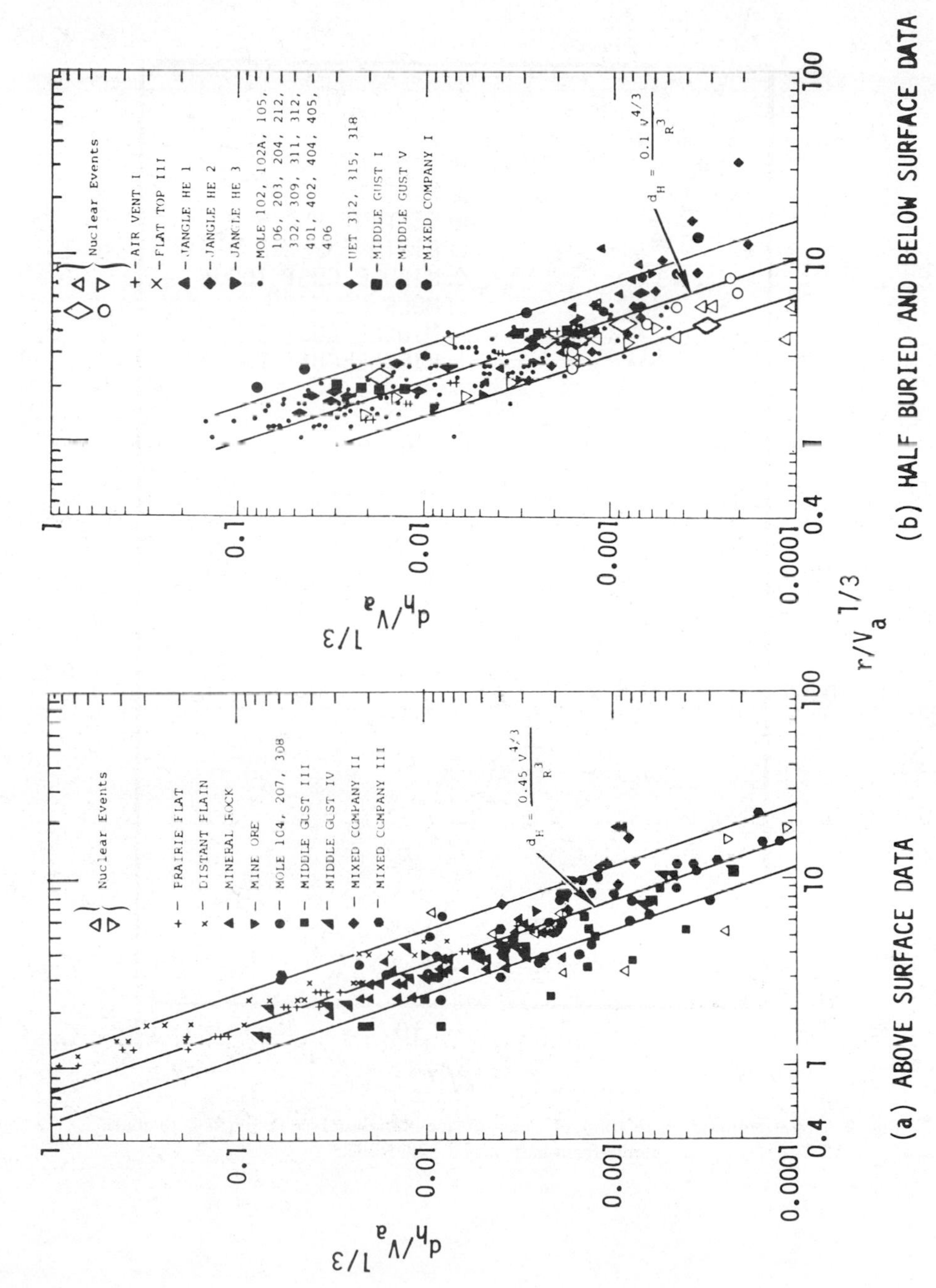

Fig. 8. Crater volume correlation for above-surface and buried explosions.

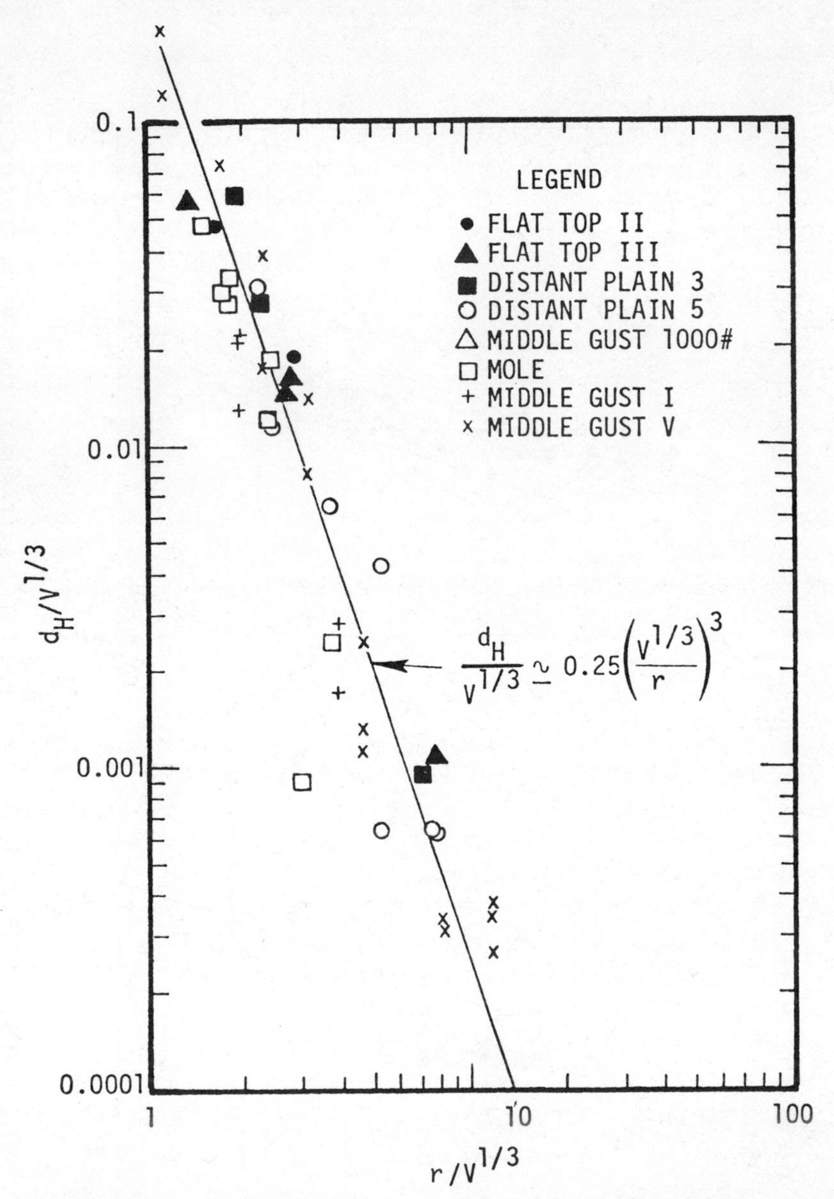

Fig. 9. Crater volume correlation of near-surface horizontal peak particle displacements from half-buried explosions.

vary greatly with depth,* so that one can interpolate between measurements at several depths.

Comparisons of existing data from surface and deep measurements on buried cratering explosions in the dry alluvium of Area 10 of the Nevada Test Site may be of particular interest to the impact cratering community. Differences between nuclear and HE cratering phenomena are minimized for such test geometries, and both should be qualitatively similar to impact cratering phenomena. First, we consider data from deep gages (on the order of the depth-of-burst) from high-explosive, near-optimum depth-of-burst cratering experiments that produced over three orders of magnitude variation in crater volume. Second, the near-surface, close-in data from buried high-explosive and nuclear explosive sources will be examined.

Buried explosion and deep gages

Project SCOOTER (500 tons of TNT detonated at a 125-ft (38.1 m) depth) produced a crater whose apparent volume was 2.64×10^6 ft^3 (7.48×10^4 m^3) (Perret, 1963). MOLE events 212 and 404 (256-lb TNT charges detonated at 2.05 and 6.84 ft (0.625 and 2.08 m) depths) produced crater volumes of 2010 ft^3 and 6000 ft^3 (57 m^3 and 170 m^3) respectively (Sachs and Swift, 1955). These experiments involved the same kind of explosive source and geology. SCOOTER and the two MOLE experiments were near-optimum depth-of-burst cratering events. On each experiment, the ground motions were measured along a single radial at depths comparable to the depth-of-burst (125 ft (38.1 m) on SCOOTER, 5 ft (1.52 m) depth on the other experiments). Figure 10 presents the $V^{1/3}$ correlation for peak horizontal particle displacements from these experiments. Also shown are data from two somewhat shallower buried JANGLE HE events. The displacements observed on identical experiments (MOLE 212 and MOLE 404) varied by a factor of 3 to 4, which is comparable to observations of the azimuthal data scatter on a well-instrumented experiment. When the displacement data are scaled with $V^{1/3}$, the other data generally fall within this scatter. An approximate fit to the logarithmic mean of the data scatter for "deep" gages and buried explosions in alluvium is

$$d_H / V^{1/3} \simeq 0.018\,(V^{1/3}/R)^2. \tag{4}$$

As indicated previously, the data appear to depart from this trend for ranges greater than about 5 $V^{1/3}$.

Buried explosions and shallow gages

Although no shallow gages were installed on SCOOTER, TEAPOT, ESS (a 1.2-kT nuclear source detonated at 67-ft (20.4 m) depth) produced a compar-

*In general, the horizontal particle velocity pulse width narrows somewhat with depth, but the qualitative features of the wave do not vary significantly for depths less than about 0.5 $V^{1/3}$.

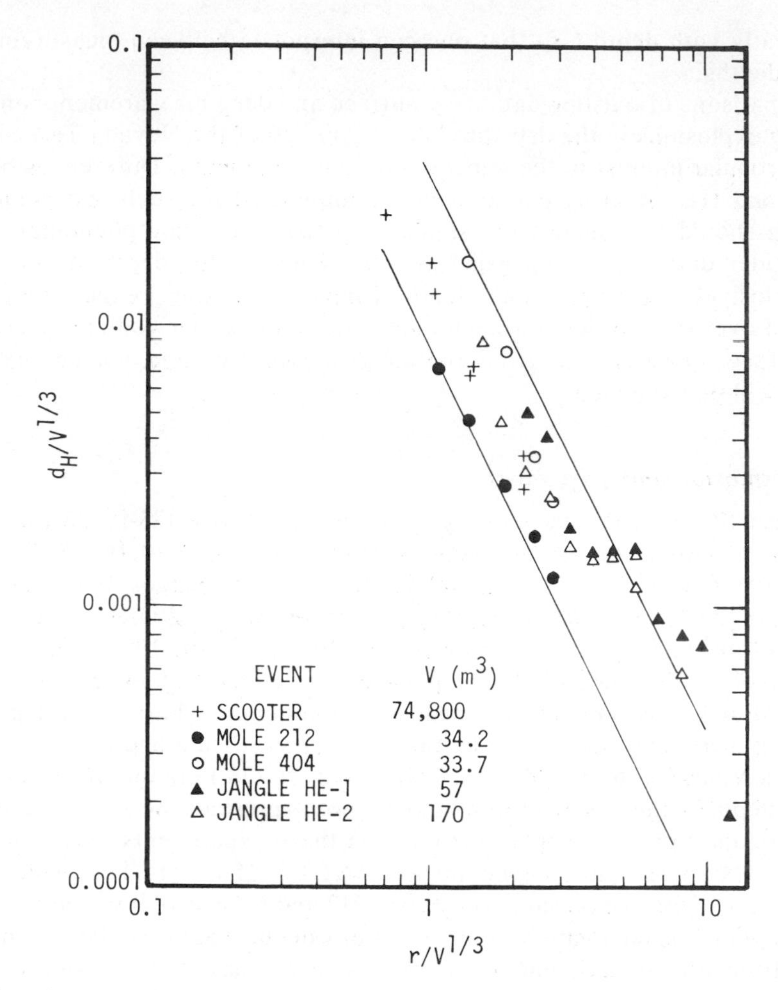

Fig. 10. Crater volume correlation of horizontal peak particle displacements for buried explosions in alluvium (deep gages).

able crater whose apparent volume was 2.6×10^6 ft^3 (7.37×10^4 m^3) (Sachs and Swift, 1958), and ground motion measurements were made at 1- and 10-ft (0.305 and 3.05 m) depths i.e., shallow depths as compared to the crater dimensions. Close-in, near-surface data from a lower yield, buried nuclear cratering event at the same scaled depth-of-burst in alluvium do not exist to establish a direct comparison of near-surface ground motions from the same kind of source. On the other hand, close-in near-surface data from several low-yield nuclear and TNT explosions at shallower scaled depths-of-burst do exist and provide some basis for examining the crater-volume correlation for near-surface motions and buried explosions in NTS alluvium. Since HE cratering phenomena are expected to be similar to NE cratering phenomena for buried bursts, we assume that any phenomenological difference caused by HE and NE sources is second order as compared to the effect of other independent variables; therefore, we have used data from several MOLE experiments (256 lb of TNT), JANGLE HE-2 (20 tons of TNT), JANGLE U (1.2 kT nuclear) and TEAPOT ESS (1.2 kT nuclear) to examine the $V^{1/3}$ correlation of near-surface ground motions.

The crater volume correlation for peak horizontal displacements from these experiments is shown in Fig. 11. The correlated displacements from TEAPOT ESS and JANGLE HE-2 are well collapsed for ranges less than about 5 $V^{1/3}$, but both appear to be lower than the MOLE data. It should be pointed out that the scatter in the correlated peak horizontal displacements is no greater than that observed on the two identical MOLE experiments in Fig. 10, and could result from a systematic azimuthal variation caused by geologic inhomogeneities.

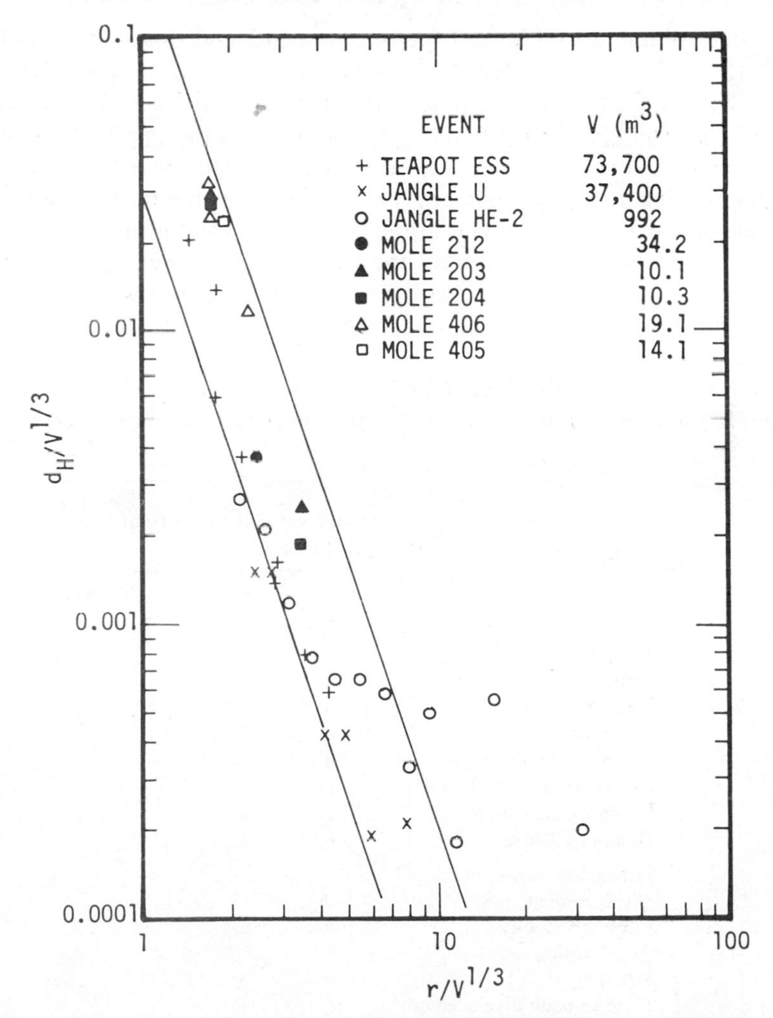

Fig. 11. Crater volume correlation of near-surface horizontal peak particle displacements from buried explosions in alluvium.

Although the data in Fig. 10 do not clearly validate the crater volume correlation, the use of $V^{1/3}$ as a characteristic length is not inconsistent with the close-in, near-surface peak horizontal particle displacements from buried cratering bursts in alluvium. An approximate fit to the logarithmic mean of the data scatter for shallow gages and buried explosions in alluvium is

$$d_H/V^{1/3} \simeq 0.08\,(V^{1/3}/r)^3, \qquad (5)$$

which is about 20% lower than Cooper's (1971) original estimate based on a more extensive data base involving media other than alluvium (Fig. 8).

4. SOME IMPLICATIONS OF DIMENSIONAL ANALYSIS

The crater volume correlation of peak crater-induced displacements in Figs. 5 and 8 through 11 suggests that $V^{1/3}$ might be used as a characteristic length associated with the late time, near-surface ground motions from cratering explosions in "uniform" geologies. This section considers the implications of using $V^{1/3}$ as a characteristic length on relationships derived by standard dimensional analysis techniques (e.g., Bridgeman, 1937). The dependent and independent variables are listed in Table 2. The crater volume is treated as an independent variable which presumably integrates the effects of independent variations in source and geology. No parameter relating to material strength is included because the crater volume is assumed to account for strength variations. A modulus value is represented by ρc^2.

Consider peak particle displacements. From dimensional analysis, we assume there exists a relation $g(d, h, r, z, V, c, g) = 0$. (Density is not included because it

Table 2. Dependent and independent variables important to crater-induced ground motions.

	Symbol	Dimension*
Independent variables		
Crater volume	V	L^3
Density	ρ	ML^{-3}
Wavespeed	c	LT^{-1}
Acceleration of gravity	g	LT^{-2}
Range from ground zero	r	L
Depth of measurement	z	L
Height of Burst	h	L
Dependent variables		
Displacement	d	L
Particle velocity	v	LT^{-1}
Acceleration	a	LT^{-2}
Stress	σ	$ML^{-1}T^{-2}$
Time to peak displacement	t_0	T

*L, M, and T are symbols for length, mass, and time.

is the only independent variable having M as a dimension.) Because there are seven quantities and two dimensions, there are five nondimensional combinations π_i, $i = 1, 2\ldots,5$ such that $q(\pi_1, \pi_2, \pi_3, \pi_4, \pi_5) = 0$. Using $V^{1/3}$ as a characteristic geometric length, we can write from inspection and usual analysis

$$\frac{d}{V^{1/3}} = f\left(\frac{h}{V^{1/3}}, \frac{z}{V^{1/3}}, \frac{r}{V^{1/3}}, \frac{gV^{1/3}}{c^2}\right). \tag{6}$$

The parameter $gV^{1/3}/c^2 = \rho gV^{1/3}/k$ (where k is a modulus equal to ρc^2) is analogous to the Froude number that has applications in fluid mechanics and which can be derived from the equations of motion as the dimensionless gravity body force term.

Peak displacement data from buried cratering bursts in alluvium, discussed in the preceding section, suggest that variation in the Froude number (via variations in V for fixed c^2 and g) has at most a weak effect on the horizontal peak displacement correlation for $10 \lesssim V \lesssim 10^5\ \mathrm{m}^3$. As shown in Fig. 5, the peak displacement correlation has the same form where variations in the Froude number were dominated by variations in c^2. Since close-in peak horizontal near-surface displacements attenuate approximately as r^{-3}, Eq. (6) may be reduced to

$$\frac{d}{V^{1/3}} \simeq K\left(\frac{h}{V^{1/3}}, \frac{z}{V^{1/3}}\right)\left(\frac{V^{1/3}}{r}\right)^3. \tag{7}$$

In examining peak particle velocity, we assume that there exists a relation where $q(v, h, r, z, V, c, g) = 0$. Again, there are seven quantities, two dimensions, and five π groups, three of which can be taken as $h/V^{1/3}$, $z/V^{1/3}$, and $r/V^{1/3}$. If we choose $gV^{1/3}/c^2$ as π_4, then two possibilities remain for π_5, v/c or $v/\sqrt{gV^{1/3}}$. Thus, two possible dimensionally consistent rules (which differ by a factor of $\sqrt{gV^{1/3}/c^2}$) for peak crater-induced particle velocity exist:

$$\frac{v}{c} = f\left(\frac{h}{V^{1/3}}, \frac{z}{V^{1/3}}, \frac{r}{V^{1/3}}, \frac{gV^{1/3}}{c^2}\right), \tag{8a}$$

or

$$\frac{v}{\sqrt{gV^{1/3}}} = f\left(\frac{h}{V^{1/3}}, \frac{z}{V^{1/3}}, \frac{r}{V^{1/3}}, \frac{gV^{1/3}}{c^2}\right). \tag{8b}$$

Figure 12 compares two alternative correlations of the near-surface peak horizontal particle velocities from buried explosions in alluvium. It appears that these data are quite reasonably collapsed independent of the Froude number by an expression of the form

$$v_H/c \simeq 0.015\,(V^{1/3}/r)^2 \tag{9}$$

whereas plotting $v_H/\sqrt{gV^{1/3}}$ as a function of scaled range appears to lead to a systematic variation with varying Froude number (i.e., varying V). Equation (9) is similar to observations from MINERAL ROCK, MINE ORE, DISTANT PLAIN 6, PRAIRIE FLAT, and DIAL PACK. As summarized in Fig. 13, the

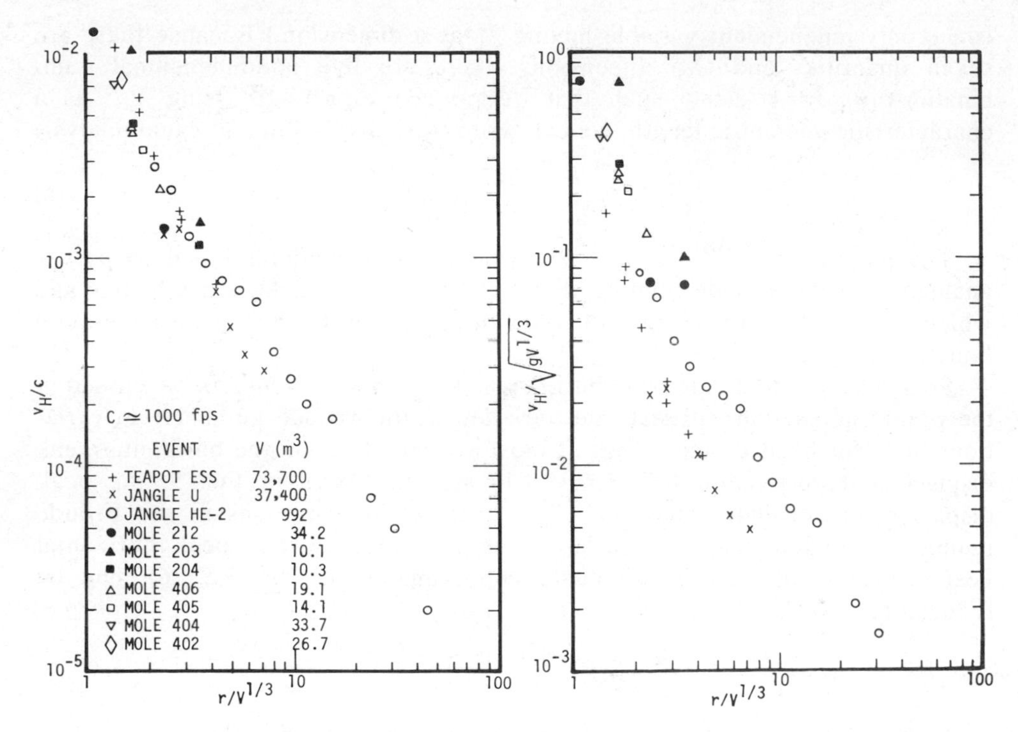

Fig. 12. Crater volume correlation of peak horizontal particle velocity from buried explosions in alluvium (near-surface gages).

soil test data fall systematically below the hard rock data, consistent with the ratio of seismic velocities of the near-surface layers. Here, the Froude number is varied by variations in both c and V. If peak wave speeds of 450 and 2000 m/sec are taken as representative of the Suffield Experiment Station silt and Cedar City tonolite, then these data from above-surface explosions could be collapsed and the logarithmic mean of the factor of 4 data scatter could be approximated by

$$v_H/c \simeq 0.02\,(V^{1/3}/r)^2. \tag{10}$$

Similar considerations as above lead to two equivalent alternatives for the manner in which characteristic times scale:

$$\frac{ct}{V^{1/3}} = f\left(\frac{h}{V^{1/3}}, \frac{z}{V^{1/3}}, \frac{r}{V^{1/3}}, \frac{gV^{1/3}}{c^2}\right), \tag{11a}$$

or

$$t\sqrt{\frac{g}{V^{1/3}}} = f\left(\frac{h}{V^{1/3}}, \frac{z}{V^{1/3}}, \frac{r}{V^{1/3}}, \frac{gV^{1/3}}{c^2}\right). \tag{11b}$$

Provided the physical phenomena are the same in experiments considered in the analysis, the manner in which distances, times and particle velocities vary is expected to be consistent with the definition $v \equiv dx/dt$. Thus, if distances and

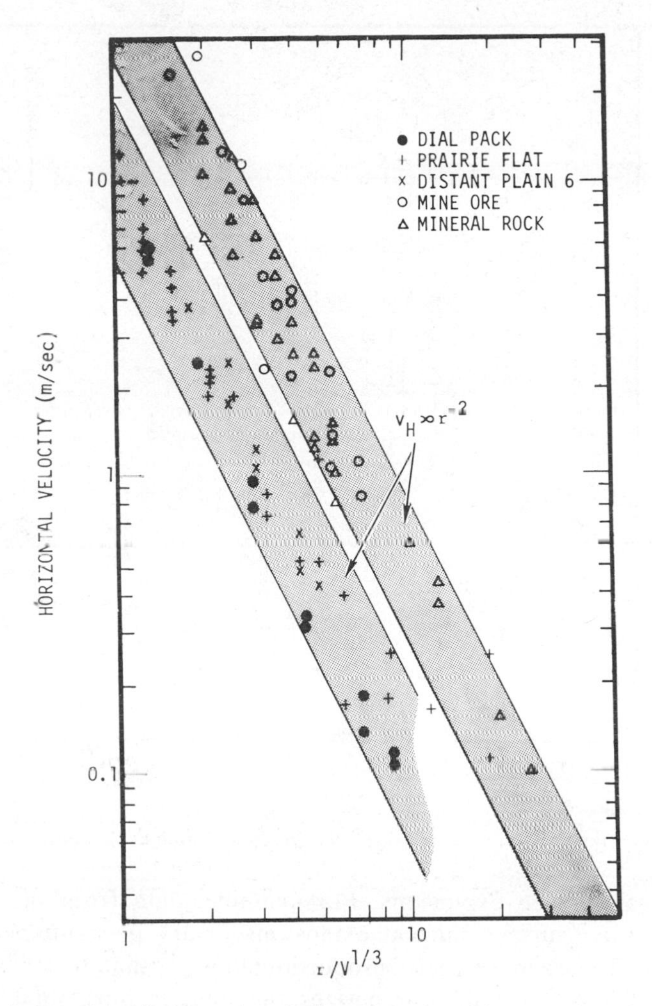

Fig. 13. Attenuation of near-surface horizontal peak particle velocity from surface-tangent HE events in "uniform" media.

displacements are normalized by $V^{1/3}$ and v at a fixed $r/V^{1/3}$ is independent of V as suggested by Eqs. (9) and (10), then t would be expected to vary as $V^{1/3}$. In this case, Eq. (11a) would be the preferred form for the correlation for characteristic times. This expectation is consistent with Fig. 14 which shows the rise time to peak horizontal particle displacement from buried cratering explosions in dry alluvium (Sachs and Swift, 1955). Thus, for buried explosions in dry alluvium, the crater-related horizontal particle velocities, displacements and characteristic times vary with the apparent crater volume in a manner consistent with the definition $v \equiv dx/dt$.

This consistent correlation of displacements, velocities and times observed for buried explosions in dry alluvium does not appear to be a general result that

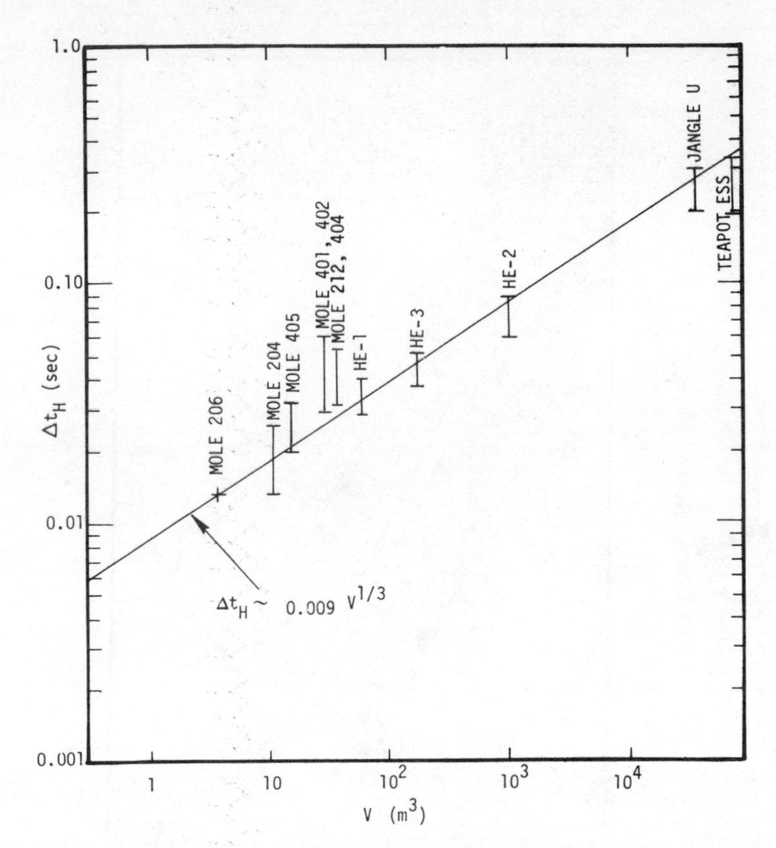

Fig. 14. Rise time to crater-related displacement from buried explosions in alluvium.

applies for near-surface explosions. For example, data from the PRE-MINE THROW IV above-surface tangent explosions in dry playa displayed characteristic times that were more nearly proportional to $V^{1/6}$ than to $V^{1/3}$ (Sauer *et al.*, 1974; Sauer, 1976). Figure 15 summarizes the time to horizontal and vertical peak displacements for surface tangent explosions. Given the definition $v \equiv dx/dt$, and the observation that distances and times vary as $V^{1/3}$ and $V^{1/6}$, peak particle velocities for similar waveforms would be expected to vary as $V^{1/6}$.

Examination of the particle velocity waveforms from the surface tangent PRE-MINE THROW IV explosions shows that airblast loading effects (which scale as $Y^{1/3}$ where Y is the explosive yield) significantly perturb the low-frequency signals and make difficult the separation of airblast-related phenomena from crater-related phenomena. The airblast-related phenomena destroy the similar nature of the vertical particle velocity waveforms, so no comment on how peak particle velocities vary is possible. The airblast-related phenomena do not appear to significantly alter the low-frequency horizontal particle velocity waveforms which appear to be similar at given "scaled" ranges ($r/V^{1/3}$) for surface tangent explosions of spherical sources with yields varying from 0.5 to 100 tons. However, the peak horizontal particle velocity at a given

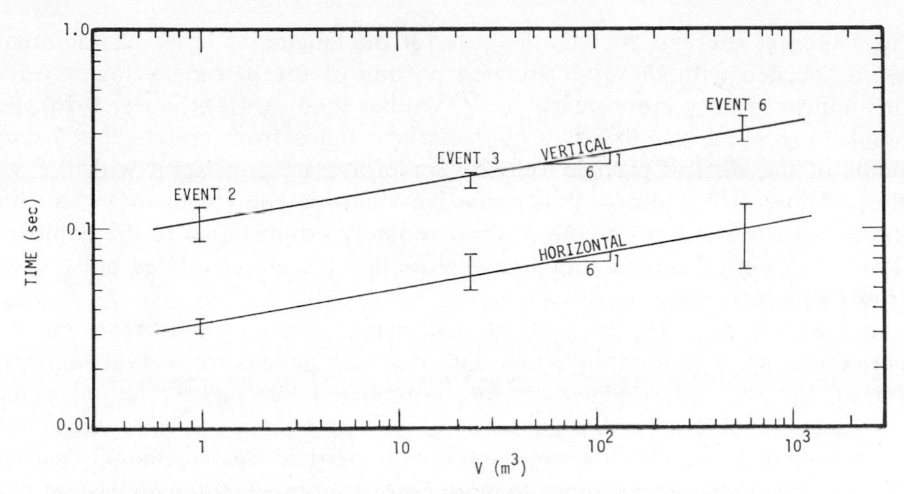

Fig. 15. Rise time to peak particle displacements from surface-tangent, high-explosive experiments on dry playa (project PRE-MINE THROW IV)

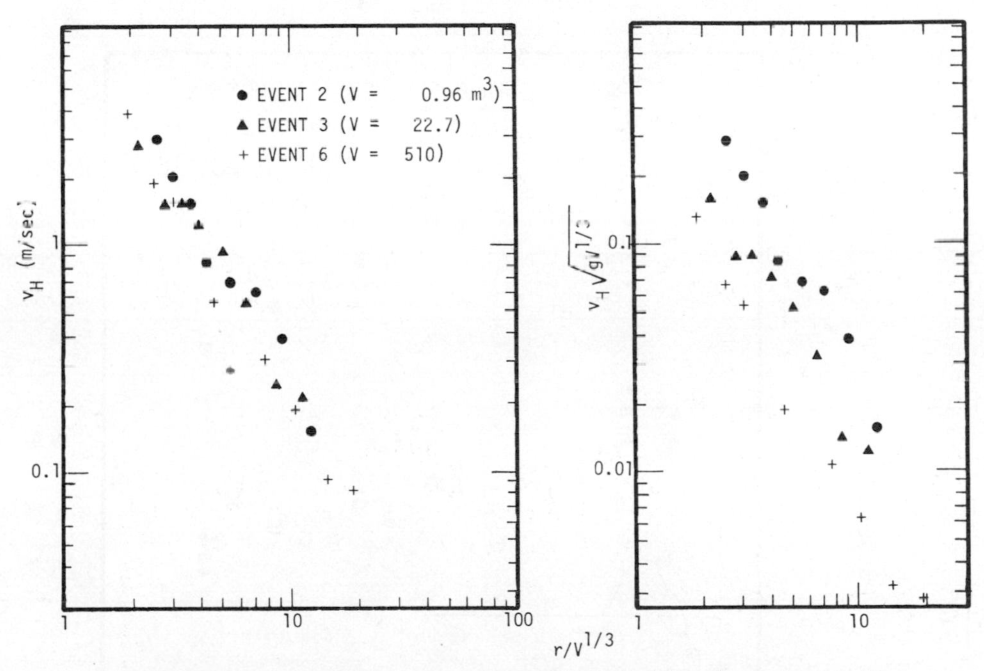

Fig. 16. Alternative correlation of the horizontal peak particle velocity from surface-tangent, high-explosive experiments on dry playa (PRE-MINE THROW IV).

$r/V^{1/3}$ from surface tangent charges appears to be nearly insensitive to variations in V, and is poorly correlated by $V^{1/6}$ (Fig. 16).

The crater-related ground motions from the PRE-MINE THROW IV 256 and 1000 lb half-buried TNT spheres appear to dominate the influence of the airblast effects that so significantly perturbed the ground motions from the surface tangent sources. As was the case for the tangent spheres, characteristic times associated with the crater-related portion of the particle velocity waveforms appear to vary more nearly as $V^{1/6}$ rather than as $V^{1/3}$ (Sauer, 1976). For example, Fig. 17 shows that two characteristic times from crater-related components of the vertical particle velocity waveform are consistent with the $V^{1/6}$ scaling. (Here; $Y^{1/6}$ scaling—Y is explosive yield—is equivalent to $V^{1/6}$ scaling of times since the crater volume is approximately proportional to the explosive yield for half-buried spherical charges.) Note that the crater-related pulse width narrows with increasing range.

As shown by Fig. 18, the vertical peak particle velocity data from the two experiments, when plotted as a function of $r/V^{1/3}$, appear to be systematically different, but they are collapsed when normalized by $\sqrt{gV^{1/3}}$. No particular preference is apparent for horizontal peak particle velocities. Thus, these data from half-buried charges are consistent with particle displacements, particle velocities and characteristic times that are correlated in a fashion consistent with $v \equiv dx/dt$, i.e., with the non-dimensional parameters $d/V^{1/3}$, $v/\sqrt{gV^{1/3}}$, and

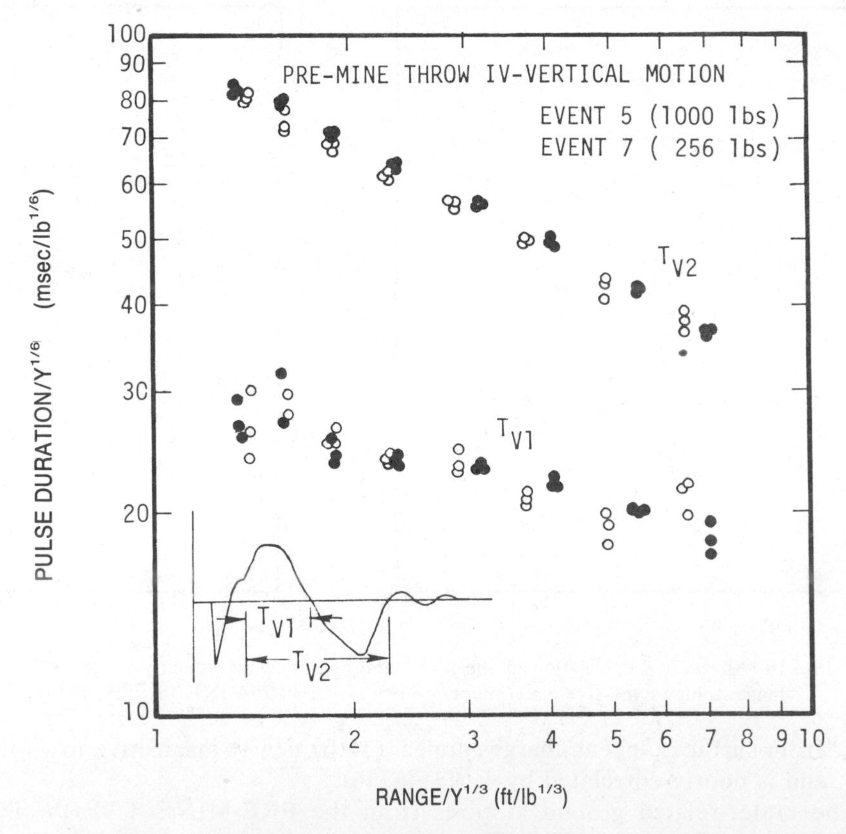

Fig. 17. Characteristic times of crater-induced vertical particle velocity waveforms from half-buried high explosive experiments.

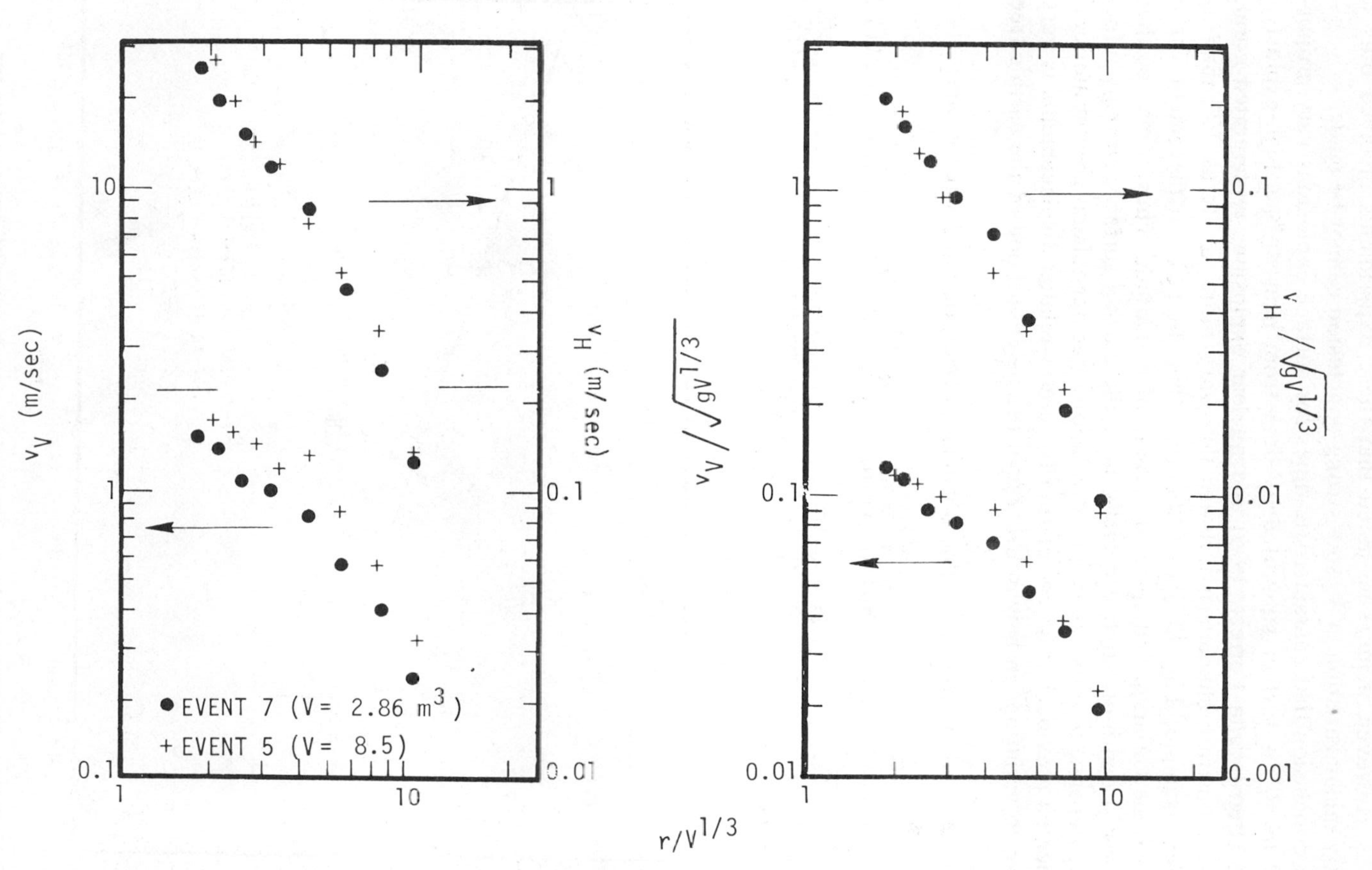

Fig. 18. Peak particle velocity from half-buried high-explosive tests in dry playa (PRE-MINE THROW IV).

$t\sqrt{g/V^{1/3}}$. However, it must be noted that these experiments involved only a relatively minor variation in V, so a strong conclusion cannot be made.

The conclusion that characteristic times for surface explosions vary proportionally to $V^{1/6}$ is further supported by data from near-surface high-explosive sources (Ferrito and Forrest, 1975) and nuclear explosions in saturated coral, which demonstrate characteristic times that vary more nearly as $V^{1/6}$ than as $V^{1/3}$. For example, Fig. 19 shows the rise time to peak displacement for a crater volume variation that spans six orders of magnitude. Thus, it may be that variations in depth-of-burst, especially near the earth's surface, strongly affect the time-dependent variables in the correlation of crater-related ground motions. As indicated previously, the correlation for crater-related displacements is very sensitive to variations in height-of-burst in the neighborhood of the earth's free surface.

Equivalent alternative formulae for peak acceleration and stresses are:

$$\frac{a}{g} = f\left(\frac{h}{V^{1/3}}, \frac{z}{V^{1/3}}, \frac{r}{V^{1/3}}, \frac{gV^{1/3}}{c^2}\right); \quad (13a)$$

$$\frac{aV^{1/3}}{c^2} = f\left(\frac{h}{V^{1/3}}, \frac{z}{V^{1/3}}, \frac{r}{V^{1/3}}, \frac{gV^{1/3}}{c^2}\right); \quad (13b)$$

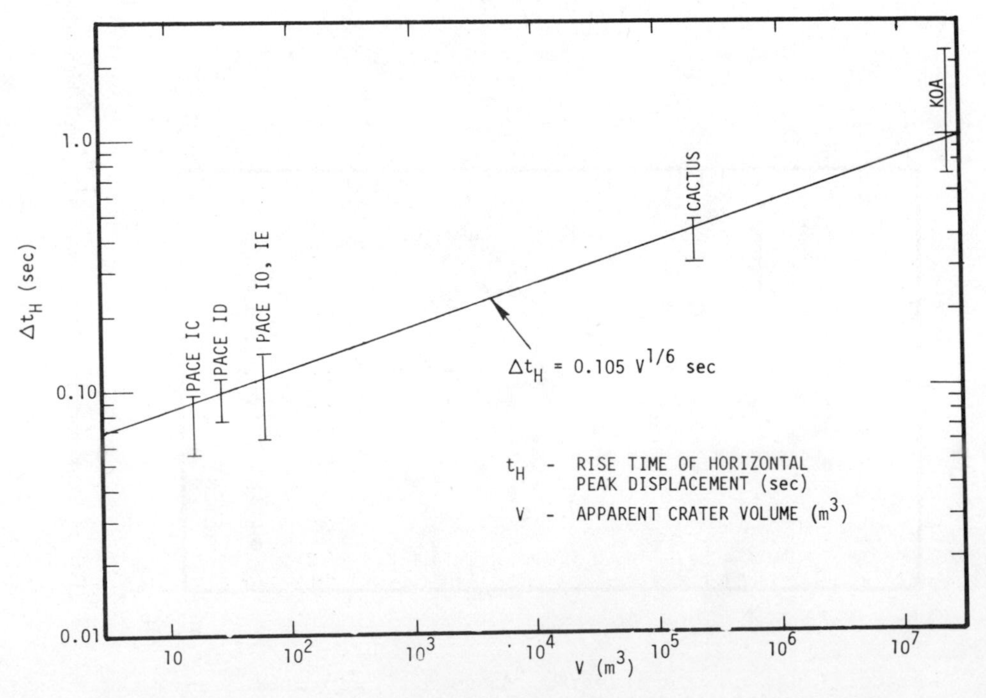

Fig. 19. Peak horizontal displacement rise time as a function of crater volume for pacific coral sand.

and

$$\frac{\sigma}{\rho c^2} = f\left(\frac{h}{V^{1/3}}, \frac{z}{V^{1/3}}, \frac{r}{V^{1/3}}, \frac{gV^{1/3}}{c^2}\right); \tag{14a}$$

$$\frac{\sigma}{\rho g V^{1/3}} = f\left(\frac{h}{V^{1/3}}, \frac{z}{V^{1/3}}, \frac{r}{V^{1/3}}, \frac{gV^{1/3}}{c^2}\right). \tag{14b}$$

Unfortunately, there are little or no data to evaluate these expressions for crater-related peak accelerations and stresses.

5. Summary and Concluding Comments

The low-frequency, near-surface ground motion data from a large number of cratering explosions in a variety of media suggest that the cube-root of the crater volume, $V^{1/3}$, may be a characteristic length common to the crater-forming process and the subsequent crater-related ground motions. The use of $V^{1/3}$ as a characteristic length integrates the effects of source parameters and geologic properties such that the peak transient particle displacements for various geologies can be correlated independently of source parameters and material properties. Illustrations of the crater volume correlation for peak displacements are given in Figs. 5 and 9 (similar sources, different media), Figs. 10 and 11 (similar sources and media, variation in V), and Fig. 8 (variety of sources and media).

For more or less uniform media, the near-surface peak transient particle displacement data are reasonably well correlated by a relationship of the form given by Eq. (3). In particular, for $r/V^{1/3} \leqslant 5$ and $z/V^{1/3} \leqslant 0.1$, approximate fits to the logarithmic mean of the data scatter are

$$\begin{aligned} d_V/V^{1/3} \simeq d_H/V^{1/3} &\simeq 0.45\,(V^{1/3}/r)^3 \quad \text{for } h/V^{1/3} > 0, \\ &\simeq 0.25\,(V^{1/3}/r)^3 \quad \text{for } h/V^{1/3} \simeq 0, \\ &\simeq 0.10\,(V^{1/3}/r)^3 \quad \text{for } h/V^{1/3} < 0, \end{aligned} \tag{15}$$

where d_V and d_H are peak transient upward and outward particle displacements, z is the depth of the observation point, r is the range from ground zero and h is the height-of-burst of the explosive source. Equation (15) should be taken as a best estimate for the peak particle displacements, and variations from these best estimates (data scatter) of up to a factor of three, even on a given experiment, would not be surprising. Permanent particle displacements vary considerably (and are sometimes downward and inward), but on the average are expected to be approximately half of the peak transient displacements (Cooper, 1971). Since the phenomena associated with impact cratering events are most like those for shallow buried explosions, the last two equations of Eq. (15) might be expected to provide good bounding estimates of the surface motions produced by meteor impacts.

Equation (15) appears to correlate reasonably well the data from all of the previous explosive cratering experiments except MIDDLE GUST III and MIDDLE GUST II, which involved detonations of 100-ton spheres of TNT at one and two charge radii above a wet clay/shale layered geology (Murrell, 1973; Pozega, 1973; Jaramillo, 1973). In these events, the observed peak transient particle displacements were systematically a factor of 2 to 3 lower than would be predicted by the first equation of Eq. (15) (see Fig. 8). On the other hand, the data from MIDDLE GUST I, which involved the detonation of a 20-ton half-buried sphere of TNT in the same wet clay/shale geology, were reasonably consistent with the second equation of Eq. (15) (see Fig. 9). Why there was such a small height-of-burst effect in the particular geology is not understood. It should be noted that the data from other experiments involving above-surface explosions over layered media (MIDDLE GUST IV in dry clay/shale and MIXED COMPANY 3 in dry sand/sandstone) are reasonably correlated by the first equation in Eq. (15). Furthermore, the data from other sites involving soils with shallow water tables (e.g., PRAIRIE FLAT, DIAL PACK, and DISTANT PLAIN 6) are consistent with Eq. (15). Thus, other tests reasonably represented by Eq. (15) were in layered media and involved wet geologies.

The crater-related ground motion data from buried explosions in dry alluvium can be collapsed reasonably well by using the consistent set of nondimensional parameters $d/V^{1/3}$, $r/V^{1/3}$, v/c and $ct/V^{1/3}$. This result appears to be independent of the Froude number $gV^{1/3}/c^2$. On the other hand, characteristic times associated with the crater-related ground motions from surface tangent and half-buried explosions in dry playa vary more nearly as $V^{1/6}$ than as $V^{1/3}$. Airblast-related effects alter the waveforms for the surface tangent events in such noncohesive materials so that the relative merits of v/c versus $v/\sqrt{gV^{1/3}}$ nondimensional velocities cannot be evaluated. (Data from cohesive soils and rocks may not be as significantly affected by the airblast, but were not considered here.) The airblast-related effects on the crater-related motions are less severe for the half-buried charges, and $d/V^{1/3}$, $r/V^{1/3}$, $v/\sqrt{gV^{1/3}}$ and $t\sqrt{g/V^{1/3}}$ appear to be a consistent set of nondimensional parameters that collapses the data in a manner independent of the Froude number. However, the variation in V was so small (only a factor of three) that no conclusion can be made with confidence.

Presumably, impact cratering events are more like shallow than deeply buried explosions. Gault (1976) has observed that the time of impact crater formation in laboratory impact experiments in dry sand varies approximately as $E^{1/6}$ where E is the kinetic energy of the impacting projectile. Assuming that V is proportional to E, this observation is consistent with our observations that characteristic times for the crater-related ground motions from near-surface explosions vary as $V^{1/6}$.

If for near-surface explosions the particle displacements and times vary as $V^{1/3}$ and $V^{1/6}$ respectively, then peak particle velocities would be expected to vary as $V^{1/6}$. Although this expectation cannot be convincingly confirmed by the data considered here, it is consistent with observations by Wisotski (1976) of the variation in the initial velocities of ejecta from surface tangent and half buried, high-explosive cratering events with yields ranging from 0.5 to 100 tons. Note that this observation is not consistent with the usual assumption that the initial ejecta velocities at geometrically similar points of origin are independent of explosive yield. On the other hand, it is consistent with observations by Post (1974) that ejecta depths normalized by $V^{1/3}$ from a wide range of crater volumes are collapsed when plotted versus $r/V^{1/3}$. (The distance ejecta travels (x) varies as v^2/g. Thus, if v varies as $V^{1/6}$, x will vary as $V^{1/3}$, and the debris depth will vary as $V^{1/3}$.)

It might be noted that if the crater volume is proportional to the explosive yield (Y), our speculations for surface burst explosions would suggest that lengths scale as $Y^{1/3}$ and velocities and times scale as $Y^{1/6}$. Although such a scaling rule is uncommon, it is recognized as the mass-gravity scaling alternative discussed by Chabai (1965).

The difficulties of developing rigorous scaling relations for explosions in a fixed gravitational field are well understood (e.g., Crowley, 1970; White, 1971; Saxe and DelManzo, 1970), therefore, we have deliberately used the terms "correlate" and "vary" rather than the term "scale" to discuss observed dimensionally consistent relationships between displacements, particle velocities, distances, times, and the crater volume. Even though the usual methods of dimensional analysis (Section 4) provide dimensionless groups of the parameters under study, we only used these dimensionless groups to guide in the correlation of ground motion data, and there is no proof that they are fundamental or complete. Whether or not the empirical correlations hold beyond the range of the data on which they are based can only be evaluated by additional experiments—or credible first-principle calculations. (In developing first-principle calculations for predicting crater-related phenomena beyond the range of the existing data, it would be helpful if they were first shown to be consistent with correlations of the type presented here.) Based on theoretical studies of the late-stage cratering phenomena by Maxwell and Seifert (1975), it does appear that features of the correlations suggested here are consistent with qualitative and quantitative features of previous calculations of cratering phenomena.

This paper focused on close-in crater-related ground motion phenomena, and the suggested correlation formulae are restricted to ranges less than about 5 $V^{1/3}$. At greater ranges, different attenuation rates and velocity/time correlations exist. In particular, in the outrunning region, near-surface ground motion waveforms have been observed by Ingram *et al.* (1975) to be relatively invariant for cratering and noncratering explosions.

References

Bratton, J. L.: 1973, Middle Gust-Mixed Company Comparisons, *Proceedings of the Mixed Company/Middle Gust Results Meeting* 13–15 *March* 1973, DNA 3151P2, C. E. Tempo, DASIAC, Santa Barbara, California.

Bratton, J. L.: 1974, The Effect of Subsurface Layers on the Simulation of Shock Waves in the Ground, *Proceedings, Fourth International Symposium on the Military Applications of Blast Simulation*, Southend-on-Sea, England.

Bridgman, P.: 1937, *Dimensional Analysis*, Yale University Press, New Haven, Connecticut.

Brode, H. L.: 1968, Review of Nuclear Weapons Effects, *Ann. Rev. of Nuc. Sci.*, **18**, 153–202.

Chabai, A. J.: 1965, On Scaling Dimensions of Craters Produced by Buried Explosions, *J. Geophys. Res.*, **70**, No. 20, p. 5075–5098.

Chisolm, S. P.: 1975, *Middle North Series, Mixed Company Event, Airblast and Ground Motion Measurements*, Air Force Weapons Laboratory, POR6630, Kirtland Air Force Base, New Mexico.
Cooper, H. F.: 1971, *On Crater-Induced Ground Motions from Near-Surface Bursts*, Air Force Weapons Laboratory, Kirtland Air Force Base, New Mexico, unpublished report.
Cooper, H. F.: 1972a, *A Review of Ground Motions from Several Russian High Explosive Cratering Experiments*, R & D Associates Marina del Rey, California, RDA-TR-062-DNA.
Cooper, H. F.: 1972b, *On Strong-Motion Seismic Ground Motions from Nuclear Airburst Explosives*, R & D Associates, Marina del Rey, California, RDA-TR-063-DNA.
Cooper, H. F.: 1972c, *Some Comments on Seismic Data from Underground Explosions at the Nevada Test Site*, R & D Associates, Marina del Rey, California, RDA-TR-077-DNA.
Cooper, H. F.: 1973, *Empirical Studies of Ground Shock and Strong Motions in Rock*, R & D Associates, Marina del Rey, California DNA 3245F.
Crawford, R. E., Higgins, C. J., and Bultman, E. H.: 1974, *The Air Force Manual for Design and Analysis of Hardened Structures*, Air Force Weapons Laboratory, Kirtland Air Force Base, New Mexico, AFWL-TR-72-102.
Crowley, B. K.: 1970, Scaling for Rock Dynamic Experiments, *Proceedings, Symposium on Engineering with Nuclear Explosives*, Las Vegas, Nevada, p. 545–559.
Ferrito, J. M. and Forrest, J. B.: 1975, *Ground Motions from Pacific Cratering Experiments*, 1000 *lb Explosive Shots*, NCEL-TR-R808, Naval Civil Engineering Laboratory, Port Hueneme, California.
Gault, D. E.: 1976, NASA-Ames Research Center, Moffett Field, California, personal communication.
Hoffman, H. V., Sauer, F. M., and Barclay, B.: 1971, *Operation Prairie Flat, Project LN-308, Strong Ground Shock Measurements*, POR-2108, Stanford Research Institute, Menlo Park, California.
Ingram, J. K., Drake, J., and Ingram, L.: 1975, *Influence of Burst Position on Airblast Ground Shock, and Cratering in Sandstone*, Miscellaneous Paper N-75-3, U.S. Army Engineer Waterways Experiment Station, Vicksburg, Mississippi.
Ingram, J. K.: 1975, *Middle North Series, Mixed Company Event, Ground Shock from a 500-ton High-Explosive Detonation on Soil over Sandstone*, U.S. Army Waterways Experiment Station, POR6613, Vicksburg, Mississippi.
Jaramillo, E. E.: 1973, *Free-Field Data Report Middle Gust III*, EG&G, Albuquerque, New Mexico, AL-831-3.
Joachim, C. E.: 1973, *MINESHAFT Series, Events Mine Under and Mine Ore*; *Subtask ss222, Ground Motion and Stress Measurements*, Technical Report N-72-1, U.S. Army Engineer Waterways Station, Vicksburg, Mississippi.
Kuhl, A. L. and Wuerker, R. F.: 1974, *Laser Cratering Research Technique*, TRW Systems Group, Redondo Beach, California, DNA 3442F.
Lockard, D. M.: 1974, *Crater Parameters and Material Properties*, Air Force Weapons Laboratory, Kirtland Air Force Base, New Mexico, AFWL-TR-74-200, October 1974.
Maxwell, D. and Seifert, K.: 1975, *Modeling of Cratering, Close-In Displacements and Ejecta*, Physics International Company, San Leandro, California, DNA 3628F.
Murrell, D. W.: 1970, *Distant Plain 6 and 1A, Project 3.02a, Earth Motion and Stress Measurements*, TR-N-70-14, U.S. Army Waterways Experiment Station, Vicksburg, Mississippi.
Murrell, D. W.: 1972, *Operation Prairie Flat, Project LN 302: Earth Motion and Stress Measurements in the Outrunning Region*, TR-N-72-2, U.S. Army Waterways Experiment Station, Vicksburg, Mississippi.
Murrell, D. W. and Carleton, H. D.: 1973, *Operation MINESHAFT: Ground Shock from Underground and Surface Explosions in Granite*, MS-2159 and MS2160, U.S. Army Waterways Experiment Station, Vicksburg, Mississippi.
Oberbeck, V. R.: 1971, Laboratory Simulation of Impact Cratering with High Explosives, *J. Geophys. Res.* **76**, 5732–5749.
Perret, W. R., Chabai, A. J., Reed, J. W., and Vortman, L. J.: 1963, *Project SCOOTER*, Sandia Corporation, Albuquerque, N.M., SC-4602 (RR).
Post, R. L.: 1974, *Ejecta Distributions from Near-Surface Nuclear and HE Bursts*, Air Force Weapons Laboratory, Kirtland Air Force Base, New Mexico, AFWL-TR-74-51.
Pozega, R. E.: 1973, *Free-Field Data Report Middle Gust II*, EG&G, Albuquerque, New Mexico.

Rooke, A. D., Carnes, B. L., and Davis L. K.: 1974, *Cratering by Explosives: A Compendium and Analysis*, U.S. Army Waterways Experiment Station, Vicksburg, Mississippi, Technical Report N-74-1.

Sachs, D. C. and Swift, L. M.: 1955, *Small Explosion Tests, Project MOLE*, Stanford Research Institute, Menlo Park, California, AFSWD-291, Vol. II.

Sachs, D. C. and Swift, L. M.: 1958, *Underground Explosion Effects, Operation TEAPOT*, Stanford Research Institute, Menlo Park, California, WT-1106.

Sauer, F. M.: 1975, *Nuclear Geoplosics*, Vol. IV, Defense Atomic Support Agency, DASA 1285.

Sauer, F. M. and Vincent, C. T.: 1967, *FERRIS WHEEL Series, FLAT-TOP Event, Project 1.2A.3A, Earth Motion and Pressure Histories*, POS3002 (WT-3002) Stanford Research Institute, Menlo Park, California.

Sauer, F. M.: 1970, *Summary Report on Distant Plain, Events 6 and 1A, Ground Motion Experiments*, DASA 2587, Stanford Research Institute, Menlo Park, California.

Sauer, F. M., Kochly, J., and Stubbs, T.: 1974, *Pre-Mine Throw IV Airblast and Ground Motion, Project No. MT*301, PRR-419, Physics International Company, San Leandro, California.

Sauer, F. M.: 1976, *Pre-Mine Throw IV, Technical Director's Report*, Physics International Company, San Leandro, California.

Saxe, H. E. and DelManzo, D. D.: 1970, A Study of Underground Explosion Cratering Phenomena, *Proceedings, Symposium on Engineering with Nuclear Explosives*, 1701–1725, Las Vegas, Nevada.

White, J. W.: 1971, Examination of Cratering Formulaes and Scaling Method, *J. Geophys. Res.*, **76**, 8599–8603.

Wisotski, J.: 1976, University of Denver Research Institute, Denver, Colorado, personal letter to H. F. Cooper.

Roddy, D. J., Pepin, R. O., and Merrill, R. B., editors.
(1977) *Impact and Explosion Cratering*, Pergamon Press (New York), p. 1165–1190.
Printed in the United States of America

Scaling of cratering experiments—an analytical and heuristic approach to the phenomenology

BARBARA G. KILLIAN and LAWERENCE S. GERMAIN

University of California, Los Alamos Scientific Laboratory, Los Alamos, New Mexico 87544

Abstract—The phenomenology of cratering can be thought of as consisting of two phases. The first phase, where the effects of gravity are negligible, consists of the energy source dynamically imparting its energy to the surroundings, rock and air. As illustrated in this paper, the first phase can be scaled if: radiation effects are negligible, experiments are conducted in the same rock material, time and distance use the same scaling factor, and distances scale as the cube root of the energy. The second phase of cratering consists of the rock, with its already developed velocity field, being "thrown out". It is governed by the ballistics equation, and gravity is of primary importance. This second phase of cratering is examined heuristically by examples of the ballistics equation which illustrate the basic phenomena in crater formation. When gravity becomes significant, in addition to the conditions for scaling imposed in the first phase, distances must scale inversely as the ratio of gravities. A qualitative relationship for crater radius is derived and compared with calculations and experimental data over a wide range of energy sources and gravities.

INTRODUCTION

SCALING CONCEPTS have been developed and used extensively both to design experiments and to provide an understanding of phenomenology. In the context of this discussion, the term "scaling" means the seeking of relationships among the variables such that one may extend the phenomenology of one observation to make reasonable predictions for a similar configuration where the numerical values of the variables are quite different (i.e., a different scale). For example, how does one predict the dimension of a crater formed under a gravitational field different from that of the earth? How can one use the crater produced by a few pounds of explosive to predict the properties of those produced by thousands of pounds of explosive?

A variety of *scaling criteria* has been proposed for cratering experiments conducted on earth, of which references (Nordyke, 1964; Chabai, 1959, 1965; Sedov, 1959; Pokrovskii and Fedorov, 1957; Vaile, 1961; Violet, 1961; Baker *et al.*, 1973; White, 1972) represent only a partial but representative list. Many of these proposed scaling criteria have been vigorously debated, and frequently confusion has existed about the role of gravity and the significance of the description of the rock.

The first part of this paper is a condensation of a previous paper by one of the authors (Killian *née* Crowley, 1970). This work uses an analytic analysis of the conservation equations to show that the scaling of two experiments performed in two different media is virtually impossible. However, scaling is possible between two experiments in the same material even with the realistically complicated descriptions for the behavior of the rock. It also shows that:

distances scale as the cube root of the energy yields and that dynamic scaling is valid, i.e., distance and time use the same scaling factor. When gravity becomes important, the conditions for scaling further require that distances must scale inversely as the gravity ratios. For two experiments conducted on earth, the gravity ratio would be unity. Hence, strictly speaking, only one-to-one scaling of dimensions (i.e., no dimensional reduction) is possible when gravity becomes important.

The second part of this paper examines the ballistics phase of the cratering process. This second phase begins after energy has been coupled to the rock surrounding the energy source. Generally, the effects of gravity are insignificant during the time in which energy is being transferred to the surrounding rock.

The ballistics equation is examined for selected cratering situations on Earth, Moon, Mars, and Jupiter. Using this equation together with heuristic arguments, the phenomenology of the ballistics phase of cratering is simply illustrated. The results obtained in Crowley (1970) are further demonstrated. A qualitative relationship is derived for crater radius as a function of the dimensionless ratio $(\text{velocity})^2/(\text{depth of energy source})(\text{gravity})$. By normalizing, this relationship is quantitatively compared with both calculations and a variety of experimental data for explosives buried near optimum cratering depths.

I. DERIVATION OF NECESSARY CONDITIONS FOR SCALING FROM THE CONSERVATION EQUATIONS

Conservation equations used to model rock mechanics experiments

The numerical simulation of rock mechanics experiments has successfully been used to model a variety of phenomena such as the generation of cavities and craters (Terhune and Stubbs, 1970; Butkovich, 1976; Bryan *et al.*, 1974; Burton *et al.*, 1975; Cherry, 1967; Cherry *et al.*, 1968) by explosive (both nuclear and chemical) sources and by the impact of projectiles (Wilkins, 1969). At present, the accuracy of such modeling appears to be limited by the accuracy of the description of the rock properties and the geologic description of the experimental site.

The one-dimensional SOC (Schatz, 1974) and the two-dimensional TENSOR (Maenchen and Sach, 1964) numerical rock dynamic codes have been extensively used to successfully perform modeling of experiments in rock and soil materials (papers cited in previous paragraph). In the following, an algebraic analysis is performed on the conservation equations used in SOC and TENSOR codes. This analysis is performed to obtain a set of necessary conditions which must be satisfied for similar solutions of the equations. The necessary conditions for similar solutions are also the necessary conditions for scaled experiments.

The SOC and TENSOR equations consider the same physics; their only differences are due to the differences in the expression of the generalized coordinate operators (divergence and gradient) in one-dimensional spherical (SOC) and two-dimensional cylindrical (TENSOR) geometries. These codes start with a set of initial conditions and descriptions of the rock properties and integrate, with respect to time, the conservation equations of continuum mechanics (Schatz, 1974; Maenchen and Sach, 1964). Since SOC and TENSOR are cast in the Lagrangian form, mass is implicitly conserved. The momentum equation considers both the stress tensor, which is composed of an isotropic and a deviatoric part, and gravity. Velocities obtained from the momentum equation are integrated with respect to time to obtain displacements. From the velocities and displacement, strain rates are calculated; and stresses are obtained from Hooke's Law. The energy equation considers work done by the total stress tensor. Equations of state for the pressure as a function of the compression and/or energy of a zone may be used for each material region under consideration. In addition, the deviatoric stresses are compared with an input failure surface, and stresses in zones which have "failed" are adjusted. For brevity, only the SOC equations are considered here.

Momentum:

$$\frac{\partial U}{\partial t} = -\frac{1}{\rho}\left[\frac{\partial P}{\partial R} + \frac{4}{3}\frac{\partial K}{\partial R} + \frac{4K}{R}\right] + g.$$

Displacement:

$$\frac{\partial R}{\partial t} = U.$$

Deviatoric strain rate:

$$\dot{e} = \frac{1}{2}\left[\frac{U}{R} - \frac{\partial U}{\partial R}\right].$$

Hooke's Law for deviatoric stress rate:

$$\dot{K} = \frac{\partial K}{\partial t} = \mu\left[\frac{U}{R} - \frac{\partial U}{\partial R}\right] = 2\mu\dot{e}.$$

Energy:

$$\frac{\partial \varepsilon}{\partial t} = \frac{4}{3}\frac{K\dot{K}}{\eta\mu} - \frac{P}{V^0}\frac{\partial V}{\partial t} = \frac{4}{3}\frac{K2\dot{e}}{\eta} - \frac{P}{V^0}\frac{\partial V}{\partial t}.$$

Equation of state:

$$P = a + a_1\eta + a_2\eta^2 + \cdots + b_1\varepsilon + b_2\varepsilon^2 + \cdots + c_1\eta\varepsilon + c_2\eta^2\varepsilon^2 + \cdots$$

where a, a_1, a_2, b_1, b_2, c_1, c_2—coefficients in equation of state; e—strain; g—acceleration of gravity; K—deviatoric stress; P—pressure; R—radial distance; t—time; v—velocity; V—volume; V^0—initial volume; ε—energy/initial volume; η—compression = current density/initial density; μ—rigidity modulus; ρ—density. Generally, the subscript zero (0) refers to dimensionless scaling parameters.

In SOC, any function or table of pressure versus compression and/or energy may be used to obtain the pressure. To determine failure, any table or function may likewise be used to specify the maximum deviatoric stress as a function of mean pressure. A simple polynomial function for pressure is considered here merely for illustrative purposes.

Derivation of the similar conditions

To obtain the necessary conditions for similar solutions, the SOC equations are now examined. The approach used here is more laborious but more systematic than either the Vaschy–Buckingham π (Birkhoff, 1960) or the dimensional analysis method (Rosenhead, 1963), to which it is equivalent (Crowley, 1967).

It is assumed that two experiments, denoted by a set of unprimed and primed variables respectively, can be adequately simulated by SOC and TENSOR. Dimensionless parameters which are denoted by the subscript 0 are formed for *all* of the variables as the ratio of unprimed to primed variables. For example, $R_0 = R/R'$, $t_0 = t/t'$.

Derivatives with respect to distance and time for an arbitary dependent variable ($\theta = \theta_0\theta'$) in the two experiments are related by

$$\frac{\partial\theta}{\partial R} = \frac{\partial\theta}{\partial R'}\cdot\frac{\partial R'}{\partial R} = \frac{1}{R_0}\cdot\frac{\partial\theta}{\partial R'},$$

$$\frac{\partial\theta}{\partial t} = \frac{\partial\theta}{\partial t'}\cdot\frac{\partial t'}{\partial t} = \frac{1}{t_0}\frac{\partial\theta}{\partial t'}.$$

With these relationships, an unprimed SOC momentum equation may be written in terms of a primed momentum equation:

$$\frac{\partial U'}{\partial t'} = \frac{-t_0}{U_0\rho_0}\frac{1}{\rho'}\left[\frac{P_0}{R_0}\frac{\partial P'}{\partial R'} + \frac{K_0}{R_0}\frac{4}{3}\frac{\partial K'}{\partial R'} + \frac{K_0}{R_0}\frac{4K'}{R'}\right] + \frac{t_0 g_0}{U_0}g'.$$

If

$$\frac{t_0}{U_0\rho_0}\frac{P_0}{R_0} = 1, \quad \frac{t_0 K_0}{U_0\rho_0 R_0} = 1, \quad \text{and} \quad \frac{t_0 g_0}{U_0} = 1,$$

the primed and unprimed momentum equations are similar. The SOC momentum equation thus imposes these three conditions on the dimensionless scaling parameters for the two experiments.

Applying the same technique to the remaining SOC equations, a set of necessary conditions for similar solutions is obtained. With minor algebraic manipulations and omitting redundant conditions, the following complete set of seven necessary conditions is obtained.

$$\frac{t_0 P_0}{U_0\rho_0 R_0} = 1. \qquad (1)$$

$P_0 = K_0 = \mu_0$ All "pressure-like" variables must use the same scaling factor. (2)

$$\frac{t_0 g_0}{U_0} = 1. \tag{3}$$

$$\frac{t_0 U_0}{R_0} = 1. \tag{4}$$

$$\dot{e}_0 = \frac{U_0}{R_0}. \tag{5}$$

$$\frac{P_0}{\varepsilon_0 \eta_0} = 1. \tag{6}$$

Polynomial equation of state: (7)

$$\frac{a_0}{P_0} = 1, \quad \frac{a_{10}\eta_0}{P_0} = 1, \quad \frac{a_{20}\eta_0^2}{P_0} = 1, \quad \text{etc.}$$

$$\frac{b_{10}\varepsilon_0}{P_0} = 1, \quad \frac{b_{20}\varepsilon_0^2}{P_0} = 1, \quad \text{etc.}$$

$$\frac{c_{10}\eta_0\varepsilon_0}{P_0} = 1, \quad \frac{c_{20}\eta_0\varepsilon_0^2}{P_0} = 1, \quad \text{etc.}$$

The two-dimensional TENSOR equations introduce the following conditions: (1) the R and Z (radial and axial) distances must use the same scaling factor; (2) the radial and axial velocities must use the same scaling factor; (3) all strains scale the same; and (4) all stresses scale the same. The complete three-dimensional equations impose the condition that the same scaling factors must be used for all three dimensions.

Scaling experiments in the same solid material

The necessary conditions (7) impose such stringent conditions on the ratios of equation-of-state parameters that scaling in two different solid materials is nearly impossible. However, if the two experiments are performed in the *same* material, a very general equation of state imposes no restrictions in addition to those imposed by an unrealistically simple equation of state. These conditions are: $\eta_0 = P_0 = \varepsilon_0 = 1$.

When the same material is being considered, $\eta_0 = 1$ implies the $\rho_0 = 1$. Then, using $t_0/R_0 = 1/U_0$ from condition (4) in condition (1) gives $U_0^2 = 1$, i.e., velocities must be the same. This criterion is physically reasonable if one considers that P and ρU^2 are both energies per unit volume. The strain rate is a change in length per unit length per unit time, $\dot{e}_0 = R_0/R_0 t_0 = 1/t_0$. Using $\dot{e}_0 = 1/t_0$ and $\eta_0 = 1$, conditions (5) and (1) are essentially the same.

Thus, for two experiments in the *same* material, the seven necessary conditions may be written as:

$$\begin{aligned} &P_0 = \eta_0 = \varepsilon_0 = U_0 = K_0 = \mu_0 = 1, \\ &R_0 = t_0, \\ &t_0 g_0 = 1 \quad \text{or} \quad R_0 g_0 = 1. \end{aligned} \tag{8}$$

The condition $R_0 = t_0$ means time and distance must scale the same, and is sometimes referred to as dynamic scaling.

When the effects of gravity are insignificant, the necessary conditions $t_0 g_0 = 1$ and $R_0 g_0 = 1$ can be neglected. However, when the effects of gravity become significant, $t_0 g_0 = 1$ and $R_0 g_0 = 1$ must be satisfied. If the two experiments are conducted on earth, $g_0 = 1$ requires that both $t_0 = 1$ and $R_0 = 1$. However, $R_0 = 1$ means that dimensions in the two experiments are identical; and *scaled experiments are not possible when gravity becomes significant.* This lack of dynamic similarity when gravity becomes significant on earth was first demonstrated by Galilei (1953).

In a manner similar to the one used to obtain Eq. (8), it can be demonstrated that when radiation is significant, radiation mean free paths and dimensions must use the *same* scaling factor (Pai, 1966). Thus, when radiation is significant, dynamic scaling is nearly impossible. For explosions at a burial depth within the radius of vaporization, about 2.2 m per $(\text{kiloton})^{1/3}$ of energy (Butkovich, 1967), radiation energy is contained within the vaporized rock. However, for shallower depths of burial, where the mean free path in air and rock are both important, dynamic scaling will be nearly impossible. Considerations such as these as well as "effective" depths of burial will be of importance in applying dynamic scaling to impact craters.

In Crowley (1970), results are reported from numerical calculations of 1 kiloton and 1 megaton energy sources at various depths of burial in cases with and without gravity. These calculations demonstrate the degeneracy of cube root dynamic scaling when gravity becomes significant. The megaton and the kiloton calculations without gravity all gave identical results when distances and times of the megaton were divided by 10. The calculations also illustrate that at explosion depths of burial near optimum for cratering (about 50 m for 1 kiloton), the effects of gravity are negligible while the stress wave is traveling to the surface.

The previously reported calculations were performed with a fairly realistic model for the rock material. It considered pressure-density loading and unloading descriptions and rock failure which included a time dependent adjustment of the principal stresses. One of the most significant results of these calculations is the demonstration that dynamic scaling holds even with such a complex description of the material. When models of rate effects and other sophistications in rock mechanics become available in the numerical codes, future investigations of their effect on dynamic scaling would be desirable.

II. HEURISTIC DESCRIPTION OF THE BALLISTICS, OR "THROWOUT" PHASE OF CRATERING

The ballistic equation

Suppose that an explosive point source is detonated in rock a distance D_* below the surface. As the stress wave propagates toward the surface, particle velocities directed radially from the source are imparted to the surrounding rock. Generally, the peak particle velocity v imparted to the surrounding rock by the stress wave at a distance D from the source can be expressed as (Crowley, 1970)

$$v = v_*\left(\frac{D}{D_*}\right)^{-n} \quad (9)$$

where v_* and D_* are a reference velocity and distance, and "n" is of the order of 2. The reference velocity, v_*, can be taken to be as the velocity of a particle at a distance D_*, directly above the source. Then, $v_i = v_*(D_i/D_*)^{-2}$ gives the velocities of other particles that are initially located a radial slant distance D_i from the source, Fig. 1a. Since $D_* = D_i \sin\theta$,

$$v_i = v_* \sin^2\theta. \quad (10)$$

From elementary ballistics, the initial velocity of a particle can be represented as separate components along the axes, Fig. 1b. One can calculate the ballistic range of the particle (x_{max}) from elementary ballistics assuming no loss of energy by the particle and a flat surface.

$$x_{max} = \frac{2V_*^2 \sin^5\theta \cos\theta}{g}. \quad (11)$$

The final position of a particle in the x direction, x_{final}, equals its initial position, x_i, plus its ballistic range, x_{max}:

$$x_{final} = x_i + x_{max}.$$

From Fig. 5a and $D_* = D_*\sin\theta$

$$x_i = D_*\frac{\cos\theta}{\sin\theta}$$

thus,

$$x_{final} = D_*\frac{\cos\theta}{\sin\theta} + \frac{v_*^2\, 2\sin^5\theta\cos\theta}{g}. \quad (12)$$

To illustrate the x_i and x_{max} functions, consider the dimensionless quantity which represents the horizontal distance

$$\frac{x_{final}}{D_*} = \frac{\cos\theta}{\sin\theta} + \frac{v_*^2}{D_*g}\, 2\sin^5\theta\cos = \frac{x_i}{D_*} + \frac{x_{max}}{D_*}. \quad (13)$$

Figure 2 shows x_i/D_* and x_{max}/D_* plotted versus θ for various values of v_*^2/D_*g.

The initial position, x_i/D_* is a monotonically increasing function. However, the ballistic range function x_{max}/D_* which is dependent on g has a maximum when $\theta = 65.9°$ for all values of v_*^2/D_*g. Assuming no energy loss, the maximum range of a projectile in a gravitational field occurs when its initial velocity is oriented at 45° with the horizontal. In the second term of Eq. (13) the rapidly decreasing velocity distribution which is assumed ($n = 2$) has been taken into account. The magnitude of the velocity at angles below 65.9° is too small to achieve maximum range.

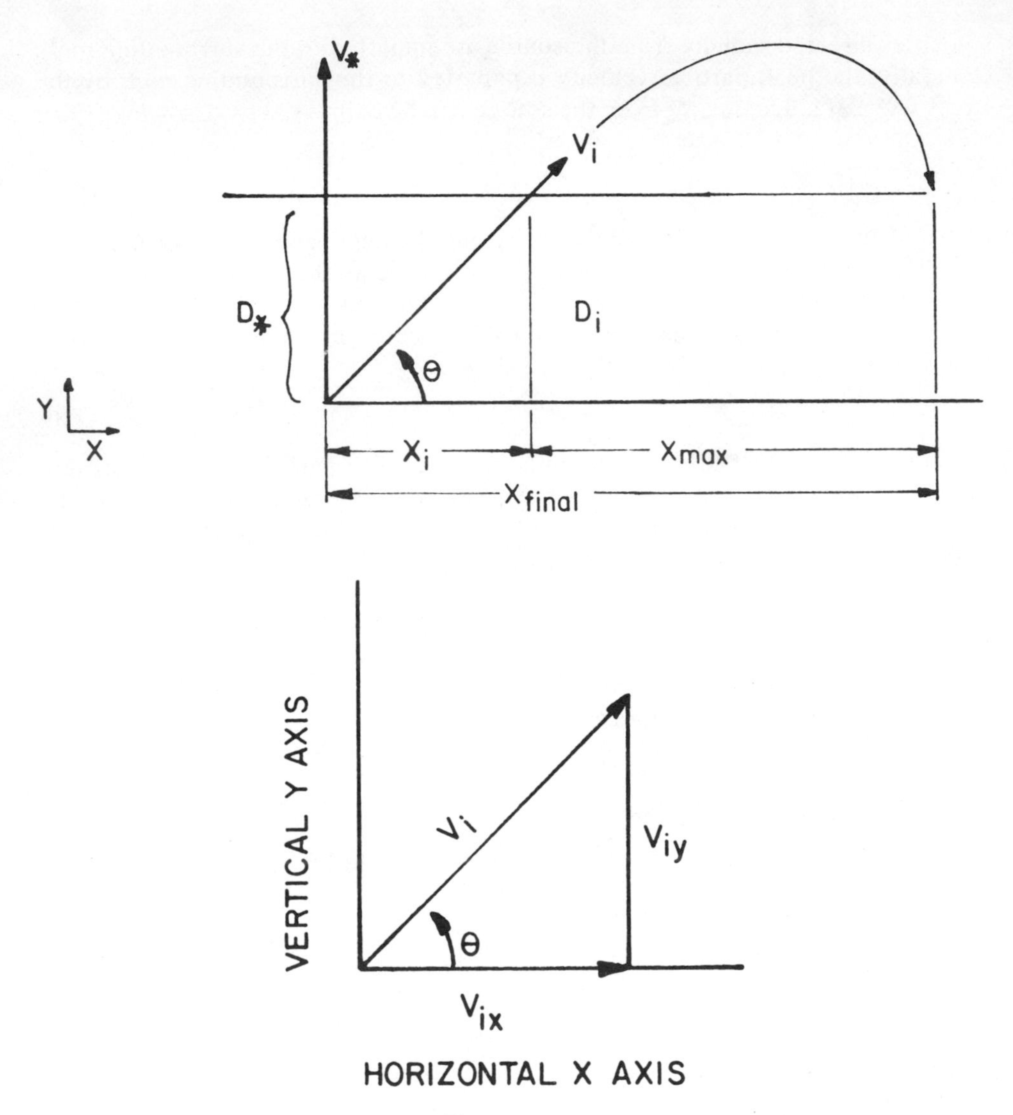

Fig. 1

Different realistically assumed values of n between 1.5 and 2.5 for different materials would slightly change this result.

The initial position and maximum range functions are added to obtain x_{final}/D_*, as shown in Fig. 3. Note that the initial position curve induces a minima in most of the x_{final}/D_* curves. As illustrated in the following, this minima can be related to the crater radius.

The factor v_*^2/D_*g can be thought of as a measure of the kinetic energy of the ballistic projectiles which are forming a crater. The quantity v_*^2 is indicative of the kinetic energy of the ground in the preballistic state and is related to the

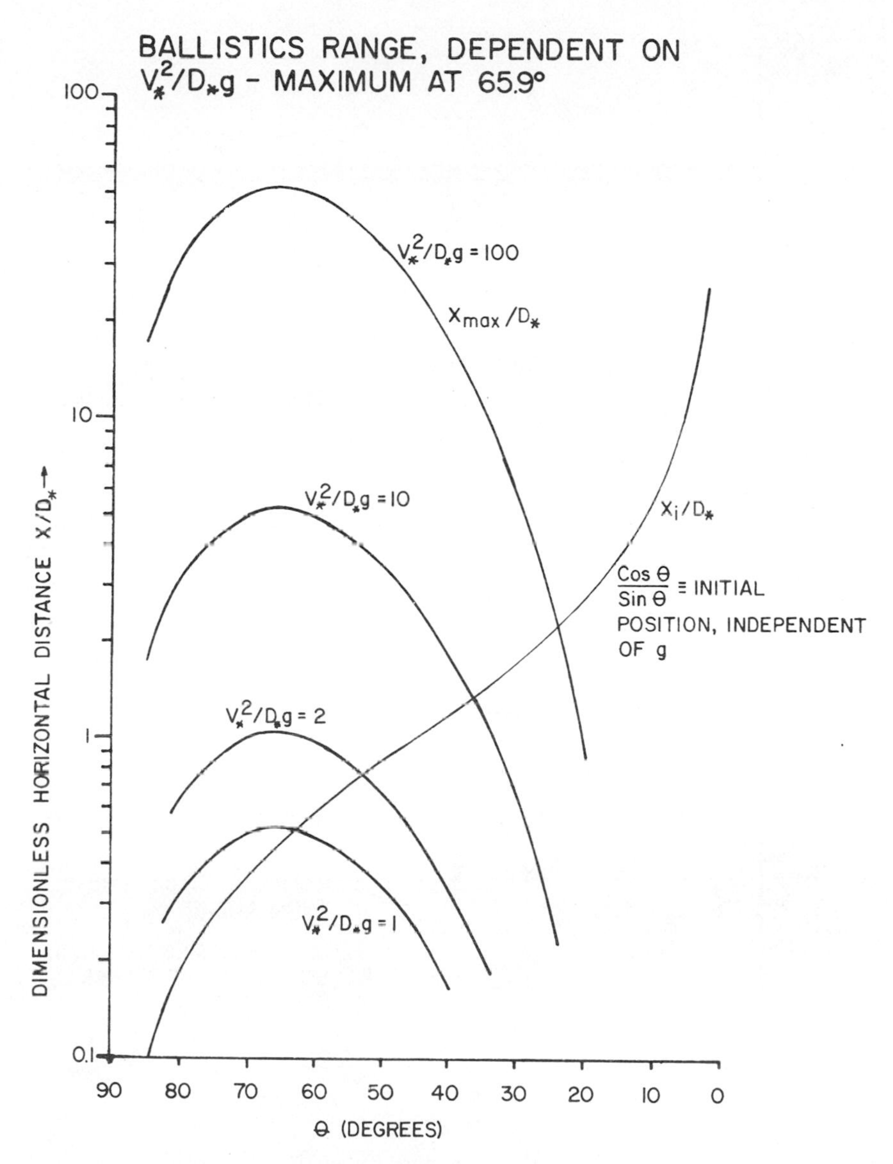

Fig. 2 Ballistics range, dependent on v_*^2/D_*g—maximum at 65.9°.

particle velocity imparted to the ground by the outgoing stress wave. A concise definition of v_* is not required by this heuristic approach since v_* is to be used as a normalizing quantity, and variations with respect to yield and depth of burial are to be recognized as illustrated in the following. Our intent here is to show that v_*^2/D_*g for a set of experiments in the same material is related to the crater radius in a systemtic manner. Thus, we intend to show that when a normalized value of v_* is obtained from one experiment, crater radii for other experiments

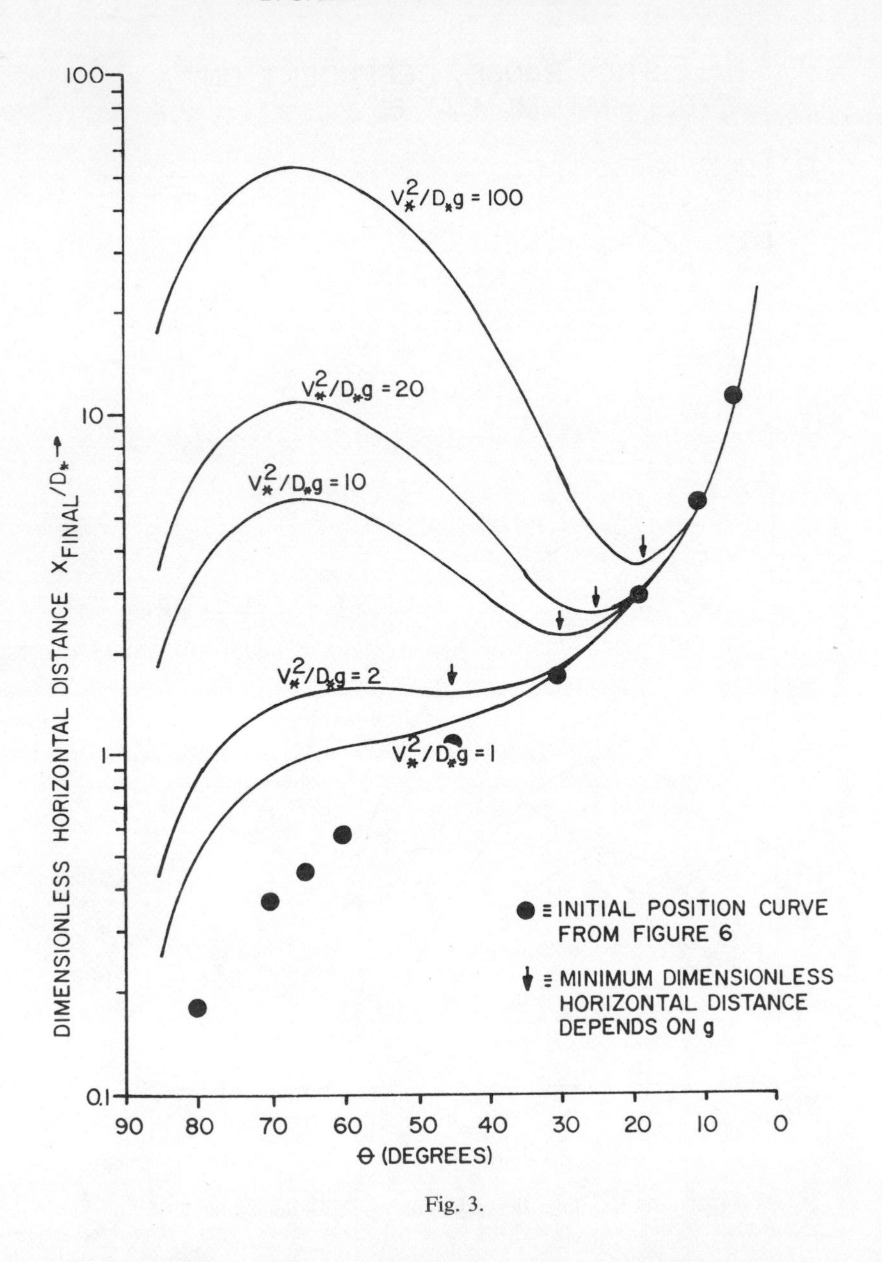

Fig. 3.

in the same material but with different yields or depths of burial can be predicted. This is not a rigorous approach, but it can be a useful one.

Examples of cratering on Moon, Mars, Earth, and Jupiter

The first set of examples uses a 20-ton point source, with $D_* = 12$ m as depicted in Fig. 4. This same point source (same yield) is considered to be

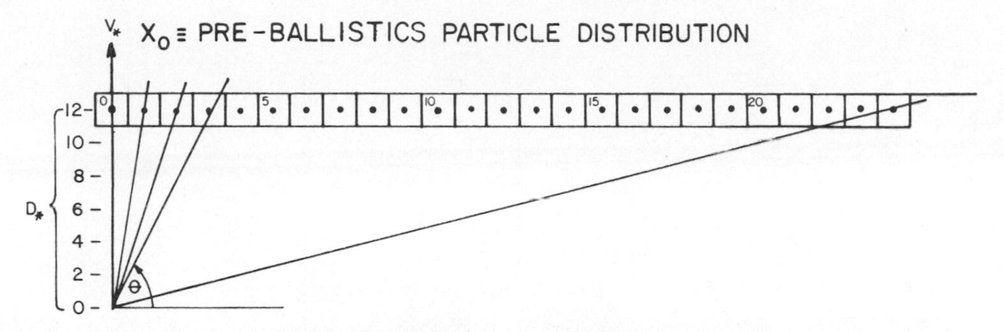

Fig. 4a. $x_0 \equiv$ pre-ballistics particle distribution.

detonated the same distance below the surface on: Moon, $g = 1.62\ \text{m/sec}^2$; Mars, $g = 3.73\ \text{m/sec}^2$; Earth, $g = 9.8\ \text{m/sec}^2$; and Jupiter, $g = 26.5\ \text{m/sec}^2$. It is assumed that the same rock material exists in all cases. As demonstrated in the last section, the velocity fields in the four cases are identical as long as gravity is insignificant. At such a shallow depth for the source, the maximum kinetic energy is imparted to the overburden before gravity becomes significant (Bryan *et al.*, 1974). Therefore, the initial conditions for the ballistic phase of cratering in the four cases are identical with v_* taken to be 30 m/sec. (This value was chosen primarily for illustrative purposes, but it represents a fairly realistic average value for the velocity of the ground for a 20-ton source.) (Bryan *et al.*, 1976).

A set of 2 m "particles" is selected (see Fig. 4). The initial position of the center of each particle, x_i, is given in Table 1 along with its initial speed, v_i, and its initial angle, θ. Calculated values of x_{max} and x_{final} for Moon, Mars, Earth, and Jupiter are also given in Table 1. Fig. 4b schematically indicates the positions of x_{final} for each particle.

As indicated in Fig. 4b, particle 0 has no horizontal component of velocity and simply lands at zero range. Particles 1, 2, and 3 land at successively larger ranges. The maximum range of particles, MR, is near particle 3 (maximum range occurs for a particle with $\theta \sim 65.9°$). Particles initially beyond 3 land at smaller horizontal ranges until a minimum range is achieved. Ranges then increase again, but the final position of the higher numbered particles is mostly influenced by their initial position. Note, that in the four cases, the particle with the minimum range varies; and in going from Moon to Jupiter, the order of particles is not preserved in the region of the minimum. This lack of order tells us that these experiments are not scaling.

Some information from Fig. 4b which will be utilized in Fig. 5 is shown in Table 2.

In Fig. 4b, note the pileup of particles around the minimum range. In the following, the minimum range is assumed to be associated with the radius of the crater. The variation in minimum range between the four cases shown in Fig. 4b and Table 2 should correspond to variations in the final crater size. That is,

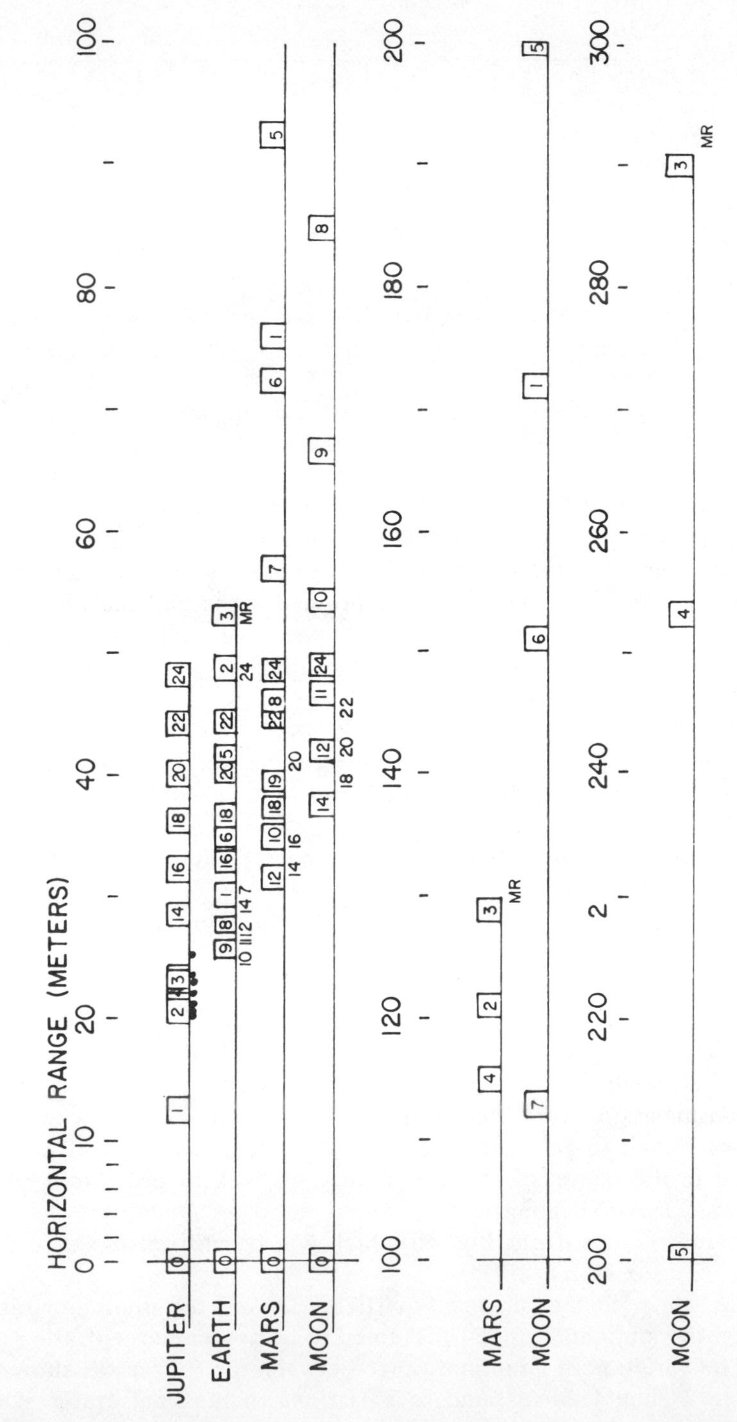

Fig. 4b. $x_{final} \equiv$ post-ballistics final particle distribution.

Table 1. Properties of the particles illustrated in Fig. 5.

Particle	x_i[1] (m)	θ	v_i[2]	x_{max}[3] (m) Moon[4]	Mars[4]	Earth[4]	Jupiter[4]	$x_{final} = x_i + x_{max}$ Moon	Mars	Earth	Jupiter
0	0	90.0	v_*	0	0	0	0	0	0	0	0
1	2	80.5	$.968v_*$	170	74.0	28.2	10.4	172	76	30.2	12.4
2	4	71.6	$.900v_*$	270	117	44.6	16.5	274	121	48.6	20.5
3	6	63.4	$.800v_*$	284	123	47.0	17.4	290	129	53.0	23.4
4	8	56.3	$.692v_*$	245	107	40.6	15.0	253	115	48.6	23.0
5	10	50.2	$.592v_*$	190	82.4	31.4	11.6	200	92.4	41.4	21.6
6	12	45.0	$.500v_*$	139	60.3	23.0	8.5	151	72.3	35.0	20.5
7	14	40.6	$.424v_*$	99	42.9	16.3	6.1	113	56.9	30.3	20.1
8	16	36.9	$.360v_*$	68.8	29.9	11.4	4.2	84.8	45.9	27.4	20.2
9	18	33.7	$.308v_*$	48.6	21.1	8.0	3.0	66.6	39.1	26.0	21.0
10	20	31.0	$.265v_*$	34.3	14.9	5.7	2.1	54.3	34.9	25.7	22.1
11	22	28.6	$.229v_*$	24.5	10.7	4.1	1.5	46.5	32.7	26.1	23.5
12	24	26.6	$.200v_*$	17.8	7.7	2.9	1.1	41.8	31.7	26.9	25.1
14	28	23.2	$.1551v_*$	9.7	4.2	1.6	.6	37.7	32.2	29.6	28.6
16	32	20.6	$.1233v_*$	5.6	2.4	.9	.3	37.6	34.4	32.9	32.3
18	36	18.4	$.100v_*$	3.3	1.4	.6	.2	39.3	37.4	36.6	36.2
20	40	16.7	$.0825v_*$	2.1	.9	.3	.1	42.1	40.9	40.3	40.1
22	44	15.3	$.0692v_*$	1.3	.6	.2	.1	45.3	44.6	44.2	44.1
24	48	14.0	$.0588v_*$	.9	.4	.1	.1	48.9	48.4	48.1	48.1
MR[5]	5.37	65.9	$.8334v_*$	287	125	47.5	17.6	292	130	52.9	23.0

[1] x_i = initial distance measured along surface.

[2] $v_i = v_* \sin^2\theta \equiv$ initial speed of particle, $v_* = 30$ m/s.

[3] $x_{max} = \frac{v_*^2\, 2\sin^5\theta\cos\theta}{g}$.

[4] g moon = 1.62 m/s² $v_*^2/g = 555$
g mars = 3.73 m/s² $v_*^2/g = 241$
g earth = 9.8 m/s² $v_*^2/g = 91.8$
g jupiter = 26.5 m/s² $v_*^2/g = 34.0$

[5] A hypothetical particle with $\theta = 65.9°$ which would travel to "Maximum Range" by ballistics throw-out.

minimum range plotted versus gravity should be a curve similar to final crater size plotted versus gravity.

Figure 5, which is taken from Bryan *et al.* (1976), shows crater radius plotted versus gravity (the ●'s in the figure). The source phase of these crater populations consisted of a numerical simulation using the TENSOR computer code of a 20-ton explosive source detonated 12 m below the surface in a weak soil. When the kinetic energy was fully coupled to the ground the source calculation was

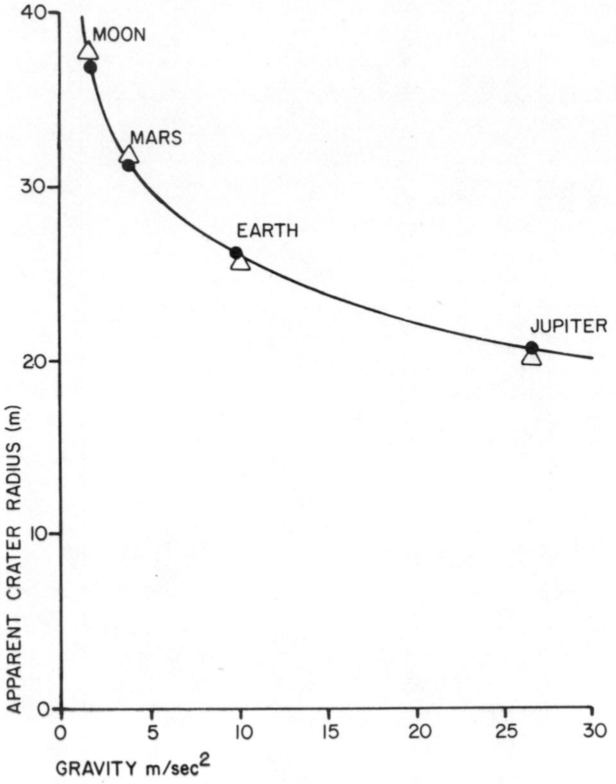

Fig. 5. Apparent crater radius vs. gravity: ● = calculations from Bryan *et al.* (1976); △—x_{min} from Fig. 4b and Table 2.

Table 2. Information from Fig. 4b.

Location	v_*^2/gD_*	#min	x_{min}	θ min	x_{min}/D_*
Moon	46.0	14&16	37.6	~22.0°	3.13
Mars	20.0	12	31.7	26.6°	2.64
Earth	7.7	10	25.7	31.0°	2.14
Jupiter	2.8	7	20.1	40.6°	1.68

#min is number of the particle with minimum range.
x_{min} is minimum range.
θ_{min} is angle of particle with minimum range.

terminated. The ballistic equation was then used on each zone of the source calculation with its directed velocity. The calculation on Earth, $g = 9.8\ \mathrm{m/sec}^2$, favorably compared with experimental data. The ballistic phase was then rerun a number of times with values for g corresponding with Moon, Mars, and Jupiter on the same zones with the same velocities. The resulting crater radii plotted versus gravity are shown in Fig. 5.

Figure 5 shows that the curve of minimum range versus gravity obtained from Fig. 4b and Table 2 is fairly similar to the final crater size plotted versus gravity as obtained from Bryan *et al.* (1976). (The fact that the two curves nearly coincide in magnitude is due to a fortuitous choice of v_* and should not be taken to mean that total crater dimensions can be determined by just one velocity, v_*.) The similarity of these two curves indicates the validity of the approach which is illustrated in Fig. 4 in describing the phenomenology of the ballistic phase of cratering, for craters produced by near optimally buried explosives. More importantly, the similarity of these two curves indicates the validity of using the minima in Fig. 3 as an indication of relative crater sizes.

Return now to the results presented in the last section, namely the necessary conditions for scaling when gravity is significant.

$$R_0 g_0 = 1 \qquad t_0 g_0 = 1 \quad \text{and} \quad R_0 = t_0 .$$

Let

$$g_0 = \frac{g_{\text{earth}}}{g_x} = \frac{9.8\ \text{m/sec}}{g_x},$$

where g_x represents the acceleration due to gravity on body x.

Then,

$$R_0 = \frac{1}{g_0} = \frac{g_x}{9.8\ \text{m/sec}} = \frac{R_{\text{earth}}}{R_x}$$

and

$$R_x = \frac{R_{\text{earth}}(9.8\ \text{m/sec})}{g_x}.$$

For a depth of burial of the explosive source of $R_{\text{earth}} = 12\ \text{m} = D_{*earth}$, $D_{*\text{moon}} = 72.6\ \text{m}$, $D_{*\text{mars}} = 31.5\ \text{m}$, and $D_{*\text{jupiter}} = 4.44\ \text{m}$. Note, that the requirements for energy scaling from the last section were:

$$\varepsilon_0 = 1 \qquad \frac{\left(\frac{\text{energy}}{\text{volume}}\right)_{\text{earth}}}{\left(\frac{\text{energy}}{\text{volume}}\right)_x} \frac{\varepsilon_{\text{earth}} R_x^3}{R_{\text{earth}}^3 \varepsilon_x} = 1,$$

or

$$\left(\frac{\varepsilon_x}{\varepsilon_{\text{earth}}}\right)^{1/3} = \frac{R_x}{R_{\text{earth}}} = \frac{g_{\text{earth}}}{g_x}.$$

Thus, in order to "scale" a 20-ton experiment on Earth at a 12 m depth, requires:

Location	Yield	Depth (D_*)	R_0	v_*
Moon	4428 tons	72.6 m	.165	30 m/sec
Mars	363 tons	31.5 m	.381	30 m/sec
Earth	20 tons	12.0 m	1.0	30 m/sec
Jupiter	1.01 tons	4.44 m	2.70	30 m/sec

To demonstrate that these experiments scale with gravity, consider Table 3. Note that the order of particles is preserved and that particle #10 has minimum range in *all* cases. Also note that the minimum values of x_{final} displayed by particle #10 are in the same ratios as R_0, $25.7/155 = .165$, $25.7/67.4 = .381$, $25.7/25.7 = 1$, $25.7/9.50 = 2.70$. This indicates that the scaling of dimensions and gravity holds, namely, $R_0 g_0 = 1$.

Derivation of a qualitative relationship for crater radius

Since the minimum of the curves in Fig. 3 can be *qualitatively* related to the radius of a throwout crater (i.e., the minimum distance at which projectiles land) a more detailed description of this minima is desired. The equation for x_{final} can be differentiated to find the initial throwout angle of the projectile which lands at the minimum horizontal distance. This angle is shown as a function of v_*^2/D_*g in Fig. 6, and it is represented by the expression

$$\frac{v_*^2}{D_*g} = [2 \sin^6 \theta \, (6 \cos^2 \theta - 1)]^{-1}. \tag{14}$$

Figure 6 indicates a minimum value of v_*^2/D_*g of about 1.64 for projectiles at about 53°. In Fig. 3, $v_*^2/D_*g = 1$ is a monotonically increasing function without a minimum, while $v_*^2/D_*g = 2$ has a minimum at about 45°. Figure 6 indicates that for v_*^2/D_*g less than 1.64, a minimum does not occur. Thus, Fig. 3 indicates that somewhere between $v_*^2/D_*g = 1$ and 2, a minimum no longer occurs. While craters still form at $v_*^2/D_*g = 1.64$, their dimensions are determined by a more direct "blowout" process rather than by a concentration of ejecta piling up at a minimum range (i.e., crater lip) as illustrated in Fig. 4b. The value of $v_*^2/D_*g = \frac{1}{64}$ or $x/D_* = 1.4$ at which cratering phenomenology changes is dependent on the assumed value for n in Eq. (9). The important point is that a phenomenological change occurs at low values of v_*^2/D_*g.

Figure 6 can be entered with a chosen value of v_*^2/D_*g, and the angle of the projectile which lands at minimum range can be found. This angle can be used in Eq. (12) to obtain the minimum value of x_{final}/D_* as a function of v_*^2/D_*g, Fig. 7. Figure 7 can be qualitatively thought of as a relationship between v_*^2/D_*g and

Table 3. Properties of particles in scaled experiments.

Particle	$x_i^{(1)} \frac{x_{earth}}{R_0}$ (m)				θ	$x_{max}^{(2)}$ (m)				$x_{final} = x_i + x_{max}$			
	Moon	Mars	Earth	Jupiter		Moon[3]	Mars[3]	Earth[3]	Jupiter[3]	Moon	Mars	Earth	Jupiter
0	0	0	0	0	90.0	0	0	0	0	0	0	0	0
1	12.1	5.25	2	.74	80.5	170	74.0	28.2	10.4	182	79.3	30.2	11.1
2	24.2	10.5	4	1.48	71.6	270	117	44.6	16.5	294	128	48.6	18.0
3	36.3	15.8	6	2.22	63.4	284	123	47.0	17.4	320	139	53.0	19.6
4	48.4	21.0	8	2.96	56.3	245	107	40.6	15.0	293	128	48.6	18.0
5	60.5	26.3	10	3.70	50.2	190	82.4	31.4	11.6	251	109	41.4	15.3
6	72.6	31.5	12	4.44	45.0	139	60.3	23.0	8.5	212	91.8	35.0	12.9
7	84.7	36.8	14	5.18	40.6	99	42.9	16.3	6.1	184	79.7	30.3	11.3
8	96.8	42.0	16	5.92	36.9	68.8	29.9	11.4	4.2	166	71.9	27.4	10.1
9	109	47.3	18	6.65	33.7	48.6	21.1	8.0	3.0	158	68.4	26.0	9.65
10	121	52.5	20	7.40	31.0	34.3	14.9	5.7	2.1	155	67.4	25.7	9.50
11	133	57.8	22	8.14	28.6	24.5	10.7	4.1	1.5	158	68.5	26.1	9.64
12	145	63.1	24	8.88	26.6	17.8	7.7	2.9	1.1	163	70.8	26.9	10.0
14	169	73.6	28	10.4	23.2	9.7	4.2	1.6	.6	179	77.8	29.6	11.0
16	194	84.1	32	11.8	20.6	5.6	2.4	.9	.3	200	86.5	32.9	12.1
18	218	99.8	38	14.1	18.4	3.3	1.4	.6	.2	221	101	36.6	14.3
20	242	105	40	14.8	16.7	2.1	.9	.3	.1	244	106	40.3	14.9
22	266	116	44	16.3	15.3	1.3	.6	.2	.1	267	117	44.2	16.4
24	290	126	48	17.8	14.0	.9	.4	.1	.1	291	126	48.1	17.9
MR	32.4	14.1	5.37	1.99	65.9	287	125	47.5	17.6	320	139	53	19.6

[1] x_i = initial distance measured along surface. R_0 for Moon = .165, Mars = .381, Earth = 1.0, Jupiter = 2.70.

[2] $x_{max} = \frac{v_*^2 \, 2 \sin^5 \theta \cos \theta}{g}$ $\quad v_* = 30$ m/sec, Moon $v_*^2/g = 555$, Mars $v_*^2/g = 241$, Earth $v_*^2/g = 91.8$, Jupiter $v_*^2/g = 34.0$.

[3] Note: θ and v_*^2/g are the same as from Table 1, therefore x_{max} is the same as Table 1.

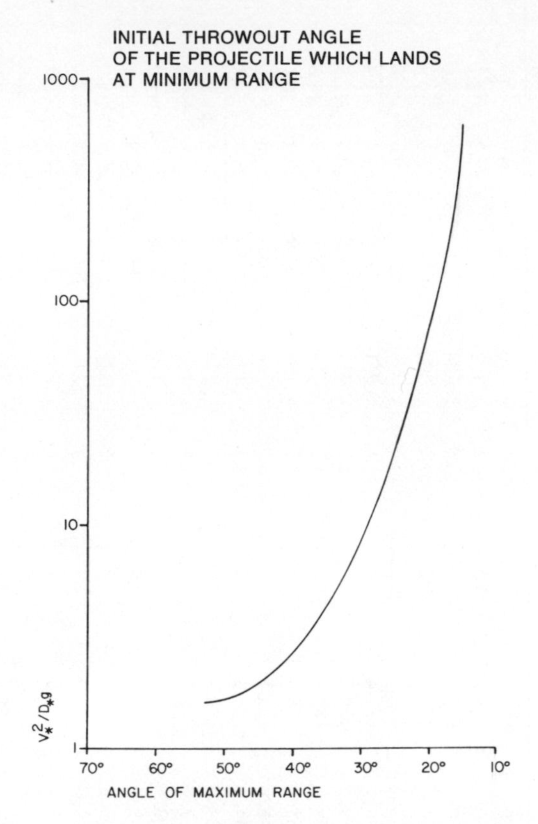

Fig. 6. Initial throwout angle of the projectile which lands at minimum range.

"crater radius/depth of source". Note that Fig. 7 should be thought of as a *qualitative* relationship since crater radius cannot be predicted by just a velocity above ground zero (v_*). But, in the following, it is shown that by normalizing a set of experimental data obtained in the same medium, a reasonable approximation to quantitative values can be obtained.

Comparison with calculational results

In order to compare Fig. 7 with calculated crater radii, a series of calculations reported by Bryan *et al.* (1976) are considered in Table 4. (It should be noted that the techniques utilized by Bryan *et al.* (1976) for calculating crater dimensions have repeatedly been compared favorably with experimental cratering data.) Four sources are considered in Table 4: (1) 20 tons of nitromethane (NM) at $D_* = 12$ m; (2) 20 tons of NM at 6 m; (3) 20 tons of NM at 17 m; and (4) 10 tons of NM at 12 m.

A value of $v_* = 30$ m/sec was selected for the 20-ton source at 12 m. This value, as discussed previously, was obtained by normalizing. To obtain v_* at

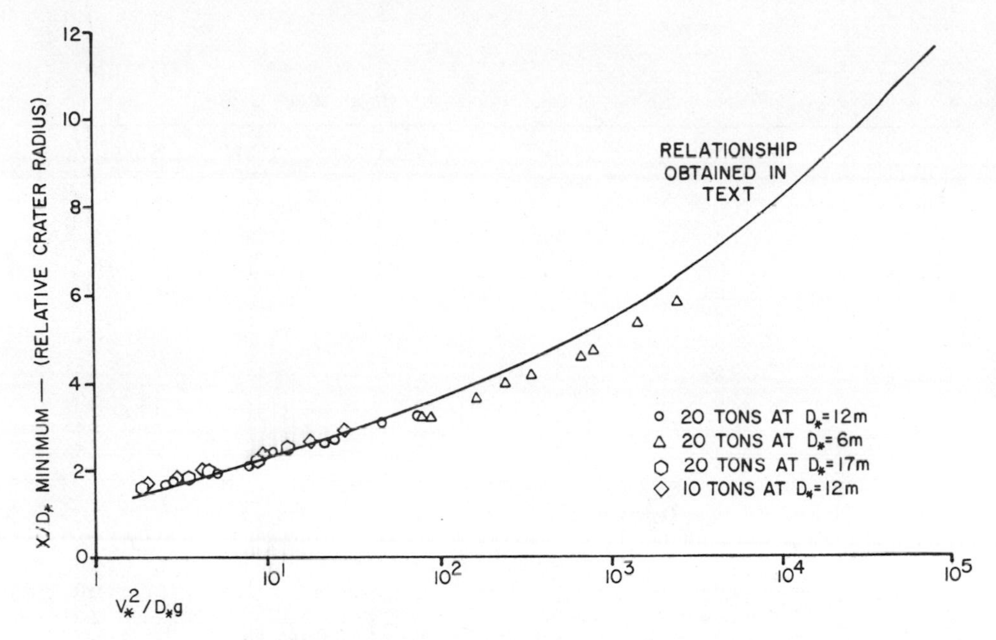

Fig. 7. Relationship between relative crater radius and v_*^2/D_*g compared with calculations from Bryan *et al.* (1976).

$D_* = 6$ m and $D_* = 17$ m, Eq. (9) was used. To obtain v_* for the 10-ton source at $D_* = 12$ m, cube root scaling of energy was used.

The results of these calculations are shown in Fig. 7 where they compare favorably with the relationship obtained for x/D_* minimum versus v_*^2/D_*g.

Comparisons with experimental data

Johnson *et al.* (1968) reported experimental cratering data at .17 g, .38 g, 1.0 g, and 2.5 g. These data were obtained at Wright-Patterson Air Force Base by flying appropriate trajectories to simulate the gravitational fields. Squibs with six grains of explosive were emplaced at various depths of burial between the surface and 12.7 cm. Both the size of the explosive source and the soil type, sand, remained constant.

Table 5 gives the data presented by Johnson *et al.* (1968) for the depth of the explosive, D_*, and the resulting crater diameter, $(2x)$. Values of x/D_* are calculated. In Table 5 values of v_i at various depths of burial are related to the value of v_i at a 5.08 cm burial, namely v_5. The same results are obtained if v_i at any burial is chosen. For the case of 1.0 g at $D_* = 5.08$, $v_*^2/D_*g = v_5^2/4978$, and $x/D_* = 3.38$. In order to normalize, Fig. 7 is used to obtain $v_*^2/D_*g = 70$ from $x/D_* = 3.38$. Thus $v_5^2 = 3.5 \times 10^5 (\text{cm/sec})^2$. Using $v_5^2 = 3.5 \times 10^5 (\text{cm/sec})^2$, the

Table 4. Results of calculations from Bryan *et al.* (1976).

	g(m/s^2)	x(m)	x/D_*	v_*(cm/s)	v_*^2/D_*g
Source: 20 tons of NM at $D_* = 12$ m				Symbol: Fig. 12⊙	
	1.0	39.9	3.33	30	75
Moon	1.62	36.9	3.08		46.3
	3.00	32.8	2.73		25.0
Mars	3.73	31.3	2.61		20.1
	7.00	28.3	2.36		10.7
Earth	9.80	26.3	2.19		7.65
	15.00	23.7	1.98		5.00
Jupiter	26.46	20.7	1.73		2.83
	30.00	20.2	1.68	↓	2.50
Source: 20 tons of NM at $D_* = 6$ m				Symbol: Fig. 12⊡	
	1.0	34.9	5.82	120	2400
Moon	1.62	32.3	5.38		1480
	3.00	28.3	4.72		800
Mars	3.73	27.8	4.63		643
	7.00	24.8	4.13		343
Earth	9.80	23.7	3.95		245
	15.00	21.7	3.61		160
Jupiter	26.46	19.7	3.28		90.7
	30.00	19.7	3.28	↓	80
Source: 20 tons of NM at $D_* = 17$ m				Symbol: Fig. 12 △	
	1.0	43.4	2.55	15.0	13.2
Moon	1.62	39.4	2.32		18.5
	3.00	34.3	2.02		4.40
Mars	3.73	32.3	1.90		3.53
	7.00	27.8	1.64		1.89
Earth	9.80	25.3	1.49		<1.64
	15.00	22.7	1.34		
Jupiter	26.46	19.7	1.16	↓	↓
Source: 10 tons of NM at $D_* = 12$ m				Symbol: Fig. 12 ×	
	1.0	34.9	2.90	18.9	29.8
Moon	1.62	31.8	2.65		18.4
	3.00	28.8	2.40		9.93
Mars	3.73	27.3	2.28		7.99
	7.00	23.7	1.98		4.26
Earth	9.80	21.7	1.81		3.04
	15.00	19.7	1.64		1.99
Jupiter	26.46	17.2	1.43		<1.64
	30.00	16.7	1.39	↓	↓

values of v_*^2/D_*g are calculated and shown in Table 5. The results of x/D_* plotted versus v_*^2/D_*g are given in Fig. 8 where good agreement with the relationship derived here is shown.

Handled in a similar manner and shown in Figs. 8 and 9 are the results of experiments reported by Chabai (1965). These experiments which are included in the next subsection, consisted of 256 lb sources of explosive at various depths.

Table 5. Experimental results from Johnson *et al.* (1968).

D_*(cm)	Diameter	Radius/D_*	v_i	v_*^2/D_*g
Experiments at .17 g = 167 cm/s^2				
0	27	∞	omit from analysis	
1.27	35.4	13.9	$16v_5$	4.24×10^5
2.54	44.0	8.66	$4v_5$	1.32×10^4
3.81	46.5	6.10	$1.78v_5$	1.74×10^3
5.08	45.3	4.46	v_5	413
7.62	40.4	2.65	$.44v_5$	53.2
10.16	34.5	1.70	$.25v_5$	12.9
12.70	23.3	.92	$.16v_5$	4.22
Experiments at .38 g = 372 cm/s^2				
0	26.3	∞	omit from analysis	
1.27	37.4	14.7	$16v_5$	1.90×10^5
2.54	39.1	7.70	$4v_5$	5.93×10^3
3.81	40.3	5.29	$1.78v_5$	782
5.08	40.2	3.96	v_5	185
7.62	36.0	2.36	$.44v_5$	23.9
10.16	30.3	1.49	$.25v_5$	5.78
12.70	19.2	.76	$.16v_5$	1.90
Experiments at 1.0 g = 980 cm/s^2				
0	22.4	∞	omit from analysis	
1.27	30.6	12.1	$16v_5$	7.20×10^4
2.54	34.1	6.71	$4v_5$	2.25×10^3
3.81	35.1	4.61	$1.78v_5$	297
5.08	34.3	3.38	v_5	70 (normalized by Fig. 7)
7.62	30.5	2.00	$.44v_5$	9.07
10.16	26.1	1.28	$.25v_5$	2.20
Experiments at 2.5 g = 2450 cm/s^2				
0	20.3	∞	omit from analysis	
1.27	26.5	10.43	$16v_5$	2.88×10^4
2.54	30.3	5.96	$4v_5$	900
3.81	31.8	4.17	$1.78v_5$	119
5.08	31.0	3.05	v_5	28.1
7.62	27.1	1.78	$.44v_5$	3.63
10.16	22.5	1.11	$.25v_5$	.88

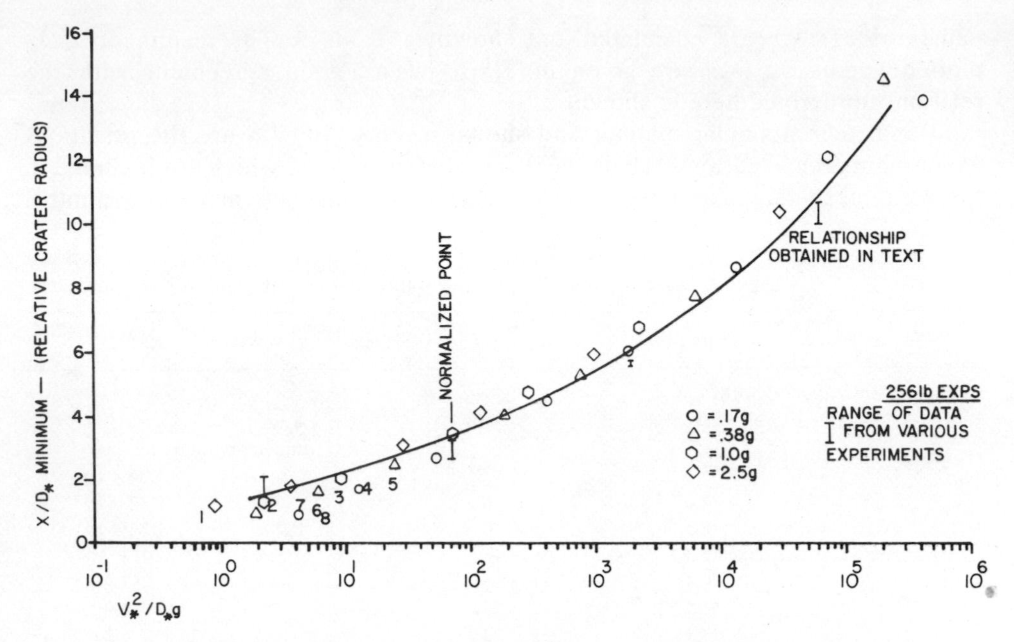

Fig. 8. Relationship between crater radius and v_*^2/D_*g compared with data: numbers 1–8, Tewes (1970); ○, △, ⬡, ◇, Johnson *et al.* (1968); 256 lbs experiments, Chabai (1965); normalized point from Johnson *et al.* (1968) data.

Normalization of velocities was based on a subsequent analysis which provided $v_* = 21.3$ ft/sec for charges located 6.35 ft below the surface.

For purposes of comparison, the following events on which some surface velocity measurements (Tewes, 1970) were obtained are also shown in Fig. 8.

Symbol Fig. 8	Name	$v_{m/sec}$	D_{*m}	R_m	R/D_*	v^2/D_*g
1	Sedan	35.1	194.0	186.0	.96	.65
2	Schooner	503.0	107.0	128.0	1.20	2.40
3	Alfa	36.6	16.1	23.2	1.40	8.50
4	Bravo	42.7	14.1	24.1	1.70	13.00
5	Charlie	533.0	13.0	24.4	1.90	23.00
6	Delta	305.0	17.3	19.8	1.10	5.50
7	Cabriolet	45.7	51.8	55.2	1.10	4.10
8	Danny Boy	45.7	33.5	32.6	.97	6.40

The v in the above are measured velocities at the surface, directly above the energy source. They are not normalized velocities which represent the kinetic energy of the ejecta. The resulting values of v^2/D_*g are too large (or to the right of the derived curve) when compared in Fig. 8. Also, note that these events mostly lie below the value $R/D_* = 1.4$ where the model formulated herein loses validity.

Additional comparisons with data

We have already shown that the model proposed herein is in good agreement with suites of data conducted in the same material with either a variable or fixed explosive strength and with variable depths of burial. However, there are a number of other experiments which are either one of a kind or part of a series of only two or three similar experiments (see the excellent summary given by Chabai, 1965).

The values of v_* were calculated directly from the model, and then scaled by cube root yield scaling and inverse square velocity scaling, to the arbitrary standard of 10 m from a source of 4×10^4 lbs of HE. Not all of the events in the Chabai listing offered a basis of comparison since this model has no meaning for values of R/D_* less than 1.4. Thirty events remained after eliminating these. They range from explosive yields of 256 lbs up to 2.4×10^6 lbs, a range of 10^4 in yield, and include nuclear as well as non-nuclear events. Figure 9 shows that the normalized scaled velocity from the model for these experiments varies by only about a factor of ± 2. Such agreement exceeds expectation considering the simplicity of the model.

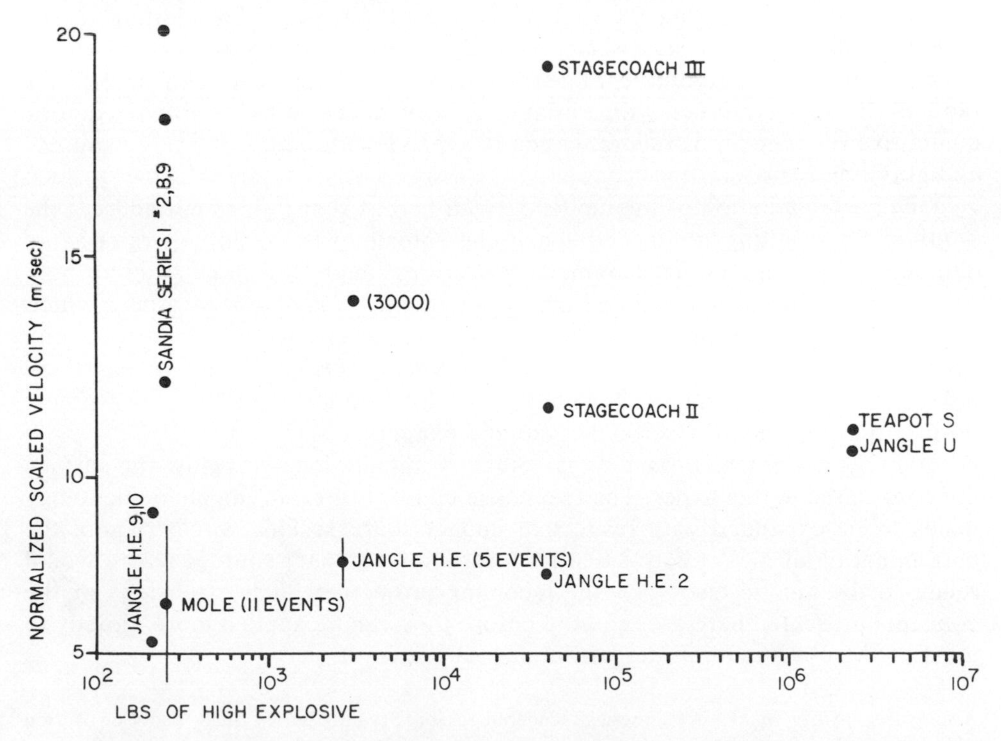

Fig. 9. Comparison of scaled and normalized velocities calculated for tests in alluvium.

Let us now reach for the moon—as it were—and try to extend the comparison to the experiments of Johnson *et al.* (1968). Their explosive was a squib with six grains of explosive giving about 4×10^{-4} lbs of HE equivalent. The same comparison gives scaled normalized velocities averaging 36 m/sec. Considering that we have scaled over an energy range of 10^{10}, and that the undefined sand media referred to in Johnson *et al.* (1968) may be substantially different than alluvium, we find these results encouraging.

CONCLUSIONS

The necessary conditions for scaling two experiments in rock materials have been derived. One may think of the cratering process as consisting of two phases. In the first phase, the effects of gravity can be considered negligible. During this phase, the energy source dynamically imparts its energy to the surrounding rock, and radiative transfer is assumed to be insignificant. In the text, it is mathematically derived, and demonstrated by calculational examples (Crowley, 1970) that two events can be scaled during this phase if: they are conducted in the same rock material, time and distance use the same scaling factor, and distances scale as the cube root of the energies.

The second phase of a cratering event is governed by the ballistic equation, and gravity is of primary importance. During this second phase, the scaling conditions imposed during the first phase are also required in addition to the condition that distances must scale inversely as the ratio of gravities.

In the text, a qualitative relationship is derived for the scaling of final crater radii of throwout craters with variable gravity fields. When normalized, this qualitative relationship is favorably compared to a wide range of energy sources and gravities.

One basic limitation of the model derived here is that it does not address the depth of the resulting crater; and hence, the volume of the resulting crater is not obtainable. Such additional parameters as critical angle for slope stability play an important role in determining the crater depth. A generalized model which includes such parameters as well as gravity variations would be very useful. More generalized relationships of v_*^2/D_*g versus X/D_* which are based on different velocity attenuations with respect to distance than the $n = 2$ power used in the text should also be derived and examined.

Only events in which the energy source is initially located below the surface are considered in this paper. The two phase concept of cratering phenomenology needs to be expanded with respect to impact sources. This would require the determination of an "effective depth of burial" for impact sources which would relate to the kinetic energy of the incoming projectile. Such extensions of the concepts presented here are required before they can be applied more directly to most of the observed extraterrestrial impact craters.

Acknowledgments—Thanks are gratefully tendered to Jon Bryan, Charles Snell, and Don Burton who generously provided the authors with the results of their calculations as well as acted as mentors for the concepts presented in the text.

References

Baker, W. E., Westine, P. S., and Dodge, F. T.: 1973, *Similarity Methods in Engineering Dynamics: Theory and Practice of Scale Modeling*, Hayden Book Co., Inc., Rochelle Park, NJ.

Birkhoff, G.: 1960, *Hydrodynamics: A Study in Logic, Fact and Similitude*, Princeton Univ. Press, Princeton, NY.

Burton, D. E., Snell, C. M., and Bryan, J. B.: 1975, Computer Design of High-Explosive Experiments to Simulate Subsurface Nuclear Detonations, *Nuclear Technology* **26**, 65–87.

Bryan, J. B., Burton, D. E., and Denny, M. D.: 1974, Numerical Studies of Cratering in Bearpaw Shale: Two-Dimensional Results, Lawrence Livermore Laboratory Report UCRL-51659, Livermore, CA.

Bryan, J. B., Burton, D. E., Snell, C. M., and Thomsen, J. M.: 1976, Numerical Simulation of Subsurface Explosion Cratering (abstract). In *Papers Presented to the Symposium on Planetary Cratering Mechanics*, p. 15–16. The Lunar Science Institute, Houston.

Butkovich, T. R.: 1967, The Gas Equation of State for Natural Materials, Lawrence Livermore Laboratory Report UCRL-14729, Livermore, CA.

Butkovich, T. R.: 1976, Correlations Between Measurements and Calculations of High-Explosive-Induced Fracture in a Coal Outcrop, *Int. J. of Rock Mech. Min. Science, and Geomech. Abstr.* **13**, 45–51.

Chabai, A. J.. 1959, Crater Scaling Laws for Desert Alluvium, Sandia Laboratories Report SC-4391, Albuquerque, NM.

Chabai, A. J.: 1965, On Scaling Dimensions of Craters Produced by Buried Explosives, *J. Geophys Res.* **70**, 5075.

Cherry, J. T.: 1967, Computer Calculation of Explosion Produced Craters, *Intern. J. Rock Mech. Min. Sci.* **4**, 1.

Cherry, J. T., Larson, D. B., and Rapp, E. G.: 1968, Computer Calculations of the Gasbuggy Event, Lawrence Livermore Laboratory Report UCRL-50419, Livermore, CA.

Crowley, B. K.: 1967, Necessary Conditions for Similar Solutions of Problems of Turbulent-Gas Dynamics, Lawrence Livermore Laboratory ReportUCRL-50211, Livermore, CA.

Crowley, B. K.: 1970, Scaling Criteria for Rock Dynamic Experiments. In *Symposium on Engineering with Nuclear Explosives 1*, 545–559, Las Vegas, NV.

Galilei, G.: 1953, Dialogue Concerning the Two Chief World Systems-Ptolemaic and Copernican, Transl. by Stillman Drake, Univ. of Calif. Press, Berkeley, CA.

Johnson, S. W., Smith, J. A., Franklin, E. G., Moraski, L. K., and Teal, D. J.: 1968, Gravity and Atmospheric Pressure Scaling Equations for Small Explosion Craters in Sand, Air Force Institute of Technology Report AFIT-TR-68-3, Wright-Patterson Air Force Base, FL.

Maenchen, O. and Sach, S.: 1964, The TENSOR Code, *Methods of Computational Physics*, Academic Press, NY.

Nordyke, M. D.: 1964, Cratering Experience with Chemical and Nuclear Explosives, Lawrence Livermore Laboratory Report UCRL-7793, Livermore, CA.

Pai, S. I.: 1966, *Radiation Gas Dynamics*, Springer-Verlag, Inc., NY.

Pokrovskii, G. I. and Fedorov, I. S.: 1957, Effect of Shock and Explosion on Deformable Media, Gos. Izd, Moscow.

Rosenhead, L.: 1963, *Laminar Boundary Layers*, Oxford at the Clarendon Press, England.

Sedov, L. I.: 1959, *Similarity and Dimensional Methods in Mechanics*, Academic Press, NY.

Schatz, J.: 1974, A One-Dimensional Wave Propagation Code for Rock Media, Lawrence Livermore Laboratory Report UCRL-51689, Livermore, CA.

Terhune, R. W. and Stubbs, T. F.: 1970, Nuclear Cratering on a Digital Computer, *Symposium on Engineering with Nuclear Explosives 1*, 334–359, Las Vegas, NV.

Tewes, H. A.: 1970, Results of the Schooner Excavation Experiment, *Symposium on Engineering with Nuclear Explosives 1*, 306–333, Las Vegas, NV.

Vaile, R. B., Jr.: 1961, Crater Survey, Defense Nuclear Agency Report WT-920, later partially reprinted in *J. Geophys. Res.* **66**, 3413.

Violet, C. E.: 1961, A Generalized Empirical Analysis of Cratering, *J. Geophys. Res.* **66**, 3461.

White, J. W.: 1971, Examination of Cratering Formulas and Scaling Methods, *J. Geophys. Res.* **76**, 8599–8603.
Wilkins, M. L.: 1969, Calculation of Elastic-Plastic Flow, Lawrence Livermore Laboratory Report UCRL-7322 Rev. 1, Livermore, CA.

Roddy, D. J., Pepin, R. O., and Merrill, R. B., editors.
(1977) *Impact and Explosion Cratering*, Pergamon Press (New York), p. 1191–1214.
Printed in the United States of America

Influence of gravitational fields and atmospheric pressures on scaling of explosion craters

ALBERT J. CHABAI

Sandia Laboratories, Albuquerque, New Mexico 87115

Abstract—Methods of obtaining scaling rules from dimensional analysis are reviewed. Scaling rules to describe the phenomena of cratering by explosions or by hypervelocity impacts are presented and discussed in terms of the requirements of similitude. Inability to perform similar cratering experiments in a gravitational field results in deviations from expected quarter-root scaling. Some sources of similarity violation are outlined and some experiments are suggested which may eliminate a suspected major source of similarity violation present in all explosion cratering experiments conducted to date in accelerated coordinate systems.

1. INTRODUCTION

IN THE STUDY of craters produced by explosives or by impacting bodies, considerable effort is devoted to the question of scaling. Scaling is the method by which information is obtained about prototype experiments from small-size or model experiments. Small charges of chemical explosives are utilized in cratering experiments in order to predict the cratering behavior of nuclear explosives having energy releases orders of magnitude greater than the chemical explosives. Laboratory experiments with hypervelocity particles a few millimeters in diameter are conducted in an effort to understand the cratering phenomena associated with meteoritic impact on lunar and planetary surfaces. Dimensional analysis is frequently invoked to establish scaling rules or the relationships by which we may predict results of a large-scale experiment from information obtained in a small-scale experiment. Comparisons of results from model and prototype experiments with predictions by scaling rules often demonstrate considerable disparity. This lack of agreement is a consequence of the fact that all conditions required to achieve similarity or similitude among model and prototype experiments are not met. Rarely is it possible in actual experiments to realize the stringent requirements placed on experiments by the necessity to conduct similar experiments. As a consequence, we must expect that scaling rules will not be obeyed. Rather, we must concern ourselves more with the deviations from expected scaling rules and their causes.

Arguments are presented to indicate that similarity among cratering experiments is rarely achieved. Data are reviewed to examine the influence of atmospheric pressure and the gravity field on crater scaling relations. Crater dimensions are found to be proportional to g^{-n} where n varies approximately from 1/8 to 1/4 and g is acceleration due to gravity. This ambiguity in the scaling rules is examined and some experiments are suggested which may remove the ambiguity.

2. Method of Dimensional Analysis

A convenient method for obtaining the scaling rules governing the phenomena of cratering is by means of dimensional analysis (Bridgman, 1949; Langhaar, 1951; Sedov, 1959; Baker *et al.*, 1973). The fundamental principle in dimensional analysis is the Pi Theorem (Buckingham, 1914; Brand, 1957). It has been stated as follows:

If an equation in n arguments is dimensionally homogeneous with respect to m fundamental units, it can be expressed as a relation between (n − r) independent dimensionless arguments. As will be discussed shortly, r is the rank of the dimensional matrix and $r \leq m$.

Consider a set of n dimensional variables, $x_1, x_2, \ldots, x_n$ which describe the phenomena of cratering and which are functionally related by

$$f(x_1, x_2, \ldots, x_n) = 0. \tag{1}$$

The function, f, in Eq. (1) is said to be dimensionally homogeneous if its form is unchanged when the dimensions of its arguments are changed. All equations describing physical phenomena are dimensionally homogeneous. The variables x_i have dimensions which are expressible in terms of arbitrarily selected units, u_1, $u_2, \ldots, u_m$, which are considered fundamental. For example, mass (M), length (L), time (T), and temperature (θ) may be taken as a set of fundamental units. Force, length, and time may also be selected as one set of fundamental units.

Let x_1 have dimensions $u_1^{a_{11}} u_2^{a_{21}} \cdots u_m^{a_{m1}}$

x_2 have dimensions $u_1^{a_{12}} u_2^{a_{22}} \cdots u_m^{a_{m2}}$

$\vdots$

x_n have dimensions $u_1^{a_{1n}} u_2^{a_{2n}} \cdots u_m^{a_{mn}}$.

A variable, x_i, representing stress, would have dimensions $M^1L^{-1}T^{-2}$, for example. Any nondimensional quantity (Pi-term) composed of these variables has the form

$$\Pi = x_1^{k_1} x_2^{k_2} \cdots x_n^{k_n}.$$

$$\Pi = (u_1^{k_1 a_{11}} u_2^{k_1 a_{21}} \cdots u_m^{k_1 a_{m1}})(u_1^{k_2 a_{12}} u_2^{k_2 a_{22}} \cdots u_m^{k_2 a_{m2}}) \cdots (u_1^{k_n a_{1n}} u_2^{k_n a_{2n}} \cdots u_m^{k_n a_{mn}}).$$

$$\Pi = u_1^{[k_1 a_{11} + k_2 a_{12} + \cdots + k_n a_{1n}]} u_2^{[k_1 a_{21} + k_2 a_{22} + \cdots + k_n a_{2n}]} \cdots u_m^{[k_1 a_{m1} + k_2 a_{m2} + \cdots + k_n a_{mn}]}.$$

Since Π must be dimensionless, the exponents of the u_i must vanish:

$$\begin{aligned} k_1a_{11}+k_2a_{12}+\cdots+k_na_{1n}&=0\\ k_1a_{21}+k_2a_{22}+\cdots+k_na_{2n}&=0\\ \vdots\qquad\qquad & \;\;\vdots\\ k_1a_{m1}+k_2a_{m2}+\cdots+k_na_{mn}&=0. \end{aligned} \tag{2}$$

In matrix form the above equations may be expressed as

$$[\mathbf{A}][\mathbf{k}]=0. \tag{3}$$

The matrix **A** is called the dimensional matrix. **A** is readily constructed by listing the variables x_i horizontally and the fundamental units u_i vertically and inserting the coefficients of **A** one column at a time by specifying the dimensions of each variable.

	x_1	$x_2 \cdots$	x_n
u_1	a_{11}	$a_{12} \cdots$	a_{1n}
u_2	a_{21}	$a_{22} \cdots$	a_{2n}
$\vdots$	$\vdots$		
u_m	a_{m1}	$a_{m2} \cdots$	a_{mn}

Thus if the stress variable considered by example earlier were x_2 and if $m=3$ with $u_1=M$, $u_2=L$ and $u_3=T$, then $a_{12}=1$, $a_{22}=-1$ and $a_{32}=-2$. Any solution of the Eqs. (2) is a set of exponents in a dimensionless product.

From the theory of homogeneous linear algebraic equations the set of Eqs. (2), is found to have $(n-r)$ linearly independent solutions, where r is the rank of the matrix **A**. The matrix **A** is of rank r when it possesses at least one nonzero determinant of order r, while all determinants of higher order are zero. We note a common case: the matrix **A** will be of rank $r=m$ provided a single nonzero determinant of order m is found for the matrix. It is frequently the case that $r=m$.

The $(n-r)$ linearly independent solutions for $[\mathbf{k}]$ of Eq. (2) provide the $(n-r)$ dimensionless quantities or Pi-terms of Buckingham's Pi Theorem. Equation (1) is then equivalently expressed as

$$\psi(\Pi_1, \Pi_2, \ldots, \Pi_{n-r})=0. \tag{4}$$

Since we have m equations with $n>m$ unknowns, the system of Eqs. (2) is underdetermined. If $r<m$, we need consider only any r of the m equations. Arbitrary values may be assigned to $(n-r)$ of the k_i and we may solve for the remaining r values of the k_i.

One convenient means of obtaining solutions to Eqs. (2) is to perform some simple operations on the matrix **A** of Eq. (3). Separate the matrix **A** into submatrices as follows

$$\underset{r\times n}{[\mathbf{A}]}=\underset{r\times(n-r)}{[\mathbf{C}_2}\;\Big|\;\underset{(r\times r)}{\mathbf{C}_1]}$$

where $\mathbf{C}_1$ is the righthandmost submatrix of $\mathbf{A}$ and is of order $(r \times r)$, i.e., having r rows and r columns. $\mathbf{C}_2$ is the remaining submatrix of $\mathbf{A}$ and is of order $[r \times (n-r)]$. Next form the solution matrix $\mathbf{B}$ of order $[(n-r) \times n]$ from which the Pi-terms are readily obtained.

$$[\ell n\, \boldsymbol{\pi}] = [\mathbf{B}][\ell n\, \mathbf{x}]$$

$$\begin{bmatrix} \ell n\, \Pi_1 \\ \ell n\, \Pi_2 \\ \vdots \\ \ell n\, \Pi_{n-r} \end{bmatrix} = [\mathbf{B}_1 \,\vdots\, \mathbf{B}_2] \begin{bmatrix} \ell n\, x_1 \\ \ell n\, x_2 \\ \vdots \\ \ell n\, x_n \end{bmatrix}$$

$\mathbf{B}_1$ is the identity matrix, $\mathbf{I}$, of order $(n-r) \times (n-r)$ and constitutes the lefthandmost submatrix of the solution matrix $\mathbf{B}$. The remaining submatrix $\mathbf{B}_2$ of order $[(n-r) \times r]$ is given by

$$\mathbf{B}_2 = -\widetilde{\mathbf{C}_1^{-1}\mathbf{C}_2}.$$

$\tilde{\mathbf{M}}$ signifies the transpose of matrix $\mathbf{M}$. The matrix $\mathbf{C}_1$ must be nonsingular and of rank r.

The form of the dimensionless Pi-terms will be determined by the arrangement of variables x_i in the dimensional matrix $\mathbf{A}$. It is often convenient to arrange the variables x_i such that the dependent variables appear first (i.e., occupy columns 1, 2, ... in $\mathbf{A}$) followed by independent variables, material properties and all dimensional constants relevant to the phenomena. All those variables which are utilized to construct $\mathbf{A}$ and are represented in $\mathbf{C}_2$ will appear in the Pi-terms only once. The variables which lead to composition of $\mathbf{C}_1$ will appear repeatedly in the Pi-terms. For this reason, it is advisable to arrange the list of variables x_i in constructing the dimensional matrix $\mathbf{A}$ such that those independent variables which can be controlled easily in experiments and those material properties considered most significant appear in $\mathbf{C}_1$ on the right-hand side of $\mathbf{A}$.

If the arrangement of variables in the dimensional matrix is such that the rank of $\mathbf{C}_1$ is less than r, the rank of $\mathbf{A}$, then a rearrangement of variables is necessary such that the rank of $\mathbf{C}_1$ is r.

To illustrate the procedure outlined, the following examples are considered.

Example 1: $(n = 5,\ m = 3,\ r = 3)$.

Assume some phenomena are described by certain variables P, ℓ, W, t, and ρ representing pressure, length, mass, time, and density, respectively. A functional relationship exists among these variables, $f(P, \ell, W, t, \rho) = 0$. Take ℓ and t to be dependent variables and P and W to be independent. Take M, L, T as fundamental units. The dimensional matrix $\mathbf{A}$, in which $m = 3$ and $n = 5$, is

	ℓ	t	W	ρ	P
M	0	0	1	1	1
L	1	0	0	−3	−1
T	0	1	0	0	−2

The rank of **A** is $r = 3$.

Two Pi-terms $(n - r = 2)$ are to be found from the solution matrix. With

$$\mathbf{C}_2 = \begin{bmatrix} 0 & 0 \\ 1 & 0 \\ 0 & 1 \end{bmatrix} \quad \text{and} \quad \mathbf{C}_1 = \begin{bmatrix} 1 & 1 & 1 \\ 0 & -3 & -1 \\ 0 & 0 & -2 \end{bmatrix},$$

$$\mathbf{B}_2 = \begin{bmatrix} -\frac{1}{3} & \frac{1}{3} & 0 \\ -\frac{1}{3} & -\frac{1}{6} & \frac{1}{2} \end{bmatrix} = -\widetilde{\mathbf{C}_1^{-1}\mathbf{C}_2}.$$

The Pi-terms are found from

$$\begin{bmatrix} \ell n\, \Pi_1 \\ \ell n\, \Pi_2 \end{bmatrix} \begin{bmatrix} 1 & 0 & -\frac{1}{3} & \frac{1}{3} & 0 \\ 0 & 1 & -\frac{1}{3} & -\frac{1}{6} & \frac{1}{2} \end{bmatrix} \begin{bmatrix} \ell n\, \ell \\ \ell n\, t \\ \ell n\, W \\ \ell n\, \rho \\ \ell n\, P \end{bmatrix}$$

$$\Pi_1 = \frac{\ell \rho^{1/3}}{W^{1/3}}, \quad \Pi_2 = \frac{t P^{1/2}}{W^{1/3} \rho^{1/6}}.$$

Thus by dimensional analysis, the functional relationship among five dimensional variables is reduced to a functional relationship among two dimensionless variables,

$$\psi\left(\frac{\ell^3 \rho}{W}, \frac{t^6 P^3}{W^2 \rho}\right) = 0$$

Example 2: $(n = 3, m = 3, r = 2)$.

Energy, E, pressure, P, and area A are the relevant variables in this example such that $f(E, A, P) = 0$. With M, L, and T as fundamental units, the dimensional matrix, **A**, is given by

	P	A	E
M	1	0	1
L	−1	2	2
T	−2	0	−2.

The determinant of **A** is zero and so **A** is not of rank 3. The rank of **A** is seen to be two. There will be one Pi-term. Because $r < m$ we may discard any $(m - r)$ of the rows of **A** in finding the Pi-term. Let us discard the bottom row of **A** and take

$$\mathbf{C}_2 = \begin{bmatrix} 1 \\ -1 \end{bmatrix} \quad \mathbf{C}_1 = \begin{bmatrix} 0 & 1 \\ 2 & 2 \end{bmatrix}.$$

$\mathbf{B}_2$ is readily determined:

$$\mathbf{B}_2 = -\widetilde{\mathbf{C}_1^{-1}\mathbf{C}_2} = [3/2 \quad -1],$$

and it is seen that $\mathbf{B}_1 = 1$. The Pi-term is obtained from

$$[\ell n\, \pi_1] = [1 \quad 3/2 \quad -1] \begin{bmatrix} \ell n\, P \\ \ell n\, A \\ \ell n\, E \end{bmatrix}$$

$$\pi_1 = PA^{3/2}/E = \text{constant},$$

or

$$\psi(PA^{3/2}/E) = 0.$$

The same result is obtained if, for example, the top rather than the bottom row of the dimensional matrix **A** had been discarded. This example serves to show that the rank of **A** can be less than m.

Example 3: ($n = 10$, $m = 4$, $r = 4$).

In this example the fundamental units are M, L, T, and θ, a temperature dimension. The dimensional matrix **A** is

	ℓ	s	r	t	W	α	ρ	P	τ	E
M	0	0	0	0	1	0	1	1	0	1
L	1	1	1	0	0	0	−3	−1	0	2
T	0	0	0	1	0	0	0	−2	0	−2
θ	0	0	0	0	0	−1	0	0	1	0.

The quantities τ and α are temperature and thermal expansion coefficient. Notice that ℓ, s, and r each have dimensions of length. The last four columns of **A** can be taken as $\mathbf{C}_1$ since, for this arrangement of variables, $\mathbf{C}_1$ is nonsingular and of rank 4, the same as the rank of **A**. Had the variables been arranged such that the last four columns of **A** were represented by the variables ℓ, s, r, and t, then $\mathbf{C}_1$ would have been of rank 2 rather than rank 4, and furthermore $\mathbf{C}_1$ would have been singular. This arrangement of variables is not permitted since $\mathbf{C}_1$ must be equal to the rank of **A** and must be nonsingular.

For this example, the matrices of interest are

$$\mathbf{C}_1 = \begin{bmatrix} 1 & 1 & 0 & 1 \\ -3 & -1 & 0 & 2 \\ 0 & -2 & 0 & -2 \\ 0 & 0 & 1 & 0 \end{bmatrix},$$

$$\mathbf{C}_2 = \begin{bmatrix} 0 & 0 & 0 & 0 & 1 & 0 \\ 1 & 1 & 1 & 0 & 0 & 0 \\ 0 & 0 & 0 & 1 & 0 & 0 \\ 0 & 0 & 0 & 0 & 0 & -1 \end{bmatrix},$$

$$\mathbf{B}_2 = \begin{bmatrix} 0 & \frac{1}{3} & 0 & -\frac{1}{3} \\ 0 & \frac{1}{3} & 0 & -\frac{1}{3} \\ 0 & \frac{1}{3} & 0 & -\frac{1}{3} \\ -\frac{1}{2} & \frac{5}{6} & 0 & -\frac{1}{3} \\ -1 & 1 & 0 & -1 \\ 0 & 0 & 1 & 0 \end{bmatrix}.$$

The relationship among the Pi-terms determined from **B** is

$$\frac{\ell P^{1/3}}{E^{1/3}} = g\left(\frac{sP^{1/3}}{E^{1/3}}, \frac{rP^{1/3}}{E^{1/3}}, \frac{tP^{5/6}}{\rho^{1/2}E^{1/3}}, \frac{WP}{\rho E}, \alpha\tau\right).$$

As has been illustrated, Pi-terms may be found from a dimensional analysis by following 8 steps:

(1) Form the dimensional matrix **A** by listing the variables x_i, $i = 1, 2, \ldots, n$ across the top and the units down the side. Order the variables such that the dependent and other more important variables generally appear as the leftmost columns of **A**.
(2) Determine the rank r of matrix **A**.
(3) Partition matrix **A** into its submatrices, $\mathbf{C}_1$ and $\mathbf{C}_2$. If $\mathbf{C}_1$ has rank less than r, rearrange variables in **A** such that $\mathbf{C}_1$ is of rank r.
(4) Find the reciprocal matrix of $\mathbf{C}_1$ and check that $\mathbf{C}_1\mathbf{C}_1^{-1} = \mathbf{I}$, the identity matrix.
(5) Form $\mathbf{B}_2$ from $\mathbf{C}_1$ and $\mathbf{C}_2$ with $\mathbf{B}_2 = -\mathbf{C}_1^{-1}\mathbf{C}_2$.
(6) Construct the solution matrix **B**.
(7) From **B** form the $(n - r)$ Pi-terms and verify that each is dimensionless.
(8) Simplify or put the Pi-terms into more useful form if necessary and write the final relationship in terms of the dimensionless Pi-terms, $\psi(\Pi_1, \Pi_2, \ldots, \Pi_{n-r}) = 0$.

In addition to the matrix method described, other methods (e.g., Killian and Germain, 1977; Kececioglu, 1960) are available for determining the Pi-terms.

At this point, it may be helpful to summarize some of the advantages and some of the shortcomings of dimensional analysis. Advantages are:

(1) A large number of (dimensional) variables is reduced to a smaller more compact number (dimensionless Pi-terms) needed to describe some physical phenomenon.
(2) Results of dimensional analysis establish the rules of similitude or scaling rules by which we learn the relationship between model and full scale tests.
(3) The Pi-terms derived from dimensional analysis are extremely helpful in planning and in graphically displaying results of experiments.
(4) Important dimensionless quantities (e.g., Froude number or Reynolds number) may be identified that describe significant physical effects.

Some shortcomings are:

(1) The functional relation among the Pi-terms is not given by dimensional analysis; it must be determined by experiment or theory.
(2) The set of variables selected as descriptive of a phenomenon must include all important variables. If an important one is omitted, the results of dimensional analysis will be incorrect.
(3) Different but equally plausible results from dimensional analysis must be differentiated by experiment.
(4) Some ambiguity exists in the manner by which the Pi-terms are formed. Once one set of Pi-terms is found, other sets can be readily obtained by forming products and quotients with the original set. Selecting the most useful set comes from experience and a real understanding of the variables involved in the problem.

3. Cube-Root Scaling, Quarter-Root Scaling, and Similitude

The methods of the previous section have been applied to 10 variables taken to describe explosive cratering. As indicated in Fig. 1, E is total energy released

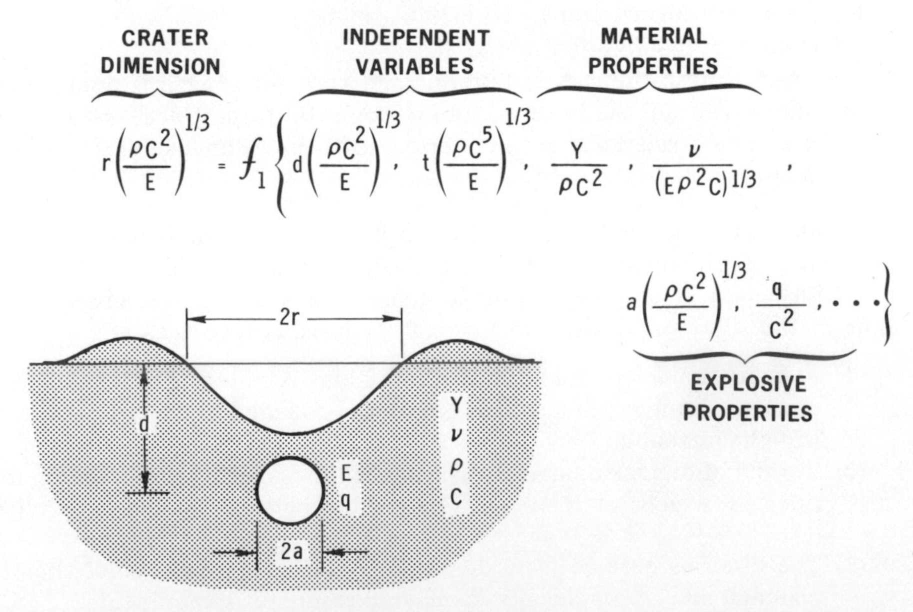

$$r\left(\frac{\rho c^2}{E}\right)^{1/3} = f_1\left\{d\left(\frac{\rho c^2}{E}\right)^{1/3}, t\left(\frac{\rho c^5}{E}\right)^{1/3}, \frac{Y}{\rho c^2}, \frac{\nu}{(E\rho^2 c)^{1/3}}, a\left(\frac{\rho c^2}{E}\right)^{1/3}, \frac{q}{c^2}, \dots\right\}$$

Fig. 1. Cube-root scaling rules derived from dimensional analysis on ten variables taken to be important in cratering by explosions.

by the explosive charge, q is charge specific energy, and a is charge radius. Material properties are represented by a strength Y, a viscosity ν, the density ρ, and the sound speed C. Crater radius is r, d is depth of charge burial, and t is time. The seven dimensionless Pi-terms shown in the equation of Fig. 1 are scaled quantities and indicate that scaled crater radius, $r(\rho C^2/E)^{1/3}$, is a function of scaled burial depth, $d(\rho C^2/E)^{1/3}$, and five other scaled variables. Dots, . . . , in the function f_1 represent other arguments for f_1 which may be inserted if additional variables, other than those initially assumed, are considered significant. For example, the explosive density, ρ_e, could enter as an argument of f_1 as ρ_e/ρ and a failure strain for the medium as ϵ_f. Because linear dimensions of scaled quantities occur in combination with $E^{1/3}$, this result is known as cube-root scaling.

The function f_1 is not given by dimensional analysis. It must be determined from theory or experiment. The arguments of the functional relationship represented by f_1 permit us to consider how model experiments can be performed which are *similar* to the prototype. If for the model experiment the arguments of f_1 have the same numerical values as for the prototype experiment, then the function f_1 will have the same value for both experiments. As a result, the scaled radius for both experiments will be identical. Employing subscripts m and p for model and prototype experiments, which scale, it is seen from the equation of Fig. 1 that

$$[r(\rho C^2/E)^{1/3}]_m = [r(\rho C^2/E)^{1/3}]_p,$$

or if both experiments are conducted in a medium where ρ and C remain constant then $r_m/r_p = (E_m/E_p)^{1/3}$. The ratio of crater radii for model and prototype experiments will be in proportion to the cube-root of the ratio of energy releases.

Similitude among experiments is achieved when all the dimensionless arguments of the function, f_1, are identical for the experiment set. This is a very strict requirement which is almost never achieved in reality. To illustrate, the Pi-term of f_1 representing scaled material viscosity, ν, is a source of similarity violation. This term demands that ν be increased as the energy of the explosion, E, is increased (ρ and C are maintained constant) if scaled experiments are to be conducted among which similitude is realized. By not scaling ν properly with E, similarity requirements are violated and to the extent that ν is an important variable, the prototype is not scaled exactly by the model experiment. Except for the Pi-term with ν it is seen that similar experiments could be conducted in the same medium without altering or scaling material properties. In particular, it is seen that the same explosive must be used when material properties are not changed.

By adding p (hydrostatic pressure) and g (acceleration due to gravity) to the list of 10 variables (in Fig. 1) considered to be important in cratering and performing a dimensional analysis on those 12 variables, the quarter-root scaling rules are obtained.

$$r\left(\frac{\rho g}{E}\right)^{1/4} = f_2\left\{d\left(\frac{\rho g}{E}\right)^{1/4}, t\left(\frac{\rho g^5}{E}\right)^{1/8}, \frac{p}{(\rho^3 g^3 E)^{1/4}},\right.$$

$$C\left(\frac{\rho}{g^3 E}\right)^{1/8}, \quad \frac{Y}{(\rho^3 g^3 E)^{1/4}}, \quad \frac{\nu}{(\rho^5 g E^3)^{1/8}},$$

$$\left. a\left(\frac{\rho g}{E}\right)^{1/4}, \quad q\left(\frac{\rho}{g^3 E}\right)^{1/4}, \ldots\right\}. \tag{5}$$

We see here for quarter-root scaling that, in order to achieve similitude among experiments in a constant gravitational field, material properties C, Y and ν must be scaled with E, hydrostatic pressure must be scaled with E and the explosive charge properties, a and q, must be altered appropriately as the explosive charge energy is increased in going from model to prototype experiment. Because material properties are not easily changed, because experiments are usually conducted in a constant gravitational field and because the same explosive charges are employed, the requirements of similarity are definitely not met and the model will not correspond to the prototype experiment. Even if the gravitational field strength can be varied in experiments, similarity cannot be achieved in experiments within a material whose properties (ρ, C, Y, ν) are not scaled but remain constant.

From dimensional analysis we obtain cube-root scaling rules or quarter-root scaling rules depending on whether or not gravity is considered unimportant or important in cratering. Results obtained here focused on explosion craters; however, the same scaling rules may be applied equally well to impact produced craters if we regard E as the kinetic energy of an impacting body and d as its diameter and if also we insert the Pi-term ρ_0/ρ for ratio of impactor density to target density. Other Pi-terms related to the impacting body may be inserted to account for other variables felt to be important, e.g., Y_0/Y, C_0/C, ν_0/ν.

4. Infractions of the Law of Similitude

Often experimenters are not cognizant of sources contributing to violation of similitude in their cratering experiments. When similitude does not exist among experiments, the scaling rules which are being employed in experiments may generate confusion and if scaling rules are being determined by experiments, strange results may be obtained. Some examples of material properties which can give rise to violation of similarity are strain rate effects (Grady *et al.*, 1976), layering (Piekutowski, 1975) and other inhomogeneities (Butters *et al.*, 1975) in the medium being cratered. Vortman (1977) has invoked strain rate effects as one mechanism possibly accounting for crater volumes scaling with energy to an exponent greater than one. Results of explosive cratering experiments in layered media are shown in Fig. 2 taken from Piekutowski (1975). In these experiments a low strength, competent medium (cemented sandstone) was overlain by a densely packed, unconsolidated Ottawa sand. It is evident that pronounced effects in size and shape of craters are obtained depending on layer thickness

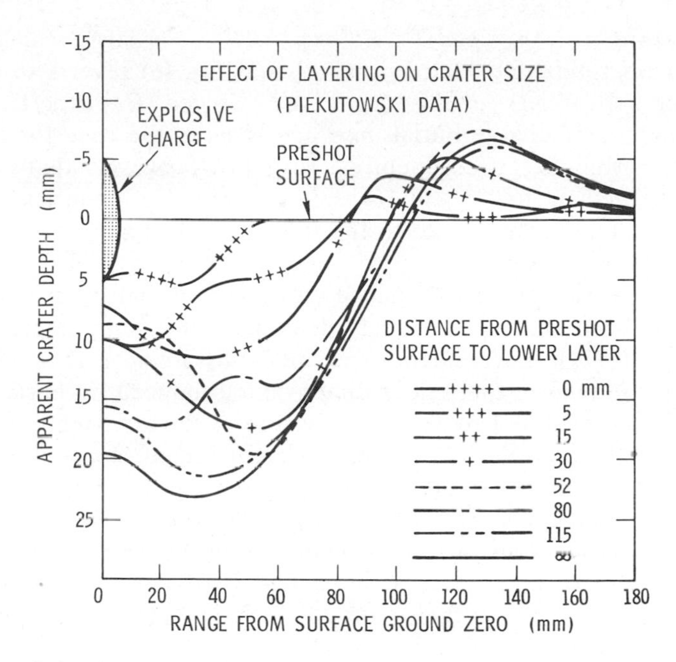

Fig. 2. Variation in apparent crater profiles as a function of top layer thickness for half-buried explosive charges.

relative to charge dimension. Attempting to scale crater dimensions resulting from one size charge to dimensions from another size charge in a medium with constant layer thickness will likely cause significant difficulties if it is meaningful at all.

A large number of cratering experiments with both chemical and nuclear explosives have been conducted in one geologic medium, Nevada desert alluvium. Several decades of energy release (10^2–10^8 kg TNT) are represented by these experiments which permit a critical examination of scaling. Neither cube-root nor quarter-root scaling is observed. Rather an empirical scaling, the 1/3.4 rule, (Chabai, 1959; Vaile, 1961; Nordyke, 1962) is found to best describe data; i.e., linear crater dimensions are proportional to the 1/3.4 power of explosive energy release as opposed to either the 1/3 or 1/4 power. It is asserted that this empirical scaling result is a consequence of similarity violation among cratering experiments conducted in Nevada alluvium. One possible explanation for the alluvium result is given by the overburden-Rodionov scaling rule (Chabai, 1965; Rodionov, 1970).

$$\frac{r}{d} = f_3\left\{\frac{(p_0 + \rho g d)^{1/3} d}{E^{1/3}}, \ldots\right\}. \tag{6}$$

Atmospheric pressure is given by p_0 and $\rho g d$ is overburden pressure at charge burial depth d. It is seen that, for small scale experiments (i.e., small E and correspondingly small d) $p_0 \gg \rho g d$, under which circumstances Eq. (6) reduces to cube-root scaling, i.e., $(p_0+\rho gd)^{1/3}d/E^{1/3} \rightarrow p_0^{1/3}d/E^{1/3}$. For large-scale experiments at correspondingly greater depth, $p_0 \ll \rho gd$, and Eq. (6) reverts to quarter-root scaling, i.e., $(p_0+\rho gd)^{1/3}d/E^{1/3} \rightarrow (\rho gd)^{1/3}d/E^{1/3} \rightarrow \rho gd^4/E \rightarrow d(\rho g/E)^{1/4}$. At intermediate depths of charge burial, $p_0 \approx \rho gd$. This is the case for many of the alluvium experiments; consequently, Eq. (6) suggests that a "scaling" exponent intermediate to 1/3 and 1/4 should apply. The empirical 1/3.4 scaling result then is largely attributable to the influence of atmospheric pressure and gravity by these arguments.

That atmospheric pressure is indeed an important and influential variable in createring has been most convincingly demonstrated by the impressive experiments of Herr (1971). Independent experiments by Johnson *et al.* (1969) and by Rodionov (1970) confirm that crater dimensions produced by buried explosions can be significantly altered by varying pressure of the atmosphere over the half space in which the cratering experiments are conducted. Some results of Herr's experiments are illustrated in the photographs of Figs. 3 and 4. For each type of material and for each group of four experiments shown, the explosive energy release and the burial depth (also scaled burial depth) are held constant. Only the atmospheric pressure was varied as indicated. It is seen that by increasing the atmospheric pressure by a factor of 20, the size of the resulting crater is strikingly reduced.

To demonstrate the influence of gravity on cratering, experiments have been conducted in accelerated frames. Viktorov and Stepenov (1960) performed explosive cratering experiments in sand at acceleration levels $g = 1g_e$, $25g_e$, $45g_e$, and $66g_e$ ($g_e = 9.8\ \mathrm{m/s^2}$). Some results from their work illustrating the influence of gravity on cratering are shown in Fig. 5. Crater size diminishes as the strength of the gravitational field, under which the crater was formed, increases. Experiments by Johnson *et al.* (1969) in sand under conditions where the gravitational acceleration was simulated to be $g = 0.17g_e$, $0.38g_e$, $1.00g_e$, and $2.50g_e$ also demonstrated significant changes in crater dimensions produced by changes in g-levels. More recent experiments (Schmidt, 1977a) with explosion craters produced in clay under g-levels up to $480g_e$ confirm earlier findings and provide information for a different medium. It should be noted that in each of the above cited experimental investigations, the atmospheric pressure was not zero but at some level (~ 0.1 MPa). Experiments to examine the influence of gravitational fields or acceleration on craters produced by hypervelocity impactors have been recently performed (Gault, 1976; Gault and Wedekind, 1976) and reveal, as in the case of explosion craters, that crater dimensions are reduced by increased acceleration levels.

The fact that acceleration noticeably affects crater size suggests that quarter-root scaling [Eq. (5)] rather than cube-root scaling (Fig. 1) should be the rule employed in modeling prototype experiments. However, as we shall see in the next section, quarter-root scaling is not confirmed by experiments. That the quarter-root rule is not experimentally observed is attributed to our inability to conduct experiments with the required similitude. As discussed in section 3, quarter-root scaling imposes severe requirements on experiments in order to

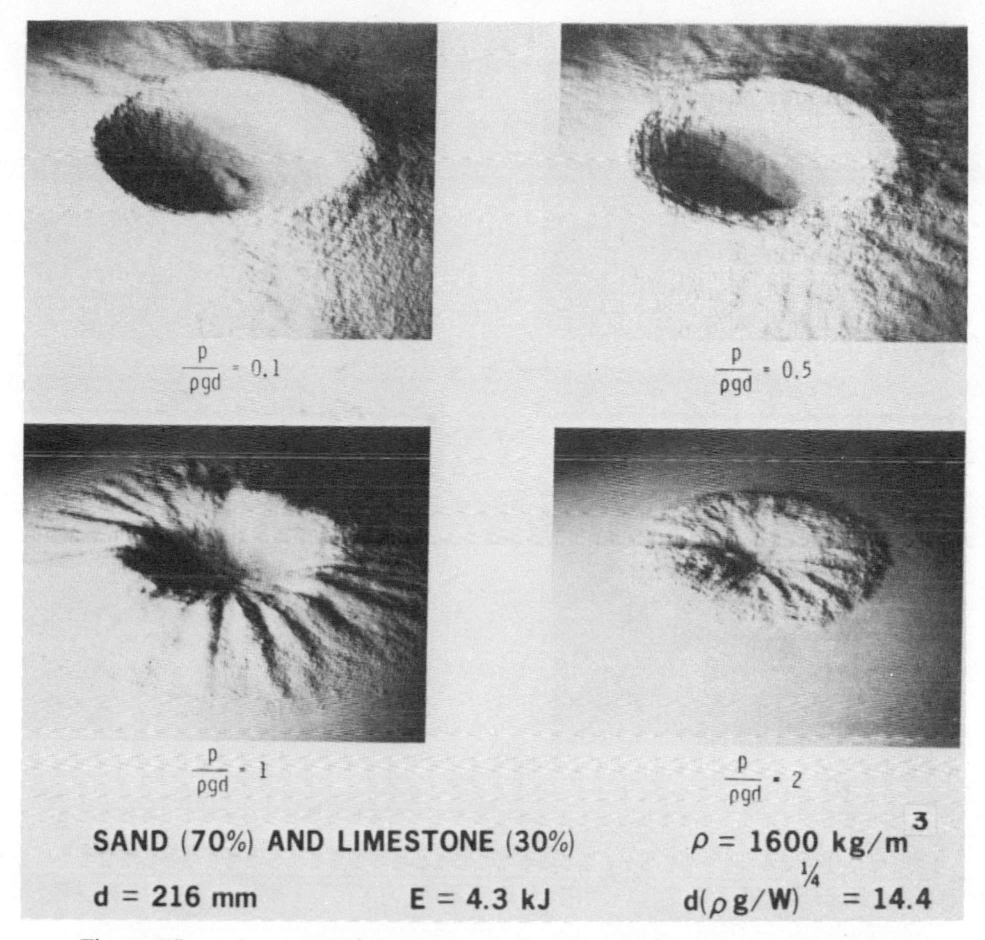

Fig. 3. Effect of atmospheric pressure on size of explosion craters produced in a sand-limestone medium.

achieve similarity. When these requirements are not met, as is usually the case, deviations from quarter-root scaling result. It should be emphasized that achieving similarity may not be just a matter of altering or properly scaling material properties and explosive characteristics. Performing similar experiments may be inherently impossible. Quarter-root scaling arises in explosive cratering when in the cratering process work is done against the gravitational potential. Late time expansion of a buried explosive fireball does work against gravity by lifting the mound of earth above it. Also still later as material is ejected ballistically to form the ejecta blanket, work is done against gravity. Early in the crater forming process, material motion and deformation are characterized by cube-root rules (Killian and Germain, 1977). Shock wave induced velocities and stresses in different scale experiments are constant at the same cube-root scaled times and distances. The radius of the explosive fireball, the extent of vaporization and

Fig. 4. Effect of atmospheric pressure on size of explosion craters in a limestone medium.

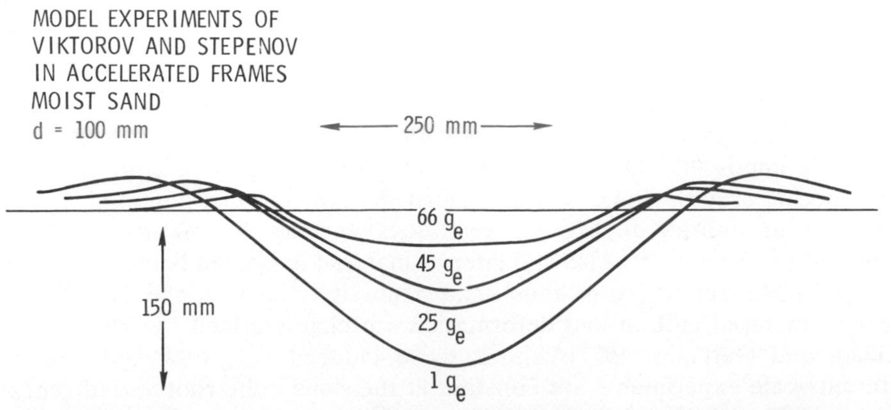

Fig. 5. Reduction in dimensions of explosion craters produced by increased frame acceleration.

melting and the distance to which material is fractured all are proportional to the cube-root of explosive energy release. As suggested by Hess and Nordyke (1961) initial velocity distributions in earth material may be established primarily by the shock wave emanated from the explosive source, an effect which is governed by cube-root scaling. Material particles which are removed from the eventual crater volume are placed in ballistic trajectories which effect is described by quarter-root scaling. Under these conditions, larger explosions will produce smaller than cube-root scaled craters. Neither cube-root nor quarter-root scaling will be observed. Crater radius and depth may each scale with different exponents of explosive energy (Vortman, 1968), indicating that crater shape is changing with explosive energy release.

5. Crater Scaling in Absence of Complete Similitude

Violations of similarity in cratering experiments cause cube-root scaling rules to overpredict crater dimensions from large explosive yields, while quarter-root rules underpredict crater sizes. Largely based on this observation a scaling relationship intermediate to cube-root and quarter-root was proposed by Saxe and DelManzo (1970) and by Westine (1970). Their result is expressed briefly as

$$\frac{r}{d} = f_4\left\{\frac{d\rho^{7/24}C^{1/3}g^{1/8}}{E^{7/24}}, \ldots\right\}, \tag{7}$$

where ... indicates the remaining terms of Saxe and DelManzo. It is seen that a linear crater dimension is proportional to the 7/24th power of explosive energy release. This scaling exponent for energy, 7/24, is, within experimental uncertainty, the same value as the empirical value, 1/3.4, established by experiment in Nevada alluvium. Equation (7) also suggests that crater dimensions vary inversely as the 1/8 power of the gravitational acceleration, g, as opposed to the 1/4 power required by quarter-root scaling [Eq. (5)].

In Fig. 6, the data of Johnson *et al.* are shown. The ratio of crater radius to depth of burst is plotted against the quantity $(g^{1/8}d)^{-1}$. A similar plot of the data of Viktorov and Stepenov is given in Fig. 7. From these figures it is readily seen that 1/8 is a better exponent, in terms of minimizing data scatter, than 1/4 to scale the data on g for both the Johnson *et al.* and the Viktorov and Stepenov data. This result is somewhat puzzling and leads one to inquire whether or not there is something significant about the $g^{1/8}$ result or whether in both sets of experiments some violations of similarity have fortuitously produced the same result. It may be recalled that both sets of experiments were conducted in sand, that in both sets of experiments atmospheric pressure was not scaled but held constant, and that the explosive charges could not be scaled to achieve similarity.

As indicated in Fig. 8, exponents, n, for g were determined by Johnson *et al.* from these data by considering the variation of crater dimension with change in acceleration level while keeping the charge burst depth constant. As the burst depth increases from 0 to 102 mm the value of n is found to increase from 0.111

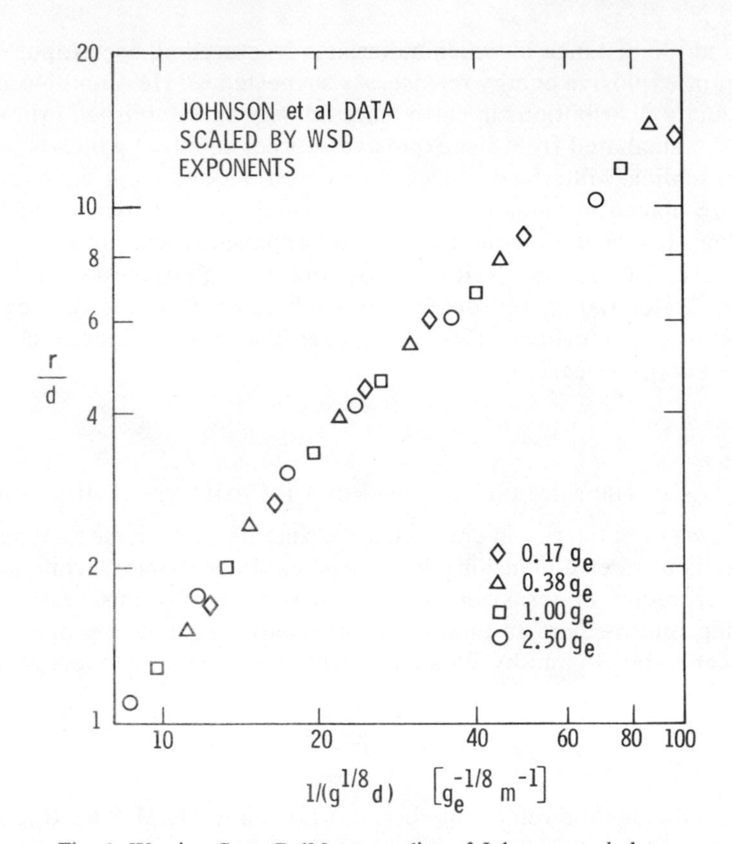

Fig. 6. Westine–Saxe–DelManzo scaling of Johnson *et al.* data.

to 0.157. The data of Viktorov and Stepenov provide a value of $n = 0.07$. Schmidt (1977b) finds an even smaller value of n from his experiments in clay.

While this n is one measure of the experimental results, it must be recognized that n determined in the manner indicated in Fig. 8 is *not* a scaling exponent for g. This is evident inasmuch as n is determined from experiments at the same actual depth of burst but *different* scaled depths of burst. From Eq. (5) it is recalled that

$$r\left(\frac{\rho g}{E}\right)^{1/4} = f_2\left\{d\left(\frac{\rho g}{E}\right)^{1/4}, \ldots\right\}.$$

At the same scaled depth of burst, if experiments are similar, we obtain

$$\frac{r_i}{r_j} = \left(\frac{g_j}{g_i}\right)^n, \quad n = \frac{1}{4},$$

since ρ and E were constant in experiments. By considering experiments of Johnson *et al.* which have the same scaled burst depth, $dg^{1/4}$, the value of n may be determined from the data with

$$n = \ell n(r_i/r_j)/\ell n(g_j/g_i). \tag{8}$$

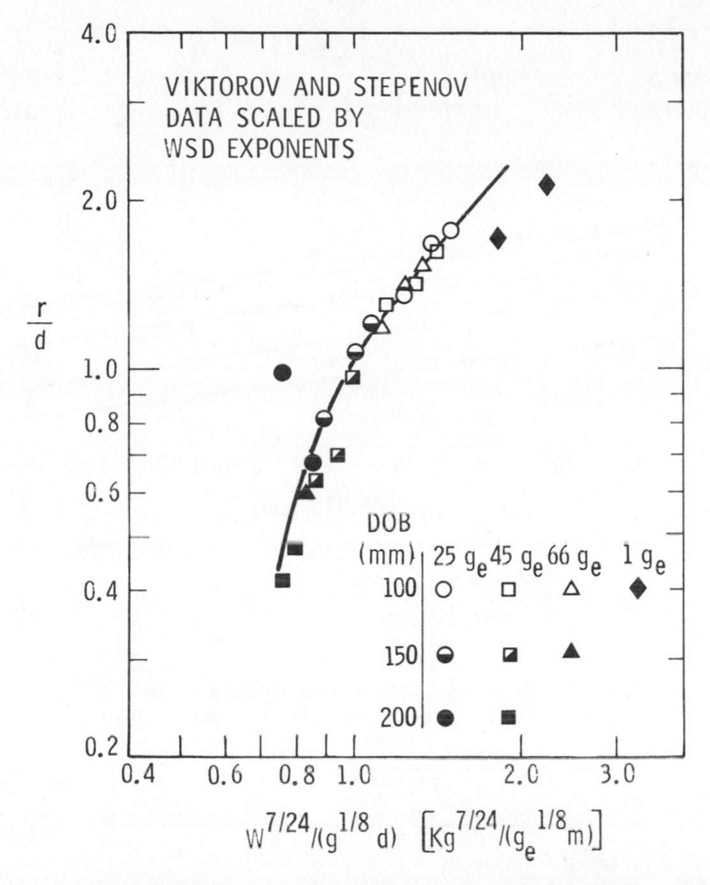

Fig. 7. Westine–Saxe–DelManzo scaling of Viktorov and Stepenov data.

Examining the data of Johnson *et al.*, five pairs of experiments are found, each pair of which has approximately the same scaled depth of burst. This set is tabulated in Table 1. From the data of Table 1 values of n were determined with Eq. (8) and are shown plotted against scaled depth of burst in Fig. 9. It is seen that the scaling exponent for g varies radically with scaled depth of burst and is in all cases less than the 1/4 value expected from quarter-root scaling. Note that the position of the Westine–Saxe–DelManzo exponent, 0.125, straddles the experimental points. Listed adjacent to each plotted point is the value of $(p_0/\rho g d)$ which prevailed in experiments. The nonconstant value for the exponent of g, as determined from the data of Johnson *et al.*, strongly suggests that similarity constraints are not being met in the experiments. It is impossible to judge which of the several nonscaling features in the experiments is most significant in violating the requirements of similarity. In addition to maintaining fixed explosive charge characteristics, atmospheric pressure and also material properties were held constant throughout experiments in understandable opposition to the demands of similitude. (Among the data of Johnson *et al.*, only

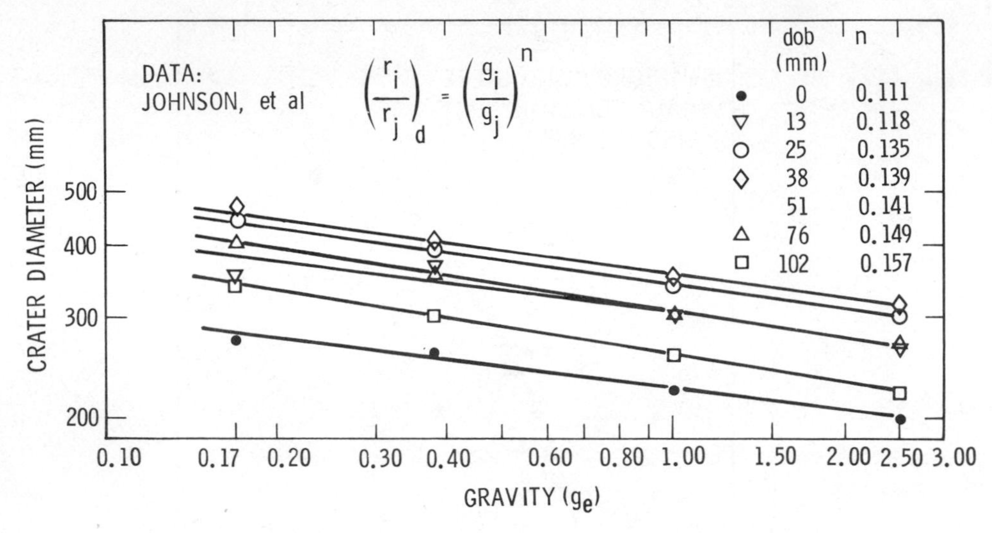

Fig. 8. Decrease in crater diameter with increasing acceleration at several actual, constant, explosion burst depths.

Table 1. Pairs of explosive cratering experiments, from data of Johnson *et al.*, having approximately common quarter-root scaled burst depths.

g	$0.17g_e$	$2.5g_e$
d	25	13
r	21.5	13.3
$dg^{1/4}$	16.1	16.3
d	51	25
r	22.8	15
$dg^{1/4}$	32.7	31.4
d	76	38
r	20.3	16
$dg^{1/4}$	48.8	47.8
d	102	51
r	17.3	15.5
$dg^{1/4}$	65.5	64.1
g	$0.38g_e$	$1.0g_e$
d	38	25
r	21	17.3
$dg^{1/4}$	29.8	25.0

(r and d in millimeters).

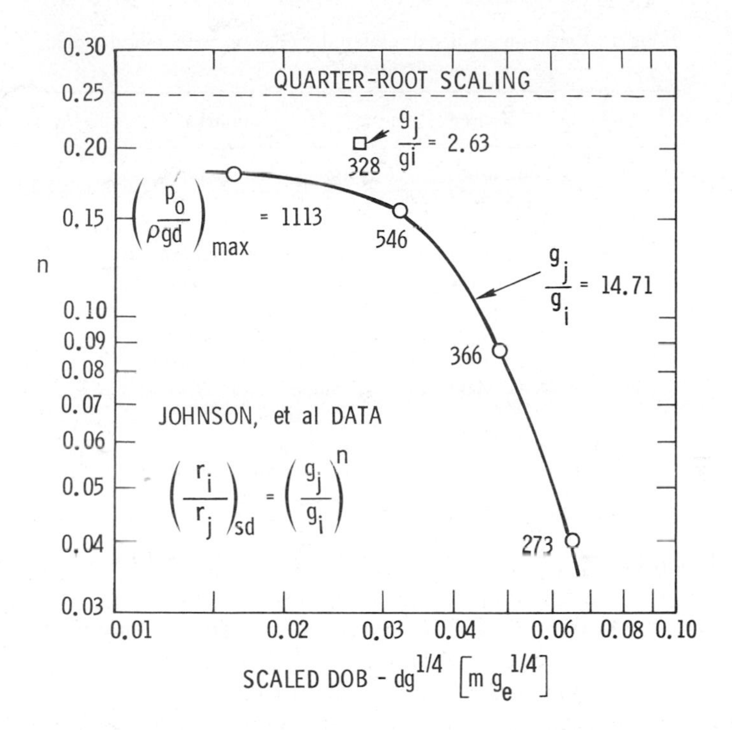

Fig. 9. Scaling exponent for g determined at constant scaled burst depths from data of Johnson *et al.*

two pairs of experiments are found which have nearly the same Westine–Saxe–DelManzo scaled burst depth, $dg^{1/8}$. Values of n determined by these pairs are 0.13 and 0.09.)

The crater radius dimension observed in experiments by Johnson *et al.* and by Viktorov and Stepenov appears to be best scaled with g^n when $n = 1/8$. From considerations of ejecta velocity distributions Ivanov (1976) argues that crater radii should be proportional to $g^{-1/7}$ or that $n = 1/7$. In an independent approach, also considering ejecta, Maxwell (1977) finds $n = 1/6.4$. The impact experiments of Gault (1976) in accelerated frames reveal $n = 1/6$. Schmidt (1977a) has obtained a best fit to crater volume data which include both accelerated frame experiments ($25g_e$ to $480g_e$) and Nevada alluvium experiments (chemical and nuclear explosions) covering a range of 12 decades in explosive energy release. Schmidt employs mass scaling (Chabai, 1965) to scale crater volume data. Converting his mass scaled crater volumes to energy scaled volumes by substituting for charge mass, W, the quantity $(E^3\rho/g^3)^{1/4}$ (Divoky, 1966), one finds that crater volumes scale according to Schmidt with g^{3n}, where $n = 5/16$. This wide range of results is summarized in Table 2.

The difference between $g^{1/4}$ and $g^{1/8}$, for example, is somewhat academic if attempts are made to predict crater sizes on other planets or the moon by scaling terrestrial craters to the gravitational fields of the lunar or planetary systems of

Table 2. Variation of linear crater dimensions with gravitational acceleration, $r \sim g^{-n}$.

n	Source	Remarks
0	Cube-root scaling	Dimensional analysis
5/16	R. M. Schmidt	Chemical, nuclear explosions, accelerated frame data
1/4	Quarter-root scaling	Dimensional analysis
1/6	D. E. Gault	Accelerated frame impact experiments in sand
1/6.4	D. E. Maxwell	*Z*-model of cratering and Ejection
1/7	B. A. Ivanov	Ejecta model
1/8	Westine, Saxe, and DelManzo	Dimensional analysis accelerated frame data

interest. Shown on the ordinate in Fig. 10 is the ratio of the crater radius, r_{WSD}, obtained with the 1/8 exponent on g, to the crater radius, r_{QRR}, obtained by the quarter-root rule, both scaled from a terrestrial crater neglecting atmospheric pressure and material property differences. It is seen that predictions of crater dimensions on the moon and planets suffer an uncertainty of 20% or less as a consequence of not being able to distinguish between 1/8 and 1/4 as the scaling exponent for g. Also shown in Fig. 10 are the frame acceleration levels covered by experiments of Johnson *et al.* and Viktorov and Stepenov. The shaded region of Fig. 10 is meant to indicate that more experiments should be conducted at acceleration levels greater than about $25g_e$ in order to obtain a measure of the scaling exponent on g with reasonable accuracy.

6. Elimination of the Similarity Violation Arising from Explosive Sources

One distinct source of similarity violation in cratering experiments at different acceleration levels is the explosive charge. In most experiments to date, the explosive charge is held fixed in cube-root fashion. As noted earlier, Eq. (5), in acknowledging the importance of gravitational acceleration, demands that the size of the explosive as well as the specific energy of the charge vary in a specific manner with g-level. From the last two Pi-terms of Eq. (5) it is seen that the constraint among these variables is given by q/ga = constant.

In Fig. 11 conditions are listed for suggested experiments at different g-levels which obey the similarity constraints placed on the explosive charges. The explosive charges are of the type employed by Rodionov (1970), being, for example, hollow glass spheres pressurized to required levels. The total energy in each charge is held constant at 100J. Craters produced will have dimensions approximately as indicated if the medium is a cohesionless, dry sand such as Ottawa quartz sand. To avoid any difficulties which may arise with scaling

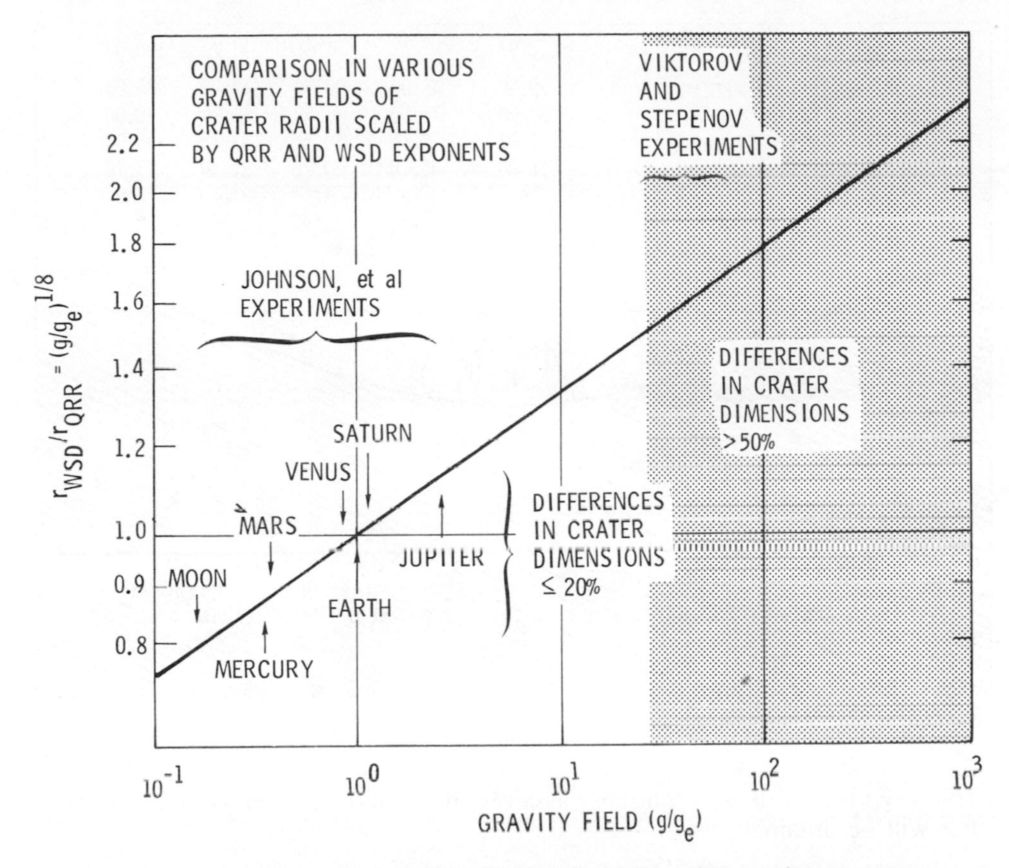

Fig. 10. Ratio of crater dimensions scaled from terrestrial craters by 1/8-exponent to that scaled by 1/4-exponent on g for various gravity fields.

atmospheric pressures, it is recommended that initial experiments be conducted in a vacuum. The different charges specific energies required at the various acceleration levels are obtained by filling the hollow glass spheres with different gases; in this example, noble gases at a temperature of 300 K. Charge specific energies must be scaled as g-level to the 3/4 power. Helium having the highest specific energy must be used at the highest acceleration level ($100g_e$), whereas xenon must be used at the lowest level ($\sim 1\ g_e$). Charge radii and also burial depths are scaled inversely as g-level to the 1/4 power. As frame acceleration is increased, charge radius and depth of burial are decreased, while charge specific energy is increased. If the experiments suggested were truly similar, they would all produce the same size crater. It should be noted that material properties C, Y and ν are not easily scaled with g and to the extent they are important, they will represent sources of similarity violation and produce different size craters. It is felt that inability to scale properties of the explosive charge is a source of similarity violation considerably greater than that associated with the medium properties. By removing the difficulties concerned with improper scaling of the

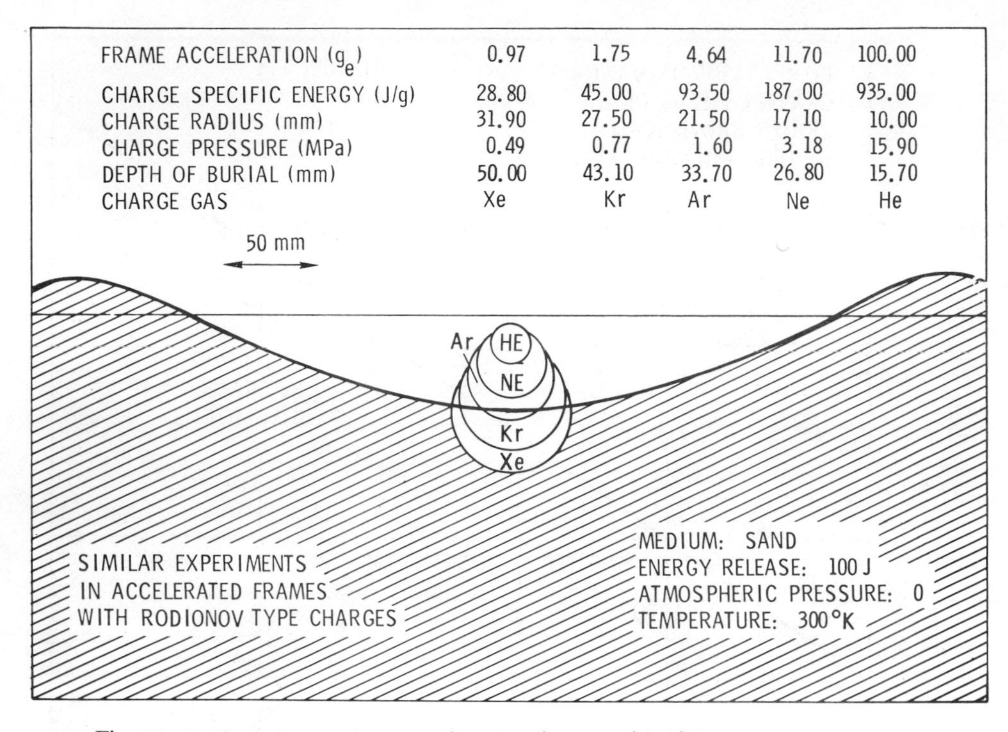

Fig. 11. Outline of explosion cratering experiments with high pressure noble gases.

explosive charges, it is hoped more accurate data on the proper scaling exponent of g will be obtained.

7. Conclusions

The influence of atmospheric pressure and gravitational field on formation and scaling of explosion and impact craters has been demonstrated experimentally. Scaling rules obtained from dimensional analysis are generally not verified by experiment. This variance is attributed to the difficulty of performing cratering experiments which adhere strictly to requirements for similitude among experiments. Recognition of the importance of the gravitational field in cratering leads to quarter-root scaling which states that crater dimensions scale directly as the 1/4 power of energy released and inversely as the 1/4 power of g, the acceleration due to gravity or the acceleration of the coordinate frame in which cratering experiments are conducted. Quarter-root scaling rules demonstrate that it is not possible to perform similar experiments in a constant gravitational field and that in general certain material properties must be scaled to achieve similarity. In addition, naturally occurring geologic materials, in which most cratering experiments are conducted, demonstrate strain rate effects, non-homogeneities, and stratification, all of which prevent realization of similarity. Various experimental scaling results such as the (1/3.4) rule established for craters in Nevada alluvium and the spread of values observed for the exponent of g (ranging from 1/8 to 5/16) are attributed to our inability to conduct similar experiments. It has been pointed out that, in the explosive cratering experiments conducted to date in accelerating frames, similarity has not been realized for at least two reasons. In these experiments, the atmospheric pressure was constant and frequently larger relative to overburden pressure at charge burial depth. Also, explosive charges employed were identical among experiments and not scaled with acceleration level as required. A set of experiments is suggested for accelerated frames in which the sources of similarity violations associated with a constant atmospheric pressure and with constant explosive charge properties are removed. From such experiments, it is hoped that the scaling exponent on g might be determined more accurately. In addition, by experiments of this type, one could begin to investigate the influence on crater scaling of similarity violations other than that due to atmospheric pressure and explosive charge properties.

Acknowledgments—The author gratefully acknowledges the photographs and other information supplied by Robert Herr and is indebted to L. J. Vortman and Frank Biggs of Sandia Laboratories for several fruitful conversations. This work was supported by the U.S. Energy Research and Development Administration, ERDA.

References

Baker, W. E., Westine, P. S., and Dodge, F. T.: 1973, *Similarity Methods in Engineering Dynamics*, Hayden Book Company, Inc., Rochelle Park, New Jersey.

Brand, L.: 1957, The Pi-Theorem of Dimensional Analysis, *Archive for Rational Mechanics and Analysis*, **I**, 1, 35–45.

Bridgman, P. W.: 1949, *Dimensional Analysis*, Yale University Press, New Haven, Connecticut.

Buckingham, E.: 1914, On Physically Similar Systems; Illustrations of the Use of Dimensional Equations, *Phys. Rev.* IV, **4**, 345–376.

Butters, S. W., Swolfs, H. S., Johnson, J. N., Hadala, P. F., and Butler, D. K.: 1975, Field, Laboratory and Modeling Studies on Mount Helen Welded Tuff for Earth Penetrator Test Evaluation, TR75-9, Terra Tek, University Research Park, 420 Wakara Way, Salt Lake City, Utah.

Chabai, A. J.: 1959, Crater Scaling Laws for Desert Alluvium, SC-4391(RR), Sandia Laboratories, Albuquerque, New Mexico.

Chabai, A. J.: 1965, On Scaling of Craters Produced by Buried Explosives, *J. Geophys. Res.* **70**, 5075–5098.

Divoky, D.; 1966, The Equivalence of Mass and Energy Scaling of Crater Dimensions: Comments on a Paper by A. J. Chabai. *J. Geophys. Res.* **71**, 2691.

Gault, D. E.: 1977, Significance of Strength and Gravity for Impact Craters (abstract). In *Papers Presented to the Symposium on Planetary Cratering Mechanics*, Lunar Science Institute, Houston.

Gault, D. E. and Wedekind, J. A.: 1977, Experimental hypervelocity impact into quartz sand: II, effects of gravitational acceleration. In *Impact and Explosion Cratering* (D. J. Roddy, R. O. Pepin, and R. B. Merrill, eds.), Pergamon Press, New York. This volume.

Grady, D. E., Hollenbach, R. E., Schuler, K. W., and Callender, J. F.: 1976, Compression Wave Studies in Blair Dolomite, SAND76-0005, Sandia Laboratories, Albuquerque, New Mexico.

Herr, R. W.: 1971, Effects of Atmospheric-Lithostatic Pressure Ratio on Explosive Craters in Dry Soil, NASA-TRR-366, NASA Langley Research Center, Hampton, Virginia.

Hess, W. N. and Nordyke, M. D.: 1961, Throwout Calculations for Explosion Craters, *J. Geophys. Res.* **66**, 3405–3412.

Ivanov, B. A.: 1976, The Effect of Gravity on Crater Formation: Thickness of Ejecta and Concentric Basins (abstract). In *Lunar Science VII*, p. 411–413.

Johnson, S. W., Smith, J. A., Franklin, E. G., Moraski, L. K., and Teal, D. J.: 1969, Gravity and Atmospheric Pressure Effects on Crater Formation in Sand, *J. Geophys. Res.* **74**, 4838–4850.

Kececioglu, D.: 1960, Dimensional Analysis in Ten Steps, *Product Engineering*, February 1, p. 54–57.

Killian, B. G. and Germain, L. S.: 1977, Scaling of Cratering Experiments—An Analytical and Heuristic Approach to the Phenomenology, In *Impact and Explosion Cratering* (D. J. Roddy, R. O. Pepin, and R. B. Merrill, eds.), Pergamon Press, New York. This volume.

Langhaar, H. L.: 1951, *Dimensional Analysis and Theory of Models*, John Wiley and Sons, Inc., New York.

Maxwell, D. E.: 1977, Simple *Z* Model of Cratering, Ejection and the Overturned Flap. In *Impact and Explosion Cratering* (D. J. Roddy, R. O. Pepin, and R. B. Merrill, eds.), Pergamon Press, New York. This volume.

Nordyke, M. D.: 1962, An Analysis of Cratering Data from Desert Alluvium, *J. Geophys. Res.* **67**, 1965–1974.

Piekutowski, A. J.: 1975, The Effect of a Layered Medium on Apparent Crater Dimensions and Ejecta Distribution in Laboratory-Scale Cratering Experiments, AFWL-TR-75-212, Air Force Weapons Laboratory, Kirtland Air Force Base, New Mexico.

Rodionov, V. N.: 1970, Methods of Modeling Ejection with Consideration of the Force of Gravity, UCRL-Trans-10476, Lawrence Livermore Laboratory, Livermore, California.

Saxe, H. C. and DelManzo, D. D.: 1970, A Study of Underground Explosion Cratering Phenomena in Desert Alluvium, *Symposium on Engineering with Nuclear Explosives*, Proceedings, Vol. 2, CONF-700101, U.S. Dept. of Commerce, Springfield, Virginia.

Schmidt, R. M.: 1977a, A Centrifuge Cratering Experiment. In *Impact and Explosion Cratering* (D. J. Roddy, R. O. Pepin, and R. B. Merrill, eds.), Pergamon Press, New York. This volume.

Schmidt, R. M.: 1977b, Personal Communication.

Sedov, L. I.: 1959, *Similarity and Dimensional Methods in Mechanics*, Academic Press, New York.

Vaile, R. B.: 1961, Pacific Craters and Scaling Laws, *J. Geophys. Res.* **66**, 3413–3433.

Viktorov, V. V. and Stepenov, R. D.: 1960, Modeling of the Action of an Explosion with Concentrated Charges in Homogeneous Ground, Translation SC-T-392, Sandia Laboratories, Albuquerque, New Mexico.

Vortman, L. J.: 1968, Craters from Surface Explosions and Scaling Laws, *J. Geophys Res.* **73**, 4621–4636.

Vortman, L. J.: 1977, Craters from Surface Explosions and Energy Dependence, A Retrospective View. In *Impact and Explosion Cratering* (D. J. Roddy, R. O. Pepin, and R. B. Merrill, eds.), Pergamon Press, New York. This volume.

Westine, P. S.: 1970, Explosive Cratering, *J. Terramech* **7**, 9–19.

Roddy, D. J., Pepin, R. O., and Merrill, R. B., editors.
(1977) *Impact and Explosion Cratering*, Pergamon Press (New York), p. 1215–1229.
Printed in the United States of America

Craters from surface explosions and energy dependence—A retrospective view*

LUKE J. VORTMAN

Field Experiments Division, Sandia Laboratories, Albuquerque, New Mexico 87110

Abstract—Treatments of the energy dependence of crater radius and depth by dimensional analysis or cratering formulas usually result in the energy dependence being the same for both dimensions. This implies that craters would not be changing shape and that the ratio of apparent radius (R_a) and apparent depth (D_a) would remain constant over a substantial range of energy. R_a/D_a has been plotted against energy for six sets of chemical explosive data, one set of nuclear explosive data, five terrestrial craters and nine lunar craters. The results show a systematic increase in the ratio over 18 orders of magnitude of energy. Unusually large rates of increase or an occasional decrease in the ratio for the chemical explosive data can be explained by nonsimilar media or geologic structure. Larger rates for the nuclear explosions are consistent with increased fireball temperatures and fireball velocity of rise with increased energy. The rate of increase of the ratio for the terrestrial impact craters is essentially the same as that for one set of chemical explosive craters (the one in the most homogeneous medium) and the smaller lunar impact craters. A larger rate of increase for the larger-energy lunar impact craters was shown to become smaller when the rise from the chord representing the crater diameter to the circumference of the moon was added to the crater depth. Instantaneous rebound of the bottom of large-energy impact craters may account for the remainder of the amount by which the rate of increase exceeds that of the smaller-energy craters. What appears to be an energy dependence greater than 1 for some of the sets of chemical explosive data is attributed to the observation that strength properties of the medium are more important for smaller than larger explosions and that the significance of rate-dependent medium response increases as shock rise times and durations decrease with decreases in energy.

1. INTRODUCTION

THE DIMENSIONS OF CRATERS produced by near-surface nuclear and chemical explosions can be related to explosive yield as follows (Vortman, 1968a)

$$R_a = C_1 W^{n_1}. \quad (1)$$

$$D_a = C_2 W^{n_2}, \quad (2)$$

$$V_a = C_3 W^{n_3}, \quad (3)$$

where W represents the energy of the explosion, R_a, D_a, and V_a are radius, depth, and volume of the apparent crater. Values of C_i and n_i ($i = 1, 2, 3$) were obtained by regression analyses of measured crater dimensions for seven sets of events, including one set of nuclear events. Those values appear in Table 1, whose main message is that (a) $n_1 \neq n_2$, (b) with one exception $n_1 > \frac{1}{3}$, (c) with two exceptions $n_2 < \frac{1}{3}$, and (d) $n_3 > 1$ for four of the chemical-explosion sets. Here we reexamine the work summarized in Table 1, and call attention to

*The work was performed under the auspices of the U.S. Energy Research and Development Administration.

Table 1. Summary of energy dependence of crater dimensions.

	Charge shape	R_a		D_a		V_a			
		C_1	n_1	C_2	n_2	C_3	n_3	$\frac{2n_1+n_2}{n_3'}$	n_1-n_2
High Explosive Experiments									
1 NTS Playa	○	0.28	0.375 ± 0.006	0.155	0.342 ± 0.013	0.0143	1.111 ± 0.018	1.092 ± 0.025	0.033
2 Watching Hill Silt	○	0.20	0.408 ± 0.005	0.37	0.259 ± 0.019	0.03	1.040 ± 0.050	1.075 ± 0.029	0.149
3 Drowning Ford Silt	⌓	0.16	0.426 ± 0.009	0.34	0.237 ± 0.012			1.089 ± 0.030	0.189
4 All Basalt and Flat Top 1	○	0.49	0.27	0.16	0.28	0.074	0.780	0.820	−0.010
5 Yakima Firing Center Only	○	0.21	0.371 ± 0.056	0.165	0.262 ± 0.042	0.0028	1.180 ± 0.117	1.004 ± 0.154	0.109
6 Yakima Firing Center, Sailor Hat	⌓	0.20	0.37	0.0265	0.47	0.0028	1.140	1.210	−0.100
Nuclear Explosive Experiments									
7 Figure 12, Vortman, 1968a		13.7*	0.443 ± 0.016	5.43*	0.256 ± 0.020	2.10×10^{3}*	1.049 ± 0.051	1.142 ± 0.052	0.187
8 Same adjusted for HOB		21.1*	0.402 ± 0.016	11.5*	0.185 ± 0.038	1.14×10^{4}*	0.885 ± 0.073	0.989 ± 0.070	0.217
9 Same as 8 adjusted for water washing		22.1*	0.387 ± 0.023	10.6*	0.215 ± 0.054	1.14×10^{4}*	0.885 ± 0.073	0.989 ± 0.100	0.172
10 Same as 9, but with adjustment using a Pacific crater HOB curve rather than NTS curve as in 8							1.000 ± 0.070		
Average of 1–6 and 9			0.372		0.295		1.023		
Average as above but excluding that value which deviates most from the above average			0.390		0.266		1.071		

Values for C for HE experiments are for W in kilograms; for NE experiments they are for W in kilotons.
*These values are corrections of those of Vortman (1968a).
Dimensions are measured with respect to the original ground surface.

dimension-modifying mechanisms not ordinarily considered in dimensional analyses. Those mechanisms may account for errors in predictions of energy dependence based on dimensional analysis or the construction of cratering formulas.

2. Considerations of Energy Dependence

2.1. *High explosive data*

The above equations treat n_1, n_2, and n_3 as constants. Chabai's (1965) dimensional analysis for buried explosions results in $n_1 = n_2$, and both approach $\frac{1}{3}$ where overburden is small and $\frac{1}{4}$ where it is large. Thus, n_1 and n_2 are functions of charge burial depth and are constant at $\frac{1}{3}$ where the charge is at the surface. The data sets examined show that $n_1 \neq n_2$.

For craters in different homogeneous media it would be expected that n_1 would be the same in all media and that medium differences would be accounted for by differences in C_1. The same would be expected for n_2 and n_3 and for C_2 and C_3. If the exponents are different for different media as in the data sets of Table 1, it is because a given medium is not sufficiently homogeneous or because the data set contains craters in similar but not sufficiently identical materials. Both these conditions occur with respect to the data sets of Table 1 except for the playa: the high water table for the explosions in silt, the inclusion of data from a limestone experiment with spherical charges in basalt, and for the hemispherical charge data the inclusion of data from a large charge (Sailor Hat) in a Hawaii basalt which was different from the basalt at the Yakima Firing Center where the other data were obtained. The set of playa data represents the best of the sets available. It is also the largest set (16 events). Other sets may constitute too few data for reliable regression analysis.

The fact that $n_3 > 1$ for four or five sets of data from chemical explosions has a plausible explanation not recognized by White (1971). A value of $n_3 > 1$ can result from large charges producing more crater volume per unit energy than small charges or by small charges producing less volume per unit energy relative to large charges. It is the thesis here that the latter is the case and the following will discuss two mechanisms which qualitatively support such a thesis.

Dillon (1972) found that "Material properties are highly important in determining the size of explosion-produced craters and some of the more important properties are unit weight, degree of saturation, shearing resistance and seismic velocity". This agrees with results reported by Gault *et al.* (1975). By way of contrast, White (1973) observes "Thus we note that the cratering results are a weak function of material properties of the cratering medium". These seemingly contradictory views are not necessarily so if the energy range for which they are applicable is defined. Dillon's data contained a large number of small-charge craters. Unit weight, degree of saturation, and seismic velocity are energy-independent. Their effects are seen in the constants of Eqs. (1), (2), and (3). Shearing resistance is not energy-independent. Craters from buried explosions in basalt (Vortman *et al.*, 1962) demonstrated that when the charges were large and the volume of material removed was large, the volume contained very many cooling cracks which constituted planes of zero (or near zero) shear and tensile strength. Here shearing resistance was relatively unimportant, supporting White's statement. But if a small charge was detonated in an uninterrupted block of the same material, the shearing resistance would be very important, as Dillon stated. The first condition is analogous to a charge fired in a medium made up of very many closely-fitting bricks without mortar and the second to a small charge fired in one of the bricks.

Thus, intrinsic strength is important in the small-energy regime with which Dillon was concerned, while pre-existing cracks may be more important in the large-energy regime with which White dealt.

The second mechanism relates to the duration and rise time of the stress pulse and its relation to the loading rate at which the medium's shearing resistance becomes strain-rate dependent. Stress pulse shapes at the boundary between where the material does or does not fail in shear (which is close to the true crater boundary within which material becomes disassociated with material on the other side of the boundary) have finite rise times and durations which are a function, probably cube-root, of the energy of the explosion. As the energy increases, the rate of rise increases and the pulse duration decreases to a point at which the strain resistance becomes more and more affected by rate-dependency, and a higher stress is required to overcome the greater strain resistance. For a given energy that higher stress exists at a smaller radial distance, the true crater boundary moves closer, and the true crater volume becomes smaller and with it the apparent crater volume. Thus, two mechanisms have been identified which could account for crater volumes at the low-energy end of the scale smaller than simple dimensional analyses would predict, and which would result in an energy dependence over a limited energy range > 1. No effort is made here to show the energy levels at which these two mechanisms are effective because the strain- or stress-rate at which a given earth medium becomes most sensitive to rate processes is not well known.

More disturbing to some investigators is that these results show n_1 to be different from n_2, whereas most dimensional analysis (Chabai, 1965) and cratering formulas (White, 1973) for buried bursts show or assume them to be equal. Inasmuch as most of the experiments in Table 1 show these differences as do the average of all and the average of all but the most deviant value in each set, the differences are believed to be real. If they are indeed real, the differences reflect a change in crater shape with increase in explosion energy, in which case n_1 and n_2 could not be equal.

Figure 1 shows the ratio of apparent crater radius to apparent crater depth (R_a/D_a) versus yield in kilotons for the data from which Table 1 was derived. Some of Baldwin's (1963) data for lunar and terrestrial impact craters (Table 2) have been added to the figure. There is an irrefutable change in crater shape over 18 decades of energy, and clearly n_1 and n_2 are not equal for surface burst explosion or impact craters. Individual data sets show differences and what may appear to be inconsistencies; however, most can be understood and are discussed below.

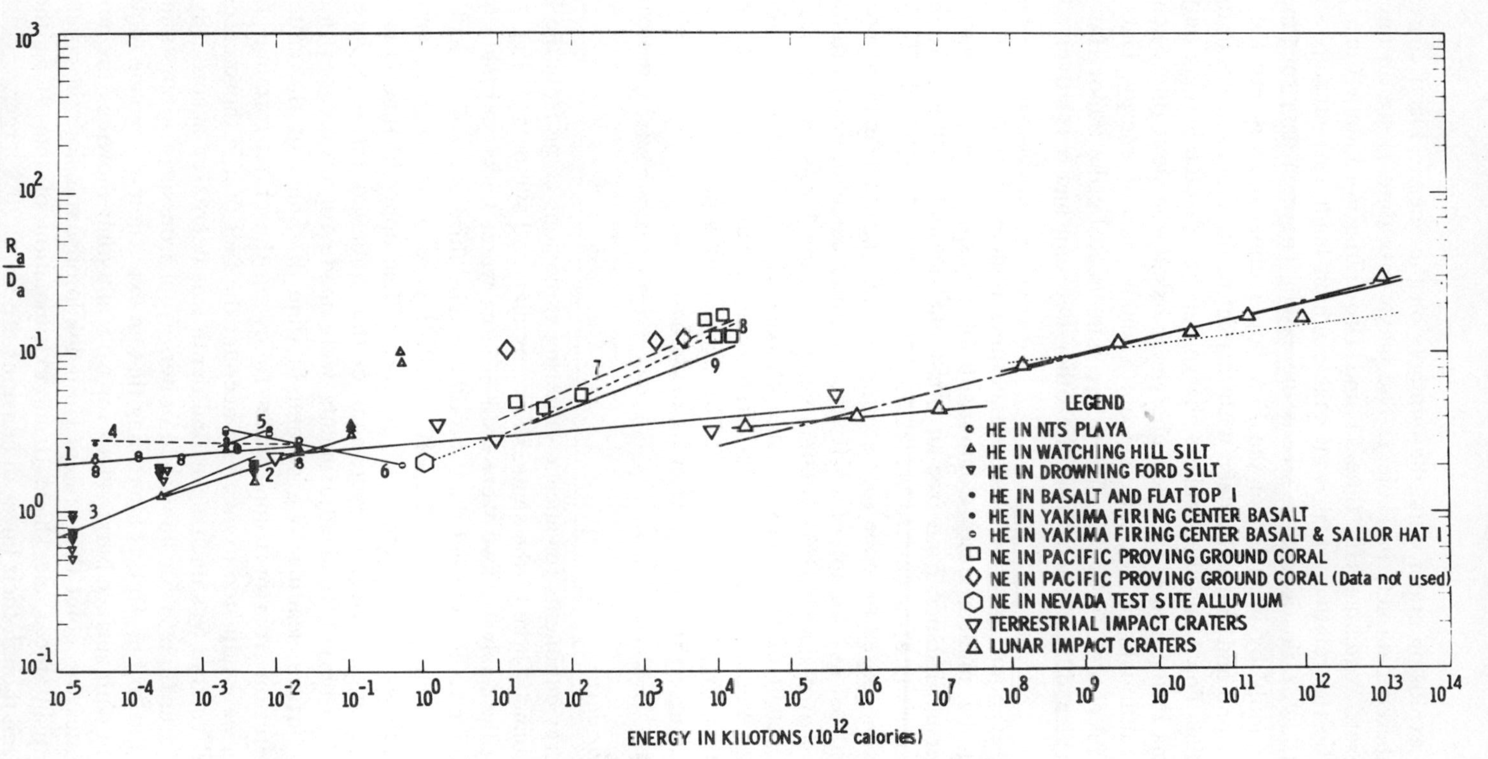

Fig. 1. Ratio of apparent crater radius to apparent crater depth versus explosion energy.

For the explosion crater data, the numbers on the lines in Fig. 1 correspond to the numbered data sets in Table 1. The playa data (line 1) are an internally consistent set. Watching Hill (line 2) and Drowning Ford data (line 3) are influenced by the relatively high water table which leads to unusually shallow craters and high ratios as charge size is increased. The next data set (line 4) is influenced unduly by inclusion of the Flat Top 1 dimensions where the crater was in limestone and unusually deep with respect to the other craters in the set. When the Flat Top 1 and the smallest-charge data are deleted leaving only data from Yakima Firing Center spherical charges in basalt, the slope (line 5) changes to positive although the set suffers from a small span of energy. Data from hemispherical charges in basalt (line 6) are influenced by the Sailor Hat crater which was at a different site from the other two and had a relatively deeper crater.

One other point must be made concerning high explosive craters before considering the nuclear case. Comparison of measured peak airblast overpressure versus distance from free-air burst nuclear and TNT explosions and similar comparison of calculated values has shown (Vortman, 1968b) that in the 5–50 psi region, based on cube-root scaling, a 1 kiloton nuclear burst produces the same airblast as $\frac{1}{2}$ kiloton of TNT. Recent crater calculations (Burton *et al.*, 1975) have shown similar HE/NE effectiveness. Figure 1 was constructed using the actual weight of the high explosive charges. If the factor of 2 in HE/NE effectiveness becomes accepted for cratering, all values for high explosive experiments in Fig. 1 should be moved to the right by a factor of two. Doing so in no way changes the arguments being made here.

2.2. *Nuclear data*

The energy exponents for nuclear cratering explosions come from the Pacific Proving Ground. Three of the shots, shown by diamond symbols on Fig. 1, were not used in the analysis. Two were fired in large water tanks and had different shock coupling to the ground than the others. The third shot, Zuni, had crater dimensions which were anomalously small. Figure 1 shows the regression analysis results for the remaining seven events. The upper fit (line 7) is for the dimensions as measured. However, none of the shots was a true surface burst (explosive half-buried), and adjustments were made for differences in height above the surface, leading to a second fit (line 8). Some of the craters experienced erosion from in-rushing water following the crater-forming process. Dimensions were adjusted to take into account the fact that in those cases radii were larger, volume was unchanged, and depth was decreased accordingly. This resulted in a fit (line 9) for unwashed craters from true surface bursts in coral rock (solid line in Fig. 1). It is interesting that an extension of this line (dotted) is in agreement with surface burst values for NTS alluvium shown by the hexagon. This illustrates the point that craters from 1 kiloton explosions in coral atolls have essentially the same shape (from extrapolation of line 9) as those in alluvium even though their linear dimensions are about 20% larger.

The most obvious observation from Fig. 1 is that the nuclear craters are changing shape at a rate ($n_1 - n_2 = 0.172$) greater than terrestrial impact craters (Table 2) and explosion craters in playa. The latter two have about the same rate of change, $n_1 - n_2 = 0.039 \pm 0.047$ and 0.033 ± 0.037, respectively. One possible explanation is that the adjustment in crater dimensions for water washing was insufficient, although the adjustment was made to be consistent with a resurvey of the craters (Circeo and Nordyke, 1964). In retrospect, the change in crater depth could have been greater simply because the crater slopes in unwashed craters would have an angle of repose characteristic of the medium, whereas washed craters would have had material normally found on the slope washed to the crater bottom, reducing the depth accordingly. The effect would be proportionately greater for the larger explosions where more of the crater periphery is exposed to open water and subject to washing. The adjustment was made solely to achieve a volume balance for the material washed in from the increased crater radius. No effort was made to achieve a particular crater profile, and it is probable that the original depth measurement took into account the change in profile brought about by water washing.

A second mechanism which may contribute to the greater rate of change of shape with increased energy is fallback of ejected material. Material from geometrically similar original positions should have similar velocities and initial trajectory angles. Thus, if unmodified by later phenomena, more of the ejecta would fall back within the crater edge as the crater radius increases with an increase in explosion energy. This tends to reduce crater depth with little change in radius, resulting in larger R_a/D_a for larger explosions.

Related modifications add still further changes. A calculation (Ganong and Roberts, 1968) of other explosion mechanisms on ejecta trajectories led to several observations.

"For 10 megatons, reflected shocks converging on ground zero reduced the ranges of all missiles ejected before 100 milliseconds. This effect diminished with 100 kilotons."

"For 10 megatons, most pebbles ejected at 10^5 cm/sec were carried up into the late time cloud. The pebbles ejected at 10^4 cm/sec were grounded by reflected shocks. However, if they were ejected as conglomerate masses later winnowed down to pebble size, or as groups of pebbles, their larger mass to drag ratio would have permitted them to remain aloft until cloud rise started."

"For 100 kilotons, most pebbles ejected at medium angles fell into the afterwinds which carried them back towards the crater but could not support them."

Thus, as energy is increased, additional material fails to escape beyond the crater edge or is brought back within it, tending to decrease the crater depth and increasing R_a/D_a.

Initial velocities and directions of material at geometrically similar locations within the volume to be cratered have been shown by calculation to be different for impact (Bjork, 1961) and nuclear explosion (Brode, 1968) sources. There is no *a priori* reason to expect the time sequence and effect of airblast and afterwinds to be similar for the two kinds of explosions.

A third mechanism which may increase fallback as energy is increased is what has come to be known as the "vacuum cleaner effect" (Ganong and Roberts, 1968), the effect of the rising fireball fed by in-rushing winds at the base of the cloud-stem which have velocities sufficient to impel material toward the explosion source. This is another mechanism which will decrease crater depth and increase R_a/D_a. The temperatures of nuclear explosions are 1 to 2 orders of magnitude greater than those of meteorite impact (Baldwin, 1963). Temperature of nuclear explosions and the rate of cloud rise increase with increased explosion energy (Brode, 1968). Thus, the contribution to decreased crater depth increases with increased energy. The data suggest that the contribution to decreased depth is small for a few kilotons and substantial for energies in the range of 1 to 10 megatons. A series of calculations similar to those of Ganong and Roberts but for impact energy sources is needed to show the relative importance of the second and third mechanisms discussed above for impact and explosion craters.

While there are several possible mechanisms to account for the crater depth scaling as a power less than $\frac{1}{3}$, the reason for crater radius to have an energy dependence greater than $\frac{1}{3}$ is not so obvious. The mechanism postulated here is airblast, which for surface bursts precedes the ground shock. The radius at which a given peak airblast overpressure occurs scales as $W^{1/3}$. Thus, if the energy dependence of crater radius is described by an exponent greater than $\frac{1}{3}$, the peak overpressure at the crater edge decreases as energy increases. However, as energy increases, the duration of the airblast pulse at the same cube-root scaled distance increases as $W^{1/3}$, as does the positive phase impulse (the area under the pressure-time curve). The net result is that if crater radius can be described by $R_a = 22.1W^{0.387}$, the positive phase impulse at the crater edge is approximately $I = 2.6W^{0.3}$ psi-sec. If the downward slap of the airblast destroys the integrity of the coral by crushing or cracking, the broken material which constitutes the upper portions of the upthrust crater edge is susceptible to a quasi-dynamic slope failure, resulting in mass at the crater edge sliding into the crater with a consequent increase in radius. The same material would be susceptible to additional water washing. This explanation for a radius dependence on energy greater than the $\frac{1}{3}$ power is more plausible than that presented in the 1968 paper.

2.3. *Terrestrial impact crater data*

The use of Baldwin's crater dimensions and energies in Fig. 1 for terrestrial and lunar craters may appear to have a certain circularity since Baldwin derived the energy based on the mass of material moved out of the crater. The circularity would exist if applied directly to R_a and D_a to determine an energy exponent. Any error in determining energy as Baldwin did would be expected to be systematic over the energy range, and as used for R_a/D_a versus energy in Fig.

1 an error would result in shifting the data in the figure to the right or left. Systematic errors would not change the slope of the fits appreciably. Others have estimated the energy involved in the Arizona impact crater. Compared with Baldwin's 8.1 megatons, Shoemaker (1963) found 1.4–1.8 megatons using cube-root scaling from the dimensions of the Teapot ESS crater. Johnson (1960) scaled crater depth as $W^{1/4}$ and crater radius as $W^{1/3.4}$ from NTS nuclear and high explosive craters to get 5 megatons. His estimates based on the brecciated zone range from 3.9 to 5.5 megatons. A hydrodynamic calculation by Bjork (1961) indicates 13.8 megatons. Roddy *et al.* (1975) got 1.6 and 18.4 megatons scaling from NTS craters by cube-root and fourth-root scaling, respectively, with intermediate energies for intermediate scaling exponents. Baldwin's estimates are between those of Johnson and Bjork and within the range cited by Roddy *et al.* and appear reasonable.

The line fit to the terrestrial impact crater ratios falls below that for high explosive craters in playa by an amount which is not statistically significant in view of the variance values. If such a discontinuity were real, it could be attributed to differences in initial velocity fields between explosion and impact craters, to a slightly subsurface burst point for the impact crater, to medium differences, or to a combination of these.

The fits to the terrestrial crater data indicate

$$R_a = 35.4 \pm 0.01\, W^{0.291 \pm 0.004}, \tag{4}$$

$$D_a = 12.5 \pm 0.05\, W^{0.251 \pm 0.017}, \tag{5}$$

$$R_a/D_a = 2.82 \pm 0.05\, W^{0.039 \pm 0.015}. \tag{6}$$

Using dimensions for Arizona Meteor Crater reported by Roddy *et al.*, $R_a/D_a = 3.46$. $R_a/D_a = 2.43$ for Jangle U. The 2.43 would scale to 3.43 using Eq. (6) and Baldwin's energy for the Arizona crater, indicating that these two craters are reasonable for comparison. If we assume that $n_3' = 2n_1 + n_2$, then for terrestrial impact craters Eqs. (4) and (5) indicate $n_3' = 0.833$. An apparent crater volume for the Arizona crater would be the total volume of ejected rock (true crater volume) minus the volume of fallback, or $73.2 \times 10^6\,\text{m}^3$. This and Jangle U volume, $V_a = 2.8 \times 10^4\,\text{m}^3$, together with $n_3' = 0.833$ leads to an energy for the Arizona crater of 15.2 megatons, slightly less than the 18.4 megatons Roddy *et al.* derived from fourth-root scaling. In addition to differences in cratering mechanics between explosive and impact craters, there is a further problem with deriving energy in the manner above. Energy is probably more nearly related to the mass of ejecta removed than to crater volume. Information on materials and dimensions for other terrestrial impact craters was not available in the detail given by Roddy *et al.* for the Arizona crater. Determining an energy dependence for mass for these craters would probably reduce the variance, since differences in material density would be taken into account. Lacking an energy dependence for mass, let us make the undefendable assumption that it, also, is the 0.833 power of energy. Then using the same two craters, and the mass of their true crater volumes, would suggest an energy for formation of the Arizona crater of 12.8 megatons.

2.4. *Lunar impact crater data*

At the overlap with lunar crater data the terrestrial crater data fit falls above the lunar data fit as it should because lunar ejecta travels further because of decreased gravity and absence of an atmosphere. Fallback is decreased accordingly, crater depth is greater, and the ratio is smaller. The one sigma variance values have been added in the accompanying equations to emphasize the recognition that with so few craters in each set, the variance is greater than the differences which are suggested here as perhaps being meaningful. Equations (4), (5), (7), and (8) indicate that for a comparable energy, 10^5 kilotons, terrestrial and lunar impact craters have nearly equal radii but that lunar craters are about 20% deeper.

Three fits have been provided for the lunar data. For the first three craters of Table 2.

$$R_a = 46.5 \pm 0.021\, W^{0.266 \pm 0.004}, \tag{7}$$

$$D_a = 20.5 \pm 0.020\, W^{0.225 \pm 0.004}, \tag{8}$$

$$R_a/D_a = 2.27 \pm 0.003\, W^{0.041 \pm 0.0005}. \tag{9}$$

Table 2. Dimensions of lunar and terrestrial impact craters

	R_a (m)	D_a (m)	$D_a^{(1)}$	R_a/D_a	$R_a/D_a^{(1)}$	W (kt)
Lunar						
In Purbach	667	195[e]	195	3.42	3.42	2.24×10^4
Piton A	1736	439	440	3.96	3.95	7.59×10^5
Bullialdus F	3339	762[e]	765	4.38	4.36	9.55×10^6
Bessel	6678	823	836	8.12	7.99	1.51×10^8
Kepler	13490	1189[e]	1241	11.35	10.87	2.75×10^9
Aristillus	23500	1798	1957	13.07	12.01	2.75×10^{10}
Tycho	36060	1860	2234	19.39	16.14	1.62×10^{11}
Pythagorus	53430	3260[e]	4081	16.39	13.09	8.91×10^{11}
Clavius	96170	3260	5922	29.50	16.24	1.17×10^{13}
Terrestrial						
Odessa 2	8.84	4		2.21		9.77×10^{-3}
Kaali Jarv	40.1[2]	11.1		3.61		1.51×10^0
Odessa 1	69.5	24.8		2.80		8.91×10^0
Arizona[3]	518.3	150		3.46		8.13×10^3
New Quebec	1389[2]	261		5.32		3.80×10^5

[e]Dimension not listed by Baldwin; estimated from dimension of other craters of comparable size.
[1]Adjusted for curvature of the moon.
[2]Apparent crater radius from Baldwin, Table 10.
[3]From Roddy *et al.* (1975).

The following six craters yield

$$R_a = 77.4 \pm 0.018\,W^{0.237\pm0.002}, \tag{10}$$

$$D_a = 71.2 \pm 0.170\,W^{0.131\pm0.016}, \tag{11}$$

$$R_a/D_a = 1.09 \pm 0.17\,W^{0.106\pm0.016}. \tag{12}$$

All nine result in

$$R_a = 60.9 \pm 0.02\,W^{0.247\pm0.003}, \tag{13}$$

$$D_a = 64.5 \pm 0.09\,W^{0.136\pm0.010}, \tag{14}$$

$$R_a/D_a = 0.945 \pm 0.080\,W^{0.111\pm0.009}. \tag{15}$$

Separating the lunar data into two separate sets was suggested by the relative position of the data in Fig. 1. It is interesting that Gault *et al.* (1975) observed that mercurian and lunar craters exhibit a discontinuity in the depth-diameter relationships. The discontinuity occurs at 7–8 km diameter for mercurian craters and at about 15 km for lunar craters. They note that this factor of approximately 2 is close to the 2.3 ratio of their respective gravitational accelerations. The 15 km rim-to-rim diameter they cite would, using Baldwin's 0.83 multiplier for apparent crater, give a radius of apparent crater of 6.225 km, just smaller than the apparent crater radius of Bessel (Table 2) and the energy level at which the lunar data of Fig. 1 are separated into two sets between Bullialdus F and Bessel. Comparison of Eqs. (7) and (10) shows that for the nine lunar craters considered here there is relatively little discontinuity in crater radius. By contrast, comparison of Eqs. (8) and (11) shows a large discontinuity in crater depth. It is recognized that the sample here is small compared with that of Gault *et al.*, however, it appears from reexamining their Fig. 15 that there too the discontinuity is in crater depth.

Head (1976) postulates a fragmental megaregolith to a depth of 2–3 km. He notes laboratory experiments which show a decrease in crater depth as an interface between incompetent and competent material becomes close to the crater bottom. He observes that crater floors become flattened as crater diameters exceed four times the thickness of the megaregolith. This would correspond to apparent crater radii of 3.3 and 5 km for 2 and 3 km megaregolith depths and R_a equal to 0.83 times the rim radius. This, too, is between Bullialdus F and Bessel. If true crater depth scales as $W^{1/4}$ from that of the Arizona Meteor Crater, a crater depth for craters the size of Bullialdus F would be about 2 km, whereas that of Bessel would be about 4 km. Thus, the onset of the discontinuity is consistent with Head's hypothesis.

Baldwin observed that taking the curvature of the moon into account in a determination of crater depth made little difference in his $\log_{10}$ plots of crater diameter and depth. However, the change in the data used in Fig. 1 and Table 2 was significant except for the three smallest lunar craters where there was essentially no change. For the following expressions, crater depths have been taken as the apparent crater depth plus the rise to the lunar circumference from a chord equal to the apparent crater diameter (twice the radius values of Table 2).

For the six largest craters

$$D_a = 27.3 \pm 0.14\, W^{0.178 \pm 0.013}, \quad (16)$$

$$R_a/D_a = 2.83 \pm 0.15\, W^{0.060 \pm 0.014}. \quad (17)$$

For all nine craters

$$D_a = 45.9 \pm 0.08\, W^{0.158 \pm 0.009}, \quad (18)$$

$$R_a/D_a = 1.38 \pm 0.09\, W^{0.089 \pm 0.009}. \quad (19)$$

With this adjustment, the slope for the six largest craters (dotted line in Fig. 1), 0.060, is still greater than that for the three smaller craters, 0.041, and the terrestrial craters, 0.039. Hodges and Wilhelms (1976) postulate that the rings in Mare Orientale represent structural interfaces thrust to the surface by an instantaneous rebound and that, momentarily, before rebound a more nearly bowl-shaped crater existed. The energy of the impact is some orders of magnitude above the largest energy shown in Fig. 1, the present depth is small, and the R_a/D_a ratio very large. Their interpretation suggests that a ratio more consistent with smaller craters could be obtained if the D_a before rebound were used. If rebound as a crater modifying mechanism begins at about 10^8 kilotons where the discontinuity is observed and increases in significance to the energy of the impact which created Mare Orientale, it might be the effect being demonstrated by a slope for the six largest craters in Fig. 1 being larger than slopes for the three smaller lunar craters and the terrestrial craters.

Despite the fact that certain of the data subsets (e.g., the Pacific data) of Fig. 1 show that there are factors other than yield influencing crater shape, the general trend of the total data array indicates R_a/D_a changing by factors of from 10 to 15 over the 18 orders of magnitude of energy shown. This is compelling evidence that the crater radius and depth should not be expected to scale by the same power law.

Equation (13) indicates that the crater radius of the nine lunar craters considered scales about as $W^{1/4}$, and it was noted earlier that there is little or no discontinuity in the dependence of radius on energy. Chabai's overburden approximate scaling leads, in the case of terrestrial explosion craters, to $W^{1/4}$ dependence where overburden pressure is large compared with atmospheric pressure. At about 40 m of overburden, this is the case. For $\frac{1}{6}$ gravity, 240 m would be required for the earth's atmospheric pressure. For the moon, essentially without atmosphere, the atmospheric term drops out, and $W^{1/4}$ dependence prevails regardless of the effective burial depth. Baldwin used an effective burial depth for In Purbach of 0.097 ft/lb$^{1/3}$ which is equivalent to an overburden depth of 242 m. All of the larger lunar craters of Table 2 have greater overburden depth if one uses Baldwin's values. It can be presumed that the three largest craters for which Baldwin shows an effective scaled burial depth of 0.000 ft/lb$^{1/3}$ are no exception and result from rounding to 0. This follows from the smallest nonzero value, that for Aristillus, 0.005 ft/lb$^{1/3}$ is equivalent to an overburden depth of 580 m. Thus, energy dependence of crater radius as $W^{1/4}$ is consistent with the overburden approximate scaling. For surface and near surface bursts the true crater radius equals or only slightly exceeds the apparent crater. If the depth of the true crater, that is, depth to the bottom of the fallback, also has an energy dependence of $W^{1/4}$, a dependence of apparent crater depth of $W^{<1/4}$ as shown in Eqs. (16) and (18) would also be consistent with the overburden approximate scaling. If rebound is a signficant mechanism, the exponent would be even smaller above the energy at which rebound is initiated.

3. Concluding Remarks

Chemical and nuclear explosive craters and terrestrial and lunar impact craters studied show a systematic change in crater shape with increase in energy. Crater-modifying mechanisms have been identified for each of the crater sources which can explain the nature of the energy dependence of R_a/D_a and for departures of some of the data sets from the overall trends. The crater-modifying mechanisms identified here (a) are not taken into account in dimensional analyses and cratering formulas, and (b) can help to explain the lack of agreement between those approaches to cratering and the results presented here. Those approaches need to be modified to reproduce the observation that $n_1 \neq n_2$, to treat true crater dimensions as fundamental dimensions, and to take proper account of dimension-modifying phenomena that affect the variation of crater size and shape with energy. As implausible as the results of the earlier paper (Vortman, 1968a) on energy dependence may have seemed to some readers, this reexamination has presented believable mechanisms to account for the results and has revealed no reason for changing them.

4. Epilogue

This elementary venture into terrestrial and lunar impact craters as an extension of explosive craters has led to several suggestions for expanding the effort begun here.

1. Baldwin's crater dimensions and energy estimates were used only because they were readily available. Topographic mapping resulting from the Ranger and Apollo programs has provided more accurate dimensions for many of the craters, including smaller craters than considered by Baldwin. It would be worthwhile to reproduce Fig. 1 with the more accurate dimensions for a larger number of craters. Impact energies could be estimated by the method used by Baldwin or by an alternate method so long as the method was consistent over the range of energies. Estimates made today could take into account the effect of substrate characteristics on crater dimensions and shape to provide greater refinement in the estimates.
2. R_a/D_a is not a good index of crater shape when considered alone. Craters

with the same R_a/D_a could have nearly conical or cylindrical shapes. Shape can more accurately be portrayed by a shape factor term such as $V_a/\pi R_a^2 D_a$ or its reciprocal. Best results would be obtained by using both the shape factor and the R_a/D_a ratio. Figure 1 could be revised using the shape factor as the ordinate. A plot of R_a/D_a ratio versus shape factor could show interesting trends in the distribution of crater shapes.

3. Impact velocities of meteorites on earth and moon average about 30 km/s with occasional excursions from as little as about 5 to as much as 70 km/s. Average material density of impactors is nearly the same for all except nickle-iron bodies. Size distribution can be estimated. It would be interesting to examine a possible relationship between the distribution of impact energy and the distribution of V_a and crater shape, especially for Class 1 craters.
4. Apparent crater volumes, V_a, determined from topographic maps would permit energy dependence to be determined from estimated energies in the same manner as the dependence for R_a and D_a was determined in the equations shown earlier. This could show whether V_a was directly proportional to energy or whether $n_3 < 1$. The effect of terraces, central mounds, and rebound could perhaps be revealed by examination of volume data.

REFERENCES

Baldwin, R. B.: 1963, The measure of the Moon. *Univ. of Chicago Press*, Chicago, IL.

Bjork, R. L.: 1961, Analysis of the formation of Meteor Crater, Arizona: a preliminary report. *J. Geophys. Res.* **66**, 3379–3387.

Brode, H. L.: 1968, Review of nuclear weapons effects. *Ann. Rev. Nucl. Sci.* **18**, 153–202.

Burton, D. E., Snell, C. M., and Bryan, J. B.: 1975, Computer design of high -explosive experiments to simulate subsurface nuclear detonations. *Nuc. Tech.* **26**, 65–67.

Chabai, A. J.: 1965, On scaling dimensions of craters produced by buried explosives. *J. Geophys. Res.* **70**, 5075–5098.

Circeo, L. J., Jr. and Nordyke, M. D.: 1964, Nuclear cratering experience at the Pacific Proving Grounds. *Rep. UCRL*-12172, Univ. of Calif., Lawrence Radiation Laboratory, Livermore, CA.

Dillon, L. A.: 1972, The influence of soil and rock properties on the dimensions of explosion-produced craters. *Rep. AFWL-TR*-71-144, Air Force Weapons Laboratory, Kirtland Air Force Base, NM.

Ganong, G. P. and Roberts, W. A.: 1968, The effect of the nuclear environment on crater ejecta trajectories for surface bursts. *Rep. AFWL-TR*-68-125, Air Force Weapons Laboratory, Kirtland Air Force Base, NM.

Gault, D. E., Guest, J. E., Murray, J. B., Dzurisin, D., and Malin, M. C.: 1975, Some comparisons of impact craters on Mercury and the Moon. *J. Geophys. Res.* **80**, 2444–2460.

Head, J. W.: 1976, Significance of substrate characteristics in determining morphology and morphometry of lunar craters. *Proc. Lunar Sci. Conf. 7th*, p. 2913–2929.

Hodges, C. A. and Wilhelms, D. E.: 1976, Formation of concentric basin rings (abstract). In *Papers presented to the Symposium on Planetary Cratering Mechanics*, p. 53–55. The Lunar Science Institute, Houston, Texas.

Johnson, G. W.: 1960, Note on estimating the energies of the Arizona and Ungava Meteorite Craters. *Rep. UCRL*-6227, Univ. of Calif., Lawrence Radiation Laboratory, Livermore, CA.

Roddy, D. J., Boyce, J. M., Colton, G. W., and Dial, A. L., Jr.: 1975, Meteor Crater, Arizona, rim drilling with thickness, structural uplift, diameter, depth, volume, and mass-balance calculations. *Proc. Lunar Sci. Conf. 6th*, p. 2621–2644.

Shoemaker, E. M.: 1963, Impact mechanics at Meteor Crater, Arizona. *The Moon, Meteorites, and Comets* (B. M. Middlehurst and G. P. Kuiper, eds.), p. 301–336, Univ. of Chicago Press, Chicago, IL.

Vortman, L. J., Anderson, E. E., Rerdinelli, R. A., Bishop, R. H., Burton, R. J., Chabai, A. J., Palmer, D. G., and Reed, J. W.: 1962, Project BUCKBOARD, 20-ton and 1/2-ton high explosive cratering experiments in basalt rock. *Rep. AC*-4675 (*RR*), Sandia Laboratories, Albuquerque, NM.

Vortman, L. J.: 1968a, Craters from surface explosions and scaling laws. *J. Geophys. Res.* **73**, 4621–4636.

Vortman, L. J.: 1968b, Air Blast from underground explosions as a function of charge burial. *Annals of the New York Academy of Sciences*, Vol. 152, Art. 1, p. 362–377.

White, J. W.: 1971, Examination of cratering formulas and scaling methods. *J. Geophys. Res.*, **76**, 8599–8603.

White, J. W.: 1973, An empirically derived cratering formula. *J. Geophys. Res.*, **78**, 8623–8633.

Roddy, D. J., Pepin, R. O., and Merrill, R. B., editors.
(1977) *Impact and Explosion Cratering*, Pergamon Press (New York), p. 1231–1244.
Printed in the United States of America

Experimental hypervelocity impact into quartz sand—II, Effects of gravitational acceleration

DONALD E. GAULT

Murphys Center of Planetology, P.O. Box 833, Murphys, California 95247

JOHN A. WEDEKIND

NASA, Ames Research Center, Moffett Field, California 94035

Abstract—We report experimental results for craters formed by aluminum spheres impacting at normal incidence against quartz sand targets in gravitational acceleration environments ranging from 0.073 to 1.0 g ($g = 980\,cm/sec^2$). Impact velocities varied from 0.4 to 8.0 km/sec. Crater dimensions and formation times are compared with results from a simplified dimensional analysis of the cratering processes. Although the comparison indicates a dominant role of gravity relative to the target strength for craters formed in sand, the results serve primarily to emphasize that both gravity and strength are variables of fundamental significance to cratering processes.

INTRODUCTION

GRAVITATIONAL ACCELERATION is one of the primary parameters that is important to impact cratering processes (together with some parameter characterizing the strength of the cratered medium). The importance of gravity stems, first, as a factor in determining the ultimate size of a crater of excavation; second, as a factor in the ballistic transportation, radial distribution, and secondary cratering effects of ejecta deposited around a crater; and third, from its influence on post-cratering modifications of very large craters of excavation by causing the formation of interior terracing and (or contributing to) central peaks and/or ringed peak complexes. Recent comparisons of lunar and Mercurian craters illustrate some effects of gravity for the latter two cases (Gault *et al.*, 1975). In this paper we present results of an experimental investigation directed toward the first subject—that is, the effects of gravity on the size of impact craters. The experiments were performed in the Vertical Gun Ballistic Range of the NASA, Ames Research Center (Gault *et al.*, 1968). Based on previous experience (e.g., Gault *et al.*, 1968; Braslau, 1970; Oberbeck, 1971; Gault, 1973; Stöffler *et al.*, 1975) dry quartz sand was used as the target medium in order to minimize strength effects and, thus, maximize gravitational effects on crater growth.

EXPERIMENTAL PROCEDURES

Aluminum spheres with diameter of 3.18 mm and mass of 0.047 g were launched from a light-gas gun at a nominal velocity of 6.64 km/sec into dry quartz sand targets. All impacts occurred from vertical trajectories into horizontal target surfaces. Ambient air pressure in the target chamber was maintained at less than 1 mm Hg. A guillotine-type shutter was explosively driven across the trajectory behind the projectile to trap propellant gases up range and prevent aerodynamic disturbances of the ejecta from the craters.

The quartz sand targets were contained in 60 cm diameter by 20.5 cm deep pans. The median grain size of the sand was approximately 0.35 mm with 70% by weight smaller than 0.5 mm. Bulk density of the unconsolidated sand was 1.65 g/cm^3. Previous experience has indicated that there are no effects of the finite size of the target containers for the maximum diameter of the largest craters produced during the experiments (35 cm rim diameter). The target containers were mounted on the end of a long cylindrical tube, which was constrained by ball bearings to only vertical motion. Springs attached to the top of the target chamber and target pans provided a constant upward force on the target-tube assembly that was independent of spring extension. When raised to the top of the target chamber and "dropped", the target experienced a transient gravitational field less than 1 "g" ($g = 980$ cm/sec^2) during which time the gun was fired, impact took place, and a crater was formed. By using springs with different spring constants, the downward acceleration of the targets was varied over a range of approximately 0.2–0.9 g, thus providing transient acceleration fields with respect to the target of approximately 0.8–0.1 g. An accelerometer (Systron-Donner Model 4310; 7.5 V output per g) embedded in the quartz sand provided measurement of target accelerations and, in addition, furnished signals of the time of impact and the termination of the constant g field. Maximum vertical travel of the targets was 120 cm.

Because the abrupt deceleration of the target system at the end of its downward stroke destroyed the craters, all impact events were stereographically recorded using 35 mm movie cameras. Framing rates were 60 per second for transient fields greater than 0.5 g, but operational problems with the cameras necessitated 30 per second for values less than 0.5 g. A xenon lamp with a 25 μsec pulse duration provided illumination for the imagery as well as serving as the effective shutter for the cameras. Both horizontal and vertical fiducial distances were placed on the target pan for calibration references for reduction of the stereo records. All measurements and distances have been referenced to the target surface. Two orthogonal pairs of crater profiles were averaged as the final data set for each cratering event. Crater profiles have been obtained for six craters formed at each of six fractional values of g (0.81, 0.67, 0.51, 0.27, 0.19, and 0.073) plus a single record for 0.12. Results for seven craters formed at 1.0 g were obtained both stereographically and by hand-profiling measurements of the craters; no systematic differences in crater dimensions are apparent within the precision of the measurements (0.5 mm), although the calculated volumes for the craters determined from the stereo profiles average 4.5% (1–6%) greater than the hand profiling measurements.

As the basis for ascertaining the functional relationship between crater diameter and projectile kinetic energy for a constant value of gravitational acceleration, profiles for another 54 craters were obtained (hand measured) at 1.0 g over a range of impact velocities from approximately 0.4 to 8.0 km/sec. In addition to the preceding results using 3.18 mm diameter spheres, 1.59 and 6.35 mm spheres (mass = 0.006 and 0.376 g) were fired into quartz sand with nominal velocities of 1.4 and 6.6 km/sec to determine the time required to form craters as a function of their size in a 1.0 g environment. These latter impacts were photographed with 16 mm movie cameras operating at approximately 8000 frames per second; precise time references were obtained from 1 msec timing marks recorded on the film.

RESULTS

The functional relationship between crater diameter and gravitational acceleration is presented in Fig. 1 for conditions when the projectile kinetic energy is maintained constant. Over the range from 1.0 to 0.073 g, an excellent empirical fit to these data is given by the exponential relationship $D \propto g^{-0.165}$ with $a \pm 0.005$ uncertainty in the value of the exponent at the 95% confidence level. No systematic changes in crater shape are evidenced over this range of gravitational environment as indicated in Table 1. The variation of crater diameter

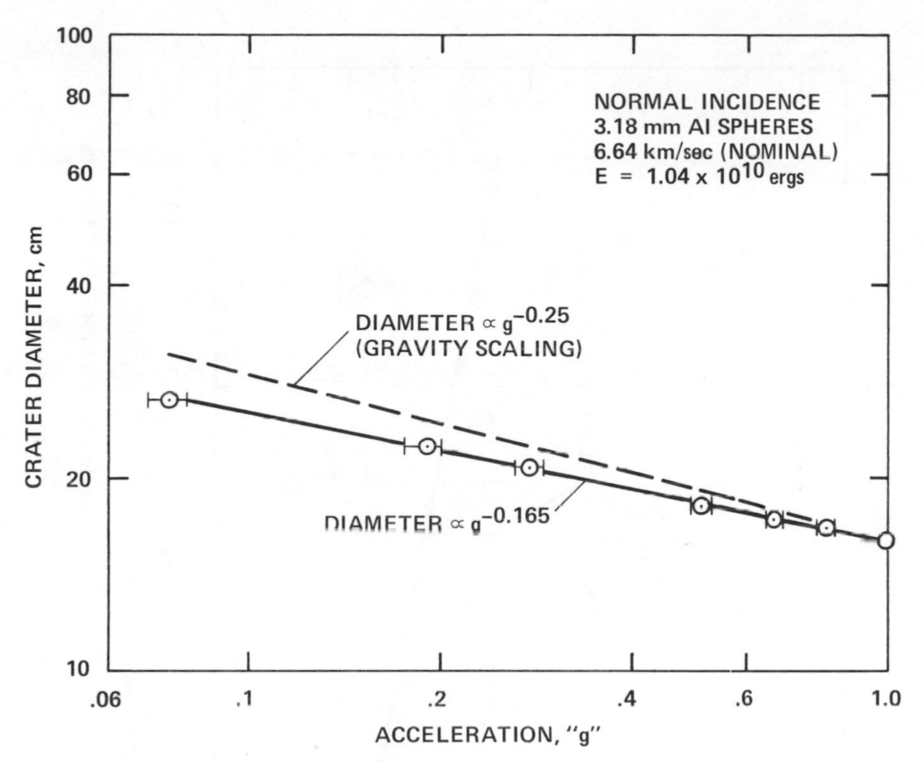

Fig. 1. Variation of crater diameter with gravitational acceleration with projectile kinetic energy constant. Diameter measured with respect to undisturbed surface. Error bars indicate maximum deviation from nominal value for g during formation of the six craters averaged for each data point. Diameter deviations less than size of data symbol.

Table 1. Crater depth-diameter ratios.*

g	d/D_a	d/D_r	D_a/D_r
1.0	0.24	0.19	1.25
0.81	0.25	0.20	1.24
0.67	0.25	0.20	1.24
0.51	0.25	0.21	1.23
0.27	0.23	0.18	1.26
0.19	0.22	0.18	1.25
0.073	0.23	0.19	1.26

*Crater depth d and apparent diameter D_a measured with respect to undisturbed surface. D_r is rim diameter.

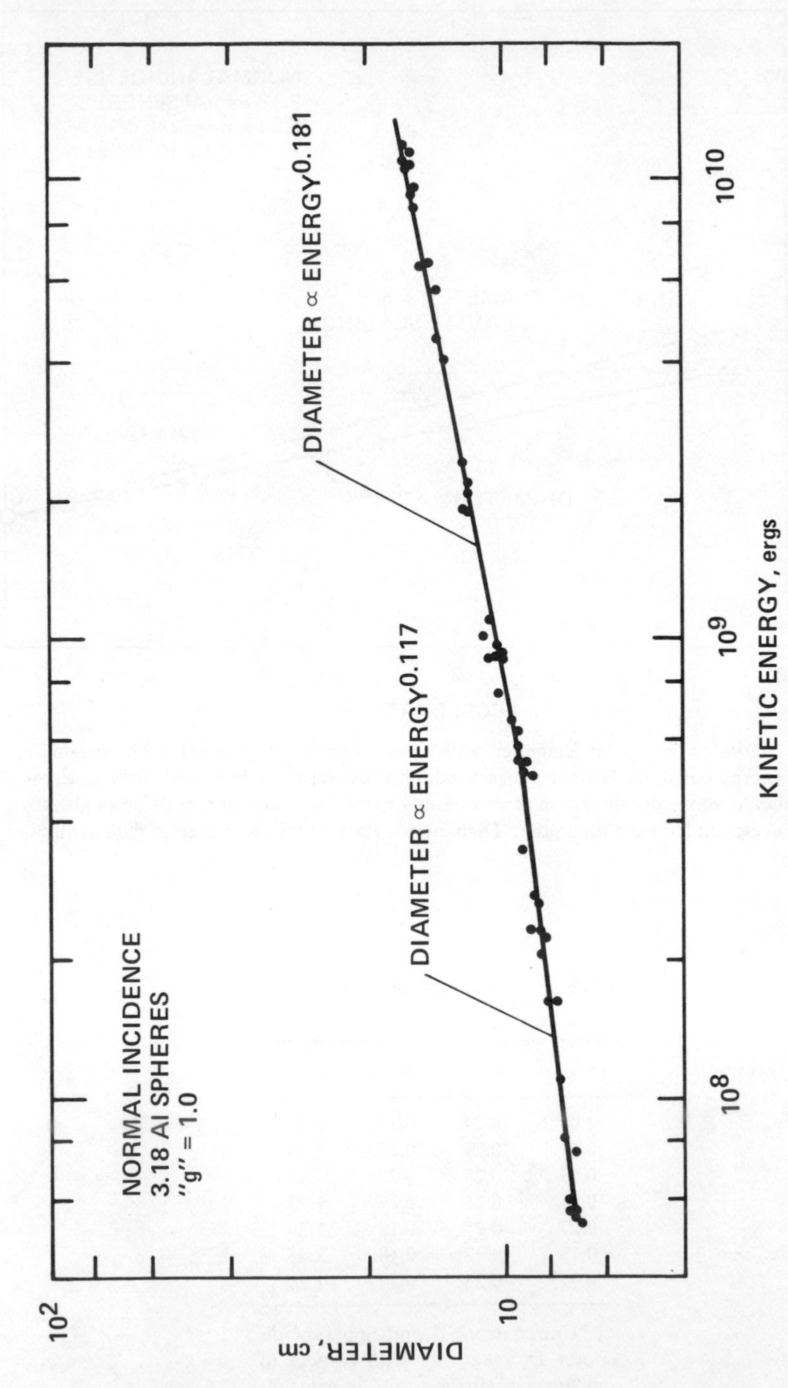

Fig. 2. Variation of crater diameter with projectile kinetic energy with gravitational acceleration constant (1.0 *g*). Diameter measured with respect to undisturbed surface.

with projectile kinetic energy in the 1.0 g environment is shown in Fig. 2. Two exponential relationships are shown; a transition in the functional constants occurs between approximately 4 to 5×10^7 erg (impact velocities of 1.3 and 1.5 km/sec, respectively). It is important to note that this transition reflects a basic change in the cratering processes. In the low-velocity regime the peak shock stresses are low relative to the strength of the projectile. As a consequence the projectile remains intact and relatively undeformed as it penetrates into the target; it comes to rest buried in the sand below the bottom of the final crater. With increasing impact velocity, however, the peak shock stresses increase and exceed the dynamic strength of the projectile. The projectile then deforms plastically, tending toward a thin, jagged-edged plate, or multiple-fragment plates, which come to rest on the bottom surface of the crater. At velocities in excess of approximately 2.0–2.5 km/sec, few, if any, fragments of projectile remain within the crater; the probability of retention of fragments and the fragment sizes decrease with increasing velocity. At still higher velocities, of course, melting and incipient vaporization of projectile occur. This change from "low-velocity" impact to "hypervelocity" impact is accompanied by a change

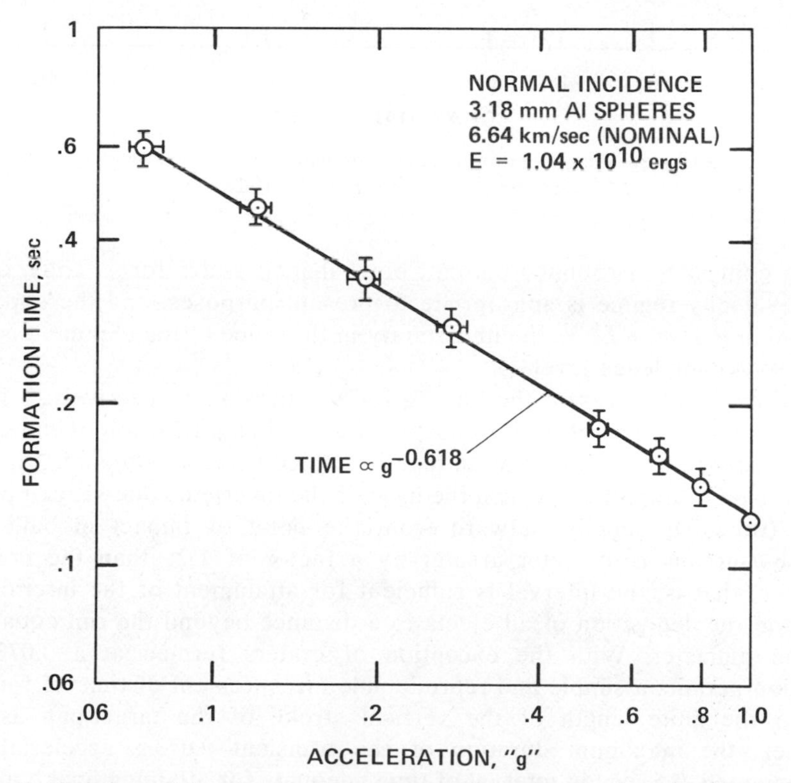

Fig. 3. Variation of the time to form crater with gravitational acceleration with projectile kinetic energy constant. Error bars on time are uncertainty due to interframe interval of stereo movie cameras.

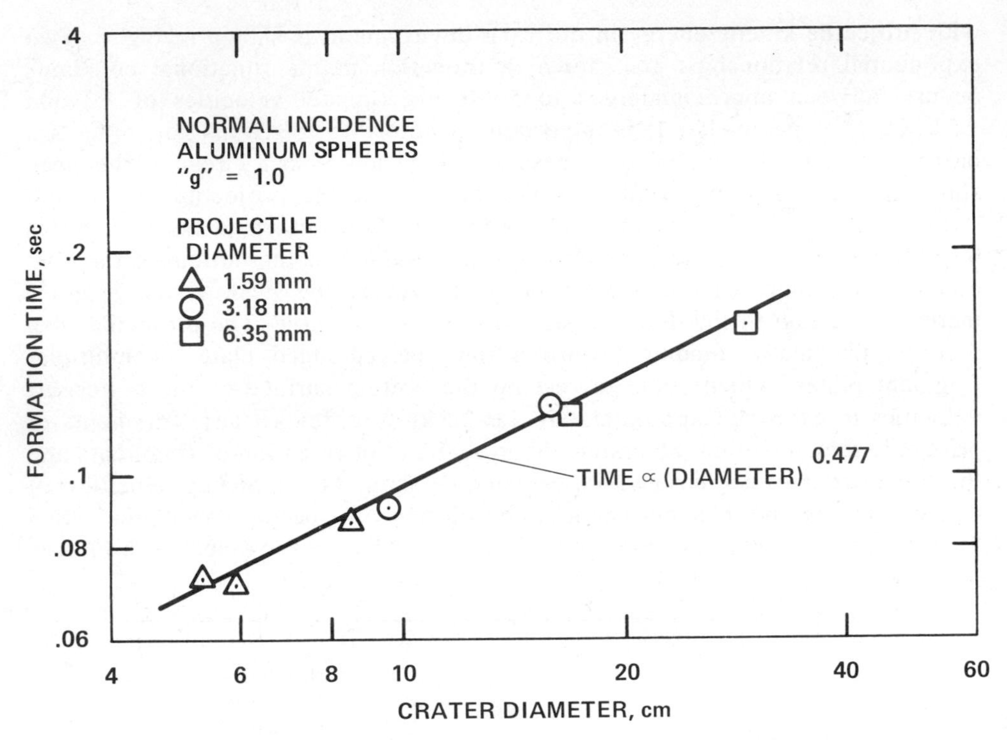

Fig. 4. Variation of the time to form crater with diameter of crater, which is measured with respect to undisturbed surface.

from a conical to a rounded-bottom, bowl-shaped crater form. Thus, only the higher velocity regime is appropriate to present purposes, and the exponential expression is $D \propto KE^{0.181}$; the uncertainty in the value of the exponent is ± 0.009 at the 95% confidence level.

Figures 3 and 4 present the time to form craters as, respectively, a function of gravitational acceleration for constant KE, and as a function of the diameter for a constant 1.0 g. The formation time is defined here to be the interval of time from impact to the time at which the base of the inverted, cone-shaped plume of ejecta (traveling radially outward from the point of impact in ballistic trajectories) attains a diameter greater by a factor of 1.25 than the crater rim diameter; that is, the interval is sufficient for attainment of the interior crater form and the deposition of all ejecta to a distance beyond the rim equal to $\frac{1}{8}$ of the rim diameter. With the exception of craters formed at a 0.073 g, this definition permits a simple and reproducible measurement of time of formation. Due to the finite length of the vertical stroke of the target-pan assembly, however, the maximum duration of the transient 0.073 g acceleration was approximately 0.5 sec, an interval of time adequate for attaining final crater form and shape, but too short to complete deposition of ejecta to the specified distance from the rim. For this single data point an extrapolated estimate for formation time is given in Fig. 3. For constant KE, $t \propto g^{-0.618}$ and with a constant 1.0 g, $t \propto D^{0.477}$; uncertainties at the 95% confidence level are, respectively, ±0.019 and ±0.032. Although not shown graphically, the relationship $t \propto KE^{0.105}$ was obtained with constant 1.0 g; uncertainty is ±0.027.

Discussion

The significance of these experimental results can be aided by their comparison with a dimensional analysis of the impact cratering processes. The approach here is not as complete as Chabai (1965) examined explosive cratering, but rather it is a simplified approach in which the number of variables is restricted to only those parameters considered to be of first-order importance. Specifically, we consider the relationships for two independent variables, crater diameter D and its time to form t, in terms of only four independent variables, the projectile kinetic energy KE, strength of the target medium s, gravitational acceleration g, and mass density of the target medium ρ. Many other parameters play a role in the cratering processes, but for present purposes they are constant or are considered of second-order importance (projectile density, porosities, etc.). Projectile kinetic energy is introduced as the primary independent variable both from first principles and experimental evidence. Energy equates directly to work which must be expended to form a crater, not the projectile momentum as advocated by Öpik (e.g., 1969), and its fundamental role is clearly indicated by experimental impact data (Gault, 1973). Moreover, the use of KE is analogous to energy yield employed in explosive cratering (equivalent weight of TNT) for which momentum has no physical meaning.

Some measure of the strength of the target medium must be considered because this physical property serves both as a dissipative agent during cratering deformations and as a factor that enters into determining the ultimate size and shape of the crater. The measure of strength that should be used is left undefined for the present (i.e., compressive, tensile, shear, or a combination thereof).

And last, from simple consideration of the energy expenditure (work) necessary to excavate (eject) mass from a crater, gravitational acceleration in company with mass density of the target medium must be included. An alternative interpretation for including g is that body forces arising from gravitational acceleration also serve to resist cratering deformations and can be considered to serve as a pseudo-strength parameter.

Thus, the dimensional analysis solutions for crater diameter and formation time are taken to start from

$$f_D(D, KE, s, g, \rho) = 0, \quad \text{(1a)}$$

$$f_t(t, KE, s, g, \rho) = 0. \quad \text{(1b)}$$

Application of the Buckingham π theorem then yields

$$F_D(\pi_1, \pi_2) = 0,$$

$$F_t(\pi_3, \pi_4) = 0,$$

where the π terms are the dimensionless ratios

$$\pi_1 = \frac{KE}{sD^3}; \quad \pi_2 = \frac{KE}{\rho g D^4}; \quad \pi_3 = \frac{KE^2\rho^3}{s^5 t^6}; \quad \pi_4 = \frac{KE}{\rho g^5 t^8}.$$

The functional relationship for diameter then becomes

$$D = k_d \left(\frac{KE}{s}\right)^{1/3} \times j(\pi_2), \tag{2a}$$

or, alternatively,

$$D = k_d \left(\frac{KE}{\rho g}\right)^{1/4} \times j(\pi_1). \tag{2b}$$

Although the dimensional analysis provides no solutions for the functions $j(\pi_1)$ and $j(\pi_2)$ or the values of the proportionality constants k_d, it is instructive to recognize the physical significance of the π_1 and π_2 dimensionless ratios. This is most easily accomplished by considering a simple physical model for cratering first suggested by Charters and Summers (1959). In their model the energy to form a crater is equated to the work expended to enlarge a cavity against the resisting forces. For present purposes the only sources of resistance are the strength and body forces. Assuming a hemispherical-shaped cavity and using a polar coordinate system (r, θ), where θ is measured from the target surface, the Charters-Summers model leads to

$$\text{Cratering energy} = \int_0^r \int_0^{\pi/2} (sr^2 \cos\theta + \rho g r^3 \sin\theta \cos\theta)\, dr\, d\theta.$$

With the strength constant, integration yields

$$\text{Cratering energy} = \frac{\pi}{12} sD^3 + \frac{\pi}{64} \rho g D^4. \tag{3}$$

The first term on the right expresses the energy expended to overcome resisting stress(es) of the target medium as it is deformed, and the second term is the energy spent to overcome the body forces arising from gravity. Note that this second term can be rewritten

$$\frac{\pi}{64} (\rho g D) D^3,$$

so that the product $(\rho g D)$ assumes an analogous role to the strength s and might be considered to represent a gravitational "strength" of the target. Comparison of these two terms with the π_1 and π_2 ratios from dimensional analysis clearly shows their similarity; the π terms are expressions representing the energy expending agents of strength and gravity.

For the limiting case of a crater formed on a gravity-free body (i.e., zero g and $j(\pi_2) = 1$), Eq. (2a) indicates that the diameter is a function of (i.e., "scales" with) the $KE^{1/3}$, or the well known Lampson cube root scaling relationship. Cube root scaling implicity implies that target strength remains constant. Any changes in the target medium will cause the crater dimensions to vary inversely with the cube root of the appropriate measure of strength. On the other hand for the opposite condition of zero strength (i.e., $j(\pi_1) = 1$) Eq. (2b) yields the diameter varying with $KE^{1/4}$ and $g^{-1/4}$, the so-called fourth root or gravity scaling relationships. The relative importance of these two scaling relationships obviously depends on the ratio $(s/\rho gD)$. For small laboratory craters formed in competent rock (Gault, 1973) strength is the dominant parameter and $(s/\rho gD) \gg 1$. For large terrestrial craters the strength ratio $(s/\rho gD) \ll 1$ and body forces must assume a dominant role. For the present experiments the quartz sand has zero strength in tension and compression (uniaxial), but has weak shear strength (Fig. 5). Accordingly, dependent on what measure of strength is introduced, $0 < (s/\rho gD) < 1$ and the possibility remains open that strength effects cannot be considered negligible for the present data set.

Comparable expressions to Eqs. (2a) and (2b) for the time to form a crater are

$$t = k_t \frac{KE^{1/3}\rho^{1/2}}{s^{5/6}} \times j(\pi_4), \quad (4a)$$

$$t = k_t \frac{KE^{1/8}}{\rho^{1/8} g^{5/8}} \times j(\pi_3), \quad (4b)$$

which can be rewritten by using π_1 and π_2 to eliminate KE

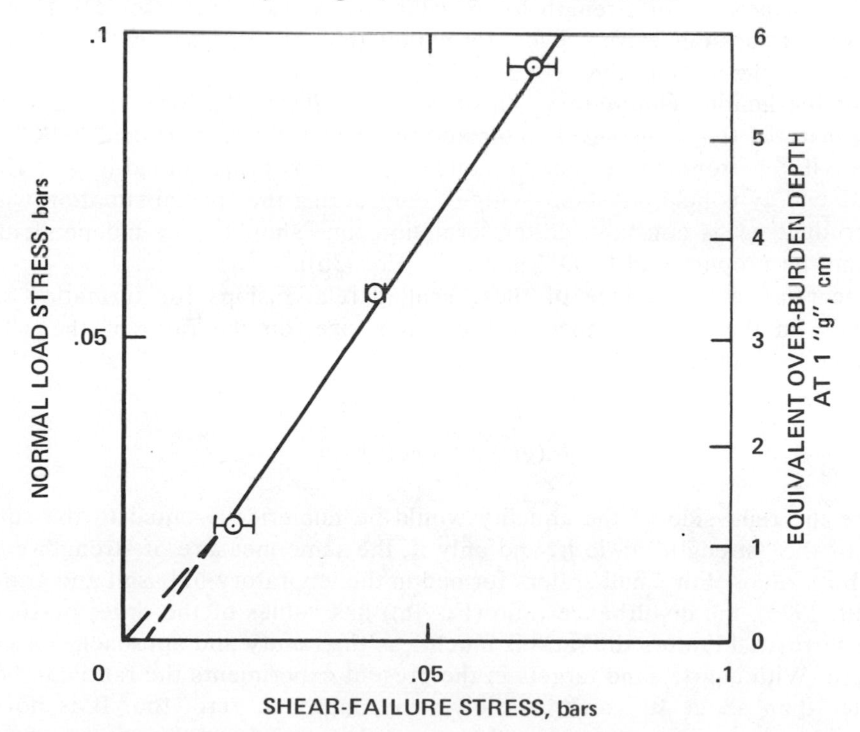

Fig. 5. Shear strength of quartz sand target material as a function of the normal load stress.

$$t = k_t \frac{D}{(s/\rho)^{1/2}} \times j(\pi_4), \tag{5a}$$

$$t = k_t (D/g)^{1/2} \times j(\pi_3). \tag{5b}$$

Because of the general correlation between bulk modulus and compressive strength for competent (elastic) materials (Farmer, 1968), the quantity $(s/\rho)^{1/2}$ in Eq. (5a) expresses a measure of velocity of a compression (dilatational) wave through the target. Similarly, the velocity excited by gravitational acceleration is proportional to $(Dg)^{1/2}$. Thus, π_3 and π_4 terms relate the time of formation of a crater to the time required to propagate a disturbance across some characteristic distance (crater diameter) of the impact event.

For the limiting case of constant strength in a gravity-free environment (i.e., $j(\pi_2) = j(\pi_4) = 1$), which has been considered before when cube root scaling is valid, Eq. (4a) indicates that formation time must also vary with $KE^{1/3}$ and, hence, is directly proportional to the size of the crater (Eq. (5a)). Changes in target material could, in principle, change formation time in two independent ways because only one measure of strength has been introduced in the dimensional analysis. The use of only one measure of strength eliminates the possibility that the term s^5 in Eq. (5a) might be better represented by the square of one measure of strength for the dilatational wave velocity (for t) and the cube of a different measure of strength for resistance to deformation (for D). In most practical circumstances, however, formation time should vary with the deformational strength approximately as $s^{5/6}$.

For the limiting condition $s = 0$, when $j(\pi_1) = j(\pi_3) = 1$), formation time in a given material (ρ = constant) is indicated to be directly proportional to $KE^{1/8}$ if the gravity environment is constant, and formation is proportional to $g^{-5/8}$ if the kinetic energy is held constant. Without considering the special situation where the product ρg^5 is constant, crater formation time should scale independent of KE and be proportional to $D^{1/2}$ and $g^{-1/2}$ (Eq. (5b)).

The relative importance of these scaling relationships for formation time depends, in the same manner as for crater size, on the ratio of their "disturbance" velocities

$$\frac{(s/\rho)^{1/2}}{(gD)^{1/2}} = \left[\frac{s}{\rho g D}\right]^{1/2}, \tag{6}$$

where the right side of the equality would be numerically equal to the square root of the "strength" ratio if, and only if, the same measure of strength enters into both ratios. For small craters formed in the laboratory in basalt and granites (Gault, 1973), the disturbance ratio (Eq. (6)) has values of the order of 10. For large terrestrial craters the ratio is much less than unity and approaches a value of zero. With quartz sand targets in the present experiments the ratio can be no greater than about 10^{-1} and, possibly, may approach zero, too. It is not unreasonable to expect, therefore, that the influence of gravity would be most pronounced on formation time in contrast to crater dimensions.

The experimental results are compared in Table 2 with the gravity scaling results from the dimensional analysis ($s = 0$, $j(\pi_1) = j(\pi_3) = 1$). For all five conditions considered the experimentally determined values of the exponents are smaller than the analytical values. Consistent with the suggestions derived from values of the strength and disturbance ratios, the greatest differences are for the crater diameters as functions of KE and g. The largest difference occurs in the variation of diameter with acceleration for constant kinetic energy where an approximately $\frac{1}{6}$ power relationship (0.165) was obtained experimentally instead of the $\frac{1}{4}$ power for gravity scaling. The three experimental exponential relationships for time of formation are in excellent agreement with those derived from the dimensional analysis, the differences being within the uncertainties in the experimental values at the 95% confidence level. This degree of agreement in the functional form for time to form a crater, in particular the $\frac{5}{8}$ power dependence on gravity for constant KE and the square-root dependence on diameter in contrast to the linear relationship for strength scaling ($s = 0$), is strong evidence for the role and importance of gravitational acceleration in these experimental impact cratering results.

The consistently smaller values of the experimental exponents, however, suggests that a small, but nevertheless significant, secondary role of a finite strength effect has been exerted by the quartz sand medium. For all five conditions in Table 2, a measure of physical strength in addition to the gravity strength would act to reduce the values of the exponents to values less than for simple gravity scaling; starting from any given reference value, as the gravity is reduced crater diameters would be smaller and require less time to form than predicted by $\frac{1}{4}$ root scaling. This suggests that that even at larger scale than the present experiments, especially in more competent materials than quartz sand, strength cannot be totally neglected and "pure" gravity scaling can never be realized—except, possibly, it might be approached when the target is water for which the effective strength approaches zero. It is cautioned, however, that subtle effects of strength cannot be solely responsible for the differences

Table 2. Comparison of experiment with gravity scaling ($s = 0$; $j(\pi_1) = j(\pi_3) = 1$).

Dependent variable	Conditions	Gravity* scaling	Experiment
$D \propto \left[\frac{KE}{g}\right]^\alpha$	KE constant	$\alpha = 0.25$	$\alpha = 0.165 \pm 0.005$†
	g constant	$\alpha = 0.25$	$\alpha = 0.181 \pm 0.009$
$t \propto \left[\frac{KE}{g}\right]^\beta$	KE constant	$\beta = 0.625$	$\beta = 0.618 \pm 0.019$
	g constant	$\beta = 0.125$	$\beta = 0.105 \pm 0.027$
$t \propto \left[\frac{D}{g}\right]^\delta$	g constant	$\delta = 0.5$	$\delta = 0.477 \pm 0.032$

*Equations (2b), (4b), and (5b).
†Uncertainty at 95% confidence level.

between the experimental results and gravity scaling. Such a singular effect of strength infers that in a transition from conditions for "pure" strength scaling to conditions for "pure" gravity scaling the value of the exponent on KE should first decrease from $\frac{1}{3}$ for strength scaling to a value less than $\frac{1}{4}$ as observed experimentally and then increase back to $\frac{1}{4}$ for gravity scaling. Such a transition involving values less than $\frac{1}{4}$ seems unacceptable on a physical basis and suggests the involvement of additional, unidentified factor(s).

It is interesting to compare these experimental results for hypervelocity impact craters with similar experimental and numerical analyses reported for explosion craters. Although the two data sets are in general agreement, the explosion crater data differ in specific details and, unfortunately, provide no basis for comparisons of formation times. The most direct experimental comparisons are with Viktorov and Stepenov (1960) and Johnson *et al.* (1969), both of whom report results of the effects of gravity on explosion craters formed also in sand; the former for moist sand for accelerations of 1, 25, 45, and 65 g: the latter for dry sand for accelerations of 0.17, 0.38, 1, and 2.5 g. In contrast to the impact data, results by Johnson *et al.* (1969) were obtained with an ambient air pressure of 600 mm Hg; Viktorov and Stepenov (1960) do not indicate this experimental condition, but presumably it was air at atmospheric pressure. Both studies report exponential relationships between acceleration and diameter that are less than $\frac{1}{4}$ root scaling (0.08 to 0.16) similar to the current results for impact craters. The smallest values are for moist sand which is consistent with a greater strength or cohesiveness (relative to dry sand) tending to negate the effects of changes in acceleration. This interpretation is also supported by Schmidt (1977) who reports results of explosion craters formed in modeling clay using a centrifuge to attain accelerations up to 480 g. The numerical values of the exponential relationships depend on the depth of burial of the explosive, the value of the exponent increasing with increasing depth of burial. This effect of burial depth also has been obtained from numerical calculations of explosion events (Bryan *et al.*, 1977; Killian and Germain, 1977). Because only the shallow depths of burial provide a reasonable simulation of an impact event (Shoemaker, 1962; Baldwin, 1963; and especially Oberbeck, 1971), the data of Viktorov and Stepenov (1960) for very deep burial depths (in addition to the use of moist sand) do not furnish a good basis for comparison here. The numerical calculations of Bryan *et al.* (1977) and Killian and Germain (1977) also correspond to burial depths too deep for impact simulation. Johnson *et al.* (1969), however, report exponential values of approximately 0.11 for surface and near-surface explosions which, at the same scale and in direct comparison with Oberbeck's (1971) experimental results, best simulates the conditions of these impact cratering experiments. Whether the difference between 0.11 and the value of 0.165 from the present investigation is attributable to (1) differences in material properties (i.e., strength), (2) differences in the ambient air pressure, or (3) fundamental differences in the cratering processes for explosions and impacts, must remain an open question at this time. It is interesting to note, however, that the rounded grains of the Ottawa sand employed by Johnson *et al.* (1969) would be expected to exert less cohesive or shear strength than the angular sand grains used in the present investigation. Less cohesiveness or shear strength should, in turn, serve to emphasize gravity effects relative to the impact data and cause the exponent to have a value greater than 0.165. Moreover, although Johnson *et al.* (1969) found a small inverse relationship between crater diameter and ambient pressure at 1.0 *g*, there is no apparent physical basis to suspect that any atmospheric effects for the given value of pressure used by Johnson *et al.* (1969) would change with the acceleration environment. Thus, the smaller value of exponent from explosions provides some speculative evidence supporting differences in the mechanics of crater formation between explosions and impacts.

Concluding Remarks

The experimental data and limited analysis presented herein furnish no general expressions for calculating the size and the time of formation of an impact crater for arbitrary initial conditions. The collective results do serve, however, to focus attention on and emphasize the significance of two primary independent variables in the impact cratering process—measure(s) of strength of the target medium and the gravitational acceleration environment. Deeper understanding and, ultimately, quantitative relationships for their roles are goals for future studies.

Acknowledgments—We thank G. Nakata and R. Jordan, U.S. Geological Survey, Flagstaff, Arizona, for their reduction of the stereo movie records. Special thanks are also due to H. F. Cooper, R. J. Port, and F. M. Sauer for their constructive reviews of the original manuscript.

References

Baldwin, R. B.: 1963, *The Measure of the Moon*, University of Chicago Press.

Braslau, D.: 1971, *J. Geophys. Res.* **75**, 3987–3999.

Bryan, J. B., Burton, D. E., Snell, C. M., and Thomsen, J. M.: 1977, In *Papers Presented to the Symposium on Planetary Cartering Mechanics*, p. 15, The Lunar Science Institute, Houston.

Chabai, A. J.: 1965, *J. Geophys. Res.* **70**, 5075–5098.

Charters, A. C. and Summers, J. L.: 1959, "Some comments on the Phenomena of High Speed Impact", NOLR 1238, U.S. Naval Ordnance Laboratory, White Oak, Silver Springs, Maryland, p. 200–221.

Farmer, I. W.: 1968, *Engineering Properties of Rocks*, E. & F. N. Spon Ltd., London.

Gault, D. E.: 1973, *The Moon* **4**, 32–44.

Gault, D. E., Guest, J. E., Murray, J. B., Dzurisin, D., and Malin, M. C.: 1975, *J. Geophys. Res.* **80**, 2444–2460.

Gault, D. E., Quaide, W. L., and Oberbeck, V. R.: 1968, *Shock Metamorphism of Natural Materials*, B. M. French and N. M. Short (eds.), p. 87–100, Mono, Baltimore, Maryland.

Johnson, S. W., Smith, J. A., Franklin, E. G., Moraski, L. K., and Teal, D. J.: 1969, *J. Geophys. Res.* **74**, 4838–4850.

Killian, B. G. and Germain, L. S.: 1977, Scaling of cratering experiments—Analytical and heuristic approach to the phenomenology. This volume.

Oberbeck, V. R.: 1971, *J. Geophys. Res.* **76**, 5732–5749.

Öpik, E. J.: 1969, *Ann. Rev. Astron. Astrophys.* **7**, 473–526.

Schmidt, R. M.: 1977, A centrifuge cratering experiment: Development of a gravity scaled yield parameter. This volume.

Shoemaker, E. M.: 1962, *Physics and Astronomy of the Moon* (Z. Kopal, ed.), p. 283–359, Academic Press.

Stöffler, D., Gault, D. E., Wedekind, J., and Polkowski, G.: 1975, *J. Geophys. Res.* **80**, 4062–4077.

Viktorov, V. V. and Stepenov, R. D.: 1960, *Inzh. Sb.* **28**, 87–96. (Translated by M. I. Weinrich, SCLT-392, Sandia Corporation, Albuquerque, New Mexico, 1961.)

Roddy, D. J., Pepin, R. O., and Merrill, R. B., editors.
(1977) *Impact and Explosion Cratering*, Pergamon Press (New York), p. 1245–1260.
Printed in the United States of America

Crater modification by gravity: A mechanical analysis of slumping

H. J. MELOSH

Division of Geological and Planetary Sciences, California Institute of Technology, Pasadena, California 91125

Abstract—This paper analyzes the stability of a crater from a mechanical point of view. The principal conclusion is that the observed slumping of lunar craters requires a perfectly plastic constituitive relation for the lunar surface rock. The angle of internal friction of this material must be less than a few degrees. Perhaps these conditions are realized for rapid shearing of the just-shocked rocks immediately after excavation of the crater. Evidence for a perfectly plastic constituitive relation is

(1) Large craters *do* slump. This would be impossible for normal angles of internal friction.
(2) Major slumping occurs only for craters larger than a definite minimum diameter, suggesting that a yield strength is exceeded.
(3) All slumped craters have about the same depth, independent of their diameter. We shall show that this is a consequence of perfect plasticity.
(4) The width of the wall terraces in large slumped craters is roughly independent of crater diameter and implies nearly the same yield strength as does the kink in the depth/diameter curve.

Accepting this evidence, a simplified model of a crater in a perfectly plastic medium is used to investigate the nature of its collapse. The stability of the crater is found to depend principally upon the dimensionless parameter $\rho gH/c$ (ρ = density, g = acceleration of gravity, H = depth, c = yield strength). Craters of all depth/diameter ratios are stable for $\rho gH/c \leq 5$. When $\rho gH/c$ is between 5 and 10 "slope failures" occur in which part of the crater wall slumps off onto the floor. When $\rho gH/c$ is larger than about 15, "floor failure" accompanies the wall slumps. Failure extends throughout the volume of rock enclosing the crater and the floor rises vertically upward as the rim subsides.

I. THE ROLE OF SLUMPING IN CRATER MODIFICATION

THE MORPHOLOGY OF FRESH CRATERS on the moon shows a regular dependence on the crater's size (Howard, 1974). Small fresh craters tend to be bowl shaped. Slump terraces and central peaks appear simultaneously in the 10–30 km diameter range. Flat hummocky floors characterize somewhat larger craters (30–100 km), while craters larger than 200 km diameter acquire multiple mountain rings.

The transition between bowl-shaped craters and craters with slumped rims and central peaks is marked by an abrupt kink in the depth/diameter curve, slumped craters tending to be shallower. This relation is well shown in Fig. 1, after Pike, 1974. The depth/diameter ratio of fresh lunar craters is close to .2 for all craters up to about 15 km rim to rim diameter. The depth of all craters larger than 15 km diameter lies between 3 and 5 km and is nearly independent of diameter. A small population of unslumped craters with diameters between 15 and 30 km lies on the continuation of the depth/diameter = .2 line.

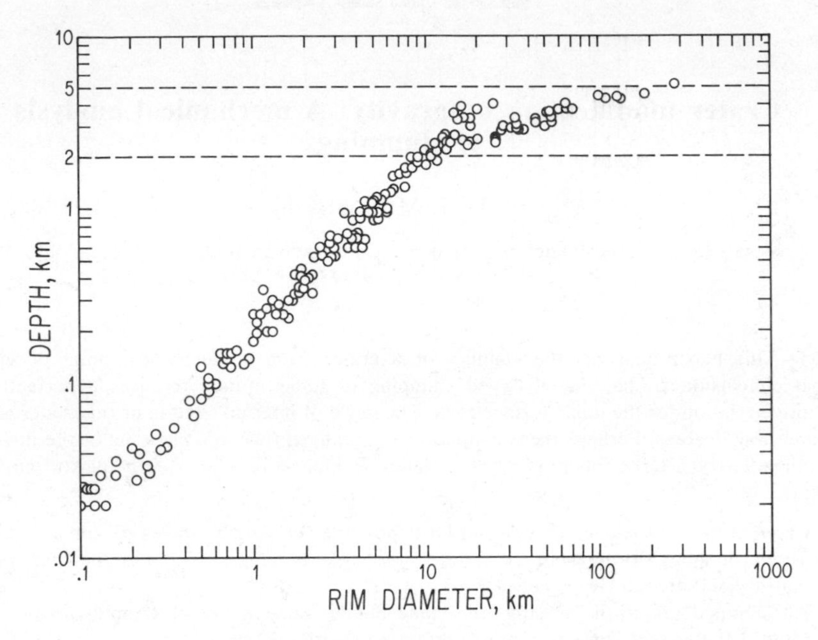

Fig. 1. Depth versus diameter for fresh lunar craters. After Pike (1974).

The dependence of the morphology of fresh lunar craters on diameter has long been attributed to gravity induced mass movements (Quaide *et al.*, 1965). Immediately after formation all craters are assumed to have the same depth/diameter ratio of about .2. Subsequent collapse of the larger craters produces the kink in the depth/diameter curve, explaining the coincidence of this kink with the appearance of wall slumps. Central peaks evidently develop as part of the collapse process, since they are correlated with wall slumps. The unslumped craters between 15 and 30 km diameter are accounted for by local variations in the strength of lunar rock and thus of the size at which slumping begins.

The purpose of the present study is to evaluate the mechanical aspects of this process in more detail. The principal result of the analysis will be that rather peculiar strength characteristics are required for the lunar surface rocks if collapse is to occur at all. In order for lunar craters to slump, the rock must fail with very low angles of internal friction (less than a few degrees). The best description of collapse is obtained if the rock fails as a perfectly plastic material, having no internal friction and a shear strength of only about 30 bars. Granting these unusual strength characteristics, a comprehensive description of crater morphology versus diameter is obtained. These results strongly suggest that lunar craters actually do fail in a plastic mode: the difficulty comes in understanding how internal friction is eliminated.

The need for low internal friction is easily demonstrated. The average internal slope of a crater with a depth/diameter ratio of .2 is only 22°. This is less than the angle of repose of cohesionless rock debris (most such materials have angles of repose in the 35–45° range, see Carson and Kirkby, 1972). Thus, even if the crater were produced in cohesionless debris, we would not expect major slumping to occur, whatever the crater's diameter might be. This argument still permits a small amount of slumping near the crater rim where slopes locally exceed 35° (thus producing a crater with straight talus-covered internal slopes like some of the smaller lunar craters), but it does not allow the observed large-scale slump features. The largest crater in Fig. 1 has a diameter of 300 km and a depth of 5 km, corresponding to a depth/diameter ratio of .017 and an average internal slope of 2°. For these proportions to have been produced by slumping, an angle of internal friction less than about 2° is required. Although the angle of repose and the angle of internal friction are not identical, they seem to be proportional to one another, and seldom differ by more than a few degrees (Carson and Kirkby, 1972, ch. 4). A low angle of repose thus implies a low angle of internal friction.

The appearance of slump features in lunar craters of about 15 km diameter suggests that a yield strength is exceeded by craters of this size. Presuming negligible internal friction from the preceding arguments, this yield strength is readily estimated. The weight of material excavated from a parabolic crater of diameter D and depth/diameter ratio $H/D = \lambda$ is given by $(\pi/8)\rho g \lambda D^3$, where ρ is the density of the excavated rock and g is the acceleration of gravity. If we suppose that this force acts over the surface of a hemisphere enclosing the crater, area $\pi D^2/2$, the shear stress τ supported by the rock near the crater is $\tau \approx \rho g \lambda D/4$. For a 15 km diameter lunar crater this shear stress is about 30 bars, which must be nearly equal to the shear strength of the rock.

These strength characteristics are decidedly peculiar for rock. Many measurements have been made of the strength of rock (Handin, 1966). No rock exhibits the properties apparently required by the slumping of lunar craters. This result is a mystery to which we have no adequate answer. Some authors are inclined to deny the reality of slumping and instead claim that the apparent slump features are a primary product of the impact process. We feel that the evidence, both photogeologic and structural (on terrestrial impact craters) is compelling in favor of slumping.

A partial answer to the strength dilemma may be that the slumping takes place under conditions for which we have little laboratory data. Slumping occurs shortly after crater excavation, as evidenced by the flat-lying sheets of what is probably impact melt on the floors of such craters as Tycho and Copernicus (Schultz, 1976). The rock involved in the slump had just previously been shocked and brecciated and might still contain considerable amounts of kinetic and internal energy at the time of the slump. Moreover, once failure begins strain rates may become very high relative to the strain rates used in conventional laboratory strength tests. Perhaps in such circumstances the strength characteristics of rock are different than in the laboratory tests. A mechanism for the reduction of the angle of internal friction is an unavoidable necessity if the dependence of crater morphology upon size is due to slumping. We have little idea of what this mechanism might be at present. Future experimental work will probably be required to resolve this question. In the remainder of this work we shall assume that the required properties are somehow attained. We shall find that this assumption leads to a self-consistent description of crater morphology, one which agrees with observation in several details.

We have shown that crater modification by gravity requires material properties closely approximated by those of a perfectly plastic substance. Standard techniques (Scott, 1963) exist for evaluating the stability of structures in a perfectly plastic medium under conditions of plane strain. In this paper we shall modify these techniques for axial symmetry and apply the resulting static stability criteria to craters. We shall find that the form of the collapse is governed almost completely by the dimensionless parameter $\rho gH/c$, where c is the yield strength of the plastic substance. When $\rho gH/c$ is less than 5 the crater is stable. For $\rho gH/c$ between 5 and about 10 slope failures occur. Material simply slumps off the crater walls and onto the crater floor. This may be associated with the "swirl textured" floor morphology of craters between 5 and 40 km diameter (Smith and Sanchez, 1973). When $\rho gH/c$ exceeds about 15 failure occurs throughout a volume of rock containing the crater. The crater floor rises vertically as the rim slumps down and inward. This mode of failure may produce well marked central peaks. In all these cases, the initial crater depth H decreases to a final depth H_f for which $(\rho gH_f/c) \cong 5$. This constant final depth explains the near independence of crater depth and diameter for slumped craters larger than 15 km and predicts the increasing extent of flat hummocky floors as crater diameter increases.

In the following pages, we shall show how these conclusions come about in more detail. The more intimate details of the modification of the plastic slip line computations for axial symmetry are relegated to an appendix. In the text we shall concentrate on the structure of the slip line calculations and describe some results.

II. A Mathematical Model of Slumping

The collapse of a lunar crater shortly after its excavation by meteorite impact is undoubtedly a very complex process. Simplifying assumptions must be made in order to make the problem tractable. The initial assumption in this section is drastic and not wholly justifiable: we assume that the collapse of a crater is a quasi-static process. Thus, we ignore the contribution of inertial forces to the stress equilibrium equations. If v is the velocity of the slumping material, then this assumption requires that the dimensionless ratio $\rho v^2/c$ be less than unity. For a yield strength c of 30 bars and a density ρ of 3 g/cm^3, the velocity of the slumping material must be less than 30 m/sec. The highest velocity which can be attained by slumping is the velocity of a body freely falling through a height equal to the depth of the crater H, $v_{\max} = \sqrt{2gH}$. If this velocity is used, $\rho v^2/c$ becomes $2\rho gH/c$ which must be greater than ten for failure to occur. Thus in making the quasi-static approximation, we are assuming that the maximum velocity of slumping is much less than the free fall velocity. This does not seem unreasonable for a first approximation, but more sophisticated calculations should include this term. Inertial forces are likely to be important when downward moving slumps meet at the center of the crater. The inward velocity of the colliding slumps is converted to a horizontal, radially inward compressive force which could be important in the process of central peak formation. The quasi-static assumption also neglects any velocity that material may have acquired from the excavation of the crater itself. This is probably not a bad approximation for the deeper seated collapses, although superficial material may still be in motion at the time of collapse.

The quasi-static assumption allows us to treat the crater as if it were a simple hole in the ground. Conventional methods may be used to analyze the static stability of the hole. The stresses in the rock surrounding the crater satisfy the usual static stress equilibrium equations

$$\frac{\partial \sigma_{rr}}{\partial r} + \frac{\partial \sigma_{rz}}{\partial z} + \frac{\sigma_{rr} - \sigma_{\varphi\varphi}}{r} = 0, \tag{1a}$$

$$\frac{\partial \sigma_{rz}}{\partial r} + \frac{\partial \sigma_{zz}}{\partial z} + \frac{\sigma_{rz}}{r} = \rho g, \tag{1b}$$

where σ_{ij} is the stress tensor in cylindrical coordinates centered around the crater (see Fig. 2 for the geometry of the problem). The stress components $\sigma_{r\varphi}, \sigma_{z\varphi}$ are zero by the assumption of axial symmetry. Compressive stress is taken to be positive.

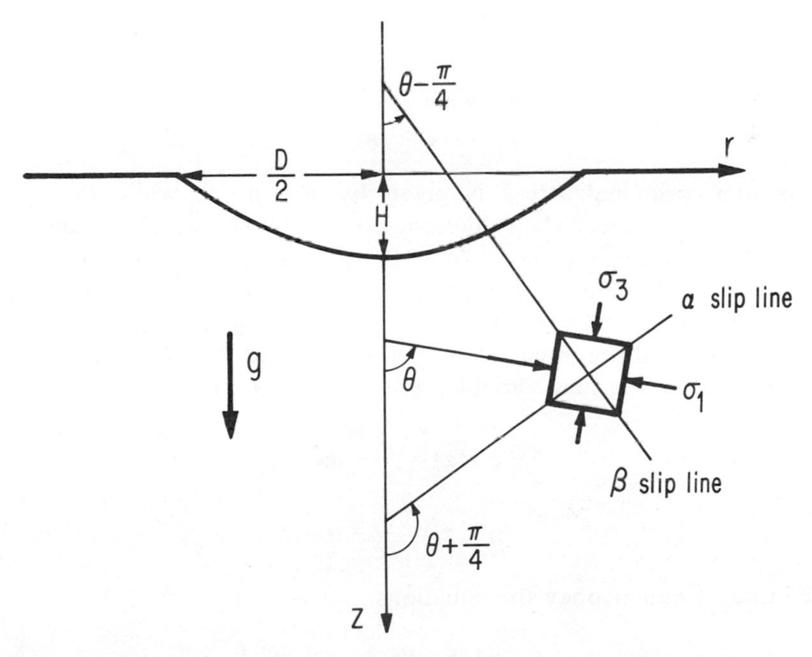

Fig. 2. Geometry of the stress field calculation. σ_1 is the maximum principal stress, σ_3 is the minimum. θ is the angle between the direction of σ_1 and the vertical. The angles $\theta \pm \pi/4$ designate the planes of maximum shear, whose traces are the α and β slip lines.

In a homogeneous, perfectly plastic substance the stress equilibrium Eqs. (1a) and (1b) are supplemented by the critical equilibrium (just failed) condition. We assume that throughout the failed mass the difference between the maximum σ_1 and the minimum σ_3 principal stresses is equal to twice the yield strength c so that the entire mass is at the point of failure (internal friction is entirely neglected in this failure criterion):

$$\frac{\sigma_1 - \sigma_3}{2} = c. \tag{2}$$

Equations (1) and (2) are not enough to compute the stresses σ_{ij} from the boundary conditions. The azimuthal stress $\sigma_{\varphi\varphi}$ is still completely undetermined. The general solution to this problem is not known. $\sigma_{\varphi\varphi}$ is fixed by equating it to the intermediate principal stress σ_2 which is in turn set equal to σ_1 (the Haar–von Kármàn Hypothesis. See Cox *et al.*, 1961).

$$\sigma_{\varphi\varphi} = \sigma_2 = \sigma_1. \tag{3}$$

This condition constitutes yet another assumption.

Equations (1), (2), and (3) may be transformed into two partial differential equations which are suitable for integration by the method of characteristics. The yield condition, Eqs. (2) and (3) are first exploited to express σ_{ij} in terms of two quantities, $p(r, z)$ and $\theta(r, z)$:

$$\sigma_{rr} = p - c \cos 2\theta, \tag{4a}$$

$$\sigma_{zz} = p + c \cos 2\theta, \tag{4b}$$

$$\sigma_{\varphi\varphi} = p + c, \tag{4c}$$

$$\sigma_{rz} = c \sin 2\theta. \tag{4d}$$

The maximum principal stress is given by $\sigma_1 = p + c$, while the minimum principal stress $\sigma_3 = p - c$. Their difference is, of course, $2c$. The quantity $p(r, z)$ plays the role of a mean pressure, while $\theta(r, z)$ is the angle between a vertical line and the direction of the maximum principal stress axis at point r, z (see Fig. 2). Substitution of expressions (4a) to (4d) into the stress equilibrium Eq. (1) leads to coupled partial differential equations for p and θ. These equations are uncoupled by using the new variables $\xi(r, z)$ and $\eta(r, z)$

$$\xi(r, z) = \frac{p}{2c} + \theta, \tag{5a}$$

$$\eta(r, z) = \frac{p}{2c} - \theta. \tag{5b}$$

The quantities ξ and η obey the equations

$$\frac{\partial \xi}{\partial z} + \tan\left(\theta + \frac{\pi}{4}\right)\frac{\partial \xi}{\partial r} = \frac{\rho g}{2c} + \frac{\cos\theta}{\sqrt{2} r \cos\left(\theta + \frac{\pi}{4}\right)}, \tag{6a}$$

$$\frac{\partial \eta}{\partial z} - \tan\left(\theta - \frac{\pi}{4}\right)\frac{\partial \eta}{\partial r} = \frac{\rho g}{2c} - \frac{\cos\theta}{\sqrt{2}r \sin\left(\theta + \frac{\pi}{4}\right)}. \tag{6b}$$

The left-hand side of Eq. (6a) may be recognized as the total derivative $d\xi/dz$ evaluated along a line (the α line) sloping at an angle $\theta + (\pi/4)$ with respect to the vertical. Similarly the left-hand side of Eq. (6b) is the total derivative $d\eta/dz$ along a line (the β line) sloping at $\theta - (\pi/4)$ and thus orthogonal to the α lines. Thus,

$$\left.\frac{d\xi}{dz}\right|_{\theta+(\pi/4)} = \frac{\rho g}{2c} + \frac{\cos\theta}{\sqrt{2}r \cos\left(\theta + \frac{\pi}{4}\right)}. \tag{7a}$$

$$\left.\frac{d\eta}{dz}\right|_{\theta-(\pi/4)} = \frac{\rho g}{2c} - \frac{\cos\theta}{\sqrt{2}r \sin\left(\theta + \frac{\pi}{4}\right)}. \tag{7b}$$

Care must be exercised in the vicinity of $r = 0$. The appendix describes how Eqs. (7a) and (7b) must be modified near $r = 0$.

Equations (7a) and (7b) can be integrated numerically. We shall illustrate the procedure by a simple but important example, that of a "crater" with a rectangular profile. The "crater" is thus a right cylindrical hole of depth H and diameter D formed in a perfectly plastic substance.

Example: Crater with a rectangular profile

Figure 3 illustrates the geometry of the ξ and η characteristic curves or "slip lines" along which information about ξ and η is transmitted. These slip lines are the traces of surfaces of maximum shear stress. It can be shown that the velocity field may be either continuous or discontinuous across these surfaces, so that the slip lines represent potential fault traces. The β lines (along which η is propagated) are seen to begin inside the crater, plunge under the rim then rise up to the surface outside the crater. The α lines (along which ξ is propagated) are orthogonal to the β lines.

There are three major divisions of the slip line field illustrated in Fig. 3. The stresses in region I, inside the crater rim, are determined by boundary conditions on the crater floor. The characteristics are straight lines for a rectangular crater profile. The principal compressive stress σ_1 is horizontal just under the crater floor so that $\theta = \pi/2$ in this region (the normal stress $\sigma_3 = 0$; Eq. (2) then requires $\sigma_1 = 2c$). Equation (7) requires that if $\theta = \pi/2$ on the crater floor, then $\theta = \pi/2$ throughout region I as shown in Fig. 3.

Region III underlies the crater rim. In this region the principal compressive stress σ_1 is normal to the surface, so that $\theta = \pi$ is the appropriate boundary condition. Region II, the "fan", lies between Regions I and III. Slip lines in the fan radiate from a singular point on the crater rim. Since θ jumps from $\pi/2$ to π

SLIP LINE FIELD, RECTANGULAR CRATER PROFILE

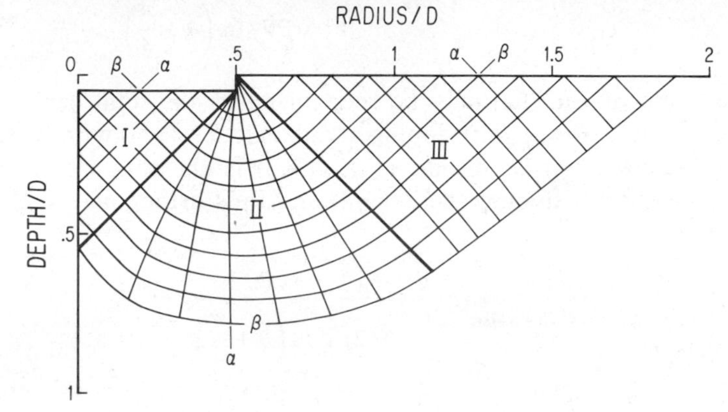

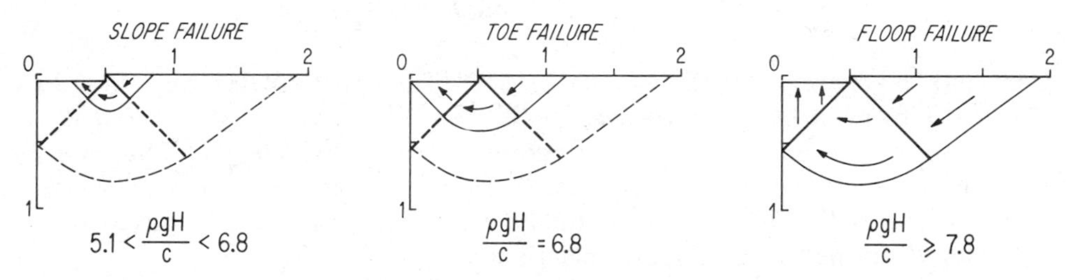

Fig. 3. Slip line field under a crater with a rectangular profile and depth/diameter ratio $\lambda = .05$. The upper part of the figure shows the entire slip line field. The form of the slip lines is not a function of $\rho gH/c$, although the extent of the failed region is. The three regions of the slip line field are labeled. Distances are normalized by the crater diameter D. Only every second line of the grid used in the numerical calculations is shown. The lower part of the figure shows how the nature of the failure changes as $\rho gH/c$ increases. The solid lines outline the failed part of the crater and the arrows schematically indicate the direction of mass movement.

across the crater rim, either η or ξ must be discontinuous. Equation (7b) shows that η must be continuous along a β line running just outside the singular point, so that ξ must be discontinuous. The α lines radiating from the singular point of the fan each transmit a different value of ξ intermediate between its values in region I and region III at the rim.

The stability of the crater is determined by integrating Eq. (7), starting at the singular point of the fan (Scott, 1963). The initial assumption (to be tested at the end of the integration) is that failure has occurred on the crater floor just inside the rim. This assumption establishes the initial values of ξ and η ($\xi = \frac{1}{2} + (\pi/2)$, $\eta = \frac{1}{2} - (\pi/2)$ on the crater floor. $\xi = \frac{1}{2} + (3\pi/2)$, $\eta = \frac{1}{2} - (\pi/2)$ just outside the rim). The β line plunges down under the rim, initially at a 45° angle. Information on ξ is propagated from the singular point so that θ can be computed at each grid point from η and ξ. The integration proceeds from grid point to grid point toward the outer rim of the crater. When the surface is reached the value of σ_{zz}

acting vertically at the end of the β line may be evaluated from ξ and η (Eqs. (4) and (5)). This stress σ_{zz} represents the normal load which must act on the crater rim in order to justify the original assumption of failure on the crater floor. If σ_{zz} is zero or negative the assumption of failure is valid. The region of the rim enclosed by the β line is in critical equilibrium and can slump. If σ_{zz} is positive, extra normal stress is required to fail the β line. Since this extra stress is not present, the line cannot have failed and the integration is terminated.

When the β line nearest the rim is found to have failed, integration is begun on a new β line closer to the center of the crater. ξ information is propagated along α lines from the previously computed β line. The value of σ_{zz} is again tested when the β line surfaces outside the crater. If the test indicates failure, a new β line is begun. The integrations are stopped only when a positive value of σ_{zz} is obtained at the end of the β line on the crater rim. The extent of failure in the crater may thus be determined for any value of the parameter $\rho gH/c$ and depth/diameter ratio λ.

Integrations like those shown in Fig. 3 have demonstrated that a crater with a rectangular profile is stable for $\rho gH/c \leq 5$ irrespective of its depth/diameter ratio. When $\rho gH/c$ is slightly larger than 5, "slope failures" occur which only involve material near the rim. When $\rho gH/c = 7$ the toe of these slope failures reaches the center of the crater, $r = 0$. Values of $\rho gH/c$ larger than 7 produce failure below the center of the crater: as the rim slumps inward, following the slip lines, a small wedge-shaped plug under the center of the crater floor rises vertically. If $\rho gH/c \gtrsim 8$ the entire floor of the crater rises vertically as the rim slumps. These modes of failure are illustrated by the inset in Fig. 3.

Surprisingly, the slip line field of a crater with a rectangular profile has little in common with the failure pattern of slumped lunar craters. Figure 3 indicates that full floor failure involves slumping throughout a region nearly 3 crater radii outside the crater rim. However, as we shall see in the next section, the exterior slip line field is highly sensitive to the detailed shape of the crater wall. For a parabolic crater profile the slip lines seem to surface much closer to the original rim than for a rectangular profile. The rectangular profile is nevertheless interesting since it exhibits several of the important features of pure plastic collapse. It is also free from the singular surfaces which appear for parabolic profiles. Another advantage of the rectangular profile is that approximate analytic solutions of the problem exist, against which the numerical code may be checked.

III. Slumping of a Parabolic Crater

Fresh, unslumped lunar craters are generally described as "bowl shaped". They are usually steeper near their rims than in their interiors, although craters with straight or even convex well profiles are known (Schultz, 1976). These characteristics suggest that a parabolic profile might approximate the form of unslumped craters. Moreover, a parabolic profile provides a good description of the Teapot Ess and Jangle U nuclear explosion craters as well as the Arizona

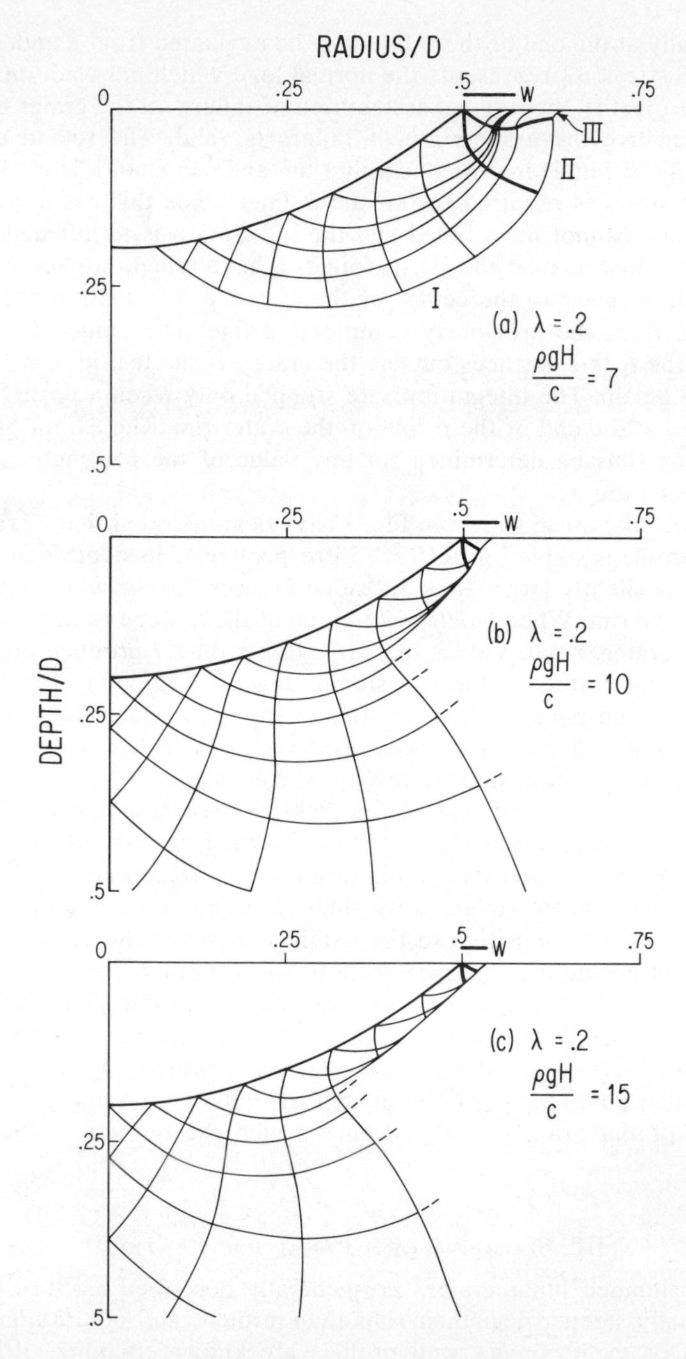

Fig. 4. Slip line field under a crater with a parabolic profile and depth/diameter ratio $\lambda = .2$. Distances are normalized by the crater diameter D. Only every fifth line of the grid used in the numerical calculations is shown. In (a) $\rho gH/c = 7$ and a deep slope failure develops. A discontinuous surface is nearly (but not quite) formed. The width of

Meteor Crater. We shall thus investigate the stability of craters with parabolic profiles where the crater depth z is related to radius r by

$$z = \lambda D\left[1 - \left(\frac{2r}{D}\right)^2\right]. \tag{8}$$

The numerical scheme described in the preceding section can be easily adapted to evaluate the stability of a crater with a parabolic profile. As before, the crater is initially assumed to have failed just inside its rim. The boundary conditions are again that the minimum principal stress σ_3 is perpendicular to the surface of the crater interior and that $\sigma_3 = 0$. Of course, the vertical stress σ_{zz} and σ_3 do not coincide for a parabolic profile; the slope of the crater wall must be taken into account. Figure 4 shows that the same three divisions of the slip line field occur for a parabolic crater as for a crater with a rectangular profile. Like the rectangular profile crater, parabolic craters begin to fail when $\rho gH/c \simeq 5$, independent of λ. The initial failure is a slope failure (Fig. 4). A new feature of the parabolic profile, however, is that the shape of the slip line field depends upon the parameter $\rho gH/c$ (as well as λ). In the case of a rectangular profile the same slip line field described the crater for all values of $\rho gH/c$.

Another new feature of the parabolic crater profile is the appearance of singular curves in the slip line field. Figure 4 shows that for $\rho gH/c \geqslant 7$ ($\lambda = .2$) the β lines from the floor and wall of the crater converge to a single line under the crater rim. This singular curve has a simple physical interpretation: the shear stress due to the weight of the overburden, resolved on the surface corresponding to the singular curve, is just equal to the yield strength c. Figure 5 shows how such a surface develops under a straight slope. At a depth h vertically beneath a point on the surface of the slope the overburden pressure is ρgh. The overburden can be resolved into a normal stress acting across a plane parallel to the slope and a shear stress $\tau = \rho gh \sin\phi \cos\phi$ acting along it (ϕ is the angle between the surface of the slope and the horizontal). The shear stress reaches the yield strength c when

$$h = c/(\rho g \sin\phi \cos\phi). \tag{9}$$

In the case of a parabolic crater the singular surface is curved, not planar; however, for sufficiently large $\rho gH/c$ the surface occurs at such shallow depths

---

the slump block calculated from Eq. (10) of the text is shown as a heavy bar over the crater rim. This is seen to be in good agreement with the width of the rim bounded by the incipient singular surface. The same three regions of the slip line field seen for a rectangular profile also occur for the parabolic profile. (b) and (c) show the slip line field for increasing $\rho gH/c$. The singular surfaces are well developed, preventing extension of the slip lines from the crater interior to the rim. We thus do not know how much of the crater interior has failed, except by analogy with the rectangular profile crater. The heavy bars near the crater rim demonstrate the agreement of Eq. (10) and the numerical computations. Regions II and III of the slip line field have shrunk to insignificance. Region I includes nearly the entire field.

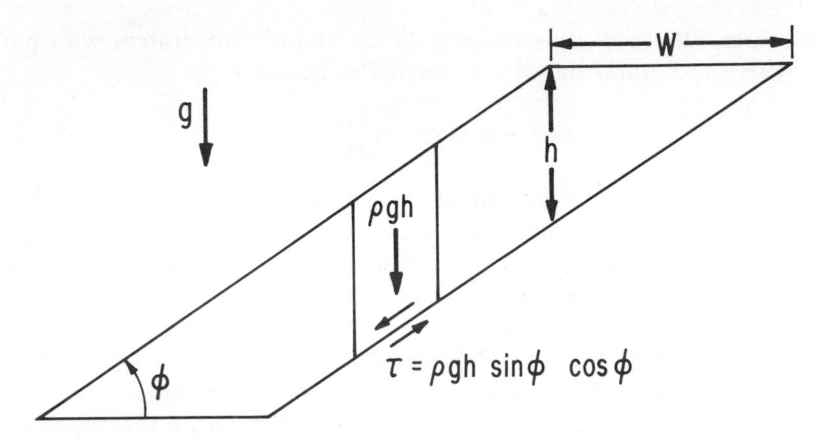

Fig. 5. Geometry for the calculation of the shear stress τ acting at depth h on a plane parallel to the surface of a slope of angle ϕ. A singular surface forms in the slip line field where this shear stress reaches the shear strength c.

that the curvature may be locally neglected. In these cases, Eq. (9) is seen to be well satisfied when the local value of ϕ is used.

These singular surfaces represent natural surfaces of detachment or faulting: the mass of material lying above the singular surface constitutes the first slump block which slides off the crater wall. When $\rho gH/c$ is sufficiently large ($\gtrsim 7$ for $\lambda = .2$) the width W of this block may be approximated by the straight slope result $W = h/\tan\phi = c/(\rho g \sin^2\phi)$. Near the rim of the crater the slope ϕ derived from Eq. (8) is $\phi_{\text{rim}} = \tan^{-1}(4\lambda)$ so that

$$W \simeq \frac{c}{\rho g}\left(\frac{1+16\lambda^2}{16\lambda^2}\right). \tag{10}$$

The length of the bar near the crater rim in Fig. 4 represents the width of the slump block computed from Eq. (10). It is clearly a good approximation to the numerical result for $\rho gH/c \gtrsim 7$. An important prediction of the plastic collapse model is that W is independent of crater size for sufficiently large craters (i.e., $\rho gH/c$ sufficiently large). This result seems to be roughly correct for lunar craters; preliminary measurements on Orbiter photographs show that for craters between 40 and 200 km diameter the slump block (wall terrace) width is in the range of 2 to 4 km. This width corresponds to a yield strength c from 40 to 80 bars, in reasonable agreement with the 30 bar figure derived from the onset of slumping at 15 km (assuming $\lambda = .2$ and using Eq. (10). A lower value of λ, as might be appropriate for slump blocks forming after the first, would result in a lower yield strength. For example, using $\lambda = .167$ Eq. (10) yields c between 30 and 60 bars).

The formation of a singular surface, however useful it may be in delineating slump blocks, constitutes a fundamental limitation of the quasi-static stability analysis. The problem is that, once a singular surface has formed, there is no way of continuing the solution to the other side of the surface. This is why the slip line fields in Fig. 4 do not continue to the rim. The β lines cannot be extended beyond the trace of the last α line to clear the singularity. The reason for this difficulty is clear: points below the singular surface are subjected to shear stresses greater than the yield strength. Equation (2) cannot be satisfied, so no solution to Eq. (7) exists. Inertial forces, neglected in the quasi-static approximation, become important in this region. The quasi-static analysis thus cannot answer questions about how wide the region of slumping is, how many slump blocks form, or even for what values of $\rho gH/c$ and λ floor failure occurs.

In spite of the limitations of the quasi-static method, some valid inferences can probably be made. Our analysis of the crater with a rectangular profile suggests that increasingly extensive failure of the crater slope and floor occurs as $\rho gH/c$ is raised (independent of λ). The value of $\rho gH/c$ at which floor failure begins may be approximately determined. On one hand, if the failure mode is not very sensitive to crater shape, floor failure could begin at $\rho gH/c = 7$, just as for the rectangular profile. On the other hand, the volume of a parabolic crater of depth H is equal to the volume of a rectangular crater of depth $H/2$. If the volumes, hence the driving forces, must be the same to initiate comparable styles of collapse, then floor failure of a parabolic crater should begin when $\rho gH/c = 14$. Similarly, full floor failure should be established for $\rho gH/c$ somewhere between 8 and 16.

IV. Conclusion

The scheme we propose for the slumping of a parabolic crater is a composite, derived partly from direct studies of the parabolic profile and partly from studies of a crater with a rectangular profile. The slip line field of the parabolic profile displays the onset of slope failure and indicates how slump blocks form. The model is, however, limited to a description of the first slump block. The deeper seated failure modes are concealed by these slope failures. The rectangular crater profile is free from singular surfaces. It thus cannot describe the more superficial slumping, but it does provide insight into the deeper seated modes of failure, especially of the floor failure mode. The two models thus complement one another. Together, they strongly suggest that a large crater ($\rho gH/c \gtrsim 16$) collapses by simultaneous slumping of the walls and floor failure, leading to a vertical rise of the crater floor as slump blocks are shed off the rim. Failure extends throughout the entire volume of rock enclosing the crater, although the exact extent of the slumped region cannot be adequately estimated from the present model. Nevertheless, comparison of the slip line field beneath the crater in Fig. 4 with the slip line field in Fig. 3 suggests that the slumped region around a parabolic profile crater is far less extensive than that around a rectangular profile crater. Moreover, if (as suggested in the introduction) the reduction of the angle of internal friction is related to the shock wave which excavated the crater, the volume of material susceptible to plastic failure may be determined more by shock dynamics than by the slip line field. At the moment, there are no data bearing on the subject.

Further work will be needed to establish the above scheme with certainty and to fill in some of the missing details. Improved numerical work, taking inertial forces into account, can eliminate a large area of uncertainty in the present quasi-static calculations. Scale model experiments, using a centrifuge to study the slumping of small craters under their own weight, will also be of great value in checking the calculations.

Nevertheless, the simplified model reported here probably describes the major features of slumping. These features are, first and foremost, the necessity of a plastic failure criterion; very little internal friction can be tolerated. Secondly, we find that sufficiently large craters should fail by a combination of slumping off the walls and a deep seated vertical rise of the crater floor. Due to the absence of internal friction, the crater depth/diameter ratio λ plays little or no role in determining the form of the collapse. The critical parameter is the crater depth H, occurring in the dimensionless combination $\rho gH/c$. Figure 6

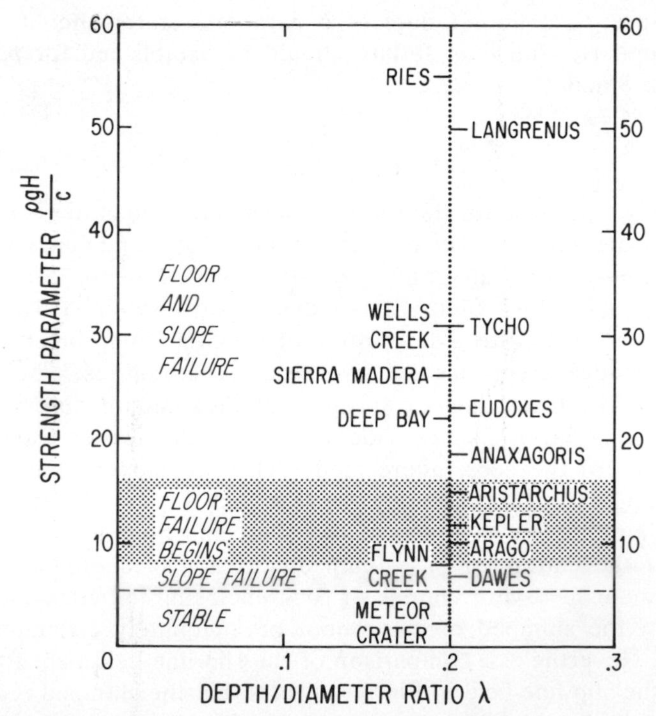

Fig. 6. Summary of the mode of collapse of a parabolic profile crater as a function of $\rho gH/c$ and λ. Floor failure probably begins for $\rho gH/c$ somewhere in the range of 8 to 16. The stippled pattern indicates uncertainty in the exact value at which this occurs. The absence of internal friction is responsible for the independence of failure mode and λ. The values of $\rho gH/c$ for various lunar (to the right of the $\lambda = .2$ line) and terrestrial (to the left of the $\lambda = .2$ line) craters are plotted for comparison. We have used the observed diameter of the crater, then assumed $\lambda = .2$ and $c = 27$ bars in computing these values (see text for fuller explanation).

summarizes these results. The mode of failure of a parabolic crater is plotted as a function of $\rho gH/c$ and λ. Included on the plot are values of $\rho gH/c$ for a variety of lunar and terrestrial craters deduced from their diameters. We have assumed $\lambda = .2$ for fresh craters and $c = 27$ bars (computed from the 15 km diameter kink point in the lunar depth/diameter plot, Fig. 1). This plot indicates a general agreement between crater morphology and theoretical $\rho gH/c$ value. Central peaks and flat floors seem to be associated with the predicted range of floor failures. Craters like Dawes are probably pure slope failures. Much more work needs to be done to fill out the details of this sort of plot. We hope to have demonstrated that it is worth the trouble to do so.

Acknowledgment—This work was supported by NASA Grant NSG-7316.

REFERENCES

Carson, M. A. and Kirkby, M. J.: 1972, *Hillslope Form and Process*, Cambridge U. Press, p. 93.

Cox, A. D., Eason, G., and Hopkins, H. G.: 1961, Axially symmetric plastic deformation in soils. *Phil Trans. Roy. Soc. London* **254A**, p. 1–45.

Handin, J. A.: 1966, Strength and ductility, *GSA Memoir* **97**, p. 223–289.

Howard, K. A.: 1974, Fresh lunar impact craters: Review of variation with size, *Proc. Lunar Sci. Conf. 5th*, p. 67–79.

Pike, R. J.: 1974, Depth/diameter relations of fresh lunar craters: Revision from spacecraft data. *Geophys. Res. Lett.* **1**, p. 291–294.

Quaide, W. L., Gault, D. E., and Schmidt, R. A.: 1965, Gravitative effects on lunar impact structures. *Annals N. Y. Acad. Sci.* **123**, p. 563–572.

Schultz, P. H.: 1976, *Moon Morphology*. U. Texas Press, Austin, Texas.

Scott, R. F.: 1963, *Principles of Soil Mechanics*, Addison-Wesley, Mass. Appendix C.

Smith, E. I. and Sanchez, A. G.: 1973, Fresh lunar craters: Morphology as a function of diameter, a possible criterion for crater origin. *Modern Geology* **4**, p. 51–59.

APPENDIX

The treatment of $r = 0$ in slip line calculations

Equations (7a) and (7b) of the text may be numerically integrated along the α and β characteristic lines to determine both a slip line field and a stress distribution consistent with the boundary conditions. This integration is straightforward except at the point $r = 0$. The presence of r in the denominator of both Eqs. (7) leads to a divergence in the derivatives of ξ and η unless other terms in (7) contribute a zero at $r = 0$. Axial symmetry requires that $\theta = \pi/2$ or π at $r = 0$. Since $\theta = \pi/2$ on the surface of the crater floor, we expect that $\theta = \pi/2$ everywhere below the floor of the crater on the $r = 0$ axis. Thus, let us expand θ in a Taylor series near $r = 0$:

$$\theta(r, z) = \frac{\pi}{2} + k(z)r + \cdots \tag{A1}$$

Substitute Eq. (A1) into Eqs. (7a) and (7b), keeping only the leading terms in r. The derivatives of ξ and η are equal to this order:

$$\left.\frac{d\xi}{dz}\right|_{\theta+(\pi/4)} = \frac{\rho g}{2c} + k(z) + \mathcal{O}(r), \tag{A2a}$$

$$\left.\frac{d\eta}{dz}\right|_{\theta-(\pi/4)} = \frac{\rho g}{2c} + k(z) + \mathcal{O}(r). \tag{A2a}$$

The function $k(z)$ may be determined by examining the stress equilibrium Eqs. (1) in the small r limit. To first order in r, Eqs. (4) become:

$$\sigma_{rr} = p + c + \mathcal{O}(r^2), \tag{A3a}$$

$$\sigma_{zz} = p - c + \mathcal{O}(r^2), \tag{A3b}$$

$$\sigma_{\theta\theta} = p + c, \tag{A3c}$$

$$\sigma_{rz} = -2ck(z)r + \mathcal{O}(r^2). \tag{A3d}$$

Substitution into Eq. (1a) yields $\partial p/\partial r = 0$, so that $p = p(z)$ is a function of z only near $r = 0$. Equation (1b) yields the desired relation

$$k(z) = \frac{1}{4c}\left(\frac{dp}{dz}\bigg|_{r=0} - \rho g\right). \tag{A4}$$

This equation may be used in the numerical integration so long as $(dp/dz)|_{r=0}$ is known. This is generally true: suppose that a new β line is begun at some depth below the crater floor. It must begin on the axis, at $r = 0$. We evaluate $(dp/dz)|_{r=0}$ by computing the difference in p between the two points on the axis immediately above the point where the new β line begins. This difference is then divided by the difference in height Δz between them. The value of the derivative $(dp/dz)|_{r=0}$ is then extrapolated to the new point. This scheme works as long as there are two points with known $p(z)$ values above the point where the new β line begins. This is not the case for the first β line to begin below the crater floor. In this case, however, $(dp/dz)|_{r=0}$ can be derived from the surface boundary conditions by Taylor expansion.

Consider a flat floored crater (a rectangular profile crater). The boundary conditions $\theta = \pi/2$ and $\sigma_3 = 0$ imply that $\xi(r, z = H) = \frac{1}{2} + (\pi/2)$ and $\eta(r, z = H) = \frac{1}{2} - (\pi/2)$ everywhere inside the crater. Consider an α line beginning at a radius $r = \Delta$ on the crater floor r and intersecting the $r = 0$ axis at $z = H + \Delta$. Equations (A2) and (A4) require

$$\xi(r = 0, z = H + \Delta) = \xi(r = \Delta, z = H) + \frac{\Delta}{4c}\left(\frac{dp}{dz}\bigg|_{r=0} + \rho g\right). \tag{A5a}$$

Similarly, since $\theta = (\xi - \eta)/2 = \pi/2$ on the axis $r = 0$,

$$\eta(r = 0, z = H + \Delta) = \eta(r = \Delta, z = H) + \frac{\Delta}{4c}\left(\frac{dp}{dz}\bigg|_{r=0} + \rho g\right), \tag{A5b}$$

but $\xi(r = \Delta, z = H) = \xi(0, H)$ and $\eta(r = \Delta, z = H) = \eta(0, H)$. Now compute $p = c(\xi + \eta)$ from (A5a) and (A5b):

$$p(r = 0, z = H + \Delta) = p(r = 0, H) + \frac{\Delta}{2}\left(\frac{dp}{dz}\bigg|_{r=0} + \rho g\right), \tag{A6}$$

or

$$\frac{p(r = 0, z = H + \Delta) - p(r = 0, H)}{\Delta} = \frac{1}{2}\left(\frac{dp}{dz}\bigg|_{r=0} + \rho g\right). \tag{A7}$$

Letting Δ approach zero, we find that at $z = H$

$$\frac{dp}{dz}\bigg|_{r=0} = \rho g \text{ (rectangular profile)}, \tag{A8}$$

which is the desired result. Similarly, for a parabolic crater with depth/diameter ratio λ and diameter D, we find that just below the crater floor at $z = H$ and $r = 0$,

$$\frac{dp}{dz}\bigg|_{r=0} = \rho g + \frac{32\lambda c}{D} \text{ (parabolic profile)}. \tag{A9}$$

These results may be combined with the computational scheme outlined in the text (and discussed in greater detail by Scott, 1963) to determine the stability and initial failure modes of an axially symmetric crater.

Roddy, D. J., Pepin, R. O., and Merrill, R. B., editors.
(1977) *Impact and Explosion Cratering*, Pergamon Press (New York), p. 1261–1278.
Printed in the United States of America

A centrifuge cratering experiment: Development of a gravity-scaled yield parameter

R. M. SCHMIDT

Shock Physics and Applied Math, M/S 42-37, Boeing Aerospace Company,
Seattle, Washington 98124

Abstract—A centrifuge was used to perform explosive cratering tests under the influence of gravitational accelerations up to 480 G*. The test matrix of 20 shots, primarily designed to demonstrate the feasibility of such experiments, provided data which suggests that increased gravitational acceleration offers a means for understanding large scale crater formation. This relevance was the primary objective of the experiment, and led to the tentative formulation of a "gravity-yield" scaling law. This proposed scaling law for crater volume was based upon NTS field data in conjunction with the centrifuge results. Good agreement is shown over a range in yield energy of 10^{12}.

1. INTRODUCTION

THE APPLICATION OF DATA obtained from a centrifuge experiment to the scaling of crater dimensions is presented. That gravity is an important parameter and can be a controlling variable in the cratering process is suggested by physical argument. The specific excavation work, energy per unit mass, required to remove material from the crater volume increases with increased gravity. In addition, ejecta range decreases as gravity is increased, the upthrust of the crater lip is resisted by an increased gravity body force, and the lithostatic confining pressure has a variable effect on material failure envelopes. As will be shown all these factors are consistent with a reduced cratering efficiency, crater volume per unit charge mass, observed with increased gravity. Relating this behavior to a comparable decrease in cratering efficiency due to increased charge size was the goal of this experiment, although the effect of gravity on crater formation is also of interest especially in the study of extra-terrestrial craters.

The objective was to utilize increased gravity to provide control of an additional crater variable on the basis of which small scale similar experiments could be performed. Comparisons among small scale laboratory experiments conducted at 1 G and larger size field experiments, such as the NTS series, show a trend in which cratering efficiency is a decreasing function of charge size. Chabai (1965) has suggested that this is a result of an inability to scale material properties, particularly material strength, contributing to non-similarity. Verification of the hypothesis that material properties become less important with

*1 G is defined as the terrestrial gravity field strength taken to be equal to 981 cm/sec^2.

increased charge size as well as with increased gravity, thereby relaxing the similarity requirement, was a primary basis of this experiment. To test this hypothesis, an experiment was performed and the results compared to large scale field data in which the only material property considered was the density.

2. EXPERIMENTAL DESCRIPTION

The centrifuge used in this study was a Gyrex Model 2133 having a 40,000 G-lb capacity with a maximum specimen weight of 500 lb and a maximum centrifugal acceleration of 500 G with an arm radius of 50 in. The centrifuge rotor with a mounted sample is shown in Fig. 1. Dynamic balance was achieved and maintained by firing two shots during the same run, thus utilizing the symmetrical design of the rotor.

The media chosen for the experiment was artist modeling clay "Permoplast" supplied by American Art Clay Co., Indianapolis, Indiana. The choice of this material was dictated by two practical constraints. One, since the centrifuge used was not equipped with a swing basket and the specimen had to start with the ground zero plane vertical, the material was required to be cohesive. Secondly, this material provided a rather simple solution to the problem of data reduction for crater dimensions. Upon the conclusion of the run, the sample was removed from the centrifuge and cut in radial sections using a 10 mil diameter wire. These sections could then be photographed and the photographs subsequently digitized to provide crater volume, radius, depth, lip dimensions, etc.

The explosive charge used in the experiment included various detonating caps supplied by Dupont. These caps, also, were chosen for convenience rather than for any desirable properties of charge similarity. The three types of charges used are shown in Fig. 2. Charge geometry as well as calculated energy release for the various configurations are compared in Table 1. By combining two or more charges in various combinations the energy ratio between shots could be controlled as well as the charge geometry.

The E106 cap was chosen as a baseline charge and, as can be seen in Table 1, the #6 cap and #8 cap have approximately two and three times the energy release respectively. Therefore an energy ratio of two can be obtained with either two E106 caps placed adjacent to each other or with a #6 cap. Similarly three adjacent E106 caps can be compared with a #8 cap for an approximate energy ratio of three. The charge energy release was calculated using both manufacturer's specified composition and configuration (Dupont, 1976) as well as explosives properties given by Johansson and Persson (1970). An equivalent weight of TNT for the various charges was calculated based upon an energy release of 1 g of TNT being equal to 4.19×10^{10} ergs.

A preliminary set of shots was performed at 1 G for the purpose of charge sizing and to confirm sample design. Other factors evaluated were charge reproducibility, sensitivity to variation in the initial temperature of the clay, and the effects of boundaries. The resulting craters were compared on the basis of diameter and depth measurements. For these preliminary tests, volumes were not determined. The samples were contained in ordinary aluminum "spring form" cake pans, 11 in. in diameter and 3 in. deep. Larger pans had no effect on crater size nor did the type of base plate material on which the pan was placed. The centrifuge has a $1\frac{1}{4}$ in. thick aluminum mounting plate which was reproduced for the 1 G shots. In addition, tests using a 1 in. thick wooden base plate confirmed that effects due to boundary reflections were negligible. Charge reproducibility did not appear to be a problem nor did clay temperature if held within a range of 67 ± 1°F. Larger variation in clay temperatures produced measurable differences in crater size.

Another factor considered was the possibility of post cratering flow due to creep of the clay at high G levels. To investigate this, samples with craters formed under 1 G conditions were mounted in the centrifuge and spun at various G levels. From these runs it was determined that negligible change in crater shape could occur during centrifuge shut down time (less than 1 min after firing). However measurable changes were observed for specimens held at 500 Gs for times greater than this. This phenomenon became more pronounced with increasing time due to heating of the air in the centrifuge.

Fig. 1. Centrifuge rotor showing sample mounted in place.

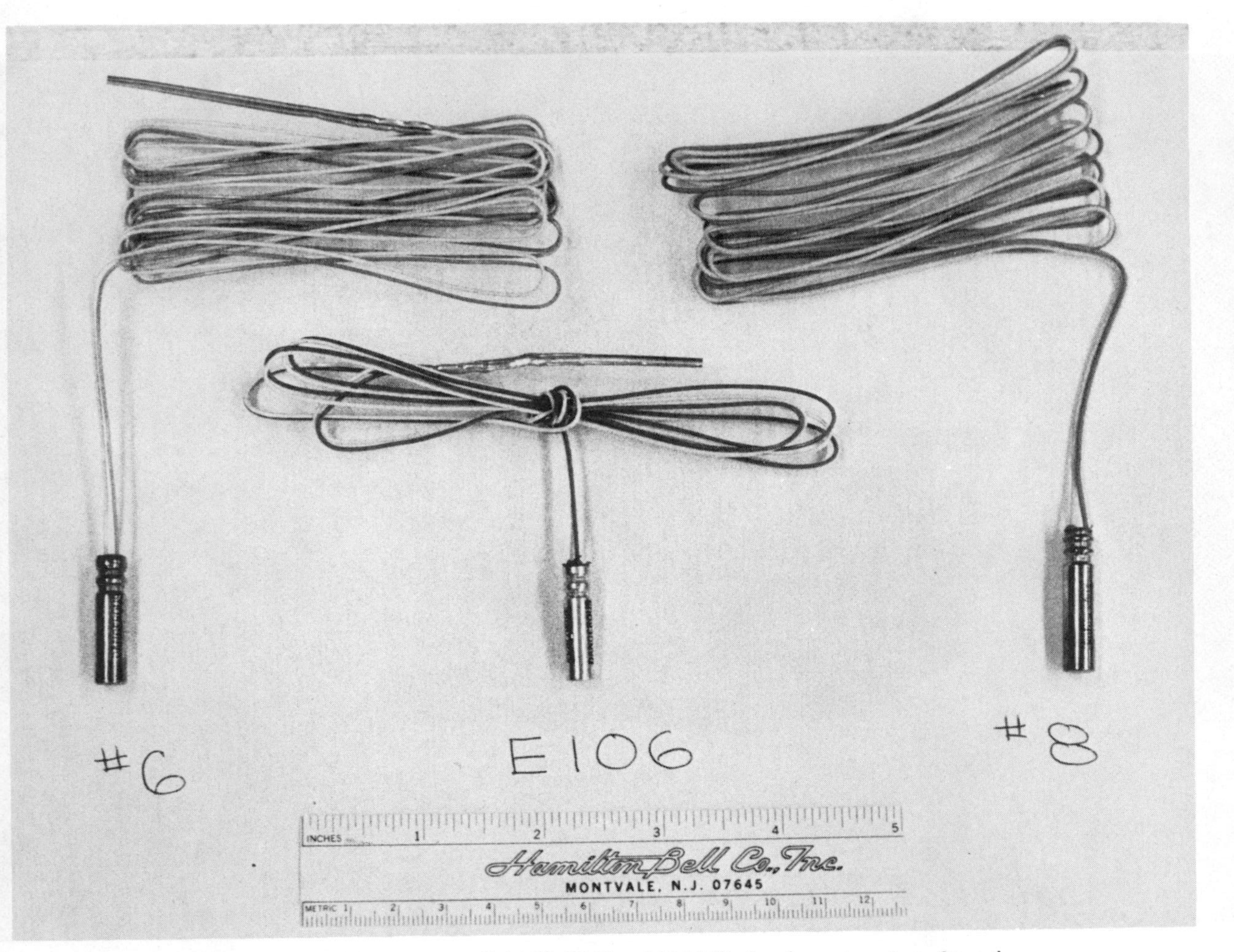

Fig. 2. Dupont detonator caps E-1A(6), E106 and E-1A(8) showing external configuration.

Fig. 3. 480 G pre-shot configuration showing installation of three E106 charges.

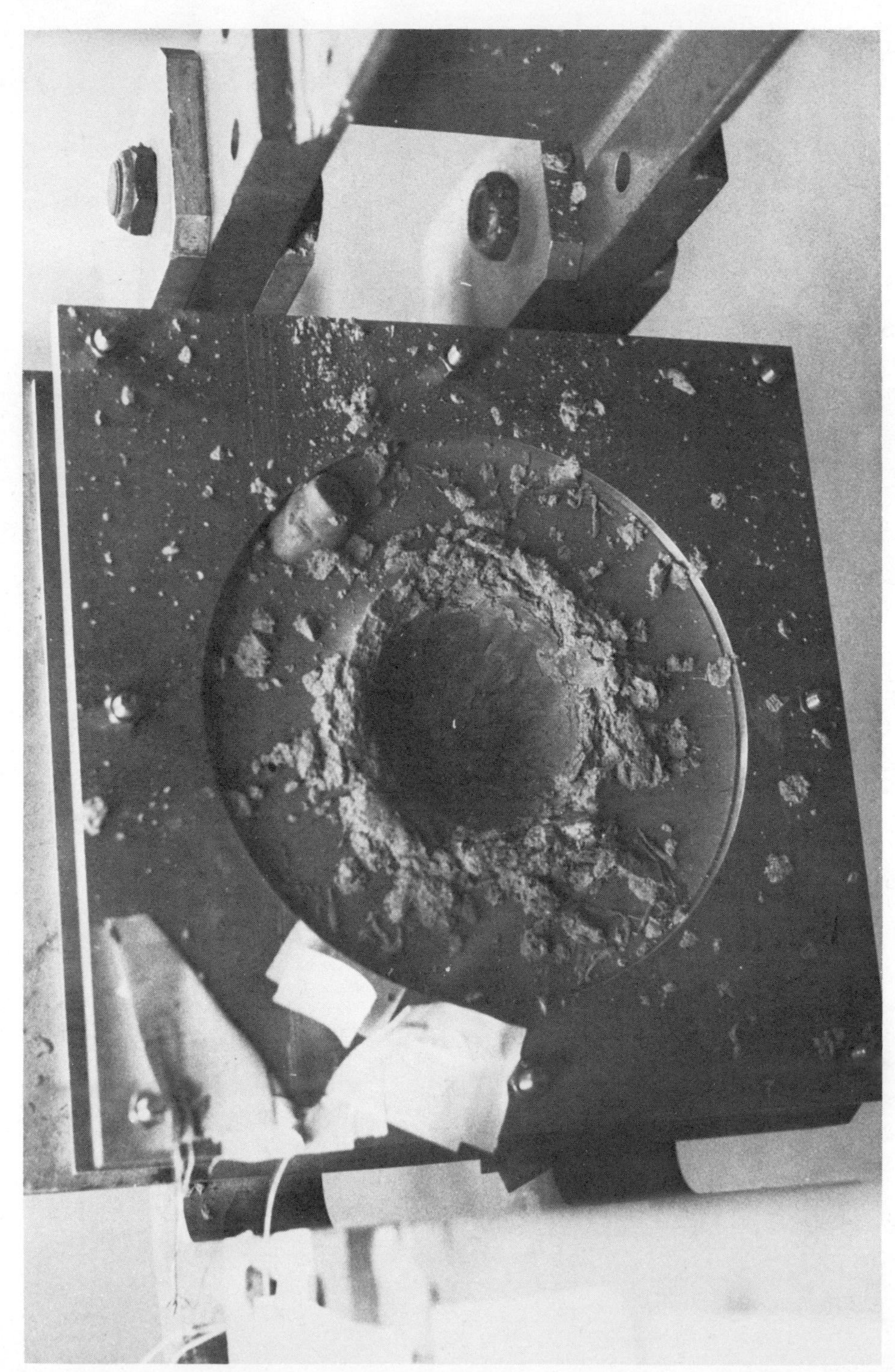

Fig. 4. 480 G post-shot configuration.

Table 1. Explosive charge configurations.

Charge type	Charge average diameter (cm)	Charge length (cm)	Length/ diameter	Calculated energy release 10^{10} erg	Equivalent weight of TNT (g)	Energy ratio (E106)
E106	.59	.43	.73	.96	.207	1.0
2-E106	.89	.43	.48	1.91	.415	2.0
3-E106	1.12	.43	.38	2.87	.622	3.0
E1A(6)	.68	.54	.79	2.10	.456	2.2
E1A(8)	.68	.76	1.12	2.97	.644	3.1

The final test series consisted of twenty shots performed at various G levels. The shots are outlined in Table 2 which describes the test configuration and the resulting crater geometry. Photographs of typical pre-shot and post-shot conditions are shown in Figs. 3 and 4 for shot number 9 fired at 480 G.

3. Dimensional Analysis

A commonly accepted rule for scaling crater dimensions is the so-called cube-root law which stems from dimensional analysis and the assumption that there are three controlling variables; a linear dimension such as crater radius (r), medium density (ρ), and charge mass (W). These three variables are comprised of only two primary dimensions, length and mass, thus giving only one non-dimensional π-group. This can be written as

$$\pi_r = r\left(\frac{\rho}{W}\right)^{1/3} = \text{constant}, \tag{1}$$

giving rise to the terminology cube-root law. This is dimensionally equivalent to a law for scaling crater volume

$$\pi_r = V\left(\frac{\rho}{W}\right) = \text{constant}. \tag{2}$$

Another example, the often quoted fourth root law suggested by Sedov (1959) sometimes referred to as gravity scaling is based upon the assumption that crater shape is determined by four variables; a linear dimension (radius or depth), medium density, charge energy (E), and gravity (g). These four variables are comprised of three independent primary dimensions and hence once again there is only one π-group which can be written

$$\pi_{rg} = r\left(\frac{\rho g}{E}\right)^{1/4} = \text{constant}, \tag{3}$$

or in terms of volume

$$\pi_{rg} = V\left(\frac{\rho g}{E}\right)^{3/4} = \text{constant}. \tag{4}$$

Table 2. Crater dimensions resulting from various test configurations.

Shot number	Clay temperature °F	Gravity G	Charge type	Depth of burial cm	Apparent crater volume cm^3	Apparent crater depth cm	Apparent crater radius cm	Aspect ratio (radius/depth)
1	67.0	120	E106	.34	24.1	2.04	2.35	1.15
2	67.0	120	E1A(8)	.60	76.4	3.75	3.45	.92
3	67.0	240	E106	.34	20.7	1.92	2.23	1.16
4	66.5	160	E106	.34	23.9	2.30	2.30	1.00
5	66.8	480	E106	.34	19.7	1.75	2.33	1.33
6	66.5	480	2-E106	1.75	106.6	3.70	3.90	1.05
7	66.7	285	E1A(6)	2.19	118.2	5.60	3.95	.71
8	66.5	320	2-E106	2.45	140.1	5.40	5.40	.73
9	67.4	480	3-E106	3.15	198.6	4.60	5.15	1.12
10	67.3	405	E1A(8)	2.89	182.5	6.00	4.30	.72
11	67.5	480	E106	1.34	48.9	2.89	2.85	.99
12	67.0	480	E106	2.34	70.3	3.85	3.27	.85
13	66.0	1	E106	.34	20.4	2.05	2.24	1.09
14	66.0	1	E106	.84	37.5	2.60	2.59	1.00
15	66.5	1	E106	1.34	63.5	3.55	3.00	.84
16	67.0	1	E106	2.34	127.4	4.80	3.40	.71
17	66.5	1	E106	3.34	166.5	6.08	3.14	.52
18	67.0	1	E106	4.34	205.4	7.18	2.65	.37
19	67.0	1	E1A(6)	.43	60.1	4.00	3.05	.76
20	67.0	1	E1A(8)	.60	71.0	3.95	3.30	.84

As more variables are introduced, dimensional analysis allows multiple π-groups, no longer insuring a unique scaling law as is given by a single π-group. In fact it does not make sense to talk about a scaling law based upon an exponent since with two or more π-groups there is no unique exponential form for a linear crater dimension (unless the functional dependence between the π-groups can be determined). Hence dimensional analysis cannot be expected to give the form of a scaling law but does reduce the total number of variables by the order of the independent primary dimensions, usually three; mass, length, and time.

In the present case, to provide a tractable analysis which can be used to compare the results of the experiment with large scale field data (Chabai, 1965), as well as with other accelerated reference frame data (Viktorov and Stepenov, 1960), only the simplest set of physical quantities which could be sufficient to describe the phenomena and could be evaluated from available data were considered. For a homogeneous non-layered geology the following set of independent variables was used:

ρ—medium density (ML^{-3})
d—depth of burial (L)
W—charge mass (M)
δ—explosive Chapman–Jouguet density (ML^{-3})
U—explosive Chapman–Jouguet particle velocity (LT^{-1})
g—acceleration of gravity (LT^{-2})

The medium is characterized only by the material density, with the depth of burial defining the geometric configuration. Charge energy was chosen to be represented by an equivalent mass of TNT. This choice is arbitrary and, as will be shown later, does not necessarily lead to the commonly used *ad hoc* cube-root law based upon mass scaling.

For buried charges, the assumption equating energy release to equivalent mass of TNT for various chemical as well as nuclear explosives is quite reasonable. Charge properties which can be used to compare various high explosives are the Chapman–Jouguet density and the Chapman–Jouguet particle velocity. In the present study in which all the charges are compared on the basis of an equivalent mass of TNT, the values chosen were those for TNT ($\delta = 2.23\ \mathrm{g/cm^3}$, $U = 1.86 \times 10^5\ \mathrm{cm/sec}$). As such they were included as dimensional constants in the analysis to maintain dimensional dependence upon given charge properties allowing a more detailed comparison of explosives on a basis other than an equivalent mass of TNT at some later time. The remaining independent variable is the acceleration of gravity and the key to this experiment.

The only dependent variable to be considered is the volume of the apparent crater*:

$$V\text{—apparent crater volume } (L^3).$$

*Crater linear dimensions (radius and depth) along with shape factor (aspect ratio, etc.) are being considered as a part of a continuing study. The work in progress includes effects of charge similarity as well as charge characteristics.

This set of seven variables, six independent, and one dependent, can be combined using the π-theorem resulting in a reduction of order for the case under consideration equal to the number of primary quantities represented (M, L, T). A consistent set of π-groups which is sufficient to describe the phenomenon is the following:

$$\pi_1 = \frac{V\rho}{W}, \tag{5.1}$$

$$\pi_2 = \frac{g}{U^2}\left(\frac{W}{\delta}\right)^{1/3}, \tag{5.2}$$

$$\pi_3 = d\left(\frac{\rho}{W}\right)^{1/3}, \tag{5.3}$$

$$\pi_4 = \frac{\rho}{\delta}. \tag{5.4}$$

If the original choice of variables contained all the relevant physical quantities, the solution for the crater volume has the folowing form:

$$\mathscr{F}_1\{\pi_1, \pi_2, \pi_3, \pi_4\} = 0. \tag{6}$$

As shown in Table 3, the medium density, ρ, is nearly constant among the three materials under consideration and δ, the Chapman–Jouguet density for TNT has the value 2.23 g/cm^3. For this collection of data π_4 can be assumed to be nearly constant since $.69 < \pi_4 < .74$ simplifying the solution to the following:

$$\mathscr{F}_2\{\pi_1, \pi_2, \pi_3\} = 0. \tag{7}$$

Table 3. Material properties comparisons.

Material	Oil base clay	Wet sand	Desert alluvium
Reference	Present centrifuge work	Viktorov and Stepenov (1960)	Chabai (1965)
Density (g/cm^3)	1.53	1.65	1.60
Cohesion	Yes	Slight	Varies with location
Angle of internal friction	None	Yes	Yes
Viscous	Yes	No	No
Remarks	Convenient test material	Drained condition	NTS geology

4. RESULTS

The surface described by Eq. (7) is plotted in Fig. 5. Data for crater volume parameter π_1 as a function of scaled charge energy π_2 and depth of burial π_3 is shown. This surface was constructed by first generating a least squares fit to volume parameter π_1 as a function of the depth of burial parameter π_3 for fixed values of gravity-scaled yield, $\pi_2 = \text{const}$. Lines connecting values of constant π_3 were then drawn as shown in Fig. 5. A cross section which represents a typical depth of burial curve for constant gravity-scaled yield is shown in Fig. 6. Here centrifuge data shows very good agreement with NTS field data (Chabai, 1965) and small scale laboratory data from accelerated reference frames (Viktorov and Stepenov, 1960).

In Fig. 6, the accelerated frame data points as well as the centrifuge data points are based upon a linear interpolation of values for π_1 vs π_3 at constant $\pi_2 = 1.15 \times 10^{-6}$. To achieve values corresponding to various constant values of π_2, all the data points were plotted in the π_1–π_2 plane. Then the data points having nearly equal values for π_3 were connected with straight lines. Thereby π_1 values for arbitrary values of π_2 could be obtained by interpolation between data

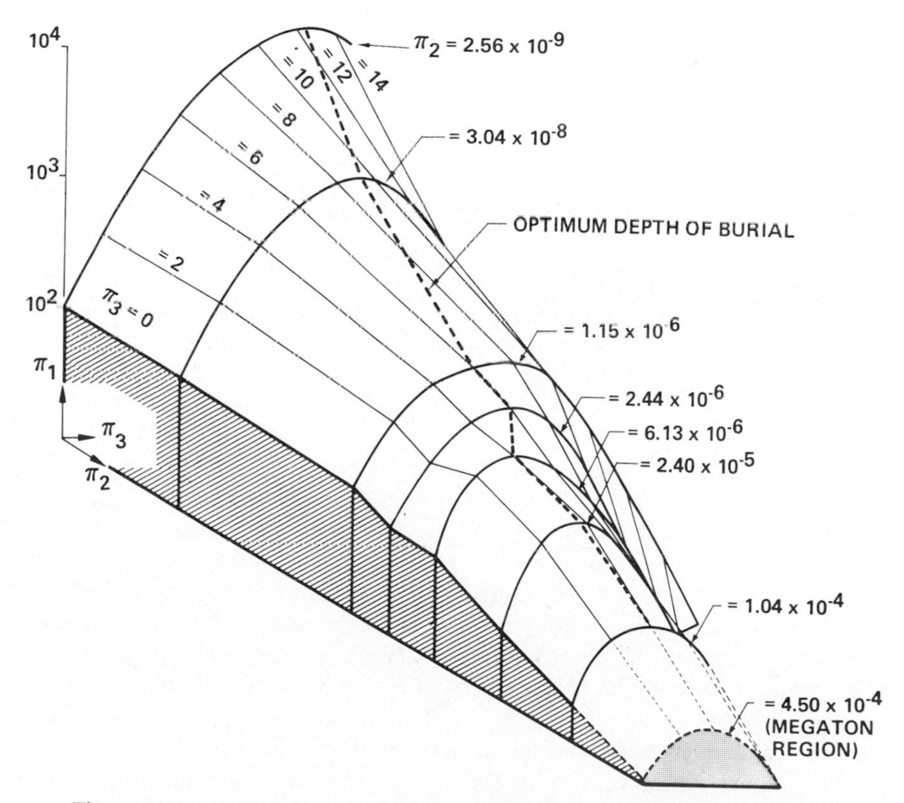

Fig. 5. $\mathscr{F}_2\{\pi_1, \pi_2, \pi_3\} = 0$, crater volume parameter (π_1) as a function of gravity-scaled yield (π_2) and depth of burial parameter (π_3).

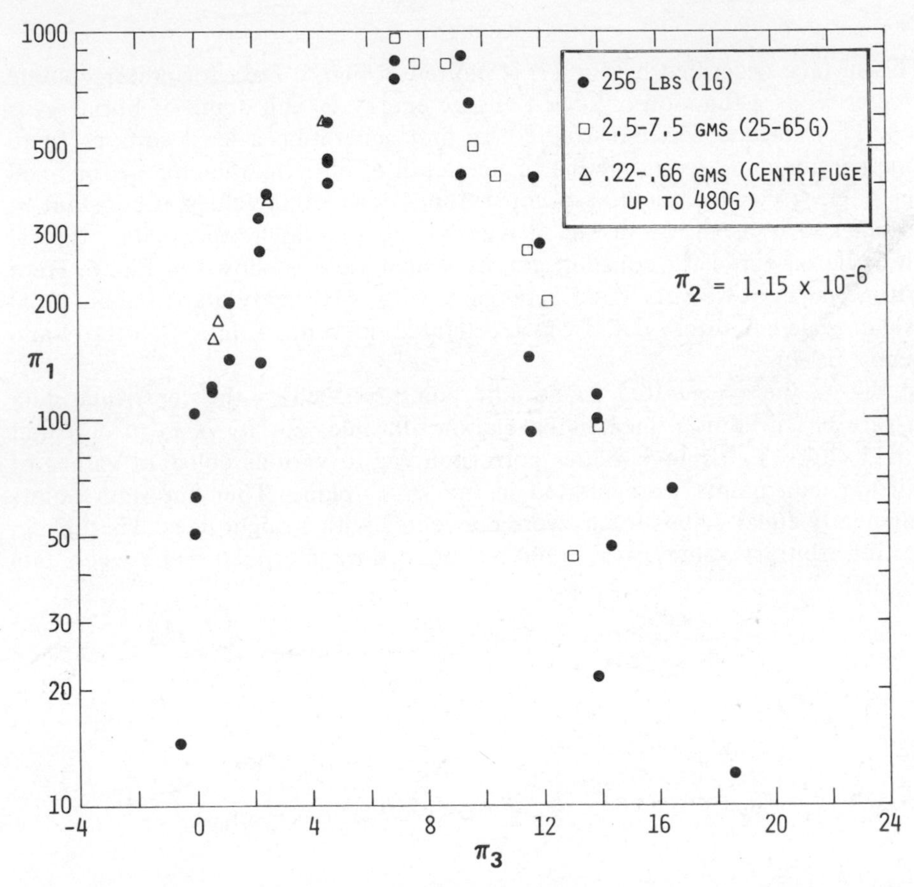

Fig. 6. Gravity parameter scaling for depth of burial effect on crater volume parameter over 10^6 range on yield.

points. Hence these points cannot be calculated directly from the reference or from Table 3. This figure is included to demonstrate how volume data over a range of 10^6 on yield energy can be correlated using the gravity scaling parameter, $\pi_2 = (g/U^2)\,(W/\delta)^{1/3}$.

The physical effect of increasing gravity is to reduce the crater volume for a given charge yield. This is illustrated in Fig. 7 which shows a comparison of a 480 G shot (number 12) with a 1 G control shot (number 16). Qualitatively this suggests that increasing gravity does indeed have the same effect as increasing charge size with regard to the cratering efficiency, represented by the π_1 parameter, a non-dimensional crater volume based upon medium density and charge mass.

A graphical technique to extend the dimensional analysis providing functional forms consisting of products of powers of the original π-groups is now described. Referring back to Fig. 5, there is a value of π_3 for each value of π_2 that maximizes the volume parameter π_1. The dotted line shown in Fig. 5 labeled

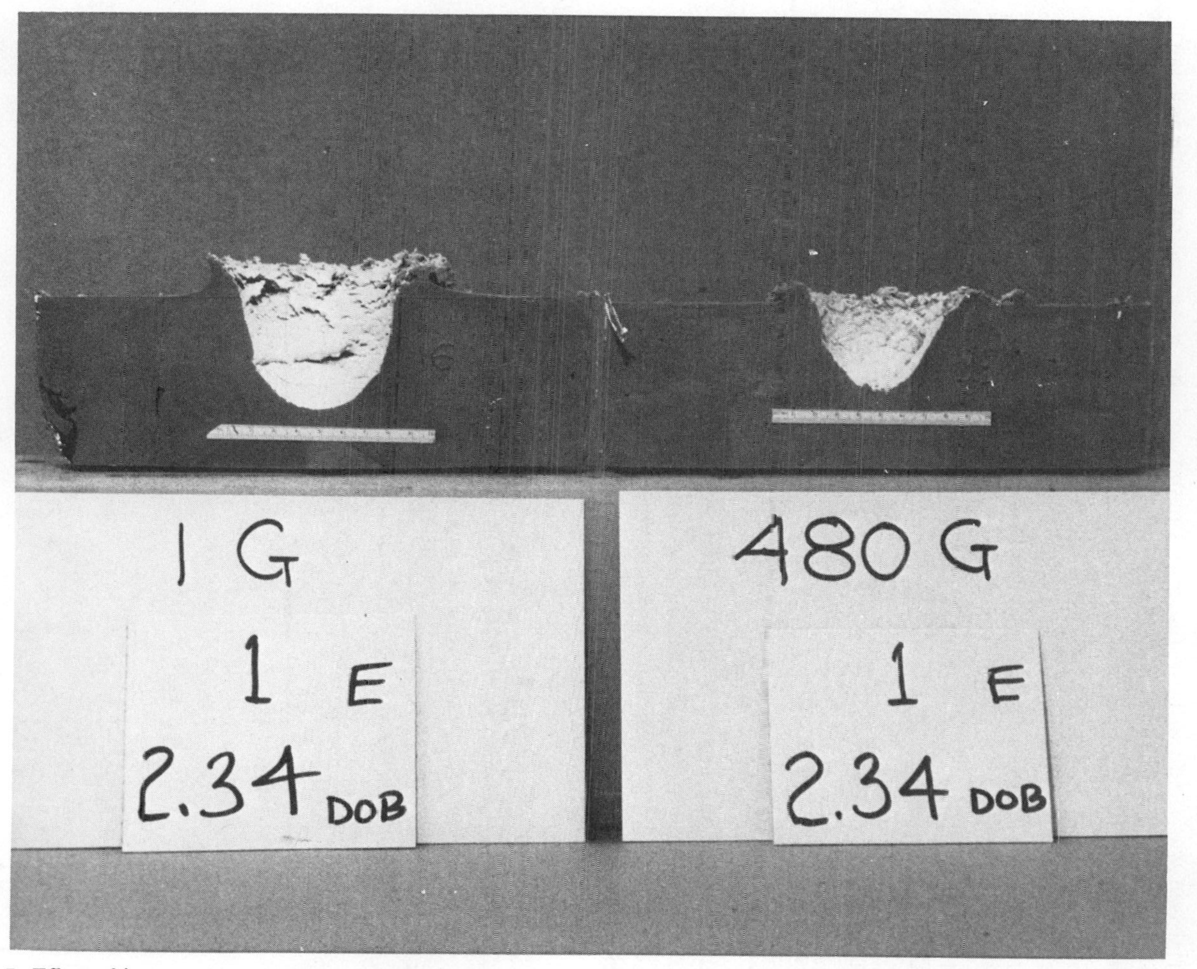

Fig. 7. Effect of increased gravitational acceleration on two otherwise identical shots. E106 charge buried at 2.34 cm at 1 G and at 480 G.

"optimum depth of burial" is the curve through these maximum values. In this case it is a measure of how the optimum burial depth varies with π_2, the gravity-scaled yield. A linearized fit to the projections of the optimum depth of burial line was made in both the π_1–π_2 plane and the π_3–π_2 plane using log–log coordinates. This was used to define a coordinate rotation which can be used to remove the explicit dependence of both π_1 and π_3 upon the variable π_2.

A schematic of this procedure is shown in Fig. 8. The two plots on the left are projections of the optimum depth of burial line onto the π_1–π_2 plane and onto the π_3–π_2 plane respectively. Multiplying the ordinate π_1, in the upper plot by the function π_2^α and similarly in the lower plot multiplying the ordinate π_3 by the function π_2^β results in a transformation of the optimum depth of burial line which is independent of the abscissa, π_2. This provides a self-consistent set of only two π-groups, $\pi_1\pi_2^\alpha$ and $\pi_2\pi_2^\beta$ derived from Eq. 7. The scaled crater volume ($\pi_1\pi_2^\alpha$) can be related directly to the scaled depth of burial ($\pi_3\pi_2^\beta$) as follows:

$$\pi_1\pi_2^\alpha = \mathscr{F}_3\{\pi_3\pi_2^\beta\}. \tag{8}$$

Values of the exponents α and β obtained from these linearized fits were used as a starting point for a regression analysis using the method of least squares to fit a third order polynomial approximation for the function $\mathscr{F}_3$. A systematic variation of these exponents produced optimum values of $\alpha = \frac{1}{4}$ and $\beta = \frac{1}{6}$ which

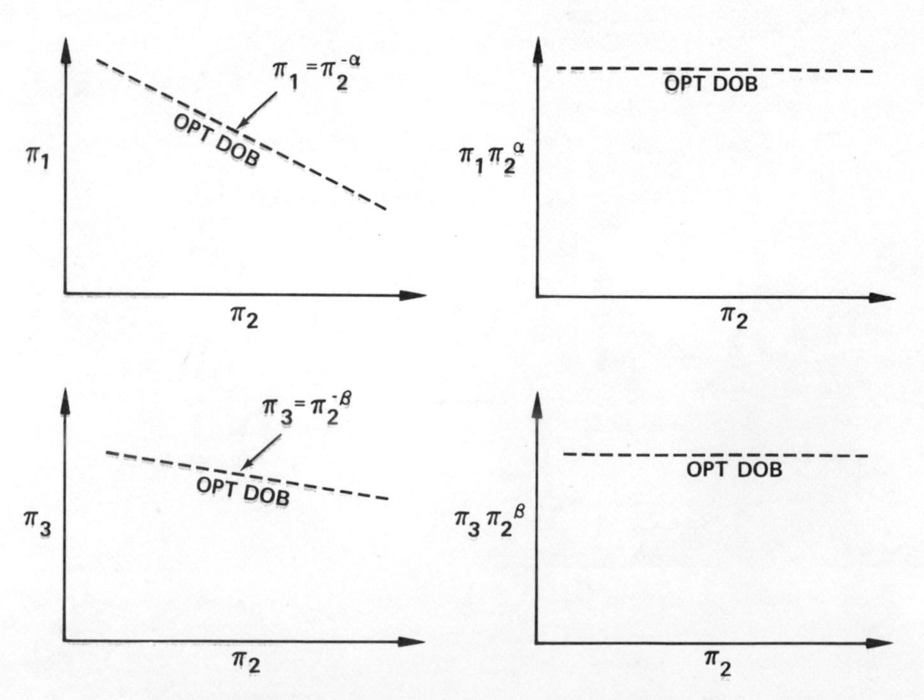

Fig. 8. Reduction in number of π-groups by coordinate transformation.

minimized the residuals providing the best fit polynomial given below.

$$\ell n\ (\pi_1\pi_2^{1/4}) = A_0 + A_1(\pi_3\pi_2^{1/6}) + A_2(\pi_3\pi_2^{1/6})^2 + A_3(\pi_3\pi_2^{1/6})^3, \tag{9}$$

where

$$A_0 = 0.3972 \qquad A_2 = -10.051$$
$$A_1 = 10.028 \qquad A_3 = 2.2524$$

A comparison of the properties of the materials used in the three data sets, i.e., NTS alluvium, damp sand, and artist's modeling clay, is shown in Table 3. No attempt has been made to quantify any property other than the density which is nearly constant among the cases considered. At 1 G comparable small charges in these three media lead to large variations in crater volume. This became evident when the 1 G shots in Table 2 were compared to 1 G control shots in damp sand performed by Viktorov and Stepenov (1960) for which the data is represented by the cross section $\pi_2 = 2.56 \times 10^{-9}$ in Fig. 5. The scatter ln π_1 along this cross section although not shown on the curve, is quite large. Attempts to include dry sand data from Johnson *et al.* (1969) for various levels of gravity from 0.2 to 2.5 G increased the scatter at this low end of the π_2 axis (small gravity, small charge). This suggests that for small charges at 1 G, material strength effects are not negligible compared to body forces due to gravity and inertia. For this reason, all the 1 G small charge data (<12 g of explosive) were omitted from the regression analysis used to determine the optimum values for the exponents α and β of the transformed variables $\pi_1\pi_2^{\alpha}$ and $\pi_2\pi_2^{\beta}$ in Eq. (5). (Based on values of the π_2 variable all data with $\pi_2 < 5 \times 10^{-8}$ were excluded.)

Noting the previously mentioned assumption that $\pi_4 = \rho/\delta \approx$ constant for these three sets of data allows δ to be replaced by ρ reducing Eq. 8 to

$$V\left(\frac{\rho}{W}\right)^{11/12}\left(\frac{g}{U^2}\right)^{1/4} = \mathscr{F}_3\left\{d\left(\frac{\rho}{W}\right)^{5/18}\left(\frac{g}{U^2}\right)^{1/6}\right\}, \tag{10}$$

where $\mathscr{F}_3$ is the best fit fifth order polynomial given by Eq. (9). The plot of $\mathscr{F}_3$ shown in Fig. 9 is a tentative form for a scaling law relating crater volume to charge energy and depth of burial, derived from accelerated frame small scale data and large scale field data. Similarity based upon the selected set of variables is achieved by any combination of ρ, d, W, δ, U, and g which gives constant values for the single independent variable $(\pi_3\pi_2^{1/6})$, fixing a point on the abscissa.

5. Discussion

It is reassuring to note that the crater volume dependence on yield is in good agreement with the "$\frac{3}{10}$ rule" based upon crater radius, i.e., $V(\rho/W)^{3n} = r(\rho/W)^n$ where $n = \frac{3}{10}$, obtained from just the large body of NTS field data for alluvium fitted separately by Chabai (1959), Vaile (1961), Saxe and Delmanzo (1970), and Baker *et al.* (1973). The derived depth of burial dependence upon yield gives an

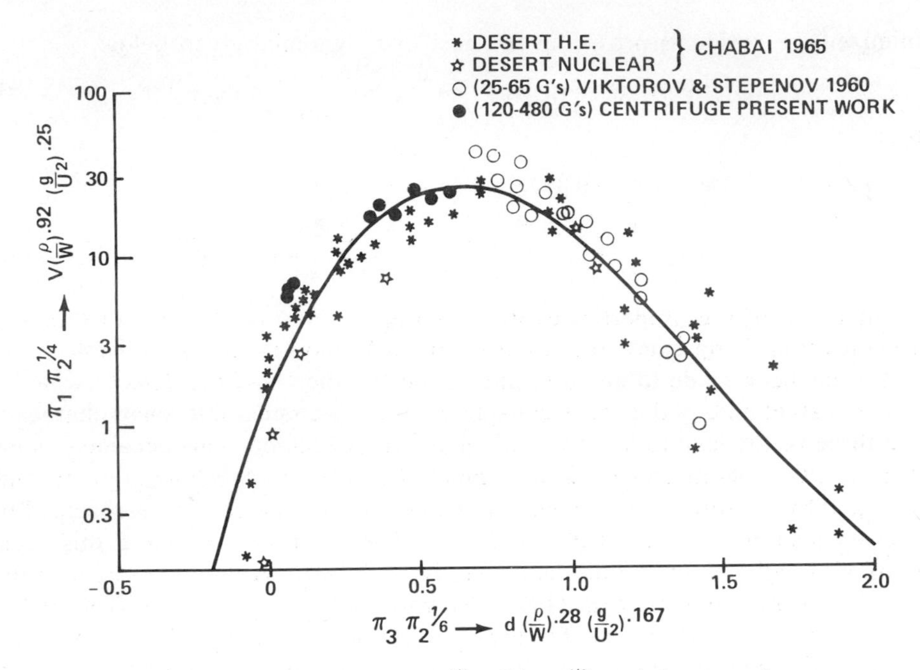

Fig. 9. Gravity-scaled yield results: $\pi_1\pi_2^{1/4} = \mathscr{F}_3\{\pi_3\pi_2^{1/6}\}$, scaled crater volume versus scaled depth of burial for 86 shots with charge yield energy range of 10^{12}.

exponent $n = 5/18 = .28$ which is also consistent with the $\frac{3}{10}$ rule for linear dimensions assuming that all linear dimensions should scale the same. It is significant that the high G small charge data points from this experiment as well as the moderate G accelerated-frame data for somewhat larger charges of Viktorov and Stepenov (1960) did not alter this dependence. This provides increased confidence in the resulting gravity dependence.

A comparison of this result with the cube-root law referred to above is merited. The inadequacy of the cube-root law ($\pi_1 = \text{constant}$) is represented by the decrease in π_1 values for increasing π_2 (i.e., projection of the surface $\mathscr{F}_2$ upon the π_1–π_3 plane results in two orders of magnitude data scatter over the same range of yields (10^{12}) as shown in Fig. 5). This demonstrates that the volume dependence upon yield is not solely determined by the π_1 parameter ($V\rho/W$) which is suggested by the cube-root law [Eq. (2)]. In fact the multiplicative function ($\pi_2^{1/4}$) obtained in this analysis can be considered a correction to the cube-root rule providing not only the correct energy dependence but also a gravity dependence. From this correlation the gravity exponent for scaling volume is $\frac{1}{4}$ (i.e. $Vg^{1/4} = \text{const.}$). Assuming $r^3 \propto V$, the gravity dependence of the radius can be written $rg^{1/12} = \text{const.}$ This value is somewhat smaller than the exponent obtained for the depth of burial dimension in the abscissa of Fig. 9, namely $dg^{1/6} = \text{const.}$ However there is no *a priori* reason that all crater linear dimensions should have the same dependence upon gravity. In fact, since the gravity field strength is a vector quantity that is parallel to the direction of the depth dimension, one could speculate that the orientation of the linear dimension term in question might be significant. This point was not addressed however and the dimensional analysis from which the π-groups (π_1, π_2, π_3, and π_4) where derived did not differentiate one linear dimension from another*. Taking another point of view that the variation is due to experimental error or uncertainties in the regression analysis the two values can be considered limits with a mean value of $n = \frac{1}{8}$.

When the number of variables is sufficiently large that more than one π-group results, dimensional analysis has a fundamental ambiguity in that an unlimited number of sets of self-consistent π-groups can be derived. None of which have any *a priori* significance unless accompanied by additional theory or experiment. Similarity can be assured with any of the sets if all π-groups are held constant, however this is many times too restrictive. If the function, $F_2(\pi_1, \pi_2, \pi_3) = 0$, relating a set of the π-groups can be determined as shown in Fig. 5 a scaling law results which does not require similarity among all π-groups, *per se*, in that each data point lies on the constructed surface which can be considered the scaling relationship.

6. Conclusions

(1) The use of a centrifuge to improve similarity between small scale laboratory tests and large scale field data has been demonstrated.

(2) A gravity-scaled yield parameter of the form $(g/U^2)(W/\delta)^{1/3}$ showed good agreement over a range of 12 orders of magnitude in charge energy.

(3) A scaling law for crater volume (given by Eq. (10) and plotted in Fig. 9)

$$V\left(\frac{\rho}{W}\right)^{11/12}\left(\frac{g}{U^2}\right)^{1/4} = \mathscr{F}_3\left\{d\left(\frac{\rho}{W}\right)^{5/18}\left(\frac{g}{U^2}\right)^{1/6}\right\}$$

was derived based upon an analysis using an equivalent weight of TNT to represent the charge energy for buried shots consisting of both chemical and nuclear explosives.

(4) Crater volume dependence upon material properties other than density diminishes with increased yield and or increased gravity.

(5) Small scale laboratory experiments in an accelerated reference frame provide a basis for improvement in very large scale prediction techniques.

(6) Based upon this investigation the dependence of a crater linear dimension (such as radius) upon gravity has the form rg^n = constant, where $\frac{1}{12} < n < \frac{1}{6}$.

---

*A procedure to consider vector lengths as fundamental units is discussed by Huntley (1952). A greater number of fundamental units can be used to distinguish between quantities which in general implies by the π-theorem a smaller number of dimensionless groups providing more useful results.

Acknowledgments—The author would like to thank Prof. K. A. Holsapple and Dr. B. M. Lempriere for many useful discussions which aided in the data interpretation. Others contributing to this effort include Mr. A. B. Zimmerschied and Mr. F. W. Davies who helped with ordnance applications and Mr. C. R. Wauchope who assisted in the data reduction.

REFERENCES

Baker, W. E., Westine, P. S., and Dodge, F. T.: 1973, *Similarity Methods in Engineering Dynamics*, Hayden Book Co., Rochelle Park, N.J.

Chabai, A. J.: 1959, *Crater Scaling Laws for Desert Alluvium*, SC-4391(RR), Sandia Laboratories, Albuquerque.

Chabai, A. J.: 1965, On Scaling Dimensions of Craters Produced by Buried Explosives, *J. Geophys. Res.* **70**, 5075–5098.

Dupont: 1976, *Detonator and Squib Guide*, E-03510, Dupont Co., Wilmington, DE.

Huntley, H. E.: 1952, *Dimensional Analysis*, MacDonald and Co., London (Dover edition 1967).

Johansson, C. H. and Persson, P. A.: 1970, *Detonics of High Explosives*, Academic Press, London and New York.

Johnson, S. W., Smith, J. A., Franklin, E. G., Moraski, L. K., and Teal, D. J.: 1969, Gravity and Atmospheric Pressure Effects on Crater Formation in Sand, *J. Geophys. Res.* **74**, 4838–4850.

Saxe, H. C. and Delmanzo, D. D.: 1970, A Study of Underground Explosion Cratering Phenomena in Desert alluvium. *Proceedings of the Symposium on Engineering with Nuclear Explosives* **2**, CONF-700101, U.S. Dept. of Commerce, Springfield, VA 22151.

Sedov, L. I.: 1959, *Similarity and Dimensional Methods in Mechanics*, Academic Press, New York and London.

Vaile, R. B.: 1961, Pacific Craters and Scaling Laws, *J. Geophys. Res.* **66**, 3413–3433.

Viktorov, V. V. and Stepenov, R. D.: Modeling of the Action of an Explosion with Concentrated Charges in Homogeneous Ground, *Inzh. Sb.* **28**, 87–96. (Translation Sandia Report SCL-T-392, 1961, Albuquerque, NM 87115).

Roddy, D. J., Pepin, R. O., and Merrill, R. B., editors.
(1977) *Impact and Explosion Cratering*, Pergamon Press (New York), p. 1279–1296.
Printed in the United States of America

Energies of formation for ejecta blankets of giant impacts

STEVEN K. CROFT

Department of Earth and Space Sciences, University of California,
Los Angeles, California 90024

Abstract—Ejection energy-diameter scaling relations for impact craters have been investigated by a simple ballistic ejection model. Scaling relations of the form $E = AD^B$, derived from least-squares fits to calculated ejection energies over a range of crater diameters, vary according to ejection angle, ejecta blanket distribution, planetary radius, and the depth-diameter ratio. Possible values of the above parameters are drawn from laboratory and field geologic crater data. Extremely broad changes in ejection angle and ejecta blanket distribution result in relatively small changes in the total energy (~2–3×). Reducing the planetary radius of curvature from infinity to lunar dimensions reduces the exponent, B, by ~0.2, for craters 100–1000 km in diameter. Simple dimensional analysis shows that this decrease from the ideal gravitational scaling relation ($B = 4$) follows from the direct proportionality between the energy necessary to travel a given distance over a planetary surface and the radius of curvature of that surface. A direct dependence of B on the exponent of the power-law depth-diameter relation is also demonstrated. Two ejection energy-diameter scaling relations explicitly including target density, surface gravity and planetary radius are presented: one for simple craters with depth-diameter ratios of ~0.2, and one for large complex craters. The wide range of proposed depth-diameter ratios for unmodified giant impact craters contributes the largest uncertainty to the energy scaling laws, amounting to a factor of ~50× for Imbrium-sized events.

INTRODUCTION

IMPACT CRATERING, an extremely complex phenomenon worthy of study in its own right, acquires great significance when studied in the context of planetary surfaces and planetary formation. The importance of cratering as a geologic process on terrestrial planets is amply demonstrated by the ubiquity of craters ranging in size from micro-pits to continent-sized basins on all ancient planetary surfaces studied to date. Its importance in planetary formation arises from the hypothesis that at least the terrestrial planets accreted from a population of small to medium-sized bodies which formed the craters visible on the moon, etc., and supplied the heat necessary for the early planetary differentiation inferred from photogeology and geochemistry (Engel *et al.*, 1974; Johnston *et al.*, 1974; Murray *et al.*, 1975). Safronov (1972) demonstrated that burying heat at depth, where it cannot escape quickly, through the process of impact cratering allows accretion times to become very long (up to 10^8 yr), in contrast to dust accretion models that require accretion times as short as 1–1000 yr (e.g., Mizutani *et al.*, 1972) to produce the heating required for differentiation. The longer accretion time is more in accord with lunar cratering rates (Hartmann, 1972). It was found, however, that by substituting different crater scaling parameters into a computer version of Safronov's model (see Safronov (1972), Ch. 15), that the significance of impact heating compared to compressional and radioactive heating could vary from negligible to dominant. Safronov also showed that impact heating is primarily due to impacting bodies larger than ~100 m in radius, which in his model corresponds to craters ≃3.6 km in diameter. The need to find an accurate scaling relation for large impacts and thereby place the most reasonable constraints possible on planetary heating by impacts was the motivation for this study.

Gault *et al.* (1975) suggest that the total kinetic energy of impact may be divided into five parts (all units in c.g.s.):

(1) Heat, $$E_h = K_1 f(V_i), \quad (1a)$$

where $f(V_i)$ is a function of impact velocity and K_1 is a constant;

(2) Comminution, $$E_c = K_2 D^{3-\alpha\delta}, \quad (1b)$$

where D = crater diameter, K_2 is a constant approximately equal to 10^9, and α, δ are parameters related to the size distribution of the crushed ejecta;

(3) Deformation, $$E_d = K_3 D^3, \quad (1c)$$

where K_3 is a constant $\cong 10^7$;

(4) Ejection, $$E_e = g K_4 D^4, \quad (1d)$$

where K_4 is a constant between 10^{-1} and 10^4, g is surface gravity; and

(5) Seismic, $$E_s = K_5 E_T, \quad (1e)$$

where E_T = total impact energy, and $K_5 \cong 10^{-4}$–10^{-5}. If, following Gault *et al.* (1975), it is assumed that for hypervelocity impacts, the heating remains a roughly constant 30–35% of E_T and seismic energy is negligibly small, then the effective power of the energy-diameter scaling law increases from $3-\alpha\delta(\sim 2.7)$ to 4 as the crater diameter increases from centimeters to kilometers. For craters larger than a few kilometers across, E_T is effectively partitioned into heat and energy of ejection only. Consequently, if a good estimate of the energy of ejection can be made, a good first-order estimate of the scaling laws for giant impacts can also be made. Unfortunately, because the estimated limits on K_4 range over a factor of 10^5, the ejection energy is the most uncertain of the five parts. To at least estimate the effects of various parameters on the ejection energy, and to possibly reduce the limits of uncertainty on K_4, a simple ejection energy model was designed and used to calculate theoretical scaling laws.

THE EJECTION MODEL

The calculation of E_e will be considered in terms of the energy necessary to eject material from a specified crater along ballistic trajectories onto a specified ejecta blanket. Figure 1 is a schematic cross section of the crater, the ejecta blanket and a sample ballistic trajectory, all of which are assumed to be axisymmetric. The initial crater form is assumed to be an inverted spherical cap of radius R_0, and depth from the original ground surface, h_s. To keep the

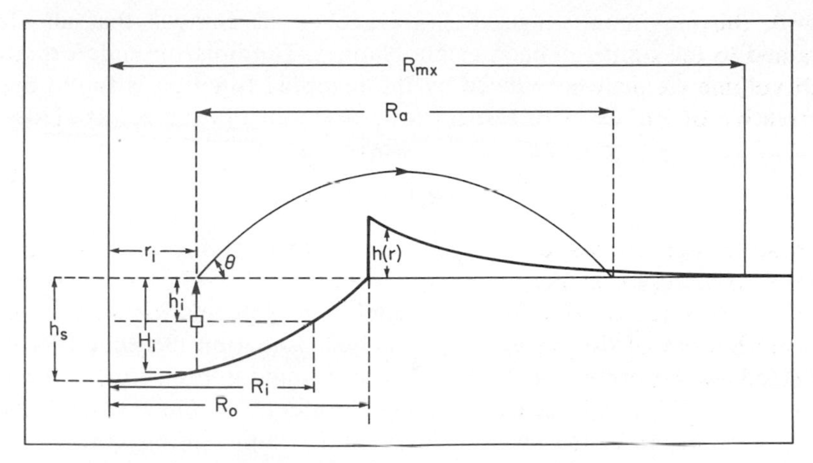

Fig. 1. Schematic diagram of ejection model: The heavy line represents the ejecta and crater profile after all material has been ejected and before post-impact modification has begun. Crater depth (h_s) is defined from the original ground surface. The true radius (R_0) is the distance from the crater center to the intersection of the crater profile with the original surface. Also shown is a volume element of material in its initial position in the crater. This element is lifted to the surface and fired along the schematic ballistic trajectory to a given range, R_a, on the ejecta blanket (which does not extend beyond R_{mx}). The crater and ballistic trajectories are defined to be axisymmetric.

diameter as the independent variable, h_s will be defined in terms of a depth-diameter relation. The crater volume is divided into annuli of radius r_i which are initially located at depth h_i in the crater. Since detailed trajectories of particles in the crater are not known, each annulus is "lifted" vertically to the surface as shown and "fired" at an elevation angle θ defined by:

$$\theta = \theta_L + \theta_D[1-(r_i/R_i)^\beta][1-(h_i/H_i)^\beta], \tag{2}$$

where R_i and H_i are defined as shown in Fig. 1, θ_L is the smallest angle matter is ejected at, and $\theta_L + \theta_D$ is the largest angle ($\sim 90°$). The terms in parentheses allow a variation of ejection angle with initial position in the crater, assigning the highest angle to material around the impact point and decaying according to the parameter β to the lowest angle for material along the final crater sides and bottom.

The range, R_a, of each annulus is given by:

$$R_a = R_{mx} - (r_i/R_i)^\alpha(R_{mx} - R_0) - r_i, \tag{3}$$

where R_{mx} is the maximum extent of the ejecta blanket which for this model was chosen to be $11R_0$. Equation (3) is essentially a mapping function that takes each horizontal layer in the crater, flips it over, and stretches it differentially onto the ejecta blanket. At the crater lip, where $r_i = R_i = R_0$, the range is seen to be zero, i.e., the volume element on the lip remains unmoved. For volume elements nearer the center of the crater, r_i decreases, resulting in an increase in R_a until, for $r_i = 0$, the maximum range, R_{mx}, is reached. It follows that all ejecta is constrained to fall on the defined ejecta blanket. The horizontal deformation (δ) of each volume element introduced by the mapping function is found by taking the derivative of Eq. (3) with respect to r_i, and substituting $R_{mx} = 11R_0$:

$$\delta = -\left(\frac{r_i}{R_i}\right)^{\alpha-1} 10\alpha \frac{R_0}{R_i}. \tag{4}$$

The negative sign causes a 180° rotation of the volume elements around a horizontal axis. Again at the crater lip, where $r_i = R_i = R_0$, the deformation is seen to be ~ 1 since $\alpha \simeq 0.1$ for this model. The deformation increases as the center and bottom of the crater is approached. Equation (3), therefore, mimics the inverted and deformed ejecta flap structure found in both small experimental (Stöffler *et al.*, 1975) and large natural impact craters (Shoemaker, 1960; Roddy *et al.*, 1975). The height of the ejecta blanket at any radius greater than R_0, $h(r)$, is found by integrating the mapping function over the volume of the crater:

$$h(r) = \left(\frac{R_{mx} - r}{R_{mx} - R_0}\right)^{(2-\alpha)/\alpha} \frac{1}{r} \frac{V_c}{\alpha\pi(R_{mx} - R_0)}, \tag{4a}$$

where V_c is the crater volume equal to $(\pi/2)R_0^2 h_s + (\pi/6)h_s^3$. The parameter α, may be roughly related to the power law decay exponent, B, of McGetchin *et al.* (1973).

From the above defined values of the ejection angle and range for each annulus, the ejection velocity may be calculated from the spherical ballistic equation:

$$V_i^2 = R_p g/[\cos^2\theta + \sin 2\theta/2 \tan(R_a/2R_p)], \tag{5}$$

where R_p is the radius of the planet on which the crater is made. The total energy required to form the ejecta blanket (E_b) is found by integrating the sum of the potential and kinetic energies over the total ejected mass (M_e):

$$E_b = \int_{M_e} \left(\frac{V_i^2}{2} + gh_i\right) dm = \pi\rho g \int_0^{h_s} \int_0^{R_i(h_i)} \left(r_i R_p \Big/ \left[\cos^2\theta + \sin 2\theta/2 \tan\left(\frac{R_a}{2R_p}\right)\right] + 2r_i h_i\right) dr_i\, dh_i, \tag{6}$$

where ρ = density and other symbols are as defined previously. It will be noted that evaluation of the ejection energy has been reduced to a problem of crater and planetary geometry only.

Finally, an energy-diameter scaling relation of the form:

$$E_b = \rho g A D_i^B, \tag{7}$$

where A and B are constants, and $D_i(=2R_0)$ is equivalent to Pike's (1967) D_t, is determined by a least-squares fit to the values of E_b calculated from Eqs. (2)–(6) for any range of crater diameters and any given set of values of the input parameters. Column 6 of Table 1 contains the logarithmic equivalents of Eq. (7)

Table 1. Calculated energy-diameter scaling relations.

(1) Model #	(2) h_s-D	(3) θ	(4) β	(5) α	(6) $\text{Log } E_b/\rho g = A + B \log D$	(7) Ratio to #1 at: 100 km	(8) Ratio to #1 at: 1000 km	(9) Notes
1	PIKEF	90–6°	0.35	0.20	$= 19.255 + 3.806 \log D$	1.00	1.00	Standard model
2	PIKEF	90–6°	0.55	0.20	$= 19.069 + 3.854 \log D$	0.82	0.91	Var. in ejection angle, θ
3	PIKEF	90–6°	0.15	0.20	$= 19.520 + 3.733 \log D$	1.31	1.11	
4	PIKEF	= 45°	—	0.20	$= 18.669 + 3.948 \log D$	0.54	0.75	
5	PIKEF	= 6°	—	0.20	$= 19.679 + 3.686 \log D$	1.53	1.16	
6	PIKEF	90–6°	0.35	0.50	$= 19.631 + 3.712 \log D$	1.54	1.24	Ej. blkt. distribution
7	PIKEF	90–6°	0.35	0.05	$= 18.895 + 3.886 \log D$	0.63	0.76	
8	PIKEM	90–6°	0.35	0.20	$= 19.964 + 2.992 \log D$	0.12	0.02	h_s-D
9	BST 1	90–6°	0.35	0.20	$= 19.789 + 3.249 \log D$	0.26	0.07	'Best' model

$R_p = 1738$ km in each case.

for nine scaling relations fit to the calculated energies of 11 diameters between 50 and 1000 km for the numerical values of the parameters (discussed in the next section) found on the same line. The r.m.s. deviation from the calculated energies over the specified diameter range for each of the nine relations was found to be <5%.

A few comments about assumptions in the model are appropriate. No post-impact modifications are included. The model crater profile shown in Fig. 1 is assumed to be the real crater profile after all ejected material has been removed and before any rim material can slump back into the crater, decreasing the interior volume and increasing the apparent diameter. This insures that all the ejected volume is accounted for. Any kinetic energy that in a real cratering event goes into moving material below the defined crater floor is neglected here. The error introduced is considered to be small because most of the material in the breccia lens never left the crater (Shoemaker and Kieffer, 1974). Thus the contribution to E_b of this material would at most be only on the order of the potential energy calculated above ($\lesssim 0.1E_b$) unless the mass of the breccia lens is ~10× that ejected from the crater, a situation that is not found in terrestrial impact structures. Also the model does not consider flowing or bouncing of ejecta after it lands, nor is deformation of subsurface strata included; the entire blanket is considered to be emplaced ballistically.

Data and Results

The four basic variables explicitly included in the model are: (1) ejection angle, (2) ejecta blanket distribution, (3) radius of curvature of the planet on which the crater is formed, and (4) the depth-diameter ratio. To obtain the most realistic results possible, the values and permissible ranges of the input parameters were defined as much as possible from laboratory cratering experiments and geologic studies of terrestrial and lunar craters. The parameter values chosen for each model $E_b - D_i$ scaling relation are listed in the second through fifth columns of Table 1. Model #1 was arbitrarily selected as the "standard" model so that comparisons of the effects of different parameter variations would all be referred to the same basis. The last two columns of Table 1 show the ratio of E_b calculated for each of the models to that calculated from model #1 at 100 and 1000 km to illustrate changes due to variation of each parameter.

Ejection angle

Measurements of the ejection angle of material from meter-sized craters in sand by Gault and Heitowit (1963), Stöffler *et al.* (1975) and others, indicate that most of the material leaves the crater at angles between 40–45°. This is in contrast to the low ejection angle of material from large lunar and terrestrial craters inferred from the impact angles of secondary craters around Copernicus (Oberbeck and Morrison, 1974), Copernican ejecta on the crater Eratosthenes (Carr, 1964), and large blocks around the Ries Basin in Germany (Chao, 1974). In both laboratory and natural craters there is evidence that relatively small amounts of high angle ejecta form the fallback layer. Compatible with these observations is an effective depth-of-burst model for Copernicus by Shoemaker (1962), in which ejecta leaves the crater at all angles between 90° and 6°, with only small amounts ejected at high angles and increasingly larger amounts at progressively lower angles. The cumulative mass-angle function calculated by Shoemaker is shown by the dots in Fig. 2. Equation (2) was arbitrarily defined to

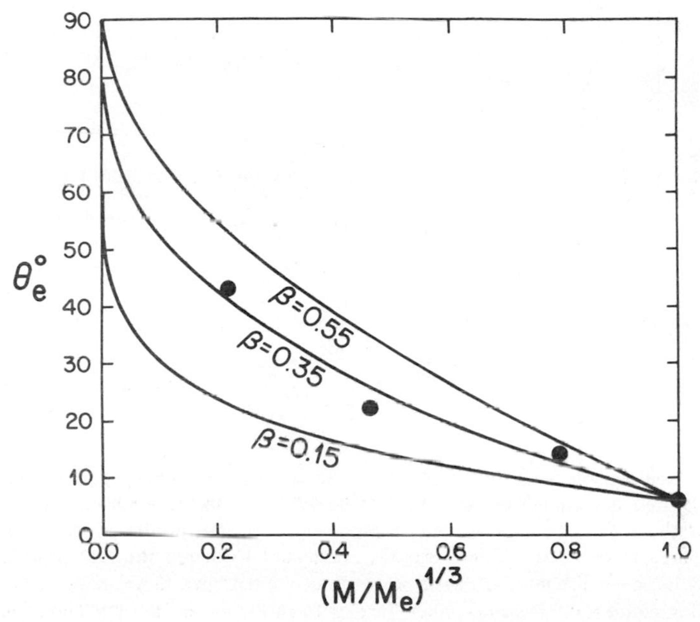

Fig. 2. Model cumulative mass-ejection angle function: Dots are derived from Shoemaker's (1962) ballistic model of Copernicus. The curves are derived from the arbitrary function Eq. (2), for the values of the free parameter β shown in the figure, to simulate the change of ejection angle for material at increasing distances in the crater from the point of impact. The abscissa variable $(M/M_e)^{1/3}$ is equivalent to the radial distance from the point of impact at the surface.

approximate Shoemaker's points, and to allow variation of the effective angle of ejection for the bulk of the material. A choice of $\theta_L = 6°$, $\theta_D = 84°$ and $\beta = 0.35$ (model #1) yields an approximate match of Shoemaker's model, which is shown in Fig. 2 along with curves calculated for $\beta = 0.15$ (model #2) and $\beta = 0.55$ (model #3). The resulting energy variations are quite small, only 10–30% (confer Table 1). Therefore, extreme models of all material ejected at the minimum energy angle of 45° (model #4), and a probably unrealistically high energy ejection angle of 6° (model #5) were calculated, with the result that the total variation in energy was still only a factor of 2–3×.

Ejecta blanket distribution

The ejecta blanket cross sections shown in Fig. 3 are calculated from Eq. (4a) for different values of the parameter α. The crater section shown is the northeastern profile of the lunar crater Timocharis taken from NASA Lunar Topographic Orthophoto maps. The profile begins at the interior radius, R_0,

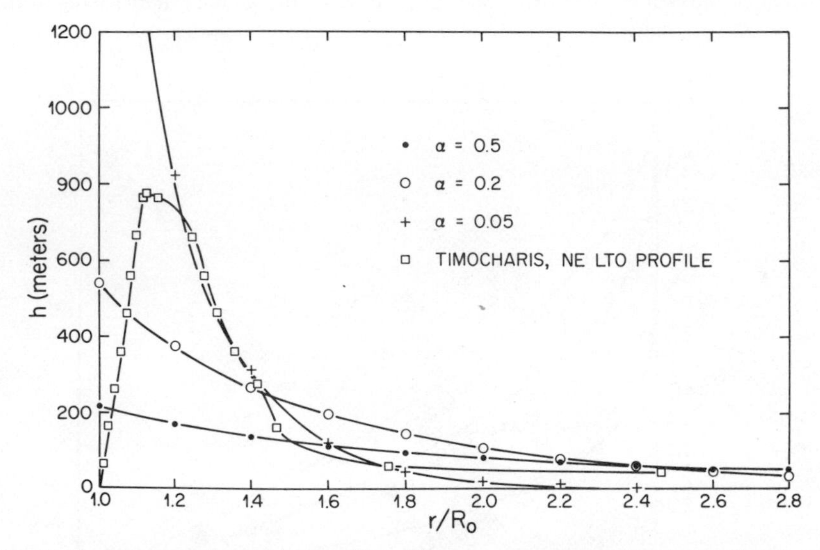

Fig. 3. Model ejecta distribution profiles: The curves corresponding to the first three symbols are calculated from Eq. (4a) for the values of the free parameter, α, shown, and for the same total ejecta volume represented by the profile of the lunar crater Timocharis. None of the model profiles are meant to match the real profile, but to bracket its mass distribution in the sense that $\alpha = 0.05$ has more mass closer to the crater lip, and $\alpha = 0.5$ has more mass farther from the crater lip than Timocharis. The profile $\alpha = 0.2$ is an intermediate case.

found by extrapolating inward the distant ($\gtrsim 2R_0$) exterior ground level, which is presumably the pre-impact ground level of the crater interior. The obvious difference in the observed and theoretical profile shape right at the crater lip is greatly lessened when the observed slumps in the interior of the crater are placed in their inferred pre-collapse positions. As is seen in the figure, $\alpha = 0.05$ (model #6) piles the ejecta close to the crater lip (~90% of the the total volume within $1.4R_0$), while $\alpha = 0.5$ (model #7) spreads the ejecta far from the crater (~70% beyond $1.4R_0$). The energy necessary to form the ejecta blanket for model #6 is smaller than that required to form the Timocharis ejecta blanket because the ballistic ranges are smaller, while the larger ranges required for model #7 result in a larger calculated energy of formation. The profile calculated from $\alpha = 0.2$, which closely approximates a McGetchin *et al.* (1973) power law profile, $h = A(R/R_0)^B$, with $B = -3$, is used in model #1 and yields an energy of formation intermediate between models #6 and #7. Interestingly enough, these rather extreme variations in the distribution of ejecta only produce a total energy variation of $\leq 2\times$. As is mentioned in the model description above, the contribution to the crater profile due to structural uplift was not included. This is primarily because: (a) the fraction of the total "ejecta" volume due to uplift is quite uncertain, and (b) the energy required to drive the material out of the crater is probably comparable to the energy required for ballistic emplacement. The only accurate estimate of the ratio of uplifted volume to total ejected volume is that of Roddy *et al.* (1975) for Meteor Crater, Arizona, which is 0.2. Estimates of the ratio of ejecta thickness to rim height for the lunar craters Linné (0.4) and Dawes (0.52) cited by Pike (1976a) are both larger than the ratio for Meteor Crater (0.33), implying that the uplifted/total ejected volume for these lunar craters is less than 0.2. Also since structurally uplifted material is concentrated around the crater rim, it would replace only low energy ejecta, and therefore contribute somewhat less than 20% to the total model energy. Since this is significantly less than the total energy variation $\sim 2\times$ due to changes in ejecta distribution, it may be safely neglected in this order-of-magnitude calculation.

Variations in surface curvature

Figure 4 shows the systematic variation of the constants A and B (confer Eq. (7)) with radius of curvature of the surface on which the given ejecta blanket is deposited. The reduction in total energy is $\leq 5\%$ for a ratio of planetary radius to crater diameter >35 ($D_i \sim 50$ km on the moon), and becomes important only for small planetary curvatures and/or large crater diameters. For example, there is a decrease of $\sim 2\times$ in the total energy required to form an Imbrium-sized ejecta blanket on the lunar surface from that required on a flat surface. The effect is also independent of the magnitude of the surface gravity, which is a constant that comes out of the integral in Eq. (6), and results from geometry alone. The "weakening" of the exponent of the flat surface gravitational scaling law ($B = 4$) of a few tenths is believed to result from dilution of the gravitational field which is everywhere constant for an infinite flat plane, but radially decreasing for a curved surface. Thus a particle of a given kinetic energy at ejection will see an effectively smaller gravitational field over its flight path in the spherical case than in the flat case, therefore, travelling farther, or in general requiring less energy to build a given ejecta blanket. Baldwin (1949) considered the effect of surface curvature on crater shape, which is different from the decrease of the energy of formation discussed here, as the crater profile and total ejecta volume for a given diameter crater in this model are defined independently of the curvature of the surface.

The regular variations of A, B seen in Fig. 4 are calculated from the BST 1 depth-diameter relation (see below) and may be fitted by least squares to an equation of the form:

$$A, B = A_\infty, B_\infty + GR_p^{-H}.$$

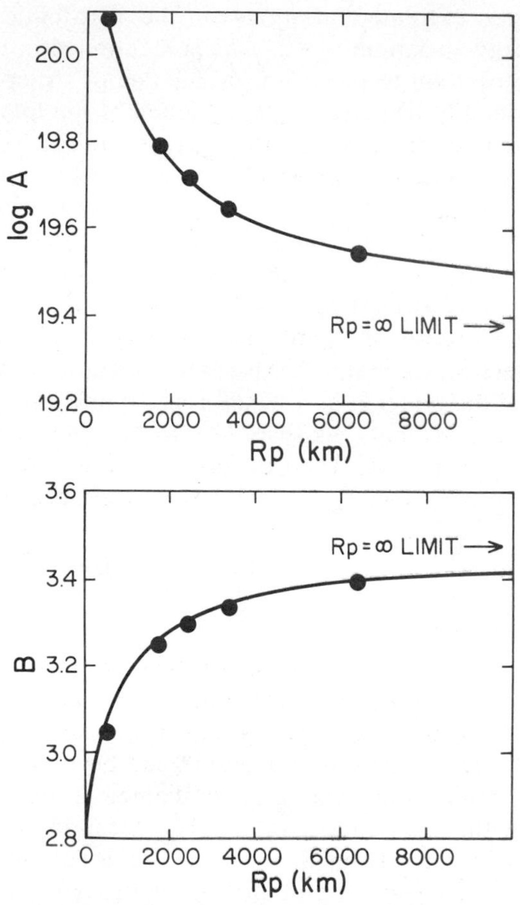

Fig. 4. Effect of planetary radius (R_p) on constants A and B of the function $E_b = \rho g A D_i^B$ (Eq. (7) in text): The dots are the values calculated for Ceres ($R_p = 500$ km), the moon (1738 km), Mercury (2440 km), Mars (3389 km), and the earth (6371 km) from the depth-diameter function designated BST 1 (see Table 2). Also calculated were the values of A and B for the case $R_p = \infty$ (indicated by the arrows). The curves are least-squares fits to the calculated points by a function of the form $A, B = A_\infty, B_\infty + GR_p^{-H}$.

This important result cautions that a scaling law developed for one planetary radius (e.g., the earth) may not be applied directly to impacts of the same size formed on another planetary surface (e.g., the moon) without due consideration for modification due to surface curvature.

Variation in depth-diameter (h_s-D) relationships

The last variable tested in this calculation determines the effect on energy resulting from varying the depth of a crater for a given diameter, which is effectively varying the volume of ejected material. To estimate the ejected volume, a spherical cap geometry was assumed with the depth derived from the diameter via an observed depth-diameter relation. This approach follows that of Short and Forman (1972). Other geometries, such as parabolic (Dence, 1973) or hemi-elliptical (Pike, 1967) of the same diameter and depth produce differences in the ejected volume of only about 10%. The depth-diameter functions used were derived from Pike's (1972, 1974) relations for lunar craters.

The upper limit for crater depths of large craters is designated PIKEF (see Table 2) which is an extrapolation of the h_s-D plot of small (<15 km) fresh lunar craters. For this case, the depth-diameter ratio remains the same, thus representing isometric growth of craters in a homogeneous medium. This extrapolation is frequently used in lunar crater studies (O'Keefe and Ahrens, 1975; DeHon, 1976; Moore *et al.*, 1974). PIKEM is the designation for the function derived (see Table 2) from the presently observed lunar h_s-D function, after all post-impact modifications have taken place, and is considered to be the lower limit.

Table 2. Definitions of depth-diameter relationships.

PIKEF:

$$h_s = 0.196D_r^{1.010} - 0.036D_r^{1.014} = 0.195D_i^{0.999}$$
$$V = 0.0802D_i^{3.000}$$

BST 1:

$$h_s = 0.547D_i^{0.501}$$
$$V = 0.218D_i^{2.498}$$

PIKEM:

$$h_s = 1.044D_r^{0.301} - 0.236D_r^{0.399} = 0.948D_i^{0.235}$$
$$V = 0.370D_i^{2.236}$$

All dimensions in km; $D_r = 1.22D_i^{0.990}$

The true crater depth and excavation volume are the quantities desired. Since Pike's functions are given in terms of depth from the rim crest, the true depth was found by subtracting the appropriate rim height function from the crater-depth function, and then converting from rim diameters to true (or interior) diameters D_i via the function given in Table 2. This function is based on unpublished data, but is similar to that found by others (Baldwin, 1963; Moore, 1976). The volume-diameter relations were then calculated for the spherical cap geometry.

The variation in energy found between PIKEF and PIKEM is seen from Table 1 to be an order of magnitude larger than the variations due to the other variables. Therefore, the dominant factor in determining an energy-diameter function is the choice of the proper depth-diameter relation. A first best guess of the proper h_s-D relation, designated BST 1, was found from the volume-diameter (D-V) relation determined by a least-squares power law fit to three points: the best estimates of the ejecta volumes of (1) Mare Imbrium, $D_i =$

970 km, $V_e = 6.5 \pm 3.5 \times 10^6\ \text{km}^3$, and (2) Mare Orientale, $D_i = 620\ \text{km}$, $V_e = 2 \pm 1 \times 10^6\ \text{km}^3$, by Head *et al.* (1975), and (3) the intersection of the volume relations of PIKEF and PIKEM, which occurs at $D_i = 7.40\ \text{km}$, $V_e = 32.5\ \text{km}^3$. The three volume-diameter relations are shown in Fig. 5 with the basin ejecta volume estimates of Head *et al.* (1975). Ejecta volume estimates for Imbrium, Orientale and other lunar basins made by other investigators occur throughout the area between PIKEF and PIKEM (though no estimates known to the author have exceeded $30 \times 10^6\ \text{km}^3$ for Imbrium, confer Head *et al.*, 1975 and Short and

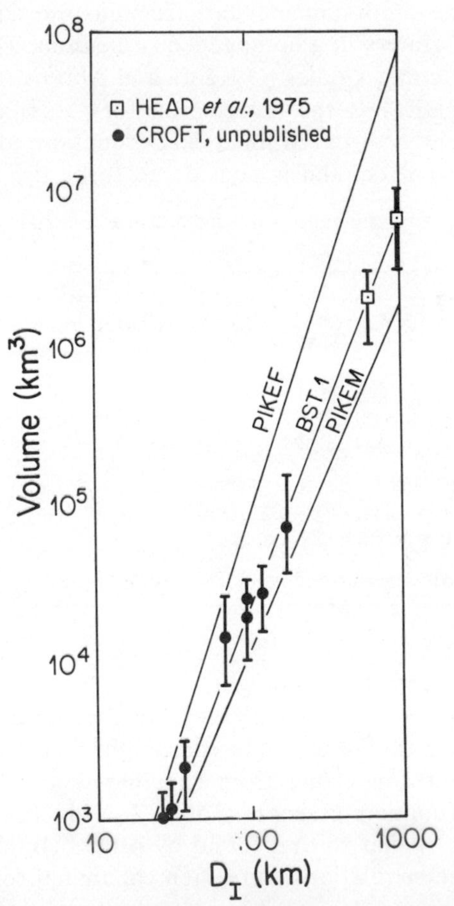

Fig. 5. Diameter-volume relations for lunar craters: The lines designated PIKEF and PIKEM are derived from the depth-diameter relations of the same name assuming spherical cap geometry. Expressions for these functions are given in Table 2. BST 1 is a least-squares fit to the estimates of Head *et al.* (1975) of ejecta volume for Imbrium and Orientale (upper and lower box, respectively) and the intersection of PIKEF and PIKEM off the bottom of the figure, at $D_i = 7.4\ \text{km}$. The dots are estimates of ejecta volumes of the following lunar craters (in order of increasing volume): Lambert, Timocharis, Plinius, King, Theophilus, Langemak, Langrenus, and Tsiolkovskii, taken from NASA Lunar Topographic Orthophoto maps.

Forman, 1972), consequently the diameters and volumes of the two lunar basins chosen to determine BST 1 are subject to dispute. In an effort to obtain a better *D-V* function, the ejecta volumes of several large fresh lunar craters were measured from Lunar Topographic Orthophoto maps, which volumes are the other points plotted in Fig. 5. The error bars are arbitrary, representing factors of two above and below the measured values. Error estimates on the actual measurements are somewhat smaller, thus the larger error bars are conservative, even allowing for post-impact modification. These volumes were not used in determining BST 1, but it is clear that the points cluster closely about that relation, and indicate it's reasonableness, at least in the 20–200 km diameter range. The energy-diameter relation for BST 1 on the moon is given as model #9 in Table 1.

Discussion

The foregoing analysis is based on the assumption that the bulk of the kinetic energy is carried away by material that ends up on the definable ejecta blanket. However, experimental data for small craters ($D \approx 4$ cm) in basalt by Gault and Heitowit (1963) indicate that such may not be the case. Measurements of the velocities of ejecta show that of the ~370 projectile masses of target material ejected from the crater, a single projectile mass, that which came out very early in the impact at very high velocities, carried 70–80% of the total ejecta kinetic energy. Breslau (1970) obtained similar results from impacts in sand ($D \cong 30$ cm), with the variation that ~3 out of 4000 projectile masses of ejecta carried off approximately half the total E_e. If this holds true for larger impacts, then E_b is significantly smaller than E_e since the possibility of small amounts of very high velocity ejecta was excluded from the model to insure a well behaved energy function near $h_i = r_i = 0$, i.e., the point of impact. A test of this effect at larger energies and crater diameters using explosives seems of limited utility since the very high energy ejecta leaves the growing crater at the very beginning of the impact, precisely at that time when equivalent high energy explosion processes differ most in detail from impact processes. There is a theoretical indication that the relative amount of the total E_e carried off by the high velocity ejecta decreases with increasing crater size. O'Keefe and Ahrens (1976) investigated the early stages of an impact of a 5 cm radius iron sphere into an anorthosite target using a computer code with more than sufficient resolution to determine the ejection energy of the first few projectile masses of ejected material. Though not tabulated directly, it appears from their figures that ~4 projectile masses (of ~40,000 total projectile masses ejected) have an ejection velocity >1 km/sec, and that this ejecta carries off ~5% of the total projectile energy. Since in their calculation $E_e/E_T = 0.06$, and these first few masses contain a large amount of thermal energy (they are partially melted), it follows that they carry away somewhat less than 5% of E_e. Assuming the importance of the initial ejecta continues to decrease as the crater diameter increases, then $E_b \cong E_e$ for the impacts considered in this paper. It would be desirable, however, to demonstrate this by experiment.

The exponent, B, of Eq. (7) may be understood in terms of a simple dimensional analysis wherein the ejection energy is compared to the potential energy necessary to lift a specified volume of material of given density to a characteristic ballistic apex height through a well defined gravitational field. The volume may be defined as the product of a characteristic depth times the square of a characteristic diameter. If the depth (h) and the vertical distance (d) through which the material is lifted are defined as simple power-law functions of the diameter:

$$h = C_1 D^l, \tag{8a}$$

and

$$d = C_2 D^m, \tag{8b}$$

then E_e may be written

$$E_e = \rho V g d = \rho g D^2 C_1 D^l C_2 D^m = C_3 \rho g D^{2+l+m}, \tag{9}$$

where $C_3 = C_1 C_2$ is a constant. Comparison with Eq. (7) shows that $B = 2 + l + m$. If over a given range of diameters, the depth-diameter ratio and the pattern of ballistic trajectories normalized to the diameter remain the same, then $l = m = 1$, and $B = 4$ as in Eq. (1d). Conversely, if either the depth-diameter ratio or the relative ballistic pattern changes systematically with diameter, then l and/or m will not equal unity, and B will be different from 4. The apex altitude (d) of a spherical ballistic trajectory is (Nelson and Loft, 1962):

$$d = \frac{R_p}{2}\left(\cos\left(\frac{R_a}{2R_p}\right) + \sin\left(\frac{R_a}{2R_p}\right) - 1\right).$$

Expanding the sine and cosine functions in power series and letting the range, R_a, of a given volume element of ejecta be directly proportional to the crater diameter, as is done in Eq. (3), results in:

$$d = C_4 D\left(1 - \frac{1}{4}\frac{D}{R_p} - \frac{1}{24}\left(\frac{D}{R_p}\right)^2 + \frac{1}{192}\left(\frac{D}{R_p}\right)^3 + \cdots\right). \tag{10}$$

For $R_p = \infty$, Eq. (10) reduces to Eq. (8b) with $m = 1$. For any finite R_p, the ratio d/D decreases as D increases for physically realistic values of D/R_p, corresponding to Eq. (8b) with $m < 1$. This is a direct demonstration of the "weakening" of B due to curvature. It is also an explanation for the variations in B for models #1–#7 in Table 1, since in these models, (a) the depth-diameter function used is PIKEF, for which $l = 0.999$, and (b) the changes of the ejection angle and ejecta distribution allows the ejecta ballistic pattern to see different relative changes in the gravitational field as the crater diameter increases. In model calculations with PIKEF and either $R_p = \infty$ or D very small, the exponent B is very nearly 4 as expected on the basis of this simple theory.

Similarly, changing the power l of the effective depth-diameter function (Eq. (8a)) changes B. This is best illustrated in Table 3 which lists the three depth-diameter functions defined in this paper (see Table 2), the value of l for each, and the calculated value of B for each when $R_p = \infty$ ($m = 1$). The

Table 3. Dependence of energy scaling on depth-diameter relation.

h_s-D function	l	B	B-m-2
PIKEF	0.999	3.999	0.999
BST-1	0.501	3.478	0.478
PIKEM	0.236	3.224	0.224

correlation of B with l is obvious, and again is in accord with this simple analysis. Also, as might be expected, the ratios in E_b derived from each of the three h_s-D relations (given in Table 2) for the 100 and 1000 km craters are equal to the ratios of the respective diameter-volume relations at those crater diameters. This underlines the point already made, that the h_s-D relation chosen for a given range of crater sizes dominates the energy scaling. It is here emphasized that PIKEF and PIKEM represent the extremes in proposed initial lunar crater volumes, and that estimates of the transient cavity volume for impacts the size of Imbrium are uncertain by at least a factor of 50×. The uncertainty decreases, of course, as the PIKEF and the PIKEM D-V relations converge with decreasing diameter, reaching a minimum where they intersect at $D_i = 7.4$ km. That is why this point (off the bottom of Fig. 5) was chosen as the third point to determine BST 1.

All fresh lunar craters less than about 10 km in diameter have the same depth-diameter ratio, (0.20) which is known quite accurately (Pike, 1974), consequently the PIKEF D-V relation probably represents the volumes of small lunar craters quite well. The same is true for Mercurian craters less than ~10 km (Gault *et al.*, 1975), and terrestrial craters smaller than ~2 km (Pike, 1976b). Therefore, the E_e-D_i scaling relation for small craters ($h_s/D = 0.2$, $R_p/D > 50$) is:

$$E_e = \rho g\, 8.24 \times 10^{18} D_i^4, \tag{11}$$

where E_e, ρ, and g are in c.g.s. units and D_i is in km. For $\rho = 3$, this corresponds to $K_4 = 0.247$ in c.g.s. units in Eq. (1d). The formal error in Eq. (11) is very small, much less than 1%, but the uncertainties in the input parameters chosen (model #1, Table 1) leads to an estimated uncertainty in E_e of ~5×.

For fresh lunar and Mercurian craters above the "knee" in the log depth–log diameter plot (Gault *et al.*, 1975) BST 1 is chosen as the appropriate D-V relation, yielding the large crater scaling law:

$$\log [E_T/(\tfrac{3}{2}\rho g)] = (19.38 - 21.47\, R_p^{-0.58}) + (3.48 - 21.58\, R_p^{-0.62}) \log D_i, \tag{12}$$

with E_T, ρ, g in c.g.s. units, R_p and D_i in km. The calculated effects of planetary curvature have been included explicitly, and account has been taken of the assumption described in the Introduction that for very large impacts at high velocity ~1/3 E_T goes into heat and the rest into E_e. Again the formal errors are small (≤10%) but the large uncertainty in the physical input parameters (model #9, Table 1) probably make Eq. (12) uncertain to ± an order of magnitude. This is still very large but significantly smaller than the original uncertainty of 10^5 in K_4.

Estimates of E_e and E_T for a large range of crater sizes are compared with those of other investigators in Table 4. The energies derived from Eqs. (11) and (12) were calculated assuming $\rho = 3\,g/cm^3$ in each case, lunar gravity (162 cm/sec^2) and radius of curvature (1738 km) for the first three craters and terrestrial gravity (980 cm/sec^2) for the last two. The energy for Imbrium calculated by O'Keefe and Ahrens (1976) is simply twice the potential energy necessary to lift the ejecta to the local surface, which in this paper was found to be only about 10% the ballistic energy. The apparent agreement in E_T stems from their use of a depth-diameter ratio of 0.2 (like PIKEF) yielding an Imbrium basin 200 km deep and containing ~10× the total volume estimated from BST 1, which was used in Eq. (12). The E_T given by Gault *et al.* (1975) is an unelaborated estimate, the question mark being theirs. The E_T for Copernicus given by Shoemaker (1962), derived from an equivalent depth-of-burst (DOB) model, agrees very well with that derived from this paper. Baldwin's (1963) E_T for Tycho is derived from extrapolations of terrestrial DOB data. Since Tycho is only slightly smaller than Copernicus, the two orders of magnitude difference between Baldwin's model and this paper applies also between Baldwin's and Shoemaker's models. A factor of 6× between the energy estimates may possibly be accounted for in the ratio of terrestrial to lunar gravities, since Baldwin apparently did not take gravity effects, which dominate in this diameter range, into account. Meteor Crater, Arizona, is probably the "most estimated" of all impact craters. Only a sample of the many energy estimates that have been made are included for comparison. See Innes (1961) and Roddy *et al.* (1975) for more complete listings. The value of E_e is shown explicitly for comparison, as Meteor Crater is still in the size range where E_d and E_c (calculated from Eq. (1))

Table 4. Comparison of calculated energies.

Crater	D_i(km)	E(ergs)	Source
Imbrium	970	2.27×10^{32}	E_T (Eq. (12))
		1×10^{34} (?)	Gault *et al.* (1975)
		$1 \rightarrow 9.4 \times 10^{32}$	O'Keefe and Ahrens (1975)
Copernicus	80	6.83×10^{28}	E_T (Eq. (12))
		7.5×10^{28}	Shoemaker (1962)
Tycho	76	5.79×10^{28}	E_T (Eq. (12))
		$1.07–6.79 \times 10^{30}$	Baldwin (1963)
Meteor Crater, Arizona	1	2.80×10^{22}	E_e (Eq. (11))
		9.26×10^{22}	E_e + Eq. (1a, b, c)
		3.4×10^{23}	Baldwin (1963)
		9.44×10^{22}	Innes (1961)
		1.8×10^{23}	Roddy *et al.* (1975), from 3.4 scaling
—	0.0104	4.60×10^{15}	E_T, O'Keefe and Ahrens (1976)
		2.76×10^{14}	E_e, estimated O'Keefe and Ahrens (1976)
		2.86×10^{14}	E_e (Eq. (11))

contribute significantly to the total energy. Lastly, E_e estimated from the theoretical calculation of O'Keefe and Ahrens (1976) is compared to E_e calculated from Eq. (11). The gratifyingly good agreement must be tempered by noting the Eq. (11) includes a specific gravity, the earth's, while the calculation of O'Keefe and Ahrens does not.

In general, energy estimates from this simple ballistic model are in satisfactory agreement with estimates of other workers. The utility of the scaling Eqs. (11) and (12) lies in the explicit inclusion of density, surface curvature, and surface gravity, allowing immediate estimates of scaling laws on different and accreting planets. Also, based on the dimensional analysis given, adjustment of Eq. (12) for different D-V relations is quite simple. The uncertainty in E_T is still quite large for major impacts, but this is mainly a result of uncertainty in true crater dimensions and the fraction of energy that goes into heating. The emphasis of this paper has been to tie energy scaling relations to geologic data as much as possible and has hopefully demonstrated what can already be inferred, as well as where much investigation has yet to be done.

Acknowledgment—This work was supported by NASA grant NGL 05-007-002. Thanks are extended to E. Luera and G. Croft who aided in preparation of the manuscript, and to S. W. Kieffer, W. M. Kaula, and R. J. Pike for helpful comments.

References

Baldwin, R. B.: 1949, *The Face of the Moon*, p. 141f, University of Chicago Press, Chicago, 239 pp.

Baldwin, R. B.: 1963, *The Measure of the Moon*, p. 147, University of Chicago Press, Chicago, 488 pp.

Bjork, R. L.: 1961, Analysis of the formation of Meteor Crater, Arizona: A Preliminary Report, *J. Geophys. Res.* **66**, 3379–3387.

Breslau, D.: 1970, Partitioning of energy in hypervelocity impact against loose sand targets, *J. Geophys. Res.* **75**, 3987–3999.

Carr, M. H.: 1964, Trajectories of objects producing Copernicus ray material on the crater Eratosthenes, *Astrogeologic Studies, Annual Progress Report, pt. A*, July1, 1963–July 1, 1964.

Chao, E. C. T.: 1974, Impact cratering models and their application to lunar studies: A geologist's view, *Proc. Lunar Sci. Conf. 5th*, p. 35–52.

DeHon, R. A.: 1976, Geologic structure of Eastern Mare Basins (abstract). In *Lunar Science VII*, p. 184–186. The Lunar Science Institute, Houston.

Dence, M. R.: 1973, Dimensional analysis of impact structures, *Meteoritics* **8**, 343–344.

Engel, A. E. J., Itson, S. P., Engel, C. G., Stickney, D. M., and Cray, E. J., Jr.: 1974, Crustal evolution and global tectonics: A petrogenic view, *Bull. Geol. Soc. Amer.* **85**, 843–858.

Gault, D. E., Guest, J. E., Murray, J. B., Dzurisin, D., and Malin, M. C.: 1975, Some comparisons of impact craters on Mercury and the Moon, *J. Geophys. Res.* **80**, 2444–2460.

Gault, D. E., and Heirowit, E. D.: 1963, The partition of energy for hypervelocity impact craters formed in rock, *Proc. Sixth Hypervelocity Impact Symposium*, Cleveland, Ohio, Vol. 2, 419–456.

Gault, D. E., Shoemaker, E. M., and Moore, H. J.: 1963, Spray ejected from the lunar surface by meteoroid impact, NASA Technical Note D-1767.

Hartmann, W. M.: 1972, Paleocratering of the moon: Review of post-Apollo Data, *Astrophys. Space Sci.* **17**, 48–64.

Head, J. W., Settle, M., and Stein, R. S.: 1975, Volume of material ejected from major lunar basins and implications for the depth of excavation of lunar samples, *Proc. Lunar Sci. Conf. 6th*, p. 2805–2829.

Innes, M. J. S.: 1961, The use of gravity methods to study the underground structure and impact energy of meteorite craters, *J. Geophys. Res.* **66**, 2225–2239.

Johnston, D. H., McGetchin, T. R., and Toksöz, M. N.: 1974, The thermal state and internal structure of Mars, *J. Geophys. Res.* **79**, 3959–3971.

McGetchin, T. R., Settle, M., and Head, J. W.: 1973, Radial thickness variation in impact crater ejecta: Implications for lunar basin deposits, *Earth Planet. Sci. Lett.* **20**, 226–236.

Mizutani, H., Matsui, T., and Takeuchi, H.: 1972, Accretion process of the moon, *The Moon* **4**, 476–489.

Moore, H. J.: 1976, Missile impact craters (White Sands Missile Range, New Mexico) and applications to lunar research, Geological Survey Professional Paper 812-B.

Moore, H. J., Hodges, C. A., and Scott, D. H.: 1974, Multi-ringed basins—Illustrated by Orientale and associated features, *Proc. Lunar Sci. Conf. 5th*, p. 71–110.

Murray, B. C., Strom, R. G., Trask, N. J., and Gault, D. E.: 1975, Surface history of Mercury: Implications of terrestrial planets, *J. Geophys. Res.* **80**, 2508–2514.

Nelson, W. C., and Loft, E. E.: 1962, *Space Mechanics*, Prentice-Hall, Inc., New Jersey, 245 pp.

Oberbeck, V. R., and Morrison, R. H.: 1974, Laboratory simulation of the Herringbone pattern associated with lunar secondary crater chains, *The Moon* **9**, 415–455.

O'Keefe, J. D. and Ahrens, T. J.: 1975, Impact cratering on the moon, Contribution #2671, Div. of Geol. Plan. Sci., California Institute of Technology.

O'Keefe, J. D. and Ahrens, T. J.: 1976, Impact ejecta on the moon, Contribution #2763, Div. of Geol. Plan. Sci., California Institute of Technology.

Pike, R. J.: 1967, Schroeter's rule and the modification of lunar crater impact morphology, *J. Geophys. Res.* **72**, 2099–2106.

Pike, R. J.: 1972, Geometric similitude of lunar and terrestrial craters, *24th Int. Geol. Cong., pt. 15*, p. 41–47.

Pike, R. J.: 1974, Depth/diameter relations of fresh lunar craters: Revision from spacecraft data, *Geophys. Res. Lett.* **1**, 291–294.

Pike, R. J.: 1976a, Ejecta thickness, rim uplift, energy type and depth of energy release (abstract), In *Papers Presented to the Symposium on Planetary Cratering Mechanics*, p. 105–107, The Lunar Science Institute, Houston.

Pike, R. J.: 1976b, Crater dimensions from Apollo data and supplemental sources, *The Moon* **15**, 463–477.

Roddy, D. J., Boyce, J. M., Colton, G. W., and Dyal, A. L., Jr.: 1975, Meteor Crater, Arizona, rim drilling with thickness, structural uplift, diameter, depth, volume, and mass-balance calculations, *Proc. Lunar Sci. Conf. 6th*, p. 2621–2644.

Safronov, V. S.: 1972, *Evolution of the Protoplanetary Cloud and Formation of the Earth and Planets*, Israel Program for Scientific Translations, Jerusalem.

Shoemaker, E. M.: 1960, Penetration mechanics of high velocity meteorites, illustrated by Meteor Crater, Arizona, *Internat. Geol. Conf. 21st Session, pt. 18*, p. 418–434.

Shoemaker, E. M.: 1962, Interpretation of lunar craters, In *Physics and Astronomy of the Moon*, Z. Kopal (ed.), Academic Press, New York.

Shoemaker, E. M., and Kieffer, S. W.: 1974, *Guidebook to the Geology of Meteor Crater*, printed for 37th Annual Meeting of the Meteoritical Society.

Short, N. M., and Forman, M. L.: 1972, Thickness of impact crater ejecta on the lunar surface, *Mod. Geol.* **3**, 69–91.

Stöffler, D., Gault, D. E., Wedekind, J., and Polkowski, G.: 1975, Experimental hypervelocity impact into quartz sand: Distribution and shock metamorphism of ejecta, *J. Geophys. Res.* **80**, 4062–4077.

Subject Index

Author Index